2019 IEEE 69th Electronic Components and Technology Conference (ECTC 2019)

Las Vegas, Nevada, USA
28-31 May 2019

Pages 785-1574

IEEE Catalog Number:	**CFP19ECT-POD**
ISBN:	**978-1-7281-1500-9**

**Copyright © 2019 by the Institute of Electrical and Electronics Engineers, Inc.
All Rights Reserved**

Copyright and Reprint Permissions: Abstracting is permitted with credit to the source. Libraries are permitted to photocopy beyond the limit of U.S. copyright law for private use of patrons those articles in this volume that carry a code at the bottom of the first page, provided the per-copy fee indicated in the code is paid through Copyright Clearance Center, 222 Rosewood Drive, Danvers, MA 01923.

For other copying, reprint or republication permission, write to IEEE Copyrights Manager, IEEE Service Center, 445 Hoes Lane, Piscataway, NJ 08854. All rights reserved.

*** This is a print representation of what appears in the IEEE Digital Library. Some format issues inherent in the e-media version may also appear in this print version.

IEEE Catalog Number: CFP19ECT-POD
ISBN (Print-On-Demand):   978-1-7281-1500-9
ISBN (Online):   978-1-7281-1499-6
ISSN: 0569-5503

Additional Copies of This Publication Are Available From:

Curran Associates, Inc
57 Morehouse Lane
Red Hook, NY 12571 USA
Phone: (845) 758-0400
Fax: (845) 758-2633
E-mail: curran@proceedings.com
Web: www.proceedings.com

Proceedings

The 2019 IEEE 69th Electronic Components and Technology Conference

Proceedings

IEEE 69th Electronic Components and Technology Conference

ECTC 2019

28 – 31 May 2019
Las Vegas, Nevada

Los Alamitos, California

Washington • Tokyo

2019 IEEE 69th Electronic Components and Technology Conference (ECTC 2019)

Las Vegas, Nevada, USA
28-31 May 2019

Pages 785-1574

IEEE Catalog Number: CFP19ECT-POD
ISBN: 978-1-7281-1500-9

2019 IEEE 69th Electronic Components and Technology Conference (ECTC)
ECTC 2019

Table of Contents

Foreword lvii
Executive Committee lix
Program Committee lx

Session 1: Wafer-Level Fan-Out Process Integration

3D-MiM (MUST-in-MUST) Technology for Advanced System Integration 1
*An-Jhih Su (Taiwan Semiconductor Manufacturing Company), Terry Ku
(Taiwan Semiconductor Manufacturing Company), Chung-Hao Tsai (Taiwan
Semiconductor Manufacturing Company), Kuo-Chung Yee (Taiwan
Semiconductor Manufacturing Company), and Douglas Yu (Taiwan
Semiconductor Manufacturing Company)*

Construction of FO-MCM with C4 Bumps Built First Using Chip Last Assembly Technology 7
*Chih-Hsun Hsu (Siliconware Precision Industries Co., Ltd), Wen-Yang Li
(Siliconware Precision Industries Co., Ltd), Chi-Jen Chen (Siliconware
Precision Industries Co., Ltd), Yih-Jenn Jiang (Siliconware Precision
Industries Co., Ltd), Jui-Feng Tai (Siliconware Precision Industries
Co., Ltd), Chang-Fu Lin (Siliconware Precision Industries Co., Ltd),
and C. Key Chung (Siliconware Precision Industries Co., Ltd)*

Feasibility Study of Fan-Out Panel-Level Packaging for Heterogeneous Integrations 14

Cheng-Ta Ko (Unimicron Technology Corporation), Henry Yang (Unimicron Technology Corporation), John H. Lau (ASM Pacific Technology Ltd), Ming Li (ASM Pacific Technology Ltd), Curry Lin (Unimicron Technology Corporation), Chieh-Lin Chang (Dow Chemical Company), Jhih-Yuan Pan (Dow Chemical Company), Hsing-Hui Wu (Dow Chemical Company), Iris Xu (Jiangyin Changdian Advanced Packaging Co., Ltd.), Tony Chen (Jiangyin Changdian Advanced Packaging Co., Ltd.), Zhang Li (Jiangyin Changdian Advanced Packaging Co., Ltd.), Kim Hwee Tan (Jiangyin Changdian Advanced Packaging Co., Ltd.), Penny Lo (ASM Pacific Technology Ltd), R. So (ASM Pacific Technology Ltd), Y. H. Chen (Unimicron Technology Corporation), Nelson Fan (ASM Pacific Technology Ltd), Eric Kuah (ASM Pacific Technology Ltd), Marc Lin (Dow Chemical Company), Y. M. Cheung (ASM Pacific Technology Ltd), Eric Ng (ASM Pacific Technology Ltd), Cao Xi (Huawei Technologies Co. Ltd.), Rozalia Beica (Dow Chemical Company), Sze Pei Lim (Indium Corporation), N. C. Lee (Indium Corporation), Mian Tao (Hong Kong University of Science and Technology), Jeffery Lo (Hong Kong University of Science and Technology), and Ricky Lee (Hong Kong University of Science and Technology)

Ultra-Thin FO Package-on-Package for Mobile Application 21

Hsiang-Yao Hsiao (Institute of Microelectronics, A*STAR), Soon Wee Ho (Institute of Microelectronics, A*STAR), Simon Siak Boon Lim (Institute of Microelectronics, A*STAR), Leong Ching Wai (Institute of Microelectronics, A*STAR), Ser Choong Chong (Institute of Microelectronics, A*STAR), Pei Siang Sharon Lim (Institute of Microelectronics, A*STAR), Yong Han (Institute of Microelectronics, A*STAR), and Tai Chong Chai (Institute of Microelectronics, A*STAR)

Development of Wafer Level Process for the Fabrication of Advanced Capacitive Fingerprint Sensors Using Embedded Silicon Fan-Out (eSiFO(R)) Technology 28

Shuying Ma (Huantian Technology (Kunshan) Electronics Co., Ltd., China), Chengqian Wang (Huantian Technology (Kunshan) Electronics Co., Ltd., China), Fengxia Zheng (Huantian Technology (Kunshan) Electronics Co., Ltd., China), Daquan Yu (Huantian Technology (Kunshan) Electronics Co., Ltd., China), Hong Xie (Filipchip International, USA), Xiaobing Yang (Huantian Technology (Kunshan) Electronics Co., Ltd., China), Li Ma (Huantian Technology (Kunshan) Electronics Co., Ltd., China), Ping Li (Huantian Technology (Xi'an) Electronics Co., Ltd., China), Weidong Liu (Huantian Technology (Xi'an) Electronics Co., Ltd., China), Jambo Yu (Synaptics, USA), and Jason Goodelle (Synaptics, USA)

Three-Dimensional Integrated Circuit (3D-IC) Package Using Fan-Out Technology 35

Jun Kyu Lee (NEPES Corporation), Sang Yong Park (NEPES Corporation), Young Ho Kim (NEPES Corporation), Jae Cheon Lee (NEPES Corporation), Sung Hyuk Lee (NEPES Corporation), Chul Hyo Lee (NEPES Corporation), Yong Tae Kwon (NEPES Corporation), Chang Woo Lee (NEPES Corporation), Jong Heon Kim (NEPES Corporation), Nam Chul Kim (NEPES Corporation), and Yun Hyun Sung (NEPES Corporation)

Ultra High Density IO Fan-Out Design Optimization with Signal Integrity and Power Integrity 41
> Keng Tuan Chang (Advanced Semiconductor Engineering, Inc.), Chih-Yi
> Huang (Advanced Semiconductor Engineering, Inc.), Hung-Chun Kuo
> (Advanced Semiconductor Engineering, Inc.), Ming-Fong Jhong (Advanced
> Semiconductor Engineering, Inc.), Tsun-Lung Hsieh (Advanced
> Semiconductor Engineering, Inc.), Mi-Chun Hung (Advanced Semiconductor
> Engineering, Inc.), and Chen-Chao Wang (Advanced Semiconductor
> Engineering, Inc.)

Session 2: Next-Generation Wirebonding and Die Attach

SB²-WB: A New Process Solution for Advanced Wire-Bonding 47
> Matthias Fettke (Pac Tech GmbH), Andrej Kolbasow (Pac Tech GmbH),
> Georg Friedrich (Pac Tech GmbH), Anna Palys (PacTech GmbH), Vinith
> Bejugam (Pac Tech GmbH), and Thorsten Teutsch (Pac Tech GmbH)

Smart Wire Bond Solutions for SiP and Memory Packages 55
> Basil Milton (Kulicke & Soffa, USA), Aashish Shah (Kulicke & Soffa,
> USA), Hui Xu (Kulicke & Soffa, USA), Odal Kwon (Kulicke & Soffa, USA),
> Gary Schulze (Kulicke & Soffa, USA), Ivy Qin (Kulicke & Soffa, USA),
> and Nelson Wong (Kulicke & Soffa, Singapore)

Preparation and Application of Cu-Ag Composite Preforms for Power Electronic Packaging 63
> Dongxiao Zhang (Wuhan University of Technology), Shengfa Liu (Wuhan
> University of Technology), Hui Xiang (Wuhan University of Technology),
> Li Liu (Wuhan University of Technology), Zhaoxia Zhou (Loughborough
> University), Stuart Robertson (Loughbrough University), Canyu Liu
> (Loughborough University), Zhiwen Chen (Wuhan University), and
> Changqing Liu (Loughborough University)

Au-Rich/Sn-Bi Interconnection in Chip-on-Module Package 69
> Jin Wang (Tsinghua University, China), Qian Wang (Tsinghua University,
> China), Xinnan Hou (GalaxyCore Inc., China), Ke Du (GalaxyCore Inc.,
> China), Lixin Zhao (GalaxyCore Inc., China), and Jian Cai (Tsinghua
> University, China)

The Properties of Cu Sinter Paste for Pressure Sintering at Low Temperature 76
> Jung-Lae Jo (Mitsui Engineered Materials Sector R&D Center, Mitsui
> Mining & Smelting Co., Ltd.), Kei Anai (Mitsui Engineered Materials
> Sector R&D Center, Mitsui Mining & Smelting Co., Ltd.), Sinichi
> Yamauchi (Mitsui Engineered Materials Sector R&D Center, Mitsui Mining
> & Smelting Co., Ltd.), and Takahiko Sakaue (Mitsui Engineered
> Materials Sector R&D Center, Mitsui Mining & Smelting Co., Ltd.)

Low Temperature Sintering of Dendritic Cu Based Pastes for Power Semiconductor Device Interconnection 81

Gang Li (Shenzhen Institutes of Advanced Technology, Chinese Academy of Sciences, Shenzhen College of Advanced Technology, University of Chinese Academy of Sciences), Jilei Fana (Shenzhen Institutes of Advanced Technology, Chinese Academy of Sciences, Shenzhen College of Advanced Technology, University of Chinese Academy of Sciences), Siyuan Liao (Shenzhen Institutes of Advanced Technology, Chinese Academy of Sciences, Shenzhen College of Advanced Technology, University of Chinese Academy of Sciences), Pengli Zhua (Shenzhen Institutes of Advanced Technology, Chinese Academy of Sciences), Baotan Zhang (Shenzhen Institutes of Advanced Technology, Chinese Academy of Sciences), Tao Zhao (Shenzhen Institutes of Advanced Technology, Chinese Academy of Sciences), Rong Sun (Shenzhen Institutes of Advanced Technology, Chinese Academy of Sciences), and Ching-Ping Wong (Shenzhen Institutes of Advanced Technology, Chinese Academy of Sciences)

A New Development of Direct Bonding to Aluminum and Nickel Surfaces by Silver Sintering in air Atmosphere 87

Ly May Chew (Heraeus Deutschland GmbH & Co. KG, Germany), Tamira Stegmann (Hochschule Aschaffenburg, University of Applied Sciences, Germany), Erika Schwenk (Hochschule Aschaffenburg, University of Applied Sciences, Germany), Monique Dubis (Hochschule Aschaffenburg, University of Applied Sciences, Germany), and Wolfgang Schmitt (Heraeus Deutschland GmbH & Co. KG, Germany)

Session 3: RDL and Additive Manufacturing

Submicron-Scale Cu RDL Pattering Based on Semi-Additive Process for Heterogeneous Integration 94
Takamasa Takano (DNP Co., Ltd.), Hiroshi Kudo (DNP Co., Ltd.), Masaya Tanaka (DNP Co., Ltd.), and Miyuki Akazawa (DNP Co., Ltd.)

Sub-Micron RDL Patterning for Advanced Packaging 101
Ken-Ichiro Mori (Canon Inc.), Yoshio Goto (Canon Inc.), Yasuo Hasegawa (Canon Inc.), Seiya Miura (Canon Inc.), and Douglas Shelton (Canon U.S.A Inc.)

Optimization of Electrolytic Plating Processes for Challenging Fan-Out Panel Level Package Designs 106
Ralph Zoberbier (Atotech Deutschland GmbH, Germany), Britta Scheller (Atotech Deutschland GmbH), and Christian Ohde (Atotech Deutschland GmbH)

3D Printed Substrates for the Design of Compact RF Systems 113
Mohd Ifwat Mohd Ghazali (Universiti Sains Islam Malaysia , Michigan State University), Saikat Mondal (Michigan State University), Saranraj Karuppuswami (Michigan State University), and Premjeet Chahal (Michigan State University)

Fully Additively Manufactured Tunable Active Frequency Selective Surfaces with Integrated On-package Solar Cells for Smart Packaging Applications 119
Syed Abdullah Nauroze (Georgia Institute of Technology), Xuanke He (Georgia Institute of Technology), and Manos M. Tentzeris (Georgia Institute of Technology)

First Demonstration of a Low Cost/Customizable Chip Level 3D Printed Microjet Hotspot-Targeted
Cooler for High Power Applications 126
> Tiwei Wei (IMEC & KU Leuven, Belgium), Herman Oprins (IMEC, Belgium),
> Vladimir Cherman (IMEC, Belgium), Ingrid De Wolf (IMEC & KU Leuven,
> Belgium), Eric Beyne (IMEC, Belgium), and Martine Baelmans (KU Leuven,
> Belgium)

Rapid Production of Customized 3D Electronics Via Hybrid Additive Manufacturing Technology 135
> Ji Li (Key Laboratory of MEMS of the Ministry of Education, Southeast
> University), Yang Wang (Key Laboratory of MEMS of the Ministry of
> Education, Southeast University), Peiren Wang (Key Laboratory of MEMS
> of the Ministry of Education, Southeast University), Jiangling He (Key
> Laboratory of MEMS of the Ministry of Education, Southeast
> University), Handa Liu (Key Laboratory of MEMS of the Ministry of
> Education, Southeast University), and Gengzhao Xiang (Key Laboratory
> of MEMS of the Ministry of Education, Southeast University)

Session 4: Advancements in Automotive and Power Devices

Solid-Liquid InterDiffusion (SLID) Bonding, for Thermally Challenging Applications 141
> Knut E Aasmundtveit (University of South-Eastern Norway, Norway),
> Thi-Thuy Luu (Zimmer and Peacock, Norway), Hoang-Vu Nguyen (University
> of South-Eastern Norway, Norway), Andreas Larsson (University of
> South-Eastern Norway, Norway), and Torleif A Tollefsen (TEGma, Norway)

Fluxless Bonding Technique of Diamond to Copper Using Silver-Indium Multilayer Structure 150
> Roozbeh Sheikhi (University of California, Irvine), Yongjun Huo
> (University of California, Irvine), and Chin C. Lee (University of
> California, Irvine)

Formulation and Processing of Conductive Polysulfide Sealants for Automotive and Aerospace
Applications 157
> Bo Song (Georgia Institute of Technology), Fan Wu (Georgia Institute
> of Technology), Kyoung-sik Moon (Georgia Institute of Technology), and
> CP Wong (Georgia Institute of Technology)

Challenges and Approaches to Developing Automotive Grade 1/0 FCBGA Package Capability 163
> Rajen Dias (Amkor Technology, USA), Mike Kelly (Amkor Technology,
> USA), Devarajan Balaraman (Amkor Technology, USA), Hideaki Shoji
> (J-Devices Corporation, Japan), Tomio Shiraiwa (J-Devices, Japan),
> KwangSeok Oh (Amkor Technology, Korea), and JoonYoung Park (Amkor
> Technology, Korea)

Advanced Substrates for GaN-Based Power Devices 168

Anthony Cibié (Univ. Grenoble Alpes, CEA, LETI, France), Julie Widiez (Univ. Grenoble Alpes, CEA, LETI, France), René Escoffier (Univ. Grenoble Alpes, CEA, LETI, France), Denis Blachier (Univ. Grenoble Alpes, CEA, LETI, France), Kremena Vladimirova (Univ. Grenoble Alpes, CEA, LETI, France), Jean-Philippe Colonna (Univ. Grenoble Alpes, CEA, LETI, France), Paul-Henri Haumesser (Univ. Grenoble Alpes, CEA, LETI, France), Stéphane Bécu (Univ. Grenoble Alpes, CEA, LETI, France), Perceval Coudrain (Univ. Grenoble Alpes, CEA, LETI, France), William Vandendaele (Univ. Grenoble Alpes, CEA, LETI, France), Jérôme Biscarrat (Univ. Grenoble Alpes, CEA, LETI, France), Charlotte Gillot (Univ. Grenoble Alpes, CEA, LETI, France), Matthew Charles (Univ. Grenoble Alpes, CEA, LETI, France), and Léa Di Cioccio (Univ. Grenoble Alpes, CEA, LETI, France)

A New Reliable, Corrosion Resistant Gold-Palladium Coated Copper Wire Material 175

Sandy Klengel (Fraunhofer Institute for Microstructure of Materials IMWS), Robert Klengel (Fraunhofer Institute for Microstructure of Materials IMWS), Jan Schischka (Fraunhofer Institute for Microstructure of Materials IMWS), Tino Stephan (Fraunhofer Institute for Microstructure of Materials IMWS), Matthias Petzold (Fraunhofer Institute for Microstructure of Materials IMWS), Motoki Eto (Nippon Micrometal Corporation), Noritoshi Araki (Nippon Micrometal Corporation), and Takashi Yamada (Nippon Micrometal Corporation)

Ultrasonic-Accelerated Intermetallic Joint Formation with Composite Solder for High-Temperature Power Device Packaging 183

Hongjun Ji (Harbin Institute of Technology at Shenzhen), Mingyu Li (Harbin Institute of Technology at Shenzhen), Weiwei Zhao (Harbin Institute of Technology at Shenzhen), and Wenwu Zhang (Harbin Institute of Technology at Shenzhen)

Session 5: Bonding Manufacturing Technologies

Comprehensive Study of Copper Nano-Paste for Cu-Cu Bonding 191

Ser Choong Chong (Institute of Microelectronics) and Pei Siang Lim Sharon (Insttitute of Microelectronics)

Enhanced Performance of Laser-Assisted Bonding with Compression (LABC) Compared with Thermal Compression Bonding (TCB) Technology 197

Kwang-Seong Choi (Electronics and Telecommunications Research Institute), Yong-Sung Eom (Electronics and Telecommunications Research Institute), Seok Hwan Moon (Electronics and Telecommunications Research Institute), Jiho Joo (Electronics and Telecommunications Research Institute), leeseul Jeong (Electronics and Telecommunications Research Institute), Kwangjoo Lee (LG Chem), Jung Hak Kim (LG Chem), Ju hyeon Kim (LG Chem), Gil-Sang Yoon (KITECH), Kwang-Hee Lee (Inha University), Chul-Hee Lee (Inha University), Geun-Sik Ahn (Protec), and Moo-Sup Shim (Protec)

A Study of 3D Packaging Interconnection Performance Affected by Thermal Diffusivity and Pressure
Transmission 204
    Jin-San Jung (Samsung Electronics Co., Ltd), Hyeong Gi Lee (Samsung
    Electronics Co., Ltd), Ji-Min Kim (Samsung Electronics Co., Ltd),
    Yong-Jin Park (Samsung Electronics Co., Ltd), Ji-In Yu (Samsung
    Electronics Co., Ltd), Yong Sung Park (Samsung Electronics Co., Ltd),
    Jun Su Lim (Samsung Electronics Co., Ltd), Hyun-Seok Choi (Samsung
    Electronics Co., Ltd), Sung-Il Cho (Samsung Electronics Co., Ltd),
    Dong wook Kim (Samsung Electronics Co., Ltd), and Sang-Ho An (Samsung
    Electronics Co., Ltd)

Vertical Laser Assisted Bonding for Advanced "3.5D" Chip Packaging 210
    Andrej Kolbasow (Pac Tech GmbH), Timo Kubsch (Pac Tech), Matthias
    Fettke (Pac Tech GmbH), Georg Friedrich (Pac Tech GmbH), and Thorsten
    Teutsch (PacTech)

Optimization of a BEOL Aluminum Deposition Process Enabling Wafer Level Al-Al Thermo-Compression
Bonding 218
    Sebastian Schulze (IHP - Innovations for High Performance
    Microelectronics), Matthias Wietstruck (IHP - Innovations for High
    Performance Microelectronics), Mirko Fraschke (IHP - Innovations for
    High Performance Microelectronics), Peter Kerepesi (EV Group, Inc.),
    Helmut Kurz (EV Group, Inc.), Bernhard Rebhan (EV Group, Inc.), and
    Mehmet Kaynak (IHP - Innovations for High Performance
    Microelectronics)

Self-Assembly Process for 3D Die-to-Wafer using Direct Bonding: A Step Forward Toward Process
Automatisation 225
    Jouve Amandine (CEA, LETI), Loïc Sanchez (CEA, LETI), Clément Castan
    (CEA, LETI), Maxence Laugier (CEA, LETI), Emmanuel Rolland (CEA,
    LETI), Brigitte Montmayeul (CEA, LETI), Rémi Franiatte (CEA, LETI),
    Frank Fournel (CEA, LETI), and Severine Cheramy (CEA, LETI)

A Single Bonding Process for Diverse Organic-Inorganic Integration in IoT Devices 235
    Tilo H. Yang (National Taiwan University), Yu-Shan Chiu (National
    Taiwan University), Hai-Yang Yu (National Taiwan University), Akitsu
    Shigetou (National Institute for Materials Science), and C. Robert Kao
    (National Taiwan University)

Session 6: Emerging Flexible Hybrid Electronics

Stretchable and Printable Medical Dry Electrode Arrays on Textile for Electrophysiological
Monitoring 243
    Yougen Hu (Shenzhen Institutes of Advanced Technology, Chinese Academy
    of Sciences), Hui Wang (Shenzhen Institutes of Advanced Technology,
    Chinese Academy of Sciences), Ommeaymen Sheikhnejad (AC2T Research
    GmbH), Yaoxu Xiong (Shenzhen Institutes of Advanced Technology,
    Chinese Academy of Sciences), Han Gu (Shenzhen Institutes of Advanced
    Technology, Chinese Academy of Sciences), Pengli Zhu (Shenzhen
    Institutes of Advanced Technology, Chinese Academy of Sciences),
    Guanglin Li (Shenzhen Institutes of Advanced Technology, Chinese
    Academy of Sciences), Rong Sun (Shenzhen Institutes of Advanced
    Technology, Chinese Academy of Sciences), and Ching-Ping Wong (Georgia
    Institute of Technology)

Screen-Printed Flexible Coplanar Waveguide Transmission Lines: Multi-physics Modeling and
Measurement 249

    *Nahid Aslani Amoli (Georgia Institute of Technology), Sridhar
Sivapurapu (Georgia Institute of Technology), Rui Chen (Georgia
Institute of Technology), Yi Zhou (Georgia Institute of Technology),
Mohamed L. F. Bellaredj (Georgia Institute of Technology), Paul A.
Kohl (Georgia Institute of Technology), Suresh K. Sitaraman (Georgia
Institute of Technology), and Madhavan Swaminathan (Georgia Institute
of Technology)*

Inkjet-Printed Filtering Antenna on a Textile for Wearable Applications 258

    *Hsuan-Ling Kao (Dept. of Electronic Engineering, Chang Gung
University, Tao-Yuan, Taiwan), Chun-Hsiang Chuang (Dept. of Electronic
Engineering, Chang Gung University, Tao-Yuan, Taiwan), and Cheng-Lin
Cho (Dept. of Engineering and System Science, National Tsing Hua
University, Hsin-Chu, Taiwan)*

Mechanical and Electrical Characterization of FOWLP-Based Flexible Hybrid Electronics (FHE) for
Biomedical Sensor Application 264

    *Yuki Susumago (Tohoku University), Qian Zhengyang (Tohoku University),
Achille Jacquemond (Tohoku University), Noriyuki Takahashi (Tohoku
University), Hisashi Kino (Tohoku University), Tetsu Tanaka (Tohoku
University), and Takafumi Fukushima (Tohoku University)*

A Wearable Fingernail Deformation Sensing System and Three-Dimensional Finite Element Model of
Fingertip 270

    *Katsuyuki Sakuma (IBM Thomas J. Watson Research Center, U.S.A.),
Bucknell Webb (IBM Thomas J. Watson Research Center, U.S.A.), Rajeev
Narayanan (IBM Thomas J. Watson Research Center), Avner Abrami (IBM
Thomas J. Watson Research Center), Jeff Rogers (IBM Thomas J. Watson
Research Center), John Knickerbocker (IBM Thomas J. Watson Research
Center), and Stephen J Heisig (IBM Thomas J. Watson Research Center)*

Heterogeneous Integration of a Fan-Out Wafer-Level Packaging Based Foldable Display on Elastomeric
Substrate 277

    *Arsalan Alam (University of California, Los Angeles), Amir Hanna
(University of California, Los Angeles), Randall Irwin (University of
California, Los Angeles), Goutham Ezhilarasu (University of
California, Los Angeles), Hyunpil Boo (University of California, Los
Angeles), Yuan Hu (University of California, Los Angeles), Chee Wei
Wong (University of California, Los Angeles), Timothy S Fisher
(University of California, Los Angeles), and S. S. Iyer (University of
California)*

A Study on the Flexible Chip-on-Fabric (COF) Assembly using Anisotropic Conductive Films (ACFs)
Materials 283

    Seung-Yoon Jung (KAIST) and Kyung-Wook Paik (KAIST)

Session 7: Advances in Flip Chip Packaging

7nm Chip-Package Interaction Study on a Fine Pitch Flip Chip Package with Laser Assisted Bonding and
Mass Reflow Technology 289

    *Ian Hsu (MediaTek), Chi-Yuan Chen (MediaTek), Stanley Lin (MediaTek),
Ta-Jen Yu (MediaTek), NamJu Cho (JCET), and Ming-Che Hsieh (JCET)*

Ultra Large Area SIPs and Integrated mmW Antenna Array Module for 5G mmWave Outdoor Applications ... 294

Pouya Talebbeydokhti (Intel Corporation), Sidharth Dalmia (Intel Corporation), Trang Thai (Intel Corporation), Raanan Sover (Intel Corporation), and Sharon Tal (Intel Corporation)

Hybrid Approach for Large Size FC-BGA to Enhance Thermal and Electrical Performance Including Power Delivery 300

Heeseok Lee (Samsung Electronics, Co. Ltd.), Yunhyeok Im (Samsung Electronics, Co. Ltd.), Junghwa Kim (Samsung Electronics, Co. Ltd.), Jisoo Hwang (Samsung Electronics, Co. Ltd.), James Jeong (Samsung Electronics, Co. Ltd.), Youngsang Cho (Samsung Electronics, Co. Ltd.), Heejung Choi (Samsung Electronics, Co. Ltd.), and Youngmin Shin (Samsung Electronics, Co. Ltd.)

Package-on-Package Micro-BGA Microstructure Interaction with Bond and Assembly Parameter 306

Pascale Gagnon (IBM Canada Limited), Clément Fortin (IBM Canada Limited), and Thomas Weiss (IBM Systems)

Low Cost Flip-Chip Stack for Partitioning Processing and Memory 314

Fabian Hopsch (Fraunhofer Institute for Integrated Circuits IIS, Division Engineering of Adaptive Systems EAS) and Andy Heinig (Fraunhofer Institute for Integrated Circuits IIS, Division Engineering of Adaptive Systems EAS)

Impact of Low Temperature Solder on Electronic Package Dynamic Warpage Behavior and Requirement 318

Wei Keat Loh (Intel Technology Sdn Bhd), Ron W. Kulterman (Flex Ltd), Haley Fu (iNEMI), and Chih Chung Hsu (CoreTech System (Moldex3D))

High Density Ultra-Thin Organic Substrates for Advanced Flip Chip Packages 325

Nokibul Islam (JCET), KH Tan (JCET), Seung Wook Yoon (JCET), and Tony Chen (JCET)

Session 8: Material and Process Trends in FOWLP and PLP

Laser Releasable Temporary Bonding Film with High Thermal Stability 330
Yong-suk Yang (3M), Kyo-sung Hwang (3M), and Robin Gorrell (3M)

Design and Demonstration of 1μm Low Resistance RDL Using Panel Scale Processes for High Performance Computing Applications 334

Bartlet DeProspo (3D Systems Packaging Research Center), Aya Momozawa (Tokyo Ohka Kogyo Co. LTD.), Atsushi Kubo (Tokyo Ohka Kogyo Co. LTD.), Chandrasekharan Nair (3D Systems Packaging Research Center), Varun Rajagoapal (3D Systems Packaging Research Center), Jenefa Kannan (Georgia Institute of Technology), Emanuel Surillo (3D Systems Packaging Research Center), Fuhan Liu (3D Systems Packaging Research Center), Mohananlingam Kathaperumal (3D Systems Packaging Research Center), and Rao Tummala (3D Systems Packaging Research Center)

Advances in Temporary Carrier Technology for High-Density Fan-Out Device Build-up 340

Arnita Podpod (IMEC), Alain Phommahaxay (IMEC), Pieter Bex (IMEC), John Slabbekoorn (IMEC), Julien Bertheau (IMEC), Abdellah Salahoueldhadj (IMEC), Erik Sleeckx (IMEC), Andy Miller (IMEC), Gerald Beyer (IMEC), Eric Beyne (IMEC), Alice Guerrero (Brewer Science Inc), Kim Yess (Brewer Science Inc), and Kim Arnold (Brewer Science Inc)

Development of Novel Low-Temperature Curable Positive-Tone Photosensitive Dielectric Materials with High Reliability 346

Yutaro Koyama (Toray Industries, Inc.), Yu Shoji (Toray Industries, Inc.), Keika Hashimoto (Toray Industries, Inc.), Yuki Masuda (Toray Industries, Inc.), Hitoshi Araki (Toray Industries, Inc.), and Masao Tomikawa (Toray Industries, Inc.)

Highly Reliable Photosensitive Negative-Tone Polyimide with Low Cure Shrinkage 352

Daisaku Matsukawa (Hitachi Chemical DuPont MicroSystems, Ltd, Japan), Hiroko Yotsuyanagi (Hitachi Chemical DuPont MicroSystems, Ltd, Japan), Shiori Sakakibara (Hitachi Chemical DuPont MicroSystems, Ltd, Japan), Noriyuki Yamazaki (Hitachi Chemical DuPont MicroSystems, Ltd, Japan), Tetsuya Enomoto (Hitachi Chemical DuPont MicroSystems, Ltd, Japan), and Takeharu Motobe (Hitachi Chemical DuPont MicroSystems, Ltd, Japan)

High Rate and Low Damage Etching Method as Pre Treatment of Seed Layer Sputtering for Fan out Panel Level Packaging 358

Tetsushi Fujinaga (ULVAC, Inc.)

Investigation and Methods Using Various Release and Thermoplastic Bonding Materials to Reduce Die Shift and Wafer Warpage for eWLB Chip-First Processes 363

Michelle Fowler (Brewer Science, Inc.), John P. Massey (Brewer Science, Inc.), Tanja Braun (Fraunhofer Institute IZM), Steve Voges (Fraunhofer Institute IZM), Robert Gernhardt (Fraunhofer Institute IZM), and Markus Wohrmann (Fraunhofer institute IZM)

Session 9: Wearables and Thin-Package Reliability and Chip Package Interaction

Effect of Charging Cycle Elevated Temperature Storage and Thermal Cycling on Thin Flexible Batteries in Wearable Applications 370

Pradeep Lall (Auburn University), Amrit Abrol (Auburn University), Ben Leever (US AFRL), and Scott Miller (NextFlex)

Bladder Inflation Stretch Test Method for Reliability Characterization of Wearable Electronics 382

Benjamin G Stewart (Georgia Institute of Technology) and Suresh K Sitaraman (Georgia Institute of Technology)

Study of BEOL Failure Mode in Flip Chip Packages at High Temperature Conditions 392

Wei Wang (Qualcomm Technologies, Inc.), Yangyang Sun (Qualcomm Technologies, Inc.), Xuefeng Zhang (Qualcomm Technologies, Inc.), Lejun Wang (Qualcomm Technologies, Inc.), Lily Zhao (Qualcomm Technologies, Inc.), Mark Schwarz (Qualcomm Technologies, Inc.), Bill Stone (Qualcomm Technologies, Inc.), and Ahmer Syed (Qualcomm Technologies, Inc.)

A Novel Metal Scheme and Bump Array Design Configuration to Enhance Advanced Si Packages CPI Reliability Performance by Using Finite Element Modeling Technique 397

Kuo-Chin Chang (Taiwan Semiconductor Manufacturing Company Ltd.), Mirng-Ji Lii (Taiwan Semiconductor Manufacturing Company Ltd.), Steven Hsu (Taiwan Semiconductor Manufacturing Company Ltd.), Hao-Chun Liu (Taiwan Semiconductor Manufacturing Company Ltd.), Yen-Kun Lai (Taiwan Semiconductor Manufacturing Company Ltd.), Sheng-Han Tsai (Taiwan Semiconductor Manufacturing Company Ltd.), and Chieh-Hao Hsu (Taiwan Semiconductor Manufacturing Company Ltd.)

Assessment of CMP Fill Pattern Effect on the Thermal Performance of Interconnects in Integrated
Circuits BEOL 405
    Assaad Helou (Southern Methodist University), Peter Raad (Southern
    Methodist University), and Archana Venugopal (Texas Instruments, Inc.)

Three-Dimensional Simulation of the Thermo-Mechanical Interaction between the Micro-Bump Joints and
Cu Protrusion in Cu-Filled TSVs of the High Bandwidth Memory (HBM) Structure 410
    Jie-Ying Zhou (South China University of Technology), Shui-Bao Liang
    (South China University of Technology), Cheng Wei (South China
    University of Technology), Wen-Kai Le (South China University of
    Technology), Chang-Bo Ke (South China University of Technology),
    Min-Bo Zhou (South China University of Technology), Xiao Ma (South
    China University of Technology), and Xin-Ping Zhang (South China
    University of Technology)

Study of Design Optimization Method for Ultra-Low Power Micro Gas Sensor 417
    Eiji Nakamura (IBM Research - Tokyo), Keiji Matsumoto (IBM Research -
    Tokyo), Andrea Fasoli (IBM Research - Almaden), Luisa Bozano (IBM
    Research - Almaden), and Hiroyuki Mori (IBM Research - Tokyo)

Session 10: Dicing and Encapsulation Technologies

A More Than Moore Enabling Wafer Dicing Technology 423
    Jeroen van Borkulo (ASM Pacific Technologies Inc), Rogier Evertsen
    (ASM Pacific Technologies Inc), and Richard van der Stam (ASM Pacific
    Technologies Inc.)

Plasma Dicing Integration Schemes for Scribe Lane Layout and the Impact on Die Strength 428
    David Parker (STMicroelectronics, France), Emmanuel Gourvest
    (STMicroelectronics, France), and Boris Bouillard (STMicroelectronics,
    France)

Advanced Dicing Technologies for Combination of Wafer to Wafer and Collective Die to Wafer Direct
Bonding 437
    Fumihiro Inoue (IMEC, Belgium), Alain Phommahaxay (IMEC, Belgium),
    Arnita Podpod (IMEC, Belgium), Samuel Suhard (IMEC, Belgium), Hitoshi
    Hoshino (Disco Hi-Tech Europe, Germany), Berthold Moeller (Disco
    Hi-Tech Europe, Germany), Erik Sleeckx (IMEC, Belgium), Kenneth June
    Rebibis (IMEC, Belgium), Andy Miller (IMEC, Belgium), and Eric Beyne
    (IMEC, Belgium)

Active Control of NCF Fillet Shape for 3D CoW by Multi Beam Laser Bonder 446
    Keiko Ueno (Hitachi Chemical Co., Ltd.), Kazutaka Honda (Hitachi
    Chemical Co., Ltd.), Tsuyoshi Ogawa (Hitachi Chemical Co., Ltd.), and
    Toshihisa Nonaka (Hitachi Chemical Co., Ltd.)

Ultrafast Laser Scribe: An Improved Metal and ILD Ablation Process 453
    Julia Chiu (Intel Corporation, USA), Aaron Gore (Intel Corporation,
    USA), Tyler Osborn (Intel Corporation, USA), Daragh Finn (Electro
    Scientific Industries, USA), Zhibin Lin (Electro Scientific
    Industries, USA), David Lord (Electro Scientific Industries, USA), and
    Jon Mellen (Electro Scientific Industries, USA)

Reliability and Benchmark of 2.5D Non-Molding and Molding Technologies 461

Yu-Hsiang Hsiao (Advanced Semiconductor Engineering, Group, Inc., Kaohsiung, Taiwan), Che-Ming Hsu (Advanced Semiconductor Engineering, Group, Inc., Kaohsiung, Taiwan), Yi-Sheng Lin (Advanced Semiconductor Engineering, Group, Inc., Kaohsiung, Taiwan), and Chien-Lin Chang Chien (Advanced Semiconductor Engineering, Group, Inc., Kaohsiung, Taiwan)

Laser-Induced Trench Design, Optimisation and Validation for Restricting Capillary Underfill Spread in Advanced Packaging Configurations 467

Gul Zeb (Université de Sherbrooke), David Danovitch (Université de Sherbrooke), and Eric Turcotte (IBM Canada)

Session 11: Automotive and Harsh-Environment Reliability

Effect of Substrate Preheating Treatment on Thermal Reliability and Micro-Structure of Ag Paste Sintering on Au Surface Finish 474

Zheng Zhang (Institute of Scientific and Industrial Research, Osaka University), Chuantong Chen (Institute of Scientific and Industrial Research, Osaka University), Katsuaki Suganuma (Institute of Scientific and Industrial Research, Osaka University), and Seigo Kurosaka (C. Uyemura & Co., Ltd.)

Package Material Selection Criteria for High Temperature Automotive Applications 479

Rene T.H. Rongen (NXP Semiconductors), A. Mavinkurve (NXP Semiconductors), G.M. O'Halloran (NXP Semiconductors), N. Owens (NXP Semiconductors), Y. Weber (NXP Semiconductors), P. Oberndorff (NXP Semiconductors), M-L Farrugia (NXP Semiconductors), E. van Olst (NXP Semiconductors), and M. van Soestbergen (NXP Semiconductors)

Solder Joint Reliability of Double-Side Mounted DDR Modules for Consumer and Automotive Applications... 486

Dongji Xie (NVIDIA), Joe Hai (Nvidia Corp.), Zhongming Wu (Nvidia Corp.), and Manthos Economou (Nvidia Corp.)

Reliability Investigation of Extremely Large Ratio Fan-Out Wafer-Level Package with Low Ball Density for Ultra-Short-Range Radar 493

P.S. Huang (MediaTek Inc.), C.K. Yu (MediaTek Inc.), W.S. Chiang (MediaTek Inc.), M.Z. Lin (MediaTek Inc.), Y.H. Fang (MediaTek Inc.), M.J. Lin (MediaTek Inc.), N.W. Liu (MediaTek Inc.), Benson Lin (MediaTek Inc.), and Ian Hsu (MediaTek Inc.)

Fatigue Behaviour of Lead-Free Solder Joints Under Combined Thermal and Vibration Loads 498

Meier Karsten (Technische Universität Dresden), Winkler Maria (Technische Universität Dresden), Leslie David (University of Maryland, Center for Advanced Life Cycle Engineering (CALCE)), Dasgupta Abhijit (University of Maryland, Center for Advanced Life Cycle Engineering (CALCE)), and Bock Karlheinz (Technische Universität Dresden)

Prognostication of Accrued Damage and Impending Failure Under Temperature-Vibration in Leadfree Electronics 505

Pradeep Lall (Auburn University), Tony Thomas (Auburn University), Jeff Suhling (Auburn University), and Ken Blecker (US Army ARDEC)

Electrochemical Impedance Spectroscopy (EIS) for Monitoring the Water Load on PCBAs Under Cycling
Condensing Conditions to Predict Electrochemical Migration Under DC Loads 515
  Simone Lauser (Robert Bosch GmbH), Theresia Richter (Robert Bosch
  GmbH), Verdingovas Vadimas (Denmarks Technical University), and Rajan
  Ambat (Denmarks Technical University)

Session 12: Advanced Photonic Devices and Packaging

Micro-Fabricated SERF Atomic Magnetometer for Weak Gradient Magnetic Field Detection 522
  Xiang Yue (Southeast University), Jintang Shang (Southeast
  University), and Chen Ye (Southeast University)

Novel Solder Pads for Self-Aligned Flip-Chip Assembly 528
  Yves Martin (IBM T.J.Watson Research Center, Yorktown Heights USA),
  Swetha Kamlapurkar (IBM T.J.Watson Research Center, Yorktown Heights
  USA), Nathan Marchack (IBM T.J.Watson Research Center, Yorktown Height
  USA), Jae-Woong Nah (IBM T.J.Watson Research Center, Yorktown Height
  USA), and Tymon Barwicz (IBM T.J.Watson Research Center, Yorktown
  Height USA)

Collective Curved CMOS Sensor Process: Application for High-Resolution Optical Design and Assembly
Challenges 535
  Bertrand Chambion (Univ. Grenoble Alpes, CEA, LETI), Christophe
  Gaschet (Univ. Grenoble Alpes, CEA, LETI), Marc Lombard (Univ.
  Grenoble Alpes, CEA, LETI), Maïlys Fernandez (Univ. Grenoble Alpes,
  CEA, LETI), Pierre Joly (Univ. Grenoble Alpes, CEA, LETI), Stéphane
  Caplet (Univ. Grenoble Alpes, CEA, LETI), Fabien Zuber (Univ. Grenoble
  Alpes, CEA, LETI), Aurélie Vandeneynde (Univ. Grenoble Alpes, CEA,
  LETI), Patrick Peray (Univ. Grenoble Alpes, CEA, LETI), Gilles
  Lasfargues (Univ. Grenoble Alpes, CEA), Marc Zussy (Univ. Grenoble
  Alpes, CEA, LETI), Jerôme Deschamps (Univ. Grenoble Alpes, CEA, LETI),
  Alexis Bedoin (Univ. Grenoble Alpes, CEA, LETI), and David Henry
  (Univ. Grenoble Alpes, CEA, LETI)

Integration and Characterization of InP Die on Silicon Interconnect Fabric 543
  Eric Sorensen (Center for Heterogeneous Integration and Performance
  Scaling (CHIPS) University of California, Los Angeles), Boris Vaisband
  (Center for Heterogeneous Integration and Performance Scaling (CHIPS)
  University of California, Los Angeles), SivaChandra Jangam (Center for
  Heterogeneous Integration and Performance Scaling (CHIPS) University
  of California, Los Angeles), Tim Shirley (Keysight Technologies), and
  Subramanian S. Iyer (Center for Heterogeneous Integration and
  Performance Scaling (CHIPS) University of California, Los Angeles)

Y-Branched Multimode/Single-Mode Polymer Optical Waveguides for Low-Loss WDM MUX Device:
Fabrication and Characterization 550
  Takaaki Ishigure (Keio University), Tomoki Nakayama (Keio University),
  Fukino Nakazaki (Keio University), and Hiroki Hama (Keio University)

Vertically Stacked and Directionally Coupled Cavity-Resonator-Integrated Grating Couplers for
Integrated-Optic Beam Steering 556
  Shogo Ura (Kyoto Institute of Technology, Japan), Junishi Inoue (Kyoto
  Institute of Technology, Japan), and Kenji Kintaka (AIST, Japan)

CiB(Chip in Board) Optical Engine Module Using Advanced Fan-Out Package Technology 563
Sang Yong Park (NEPES Corporation), Ju Hyun Nam (NEPES Corporation),
Ji Ni Shim (NEPES Corporation), Jun Kyu Lee (NEPES Corporation), Yong
Tae Kwon (NEPES Corporation), Chang Woo Lee (NEPES Corporation), Jong
Heon Kim (NEPES Corporation), and Nam Chul Kim (NEPES Corporation)

Session 13: Technologies Enabling 3D and Heterogeneous Integration

Active Interposer Technology for Chiplet-Based Advanced 3D System Architectures 569
Perceval Coudrain (Univ. Grenoble Alpes, CEA, LETI, France), Jean
Charbonnier (Univ. Grenoble Alpes, CEA, LETI, France), Arnaud Garnier
(Univ. Grenoble Alpes, CEA, LETI, France), Pascal Vivet (Univ.
Grenoble Alpes, CEA, LETI, France), Rémi Vélard (Univ. Grenoble Alpes,
CEA, LETI, France), Andrea Vinci (Univ. Grenoble Alpes, CEA, LETI,
France), Fabienne Ponthenier (Univ. Grenoble Alpes, CEA, LETI,
France), Alexis Farcy (STMicroelectronics, France), Roselyne Segaud
(Univ. Grenoble Alpes, CEA, LETI, France), Pascal Chausse (Univ.
Grenoble Alpes, CEA, LETI, France), Lucile Arnaud (Univ. Grenoble
Alpes, CEA, LETI, France), Didier Lattard (Univ. Grenoble Alpes, CEA,
LETI, France), Eric Guthmuller (Univ. Grenoble Alpes, CEA, LETI,
France), Giovanni Romano (Univ. Grenoble Alpes, CEA, LETI, France),
Alain Gueugnot (Univ. Grenoble Alpes, CEA, LETI, France), Frédéric
Berger (Univ. Grenoble Alpes, CEA, LETI, France), Jérôme Beltritti
(STMicroelectronics), Therry Mourier (Univ. Grenoble Alpes, CEA, LETI,
France), Mathilde Gottardi (Univ. Grenoble Alpes, CEA, LETI, France),
Stéphane Minoret (Univ. Grenoble Alpes, CEA, LETI, France), Céline
Ribière (Univ. Grenoble Alpes, CEA, LETI, France), Gilles Romero
(Univ. Grenoble Alpes, CEA, LETI, France), Pierre-Emile Philip (Univ.
Grenoble Alpes, CEA, LETI, France), Yorrick Exbrayat (Univ. Grenoble
Alpes, CEA, LETI, France), Daniel Scevola (Univ. Grenoble Alpes, CEA,
LETI, France), Didier Campos (STMicroelectronics), Maxime Argoud
(Univ. Grenoble Alpes, CEA, LETI, France), Nacima Allouti (Univ.
Grenoble Alpes, CEA, LETI, France), Raphaël Eleouet (Univ. Grenoble
Alpes, CEA, LETI, France), César Fuguet Tortolero (Univ. Grenoble
Alpes, CEA, LETI, France), Christophe Aumont (Univ. Grenoble Alpes,
CEA, LETI, France), Denis Dutoit (Univ. Grenoble Alpes, CEA, LETI,
France), Corinne Legalland (Univ. Grenoble Alpes, CEA, LETI, France),
Jean Michailos (STMicroelectronics, France), Séverine Chéramy (Univ.
Grenoble Alpes, CEA, LETI, France), and Gilles Simon (Univ. Grenoble
Alpes, CEA, LETI, France)

Process Development of Power Delivery Through Wafer Vias for Silicon Interconnect Fabric 579
Meng-Hsiang Liu (University of California, Los Angeles), Boris
Vaisband (University of California, Los Angeles), Amir Hanna
(University of California, Los Angeles), Yandong Luo (University of
California, Los Angeles), Zhe Wan (University of California, Los
Angeles), and Subramanian S. Iyer (University of California, Los
Angeles)

Active Through-Silicon Interposer Based 2.5D IC Design, Fabrication, Assembly and Test 587
Jayasanker Jayabalan (Institute of Microelectronics), Vivek
Chidambaram Nachiappan (Institute of Microelectronics), Sharon Lim Pei
Siang (Institute of Microelectronics), Wang Xiangyu (Institute of
Microelectronics), Jong Ming Chinq (Institute of Microelectronics),
and Surya Bhattacharya (Institute of Microelectronics)

xviii

System on Integrated Chips (SoIC(TM) for 3D Heterogeneous Integration 594

    Ming-Fa Chen (Taiwan Semiconductor Manufacturing Company (TSMC)),
    Fang-Cheng Chen (Taiwan Semiconductor Manufacturing Company (TSMC)),
    Wen-Chih Chiou (Taiwan Semiconductor Manufacturing Company (TSMC)),
    and Doug C.H. Yu (Taiwan Semiconductor Manufacturing Company (TSMC))

Die-to-Wafer (D2W) Processing and Reliability for 3D Packaging of Advanced Node Logic 600

    Luke England (GLOBALFOUNDRIES), Daniel Fisher (GLOBALFOUNDRIES), Katie
    Rivera (GLOBALFOUNDRIES), Bill Guthrie (GLOBALFOUNDRIES), Ping-Jui Kuo
    (Advanced Semiconductor Engineering (ASE)), Chang-Chi Lee (Advanced
    Semiconductor Engineering), Che-Ming Hsu (Advanced Semiconductor
    Engineering), Fan-Yu Min (Advanced Semiconductor Engineering),
    Kuo-Chang Kang (Advanced Semiconductor Engineering), and Chen-Yuan
    Weng (Advanced Semiconductor Engineering)

Enabling Ultra-Thin Die to Wafer Hybrid Bonding for Future Heterogeneous Integrated Systems 607

    Alain Phommahaxay (IMEC), Samuel Suhard (IMEC), Pieter Bex (IMEC),
    Serena Iacovo (IMEC), John Slabbekoorn (IMEC), Fumihiro Inoue (IMEC),
    Lan Peng (IMEC), Koen Kennes (IMEC), Erik Sleeckx (IMEC), Gerald Beyer
    (IMEC), and Eric Beyne (IMEC)

The Thermal Dissipation Characteristics of The Novel System-In-Package Technology (ICE-SiP) for
Mobile and 3D High-end Packages 614

    Taejoo Hwang (Samsung Electronics Co., Ltd., Republic of Korea),
    Dan(Kyung Suk) Oh (Samsung Electronics Co., Ltd., Republic of Korea),
    Jaechoon Kim (Samsung Electronics Co., Ltd., Republic of Korea),
    Euseok Song (Samsung Electronics Co., Ltd., Republic of Korea), Taehun
    Kim (Samsung Electronics Co., Ltd., Republic of Korea), Kilsoo Kim
    (Samsung Electronics Co., Ltd., Republic of Korea), Joungphil Lee
    (Samsung Electronics Co., Ltd., Republic of Korea), and Taehwan Kim
    (Samsung Electronics Co., Ltd., Republic of Korea)

Session 14: Fine-Pitch Solderless Bonding

Fine-Pitch (≤10 μm) Direct Cu-Cu Interconnects Using In-Situ Formic Acid Vapor Treatment 620

    SivaChandra Jangam (University of California Los Angeles), Adeel Ahmed
    Bajwa (Kulicke & Soffa Industries Inc), Umesh Mogera (University of
    California Los Angeles), Pranav Ambhore (University of California Los
    Angeles), Tom Colosimo (Kulicke & Soffa Industries Inc), Bob Chylak
    (Kulicke & Soffa Industries Inc), and Subramanian Iyer (University of
    California Los Angeles)

Low Temperature Cu Interconnect with Chip to Wafer Hybrid Bonding 628

    Guilian Gao (Xperi Corporation, USA), Laura Mirkarimi (Xperi
    Corporation, USA), Thomas Workman (Xperi Corporation, USA), Gill
    Fountain (Xperi Corporation, USA), Jeremy Theil (Xperi Corporation,
    USA), Gabe Guevara (Xperi Corporation, USA), Ping Liu (Xperi
    Corporation, USA), Bongsub Lee (Xperi Corporation, USA), Pawel Mrozek
    (Xperi Corporation, USA), Michael Huynh (Xperi Corporation, USA),
    Catharina Rudolph (Fraunhofer Institute for Reliability and
    Micro-Integration, IZM – ASSID, Germany), Thomas Werner (Fraunhofer
    Institute for Reliability and Micro-Integration, IZM – ASSID,
    Germany), and Anke Hanisch (Fraunhofer Institute for Reliability and
    Micro-Integration, IZM – ASSID, Germany)

Cu Microstructure of High Density Cu Hybrid Bonding Interconnection 636

Seokho Kim (Samsung Electronics Co., Ltd., Korea), Pilkyu Kang
(Samsung Electronics Co., Ltd., Korea), Taeyeong Kim (Samsung
Electronics Co., Ltd., Korea), Kyuha Lee (Samsung Electronics Co.,
Ltd., Korea), Joohee Jang (Samsung Electronics Co., Ltd., Korea),
Kwangjin Moon (Samsung Electronics Co., Ltd., Korea), Hoonjoo Na
(Samsung Electronics Co., Ltd., Korea), Sangjin Hyun (Samsung
Electronics Co., Ltd., Korea), and Kihyun Hwang (Samsung Electronics
Co., Ltd., Korea)

Low-Resistance and high-Strength Copper Direct Bonding in no-Vacuum Ambient Using Highly
(111)-Oriented Nano-Twinned Copper 642

Jing Ye Juang (National Chiao Tung University, Taiwan), Kai Cheng Shie
(National Chiao Tung University, Taiwan), Po-Ning Hsu (National Chiao
Tung University, Taiwan), Yu Jin Li (National Chiao Tung University,
Taiwan), K N Tu (University of California at Los Angeles), and Chih
Chen (National Chiao Tung University, Taiwan)

Sub-10μm Pitch Hybrid Direct Bond Interconnect Development for Die-to-Die Hybridization 648

John P. Mudrick (Sandia National Laboratories, USA), Jonatan A.
Sierra-Suarez (Sandia National Laboratories, USA), Matthew B. Jordan
(Sandia National Laboratories, USA), T. A. Friedmann (Sandia National
Laboratories, USA), Robert Jarecki (Sandia National Laboratories,
USA), and M. David Henry (Sandia National Laboratories, USA)

Cu Pillar with Nanocopper Caps: The Next Interconnection Node Beyond Traditional Cu Pillar 655

Ramón A. Sosa (Georgia Institute of Technology - Packaging Research
Center), Kashyap Mohan (Georgia Institute of Technology - Packaging
Research Center), Luu Nguyen (Texas Instruments), Rao Tummala (Georgia
Institute of Technology - Packaging Research Center), Antonia Antoniou
(Georgia Institute of Technology), and Vanessa Smet (Georgia Institute
of Technology - Packaging Research Center)

Cu-Cu Bonding by Low-Temperature Sintering of Self-Healable Cu Nanoparticles 661

Junjie Li (Huazhong University of Science and Technology, China), Qi
Liang (Huazhong University of Science and Technology, China), Chen
Chen (Huazhong University of Science and Technology, China), Tielin
Shi (Huazhong University of Science and Technology, China), Guanglan
Liao (Huazhong University of Science and Technology, China), and
Zirong Tang (Huazhong University of Science and Technology, China)

Session 15: High-Bandwidth Packaging

Electrical Performance Limits of Fine Pitch Interconnects for Heterogeneous Integration 667

Ahmet C. Durgun (Assembly and Test Technology Development Intel
Corporation), Zhiguo Qian (Assembly and Test Technology Development
Intel Corporation), Kemal Aygun (Assembly and Test Technology
Development Intel Corporation), Ravi Mahajan (Assembly and Test
Technology Development Intel Corporation), Tim Tri Hoang (Programmable
Solutions Group Intel Corporation), and Sergey Yuryevich Shumarayev
(Programmable Solutions Group Intel Corporation)

A High-Bandwidth Fine-Pitch 2.57Tbps/mm In-package Communication Link Achieving 48fJ/bit/mm
Efficiency 674

Nicolas Pantano (IMEC, KULeuven), Geert Van der Plas (IMEC), Pieter
Bex (IMEC), Philip Nolmans (IMEC), Dimitrios Velenis (IMEC), Marian
Verhelst (KU Leuven), and Eric Beyne (IMEC)

A New SI-PI co-Simulation Approach for Efficient Consideration of Coupling Between PDN and SDN 682

Heesok Lee (Samsung Electronics, Co. Ltd.), Jisoo Hwang (Samsung
Electronics, Co. Ltd.), Hoi-jin Lee (Samsung Electronics, Co. Ltd.),
and Youngmin Shin (Samsung Electronics, Co. Ltd.)

Signal Integrity of Submicron InFO Heterogeneous Integration for High Performance Computing
Applications 688

Chuei-Tang Wang (Taiwan Semiconductor Manufacturing Company Ltd.),
Jeng-Shien Hsieh (Taiwan Semiconductor Manufacturing Company Ltd.),
Victor C. Y. Chang (Taiwan Semiconductor Manufacturing Company Ltd.),
Shih-Ya Huang (Taiwan Semiconductor Manufacturing Company Ltd.), T. Ko
(Taiwan Semiconductor Manufacturing Company Ltd.), Han-Ping Pu (Taiwan
Semiconductor Manufacturing Company Ltd.), and Douglas Yu (Taiwan
Semiconductor Manufacturing Company Ltd.)

28GHz Through Glass Via (TGV) Based Band Pass Filter Using Through Fused Silica Via (TFV) Technology . 695

Renuka Bowrothu (University of Florida), Seahee Hwangbo (University of
Florida), Todd Schumann (University of Florida), and Yong-Kyu Yoon
(University of Florida)

Innovative Packaging Solutions of 3D Double Side Molding with System in Package for IoT and 5G
Application 700

Mike Tsai (Siliconware Precision Industries Co., Ltd., Taiwan), Ryan
Chiu (Siliconware Precision Industries Co., Ltd., Taiwan), Dick Huang
(Siliconware Precision Industries Co., Ltd., Taiwan), Feng Kao
(Siliconware Precision Industries Co., Ltd., Taiwan), Eric He
(Siliconware Precision Industries Co., Ltd., Taiwan), J. Y. Chen
(Siliconware Precision Industries Co., Ltd., Taiwan), Simon Chen
(Siliconware Precision Industries Co., Ltd., Taiwan), Jensen Tsai
(Siliconware Precision Industries Co., Ltd., Taiwan), and Yu-Po Wang
(Siliconware Precision Industries Co., Ltd., Taiwan)

Enhancing Efficiency of Antenna-in-Package (AiP) by Through-Silicon-Interposer (TSI) with Embedded
Air Cavity and Polyimide Dielectric Micro-Substrate 707

Yunna Sun (Shanghai Jiao Tong University), Yunting Sun (Shanghai Jiao
Tong University), Jiangbo Luo (Shanghai Jiao Tong University), Huiying
Wang (Shanghai Jiao Tong University), Zhuoqing Yang (Shanghai Jiao
Tong University), Yan Wang (Shanghai Jiao Tong University), Guifu Ding
(Shanghai Jiao Tong University), and Kwangwoo Han (Samsung Electronics
Co.)

Session 16: Advanced Materials for High-Speed Electronics

Low-Loss Glass Substrates Formulated with a Variety of Dielectric Characteristics for
Millimeter-Wave Applications 712

Kazutaka Hayashi (AGC Inc.), Nobutaka Kidera (AGC Inc.), and Yoichiro
Sato (AGC Inc.)

Evaluation of Fine-Pitch Routing Capabilities of Advanced Dielectric Materials for High Speed
Panel-RDL in 2.5D Interposer and Fan-Out Packages 718
Shreya Dwarakanath (Georgia Institute of Technology), P. Markondeya
Raj (Florida International University), Amit Agarwal (Microchips,
USA), Daichi Okamoto (TAIYO INK MFG. CO., LTD. Japan), Atsushi Kubo
(Tokyo Ohka Kogyo Co., Ltd., Japan), Fuhan Liu (Georgia Institute of
Technology), Mohan Kathaperumal (Georgia Institute of Technology), and
Rao R. Tummala (Georgia Institute of Technology)

Attenuation of high Frequency Signals in Structured Metallization on Glass: Comparing Different
Metallization Techniques with 24 GHz, 77 GHz and 100 GHz Structures 726
Letz Martin (SCHOTT AG, Germany), Jost Matthias (TU Darmstadt),
Brandon T. Gore (Samtec, Colorado Springs), William J. Kozlovsky
(Samtec, Colorado Springs), Romeo Premerlani (Varioprint AG), Alex
Bruderer (Varioprint AG), Manuel Martina (Schweizer Electronic AG),
Thomas Gottwald (Schweizer Electronic AG), Tetsuya Onishi (Grand Joint
Technology Ltd.), Shigeo Onitake (KOTO Electric Co.), Siddharth
Ravichandran (Packaging Research Center), Holger Maune (Institute for
microwave engineering and photonics), and Mathias Mydlak (SCHOTT AG)

The Highly Effective EMI Shielding Materials for Electric and Magnetic Fields Over the Wide Range of
Frequency in Near-Field Region 733
Yoon-Hyun Kim (Ntrium Inc.), Kisu Joo (Ntrium Inc.), Kyu Jae Lee
(Ntrium Inc.), Jung Woo Hwang (Ntrium Inc.), Seung Jae Lee (Ntrium
Inc.), Se Young Jeong (Ntrium Inc.), and Hyun Ho Park (Ntrium Inc.)

Low Loss NCF Material for High Frequency Device 740
Kazutaka Honda (Hitachi Chemical Co., Ltd.), Keiko Ueno (Hitachi
Chemical Co., Ltd.), Tsuyoshi Ogawa (Hitachi Chemical Co., Ltd.), and
Toshihisa Nonaka (Hitachi Chemical Co., Ltd.)

In-Situ Redox Nanowelding of Copper Nanowires with Surficial Oxide Layer as Solder for Flexible
Transparent Electromagnetic Interference Shielding 746
Xianwen Liang (Chinese Academy of Sciences), Jianwen Zhou (Chinese
Academy of Sciences), Gang Li (Chinese Academy of Sciences), Tao Zhao
(Chinese Academy of Sciences), Pengli Zhu (Chinese Academy of
Sciences), Rong Sun (Chinese Academy of Sciences), and Ching-ping Wong
(Georgia Institute of Technology)

Compartmental EMI Shielding with Jet-Dispensed Material Technology 753
Xuan Hong (Henkel Corporation), Qizhuo Zhuo Zhuo (Henkel Corporation),
Xinpei Cao (Henkel Corporation), Dan Maslyk (Henkel Corporation), Noah
Ekstrom (Henkel Corporation), Juliet Sanchez (Henkel Corporation),
Selene Hernandez (Henkel Corporation), and Jinu Choi (Henkel
Corporation)

Session 17: Materials and Design for Reliability of Next-Generation Packages:

Highly (111)-Oriented Nanotwinned Cu for High Fatigue Resistance in Fan-Out Wafer-Level Packaging 758
Yu-Jin Li (National Chiao Tung University), Chih-Han Theng (National
Chiao Tung University), I-Hsin Tseng (National Chiao Tung University),
Chih Chen (National Chiao Tung University), Benson Lin (MediaTek Inc),
and Chia-Cheng Chang (MediaTek Inc)

WLCSP Package and PCB Design for Board Level Reliability 763

Jason Chiu (Taiwan Semiconductor Manufacturing Company, Ltd.), K.C. Chang (Taiwan Semiconductor Manufacturing Company, Ltd.), Steven Hsu (Taiwan Semiconductor Manufacturing Company, Ltd.), Pei-Haw Tsao (Taiwan Semiconductor Manufacturing Company, Ltd.), and M.J. Lii (Taiwan Semiconductor Manufacturing Company, Ltd.)

Assessing the Reliability of Highly Stretchable Interconnects for Flexible Hybrid Electronics 768

Rajesh Sharma Sivasubramony (Binghamton University), Ashwin Varkey Zachariah (Binghamton University), Mohammed Alhendi (Binghamton University), Manu Yadav (Binghamton University), Peter Borgesen (Binghamton University), Mark D. Poliks (Binghamton University), Nancy C. Stoffel (GE Global Research), David M. Shaddock (GE Global Research), and Liang Yin (GE Global Research)

The How and why of Biased Humidity Tests with Copper Wire 777

Amar Mavinkurve (NXP Semiconductors), R.T.H. Rongen (NXP Semiconductors), L. Goumans (NXP Semiconductors), M-L Farrugia (NXP Semiconductors), E. van Olst (NXP Semiconductors), Orla O'Halloran (NXP Semiconductors), and M. van Soestbergen (NXP Semiconductors)

Twist Testing for Flexible Electronics 785

Justin H. Chow (Georgia Institute of Technology), Jeffrey Meth (DuPont Electronics and Imaging), and Suresh K. Sitaraman (Georgia Institute of Technology)

Mechanical Properties and Microstructural Fatigue Damage Evolution in Cyclically Loaded Lead-Free Solder Joints 792

Sinan Su (Auburn University), Mohd Aminul Hoque (Auburn University), Md Mahmudur Chowdhury (Auburn University), Sa'd Hamasha (Auburn University), Jeffrey C. Suhling (Auburn University), John L. Evans (Auburn University), and Pradeep Lall (Auburn University)

Reliability Studies of Silicon Interconnect Fabric 800

Niloofar Shakoorzadeh (UCLA), Siva Chandra Jangam (UCLA), Kaysar Rahim (GlobalFoundries), Pranav Ambhore (UCLA), Han Chien (National Chiao Tung University), Amir Hanna (UCLA), and Subramanian S. Iyer (UCLA)

Session 18: Warpage and Material Performance

Improved Finite Element Modeling of Moisture Diffusion Considering Discontinuity at Material Interfaces in Electronic Packages 806
Lulu Ma (Lamar University), Rahul Joshi (AMD), Keith Keith Newman (AMD), and Xuejun Fan (Lamar University)

Study of Thermal Aging Behavior of Epoxy Molding Compound for Applications in Harsh Environments 811
Adwait Inamdar (Robert Bosch GmbH), Alexandru Prisacaru (Robert Bosch GmbH), Martin Fleischman (Robert Bosch GmbH), Erick Franieck (Robert Bosch GmbH), Przemyslaw Gromala (Robert Bosch GmbH), Agnes Veres (Robert Bosch Kft), Csaba Nemeth (Robert Bosch Kft), Yu-Hsiang Yang (University of Maryland), and Bongtae Han (University of Maryland)

xxiii

Warpage Variation Analysis and Model Prediction for Molded Packages 819
Yuling Niu (Qualcomm Technologies, Inc., USA), Wei Wang (Qualcomm
Technologies, Inc., USA), Zhijie Wang (Qualcomm Technologies, Inc.,
USA), Karthik Dhandapani (Qualcomm Technologies, Inc., USA), Mark
Schwarz (Qualcomm Technologies, Inc., USA), and Ahmer Syed (Qualcomm
Technologies, Inc., USA)

Peridynamics for Predicting Thermal Expansion Coefficient of Graphene 825
Erdogan Madenci (University of Arizona), Atila Barut (University of
Arizona), and Mehmet Dorduncu (University of Arizona)

Machine Learning Approach to Improve Accuracy of Warpage Simulations 834
Cheryl Selvanayagam (Singapore University of Technology and Design),
Pham Luu Trung Duong (Singapore University of Technology and Design),
Rathin Mandal (Advanced Micro Devices Inc.), and Nagarajan Raghavan
(Singapore University of Technology and Design)

Study on Warpage of Fan-Out Panel Level Packaging (FO-PLP) Using Gen-3 Panel 842
Fa Xing Che (Institute of Microelectronics A*STAR), Kazunori Yamamoto
(Institute of Microelectronics A*STAR), Vempati Srinivasa Rao
(Institute of Microelectronics A*STAR), and Vasarla Nagendra Sekhar
(Institute of Microelectronics A*STAR)

Mechanical Properties of Intermetallic Compounds at Elevated Temperature by Nanoindentation 850
Fan Yang (The Institute of Technological Sciences, Wuhan University,
Wuhan, China), Sheng Liu (Wuhan University), Zhaoxia Zhou
(Loughborough University), Zhiwen Chen (Wuhan University), Li Liu
(Wuhan University of Technology), Canyu Liu (Loughborough University),
and Changqing Liu (Loughborough University)

Session 19: MEMS, Sensors, and IoT

A MEMS Microphone in a FOWLP 855
Horst Theuss (Infineon Technologies AG), Christian Geissler (Infineon
Technologies AG), Franz-Xaver Muehlbauer (Infineon Technologies AG),
Claus von Waechter (Infineon Technologies AG), Thomas Kilger (Infineon
Technologies AG), Juergen Wagner (Infineon Technologies AG), Thomas
Fischer (Infineon Technologies AG), Ulf Bartl (Infineon Technologies
AG), Stephan Helbig (Infineon Technologies AG), Alfred Sigl (Infineon
Technologies AG), Dominic Maier (Infineon Technologies AG), Bernd
Goller (Infineon Technologies AG), Matthias Vobl (Infineon
Technologies AG), Matthias Herrmann (Infineon Technologies AG),
Johannes Lodermeyer (Infineon Technologies AG), Ulrich Krumbein
(Infineon Technologies AG), and Alfons Dehe (Hahn-Schickard)

Fan-Out Wafer Level Packaging - A Platform for Advanced Sensor Packaging 861
  Tanja Braun (Fraunhofer IZM), Karl-Friedrich Becker (Fraunhofer
  Institute for Reliability and Microintegration), Ole Hoelck
  (Fraunhofer Institute for Reliability and Microintegration), Steve
  Voges (Fraunhofer Institute for Reliability and Microintegration),
  Ruben Kahle (Fraunhofer Institute for Reliability and
  Microintegration), Pascal Graap (Fraunhofer Institute for Reliability
  and Microintegration), Markus Wöhrmann (Fraunhofer Institute for
  Reliability and Microintegration), R. Aschenbrenner (Fraunhofer
  Institute for Reliability and Microintegration), Tanja Braun
  (Technical University Berlin), Marc Dreissigacker (Technical
  University Berlin), Martin Schneider-Ramelow (Technical University
  Berlin), and Klaus-Dieter Lang (Technical University Berlin)

3D-MID Evaluation and Validation for Space Applications 868
  Etienne Hirt (Art of Technology AG), Klaus Ruzicka (Art of Technology
  AG), Benedikt Wigger (Hahn – Schickard, Mikromontage), Maximilian
  Barth (Hahn – Schickard), Rafat Saleh (Hahn – Schickard), Florian
  Janek (Hahn – Schickard), and Ernst Müller (University Stuttgart)

High-Temperature Pressure Sensor Package and Characterization of Thermal Stress in the Assembly up
to 500 °C 878
  Nilavazhagan Subbiah (University of Freiburg, Germany), Qingming Feng
  (University of Freiburg, IMTEK), Juergen Wilde (University of
  Freiburg, Germany), and Gudrun Bruckner (CTR AG, Austria)

Development of 3D WLCSP with Black Shielding for Optical Finger Print Sensor for the Application of
Full Screen Smart Phone 884
  Daquan Yu (Huantian Technology (Kunshan) Electronics Co., Ltd), Yichao
  Zou (Huantian Technology (Kunshan) Electronics Co., Ltd), Xirui Xu
  (Huantian Technology (Kunshan) Electronics Co., Ltd), Aihua Shi
  (Huantian Technology (Kunshan) Electronics Co., Ltd), Xiaobing Yang
  (Huantian Technology (Kunshan) Electronics Co., Ltd), and Zhiyi Xiao
  (Huantian Technology (Kunshan) Electronics Co., Ltd.)

Micro Fountain-Like Resonators 890
  Jianfeng Zhang (Southeast University, China), Jintang Shang (Southeast
  University, China), Bin Luo (Southeast University, China), and Zhaoxi
  Su (Southeast University, China)

Novel Additively Manufactured Packaging Approaches for 5G/mm-Wave Wireless Modules 896
  Tong-Hong Lin (Georgia Institute of Technology), Aline Eid (Georgia
  Institute of Technology), Jimmy Hester (Georgia Institute of
  Technology), Bijan Tehrani (Georgia Institute of Technology), Jo Bito
  (Texas Instrument), and Manos M. Tentzeris (Georgia Institute of
  Technology)

Session 20: Fanout and Heterogeneous Integration

Feasibility Study of Fan-Out Wafer-Level Packaging for Heterogeneous Integrations 903
John Lau (ASMPT), Ming Li (ASMPT), Iris Xu (Jiangyin Changdian
Advanced Packaging Co., Ltd.), Tony Chen (Jiangyin Changdian Advanced
Packaging Co., Ltd.), Kim Hwee Tan (Jiangyin Changdian Advanced
Packaging Co., Ltd.), Zhang Li (Jiangyin Changdian Advanced Packaging
Co., Ltd.), Nelson Fan (ASMPT), Eric Kuah (ASMPT), Raymond So (ASMPT),
Penny Lo (ASMPT), Y. M. Cheung (ASMPT), Cao Xi (Huawei Technologies
Co. Ltd.), Rozalia Beica (Dow Chemical Company), Sze Pei Lim (Indium
Corporation), NC Lee (Indium Corporation), Cheng-Ta Ko (Unimicron
Technology Corporation), Henry Yang (Unimicron Technology
Corporation), YH Chen (Unimicron Technology Corporation), Mian Tao
(Hong Kong University of Science and Technology), Jeffery Lo (Hong
Kong University of Science and Technology), and Ricky Lee (Hong Kong
University of Science and Technology)

Experiment of 22FDX(R) Chip Board Interaction (CBI) in Wafer Level Packaging Fan-Out (WLPFO) 910
Jae Kyu Cho (GLOBALFOUNDRIES, USA), Jens Paul (GLOBALFOUNDRIES,
Germany), Simone Capecchi (GLOBALFOUNDRIES, Germany), Frank
Kuechenmeister (GLOBALFOUNDRIES, Germany), and Ta-Chien Cheng
(GLOBALFOUNDRIES, Singapore)

FOWLP Design for Digital and RF Circuits 917
Teck Guan Lim (Institute of Microelectronics, Singapore), David Soon
Wee Ho (Institute of Microelectronics, Singapore), Eva Wai Leong Ching
(Institute of Microelectronics, Singapore), Zihao Chen (Institute of
Microelectronics, Singapore), and Surya Bhattacharya (Institute of
Microelectronics, Singapore)

Next Generation of 2-7 Micron Ultra-Small Microvias for 2.5D Panel Redistribution Layer by Using
Laser and Photolithography Technologies 924
Fuhan Liu (Georgia Institute of Technology), Chandrasekharan Nair
(Georgia Institute of Technology), Gaurav Khurana (Georgia Institute
of Technology), Atom Watanabe (Georgia Institute of Technology),
Bartlet H. DeProspo (Georgia Institute of Technology), Atsushi Kubo
(Tokyo Ohka Kogyo Co. Ltd., Japan), Cheng Ping Lin (Panasonic
Corporation, Japan), Toshiyuki Makita (Panasonic Corporation, Japan),
Naoki Watanabe (Panasonic Industrial Devices Sales Company of America,
USA), and Rao R. Tummala (Georgia Institute of Technology)

Multilayer RDL Interposer for Heterogeneous Device and Module Integration 931
Yi-Hang Lin (Taiwan Semiconductor Manufacturing Company Ltd.), M.C.
Yew (Taiwan Semiconductor Manufacturing Company Ltd.), S.M. Chen
(Taiwan Semiconductor Manufacturing Company Ltd.), M.S. Liu (Taiwan
Semiconductor Manufacturing Company Ltd.), Pravin Kavle (Taiwan
Semiconductor Manufacturing Company Ltd.), T.M. Lai (Taiwan
Semiconductor Manufacturing Company Ltd.), C.T. Yu (Taiwan
Semiconductor Manufacturing Company Ltd.), F.C. Hsu (Taiwan
Semiconductor Manufacturing Company Ltd.), C.S. Chen (Taiwan
Semiconductor Manufacturing Company Ltd.), T.J. Fang (Taiwan
Semiconductor Manufacturing Company Ltd.), C.K. Hsu (Taiwan
Semiconductor Manufacturing Company Ltd.), K.C. Lee (Taiwan
Semiconductor Manufacturing Company Ltd.), C.H. Lin (Taiwan
Semiconductor Manufacturing Company Ltd.), P.Y. Lin (Taiwan
Semiconductor Manufacturing Company Ltd.), and Shin-Puu Jeng (Taiwan
Semiconductor Manufacturing Company Ltd.)

Effects of Dielectric Curing Conditions on the Interfacial Adhesion of Cu RDL for Fan-Out Wafer Level Packaging 937

> Gahui Kim (Andong National University, Korea), Kirak Son (Andong
> National University, Korea), Dogeun Kim (Korea Institute of Materials
> Science, Korea), Seok-hyun Lee (SAMSUNG ELECTRONICS CO., LTD, Korea),
> and Young-Bae Park (Andong National University, Korea)

Al-Al Direct Bonding with Sub-μm Alignment Accuracy for Millimeter Wave SiGe BiCMOS Wafer Level Packaging and Heterogeneous Integration 942

> Matthias Wietstruck (IHP - Leibniz Institut für innovative
> Mikroelektronik), Sebastian Schulze (IHP - Leibniz Institut für
> innovative Mikroelektronik), Bernhard Rebhan (EV Group E. Thallner
> GmbH), Peter Kerepesi (EV Group E. Thallner GmbH), Helmut Kurz (EV
> Group E. Thallner GmbH), Gerald Silberer (EV Group E. Thallner GmbH),
> Josef Meiler (EV Group E. Thallner GmbH), Selin Tolunay Wipf (IHP -
> Leibniz Institut für innovative Mikroelektronik), Christian Wipf (IHP
> - Leibniz Institut für innovative Mikroelektronik), and Mehmet Kaynak
> (IHP - Leibniz Institut für innovative Mikroelektronik, Sabanci
> University)

Session 21: 5G, mm-Wave, and Antenna-in-Package

Vivaldi Antenna Array Fabricated Using a Hybrid Process 948

> Vincens Gjokaj (Michigan State University), Cameron Crump (Michigan
> State University), John Papapolymerou (Michigan State University),
> John Albrecht (Michigan State University), and Premjeet Chahal
> (Michigan State University)

Novel Multicore PCB and Substrate Solutions for Ultra Broadband Dual Polarized Antennas for 5G Millimeter Wave Covering 28GHz & 39GHz Range 954

> Trang Thai (Intel Corporation), Sidharth Dalmia (Intel Corporation),
> Josef Hagn (Intel Corporation), Pouya Talebbeydokhti (Intel
> Corporation), and Yossi Tsfati (Intel Corporation)

3D Glass Package-Integrated, High-Performance Power Dividing Networks for 5G Broadband Antennas 960

> Muhammad Ali (Georgia Institute of Technology), Atom Watanabe (Georgia
> Institute of Technology), Tong-Hong Lin (Georgia Institute of
> Technology), Markondeya Raj Pulugurtha (Florida International
> University), Manos M. Tentzeris (Georgia Institute of Technology), and
> Rao R. Tummala (Georgia Institute of Technology)

Advanced Wafer Level PKG Solutions for 60GHz WiGig (802.11ad) Telecom Infrastructure 968

> Dapeng Wu (Sivers IMA AB), Robin Dahlbäck (Sivers IMA AB), Erik
> Öjefors (Sivers IMA AB), Mats Carlsson (Sivers IMA AB), Francis Chee
> Peng Lim (STATS ChipPAC Pte. Ltd.), Yew Kheng Lim (STATS ChipPAC Pte.
> Ltd.), Aung Kyaw Oo (STATS ChipPAC Pte. Ltd.), Won Kyung Choi (STATS
> ChipPAC Pte. Ltd.), and Seung Wook Yoon (STATS ChipPAC Pte. Ltd.)

xxvii

Low-Loss Additively-Deposited Ultra-Short Copper-Paste Interconnections in 3D Antenna-Integrated
Packages for 5G and IoT Applications 972

*Atom O. Watanabe (Georgia Institute of Technology), Yiteng Wang
(Georgia Institute of Technology), Nobuo Ogura (Nagase & Co., LTD.,
Japan), P. Markondeya Raj (Florida International University), Vanessa
Smet (Georgia Institute of Technology), Manos M. Tentzeris (Georgia
Institute of Technology), and Rao R. Tummala (Georgia Institute of
Technology)*

Advanced Thin-Profile Fan-Out with Beamforming Verification for 5G Wideband Antenna 977

*Ricky Hsieh (Advanced Semiconductor Engineering (ASE), Inc., Taiwan),
Fu-Cheng Chu (Advanced Semiconductor Engineering (ASE), Inc., Taiwan),
Cheng-Yu Ho (Advanced Semiconductor Engineering (ASE), Inc., Taiwan),
and Chen-Chao Wang (Advanced Semiconductor Engineering (ASE), Inc.,
Taiwan)*

Integrated Compact Planar Inverted-F Antenna (PIFA) with a Shorting Via Wall for Millimeter-Wave
Wireless Chip-to-Chip (C2C) Communications in 3D-SiP 983

*Seahee Hwangbo (University of Florida), Renuka Bowrothu (University of
Florida), Hae-in Kim (University of Florida), and Yong-Kyu Yoon
(University of Florida)*

Session 22: Advanced Substrates and Interconnect Technology

Temporary SiC-SiC Wafer Bonding Compatible with High Temperature Annealing 989
*Fengwen Mu (The University of Tokyo), Tadatomo Suga (The University of
Tokyo), Miyuki Uomoto (Tohoku University), and Takehito Shimatsu
(Tohoku University)*

Ultrathin Glass to Ultrathin Glass Bonding Using Laser Sealing Approach 995
*Messaoud Bedjaoui (Univ. Grenoble Alpes, CEA-LETI, France), Johnny
Amiran (Univ. Grenoble Alpes, CEA-LETI, France), and Jean Brun (Univ.
Grenoble Alpes, CEA-LETI, France)*

Development of Resins for Bumpless Interconnects and Wafer-On-Wafer (WOW) Integration 1002
*Naoko Araki (Daicel Corporation), Shinji Maetani (Daicel Corporation),
Kim Young Suk (Disco Corporation), Shoichi Kodama (Disco Corporation),
and Takayuki Ohba (Tokyo Institute of Technology)*

Development of Novel Photosensitive Dielectric Material for Reliable 2.1D Package 1009
*Yune Kumazawa (Mitsubishi Gas Chemical Company, Inc.), Seiji Shika
(Mitsubishi Gas Chemical Company, Inc.), Shunsuke Katagiri (Mitsubishi
Gas Chemical Company, Inc.), Takuya Suzuki (Mitsubishi Gas Chemical
Company, Inc.), Tsuyoshi Kida (Mitsubishi Gas Chemical Company, Inc.),
and Shu Yoshida (Mitsubishi Gas Chemical Company, Inc.)*

High Reliability Solder Resist with Strong Adhesion and High Resolution for High Density Packaging 1015
*Sawako Shimada (TAIYO INK MFG. CO., LTD.), Kazuya Okada (TAIYO INK
MFG. CO., LTD.), Tomoya Kudo (TAIYO INK MFG. CO., LTD.), Chiho Ueta
(TAIYO INK MFG. CO., LTD.), and Yuya Suzuki (TAIYO INK MFG. CO., LTD.)*

Method for Mitigating the Warpage of Ultra-Thin FC-CSPs by Controlling of EMC Properties 1022
*Chika Arayama (Panasonic Corporation), Takahiro Akashi (Panasonic
Corporation), Yasunari Tomita (Panasonic Corporation), and Naoki
Kanagawa (Panasonic Corporation)*

Innovative Socketable and Surface-Mountable BGA Interconnections 1028

Omkar Gupte (Georgia Institute of Technology - Packaging Research Center), Kristie Teoh (Georgia Institute of Technology - Packaging Research Center), Rao Tummala (Georgia Institute of Technology - Packaging Research Center), Gregorio Murtagian (Intel Corporation), and Vanessa Smet (Georgia Institute of Technology - Packaging Research Center)

Session 23: High-Bandwidth 3D and Photonic Integration

A Highly Reliable 1.4μm Pitch Via-Last TSV Module for Wafer-to-Wafer Hybrid Bonded 3D-SOC Systems . 1035

Stefaan Van Huylenbroeck (IMEC), Joeri De Vos (IMEC), Zaid El-Mekki (IMEC), Geraldine Jamieson (IMEC), Nina Tutunjyan (IMEC), Karthik Muga (IMEC), Michele Stucchi (IMEC), Andy Miller (IMEC), Gerald Beyer (IMEC), and Eric Beyne (IMEC)

Nanoscale Topography Characterization for Direct Bond Interconnect 1041

Bongsub Lee (Xperi Corporation), Pawel Mrozek (Xperi Corporation), Gill Fountain (Xperi Corporation), John Posthill (Xperi Corporation), Jeremy Theil (Xperi Corporation), Guilian Gao (Xperi Corporation), Rajesh Katkar (Xperi Corporation), and Laura Mirkarimi (Xperi Corporation, USA)

Fully-Filled, Highly-Reliable Fine-Pitch Interposers with TSV Aspect Ratio >10 for Future 3D-LSI/IC Packaging 1047

Murugesan Murugesan (Tohoku University, Japan), Takafumi Fukushima (Tohoku University, Japan), Kiyoharu Mori (T-Miro, Japan), Ai Nakamura (T-Micro, Japan), Yisang Lee (T-Micro, Japan), Makoto Motoyoshi (T-Micro, Japan), J.C Bea (Tohoku University), Shigeru Watariguchi (Meltex, Japan), and Mitsumasa Koyanagi (Tohoku University, Japan)

3D Silicon Photonics Interposer for Tb/s Optical Interconnects in Data Centers with Double-Side Assembled Active Components and Integrated Optical and Electrical Through Silicon Via on SOI 1052

Bogdan Sirbu (Fraunhofer Institute for Reliability and Microintegration IZM Berlin), Yann Eichhammer (Fraunhofer Institute for Reliability and Microintegration IZM Berlin), Hermann Oppermann (Fraunhofer Institute for Reliability and Microintegration IZM Berlin), Tolga Tekin (Fraunhofer Institute for Reliability and Microintegration IZM Berlin), Jochen Kraft (ams AG, Austria), Victor Sidorov (ams AG, Austria), Xin Yin (IMEC, Belgium), Johan Bauwelinck (IMEC, Belgium), Christian Neumeyr (Vertilas GmbH, Germany), and Francisco Soares (Fraunhofer Heinrich-Hertz-Institut HHI Berlin)

Flip-Chip III-V-to-Silicon Photonics Interfaces for Optical Sensor 1060
Yves Martin (IBM T. J. Watson Research Center, Yorktown Heights USA),
Jason S. Orcutt (IBM T. J. Watson Research Center, Yorktown Heights
USA), Chi Xiong (IBM T. J. Watson Research Center, Yorktown Heights
USA), Laurent Schares (IBM T. J. Watson Research Center, Yorktown
Heights USA), Tymon Barwicz (IBM T. J. Watson Research Center,
Yorktown Heights USA), Martin Glodde (IBM T. J. Watson Research
Center, Yorktown Heights USA), Swetha Kamlapurkar (IBM T. J. Watson
Research Center, Yorktown Heights USA), Eric J. Zhang (IBM T. J.
Watson Research Center, Yorktown Heights USA), William M.J. Green (IBM
T. J. Watson Research Center, Yorktown Heights USA), Victor
Dolores-Calzadilla (Fraunhofer Heinrich-Hertz Institute, Germany),
Ariane Sigmund (Fraunhofer Heinrich-Hertz Institute, Germany), and
Martin Moehrle (Fraunhofer Heinrich-Hertz Institute, Germany)

Extremely Low-Profile Single Mode Fiber Array Coupler Suitable for Silicon Photonics 1067
Mitsuharu Hirano (Sumitomo Electric Industries, Ltd., Japan), Akira
Furuya (Sumitomo Electric Industries, Ltd., Japan), Hideki Machida
(Sumitomo Electric Industries, Ltd., Japan), Koichi Koyama (Sumitomo
Electric Industries, Ltd., Japan), Yasunori Murakami (Sumitomo
Electric Industries, Ltd., Japan), and Kazunori Tanaka (Sumitomo
Electric Industries, Ltd., Japan)

Micro Lens Array Assembly for Optical Organic Substrate 1074
Patrick Jacques (IBM Bromont, Canada), Richard Langlois (IBM Bromont,
Canada), Élaine Cyr (IBM Bromont, Canada), Alexander Janta-Polczynski
(IBM Bromont, Canada), Paul Fortier (IBM Bromont, Canada), Koji Masuda
(IBM Tokyo Research, Japan), Masao Tokunari (IBM Tokyo Research), and
Hsiang-Han Hsu (IBM Tokyo Research)

Session 24: Advancements in Solder Joint Characterization and Reliability Evaluation

Effects of In and Zn Double Addition on Eutectic Sn-58Bi Alloy 1081
Shiqi Zhou (Osaka University, Japan), Yu-An Shen (Osaka University,
Japan), Tiffani Uresti (Texas A&M University at Qatar, Qatar), Vasanth
Shunmugasamy (Texas A&M University at Qatar, Qatar), Bilal Mansoor
(Texas A&M University at Qatar, Qatar), and Hiroshi Nishikawa (Osaka
University, Japan)

Microstructural Evolution in SAC+X Solders Subjected to Aging 1087
Jing Wu (Auburn University), Jeffrey C. Suhling (Auburn University),
and Pradeep Lall (Auburn University)

Microstructure Signature Evolution in Solder Joints, Solder Bumps, and Micro-Bumps Interconnection
in A Large 2.5D FCBGA Package During Thermo-Mechanical Cycling 1099
Arman Ahari (Portland State University), Andy Hsiao (Portland State
University), Greg Baty (Portland State University), Peng Su (Juniper
Networks, USA), and Tae-Kyu Lee (Portland State University)

Long-Term Reliability of Solder Joints in 3D ICs Under Near-Application Conditions 1106
Omar Ahmed (University of Central Florida), Golareh Jalilvand
(University of Central Florida), Hector Fernandez (University of
Central Florida), Peng Su (Juniper Networks), Tae-Kyu Lee (Portland
State University), and Tengfei Jiang (University of Central Florida)

Experimental Investigation of the Correlation between a Load-Based Metric and Solder Joint
Reliability of BGA Assemblies on System Level 1113
Fabian Schempp (Robert Bosch GmbH, University of Freiburg - IMTEK),
Marc Dressler (Robert Bosch GmbH), Daniel Kraetschmer (Robert Bosch
GmbH), Friederike Loerke (Robert Bosch GmbH), and Juergen Wilde
(University of Freiburg - IMTEK)

Fatigue Life Prediction Model Development for Decoupling Capacitors 1121
Krishna Tunga (IBM Corporation, USA), Joseph Ross (IBM Corporation,
USA), Kamal Sikka (IBM Corporation, USA), and Bakul Parikh (IBM
Corporation, USA)

A Study of Substrate Models and Its Effect On Package Warpage Prediction 1130
Van-Lai Pham (Binghamton University), Huayan Wang (Binghamton
University), Jiefeng Xu (Binghamton University), Jing Wang (Binghamton
University), Chrandeep Singh (Corning Inc.), and Seungbae Park
(Binghamton University)

Session 25: Wafer Level Packaging and Fan-In/Fan-Out Structures & Materials

3D Fan-Out Package Technology with Photosensitive Through Mold Interconnects 1140
Kentaro Mori (Toshiba Electronic Devices and Storage Corporation),
Soichi Yamashita (Toshiba Electronic Devices and Storage Corporation),
Takafumi Fukuda (Toshiba Development & Engineering Corporation),
Masahiro Sekiguchi (Toshiba Electronic Devices and Storage
Corporation), Hirokazu Ezawa (Toshiba Memory Corporation), and Shuzo
Akejima (Toshiba Electronic Devices and Storage Corporation)

Effects of the Materials Properties of Epoxy Molding Films (EMFs) on Fan-Out Packages (FOPs)
Characteristics 1146
Sangmyung Shin (KAIST), Hanmin Lee (KAIST), JunMo Kim (KAIST), Tae-Ik
Lee (KAIST), Taek-Soo Kim (KAIST), Youjin Kyung (LG Chem), Minsu Jeong
(LG Chem), Kwangjoo Lee (LG Chem), and Kyung-Wook Paik (KAIST)

Mechanism of Moldable Underfill (MUF) Process for RDL-1st Fan-Out Panel Level Packaging (FOPLP) ... 1152
Lin Bu (Institute of Microelectronics A*STAR), F. X. Che (Institute of
Microelectronics A*STAR), Vempati Srinivasa Rao (Institute of
Microelectronics A*STAR), and Xiaowu Zhang (Institute of
Microelectronics A*STAR)

Study of the Board Level Reliability Performance of a Large 0.3 mm Pitch Wafer Level Package 1159
Bernd Waidhas (Intel Deutschland GmbH), Jan Proschwitz (Intel
Deutschland GmbH), Christoph Pietryga (Intel Deutschland GmbH), Thomas
Wagner (Intel Deutschland GmbH), and Beth Keser (Intel Deutschland
GmbH)

Study of Board Level Reliability of eWLB (embedded Wafer Level BGA) for 0.35mm Ball Pitch 1165
Kang Hai Lee (STATS ChipPAC Pte. Ltd.), Yeow Kheng Lim (STATS ChipPAC
Pte. Ltd.), Seng Guan Chow (STATS ChipPAC Pte. Ltd.), Kang Chen (STATS
ChipPAC Pte. Ltd.), Won Kyung Choi (STATS ChipPAC Pte. Ltd.), Seung
Wook Yoon (STATS ChipPAC LTD PTE), NW Liu (Advanced Package
Technology, Mediatek Inc.), Yenyao Chi (Advanced Package Technology,
Mediatek Inc.), and Benson Lin (Advanced Package Technology, Mediatek
Inc.)

Board Level Reliability Study of Fan-Out Single Die Package with 350um Bump Pitch 1170
Chieh-Lung Lai (Siliconware Precision Industries Co., Ltd.), Gu-Yan
Lin (Siliconware Precision Industries Co., Ltd.), Tz-Yuan Chao
(Siliconware Precision Industries Co., Ltd.), Yih-Sin Chen
(Siliconware Precision Industries Co., Ltd.), and Feng-Lung Chien
(Siliconware Precision Industries Co., Ltd.)

The Analysis for Bump Resistance Improvement by Optimizing the Sputter Condition 1175
Ming-Sin Su (Taiwan Semiconductor Manufacturing Company Ltd.),
Chang-Ning Wang (Taiwan Semiconductor Manufacturing Company Ltd.),
Clair Tsai (Taiwan Semiconductor Manufacturing Company Ltd.), T. L.
Yang (Taiwan Semiconductor Manufacturing Company Ltd.), Rolance Yang
(Taiwan Semiconductor Manufacturing Company Ltd.), W. C. Wu (Taiwan
Semiconductor Manufacturing Company Ltd.), C. S. Liu (Taiwan
Semiconductor Manufacturing Company Ltd.), J. M. Chiu (Taiwan
Semiconductor Manufacturing Company Ltd.), Y. F. Chen (Taiwan
Semiconductor Manufacturing Company Ltd.), Ponder Pang (Taiwan
Semiconductor Manufacturing Company Ltd.), Harry Ku (Taiwan
Semiconductor Manufacturing Company Ltd.), Kirin Wang (Taiwan
Semiconductor Manufacturing Company Ltd.), C.H. Su (Taiwan
Semiconductor Manufacturing Company Ltd.), Steven Hsu (Taiwan
Semiconductor Manufacturing Company Ltd.), Calvin Lu (Taiwan
Semiconductor Manufacturing Company Ltd.), K. C. Liu (Taiwan
Semiconductor Manufacturing Company Ltd.), and Marvin Liao (Taiwan
Semiconductor Manufacturing Company Ltd.)

Session 26: High-Speed Signaling for High-Performance Computing and Memory

Hybrid Prepreg Conventional Build-Up Laminate for 112Gbit/s SerDes 1179
Kwang Won Choi (GLOBALFOUNDRIES US Inc.), Edmund Blackshear
(GLOBALFOUNDRIES US Inc.), Eric Tremble (GLOBALFOUNDRIES US Inc.),
David Stone (GLOBALFOUNDRIES US Inc.), Jean Audet (IBM Corporation,
Canada), and Keiichi Hirabayashi (Shinko Electric Industries Co.,
LTD., Japan)

PI/SI Analysis and Design Approach for HPC Platform Applications 1188
Sungwook Moon (Samsung Electronics Co. Ltd.), Chanmin Jo (Samsung
Electronics Co. Ltd.), and Seungki Nam (Samsung Electronics Co. Ltd.)

PoP LPDDR5 (6.4 Gbps) NTODT and 1-Tap DFE for Signal Integrity Enhancement 1194
Sunil Gupta (Qualcomm Technologies, Inc.)

OpenCAPI Memory Interface Signal Integrity Study for High-Speed DDR5 Differential DIMM Channel with Standard Loss FR-4 Material and SNIA SFF-TA-1002 Connector 1200
>Biao Cai (IBM), Jose Hejase (IBM), Kyle Giesen (IBM), Junyan Tang
(IBM), Brian Connolly (IBM), KyuHyoun Kim (IBM), Daniel Dreps (IBM),
Zhineng Fan (Amphenol ICC), Rocky Huang (Amphenol ICC), Luyun Yi
(Amphenol ICC), Qiaoli Chen (Amphenol ICC), Yifan Huang (Amphenol
ICC), and Stephen Smith (Amphenol ICC)

Effectiveness of Equalization and Performance Potential in DDR5 Channels with RDIMM(s) 1208
>Nanju Na (Xilinx) and Hing "Thomas" To (Xilinx)

Inductive Links for 3D Stacked Chip-to-Chip Communication 1215
>Xiao Sun (IMEC, Belgium), Nicolas Pantano (IMEC, Belgium), Kim
Soon-Wook (IMEC, Belgium), Geert Van der Plas (IMEC, Belgium), and
Eric Beyne (IMEC, Belgium)

System Co-design of a 600V GaN FET Power Stage with Integrated Driver in a QFN System-in-Package
(QFN-SiP) 1221
>Jie Chen (Texas Instruments, Inc), Yong Xie (Texas Instruments, Inc),
Django Trombley (Texas Instruments Incorporated), and Rajen Murugan
(Texas Instruments Incorporated)

Session 27: Advanced Biosensors and Bioelectronics

Flexible Probe for Electrical Neural Signal Recording 1227
>Sajay Bhuvanendran Nair Gourikutty (Institute of Microelectronics,
A*STAR, Singapore) and Ruiqi Lim (Institute of Microelectronics,
A*STAR, Singapore)

Stretchable, Implantable Nanomembrane Biosensor for Wireless, Real-Time Monitoring of Hemodynamics .. 1233
>Robert Herbert (Georgia Institute of Technology) and Woon-Hong Yeo
(Georgia Institute of Technology)

A Wearable Passive pH Sensor for Health Monitoring 1240
>Saikat Mondal (Michigan State University), Saranraj Karuppuswami
(Michigan State University), Rachel Steinhorst (Michigan State
University), and Premjeet Chahal (Michigan State University)

Novel Packaging Structure and Processes for Micro-TFB (Thin Film Battery) to Enable Miniaturized
Healthcare Internet-of-Things (IoT) Devices 1246
>Bing Dang (IBM Research), Qianwen Chen (IBM Research), Leanna Pancoast
(IBM Research), Yu Luo (IBM Research), Hongqing Zhang (IBM Systems),
Jae-woong Nah (IBM Research), John Knickerbocker (IBM Research), Andy
Shih (Front Edge Technologies Inc.), Po Wen Cheng (Front Edge
Technologies Inc.), Kai Liu (Front Edge Technologies Inc.), Mengnian
Niu (Front Edge Technologies Inc.), and Simon Nieh (Front Edge
Technologies Inc.)

Screen Printed Temporary Tattoos for Skin-Mounted Electronics 1252
>Samuli Tuominen (Tampere University) and Matti Mantysalo (Tampere
University)

Thermoset Polymers for Bioelectronic Interfaces - Engineering of Thermomechanical Properties 1258

Adriana Carolina Duran-Martinez (The University of Texas at Dallas),
Seyedmahmoud Hosseini (The University of Texas at Dallas), Daniel Del
Nero (The University of Texas at Dallas), Alexandra Joshi-Imre (The
University of Texas at Dallas), Walter E. Voit (The University of
Texas at Dallas), and Melanie Ecker (The University of Texas at
Dallas)

Direct Heterogeneous Bonding of SiC to Si, SiO2, and Glass for High-Performance Power Electronics
and Bio-MEMS 1266

Jikai Xu (Harbin Institute of Technology), Chenxi Wang (Harbin
Institute of Technology), Qiushi Kang (Harbin Institute of
Technology), Shicheng Zhou (Harbin Institute of Technology), and
Yanhong Tian (Harbin Institute of Technology)

Session 28: Embedded and Integrated Technologies

Development of Flexible Hybrid Electronics Using Reflow Assembly with Stretchable Film 1272

Weifeng Liu (Flex), William Uy (Flex), Alex Chan (Flex), Dongkai
Shangguan (Flex), Andy Behr (Panasonic), Takatoshi Abe (Panasonic),
and Fukao Tomohiro (Panasonic)

Highly Compact RF Transceiver Module Using High Resistive Silicon Interposer with Embedded Inductors
and Heterogeneous Dies Integration 1279

G. Pares (Univ. Grenoble Alpes, CEA), Michel Jean-Philippe (CEA),
Deschaseaux Edouard (CEA), Ferris Pierre (CEA), Serhan Ayssar (CEA),
and Giry Alexandre (CEA)

Process Induced Wafer Warpage Optimization for Multi-chip Integration on Wafer Level Molded Wafer 1287

Chen-Yu Huang (Siliconware Precision Industries Co., Ltd, Taiwan),
Daniel Ng (Siliconware Precision Industries Co., Ltd, Taiwan), Hung-Ho
Lee (Siliconware Precision Industries Co., Ltd, Taiwan), Vito Lin
(Siliconware Precision Industries Co., Ltd, Taiwan), Chang-Fu Lin
(Siliconware Precision Industries Co., Ltd, Taiwan), and C. Key Chung
(Siliconware Precision Industries Co., Ltd, Taiwan)

Improved Structure for Package Substrates with Embedded Thin-Film Capacitor 1294

Tomoyuki Akahoshi (Fujitsu Laboratories Ltd.), Daisuke Mizutani
(Fujitsu Laboratories Ltd.), Kei Fukui (Fujitsu Interconnect
Technologies Ltd.), Seigo Yamawaki (Fujitsu Interconnect Technologies
Ltd.), Hidehiko Fujisaki (Fujitsu Interconnect Technologies Ltd.),
Manabu Watanabe (Fujitsu Advanced Technologies Ltd.), and Masateru
Koide (Fujitsu Advanced Technologies Ltd.)

3D Packaging with Embedded High-Power-Density Passives for Integrated Voltage Regulators 1300

Teng Sun (Georgia Institute of Technology), Robert G. Spurney (Georgia
Institute of Technology), Atom Watanabe (Georgia Institute of
Technology), P. Raj Pulugurtha (Florida International University),
Himani Sharma (Georgia Institute of Technology), Rao Tummala (Georgia
Institute of Technology), and Furukawa Yoshihiro (Nitto Denko
Corporation)

A Novel Panel Level Double Side Embedded Package for Small Size Power Devices 1306

Kunpeng Ding (Shenzhen Siptory Technologies Co., Ltd, China), Zhichao
Wu (Institute of Microelectronics, Tsinghua University, China), Mian
Huang (Shenzhen Siptory Technologies Co., Ltd, China), Bowei Zhang
(Wuxi Sky Chip Interconnection Technology Co., Ltd, China), and Jian
Cai (Institute of Microelectronics, Tsinghua University, China)

Chiplet Micro-Assembly Printer 1312

Bradley B. Rupp (PARC), Anne Plochowietz (PARC), Lara S. Crawford
(PARC), Matthew Shreve (PARC), Sourobh Raychaudhuri (PARC), Sergey
Butylkov (PARC), Yunda Wang (PARC), Ping Mei (PARC), Qian Wang (PARC),
Jamie Kalb (PARC), Yu Wang (PARC), Eugene M. Chow (PARC), and JengPing
Lu (PARC)

Session 29: Electromigration and Innovative Reliability Test Methods

Effect of Intermetallic Compound Growth on Electromigration Failure Mechanism in Low-Profile Solder
Joints 1316

Hossein Madanipour (University of Texas at Arlington), Yi-Ram Kim
(University of Texas at Arlington), Choong-Un Kim (University of Texas
at Arlington), Ninad Shahane (Texas Instruments, Inc.), Dibyajat
Mishra (Texas Instruments, Inc.), and Luu Nguyen (Texas Instruments,
Inc.)

Effect of Grain Orientation and Microstructure Evolution on Electromigration in Flip-Chip Solder
Joint 1324

Xing Fu (Science and technology on reliability physics and application
of electronic component laboratory, China), Bin Zhou (Science and
technology on reliability physics and application of electronic
component laboratory), Ruohe Yao (University of Technology), Yunfei En
(Science and technology on reliability physics and application of
electronic component laboratory), and Si Chen (Science and technology
on reliability physics and application of electronic component
laboratory)

High Electromigration Lifetimes of Nanotwinned Cu Redistribution Lines 1328

I-Hsin Tseng (National Chiao Tung University), Yu-Jin Li (National
Chiao Tung university), Benson Lin (MediaTek Inc), Chia-Cheng Chang
(MediaTek Inc), and Chih Chen (National Chiao Tung University)

Non-destructive Failure Analysis of Various Chip to Package Interaction Anomalies in FCBGA Packages
Subjected to Temperature Cycle Reliability Testing 1333

Vishnu V. B. Reddy (Georgia Institute of Technology), I. Charles Ume
(Georgia Institute of Technology), Jaimal Williamson (Texas
Instruments Inc., USA), and Luu Nguyen (Texas Instruments Inc., USA)

Assessment of Accelerometer Versus LASER for Board Level Vibration Measurements 1339

Varun Thukral (NXP Semiconductors), M. Cahu (NXP Semiconductors),
J.J.M. Zaal (NXP Semiconductors), J. Jalink (NXP Semiconductors), R.
Roucou (NXP Semiconductors), and R.T.H. Rongen (NXP Semiconductors)

Effect of Process Parameters on the Long-Run Print Consistency and Material Properties of Additively Printed Electronics 1347

Pradeep Lall (Auburn University), Nakul Kothari (Auburn University),
Amrit Abrol (Auburn University), Jeff Suhling (Auburn University),
Sudan Ahmed (Auburn University), Ben Leever (US AFRL), and Scott
Miller (NextFlex)

A Viscoplastic-Based Fatigue Reliability Model for the Polyimide Dielectric Thin Film 1359

Yu-Chen Chang (National Cheng Kung University), Tz-Cheng Chiu
(National Cheng Kung University), Yu-Ting Yang (Advanced Semiconductor
Engineering Group, Inc.), Yi-Hsiu Tseng (Advanced Semiconductor
Engineering Group, Inc.), and Xi-Hong Chen (Advanced Semiconductor
Engineering Group, Inc.)

Session 30: Assembly and Process Modeling

Explicit FE Failure Prediction of Interfaces and Interconnect in Potted Electronics Assemblies
Subject to High-g Acceleration Loads 1366

Pradeep Lall (Auburn University), Kalyan Dornala (Auburn University),
Ryan Lowe (ARA Associates), and John Deep (US AFRL)

Numerical Simulation on the Formation Process of Metal Droplets by Pneumatic Diaphragm Drop-on
Demand Technology 1377

Kun Ma (Wuhan University), Sheng Liu (Wuhan University), Zhiwen Chen
(Wuhan University), Li Liu (Wuhan University of Technology), Hao Zheng
(China Ship Development and Design Center), and Yao Zhang (China Ship
Development and Design Center)

On Curing-Induced Residual Stresses After Molding Processes: Mold Shrinkage, Chemical Shrinkage or
Both? 1382

Changsu Kim (University of Maryland), Sukrut Phansalkar (University of
Maryland), Hyun-Seop Lee (University of Maryland), and Bongtae Han
(University of Maryland)

Realistic Solder Joint Geometry Integration with Finite Element Analysis for Reliability Evaluation
of Printed Circuit Board Assembly 1387

Chun Sean Lau (Western Digital Corporation), Ning Ye (Western Digital
Corporation), and Hem Takiar (Western Digital Corporation)

Multi-physics Modelling and Experimental Investigation – An Original Approach for
Laser-Dicing/Grooving Process Optimization 1396

Jeff Moussodji Moussodji (3IT-UdeS/C2MI/IBM), Oswaldo Chacon (IBM
Canada Ltd), Francis Santerre (IBM Canada Ltd), and Dominique Drouin
(3IT-Université de Sherbrooke)

Thermal Characteristics of Vertically-Integrated GaN/SiC-on-Si Assemblies: A Comparative Study 1405

Kimmo Rasilainen (Chalmers University of Technology, Sweden), Per
Ingelhag (Ericsson AB, Sweden), Peter Melin (Ericsson AB, Sweden),
Torbjörn M. J. Nilsson (Saab AB, Sweden), Mattias Thorsell (Chalmers
University of Technology, Sweden and Saab AB, Sweden), and Christian
Fager (Chalmers University of Technology, Sweden)

Comprehensive Investigation on Warpage Management of FOPLP with Multi Embedded Ring Designs 1413

> Chang-Chun Lee (National Tsing Hua University), Yan-Yu Liou (National
> Tsing Hua University), Pei-Chen Huang (National Tsing Hua University),
> Fussen Hsu (Unimicron Technology Corporation), Puru Bruce Lin
> (Unimicron Technology Corporation), Cheng-Ta Ko (Unimicron Technology
> Corporation), and Yu-Hua Chen (Unimicron Technology Corporation)

Session 31: Automotive and Power Packaging

Development of High Power and High Junction Temperature SiC Based Power Packages 1419

> Gongyue Tang (Institute of Microelectronics, A*STAR), Leong Ching Wai
> (Institute of Microelectronics, A*STAR), Teck Guan Lim (Institute of
> Microelectronics, A*STAR), Yong Liang Ye (Institute of
> Microelectronics, A*STAR), Pal Singh Ravinder (Institute of
> Microelectronics, A*STAR), Lin Bu (Institute of Microelectronics,
> A*STAR), Boon Long Lau (Institute of Microelectronics, A*STAR), Tai
> Chong Chai (Institute of Microelectronics, A*STAR), Kazunori Yamamoto
> (Institute of Microelectronics, A*STAR), and Xiaowu Zhang (Institute
> of Microelectronics, A*STAR)

New Developments of Copper Plating Technology for Embedded Power Chip Packages Challenges 1426

> Yung-Da Chiu (Advanced Semiconductor Engineering (ASE) Inc.),
> Shiu-Chih Wang (Advanced Semiconductor Engineering (ASE) Inc.), David
> Tarng (Advanced Semiconductor Engineering (ASE) Inc.), An-Tai Wu
> (Advanced Semiconductor Engineering (ASE) Inc.), Allenyl Chen
> (Advanced Semiconductor Engineering (ASE) Inc.), Louis Chen (Advanced
> Semiconductor Engineering (ASE) Inc.), and Chi-Tsung Chiu (Advanced
> Semiconductor Engineering (ASE) Inc.)

Innovative Flip Chip Package Solutions for Automotive Applications 1432

> Tom Tang (Siliconware Precision Industries Co., Ltd. Taiwan), Bo-Siang
> Fang (Siliconware Precision Industries Co., Ltd. Taiwan), David Ho
> (Siliconware Precision Industries Co., Ltd. Taiwan), B.H. Ma
> (Siliconware Precision Industries Co., Ltd. Taiwan), Jensen Tsai
> (Siliconware Precision Industries Co., Ltd. Taiwan), and Yu-Po Wang
> (Siliconware Precision Industries Co., Ltd. Taiwan)

Reliability of Laminated Bond Structure Using (Cu, Ni)/Sn TLP Bonding with Al Interlayer for High
Temperature Power Electronics Packaging 1437

> Yanghe Liu (Toyota Research Institute of North America), Shailesh N.
> Joshi (Toyota Research Institute North America), and Ercan M. Dede
> (Toyota Research Institute North America)

Silver Sintering on Organic Substrates for the Embedding of Power Semiconductor Devices 1443

> Alexander Schiffmacher (IMTEK University of Freiburg, Germany),
> Juergen Wilde (IMTEK University of Freiburg, Germany), Lorenz
> Litzenberger (IMTEK University of Freiburg, Germany), Till Huesgen
> (University of Applied Science Kempten, Germany), and Vladimir
> Polezhaev (University of Applied Science Kempten, Germany)

High Temperature Resistant Packaging Technology for SiC Power Module by Using Ni Micro-Plating
Bonding 1451

Kohei Tatsumi (Waseda University), Isamu Morisako (Waseda University),
Keiko Wada (Waseda University), Minoru Fukuomori (Waseda University),
Tomonori Iizuka (Waseda University), Nobuaki Sato (Mitsui High-tec
Inc.), Koji Shimizu (Mitsui High-tec Inc.), Kazutoshi Ueda (Mitsui
High-tec Inc.), Masayuki Hikita (Kyushu Institute of Technology),
Rikiya Kamimura (Kitakyushu Foundation for the Advancement of
Industry, Science and Technology), Naoki Kawanabe (WALTS Co., LTD.),
Kazuhiko Sugiura (DENSO Corporation), Kazuhiro Tsuruta (DENSO
Corporation), and Keiji Toda (TOYOTA Motor Corporation)

Pb-Free, High Thermal and Electrical Performance Driven Die Attach Material Development for Power
Packages 1457

Kim Byong Jin (AMKOR), Dong Su Ryu (AMKOR), HyeongIl Jeon (AMKOR),
Muhammad Hadhari Hazellah (AMKOR), Weng Tuck Chim (AMKOR), and Jin
Young Khim (AMKOR)

Session 32: Power and Panel Assembly

An RDL-First Fan-Out Panel-Level Package for Heterogeneous Integration Applications 1463
Yu-Min Lin (Industrial Technology Research Institute (ITRI), Taiwan),
Sheng-Tsai Wu (Industrial Technology Research Institute (ITRI),
Taiwan), Chun-Min Wang (Unimicron Technology Corporation, Taiwan),
Chia-Hsin Lee (Brewer Science, Taiwan), Shin-Yi Huang (Industrial
Technology Research Institute (ITRI), Taiwan), Ang-Ying Lin
(Industrial Technology Research Institute (ITRI), Taiwan), Tao-Chih
Chang (Industrial Technology Research Institute (ITRI), Taiwan), Puru
Bruce Lin (Unimicron Technology Corporation, Taiwan), Cheng-Ta Ko
(Unimicron Technology Corporation, Taiwan), Yu-Hua Chen (Unimicron
Technology Corporation, Taiwan), Jay Su (Brewer Science, Taiwan), Xiao
Liu (Brewer Science, USA), Luke Prenger (Brewer Science, USA), and
Kuan-Neng Chen (National Chiao Tung University, Taiwan)

High Yield Precision Transfer and Assembly of GaN μLEDs Using Laser Assisted Micro Transfer Printing.... 1470
Goutham Ezhilarasu (University of California Los Angeles), Amir Hanna
(University of California Los Angeles), Ajit Paranjpe (Veeco
Instruments Inc., USA), and Subramanian Iyer (University of California
Los Angeles)

High-Density Flexible Substrate Technology with Thin Chip Embedding and Partial Carrier Release
Option for IoT and Sensor Applications 1475
Kai Zoschke (Fraunhofer IZM), Piotr Mackowiak (Fraunhofer Institute
for Reliability and Microintegration), Ha-Duong Ngo (Fraunhofer
Institute for Reliability and Microintegration), Christian Tschoban
(Fraunhofer Institute for Reliability and Microintegration), Carola
Fritsche (Fraunhofer Institute for Reliability and Microintegration),
Kevin Kröhnert (Fraunhofer Institute for Reliability and
Microintegration), Thorsten Fischer (Fraunhofer Institute for
Reliability and Microintegration), Ivan Ndip (Fraunhofer Institute for
Reliability and Microintegration), and Klaus-Dieter Lang (Technical
University of Berlin)

Advance Embedded Packaging for Power Discrete Device 1485

Jia Ren Huo (Wuxi Sky Chip Interconnection Technology co., LTD), Song Guan Qiang (Wuxi Sky Chip Interconnection Technology co., LTD), Jing Jiang (Wuxi Sky Chip Interconnection Technology co., LTD), Wang Jun Tao (Wuxi Sky Chip Interconnection Technology co., LTD), and Ling Wen Kong (Wuxi Sky Chip Interconnection Technology co., LTD)

Large Panel Size Bonder with High Performance and High Accuracy 1492

Hubert Selhofer (Besi Austria GmbH), Andreas Mayr (Besi Austria GmbH), and Hugo Pristauz (Besi Austria GmbH)

Advances in high Speed Plating for Vertical Glass Panel Fine-Line Plating 1498

Christian Dunkel (Semsysco), Herbert Ötzlinger (Semsysco), Onishi Tetsuya (GJTech / Semsysco), and Raoul Schroeder (Semsysco)

Study of the Properties of AlN PMUT used as a Wireless Power Receiver 1503

Dan Gong (Xiamen University), Shenglin Ma (Xiamen University), Yihsiang Chiu (Peking University), Hungping Lee (J-Metrics Technology, Shenzhen), and Yufeng Jin (Peking University)

Session 33: Fan-Out, Flip Chip, and WLCSP

A Sequential Finite Volume Method / Finite Element Analysis of a Power Electronic Semiconductor Chip..... 1509

Mario Gschwandl (Polymer Competence Center Leoben GmbH, Austria), Peter Filipp Fuchs (Polymer Competence Center Leoben GmbH, Austria), Thomas Antretter (Montanuniversitaet Leoben, Austria), Martin Pfost (TU Dortmund University), Ivaylo Mitev (Polymer Competence Center Leoben GmbH, Austria), Tao Qi (Austria Technologie & Systemtechnik Aktiengesellschaft), Thomas Krivec (Austria Technologie & Systemtechnik Aktiengesellschaft), Angelika Schingale (CPT Group GmbH, Germany), and Michael Decker (CPT Group GmbH, Germany)

Failure Life Prediction of Wafer Level Packaging using DoS with AI Technology 1515

P. H. Chou (National Tsing Hua University), H. Y. Hsiao (National Tsing Hua University), and K.N. Chiang (National Tsing Hua University)

Thermal Cycling Simulation and Sensitivity Analysis of Wafer Level Chip Scale Package with Integration of Metal-Insulator-Metal Capacitors 1521

Yi Zhou (Georgia Institute of Technology), Liangbiao Chen (ON Semiconductor), Yong Liu (ON Semiconductor), and Suresh Sitaraman (Georgia Institute of Technology)

Effect of Time-Dependent Bulk Modulus on Reliability Assessment of Automotive Electronic Control Unit 1529

Hyun Seop Lee (University of Maryland), Bongtae Han (University of Maryland), and Przemyslaw Gromala (Robert Bosch GmbH, Germany)

Thermal and Mechanical Simulations for Fan-Out Wafer-Level Packaging Technology: Introduction of a "Solder Heatsink" 1535

Jean-Philippe Colonna (CEA-Leti, Université Grenoble Alpes, France), Loic Marnat (Université Grenoble Alpes), Mathilde Cartier (Université Grenoble Alpes), Gabriel Pares (Université Grenoble Alpes), and Dominique Noguet (Université Grenoble Alpes)

Wafer Level Warpage Modelling and Validation for FOWLP Considering Effects of Viscoelastic Material
Properties Under Process Loadings 1543

zhaohui chen (Institute of Microelectronics, A*STAR (Agency for
Science, Technology and Research)), Xiaowu Zhang (Institute of
Microelectronics, A*STAR (Agency for Science, Technology and
Research)), Sharon Pei Siang Lim (Institute of Microelectronics,
A*STAR (Agency for Science, Technology and Research)), Simon Siak Boon
Lim (Institute of Microelectronics, A*STAR (Agency for Science,
Technology and Research)), Boon Long Lau (Institute of
Microelectronics, A*STAR (Agency for Science, Technology and
Research)), Yong Han (Institute of Microelectronics, A*STAR (Agency
for Science, Technology and Research)), Ming Chinq Jong (Institute of
Microelectronics, A*STAR (Agency for Science, Technology and
Research)), Songlin Liu (A*STAR (Agency for Science, Technology and
Research)), Xiaobai Wang (A*STAR (Agency for Science, Technology and
Research)), and Yosephine Andriani (A*STAR (Agency for Science,
Technology and Research))

Ultra-Thin Package Board Level Drop Impact Modeling and Validation 1550

Shu-Shen Yeh (Taiwan Semiconductor Manufacturing Company (TSMC)), P.
Y. Lin (Taiwan Semiconductor Manufacturing Company (TSMC)), M. C. Yew
(Taiwan Semiconductor Manufacturing Company (TSMC)), W. Y. Lin (Taiwan
Semiconductor Manufacturing Company (TSMC)), K. C. Lee (Taiwan
Semiconductor Manufacturing Company (TSMC)), C. C. Yang (Taiwan
Semiconductor Manufacturing Company (TSMC)), J. H. Wang (Taiwan
Semiconductor Manufacturing Company (TSMC)), P. C. Lai (Taiwan
Semiconductor Manufacturing Company (TSMC)), C. K. Hsu (Taiwan
Semiconductor Manufacturing Company (TSMC)), and Shin-Puu Jeng (Taiwan
Semiconductor Manufacturing Company (TSMC))

Session 34: Emerging Materials and Processing

Flexible Graphene-Glass Fiber Composite Film with Ultrahigh Thermal Conductivity and Mechanical
Strength as Highly Efficient Thermal Spreader Materials 1556

Xiaoliang Zeng (Center for Advanced Material Research Shenzhen
Institutes of Advanced Technology, Chinese Academy of Sciences),
Linlin Ren (Center for Advanced Material Research Shenzhen Institutes
of Advanced Technology, Chinese Academy of Sciences), Rong Sun (Center
for Advanced Material Research Shenzhen Institutes of Advanced
Technology, Chinese Academy of Sciences), Jianbin Xu (Center for
Advanced Material Research Shenzhen Institutes of Advanced Technology,
Chinese Academy of Sciences), and Ching-Ping Wong (School of Materials
Science and Engineering Georgia Institute of Technology Atlanta, USA)

Highly Thermal Conductive and Electrically Insulated Graphene Based Thermal Interface Material with
Long-Term Reliability 1564

Nan Wang (SHT Smart High Tech AB), Ya Liu (Chalmers University of
Technology), Shujing Chen (Shanghai University), Lilei Ye (SHT Smart
High Tech AB), and Johan Liu (Chalmers University of Technology)

Further Enhancement of Thermal Conductivity through Optimal Uses of h-BN Fillers in Polymer-Based Thermal Interface Material for Power Electronics 1569

Han Jiang (Loughborough University), Han Zhou (Loughborough University), Stuart Robertson (Loughborough University), Zhaoxia Zhou (Loughborough University), Liguo Zhao (Loughborough University), and Changqing Liu (Loughborough University)

Wafer Level Integration of Thin Silicon Bare Dies Within Flexible Label 1575

Jean-Charles Souriau (Univ. Grenoble Alpes, CEA, LETI), Ahmad Itawi (Univ. Grenoble Alpes, CEA, LETI), and Laetitia Castagné (Univ. Grenoble Alpes, CEA, LETI)

Laser Sintering of Aerosol Jet Printed Conductive Interconnects on Paper Substrate 1581

Mohammed Alhendi (Binghamton University), Rajesh S. Sivasubramony (Binghamton University), Jack Lombardi (Binghamton University), Darshana L. Weerawarne (Binghamton University), Peter Borgesen (Binghamton University), Mark D. Poliks (Binghamton University), and Azar Alizadeh (General Electric Global Research)

In-Situ Investigation of Organic Additive Interactions in Copper Electroplating Solutions with Surface Enhanced Raman Spectroscopy (SERS) 1588

Nithin Nedumthakady (Georgia Institute of Technology), Bartlet DeProspo (Georgia Institute of Technology), Himani Sharma (Georgia Institute of Technology), Rahul Manepalli (Intel Corporation), Sashi Kandanur (Intel Corporation), Sajanlal Panikkanvalappil (Georgia Institute of Technology), Nasrin Hooshmand (Georgia Institute of Technology), and Rao Tummala (Georgia Institute of Technology)

C4 Compatible Ultra-Thick Cu On-chip Magnetic Inductor Architecture Integrated with Advanced Polymer/Cu Planarization Process 1595

C.H. Kuo (Taiwan Semiconductor Manufacturing Company, Ltd.), S.B. Yang (Taiwan Semiconductor Manufacturing Company, Ltd.), C.C. Kuo (Taiwan Semiconductor Manufacturing Company, Ltd.), Y.N. Chen (Taiwan Semiconductor Manufacturing Company, Ltd.), K.S. Yuan (Taiwan Semiconductor Manufacturing Company, Ltd.), G.C. Huang (Taiwan Semiconductor Manufacturing Company, Ltd.), C.N. Ke (Taiwan Semiconductor Manufacturing Company, Ltd.), Grace Chang (Taiwan Semiconductor Manufacturing Company, Ltd.), C.C. Hsu (Taiwan Semiconductor Manufacturing Company, Ltd.), H.L. Huang (Taiwan Semiconductor Manufacturing Company, Ltd.), Kirin Wang (Taiwan Semiconductor Manufacturing Company, Ltd.), Harry Ku (Taiwan Semiconductor Manufacturing Company, Ltd.), C.S. Chen (Taiwan Semiconductor Manufacturing Company, Ltd.), K.C. Liu (Taiwan Semiconductor Manufacturing Company, Ltd.), Alex Kalnitsky (Taiwan Semiconductor Manufacturing Company, Ltd.), and Marvin Liao (Taiwan Semiconductor Manufacturing Company, Ltd.)

Session 35: New Interconnects for Package Scaling

Development of 2.3D High Density Organic Package using Low Temperature Bonding Process with Sn-Bi Solder 1599
> Shota Miki (SHINKO ELECTRIC INDUSTRIES CO., LTD.), Hiroshi Taneda
> (SHINKO ELECTRIC INDUSTRIES CO., LTD.), Naoki Kobayashi (SHINKO
> ELECTRIC INDUSTRIES CO., LTD.), Kiyoshi Oi (SHINKO ELECTRIC INDUSTRIES
> CO., LTD.), Koji Nagai (SHINKO ELECTRIC INDUSTRIES CO., LTD.), and
> Toshinori Koyama (SHINKO ELECTRIC INDUSTRIES CO., LTD.)

PowerTherm Attach Process for Power Delivery and Heat Extraction in the Silicon-Interconnect Fabric Using Thermocompression Bonding 1605
> Pranav Ambhore (University of California, Los Angeles), Umesha Mogera
> (University of California, Los Angeles), Boris Vaisband (University of
> California, Los Angeles), Ujash Shah (University of California, Los
> Angeles), Timothy Fisher (University of California, Los Angeles), Mark
> Goorsky (University of California, Los Angeles), and Subramanian S.
> Iyer (University of California, Los Angeles)

Interconnect Scheme for Die-to-Die and Die-to-Wafer-Level Heterogeneous Integration for High-Performance Computing 1611
> Rabindra Das (MIT Lincoln Laboratory), Vladimir Bolkhovsky (MIT
> Lincoln Laboratory), Christopher Galbraith (MIT Lincoln Laboratory),
> Daniel Oates (MIT Lincoln Laboratory), Jason Plant (MIT Lincoln
> Laboratory), Renée Lambert (MIT Lincoln Laboratory), Scott Zarr (MIT
> Lincoln Laboratory), Ravi Rastogi (MIT Lincoln Laboratory), Dmitri
> Shapiro (MIT Lincoln laboratory), Manuel Docanto (MIT Lincoln
> Laboratory), Terence Weir (MIT Lincoln laboratory), and Leonard
> Johnson (MIT Lincoln Laboratory)

Ultra Wide Micro Bumps Interconnection Matrix for High Energy Particle Detection: Process and Assembly 1622
> Jean Charbonnier (CEA Leti), Myriam Assous (CEA-Leti), Thierry Mourier
> (CEA-Leti), Céline Ribière (CEA-Leti), Stéphane Minoret (CEA-Leti),
> Sophie Verrun (CEA-Leti), Pierre Tissier (CEA-Leti), Rémi Coquand
> (CEA-Leti), Mehmet Bicer (CEA-Leti), Fabienne Allain (CEA-Leti), Rémi
> Franiatte (CEA-Leti), and Gabriel Pares (CEA-Leti)

Growth Behavior and Orientation Evolution of Cu6Sn5 Grains in Micro Interconnect During Isothermal Reflow 1629
> S. Chen (Dalian University of Technology), N. Zhao (Dalian University
> of Technology), Y.Y. Qiao (Dalian University of Technology), Y.P. Wang
> (Dalian University of Technology), H.T. Ma (Dalian University of
> Technology), and C.M.L. Wu (City University of Hong Kong)

Development of a no Reflow Cu Pillar Bump to Improve Chip/Package Interactions (CPI) Process and Reliability Performance 1635
> Kuei Hsiao (Frank) Kuo (Siliconware Precision Industries Co., Ltd.
> (SPIL)), Jiunn Jie Wang (Siliconware Precision Industries Co., Ltd.
> (SPIL)), Yen Neng Wang (Siliconware Precision Industries Co., Ltd.
> (SPIL)), Feng Lung Chien (Siliconware Precision Industries Co., Ltd.
> (SPIL)), and Rick Lee (Siliconware Precision Industries Co., Ltd.
> (SPIL))

A Novel Interconnection Technology Using Ultra-Thin Under Barrier Metal for Multiple Chip-on-Chip
Stacking Structure 1641
> Takuya Nakamura (Sony Semiconductor Solutions), Kan Shimizu (Sony
> Semiconductor Solutions), Masataka Maehara (Sony Semiconductor
> Solutions), Toshihiko Hayashi (Sony Semiconductor Solutions), Kentaro
> Akiyama (Sony Semiconductor Solutions), Junichiro Fujimagari (Sony
> Semiconductor Solutions), Tomohiro Ohkubo (Sony Semiconductor
> Manufacturing), Atsushi Fujiwara (Sony Semiconductor Manufacturing),
> and Hayato Iwamoto (Sony Semiconductor Solutions)

Session 36: RF & Power Components and Modules

Multilayer Decoupling Capacitor using Stacked Layers of BST and LNO 1647
> Todd Schumann (University of Florida), Sheng-Po Fang (University of
> Florida), Yong-Kyu Yoon (University of Florida), Jongmin Yook (Korea
> Electronics Technology Institute), and Dongsu Kim (Korea Electronics
> Technology Institute)

System Co-Design of a High Current (40A) Synchronous Step-Down Converter in an Innovative Multi-chip
Module (MCM) LQFN-Type Packaging Technology 1653
> Todd Harrison (Texas Instruments, Inc.), Jie Chen (Texas Instruments,
> Inc.), and Rajen Murugan (Texas Instruments, Inc.)

Integrating Solid State Protection with a RF-MEMS Switch for Achieving ESD Robustness 1660
> Srivatsan Parthasarathy (Analog Devices, USA), Padraig Fitzgerald
> (Analog Devices, Ireland), Javier Salcedo (Analog Devices, USA), Ray
> Goggin (Analog Devices, Ireland), and Jean-Jacques Hajjar (Analog
> Devices, USA)

A Zero Height Small Size Low Cost RF Interconnect Substrate Technology For RF Front Ends For M.2
Modules And SiP 1666
> Sidharth Dalmia (Intel Corporation), Kirthika Nahalingam (Intel
> Corporation), Swathi Vijayakumar (Intel Corporation), and Pouya
> Talebbeydokhti (Intel Corporation)

Open and Closed Loop Inductors for High-Efficiency System-on-Package Integrated Voltage Regulators 1672
> Claudio Alvarez (Georgia Institute of Technology), Mohamed Bellaredj
> (Georgia Institute of Technology), and Madhavan Swaminathan (Georgia
> Institute of Technology)

RF Inductors Integrated in Organic Packaging 1680
> Denis Mercier (CEA-Leti), Jean-Philippe Michel (CEA-Leti), Christine
> Raynaud (CEA-Leti), and Christophe Billard (CEA-Leti)

3D Printed Interposer Layer for High Density Packaging of IoT Devices 1687
> Saikat Mondal (Michigan State University), Mohd. Ifwat Mohd. Ghazali
> (Universiti Sains Islam Malaysia), Kanishka Wijewardena (Michigan
> State University), Deepak Kumar (Michigan State University), and
> Premjeet Chahal (Michigan State University)

xliii

Session 37: Interactive Presentations 1

Comprehensive Solution for Micro Bump Coplanarity Control 1693
Chun-Chen Liu (Taiwan Semiconductor Manufacturing Company), J.H. Chen
(Taiwan Semiconductor Manufacturing Company), Y.N. Hsu (Taiwan
Semiconductor Manufacturing Company), Rung-De Wang (Taiwan
Semiconductor Manufacturing Company), Yu-Cheng Wang (Taiwan
Semiconductor Manufacturing Company), Bin-En Ho (Taiwan Semiconductor
Manufacturing Company), Y.H. Wu (Taiwan Semiconductor Manufacturing
Company), Ponder Pan (Taiwan Semiconductor Manufacturing Company),
Harry Ku (Taiwan Semiconductor Manufacturing Company), Kirin Wang
(Taiwan Semiconductor Manufacturing Company), Calvin Lu (Taiwan
Semiconductor Manufacturing Company), K.C. Liu (Taiwan Semiconductor
Manufacturing Company), and Marvin Liao (Taiwan Semiconductor
Manufacturing Company)

Structural Enhancement for a CMOS-MEMS Microphone Under Thermal Loading by Taguchi Method 1697
Chun-Lin Lu (National Tsing Hua University) and Meng-Kao Yeh (National
Tsing Hua University)

A Methodology to Correct in-Fixture Measurement of Impedance by a Machine Learning Model 1704
Bo-Siang Fang (Siliconware Precision Industries Co., Ltd. (SPIL),
Taiwan), Chia-Chu Lai (Siliconware Precision Industries Co., Ltd.
(SPIL), Taiwan), Ying-Wei Lu (Siliconware Precision Industries Co.,
Ltd. (SPIL), Taiwan), Kuan-Ta Chen (Siliconware Precision Industries
Co., Ltd. (SPIL), Taiwan), Mike Tasi (Siliconware Precision Industries
Co., Ltd. (SPIL), Taiwan), and Don-Son Jiang (Siliconware Precision
Industries Co., Ltd. (SPIL), Taiwan)

Material and Structure Design Optimization for Panel-Level Fan-Out Packaging 1710
Dao-Long Chen (Advanced Semiconductor Engineering, Inc.), Ian Hu
(Advanced Semiconductor Engineering, Inc.), KarenYU Chen (Advanced
Semiconductor Engineering, Inc.), Meng-Kai Shih (Advanced
Semiconductor Engineering, Inc.), David Tarng (Advanced Semiconductor
Engineering, Inc.), Dinos Huang (Advanced Semiconductor Engineering,
Inc.), and JY On (Advanced Semiconductor Engineering, Inc.)

The Microstructure and Mechanical Property of the High Entropy Alloy as a low Temperature Solder 1716
Li Pu (Beijing Institute of Technology), Quanfeng He (City University
of Hong Kong), Yong Yang (City University of Hong Kong), Xiuchen Zhao
(Beijing Institute of Technology), Zhuangzhuang Hou (Beijing Institute
of Technology), K. N. Tu (University of California, USA), and Yingia
Liu (Beijing Institute of Technology)

A Versatile Fan-Out Infrastructure Based on Die-Stencil Substrate Promoted by an Advanced
Multifunctional Temporary Bonding Material 1722
Xiao Liu (Brewer Science, Inc.), Baron Huang (Brewer Science, Inc.),
Hong Zhang (Brewer Science, Inc.), Lisa Kirchner (Brewer Science,
Inc.), Arthur Southard (Brewer Science, Inc.), Rama Puligadda (Brewer
Science, Inc.), and Tony Flaim (Brewer Science, Inc.)

Low Temperature and Pressureless Microfluidic Electroless Bonding Process for Vertical
Interconnections 1729
Han-Tang Hung (National Taiwan University), Sean Yang (National Taiwan
University), I-An Weng (National Taiwan University), Yan-Hao Chen
(Unimicron Technology Corporation, Taiwan), and C. Robert Kao
(National Taiwan University)

3D Integration of CMOS-Compatible Surface Electrode Ion Trap and Silicon Photonics for Scalable Quantum Computing 1735

Jing Tao (Nanyang Technological University), Yu Dian Lim (Nanyang Technological University), Hong Yu Li (Agency for Science, Technology and Research (A*STAR)), Nam Piau Chew (Nanyang Technological University), Anak Agung Alit Apriyana (Agency for Science, Technology and Research (A*STAR)), Lin Bu (Agency for Science, Technology and Research (A*STAR)), Peng Zhao (Nanyang Technological University), Luca Guidoni (Université Paris Diderot), and Chuan Seng Tan (Nanyang Technological University)

Integrated RTD Sensors for Maintaining Thermal Uniformity During TCB Process 1744

Salwa Ben Jemaa (Interdisciplinary Institute for Technological Innovation (3IT) Sherbrooke University), Julien Sylvestre (Interdisciplinary Institute for Technological Innovation (3IT) Sherbrooke University), and Pascale Gagnon (IBM Canada Bromont, QC, Canada)

Wireless Transfer of Power and Data Via a Single Resonant Inductive Link 1751

Shiang-Hwua Yu (National Sun Yat-sen University), Yi-Chen Hsieh (National Sun Yat-sen University), Chin-Wei Chan (National Sun Yat-sen University), I-Fang Lo (National Sun Yat-sen University), Heri Suryoatmojo (Institut Teknologi Sepuluh), and Lih-Tyng Hwang (National Sun Yat-sen University)

Adaptive Patterning of Optical and Electrical Fan-Out for Photonic Chip Packaging 1757

Ahmed Elmogi (Centre for Microsystems Technology, imec and Ghent University), Andres Desmet (Centre for Microsystems Technology, imec and Ghent University), Jeroen Missinne (Centre for Microsystems Technology, imec and Ghent University), Hannes Ramon (Ghent University-imec), Joris Lambrecht (Ghent University-imec), Peter De Heyn (imec), Marianna Pantouvaki (imec), Joris Van Campenhout (imec), Johan Bauwelinck (Ghent University-imec), and Geert Van Steenberge (imec and Ghent University)

Low Surface Reflectance Structure at Near Infrared Wavelength by Injection Molding 1764

Sho Yakabe (Sumitomo Electric Industries, Ltd.), Takuro Watanabe (Sumitomo Electric Industries, Ltd.), Takayuki Shimazu (Sumitomo Electric Industries, Ltd.), Ryohei Hokari (National Institute of Advanced Industrial Science and Technology), and Kazuma Kurihara (National Institute of Advanced Industrial Science and Technology)

A Novel Design of a Bandwidth Enhanced Dual-Band Impedance Matching Network with Coupled Line Wave Slowing 1770

Deepayan Banerjee (IIIT Delhi, India), Antra Saxena (IIIT Delhi, India), and Mohammad Hashmi (Nazarbayev University, Kazakhstan)

Effects of Electromigration on Microstructural Evolution and Mechanical Properties of Preferential Growth Intermetallic Compound Interconnects for 3D Packaging 1774

Mingliang L. Huang (Dalian University of Technology) and Lin Zou (Dalian University of Technology)

Telemetry for Implantable Biosensors 1782

Ryan B. Green (Virginia Commonwealth University) and Erdem Topsakal (Virginia Commonwealth University)

Ultra-Thin QFN-Like 3D Package with 3D Integrated Passive Devices 1789

Ayad Ghannam (3DiS Technologies S.A.S, France), Niek van Haare (Besi Netherlands, B.V., Netherlands), Julian Bravin (EV Group E.Thallner GmbH, Austria), Elisabeth Brandl (EV Group E.Thallner GmbH, Austria), Birgit Brandstätter (Besi Austria GmbH, Austria), Hannes Klingler (Besi Austria GmbH, Austria), Benedikt Auer (Besi Austria GmbH, Austria), Philippe Meunier (NXP Semiconductors, France), and Sebastiaan Kersjes (Besi Netherlands, B.V., Netherlands)

Low-Cost Non-TSV Based 3D Packaging Using Glass Panel Embedding (GPE) for Power-Efficient, High-Bandwidth Heterogeneous Integration 1796

Siddharth Ravichandran (Georgia Institute of Technology), Shuhei Yamada (Murata Manufacturing Co. Ltd, Kyoto, Japan), Fuhan Liu (Georgia Institute of Technology), Vanessa Smet (Georgia Institute of Technology), Mohanalingam Kathaperumal (Georgia Institute of Technology), and Rao Tummala (Georgia Institute of Technology)

Polylithic Integration of 2.5D and 3D Chiplets Using Interconnect Stitching 1803

Paul K. Jo (Georgia Institute of Technology), Ting Zheng (Georgia Institute of Technology), and Muhannad S. Bakir (Georgia Institute of Technology)

Characterization of the Current Mechanisms and Improved Leakage Current in Silver Doped Barium Strontium Titanate 1809

Todd Schumann (University of Florida), Kyoung-Tae Kim (University of Florida), Sheng-Po Fang (University of Florida), and Yong-Kyu Yoon (University of Florida)

High Temperature Aging Effects in SAC and SAC+X Lead Free Solders 1815

Mohammad S. Alam (Auburn University), KM Rafidh Hassan (Auburn University), Jeffrey C. Suhling (Auburn University), and Pradeep Lall (Auburn University)

Session 38: Interactive Presentations 2

Laundering Reliability of Electrically Conductive Fabrics for E-Textile Applications 1826

Jeffrey ChangBing Lee (iST-Integrated Service Technology Inc.), Weifeng Liu (FLEX Ltd.), ChangHo Lo (iST-Integrated Service Technology Inc.), and Cheng-Chih Chen (iST-Integrated Service Technology Inc.)

Preconditioning Technologies for Sputtered Seed Layers in FOPLP 1833

Johannes Weichart (Evatec AG), Jüergen Weichart (Evatec AG), Andreas Erhart (Evatec AG), and Kay Viehweger (Fraunhofer IZM ASSID)

Impact of Thermal Boundary Resistance on the Thermal Design of GaN-on-Diamond HEMTs 1842

Huaixin Guo (Nanjing Electronic Devices Institute), Yuechan Kong (Nanjing Electronic Devices Institute), and Tangsheng Chen (Nanjing Electronic Devices Institute)

Measuring the Electric Properties of Thin Film Shape Memory Polymers in Simulated Physiological Conditions 1848

Daniel Del Nero (The University of Texas at Dallas), Alexandra Joshi-Imre (The University of Texas at Dallas), and Walter Voit (The University of Texas at Dallas)

xlvi

Evaluation of WLP Dielectrics for High Voltage Applications 1853
    Markus Wöhrmann (Fraunhofer IZM), Michael Toepper (Fraunhofer IZM),
    Marcus Paeck (Fraunhofer IZM), and Klaus-Dieter Lang (Technical
    University Berlin)

Mitigating the Effects of Microvortices in high-Re Deterministic Lateral Displacement by Using
Symmetric Airfoil-Shaped Pillars .... N/A
    Brian Dincau (Washington State University Vancouver), Kawkab Ahasan
    (Washington State University Vancouver), and Jong-Hoon Kim (Washington
    State University Vancouver)

Plasma Dry Process Technology Development of Glass-Epoxy Film on the Silicon Substrate to Fabricate
RDL for Future GPU/AI Application 1865
    Takahide Murayama (ULVAC, Inc.), Muneyuki Sato (ULVAC, Inc.), Akiyoshi
    Suzuki (ULVAC, Inc.), Atsuhito Ihori (ULVAC, Inc.), Tetsushi Fujinaga
    (ULVAC, Inc.), and Yasuhiro Morikawa (ULVAC, Inc.)

Fully Solid-State Integrated Capacitors Based on Carbon Nanofibers and Dielectrics with Specific
Capacitances Higher Than 200 nF/mm2 1870
    Amin Saleem (Smoltek AB, Sweden), Rickard Andersson (Smoltek AB,
    Sweden), Maria Bylund (Smoltek AB, Sweden), Charlotte Goemare (Smoltek
    AB, Sweden), Guilhem Pacot (Smoltek AB, Sweden), Mohammed Kabir
    (Smoltek AB, Sweden), and Vincent Desmaris (Smoltek AB, Sweden)

Application of Fan-Out Panel Level Packaging Techniques for Flexible Hybrid Electronics Systems 1877
    Wei-Yuan Cheng (ITRI), Shau-Fei Cheng (ITRI), Chen-Tsai Yang (ITRI),
    Shau-Fei Cheng (ITRI), Wei-Han Chen (ITRI), Hsin-Cheng Lai (ITRI),
    Tai-Jui Wang (ITRI), and Yuh-Zheng Lee (ITRI)

Structuring of Laser Activated Polymers for Sensor Applications 1883
    Sebastian Bengsch (University Hanover), Marc Christopher Wurz
    (University Hanover), Kevin Cromwell (University Hanover), and
    Maximilian Aue (University Hanover)

A Deep Learning Approach for Volterra Kernel Extraction for Time Domain Simulation of Weakly
Nonlinear Circuits 1889
    Thong Nguyen (University of Illinois at Urbana Champaign), Xinying
    Wang (University of Illinois at Urbana Champaign), Xu Chen (University
    of Illinois at Urbana Champaign), and Jose Schutt-Aine (University of
    Illinois at Urbana Champaign)

224G Package Interconnect Design Study - Based on Artificial Neural Network Modeling Approach 1897
    Hui Liu (Intel Corporation), Qian Ding (Intel Corporation), and
    Penglin Liu (Intel Corporation)

Enhanced Reliability of a RF-SiP with Mold Encapsulation and EMI Shielding 1902
    Chan-Yuan Liu (Advanced Semiconductor Engineering, Inc., Taiwan),
    Jason Chien (Advanced Semiconductor Engineering, Inc., Taiwan),
    Yu-Chou Tseng (Advanced Semiconductor Engineering, Inc., Taiwan),
    Kuo-Hsien Liao (Advanced Semiconductor Engineering, Inc., Taiwan),
    Alex Chan (Advanced Semiconductor Engineering, Inc., Taiwan), Dao-Long
    Chen (Advanced Semiconductor Engineering, Inc., Taiwan), Meng-Kai Shih
    (Advanced Semiconductor Engineering, Inc., Taiwan), and Mark Gerber
    (Advanced Semiconductor Engineering, Inc., U.S.)

Study of the Effect and Mechanism of a Cap Layer in Controlling the Statistical Variation of Via Extrusion 1909
  Golareh Jalilvand (University of Central Florida) and Tengfei Jiang
  (University of Central Florida)

Three Dimensional Copper Foam-Filled Elastic Conductive Composites with Simultaneously Enhanced Mechanical, Electrical, Thermal and Electromagnetic Interference (EMI) Shielding Properties 1916
  Tan Lu (Shenzhen Institutes of Advanced Technology, Chinese Academy of
  Sciences), Han Gu (Shenzhen Institutes of Advanced Technology, Chinese
  Academy of Sciences), Yougen Hu (Shenzhen Institutes of Advanced
  Technology, Chinese Academy of Sciences), Tao Zhao (Shenzhen
  Institutes of Advanced Technology, Chinese Academy of Sciences),
  Pengli Zhu (Shenzhen Institutes of Advanced Technology, Chinese
  Academy of Sciences), Rong Sun (Shenzhen Institutes of Advanced
  Technology, Chinese Academy of Sciences), and Ching-Ping Wong (Georgia
  Institute of Technology)

Vertical Interconnect Technology for Enlarging Capacity on Micro Solid Thin Film Rechargeable Battery 1921
  Akihiro Horibe (IBM Research - Tokyo), Kuniaki Sueoka (IBM Research -
  Tokyo), Takahiro Mori (IBM Research - Tokyo), Risa Miyazawa (IBM
  Research - Tokyo), and Hiroyuki Mori (IBM Research - Tokyo)

Characterization of Fine Pitch Hybrid Bonding Pads using Electrical Misalignment Test Vehicle 1926
  Imed Jani (CEA, LETI), Didier Lattard (CEA, LETI), Pascal Vivet (CEA,
  LETI), Lucile Arnaud (CEA, LETI), Severine Cheramy (CEA, LETI), Edith
  Beigné (CEA, LETI), Alexis Farcy (STMicroelectronics), Joris Jourdon
  (STMicroelectronics), Yann Henrion (STMicroelectronics), Emilie
  Deloffre (STMicroelectronics), and Halim Bilgen (STMicroelectronics)

Dynamic Characteristics Evaluation on NCF Under Challenging Conditions and Its Application 1933
  Tomonori Nakamura (Shinkawa LTD, Japan), Hiromi Shibahara (Shinkawa
  LTD, Japan), Osamu Watanabe (Shinkawa LTD, Japan), Tetsuya Utano
  (Shinkawa LTD, Japan), Daisuke Tani (Shinkawa LTD, Japan), Sung
  Chenhsiu (Shinkawa LTD, Japan), Toru Maeda (Shinkawa LTD, Japan), Doug
  Day (Shinkawa LTD, Japan), Hidekazu Yagi (Dexerials Corporation,
  Japan), Ryoji Kojima (Dexerials Corporation, Japan), Daichi Mori
  (Dexerials Corporation, Japan), Tatsuo Nagamatsu (Dexerials
  Corporation, Japan), and Junichi Kaneko (Dexerials Corporation, Japan)

Study of Electrical and Mechanical Characteristics of Inkjet-Printed Patch Antenna Under Uniaxial and Biaxial Bending 1939
  Yi Zhou (Georgia Institute of Technology), Sridhar Sivapurapu (Georgia
  Institute of Technology), Rui Chen (Georgia Institute of Technology),
  Nahid Aslani Amoli (Georgia Institute of Technology), Mohamed
  Bellaredj (Georgia Institute of Technology), Madhavan Swaminathan
  (Georgia Institute of Technology), and Suresh K. Sitaraman (Georgia
  Institute of Technology)

Effects of Oven and Laser Sintering Parameters on the Electrical Resistance of IJP Nano-Silver Traces on Mesoporous PET Before and During Fatigue Cycling 1946
  G.S. Khinda (SUNY Binghamton), M.Z. Kokash (SUNY Binghamton), M.
  Alhendi (SUNY Binghamton), M. Yadav (SUNY Binghamton), J.P. Lombardi
  (SUNY Binghamton), D.L. Weerawarne (SUNY Binghamton), Mark D. Poliks
  (SUNY Binghamton), P. Borgesen (SUNY Binghamton), and Nancy C. Stoffel
  (General Electric Global Research Center)

Multilayer Glass Substrate with High Density Via Structure for All Inorganic Multi-chip Module 1952
  Toshiki Iwai (FUJITSU LABORATORIES LTD.), Taiji Sakai (FUJITSU
  LABORATORIES LTD.), Daisuke Mizutani (FUJITSU LABORATORIES LTD.),
  Seiki Sakuyama (FUJITSU LABORATORIES LTD.), Kenji Iida (FUJITSU
  INTERCONNECT TECHNOLOGIES LIMITED), Takayuki Inaba (FUJITSU
  INTERCONNECT TECHNOLOGIES LIMITED), Hidehiko Fujisaki (FUJITSU
  INTERCONNECT TECHNOLOGIES LIMITED), Akira Tamura (FUJITSU INTERCONNECT
  TECHNOLOGIES LIMITED), and Yoshinori Miyazawa (FUJITSU INTERCONNECT
  TECHNOLOGIES LIMITED)

The Poisson's Ratio of Lead Free Solder - The Often Forgotten But Important Material Property 1958
  KM Rafidh Hassan (Auburn University), Mohammad S. Alam (Auburn
  University), Jeffrey C. Suhling (Auburn University), and Pradeep Lall
  (Auburn University)

Additive Laser Metal Deposition Onto Silicon for Enhanced Microelectronics Cooling 1970
  Arad Azizi (Binghamton University (SUNY)), Matthias A. Daeumer
  (Binghamton University (SUNY)), Jacob C. Simmons (Binghamton
  University (SUNY)), Bahgat G. Sammakia (Binghamton University (SUNY)),
  Bruce T. Murray (Binghamton University (SUNY)), and Scott N. Schiffres
  (Binghamton University (SUNY))

Moisture Barrier, Mechanical, and Thermal Properties of PDMS-PIB Blends for Solar Photovoltaic (PV)
Module Encapsulant 1977
  Jinho Hah (Georgia Institute of Technology), Michael Sulkis (Georgia
  Institute of Technology), Chao Ren (Georgia Institute of Technology),
  Minsoo Kang (Georgia Institute of Technology), Kyoung-sik Moon
  (Georgia Institute of Technology), Samuel Graham (Georgia Institute of
  Technology), and C. P. Wong (Georgia Institute of Technology)

Session 39: Interactive Presentations 3

Modeling and Design of Power Distribution Network for a Heterogeneous Integrated Active Interposer
with Neuromorphic Computing Circuits 1983
  Min Miao (Beijing Information Science and Technology University),
  Tianfang Chen (Beijing Information Science and Technology University),
  Yang Yang (Peking University Shenzhen Graduate School), Jincan Zhang
  (Beijing Information Science and Technology University), Na Li
  (Beijing Information Science and Technology University), Kunkun Li
  (Beijing Information Science and Technology University), Liyuan Wang
  (Beijing Information Science and Technology University), Huan Liu
  (Peking University), Xiaole Cui (Peking University Shenzhen Graduate
  School), and Yufeng Jin (Peking University Shenzhen Graduate School)

PCB Microstrip Line Far-End Crosstalk Mitigation by Surface Mount Capacitors 1989
  Zhaoqing Chen (IBM Corporation)

New Cost-Effective Via-Last Approach by "One-Step TSV" After Wafer Stacking for 3D Memory
Applications 1996
  Masaya Kawano (Institute of Microelectronics, A*STAR), Xiangy-Yu Wang
  (Institute of Microelectronics, A*STAR), and Qin Ren (Institute of
  Microelectronics, A*STAR)

xlix

Microstructure and Property Changes in Cu/Sn-58Bi/Cu Solder Joints During Thermomigration 2003
  Yu-An Shen (Joining and Welding Research Institute (JWRI), Osaka
  University), Shiqi Zhou (Osaka University), Jiahui Li (City University
  of Hong Kong), K. N. Tu (UCLA), and Hiroshi Nishikawa (Joining and
  Welding Research Institute (JWRI), Osaka University)

Simulation and Experimental Validations of EM/TM/SM Physical Reliability for Interconnects Utilized
in Stretchable and Foldable Electronics 2009
  Chang-Chun Lee (National Tsing Hua University), Oscar Chuang (National
  Tsing Hua University), Chia-Ping Hsieh (National Taiwan University),
  Wei-Yuan Cheng (Industrial Technology Research Institute), and Steve
  Chiu (Industrial Technology Research Institute)

A Complex Integrated Circuit Structure Transformation, Modeling and Simulation Method 2016
  Daixing Wang (Peking University Shenzhen Graduate School), Wei Wang
  (Institute of Microelectronics Peking University), and Yufeng Jin
  (Peking University Shenzhen Graduate School)

A Study on the Oxygen Plasma Treatment on the Peel Adhesion Strength and Solder Wettability of
SnBi58 Based Anisotropic Conductive Films 2022
  Shuye Zhang (Harbin Institute of Technology), Mingliang Huang (Dalian
  University of Technology), Yang Wu (Dalian University of Technology),
  Ming Yang (Hisilicon Optoelectronics Co., Ltd), Tiesong Lin (Harbin
  Institute of Technology), Peng He (Harbin Institute of Technology),
  and Kyung-Wook Paik (Nano-Packaging and Interconnection Laboratory)

Numerical Analysis of the Influence of Polymeric Materials on a MEMS Package Performance Under
Humidity and Temperature Loads 2029
  Mahesh Yalagach (Polymer Competence Center Leoben GmbH, Leoben,
  Austria.), Peter Filipp Fuchs (Polymer Competence Center Leoben GmbH,
  Leoben, Austria.), Archim Wolfberger (Polymer Competence Center Leoben
  GmbH, Leoben, Austria.), Mario Gschwandl (Polymer Competence Center
  Leoben GmbH, Leoben, Austria.), Thomas Antretter (Montanuniversitaet
  Leoben, Institute of Mechanics, Leoben, Austria.), Michael Feuchter
  (Montanuniversitaet Leoben, Institute of Material Science and Testing
  of Polymers, Leoben, Austria.), Coen Tak (ams AG, Premstaetten,
  Austria), and Qi Tao (Austria Technologie & Systemtechnik
  Aktiengesellschaft, Leoben, Austria.)

Electromigration-Induced -Sn Grain Rotation in Lead-Free Flip Chip Solder Bumps 2036
  Mingliang L. Huang (Dalian University of Technology), Jiameng M. Kuang
  (Dalian University of Technology), and Hongyu Y. Sun (Dalian
  University of Technology)

Low-Cost MT-Ferrule-Compatible Optical Connector for Co-packaged Optics Using Single-Mode Polymer
Waveguide 2042
  Akihiro Noriki (National Institute of Advanced Industrial Science and
  Technology (AIST)), Takeru Amano (National Institute of Advanced
  Industrial Science and Technology (AIST)), Masatoshi Tsunoda (Kyocera
  Corporation), and Toshiaki Michihiro (Kyocera Corporation)

Characterization of Coated Silver Wire Bond Interface Using TEM 2048
  Murali Sarangapani (Heraeus Materials Singapore Pte. Ltd.,), Eric Tan
  Swee Seng (Heraeus Materials Singapore Pte. Ltd.,), and Jason Wong
  Chin Yeung (Heraeus Materials Singapore Pte. Ltd.,)

Research on Applied Reliability of BGA Solder Balls in Extreme Marine Environment 2054
  Liyuan Liu (China Electronic Product Reliability and Environmental
  Testing Research Institute), Tao Lu (China Electronic Product
  Reliability and Environmental Testing Research Institute), Daojun Luo
  (China Electronic Product Reliability and Environmental Testing
  Research Institute), and Hui Xiao (China Electronic Product
  Reliability and Environmental Testing Research Institute)

Influence of Single/Double Sweeping Mode and Sweeping Voltage Increment/Polarity on Measurement of
TSV Leakage Current 2061
  Qinghua Zeng (Peking University), Jing Chen (Peking University), and
  Yufeng Jin (Peking University)

Improving the Solder Wettability Via Atmospheric Plasma Technology 2067
  Sagung Dewi Kencana (National Taiwan University of Science and
  Technology), Yu-Lin Kuo (National Taiwan University of Science and
  Technology), Yee-Wen Yen (National Taiwan University of Science and
  Technology), Eckart Schellkes (Robert Bosch Taiwan Co., Ltd), and
  Wallace Chuang (Robert Bosch Taiwan Co., Ltd)

Orthogonal Quilt Packaging 3D Integration for High-Energy Particle Detectors 2072
  Jason Kulick (Indiana Integrated Circuits, LLC), Tian Lu (Indiana
  Integrated Circuits, LLC), Edit Varga (Indiana Integrated Circuits,
  LLC), Gary H. Bernstein (Indiana Integrated Circuits, LLC), Carlos
  Ortega (Indiana Integrated Circuits, LLC), Christopher Kenney (SLAC
  National Accelerator Laboratory), and Julie Segal (SLAC National
  Accelerator Laboratory)

Carbonized Electrodes for Electrochemical Sensing 2073
  Mohammad Aminul Haque (The University of Tennessee, Knoxville),
  Nickolay V. Lavrik (Oak Ridge National Laboratory), Dale Hensley (Oak
  Ridge National Laboratory), and Nicole McFarlane (The University of
  Tennessee, Knoxville)

Moldability Challenges Associated with the Assembly of Thicker IC Packages for High Voltage and
Power Applications 2079
  Sadia Naseem (Texas Instruments Inc.), Jack Chiang (Texas Instruments
  Inc.), Megan Chang (Texas Instruments Inc.), Bob Lee (Texas
  Instruments Inc.), and Jason Chien (Texas Instruments Inc.)

Highly Compact, Multiband Composite-Right/Left-Handed(CRLH) Transmission Line Based Stub for GPS
Applications 2085
  Hae-In Kim (University of Florida), Seahee Hwangbo (University of
  Florida), Renuka Bowrothu (University of Florida), and Yong-Kyu Yoon
  (University of Flordia)

Session 40: Interactive Presentations 4

Die Thickness Optimization for Preventing Electro-Thermal Fails Induced by Solder Voids in Power
Devices 2091
  Dario Vitello (STMicroelectronics), Andrea Albertinetti
  (STMicroelectronics), and Marco Rovitto (STMicroelectronics)

3-T (8-T) Decoupling Capacitors for Improved PDN in LPDDR4/4X/5 System 2097
  Sunil Gupta (Qualcomm Technologies, Inc.)

Improved Correlation Between Accelerated Board Level Reliability (BLR) Testing and Customer BLR
Results Using a Hybrid Closed-Form/Finite Element Methodology 2103

Maxim Serebreni (DfR Solutions), Natalie Hernandez (DfR Solutions),
Gil Sharon (DfR Solutions), Nathan Blattau (DfR Solutions), Craig
Hillman (DfR Solutions), and Ken Symonds (Western Digital)

Fabrication and Reliability Demonstration of 3 μm Diameter Photo Vias at 15 μm Pitch in Thin
Photosensitive Dielectric Dry Film for 2.5 D Glass Interposer Applications 2112

Daichi Okamoto (TAIYO INK MFG. CO., LTD.), Yoko Shibasaki (TAIYO INK
MFG.CO.LTD), Daisuke Shibata (TAIYO INK MFG.CO.LTD), Tadahiko Hanada
(TAIYO INK MFG.CO.LTD), Fuhan Liu (Georgia Institute of Technology),
Mohanalingam Kathaperumal (Georgia Institute of Technology), and Rao
R. Tummala (Georgia Institute of Technology)

Pre-Cure Modification of Electrically Conductive Adhesive for Low Temperature Interconnection 2117

Jinto George (University of Sherbrooke, Bromont, QC, Canada), David
Danovitch (University of Sherbrooke, Bromont, QC, Canada), Alexandre
Leblanc (IBM Canada Ltd, Bromont, QC, Canada), Eric Savage (IBM Canada
Ltd, Bromont, QC, Canada), Michael Ayukawa (Redlen Technologies,
Saanichton, BC, Canada), and Dexter Macaisa (Redlen Technologies,
Saanichton, BC, Canada)

RDL-1st Fan-Out Panel Level Packaging (FOPLP) for Heterogeneous and Economical Packaging 2126

Nagendra Sekhar Vasarla (Institute of Microelectronics, A*STAR (Agency
for Science, Technology and Research)), Vempati Srinivasa Rao
(Institute of Microelectronics, A*STAR (Agency for Science, Technology
and Research)), F. X. Che (Institute of Microelectronics, A*STAR
(Agency for Science, Technology and Research)), Chong Ser Choong
(Institute of Microelectronics, A*STAR (Agency for Science, Technology
and Research)), and Kazunori Yamamoto (Institute of Microelectronics,
A*STAR (Agency for Science, Technology and Research))

Epoxy Composites with Surface Modified Silicon Carbide Filler for High Temperature Molding Compounds. 2134

Fan Wu (Georgia Institute of Technology), Nicholas C Mitchell
(Nicholas C), Bo Song (Georgia Institute of Technology), Kyoung-sik
Moon (Georgia Institute of Technology), and CP Wong (Georgia Institute
of Technology)

Ultra Low Resistivity and High Electrical Stability Silo-Ag ECAs Produced from Curing Chemistry
Optimization for Flexible Electronics 2140

Xueqiao Wang (Georgia Institute of Technology), Bo Song (Georgia
Institute of Technology), Kyoung-Sik Moon (Georgia Institute of
Technology), and C. P. Wong (Georgia Institute of Technology)

Physics of Failure Based Simulation and Experimental Testing of Quad Flat No-Lead Package 2144

Jia-Shen Lan (National Sun Yat-sen University) and Mei-Ling Wu
(National Sun Yat-sen University)

lii

An Assessment of Electromigration in 2.5D Packaging 2150
Jiefeng Xu (The State University of New York at Binghamton), Scott McCann (The State University of New York at Binghamton), Huayan Wang (The State University of New York at Binghamton), Jing Wang (The State University of New York at Binghamton), VanLai Pham (The State University of New York at Binghamton), Stephen R. Cain (The State University of New York at Binghamton), Gamal Refai-Ahmed (The State University of New York at Binghamton), and S.B. Park (The State University of New York at Binghamton)

Diffusion Enhanced Drive Sub 100 °C Wafer Level Fine-Pitch Cu-Cu Thermocompression Bonding for 3D IC Integration 2156
Asisa Kumar Panigrahy (Gokaraju Rangaraju Institute of Engineering & Technology, Hyderabad, India), Satish Bonam (Indian Institute of Technology Hyderabad, India), Tamal Ghosh (Indian Institute of Technology Hyderabad, India), Siva Rama Krishna Vanjari (Indian Institute of Technology Hyderabad, India), and Shiv Govind Singh (Indian Institute of Technology Hyderabad)

Development of Sheet Type Molding Compounds for Panel Level Package 2162
Kenichi Ueno (Company), Kazuhiro Dohi (SANYU REC CO., LTD.), Yui Suzuki (SANYU REC CO., LTD.), and Masakazu Hirose (SANYU REC CO., LTD.)

Defect Detection for the TSV Transmission Channel Using Machine Learning Approach 2168
Huan Liu (Peking University), Runiu Fang (Peking University), Min Miao (Beijing Information Science and Technology University), Yang Yang (Shenzhen Graduate School, Peking University), and Yufeng Jin (Shenzhen Graduate School, Peking University)

Direct Printing of Heat Sinks, Cases and Power Connectors on Insulated Substrate Using Selective Laser Melting Techniques 2173
Rabih Khazaka (Safran SA), Donatien Martineau (Safran SA), Toni Youssef (Safran SA), Thanh Long Le (Safran SA), and Stephane Azzopardi (Safran SA)

Server CPU Package Design Using PoINT Architecture 2180
Arun Chandrasekhar (Intel), Vijaya Boddu (Intel), Erich Chuh (Intel), Krishna Bharath (Intel), Farzaneh Yahyaei-Moayyed (Intel), Srikrishnan Venkataraman (Intel), Sriram Srinivasan (Intel), Ram Viswanath (Intel), Huthasana Kalyanam (Intel), and Ritesh Jain (Intel)

Highly Reliable Die-Attach Silver Joint with Pressure-Less Sintering Process 2186
Sihai Chen (Indium Corporation, USA), William Shambach (Rochester Institute of Technology), Jordan Palmer (Rochester Institute of Technology), Christine Labarbera (Indium Corporation), Xuanyi Ding (Cornell University), and Ning-Cheng Lee (Indium Corporation)

3D Power Packaged Device Thermo-Mechanical Modeling and Stress Analysis After Reliability Trials 2194
Lucrezia Guarino (STMicroelectronics, Italy), Lucia Zullino (STMicroelectronics, Italy), Luca Cecchetto (STMicroelectronics, Italy), Fiorella Pozzobon (STMicroelectronics, Italy), and Antonio Andreini (STMicroelectronics, Italy)

Millimeter Wave Dual Polarization Design Using Frequency Selective Surface (FSS) for 5G Base-Station Applications 2200

Chi-Hau Yang (National Sun Yat-Sen University), Chung-Yi Hsu (National Sun Yat-Sen University), and Lih-Tyng Hwang (National Sun Yat-Sen University)

Direct Bonding of low Temperature Heterogeneous Dielectrics 2206

Serena Iacovo (imec), Ian Peng (imec), Alain Phommahaxay (imec), Fumihiro Inoue (imec), Patrick Verdonck (imec), Soon-Wook Kim (imec), Erik Sleeckx (imec), Andy Miller (imec), Gerald Beyer (imec), and Eric Beyne (imec)

Session 41: Student Interactive Presentations

Low Temperature Transient Liquid Phase (TLP) Bonding using Eutectic Sn-In Solder Anisotropic Condctive Films (ACFs) for Flexible Ultrasound Transducer 2213

Jae-Hyeong Park (KAIST), Jongcheol Park (NanoFab Center), and Kyung-Wook Paik (KAIST)

Room-Temperature Bonding with Pd Coated Cu Wire on Al Pads: Ball Bond Optimization with 2-Stage Methodology 2219

Nicholas Kam (University of Waterloo), Michael David Hook (University of Waterloo), Celal Con (KA Imaging), Karim S. Karim (University of Waterloo), and Michael Mayer (University of Waterloo)

On-Chip ESD Monitor 2225

Kannan Kalappurakal Thankappan (University of California, Los Angeles), Boris Vaisband (University of California, Los Angeles), and Subramanian S. Iyer (University of California, Los Angeles)

Preparation and Characterization of Electroplated Cu/Graphene Composite 2234

Xin Wang (Tsinghua University), Qian Wang (Tsinghua University), Jian Cai (Tsinghua University), Changming Song (Tsinghua University), Yang Hu (Tsinghua University), Yang Zhao (University of Science and Technology of China), and Yu Pei (University of Science and Technology of China)

Quantifying the Impact of RF Probing Variability on TRL Calibration for LTCC Substrates 2240

Ömer Faruk Yildiz (Hamburg University of Technology, Germany), David Dahl (Hamburg University of Technology, Germany), and Christian Schuster (Hamburg University of Technology, Germany)

Effects of NCF and UBM Materials on Electromigration Reliabilities of Sn-Ag Microbumps for Advanced 3D Packaging 2246

Kirak Son (Andong National University), Gahui Kim (Andong National University), Hyodong Ryu (Andong National University), Gyu-Tae Park (Amkor Technology Korea Inc.), Ho-Young Son (SK hynix Inc.), Nam-Seog Kim (SK hynix Inc.), Cheol-Woong Yang (Sungkyunkwan University), Young-Cheon Kim (Andong National University), Jeong Sam Han (Andong National University), and Young-Bae Park (Andong National University)

Ag Diffusion Control Through Sn on a Sequential Plating-Based Bumping Process 2252

Abderrahim EL Amrani (Université de Sherbrooke), Etienne Paradis (Université de Sherbrooke), David Danovitch (Université de Sherbrooke), and Dominique Drouin (Université de Sherbrooke)

Mechanical Reliability Assessment of Cu_6Sn_5 Intermetallic Compound and Multilayer Structures in Cu/Sn Interconnects for 3D IC Applications 2258
    Jui-Yang Wu (National Taiwan University), C. Robert Kao (National
    Taiwan University), and Jenn-Ming Yang (University of California, Los
    Angeles)

A Study on the Anchoring Polymer Layer (APL) Anisotropic Conductive Films (ACFs) with Self-Exposed Conductive Particles Surface for Ultra-Fine Pitch Chip-on-Glass (COG) Applications 2266
    Dal-Jin Yoon (KAIST) and Kyung-Wook Paik (Korea Advanced Institute of
    Science and Technology)

Bending Properties of Fine Pitch Flexible CIF (Chip-in-Flex) Packages Using APL (Anchoring Polymer Later) ACFs (Anisotropic Conductive Films) 2272
    Ji-Hye Kim (KAIST, Korea), Dal-Jin Yoon (KAIST, Korea), and Kyung-Wook
    Paik (KAIST, Korea)

Effects of the Curing Properties and Viscosities of Non-Conductive Films (NCFs) on the Sn-Ag Solder Bump Joint Morphology and Reliability 2278
    HanMin Lee (KAIST, South Korea), SeYong Lee (KAIST, South Korea),
    SangMyung Shin (KAIST, South Korea), TaeJin Choi (Doosan Corporation
    Electro-Materials BG, South Korea), SooIn Park (Doosan Corporation
    Electro-Materials BG, South Korea), and Kyung-Wook Paik (KAIST, South
    Korea)

Experimental Investigations on Vertical Ultrasonic Assisted Low Temperature Sintering Process 2284
    Henning Seefisch (Leibniz Universität Hannover) and Jens Twiefel
    (Leibniz Universität Hannover)

Pressureless Transient Liquid Phase Sintering Bonding of Sn-58Bi with Ni Particles for High-Temperature Packaging Applications 2290
    Kyung Deuk Min (Sungkyunkwan University, Republic of Korea), Kwang-Ho
    Jung (Sungkyunkwan University, Republic of Korea), Choong-Jae Lee
    (Sungkyunkwan University, Republic of Korea), and Seung-Boo Jung
    (Sungkyunkwan University, Republic of Korea)

Epoxy/ Triazine Copolymer Resin System for High Temperature Encapsulant Applications 2296
    Jiaxiong Li (Georgia Institute of Technology), Chao Ren (Georgia
    Institute of Technology), Kyoung-sik Moon (Georgia Institute of
    Technology), and Ching-ping Wong (Georgia Institute of Technology)

Low Temperature Ag-Ag Direct Bonding Technology for Advanced Chip-Package Interconnection 2302
    Jiaqi Wu (University of California Irvine) and Chin C. Lee (University
    of California Irvine)

Reliability of Micro-Alloyed SnAgCu Based Solder Interconnections for Various Harsh Applications 2309
    Sinan Su (Auburn University), Francy John Akkara (Auburn University),
    Anto Raj (Auburn University), Cong Zhao (Auburn University), Seth
    Gordon (Auburn University), Sharath Sridhar (Auburn University),
    Sivasubramanian Thirugnanasambandam (Auburn University), Sa'd Hamasha
    (Auburn University), Jeffery Suhling (Auburn University), and John
    Evans (Auburn University)

Wideband Low-Profile Ka-Band Microstrip Antenna with Low Cross Polarization Using Asymmetry AMC Structure 2318

Mei Xue (Institute of Microelectronics of the Chinese Academy of Sciences), Weikang Wan (Institute of Microelectronics of the Chinese Academy of Sciences), Qidong Wang (Institute of Microelectronics of the Chinese Academy of Sciences), and Liqiang Cao (Institute of Microelectronics of the Chinese Academy of Sciences)

Automatic Transient Thermal Impedance Tester for Quality Inspection of Soldered and Sintered Power Electronic Devices on Panel and Tile Level 2324

Maximilian Schmid (Technische Hochschule Ingolstadt), Bhogaraju Sri Krishna (Technische Hochschule Ingolstadt), and Gordon Elger (Technische Hochschule Ingolstadt)

Time 0 Void Evolution and Effect on Electromigration 2331

Jiefeng Xu (The State University of New York at Binghamton), Scott McCann (Xilinx, Inc.), Huayan Wang (The State University of New York at Binghamton), VanLai Pham (The State University of New York at Binghamton), Stephen R. Cain (The State University of New York at Binghamton), Gamal Refai-Ahmed (Xilinx, Inc.), and S.B. Park (The State University of New York at Binghamton)

Quintuple Band Lambda/4 Stub by using Unbalanced Bridged CRLH Transmission Lines 2337

Renuka Bowrothu (University of Florida), Seahee Hwangbo (University of Florida), Haein Kim (University of Florida), and Yong-Kyu Yoon (University of Florida)

Product Level Design Optimization for 2.5D Package Pad Cratering Reliability During Drop Impact 2343

Huayan Wang (State University of New York at Binghamton), Jing Wang (State University of New York at Binghamton), Jiefeng Xu (State University of New York at Binghamton), Vanlai Pham (State University of New York at Binghamton), Ke Pan (State University of New York at Binghamton), Seungbae Park (State University of New York at Binghamton), Hohyung Lee (Xilinx Inc, USA), and Gamal Refai-Ahmed (Xilinx Inc, USA)

Microstructures of Pb-Free Solder Joints by Reflow and Thermo-Compression Bonding (TCB) Processes 2349

Youngja Kim (Samsung Electronics), Jinho Hah (Georgia Institute of Technology), Patxi Fernandez-Zelaia (Georgia Institute of Technology), Sangil Lee (Georgia Institute of Technology), Leroy Christie (ASM Pacific Technology), Paul Houston (Engent Inc.), Shreyes Melkote (Georgia Institute of Technology), Kyoung-Sik Moon (Georgia Institute of Technology), and Ching-Ping Wong (Georgia Institute of Technology)

Reduction of Ag Corrosion Rate During Decapsulation of Ag Wire Bond Packages 2359

Young-Ja Kim (Samsung Electronics, Korea), Jinho Hah (Georgia Institute of Technology), Kyoung-Sik (Jack) Moon (Georgia Institute of Technology), and C. P. Wong (Georgia Institute of Technology)

Author Index

Foreword

On behalf of the Program Committee and Executive Committee, it is our pleasure to welcome you to the 69th Electronic Components and Technology Conference (ECTC) which will be held at The Cosmopolitan of Las Vegas in Las Vegas, Nevada from May 28-31, 2019. This premier international conference is sponsored by the IEEE Electronic Packaging Society (EPS). The ECTC Program Committee has selected over 350 papers which will be presented in 36 oral sessions and five interactive presentation session including one interactive presentation session exclusively featuring papers by student authors. The oral sessions will feature selected papers on key topics such as fan-out packaging, wafer-level packaging, flip-chip packaging, 3D/TSV technologies, design for RF performance and signal/power integrity, thermal and mechanical modeling, optoelectronics packaging, materials and reliability. Interactive presentation sessions will showcase papers in a format that encourages more in-depth discussion and interaction with authors about their work.

Authors from over twenty countries are expected to present their work at the 69th ECTC, covering ongoing technology development within established disciplines or emerging topics of interest for our industry such as additive manufacturing, heterogeneous integration, flexible and wearable electronics.

ECTC will also feature six special sessions with invited industry experts covering several important and emerging topic areas. On Tuesday, May 29 at 10 a.m., W. Hong Yeo and Mikel Miller will chair a special session covering "Transient Electronics: A Green Revolution for Packaging". On the same day at 2 p.m., Rena Huang and Soon Jang will chair a session focused on Photonics on the Cutting-Edge of Technology Evolution. Tuesday evening will also include the ECTC Panel Session at 7:30 p.m. chaired by IEEE EPS President Avi Bar-Cohen and Karlheinz Bock, where young researchers will share their visions of future packaging technologies and participate in discussions with experts in the field.

This conference will also feature a Women's Panel and Reception jointly organized by ECTC and ITherm on Wednesday, May 29 at 6:30pm. This year, panelists from around the globe will share their perspectives on efforts to enhance the participation of women in engineering, and the panel will be chaired by Kristina Young-Fisher and Cristina Amon. On the same day at 7:30 p.m., Tanja Braun will chair the ECTC Plenary Session titled "Sensors and Packaging for Autonomous Driving". In this plenary session, experts will address the challenges and demands for sensors and packages for autonomous driving along the value chain. On Thursday, May 31 at 8 p.m., the IEEE EPS Seminar titled "Roadmap of IC Packaging Materials to Meet Next-Generation Smartphone Performance Requirements" will be moderated by Yasumitsu Orii and Sheigenori Aoki from the High-Density Substrates & Boards Technical Committee of the IEEE EPS Society.

Supplementing the technical program, ECTC also offers Professional Development Courses (PDCs) and the Technology Corner exhibits. Co-located with the IEEE iTHERM Conference this year, the 69th ECTC will offer eighteen PDCs, organized by the PDC Committee chaired by Kitty Pearsall and Jeffrey Suhling. The PDCs will take place on Tuesday, May 28 and are taught by distinguished experts in their respective fields. The Technology Corner will showcase the latest technologies and products offered by leading companies in the electronic components, materials, packaging and services fields. More than one hundred Technology Corner exhibits will be open Wednesday and Thursday starting at 9 a.m. ECTC also offers attendees numerous opportunities for networking and discussion with colleagues during coffee breaks, daily luncheons and nightly receptions.

Whether you are an engineer, a manager, a student or an executive, ECTC offers something unique for everyone in the microelectronics packaging and components industry. I invite you to make your plans now to join us for the 69th ECTC and be a part of all the exciting technical and professional opportunities. I also take this opportunity to thank our sponsors, exhibitors, authors, speakers, PDC instructors, session chairs, and program committee members, as well as all the volunteers who help make the 69th ECTC a success. We look forward to meeting you in Las Vegas, Nevada May 28 –31, 2019.

Nancy Stoffel
69th ECTC Program Chair
General Electric Research
stoffel@ge.com

Mark Poliks
69th ECTC General Chair
Binghamton University
mpoliks@binghamton.edu

2019 Executive Committee

General Chair
Mark Poliks
Binghamton University
mpoliks@binghamton.edu

Vice-General Chair
Christopher Bower
X-Celeprint Inc.
cbower@x-celeprint.com

Program Chair
Nancy Stoffel
GE Research
nstoffel1194@gmail.com

Assistant Program Chair
Rozalia Beica
DuPont
rozalia.beica@dupont.com

Web Administrator
Ibrahim Guven
Virginia Commonwealth University
iguven@vcu.edu

Jr. Past General Chair
Sam Karikalan
Broadcom Inc.
sam.karikalan@broadcom.com

Sr. Past General Chair
Henning Braunisch
Intel Corporation
braunisch@ieee.org

Sponsorship Chair
Wolfgang Sauter
GLOBALFOUNDRIES
wolfgang.sauter@globalfoundries.com

Finance Chair
Patrick Thompson
Texas Instruments, Inc.
patrick.thompson@ti.com

Publications Chair
Steve Bezuk
sbezuk@gmail.com

Publicity Chair
Eric Perfecto
eric.perfecto.us@ieee.org

Treasurer
Tom Reynolds
T3 Group LLC
t.reynolds@ieee.org

Exhibits Chair
Joe Gisler
Vector Associates
gisler.h.dr@ieee.org

Exhibits Co-Chair
Alan Huffman
Micross Advanced Interconnect Technology
alan.huffman@micross.com

Arrangements Chair
Lisa Renzi Ragar
Renzi & Company, Inc.
lrenzi@renziandco.com

EPS Representative
C. P. Wong
Georgia Institute of Technology
cp.wong@mse.gatech.edu

2019 Program Committee

Applied Reliability

Chair
Deepak Goyal
Intel Corporation
deepak.goyal@intel.com

Assistant Chair
Darvin R. Edwards
Edwards Enterprise Consulting, LLC
darvin.edwards1@gmail.com

Tim Chaudhry
Amkor Technology, Inc.

Tz-Cheng Chiu
National Cheng Kung University

Vikas Gupta
Texas Instruments, Inc.

Sandy Klengel
Fraunhofer Institute for Microstructure of Materials and Systems

Pilin Liu
Intel Corporation

Varughese Mathew
NXP Semiconductors

Toni Mattila
Aalto University

Keith Newman
AMD

Donna M. Noctor
Nokia

S. B. Park
Binghamton University

Lakshmi N. Ramanathan
Microsoft Corporation

René Rongen
NXP Semiconductors

Scott Savage
Medtronic Microelectronics Center

Jeffrey Suhling
Auburn University

Pei-Haw Tsao
Taiwan Semiconductor Manufacturing Company, Ltd.

Dongji Xie
NVIDIA Corporation

Assembly & Manufacturing Technology

Chair
Mark Gerber
Advanced Semiconductor Engineering Inc.
mark.gerber@aseus.com

Assistant Chair
Jin Yang
Intel Corporation
jin1.yang@ieee.org

Sai Ankireddi
Soraa, Inc

Christo Bojkov
Qorvo

Garry Cunningham
NGC

Habib Hichri
Suss Microtech Photonic Systems Inc.

Paul Houston
Engent

Li Jiang
Texas Instruments

Chunho Kim
Medtronic Corporation

Wei Koh
Pacrim Technology

Ming Li
ASM Pacific Technology

Debendra Mallik
Intel Corporation

Jae-Woong Nah
IBM Corporation

Valerie Oberson
IBM Canada Ltd

Chandradip Patel
Schlumberger Technology Corporation

Shichun Qu
Intersil, a Renesas Company

Paul Tiner
Texas Instruments

Andy Tseng
JSR Micro

Jan Vardaman
Techsearch International

Yu Wang
Sensata Technologies

Shaw Fong Wong
Intel Corporation

Wei Xu
Huawei

Tonglong Zhang
Nantong Fujitsu Microelectronics Ltd.

Emerging Technologies

Chair
Florian Herrault
HRL Laboratories, LLC
fgherrault@hrl.com

Assistant Chair
Benson Chan
Binghamton University
chanb@binghamton.edu

Isaac Robin Abothu
Siemens Healthineers

Meriem Akin
Robert Bosch GmBH

Vasudeva P. Atluri
Renavitas Technologies

Karlheinz Bock
Technische Universitat Dresden

Vaidyanathan Chelakara
Acacia Communications

Rabindra N. Das
MIT Lincoln Labs

Dongming He
Qualcomm Technologies, Inc.

TengFei Jiang
University of Central Florida

Jong-Hoon Kim
Washington State University Vancouver

Ahyeon Koh
Binghamton University

Ramakrishna Kotlanka
Analog Devices

Santosh Kudtarkar
Analog Devices

Kevin J. Lee
Qorvo Corporation

Zhuo Li
Fudan University

Chukwudi Okoro
Corning

Bharat Penmecha
Intel Corporation

C. S. Premachandran
GLOBALFOUNDRIES

Jintang Shang
Southeast University

Rohit Sharma
IIT Ropar

Nancy Stoffel
GE Research

Liu Yang
IBM

Jimin Yao
Intel Corporation

W. Hong Yeo
Georgia Institute of Technology

Hongqing Zhang
IBM Corporation

High-Speed, Wireless & Components

Chair
Wendem Beyene
Intel Corporation
wendem.beyene@intel.com

Assistant Chair
Lianjun Liu
NXP Semiconductor, Inc.
lianjun.liu@NXP.com

Amit P. Agrawal
Microsemi Corporation

Kemal Aygun
Intel Corporation

Eric Beyne
IMEC

Prem Chahal
Michigan State University

Zhaoqing Chen
IBM Corporation

Charles Nan-Cheng Chen
HiSilicon Technologies

Craig Gaw
NXP Semiconductor

Abhilash Goyal
Velodyne LIDAR, Inc.

Xiaoxiong (Kevin) Gu
IBM Corporation

Rockwell Hsu
Cisco Systems, Inc.

Lih-Tyng Hwang
National Sun Yat-Sen University

Bruce Kim
City University of New York

Timothy G. Lenihan
TechSearch International

Rajen M Murugan
Texas Instruments

Nanju Na
Xilinx

Dan Oh
Samsung

P. Markondeya Raj
Florida International University

Hideki Sasaki
Renesas Electronics Corporation

Li-Cheng Shen
Wistron NeWeb Corporation

Jaemin Shin
Qualcomm Corporation

Manos M. Tentzeris
Georgia Institute of Technology

Maciej Wojnowski
Infineon Technologies AG

Yong-Kyu Yoon
University of Florida

Interconnections

Chair
Wei-Chung Lo
ITRI
lo@itri.org.tw

Assistant Chair
Dingyou Zhang
Broadcom Inc.
dingyouzhang.brcm@gmail.com

Thibault Buisson
Yole Développement

Jian Cai
Tsinghua University

William Chen
Advanced Semiconductor Engineering, Inc.

David Danovitch
University of Sherbrooke

Rajen Dias
Amkor Technology, Inc.

Bernd Ebersberger
Infineon Technologies

Takafumi Fukushima
Tohoku University

Tom Gregorich
Zeiss Semiconductor Manufacturing Technology

Kangwook Lee
Amkor Technology Korea

Steward Lee

Li Li
Cisco Systems, Inc.

Changqing Liu
Loughborough University

Nathan Lower
Rockwell Collins, Inc.

James Lu
Rensselaer Polytechnic Institute

Voya Markovich
Microelectronic Advanced Hardware Consulting, LLC

Lou Nicholls
Amkor Technology, Inc.

Peter Ramm
Fraunhofer EMFT

Katsuyuki Sakuma
IBM Corporation

Lei Shan
IBM Corporation

Ho-Young Son
SK Hynix

Jean-Charles Souriau
CEA Leti

Chuan Seng Tan
Nanyang Technological University

Matthew Yao
GE Energy Management

Materials & Processing

Chair
Mikel Miller
EMD Performance Materials
mikel.miller@emdgroup.com

Assistant Chair
Tanja Braun
Fraunhofer IZM
tanja.braun@izm.fraunhofer.de

Yu-Hua Chen
Unimicron

Qianwen Chen
IBM Research

Bing Dang
IBM Research

Yung-Yu Hsu
Apple Inc.

Lewis Huang
Senju Electronic

C. Robert Kao
National Taiwan University

Chin C. Lee
University of California, Irvine

Alvin Lee
Brewer Science

Yi (Grace) Li
Intel Corporation

Ziyin Lin
Intel Corporation

Yan Liu
Medtronic Inc. USA

Daniel D. Lu
Henkel Corporation

Joon-Seok Oh
Samsung Electro-Mechanics

Praveen Pandojirao-S
Johnson & Johnson

Mark Poliks
Binghamton University

Dwayne Shirley
Inphi

Ivan Shubin
Oracle

Bo Song
HP Inc.

Yoichi Taira
Keio University

Lejun Wang
Qualcomm Technologies, Inc.

Frank Wei
Disco Japan

Kimberly Yess
Brewer Science

Myung Jin Yim
Apple

Hongbin Yu
Arizona State University

Packaging Technologies

Chair
Dean Malta
Micross Advanced Interconnect Technology
Dean.Malta@micross.com

Assistant Chair
Luke England
GLOBALFOUNDRIES
luke.england@globalfoundries.com

Daniel Baldwin
H.B. Fuller Company

Bora Baloglu
Amkor Technology

Jie Fu
Apple

Mike Gallagher
DuPont

Ning Ge
Consultant

Allyson Hartzell
Veryst Engineering

Kuldip Johal
Atotech

Beth Keser
Intel Corporation

Young-Gon Kim
Integrated Device Technology, Inc.

Andrew Kim
Intel Corporation

John Knickerbocker
IBM Corporation

Albert Lan
Applied Materials

John H. Lau
ASM Pacific Technology

Jaesik Lee
Nvidia

Markus Leitgeb
AT&S

Luu Nguyen
Texas Instruments Inc.

Deborah S. Patterson
Harbor Electronics, Inc.

Raj Pendse
Facebook FRL (Facebook Reality Labs)

Subhash L. Shinde
Notre Dame University

Joseph W. Soucy
Draper Laboratory

Peng Su
Juniper Networks

Kuo-Chung Yee
Taiwan Semiconductor Manufacturing Corporation, Inc.

Christophe Zincke
Advanced Semiconductor Engineering, Inc.

Photonics

Chair
Ping Zhou
LDX Optronics, Inc.
pzhou@ldxoptronics.com

Assistant Chair
Z. Rena Huang
Rensselaer Polytechnic Institute
zrhuang@ecse.rpi.edu

Mark Beranek
Naval Air Systems Command

Stephane Bernabe
CEA Leti

Fuad Doany
IBM Research

Gordon Elger
Technische Hochschule Ingolstadt

Takaaki Ishigure
Keio University

Ajey Jacob
GLOBALFOUNDRIES

Soon Jang
ficonTEC USA

Harry G. Kellzi
Teledyne Microelectronic Technologies

Richard Pitwon
Resolute Photonics Ltd

Alex Rosiewicz
A2E Partners

Henning Schroeder
Fraunhofer IZM

Andrew Shapiro
JPL

Masato Shishikura
Oclaro Japan

Masao Tokunari
IBM Corporation

Shogo Ura
Kyoto Institute of Technology

Stefan Weiss
II-VI Laser Enterprise GmbH

Feng Yu
Huawei Technologies Japan

Thomas Zahner
OSRAM Opto Semiconductors GmbH

Thermal/Mechanical Simulation & Characterization

Chair
Przemyslaw Gromala
Robert Bosch GmbH
Przemyslawjakub.gromala@de.bosch.com

Assistant Chair
Ning Ye
Western Digital
ning.ye@wdc.com

Christopher J. Bailey
University of Greenwich

Kuo-Ning Chiang
National Tsinghua University

Xuejun Fan
Lamar University

Nancy Iwamoto
Honeywell Performance Materials and Technologies

Pradeep Lall
Auburn University

Chang-Chun Lee
National Tsing hua University (NTHU)

Yong Liu
ON Semiconductor

Sheng Liu
Wuhan University

Erdogan Madenci
University of Arizona

Tony Mak
Wentworth Institute of Technology

Karsten Meier
Technische Universität Dresden

Erkan Oterkus
University of Strathclyde

Sandeep Sane
Intel Corporation

Suresh K. Sitaraman
Georgia Institute of Technology

Wei Wang
Qualcomm Technologies, Inc.

G. Q. (Kouchi) Zhang
Delft University of Technology (TUD)

Tieyu Zheng
Microsoft Corporation

Jiantao Zheng
Hisilicon

Interactive Presentations

Chair
Michael Mayer
University of Waterloo
mmayer@uwaterloo.ca

Assistant Chair
Pavel Roy Paladhi
IBM Corporation
Pavel.Roy.Paladhi@ibm.com

Swapan Bhattacharya
Engent Inc.

Rao Bonda
Amkor Technology

Mark Eblen
Kyocera International SC

Ibrahim Guven
Virginia Commonwealth University

Alan Huffman
Micross Advanced Interconnect Technology

Jeffrey Lee
iST-Integrated Service Technology Inc.

Nam Pham
IBM Corporation

Mark Poliks
Binghamton University

Patrick Thompson
Texas Instruments, Inc.

Kristina Young-Fisher
GLOBALFOUNDRIES

Professional Development Courses

Chair
Kitty Pearsall
Boss Precision, Inc.
kitty.pearsall@gmail.com

Assistant Chair
Jeffrey Suhling
Auburn University
jsuhling@auburn.edu

Vijay Khanna
IBM Corporation

Eddie Kobeda
Nypro, A Jabil Company

Lakshmi N. Ramanathan
Microsoft Corporation

Twist Testing for Flexible Electronics

Justin H. Chow[1], Jeffrey Meth[2] and Suresh K. Sitaraman[1]

[1] Computer-Aided Simulation of Packaging Reliability (CASPaR) Lab
G. W. Woodruff School of Mechanical Engineering
Flexible Wearable Electronics Advanced Research
Georgia Institute of Technology
Atlanta, GA 30332-0405, USA
suresh.sitaraman@me.gatech.edu

[2] DuPont Electronics & Imaging
Wilmington, DE 19803

Abstract— **Electronic devices have found use in a variety of flexible form factors such as wearable health monitoring devices, fitness trackers integrated into garments, etc. These form factors expose the electronic components (printed traces, passives, active devices, etc.) to a wide range of mechanical strains that are not typically seen in traditional rigid devices. In fact, the utility of these devices depends on its ability to stretch, bend, fold, and/or twist. As such, there is a need for methods and standards for testing these devices under these mechanical loads. In our ongoing work, we have conducted extensive stretch and bend testing of conductors, co-planar waveguides, antennas, and other elements. Both monotonic and fatigue testing of some of these components have also been carried out.**

In this paper we describe the development of a test apparatus that can subject flexible electronic test coupons to controlled twists while monitoring in-situ the electrical resistance of the printed conductor. Unlike in stretch testing, in twist testing buckling of the samples is a concern, and thus, appropriate models have been used to ensure that the substrates are subjected to pure twisting without any apparent buckling of the sample. Samples were subjected to twist angles ranging from ± 30° to ± 135° and the resulting resistance was measured in-situ. Both polyimide and PET substrates with screen-printed silver conductors have been tested and the results have been presented for both monotonic and cyclic loadings. The thermal coefficient of resistance (TCR) of screen-printed silver ink has been measured to be 0.00233 per °C and accounted for in the twist resistance measurements to reduce the temperature effect on resistance during long-term cyclic twist testing. Cyclic twisting to ±135° or shear strains of ± 2.99 x 10-3 induced minimal changes in resistance (less than 1.5%) over 5000 cycles.

Keywords- flexible electronics, mechanical testing, reliability, twist, torsion, thermal coefficient of resistance

I. INTRODUCTION

Electronic devices have found use in a variety of flexible form factors such as wearable health monitoring devices, fitness trackers integrated into garments, etc. These form factors expose the electronic components (printed traces, passives, active devices, etc.) to a wide range of mechanical loads that are not typically seen in traditional rigid devices. In fact, the utility of these devices depends on their ability to

stretch, bend, fold, and/or twist. As such, there is a need for methods and standards for testing these devices under these mechanical loads. While there are several existing studies that focus on stretch and bend testing of flexible electronics, there is comparatively little or no systematic study of flexible electronics under twist [1], especially with in-situ electrical monitoring. Twist is a mode of mechanical loading that is common to wearables. For example, a fitness shirt with embedded traces needs to function even if the wearer twists his torso. On average, a human back can twist up to 30° in either direction [2]. When subjected to machine washing, wearable may experience even more extreme twisting. Thus the reliability of flexible electronics under twisting is important.

II. THEORY

A. Strains under Torsion

An approximation of the shear strain distribution within the narrow rectangular cross-section of a ribbon-like film subjected to torsion (Figure 1) is

$$\gamma_{zx}(x, y) = -2 * \theta_z * y \qquad (1)$$

$$\gamma_{zy}(x, y) = 0 \qquad (2)$$

where $\gamma_{zx}(x,y)$ and $\gamma_{zy}(x,y)$ are the shear strains as a function of x and y in the zx and zy directions respectively. θ_z is the twist angle per length (radians/mm) about the axial (z) direction. This approximation is well known and is included in many textbooks on mechanics of materials (e.g. [3]). This equation will be used to approximate the strain even though it is limited by some simplifying assumptions such as small deformations/angles. Eq. 1 assumes that there is no applied axial force, and that the deformations are sufficiently small so that no axial strains are induced. With increasing twist angle, the overall length of the strip will decrease, and unless the ends are allowed to move, there will be axial strain in the z direction. These equations are not valid near $x = \pm b$ as the shear strains must go to zero at the corners. One can see from Eq. 1 that the shear strain is constant along the top and bottom surfaces ($y=\pm h$) of the substrate. Thus, conductive traces of the different widths should experience the same shear strain

978-1-7281-1500-9/19 $31.00 © 2019 IEEE

for a given twist. For this paper, it will be assumed that the conductive trace experiences the same shear strain as the top surface of the substrate.

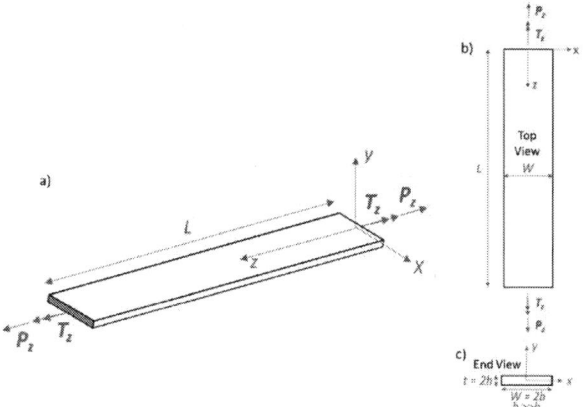

Figure 1. a) Ribbon-like film subjected to axial torsion and force b) A top view c) An end view of the narrow rectangular cross-section.

Further analysis of strain using large deformation theory and finite-element analysis is ongoing for substrates with traces on them.

B. Buckling Shapes and Twist Limits

When thin ribbon-like films are twisted, various shapes can be achieved depending on the magnitude of the applied twist angle and axial tension forces. Figure 2 shows photographs of five of the possible mode shapes using a PET substrate. Chopin and Kudrolli [4] created a phase diagram for the different mode shapes based on the unitless parameters of normalized twist angle and normalized axial tension. The normalized twist angle is defined as

$$\eta = \frac{\alpha}{L/W} \qquad (3)$$

where α is the angle of twist, and L and W are the length and width of the sample respectively. The normalized axial force is defined as

$$F = \frac{P}{E \cdot t \cdot W} \qquad (4)$$

where P is the applied axial force, E is the sample's Young's modulus, and t is the thickness of the sample.

At low twist angles, the strips take on the smooth, well-behaved helicoid shape as seen in Figure 2b. When tension forces are kept low and the twist angle is increased, the strip buckles longitudinally (axial or z direction) (Figure 2c) when η reaches a critical angle of η_L. With further twisting above η_L, the strip takes on a creased helicoid shape (Figure 2d). The longitudinal buckling shapes and creased helicoid shapes are only possible when F is low (approximately below 4 x 10⁻³). Chopin and Kudrolli describe the region of longitudinal buckling with the following equation that matches their empirical results:

$$\eta_L = \sqrt{24F} + 6\frac{t}{W} \qquad (5)$$

At higher values of F, the strip transitions from helicoid shapes to a transverse (width or x direction) buckling shape Figure 2e.

For flexible electronic applications, the helicoid shape (Figure 2b) would seem to be preferred over the creased helicoid shape. The creased helicoid shape concentrates the strain into linear regions to form creases and sharp corners are formed where two creases meet. Over several cycles, these corners can become initiation sites for tears in the substrate.

III. SAMPLE AND APPARATUS DESIGN AND FABRICATION

A. General Sample Design

A standard set of samples was designed for use in various mechanical tests. This sample design has been used for both stretch testing [5] and bend testing [6]. The sample design is shown in Figure 3 and consists of six different samples with conductor line-widths ranging from 0.25 mm to 10 mm to investigate any dependency on line-width. Each sample has a substrate width of 25 mm and a total sample length of 175 mm. These samples are intended to be tested with an initial grip separation of 100 mm. These overall dimensions were selected in part to be compatible with ASTM D882-12 which is often used to test plain substrates. Evenly spaced tick marks on both edges of each sample provide for easy alignment as well as a means of optically measuring local strain. Electrical test pads are configured for continuous, in-situ 4-wire resistance measurement.

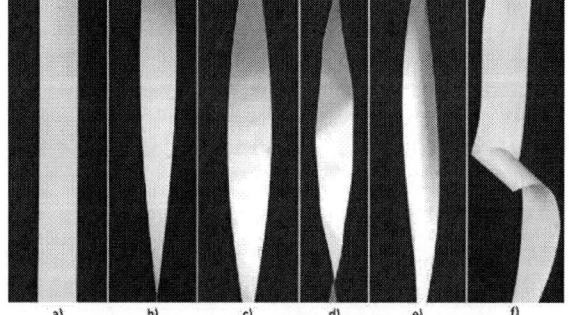

Figure 2. Potential shapes of a thin twisted ribbon: a) the initial untwisted, painted PET film with L x W x t dimensions of 200 mm x 20 mm x 0.127 mm, b) helicoid shape, c) longitudinal buckling, d) creased helicoid shape, e) transverse buckling, and f) looped shape

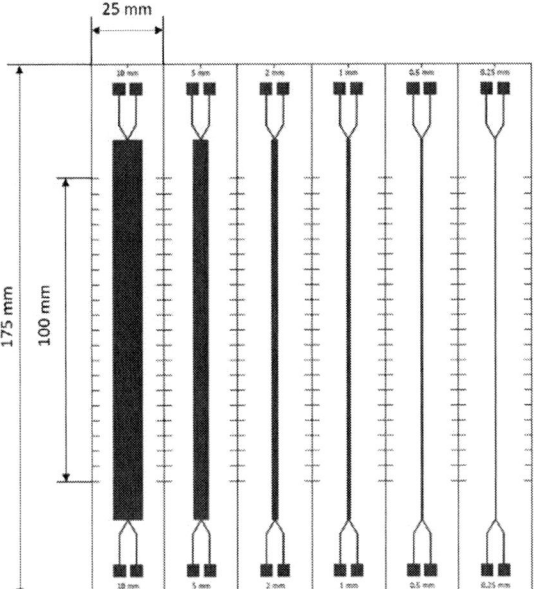

Figure 3. Standard sample design for mechanical testing of flexible electronics.

This design was screen-printed at DuPont using their 5025 silver ink on both Kapton 500HN (polyimide) and Melinex ST506 (PET) substrates as shown in Figure 4. After printing and curing the conductive ink, an additional dielectric material (DuPont 5018 UV Dielectric) was used to encapsulate the ink. Substrates are approximately 127 µm thick and the 5025 ink is approximately 10-12 µm thick.

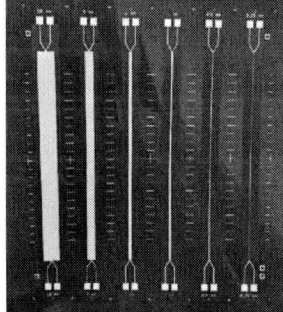
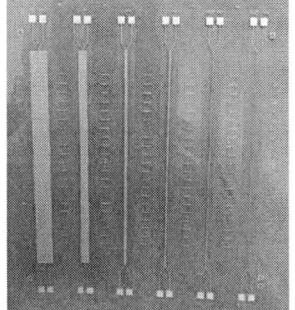

Kapton 500HN (Polyimide)
Silver Ink: 5025
UV Encapsulant: 5018

Melinex ST506 (PET)
Silver Ink: 5025
UV Encapsulant: 5018

Figure 4. Kapton and Melinex samples with screen-printed silver inks of different widths

B. Twist-Specific Samples

In order to evaluate whether the general samples shown above would be adequate for twist testing, the helicoid twist limits of the Kapton and PET samples were first determined using Eq. 5 assuming there was no applied tension force. With no applied tension force, the twist limit is material independent. The maximum twist angles are shown in Table 1 as a function of sample width. For both substrates the maximum twist angle to remain in the helicoid shape is about

7° for the general sample design with widths of 25 mm. This twist limit is rather low, so in order to allow for a larger twist range, the general samples were cut down to widths of 10 mm.

Table 1. Maximum twist angle for a given substrate width with no axial force

	Sample Width (mm)					
	1	5	10	15	20	25
Max Angle (deg)	4366	175	44	19	11	7

C. Design and Fabrication of a Twist-Testing Apparatus

While instruments for stretch testing materials are readily available on the commercial market, twist or torsion equipment is more specialized. Twist testers that allow for sample shortening and in-situ electrical measurements are not available commercially. Thus a twist testing apparatus was designed that could apply controlled axial twists up to an infinite number of rotations in either direction in both a monotonic and cyclic manner. Provisions were made for in-situ monitoring of 4-wire resistance, angle of twist and reaction torque. The apparatus allowed for the sample to shorten as it twisted to reduce any axial strains in the sample.

Figure 5 shows the actual experimental setup that was fabricated in-house and used for testing. The setup consists of a computer-controlled servo motor with an absolute magnetic rotary encoder to measure twist angle. The servo motor is connected to a hollow drive shaft via a timing belt and gears. The rotating end of the sample is fixed to the hollow drive shaft which in turn is supported by combination radial/axial thrust bearings. The electrical resistance of the printed conductor is probed via a set of pogo pins that are clamped on both ends of the sample. The lead wires from the pogo pins on the rotating side are fed through the hollow shaft and are connected to a slip ring connector. This hollow shaft and slip ring connector allows for an infinite number of rotations without the lead wires being twisted or tangled up. On the fixed end of the setup, the sample is attached to a reaction torque sensor that measures the torque experienced by the sample. The reaction torque sensor is mounted to a linear rail system that allows the sample to shorten during twisting and thus, ensures minimal axial loads are induced. A constant, low-force (~4N) return spring is used to return the fixed end to its initial position after untwisting.

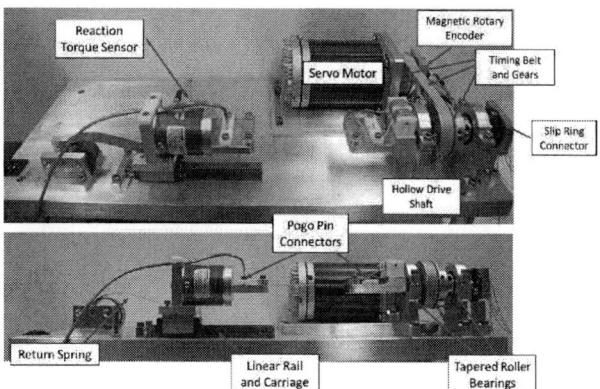

Figure 5. Fabricated experimental apparatus for twist testing.

IV. MONOTONIC TESTING

With the axial load of 4N from the return spring, the allowable twist angle before buckling increases significantly and is presented in Table 2. The Young's modulus of Kapton and Melinex is 2.5 GPa and 3.5 GPa respectively and these values were used for the calculation F in Eq. 5. With the addition of the axial force, 10 mm wide PET and Kapton samples can twist up to 128° and 144° respectively before buckling.

Table 2. Maximum twist angle before buckling occurs for various substrate widths with a 4N axial force

		Sample Width (mm)					
		1	5	10	15	20	25
Max Angle (deg)	Kapton	7535	458	144	74	46	32
	PET	7045	414	128	66	41	28

Some initial monotonic testing was performed on Kapton samples. Two samples with a 0.25 mm and a 10 mm wide trace, both on a 25 mm wide substrate were twisted between ±30°. Samples with 0.5 mm and 1 mm wide traces on 10 mm wide substrates were subsequently tested between ±135°. In both cases, the shape remained helicoid without any apparent buckling, as designed. The resulting normalized resistance vs angle of twist and shear strain curves are plotted Figure 6. The shear strains were calculated using Eq. 1 neglecting the effects of the small 4N axial load. In Figure 6, the shear strains are shown along the top horizontal axis of the plot. For all four cases studied, it is seen that there is practically no change in resistance with the twist angle (or the shear strain).

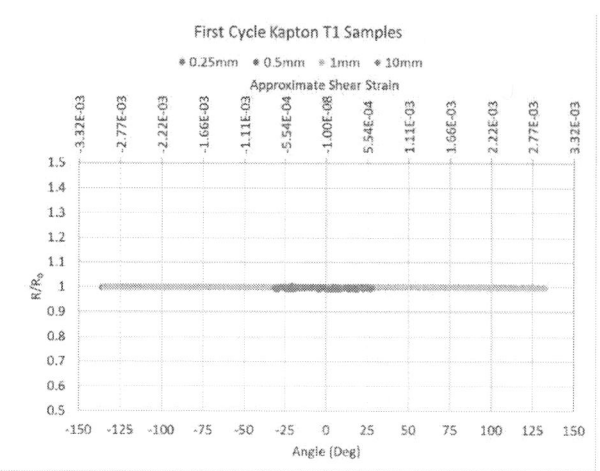

Figure 6. R/R_0 vs twist angle and shear strain for four Kapton samples of various trace and substrate widths. 0.25 mm and 10 mm trace width samples had a 25 mm substrate and were twisted to ±30°. 0.5 mm and 1 mm trace width samples had a 10 mm substrate and were twisted to ±135°.

The experiment was repeated with PET samples of substrate width of 10 mm and an ink width of 0.25, 0.5, 1, 2, and 5 mm through a twist angle of ±135°. The normalized electrical resistance results are shown in Figure 7, and it is seen that there is little or no change in the resistance value, which is consistent with the Kapton results. Upon closer examination of the results, the change in resistance at 135° was found to be under 0.5% for both Kapton and PET. This is because the approximate shear strain at a twist of 135° is only 2.99 x 10^{-3}, which is very low to create resistance change.

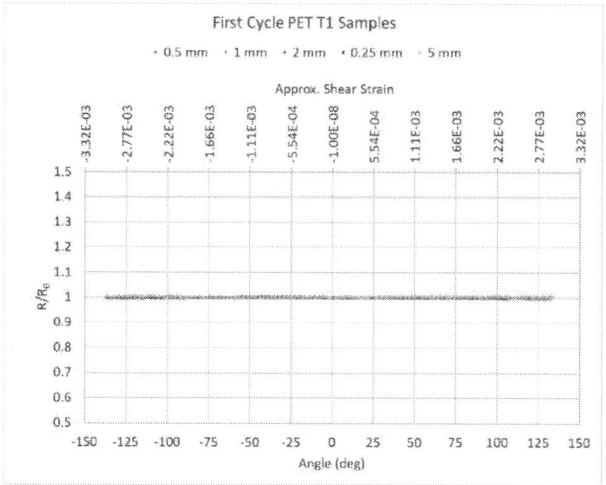

Figure 7. R/R_0 vs twist angle and shear strain for five PET samples of various trace widths subjected to a ±135° twist.

V. CYCLIC TESTING

A. Cyclic Testing Without Temperature-Control

After monotonic testing, the samples were subjected to further cyclic testing. The 25 mm wide Kapton sample with a 0.25 mm trace width was twisted to ±30° for several thousand cycles over the course of a couple days. Figure 8 shows a plot of the absolute resistance at the maximum twist angle vs cycle. The plotted resistances are taken at the maximum twist angle during each cycle. Figure 8a shows that overall, the resistance remains flat over 4500 cycles. However, a closer look at the results (Figure 8b) reveals obvious, non-random variations in resistance over time. These changes in resistance did not appear to be due to twist effects, as the twist amplitude and speed remained constant. One would expect a monotonic change in resistance under these conditions and not a stepped change in resistance in both the positive and negative directions.

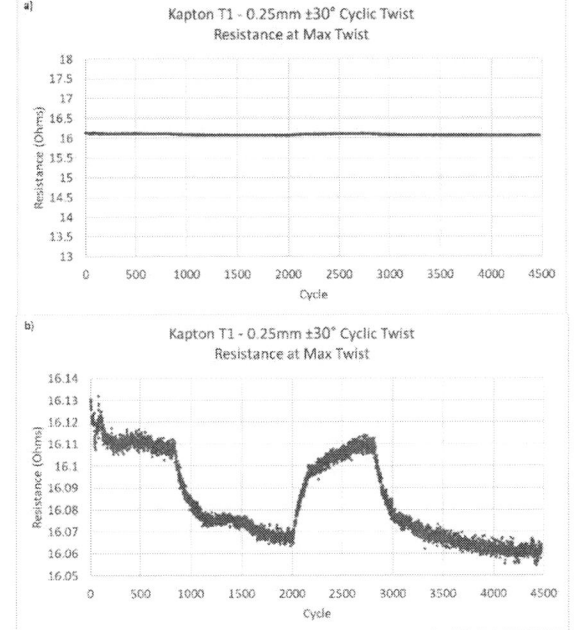

Figure 8. a) Absolute resistance at max twist vs cycle for a 25 mm wide Kapton sample with a 0.25 mm trace width twisted to ±30°. b) The same plot as in (a), but with a finer resistance scale.

The reason behind the stepped change in resistance became evident when the resistance data was plotted as a function of absolute time as shown in Figure 9 rather than cycle. The elevated portions of the plot were during normal work hours while the lower portions of the plot were after normal work hours when the building HVAC system was shut off to conserve energy. Temperature effects appeared to be masking any effects from twisting.

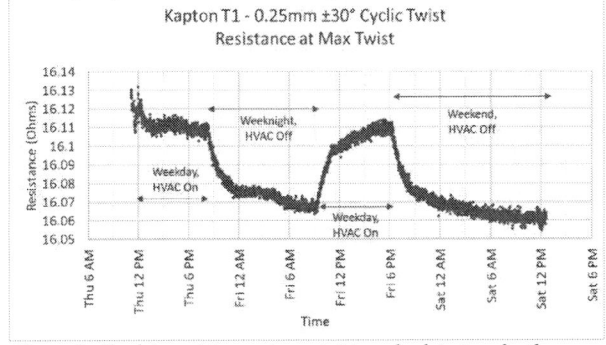

Figure 9. Absolute resistance at max twist vs absolute time for the same sample shown in Figure 8.

B. Cyclic Testing with Temperature-Control

To minimize temperature effects on the twist testing, the entire twist testing setup was enclosed within an insulated box and a PID-controlled thermoelectric cooler was installed to control the temperature as shown in Figure 10. A thermistor was also added to the twist apparatus to monitor sample temperature.

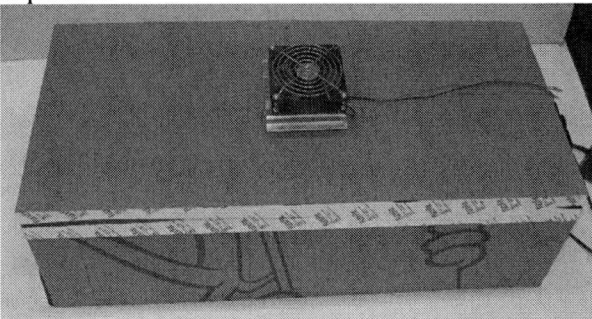

Figure 10. An insulated box and PID controlled thermo-electric cooler for temperature control.

Using this temperature-controlled setup, a 10 mm wide PET sample with a 0.5 mm trace width was cycled to ±135° for about 2500 cycles while both the resistance and temperature were monitored. The results are shown in Figure 11 with resistances measured at the maximum twist angle during each cycle. Overall, there is very little change in absolute resistance over the course of cycling. The maximum change in resistance was less than 0.3%. However, even with temperature control to ±0.5°C, temperature effects still appeared to dominate the resistance behavior. The trend in resistance still closely matched the trend in temperature change.

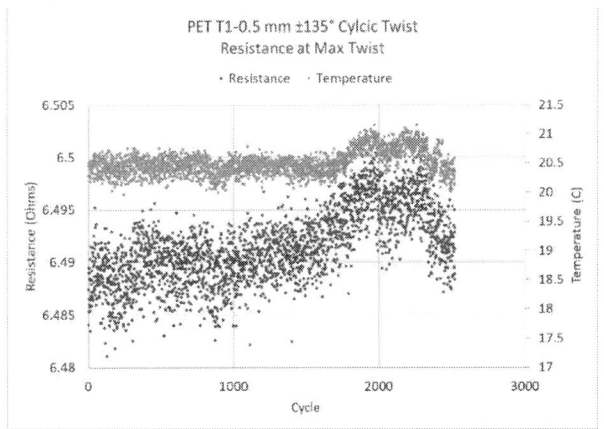

Figure 11. Absolute resistance and temperature vs number of cycles for a 10 mm wide PET sample with a 0.5 mm trace width twisted to ±135°. Absolute resistance values are plotted in blue and are shown on the left axis. Temperature values are plotted in gray above the resistance values and are shown on the right axis.

C. TCR Measurements and Results

Based on the above analysis, it became clear that although the resistance change due to twisting alone was negligible, minor changes in room temperature could affect the resistance values, and thus, such thermal effects on resistance needed to be addressed. Therefore, to mitigate the effects of temperature on twist testing measurements, the thermal coefficient of resistance (TCR) was determined to correct for temperature changes. An entire sheet of uncut PET samples with the six different trace widths was placed in a forced-air oven and the 4-wire resistance of each trace was recorded continuously. The oven's temperature was ramped from approximately 30°C to 70°C in 5°C increments, and the temperature profile is shown in Figure 12. Each temperature set point was held until the standard deviation of the temperature remained within 0.5% for 5 mins. The temperature was monitored with a 10k NTC thermistor with a tolerance of ±0.1°C. The change in absolute resistance was normalized by the starting resistance and plotted as a function of temperature in Figure 13. A linear least squares fit was used to determine the TCR (the slope of the fit) and is shown in Table 3. The fit was excellent with all of the coefficients of determination or R^2 values for each trace width exceeding 0.999. The average of all the TCR values is 0.00233 per °C. This TCR value for DuPont 5025 Ag ink, which is a composite silver material, is approximately 60% of that of pure bulk silver. The TCR of pure bulk silver is 0.0038 per °C [7].

Table 3. TCR values of 5025 ink on PET for the 6 different trace widths

Trace Width (mm)	TCR (per °C)
0.25	0.00227950
0.5	0.00233529
1	0.00236773
2	0.00237705
5	0.00232219
10	0.00227077
Average	0.00232542
Standard Deviation	4.3952E-05

D. Cyclic Testing with Temperature-Control and Correction

Using the measured TCR, the absolute resistances from Figure 11 were corrected to a constant temperature of 25°C. These temperature-controlled and corrected results are shown in Figure 14. Once temperature-corrected, the resistance is almost completely flat and completely independent of temperature.

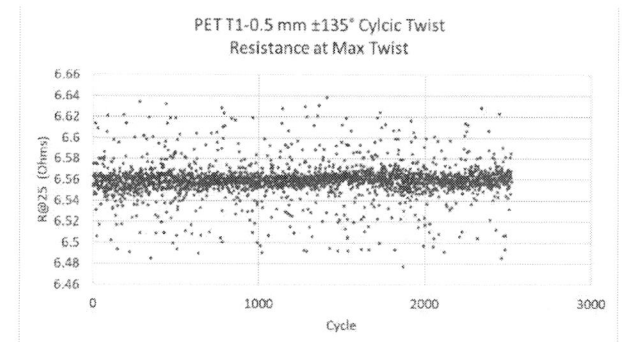

Figure 14. Temperature-corrected absolute resistance and temperature vs number of cycles for a 10 mm wide Kapton sample with a 0.5 mm trace width twisted to ±135°. See Figure 11 for uncorrected results.

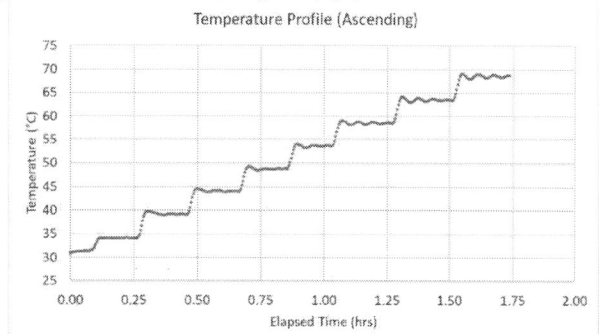

Figure 12. TCR measurement temperature profile from 30°C to 70°C in 5°C increments.

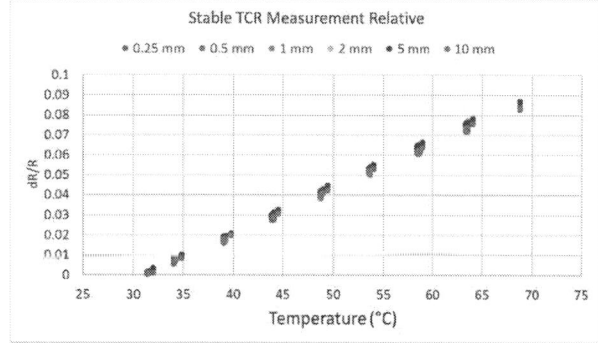

Figure 13. Relative resistance change at each temperature set point for all six traces widths from 0.25 mm to 10 mm.

Additional 10 mm wide PET samples were cycled between ±135° at a twist rate of 2 rpm or 12°/s for several thousand cycles. The temperature-corrected, normalized resistance results of the twist cycling are shown in Figure 15a. From the figure, one can see that the resistance stays essentially flat through at roughly 5500 cycles. Figure 15b provides a more detailed look at the results. There appears to be a slight increase in normalized resistance for some of the trace widths but overall the change in resistances is less than 1.5%. As in monotonic testing, cyclic twist effects appear to be small in comparison to other modes of mechanical loading. However, resistance alone is not an indication of potential damage in the ink. Future SEM imaging may reveal cracks or other damage to the ink structure. Subsequent stretch testing of these samples after twist testing may reveal faster resistance increase than in samples without prior twist testing.

978-1-7281-1500-9/19 $31.00 © 2019 IEEE

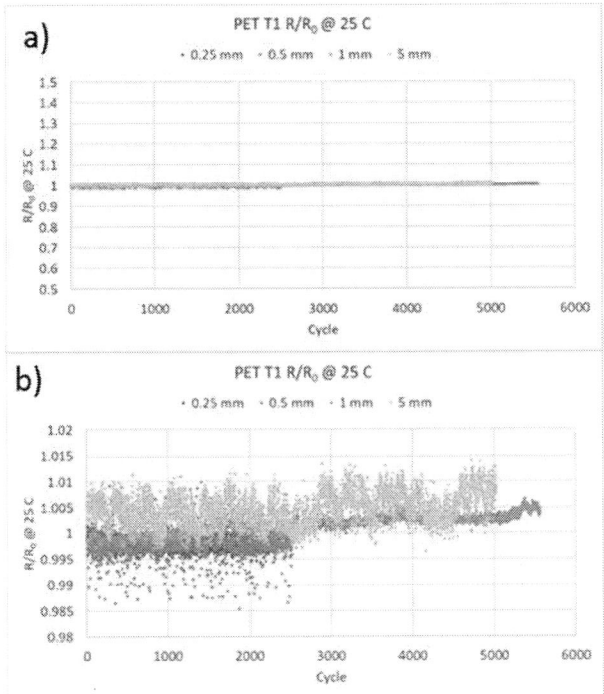

Figure 15. a) R/R₀ at max twist corrected to 25°C vs number of ±135° twist cycles. All samples were on 10 mm wide PET substrates with various trace widths. b) The same plot with a finer scale.

VI. CONCLUSIONS

A twist testing apparatus was successfully designed and fabricated to characterize different flexible electronic components under almost pure torsion. Twist buckling modes and their associated twist limits are an important consideration when designing devices or samples subjected to twist. Monotonic twisting of conductive inks causes minimal changes in measured resistance (less than 0.5%) at a maximum twist of 135° or a shear strain of 2.99×10^{-3}. Temperature effects on resistance dominate over twist effects under long-duration twist cycling, and thus should be controlled and corrected for. The TCR of 5025 ink on PET was measured to be 0.00233 per °C. Cyclic twisting to ±135° induced minimal changes in resistance (less than 1.5%) over 5000 cycles.

ACKNOWLEDGMENT

This material is based, in part, on research sponsored by Air Force Research Laboratory under agreement number FA8650-15-2-5401, as conducted through the flexible hybrid electronics manufacturing innovation institute, NextFlex. The U.S. Government is authorized to reproduce and distribute reprints for Governmental purposes notwithstanding any copyright notation thereon. The views and conclusions contained herein are those of the authors and should not be interpreted as necessarily representing the official policies or endorsements, either expressed or implied, of Air Force Research Laboratory or the U.S. Government. The authors would also like to thank various colleagues at Georgia Tech and DuPont as well as collaborators at Binghamton University for their valuable input and discussion during the ongoing project.

REFERENCES

[1] K. Harris, A. Elias, and H.-J. Chung, "Flexible electronics under strain: a review of mechanical characterization and durability enhancement strategies," *Journal of materials science,* vol. 51, no. 6, pp. 2771-2805, 2016.

[2] R. J. Hindle, M. J. Pearcy, A. T. Cross, and D. H. T. Miller, "Three-dimensional kinematics of the human back," *Clinical Biomechanics,* vol. 5, no. 4, pp. 218-228, 1990/11/01/ 1990.

[3] A. Boresi and R. Schmidt, *Advanced Mechanics of Materials*, 6th ed. John Wiley & Sons, 2003.

[4] J. Chopin and A. Kudrolli, "Helicoids, wrinkles, and loops in twisted ribbons," *Physical Review Letters,* vol. 111, no. 17, p. 174302, 2013.

[5] J. H. Chow *et al.*, "Stretch Testing of Flexible Electronics," in *Flex 2018*, Monterey, CA, 2018.

[6] R. Chen, J. Chow, C. Taylor, J. Meth, and S. Sitaraman, "Adaptive Curvature Flexure Test to Assess Flexible Electronic Systems," in *2018 IEEE 68th Electronic Components and Technology Conference (ECTC)*, 2018, pp. 236-242: IEEE.

[7] R. A. Serway, *Principles of Physics*, 2nd ed. Fort Worth, TX: Saunders College Pub., 1998.

Mechanical Properties and Microstructural Fatigue Damage Evolution in Cyclically Loaded Lead-Free Solder Joints

Sinan Su[1], Mohd Aminul Hoque[2], Md Mahmudur Chowdhury[2], Sa'd Hamasha[1],
Jeffrey C. Suhling[2], John L. Evans[1], Pradeep Lall[2]
[1] Department of Industrial and Systems Engineering
[2] Department of Mechanical Engineering
Center for Advanced Vehicle and Extreme Environment Electronics (CAVE3)
Auburn University, Auburn, AL 36849
Email: smh0083@auburn.edu

Abstract—Solder joints in electronic assemblies are typically subjected to cyclic loadings, either in actual applications or in accelerated life tests. Mismatches in the thermal expansion coefficients of the assembly materials lead to the solder joints being subjected to cyclic (positive/negative) mechanical shear strains and stresses when they are exposed to repetitive thermal cycles with fixed temperature extremes. This cyclic loading leads to thermomechanical fatigue damage that involves damage accumulation, crack initiation, crack propagation, and failure/fracture. The material damage that occurs during cyclic loading becomes immediately evident through the "load drop" and "widening" that occurs in the cyclic stress-strain curves as cycling progresses. Eventually, this damage leads to microcrack formation and recrystallization of the Sn grain structure.

In the current work, a novel technique has been developed to characterize the fatigue damage evolution occurring in the lead free SAC305 and SAC_Q (SAC+Bi) solder joints subjected to shear mechanical cycling. The prime objective of this study was to better understand the damage accumulation and microstructural evolution in lead free solder joints subjected to cyclic fatigue loading. Specifically, both SAC305 and SAC_Q solder joints were fabricated, and then they were cycled for various durations up to values near failure. The fatigue life and accumulated inelastic work were also measured and correlated to the microstructural damage. In addition, nanoindentation tests were performed on the shear cycled joints to study the evolution of the joint creep properties occurring as a function of the duration of cycling and microstructural damage.

Keywords: Solder Joint; Fatigue; Doping; Microstructure

I. INTRODUCTION

Solder joints in realistic applications are typically exposed to cyclic loading, either through thermal cycling or mechanical cycling. In general, the failure modes for mechanical cycling are completely different than from thermal cycling. For thermal cycling, the failure is mainly due to the mismatches of the coefficients of thermal expansion (CTE) within the assembly, where precipitates are coarsened and global recrystallization occurs at the corners of the solder joint, followed by the crack initiation and propagation in the recrystallization region [1, 2]. In contrast, mechanical cycling normally does not lead to major recrystallization, and transgranular failure is observed to be the dominant failure mode. During mechanical cycling, cracks nucleated at the defects or voids created by the plastic deformation induced by cyclic loading, and then propagate within the bulk solder [3]. As mentioned by Anderson, et al. [4], the crack propagation path under transgranular failure is strain range dependent, where the cracks go through the bulk solder under the high strain range, and are directed along the interfacial area at lower strain range.

Moreover, the accelerated life test (ALT) has been utilized as one of the most effective methods in predicting the service life of microelectronic assemblies and identifying the potential failure mechanisms in a short period of time. However, as stated by Borgesen, et al. [5], "results of accelerated testing would be misleading, even useless if the testing could not reflect the performance in service." Bulk samples have been used for decades in testing of the mechanical properties of solder materials. However, studies have revealed that mechanical properties exacted from bulk samples may not reflect those from solder joints implemented in the realistic application, since the solder joints often demonstrate a much more complex structure than bulk samples [4, 6].

When testing bulk samples, several important factors are omitted, which may have significant impacts on the fatigue behaviors of solder joints. For example, an individual solder joint typically consists of either one single oriented anisotropic grain or three grains in 'beach ball' structure, and the shear properties of solder joints are largely depended on the number of grains and their orientation [7]. The Intermetallic compound layer (IMC) formed during the reflow process has also been proven to impact the fatigue life of solder joints, and IMC composition and thickness demonstrated the largest effects [8]. Moreover, the printed circuit board (PCB) surface finish also can have a large influence on the reliability of solder joints because it strongly affects solderability and wettability [9]. By applying cyclic shear mechanical forces directly to solder joint samples, the above effects can be investigated. In addition, the the in-situ load vs. displacement behavior

978-1-7281-1500-9/19 $31.00 © 2019 IEEE

during the test can be monitored, and the evolution of several parameters such as plastic strain, loading slope, and energy dissipated per cycling can be quantitively measured [10-14].

Solder doping or micro-alloying has been documented to improve solder joint mechanical stiffness, strength, and thermal cycling reliability [15-21]. Specifically, alloying Bi to the ternary Sn-Ag-Cu (SAC) solder matrix has shown some promising results such as improved mechanical properties and reduction of aging effects [22-26]. Previous cyclic loading studies have included measurement of (1) damage accumulation during fatigue testing for both lead-free SAC and doped SAC+X alloys [27-28], (2) changes in the constitutive properties due to the mechanical cycling of SAC305 and SAC_Q solder alloys [29-30], and (3) microstructural evolution in a fixed region for solder alloys subjected to cycling [29, 31]. However, all of these studies were carried out on miniature bulk tensile samples (circular or rectangular in cross-section), and only limited studies have considered using solder joints themselves as test specimens.

In the current work, a novel technique has been developed to characterize the fatigue damage evolution occurring in the lead free SAC305 and SAC_Q (SAC+Bi) solder joints subjected to shear mechanical cycling. The prime objective of this study was to better understand the damage accumulation and microstructural evolution in lead free solder joints subjected to cyclic fatigue loading. Specifically, both SAC305 and SAC_Q solder joints were fabricated, and then they were cycled for various durations up to values near failure. The fatigue life and accumulated inelastic work were also measured and correlated to the microstructural damage. In addition, nanoindentation tests were performed on the shear cycled joints to study the evolution of the joint creep properties occurring as a function of the duration of cycling and microstructural damage.

II. EXPERIMENTAL PROCEDURES

In this study, SAC305 and SAC_Q (SAC+Bi) lead free solders were studied using cyclic shear testing. Table 1 summarizes the solder joint compositions. The test board and test vehicles are shown in Figure 1. The test board is made of a two-layer FR-4 glass-epoxy material. Each test board consisted of 144 square test vehicles with dimensions of 10 x 10 mm and separated by v-scoring. Each test vehicle contained a 3 x 3 square array of 9 individual solder joints (0.75 mm in diameter) with a pitch size of 3 mm. Individual solder joints were formed on the PCBs by placing solder spheres directly on tacky solder flux through the apertures of a 'bumping' stencil, and then the whole assembly was reflowed using a full convection reflow oven with a thermal profile having a peak temperature of 245 °C and time of 45-60 sec above liquidus.

Table 1. Solder Alloy Compositions

Solder Alloy	Composition
SAC305	96.5Sn – 3.0Ag - 0.5Cu
SAC_Q	92.77Sn - 3.41Ag - 0.52Cu - 3.3Bi

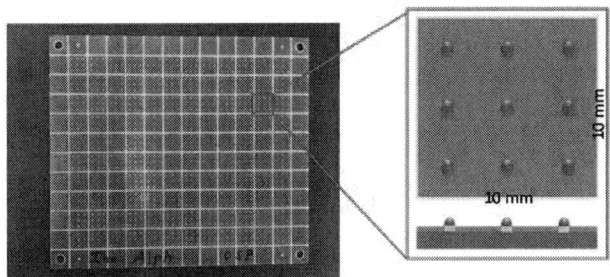

Figure 1. Test Board with Solder Joint Test Vehicles

Cyclic shear testing was conducted using a precision Instron micro-mechanical tester as shown in Figure 2. During a test, an individual solder joint was first placed in the center of the shear probe, and shear force was provided by pressing against the solder joint with the wall of the probe. The cyclic testing was performed in a load (stress) controlled manner. Referring to the detailed testing schematic in Figure 3, the shear probe first moved to the right and contacted the solder joint. The motion continued to the right until the pre-selected force (stress) level was reached, and then the shear probe motion was stopped and the motion of the probe was subsequently reversed to move towards the left until the same force (stress) magnitude was reached. Such a pattern was then repeated over and over to perform the stress-controlled shear cycling of the joints.

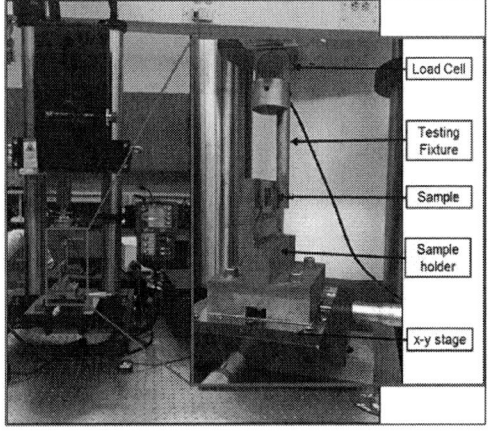

Figure 2. Micromechanical Testing Machine Setup

Each type solder joint alloy was testing using four different stress amplitudes. These included maximum applied stresses of 16, 20, 24, and 28 MPa for SAC305; and maximum applied stresses of 24, 28, 32, and 36 MPa for

SAC_Q. All of these levels are below the ultimate shear strengths of the two solder materials. As the cycling progressed, fatigue failure of a joint was taken to have occurred when a dramatic drop of the applied stress was detected, and the fatigue life of the solder joint was defined to be the number of total cycles that had occurred at the time of joint failure.

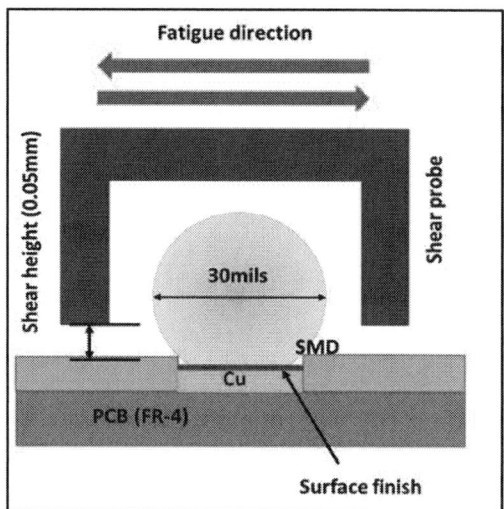

Figure 3. Detailed Cyclic Loading Test Schematic

The solder joints were subjected to cyclic shear loading at a constant strain rate of 0.01 sec^{-1}. Reversal of the shear direction was performed when the assigned stress amplitude was reached. The cyclic stress vs. strain data were systematically recorded, and the hysteresis loops were extracted. A typical stress vs. strain hysteresis loop for one cycle is shown in Figure 4. Two important parameters can be extracted from the hysteresis loop: (1) the plastic strain range, which is the width of the loop at 0 MPa stress; and (2) the plastic work dissipated in the cycle, which is the total area within the hysteresis loop (grey shaded). The work dissipation in each cycle was calculated using the numerical integration method and Matlab. Cycling was continued until joint failure occurred, and the evolutions of the hysteresis loop area and plastic strain with the number of cycles were monitored.

The results from all of the fatigue tests for a particular alloy were used to construct plots of fatigue life versus stress amplitude data. The data were modeled using a two-factor power equation as proposed by Hamasha, et al. [13]:

$$N = aP^{-b} \qquad (1)$$

where N is fatigue life, P is stress amplitude, and a, b are material constants.

The creep behavior of the solder joints was measured using a nanoindenter as shown in Figure 5. To characterize the evolution of the creep properties with cycling, the top of

the solder joint sample was polished prior to the mechanical shear cycling, and then it was cycled for various durations (e.g. 50, 100, 150, 200, and 300 cycles) using a stress amplitude of 24 MPa. After each duration of cycling, the sample was removed from the micromechanical testing machine and placed in the nanoindenter. A nanoindentation creep test was subsequently performed to characterize the creep strain rate vs. stress behavior of the solder material near the center of the joint. The load vs. displacement response was measured by a Berkovich tip in the direction that is normal to the cross-sectional face for each indentation test as indicated in the test schematic in Figure 6.

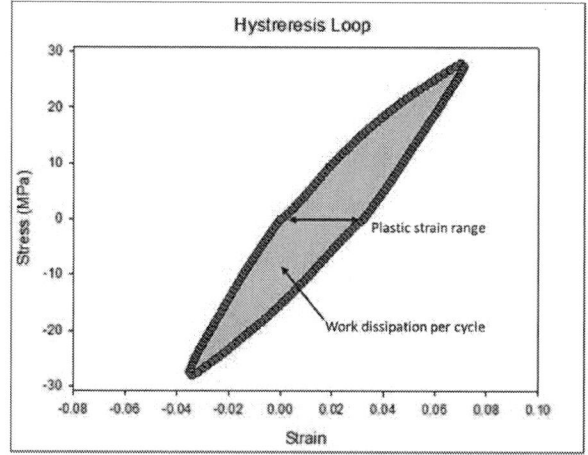

Figure 4. Typical Hysteresis Loop for SAC305

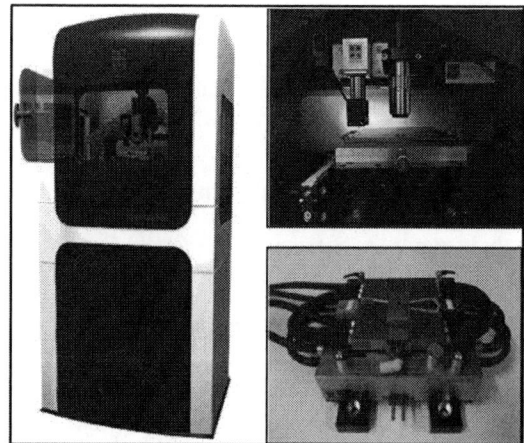

Figure 5. Nanoindentation System

For the indentation creep tests, a constant load of 30 mN was applied for 900 sec at the holding stage of the loading profile. The indentation depth versus time behavior was then recorded, and modeled using a hyperbolic tangent and log empirical model proposed by Chhanda, et al. [32]:

$$h = C_1 \ln(1+t) + C_2 \tanh(C_3 t) + C_4 t + C_5 \qquad (2)$$

where h is the indentation depth, t is the test time, and C_1 to C_5 are fitting constants. The creep strain rate was determined using the model proposed by Mayo and Nix [33]:

$$\dot{\varepsilon} = \frac{1}{h}\frac{dh}{dt} \qquad (3)$$

where $\dot{\varepsilon}$ is the creep strain rate. The extracted creep rate vs. stress response was then fit using the following exponential power law model:

$$\dot{\varepsilon} = C_1 e^{C_2 \sigma} \qquad (4)$$

where σ is the applied stress, and C_1 to C_5 are the fitting constants. Reference [15] gives further details of the nanoindentation creep test procedure.

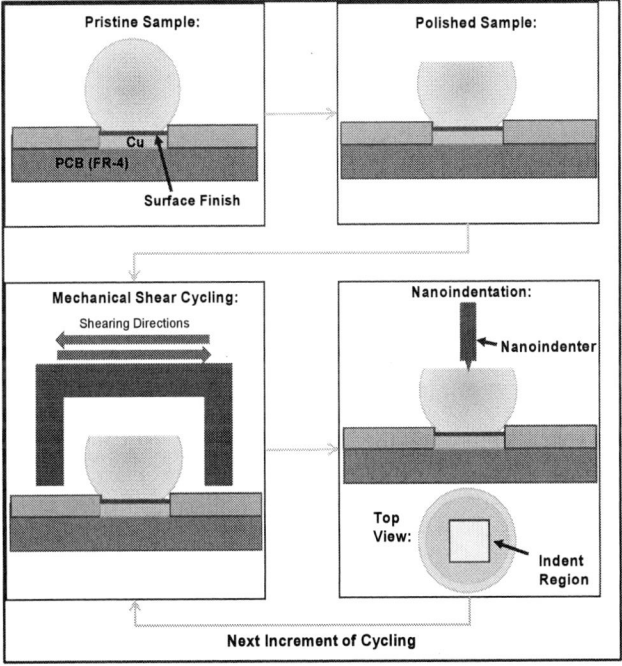

Figure 6. Test Schematic for Nanoindentation Creep Testing

III. TESTING RESULTS & DISCUSSION

Effect of Stress Amplitude on the Hysteresis Loop and Life

Once a solder joint was cycled until complete failure, the evolution of the work per cycle (area within the hysteresis loop) occurring during the shear fatigue test was plotted as a function of the cycle number as shown in Figure 7. From this plot, we can observe that there are three distinct regions of response labelled R1, R2, and R3. In region R1, there is

an initial drop of the work per cycle, which is due to the initial strain hardening of the solder joint. During the first few cycles of the performed tests, a limited flattening of the solder joint occurs in the regions of the probe contact areas. After this initial plastic deformation takes place, no additional significant flattening is observed. In region R2, the work per cycle becomes essentially constant (steady state). In the following discussions, all of the work per cycle comparisons will be based on values extracted from the steady-state region. In region R3, the work per cycle increases dramatically from the steady-state value, and complete failure occurs. This region features major crack initiation and propagation within the solder, and finally the fatigue failure of the solder joint.

As mentioned in the experimental procedure section, solder joints from each alloy were exposed to 4 different stress amplitudes, and the effects of the maximum stress level on the solder material fatigue properties were determined. Figure 8 shows hysteresis loop (half loop) comparisons for SAC305 solder joints under various stress amplitudes (16, 20, 24, and 28 MPa). As expected, both the plastic strain range and the area within the hysteresis loop were increased with increasing stress amplitude. Thus, larger stress amplitudes create more 'damage' per cycle relative to lower stress amplitudes. Figures 9 and 10 further expound this phenomenon by showing the work per cycle and plastic strain range evolution occurring for the different stress amplitudes as a function of the number of cycles (regions R1 and R2). Examining Figure 9, the same trend is observed for all four of the different stress amplitudes. In particular, the work per cycle dropped during early cycling and then became nearly steady state. Higher stress amplitudes resulted in larger work per cycle values in the steady state region. Similar trends were observed for the plastic strain range evolution.

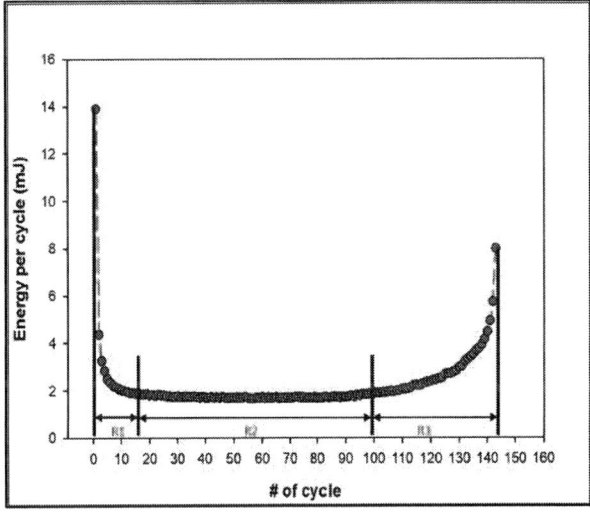

Figure 7. Typical Energy Dissipated per Cycle During a Shear Fatigue Test (SAC305)

978-1-7281-1500-9/19 $31.00 © 2019 IEEE

The stress amplitude was not the only factor that impacted the hysteresis loop. Differences in solder material composition can also influence the size and shape of the hysteresis loop as shown in Figure 11. At the same stress amplitude, the SAC305 demonstrated a more ductile response with a larger plastic strain range, wider hysteresis loop, and higher plastic work per cycle relative to that for SAC_Q. The SAC_Q alloy has more than 3.0% Bi, which causes solid solution hardening and thus increases the strength and fatigue resistantance when compared to SAC305 that has no Bi.

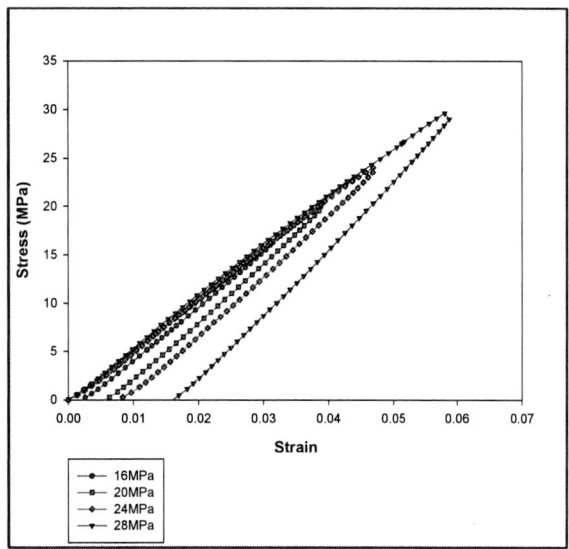

Figure 8. Hysteresis Loops for SAC305 under Various Stress Amplitudes

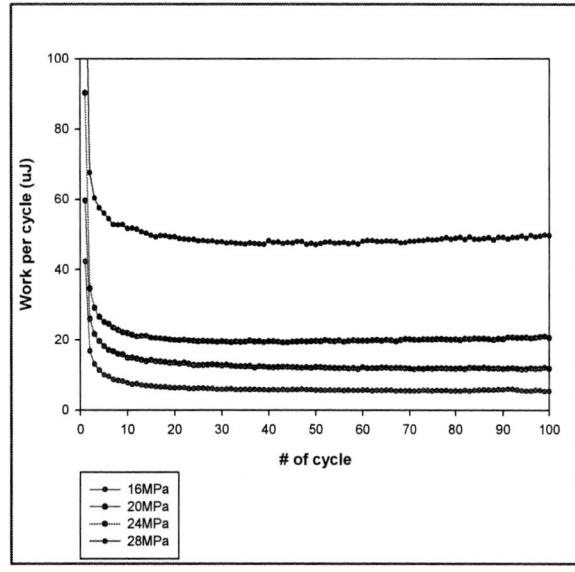

Figure 9. Plastic Work per Cycle Evolution for Various Stress Amplitudes (SAC305)

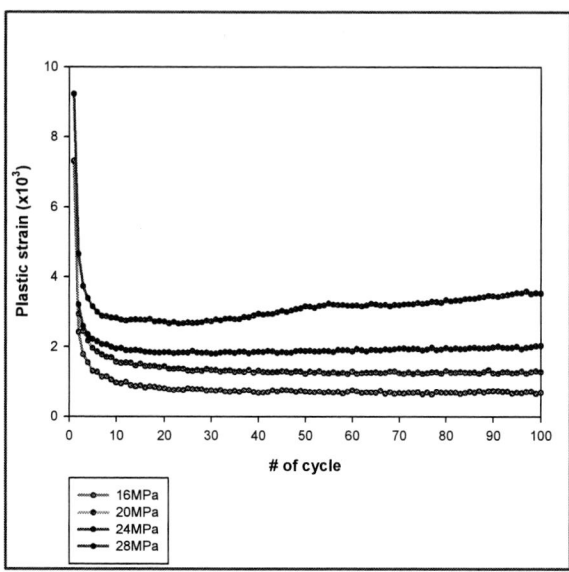

Figure 10. Plastic Strain Range per Cycle Evolution for Various Stress Amplitudes (SAC305)

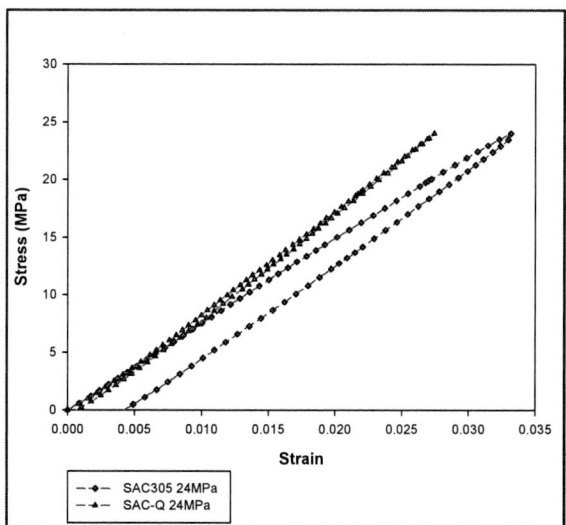

Figure 11. Hysteresis Loop Comparison between SAC305 and SAC_Q at a Stress Amplitude of 24MPa

The fatigue life (cycles per failure) is plotted as a function of the applied stress amplitude for the two solder alloys in Figure 12. For each alloy, the two-factor power law equation in eq. (1) was used to fit the data. The results show that SAC_Q demonstrated a longer fatigue life than SAC305 at all stress amplitudes. However, the fitting lines are converging at higher stress amplitudes. Similarly, the accumulative plastic work leading to a major crack is plotted as a function of stress amplitude for the two solder alloys in 13. The accumulated work need for a major failure tends to decrease with the increase of stress amplitude for both solder alloys. This is because cycling at lower stress amplitudes requires more damage accumulation to reach the point that the solder can not handle the applied stress. Moreover, SAC305 demonstrated slightly higher total accumulated work than SAC_Q at the time of failure, which is due to the observation that SAC305 is more ductile and can deform and dissipate more energy before reaching the failure point.

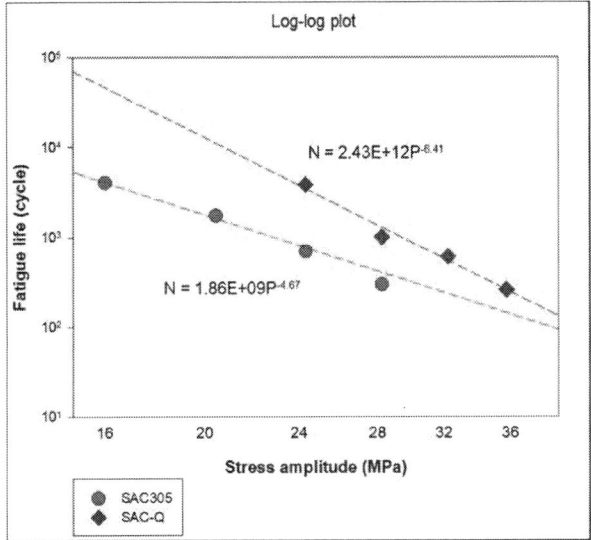

Figure 12. Variation of Fatigue Life with Stress Amplitude

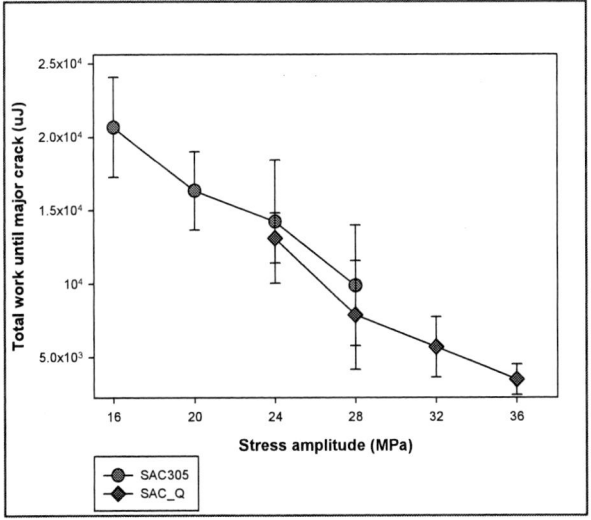

Figure 13. Variation of Accumulated Plastic Work at Failure with Stress Amplitude

Evolution of Creep Properties

The change of creep resistance with mechanical shear cycling was studied for a set of SAC305 solder joints using nanoindentation as described above. A 30 mN load with a dwell time of 900 sec was used to initiate the nanoindentation creep experiments. Before performing the creep indentation experiments, the solder joints were cycled for various durations (e.g., 0, 50, 100, 150, 200, 300 cycles). Specifically, the initial indentations were performed on virgin samples with no cycling. The same samples were then mechanically cycled in shear for the 50 cycles using the Instron micro-mechanical tester with a constant peak load of 24 MPa. After this cycling was completed, the sample was indented again in a nearby location to re-measure the creep behavior of the solder joint. This process was then repeated for total cycling durations of 100, 150, 200, and 300 cycles. Example creep displacement versus time curves for a typical joint with different durations of prior cycling are shown in Figure 14. It is observed that with the increased number of cycles, the nanoindentation creep displacement at any time increased with the amount of prior cycling, indicating material property degradation due to the cycling induced fatigue damage.

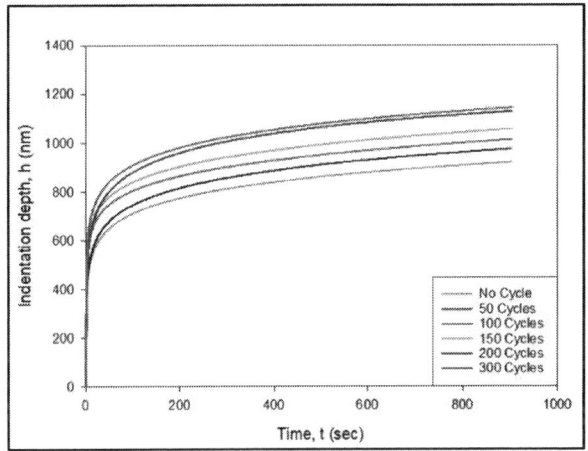

Figure 14. Dependence of Creep Indentation Displacement on Time for Various Prior Mechanical Cycling Durations

Figure 15 illustrates the measured variations of the creep strain rate with the applied stress for the various durations of prior cycling. As can be seen, the shear mechanical cycling demonstrated a strong effect on the creep strain rate. For example, the creep strain rate increased by a factor of 28X after 300 cycles relative to the solder material in the initial state (0 cycles).

The mechanical property evolutions exhibited by the cycled solder joints are due to microstructure changes and damage caused by the cycling. An example of such changes is shown in Figure 16 for SAC305 [29]. It can be observed that as the cycling progresses, microcracks began to develop along the subgrain boundaries of the β-Sn dendrites. With further cycling, those microcracks grew in length, connected and entangled with each other, and eventually formed larger transgranular cracks. Therefore, the softening and degradation of the mechanical properties of the solder joints were due to this damage accumulation and microcracks formation.

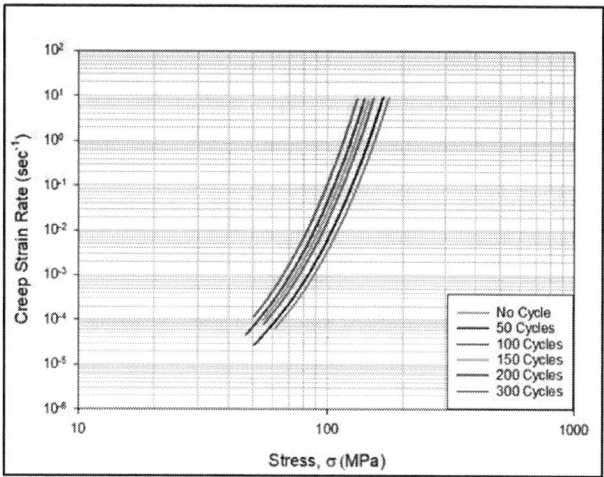

Figure 15. Variation of Creep Strain Rate with Applied Stress for Various Prior Cycling Durations

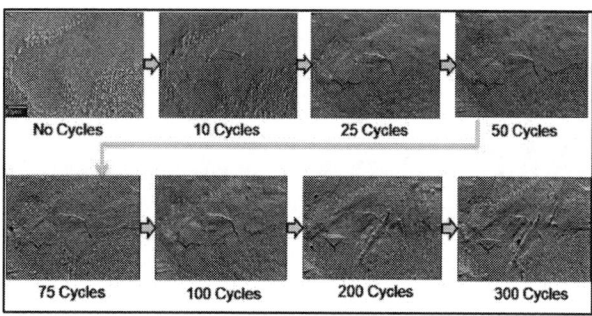

Figure 16. Evolution of SAC305 Solder Microstructure During Mechanical Cycling [29]

IV. SUMMARY

In this investigation, the effect of stress on the evaluation of fatigue life and hysteresis loop area during mechanical cycling with two types of solder alloys was studied, and then the effects of mechanical cycling on the damage accumulation were explored. Such studies are necessary to help improve the fatigue models for lead-free solder joints. The innovative part of this study was applying cyclic shear loading directly on the solder joint (rather than bulk specimens), so that the results acquired would be a reflection of realistic conditions. Individual solder joints were subjected to various stress amplitudes from 16 to 36 MPa until complete failure. Other individual solder joints were cycled for various durations (e.g. 50, 100, 150, 200, 300 cycles), and then the creep properties were measured using nanoindentation.

It was observed that work per cycle and plastic strain range increased with the increasing of stress amplitudes,

while fatigue life decreased with the increase of stress amplitudes. The SAC_Q solder joints with Bi demonstrated a higher fatigue life than the SAC305 solder joints. The presence of Bi can improve the mechanical properties by distorting the Sn crystalized lattice and blocking the relative movement of the dislocation cell structure by 'solid solution hardening.' The work per cycle and plastic strain range in SAC_Q were seen to be less than those of SAC305 for the same applied stress amplitude.

Furthermore, the creep displacement and creep rate of SAC305 solder joints increased with the progression of mechanical cycling. At the same stress level, the creep strain rate increased by a factor of 28X after 300 cycles. We also observed a transgranular failure mechanism associated with the cyclic fatigue testing. Cracks were initiated within the β-Sn dendrites, tended to grow along the subgrain boundaries, and increased in length with the cycling. Finally, the cyclic loading led to the creation of the transgranular failure, weakened the β-Sn dendrites, and attributed to the degradation of solder joint mechanical properties.

V. REFERENCES

[1] L. Yin, L. Wentlent, L. Yang, B. Arfaei, A. Oasaimeh, and P. Borgesen, "Recrystallization and precipitate coarsening in Pb-free solder joints during thermomechanical fatigue," *Journal of electronic materials,* vol. 41, pp. 241-252, 2012.

[2] A. Qasaimeh, Y. Jaradat, L. Wentlent, L. Yang, L. Yin, B. Arfaei, *et al.,* "Recrystallization behavior of lead free and lead containing solder in cycling," in *Electronic Components and Technology Conference (ECTC), 2011 IEEE 61st,* 2011, pp. 1775-1781.

[3] A. Qasaimeh, S. Hamasha, Y. Jaradat, and P. Borgesen, "Damage evolution in lead free solder joints in isothermal fatigue," *Journal of Electronic Packaging,* vol. 137, p. 021012, 2015.

[4] C. Andersson, Z. Lai, J. Liu, H. Jiang, and Y. Yu, "Comparison of isothermal mechanical fatigue properties of lead-free solder joints and bulk solders," *Materials Science and Engineering: A,* vol. 394, pp. 20-27, 2005.

[5] P. Borgesen, S. Hamasha, M. Obaidat, V. Raghavan, X. Dai, M. Meilunas, *et al.,* "Solder joint reliability under realistic service conditions," *Microelectronics Reliability,* vol. 53, pp. 1587-1591, 2013.

[6] S. Su, N. Fu, F. J. Akkara, and S. Hamasha, "Effect of Long-Term Room Temperature Aging on the Fatigue Properties of SnAgCu Solder Joint," *Journal of Electronic Packaging,* vol. 140, p. 031005, 2018.

[7] B. Arfaei, Y. Xing, J. Woods, J. Wolcott, P. Tumne, P. Borgesen, *et al.,* "The effect of Sn grain number and orientation on the shear fatigue life of SnAgCu solder joints," in *2008 58th Electronic Components and Technology Conference,* 2008, pp. 459-465.

[8] F. Akkara, S. Su, S. Thirugnanasambandam, A. Dawahdeh, A. Qasaimeh, J. Evans, *et al.,* "Effects of Long-Term Aging on SnAgCu Solder Joints Reliability in Mechanical Cycling Fatigue," in *SMTA International Conference, Rosemont, IL, Sept,* 2017, pp. 17-21.

[9] F. Akkara, M. Abueed, M. Rababah, C. Zhao, S. Su, J. Suhling, *et al.*, "Effect of Surface Finish and High Bi Solder Alloy on Component Reliability in Thermal Cycling," in *2018 IEEE 68th Electronic Components and Technology Conference (ECTC)*, 2018, pp. 2032-2040.

[10] S. Hamasha, F. Akkara, S. Su, H. Ali, and P. Borgesen, "Effect of Cycling Amplitude Variations on SnAgCu Solder Joint Fatigue Life," *IEEE Transactions on Components, Packaging and Manufacturing Technology*, 2018.

[11] S. Hamasha and P. Borgesen, "Effects of Strain Rate and Amplitude Variations on Solder Joint Fatigue Life in Isothermal Cycling," *Journal of Electronic Packaging*, vol. 138, p. 021002, 2016.

[12] S. Hamasha, Y. Jaradat, A. Qasaimeh, M. Obaidat, and P. Borgesen, "Assessment of Solder Joint Fatigue Life Under Realistic Service Conditions," *Journal of electronic materials*, vol. 43, 2014.

[13] S. Hamasha, A. Qasaimeh, Y. Jaradat, and P. Borgesen, "Correlation between solder joint fatigue life and accumulated work in isothermal cycling," *IEEE Transactions on Components, Packaging and Manufacturing Technology*, vol. 5, pp. 1292-1299, 2015.

[14] S. Hamasha, S. Su, F. Akkara, A. Dawahdeh, P. Borgesen, and A. Qasaimeh, "Solder joint reliability in isothermal varying load cycling," in *Thermal and Thermomechanical Phenomena in Electronic Systems (ITherm), 2017 16th IEEE Intersociety Conference on*, 2017, pp. 1331-1336.

[15] Hasnine, M., Suhling, J. C., Prorok, B. C., Bozack, M. J., and Lall, P., "Nanomechanical Characterization of SAC Solder Joints - Reduction of Aging Effects Using Microalloy Additions," *Proceedings of the 65th IEEE Electronic Components and Technology Conference*, pp. 1574-1585, San Diego, CA, May 27-29, 2015.

[16] A. Raj, S. Gordon, J. Evans, M. Bozack, W. Johnson, "Long Term Isothermal Aging of BGA Packages Using Doped Lead Free Solder Alloys," in *SMTA International Proceedings*, 2018.

[17] A. Raj, T. Sanders, S. Sridhar, J. Evans, M. Bozack, W. Johnson "Thermal Shock Reliability Test on Multiple Doped Low Creep Lead Free Solder Paste and Solder Ball Grid Array Packages," in *SMTA International Proceedings*, 2015.

[18] C. Zhao, "Board Level Reliability of Lead-Free Solder Interconnections with Solder Doping Under Harsh Environment," 2017.

[19] C. Zhao, T. Sanders, Z. Hai, C. Shen, and J. L. Evans, "Reliability Analysis of Lead-Free Solder Joints with Solder Doping on Harsh Environment," in *International Symposium on Microelectronics*, 2016, pp. 000117-000122.

[20] S. Su, F. J. Akkara, M. Abueed, M. Jian, J. Suhling, and P. Lall, "Fatigue Properties of Lead-free Doped Solder Joints," in *2018 17th IEEE Intersociety Conference on Thermal and Thermomechanical Phenomena in Electronic Systems (ITherm)*, 2018, pp. 1243-1248.

[21] F. J. Akkara, C. Zhao, S. Su, S. Hamasha, J. Suhling, "Effects of Mixing Solder Sphere Alloys with Bismuth-Based Pastes on The Component Reliability in Harsh Thermal Cycling," *Proceedings of SMTA International*, 2018.

[22] Cai, Z., Zhang, Y., Suhling, J. C., Lall, P., Johnson, R. W., Bozack, M. J., "Reduction of Lead Free Solder Aging Effects Using Doped SAC Alloys," *Proceedings of the 60th IEEE Electronic Components and Technology Conference*, pp. 1493-1511, Las Vegas, NV, June 2-4, 2010.

[23] Ahmed, S., Basit, M., Suhling, J. C., Lall, P., "Effects of Aging on SAC-Bi Solder Materials" *Proceedings of ITherm 2016*, pp. 746-754, Las Vegas, NV, May 30 - June 3, 2016.

[24] Ahmed, S., Suhling, J. C., Lall, P., "The Anand Parameters of Aging Resistant Doped Solder Alloys" *Proceedings of ITherm 2017*, pp. 1416-1424, Orlando, FL, May 30 - June 2, 2017.

[25] Alam, M. S., Hassan, R., Suhling, J. C., and P. Lall, "High Temperature Mechanical Behavior of SAC and SAC+X Lead Free Solders," *Proceedings of the 68th IEEE Electronic Components and Technology Conference*, pp. 1781-1789, San Diego, CA, May 29 - June 1, 2018.

[26] Alam, M. S., Hassan, R., Suhling, J. C., and P. Lall, "A Comparative Study of the High Temperature Mechanical Behavior of Lead Free Solders," *Proceedings of ITherm 2018*, pp. 1314-1323, San Diego, CA, May 29 - June 1, 2018.

[27] Chowdhury, M. M. R., Ahmed, S., Fahim, A., Suhling, J. C., Lall, P., "Mechanical Characterization of Doped SAC Solder Materials at High Temperature" *Proceedings of ITherm 2016*, pp. 1202-1208, Las Vegas, NV, May 30 - June 3, 2016.

[28] Chowdhury, M. M. R., Fu, N., Suhling, J. C., and Lall, P., "Evolution of the Cyclic Stress-Strain Behavior of Doped SAC Solder Materials Subjected to Isothermal Aging," *Proceedings of ITherm 2017*, pp. 1369-1379, Orlando, FL, May 30 - June 2, 2017.

[29] Chowdhury, M. M. R., Hoque, M. A., Fu, N., Suhling, J. C., Hamasha, S., and Lall, P., "Characterization of Material Damage and Microstructural Evolution Occurring in Lead Free Solders Subjected to Cyclic Loading," *Proceedings of the 68th IEEE Electronic Components and Technology Conference*, pp. 865-874, San Diego, CA, May 29 - June 1, 2018.

[30] Hoque, M. A. Chowdhury, M. M., Fu, N., Suhling, J. C., Hamasha, S., and Lall, P., "Evolution of the Cyclic Stress-Strain and Constitutive Behaviors of Doped Lead Free Solder During Fatigue Testing," *Proceedings of ITherm 2018*, pp. 1387-1395, San Diego, CA, May 29 - June 1, 2018.

[31] Chowdhury, M. M. R., Hoque, M. A., Ahmed, S., Suhling, J. C., Hamasha, S., and Lall, P., "Effects of Mechanical Cycling on the Microstructure of SAC305 Lead Free Solder," *Proceedings of ITherm 2018*, pp. 1324-1332, San Diego, CA, May 29 - June 1, 2018.

[32] N. J. Chhanda, J. C. Suhling, and P. Lall, "Experimental characterization and viscoplastic modeling of the temperature dependent material behavior of underfill encapsulants," in *ASME 2011 Pacific Rim Technical Conference and Exhibition on Packaging and Integration of Electronic and Photonic Systems*, 2011, pp. 749-761.

[33] M. Mayo and W. Nix, "A micro-indentation study of superplasticity in Pb, Sn, and Sn-38 wt% Pb," *Acta Metallurgica*, vol. 36, pp. 2183-2192, 1988.

Reliability Studies of Silicon Interconnect Fabric

Niloofar Shakoorzadeh[1], SivaChandra Jangam[1], Kaysar Rahim[2], Pranav Ambhore[1], Han Chien[1], Amir Hanna[1], and Subramanian S. Iyer[1]

[1]Center for Heterogeneous Integration and Performance Scaling (CHIPS),
Samueli School of Engineering, University of California Los Angeles
[2]Global Foundries, Malta, USA

nellysh@ucla.edu, md.rahim@globalfoundries.com

Abstract— The silicon interconnect fabric (Si-IF) is a heterogeneous integration platform where unpackaged dies are assembled with fine-pitch interconnects (< 10 μm) using solderless metal-metal thermal compression bonding. This assembly process does not utilize any underfill or molding compound and therefore, requires alternative passivation approaches. We developed a novel passivation technique comprising two processes. First, we passivate sidewalls of the copper (Cu) pillars on Si-IF by controlling the recess profile of inter layer dielectric (SiO₂). In the second process, we passivate the assembled dies on the Si-IF post bonding using thin films. To achieve this encapsulation, we used a multilayer thin film structure consisting of an inorganic (SiN$_x$) layer (500 nm) and an organic (Parylene C) layer (3 μm). Humidity testing was done in accordance with "85/85" Steady-State Humidity Life Test standard on passivated blanket Cu samples. X-ray powder diffraction (XRD) scans of samples subjected to humidity testing showed no signs of Cu oxide peaks even after 168 hours. This multilayer structure was used to encapsulate bonded dies on Si-IF and humidity testing was performed for 120 hrs. There was no measurable change (<0.4%) in the average shear force of the samples before and after testing respectively, demonstrating the effectiveness of this passivation technique. Further, the resistance of links on the passivated Si-IF didn't change (<0.1%) after 120 hrs of humidity testing. Moreover, we present results of temperature cycling of Si-IF assembly. The results show that links with low initial resistance didn't fail after 1000 cycles of testing.

Keywords: Silicon interconnect fabric, heterogeneous integration, Reliability, Humidity testing, Temperature cycling

I. INTRODUCTION

The scaling of package and PCB dimensions is crucial for enabling integration of high-performance systems. At UCLA, we have developed the Silicon Interconnect Fabric (Si-IF), which is a platform for heterogeneous integration of unpackaged dies at fine interconnect pitch (≤10 μm) and close spacings (<100 μm). We have fabricated 10 μm pitch Cu pillars with 5 μm diameter and 1.5 μm height on the Si-IF. These pillars are bonded to corresponding pads on the die. Both the pillars and pads are fabricated using a damascene process [1]. The Si-IF fabrication process is compatible with conventional Si back end of the line (BEOL) processes which makes this platform very attractive in terms of cost and feasibility. The dies are bonded to the Si-IF using a solderless metal-metal thermal compression bonding (TCB). Further,

due to the fine-pitch interconnects and short links on the Si-IF, we achieve high data-bandwidth with low latency and low power consumption [1,2]. Fig. 1 (a) is a schematic of Si-IF showing pillars and pads and thermocompression bonding process.

Our earlier demonstrations did not involve any passivation of the system. However, to ensure reliability of the assembled Si-IF against various environmental stressing, we need to passivate the system. Typically, in packages today, underfill or molding compounds are used not only to passivate but provide mechanical robustness. Since we use fine-pitch interconnects with only ~1.5 μm of exposed Cu pillars with no underfill, we need alternative passivation techniques. Further these passivation techniques should have low thermal budget because of functional dies are already assembled on the Si-IF.

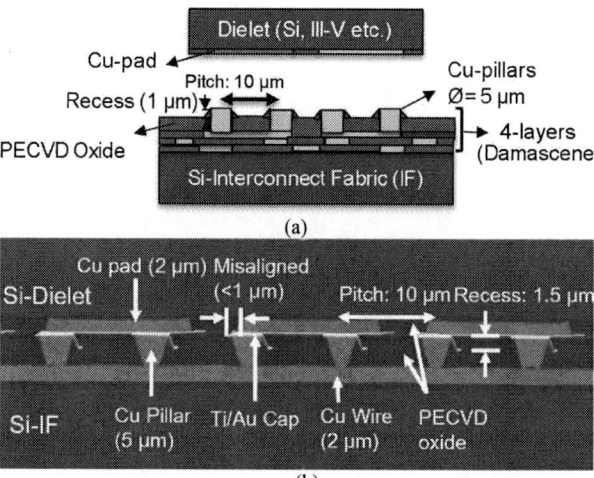

Figure. 1 (a) A schematic of an Si-IF and a dielet showing pillars and pads. Si-IF has 4 layers of metallization that's been fabricated using damascene process. (b) a cross section of a bonded assembly, showing pillars and pads bonded. Cu pillars are protected by sidewall SiO₂ [3]. Some excess Au is visible on the right side of the pillars. That is due to misalignment in Au lift off step.

We have developed a two-step method for passivating the system on Si-IF. The first step involves tapered etching of the side walls of the inter layer dielectric (ILD) to expose the Cu pillars for TCB. This ensures the Cu pillars and corresponding are encapsulated with the dielectric and therefore are not exposed to the environment. The second step is to passivate

978-1-7281-1500-9/19 $31.00 © 2019 IEEE

the entire system after assembling dies. This passivation is needed to ensure protection of any misaligned pillars and pads from degradation. Fig. 1 (b) is a cross section of a bonded die on the Si-IF, showing the Cu pillars, traces and pads. The side walls of individual Cu pillars have SiO$_2$ that has a tapered profile [3]. We present more details on the encapsulation, as well as, verification tests done on bonded samples in section II.

In conventional packaging, coefficient of thermal expansion (CTE) mismatch between different components of the package, i.e. dies, molding material, FR4 board and so on is the primary cause of failure during temperature cycling [4]. Moreover, formation of brittle intermetallic compounds (IMC) in solder interconnects is another reason for failure that occurs during temperature cycling. The IMC formation results in crack initiation due to CTE mismatch [5,6]. This crack formation along with brittleness of IMCs and subsequent crack propagation through the interconnect results in ultimate failure of the joints under extreme temperature variations [5,6].

Since Si-IF utilizes direct solderless TCB, IMC formation is eliminated in the system and there is no concern of crack formation due to variation in mechanical properties of the joint. Although the effect of CTE mismatch is reduced in the Si-IF due to use of fewer materials, there remains concerns of mismatch between Cu and SiO$_2$. Further, there may be crack formation at the bond interface due to existence of micro voids. Therefore, we studied effects of temperature cycling and report the results in section III. Finally, the conclusion and future work is presented in section IV.

II. ENCAPSULATION AND HUMIDITY TESTING

As mentioned earlier, encapsulation in electronic packaging is a crucial step to ensure protection of integrated circuit from degradation due to environmental factors. There are two different types of encapsulation used in electronic packaging, namely, hermetic (metallic or ceramic) and non-hermetic (plastic) [7]. Currently, most of the encapsulation in microelectronics is non-hermetic and a lot of effort has been put into incorporating new polymer-based encapsulations in the recent years [7]. Different criteria must be considered when choosing a new encapsulation. These include CTE, glass transition temperature (T$_g$) and mechanical properties of the material, its adhesion to the substrate, and conformality of the encapsulation [8].

A. Parylene C as encapsulation

Even though the Cu pillars in Si-IF are protected by sidewall dielectric, there is still a need for a moisture barrier to protect interconnects and the Si-IF assembly from degradation due to moisture ingress. Also, Cu pads and Cu pillars may still be exposed due to misalignment during the bonding process. We have investigated use of Parylene C as an encapsulation for these fine-pitch Cu interconnects.

Poly-(para-xylylene), known as Parylene, is a class of semi crystalline polymers that are deposited through chemical vapor deposition (CVD) process. There are different types of Parylene such as Parylene C, N and D [9]. Fig. 2 is the chemical structure of three types of Parylene [9]. Parylene N

is mainly used in applications where high temperature stability is required whereas Parylene C is the most common type encountered in MEMS and has the best moisture barrier property compared to the other types [9].

Parylene C deposition by CVD process uses a dimer which is heated in vacuum to form a gas. Pyrolyzing the gas creates the monomers. Monomers are transferred to the deposition chamber where they sublimate and form a conformal thin film, free of pin holes, at room temperature [10]. Aside from forming a defect free thin film at room temperature, Parylene C has other advantages as an encapsulation material including mechanical flexibility, electrical insulation and low intrinsic stress. As a result, it has been used as a water-resistant encapsulation in electronics for the past several decades [9].

To evaluate the robustness of Parylene C as moisture

Figure. 2 Chemical structure for Parylene N, C and D [9].

barrier, humidity testing was done on blanket Cu samples passivated with 1 and 3 μm of Parylene C. Test coupons were prepared by growing 500 nm of thermal SiO$_2$ on Si wafers followed by sputtering of 20nm of Ti as a diffusion barrier and 200 nm of Cu as the seed layer. Then, 1 μm of Cu was electroplated on the samples. These test coupons were then transferred to the deposition chamber and Parylene C was deposited.

Humidity test condition was 85°C and 85% relative humidity (RH) which is the Steady-State Humidity Life Test standard. Passivated samples were subjected to above testing conditions for 72 hours and then x-ray powder diffraction scan (XRD) was done on the samples to measure amount of Cu oxidation. Fig. 3 shows the schematic of the blanket samples with passivation. Fig. 4 illustrates the XRD scans for two

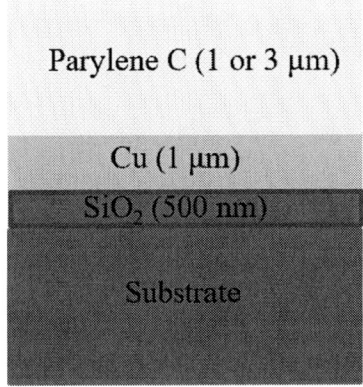

Figure. 3 A schematic of the blanket Cu test coupon. Test coupons were encapsulated by 1μm or 3μm of Parylene C and humidity test was performed for 72 hours

samples with 1 and 3 μm thick layers of Parylene C after 72 hours of humidity testing.

In both the samples, Cu$_4$O$_3$ was detected. Scherrer equation (1) was used to estimate average grain size of oxide in each sample shown below.

$$\tau = \frac{k\lambda}{\beta \cos \theta} \qquad (1)$$

Where τ is the mean size of crystalline domain, k is Shape Factor and is close to unity, λ is the x-ray wavelength, β is peak broadening at half the maximum intensity (FWHM) after subtracting instrumental peak broadening in radians and θ is Bragg angle.

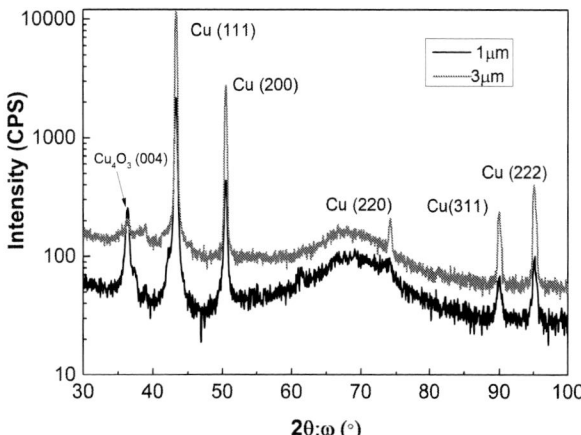

Figure. 4 XRD scans of samples passivated with 1μm and 3 μm of Parylene C after 72 hours of humidity testing. Copper oxide is present in both samples.

Table I summarizes mean grain size of Cu oxide and Cu grain size in each sample.

Table I. Summary of average grain size in sample passivated with 1 and 3 μm of Parylene C after 72 hours of humidity testing.

item	1μm Parylene C	3 μm Parylene C
Cu$_4$O$_3$ Average grain size	15.53 nm	1.5 nm
Cu Average grain size	22.5 nm	22.5 nm

B. Multi-layer encapsulation

Another encapsulation methodology was implemented using a multi-layer thin films consisting of plasma enhanced chemical vapor deposition (PECVD) SiN$_x$ and CVD Parylene C. SiN$_x$ is a common material in microfabrication that is used as dielectric, polish stop, and as a thin film passivation. SiN$_x$ is inert to moisture ingress; however, due to the inherent defects existing in the PECVD process (pin holes), water molecules can penetrate through it and reach the substrate [11]. It has been shown that combining SiN$_x$ and Parylene C

as a multi-layer passivation will form an excellent barrier

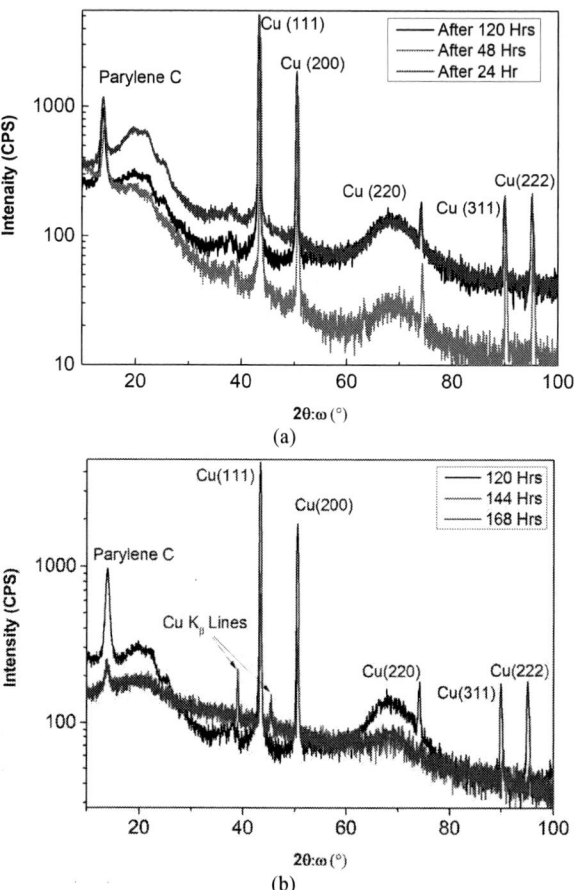

Figure. 5 (a) XRD scans for samples after 24, 48 and 120 hours of humidity testing and (b) after 120,144 and 168 hours of testing. No oxide peaks are detected in the samples.

against water molecules in [12].

Since passivation is the last step in the integrated Si-IF assembly, it's crucial to develop a low-temperature deposition process to ensure the integrity of any temperature sensitive dies. Hence, a low temperature (100°C) PECVD process was used for SiN$_x$ deposition.

To evaluate the effectiveness of the multi-layer passivation, humidity testing like the one mentioned earlier was performed on passivated Cu blanket samples. Test coupons were fabricated similar to previously mentioned process and were passivated with 500nm of PECVD SiN$_x$ (100°C). Parylene C adhesion promoter was spin coated on the samples and annealed. We studied the effect of annealing after application of adhesion promoter previously [3]. We had shown that annealing at 130 °C for 5 min significantly improved Parylene C adhesion and the adhesion strength does not degrade after exposure to humidity testing for up to 144 hours. Next, Parylene C is deposited on test coupons (3μm) and finally, another layer of SiN$_x$ is deposited (300 nm).

These test coupons were tested for up to 168 hours at 85°C and 85% RH and XRD scan was performed on samples after every 24 hours. Fig. 5 (a) shows the XRD scan of sample after

24, 48 and 120 hours of testing and Fig. 5 (b) shows the scan for 120, 144 and 168 hours of testing.

One of the challenges in using low temperature PECVD SiN$_x$ for passivation of bonded samples is step coverage. To ensure the multi-layer passivation is in fact covering the entire assembled Si-IF, it is necessary to study the side wall coverage of PECVD SiN$_x$. Focused ion beam (FIB) cross sectioning was used to examine the step coverage. After bonding, 500 nm of PECVD SiN$_x$ was deposited and FIB was done on top of the die, and in the middle and bottom of the die side wall to examine the thickness of the deposited SiN$_x$, shown in Fig. 6. These thicknesses indicate the quality of step coverage. At top the die thickness of SiN$_x$ is 522.9 nm this thickness is reduced to 55.2 nm in the middle of the side wall and at the bottom, no SiN$_x$ is deposited.

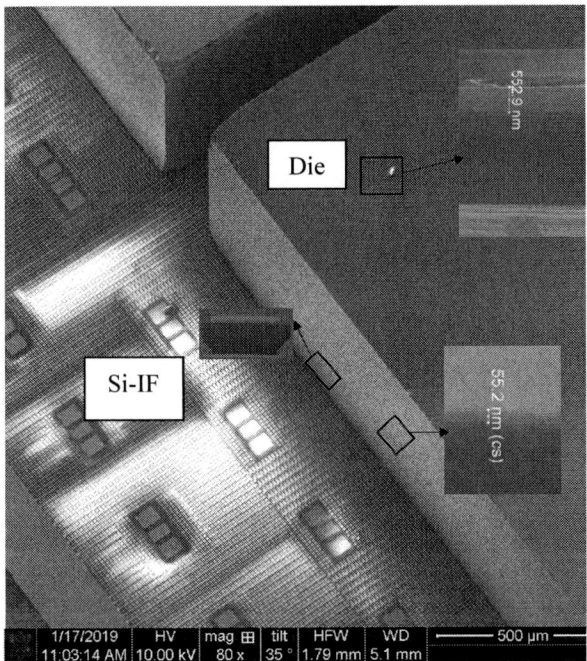

Figure. 6 FIB cross section image of the thickness of deposited PECVD SiN$_x$ at the top of the die (529 nm), middle of the die (55.2 nm) and bottom of the die (no SiN$_x$)

To improve step coverage, PECVD recipe was modified to include wafer bias. This significantly improved SiN$_x$ conformality. Fig. 7 shows the FIB cross section of the die with SiN$_x$ thicknesses at the top (283 nm), middle (125.3) and the bottom (44.1 nm) of the side wall. This modified recipe was used for passivation of the Si-IF assemblies.

We fabricated and assembled dies on the Si-IF with daisy chain structures. These samples had no Au capping of the Cu pillars and Cu pads on contrary to the previously demonstrated assemblies in [1]. These dies were bonded using a direct Cu-Cu TCB with in-situ formic acid vapor treatment. The details of this Cu-Cu TCB process is explained elsewhere [13]. After bonding, the passivation of the Si-IF assembly was done by deposition of 500 nm SiN$_x$ followed by CVD deposition of 3μm of Parylene C and finally, deposition of 300nm SiN$_x$.

The effectiveness this passivation was investigated. The bonded samples were used to evaluate change in the shear strength after humidity testing. The average shear force of the bonded samples before humidity testing was 135 N. After 120 hours of humidity testing, the average shear force of the assembled samples was 134.58 N, which shows that the bond was intact before and after the test. This change is <0.4% which is within the experimental errors.

Further, we performed humidity testing and measured the change in wiring resistance of the earlier demonstrated Si-IF [1], after passivation. We did not observe any measurable change in the Si-IF wiring resistance even after 120 hours of humidity testing.

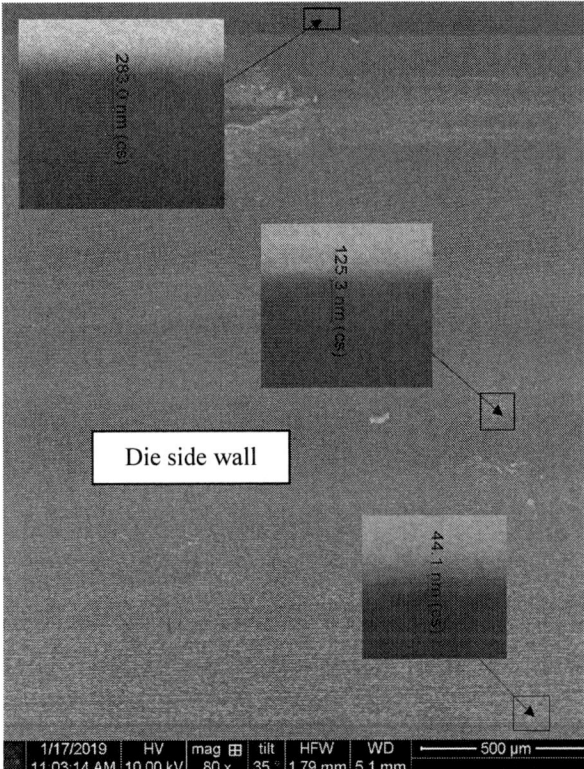

Figure. 7 Optical image of the side wall of a die showing the thickness of deposited SiN$_x$ at top of the die (283nm), middle section of the sidewall (125.3 nm) and bottom at the interface with Si-IF (44.1 nm).

III. TEMPERATURE CYCLING

Characterizing interconnects under severe temperature change is a crucial aspect of the reliability of electronic packaging. Regardless of the type of interconnects, it is important to ensure they don't fail as a result of the induced thermo-mechanical stresses [14].

To evaluate the failure mechanism in the Si-IF assemblies, temperature cycling was done on daisy chain samples where the dies were bonded to the Si-IF using 10 μm fine-pitch Au-capped Cu pillars, demonstrated in [1]. Temperature cycling testing condition was in accordance with JESD22-A104E test condition G [15]. Temperature was varied between T_{max} of 125°C and T_{min} of -40°C. Soak time at each temperature was 10 min. No overmold or encapsulation was utilized in order to

978-1-7281-1500-9/19 $31.00 © 2019 IEEE

eliminate their effects on failure. Moreover, no underfill or overmold is required in Si-IF integration scheme as explained earlier.

Temperature cycling was done for up to 1000 cycles and the resistance of the daisy chains were measured before and after the test. Table II is the summary of the resistance before and after the temperature cycling which shows only one of the five tested links showed high change in resistance. After identifying the high resistance link, infrared camera (IR) camera was used to sense heat signatures and detect the hot spots when current was passed. The temperature map of the sample being is shown Fig. 8. The bright spot represents the failed interconnect in the daisy chain.

Consequently, FIB cross sectioning was done at the interconnect with high resistance to evalute the failure mechanism of the bonding. Fig. 9 shows the SEM image of the cross-section of the high resistance link with the failed interconnect, Fig. 9(b). The failure mechanism during temperature cycling was due to both the delamination at Au-Au interface (bond interface) and the delamination at Cu-Au interface. Delamination at Au-Au interface is due to the micro voids that remain after bonding. These micro voids work as stress concentrators and during temperature cycling, due to the induced thermo-mechanical stresses, microcracks are formed. These microcracks propagate through the bond and result in the delamination and ultimate failure of the link. The delamination of the Cu-Au interface is due to the poor adhesion of the Au thin films to underlying Cu pillars despite using Ti as the adhesion layer. This will be addressed elsewhere.

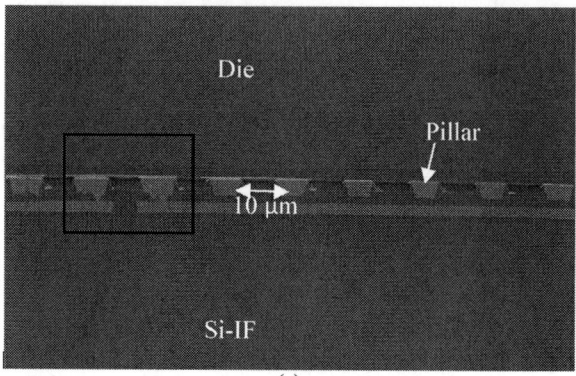

(a)

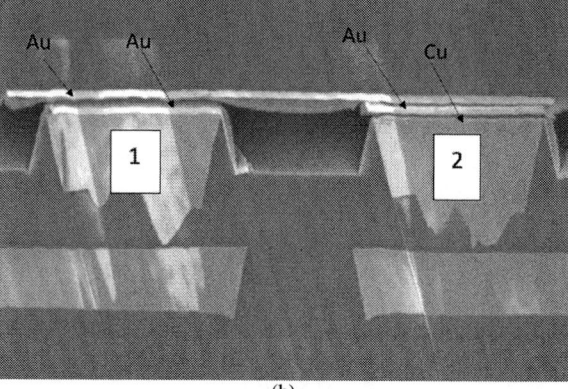

(b)

Figure. 9 (a) SEM image of the link with high resistance, two pillars shown delamination. (b) a zoomed in image of the two pillars, delamination happened at Au-Au interface for pillar 1 and for pillar 2, delamination happened at Au/Cu interface

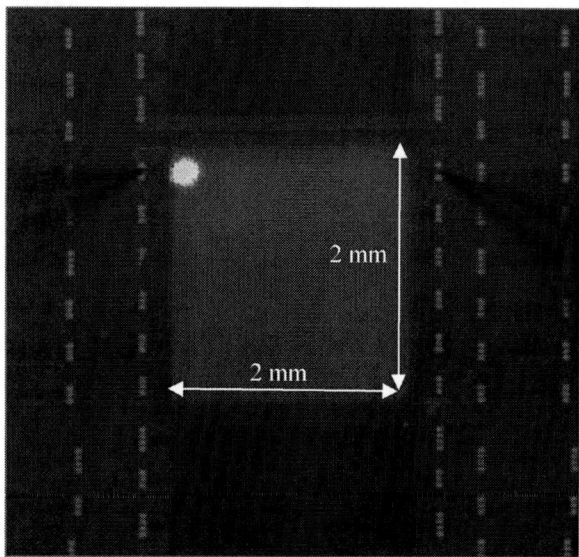

Figure. 8 Hot spot is detected in the thermal image on the upper left of the die.

Table II. Summary of the daisy chains resistance before and after 1000 cycles of humidity testing

Pad No	Resistance (Ω) before test	Resistance (Ω) after 1000 cycles
1-2	14.056	443
3-4	7.333	7.598
5-6	7.092	7.281
7-8	7.459	7.684
9-10	7.374	7.628

IV. CONCLUSION AND FUTURE WORK

We observe that oxidation happens in the samples passivated with only Parylene C, shown in Fig. 4 and TableI1. Therefore, Parylene C by itself is not an adequate encapsulation. However, increasing the thickness of Parylene C from 1 to 3 μm significantly reduced the amount of oxidation as shown in Table I. We observed combining Parylene C with an inorganic thin film such as SiN_x significantly reduced the penetration of water molecules to the substrate. Therefore, our multi-layer passivation scheme is an effective moisture barrier. In both Fig. 5 (a) and (b), no copper oxide peaks were detected on the passivated samples even after humidity testing for up to 168 hours.

For the Si-IF assemblies, PECVD step coverage is an important parameter to ensure proper passivation at the bond interface. The PECVD recipe was modified to include wafer bias. Biasing the wafer prevented charge build up on the sample which improved the conformality and step coverage of the deposited thin film (Fig. 6 and 7).

The average shear force before the test was 135 N and after 120 hours of humidity testing the average shear force didn't change. This suggests that multilayer encapsulation was an effective barrier for moisture ingress. Resistance of the passivated Si-IF didn't change after the period of humidity testing as well, which also shows that this encapsulation is in fact effective. Further investigation is needed on the failure mechanisms due to moisture ingress as well as identifying the optimum thickness of Parylene C and SiN_x layers.

Table II. shows the link resistances of a bonded samples before and after 1000 temperature cycles. The link that failed had a higher initial resistance which is the result of a poor bond (due to misalignment and/ or contamination of pads or pillars). The FIB cross section images showed delamination at the Au-Au interface as well as Cu-Au interface (Fig. 9 (a) and (b)). Defects in the bonding interface acted as stress concentrators during thermo-mechanical stressing of the sample which ultimately resulted in delamination at Au/Au interface. No failure was detected in other links that had low initial resistance. More process optimization is needed to avoid delamination at Au/Cu interface. Moreover, elimination of Au from Si-IF process and utilizing Cu-Cu bonding will eliminate this issue.

ACKNOWLEDGMENT

This work was supported in part by the Semiconductor Research Corporation (SRC), DARPA, and the CHIPS consortium.

REFERENCES

[1] S. Jangam, A. Bajwa, K. Thankkappan, P. Kittur and S. Iyer, "Electrical Characterization of High Performance Fine Pitch Interconnects in Silicon-Interconnect Fabric", 2018 IEEE 68th Electronic Components and Technology Conference (ECTC), 2018.

[2] S. Iyer, "Heterogeneous Integration for Performance and Scaling", IEEE Transactions on Components, Packaging and Manufacturing Technology, vol. 6, no. 7, pp. 973-982, 2016

[3] N.Shakoorzadeh, A. Hanna, and S. Iyer, "Bilayer Encapsulation for Silicon Interconnect Fabric", 2019 IEEE International Reliability Physics Symposium, 2019 (In press)

[4] D. Shnawah, M. Sabri and I. Badruddin, "A review on thermal cycling and drop impact reliability of SAC solder joint in portable electronic products", Microelectronics Reliability, vol. 52, no. 1, pp. 90-99, 2012. Available: 10.1016/j.microrel.2011.07.093.

[5] R. Tian, C. Hang, Y. Tian and L. Zhao, "Growth behavior of intermetallic compounds and early formation of cracks in Sn-3Ag-0.5Cu solder joints under extreme temperature thermal shock", Materials Science and Engineering: A, vol. 709, pp. 125-133, 2018. Available: 10.1016/j.msea.2017.10.007.

[6] R. Tian, C. Hang, Y. Tian and J. Xu, "Brittle fracture of Sn-37Pb solder joints induced by enhanced intermetallic compound growth under extreme temperature changes", Journal of Materials Processing Technology, vol. 268, pp. 1-9, 2019. Available: 10.1016/j.jmatprotec.2019.01.006.

[7] H. Ardebili, J. Zhang and M. Pecht, Encapsulation Technologies for Electronic Applications. [S.l.]: WILLIAM ANDREW PUBLISHING, 2018.

[8] Y. Yao, G. Lu, D. Boroyevich and K. Ngo, "Survey of High-Temperature Polymeric Encapsulants for Power Electronics Packaging", IEEE Transactions on Components, Packaging and Manufacturing Technology, vol. 5, no. 2, pp. 168-181, 2015. Available: 10.1109/tcpmt.2014.2337300.

[9] J. Ortigoza-Diaz et al., "Techniques and Considerations in the Microfabrication of Parylene C Microelectromechanical Systems", Micromachines, vol. 9, no. 9, p. 422, 2018.

[10] "Parylene Deposition Process – Specialty Coating Systems", Scscoatings.com.[Online].Available: https://scscoatings.com/parylene-coatings/parylene-expertise/parylene-deposition/. [Accessed: 09- Feb- 2019].

[11] J. Lee, B. Sahu and J. Han, "Simple realization of efficient barrier performance of a single layer silicon nitride film via plasma chemistry", Physical Chemistry Chemical Physics, vol. 18, no. 47, pp. 32198-32209, 2016. Available: 10.1039/c6cp06722k.

[12] N. Kim, S. Graham and K. Hwang, "Enhancement of the barrier performance in organic/inorganic multilayer thin-film structures by annealing of the parylene layer", Materials Research Bulletin, vol. 58, pp. 24-27, 2014. Available: 10.1016/j.materresbull.2014.03.022.

[13] S. Jangam, A. Bajawa, U. Mogera, P. Ambhore, T. Colosimo, T. Paulumbo, D. Deangelis, B. Chylak and S.S. Iyer, "Fine Pitch (< 10μm) Direct Cu-Cu Interconnects using In- Situ Formic Acid Vapor Treatment", 2019 IEEE 69th Electronic Components and Technology Conference (ECTC), (in press).

[14] E. Dalton, G. Ren, J. Punch and M. Collins, "Accelerated temperature cycling induced strain and failure behaviour for BGA assemblies of third generation high Ag content Pb-free solder alloys", Materials & Design, vol. 154, pp. 184-191, 2018. Available: 10.1016/j.matdes.2018.05.030.

[15] "TEMPERATURE CYCLING | JEDEC", Jedec.org. [Online]. Available:https://www.jedec.org/standards-documnts/docs/jesd-22-a104e.

978-1-7281-1500-9/19 $31.00 © 2019 IEEE

Improved Finite Element Modeling of Moisture Diffusion Considering Discontinuity at Material Interfaces in Electronic Packages

Lulu Ma[1,3], Rahul Joshi[2], Keith Newman[2], Xuejun Fan[1]

[1] Department of Mechanical Engineering, Lamar University, Beaumont TX 77710, USA
[2] AMD, 7171 Southwest Parkway, Austin, TX 78735 USA
[3] Department of Mechanics, Tianjin University, Tianjin 300350, China
xuejun.fan@lamar.edu

Abstract—**The modeling of moisture diffusion plays an important role for the integrity and reliability of electronic packages. In this paper, a new normalization approach and its implementation using ANSYS finite element analysis software are presented. Such an approach can solve the diffusion problem with varying temperature and humidity. Two different options in moisture diffusion modeling provided by ANSYS are discussed. As a validation, the numerical results are compared to that using the conventional normalization approach.**

Keywords: electronic package; finite element modeling; generalized solubility; moisture diffusion; normalization.

I. INTRODUCTION

Polymer-based materials play an important role in microelectronic and optoelectronic packaging. Polymeric materials are capable of absorbing moisture, retaining moisture, and transporting moisture [1-2]. The physical and chemical interactions of moisture with polymers can lead to physical, mechanical, and chemical changes of the polymers [3]. The moisture diffusion can lead to many deleterious effects, including material aging [4], hygroscopic swelling, electrochemical migration [5], and even "popcorn" failures [6-10]. Accurate modeling of moisture diffusion is crucial for assessing the reliability of electronic packages comprised of moisture sensitive materials such as substrate, adhesive, and encapsulation [11].

Fick's diffusion theory has been widely used in the field of moisture diffusion. The mathematical equation for Fickian moisture diffusion can be written as

$$\frac{\partial C}{\partial t} = -\nabla \cdot (-D_0 \nabla C) \qquad (1)$$

where C (kg/m^3) is moisture concentration, D_0 (m^2/s) is moisture diffusivity, which is temperature-dependent. However, moisture concentration is discontinuous at multi-material interfaces, as shown in Fig. 1a, which makes the problem unconventional in standard finite element modeling. As a result, (1) cannot be solved directly using the field variable: moisture concentration C. To overcome the limitation, a common practice is to adopt a continuous field variable through normalization, as shown in Fig. 1b. Different normalizing approaches have been developed in the literature, including the wetness theory [12], partial pressure [13], advanced normalization concentration [14], and recently, surface humidity potential (SHP) method [15]. These methods, however, all have limitations in the actual applications. A water activity-based diffusion theory has been developed to consider both linear and non-linear diffusion behaviors [16]. The theory adopts a field variable called water activity, which is a measure of water energy state and is always continuous. Unfortunately, the existing commercial finite element analysis software does not have the capability of using water activity. In the case of Henry's law, which assumes a linear relationship between the saturated moisture concentration and ambient humidity (*RH*), it has been shown that water activity-based diffusion theory is reduced to an established "wetness" approach [16]. Consequently, it may not be necessary to learn the water activity theory for diffusion modeling for many packaging challenges where Henry's law is applicable [17]. Even so, incorrect modeling results may still be obtained in diffusion modeling with finite element software [18].

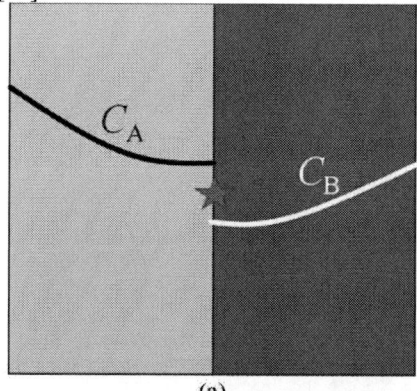

(a)

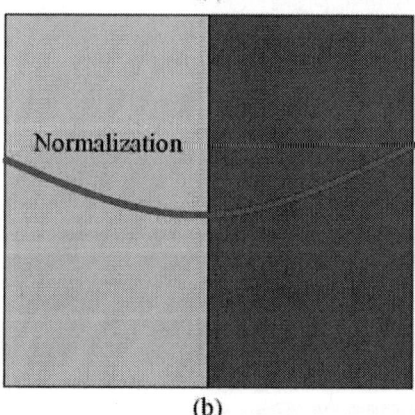

(b)

Figure 1.   (a) Illustration of a multi-material system with discontinuous moisture concentration C_A for material A and C_B for material B; (b) graph of normalized solution.

In this paper, the methodology of finite element moisture diffusion modeling using ANSYS will be studied. A new normalization approach is introduced, which can use the current capability of ANSYS moisture diffusion to solve much broader diffusion problems. In addition, the paper will discuss two different options in moisture diffusion provided by ANSYS – namely, "diffusion element" and "coupled element". We will present a benchmark example to demonstrate how moisture diffusion modeling can be conducted with accuracy in ANSYS. Specific examples include both humidity and temperature dependent material properties and actual applications.

II. CURRENT MOISTURE DIFFUSION THEORY IN ANSYS

ANSYS has extended its capability to solve moisture diffusion with the wetness approach, in which the moisture concentration is normalized by saturated concentration. The normalized diffusion model implemented in ANSYS is

$$\frac{\partial(C_{sat}\overline{C})}{\partial t} = \nabla \cdot (D\nabla(C_{sat}\overline{C})) + G \tag{2}$$

where $\overline{C}$ is the normalized concentration, similar to wetness, G is diffusing substance generation rate per unit volume, ∇ is gradient operator $= \{\frac{\partial}{\partial x} \frac{\partial}{\partial y} \frac{\partial}{\partial z}\}$, and $\nabla\cdot$ is divergence operator. Equation (2) can be rewritten as

$$C_{sat}\frac{\partial\overline{C}}{\partial t} + \overline{C}\frac{\partial C_{sat}}{\partial T}\frac{\partial T}{\partial t} = \nabla \cdot ([D]C_{sat}\nabla\overline{C} \\ + \overline{C}\frac{\partial C_{sat}}{\partial T}\nabla T) + G \tag{3}$$

In the above formulation, C_{sat} is considered as temperature-dependent only. However, C_{sat}, in general, is both humidity and temperature dependent. According to the Henry's law, the C_{sat} can be written as

$$C_{sat} = p_{amb}S \tag{4}$$

where p_{amb} is ambient partial vapor pressure and S is solubility. And the p_{amb} is given below

$$p_{amb} = RH\, p_g(T) \tag{5}$$

with $p_g(T)$ as saturated water vapor pressure, which has a form as

$$p_g(T) = p_0 e^{(-\frac{E_p}{RT})} \tag{6}$$

where $E_p=4.01\times10^{-4}$ (J/mol)=0.415 (ev), and $p_0=3.82\times10^{10}$ (Pa).
Henry solubility S is defined as

$$S(T) = S_0 e^{(\frac{E_s}{RT})} \tag{7}$$

Combining (4), (5), (6) and (7) yields

$$C_{sat} = RH\, p_0 S_0 e^{(\frac{E_s - E_p}{RT})} \tag{8}$$

The generalized solubility K, as defined by [16], may be introduced

$$K = K(T) = K_0 \exp(-\frac{E_k}{RT}) \tag{9}$$

where $K_0 = p_0 S_0$, $E_k = E_s - E_p$. With (9), the (8) simplifies into

$$C_{sat} = RH(t)K(T) \tag{10}$$

Equation (10) states that $C_{sat}(RH(t), T(t))$ is **both** time- and temperature-dependent. However, the current moisture diffusion theory in ANSYS assumes that C_{sat} is temperature-dependent only. Therefore, ANSYS cannot solve the problem with varying RH as function of time.

III. IMPLEMENTATION OF NEW NORMALIZATION THEORY

A. New Normalization Theory

We adopt $\overline{C}_K$ as the field variable, which can be written as

$$\overline{C}_K = \frac{C}{K} \tag{11}$$

where $\overline{C}_K$ is normalized concentration, and K is general solubility, as defined in (9). The moisture diffusion equation can be written as

$$\frac{\partial(K\overline{C}_K)}{\partial t} = \nabla \cdot (D\nabla(K\overline{C}_K)) \tag{12}$$

As K is only a function of temperature only, (12) is expanded to as follows

$$K\frac{\partial\overline{C}_K}{\partial t} + \overline{C}_K\frac{\partial K}{\partial T}\frac{\partial T}{\partial t} = \nabla \cdot ([D]K\nabla\overline{C}_K + \overline{C}_K\frac{\partial K}{\partial T}\nabla T) \tag{13}$$

Comparing (13) with (3), the governing equation using $\overline{C}_K$ is exactly identical to the diffusion equation in ANSYS. And $\overline{C}_K$ turns out to be the water activity a_w, which has been proved to be continuous at interface [16].

B. Analogy using ANSYS

To implement the new normalization theory, one can make analogy between (13) and (3). The analogy is summarized in Table I. In using ANSYS, the field variable $\overline{C}$ is replaced by $\overline{C}_K$. Temperature-dependent material property C_{sat} is replaced by general solubility K. From (10), the general solubility K can be rewritten as

$$K(T) = \frac{C_{sat}}{RH(t)} \qquad (14)$$

Substituting (14) into (11), the field variable $\overline{C}_K$ can be rewritten as

$$\overline{C}_K = \frac{C}{C_{sat}} RH(T) \qquad (15)$$

For the analogy of the two field variables $\overline{C}$ and $\overline{C}_K$, the initial and boundary conditions are summarized in Table II.

TABLE I.   ANALOGY BETWEEN NEW NORMALIZATION THEORY AND MOISTURE DIFFUSION WITH THE WETNESS APPROACH IN ANSYS

Equation	Analogy	
	Field Variable	Material Property
(3)	$\overline{C}$	C_{sat}
(13)	$\overline{C}_K$	K

TABLE II.   ANALOGY OF THE INITIAL AND BOUNDARY CONDITIONS FOR THE TWO FIELD VARIABLES

Field Variable	Analogy			
	Boundary condition		Initial condition	
	Absorption	Desorption	Absorption	Desorption
$\overline{C}$	1	0	C_0/C_{sat} (=0 if dry)	C_0/C_{sat} (=1 if fully saturated initially)
$\overline{C}_K$	$RH(t)$	0	C_0/K (=0 if dry)	C_0/K

IV.   CASE STUDY

A. Benchmark Problem

A 1-D bi-material absorption problem with varying RH and temperature is studied. The benchmark problem is shown in Fig. 2. The initial relative humidity $RH_0 = 0\%$ and the temperature $T = 30°C$. Ambient temperature profile and relative humidity profile are given below,

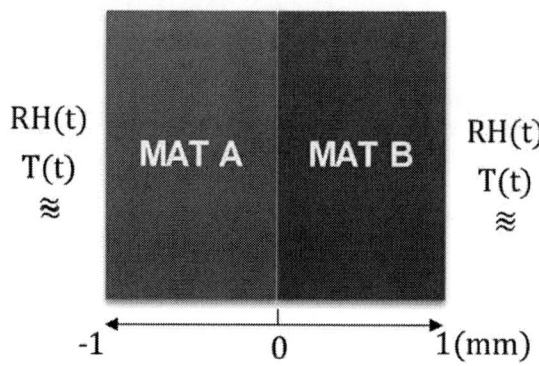

Figure 2.   A benchmarking problem for moisture diffusion in a bi-material system.

$$\begin{cases} T(t) = 30 + \dfrac{50t}{200} & 0 < t \le 200 \text{ (min)} \\ RH(t) = 0.3 + \dfrac{0.7t}{200} & 0 < t \le 200 \text{ (min)} \end{cases} \qquad (16)$$

From (16), C_{sat} is both time and temperature-dependent in this problem. Each material in the benchmarking problem has a thickness of 1 mm. The material properties are given in Table III. Based on (9), the generalized solubility K can be evaluated from the material properties given in the table.

TABLE III.   MATERIAL PROPERTIES FOR BENCHMARKING THE NEW NORMALIZATION THEORY

Properties	Material	
	Mat A	Mat B
D_0 (m²/s)	5.0×10^{-3}	4.0×10^{-3}
E_D (ev)	0.518	0.518
S_0 (Kg/m/Pa)	6.0×10^{-10}	2.0×10^{-10}
E_S (ev)	0.383	0.362

B. Results

The analysis was performed using the field variable $\overline{C}_K$. The thermal-diffusion coupled element was selected.

The calculated distributions of moisture concentration at $t = 100$ min and $t = 200$ min are plotted in Fig. 3. To verify the FEA results, a finite difference method (FDM) is also applied to solve the same problem with MATLAB programming. The numerical results from both FDM and FEA are in excellent agreement, as shown in Fig. 3. This confirms that the FEA presents correct results using the coupled thermal-diffusion element with $\overline{C}_K$.

To further demonstrate the validity of field variable $\overline{C}_K$, we calculated the distributions of moisture concentration at $t = 200$ min by using the field variables $\overline{C}$ and $\overline{C}_K$. The results are shown in Fig. 4. It can be seen from the figure that the distribution results of moisture concentration are different. Since C_{sat} is temperature-dependent and time-dependent, the moisture diffusion equation can be written as

$$C_{sat}\frac{\partial \overline{C}}{\partial t}+\overline{C}\frac{\partial C_{sat}}{\partial T}\frac{\partial T}{\partial t}+\overline{C}\frac{\partial C_{sat}}{\partial t}=\nabla \cdot ([D]C_{sat}\nabla \overline{C}$$
$$+\overline{C}\frac{\partial C_{sat}}{\partial T}\nabla T)+G \tag{17}$$

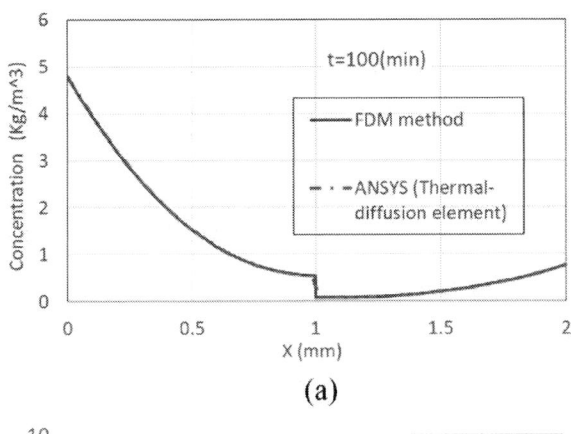

(a)

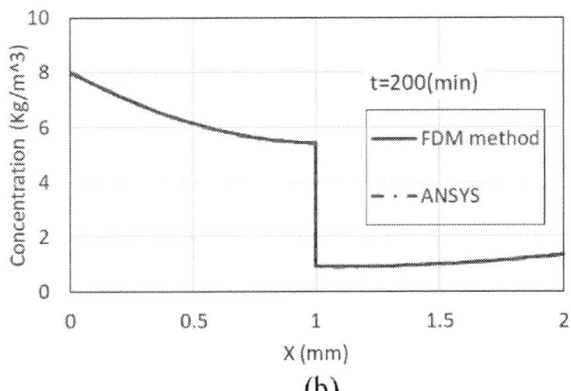

(b)

Figure 3. Results of moisture concentration by using the field variable $\overline{C}_K$ in ANSYS and the finite difference method (a) $t = 100$ (min) (b) at $t = 200$ (min).

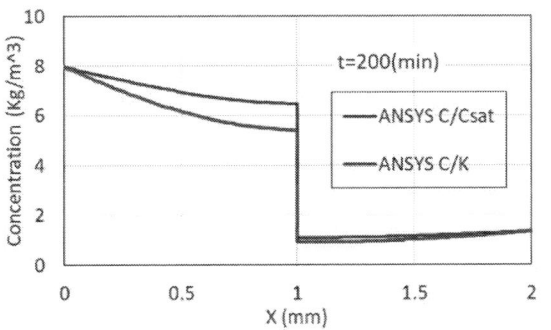

Figure 4. The distributions of moisture concentration at $t = 200$ min by using the field variable $\overline{C}$ and $\overline{C}_K$

Comparing (17) with the normalized diffusion model in (3), it is found that the actual moisture diffusion equation is different from the diffusion model implemented in ANSYS, so the results obtained by $\overline{C}$ are incorrect.

V. ELEMENT OPTIONS IN ANSYS

There are two different types of element option in ANSYS 17.2, namely "diffusion element" and "coupled element". In the "diffusion element" approach, however, the saturated moisture concentration must be temperature independent. In other words, C_{sat} must be a constant throughout the temperature range to be studied. Alternatively, ANSYS provides a "coupled element" option for moisture diffusion modeling, in which the saturated moisture concentration can be temperature-dependent. Essentially, coupled elements in ANSYS features the ability to solve coupled field simultaneously, such as heat conduction, moisture diffusion, and structural analysis.

In order to make the comparison among different element options in ANSYS, we analyzed the benchmark problem with different elements, and the results are shown in Figure 5.

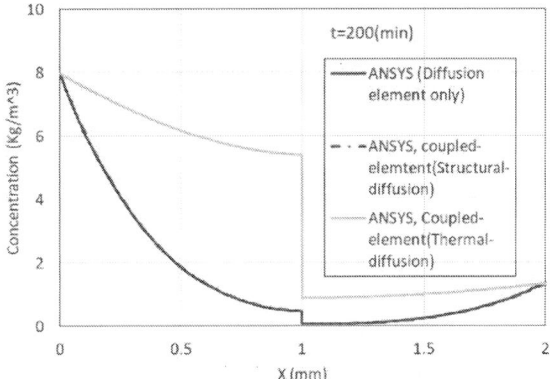

Figure 5. Results of moisture concentration calculated by different element options in ANSYS

In diffusion element only and the coupled element with the structure-diffusion approach, in which the saturated moisture concentration must be constant, the moisture diffusion equation can be written as

$$K\frac{\partial \overline{C}_k}{\partial t}=\nabla \cdot ([D]K\nabla \overline{C}_k) \tag{18}$$

which is different from (13). This is the reason why the diffusion element only and the coupled element with structural-diffusion option present the same incorrect results. Therefore, when materials behave as temperature-dependent saturated concentration, the coupled element with thermal-diffusion or thermal-diffusion structural element must be used.

VI. CONCLUSIONS

In this paper, a new normalization approach with the field variable $\overline{C}_K$ for moisture diffusion has been developed.

ANSYS built-in $\overline{C}$ approach cannot solve the problem with varying RH correctly. For a general moisture diffusion problem with temperature-dependent C_{sat} and varying ambient RH and temperature with time, $\overline{C}_K$ must be used and the coupled element with thermal-diffusion or thermal-structural-diffusion option must be applied at the same time.

This study concluded that the field variable $\overline{C}_K$ is applicable and effective with no restriction for modeling any moisture diffusion in electronic packages problems involving multi-material systems. Different options in moisture diffusion modeling provided by ANSYS were discussed in this paper. Caution must be made in the selection of different options for diffusion problems

ACKNOWLEDGMENT

Dr. Liangbiao Chen of On Semiconductor provided the numerical results using FDM. Ms. Leila Saberi helped to run the simulations and made plots for the results. Their efforts and contributions were greatly appreciated.

REFERENCES

[1] L. Chen, J. Zhou, H. Chu, G. Zhang, and X. Fan, "A Review on Water Vapor Pressure Model for Moisture Permeable Materials Subjected to Rapid Heating," Applied Mechanics Reviews, vol. 70, 2018, pp. 020803.

[2] X. Fan, "Moisture related reliability in electronic packaging," ECTC Professional Development Course Notes, 2008.

[3] E.-H. Wong and S. Park, "Moisture diffusion modeling–A critical review," Microelectronics Reliability, vol. 65, 2016, pp. 318-326.

[4] X. Q. Shi, Y. L. Zhang, W. Zhou, and X. J. Fan, "Effect of hygrothermal aging on interfacial reliability of silicon/underfill/FR-4 assembly," IEEE Transactions on Components and Packaging Technologies, vol. 31, 2008, pp. 94-103.

[5] X. J. Fan and E. Suhir, Moisture sensitivity of plastic packages of IC devices: Springer, 2010.

[6] A. A. Gallo and R. Munamarty, "Popcorning: a failure mechanism in plastic-encapsulated microcircuits," IEEE Transactions on Reliability, vol. 44, 1995, pp. 362-367.

[7] L. Chen, J. Adams, H. Chu, and X. Fan, "Modeling of moisture over-saturation and vapor pressure in die-attach film for stacked-die chip scale packages," Journal of Materials Science: Materials in Electronics, vol. 27, 2016, pp. 481-488.

[8] L. Zhu, J. Zhou, and X. Fan, "Rupture and instability of soft films due to moisture vaporization in microelectronic devices," Comput. Mater. Continua, vol. 39, 2014, pp. 113-134.

[9] X. Fan, G. Zhang, W. D. van Driel, and L. J. Ernst, "Interfacial delamination mechanisms during soldering reflow with moisture preconditioning," IEEE Transactions on Components and Packaging Technologies, vol. 31, 2008, pp. 252-259.

[10] C. G. Shirley, "Popcorn cavity pressure," IEEE Transactions on Device and Materials Reliability, vol. 14, 2014, pp. 426-431.

[11] L. Chen, H.W. Chu, X. Fan, "A convection–diffusion porous media model for moisture transport in polymer composites: model development and validation," Journal of Polymer Science Part B: Polymer Physics, vol 53, 2015, pp. 1440–1449.

[12] E. H. Wong, Y. C. Teo, and T. B. Lim, "Moisture diffusion and vapor pressure modeling of IC packaging," IEEE 48th Electronic Components and Technology Conference (ECTC), 1998, pp. 1372-1378.

[13] J. E. Galloway and B. M. Miles, "Moisture absorption and desorption predictions for plastic ball grid array packages," IEEE Transactions on Components, Packaging, and Manufacturing Technology: Part A, vol. 20, 1997, pp. 274-279.

[14] C. Jang, S. Park, B. Han, and S. Yoon, "Advanced thermal-moisture analogy scheme for anisothermal moisture diffusion problem," Journal of Electronic Packaging, vol. 130, 2008, pp. 011004.

[15] D. Markus, M. Schmidt, K. Lunz, and U. Becker, "A new method to model transient multi-material moisture transfer in automotive electronics applications," 17th International Conference on Thermal, Mechanical and Multi-Physics Simulation and Experiments in Microelectronics and Microsystems (EuroSimE),, 2016, pp. 1-8.

[16] L. Chen, J. Zhou, H. Chu, G. Zhang, and X. Fan, "Modeling nonlinear moisture diffusion in inhomogeneous media," Microelectronics Reliability, vol. 75, ,2017, pp. 162-170.

[17] L. Chen, Y. Liu, and X. Fan, "Application of water activity-based theory for moisture diffusion in electronic packages using ANSYS," 19th International Conference on Thermal, Mechanical and Multi-Physics Simulation and Experiments in Microelectronics and Microsystems (EuroSimE), 2018, pp. 1-6.

[18] C. Diyaroglu, S. Oterkus, and E. Oterkus, "A novel finite element technique for moisture diffusion modeling using ANSYS," IEEE 68th Electronic Components and Technology Conference (ECTC), 2018, pp. 227-235.

Study of thermal aging behavior of epoxy molding compound for applications in harsh environments

Adwait Inamdar[1], Alexandru Prisacaru[1], Erick Franieck[1], Martin Fleischmann[1], Przemyslaw Gromala[1],
Agnes Veres[2], Csaba Nemeth[2], Yu-Hsiang Yang[3], Bongtae Han[3]
[1]Robert Bosch GmbH, Automotive Electronics, Reutlingen, Germany
[2]Robert Bosch Kft., Automotive Electronics, Budapest, Hungary
[3]University of Maryland, College Park, USA
PrzemyslawJakub.Gromala@de.bosch.com

Abstract—In the automotive industry, epoxy-based molding compounds (EMCs) are often used to protect not only single IC packages but also entire electronic control units (ECUs). The EMC undergoes thermal aging during the operation-lifecycle of its parent electronic package. The thermal aging oxidizes the EMC, which can alter its mechanical properties significantly. Understanding the oxidation phenomenon of EMC and its effect on property changes is critically required to predict the reliability of ECUs subjected to harsh environments. In this study, the oxidation phenomenon of EMC is characterized experimentally by measuring oxidation growth rate and the mechanical properties of oxidized EMC. In the first task, EMC samples are subjected to three different high temperature storage (HTS) conditions – 170 °C, 200 °C and 230 °C. The thicknesses of the oxidized layers are measured as a function of storage time (from 0 to 1500 hours) using a fluorescent microscope. The oxidation growth rates at the storage temperatures are determined from the thickness measurements, and they are subsequently used to determine the activation energy of the growth rate. In the second task, thin samples (300 µm thick) are subjected to a HTS condition until they are fully oxidized. Then, critical thermo-mechanical properties of oxidized EMC, including coefficient of thermal expansion (CTE), glass transition temperature, and modulus of elasticity, are measured using digital image correlation (DIC) and dynamic mechanical analysis (DMA), respectively. The detailed procedures of the experimental characterization are presented together with the test results. Their implications on the ECU reliability is also discussed.

Keywords- Epoxy molding compound (EMC); thermal aging; oxidation; fluorescence microscopy; activation energy; material characterization; dynamic mechanical analysis (DMA); digital image correlation (DIC); moiré interferometry; coefficient of thermal expansion (CTE); glass transition temperature

I. INTRODUCTION

Epoxy molding compound (EMC) is a composite material consisting of epoxy resin as matrix, silica as filler material, hardeners, and other additives [1]. EMC is a thermosetting polymer formed by cross-linking reaction between epoxy resin and hardener.

A typical EMC contains very high content of silica filler, which helps in attaining desired thermo-mechanical properties. Hardeners contribute to the heat resistance and storage stability. Additives like cure promoter and flame-

retardants provide increased reaction between epoxy and hardener and lower risk of flammability, respectively.

Due to all of the constituent materials, EMC possesses high adhesion strength, low shrinkage, excellent chemical resistance and heat resistance [2]. Additionally, its ability to be molded and relatively low cost compared to traditional ceramic packages make EMC a better and very commonly used alternative as an encapsulating material for electronic packages. EMC protects the electronic chips from various environmental conditions including shock, chemical loads, heat, and moisture [3].

In the automotive industry, EMCs are often used to protect not only single IC packages but also entire electronic control units (ECUs). EMC undergoes thermal aging during the ECU operation. Thermal aging causes oxidation of EMC, which can significantly change its mechanical properties [4] [5].

Understanding the oxidation phenomenon of EMC and its effect on property changes is critically required to predict the reliability of ECUs subjected to harsh environments. This study aims at experimental characterization of the oxidation phenomenon of EMC. The study is carried out by conducting two separate tasks: (1) measurement of oxidation growth rate, and (2) measurement of mechanical properties of oxidized EMC.

II. BACKROUND: OXIDATION OF EMC

Oxidation of EMC is a combination of two processes – diffusion of oxygen and reaction of oxygen with EMC [6]. Oxygen present in the vicinity diffuses into the porous structure of EMC and it reacts with the EMC to form the oxide layer. The slower of these two simultaneous processes determines the rate of oxidation. Depending upon which process is slower, the type of resulting oxidation layer also varies. Since the reaction rates are governed by temperature, the temperature is a main factor affecting the rate of oxidation as well as the type of oxidation layer.

When the reaction process is much slower than the diffusion, oxygen can reach into deeper areas before getting fully consumed in the reaction process [7]. As a result, the oxidation layer does not have a clear separation from the bulk-EMC, and the diffused but unreacted oxygen could be found into the bulk-EMC region [2].

On the other hand, when the reaction process is much faster than the diffusion, the diffused oxygen is consumed rapidly by the reaction, and thus it cannot go into deeper areas. This results in a darker oxide layer with distinguishable boundary from bulk-EMC. Temperature being an important factor affecting this process, experimental trials of thermal aging are carried out at different temperatures to verify these different kinds of oxidation layers. Details about the experiments are discussed in the next section.

III. EXPERIMENTAL STUDY OF OXIDATION GROWTH

In order to study and model the oxidation growth, experimental data of the oxidation layer thickness at different temperatures was required.

A. Specimen and Testing Condition

For the experimental study, EMC samples of rectangular cross section 10 mm × 4 mm were subjected to thermal aging by means of high temperature storage (HTS) at the selected temperatures – 170 °C, 200 °C and 230 °C. Samples were stored for up to 1500 hours with selected reading time of 2, 4, 8, 24, 100, 500, 1000, and 1500 hours. After the aging process, each sample was embedded in clear resin as shown in Fig. 1. Cross sections of the samples were then prepared for the measurement by grinding with three different grain sizes – 120, 500, and 1200 and later polishing with two different roughness stages – 9 μm and 1 μm.

Figure. 1   EMC sample embedded in clear resin.

B. Measurement of Oxidation Layer

Measurement of oxidation layer thickness was done with different methods. Starting with digital microscopy, thickness values were measured at three different locations each on three different boundaries of the cross section, as shown in Fig. 2. Hence a set of 9 readings was obtained for each sample. At higher temperatures, thickness boundary is quite clearly visible and is easy to measure. On the contrary at the lower temperatures, the reaction of oxygen with EMC is slower and hence, the oxide layer boundary is difficult to

define and measure. Due to this challenge, especially at lower temperature and lower storage time, software based image enhancement was done, followed by re-measurement of the thickness.

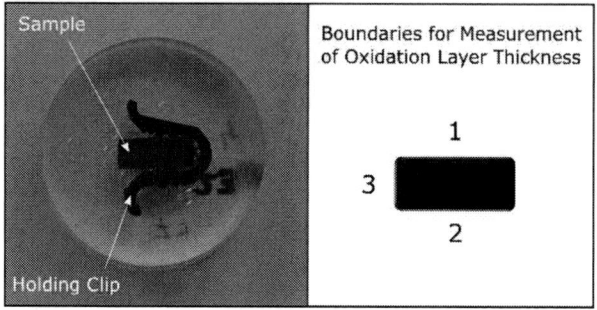

Figure. 2   Boundaries for measurement of oxidation layer thickness.

This still did not provide consistent data and hence, measurement using a fluorescence microscope was implemented. The illumination used in this microscopy is in the ultra-violate range. This results into more detailed images, since UV rays have smaller wavelength than visible light [8]. The same approach for number of measurements was followed to have 9 readings per sample. More consistent data with low standard deviation were obtained from fluorescence microscopy and it is represented in the plots in Fig. 3.

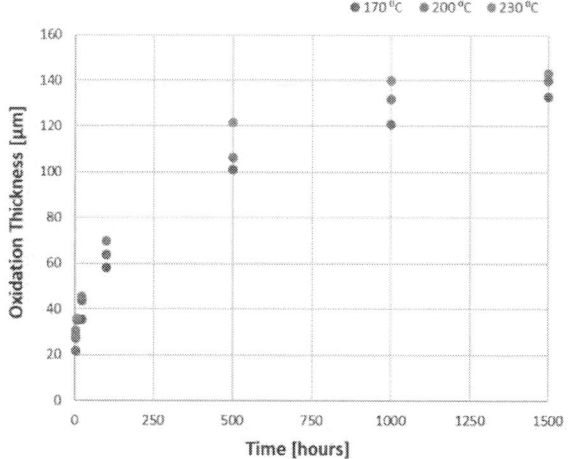

Figure. 3   Data of oxidation thickness measurement using fluorescence microscopy.

Fluorescence microscopy lead to some interesting observations. Similar to the previous measurements, images of samples kept at higher temperature show distinct, dark, and brown colored oxide layer, which is clearly separated from the green colored bulk-EMC. This also made determining the boundary between oxidation layer and bulk-EMC easier. However, samples at lower temperature show a prominent gradient from brown to green color. Fig. 4 shows the images of samples kept at 170 °C, 200 °C and 230 °C for 1000 hours.

In such cases, the gradient was closely observed and the boundary was defined at the point where the brown portion of gradient ends and the green colored region begins. In this way, the oxidation layer thickness was recorded.

Figure. 4   Comparison of oxide layer in EMC samples kept at different temperatures for 1000 hours.

IV.   OXIDATION GROWTH PREDICTION MODEL

An empirical model of the oxidation layer thickness is proposed based on the data collected from the measurements. The recorded data of oxidation layer thickness show faster growth in early hours of storage for all three temperatures. A trend of saturation is also observed at higher values of storage time. Taking these trends into consideration, a suitable time and temperature dependent mathematical model to predict the oxidation layer thickness was worked upon.

A.   Mathematical Model

Equitation represented in (1) indicates the proposed model. The oxidation thickness 'h' as a function of time 't' depends on the parameters 'A' and 'B'. Parameter 'A' is a constant and independent of temperature, while 'B' is temperature dependent parameter. Arrhenius equation, which is an empirical relationship for modelling temperature dependent processes, was considered for defining the parameter 'B'. The relation is indicated in (2), where 'C' is a constant, 'E_B' is activation energy, 'T' is temperature and 'R' is gas constant.

$$h(t) = A\left(1 - e^{-Bt}\right) \quad (1)$$

$$B(T) = C\, e^{-\frac{E_B}{RT}} \quad (2)$$

$$\ln B = \ln C - \frac{E_B}{R} \cdot \frac{1}{T} \quad (3)$$

When the equation (2) is manipulated by taking logarithm on both sides, an equation of straight line is obtained with '$1/T$' as abscissa and '$\ln B$' as the ordinate, as indicated in (3).

B.   Procedure to Obtain the Parameters 'A' and 'B'

In order to obtain the activation energy, the straight line indicated in (3) must be defined. Although 2 points are sufficient, 3 points define a straight line better. Hence, to obtain the three points – ($ln\, B_1$, $1/T_1$), ($ln\, B_2$, $1/T_2$), and ($ln\, B_3$, $1/T_3$), values of the parameter 'B' should be obtained at three different temperatures. Additionally, the temperature values T_1, T_2, and T_3 should be such that their reciprocal values are distant enough to define the straight line more correctly. Hence, the temperature values of 170 °C, 200 °C, and 230 °C were selected for the experiment to record the data of the oxidation layer thickness.

In order to obtain the complete mathematical model, MATLAB curve fitting tool was used. Curves were fit to the experimental data at three different temperatures, by adjusting the ranges of parameters 'A' and 'B'. Procedure was continued until a constant value of 'A' was obtained and corresponding three different values of 'B' were noted. They were then used to fit the straight line shown in (3) and to obtain the activation energy 'E_B' and constant 'C'. In this way, Arrhenius equation for parameter 'B' was fully defined. Hence, with the knowledge of parameters 'A' and 'B', the complete time and temperature dependent mathematical model of the oxidation layer thickness 'h' was obtained.

V.   EFFECT OF OXIDATION ON MECHANICAL PROPERTIES

There are three material properties of primary importance for an EMC – modulus of elasticity (E), coefficient of thermal expansion (CTE), and glass transition temperature (T_g). These properties define the thermo-mechanical behavior of EMC. Specific experimental tests were performed to record these properties. Dynamic mechanical analysis (DMA) is used to get the temperature variation of elasticity modulus, while thermo-mechanical analysis (TMA) or digital image correlation (DIC) is used to obtain the strain vs. temperature plot. Slope of the strain vs. temperature graph defines the CTEs of EMC, and the point of change in slope (the 'kink' point) defines the T_g. Along with the E vs. temperature plot, DMA also provides temperature variation of $tan\,\delta$, which is the ratio of loss modulus (E'') and storage modulus (E'). The single peak in this plot also indicates the positions of T_g. Hence, T_g of EMC can be defined from either of two experimental methods. Fig. 5 shows typical plots obtained from DMA and TMA/DIC and also indicates above mentioned material properties.

A.   Study of Partially oxidized EMC Samples

Effect of thermal aging on mechanical properties of EMC was studied by subjecting EMC samples to various sets of standard thermal aging processes, e.g. reflow soldering, thermal cycling, high temperature storage etc. These processes lead to partial oxidation of the samples. Later, standard tests – DMA and TMA (or DIC) were performed to observe changes in material properties. Fig. 6 [2] shows one example of comparison between results of non-aged and aged samples. Typical observation was that, in case of aged samples, more than one peaks in tan δ vs. temperature plot, more than one sudden drops in value of elasticity modulus, and more than one kinks in strain vs. temperature plot were observed. Since the non-aged EMC samples show all of the above features in respective plots only once, this leads to the conclusion that an aged-sample no longer consists of a single material, but is a multi-material system [9].

In other words, this suggests that the oxidation layer should have comparably different material properties than the bulk-EMC, which makes the plots of aged samples show multiple features instead of one. This observation highlights the requirement of characterizing material properties of the oxide layer separately.

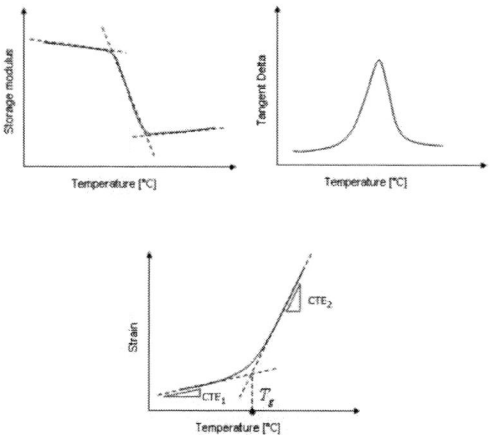

Figure. 5   Typical plots obtained from DMA and TMA / DIC tests.

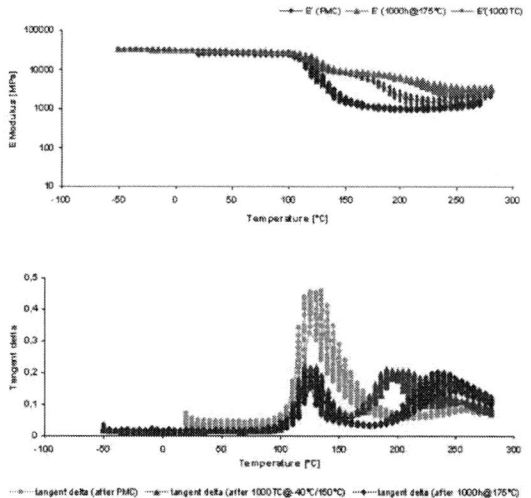

Figure. 6   Experimental plots obtained from the partially-oxidized EMC samples.

B. Characterization of Fully-Oxidized EMC Samples

In order to characterize the oxidized EMC using standard experimental tests, samples of fully-oxidized EMC were required. Since the regular samples used in the tests are too thick to be fully oxidized, special samples with 300 µm thickness were prepared.

1)   Sample Preparation

The thin samples were prepared from the regular samples, with about 2mm thickness, by a specially designed grinding process. In order to perform the sample thickness reduction, grinding blocks were designed and constructed such that the sample holders can be fixed onto them. The sample holders have cavities matching the dimensions of the samples, as shown in Fig. 7. The samples were placed inside the cavities and then were fixed into position using the venturi jet in combination with compressed air.

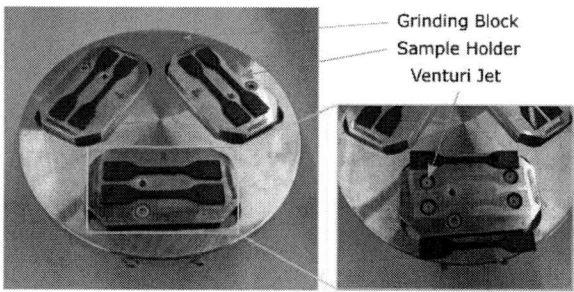

Figure. 7   Specially designed sample holders for the grinding process to produce 300 µm thin samples.

After the fixing process, grinding was carried out to attain the desired sample thickness. This was followed by polishing to guarantee a reproducible surface of the ground EMC, later for the temperature storage. Samples were ground and polished equally on both sides. As a result, samples with 300 µm thickness with a shiny and reflective surface were obtained. Fig. 8 shows the thin and the regular sample, marked with the numbers 1 and 2 respectively.

Figure. 8   EMC samples: 300 µm thin (1) and 2 mm thick (2).

Thin samples were then stored at a constant temperature of 150 °C for over 2000 hours to ensure complete oxidation. It was also verified by observing the cross sections under fluorescence microscope, as shown in Fig. 9. Similar procedure to the oxidation thickness measurement was followed to prepare the cross section. CTE, glass transition temperature, and modulus of elasticity of the fully-oxidized EMC were then characterized using these fully-oxidized thin samples.

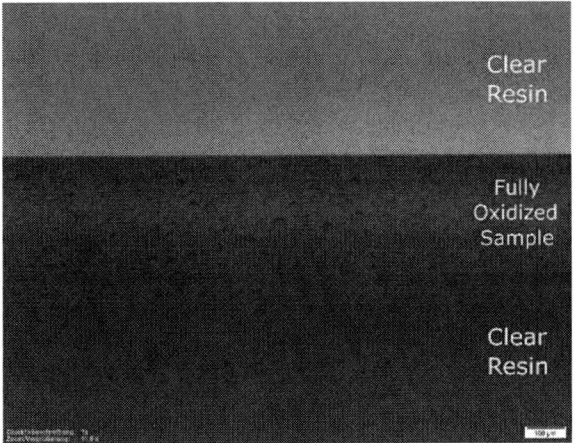

Figure. 9 Cross section of fully-oxidized EMC sample observed under fluorescence microscope.

2) Experimental Tests
a) Coefficient of Thermal Expansion

The CTE of fully oxidized and non-aged thin samples (300 μm thickness) was measured using the digital image correlation (DIC) technique. This method records images of samples at different temperatures with the help of two cameras angularly directed towards them. The strain is calculated along a user-defined path by comparing pixels of images at different temperatures [10]. The DIC setup is schematically represented in Fig. 10.

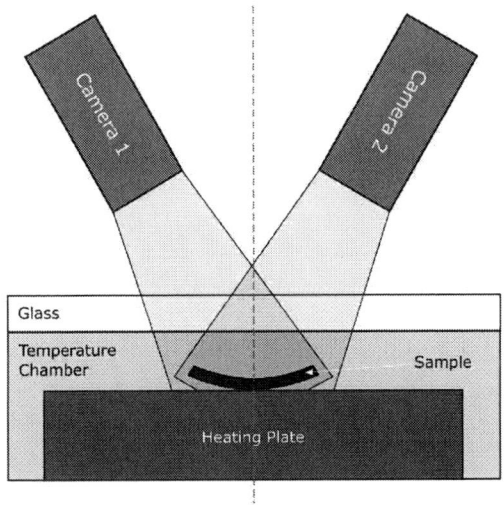

Figure. 10 Schematic illustration of the DIC setup.

Fig. 11 shows the temperature profile used in one of the trials of this experiment. It is important to note that the initial two temperature cycles are required as pre-conditioning for the samples. This eliminates unnecessary multiple kinks in the strain vs. temperature graph. Actual measurements were carried out during the last heating cycle with a low rate of temperature change.

Evaluation of strain was carried out at different temperatures for both fully-oxidized and non-aged samples of EMC. A straight line was used as the path for strain measurement. Fig. 12 shows an example of evaluation image of samples at 120 °C. The horizontal lines in black color are the paths used to evaluate the strain.

The strain versus temperature plot obtained from this test is shown in Fig. 13 while the Table 1 summarizes the values of CTE for both samples. The fully-oxidized EMC shows significantly higher value of T_g, which is determined from the sudden change of slope or the position of 'kink-point'. The CTE values of oxidized EMC are lower than corresponding values of non-aged EMC for the entire range of temperatures.

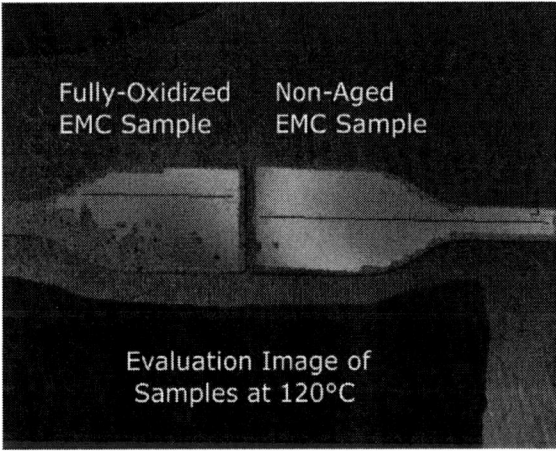

Figure. 11 Temperature profile used for the DIC test.

Figure. 12 Evaluation image of samples at 120 °C during the DIC test.

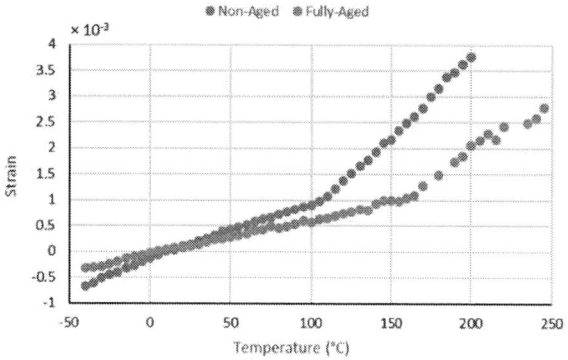

Figure. 13   Strain vs. temperature plot obtained from the DIC test.

TABLE I.   CTE VALUES FOR NON-AGED AND FULLY-OXIDIZED EMC SAMPLES

	T_g (°C)	CTE1 (ppm/K)	CTE2 (ppm/K)
Non-Aged	100	11	30
Fully-Oxidized	160	7	20

Before T_g of the non-aged EMC (100 °C), the difference in CTE values is 3 ppm/K, while after T_g of the oxidized EMC (160 °C), the difference rises to 10 ppm/K. In the range between 2 T_g's, this difference is 23 ppm/K and is the highest. Clearly, both materials exhibit noticeably different thermal properties.

b) Modulus of Elasticity

Modulus of elasticity, more precisely the storage modulus (E'), was characterized using dynamic mechanical analysis (DMA). Samples of dimension 25 mm × 10 mm × 0.3 mm were used for the experiment. A sinusoidal stress was applied at a frequency of 10 Hz and strain in the sample was measured as a response. As a result, the complex modulus (E), which has storage modulus E' as the real coefficient and loss modulus E'' as the imaginary coefficient, was obtained. This process was carried out at different temperatures in the range -40 °C to 300 °C. Fig. 14 shows the storage modulus versus temperature plot, obtained from this test. The y-axis values are normalized by storage modulus of non-aged EMC at room temperature.

The plot clearly indicates that the oxidized EMC has higher value of modulus for the entire range of temperature. At the room temperature (20 °C), it possesses nearly 20% higher value of storage modulus than the non-aged EMC. The beginning of the transition zone of oxidized EMC is also shifted by about 80 °C. The transition zone of oxidized EMC is also spread over a bigger temperature range than that of non-aged EMC. These results highlight that the oxidized and non-oxidized EMC have significantly different mechanical properties too.

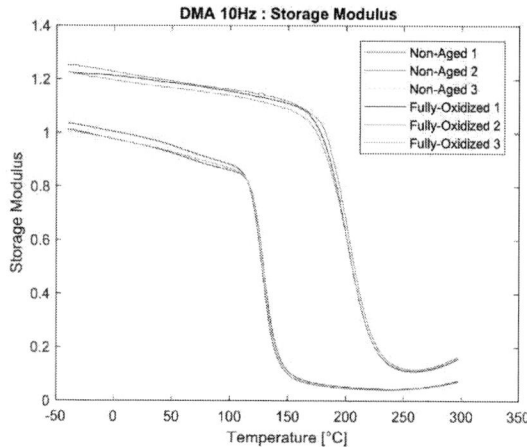

Figure. 14   Storage modulus vs. temperature plot obtained from the DMA test.

c) Thermal Deformation Analysis by Moiré Interferometry

Real-Moiré interferometry was utilized to investigate the effect of the oxidized layer on the package deformations. The method is a full-field optical technique to measure the in-plane deformations with high sensitivity, high signal-to-noise ratio, and excellent clarity [11]. The outputs are the contour maps of in-plane displacements. It has been used widely for electronic packaging design and reliability assessment [12].

The optical / mechanical configuration used in the study is illustrated in Fig. 15. It consists of (1) a portable engineering moiré interferometer that provides two sets of virtual reference gratings and (2) a conduction chamber built on a high performance thermo-electric cooler that provides accurate temperature control. The thermal conduction chamber is mounted on an x-y-z translation stage, which allows positioning as well as focusing the specimen. More details of the system can be found in [13].

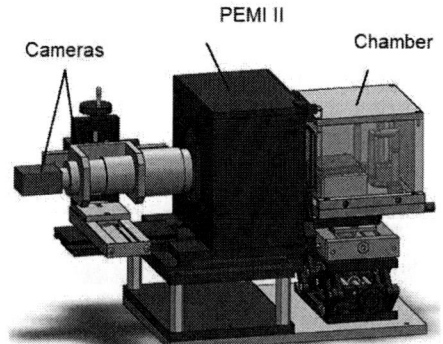

Figure. 15   Schematic illustration of the optical/mechanical configuration.

Two packages – a non-aged package and a package subject to an aging temperature of 150 °C for over 2000 hours – were prepared for moiré experiments. The cross-section of the aged package is shown in Fig. 16.

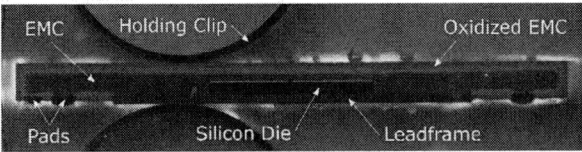

Figure. 16 Cross-section of aged package used for moiré experiments.

The specimen gratings with a frequency of 1200 lines/mm were replicated on the cross sections at room temperature (20 °C). The details about the grating and replication procedures used in this study can be found in [11] and [14]. The packages with the gratings were then heated to 85 and 150°C, and the thermal deformations were documented by the moiré system. It is to be noted that the first measurement temperature is below T_g of both non-aged and fully-oxidized EMC while the second temperature lies between T_g's of non-aged and fully-oxidized EMC.

The U and V fringe patterns obtained at the two temperatures, representing the in-plane displacements in the x and y directions, respectively, are shown in Fig. 17.

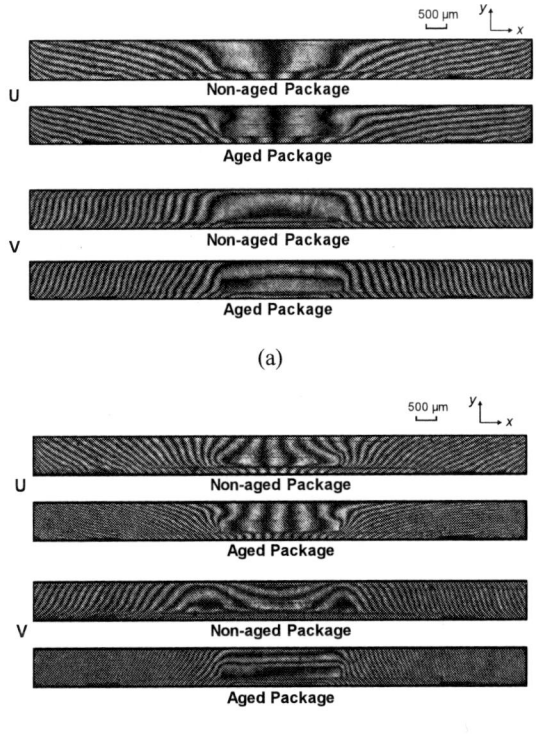

Figure. 17 Fringe patterns representing in-plane displacements obtained at (a) 85°C and (b) 150°C, where contour interval is 407 nm.

Both packages at 85°C show similar behaviors, but at 150°C, the aged package shows much stiffer behavior than the non-aged package. It is attributed to the higher T_g and thus higher modulus and lower CTE of the aged part of EMC at 150°C.

A more quantitative analysis after measuring the thickness of the oxidized layer of the aged package will be conducted, and the results will be reported in the future publication.

VI. CONCLUSION

Thermal aging of EMC is an important phenomenon with respect to predicting the reliability of an electronic package. Hence, a thorough study of the effect of thermal aging on EMC was carried out. It has two key aspects – (1) study of the oxidation growth and (2) study of thermo-mechanical properties of the oxidized-EMC.

In the first aspect, EMC samples were thermally aged at different temperatures and for different storage periods. Measurement of oxidation layer thickness was carried out by fluorescence microscopy. Collected data was then plot and a suitable mathematical model was proposed based on the observed trends. A detailed procedure with curve fitting, in order to obtain the time and temperature dependent model of oxidation layer growth, is also described.

The second aspect deals with the effect of thermal aging on thermo-mechanical properties of EMC. Initially, a comparison of experimental results from DMA and TMA / DIC for the partially-aged and non-aged EMC samples is presented. The results highlight the necessity of characterizing material properties of the oxide layer solely. Hence, a novel method to prepare 300 µm thin samples was designed. Prepared thin samples were fully-oxidized and then used for experimental tests.

Coefficient of thermal expansion and glass transition temperature were characterized using DIC, while the modulus of elasticity was evaluated using DMA. Moiré interferometry was also carried out on aged and non-aged packages. The moiré interferometry results reflect the behavior based on the quantitative differences in the material properties of oxidized and non-aged EMC. Collectively, all of the experimental tests show a significant difference in all thermo-mechanical properties of the oxide layer and bulk-EMC.

This study presents a mathematical model, which can be used to predict the oxide layer thickness inside EMC, after an exposure to a particular temperature for a certain time. Material characterization provides values of thermo-mechanical properties of the oxide layer. Combination of both can be incorporated into a finite element simulation of an electronic package, and its reliability can be predicted more realistically and accurately.

ACKNOWLEDGMENT

The authors would like to thank Dr. Eric Nguegang for his guidance and support.

REFERENCES

[1] Eric Nguegang, Jochen Franz, André Kretschmann, and Hermann Sandmaier, "Aging effects of epoxy moulding compound on the long-term stability of plastic package", 11th. International Conference on Thermal Mechanical and Multiphysics Simulation and Experiments in Micro-Electronics and Micro-Systems EuroSimE 2010, pp. 1-6, April 2010.

[2] Eric Nguegang, "Experimental and simulation-based investigations on the influence of thermal aging and humidity on the warpage of molded plastic packages", 2013.

[3] K.M.B. Jansen, J. de Vreugd, L.J. Ernst, and C. Bohm, "Thermal aging of molding compounds", 11th International Conference on Electronic Packaging Technology & High Density Packaging 2010, pp. 778-780, August 2010.

[4] J. de Vreugd, K.M.B. Jansen, L.J. Ernst, C. Bohm, and R. Pufall, "High temperature storage influence on molding compound properties", 11th International Conference on Thermal Mechanical and Multiphysics Simulation and Experiments in Micro-Electronics and Micro-Systems EuroSimE 2010, pp. 1-6, April 2010.

[5] J. de Vreugd, A. Sánchez Monforte, K.M.B. Jansen, L.J. Ernst, C. Bohm et al., "Effect of postcure and thermal aging on molding compound properties", 11th Electronics Packaging Technology Conference 2009, pp. 342-347, December 2009.

[6] L. Olivier, C. Baudet, D. Bertheau, J. C. Grandidier, M. C. Lafarie-Frenot, "Development of experimental, theoretical and numerical tools for thermo-oxidation of CFRP composites",Composites part A, vol. 40, issue 8 (2009), pp. 1008-1016.

[7] P. Dole, J. Cauchard, "Determination of oxidation profiles of elastomeric materials. Part I. microscopic approach: pinpoint DMA", Polymer Degradation and Stability, vol.47, issue 3 (1995), pp. 441-448.

[8] Fred Rost, "Fluorescence microscopy, applications", Encyclopedia of Spectroscopy and Spectrometry (Third Edition), 2017, pages 627-631.

[9] J. de Vreugd, "The Effect of Aging on Molding Compound Properties", 2011.

[10] Bing Pan, Kemao Qian, Huimin Xie, and Anand Asundi, "Two-dimensional digital image correlation for in-plane displacement and strain measurement: a review", Measurement Science and Technology, vol. 20, number 6, April 2009, doi:10.1088/0957-0233/20/6/062001.

[11] D. Post, B. Han, and P. Ifju, "High Sensitivity Moiré: Experimental Analysis for Mechanics and Materials, Mechanical Engineering Series," Springer-Verlag, NY, 1994. (Student edition, 1997).

[12] B. Han, "Thermal stresses in microelectronics subassemblies: quantitative characterization using photomechanics methods," Journal of Thermal Stresses 26:6 (2003), pp. 583-613, doi:10.1080/713855954.

[13] B. Wu and B. Han, "Advanced mechanical/optical configuration of real-time moiré interferometry for thermal deformation analysis of fan-out wafer level package," IEEE Transactions on Components, Packaging and Manufacturing Technology, vol. 8, number 5, March 2018, doi: 10.1109/TCPMT.2018.2805873.

[14] B. Han, "Recent advancements of moiré and microscopic moiré interferometry for thermal deformation analyses of microelectronics devices," Experimental Mechanics, vol. 38, issue 4, pp. 278-288, December 1998, doi: 10.1007/BF02410390.

Warpage Variation Analysis and Model Prediction for Molded Packages

Yuling Niu, Wei Wang, Zhijie Wang, Karthik Dhandapani, Mark Schwarz, Ahmer Syed

Qualcomm Technologies, Inc.
5775 Morehouse Drive
San Diego, CA 92121
E-mail: niu@qti.qualcomm.com

Abstract—Mechanical simulation model is an effective way to analyze warpage of electronic packages. To establish a reliable numerical model, simulation results are usually calibrated by warpage measurement results. In general, the measured warpage data are collected through averaging the warpage of several samples. Some categories of electronic packages show relatively stable warpage behavior so that the average warpage can eliminate the measurement error. However, in this study, it was observed that samples at different locations of substrate displayed quite different warpage values and shapes, which means the average warpage cannot represent the actual deformation situation. The warpage variation and location dependency were attributed to the shrinkage impact of epoxy molding compound. Therefore, molding shrinkage needs to be considered in simulation models to better predict molded packages. In this work, the molded packages are simulated during reflow profile. The effective strain of molding material, consisting of chemical shrinkage and cooling strain at gel point, is calculated and included in numerical model. By employing different chemical shrinkage strains from different directions, the warpage variation curve was accurately described and simulated by numerical model.

Keywords- Warpage Variation; Epoxy Molding; Shrinkage Strain; Simulation; Finite Element Method; Molded Package

I. INTRODUCTION

Warpage, generated by the coefficient of thermal expansion (CTE) mismatch among different material layers and chemical shrinkage of film materials, is always the principal concern during fabrication process of electronic packages. Nowadays, there are effective tools like Shadow Moiré [1] and Digital Image Correlation assisting in real-time warpage measurement of electronic packages during reflow process [2], [3]. Besides directly building test vehicles and conducting measurements, finite element warpage simulation and prediction is another efficient and robust way to understand the effect of different design factors and to compare the advantage and disadvantage among various package structures or material selections [4]–[6]. It is essential to calibrate the finite element model with experiment data for precise warpage prediction [7]–[9]. The conventional numerical model calibration utilized average warpage of samples as reference. However, in this study, when film materials were applied to electronic package, the warpage showed clear variation. Rather than average warpage, the maximum and the minimum warpage are the real risk endangering the yield rate during mass production.

Organic substrate, Silicon chip and Copper lid are relatively steady in terms of material properties, which means they will not cause evident warpage variation [10]. In contrast, film material is an unstable factor during curing process [11]–[13]. The epoxy molding curing process involves chemical shrinkage strain and cooling strain from curing temperature to gel point. Warpage variation emerged due to various shrinkage strains of epoxy molding from different directions. It's difficult to perform quantitative calculation of these two strains individually. It is commonly believed that the chemical shrinkage and cooling strain function together to generate a permanent effective strain at gel point.

In this study, warpage variation curves and diverse warpage shapes were designed to be learned and modeled. At beginning, bilayer samples (organic substrate and epoxy molding) were built and measured to understand molding effect on warpage variation. After selecting the representative warpage measurement results of several samples, which stands for the maximum and minimum warpage curves and all varieties of warpage shapes, their effective shrinkage and cooling strains were calculated on the basis of warpage curvature [14]–[17]. Because chemical shrinkage is a type of deformation effect, effective strain was preferred to be applied to simulation model instead of built-in stress [18]. In the finite element model, different effective strains were applied to the molding material and then the bilayer sample models duplicated different warpage shapes and curves. The results were compared with linear model and experiment data. After verifying the shrinkage strain induced warpage model for bilayer structure, the complete molded package was modeled [19], and the potential warpage shapes and results were predicted. In the end, experiment results of molded package were collected to discuss the accuracy and effectiveness of the warpage prediction method, that is applying molding strain into the warpage variation model.

II. EXPERIMENT OF BYLAYER SAMPLES

To understand the molding deformation at curing temperature, test vehicles with two layers were built. In Fig. 1(a), the bilayer sample consists of 180μm epoxy molding and 136μm substrate with a dimension of $12.4 \times 12.4mm^2$ (Fig. 1(b)). The samples were fabricated on a large substrate strip while the molding was flowed on the substrate strip at 175°C. Total nine samples were selected at different locations of the substrate strip and baked at 125°C for 24 hours. Shadow Moiré was adopted to perform warpage measurement of these samples in a radiation chamber from 25°C to 260°C. Shadow Moiré was equipped with 100 LPI glass grating to achieve 2.5μm warpage resolution.

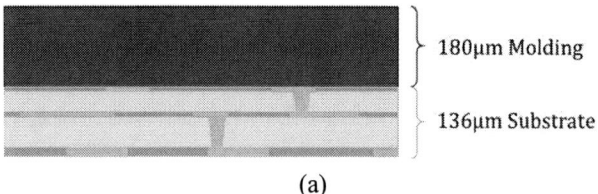

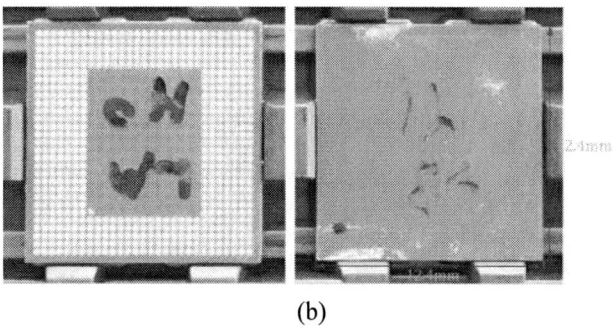

Figure 1. (a) Cross-section schematic and thickness of the bilayer (molding and substrate) sample; (b) Dimension of the bilayer sample and its top and bottom surface view.

The signed absolute warpage of nine bilayer samples was shown in Fig. 2. During 25°C - 260°C - 25°C temperature profile, the plot displayed symmetrical warpage curve, which proved the repeatability of Shadow Moiré measurement. At room temperature, the samples exhibited the smiling shape with negative warpage. Then the negative to positive warpage transition occurred between 183°C and 217°C. Whereas, the epoxy molding was cured at 175°C, where the flat shape or the warpage transition was ideally expected to be. It is recognized that the chemical shrinkage and cooling compression strain of molding at 175°C are the real reason to the samples' negative warpage.

In addition, the warpage variation was clearly expressed by three warpage curves. At 25°C, the minimum warpage is twice the maximum warpage. As temperature increases to 260°C, the warpage variation started with 133μm, then decreased to 60μm and end up with 85μm, an evident warpage variation persisted. Based on this warpage curve, at room temperature, the minimum warpage is the worst case, while at 260°C, the maximum warpage became the worst. In this study, the warpage variation range was set to be predicted to make the numerical model more useful instead of exclusively correlating with the average warpage.

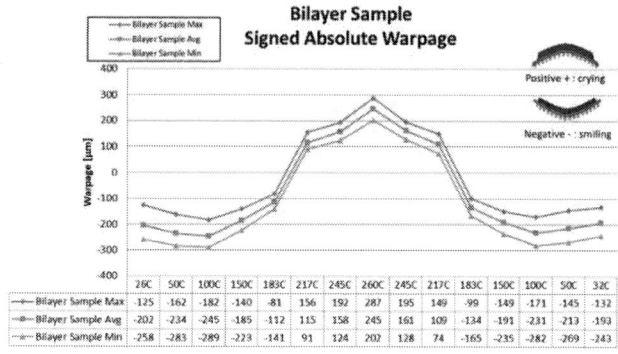

Figure 2. Signed absolute maximum, average, and minimum warpage curves of nine bilayer samples during reflow tempearture profile.

III. CURVATURE CALCULATION

Among nine bilayer samples, it is noticed that the warpage curve of sample #1 is the same with the minimum warpage curve, and the warpage curve of sample #6 is the same with the maximum warpage curve. These two samples were chosen as the curvature calculation reference to simulate the warpage range. In addition to the maximum and minimum warpage curvature, the warpage curve of sample #5 showed similar trend and value with the average warpage curve. Therefore, three samples were utilized to simulate the warpage variation curve and their warpage contours were shown in Fig. 3. The warpage contours of sample #1 and #6 showed the phenomenon of cylindrical shape with a principal warpage curvature, i.e. curvature of bifurcation. By checking the location of sample #1 and #6, it was noticed that the dominant curvature is dependent on molding flow direction. Meanwhile, sample #5 was located at the center of the substrate strip with regular spherical shape and quarter symmetric out-of-plane deformation, which is a common assumption in numerical simulation.

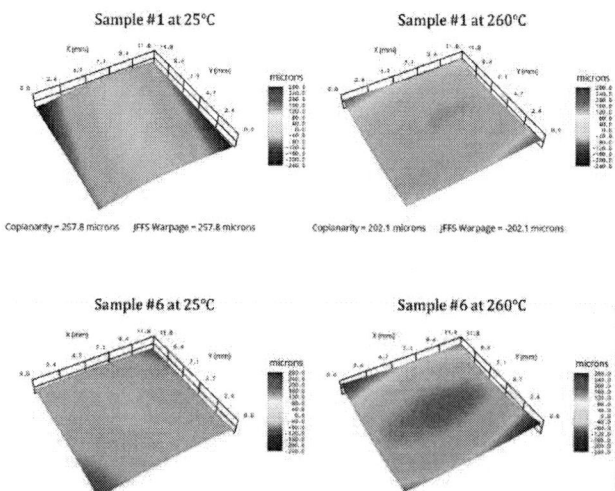

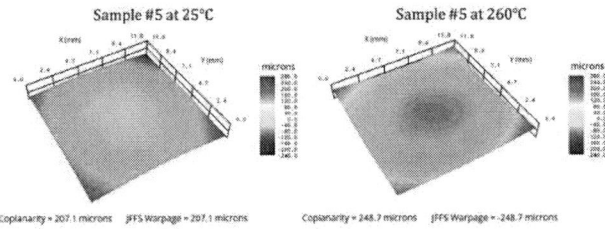

Figure 3. Absolute Shaodow Moiré substrate warpage contours of the bilayer sample #1, #6 and #5 at 25°C and 260°C.

In order to calculate the effective shrinkage strain at 175°C, linear interpolation was employed to compute the warpage and curvature of three samples. Given that the square of the displacement slope is small compared to the package size, the relationship between curvature and warpage can be defined as

$$\kappa_x = \frac{\partial^2 \omega}{\partial x^2} \qquad (1)$$

$$\kappa_y = \frac{\partial^2 \omega}{\partial y^2} \qquad (2)$$

$$\kappa_{xy} = \kappa_{yx} = \frac{\partial^2 \omega}{\partial x \partial y} \qquad (3)$$

where $z = \omega(x, y)$ is the sample warpage. κ_x and κ_y indicate the orthogonal x and y direction, respectively. κ_{xy} and κ_{yx} represent the twist curvature, meaning the rate of slope change along the perpendicular direction. According to the warpage shapes of three samples, the equation $(1) - (3)$ have different conditions:

Sample #1 (cylindrical shape): $\kappa_x \neq 0, \kappa_y \neq 0, \kappa_{xy} = 0$
Sample #6 (cylindrical shape): $\kappa_x \neq 0, \kappa_y \neq 0, \kappa_{xy} = 0$
Sample #5 (spherical shape): $\kappa_x = \kappa_y, \kappa_{xy} = 0$

After defining the curvature for each sample, the next step is to convert the curvature into shrinkage strain. Regarding that the thickness of molding is close to bottom substrate, the Stoney equation is not applicable any more [18]. In this scenario, the thickness of molding is considered, so the strain energy of the epoxy molding must be included in the calculation of total potential energy. Hence, the molding effective shrinkage strain was expressed as [20]

$$\varepsilon_m = \frac{\kappa E_s^*}{6 t_m E_m} \left(1 + \frac{t_m}{t_s}\right)^{-1} \left[1 + 4 \frac{t_m E_m^*}{t_s E_s^*} + 6 \frac{t_m^2 E_m^*}{t_s^2 E_s^*} + 4 \frac{t_m^3 E_m^*}{t_s^3 E_s^*} \right.$$

$$\left. + \frac{t_m^4 E_m^*}{t_s^4 E_s^*}\right] \qquad (4)$$

$$E_s^* = \frac{E_s}{1 - v_s} \qquad (5)$$

$$E_m^* = \frac{E_m}{1 - v_m} \qquad (6)$$

where t_m and t_s are the thickness of molding and substrate, respectively; E_s^* and E_m^* represent the substrate and molding biaxial elastic modulus; E_m, v_m, E_s, and v_s imply Young's modulus and Poisson's ratio of molding and substrate.

IV. RESULT AND DISCUSSION

A. Bilayer Sample Model Calibration

With the help of finite element method, two simulation models were built with solid hexahedral elements. Quarter symmetrical model is used for sample #5, the spherical shape warpage simulation. At the same time, the half model with diagonal symmetrical boundary condition is prepared for mimicking the cylindrical warpage shapes of sample #1 and #6 (Fig. 4).

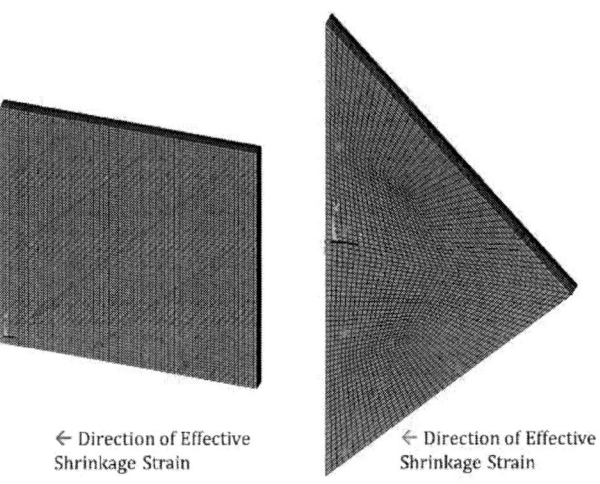

← Direction of Effective Shrinkage Strain

← Direction of Effective Shrinkage Strain

Figure 4. Finite element model mesh contours of the bilayer samples with quarter symmetrical boundary condition for spherical shape model, and half symmetrical boundary condition along diagonal for two cylindrical shape models.

Before running 25°C - 260°C - 25°C profile, shrinkage strain was applied to molding material at 175°C. After the model completing the thermal cycle, simulation results were collected and plotted in Fig. 5. The solid blue lines represent simulation results, in which light blue line indicates minimum warpage and dark blue line marks maximum warpage. Meanwhile, yellow dashed line and dark orange dashed line describe the experiment data of maximum and minimum warpage. It should be clarified that sample #5 was chosen to simulate the average warpage. It has similar but not the same warpage value with average warpage curve, which was marked as "Exp_#5" and to compare with simulation average warpage curve.

The trend of warpage transition and its peak and valley values in numerical models were consistent with experiment result. During temperature rising process, the simulation warpage curve predicted warpage accurately with the correlation error less than ±8% approximately. In most temperature stages, the correlation error is within ±5%. It is known that molding shrinkage strain dominates sample deformation between 150°C and 183°C. The analytical solution of the applied shrinkage strain lays several ideal assumptions according to the Freud's Equation. The calculation of effective strain can be improved in coming study. Through cooling process, Shadow Moiré has to cool

978-1-7281-1500-9/19 $31.00 © 2019 IEEE

down the chamber with wind blower, which may generate the temperature gradient and affect the accuracy of temperature detection. The simulation and experiment curves cannot match perfectly, but simulation results still show correct tendency and warpage range. Fig. 6 exhibited the examples of room temperature warpage contours of simulation and experiment. By extracting different curvatures of x and y axis and applying corresponding initial shrinkage strain to molding material, the cylindrical shape and the spherical shape were imitated in numerical model. Instead of the general quarter symmetrical boundary condition, the models of sample #1 and #6 portrayed the cylindrical warpage from one dominating direction. With the same molding shrinkage method, saddle shape and twisted shape can be modeled as well. In summary, warpage curves and contours of numerical model succeed to replicate the actual cylindrical out-of-plane deformation and the warpage transition tendency.

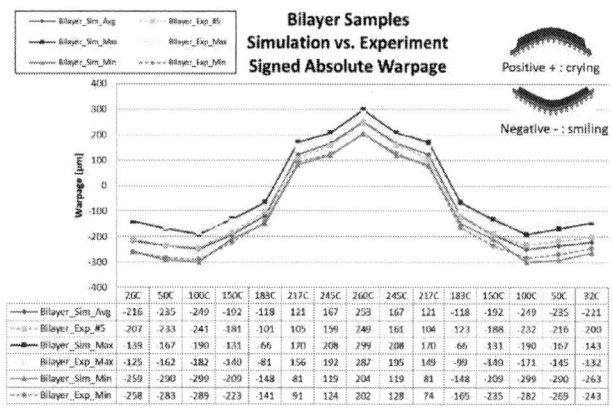

Figure 5. The comparsion of bilayer samples' average, maximum and minimum warpage variation curve of simulation and experiment during reflow temperature profile.

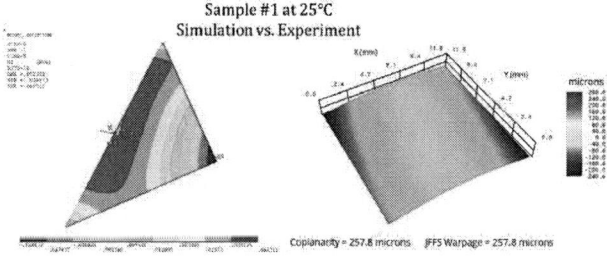

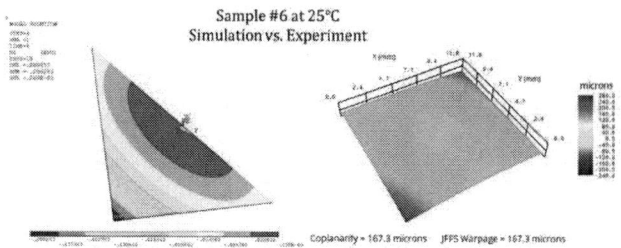

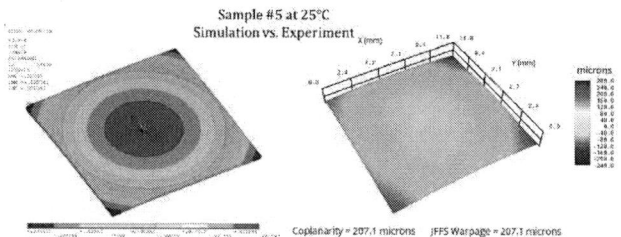

Figure 6. The comparsion of absolute warpage shapes and contours between simulation and experiment of bilayer sample #1, #6 and #5 at room temperature.

B. Molded Package Warpage Prediction

After confirming the molding shrinkage strain method on warpage variation simulation, this work moved on to the real package sample, the molded package, which has a piece of silicon chip inserting into molding (Fig. 7). It has the same dimension with bilayer sample. The approximative three-layer structure was more stable than bilayer. Moreover, the complex structure means that it is challenging to characterize the behavior of epoxy molding. Therefore, efforts were made in this study to build and analyze the bilayer sample at first. When calibrated, the material properties and effective strain were applied to the molded package model to learn warpage variation and compare to measurement result.

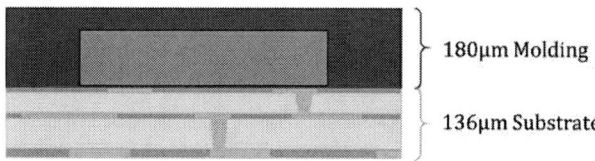

Figure 7. Schematic of the molded package's thickness cross-section with silicon die inserted.

Fig. 8 plotted both simulation and experiment warpage curves of molded package. Blue dashed lines describe experiment result, while solid orange and yellow line signify maximum and minimum warpage in simulation model. Due to the existence of die at package center, the maximum and minimum out-of-plane deformation were less than bilayer sample, and the warpage shape and trend were reversed, which revealed crying at low temperature and smiling at high temperature. For bilayer sample, there is a warpage turning point at 100°C, which is related to the material property adjustment of molding at glass transition temperature. Nevertheless, the molded package did not display the similar warpage turning phenomenon. In the aspect of model correlation, the numerical model was implemented to predict warpage transition and precise warpage variation of molded package.

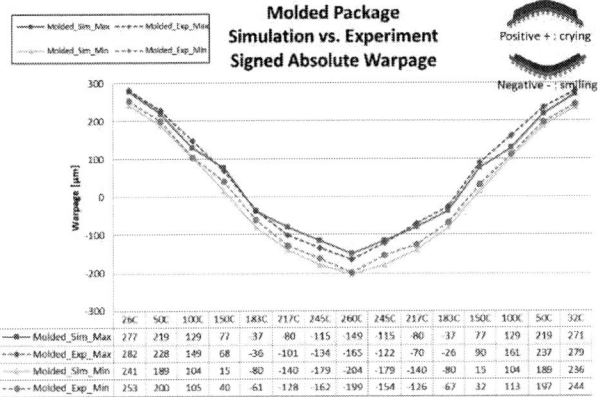

Figure 8. The comparison of molded package's maximum and minimum warpage cruves between simulation and experiment through reflow temperature profile.

	26C	50C	100C	150C	183C	217C	245C	260C	245C	217C	183C	150C	100C	50C	32C
Molded_Sim_Max	277	219	129	77	37	-80	-115	-149	-115	-80	-37	77	129	219	271
Molded_Exp_Max	282	228	149	68	-36	-101	-134	-165	-122	-70	-26	90	161	237	279
Molded_Sim_Min	241	189	104	15	-80	-140	-179	-204	-179	-140	-80	15	104	189	236
Molded_Exp_Min	253	200	105	40	-61	-128	-162	-199	-154	-126	67	32	113	197	244

C. Linear Model Discussion

The shrinkage strain method assisted the numerical model to simulate warpage variation effectively. At curing temperature, if the finite element model did not incorporate the molding shrinkage and diametrically utilized the material property, the linear model would predict warpage falsely. In Fig. 9, the green dotted line illustrates the average warpage curve with linear model. Without counting effective shrinkage strain, the sample was assumed to be stress-free at 175°C. The maximum and minimum warpage of linear model are smaller than actual measurement data. Though the warpage trend was consistent with measurement, the linear model was expected to generate distinct bias on warpage prediction of molded package. Meanwhile, the linear model is inadequate to compromise different shrinkage conditions to depict warpage variation.

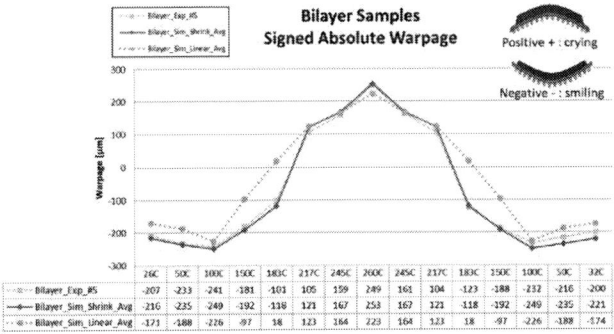

Figure 9. Average warpage comparison among experiment, shirnakge strain applied model and linear elastic model during reflow process.

	26C	50C	100C	150C	183C	217C	245C	260C	245C	217C	183C	150C	100C	50C	32C
Bilayer_Exp_#5	-207	-233	-241	-181	-101	105	159	249	161	304	-123	-188	-232	-216	-200
Bilayer_Sim_Shrink_Avg	-216	-235	-249	-192	-118	171	167	253	167	121	-118	-192	-249	-235	-221
Bilayer_Sim_Linear_Avg	-171	-188	-226	-97	18	123	164	223	164	123	18	-97	-226	-188	-174

V. CONCLUSION

This work laid a foundation of warpage variation prediction with numerical model. The simulation results correlated with measured warpage result for both bilayer sample and the molded package. This method pushed warpage prediction from a single value to a comparatively reasonable range. The prognostication of warpage range could be a good guidance to assist the electronic package developers in understanding the potential risks. Furthermore, the effective molding shrinkage strain method provides a novel opinion to define the shrinkage phenomenon of film and polymer materials. Similar concept can be extensively utilized in numerical models for various categories of electronic packages. This study will also be carried forward in model calibration and methodology verification on more electronic packages and warpage shapes in the future.

ACKNOWLEDGMENT

This research presented in this paper was highly supported by the electronic packaging team of Qualcomm Technologies, Inc. The authors appreciate the support.

REFERENCES

[1] Y. Y. Wang and P. Hassell, "Measurement of thermally induced warpage of BGA packages/substrates using phase-stepping shadow moire," in *Proceedings - Electronic Components and Technology Conference*, 1997, pp. 283–289.

[2] L. Xue, Y. Niu, H. Lee, D. Yu, S. Chaparala, and S. Park, "An Experimental and Numerical Study of the Dynamic Fracture of Glass," in *ASME 2013 International Technical Conference and Exhibition on Packaging and Integration of Electronic and Photonic Microsystems*, 2013, pp. 1–9.

[3] Y. Niu, J. Wang, and S. B. Park, "The Complete Packaging Reliability Studies through One Digital Image Correlation System," in *Proceedings - Electronic Components and Technology Conference*, 2017, pp. 1916–1921.

[4] W. Lin, "A feasible method to predict thin package actual warpage based on an FEM model integrated with empirical data," in *Proceedings of Electronic Components and Technology Conference*, 2015, pp. 1985–1990.

[5] S. Shao et al., "Comprehensive study on 2.5D package design for board-level reliability in thermal cycling and power cycling," in *Proceedings - Electronic Components and Technology Conference*, 2018, pp. 1668–1675.

[6] S. Shao, D. Liu, Y. Niu, and S. Park, "Die stress in stealth dicing for MEMS," in *Proceedings of the 15th InterSociety Conference on Thermal and Thermomechanical Phenomena in Electronic Systems, ITherm 2016*, 2016, pp. 539–545.

[7] J. Wang, Y. Niu, and S. Park, "An Investigation of Moisture-Induced Interfacial Delamination in Plastic IC Package During Solder Reflow," in *ASME 2017 International Technical Conference and Exhibition on Packaging and Integration of Electronic and Photonic Microsystems*, 2017, pp. 1–8.

[8] Y. Niu, H. Lee, and S. Park, "A new in-situ warpage measurement of a wafer with speckle -free digital image correlation (DIC)

method," in *Proceedings of the 2015 IEEE 65th Electronic Components and Technology Conference*, 2015, pp. 425–431.

[9] Y. Niu, J. Wang, S. Shao, H. Wang, H. Lee, and S. B. Park, "A comprehensive solution for electronic packages' reliability assessment with digital image correlation (DIC) method," *Microelectron. Reliab.*, vol. 87, no. March, pp. 81–88, 2018.

[10] Y. Niu, H. Wang, and S. B. Park, "A general strategy of in-situ warpage characterization for solder attached packages with digital image correlation method," *Opt. Lasers Eng.*, vol. 93, pp. 9–18, 2017.

[11] Y. Niu, H. Wang, S. Shao, and S. B. Park, "In-Situ Warpage Characterization of BGA Packages with Solder Balls Attached During Reflow with 3D digital image correlation (DIC)," in *Proceedings of the 2016 IEEE 66th Electronic Components and Technology Conference*, 2016, pp. 782–788.

[12] C. Zhu, W. Ning, H. Lee, J. Ye, G. Xu, and L. Luo, "Experimental identification of warpage origination during the wafer level packaging process," in *Proceedings - Electronic Components and Technology Conference*, 2014, pp. 815–820.

[13] S. Park, D. Liu, Y. Kim, H. Lee, and S. Zhang, "Stress evolution in an encapsulated MEMS package due to viscoelasticity of packaging materials," in *Proceedings of Electronic Components and Technology Conference*, 2012, pp. 70–75.

[14] T. L. Chou, S. Y. Yang, and K. N. Chiang, "Overview and applicability of residual stress estimation of film-substrate structure," *Thin Solid Films*, vol. 519, no. 22, pp. 7883–7894, 2011.

[15] G. G. Stoney, "The Tension of Metallic Films Deposited by Electrolysis," in *Proceedings of the Royal Society of London.*, 1909, vol. 82, no. 553, pp. 172–175.

[16] H. Gleskova, I. C. Cheng, S. Wagner, J. C. Sturm, and Z. Suo, "Mechanics of thin-film transistors and solar cells on flexible substrates," *Sol. Energy*, vol. 80, no. 6, pp. 687–693, 2006.

[17] Y. Niu, V. Pham, J. Wang, and S. Park, "An Accurate Experimental Determination of Effective Strain for Heterogeneous Electronic Packages With Digital Image Correlation Method," *IEEE Trans. Components, Packag. Manuf. Technol.*, no. Dic, pp. 1–11, 2018.

[18] C. Kim, T. Lee, H. Choi, M. S. Kim, and T. S. Kim, "Methodology development of warpage analysis of polymer based packaging substrate," in *Proceedings - Electronic Components and Technology Conference*, 2014, pp. 1004–1009.

[19] M. Brunnbauer *et al.*, "An embedded device technology based on a molded reconfigured wafer," in *Proceedings - Electronic Components and Technology Conference*, 2006, pp. 547–551.

[20] L. B. Freund and S. Subra, *Thin film materials stress, defect formation, and surface evolution.* Cambridge University Press, 2004.

2019 IEEE 69th Electronic Components and Technology Conference (ECTC)

Peridynamics for predicting thermal expansion coefficient of Graphene

E. Madenci, A. Barut and M. Dorduncu
Department of Aerospace and Mechanical Engineering
University of Arizona, Tucson, Arizona, USA

Abstract

This study presents an investigation of thermal fluctuations of a graphene layer by using peridynamics (PD). The stored energy in the graphene layer due to the fluctuations is expressed in a quadratic form in terms of the stiffness matrix under von Karman assumptions. The Gibbs free energy of the graphene layer related to the partition function is calculated using the Gaussian integrals. However, the partition function requires the evaluation of the determinant of stiffness matrix appearing in the energy expression. Although conceptually very attractive, computing the determinant of an extremely large stiffness matrix whose size is dictated by the characteristic length scale poses computational challenges. Therefore, the PD form of the stiffness matrix is constructed by using two levels of discretization in order to evaluate its determinant accurately without any computational challenges. The derivatives of the partition function permit the determination of several thermodynamic quantities such as the thermal expansion coefficient and its dependence on temperature. This approach enables the exploration of the effect of different geometries, boundary conditions and the nature of loading conditions as well as heterogeneous material properties on the fluctuations.

Introduction

Graphene is a promising material for electronics industry due to its unique mechanical, thermal and electronic properties. Therefore, its excellent thermal, mechanical, and electrical properties [1] have received a great deal of attention. However, its thermal fluctuations received relatively less attention [2]. A graphene layer has a single-atom thickness, and in-plane carbon atoms are covalently bonded resulting in very strong connections between the carbon atoms. Also, there exist many experimental investigations on single- and multi-layer graphene in order to control the wrinkle amplitude, wavelength and orientation. These wrinkles may have some influence on its physical properties [2-5] including the negative thermal expansion coefficient based on experiments and simulations [6].

This study focuses on the thermal fluctuations of a graphene layer. It specifically concerns the determination of the thermal expansion coefficient and its dependence on temperature. The thermal fluctuations of extremely thin material layers were investigated experimentally [7], analytically [8] and through simulations [9] based on the relationship between geometry and statistical mechanics [10]. The majority of the simulations employ triangulation of the geometry and Monte Carlo and time-dependent Ginzburg methods [11].

Recently, Liang and Purohit [12] constructed the energy stored in the fluctuating material layer in terms of a quadratic function of out-of-plane deflections under von Karman assumptions after discretizing the geometry through equilateral triangle elements. The quadratic form of the energy enables the computation of the partition function by using the Gaussian integrals rather than commonly accepted simulation methods [13]. Also, it enables the exploration of the effect of different geometries, boundary conditions and the nature of loading conditions on the fluctuations.

Within the realm of finite element method, the partition function requires the evaluation of the determinant of stiffness matrix appearing in the quadratic form of the energy. Although conceptually very attractive, computing the determinant of an extremely large stiffness matrix whose size is dictated by the characteristic length scale for sufficient geometric accuracy poses computational challenges.

In order to avoid such computational challenges, this study employs the peridynamic differential operator (PDDO) [14] at two levels of discretization: level-1 is a course grid and level-2 is a fine grid. The size of the stiffness matrix in the energy expression is dictated by discretization in level-1rst level. However, the accuracy of the calculation is achieved by the degree of discretization in level-2. The unknowns associated with the fine grid are expressed in terms of the unknowns of the course grid by employing the concept of PD interpolation [14]. Also, the PD approach readily enables the inclusion of the effects of heterogeneity due to variation in material properties and the presence of defects on the thermal fluctuations.

Energy in a fluctuating graphene layer under hydrostatic tension

Although the structure of a graphene layer can be in the form of zigzag or armchair patterns, it can be considered as a homogenous and isotropic membrane [15]. Also, the energy stored due to thermal fluctuations in a graphene layer in the presence of hydrostatic tension, shown in Fig 1, can be approximated under the assumption that it experiences moderately large deflections from its flat (undeformed) state [12, 16-18].

Employing the von Karman plate theory, the work done by internal forces, U in a graphene layer can be expressed as a combination of the stretching and bending energies, U_s and U_b, respectively.

978-1-7281-1500-9/19 $31.00 © 2019 IEEE

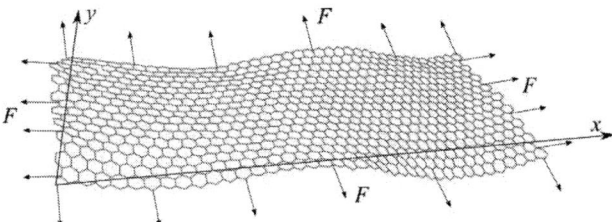

Fig. 1. Fluctuating graphene layer under hydrostatic tension

As suggested by Liang and Purohit [12, 18], they can be expressed in the form

$$U_s = \frac{h}{2}\int_A K\left(\varepsilon_x + \varepsilon_y\right)^2 + G\left[\left(\varepsilon_x - \varepsilon_y\right)^2 + \gamma_{xy}^2\right]dA \tag{1a}$$

and

$$U_b = \frac{1}{2}\int\int_A\left[K_b\left(w_{,xx} + w_{,yy}\right)^2 - 2K_G\left(w_{,xx}w_{,yy} - w_{,xy}^2\right)\right]dA \tag{1b}$$

where $\varepsilon_x, \varepsilon_y$ and ε_{xy} are the in-plane strain components and w is the out-of-plane deflection of graphene on its mid-plane. Its area is $A = L \times W$ with thickness, h. The subscripts after comma indicate differentiation with respect to the spatial variables, (x, y). The material constants K and G represent the area bulk and shear moduli, respectively. They are defined as

$$G = \frac{E}{2(1+v)} \quad \text{and} \quad K = \frac{E}{2(1-v)} \tag{2a}$$

in which E and v represent the Young's modulus and Poisson ratio, respectively. The bending moduli, K_b and K_G are defined as

$$K_b = \frac{Eh^3}{12(1-v^2)} \quad \text{and} \quad K_G = (1-v)K_b \tag{2b}$$

The energy associated with the hydrostatic tension, F in the graphene layer can be expressed as

$$V = -\int_A\left[F(u_{,x} + v_{,y})\right]dA \tag{3}$$

where u and v are the in-plane displacement components in the x- and y- directions, respectively, with F representing the hydrostatic tension per unit length.

Under von Karman assumptions, the in-plane strain components can be expressed as

$$\varepsilon_x = u_{,x} + \frac{w_{,x}^2}{2} \tag{4a}$$

$$\varepsilon_y = u_{,x} + \frac{w_{,y}^2}{2} \tag{4b}$$

and

$$\varepsilon_{xy} = \frac{u_{,x} + u_{,y}}{2} + \frac{w_{,x}w_{,y}}{2} \tag{4c}$$

Thus, the energy due to hydrostatic tension becomes

$$V = -\int_A F\left[\left(\varepsilon_x - \frac{w_{,x}^2}{2}\right) + \left(\varepsilon_y - \frac{w_{,y}^2}{2}\right)\right]dA \tag{5a}$$

or

$$V = -\int_A F\left[\left(\varepsilon_x + \varepsilon_y\right) - \frac{1}{2}\left(w_{,x}^2 + w_{,y}^2\right)\right]dA \tag{5b}$$

As discussed by Ling and Prohit [18, 19], the stretching and bending energies are proportional to the strain, ε and curvature, κ in the fluctuating graphene as

$$U_s \sim EA\varepsilon^2/2 \tag{6a}$$

and

$$U_b \sim K_bA\kappa^2/2 \tag{6b}$$

In accordance with the equipartition theorem of statistical mechanics, their mean square fluctuations can be estimated as

$$\langle\varepsilon^2\rangle = k_BT/EA \tag{7a}$$

and

$$\langle\kappa^2\rangle = k_BT/K_bA \tag{7b}$$

in which k_B is the Boltzmann constant, T is the absolute temperature. The ratio of mean-square fluctuations in the in-plane strain to the bending strain can be written as

$$\langle\varepsilon^2\rangle/\langle\kappa^2\rangle h^2 = K_b/Eh^2 \tag{8}$$

This ratio has a typical value of

$$K_b/Eh^2 \approx 10^{-3} \tag{9}$$

for a graphene sheet has $E \approx 1\text{TPa}$, $K_b \approx 10^{-19}\text{Nm}$ and $h = 0.3\text{nm}$ [18]. The in-plane fluctuations are negligible with respect to the bending fluctuations. Thus, the contribution of the in-plane deformation due to the hydrostatic tension can be disregarded. This assumption simplifies the expressions for U_s and V as

$$U_s(x,y) \approx 0 \tag{10a}$$

and

$$V = \frac{1}{2} \int_A F\left(w_{,x}^2 + w_{,y}^2\right) dA \tag{10b}$$

The final form of the total energy, U in a fluctuating graphene layer becomes

$$U = \frac{1}{2} \int_A \left[K_b \left(w_{,xx} + w_{,yy}\right)^2 - 2K_G \left(w_{,xx} w_{,yy} - w_{,xy}^2\right) \right] dA + \frac{1}{2} \int_A F\left(w_{,x}^2 + w_{,y}^2\right) dA \tag{11}$$

Discrete form of energy state

The potential energy, Π of a graphene layer is composed of strain energy, U and work done by external loads, Ω as

$$\Pi = U - \Omega \tag{12}$$

The strain energy and the external work can be evaluated in discrete form as

$$U = \frac{1}{2} \mathbf{w}^T \mathbf{K} \mathbf{w} \tag{13a}$$

and

$$\Omega = \mathbf{w}^T \mathbf{P} \tag{13b}$$

where $\mathbf{w}$ is a vector of out-plane deflections at M discrete points defined as

$$\mathbf{w}^T = \{w_1, w_2, \ldots\ldots, w_{M-1}, w_M\} \tag{14}$$

The symmetric and positive definite matrix, $\mathbf{K}$ of size ($M \times M$), representing the stiffness of a graphene layer depends on the mean curvature bending modulus, K_b, the hydrostatic tension, F, geometry, L, and the spacing, Δ between the discrete points. The vector, $\mathbf{P}$ represents the applied transverse loads. Enforcing the principle of virtual work, $\delta\Pi = 0$ leads to the equilibrium equation in the form

$$\delta\mathbf{w}^T \left(\mathbf{K}\mathbf{w} - \mathbf{P}\right) = 0 \tag{15}$$

In order to apply the statistical mechanics, Su and Purohit [20-22] introduced the fluctuation, $\Delta\mathbf{w}$ away from the static state as

$$\Delta\mathbf{w} = \mathbf{w} - \mathbf{w}_{min} \tag{16}$$

in which $\mathbf{w}_{min}$ represents the solution to the static state without the effect of thermal fluctuations. Substituting for $\mathbf{w} = \mathbf{w}_{min} + \Delta\mathbf{w}$ in Eq. (12) along with Eq. (13) results in

$$\Pi = \frac{1}{2} \mathbf{w}_{min}^T \mathbf{K} \mathbf{w}_{min} + \Delta\mathbf{w}^T \left(\mathbf{K}\mathbf{w}_{min} - \mathbf{P}\right) + \frac{1}{2} \Delta\mathbf{w}^T \mathbf{K} \Delta\mathbf{w} - \mathbf{w}_{min}^T \mathbf{P} \tag{17}$$

After invoking Eq. (15), this expression for the total potential becomes

$$\Pi = \frac{1}{2} \mathbf{w}_{min}^T \mathbf{K} \mathbf{w}_{min} + \frac{1}{2} \Delta\mathbf{w}^T \mathbf{K} \Delta\mathbf{w} - \mathbf{w}_{min}^T \mathbf{P} \tag{18}$$

Its comparison to Eqs. (12) and (13) results in the expression for the strain energy due to fluctuations as

$$U = U_{min} + \frac{1}{2} \Delta\mathbf{w}^T \mathbf{K} \Delta\mathbf{w} \tag{19a}$$

where U_{min} is the energy of the static state defined as

$$U_{min} = \frac{1}{2} \mathbf{w}_{min}^T \mathbf{K} \mathbf{w}_{min} \tag{19b}$$

Partition function

Based on the discrete form of the strain energy stored in a graphene layer due to fluctuations, its partition function Z can be expressed in the form of a multi-dimensional Gaussian integration as suggested by Su and Purohit [20-22]

$$Z = \int e^{-U(\mathbf{w})/k_B T} d\mathbf{w} \tag{20}$$

The partition function is simply an integration of $e^{-U(\mathbf{w})/k_B T}$ over all possible deformation states described by the vector, $\mathbf{w}$. At zero temperature in the absence of fluctuations, the graphene is flat ($\mathbf{w} = 0$) and has the lowest energy state. It fluctuates about this flat state for $T > 0$. As suggested by [8, 20], this integration can be carried out for the stored energy of a quadratic form as

$$Z = \int e^{-U(\mathbf{w})/k_B T} d\mathbf{w} = e^{-\frac{E_{min}}{k_B T}} \sqrt{\frac{\left(\pi k_B T\right)^M}{\det \mathbf{K}}} \tag{21}$$

where M is the number of unknowns in the vector of $\mathbf{w}$.

The evaluation of this expression requires the determination of the determinant of stiffness matrix, $\mathbf{K}$. However, its size becomes extremely large for accurate evaluation due to the characteristic length scale which may be on the order of a few nm. Thus, it may become computationally challenging. In order to reduce the storage requirements, and decrease the computational time, the skyline method can be employed to construct and decompose it in the form

$$\mathbf{K} = \mathbf{LDL}^T \tag{22}$$

where $\mathbf{L}$ is a lower triangular matrix whose diagonals are equal to unity and $\mathbf{D}$ is a diagonal matrix with positive entries. Hence, the determinant of matrix, $\mathbf{K}$ can be evaluated as

$$\det \mathbf{K} = \det \mathbf{L} \det \mathbf{D} \det \mathbf{L}^T \tag{23}$$

Note that $\det \mathbf{L} = 1$; thus, Eq. (23) reduces to

$$\det \mathbf{K} = \det \mathbf{D} = D_1 D_2 \cdots D_M \tag{24}$$

where D_i $(i = 1, \ldots, M)$ are the diagonal entries of the diagonal matrix $\mathbf{D}$.

The free energy function, G, is expressed in terms of

$$G = -k_B T \ln Z \tag{25}$$

Substituting for the partition function, Z, from Eq. (21) and expanding the terms, the free energy function can be evaluated as

$$G = E_{\min} - \frac{k_B T M}{2} \ln(\pi k_B T) + \frac{k_B T}{2} \ln\left(\det \mathbf{K}\right) \tag{26}$$

When the graphene is subjected to thermal fluctuations, it bends and its projected area reduces from the area of flat configuration. Furthermore, applying tension while keeping the thermal loading changes the projected area. The reduction in projected area can be established as [8]

$$\Delta A = A(\infty, T) - A(F, T) = -\frac{\partial G}{\partial F} \tag{27}$$

where $A(\infty, T)$ is the area at very large tension such that all fluctuations vanish and it is flat. Hence, when $F = 0$ (unstretched), the reduction in projected area is only due to thermal loading of T, and $\Delta A \rightarrow 0$ as $F \rightarrow \infty$. In Eq. (26), the only term that varies as a function of hydrostatic tension, F is the stiffness matrix, $\mathbf{K}$. Therefore, the reduction in projected area can be expresses as

$$\Delta A = -\frac{k_B T}{2} \frac{d}{dF} \ln\left(\det \mathbf{K}\right) \tag{28}$$

Substituting for $\det \mathbf{K}$ from Eq. (24) into Eq. (28) yields

$$\Delta A = -\frac{k_B T}{2} \frac{d}{dF} \left(\ln D_1 + \ln D_2 + \cdots + \ln D_M\right) \tag{29}$$

The ratio of reduction in projected area, ΔA to the original area, A at zero tension and temperature becomes

$$\frac{\Delta A}{A} = -\frac{k_B T}{2A} \frac{d}{dF} \ln\left(\det \mathbf{K}\right) \tag{30}$$

Also, the entropy of a graphene layer can be evaluated as

$$S = -\frac{\partial G}{\partial T} \tag{31}$$

Substituting from Eq. (26) leads to the explicit form of the entropy as

$$S = \frac{k_B M}{2} \left(1 + \ln\left(\pi k_B T\right)\right) - \frac{k_B}{2} \left(\ln\left(\det \mathbf{K}\right) + T \frac{\partial}{\partial T} \ln\left(\det \mathbf{K}\right)\right) \tag{32}$$

The temperature dependent bending modulus $K_b(T)$ can also contribute to the thermal expansion coefficient. Differentiating Eq. (28) with respect to temperature T results in the expression for thermal expansion coefficient as

$$\alpha = -\frac{\partial^2}{\partial T \partial F}\left[\frac{k_B T}{2A} \ln\left(\det \mathbf{K}\right)\right] = \frac{\partial}{\partial T}\left(\frac{\Delta A}{A}\right) \tag{33a, b}$$

For a specified tension, F, the slope of α with respect to T can be evaluated as

$$\frac{\partial \alpha}{\partial T} = -\frac{1}{A} \frac{\partial^3 G}{\partial F \partial T^2} = \frac{1}{AT} \frac{\partial C_F}{\partial F} \tag{34}$$

in which C_F, representing the heat capacity, is obtained as

$$C_F = -T \frac{\partial^2 G}{\partial T^2} \tag{35}$$

Peridynamic evaluation of partition function

The PD evaluation of the partition function due to thermal fluctuations is achieved by two levels of structured discretization as shown in Fig 2. The level-1 discretization is coarse and it dictates the number of unknowns in the energy expression. i.e., size of the stiffness matrix. The level-2 discretization is fine and controls the accuracy of integration.

As denoted by green points, the spacing is defined by $\Delta x_1 = L/(m-1)$ where m represents the number of points in x- and y- directions in the course grid. It results in $M = m \times m$ points in the discretization of the domain, and defines the size of the stiffness matrix, $\mathbf{K}$ as $(M \times M)$. As shown in Fig 2, the position vector, the out-of-plane displacement, and the volume of j-th point in level-1 grid are designated as $\mathbf{x}_{(j)}$, $w_{(j)}$, and $V_{(j)}$, respectively.

978-1-7281-1500-9/19 $31.00 © 2019 IEEE

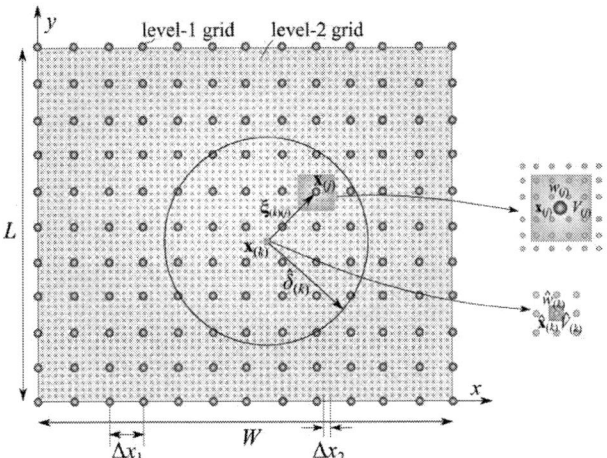

Fig. 2. Discretization of graphene with level-1 (coarse) and level-2 (refined) grids

In level-2 grid, the spacing is defined by $\Delta x_2 = L/n$ where $N = n \times n$ represents the number of PD points denoted by blue points. In this fine grid, the position vector, the out-of-plane displacement and the volume of k-th point are denoted by $\hat{\mathbf{x}}_{(k)}$, $\hat{w}_{(k)}$ and $\hat{V}_{(k)}$, respectively.

As shown in Fig. 2, the horizon (radius) of the k-th PD point in level-2 grid is denoted by $\hat{\delta}_{(k)}$. The distance between k-th PD point of level-1 grid and any other PD point in the level-2 grid is represented by $\xi_{(k)(j)} = \mathbf{x}_{(j)} - \hat{\mathbf{x}}_{(k)}$. The unknown displacements of the points in the fine grid, $w_{(j)}$ are expressed in terms of the unknown displacements of the points in the course grid $\hat{w}_{(k)}$ by using the PDDO [14] while preserving the total area of the points in both grids as

$$\sum_{j=1}^{M} A_{(j)} = \sum_{k=1}^{N} \hat{A}_{(k)} \tag{36}$$

As shown in Fig 2, the area of each PD point in the course and fine grids can be calculated as

$$A_{(j)} = LW/M \quad j = 1, \ldots, M \tag{37a}$$

and

$$\hat{A}_{(k)} = LW/N \quad k = 1, \ldots, N \tag{37b}$$

Using the concept of PD interpolations introduced by Madenci et al. [14], the out-of-plane displacement and its derivatives at each point in the fine grid can be expressed as

$$\hat{w}_{(k)} = \sum_{j=1}^{N_{(k)}} w_{(j)} g_2^{00} (\xi_{(k)(j)}; \rho_{(k)(j)}) A_{(j)} \tag{38a}$$

$$\hat{w}_{(k),x} = \sum_{j=1}^{N_{(k)}} w_{(j)} g_2^{10} (\xi_{(k)(j)}; \rho_{(k)(j)}) A_{(j)} \tag{38b}$$

$$\hat{w}_{(k),y} = \sum_{j=1}^{N_{(k)}} w_{(j)} g_2^{01} (\xi_{(k)(j)}; \rho_{(k)(j)}) A_{(j)} \tag{38c}$$

$$\hat{w}_{(k),xx} = \sum_{j=1}^{N_{(k)}} w_{(j)} g_2^{20} (\xi_{(k)(j)}; \rho_{(k)(j)}) A_{(j)} \tag{38d}$$

$$\hat{w}_{(k),yy} = \sum_{j=1}^{N_{(k)}} w_{(j)} g_2^{02} (\xi_{(k)(j)}; \rho_{(k)(j)}) A_{(j)} \tag{38e}$$

and

$$\hat{w}_{(k),xy} = \sum_{j=1}^{N_{(k)}} w_{(j)} g_2^{11} (\xi_{(k)(j)}; \rho_{(k)(j)}) A_{(j)} \tag{38f}$$

where g_2^{pq} with ($p, q = 0, 1, 2$) represents the PD functions by enforcing the orthogonality condition of PDDO [14]. In these equations, $N_{(k)}$ represents the number of level-1 points within the horizon of level-2 points, $\hat{\mathbf{x}}_{(k)}$ and $\rho_{(k)(j)}$ represents the weight function. It is defined as

$$\rho_{(k)(j)} = e^{-4\xi_{(k)(j)}/\hat{\delta}_{(k)}} \tag{39}$$

In matrix form, Eq. (38) is rewritten as

$$\hat{w}_{(k)} = \mathbf{h}_{(k)}^T \mathbf{w}_{(k)} \tag{40a}$$

$$\hat{w}_{(k),x} = \mathbf{h}_{(k),x}^T \mathbf{w}_{(k)} \tag{40b}$$

$$\hat{w}_{(k),y} = \mathbf{h}_{(k),y}^T \mathbf{w}_{(k)} \tag{40c}$$

$$\hat{w}_{(k),xx} = \mathbf{h}_{(k),xx}^T \mathbf{w}_{(k)} \tag{40d}$$

$$\hat{w}_{(k),yy} = \mathbf{h}_{(k),yy}^T \mathbf{w}_{(k)} \tag{40e}$$

and

$$\hat{w}_{(k),xy} = \mathbf{h}_{(k),xy}^T \mathbf{w}_{(k)} \tag{40f}$$

where the coefficient vectors, $\mathbf{h}_{k,x}$, $\mathbf{h}_{k,y}$, $\mathbf{h}_{k,xx}$, $\mathbf{h}_{k,yy}$ and $\mathbf{h}_{k,xy}$ and the unknown displacement vector, $\mathbf{w}_k$, are defined as

$$\mathbf{h}_{(k)} = \left\{ g_2^{00} (\xi_{(k)(1)}; \rho_{(k)(1)}) A_{(1)}, \ldots, g_2^{00} (\xi_{(k)(N_{(k)})}; \rho_{(k)(N_{(k)})}) A_{(N_{(k)})} \right\} \tag{41a}$$

$$\mathbf{h}_{(k),x} = \left\{ g_2^{10} (\xi_{(k)(1)}; \rho_{(k)(1)}) A_{(1)}, \ldots, g_2^{10} (\xi_{(k)(N_{(k)})}; \rho_{(k)(N_{(k)})}) A_{(N_{(k)})} \right\} \tag{41b}$$

978-1-7281-1500-9/19 $31.00 © 2019 IEEE

$$\mathbf{h}_{(k),y} = \left\{ g_2^{01}(\xi_{(k)(1)}; \rho_{(k)(1)}) A_{(1)}, \ldots, g_2^{01}(\xi_{(k)(N_{(k)})}; \rho_{(k)(N_{(k)})}) A_{(N_{(k)})} \right\} \quad (41c)$$

$$\mathbf{h}_{(k),xx} = \left\{ g_2^{20}(\xi_{(k)(1)}; \rho_{(k)(1)}) A_{(1)}, \ldots, g_2^{20}(\xi_{(k)(N_{(k)})}; \rho_{(k)(N_{(k)})}) A_{(N_{(k)})} \right\} \quad (41d)$$

$$\mathbf{h}_{(k),yy} = \left\{ g_2^{02}(\xi_{(k)(1)}; \rho_{(k)(1)}) A_{(1)}, \ldots, g_2^{02}(\xi_{(k)(N_{(k)})}; \rho_{(k)(N_{(k)})}) A_{(N_{(k)})} \right\} \quad (41e)$$

$$\mathbf{h}_{(k),xy} = \left\{ g_2^{02}(\xi_{(k)(1)}; \rho_{(k)(1)}) A_{(1)}, \ldots, g_2^{11}(\xi_{(k)(N_{(k)})}; \rho_{(k)(N_{(k)})}) A_{(N_{(k)})} \right\} \quad (41f)$$

and

$$\mathbf{w}_{(k)} = \left\{ w_{(1)}, w_{(2)}, \ldots, w_{(N_{(k)})} \right\} \quad (41g)$$

Using Eq. (11), the strain energy of each point $\hat{\mathbf{x}}_{(k)}$ in the fine grid is expressed in the form

$$U_{(k)} = \frac{1}{2} K_b \left(\hat{w}_{(k),xx} + \hat{w}_{(k),yy} \right)^2 \hat{A}_{(k)} \\ - K_G \left(\hat{w}_{(k),xx} \hat{w}_{k,yy} - \hat{w}_{(k),xy}^2 \right) \hat{A}_{(k)} \\ + \frac{1}{2} F \left(\hat{w}_{(k),x}^2 + \hat{w}_{(k),y}^2 \right) \hat{A}_{(k)} \quad (42)$$

with $A_{(k)} = WL/N$ representing the area of each point. After expanding this equation and substituting for the derivatives of $\hat{w}_{(k)}$ from Eq. (40), the strain energy at point $\hat{\mathbf{x}}_{(k)}$ is expressed in matrix form as

$$U_{(k)} = \frac{K_b A_{(k)}}{2} \begin{bmatrix} \mathbf{w}_{(k)}^T \mathbf{h}_{(k),xx} \mathbf{h}_{(k),xx}^T \mathbf{w}_{(k)} + \mathbf{w}_{(k)}^T \mathbf{h}_{(k),yy} \mathbf{h}_{(k),yy}^T \mathbf{w}_{(k)} \\ + \mathbf{w}_{(k)}^T \mathbf{h}_{(k),xx} \mathbf{h}_{(k),yy}^T \mathbf{w}_{(k)} + \mathbf{w}_{(k)}^T \mathbf{h}_{(k),yy} \mathbf{h}_{(k),xx}^T \mathbf{w}_{(k)} \end{bmatrix} \\ - K_G A_{(k)} \begin{bmatrix} \frac{1}{2} \mathbf{w}_{(k)}^T \mathbf{h}_{(k),xx} \mathbf{h}_{(k),yy}^T \mathbf{w}_{(k)} + \frac{1}{2} \mathbf{w}_{(k)}^T \mathbf{h}_{(k),yy} \mathbf{h}_{(k),xx}^T \mathbf{w}_{(k)} \\ - \mathbf{w}_{(k)}^T \mathbf{h}_{(k),xy} \mathbf{h}_{(k),xy}^T \mathbf{w}_{(k)} \end{bmatrix} \quad (43) \\ + \frac{F A_{(k)}}{2} \begin{bmatrix} \mathbf{w}_{(k)}^T \mathbf{h}_{(k),x} \mathbf{h}_{(k),x}^T \mathbf{w}_{(k)} + \mathbf{w}_{(k)}^T \mathbf{h}_{(k),y} \mathbf{h}_{(k),y}^T \mathbf{w}_{(k)} \end{bmatrix}$$

This equation can be simplified to a compact form as

$$U_{(k)} = \mathbf{w}_{(k)}^T \mathbf{K}_{b(k)}^* \mathbf{w}_{(k)} + \mathbf{w}_{(k)}^T \mathbf{K}_{G(k)}^* \mathbf{w}_{(k)} + \mathbf{w}_{(k)}^T \mathbf{K}_{F(k)}^* \mathbf{w}_{(k)} \quad (44)$$

where the stiffness matrices $\mathbf{K}_{b(k)}^*$, $\mathbf{K}_{G(k)}^*$, and $\mathbf{K}_{F(k)}^*$ are defined as

$$\mathbf{K}_{b(k)}^* = \frac{K_b A_{(k)}}{2} \begin{pmatrix} \mathbf{h}_{(k),xx} \mathbf{h}_{(k),xx}^T + \mathbf{h}_{(k),yy} \mathbf{h}_{(k),yy}^T \\ + \mathbf{h}_{(k),xx} \mathbf{h}_{(k),yy}^T + \mathbf{h}_{(k),yy} \mathbf{h}_{(k),xx}^T \end{pmatrix} \quad (45a)$$

$$\mathbf{K}_{G(k)}^* = K_G A_{(k)} \begin{pmatrix} \frac{1}{2} \mathbf{h}_{(k),xx} \mathbf{h}_{(k),yy}^T + \frac{1}{2} \mathbf{h}_{(k),yy} \mathbf{h}_{(k),xx}^T \\ - \mathbf{h}_{(k),xy} \mathbf{h}_{(k),xy}^T \end{pmatrix} \quad (45b)$$

$$\mathbf{K}_{F(k)}^* = \frac{F A_{(k)}}{2} \left(\mathbf{h}_{(k),x} \mathbf{h}_{(k),x}^T + \mathbf{h}_{(k),y} \mathbf{h}_{(k),y}^T \right) \quad (45c)$$

The strain energy of the graphene layer given by Eq. (11) can be evaluated as the sum of the strain energies of the PD points of level 2 grid as

$$U = \sum_{k=1}^N U_{(k)} \quad (46)$$

Substituting for $U_{(k)}$ in this equation from Eq. (44) results in the strain energy of a graphene layer due to the fluctuations in matrix form as

$$U = \sum_{k=1}^N \left(\mathbf{w}_{(k)}^T \mathbf{K}_{b(k)}^* \mathbf{w}_{(k)} + \mathbf{w}_{(k)}^T \mathbf{K}_{G(k)}^* \mathbf{w}_{(k)} + \mathbf{w}_{(k)}^T \mathbf{K}_{F(k)}^* \mathbf{w}_{(k)} \right) \quad (47)$$

which can be rewritten in compact form

$$U = \mathbf{w}^T \mathbf{K}_b^* \mathbf{w} + \mathbf{w}^T \mathbf{K}_G^* \mathbf{w} + \mathbf{w}^T \mathbf{K}_F^* \mathbf{w} \quad (48)$$

where

$$\mathbf{w}^T = \left\{ w_1, w_2, \ldots, w_M \right\} \quad (49a)$$

$$\mathbf{K}_b^* = \text{Assemble} \left[\mathbf{K}_{b(1)}^*, \mathbf{K}_{b(2)}^*, \cdots, \mathbf{K}_{b(M)}^* \right] \quad (49b)$$

$$\mathbf{K}_G^* = \text{Assemble} \left[\mathbf{K}_{G(1)}^*, \mathbf{K}_{G(2)}^*, \cdots, \mathbf{K}_{G(M)}^* \right] \quad (49c)$$

$$\mathbf{K}_F^* = \text{Assemble} \left[\mathbf{K}_{F(1)}^*, \mathbf{K}_{F(2)}^*, \cdots, \mathbf{K}_{F(M)}^* \right] \quad (49d)$$

Finally, the strain energy expression in Eq. (48) can be rewritten as

$$U = \mathbf{w}^T \mathbf{K} \mathbf{w} \quad (50a)$$

with

$$\mathbf{K} = \mathbf{K}_b^* + \mathbf{K}_G^* + \mathbf{K}_F^* \quad (50b)$$

in which the matrix $\mathbf{K}$ represents the overall stiffness matrix that accounts for stretching effects on bending and shear energy of a graphene layer. The stiffness matrix resulting from the PD interactions, $\mathbf{K}$ is symmetric and positive definite. However, it is a sparsely populated large matrix, and band width is dictated by the horizon size.

Evaluation of reduction in projected area and thermal expansion coefficient

The reduction in normalized projected area, $\Delta A/A$, and the thermal expansion coefficient, α are evaluated for specified hyro-static tension values of $F - \Delta F$, F and $F + \Delta F$, and temperature values of $T - \Delta T$, T and $T + \Delta T$ with ΔF and ΔT representing the incremental values for numerical differentiation. Based on central difference approximation, normalized projected area, $\Delta A/A$ at point (F,T) is expressed as

$$\frac{\Delta A}{A}(F,T) = -\frac{k_B T}{4A\Delta F}\ln\frac{(\det \mathbf{K}(F+\Delta F,T))}{(\det \mathbf{K}(F-\Delta F,T))} \quad (51)$$

Applying the central difference approximation to Eq. (33a), the thermal expansion coefficient, $\alpha(F,T)$ can be expressed as

$$\alpha(F,T) = \frac{\dfrac{\Delta A}{A}(F,T+\Delta T) - \dfrac{\Delta A}{A}(F,T-\Delta T)}{2\Delta T} \quad (53)$$

in which $\dfrac{\Delta A}{A}(F,T+\Delta T)$ and $\dfrac{\Delta A}{A}(F,T+\Delta T)$ are approximated by also using the central difference approximation as

$$\frac{\Delta A}{A}(F,T+\Delta T) = -\frac{k_B(T+\Delta T)}{4A\Delta F}\ln\frac{\det \mathbf{K}(F+\Delta F,T+\Delta T)}{\det \mathbf{K}(F-\Delta F,T+\Delta T)} \quad (54a)$$

and

$$\frac{\Delta A}{A}(F,T-\Delta T) = -\frac{k_B(T-\Delta T)}{4A\Delta F}\ln\frac{\det \mathbf{K}(F+\Delta F,T-\Delta T)}{\det \mathbf{K}(F-\Delta F,T-\Delta T)} \quad (54b)$$

Numerical results

In order to verify the accuracy of the PD representation of the energy expression, a simply supported membrane is subjected to hydrostatic tension, F varying from 0.1pN/nm to 1pN/nm. The membrane is inextensible, i.e., $\varepsilon_x + \varepsilon_y = 0$, and lacks shear rigidity, i.e., $K_G \approx 0$. For a simply supported membrane, the analytical expressions for the reduction in projected area and the entropy of the membrane are of the form [8, 10]

$$\frac{\Delta A}{A} = \frac{k_B T}{8\pi K_b}\left(\ln\left(\frac{\pi^2}{b^2}+\frac{F}{K_b}\right) - \ln\left(\frac{\pi^2}{A}+\frac{F}{K_b}\right)\right) \quad (55a)$$

and

$$\Delta S = S - S_0(T) = \frac{Ak_B}{8\pi}\left(\frac{\pi^2}{A}+\frac{F}{K_b}\right)\ln\left(\frac{\pi^2}{A}+\frac{F}{K_b}\right)$$
$$-\frac{Ak_B}{8\pi}\left(\frac{\pi^2}{b^2}+\frac{F}{K_b}\right)\ln\left(\frac{\pi^2}{b^2}+\frac{F}{K_b}\right) \quad (55b)$$

in which b is the radius of a lipid head group and is on the order of 1 nm. The geometry is defined by $L = W = 1000$ nm, and its flexural rigidity is specified as $K_b = 40$ (pN)(nm)K^{-1}T with T representing temperature in

K. The determinant of the membrane stiffness matrix is computed for three different level-1 grid sizes specified by $m = 501$, $m = 301$ and $m = 101$ points in the x- and y-directions. The corresponding level-2 grid division is achieved by $n = 4(m-1)$ points. The horizon of point, $\hat{\mathbf{x}}_{(k)}$ in the refined grid is specified as $\hat{\delta}_{(k)} = 4\Delta x_1$. The ratio of reduction in projected area to the original area, $\Delta A/A$ from Eq. (27) and the change in entropy, ΔS from Eq. (32) are computed for each discretization at $T = 300$ K. As shown in Figs. 3 and 4, the PD predictions converge for decreasing grid spacing, and recover the analytical solutions by Helfrich [10] remarkably well.

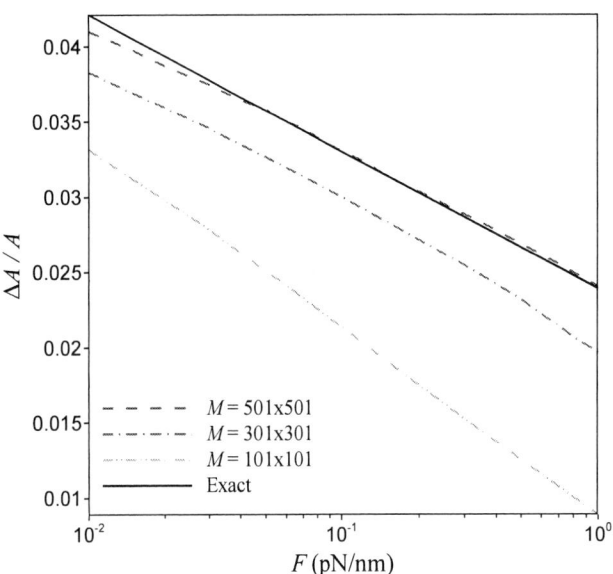

Fig. 3. Reduction in projected area of a fluctuating membrane at $T = 300$ K

In the case of thermal fluctuations of a graphene layer, its geometry is defined by $L = W = 100$ nm. It has a Young's modulus and Poisson's ration of $E \approx 1$TPa, $v \approx 0.35$, respectively, and its thickness is $h = 0.3$nm [1]. Based on the data given in [2], its flexural rigidity is specified as [18]

$$K_b = 131.36 + 200\tanh\frac{T}{1500} \quad (56)$$

where the units of K_b and T are pN nm and K, respectively.

The thermal expansion coefficient of graphene is obtained by using Eq. (33b) along with Eq. (53). The reduction in projected area due to thermal fluctuations is first computed as a function of temperature at two neighboring temperatures and the thermal expansion coefficient is obtained from the partial derivative of $\Delta A/A$ with respect to temperature using Eq. (53). The neighboring values are specified as $\Delta F = 0.001$ pN/nm and $\Delta T = 1$ K.

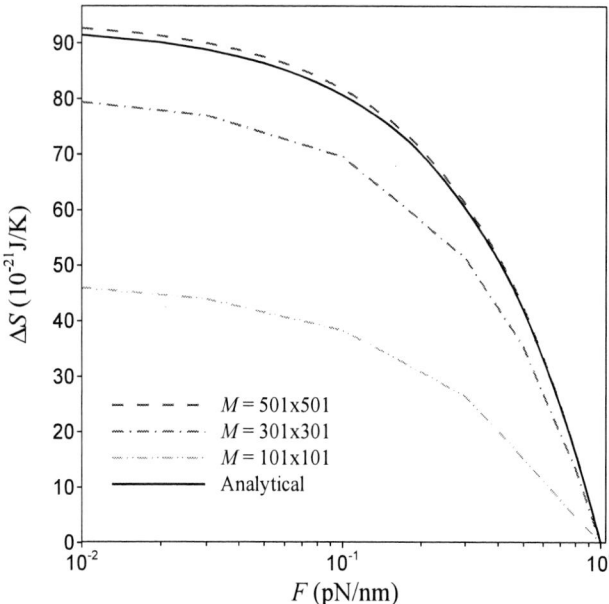

Fig. 4. Entropy change in a fluctuating membrane at $T = 300\ °K$

In order to ensure the convergence of level-1 grid size, the variation of reduction in projected area of the graphene layer as a function of hydrostatic tension from 0.001 to 1 pN/nm for level-1 grid intervals of $m = 101$, 201, 301, 401, 501, and 601 are computed as shown in Fig. 5 . The convergence is achieved with less than 2% difference for a grid size of $M = 601 \times 601$ with $m = 601$ and $n = 2400$, respectively. The horizon size is specified as $\hat{\delta}_{(k)} = 3\Delta x_1$.

The reduction in projected area at a minimum tension of $F = 0.001$ pN/nm is around 0.0121, and it slightly reduces to 0.011 when the applied tension is $F = 1.0$ pN/nm The negative thermal expansion coefficient of the graphene layer is due to the increase in fluctuations of the graphene at higher temperatures resulting in reduced projected area. The expression for the thermal expansion coefficients given Eq. (53) is computed at $F = 0.001$ pN/nm for temperatures varying from $T = 200$ to 400K using the level-1 grid with $m = 601$.

Figure 6 shows the comparison of the PD predictions with those experimentally measured and analytically predicted by Mounet and Marzari [23]. In this figure, the experimental measurements are shown with bands of curves representing the results of several measurements. In these experiments, the measurements were conducted at temperatures varying from $T = 50$ to $300\ K$. Also, the analytical prediction by Mounet and Marzari [23] is denoted by the solid red line. The PD prediction is represented by the dark solid line. It is observed that the PD predictions agree reasonably well against the second experimental set by Mounet and Marzari [23]. In particular, it is observed that the thermal expansion coefficient decreases with increasing temperature, even though the bending stiffness of the graphene, K_b, increases.

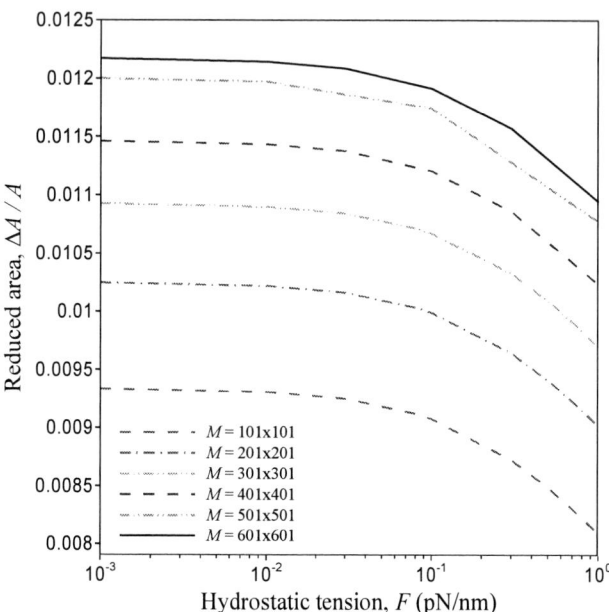

Fig. 5. Reduction in projected area of a fluctuating graphene at $T = 300\ °K$

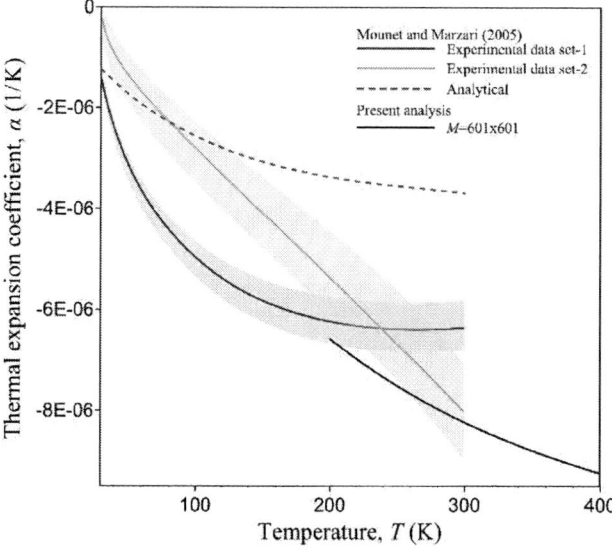

Fig. 6. Influence of temperature on the thermal expansion coefficient of graphene at $F = 0.001$ pN/nm

Conclusions

This study presents a new modeling approach for thermal fluctuations of thin layers by using peridynamics. It specifically employs the PD differential operator for accurate evaluation of the determinant of the membrane stiffness matrix which is a key step in the determination of the partition function. The simultaneous use of fine and course grids along with PD interpolations eliminates the computational challenges for decreasing grid spacing. In the case of membrane, the PD predictions converge and approach the analytical solution as the grid spacing decreases. The PD predictions exhibit the expected experimental observations and agree reasonably well with certain data published in the literature. This approach is applicable to any thin material layer with inclusions, defects, internal cutouts, heterogeneous material properties, and different type of boundary conditions.

References

1. Lee, C., Wei, X., Kysar, J. W. and Hone, J., 2008, "Measurement of the Elastic Properties and Intrinsic Strength of Monolayer Graphene," Science, 321, pp. 385–388.

2. Fasolino, A., Los, J. H. and Katsnelson, M. I., 2007, "Intrinsic Ripples in Graphene," Nat. Mater., 6, pp. 858–861.

3. Los, J. H., Katsnelson, M. I., Yazyev, O. V., Zakharchenko, K. V. and Fasolino, A., 2009, "Scaling Properties of Flexible Membranes From Atomistic Simulations: Application to Graphene," Phys. Rev. B, 80, p. 121405.

4. Zakharchenko, K. V., Los, J. H., Katsnelson, M. I. and Fasolino, A., 2010, "Atomistic Simulations of Structural and Thermodynamic Properties of Bilayer Graphene," Phys. Rev. B, 81, p. 235439.

5. He, Y. Z., Li, H., Si, P. C., Li, Y. F., Yu, H. Q., Zhang, X. Q. and Liu, X. F., 2011, "Dynamic Ripples in Single Layer Graphene," Appl. Phys. Lett., 98, p. 063101.

6. Yoon, D., Son, Y. W. and Cheong, H., 2011, "Negative Thermal Expansion Coefficient of Graphene Measured by Raman Spectroscopy," Nano Lett., 11, pp. 3227–3231.

7. Monzel, C and Sengupta, K., 2016, "
Measuring shape fluctuations in biological membranes," J. Phys. D: Appl. Phys., 49, p. 243002

8. Boal, D., 2002, Mechanics of the Cell, Cambridge University Press, Cambridge, UK.

9. Kosmrlj, A. and Nelson, D. R., 2014, "Thermal Excitations of Warped Membranes," Phys. Rev. E, 89, p. 022126.

10. Helfrich, W., 1975, "Out-of-Plane Fluctuations of Lipid Bilayers," Z. Naturforsch. Sect. C: Biosci., 30, p. 841.

11. Ramakrishnan, N., Kumar, P. S. and Ipsen, J. H., 2010, "Monte Carlo Simulations of Fluid Vesicles With In-Plane Orientational Ordering," Phys. Rev. E, 81, p. 041922.

12. Liang, X. and Purohit, K. P., 2015, "A Fluctuating Elastic Plate and a Cell Model for Lipid Membranes," J. Mech. Phys. Solids, 90, pp. 29–44.

13. Hanlumyuang, Y., Liu, L. and Sharma, P., 2014, "Revisiting the Entropic Force Between Fluctuating Biological Membranes," J. Mech. Phys. Solids, 63, pp. 179–186.

14. Madenci, E., Barut A. and Dorduncu, M., 2019, Peridynamic differential operator for numerical analysis, Springer, NY.

15. Min, K. and Aluru, N. R., 2011, "Mechanical Properties of Graphene Under Shear Deformation," Appl. Phys. Lett., 98, p. 013113.

16. Wei, X., Fragneaud, B., Marianetti, C. A. and Kysar, J. W., 2009, "Nonlinear Elastic Behavior of Graphene: Ab Initio Calculations to Continuum Description," Phys. Rev. B, 80(20), p. 205407.

17. Lee, G. H., Cooper, R. C., An, S. J., Lee, S., van der Zande, A., Petrone, N. and Kysar, J. W., 2013, "High-Strength Chemical-Vapor-Deposited Graphene and Grain Boundaries," Science, 340, pp. 1073–1076.

18. Liang, X. and Purohit, K. P., 2016, "A Fluctuating Elastic Plate Model Applied to Graphene," J. Applied Mechanics, 83, p. 081088

19. X. Liang and P. K.Purohit, 2018, A method to compute elastic and entropic interactions of membrane inclusions, ExtremeMechanics Letters, Vol. 18, pp. 29-35

20. Su, T. and Purohit, P. K., 2010, "Thermomechanics of a Heterogeneous Fluctuating Chain," J. Mech. Phys. Solids, 58, pp. 164–186.

21. Su, T. and Purohit, P. K., 2011, "Fluctuating Elastic Filaments Under Distributed Loads," Mol. Cell. Biomech., 8, pp. 215–232.

22. Su, T. and Purohit, P. K., 2012, "Semiflexible Filament Networks Viewed as Fluctuating Beam-Frames," Soft Matter, 8, pp. 4664–4674.

23. Mounet, N. and Marzari, N., 2005, "First-Principles Determination of the Structural, Vibrational and Thermodynamic Properties of Diamond, Graphite, and Derivatives," Phys. Rev. B, 71, p. 205214.

Machine Learning Approach to Improve Accuracy of Warpage Simulations

Cheryl Selvanayagam[1,2,*], Pham Luu Trung Duong[1], Rathin Mandal[2] and Nagarajan Raghavan[1]

[1] Engineering Product Development, Singapore University of Technology and Design, Singapore 487372
[2] Advanced Micro Devices, Inc. (AMD), Package Engineering, Singapore 469032
*Corresponding Author: cheryl_selavanayagam@mymail.sutd.edu.sg

Abstract— Warpage control of electronic packages has become a critical challenge given the requirement of thinner packaging solutions for the future. While modeling warpage using finite element models is a good way to predict stresses and warpage, the analysis results are only as good as the assumed model inputs such as material properties. This is especially true with warpage simulations and the anisotropic and temperature - dependent material properties of the electronic substrate. With the varying metal line patterns and densities across the substrate, the substrate material properties can be spatially varying in ultra-thin packages resulting in complex warpage profiles. To aid in better design for reliability of future ultra-thin packages, we propose here the use of the Markov Chain Monte Carlo (MCMC) approach (Bayesian inference) combined with finite element simulations to identify the sensitive material parameters that most affect warpage and learn the localized material properties based on warpage contours that have been measured using digital image correlation (DIC). The proposed technique can enable us to design better packages with locally tailored material properties (by tuning metal layer densities, for example) to enable us to stay within an acceptable warpage threshold.

Keywords – Bayesian inference, Digital Image Correlation, Markov Chain Monte Carlo (MCMC); Ultra-thin package; Warpage.

I. INTRODUCTION

As electronic packages get thinner, controlling the warpage of these parts becomes more challenging. Highly warped packages can encounter yield losses due to bump bridging and solder extrusion [1, 2]. Modeling the warpage using finite element method (FEM) is a good way to understand the various phenomena that affect it such as process temperature and initial bare substrate warpage. However, a good FEM model requires accurate material property inputs for the simulation results to be useful. Electronic substrates with their anisotropic metal lines and temperature-dependent core and Ajinomoto Build-up Film (ABF) are quite cumbersome to characterize [3]. Often, at the design stage, substrate samples are not available for characterization. Several authors have attempted to use trace import methods to model the substrate [4] – [7] or printed circuit boards (PCBs) [8] in their FEM models to varying degrees of success. Unfortunately, this method only accounts for the copper line anisotropy while neglecting the temperature-dependence of the other materials that comprise the substrate. In addition, the trace import method is time-consuming to implement in the design stage as it would require a cycle of trace import, mechanical simulation, iterative updates to the electrical artwork and then back to trace import simulations yet again. One other paper proposed an analytical approach to calculate the substrate properties

[9]; however, this method also requires that the electronic artwork is available. As such, it is not feasible for implementation in the package design stage.

In this study, the aim is to develop an efficient simulation methodology by which we could learn the localized material properties of the substrate and the stiffener based on the warpage contours obtained through digital image correlation (DIC) characterizations. The objective is to make the methodology easily applicable during the package design stage for smarter material selection. One method to do this is to use Bayes' Theorem which refines the probability of a hypothesis based on each new evidence or experimental observation collected. This *Bayesian Inference* technique paired with a physics-based model has been successfully used recently in photovoltaic research to determine material properties of different layers and interfaces in the solar cell [10]. One drawback of this implementation, as acknowledged by the authors, is that it utilizes predetermined discrete values for the parameters which does not give the model the freedom to explore the entire design space.

In this work, a Bayesian framework incorporating Markov Chain Monte Carlo (MCMC) is used. This method is a subset of the Bayesian Inference approach where the posterior distribution of the parameter is determined by random sampling of the probabilistic space [11, 12]. Evidence for this analysis is generated through FEM. This approach can be used to determine the material properties of different regions of the substrate and at different temperatures. The enhancement in accuracy of this approach will be validated by comparing the experimental (from DIC) and simulated warpage trends after the reflow process. With the learned material properties over the different regions of the substrate and over a range of temperatures, the reflow FEM simulations of the flip chip are expected to be more accurate and realistic. The proposal here is a good illustration of how data-driven and physics-driven modeling approaches can come together to enable an accurate simulation of the warpage evolution in a complex ultra-thin package.

II. METHODOLOGY

The MCMC toolbox for Matlab® developed by Haario *et al.* [13] was used for this analysis. A flow chart of the steps involved is shown in Figure 1. First, the parameters of interest are identified. Then, the prior distribution of these parameters are specified, this includes the range of values for each parameter (maximum and minimum), the mean and the standard deviation. The warpage contour from DIC, measured at room temperature is used as the likelihood distribution. The code was then run to generate $x = 1000$ MCMC samples, where x is the number of iterations of the analysis. Parameter convergence is gauged based on whether each parameter converges to a narrow band of values over the 1000 interations of the analysis. If the parameters have not converged satisfactorily, prior distributions can be tweaked and the MCMC analysis rerun. The MCMC samples generated represent the posterior distributions of the parameters which can be visualized in plots of probability density functions. Besides looking at probability density plots to determine the range of acceptable values from the analysis, the latter values from the MCMC samples can be averaged to determine the values of the parameters.

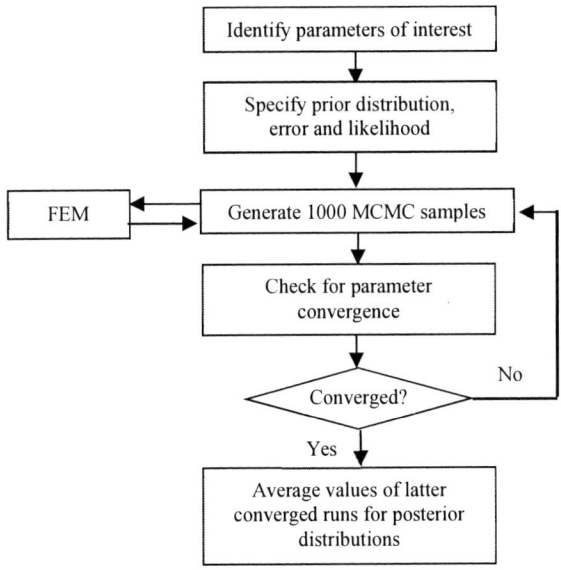

Figure 1. Flow chart showing the steps in the execution of the MCMC for warpage simulations and model parameter probability density estimation.

III. FEM AND MCMC MODEL DETAILS

A. Finite Element Model (FEM)

A finite element model of a package was built on ANSYS® v19 to evaluate the warpage of the part. This model was fully parameterized to allow easy tweaking of the parameters such as the material properties and the model geometry. Post-processing of the model was automated such that the out-of-plane displacement along the diagonal of the package was written to a text file.

The quarter model built is shown in Figure 2(a). Typical warpage contours from the simulations are shown in Figure 2(b). Displacement plots of DIC and FEM along the diagonal of the package are shown in Figure 2(c). From this figure, we notice a large discrepancy in terms of both shape and warpage magnitude even when using measured substrate properties in the FEM models. The squared sum of the difference between the experimental and simulated curves is 0.036, which is relatively large. The ideal situation would be to have overlapping experimental and simulation warpage curves.

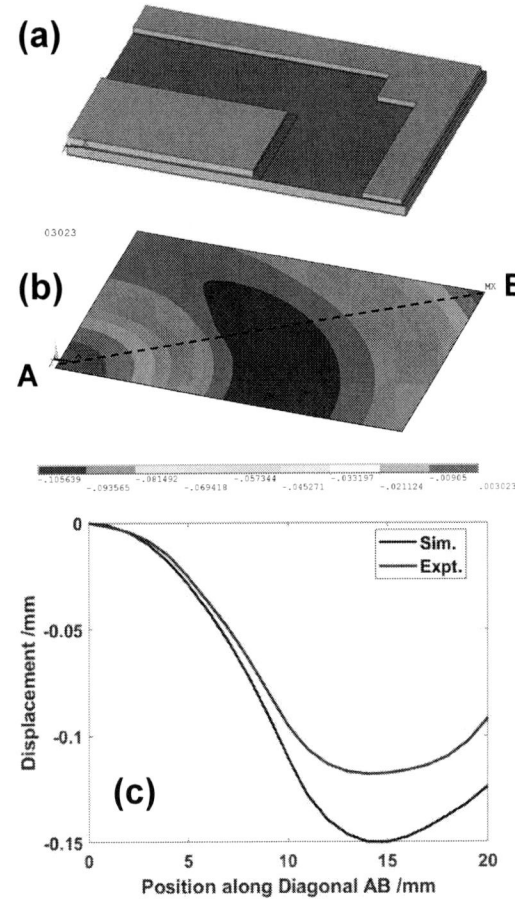

Figure 2. (a) Quarter FEM model of the package for warpage analysis and the corresponding (b) warpage contour plot. (c) Displacement plot along package diagonal (AB) comparing experimental and simulation results.

B. Markov Chain Monte Carlo (MCMC)

The FEM model gives the relationship between displacements y_i and the positions p_i along the diagonal as:

$$y_i = f_i(p_i, \mathbf{x}) \quad , i = 1, ..., N \qquad (1)$$

where $\mathbf{x} = (x_1, ..., x_{n_p})$ is a vector of n_p parameters of the model. Let us assume that the measurement of displacement data along the package diagonal AB: $\Upsilon = \{z_i = y_i + \varsigma_i\}$ $i = 1, ..., N$ is available. Let us define the likelihood function from the Bayesian perspective as:

$$L(\mathbf{x} \mid \Upsilon) = \frac{e^{-SS/(2v^2)}}{(2\pi v^2)^{n/2}} \qquad (2)$$

where v is the variance of the errors, ς_i, and the sum of square error between measurement and model output is :

$$SS(\mathbf{x}) = \sum_{i=1}^{N} \left(z_i - f_i(p_i, \mathbf{x}) \right)^2 \qquad (3)$$

978-1-7281-1500-9/19 $31.00 © 2019 IEEE

The Bayesian framework considers the model parameters as a random vector with a prior distribution: $p_0(\mathbf{x})$, which encodes our prior knowledge about the parameters. The Bayes' theorem fundamentally states that the posterior distribution of parameters given the observation is

$$p(\mathbf{x}\,|\,\Upsilon) \propto L(\mathbf{x}\,|\,\Upsilon)p_0(x) \tag{4}$$

Thus, the posterior takes into account the observed data and refines our knowledge about the parameters. However, the main technical difficulty for evaluating the posterior distribution in Eqn. (4) is the requirement of a normalization constant. An alternative is the Markov Chain Monte Carlo (MCMC) method, where one can sample any process whose density is known up to a normalization constant. The delay rejection adaptive metropolis (DRAM) [13] is one popular adaptive MCMC methods. Its algorithm is briefly described below:

- Define the number of the chain iterations: M, the length of adaption: k_0 and $p_0(\mathbf{x})$: the prior distribution incorporating our current knowledge about the parameters.
- Define a starting point for estimation $\mathbf{x}_0$ and compute: The initial variance of output:

$$(s_0)^2 = \mathrm{SS}(x_0)/(N - n_p) \tag{5}$$

Initial covariance:

$$V = (s_0)^2[\boldsymbol{\chi}^T(\mathbf{x}_0)\boldsymbol{\chi}^T(\mathbf{x}_0)]^{-1} \tag{6}$$

where the sensitivity matrix is defined as:

$$\boldsymbol{\chi}(\mathbf{x}_0) = \frac{\partial f_i}{\partial x_{0j}}, \ i = 1,...,N, j = 1,...,n_p \tag{7}$$

- For $k = 1, ..., M$

  (a) Decompose the covariance matrix using the Cholesky decomposition: $R = chol(V)$

  (b) Sample a random vector from the standard Gaussian distribution: $\eta \sim N(\mathbf{0}, I_{n_p})$

  (c) New sampling candidate can be expressed by $\mathbf{x}_{new} = \mathbf{x}_{k-1} + R\eta$ and compute $SS_{new} = SS(\mathbf{x}_{new})$, $SS_{k-1} = SS(\mathbf{x}_{k-1})$

  (d) Sample a random number u_α uniformly distributed in $[0,1]$ and $\alpha(\mathbf{x}_{new}\,|\,\mathbf{x}_{k-1}) = \min(1, e^{-|SS_{new} - SS_{k-1}|/(2s^2_{k-1})})$

  (e) If $u_\alpha < \alpha(x_{new}\,|\,\mathbf{x}_{k-1})$:
      Set $\mathbf{x}_k = \mathbf{x}_{new}$, $SS_k = SS_{new}$

  Else (delay rejection step):

      Sample $\eta \sim N(\mathbf{0}, I_{n_p})$
      Construct $\mathbf{x}_{new2} = x_{k-1} + (1/5)R\eta$

Sample u_{α_2} within $[0,1]$
Compute

$$\alpha_2 = \min(1, \frac{L(\mathbf{x}_{new2})[1 - \alpha(\mathbf{x}_{new}\,|\,\mathbf{x}_{new2})]}{L(\mathbf{x}_{k-1})[1 - \alpha(\mathbf{x}_{new}\,|\,\mathbf{x}_{k-1})]})$$

If $u_{\alpha_2} < \alpha_2$:
    Set $\mathbf{x}_k = \mathbf{x}_{new2}$, $SS_k = SS(\mathbf{x}_{new2})$

Else:
    Set $\mathbf{x}_k = \mathbf{x}_{k-1}$

End if

End if

  (f) If $\mathrm{mod}(k, k_0) = 1$:

      Update
      $$V = (2.38^2 / n_p)\mathrm{cov}(\mathbf{x}_0,...,\mathbf{x}_{k-1})$$
      End if

End For

The critical part about running an MCMC analysis is selecting the parameters and providing the prior distributions. Selecting fewer parameters allows the analysis to converge more quickly, at the risk of oversimplification. Similarly, a wider range for the prior distribution will require a longer time for convergence. In the next section, three different MCMC studies are presented in a sequential attempt to improve the accuracy of warpage simulations by learning the material properties more accurately.

IV. Results and Discussion

A. Two-Parameter Model: In-plane and Out-of-plane CTE

Since the objective of this work is to improve the accuracy of warpage simulations, it is natural to focus on the substrate material properties first. The first study here considers two parameters - the in-plane and out-of-plane substrate CTEs, for simplicity. The in-plane CTE and out-of-plane CTE were assumed to have normal distributions with means of 14 ppm/^{0}C and standard deviations of 2 ppm/^{0}C. These simulations were run for 1000 iterations.

The convergence plot for the analysis is shown in Figure 3. In-plane CTE converges to a narrow band between 11.5 to 12.5 ppm/^{0}C while the out-of-plane CTE still has considerable spread after 1000 iterations. This indicates that out-of-plane CTE has little effect on the final warpage shape within the range of 12 to 16 ppm/^{0}C. The probability density plots for each parameter obtained from the simulations are shown in Figure 4. The in-plane CTE is determined to be 12.1 ppm/^{0}C, while the out-of-plane CTE is 14.1 ppm/^{0}C, as obtained by averaging the latter runs of the MCMC samples.

A comparison of the warpage simulation results using these optimized stochastic material properties is shown in Figure 5. As can be seen, a significant reduction in the model-experiment discrepancy is observed. The squared sum of the difference between the experimental and simulated curves has decreased from 0.036 to 9.5×10^{-4}.

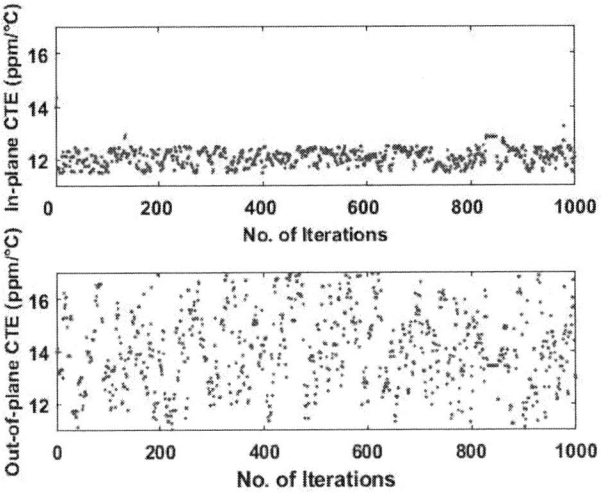

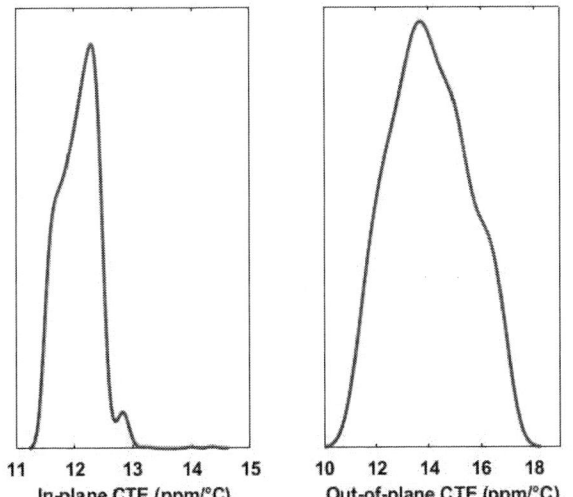

Figure 3. Convergence plot of the in-plane and out-of-plane CTEs for the ultra-thin package considered in this study for 1000 iterations.

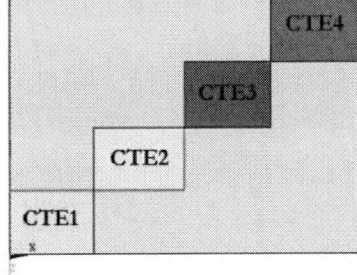

Figure 4. Extracted probability density functions of the in-plane and out-of-plane CTEs for the ultra-thin package considered in this study.

B. Four-Parameter Model: In-plane CTE at Different Locations Along the Diagonal

In this sub-section, the number of parameters selected was increased from two to four for an even better fit to the experimental warpage data. As out-of-plane CTE was determined to be non-critical to the warpage, the parameter was fixed at 14.1 ppm/°C, as determined in the previous section and dropped from the MCMC analysis. Instead, the in-plane CTE was varied along the diagonal of the substrate. Figure 6 shows the FEM model of the substrate, divided such that regions along the diagonal can be assigned different values of CTE. The rationale for this is that the copper density and volume fraction would vary across the substrate,

causing variations in the in-plane CTE. Capturing the variation of the CTE would increase the warpage simulation accuracy.

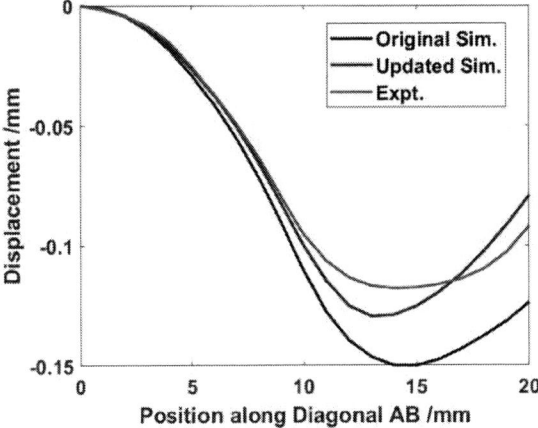

Figure 5. Displacement plot along the package diagonal (AB) for three different cases → experimental measurement (using DIC), simulation with initial and optimized set of parameter values. Note that the warpage profile (blue line) is now much closer to the experimental result (green line). The squared sum of the difference between the experimental and updated simulated curves is 9.5×10^{-4}.

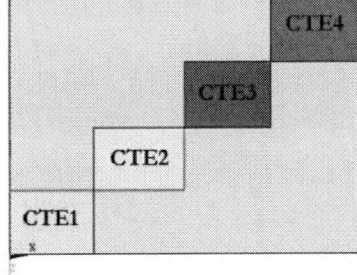

Figure 6. Top view of FEM model of the substrate, showing the different sections with different CTEs. (Colors indicate different material property).

This analysis was run for 1000 iterations as well. The convergence plot is shown in Figure 7. CTE1 and CTE3 seem to converge to about 10 and 19 ppm/°C respectively, while CTE2 and CTE4 oscillate within certain ranges. The corresponding probability density plots are shown in Figure 8. The best estimate for CTE1 through CTE4 have been determined to be 9.7, 13.0, 18.9 and 18.2 ppm/°C respectively by averaging the latter values of the MCMC trials. It is interesting to note here that the CTE increases from the center to the corner of the substrate. Note that the values of CTE3 and CTE4 are similar, indicating that these regions could be represented by the same value for further analysis. The squared sum of the difference between the experimental and simulated curves has now decreased further from 9.5×10^{-4} in the previous section to 2.4×10^{-4}. As we believe there is still room for improvement in terms of the shape of the curve, in the next section, the stiffener modulus and CTE are considered in addition to the substrate CTE.

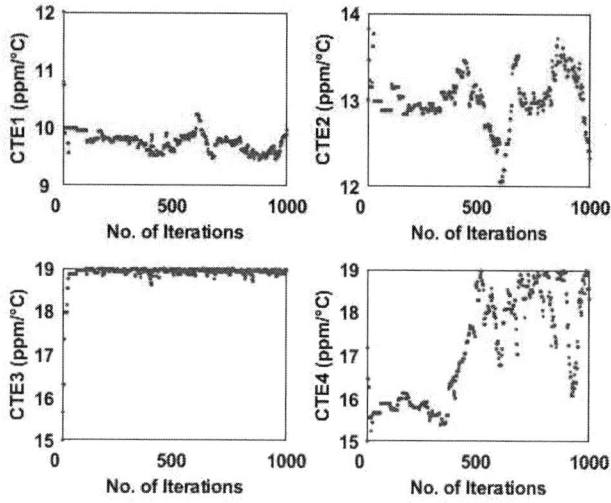

Figure 7. Convergence plot of the four in-plane CTEs along the diagonal AB during the 1000 iterations.

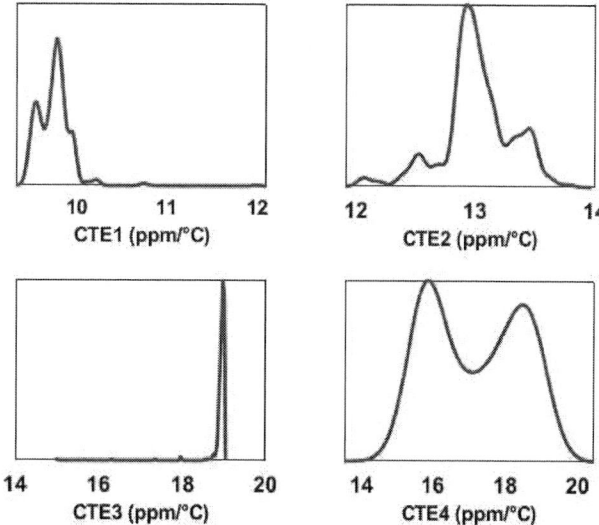

Figure 8. Extracted probability density functions for the four in-plane CTEs along the diagonal AB.

C. Four-Parameter Model: In-plane CTE at Two Locations along Diagonal + Modulus and CTE of Stiffener

To further revise and improve our model predictions, in this sub-section, the parameters selected for analysis are the two substrate CTEs along the diagonal, and the modulus and CTE of the stiffener. Though the stiffener is of a known material composition, material processing for the fabrication of the geometry could change the stiffener material properties. This analysis considers the additional effect of stiffener modulus and CTE on the warpage magnitude and shape. Results are shown in Figures 10 to 12. Figure 10 shows the convergence plot, Figure 11 is the extracted probability density function and Figure 12 is the final displacement plot. The in-plane CTE of the substrate at the center and corner have been determined to be 10.6 and 13.8

ppm/^{0}C respectively. Initial values for the stiffener modulus and CTE were 193 GPa and 16.9 ppm/^{0}C respectively. After the MCMC analysis, the values for the stiffener modulus and CTE have been determined to be 174GPa and 16.1 ppm/^{0}C respectively. Though the revision in these parameter values does not seem significant, they affect the warpage output significantly. The squared sum of the difference between the experimental and simulated curves has decreased further from 2.4×10^{-4} in the previous section to 1.1×10^{-4}. This set of parameters results in the closest replication of the experimental warpage plot. Note that varying the stiffener material properties allows for a simulation envelope (dashed black lines on Figure 12) that better encompasses the experimental curve.

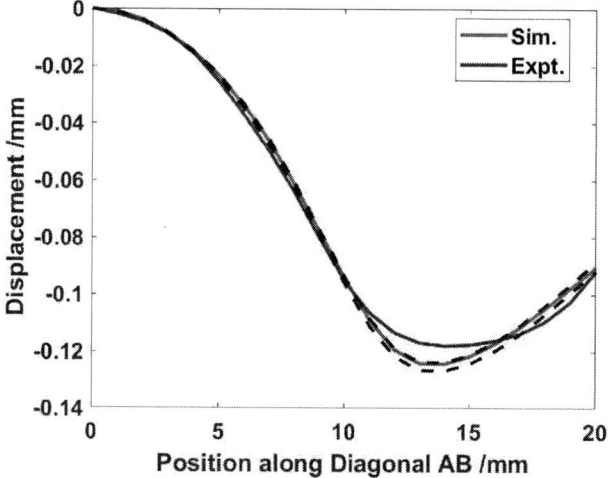

Figure 9. Displacement plot along the package diagonal (AB) for two cases → experimental measurement (blue) and simulation with optimized set of four in-plane CTE values. Dashed black lines denote are the 5% and 95% quantiles of the 1000 simulation runs. As can be seen, the confidence intervals are very close to the mean value, which indicate the robustness of our predictions.

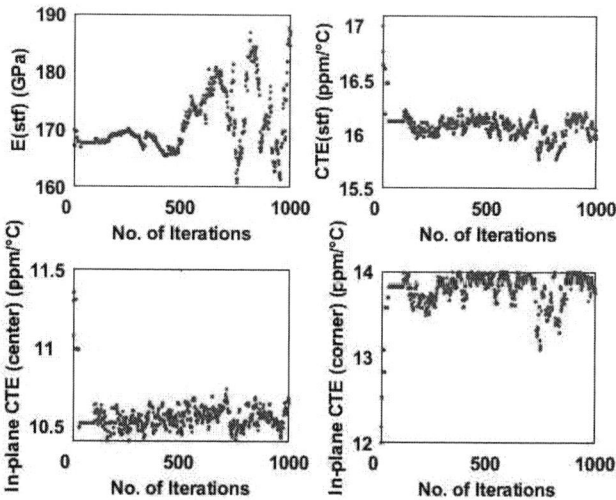

Figure 10. Convergence plot of the four parameters – stiffener modulus, CTE and in-plane substrate CTE at the center and corner.

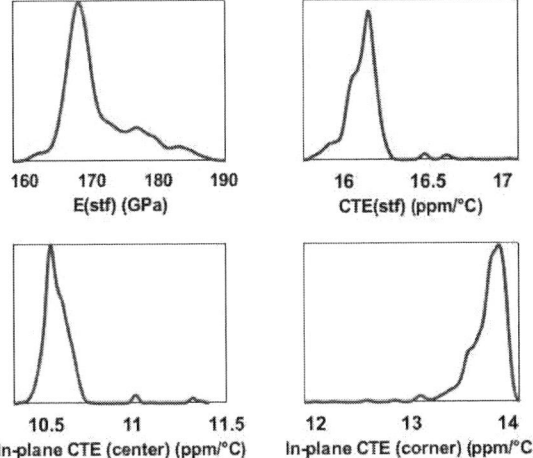

Figure 11. Extracted probability density functions of the four parameters – stiffener modulus and CTE and in-plane substrate CTE at center and corner.

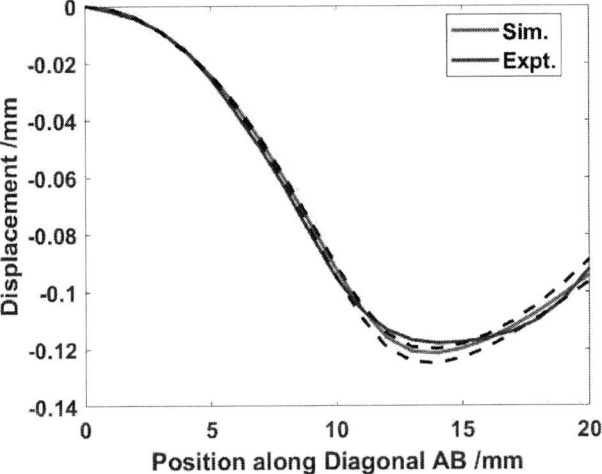

Figure 12. Displacement plot along the package diagonal (AB) for two cases → experimental measurement (blue) and simulation with optimized set of four parameters – stiffener modulus and CTE and in-plane substrate CTE at center (CTE2) and corner (CTE4). Dashed black lines are the 5% and 95% quantiles of the 1000 simulation runs.

D. Using MCMC Approach for Package Design with 20% Reduced Warpage

The MCMC approach could also be used during the design phase of a package to ensure that warpage shape and magnitude are within expected ranges. This can be done by specifying the desired warpage magnitude and shape in terms of a modified warpage contour as the likelihood input to the MCMC.

This is demonstrated in this section where the warpage magnitude along the diagonal AB is reduced by 20% and the resulting material parameters needed to achieve this are back-calculated. The results are shown in Figures 13 to 15. The needed values for each parameter are determined by averaging the latter runs of the parameter in the MCMC analysis. Stiffener modulus and CTE have been determined

as 175GPa and 14.9 ppm/°C. The corresponding desired in-plane CTE of the substrate at the center and corner have been determined to be 8.1 and 13.3 ppm/°C respectively. Once the range of values for each material parameter to be designed has been identified, material selection can then be carried out to meet these new requirements. In this manner, the part can be designed to any required warpage specification.

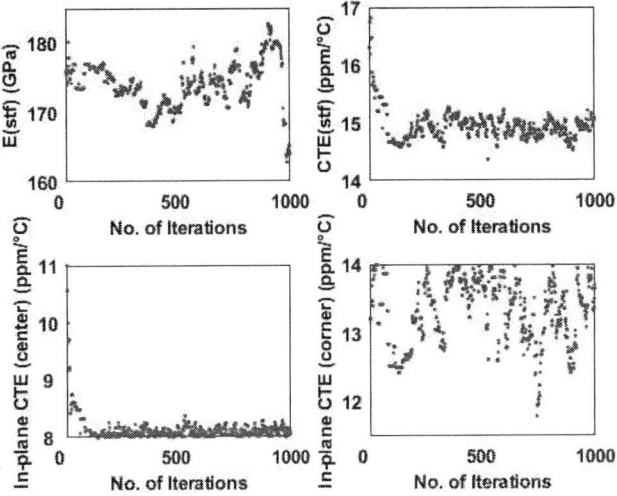

Figure 13. Convergence plot of the material parameter values to be attained so as to enable a 20% reduction in warpage design requirement.

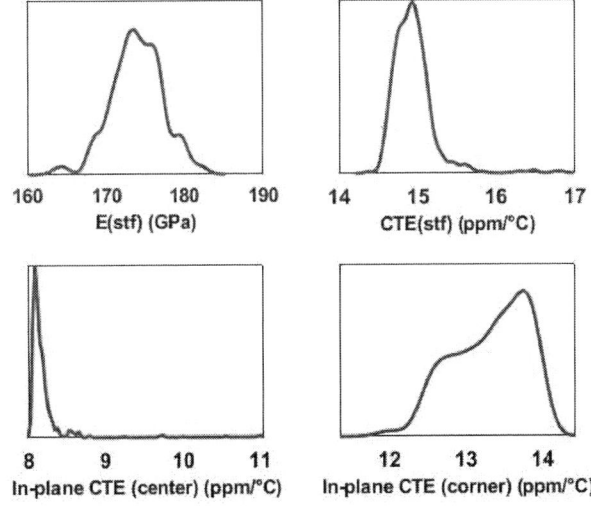

Figure 14. The desired (needed) probability density functions of the material design parameters in order to achieve 20% reduction in the package warpage.

E. How is this approach better than the conventional FEM parametric studies?

The parameter estimation problem is the problem of finding the optimal set of values for the parameters for which the model prediction best matches with our experimental observations. The standard non-linear least squares approach can give only a unique best vector of parameters and it does not take into account the errors and uncertainties inherent in the model and the measurements. Unlike the nonlinear least square technique, the Bayesian approach provides a means to account for the model and measurement uncertainties, which can then be translated into uncertainties in the estimated parameter values. The posterior density function provides us with additional important information: the probability that the displacement at a given position belongs to a given interval, which is useful to initiate design for variability and ensure a robust design of the package.

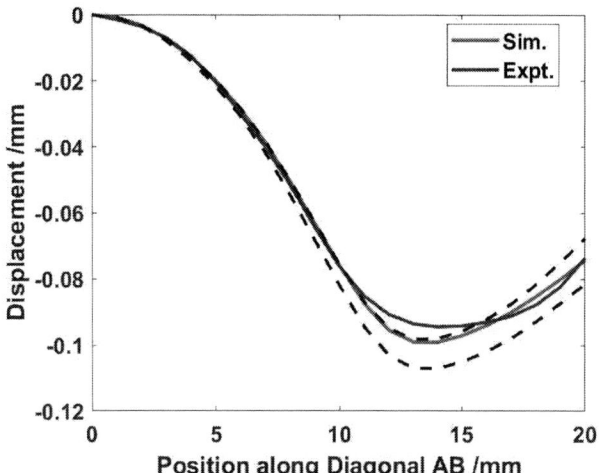

Figure 15. Displacement plot along the package diagonal for two cases → experimental measurements artificially reduced by 20% (blue), simulation with optimized set of four parameters – stiffener modulus and CTE and in-plane substrate CTE at the center and corner. The dashed black lines are the 5% and 95% quantiles for the 1000 simulation runs.

V.   CONCLUSIONS AND FUTURE WORK

In this study, we presented the MCMC-based machine learning approach to improve the accuracy of warpage simulations quite significantly. Using this prescribed approach, unknown material parameters can be learnt such that the simulation output matches with the experimental data as closely as possible. In particular, the values for the CTE of the substrate at the different regions and the modulus and CTE of the stiffener have been successfully learnt, which in turn helped us reduce warpage simulation error by more than two orders of magnitude.

The results in this study will be useful for the following design for reliability (DfR) initiatives:

1. Learning material properties over different regions in the substrate for more accurate warpage simulations.

2. Tracking material property evolution over time and temperature provided that warpage data over a range of temperatures and over time are available.

3. Determining unknown material properties in electronic packaging simulations, including mold compound properties for wafer level warpage modeling and interlayer dielectric (ILD) properties for stress analysis, where warpage data is available.

4. And finally, designing thin packages with lower warpage through smarter material selection using the proposed hybrid data-driven and physics-driven modeling approach.

This work clearly illustrates the power of Bayesian inference using the FEM model as the basis. While the direct use of the pure FEM model can be very time consuming for material design and optimization, the Bayesian approach here is found to be very computationally efficient. This is because each iteration using the Bayesian approach moves the parameter selection in a direction for reduced error whereas a purely FEM-based non-linear optimization would process all predetermined discrete variables in the range for each parameter. For our future work, we will consider the surrogate models such as Gaussian process regression (GPR) or polynomial chaos methods in place of FEM model so as to further reduce the computational time and also study the sensitivity of the warpage profile to the different model parameters [**14**, **15**].

ACKNOWLEDGMENT

The authors would like to acknowledge the Economic Development Board of Singapore (EDB) and Advanced Micro Devices, Inc. (AMD) for funding this work under Grant No. IGIPAMD1801.

REFERENCES

[1]   C. C. Meng, S. Stoeckl, H. Pape, F. M. Yee, and T. A. Min, "Effect of substrate warpage on flip chip BGA thermal stress simulation," *2010 12th Electronics Packaging Technology Conference*, Singapore, 2010, pp. 500-504.

[2]   M. Kurashina, D. Mizutani, M. Koide, M. Watanabe, K. Fukuzono, and H. Suzuki, "Low warpage coreless substrate for large-size LSI packages," *2012 IEEE 62nd Electronic Components and Technology Conference*, San Diego, CA, 2012, pp. 1378-1383.

[3]   C. Selvanayagam, R. Mandal and N. Raghavan, Comparison of experimental, analytical and simulation methods to estimate substrate material properties for warpage reliability analysis, *Microelectronics Reliability*, Vol. 88-90 , pp. 817–823, (2018).

[4]   B. Kim and B. Han, "Numerical/Experimental Hybrid Approach to Predict Warpage of Thin Advanced Substrates," *2018 IEEE 68th Electronic Components and Technology Conference (ECTC)*, San Diego, CA, 2018, pp. 267-272.

[5]   P. Chen, Z. Ji, Y. Liu, C. Wu, N. Ye and H. Takiar, "Warpage Prediction Methodology of Extremely Thin Package," *2017 IEEE 67th Electronic Components and Technology Conference (ECTC)*, Orlando, FL, 2017, pp. 2080-2085.

[6]   L.O. McCaslin, S. Yoon, H. Kim, S.K. Sitaraman, Methodology for modeling substrate warpage using copper trace pattern implementation, *IEEE Trans. Adv. Packag.*, Vol. 32 (4), (2009), pp. 740–745.

[7]   M. Wang and B. Wells, "Substrate Trace Modeling for Package Warpage Simulation", *2016 IEEE 66th Electronic Components and Technology Conference (ECTC)*, Las Vegas, NV, 2016, pp. 516-523.

[8]   V.K. Yaddanapudi, S. Krishnaswamy, R. Rath and R. Gandhi, "Validation of New Approach of Modelling Traces by Mapping Mechanical Properties for a Printed Circuit Board Mechanical

Analysis", *2015 IEEE 17th Electronics Packaging and Technology Conference (EPTC)*, Singapore, 2015, pp. 1-6.

[9] L. Valdevit, V. Khanna, A. Sharma, S. Sri-Jayantha, D. Questad and K. Sikka, Organic substrates for flip-chip design: A thermo-mechanical model that accounts for heterogeneity and anisotropy, *Microelectronics Reliability*, 48 (2) 2008 pp. 245-260.

[10] R.E. Brandt, R.C. Kurchin, V. Steinmann, D. Kitchaev, C. Roat, S. Levcenco, G. Ceder, T. Unold and T. Buonassisi, Rapid Photovoltaic Device Characterization through Bayesian Parameter Estimation, *Joule*, 1 (4) 2017, 843-856.

[11] Gamerman, D., Lopes, H. (2006). *Markov Chain Monte Carlo*. New York: Chapman and Hall/CRC.

[12] D.P. Kroese, T. Taimre, Z.I. Botev (2011). *Handbook of Monte Carlo Methods*, Wiley Series in Probability and Statistics, John Wiley and Sons, New York.

[13] H. Haario, M. Laine, A. Mira and E. Saksman, 2006. DRAM: Efficient adaptive MCMC, *Statistics and Computing* 16, pp. 339-354. doi: 10.1007/s11222-006-9438-0.

[14] L. Q. Minh, P.L.T Duong, M. Lee, 2018. Global Sensitivity Analysis and Uncertainty Quantification of Crude Distillation Unit Using Surrogate Model Based on Gaussian Process Regression, *Industrial & Engineering Chemistry Research* 57 (14), pp. 5035-5044.

[15] P.L.T. Duong, W. Ali, E. Kwok, M. Lee, 2016. Uncertainty quantification and global sensitivity analysis of complex chemical process using a generalized polynomial chaos approach, *Comput. Chem. Eng.*, Vol. 90, pp. 23–30.

Study on Warpage of Fan-Out Panel Level Packaging (FO-PLP) using Gen-3 Panel

F. X. Che*, Kazunori Yamamoto, Vempati Srinivasa Rao, and Vasarla Nagendra Sekhar

Institute of Microelectronics, A*STAR (Agency for Science, Technology and Research)

2 Fusionopolis Way, #08-02, Innovis Tower, Singapore 138634

*email: chef@ime.a-star.edu.sg; chefaxing@pmail.ntu.edu.sg

Abstract— In this study, fan-out panel level packaging (FO-PLP) technology using redistribution layer (RDL) first approach is demonstrated using Gen-3 glass substrate (550mm × 650mm size). Investigation on panel level warpage during process is carried out through finite element analysis (FEA) to address warpage issue with considering material selection, structure design and process optimization. Taguchi method helps to identify important parameters for each process and then further parametric study is conducted with detailed FEA simulation. The effect of gravity on panel warpage has been simulated and analyzed by considering panel process and panel size. Modelling gravity effect reduces panel warpage significantly, which is much close to the real case and need to be considered in the large panel warpage simulation.

Keywords-fan-out technology; finite element modelling; FO-PLP; panel level packaging; warpage; gravity effect.

I. INTRODUCTION

With multiple chips and multiple redistribution layers (RDLs) developed in fan-out wafer level packaging (FO-WLP) technology, FO-WLP technology has a wide range application in high I/O integrated packaging and heterogeneous integrated packaging with its cost effective solution [1-3]. To further improve throughput of fan-out packaging technology, fan-out panel level packaging (FO-PLP) has been investigated by several industries such as ASE, NEPES, and IZM Fraunhofer [4-6]. Panel size can be from 300mm × 300mm to 600mm × 600mm. Currently, FO-PLP is mostly focusing on low-end or mid-end application with low I/O density and coarse RDL line/space. In this study, FO-PLP technology is developed using Gen-3 glass substrate (size of 550mm × 650mm), which follows the standard for flat panel display manufacturing industry. Based on the used panel size, the packaging area per panel is 5 times area per 12" wafer, so the throughput can be significantly increased. It is known that wafer warpage is one of critical challenges for FO-WLP technology [1,2,7]. Such challenge will become more severe for FO-PLP technology due to large panel size [6]. There are several approaches to overcome and reduce panel warpage during processes including material selection, structure optimization design, process improvement, which are discussed in this paper.

Finite element analysis (FEA) modelling has been widely used for warpage investigation and reduction in advanced packaging technology [7-10] and is established for process sequence modelling for panel warpage analysis in this study to optimize structure, process condition and material selection. Due to huge practices needed in full DOEs,

Taguchi method is implemented by considering 6 input parameters including glass CTE and thickness, die thickness, overmold thickness, molding compound and dielectric materials. Each parameter has 3 levels. From simulation results and Taguchi analysis, critical parameters can be identified for each process in terms of warpage. Further parametric studies focusing on structure and glass carrier CTE have been conducted for each process to deeply understand and reduce panel warpage through optimization. In addition, the effect of gravity on panel warpage has been simulated and analyzed by considering panel process and panel size because it is important to simulate gravity effect on panel warpage for thin large panel. Results show that considering gravity reduces panel warpage significantly, which is much close to the real case. Panel warpage simulation result with considering gravity effect is successfully validated by panel warpage measurement data.

II. TEST VEHICLE AND PROESS FLOW

In this study, RDL-first approach, also called die last approach, is implemented for FO-PLP technology with Gen-3 glass substrate as a carrier panel. Gen 3 panel has a geometry size of 650mm × 550mm, which is 5 times area of 12 inch wafer as shown in Fig. 1. Therefore, high throughput can be significantly increased when fan-out packaging technology uses Gen 3 panel as a supporting carrier. In this study, package size is designed as 15mm × 15mm with embedding single chip with size of 10mm × 10mm. Packages are fully populated in Gen 3 panel with array of 34 rows by 40 columns, so there are total of 1360 packages in one panel. Fig. 2 shows process flow of FO-PLP packaging technology. One very thin release layer used for debonding process with laser release is coated onto glass panel carrier. Then dielectric coating, lithographical patterning, seed layer deposition, photo-resistance coating and development are conducted for preparing RDL layer plating process. Two RDL layers are fabricated on a glass carrier panel according to above processes. Then chips are bonded to panel with Cu pillar connections and reflow process. After that, panel level compression molding using epoxy molding compound (EMC) is conducted. Two types of EMC material, i.e. liquid or granular, can be used in compression molding for FO-PLP packaging technology. According to designed package thickness, back grinding process on EMC layer is carried out. After panel back grinding, glass carrier is removed through release layer using laser debonding process. Solder ball attachment process can be carried out on the partitioned panel. Finally, separate package can be achieved after singulation process. To ensure the successful processing, it is critical to control panel warpage during whole processes. In this study, FEA modelling and simulation are implemented to investigate panel level warpage and provide guideline for reducing panel warpage.

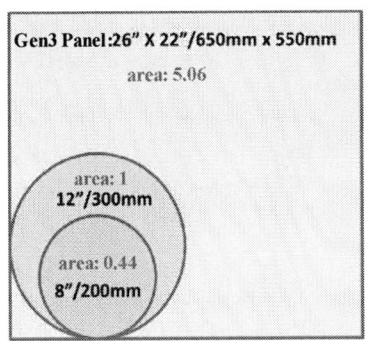

Figure 1.   Area comparion among different carriers.

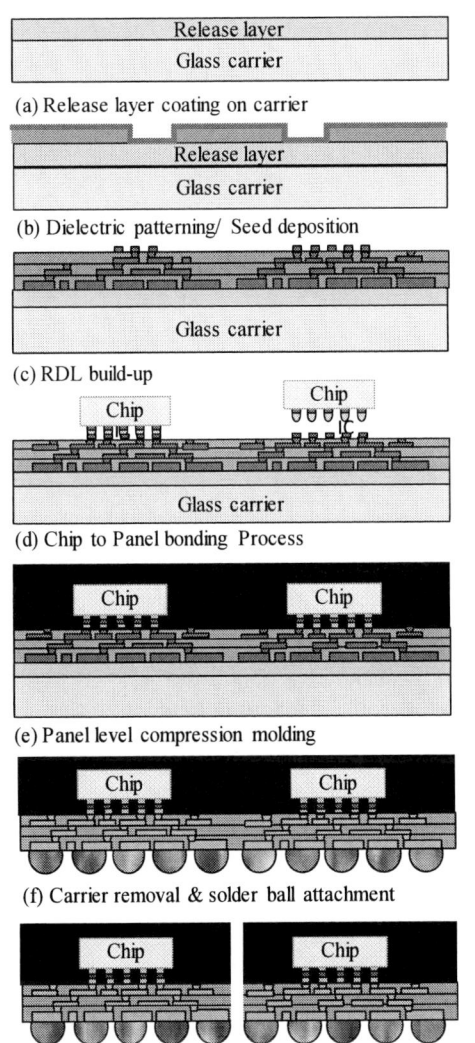

Figure 2.   Process flow of RDL-first approach for FO-PLP packaging.

III.   FINITE ELEMENT MODELLING

Investigation on panel level warpage during process is carried out through FEA modelling and simulation to address warpage issue with considering material selection, structure design and process optimization. Due to symmetric layout, A 3D quarter FEA model is used for panel level packaging technology as shown in Fig. 3. Usually, researchers only simulate CTE mismatch induced warpage during different process conditions. In this study, CTE mismatch induced panel warpage is also modeled considering different process conditions and FEA model is shown in Fig. 3(a). Symmetric boundary condition is applied properly. We also consider gravity effect on panel warpage. FEA model with gravity effect is shown in Fig. 3(b) by considering table support. Gravity is one type of loading added onto to panel in vertical z direction. Contact pair is defined between the bottom of the panel and the top of the table. Table support is modelled as a rigid body. Two steps of modelling are carried out. Firstly, normal modelling only considering CTE mismatch induced warpage is conducted. Then gravity is added as a static loading.

Material properties are very important for accurate modelling results. Table I lists material properties used in FEA modelling. Three photo-dielectric (PD) and EMC materials and 4 types of glass carrier are modelled for material selection purpose. Temperature-dependent properties are simulated for PD and EMC materials. Curing temperature is defined as stress free temperature for PD and EMC materials. Element birth and death technique is used to simulate process dependent warpage.

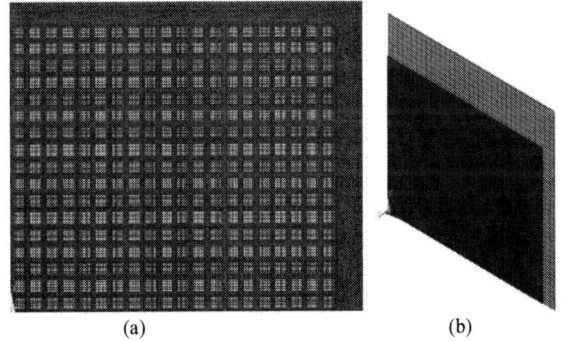

Figure 3.   FEA models for FO-PLP packaging technology, (a) modelling CTE effect on warpage, (b) adding weight effect on warpage.

TABLE I.   MATERIAL PROPERTIES USED IN FEA MODELLING

Materials	Young's modulus (GPa)	CTE (ppm/°C)	Poisson's ratio	Tg (°C)	Curing Temp. (°C)
Die	131	2.8	0.3		
Photodielectric 1 (PD1)	3	55	0.3	276	220
Photodielectric 2 (PD2)	4.4	40/128*	0.3	240	230
Photodielectric 3 (PD3)	3.83	52/440*	0.3	215	230
EMC1 (Granule)	6/0.2*	10/58*	0.3	195	150
EMC2 (Liquid)	23/2*	7.3/32*	0.3	165	125
EMC3 (Liquid)	14/0.2*	12/45*	0.3	165	130
Glass1	69	3	0.3		
Glass2	70	5	0.3		
Glass3	69.3	7.6	0.3		
Glass4	71.7	9.6	0.3		

*. Data for Below Tg/Above Tg.

IV. RESULTS AND DISCUSSION

A. Process Dependent Panel Warpage Analysis

Process dependent warpage simulation results at room temperature for reference model is shown in Fig. 4 for FO-PLP technology with RDL first approach. Two layer RDLs are fabricated so there are total three photo-dielectric layers with 7μm thickness of each layer are modelled. In the reference model, the used materials are EMC1, PD2, and glass2 and their properties are listed in Table I. Thickness of glass carrier, chip and overmold is 1.1mm, 250μm and 100μm, respectively. Panel warpage increases approximately linearly with PD layer thickness. For convenience, total thickness of 21μm for three PD layers will be modeled in one step in the following simulation. After die attach, panel warpage changes from positive value to negative value mainly due to larger CTE of glass than silicon chip. Panel warpage reduces to a small value after compression molding because overall CTE of EMC2 and Si matches with glass CTE. Large panel warpage occurs after debonding process due to very thin and large panel structure and CTE mismatch among PD, Si and EMC materials. Such large warpage arises challenges for the following solder ball attachment and singulation. Potential solution is to reduce warpage or partition large panel to several small panels.

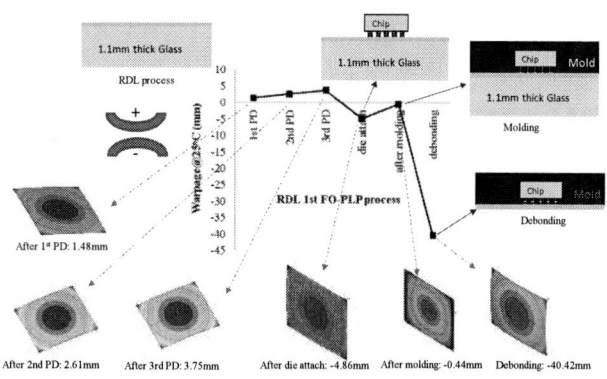

Figure 4.   Process dependnet panel warpage simulation results.

B. Parametric Study on Panel Warpage

In order to further reduce panel warpage and understand which factors having significant effect on panel warpage in different processes, parametric study is conducted by considering glass carrier CTE and thickness, chip thickness, overmold thickness, PD and EMC materials. Geometry and materials used in the reference model is considered as a benchmark and only one parameter is investigated with other parameters keeping the same as ones in the reference model in parametric study.

Figs. 5 and 6 show the effect of glass thickness and CTE on panel warpage, respectively. Panel warpage decreases with glass carrier thickness in RDL, die attach and molding processes. Effect of glass CTE on panel warpage is complex in different processes. Panel warpage slightly decreases with increasing glass CTE due o low CTE mismatch between PD and glass when increasing glass CTE. After die attach and

molding process, panel may show different warpage direction with using different CTE glass. Glass carrier with 3ppm/K CTE leads to positive warpage but glass carrier with 7ppm/K CTE results in negative warpage. Warpage direction and value depend on CTE mismatch among materials in different processes. General speaking, glass CTE should match EMC CTE for low panel warpage after molding process.

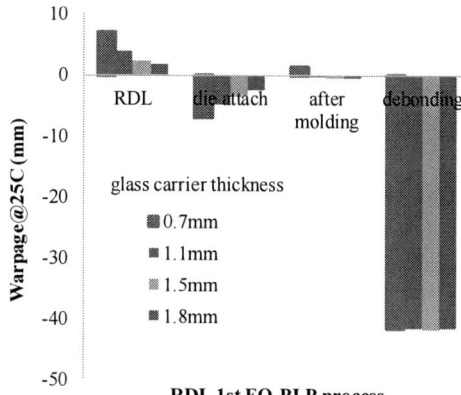

Figure 5.   Effect of galss carrier thickness on panel warpage.

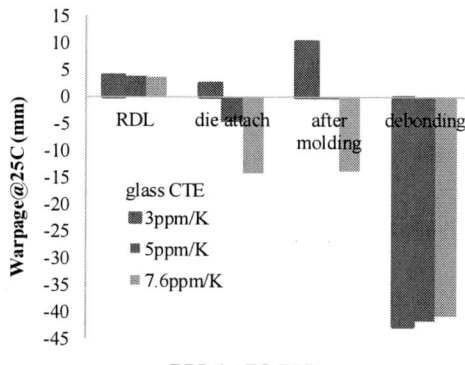

Figure 6.   Effect of glass CTE on panel warpage.

Figs. 7 and 8 show the effect of chip thickness and overmold thickness on panel warpage, respectively. Panel warpage after die attach process slightly increases with chip thickness. Effect of chip thickness on panel warpage after molding is not significant due to matched CTE between glass and EMC. Panel warpage after debonding decreases with chip thickness due to much stiffer structure for panel with thick chip. Similarly, panel warpage after debonding decreases with overmold thickness due to much stiffer structure for panel with thick overmold. Thick overmold also helps to reduce panel warpage after molding process due to stiffer structure. Chip thickness and overmold thickness designs are important for reducing final panel warpage after debonding because these 2 factors determine the structure of test vehicle. While glass carrier thickness and CTE affect panel warpage in the intermediate process, not on warpage after debonding.

978-1-7281-1500-9/19 $31.00 © 2019 IEEE

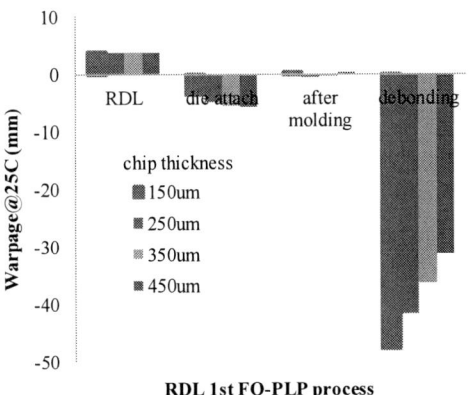

Figure 7. Effect of chip thickness on panel warpage.

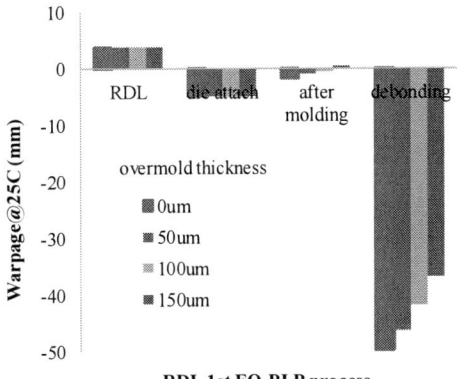

Figure 8. Effect of overmold thickness on panel warpage.

Figs. 9 and 10 show the effect of EMC and PD materials on panel warpage, respectively. EMC has significant effect on panel warpage after molding and debonding process. EMC3 leads to large warpage after molding due to large CTE compared to glass carrier. EMC1 leads to large warpage after debonding due to low Young's modulus which makes thin panel softer and easy for deformation. PD3 leads to larger warpage compared to PD1 and PD2 after RDL and debonding process due to its low glass transition temperature, Tg, which is lower than its curing temperature.

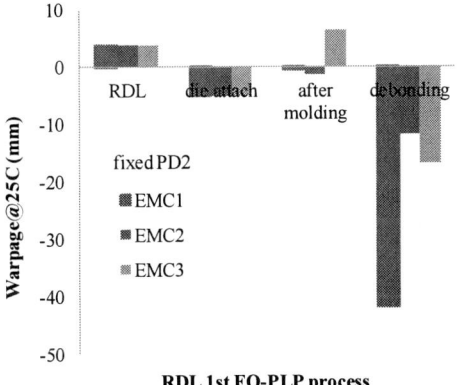

Figure 9. Effect of EMC material on panel warpage.

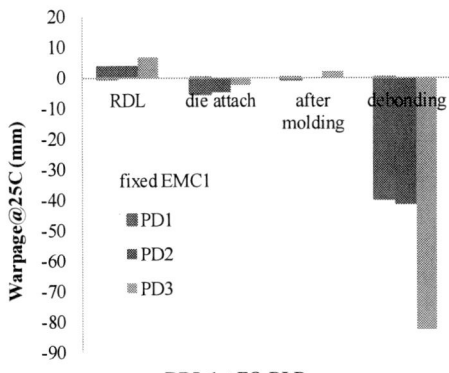

Figure 10. Effect of photo-dielectric material on panel warpage.

C. Taguchi Statistic Method for Panel Warpage

It is difficult to optimize overall design and materials using above parametric study which cannot capture interaction of different parameters in the different processes. It is also not reality to conduct full DOEs simulation considering all variables. Taguchi statistic method is one powerful technique to use limited DOE designs to predict results for overall parameter combinations. Table II lists total 27 runs by considering 6 parameters discussed earlier with each one having 3 levels. Total 729 combinations for full DOEs are cut down to 27 Taguchi runs, which saves huge time and efforts. Each Taguchi run will be simulated with considering 6 parameters as input and panel warpage as a response will be output in following process: RDL, die attach, molding and debonding process. Through Taguchi method, design and material optimization will be provided for low panel warpage requirement in different processes.

TABLE II. DOEs OF TAGUCHI METHOD FOR WARPAGE STUDY

Run No.	Glass CTE (ppm/K)	Glass thickness (mm)	Die thickness (mm)	Overmold thickness (mm)	EMC	Photodielectric
1	3	1.1	0.15	0.05	EMC1	PD1
2	3	1.1	0.15	0.05	EMC2	PD2
3	3	1.1	0.15	0.05	EMC3	PD3
4	3	1.5	0.25	0.1	EMC1	PD1
5	3	1.5	0.25	0.1	EMC2	PD2
6	3	1.5	0.25	0.1	EMC3	PD3
7	3	1.8	0.35	0.15	EMC1	PD1
8	3	1.8	0.35	0.15	EMC2	PD2
9	3	1.8	0.35	0.15	EMC3	PD3
10	5	1.1	0.25	0.15	EMC1	PD2
11	5	1.1	0.25	0.15	EMC2	PD3
12	5	1.1	0.25	0.15	EMC3	PD1
13	5	1.5	0.35	0.05	EMC1	PD2
14	5	1.5	0.35	0.05	EMC2	PD3
15	5	1.5	0.35	0.05	EMC3	PD1
16	5	1.8	0.15	0.1	EMC1	PD2
17	5	1.8	0.15	0.1	EMC2	PD3
18	5	1.8	0.15	0.1	EMC3	PD1
19	7.6	1.1	0.35	0.1	EMC1	PD3
20	7.6	1.1	0.35	0.1	EMC2	PD1
21	7.6	1.1	0.35	0.1	EMC3	PD2
22	7.6	1.5	0.15	0.15	EMC1	PD3
23	7.6	1.5	0.15	0.15	EMC2	PD1
24	7.6	1.5	0.15	0.15	EMC3	PD2
25	7.6	1.8	0.25	0.05	EMC1	PD3
26	7.6	1.8	0.25	0.05	EMC2	PD1
27	7.6	1.8	0.25	0.05	EMC3	PD2

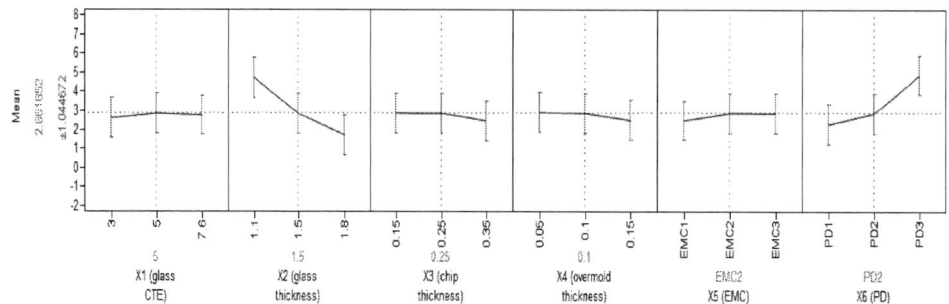

Figure 11. Panel warpage prediction after RDL process using Taguchi method.

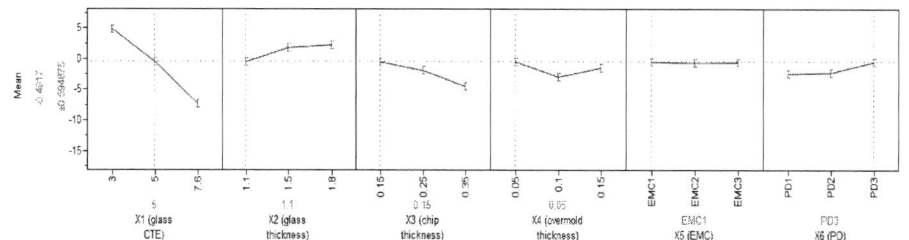

Figure 12. Panel warpage prediction after die attach process using Taguchi method.

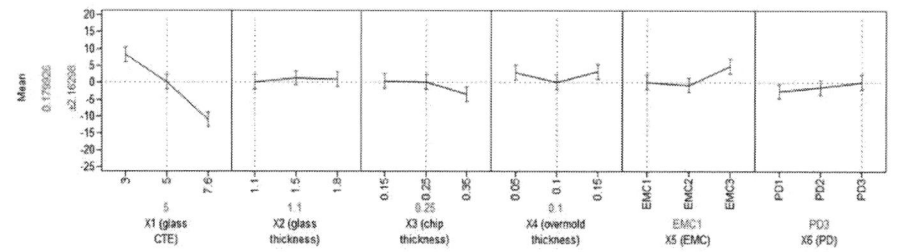

Figure 13. Panel warpage prediction after molding process using Taguchi method

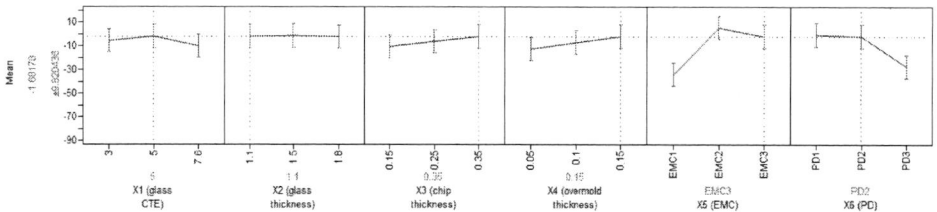

Figure 14. Panel warpage prediction after debonding process using Taguchi method

Fig. 11 shows the panel warpage prediction results based on Taguchi method for FO-PLP after RDL process. Panel warpage of all the combinations of 6 parameters can be obtained based on Fig. 11. Glass thickness and PD material are 2 important parameters for panel warpage after RDL process. Thick glass leads to low warpage. PD3 leads to large warpage compared to PD1 and PD2. Glass CTE is not significant factor for panel warpage after RDL process. Panel warpage changes from 1.14mm (lowest) to 6.74mm (highest) among all combinations. Fig. 12 shows the panel warpage prediction results after die attach process. Glass CTE, glass thickness and chip thickness are 3 important parameters for controlling panel warpage after die attach. Warpage direction changes when glass CTE changes from low to high value.

Thicker glass increases panel warpage when glass has low CTE. Thicker chip reduces panel warpage when glass has low CTE. However, when glass has high CTE, e.g. more than 5ppm/K, thicker glass and thin chip help to reduce panel warpage. To control panel warpage after die attach, first consideration is to chose suitable glass CTE, e.g. 3ppm/K to 5ppm/K in this study. The optimal case for panel warpage after die attach is glass carrier having 5ppm/K CTE and 1.1mm thickness and 0.15mm thick chip, which leads to only 0.5mm panel warpage after die attach. Fig. 13 shows the panel warpage prediction results after compression molding process. Fig.13 provides an optimal case with low warpage of 180μm when using EMC1 material. The vary important parameter for controlling warpage after molding is glass

978-1-7281-1500-9/19 $31.00 © 2019 IEEE   846

carrier CTE. Other parameters have insignificant effect on warpage. Fig. 14 shows the panel warpage prediction results after debonding process. Warpage after debonding is mainly determined by EMC and PD materials. Chip thickness and ovemold thickness also affect warpage after debonding. Usually, thick chip helps to reduce warpage after debonding due to stiffer structure. When EMC2 or EMC3 and PD1 and PD2 are used, warpage can be controlled to less than 5mm after debonding process. When EMC1 is used, warpage is more than 30mm after debonding. Therefore, it needs to systematically consider material selection and geometry design to make sure achieving low warpage and manufacturable through whole panel processes. Table III lists panel warpage under different combinations. Priority 1 is the first choice to achieve low warpage to make sure warpage less than 7mm through whole processes.

TABLE III.   PANEL WARPAGE UNDER DIFFERENT COMBINATIONS

| Glass CTE (ppm/K) | Chip thickness (mm) | Overmold thickness (mm) | EMC | PD | Panel warpage (mm) | | | | Priority |
					RDL	Die attach	Molding	Debonding	
3	0.15	0.05	EMC2/3	PD1/2		2.9	0.5 to 4.8	-12 to -20	
		0.1						-6.8 to -14	
		0.15						-1.5 to -10.2	
	0.25	0.05	EMC2/3	PD1/2	3.6 to 4.8	1.5	1 to 3.5	-7.7 to -9.1	2
		0.1						-2.4 to -11.1	
		0.15						1.4 to -6	1
	0.35	0.05	EMC2/3	PD1/2		-1	-3.5 to 1	-3.5 to -12.2	
		0.1						1.9 to -6.9	
		0.15						7 to -2	1
5	0.15	0.05	EMC2/3	PD1/2		-2.4	-4.8 to -0.5	-12 to -20	
		0.1						-6.8 to -14	
		0.15						-1.5 to -10.2	
	0.25	0.05	EMC2/3	PD1/2	3.9 to 4.4	-3.8	-6.3 to -2	-7.7 to -9.1	2
		0.1						-2.4 to -11.1	
		0.15						1.4 to -6	1
	0.35	0.05	EMC2/3	PD1/2		-6.4	-8.8 to -4.5	-3.5 to -12.2	
		0.1						1.9 to -6.9	2
		0.15						7 to -2	2

D. Effect of Weight Effect on Panel Warpage

Above FEA simulation only considers CTE mismatch induced warpage. However, gravity effect on panel warpage will become significant for large thin panel when put it onto a supporting table. Therefore, gravity should be considered for accurate warpage simulation. Fig. 15 shows panel warpage measurement results using a rule to measure four corners. Total of 4 panels are used and the averaged warpage is 1.3mm. Total 2 RDL layers are used for the panel in Fig. 15 and each layer has thickness of 8.4μm. In the FEA

simulation, total RDL thickness of 17μm is modelled for comparison. Fig. 16 shows panel warpage simulation results. Traditional method with only modelling CTE effect leads to 3mm warpage. When adding gravity effect on warpage, panel warpage is reduced to 1.5mm, which is consistent with measurement results. Therefore, modelling gravity effect is necessary and provides much accurate simulation results.

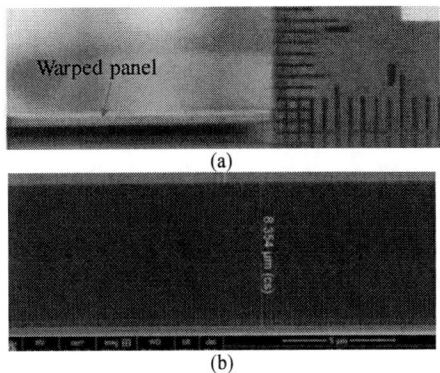

Figure 15. Fabricated panel, (a) warpage measurement, (b) PD layer thickness.

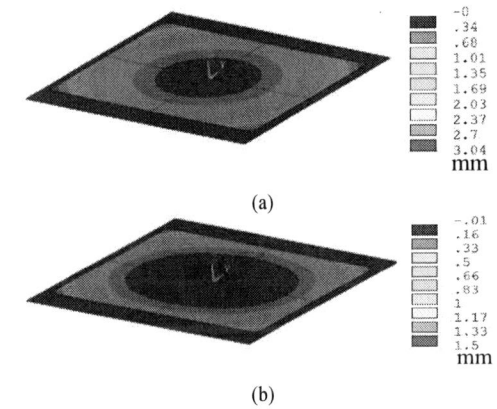

Figure 16. Panel warpage simulation results, (a) only model CTE miamatch, (b) add gravity effect.

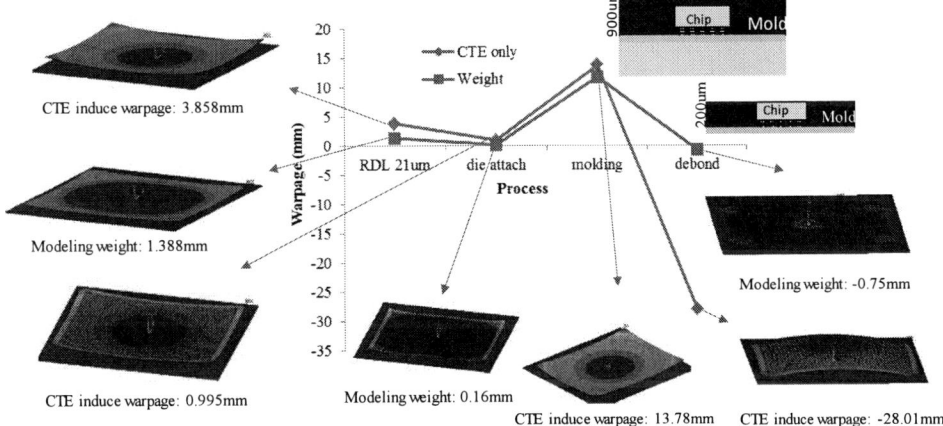

Figure 17. Panel warpage result comparison between FEA simulation with and without modelling gravity.

Fig.17 shows warpage comparison between FEA simulation with and without modelling gravity effect. In the FEA simulation, chip thickness is 775μm, molding thickness is 900μm, final panel thickness is 200μm after back grinding, glass carrier has 1.1mm thickness and CTE of 5ppm/K, EMC3 and PD2 are modelled. For all processes, gravity helps to reduce panel warpage when panel is sitting onto a table. After debonding process, the simulated panel warpage is 28mm when only modelling CTE effect. Modelling gravity effect makes warpage significantly reduce to about 1mm. Gravity must be modelled for large panel warpage study, especially for thin panel. Table IV lists warpage reduction by weight for each process step. Weight helps to reduce warpage by 14.4% after molding process in which panel has very thick structure (including glass carrier) so that weight effect is not significant. Weight helps to reduce warpage by 64%, 83.9% and 97.3% for panel after RDL process, die attach and debonding process, respectively. After debonding process, 200μm thin large panel becomes flat when sitting onto the table top due to weight effect. Therefore, large thin panel warpage may not be big challenging, which usually cannot be predicted using the traditional simulation method without considering weight effect. Handling thin panel is one of big challenges.

TABLE IV.   PANEL WARPAGE REDUCTION DUE TO WEIGHT EFFECT

Process	RDL 21um	die attach	molding	debond
CTE only (mm)	3.858	0.995	13.78	-28.01
Adding weight (mm)	1.388	0.16	11.79	-0.75
Reduced by	64.0%	83.9%	14.4%	97.3%

Effect of panel size on panel warpage is shown in Fig. 18 with only modelling CTE effect. Warpage increases with panel size. Effect of weight on panel warpage is carried out for different panel sizes and the results are shown in Fig. 19. Weight helps to reduce panel warpage at each process step. Significant reduction on panel warpage due to weight happens for the thin panel after debonding. Interesting thing is that larger panel has lower warpage than smaller panel after debonding when simulating weight effect, which usually never happens using the traditional FEA method.

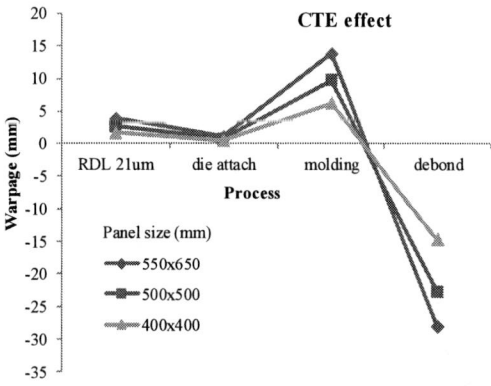

Figure 18. Effect of panel size on warpage when only modelling CTE.

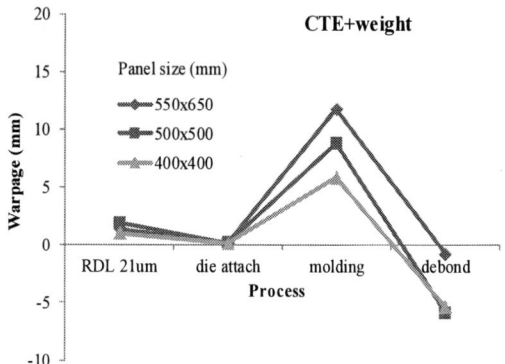

Figure 19. Effect of panel size on warpage when modelling weight effect.

V.   CONCLUSIONS

Based on FEA simulation results and experimental correlation of panel warpage for FO-PLP technology, some important findings can be drawn below:

1. Process dependent panel warpage modelling methodology is established for FO-PLP technology using RDL-first approach. Using thick glass carrier helps to reduce panel warpage significantly for RDL process. Glass carrier should have matched CTE with EMC material to reduce panel warpage after molding process. Chip and overmold thicknesses, PD and EMC materials are important parameters affecting panel warpage after debonding.

2. Taguchi method helps to identify important parameters for each process. Based on optimized structure and materials, panel warpage can be controlled less than 7mm in each process step for 550mm × 650mm panel.

3. Gravity helps to significantly reduce panel warpage of the thin large panel. Gravity effect on warpage is not significant for panel after molding due to thick and stiff structure. Considering gravity effect for large panel provides accurate simulation result which is validated by warpage measurement.

4. Panel warpage increases with panel size based on traditional FEA simulation method with only modelling CTE mismatch induced warpage. When modelling weight effect, larger panel may have lower warpage, such as thin large panel after debonding in this study.

ACKNOWLEDGMENT

This work is the result of a project initiated by *Fan-Out Panel Level Packaging (FO-PLP) Consortium*. The authors greatly appreciate the members' participation in discussions and encouragement throughout the course of the project which makes this research possible.

REFERENCES

[1] F.X. Che, David Ho, M.Z. Ding and Daniel Rhee, "Study on Process Induced Wafer Level Warpage of Fan-Out Wafer Level Packaging," Proc. 66th Electron. Compon. Technol. Conf., 2016, pp. 1879–1885.

[2] J. H. Lau, M. Li, and D. Tian et al, "Warpage and Thermal Characterization of Fan-Out Wafer-Level Packaging," IEEE Trans. Compon. Packag. Manuf. Technol., vol. 7, no. 10, pp. 1729–1738, 2017.

[3] F.X. Che, M. Kawano, M.Z. Ding, Y. Han, and S. Bhattacharya, "Study on Low Warpage and High Reliability for Large Package using TSV-Free Interposer Technology through SMART Co-Design Modelling," IEEE Trans. Compon. Packag. Manuf. Technol., vol. 7, no. 11, pp. 1774–1785, 2017.

[4] T. Braun, K.F. Becker, S. Voges, J. Bauer, R. Kahle, V. Bader, et al., "24"x18" fan-out panel level packing," Proc. 64th Electron. Compon. Technol. Conf., 2014, pp.940–946.

[5] H.D. Chang, D. Chang, K. Liu, H.S. Hsu, R.F. Tai, and H.C. Huang, "Development and characterization of new generation panel fan-out (P-FO) packaging technology," Proc. 64th Electron. Compon. Technol. Conf., 2014, pp. 940–946.

[6] H.W. Liu, Y.W. Liu, J. Ji, J. Liao, A. Chen, Y.H. Chen, et al., "Warpage characterization of panel fan-out (P-FO) package," Proc. 64th Electron. Compon. Technol. Conf., 2014, pp. 1750–1754.

[7] F.X. Che, David Ho, Mian Zhi Ding, Xiaowu Zhang, "Modelling and Design Solutions to Overcome Warpage Challenge for Fan-Out Wafer Level Packaging (FO-WLP) Technology," Proc. 17th Electron. Packag. Technol. Conf., 2015, P203.

[8] F.X. Che, H.Y. Li, Xiaowu Zhang, S. Gao, and K.H. Teo, "Wafer Level Warpage Modelling Methodology and Characterization of TSV Wafers," Proc. 61st Electron. Compon. Technol. Conf., 2011, pp. 1196–1203.

[9] S. Zhang, Q. Dan, C. Liu, Y. Xu, X. Wu, S. Liu, et al., "Simulations for the impact of warpage on the accuracy of attitude and heading reference system," Proc. 64th Electron. Compon. Technol. Conf., 2014, pp. 1010–1015.

[10] F.X. Che, David Ho, and T.C. Chai, "Co-design for Extreme Large Package Solution with Embedded Fine Pitch Interposer (EFI) Technology," Proc. 68th Electron. Compon. Technol. Conf., 2018, pp. 1030–1038.

Mechanical properties of intermetallic compounds
at elevated temperature by nanoindentation

Fan Yang[a], Sheng Liu[a], Zhaoxia Zhou[b], Zhiwen Chen[a,*], Li Liu[c], Canyu Liu[d], Changqing Liu[d,*]

[a]The Institute of Technological Sciences, Wuhan University, Wuhan, China
[b]Loughborough Materials Characterisation Center, Loughborough University , Leicestershire, the United Kingdom
[c]School of Materials Science and Engineering, Wuhan University of Technology, Wuhan, People's Republic of China
[d]Wolfson School of Mechanical, Electrical and Manufacturing Engineering, Loughborough University, Leicestershire, the United Kingdom
zwchen_lu@163.com,C.Liu@lboro.ac.uk

Abstract—**With the high operating temperature in power electronics devices, the reliability of lead-free solder joints in electronic packages at high temperatures is important. Solder joint reliability depends on a large extent on the mechanical properties of the thin intermetallic compound layer, especially hardness and Young's modulus. In this paper, the mechanical properties of Cu6Sn5, the main component of IMC, at high temperature are measured by nanoindentation. After being processed and analyzed, data are compared with the room temperature one. The results show that temperature changes have effect on the mechanical properties of lead-free solder joints.**

Keywords- Cu6Sn5; Nanoindentation; Mechanical property; Young's Modulus; Hardness

INTRODUCTION

With the development of power electronic devices, the reliability of electronic packaging materials for connecting chips and substrates at high temperatures is worth studying [1-3]. The brittle intermetallic compound (IMC) formed between the solder and the substrate is a key factor determining the reliability of the electronic package [4,5]. Studying the mechanical properties of IMC materials at high temperatures is a premise for ensuring the mechanical integrity of the device. The hardness and Young's modulus of the IMCS layer are two important element [6,7]. Previous researchers focused on the mechanical properties of IMCS at room temperature. Evidently, the properties of materials vary greatly with the operating temperature of the device [8]. Lead free solders include SnCu, SnAg, SnBi, and SnAgCu [9]. The solder is usually reflowed on the copper substrate, and the copper and tin atoms react with each other to form IMCs composed of Cu6Sn5 and Cu3Sn. The former accounts for a larger proportion. This paper is to study the performance of Cu6Sn5 in high temperature. The experimental results are then compared to materials under room conditions.

This paper introduces the experimental process and then analyzes the experimental results. To simplify the indentation test material, IMCs were prepared by reflowing. It's difficult to test the mechanical properties of micro scale IMCs by macro scale method , so they are characterized by nanoindentation [10-12].

EXPERIMENT

Material preparation

To simplify the experimental materials and reduce unnecessary contamination, the indentation material IMCS is prepared by reflowing a Sn99Cu1 solder ball on a polished Cu substrate. Figure 1 shows the reflow diagram. Qn is a parameter related to reflow, and it is defined as the integral of the temperature T ℃ during the solder alloy is melting, which can be approximated by t × ΔT [13]. Value of Qn of Cu6Sn5 is 2500 s·℃. The larger Qn will make the IMCs thicker. The thick IMCS layer makes it easy to find the correct indentation position, and the experimental results are better. The samples are then aged in a vacuum oven at 170 ° C for 100 hours. Finally, the solder joint is cut to expose the cross section and polished, which containing the thick IMCs layer [14]. The nanoindentation experiment require a smooth and flat surface [15]. Before experiment at 100 ℃, it is necessary to wait for the sample temperature being stable, which takes about two hours.

Hystron TI 950 is employed to perform nanoindentation test on the cross section of the sample. The Berkovich indenter is used. Considering the melting point of the material, the actual environment of the solder joint and the application procedure, nanoindentation experiments were carried out at room temperature and 100 ° C [16].

Among them, creep on unloading characteristics has a great influence on data, especially at high temperature [17]. To eliminate the effect, a ten seconds of holding period is set each time. And the indenter was in the deepest position [18]. To reduce errors, each experimental group with different temperature and loading speed have five indentations. Per indentation interval is 20 um. In this experiment, the loading speed is controlled by adjusting the loading time. Maximum load and maximum displacement are 5 mN and 220 nm, respectively. The main parameters of the test are listed in the table I.

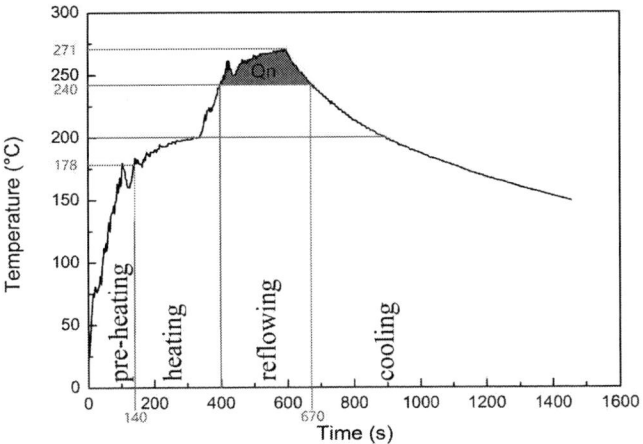

Fig. 1 The measured temperature profile for IMCs preparation.

TABLE 1 MAJOR PARAMETERS OF THE NANOINDENTATION TESTS

Test mode	load controlled
Maximum load (μN)	5000
Loading rate (μN/s)	2000,500,50
Unloading rate (μN/s)	2000,500,50
Dwelling time (s)	10
Test temperature	25°C, 100°C

After the experiment, the indentation of the Sn99Cu1/Cu sample is observed by an electron microscope. The experimental data are thousands of points in relation to displacement and load, and the indentation curves are collected by data processing software. According to the method proposed by Oliver and Pharr, properties of materials like Young's modulus and hardness at each loading rate and two temperature can be obtained [19,20].

Result

The surface morphology of the sample is investigated by scanning electron microscopy (SEM, Zeiss Ultra Plus). At the same time, the indentation element is detected by scanning electron microscopy and energy dispersive spectroscopy(EDS). Figure 2 (a) shows the surface of the IMCs under electron microscopy. An IMCs layer with a width of approximately 13μm can be seen. There is a slight pile-up in the indentation.

The results of the EDS in Figure 2(b) show that copper element occupies 39.66% of the total weight. The main ingredient is tin, which weight percentage is 60.34%. It further proves that Cu6Sn5 mainly consist of the IMCs layer. It also demonstrates that oxidation of Sn-rich sample does not pose identifiable influence on test results.

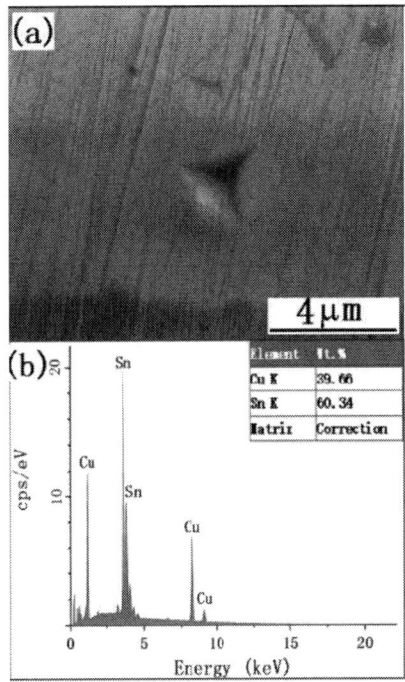

Fig. 2 The (a) morphology and (b) composition of Cu6Sn5 at 100°C.

TABLE 2 MAJOR PARAMETERS OF THE NANOINDENTATION TESTS

symbol	Meaning of the symbol
h_{max}	maximum depth of nanoindents
ε	constant related to the geometry of indenter
S	initial unloading contact stiffness of the material

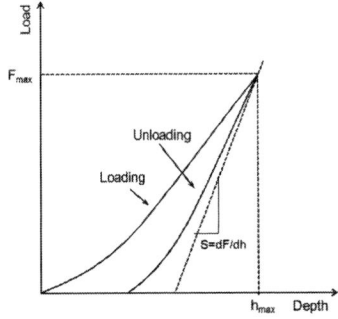

Fig. 3 The schematic of elasto-plastic stress-strain curve of metals.

Figure 3 is a schematic diagram of the nanoindentation curve. The meaning of each symbol is shown in Table 2.

Figure 4 is the comparison of room temperature and 100 °C indentation curves at three loading speed. There are serrations for each curve. At high temperature the curve is smoother and the displacement during holding period is bigger. Curves at room temperature show multiple pop-ins, characteristic of cracking, which symbolize the formation of shear band [21]. The crack shown by Figure 5 can verifies this inference. It's a example of indenter at 25 ℃. This result suggests a brittle-to-ductile transition in the deformation behavior of the IMCs [22]. Under the minimum loading rate, the displacement at 100 ℃ has a maximum depth, which is more than that at 25 ℃. Under the maximum loading rate the maximum depth of 100 ℃ is 52 nanometers deeper than that of 25 ℃. As the loading rate going down, depth disparity between two temperature increases. With the temperature rising, the speed of molecular movement increase, therefore the creep phenomenon is more intense [23].

Many results can be obtained from the nanoindentation curves [24]. The two most important result of these are Young's modulus and hardness. The reduced modulus E* is determined by the indenter and the sample material. The simple equation is given by

$$\frac{1}{E^*} = \frac{(1 - v^2)}{E} + \frac{(1 - v_i^2)}{E_i}$$

Where Ei and E are the elastic moduli of the indenter and the specimen, respectively, and vi and v are their Poisson's ratios. Because the indenter material is generally diamond in the nanoindentation test, E* is mainly determined by the mechanical properties of the sample.

H is the hardness of the sample,which is defined as the average stress at the maximum contact area A (contact projection area) or the maximum load F (maximum force applied in the indentation test).

$$H = F/A \qquad (1)$$

The theories proposed by Oliver and Pharr are often used to derive the relationship of the parameters in the nanoindentation method [25].

$$h_c = h_{max} + \varepsilon \frac{F_{max}}{S} \qquad \textbf{(2)}$$

Figure 6 shows the comparison of Young's modulus and hardness exhibited by the two experimental temperature conditions at each loading rate.

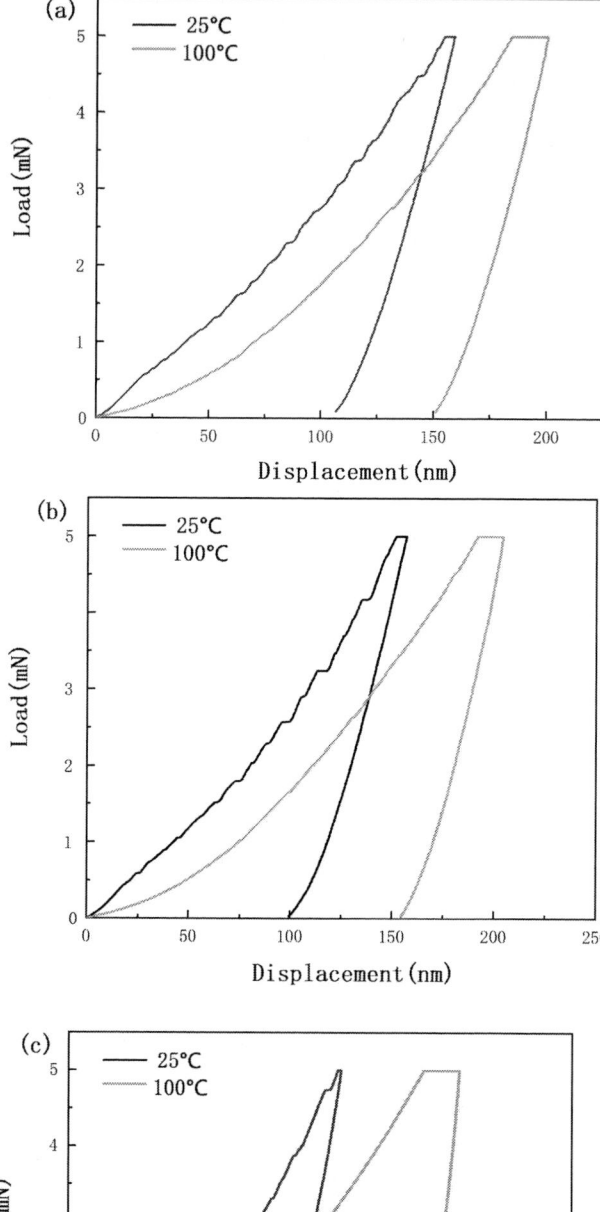

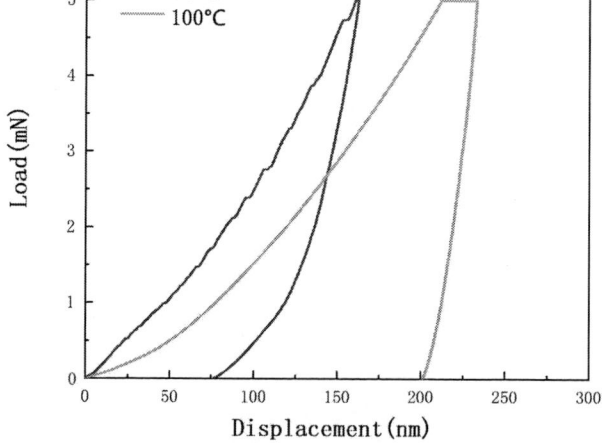

Fig. 4 Examples of load-displacement curves at different loading speed of (a)2000μN/s, (b)500μN/s, (c)50μN/s.

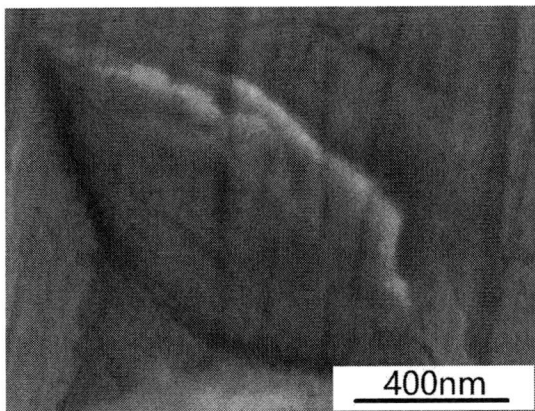

Fig. 5 Examples of nanoindents at 25°C.

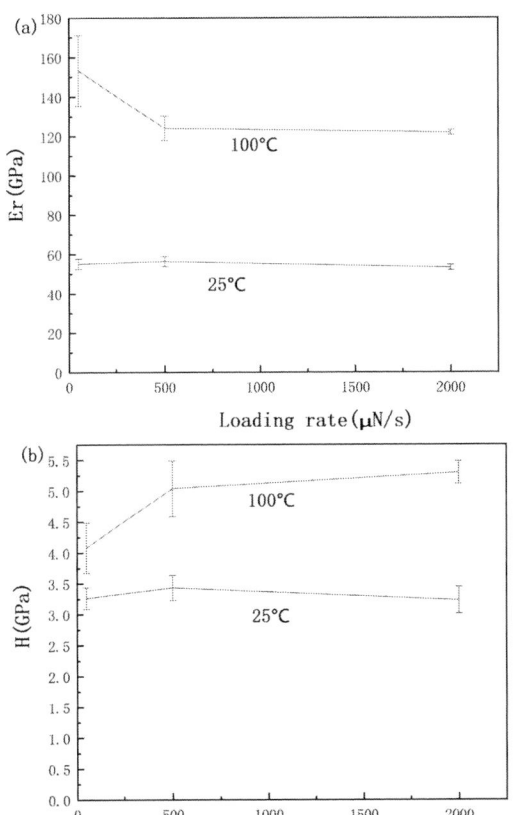

Fig. 6 Relation between (a) reduced modulus and (b) hardness of Cu6Sn5, and loading rates at room temperature and 100°C.

The results show that the hardness value of Cu6Sn5 increases with the increasement of temperature. When loading rate ranges from 50 μN/s to 2000 μN/s, Er at room temperature ranges from 53.22 GPa to 55.13 GPa. When the temperature rises to 100 °C, Er is 121.89 ~153.32 GPa. The hardness of the sample at 25 ° C is in the range of 3.23 GPa to 3.26 GPa. At 100 ° C, the hardness of the sample may vary from 5.30 GPa to 4.08

GPa. Noteworthily, When every condition except temperature is the same, the Young's modulus at 100 °C is twice the Young's modulus at room temperature. For example, when the loading rate is 500 uN per second. During the heating process, the hardness and Young's modulus of Cu6Sn5 increased from 3.43 GPa to 5.04 GPa, respectively, and the Young's modulus increased from 56.37 GPa to 124.10 GPa. However, most studies show that the higher the temperature, the smaller the hardness of the material, which is contrary to the experimental results. This maybe due to experimental errors caused by pile-up and sink-in. Sink-in is more likely to occur when material properties are more brittle [26]. And that lead to an underestimation of hardness at 25°C. The strain hardening theory can partly explained that pile-up happened at high temperature [27]. So the contact area is less calculated at 100 °C, causing the hardness to be overestimated.

The temperature change has a great influence on the mechanical properties of Cu6Sn5. The mechanism needs further research. Studying the mechanical properties of IMCS materials at different temperature is a premise for improving the effective life. In this paper, only the effect of temperature changes on a small fraction of the properties of the material is discussed.

CONCLUSIONS

By studying and observing the mechanical properties of IMCs dominated by Cu6Sn5, the following conclusions can be drawn:

1) The load-displacement curve is smoother at high temperatures and the displacement during holding period is greater. This due to brittle-to-ductile transition and creep phenomenon.

2) Under the minimum loading rate, the displacement at 100 °C has a maximum depth, which is more than that at 25 °C. The maximum depth of 100 °C is 52 nanometers deeper than that of 25 °C. As the loading rate going down, depth disparity between two temperature increases.

3) When loading rate ranges from 50 μN/s to 2000 μN/s, Er at room temperature ranges from 53.22 GPa to 55.13 GPa. When the temperature rises to 100 °C, Er is 121.89 GPa~153.32 GPa.

4) The hardness of the sample at 25 ° C is in the range of 3.23 GPa to 3.26 GPa. When the temperature reaches 100 ° C, the hardness of the sample vary from 5.30 GPa to 4.08 GPa. This may be the due to the errors caused by pile-up and sink-in.

5) The change of hardness of Cu6Sn5 is not obvious with the change of loading rate at room

temperature. But the hardness of Cu6Sn5 decreases slightly, when the loading rate increases at high temperature. The reasons and mechanisms behind this require more work.

ACKNOWLEDGMENT

This work was supported by The Startup Foundation for Introducing Talent of Wuhan University (NO. 413100011).

REFERENCES

[1] Plieninger, R., M. Dittes, and Klaus Pressel. "Modern IC packaging trends and their reliability implications." Microelectronics Reliability 2006, 1868-1873.

[2] Shiue, R. K., et al. "The reliability study of selected Sn–Zn based lead-free solders on Au/Ni–P/Cu substrate." Microelectronics reliability 2003, 453-463.

[3] Sun, Lei, et al. "Reliability study of industry Sn3. 0Ag0. 5Cu/Cu lead-free soldered joints in electronic packaging." Journal of Materials Science: Materials in Electronics 26.11 (2015): 9164-9170.K. Elissa, "Title of paper if known," unpublished.

[4] Chen, Wen-Hwa, et al. "IMC growth reaction and its effects on solder joint thermal cycling reliability of 3D chip stacking packaging." Microelectronics Reliability 53.1 (2013): 30-40.

[5] Chan, Yan Cheong, and Dan Yang. "Failure mechanisms of solder interconnects under current stressing in advanced electronic packages." Progress in Materials Science 55.5 (2010): 428-475.

[6] Rao, BSS Chandra, et al. "Morphology and mechanical properties of intermetallic compounds in SnAgCu solder joints." Microelectronic engineering 87.11 (2010): 2416-2422

[7] Yang, Ping-Feng, et al. "Nanoindentation identifications of mechanical properties of Cu6Sn5, Cu3Sn, and Ni3Sn4 intermetallic compounds derived by diffusion couples." Materials Science and Engineering: A 485.1-2 (2008): 305-310.

[8] Burward-Hoy, Trevor. "Method and apparatus for rapidly varying the operating temperature of a semiconductor device in a testing environment." U.S. Patent No. 5,977,785. 2 Nov. 1999.

[9] Yang, Ming, et al. "Effects of Ag content on the interfacial reactions between liquid Sn – Ag – Cu solders and Cu substrates during soldering." Journal of Alloys and Compounds 679 (2016): 18-25.

[10] Doener,M.F. and Nix,W.D.,J. Mater. Res.,1986,1,601

[11] Oliver, Warren Carl, and George Mathews Pharr. "An improved technique for determining hardness and elastic modulus using load and displacement sensing indentation experiments." Journal of materials research 7.6 (1992): 1564-1583.

[12] Cheng, Yang-Tse, and Che-Min Cheng. "Scaling approach to conical indentation in elastic-plastic solids with work hardening." Journal of Applied Physics 84.3 (1998): 1284-1291.

[13] Tu, P. L., et al. "Comparative study of micro-BGA reliability under bending stress." IEEE transactions on advanced packaging 23.4 (2000): 750-756.

[14] Tu, P. L., Y. C. Chan, and Joseph KL Lai. "Effect of intermetallic compounds on vibration fatigue of/spl mu/BGA solder joint." IEEE Transactions on Advanced Packaging 24.2 (2001): 197-205.

[15] McElhaney KW, Vlassak ‖. Nix WD. "Determination of indenter tip geometry and indentation contact area for depth-sensing indentation experiments. "J Mater Res 1998;13:1300-6.

[16] Suh, J. O., et al. "Size distribution and morphology of Cu6Sn5 scallops in wetting reaction between molten solder and copper." Acta Materialia 56.5 (2008): 1075-1083.

[17] Chudoba, T., and F. Richter. "Investigation of creep behaviour under load during indentation experiments and its influence on hardness and modulus results." Surface and Coatings Technology 148.2-3 (2001): 191-198.

[18] Chudoba, T., and F. Richter. "Investigation of creep behaviour under load during indentation experiments and its influence on hardness and modulus results." Surface and Coatings Technology 148.2-3 (2001): 191-198

[19] Oliver, W.C. and G.M. Pharr, An improved technique for determining hardness and elastic modulus using load and displacement sensing indentation experiments. Journal of materials research, 1992. 7(06): p. 1564-1583.

[20] Li, X. and B. Bhushan, A review of nanoindentation continuous stiffness measurement technique and its applications. Materials Characterization, 2002. 48(1): p. 11-36.

[21] Chen, Kuan-Wei, et al. "A study of the relationship between semi-circular shear bands and pop-ins induced by indentation in bulk metallic glasses." Intermetallics 18.8 (2010): 1572-1578.

[22] Lotfian, S., et al. "Mechanical characterization of lead-free Sn-Ag-Cu solder joints by high-temperature nanoindentation." Journal of electronic materials 42.6 (2013): 1085-1091.

[23] Liu, Y. C., et al. "High-temperature creep and hardness of eutectic 80Au/20Sn solder." Journal of Alloys and Compounds 448.1-2 (2008): 340-343.

[24] Ebisu, T., and S. Horibe. "Analysis of the indentation size effect in brittle materials from nanoindentation load – displacement curve." Journal of the European Ceramic Society 30.12 (2010): 2419-2426.

[25] Zysset, Philippe K., et al. "Elastic modulus and hardness of cortical and trabecular bone lamellae measured by nanoindentation in the human femur." Journal of biomechanics 32.10 (1999): 1005-1012.

[26] Marques, V. M. F., C. Johnston, and P. S. Grant. "Nanomechanical characterization of Sn−Ag−Cu/Cu joints—Part 1: Young′s modulus, hardness and deformation mechanisms as a function of temperature." Acta materialia 61.7 (2013): 2460-2470.

[27] Huang, W. M., et al. "Pile-up and sink-in in micro-indentation of a NiTi shape-memory alloy." Scripta Materialia 53.9 (2005): 1055-1057.

A MEMS Microphone in a FOWLP

Horst Theuss, Christian Geissler, Franz-Xaver
Muehlbauer, Claus von Waechter, Thomas Kilger,
Juergen Wagner, Thomas Fischer, Ulf Bartl,
Stephan Helbig, Alfred Sigl, Dominic Maier,
Bernd Goller, Matthias Vobl, Matthias Herrmann,
Johannes Lodermeyer, Ulrich Krumbein

Infineon Technologies
Munich, Germany
Horst.Theuss@infineon.com

Alfons Dehé

Hahn-Schickard
Villingen-Schwenningen, Germany
Alfons.Dehe@Hahn-Schickard.de

Abstract —**This work presents a fully functional miniaturized MEMS microphone demonstrator assembled within a modified Fan Out Wafer Level Package (FOWLP) process chain. Core of the development is the adaption of the FOWLP-process to MEMS-microphones, which have not yet been fully processed in the wafer fab. Instead, these microphone chips contain non-released membranes making them sufficiently robust to withstand backend processes, such as laminating and molding. The membrane release itself is performed on the reconstituted mold-wafer and postponed to a later process step. Mechanical stress effects induced by the package onto the MEMS are limited to a minimum by the implementation of stress decoupling suspension structures on the MEMS die.**

Keywords – MEMS; microphone; FOWLP; eWLB; Wafer Level Packaging

I. INTRODUCTION

Making use of its superior routing capability, developments on FOWLPs typically focus on high pin count applications, fine pitches or superior RF-properties. Moreover, FOWLP represents an elegant technology for interconnecting different chip technologies within a System-in-Package (SiP). In order to fully utilize the potential of FOWLP regarding SiP-devices, the technology needs to be extended to semiconductors containing MEMS structures, for instance membranes. Particular barriers to apply FOWLP technology to MEMS microphones are the incompatibility of the thin sensitive chip membranes with processes such as laminating, molding, backgrinding and dicing. Moreover, package induced mechanical stress deteriorates the oscillation quality of the MEMS membrane. These stress phenomena are mainly caused by mold encapsulation and thermomechanical incompatibilities of interfacing materials. The obvious packaging challenge, however, results from the need for cavities on both sides of the membrane in order to enable accurate acoustic oscillations. In addition, the package has to provide a certain acoustic back volume opposite to the signal port. This back volume directly impacts the acoustic device performance by strongly

contributing to the damping properties of the oscillating system.

Figure 1 shows a state of the art MEMS microphone chip with an illustration of the capacitive structure. A standard package solution for a digital device with a bottom acoustic port is shown as well. Dimensions shown are typical but can vary within a certain range. A comprehensive description of such a device including design considerations and acoustic characterization is given in [1]. Subject of this paper is to describe a packaging process within a FOWLP flow.

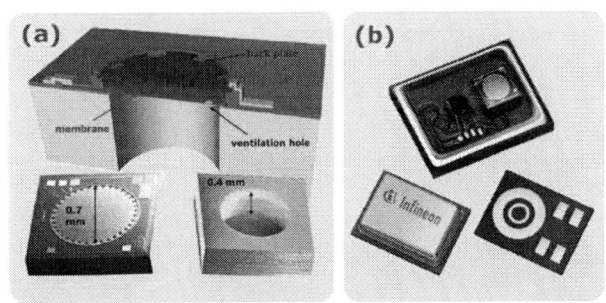

Figure 1: (a) Sketch of a simple MEMS microphone chip. The membrane oscillates relative to the thicker and stiffer backplate. Membrane and backplate represent the capacitor electrodes, while ventilation holes enable pressure equalization between acoustic front and back volume.
(b) Standard microphone package: Die-/wirebonding onto organic substrate, optionally followed by globtop coverage of the ASIC with subsequent lid attach.

II. MEMS ROBUSTNESS DURING PROCESSING

In contrast to the MEMS shown in Fig. 1(a), the microphones used in this study are designed with the membrane on top of the backplate (Figure 2). This is beneficial with respect to the overall process flow comprising various etching processes for creating membrane, backplate and stress decoupling suspension structures. These stress decoupling features are discussed in more detail in section V.

978-1-7281-1500-9/19 $31.00 © 2019 IEEE

How can a MEMS microphone chip survive the packaging processes involving harsh processes such as lamination and mold encapsulation? The fully processed chip (Figure 2, bottom) with released membrane and backplate is obviously far too sensitive to withstand most of the backend processes. In particular the die attach with the membrane facing an adhesive layer as well as its removal after compression molding are processes which destroy the MEMS structures irreversibly.

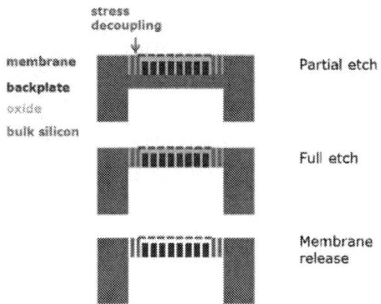

Figure 2: Nomenclature for the different evaluated MEMS conditions:
Partial etch: Incomplete removal of the bulk silicon (DRIE-etching, Bosch-process) from the wafer backside
Full etch: The Bosch process is completed. Backplate and membrane are still connected by an oxide layer.
Membrane release: The oxide layer is removed by wet etching. The MEMS microphone is fully processed.

Thus, the MEMS processing was stopped after full etching from the backside using standard DRIE processing (Deep Reactive Ion Etching, also known as Bosch process), but prior to the wet etch release of the membrane from the backplate (Figure 2, middle). This condition was named "full etch". A nano-indenter test using a sphere with 50μm diameter pressed onto the membrane served for the characterization of its mechanical stability:

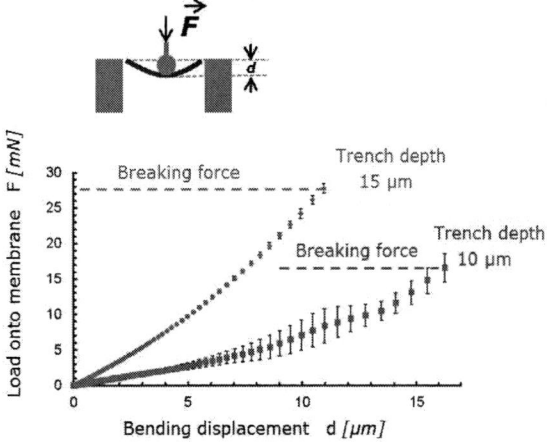

Figure 3: Mechanical stability of membrane after full etching from the backside (indenter test). Deeper trenches lead to increased stability (breaking forces) of the membrane - a firm requirement to survive the assembly processes.

Force versus displacement until breakage is shown in Figure 3 for two different trench depths which equal the thicknesses of the stress decoupling grid structure (details of this structure are shown in Figure 8). As expected, deeper trenches were able to withstand higher bending forces (Figure 3). However, none of these "partial etch" samples survived assembly test runs in the FOWLP line. Apparently process stability requires far higher mechanical robustness levels. This led to the introduction of the "partial etch" version (Figure 2, top), where the Bosch-process is disrupted prior to reaching the backplate. A bulk silicon layer of approx. 50-80 μm thickness remains and ensures the required mechanical stability of the MEMS during further processing. Further test runs in the line included samples in the version "partial etch". These evaluation allowed the correlation of the minimum required breaking force with process stability. The respective minimum could be estimated to be approximately 75 mN.

III. PROCESS FLOW

The various known FOWLP process chains differ in minor modifications. Basis for the present study is the so-called eWLB process (embedded Wafer Level Ball Grid Array) [2]. Figure 4 schematically illustrates how the eWLB flow can be adapted to the MEMS specific constraints and boundaries:

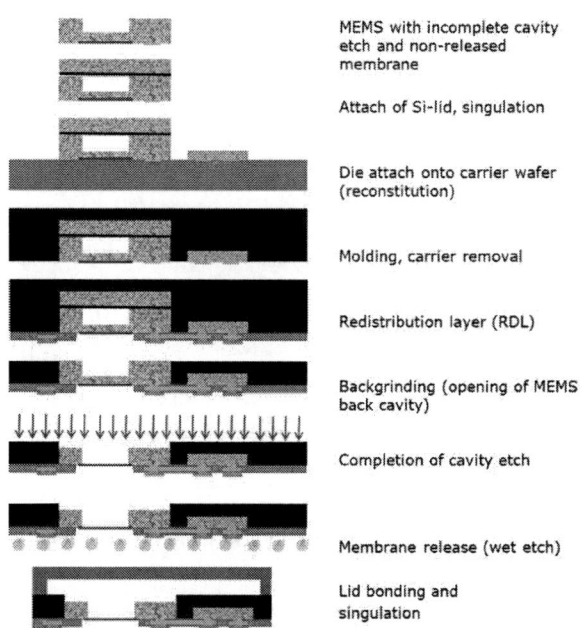

Figure 4: Selected process flow out of various options. The Figure highlights novel MEMS-specific processes, while well-known standard eWLB processes were partially combined into one sketch.

The MEMS in condition "partial etch" is covered with a silicon lid on wafer level prior to package singulation. The lid protection enables the application of a standard blade dicing process. MEMS and ASIC are bonded face down onto a standard eWLB carrier wafer, followed by compression mold encapsulation and deposition of a 2 layer RDL (redistribution layer). Mechanical backgrinding removes the silicon cover of the MEMS and opens the back cavity. Now the membrane needs to be released by completing the Bosch etch process followed by a wet etch process removing the oxide in between membrane and backplate. The Bosch DRIE process will certainly thin down all exposed silicon areas, as indicated in Figure 4. Remarkably the Bosch process can be applied using the mold compound itself as a masking feature. DRIE will also interact with the surface of the mold compound. Glas filler particles might partially be removed and potentially complicate subsequent processes. This could be avoided by depositing an organic backside protection layer after the grinding process. Being just an optional process, backside protection is not shown in Figure 4. In this basic study the backside protection was skipped but it may have to be reconsidered for quality and yield improvement. A lid bonding represents the last process on wafer level. Here, the lids were still individually glue attached within a pick and place step. However, a high volume manufacturing process would certainly strive for a wafer level bonding option prior to singulation. The choice of thermoset lid material ensures a good thermomechanical match with the mold encapsulation. The lid creates the acoustic back volume. Blade dicing singulates the devices and completes the packaging process flow.

Figure 5: eWLB-wafer after membrane release and lid attach. The pictures show a developmental wafer with additional large bare silicon chips (right, bright areas). Their purpose is to limit warpage effects to a minimum.

MEMS microphone devices typically require a grounded conductive shielding – e. g. a metal or metallized lid, which reduces electromagnetic interferences. This topic has not yet been considered in this study, but the feature can be implemented based upon published FOWLP technology: A vertical interconnect has to be introduced to establish the necessary ground interconnect to the lid. Various options for such interconnects are conceivable. These comprise Through Silicon Vias (TSVs) implemented into MEMS or ASIC but also interconnects through the mold encapsulation, [3, 4]. An

alternative to the conductive lid could be the use of a conductive backside protection as mentioned in section III.

IV. PROCESS INDUCED MECHANICAL STRESS

Being mechanical systems, MEMS are sensitive to mechanical distortion. Package induced stress is a major source for the phenomenon. Package induced stress can for instance be caused by encapsulation processes, shrink of mold compounds and other polymer materials, but also by board assembly and thermal mismatch effects during operation. A case study of package induced stress impact on the performance of a MEMS device has been reported in [5] on the example of a pressure sensor.

Within a thermomechanical simulative model, the impact of the assembly processes were investigated. The entire process flow with its associated temperature profile was taken into account followed by temperature cycling (-40°C to 125°C) and cooling down to room temperature. For simplicity reasons, the membrane/backplate-system was modeled by just one layer using average effective properties. In the final state the backplate showed an upward bending with a deflection of approx. 1 μm. An according upward bending was confirmed by profilometer measurements of processed samples. In the presence of stress decoupling structures, this deflection seems to be surprising. However, it is caused almost exclusively by intrinsic stresses and not by package induced stress. These intrinsic stress mainly originates from the doping profile of the silicon substrate and from the deep ion etching process. In fact such a defined reference state has turned out to be beneficial for achieving reproducible acoustic performances.

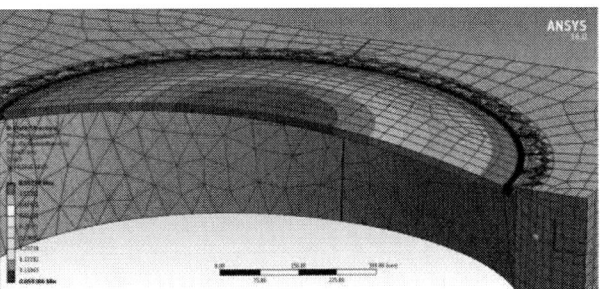

Figure 6: Simulative evaluation of the backplate deflection including manufacturing-specific T-profile as well as board assembly and T-cycle between -40°C to 125°C. A residual upward deflection of approx. 1 μm remains.

V. STRESS DECOUPLING

In standard microphone packages (Figure 1(b)), the MEMS is typically attached using a soft glue with a very low Youngs modulus. This way, the stress level can be kept sufficiently low. Mold encapsulation however, exerts a stress level onto a microphone MEMS which is by no means acceptable. This led to the introduction of stress decoupling suspension structures on silicon wafer level. These structures consist of trench-based silicon-springs, which encompass the whole membrane. Figure 7 shows pictures of the specific suspension structures used. MEMS with 16, 24 and 32 springs were manufactured and investigated, using trench depths of 10 μm and 15 μm.

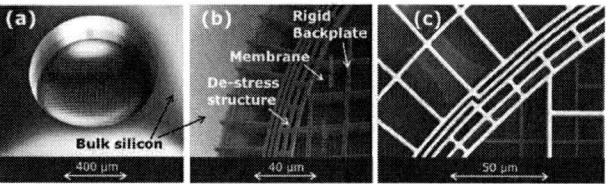

Figure 7: De-stress structures decoupling backplate and membrane from the bulk silicon.
(a) MEMS chip, bottom view
(b) and (c): Detailed view of the deep trench based suspension structures. The brighter area behind the backplate in picture (c) is the electrical connection of the membrane to the chip area outside the membrane.

The simulation model in Figure 8 illustrates more details of the spring structures as well as their deflection under mechanical load. Further advanced membrane suspension approaches are discussed in ref. [6] within simulative analysis. The stress decoupling quality of a specific structure can nicely be characterized by means of the so-called acoustic center compliance – determined while the sample is exposed to acoustic load. This value equals the deflection of the membrane center during exposure of acoustic pressure, typically determined at 1 kHz and 1Pa acoustic pressure.

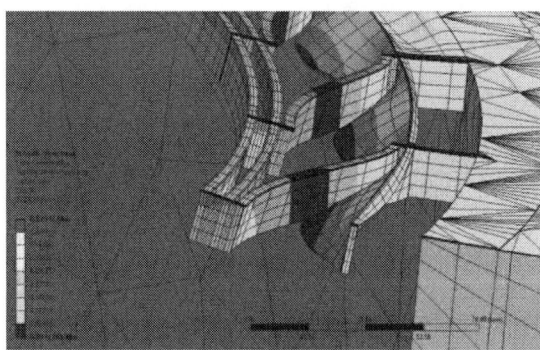

Figure 8: Deflection of the stress-decoupling spring-structures after applying a tensile stress to the chip.

Table 1 compares simulated compliance values of samples with and without stress decoupling features. In the model, samples were exposed to external stress leading to a bending strain of 200 nm. Tensile stress is given by positive values, negative values translate into compressive stress and "zero"

represents the relaxed state without external forces. While the external stress has almost no impact onto the stress decoupled MEMS, the reference sample changes its compliance by a significant factor making it unacceptable for a highly performing microphone device. In fact, it seems comprehensible, that a strained membrane experiences some kind of stiffening and becomes less sensitive to acoustic excitation.

External Strain [nm]	Center compliance [nm/Pa]	
	stress decoupling	No stress decoupling
-200	8.614	2.90
0	8.605	7.70
+200	8.596	2.78

Table 1: Center compliance of MEMS with and without stress decoupling after exposure to different mechanical loads. These loads lead to the mentioned "external strain". Values refer to samples with membrane diameters of 1000 μm. The stress decoupling version is based on 32 silicon springs and a trench depth of 15 μm.

VI. DEMONSTRATORS

Optimum results were achieved using 32 suspension springs and a trench depth of 15 μm. Pictures in Figure 9 document intermediate states after carrier removal and RDL deposition. Figure 10 presents the final samples which were assembled as described in section III. With reference to the eWLB process, the devices were named "eMic". Dimensions are 2.0 x 3.0 x 0.8 mm³, but they can easily be adapted to other chip sizes. Depending on lateral dimensions, height and layout of the lid, the acoustic back volume in the presented eMic demonstrators amounts to 1.78 mm³.

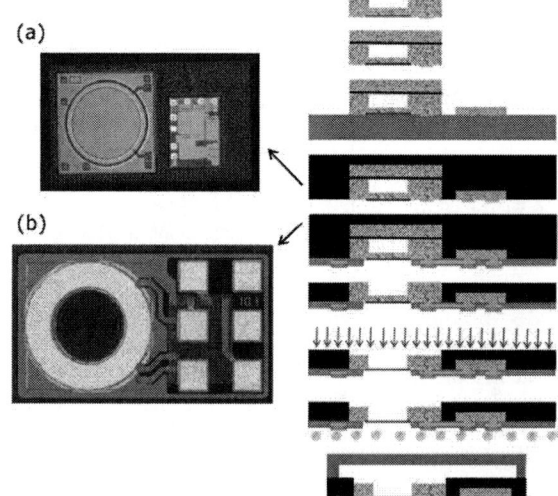

Figure 9: eMic demonstrator after
(a) molding and carrier removal
(b) application of RDL

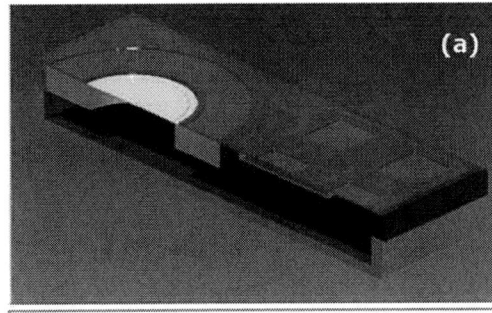

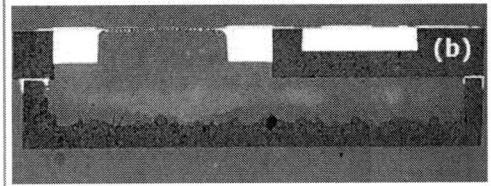

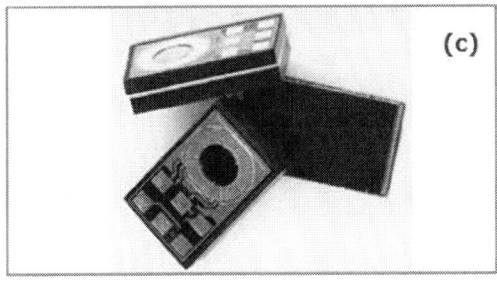

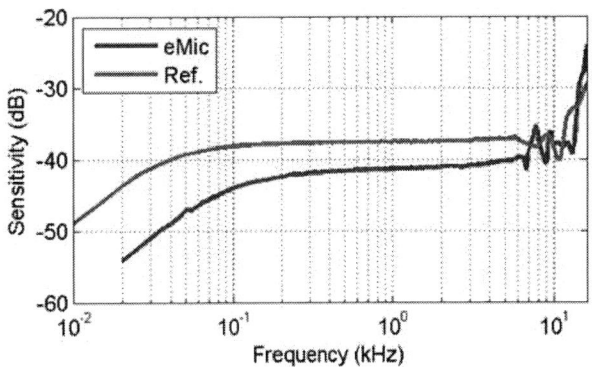

Figure 11: Frequency response of the eMic in comparison to a reference microphone in a standard package.

Again, these prototypes refer to stress decoupled 700 µm membranes, while the simulated curves correspond to the reference type with 850 µm membranes. This range is completely acceptable for commercial use and comes already very close to high performance devices (Figure 12) - in particular, if we consider the smaller membrane diameter of the eMics. Since the simulated curves represent optimized microphone models, the somewhat lower value of the eMics is certainly no surprise. The MEMS have not yet been optimized with respect to electro-acoustic parameters.

Figure 10: Final eMic demonstrator (dimensions are 2.0 x 3.0 x 0.8 mm³)
(a)   Cross section, schematic
(b)   Cross section of sample
(c)   Appearance of final device

The noticeable roughness of the inner lid surface is due to the lab based manufacturing process: Samples were laser milled out of bulk glass-filled thermoset material.

VII.   ACOUSTIC CHARACTERIZATION

Functionality of the demonstrators was confirmed by acoustic characterization. The sensitivity response measured versus frequency showed the expected course which is typical for MEMS microphones. The reference frequency response in Figure 11 corresponds to microphones with 850 µm membranes on chip size 1.2 by 1.2 mm². The eMic-curve refers to stress decoupled membranes of diameter 700 µm on a chip with dimensions 1.6 by 1.6 mm². The reference microphone is certainly optimized with respect to parameters such as back volume, parasitic effects, ventilation features and membrane dimension. However, it is remarkable, that even the lab based eMic already shows a good acoustic performance with a nice flat course of the sensitivity level in the relevant audio range.

The SNR (Signal to Noise Ratio) is a well-known indicator for the acoustic performance of MEMS microphones. With a back volume of 1.78 mm³ the SNR of two representative eMics was determined to be 63.8 dB(A) and 64.4 dB(A), respectively.

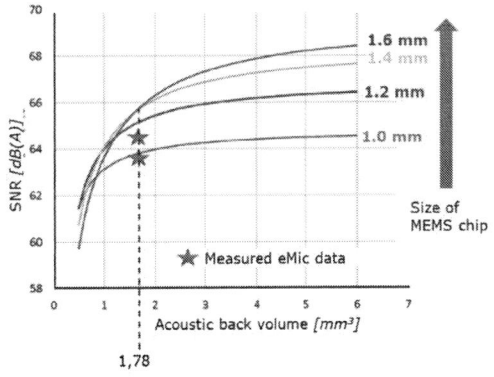

Figure 12: SNR values for various sizes of MEMS microphones and their dependence on the acoustic back volume. Lines represent numerical simulations for reference devices with membrane diameters of 850 µm. Red stars represent measurements on two eMic samples with membrane diameters of 700 µm.

VIII.   RELIABILITY

Lateral dimensions of the eMic demonstrators are smaller than most standard commercial eWLB devices. This is why the board level reliability is considered to be of a minor problem, even though the standoff will be lower – due to the missing solder balls. Board assembly tests and TCoB runs (-40°C – 125°C) were performed on samples with different RDL-layout containing non-functional testchips. It was shown, that using a non-soldermask defined layout is beneficial: No failures for 44 samples were detected after 1200 cycles.

Due to the mechanical nature of the MEMS, a higher focus was put onto mechanical stress tests. Particularly drop tests were carried out. These are not standardized drop tests but specific mechanical stress tests used within Infineon for MEMS characterization: The impact of a defined drop within a heavy metal fixture creates a pressure pulse hitting the membrane. This pressure pulse rather than inertia phenomena cause the failure.

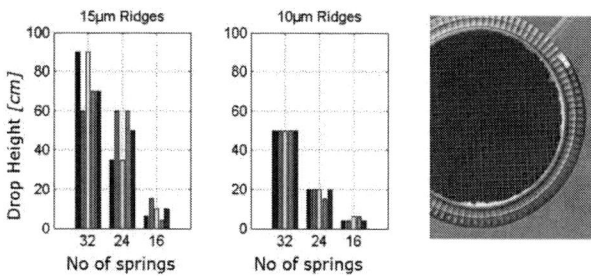

Figure 13: Drop test results of eMic devices for two different thicknesses of the backplate ridges. Each bar represents one sample. Values denote maximum drop heights the devices can withstand without failure. Typical failures are ruptures of the membrane (right).

For maximum robustness, test results advise to choose thick backplates and a high number of stress decoupling silicon springs (Figure 13). Typical failures are ruptures at the edge of the membrane. Ruptures of the stress decoupling system were also observed, but only for the least robust version: Lowest number of springs in combination with 10 μm ridges.

IX. SUMMARY

The processing of a MEMS microphone chip within a modified FOWLP flow was successfully described and performed. Key modifications from standard flows are

a) Creation of a robust MEMS chip which can survive the harsh backend processes. This was achieved by early interrupting the bulk silicon DRIE etch in the wafer fab.

b) Introduction of stress decoupling spring-like suspension structures on wafer level. These enable the microphone to acoustically perform, despite the increased package induced stress.

c) Performing the final membrane release *after* encapsulation, RDL-deposition and backgrinding.

Part of the processes were still carried out in a laboratory environment. However, proposals for increasing the efficiency of the manufacturing process do exist.

Demonstrators were acoustically characterized and confirmed to be fully functional microphone devices. Mechanical stability of the membrane suspension might need further improvement to reach a safe robustness level. The proposed process flow is a further example for dissolving barriers between wafer fab and backend processes.

ACKNOWLEDGMENT

This project has received funding from the Electronic Component Systems for European Leadership Joint Undertaking under grant agreement No 692480. This Joint Undertaking receives support from the European Union's Horizon 2020 research and innovation program and Germany, Saxony, Spain, Austria, Belgium, Slovakia. The project IoSense is co-funded by Federal Ministry of Education and Research.

Particularly we would like to thank Robert Wieland and Thi Xuan Anh Bui, Fraunhofer EMFT, Munich, for their support in plasma etching.

REFERENCES

[1] A. Dehé, M. Wurzer, M. Füldner and U. Krumbein, "The Infineon Silicon MEMS Micropohone", AMA Conferences 2013, Nuernberg, Proc. Sensor 2013, DOI 10.5162/sensor2013/A4.3

[2] M. Brunnbauer, E. Fuergut, G. Beer, T. Meyer, H. Hedler, J. Belonio, E. Nomura, K. Kiuchi, and K. Kobayashi, "An Embedded Device Technology Based on a Molded Reconfigured Wafer", Proc. 56th Electronic Components and Technology Conference (ECTC 2006), San Diego, USA, pp 547-551

[3] M. Wojnowski, G. Sommer, K. Pressel, G. Beer, "Horizontal and Vertical Interconnects for Integration of Passive Components", Proc. 63rd Electronic Components Technology Conference (ECTC 2013), Las Vegas, USA

[4] M.Wojnowski, K.Pressel, and G.Beer, "Novel Embedded Z Line (EZL) Vertical Interconnect Technology for eWLB". Proc. 65th Electronic Components and Technology Conference (ECTC 2015), San Diego, USA, pp 1071-1076

[5] H. Theuss, K. Elian, M. Fink, M. Fries, F.-P. Kalz, B. Schwabe, M. Vaupel, C. Jaroschek1, A. Schubert, H. Hummel, O. Steffens, "Mechanical Stress Impact of Assembly Processes onto a Stress-Sensitive Testchip", Proc. 44th International Symposium on Microelectronics, (IMAPS 2011), Long Beach, USA p. 619

[6] M. Füldner, A. Dehé and R. Lerch, "Analytical Analysis and Finite Element Simulation of Advanced Membranes for Silicon Microphones", IEEE Sensors Journal, October 2005, vol. 5, no. 5, pp. 857-863

Fan-out Wafer Level Packaging - A Platform for Advanced Sensor Packaging

Tanja Braun, Karl-Friedrich Becker, Ole Hoelck,
Steve Voges, Ruben Kahle, Pascal Graap, Markus
Wöhrmann, R. Aschenbrenner
Fraunhofer Institute for Reliability and
Microintegration
Berlin, Germany
e-mail: tanja.braun@izm.fraunhofer.de

Tanja Braun, Marc Dreissigacker, Martin
Schneider-Ramelow, Klaus-Dieter Lang
Technical University Berlin,
Berlin, Germany

Abstract—As mobile devices have become ubiquitous, also a large number of integrated sensors have entered our daily life. And with every new generation of mobiles the number of sensors increases, forming a growing market. Applications range from consumer to automotive and industrial up to biomedical. Especially in the era of Internet of Things (IOT), of smart manufacturing or of autonomous driving highly integrated and low cost sensor packaging solutions are required. Fan-out Wafer Level Packaging (FOWLP) is currently the hottest packaging trend in microelectronics - offering opportunities for a miniaturized system integration with multiple die packaging and 3D integration as well as packaging of stress sensitive components or sensors where a direct access to the sensing surface is needed.

MEMS pressure sensors are used in a wide variety of applications with a still growing market share. Packaging challenges come from the stress sensitive and fragile membrane and the necessity for direct media access to this membrane. Low pressure compression molding as used for FOWLP seems to be a suitable choice for encapsulation. The access to the sensor membrane then can be managed by removing the RDL above the membrane. However, depending on membrane type, a protection of the membrane during processing might be mandatory, especially during assembly, molding and debonding. Here, different approaches to protect such delicate structures will be introduced. Gas sensors used for air quality measurement as well as early fire detection have similar demands as MEMS sensors. The integration of particle and humidity barrier protection layers will be also discussed here.

Another challenging market are bio-medical sensors. Point of Care devices for medical applications are becoming more and more widespread. Besides the described challenges for biomedical sensor also the question arises when the sensor die will be functionalized for detection of e.g. biomarkers as this functionalization is typically not temperature stable. The planar FOWLP packaging approach offers also the direct integration of a microfluidics on top.

Optical sensors as e.g. proximity or image sensor are also a suitable application for FOWLP also due to the planar packaging and direct accessibility of component surfaces. 77 GHz automotive radar modules are another example which is already introduced in volume manufacturing using the advantage of very short and low inductance interconnects in FOWLP as well as the possibility of integrating antennas on package.

Packaging cost could be lowered when moving from wafer to large area panel level packaging. One example here is the fingerprint sensor packaged on panel from Nepes which is ready for high volume manufacturing. An example of gas sensor packaging on panel will also be given in the paper.

In summary the paper will present the potential of Fan-out Wafer Level Packaging for advanced sensor packaging and show FOWLP examples for a wide variety of sensors including MEMS based acceleration and pressure sensors, capacitive micro-machined ultrasonic transducers (CMUTs), gas sensor and bio-medical sensors.

Keywords: fan-out wafer level packaging, sensor packaging

I. INTRODUCTION

Fan-out Wafer Level Packaging (FOWLP) is one of the latest packaging trends in microelectronics. FOWLP has a high potential in significant package miniaturization concerning package volume but also in thickness. Main advantages of FOWLP are the substrate-less package, lower thermal resistance, higher performance due to shorter interconnects together with direct IC connection by thin film metallization instead of wire bonds or flip chip bumps and thus lower parasitic effects. Especially the inductance of the FOWLP is much lower compared to FC-BGA packages.

For Fan-out Wafer Level Packaging two basic process flows are under discussion the "Mold first" and the "RDL first" approach. Where for the "Mold first" process meanwhile a face-down and a face-up option exist. Process flows are depicted in Figure 1.

978-1-7281-1500-9/19 $31.00 © 2019 IEEE

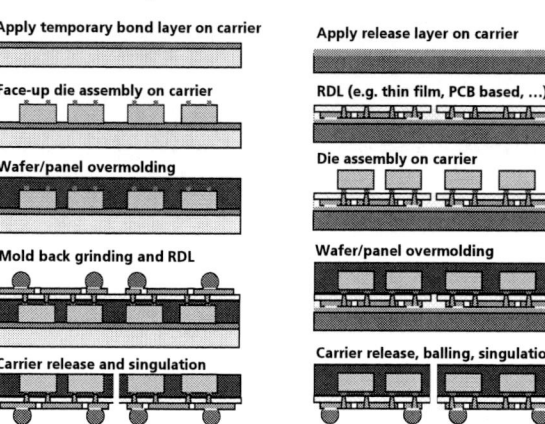

Figure 1. Fan-out Wafer/Panel Level Packaging process flow options.

"Mold first" face-down starts with die assembly on an intermediate carrier followed by overmolding and debonding of the molded wafer/panel from the carrier. The redistribution layer typically based on thin film technology is finally applied on the reconfigured molded wafer/panel. A new face-down panel approach was recently introduced by Samsung. Here dies are assembled on a carrier prepared with a PCB frame and then overmolded. The panel is released from the carrier and an RDL is applied on the active die side. Vias are drilled through the mold on the backside down to the PCB with integrated vias. A backside RDL is then applied allowing package-on-package stacking.

The face-up approach also starts with die assembly on a carrier with a temporary adhesive layer. But here dies have a Cu-bump and are placed face-up on the carrier. After overmolding a backgrinding step opens access to the Cu-bumps of the dies again. The redistribution is applied and finally the wafer is released from the carrier and diced for package singulation.

Fan-out Wafer Level Packaging offers many opportunities for sensor packaging. First it is a substrateless approach for miniaturized packaging and heterogeneous 3D integration. Multiple sensors in one package, integration of ASICs and or passive components as well as stackable configurations could be realized. The very short electrical interconnects and therewith the low inductance package is well suited for high frequency and mm-wave packaging as used for current 77 GHz automotive radar modules. By using through mold vias (TMV) not only stackable packages could be manufactured but also packages where the sensing surface is "looking" in a defined direction which is not the SMD board side. For biosensors, e.g. a clear separation between the wet microfluidic side and the dry electrical connection can be achieved by integrating TMVs.

The "Mold first" face-down process flow fits well to sensor packaging needs. No special sensor die preparation as bumping or component thickness adaptation is needed. Direct access to sensor surfaces could be also achieved by

opening the RDL in that area with the same lithography step where the µvias were formed.

Also process-wise compression molding works at lower pressures as transfer molding used typically in microelectronics packaging for e.g. BGA or QFN molding and is therefore a better choice for the encapsulation of stress sensitive components.

II. CHALLENGES FOR SENSOR PACKAGING

Sensors own special packaging demands and are often application specific. In general they have to interact with their environment to sense e.g. pressure, acceleration, yaw, sound, light, RF frequencies, gas or target molecules.

MEMS sensors are used are used in a wide variety of applications with a still growing market share [1], [2]. Packaging challenges come from the stress sensitive and fragile membrane and the necessity for direct media access to this membrane especially in the case of e.g. pressure sensors. Low pressure compression molding as used for FOWLP seems to be a suitable choice for encapsulation. The access to the sensor membrane then can be created by removing the RDL above the membrane. However, depending on membrane type, the stress on the membrane during processing might be challenging, especially during assembly, molding and debonding. Temporary protection layers or optimized materials and processing are mandatory to avoid membrane cracking.

Micro hotplate gas sensors used e.g. for air quality measurement as well as early fire detection may have similar demands as MEMS sensors. Here also the integration of particle and humidity barrier protection layers would give additional benefit. However for all the stress sensitive sensors the overall package construction has to transfer only a low amount of stress during the manufacturing but also over the lifetime to guarantee full functionality.

Another challenging market are bio-medical sensors. Point of Care devices for medical applications are becoming more and more widespread. Besides the already described challenges for biomedical sensors the question arises when

978-1-7281-1500-9/19 $31.00 © 2019 IEEE

the sensor die will be functionalized for detection of e.g. biomarkers as this functionalization is typically not very stable at elevated temperatures.

Fan-out wafer level packaging includes a lot of process steps with a mechanical stress as molding and debonding to the sensor surface as well as steps where different chemistry as solvents or acids are involved especially during RDL application. This may also require temporary protection layers to avoid damage of the sensing surface.

Within the next sections examples of successfully developed sensor packages based on FOWLP are described in detail.

III. MEMS SENSOR PACKAGING

FOWLP technologies have been also used to manufacture a multi-sensor stack consisting of a pressure sensor and ASIC package assembled on the backside of an acceleration sensor and ASIC package with through mold vias (TMVs) for 3D routing (see Figure 2).

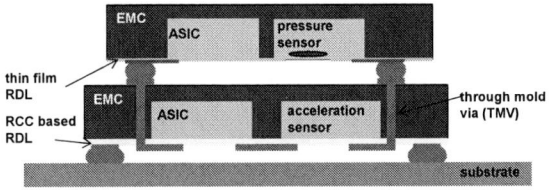

Figure 2.   Multi-sensor stack consisting of a pressure sensor-ASIC package and an acceleration sensor-ASIC package manufactured in wafer level embedding technology.

For pressure sensor packaging FOWLP technology with a chip first approach and thin film based redistribution layer (RDL) has been chosen, as a photosensitive thin film dielectrics allows direct access to the Si membrane of the pressure by an opening in the RDL [1]. For acceleration sensor and ASIC packaging a FOWLP technology with a double sided resin coated copper (RCC) based redistribution has been selected [5]. Advantage of this technology option is the direct implementation of laser drilled vias for 3D routing from package bottom to top side.

Reconfigured wafer assembly and compression molding was done on 6" wafer size. For compression molding a liquid compound with a maximum filler size of 55 µm has been selected. Wafer thickness was set to 670 µm due to sensor thickness of 550 µm to allow homogeneous overmolding of the components without damaging the stress sensitive MEMS components by the filler particles. The embedded sensor and ASIC dies need to be prepared with an under bump metallization (UBM) to avoid incompatibilities with the PCB processing used. Typically, Cu or NiPd pad reinforcement is applied to the silicon wafer or to the reconfigured mold wafer.

A filled RCC material was laminated on both sides of the reconfigured wafer and µvias to the sensor and ASIC pads as well as the trough mold vias (TMV) were lased drilled in one step. Next process steps were via cleaning, palladium activation and copper plating. By plating, vias are filled, die pads are connected to the copper layer and connection of top

copper layer to bottom copper layer is achieved. Conductor line formation was done by laser direct imaging (LDI) in combination with a dry film resist and copper etching. Finally, a solder mask and solderable surface finish as NiAu and solder balls can be applied. Figure 3 shows the final acceleration sensor package with pads on the package top side for stacking of the pressure sensor package and TMV for 3D routing to the package bottom to connect ASIC and acceleration sensor as well as the substrate.

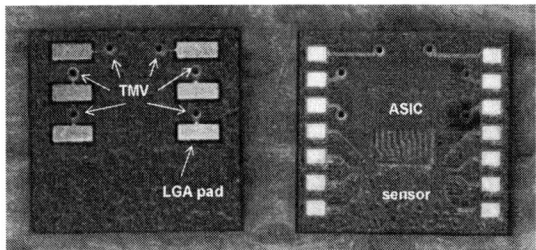

Figure 3.   Photograph of an acceleration sensor with ASIC package, package top (left) and bottom (right).

ASIC and sensor show a minimum contact pitch of 110 µm and a pad size of 80x80 µm². Die positions have been measured after molding and automatically used to adapt the µvia drill position to the die pads and the wiring of the conductor lines to compensate the die shift during assembly and molding. Therewith, a good alignment between µvias and die pads could be achieved without shorts or off target positions. Figure 4 depicts an X-Ray image of a manufactured demonstrator package showing the interconnection of the top and bottom package metallization by the through mold vias.

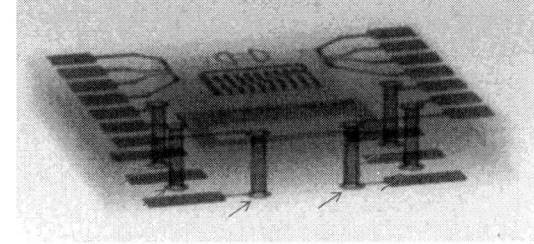

Figure 4.   X-Ray image of an acceleration sensor and ASIC package with through mold vias (TMV) marked by red arrows.

Pressure sensor and acceleration sensor packages were stacked and assembled on board by soldering. The demonstrator as shown in Figure 5 allows the functional testing of the entire sensor stack and therewith the proof of technology. Both sensors performed properly within their specification range. Hence, it can be derived that the FOWLP technology is well suited also for packaging of stress sensitive components as MEMS devices and sensors.

Figure 5. Photograph of a demonstrator stack with acceleration and pressure sensor package.

However, one of the major challenges for embedding MEMS components is the combination of the process with fragile or fracture sensitive sensor structures which are not covered. For the case of the mold first approach the active die side is placed on the thermal release tape (TRT). The TRT fixes the dies during the molding process. During the TRT release the remaining adhesion of the tape to the dies could lead to a damage of unprotected MEMS structures.

To ensure integrity of delicate structures (e.g. membranes), a variety of temporary sacrificial layers exist. These protect sensitive MEMS e.g. with a thin film coating on wafer-level by lamination or spin-coating. On individual MEMS, application of the protective layer prior to assembly, as well as during assembly by dipping the MEMS into a material depot right before placement, is possible. Typically, photosensitive polymers or thermo-plastic materials are used. The main advantage of this is that established and proven assembly strategies can be used.

Another approach is to substitute the commonly used TRTs with a thin polymer layer. The MEMS is then heated by the assembly head, while the substrate is heated as well. This lowers the viscosity of the temporary adhesive and allows for gentle placement under careful control of bonding force and time, as is described in Figure 6.

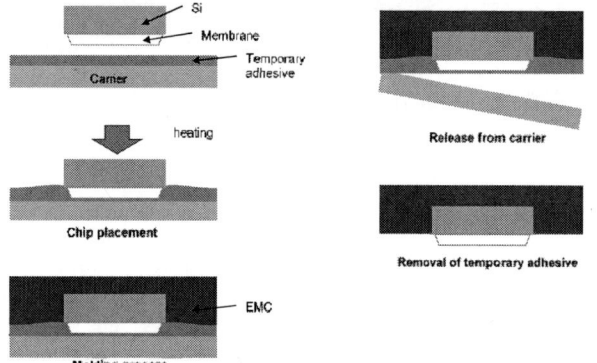

Figure 6. Process flow for placing heated MEMS into temporary adhesive with subsequent release.

For fracture sensitive MEMS components a new embedding technology was developed like it is shown in Figure 5. The hybrid Fan-out process flow combines the mold-first and RDL-first approach and is already described elsewhere [6].

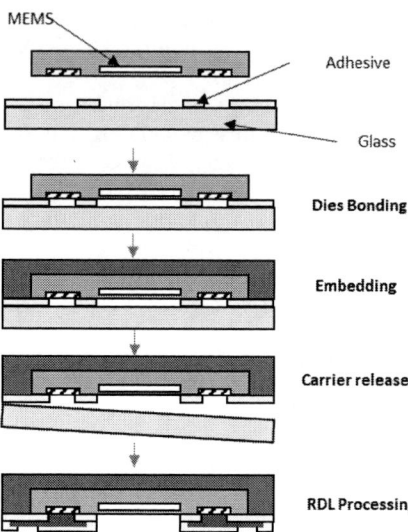

Figure 7. Fan-out Wafer Level Packaging process flow options with pre-structured dielectric layer.

The assembly process is optimized for handling MEMS. Therefore a glass carrier with a structured adhesive layer is used instead of the TRT. The adhesive is removed in the area where contact pads and MEMS structures on the die are located (see Figure 8). The bonding process of the die on the glass carrier is done at temperature in the range of 80°C where the adhesive becomes sticky. The structured adhesive avoids contamination on or damaging of the fracture sensitive structures. After placing the dies the adhesive is cured where the thermosetting material becomes solid and any die shifting during the overmolding process could be avoided. The glass carrier is removed by laser release. A laser of 248 nm is transmitted from the backside through the glass carrier and is absorbed in the first hundred nanometer of the adhesive layer which leads to a brake of the polymer backbone structure and allows a release of the carrier at room temperature. The adhesive is a thermoset material in thermal stable after curing. Therefore the adhesive can be already used as the first passivation layer following by the generation of the RDL layer.

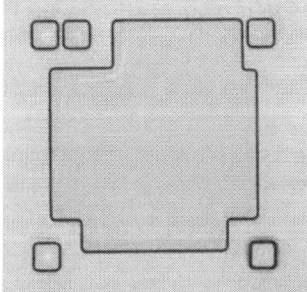

Figure 8. Structured adhesive layer for large MEMS structure in the center of die and openings of for the contact pads.

The overall proof of concept for MEMS packaging with the described hybrid fan-out approach is ongoing and will be reported later.

IV. GAS SENSOR PACKAGING

A FET (Field-Effect Transistors) based gas sensor was chosen for fan-out wafer and panel level packaging. Two packaging options have been evaluated with channels and a cover layer for sensor protection and with small channel opening formed by lithography (see Figure 9). Laser drilled through mold vias (TMV) have been chosen for electrical routing of the sensor interconnects to the package backside. Therewith the package can be reflow soldered without high risk of flux contamination and the sensor has open access to the environment and not only to the small gap between package and printed circuit board.

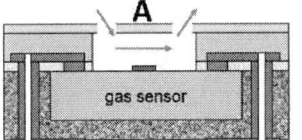 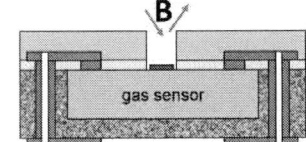

Figure 9. Gas sensor packaging options; A: with channels and a cover layer for sensor protection, B: with small channel opening realized by lithography.

Different materials for channel and cover layer manufacturing have been evaluated for option A. No successful combination could be found due to air entrapments, cover layer stability & overall material compatibility (see Figure 10). Therefore, approach B has been chosen for panel level demonstrator manufacturing. Even if these first trials were not successful the idea of a cover layer with a separate in- and outlet integrated in the RDL to protect the sensitive die surface is attractive for various application and should be further investigated and developed.

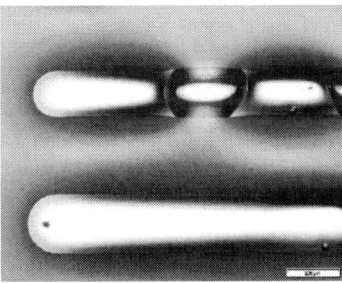

Figure 10. Cover layer structure of option A, collapsed lid layer.

Technology has been developed on 200 mm wafer size with test dies first and then transferred to large 457x305 mm^2 panel size (see Figure 11). When moving from round to panel formats also different materials, equipment and processes can be considered to further lower cost. In this study dry film materials and maskless laser direct imaging technology have been used for redistribution layer manufacturing. Both processes are typically used in a PCB manufacturing environment. Based on this approach a direct upscaling from wafer level processing to larger panel sizes was possible [7].

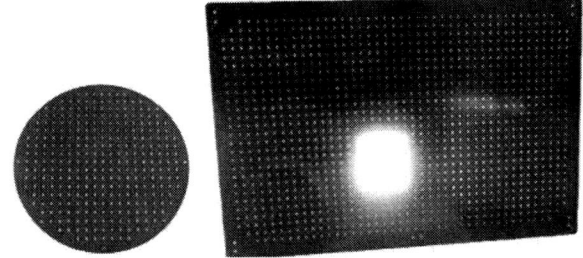

Figure 11. Gas sensor packaging left: 200 mm reconfigured wafer; right: 457x305 mm^2 reconfigured panel.

Packing option B could be demonstrated successfully on panel level using functional sensor dies as well as test dies to fill the entire panel area. Figure 12 shows a detail of the processed panels with functional dies in the center as well as a cross cut of the TMV.

 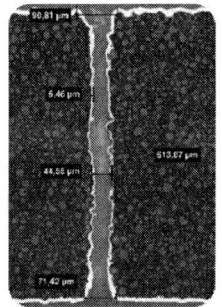

Figure 12. Gas sensor packaging left: fan-out panel with functional sensors and test dies, right: cross section of through mold via (TMV).

V. BIOSENSOR PACKAGING WITH INTEGRATED MICROFLUIDICS

FOWLP could not only be a packaging platform for MEMS sensor, also other sensors where direct media access is needed as e.g. biosensors attached to microfluidics can benefit from this technology approach. Within the European funded project PoC-ID a microelectronic biosensor platform targeting the diagnostic respiratory syncytial virus (RSV) was developed. The core of the system is a BioGrFET sensor based on a graphene field effect transistor (GrFET) [8], [9]. The liquid sample containing biomarkers flows over the sensor's surface with probe molecules, where the target molecules (specific biomarkers) of the fluid can be immobilized. The charge of the biomarker on the surface changes the charge carrier density inside the graphene which can be detected by measuring the graphene field effect transport characteristic. Specific packaging challenge for the sensor is to develop a packaging technology process flow that allows to add the sensor functionalization during packaging and leaves this functionalization intact until the packaging processes are finalized, which implies a process selection with reduced thermal and mechanical load on the delicate functionalized sensor [10].

FOWLP of the BioGrFET chips compromises five steps (see Figure 13). The first step is the creation of the reconfigured mold wafer. Electrical characterization of functional BioGrFET chips before and after compression molding showed their functionality even after these process steps.

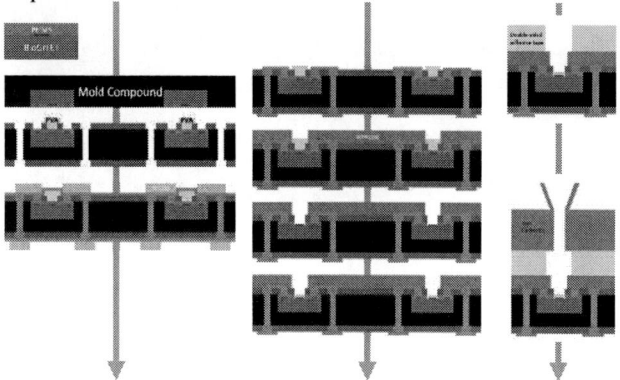

Figure 13. Process flow of the Fan-Out Wafer Level Process. 1. Manufacturing of the reconfigured mold wafer. 2. Adding the redistribution layer with through mold vias. 3. Applying the microfluidic layer. 4. Singulation. 5. Covering the microfluidics channels.

The second step is the application of the redistribution layer (RDL). Through mold vias are laser-drilled into the mold for electrical routing to the wafer back side. Several steps, comprising vacuum resist lamination and photolithographic structuring of the layers, sputtering of Ti:W as a seed layer as well as copper plating and etching, follow.

The third step is the creation of the microfluidics. The planar surface of the mold wafer allows the vacuum lamination of photoresists. Channels and openings for the sensor devices are created by photolithography. Afterwards, the packages are singulated into single packages.

Technology has been developed and evaluated using test dies with daisy structures to prove the concept of adding microfluidic layers on FOWLP wafers. In Figure 14 test vehicles with two different microfluidic channel structures are shown.

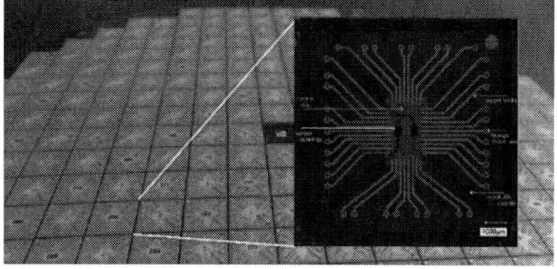

Figure 14. Reconfigured test dies wafer with RDL and microfluidic channels with two different alternating microfluidic structures (large image), hybrid microfluidic-microelectronic package layout (detail).

The successful structuring of the RDL with sensor opening and microfluidic channel structure on top could be verified with the cross sections depicted in Figure 15.

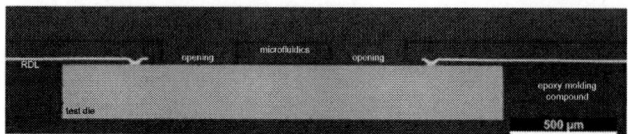

Figure 15. Cross section test die with RDL opening and microfluidic channels

Transferring the process flow to functional BioGrFET sensors also the protection of sensor surface during processing has to be considered. The first three FOWLP process steps involve methods that could attack the sensing surface with amino groups. A PMMA layer turned out to be a good protection layer. It protects the sensor surface against contamination from the thermal release film used for compression molding and avoids contaminations from different solvents during the RDL process and the application of the microfluidic part. It is applied by spin coating before the compression molding process and is removed with acetone right before functionalization of the sensor surface. Nevertheless there is still the bombardment with the high energetic particles from sputtering. Tests showed that the thin PMMA layer (around 500 nm thick) did not protect the sensor structure against the high energetic particles from sputtering. Fluorescence markers usually docking to sensor surface could not be captured on the surface after sputtering. An additional PVA (polyvinyl acetate) layer leads to an adequate protection. This PVA can easily be removed with DI-water after the RDL processing. The fourth step is the application of different biomimetic recognition molecules (aptamers) on the devices, called multiplexing. This is done by spotting an aptamer solution on the devices with a jetting process (using a GESIM nanoplotter). Afterwards the open channels are covered. The luer connector is attached to the package using a laser-cut double-sided adhesive tape, placed with a standard pick and place tool (see Figure 16). With that configuration the sensor system could be successfully evaluated.

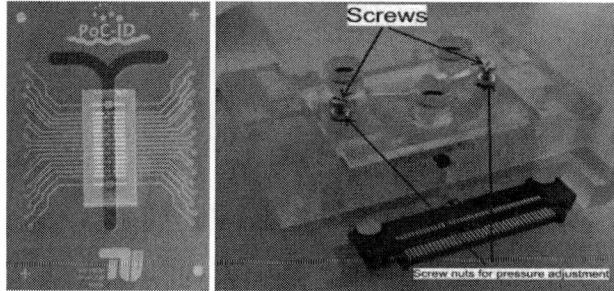

Figure 16. FOWLP BioGrFET package with integrate microfluidic channel in RDL (left) and embedded system with electrical socket with screw nuts for pressure adjustment (right).

VI. CONCLUSIONS

Fan-out Wafer Level Packaging offers many opportunities for sensor packaging as e.g. high degree of miniaturization, suitable for heterogeneous 3D integration or access to sensor surface through RDL structuring. Packaging examples has been presented for MEMS components, gas sensors and a bio-sensor with integrated microfluidics.

However, besides a multitude of advantages sensor packaging by FOWLP holds also some challenges. Sensors with fragile membranes are sensitive to stress during packaging and use and low-stress solutions have to be developed. During FOWLP processing also a high variety of different chemistry is involved and may harm sensor surfaces. Here, suitable temporary sensor protection layers have to be applied and removed without damaging the sensor.

Finally, moving from wafer to panel level packaging offers a fan-out packaging platform with advantages mentioned above with a high potential for lowering cost. This has been also already demonstrated for the gas sensor package described.

ACKNOWLEDGMENT

The authors would like to acknowledge the support for the projects
- MST SmartSense by German ministry for education and research [BMBF] and VDI/VDE Innovation + Technik GmbH (Förderkennzeichen: 16SV3672)
- SAFESENS by European Commission and German ministry for education and research [BMBF] and VDI/VDE Innovation + Technik GmbH (ENIAC-JU-2013-1(621272))
- PoC-ID project by European Commission (EU Horizon2020 Program under Grant Agreement number 634415)

The authors would also like to thank Leopold Georgi, Patrick Reinecke, Volker Bader, Oliver Nallaweg, Jasmin Zühlke and Damian Freimund for supporting the work presented.

REFERENCES

[1] M. Tilli, M. Paulasto-Krockel, T. Motooka, V. Lindroos; Handbook of Silicon Based MEMS Materials and Technologies; Elsevier, William Andrew; 2nd edition, October 2015.

[2] J. H. Lau , C. K. Lee, C. S. Premachandran, Y. Aibin; Advanced MEMS Packaging; International Marine/Ragged Mountain Press; October 2009.

[3] J. Kim, I. Choi, JH. Park, J.-E. Lee, TS. Jeong, J. Byun, YG. Ko, K. Hur, D.-W. Kim, K. S. Oh; Fan-out Panel Level Package with Fine Pitch Pattern; Proceedings of ECTC 2018, San Diego, Ca, USA.

[4] M. Bründel, U. Scholz, F. Haag, E. Graf, T. Braun, K.-F. Becker; Substratless sensor packaging using wafer level fan-out technology; Proceedings of EPTC 2012, Singapore.

[5] T. Braun, K.-F. Becker, L. Böttcher, J. Bauer, T. Thomas, M. Koch, R. Kahle, A. Ostmann, R. Aschenbrenner, H. Reichl, M. Bründel, J.F. Haag, U. Scholz; Large Area Embedding for Heterogeneous System Integration; Proceedings of ECTC 2010, Florida, USA.

[6] M. Woehrmann, T. Braun, M. Toepper, K.-D. Land; Ultra-thin 50 um Fan-Out Wafer Level Package: Development of an Innovative Assembly and De-bonding Concept, Procedings of ECTC 2018, San Diego, USA.

[7] T. Braun; K.-F. Becker; O. Hoelck; R. Kahle; M. Wöhrmann; L. Boettcher; M. Töpper; L. Stobbe; H. Zedel; R. Aschenbrenner; S. Voges; M. Schneider-Ramelow; K.-D. Lang; Panel Level Packaging - A View Along the Process Chain; 2018 IEEE 68th Electronic Components and Technology Conference (ECTC); San Diego, Ca., USA.

[8] A. Turchanin, A. Gölzhäuser; Carbon nanomembranes. Advanced Materials, 28(29), pp. 6075-6103, 2016.

[9] M. Woszczyna, A. Winter, M. Grothe, A. Willunat, S. Wundrack, R. Stosch, A. Turchanin; All-carbon vertical van der Waals heterostructures: Non-destructive functionalisation of graphene for electronic applications. Adv. Mater., 26, pp. 4831-4837, (2014, July 23.

[10] P. Reinecke, M.-T. Putze, L. Georgi, R. Kahle, D. Kaiser, D. Hüger, P. Livshits, J. Weidenmüller, T. Weimann, A. Turchanin, T. Braun, K.-F. Becker, M. Schneider-Ramelow, K.-D. Lang; Scalable hybrid microelectronic-microfluidic integration of highly sensitive biosensors; Proc. of 51st Symposium on Microelectronics, IMAPS 2018, Passadena, Ca, USA.

3D-MID Evaluation and Validation for Space Applications

Etienne Hirt
Art of Technology AG
Zurich, Switzerland
hirt@art-of-technology.ch

Klaus Ruzicka
Art of Technology AG
Zurich, Switzerland
ruzicka@art-of-technology.ch

Benedikt Wigger
Hahn – Schickard, Mikromontage
Stuttgart, Germany
Benedikt.Wigger@Hahn-Schickard.de

Maximilian Barth, Rafat Saleh,
Florian Janek
Hahn – Schickard
Stuttgart, Germany

Ernst Müller
University Stuttgart, IFM
Stuttgart, Germany
ernst.mueller@ifm.uni-stuttgart.de

Abstract— **The potential of 3D-MID technologies for Space applications is investigated within the ESA Artes 5.1 programme. Several test samples and test vehicles have been built to demonstrate assembly techniques, weight gain and have been characterised to achieve TRL 5. In this paper possible assembly processes for 3D-MID are shown and the results of environmental testing are presented.**

Keywords - 3D-MID, Space, metallised plastics, technology

I. INTRODUCTION

3D MID Technology combines electrical and mechanical functionality. Circuit traces are embedded onto plastic parts together with pads for component assembly. In order to investigate and demonstrate the capabilities the following tasks have been performed by Art of Technology within the ESA Artes 5.1 program (Contract 4000117360/16/UK/ND):

In a first step, an overview of the vast number of commercial 3-MID techniques was given together with a preselection on base materials and most promising technologies see [1].

Dedicated test plates have been developed to characterize mechanical, electrical and assembly properties like metal adhesion, shear strength and wire bond pull strength as well as track resistance with and without vias and RF properties.

Two space products have been selected as test vehicles to demonstrate design and assembly processes. The focus is on the integration (or better use) of mechanical, thermal, optical and electrical functions, to achieve small, lightweight and reliable solutions for space applications.

II. TECHNOLOGY AND MAterial Selection

Laser Direct Structuring (LPKF-LDS) [2] was identified as the most promising technology. It is a mature technology which allows fully three-dimensional parts. Several modern base materials including LCP and PEEK can be used, see

TABLE I. This is important as these materials are already in use within space environments. Besides that also low volumes can be produced cost-effectively. Commonly used metallisation like Cu/Ni/Au or other plating combinations such as Cu/Ag, Cu/Sn or Cu/Pd/Au can be realised. Typically, the layer system with an overall thickness of approximately 10 µm is used, which is significantly lower than on standard PCB's. For high power, electroplating can be used. Vias are holes formed in the injection moulding process, or laser drilled, followed by electroless plating. The main characteristics of this technology are:

- Lots of base material polymers from several suppliers
- Economic solutions for low to high production lot sizes by selecting proper mould tool materials (aluminium, steel). Reduced overall effort due to the integration of mechanics with electronics.
- Very flexible electrical design of circuitry on 3D substrates
- A high degree of miniaturization

TABLE I. TYPICAL LDS MATERIALS *OVERVIEW*

Material (Chem. Nature)	CTE XY/Z (ppm/°C)	Thermal Conductivity (W/mK)	Mass Density (kg/m³)
TECACOMP® LCP LDS black 4107 (LCP)	16/32 (50-100°C)	1.61/0.76	1730
TECACOMP® PEEK LDS black 3980 (PEEK)	15/25 (23-100°C)	0.95	1700
Al2O3 (reference)	8.4	35	3940
Arlon 45N (reference)	15 / 55	0.25	1750

III. ENVIRONMENTAL CONDITIONS

In order to assess the viability of the selected manufacturing process and base materials all test samples and demonstrators have been tested and validated to the space norms (ECSS) for reliability and electrical performance. Generic environmental conditions have been used for the characterisation as no specific mission requirements were specified.

- Accelerated ageing (Life Test). Storage at 125°C for 1000 hours with measurements taken at 0/250/500/1000h
- Thermal Cycle Test: 500 cycles between -55°C and +100°C, rate of temperature change 4 K/min with a dwell time of 30 min measurements taken at room temperature after 0/100/250/500 cycles.
- Sine Vibration (Demonstrators only): 25 to 100Hz; 20g and 100 to 200Hz, 15g; 1 octave/min.
- Random Vibration (Demonstrators only): Perpendicular: 20 to 100Hz, +6dB/oct; 100 to 500Hz, $1.0g^2$/Hz; 500 to 2000Hz -6dB/oct; Parallel: 20 to 100Hz, +6dB/oct; 100 to 800Hz, $0.5g^2$/Hz; 800 to 2000Hz -3dB/oct;; 5 minutes per axis
- Shock (Demonstrators only): Mechanical shock 1000g – 0.5ms.
- Radiation: Total Ionisation Dose (TID), Heavy Ion Test, Proton Test. Tests omitted as base materials known to withstand radiation at higher levels than electronic components.

IV. TEST STRUCTURES: TESTS AND RESULTS

This section describes the tests and the corresponding results performed on the test samples during applying the environmental conditions.

A. Mechanical Properties

As all moulded parts shrink, a manufacturer with process heritage is required in order to ensure geometric accuracy. Material flow and shrinkage can be simulated especially for designs with "constant" wall thickness. Once correctly designed the repeatability of moulded parts is excellent.

Material density and porosity is defined by the plastic used, by the parts geometry and the injection positions.

As shown in TABLE II. no degradation tendencies could be detected on material properties due to environmental testing.

TABLE II. TENSILE BAR TEST RESULTS

Measurement Average=4	Sample	Elastic modulus [MPa]	Rupture point [%]	Yield strength [MPa]	Tensile strength [MPa]
LCP, Life Test 0h	Avg	13753	1.09	88.9	28
	Std.Dev.	376	0.06	3.3	6.0
LCP, Life Test, 250h	Avg	12974	1.01	86.7	23
	Std.Dev.	565	0.10	6.3	6.8
LCP, Life Test, 500h	Avg	13084	0.99	87.4	25
	Std.Dev.	126	0.04	1.5	3.5
LCP, Life Test, 1000h	Avg	12783	1.07	92.7	27
	Std.Dev.	170	0.04	1.4	4.0
LCP, Thermal cycle test, 0cycles	Avg	13753	1.09	88.9	28
	Std.Dev.	376	0.06	3.3	6.0
LCP, Thermal cycle test, 100cycles	Avg	12923	0.95	83.8	19
	Std.Dev.	305	0.04	1.5	2.0
LCP, Thermal cycle test, 250cycles	Avg	13172	0.95	84.0	24
	Std.Dev.	403	0.08	1.4	3.7
LCP, Thermal cycle test, 500cycles	Avg	13317	0.94	86.7	22
	Std.Dev.	463	0.02	1.5	4.0

Surface roughness is a locally controllable parameter for LPKF-LDS technology, this is achieved by optimising the laser parameters. For tracks and SMD pads a rougher surface is desirable to increase metal adhesion, while for bond pads and RF tracks a smoother surface is preferable.

The layer thickness of the applied metallisation (usually Cu / Ni / Au) is in the range of 8 to 10 µm, with thicker tracks for high current applications being achievable via galvanic metallisation. Reliability of assembled "Boards" depends on the metallisation adhesion to the base material. A hot pin pull test is used instead of a peel test due to the thin metallisation which is not compatible with peel tests.

The adhesion on MID parts is slightly lower than on classical PCB materials as it can be seen in TABLE III. The most common base material for space grade PCB's, Arlon 45N was chosen for comparison. LCP values are lower but very well defined as this material provides smoother surfaces than PEEK. No degradation of hot pin pull test was found due to environmental tests.

TABLE III.   HOT PIN PULL TEST RESULTS

Measurement Average=20	Adhesive Strength (N/mm²)		
	Average	*Min*	*Std. Dev.*
Arlon 45N, untreated (avg=10)	35.05	---	4.97
PEEK, untreated (avg=10)	20.02	---	0.30
LCP, Life Test 0h	6.26	5.46	0.54
LCP, Life Test, 250h	7.15	5.52	0.71
LCP, Life Test, 500h	6.96	5.99	0.41
LCP, Life Test, 1000h	6.53	5.32	0.56
LCP, Thermal cycle test, 0cycles	6.44	4.34	0.58
LCP, Thermal cycle test, 100cycles	6.78	5.39	0.41
LCP, Thermal cycle test, 250cycles	6.89	5.95	0.50
LCP, Thermal cycle test, 500cycles	5.85	4.35	0.64

SMD shear strength is also related to metal adhesion. These results have more practical significance as space applications have to cope with shock and vibration. The results are compared to Standard 0603 and 0805 resistors documented in [3].

The shear test results in TABLE IV. before, during and after environmental tests match well with the hot pin pull test. Shear forces on PEEK are in the same range as for standard PCB's. This is expected as the show the failure mechanism of breaking of the solder joint. On LCP we found removal of pads, which corresponds to lower shear forces being measured and is coherent with the smoother surface compared to PEEK. There was no degradation of the shear forces during environmental tests.

TABLE IV.   SMD SHEAR TEST RESULTS

Measurement Average=10	Force (N)			
	0805		0603	
	Avg	*Std.Dev*	*Avg*	*Std.Dev*
PCB [3]	35 ~ 75		25 ~ 35	
PEEK	48.5	9.0	23.50	3.0
LCP, Life Test 0h	11.75	3.15	22.02	8.08
LCP, Life Test, 250h	11.65	3.54	14.95	2.05
LCP, Life Test, 500h	10.33	2.62	28.33	8.68
LCP, Life Test, 1000h	10.74	3.07	14.71	2.15
LCP, Thermal cycle test, 0cycles	10.33	2.68	21.77	5.44
LCP, Thermal cycle test, 100cycles	10.33	1.73	16.48	2.46
LCP, Thermal cycle test, 250cycles	13.12	4.97	21.13	6.80
LCP, Thermal cycle test, 500cycles	10.84	2.16	19.35	3.37

Wire bonding strength depends on the quality of the bond pads that were optimized by trimming the laser parameters. Pull test according to DVS 2811 were used to check the quality. The required minimum separation load is 40mN without lift-offs. As shown in TABLE V. all test bonds passed the requirements. LCP has a smoother surface than PEEK leading to higher separation loads.

While Al bonding is the state-of-the-art process, Au bonding was used for environmental testing as this is preferred for space applications in order to avoid issues with purple plague. Due to Al bonding heritage, the separation load standard deviation is quite low with values below 8%.

Although Au bonding is an uncommon process on 3D-MID, bonds showed higher separation load values but with a higher standard deviation of up to 25%, indicating the lower maturity of the process.

No degradation of shear force was caused during the environmental tests.

TABLE V.   WIRE BOND PULL TEST RESULTS (AU BONDING UNLESS MARKED OTHERWISE)

Value Average=25 each	Orient.	Avg Spearation Load (mN)	Std Dev	Min separation load (mN)	Lift Off
PEEK (Al)	↕	86.7	11.9	60.3	0
	↔	86.9	8.1	71.3	0
LCP (Al)	↕	110.3	7.8	74.4	0
	↔	93.7	5.8	77.6	0
LCP, Life Test 0h	?	129.17	18.39	74.86	0
	?	123.1	13.56	93.26	0
LCP, Life Test, 250h	↕	149.94	15.32	115.13	0
	↔	142.68	20.38	92.42	0
LCP, Life Test, 500h	↕	147.06	23.49	65.46	0
	↔	173.08	13.15	143.29	0
LCP, Life Test, 1000h	↕	125.69	22.89	72.09	0
	↔	137.92	7.84	116.64	0
LCP, Thermal cycle test, 0cycles	?	139.09	16.85	125.26	0
	?	168.89	22.91	112.69	0
LCP, Thermal cycle test, 100cycles	↕	136.50	27.43	68.01	0
	↔	202.70	22.07	125.95	0
LCP, Thermal cycle test, 250cycles	↕	300.82	74.09	142.47	0
	↔	184.51	21.12	147.94	0
LCP, Thermal cycle test, 500cycles	↕	178.25	44.72	90.67	0
	↔	148.65	16.82	107.55	0

B. Electrical Properties

LPKF-LDS allows fine structures, but together with the relatively thin metallisation track resistances can get significant. The resistance is minimized with a very uniform metallisation. The performance before, during and after environmental test is checked on the test plates Figure 1. featuring three different track widths directly connected or as via chain.

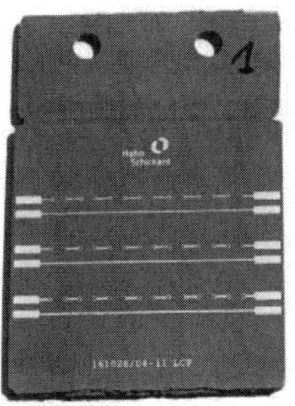 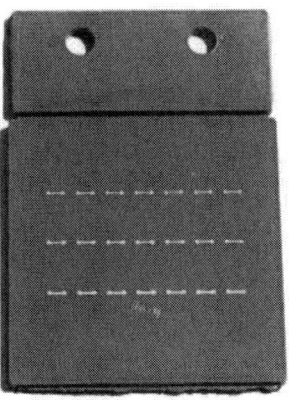

Figure 1. Ohmic Test Plate

Track resistance on LCP shows slightly lower values than on PEEK, but still higher than a conventional 8µm copper foil. This is an expected result as the metallisation roughness on 3D-MID is higher than on copper foil of a standard PCB. PEEK surfaces are less smooth than LCP, so the resistance is higher. Vias decrease the resistance marginally because of increase of metallisation cross-section.

As shown in TABLE VI. some vias degraded during thermal cycle testing. The cross-section in Figure 2. indicates possible reasons. The metallisation in the edge of the vias is thinner than the normal tracks and there is a relatively large CTE mismatch between copper (~16ppm/K) and the LCP in the out-of-plane direction (~23ppm/K) causing higher stress during thermal cycling.

TABLE VI.   AVERAGE TRACK RESISTANCE [OHM] TEST RESULTS

Structure Track length 29mm 14 Vias (mm) Average resistance [Ohm] of 2 test plates (max) values in brackets	200 µm		250 µm		300 µm	
	No	0.30 / 0.50	No	0.35 / 0.55	No	0.40 / 0.60
8 /35 µm thick PCB track calculation	0.31 0.07	--	0.25 0.06	--	0.21 0.05	--
LCP, Life Test 0h	0.55	2.10	0.45	1.00	0.4	0.50
LCP, Life Test, 250h	0.55	2.15	0.45	1.00	0.4	0.55
LCP, Life Test, 500h	0.55	2.25	0.45	1.00	0.4	0.60
LCP, Life Test, 1000h	0.55	2.45	0.50	1.05	0.4	0.65
LCP, Thermal cycle test, 0cycles	0.60	2.40	0.5	0.80	0.5	0.55
LCP, Thermal cycle test, 100cycles	0.60	2.60	0.5	1.05	0.4	0.65
LCP, Thermal cycle test, 250cycles	0.60	2.70	0.5	1.25	0.4	2.10 (3.6)
LCP, Thermal cycle test, 500cycles	0.60	2.75	0.5	1.8 (2.1)	0.5	4.70 (7.8)

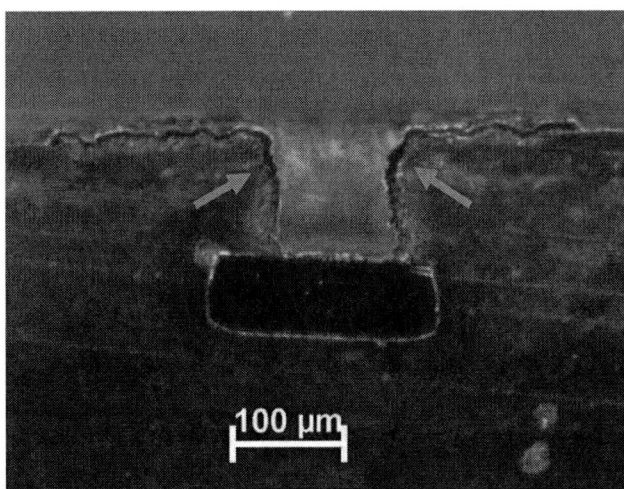

Figure 2.   Sun Sensor cross-section: Via to Connector Pin.

RF properties were tested using a non-grounded co-planar waveguide structure. For the measurements a Keysight N9917A was used. The spatial resolution of 3mm is relatively low but still sufficient to demonstrate usability. The measurement results show a distance between instrument and connector of the test sample of ~55mm and an RF track of about 36mm. These results correspond to the physical setup as shown in Figure 3.

Figure 3. RF / Isolation Test Plates

The impedance matching at the border between the connector and MID still shows a return loss of 10dB after the environmental tests which should be optimised for a RF design. This is not a 3D-MID specific or degradation issue. Figure 4. shows a typical result after 500 thermal cycles between -55°C and +100°C.

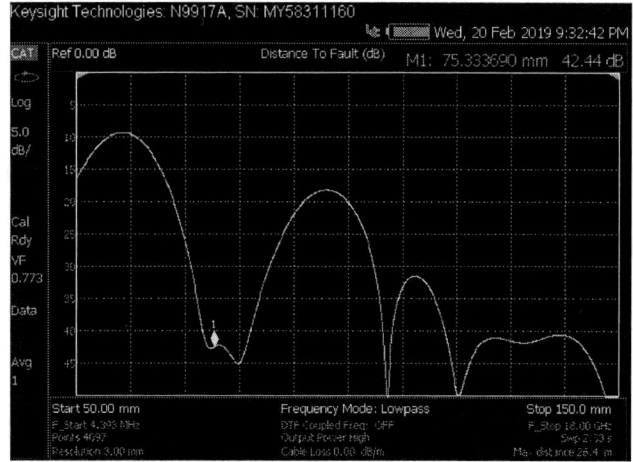

Figure 4. Distance to fault (Return Loss) for RF Test Plate, after 500 thermal cycles

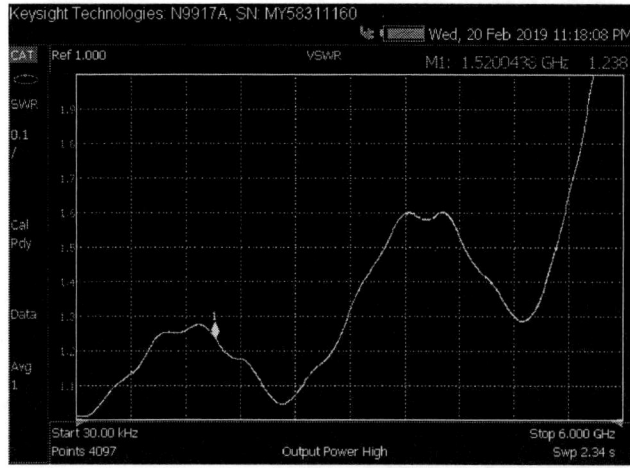

Figure 5. RF Test Plate VSWR up to 6GHz

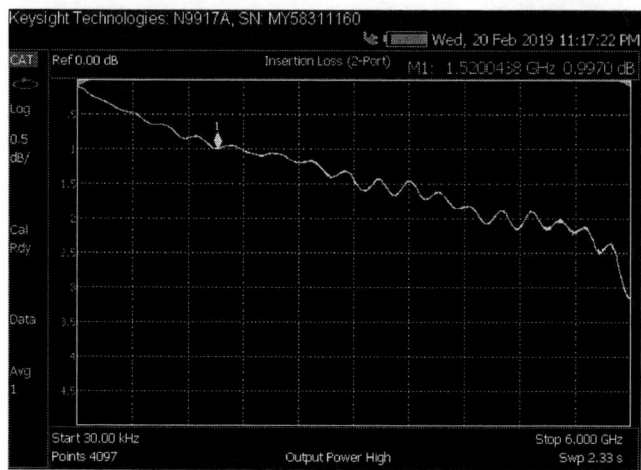

Figure 6. RF Test Plate Insertion Loss up to 6GHz

The VSWR is excellent, < 1.35 up to 3GHz and < ~1.8 from 3 to 5GHz, which is a very good result especially for the design based only on datasheet properties. The insertion loss at 3GHz for two connectors and the co-planar waveguide in the MID is between 0.85dB and 1dB.

The RF structure was also used for isolation measurements. 50V has been applied to the structure with a 150µm gap. All results showed isolation resistance greater than 1TΩ which is well above requirements. No degradation of isolation found during environmental tests.

C. Other relevant Properties and Tests

1) Radiation
LCP and especially PEEK are highly resistant against radiation up to 5000 Mrad, which is much higher than acceptable for electronic components.

2) Outgassing
LCP and PEEK are already used in space applications, the required additive is non-organic and is not expected to contribute to the outgassing performance. Literature values are well below requirements. Fully processed samples will be tested at the ESTEC (ESA) facility as soon as a time-slot is available.

3) Thermal Conductivity
PEEK shows a thermal conductivity of about 0.95 W/mK and LCP is with up to 1.61 W/mK even better. Compared to 0.25 W/mK for the standard PCB material Arlon 45 this is an improvement by a factor of 2 ~ 5.

978-1-7281-1500-9/19 $31.00 © 2019 IEEE

4) CTE / Thermal Cycling

The in-plane CTE is in the same range or slightly better than standard PCB material compared to ceramic materials which are used as carrier for electronic components. This reduces thermally induced stress for mounted components during temperature changes while the better thermal conductivity keeps the temperature change lower.

Out of plane CTE for LCP differs in comparison to the CTE of copper causing higher stress for vias.

5) UV resistivity

UV resistivity is an important material property as exposure to UV radiation may make parts brittle. LCP and PEEK (in absence of oxygen) have good UV performance.

V. DEMONSTRATORS

Two test vehicles were selected to demonstrate the technology, production methods and assembly features. Only LCP was selected to reduce cost and processing time for demonstration. LCP was preferred to PEEK as it features smoother surfaces and fine structures.

A. Sun Sensor

The sun sensor is based on a design from Lens R&D, Netherlands. The conventional design has an aluminium case with press-fit connectors. Its components are mounted on thick film ceramic substrate that is bonded into the housing. The field of view is defined by a patterned cover window requiring optical alignment. Weight is approximately 22 grams.

Figure 7. Sun Sensor Conventional Design.

The 3D-MID design was produced using a simple two-piece mould designed to ensure a "constant" wall thickness. The test vehicle was designed allowing the laser writing without workpiece position change.

A standard SMT connector is inserted upside down into the moulding tool; the housing and solder pins are completely integrated into the thermoplastic by the moulding process. For the electrical connection, two laser drilled vias per pin connect to the overmoulded SMT pins. As the laser activates the material during drilling, the electrical connection is created during electroless metallisation. The connector is locked by moulded hooks instead of screws. During testing, the hooks were removed to ease mating and de-mating of test cables.

The body is a single plastic part with a weight of 0.5 gram and the total weight is only 5.6 grams.

The photodiode is mounted with conductive glue and Au wire bonding is used for quadrant connections. Conductive bonding is used for the SMD components. Consequently, the demonstrator allows the investigation of several assembly technologies.

The field of view is uniquely defined by the moulded housing walls around the photodiode and not by the cover glass, so the accuracy is defined exclusively by the placement of the photodiode onto the moulded part and avoids the need for window alignment and structuring.

Figure 8. 3D-MID Sun Sensor.

All Sun Sensors were operational after environmental testing without functional degradation. The normalised photocurrent ratio between the illuminated and the shadowed areas on the photodiodes (measurement principle) did not degrade. Same is valid for the angle where the illumination ratio is at a maximum (equal to the field of view) as well as the shape of the peak.

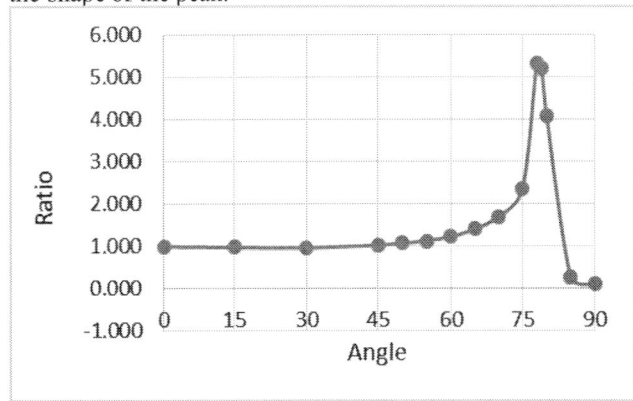

Figure 9. 3D-MID Sun Sensor: Typical Normalised Photocurrent Ratio between Illuminated and Shadowed Areas vs Angle to Light Source

Vibration and shock tests indicated that the mounting structures are critical as both parts showed small cracks in the walls around the mounting holes.

TABLE VII.   SUN SENSOR FUNCTIONAL TEST RESULTS

	Ratio	Angle [°]	-3dB [°]
Life Test 0h	5.25	78.5	1.86
Life Test, 250h	5.24	78.4	1.94
Life Test, 500h	5.36	78.7	1.84
Life Test, 1000h	5.44	78.9	1.73
Life Test Median	**5.39**	**78.6**	**1.85**
Life Test Std. Deviation	**0.085**	**0.222**	**0.087**
Thermal cycle test, 0cycles	5.48	78.7	1.75
Thermal cycle test, 100cycles	5.24	78.8	1.58
Thermal cycle test, 250cycles	5.39	78.7	1.62
Thermal cycle test, 500cycles	5.36	78.7	1.79
Thermal cycle test Median	**5.375**	**78.7**	**1.685**
Thermal cycle test Std. Deviation	**0.062**	**0.05**	**0.101**
Vibration Initial	5.09	79.0	3.61
Vibration After Sine Survey	4.88	78.9	3.46
After Sine Vibration	4.67	78.9	3.31
After Random Vibration	4.61	78.9	3.27
Vibration Median	**4.775**	**78.90**	**3.385**
Vibration Std. Deviation	**0.218**	**0.05**	**0.155**
Shock Initial	5.31	78.5	1.73
After Shock	5.67	78	1.75
Shock Median	**5.490**	**78.25**	**1.74**
Shock Std. Deviation	**0.255**	**0.354**	**0.014**

The suns sensors functionally also survived vibration and shock tests. Especially all bond wires are in place, see green marks in Figure 10. The first Eigenfrequency is above 1000Hz as shown in Figure 11. However, some structural issues have been observed. Washers to protect the mounting area are mandatory, otherwise the screw head will damage the structure. But even with that protection one crack occurred in the material around the mounting hole during the shock tests, see the red mark in Figure 10.

Figure 10. 3D-MID Sun Sensor after Shock Tests.

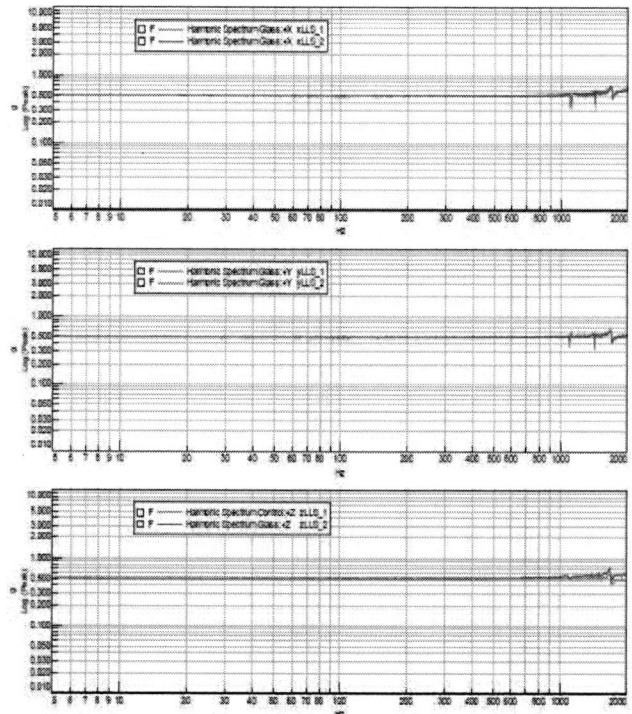

Figure 11. Sun Sensor Comparison of Sine Survey Results, red: before Shock, blue: after Shock

A sine survey before and after vibration and shock tests indicates the mechanical issue, best seen in Figure 11. Z-axis curve right above 1000Hz.

B. Helical Antenna

The helical antenna is based on a design from Tryo Aerospace, Spain. As shown in Figure 12. their original design uses an extruded tube screwed to an aluminium base plate. They bonded the helix conductor onto the tube or melted it into the tube. As this is an experimental antenna, no matching network is provided. Total mass is 450 grams.

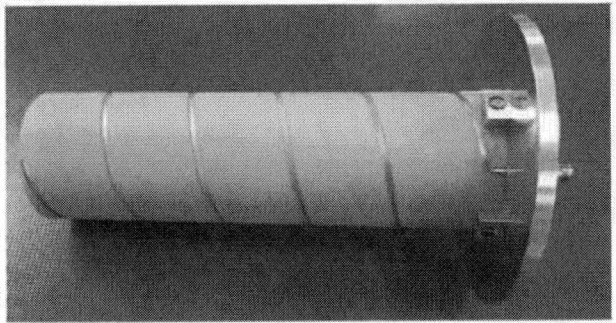

Figure 12. Monofilar Helix based on Peek (ESA Contract No. 4000114254/15/NL/CBi/fk)

Figure 13. 3D-MID Helical Antenna.

The 3D-MID solution consists of two parts. A tube with a draft angle of 1 - 2° which is soldered to a small plate to ensure a closed ground plane. The helix is metallised directly to the tube. To be equivalent to the reference design no matching network is implemented but it would be possible by updating the laser structuring and placing the required components on the shoulder where the logos are located. The solution reduces the mass to 114 grams.

The mechanical construction provides "constant" wall thickness and avoids mould seams in the metallised area. The injection point is on top of the structure, therefore weld lines in the metallised area cannot be avoided. The part is quite large for MID processing and helix writing with the laser is challenging. To minimise laser time and reduce the risk of unwanted debris deposition, the ground plane could be implemented with a hatched structure.

The first part consists of a cone with ~2° draft angle with the brim, the second part is a plate only. The closed bottom metallisation is important for the antenna characteristics. All other approaches discussed need either more complex mould tools, have seams in the metallised area, or the differences in the wall thickness are too high.

This intention of this test vehicle is to demonstrate the processing of big components and the RF performance in the Gigahertz range.

VI. SUMMARY AND CONCLUSION

3D-MID is a very interesting technology for use in space applications.

A. Technology

- Extreme lightweight products are possible compared to state-of-the-art designs. Up to 75% weight savings demonstrated.

- LPKF-LDS is a mature technology and fits all requirements for RF structures and sensor electronics.

- Several high-end plastics can be used with the LDS technology, LCP and PEEK are the most promising as there is already an existing space heritage.

- Assembly effort can be reduced as mechanical and electrical functions are combined. State-of-the-art sun sensor requires three function relevant alignment steps – Photodiode to the carrier, carrier to the case, structured window to the case (photodiode). The MID sun sensor requires only one alignment step – photodiode to 3D-MID body, the window is only a protective cover and not function relevant. Even the assembly of the helical antenna is simpler as no base plate has to be mounted.

B. Electronics

- 3D-MID is best suited for low complexity electronics. Single layer or double layer assembly is possible if restrictions on via geometry are followed. In case of via use, double vias are recommended to enhance reliability during thermal cycles.

- The frequency range is low power DC up to several Gigahertz. High power DC tracks are also possible with electrochemical plating.

- All commonly used assembly technologies such as soldering, glueing and wire bonding can be used:

  o Compatible with standard surface mount and through-hole components

  o Component placement can be done either manually or by means of a 6-axes placement machine.

  o For wire bonding the Al wire is preferred on 3D-MID due to the lower process temperatures: However, Au bonding successfully passed all of the environmental tests. Some effort is still required to improve the stability of the Au bond process (much higher variation on separation load, TABLE V.).

C. Injection Molding and Laser Structuring

- Optimise the mechanical design such that a two-piece moulding tool is sufficient in order to minimize production cost.

- The mechanical restrictions on size (for moulding tool and laser process) depend on the manufacturer's capabilities.

- Shrinking of the moulded parts needs to be calculated and/or simulated by the manufacturer, but the repeatability between the parts is very high.

- To obtain clean moulded parts, no "sharp" edges allowed. 1° - 2° draft angle is required to allow the moulded parts to be ejected from the moulding tool.

- "Constant" wall thickness is required for clean moulds. Otherwise, depressions in surfaces will occur when parts cool down after moulding.

- Complex (e.g. over the edge - helical antenna) require high effort on laser structuring as the part has to be repositioned to reach all areas to be metallised. Very large areas increase the laser processing time.

- Avoid injection moulding seams where metallisation is required. A clean and continuous metallised track cannot be guaranteed in such areas. This is the reason for the conical form of the helical antenna. No mould tool split is required in this area and therefore no disturbing weld lines.

- Two electrical layers are possible if the substrate thickness is less than ~1.5mm. This is a restriction for laser-drilled vias only. Alternatively, mould tool defined vias or through-hole components can be used for plane change of tracks for thicker substrates.

- Electroless metallisation is typically 8~10μm. High current tracks need an additional electrochemical copper process.

- Large metallised areas in combination with fine structures should be avoided. They might be distorted by the debris of the large areas.

- Large areas (ground planes) can be hatched to reduce laser time, there is no significant weight gain due to the thin metallisation. Adhesion is better for non-structured planes.

- Mechanically stressed sections are more critical than those in conventional designs. Mounting structures (holes) have to be designed to minimise the risk of damage during integration.

D. Change Possibilities

Electrical tracks can be individual for every part as no masks are required. Mechanical changes have to be small, otherwise a new mould tool has to be designed and built.

VII. RECOMMENDATIONS

- Adherence to the design requirements for moulded parts is essential. Additionally, weld lines should be avoided in metallised areas

- Vias may degrade due to the thin metallisation especially due to thermal cycling. Use more than one in critical paths. This is not a severe issue as the complexity of electronics is usually low and, in most cases, only single sided assembly is used. To add redundancy, the sun sensor uses 2 vias each for contacting the connector pins.

- As the metallisation is usually relatively thin, the track resistance needs to be considered in electronic design.

- The material around mounting holes is critical. Weld lines with lower mechanical stability occur due to the material flow around the hole during the moulding process. This can cause cracks under mechanical stress, the profile of 2 x 2mm around the holes was not sufficient. Either increase the profile or use an insert applied directly to the mould.

- The design goal was technology readiness level (TRL) 5 - component and/or breadboard validated in simulated or real-space environment. All relevant tests have been performed, acceptance currently under negotiation.

ACKNOWLEDGEMENT

In addition to the co-authors, we would like to extend our gratitude to the following companies for their support with the test vehicle design, production and verification.

- Multiple Dimensions, Switzerland
  - 3D-MID production in Biel, Switzerland
  - Technology research
  - http://www.multipledim.com/

- Lens R&D, Netherlands
  - Optimising high reliability (electro-optical) sensors and systems for volume fabrication
  - Requirements and support for the Sun Sensor Demonstrator
  - https://lens-rnd.com/

- Tryo Aerospace, Spain
  - Design and manufacturing of communication equipment for satellite communication
  - Requirements of the Helical Antenna
  - https://www.tryo.es/

GLOSSARY

3D-MID   Three-Dimensional Moulded Interconnect Device

CTE   Coefficient of Thermal Expansion

ECSS   European Cooperation for Space Standardization

LCP   Liquid Crystal Polymer

LDS   Laser Direct Structuring

LPKF   LPKF Laser & Electronics AG

PEEK   PolyEther Ether Ketone

RF   Radio Frequency

TDR   Time Domain Reflectometry

TRL 5   Technology Readiness Level 5: Component and/or breadboard validation in relevant environment,

VSWR   Voltage Standing Wave Ratio

REFERENCES

[1] Etienne Hirt, Ruzicka, Klaus. 3D-MID for Space. 1-8. 10.1109/ESTC.2018.8546449, 2018.

[2] LPKF LDS: Laser Direct Structuring for 3D Moulded Interconnect Devices *https://www.lpkf.com/en/industries-technologies/electronics-manufacturing/3D-MIDs-with-laser-direct-structuring-lds/ (fetched Feb 2019).*

[3] Transactions on Engineering Technologies: World Congress on Engineering 2014, p. 43

High-Temperature Pressure Sensor Package and Characterization of thermal stress in the assembly up to 500 °C

Nilavazhagan Subbiah, Qingming Feng,
Juergen Wilde
Laboratory for Assembly and Packaging Technology
Department of Microsystems Engineering
University of Freiburg, Germany
subbiah@imtek.uni-freiburg.de

Gudrun Bruckner
CTR AG, HIT
Villach, Austria
Gudrun.Bruckner@CTR.at

Abstract— High temperature stable (500 °C) pressure sensors are required in various fields like aerospace, automobile and manufacturing industries. However, reliable sensors working at such high temperatures are still not sufficiently developed [3]. Mainly, developing a high temperature stable package imposes new challenges due to thermal cross-sensitivity and temperature induced stresses [2]. Important challenges are to identify stable materials at high temperatures and stress-tolerant sensor mounting techniques.

This current research work will mainly focus on the implementation of a high temperature (HT) concept for applications up to 500 °C as follows: A micro strain gauge is deposited and patterned on a Langasite (LGS) crystal. It is attached to a ceramic substrate (Al_2O_3) like a cantilever by flip-chip interconnection and glass solder underfill. The ceramic substrate has a membrane structure that was fabricated by ultrasonic machining. The deflection of the deforming membrane is transferred pointwise to the free end of the crystal inside the package using a quartz spacer. The strain produced on the cantilever is measured by the change of resistance of a microstrain gauge. This special design concept aims at the elimination of thermal stresses between membrane and the sensing element.

Despite the effort taken to minimize stresses in the assembly, the die attachment area will however experience thermal stress due to the mismatch of Thermal Coefficient of Expansion (TCE) of the different materials employed. Hence, it is necessary to evaluate the development of stress along the sensing element to prove the stress tolerant design of this sensor. Characterizing thermal stresses could be used to compensate for the offset signal and to optimize the sensor design. This research work will establish a method to develop a pressure sensor stable for operations up to 500 °C and also a final implementation of this concept is shown. This concept has potential for the generation of very compact wireless sensors working autonomously in high-temperature environments.

Keywords-flip-chip, ceramic, high-temperature, glass-solder, Pressure-sensor

I. INTRODUCTION

A concept for the stress tolerant pressure sensor design is already investigated and a low temperature stable (200 °C)

version of this concept was implemented in previous research work [2]. To make the die attachment stable at high temperatures up to 500 °C, modifications are performed on the first implementation by changing the sensing element (die) material to LGS and under filling material to glass solder [3]. This stress tolerant design is realized by the LGS cantilever structure flip-chip die attached to the ceramic substrate using gold stud bumps produced on the LGS sensing element. This die attachment is under filled with glass solder to improve the strength of the interconnection which also relieves stress under thermal load.

The cantilever sensing element is free to expand along its length; this should compensate for the stresses that arise in the assembly. Such high operating temperatures (500 °C) impose challenges in terms of materials and manufacturing technology.

The main focus of this research work is to combine the earlier developments and to produce a final version of the sensor that is stable for operations at 500 °C. Additionally, the influence of thermal stresses in the sensor assembly must be characterized.

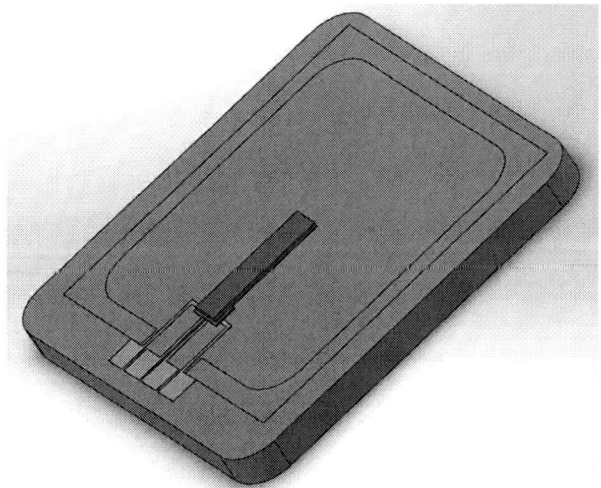

Figure 1. CAD model of the HT pressure sensor showing cantilever based sensing element die attached to the contact pads on the substrate

This paper is divided into three major sections. In section II, the concept of the die attachment and the whole package is discussed with the relevant material choices and fabrication techniques. In section III, implementation of this high temperature assembly technique is summarized and the results are discussed. In section IV, the characterization of thermally induced deformation in the die attachment area is presented. Additionally, the sensor assembly is subjected to heat cycle treatments and the stability of the interconnections is tested after several cycles and the strength of the die attachment is evaluated using shear tests.

II. CONCEPT OF THE HT SENSOR

The main aim of this pressure sensor concept was to minimize the influence of thermally induced stresses. In most sensors the cross sensitivity is introduced by the stress developed in the die attachment area. In this design the attachment area of the sensor was kept to a minimum. Therefore, a bending cantilever with a fixed end and a loaded movable end is used for the transducer. Since the cantilever is free to expand along its length, this should compensate for the stresses that arise in the assembly. A substrate with a deformable membrane is used as pressure transfer mechanism. Force that is generated by the deformable membrane under external pressure is transferred to the sensing element by a spacer attached on the lower side of the membrane. This spacer has the shape of a cylindrical rod. It is attached under the membrane and touches the top surface of the sensing element. When the sensing element is strained by bending, we measure a change in the sensor signal [2].

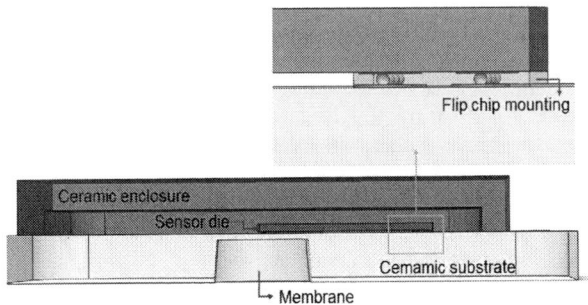

Figure 2. Assembly concept of the HT pressure sensor.

The selection of materials for each individual part of the assembly was established in previous research work [3]. These materials should be closely matched with respect to the Thermal Coefficient of Expansion (TCE) to minimize the thermal stresses. An alumina ceramic is used as the base substrate as well as for the cap with the membrane structure. Langasite (LGS) is chosen for the sensing element as it is well established for operations at higher temperatures. A flip-chip die attachment with gold stud bumps and glass solder under-fill material were investigated for reliable

operations up to 500 °C under similar conditions [4]. A quartz fiber rod is employed as spacer between the

deforming membrane and the free end of the sensing element. The glass solder is used to relieve the thermal stresses and increases the strength of the die attachment. Glass solder is also available with varying TCE and can be conveniently chosen based on the TCE of the substrate and the die. Glass solder is also used as a high temperature stable adhesive to fix the ceramic enclosure and the quartz spacer under the membrane. These material choices are well tested for stable operations at high temperature and in corrosive environments.

III. IMPLEMENTATION OF THE HT PRESSURE SENSOR

A. Manufacturing of Assembly Parts

To successfully realize the complete assembly of the sensor, the individual parts of the assembly must first be fabricated. This assembly consists of the following parts as shown in Figure 2: A ceramic base plate, a ceramic enclosure, a sensing element and a quartz fiber spacer. The ceramic base plate and the enclosure are structured using an ultrasonic machining process. This is a mechanical ablation process with a tool vibrating at ultrasonic frequency and diamond abrasive which removes the ceramic materials. The base plate is structured with the membrane and the markings for the placement of the spacer on the inner side of the substrate. Additionally, markings for the location of the external contact electrodes and the placement of the bottom enclosure are done using an UV marking laser. The sensing element is patterned with strain gauge structures using standard thin-film and photolithography processes [3]. The strain gauge is formed by depositing three metal layers of Cr-Au-Pt (20 nm - 30 nm - 30 nm) and structuring them using a lift-off process.

The quartz spacer is produced from commercially available optical fibers. The fibers are stripped from their acrylic encapsulation. The resulting fiber is left with standard cladding diameter of 120 µm. This cylindrical fiber cables are then cut into rods of 2 mm length and used as spacer in the assembly.

B. Integration of the Assembly Parts

Once the parts are manufactured, the assembly is performed by attaching the parts together using appropriate techniques. The ceramic base plate is first printed with contact electrodes using a thick-film screen printing process. Ag-Pt thick-film paste C1076SD (LPA609-022) from *Heraeus Electronics* is used to print these tracks. This paste is specified to withstand temperatures up to 800 °C. The position of the printing is set using the alignment marks on the substrate. After the paste is printed on the alumina, it is dried at 150 °C for 15 minutes, and then successively cured at 850 °C for another 15 min [3].

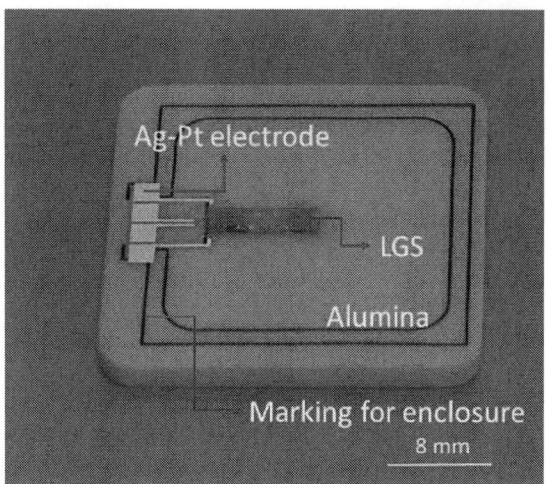

Figure 3.   Flip-chip die attached LGS based strain gauge on Al₂O₃.

To attach the quartz rod onto the substrate, glass solder is dispensed in the area of the marking made for the spacer. Then the quartz rod is placed on top of the dispensed glass solder. It is then cured at 780 °C for 15 min [3].

Figure 4.   Cross section of the flip-chip die attachment showing Au stud bumps

Once the substrate is ready with the electrical contact pads and the spacer, the flip-chip die attachment is done on the substrate to mount the sensing element. This flip-chip die attachment defines the space between crystal and the membrane and also acts as the electrical interconnection to the sensing element on the chip. A bump with approximately 50 μm height and very small tail length is desired. A 50 μm thick gold wire is used to create the bumps in a Delvotec 5410 ball wedge bonder. The stud bumps are generated on the contact pads of the sensing element through ball-wedge wire bonding. Thermosonic pressure is applied to die attach the LGS sensing element with stud bumps to the Ag-Pt contact pads printed on the ceramic plate. To provide strength to the die attachment, after flip-chip bonding glass solder is used as under fill.

A ceramic enclosure that was manufactured by Ultra-sonic machining is then attached to the substrates using glass solder. The lid acts as the protection against harsh environment and also as a mechanical support for the deforming top plate.

C.   Final sensor package

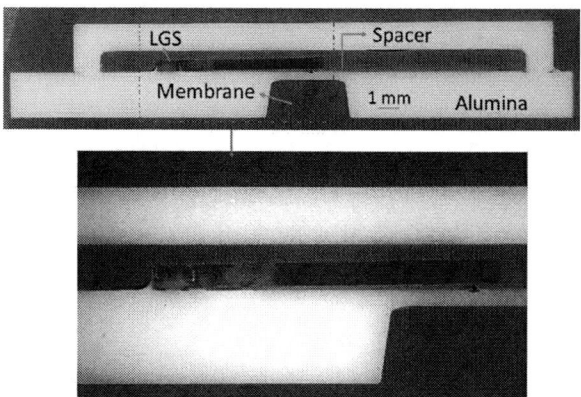

Figure 5.   Cross section of the integrated assemlby showing all assembly parts.

A cross section of the encapsulated assembly is shown in Figure 5. It shows the flip-chip die attached LGS cantilever slightly having a slope towards the free end of the cantilever due to the difference of height of the flip-chip stud bumps and the quartz spacer. However, this slight elevation at the free end helps the spacer to be always in contact with the cantilever. In this package the positioning of the different parts of the assembly is important in order to realize this concept. The cross section shows that the spacer is attached in the middle of the membrane structure and has contact to the cantilever as per the design. Figure 4 shows the cross section of the flip-chip die attached area. The glass solder under-filler is pulled into the small gap between the substrate and the LGS sensing element. Although the distribution of glass solder is not even, it can still help to improve the die attachment strength. It also shows that the stud bumps are in good contact with the chip and substrate metallization.

IV.   CHARACTERIZATION

In this sensor, the LGS cantilever is attached on one end using flip-chip interconnection. This die attached area is of high interest to estimate the influence of thermal stresses. In the previous study, a Digital Image Correlation technique (DIC) was used to estimate the deflection of the free end of the cantilever [3]. This is designed to expand unconstrained at elevated temperature, thereby minimizing the stress on the sensing element. The DIC measurements confirmed that the cantilever is expanding freely and its stress tolerant behaviour was proven [3]. However, the die attachment area at one end of the cantilever is fixed and it cannot freely expand.

In this research work, the deformation on the surface of the die attach area is measured using DIC technique. This die attachment is approximately 2 x 1.5 mm². The expansion of the LGS crystal along its width in Y direction and expansion in the Z direction (out of plane) is measured. Any strain on the die attachment area is an indicator of stress. In order to perform this experiment, samples are patterned with a stochastic pattern as shown in Figure 6. The yellow box represents the die attach area and the red dots mark the segment taken for investigation along the Y direction.

A. Deformation along Y-axis

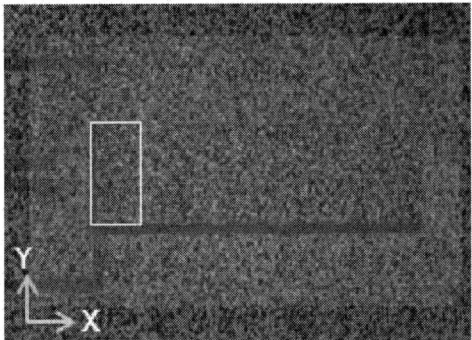

Figure 6. Flip-chip sample coated with stochoistic pattern for DIC measurements.

The LGS crystals used as sensing elements have anisotropic thermal expansion properties. A cut with an Euler angle (0°, 138.5°, 26.6°) is used in this assemlby design. The dimensions of the LGS crystal are 8 x 2 mm². This sample is placed in a heating chamber under vacuum. The temperature is increased in 50 K steps and the images are captured in the DIC measurement system. For the analysis, the expansion of the chosen segment along the Y-axis is measured and recorded in Figure 7. There is a total expansion of ~ 7 μm in this direction at 500 °C. It also shows that the expansion increases linearly with the temperature.

B. Deformation along Z-axis (Out of plane)

The out of plane deformation along the analysis segment in the Y-axis is measured. During this measurement the movement of the substrate is fixed and only the expansion of the crystal alone is considered. Figure 8 shows that the LGS crsytal expands in the z-direction (out of plane) along the segment taken in Y-axis. The measurement also shows that the expansion increases with temperature and a maximum of ~3 μm is recorded at 500 °C.

In both these measurements, no warpage is observed. It proves that there is no bending stress on the surface of the die attachment. However, further investigations have to be performed with white light interferometer to confirm whether there is any residual stress in the assembly.

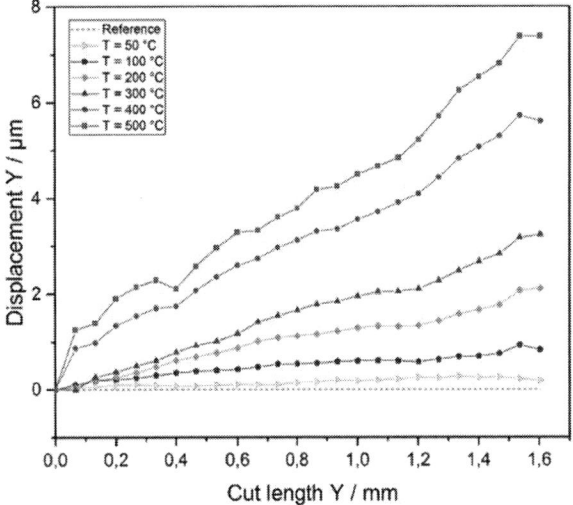

Figure 7. Expansion of the die attachment measured in the Y - direction. strength of the die attachment

C. Strength of the die attachment

The concept of bending cantilever is designed to have a small fixed area and a larger movable area. This small attachment area has to be stable to take the necessary load. The strength of the flip-chip die attachment can be characterized using shear tests. To perform these experiments, test samples were prepared resembling the original die attachment. The objects are divided into two groups, and each group had four samples. The four samples in Group 1 were directly subjected to the shear test. The other four samples in Group 2 were firstly treated with 10 heating cycles from room temperature to 500 °C before the shear test.

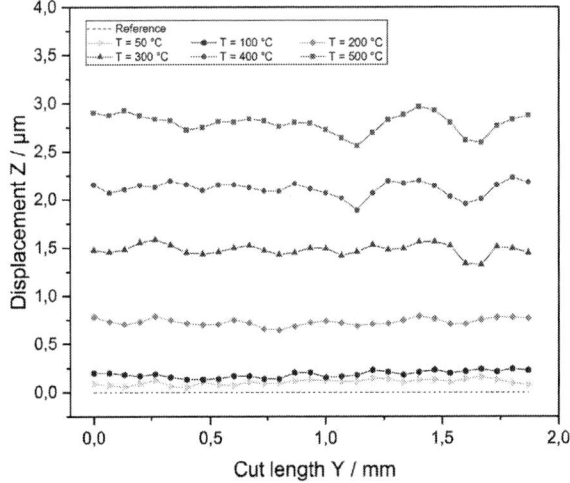

Figure 8. Expansion of the die attachment measured along the Z-direction. The off-set was subtracted at the surface of the substrate

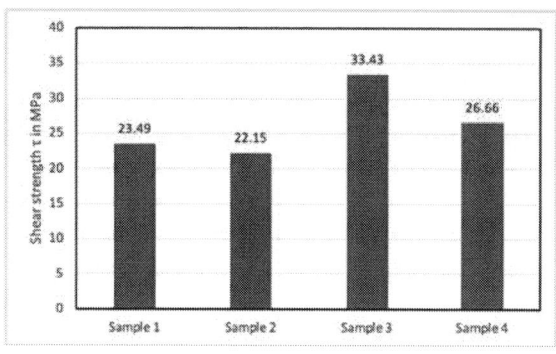

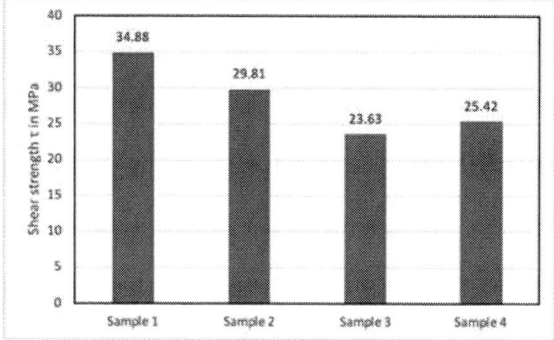

Figure 9. Measured shear strength for the flip-chip die attached with glass solder underfill before and after the heat treatment

The shear strength τ was determined by the following equation

$$\tau = F / A$$

Where, F was the loaded shear force recorded by the data logging system, A is the area of the die attachment.

$$A = W * L$$

W=2mm was the width of the LGS chip, L was defined as the length of the die attached area, which was measured on the microscopic image of the shear test broken sample. The shear test results are shown in Figure 9 . The mean value of the shear strength of Group 1 was 26.5 MPa, and the mean value of the shear strength of Group 2 was 28.4 MPa.

These values are sufficient for MEMS based die attachment technique. The investigation after shearing showed that the samples broke in the glass solder - stud bump layer in all cases. As observed, the adhesion of the screen printed electrode remained intact during the shear tests. There is no significant change in the mean value of shear strength before and after the heat treatment. This die attachment has good resistance to thermal load and it could be used realibly for operations at 500 °C.

V.   CONCLUSION & OUTLOOK

This research work presents the concept of a pressure sensor package for applications up to 500 °C. In this approach, ceramic substrates are used as active membrane structure and enclosure. Glass solder is used to attach the ceramic parts and the spacer between the membrane and the cantilever. High temperature resistant metals like gold, platinum and silver are used for interconnection and metallization. All these parts are integrated and a first prototype assembly is produced.

This sensor assembly must be characterized further for its behavior under operation up to 500 °C under pressure load. The effects of temperature on the sensor signal must be separately determined and compensated. The influence of temperature on the assembly over time must be estimated by subjecting the assembly to longer hours and by thermal cycles. In the future, resistive based sensing element can also be replaced with SAW based delay line element. This will enable us to measure the pressure wirelessly.

ACKNOWLEDGMENT

"This project has been supported by the COMET K1 centre ASSIC - Austrian Smart Systems Integration Research Center. The COMET – Competence Centers for Excellent Technologies-program is supported by BMVIT, BMWFW and the federal provinces of Carinthia and Styria."

REFERENCES

[1] Subbiah N, Ghosh S, Zeiser R, Wilde J, Comparison of Packaging Concepts for High-Temperature Pressure Sensors at 500 °C. 2017 IEEE 67th Electronic Components & Technology Conference (ECTC), Orlando, FL, USA, DOI: 10.1109/ECTC.2017.255.

[2] Subbiah N, Ramirez K.A.B, Wilde J and Bruckner G, Concept and Implementation of High-Temperature Pressure Sensor Package up to 500 °C, 2018 19th International Conference on Sensors and Measurement Technology (AMA Conferences - SENSOR 2018), Nuremberg, Print ISBN: 978-3-8007-4683-5

[3] Subbiah N, Feng Q, Ramirez K.A.B, Wilde J and Brucker G. Flip-chip Die Attachment for High-temperature Pressure Sensor Packages up to 500 °C. 2018 IEEE Electronic System-Integration Technology Conference (ESTC), Dresden, Germany, DOI: 978-1-5386-6814-6/18.

[4] Binder A, Bardong J, and Bruckner G, Design and Characterisation of a Combined Pressure, Temperature, ID Sensor for Harsh environment, Proc. IEEE International Ultrasonics Symposium, IEEE Press, 2014, DOI: 10.1109/ULTSYM.2014.0374.

[5] Zeiser R, Mikromechanische Drucksensoren mit spezifischer Aufbautechnik für den Einsatz bei 500 °C, Dissertation, University of Freiburg, 2015, ISBN-10: 3862475441 (3862475441), ISBN-13: 978-3-86247-544-5 (9783862475445).

[6] Zeiser R, Ayyub S, Hempel J, Berndt M and Wilde J, Mechanical Stress Analyses of Packaged Pressure Sensors for Very High Temperatures, Journal of Microelectronics and Electronics Packaging, (2014), pp.30-35, DOI:10.4071/imaps.399.

Development of 3D WLCSP with Black Shielding for Optical Finger Print Sensor for the Application of Full Screen Smart Phone

Daquan Yu*, Yichao Zou, Xirui Xu, Aihua Shi, Xiaobing Yang, Zhiyi Xiao

Huantian Technology (Kunshan) Electronics Co., Ltd., 112 LongTeng RD, Kunshan, Jiangsu Province, China

Email: daquan.yu_ks@ht-tech.com

Abstract—As an important biometric technique, fingerprint verification is widely used for smartphone, IoT and payment due to the small size and high security. To meet full screen trend of smartphone, optical finger print sensor (FPS) using CIS received great interest and gradually replaced capacitive FPS. For such a device, a thin package size with better image performance is required. To meet these critical requirements, 3D wafer-level chip scale package (WLCSP) with via last TSV (through silicon via) was the best solution for CIS.

In this paper, 3D WLCSP for FPS based on CIS with via last TSV interconnects was developed. For the optical FPS, the package size is 920×656μm with 580μm height, which includes 200μm thick glass, ~100μm thick optical layer, ~100μm thick CIS chip and BGAs of 80μm height. Electrical interconnections are implemented by via last TSVs, where big silicon trenches were formed on the backside of the two-row pads on the adjacent chips, then the small TSVs were etched for contacting with the pads of the backside. Dry etch was used to remove the oxide of the pads and PVD, electroplating and electroless plating were used to form redistribution layer (RDL).The light reflection and scattering at the sidewalls of the CIS chip and optical layer will affect the sensor performance greatly. Therefore, the coating of black materials on the sidewalls was developed to improve the performance of the optical FPS for high-end applications. The precut on dicing street was used to form the deep trench on CIS wafer and optical layer for black shielding. To form a void-free filling, the wetting property of the black glue is very important. Three different materials with required optical shielding properties were used. The process of dispensing of black glue into the trenches was developed. The droplet diameter of 50μm with a jetting speed of 70mm/s was used for black material filling. The results showed that one of the materials with 15μm thickness on the sidewalls with 99.9% optical shielding from 430-560nm light wavelength and void-free filling was obtained which is good enough for the application. Reliability of the 3D WLCSP with black shielding was characterized by thermal cycling (TC) and highly accelerated stress test temperature storage (HAST) tests and the results show good package level reliability. The results indicate that the 3D WLCSP can provide a low cost and reliable solution for optical FPS.

Keywords-Finger Print Sensor; CMOS image sensor, WLP; TSVs; Biometric Technique

I. INTRODUCTION

The CMOS image sensor (CIS) industry is able to grow at the speed of the global semiconductor industry, which has become a key segment of the broader semiconductor industry. Mobile, security and automotive markets are all in the middle of booming expansion. Recently, there are a number of new applications for CIS, such as artificial intelligence, bio-device like PillCam, FPS and light sensor based on CIS. Technology advancement and the switch from imaging to sensing is fostering innovation at multiple levels: pixel, chip, wafer, all the way to the system. CIS sensors are also at the forefront of 3D semiconductor approaches. The number of cameras increases at the same time as quality increases, which has a multiplier effect on revenue generation. The CIS industry sector is clearly in the middle of a golden age [1].

As an important biometric technique, fingerprint verification is widely used for smart phone, IoT, and payment due to small size and high security. The fingerprint sensor market is expanding significantly with the emergence of new applications and technologies. These sensors are being used in many applications such as consumer electronics; banking and finance. Nowadays there are mainly three kinds of FPS, i.e., capacitive, optical and ultrasonic [2]. To meet full screen trend of smart phone, optical FPS using CIS aroused great interest and gradually replaced capacitive FPS. For such a device, a thin package size with better image performance is required.

It is well known that through silicon via (TSV) is the key packaging technology for 3D IC [3-4], 2.5D interposer integration [5-6] and 3D WCSLP for MEMS and sensor integration [7-8]. To meet these critical requirements for FPS packaging, 3D WLCSP with via last TSV was the best solution for CIS. The 3DWLCSP for CIS was capped with glass with 300~400μm thickness with cavity wall of ~40μm.Forultra-small package or high-end applications, the light reflection and scattering at the edge of glass sidewall will affect the sensor performance greatly. Therefore, the coating of black materials on the glass or package sidewalls to absorb the light is needed. In this paper, the 3D WLCSP for optical FPS was developed and the process for sidewall coating for light shielding was studied and reported.

II. PACKAGE AND PROCESS

A. Package information

The FPS package size is 920×656μm. The main package dimensions are listed in Table I. The total package thickness is about 580μm, which includes 200μm glass with thin IR layer, 130μm optical layer, 100μm CIS chip and 80μm height BGAs. In the package, electrical interconnections are implemented by via last TSVs. There is a slope at the edge of the chip and a thinner silicon sidestep is formed where TSVs are fabricated to connect with the pads. Such a structure is easier for manufacturing with simple process comparing with vertical TSVs. RDL is built to connect the pads and the BGAs on the backside of the chip.

For the present FPS package, the complex layers with different CTE will result in large wafer warpage during packaging process. The select of suitable glass with IR coating is important. To meet optical performance, the coating of thin black material on the sidewall of the chip and optical layer should prevent light reflection and scattering. Specifically, a material with thin thickness coating on the sidewalls should have the capability of 99.9% optical shielding for the light with wavelength of 430-560nm. Furthermore, there should have no voids inside the material. Therefore, the coating of the black material is very challenging since it needs to find out the suitable material and develop the coating process.

Fig. 1 Cross-sectional structure of 3D WLCSP package

TABLE I THICKNESS OF EACH LAYER FOR THE 3D WLCSP

Parameters	Dimensions
Glass thickness	200μm
IR film thickness	6-7μm
Non-cavity glue thickness	20μm
Optical layer thickness	130μm
Silicon chip thickness	100μm
Passivation thickness	10μm
Solder mask thickness	30μm
BGA height	80μm
Total package thickness	~580μm

B. 3D WLCSP Process flow

The process flow of the proposed 3D TSV-based WLCSP is illustrated in Fig. 2. In step (1), incoming IR glass with 130μm optical layer was bonded with CIS wafer. For the wafer bonding process, high transparent non-cavity glue was applied. In step (2), the wafer was thinned to 110μm using coarse grinding and fine grinding after bonding. Dry etching was followed to reduce the wafer to 100μm to remove subsurface damage and release the residual stress caused by mechanical grinding [8]. In step (3), trenches were formed by photolithography and isotropic etching. In step (4), tapered TSVs were formed at first. Then the etching of the oxide on backside of the pads was performed directly after photoresist stripping. In step (5), the first passivation layer

was deposited by spray coating. In step (6), Cu filling and RDL forming were finished by multi-step depositing process. Ti barrier layer and Cu seed layer were sputtered through physical vapour deposition with thickness of 0.3 and 0.5μm respectively. A Cu layer with about 3.5μm thickness was then deposited on the whole wafer surface by electroplating, thus the trenches and TSVs was partially filled with the Cu layer. After Cu electroplating, RDL were formed by photolithography and wet etching. Subsequently, electroless plating of Ni/Au with thickness of 3.5/0.1μm was performed on the Cu layer. Precutting was used to form deep trench on silicon and optical where black coating material will be filled. In step (7), black photoresist material was used as the final passivation. The openings for BGA were formed after photoresist exposure and development. In step (8), the bonded wafer was separated into single packages by blade dicing.

In this process flow, for step (8) the performance of the black photeresis is not proved. If the optical shielding cannot meet the requirement, new materials and process will be studied.

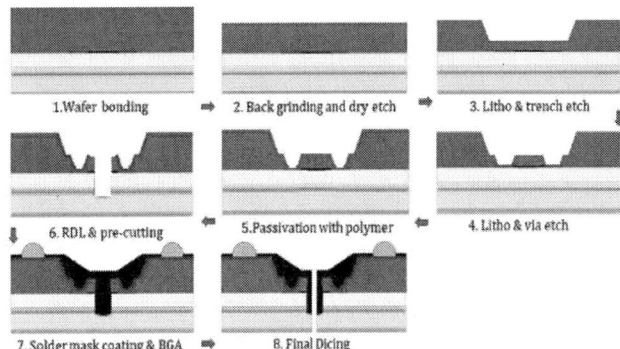

Fig. 2 The schematic flow of the 3D WLCSP

III. RESULTS

A. Selection of glass for wafer bonding

Three kinds of glass, named D236T, AF32 and D263T with thickness of 145, 200 and 210μm respectively were used for bonding evaluation. As shown in Table II, The warpage values of the three glasses before bonding were about 10, 3 and 6mm respectively. As shown in Fig. 3, after wafer bonding and CIS wafer thinning, bonding using AF32 glass showed good results. The warpage is less than 3μm which is accepted for wafer level process. Therefore, AF32 was selected for 3D WLCSP.

TABLE II THE EFFECT OF GLASS ON THE WARPAGE OF BONDED WAFERS

Glass	Thickness (μm)	Warpage (mm)		
		Before bonding	After bonding	After thinning
D236T	145	10	5	NA
AF32	200	3	3	3
D263T	210	6	3.5	7

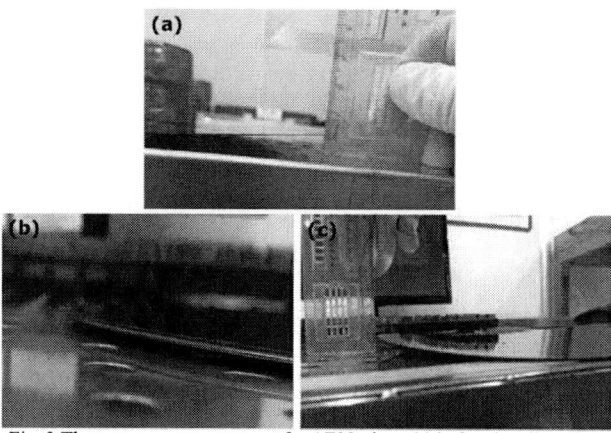

Fig. 3 The warpage measurement for AF32 glass: (a) before bonding, (b) after bonding, and (c) after CIS wafer thinning

B. Process results

As shown in Fig.2, in step (3), trenches were formed by isotropic etching. Along the X direction of the package, the specs of trench opening bottom width are 523 and 470μm. For Y direction, the specs of trench opening bottom width are 170 and 120μm respectively. The results after etching were within the specs as listed in Table III.

For the tapered vias, the specs for top opening, bottom width and depth are 82, 55 and 45μm respectively. The results were within the specs. As shown in Fig. 4(a), after oxide etching, the Al pad was exposed. As shown in Fig. 4(b), ~30μm opening of the first passivation layer on the bottom of the vias was achieved. RDL with 42±2μm width was built for connecting vias and BGAs as shown in Fig. 4(c).

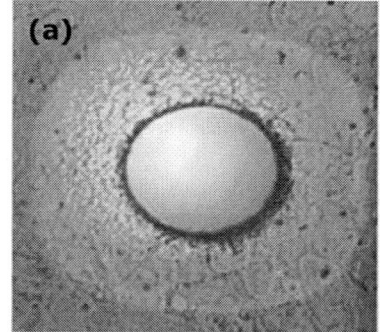

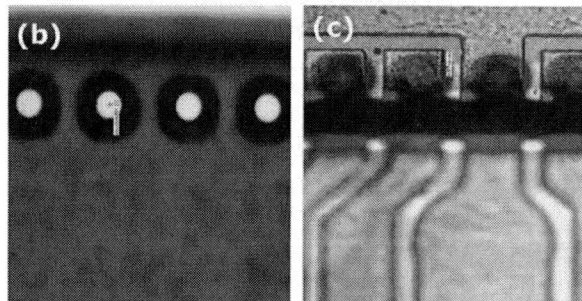

Fig. 4 Process results for (a) vias after oxide etch, (b) passivation opening in the vias, and (c) RDL formation on vias and backside of the wafer

TABLE III TRENCH ETCH RESULTS

Direction	Items	Spec(μm)	Positions				
			1	2	3	4	5
X-Axis	Top	523±20	525	530	527	533	533
	Bottom	470±15	468	471	470	470	468
	Depth	60±5	58	57	61	58	58
Y-Axis	Top	170±15	182	181	182	184	182
	Bottom	120±15	118	121	119	119	122

Spin coating of black photoresist was performed to form optical shielding. As shown in Fig. 5, the thickness of black photoresist on the sidewall was larger than 15μm. However, for the application of FPS, the coating without voids is required. For the present process and material, voids and bubbles were found. Even after optimization, the yield is less than 10%. Therefore, new process and materials are needed.

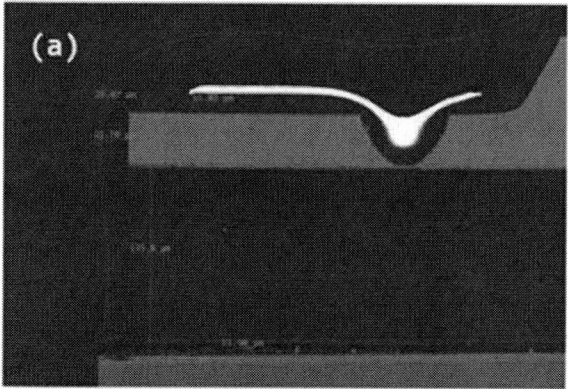

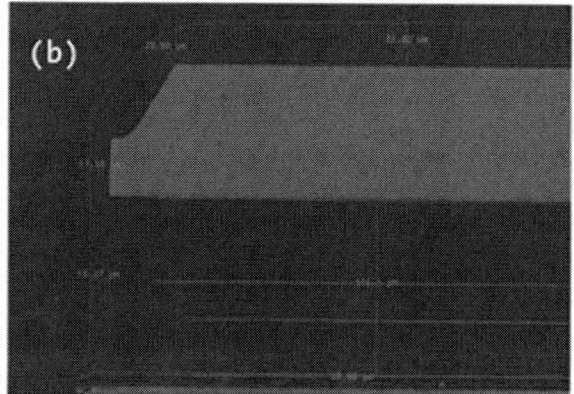

Fig. 5 Cross-section views of the package edge: (a) the side with TSV, (b) the another side without TSVs

C. Development of black shielding technology

1) Printing method

The printing of black glue from DOW was tried. The blade printing speed is 2mm/s with 45° and 90°was used. Since the trench is quite deep, double printing was performed. After printing, vacuum placement was applied under 13Kpa for 30min followed 13pa for another 30min. Fig. 6(a) shows

978-1-7281-1500-9/19 $31.00 © 2019 IEEE

the wafer after black glue printing and Fig. 6(b) shows the cross-section view of the sidewall on the trench.The results showed that long strip shape bubbles were formed inside the glue on the sidewall as shown in Fig. 6(b). The yield is nearly zero. We tried different printing parameters. However, there is no improvement. Therefore, it can be concluded that the poor flowability of the black glue led to bad results.

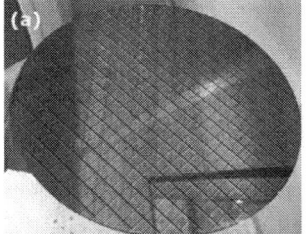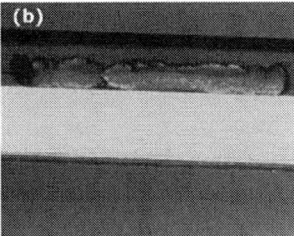

Fig. 6 Printing results of black glue:(a) wafer after black glue printing, (b) cross-section view of the sidewall on the trench

2) Spin method

Spin coating was used to fill the trench with black photoresist. The rotation rate of the spin is 200r/min first and then 300r/min. As shown in Fig 7, the wafer surface was uneven and the round bubbles were found on the sidewall. The rejection rate is about 80%.

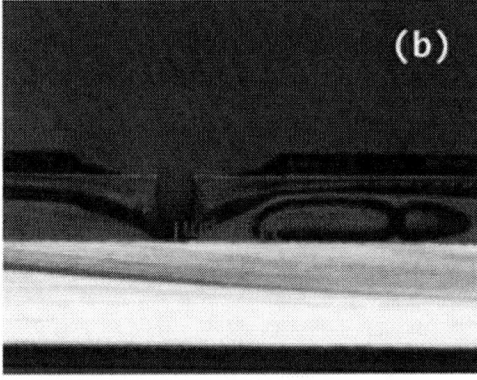

Fig. 7 Spin coating of black photoresist: (a) wafer surface, (b) cross-section view on the crossing of the dicing street

Spin coating plus vacuum placement was applied under 13Kpa for 30min followed 13pa for another 30min. The cross-section view of the black photoresist on sidewall was shown in Fig. 8. Large bubbles were observed and the yield is quite poor.

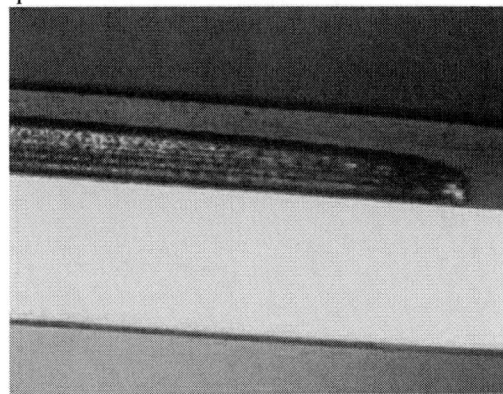

Fig. 8 Cross-section view of the black photoresist on sidewall using spin coating plus vacuum placement

3) Dispensing with new materials

We realized that to form void-free filling, the wetting property of the black glue is very important. If the material like underfill, it will fill the trench easily. Then, we try to find out black glue which has good flowability and optical shielding. Three different materials from Henkel and Panacol with required optical shielding properties were used for evaluation.

The jetting model was used for dispensing, moving speed of the nozzle is 70mm/s, the droplet diameter is 50μm. The temperature of the wafer was 55°C. The baking temperature after dispensing is 130 °C for 10min. Since these three black materials have better flowability, the trenches were easy to get void-free filling result. Fig.9 shows the surface of the dummy wafer with the same trench structure of the 3D WLCSP where the trenches were successful filled.

Fig. 9 Wafer morphology after black glue dispensing

As shown in Fig. 10(a), after dispensing for the dummy wafer, the trench was filled with black glue and there are small droplets on the edge of the wafer which is not an issue since the BGA is far away from the edge. As shown in Fig. 10(b) and (c), black material on the sidewall for X and Y directions has no voids or bubbles. It means the idea using

978-1-7281-1500-9/19 $31.00 © 2019 IEEE   887

better flowability material with the dispensing method can be used for edge optical shielding of the 3D WLCSP.

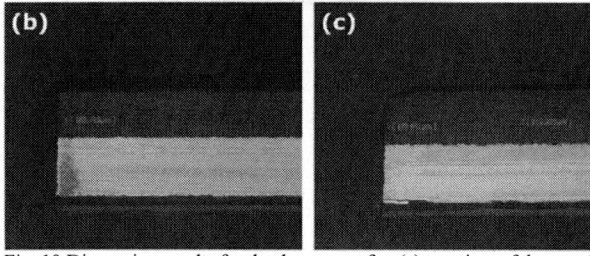

Fig. 10 Dispensing results for the dummy wafer: (a) top view of the trench before dicing, (b) material on the sidewall of the trench along X direction, (c) material on the sidewall of the trench along Y direction after dicing

According to the performance requirement, the material shield rate should be larger than 99.9% with 15μm coating thickness. The optical shielding property of the three materials was tested. As listed in Table IV, one of the materials with 15μm thickness on the sidewalls with 99.9% optical shielding for 430-560nm light wavelength which fully meet the performance requirement. Therefore, this material is selected for the final package.

TABLE IV TEST RESULT OF OPTICAL SHIELDING PROPERTY

Material	Thickness (μm)	Optical shielding rate % for 430-560nm light wavelength		
		Min	Max	Avg
A	27	98.20	99.60	98.90
B	20	99.64	99.88	99.74
C	15	99.80	100	99.90

D. 3D WLCSP with optimized black shielding process

According to the developed black material process, the 3D WLCSP process flow was optimized. For the step (7) showed in Fig.2, instead of black photoresist material, solder mask material was used for passivation. The precut was performed to make the trenches for shielding. After forming BGA, the selected material was dispensed into the trench by the process developed. After baking of the black material, finial dicing was performed to get the final package with black shielding on the sidewalls of the optical layer and CIS chip. As shown in Fig. 11, about 20μm thick black material was formed at the edge of the silicon chip and optical layer. The yield is greater than 90% according to voids free criteria.

E. Performance and reliability of 3D WLCSP

After packaging, the electrical and optical properties of the FPS were tested. The results showed good performance for mobile applications.

Reliability of the 3D WLCSP with black shielding was characterized by thermal cycling (TC) and highly accelerated stress test temperature storage (HAST) tests and the results show good package level reliability.

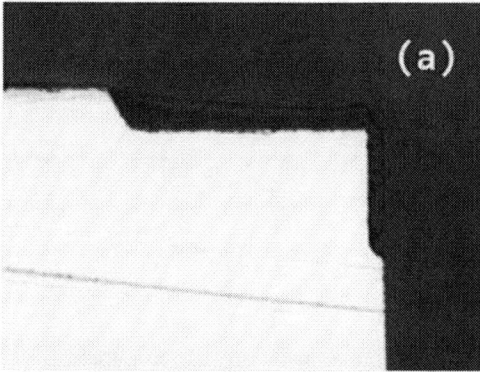

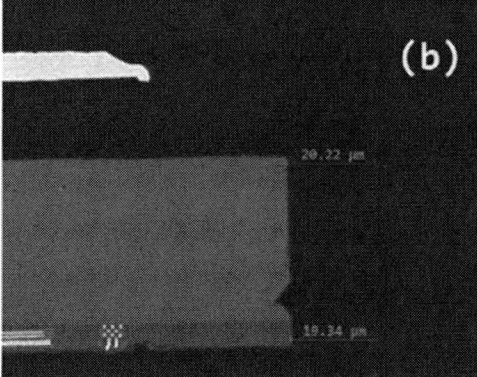

Fig. 11 The cross-section view of the edge of the final FPS package using suitable material: (a) optical picture and (b) SEM picture of the sidewall

IV. CONCLUSION

In this paper, a 3D WLCSP for FPS based on CIS with via last through silicon via (TSV) interconnects and black shielding on package sidewalls was developed. For the optical FPS, the package size is 920×656μm with 580μm height, which includes 200μm glass, ~130μm optical layer, ~100μm CIS chip and 80μm BGAs.

The black shielding on the sidewalls of the package is crucial for the optical FPS performance and it is the most challenging process. Several processes were evaluated. Spin coating and printing methods did not get a good yield. Voids and bubbles were found on the sidewalls. The dispensing process using good flowability black glue was proposed and developed to fill the trenches. The droplet diameter of 50μm with a jetting speed of 70mm/s was used for black material filling. To form a void-free filling, the wetting property of the black glue is very important. Three different materials with required optical shielding properties were used. And one of the materials with 15μm thickness on the sidewalls with 99.9% optical shielding from 430-560nm light

978-1-7281-1500-9/19 $31.00 © 2019 IEEE

wavelength was obtained. The yield for black shielding is greater than 90% according to voids free criteria.

Performance and reliability of the 3D WLCSP with black shielding was characterized. The results show good performance and package level reliability. The results indicate that the present 3D WLCSP of the FPS can provide a low cost and reliable solution for optical FPS.

ACKNOWLEDGEMENT

The authors gratefully acknowledge the support from Panacol and Henkel for material development. The support from H&H Technology Co., Ltd for jetting process and tool demo is highly appreciated. Present study was financially supported by National Science and Technology Major Project No. 2017ZX02519.

REFERENCES

[1] https://www.i-micronews.com/imaging-report/product/status-of-the-cmos-image-sensor-industry-report.html

[2] L.Liu, H.Jiang, W.Ying, et al.,"Finger print sensor molding thickness none destructive measurement with Terahertz technology," Semiconductor Technology International Conference,2017,pp.1-3.

[3] M.Motoyoshi,"Through-Silicon Via (TSV). Proceedings of the IEEE," 2009, 97(1):43-48.

[4] M. Yao, N. Zhao, T. Wang, D, Yu, Z. Xiao, H. Ma, "Study of three-dimensional small chip stacking using low cost wafer-level micro-bump/dry film adhesive hybrid bonding and via-last TSVs,"Journal of Electronic Materials, 2018, 47(12): 7544–7557.

[5] B.Banijamali, S.Ramalingam, K.Nagarajan, et al.,"Advanced reliability study of TSV interposers and interconnects for the 28nm technology FPGA,"Proc. 61th Electronic Components and Technology Conference (ECTC), IEEE Press, Jun 2011, pp.286-290.

[6] H.Wang, X.Ren, Z.Jing, et al. "High frequency characterization and analysis of through silicon vias and coplanar waveguides for silicon interposer," Microsystem Technologies, 2016, 22(2):337-347.

[7] S. Zhao, D. Yu, Y. Zou, C. Yang, X. Yang, Z. Xiao, et al."Integration of CMOS image sensor and microwell array using 3D WLCSP technology for bio-detector application," IEEE Transactions on Components, Packaging and Manufacturing Technology, accepted.

[8] Z. Xiao, J. Fan, Y. Ren, Y. Li, X. Huang, D. Yu, W. Zhang, "Development of 3D Thin WLCSP Using Vertical Via Last TSV Technology with Various Temporary Bonding Materials and Low Temperature PECVD Process," Proc. 66th Electronic Components and Technology Conference (ECTC), IEEE Press, Jun. 2016, pp. 302-309.

Micro Fountain-Like Resonators

Jianfeng Zhang[1,2], Jintang Shang*[1,2], Bin Luo[1,2], Zhaoxi Su[1,2]

[1] Key Lab of MEMS of Ministry of Education
[2] Quantum Information Research Center
Southeast University
Nanjing, China
*E-mail: jshang@seu.edu.cn

Abstract—**This paper presents a novel shell resonator, fountain like resonator (FLR). The FLR is of doughnut shape, anchored at the outer rim to enlarge the bonding or support area. The working mode, wineglass n=2 mode, is generated by the inner rim. Finite element mechanical (FEM) simulation is utilized to analyze the influence of shell size on the resonant frequencies. The foaming process is adopted to fabricated the FLR. The frequency response is obtained by Laser Doppler Vibrometer (LDV), and the wineglass n=2 frequency of this resonator is 50.1kHz. The novel shell resonator with doughnut shape shows potential for gyroscopic application.**

Keywords：resonator; frequency; characteristic mode

I. INTRODUCTION

Hemispherical shell resonator (HSR) is at the heart of hemispherical resonator gyroscopes (HRG), which have played an increasingly important role in the aviation for their high performance, low noise and high stability [1]. In general, there are three directions for development of HRG--scaling down the volume, reducing the cost and improve the performance. With technology of microelectromechanical system (MEMS), micro hemisphere resonators with small volume and low price can be fabricated at a wafer level [2, 3], so that micro fabricated hemispherical resonators have attracted recent attention around world. However, HRGs with resonator fabricated by MEMS process have not been commercialized at present. To make micro HRGs practical, it is essential to fabricate resonators of high symmetry, high quality factor and high reliability. For applications including spacecraft stabilization and aircraft navigation, the reliability is especially important.

Resonators with different geometric structures using different fabrication methods have sprung up, as summarized in Fig. 1. Solid anchored wineglass resonator is shown in Fig. 1 (a), and the resonator was fabricated by glassblowing [4]. This wineglass-shaped shell resonator with quality factor of 1 million was demonstrated in [5]. However, insufficient air results in small height–to-radius ratio of these resonators. As shown in Fig. 1 (b), long anchored HSR was demonstrated using glassblowing and the effect of annealing on quality factor of HSR was investigated [6]. It was found that the quality factor was improved from 6.66×10^5 to 9.32×10^5 [6] by annealing at 800°C and slow cooling down for a long time. As shown in Fig. 1 (c), blowtorch molding was utilized

to shape the hollow anchored birdbath resonator [7]. This kind resonator with a working frequency of 10kHz achieved quality factor of 9.8 million [8]. As shown in Fig. 1 (d), a new method was reported to fabricate fused silica resonator with long strong stem and thin shell rim [9]. The quality factor of this resonator was up to 2.55 million [9]. In order to acquire shell resonator with small frequency split and high quality factor, this method requires molds with extremely high-degree accuracy and low roughness. As shown in Fig. 1 (e), the distribution of effective mass and stiffness along the radial direction was adjusted by wet etching of silica wafer before blowing [10]. By this way, the frequency of working mode can be accommodated to the desirable range. In addition, other possible parasitic frequencies can be kept away from the working frequency. In order to ensure symmetry of resonators, this process has stricter requirement in aligning to avoid the mismatch between the etched silica wafer and the graphite mold. A bell resonator [11] was proposed with large height-to-radius ratio, as shown in Fig. 1 (f). By adjusting the structural parameters of graphite mold, the geometry of bell resonators can be accommodated and resonators with more complex structure can be obtained. The frequency of wineglass n=2 mode of the bell resonator is 12.9kHz and the quality factor is 271k [11]. Fig. 1 (g) shows a short anchored SiO_2 HSR fabricated with pop-up-rings mask and deep isotropic chemical etching [12]. 650nm thick HSR with its diameter of 160μm was demonstrated with this method. This SiO_2 HSR was fabricated in wafer level and achieved frequency split of 512Hz and quality factor of 31542 [12]. Compared with SiO_2 resonators mentioned above, this HSR is exceedingly small and its distribution of thickness is uniform. However, using this technology to control the symmetry and aspect ratio of hemispherical shells requires a lot of research on process details.

In our previous work [13], central-anchored umbrella shell with large height-to-radius ratio was fabricated by foaming process. Currently, most of micro shell resonators adopt structures such as birdbath, mushroom, wineglass and bell, which are all anchored by central support. Small stems result in small bonding area or fixing area, so that these resonators are subject to external acceleration or shock. In addition, small bonding area may bring negative impact on the reliability. This paper proposes a novel outer-rim-anchored resonator to enlarge the support area, as shown in Fig. 1 (h).

978-1-7281-1500-9/19 $31.00 © 2019 IEEE

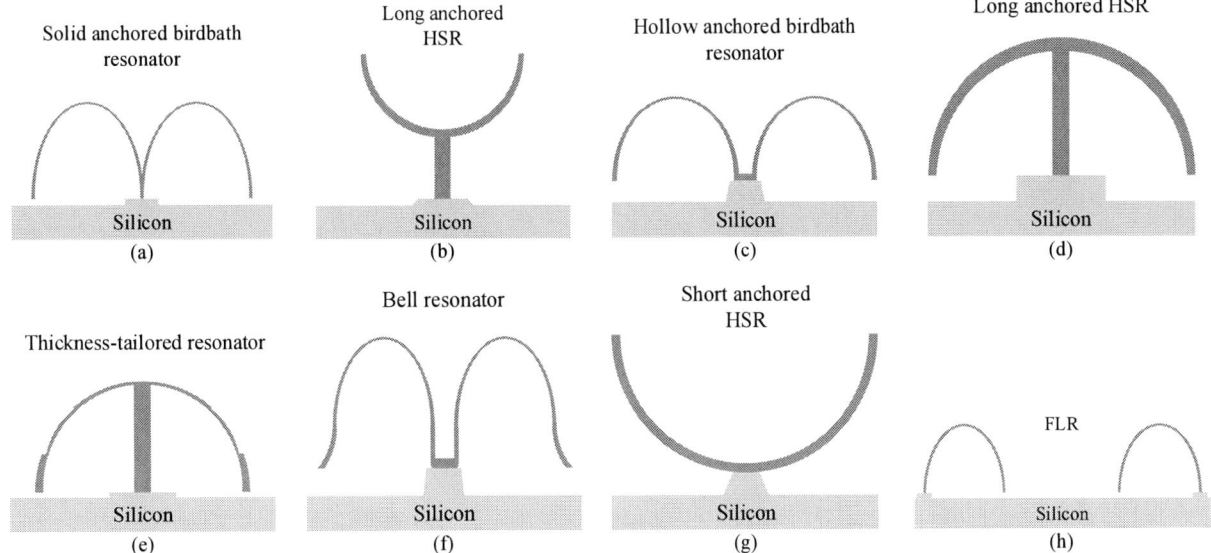

Figure. 1 Cross-section of different shell structures: (a) solid anchored wineglass resonator [4]; (b) long anchored HSR [6]; (c) hollow anchored birdbath resonator [7]; (d) long anchored HSR [9]; (e) thickness-tailored resonator [10]; (f) bell resonator [11]; (g) short anchored HSR [12]; (h) FLR

II. DESIGN AND SIMULATION

The FEM simulations are conducted in COMSOL to analyze characteristic frequencies of FLRs of different sizes. Because of the structural complexity of FLR, a simplified model is utilized to analyze the relationship between resonant frequencies and the structural dimension of FLR. As shown in Fig. 2, the model is a 3D axisymmetric shell with its cross-section is a regular semi elliptical ring. Dashed line o-a is the symmetry axis of the model. Four geometric parameters used in the simulation, namely: R_{outer}, R_{inner}, Height and Thickness. R_{outer} is the distance from the outer rim of the shell to the symmetry axis. R_{inner} is the distance from the inner rim of the shell to the symmetry axis. Height is the distance from top of shell to its bottom. Thickness is the thickness of the shell. Fig. 2 also shows the two main differences between the simplified model and real FLR. First, the thickness of the simplified model is uniform, whereas the thickness of FLR decreases with its height. Second, influenced by the surface tension, the cross section of FLR tilts toward the symmetry axis. Under this circumstance, the characteristic frequencies of FLR will be lower than the result of simulation.

Fig. 3 shows the first eight simulated resonance modes of simplified model. Fig.4 shows the relationship between the first three resonant frequencies and geometric parameters, which indicates that the order of characteristic modes may vary with the structural parameters. As shown in Fig. 4 (a), the frequencies of tilting mode and vertical mode decrease with R_{inner}. But for frequency of wineglass n=2 mode, there is a specific R_{inner} corresponding to the its minimum. As shown in Fig. 4 (b), the first three characteristic frequencies of the model decrease with R_{outer}. Fig. 4 (c) shows that as the height of the shell increases, its first three resonant frequencies rise first and then decreases. Moreover, the wineglass n=2 mode is more susceptible to the changing of

height than other modes. As shown in Fig. 4 (d), the resonant frequencies of the shell increase with its thickness as a result of growing stiffness.

Based on the above results, FLRs of four different dimensions is designed as shown in Tab. 1. The fabrication process of FLR is simulated according to the designed dimensions to optimize FLR by tailoring its height. In this process, the height of FLR is controlled by the adjusting the pressure difference and the initial thickness is 300μm. This simulation results are considered to be consistent with the real results. Low frequency of working mode reduces difficulty in design of detection circuit. Keeping parasitic mode frequency far from working mode can improve the performance of resonator. Based on the simulation result, the dimension of FLR is appropriate when the height of FLR is slightly greater than half of difference of R_{outer} and R_{inner}.

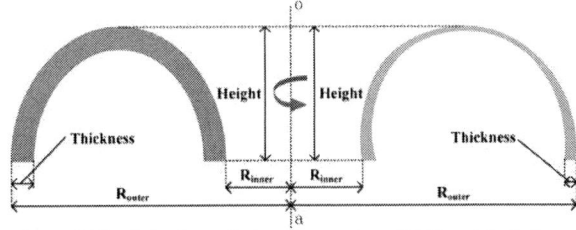

Figure. 2 The right picture is the cross section of FLR and the left picture is the cross section of the simplified model. Dashed line o-a is the symmetry axis of the shell.

TABLE. I GEOMETRIC DIMENSION OF FLR

R_{inner}(mm)	R_{outer}(mm)
2.5	5.5
2.5	6
3	5.5
3	6

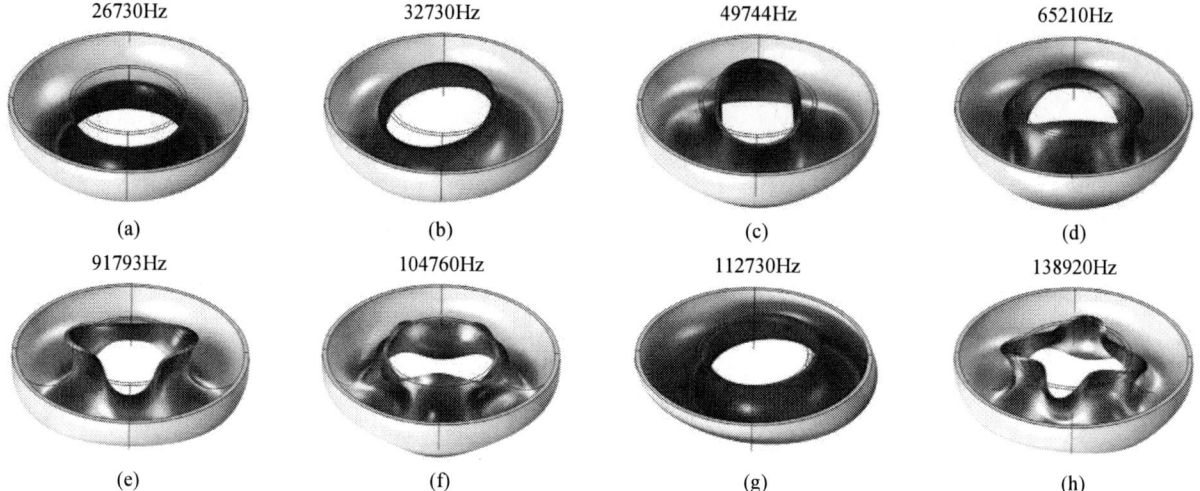

Figure. 3: First eight resonant modes of FLR: (a) vertical mode; (b) tilting mode; (c) wineglass n=2 mode; (d) deformed wineglass n=2 mode;(e) wineglass n=3 mode; (f) deformed wineglass n=3 mode; (g) deformed tilting mode; (h) wineglass n=4 mode

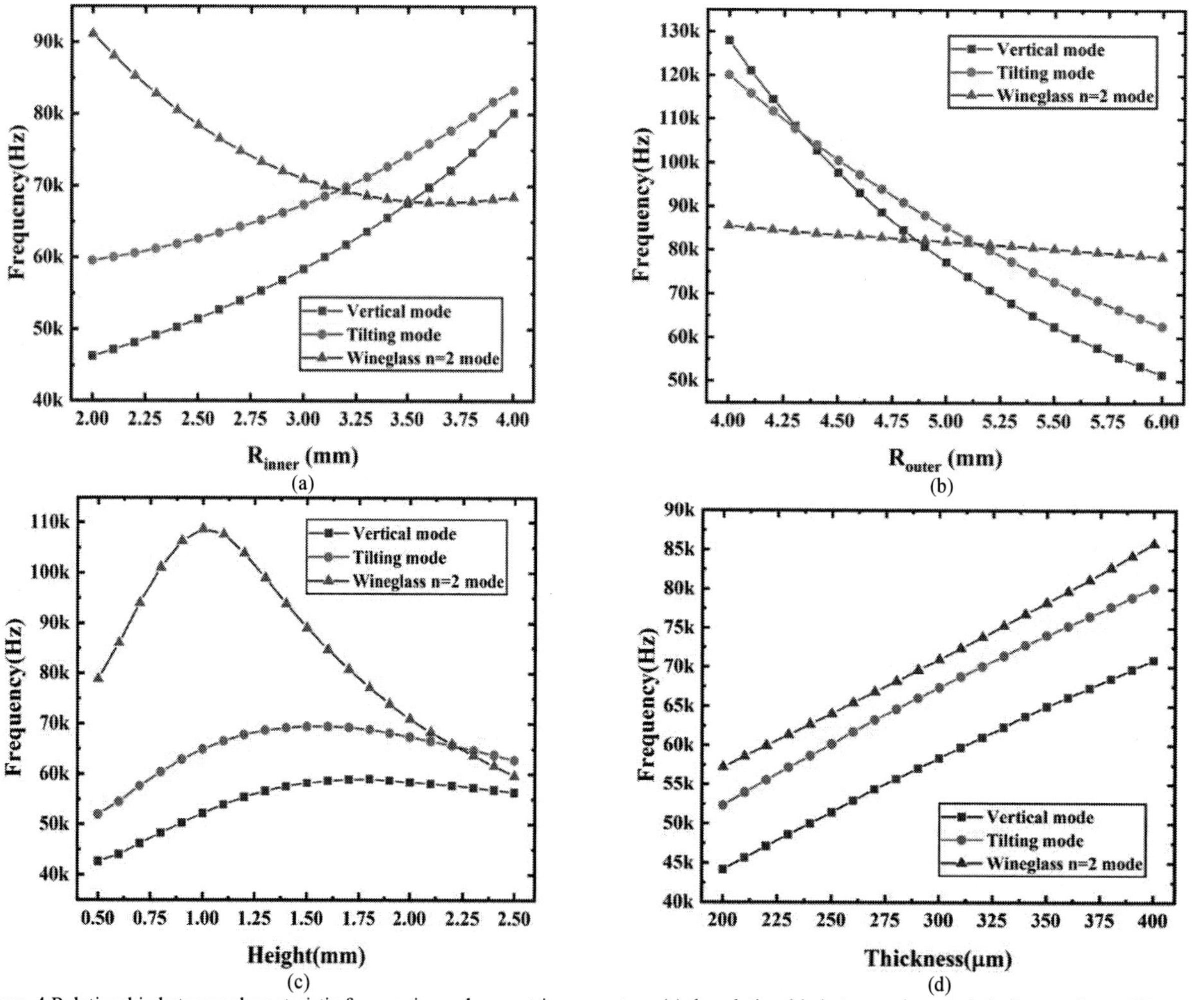

Figure. 4 Relationship between characteristic frequencies and geometric parameters: (a) the relationship between characteristic frequencies and R_{inner}; (b) the relationship between characteristic frequencies and R_{outer}; (c) the relationship between characteristic frequencies and height; (d) the relationship between characteristic frequencies and thickness. The default values for R_{inner}, R_{outer}, height and thickness are 2.5mm, 6mm, 2mm and 300μm, respectively.

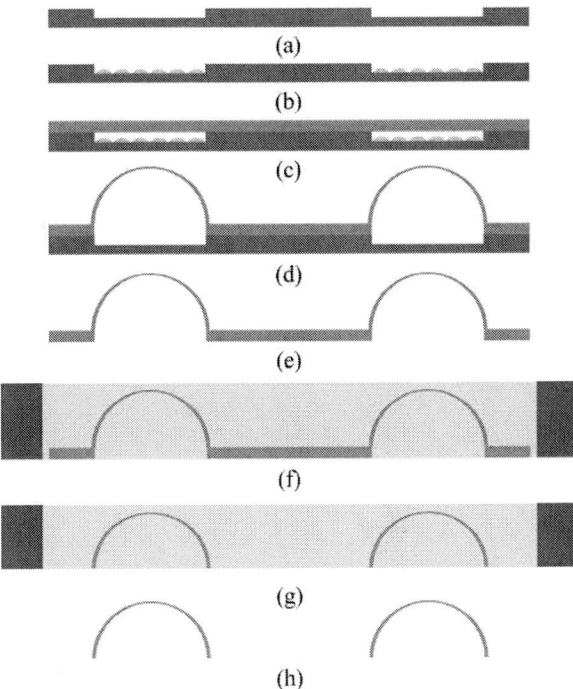

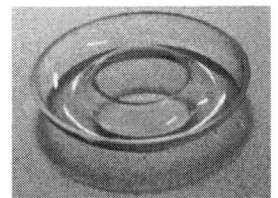 

Figure. 6 Released FLR from different point of view

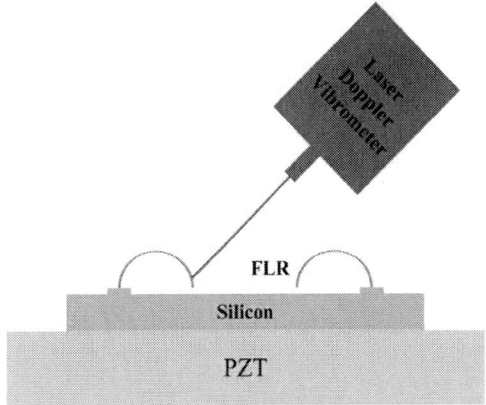

Figure. 7 Experimental setup

Figure. 5 Fabrication process of FLR: (a) cavities are graved with laser; (b) foaming agent is dropped into the cavities; (c) bonding a glass wafer to silicon; (d) resolve the foaming agent to mold FLR; (e) remove the silicon wafer; (f) put the FLR with substrate into a mold and fill the interspace with polymer; (g) remove the glass substrate; (h) release the FLR

III. FABRICATION

The fabrication procedure for FLR is shown in Fig.5, namely: (1) graving cavities on the silicon wafer with laser; (2) dropping foaming agent into the cavities and drying the water; (3) bonding the silicon wafer to a glass wafer; (4) heating the bonded wafer to mold the FLRs; (5) removing silicon wafer with TMAH; (6) releasing the FLRs with grinding.

In detail, first of all, oxide layer is generated on a 500μm silicon wafer to protect the silicon wafer from contamination of silicon powder generated during laser graving. Then cavities are graved with laser on the silicon wafer. Remove the silicon powder with tetramethy-lammonium hydroxide (TMAH) with the oxide layer acting as a mask. Secondly, the foaming agent (e.g. $CaCO_3$) is dropped into the cavities in the form of a suspension with water. In this way, the amount of foaming agent could be controlled precisely. Then the suspension is heated to 50 °C for drying the water and leaving the foaming agent. The heating temperature should not be too high to prevent powder left behind from splashing. Thirdly, the silicon wafer with cavities is bonded to a glass wafer of 300μm. Fourthly, the bonded wafers are heated for resolving the foaming agent and creating FLR. This is followed by wet etching to remove silicon wafer with TMAH. Then the FLR with the substrate is put into a plat and firm silicon mold and polymer is filled into the

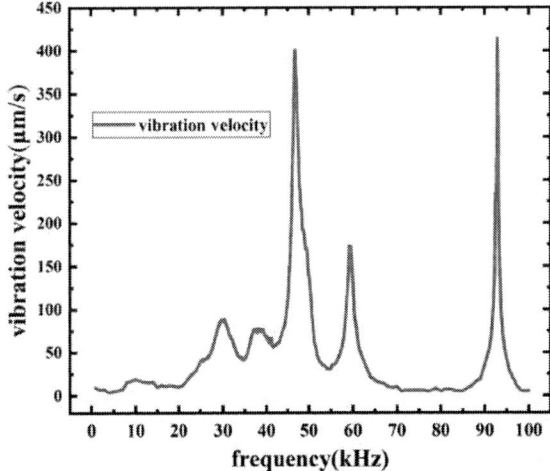

Figure. 8 The frequency response of FLR

interspace between the FLR and the mold. After the polymer solidifies, the glass substrate is eliminated by grinding. To avoid damage of the edge of FLR, the emery papers with small granularity are preferred and the grinding speed should be slow. Finally, the polymer is removed by immersed in acetone and the FLR are released as shown in Fig.6.

IV. EXPERIMTENATL

As shown in Fig.7, the FLR is fixed on a silicon bulk and driven by a piezoelectric ceramic transducer (PZT). The vibration velocity of the inner rim of FLR was measured with Laser Doppler Vibrometer (LDV).

Fig.8 shows the frequency response of a FLR, in which we can find three obvious peak, namely: 46.4kHz, 59.6kHz,

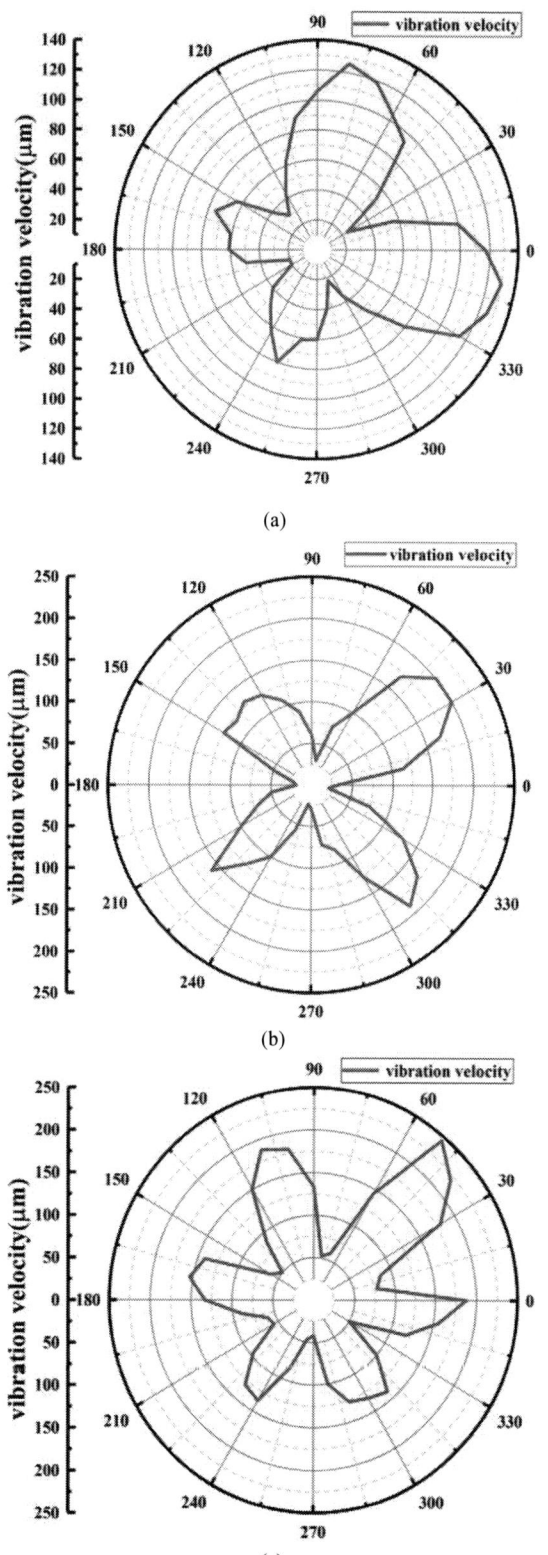

(a)

(b)

(c)

Figure. 9 Vibration velocity of different part of the rim of FLR working at 50.1kHz, 59.6kHz and 93.1kHz, respectively

93.1kHz. In another testing process, we find that there is also an obvious peak at 50.1 kHz, which may result from that the region of rim of FLR detected in the first time is the wave node of this mode.

Then vibration velocity of different regions of rim of FLR is detected for determine the order of renounce modes. Fig. 9 shows the vibration velocity of different regions of the rim of FLR driven by different frequencies. Depend on Fig. 9, we can ensure that the frequencies of wineglass n=2 mode, deformed wineglass n=2 mode and wineglass n=3 mode are 50.1kHz, 59.6kHz and 93.1kHz, respectively. This result basically corresponds to the simulation.

V. CONCLUSION

In this paper, a novel micro resonator--FLR is demonstrated for working at strong external acceleration or shock. The relationship between characteristic frequencies and geometrical dimensions of FLR is investigated with FEM simulation. The characteristic frequencies of FLR are correspond to the simulation result. The frequency of working mode of FLR is 50.1kHz. Future work includes improvement of symmetry and quality factor of the FLR and reduction of frequency of working mode.

ACKNOWLEDGMENT

This work is supported by National Science Foundation of China (No. 51675102 and No. 51275091). The authors also acknowledge funding supplied by Specially-Appointed Professors by Universities in Jiangsu Province and "333 Projects" of Jiangsu Province. Xiang Yue is acknowledged for his support in fabrication of samples and assembly of instruments. Chen Ye is thanked for his help on simulation and suggestion.

REFERENCES

[1] D. M. Rozelle, "The hemispherical resonator gyro: From wineglass to the planets (AAS 09-176)," in *Proc. 19th AAS/AIAA Space Flight Mechanics Meeting,* 2009, pp. 1157–1178.

[2] C. L. Fegely, D. N. Hutchison and S. A. Bhave, "Isotropic etching of 111 SCS for wafer-scale manufacturing of perfectly hemi-spherical silicon molds," in *Proc. IEEE Transducers*, Beijing, China, June 5–9, 2011, pp. 2595–2598.

[3] E. J. Eklund, A. M. Shkel, "Glass Blowing on a Wafer Level," *J. Microelectromech. Syst.*, vol. 16, no. 2, pp. 232–239, Apr. 2007.

[4] D. Senkal, C. R. Raum, A. A. Trusov, and A. M. Shkel, "Titania silicate/fused quartz glassblowing for 3-D fabrication of low internal loss wineglass microstructures," in *Proc. Solid-State Sensors, Actuat., Microsyst. Workshop*, Hilton Head Island, SC, USA, June 3–7, 2012, pp. 267–270.

[5] D. Senkal, M. J. Ahamed, S. Askari, A. M. Shkel, "1 million Q-factor demonstrated on micro-glassblown fused silica wineglass resonators with out-of-plane electrostatic transduction" in *Hilton Head*, Hilton Head Island, SC, USA, pp. 68-71, 2014.

[6] M. J. Ahamed, D. Senkal and A. M. Shkel, "Effect of annealing on mechanical quality factor of fused quartz hemispherical resonator," *2014 International Symposium on Inertial Sensors and Systems (ISISS)*, Laguna Beach, CA, 2014, pp. 1-4.

978-1-7281-1500-9/19 $31.00 © 2019 IEEE

[7] J. Y. Cho, J. Yan, J. A. Gregory, H. W. Eberhart, R. L. Peterson and K. Najafi, "3-Dimensional Blow Torch-Molding of Fused Silica Microstructures," in *Journal of Microelectromechanical Systems*, vol. 22, no. 6, pp. 1276-1284, Dec. 2013.

[8] Singh, T. Nagourney, J. Y. Cho, A. Darvishian, K. Najafi and B. Shiari, "Design and fabrication of high-Q birdbath resonator for MEMS gyroscopes," *2018 IEEE/ION Position, Location and Navigation Symposium (PLANS)*, Monterey, CA, 2018, pp. 15-19.

[9] J. Y. Cho and K. Najafi, "A high-q all-fused silica solid-stem wineglass hemispherical resonator formed using micro blow torching and welding," *2015 28th IEEE International Conference on Micro Electro Mechanical Systems (MEMS)*, Estoril, 2015, pp. 821-824.

[10] T. Nagourney, S. Singh, B. Shiari, J. Y. Cho and K. Najafi, "Fabrication of hemispherical fused silica micro-resonator with tailored stiffness and mass distribution," *2018 IEEE Micro Electro Mechanical Systems (MEMS)*, Belfast, 2018, pp. 1000-1003.D.

[11] T. Nagourney, J. Cho, A. Darvishian, B. Shiari and K. Najafi, "Micromachined high-Q fused silica bell resonator with complex profile curvature realized using 3D micro blowtorch molding," *2015 Transducers - 2015 18th International Conference on Solid-State Sensors, Actuators and Microsystems (TRANSDUCERS)*, Anchorage, AK, 2015, pp. 1311-1314.

[12] M. M. Torunbalci, S. Dai, A. Bhat and S. A. Bhave, "Acceleration insensitive hemispherical shell resonators using pop-up rings," *2018 IEEE Micro Electro Mechanical Systems (MEMS)*, Belfast, 2018, pp. 956-959.

[13] B. Luo, J. Shang, Z. Su Z, and C.-P. Wong, "Height Adjustment of 3-D Axisymmetric Micro Umbrella Shells for Tailoring Wineglass Frequency," *IEEE Transactions on Components, Packaging and Manufacturing Technology*, 2018.

Novel Additively Manufactured Packaging Approaches for 5G/mm-Wave Wireless Modules

Tong-Hong Lin*, Aline Eid*, Jimmy Hester*, Bijan Tehrani*, Jo Bito†, and Manos M. Tentzeris*

*School of Electrical and Computer Engineering
Georgia Institute of Technology, Atlanta, Georgia 30332–0250
Email: tlin97@gatech.edu
†Texas Instrument, Dallas, Texas

Abstract—Additive packaging has garnered an increasing amount of interest due to its promising industrial scalability, and its materials and topological patterning versatility. The effort described in this paper benchmarks the performance of such a packaging approach relative to its commercial counterpart for the integration of devices operating in the mm-wave 5G bands. The approach first demonstrates the fully-inkjet-printed integration of a tunnel diode—a component empowering a plethora of emerging applications in these frequency bands—before comparing it to its pre-packaged counterpart. In an effort to enable an accurate and quantifiable comparison, the measured properties of the two integrated diodes were de-embedded using a TRL extraction method before their lumped element models were determined and compared. The comparison demonstrates a general and often significant reduction in all the packaging parasitics, up to 53%, using the additive approach. This result adds upon the growing evidence supporting the superior electrical performance of additive packaging for the integration of mm-wave dies and strengthens the appeal for the development of ultra-compact and ultra-high-performance fully-additively-manufactured mm-wave Systems in Package (SiP).

Index Terms—inkjet printing; packaging; 5G; mm-wave; modeling;

I. INTRODUCTION

The fifth generation (5G) communication protocol has been proposed to accommodate the continuously increasing desire for faster data transmission. One of the key reasons to realize a faster data link for 5G communication is the use of high mm-wave frequency bands which open new possibilities as well as new challenges. For starters, the small wavelength due to high frequencies is helpful to reduce the overall size for radio frequency (RF) components making it possible to integrate RF components with active components inside a single package to achieve system-on-package (SoP) designs [1]–[3]. In order to integrate the dies with peripheral RF components, inter-connections are used to connect the pins of a die to package body where RF components are located. Conventionally, wire-bonding interconnections are applied. However, the parasitics due to wire bonds are large and will significantly degrade the performances especially in the high frequencies like 5G and mm-wave.

In order to alleviate the parasitics due to the intercon-nections, the inkjet printed ramp structure is proposed [4]. To overcome the height difference between the die and the package, ramp structures are printed around the die and then

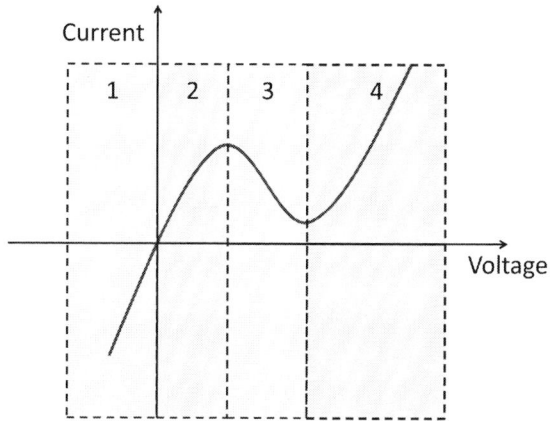

Fig. 1: The general I-V curve for a tunnel diode.

the silver traces are printed on the ramp to serve as the interconnections between the die and the package. However, performances comparisons between conventional packaging structure and the ramp interconnections are not included. Thus, to establish the comparison between the conventional packaging interconnections and the additively manufactured ones, the same functional die has to be packaged using both ways.

Tunnel diodes are chosen for packaging since efforts have indicated the high potential of tunnel diodes for 5G and mm-wave applications [5], [6]. A tunnel diode can find different applications depending on its biasing condition. A general I-V curve for a tunnel diode is shown in Fig. 1 and it can be separated into four different zones. In reverse biasing, zone 1, a heterostructure backward tunnel diode which can be turned on with ultra-low voltages is used for energy harvesting applica-tions. The ultra-low turn-on voltages can be useful to improve the existing energy harvesting techniques and applications [7]–[9]. In zone 2, it behaves like a Schottky diode and can be used in mixer applications. Finally, in zone 3, a decreasing current is observed with the increasing voltage which gives to the tunnel diode a natural negative differential resistance. In that negative slope region of its I-V curve, the tunnel diode can be used as a reflection amplifier and an oscillator. The energy harvester, oscillator, and the reflection amplifier are all key elements for backscattering sensing techniques. The energy harvester built using the tunnel diodes can operate with high efficiency at low power conditions [5]. The oscillator and the reflection amplifier which can be operated with low biasing power can be also great advantages for backscattering systems whose power budget is typically very low [6].

In this paper, the performance comparisons between the conventional packaging techniques and additively manufactured packaging are done for the first time. The tunnel diode die is packaged using inkjet printed technique. The same die is also packaged utilizing commercially available methods. The complete circuit models are built for both test samples. The through, reflect, and line (TRL) calibration method is used to calibrate out the effects of feeding lines and bias-tees. The parasitic inductors and capacitors due to packaging are extracted and used to prove that the additively manufactured packaging outperforms the conventional packaging techniques.

II. FABRICATION PROCESS

A bare-die tunnel diode from Aeroflex / Metelics (model no. MBD2057-C18) is interconnected with a laminate substrate using a multi-material inkjet printing fabrication process as a comparison to a traditional diode package (model no. MBD2057-E28X). Inkjet printing is performed using a Dimatix DMP-2831 materials printer with 10 pL drop-volume cartridges. The host substrate chosen for this evaluation is 32 mil Rogers RO4003C with 1 oz metalization. Printed metallic features, such as the conductive die attach and interconnect, are patterned with Sun Chemical EMD-5730 silver nanoparticle (SNP) ink, capable of achieving a resistivity of 5–30 $\mu\Omega$ cm after thermal sintering. The dielectric ramp is patterned using a MicroChem SU-8 polymer-based ink formulated to achieve a printable viscosity for the Dimatix system [10]. A test board with a microstrip topology is fabricated using an inkjet-printed lithography technique with the Rogers RO4003C laminate. First, the microstrip pattern is printed using 3 layers of the SU-8 polymer ink. The printed SU-8 mask is then crosslinked with a 600 mJ/cm^2 exposure of 365 nm ultraviolet (UV) light. Etching of the board is then performed using FeCl$_3$ at a temperature of 55 °C. Finally, acetone is used to strip the printed SU-8 mask from the board.

A general outline of the fabrication process for the inkjet-printed bare-die interconnects is shown in Fig. 2. First, a die attach is printed using SNP ink (2 layers, 20 μm drop spacing) onto the copper microstrip trace. The 375×375×125 μm diode die is then manually placed onto the wet SNP and then dried and sintered at 150 °C for 2 h. Next, SU-8 is printed (6 layers, 10 μm drop spacing) to pattern the 3D dielectric ramp from the PCB microstrip trace to the top of the diode die. The three-step curing profile for the SU-8 ink is as follows: thermal soft bake ramping from 60–95 °C over 10 min on a hotplate, 300 mJ/cm^2 UV crosslinking exposure, and finally a thermal hard bake at 95 °C for 7 min on a hotplate. After the SU-8 ink is cured, a 2.5 min exposure to UV O$_3$ is performed to facilitate proper ink wetting for the subsequent SNP printing. Finally, SNP is printed (5 layers, 20 μm drop spacing) to pattern the

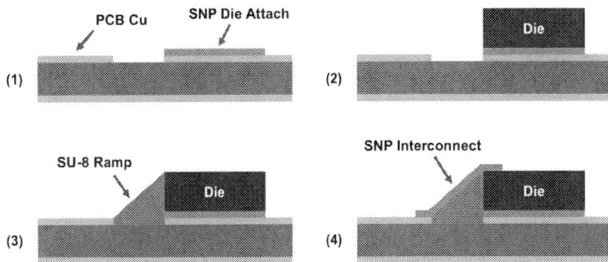

Fig. 2: Side-view schematic of inkjet-printed interconnect fabrication process: (1) inkjet print SNP die attach, (2) attach diode die, (3) inkjet print SU-8 dielectric ramp, (4) inkjet print SNP interconnect.

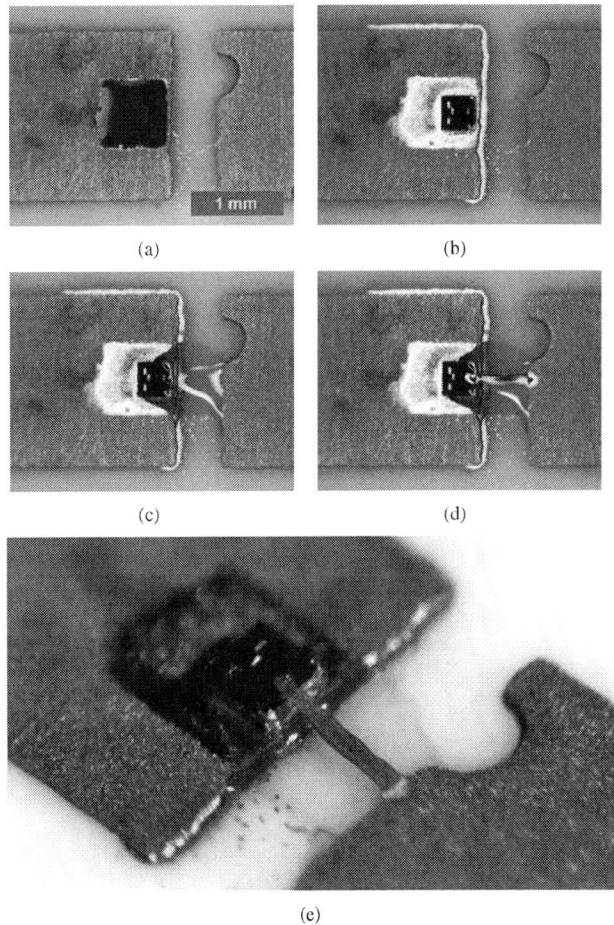

Fig. 3: Inkjet-printed interconnect fabrication process with tunnel diode die: (a) SNP die attach patterning, (b) die placement, (c) SU-8 ramp patterning, (d) final SNP interconnect patterning, and (e) perspective micrograph.

interconnect to the anode pad of the diode die, followed by final sintering at 150 °C for 2 h. Micrographs of the fabrication process with a tunnel diode sample are presented in Fig. 3.

The printed transition between the copper microstrip topol-

978-1-7281-1500-9/19 $31.00 © 2019 IEEE

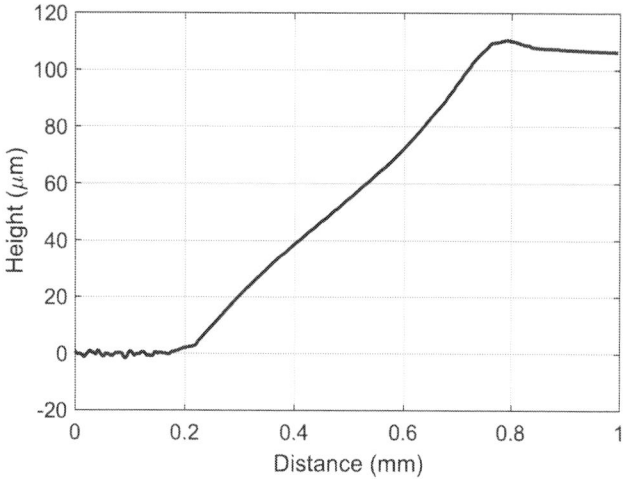

Fig. 4: Profilometer scan of inkjet-printed ramp interconnect to diode die.

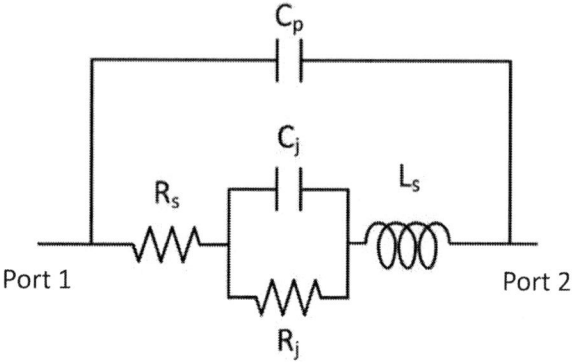

Fig. 5: The equivalent circuit model for the tunnel diode with packaging.

ogy of the PCB and the top surface of the diode die is measured using an Alpha-Step D-500 surface profilometer from KLA Tencor. Fig. 4 shows the measured profile of the ramp interconnect, exhibiting a slope of approximately 10°. As seen in the measured profile, the printed transition is smooth and lacks abrupt discontinuities at the interfaces between the PCB, ramp, and die.

From the mechanical viewpoint, the proposed additively manufactured interconnections are more reliable. Since the wire bonds are floating on the air, they are vulnerable to shocks or vibrations and it is easier to cause breaking wires or delamination. On the contrary, the printed silver traces are placed on the SU8 substrate instead of floating, and thus can resist shocks and vibrations. The property makes additively manufactured packaging suitable for wearable applications which often embrace shocks and vibrations due to constant human movements. From the electrical standpoint, the additively manufactured interconnections are shorter compared with the wire bonds. Therefore, the parasitic due to the interconnections is also expected to be smaller. This is critical for high-frequency applications such as 5G and mm-wave where the effects of parasitic become more significant.

III. TUNNEL DIODE CIRCUIT MODELING

The equivalent circuit model for the tunnel diode with packaging is shown in Fig. 5. The circuit model is composed of 5 components. The C_j is used to model the junction capacitance of the tunnel diode. The R_j is used to model the non-linear I-V relation of the tunnel diode. The R_s is for the exterior series resistance. Finally, the L_s and C_p are used to model the capacitance and inductance induced by packaging structures.

The junction capacitance, C_j, is 0.3 pF from the datasheet. The R_s and R_j can be extracted from the measured I-V curve of the tunnel diode. The L_s and C_p are extracted by measured 2-port scattering parameters of the tunnel diode.

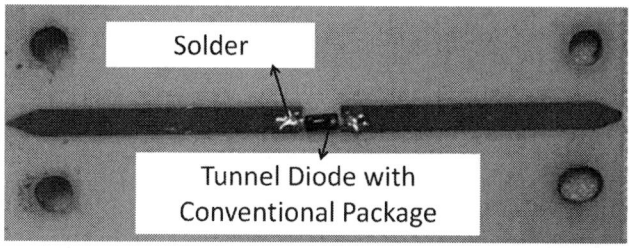

(a)

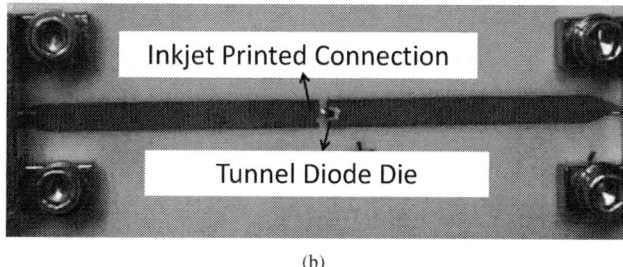

(b)

Fig. 6: The photo of the (a) die with inkjet printed package, and (b) die with traditional package.

Since the 2-port scattering parameters are used, the TRL calibration method is used to de-embedded the tunnel diode to get higher accuracy of the extracted values for the equivalent models.

IV. TRL CALIBRATION

The fabricated samples of the tunnel diode with inkjet printed package and with the traditional package are shown in Fig. 6. As shown in the figure, there are two 20 mm long feeding line with tapering structures at the end. This feeding line will induce additional loss and mismatch to the entire system. Thus, in order to obtain equivalent models with higher accuracy, the TRL calibration method is adopted. The TRL calibration circuits are shown in Fig. 7. The through circuit is composed of two feeding lines connect together without the tunnel diode. The reflect circuit is the two feeding lines with open circuits. The line circuit is the two feeding lines

978-1-7281-1500-9/19 $31.00 © 2019 IEEE   898

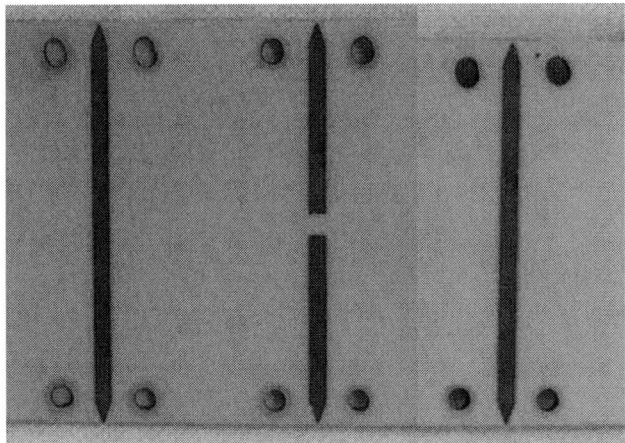

Line   Reflect   Through

Fig. 7: The TRL calibration circuits.

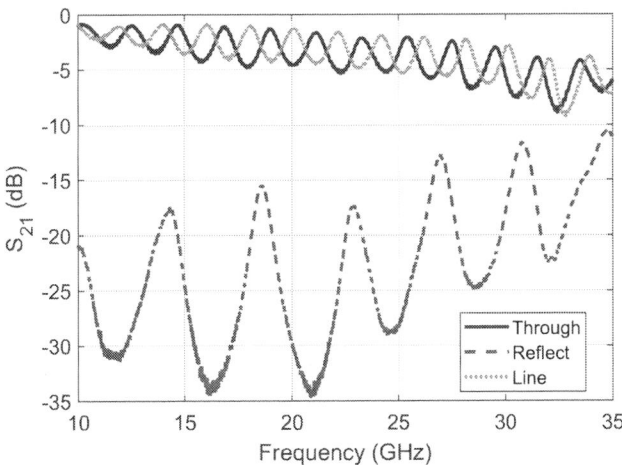

Fig. 8: The measured S21 for the TRL calibration.

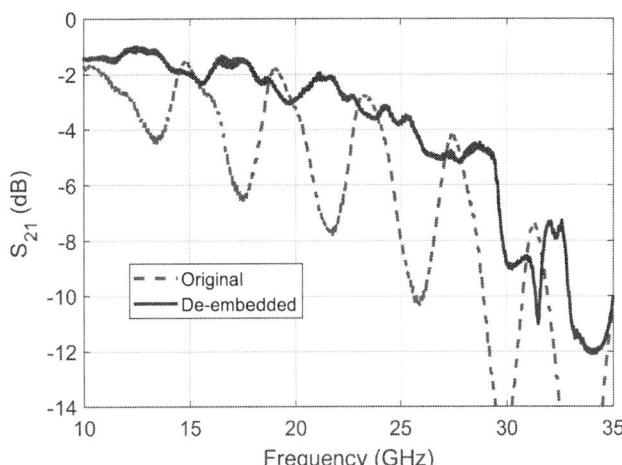

Fig. 9: The measured and TRL de-embedded tunnel diode with inkjet printed package.

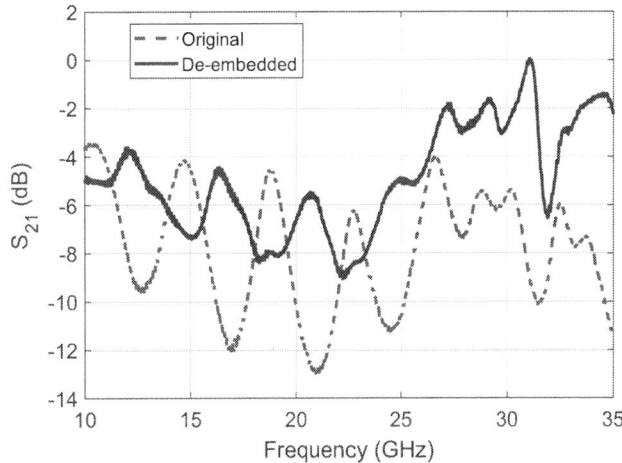

Fig. 10: The measured and TRL de-embedded tunnel diode with traditional package.

with one additional length inserted between them. The inserted line length has to be $\lambda/4$ at the geometric mean of the highest and lowest frequencies. Furthermore, the ratio between the lowest and highest frequencies has to be smaller than 1:8. For example, the lowest frequency here is 10 GHz while the highest frequency is 35 GHz. Thus, the inserted $\lambda/4$ line is calculated at 18.7 GHz. Besides, the ratio between the lowest and highest frequencies is 1:3.5 which is smaller than the maximum of 1:8.

The measured S_{21} of the TRL circuits are shown in Fig. 8. As depicted in the figure, there are ripples and losses due to the taper structure and the microstrip feeding lines. These effects will exist when measuring the 2-port scattering parameter of the tunnel diode test samples, and thus has to be calibrated out using TRL before extracting the parameters for the equivalent model. The measured S_{21} of the tunnel diode with inkjet printed package and the tunnel diode with the traditional package are shown in Fig. 9 and Fig. 10, respectively. The de-embedded results calculated using TRL calibration are also included for comparisons. As shown in the figures, The ripples can be alleviated significantly and the de-embedded S_{21} is higher because the loss of the feeding lines is calibrated out. The measured S_{21} after TRL calibration will be used to extract the equivalent models of the test samples.

V. MODEL PARAMETERS EXTRACTION

As shown in Fig. 5, in addition to the junction capacitance (C_j) which is already provided in the datasheet, there are four remaining parameters including junction resistance (R_j), series resistance (R_s), and packaging parasitics (L_s and C_p) which have to be extracted by the measured results.

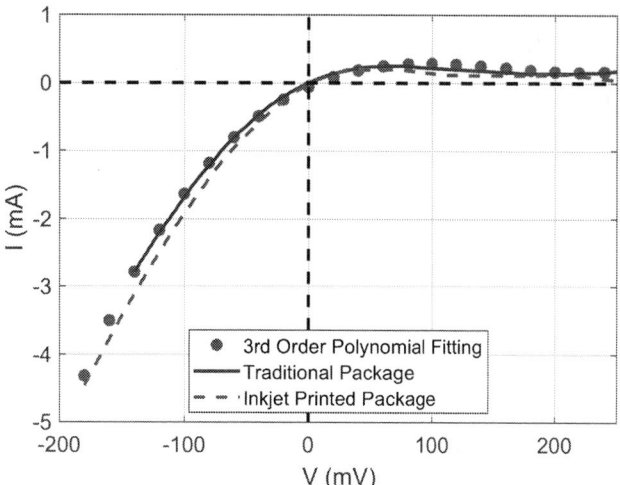

Fig. 11: The measured DC I-V Curves of the tunnel diodes.

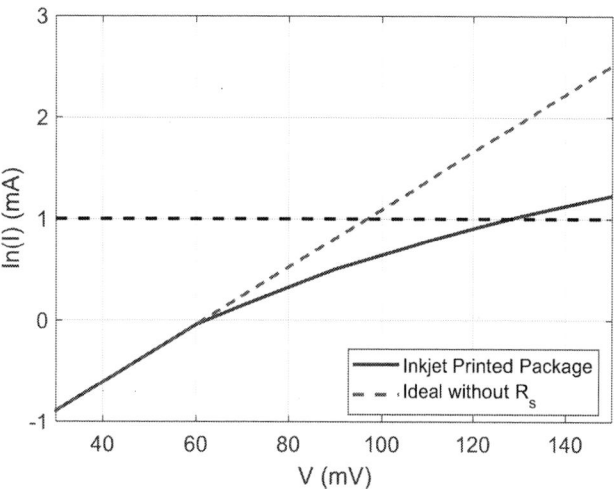

Fig. 12: The extraction of R_s for the tunnel diode with inkjet printed package.

A. Junction Resistance

The non-linear effects of the tunnel diode are modeled using the junction resistance. The measured DC I-V curve is shown in Fig. 11. The measured DC I-V curve has similar shape as the general tunnel diode as shown in Fig. 1. The region 2 is ranging from 0 mV to 60 mV and the region 3 is between 60 mV and 210 mV. Since it is measured in DC, it is only affected by the junction and series resistance. The overall resistance can be approximated with 3rd order polynomial fitting as the following

$$I(V) = 1.3 * 10^{-7}V^3 - 6.2 * 10^{-5}V^2 \\ + 8.3 * 10^{-3}V - 0.058 \quad (1)$$

The resulting curve is also included in Fig. 11 for comparisons and a good agreement can be observed. The junction resistance is the overall resistance minus the series resistance.

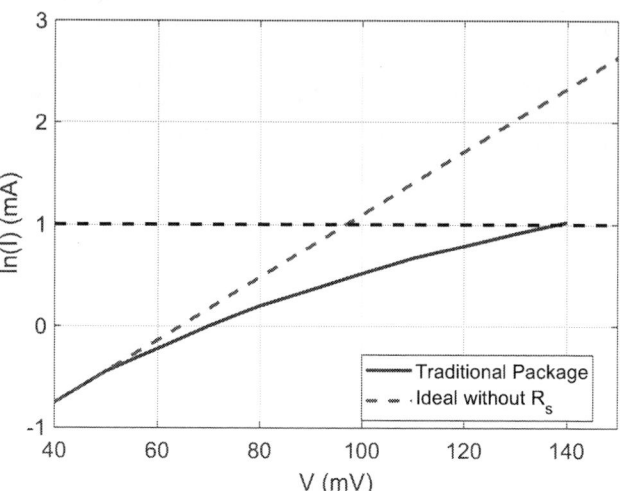

Fig. 13: The extraction of R_s for the tunnel diode with traditional package.

B. Series Resistance

The current through the diode can be expressed as [11]

$$I \approx I_s \exp(\frac{V - IR_s}{V_T}) \quad (2)$$

$$V_T = \frac{nkT}{q} \quad (3)$$

where I_s is the saturation current. Thus, by (2), we can know

$$ln(I) \approx ln(I_s) + \frac{V - IR_s}{V_T} \quad (4)$$

Therefore, if the series resistance is zero, the relationship between the natural log of current and the voltage is linear. Hence, we know that

$$ln(I) \approx ln(I_s) + \frac{V_1 - IR_s}{V_T} \quad (5)$$

$$ln(I) \approx ln(I_s) + \frac{V_2}{V_T} \quad (6)$$

where (5) is for the measured curve and (6) is the ideal curve without series resistance. Thus, for the same value of current and by (5) and (6), we can obtain

$$\frac{V_1 - IR_s}{V_T} = \frac{V_2}{V_T} \quad (7)$$

$$R_s = \frac{V_1 - V_2}{I} \quad (8)$$

The resulting figures for both the tunnel diode with inkjet printed package and the traditional package is shown in Fig. 12 and Fig. 13, respectively. The ideal linear tangential curve is created to calculate the series resistance. For the tunnel diode with inkjet printed package (Fig. 12), the V_1 and V_2 are 127.6 mV and 96.83 mV, respectively where the current is

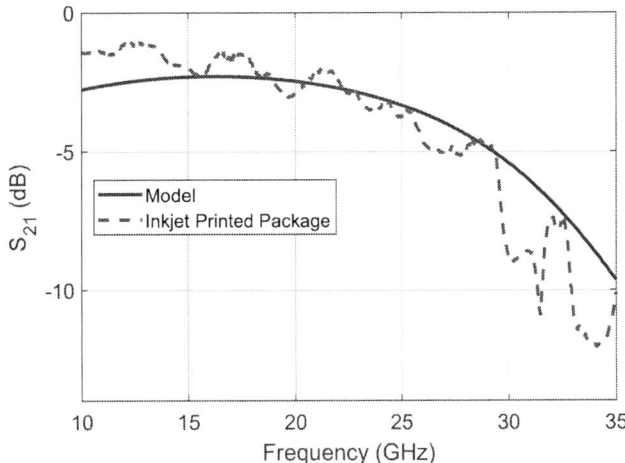

Fig. 14: The comparisons between the equivalent model and the measured S21 of the tunnel diode with inkjet printed package.

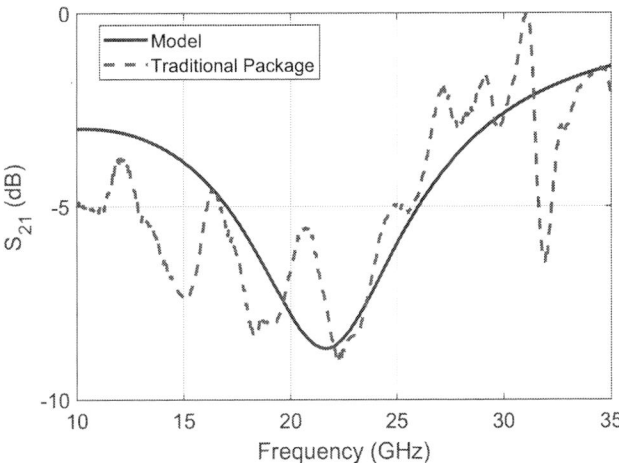

Fig. 15: The comparisons between the equivalent model and the measured S21 of the tunnel diode with traditional package.

e(2.71828) mA. Therefore, by (8), the series resistance for the tunnel diode with inkjet printed package is 11.3 Ω. Similarly, for the tunnel diode with traditional package (Fig. 13), the V_1 and V_2 are 138 mV and 97 mV, respectively where the current is e(2.71828) mA. Thus, the series resistance for the tunnel diode with the traditional package is 15 Ω. The series resistance for inkjet printed package is smaller than the traditional package, and thus the loss induced by the package is smaller.

C. Package Parasitics

The final step toward the extraction of the equivalent model is the package parasitics (C_p and L_s). These parameters are extracted by placing the equivalent model in Fig. 5 with the values of R_s and R_j from the previous sessions in the simulation software, advanced design system, and performs

Parameter	Inkjet Printed Package	Traditional Package
C_j	0.3 pF	0.3 pF
R_s	11.3 Ω	15 Ω
C_p	0.056 pF	0.12 pF
L_s	0.351 nH	0.61 nH

TABLE I: Parameter comparison between the inkjet printed package and the traditional package

the curve fitting to fit the measured S21 which is calibrated using TRL calibration.

The resulting S21 for the model and the tunnel diode with inkjet printed package is shown in Fig. 14. The good agreement between the equivalent model and the measured results can be observed and validate the correctness of the equivalent model. The same process is applied for the tunnel diode with the traditional package and the results are shown in Fig. 15. The good agreement can also be seen in the figure.

The extracted parameters for the equivalent models for both the inkjet printed package and traditional package are summarized in Table I. The inkjet printed package can support lower series resistance. Furthermore,the capacitance due to the traditional package is 0.12 pF which is 2.14 times of the inkjet printed package one, 0.056 pF. The inductance due to the traditional package is 0.61 nH which is 1.74 times of the inkjet printed package one, 0.351 nH. In summary, by extraction the parameters of the equivalent models for the tunnel diode with different packaging method, the inkjet printed package method outperforms the traditional method since both the resistance and the parasitics are lower. Thus, inkjet printed package is more suitable for high frequencies applications such as 5G and mm-wave.

VI. CONCLUSION

Adding to the foundation built by the results presented in [5] and [6], the work presented in this paper adds to the evidence supporting the superior electrical performance of additive packaging for systems integration, through the example of the interfacing of an mm-wave tunnel diode die. Using this scalable approach—a packaging scheme expandable to the assembly of ultra-compact and complex systems—the parasitics were shown to be reduced by up to 53%. While the testing and improvement of the mechanical and thermal properties required for the industrial scaling of this additive scheme (and for the guarantee of its long-term reliability) are still in need of attention and development, the remarkable electrical properties obtained in mm-wave regimes using low-cost additive packaging approaches should provide a strong motivator for these endeavors, and opens the prospect for the emergence of ultra-low-cost, ultra-compact, and ultra-high-performance mm-wave additive SiPs.

REFERENCES

[1] K. Lim, S. Pinel, M. Davis, A. Sutono, C.-H. Lee, D. Heo, A. Obatoynbo, J. Laskar, E. M. Tantzeris, and R. Tummala, "Rf-system-on-package (sop) for wireless communications," *IEEE Microwave Magazine*, vol. 3, no. 1, pp. 88–99, March 2002.

[2] T. Lin, R. Bahr, M. Tentzeris, R. Pulugurtha, V. Sundaram, and R. Tummala, "Novel 3D-/Inkjet-Printed Flexible On-package Antennas, Packaging Structures, and Modules for Broadband 5G Applications," in *2018 IEEE 68th Electronic Components and Technology Conference (ECTC)*, May 2018, pp. 214–220.

[3] T. Lin, P. M. Raj, A. Watanabe, V. Sundaram, R. Tummala, and M. M. Tentzeris, "Nanostructured miniaturized artificial magnetic conductors (AMC) for high-performance antennas in 5G, IoT, and smart skin applications," in *2017 IEEE 17th International Conference on Nanotechnology (IEEE-NANO)*, July 2017, pp. 911–915.

[4] B. K. Tehrani, B. S. Cook, and M. M. Tentzeris, "Inkjet-printed 3d interconnects for millimeter-wave system-on-package solutions," in *2016 IEEE MTT-S International Microwave Symposium (IMS)*, May 2016, pp. 1–4.

[5] C. H. P. Lorenz, S. Hemour, W. Li, Y. Xie, J. Gauthier, P. Fay, and K. Wu, "Breaking the efficiency barrier for ambient microwave power harvesting with heterojunction backward tunnel diodes," *IEEE Trans. Microw. Theory Techn.*, vol. 63, no. 12, pp. 4544–4555, Dec 2015.

[6] F. Amato, C. W. Peterson, M. B. Akbar, and G. D. Durgin, "Long range and low powered rfid tags with tunnel diode," in *2015 IEEE International Conference on RFID Technology and Applications (RFID-TA)*, Sep. 2015, pp. 182–187.

[7] T. Lin, J. Bito, J. G. D. Hester, J. Kimionis, R. A. Bahr, and M. M. Tentzeris, "On-Body Long-Range Wireless Backscattering Sensing System Using Inkjet-/3-D-Printed Flexible Ambient RF Energy Harvesters Capable of Simultaneous DC and Harmonics Generation," *IEEE Trans. Microw. Theory Techn.*, vol. 65, no. 12, pp. 5389–5400, Dec 2017.

[8] T. Lin, W. Su, and M. M. Tentzeris, "Expand Horizons of Microfluidic Systems: An Inkjet Printed Flexible Energy Autonomous Micropump System for Wearable and IoT Microfluidic Applications," in *2018 IEEE/MTT-S International Microwave Symposium - IMS*, June 2018, pp. 812–815.

[9] T. Lin, J. Bito, and M. M. Tentzeris, "Wearable inkjet printed energy harvester," in *2017 IEEE International Symposium on Antennas and Propagation USNC/URSI National Radio Science Meeting*, July 2017, pp. 1613–1614.

[10] B. K. Tehrani, C. Mariotti, B. S. Cook, L. Roselli, and M. M. Tentzeris, "Development, characterization, and processing of thin and thick inkjet-printed dielectric films," *Organic Electronics*, vol. 29, pp. 135 – 141, 2016.

[11] T. Kiuru, J. Mallat, A. V. Raisanen, and T. Narhi, "Schottky diode series resistance and thermal resistance extraction from *s*-parameter and temperature controlled iv measurements," *IEEE Trans. Microw. Theory Techn.*, vol. 59, no. 8, pp. 2108–2116, Aug 2011.

Feasibility Study of Fan-Out Wafer-Level Packaging for Heterogeneous Integrations

John Lau[1], Ming Li[1], Iris Xu[2], Tony Chen[2], Kim Hwee Tan[2], Zhang Li[2], Nelson Fan[1], Eric Kuah[1], Raymond So[1], Penny Lo[1], Y. M. Cheung[1], Cao Xi[3], Rozalia Beica[4], Sze Pei Lim[5], NC Lee[5], Cheng-Ta Ko[6], Henry Yang[6], YH Chen[6], Mian Tao[7], Jeffery Lo[7], and Ricky Lee[7]

[1]ASM Pacific Technology Ltd
[2]Jiangyin Changdian Advanced Packaging Co., Ltd.
[3]Huawei Technologies Co. Ltd.
[4]Dow Chemical Company
[5]Indium Corporation
[6]Unimicron Technology Corporation
[7]Hong Kong University of Science and Technology.
Ph: 852-2619-2757, Email: john.lau@asmpt.com

ABSTRACT

The design, materials, process, and fabrication of a heterogeneous integration of 4 chips by a FOWLP (fan-out wafer-level packaging) with chip-first and dies face-down method are investigated in this study. Emphasis is placed on the application of a new assembly process and materials for fabricating the RDLs (redistribution layers) of the heterogeneous integration. The size of the reconstituted wafer is 300mm. The epoxy molding compound (EMC) is a dry-film material and is molded by lamination method. The minimum metal line width and spacing of the RDLs is 5µm.

(1) Introduction

Heterogeneous integration uses packaging technology to integrate dissimilar chips, photonic devices, and/or components (side-by-side, stack, or both) with different materials and functions, and from different fabless design houses, foundries, wafer sizes, feature sizes and companies into a system or subsystem [1]. In general, heterogeneous integrations can be classified [1] as heterogeneous integrations on organic substrates [2-4], on silicon substrates with TSV (through-silicon via)-interposers [5-7], on silicon substrates with TSV-less interposers such as bridges [8-11], on fan-out RDL (redistribution-layer) substrates [12-16], and on ceramic substrates [17]. For the next few years, there will be more implementations of a higher level of heterogeneous integrations on these various substrates, whether it is for performance, form factor, power consumption, signal integrity, or cost [1]. In this study, heterogeneous integration on organic substrate will be investigated.

There are many fan-out wafer-level packaging (FOWLP) formations [18-33]. However, basically there are three different kinds, namely chip-first (die face-down) [19-24], chip-first (die face-up) [25-31], and chip-last or RDL-first [32, 33]. In this study, chip-first (die face-down) formations will be employed.

In [22], we had demonstrated the chip-first and die face-down fan-out wafer-level packaging for heterogeneous integration. The liquid EMC (epoxy molding compound) was compression molded. The metal line width and spacing of the RDLs (redistributed layers) were 10µm and 15µm. In this study, instead of using compression molding of the liquid

EMC reported in [22], herein we use a lamination method of a new dry-film EMC. Also, the metal line width and spacing of the RDLs will be reduced from 10µm to 5µm. Furthermore, in order to save the expensive EMC materials and achieve low-profile (thin) package, a special assembly process will be developed.

The dimensions of the package are 10mm x 10mm x 300µm, which consists of one 5mm x 5mm chip, and three 3mm x 3mm chips. The spacing between the large chip and the small chip is only 100µm. One practical application of this kind of package is for the application processor chipset, i.e., the large chip could be a processor and the small chips could be memories. The temporary carrier is a 300mm glass wafer, which supports the chips with a 2-side thermal release tape. In order to have a very low-profile package and save the EMC material, the reconstituted wafer is attached to another temporary glass wafer with a coated light-to-heat-conversion (LTHC) layer. Then, remove the first glass carrier and peel off the tape. It is followed by making the RDLs with a photosensitive polyamide for the dielectric layers and electroplating Cu for the conductor layers. Then, mount the solder ball and debond the second glass carrier by scanning a laser. SEM (scanning electron microscopy) shows that the 5µm metal line width and spacing RDL are properly made. C-mode SAM (scanning acoustic microscope) demonstrates that there are not voids in the EMC even in the 100µm gap between the large and small chips. This is a very high-throughput process. In one shot, 629 (10mm x 10mm) heterogeneous integration packages can be made.

(2) Test Chips

Figure 1 shows the test chips under consideration. The layout and the fabricated large test chip are shown in Figure 1(a). It can be seen that the large chip sizes are 5mm x 5mm x 150µm and there are 160 pads with a pitch = 100µm (the inner rows). The SiO_2 passivation opening of the Al-pad is 50µm x 50µm and the size of the Al-pad is 70µm x 70µm. The dimensions of the small chip are 3mm x 3mm x 150µm and the fabricated chip is shown in Figure 1(b). It can be seen that there are 80 pads and are on 100µm-pitch (inner rows). The cross section and dimensions of the pads of the small chip are the same as those of the large chip.

978-1-7281-1500-9/19 $31.00 © 2019 IEEE

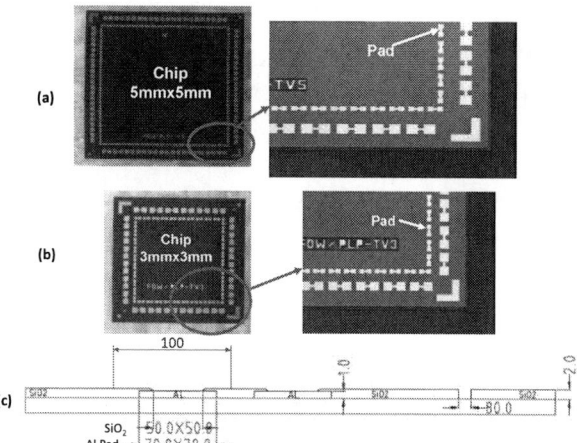

Figure 1 Layout and fabricated test chips. (a) 5mm x 5mm with 160 pads on 100μm-pitch (inner rows). (b) 3mm x 3mm with 80 pads on 100μm-pitch (inner rows).

(3) Test Package

Figure 2 schematically shows the test package under consideration. The dimensions of the test package are: 10mm x 10mm and it consists of one large chip (5mm x 5mm) and three small chips (3mm x 3mm). The spacing (gap) between the large chip and the small chip is 100μm. There are two RDLs of the test package. The RDLs between the chips and RDL1 is shown in Figure 3(a), between the RDL1 and RDL2 is shown in Figure 3(b), and between the RDL2 and the PCB is shown in Figure 3(c). Figure 3(d) shows the footprint of the test package. These packages are to be made from a 300mm glass reconstituted wafer and the pitch of the test package on the wafer is 10.2mm. In real applications, the large chip could be an application processor and the small chips could be memories.

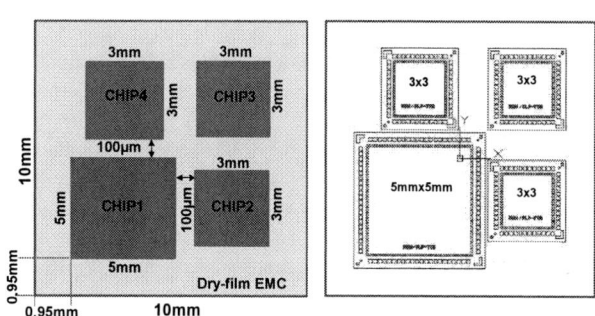

Figure 2 Schematic of the heterogeneous integration of 4 chips in a package. The dimensions of the package are 10mm x 10mm.

Figure 4 schematically shows the cross-sectional view of the test package. It can be seen that there are 2 RDLs and the thickness of RDL1 is 3μm and RDL2 is 7.5μm. The metal line width and spacing of RDL1 are 5μm and those of RDL2 are 15μm, respectively. The dielectric layer thickness of DL1 and DL2 is 5μm, and DL3 is 10μm.

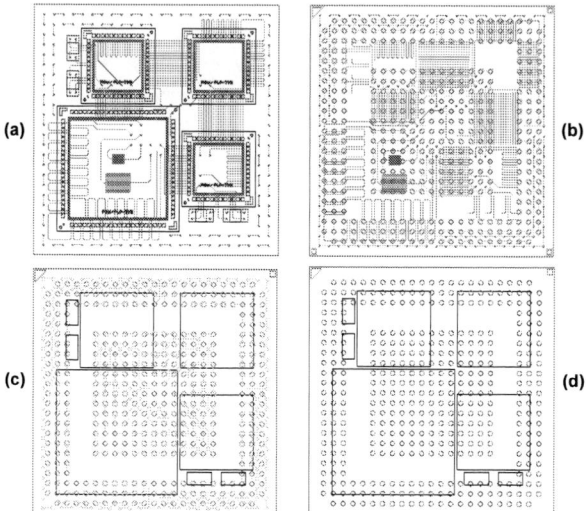

Figure 3 Test package layouts. (a) RDLs between test chips and RDL1. (b) RDLs between RDL1 and RDL2. (c) RDLs between RDL2 and PCB. (d) Footprint of the test package bottom.

The via (V_{C1}), through the first dielectric layer (DL1), connecting the Cu contact-pad of the test chip to the first RDL (RDL1) is 20-30μm in diameter. The pad-diameter on the RDL1 is 55μm, which is connected to RDL2 through the via (V_{12}) with a diameter of 30-40μm. Similarly, the pad-diameter on the RDL2 is 65μm. Finally, 220μm solder-ball Cu pads are formed on RDL2. The opening of the passivation (DL3) is 180μm. The solder ball size is 200μm and ball pitch is 0.4mm.

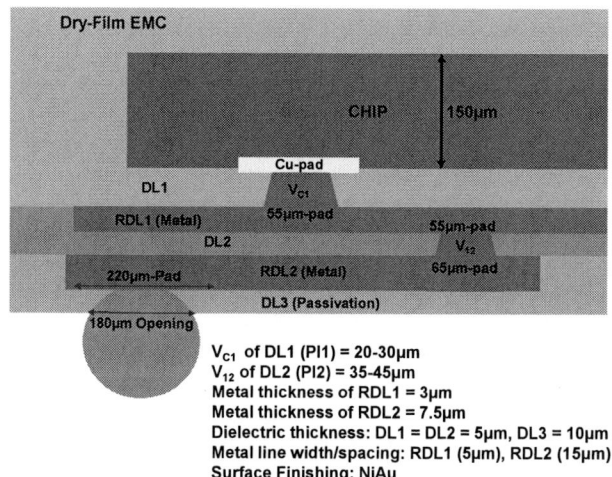

Figure 4 Schematic of the cross section of the test package

(4) Conventional Chip-First (Face-Down) Wafer Process

Figure 5 shows, in general, the conventional process flow of chip-first with die face-down FOWLP [18]. First, the device wafer is tested for known-good dies (KGDs) and then singulated into individual dies. This is followed by picking up the KGDs and placing them face-down on a temporary carrier (which can be metal, silicon, or glass) that can be round

978-1-7281-1500-9/19 $31.00 © 2019 IEEE   904

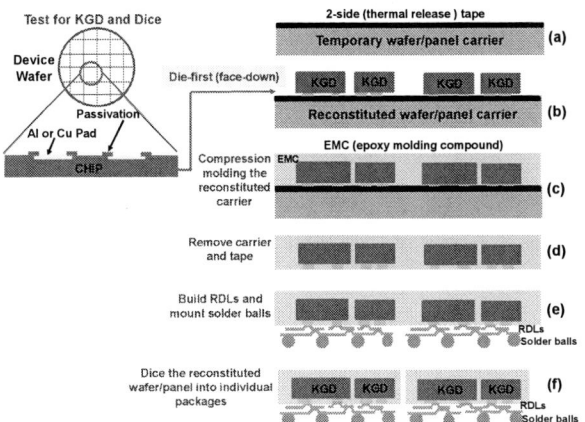

Figure 5 Conventional FOWLP with chip-first face-down.

(wafer) or rectangular (panel), Figure 5(b), with a double-sided thermal release tape, Figure 5(a). Then, the reconstituted (temporary) carrier with the KGDs are molded with EMC, Figure 5(c), using the compression method + PMC (post mold cure) before removing the carrier and the peeling off the double-sided tape, Figure 5(d). Next comes building the RDLs 5(e). Finally, solder balls are mounted and the whole reconstituted carrier (with KGDs, RDLs, and solder balls) is diced into individual packages, Figure 5(f).

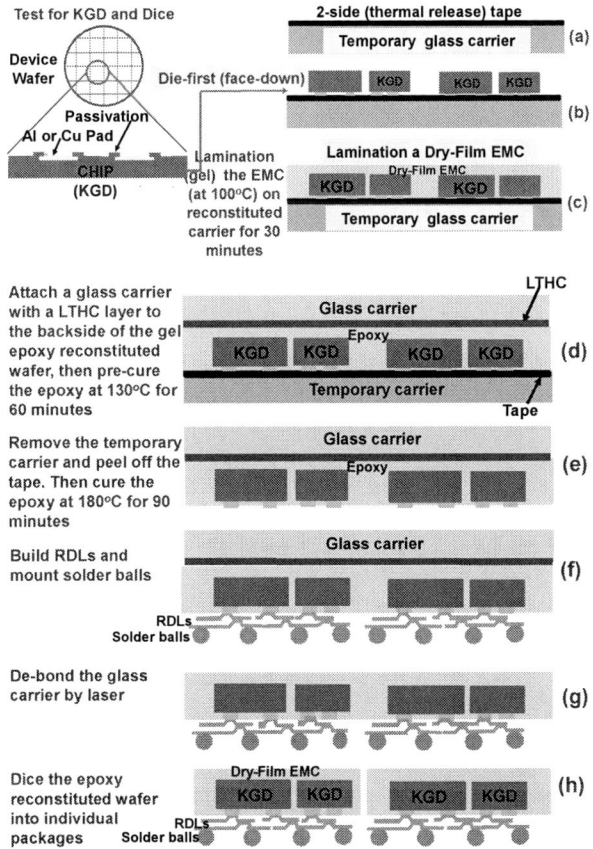

Figure 6 New FOWLP with chip-first and dies face-down.

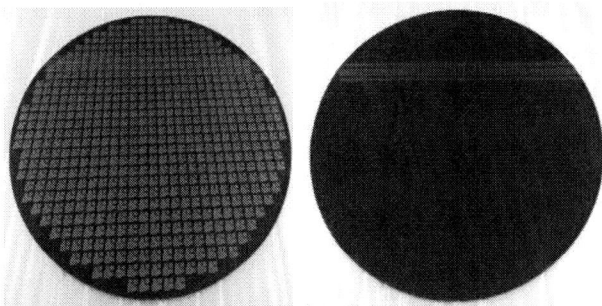

Figure 7 Right after the step of Figure 6(c). (L) Glass-side. (R) Dry-film EMC-side.

(5) New Process for Heterogeneous Integration Package

In the present study, we will not use the conventional method such as Figure 5(c) to compression molding the EMC and Figure 5(e) to build the RDLs. Instead, we will use the new process shown in Figure 6.

(5A) Dry-Film EMC Lamination

In [22], we had used a liquid EMC and compression molding method. In this study, we will use a new dry-film EMC with material properties shown in Table 1 and the EMC molding method is by lamination. After pick and place the KGDs on the carrier, Figures 6(a) and 6(b), the 200μm-thick dry-film EMC is laminated on the reconstituted wafer at 100°C for 30 minutes as shown in Figure 6(c) and Figure 7.

(5B) Temporary Bonding Another Glass Carrier

In the conventional FOWLP, after EMC molding, then debond the carrier and peel off the tape, Figure 5(d), the total thickness of the reconstituted wafer is usually ≥ 450μm. It is followed by building the RDLs and mounting the solder balls, Figure 5(e). However, in this study, in order to save the expensive EMC materials and have a very low profile (thin) package, the total thickness of our reconstituted wafer without the carrier is only 300μm. Thus, the reconstituted wafer is too fragile to fabricate the RDLs and mount the solder balls.

TABLE 1 Material Properties of Dry-Film EMC

Min. Melting Viscosity (ps)		3590 (120μm)
Cure condition (degC x min.)		180 x 90
Tg (degC)	DMA	189
	TMA	171
CTE (30-150degC)(ppm/K)	TMA	10
CTE (150-240degC)(ppm/K)		26
Young's modulus (MPa)	Tensile mode	13,000
Breaking strength (MPa)		82
Elongation (%)		1.0

One of the solutions is to attach the thin reconstituted wafer on another 1mm-thick glass (CTE = 6.4x10^{-6}/°C) wafer with a coated LTHC (light-to-heat-conversion) layer as shown in Figure 6(d). Then, pre-cure the dry-film EMC at 130°C for 60 minutes. Next, remove the temporary glass carrier and peel off the tape, Figure 6(e). It is followed by curing the dry-film EMC at 180°C for 90 minutes.

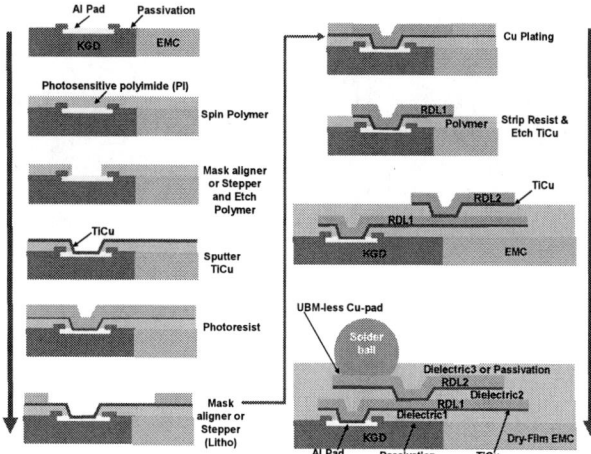

Figure 8 RDLs key process steps.

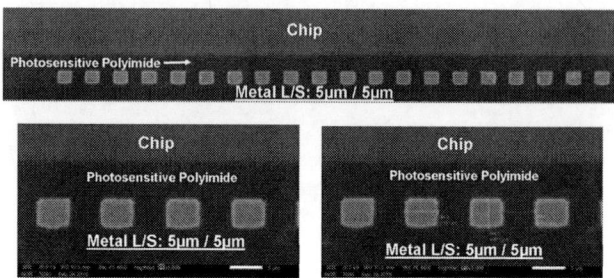

Figure 10 Cross section images of the chip, photosensitive polyimide, and RDL1s. The metal line width and spacing of RDL1 are 5μm.

Figure 10 shows the cross section images of the RDL1. It can be seen that the chip, the photosensitive polyimide, and the RDL1 with metal line width and spacing = 5μm.

(5D) Solder Ball Mounting

There are two different stencils for the solder ball mounting, Figure 6(f), one is for stencil printing the flux and the other is for stencil mounting the solder balls. The solder (Sn3wt%Ag0.5wt%Cu) balls (200μm-diameter) used are on a 0.4mm-pitch. The peak temperature for solder reflow is 245°C.

(5C) RDLs

Now, it is ready to build the RDLs as shown in Figure 6(f). Figure 8 shows the key process steps in making the RDLs. First, spin coat a photosensitive polyimide (PI) on the reconstituted wafer. Then apply a stepper (every 4 test packages as a unit) and use photolithography techniques to align, expose, and develop the vias of the PI. Finally, cure the PI at 200°C for one hour. This will form a 4 to 5-μm-thick PI layer. (PI development.) It is followed by Sputtering Ti and Cu by physical vapor deposition (175 to 200°C) over the entire wafer. Apply a photoresist and a stepper and then use photolithography techniques to open the redistribution-traces locations. Then electroplate Cu by electrochemical deposition (room temperature) on Ti/Cu in photoresist openings. Strip off the photoresist and etch off the Ti/Cu. (RDL1 is obtained.) Repeat the above steps to obtain RDL2. Figure 9 shows the metal line width and spacing of RDL1. It can be seen that the metal line width and spacing of RDL1 are very close to the design value (5μm).

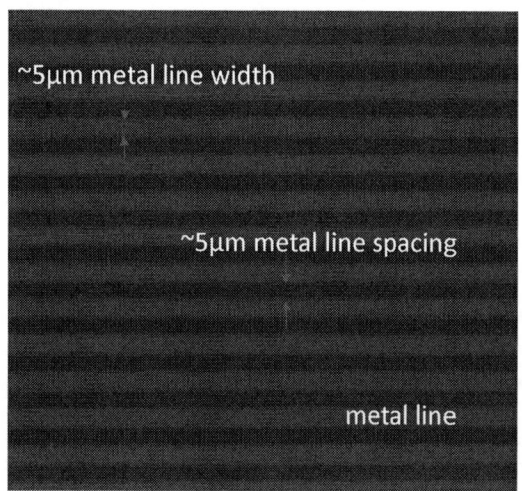

Figure 9 Top view of the metal line width and spacing of RDL1.

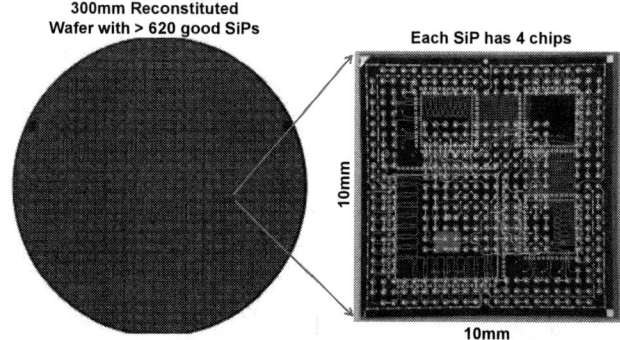

Figure 11 300mm reconstituted wafer with > 620 good SiPs. Each SiP (10mm x 10mm) has one large chip (5mm x 5mm) and three small chips (3mm x 3mm).

(5E) Final De-Bonding

The de-bonding of the glass carrier as shown in Figure 6(g) is by scanning a laser (355nm DPSS Nd: YAG UV laser source is used) from the glass carrier side. The laser spot-size is 240μm, the scanning speed is 500mm/s and the scanning pitch is 100μm. When the LTHC layer "sees" the laser light, it converts into powders and the glass carrier is easily removed. It is followed by chemical cleaning. Figure 11 shows the reconstituted wafer without any carrier and a closed-up on one of the package. It can be seen that there are 4 chips in a package and they are properly fabricated. The reconstituted wafer is diced, as shown in Figure 6(h), into individual packages as shown in the x-ray image in Figure 12, where also shows the RDL1 and RDL2 of the package.

978-1-7281-1500-9/19 $31.00 © 2019 IEEE

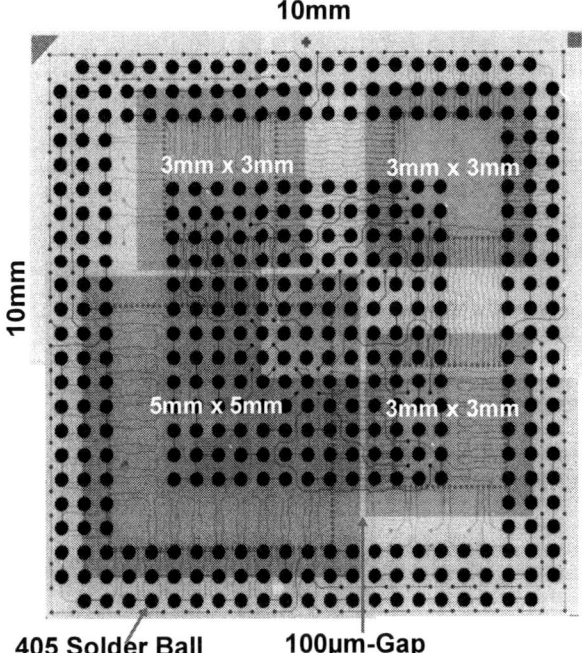

Figure 12 X-ray image of the heterogeneous integration of one large chip (5mm x 5mm) and three small chips (3mm x 3mm). The dimensions of the SiP are 10mm x 10mm and there're 405 200µm-diameter solder balls on 0.4mm-pitch.

Figure 13 shows a typical cross section of the heterogeneous integration package. It can be seen that there are two RDLs. The thickness of RDL2 (7.5µm) is thicker than that of RDL1 (3µm). The thicker RDL2 is for the UBM-less thicker Cu pads to "resist" the Cu consumption from the solder ball reflow and during operation.)

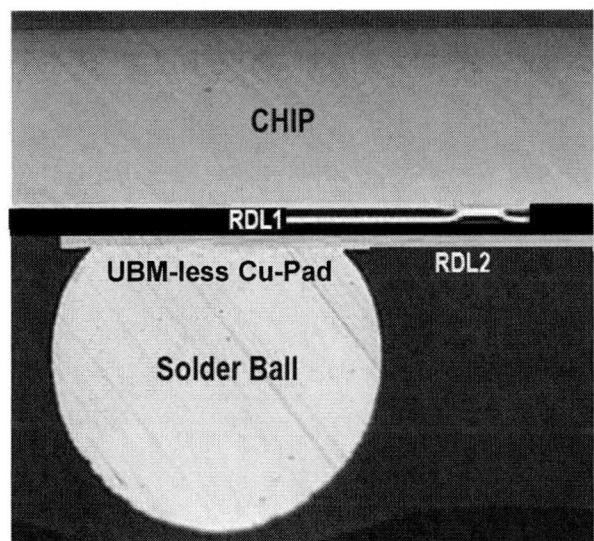

Figure 13 Cross section of the individual package

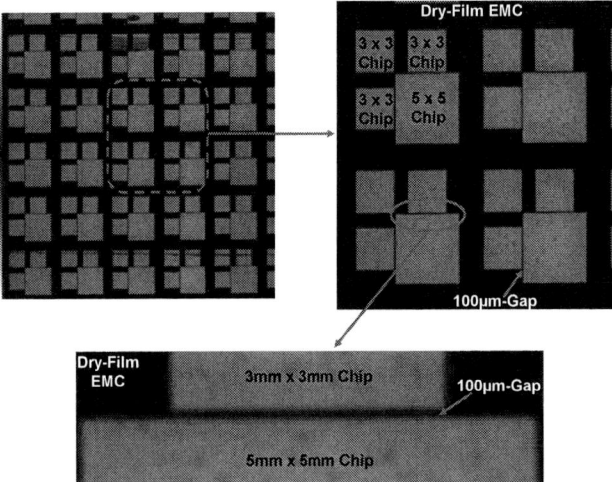

Figure 14 C-mode SAM image showing there is not any obvious void in the dry-film EMC even in the 100µm-gap between the large chip and small chips.

Inspections for dry-film EMC lamination molding voids are carried out by C-SAM (C-mode scanning acoustic microscopy). In order to balance the resolution and signal penetration depth, a transducer of 75MHz is selected for the voids observation. After a couple of parametric studies, there is not any void in the optimal wafers (EMC as well as gap) as shown in Figure 14.

(8) Summary and Recommendations

The design, materials, process, and fabrication of a thin heterogeneous integration of 4 chips by a new FOWLP method have been investigated. Some important results and recommendations are summarized as follows.

➢ The minimum metal line width and spacing of the RDLs = 5µm have been fabricated. The accuracy has been demonstrated by both optical and SEM images.

➢ The gap between the large chip (5mmx5mm) and the small chip (3mmx3mm) is 100µm. It has been found by the C-SAM method that there is not any void in the dry-film EMC and the gap.

➢ A dry-film EMC has been chosen and a lamination method of the dry-film EMC has been developed.

➢ Both x-ray and optical microscope images verified that the 4 chips have been assembled properly.

➢ Both x-ray and optical microscope images demonstrated that the RDLs have been properly fabricated.

➢ The heterogeneous integration package thickness is 300µm (not including the solder balls). It is not only save expensive EMC cost but leads to a very low-profile (thin) package.

➢ The proposed assembly process, i.e., by attaching the thin reconstituted molded wafer on another temporary glass wafer with a coated LTHC layer, provided a better warpage control, an opportunity for making supper low-profile package, and a way to save the EMC material.

➢ The thin heterogeneous integration package has been assembled on a PCB and through a shock (drop) test and thermal cycling test. The reliability results will be presented at the conference.

(9) Acknowledgements

The authors would like to thank their upper managements from ASM, DOW, Huawei, Indium, JCAP, and Unimicron for their strong support of this consortium project. The constructive contributions from TJ Tseng, CM Lai, Marc Lin, Casper Tsai, YM Chan, Leslie Chang, TW Lam, JW Dong, and Jiang Leon are greatly appreciated.

References

[1] Lau, J. H., *Heterogeneous Integrations*, Springer, 2019.

[2] Shimizu, N., Kaneda, W., Arisaka, H., Koizumi, N., Sunohara, S., Rokugawa, A., and Koyama, T., "Development of Organic Multi Chip Package for High Performance Application," *IMAPS International Symposium on Microelectronics*, Orlando, FL, Sep. 30–Oct. 3, 2013, pp. 414–419.

[3] Oi, K., Otake, S., Shimizu, N., Watanabe, S., Kunimoto, Y., Kurihara, T., Koyama, T., Tanaka, M., Aryasomayajula, L., and Kutlu, Z., "Development of New 2.5D Package With Novel Integrated Organic Interposer Substrate With Ultra-Fine Wiring and High Density Bumps," *IEEE 64th Electronic and Components Technology Conference*, Orlando, FL, May 27–30, 2014, pp. 348–353.

[4] Li, L., P. Chia, P. Ton, M. Nagar, S. Patil, J. Xue, J. DeLaCruz, M. Voicu, J. Hellings, B. Isaacson, M. Coor, and R. Havens, "3D SiP with organic interposer of ASIC and memory integration," *Proceedings of IEEE/ECTC*, May 2016, pp. 1445-1450.

[5] Hou, S., W. Chen, C. Hu, C. Chiu, K. Ting, T. Lin, W. Wei, W. Chiou, V. Lin, V. Chang, C. Wang, C. Wu, and D. Yu, "Wafer-Level Integration of an Advanced Logic-Memory System Through the Second-Generation CoWoS Technology", *IEEE Transactions on Electron Devices*, October 2017, pp. 4071-4077.

[6] Yu, A., J. H. Lau, S. Ho, A. Kumar, W. Hnin, W. Lee, M. Jong, V. Sekhar, V. Kripesh, D. Pinjala, S. Chen, C. Chan, C. Chao, C. Chiu, C. Huang, and C. Chen, "Fabrication of High Aspect Ratio TSV and Assembly with Fine-Pitch Low-Cost Solder Microbump for Si Interposer Technology with High-Density Interconnects", *IEEE Transactions on CPMT*, Vol. 1, No. 9, September 2011, pp. 1336-1344.

[7] Selvanayagam, C., J. H. Lau, X. Zhang, S. Seah, K. Vaidyanathan, and T. Chai, "Nonlinear Thermal Stress/Strain Analyses of Copper Filled TSV (Through Silicon Via) and Their Flip-Chip Microbumps", *IEEE Transactions on Advanced Packaging*, Vol. 32, No. 4, November 2009, pp. 720-728.

[8] Chiu, C., Z. Qian, and M. Manusharow, "Bridge interconnect with air gap in package assembly," *US Patent No. 8,872,349*, 2014.

[9] Mahajan, R., R. Sankman, N. Patel, D. Kim, K. Aygun, Z. Qian, et al., "Embedded multi-die interconnect bridge (EMIB) – a high-density, high-bandwidth packaging interconnect," *IEEE/ECTC Proceedings*, May 2016, pp. 557-565.

[10] Podpod, A., J. Slabbekoorn, A. Phommahaxay, F. Duval, A. Salahouedlhadj, M. Gonzalez, K. Rebibis, R.A. Miller, G. Beyer, and E. Beyne, "A Novel Fan-Out Concept for Ultra-High Chip-to-Chip Interconnect Density with 20-μm Pitch", *IEEE/ECTC Proceedings*, May 2018, pp. 370-378.

[11] Lau, J. H., C. Lee, C. Zhan, S. Wu, Y. Chao, M. Dai, R. Tain, et al., "Through-Silicon Hole Interposers for 3D IC Integration," *IEEE Transactions on CPMT*, Vol. 4, no. 9, September 2014, pp. 1407–1418.

[12] Pendse, R., "Semiconductor device and method of forming extended semiconductor device with fan-out interconnect structure to reduce complexity of substrate," filed on Dec. 23, 2011, US 2013/0161833 A1, pub. date: June 27, 2013.

[13] Yoon, S. W., P. Tang, R. Emigh, Y. Lin, P. C. Marimuthu, and R. Pendse, "Fan-out flip-chip eWLB (embedded wafer-level ball grid array) technology as 2.5D packaging solutions," *Proc. of IEEE/ECTC*, May 2013, pp. 1855–1860.

[14] Lin, Y., W. Lai, C. Kao, J. Lou, P. Yang, C. Wang, et al., "Wafer warpage experiments and simulation for fan-out chip-on-substrate," *Proc. of IEEE/ECTC*, May 2016, pp. 13-18.

[15] Lee, Y., W. Lai, I. Hu, M. Shih, C. Kao, D. Tarng, and C. Hung, "Fan-Out Chip on Substrate Device Interconnection Reliability Analysis", *Proceedings of IEEE/ECTC*, May 2017, pp. 22-27.

[16] Suk, K., S. Lee, J. Kim, S. Lee, H. Kim, S. Lee, et al., "Low-cost Si-less RDL interposer package for high-performance computing applications," *Proc. of IEEE/ECTC*, May 2018, pp. 64-69.

[17] Garrou, P. E., and I. Turlik, *Multichip Module Technology Handbook*, McGraw-Hill, 1997.

[18] Lau, J. H., *Fan-Out Wafer-Level Packaging*, Springer, 2018.

[19] Brunnbauer, M., E. Furgut, G. Beer, T. Meyer, H. Hedler, J. Belonio, E. Nomura, K. Kiuchi, and K. Kobayashi, "An Embedded Device Technology Based on a Molded Reconfigured Wafer", *IEEE/ECTC Proceedings*, May 2006, pp. 547-551.

[20] Brunnbauer, M., E. Furgut, G. Beer, and T. Meyer, "Embedded Wafer Level Ball Grid Array (eWLB)", *IEEE/EPTC Proceedings*, December 2006, pp. 1-5.

[21] Ko, CT, H. Yang, J. H. Lau, M. Li, M. Li, C. Lin, J. W. Lin, T. Chen, I. Xu, C. Chang, J. Pan, H. Wu, Q. Yong, N. Fan, E. Kuah, Z. Li, K. Tan, Y. Cheung, E. Ng, K. Wu, J. Hao, R. Beica, M. Lin, Y. Chen, Z. Cheng, S. Koh, R. Jiang, X. Cao, S. Lim, N. Lee, M. Tao, J. Lo, and R. Lee, "Chip-First Fan-Out Panel-Level Packaging for Heterogeneous Integration", *IEEE Transactions on CPMT*, September 2018, pp. 1561-1572.

[22] Lau, J. H., M. Li, M. Li, T. Chen, I. Xu, X. Qing, Z. Cheng, N. Fan, E. Kuah, Z. Li, K. Tan, Y. Cheung, E. Ng, P. Lo, K. Wu, J. Hao, S. Koh, R. Jiang, X. Cao, R. Beica, S. Lim, N. Lee, C. Ko, H. Yang, Y. Chen, M. Tao, J. Lo, and R. Lee, "Fan-Out Wafer-Level Packaging for

Heterogeneous Integration", *IEEE Transactions on CPMT*, 2018, September 2018, pp. 1544-1560.

[23] Lau, J. H., M. Li, Y. Lei, M. Li, I. Xu, T. Chen, Q. Yong, Z. Cheng, K. Wu, P. Lo, Z. Li, K. Tan, Y. Cheung, N. Fan, E. Kuah, X. Cao, J. Ran, R. Beica, S. Lim, NC Lee, C. Ko, H. Yang, Y. Chen, M. Tao, J. Lo, and R. Lee, "Reliability of Fan-Out Wafer-Level Heterogeneous Integration", *IMAPS Transactions, Journal of Microelectronics and Electronic Packaging*, Vol. 15, Issue: 4, October 2018, pp. 148-162.

[24] Ko, CT, H. Yang, J. H. Lau, M. Li, M. Li, C. Lin, J. W. Lin, C. Chang, J. Pan, H. Wu, Y. Chen, T. Chen, I. Xu, P. Lo, N. Fan, E. Kuah. Z. Li, K. Tan, C. Lin, R. Beica, M. Lin, X. Cao, S. Lim, NC Lee, M. Tao, J. Lo, and R. Lee, "Design, Materials, Process, and Fabrication of Fan-Out Panel-Level Heterogeneous Integration", *IMAPS Transactions, Journal of Microelectronics and Electronic Packaging*, Vol. 15, Issue: 4, October 2018, pp. 141-147.8-162.

[25] Yu, D., "Wafer-level system integration (WLSI) technologies for 2D and 3D system-in-package," *SEMIEUROPE* 2014.

[26] Lin, J., J. Hung, N. Liu, Y. Mao, W. Shih, and T. Tung, "Packaged semiconductor device with a molding compound and a method of forming the same," *US Patent 9,000,584*, Filed on December 28, 2011, Patented on April 7, 2015.

[27] Tseng, C., Liu, C., Wu, C., and D. Yu, "InFO (Wafer Level Integrated Fan-Out) Technology", *IEEE/ECTC Proceedings*, 2016, pp. 1- 6.

[28] Lau, J. H., M. Li, D. Tian, N. Fan, E. Kuah, K. Wu, M. Li, J. Hao, Y. Cheung, Z. Li, K. Tan, R. Beica, T. Taylor, C.T. Lo, H. Yang, Y. Chen, S. Lim, N.C. Lee, J. Ran, X. Cao, S. Koh, and Q. Young, "Warpage and thermal characterization of fan-out wafer-level packaging," *IEEE Transactions on CPMT*, Vol. 7, No. 10, pp. 1729-1738, 2017.

[29] Lau, J. H., M. Li, N. Fan, E. Kuah, Z. Li, K. Tan, T. Chen, I. Xu, M. Li, Y. M. Cheung, Wu Kai, Ji Hao, R. Beica, T. Taylor, C. Ko, H. Yang, Y. Chen, S. Lim, N. Lee, J. Ran, K. Wee, Q. Yong, C. Xi, M. Tao, J. Lo, and R. Lee, "Fan-out wafer-level packaging (FOWLP) of large chip with multiple redistribution layers (RDLs)," *IMAPS Transactions, Journal of Microelectronics and Electronic Packaging*, Vol. 14, No. 4, pp. 123-131, 2017.

[30] Lau, J. H., M. Li, Y. Li, M. Li, I. Au, T. Chen, S. Chen, Q. Yong, J. Madhukumar, K. Wu, N. Fan, E. Kuah, Z. Li, K. Tan, W. Bao, S. Lim, R. Beica, C. Ko, and X. Cao, "Warpage measurements and characterizations of FOWLP with large chips and multiple RDLs," *IEEE Transactions on CPMT*, Vol. 8, No. 10, pp. 1729-1737, 2018.

[31] Lau, J. H., M. Li, Q. Li, I. Xu, T. Chen, Z. Li, K. Tan, X. Qing, C. Zhang, K. Wee. R. Beica, C. Ko, S. Lim, N. Fan, E. Kuah, K. Wu, Y. Cheung, E. Ng, X. Cao, J. Ran, H. Yang, Y. Chen, N. Lee, M. Tao, J. Lo, and R. Lee, "Design, Materials, Process, and Fabrication of Fan-Out Wafer-Level Packaging", *IEEE Transactions on CPMT*. June, 2018, pp. 991-1002.

[32] Kurita, Y., T. Kimura, K. Shibuya, H. Kobayashi, F. Kawashiro, N. Motohashi, et al., "Fan-out wafer-level packaging with highly flexible design capabilities," *IEEE/ESTC Proceedings*, September 2010, pp. 1-6.

[33] Motohashi, N., T. Kimura, K. Mineo, Y. Yamada, T. Nishiyama, K. Shibuya, et al., "System in wafer-level package technology with RDL-first process," *IEEE/ECTC Proceedings*, 2011, pp. 59-64.

Experiment of 22FDX® Chip Board Interaction (CBI) in Wafer Level Packaging Fan-Out (WLPFO)

Jae Kyu Cho[1], Jens Paul[2], Simone Capecchi[2], Frank Kuechenmeister[2], Ta-Chien Cheng[3]

1. GLOBALFOUNDRIES Fab8, 400 Stone Break Road Extension, Malta, NY 12020, USA
e-mail: jaekyu.cho@globalfoundries.com

3. GLOBALFOUNDRIES Fab7, 60 Woodlands Industrial Park D Street 2, Singapore, Singapore

2. GLOBALFOUNDRIES Fab1, Wilschdorfer Landstr. 101, Dresden, Saxony, Germany 01109

Abstract—There is growing demand for wafer level packaging, which enables thinner, lighter, and more cost effective packaging solutions. However, these demands come with a penalty; reduced package footprint limits the number of I/Os that can be realized in the smaller formfactor. In order to compensate this drawback, wafer level packaging fan-out has been drawing great attention, since fanning out of RDL beyond the die domain not only allows a higher number of I/Os without increasing the die size, but also allows to mount other passives or chips within the package. Compared to conventional packages, a significantly thinner formfactor is thereby achieved. Without a substrate, the wafer level packaging fan-out interacts directly with the Printed Circuit Board (PCB). Like conventional flip chip packages, which are designed and manufactured to mitigate chip package interaction risks, wafer level packaging fan-out needs to be designed and manufactured to mitigate chip board interaction risks. To address this challenge, a test vehicle was designed and manufactured based on silicon on insulator technology. Drop and temperature cycling test on board were performed on the assembled test vehicle. To understand the failure mechanisms, failure mode analysis and Weibull analysis were performed.

Keywords-Wafer Level Packaging Fan-Out, Chip Board Interaction, Board level reliability test, Drop test, Temperature Cycle on Board

I. INTRODUCTION

The increasing demand for miniaturization and higher functionality at lower cost drove the adoption of smallest form-factor packaging architectures, Wafer level Chip Scale Packaging (WLCSP). The fact that WLCSP can be continuously manufactured on uncut wafers without disruption from silicon fabrication processes helps to simplify the supply chain. Since no additional substrate is required, WLCSP has become very popular. Although it cannot compete in high performance computing applications, it has been adopted for wireless connectivity,

analog/mixed signals, and image sensors with reasonable performance at a lower cost. However, the application of WLCSP has been limited to lower I/O counts as the package size is confined by the size of the chip area. To overcome this limitation of WLCSP, Wafer Level Packaging Fan-Out (WLPFO) has been introduced. WLPFO has also been drawing significant attention as an emerging packaging platform for System in Package (SiP) designs, as WLPFO enables a high degree of integration. By expanding the Fan-Out area of the molding compound through re-distribution layer (RDL), WLPFO can accommodate higher I/O counts beyond the chip area and passives or other devices can be paired on the same packages. Compared to flip chip, additional benefits such as better thermal dissipation and electrical performance with less parasitic noise can be achieved through WLPFO.

In contrast to WLCSP, WLPFO provides better environmental protection due to the fan-out molding compound area. However, the lack of a PCB substrate in WLPFO technology still faces chip-board-interaction (CBI) challenges, which disrupt the structural integrity of the device and eventually results in reliability issues. It is well understood that CBI originates from extreme dissimilarities of material properties of the chip and the PCB board. The mismatch of thermal expansion coefficients and modulus can be overcome by various engineering solutions. The introduction of mechanically weaker ultra low-k (ULK) dielectric layers in the silicon Back-End-of-Line (BEOL) makes CBI control even more demanding.

In this work, to address these CBI-related challenges systematically, a test chip was fabricated based on 22FDX® technology. The test chip was equipped with various CBI sensors and used in the chip-first WLPFO processing technology. Based on JEDEC standard, a PCB board was fabricated and subsequently CBI reliability testing was carried out at the board level and *in situ,* electrical read-outs were measured. Die thickness variations and the application of an underfill were incorporated as part

978-1-7281-1500-9/19 $31.00 © 2019 IEEE

of the CBI testing and demonstrated good reliability performance throughout temperature cycles in the board test beyond 500x and in the drop test beyond 100x. Weibull analysis showed that failure is based on a wear-out mechanism indicated by solder fatigue, RDL cracks, and pad fracture in the board.

II. TEST VEHICLE DESCRIPTION

In order to investigate chip package interactions, a test chip was designed based on GLOBALFOUNDRIES 22nm Fully Depleted - Silicon On Insulator (FD-SOI). Key attributes of the test vehicle are summarized in Table I. As was described in other studies [1], the test vehicle was equipped with various CPI sensory macros. The sensory macros were constructed in BEOL of 8 metal layers in the silicon die level and then re-routed through 1 Cu RDL layer up to BGAs. Through the Cu RDL layer, the CPI macros were fanned out from 7×7 mm^2 to 10×10 mm^2 on the WLP level. To date, WLP has not been adopted in high computing application, so a 0.35 mm BGA pitch provides enough I/O counts and was selected for this study.

TABLE I. Test Vehicle Key Attributes

ATTRIBUTES	*CONTENT*
Package	WLP-FO
Die Size	7×7 mm^2
Package Size	10×10 mm^2
BEOL stack	8 metal layer
Cu RDL	1 layer
BGA pitch	0.35 mm
Number of I/Os	707

(A)

(B)

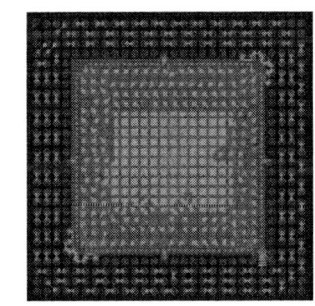

(C)

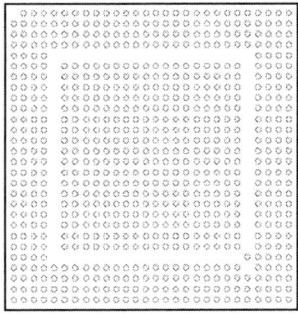

Figure 1. Pictorial description of WLP test vehicle: (A) A picture of WLP test vehicle, showing backside and front side view. (B) A Schematic of CPI TV with terminal layers view, (C) BGA layout.

A large group of macros is located in the peripheral area where the Distance to Neutral Point (DNP) is high, other macros were placed in the middle of the die. Table II shows a list of macros and their purpose. These are designed to monitor any degradation or changes of structural integrity upon moisture ingress or thermomechanical stress during reliability test.

TABLE II. WLP CPI sensory macros

TEST SENSORS	*PURPOSE*
Perimeter Line Ring	Check mechanical integrity for die seal (Resistance / Leakage)
Perimeter Line Stitch	
Perimeter Line Top layer	
Delamination Sensor	Check mechanical integrity for die seal with better local resolution than perimeter line (Resistance)
WLP stitch	Stitch for test of adhesion between LB and RDL and ball integrity during board level reliability
Edge stitch	Assessment of interconnect die – RDL based on chains
THB Sensor	Sensitive humidity sensor in die edge (Leakage)
Serpentine via chain + comb	Assessment of damage in typical BEOL structures (Resistance / Leakage)

III. 22FDX BEOL STRUCTURAL INTEGRITY

Back-End-Of-Line (BEOL) is where devices are wired through various metal layers. While Moore's law drove scaling and functionality improvement in Front-End_of-Line (FEOL), similar efforts have been made in BEOL to decrease the RC delay and improve funcionality. These efforts involve not only BEOL feature scaling but also chemical/mechanical modifications in dielectric materials and barrier/interface layers. Especially, the low k (LowK) dielectric material is well known for it's brittle nature and 22FDX also includes this material in BEOL. In general, advanced silicon nodes provide various BEOL integration schemes, thus it is essential to ensure that there is no adverse effect with different integrations. GLOBALFOUNDRIES investigates the BEOL structural integrity via a Double Cantilever Beam (DCB) test and a modified Edge Liftoff Test (MELT) [2]. These tests are destructive tests and Fig. 2 briefly describes how the tests are performed.. The DCB test requires fully integrated BEOL layers in a strip form. A pre-crack is introduced and mulitple loads are applied on the sample until complete sample fracture is achieved. Based on the load-displacement curve, the critical energy release rate (G_c) can be extracted and it provides useful information such as BEOL integrity strength, crackstop strength, or adhesive and cohesive failure. In the MELT test, the sample is coated with a polymeric layer and immersed under cryogenic conditions. Based on the fracture temperature, the racture toughness (K_{1c}) can be calculated.

The results from the DCB and MELT are shown in Fig. 3. 22FDX offers various BEOL stack integration schemes for different applications and a couple of those were investigated by a wafer level analytical test. Stack #1 was chosen to proceed further with the WLP reliability test. Fig. 3 shows that both G_C and K_{1C} from various BEOL stack options with passing cut-off criteria. It is important to note that there is no significant difference between the SOI technology vs. 14nm Fin Field Effect Transitor (FINFET) or conventional 28 nm bulk complementary metal-oxide-semiconductor (CMOS) technology using planar devices.

(A)

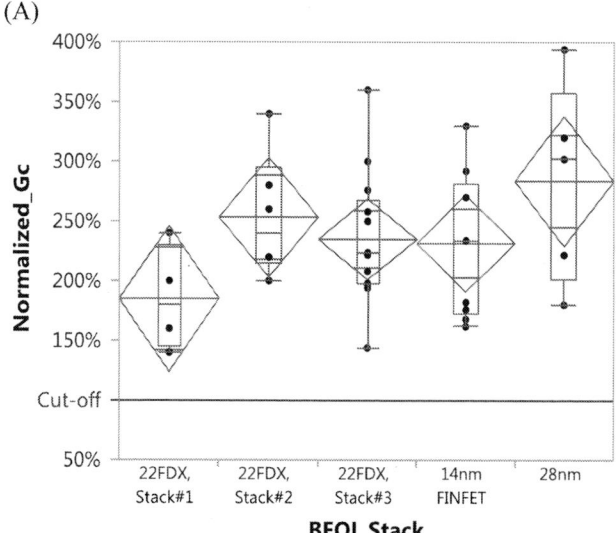

(B)

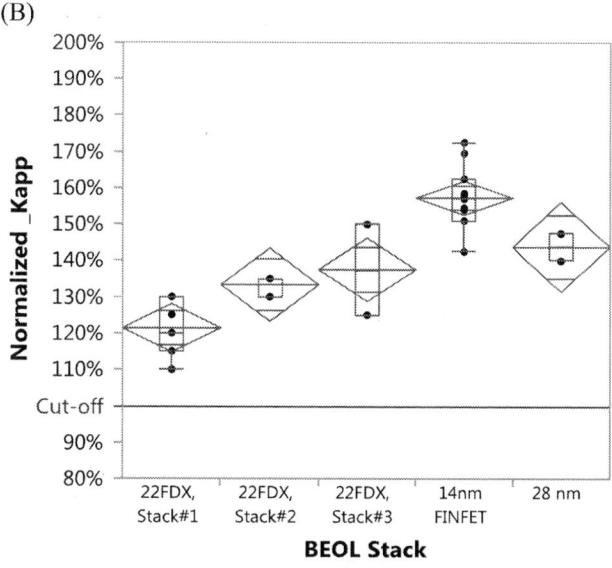

Figure 3. Result from structural integrity analytical test. The raw data has been normalized based on internal cut-off specifications: (A) result from DCB test and (B) result from MELT test.

(A)

(B)

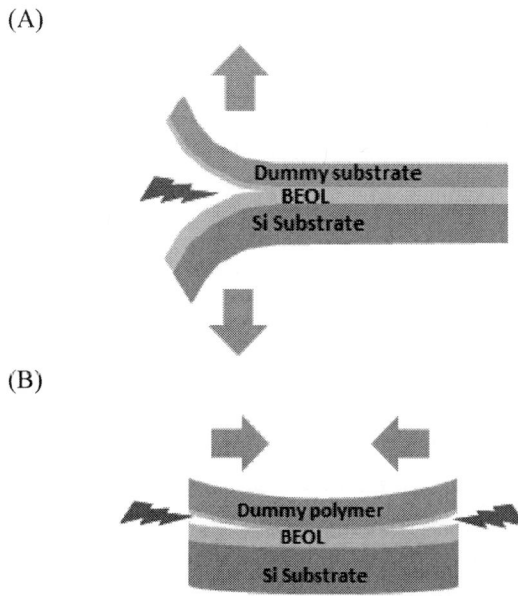

Figure 2. Structural integrity analyical test: (A) Double Cantilever Beam (DCB) test and (B) Modified Edge Liftoff Test (MELT)

IV. WAFER LEVEL PACKAGING PROCESSING

Wafer level packaging was performed based on a conventional chip-first approach. 22FDx wafers, which were terminated by an aluminum pad layer, were continuously processed at an OSAT beginning with backgrinding to the desired silicon die thickness, as described in Fig. 4. Each re-constituted wafers provided ~ 600 WLP-FO dies post singulation. Fig. 5 shows the key features in a final WLP-FO die.

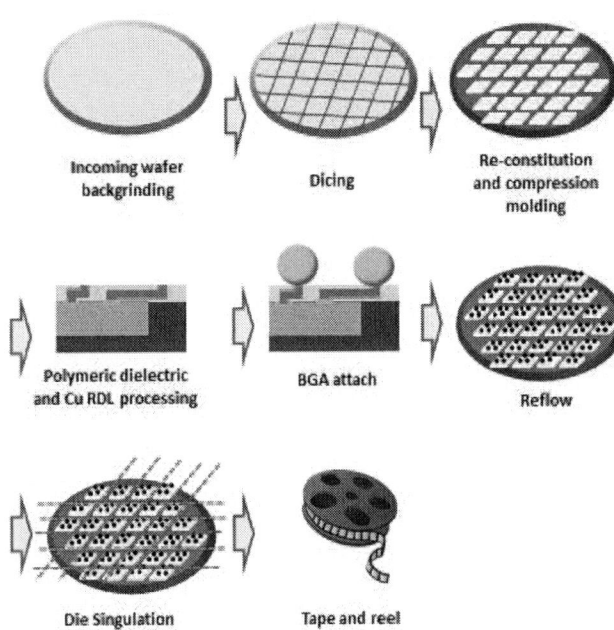

Figure 4. Wafer level packaging processing.

(A)

(B)

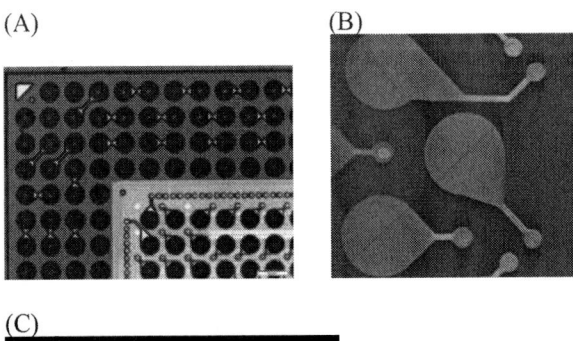

(C)

Figure 5. Construction analysis: (A) Top-down optical image at a die corner, showing good transition between die and fanned out molded area, (B) Top down 2D X-ray focusing on RDL construction, and (C) 3D X-ray over a BGA, RDL, and Si BEOL.

V. COMPONENT LEVEL RELIABILITY TEST

WLP-FO dies were exposed to a JEDEC standard reliability test without surface mount. The main focus of the component level reliability (CLR) test was to investigate the effect of moisture ingress, which is not well captured with the board level reliability test. Table III shows the passing CLR results.

TABLE III. Component Level Reliability (CLR) test result

Environmental Stress	Condition	Final Readout	Pass/ Fail
Pre-conditioning (MSL1)	85 °C / 85% RH soak for 168 hr	168 hr	Pass
Unbiased HAST (with pre-conditioning)	130 °C, 85% RH	96 hr	Pass
Temperature Cycling (with Pre-conditioning)	B: - 55 °C to 125 °C	1000 x	Pass
High Temperature Storage	150 °C	1000 hr	Pass

VI. CHIP BOARD INTERACTION

As WLP does not require assembly on a laminate level, more rigorous requirements are imposed on the board level. Thus, it is essential to investigate the Chip Board Interaction (CBI) to reduce any risk to the final form factor. To achieve this goal, a PCB board has been designed and fabricated based on JEDEC recommendation [3, 4]. JEDEC defined number of layers and recommended material properties. Fig. 6 (A) gives an overview of the material set used to manufacture the PCB.

(A)

Board Layer	Thickness (microns)	Copper Coverage (%)	Material
Solder Mask	20		LPI
Layer 1	35	Pads + traces	Copper
Dielectric 1-2	65		RCC*
Layer 2	35	40% including daisy chain links	Copper
Dielectric 2-3	130		FR4†
Layer 3	18	70%	Copper
Dielectric 3-4	130		FR4†
Layer 4	18	70%	Copper
Dielectric 4-5	130		FR4†
Layer 5	18	70%	Copper
Dielectric 5-6	130		FR4†
Layer 6	18	70%	Copper
Dielectric 6-7	130		FR4†
Layer7	35	40%	Copper
Dielectric 7-8	65		RCC*
Layer 8	35	Pads + Traces + daisy chain links	Copper
Solder Mask	20		LPI

* Suggested RCC Material: Polyclad PCL-CF-400 12/35/35
† Suggested FR4 Material: NELCO N-4000-6 or equivalent

(B)

Figure 6. (A) PCB construction requirement per JEDEC (B) Picture of a PCB board with 15 WLP-FO dies surface mounted.

To investigate CBI, three selective formfactors were chosen and tested by the board level reliability (BLR) test. A PCB board can surface mount 15 dies in a 3 × 5 array configuration, as shown in Fig. 6B. Two PCB boards were used per form factor to confirm reproducibility. A silicon die thickness of 300 μm and 360 μm was adapted (Table IV), since these are the common formfactors in wafer level chip scale packaging and a board level underfill (also known as, secondary underfill) was selectively tested to understand its impact on CBI (Table V).

TABLE IV. Summary of BLR formfactor. Thickness was defined without BGA placement.

Group	Si Die thickness	Package thickness	Board level underfill
1	300 μm	480 μm	No
2	360 μm	480 μm	No
3	360 μm	480 μm	Yes

TABLE V. Key material properties of board level underfill.

Properties	Contents
Tg (°C)	135
CTE (ppm/ °C)	32
M (GPa)	8

There are several options to investigate CBI. In this study, drop test and temperature cycling on the board (TCoB) were explored and procedures were followed as defined in JEDEC.

In the drop test, 15 dies were surface-mounted on a PCB board and then the board was fastened via four screws at each corner in a horizontal way. The surface mounted PCB side faced downward to ensure the most significant board deflection during the drop. The mounted PCB board then traveled vertically through guided rails and pounded toward the strike surface of at the rate of 1500 Gs within 0.5 ms as defined by JEDEC condition B. The PCB board was connected to a data logger and the resistances of each individual unit were monitored *in-situ*. The results are shown in Fig. 7 (A), in a Weibull format. Although there is a difference in the silicon die thickness, Group #1 and group #2 showed no significant difference. In flip chip chip scale package (FcCSP) case, it was often observed that thinner dies tend to exhibit better reliability performance. This behavior was not observed in this measurement. It is also worthwhile to highlight that these experiments were performed without using a board level underfill . When an underfill was used, (as shown in Table V), Group#3), failure did not occur over the entire measurement range. This is a significant performance improvement compared to the groups without an underfill. Since all three groups performed well beyond GLOBALFOUNDRIES' internal cut-off specification (100 cycles), the test was stopped at 2000 cycles. Table VI shows the Weibull analysis results. The shape parameter ($\beta > 1$) indicates that the main failure mechanism in both, group #1 and #2 is a wear-out mechanism, which is proof of the silicon and WLP processing maturity.

For the temperature cycling test, same PCB board design was used as for the drop test. All surface-mounted PCB boards were placed in a temperature cycling chamber and any resistance change of the individual units were recorded and monitored *in-situ*. The entire measurement consisted of up to 2300 cycles thereby exceeding GLOBALFOUNDRIES' requirement of 500 cycles. It was surprising to observe that only group #1 showed failures within the measurement range, whereas groups #2 and #3 showed no failure. However, group #1 has a thicker molding compound compared to groups #2 and #3; thus, it experiences a greater thermomechanical stress/strain in the CBI. Likewise, in the drop test case, a shape parameter greater than 1 indicates a wear-out phenomenon as a failure mechanism.

(A)

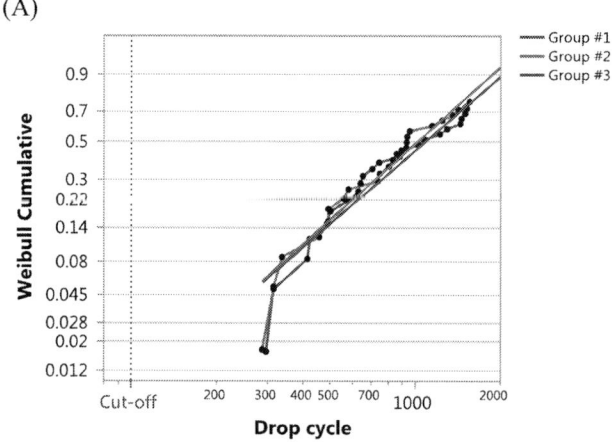

(B)

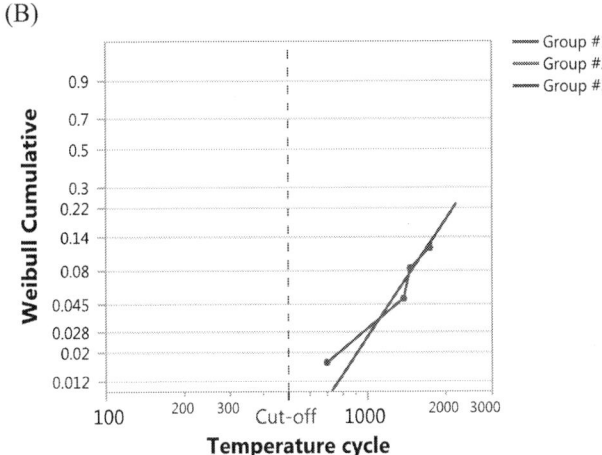

Figure 7. Weibull plot of (A) drop test and (B) TCoB test

TABLE VI. Weibull parameter estimates from drop test and TCoB test

Test	Contents	Group #1	Group #2	Group #3
Drop test	First Time to Failure (FTTF) cycles	297	288	N.A.
	α (scale parameter)	1312	1221	N.A.
	β (shape parameter)	1.88	1.96	N.A.
TCoB test	First Time to Failure (FTTF) cycles	700	N.A.	N.A.
	α (scale parameter)	3409	N.A.	N.A.
	β (shape parameter)	2.98	N.A.	N.A.

After the BLR test, the failure mode was further inspected by physical analyses. Fig. 8 shows major failure modes during BLR. In the drop test, a BGA interface fracture and pad crater were common failure modes, as shown in Fig. 8. (A). These are the common failures and agree well with other studies [5]. Although both group #1 and #2, satisfied cut-off criteria, the interface fracture thoughness can be improved by a proper pad finish selection and the pad crater failure mechanism can be improved with better build-up material. Since group #3, which has an underfill did not provide any failure, no failure analysis was performed for this group. It is important to highlight that the silicon BEOL integrity was maintained without any interruption during the Drop Test. Fig. 8 (B) shows failure images post the TCoB test. Only group #1 showed failures and were investigated by a cross-sectional image analysis. As for the drop test case, a BGA interface fracture was the most common failure mode and sometimes an RDL crack was also observed. A crack propagation into the BGA bulk was often observed, which can be attributed to intermetallic compound (IMC) growth.

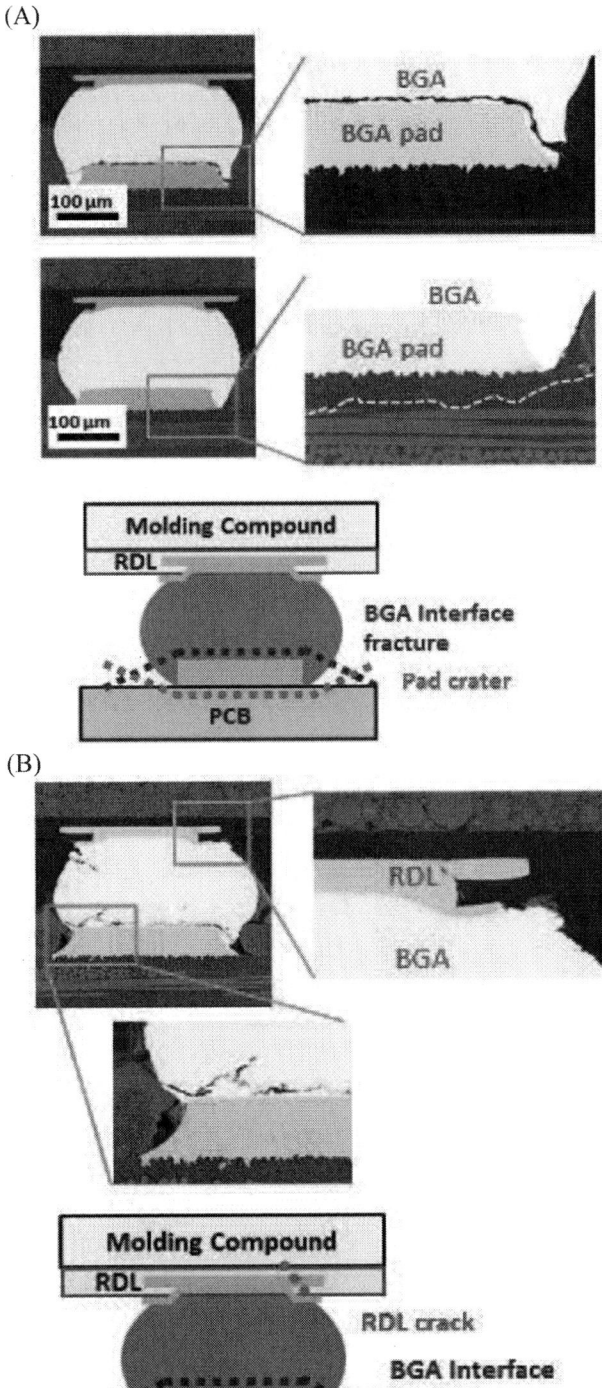

Figure 8. Post BLR, representative cross-sectional images along with schematic description on failure modes: (A) failures post drop test. (B) failures post TCoB .

VII. CONCLUSION

This paper described chip board interactions for wafer level packaging fan-out. A silicon test vehicle was fabricated by SOI technology. Structure integrity analytical tests showed stable and mature silicon BEOL. Subsequently, the test vehicle was surface-mounted on a PCB and subject to drop and TCoB reliability tests. The results showed that both tests successfully passed cut-off criteria without using a board level underfill. It is important to point out that a significantly higher reliability was observed when an underfill was used, as indicated by no fail during drop and TCoB test within the measurement range. In the drop test, the BGA interface fracture and PCB pad crater were major failure modes. Whereas in the TCoB test, BGA interface fractures and RDL cracks were common failure modes.

22FDX® is a registered trademark of GLOBALFOUNDRIES' 22-nm fully depleted silicon on insulator technology.

ACKNOWLEDGEMENT

The authors would like to thank Advanced Semiconductor Engineering, Inc (ASE/Taiwan), in particular, Chiuyuan Lu, Richie Kuo, Justin Lin. Daniel Chang, and Nick Hsiao for their WLP support in this work. The authors also would like to thank Christian Klewer and Jeannette Koernert (GLOBALFOUNDRIES) for their support throughout this work.

REFERENCES

[1] J. Cho, J. Paul, S. Capecchi, D. Breuer, F. Kuechenmeister, D. Scott, J. Choi, and W. Kang, "Chip Board Interaction Analysis of 22-nm Fully Depleted Silicon on Insulator (FD-SOI) technology in Wafer Level Packaging (WLP)", *International Wafer-Level Packaging Conference (IWLPC 2018)*, San Jose, CA, USA, 2018.

[2] F. Kuechenmeister, D. Breuer, H. Geisler, C. Klewer, B. Boehme, K. Machani, M. Hecker, C. Goetze, J. Cho, H. Kamineni, J. Paul and M. Thiele, "A Generic Strategy to Assess and Mitigate Chip Package Interaction Risk Factors for Semiconductor Devices with Ultra-low k Dielectric Materials in Back End of Line", *The 50th International Microelectronics Assembly and Pacakging Society (IMAPS 2017)*, Raleigh, NC, USA, 2017.

[3] JEDEC Standard JESD22-B111, Board Level Drop Test Method of Components for Handheld Electronic Products, July, 2003.

[4] JEDEC Standard JESD22-A104, Temperature Cycling, March, 2009.

[5] A. Farris, J. Pan, A. Liddicoat, M. Krist, N. Vickers, B.J. Toleno, D. Maslyk, D. Shangguan, J. Bath, D. Willie, and D.A. Geiger, "Drop impact reliability of edge-bonded lead-free chip scale packages," *Microelectronics Reliability,* Vol. 49, Issue 7, p 761-770, 2009.

FOWLP Design for Digital and RF Circuits

Lim Teck Guan, David Soon Wee Ho, Eva Wai Leong Ching, Chen Zihao*, Surya Bhattacharya

Institute of Microelectronics, Singapore

A*STAR (Agency for Science, Technology and Research)

Abstract— Fan-out Wafer Level Packaging (FOWLP) is a versatile semiconductor packaging technology which can be used for various RF and Digital applications. In the FOWLP technology, the bare die IC is embedded in a molding compound for protection, and a Redistribution Layer (RDL) is formed to route the electrical input/output (I/O) of the IC to the solder ball pads which are distributed on the surface area of the package. Besides fanning out of the electrical I/O, it can also be used for multi-chip 2.5D integration. In addition, vertical interconnect can be formed in the molding compound. Together with additional RDL on the top of the package, the FOWLP allows the electrical circuit connection on both sides of the package to form a much compact Package-on-Package or 3D integrated circuit. Furthermore, the RDL can be used to form high performance RF passive circuit to realize an RF System in Package module.

Keywords-component; FOWLP, RF integration, Digitial Integration, PoP, AiP, HBM, Advacned Packaging, SiP

I. INTRODUCTION

FOWLP is a cost effective packaging technology which the electronic IC is embedded in the molding compound. The epoxy molded compound (EMC) besides providing the mechanical protection, together with the Retribution Layer (RDL) and Through Mold Interconnect (TMI), it provides electrical routing. Currently, there are two main variances of the FOWLP process, they are the Mold-first and the RDL first [1]. The Mold-first is more suitable for integration application which does not require fine RDL width and spacing. The typical minimum line width spacing for the Mold-first process is about 10μm. This is due to the inherent die shift which occurs during the molding process. On the other hand, the RDL-first can achieve finer RDL width and spacing of 2μm, but it needs the additional under-fill process which leads to higher cost compares with the Mold-first. The RDL-first is targeted for high end applications which need very high density routing. The FOWLP is also a very versatile packaging platform for multi-chip integration. Currently, there are many developments of the FOWLP for the 2.5D and 3D integration schemes to realize System-in-package (SiP) and Package-on-Package (PoP) integration. In this paper, the electrical performance of the three main FOWLP integration designs for different digital and RF integration are presented.

II. ULTRA-THIN PACKAGE ON PACKAGE

The targeted application of this PoP design is for the integration of Low Power Double Data Rate (LPDDR) memory chip with a connection of speed of 3.2Gbps to the ASIC in mobile devices. For this PoP design, one or multiple LPDDR memory chips are stacked on top of the ASIC chip. Due to the current trend of the mobile handset thin design, a low package profile is critical [[1]-[4].

In this FOWLP integration, the memory dies are stacked directly on top of the ASIC chip instead of using a support substrate, as shown in Figure 1. The ASIC chip is first embedded face-down and its electrical I/O to the memory chips are routed to the opposite side of the chip using the RDL and the TMI. The memory dies are then assembled face-up on the top of the ASIC chip, and all the memory electrical I/O are connected to the top side of the TMI pads using low loop reverse stitch fine pitch wire bond. A final molding is then applied to encapsulate the top memory die and the wire bond. In this integration scheme, there is no substrate between the ASIC and memory die. In this way, it helps to achieve better electrical performance as there are less discontinuity, and at the same time, it provides a very low profile of less than 0.8mm as compared to the typical PoP with a substrate.

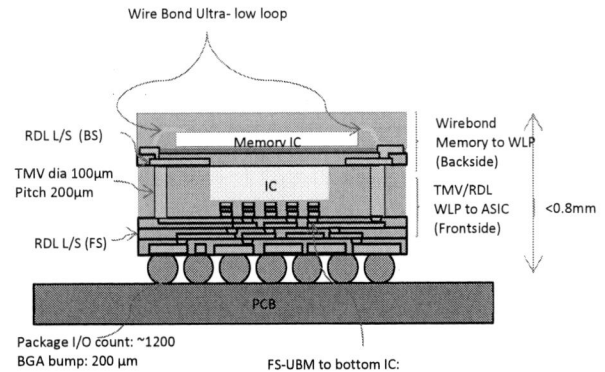

Figure 1: PoP stack up design

Figure 2 shows the 3D layout of the package with 2 memory dies place side by side and are on top of the ASIC die. The memory dies I/Os are located only at both the long sides of the chips. The ASIC and memory I/O are expected to be also located on the same side so that they can be routed directly.

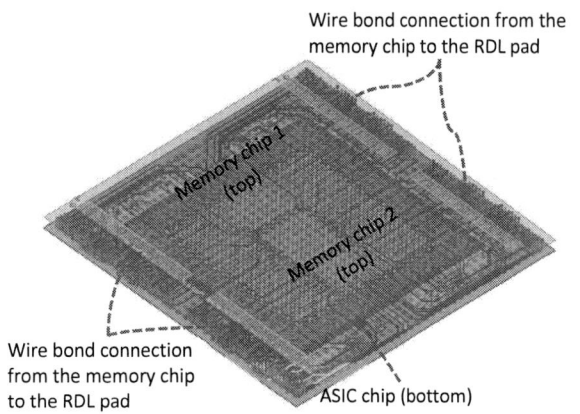

Figure 2: PoP design with 2 lateral memory chip on top of the ASIC chip

For high speed and high density integration, flip chip design has always been the preferred choice as compared with the wire bond. This is because the interconnect bandwidth of the wire bond is limited by the high inductance of the wire. In addition, there is a pad pitch mismatch between the die and the substrate. The typical pad pitch of the die is around 70μm while the minimum pad pitch of the substrate is around 150μm. This causes the fan out of the wire bond for a Ground-Signal-Ground (GSG) interconnect as shown in Figure 3. The fan out GSG wire bond reduces the interconnect bandwidth. Figure 4 shows the simulated frequency response of the GSG wire bond interconnects with various substrate pad pitch. During the simulation, the center signal wire bond length is kept constant. The simulated result shows that the insertion loss increases as the substrate pad pitch increases. Besides causing impedance discontinuities, the fan out wire bond causes the length for the various signal bond wire to vary. Those signal I/O which are located at the edge of the chip will have a longer wire length as compared those signal I/O located at the center of the chip. As the inductance increases with its wire length, those signal I/Os will have smaller interconnect bandwidth. The various different signal wire bond length will also create the undesirable skew in the I/O timing.

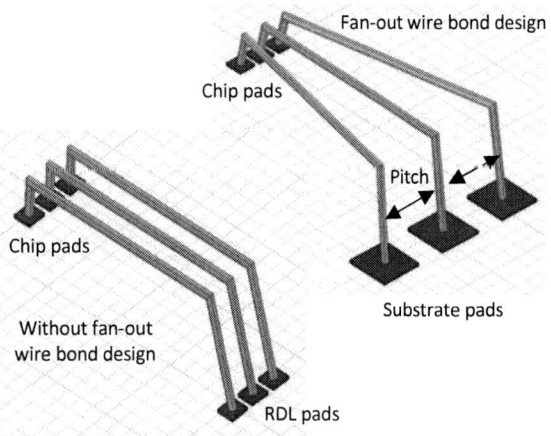

Figure 3: Wire bond GSG design from die to substrate with ultra-low profile

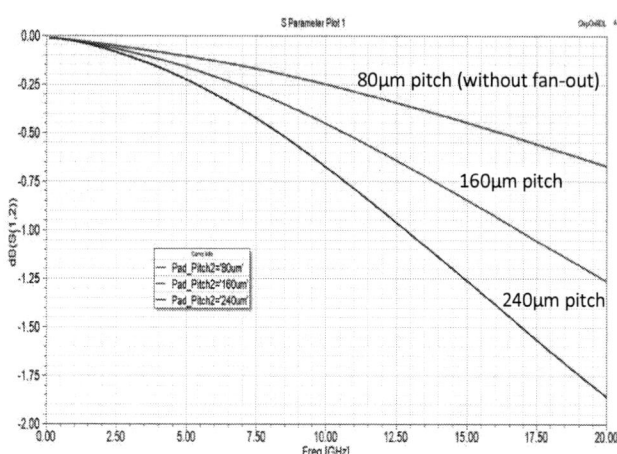

Figure 4: Insertion loss of the wire bond with various substrate pitch

On the other hand, the minimum pad size and the spacing of the RDL in the FOWLP can be fabricated as small as the die pad design. This allows the GSG wire bond to be designed having the same length and constant pitch as shown in Figure 2. In addition, all the signal wire bond connections will have the same length as no fan out is required. Figure 5 shows the X-ray pictures of the fabricated test vehicle. It shows the constant pitch wire bonding connection from the top memory chip to the middle layer RDL pads for all the I/O. The wire bond used is a 0.8mil Au wire and is formed using a low loop with reveres stitch bonding. The signal lines are then connected to the TMI to the bottom RDL. The fan out routing to the bottom chip is designed using the bottom RDL. The RDL allows more flexibility in the routing and does not suffer from the high inductance issue as the wire bond.

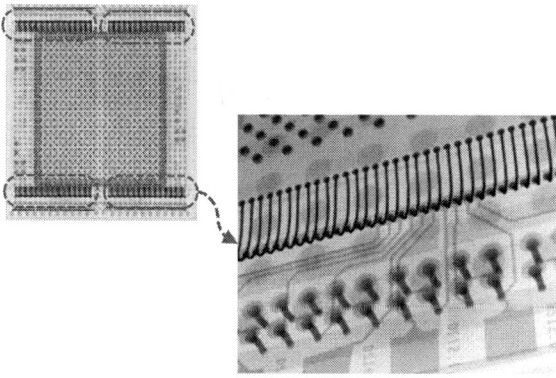

Figure 5: X-ray picture of the PoP wire bond with constant pitch connecting the memory die to the RDL pad

The frequency response of this GSG wire bond and TMI interconnect is modelled as shown in Figure 6. The TMI has a diameter of 100μm and a pitch of 200μm. In order to minimize the package size, the GSG TMI are designed with an alternate configuration. The wire bond and the TMI are connected using the RDL. The simulated frequency response of the wire

978-1-7281-1500-9/19 $31.00 © 2019 IEEE

bond with TMI interconnect is as shown in Figure 7. Due to the additional TMI and RDL connection, the insertion loss is increased to ~1dB at 20GHz while the -10dB Return Loss bandwidth is about 12GHz. Nevertheless, this still meets the requirement of LPDDR integration.

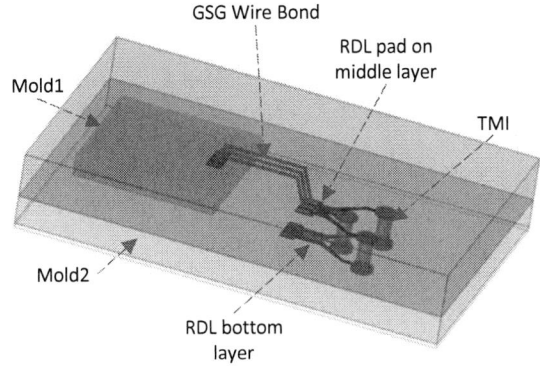

Figure 6: Simulation model showing the wire bond from memory chip to the middle RDL pads, to the TMV, and the RDL to the bottom ASIC chip

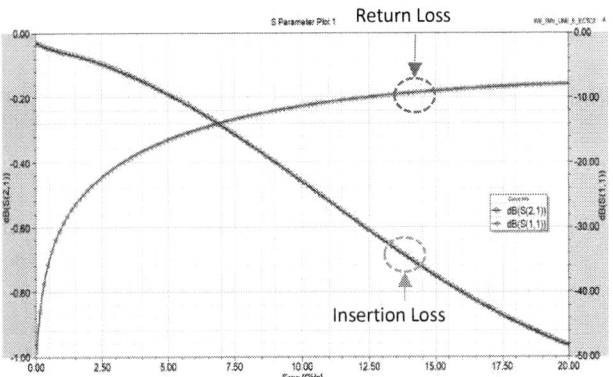

Figure 7: Insertion Loss (dB) and Return Loss (dB) of the wire bond and TMI interconnect

A set of the back-to-back GSG TMI connected with the constant gap wire bond using the proposed FOWLP stack up is designed and fabricated. Figure 8 shows the side view of the 3D Electromagnetic (EM) simplify model uses for the analysis. The direct straight length of the wire bond is approximately 600µm and the gap is 80µm. The wire bonds are connected to the RDL pads and routed to the TMI of height 200µm. The diameter and the pitch of the TMI is 100µm and 200µm, respectively. The RF probe pads are designed at the bottom of the test vehicle. Figure 9 shows the fabricated test vehicle before the over-molding. The wire bonding is done using the low loop reverse stitch bonding. Figure 9 shows the simulation and preliminary measurement results of the wire bond to TMI back-to-back test vehicle. The measurement and the simulation result match well, the differences are probably due to the simplified model use which does not include the various layer of contact parasitic

components. In addition, due to the fabrication tolerance, there are some variations in the pitch of the wire bond along which may contribute to this mismatch. Nevertheless, the insertion loss for the measurement and simulation is less than 0.5dB up to 20GHz. The results show that the low loop reverses wire bond on RDL, which enable a constant gap and short GSG wires can be used to support high speed signal up to the next generation memory. In addition, the lower inductance wire bond design which gives a high bandwidth can also be used to support RF integration, especially the RF front end module.

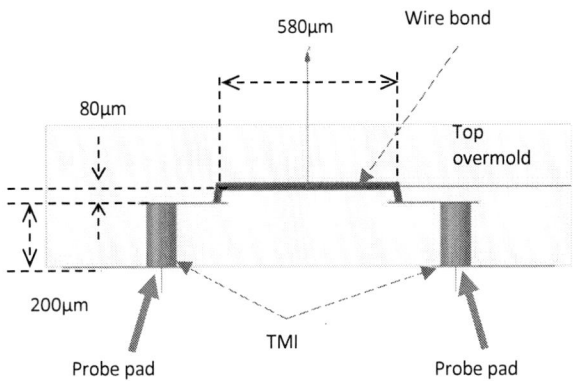

Figure 8: Side view of the FOWLP Wire bond to TMI back to back test vehicle design

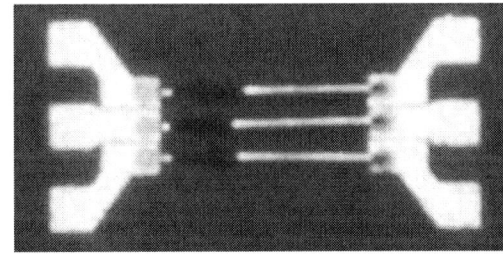

Figure 9: Top view of the fabricated test vehicle before top over mold

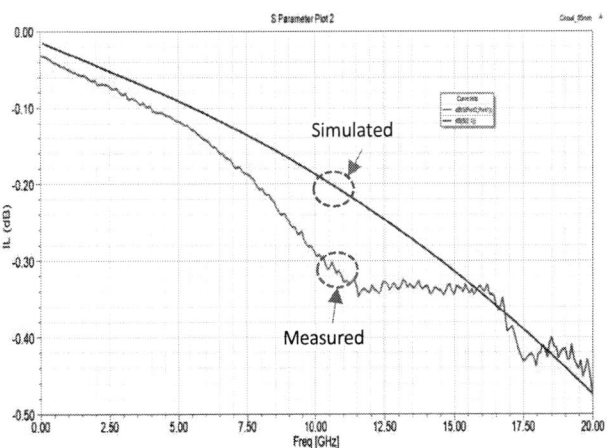

Figure 10: Insertion loss (dB) of the measured and simulated back-to-back wire bond TMI test vehicle.

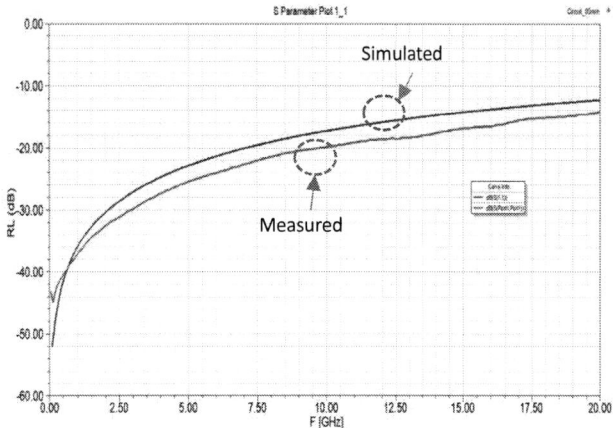

Figure 11: Return loss (dB) of the measured and simulated back-to-back wire bond TMI test vehicle.

III. HIGH DENSITY 2.5D

The electrical integration design is based on the Wide I/O connection standard, where multiple fine signal lines of about 2μm are used for the lateral integration of two ICs, This integration design is mainly for the high end application such as the GPU and HBM as shown in Figure 12. Currently, there are several 2.5D integration technologies being developed, these are Silicon Interposer [8], Organic interposer [9], low-temperature co-fired ceramics (LTCC) [10] and Embedded Multi-Die Interconnect Bridge (EMIB) [11]. The current bit rate of each line or channel is about 2Gbps. Depending on the number of the line used, the aggregated data rate of the Wide I/O connection can reach several Terabit/sec. For this high density signal line, there is no impedance match and termination. Although this help to improve power efficiency, it leads to signal integrity challenge, especially crosstalk. The problem is elevated when the signal speed increases or the length of the connection increases.

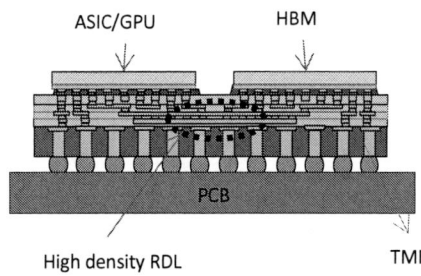

Figure 12: 2.5D High density FOWLP for GPU HBM integration

Currently, there are two methods of realizing the high density connection through the FOWLP being pursued in [5]. In the first method, a thin Si substrate with fine line width and spacing line of down to minimum 0.4μm width prefabricated using the BEOL technology is embedded (EFI) in the package to provide the high density routing. In this design, the 0.4μm

is used for shielding between two adjacent channels which helps to reduce the crosstalk level. The signal line width is approximately 2μm, and the metal thickness is limited to around 1μm due to the BEOL process. For long interconnect length of 8mm, the line resistance is the major issue affecting the signal quality.

In the other design, the high density lines are realized directly using the RDL which has a minimum width of 2μm. Figure 13 shows the fabricated high density portion of the 2μm line of the interconnecting lines for the HBM to ASIC test vehicle. The RDL process has a higher metal thickness of more than 2μm, and this helps to reduce the line resistance effect. However, due to the cost and process complexity issue, the number of fine RDL that could be used for the signal routing in this design is restricted to 2 layers. With these minimum 2μm line width spacing and the 2 RDL layers constrain, it is not possible to implement a shielding line between the 2 adjacent channels while meeting HBM routing requirement.

The cross section design of this stack up is shown in Figure 14. The design of the signal lines in those two RDLs are such that they are arranged alternately. In this way, the crosstalk or interference between the signals between the two RDL is minimized. For the crosstalk reduction between the adjacent channels in the same RDL, it is proposed to reduce RDL dielectric thickness between the signal and ground [5]. This is expected to increase the coupling electromagnetic field strength between the signal and the ground, and reduces the adjacent channel electromagnetic coupling field strength.

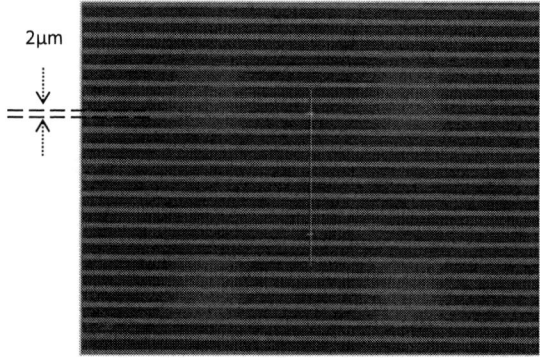

Figure 13: Constant routing section with 2μm line width with 4μm gap spacing

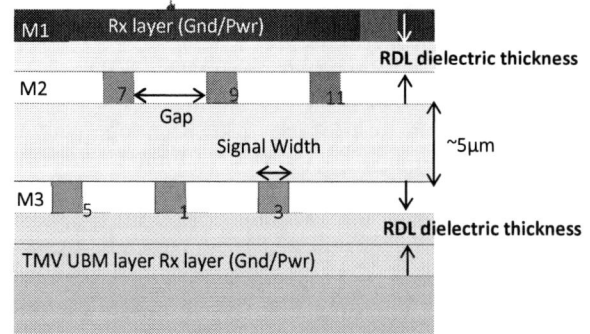

Figure 14: Cross Section design of the 2 routing high density RDL FOWLP for HBM and GPU applications

978-1-7281-1500-9/19 $31.00 © 2019 IEEE

The simulated and measured crosstalk frequency response between the two adjacent channels on the same metal layer of reducing RDL dielectric thickness is shown in Figure 15. The measured crosstalk shows that the RDL with a ~6μm thick dielectric thickness has a crosstalk level of about -15dB at 2GHz. The crosstalk level is reduced to less than -20dB at 2GHz when the dielectric thickness is reduced to ~2μm. This shows that the RDL dielectric thickness can be used to improve the electrical performance of the interconnect for the HBM to ASIC integration.

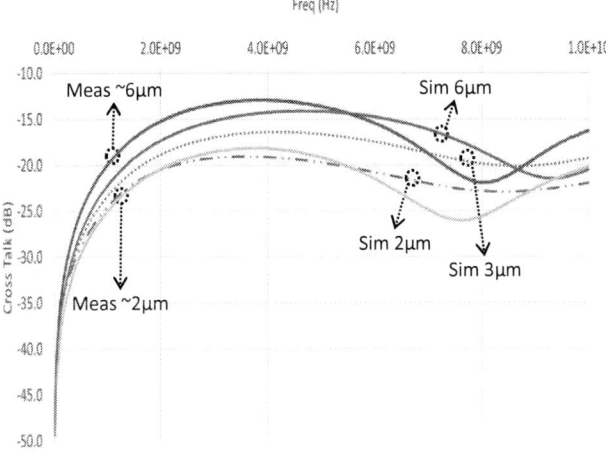

Figure 15: Cross talk measurement between two adjacent channels for different RDL dielectric thickness

IV. ANTENNA IN PACKAGE

The last targeted application is for the mmWave circuit packaging. At mmWave frequency, the typical transition of the BGA package to the PCB has led to high insertion loss and limited bandwidth. The FOWLP AiP offers an elegant solution to integrate the antenna with the chip embedded to improve the RF performance by reducing the interconnect length and avoids the BGA transition to the PCB board. However, for the antenna design in the package, its performance is affected by the small size of the package. Although at a higher frequency, the size of the antenna is scaled down accordingly, the free space path loss increases at the function to the square of the frequency. As a result, a higher antenna gain is required and this is usually overcome by using antenna array to provide beam forming and beam steering. Beside electrical performance, the thermal performance is also critical as the chip is embedded inside the package.

In order to meet these requirements, a double EMC FOWLP AiP was proposed in [12]. In this 2 EMC layers design, the MMIC is embedded in EMC 1. The MMIC is made to face up so that the thermal solution can be designed on the backside of the MMIC. The face up design allows the RF feedline to be connected directly from the MMIC to the antenna using Metal 1 (M1), as shown in Figure 16. M1 is formed between EMC 1 and EMC 2. The other electrical I/Os

are then connected to the external circuit through the TMV formed on EMC 1. The patch antenna is designed with capacitively coupled so that there is no TMV required for EMC 2. The radiating elements of the antenna are designed on Metal 2 (M2) which is on top of EMC 2. Below EMC 1, Metal 3 (M3) is used for the antenna ground as well as the routing of the electrical signal to the solder bump. The exploded view of this AiP design is as shown in Figure 17.

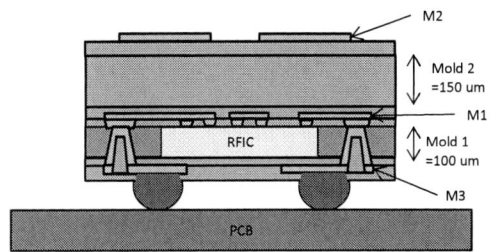

Figure 16: Antenna in Package stack up design

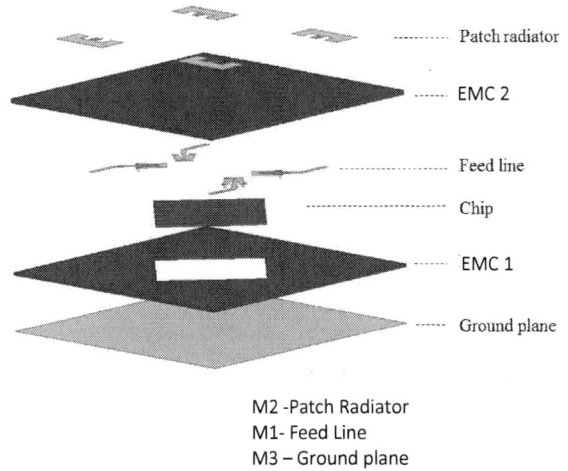

M2 -Patch Radiator
M1- Feed Line
M3 – Ground plane

Figure 17: Exploded view of the proposed AiP stack up

In order to evaluate the antenna performance of this AiP for 60GHz WLAN, a test vehicle which consists only the antenna is designed and fabricated. The antenna array consists of 2x2 patch elements and they are connected together through a power divider feed instead of individually feed. The design of the antenna test vehicle is as shown in Figure 18 and it uses the same stack up as shown in Figure 17. Figure 19 shows the front side of the fabricated AiP test vehicle.

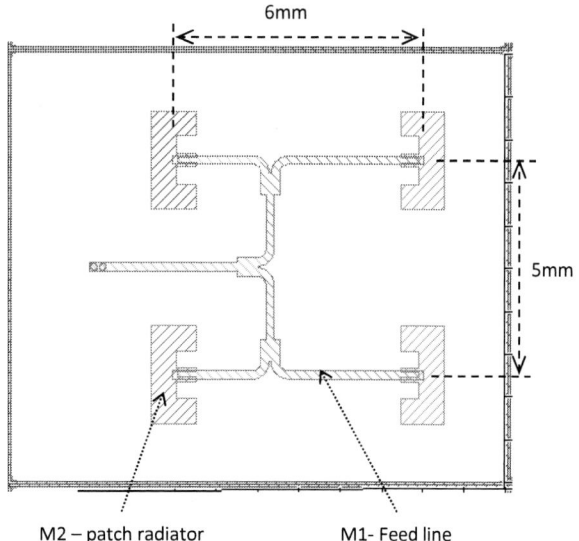

Figure 18: 2x2 array AiP test vehicle design

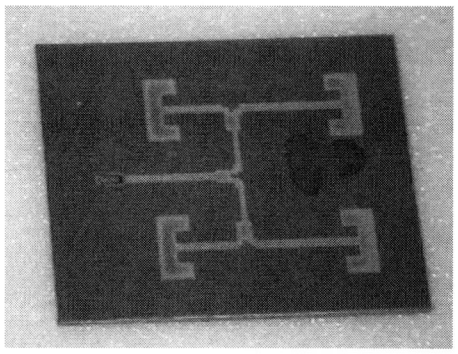

Figure 19: Fabricated 2x2 array AiP test vehicle

Figure 20 shows the measured and simulated Return Loss (RL). The results show that there is some mismatch in the RL between the measured and simulated. The measured resonance frequencies are shifted by about 1% and the RL is slightly higher. The slight mismatch in the response is probably due to the EMC material properties. In addition, it is noted that the height of the EMC can also lead to changes in the RL response as shown in Figure 20. The H-plane of the 60GHz radiation patterns of the AiP were also measured and compared with the simulation as shown in Figure 21. The radiation pattern matches well with the simulation results.

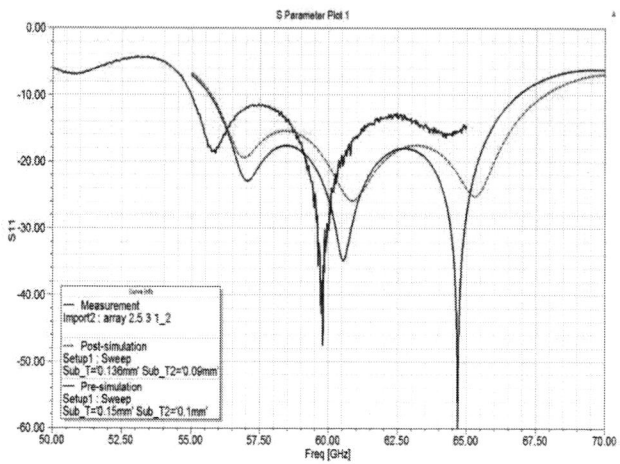

Figure 20: Measured and simulated RL of 2x2 array AiP test vehicle

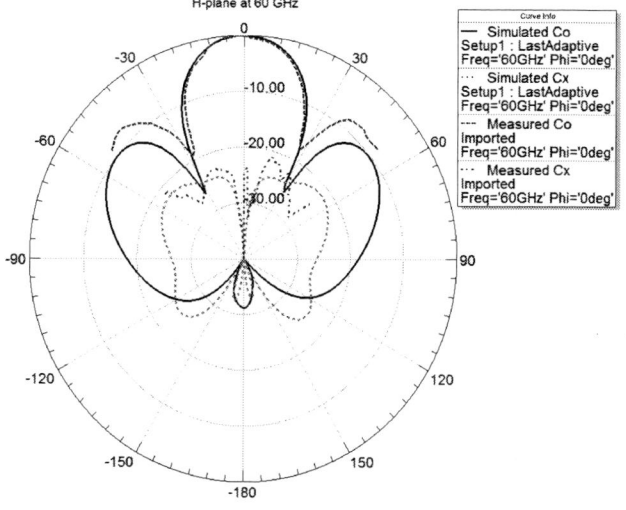

Figure 21: Measured and simulated H-plane radiation pattern at 60GHz of 2x2 array AiP test vehicle

CONCLUSION

The preliminary electrical results of the Ultra-thin PoP, 2.5 High density Fan-out and the AiP test vehicles show that the FOWLP technology can be used to realize various digital and RF modules. The FOWLP 3D integration with ultra-low loop wire bond is well suit for multi-die stacking with high speed connection in a small form factor for mobile ASIC and memory integration. It is a good alternative for substrate PoP. The FOWLP RDL can achieve fine line width spacing of 2μm to support the 2.5D high density Wide I/O high end HBM and GPU integration, which is currently realized using the costly Si interposer. Finally, FOWLP is also an attractive

978-1-7281-1500-9/19 $31.00 © 2019 IEEE

solution to realized mmWave AiP or integration for the 5G communication and mmWave radar applications.

ACKNOWLEDGMENT

This project is part of the Multi-Chip Fan-Out Wafer Level Packaging Development Line Consortium (DLC). The author will like to thanks the IME FOWLP team for the process discussion, and project members for their support in this work. Chen Zihao has left IME and is now with Harbin Institute of Technology (Shenzhen), Shenzhen, China.

REFERENCES

[1] "Development of High Density Fan Out Wafer Level Package (HD FOWLP) with Multi-layer Fine Pitch RDL for Mobile Applications", Vempati Srinivasa Rao ; Chai Tai Chong ; David Ho ; Ding Mian Zhi ; Chong Ser Choong ; Sharon Lim PS ; Daniel Ismael ; Ye Yong Liang, 2016 IEEE 66th Electronic Components and Technology Conference (ECTC), Year: 2016, Pages: 1522 - 1529

[2] "Advanced flip chip package on package technology for mobile applications", Ming-Che Hsieh, 2016 17th International Conference on Electronic Packaging Technology (ICEPT), Year: 2016, Pages: 486 – 491

[3] "Chip Stackable, Ultra-thin, High-Flexibility 3D FOWLP (3D SWIFT® Technology) for Hetero-Integrated Advanced 3D WL-SiP", WonMyoung Ki ; WonGeoL Lee ; InSu MoK ; IlBok Lee ; WonChul Do ; Moh Kolbehdari ; Alex Copia ; Suresh Jayaraman ; Curtis Zwenger ; KangWook Lee, 2018 IEEE 68th Electronic Components and Technology Conference (ECTC), Year: 2018, Pages: 580 – 586

[4] "Design and electrical analysis for adavanced fan-out package-on-package", Po-Chih Pan ; Tsun-Lung Hsieh ; Chih-Yi Huang ; Ming-Fong Jhong ; Chen-Chao Wang, 2017 IEEE 19th Electronics Packaging Technology Conference (EPTC), Year: 2017, Pages: 1 - 4

[5] "Electrical Design for the Development of FOWLP for HBM Integration", Teck Guan Lim ; David HO Soon Wee, 2018 IEEE 68th Electronic Components and Technology Conference (ECTC), Year: 2018, Pages: 2142 – 2148

[6] "Passive Devices Fabrication on FOWLP and Characterization for RF Applications", Chunmei Wang ; King Jien Chui ; Xiangyu Wang ; Teck Guan Lim ; Mingbin Yu ; Gilbert See ; Gu Yu, 2017 IEEE 67th Electronic Components and Technology Conference (ECTC), Year: 2017, Page s: 312 – 318

[7] "Comparison of Package-on-Package Technologies Utilizing Flip Chip and Fan-Out Wafer Level Packaging", Amy Lujan, 2018 IEEE 68th Electronic Components and Technology Conference (ECTC), Year: 2018, Page s: 2089 – 2094

[8] "An Overview of the Development of a GPU with integrated HBM on Silicon Interposer" Chang-Chi Lee; Cp Hung; Calvin Cheung; Ping-Feng Yang; Chin-Li Kao; Dao-Long Chen; Meng-Kai Shih; Chien-Lin Chang Chien; Yu-Hsiang Hsiao; Li-Chieh Chen; Michael Su; Michael Alfano; Joe Siegel; Julius Din; Bryan Black, IEEE 66th Electronic Components and Technology Conference (ECTC), Year: 2016, Pages: 1439 – 144

[9] "3D SiP with Organic Interposer for ASIC and Memory Integration" Li Li; Pierre Chia; Paul Ton; Mohan Nagar; Sada Patil; Jie Xue; Javier Delacruz; Marius Voicu; Jack Hellings; Bill Isaacson; Mark Coor; Ross Havens, IEEE 66th Electronic Components and Technology Conference (ECTC), Year: 2016, Pages: 1445 - 1450

[10] "LTCC Package for High-bandwidth Logic to Memory Interconnection", Norio Chujo; Yutaka Uematsu; Toshiaki Takai; Masahiro Toyama; Junichi Masukawa; Hiroyuki Nagatomo Yamazaki, IEEE Electrical Design of Advanced Packaging and Systems Symposium (EDAPS), Year: 2015, Pages: 5 – 8

[11] "Embedded Multi-Die Interconnect Bridge (EMIB) – A High Density, High Bandwidth Packaging Interconnect" Ravi Mahajan; Robert Sankman; Neha Patel; Dae-Woo Kim; Kemal Aygun; Zhiguo Qian; Yidnekachew Mekonnen; Islam Salama; Sujit Sharan; Deepti Iyengar; Debendra Mallik, IEEE 66th Electronic Components and Technology Conference (ECTC), Year: 2016, Pages: 557 – 565

[12] "Millimeter-Wave Antenna in Fan-Out Wafer Level Packaging for 60 GHz WLAN Application".Zihao Chen ; Lim Teck Guan ; David Soon Wee Ho ; Surya Bhattacharya, 2018 IEEE 68th Electronic Components and Technology Conference (ECTC), Year: 2018, Page s: 331 – 336

[13] "Array antenna integrated fan-out wafer level packaging (InFO-WLP) for millimeter wave system applications", Chung-Hao Tsai ; Jeng-Shien Hsieh ; Monsen Liu ; En-Hsiang Yeh ; Hsu-Hsien Chen ; Ching-Wen Hsiao ; Chen-Shien Chen ; Chung-Shi Liu ; Mirng-Ji Lii ; Chuei-Tang Wang ; Doug Yu, IEEE International Electron Devices Meeting, Year: 2013, Pages: 25.1.1 - 25.1.4

Next Generation of 2-7 Micron Ultra-small Microvias for 2.5D Panel Redistribution Layer by using Laser and Photolithography Technologies

Fuhan Liu, Chandrasekharan Nair, Gaurav Khurana, Atom Watanabe, Bartlet H. DeProspo, Atsushi Kubo*, Cheng Ping Lin**, Toshiyuki Makita**, Naoki Watanabe+, Rao R. Tummala

3D Systems Packaging Research Center, Georgia Institute of Technology, Atlanta, GA, USA
*Tokyo Ohka Kogyo Co. Ltd., Japan
** Panasonic Corporation, Osaka 571-8686, Japan
+ Panasonic Industrial Devices Sales Company of America, USA
Email: fliu@ece.gatech.edu

Abstract— Microvia is the vertical interconnect structure for multi-layer redistribution layers (RDLs) in high-density interconnect (HDI) printed circuit boards (PCBs), HDI package substrates, 2.5D interposers and fan-out packages. Three technologies such as photolithography, UV laser and excimer laser have been used to form small microvias (≤ 20 µm diameter) in polymer dielectrics. All the three above mentioned technologies are studied and compared in the work presented in this paper. Photovia was first introduced by IBM for Surface Laminar Circuit technology and it has scaled down from 125 µm then to below 10 µm today. The smallest photovia demonstrated is 2 µm in diameter by using 365 nm photolithography in 5 µm thick TOK photo-imageable dielectric (PID) (IF4605) film. Photovias of 3 µm diameter were also demonstrated in 5 µm thick Taiyo Ink dielectric dry film material (PDM) which passed 1,500 thermal cycles (-55 °C to 125 °C). The limitation of photovia technology is the availability and cost of photo-sensitive dielectric materials with the required electrical, mechanical, thermal and chemical properties. The state-of-the-art microvia diameter is 20 µm by using conventional high-speed UV laser technologies. Multi-layer RDL with microvias and trenches of 4 µm feature sizes are simultaneously fabricated in a 7 µm thick Ajinomoto Build-up Film (ABF) with small fillers by using excimer laser and passed 1,000 thermal cycles (-55 °C to 125 °C). This paper demonstrates a novel picosecond UV laser technology to push the limits of low-cost UV laser technology by optimizing laser parameters and dielectric materials. The Cornerstone picosecond UV laser tool from ESI is capable of producing output power of 16W at 355 nm wavelength. The pulse duration is 5 ps which minimizes the heat-affected zone of polymer dielectric and the high (80 MHz) repetition rate enables this laser to be used in high throughput manufacturing processes. Microvias with minimum diameter of < 7 µm were fabricated in 5 µm thick ABF with small fillers and in 7 µm thick novel Panasonic low stress dielectric film-S (PLS-S), by using 355 nm picosecond UV laser tool. These ABF and PLS-S films are non-photosensitive dielectric materials. This is the first demonstration of very small microvias (< 7 µm) in polymer dielectrics using UV laser ablation. The motivation of this work is to address the high RDL interconnect density requirements for 2.5D interposer and high density (HD) fan-out packages. The next generation of low-cost, ultra-small microvias will (1) Increase the RDL I/O density, (2) Meet fine bump pitch requirements, (3) Reduce the metal layer count for package substrate RDL, (4) Fill the gap between semiconductor back-end-of-line (BEOL) process and semi-

additive process (SAP) and thereby (5) Improve the packaging performance at lower costs.

Keywords – Microvia, RDL Interconnect, UV laser, HDI, IO Density, 2.5D interposer, Fan-Out packaging

I. INTRODUCTION

The fast-growing demands for increasing speed, bandwidth, and functionality of electronic devices and systems pose several challenges for packaging technologies. The increase in bandwidth and functionality drives a larger number of signal input and output (I/Os) numbers. However, at the same time, the die size is becoming smaller and smaller due to scaling down of CMOS semiconductor technology. Thus, the large number of I/Os coupled with reduction of die size pushes toward finer die interconnect pitches. Development of small and ultra-small microvia technologies becomes critical and vital. Figure 1 (a) shows the state-of-the-art organic RDL with 6 µm L/S and 20 µm via at 50 µm pitch. The I/O density is 40 IOs/mm/layer. [1] Figure 1 (b) shows the improved RDL on glass substrate with 2 µm L/S and 2 µm via. The I/O density will have 5X increment and the pitch can be reduced from 50 µm to 20 µm or less.

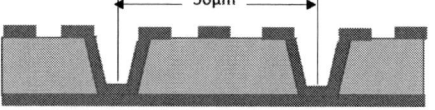

(a) State-of-the-art organic 2.5D interposer RDL w/20µm Via, 50µm Pitch and 6µm L/S

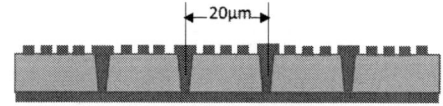

(b) Novel glass 2.5D interposer RDL w/2µm via, 20µm Pitch and 2um L/S and 5X wiring incensement

Figure 1. Comparison of state-of-the-art organic 2.5D interposer and novel glass interposer RDL technologies with small microvia interconnects

In the early 1980s, IBM developed C4 bumping process with a bump pitch of 250 µm. Later, IBM Japan developed Surface Laminar Circuit (SLC) technology using photovia process. [2] Microvia was defined as via with diameter less than or equal to 150 μm. The microvia technology was quickly used in chip scale packages (CSP) and flip chip ball grid array (FCBGA) package substrates. In the year 2000, the chip pitch was scaled down to 200 μm. The minimum diameter of microvia demonstrated was 25 μm in a 25 μm thick Vialux PID film. [3] However, due to the limited availability of photosensitive dielectric materials, laser technologies were introduced for microvia formation. The bump pitch was scaling but slowly down to about 160-180 μm till 2010. During the period of 2010-2015, with the development of copper pillar bumping technology, the flip chip BGA bump pitch quickly scaled down to 100 μm and even 80 μm. At the same time, 2.5D interposer and fan-out technologies emerged and significant progress was made in RDL. As a leading-edge organic interposer, Kyocera demonstrated 50 μm bump pitch organic laminate substrate by using advanced semi-additive process (SAP) process. [1] On the other hand, Xilinx used semiconductor back-end-of-line (BEOL) process and demonstrated the first 2.5D silicon interposer. The bump pitch was 45 μm. Shinko pushed the bump pitch to 40 μm in 2.1D organic interposer using hybrid RDL technology. [4] With the silicon moving to 14 nm node, more companies reduce their bump pitch to 50 μm and below in their roadmaps. With the development of sub-5 μm microvia technologies, the interposer bump pitch in the range of 20-30 μm will be feasible in the near-future. Figure 2 shows the progress of IC bump pitch during the past three to four decades and the trends in the near future.

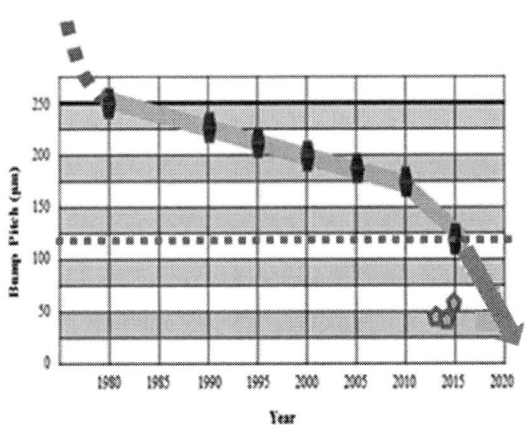

Figure 2. IC bump pitch improvement in the past decades and near future trends

The reduction of bump pitch requires scaling down of diameter of microvias. Laser ablation has been the primary method of microvia formation. CO_2 laser is widely used in PCBs but via size is limited to 40 μm. UV laser is used in HDI substrates and has the advantages of ablating small vias in various polymer dielectrics. However, the conventional UV lasers have challenges to scale down to sub 10 μm microvia diameters. Microvias with diameters of 20 μm in 8 μm thick polymer dielectrics were reported in 2.5D organic interposer Advanced Package X (APX). [1] In order to further reduce the microvia diameter, Shinko used thin

film photosensitive dielectric to fabricate 10 μm diameter microvias in their 2.1D organic interposer iTHOP. [4] SEMCO demonstrated 6 μm microvias using photolithography and Fraunhofer and Suss Microtec reported 5 μm microvias generated by using Excimer laser ablation. [5] As 2.5D interposer and high-density fan-out packaging RDL is emerging, smaller diameter microvias become critical. Small microvia diameters meet the fine pitch IC needs and allow high density routing between the fine pitch pads. The pros and cons among the key technologies will be discussed in the paper. The design engineers will have the flexibility for designing package architectures based on these most advanced microvia technologies.

II. TEST STRUCTURE AND PROCESS FLOW

Optical microscope, scanning electron microscope (SEM) techniques are critical for imaging of microvias. Normal mechanical polishing and focused ion beam (FIB) techniques are used for cross-sectioning of microvias. Optical microscope is a simple and quick non-destructive method of inspection and it is convenient for process development and optimization. However, when via diameter gets below 10 μm, it becomes difficult to inspect the vias using a microscope. A test structure for small via process development is shown in Figure 3. There is a thin layer copper between the dielectric film and substrate. The substrate can be an organic laminate or an inorganic silicon wafer or glass panel. The thin copper layer can be fabricated by means of lamination of copper foil, plating, sputtering on the surface of the substrate or it can be a copper cladded substrate. After cleaning and surface treatment of copper, the dielectric film to be tested was laminated on the substrate. The non-photo-imageable dielectric film was cured for laser ablation and the PID film was patterned by UV lithography and cured further. After microvia formation and a short O_2 plasma desmear process to remove any residues from copper surface, electroplating using the copper layer beneath the dielectric film was done for bottom up filling of copper in the via. High magnification (50X-1000X) optical microscope was used for via inspection. With the bottom-up plated copper, it is easier to judge whether the bottom of via is open or not by using optical microscope.

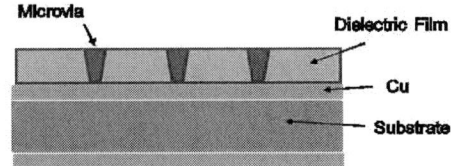

Figure 3. Test Structure for small microvia process development.

III. MICROVIA FORMATION TECHNOLOGIES

There are three primary technologies used for microvia formation in PCB and package substrate fabrication: (A) Photolithography, (B) Plasma Etch and (C) Laser ablation (CO_2, UV and Excimer). Photolithography was the first method used to create microvia in a photo-imageable dielectric (PID) polymer that initiated the high-density interconnect (HDI) substrate for flip chip applications. All vias, as per the mask design, are formed at the same time in the photolithography defined microvia formation. As mentioned before, the photo-defined microvia could achieve reliable 25 μm very small microvias in earlier time. [3] The drawback of this method was limited to the availability of materials. Same as photo defined via, plasma etch makes all microvia simultaneously. However, it is difficult to make small vias due to the nature of isotropic etching. Laser ablation technologies were rapidly developed in late 1990s and dominate the microvia formation since then. CO_2 laser emits infrared light with wavelengths between 9.3 μm to 10.6 μm. The mechanism of CO_2 laser ablation is based on photo thermal ablation by burning the polymer material away from the desired areas to form vias. The relative long wavelength limits the minimal focus diameter of the laser beam. The via diameter size drilled by CO_2 laser is limited to 40 μm. UV laser has shorter wavelengths of 355 nm and 266 nm. The quantum energy hv of a short wavelength photon is much higher than CO_2 laser. The mechanism of UV laser ablation is based on a combination of photo chemical and thermal ablations. The UV laser beam can be focused into a very small spot resulted in very high light intensity. These beams can break molecular bonds and ablate most of the dielectric polymers and metals. The minimum via diameter as mentioned was 20 μm. [1, 6] Due to strong demands of smaller microvias for the 2.5D interposer and fan-out package, photo-define via was back due to its excellent photolithographic capability. With the development of thin high-resolution photo-imageable dielectric materials, as small as 2 um diameter microvias are feasible. The excimer is commonly used in semiconductor-based ICs fabrication. In the recently, the excimer laser was interested to the packaging Engineers. It is used for embedded trench RDL fabrication and as well as for microvia formation in package substrates. Excimer laser can emit wavelengths from 126 nm to 351 nm light depending on the gases used. Wavelength of 308 nm using XeCl excimer laser is typical for packaging applications. A reticle mask is required for projecting the beam onto the substrate to create microvias. Diameter of 4 to 5 μm microvias were reported. [7]

IV. NEW GENERATION ULTRA-SMALL PHOTOVIA

A. Materials and Tools

In the recent years, panel level RDL with 1.5 μm ultra-fine line and space was demonstrated by using embedded photo trench method. The key material used in the photolithography was ultra-thin high resolution photosensitive dielectric film, IF4605. The film IF is a negative tone resist sensitive to wavelength of i-line 365 nm. Its dielectric constant is 3.5, glass transition temperature is 250 ⁰C and coefficient of thermal expansion is 45 ppm/K. Vacuum laminator was used for laminating the film to the substrate at 90 ⁰C and cured at 200 ⁰C. Another new photosensitive dielectric film reported in this paper is a

photosensitive dielectric dry film material (PDM) with a low CTE and low stress. An i-line projection stepper and low NA Ushio tool was used for the photo-defined via formation.

B. Results

Microvias with diameters of 2-5 μm at via pitches of 4-10 μm was successfully demonstrated in this paper. After microvias were formed, the vias were filled by using Ti-Cu sputtering followed by panel copper plating. Chemical wet etching was used to remove the overburden copper. Figure 4 shows the top view of copper filled 2-5 μm diameter microvias. Figure 5 (A) shows SEM images of 2 μm diameter area array microvias and (B) a magnified SEM image of 2 μm diameter.

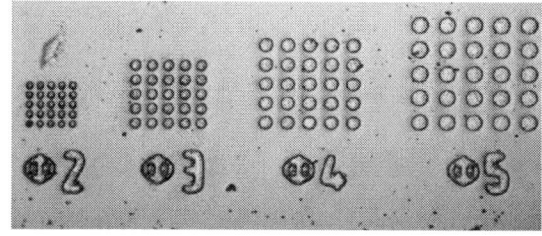

Figure 4. Micrograph of 2 to 5μm copper filled microvias in a 5 μm thick photosensitive dry film IF4605. The minimum via diameter opened was 2 μm at 4 μm pitch.

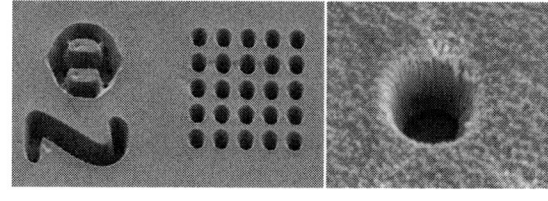

(A)	(B)

Figure 5. (A) Top view of 2 μm microvia array and (B) SEM image of a 2 μm diameter via

Another photosensitive dielectric film (PDM) from Taiyo Ink was also developed as a part of this work. Figure 6 shows cross section of a via with 3 μm diameter and microvia in 5 μm thick PDM film by using Focused Ion Beam (FIB) method. Test vehicle consists of a 400 microvia daisy chain consiting of 3 μm microvias was fabricated and the test vehicle succesfully passed 1,500 thermal cycle tests (TCT). The latest results will be reported in the 2019 ECTC conference in Las Vegas, NV, USA. [8]

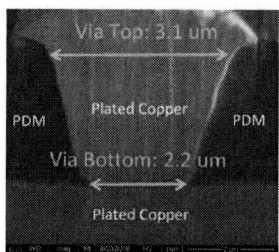

Figure 6. Image of FIB cross section of 3 μm diameter microvia formed in a 5 μm thick PDM film

V. EXCIMER LASER

A 308 nm XeCl excimer laser process was used for patterning the polymer dielectric. XeCl laser is inert to air and causes much less damage to the optics while having high processability for polymer dielectrics. Excimer laser uses a mask to ablate polymer dielectric in scan areas as large as 50 mm × 50 mm. The step and repeat mode helps to align vias and trenches in the RDL with +/- 1 µm precision alignment over a large dimensionally stable panel like glass substrate. Due to this precision alignment feature, simultaneous ablation of trenches followed by vias is possible in polymer dielectrics, resulting in an organic dual damascene process with reduced number of steps compared to conventional SAP. The process flow used is shown in Figure 7. Laser trenches were patterned to 4-5 µm depths in the polymer dielectric. The fluence used was 800 mJ/cm^2 and the total pulse count employed was 10. A daisy chain structure was fabricated in ABF having fine fillers with 5 µm trench widths and 4 µm microvia diameters. In this study, Disco Surface Planar tool was used to remove the copper overburden and planarize the metallized RDL. The diameter of via increased to 5 µm after the electroless copper seed desmear process to etch and roughen the polymer dielectric surface. The fabricated structure is shown in Figure 8. Thermal cycling reliability studies were performed on the samples according to JEDEC standards with proper pre-conditioning of samples. The thermal cycling conditions were as follows: (A) Hold at -55 ^{0}C for 15 mins, (B) Ramp from -55 ^{0}C to 125 ^{0}C in 15 mins and (C) Hold at 125 ^{0}C for 15 mins. The electrical resistance of the daisy chain structure was measured for 100 coupons using a four-probe method. The values of the electrical resistance of some of the coupons before and after 1000 thermal cycles are reported in Table 1.

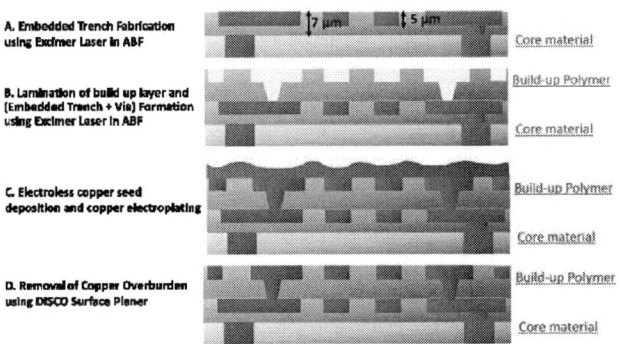

Figure 7. Process flow of dual damascene process using excimer laser for polymer RDL

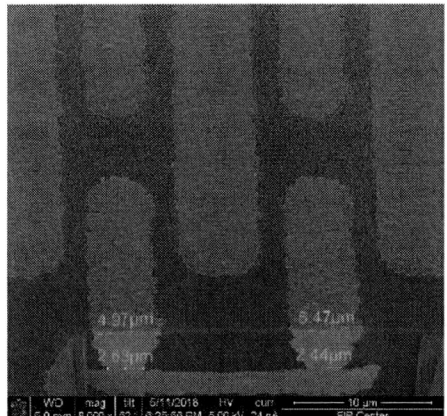

Figure 8. Multi-layer RDL with 5 µm diameter microvias and trenches fabricated using excimer laser with ABF dielectric

Table 1. Electrical resistance values of daisy chain structures fabricated in ABF dielectric before and after thermal cycling

Coupon #	Initial Resistance (After Fabrication)	Final Resistance (After 1000 cycles)	Resistance Change (%)
1	16.8 Ω	17.2 Ω	2.4 %
2	16.3 Ω	16.6 Ω	1.8 %
3	15.4 Ω	16.2 Ω	5.2 %
4	17.4 Ω	17.8 Ω	2.3 %
5	19.4Ω	19.7 Ω	1.5 %

VI. PICOSECOND UV LASER AND SMALL VIA FORMATION

A. Picosecond UV Laser

The system used in this study is a picosecond pulsed UV laser system, "Corner Stone", developed by ESI. The system is equipped with a coherent solid-state laser that generates picosecond pulses at 80 MHz repetition rate with an output power of 16 Watts. The wavelength of the UV laser is 355 nm which is the third harmonic of the original wavelength of 1,064 nm with a pulse width of 5 ps. The output energy of each pulse is 200 nJ. The image of the Corner Stone Laser system is shown in Figure 9. Compared to the conventional nanosecond UV laser, picosecond laser system has smaller pulse width (1/1000th) of the width of a nanosecond laser), resulting in very high pulse power (1000X power of a nanosecond laser). This high-power beam when focused in a very small spot will result in extremely high intensity. When the beam hits the dielectric material with intensity that exceeds the threshold of the material, the interaction in the area will be enhanced and completed in very short time. In the area of illumination, the material will be removed rapidly. Only very less energy will be left and escape into adjacent areas, minimizing the heat-affected zones in polymer dielectrics. Small and high quality microvia can be generated using this method. Figure 10 illustrates the comparison of nanosecond pulse and picosecond pulse laser ablation. The primary factors influencing the microvia fabrication are pulse power, width, beam profile, material absorption, density and composition, thickness, and the reflective properties of the surface of the substrate. Therefore, the process optimization in terms of laser power control, the amount of the laser energy and beam focus are important to achieve the required quality and size of the microvia.

Figure 9. Photo of new generation Picosecond Ultra-fast UV laser "Corner Stone" by ESI

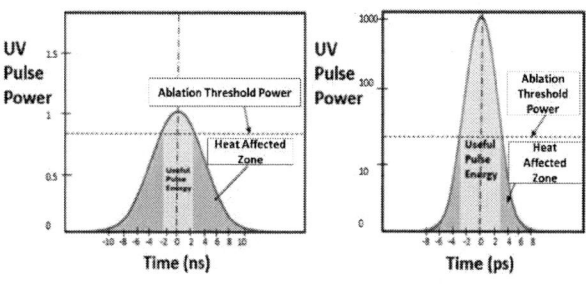

Figure 10. Comparison of nanosecond and picosecond lasers. Short pulse picosecond has higher power.

B. Materials

Two types of ultra-thin dielectric films were used in the present study: PLS (S) film from Panasonic and fine filler containing ABF film from Ajinomoto. The PLS (S) which is a newly developed thin dielectric dry film by Panasonic has the advantages of low D_k (2.99), high T_g and low residual stress. Two kinds of the PLS films with different thickness and filler sizes were used. The thicker one, PLS, is 15 μm with standard filler size. The thinner one is 7 μm thick, PLS-S, has smaller fillers than PLS. The residual stress of 15 μm PLS film was 8 MPa, which is 50% lower than that of conventional dielectric film, measured by BowOptic 208 Wafer Stress Curvature Tool. The lower stress allows the substrate with lower warpage that is greatly beneficial to the stability of fabrication process. The properties of the material are listed in Table 2. The ABF is a commonly used material in package substrates. The ABF film used in this study was 5 μm thick with fine fillers.

Table 2. Properties of PLS and PLS-S

Item	Condition / Method	Unit	PLS	PLS-S
Elastic modulus	DMA	GPa (30℃)	1.9	2.1
Tg	DMA	℃	277	270
Dk (@10GHz)	Empty Cavity Resonator	–	2.99	2.99
Df (@10GHz)	Empty Cavity Resonator	–	0.02	0.02

C. UV laser Microvia Formation

Spiral raster and pulse punch methods were used in this study. In spiral raster method, the UV beam is first focused in the center of the via and moves outward from a smaller ring to a larger ring in a spiral motion with a set of radii (r) as shown in Figure 11. The final diameter of via is determined by the outermost circle. Whereas, in pulse punch method, UV beam position is fixed and via is drilled by controlling the number of pulses and power level, shown in Figure 12. In the selected drill test, pulse numbers and power of the pulses were varied. The hypothesis is that smaller diameter microvias can be made by punch method.

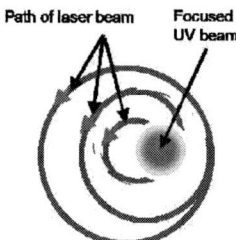

Figure 11. Sketch of spiral via ablation starting at the center. The final via diameter is determined by the outermost circle.

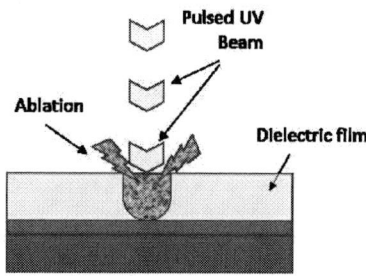

Figure 12. Schematic of pulsed UV beam punch process for small microvia formation. The via diameter is determined by the beam profile of the focused laser beam

D. Results

1) Spiral raster ablation in 15 μm PLS film

A thin FR-4 laminate with copper foil was used as core. After surface treatment the PLS was laminated on both sides of the laminate and cured at 200°C for 1 hour. The sample was ablated by focused laser beam using 1 watt and the spiral raster method. The results of spiral raster microvia ablation are shown in Figure 13. The tested material was 15 μm thick dielectric PLS film. A group of four vias were drilled using 1, 3, 5 and 7 circles respectively. The largest microvia was obtained by 7 circles and the smallest microvia was achieved by only one circle. An optical microscope was used for the inspection of the fabricated vias. All the vias can be seen from top to the bottom. Chemical desmear and electroless copper plating processes were used for via metallization. Figure 13(a) shows the images of via top and 13(b) via bottom after UV laser ablation. Figure 13(c) shows via cross sections after copper plating. The diameter of the four groups were 10, 20, 30 and 40 μm respectively. The thickness of the film measured was 13 μm. The bottom diameter of the smallest microvia is 7-8 μm. The aspect ratio of the smallest microvia is 1.3:1.

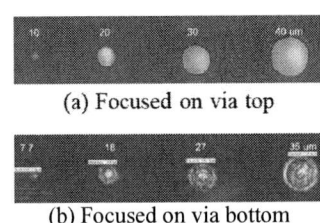

(a) Focused on via top

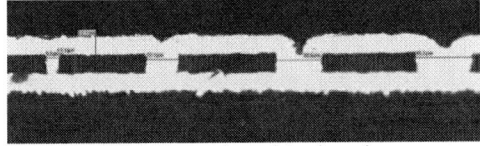

(b) Focused on via bottom

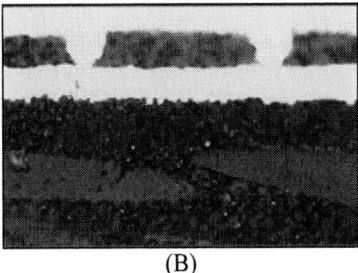

(c) Cross section of 4 microvias

Figure 13. Optical micrographs of microvias drilled by picosecond ultra-fast laser with various circles at different radii

2) Pulse punch ablation in 7 μm PLS-S

PLS-S is a modified version of PLS with small size fillers. The thickness is 7 μm. An organic laminate with a smooth surface profile, thin copper foil was employed as the core substrate. The dielectric film was laminated by a vacuum laminator and cured at 200 ^{0}C for 1 hour. A matrix was designed for the pulse punch test. The pulse power was varied from 0.1 to 1 Watt with an increment of 0.1 W and with punch numbers ranging from 1 to 10. Experiments showed that the minimum power required for via opening was 0.2 W with respect to 1 punch and 0.1 W with respect to 2 punches. It was found that the inspection is difficult when via size is less than 10 μm. Bottom up plating was done to plate copper into the vias. Figure 14 shows the optical images of the small microvias in the PLS-S by pulse punch method after bottom-up copper plating. The diameter of via was 5-7 μm.

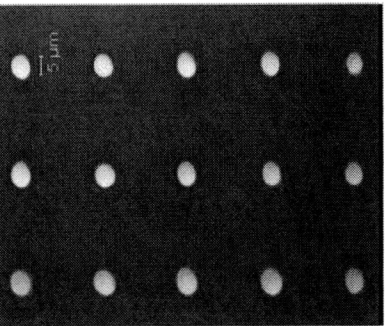

Figure 14. Top view of bottom-up copper plated microvias in 7 μm thick PLS-S film drilled by picosecond UV laser. The via diameter is 5-7 μm

3) Pulse Punch ablation in 5μm ABF

ABF film is a standard build up film widely used in package substrate. The ABF used in the test is 5 μm thick with small fillers. Similar to PLS-S, the ABF film was laminated on an organic laminate with low profile thin copper foil on both sides and cured at 130 ^{0}C for 30 minutes plus 180 ^{0}C for another 30 minutes. The result of pulse punch process on 5 μm thick ABF film is shown in Figure 15. The minimum diameter of microvia was 5-6 μm. Experimental test on different power level showed that the higher power resulted in larger microvias. Figure 16 shows the relationship of via diameter of power levels.

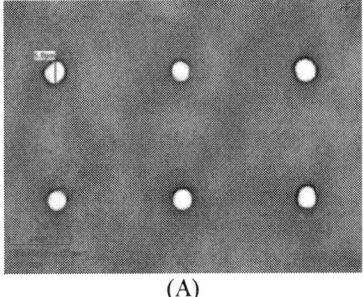

(A)

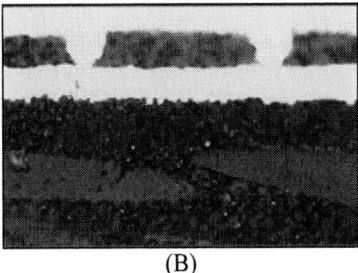

(B)

Figure 15. Optical images of microvias in ABF (5 μm thick, fine fillers) (A) Top view of bottom up plated copper and (B) Cross section of fully copper filled

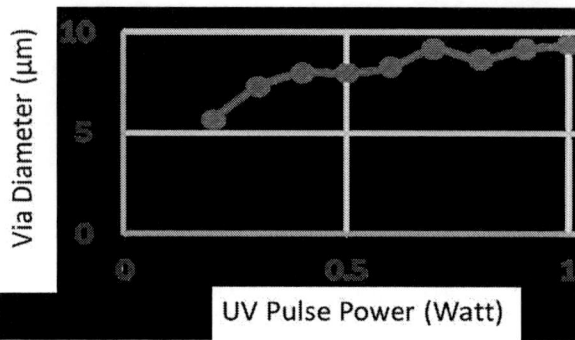

Figure 16. Experimental result of via diameter with UV laser power

The status of today's advanced technologies for generating small and ultra-small microvia technologies are summarized in Table 3.

Table 3. Summary of small diameter microvia technologies

Method		Λ	Via Dia, um	Advantages	Drawback
Laser	CO_2	9.3-10.6 um	40-60	Fast Low Cost	Beam size limits Heat effects
	UV	355 nm	7	Various materials Maskless	Point-by-point Position accuracy
	Excimer	308 nm	3-5	Various materials Position accuracy	Maintenance Mask required
Photo		365 nm	2	Simultaneously formation Position defined by mask	Limited material avaiablity

VII. SUMMARY

This paper studied and compared today's most advanced microvia generation technologies and reported the latest progress. We have demonstrated fabrication of microvias with 4 μm diameter through ablation of 7 μm thick ABF by excimer laser which successfully passed 1,000 thermal cycle test. Diameter of 2 μm ultra-small microvia was generated using a 5 μm thick IF film. Diameter of 3 μm microvia in PDM by using photolithographic method also have been shown to complete 1,500 TCT. We have also shown that a UV laser with picosecond pulse width is capable of generating 5-7 μm microvia. Further, this paper illustrates that the design engineers will have flexibility for designing their architectures based on these most advanced microvia technologies. Finally, with the development of sub-5 μm microvia technologies using various methods shown in this paper, fabrication of system architectures consisting of an interconnect bump pitch of 20 to 30 μm will be feasible in the near future.

VIII. ACKNOWLEDGEMENT

The authors would like to acknowledge PRC's consortium members for their support. In particular, we are very thankful to Electro Scientific Industry Inc for Ultra-fast UV laser, Habib Hichri from Suss MicroTec for Excimer laser, Ushio Inc. for Projection Stepper, Ajinomoto Co. Inc. for ABF films, Corning Glass and Asahi Glass companies for their ultra-thin glass substrates, AtoTech Corp. for copper plating chemicals and processes. The authors would also like to specially thank Mr. Martin Orrick and Mr. Lamar McDonald from Electro Scientific Industry Inc for their kind support and useful discussions.

IX. REFERENCES

[1] M. Ishida, "APX (Advanced Package X) - Advanced Organic Technology for 2.5D Interposer," in 2014 CPMT Seminar, Latest Advances in Organic Interposers, Lake Buena, Vista, Florida, USA, May 27-30, 2014

[2] Y. Tsukada, S. Tsuchida, Y. Mashimoto, "Surface Laminar Cicuit Packaging," in Proc. 42th IEEE Electronic Components and Technol. Conf. (ECTC), San Diego, CA, USA, May 18-20, 1992, pp. 22-27

[3] F. Liu, J. Lu, V. Sundaram, D. Sutter, G. White, D. Baldwin, R.R. Tummala, "Reliability Assessment of Microvias in HDI Printed Circuit Boards", IEEE Transactions on Components And Packaging Technologies, 254-259, Vol. 25, No. 2, June 2002

[4] K. Oi, S. Otake, N. Shimizu, S. Watanabe, Y. Kunimoto, T. Kurihara, T. Koyama, M. Tanaka, L. Aryasomayajula and Z. Kutlu, "Development of New 2.5D Package with Novel Integrated Organic Interposer Substrate with Ultra-fine Wiring and High-Density Bumps," in Proc. IEEE Electronic Components and Technol. Conf. (ECTC), Lake Buena, Vista, Florida, USA, May 27-30, 2014, pp. 348-353

[5] Woehrmann, M., Hichri, H., Gernhardt, R., Hauck, K., Braun, T., Toepper, M., Arendt, M.; Lang, K.-D, "Innovative excimer laser dual damascene process for ultra-fine line multi-layer routing with 10 μm pitch micro-vias for wafer level and panel level packaging," ECTC 2017, the 67th Electronic Components and Technology Conference, 30 May-2 June 2017, Lake Buena Vista, Florida.

[6] W-C Chen, C-W Lee, M-H Chung, C-C Wang, S-K Huang, Y-S Liao, H-C Kuo, C-C Wang and D Tarng, "Development of novel fine line 2.1 D package with organic interposer using advanced substrate-based process", in Proc. IEEE Electronic Components and Technol. Conf. (ECTC), San Diego, CA, USA, May 29-June 1, 2018, pp. 601-606.

[7] Nair, Chandrasekharan, Bartlet DeProspo, Habib Hichri, Markus Arendt, Fuhan Liu, Venky Sundaram, and Rao Tummala. "Reliability Studies of Excimer Laser-Ablated Microvias below 5 Micron Diameter in Dry Film Polymer Dielectrics for Next Generation, Panel-Scale 2.5 D Interposer RDL." In 2018 IEEE 68th Electronic Components and Technology Conference (ECTC), pp. 1005-1009. IEEE, 2018.

[8] D. Okamoto, Y. Shibasaki, Daisuke, F. Liu, M, Kathaperumal, R. R. Tummala, "Fabrication and Reliability Demonstration of 3 μm Diameter Photo Vias at 15 μm Pitch in Thin Photosensitive Dielectric Dry Film for 2.5 D Glass Interposer Applications", to be published in The 2019 IEEE 60th ECTC, Las Vagas, NV, USA

Multilayer RDL Interposer for Heterogeneous Device and Module Integration

Yi-Hang Lin, M.C.Yew, M.S. Liu ,S.M. Chen, T.M. Lai, P.N. Kavle, C.H. Lin, T.J. Fang, C.S. Chen, C.T. Yu, K.C. Lee, C.K. Hsu, P.Y. Lin, F.C Hsu and Shin-Puu Jeng*

Taiwan Semiconductor Manufacturing Company, No.6, Creation Rd. II,
Hsinchu Science Park, Hsinchu, Taiwan (R.O.C.) 30077
Email: *spjeng@tsmc.com

Abstract—in this paper, we demonstrate a high density heterogeneous large package using a RDL interposer with six interconnection layers. Four Si chiplets and two HBM modules are connected with fine pitch copper lines to deliver a complete system-in-package solution for high performance computation. The multilayer interconnections provide excellent design flexibility to optimize signal, power, and ground planes. The RDL interposer has generic structural advantages in interconnection integrity and bump joint reliability, which allows further scaling up of the package size for more complicated functional integration.

Keywords- Fanout and Heterogeneous Integration; Interconnections; Chiplets; System in package

I. INTRODUCTION

In high performance computing applications, one of the key enabling components is the fine-pitch RDL, which provides connection between logic and high-bandwidth memory (HBM) or between chiplets. The interconnection density determines the electrical performance of the packages. As the connection length between logic and JEDEC standard HBM memory is over 4mm, long interconnects of this type require not only insertion loss reduction with low impedance, but also strong crosstalk protection in both horizontal and vertical directions. Fine pitch Cu vias and traces allow finer power mesh, which reduces power delivery network (PDN) impedance, and reduces noise. In the chiplet scheme, multiple RDL interconnections are required to connect electrical interfaces, especially for high pin counts and to enhance design flexibility.

Si interposers have been successfully adopted for chiplets and HBM integration. [1] These packages exhibit excellent performance, which meet increasing bandwidth demands and unveil various important applications in network and artificial intelligence computation. In this paper, we demonstrate a large package that integrates four Si chiplets and two HBM modules on a RDL interposer with six layers of interconnections. The benefits of electrical performance using six RDL interconnections are analyzed. Furthermore, the generic mechanical advantages in RDL integrity and bump joint reliability of the new type of package are presented.

A. HBM integration

High-performance computing requires high density on-package integration with a high data rate. For example, the

packages in the references [2,3] integrate three FPGA chiplets and two HBMs and one big GPU and four HBMs, respectively. Higher computing efficiency can be achieved with more HBMs – for example, the package in reference [4] uses four HBMs to gain >2.7 TFLOPS and reference [5] employs six HBMs. The data rate of HBM continues to increase: the HBM1 data rate is 1 Gbps, HBM2 goes up to 2.4Gbps, and HBM3 plans to reach about 3.2Gbps [6, 7]. The high-performance package needs to provide good SI/PI performance to support such high data rates.

B. CPU cores integration :

To achieve higher performance computing, thread numbers and the number of CPU cores increase year by year. In the HPC processor in reference [8], the main blocks are divided into 3 parts: the system agent part for I/O, the multicore part for central computing, and the GPU part. The size of the central computing part expands for high performance computing. In order to increase core numbers and reduce cost, the package in references [9,10] divides one big processor into four chiplets.

C. System-level heterogeneous integration

From a system computation efficiency point of view, the seamless integration between FPGA, CPU, GPU, NPU, IO interfaces, SRAM, and HBM is critical. It is a challenge for package technology to integrate such diverse functional components. The package in reference [11], which integrates CPU, GPU and HBM, is likely the beginning of such a trend.

Here, we will demonstrate the potential of our multilayer RDL interposer for system level heterogeneous integration.

II. MULITIPLE RDL INTERPOSER FABRICATION

As shown in Figure 1, the basic integration scheme of the RDL fan-out interposer resembles that of the familiar Si interposer. Si chips and memory modules are attached to the interposer with protective molding compound, and the "chip-on-RDL interposer" structure is then jointed unto a PCB substrate with C4 bumps. Figure 2 shows the detailed constituents of the structure, including Si chips, memory modules, micro bumps, RDL interposer, C4 bumps, PCB substrate and BGA. The key components of RDL interposer are listed in Table 1. The line width/spacing of baseline RDL interposer is 2/2 um, which is larger than that of a

---

978-1-7281-1500-9/19 $31.00 © 2019 IEEE

typical Si interposer. The vertical interconnection is composed of fine pitch stacking vias and stagger vias, which allows flexible routing design without extra parasitic capacitance.

Figure 1 Schematic cross-section of SOC and HBM modules on multilayer RDL interposer. The interposer stack is attached to a PCB substrate.

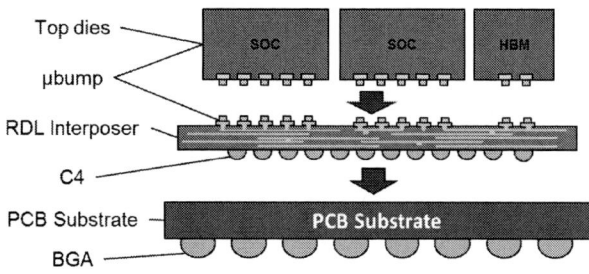

Figure 2 Schematic drawing showing the details and process sequence of heterogeneous integration of SOC and HBM on RDL interposer.

TABLE 1 KEY COMPONETS IN RDL INTERPOSER INTEGRATION

		RDL interposer
Heterogeneous integration		Yes
Si chip, module I/O		Cu, solder bumps
Interposer	Dielectric	Organic (Polyimide)
	RDL	Fine pitch Cu lines
	Vertical Interconnect	Cu via (staggered, stacking or mixture of both)
	C4	Cu, solder bumps

Figure 3 shows a fully assembled RDL interposer package. There are four Si chips and two HBMs in this package. The sizes of RDL interposer and PCB substrate are 32x35mm^2 and 55x55mm^2, respectively. Figure 4 shows the X-ray images of good self-aligned micro bump joints. The minimal bump pitch is 55um here.

Figure 5(a) shows the cross-sectional view of the package. Both Si chips and HBM are jointed onto a thin RDL interposer. The molded interposer structure is assembled to a PCB substrate with C4 bumps. Figure 5(b) shows the 6 RDL structure. The high density fine-pitch Cu RDL is for the connection between the PHY of chiplets and HBM. Ground RDL mesh is used as shielding for good SI/PI performance. Figure 5 (c), (d) and (e) show the SEM pictures of stagger vias, two stacking vias, and four stacking vias, respectively. The use of stacking via can reduce the RDL routing distance and increase design flexibility. Finally, Figure 6 shows the OM and SEM images of fine pitch Cu-lines with minimal 2um width.

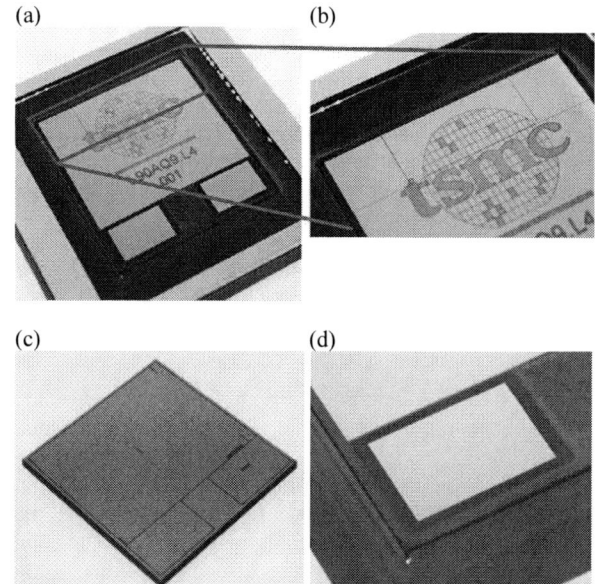

Figure 3 The interposer and PCB substrate dimensions are 32 x35 mm^2 and 55x55mm^2, respectively. (a), (b) The package has four Si chiplets and two HBM modules, (c) the backside of interposer with C4 bumps, (d) HBM module on interposer.

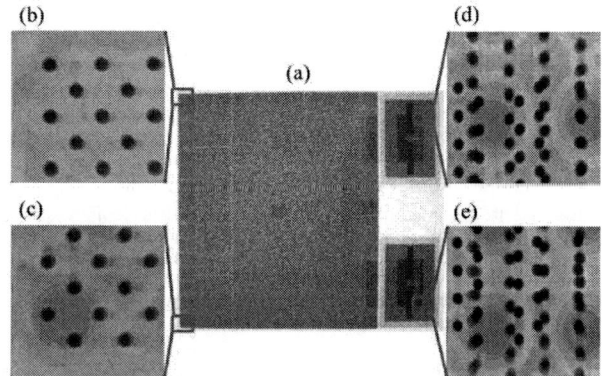

Figure 4 X-ray images that shows good micro-bump joints. (d), (e) are the micro-bump joints of HBM. The images of the bumps inside HBM overlap with the ones on interposer.

978-1-7281-1500-9/19 $31.00 © 2019 IEEE

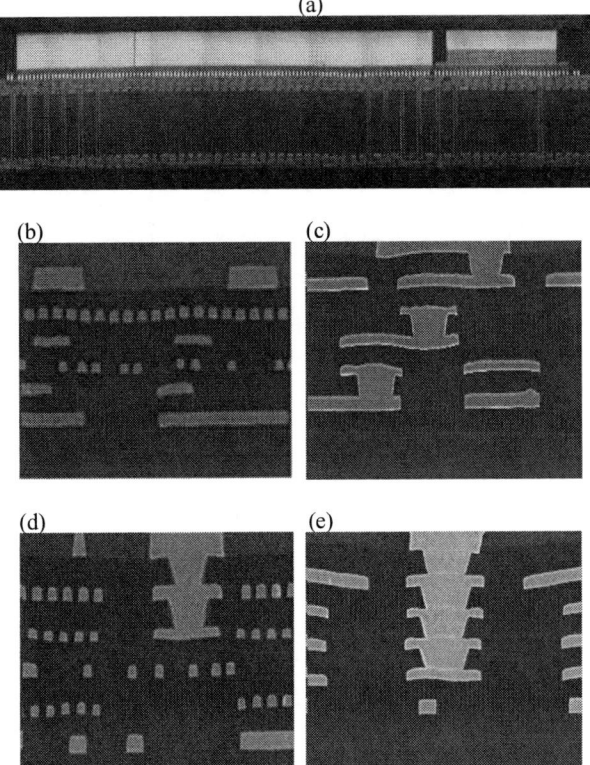

Figure 5 Cross-sectional views of (a) RDL interposer package on a PCB substrate, (b) six layers of Cu interconnections, (c) six layers of Cu interconnections with stagger vias, (d) Cu interconnections with two stacking vias, (e) Cu interconnections with four stacking vias.

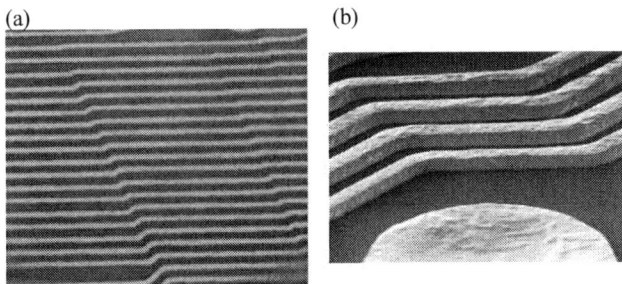

Figure 6 (a) Optical microscope and (b) SEM images of 2 um Cu lines.

III. ELECTRICAL PERFORMANCE OF MULTILAYER RDL

The electrical performance of eye diagrams and the insertion loss (S parameter) of three different RDL arrangements, co-planar GSSG structure in the three RDL scheme [Fig 7(a)], a co-planar GSGSG structure in the three RDL scheme [Fig. 7(b)], and the co-planar GSGSG structure shielded by three extra ground traces in the six RDL scheme [Fig 7(c)], is studied. The signal integrity performance, i.e.,

the height and jitter noise of eye diagrams, is simulated using HFSS and ADS. Figure 8 (a), (b) and (c) compare the eye diagrams of these different isolation configurations. Both the eye height and jitter noise of signal lines are significantly improved with additional ground isolation. The electrical performance of the six-RDL scheme has a superior performance as compared to that of the three-RDL scheme. Table 2 summarizes the values of improvement: eye height of six-RDL design is 6%-16% better than that of three-RDL scheme. For jitter noise, the six-RDL design is 2-3 times better than the three-RDL design.

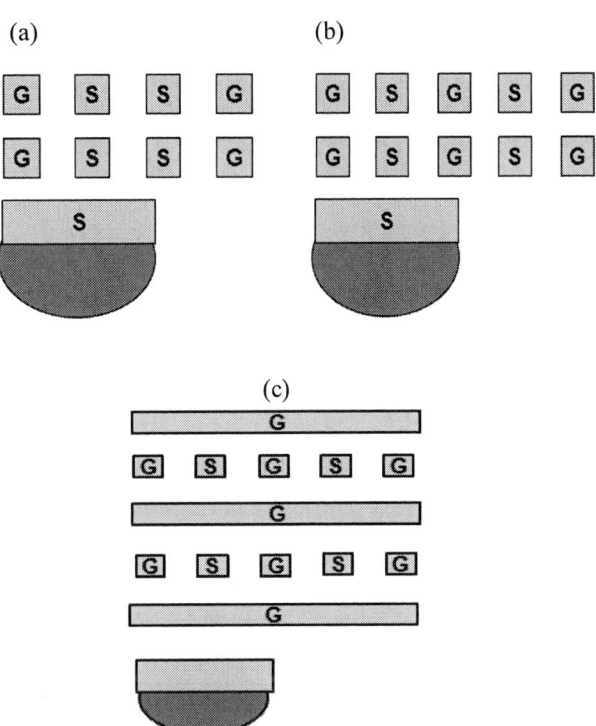

Figure 7 Signal routing arrangements of three-RDL and six-RDL (a) coplanar GSSG with three RDL interconnections, (b) coplanar GSGSG with three RDL interconnections, (c) coplanar GSGSG and interlayer ground shielding with six RDL interconnections.

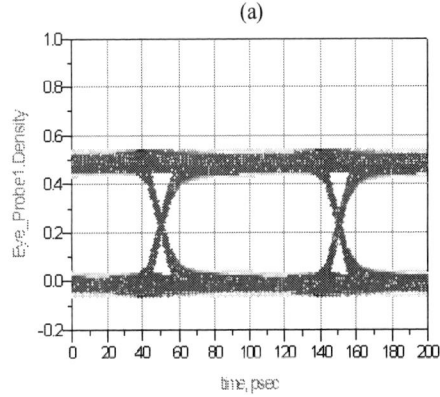

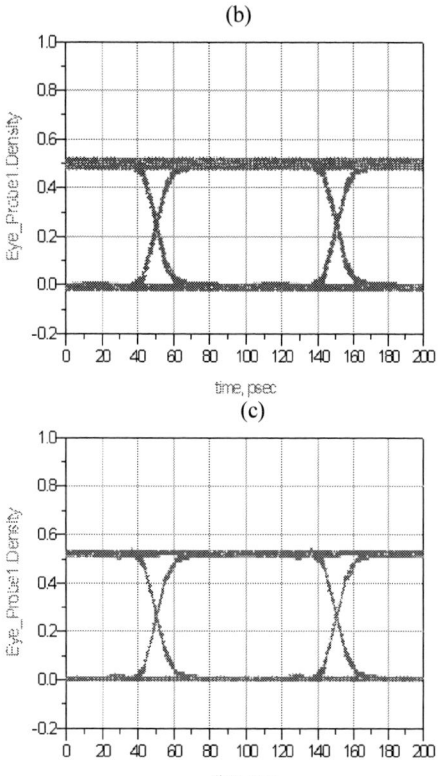

Figure 8 Simulated eye diagrams of (a) coplanar GSSG with three RDL interconnections, (b) coplanar GSGSG with three RDL interconnections, (c) coplanar GSGSG and interlayer ground shielding with six RDL interconnections.

Table 2 the Signal integrity

	RDL type	3RDL-GSSG	3RDL-GSGSG	6RDL-GSGSG
Signal Integrity	Eye height	0.84x	0.94x	1x
	Jitter(rms)	1x	0.49x	0.34x

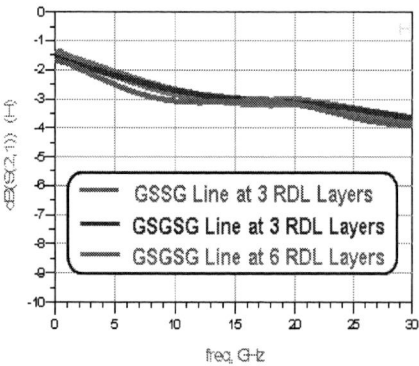

Figure 9 Simulated insertion loss of different configurations is compared for HBM-SOC PHY connections.

The insertion loss performance is important for high frequency operation. High insertion loss degrades the signal intensity and increases the operational power. The insertion loss (S21 parameter) performance of these three signal routing structures is compared in Figure 9. Comparable performance is observed due to equal line width and thickness for all signal routings.

The crosstalk performance of the two adjacent co-planar signal lines is compared in Figure 10(a). The coplanar GSSG and GSGSG structures with three RDL interconnections exhibit larger crosstalk than that of coplanar GSGSG with six RDL interconnections. Additional interlayer ground shielding with six RDL interconnections provides significant performance improvement. For the layer-to-layer crosstalk, the inserted ground plane in the six-RDL scheme is capable of completely isolating the signal lines, and produces nearly zero crosstalk, as shown in Fig. 10 (b).

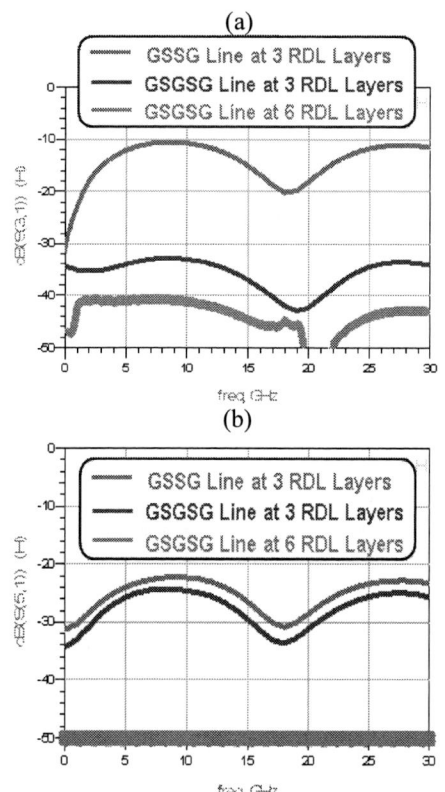

Figure 10 (a) Simulated crosstalk of two adjacent co-planar signal lines, (b) simulated layer-to-layer crosstalk of two adjacent signal lines in vertical direction.

IV. STRUCTURAL ADVANTAGES OF RDL INTERPOSER AND RELIABILITY ASSESSMENT

The four-chips-plus-two-HBM RDL interposer package successfully passes the stringent reliability torture without failures. There are generic structural advantages of the RDL interposer, particularly in RDL integrity and bump joints reliability.

A. Mutilayer RDL Integrigy

Compared to the RDL layer elsewhere on the package, the fine pitch Cu lines underneath the gaps between Si chip and HBM have a relatively lower structural stiffness support. These lines can be deformed and broken during the reliability test.

The stress on RDL interposer with a temperature loading from room temperature to 250C is characterized with finite element analysis. Due to its shortest distance to Si and HBM, RDL1 has the highest stress from CTE mismatch as shown in the contour plot in Figure 11. Fortunately, the underfill material between RDL and Si chip/HBM serves as good stress buffer layer, which significantly reduces the stress to below the risk level, as shown in Figure 12.

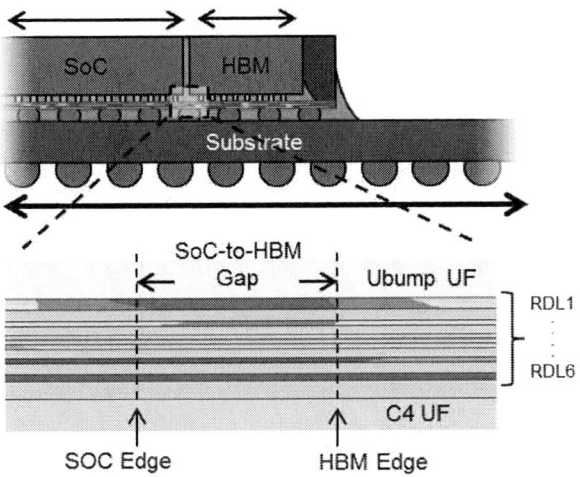

Figure 11 Cross sectional schematic of RDL interposer, and P1 stress contours of the RDL below the SoC-to-HBM gap.

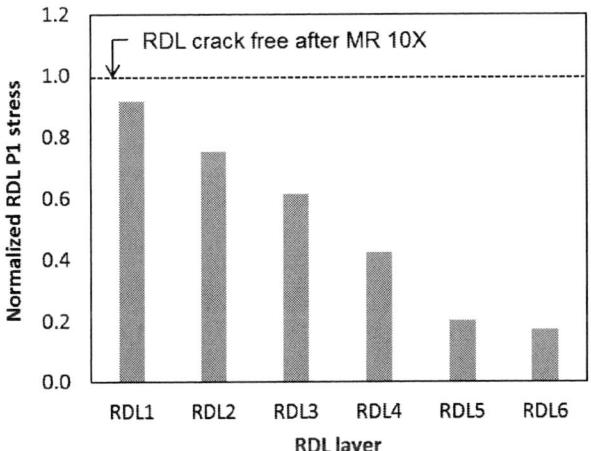

Figure 12 Normalized P1 stress each RDL layers.

B. Micro bump, C4 Joint Reliability

The micro solder joint reliability is investigated through mechanical stress simulation. The temperature cycling (TC) loading ranges from -40°C to 125°C with a 1-hour cycle duration. Figure 13 shows the accumulated strain energy density (SED) of the corner micro bump on Si chip for both RDL interposer and flip chip packages. The SED on micro bump is significantly reduced by the RDL layer and the underfill layer for C4 bumps. The normalized maximum delta SED within 1 TC of corner micro bump is 0.52, which is lower than experimentally proven safe delta SED level.

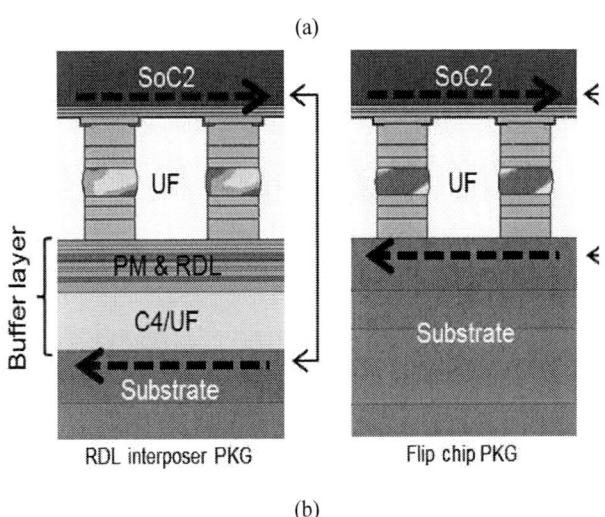

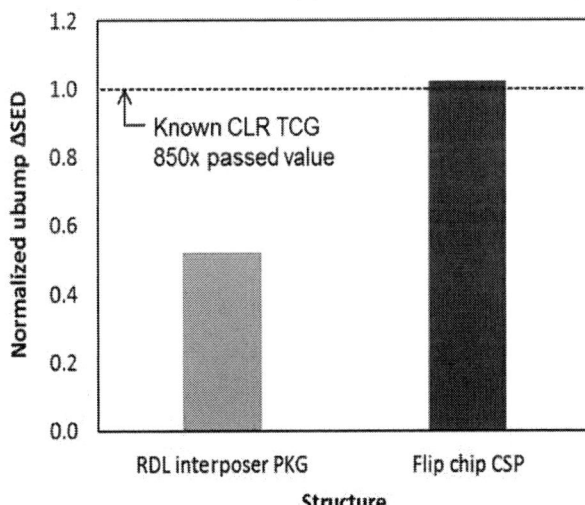

Figure 13 Comparison of (a) micro-bump strain contour, (b) normalized micro-bump strain energy density between RDL interposer package and flip chip package with same boundary condition.

Similarly, the accumulated SED of the corner C4 bump on Si chip can be reduced by the flexible RDL layer, as shown in Figure 14. The C4 joint reliability, i.e., the chip-package-interaction (CPI), window is substantially larger than the

978-1-7281-1500-9/19 $31.00 © 2019 IEEE

typical flip chip type package. This is the primary reason why the RDL interposer is scalable to large sizes.

(a)

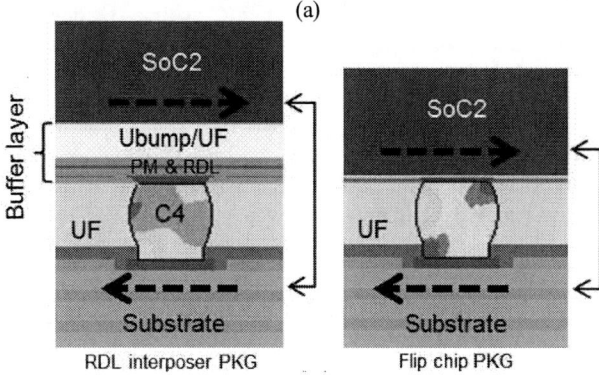

(b)

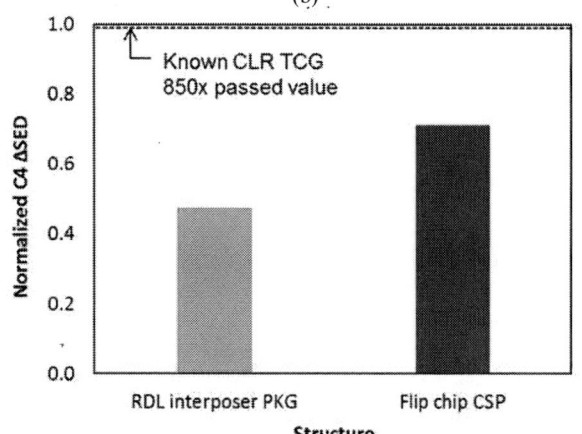

Figure 14 Comparison of (a) C4 strain contour, (b) normalized C4 strain energy density between RDL interposer package and flip chip package with same boundary condition.

V. CONCLUSION

The multilayer RDL interposer package is an excellent heterogeneous integration platform. Six layers of interconnection provide design flexibility for chiplets and HBM integration with good electrical performance, such as large eye height, low jitter, and nearly zero layer-to-layer crosstalk performance. This unique scheme, due to the flexible organic RDL layers used as a stress buffer layer to protect fine pitch Cu lines and bump joints, offers good package reliability and scalability to larger package sizes.

REFERENCES

[1] Suresh Ramalingam, "HBM package Integration: Technology Trends, Challenges and Applications", IEEE Hot Chip Symposium, 2016.
[2]Gaurav Singh et al., "Xilinx 16nm Datacenter Device Family with In-Package HBM and CCIX Interconnect ", IEEE Hot Chips Symposium, 2017.
[3] Jack Choquette, "Volta: Performance and Programmability", IEEE Hot Chips Symposium, 2017.
[4] Toshio Yoshida, "Fujitsu High Performance CPU for the Post-K Computer", IEEE Hot Chips Symposium, 2018.
[5] Yohei Yamada, "Vector Engine Processor of NEC's Brand-New Supercomputer SX-Aurora TSUBASA", IEEE Hot Chips Symposium, 2018.
[6] Jin Hee Cho et al., "A 1.2V 64Gb 341GB/s HBM2 Stacked DRAM with Spiral Point-to-Point TSV Structure and Improved Bank Group Data Control", ISSCC, 2018.
[7] Hongshin Jun et al.,"HBM (High Bandwidth Memory) DRAM Technology and Architecture", International Memory Workshop, 2017.
[8] https://en.wikipedia.org/wiki/Intel_Core
[9]Kevin Lepak et al., "The next generation amd enterprise server product architecture", IEEE Hot Chips Symposium, 2017.
[10] Noah Beck et al., "Zeppelin: An SoC for Multichip Architectures", ISSCC, 2018.
[11] Srinivas Chennupaty, "Thin & Light & high performance graphics", IEEE Hot Chips Symposium, 2018.

Effects of Dielectric Curing Conditions on the Interfacial Adhesion of Cu RDL for Fan-Out Wafer Level Packaging

Gahui Kim, Kirak Son, Young-Bae Park
School of Materials Science and Engineering
Andong National University
Andong -si, Korea
e-mail: ybpark@anu.ac.kr

Dogeun Kim
Principal Researcher of Advanced Nano-Surface
Department, Head of Research Planning Department
Korea Institute of Materials Science (KIMS)
Changwon, Korea
e-mail: dogeunkim@kims.re.kr

Seok-hyun Lee
Advanced Package Development Team
Test & System Package
SAMSUNG ELECTRONICS CO.,LTD
Hwasung-city, Korea
e-mail: sh1923.lee@samsung.com

Abstract— The effect of polybenzoxazole (PBO) dielectric curing and post-annealing treatment conditions at 150°C on the interfacial adhesion energies of Ti barrier and PBO dielectric layers for fan-out wafer level packaging applications were systematically investigated. The initial interfacial adhesion energies were 16.63, 25.95, and 16.58 J/m^2 under PBO curing conditions at 175, 200, and 225°C, respectively. The interfacial adhesion energies were 25.95, 28.79, and 8.86 J/m^2 at post-annealing times of 0, 24, and 100 h, respectively. The increased interfacial adhesion energy during PBO curing and post-annealing treatment for 24h seems to be closely related to M-O-C species at Ti/PBO interface, while decreased interfacial adhesion energy during post-annealing treatments for 100h seems be closely related to degradation of the PBO.

Keywords-FOWLP, PBO, 4-point bending test, interfacial adhesion energy

I. INTRODUCTION

In recent years, electronic devices have been increasingly expected to meet the demand for higher performance, small size, weight reduction, and high input/output (I/O) density [1,2]. Fan-out wafer level packaging (FOWLP) is one of the latest trends in microelectronic packaging. FOWLP technology is a promising advanced packaging solution with many applications due to its advantages such as high I/O density, cost-effectiveness and small form factor [3]. A key ingredient enabling the FOWLP technology is the redistribution layer (RDL), which routes high density connections on the chip to lower density connections on the substrate [4]. The RDL of FOWLP is configured into an electroplated Cu/polymer dielectric structure. Currently, the polymer dielectric materials for FOWLP are cured at around 230 ° C [5]. However, low temperature curing requirements

for dielectric materials are required low heat resistance, strong adhesion to the Cu RDL, and high reliability in the multi-layered dielectric materials after the reliability test [6]. These requirements are essential for FOWLP technology since these dielectric materials pass through thermal humidity test, high temperature storage, a thermal cycle test, and a high-impact test [6].

Previous reports were studied the interfacial adhesion in molding resin/polyimide systems on the polyimide curing conditions. Poor stability of the molding resin/polyimide interface at the lower cure temperatures is caused by residual polyimide contamination [7]. In addition, the effects of conversion from polyhydroxyamide (PHA) to polybenzoxazole (PBO) on the interfacial adhesion energy between PBO and Cu films were systemically analyzed [8]. The lap-shear strength of PBO/ Cu systems increased with the conversion of PHA to PBO [8]. Also, previous reports were studied the effects of interfacial adhesion on metal film/polyimide systems after post-annealing treatment [9]. The effect of annealing conditions on the interfacial adhesion energy between electroless-plated Ni and polyimide film was systematically analyzed by a 180-degree peel test. It is reported that degradation of the polyimide structure due to oxygen diffusion through the interface between Ni and alkali wet-treated polyimide [9].

In this study, the effects of dielectric curing and post-annealing treatment conditions on the interfacial adhesion energy between the Ti barrier and PBO dielectric layers was quantitatively measured using a 4-point bending test. The delaminated surfaces of the metal and the PBO dielectric were analyzed with scanning electron microscopy (SEM), energy dispersive spectroscopy (EDS), and X-ray photoemission spectroscopy (XPS).

II. EXPERIMENTAL

The dielectric was based on a low-temperature curable photo-sensitive PBO (HD-8930) provided by Hitachi Chemical and DuPont Microsystems Ltd [10]. The PBO dielectric was formed on a 500μm thick 4-inch Si wafer using a spin coater to a thickness of 3μm followed by curing at 175, 200, and 225 ° C for 2 h in air. The 50nm thick Ti barrier and 100nm Cu seed layer were sequentially deposited using DC magnetron sputtering. Cu was electroplated at room temperature at a current density of 20 mA/cm^2 in an electroplating Cu solution containing $CuSO_4 \cdot 5H_2O$ with a mixture of additives. A 2μm thick electroplated Cu layer was finally obtained.

The quantitative interfacial adhesion energy was measured using a 4-point bending test. Figure 1 shows a schematic of the 4-point bending geometry and the sample structure. A 4-point bending test can measure the interfacial adhesion energy between very weakly bonded brittle material [11,12]. The energy release rate, G, which is required for a crack to propagate at the inner homogeneous material, can be expressed follows [12, 13]

$$G = \frac{21(1 - v^2)M^2}{4E b^2 h^3} = \frac{21(1 - v^2)P^2 L^2}{16E b^2 h^3} \qquad (1)$$

Here, v represents the Poisson's ratio for a Si wafer (0.28), E denotes the elastic modulus of the elastic material (139 GPa), b is the specimen width (3mm), h is one wafer thickness (500μm). L is the distance between inner and outer loading points, and M is the moment defined as $PL/2$. A 4-point bending tests were conducted with a loading rate of 0.08 μm/sec with 20N load cell.

To identify the location of the debonded path, the morphology of the resulting fracture surfaces was observed by SEM and EDS. XPS measurements were performed to investigate the interface chemistry on delaminated surfaces in a Thermo Fisher Scientific with monochromatic Al Kα X-ray radiation at 1486.6 eV, and the C 1s peak at 284.7 eV corresponding to surface contamination was used as a reference peak[12].

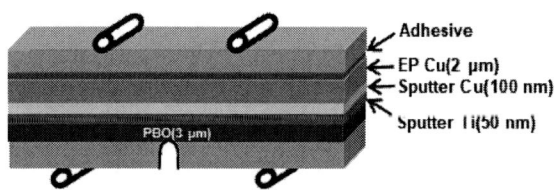

Figure 1.   Schematic sample structure of RDL/dielectric for 4-point bending test .

III. RESULTS AND DISCUSSION

The effects of dielectric curing and post-annealing treatment conditions on the interfacial adhesion energy between the Ti barrier and PBO dielectric layers were evaluated quantitatively. Load–displacement graphs were obtained, and the interfacial adhesion energy was calculated using Eq. (1)[12,14]. When the crack propagates on a weak interface, the load decreases abruptly and then shows a steady-state value on the load versus displacement graph [12,15]. The value of the constant load is used to calculate the interfacial adhesion energy [12,16,17]. Figure 2 shows the quantitative measured interfacial adhesion energy in the sample specimens for the PBO curing conditions based on 4-point bending tests. The interfacial adhesion energy was 16.63 ± 7.55, 25.95 ± 5.59, and 16.58 ± 2.78 J/m^2 for PBO curing conditions of 175, 200, 225°C, respectively. The interfacial adhesion energy increased or decreased with the PBO curing conditions.

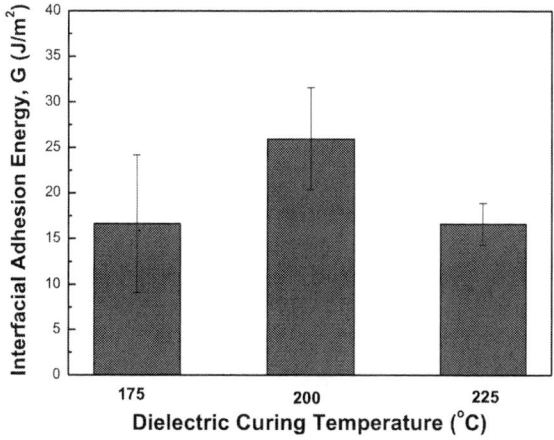

Figure 2.   Effect of dielectric curing temperatures on the measured interfacial adhesion energies.

After the 4-point bending test, both sides of the delaminated surfaces were analyzed using SEM and EDS. Figure 3 shows SEM images of the delaminated upper and lower sides at PBO curing conditions. As shown in Fig. 3, the delaminated upper and lower sides had relatively smooth surface morphologies. In the EDS results of the delaminated upper and lower sides, C, O, F, and Si were detected on the upper side, and C, O, F, Ti, and Cu were detected on the lower side. This indicates that the main path of delamination in the EP Cu/Cu seed/Ti barrier/PBO dielectric/Si multilayer is the Ti/PBO interface.

978-1-7281-1500-9/19 $31.00 © 2019 IEEE

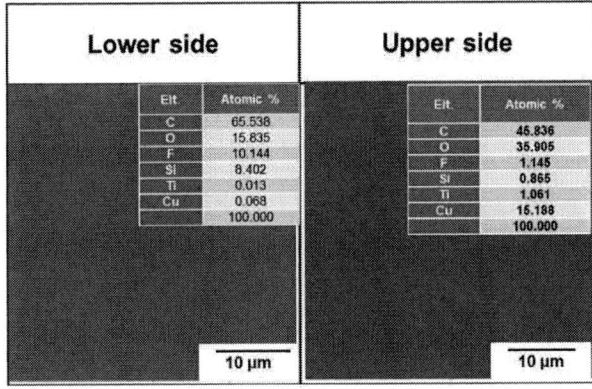

Figure 3. SEM images of the upper and lower sides after the 4-point bending test at the Ti/PBO interface after PBO curing condition.

XPS analysis of the delaminated PBO surface was used to investigate the changes in chemical bonding resulting from the PBO curing conditions. The XPS wide-scan spectra of the PBO surface detected C 1s, N 1s, F 1s, and O 1s in all samples. The XPS C 1s peak of the delaminated PBO surface was decomposed into the component peaks by Gaussian curve fitting for the PBO curing conditions as shown in Fig. 4. The resulting peaks were assigned to C-O (286.3 eV) and C=O (288.6 eV) bonding[9]. The C-O bonding increased from 175 to 200 °C, while the C-O bonding was decreased from 200 to 225 °C. The improved adhesion has been attributed to the interaction between the metal and the surface carbonyl groups resulting in the formation of M-O-C species [18]. Therefore, we speculate that the formation of the functional groups, such as M-O-C species, is closely related to the interfacial bonding mechanism between Ti and PBO.

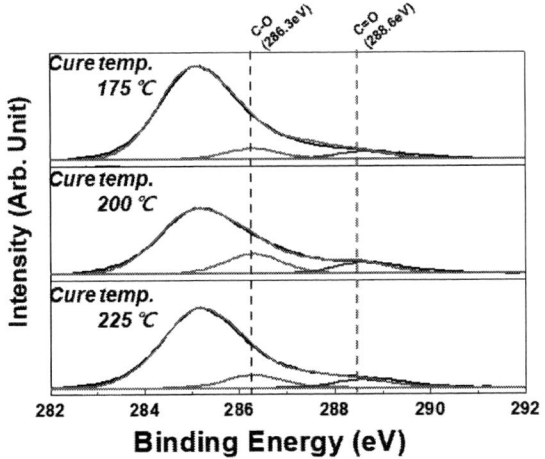

Figure 4. XPS analysis of C 1s peak at the delaminated lower side after the 4-point bending test for PBO curing conditions.

Figure 5 shows the interfacial adhesion energy of sample measured using the 4-point bending test after post-annealing treatment at 150°C after PBO dielectric curing at 200°C. The interfacial adhesion energies were 25.95 ± 5.59, 28.79 ± 2.23, and 8.86 ± 4.89 J/m^2 at post-annealing times of 0, 24, and 100 h, respectively. The interfacial adhesion energy was increased after a post-annealing treatment of 24 h while the interfacial adhesion energy was decreased after post-annealing treatment of 100h.

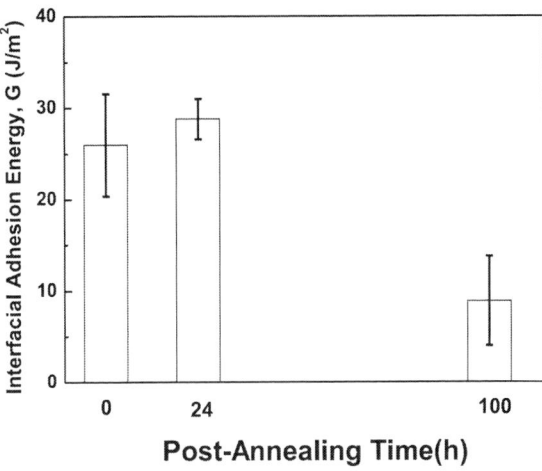

Figure 5. Effect of post-annealing time on the measured interfacial adhesion energies at 150°C

Figure 6 represents an SEM image of a delaminated surface of metal film and PBO dielectric after post-annealing treatment. As shown in Fig. 6, the delaminated upper and lower sides had relatively smooth surface morphologies. Most of the Ti/PBO interface was delaminated after 24 h, and partial delamination occurred at the interface between Cu and adhesive. Delaminated interfaces occurred at the Ti/PBO interface after 100 h.

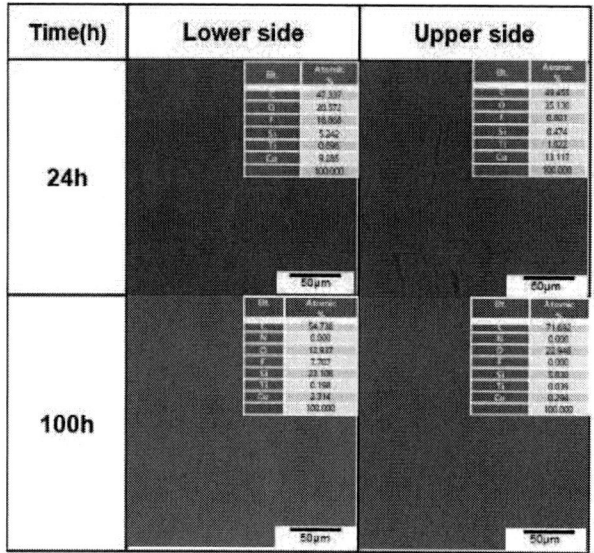

Figure 6. SEM images and EDS results of the delaminated interfaces after post-annealing treatment.

The XPS analysis of the delaminated PBO surface revealed changes in chemical bonding arising from the post-annealing treatment as shown in Fig. 7. The C-O bonding increased from 0 to 24 h while the C-O bonding decreased from 24 to 100 h. The increase in interfacial adhesion energy during 24 h appears to be closely related to the metal carrying surface carbonyl groups to form M-O-C species[18].

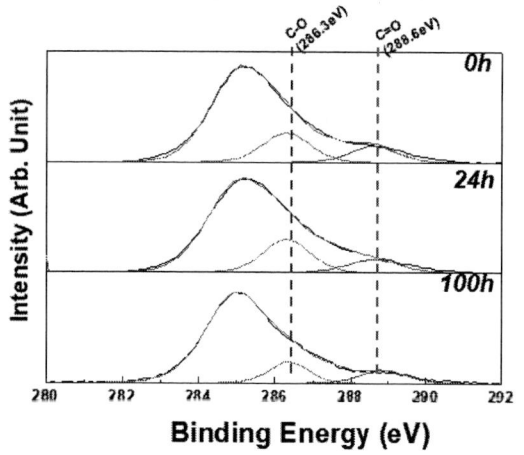

Figure 7. XPS analysis of C 1s peak on the lower side after the 4-point bending test at the Ti/PBO interface after annealing at 150 °C.

IV. CONCLUSION

The effect of PBO dielectric curing conditions and post-annealing treatment on the interfacial adhesion energy between Ti film and a PBO dielectric was evaluated using a 4-point bending test. Measured interfacial adhesion energy increased or decreased during both PBO curing condition and post-annealing treatment. The increased interfacial adhesion energy following PBO curing and 24 h of post-annealing treatment appears to be closely related to the M-O-C species at Ti/PBO interface, while the decreased interfacial adhesion energy following 100 h of post-annealing treatments appears be closely related to PBO degradation.

ACKNOWLEDGMENT

This research was supported by Basic Science Research Program through the National Research Foundation of Korea(NRF) funded by the Ministry of Education(2016R1D1A3B03933937), MOTIE(Ministry of Trade, Industry & Energy (10067804) and KSRC(Korea Semiconductor Research Consortium) support program for the development of the future semiconductor device.

REFERENCES

[1] K. Nishido, H. Onozeki, N. Suzuki, and T. Nonaka, "Study of the die potion accuracy in the fabrication process of a die first type FO-PLP", Proc. 2018 20th Electronics Packaging Technology Conference (EPTC), Dec., 2018.

[2] Y. Andriani, X. Wang, S. Liu, Z. Chen, X. Zhang, "Thermomechanical and Vixcoelasic Properties of Dielectric Materials Used in Fan-Out Wafer-Level Packaging", Proc. 2018 20th Electronics Packaging Technology Conference (EPTC), Dec., 2018.

[3] F. X. Che, Z. Chen, "Study on Electrical Performance Mechanical Reliability of Antenna in Package (AIP) with Fan-Out Wafer Level Packaging Technology", Proc. 2018 20th Electronics Packaging Technology Conference (EPTC), Dec., 2018.

[4] W. W. Flack et al., "One Micron Damascene Redistribution for Fan-Out Wafer Level Packaging using a Photosensitive Dielectric Material", Proc. 2018 20th Electronics Packaging Technology Conference (EPTC), Dec., 2018.

[5] T. Enomoto, S. Abe, D. Matsaku, T. Nakamura, "Recent Progress in Low Temperature Curable Photosensitive Dielectrics", Proc. International Conference on Electronics Packaging (ICEP), pp. 498-501, Apr., 2017.

[6] T. Fujiwara, Y. Shoji, Y. Masuda, K. Hashimoto, Y. Koyama, K. Isobe, H. Araki, R. Okuda, and M. Tomikawa, "Higher Reliability for Low-temperature Curable Positive-Tone Photodefinable Dielectric Materials", Proc. 19th Electronics Packaging Technology Conference (EPTC), Dec., 2017

[7] M. Amagai, M. Ohsumi, E. Kawasaki, R. Baumann, and H. Kitagawa, " The Effect of Polyimide Surface Morphology and Chemistry on Package Cracking Induced by Interfacial Delamination", Proc. of 1994 IEEE International Reliability Physics Symposium (IRPS), Apr., 1993, pp.101-107 doi: 10.1109/RELPHY.1994.307850.

[8] D. H. Baik, W. O. Lee, Y. H. Park, "INTERFACIAL CHARACTERIZATION OF POLYBENZOLE/COPPER SYSTEM", Mol. Cryst. Liq. Cryst., vol. 424, Oct., 2004, pp. 265–271, doi: 10.1080=15421400490506270.

[9] S.-C. Park, K. -J. Min, K. H. Lee, Y. Jeong, and Y. -B. Park, "Effect of Annealing on the Interfacial Adhesion Energy between Electroless-Plated Ni and Polyimide" Met. Mater. Int., vol. 17, Feb. 2011, pp. 111-115, doi: 10.1007/s12540-011-0215-z.

[10] S. Kang, Y. Kim, A. Moon, S. Lee, S. E. Kim, and S. Kim, "Surface Planarization of Polymeric Interlayer Dielectrics for FOWLP Applications", Proc. 2018 20th Electronics Packaging Technology Conference (EPTC), Dec., 2018.

[11] R. H. Dauskardt, M. Lane, Q. Ma, N. Krishna, "Adhesion and debonding of multi-layer thin film structures" Engineering Fracture Mechanics vol. 61, Agu., 1998, pp. 141-162, doi: 10.1016/S0013-7944(98)00052-6.

[12] M. Jeong, B. -H. Bae, H. Lee, H. -O. Kang, W. -O. Hwang, J. -M. Yang, and Y. -B. Park, "Effects of post-annealing and temperature/humidity treatments on the interfacial adhesion energy of the Cu/SiNx interface for Cu interconnects", Jpn. J. Appl. Phys., vol. 55, May, 2016, pp. 06JD01- 06JD01-5, doi: 10.7567/JJAP.55.06JD01.

[13] P. G. Charalambides, J. Lund, A. G. Evans and R. M. McMeeking, "A Test Specimen for Determining the Fracture Resistance of Bimaterial Interfaces", J. Appl. Mech., vol. 56, Mar., 1989, pp. 77-82, doi:10.1115/1.3176069

[14] Q. Ma, J. Bumgarner, H. Fujimoto, M. Lane and R. H. Dauskardt, "Adhesion Measurement of Interfaces in Multilayer Interconnect Structures," in Materials Reliability in Microelectronics VII, Pro. of MRS Annual Meeting, Feb., 1997, pp. 3-14, doi: 10.1557/PROC-473-3.

[15] J. -W. Kim, K. -S. Kim, H. -J. Lee, H. -Y. Kim, Y. -B. Park, and S. Hyun, "The effect of plasma pre-cleaning on the Cu-Cu direct bonding for 3D chip stacking", proc. 18th IEEE International Symposium on the Physical & Failure Analysis of Integrated Circuits (IPFA), Jul., 2011

[16] Z. Huang, Z. Suo, G. Xu, J. He, J. H. Prévost, and N. Sukumar, "Initiation and arrest of an interfacial crack in a four-point bend test", Eng. Fracture Mech., vol. 72, Nov., 2005, pp 2584-2601, doi: 10.1016/j.engfracmech.2005.04.002.

[17] M. Damayanti, J. Widodo, T. Sritharan, S. G. Mhaisalkar, W. Lu, Z. H. Gan, K. Y. Zeng, and L. C. Hsia, "Adhesion study of low-k/Si system using 4-point bending and nanoscratch test", Mater. Sci. Eng. B, vol. 121, Aug., 2005, pp 193-198, https://doi.org/10.1016/j.mseb.2005.03.030.

[18] J. E. Graya, P. R. Norton, K. Griffiths, "Mechanism of adhesion of electroless-deposited silver on poly(ether urethane)", Thin Solid Films, vol. 484, Mar., 2005, pp.196-207, doi: 10.1016/j.tsf.2005.01.090.

Al-Al Direct Bonding with Sub-µm Alignment Accuracy for Millimeter Wave SiGe BiCMOS Wafer Level Packaging and Heterogeneous Integration

M. Wietstruck[1], S. Schulze[1], B. Rebhan[2], P. Kerepesi[2], H. Kurz[2], G. Silberer[2], J. Meiler[2],
S. Tolunay Wipf[1], C. Wipf[1], M. Kaynak[1,3]

[1]IHP – Leibniz Institut für innovative Mikroelektronik, Im Technologiepark 25, 15236 Frankfurt (Oder), Germany
[2]EV Group E. Thallner GmbH, DI Erich Thallner Strasse 1, A-4782 St. Florian am Inn, Austria
[3]Sabanci University, Orta Mahalle, Tuzla 34956 Istanbul, Turkey

Abstract— In this work, we demonstrate an Al-Al direct bonding process for advanced wafer-level packaging and heterogeneous integration. The wafer bonding process is performed at 300°C with an alignment accuracy of <1 µm enabling fine-pitch wafer-to-wafer interconnections. The electrical performance of the Al-Al direct bonding is analyzed using different types of DC and RF test structures. Low resistance Al-Al bonding interconnections with mΩ-range contact resistance and insertion loss values <0.1 dB per interconnection are realized. An interposer based WLP process flow is developed which enables high performance mm-wave SiGe BiCMOS wafer level packaging applications.

Keywords-Wafer-level packaging; Waferbonding; BiCMOS; Interposer.

I. INTRODUCTION

Wafer-level packaging (WLP) is one of the key enabling technologies for heterogeneous integration. In the past, a wide range of WLP technologies have been developed. WLP technologies need to fulfill the requirements for both an electrical redistribution of ICs using Fan-In and Fan-Out Wafer-Level Packaging (FOWLP) as well as adding new components and functionalities into the systems using 2.5/3D integration [1]. A general overview about different types of advanced WLP technologies and application-dependent packaging approaches is given in Figure 1.

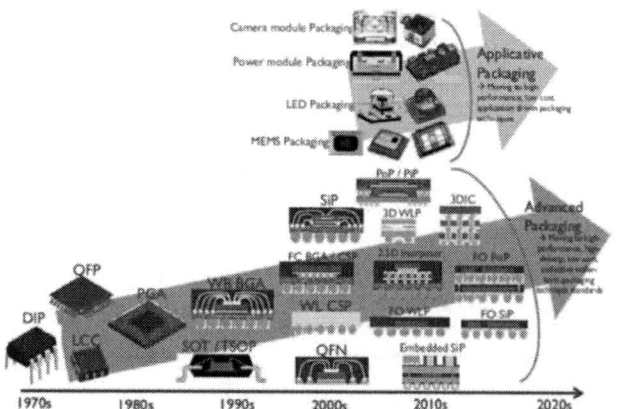

Figure 1. Evolution of advanced WLP technologies and applications showing the wide variation of packaging integration concepts [1].

Among the general electrical, thermal and reliability requirements, the interconnections are one of the major packaging concerns. The continuous demand for higher interconnection densities requires ultra-fine pitch interconnections with smaller dimensions and pitches for die-to-wafer and wafer-to-wafer (W2W) integration techniques. Beside conventional flip-chip technologies based on µ-bumps and Cu pillars, different metal bonding techniques e.g. thermo-compression bonding and solid-liquid inter-diffusion bonding have been developed. In the last decade, Cu-Cu direct and hybrid bonding are considered as key wafer bonding technologies providing ultra-fine pitch interconnections [2]. Although the Cu dual Damascene process is a well-established process for standard CMOS metallization, a significant process integration effort is required to fulfill the requirements for wafer bonding applications. The requirements in terms of surface roughness is one of the main limitation whereas advanced diffusion barriers need to be also considered [3]. As a counterpart for Cu-Cu bonding, Al-Al direct bonding is considered as a promising alternative for direct metal bonding applications providing low-cost W2W interconnections with minimum additional process effort and complexity [4]. Conventional Al-Al thermo-compression bonding mainly applied for device encapsulation requires high process temperatures of ~450°C together with high bond forces to provide reliable bonding interfaces [5]. Both the high temperatures and high bond forces are strongly limiting the CMOS compatibility thus conventional Al-Al bonding has a limited range of potential applications. Recently, new wafer bonding equipment and process technologies have proven that Al-Al bonding with bonding temperatures down to 150°C is feasible making this technology very attractive for CMOS compatible wafer-level packaging applications [6].

In this work, we demonstrate an Al-Al direct bonding process for advanced WLP and heterogeneous integration. The wafer bonding process is performed at 300°C with an alignment accuracy of <1 µm enabling fine-pitch W2W interconnections. The electrical performance of the Al-Al direct bonding is analyzed and low resistance Al-Al bonding interconnections with mΩ-range contact resistance and insertion loss <0.1 dB per interconnection are realized. An interposer based WLP process is developed which enables a high performance mm-wave SiGe BiCMOS wafer level packaging.

978-1-7281-1500-9/19 $31.00 © 2019 IEEE

II. AL-AL WAFER BONDING PROCESS TECHNOLOGY

A. Al-Al Direct Wafer Bonding Process

The Al-Al direct wafer bonding process is realized using the ComBond® - Automated High-Vacuum Bonding System from EVG (Figure 2) [7]. By the use of these advanced wafer bonding equipment and technologies, a fully CMOS compatible Al-Al direct wafer bonding process with low bonding temperature and high alignment accuracy becomes feasible. The unwanted oxide formation of the Al surface is considered as one of the main limitation to realize reliable electrical interconnections with low contact resistance. A novel pre-bond surface treatment which is based on a plasma cleaning process is one of the key features to remove unwanted oxide enabling an oxide-free Al surface for the wafer bonding process. In addition, the operation and handling under ultra-high vacuum prevents from any re-oxidation of the surface in between the pre-bond surface treatment and the actual bonding process.

Figure 2. EVG ComBond® High-Vacuum Bonding System [7].

The Al-Al wafer bonding process is started by applying the ComBond activation process on both wafers to remove unwanted oxide at the Al surface. After cooling down the wafers to room temperature, the wafers are aligned to each other using a high accuracy reflective infrared (RIR) live alignment. The wafers are clamped and a pre-bond process is initiated to prevent from any handling-induced misalignment during the subsequent wafer handling and bonding process. After declamping the wafers, the bonding process is applied. The Al-Al bonding process is realized using a bond force of 60 kN together with bonding temperatures of 300°C. A subsequent annealing at 300°C for 1 h is realized to enable a sufficient Al-Al bonding interface formation. Finally the wafers are cooled down to room temperature and the overall wafer bonding process is finished.

Based on subsequent infrared and C-mode scanning acoustic microscopy (C-SAM) measurements, a successful void-free Al-Al bonding with a post bond alignment accuracy of <1 μm is achieved. The result of the C-SAM measurement is shown in Figure 3, which clearly indicates a successful Al-Al bonding.

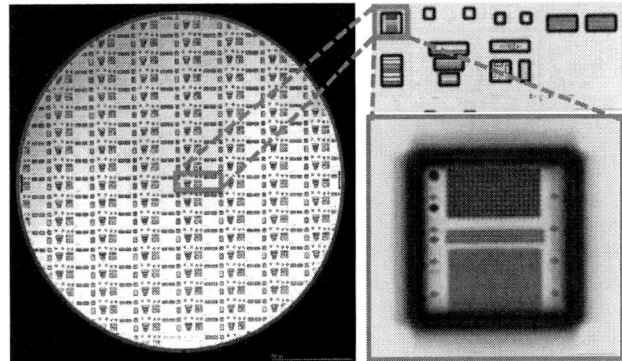

Figure 3 C-mode scanning acoustic microscope images with different magnifications showing the complete 200 mm wafer, a full reticle and a selected test chip.

After the successful wafer bonding process is accomplished, the Al-Al bonding interface is analyzed using focused-ion-beam (FIB) measurements. A SEM cross-section image of an Al-Al bonding pad with a pad size of 20x20 μm² is shown in Figure 4. In addition, a detailed image of the Al-Al bonding pad interface is also shown in the same figure.

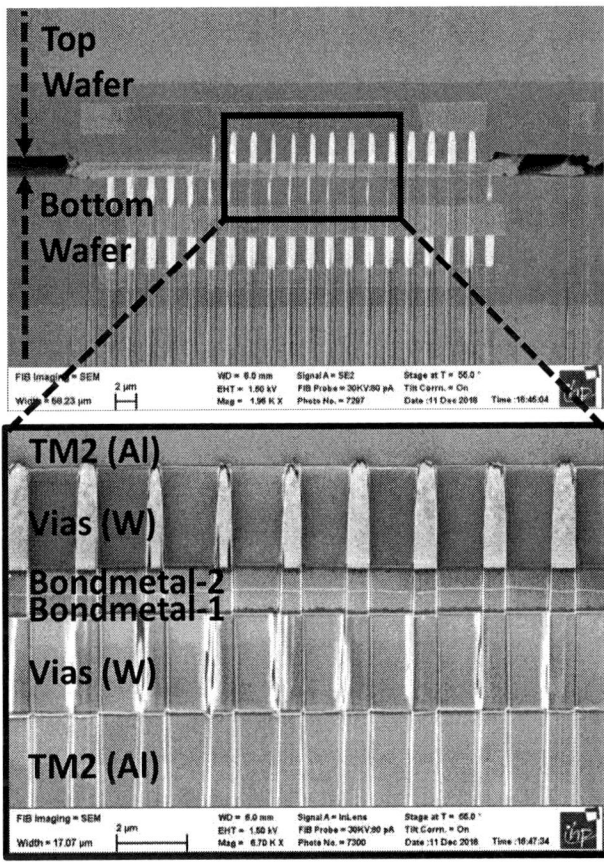

Figure 4 FIB cross section image of an Al-Al bonding pad (top) and a zoom-in image of the bonding area showing the Al-Al bonding interface (bottom).

B. Development of Interposer Packaging Platform

The potential of Al-Al direct bonding is demonstrated by a 200 mm interposer packaging platform which can be applied as a SiGe BiCMOS WLP platform including a high-resistive silicon (Si) interposer bonded together with a SiGe BiCMOS wafer. The interposer comprises a two layer Al metallization stack enabling integrated mm-wave high-Q passives whereas the BiCMOS wafer consists of a standard 7-layer Al metallization stack. For the initial process development, the BiCMOS wafer is emulated by a silicon wafer with a single-layer BEOL metallization and in addition low-resistive silicon substrates with 10 Ωcm resistivity are used both for the interposer and the BiCMOS wafer. The overall process flow is realized in a way that it can be directly transferred to a real interposer and BiCMOS wafer without any additional process adaptations. The electrical interconnection in between both wafers is realized using the developed Al-Al direct wafer bonding process. The overall process flow is shown in Figure 5.

At the beginning, both the interposer and the BiCMOS wafers are fabricated (a). The second process step comprises the deposition of the Al wafer bonding metallization layer using a PVD deposition process (b). The additional Al bonding layer with a 1 µm thickness is added on top of each wafer enabling a separate wafer bonding interconnection pad optimization without any negative influence on the qualified BiCMOS or interposer technology. In case of the BiCMOS wafer, a plasma dicing process using a silicon deep-reactive ion etching is realized to provide deep trenches with a width of 60 µm and a depth of >250 µm (c). Thanks to the plasma dicing process, very fine and deep trenches with high aspect ratios >20 can be realized. This process principle is similar to the well-known Grinding-before-Dicing (DBG) process enabling dicing of ultra-thin silicon chips [8]. In general, a blade dicing process can be also applied but due to the high risk for particle contamination, a plasma dicing approach is preferred. In the next step, the aforementioned Al-Al direct wafer bonding process is applied to realize an electrical interconnection between both the interposer and the BiCMOS wafer (d). The process flow and the process parameters are explained in the aforementioned chapter. The bonded dies are released by a subsequent wafer grinding step from top of the wafer stack to separate the bonded dies and the residual silicon frame (e). The residual Si frame is released and can be removed using a tape-supported debonding step (f).

In terms of the development of the Al-Al wafer bonding process, this process flow is applied to provide a direct access of the electrical probe pads on the interposer after the wafer bonding process for electrical evaluation of the Al-Al wafer bonding. As mentioned before, this process can be directly applied also for different types of WLP e.g. as an interposer based SiGe BiCMOS FOWLP platform.

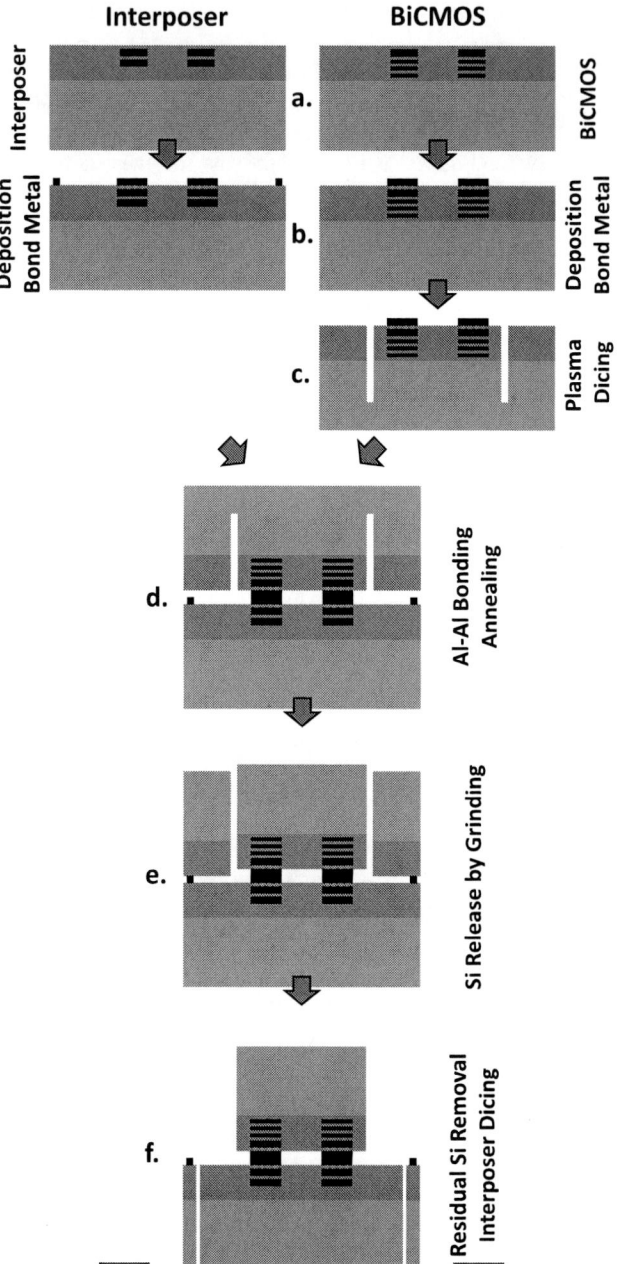

Figure 5 Process flow for the fabrication of the interposer wafer-level packaging platform using Al-Al wafer bonding.

A microscope image of the bonded wafer after silicon frame release by grinding (e) is shown in in Figure 6. The active die, the residual silicon frame and the plasma dicing trenches are highlighted. In addition, an interposer die with the bonded active die on top is also shown after removal of the Si frame by tape-supported debonding in Figure 6.

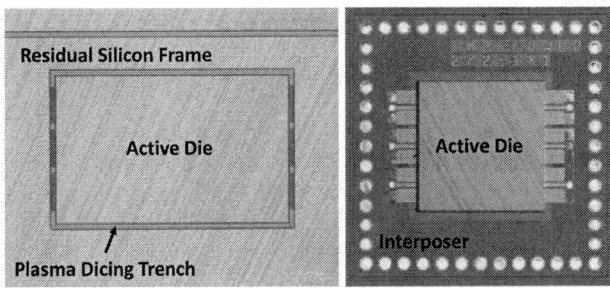

Figure 6. Microscope images after release of residual silicon frame using the grinding process (left) and final image of a silicon interposer die after removal of the residual silicon frame (right).

During the bond metal deposition process in step b, additional metal structures are realized outside of the active interposer area. These structures work as mechanical stoppers preventing from a direct mechanical contact between the residual silicon frame and the interposer wafer surface in process step e, thus any mechanical damage of the surface is prevented. A microscope image and a white-light interferometer (WLI) surface topography measurement is shown in Figure 7 and a height difference of 1.2 µm can be extracted. These mechanical stoppers are distributed over the interposer wafer.

Figure 7 Mechanical stopper based on interposer bond metal pad preventing from unwanted contact between interposer surface and BiCMOS wafer.

After finalizing the overall process, the 200 mm silicon interposer wafer and an additional reticle-level zoom-in image of different bonded dies can be seen in Figure 8.

Figure 8 Microphotograph of fabricated interposer wafer and zoom-in image showing different dies on top of the interposer wafer based on the Al-Al direct bonding process.

III. EXPERIMENTAL RESULTS

A. DC Characterization

The Al-Al wafer bonding process is analyzed using a DC test chip including both kelvin-type single electrical contacts for the extraction of the contact resistance and the related specific contact resistivity as well as daisy chains to evaluate the Al-Al wafer bonding process stability and reliability.

The kelvin-type DC test structures are realized for three different contact sizes ranging from 80x80 to 20x20 µm² bonding pads. The overall test structure and a zoom-in image of the specific Al-Al bond pad area is shown in Figure 9 for the case of a 20x20 µm² bonding pad. The purple colored interconnection lines are realized on the interposer wafer whereas the orange colored interconnection lines are realized on the BiCMOS wafer. Due to the integration concept of a kelvin-type test structure, the top wafer cannot be directly electrically connected via electrical probes therefore both the related current and voltage interconnection lines require an additional Al-Al electrical interconnection bonding pad to the interposer wafer and the final measurement pads.

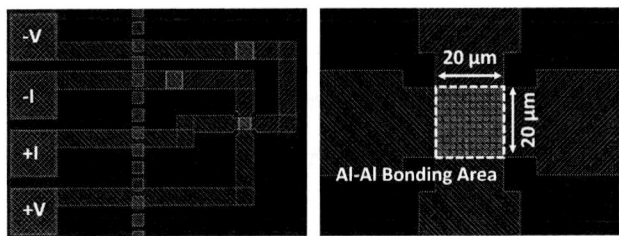

Figure 9 Layout of a kelvin-type test structure with a 20x20 µm² Al-Al bonding pad for the extraction of the contact resistance and specific contact resistivity.

The current-voltage (I-V) curves are measured at wafer-level ranging from -40 to +40 mA using a Keysight precision measurement unit B2912A and a semi-automatic probe station PA200IS from Cascade. The differential contact resistances are extracted and the measurement results are shown in Figure 10 for all functional dies.

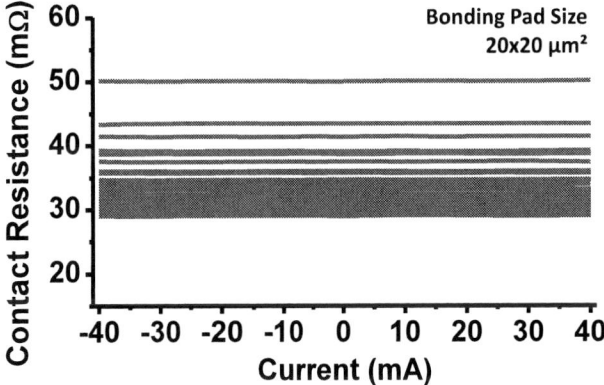

Figure 10 Differential contact resistance for a bond pad size of 20x20 µm² for all functional dies on the 200 mm wafer.

In case of the 20x20 μm² bond pad, a wafer map of the contact resistance for a fixed current of 40 mA is shown in Figure 11. The non-functional electrical dies which are mainly located at the edge of the wafer can be correlated with residual silicon which could not be properly removed during the tape-supported debonding step thus the electrical pads could not be accessed.

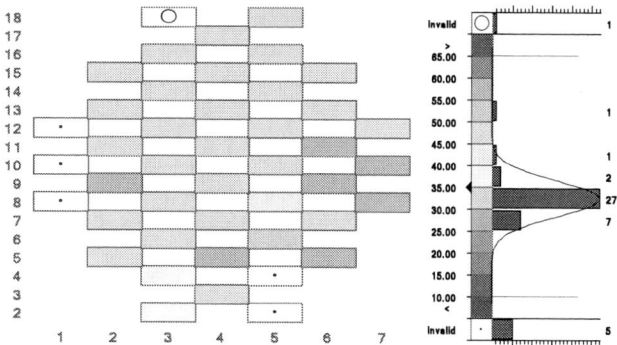

Figure 11 Wafer map of the extracted differential contact resistance of the 20x20 μm² bonding pads with at a fixed current of 40 mA.

Beside the single kelvin-type contacts, daisy chains with a fixed contact size of 20x20 μm² with 480 series electrical contacts are realized to analyze the stability and reliability of the Al-Al direct wafer bonding process. The extracted resistance per interconnection is shown in Figure 13. The difference of the extracted contact resistance for the kelvin-type and daisy chain test structure with a fixed 20x20 μm² contact size can be explained by the additional interconnection lines between each of the 480 series contacts adding an additional parasitic resistance.

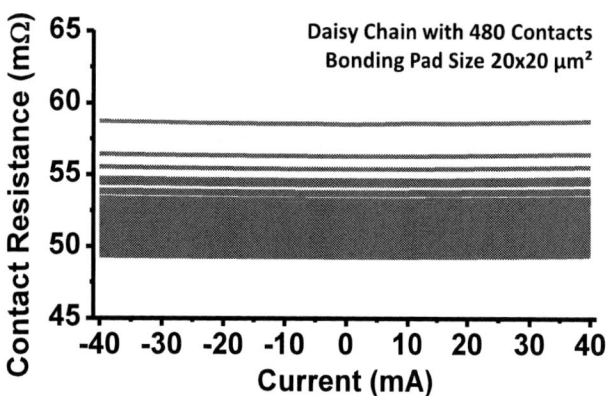

Figure 12. Extracted contact resistance divided by the number of series contacts for a daisy chain with 20x20 μm² bonding pads.

A summary of the DC measurement results is given in Table 1 for the different contact sizes ranging from 20x20 to 80x80 μm² and the different types of test structures. The following electrical and process parameters are summarized: the mean contact resistance R_{mean}, the standard deviation SD, the specific contact resistivity ρ_C, and the process yield which is based on a defined contact resistance threshold value of 200 mΩ per contact independent from the contact size. Obviously Al-Al bonding interconnections with very

low contact resistance in the mΩ-range can be realized. The extracted R_{mean} values multiplied with the contact area leads to a specific contact resistivity of <1.3e-7 Ωcm² which is comparable to standard electrical interfaces e.g. in metal-semiconductor technologies [9]. Considering the aforementioned limit of a maximum contact resistance of 200 mΩ/contact, a yield of 80-90% is achieved. It is worth to mention, that an optimized removal of the residual silicon frame at the wafer edges will help to increase the bonding yield to values above 95%.

Table 1. Summary of DC Measurement Results.

Parameter	Al-Al Bonding Contact Size			
	Kelvin 20x20 μm²	Kelvin 40x40 μm²	Kelvin 80x80 μm²	Daisy Chain 20x20 μm²
R_{Mean} (mΩ)	31.98	7.58	1.75	51.48
SD (mΩ)	3.47	1.02	0.20	2.16
ρ_C (Ω*cm²)	1.29e-7	1.21e-7	1.12e-7	---
Yield (%)	84.6	81.3	82.4	86.8

B. RF Characterization

The potential of using Al-Al bonding and the related interposer packaging platform at mm-wave frequencies is analyzed using an additional RF test chip. This chip consists of coplanar waveguides (CPW) both on the interposer and the BiCMOS wafer including 18 series transitions in between these two wafers with a fixed Al-Al bond pad size of 15x15 μm². The 3D EM simulation model with different perspectives of the overall chip and the RF test structure is shown in Figure 13.

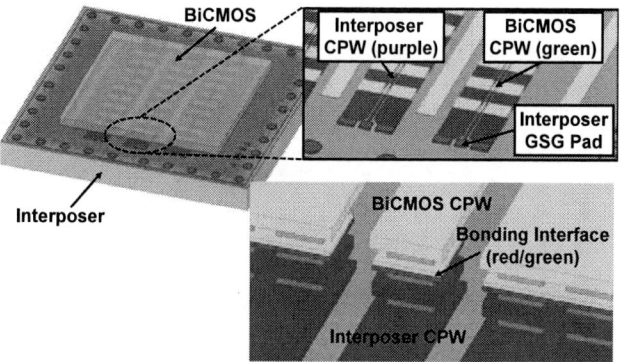

Figure 13 3D EM simulation model of the complete RF test chip together with additional zoom-in images of the test structure and the Al-Al bonding interconnection pads.

The S-parameter measurement results from 1-67 GHz are given in Figure 14. The overall insertion loss of the full test structure is around 4 dB at 67 GHz and a very good input matching with reflection better than -15 dB is achieved for the full frequency band. After deembedding of the GSG pads, an insertion loss of ~0.2 dB per transition can be extracted including the peripheral additional transmission lines on the BiCMOS and interposer wafers.

978-1-7281-1500-9/19 $31.00 © 2019 IEEE

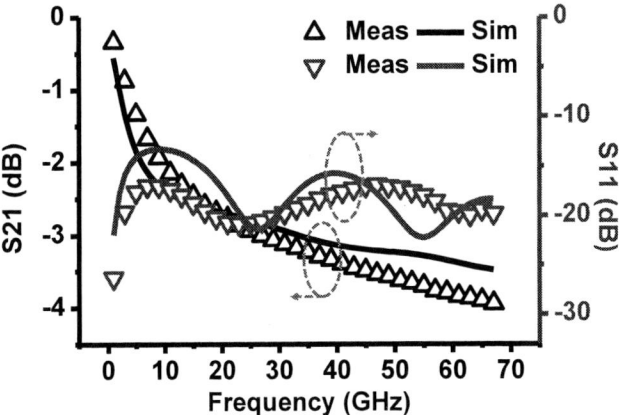

Figure 14 S-parameter measurement results of full test structure including the Al-Al interconnections, the CPW transmission lines and the GSG pads.

As mentioned before, low-resistivity Si substrate with a resistivity of 10 Ωcm is used for the current process development. Obviously, this can lead to an increase of the insertion loss of the peripheral transmission lines itself. To analyze this effect, an additional EM simulation is performed to evaluate the influence of the Si resistivity. In total, four different substrate resistivity combinations are analyzed: 10 Ωcm for both wafers, 50 Ωcm for both wafers, 50 Ωcm for the BiCMOS and 1000 Ωcm for the interposer wafer and finally 50 Ωcm for the BiCMOS wafer and 4000 Ωcm for the interposer wafer. The S-parameter simulation results are shown in Figure 15. With increasing substrate resistivity, the overall loss of the test structure can be strongly reduced. In case of the interposer, a substrate resistivity of ~1k Ωcm should be selected whereas a further increase will not lead to any significant improvement. Obviously, the main loss of the currently fabricated interposer WLP platform can be correlated to the low-resistive silicon substrates. In case of high-resistivity interposer, the overall loss can be reduced to <2 dB which proves that the loss of the Al-Al interconnections is neglectable with <0.1 dB per transition.

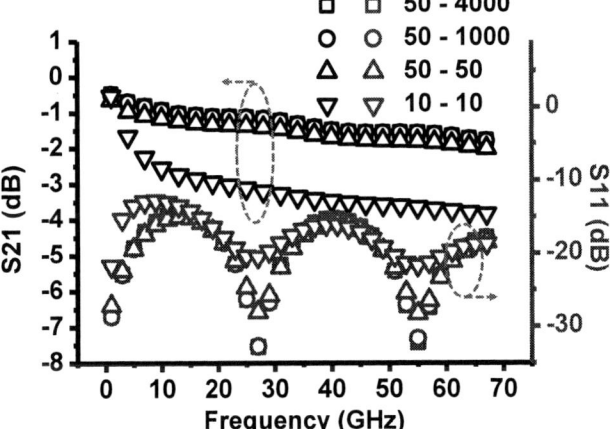

Figure 15 S-parameter measurement results of full test structure including Al-Al interconnections, CPW lines and GSG pads for four different substrate resistivity combinations for the interposer and BiCMOS wafer.

IV. CONCLUSION

In this work, the development of an Al-Al direct bonding process for advanced WLP and heterogeneous integration is demonstrated. The wafer bonding process is performed at 300°C with an alignment accuracy of <1 μm enabling fine-pitch W2W interconnections. The electrical performance of the Al-Al direct bonding is analyzed and thanks to the pre-bond oxide removal process and high vacuum wafer handling, low resistance Al-Al bonding interconnections with a specific contact resistivity of ~1.2e-7 Ωcm² are realized which enables interconnections with a mΩ-range contact resistance and RF interconnections with insertion loss values <0.1 dB at 67 GHz. An interposer based WLP process is developed which enables e.g. a high performance mm-wave SiGe BiCMOS FOWLP platform. This process module enables a wide range of new applications by adding additional functionalities to conventional BiCMOS and interposer substrate technologies.

ACKNOWLEDGMENT

The authors thanks to the team of IHP pilot line for the excellent wafer fabrication. Furthermore special thanks to Jens Katzer for the FIB analysis and Gerald Strzalka from DISCO HI-TEC Europe for the back grinding process.

REFERENCES

[1] Yole Developpement, "Status of the Advanced Packaging Industry 2018," 2018

[2] A. Jouve et al., "1μm Pitch direct hybrid bonding with <300nm wafer-to-wafer overlay accuracy," 2017 IEEE SOI-3D-Subthreshold Microelectronics Technology Unified Conference (S3S), Burlingame, CA, 2017, pp. 1-2.

[3] S. Kim et al., "Ultra-Fine Pitch 3D Integration Using Face-to-Face Hybrid Wafer Bonding Combined with a Via-Middle Through-Silicon-Via Process," 2016 IEEE 66th Electronic Components and Technology Conference (ECTC), Las Vegas, NV, 2016, pp. 1179-1185.

[4] Ji Fan and Chuan Seng Tan (September 19th 2012). Low Temperature Wafer-Level Metal Thermo-Compression Bonding Technology for 3D Integration, Metallurgy Yogi-raj Pardhi, IntechOpen, DOI: 10.5772/48216. Available from: https://www.intechopen.com/books/metallurgy-advances-in-materials-and-processes/low-temperature-wafer-level-metal-thermo-compression-bonding-technology-for-3d-integration

[5] N. Malik, K. Schjølberg-Henriksen, E. Poppe and T. G. Finstad, "Al-Al thermocompression bonding for wafer-level MEMS packaging," 2013 Transducers & Eurosensors XXVII: The 17th International Conference on Solid-State Sensors, Actuators and Microsystems (TRANSDUCERS & EUROSENSORS XXVII), Barcelona, 2013, pp. 1067-1070.

[6] B. Rebhan et al., "Low-Temperature Aluminum-Aluminum Wafer Bonding," ECS Trans. 2016, volume 75, issue 9, pp.15-24

[7] https://www.evgroup.com/en/products/bonding/integrated_bonding/evg_combond/, (21.02.18)

[8] Shinya Takyu, Junya Sagara and Tetsuya Kurosawa, "A study on chip thinning process for ultra thin memory devices," 2008 58th Electronic Components and Technology Conference, Lake Buena Vista, FL, 2008, pp. 1511-1516.

[9] Y.K. Fang, C.Y. Chang, Y.K. Su, "Contact resistance in metal-semiconductor systems," Solid-State Electronics, Volume 22, Issue 11, 1979, Pages 933-938.

Vivaldi Antenna Array Fabricated Using A Hybrid Process

Vincens Gjokaj, Cameron Crump, John Papapolymerou, John Albrecht, and Premjeet Chahal
Department of Electrical and Computer Engineering
Michigan State University
East Lansing, MI 48824
chahal@egr.msu.edu

Abstract—**Additive manufacturing (AM) has recently gained much attention from the electromagnetic community touted as a replacement for traditional manufacturing and a potential to change the way high-frequency components are designed and manufactured. It promises to produce complex components that are difficult to realize using conventional techniques and provides new opportunities for novel system designs along with reduced time to prototype. There is a range of AM print technologies that are commercially available, each with its specific print properties and reliance on the materials adapted to the print process. Polymer composite inks can readily be deposited using aerosol-jet printing while achieving high print resolution ($<10\mu$m) and good repeatability ($<2\mu$m). Due to the large working distance of the spray nozzle, aerosol printing readily allows for the fabrication of components on non-planar structure which is desirable for the next generation of high functional density systems. Polyjet printing allows for the fabrication of thick structures with print resolution below 100μm. Ability to print conductive, dielectric and magnetic materials in thick and thin film form allows for the design a range of EM structures.**

This paper demonstrates the use of aerosol-jet and poly-jet printing, and conventional lithography in the fabrication of a wideband (10-25 GHz) Vivaldi antenna array with high gain (>10 dBi). Polyjet printing is utilized to fabricate the Vivaldi structure and aerosol-jet printing is used to fabricate the resistor for the Wilkinson power divider. A Vivaldi antenna array fabricated using conventional technique, clean room process and surface mount resistors, is also designed and fabricated to benchmark the proposed new technology. The printed conductive polymer based resistor is characterized over a wide frequency range. Overall, this paper presents the details of design, fabrication, and measurement of a 1x2 Vivaldi antenna array operating in the 10-25 GHz fabricated through the use of this hybrid process which combines AM and standard lithography.

Index Terms—**Additive manufacturing (AM), 3D Printing, Antennas, Aerosol Jet, Resonators, Rigid-Flex.**

I. INTRODUCTION

Additive manufacturing (AM) has attracted significant interest for rapid prototyping and low-cost manufacturing of RF electronic components. Many RF components have been demonstrated based on plastic printed parts with blanket metal coatings [1]–[6] and showed promising results. Recently, a new fabrication technique has been demonstrated that allows direct patterning of metal thin films onto printed plastic parts using a mechanical process instead of chemical etching, resulting in reduced fabrication cost and ability to pattern on non-planar structures [7]. To reduce dielectric loss, an air-like

This work was supported by Honeywell Federal Manufacturing and Technologies under contract DE-NA0002839.

substrate is fabricated using AM [8]. Other techniques are also being investigated to achieve metal patterning including deposition of silver inks using AM processes [9]. However, silver inks have poor electrical conductivity, especially at high frequencies. Although plastic based AM has been demonstrated to be an effective method of fabricating large 3D RF components, it is still limited with respect to fine line patterning owing to the resolution control and limited material availability. Furthermore, thin-film deposition in free form of AM materials is still a challenge as the plastic materials are brittle and also difficult to deposit in thin-film form. On the other hand, flexible or thin-film materials have long been used to fabricate planar RF components with fine line features [10][11]. Both of these methods have their unique properties that make them attractive for RF component design and fabrication, and combining these methods provides complementary advantages that have not been previously exploited. Building upon the work presented in ref. [12], this rigid-flex coax-like structure is used in the implementation of a foldable Vivaldi antenna array. This hybrid process uses AM to fabricate the larger features of the array such as the antennas and standard lithography to create the fine features like the Wilkinson power divider. In addition, AM is used to print the 100 Ω resistor needed for the Wilkinson power divider using a conductive polymer material.

This rigid-flex structure is similar to higher cost air dielectric coax cables in terms of being low loss, but enables far greater levels of system integration. This unique structure bridges the gap between standard PCB technologies (planar) and 3D printed structures. A novel combination of LCP processing and 3D printing is used in a hybrid process flow. Building towards a foldable Vivaldi antenna array, key components are individually evaluated first followed by the array which utilizes these components. In particular, a rigid-flex coax transmission line, rigid-flex UWB power dividers, and rigid-flex antenna feeds for both single antennas and multiple element folded arrays are simulated (in ANSYS HFSS®), fabricated and characterized.

II. AEROSOL PRINTED RESISTOR

With surface mount components, there are physical and electrical limitations that could potentially be improved by using additive manufacturing (AM). Previous work has been done with static (DC) circuits to fabricate passive components such as resistors, capacitors, and inductors using injet printing techniques [13]–[16]. However, to the best of our knowledge, resistors for microwave applications. By fabricating the passive elements directly onto a circuit there is a potential for reducing undesired parasitics. There are many materials available for AM that can be used individually or in combination to fabricate novel passive components. A common material used often in the fabrication of flexible circuits and solar cells for electrode connection is conductive polymers. For the purpose of replacing a chip-resistor with an aerosol printed thin-film resistor, we chose to use of PEDOT:PSS (Poly(3,4-ethylenedioxythiophene)-poly(styrenesulfonate)). PEDOT:PSS is a conjugated polymer [17] whose conductivity is low compared to copper and can readily be used in the design of a resistor. The geometry of the printed resistor can be used to achieve desired resistance values. In order to replace the 100 Ω chip-resistor with a printed thin-film conductive polymer resistor, the same footprint was used. By using the same footprint for the resistor, a one-to-one comparison can be made between the commercially available and the customized aerosol-jet printed resistor. Using an Optomec 5X aerosol-jet printer with ultrasonic atomizer, we were able to deposit thin-film conductive polymers onto a liquid crystal polymer (LCP) substrate with direct contact to etched copper pads and traces. Since PEDOT:PSS is a low viscosity material, it can easily be aerosolized and deposited. The area taken up by the chip-resistor was used as the pattern area of the printed thin-film resistor, with feature sizes as small as 40μm. By adjusting the amount of material deposited and successive layers of PEDOT:PSS, the resistance of the film can be precisely controlled. With the use of an Optomec Aerosol Jet 5X (AJ5X) aerosol jet printer, this paper presents the use of a conductive conjugated polymer. Atomization is the process of agitating nanoparticles disperse in a liquid solvent into small particles in a flow of inter gas. Aerosolization in the ultrasonic atomizer on the AJ5X printer is achievable for viscosity less than 10 cP for materials of volume less than 5 mL. Liquid ink is placed inside a thin-walled glass vial which is submersed into a bath of chilled water containing an ultrasonic transducer. Aerosol droplets of ink particles in the vial are created by ultrasonic agitation, then transported out of the atomizer by nitrogen carrier gas. The flow rate of nitrogen through the atomizer determines the amount of aerosolized particles to leave the atomizer and travel through the delivery tube leading to the deposition head (print head).

In the print head, the aerosolized gas combines with a focusing jet of nitrogen known as the sheath gas. The aerosolized mixture is then exited through a small nozzle orifice ranging from 100 μm. to as large as 300 μm. in diameter. The 300 μm. diameter nozzle was used in the printing of this resistor with line dimensions as small as 40 μm.. The flow rate of sheath gas will determine the focus or resolution achieved by the nozzle. Controlling speed of the 2D axis will determine how thick a deposited feature is, as well as the uniformity of a line. A fast process speed will create narrow and thin line features, and a slow process will create wide features that can also form large deposits of puddles or overspray around

the desired feature. For the PEDOT:PSS conductive polymer, ambient room temperatures proved to be a good atmosphere for both adhesion and curing of the deposition. A major benefit of using an aerosol-jet printed thin-film is the fine feature and high resolution capability. Injket printing and screen printing do not currently allow for features of 10μm to 60μm with high repeatability. A recent publication showed aerosol jet printing of feature sizes as small as 8μm[18]. Aerosol jet printing also allows for successive layering of material and overcomes many geometrical challenges. If a printed component such as a thin-film resistor is not within a margin of error, then additional geometric structures can be added to increase or reduce the resistivity. This is a major improvement over the current processes of laser trimmed chip-resistors.

III. FLEXIBLE UWB POWER DIVIDER

A major advantage of the Rigid-Flex structure is that it uses a standard substrate material and with this advantage a power divider can be made in a straightforward way and can be used for various circuits incorporated with the flex-rigid coax. Here, the Wilkinson divider was chosen because of its large operational bandwidth [19] and low insertion loss. The center frequency was chosen to be 20 GHz which allowed the Wilkinson power divider to operate between 10-25 GHz[20]. The layout of the Wilkinson power divider is shown in Fig. 1a. The 100 Ω resistor was modeled as a 100 Ω impedance for the simulation. The dimensions of the modeled resistor were chosen to be 1.6 mm long and 0.80 mm wide which would make it the size of a standard 0603 surface mount component. The fabricated Wilkinson power divider with the printed resistor can be seen in Fig. 1c. It was fabricated on a 7 mil Liquid Crystal Polymer (LCP)(180 μm) Rogers ULTRALAM 3850HT (ϵ = 2.9, tanδ =.001) substrate. LCP was chosen for this paper as it offers flexibility aswell as low loss for rf applications [21]. was The copper cladding on the LCP was patterned using a positive photolithography process. An identical Wilkinson power divider was fabricated using the same process as discussed above and it used a surface mount 0603 resistor rated to 50 GHz and is shown in Fig. 1b. The measured results for the Wilkinson power divider with the printed resistor compared to the Wilkinson power divider that utilizes a standard surface-mount resistor are shown in Fig. 1d. As can be seen, the results for both Wilkinson power dividers matches closely, and shows that the printed resistor can provide similar performance to a commercially available high frequency resistor.

IV. 3D PRINTED VIVALDI

AM has seen a huge influx on interest for RF applications specifically for fabrication of thick complex antenna designs. Recently, a 3D printed substrate integrated waveguide fed Vivaldi was demonstrated by our group [22], and the antenna showed good results. For this paper, a Vivaldi antenna was designed to operate between 10-25 GHz to match the operation of the Wilkinson power divider described above. The designed Vivaldi with dimensions is shown in Fig. 3a. The simulated

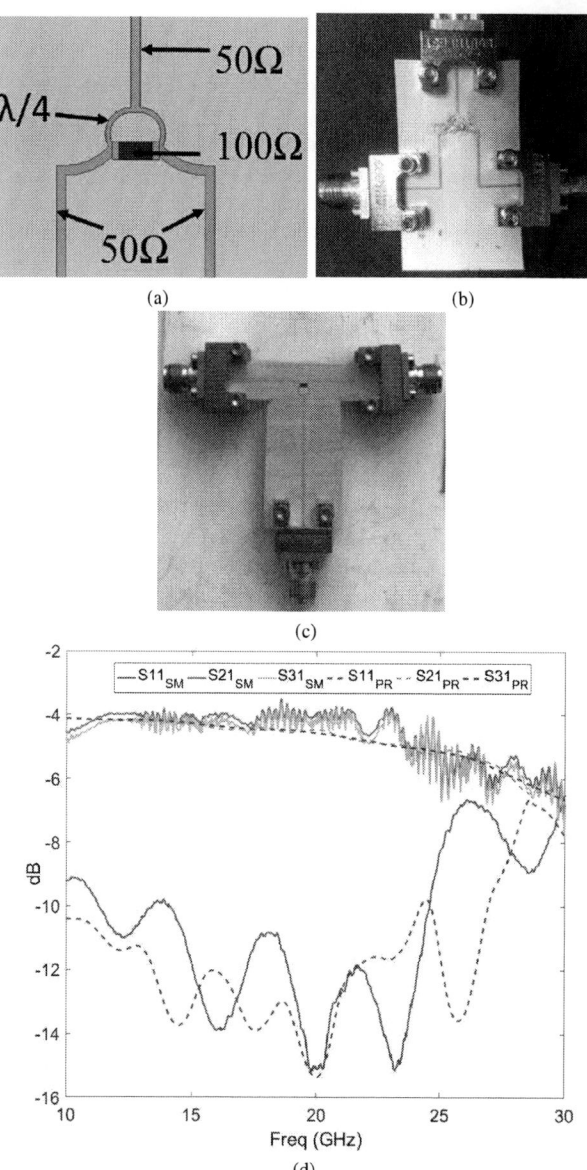

Fig. 1: (a) Wilkinson power divider design. (b) Wilkinson power divider fabricated with surface mount resistor. (c) Wilkinson Power divider fabricated with printed resistor (d) Measured results of the Wilkinson power divider.PR- Printed Resistor, SM- Surface Mount

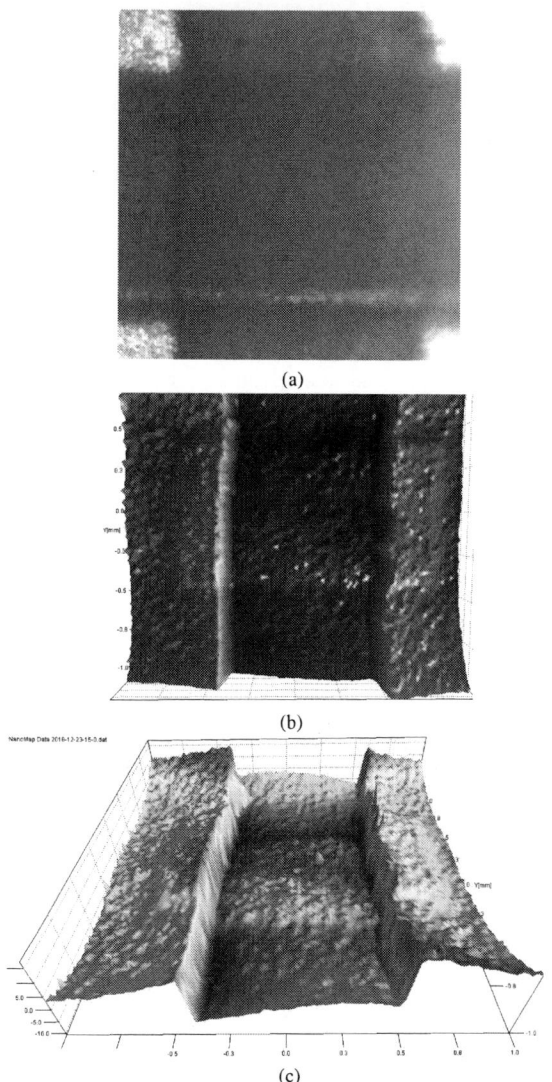

Fig. 2: (a) Microscope image of printed resistor. (b) Surface profile of the printed resistor. (c) Surface profile of the printed resistor.

S_{11} for the Vivaldi is shown in Fig. 4a. the simulation results show an S_{11} of -10 dB or better over the entire 10-25 GHz band. The antenna was 3D printed using an Objet Connex 350 printer. Following 3D printing the antenna was then blanket metalized up to 6 μm of copper. This was done by first sputter depositing a seed layer of Titanium/Copper (Ti/Cu: 200nm/500nm) followed by electroplating of copper up to 6 μm. The fabricated Vivaldi antenna can be seen in Fig. 3b and the measured results are shown in Fig. 4a. The simulated gain is 9 dBi with a maximum gain over 12 dBi which is shown in Fig. 4b. In comparison, the antenna had a measured gain of 8.7 dBi with a maximum gain of 11 dBi.

978-1-7281-1500-9/19 $31.00 © 2019 IEEE

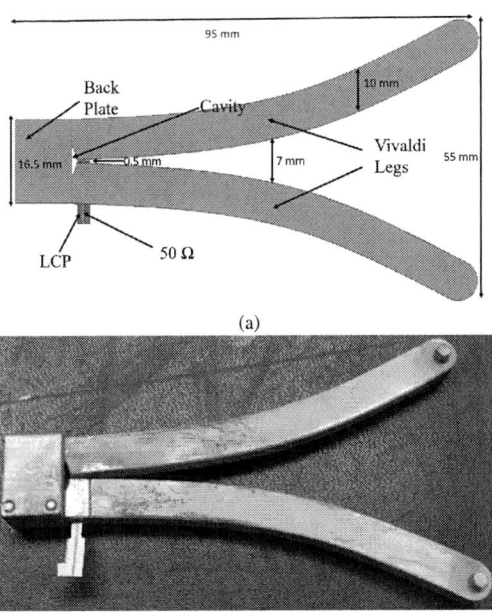

(a)

(b)

Fig. 3: (a) Designed Vivaldi antenna . (b) Fabricated Vivaldi antenna.

V. RIGID-FLEX UWB VIVALDI ARRAY

Next, the Wilkinson power dividers described above are combined with the 3D printed Vivaldi to create a 1x2 element antenna array. The Wilkinson power dividers from above will have their connectors removed and used with the printed Vivalid structure. The fabricated 1x2 array that uses a surface mount resistor is shown in Fig. 5a and the array that uses a printed resistor is shown in Fig. 5b. The array S_{11} results are shown in Fig. 6a. There is good correlation between simulation and measured results for the arrays. As shown in Fig. 6a, the array that utilized a 0603 surface mount resistor had a resonance between 16-19 GHz; through simulation it was found that the resonance was caused from the parasitic inductance and capacitance of the 0603 package and soldering to the board. The PEDOT:PSS based resistor did not show any resonance over the measured frequency band. Fig. 6b shows the gain of the array. The array had a simulated gain of 12 dBi at 10 GHz and a max gain of 15 dBi at 18 GHz. The measured gain for the surface mount resistor array was approximately 10 dBi at 10 GHz with a maximum gain of approximately 11.75 dBi at 16 GHz; however, the gain begins to drop where the resonance occurred. For the printed resistor array the gain started at approximately 10 dBi and with a maximum gain of 12.5 dBi at 17 GHz.

VI. CONCLUSION

In this paper, an aerosol jet printed resistor was used on a Wilkinson power divider fabricated using standard pho-tolithography and compared to the same Wilkinson power divider that used a surface mount 0603 resistor. This paper for the first time utilizes PEDOT:PSS as a high frequency

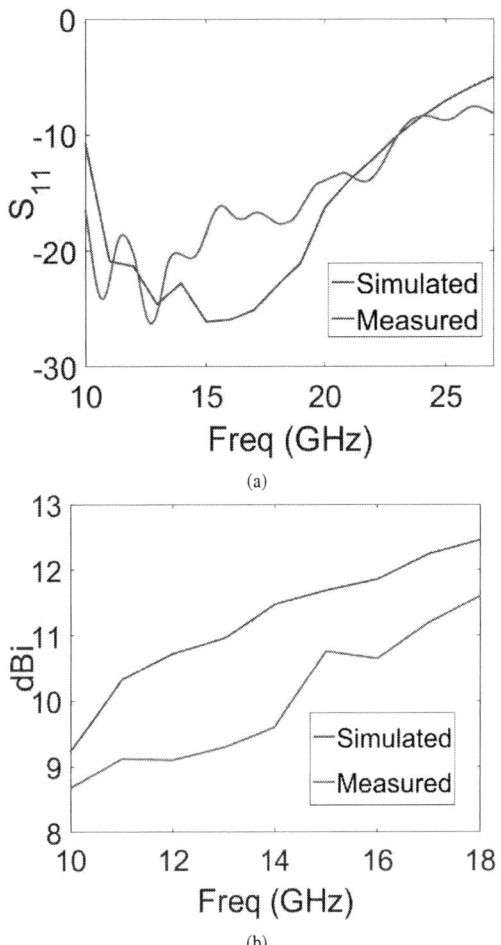

(a)

(b)

Fig. 4: (a) Simulated and Measured S_{11} . (b) Simulated and Measured gain results.

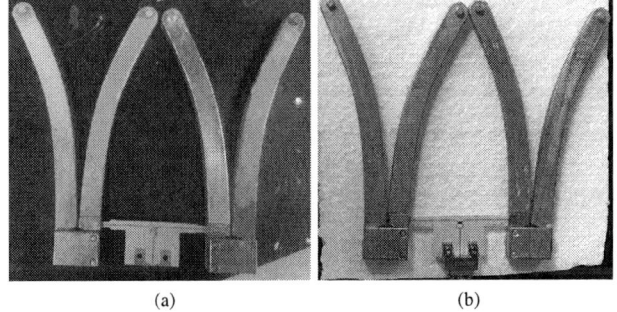

(a)   (b)

Fig. 5: (a) 1x2 Vivaldi array with surface mount resistor. (b) 1x2 Vivaldi array with printed resistor.

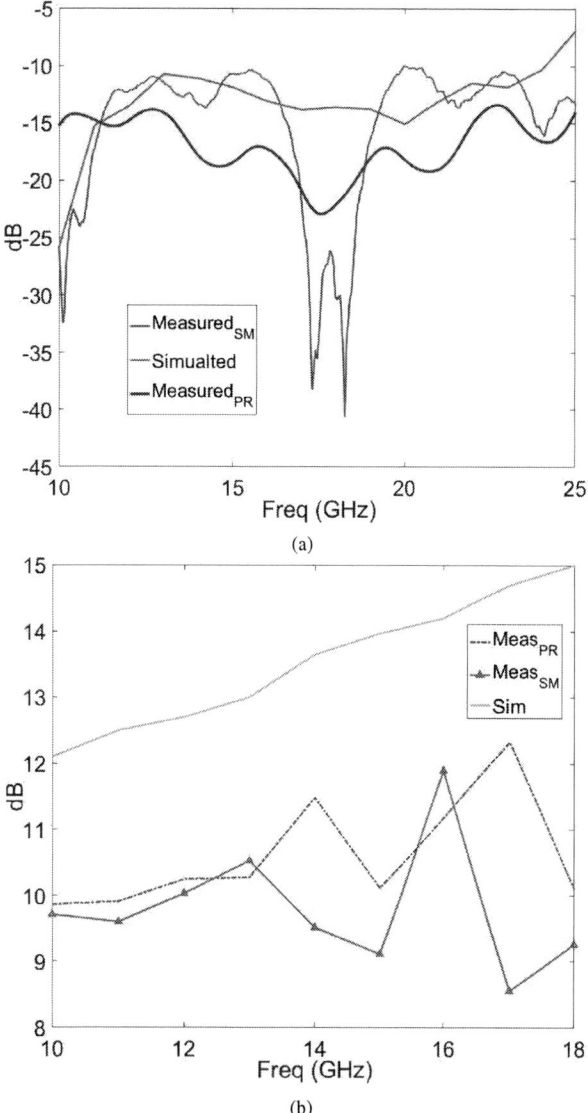

Fig. 6: (a) Simulated and Measured S_{11} . (b) Simulated and Measured gain results. PR- Printed Resistor, SM- Surface Mount

resistor for the Wilkinson power divider and shows measured results up to 25 GHz. This paper showed an effective way of creating RF resistors through additive manufacturing. This resistor was used in an application and showed good results compared to traditional surface mount RF resistors. Overall, a Vivaldi antenna array is demonstrated through the use of hybrid process, which combines the benefits of AM to create large features fast and extremely cost effectively and photolithography to create the fine features of a Wilkinson power divider.

REFERENCES

[1] V. Gjokaj and P. Chahal, "A design study of 3d printed reduced height waveguide structures," in *2018 IEEE 68th Electronic Components and Technology Conference (ECTC)*, pp. 125–130, May 2018.

[2] F. Bongard, M. Gimersky, S. Doherty, X. Aubry, and M. Krummen, "3d-printed ka-band waveguide array antenna for mobile satcom applications," in *2017 11th European Conference on Antennas and Propagation (EUCAP)*, pp. 579–583, March 2017.

[3] M. I. M. Ghazali and P. Chahal, "A 3d printed cavity backed 24 slotted waveguide antenna array," in *2018 IEEE International Symposium on Antennas and Propagation USNC/URSI National Radio Science Meeting*, pp. 1435–1436, July 2018.

[4] M. Szymkiewicz, Y. Konkel, C. Hartwanger, and M. Schneider, "Kuband sidearm orthomode transducer manufactured by additive layer manufacturing," in *2016 10th European Conference on Antennas and Propagation (EuCAP)*, pp. 1–4, April 2016.

[5] A. I. Dimitriadis, M. Favre, M. Billod, J. Ansermet, and E. de Rijk, "Design and fabrication of a lightweight additive-manufactured ka-band horn antenna array," in *2016 10th European Conference on Antennas and Propagation (EuCAP)*, pp. 1–4, April 2016.

[6] J. Thornton, B. Dalay, and D. Smith, "Additive manufacturing of waveguide for ku-band satellite communications antenna," in *2016 10th European Conference on Antennas and Propagation (EuCAP)*, pp. 1–4, April 2016.

[7] J. A. Byford, M. I. M. Ghazali, S. Karuppuswami, B. L. Wright, and P. Chahal, "Demonstration of RF and microwave passive circuits through 3-D printing and selective metalization," *IEEE Transactions on Components, Packaging and Manufacturing Technology*, vol. 7, pp. 463–471, March 2017.

[8] M. I. M. Ghazali, E. Gutierrez, J. C. Myers, A. Kaur, B. Wright, and P. Chahal, "Affordable 3D printed microwave antennas," in *2015 IEEE 65th Electronic Components and Technology Conference (ECTC)*, pp. 240–246, May 2015.

[9] C. Oakley, A. Kaur, J. A. Byford, and P. Chahal, "Aerosol-Jet Printed Quasi-Optical Terahertz Filters," in *2017 IEEE 67th Electronic Components and Technology Conference (ECTC)*, pp. 248–253, May 2017.

[10] L. Hepburn and J. Hong, "On the development of compact lumped-element LCP filters," in *2014 44th European Microwave Conference*, pp. 544–547, Oct 2014.

[11] H. H. Ta and A.-V. Pham, "Compact wilkinson power divider on multilayer org anic substrate," in *2012 IEEE/MTT-S International Microwave Symposium Digest*, pp. 1–3, June 2012.

[12] V. Gjokaj and P. Chahal, "3d printed hybrid rigid-flex coaxial-like transmission line structures," in *2018 IEEE International Symposium on Antennas and Propagation USNC/URSI National Radio Science Meeting*, pp. 1425–1426, July 2018.

[13] C. Ionescu, P. Svasta, A. Vasile, and D. Bonfert, "Investigations on organic printed resistors based on pedot:pss," in *2012 IEEE 18th International Symposium for Design and Technology in Electronic Packaging (SIITME)*, pp. 85–89, Oct 2012.

[14] C. Sriprachuabwong, C. Srichan, T. Lomas, and A. Tuantranont, "Simple rc low pass filter circuit fabricated by unmodified desktop inkjet printer," in *ECTI-CON2010: The 2010 ECTI International Confernce on Electrical Engineering/Electronics, Computer, Telecommunications and Information Technology*, pp. 929–932, May 2010.

[15] B. J. Kang, C. Kyu Lee, and J. Hoon Oh, "All-inkjet-printed electrical components and circuit fabrication on a plastic substrate," *Microelectronic Engineering*, vol. 97, p. 251254, 09 2012.

[16] V. Correia, K. Mitra, H. Castro, J. Rocha, E. Sowade, R. Baumann, and S. Lanceros-Mndez, "Design and fabrication of multilayer inkjet-printed passive components for printed electronics circuit development," *Journal of Manufacturing Processes*, vol. 31, pp. 364–371, 01 2018.

[17] J. A. Paulsen, M. Renn, K. Christenson, and R. Plourde, "Printing conformal electronics on 3D structures with aerosol jet technology," in *2012 Future of Instrumentation International Workshop (FIIW) Proceedings*, pp. 1–4, Oct 2012.

[18] C. Oakley and P. Chahal, "Aerosol jet printed quasi-optical terahertz components," *IEEE Transactions on Terahertz Science and Technology*, vol. 8, pp. 765–772, Nov 2018.

[19] D. De, A. Prakash, N. Chattoraj, P. K. Sahu, and A. Verma, "Design and analysis of various Wilkinson Power Divider Networks for L band applications," in *2016 3rd International Conference on Signal Processing and Integrated Networks (SPIN)*, pp. 67–72, Feb 2016.

[20] D. Pozar, *Microwave Engineering*. Wiley, 2004.

[21] D. C. Thompson, O. Tantot, H. Jallageas, G. E. Ponchak, M. M. Tentzeris, and J. Papapolymerou, "Characterization of liquid crystal polymer (lcp) material and transmission lines on lcp substrates from 30

to 110 ghz," *IEEE Transactions on Microwave Theory and Techniques*, vol. 52, pp. 1343–1352, April 2004.

[22] V. Gjokaj, P. Chahal, J. Papapolymerou, and J. D. Albrecht, "A novel 3D printed vivaldi antenna utilizing a substrate integrated waveguide transition," in *2017 IEEE International Symposium on Antennas and Propagation USNC/URSI National Radio Science Meeting*, pp. 1253–1254, July 2017.

Novel Multicore PCB and Substrate Solutions for Ultra Broadband Dual Polarized Antennas for 5G Millimeter Wave Covering 28GHz & 39GHz range

Trang Thai, Sidharth Dalmia, Josef Hagn
Intel Component and Devices Group (iCDG)
Intel Corporation
Santa Clara, CA USA
trang.thai@intel.com
sidharth.dalmia@intel.com
josef.hagn@intel.com

Pouya Talebbeydokhti, Yossi Tsfati
Intel Component and Devices Group (iCDG)
Intel Corporation
Santa Clara, CA USA
pouya.talebbeydokhti@intel.com
yossi.tsfati@intel.com

Abstract— 5G millimeter wave (mmWave) communications that enables Gbps data channels often require broadband operation with dual polarization in multi frequency ranges (27-30, and 37-40GHz). Additionally, access points and base stations need large high gain antenna arrays and consistent radiation patterns across operation frequencies for ease of beam scanning control. Many stacked patch antennas were proposed as the solution. However, broadband patch antennas require thick substrate and multiple layers. For high volume production in standard substrate technology with acceptable yield, such solution targeting the consumer market has not been attempted previously. In this work, we present a novel multicore packaging approach to realize a unique and cost effective solution based on standard PCB/ Substrate technology. We demonstrate this with a complete mmWave RF module board with dimensions of 30mm x 50mm x 1.5mm. It consists of a large area dual polarization dual frequency (28GHz and 39GHz) stacked patch antenna array of 4x8 elements and fully integrated Systems-In-Package (SIPs) along with all the thermal and mechanical functions. This solution achieves a very high cross polarization discrimination ratio (>25dB), which is critical in the MIMO operation of the dual polarization network. The work includes measurements to validate the design and the packaging concept suitable for low cost mmWave 5G CPE (Customer Premises Equipment) and base station applications.

Keywords-component; 5G; Ultra Large SIP; mmW Antenna Array; Outdoor application

I. INTRODUCTION

5G technology is the fifth generation mobile communications system that promises wireless channels with data rates above a Gigabit per second. The high speed channels are enabled by both the sub 6 GHz standard, in which MIMO techniques is key, and the millimeter wave (mmWave) frequency range, in which ultra-high bandwidth of the carrier frequencies is key. 5G networks aim to achieve ubiquitous communication between anybody and anything, anywhere and at any time. However, mmWave frequencies suffer from high loss in free space propagation and function most efficiently in line of sight only, therefore small cell networks are introduced instead of the current large high power cell towers. Customer Premises Equipment is analagous to a cable modem that interfaces between the small cell network nodes and the users' home. Today, the

only commercial Gbps internet service to buildings or home is fiber optics, which is expensive and limited in residential coverage. On the other hand, cellular providers already have cell towers everywhere, once deployed with 5G technology, multi Gbps links can be delivered to virtually any location at low cost with greater scalability. A small cell network is illustrated in Fig. 1, where the outdoor unit functions as the interface to connect the signals from the small cell tower to buildings and homes, and the CPE as an indoor unit to deliver multi Gbps to devices, all completely wire-free.

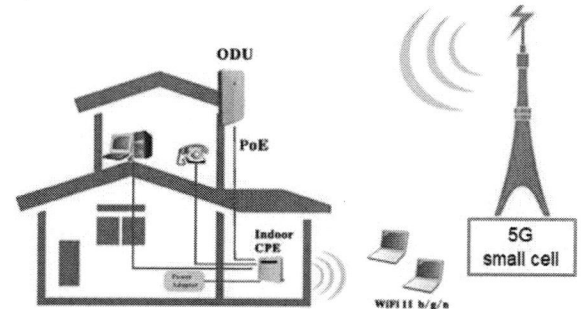

Figure 1. Small cell Network.

Consequently, those CPE and PDU devices, which would be deployed to every home, building, and public space, would benefit significantly from a high performance antenna array module that can be built with high yield and commercially low cost. The engineering cost of such a highly complex and highly integrated module is often high, therefore 5G carriers and customers have great demand for a complete plug-n-play solution. In this work, we present a novel multicore packaging approach to realize a unique and cost effective solution to mass produce such complex RF mmWave modules, based on standard PCB/Substrate technology. We demonstrate with a complete mmWave RF module board whose size is 30mm x 50mm x 1.5mm, which will deliver linear dual polarization and dual frequencies (28GHz and 39GHz) based on a stacked patch antenna array of 4x8 elements and fully integrated Systems-In-Package (SIPs).

II. RF MODULE ARCHITECTURE

Previously Intel developed 802.11ad products that operate also in the mmWave frequencies (60GHz) for communications [1]. However, the challenges for 5G is the balance between size, performance, cost, and power efficiency. In order to keep the module cost low, our team at Intel developed a novel module architecture utilizing standard PCB/ substrate technology. In this architecture, the module would need to support both the SIP PCB and the complete hardware architecture that integrates all thermal, mechanical, and electrical functionalities. As a result, we partitioned the design into two parts. The first part consists of large area super SIP, each is about 21mm x 18mm. And the second part consists of the ultra-rigid board which is 1.5mm thick. The rigid board contains interconnections to all the nets on the SIP including more than 500 nets, 3000 BGA connections, locations for housing heat sinks, board connectors, and all 32 antenna elements from the 4x8 antenna array. Each antenna element has four inputs, two of which support the low band around 28GHz frequencies, and the other two inputs which support the high band around 39GHz frequencies. Each frequency band has two orthogonal polarizations generated by two separate antenna inputs. The architecture of the module is illustrated in Fig. 2. In the first part, the two super SIPs are packaged separately, with the complete BOM (Bill Of Materials) and conformal shield utilizing a single laminate for routing. In the second part, the large area multi-core substrate consists of 10 layers that contain all the antenna elements, RF and DC connectors, mechanical screws, and additional routing between all of these elements and the SIP input/output ports.

Figure 2. Architecture of the 5G mmWave module.

In order to accommodate both the high density circuit layers and the low density high volume antenna structure, we developed a stackup that is composed of three portions with different core layers with different thicknesses, which is illustrated in Fig. 3. The first portion consists of the top four metal layers that support the ultra-high density circuit with hundreds of micro vias. The second portion is a single core to introduce volume (without metal) that is necessary

for most broadband patch antennas. Additionally, it also provides stiffness and mechanical symmetry for the stack up, holding the top portion and the bottom portion together. The third portion consists of the bottom four metal layers that contain the 32 antenna patches. There are no vias in the low density layers. They consist of only metal patches that are parasitically coupled to each other. This third portion employs two cores to enable the flexibility in antenna design as well as to maintain stiffness for the required volume. Note that each portion maintains its own layer thickness symmetry as well as the balance of its copper distribution in different layers. As a result, the entire stackup only needs 10 layers to support all the required functionalities and components of the module board. Traditionally, such complex boards would be built with a single thick core with multiple prepregs added on both sides [2 – 3], which often results in 16-20 layers. Such a process is expensive and requires longer lead time.

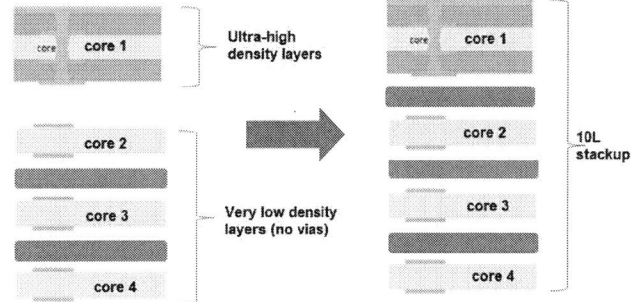

Figure 3. Multi-core PCB/substrate low cost stackup of the mmWave module.

The module board is constructed using 2-2-2 + 1-2-1 boards laminated together as illustrated in Fig 4. All the core stackups were laminated in 2 steps, and aligned by the through vias. These are the stitching vias on the outline of the module board (more than five hundred stitching vias were used). The stitching vias do not only stitch all the cores together in a low cost high yield process, but also align all layers in a very tight tolerance for extraordinarily thick board in mmWave. These have highly asymmetrical dielectric thicknesses and copper distribution. The final module is 50mm x 32.5mm x 1.5mm, the first and one of the largest mmWave RFEM implementations of its kind ever built for high volume 5G technology in the consumer market. As a result, we have filed multiple patents (pending) in this area. In the next section, the antenna topology and operation principles will be discussed.

978-1-7281-1500-9/19 $31.00 © 2019 IEEE

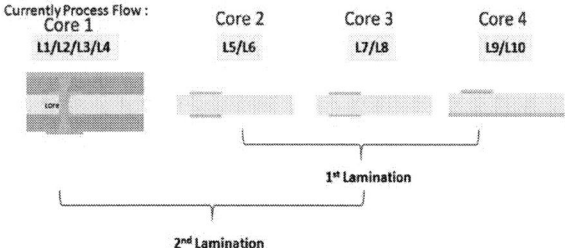

Figure 4. Lamination steps for the multi-core layers.

III. ANTENNA DESIGN

5G mmWave standards offer two operating frequency ranges in the US. The lower band is from 27GHz – 29.5GHz, and the higher band from 37GHz – 40GHz. Each frequency range (8 – 10% of relative bandwidth) is allowed to operate two orthogonal polarizations concurrently (linear polarizations) to increase the channel capacity. This feature is translated as highly isolated vertical and horizontal polarizations, i.e. high isolation between antenna ports of the same operating frequencies (>20dB) and high cross polarization discrimination in their radiation characteristics (XPD >20dB). In order to accommodate all operation frequency bands for an all-in-one solution as well as all the performance requirements of the antennas, we have utilized the stacked patch antenna concept to achieve dual frequency and orthogonal dual linear polarization. There are three patches to form two wideband resonant frequency bands [4]. A capture of the antenna unit is displayed in Fig. 5.

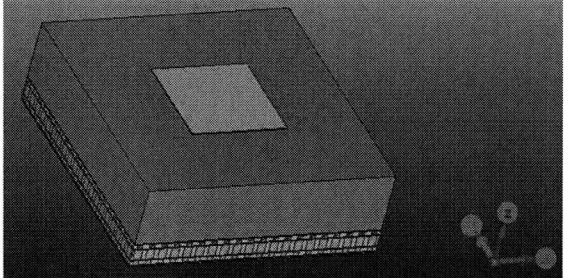

Figure 5. Topology of the tual band dual polarization antenna concept (exclude substrate in the view)

The patch antenna can be excited by a coupling strip feed line [4]. Furthermore, to off load the burden of switches required to implement in the RFIC, a divider is embedded into the antenna feed structure. All of which are embedded in the four layers of high density circuits contained within the high density circuit layers. This divider splits each polarization into high band and low band ports. As a result, we have a quad-fed antenna with each port responsible for one polarization and one frequency band. The divider implemented in this approach can be matched better to the complex input impedance at the slot for an ultra-wideband frequency range, as each frequency and each polarization is excited by a separate antenna port, we allow much greater flexibility in the operation of beam forming, scanning, and different MIMO schemes.

The simulation results of the antenna are shown in Figs. 6 – 7. Because the antenna is designed for a large array (4x8), each antenna input impedance needs to be optimized as a unit cell, in which the antenna ports effectively see the impedance of its neighbors while the neighbor elements are also excited and in operation (each has 8 neighbors except for those at the edges of the board). Such impedance in an array is referred to as active impedance, as opposed to passive impedance in which all other ports are terminated with match impedance. Reflection coefficients of each port in the unit cell design are shown in Fig. 6, where the low band (LB) ports and high band (HB) ports show excellent bandwidth coverage at the -10dB reflection level. The coupling level between the ports of each frequency range are shown in Fig. 7. Both LB port isolation and HB port isolation are higher than 15dB within the required bandwidth of 27GHz – 29.5GHz and 37GHz – 40GHz respectively. These highly challenging demands of the antenna module performance were satisfied by the unique antenna concept and the novel stackup approach that uses combination of very thin and thick laminate layers for antenna performance.

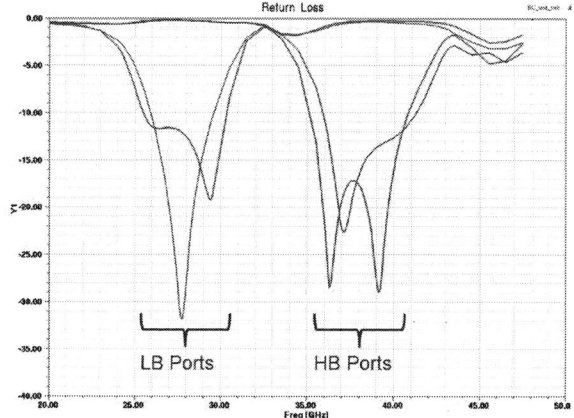

Figure 6. Simulation results of S11 (return loss) of the antenna ports (optimized for a unit cell in the large array).

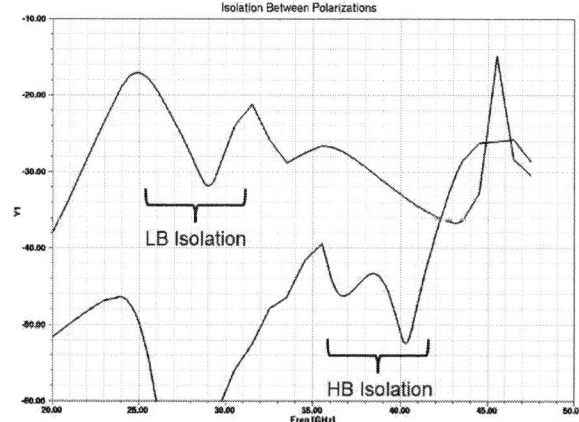

Figure 7. Simulation results of the isolations between V-pol and H-pol ports of the unit cell antenna.

In terms of XPD, we observed a wide solid angle (relative to the propagation direction) in which operation frequencies show an XPD level of at least 30dB. This performance allows for 2x channel capacity when both polarizations are operated concurrently. The gain of the 4x8 antenna array is about 16.2dB and 17.6dB at LB and HB respectively.

IV. FABRICATION AND MEASURMENTS RESULTS

The outcome of this innovation is the fully built module with complete SIPs, antennas, connectors, mechanical interfaces, and shield integration as demonstrated in Figs. 8 – 9. The capture in Fig. 8 shows the compact assembly of the SIPs with their conformal shield and RF/ DC connectors. Furthermore, several hundreds of the stitching vias surrounding the outline of the module board can be observed in Fig.9 along with the 4x8 antenna patches. Here, the stitching vias are also designed into a configuration of a hard surface structure [5] (frequency selective impedance surface) to couple the radiation of the array toward end-fire direction so as to improve the spatial coverage of beam scanning. The X-ray picture in Fig. 10 reveals more than 3000 connections, 1500 per SIP.

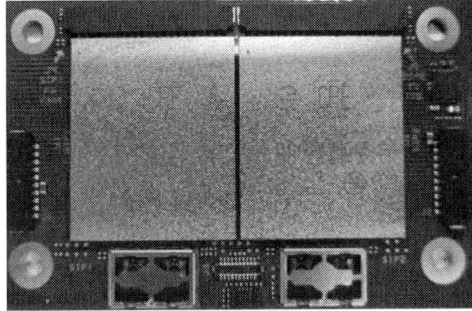

Figure 8. Backside view of the fully assembled mmWave antenna module and its corresponding X-ray image.

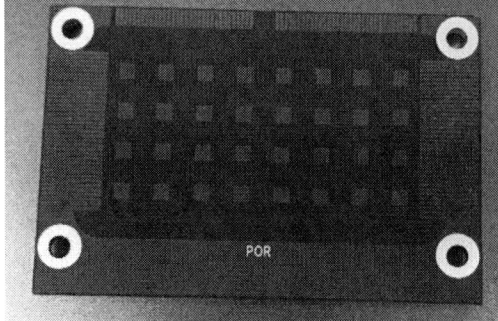

Figure 9. Frontside view of the fully assembled mmWave antenna module revealing the antenna patches.

Figure 10. X-ray image of the mmWave module board.

We observed no warpage issue for this board. In validation of the alignment of 10 layers in this thick board, the cross-section in Fig. 11 shows excellent through-hole alignment for the entire stackup with no issue. The through vias appear to fit well within all the via pads throughout the stackup with sufficient margin by design. This cross-section was taken at the top left corner of the board where ultra-high density circuits occur. Furthermore, the traces in the high density layers run very close to the vias, whose gaps are verified and shown in Fig. 12, where we performed the X-ray/ burnish cross-section analysis for the board in the ultra-high density layers. This parameter is particularly important for mmWave high density circuits, because it can easily detune the matching for the divider and compact routing. The differences between the target (text in blue) and the actual values (text in red) of the gaps between traces themselves and between traces and vias are demonstrated to be within 10 – 15um, an exceptional margin tolerance considering the number of layers and the stackup complexity. Consequently, we have achieved excellent margin tolerance for this novel multi-core approach.

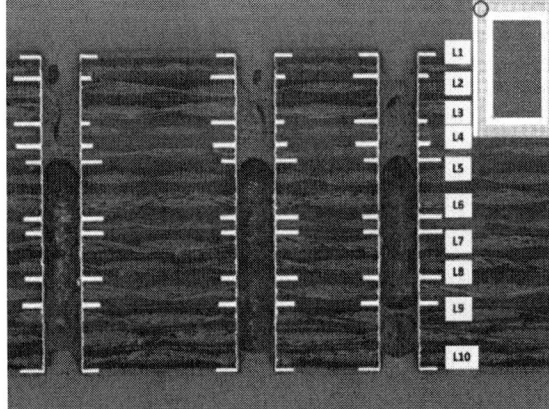

Figure 11. Cross-section of the mmWave module stackup.

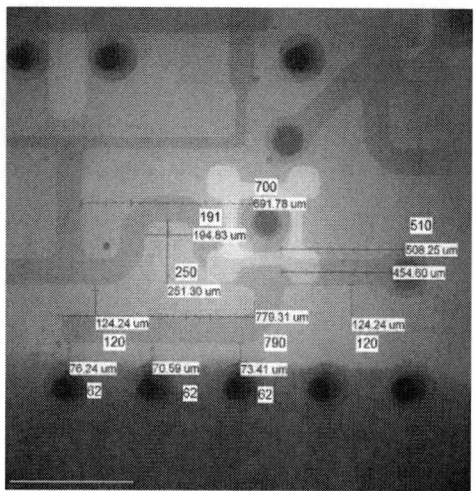

Figure 12. X-ray/burnish cross-section analysis of the ultra-high density layers of the mmWave board.

Finally, to validate the performance and fabrication of the antennas in this complex module board, the antenna measurement of a representative element in the array (element 15 in Fig. 13) is presented in Figs. 14 – 15. This antenna element is situated in the middle of the array, surrounded by 8 neighboring elements, making it an excellent validation for the unit cell design. The reflection coefficients of the two HB ports (for two polarizations) of two different samples are shown in Fig. 14 with overlay of simulation results of the same antenna element. Similarly, the measured reflection level of the two LB ports are as shown in Fig. 15. Note that in these measurements and simulations of a single element, all other RF ports on the board are left open. This configuration presents a different impedance environment than the one in simulations that were previously discussed in Sect. III. Although this impedance environment is not the same as the operation-environment of the array where all ports are excited, it is valid for the verification of the measurements against the simulations. The results in Figs. 14 – 15 show an excellent agreement between measurements and simulations, exhibiting the expected resonant peaks. Additionally, the variation between two random samples is virtually none. The coupling level between the orthogonal polarizations ports are presented in Fig.16. Again we observe excellent agreement between the measurements and simulations with a high isolation level of more than 30dB across the operation frequencies.

Figure 13. The mmWave module board with bare copper exposed for antenna ports measurements where element 15 was investigated

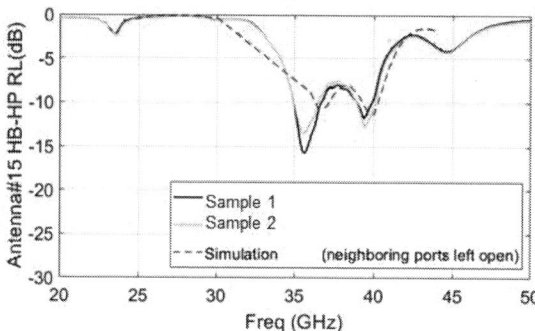

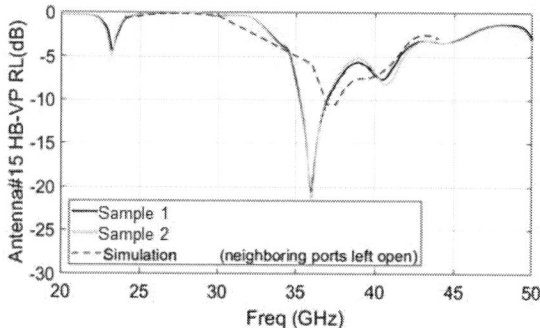

Figure 14. Reflection coefficients of two samples at the HB ports of antenna element 15 for a) horizontal polarizations in the top graph; b) vertical polarizations in the bottom graph.

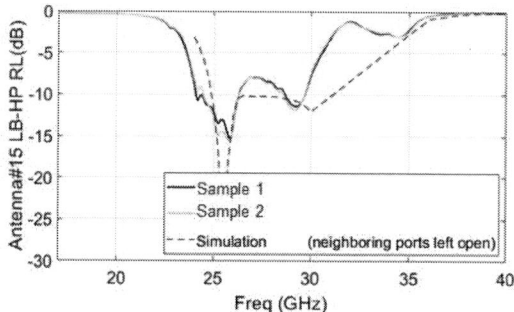

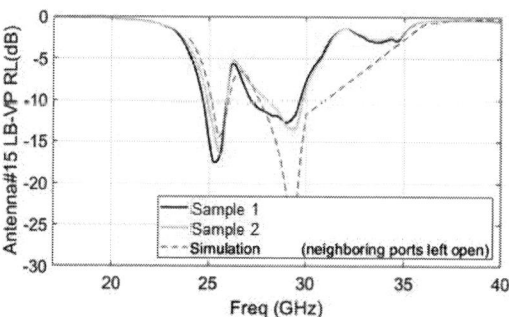

Figure 15. Reflection coefficients of two samples at the LB ports of antenna element 15 for 1) horizontal polarizations in the top graph; b) vertical polarizations in the bottom graph.

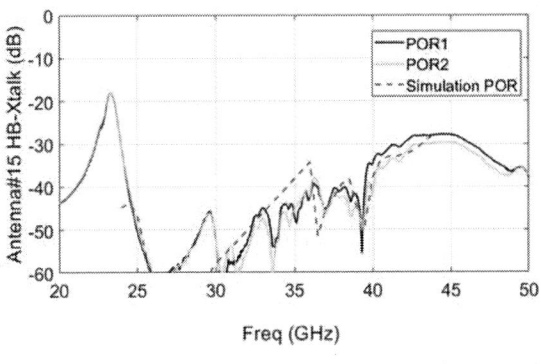

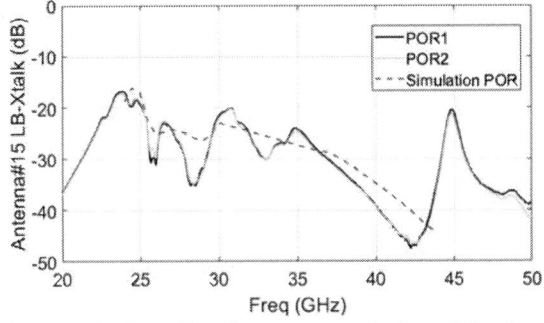

Figure 16. Coupling between vertical and horizontal polarizations of two samples of antenna element 15 for 1) HB ports in the top graph; b) LB ports in the bottom graph.

V. CONCLUSION

In this work, we presented a novel packaging approach using multicore PCB and substrate solutions to realize an ultra-broadband dual polarized antenna array module for 5G mmWave covering both 28GHz and 39GHz frequency ranges. We demonstrated the success of our innovative techniques with cross-section analysis and measurements of the antenna module board. The packaging approach was able to address highly challenging demands of the unique antenna concept with the combination of very thin and thick laminate layers for antenna performance. Stacked patch antennas provide many advantages in large beam steering array so they are ubiquitous in mmWave applications. Their

bandwidth performance is corresponding to the thickness of the antenna layers. Additionally, aperture fed antennas also provide high isolation between ports and excellent XPD only if the manufacturing techniques can accommodate various layer thicknesses and highly asymmetrical stackup and copper density. Our approach provides these degrees of freedom in the design with low cost and high yield, validated with a complete design, fabrication, and measured mmWave 5G module, making commercial deployment of the 5G mmWave technology easier. For the largest mmWave module – a first of its kind - we have demonstrated exceptional yield for all parameters from alignment to trace/feature size to trace gap. The result is a high performance RFEM for a large array (4x8 elements) that can support all 5G millimeter wave communication standards with dual polarization and a high cross polarization discrimination ratio (>25dB) to enable excellent MIMO operation for mobile and base-station applications.

ACKNOWLEDGMENT

We would like to acknowledge the support of William Lambert, Zhenguo Jiang and the rest of the ATTD team & Intel Lab as well as the collaboration of our OSAT partners.

REFERENCES

[1] Intel Wireless Gigabit Wireless Solutions https://www.intel.com/content/www/us/en/products/wireless/ad-products/wireless-gigabit-11100vr.html

[2] X. Gu *et al.*, "A multilayer organic package with 64 dual-polarized antennas for 28GHz 5G communication," *IEEE International Microwave Symposium (IMS)*, pp. 1899-1901, Honolulu, HI, 2017

[3] D. Liu, J. A. G. Akkermans, H.-C. Chen, and B. Floyd, "Packages with integrated 60-GHz aperture-coupled patch antennas," in *IEEE Transactions on Antennas and Propagation*, vol. 59, no. 10, pp. 3607-3616, Oct. 2011.

[4] B. Lindmark, "A dual polarized dual band microstrip antenna for wireless communications," in IEEE Aerospace Conference, Mar, 1998.

[5] Per-S. Kildal, "Artificially soft and hard surfaces in electromagnetics and their application to antenna design," in IEEE European Microwave Conference, Sept 1993.

2019 IEEE 69th Electronic Components and Technology Conference (ECTC)

3D Glass Package-Integrated, High-Performance Power Dividing Networks for 5G Broadband Antennas

Muhammad Ali[1], Atom Watanabe[1], Tong-Hong Lin[1], Markondeya R. Pulugurtha[2],
Manos M. Tentzeris[1] and Rao. R. Tummala[1]

[1]3D System Packaging Research Center, Georgia Institute of Technology, Atlanta, GA, USA
[2]Department of Biomedical Engineering, Florida International University, Miami, FL, USA
Email: ali_cmi@gatech.edu

Abstract—This paper demonstrates package-integrated power dividers with footprint smaller than the free-space wavelength corresponding to the operating frequency of 28 GHz band for 5G Antenna-in-Package (AiP), by utilizing precision low-loss redistribution layers (RDL) on glass substrates for highly-integrated mixed-signal systems. Two configurations of power dividers with two-way and three-way equal power split are modeled, designed and fabricated on glass substrates with thin-film build-up layers. This approach combines the benefits of ceramic and low-loss polymers for electrical performance, and silicon-like dimensional stability of glass for precision panel-scale patterning. Multilayered RDL with sub-20 micron features are utilized to design innovative power divider topologies with benefits in terms of low added insertion loss (<0.8-dB) and minimal phase-shift between the output ports, due to high precision of distributed transmission lines and through panel vias (TPVs). These power dividing networks depict upto 25% lower added insertion loss as compared to similar structures on fine pitch InFO RDL. The power dividers are also configured as 2×1 and 3×1 antenna arrays using Yagi-Uda antennas which cover the entire 28 GHz 5G band. The performance of power dividers as well as corresponding antenna arrays shows an excellent correlation between simulated and measured results.

Index Terms—5G and mm-wave; small-cell; RF; power divider; T-junction; Yagi-Uda; antenna array; semi-additive process

I. INTRODUCTION

The millimeter-wave (mm-wave) packaging technologies along with embedded antenna-in-package (AiP) is one of the vital contributors for the realization of high data-rate wireless communication. Significant momentum has been building up around the implementation of next generation of networks, namely 5G mobile communication systems, which are rumored to be commercially available to the end-users by 2020 [1–4]. For developing 5G wireless networks, the objectives include higher data-rate, higher capacity, high quality-of-service (QoS) low end-to-end latency, massive multiple-input multi-output (MIMO) capability and reduced cost [5], [6]. Millimeter-wave beam-forming is one of the most active areas in pursuit of compact, efficient and low-cost end-user devices as it is one of the key enabling technologies for 5G infrastructure such as small-cell and base-stations [7–10].

Due to higher free-space loss of mm-waves, high directivity and beam-steering of antennas becomes an inevitable requirement to enable a stronger and reliable wireless 5G connection. With the trend shifting towards package-integrated antennas, integrating a power dividing network with the antenna array in a package becomes extremely important as the power delivery to antennas operating at different bandwidths need to have minimal phase-shift [10]. Power dividers can be integrated either in the same metal layer or in the metal layer underneath the antenna array, depending primarily upon the complexity of the system as well as the chosen structure for power divider. Typical structures are based on microstrip, conductor-backed coplanar waveguide (CBCPW) or stripline [11–13]. Since the target layer is the top metal, microstrip structure is chosen for this demonstration as shown in Fig. 1.

A range of innovative substrate technologies have been explored for mm-wave antenna design. Antenna arrays have been demonstrated on state-of-the-art substrate technologies which include organic laminates, low-temperature co-fired ceramic (LTCC), integrated fan-out redistribution layer (InFO-RDL) and glass [14–18]. Antenna performance need to be optimized regardless of the substrate type in terms of desired bandwidth, gain and radiation pattern. Antenna-chip-package co-design also dictates the size of the antenna array, optimum material properties for interconnects as well as thermal and mechanical considerations. LTCC is limited by its high cost and scalability to large panels whereas organic substrates take

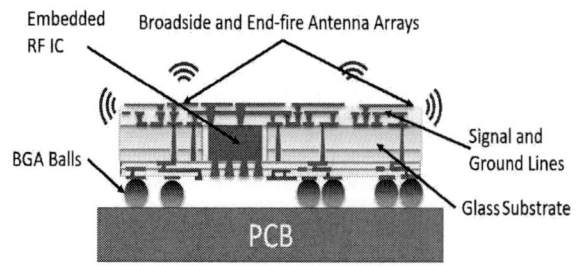

Figure 1. Illustration of an antenna-in-package module

978-1-7281-1500-9/19 $31.00 © 2019 IEEE

advantage of their low cost and scalability. However, they fall short in processability due to panel-scale warpage and reliability issues. InFO has challenges in use of epoxy molding compound and co-efficient of thermal expansion (CTE) mismatch with silicon dies. Glass-based packages are emerging as competitors to realize mm-wave technologies because of superior dimensional stability, availability in large-area low-cost panels, ability to form fine-pitch through-vias, stability to temperature and humidity, and matched coefficient of thermal expansion (CTE) with devices along with low dielectric loss compared to silicon and mold compounds used in fan-out packages.

In this paper, 3D package-integrated, equal-split power dividing networks with footprint smaller the free-space wavelength at the operating frequency of 28 GHz (24.5-29.5 GHz, FBW=18.51%) bands for 5G applications. Two power divider configurations: two- and three-way, are modeled, designed and fabricated on thin-film build-up layers on ultra-thin laminated glass substrate. The power dividers are connected with Yagi-Uda antennas to demonstrate their effectiveness. Utilizing glass core combines the benefits of ceramic or low-loss polymers for electrical performance, and silicon-like dimensional stability of glass for precision panel-scale patterning. Multilayered RDL are utilized to design innovative power divider topologies with benefits in terms of low added insertion loss, wide-bandwidth and minimal phase-shift. These power dividing networks depict upto 25% lower added insertion loss as compared to similar structures on fine pitch InFO RDL [16].

This paper is organized as follows: An introduction to the motivation and applications of this research is given in Section-1. The materials stackup and design techniques for the demonstration of power dividers is discussed in Section-2 followed by the detail of fabrication process in the Section-3. Finally, a comparison of simulated and measured results is given in Section-4.

II. MATERIAL STACKUP, MODELING AND DESIGN PROCEDURE OF POWER DIVIDERS

A. Material Stackup

In this section, the material stackup and design procedure of power dividers is discussed. The microstrip structure is chosen for power dividers with the intent of integrating them into a module on the top metal layer, along with the antennas. The material stackup is shown in Fig. 2.

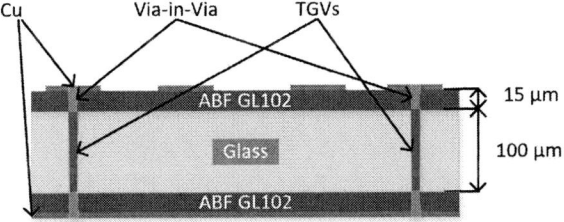

Figure 2. Material stackup for demonstration of power dividers

The material stackup consists of ultra-thin 100-μm EN-A1 glass substrate from Asahi Glass Co., (AGC) laminated with 15-μm Ajinomoto ABF GL102 on both sides. The electrical properties of the glass substrate are as follows: dielectric constant (Dk) = 5.4 and loss tangent (Df): 0.005 at 10 GHz. Similarly, the electrical properties of ABF GL102 are: Dk = 3.3 and Df = 0.0044 at 5.8 GHz and it has stable properties upto 50 GHz [19–21]. The desired copper thickness for the microstrip power divider structures is targeted to be 8-μm. The skin-depth at the highest operating frequency of 29.5 GHz is approximately 0.38-μm. The copper thickness of traces gives enough room to conveniently fabricate the microstrip transmission lines with decent power handling capabilities while keeping the conductor thickness more than five times of skin-depth for low-loss applications. The through-glass via (TGV) diameter is set to 100-μm, whereas the via-in-via diameter is adjusted accordingly. The minimum center-to-center via pitch is 450-μm. This material stackup is used in simulation using ANSYS High Frequency Structure Simulator (HFSS) for both power dividers in this demonstration.

B. Power Divider Design

The transmission line modeling of a power divider is based on a T-junction, as shown in the Fig. 3. Another usual choice of power divider is Wilkinson power divider as it is advantageous in terms of providing isolation between the output ports [22]. However, it can be disadvantageous for small packages as it requires a lumped resistor. Generally, there can be fringing

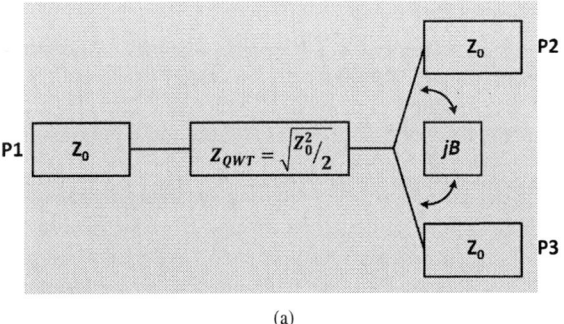

(a)

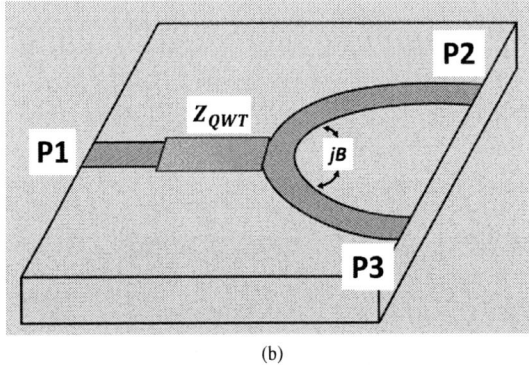

(b)

Figure 3. A T-junction power divider with a matching quarter-wave transformer (a) schematic, and (b) an example microstrip realization

978-1-7281-1500-9/19 $31.00 © 2019 IEEE

fields in a T-junction, especially in a microstrip configuration, and higher order modes that are associated with the discontinuity at the junction of the three transmission lines. This can lead to stored energy, donated by lumped susceptance, B. Since the divider is lossy, it can be matched to all ports using a quarter wave transformer. Looking into the junction, the input admittance is seen as:

$$Y_{in} = jB + \frac{1}{Z_2} + \frac{1}{Z_3} = \frac{1}{Z_1} \qquad (1)$$

where Z_n indicate impedance of the respective transmission line attached to the port n (n=1, 2, 3,...). In case of low-loss transmission lines, the characteristic impedance (Z_0) of transmission lines can be assumed to be real, equating B to zero. Thus, (1) transforms into (2):

$$\frac{1}{Z_1} = \frac{1}{Z_2} + \frac{1}{Z_3} \qquad (2)$$

In practicality, B is not negligible but it can be compensated by either some discontinuity compensation or a reactive tuning element. Often, the impedance of output networks is adjusted as such to cancel out B over a narrow or a wide frequency range.

The power dividing networks designed for this demonstration are equal-split two- and three-way power dividers. For both cases, the characteristic impedance of the input and output lines is set at 50 Ω. Looking into the transmission line junction of the two-way network, two 50 Ω impedances are seen in parallel, making an effective impedance of 25 Ω. The input transmission line has an impedance of 50 Ω which is matched to the 25 Ω impedance using a quarter-wave transformer. Similarly, the three-way equal-split power divider has an impedance of 16.67 Ω looking into the junction of transmission lines and it is matched to the input impedance of 50 Ω using an appropriate quarter-wave transformer. This transformer also helps in countering the lumped susceptance B at the junction of the power divider. The distance between output ports is adjusted in reference to the dimensions of Yagi-Uda antennas and the required spacing between them for antenna arrays. The ideal spacing between array elements is usually half of the free-space wavelength (λ_0).

Since the material stackup for this demonstration consists of three dielectric layers sandwiched together as shown in Fig. 4, it is imperative to find the effective dielectric permittivity (ε_{rc}) of this stackup. The dielectric constants of each layer are donated by ε_n where n is the number of layers (n=1, 2, 3,...). It is expressed using the following set of equations in which h_n, h_{n-1},...h_1 represent the height of individual layer starting from the top and ε_n, ε_{n-1},...ε_1 correspond to the complex relative permittivity of the respective dielectric layer [23], [24]:

$$\varepsilon_{rc} = \frac{|d_1| + |d_2| + |d_3|}{|\frac{d_1}{\varepsilon_1}| + |\frac{d_2}{\varepsilon_2}| + |\frac{d_3}{\varepsilon_3}|} \qquad (3)$$

for $h_n + h_{n-1} + ... + h_1 \simeq \lambda/10$.

$$d_1 = \frac{K(k_1)}{K'(k_1)} \qquad (4)$$

$$d_2 = \frac{K(k_2)}{K'(k_2)} - \frac{K(k_1)}{K'(k_1)} \qquad (5)$$

$$d_3 = \frac{K(k_3)}{K'(k_3)} - \frac{K(k_2)}{K'(k_2)} - \frac{K(k_1)}{K'(k_1)} \qquad (6)$$

$$d_n = \frac{K(k_n)}{K'(k_n)} - \frac{K(k_n - 1)}{K'(k_n - 1)} - ... - \frac{K(k_1)}{K'(k_1)} \qquad (7)$$

Generally, k_n is defined as [25]:

$$k_n = \frac{1}{cosh(\frac{w\pi}{h_n + h_{n-1} + ... + h_1})} \text{ for n=1, 2, 3,...} \qquad (8)$$

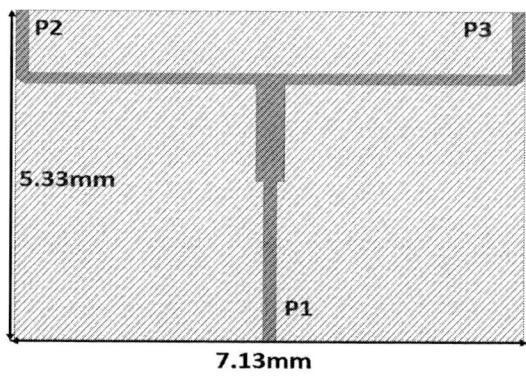

7.13mm

(a)

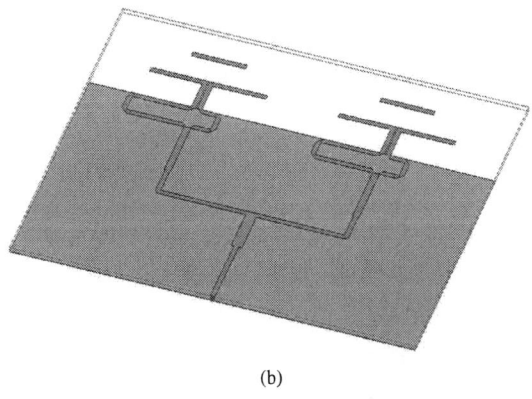

(b)

Figure 5. A two-way power divider with a matching quarter-wave transformer (a) simulation layout, and (b) configured as 2×1 antenna array

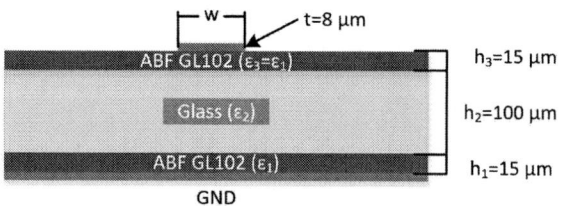

Figure 4. Multilayered microstrip line for power dividers

The value of $\frac{K()}{K'()}$ is given by (9).

$$\frac{K(k_n)}{K'(k_n)} = \frac{1}{\pi} ln(2\frac{1+\sqrt{k_n}}{1-\sqrt{k_n}}) \text{ for } 0.7 \leq k_n \leq 1 \quad (9)$$

The frequency-independent relative dielectric constant of the multilayer substrate is obtained using (3)-(9). After obtaining this parameter, the next objective is to find the frequency-dependent behavior of the material stackup. The concept transforms to a single substrate with permittivity ε_{rc} and thickness of h=h_n+h_{n-1}+...+h_1 and it is discussed in detail in [26], [27]. It is assumed in this model that the total substrate thickness is approximately equal to $\lambda_0/10$ to obtain quasi-TEM characteristics.

Although calculating frequency-dependent behavior of the multilayer stackup is not absolutely critical below 100 GHz, the frequency-independent ε_{rc} is important as it helps in calculating the width of transmission line corresponding to the characteristic impedance of 50 Ω. The frequency-dependent behavior of the multilayer stackup is useful in accurately finding the electrical length associated with the physical length of the transmission line. Moreover, ε_{rc} can also be estimated intuitively by looking at the contribution of each dielectric to the stackup based on the thickness of individual layer. In this specific case, the ε_1 and ε_3 both equal to Dk of ABF GL102

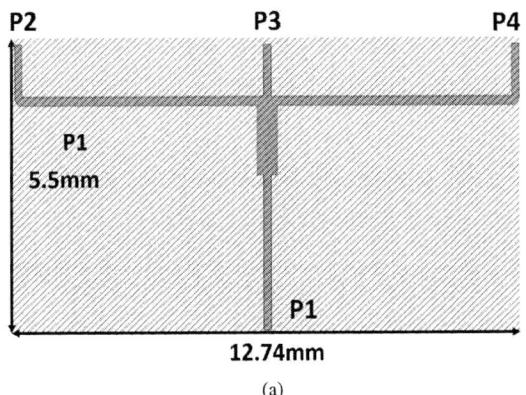

(a)

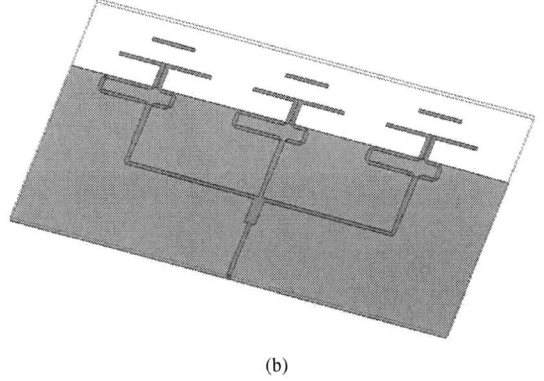

(b)

Figure 6. A three-way power divider with a matching quarter-wave transformer (a) simulation layout, and (b) configured as 3×1 antenna array

which makes 30-μm of the total stackup height. The remaining 100-μm is the Dk of glass substrate. Using this data, the Dk of the entire stackup can be estimated to be slightly less than the Dk of glass substrate. Utilizing this technique, width of the microstrip transmission line corresponding to the characteristic impedance of 50 Ω is calculated.

Using this method, the two- and three-way power dividers are simulated. They are also configured as 2×1 and 3×1 using Yagi-Uda antennas which cover the entire 28 GHz 5G band. The layout of both power dividers along with corresponding antenna arrays is shown in Fig 5 and Fig 6, respectively.

III. FABRICATION PROCESS OF POWER DIVIDERS

The fabrication process uses the semi-additive patterning (SAP) process to pattern polymers. SAP enables the fabrication of sub-5 micron features for glass substrates due to their surface planarity and dimensional stability. The process is illustrated with cross-sections in Fig. 7. Proper handling procedures are advised to administer glass fabrication process to address its fragility and brittleness [28], [29]. Polymer lamination on glass addresses the challenge of glass handling and acts a buffer layer that reduces the effect of high CTE copper on glass. It also enables metallization as well as prevents copper migration to glass surface under high electrical voltage during plating.

Through-Glass Vias (TGVs) of 100-μm diameter are drilled by AGC. The glass panels are treated with silane to promote adhesion to polymers and mitigate delamination during the whole fabrication process. Polymer lamination is performed on both sides of glass panel using a roll laminator and 15-μm ABF GL102. This step is followed by polymer curing. The fabrication process continues to via-in-via ablation. The power, power density and number of repetitions are optimized for the UV laser to drill vias in the polymer. This step is performed on both sides of the panel. Next, a 0.2-μm copper seed layer is deposited uniformly on the polymer using electroless plating method. The electroless plating method has a surface roughening step utilizing a permanganate etch which creates mechanical anchor sites on the polymer to enable latching of copper particles. It also helps in removal of residual polymer left after via-in-via ablation.

Afterwards, both sides of the panel are laminated with a 15-μm dry-film negative photoresist. The panel is photolithographically patterned with optimized dose time for the most critical feature of the design. Followed by photoresist development, the panel is subjected to O_2 plasma which removes photoresist residue and improves the surface conditions for copper plating. The metallization of traces and vias is performed using electrolytic plating. SAP yields better dimensions and sidewall control of the deposited copper unlike subtractive etching techniques which have limited control. The target copper thickness after fabrication process is 8-μm so the authors strive to achieve plated copper thickness of 8.2-μm to account for removal of copper seed layer. After electrolytic plating, the photoresist is stripped off using a stripper solvent and seed layer etching is performed. The measured copper thickness

Glass Panel with TGVs

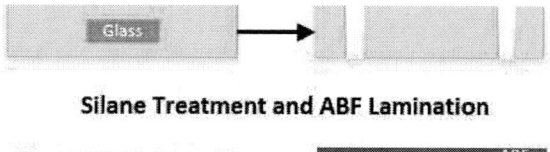

Silane Treatment and ABF Lamination

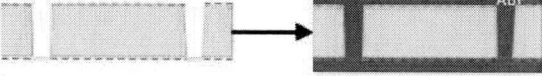

Via-in-Via Ablation and Cu Seed Layer Deposition

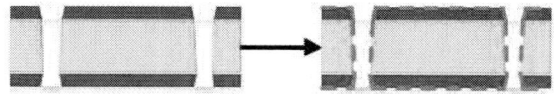

Photoresist Lamination and Photolithography

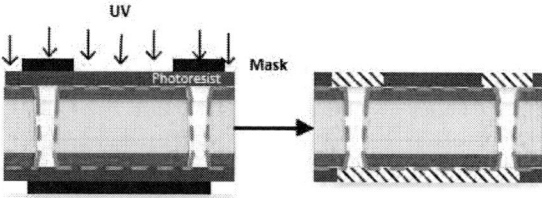

Electrolytic Cu Plating

Photoresist Stripping and Seed Layer Etching

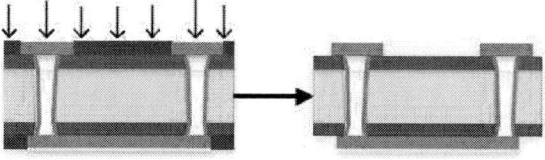

Figure 7. Cross-sectional illustration of SAP

after fabrication process is 8.5±0.5-μm. The fabricated two- and three-way power dividers and antenna arrays are shown in Fig. 8a and Fig. 8b, respectively.

IV. CHARACTERIZATION RESULTS AND ANALYSIS

In this section, the characterization results of fabricated power dividers are discussed in detail. The measurements are performed using a VNA in the frequency range of 20-32 GHz using ACP50 GSG probes and Short-Open-Load-Through (SOLT) calibration. For the normalized antenna radiation pattern measurements, the panels are diced into individual coupons and 2.92 mm SMA connectors are soldered onto each coupon as shown in Fig. 9. The measurement results

are compared with simulation results to perform a model-to-hardware correlation study. Moreover, dimensional analysis is performed to analyze the fabricated dimensions and compare them with the dimensions in simulation.

A. Two-Way Power Divider

The s-parameters of fabricated two-way power divider are shown in Fig 10. The ideal insertion loss of a two-way power divider is 3.01-dB. As evident from Fig. 10, the power divider has return loss less than 15-dB in the entire 28 GHz 5G band and the insertion loss is fairly constant (~3.46-dB) as well. The added insertion loss of the two-way power divider is 0.45-dB. Comparison of simulated and measured return loss of 2×1 Yagi-Uda antenna array is shown in Fig. 11. The 10-dB return loss bandwidth of antenna array ranges from 24.3 to 31.4 GHz (FBW=25.5%). The simulated and measured radiation patterns of the antenna array are compared in Fig. 12. It is to be noted that the individual element realized gain of Yagi-Uda antenna

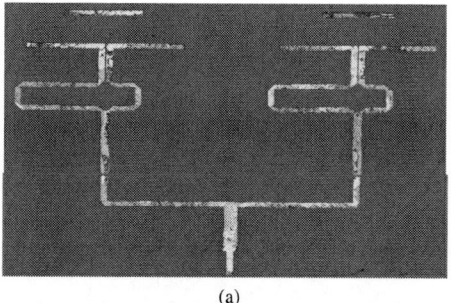

(a)

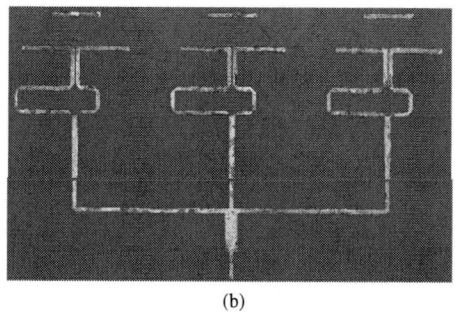

(b)

Figure 8. Fabricated power divider with Yagi-Uda antennas (a) two-way, and (b) three-way

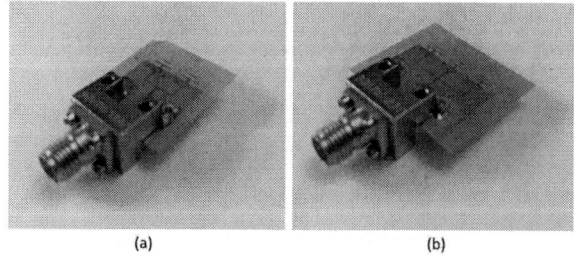

(a)   (b)

Figure 9. Antenna arrays with SMA connectors

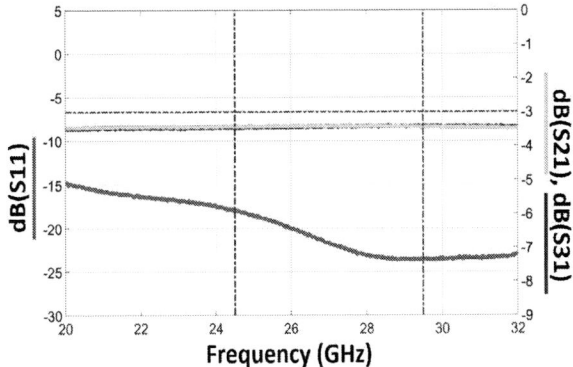

Figure 10. S-parameters of the fabricated two-way power divider

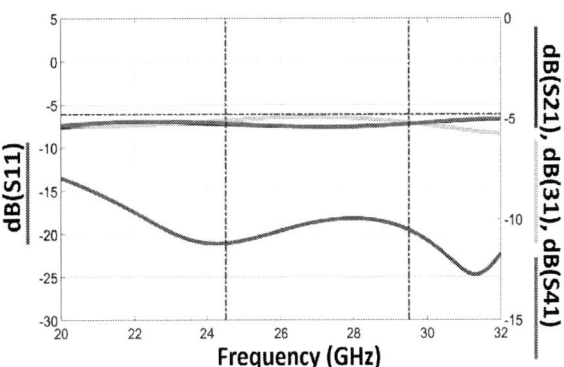

Figure 13. S-parameters of the fabricated three-way power divider

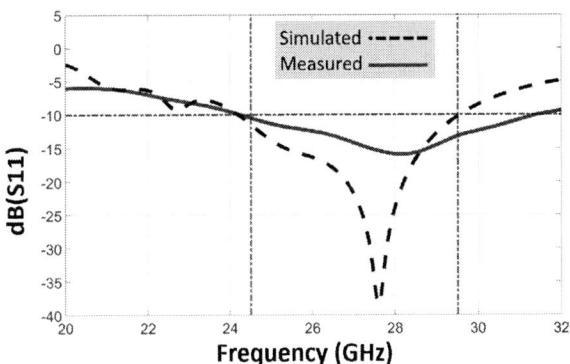

Figure 11. Comparison of simulated and measured S11 of the fabricated 2×1 Yagi-Uda antenna array

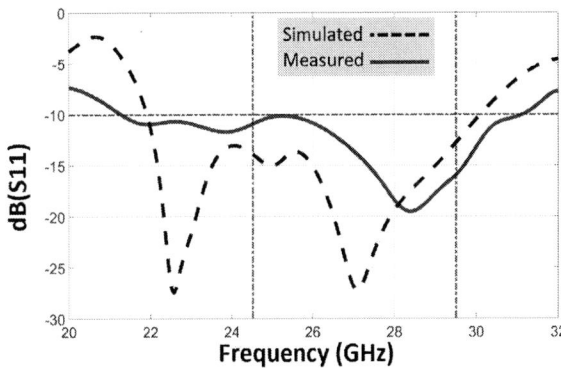

Figure 14. Comparison of simulated and measured S11 of the fabricated 3×1 Yagi-Uda antenna array

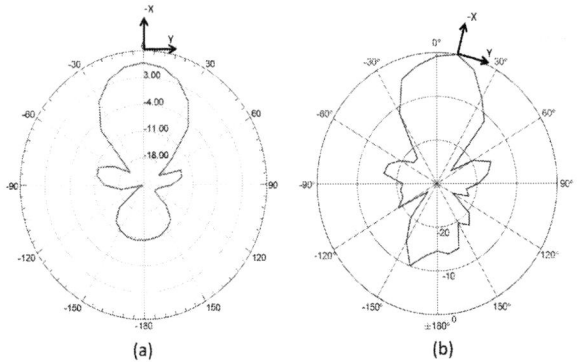

Figure 12. Comparison of simulated and measured radiation pattern of 2×1 Yagi-Uda antenna array

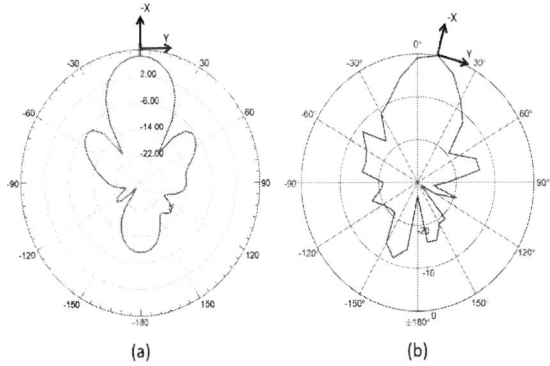

Figure 15. Comparison of simulated and measured radiation pattern of 3×1 Yagi-Uda antenna array

is 4-dB. When configured as a 2×1 array, the realized gain increases to 7-dB with >78% radiation efficiency.

B. Three-Way Power Divider

The s-parameters of the fabricated three-way power divider are shown in Fig 13. The ideal insertion loss of a three-way power divider is 4.77-dB. As depicted in Fig. 13, the power divider has return loss less than 15-dB and a maximum insertion loss 5.42-dB in the entire 28 GHz 5G band. This

gives the added insertion loss of the three-way power divider to be 0.65-dB. Comparison of simulated and measured return loss of 3×1 Yagi-Uda antenna array is shown in Fig. 14. The 10-dB return loss bandwidth of antenna array ranges from 21.3 to 31.1 GHz (FBW=37.4%). The simulated and measured radiation patterns of the antenna array are compared in Fig. 15. The realized gain of the 3×1 Yagi-Uda antenna array is 8.24-dB with >85% radiation efficiency.

978-1-7281-1500-9/19 $31.00 © 2019 IEEE

TABLE I. Physical and Electrical Dimensions of Fabricated Power Dividers and Antenna Arrays

Structure	Physical Dimensions (mm^2)	Electrical Dimensions (λ_0)^2
2-Way Power Divider	5.44×7.13	0.51×0.67
2×1 Antenna Array	14.95×12.33	1.39×1.15
3-Way Power Divider	5.5×12.74	0.51×1.19
3×1 Antenna Array	12.5×20.6	1.17×1.92

C. Dimensional Analysis

The physical dimensions along with the corresponding electrical dimensions in terms of λ_0 of each fabricated power divider and corresponding antenna array is given in Table I. The physical dimensions of the power dividers are normalized by the corresponding free-space wavelength of 28 GHz 5G band: 10.71 mm.

To the best understanding of the author, the fabricated two-way power divider is one of the smallest in terms of x-y dimensions as well as z-height. The total height of the demonstrated power dividers is 147-μm. Moreover, they depict superior performance in terms of low-added insertion loss and minimal phase-shift between the output ports.

V. Conclusions

This paper presents the detailed design, fabrication and analysis of 3D glass package-integrated, equal-split power dividers with footprint smaller than the unit free-space wavelength corresponding to the operating frequency of 28 GHz 5G band. The power dividers are fabricated using SAP process to meet the dimensional accuracy requirement for such structures on ultra-thin glass. The demonstrated power dividers are also configured as antenna arrays and they can be used in the top metal layers of RF front-end packages where size in all dimensions is a critical requirement The power dividers exhibit low added insertion loss and minimal phase-shift between output ports, making them ideal for strict-footprint 5G and mm-wave module specifications.

Acknowledgment

The authors wish to acknowledge the industry sponsors of the consortia program at GT-PRC for their technical guidance and support.

References

[1] J. Thompson, X. Ge, H. Wu, R. Irmer, H. Jiang, G. Fettweis, and S. Alamouti, "5g wireless communication systems: prospects and challenges [guest editorial]," *IEEE Communications Magazine*, vol. 52, no. 2, pp. 62–64, February 2014.

[2] C. Huang and W. Lin, "A radio transceiver architecture for coexistence of 4g-lte and 5g systems used in mobile devices," in *2017 IEEE MTT-S International Microwave Symposium (IMS)*, June 2017, pp. 2056–2058.

[3] E. Dahlman, G. Mildh, S. Parkvall, J. Peisa, J. Sachs, and Y. Selén, "5g radio access," https://www.ericsson.com/en/ericsson-technology-review/archive/2014/5g-radio-access, June 2014.

[4] J. Nightingale, P. Salva-Garcia, J. M. A. Calero, and Q. Wang, "5g-qoe: Qoe modelling for ultra-hd video streaming in 5g networks," *IEEE Transactions on Broadcasting*, vol. 64, no. 2, pp. 621–634, June 2018.

[5] A. Gupta and R. K. Jha, "A survey of 5g network: Architecture and emerging technologies," *IEEE Access*, vol. 3, pp. 1206–1232, 2015.

[6] Y. Huo, X. Dong, and W. Xu, "5g cellular user equipment: From theory to practical hardware design," *IEEE Access*, vol. 5, pp. 13 992–14 010, 2017.

[7] X. Gu, D. Liu, C. Baks, A. Valdes-Garcia, B. Parker, M. R. Islam, A. Natarajan, and S. K. Reynolds, "A compact 4-chip package with 64 embedded dual-polarization antennas for w-band phased-array transceivers," in *2014 IEEE 64th Electronic Components and Technology Conference (ECTC)*, May 2014, pp. 1272–1277.

[8] F. Boccardi, R. W. Heath, A. Lozano, T. L. Marzetta, and P. Popovski, "Five disruptive technology directions for 5g," *IEEE Communications Magazine*, vol. 52, no. 2, pp. 74–80, February 2014.

[9] W. Roh, J. Seol, J. Park, B. Lee, J. Lee, Y. Kim, J. Cho, K. Cheun, and F. Aryanfar, "Millimeter-wave beamforming as an enabling technology for 5g cellular communications: theoretical feasibility and prototype results," *IEEE Communications Magazine*, vol. 52, no. 2, pp. 106–113, February 2014.

[10] D. Liu, X. Gu, C. W. Baks, and A. Valdes-Garcia, "Antenna-in-package design considerations for ka-band 5g communication applications," *IEEE Transactions on Antennas and Propagation*, vol. 65, no. 12, pp. 6372–6379, Dec 2017.

[11] B. Mishra, A. Rahman, S. Shaw, M. Mohd, S. Mondal, and P. P. Sarkar, "Design of an ultra-wideband wilkinson power divider," in *2014 First International Conference on Automation, Control, Energy and Systems (ACES)*, Feb 2014, pp. 1–4.

[12] O. D. Gürbüz, K. Topalli, M. Ünlü, . Demir, and T. Akin, "A 35 ghz coplanar waveguide power divider," in *2010 10th Mediterranean Microwave Symposium*, Aug 2010, pp. 22–24.

[13] A. Hirota, Y. Tahara, H. Yukawa, T. Owada, Y. Yamaguchi, and H. Miyashita, "A stripline power divider with insensitive to resistance variations using a parallel resistor pair," in *2014 IEEE MTT-S International Microwave Symposium (IMS2014)*, June 2014, pp. 1–3.

[14] C. Ho, M. Jhong, P. Pan, C. Huang, C. Wang, and C. Ting, "Integrated antenna-in-package on low-cost organic substrate for millimeter-wave wireless communication applications," in *2017 IEEE 67th Electronic Components and Technology Conference (ECTC)*, May 2017, pp. 242–247.

[15] Y. P. Zhang and D. Liu, "Antenna-on-chip and antenna-in-package solutions to highly integrated millimeter-wave devices for wireless communications," *IEEE Transactions on Antennas and Propagation*, vol. 57, no. 10, pp. 2830–2841, Oct 2009.

[16] C. Hsu, C. Tsai, J. Hsieh, K. Yee, C. Wang, and D. Yu, "High performance chip-partitioned millimeter wave passive devices on smooth and fine pitch info rdl," in *2017 IEEE 67th Electronic Components and Technology Conference (ECTC)*, May 2017, pp. 254–259.

[17] M. Wojnowski and K. Pressel, "Embedded wafer level ball grid array (ewlb) technology for high-frequency system-in-package applications," in *2013 IEEE MTT-S International Microwave Symposium Digest (MTT)*, June 2013, pp. 1–4.

[18] T. Kamgaing, A. A. Elsherbini, T. W. Frank, S. N. Oster, and V. R. Rao, "Investigation of a photodefinable glass substrate for millimeter-wave radios on package," in *2014 IEEE 64th Electronic Components and Technology Conference (ECTC)*, May 2014, pp. 1610–1615.

[19] M. Ali, F. Liu, A. Watanabe, P. M. Raj, V. Sundaram, M. M. Tentzeris, and R. R. Tummala, "First demonstration of compact, ultra-thin low-pass and bandpass filters for 5g small-cell applications," *IEEE Microwave and Wireless Components Letters*, vol. 28, no. 12, pp. 1110–1112, Dec 2018.

[20] M. Ali, F. Liu, A. Watanabe *et al.*, "Miniaturized high-performance filters for 5g small-cell applications," in *2018 IEEE 68th Electronic Components and Technology Conference (ECTC)*, May 2018, pp. 1068–1075.

[21] A. O. Watanabe, M. Ali, B. Tehrani, J. Hester, H. Matsuura, T. Ogawa, P. M. Raj, V. Sundaram, M. M. Tentzeris, and R. R. Tummala, "First demonstration of 28 ghz and 39 ghz transmission lines and antennas on glass substrates for 5g modules," in *2017 IEEE 67th Electronic Components and Technology Conference (ECTC)*, May 2017, pp. 236–241.

[22] D. Pozar, *Microwave Engineering*. Wiley, 2005. [Online]. Available: https://books.google.com/books?id=LwisngEACAAJ

[23] Y. J. Yoon and B. Kim, "A new formula for effective dielectric constant in multi-dielectric layer microstrip structure," in *IEEE 9th Topical Meeting on Electrical Performance of Electronic Packaging (Cat. No.00TH8524)*, Oct 2000, pp. 163–167.

[24] K. R. Jha and G. Singh, "Analysis of dielectric permittivity and losses of two-layer substrate materials for microstrip antenna at thz frequency," in *2009 International Conference on Advances in Recent Technologies in Communication and Computing*, Oct 2009, pp. 672–675.

[25] R. Collin, *FOUNDATIONS FOR MICROWAVE ENGINEERING, 2ND ED*. Wiley India Pvt. Limited, 2007. [Online]. Available: https://books.google.com/books?id=coBpP2SLiZQC

[26] M. Kobayashi, "A dispersion formula satisfying recent requirements in microstrip cad," *IEEE Transactions on Microwave Theory and Techniques*, vol. 36, no. 8, pp. 1246–1250, Aug 1988.

[27] E. O. Hammerstad, "Equations for microstrip circuit design," in *1975 5th European Microwave Conference*, Sep. 1975, pp. 268–272.

[28] P. M. Raj, C. Nair, H. Lu, F. Liu, V. Sundaram, D. W. Hess, and R. Tummala, ""zero-undercut" semi-additive copper patterning - a breakthrough for ultrafine-line rdl lithographic structures and precision rf thinfilm passives," in *2015 IEEE 65th Electronic Components and Technology Conference (ECTC)*, May 2015, pp. 402–405.

[29] H. Lu, R. Furuya, B. M. D. Sawyer, C. Nair, F. Liu, V. Sundaram, and R. R. Tummala, "Design, modeling, fabrication and characterization of 2–5-μm redistribution layer traces by advanced semiadditive processes on low-cost panel-based glass interposers," *IEEE Transactions on Components, Packaging and Manufacturing Technology*, vol. 6, no. 6, pp. 959–967, June 2016.

Advanced Wafer Level PKG solutions for 60GHz WiGig (802.11ad) Telecom Infrastructure

Dapeng Wu, Robin Dahlbäck, Erik Öjefors and Mats Carlsson
Sivers IMA AB, Torshamnsgatan 9, SE16440, Kista, Sweden

Francis Chee Peng Lim, Yew Kheng Lim, Aung Kyaw Oo, Won Kyung Choi and Seung Wook Yoon
STATS ChipPAC Pte. Ltd., JCET Group, 5 Yishun Street 23, Singapore 768442
seungwook.yoon@statschippac.com

Abstract

The continued growth in mobile data traffic is pushing for new and innovative solutions. Over the next few years, cellular phones, tablets, and computers will switch to 5G wireless technology. In this work, an highly integrated WiGig/802.11ad compliant 16+16 beam forming transceiver RFIC with advanced FOWLP (fan out wafer level packaging) technology known as eWLB (embedded wafer level BGA) is presented, which supports the need for the increased demand of data traffic. The 0.5mm pitch eWLB package measures 12.6x12.6x0.8mm and is using advanced dielectric materials and 2 metal layers for the redistribution layers (RDL). Full-wave electromagnetic simulations were performed including models of the chip layout, interconnects on the RDL, transitions and PCB. JEDEC level reliability was tested for component level reliability including MSL, TC, uHAST and HTS. The measurement results of compressed output power and noise figure for both bare die and packaged chip are presented.

Keywords-component; eWLB, FOWLP, 60Ghz mmWave, 5G, Telecom Infra, beam forming tranceiver, component level reliability, mmWave electrical simulation and characterization.

I. INTRODUCTION

60GHz (mmWave) Communication

In recent years millimeter wave systems have started to gain traction in the communications area due to advances in process technologies and integration solutions allowing for more cost effective products[1]. Today millimeter wave technology is used extensively in point-to-point communications in backhaul applications [1], as an alternative to running optical fiber. Millimeter wave outdoor fixed point-to-point radio links can allow for a low installation cost compared to optical fiber installations, whilst still allowing for a high capacity. Millimeter waves will be a part of the next generation cellular standard, 5G. The 60 GHz band is as of today used in communication for both backhaul applications, as well as WLAN applications. The IEEE 802.11ad WLAN communication standard uses the 60 GHz band[2]. The focus of future applications is often in the access network, providing connectivity to the end user devices such as smartphones and laptops. This is made possible due to the decrease in cost[3]. Most modern access network communication systems such as LTE or Wi-Fi utilize sub 6 GHz frequencies. There are several important differences between sub 6 GHz systems and millimeter wave systems. One difference is the physical size of the actual components. The size of an antenna is related to the wavelength. The fact that antennas in general are very small for 60 GHz systems allows for much more compact receiver and transmitter systems. At sub 6 GHz bands, a few hundred megahertz of unlicensed spectrum is available in most regions of the world, however at around 60 GHz several gigahertz of unlicensed spectrum is available. This gives the possibility of a very high throughput since the capacity is proportional to the available spectrum for high values of signal-to-noise ratio [4]

mmWave for 5G – FWA[5]

The total number of units that have one antenna or multiple RF transceivers, will in 2021 be 5.9 million and in 2028 60 million. Of which in 2021, the total of 15% are "base stations" [6]. The traditional macro bas station (BTS) will only cover 0.8% of the total market, while remote radio heads (RRH) cover 10% and small cells 3.6%. The main bulk will be CPEs with a market share of 85% (Customer Premises Equipment). A similar distribution is estimated for 2028.

This means that approximately 85% of the total number of RF modules that will be sold for the 5G mmWave market will be located in indoor or outdoor CPEs.

A 5G system includes multiple sub-systems which should be optimized and operate together. The RF subsystem transfers the radio signal all the way from the baseband signal (from the modem) into the air. How the RF solution is architected is very dependent on the use case and what you like to achieve, e.g. if you want to reach 150 meters or 1000 meters. There are four parts that sums up the total effective isotropic radiated power (EIRP) in an RF system (i.e the total "power" or "signal strength"):

- The RF power amplifier itself, i.e. its output power and linearity performance

- The number of power amplifiers tiled together in the antenna array
- The gain achieved from the antenna. For a given frequency, the antenna's effective area is proportional to the power gain, i.e. the size of the antenna is the main driver of the total gain that the antenna can achieve [7]
- Losses in the substrate, wave guides and packages

5G mmWave FWA infrastructure will be built mostly on small cells/RRHs connecting to CPEs. The main RF component volume will be on the CPE side which is close to 90% of all units sold. Using outdoor CPEs will probably be the most environmental friendly and the most cost-effective solution and the best fit for mmWave solutions, which is mainly a line of sight technology. According to Samsung[8], the size of the 5G system will be important, hence by bringing highly integrated 60 GHz technology to mmWave 5G, it would offer low cost, lower power consumption and the right size and solution to 5G marketplace.

II. TECHNOLOGY

Emerging WLCSP market of mmWave Applications

Market trends as experienced by end applications drive the emergence and evolution of any package technology. Currently, the primary automotive packaging solution is leaded or laminate wirebonding which account for more than 80% of the total assembly market.

The smallest possible package size is the Wafer Level Chip Scale Package (WLCSP), since the final package is no larger than the required circuit area. Since its introduction, WLCSP has experienced significant growth due to the small form factor, lower cost and high performance requirements of mobile and portable applications.

For RF and high-frequency applications, advanced wafer level packaging, eWLB (embedded wafer level BGA) showed less parasitic electrical interference, therefore, significantly improving overall device electrical performance. In one example, a 77GHz SiGe mixer packaged with eWLB achieved excellent high-frequency electrical performance due to the small contact dimensions and short signal pathways or interconnection length, which decreased parasitic effects [9,10].

The list below demonstrates the advantages of eWLB packaging solution for the mmWave device or high-frequency applications as compared to substrate or laminate-based packaging, such as flipchip or wirebonding.

1. **Interconnection length**: eWLB enables integration where the distance has to be as short as possible (loss increases with distance) to minimize loss (assuming both technologies have the same material loss).

2. **Conductance loss**: Plated Cu in organic substrate materials have large surface roughness because of the process used to improve adhesion and plating process control. eWLB uses a thin-film fab process for seed-layer and a well-controlled Cu plating to achieve a smooth Cu RDL surface which is more effective for skin effect in high-frequency ranges (At 100GHz, Cu skin depth is ~0.2µm).

3. **Dielectric loss**: Organic substrate materials have high losses in mm-Wave range and also heterogeneous material sets bring complexity in terms of electrical behaviors. eWLB uses molding compound and low-loss dielectric materials enabling achievement of less dielectric loss.

4. **Design flexibility**: eWLB provides more design flexibility for less routing interference with fine line width (LW) and line spacing (LS) capability (less than 10/10um LW/LS).

Traditionally, bare die technology is used where the die is attached with adhesive to the printed circuit board (PCB) and electrical contact is performed by wire bonding on the board. This challenging assembly has to endure several critical process steps: from bare die handling to shaping wire bonds in a way that RF requirements are met.

One key element for the change from a rather complex and expensive solution to an easy-to-use and, therefore, inexpensive and affordable product is the use of standard surface mount device (SMD) packaging technology. The eWLB package is SMD attached thereby simplifying the upstream assembly process and has already been proven in a few mmWave applications.

III. EXPERIMENTAL RESULTS

A. 60GHz WiGig eWLB assembly

eWLB package was designed in 12.6x12.6mm with 2-L RDL design utilizing low temperature curable advanced dielectric materials (ADM), which provide robust package reliability [11]. ADM is low temperature curable dielectric materials in eWLB process and it would be good for thermos-sensitive devices (embedded memory etc.) and less stress on the device with lower thermal budget.

As device performance is increased with high frequency, there is needs of multi-layer RDL for design flexibility in package level for system integration.

The specification details of each test vehicle are shown below in Table 1.

Table 1. Device specification of 60GHz eWLB.

Item	Description
PKG Type	eWLB FOWLP
PKG size	12.6x12.6mm
Lead count	314
Solder ball pitch	0.5mm
Ball size	0.30mm
Die Size	5.0x5.0mm
RDL Layer	2L
Fab node	SiGe
Die Thickness	0.40mm
PKG height	0.82mm
UBM	With UBM

 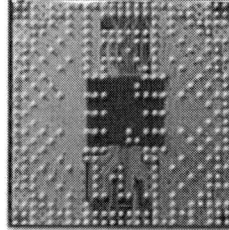

Figure 1. Beamforming transceiver for 60GHz WiGig applications.

B. Component Level Reliability

eWLB test vehicles were assembled and prepared for component level reliability tests according to Table 1 spec. Table 2 shows the package level reliability test conditions in this study. eWLB test vehicles is currently on-going with JEDEC standard package level reliability tests of MSL1, Temperature cycle (TC), high temperature storage life (HTS) and un-biased temperature humidity storage (uHAST).

Table 2. Package Level Reliability Results of eWLB with advanced dielectric materials (ADM).

Test	Test Condition	Test Conditions
Pre-Cond	JEDEC J-STD-020	MSL1
TC Temp. Cycling	JESD22-A104	Ta = -55/+125°C 1000 cycles
HTSL, High Temp. Storage Life	JESD22-A103	Ta=150°C 1000h
uHAST (w/o bias) after Precon	JESD22-A101	Ta=130°C, 85%RH 192h without bias

C. Electrical Performance and Characterization

The RF transitions in the package was designed in a 3D electromagnetic simulation software. This includes the chip-pad to RDL transition, the CPW type transmission-line structure within the package and the package to board transition. The RF design was first drafted in the three steps mentioned and then the complete chain, from microstrip on chip to microstrip on PCB was simulated for one channel. Finally the complete RF layout of the package was imported to the simulation software and a verification simulation with all paths and their interactions were performed. A picture of the RDL to PCB transition model can be seen in Fig. 2. The simulated package loss is approximately 2 dB over the band of operation.

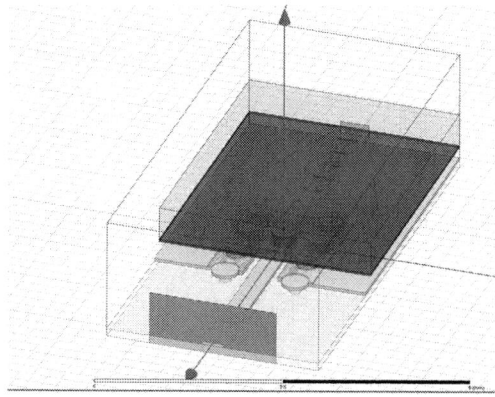

Figure 2. Sub-set of the 3D package model including coplanar waveguide (CPW) port, transition from package to PCB and microstrip on the PCB

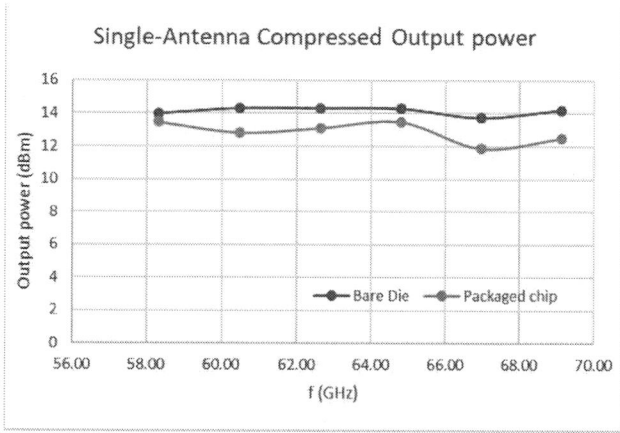

Figure 3. Comparison of single-antenna compressed output power between bare die and packaged chip

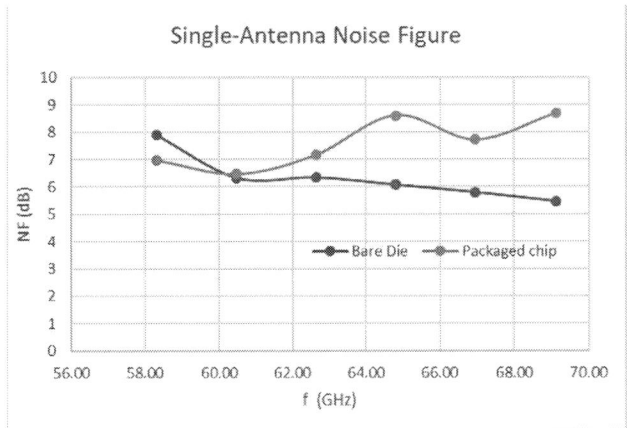

Figure 4. Comparison of single-antenna noise figure between bare die and packaged chip

To evaluate the performance of eWLB package, the compressed output power and noise figure of one single-antenna element is measured at the center frequency of each WiGig (802.11ad) channel for both bare die and package chip at board level with probe, as shown in Fig. 3 and 4. The difference in output power between bare die and packaged chip represents the actual loss of eWLB package including the PCB transition. Due to different source and load impedances when the bare die is probed compared to when it is packaged the difference shown in the figures varies slightly.

IV. CONCLUSION

60GHz beamforming transceiver eWLB was assembled and characterized for its reliability and electrical performance of eWLB FOWLP with 0.5mm ball pitch and 2-L RDL of ADM.

1. eWLB test vehicles passed JEDEC standard reliability tests, MSL-1, TC, HTS, uHAST with 2-L RDL structure.
2. Simulation of the complete RF transitions in the package for one channel shows 2dB loss over the WiGig (802.11ad) band. The compressed output power and noise figure for both bare die and packaged chip are measured and compared.

Furthermore, factors such as superior high frequency electrical performance and the ability to enable heterogeneous integration such as the integration of Antenna, ie. AiP (Antenna-in-Package). into the various thin-film layers, active/passive devices into the mold compound or encapsulation, and achieve 3D vertical interconnections for new 3D SiP and 2.5D/3D packaging solutions, differentiate eWLB from other packaging technologies. eWLB technology provides a more holistic performance relevant to an increasingly broad range of emerging applications including 5G mmWave applications with AiP.

ACKNOWLEDGMENT

The authors would like to express their sincere appreciation to the eWLB NPI and R&D team of STATS ChipPAC Pte Ltd for their helpful advice and support of this work.

REFERENCES

[1] K.-C. Huang and D. J. Edwards, "Gigabit wireless communications", in Millimetre Wave Antennas for Gigabit Wireless Communications:A Practical Guide to Design and Analysis in a System Context. Wiley Telecom, 2008, pp. 286–, isbn: 9780470712467. doi: 10.1002/9780470712467.ch1.]

[2] "ISO/IEC/IEEE international standard for information technology–telecommunications and information exchange between systems–local and metropolitan area networks–specific requirements-part 11: Wireless LAN medium access control (MAC) and physical layer (PHY) specifications amendment 3: Enhancements for very high throughput in the 60 GHz band (adoption of IEEE std 802.11ad-2012)", ISO/IEC/IEEE 8802-11:2012/].

[3] K. Nguyen, M. G. Kibria, K. Ishizu, and F. Kojima, "Empirical investigation of IEEE 802.11ad network", in 2017 IEEE International Conference on Communications Workshops (ICC Workshops), May 2017, pp. 192–197. doi: 10.1109/ICCW.2017.7962656], which allows the communication modules to be implemented in devices.

[4] D. Tse and P. Viswanath, Fundamentals of Wireless Communication. Cambridge

[5] https://www.siversima.com/news/mmwave-for-5g-fixed-wireless-access-a-review/

[6] SNS research report "5G for FWA (Fixed Wireless Access): 2017 – 2030"

[7] https://en.wikipedia.org/wiki/Antenna_gain

[8] https://www.fiercewireless.com/tech/samsung-touts-breakthrough-5g-ready-antenna-power-amplifier-at-28-ghz

[9] M. Wojnowski1, M. Engl1, B. Dehlink:; G. Sommer', M. Brunnbauer, K. Pressed and R. Weigel, "A 77 GHz SiGe mixer in an embedded wafer level BGA package", Proceedings of 58th Electronic Components and Technology Conference, 2008. ECTC 2008. (2008)

[10] G. Haubner, W. Hartner, S. Pahlke, M. Niessner, "77 GHz automotive RADAR in eWLB package: From consumer to automotive packaging", Microelectronics Reliability, September 2016, DOI: 10.1016/j.microrel.2016.07.104. (2016)

[11] S.W. Yoon, Tom Strothmann, Yaojian Lin and Pandi C. Marimuthu, "Robust Reliability Performance of Large size eWLB (Fan-out WLP),' iMAPS Device Packaging Conferences 2013, Pheonix, Arizona (2013).

978-1-7281-1500-9/19 $31.00 © 2019 IEEE

Low-Loss Additively-Deposited Ultra-Short Copper-Paste Interconnections in 3D Antenna-Integrated Packages for 5G and IoT Applications

Atom O. Watanabe[1], Yiteng Wang[1], Nobuo Ogura[2], P. Markondeya Raj[1,3],
Vanessa Smet[1], Manos M. Tentzeris[1], and Rao R. Tummala[1].
[1]3D Systems Packaging Research Center, Georgia Institute of Technology, Atlanta, GA USA.
[2]Nagase & Co., LTD., Tokyo, Japan.
[3]Florida International University, Miami, FL USA.

Abstract—High-bandwidth 5G and 6G communication systems will inevitably migrate to 3D package architectures with backside or embedded dies and antenna-integrated packages for ultra-low losses and smaller footprints. With the trend to such 3D millimeter-wave (mm-wave) packages, the losses from the assembly and through-vias tend to dominate the overall losses. Traditional wirebond and thick solder interconnections lead to large mm-wave interconnect losses that are not acceptable for emerging 5G and 6G communications. This paper focuses on the material syntheses and process development of nanocopper interconnections with ultra-low interconnect losses for chip-last or flip-chip assembly in packages. The first part of the paper introduces the material synthesis of an innovative copper paste with shorter sintering times and temperatures. Optimized conditions are obtained to attain a conductivity of 1.4×10^7 S/m. This is equivalent to 82% increase in conductivity compared to that of solder. The surface roughness is also measured through atomic-force microscopy. Results suggest that the copper paste features higher roughness than that of solders. The second part of this paper discusses the potential of novel nanocopper paste to replace solders as a package assembly material, focusing on the effect of the conductivity and surface roughness with regard to the insertion loss in interconnection bumps. Based on the improved material properties of nanocopper paste, the model shows a 53% reduction in the dB scale at 28 GHz, by employing nanocopper paste. Die shear test for copper paste is also performed to show a high potential to replace solders as a flip-chip assembly material in both printed-circuit-board and mm-wave packaging technologies.

Index Terms—5G communications, Millimeter wave, Low loss interconnect, Interconnect bump, Copper paste, Antenna in package.

I. INTRODUCTION

Ongoing changes in the way humans interact with and consume data is anticipated to create an explosive growth in the number of devices connected to each other as well as to the internet. The bandwidth in electronic communications and wireless sensing in automotive safety, intelligent navigation, wireless sensing and terahertz (THz) communication, and smartphone-like infotainment, communication data rates are projected to be at least 10-100X higher than existing 4G LTE connections [1], [2]. With emerging 5G and 6G communications, trillions of devices with hundreds of radios per person are projected, leading to THz communications

with 100 Gbps of bandwidth. In order to realize such systems, heterogeneous active devices and passive components need to communicate within the package and with the outside world in a seamless manner. This trend demands innovative assembly materials or interconnection methods to enable such applications. Conventional methodologies employ wire bonding to form interconnections between the chips and package substrate. However, wire bonds entail more difficult designs in higher frequencies in both IC (e.g., a switch and low-noise amplifier) and package designs because of the increasing parasitic inductance, which causes higher noise figure due to more discontinuity in signal paths and degrades the overall performance. Additionally, wire-bonding methods require more space for landing pads, which makes the miniaturization of IC packages more difficult. Flip-chip technology using solder paste, copper pillar interconnections [3], and all copper [4] addresses the challenges with short interconnection lengths and smaller parasitic inductance [5]. Fan-out packaging is emerging as an alternative approach but flip-chip on low-loss laminate still prevails as the mainstream approach [6] to realize 5G and 6G modules because of its design, material, and process flexibility and simplicity, as shown in Figure 1.

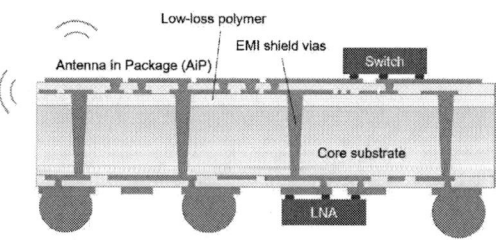

Fig. 1. Antenna-integrated packages with mm-wave chips assembled.

Because of the relatively-low conductivity of solder and process complexity associated with planarization and surface smoothness requirements of direct copper-to-copper interconnections, copper paste is emerging [7], [8] as an ideal material candidate in high-frequency low-loss interconnection assembly. Copper paste shows higher conductivity than solder, given oxidation of copper particles is prevented. The viscosity of copper paste also enables various assembly methods such as stencil printing and pin transfer. However, copper paste has not been investigated as a chip-assembly material for 5G or millimeter-wave (mm-wave) applications..

This paper discusses the potential of copper paste as an off-chip interconnection material for high-frequency applications such as 5G and mm-wave communications and RADAR systems. Section II introduces the significance of highly-conductive materials through electrical simulations of flip-chip interconnect bumps between a chip and package. Section III discusses the materials syntheses of various copper pastes, evaluating electrical conductivity, the size of copper particles, and surface roughness, optimizing the sintering conditions. In order to quantify the signal losses caused by interconnect bumps with the synthesized materials, Section IV focuses on the potential of copper paste using fabricated test vehicles with the synthesized materials.

II. MODELING AND DESIGN OF C4 BUMPS EMPLOYED FOR CHIP ASSEMBLY

The objective of this task is to quantify signal losses in interconnect bumps in the frequency bands mainly used for mobile and radar communications. High-frequency signals were designed, as illustrated in the inset of Figure 2. The signal travels from the chip to the package-integrated antenna arrays, through the 50-Ω conductor-backed coplanar waveguide (CPWG) and landing pads. The diameter of the bumps is designed to be 100 μm. In the 3D full-wave electromagnetic (EM) simulation, it is assumed that three frequency bands were chosen and the conductivity of the interconnect bumps was swept from 7.69×10^6 S/m (lead-free solder: SAC305 [9]) through 5.81×10^7 S/m (bulk copper) for twenty bumps in a chain. The signal losses were plotted in Figure 2, implying that high conductivity leads to lower signal losses in interconnect bumps because of the lower signal attenuation and lower heat generation. The results also show that the signal losses at higher frequencies (i.e., 28 and 60 GHz) are more critical than those in the lower frequencies (i.e., 2.4 GHz). Therefore, higher conductivity is essentially required to reduce signal losses and noise figure from the electrical standpoint. The reduction in signal losses lowers the link budget and allows design flexibility in other elements in mm-wave antenna-in-packages.

III. MATERIAL SYNTHESIS OF COPPER PASTE

Higher conductivity with copper paste makes it a compelling alternative to solder in the form of Sn/Ag/Cu (SAC) [9]. However, the major challenge with copper paste is the oxidation occurring on surface of copper particles during the sintering process. This section discusses the design of copper paste interconnections to obtain high conductivity and smooth surface to lower insertion losses in interconnection bumps.

Copper paste (A) with large copper particles with diameters of approximately 1 μm was chosen. After sintering the copper paste A, the electrical conductivity was measured using the four-point probe method, also known as Kelvin method. The copper paste A was annealed at sintering

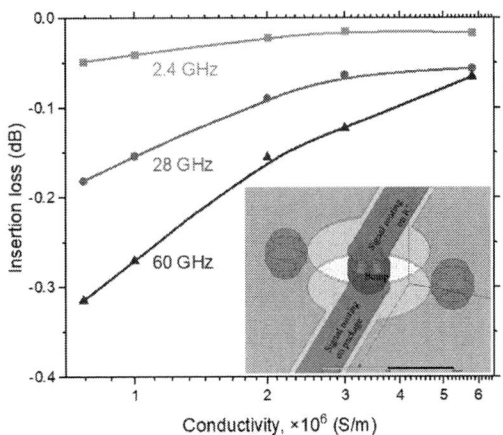

Fig. 2. Insertion loss in interconnect bumps at widely-used frequency bands for RF and 5G/mm-wave applications.

temperatures, T_s, of 200°C and 230°C for half an hour to evaluate the conductivity. Figure 3 depicts the scanning-electron-microscope (SEM) images of the annealed copper paste A at the different temperatures. It is observed in the SEM images that copper particles are not fused with adequate neck formation, which resulted in low conductivity, as listed in Table I. The conductivities under the sintering temperatures were lower than that of solder in the form of SAC305.

TABLE I

MEASURED CONDUCTIVITY OF THE COPPER PASTE A AT TWO SINTERING TEMPERATURES.

	Sintering temperature		
	200°C	230°C	260°C
Conductivity (S/m)	2.14×10^6	3.41×10^6	5.90×10^6

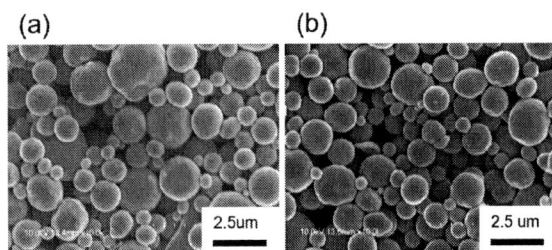

Fig. 3. SEM images of the copper paste A after sintering for 30 minutes at (a) 200°C and (b) 230°C.

To obtain higher conductivity than solder, another copper paste was selected; this copper paste B is bimodal paste, where large and small copper particles coexist and promote high green density in addition to an organic reducing agent. The reducing agent prevents copper from oxidation below 100°C and decomposes at around 100°C. It also reduces the viscosity of copper paste and enables various assembly methodologies such as pin transfer. Sintering under nitrogen also prevents copper particles from oxidation. For the optimization of sintering conditions, two parameters (i.e., time t_s and temperature T_s) were varied. To evaluate the copper paste B sintered under each condition, electrical conductivity was measured through the four-point probe method. The results are summarized in Figure 4, comparing to the conductivity of solder in the form of SAC305. The results indicate that the sintering temperature is more critical to the conductivity than the sintering time. The copper paste sintered at 260°C showed higher conductivity than the solder (7.69×10^6 S/m) regardless of the sintering time. The copper paste B sintered for 30 minutes at 260°C showed a conductivity of 1.4×10^7 S/m, which is 82% higher than that of solder. It is found that, controlling the sintered temperature is more significant than the time, implying that the reduction of sintering time with higher sintering temperatures will decrease the time of the total process of chip assembly.

Fig. 5. SEM images of the copper paste B sintered for 30 minutes at (a) 200°C, (b) 230°C, and (c) 260°C.

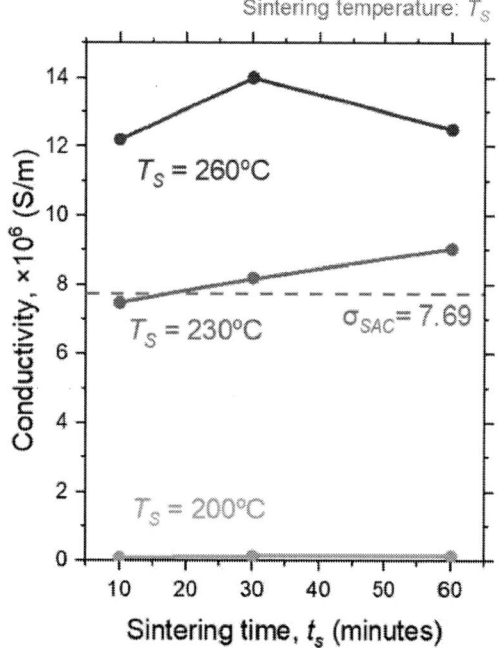

Fig. 4. Measured conductivity of the copper paste B with various sintering time and temperature annealed in the nitrogen gas.

Since the increase in the sintering temperature provided a remarkable improvement of conductivity, we also observed the necking of the copper paste B at each sintering temperatures (200°C, 230°C, 260°C) through SEM. The SEM pictures, as illustrated in Figure 5, show that copper particles at 230°C and 260°C have more interparticle neck formation than the one sintered at 200°C. Copper nano-particle necking is especially observed in Figure 5 (c). Compared to Figure 3, the areas of copper particles sintering are much higher in the copper paste B (Figure 5).

Signal losses are attributed to the ratio of the surface roughness (Δ) to the skin depth (δ). As the surface roughness has a negative impact on the signal loss, the surface

roughness analysis was performed through atomic-force-microscope (AFM), which provides a very high resolution on the order of fractions of nanometer, more than 10^3 times better than the optical diffraction limit. The roughness causes the conductor loss. Liang *et. al.* [10] identified a good agreement between the attenuation factor between a conductor loss (α_c) and surface roughness (Δ) with a positive correlation in the form of $1 + \frac{2}{\pi} \arctan \left[1.4 \left(\frac{\Delta}{\delta} \right)^2 \right]$. The AFM images are illustrated in Figure 6, showing in the roughness average, R_a, of 162 nm and 81 nm for the copper paste B sintered for 30 minutes at 260°C and the re-flowed solder paste, respectively. While the solder paste offers relatively smooth surface due to the melted and re-formed solder, the roughness of the sintered copper paste B is primarily attributed to the copper bimodal particles with the size of roughly 100 nm, as depicted in Figure 5. The impact of these surface roughness is discussed in Section IV.

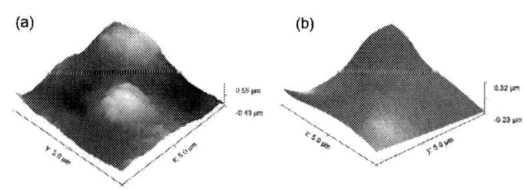

Fig. 6. AFM images of (a) the copper paste B sintered for 30 minutes at 260°C and (b) a reflowed solder paste.

IV. CHARACTERIZATION OF COPPER PASTE FOR CHIP ASSEMBLY

This section discusses the demonstration of chip-last flip chip assembly using the copper paste B. To verify the potential as the chip assembly material, die shear test is also performed for reliability.

A. Die assembly with copper paste

The daisy-chain structure with a test-chip, utilized in the simulation depicted in the inset of Figure 2, was also employed as the test vehicle. The test vehicles are fabricated through the semi-additive patterning (SAP) process, which provides sub-10μm patterning. The fabricated test vehicles, shown in Figure 7, consist of conductor-backed coplanar waveguides (CPWG) with a spacing of 15 μm and landing pads with a diameter of 100 μm.

Fig. 7. Top view of the fabricated test vehicle with CPWG and landing pads for flip-chip assembly.

The fabricated test vehicles are assembled through the pin-transfer method, where the copper paste B is transferred from a diameter-controlled pin (80 μm) onto 100-μm landing pads, as illustrated in Figure 8. Once copper paste is provided onto landing pads on the fabricated package, the test-chip is assembled, aligned with the landing pads of both the package and die. This assembly was performed with a pick-and-place assembly machine with an accuracy below 5 μm. The pair of the package and die was annealed under the optimized condition (t_s=30 minutes and $T_s = 260°$C) discussed in Section III.

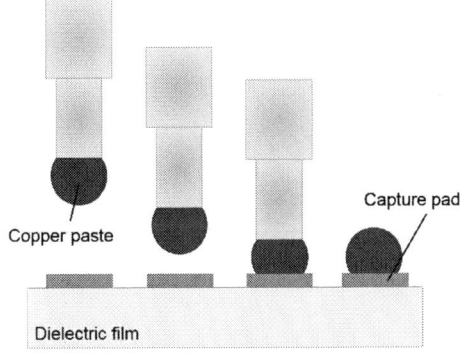

Fig. 8. Pin transfer assembly with a 80- pin for transferring copper paste onto 100-μm landing pads.

The cross-sectional images of the assembled test vehicle at two different locations are shown in Figure 9. The copper paste is observed to deposit within the 100-μm copper pads with diameters of approximately 90 μm. The height of the copper-paste interconnects (4–6 μm) is much shorter than other conventional methods such as C4 and C2 bumps.

These shorter interconnects will suppress the parasitic inductance and unwanted electromagnetic interference occurring in bumps. In addition, the accuracy of copper-paste interconnect size is enabled by the control of the amount of copper paste deposited during the pin transfer (Figure 8).

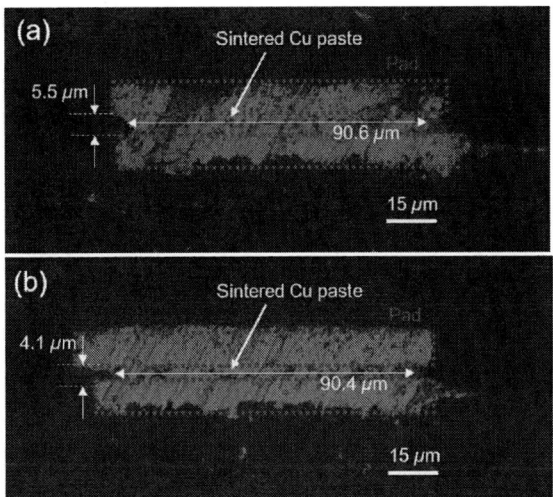

Fig. 9. Cross sectional images through optical microscope at two different locations of the assembled test vehicle with the copper paste B sandwiched by 100-μm copper landing pads.

To quantify the insertion losses in the bumps with the copper paste, the fabricated test vehicle, which includes twenty bumps in a chain, was modeled in the 3D full-wave EM simulator (HFSS). In these simulations, the measured conductivity and surface roughness (Section III) were considered. To verify the chip assembly performance using copper paste, the signal loss when employing solder-paste was also computed. The simulations, summarized in Figure 10, showed an insertion loss of -0.118 dB with copper paste at 28 GHz while the use of solder paste led to an insertion loss of -0.181 dB. Although copper paste features higher surface roughness, this result indicates that the improvement of conductivity reduced signal losses by 53% by employing nanocopper paste.

B. Die Shear Test for Reliability

As an interconnection assembly material in packages, reliability is one of the most significant parameters. Reliability is assessed through a die-shear test, which determines the strength of the system component assembly when subjected to force. A controlled amount of the copper paste B was deposited with pressure, onto a gold surface-finished silicon die, so that the copper paste forms a 6-mm bump shape. A force was applied in the transverse direction using a tensile strength tester, and the material failure occurred at 6.24 MPa before the separation of copper paste from the gold deposited silicon die, as shown in Figure 11. Dark-color circles are the sintered copper paste after the die shear test. This result demonstrates the strong adhesion between the copper paste and gold.

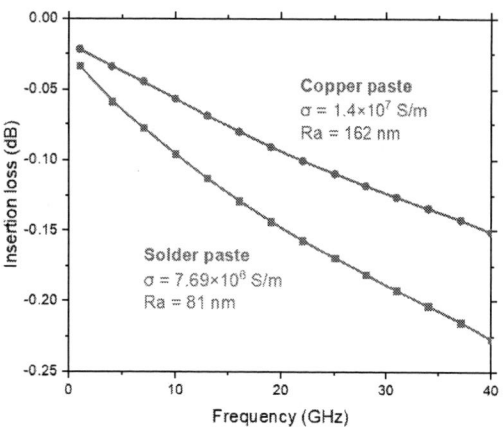

Fig. 10. Insertion loss as a function of frequency, computing the measured conductivity and surface roughness of the copper pate B and solder paste in the form for SAC305.

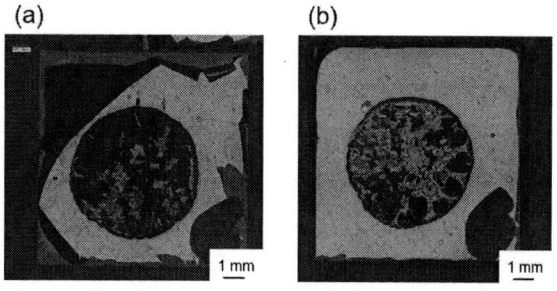

Fig. 11. Microscope pictures of (a) the top die and (b) the bottom die. The black circles

V. CONCLUSIONS

There is increasing need for low-loss ultra-wideband interconnects in the 5G and 6G mm-wave frequency bands to attain seamless chip-to-antenna transitions. This paper focuses on the modeling, material syntheses and process development of off-chip nanocopper interconnections as an alternative to solders. The benefits in conductivity, interconnection height and equivalent surface roughness resulted in a lower insertion loss in the chip-to-package interconnections compared to solder. The first part of the paper discussed the material synthesis of copper paste, optimizing the sintering time and temperature. The optimized conditions with the current nanocopper paste were found with a sintering temperature of 260°C for 30 minutes, which led to a conductivity of 1.4×10^7 S/m. This is equivalent to 82% increase in conductivity compared to that of solder in the form of SAC305. The surface roughness was also measured through atomic-force microscopy. The copper paste showed twice higher roughness than that of solder. The second part of this paper focuses on the potential of copper paste to replace solder as a package assembly material. Computing the material properties of copper paste and solder, the model showed a 53% reduction in the dB scale, at 28 GHz, by employing copper paste. Mechanical reliability was also assessed through a die-

shear test. Copper paste was shown to have a high potential to replace solder as a flip-chip assembly material in printed-circuit-board and packaging technologies.

REFERENCES

[1] A. Rashidian, S. Jafarlou, A. Tomkins, K. Law, M. Tazlauanu, and K. Hayashi, "Compact 60 ghz phased-array antennas with enhanced radiation properties in flip-chip bga packages," *IEEE Transactions on Antennas and Propagation*, 2018.

[2] A. Watanabe, T. Lin, P. M. Raj, V. Sundaram, M. M. Tentzeris, R. R. Tummala, and T. Ogawa, "Leading-edge and ultra-thin 3d glass-polymer 5g modules with seamless antenna-to-transceiver signal transmissions," in *2018 IEEE 68th Electronic Components and Technology Conference (ECTC)*, May 2018, pp. 2026–2031.

[3] I. Panchenko, M. Kunz, J. M. Mathias Boettcher, L. Lehmann, T. Atanasova, and M. Wieland, "In-line metrology for cu pillar applications in interposer based packages for 2.5 d integration," 2017.

[4] K. Mohan, N. Shahane, R. Sosa, S. Khan, P. M. Raj, A. Antoniou, V. Smet, and R. Tummala, "Demonstration of patternable all-cu compliant interconnections with enhanced manufacturability in chip-to-substrate applications," in *2018 IEEE 68th Electronic Components and Technology Conference (ECTC)*, May 2018, pp. 301–307.

[5] N. Li, Y. Wu, L. Chen, B. Wu, C. Zhao, Y. Wang, and Y. Sun, "A novel bump-cpw-bump structure for interconnection/transition of rf mems packaging," in *Electronic Packaging Technology (ICEPT), 2015 16th International Conference on*. IEEE, 2015, pp. 845–847.

[6] L. Del Carro, M. Kossatz, L. Schnackenberg, M. Fettke, I. Clark, and T. Brunschwiler, "Laser sintering of dip-based all-copper interconnects," in *2018 IEEE 68th Electronic Components and Technology Conference (ECTC)*. IEEE, 2018, pp. 279–286.

[7] A. A. Zinn, R. M. Stoltenberg, J. Chang, Y.-L. Tseng, S. M. Clark, and D. A. Cullen, "A novel nanocopper-based advanced packaging material," in *Electronics Packaging Technology Conference (EPTC), 2016 IEEE 18th*. IEEE, 2016, pp. 1–6.

[8] Y. Kamikoriyama, H. Imamura, A. Muramatsu, and K. Kanie, "Ambient aqueous-phase synthesis of copper nanoparticles and nanopastes with low-temperature sintering and ultra-high bonding abilities," *Scientific reports*, vol. 9, 2019.

[9] S. Nai, J. Wei, and M. Gupta, "Effect of carbon nanotubes on the shear strength and electrical resistivity of a lead-free solder," *Journal of Electronic Materials*, vol. 37, no. 4, pp. 515–522, Apr 2008.

[10] T. Liang, S. Hall, H. Heck, and G. Brist, "A practical method for modeling pcb transmission lines with conductor surface roughness and wideband dielectric properties," in *2006 IEEE MTT-S International Microwave Symposium Digest*. IEEE, 2006, pp. 1780–1783.

Advanced Thin-Profile Fan-Out with Beamforming Verification for 5G Wideband Antenna

Sheng-Chi Hsieh, Fu-Cheng Chu, Cheng-Yu Ho and Chen-Chao Wang

Electrical Laboratory, Corporation Design Division, Corporate Research and Development, Advanced Semiconductor

Engineering (ASE), Inc., No. 26, Chin 3rd Road Nantze Export Processing Kaohsiung 811, Taiwan

Email: Ricky_Hsieh@aseglobal.com

Abstract— In this paper, the fabrication process of AiP is based on Fan-Out chip on substrate. This solution brings short interconnection between die to die and die to antenna for excellent electrical performance. They tend to offer the lowest transition loss due to millimeter-wave front-end circuitry directly connected to the antenna. The packaging technology has a number of important features. Package height is slightly less than 0.75 mm, with three RDL layers, two mold layers for antenna design, mold thickness of 250um, and ball height of 200um. Its benefits include a smaller package footprint compared to conventional organic substrate or laminate packages, higher maximum connection density, as well as desirable electrical and thermal performance. In this work, the coupling patch antenna is implemented to Fan-out package. Based on design theory of antenna array, proper design requires careful optimization of the spacing, length, and width of antenna array. By choosing the antenna parameters properly, the proposed coupling patch antenna on fan-out package has better than 10dB return loss in 26-33GHz range, with ~7GHz bandwidth and provides a high-gain (above ~10.3 dBi) radiation pattern with 2x2 patch antenna array. Beamforming array is a very important technique in 5G application. There are many behaviors needed to be figured out in a realistic beamforming model, such as mutual coupling effect, which will affect data transmission quality and even degrade the expected beamforming pattern. Mutual coupling effect is analyzed in this study. The proposed thin profile AiP solution can achieve both broadband and high directivity characteristics, which meet the requirement of 5G systems in the allocated frequency bands at 28 GHz applications.

Keywords—28 GHz 5G systems; flip-chip ball grid array (FCBGA) package; flip chip chip scale (FCCSP) package; Fan out wafer level chip scale package (Fan-out WLP); Advanced single sided substrates (fan-out package); Antenna in package (AiP).

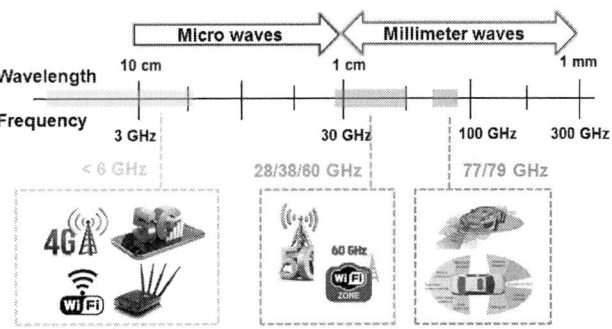

Fig. 1. Comparison of wavelength in air at 28 GHz, 60GHz and 77 GHz mmWave applications.

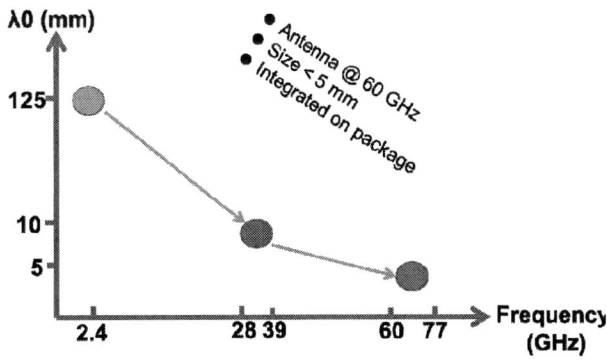

Fig. 2. Comparison of wavelength in air at 28 GHz, 60GHz and 77 GHz mmWave applications

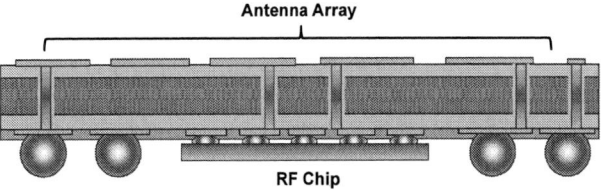

Fig.3. Cross-section view of the mmWave transceiver on flip chip ball grid array (FCBGA) AiP.

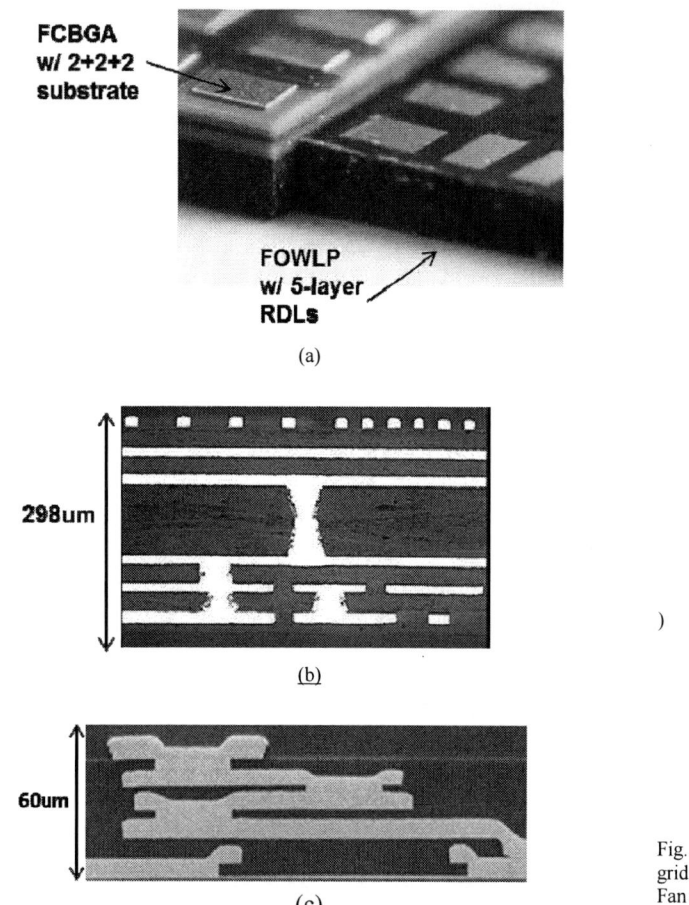

Fig. 4 (a) Photograph of FCBGA and Fan-out WLP (b) SEM cross-section of FCBGA (c) SEM cross-section of Fan-out WLP

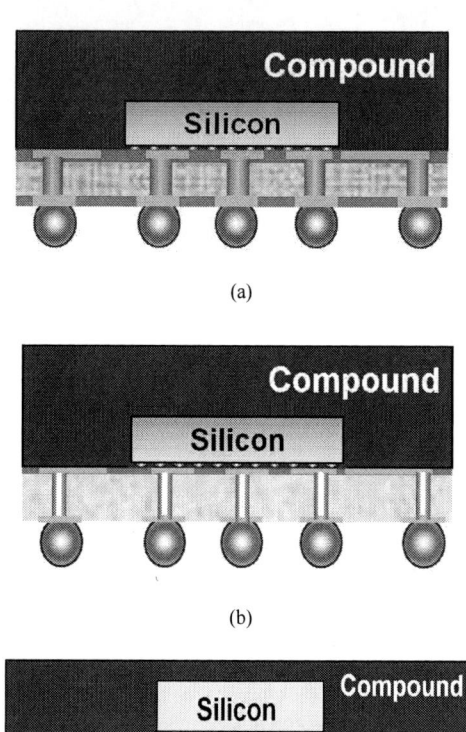

Fig. 5 Cross-section schematic of different package type. (a) Flip-chip ball grid array (FCBGA) package. (b) Flip chip chip scale (FCCSP) package. (c) Fan out wafer level chip scale package (Fan-out WLP)

I. INTRODUCTION

The recent trends have significantly increased in demands for millimeter-wave wireless communication systems as shown in Fig. 1. The potential applications include 28/39-GHz 5th mobile generation 60-GHz high-speed wireless data link, 77-GHz automotive radar and 94-GHz radar imaging, etc[1]-[2]. In order to enable low-cost and high performance millimeter-wave solution, integration for antennas and ASIC, which requires short interconnection and low transmit loss, are highly desired. For example, both LTE and sub-6 GHz within 1-6 GHz frequency range can coexist with large distance between antenna and multi-mode RF transceiver. Usually, the antenna has been designed separately from the OEM and manufactured in a different technology. In the microwave frequency range, the physical wavelength is typically too large to allow antennas be placed on PCB substrate or phone case of reasonable sizes. Hybrid circuits with the antennas on PCB also often are used at low frequencies due to low conductive loss. Connectors and RF cables have generally been used to interface the antenna and RF module. However, as the operating frequency of the mobile systems increases to millimeter-wave, it is often severely degraded. Thus, the

complexities of the packages and interconnections are raised since it is required to integrate antenna and RF chips to improve performance by reducing interconnected distance. One solution is the millimeter wave antenna in package (AiP) to integrate antenna with RF ICs into chip-scale package devices. As shown in Fig. 2, the physical wavelength in millimeter-wave range is smaller than current commercial systems including GSM, LTE, and Wi-Fi. The more compact antenna size makes it suitable for further system in package integration. AiP offers a solution for interconnect and antenna implementation problem by integrating the adaptive antennas to the RF chips in the package. It can reduce the path loss and allow miniaturization of wireless systems. Furthermore, as mobile and wearable devices are transferred to 5G platforms, they are required to obtain high data rate communication, while still needing to meet high performance requirements for mobile devices. Even if only one antenna element isn't fulfilled, such as not having enough gain, it will be difficult to provide high quality transmission. A possible high gain antenna can be achieved by antenna array designed with directivity radiators. An increased number of antenna elements

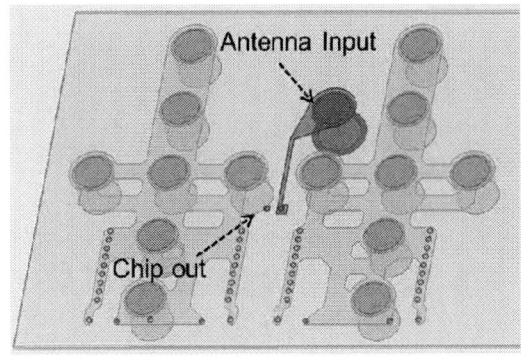

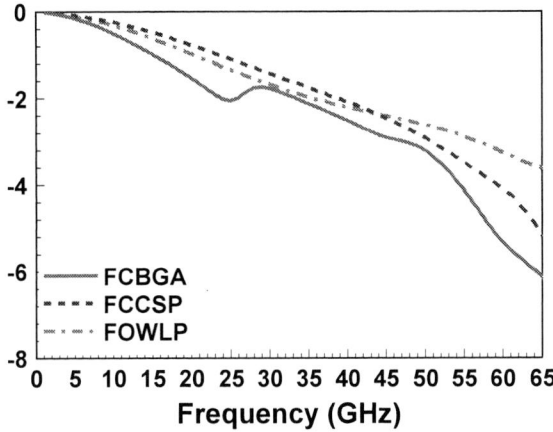

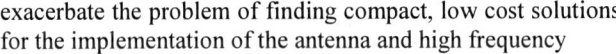

Fig. 6. Comparison of insertion loss for single-end transmission line between FCBGA package, FCCSP package and Fan-out WLP

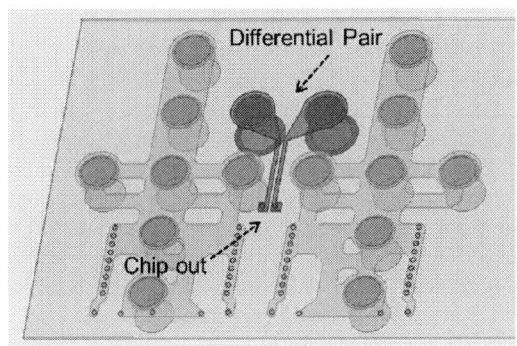

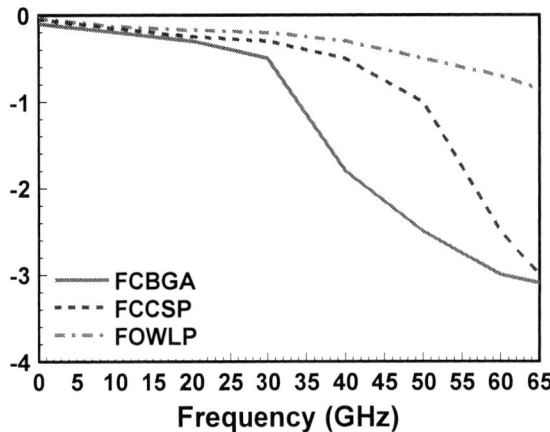

Fig. 7. Compairson of insertion loss for differential-pair transmission line between FCBGA package, FCCSP package and Fan-out WLP

exacerbate the problem of finding compact, low cost solutions for the implementation of the antenna and high frequency

interconnects. Although AiP is highly integrated with the front-end circuitry, this technology essentially takes advantage of high performance material, such as low- temperature co-fired ceramic (LTCC) [3], liquid crystal polymer (LCP), Teflon polytetrafluoroethylene (PTFE) [4],
ceramic-filled PTFE (Rogers RO3003) [5], and glass-reinforced ceramic (Rogers RO4000) to realize broadband highly efficient antennas because of their low-k and low-loss properties at millimeter-wave frequencies. Low-cost AiP substrate solutions are very important for market success. Little work has been published on cost-effective antenna array implemented in packages. One of the effective solutions is implemented on organic substrate[6]-[8]. It is a low-cost and low loss multilayer organic substrate for flip chip ball grid array (FCBGA) package as shown in Fig. 3. However, the thick substrate thickness is not easy to mount in thin mobile phone case. In addition, 5G mmWave systems create significant challenges for packaging engineers since the power consumption caused by high date rate at mmWave is obviously from active device. The thermal issue of interface on PCB is a very serious challenge to the 5G millimeter-wave

systems. A novel solution for advanced packaging is the Fan out wafer level chip scale package (Fan-out WLP). Fan-out WLP is a new generation platform that will support die to die integration, particularly for heterogeneous high-density IO system in a package [9]-[11]. Its benefits include a smaller package footprint compared to conventional lead-frame or laminate packages, maximum connection density, as well as desirable electrical and thermal performance. Fig. 4 shows the comparison of package thickness between FCBGA and FOWLP. After that, fan-out WLP offers small form factor, excellent electrical and thermal performance for mmWave AiP into mobile devices. In addition, foundry-like process is another approach to reduce process variation and increase yield. A stacking patch antenna with broadband frequency has been widely used in wireless application. A slot tuning has been proposed in the antenna performance improvement. This work proposes a stacking patch antenna with slot structure to increase the bandwidth and directivity characteristics, which meets the requirement of 5G millimeter-wave application.

II. LOSSES OF TRANSITION FROM CHIP TO PACKAGE

As you know, millimeter-wave system has a larger interconnection loss than sub-6 GHz systems. Thus millimeter-wave systems need to overcome the significant challenges for packaging engineers since the interconnection of the package may induce more losses of transition from antenna to chip side. The losses of transition from chip to package's pin out and chip to antennas are a serious issue to the millimeter-wave systems for commercial applications. In this paper, we study losses of the transition from chip to package including flip-chip ball grid array (FCBGA) package, flip chip scale (FCCSP) package and Fan out wafer level chip scale package (Fan-out WLP). Schematically depicting the cross-section of FCBGA package, FCCSP package and Fan-out WLP are show in Fig. 5 (a)-(c), respectively. Fig. 6 shows the 3D electrical modeling and EM simulation for single-end transmission line on millimeter-wave package systems. Based on the EM model, we employ a single-end transmission line to study the loss that is fromRF chip to patch antenna input port in millimeter wave AiP system. Figure 6 is also shows a comparison of insertion loss results by HFSS simulation with single-end transmission line for FCBGA package, FCCSP package and Fan-out WLP, respectively. Figure 7 shows the EM simulation model with differential-pair transmission line, it is a RF chip to dipole antennas in millimeter wave AiP system. Because of the shortest interconnection in Fan-out WLP, they have lower loss of transition from chip to package than FCBGA and FCCSP. As shown in Fig. 6 and Fig. 7, the single-end and differential-pair transmission line implemented on Fan-out WLP have lower insertion loss for chip to antenna array. Based on this study, the fan-out WLP not only can offer small form factor but also obtain excellent electrical performance for mmWave AiP into future 5th Gen. mobile devices. Its small form factor of AiP solutions are important prerequisites for millimeter-wave market.

III. FANOUT ANTENNA IN PACKAGE (AiP) DESIGN

The proposed fan-out structure is shown in Fig 8. The packaging technology has a number of important features. Package height is slightly less than 0.75 mm, with three RDL layers. It has two mold layers for antenna design, with a mold thickness of 250um, and ball height of 200um. Based on this structure, a 5G AiP implemented on fan-out package has a thin substrate thickness than when implemented on FCBGA and FCCSP. This work proposed a patch antenna design in fan-out package for 28GHz band. Patch antenna consists of a patch of metal layer with a large ground. There are several advantages to employ patch antenna for mobile applications including low profile, high gain and high compatibility for packaging with IC. However, there is a problem in patch antenna for wireless application. It is well-known that patch antenna have very narrow band. In this paper, we propose stacking patch to obtain broadband antenna design as shown in Fig. 9. In the past years, several ideas have been published to increase the bandwidth. One effective idea to improve this issue is by adding a parasitic element above the main patch. In general theory of stacking patches, it is realized by electromagnetic coupling to add bandwidth. In this design, the parasitic element is located on top mold layer above main patch with a thickness of mold layer 250 μ m. The large ground is located at bottom of

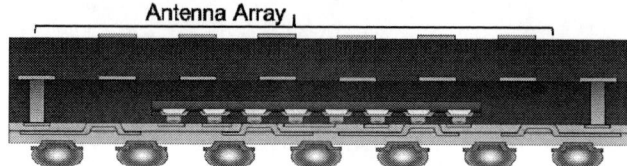

Fig. 8. Geometry of the fan-out Antenna in Package

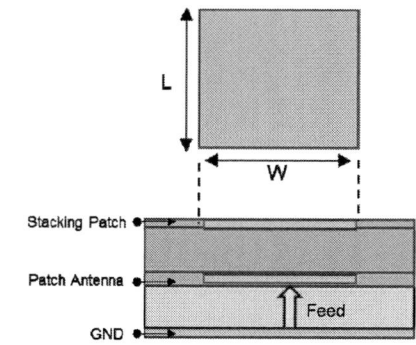

Fig. 9. The Schematic of propose stacking patch antenna

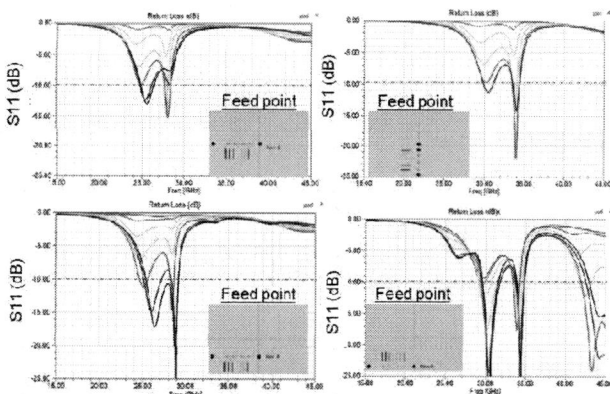

Fig. 10. Varrying return loss by adjusting the position of the feed point

mold layer. Based on design theory of a patch antenna, proper design requires careful optimization of the feeding point, and width of antenna. The design of the feeding structure is directly connected to main patch. However, the feeding point has to match the antenna impedance. The input impedance varies obviously with resonance frequency depending on different feeding point of patch antenna. A rectangular patch of dimensions is around 2.5mm by 2.5mm. The dielectric constant of mold layers is εr=3.6. A well feeding point can obtain required bandwidth to cover operating frequency ranges. The return loss of our antenna is a function of frequency which we simulated from HFSS simulation. Fig. 10 shows varying return loss by adjusting the position of the feed point. The return loss depends on feeding location which can achieve good impedance matching between the plate and the feeding point. The current loop will introduce the inductance to match the antenna impedance for capacitive loading. Optimization of feeding point for patch antenna is an important initial step to

978-1-7281-1500-9/19 $31.00 © 2019 IEEE

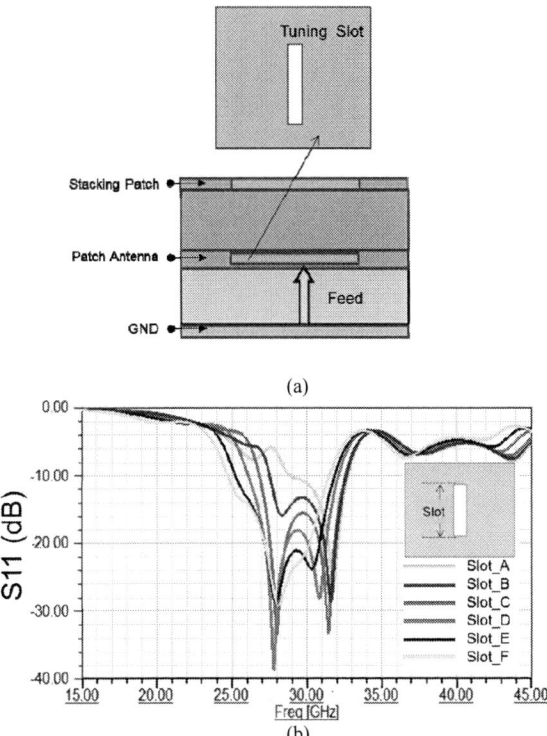

(a)

(b)

Fig. 11. Simulation result of various return loss by adjusting the dimension of notching slot structure.

determine antenna structure and to improve impedance bandwidth. In this case, the antenna bandwidth can achieve up to 3.6-GHz after optimization by adjusting the position of the feed point.

IV. BROADBAND SLOTED PLATE DESIGN

In order to meet the system requirement, the antenna bandwidth still needs to be enhanced for 28-GHz application. In this paper, two techniques are employed to improve the bandwidth for patch antenna. One is adding a stacking patch to obtain the broadband antenna design. The other way of enhancing the bandwidth of patch antenna is to add a tuning slot. The proposed tuning slot for patch antenna is as shown in Fig 11(a). The tuning slot is designed on main patch near the feeding strip that can be used to achieve good matching with the inductance. By having the slot design being closed to resonance allows an impedance response similar to adding parasitic to improve the bandwidth. In fact, to increase the bandwidth of a patch antenna, thick mold layers are required. However, thick mold layer will cause warpage issue that is hard for manufacturing and suffer yield loss. In this case, we choose the simple way of enhancing the bandwidth without increasing the complexity of fan-out structure by adding a tuning slot for antenna impedance matching. Fig. 10 shows varrying return loss by adjusting the dimension of notching slot structure. As shown in Fig. 12, the return loss of < - 10 dB bandwidth is from 26.8 to 32.5 GHz. By combining both

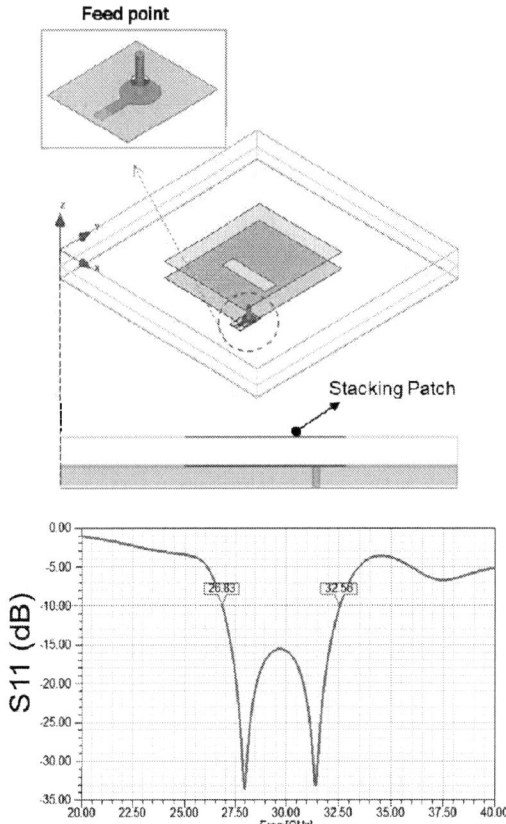

Fig. 12. Simulation result of stacking patch antenna

techniques, the patch antenna can achieve 6.7 GHz bandwidth which can fit the requirements of 28-GHz for wireless communication applications.

V. PATCH ANTENNA ARRAY DESIGN

The antenna array is generally used to increase antenna directivity for high speed and high data-rate transmission. They provide a solution to increase the antenna gain and improve the performance over a single antenna. The patch antenna array allows better control of direction and can be easily controlled for beam direction with phase array technology. In order to obtain best directivity and linearity, the linear arrays are designed to half-wave spacing to give identical results. In this work, we performed a full EM simulation of the 1x2 and 2x2 wideband antenna arrays with identical feeding network. The 3D schematic model of a 1x2 and 2x2 patch antenna is shown in Fig. 13. The 2x2 antenna array has dimensions of around 10mm by 10mm with 3mm spacing for each antenna. The four antenna elements are made of stacked patch antenna structure with same feeding location.

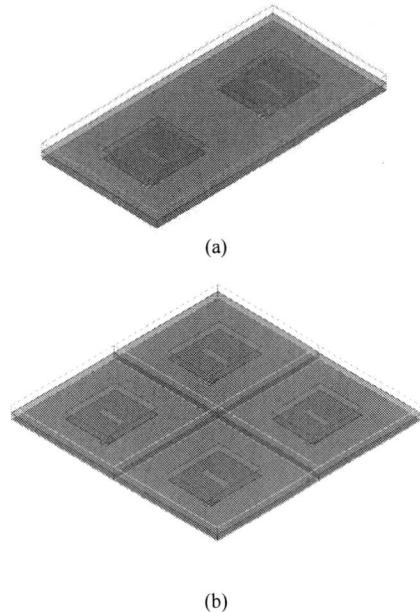

(a)

(b)

Fig. 13. Geometry of the 1x2 and 2x2 patch antenna array

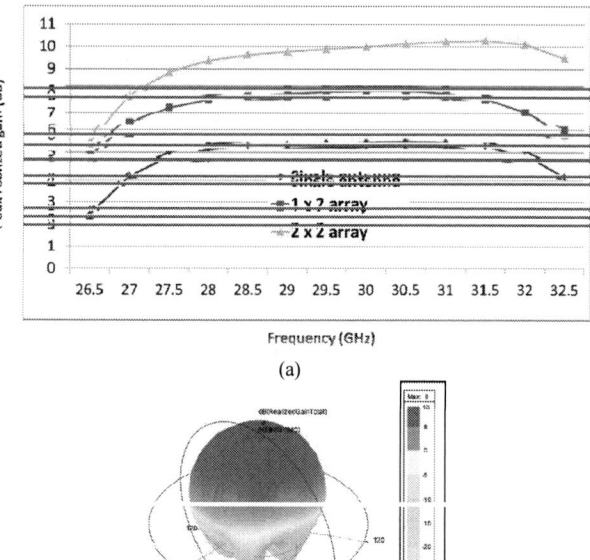

(a)

(b)

Fig. 14. Simulation result of antenna performance(a) antenna gain (b) 3D radiation pattern

The antenna gain simulation results are shown in Fig. 14(a). The maximum gain of the proposed 1x2 and 2x2 antenna arrays are 8.1 dBi and 10.3 dBi. The simulated 3D radiation patterns in free space are shown in Fig. 14(b). The antenna array combined with phase shift is widely used to achieve the beamforming characteristic in wireless application. In this case, adjusting phase difference of each antenna, the main beam of the radiation pattern can change the direction and give the best communication link.

VI. CONCLUSION

Fan-out WLP minimizes 28-GHz interconnection losses of transition from chip to fan-out package to PCB. Small form factor AiP solutions are important prerequisites for 5G systems market. The proposed AiP solution on fan-out package has a broadband and high-gain (above 5 dBi w/ single antenna) performance. The return loss of < - 10 dB bandwidth is from 26 to 33 GHz, which covers the requirements of 28GHz systems. Finally, this work also demonstrates the simulation for antenna array and radiation pattern of AiP on fan-out package. The maximum gain of the proposed 1x2 and 2x2 antenna arrays are 8.1 dBi and 10.3 dBi.. This work provides a thin AiP approach and is suitable for mmWave applications.

REFERENCES

[1] A. Fischer, Z. Tong, A. Hamidipour, L. Maurer and A. Atelzer, "77-GHz Multi-Channel Radar Transceiver With Antenna in Package," IEEE Trans Antennas Propag., vol. 62, no. 3, pp. 1386 – 1394, March 2014.

[2] J. Hasch, E. Topak, R. Schnabel, T. Zwick, R. Weigel and C. Waldschmidt, "Millimeter-Wave Technology for Automotive Radar Sensors in the 77 GHz Frequency Band, IEEE Trans. Microw. Theory Techn., vol. 62, no. 3, pp. 1386 – 1394, March 2014.

[3] O. Kramer, T. Djerafi, and W. Ke, "Very small footprint 60 GHz stacked Yagi antenna array," IEEE Trans Antennas Propag., vol. 59, no. 9, pp. 3204 – 3210, Sep.2011.

[4] A. L. Amadjikpe, D. Choudhury, G. E. Ponchak, and J.Papapolymerou, "High gain quasi-Yagi planar antenna evalution in platform material environment for 60 GHz wireless applications," in Proc. IEEE MTT-S Int. Microw.Symp. Dig, June 2009, pp. 385 – 388.

[5] R. Kulke et al., "24 GHz radar sensor integrates patch antenna and frontend module in single multilayer LTCC substrate," in Proc. Eur. Microelectronics and Packaging Conf., Jun. 2005, pp. 239–242.

[6] A. Fischer, Z. Tong, A. Hamidipour, L. Maurer, and A. Stelzer, "A 28-GHz antenna in package," in 41st Europ. Microwave Conference, vol. 57, no. 11, pp. 1316–1319, Oct. 2011.

[7] C.-Y. Ho, S.-C. Hsieh, M.-F. Jhong, C.-C. Wang and C.-Y. Ting, "A 77GHz Antenna-in-Package with Low-Cost Solution for Automotive Radar Applications," in Proc. IEEE Electronic Components and Technology Conference (ECTC), pp. 191-196, 2018

[8] C.-Y. Ho, M.-F. Jhong, P.-C Pan, C.-Y. Huang, C.-C. Wang and C.-Y. Ting, "Integrated Antenna-in-Package on Low-Cost Organic Substrate for Millimeter-Wave Wireless Communication Applications," in Proc. IEEE Electronic Components and Technology Conference (ECTC), pp. 242-247, 2017

[9] C.-W. Hsu, C.-H. Tsai, J.-S. Hsieh, K.-C. Yee, C.-T.Wang, and Douglas Yu, "High performance chippartitioned millimeter wave passive devices on smooth and fine pitch InFO RDL," in Proc. IEEE Electronic Components and Technology Conference (ECTC), pp. 254-259, 2017.

[10] C.-T. Wang, T.-C. Tang, C.-W. Lin, C.-W. Hsu, J.-S. Hsieh, C.-H.Tsai, K.-C. Wu, H.-P. Pu, and Douglas C.-H. Yu, "InFO_AiP Technology for High Performance and Compact 5G Millimeter Wave System Integration," in Proc. IEEE Electronic Components and Technology Conference (ECTC), pp. 202-207, 2018

[11] C.-T. Wang, and Douglas C.-H. Yu, "Signal and Power Integrity Analysis on Integrated Fan-out PoP(InFO_PoP) Technology for Next Generation Mobile Applications," in Proc. IEEE Electronic Components and Technology Conference (ECTC), pp. 380-385, 2016

Integrated Compact Planar Inverted-F Antenna (PIFA) with a Shorting Via Wall for Millimeter-wave Wireless Chip-to-chip (C2C) Communications in 3D-SiP

Seahee Hwangbo, Renuka Bowrothu, Hae-in Kim and Yong-Kyu Yoon
University of Florida, Gainesville, FL, USA
FL, USA
shhwangbo55@gmail.com, ykyoon@ece.ufl.edu

Abstract— W-band (75 GHz – 110 GHz) wireless in-plane chip-to-chip (C2C) communications in 3D System-in-Packaging (SiP) are demonstrated using an integrated compact planar inverted-F antenna (PIFA) with a shorting via wall. As a test vehicle, a 77 GHz compact PIFA with a shorting via has been designed in a PCB substrate and their RF parameters such as reflection coefficients (S11s), antenna gains, radiation patterns have been studied. Milling machine and micro-fabrication has been utilized to implement the proposed 77 GHz PIFA with a shorting via and their RF characterization has been performed. The measured S11 results agree well with the simulated ones. The simulated efficiency and antenna peak gain is 96 % and 1.77 dBi, respectively. An antenna footprint of 0.19 mm^2 excluding the feeding line is demonstrated, showing more than 50 % area reduction compared with a disc-loaded monopole antenna.

Keywords-Braodside directional antennas; RF/millimeter-wave antennas; Through Glass Vias (TGVs); 3D System-in-Packaing (SiP); Wireless C2C communications; Wireless interconnects;

I. INTRODUCTION

Recently, wireless chip-to-chip (C2C) communications using substrate integrated millimeter-wave antennas have been actively investigated to realize far distance high rate data transmission between multiple chips in 3D-SiP [1], [2], [3], [4]. It has been observed that such a wireless interconnect allows us to achieve superior performance such as low power consumption compared to the conventional wired interconnects such as bonding wires and TSV/TGV [5], [6], [7]. To realize far-field wireless C2C communications, a variety of millimeter-wave antenna structures have been utilized for wireless interconnects in 3D-SiP [8], [9], [10], [11]. For instance, a 60 GHz cavity-backed array antenna has been demonstrated in [8], showing a gain of 4 dBi and a large size of 19.36 mm^2, being less practical to be integrated within a few centimeter-long chip.

On the other hand, our group has demonstrated several millimeter-wave TGV-integrated antennas with a gain of 5.36 dBi (Max.) and a compact size of 9 mm^2 [12] - [16]. Various lateral/vertical wireless intra-/inter-chip communications have been achieved by investigating a variety of millimeter-wave TGV-integrated antennas. One of the advantages of the wireless chip-to-chip communications using TGV-integrated antennas is that the

time delay and communication power consumption caused from long wired interconnects and lossy substrate is significantly decreased. As a result, high-rate wireless data transmission will be realized within many integrated circuits and chips. The previously reported disc-loaded monopole antennas consist of a circular shape loading disc and a through substrate via, where the most area is occupied by the circular shape disc.

In this work, a planar inverted-F antenna (PIFA) with a shorting via is demonstrated for millimeter-wave wireless chip-to-chip communications in 3D-SiP for the first time. The advantages of using a PIFA with a shorting via for wireless C2C communications are as follows: First, a vertical via (a shorting via wall) is utilized to connect a ground plane on the top and a patch on the bottom, making it possible to reduce the patch size approximately by 50 %. Also, the proposed PIFA structure can be integrated with any TSV and TGV in 3D-SiP as the via structures are used as a part of the PIFA such as a feeding line and a shorting via wall. Different from the conventional air-lifted or extruded PIFA antenna, it is embedded in the substrate, package-compatible, and stackable in 3D-SIP. As a test vehicle, a compact 77 GHz (W-band) PIFA with a shorting via wall is implemented and its performance on return losses, antenna gain, and radiation patterns is investigated. RF characterization of the fabricated antenna is performed to measure its S-parameter and the experimental results are compared with the simulated results.

II. ANTEANNA DESIGN

A 77 GHz integrated compact planar inverted-F antenna (PIFA) with a shorting via wall was designed on a Rogers TMM4 substrate with a dielectric constant of 4.54, a loss tangent of 0.002, and a thickness of 510 μm. High Frequency Structure Simulator (HFSS, Ansys Inc.) is utilized to simulate a full 3D structure of the proposed PIFA as shown in Fig. 1. Fig. 1 (a) and (b) show the perspective view of the antenna and the zoomed-in view of vias, respectively. In order to excite the proposed 77 GHz PIFA, a 50 Ω ground/signal/ground (GSG) coplanar waveguide (CPW) feeding planar line is designed and connected to the top side of the PIFA. The top view, bottom view, and cross-section view of the proposed PIFA with a shorting via wall are shown in Fig. 1 (c), (d), and (e), respectively.

Table 1 represents the optimized parameters of the 77 GHz integrated compact PIFA with a shorting via wall. The width

978-1-7281-1500-9/19 $31.00 © 2019 IEEE

(w) of CPW signal line and the gap (g) between the signal and ground is 130 µm and 20 µm, respectively. The diameter (d) of feeding via and shorting via wall is 200 µm. The width (b) and length (a) of the patch antenna bar on the bottom side of the substrate is 110 µm and 645 µm, respectively.

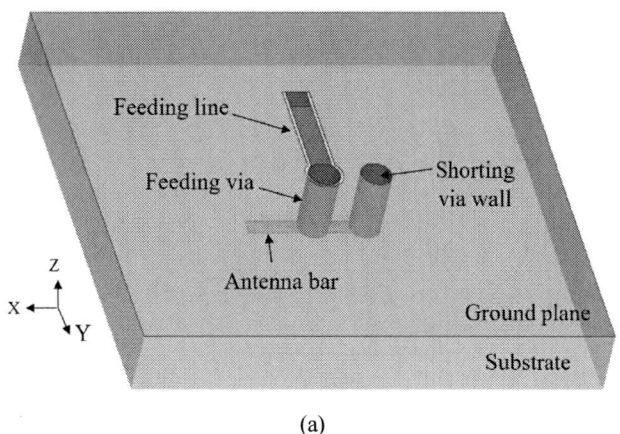

(a)

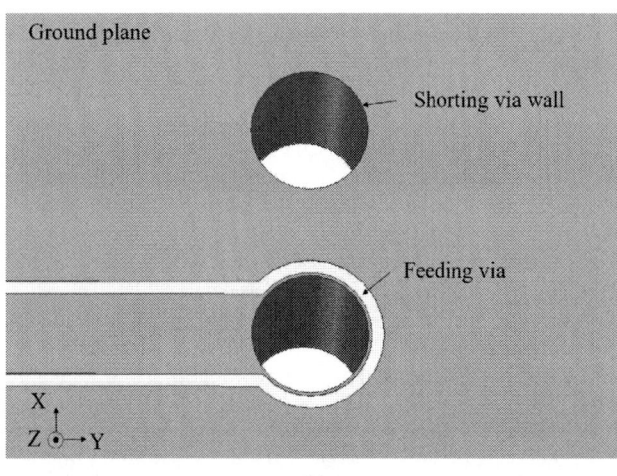

(b)

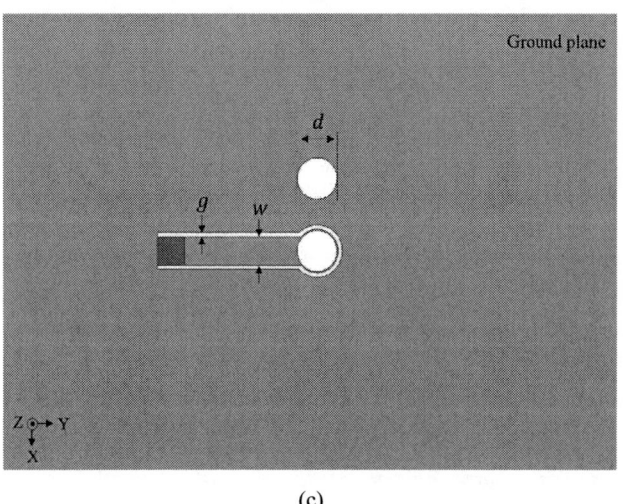

(c)

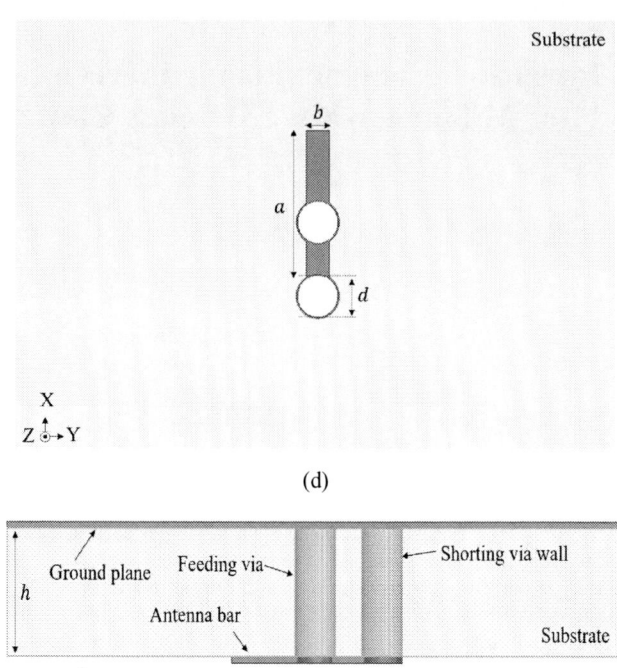

(d)

(e)

Figure 1. Schematic of the proposed 77 GHz integrated compact planar inverted-F antenna (PIFA) with a shorting via wall: (a) perspective view of the antenna, (b) zoomed-in view of vias, (c) top view, and (d) bottom view, and (e) cross-section view of the antenna.

TABLE I.   GEOMETRICAL PARAMETERS OF THE PROPOSED 77 GHZ INTEGRATED COMPACT PLANAR INVERTED-F ANTENNA (PIFA) WITH A SHORTING VIA WALL

Parameters	Values [mm]
a	0.645
b	0.11
d	0.2
w	0.13
g	0.02
h	0.51

III. FABRICATION

A Rogers substrate has been utilized to fabricate the proposed 77 GHz integrated compact planar inverted-F antenna (PIFA) with a shorting via wall. Fig. 2 illustrates the fabrication process of the PIFA with a shorting via wall. The substrate has been cleaned using the piranha cleaning (hydrogen peroxide : sulfuric acid = 1 : 3) and the organic cleaning (acetone/isopropanol/deionized water) (A). Two holes for the PIFA antenna and the shorting via wall have been made using milling machine (B). After laminating a negative photoresist dry film, MX5020 (DuPont, Inc.) on top side of the substrate, UV exposure has been performed with an optical dose of 100 mJ/cm^2 and the dry film has been developed in 0.8 wt. % potassium carbonate (K_2CO_3) solution

A) Etching copper on a substrate

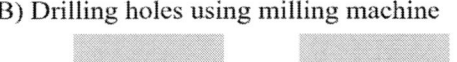

B) Drilling holes using milling machine

C) Lamination of dry film MX5020, UV exposure, and development

D) Sputtering Ti/Cu/Ti (30nm/2μm/30nm)

E) Lift-off dry film on top layer

F) Lamination of dry film MX5020, UV exposure, and development

G) Sputtering Ti/Cu/Ti (30nm/2μm/30nm)

H) Lift-off dry film on bottom layer and cleaning

Figure 2. Fabrication process of 77 GHz integrated compact planar inverted-F antenna (PIFA) with a shorting via wall.

for 2 minutes (C). Titanium (Ti)/Copper (Cu)/Titanium (Ti) (30 nm/2 μm/30 nm) has been deposited for the 50 Ω antenna feeding lines, TGVs, and ground plane (D). Lift-off of the remaining dry film has been performed to complete the top layer of the substrate (E). For the antenna bar on the bottom side of the substrate, the dry film has been laminated, baked on a hot plate at 95 °C for 10 minutes, and patterned using the

ultraviolet (UV) light exposure (F). The potassium carbonate developer has been utilized to remove the remaining dry film. After that, Titanium (Ti)/Copper (Cu)/Titanium (Ti) (30 nm/2 μm/30 nm) has been sputtered (G), and the remaining dry film photoresist on the bottom side of the substrate was cleaned utilizing acetone (H).

Fig. 3 (a) and (b) show the top view of a fabricated 77 GHz integrated compact planar inverted-F antenna (PIFA) with a shorting via wall after step (C) and step (E) in Fig. 2, respectively. The exposed dry film photoresist is remaining on top of the substrate after dry film development as shown in Fig. 3 (a). Also, Fig. 3 (b) represents the Ti/Cu/Ti layers sputtered on top of the substrate, showing that the CPW feeding line and ground plane is successfully fabricated to excited the 77 GHz PIFA with a shorting via wall.

Fig. 4 show the images of the completed antenna: (a) zoom-out view, (b) magnified top view, and (c) magnified bottom view.

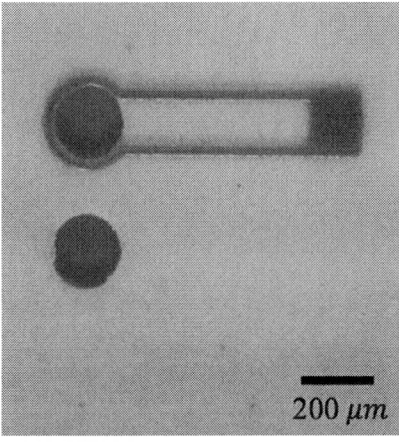

(a)

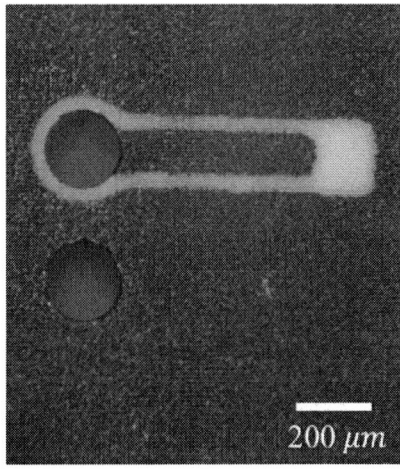

(b)

Figure 3. Photos of the 77 GHz integrated compact planar inverted-F antenna (PIFA) with a shorting via wall: top view of the antenna after (a) step (C) and (b) step (E) in Fig. 2

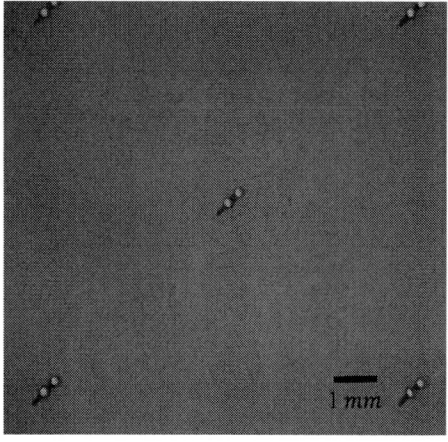

(a)

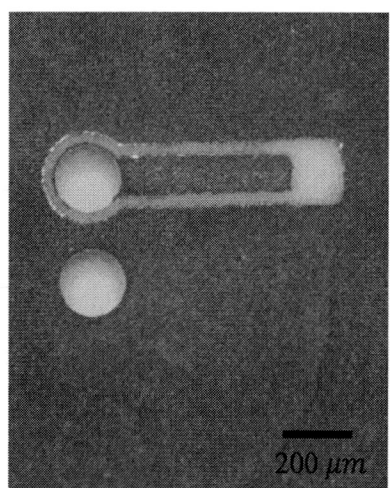

(b)

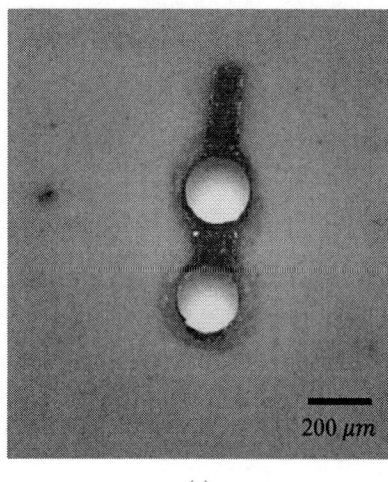

(c)

Figure 4. Photos of the fabricated 77 GHz integrated compact planar inverted-F antenna (PIFA) with a shorting via wall: (a) zoom-out view, (b) top view, and (c) bottom view.

IV. MEASUREMENT AND CHARACTERIZATION

The fabricated antenna shows similar dimensions as designed in Table 1. It has a length of approximately 850 μm and a width of 220 μm, resulting in a footprint of 0.19 mm^2.

High frequency measurement has been performed to characterize the fabricated 77 GHz integrated compact planar inverted-F antenna (PIFA) with a shorting via wall. A vector network analyzer (HP E8361A, Agilent, Inc.), a millimeter wave module (N5260-60003, Agilent, Inc.), and a ground-signal-ground (GSG) probe (110H, GGB Industries, Inc.) have been used for the RF characterization including return losses after standard one port short-open-load (SOL) calibration between 67 GHz and 87 GHz. The measured S11 of 77 GHz PIFA with a shorting via wall is compared with the simulated one as shown in Fig. 5. The measured S11 shows a return loss of 12 dB at 77 GHz with a 10 dB bandwidth (BW) of 13.88 GHz from 67 GHz to 80.88 GHz. The simulated one shows a return loss of 28.3 dB at 77 GHz with a 10 dB bandwidth (BW) of 14.3 GHz from 67 GHz to 81.3 GHz, concluding that the measurement results of the fabricated 77 GHz PIFA with a shorting via wall agree well with the simulated ones. Also, the measured S11 of the fabricated PIFA shows degraded impedance matching between the antenna and input port, which is attributed to the milling machine and micro-fabrication tolerance.

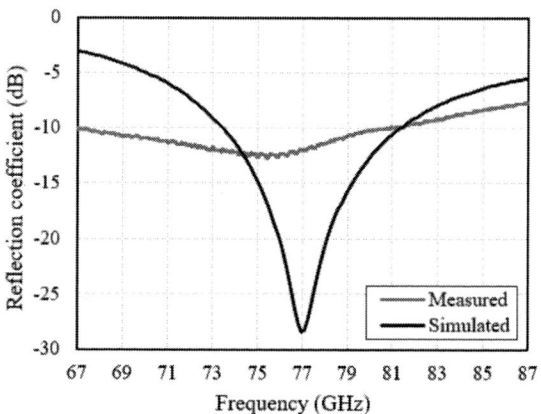

Figure 5. Comparison of simulated and measured reflection coefficients (S11s) of the proposed 77 GHz integrated compact planar inverted-F antenna (PIFA) with a shorting via wall.

Fig. 6 (a) and (b) show the simulated radiation patterns of the 77 GHz integrated compact planar inverted-F antenna (PIFA) with a shorting via wall on the XZ plane and YZ plane. Broadside radiation patterns on the XZ plane (phi = 180°) and YZ plane (phi = 180°) have been observed for wireless C2C communications. The calculated antenna parameters including efficiency and peak gain have been tabulated in Table 2. The simulated antenna peak gain and efficiency is 1.77 dBi and 96 %, respectively.

978-1-7281-1500-9/19 $31.00 © 2019 IEEE

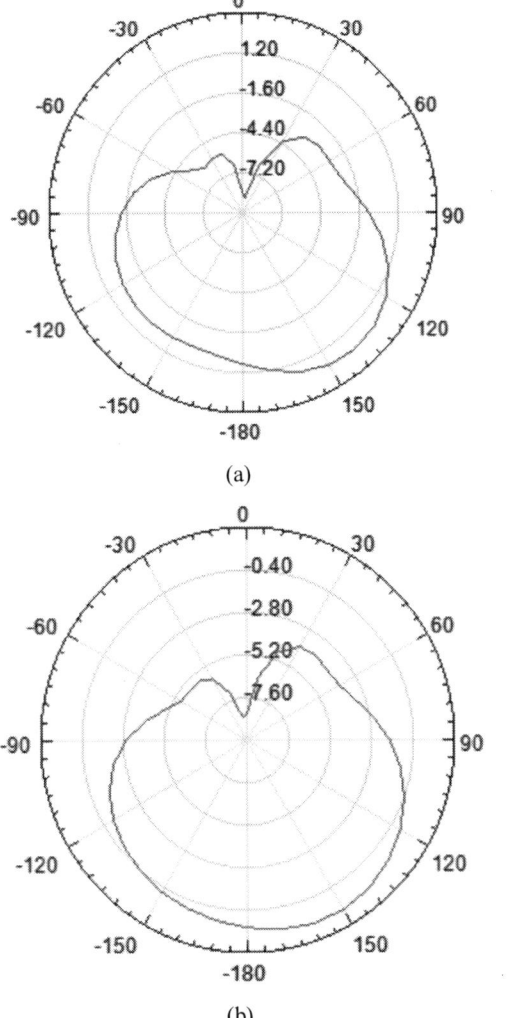

(a)

(b)

Figure 6. Simulated radiation patterns of the proposed 77 GHz integrated compact planar inverted-F antenna (PIFA) with a shorting via wall: (a) XZ plane and (b) YZ plane.

V. CONCLUSION

A 77 GHz integrated compact planar inverted-F antenna (PIFA) with a shorting via wall was proposed for W-band (75 GHz - 110 GHz) wireless C2C chip communications in 3D-SiP. 77 GHz integrated compact PIFAs were designed in a PCB and utilized to demonstrate wireless out-of-plane (vertical) communications among multiple chips and integrated circuits. The designed 77 GHz PIFA was fabricated using milling machine and micro-fabrication and characterized by investigating S-parameters. The measured reflection coefficient (S11) of the 77 GHz PIFA agrees with the simulated one and the simulated results show a broad-side radiation pattern. It is concluded that highly compact, energy efficient wireless chip-to-chip (C2C) communications in 3D-SiP can be realized by using the proposed 77 GHz PIFA with a shorting via wall.

TABLE II. SIMULATED ANTENNA PERFORMANCES OF 77 GHZ INTEGRATED COMPACT PLANAR INVERTED-F ANTENNA (PIFA) WITH A SHORTING VIA WALL

Parameters	Simulated results
Peak gain	1.77 dBi
Efficiency	96 %

ACKNOWLEDGMENT

A negative photoresist dry film (MX5020) is donated by DuPont Inc. Milling machine, micro-fabrication, and RF experiment/characterization have been performed at the University of Florida.

REFERENCES

[1] Md Shahriar Shamim, Naseef Mansoor, Rounak Singh Narde, Vignesh Kothandapani, Amlan Ganguly, and Jayanti Venkataraman, "A Wireless Interconnection Framework for Seamless Inter and Intra-Chip Communication in Multichip Systems," IEEE Transactions on Computers, vol. 66, no. 3, March 1, 2017, pp. 389 - 402.

[2] Xinmin Yu, Joe Baylon, Paul Wettin, Deukhyoun Heo, Partha Pratim Pande, and Shahriar Mirabbasi, "Architecture and Design of Multichannel Millimeter-Wave Wireless NoC," IEEE Design & Test, vol. 31, no. 6, May 2014, pp. 19 - 28.

[3] M.F. Chang, V.P. Roychowdhury, Liyang Zhang, Hyunchol Shin, and Yongxi Qian, "RF/wireless interconnect for inter- and intra-chip communications," Proceedings of the IEEE, vol. 89, no. 4, Apr 2001, pp. 456 - 466.

[4] John H. Lau, "3D IC Packaging 3D IC Integration," [Online]. Available:http://s3.amazonaws.com/sdieee/1817-SanDiegoCPMTDL_Lau_advancedpackaging.pdf

[5] Amlan Ganguly, et al. "The Advances, Challenges and Future Possibilities of Millimeter-Wave Chip-to-Chip Interconnections for Multi-Chip Systems," Journal of Low Power Electronics and Applications vol. 8, no. 1, 2018.

[6] Rozalia Beica, "3D Integration: Applications and Market Trends," IEEE 2015 International 3D Systems Integration Conference, Sendai, Japan, 31 Aug.-2 Sept. 2015.

[7] Iftekhar Ibne Basith, Rashid Rashidzadeh, "Contactless Test Access Mechanism for TSV-Based 3-D ICs Utilizing Capacitive Coupling," IEEE Transactions on Instrumentation and Measurement, vol. 65, no. 1, Oct 2015, pp. 88 - 95.

[8] L. Dussopt, et al. "Silicon interposer with integrated antenna array for millimeter-wave short-range communications," IEEE/MTT-S International Microwave Symposium Digest, Montreal, QC, Canada, June 2012.

[9] Aman Samaiyar, Shobha Sundar Ram, and Sujay Deb, "Millimeterwave planar log periodic antenna for on-chip wireless interconnects," 8th European Conference on Antennas and Propagation (EuCAP), April 6-11, 2014.

[10] Pierre Marie Martin, Thierry Le Gouguec, and Najib Mahdi, "Wireless interconnects by using printed antennas for inter-chip communications in PCB context," 11th European Radar Conference (EuRAD), Oct. 8- 10, 2014.

[11] Jack W. Holloway, Luciano Boglione, Timothy M. Hancock, Ruonan Han, "A Fully Integrated Broadband Sub-mmWave Chip-to-Chip Interconnect," IEEE Transactions on Microwave Theory and Techniques, vol. 65, no. 7, July 2017, pp. 2373 - 2386.

978-1-7281-1500-9/19 $31.00 © 2019 IEEE

[12] Seahee Hwangbo, et al. "Through Glass Via (TGV) disc loaded monopole antennas for millimeter-wave wireless interposer communication." *IEEE 65th Electronic Components and Technology Conference (ECTC)*, 2015.

[13] Seahee Hwangbo, Aric B. Shorey, and Yong-Kyu Yoon. "Millimeter-Wave Wireless Intra-/Inter Chip Communications in 3D Integrated Circuits Using Through Glass Via (TGV) Disc-Loaded Patch Antennas." *IEEE 66th Electronic Components and Technology Conference (ECTC)*, 2016.

[14] Seahee Hwangbo, Yong-Kyu Yoon, and Aric Shorey. "Glass interposer integrated dual-band millimeter wave TGV antenna for inter-/intra chip and board communications." *IEEE International Symposium on Antennas and Propagation (APSURSI)*, 2016.

[15] Seahee Hwangbo, A. B. Shorey, and Y.-K. Yoon, "Directional Through Glass Via (TGV) Antennas for Wireless Point-to-point Interconnects in 3D Integration and Packaging," *IEEE Electronic Components and Technology Conference (ECTC)*, May 30 - June 2, 2017.

[16] Seahee Hwangbo, Yong-Kyu Yoon, Aric B. Shorey, "Millimeter-Wave Wireless Chip-to-Chip (C2C) Communications in 3D System-in-Packaging (SiP) Using Compact Through Glass Via (TGV)-Integrated Antennas," *IEEE Electronic Components and Technology Conference (ECTC)*, 29 May-1 June 2018.

Temporary SiC-SiC wafer bonding compatible with high temperature annealing

Fengwen Mu, Tadatomo Suga
Department of Precision Engineering, School of
Engineering, The University of Tokyo,
Tokyo, Japan

Miyuki Uomoto, Takehito Shimatsu
Research Institute of Electrical Communication,
Tohoku University,
Sendai, Japan

Abstract—**A temporary wafer bonding of SiC-SiC compatible with rapid thermal annealing at ~1000 °C has been developed. An intermediate Ni nano-layer was employed to realize a strong and seamless bonding of two SiC wafers at room temperature without additional pressure. By the rapid thermal annealing process, the interface strength was remarkably decreased and the de-bonding could be achieved at the annealed interface. Interface analyses were carried out to investigate the mechanisms of both bonding and de-bonding. Further development of this temporary bonding technology is expected to be able to make the fabrication of thin SiC device compatible with the common rapid thermal annealing process.**

Keywords- wafer bonding; temporary; SiC; interface; thin SiC device

I. INTRODUCTION

On account of the intrinsic properties of Si such as low bandgap energy and low thermal conductivity, Si power devices are meeting some limitations. SiC power devices have been drawing much attention due to their superior characteristics such as low power losses, high speed switching, high working temperature and fast heat dissipation [1-2]. In recent years, many technologies of SiC material and device have been developed, such as production of large-size high-quality SiC wafer, fabrication of diverse SiC power devices and so on [3-10]. Although many kinds of SiC devices have been successfully commercial, the potential of SiC device can be further developed. One of the new trends is to make the SiC device thinner for further reduction of losses and increase of heat dissipation [11-12]. However, different from that of Si device, the ohmic contact formation of SiC usually needs a rapid thermal annealing up to ~1000 °C [11-13]. Given the possible thermal stress in the thin wafer, it is very challenging to fabricate a thin device. To overcome this challenging, a laser annealing process has been developed for the ohmic contact formation of SiC [11-12].

Another solution is to employ a temporary wafer bonding technique compatible with high temperature annealing [14-15]. Until now, polymer adhesive is most widely used for temporary wafer bonding [16-18]. Recently, temporary bonding using Si nanolayer [19] and diamond-like-layer (DLC) [20] have been developed. Unfortunately, all of them are not suitable for thin SiC device fabrication, because they are not compatible with rapid thermal annealing at ~1000 °C.

In this research, we attempted to develop a new temporary wafer bonding for thin SiC device fabrication, which is compatible with high temperature annealing.

II. IDEA OF TEMPORARY BONDING VIA Ni-SiC REACTION

Recently, much progresses have been achieved in terms of permanent wafer bonding of SiC [4-5, 20-25], which, however, cannot be easily de-bonded.

Our idea is firstly bonding two SiC wafers by an intermediate Ni nano-layer at room temperature and then de-bonding through enough interfacial carbon precipitation from Ni-SiC reaction during the rapid thermal annealing for ohmic contact formation. The Ni-SiC reaction at high temperature has been researched in many publications about the graphene formation [26-30] and the Ni-based ohmic contact formation on SiC [13]. The carbon precipitation from the Ni-SiC reaction during cooling has been well demonstrated. This is due to the low carbon solubility in Ni-silicide at a low temperature. Besides, when the thickness of Ni film is at nanometer level, the carbon diffusion in perpendicular direction may be limited, as a result, the carbon precipitation is likely to be parallel to SiC surface. In case of the structure of SiC/Ni nano-film/SiC, the carbon precipitation is likely to be enhanced at interface, because of none of exposed surfaces, different from the Ni nano-film/SiC in previous researches [26-30]. The enhanced carbon precipitation at the interface is expected to be helpful for the de-bonding after high temperature process.

III. EXPERIMENT METHODS

A. Bonding by Ni nano-film at room temperaute and annealing for de-bonding

The wafer bonding experiments were carried out in a machine with an ultra-high vacuum sputtering-bonding chamber by a room temperature bonding method, which has been named as atomic diffusion bonding. In the chamber, two Ar fast-atom-beam (FAB) sources were set for surface cleaning and a dc magnetron sputtering cathode was set for deposition of Ni films. Working pressure of Ar gas and current for FAB sources were 0.025 Pa and 15 mA, respectively. This condition resulted in the acceleration voltage of ~1.8 kV for one FAB source, and ~2.1 kV for the other. Working pressure of Ar gas during the sputtering deposition of Ni nano-film was 0.5 Pa. After loading two SiC wafers into the chamber, the chamber was vacuumed to be ~5×10^{-7} Pa. Then, both of the SiC wafers were cleaned by their respective FAB irradiation for 90 s, which is enough to completely remove the surface oxide layer on common Si wafer. Immediately after the surface cleaning, working Ar gas pressure was increased to 0.5 Pa, then, Ni films with a thickness of ~15 nm were deposited one by one on the cleaned SiC wafer surfaces by sputtering deposition. Thereafter, the two deposited Ni films were contacted immediately for bonding. The Ni-film has a total thickness of ~30 nm. During bonding process, neither annealing nor pressure was applied.

A rapid thermal annealing was conducted using an infrared lamp heating system at 1273 °C for 120 s in flowing Ar gas. The heating rate is ~25 °C/s and the cooling rate from 1273 °C to 873 °C is also ~25 °C/s. From 873 °C to room temperature, the sample was cooled as fast as possible at a not specified rate.

B. Specimens

The wafers used are n-type, 2-in., on-axis 4H-SiC wafers with the thickness of ~330 μm. The Si-face of 4H-SiC wafers, smoothed by chemical mechanical polishing, was chosen as the bonding surface. The root-means-square roughness of the original bonding surface is ~0.25 nm, confirmed by a dynamic force microscopy (DFM, NanoNavi/L-trace II; Hitachi High Tech Science).

C. Anlysis methods

After bonding, the bonded wafer was immersed in distilled water and observed by scanning acoustic microscopy (SAM, Hitachi FineSAT FS300) to examine the un-bonded areas. The acoustic pixel size is set as 50 μm. Then, the bonded wafer was diced into many chips with the size of 10×10 mm² for multiple measurements. The tensile strengths of the bonding interface before and after RTA were evaluated by a tensile pulling tester (Shimadzu AG-X) using the diced chips. A high-resolution transmission electron microscope (HRTEM, Tecnai F20, FEI Corp.) equipped with an electron energy loss spectroscopy (EELS) was employed to investigate the bonding interfaces of SiC/Ni nano-film/SiC before and after RTA, which was prepared by a focused ion beam unit (FIB, FEI Helios).

IV. RESULTS AND DISCUSSION

The photo and SAM image of the bonded wafer are shown in Fig. 1. None of voids can be found. Fig. 2 shows the comparison of the tensile strength of bonded interface before and after the rapid thermal annealing at ~1000 °C. The tensile strength of the interface without the rapid thermal annealing is out of evaluation since the bonding is stronger than the adhesive used for fixing or the interface between adhesive and bonded samples. The fractures always happened without the separation of bonded chips, as shown in Fig. 3(a). The measured strength, not the actual strength, is from 20.0 MPa to 21.8 MPa, which has an average value of 20.8 MPa. The real interfacial tensile strength should be, at least, greater than the minimum value of 20.0 MPa. This is strong enough for most of the SiC processing steps. In contrast with the interfacial strength before annealing, the strength of the annealed interface is dramatically weakened

to the range of 6.4~10.6 MPa, which has an average value of ~8.7 MPa. Besides, the separation always happened at the annealed bonding interface, as shown in Fig. 3(b). This result demonstrates the feasibility of the de-bonding after a rapid thermal annealing at ~1000 °C.

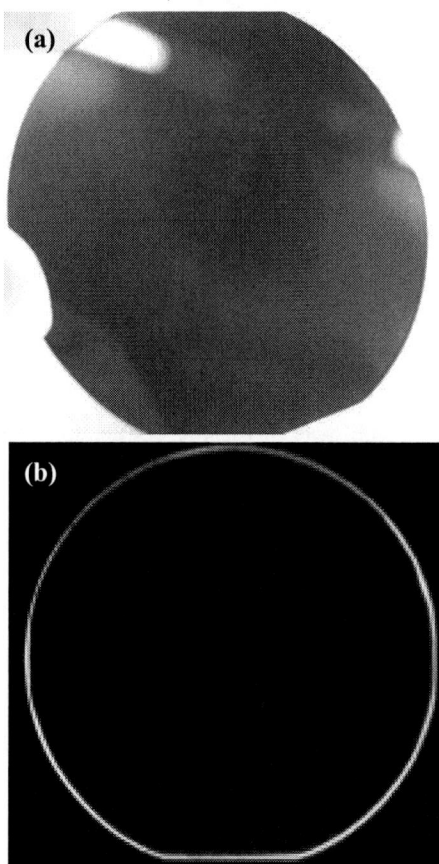

Figure 1. (a) Photograph and (b) SAM image of the 2" SiC-2" SiC wafer bonded via ~30nm Ni layer.

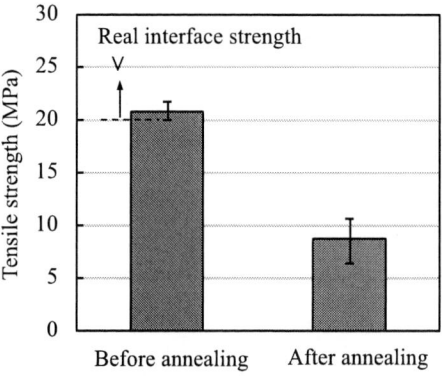

Figure 2. Comparison of the tensile strength of the bonded samples before and after rapid thermal annealing.

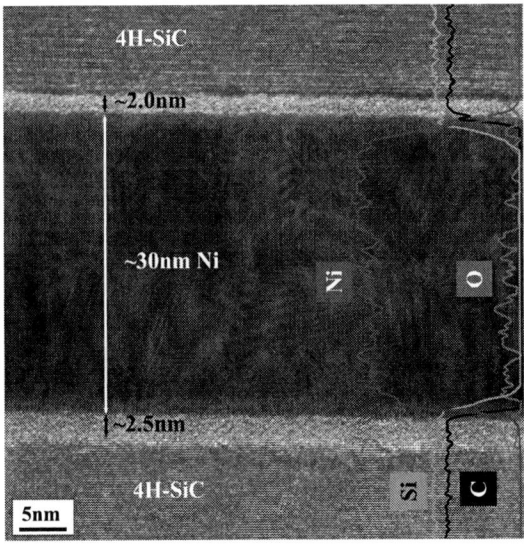

Figure 3. Fractures of the chips after pulling test: (a) before and (b) after rapid thermal annealing.

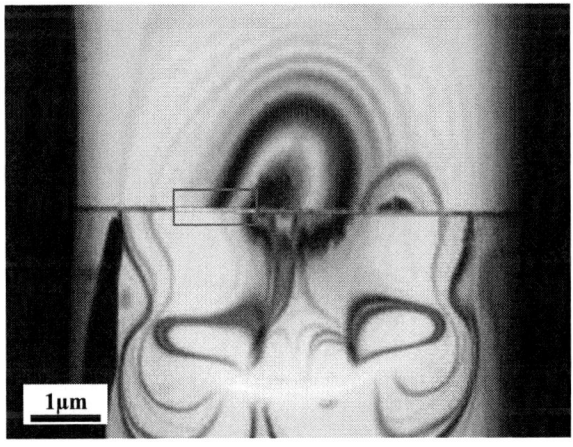

Figure 4. TEM cross-section image of the SiC/Ni nano-film/SiC bonding interface before rapid thermal annealing with element profiles of EELS across the interface.

Fig. 4 shows the TEM cross-section of bonded interface before rapid thermal annealing. The interface corresponding to the Ni-film surface after deposition cannot be distinguished, which indicates that recrystallization may occur across the bonding interface due to the sufficient interfacial self-diffusion, as discussed in previous publications [31-33]. Even in a low-magnification, none of

voids or cracks can be found. This confirmed the seamless feature of bonding via Ni nano-film.

The thickness of the intermediate Ni layer is about ~30 nm, which agrees well with the nominal thickness. Besides, two amorphous layers between deposited Ni nano-film and each SiC substrate caused by surface activation could be confirmed, which have a thickness of ~2.0 nm and ~2.5 nm, respectively. The difference in thickness is probably caused by the difference in surface activation process described in the above experiment methods. EELS line-scanning was employed to investigate the atomic distribution of the elements of Si, O, C and Ni across the interface. As shown in Fig. 4, the line-scanning results of Si, O, C and Ni were represented in green, red, black and blue, respectively. According to the results of line-scanning, it can be confirmed that the surface oxide layer of SiC substrate has been completely removed by surface activation process. At the interfacial region between Ni nano-film and SiC amorphous layer, a mixing layer consisting of Ni, Si and C appeared. This interfacial mixing is consider to be caused by the Ni sputtering deposition. In addition, it was found that the Si diffusion into Ni nano-film is deeper than that of C. It is assumed that both of the sufficient interfacial atomic diffusion together with the interfacial mixing between amorphous SiC and Ni layer contribute to the strong bonding of SiC-SiC at room temperature without any pressure.

The bonded interface after rapid thermal annealing was also analyzed by TEM and EELS. Fig. 5 shows a TEM cross-section of the interface after annealing in a relative low magnification. It was found that there is a ~1.2 μm void in the ~6.5 μm observed area. The formation reason of voids is still unclear. Three typical areas of the interface after annealing were illustrated in Fig. 6. One part of the void areas is shown in Fig. 6(a) and two typical contacted areas are shown in Fig. 6(b) and 6(c).

Figure 5. Bonding interface of the SiC/Ni nano-film/SiC after rapid thermal annealign at a low magnification. The red rectangular represented the voids area.

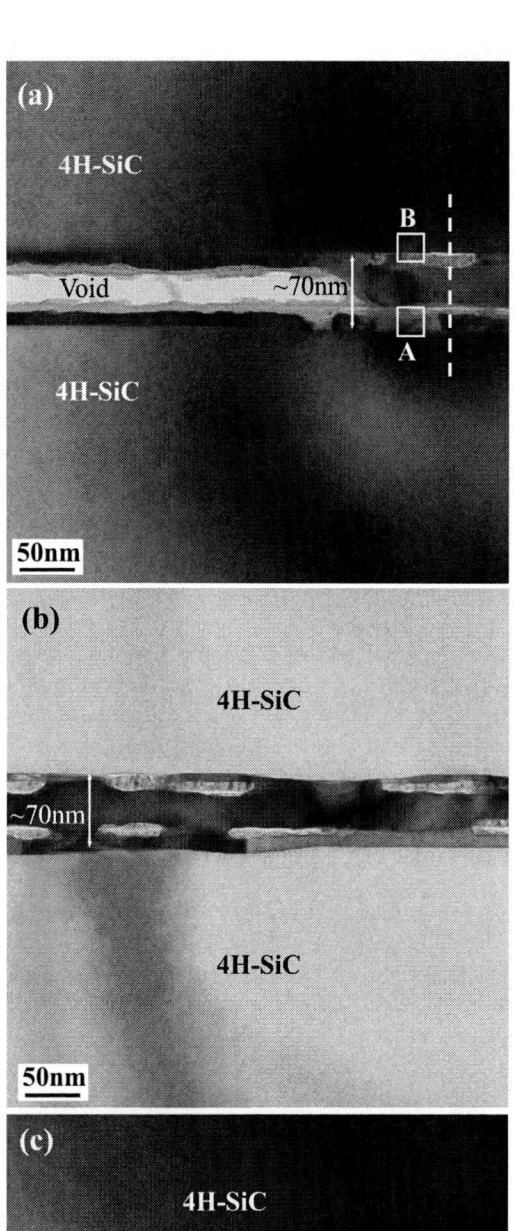

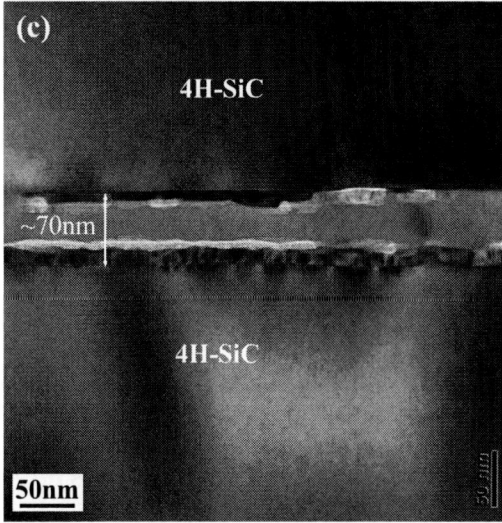

Figure 6. Typical areas of the SiC/Ni nano-film/SiC bonding interface after rapid thermal annealing at high magnification: (a) void area; (b), (c) two typical contacted areas

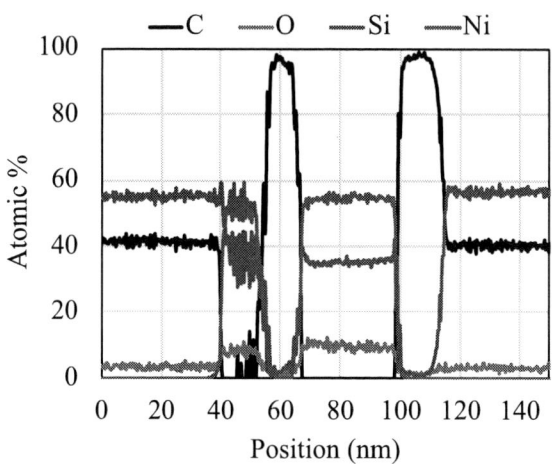

Figure 7. Element profiles of EELS across the annealed interface. The carbon peak areas correspond to the white strips at the annealed interface. The dash line in Fig. 6(a) represent the position of EELS line scanning.

In accordance with the Fig. 6, the thickness of the interfacial layer became ~70 nm from ~30nm. EELS line-scanning was carried out to investigate the interface composition. The scanning position is as indicated by the dash line in Fig. 6(a). According to the EELS results shown in Fig. 7, the annealed interfacial layer is composed of Ni-silicide and carbon material. The white strips at the interface are corresponding to the carbon material precipitated during cooling and the other parts are Ni-silicide. As we assumed, the precipitation of the carbon material seems almost parallel to the SiC surface. Since the interface before annealing is almost close to symmetrical, the precipitation of carbon seems also at two parallel planes, close to SiC substrate, during cooling. Very interestingly, the length of the carbon material in Fig. 6(b) and Fig. 6(c) is very different. The length of the carbon material in Fig. 6(b) is in the range of ~50 to ~150 nm, while that in Fig. 6(c) is even longer than 300 nm. The detailed reason needs further investigation.

Two rectangular positions in Fig. 6(a) were further magnified to confirm the structure of the precipitated carbon material, as shown in Fig. 8. A layered structure could be observed, which indicates the precipitated carbon material is graphite or multilayer graphene. Since the precipitation is parallel to the interface, this is assumed to be one of the mechanisms for de-bonding, considering the adhesion in the layered carbon is relatively weak. Moreover, the enhancement of the carbon precipitation at the interface because of none of exposed surface is also assumed to be very helpful for de-bonding [26-30].

Further study on the affect of Ni-film thickness and other parameters will be proceeded. As a final target, a temporary bonding in wafer size will be conducted.

978-1-7281-1500-9/19 $31.00 © 2019 IEEE

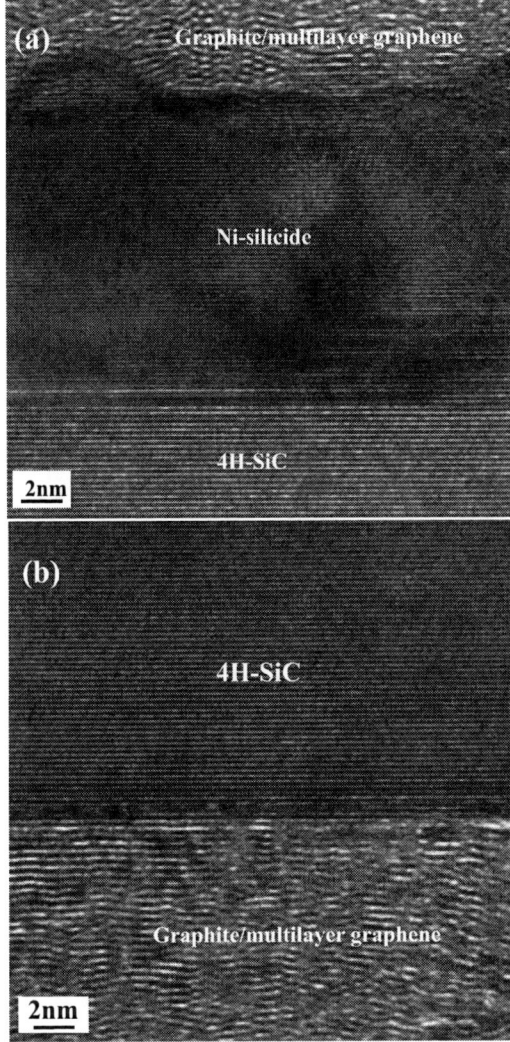

Figure 8. Magnified images of the two rectangular areas (a) A and (b) B in Fig. 6(a).

V. SUMMARY AND OUTLOOK

As a conclusion, a temporary SiC-SiC wafer bonding, compatible with rapid thermal annealing at ~1000 °C, has been developed via a ~30 nm Ni layer. The bonding was obtained at room temperature without any pressure, which has a seamless interface and a tensile strength higher than 20.0 MPa. Both of the interfacial atomic diffusion and the interfacial mixing during sputtering deposition contribute to the robust bonding of SiC-SiC at room temperature. After the annealing at ~1000 °C, the de-bonding at the annealed interface can be realized by pulling test. In accordance with the analyses of the annealed interface, the precipitation of layered carbon material parallel to the SiC surfaces is assumed to be able to weaken the interface. We expect that this temporary bonding technology after further development will advance the fabrication of thin SiC device.

ACKNOWLEDGMENT

This research was conducted under a contract of R&D for Expansion of Radio Wave Resources, organized by the Ministry of Internal Affairs and Communications, Japan.

REFERENCES

[1] T. Kimoto, and J. A. Cooper, Fundamentals of Silicon Carbide Technology: Growth, Characterization, Devices and Applications, John Wiley & Sons, Singapore, 2014.

[2] H. Okumura, "A roadmap for future wide bandgap semiconductor power electronics," MRS Bull. Vol. 40, May 2015, pp. 439-444, https://doi.org/10.1557/mrs.2015.97.

[3] M. Kracica, E. L. H. Mayes, H. N. Tran, A. S. Holland, D. G. McCulloch, and J. G. Partridge, "Rectifying electrical contacts to n-type 6H–SiC formed from energetically deposited carbon," Carbon Vol. 102, Jun. 2016, pp. 141-144, doi:https://doi.org/10.1016/j.carbon.2016.02.038.

[4] J. Liang, S. Nishida, M. Arai, and N. Shigekawa, "Effects of thermal annealing process on the electrical properties of p+-Si/n-SiC heterojunctions," Appl. Phys. Lett. Vol. 104, April 2014, pp. 161604, doi:https://doi.org/10.1063/1.4873113.

[5] H.-P. Phan, H. H. Cheng, T. Dinh, B. Wood, T. K. Nguyen, F. Mu, H. Kamble, R. Vadivelu, G. Walker, L. Hold, A. Iacopi, B. Haylock, D. V. Dao, M. Lobino, T. Suga, and N.-T. Nguyen, "Single-Crystalline 3C-SiC anodically Bonded onto Glass: An Excellent Platform for High-Temperature Electronics and Bioapplications," ACS Appl. Mater. Inter. Vol. 9, Aug. 2017, pp. 27365-27371, doi: 10.1021/acsami.7b06661.

[6] R. Maboudian, C. Carraro, D. G. Senesky, and C. S. Roper, "Advances in silicon carbide science and technology at the micro-and nanoscales," J. Vac. Sci. Technol. A Vol. 31, Jun. 2013, pp. 050805, doi: https://doi.org/10.1116/1.4807902.

[7] R. S. Okojie, D. Lukco, V. Nguyen, and E. Savrun, "4H-SiC piezoresistive pressure sensors at 800°C with observed sensitivity recovery," IEEE Electron Device Lett. Vol. 36, Feb. 2015, pp. 174-176, doi:10.1109/LED.2014.2379262.

[8] M. B. J. Wijesundara, and R. G. Azevedo, Silicon Carbide Microsystems for Harsh Environments, Springer, New York, 2011.

[9] F. Zhao, M. M. Islam, and C. F. Huang, "Photoelectrochemical etching to fabricate single-crystal SiC MEMS for harsh environments," Mater. Lett. Vol. 65, Feb. 2011, pp. 409-412, doi: https://doi.org/10.1016/j.matlet.2010.10.034.

[10] D. H. van Dorp, J. L. Weyher, and J. J. Kelly, "Anodic etching of SiC in alkaline solutions," J. Micromech. Microeng. Vol. 17, Mar. 2007, pp. S50, doi:https://doi.org/10.1088/0960-1317/17/4/S04.

[11] Y. Nakanishi, T. Tominaga, H. Okabe, Y. Suehiro, K. Sugahara, T. Kuroiwa, Y. Toyoda, S.Yamakawa, H. Murasaki, K. Kobayashi, and H. Sumitani, "Properties of a SiC Schottky Barrier Diode Fabricated with a Thin Substrate," Mater. Sci. Forum, Vol. 820, Feb. 2014, pp. 778-780, doi: https://doi.org/10.4028/www.scientific.net/MSF.778-780.820.

[12] R. Rupp, R. Gerlach, U. Kirchner, A. Schlögl, and R. Kern, "Performance of a 650V SiC diode with reduced chip thickness," Mater. Sci. Forum, Vol. 717-720, Jun. 2012, pp. 921-924, https://doi.org/10.1016/j.carbon.2016.02.038.

[13] A. V. Kuchuk, P. Borowicz, M. Wzorek, M. Borysiewicz, R. Ratajczak, K. Golaszewska, E. Kaminska, V. Kladko, and A. Piotrowska, "Ni-Based Ohmic Contacts to n-Type 4H-SiC: The Formation Mechanism and Thermal Stability," Adv. Cond. Matter Phys. Vol. 2016, Feb. 2016, pp. 9273702, doi: http://dx.doi.org/10.1155/2016/9273702.

[14] S. H. Christiansen, R. Singh, and U. Gösele, "Wafer direct bonding: From advanced substrate engineering to future applications in micro/nanoelectronics," Proc. IEEE Vol. 94, Dec. 2006, pp. 2060-2106, doi: 0.1109/JPROC.2006.886026.

[15] H. Takagi, K. Kikuchi, R. Maeda, T.-R. Chung, and T. Suga, "Surface activated bonding of silicon wafers at room temperature," Appl. Phys. Lett. Vol. 68, Feb. 1996, pp. 2222-2224, doi: https://doi.org/10.1063/1.115865.

[16] J. Hermanowski, "Thin wafer handling-Study of temporary wafer bonding materials and processes," IEEE Int. Conf. on 3D Syst. Integr. (3DIC), Spet. 2009, pp. 1-5. doi:10.1109/3DIC.2009.5306550

[17] B. Dang, B. Webb, C. Tsang, and P. Andry, J. Knickerbocker, "Factors in the selection of temporary wafer handlers for 3D/2.5 D integration," 64th Electron. Compon. Technol. Conf., May 2014, pp. 576-581, doi: 10.1109/ECTC.2014.6897343.

[18] S.-W. Lee, J.-W. Park, C.-H. Park, D.-H. Lim, H.-J. Kim, J.-Y. Song, and J.-H. Lee, "UV-curing and thermal stability of dual curable urethane epoxy adhesives for temporary bonding in 3D multi-chip package process," Int. J. Adhes. Adhes. Vol. 44, 2013, pp. 138-143.

[19] K. Takeuchi, M. Fujino, Y. Matsumoto, and T. Suga, "Room temperature bonding and debonding of polyimide film and glass substrate based on surface activate bonding method," Jpn. J. Appl. Phys. Vol. 57, Jan. 2018, pp. 02BB05, doi: https://doi.org/10.7567/JJAP.57.02BB05.

[20] R. Bellmana, P. Mazumdera, R. G. Manleya, K. Adiba, and S. Liu, "Temporary Bonding for High Temperature Processing of Thin Glass Using Plasma Activated DLC Layer," AiMES 2018 ECS and SMEQ Joint International Meeting, Abstract MA2018-02 951, Oct. 2018.

[21] T. Suga, F. Mu, M. Fujino, Y. Takahashi, H. Nakazawa, and K. Iguchi, "Silicon carbide wafer bonding by modified surface activated bonding method," Jpn. J. Appl. Phys. Vol. 54, Jan. 2015, pp. 030214, doi: https://doi.org/10.7567/JJAP.54.030214.

[22] F. Mu, K. Iguchi, H. Nakazawa, Y. Takahashi, M. Fujino, and T. Suga, "Room-temperature wafer bonding of SiC–Si by modified surface activated bonding with sputtered Si nanolayer," Jpn. J. Appl. Phys. Vol. 55, Mar. 2016, pp. 04EC09, doi:https://doi.org/10.7567/JJAP.55.04EC09.

[23] E. Higurashi, K. Okumura, K. Nakasuji, and T. Suga, "Surface activated bonding of GaAs and SiC wafers at room temperature for improved heat dissipation in high-power semiconductor lasers," Jpn. J. Appl. Phys. Vol. 54, Jan. 2015, pp. 030207, doi: https://doi.org/10.7567/JJAP.54.030207.

[24] F. Mu, K. Iguchi, H. Nakazawa, Y. Takahashi, M. Fujino, and T. Suga, "Direct Wafer Bonding of SiC-SiC by SAB for Monolithic Integration of SiC MEMS and Electronics," ECS J. Solid State Sci. Technol. Vol. 5, Jun. 2016, pp. P451-P456, doi: 10.1149/2.0011609jss.

[25] G. N. Yushin, and Z. Sitar, "Influence of relative wafer rotation on the electrical properties of the bonded SiC/SiC interface," Appl. Phys. Lett. Vol. 84, Mar. 2004, pp. 3993-3995, doi: https://doi.org/10.1063/1.1753065.

[26] A. A. Woodworth, and C. D. Stinespring, "Surface chemistry of Ni induced graphite formation on the 6H–SiC (0001) surface and its implications for graphene synthesis," Carbon Vol. 48, Jun. 2010, pp. 1999-2003, doi: 10.1016/j.carbon.2010.02.007.

[27] Z.-Y. Juang, C.-Y. Wu, C.-W. Lo, W.-Y. Chen, C.-F. Huang, J.-C. Hwang, F.-R. Chen, K.-C. Leou, and C.-H. Tsai, "Synthesis of graphene on silicon carbide substrates at low temperature," Carbon Vol.47, July 2009, pp. 2026-2031, doi: https://doi.org/10.1016/j.carbon.2009.03.051.

[28] J. Hofrichter, B. N. Szafranek, M. Otto, T. J. Echtermeyer, M. Baus, A. Majerus, V. Geringer, M. Ramsteiner, and H. Kurz, "Synthesis of graphene on silicon dioxide by a solid carbon source," Nano Lett. Vol. 10, Dec. 2009, pp. 36-42, doi:10.1021/nl902558x.

[29] E. Escobedo-Cousin, K. Vassilevski, T. Hopf, N. Wright, A. O'Neill, A. Horsfall, J. Goss, P. Cumpson, "Local solid phase growth of few-layer graphene on silicon carbide from nickel silicide supersaturated with carbon," J. Appl. Phys. Vol. 113, March. 2013, pp. 114309, https://doi.org/10.1063/1.4795501.

[30] A. Hähnel, V. Ischenko, and J. Woltersdorf, "Oriented growth of silicide and carbon in SiC-based sandwich structures with nickel," Mater. Chem. Phys. Vol. 110, Aug. 2008, pp. 303-310, doi: https://doi.org/10.1016/j.matchemphys.2008.02.009.

[31] T. Shimatsu, and M. Uomoto, "Atomic diffusion bonding of wafers with thin nanocrystalline metal films," J. Vac. Sci. Technol. B Vol. 28, May 2010, pp. 706-714, doi: https://doi.org/10.1116/1.3437515.

[32] T. Shimatsu, and M. Uomoto, "Room temperature bonding of wafers with thin nanocrystalline metal films," ECS Trans. Vol. 33, 2010, pp. 61-72, doi:10.1149/1.3483494.

[33] T. Shimatsu, M. Uomoto, and H. Kon, "Room Temperature Bonding Using Thin Metal Films (Bonding Energy and Technical Potential)," ECS Trans. Vol. 64, 2014, pp. 317-328, doi: 10.1149/06405.0317ecst.

Ultrathin Glass to Ultrathin Glass Bonding Using Laser Sealing Approach

Messaoud Bedjaoui*, Johnny Amiran, Jean Brun
Univ. Grenoble Alpes, CEA, LETI, DCOS
Minatec Campus
Grenoble F-38000, France
*messaoud.bedjaoui@cea.fr

Abstract—In this paper, a vacuum laser-assisted glass frit bonding approach and its use for the encapsulation of thin film battery devices is reported. This sealing method is evaluated using alkali free glass (50µm) as a substrate as well as a cover. A detailed description of the experimental process (thin film battery deposition, cover preparation, localized laser sealing) is carried out. Specific focus is dedicated to process adaptation regarding the particular case of ultrathin glasses. A suitable tool equipped with mechanical pressure and vacuum facilities is proposed to prevent the thermal stress within the ultrathin glasses. The controllability of the mechanical force bonding between the glass frit layer and the UTG substrate during the laser sealing ensures a full hermetic seal. Moreover, the fine-tuning of the heating time by increasing sealing speed (up to 20mm/s) and increasing the number of passes (> 10) can significantly improve the cooling rate of the glass and helps to reduce the ultrathin glass cracks. Preliminary demonstration experiments illustrate potential applications of this technique for hermetic sealing of temperature sensitive electrochemical devices like thin film batteries.

Keywords-Ultrathin glass; Laser sealing; Thin film batteries; Vacuum packaging; Glass frit

I. INTRODUCTION

The advent of ultrathin glass UTG (<100µm) such as AF32® Eco Thin Glass (Schott AG), Willow® Glass (Corning) and G-Leaf™ (Nippon Electric Glass), have generated significant interest as a flexible substrate materials as well as packaging materials [1, 2, 3]. In microelectronic applications, ultrathin glass exhibits a plethora of properties; it is highly transparent, smooth and planar with extraordinarily good mechanical stiffness. These characteristics are leading to a number of promising thin-film devices in the area of energy sources and advancing displays (e.g lithium microbatteries, solid and organic state lighting, liquid crystals displays…). In practice, these kinds of miniature electronic devices require hermetic packaging as they undergo rapid degradation upon atmospheric elements exposition. For example, lithium-based thin film batteries require low water permeability ($\leq 10^{-4}$ g/m^2/day) and low oxygen permeability ($\leq 10^{-4}$ cc/m^2/day) due to the very great sensitivity of lithium with water and oxygen [4]. In this way, there are several solutions developed to provide hermetic encapsulation of thin-film batteries (TFB). The most common, easier and less costly approach uses lamination of adhesive barrier films [5, 6]. The weakness of this method is lies to the fact that adhesive layers allows the penetration of moisture and oxygen from the lateral side. However, this fact is greatly severe for small devices as the diffusion paths of oxidants species are extremely shorts which increasing the vulnerability of lithium layers. To overcome these problems and ensure a real encapsulation, an ideal scheme preconizes the hermetic joining of two glass panes. Apparently, the hermeticity is assured by the encapsulation capping as well as by the substrate material. In this regard, ultrathin glass should be ideal for producing excellent diffusion barrier [1]. The encapsulating system in turn should be thin, robust and offer seal with long-term stability. However, the relative low melting temperature point of lithium (180°C) and the need of high integration ratio (packaging surface vs active surface) considerably restrict the choice of encapsulating process.

Amongst various methods reported in literature, glass frit bonding is a common and easier thermo-compressive process in wafer level encapsulation and packaging [7]. It's a simple and robust process based on the use of an intermediate layer positioned between two rigid substrates (>200 µm) and subjected to pressure and heating. The packaging principle is achieved by the creation of a cavity housing the microelectronic device using the intermediate layer defined by the glass frit, the substrate and the cover. Traditionally, the frit glass process consists of three main steps: screen-printing of a glass paste, its thermal conditioning, and thermo-compressive bonding. However, the thermo-compressive bonding generally requires the heating of the whole device at high temperature (several hundred of degrees). Such processes, hence, do not allow the encapsulation of temperature sensitive devices (e.g. lithium thin film batteries) within standard glass frit bonding. To cope with this issue, the ideal solution combines glass frit bonding and laser joining techniques. The laser sealing is an alternative localized joining technique typically adapted for packaging of temperature-sensitive devices. This technique exclusively allow localized heating of the frit glass. Moreover, the choice of the laser source is of a special interest in achieving hermetic seals. The laser-based process is advantageously compatible with similar materials as well as with dissimilar materials, such as glass to glass or silicon to glass.

Against the background of laser-based glass frit sealing very little research has been properly focused on the use of ultrathin glasses (<100μm). Overall, it is obvious that the employment of ultrathin glass for the packaging accentuates the warpage challenge during cover-substrate assembly. In addition, the ultrathin glasses assembly imposes mechanical pressure during the laser sealing process to ensure an appropriate sealing exacerbating, thereby, the glass cracking. Accordingly, an adapted and reliable process play a critical role for the requirements of thinner packages using ultrathin substrates and ultrathin covers. In this paper, we report preliminary engineered experiences on the use of alkali-free ultrathin glass materials for the packaging of lithium thin film batteries. More particularly, we propose a sealing cell for vacuum laser-assisted glass frit bonding avoiding the thermal and mechanical stress of the ultrathin glasses during the sealing process. The study includes an initial evaluation of the glass frit deposition on ultrathin glasses in peripheral cordon shape, thus enabling the fabrication of cover panes. We further show that the process parameters adjusting prevents the major issues generated by the laser on the frit glass (stress, cracks). In particular, we investigate the benefit of the fabricated vacuum cell to improve the bonding force on the upper face of the frit glass cords, thereby reducing the stress risk on the frit glass and the ultrathin glass panes. Finally, we evaluates the proposal concept through the realization and the encapsulation of lithium thin-film structures (<20μm) using alkali free glass (50μm) as a substrate as well as a cover.

II. EXPERIMENTAL PROCEDURES

A. UTG Materials

In this work, we primarily focused on the bonding of symmetrical ultrathin glass. From practical point of view, the sealing of glasses with similar properties helps to reduce thermo-mechanical stress in the bonding interface especially when the glass materials are ultrathin (<100μm). The host substrate and cover materials used in all bonding experiments were AF32 alkaline-free borosilicate glass with a thickness of 50 μm manufactured by Schott. The outer dimensions of the substrate and cover glass are typically 25.4mmx55.8mm. Experimentally, AF32 substrate materials characterized by relatively high glass transformation temperature (typically Tg>500°C) are suitable for the fabrication of lithium thin film batteries (TFB). Indeed, the active layers of TFB such as lithium cobalt oxide electrodes ($LiCoO_2$), in general, requires a post annealing treatment at high temperature (500°C-800°C) to improve the crystallinity of $LiCoO_2$ deposit [8]. Furthermore, the UTG is a high barrier diffusion material [1] that acts as a hermetic encapsulation of lithium layers from various oxidants (humidity, oxygen…). Finally, the AF32 UTG material is highly transparent in the visible range allowing, hence, a large choice of appropriate laser sources for the laser-assisted frit glass process. Table 1 resumes the main characteristics of AF32 UTG materials used as a host substrate for lithium-based layers as well as a cover system.

TABLE I. GENERAL CHARACTERISTICS OF EMPLOYED MATERIALS (UTG AND FRIT GLASS)

Material	UTG substrate and UTG cover	Corning Vita® Frit glass
Transformation temperature Tg	715°C	303°C
Coeficient of Thermal Expansion CTE (0 to 300°C)	3-5 ppm/°C	7 ppm/°C
Transmission (λ=550nm-1000nm)	92%	15%

B. Glass Frit

For a successful glass frit bonding, it is important to select the suitable glass frit material for various reasons. For instance, the matching of Coefficient of Thermal Expansion (CTE) of glass frit and UTG materials is a critical factor for the process achievement. In fact, the glass frit bonding of materials with different mechanical properties can generates significant cracking and delamination of the frit from the glass during the sealing process. In the present paper, the glass paste used is a lead-free borosilicate-based (Vita®, a viscosity of 57 Pa.s) commercialized by Corning. According to the supplier specifications, the thermal expansion coefficient (Table 1) is in the same range to that of the UTG materials. In general, the glass paste used is a mixture of powders (frits), organic binders and solvents to form a paste compatible with printing deposit techniques such as screen printing, doctor-blading or dispensing. This fact supposes the implementation of a burn out stage before bonding in order to remove completely the organic components. Indeed, insufficient heat treatment of the frit matrix could lead to the void formation inside the bonded glass [7] degrading the reliability of the packaging. Therefore, adequate temperature firing profile is mandatory without exceeding the glass transition temperature of the glass frit.

To define the sintering schedule of the Vita® paste, thermal properties are investigated using TG thermal analysis. Fig. 1 shows the TG measurement results of 14 mg glass frit heated at 10°C/min from 25°C to 1200°C in air atmosphere using an Al_2O_3 crucible. At 139°C, TG curve shows large weight loss (~39.5%) generally caused by the solvents volatilization. A second weight loss less important (~3.8%) is observed in the temperature range 220°C-320°C related to the organic burn-out [9]. The third weight loss (<2%) in the temperature range 500°C-600°C corresponds to the glazing of the frit glass [9]. More generally, the weight loss value in the TG curve almost remains invariably when the temperature is higher than 300 °C, i.e. near the glass transition temperature Tg (Table 1). To deep investigate the thermal characteristics, supplementary TG measurements (not shown) of the glass frit paste are conducted at different temperature stage (150°C for 30 min; 280°C for 60 min and 380°C for 60 min; 520°C for 10 min in air atmosphere). These results confirm that 380°C stage is sufficient for the complete removing of the organic binder. As a first assessment, the thermal conditioning of the Vita® frit paste depicted in Fig. 2 includes two successive steps: drying at 150°C for 30 min and burnt out at 380°C for 60 min.

978-1-7281-1500-9/19 $31.00 © 2019 IEEE

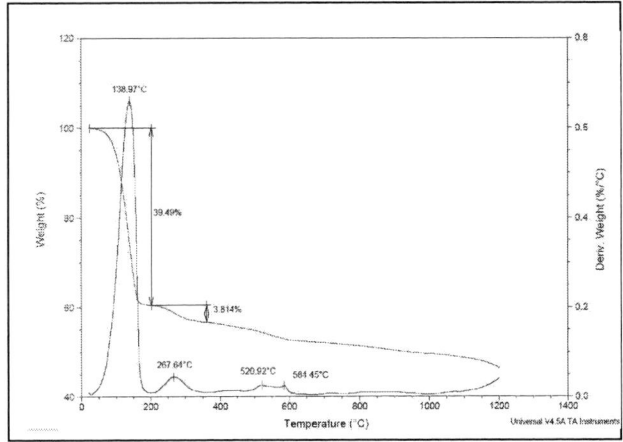

Figure 1. TG measurement of the Vita® glass frit under air atmosphere with a heating rate of 10°C/min.

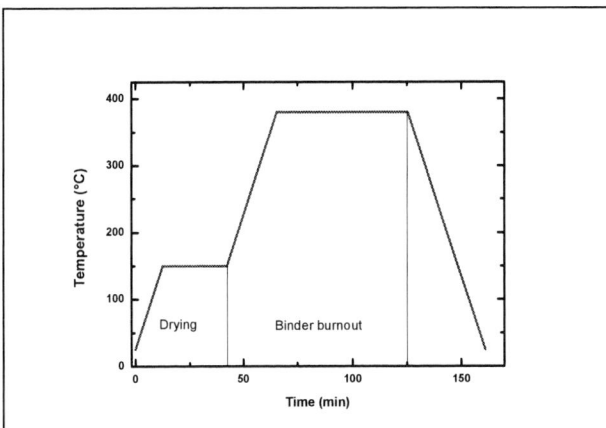

Figure 2. Temperature profile of the thermal conditionning of Vita® frit glass before laser process.

In this paper, the Vita® glass frit was screen-printed onto the 50µm UTG cover using a screen printer with a 300 mesh/in stainless steel screen. The vehicle test design of this work is a pseudo-square glass frit cords with dimensions of 23x23 mm² and two different widths (0.4 mm and 0.8 mm). The screen printing parameters were 50 N necessary for forcing the paste through the slits in the mask and 350mm/s for the razor blade speed. After glass frit deposition, the UTG cover were then submitted to thermal treatment under air ambient with firing profile as shown in Fig. 2 to drive out the solvents and the organics elements. The weight loss caused by the thermal treatment results in a reduction of the glass paste thickness between the screen printing stage and the thermal conditioning stage. Using the experimental conditions of this work, the thickness reduction value is around 30%. After the thermal treatment, the frit glass cords normally consist of glass powder compatible with melting and flowing under higher temperature budget. Typical example of the frit glass cords obtained on 50 µm UTG cover are shown in Fig. 3. For this example, the measured height of the glass frit cord is 14-15 µm thick after screen-printing and 9-10 µm thick after annealing treatment.

Figure 3. Photograph of 50µm UTG cover with Vita® glass frit cord after screen printing and thermal treatment.

C. Test vehicle

In this work, the test vehicle consists of lithium-based layers deposited on 50µm UTG substrate and packaged by 50µm UTG cover using Vita® glass frit as an intermediate layer. The experimental procedure, depicted by Fig. 4, mainly comprises three steps. The process starts with screen-printing of the glass frit cord on UTG cover followed by the thermal conditioning as described in Section II-B. The UTG cover was placed on the top of UTG substrate containing lithium layers before laser sealing. The positioning of the UTG substrate and the UTG cover plays a major role in the sealing quality. The main objective of the laser-bonding tool is to facilitate the handling of the UTG materials and, hence, proposing an easier and flexible support for the laser-assisted frit glass bonding. It must fulfill the requirement for laser sealing of UTG materials and the particular sensitivity criteria of microbattery lithium-based components. Indeed, it is mostly critical to protect the lithium layers in fine as well as during the laser sealing process itself. For instance, as the lithium deposit equipment are systematically connected to argon glove box the laser-bonding tool had to be versatile ensuring, inter alia, a use in glove box area. By the way, the laser-bonding tool ensure a hermetic transfer of the UTG samples with lithium layers from the lithium deposit equipment to the laser sealing equipment.

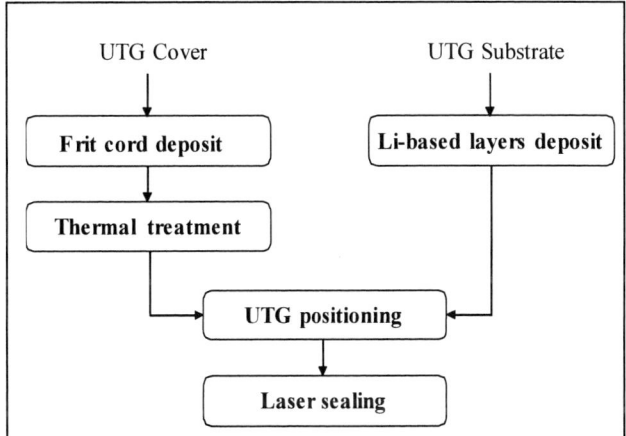

Figure 4. Elementary steps of the process flow applied for the packaging of lithium based layers using laser-assited glass frit technique.

D. Laser-Bonding Tool

The design of the laser-bonding tool allows bonding in inert gas, in ambient atmosphere as well as under vacuum conditions. Fig. 5 shows the schematic of the laser-bonding tool approved for this work before (a) and after (b) vacuum pumping, respectively. The proposed setup is a sealing cell consisting of two main parts. The first one is a copper block with a circular sidewall configured to support an upper cap window, in addition to a receiving zone dedicated to hold the UTG substrate (25.4mmx55.8mmx120µm). On the one hand, the bottom basis of the copper block is equipped with a vacuum port connected to a pumping system. On the other hand, the upper face of the circular sidewall contains a groove with an elastomer seal allowing the accommodation of the second part of the sealing cell. The glass window of the second part shown in Fig. 5 (a) serves as an upper capping to close hermetically the sealing cell. The choice of this window must meet a number of criteria including the flexion ability under vacuum pressure and the total transparency vs laser wavelength. For our experiments, it is preferably an AF32 glass wafer (200 mm) with a thickness of 500µm. The physical, optical and mechanical properties are closely similar to UTG materials.

In the present approach, the UTG substrate and the UTG cover are brought into the laser-bonding tool. The design of the receiving zone guarantees an accurate positioning of the UTG cover with frit glass vs the UTG substrate with lithium-based layers. Prior to the effective laser sealing process, the sealing cell is closed under argon atmosphere (Fig. 5(b)) by evacuating the chamber to a moderate vacuum. This process is suitable with lithium-based layers fabrication that employ glove boxes with argon ambient. For instance, a vacuum around 3kPa leads to a glass window deflection around 70µm over the UTG substrate dimension. Advantageously, the cap glass window is in close contact with the UTG cover while applying light pressure to the glass frit cord, generally required for high quality sealing. After that, the sealing chamber is ready for transfer to be loaded on the workpiece of the laser equipment as shown by the photograph in Fig. 6. Then, the laser assisted sealing can takes place.

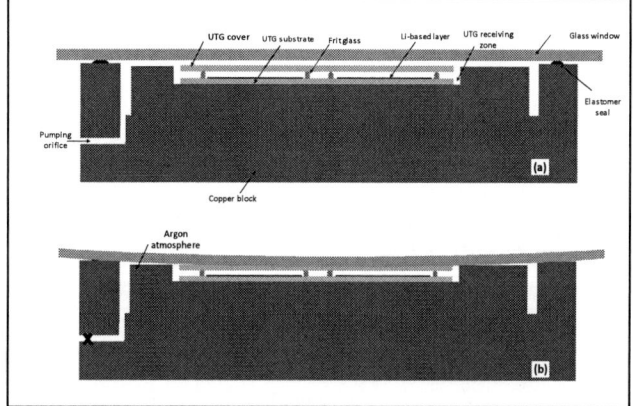

Figure 5. 2D schematic sketch of the laser-bonding tool approved in this work (a) after positionning of the UTG samples and the glass window, (b) after vacuum pumping on the sealing cell. The geometrical dimensions of the various elements are not in scale.

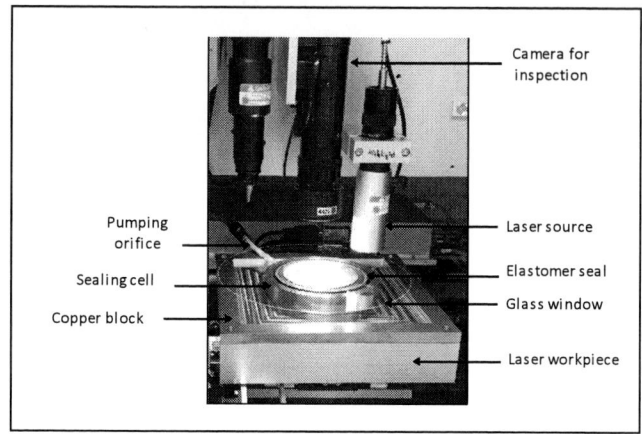

Figure 6. Photograph of the laser-bonding tool.

The laser source used in all our experiments is a laser diode supplied by Coherent with an optical fiber delivery. The wavelength emission range is 800-820nm permitting a total transmission through glass window and UTG materials and a high absorption in contact with frit glass with respect of the characteristics given in Table 1. The laser power output was up to 100 W at CW mode. The laser spot has a circular shape with a diameter of approximately 800 µm. The laser-bonding tool is fixed on workpiece equipped by XY motion stage (200 mm x 200 mm) with speed capabilities up to 300mm/s.

III. LASER SEALING RESULTS

A. Preliminary results

The principle of laser-assisted frit glass approach is driven by the heat conduction resulted from the optical energy. In this process, the UTG materials and the glass window are virtually transparent for the laser wavelength (800-820nm). Thus, the laser radiation is mainly absorbed in the glass frit. It is obvious that sealing process requires close contact between glass frit and glass parts, which have to be sealed. Using the standard process by applying external mechanical force on the top of the glass window without pumping in the sealing tool, the glass frit partly reflows along the patterns as shown in Fig. 7. Therefore, the UTG cover and the UTG substrate are easily separated which indicates that no reliable seal is achieved between the various elements. The potential root cause of this fact could be a break contact between the UTG cover and the glass frit cord from one side and the UTG substrate from the other side during the laser scanning movement. Indeed, the glass frit melting under laser beam irradiation implies a local compression of the irradiated zones in comparison to other zones of the frit pattern. Thus, the UTG cover might flex slightly resulting in a change of pressure applied to the non-sealed parts of UTG samples without compensation by the external force on the glass window. Additionally, the unequal pressure on the entire surface of the glass frit cord throughout the laser process even leads to void generation in the bonding layer.

978-1-7281-1500-9/19 $31.00 © 2019 IEEE

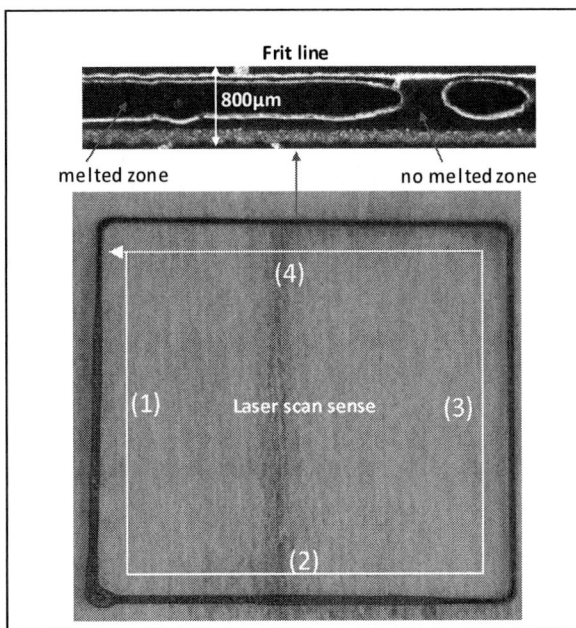

Figure 7. Photograph of unsuccessful UTG to UTG bonding laser using standard method: bonding in air with external force applied on the glass cover window. The Vita® frit glass is 15µm hight and 800µm width. The numbers (1, 2, 3, 4) refers to the laser scan sequence.

A possible solution of the fact shown in Fig.7 is the use of the laser-bonding tool proposed in our approach. In this way, the bending of the closing cap under vacuum pumping generates the bonding force required for the laser sealing. As the sealing cell can be permanently connected to the pumping source, the proposed approach allows stable force independently of the laser movement. Thus, this method enables homogeneous distribution of the bearing force at the level of glass frit cord. From a design point of view, the pressure in the sealing chamber ranges from 1 kPa to 3 kPa. The sealing cell has a circular form with a diameter of 90 cm able to embark UTG substrates 25.4mmx55.8mm.

B. Effect of laser parameters

In the rest of this document, we mainly concentrate on the bonding trials of the UTG materials using the intermediate Vita® frit glass in order to test and to optimize the potential use of the proposed laser-bonding tool. The quality of the bonding was assessed on visual inspection of the seal using optical microscope. In present work, the sealing method is a multi-pass contour bonding protocol that implies the movement of the laser along the frit cord sandwiched between the two UTG materials. This approach uses fast scan of the entire sample for several loops. The negative rating corresponds to delamination; crack on the UTG materials, or glass frit, while positive assessment concerns the samples with no crack or delamination. In addition, the good noted specimens are submerged in ethanol liquid for leak test. This easiest and fast test is very informative regarding the sealing assessment.

Fig.8 shows a typical seal of 50 µm UTG substrate and 50 µm UTG cover with Vita® frit glass elaborated according to the experimental conditions mentioned in section II.B. In present document, the main operating variables of the laser-assisted sealing process considered are laser power (from 10W to 40W), scanning velocity (from 10mm/s to 40mm/s) and loops repetition (from one loop to 40 loops). The optimized operating parameters of the sample shown in Fig.8 was a scanning velocity of 20mm/s and a laser power of 25W for 10 complete loops. These conditions allow free-defect seals, such as possible cracking or delamination in the frit cord or cracking in the UTG materials. More particularly, the bonded samples are void free and this clearly demonstrates that the experimental conditions are suitable for UTG to UTG bonding. Furthermore, after the ethanol injection test no leakage was observed for this specimen. These first bonding experiments demonstrated that it is possible to achieve full hermetic seals using the proposed laser-bonding tool.

The joint quality of the bonding was analyzed by Scanning Electron Microscope (SEM) after sectioning and polishing of the bonded sample prepared using the optimized parameters (20mm/s, 25W, 10 loops). Fig. 9 shows a SEM photo of a transversal cross-section of the Vita® glass frit-bonding layer. This image reveals that glass frit has excellent contact with both UTG materials with no visible gas inclusions or voids, an indication of an effective sealing. The obtained glass frit cord is approximately 9 µm thick, slightly different in comparison to the value obtained after the burn out step. This minor shift is probably related to the thermal effect of the glazing occurred during the laser beam impact, that is coherent with the TG measurement of Fig. 1. However, the UTG to UTG bonding approach is consistent with thick frit glass as the final thickness of the frit cord can be increased by adjusting the screen printing parameters. Despite this fact, the morphology of the bonding cord obtained in our approach is highly comparable to the laser-assisted glass frit bonding of thick glass [10, 11] as well as to the traditional glass frit thermo-compressive process [7, 12].

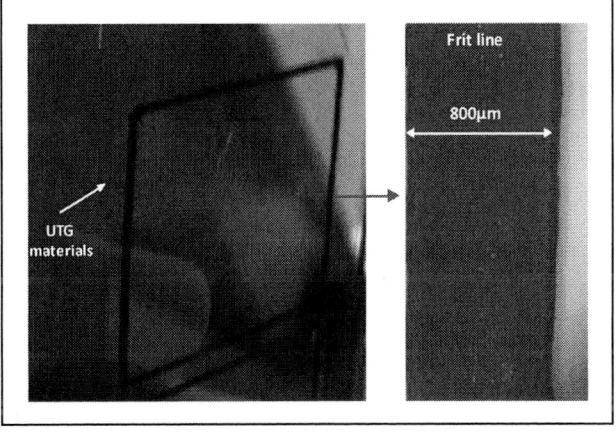

Figure 8. Photograph of UTG to UTG bonding after laser sealing using the laser-bonding tool.

978-1-7281-1500-9/19 $31.00 © 2019 IEEE

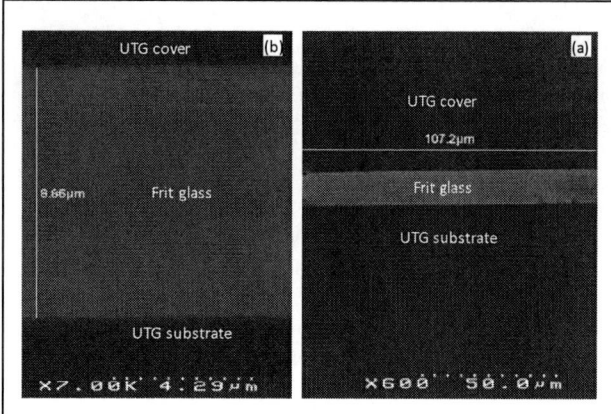

Figure 9. SEM image (a and b refer to two magnifications) of a cross section of Vita® glass frit sandwiched between UTG cover and UTG substrate using laser-assited frit glass approach as proposed in this work.

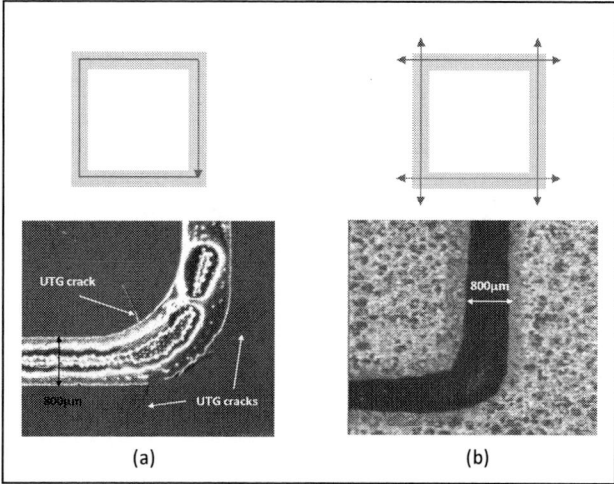

Figure 10. Schematic and microscopic photgraphs illustration of different laser approach employed with laser-bonding tool. (a) quasi-simultaneous process; (b) separate laser scan segments.

In present work, UTG to UTG bonding based on frit glass laser-assisted employs several closed loops to achieve the required temperature of bonding. This approach well known as quasi-simultaneous process using fast scan is obviously suited for 23mmx23mm patterns [13]. Using this process scheme, bonding experiments duplication have shown that it is hard to obtain reliable seals with high repeatability and crack-free defects. The cracks seems usually originates from the starting point which is most probably related to the cooling rate during sealing process. Indeed, one of the parametrs affecting the residual stress in the UTG and the frit glass is the smooth cooling. In fact, the laser beam starts and finishes the scan at the same corner of the pseudo-square pattern. We clearly see the impact of laser beam on the UTG crack at the starting/end point. The effective heating time of the starting/end point is obviously

very important in comparison to the entire bond line yielding fractured samples. Thus, this work introduces a modified scheme to avoid corner overexposure to laser beam and reduce the effective heating time of the critical points. Generally, it was observed that crack-free samples were obtained by applying fast separate laser-scan segments. The scan is systematically repeated more than 10 time in back and forth direction where the starting point is external to the frit cord. This principle as well as microscopic photography observations are depicted in Fig. 10.

C. Lithium test

As mentioned above, the laser-assisted frit glass-bonding process exploits the heat flow derived from laser radiation to melt the glass frit. The laser beam locally heats the glass paste cord to its melting point in few seconds. Indeed, a major advantage of laser heating is that in contrast to furnace heating only a localized zone is exposed to elevated temperature for a time greatly reduced. This is notably appreciated for the packaging of temperature sensitive components like lithium batteries. As such, lithium would be oxidized very quickly in contact with ambient atmosphere. A completely transparent aspect of the lithium material is the typical manifestation of this degradation after few seconds. As for calcium test for organic light emitting diode devices, lithium test is a screening approach, which uses a unique lithium layer deposit without recourse to the whole process of lithium batteries. Despite the simplicity of the lithium test, it is still an effective method to find very small leaks of the sealed UTG samples.

Preliminary hermeticity tests of encapsulated lithium layers were performed by using the process flow and the laser-bonding tool described in sections II-C and II-D, respectively. Typical seals capable of surviving a hermeticity test (650 hours in 60°C/ 85% RH) are shown in Fig. 11. In this example, the Li layer (1 μm thick) can be assimilated to 19mmx19mm square, the UTG size is 25.4mmx55.8mm, and the Vita® glass frit is 23mmx23mm. Both UTG are 50μm thickness, the frit height is 10μm while the frit width is 800μm. The voluntary adjustment of the laser parameters in Fig.11 provides two distinguish cases: high efficiency sealing for the cell 1 (laser power = 25W) and low efficiency sealing for the cell 2 (laser power = 15W). The other laser parameters are constant; scan velocity 20mm/s, 10 back and forth by segment. The surface appearance of lithium layer of the cell 1 shown in Fig. 11 (b) exhibits a slightly dimpled form after 650 hours in 60°C/ 85% RH while the cell 2 presents a heavily impacted surface. In addition, the contour of the lithium patch of the cell 2 reveals a cloudy aspect indicating clearly an oxidation reaction of lithium with eventual penetrant oxidant through the frit seal. The given results confirm that the laser-assisted glass frit bonding process is a promising approach for hermetic sealing criteria of temperature sensitive electrochemical devices like thin film batteries.

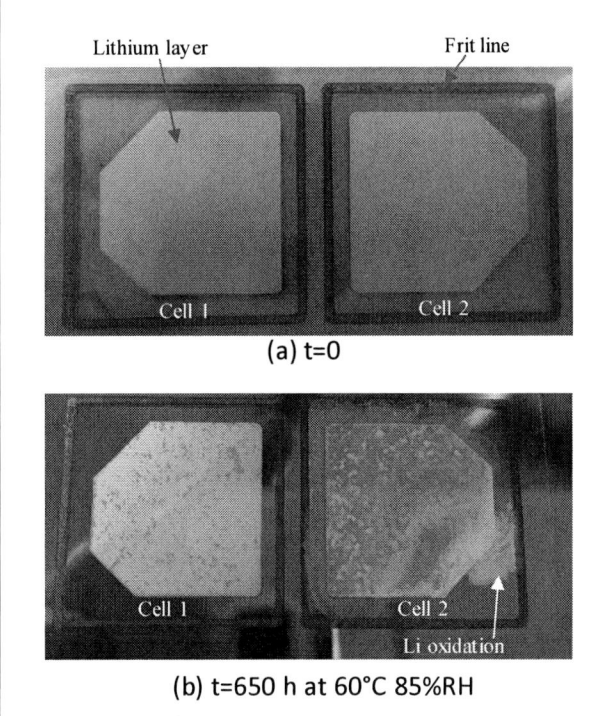

Figure 11. Photograph of lithium layers encapsulated by UTG to UTG bonding approach described in this work. (a) after sealing, (b) after acclerate ageing at harsh conditions 60°C 85%RH. Cell 1 (laser power =25W); Cell 2 (laser power =15W).

IV. SUMMARY

We presented a new promising approach for ultrathin glasses bonding using glass frit cord as intermediate layer and adapted laser-bonding tool as a sealing cell. The developed process can be used for the encapsulation and the packaging of temperature-sensitive devices, established on UTG substrates and UTG covers in well-controlled manner. This process is based on the transfer of UTG cover with frit glass patterns on UTG substrate using a vacuum closed laser-bonding tool. The bonding step is carried out by a laser beam directly fired onto the frit pattern through the glass cap and the UTG cover. The main key insight in the process development was the adaptation of the laser parameters (power, scan velocity, scanning manner) to accommodate the thermal shocks, thus avoiding the formation of micro-cracks across the bonding cord and the UTG materials.

The proposed approach originates inclusions (voids, gas)-free glass frit bonding layers with excellent contact with UTG cover and UTG substrate indicating an effective sealing. As such, hermeticity tests using lithium layers corroborate the suitability of the proposed approach for obtaining hermetic sealing of lithium microbatteries from external exposures. Next investigations should verify the feasibility of the laser-assisted glass frit bonding in the presence of electrical contacts of microbatteries. Additional experiments will be extended to study the metal influences on the process window.

ACKNOWLEDGMENT

The authors thank EnSO project for financial support. EnSO has been accepted for funding within the Electronic Components and Systems for European Leadership Joint Undertaking in collaboration with the European Union's H2020 Framework Program (H2020/2014-2020) and National Authorities, under grant agreement n° 692482.

REFERENCES

[1] S. M. Garner, "Flexible glass: Enabling thin, Lightweight, and Flexible Electronics," Hoboken, NJ : John Wiley & Sons, Inc. ; Beverly, MA : Scrivener Publishing LLC, 2017.

[2] R. Sprengard, M. Jotz, M. Letz, L. Parthier, F. Prince, M. Woehrmann, J. U. Thomas and M. Toepper, "Ultrathin glasses for semiconductor packaging," International Symposium on Microelectronics: Fall 2016, Vol. 2016 N°1, 2016, pp. 000293-000298, doi: 10.4071/isom-2016-wp24.

[3] V. Sundaram, Y. Sato, T. Seki, Y. Takagi, V. Smet, M. Kobayashi and R. R. Tummala, "First demonstration of a surface mountable, ultra-thin glass BGA package for smart mobile logic devices," Proc. Electronic Components and Technology Conference (ECTC 64th), IEEE Xplore, Sep. 2014, pp. 365-370, doi: 10.1109/ECTC.2014.6897313.

[4] R. Salot, S. Martin, S. Oukassi, M. Bedjaoui, J. Ubrig, "Microbattery technology overview and associated multilayer encapsulation process," Applied Surface Science 256S, pp. S54-S57, 2009.

[5] F. Gasco and P. Feraboli, "Manufacturability of composite laminates with integrated thin film Li-ion batteries," Journal of composite materials, vol.48(8), No 1-2, pp. 899-910, 2014.

[6] M. Bedjaoui and S. Poulet, "Direct bonding and debonding approach of ultrathin glass substrates for high temperature devices," Proc. Electronic Components and Technology Conference (ECTC 67th), IEEE Xplore, Aug. 2017, pp. 725-732, doi: 10.1109/ECTC.2017.79.

[7] R. Knechtel, "Glass frit bonding: an Universal technology for wafer level encapsulation and packaging," Microsystem Technologies, vol.12, No 1-2, pp. 63-68, 2005.

[8] S. Tintignac, R. Baddour-Hadjean, J-P. Periera-Ramos and R. Salot, "High performance spputtered LiCoO₂ thin films obtained at a moderate annealing treatment combined to a bias effect," Electrochimica acta, vol.60, pp. 121-129, 2012.

[9] G. Wu, D. Xu, X. Sun, B. Xiong and Y. Wang, "Wafer-level vacuum packaging for microsystems using glass frit bonding," IEEE Transactions on Components, Packaging and Manufacturing Technology, vol.3, No 10, pp. 1640-1646, 2013.

[10] R. Cruz, J. Ranita, J. Maçaira, F. Ribeiro, A. da Silva, J. M. Oliveira, M. Fernandes, H. Ribeiro, J. Mendes and A. Mendes "Glass-Glass Laser-assisted glass frit bonding," IEEE Transactions on Components, Packaging and Manufacturing Technology, vol.2, No 12, pp. 1949-1956, 2012.

[11] S. Emami, J. Martins, L. Andrade, J. Mendes and A. Mendes "Low temperature hermetic laser-assisted glass frit encapsulation of soda-lime glass substrates," Optics and Lasers in Engineering Technologies, vol.96, pp. 107-116, 2017.

[12] R. Knechtel, M. Wiemer and J. Frömel "Wafer level encapsulation of microsystems using glass frit bonding," Microsystem Technologies, vol.12, pp. 468-472, 2006.

[13] H. Kind, E. Gehlen, M. Aden, A. Olowinsky and A. Gillner "Laser glass frit sealing of vacuum insulation glasses," Physics Procedia, vol.56, pp. 673-680, 2014.

Development of Resins for Bumpless Interconnects and Wafer-On-Wafer (WOW) Integration

N. Araki[1,3], S. Maetani[1,3], Y. S. Kim[2,3], S. Kodama[2,3], and T. Ohba[3]

[1] DAICEL Corp., 1239, Shinzaike, Abosho-ku, Himeji, Hyogo 671-1283, Japan
[2] DISCO Corp., 13-11 Omori-Kita 2-chome, Ota-ku, Tokyo 143-8580, Japan
[3] WOW alliance, Tokyo Institute of Technology, 4259 Nagatsuda, Midori-ku, Yokohama 226-8503, Japan
na_araki@jp.daicel.com

Abstract— Our permanent and temporary adhesives are suitable for wafer-on-wafer (WOW) technology using ultra-thin wafer multilayer stacks with bumpless dual damascene interconnects.. Because our temporary adhesive is a reactive hot-melt type, the bonding temperature is around 130 °C and the de-bonding temperature is higher than the bonding temperature (>200 °C). After de-bonding and washing processes, the temporary adhesive residue was uniformly distributed on the de-bonding wafer, and the thickness of the residue was estimated to be 1–2 nm in a secondary ion mass spectrometry (SIMS) profile. A permanent adhesive with high heat resistance (maximum operating temperature, 300 °C) was prepared and could be cured at 135–170 °C, which is lower than the de-bonding temperature of the temporary adhesive. The permanent adhesive exhibited adhesion to SiO_2 and Si with highly reliable thermal cycle test (TCT) resistance, and a good etching profile. The solid state of the permanent adhesive coated on an Si substrate exhibited no diffusion in a SIMS profile. The electrochemical ability of the permanent adhesive is indicated by its current density of 10^{-9} to 10^{-8} A/cm^2 at -400 to 400 MV/cm, indicating that it is a near-perfect electrical insulator. Vertical through-silicon via interconnects were demonstrated by optimizing the WOW process using our two adhesives.

Keywords- permanent and temporary adhesives; 3D integration; wafer-on-wafer (WOW); TSV; bumpless dual-damascene interconnects

I. INTRODUCTION

3D integration, which allows devices on stacked silicon wafers to be vertically connected by through-silicon vias (TSVs), is an essential technology for enhancing system performance, such as achieving low-energy data transfer and increasing memory capacity [1-3]. Chip-on-Chip (COC) and Chip-on-Wafer (COW) processes with TSVs and microbumps are commonly used in conventional 3D integration [4]; however, the chip-based processes have limitations in the flip-chip process attributed to the chip thickness, the maximum number of stacked chips and, in particular, the throughput. These result in a significant bottleneck to achieving low-cost and high-volume production. Moreover, the extremely small bump size is a major barrier to vertical interconnect scaling, and the mechanical strength causes reliability problems due to thermal mismatch. To overcome these problems, we have

developed permanent and temporary adhesives for a wafer-on-wafer (WOW) technology that enables multilevel stacking using several-micrometer-thick wafers [1-3, 5-10].

Our WOW technology features bumpless dual-damascene interconnects with a via-last after bonding scheme. Figure 1 shows the WOW process flow for the case of two stacked wafers. Thermal stability in the stacked wafers and thermal budget controls for the temporary and permanent adhesives are essential to achieve WOW stacking. Most temporary adhesives are de-bondable by heat, ultraviolet light, and/or mechanical force and are necessary for the conventional manufacture of 3D devices, such as COW and WOW. In order to release the device wafer from the carrier with low stress, thermal release adhesives, referred to as hot melt adhesives, are preferred. However, hot-melt adhesives having a wide range of operating temperatures, such as from room temperature to temperatures above 200 °C, have not been developed yet. On the other hand, many compounds useful for permanent adhesives with high thermal stability require high temperatures for the curing process, and low temperature curable compounds generally have poor thermal stability. For example, benzocyclobutene (BCB) resin, which has high thermal stability, needs a high curing temperature of about 250 °C, and epoxy compounds which can be cured at a low temperature under 200 °C do not have good thermal stability. Therefore, these materials are not useful for permanent adhesives.

This article describes the properties of these permanent and temporary adhesives for WOW total integration. A hot-melt adhesive was prepared for use in combination with a permanent adhesive having high heat resistance. For the temporary adhesive residue, which was considered to consist of hydrocarbon compounds, a uniform distribution on the de-bonding wafer washed with PGME was observed, and the thickness of the residue was estimated to be 1–2 nm. A permanent adhesive with lower curing temperature (about 135-170 °C) than BCB and high heat resistance (maximum operating temperature, 300 °C) was successfully synthesized from a specific molecular structure. The permanent adhesive could be cured at 135–170 °C, which is lower than the de-bonding temperature of the temporary adhesive (over 200 °C). In addition, The permanent adhesive exhibited insulating properties, adhesion to SiO_2 and Si with highly reliable thermal cycle test (TCT) resistance, and a good

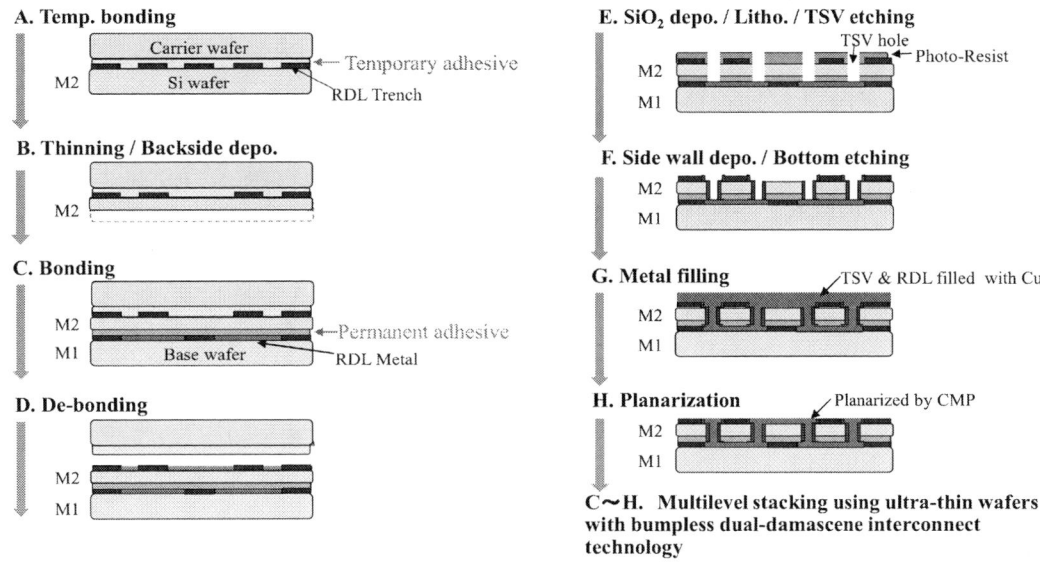

Figure 1. Process flow of WOW stack.

etching profile. The diffusion profiles of the permanent adhesive in the solid state were observed using secondary ion mass spectrometry (SIMS). The contact resistivity in stacked 300 mm wafers using the permanent adhesive indicated that its properties were suitable for WOW total integration.

II. EXPERIMENTAL

A. Chemicals

A reactive hot-melt type temporary adhesive (DTB-TP005), whose glass transition temperature (Tg) was controlled at the bonding and de-bonding temperatures, was developed by chemical design. Recommended temperatures for bonding and de-bonding of the temporary adhesive are around 130 °C and over 200 °C, respectively. The main structure of the permanent adhesive (DPAS100) was synthesized from organic-inorganic hybrid components and the permanent adhesive exhibited high thermal stability (maximum operating temperature, 300 °C). The functional groups and curing temperature of the permanent adhesive were optimized since the permanent adhesive needs to be cured within the operating temperature range of the temporary adhesive. To render a Si surface more compatible with an organic polymer, based on organosilane primer chemistry, the surface needs to be covered with an organosilane and the organic portion needs to face outward. In our process, DSM003-2 was supplied as an adhesion promoter made of a solution of organosilane compounds in PGMEA. One of the driving forces of adhesion between Si and the permanent adhesive is condensation of silanols with surface oxygen atoms or hydroxyl groups. TABLE I shows the properties and process conditions of the materials developed in this study.

TABLE I. PROPERTIES AND PROCESS CONDITIONS.

	Permanent adhesive on adhesion promoter	Adhesion promoter	Temporary adhesive
Thickness	0.5–5 µm	~10 nm	2–20 µm
Softening temperature	over 40 °C	-	over 130 °C
Solidification	135 ± 5 °C, 30 min and 170–195 °C, 30 min	100–120 °C, 5 min	200–230 °C, 5 min
De-bonding	-	-	200–235 °C, 1~10 mm/s
Cleaning solvent	-	-	PGME
Modified tape peel test of stacked wafers after TCT 1000 cycles	No delamination	-	-

978-1-7281-1500-9/19 $31.00 © 2019 IEEE

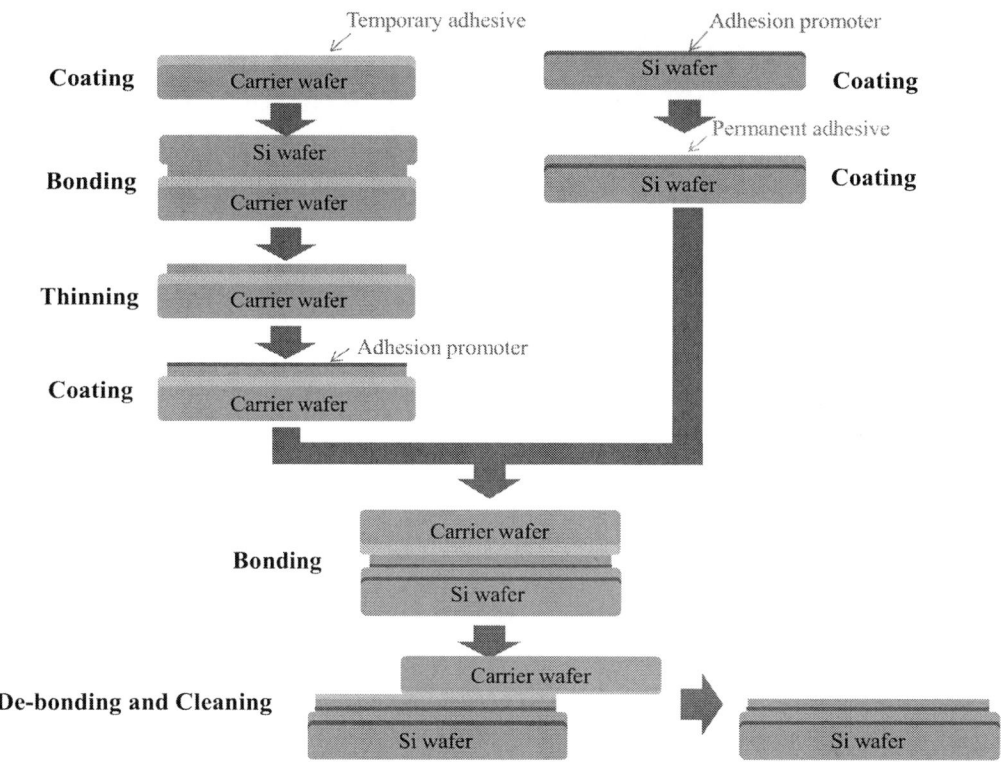

Figure 2. Process flow for preparation of stacked wafers.

B. Preparation of stacked wafers

The process of preparing stacked wafers is shown in Figure 2. On a carrier wafer, a temporary adhesive layer about 10 μm thick was formed, and a device wafer was stacked on top. After the device wafer was thinned to about 10 μm by grinding and polishing, an adhesion promoter was coated on the thinned wafer surface. In the next process, an adhesion promoter and a permanent adhesive layer were formed in this order on the device wafer surface for permanent adhesion, and the adhesion promoter-coated thinned wafer was stacked on top. The thickness of the permanent adhesive layer was about 2.5 μm. After a curing process at 135 °C for 30 min and then at 170 °C for 30 min, there was no void between the two stacked wafers. The carrier was de-bonded by sliding at a speed of 10 mm/s at 235 °C, and then the residual temporary adhesive was able to be removed with PGME.

C. Adhesion test methods

A number of different methods have been used to measure the adhesion of thin films. In previous reports, the results obtained by three different test methods, namely tape peeling, improved edge lift-off, and dicing, have been described in the case of a permanent adhesive coating on Si and SiO_2 coated with an adhesion promoter [2]. In this paper,

the results obtained in a modified tape peeling test after a thermal cycle test (TCT) are described for the case of stacked wafers in which a permanent adhesive was coated on Si and SiO_2 coated with an adhesion promoter.

The stacked wafers were diced as shown in Figure 3. After the dicing, a modified tape peeling test based on JIS K 5400 was performed before and after TCT 1000 cyclesfrom a minimum temperature of -55 °C to a maximum temperature of 150 °C. In this test, a semi-quantitative rate of damage to the diced wafers after removal of the tape was obtained.

D. Measurement of coating layers

Measurement of the permanent and temporary adhesive coating thicknesses were obtained from contact/non-contact measurement techniques. The *I-V* curve of the permanent adhesive coating was measured as a basic parameter of electrochemical ability. For the elements constituting the adhesives, the diffusion profiles of the elements that greatly affect the device characteristics were observed using SIMS.

E. TSV formation

In the TSV formation process for forming a bumpless interconnect, SiO_2 deposition, lithography and etching processes were carried out after thinning the Si wafer. In the TSV etching process, three steps were performed: SiO_2, Si,

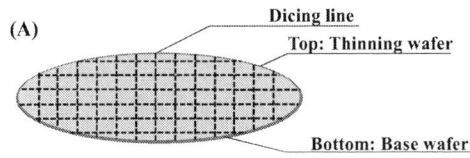

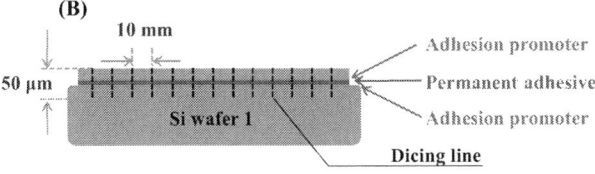

Figure 3. Diced wafers for a modified tape peeling test after TCT :
(A) overhead view, (B) cross-sectional view.

and permanent adhesive etching. Bottom cleaning processes, such as O_2 plasma, wet cleaning, and Ar sputtering, were performed to improve the contact resistance of the TSV by removing the contamination. Finally, metallization and planarization were carried out from the front, as a TSV last process.

III. RESULTS AND DISCUSSION

A. Adhesion of the permanent bonded wafers after 1000 cycles of TCT

In the modified tape peeling test described above, perfect bonding by the permanent adhesive coating on the Si and SiO_2 coated with the adhesion promoter was observed. In the modified tape peeling test, as shown in Figure 4, no chipping or delamination of the diced thinned wafers was observed before and after 1000 cycles of TCT from -55 °C to 150 °C. This result indicates that adhesion of the permanent adhesive had highly reliable TCT resistance.

B. Electrochemical ability of the permanent adhesive layer

The *I-V* curves of a 0.25 μm-thick permanent adhesive coating obtained at ten arbitrary locations are shown in Figure 5. The current density of the permanent adhesive coating, 10^{-9} to 10^{-8} A/cm^2, indicates that it is a near-perfect electrical insulator at -400 to 400 MV/cm.

C. Diffusion profiles from adhesive coatings to wafers

Among the elements constituting the permanent adhesive, the elements that greatly affected the device characteristics, namely, Sb and P, were observed using SIMS. In the SIMS profiles in Figure 6, the left axis shows P and Sb concentrations (atoms/cm³), and the right axis shows C and Si secondary ion intensities (count). As shown in Figure 6, no diffusion of the elements constituting the permanent adhesive into the wafers was observed. Although P and Sb were slightly detected in the vicinity of the Si substrate/adhesive layer interface in each sample, since it is equivalent to the rising edge of C in the adhesive component, clear diffusion into the Si substrate was not observed. This

Figure 4. Stacked wafers which were tested at 1000 thermal cycles from -55 °C to 150 °C and then attached and removed tape.

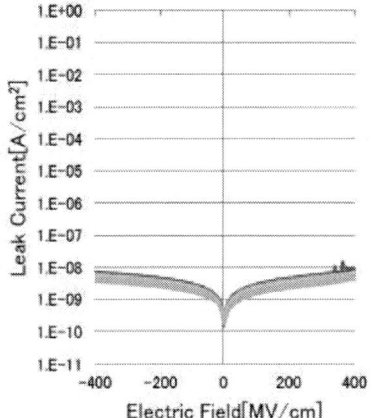

Figure 5. *I-V* curves of the permanent adhesive coating at ten optional locations.

means that the permanent adhesive components do not significantly affect the characteristics of the semiconductor device. For the temporary adhesive residue, which was considered to consist of hydrocarbon compounds, a uniform in-plane distribution was observed, and the thickness of the residue was estimated to be 1-2 nm from the SIMS profile.

D. Etching property and electrical characteristics of TSVs

The ideal etch profile is one in which there is no undercut at the interface between Si and the adhesive layer and no bowing shape, in order to suppress leakage paths into Si. As shown in Figure 7, the required profile was achieved by using the permanent adhesive.

In this work, the lengths of the TSV interconnects, which were equal to the total thickness of the thinned wafer and the adhesive layer, were 12 μm, and the diameter was 10 μm. Bottom cleaning processes such as O_2 plasma, wet cleaning, and Ar sputtering removed contaminants, including residues, byproducts, and oxides on the copper surface. Figure 8 shows the low TSV resistance of the WOW stack optimized using our adhesives.

In addition, the excellent uniformity of the TSV resistance in Figure 8 indicates low warpage and low curing shrinkage of the stacked wafers. The reason is that for stacked wafers such as those produced by the WOW process, wafer-to-wafer parallelism affects the line resistance,

978-1-7281-1500-9/19 $31.00 © 2019 IEEE 1005

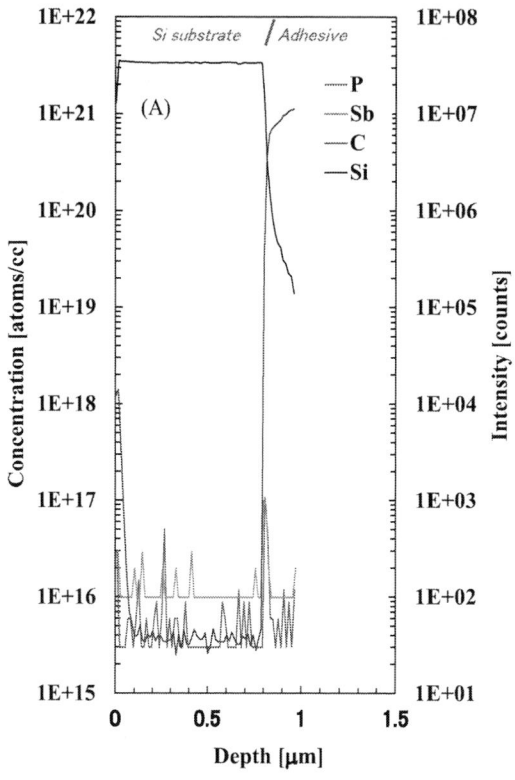

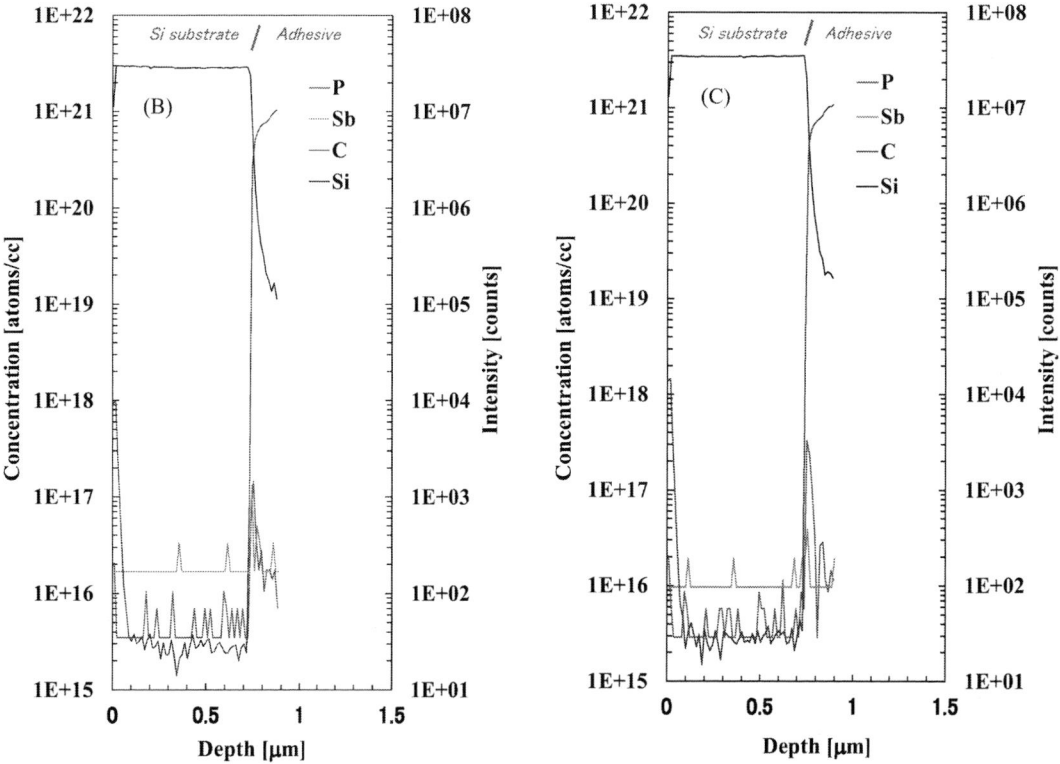

Figure 6. Diffusion profile from permanent adhesive to wafer using SIMS: (a) before heating, (b) center of wafer after 8 heating cycles at 280 °C for 30 minutes, (c) edge of wafer after 8 cycles of heating at 280 °C for 30 minutes. The left axis shows P and Sb concentrations (atoms/cm^3) and the right axis shows C and Si secondary ion intensities (count).

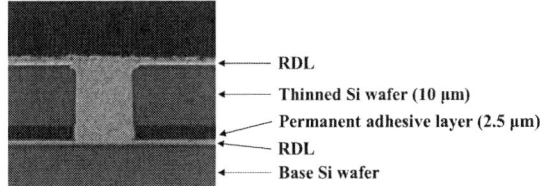

RDL
Thinned Si wafer (10 μm)
Permanent adhesive layer (2.5 μm)
RDL
Base Si wafer

Figure 7. A typical cross-sectional SEM image after etching, Cu metal filling, and planarization by CMP.

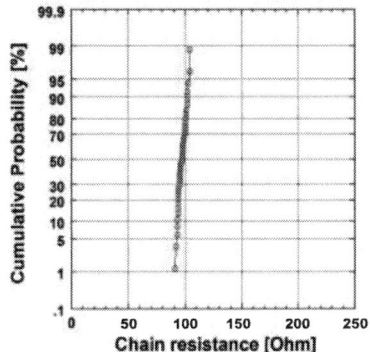

Figure 8. TSV daisy chain resistance of the optimized WOW stacking using our two adhesives.

contact resistance, and vertical TSV interconnect uniformity.

IV. CONCLUSION

Our permanent and temporary adhesives are suitable for fabricating multilevel stacks using ultra-thin wafers for wafer-on-wafer (WOW) technology with bumpless dual-damascene interconnects (via-last after bonding).

In the experiments described here, the reactive hot-melt type temporary adhesive had a bonding temperature of around 130 °C and a de-bonding temperature (>200 °C) higher than the bonding temperature. After de-bonding and washing with PGME, the temporary adhesive residue was uniformly distributed on the de-bonding wafer. The thickness of the residue, which is considered to consist of hydrocarbon compounds, was estimated to be 1–2 nm in the SIMS profile.

A permanent adhesive with lower curing temperature than that of BCB and heat resistance (maximum operating temperature, 300 °C) was successfully synthesized from specific organic-inorganic hybrid components. The permanent adhesive was able to be cured at 135–170 °C, which is lower than the de-bonding temperature (over 200 °C) of the temporary adhesive. The adhesion strength of the permanent adhesive coating to SiO_2 and Si showed highly reliable TCT resistance. In the SIMS results, the lack of a diffusion profile from the permanent adhesive coating on the wafer means that the adhesive components did not significantly affect the characteristics of the semiconductor device. The *I-V* curve of the permanent adhesive indicated that the permanent adhesive was a near-perfect electrical insulator. The permanent adhesive coating had a good etching profile and low TSV resistance with excellent uniformity using 300 mm stacked wafers. In addition, the low warpage and shrinkage of the stacked wafers were estimated from the TSV resistance with excellent uniformity.

In conclusion, our permanent adhesive with high heat resistance (maximum operating temperature, 300 °C), low curing temperature (<200 °C) and highly reliable TCT resistance will be useful for WOW processes.

ACKNOWLEDGMENT

This study was carried out in the WOW Alliance program of the Tokyo Institute of Technology.

REFERENCES

[1] Y. S. Kim, S. Kodama, Y. Mizushima, N. Araki, C. Hsiao, H. Chang, C. Lin, and T. Ohba, "Optimization of Via Bottom Cleaning for Bumpless Interconnects and Wafer-On-Wafer (WOW) Integration", Proceedings of the 2018 IEEE 68th Electronic Components and Technology Conference (ECTC), pp.1962, 2018.

[2] N. Araki, Y. S. Kim, S. Kodama, C. Hsiao, H. Chang, C. Lin, and T. Ohba, "Development of Micrometer-Thick Bonding Material for Wafer-On-Wafer (WOW) Applications", Proceedings of 13th International Microsystems, Packaging, Assembly and Circuits Technology (IMPACT), pp214-217, 2018.

[3] T. Ohba, N. Maeda, H. Kitada, K. Fujimoto, K. Suzuki, T. Nakamura, A. Kawai, K. Arai, "Thinned Wafer Multi-stack 3DI Technology", Microelectronic Eng., 87 p. 485 (2010).

[4] A. Klumpp, R. Merkel, R. Wieland and P. Ramm, "Chip-to-Wafer Stacking Technology for 3D System Integration", Proceedings of the Electronic Components and Technology Conference (ECTC), pp.1080, 2003.

[5] Y. S. Kim, N. Maeda, H. Kitada, K. Fujimoto, S. Kodama, A. Kawai, K. Arai, K. Suzuki, T. Nakamura, and T. Ohba, "Advanced Wafer Thinning Technology and Feasibility Test for 3D Integration", Microelectronic Eng., Elsevier, 107 pp. 65, 2013.

[6] Y. S. Kim, S. Kodama, Y. Mizushima, N. Maeda, H. Kitada, K. Fujimoto, T. Nakamura, D. Suzuki, A. Kawai, K. Arai, and T. Ohba, "Ultra Thinning down to 4-μm using 300-mm Wafer proven by 40-nm Node 2Gb DRAM for 3D Multi-stack WOW Applications", IEEE Symp. on VLSI Technol., pp.26, 2014.

[7] Y. S. Kim, S. Kodama, Y. Mizushima, T. Nakamura, N. Maeda, K. Fujimoto, A. Kawai, K. Arai, and T. Ohba, "A Robust Wafer Thinning down to 2.6-μm for Bumpless Interconnects and DRAM WOW Applications", IEEE IEDM Tech. Dig. pp.189, 2015.

[8] Y. S. Kim, S. Kodama, M. Maeda, K. Fujimoto, Y. Mizushima, A. Kawai, T.C. Hsu, P. Tzeng, T.K. Ku and T. Ohba, "Electrical Characteristics of Bumpless Interconnects for Through Silicon Via (TSV) and Wafer-On-Wafer (WOW) Integration", International Conference on Electronics Packaging (ICEP), pp 74., 2016.

[9] Y. S. Kim, S. Kodama, Y. Mizushima,, T. Nakamura, M. Maeda, K. Fujimoto, A. Kawai, and T. Ohba, "Warpage-free Ultra-Thinning ranged from 2 to 5-um for DRAM Wafers and Evaluation of Devices Characteristics", Proceedings of the Electronic Components and Technology Conference (ECTC), pp.1471, 2016.

[10] Y. Mizushima, Y. S. Kim, T. Nakamura, A. Uedono, T. Ohba, "Behavior of Copper Contamination on Backside Damage for

978-1-7281-1500-9/19 $31.00 © 2019 IEEE

Ultra-thin Silicon Three Dimensional Stacking Structure", Microelectronic Eng., Elsevier, 167 pp. 23, 2017.

[11] M. E. Mills et al., Microelectronic Eng., 33 p. 327 (1997).

Development of novel photosensitive dielectric material for reliable 2.1D package

Yune Kumazawa, Seiji Shika, Shunsuke Katagiri, Takuya Suzuki, Tsuyoshi Kida, Shu Yoshida
Mitsubishi Gas Chemical Company, Inc.
TOKYO TECHNOPARK 1-1, Niijuku 6-chome, Katsushika-ku,Tokyo, Japan
E-mail: yune-kumazawa@mgc.co.jp

Abstract— **Photo-sensitive dielectric materials applicable to advanced organic interposers are desired to realize cost-effective semiconductor packages equipped with fine patterning capability. In this work, the novel photo-polymerizable compound expected to be a major component of the photo-sensitive dielectric materials for 2.1D was developed and studied. The new compound has some unique features such as high transparency or enough reactivity, which are considered to lead to excellent process-ability for fine patterns, appropriate mechanical properties and high insulation reliability. Firstly, dependency of light transmittance on photo-polymerizability was investigated with several photo-polymerizable compounds, and it was confirmed that the new photo-polymerizable compound had enough capability of polymerization under typical UV exposure conditions. Secondly, the feasibility study on fine patterning was conducted, and it was confirmed that 10um via holes were formed on the composition based on the new compound without any technical issue. Thirdly, several sets of HAST (High Accelerated Stress Test) were conducted with 40um/40um or 10um/10um line and space test vehicles to compare insulation performances between the new compound and other conventional polymerizable compounds (epoxy and acrylate). Finally, the insulation reliability of fine wirings was studied with 5um/5um and 4um/2um test vehicles, and it was confirmed that the new compound would be applicable to the photo-sensitive dielectric material for 2.1D as the major component. Thus the effectiveness of the new photo-polymerizable compound for next generation photo-sensitive dielectric materials was successfully demonstrated.**

Keywords-2.1D; organic interposer; photosensitive dielectric materials; via hole; insulation reliavility;

I. INTRODUCTION

2.1D package capable of integrating wide band memory units (ex. HBM: High Bandwidth Memory) and logic dies (ex. GPU: Graphics Processing Unit) directly on an organic interposer has been considered to take the place of 2.5D. It is also expected that 2.1D will contribute to progresses of several technology drivers such as 5G communications and AI (Artificial Intelligence); however, 2.5D with Si interposer is still regarded as the only solution for real products in spite of too high production cost. Actually, most of the organic interposers being developed for 2.1D still have some technical issues, especially in the process-ability for fine patterning including small sized via holes and the insulation reliability of high density wirings.

Table 1 2.1D and 2.5 D PKGs

Item	2.5D	2.1D
Schematic diagram	Si interposer HBM GPU Substrate	Photosensitive Dielectric Material HBM GPU Substrate
Interposer	Silicon + Organic	Organic
Package profile	High	Low
Electrical properties	Poor	Good
Material cost	High	Low
Assembly cost	High	Low
Lead time	Long	Short
Availability	Mass Production	Under development

So, this work aimed to enhance performances of the organic interposers by providing the new photo-sensitive dielectric material to the semiconductor market on the basis of experiences of BT resins so that reliable 2.1D would be realized with the reasonable production cost (Table 1) [1].

As already known well, an interposer targeted for advanced SiP (System in Package) is normally required to have fine pattering capability which realizes high density wirings (Line and Space: 2um/2um and below) with small via holes (Via size: 5um-10um). Although it is often said that the target via size should be less than 5um, the conventional via forming process with CO_2 laser beam for standard package substrates is approaching to the theoretical limitation around 20um. In addition to the technical limitation, large number of via holes required in 2.1D will be supposed to cause drastic increase of via forming cost. That is why this work focused on the photo-sensitive dielectric material for the organic interposers, especially on the film type material which can accomplish the excellent process-ability of via holes with low production cost.

Moreover, it is becoming more difficult for the conventional dielectric materials to ensure the insulation reliability for fine wirings. Generally, the photo-sensitive dielectric materials targeted for the organic interposers are designed to contain two major components, which are a photo-polymerizable compound for fine patterning under UV exposure and a thermosetting compound to ensure the insulation reliability with its strong adhesion to combined materials [2]. For example, photo-polymerizable acrylate compounds and thermosetting epoxy compounds are often adopted in the typical photo-sensitive dielectric material. In this case, the epoxy compounds cover the lack of adhesion of the acrylate compounds, however, it is becoming severe to achieve sufficient adhesion to keep insulation reliability only by the epoxy compounds. To realize the advanced interposers associated with fine wirings, the photo-polymerizable compound should also fulfill the role to enhance the insulation properties of the photo-sensitive materials.

In this work, some types of photo-polymerizable compounds were newly developed and studied to get some clues leading to the reliable photo-sensitive dielectric materials for 2.1D interposers.

II.   EXPERIMENTS

A.   Feasibility study on photo-polymerizable compound
A-1   Light transmittance

Normally, the photo-sensitive dielectric materials contain one or several photo-initiators to start their polymerization. The photo-initiator generates active species under UV exposure, and followed by the polymerization as shown in Fig.1. The photo-initiator has to absorb sufficient energy from UV radiation through the composition to proceed polymerization of the photo-polymerizable compound by created active species. So, high transparency of the photo-polymerizable compound is an important parameter to ensure the patterning process by UV exposure.

Generally, the light transmittances of typical acrylate compounds used for photo-sensitive materials dedicated to package substrate are said to be around 90-98% at 405nm.

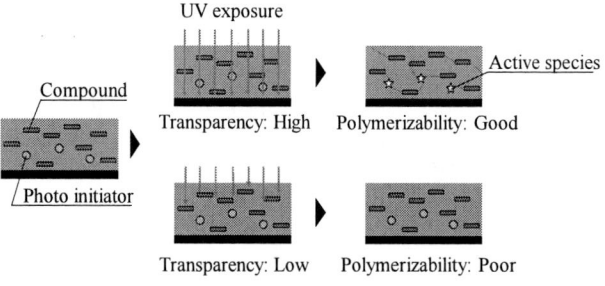

Fig.1 Polymerization process

Table 2 shows the light transmittances of the compound A (newly developed in this work) and 2 types of acrylate compounds which are typically used as photo-polymerizable components. The measurement was conducted by UV-vis spectroscopy this time.

The light transmittances of Acrylate 1 and 2 were 99% and 98% respectively and that of the compound A was 88%.

This result suggests that the compound A has little bit lower transparency than the other acrylate compounds, but it does not affect the behavior of the photo-initiator.

In other words, the photo-polymerizable compound A has enough potential to be polymerized under normal UV exposure and would be able to become a suitable component for the photo-sensitive dielectric materials.

Table 2 Light transmittance at 405nm

	Light Transmittance [% T]
Compound A	88
Acrylate 1	99
Acrylate 2	98

A-2   Polymerization under UV exposure

Next, photo-polymerizability under UV exposure was investigated to understand the capability of the compound A as a component of the photo-sensitive dielectric materials. When the photo-sensitive dielectric materials are applied as insulation layers in 2.1D interposers, they should have enough reactivity under moderate UV radiation, because the excessive UV exposure brings some side effects such as high manufacturing cost due to long process time or uncontrollable diffraction leading to insufficient resolution.

To evaluate the performances of several combinations of some photo-polymerizable compounds and photo-initiators, FT-IR was conducted in this work. Peak intensity ratios at specific wave numbers appeared in each compound were compared before and after UV exposure.

In this evaluation, each varnish composed of each photo-polymerizable compound (compound A or Acrylate 1) and suitable photo-initiator (5wt%) was prepared and coated on CCL (Copper-Clad Laminate). Each CCL coated with each varnish was dried for 5 minutes in a clean oven. Measurement conditions of FT-IR were as follows.

UV Exposure time: 60sec
Intensity of UV: 10mW/cm^2
Wave length: 405nm

Fig.2 shows a typical spectrum of an acrylate compound obtained by FT-IR. In the case of a typical acrylate compound, it is known well that the intensity at 833cm^{-1} decreases along with the progress of polymerization during UV exposure. So, the consumption of reactive functional groups contributing to polymerization can be analyzed easily by tracing the changes at their specific wave numbers.

Fig.3 shows the consumption rates of reactive functional groups of compound A and Acrylate 1, and it was confirmed that the difference between their consumption ratios was very small and ignorable. This result means that the polymerizable performances of compound A and Acrylate 1 under UV exposure are almost same in the viewpoint of practical use.

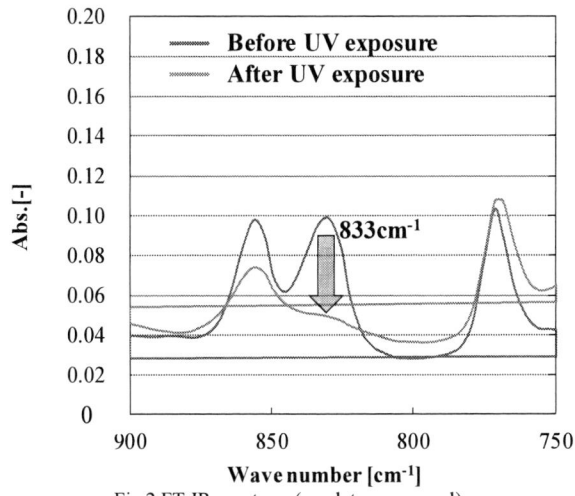

Fig.2 FT-IR spectrum (acrylate compound)

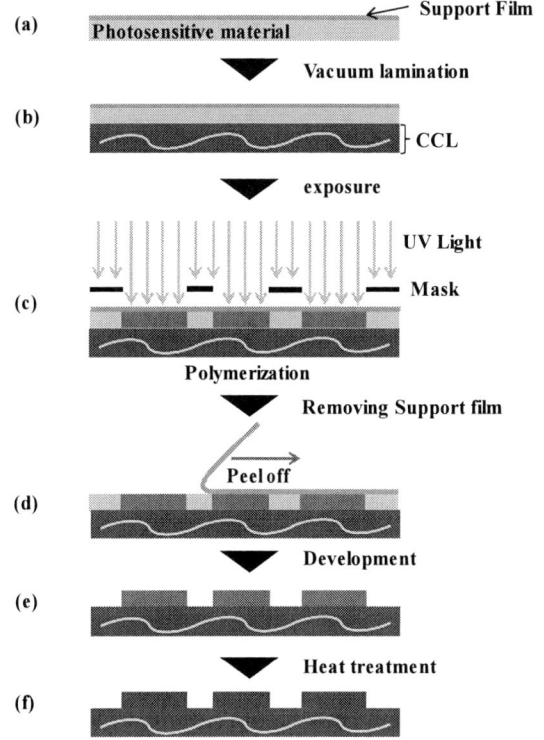

Fig.4 Patterning test flow

Fig.4 shows the patterning test flow in this work. Firstly, the composition composed of compound A and 5wt% photo-initiator is coated manually on support film (ex. polyethylene terephthalate film) and dried in a clean oven (a). The dried composition is laminated on CCL by vacuum lamination (b). The composition is exposed to 300mJ/cm^2 of 405nm-UV light (17mW/cm^2) (c). The support film is peeled off from the composition (d). Development is conducted for 180sec with basic solution. (ex. Tetra-methyl-ammonium hydroxide) (e). Heat treatment is conducted at 180degC for 1 hour (f).

Fig. 5 (a), (b) show images of 20um and 10um via holes. In this work, 20um and 10um via holes were formed in 10um thickness and 5um thickness compositions respectively.

Fig.3 Consumption of reactive functional groups

A-3 Patterning test on CCL

As mentioned before, the compound A had enough transparency for UV exposure and would have practical reactivity for polymerization. Then patterning test on CCL was performed as the next feasibility study.

As already known, two forms of photo-sensitive dielectric materials, liquid and film types have been proposed for the 2.1D interposer. Although each type has each pros and cons, the film type was adopted in this work because the film type has excellent flexibility and good compatibility with current manufacturing processes in substrates makers.

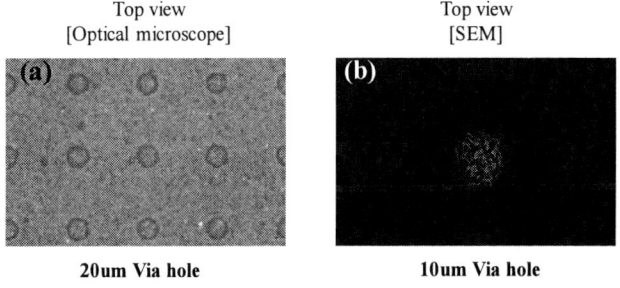

Fig.5 Via holes (a) 20um, (b) 10um

B. Insulation Reliability

Along with the shrinkage of wiring L/S on the interposers, the insulation reliability are apparently becoming more important. Especially, 2.1D is considered to have finer wiring than any other conventional substrates, therefor, more reliable insulation properties should be indispensable.

Photo-sensitive dielectric materials used for semiconductor devices often encounter critical insulation troubles caused by moisture or other environmental factors. Particularly, absorbed moisture in the package accelerates the creation of electro migration (Cu^{2+}, Sn^{2+} and so on). Moreover, it is often said that it is impossible for the photo-sensitive dielectric materials to ensure insulation reliabilities as same as thermosetting resin does.

In this work, HAST was conducted to evaluate the performance of each photo-polymerizable compound in the form of composition containing each photo-initiator. The insulation properties derived from the compound A and typical acrylate compounds were also compared to understand the effectiveness of the compound A.

B-1   HAST with 40um/40um L/S test vehicle

HAST was conducted with the test vehicles shown in Fig.7. The test vehicles were prepared by laminating dried compositions comprising of the compound A and the photo-initiator on the patterned CCL made of BT resin. Detailed conditions for sample preparation are as follows.

Thickness of composition layer: 30um
Line and space of CCL: 40um/40um
UV exposure: 500mJ/cm^2 (17mW/cm^2)
Wave length: 405nm
Heat treatment after UV exposure: 180degC, 1hour

Prior to HAST, pre-treatment was conducted to the test vehicles by following JEDEC MSL2 and 260degC reflow at three times. HAST conditions were 130degC/ 85%RH and 5 VDC.

For reference, an acrylate compound and an epoxy compound were also evaluated in a same manner to compare with the compound A.

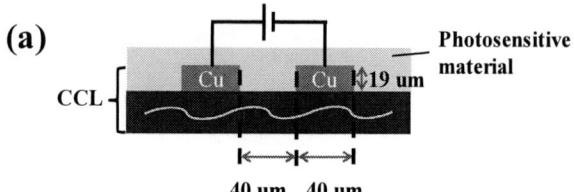

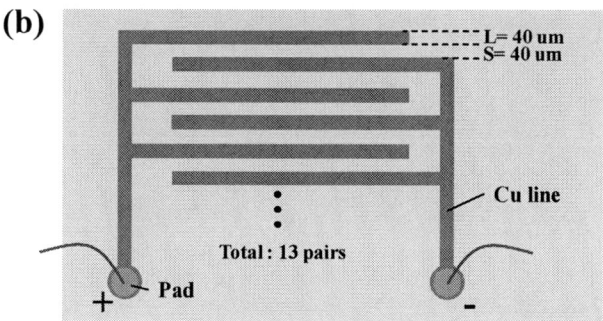

Fig.6 Test vehicle (L/S: 40um/40um)

Fig.7 shows the transitions of electric resistances of the compositions based on photo-polymerizable compounds. The resistances of all samples were normalized by each initial value to make it easy to compare each behavior. In this step, the compound A, an epoxy compound (Epoxy) and an acrylate compound (Acrylate) were compared.

The resistance of Epoxy was dropped rapidly around 50h and failed after 55h. After HAST, the resistance of Epoxy showed less than 1.00E-06 and did not recover to high resistance even after drying. The resistance of Acrylate decreased gradually and dropped to 1.08E-02 after 400h.

On the other hand, the resistance of the compound A showed stable transition and maintained over 1.0E-01 even after 400h.

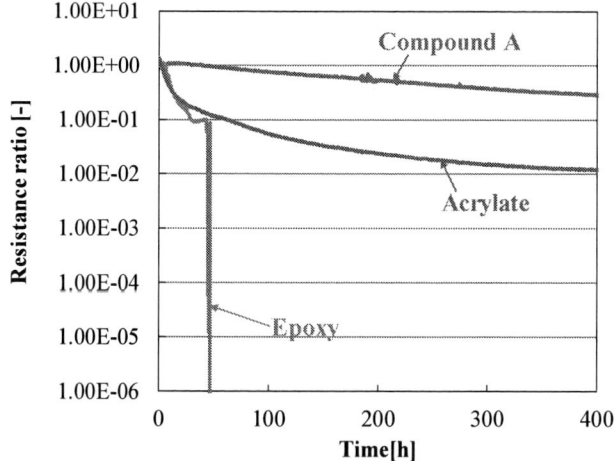

Fig.7 Transitions of Electrical resistance in HAST
(L/S=40um/40um)

To clarify failure modes in Epoxy and Acrylate, samples after HAST were analyzed optically by microscope. As

Table 3 shows, it was confirmed that electro-migration (Cu dendrite) were occurred between adjacent Cu lines of Epoxy and Acrylate samples. In contrast, no Cu migration was observed in any samples based on the compound A.

From the details collected in this analysis, it was assumed that de-laminations between the composition and CCL occurred firstly, then Cu migration proceeded in succession. In addition, cationic ions brought by the photo-initiator accelerated the creation of Cu ions leading to Cu migration in Epoxy samples.

The mechanism explaining that the compound A did not show any migration is still under investigation, but it can be said that the adhesion strength between the composition and CCL might be strong enough to prevent de-laminations in only samples based on the compound A.

Table 3 Sample analysis after HAST (400h)

Type	Compound A	Acrylate	Epoxy
Image ×1000			
Electro-migration	None	Observed	Observed

B-2 Molecular structure vs. Insulation reliability

To study the relation between the insulation reliability and the molecular structure of the photo-polymerizable compound, 2 types of compounds (compound B, C) were newly prepared by modifying the compound A (Fig.8). The compound B has acrylate groups, and the compound C has epoxy groups as reaction points for polymerization. All compounds has the same structure in their molecules except for respective functional groups (acrylate, epoxy, X) for polymerization.

For this evaluation, 10um/10um line and space test vehicles were prepared and compositions based on compound A, B, and C (Table 4) were laminated on the patterned CCL. HAST conditions were same as the abovementioned 40um/40um L/S test.

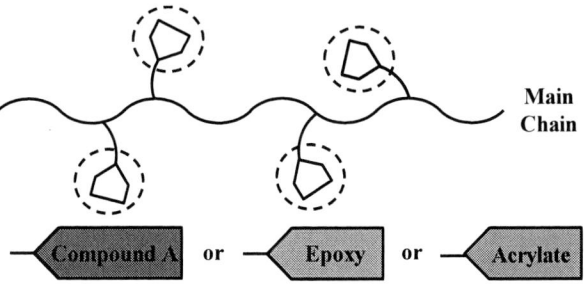

Fig.8 Molecular structure image

Table 4 Compositions for HAST

	Composition A	Composition B	Composition C
Major Compound (Functional group)	Compound A (Group X)	Compound B (Acrylate)	Compound C (Epoxy)
Initiator	Initiator 1	Initiator 2 (radical)	Initiator 3 (cation)

Fig.9 shows the transitions of normalized resistances during HAST. The resistance of the composition C containing epoxy groups dropped rapidly around 60h. In case of the composition B (acrylate groups), the resistance decreased gradually and reached to 1.41E-01 after 400h. On the other hand, the resistance of the composition A (X groups) maintained over 1.0E+00 during HAST. The composition A based on the compound A showed the most stable resistance.

For further investigation, samples after HAST were analyzed by microscope. As the result, Cu migrations between Cu lines were observed in the composition B and C, but any migration was not seen in the composition A. So it was found that the insulation reliability of the photo-sensitive dielectric material strongly depended on the type of functional group in the photo-polymerizable compound.

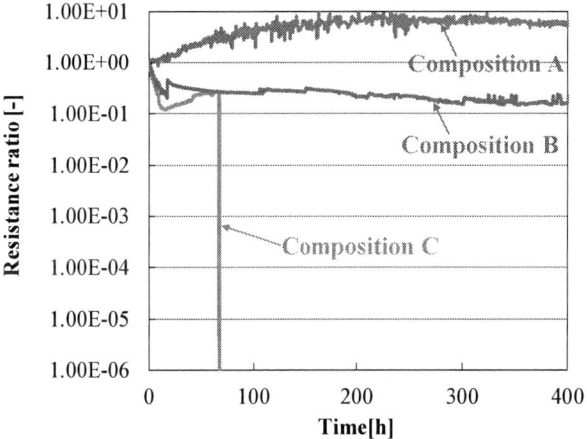

Fig. 9 Transitions of Electrical resistance in HAST (L/S=10um/10um)

B-3 Insulation reliability of fine wiring

As the final step in this work, insulation reliability tests with 5um/5um and 4um/2um L/S vehicles were performed to confirm the insulation capability of the compound A for 2.1D interposers.

For this evaluation, another test vehicle was designed as shown in Fig.10 and Fig.11. Both of 4um/2um and 5um/5um vehicles were laminated with the composition A by vacuum lamination. Assembly conditions (lamination, UV radiation, etc.) for test vehicles were same as the previous evaluations.

Samples were treated by JEDEC MSL2 and 260degC reflow at three times prior to HAST. HAST conditions were 130degC/ 85%RH and 3.3VDC.

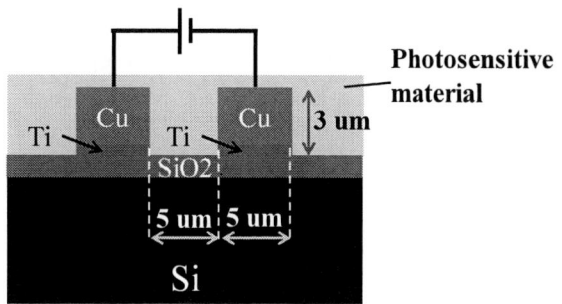

Fig. 10 Test vehicle (L/S: 5um/5um)

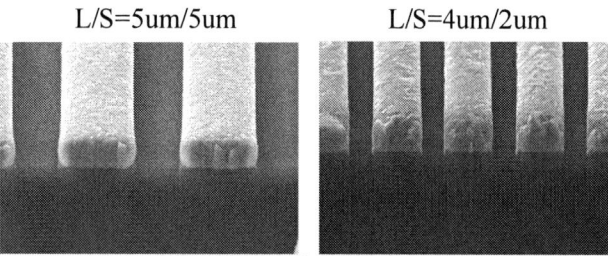

Fig.11 Cu lines (before lamination)

Fig.12 shows the transitions of normalized resistances of 4um/2um and 5um/5um test vehicles. It was confirmed that both of the resistances were very stable during HAST and showed similar trend. The compound A has enough capability to be applied to the photo-sensitive dielectric materials for 2.1D.

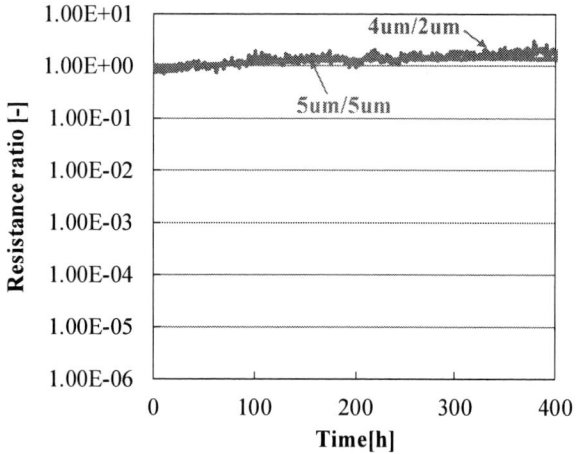

Fig.12 Electrical resistance in HAST
(L/S: 5um/5um and 4um/2um)

III. CONCLUSION

In order to realize the photo-sensitive dielectric materials for 2.1D organic interposer, the photo-polymerizable compound equipped with high insulation reliability must be the key component. In this work, the process-ability and the insulation reliability of the developed compound A was successfully demonstrated. In addition, it was confirmed that

the compound A has the superior properties in the comparison with the typical photo-polymerizable compounds.

IV. NEXT CHALLENGE

As the next challenge of the development, several evaluations in the form of multi layered vehicles emulated for actual devices will be performed to investigate the capability of the developed photo-sensitive material. Process-ability and reliability on smaller via (less than 5um) and finer wirings (less than 1um) will be also reported in the next report.

REFFERENCES

[1] W. Chung Chen, C. Wen Lee, M. Hua Chung, C. Chi Wan, S. Kun Huang, Y. Sen Liao, H. Chun Kuo, C. Chao Wang and D. Tarng, "Development of novel fine line 2.1 D package with organic interposer using advanced substrate-based process" , ECTC 2018

[2] X. Wei, Y. Shibasaki, "A novel photosensitive dry-film dielectric material for high density package substrate, interposer and wafer level package", ECTC 2016

High Reliability Solder Resist with Strong Adhesion and High Resolution for High Density Packaging

Sawako Shimada, Kazuya Okada, Tomoya Kudo, Chiho Ueta, Yuya Suzuki
TAIYO INK MFG. CO., LTD.
900 Hirasawa, Ranzan-machi, Hiki-gun, Saitama, 355-0215, Japan
Phone : +81-493-61-2728, Fax : +81-493-61-2729
E-mail : shimada.sawako@taiyoink.co.jp

Abstract—**This paper describes material analysis and material design of high reliability solder resist (SR) with excellent performance for high density packaging. There is growing demand for higher speed and higher data bandwidth signal transmission for many applications, such as 5G communication, artificial intelligence (AI), and advanced driver-assistance systems (ADAS). Such applications require high density and high performance IC packaging with fine Cu wiring and high frequency signal transmission. Solder resist materials for such packaging need to satisfy many special properties, such as high resistance to Cu electrochemical migration, strong adhesion to low profile Cu layers, and accurate photo-lithography resolution. However, development of a solder resist material that has all the excellent properties above is highly challenging, because many of these properties are trade-off. Indeed, adhesion of conventional SR to low profile Cu layer dropped more than 80% after high temperature and moisture HAST condition. Additionally, photolithography resolution below 50 μm was highly challenging due to light scattering. To overcome the trade-offs, this research began with the detail material analysis of the organic and inorganic components in SR materials. First analysis in polymer structures showed that resin with less shrinkage and less hydrolysis increased the initial adhesion, as well as adhesion after high temperature and high moisture condition. Next study on filler type and surface treatment revealed that the organic and inorganic surface treatment were effective to improve adhesion stability and resolution. This can be explained by the higher electrical affinity and less light scattering. By integrating the fundamental analyses, a new SR with excellent adhesion stability (85% of initial adhesion), high photolithography resolution below 40 μm, and excellent Cu migration resistance below 8 μm L/S.**

Keywords – Solder resist, Reliability, High density packaging, Advanced packaging

I. INTRODUCTION

Needs for the high performance electronic devices are increasing due to the evolution in new applications, such as 5G, ADAS, and AI. These devices require high performance packaging with high Cu wiring density for large number of I/O routing and low Cu surface roughness for reduced signal transmission loss at high frequencies [1]. SR for such packaging should satisfy many properties, including excellent Cu migration resistance and strong adhesion to surrounding materials for better packaging reliability, as well as high and precise photo lithography resolution for narrow bump pitch interconnections.

Our previous research revealed that the high loading of inorganic filler in SR has contributed to excellent Cu migration resistance under biased highly accelerated test (BHAST) and robust crack performance under thermal cycling test (TCT) [2,3]. However, large amount of filler inclusion in SR results in weak adhesion to surrounding materials, and poor photo lithography resolution. To overcome the challenges, this research focuses on material analysis and design to develop a new SR with strong adhesion and excellent resolution even under high filler loading.

Strong bonding of SR to surrounding materials, especially to Cu layer is critical, because delamination at the interface induces high risk of insulation failure. To enhance the adhesion between different materials, surface roughening processes are typically used for mechanical anchoring [4]. Conventional packaging utilizes chemical roughening of Cu surface in the order of micron level to improve adhesion between SR and Cu. However, such high Cu roughness creates high signal transmission loss, which is not desirable for high performance packaging. Recently, low profile Cu treatments with less surface roughening have been developed, and applied for high performance packaging. However, conventional SR has poor adhesion to the Cu with such processes due to poor mechanical anchoring. Figure 1 shows peel strength of conventional SR to Cu treated with a traditional process (Ra=0.5 μm) and a new process (Ra=0.05 μm). As described in the graph, adhesion of SR to Cu with the new process dramatically decreased after high temperature and high moisture condition. In addition to the lower Cu surface roughness, there are other factors to make it challenging to achieve high SR adhesion; 1) low polarity resin in the SR with high hydrophobicity for high Cu migration resistance, 2) low material flow during the SR coating process due to high filler loading. Lower intermolecular interactions due to these factors results in interfacial delamination failures.

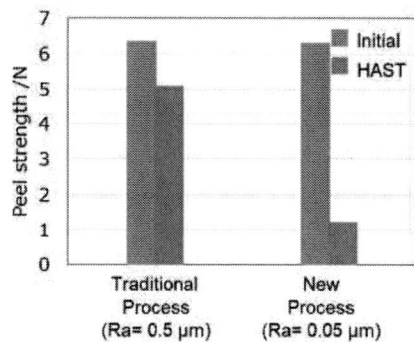

Figure 1. Cu peel strength of conventional SR with different roughening process

High photo lithography resolution of small solder resist opening (SRO) is another important criterion for next generation SR, because high density packaging has a lot of I/Os with fine bump pitch. Tight tolerance of SRO size is also critical, because it has influence on the solder joint reliability. However, large amount of filler inclusion results in poor resolution and tolerance, due to increased light scattering at the interface between the resin and filler particles. Since most of the SR materials are negative tone, light scattering at the edge of unexposed part results in narrowing of SRO size (figure 2).

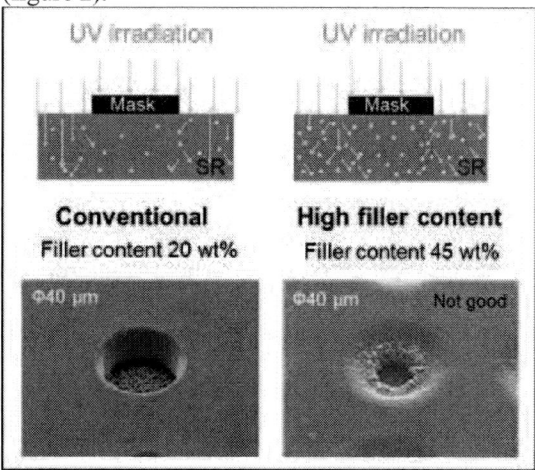

Figure 2. SR resolution with different filler content

This research focuses on the design of SR components, namely resin and filler systems to achieve strong adhesion with excellent resolution and tolerance under high filler loading. By integrating the findings from the fundamental material analyses, a new SR, with high reliability, strong adhesion, and excellent resolution and tolerance, was successfully developed. Following sections describe detail analyses and evaluations.

II. MATERIAL ANALYSIS

2-1. Analysis on polymer resin

This section explains the analysis on the alkali-soluble photosensitive resin for better adhesion. Factors below are expected to have impacts on the adhesion of SR to other materials.

1) *Amount of photo-reactive group*
2) *Amount of developable group*
3) *Molecular weight.*
4) *Apolar polymer backbone*

In this study, adhesion and properties of SR with various parameters above were evaluated, and the best suitable resin system was defined for the new SR.

2-1-1. Photo-reactive group

Amount of photo-reactive group was examined. Larger amount of reactive group generally contributes to better photo-lithography resolution, while it is expected to weaken the adhesion because of larger cure shrinkage. First, resin with various amount of photo-reactive group were prepared. Then they were coated on low-profile Cu foils with adhesion promotor (Ra=0.05 μm). After that, the samples were placed under highly accelerated stress test (HAST) condition, 130 °C and 85%RH, for 100 hours. Peel strengths of the Cu layers from resin surfaces were measured with Autograph AG-X (SHIMADZU CORPORATION) before and after HAST. As shown in figure 3, resin with lower photo-reactive group had initially higher adhesion as expected, but no specific correlation after HAST was observed.

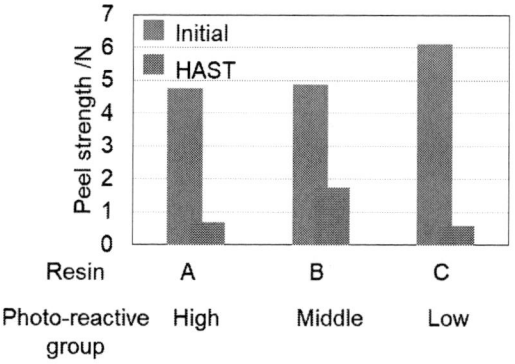

Figure 3. Cu peel strength of resins with different amount of photo-reactive group

2-1-2. Developable group

For more detail analysis on adhesion after HAST, hydrolysis rate of resin was examined, because resin generally degrades under high temperature and moisture by ester hydrolysis. Ester bonding exists in the developable group of resin system, so two resin systems with different amount of developable group were prepared and tested. First is resin B that had the highest adhesion after HAST in the previous test, and another is resin D with lower amount of developable group. First, hydrolysis rate after HAST condition of each material was tested by FT-IR spectrum 100 (PerkinElmer Japan Co., Ltd.). As expected, hydrolysis rate after HAST in resin D was 10-15% lower than that in resin B (table 1).

Table 1. Hydrolysis rate of resin B and D

Resin	B	D
Hydrolysis rate (%)	70~90	60~85

Figure 4 is the result of Cu peeling test, showing that the Cu peel strength of resin D after HAST was higher than that of resin B, which can be due to less hydrolysis in resin D.

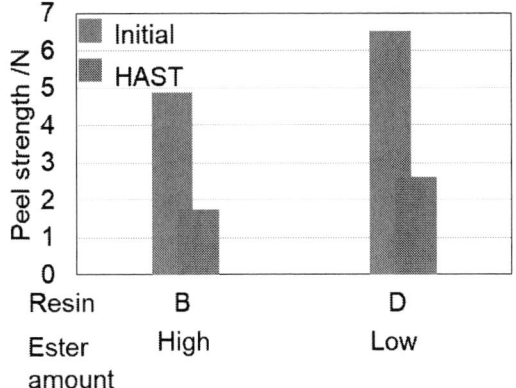

Figure 4. Cu peel strength of resins with different amount of developable group

Furthermore, initial Cu peel strength of resin D was found to be higher than resin B, which is unexpected since less polarity with smaller developable group could have led to lower initial adhesion. This could be attributed to 1/3 lower molecular weight of resin D, and better resin flow and coverage on the Cu surface.

2-1-3. Apolar polymer backbone

Next, analysis on polymer backbone structure was investigated. Incorporation of apolar and rigid cyclic group (such as benzene ring) into resin system is expected to reduce cure shrinkage as well as water absorption [5]. On the other hand, less polarity can cause decline in initial adhesion. Therefore, resin with heterocyclic ring group (resin E) was also tested, expecting larger molecular interaction with Cu surface while keeping low shrinkage and hydrolysis. Figure 5 shows the result of Cu adhesion before and after HAST, and Table 2 summarizes the rate of polymer hydrolysis. Resin E and F, which have cyclic group into the polymer backbone showed lower initial adhesion than resin D, while having higher bonding strength after HAST than D. Reduced initial adhesion can be explained by the reduced polarity, and enhanced adhesion after HAST can be due to the improved resistance to hydrolysis. Highly stable adhesion after HAST was achieved by incorporation of heterocyclic ring group.

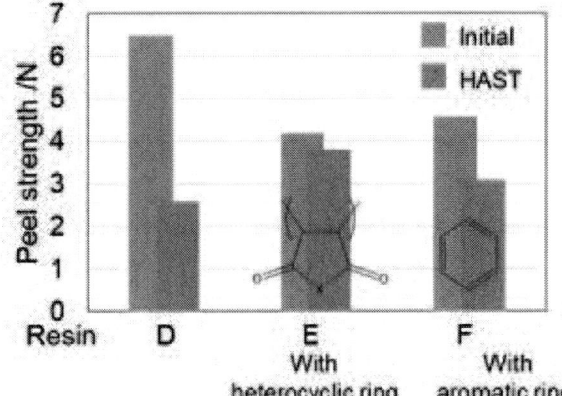

Figure 5. Cu peel strength of resins with various polymer backbone structures

Table 2 Hydrolysis rate of resin D, E and F under HAST

Resin	D	E	F
Hydrolysis rate (%)	60~85	50~70	60~85

As the final step of resin analysis, thermal material property was tested. As one of the important properties, glass transition temperature (Tg) of resin was measured with TMA-Q400EM (TA Instruments Japan Inc.) and values are summarized in table 3. Resin A, B and E have relatively high Tg, which is beneficial for high thermal stability, especially under reflow process, and high temperature storage test for automotive applications. High Tg of resin A and B may be due to large amount of photo-reactive group and their highly cross-linked polymer network after cure, while high Tg of resin E potentially because of its rigid polymer structure.

Table 3 Tg of various resins

Resin	A	B	C	D	E	F
Tg/deg.	179	179	157	156	181	158

To verify the hypothesis above, an additional test of the three resin materials (A, B, and E) to check substrate warpage was conducted. The materials were coated on 5 mm × 5mm × 36 μm size of Cu foils and warpage of the sample after full curing was evaluated as average of warp height of the four corners (table 4). Resin A and B showed larger substrate warpage, that indicates their large cure shrinkage associated with large amount of photo-reactive group. On the other hand, resin E has less warpage, which implies small cure shrinkage because of the rigid structure [6, 7].

Table 4. Substrate warpage after cure for resin A, B, and E

Resin	A	B	E
Warpage/mm	6.2	6.0	3.5

Analyzed results in this section are summarized as followings;

- Lower amount of photo-reactive group results in higher initial adhesion to Cu, and reduced substrate warpage after cure
- Reduction in developable group suppresses hydrolysis and adhesion degradation under HAST condition
- Resin with small molecular weight leads to higher adhesion
- Incorporation of heterocyclic ring group increases polymer Tg with stable adhesion under HAST

From these analyses, resin E was selected as a candidate organic part for our new SR material.

2-2. Analysis on filler

2-2-1. Type of filler

Filler materials are inorganic elements with high thermal stability and hydrophobicity, therefore higher loading of filler materials into SR increases Cu migration resistance and crack resistance, as reported previously [2,3]. Filler materials also largely influence material properties of SR, therefore SR with various filler materials were prepared and their properties were measured.

First, coefficient of thermal expansion (CTE) of SR with various filler materials were measured, since CTE is one of the most important properties for packaging stress, and crack performance. Table 5 summarizes the CTE of SR below Tg with 20 Vol% filler loading (SiO$_2$: 35 wt%, Others: 40~45 wt%) . SR with BaSO$_4$ had higher CTE compared to other SR materials.

Table 5. CTE of SR with various filler materials

Filler	SiO$_2$	BaSO$_4$	Al$_2$O$_3$	TiO$_2$
CTE α1 (ppm/K)	40~45	50~55	42~47	40~45

Table 6 summarizes the electrical properties of filler materials. Low dielectric constant is favorable for high speed and low loss signal transmission.

Table 6. Dielectric constant of filler materials

Filler	SiO$_2$	BaSO$_4$	Al$_2$O$_3$	TiO$_2$
Dielectric constant	3.5~4.5	7.9~11	8.5~10	40~80

Table 7 shows refractive indexes of filler materials at 350-450 nm, which is the range of wavelength used for photolithography processing. Large difference in refractive index between filler and resin causes poor SR resolution, because it leads to larger light reflection and scattering at the interfaces. Therefore, fillers with similar refractive index to

resin are preferred. From analyses on the mechanical, electrical, and optical properties, SiO$_2$ was selected for the outstanding properties.

Table 7. Refractive index of filler materials at 350-450 nm

Filler	SiO$_2$	BaSO$_4$	Al$_2$O$_3$	TiO$_2$	Resin
Refractive index	1.46~1.48	1.64~1.65	1.77~1.78	3.14~4.47	1.50~1.60

Finally, Cu peel strengths of SR with various filler materials were measured, following the same procedure in the subsection 2-1-1. As seen in figure 6, SR with SiO$_2$ had lower adhesion than other SR materials. To improve the adhesion of the SR with SiO$_2$, surface treatment of SiO$_2$ filler was conducted and the result is described in the next subsection.

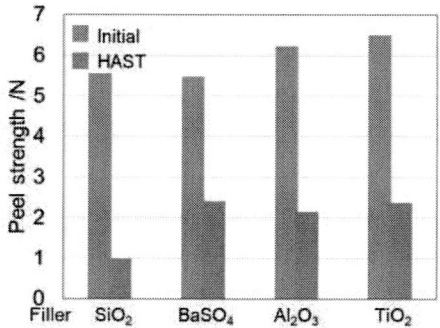

Figure 6. Cu peel strength of SR with different kind of filler

2-2-2. Filler surface treatment

As discussed, SiO$_2$ filler has various benefit in SR formulation, while it poses lower adhesion to Cu layer. Therefore, various surface treatments on SiO$_2$ were applied, and SR performance with the filler materials was evaluated.

First, as typical surface treatment, organic silanization processes were applied to SiO$_2$. Four different silane treatment agents were selected; a photo-reactive agent and three thermal-reactive agents A, B, and C. SR materials with the four type of filler were prepared and their mechanical properties were measured by nominal tensile test. Nominal tensile test was conducted with RSA-G2 (TA Instruments Japan Inc.), before and after HAST condition. Figure 7 shows the tensile strength and elongation values of the SR materials. All the SR with treated filler had higher tensile strength and elongation, especially SR with photo-reactive treatment had outstanding increase. This can be explained by suppressed interfacial fracture between filler and resin, due to enhanced chemical bonding.

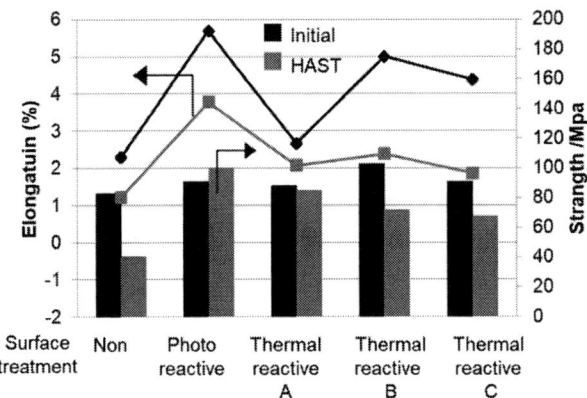

Figure 7. Elongation (line graph) and tensile strength (bar graph) of SR with various type of organic treated SiO_2 filler

Next, Cu peel strengths of these SR materials were measured and shown in figure 8. It was found that no specific difference in peel strength was observed between with or without organic silanization. Especially, adhesion after HAST condition decreased dramatically for all the SR.

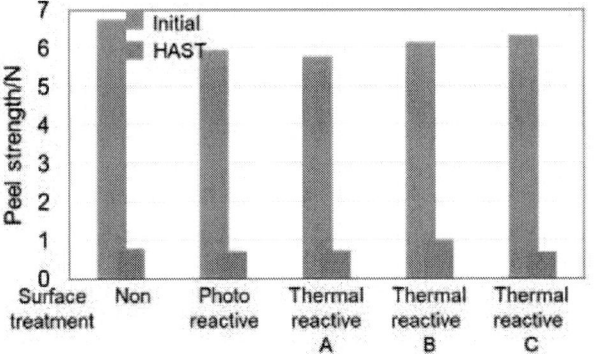

Figure 8. Cu peel strength of SR with various type of organic treated SiO_2 filler

To improve the Cu peel strength, inorganic surface treatment was examined. As discussed in the previous subsection, SR with Al_2O_3 filler maintained high peel strength after HAST, indicating their effectiveness in adhesion. Therefore, Al_2O_3 surface treatment was applied on the SiO_2 filler, modifying the adhesion performance of SR while keeping basic mechanical and electrical properties of SiO_2. Figure 9 shows the Cu peel strength of SR with non-treated SiO_2 filler, and SR with Al_2O_3 treated SiO_2 filler. By adding Al_2O_3 treatment, Cu peel strength after HAST condition were stabilized.

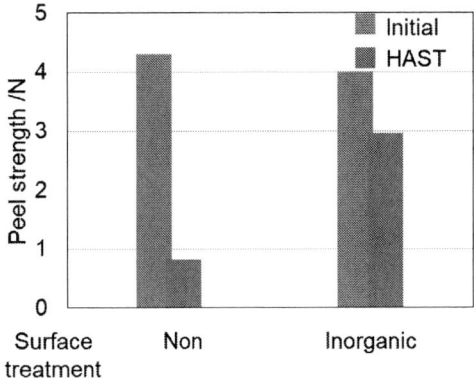

Figure 9. Cu peel strength of SR with inorganic treated or non-treated SiO_2 filler

SEM surface analysis on Cu foil after HAST and peel testing was conducted with JSM-7600F (JEOL Ltd.), shown in figure 10. Exposed Cu grains for non-treated SiO_2 indicate interfacial delamination, while Cu surface for treated SiO_2 is fully covered by SR that indicates cohesive failure. This change in fracture mode between Cu foil and SR explains adhesion improvement by applying Al_2O_3 treatment on SiO_2 filler.

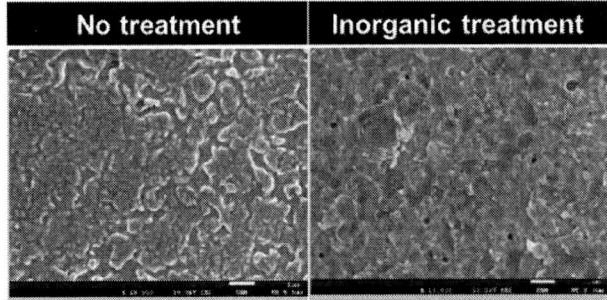

Figure 10. SEM images of Cu foil surface after HAST and peel testing

For more detail analysis, zeta potentials of filler materials were evaluated. The zeta potentials represent the electrical potential on the sliding surface of fine particles in colloid solutions. From the magnitude of zeta potential, affinity between the fine particles and other materials can be predicted, therefore it is adopted to the evaluation of dispersion stability of colloidal solutions and adhesion stability of inkjet ink on paper [8,9]. Therefore, zeta potentials of filler materials can be indicator of the SR adhesion. Zeta potentials of SiO_2 filler materials, without surface treatment, with photo-reactive organic treatment, and with Al_2O_3 inorganic treatment, were measured. Measurement was conducted with ELSZ-2000 (Otsuka Electronics Co., Ltd.) and shown in figure 11. Dramatical decrease in magnitude of zeta potential, from -70 mV to 15 mV by inorganic treatment, was observed.

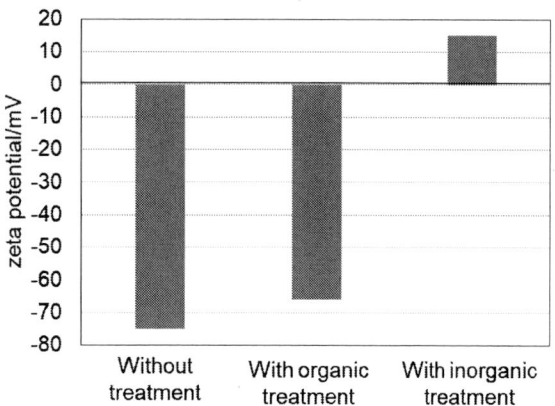

Figure 11. Zeta potential of SiO$_2$ filler with different treatments

In addition, zeta potentials of SiO$_2$, BaSO$_4$, Al$_2$O$_3$, and TiO$_2$ filler surfaces were measured (figure 12). Tendency of much smaller zeta potential of Al$_2$O$_3$ filler compared to SiO$_2$ filler is in accordance with the previous observation on Al$_2$O$_3$ treated SiO$_2$ filler. Large magnitude of zeta potential generates repulsion force between filler particles in SR, potentially causing less adhesion onto surrounding materials. Furthermore, electrostatic attraction is less susceptible to high temperature and moisture condition, which could be the reason for higher adhesion stability of SR with Al$_2$O$_3$ treated SiO$_2$ after HAST.

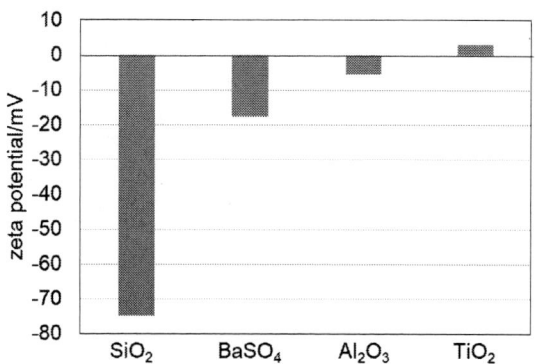

Figure 12. Zeta potential of filler materials

In addition to the adhesion performance, photolithography resolution of SR with and without Al$_2$O$_3$ treatment was tested. Designed opening size for the SR opening (SRO) was 50 μm, and SEM pictures are shown in figure 13. SRO size in SR with Al$_2$O$_3$ treatment was 50 μm as designed. However, SR without treatment had smaller SRO than designed, which indicates larger light scattering in SR at the edge of photomask as discussed in figure 2.

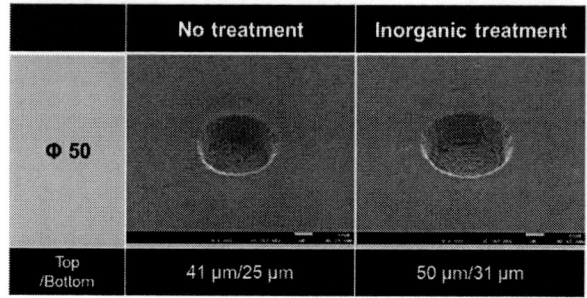

Figure 13. SEM images of SRO (50 μm design)

This could be attributed to the change in refractive index of filler particles by Al$_2$O$_3$ surface treatment. As discussed earlier, Al$_2$O$_3$ has higher index (1.77-1.78) than SiO$_2$ (1.46-1.48) at 350-450 nm. By applying Al$_2$O$_3$ surface treatment, refractive index of SiO$_2$ can be increased to that of resin (1.5-1.6). As a result, lower light scattering at the filler-resin could contributed to the higher resolution accuracy of SR.

The analyses in this subsection are summarized as followings;

- SiO$_2$ filler effectively enhances mechanical, electrical properties of SR, while adhesion to Cu is low
- Mechanical properties of SR with SiO$_2$ filler can be enhanced by photo-reactive organic silanization
- Adhesion of SR to Cu was increased by Al$_2$O$_3$ surface treatment on SiO$_2$ filler
- Photolithography resolution accuracy of SR was improved by Al$_2$O$_3$ surface treatment on SiO$_2$ filler

From the results, SiO$_2$ filler with Al$_2$O$_3$ and photo-reactive silane surface treatments was selected for our new SR material design.

III. DEVELOPMENT OF NEW SR

This chapter describes development of new SR that has stable adhesion, excellent resolution, high Cu migration resistance, and robust crack resistance. From the analyses in chapter 2, two main components in the new SR were selected.

1. Resin E with heterocyclic ring and proper amount of functional groups
2. SiO$_2$ filler treated with Al$_2$O$_3$ and photo-reactive silane agents

A new SR was prepared by incorporating these components with high filler loading around 50 wt%. Figure 14 shows Cu peel strengths of reference SR (conventional resin and filler without surface treatment) and the newly formulated SR on low-profile Cu (Ra: 0.05 μm). Peel strength of reference SR decreased after HAST condition down to 20% of initial value. On the other hand, the new SR had 20% higher initial adhesion and kept 85% of the initial value even after HAST condition.

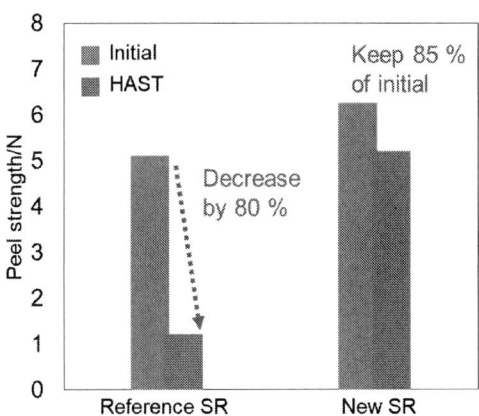

Figure 14. Cu peel strength of reference SR and new SR

Photolithography resolution of the new SR was evaluated, and SEM pictures are shown in figure 15. Small SRO size down to 40 μm was successfully formed in the new SR, which could not be achieved in conventional SR with high filler loading (figure 2). SRO size in the new SR was as designed with high accuracy, without undercut or halation issues. This is because of suppressed light scattering at filler-resin interface even with high filler loading.

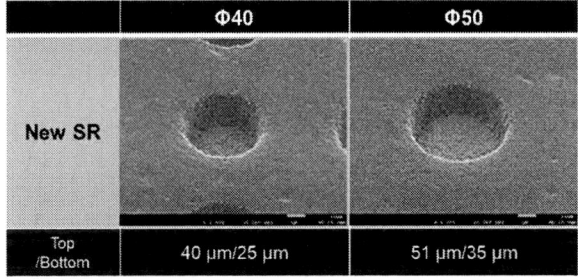

Figure 15. SEM images of SRO in new SR (left: 40 μm design, right: 50 μm design, SR thickness: 20 μm)

Cu migration resistance was tested with reference and new SR. These SR materials were first coated on top of fine pitch Cu wiring with 8 μm/8 μm line and space structures. Then 3.3 V voltage bias was applied in between the wirings, under HAST condition. Failure criteria was 63.2% reduction of initial resistance. Figure 16 shows the Weibull plot of the failure rate for reference SR and new SR. New SR showed longer life to failure, because of less amount of organic component where Cu ion migration takes place. Additionally, less polarity and high rigidity in the new resin system, with less moisture uptake and hydrolysis, may have contributed to the improvement in Cu migration resistance.

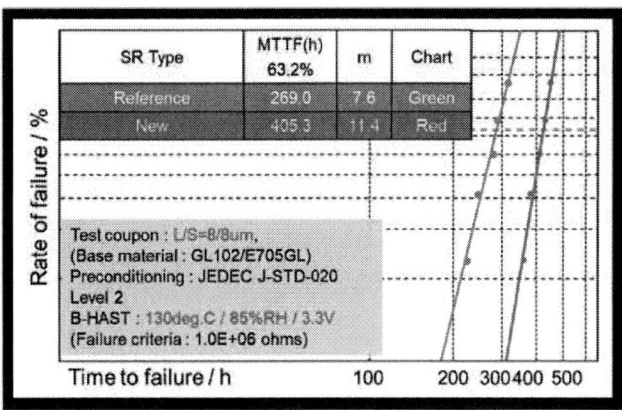

Figure 16. Weibull plot from Cu migration test

IV. CONCLUSION

Analyses of organic and inorganic components provided design guideline for material development. Chemical structure of resin was tailored to realize excellent SR adhesion. Organic and inorganic mixed filler surface treatment was found to be effective to improve mechanical properties of SR, adhesion stability, and photolithography resolution. By integrating the best components, a new SR with high reliability, strong adhesion to low profile Cu, and excellent resolution, was successfully developed. This material is beneficial for high density packaging that is vital need for various applications, such as 5G, ADAS, and AI.

[1] C. H. Yu, L. J. Yen, C. Y. Hsieh, J. S. Hsieh, Victor C. Y. Chang, C. H. Hsieh, C. S. Liu, C. T. Wang, KC Yee and Doug C. H. Yu"High Performance, High Density RDL for Advanced Packaging" IEEE Electronic Components and Technology Conference(2018) pp.587-593

[2] Chiho Ueta, Kazuya Okada, Toko Shiina, Tadahiko Hanada, Nobuhito Ito "Development of Solder Resist with Improved Adhesion at HTSL (175 deg C for 3000 Hours) and Crack Resistance at TST for Automotive IC Package" IEEE Electronic Components and Technology Conference(2017) pp156-165

[3] Kazuya Okada, Toko Shiina "Development of Photosensitive Solder Resist with High Reliability for Semiconductor Package" IEEE Electronic Components and Technology Conference(2017)

[4] Yasuaki Suzuki "Mechanism of Bonding of Dissmilar Materials and Applied Technology" Journal of the Adhesion Society of Japan ,Vol.54 No.5, 2018, pp 169-188

[5] Masashi Kaji "Structure and Properties of Epoxy Resins Containing Aromatic Unit in the Main Chain" Jounal of Network Polymer, Vol.32 No.1, 2011, pp35-40

[6] Kazuo Arita "Fundamental Study on Development of High Heat Resistant Epoxy Resins" Jounal of Network Polymer,Vol.36,No.5,2015, pp255-264

[7] Miyuki Harada ,Adhesion Technology, Japan, Vol.37, No.1, 2017,pp-1-5

[8] Shinnosuke Usui, Hiroshi Sasaki "Dispersion and Congulation of Fine Particles – Fundementls and aplocations –" Journal of the mining and materials processing institute of japan, 107,No.9, 1991, pp 585-591

[9] Shoichi Nakamura, "Surface Zeta-Potential Analyzer Analysis of cllarge on Polymer Membrane using Zeta-potential" MEMBRANE,30,(6),2005, pp344-347

Method for mitigating the warpage of Ultra-thin FC-CSPs by controlling of EMC properties

Chika Arayama, Takahiro Akashi, Yasunari Tomita, Naoki Kanagawa
Panasonic Corporation
Automotive & Industrial Systems Company
2-3 Tomarikoyanagi-cho, Yokkaichi city, Mie, Japan
e-mail: arayama.chika@jp.panasonic.com

Abstract— Recent developments in high-speed communication technology have accelerated reductions in the size and thickness of semiconductor packages. Fan-out wafer level packaging (FO-WLP) is an emerging packaging trend. Considering cost and scalability, there are also high expectation for ultra-thin Flip chip-Chip scale package (FC-CSP) technology which has evolved into a structure that is an approaching die-sized dimensions. As the packaging structure changes to Ultra-thin FC-CSP, it is necessary to control package warpage with smaller quantity of epoxy molding compound (EMC.) This study evaluated strip warpage of molded array package (MAP) and singulated unit warpage after dicing to investigate the effect of EMC properties on the package warpage. It has been understood that the mold shrinkage of EMC is the most dominant factor in warpage control of packages. However, the degree of warpage amount is not proportional to various properties in a packaging structures with a high ratio of silicon die to EMC, such as Ultra-thin FC-CSP when modelling low strip and unit warpage. For strip warpage, high mold shrinkage and low 25°C flexural modulus showed a proportional relationship. While high Tg demonstrated some effectiveness for unit warpage, clear correlation was not established. This investigation developed a "molding stress index" which is calculated by integrating the storage modulus and the coefficient of thermal expansion (CTE) multiplier over a temperature range from temperature (35 °C) to molding temperature (175 °C.) The results demonstrate that this parameter can be used to evaluate the changes in properties such as CTE, modulus and Tg in accordance with the change of temperature comprehensively. There is a tendency toward strip and unit warpage when the molding stress index is high. This study suggests that the molding stress index is an effective evaluation method to lower the unit warpage. This evaluation also confirmed that both lower strip and unit warpage for various substrates were realized with EMC having high mold shrinkage and low modulus at normal temperatures while maintaining a high molding stress index. In addition, further warpage reduction is possible by jointly optimizing the properties of EMC and substrate.

Keywords- Ultra-thin FC-CSP, Warpage control technology, Epoxy molding compound, Substrate

I. INTRODUCTION

Recent developments in high-speed communication technology are driving semiconductor device performance improvements. As the Internet of Things (IoT) adoption accelerates, the demand for small form factor, high density packaging continues to increase. Even as flip chip semiconductor packages are getting smaller, their electrical performance is improving [1-4]. For example, there are high expectations for fan-out wafer level packaging (FO-WLP) semiconductor packaging technology. As a result of cost and scalability advantages, there are also high expectations for ultra-thin Flip chip-Chip scale package (FC-CSP) (Fig.1) which has evolved into a structure that is extremely thin as and approaches wafer-scale dimensions [5].

A feature of Ultra-thin FC-CSP is high volumetric ratio of silicon wafer (die) in the molded package (Fig. 2). Warpage control of Ultra-thin FC-CSP becomes more difficult than normal FC-CSP form factors because the smaller amount of EMC available to provide rigidity to the package.

To increase throughput and increase material efficiency, thin packages can be manufactured by mold array package (MAP) technology. However, there have been problems in controlling both the strip (after molding) and unit (after dicing) warpage. Solutions to these issues have been reported by some papers, but these studies were mainly conducted using normal type FC-CSP devices with thick substrates and mold caps. Limited research has been reported on Ultra-thin type package [6-8].

In this study, MAP form test vehicles simulating ultra-thin FC-CSP were prepared and evaluated for strip and unit warpage after molding to examine the effect of EMC properties on the package warpage.

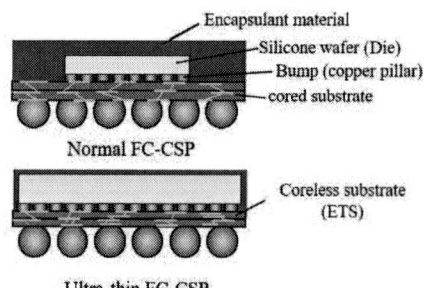

Figure 1. Typical structure of FC-CSP and Ultra-thin FC-CSP.

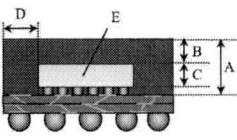

	Normal FC-CSP	Ultra-thin FC-CSP
A: Mold thickness (um)	300~500	< 250
B: Mold clearance (um)	< 100	< 100
C: Die thickness (um)	100 ~ 200	100 ~ 200
D: Mold edge clearance (um)	500 <	< 200
E: Die volume ratio (per 1unit)	20 ~35	50 <

Figure 2. Typical package structure design of Ultra-thin FC-CSP.

II. MOLDED SAMPLE USED IN THE EXPERIMENT

Molded array package (MAP) device construction was used in this experiment. The strip size of the test vehicle (TV) and molding conditions are shown in the Fig. 3 and Table 1. The test vehicle, composed of a substrate and die attach film (DAF) mounted die, was molded with EMC. A post-molding cure process was performed to complete the polymerization. Following curing, the strip of MAP devices was singulated into discrete package for evaluation.

Strip warpage was measured at the four corners of the strip after molding, post mold cleaning (PMC) and solder reflow (from 160 °C to 200 °C for 90 seconds, 260 °C for 10 seconds). Singulated package coplanarity was determined by shadow moiré over the temperature range of 30 °C to 260 °C. The result of warpage value was the largest coplanarity value in the observation temperature range.

Table 1. Test vehicle design and molding condition.

MAP design	Frame size	60.0 mm x 220.0 mm
	Molding area	53.6 mm x 213.6 mm
	Distance of chip	600 um
UNIT design	PKG size	11.3 mm x 12.8 mm
	Mold thickness	270 um
	Die size	10.9 mm x 12.4 mm
	Die thickness	150 um
	DAF thickness	15 um
	Ratio of die volume	52 %
Substrate design	Layer	3L coreless
	Substrate thickness	110 um
	Residual copper rate	55%, 70%, 60%
Molding condition	Molding method	Compression molding
	Injection time	14 s
	Mold cure time	180 s
	Molding pressure	8.6 MPa
	Post mold cure	175 °C, 6 h

■ Frame sample (before molding)

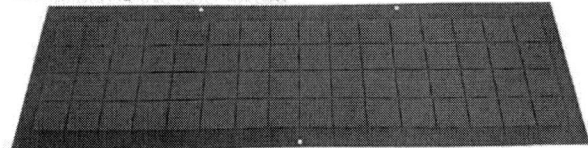

■ Molded sample

■ Cross section picture

Figure 3. Pictures of STRIP and UNIT flame sample.

III. RESEARCH FINDING NEW EVALUATION INDEX FOR CONTROL OF STRIP WARPAGE AND UNIT WARPAGE.

The TV shown in Fig. 3 was manufactured using seven types of EMC in shown in Table 2 and two types of substrates. Fig.4 illustrates the correlation between strip warpage and EMC properties. Fig. 5 shows the correlation between the singulated unit warpage results and properties of EMC.

From the results in Fig. 4, it can be observed that strip warpage was positively correlated with the 25 °C flexural modulus property and negatively correlated with mold shrinkage property. In contrast, there did not appear there was a correlation between Glass-Transition temperature (Tg) and 260 °C flexural modulus properties. These results suggest that lowering 25 °C Flexural modulus and increasing Mold shrinkage should reduce strip warpage.

The effect high Tg EMC in decreasing unit warpage has been reported previously [9]. The results shown in Fig.5, again confirmed that unit warpage was reduced by high Tg EMC. The effects of decreases in unit warpage by mold shrinkage and CTE are generally accepted. [2, 8]. However, these results of this evaluation did not indicate a clear correlation between the mold shrinkage and unit warpage in this TV with a high die volume occupation rate.

Warpage in FO-WLPs tend to result from large CTE differences between the silicon die and EMC. Various companies are currently conducting studies in this field [10, 11]. One study identified a method to decrease warpage by the EMC with an index incorporating both the room temperature CTE and the room temperature modulus [12].

The current study developed a complex index termed the "Molding Stress Index" by multiplying the CTE by the storage modulus as shown in equation (1.) This index was evaluated from room temperature to molding temperature. Fig. 6 shows the correlation between the molding stress index and unit warpage.

$$\text{Molding stress index} = \int_{35°C}^{175°C} [E'(T)Resin \times CTE(T)]dT \quad (1)$$

This index contains three EMC material properties; CTE, storage modulus (E ') and Tg. These properties are known to be effective in controlling unit warpage were therefore selected for this evaluation. In this study, a sampling interval of 5 °C was used to calculate the integral value. Fig. 6 shows decrease of unit warpage with a high molding stress index value.

In this phase of the study it was found that lowering the 25 °C flexural modulus and increasing the mold shrinkage properties were found to be an effective method of reducing strip warpage. On the other hand, increasing the value of molding stress index which comprehensively evaluating CTE, storage modulus and Tg was found to be an effective method of reducing the unit warpage.

However, the actual result values of strip warpage and unit warpage were large. In a mass production environment, this result would be difficult to apply. Based on the results in the previous phase, the development of materials which achieve both low strip and unit warpage were investigated.

Table 2. Property of EMC candidates.

		Unit	EMC A	EMC B	EMC C	EMC D	EMC E	EMC F	EMC G
Mold Shrinkage (PMC)		%	0.35	0.22	0.52	0.39	0.44	0.68	0.66
Tg (tanδ)		°C	190	151	151	186	190	153	170
Storage modulus E'(25 °C)		GPa	23	25	12	21	19	15	14
Storage modulus E' (260 °C)		GPa	1.3	1.3	0.44	1.2	1.1	0.23	0.22
CTE 1		ppm / °C	20	11	20	24	24	28	28
CTE 2		ppm / °C	52	35	64	55	55	77	76
Flexural modulus 25 °C		GPa	15	23	13	16	16	12	11
Flexural modulus 260 °C		GPa	0.67	0.63	0.35	0.74	0.74	0.25	0.32
Molding stress index		-	12597	6334	6404	12871	12211	11173	11011
STRIP	Substrate A	mm	28	32	33	27	26	25	24
Warpage	Substrate B	mm	30	34	35	33	34	17	20
UNIT	Substrate A	um	139	145	154	126	116	148	128
Warpage	Substrate B	um	142	159	156	136	137	140	134

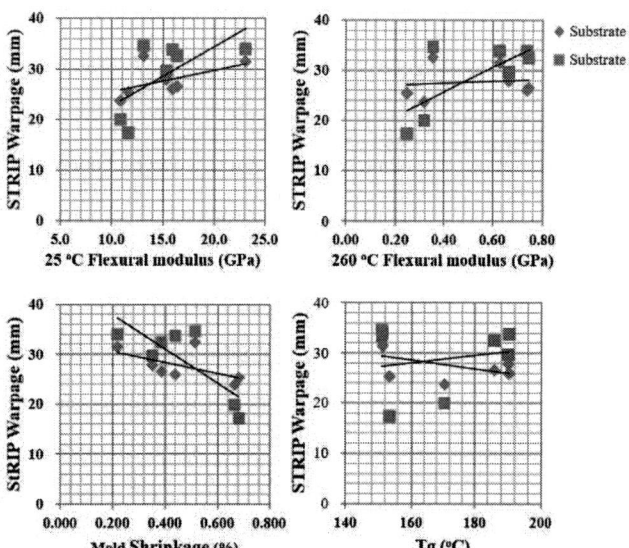

Figure 4. Correlation between STRIP warpage and EMC property.

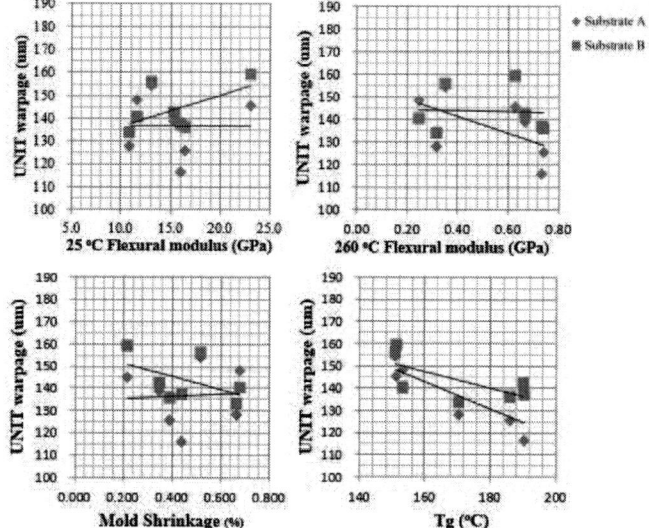

Figure 5. Correlation between UNIT warpage and EMC property.

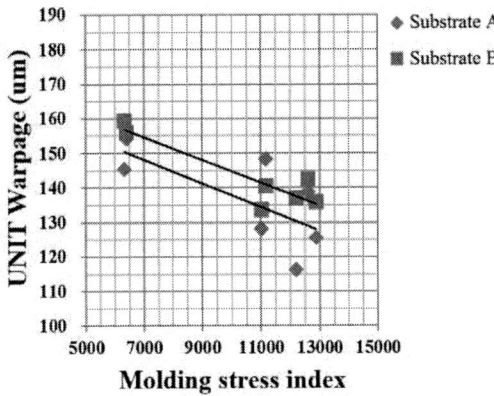

Figure 6. Correlation between UNIT warpage and Molding stress index.

IV. THE DEVELOPMENT OF ENCAPSULANT MATERIALS WHICH ACHIEVE BOTH LOW STRIP AND UNIT WARPAGE

The purpose of this section of the study was to achieve both low strip and unit warpage. The target value was set in consideration of mass production usability and secondary mounting. The target was a strip warpage value of <5 mm and a unit warpage value of <80 um.

Based on the results in the first section, an EMC was formulated targeting a 25 °C flexural modulus of <11 GPa, mold shrinkage <0.6% and a molding stress index < 10,000.

Molded products were produced using four types of EMC designated X, Y, Z and W (Table 3.) and three types of substrates A, C and D (Table 4.). The strip and unit warpage behavior of these combinations was measured and the results are shown in Fig. 7.

The results suggest that lowering the 25 °C flexural modulus and increasing mold shrinkage can achieve low strip warpage, similar to the trends discussed in the previous section. Notably, EMC formulation W showed the lowest strip warpage in this study. EMC W exhibited the largest mold shrinkage value and lowest 25 °C flexural modulus. The unit warpage behavior exhibited the same tendencies as discussed in the previous system in that the high molding stress index EMC showed the lowest unit warpage.

For substrate C, the unit warpage with EMC W showed significantly better results than the trend line. EMC W had the highest mold shrinkage property than the other EMC formulations evaluated. It was assumed that this high mold shrinkage caused the lowest unit warpage result with substrate C. However, a clear cause of this result was not determined. Further studies are needed.

Fig. 8 and 9 summarize the results of substrate and strip combinations unit warpage. When Substrate A was changed to Substrate C or Substrate D, these substrates reduced and unit warpage regardless of which EMC was applied.

In the case of strip warpage, it was assumed that substrate low modulus causes lower warpage. The substrate C with the lowest modulus was the best result for strip warpage. For lowering unit warpage, the high Tg of the substrate is apparently the primary factor.

The warpage behavior in the structure having high die volume occupation ratio may be controlled by using a material which satisfies the high molding stress index, high mold shrinkage and low 25℃ flexural modulus like EMC W.

This study also confirmed that warpage behavior was affected by different combinations of substrate and EMC. These results suggest that further warpage reduction is possible by optimizing the properties of EMC and substrate.

Table 3. Property of EMC candidates.

		Unit	EMC X	EMC Y	EMC Z	EMC W
Mold shrinkage (PMC)		%	0.47	0.32	0.65	1.1
Tg (tanδ)		℃	155	203	187	148
Storage modulus E'(25 ℃)		GPa	17	22	14	12
Storage modulus E' (260 ℃)		GPa	0.71	0.97	1.3	0.74
CTE 1		ppm / ℃	20	19	28	37
CTE 2		ppm / ℃	60	55	83	110
Flexural modulus 25 ℃		GPa	13	17	11	9.2
Flexural modulus 260 ℃		GPa	0.36	0.95	0.31	0.15
Molding stress index		-	8383	11299	10834	10300
STRIP Warpage	Substrate A	mm	21	20	15	11
	Substrate C	mm	15	15	13	9
	Substrate D	mm	15	16	16	10
UNIT Warpage	Substrate A	um	92	71	87	80
	Substrate C	um	77	65	65	54
	Substrate D	um	67	52	62	58

Table 4. Property of substrate candidates.

		Substrate A	Substrate C	Substrate D
Tg (tanδ)	℃	212	239	257
CTE (X,Y) 1	ppm / ℃	14	13	8.5
Storage modulus E' (25 ℃)	GPa	14	12	14
Storage modulus E' (260 ℃)	GPa	1.1	2.2	6.5

■ STRIP warpage

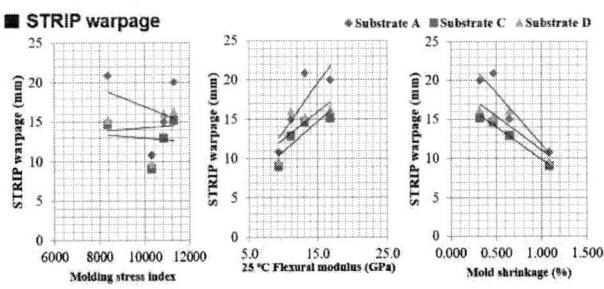

■ UNIT warpage

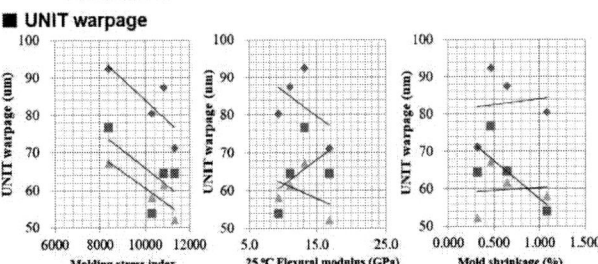

Figure 7. Correlation between warpage result and EMC property.

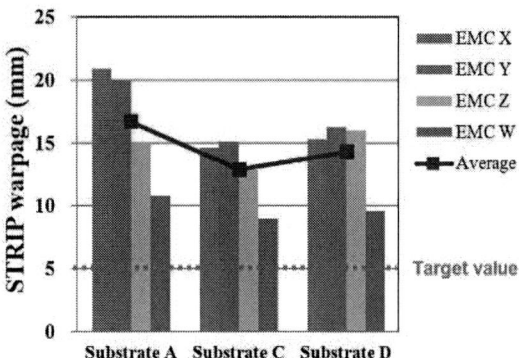

Figure 8. STRIP warpage result of different substrates.

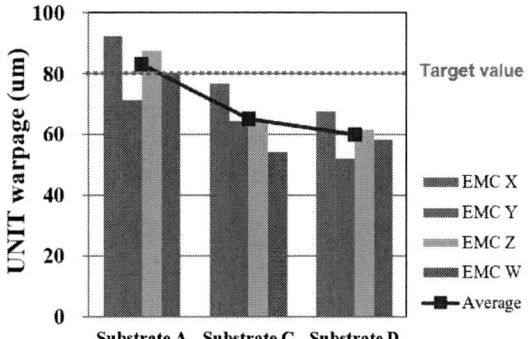

Figure 9. UNIT warpage result of different substrates.

V. CONCLUSION

The first section of this study confirmed the differences in warpage behavior between standard FC-CSP and ultra-thin FC CSP. In constructions with high die volume occupation rate such as ultra-thin FC CSP (exceeding 50 %,) it was difficult to correlate EMC basic property values (mold shrinkage, Tg, etc.) with warpage. A new evaluation method, the molding stress index, was developed. This index showed a positive correlation with unit warpage measurements.

In the second portion of the study, EMC W which exhibited a high molding stress index, low 25 °C modulus and high mold shrinkage showed low strip and unit warpage regardless of the substrate. On the other hand, it was confirmed that warpage behavior depended on the substrate properties even when the same EMC was used.

This study concluded that it is effective to mitigate ultra-thin FC-CSP warpage via the EMC properties of high molding stress index, low 25 °C flexural modulus and high mold shrinkage. Further warpage reduction is potentially possible by managing the properties governing the interaction of EMCs and substrates.

REFERENCES

[1] T. Tsuji, T. Akashi, N. Watanabe, K. Nishidono, M. Nakamura, "Development of encapsulant material for molded underfill for fine pitch flip chip packages", 66 th IEEE Electron. Compon. Technol. Conf. (ECTC 2016), p.750-753.

[2] M. Nakamura, "Technological Trends of Semiconductor Encapsulation Material", Panasonic Electric Works Technical Report, vol.59 (2011), p.10-16.

[3] T. Akashi, T. Tsuzuki, T. Tsuji, M. Nakamura, "Encapsulation Materials for Mold Underfill with High Filling Ability", Panasonic Electric Works Technical Report, vol.59 (2011), p.50-54

[4] R. L. Hubbard, Z. Pierino, Z. Pukun, "Flip-chip process improvements for low warpage", 60th IEEE Electron. Compon. Technol. Conf. (ECTC 2010), pp. 25-30.

[5] Electronic Publication: "Fan-Out Packaging Technologies and market trends", YOLE Development, 2016.

[6] Y.Nakamura, S. Katogi, "History of semiconductor mounting substrate materials and future technology trends", Hitachi chemical Technical Report, No.55 (2013), p.25-30.

[7] M.S. Chae, E. Ouyang, "Strip Warpage Analysis of a Flip Chip Package Considering the Mold Compound Processing Parameters", Proc. 63th IEEE Electron. Compon. Technol. Conf. (ECTC 2013), pp. 441-448.

[8] Gu Bin, Jun Dimaano, Richen Chen et al., "Unit Warpage Control with Universal Die Thickness", Proc. IEEE Symp. Electronics Packaging Technology Conference (EPTC 2014), pp. 303-306.

[9] K. Miyake, "Warpage Thermoelastic Analysis Considering Cure Shrinkage of BGA Package", The Japan Institute of Electronics Packaging Report, vol.7 No.1(2004), P54-61.

[10] Y. Oi et al., "Warpage Control of Liquid Molding Compound for Fan-Out Wafer Level Packaging", 2018 IEEE 68th Electronic Components and Technology Conference (ECTC 2018), pp. 967-972.

[11] K. Kwon et al., "Compression molding encapsulants for wafer-level embedded active devices: Wafer warpage control by epoxy molding compounds", Proc. 67th IEEE Electron. Compon. Technol. Conf. (ECTC2017), pp. 319-323.

[12] H. Ito, K. Sato, K. Arai, Japan patent: Patent No. 6423119.

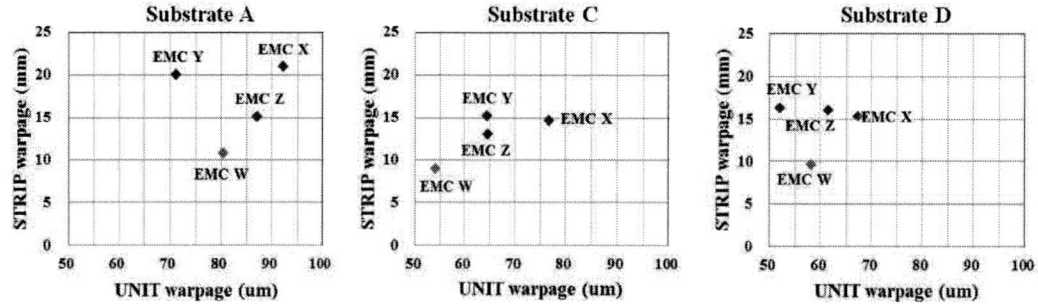

Figure 10. The results of STRIP warpage and UNIT warpage.

978-1-7281-1500-9/19 $31.00 © 2019 IEEE

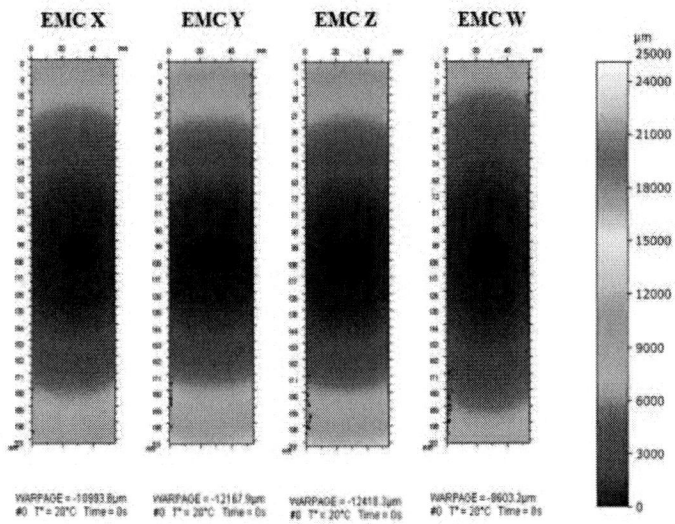

Figure 11. Moire result of STRIP warpage at 25 °C (substrate D).

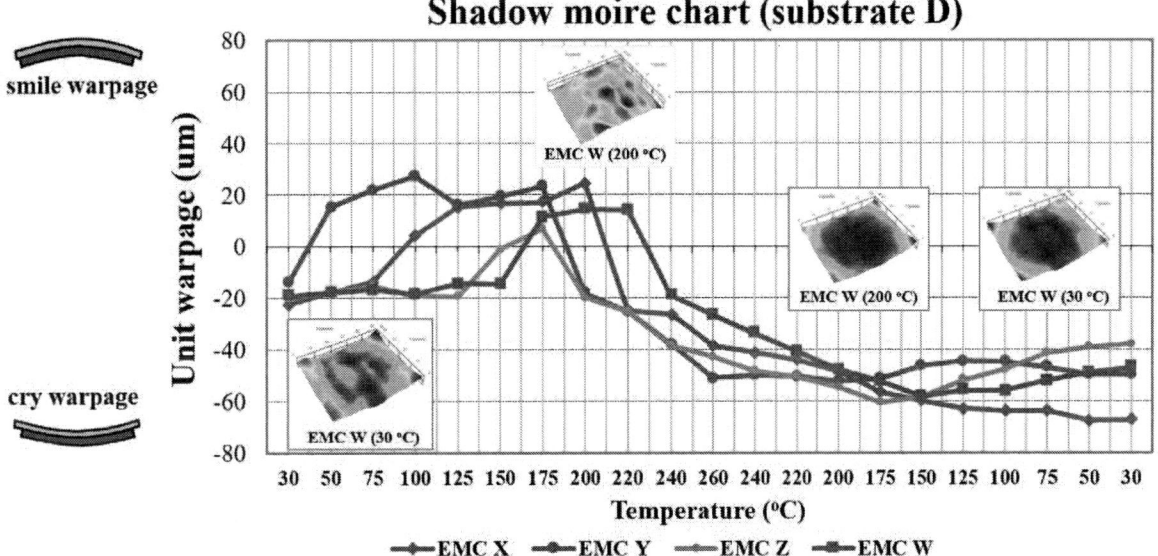

Figure 12. Moire result of UNIT warpage (substrate D).

Innovative Socketable and Surface-Mountable BGA interconnections

Omkar Gupte, Kristie Teoh, Rao Tummala,
Vanessa Smet
3D systems Packaging Research Center
Georgia Institute of Technology
Atlanta GA, USA
Email: vanessa.smet@prc.gatech.edu

Gregorio Murtagian
Intel Corporation
Chandler AZ, USA
Email: gregorio.r.murtagian@intel.com

Abstract— This paper introduces a universal BGA technology that can reliably be used both in sockets and SMT applications. Socketing and SMT have been driving two different board-level interconnection technologies: LGA for socketing, to provide a stable and rewordable mechanical contact, and BGA for SMT, for low-temperature metallurgical bonding to the board. Enabling socketable BGAs has become critical to simplify microprocessor package designs and converge towards a unique product. The approach presented in this research involves the use of multi-layered coatings on solder spheres. These coatings consist of a diffusion barrier/noble metal combination designed based on diffusion models to meet the requirements of both applications. Standard surface finish Electroless Ni Immersion Au (ENIG) has been studied in detail as a first approach. This paper focuses on the challenges faced in plating ENIG on traditional SAC solder balls and a new process combining sputtering and electrodeposition is demonstrated as a promising alternative to uniformly coat solder spheres with a controlled thickness of ENIG without corrosion. A first demonstration of attaching these coated spheres to the package is also presented.

Keywords- BGA, ENIG, intermetallics, socketing, solder, surface mount

I. INTRODUCTION

OEM Microprocessors have conventionally been packaged using Land Grid Array (LGA) designs, press-fitted into sockets for ease of reworkability. However, Ball Grid Array (BGA) packages have recently become mainstream for surface mount (SMT) applications, driven by the need for miniaturization of electronic systems. While SMT processes and applications with BGA are becoming more widespread, the market need for sockets is also expected to increase significantly over the next 5 years [1]. While microprocessor companies would benefit from producing a single BGA package design, this raises challenges for the OEM supply-chain as no BGA-compatible socket is currently available. Enabling universal BGA packages compatible with both socketing and SMT processes is, therefore, critical to this industry transition.

In LGA sockets, electrical contact is initiated between Au-plated LGA pads and Au-plated socket paddles, forming stable mechanical contact interfaces at up to 100-120°C with no risk of self-diffusion and formation of a metallurgical bond. This renders the sockets and packages to be rewordable over multiple socketing cycles. However, BGA sockets face major reliability challenges that have hindered their adoption to date.

When BGA packages are placed into sockets, the solder balls come directly into contact with the Au paddle of the socket. The BGA balls are typically made of a soft alloy Sn-Ag-Cu (SAC) with various Ag content (SAC105, SAC305, SAC405) with relatively low melting points in the $217 - 225°C$ range. Upon being latched in the socket, BGA packages face three primary challenges: (1) damage to the solder balls, (2) surface oxidation, and (3) intermetallics formation between solder balls and paddles. The typical latching forces used in socketing (30-50 g/pin) are sufficient to initiate significant damages to the soft solder balls [2]. Along with this, the surface of the solder balls is easily oxidized in operating conditions, which leads to an increase in contact resistance over time as the oxide layer grows. Another major challenge faced by BGAs in sockets is the formation of intermetallics (IMCs) at the interface of the solder ball and the socket paddle. Solid-state reaction of Au (from the socket paddle) and Sn (from the solder ball) leading to the rapid formation of Au-Sn IMCs has been reported [3]. The IMCs formed at the contact interface not only increase the interconnects' resistance, but also leave a residue on the paddle after every socketing cycle, degrading reworkability over multiple socketing cycles [2].

To overcome these challenges, most research efforts have focused on modifying the contact technology of the socket paddle with solutions like cantilever springs [2] which provides compliance, Dendriplates [2] or tweezer contacts [4]. While these solutions limit damage to the solder balls, and to some extent, address the oxidation issue by penetrating into the solder, IMC formation is still very much a challenge. To address this last challenge, re-balling the package by removing the solder BGAs and replacing them with Au-coated Cu-cored balls has been demonstrated [5]. These Cu-cored balls are attached to the package using solder paste. This provides the stable Au-Au interface between the balls and the socket paddle for socketing. However, this process requires rework which increases the cost of production.

To comprehensively address all three challenges, Georgia Tech PRC is developing a novel interconnection system that is compatible with both SMT and socketing. In this system, solder balls are coated with multi-layered (diffusion barrier/noble metal) coatings designed based on theoretical diffusion models. The multi-layered coating is directly applied on the solder ball with controlled thickness and the coated balls are then attached to the package using a printed solder paste. This package can now be used for socketing as the outermost noble metal layer provides the required oxide-

978-1-7281-1500-9/19 $31.00 © 2019 IEEE

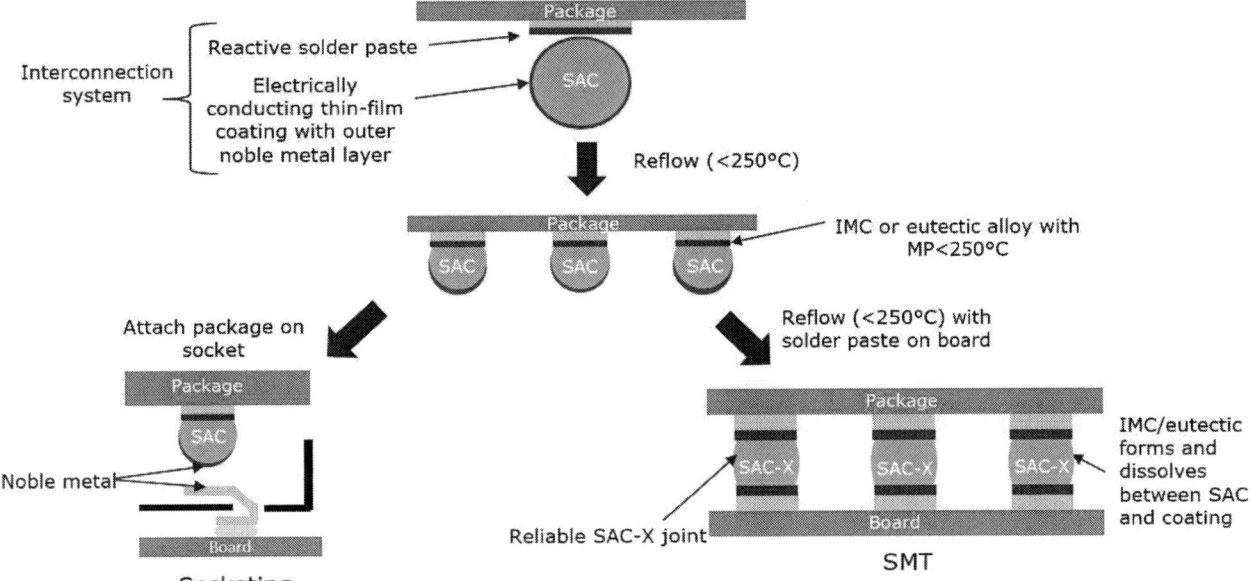

Figure 1: Process flow for socketable BGAs using solder balls with multi-layered coatings.

free, stable contact interface with the socket paddle while the diffusion barrier layer prevents migration of solder to the surface of the ball and subsequent intermetallic formation.

Simultaneously, this package with coated balls can be attached to the board with another solder paste for SMT applications. During this second reflow cycle for board attach, the multi-layered coating can be dissolved into the bulk solder by physical forces and chemical reactions occurring at the interface at the reflow temperature. This approach is illustrated in Fig. 1. The major challenge in this new concept is the contradicting requirements on the diffusion barrier layer design imposed by both applications. Socketing applications need the diffusion barrier layer to be intact throughout the lifetime of the product to prevent diffusion of Sn to the surface. On the other hand, SMT applications require the entire coating to break down and dissolve in the solder during the reflow of the package on the board. Hence, the barrier layer needs to be designed according to the minimum required thickness. The use of standard electroless Ni/immersion Au as surface modification was first explored in this research. Ni is a well-known diffusion barrier for Sn, which is the main component of the solder ball used [6]. Reaction of Ni with Sn forms hard IMC phases at the interface, which then need to be dissolved during SMT. This paper focuses on the kinetic modeling for the IMC formation at the interface and the diffusion modeling of the Ni in Au to design the coating. The electroless Ni immersion Au system is then discussed followed by the processing challenges for the metal deposition on solder and the process modifications for ENIG coating on solder spheres. This is followed by a first demonstration of attaching the coated spheres to the package using solder paste.

II. DIFFUSION MODELING FOR BARRIER LAYER DESIGN

Socketing applications require the noble metal contact between interconnections and socket paddles to be thermally stable at 100-120°C for at least 20 rework cycles. This requires the diffusion barrier layer to be stable and prevent the diffusion of Sn to the Au surface through the ball-attach reflow during which solder may be molten, and the operating life. The reactions of Ni barrier layer with the solder were studied at different temperatures to determine the phases that are formed at the interface of solder ball and barrier layer. It was understood that at all temperatures, $(CuNi)_6Sn_5$ IMC phase forms first, followed by $(NiCu)_3Sn_4$ [7]. Parabolic growth law, as seen in Equation 1, is followed for the IMC formation when solder is in either solid or molten states. 'x' is the IMC thickness in µm, 'k' is a growth constant in $\mu m/s^{1/2}$ and 't' is the time in s. The rate constants for the reaction in both conditions follow the Arrhenius equation and were calculated from literature data. At 125°C for the solid-state reaction, 'k' was calculated to be 7.91E-4 $\mu m/s^{1/2}$ whereas at 260°C for the molten state reaction, 'k' was calculated as 0.37 $\mu m/s^{1/2}$ [8][9].

$$x = k\sqrt{t} \qquad (1)$$

The thickness of intermetallic formed was calculated at different temperatures for varying times, representative of the conditions of use of the socketable BGA package. The shelf life of the coated spheres before ball-attach was simulated with accelerated conditions of 100°C for 1000 h. Two different solder pastes were considered for ball-attach, SnBiAg (SBA) and SAC305 with melting temperatures of 138°C and 217°C, respectively, giving reflow peak temperatures of 170°C and 250°C, respectively. Lastly, the operating conditions for the package in socketing were represented through thermal aging at 125°C for 1000 h. The thicknesses of the IMCs formed during all these process steps are reported in Table I.

Based on the total thickness of IMC that is formed, the thickness of the Ni layer consumed for this reaction was

978-1-7281-1500-9/19 $31.00 © 2019 IEEE

Table I: Calculations for IMC thickness at SAC-Ni interface

Solder paste	Shelf life thermal aging	IMC thickness	Condition	Temperature, time	IMC reaction	IMC thickness	Total IMC thickness	Ni layer consumed
SAC 305	100°C, 1000 hr	1 μm	1st reflow	250°C	Solid-liquid	2 μm	4.5 μm	3.8 μm
			Thermal aging	125°C, 1000 hr	Solid state	1.5 μm		
SBA	100°C, 1000 hr	1 μm	1st reflow	170°C	Solid state	~0	2.5 μm	2.2 μm
			Thermal aging	125°C, 1000 hr	Solid state	1.5 μm		

estimated. This calculation was done by considering the mass of Ni is conserved during the reactions. This relation is shown in Equation 2 where ρ represents the density of the material and V represents the volume of the material. The density of Ni was assumed as 8.912 g/cc and that of the IMC was considered as 8.017 g/cc [10].

$$\rho_{Ni}\ V_{Ni} = \rho_{IMC}\ V_{Ni\ in\ IMC} \qquad (2)$$

Diffusion modeling was also done for the Ni-Au system to estimate the thickness of the Au layer to prevent the diffusion of Ni to the outer surface of Au during the socketing lifetime. The initial thickness of the Au layer was considered to be 100-200 nm. Since the Ni layer was more than 200 times thicker than the Au layer, a semi-infinite diffusion model was applied to the system by assuming that the concentration of Ni remains constant at the interface.

Equation 3 shows the model established for the Ni-Au system. C represents the concentration of Ni in the Au layer at distance 'x' from the interface and at time 't'. C_s represents the concentration of Ni at the interface. C_0 stands for the initial concentration of Ni at the outer surface of Au and D is the diffusion coefficient of Ni in Au. The diffusion coefficient D was considered as 7.76E-13 μm²/s [11]. The predicted time for the concentration of Ni on the Au surface to increase by more than 1% was more than 10,000 h.

$$\frac{C(x,t)-C0}{Cs-C0} = 1 - \mathrm{erf}\left(\frac{x}{2\sqrt{Dt}}\right) = \mathrm{erfc}\left(\frac{x}{2\sqrt{Dt}}\right) \qquad (3)$$

In addition to the semi-infinite model assumption, a geometrical assumption was made during the model construction. Due to the large difference in the radius of the solder ball in consideration (125μm) and the thickness of Ni layer (4.5μm), the geometry for the diffusion was considered as a parallel plate geometry instead of a spherical geometry.

III. PLATING PROCESS FOR MULTI-LAYERED COATINGS

A. Electroless Ni immersion Au (ENIG) coating

Electroless Ni-Immersion Au (ENIG) is a standard terminal metal plating couple used as surface finish in the microelectronics industry. The plating is usually applied on Cu pads on substrates and boards and plated using an acidic bath with hypophosphite according to the reaction described in Equation 4. The plated Ni layer contains a small amount of phosphorous due to the presence of hypophosphite in the solution.

$$Ni^{+2} + 4H_2PO_2^- + H_2O \rightarrow Ni^0 + P + 2HPO_3^{-2} + H_2PO_3^- + 3H^+ + 3/2H_2 \quad (4)$$

Immersion Au is plated on Ni using a simple atomic exchange reaction between less noble Ni atoms and the more noble Au atoms from a solution of potassium gold cyanide. Since this is an atomic exchange process, it is a slow process. The thickness of the Au layer deposited is limited as once the entire surface is covered with Au, Ni and Au atoms need to interdiffuse through the Au layer to continue the deposition. Hence, deposition of Au layer thicker than a few hundred nanometers takes a significantly longer time.

B. Challenges of ENIG plating on solder

A series of pre-treatment baths are required to prepare the sample for ENIG plating. The process flow for the ENIG plating is shown in Fig. 2. The baths with the required chemistries were provided by Atotech GmbH. The pre-treatment baths include microetch and acid dip (5% H_2SO_4) to remove the surface oxides, and an activator to activate the surface for Ni plating. These baths are acidic with pH less than 1 and are established for plating on Cu substrates. The pH and temperature of the Ni bath is maintained at 4.8 and 85°C, respectively. The solder balls were first reflowed on the board and the ENIG process with all the pre-treatment steps was carried out. Severe corrosion was observed on the surface of the solder balls as seen in Fig. 3.

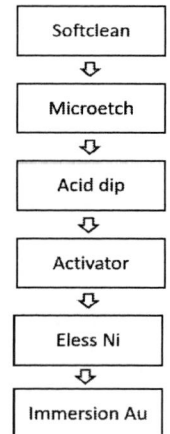

Figure 2: Process flow for ENIG (Courtesy Atotech GmbH)

This observation was confirmed in literature and the exposure of solder to severe acidic conditions was confirmed as the reason for the corrosion [12]. Another major challenge was to evenly coat the entire surface of free-standing spheres not attached to a substrate. This requires the spheres to rotate continuously in the solution to ensure uniform coating.

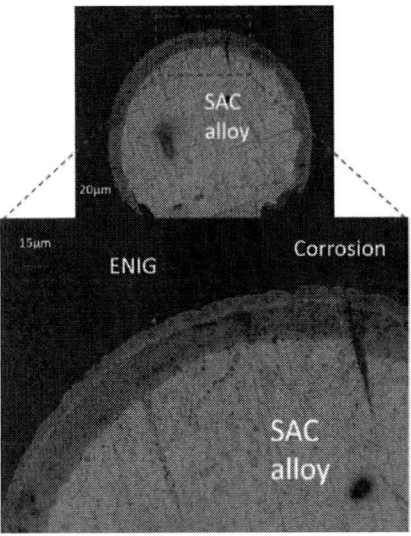

Figure 3: Corrosion on solder ball surface after ENIG plating

C. Plating process development

To address the challenge of corrosion, the solder spheres attached to the board were treated with the Electroless Ni solution without the pretreatment steps. It was observed that the plating rate was slow, and the plating was not uniform on the entire surface of the spheres. Some corrosion was also observed on parts of the surface. This was because the bare solder was exposed to acidic conditions until a thin layer of Ni could form on the surface of the sphere. Subsequently, the pH of the bath was increased to reduce corrosion. On increasing the pH from 4.8 to 5.5, it was observed that a thicker Ni layer was deposited for the same plating time. It was concluded that the redox reaction is kinetically favorable under weak acidic conditions. However, the non-uniformity in the plated layer persisted and some corrosion was still observed on the surface of the sphere as can be seen in Fig. 4

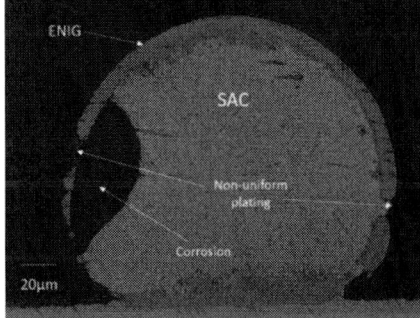

Figure 4: Corrosion and non-uniform plating after eliminating pre-treatment steps

Fig. 5 shows the solder on the outside of the plated ENIG shell after reflowing on the board. This is caused by the solder flowing out through the discontinuities in the Ni shell during reflow and solidifying on the outer surface of the sphere.

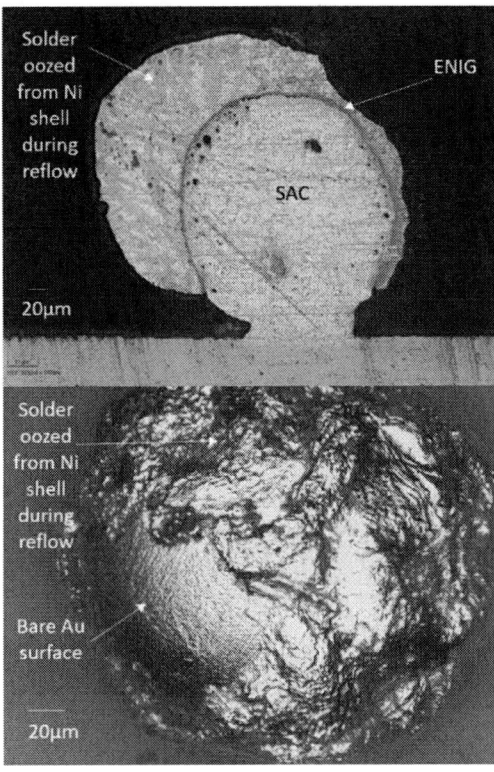

Figure 5: Solder flowing out through the discontinuities in the ENIG shell: cross section view (top) and top view (bottom)

To achieve uniform plating, a thin seed layer of Ni was sputtered using the Unifilm sputtering tool. Prior to sputtering, the surface oxides formed on the solder sphere surface were etched by treating the spheres with hydrogen plasma. The etching was done on the Plasmatherm SLR RIE at 300 W power with a hydrogen flow rate of 30 sccm. The solder spheres were placed directly on the sample holder plate for easy rotation to ensure uniform exposure of the surface to plasma. A 200 nm-thick seed layer of Ni was then sputtered on the solder spheres. Similar to the plasma tool, the treated solder balls were placed directly on the sample holder to ensure uniform coverage. The solder balls with a Ni seed layer were then treated with all the pre-treatment steps before exposing them to the Ni bath. The pre-treatment baths remove any oxide formed on the seed layer before plating Ni. To achieve uniform plating of Ni on the entire surface of the sphere, the baths were prepared in small 100mL beakers and were continuously stirred using a magnetic stirrer. The solder balls were dropped directly in the baths and were stirred continuously in the bath due to their proximity to the stirrers. The pH and temperature for the Ni bath was maintained at 5.5 and 85°C, respectively, while that of the Au bath was maintained at 4.8 and 85°C, respectively. With these process modifications, a uniform ENIG coating was obtained on the surface of the solder spheres as shown in Fig. 6. The process was started with approximately 0.1 g solder spheres and ~75% yield was achieved for the modified coating process. Fig. 7 shows the entire modified process flow for ENIG coating.

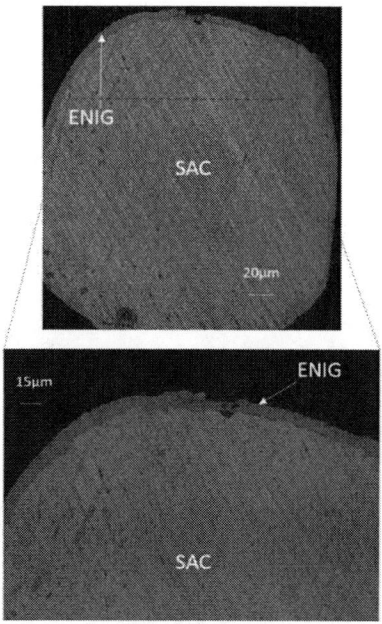

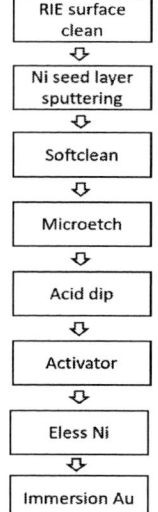

Figure 6: Uniformly ENIG coated solder ball with optimized plating process

RIE surface clean
⇩
Ni seed layer sputtering
⇩
Softclean
⇩
Microetch
⇩
Acid dip
⇩
Activator
⇩
Eless Ni
⇩
Immersion Au

Figure 7: Modified process flow for coating solder spheres with ENIG

IV. COATED BALL ATTACH

The coated solder balls are attached to the package and board using a paste printing and ball drop process. For attaching the coated solder balls to the package during the first reflow, minimal reaction between the solder core and the diffusion barrier layer is desired. This drives the need to use a low melting point solder paste such as SBA, with a melting point <140°C. SBA has shown to have moderate creep resistance and high modulus but is also known to be brittle [13]. This would enable the use of SBA at 100-120°C. However, if the working temperature of the socketing application is very close to or higher than the melting temperature of the paste, a solder paste with higher melting point (>200°C) needs to be considered for attaching the coated spheres. Solder paste is printed on the pads using a stainless steel stencil, and the coated balls are then dropped using a second stencil as shown in Figure 8a, prior to reflow using a 5-zone Electrovert Omniflo 5 oven.

A major challenge observed during the ball-attach process was the wicking of solder paste on the entire surface of the sphere during reflow as shown in Figure 8b. This was addressed by controlling the volume of solder paste printed, as well as the time above liquidus (TAL). The volume of solder paste was theoretically calculated such that it would wick to half the height of the solder ball as shown in Figure 9 and Equation 5.

$$Vsp = \pi(r^3 + (r')^2 h) - \frac{2}{3}\pi r^3 \qquad (5)$$

Single ball shear test was performed on the attached balls and it was understood that the shear strength of the solder joint depends not only on the initial printed volume of the paste, but also on the time for which the solder paste is in molten state during reflow. Longer time above liquidus (TAL) leads to a weaker joint as more amount of paste wicks along the ball surface leaving lesser unreacted solder in the joint. Figure 10 shows the shear strength of the joint formed using SBA paste for the attach of ENIG-coated balls, 225 μm in diameter, with printed paste diameters of 205 μm, 210 μm and 220 μm and three reflow speeds of 12in/min, 14in/min and 16 in/min, which lead to three TAL values of 150 s, 123 s and 111 s, respectively. Figure 8c shows the top view of ENIG-coated BGAs with controlled volume of printed paste and reflow profile, demonstrating the feasibility of this approach to achieve socketable BGA packages.

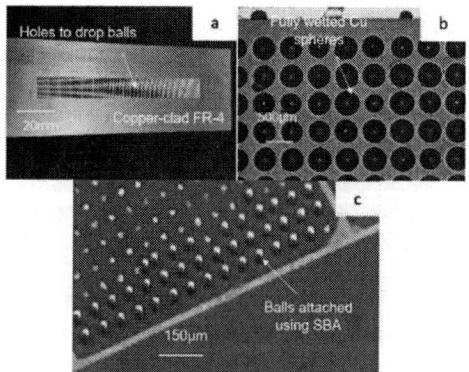

Figure 8: (a) Stencil to drop balls (b) Solder paste fully wicked on Cu spheres (c) Cu spheres attached with controlled solder paste wicking

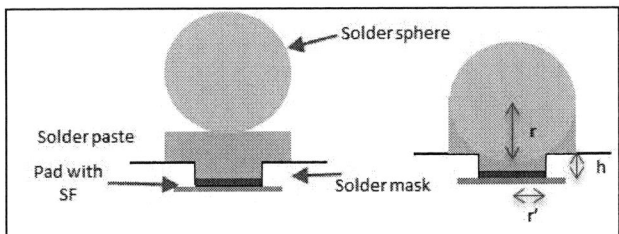

Figure 9: Schematic for controlled solder paste wicking calculations

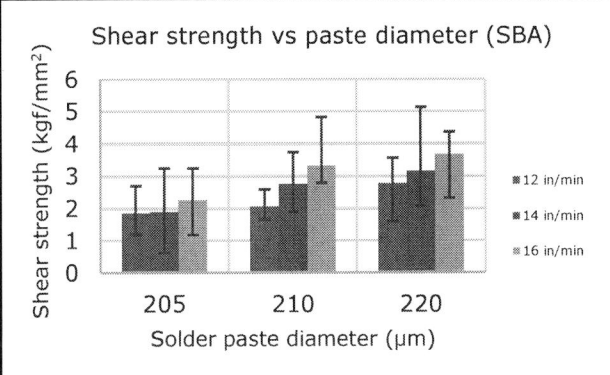

Figure 10: Shear strength relation of SBA solder joint to printing diameter and TAL

V. CONCLUSIONS

Enabling socketable BGAs is critical to the semiconductor industry's transition to a universal package design, compatible with both OEM socketing and SMT applications. A new approach was proposed, relying on the surface modification of solder spheres using thin-film metallic coatings to control reactions at every step of the processes. A bi-layer coating was explored consisting of an outer layer of Au to provide a stable electrically-conductive contact with excellent wettability for ball attach and socketing, and a diffusion barrier layer to prevent migration of Sn to the surface of the balls and subsequent intermetallic formation with the substrate paddle. The contradicting requirements the barrier layer faced in socketing and SMT applications were first highlighted in this study. Based on these requirements, the minimum thickness of Ni barrier layer was calculated for the ENIG system evaluated as a first proof-of-concept using theoretical diffusion models. This thickness is also dependent on the reflow temperature used for ball attach to the package. For a solder paste with melting points in the range of 120–140°C, the thickness of the Ni layer is 2.5 μm whereas that for a solder paste with melting point in the range of 210–230°C, it is 3.8 μm. The issue of corrosion of solder in acidic conditions was observed using standard ENIG plating chemistries to coat the balls and addressed by sputtering a thin seed layer of Ni on the sphere surface prior to surface finish plating. The ENIG plating conditions were optimized to achieve complete coverage on the balls with a process yield of ~75% at lab scale. Conditions for ball attach to the board were investigated and the dependency of ball shear strength on multiple process parameters was established. This work has introduced an innovative method of enabling socketable BGAs by surface modification. It has established the pathway for exploring new materials and methods to tackle the contradicting requirements of this technology and develop new BGA spheres with solder cores that can be universally used in many applications.

ACKNOWLEDGMENT

This research was sponsored by the Semiconductor Research Corporation (SRC). The authors would like to thank Atotech GmbH for providing the ENIG chemicals, Marlee Newman for her help with experiments, Colin Holmes for discussions and the PRC staff for tools and infrastructure support.

REFERENCES

[1] https://www.mordorintelligence.com/industry-reports/ic-socket-market

[2] Chan, Benson, and Pratap Singh. "BGA sockets-a dendritic solution." In *1996 Proceedings 46th Electronic Components and Technology Conference*, pp. 460-466. IEEE, 1996.

[3] Lee, Teck Kheng, Sam Zhang, C. C. Wong, A. C. Tan, and Davin Hadikusuma. "Interfacial microstructures and kinetics of Au/SnAgCu." *Thin Solid Films* 504, no. 1-2 (2006): 441-445.

[4] Liu, Weifeng, and Michael Pecht. *IC component sockets*. John Wiley & Sons, 2004, pp.106-108

[5] Al-Momani, Emad S. "Re-balling BGA with gold-plated copper spheres, the need and the SMT challenges." In *2016 11th International Microsystems, Packaging, Assembly and Circuits Technology Conference (IMPACT)*, pp. 1-4. IEEE, 2016.

[6] Islam, M. N., Y. C. Chan, Ahmed Sharif, and M. O. Alam. "Comparative study of the dissolution kinetics of electrolytic Ni and electroless Ni–P by the molten Sn3. 5Ag0. 5Cu solder alloy." *Microelectronics Reliability* 43, no. 12 (2003): 2031-2037.

[7] Laurila, Tomi, and Vesa Vuorinen. "Combined thermodynamic-kinetic analysis of the interfacial reactions between Ni metallization and various lead-free solders." *Materials* 2, no. 4 (2009): 1796-1834.

[8] Choubey, Anupam, Hao Yu, Michael Osterman, Michael Pecht, Fu Yun, Li Yonghong, and Xu Ming. "Intermetallics characterization of lead-free solder joints under isothermal aging." *Journal of Electronic Materials* 37, no. 8 (2008): 1130-1138.

[9] Dariavach, N., P. Callahan, J. Liang, and R. Fournelle. "Intermetallic growth kinetics for Sn-Ag, Sn-Cu, and Sn-Ag-Cu lead-free solders on Cu, Ni, and Fe-42Ni substrates." *Journal of Electronic Materials* 35, no. 7 (2006): 1581-1592.

[10] Yang, Jian, Jihua Huang, Dongyu Fan, Shuhai Chen, and Xingke Zhao. "Structural, mechanical, thermo-physical and electronic properties of η′-(CuNi) 6Sn5 intermetallic compounds: First-principle calculations." *Journal of Molecular Structure* 1112 (2016): 53-62.

[11] Abdul-Lettif, Ahmed M. "Determination of diffusion coefficients in Au/Ni thin films by Auger electron spectroscopy." *physica status solidi (a)* 201, no. 9 (2004): 2063-2066.

[12] Nordarina, J., H. Z. Mohd, A. M. Ahmad, and F. M. N. Muhammad. "Corrosion Behaviour of Sn-based Lead-Free Solders in Acidic Solution." In *IOP Conference Series:*

Materials Science and Engineering, vol. 318, no. 1, p. 012003. IOP Publishing, 2018.

[13] Hwang, J. S., and R. M. Vargas. "Solder joint reliability—Can solder creep?." *Soldering & Surface Mount Technology* 2, no. 2 (1990): 38-45.

2019 IEEE 69th Electronic Components and Technology Conference (ECTC)

A Highly Reliable 1.4µm pitch Via-last TSV Module for Wafer-to-Wafer Hybrid Bonded 3D-SOC Systems

Stefaan Van Huylenbroeck, Joeri De Vos, Zaid El-Mekki, Geraldine Jamieson, Nina Tutunjyan, Karthik Muga, Michele Stucchi, Andy Miller, Gerald Beyer and Eric Beyne

Imec vzw
Kapeldreef 75, B-3001 Leuven, Belgium
e-mail: Stefaan.VanHuylenbroeck@imec.be

Abstract—This paper demonstrates the fabrication of a reliable 0.7µm diameter and 5µm deep (0.7x5µm) via-last module, fitting a 1.4µm TSV pitch. Enabling sub-micron TSV diameters requires a thinner photo resist, however still withstanding the top passivation dielectric etch, the deep silicon etch and the bottom dielectric etch. The actual TSV silicon diameter is 0.8µm just below the top dielectric hard mask, but reduces to 0.7µm in the middle and to 0.65µm at the bottom of the via. The bottom dielectric tri-layer, consisting of an STI oxide, a thin SiN and a PMD oxide layer, is etched using a dedicated three step selective etch recipe. A thin ALD TiN embedded barrier is implemented, assuring good TSV reliability. An alternative and scalable protection of the oxide liner at the top of the TSV during bottom liner etch is worked out. It makes use of an APF strippable amorphous carbon film. Despite the sub-micron TSV diameter, a conventional PVD Ta barrier and PVD Cu seed is still maintained. Discontinuities in the PVD Cu seed are repaired by using a 30nm thin alkaline ECD seed layer enhancement (SLE), resulting in a conformal copper seed all over the TSV and ensuring void less ECD copper fill.

Electrical results prove the maturity of this 0.7µm diameter, 1.4µm pitch via-last module. The connectivity of the TSV, from wafer front to back side, has been checked by means of kelvin and daisy chain structures, showing 100% yield and low spread on the measured resistance values. High breakdown voltage of the TSVs is obtained. The integrity of the oxide liner all over the TSV sidewall is proven by means of IV-controlled reliability measurements (IV_{CTRL}). The breakdown voltage V_{bd} has very little dependence on the applied stress voltage ramp rate, resulting in high field accelerating factor γ, confirming the high TSV liner/barrier reliability.

Keywords-component; TSV, Via-last, 3D Integration, hybrid bonding

I. INTRODUCTION

One of the approaches to realize small pitch 3D interconnects is hybrid wafer-to-wafer bonding. The metal layer of the top wafer can be fed through to the wafer backside by means of via-last TSVs. Imec already succeeded to reduce the pitch in hybrid wafer-to-wafer bonding to 1.4µm [1]. In this paper, we will demonstrate the fabrication of a reliable 0.7µm diameter and 5µm deep (0.7 x 5µm) via-last module that can be realized on the same 1.4µm pitch.

II. 1 X 5 UM VIA-LAST REFERENCE MODULE

A 1x5µm and 2µm pitch via-last module was presented in [2]. Via-last TSVs are implemented after thinning the top wafer to a Si thickness of 5µm. After TSV litho and deep silicon etch, the STI/PMD oxide is etched till just above the metal landing pad, followed by the resist removal (Fig. 1). A conformal PEALD liner oxide and a non-conformal PECVD nitride layer are deposited next. The latter protects the liner at the top of the TSV during bottom liner opening.

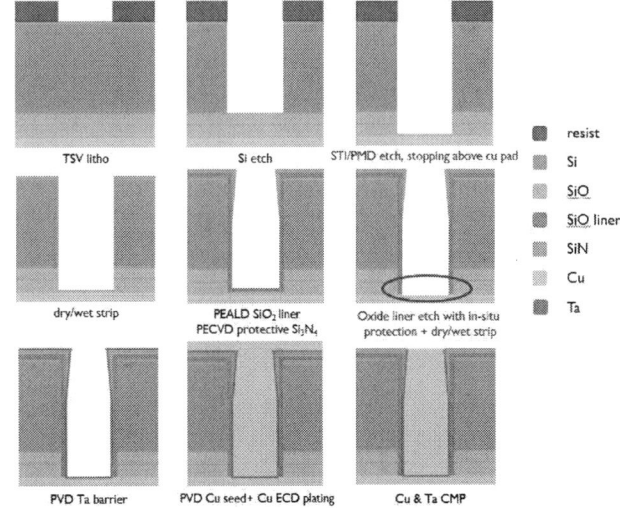

Figure 1. 1 x 5µm via-last reference flow.

The bottom liner etch exposes the metal landing pad, ensuring electrical contact with the TSV. A conventional PVD Ta barrier and PVD copper seed system is implemented. ECD Cu fill and CMP steps, removing the Cu overburden and Ta barrier material from the surface, complete the via-last TSV module.

It is also reported in [2] that copper of the metal landing plate gets re-sputtered onto the TSV liner oxide, as the liner etch is not indefinitely selective to the Ta barrier and the copper of this landing pad. The introduction of an additional – embedded – barrier layer prior to this critical liner opening dry etch step improves the TSV reliability (Fig. 2). A highly conformal thin ALD TiN layer is therefore embedded in the oxide liner. This embedded barrier stops the diffusion

978-1-7281-1500-9/19 $31.00 © 2019 IEEE   1035

through the liner oxide of any re-sputtered copper on the TSV sidewalls and protects as well the liner oxide during bottom liner opening etch.

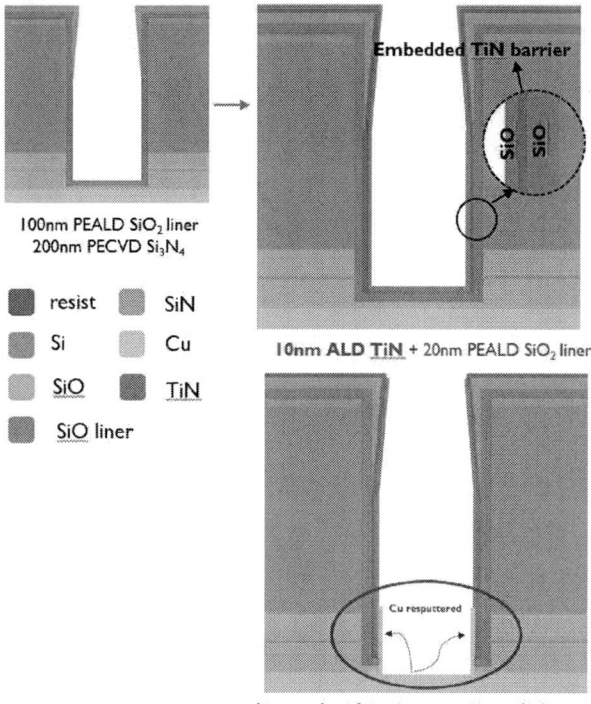

Figure 2. Embedded barrier concept. The 10nm ALD TiN barrier is embedded into the oxide liner and deposited prior to the liner opening etch, as such stopping the diffusion of re-sputtered copper through the oxide liner.

III. 0.7 X 5 UM VIA-LAST PROCESS MODULE

A. TSV lithography

Because of the TSV diameter scaling from 1μm to 0.7μm, a further reduction of the photo resist thickness is needed in order to keep the ratio of resist thickness to TSV diameter close to 3. The resist type is therefore changed from a 3.6μm thick AZ® TX 1311 DUV (248nm) to a 2.0μm thick Fujifilm KRF DUV (248nm) positive tone photo resist. There is no BARC layer used.

The TSV resist has to withstand the top passivation dielectric etch, the deep silicon etch as well as the bottom dielectric etch. Blanket resist consumption tests were performed for each individual etch step (Fig. 3) with respectively a 60 sec etch time for hard mask and STI/PMD bottom dielectric etch and 50 Bosch etch cycles for the deep-Si etch step. After real via-last processing, there still remains around 700nm resist.

The resist profile after litho exposure and development was verified with cross-sectional SEM (Fig. 4). A bottom CD close to 700nm is obtained.

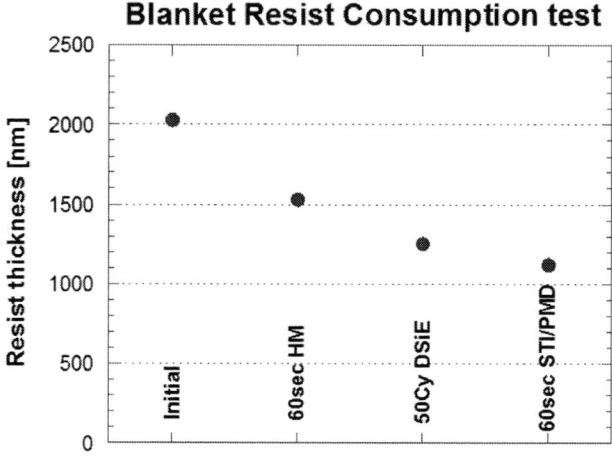

Figure 3. Blanket resist cosumption test.

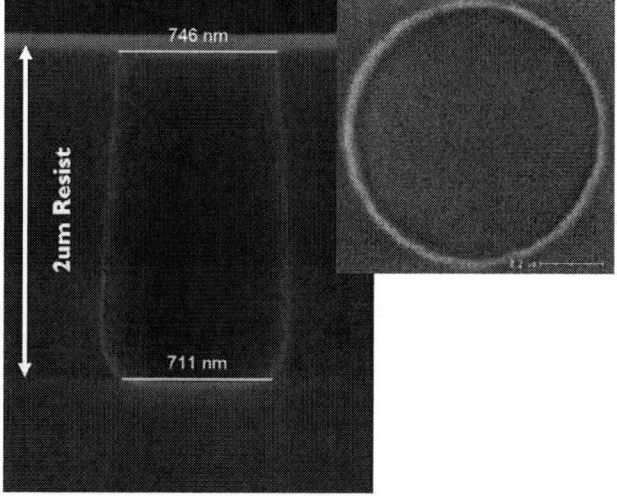

Figure 4. TSV litho resist profile.

B. Hard Mask, deep-Si and bottom dielectric etch

The etch of the top dielectric hard mask stack, which consists of a 150nm silicon oxide and 250nm silicon nitride layer, is done in-situ on a Lam research 2300® Syndion® V2 chamber prior to the deep silicon etch, providing a bottom hard mask diameter of 0.85μm. After the Rapid Alternating Process (RAP), mimicking the Bosch type deep silicon etch, the actual TSV silicon diameter remains around 0.8μm just below the top dielectric, eliminating any re-entrant top profile. The silicon diameter reduces to 0.7μm in the middle and to 0.65μm at the bottom of the via (Fig. 5). With these TSV dimensions, a pitch of 1.4μm can be assured. The time-ramped approach has been implemented in both deposition and etch phase, in order to be able to better control the TSV profile during the RAP process.

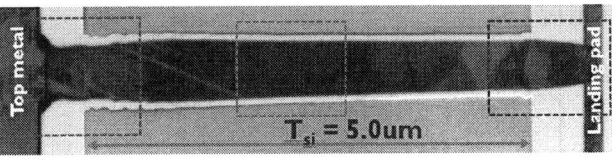

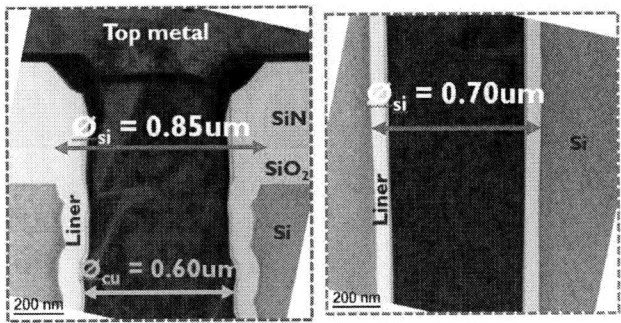

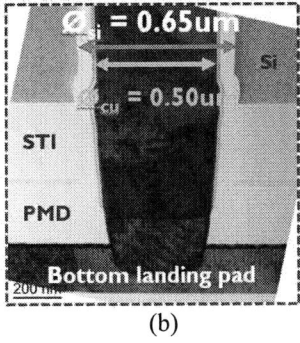

(a)

(b)

Figure 5. Copper filled 0.7 x 5 µm TSV TEM pictures showing respectively effective Si and Cu diameter
(a) TSV overview ; (b) Zoom-in on top, center and bottom part.

For the bottom dielectric tri-layer, consisting of an STI oxide, a thin SiN and a PMD oxide layer, a dedicated three step selective etch recipe has been developed, as such obtaining a uniform process that leaves a fixed amount of oxide above the metal landing pad. In the first step, the STI oxide is etched with a 4:1 selectivity towards the thin SiN layer. The SiN etch step is highly selective to the underlying PMD oxide layer. The final etch step consists of a timed PMD oxide etch. About 100nm of oxide is remaining on top of the metal landing pad. The etch front is also very flat all over the TSV (Fig. 6), another advantage of this three step etch recipe.

C. Liner deposition and bottom liner opening

As explained in § II and in [2], an embedded barrier concept is also applied to this 0.7 x 5µm via-last module. Compared to the reference 1 x 5µm TSV module, the thickness of the oxide liner and the embedded barrier capping oxide are halved to respectively 50nm and 10nm. For the ALD TiN embedded barrier, wafers were processed with both the reference 10nm embedded barrier thickness and with a reduced 5nm one.

The liner oxide at the top of the TSV is protected by a 150nm non-conformal PECVD low temperature SiN layer,

implemented before the embedded barrier deposition [3]. This SiN layer deposits only on the top of the TSV, as such protecting the oxide liner in that area against recess during the bottom liner opening etch.

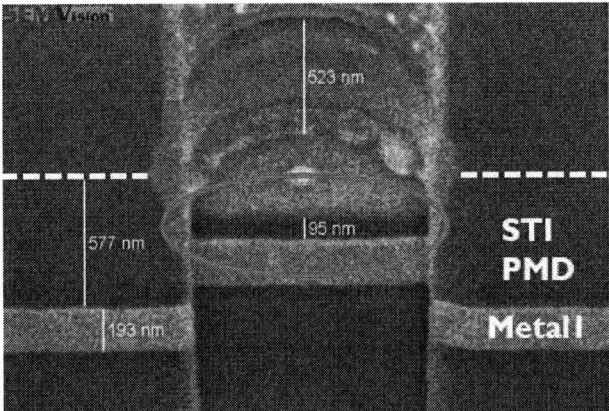

Figure 6. Inline FIB picture (taken under 45°) after bottom dielectric etch. Polymers on TSV sidewall are from FIB preparation.

Although this SiN protection concept is successfully demonstrated on this 0.7 x 5µm TSV-last module, an alternative liner protection at the top of the TSV is also developed. In this case, the PECVD SiN is replaced by a 100nm strippable amorphous carbon layer (APF, Advanced Patterning Film), deposited after the capping oxide. This film has, just as the PECVD SiN, a low step coverage and will protect the liner oxide only at the top of the TSV. However, the APF layer is removed during the O_2 strip after bottom liner etch. This alternative liner protection approach is therefore more scalable than the PECVD SiN protection scheme, as there is no extra CD loss at the top part of the TSV at the time of the barrier/seed deposition and TSV copper plating.

D. Barrier/seed and TSV copper fill

In order to keep this scaled sub-micron TSV-last module cost effective, we investigated a further usage of a conventional PVD Ta barrier and PVD copper seed system to enable void free copper filling by means of bottom-up electroplating. When keeping the PVD copper thickness to 400nm, the reference thickness as used in the 1 x 5µm module, we observe early pinch-off at the top of the TSV and this both in the center and at the edge of the wafer (Fig. 7). A reduction of the copper seed to 300nm enables void free filling in the center of the wafer, but still results in a fully voided TSV at the wafer edge due to this early pinch off at the TSV top. Further reducing the PVD copper seed to 250 and even 200nm shows another behavior. For these thinner copper seeds, discontinuities in the seed layer causes typical defects such as bottom-sidewall voids as it becomes challenging to reliably deposit such a thin PVD seed layer on the sides and bottom of a high aspect ratio via. The cross-sectional FIB pictures in Fig. 6 reveal that this issue pops up rather in the center of the wafer and not at the wafer edge.

A 30nm thin alkaline ECD seed layer enhancement (SLE) is therefore introduced. This layers "repairs" the PVD

978-1-7281-1500-9/19 $31.00 © 2019 IEEE

copper seed at the bottom of the TSV, therefore resulting in a conformal copper seed layer all over the TSV. It ensures void free ECD copper fill of 0.7 x 5µm TSVs, as it is shown and demonstrated in Fig. 5 for an end-of-line scaled via-last TSV.

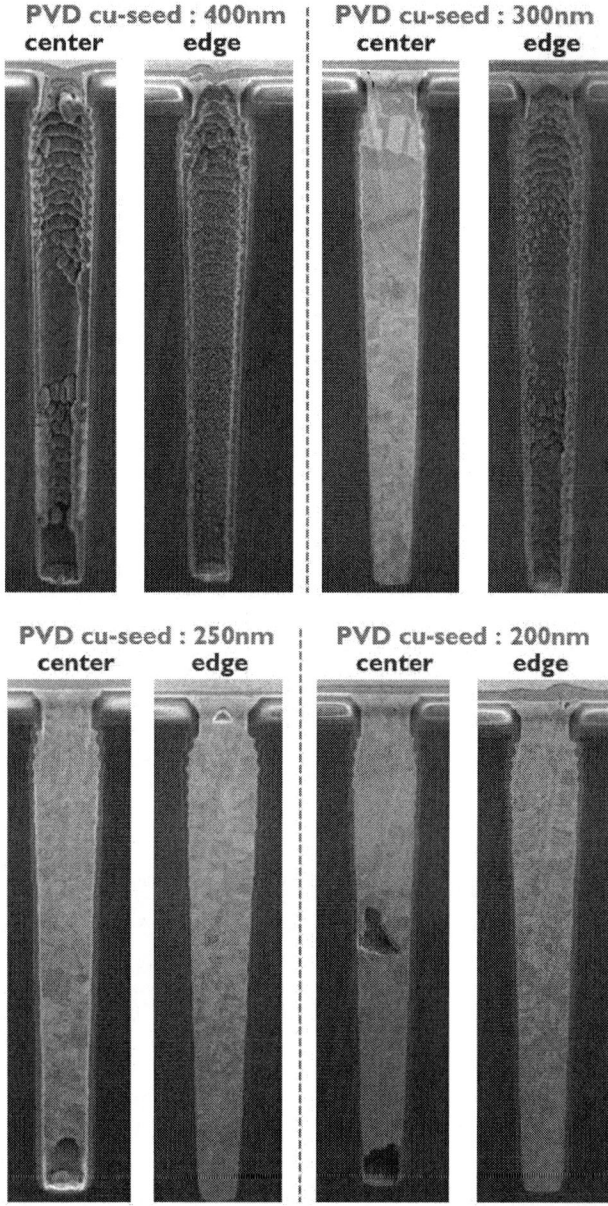

Figure 7. Cross-sectional FIB pictures after ECD copper fill, taken in center and edge of a wafer for respectively a 400nm, 300nm, 250nm and 200nm PVD cu-seed.

E. TSV copper and barrier CMP

The via-last module classically finishes with a 2-step chemical-mechanical polishing (CMP) removing first the copper overburden from the electroplated TSV filling, followed by the removal of the Ta TSV barrier and the TiN embedded barrier. No post-plate anneal is foreseen in this

module. Backside RDL metallization, by using a semi-additive process, provides the electrical access to the TSV at the thinned wafer back side.

To conclude this process integration paragraph, detailed TEM pictures of the TSV top part are shown in Fig. 8 and 9 for respectively a via-last TSV with PECVD SiN protection and APF liner protection. Both wafers are processed with the embedded ALD TiN concept, 5 or 10nm thick. The barrier CMP was performed equal on both wafers.

In the case of a PECVD SiN protection, the barrier CMP has removed around 100nm of this SiN liner protection layer at the surface. The 50nm thick liner oxide is not attacked, neither is the hard mask stack consisting of a 250nm SiN and a 150nm SiO₂ layer. The PECVD SiN remains present at the TSV shoulder (protecting the liner oxide) and is also visible all over the TSV sidewall in this TSV top part picture.

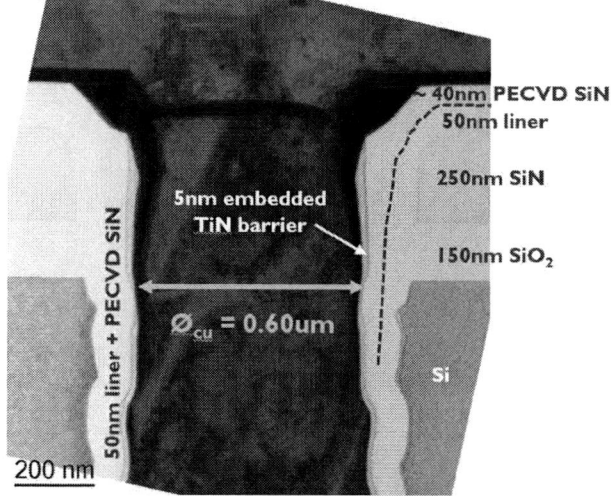

Figure 8. TEM picture (final process) of the 0.7 x 5µm TSV top part for a wafer with PECVD SiN liner protection.

For the wafer with an APF liner protection, and thus without PECVD SiN, the barrier CMP removes at the wafer surface the 50nm liner oxide and about 100nm of the originally 250nm thick SiN hard mask layer. The liner oxide however remains intact all over the TSV sidewall. In fact, the ALD TiN embedded barrier is nicely protecting the liner oxide during the bottom liner opening etch. The effective copper diameter at the TSV top part increased from ~0.60µm in the case of PECVD SiN liner protection to close to 0.7µm for the APF protected wafer.

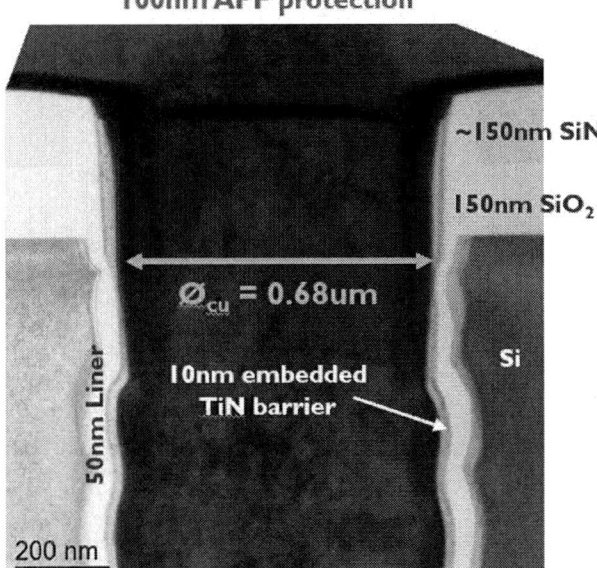

Figure 9. TEM picture (final process) of the 0.7 x 5μm TSV top part for a wafer with APF liner protection.

IV. ELECTRICAL RESULTS

A. Via-last Kelvin and Chain Resistance

A cumulative plot of the single TSV kelvin resistance is shown in Fig. 10 for respectively a 1 x 5μm and the scaled 0.7 x 5μm via-last structure. The expected increase in single TSV resistance, when scaling the TSV diameter, is limited as we also halved the thickness of the liner oxide and embedded barrier capping oxide when moving to the 0.7 x 5μm TSV, as such increasing the effective copper diameter.

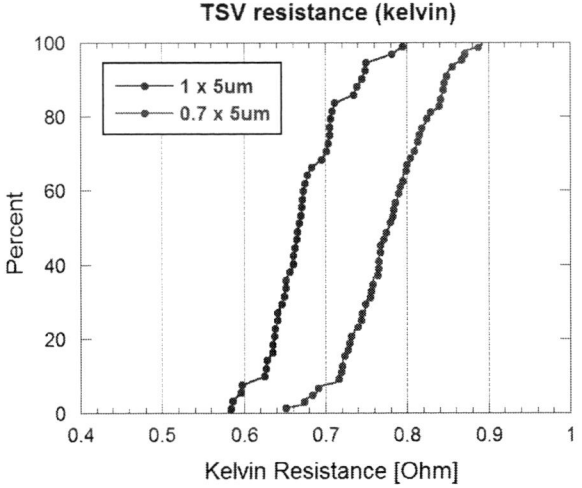

Figure 10. 1 x 5μm and 0.7 x 5μm kelvin resistance for a single TSV via-last structure.

Fig. 11 compares the single TSV kelvin resistance with the unit cell resistance obtained from a daisy chain containing 78 TSVs in series. 100% yield is obtained on this chain structure.

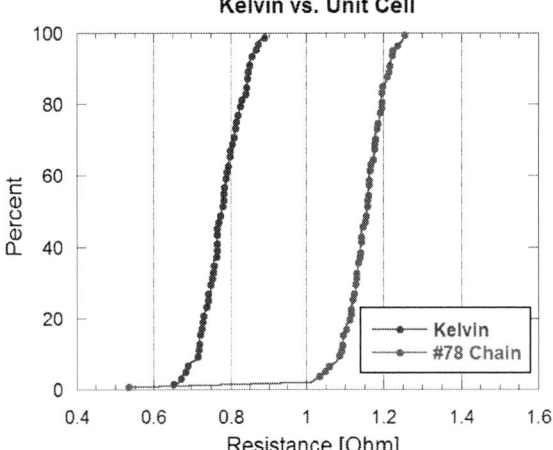

Figure 11. Kelvin and unit cell resistance (chain of 78 TSVs) for a 0.7 x 5μm via-last TSV.

B. Via-last Capacitance

The via-last capacitance is measured on array structures with varying number of TSVs, ranging from 0 (to measure the parasitic capacitance of the measurement bond pad and metal connections to the array) to respectively 50, 75 and 100 parallel connected TSVs. The obtained accumulation capacitance can then be plotted as a function of the number of TSVs. The slope of the linear fit represents the actual single via-last capacitance value. Fig. 12 shows the data for respectively a 1 x 5μm via-last reference and the scaled 0.7 x 5μm via-last. The via-last capacitance almost doubles from 5.4fF/TSV (1 x 5μm) to 10.1fF/TSV for the 0.7 x 5μm via-last, due to a twice as thin oxide liner and embedded barrier capping oxide used in the 0.7μm diameter scaled via-last module.

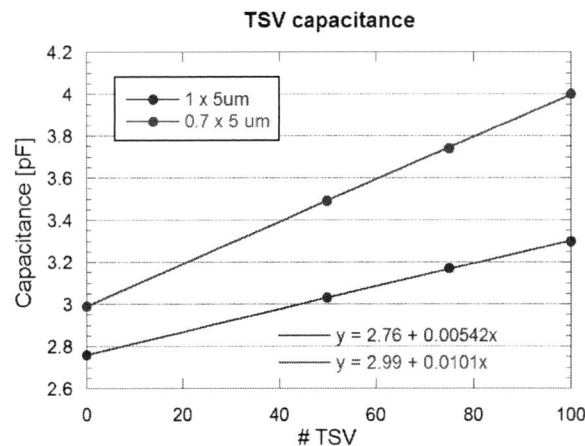

Figure 12. TSV capacitance as a function of the number of parallel connected TSVs, for respectively 1 x 5μm and 0.7 x 5μm via-last.

C. Via-last Reliability

The TSV liner/barrier integrity is verified by using the controlled I-V method (IV_{CTRL}) [4]. This method is a fast and quantitative characterization alternative for the time dependent dielectric breakdown (TDDB) analysis and detects whether or not copper is involved in the dielectric degradation or whether the observed breakdown is intrinsic. The TSVs are tested in both accumulation and depletion mode. In accumulation mode, also called copper-confined regime, the intrinsic quality of the oxide liner is tested. In depletion mode (copper-driven regime) the defectivity of the barrier is tested as the positively charged copper ions are repelled from the TSV and can diffuse through the oxide into the silicon in the case the barrier is not integer all over the TSV sidewall.

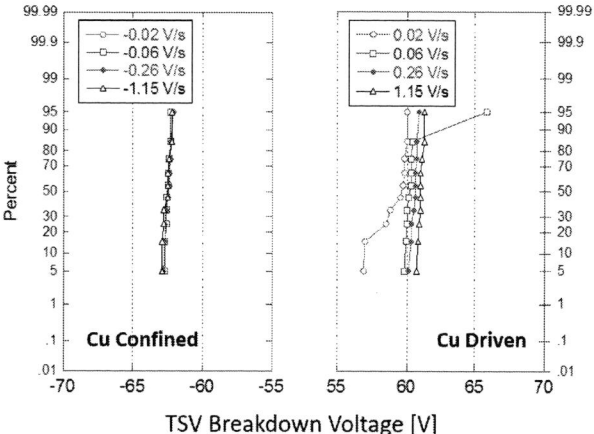

Figure 13. Cumulative plot of the TSV breakdown voltage for different voltage ramp rates, measured in both cu-confined and cu-driven mode. 0.7 x 5μm via-last TSV with 5nm ALD TiN embedded barrier.

Fig. 13 shows, for different voltage ramp rates, the cumulative distribution of the measured breakdown voltage (V_{bd}) of the 0.7 x 5μm via-last TSV with a 5nm ALD TiN embedded barrier. A V_{bd} around 60V is the expected value for a 50nm thick liner oxide.

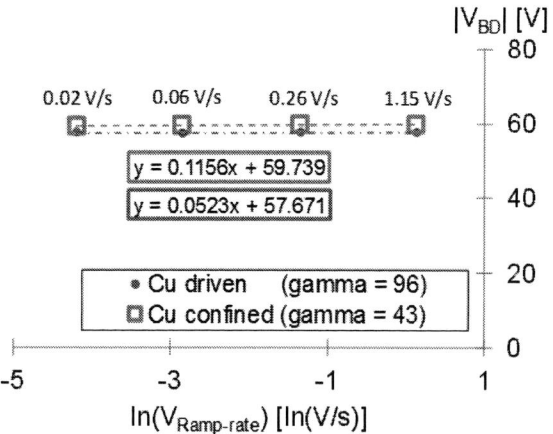

Figure 14. TSV breakdown voltage as a function of applied IV_{CTRL} voltage ramp rate. 0.7 x 5μm via-last TSV with 10nm ALD TiN embedded barrier.

The dependence of the breakdown voltage as a function of the voltage ramp rate is represented in Fig. 14 and 15 for a 0.7 x 5μm via-last with respectively a 10nm and 5nm TiN embedded barrier thickness. From these plots, the field acceleration factor γ can be extracted. Very high γ absolute values are obtained for either ALD TiN thickness and this at both cu-driven and cu-confined polarities. This illustrates the high TSV reliability for this scaled 0.7 x 5μm via-last module.

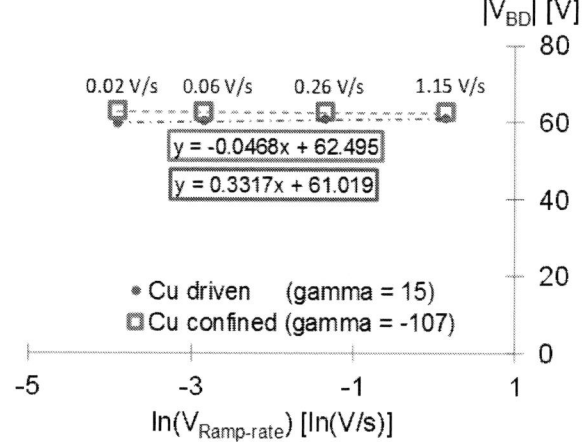

Figure 15. TSV breakdown voltage as a function of applied IV_{CTRL} voltage ramp rate. 0.7 x 5μm via-last TSV with 5nm ALD TiN embedded barrier.

V. CONCLUSIONS

We demonstrated the realization of a reliable 0.7 x 5μm via-last module, fitting a 1.4μm TSV pitch. The individual process step of this TSV module have been extensively discussed. Electrical measurements show the potential of the module in terms of TSV continuity, yield and capacitance. Furthermore, liner/barrier integrity is proven to be very good.

REFERENCES

[1] E. Beyne, S-W Kim, L. Peng, N. Heylen, J. De Messemaeker, O.O. Okudur, A. Phommahaxay, T-G Kim, M. Stucchi, D. Velenis, A. Miller, and G. Beyer, "Scalable, sub 2μm Pitch, Cu/SiCN to Cu/SiCN Hybrid Wafer-to-Wafer Bonding Technology," IEEE International Electron Devices Meeting - IEDM, Dec 2017, pp. 32.4.1–32.4.4.

[2] S. Van Huylenbroeck, Y. Li, J. De Vos, G. Jamieson, N. Tutunjyan, A. Miller, G. Beyer and E. Beyne, "A Highly Reliable 1x5μm Via-last TSV Module," IEEE International Interconnect Technology Conference - IITC, May 2018, pp. 94–96.

[3] S. Van Huylenbroeck, M. Stucchi, Y. Li, J. Slabbekoorn, N. Tutunjyan, S. Sardo, N. Jourdan, L. Bogaerts, F. Beirnaert, G. Beyer and E. Beyne, "Small Pitch, High Aspect Ratio Via-last TSV Module," IEEE Electronic Components and Technology Conference - ECTC, May 2016, pp. 43–49.

[4] Y. Li, D. Velenis, T. Kauerauf, M. Stucchi, Y. Civale, A. Redolfi and K. Croes, "Electrical Characterization Method to Study Barrier Integrity in 3D Through-Silicon Vias," IEEE Electronic Components and Technology Conference - ECTC, May 2012, pp. 304–308.

Nanoscale Topography Characterization for Direct Bond Interconnect

Bongsub Lee, Pawel Mrozek, Gill Fountain, John Posthill,
Jeremy Theil, Guilian Gao, Rajesh Katkar, and Laura Mirkarimi
Xperi Corporation
San Jose, California, USA
bongsub.lee@xperi.com

Abstract—**Hybrid bonding achieves mechanical and electrical connection between device wafers or dies, by directly joining dielectric and metal surfaces to form an all-inorganic interface. This direct bond interconnect (DBI) technology enables very fine pitch interconnects for high bandwidth interfaces. DBI is currently used for mass production of image sensors and is actively investigated for NAND, DRAM and MEMS applications. Characterizing and controlling nanoscale topography are essential for this type of bonding. After chemical mechanical polishing (CMP), the dielectric surface (usually SiO$_2$) should have high planarity and sub-nm roughness, and the metal surface (usually Cu) should be slightly recessed below the dielectric surface in general. Atomic force microscopy (AFM) is a critical technique required to monitor the CMP process module and ensure a robust manufacturing process. While AFM and related techniques have been known for decades, nanoscale or sub-nm scale characterization for DBI requires careful choice of the analysis configurations and parameters to avoid misinterpretation. Here we discuss key considerations for AFM analysis, extraordinary AFM artifacts in the relative heights of the Cu and SiO$_2$ areas, and topographic characteristics of Cu/SiO$_2$ surface for successful hybrid bonding. The force between an AFM tip and a sample should be sufficiently low for consistent roughness measurement but sufficiently high for minimizing the effects of surface contamination or artifacts. Proper data processing such as flattening should be done to make realistic images. Occasionally we observed artifacts that produced an incorrect Cu height, which could render an actually recessed Cu area as protruding. This artifact tends to occur more if the tip is not fresh or the tapping force is low. If a data image is unusually blurry and the oxide roughness is much smaller than usual, it may be a sign of this artifact. Replacing the tip or scanning in contact mode can usually demonstrate if there was an artifact. AFM analysis revealed that a curved SiO$_2$ surface tends to occur in the vicinity of Cu interconnect areas. Optimized CMP conditions can reduce the size of seams and eliminate them.**

Keywords—Hybrid bonding; nanoscale topography; 3D-IC; atomic force microscopy

I. INTRODUCTION

Hybrid bonding, or Direct Bond Interconnect (DBI), is currently used for mass production of CMOS image sensors for mobile phones [1] and is actively investigated for other 3-dimensional integrated circuit (3D-IC) applications such as 3D NAND [2] and die-to-wafer stacked DRAM [3][4].

Wafer or chip surfaces are polished, chemically activated, and then bonded at room temperature with almost no force. This room-temperature bonding already provides sufficient strength for normal handling. After many chips are bonded, they can be batch-annealed to achieve full mechanical strength and electrical connection [5] [6] [7]. This bonding technique forms a completely inorganic interface, without involving solder bumps or organic underfill between layers. Compared to the current solder-based stacking technology, hybrid bonding provides much finer pitch, higher bandwidth, better thermal performance, improved RLC characteristics, and higher assembly throughput [3].

Characterizing and controlling nanoscale topography is critical for the direct bond interconnect technology. In the case of simple dielectric-to-dielectric bonding (called direct bonding or ZiBond), the dielectric surface should be extremely smooth before activation and bonding. To achieve electrical connection between stacked layers (called hybrid bonding or DBI), the surface should have dielectric background areas (usually SiO$_2$) as well as metal pad areas (usually Cu). Chemical mechanical polishing (CMP) should achieve a very low dielectric roughness and also a certain recess of metal areas below the dielectric surface [8], which is illustrated in Fig. 1(a). Upon contact, the plasma-activated dielectric surfaces bond together instantaneously (Fig. 1(b)). Metal-to-metal bond occurs during a subsequent batch annealing. The coefficient of thermal expansion of metals are typically far larger than dielectrics. The metal expands to fill the gap and then build up internal pressure (Fig. 1(c)). It is under this internal pressure and annealing temperature that metal atoms diffuse across the interface, making good metal-to-metal bond and hence electrical connection. External pressure is not required for this type of bonding.

Atomic force microscopy (AFM) is a surface characterization tool that can monitor the nanoscale topography resulting from CMP. AFM analyzes topography by tracking the interaction between an ultra-sharp tip and surface atoms. While AFM and related techniques have been known for decades, we find that AFM characterization for DBI requires careful consideration of the analysis procedure and parameters to avoid misinterpretation. In this study, we discuss key considerations for AFM analysis and demonstrate surface topography characteristics of Cu/SiO$_2$ DBI surfaces. During CMP, relative removal rate of metal, dielectric, and barrier materials should be controlled to achieve the required topography for DBI. CMP may

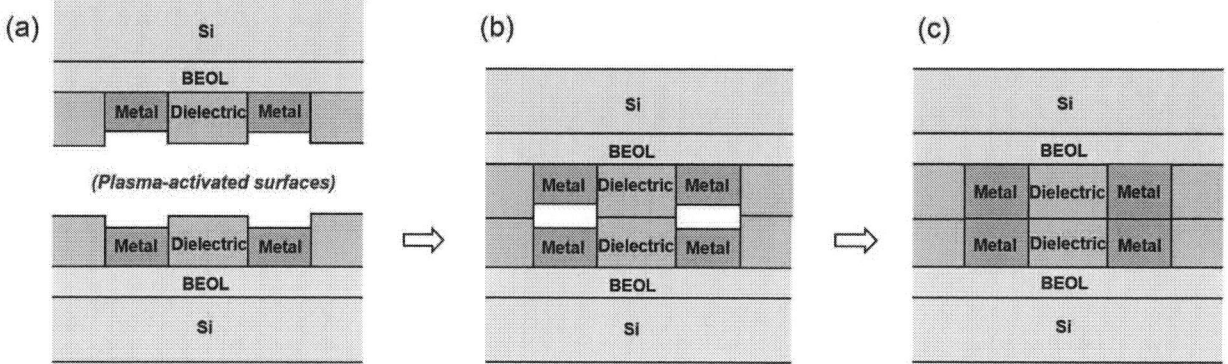

Figure 1. Schematic flow for Direct Bond Interconnect (DBI) process. (a) CMP prepares flat dielectric surfaces as well as moderately recessed metal surfaces. The surfaces are cleaned and plasma-activated (or activated by wet chemistry). (b) The surfaces are brought into contact. Dielectric surfaces bond together instantaneously at room temperature. (c) Upon batch annealing, the metal expands more than the dielectric and becomes internally pressurized. Metal-to-metal bonding is accomplished under this temperature and internal pressure (without external pressure). The initial metal recess has to be carefully controlled to make this process possible at a moderate annealing temperature.

produce a curved profile of oxide. We illustrate the effect of such curved profiles on the final bond interface. We also report occurrence of extraordinary AFM artifact from DBI structures. It can falsely indicate significant protrusion of Cu over SiO_2 and excessively smooth SiO_2 surface. We present examples of this artifact and discusses how to identify such problems.

II. EXPERIMENTS: CONSIDERATIONS FOR AFM TO CHARACTERIZE NANOSCALE TOPOGRAPHY FOR DBI

A. Scanning Process for AFM

Fig. 2(a) is one example of 3-dimensional representation of a DBI sample surface, which can be constructed by AFM. It consists of a dielectric area (SiO_2) and evenly distributed metal pads (Cu). The metal pads are slightly recessed from the dielectric background, as depicted in Fig. 1(a). Fig. 2(b) is a magnified view of a small dielectric area. The root-mean-square roughness of the dielectric surface can be obtained by calculating the standard deviation of the z-values from each pixel. Note that it is popular to use μm as the x/y-axis unit and use nm or Å as the z-axis unit to show AFM results. The appearance of surface features is hence highly magnified along the z-axis in this kind of visual representation.

To properly understand the topography of a DBI structure by AFM, it is important to understand how to obtain height information and how to post-process the acquired data. The AFM tip scans the sample in one direction (usually in the x-direction) and obtain the height (z) information along that direction, as depicted in Fig. 3(a). Then the tip or the sample shifts in the orthogonal direction (usually in the y-direction) and repeats the x-scan there. Fig. 3(b) shows schematic examples of such x-z profiles. By stacking many x-z profiles together in the y-direction, one can obtain a 3-dimensional representation such as Fig. 3(c), which is an equivalent of Fig. 2(a). In Fig. 3(c), darker

contrast represents lower points, and brighter contrast shows higher points.

B. Flattening Operation and Height Measurement

Combining AFM line profiles, as illustrated in Figs. 3(b) and 3(c), is usually accomplished by a "flattening" operation included in AFM analysis software [9]. If the AFM instrument can keep the relative height information from all points of the x-y plane, we will be able to see the entire 3-dimensional surface without distortion. One should adjust the sample tilt, since the sample plane cannot be perfectly perpendicular to the z-axis for measurement. This process is usually done by selecting the "whole" or "whole plane" option for flattening. However, while the relative height information is quite well preserved in the scanning direction (x-direction in this example), it is not always the case in the other direction (y-direction). For example, after

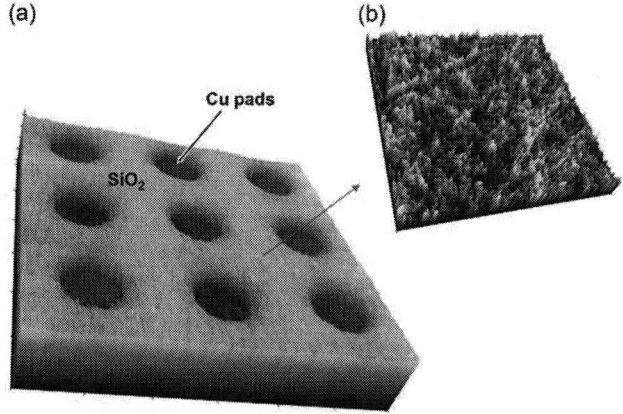

Figure 2. Examples of 3-dimensionally represented AFM data from a DBI sample. (a) Topography of an area of a DBI sample, which consists of SiO_2 background and recessed Cu pads. The size and pitch of Cu pads can vary widely depending on applications. (b) Magnified view of a small SiO_2 area. It is common to use μm as the x/y-axis unit and use nm or Å as the z-axis unit.

978-1-7281-1500-9/19 $31.00 © 2019 IEEE

a tip spends time to obtain many profiles, the sample temperature may change slightly. Even infinitesimally small temperature fluctuation (or other kinds of fluctuation) may effectively shift the measured height in the nanometer scale. Such difference may happen between the times to obtain the first profile and the last profile of Fig. 3(b); the height relationship between the profiles (i) and (vi) is not always clear.

A popular method of combining profiles is "line-by-line" flattening. In general, it takes each x-z profile and corrects its tilt (line-by-line) and, when combining those x-z profiles, it relies on the user's definition of the relationship between the profiles. Most AFM operators are accustomed to performing line-by-line flattening after selecting the entire area. It fits all x-z profiles into a single plane by least squares fitting. In other words, the software obtains the average height in each profile and aligns all of the profiles at their averages. It is indeed a good way to obtain results like Fig. 2(b); it is reasonable to assume that all x-z profiles have the same average height. On the other hand, this simple method was not adequate for the DBI profiles in Fig. 3(b). The average of heights in the profile (iii) should be clearly different from that of (v). If one aligns all profiles at their average values (i.e. tries to fit all profiles into one plane by least squares fitting), the result will look like Fig. 3(d), which is clearly unrealistic with "leveling artifacts [9]". Instead, we typically aligned the profiles by selecting only a part of SiO_2 areas as shown in Fig. 3(e) and performing line-by-line flattening. It fits only the selected areas into one plane by least squares fitting, to construct a result like Fig. 3(c). This is a robust way to make a visually reasonable AFM image. This is still not completely accurate in the y-direction but does not include the effects of uncontrollable fluctuation of conditions. Whichever flattening method is used, the height relationship between different profiles (y-z information) is not very dependable.

To obtain Cu recess below the SiO_2 surface, one should measure the height difference between points along the scanning direction (e.g. within one x-z profile such as the profile (iii) in Fig. 3(b)), instead of trying to compare points at different y-locations. If one must compare points at different y values, the operator should align those points along the scan direction by rotating the sample before scanning. Some machines also allow users to scan along an arbitrary direction. In this work, we consistently scanned along one direction that Cu pads are aligned to.

We employed an AFM with a flexure scanner for this study and always used first-order data flattening. It only corrects the tilting (slope) of the sample. Many traditional AFM systems have a tube scanner, which makes a pendulum-like motion while scanning. With that, a scanned profile along a perfectly flat surface would look like a parabola instead of a straight line. While second-order flattening method has been commonly employed to correct this effect, this method is complicated when the sample surface has some curvature. Flexure scanners do not have this curved background issue, while they may still show a slightly uneven background from a flat sample due to mechanical imperfection (out-of-plane motion), e.g. about

a couple of nm in z out of 100 μm in x. A proper flattening protocol should be chosen depending on the machine type. Good practice includes scanning a flat sample, such as a blank silicon wafer or an optical flat, to check the inherent background profile from the machine on a routine basis.

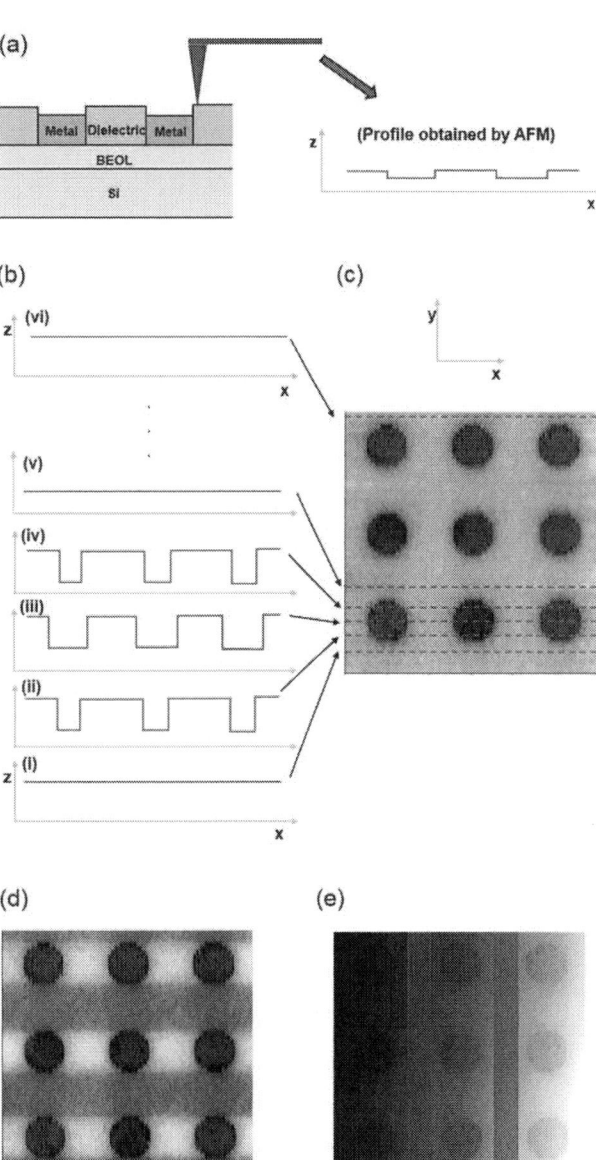

Figure 3. Method to construct a 3-dimensional representation of a DBI sample surface from AFM scans. (a) Obtaining a surface profile by scanning a sample surface. After taking one x-z profile, scanning can be repeated at different y-positions. (b) Schematic examples of surface profiles (x-z profiles). In actual AFM scanning, (c) Collection of many x-z profiles that shows the x, y, z information together. This is an equivalent 2D representaion of Fig. 2(a). (d) Result of a common line-by-line flattening, which selected the entire area and fitted them into a flat plane by least squares fitting. It gave visual artifacts (horizontal lines) for a typical DBI case. (e) Line-by-line flattening that was used to produce the result in c. Only the shaded SiO_2 areas were selected and fitted to a flat plane.

C. Imaging Mode for DBI Samples

We employed tapping, non-contact, and contact modes of AFM for this study. Tapping mode is the most popular imaging mode for AFM. The cantilever oscillates near its resonance frequency, and the tip "taps" on the surface [9]. Non-contact mode (or "true non-contact mode") operates at a larger distance where attractive force is dominant instead of repulsive force. Since the tip does not effectively touch the surface, this mode allows a very long tip life without mechanical wear. Note that an AFM tip has nanometers of radius, which can increase by mechanical wear. When a tip becomes less sharp, it cannot track very fine surface features, often measuring a lower roughness from the same surface. Non-contact mode was sometimes useful to measure dielectric roughness of many DBI samples consistently. However, this mode was often too sensitive to the surface conditions including slight contamination by organic substances and was prone to the interaction-force-related artifacts discussed in Section IV. The sensitivity of non-contact mode may be advantageous for very soft samples, but it is unnecessary for DBI sample surfaces that consist of relatively hard materials (SiO_2 and Cu). On the other hand, contact mode does not use oscillation. The tip "drags" on the surface with a relatively large force. The tip wears out quickly in contact mode, so it was not suitable for measuring roughness from multiple areas. It was robust for measuring Cu recess and especially useful to minimize the effects of organic contamination or force-related artifacts. When the data from contact and tapping modes were obviously different, the issue was due to degraded tip or surface condition.

For the purpose of DBI surface characterization, tapping mode AFM was a good compromise to obtain recess, roughness, and other surface characteristics from multiple areas with relatively good consistency. We replaced the tip after a certain amount of use before the measured roughness became too low, i.e. before the tip was worn out. When an artifact was suspected, we replaced the tip or tested the sample in contact mode.

III. EXTRAORDINARY ARTIFACTS IN AFM DATA IN DBI STRUCTURES

The validity of AFM data is critical in developing the CMP process for DBI. It is therefore important to understand all possible AFM artifacts. For example, it is well known that a worn-out tip may measure a low roughness value from an unacceptably rough sample.

In this study, we occasionally observed AFM artifacts that produced incorrect Cu recess values. It could even reverse the contrast, making recessed Cu areas appear as protruded. This type of artifact is not widely known with most AFM users and may cause significant misinterpretation. Fig. 4(a) is an example of normal AFM result from a DBI sample, showing recessed Cu areas. In some cases, the same sample showed completely different results as shown in Fig. 4(b). The Cu areas appear to be much higher than in the normal result, showing artificial protrusion in this case. The oxide roughness also appeared

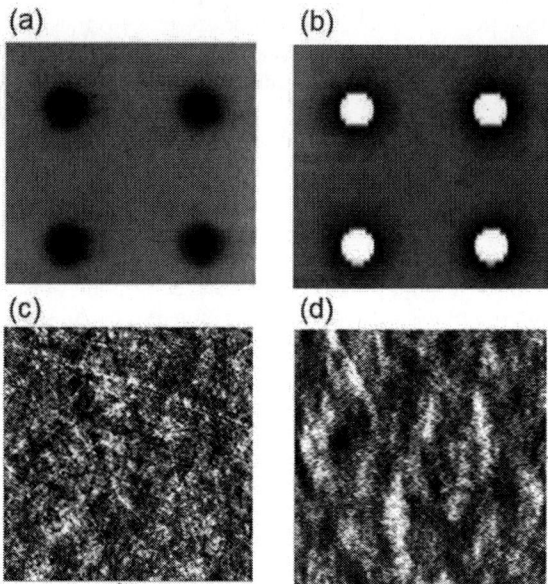

Figure 4. AFM artifacts due to abnormal tip-to-sample interaction. (a) Normal AFM result from a DBI sample with recessed Cu areas. (b) Erroneous AFM result from the sample sample, showing protruding Cu areas. (c) Normal AFM result from a small SiO_2 area of the same sample. (d) AFM result from a small SiO_2 area taken after the artifact was observed. The image is blurry, and the resulting roughness is lower than normal.

lower when this artifact occurred. Fig. 4(c) is a regular AFM result from a small SiO_2 area on this sample, and Fig. 4(d) is an erroneous result. Fig. 4(d) appears blurry and shows a lower roughness, while Fig. 4(c) is crisp and reveals actual surface features such as nanoscale scratches from CMP.

We have also observed a reversible behavior with such artifact. Fig. 5 shows two AFM results from the same area. In the case of Fig. 5(a), the scan started from the bottom and showed artificial protrusion. After the tip hit a large particle (or particles, shown with an arrow), the artifact suddenly disappeared and Cu recess was shown. After that, the same area was scanned again from the bottom for Fig. 5(b). Normal Cu recess was shown at first but, after the tip hit the particle again, artificial protrusion reappeared. This repetition occurred also from additional scans on this area.

Occurrence of this artifact was related to tip freshness and tip-sample interaction force. This artifact did not occur with an unused tip. It occurred with some tips after used for a certain amount, but it was difficult to predict how soon it could occur. Non-contact mode appeared more prone to this artifact. In one case, the artifact occurred only after scanning a couple of areas in a non-contact mode. When we switched to tapping mode with the same tip, the artifact disappeared. The artifact could occur also in tapping mode after some use. Contact mode was the most robust against this type of artifact.

The origin of this artifact is not fully known to us, but it should be related to the difference between the tip-Cu interaction and the tip-SiO_2 interaction in certain circumstances. One explanation is electrical charging near

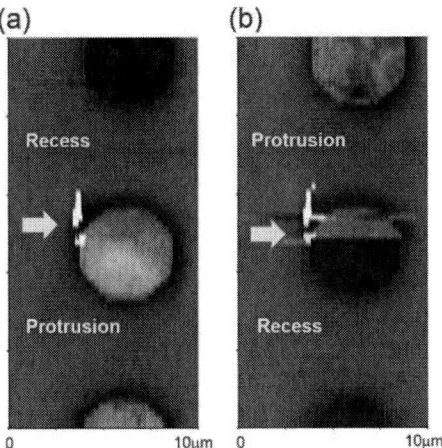

Figure 5. A case of reversible artifact. (a) AFM result that first showed artificial protrusion (bottom) then normal Cu recess (top) after the tip hit a large particle. (b) AFM result taken immediately afterward. The scan first showed normal recess at the bottom then artificial protrusion after the tip hit the same particle.

the AFM tip. A charged tip will interact differently on metal and dielectric areas on a DBI sample and have more difficulty in correctly analyzing roughness or nanoscale scratches (Fig. 4(c) and (d)). Charging may occur more easily with a used tip as it collects contaminants or become damaged after multiple scans. Non-contact mode should be more affected to the electrostatic force than contact mode. We tried using an ionizing air blower to remove static charges and observed limited improvement. It reduced the amount of artificial protrusion but did not completely remove the problem even after many minutes of exposure. As discussed in the previous section, tapping mode is the preferred method for general analysis of DBI structures, even though this artifact may occur. A good practice is to observe dielectric roughness over the course of measurements. If the data image is unusually blurry and the oxide roughness is much smaller than normal, it may be a sign of the artifact. Replacing the tip or testing with contact mode can usually demonstrate whether the feature is real or an artifact.

IV. NANOSCALE CHARACTERISTICS OF CU/SIO$_2$ DBI SURFACES AND THEIR EFFECTS ON BONDING

For DBI, large deviations from flatness can affect the bond quality and bond strength. Curvature of the dielectric surface can affect the bond result but has not been discussed in detail in previous studies. Height change from SiO$_2$ to Cu may not usually be a strict step function especially when interacting with the CMP slurry. Therefore, the dielectric surface may become curved in the vicinity of metal interconnect areas. After initial bonding and annealing, this non-flatness may result in confined non-bonded areas or seams.

We developed multiple DBI designs and the corresponding CMP processes [3]. The dependence of layout and design on the resulting topography is well known

in the CMP industry [10]. AFM measurements must be performed in conjunction with the CMP development. Die bonded with non-optimized CMP leave seams as shown in Fig. 6(a). The seam is described as a non-bonded SiO$_2$-to-SiO$_2$ area. In Fig. 6, the seams are observed near Cu areas. In the case of Fig. 6(b), the CMP condition was tuned to make much flatter oxide. It was also possible to minimize the occurrence of seams, depending on CMP conditions and DBI design. The example shown in Fig. 6(c) has no visible seams at the bond interface. Optimizing the CMP condition is the key to produce the right amount of surface characteristics such as metal recess, dielectric roughness, and dielectric curvature for DBI.

V. CONCLUSIONS

AFM is a critical technique to characterize the nanoscale topography for hybrid bonding. Careful protocols should be followed to ensure that the AFM characterization is accurate without artifacts that result in misleading interpretation. The force between an AFM tip and a sample should be sufficiently low for consistent roughness measurement but sufficiently strong for minimizing the effects of surface contamination or artifacts. Tapping mode was usually a good compromise for the purpose of DBI characterization. It is possible to produce a 3-dimensional representation of a surface by stacking many profiles. One should note that, in such a 3D representation, the height relationship is far more accurate within one profile than between different profiles. Accordingly, comparison of

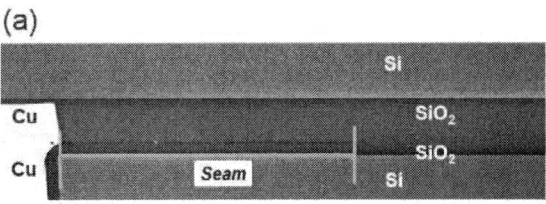

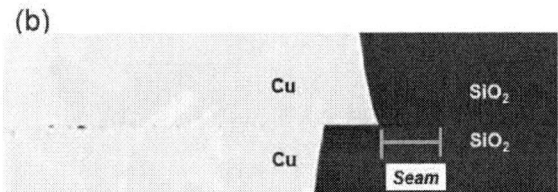

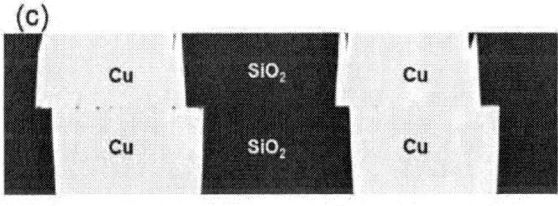

Figure 6. Cross-sectional scanning electron microscope images of DBI pairs. (a) Significant non-bonded SiO$_2$ area (seam) near the Cu pads. (b) Minimal seam during bond. (c) DBI pair without visible seams due to optimized CMP conditions.

heights should be done by comparing points along the scanning direction.

Occasionally we observed AFM artifacts that produced an incorrect Cu height, often showing artificial Cu protrusion from a truly recessed region. The artifact tends to occur more if the tip is not fresh or the tapping force is low. It is possibly due to local charging of the tip. Using an ionizer could reduce the degree of this artifact but did not completely remove it in our case. This artifact usually accompanies blurriness and excessively low roughness, so it is a good practice to watch the roughness in the course of scanning many areas. Replacing the tip or testing in contact mode can usually demonstrate whether there was an artifact.

Topographic metrology and AFM analysis are critical to characterize the CMP process which ultimately impacts that quality of the bond. By optimizing CMP conditions with the topographic metrics of roughness, flatness, and recess, we could eliminate the bond seams.

ACKNOWLEDGMENT

We thank Michael Huynh and Benny Fuentes for their help in sample preparation and SEM imaging, and Kang-Deuk Choi, Ardavan Zandiatashbar, Phani Kondapani, and David Braunstein for discussions.

REFERENCES

[1] P. Garrou, "IFTLE 303 Sony Introduces Ziptronix DBI Technology in Samsung Galaxy S7," Solid State Technology, 2016. http://electroiq.com/insights-from-leading-edge/2016/09/iftle-303-sony-introduces-ziptronix-dbi-technology-in-samsung-galaxy-s7/.

[2] R. Merritt, "YMTC Adds Detail to NAND Plans, EE Times," 2018. https://www.eetimes.com/document.asp?doc_id=1333563.

[3] G. Gao, L. Mirkarimi, G. Fountain, L. Wang, C. Uzoh, T. Workman, G. Guevara, C. Mandalapu, B. Lee and R. Katkar, "Scaling Package Interconnects Below 20µm Pitch with Hybrid Bonding," in IEEE 68th Electronic Components and Technology Conference, San Diego, 2018.

[4] B. Lee, R. Katkar, G. Gao, G. Fountain, S. Lee, L. Wang, C. Mandalapu, C. Uzoh, L. Mirkarimi, B. Sykes, M. Litjens, Y. Niu, S. Shao, J. Wang and S. Park, "Mechanical Strength Characterization of Direct Bond Interfaces for 3D-IC and," in Proc. IEEE 68th Electronic Components and Technology Conference, San Diego, 2018.

[5] V. Masteika, J. Kowal, N. S. J. Braithwaite and T. Rogers, "A Review of Hydrophilic SiliconWafer Bonding," ECS J. Solid State Sci. and Technol., vol. 3, no. 4, pp. Q42-Q54, 2014.

[6] P. Enquist, "Scalability and Low Cost of Ownership Advantages of Direct Bond Interconnect (DBI®) as Drivers for Volume Commercialization of 3-D Integration Architectures and Applications," in Materials Research Society Fall Meeting, 2008.

[7] P. Enquist, G. Fountain, C. Petteway, A. Hollingsworth and H. Grady, "Low Cost of Ownership Scalable Copper Direct Bond Interconnect 3D IC Technology for Three Dimensional Integrated Circuit Applications," in IEEE International Conference on 3D System Integration, 2009.

[8] P. Enquist, "3D Technology Platform – Advanced Direct Bond Technology," in 3D Integration for VLSI Systems, Pan Stanford Publishing, 2011, p. 175.

[9] P. Eaton and P. West, Atomic Force Microscopy, Oxford University Press, 2010.

[10] T.-C. Chen, M. Cho, D. Z. Pan and Y.-W. Chang, "Metal-Density-Driven Placement for CMP Variation and Routability," IEEE Trans.

on Computer-Aided Design of Integ. Circ. and Sys., vol. 27, no. 12, pp. 2145-2155, 2008.

Fully-Filled, Highly-Reliable Fine-Pitch Interposers with TSV Aspect Ratio >10 for Future 3D-LSI/IC Packaging

M. Murugesan[1], T. Fukushima[1], K. Mori[2], A. Nakamura[2], Y. Lee[2], M. Motoyoshi[2], J.C. Bea[1], S. Watariguchi[3] and M. Koyanagi[1],

[1]GINTI, 6-6-10, Aza-Aoba, Aramaki, Aoba-ku, Sendai, 980-8579, Japan
[2]T-Micro, 6-6-10, Aramaki, Aoba-ku, Sendai, 980-8579, Japan
[3]MELTEX INC., 2-3-1, Yoshino-cho, Kita-ku Saitama 331-0811, Japan
E-mail: murugesh@bmi.niche.tohoku.ac.jp

Abstract —Si interposer with 10 μm-width, 100 μm-deep through-silicon via (TSV) has been fabricated using electroless (EL) Ni as barrier and seed layers, and characterized for their electrical resistance. The chemistry of electroless-Ni plating bath was meticulously adjusted for the conformal formation of Ni along the TSV side wall. From the resistance value of 36 mΩ per TSV obtained from the Kelvin measurement of these Cu-TSV chain showed that the electroless Ni layer well acts as a good seed layer for completely filling the high aspect ratio TSVs by Cu-electroplating.

Keywords - Electroless Ni; Barrier/Seed layer; Cu-TSV; Si interposer; Cu-diffusion

I. INTRODUCTION

The transistors inside the chip continually improve their performance and energy consumption, thanks to the highly scaled down integrated circuits with tens of nanometers. Though there is an enormous improvement in the on-chip performance, the improvement in the input/output (I/O) performance is yet to keep in pace. This renders to a severe degradation in performance of system-on-chip (SOC), system-in-package (SIP), etc. owing to only a limited number of I/O availability. The integration of multiple functional chips on a Si or glass interposers paves the way for enhancing the number of I/Os in SOC or SIP. The Si interposer technology not only eases the I/O bottleneck problem, but also helps to minimize the thermal issues of 3D-ICs.

In recent days through-silicon-via (TSV) interposer technology has been widely used to improve the miniaturization and product performance by vertically stacking not only for FPGA, HBM, CIS chips, etc. [1-5], but also there is an growing demand for different RF applications such as active and passive interposers, high-Q passives and filters[6-8]. In contrast to 3D-ICs where the smaller diameter (<5 μm) TSVs with less than a few tens of microns in depth are included in large numbers in ultra-thin IC stacks [9-12], the TSVs for 3D heterogeneous integration of ICs with MEMS or RF to the Si interposers are less in number and their diameter is in the tens of microns with hundreds of microns in depth.

It is conventional to deposit the barrier and seed layers inside the TSV sidewall by using sputtering tool. However, it is increasingly becoming difficult to conformally deposit barrier and seed layers inside the TSV as the aspect ratio (AR) increases. This leads to a very thick deposition of barrier and seed layers at the TSV top (around one order

higher) as against the thickness of the respective layers at the bottom corner of the TSV. Even the sophisticated sputter machine currently available at the market do only deposit less than 10% of barrier and seed material at the bottom corner of the TSV. This leads to three different serious reliability problems. (1) as the thickness of the barrier layer getting thicker and thicker at the top of TSV, the growth of grain also proceeds. This results into the formation of grain boundaries in the barrier layer that degrades the Cu diffusion barrier ability, since the diffusion of Cu primarily occurs via grain boundary. (2) Since the barrier layer is very thin at the bottom corner of the TSV, there may be a discontinuity in the barrier layer. This will also leads to diffusion of Cu into the SiO_2 and then into the Si. The above mentioned two issues become more severe with not only the TSV size is scaled down but also with the increase in the TSV AR. (3) As both the barrier and seed layers becoming thicker and thicker, an overhang structure formed by barrier and seed layers at the top corner of the TSV. This results into earlier closure of the TSV top, leaving behind the void inside the TSV middle and/or bottom. This void formation could be explained by two ways. (i) Just because of the presence of physical barrier (i.e. overhang structure), the diffusion of plating solution into the TSV bottom is hindered, and thus less plating of metal at the bottom of the TSV. (ii) The non-uniform resistance value of seed layer [i.e lower resistance at the top (owing to thick seed layer) and higher resistance at the TSV bottom (due to very thin layer of seed material, caused by smaller coverage of seed layer during deposition)] itself leads to faster electroplating at the TSV top as against the TSV bottom, and hence an earlier closure of TSV at the top.

To get rid of the non-uniformity issue associated with the conventional sputtering of barrier and seeds layers inside the high AR TSV, chemical vapor deposition (CVD) process is recently used to conformal deposition of manganese (Mn) or its oxides as barrier layer inside the high AR TSV. Thin CVD Mn oxide acts as better barrier layer to arrest Cu diffusion compared to sputtered Ta or Ti barrier layer [13, 14]. However, the CVD technique is relatively costlier process than sputtering, and it involves high temperature which may not be suitable for the TSV-last approach.

Thus an alternative approach to both sputtering as well as the CVD techniques to deposit barrier and seed layers, a low-cost and low-temperature process is immensely needed, especially for Si interposer application. An electroless metal (EL) plating process is now widely used to form conformal

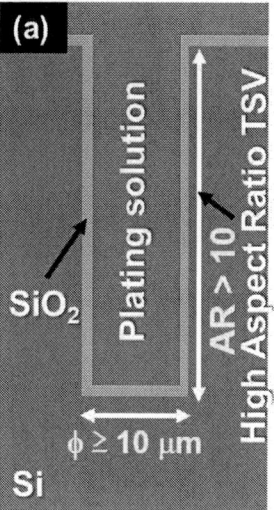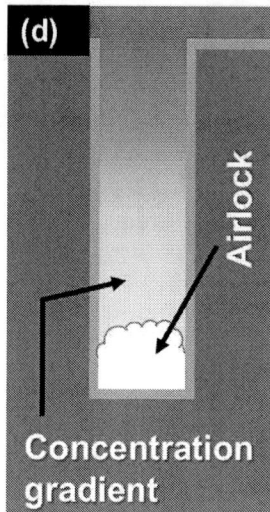

Figure 1. Schematic illustration of possible scenarios during electroless plating of metal inside high aspect ratio TSV (a) ideal; (b) presence of airlock inside TSV; (c) existence of concentration gradient of metal ions; (d) combination of both (c) and (d).

barrier and seed layers inside the high AR TSV. Previously it has been shown that conformal EL plating of Ni, Co, Cu, Au, etc. inside the TSV for 3D-LSI/IC fabrication [15-18]. In this work, we propose to use the EL plating technique for the conformal formation of Ni over the SiO₂ dielectric inside the TSV for interposer applications. Although EL plating process relatively forms barrier and seed layer deep inside the TSV conformally, the chemistry of EL plating bath needs to be meticulously adjusted especially in the case of TSVs with AR greater than 10.

II. Experiment

10 μm-width, 100 μm-deep TSV structures on 12-inch Si wafer were formed by Bosch etch method. 500 nm-thick SiO₂ was deposited along the TSV side wall using plasma enhanced chemical vapor deposition. Followed by El-Ni plating was carried out using the plating solutions provided by Meltex corp. In this work Pd was used as a catalyst to initiate the Ni deposition. EL-Ni plating was optimized for the temperature, plating period, and the concentration of EL-Ni plating bath. Cu electro plating was employed to completely fill the TSVs using the EL-Ni layer as seed layer. In order to measure the resistivity of these TSVs, we have also fabricated the daisy chain Cu-TSV structure. Kelvin measurements were carried out using Frame Transfer Prober FP3000 from Accretech, which automatically performs transfer and probing thin wafers, singulated wafers, and equipped with die position correction software, automatic wafer alignment, and automatic probe needle to pad alignment.

III. Results and Discussion

Shown in Fig. 1 is the schematic view of TSV structures charged with the EL-Ni plating solution. Fig. 1(a) is the ideal condition of EL-Ni plating solution inside the TSV. Fig. 1(a) to (c) reveals the possible scenarios of defective plating processes. As the AR of the TSV increases, it is extremely difficult not only to maintain the concentration of metal ion deep inside the TSV as well as free from airlock at the bottom of the TSV. Fig. 1(b) reveals the presence of airlock at the bottom of TSV, which results into a void at the bottom of the TSV. A conventional outgassing process has little effect in removing the airlock present in the bottom of the TSV with AR > 10. Fig. 1(c) reveals the non-uniform plating solution concentration as the time passes by. This plating condition results into very poor coverage ratio of Ni at the bottom of TSV, which also poses severe reliability issue. Fig. 1 (d) is worst possible scenario during the EL plating.

Fig. 2 reveals the SEM images taken after the El-Ni plating inside the TSV with AR 10. Fig. 2(a), 2(b) and 2(c) are for similar to the EL-Ni plating conditions shown in Fig. 1(b), 1(c) and 1(d), respectively. Although the EL-Ni coverage ratio is conformal (close to 1) in fig. 2(a), the

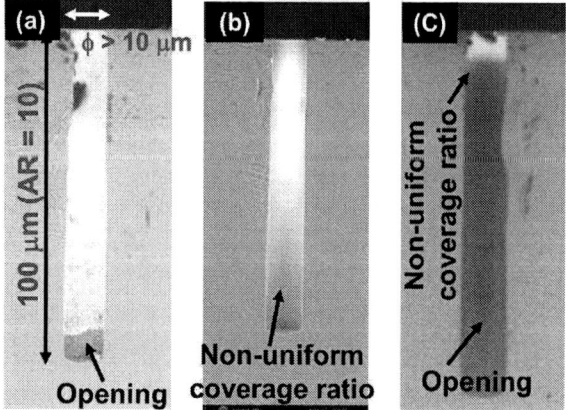

Figure 2. X-sec. (hand cleaved) SEM images obtained after the EL-Ni plating inside the high AR TSV using EL-Ni chemistry 1 and 2..

978-1-7281-1500-9/19 $31.00 © 2019 IEEE   1048

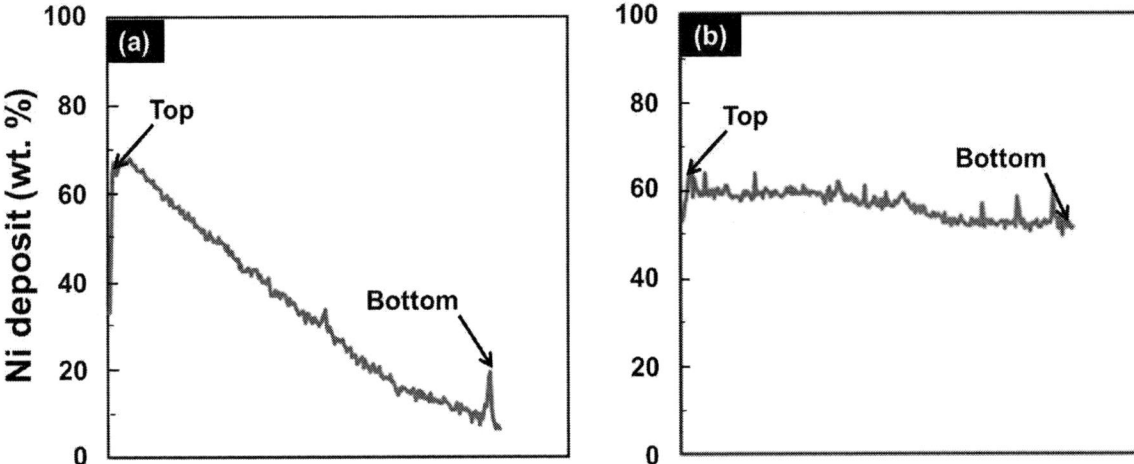

Figure 3. EDS line profile for Ni content in the deposited EL-Ni inside the high AR TSV using (a) EL-Ni chemistry 1 and (b) El-Ni-chemistry 2.

presence of airlock during the plating process leads to an open part at the bottom of the TSV.

Fig.3 is the plot of Ni content (in wt. %) deduced from EDS analysis of the Ni deposit formed inside the TSV, using (a) EL-Ni chemistry 1 and (b) EL-Ni chemistry 2. The major difference between the EL-Ni chemistry 1 and 2 are the difference in the $Ni2+$ concentrations in the plating bath and the amount of stabilizer. The Si chips were cleaved after the EL-Ni plating to get the X-sec. sample. Obtained EDS data were calibrated by considering the total wt. % of the Ni, P, O and Si as 100. It is that the Ni content continuously varied from the top of the TSV to the bottom of the TSV, by using the EL-Ni chemistry 1. The difference in Ni coverage between the TSV top and the TSV bottom is around 6x~7x, and it is thinner at the bottom of the TSV. To overcome this

issue we have changed the $Ni2+$ content of the plating bath in EL-Ni chemistry 2, in addition to the amount of stabilizer. Further to facilitate the EL-Ni plating at the bottom of the TSV we have carefully controlled the plating temperature and time, and the EDS result is shown in fig. 3(b). Upon comparing it with the results of fig. 3(a), it is clear the coverage ratio of Ni all along the TSV has been tremendously improved for EL-Ni chemistry 2, and it is close to 100%.

Shown in fig. 4 is the plot revealing the dependency of Ni coverage ratio with growth rate for both the EL-Ni chemistry of 1 and 2. For the EL-Ni chemistry 2, the coverage ratio of Ni at the middle of the TSV is nearly independent of the growth rate from anywhere between 30 nm/min 140 nm/min. However, even for the EL-Ni chemistry 2, the coverage ratio drops from 88% to 10 %, as the growth

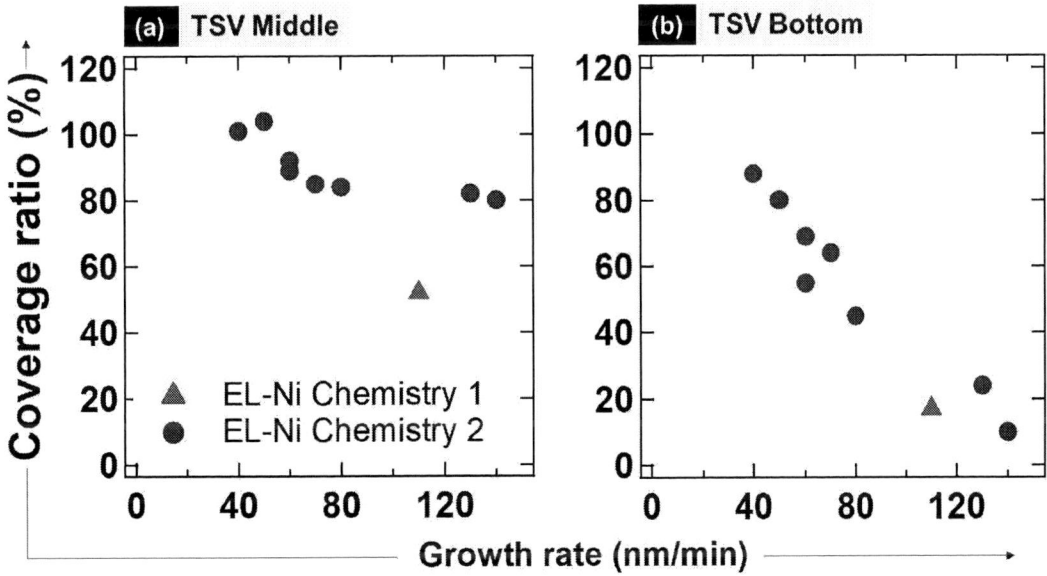

Figure 4. Dependency of coverage ratio of Ni inside the high AR TSV (a) TSV middle and (b) TSV bottom on growth rate of EL-Ni.

978-1-7281-1500-9/19 $31.00 © 2019 IEEE

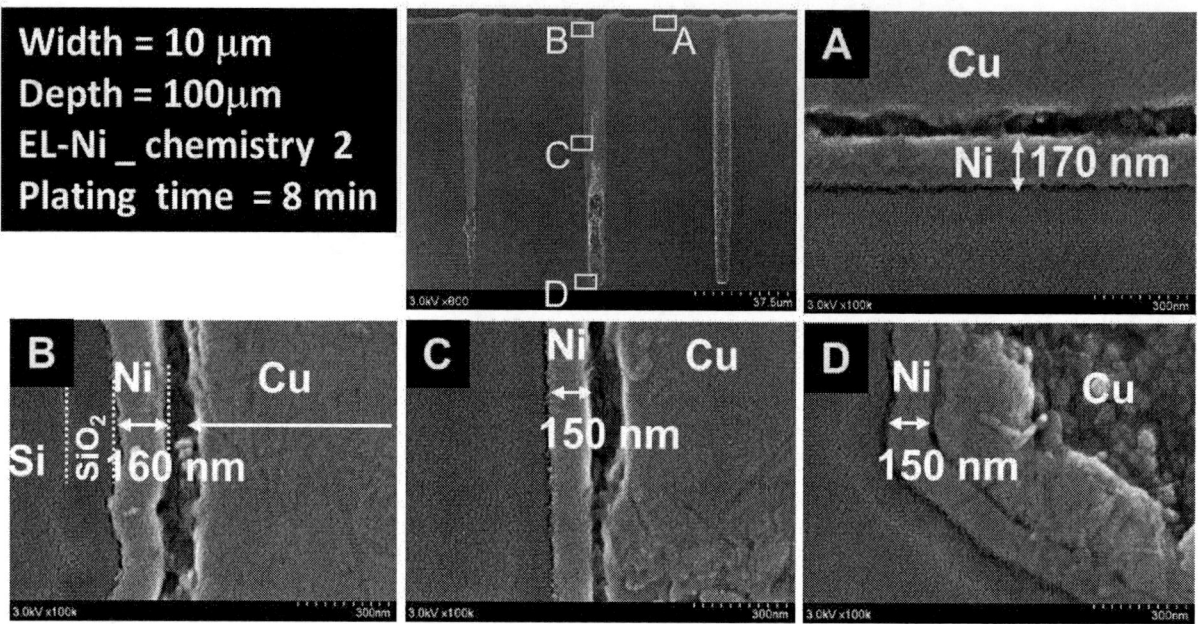

Figure 5. X-sec. SEM images obtained after the conformal EL-Ni plating upto the bottom of high AR TSV. EL-Ni thickness (a) on the field, (b) TSV top, (c) TSV middle, and (d) TSV bottom.

rate increased from 30 nm/min to 140 nm/min. It is imperative that in order to enhance the Ni deposit formation at the bottom of the TSV, one need to slow down the overall Ni growth rate.

Having obtained the conformal deposition of EL-Ni inside the TSV with diameter 10 μm and AR 10, we have attempted to completely fill the TSV with Cu by electroplating. As deposited EL-Ni layer forms a seed layer for Cu electroplating, and we were able to successfully fill the TSV with Cu. Shown in Fig. 5 are the X-sec. SEM images taken after Cu electroplating over the EL-Ni seed layer. The EL-Ni seed layer was formed by using the EL-Ni chemistry 2, and the growth rate of Ni to 30 nm/min. As can be seen from the SEM images, the EL-Ni seed layer thicknesses were

around 160 nm, 150 nm and 150 nm, respectively at the TSV top, middle and bottom.

To measure the TSV resistance, we have also fabricated the daisy chain Cu-TSV structure, and the X-sec. SEM image is shown fig. 6. Shown in fig. 7 is the results of Kelvin measurement carried over the sample shown in fig. 6. The daisy chain contains 64 TSVs. The cumulative resistance value of the daisy chain was 2.35 Ω. We have also confirmed the repeatability of the data. The deduced single TSV resistance value is around 36.7 mΩ, which is on par with Cu-TSV fabricated from Cu seed layer. This reveals that using EL-Ni plating method which is a low cost process compared to sputtering and CVD processes, one can easily fabricate Cu-TSV with high AR for Si interposer application.

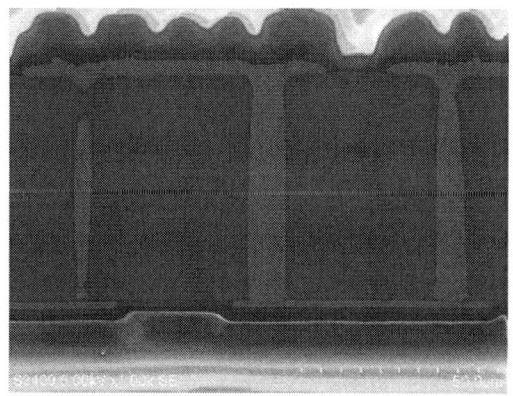

Figure 6. X-sec. SEM images revealing the fully filled Cu-TSV daisy-chain structure using EL-Ni as barrier and seed layers.

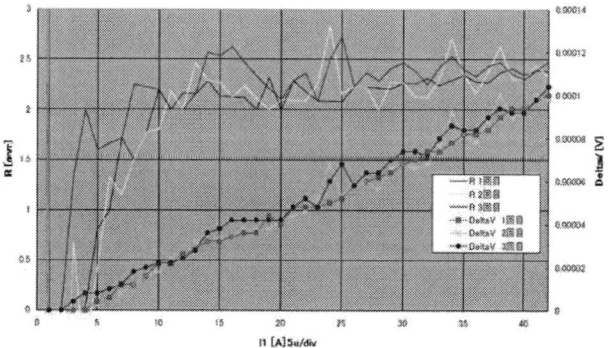

Figure 7. Resistance value deduced from I-V measurements for the fully filled 10 μm-width Cu TSV daisy chain structure containing 64 TSVs.

978-1-7281-1500-9/19 $31.00 © 2019 IEEE

IV. SUMMARY

We have demonstrated the fabrication of high aspect ratio Cu-TSVs using low-cost, highly scalable, CMOS-compatible EL-Ni plated Ni as seed layer for Si interposer applications. By modifying the EL-Ni chemistry of the plating solution, we have obtained highly conformal Ni seed layer growth inside the high aspect ratio TSV, with the coverage ratio of 100%. The resistance value of 36 mΩ for 10 μm-width Cu-TSV is on par with the best reported values and suggests that EL-Ni plating method can readily be employed to fabricate high aspect ratio Cu-TSVs for Si interposer application.

ACKNOWLEDGMENT

Part of this work was performed at the Junichi Nishizawa Research Center and Micro/Nano-machining research and education Center (MNC) at Tohoku University, Japan. We would like to thank all the staff members of GINTI (Global INTegration Initiative) for their great help in carrying out 3D-integaration processes.

REFERENCES

[1] M. Wiestruck, S. Marschmeyer, P. Kulse, T. Vob, M. Lisker, A. Kruger, D. Wolansky, M. Fraschke, and M. Kaynak, "Development of a Through-Silicon Via (TSV) Process Module for Multi-project Wafer SiGe BiCMOS and Silicon Interposer", Proceedings of IEEE 68th Electronic Components and Technology Conferecne, pp. 2267-2274, 2018.

[2] B, Banijamali, S. Ramalingam, K. Nagarajan, and R. Chaware, "Advanced Reliability Study of TSV Interposers and Interconnects for the 28 nm Technology FPGA", Proceedings of IEEE 61st Electronic Components and Technology Conference, pp. 285-290, 2011.

[3] V.N. Sekhar, J.S. Toh, J. Cheng, J. Sharma, S. Fernando, C. Bangtao, "Wafer Level Packaging of RF MEMS Devices using TSV Interposer Technology", Proceedings of IEEE 14th Electronics Packaging Technology Conference, pp. 231-235, 2012.

[4] S. McCann, H. H. Lee, G. R-Ahmed, T. Lee, and S. Ramalingam, "Warpage and Reliability Challenges for Stacked Silicon Interconnect Tehnology in large Packages", Proceedings of IEEE 68th Electronic Components and Technology Conference, pp. 2345-2350, 2018.

[5] J. Lee, C.Y. Lee, C. Kim, S. Kalchuri, "Micro Bump System for 2nd Generation Silicon Interposer with GPU and High Bandwidth Memory (HBM) Concurrent Integration", Proceedings of 68th Electronic Components and Technology Conference, pp. 607-612, 2018.

[6] J. Onohara, F. Takagi, T. Kizu, K. Imayoshi, H. Nomura, H. Yun, "Development of the Integrated Passive Device using Through-Glass-Via Substrate", Proceedings of International Conference on Electronics Packaging and iMAPS All Asia Conference, pp.19-22, 2019.

[7] D. Malta, C. Gregory, M. Lueck, J. Lannon, J. Lewis, D. Temple, P. DiFionzo, F. Naumann, and M. Petzold, "Characterization and Modeling of Copper TSVs for Silicon Interposers", Proceedings of IEEE 63rd Electronic Components and Technology Conference, pp. 2235-2242, 2013.

[8] T. Ebefors, J. Fredlund, d. Perttu, R.V. Dijk, L. Cifola, M. Kaunisto, P. Rantakari, T.V. Keikkila, "The Development and Evaluation of RF TSV for 3D IPD Applications", Proceedings of IEEE International 3D Systems Integration Conference, pp. 1-8, 2013.

[9] D. Temple, D. malta, J.M. Lannon, M. Lueck, A. Huffman, C.W. Gregory, J.E. Robinson, P.R. Coffman, T.B. Welch, and M.R. Skokan, "Bonding for 3-D Integration of Heterogeneous Technologies and materials", ECS Transactions, 16 (8), pp. 3-13, 2008.

[10] K.W. Lee, Y. Ohara, K. Kiyoyama, J.C. Bea, M. Murugesan, T. Fukushima, T. Tanaka, and M. Koyanagi, "Die-Level 3-D Integration Technology for Rapid Prototyping of High-Performance Multifunctionality Hetero-Integrated Systems", IEEE Transactions on Electron Devices, 60 (11), pp. 3842-3848, 2013.

[11] M. Murugesan, T. fukushima, J.C. Bea, H. Hashimoto, and M. Koyanagi, "Intra- and Inter-chip Electrical Interconnection Formed by Directed Self Assembly of Nanocomposite Containing Diblock Copolymer and Nanometal", Proceedings of IEEE International Reliability Physics Symposium, pp. 4D.2_1-7, 2018.

[12] M. Murugesan, T. Fukushima, and M. Koyanagi, "500 nm-sized Ni-TSV with Aspect Ratio 20 for Future 3D-LSIs _ A Low-cost Electroless-Ni Plating Approach", Proceedings of Advanced Semiconductor Manufacturing Conference, in press, 2019.

[13] K.W. Lee, H. Wang, J.C. Bea, M. Murugesan, Y. Sutou, T. Fukushima, J. Koike, and M. Koyanagi, "Barrier Properties of CVD Mn Oxide Layer to Cu Diffusion for 3-D TSV", Proceedings of IEEE Electron Device Letters, Vol. 35 (1), pp. 114-116, 2014.

[14] M. Murugesan, J.C. Bea, K.W. Lee, T. Fukushima, T. Tanaka, M. Koyanagi, Y. Sutou, H. Wang, and J. Koike, "Effect of CVD Mn Oxide Layer as Cu Diffusion Barrier for TSV", Proceedings of IEEE 3D System Integration Conference, pp. 1-4, 2013.

[15] M. Kawano, N. Takahashi, M. Komuro, S. Matsui, "Low –cost TSV Process Using Electroless Ni Plating for 3D Stacked DRAM", Proceedings of IEEE 60th Electronic Components and Technology Conference, pp. 1094-1099, 2010.

[16] F. Inoue, T. Shimizu, T. Yokoyama, H. Miyake, K. Kondo, T. Saito, T. Hayashi, S. Tanaka, T. Terui, and S. Shingubara, "Formation of electroless barrier and seed layers in a high aspect ratio through-Si vias using Au nanoparticle catalyst for all-wet Cu filling technology", Electrochemica Acta, 56, pp. 6245-6250, 2011.

[17] F. Inoue, H. Philipsen, M. H. van der Veen, S. Huylenbroeck, S. Armini, H. Struyf, and T. Tanaka, "Role of Bath Composition in Electroless Cu Seeding on Co Liner for through-Si Vias", J. of Solid State Science and Technology, 4 (1), pp. N3108-N3112, 2015..

[18] K.W. Lee, C. Nagai, A. Nakamura, J.C. Bea, M. Murugesan, T. Fukushima, T. Tanaka, and M. Koyanagi, "Effects of Electro-less Ni Layer as Barrier/Seed Layers for High Reliable and Low Cost Cu TSV", Proceedings of IEEE 3D Systems Integration Conference, pp. 1-4, 2014.

3D Silicon Photonics Interposer for Tb/s Optical Interconnects in Data Centers with double-side assembled active components and integrated optical and electrical Through Silicon Via on SOI

Bogdan Sirbu[1], Yann Eichhammer[2], Hermann Oppermann[2], Tolga Tekin[1]

[1]System Integration & Interconnection Technologies Dept.
[2]Wafer Level System Integration Dept.
Fraunhofer Institute for Reliability and Microintegration IZM
Berlin, Germany
marian.bogdan.sirbu@izm.fraunhofer.de

Jochen Kraft, Victor Sidorov
ams AG
Graz, Austria
jochen.kraft@ams.com

Xin Yin, Johan Bauwelinck
IMEC
Gent, Belgium
xin.yin@ugent.be

Christian Neumeyr
Vertilas GmbH
München, Germany
neumeyr@vertilas.com

Francisco Soares
Fraunhofer Heinrich-Hertz-Institut HHI
Berlin, Germany
francisco.soares@hhi.fraunhofer.de

Abstract—**In this paper, we present the concept, fabrication, process and packaging of a 3D Si photonics interposer. This Si photonics interposer merges passive photonic and electronic functionalities within a single chip. The interposer is populated with active optical and electronic add-ons, which are flip-chip bonded to the interposer using thermo-compression bonding. The interposer itself is then flip-chip bonded to a glass and Si carrier for further testing purposes. This integration concept enables a high connection density (Gb/s/mm²) by assembling 40Gb/s per channel opto-electrical components on both sides of the interposer. Communication between components on both sides of the interposer is enabled by optical and electrical TSVs with a 3dB bandwidth >28GHz. A single mode photonic layer, designed for 1.55µm wavelength is integrated within the interposer to be used for routing and switching of the optical signals. Main fabrication and packaging steps are described here, together with some demonstrator evaluation results.**

Keywords- Silicon photonic interposer, through silicon via (TSV), electronic-photonic integrated circuits (EPIC), packaging, Cu pillars, thermocomompression bonding, data center, optical switch, optical interconnects, heterogeneous integration

I. INTRODUCTION

In order to support the increasing presence of connected devices in a broad range of applications in modern societies, data processing infrastructures such as data centers and high performance computers have to continuously increase their capacities and performances. This enhancement in data processing performances should be accompanied by a concurrent decrease in power consumption of these infrastructures as well as by a reduction of manufacturing costs. In our vision, these goals could be achieved by exploiting existing photonic technologies in a holistic way, synergizing the different fabrication platforms in order to deploy the optimal "mix&match" technology and tailor this to each interconnect layer.

Silicon photonics is one of the most promising technologies to reach these ambitious goals [1]. The manufacturing of silicon photonic devices can rely on existing CMOS foundries and their well-established processes and low production costs. Important hurdles still need to be overcome to move towards a massive adoption of silicon photonic devices, as silicon shows a certain number of shortcomings in terms of the performance of photonic components (light sources, modulators, detectors). Integration of III-V light sources and detectors is therefore still the best option in order to obtain satisfactory performances.

Several approaches to integrate III-V components and silicon photonic integrated circuits together are currently followed: either an active III-V layer is attached to the silicon wafer using direct bonding or epitaxy [2]-[3], or III-V components are bonded on or next to the silicon photonic chip [4]. Furthermore, CMOS electronics still have to be integrated as well, as driving electronics are necessary for functional optical components.

We have followed the latter mentioned method, which is usually referred to as hybrid photonic integration. In this paper, we present a 3D Silicon Photonics Interposer as a true 3D integration solution for Tb/s Photonic Interconnects in data centers, developed in the European flagship research project PhoxTroT [5]. The 3D Silicon Photonics Interposer consists of active electro-optical and RF components, such as III-V VCSEL arrays, III-V PD arrays, TIA arrays or Driver arrays, which are assembled to both sides of the Interposer,

978-1-7281-1500-9/19 $31.00 © 2019 IEEE

allowing an increase of the connection density (Gb/s/mm2) as well as an increase of the overall throughput. In order to enable communication between components assembled to both sides of the interposer, electrical TSVs (bandwidth >28GHz) as well as optical TSVs are integrated to the Silicon Photonics Interposer [6]. The Silicon Photonics Interposer includes also a Silicon photonic layer for routing and switching of optical signals. The optical coupling into the Silicon photonic layer is done using grating couplers from the active components (VCSEL and PD arrays) based on patented tilted assembly solution in order to cope with the acceptance of the grating coupler[7].

Fabrication and assembly results are reported in detail, as well as electrical and optical measurements. This 3D-integration concept would be compatible with a further assembly of the interposer to a single mode optical printed circuit board (OPCB) [8] (either polymer or glass based), as the 3D Silicon Photonics Interposer includes about 400 electrical I/Os as well as 96 optical I/Os with a pitch of 250μm.

The 3D Silicon Photonics Interposer supports a data throughput of 1.28Tb/s as on-board photonic interconnection for data centers.

This paper is organized as follows. In Section 2 the concept of the photonic interposer is described. Section 3 deals with the fabrication process of the interposer itself, done at the ams AG facilities, whereas Section 4 describes the assembly and packaging procedures. This is followed by the characterization results, presented in Section 5 and, finally, the conclusion is included in Section 6.

II. CONCEPT

The aim of the present work was the development and fabrication of a 3D single mode optoelectronic router. This single mode router comprises a silicon photonic layer, with bonded PD and VCSEL arrays driven by electronics TIA and Driver chips respectively bonded to the router chip as well.

In detail, the Silicon Photonics Interposer with router functionality comprises single mode silicon photonics layer designed for 1.55μm wavelength, assembled 40 Gb/s VCSELs and PD arrays driven by respective RF Driver/TIAs 3D integrated on the active interposer. The Silicon Photonics Interposer is assembled on a glass carrier, providing the electrical signals driving the optoelectronic components and chip-to-board optical interfaces. The silicon photonics layer itself implements a 4x4 switching matrix following the Benes topology with six electro-optical Mach-Zehnder Interferometer (MZI) switches that exploit plasma dispersion phenomena as switching function [9]. At each one of the four inputs and outputs of the switch, a 12 channel arbitrary waveguide grating (AWG) is inserted for multiplexing and de-multiplexing of the 12 optical wavelengths into one data stream, to handle the coarse routing functionalities. The VCSELs/PDs and the Driver/TIA electronic chips are placed at the inputs of the switch, being responsible for o/e/o conversion and fine wavelength routing functionalities.

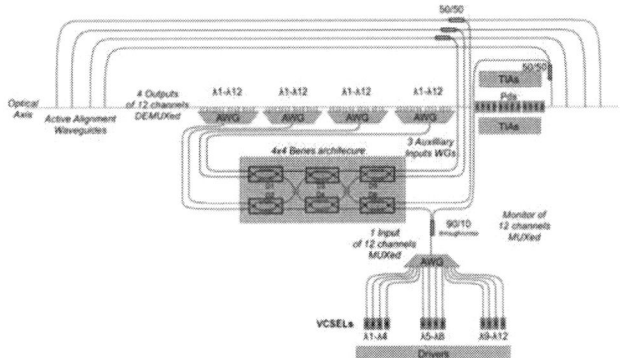

Figure 1. Conceptual layout of the single mode router with 4x4 switching matrix

Figure 1 shows the conceptual layout of the single mode router. The interposer was intended to be assembled on an glass based OPCB, where the interconnectivity between the optical layer of the OPCB and the Si Interposer was done vertically by means of one mirror coupling element, as shown in Figure 2. This is the reason why all optical inputs and outputs of the interposer are aligned on a common optical axis. For the routing operation, each one of the 12 optical channels streamed from the OPCB is fed to a separate PD and its respective electronic TIA. The TIA can then perform opto-electronic conversion of the incoming signals, while the received electrical signal is then transmitted to an electronic Driver amplifier prior to driving the modulation operation of a VCSEL. Each VCSEL is then tuned to a different wavelength through current injection to match the channel spacing of the AWG multiplexer on the silicon layer.

At this point it should be noted, that this o/e/o conversion is envisioned to control the fine routing functionality at a wavelength granularity through electronic control logic circuitry. This implies that a certain optical input channel of the 12 wavelengths (e.g. channel #1), after the first opto-electronic conversion, can be fed to any of the 12 VCSELs, which are tuned to emit at different wavelengths. The electrical signal is converted back to the optical domain at any of the different wavelengths within the group of 12 channels (e.g. channel #5). The 12 VCSEL outputs are then wavelength multiplexed through the AWG before being fed to the 4x4 switching matrix, where they are coarsely routed to one of the four outputs, and subsequently de-multiplexed through the AWG and vertically coupled back to the OPCB through 12x grating couplers and mirroring element.

From the sketch in Figure 2, we can deduce the need of employing 3D-Integration techniques for the packaging of the interposer. However, in this paper we focus on the fabrication, assembly and characterization of the Si interposer itself. The packaging demonstration of the Si Interposer on a single-mode OPCB is out of the scope of this paper. For the assembly and characterization of the Si Interposer, glass and Silicon substrates without any optical waveguides have been employed instead of the OPCB shown in Figure 2. In this case the optical in- and out optical

978-1-7281-1500-9/19 $31.00 © 2019 IEEE

coupling of the Si Interposer has been done by means of a fiber array.

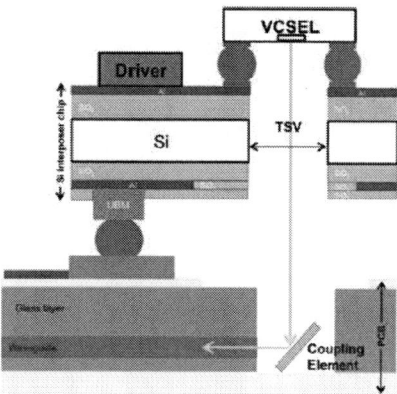

Figure 2. Sketch of the 3D router optoelectronic chip mounted to an OPCB

III. INTERPOSER FABRICATION

The starting material for wafer runs with wave guides was SOI (Silicon on Isolator) with an isolation oxide layer with 3μm thickness. The wafers were processed on both sides and process flow schematics are represented in Figure 3.

Front side process:
Alignment marks generation
Active photonics
- pn-junctions implantation
- waveguides etch (70nm)
- p+, n+ contacts implantations
- SiO2 cladding deposition
Metals 1,2,3
SiO2 stress compensation layer on wafer backside deposition

Bonding of Handling wafer
Thinning wafer (100μm or 200μm)

Back side process:
TSV formation
TSV metal
Si oxide stress compensation layer (on top of passivation layer) deposition
Passivation (TSV side) etch

Handling wafer Removal

Front side process: Opening of passivation windows for electrical contacts

Figure 3. Schematics of front- and back-sides process flow

The first mask level was used for the generation of alignment patterns. As the active photonic devices, the modulators use a vertical pn junction in the wave guide region. The mask for the low dose p- and n-implant needs an

alignment mark, as the n- and the p–implant itself do not leave a visible pattern. For the next mask level, the wave guide mask which is responsible for the structuring of the SOI layer, the alignment patterns of the first mask are used. Only wave guides of the rib type were used. The target depth of the etching is 70nm. After finishing the passive and active wave guide structures, a cladding layer was deposited and planarized. On top of these wave guide structures, a succession of structured metal layers with vertical electrical contacts was deposited, similar to a metal back-end of a CMOS chip.

Above the metal stack a handling wafer was mounted by direct bonding, which enabled the thinning of the wafers to a substrate thickness of 100μm. For reference purposes, a few wafers were thinned to a thickness of 200μm. For further processing on the backside, a wafer mask was aligned to marks on the frontside of the wafer by the aid of infrared light. In a next step, TSVs were etched through the thinned substrate. Isolation oxide layers and a metallization inside the TSVs were processed. Figure 2 demonstrates TSVs used for electrical connection of frond-and backsides of the wafer using underlying metals stack (left) and so-called optical TSV without metal layers at the TSV bottom (right). Only transparent Si oxide remained at the optical TSV bottom (Figure 4, right).

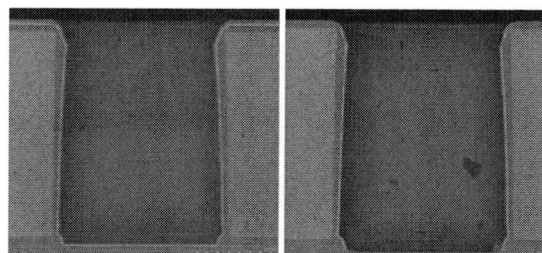

Figure 4. Left: electrical TSV, right: optical TSV

After the TSVs were formed, metal was deposited and patterned, passivation and stress-compensating dielectrics were completed, and finally the handling wafer was removed.

Cleaning results controlled on wafer level by both macroscopic visual and optical microscope inspection on both TSV and CMOS sides of the wafers. This inspection showed no adhesive residuals or other kinds of surface damages (aluminium pads corrosion, etc.). Each optical inspection (macro and micro) included both front side illumination for surface inspection and backside illumination in order to indicate optical TSVs transparency. Figure 5 depicted clear difference between optical and electrical TSVs with light comes through although they look similarly in front-side illumination.

Figure 5 demonstrates the same pattern on opposite sides of the wafer – front side (top, a and b) and back (TSV) side (bottom, s and d).

978-1-7281-1500-9/19 $31.00 © 2019 IEEE

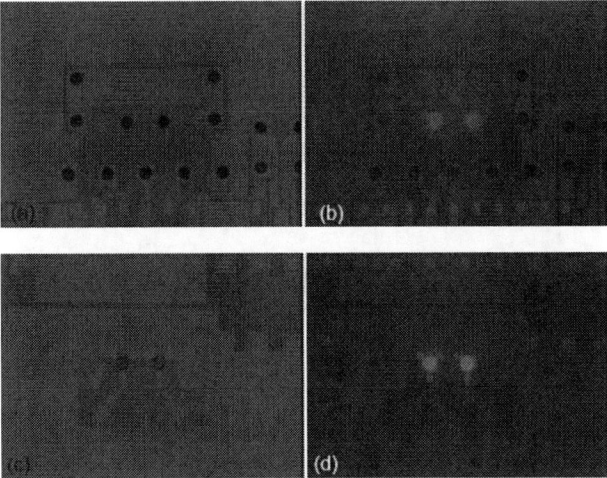

Figure 5.   Optical micrographs. (a) – TSV side, (b) – back-side illumination. Light comes through optical TSVs. (c) – The same patterns on CMOS side, (d) - back-side illumination. Light comes through optical TSVs

IV.   PACKAGING/ASSEMBLY

Following completion of their fabrication, photonic inteposer wafers have been post-processed in order to deposit Cu pillars with a finish metallization suitable for stud bumping and thermocompression bonding. After these post-processing steps, wafer level processes and stud bumping were completed as well as assembly of electronic and optical components, which took place at chip level. Once components had been assembled to the interposer, the packaged interposer was assembled to a glass or silicon substrate, for further testing purposes.

A.   Post-processing of interposer wafers

Prior to assembly of the add-ons to the photonic interposers, Cu pillars have been deposited on the interposers in order to enable the bonding of the interposer to substrates for test purposes.

Prior to processing, the wafers were glued to a glass carrier wafer in order to undergo further processing. This is because they are only 100µm thin and rather fragile due to the multiple metal layers that have already been deposited. The pillars were processed using common microfabrication techniques i.e. sputtering, lithography and electrodeposition. In Figure 6, a cross-section of the interposer with Cu pillars can be seen.

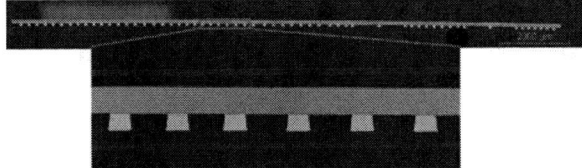

Figure 6.   Cross-section of the photonic interposer, with Cu pillars

Deposition of the pillars itself did not lead to any specific difficulty, as common microfabrication techniques have been employed. Nevertheless, singulation of the wafers was difficult. Glass handling wafers have obviously been glued to the pillar-less side of the interposers: therefore, interposer wafers have been transferred to the dicing frame with the pillar-side being glued to the frame. Due to the design of the wafers, the distribution of the pillars across the wafer was not homogeneous: the dicing frame therefore tended to wrap around the pillars, exerting stress on the fragile wafers, which happened to break during the first dicing trials. It was consequently decided that prior to debonding the wafers from the glass carrier and transferring them to the dicing frame, a protective resist would be spin coated onto the pillar-side of the interposer. The wafers could then be mounted to the frame on their pillar-less side, and dicing could be carried out without damaging the wafers.

B.   Assembly of optical and electronic add-ons

After deposition of the Cu pillars and singulation of the interposer wafers, electronic and optical add-ons were assembled to the interposer chips. Both sides of the interposer have been populated with components, which raises challenges in terms of handling the interposers and the components during the bonding processes at the flip-chip bonders.

The method that was chosen for assembling the components to the photonic interposer is Au-Au thermo-compression flip-chip bonding. The following constraints led to the choice of this bonding method: first, the electronic components were available only as chips and could therefore not be post-processed using wafer-level methods, and secondly, optical components should be assembled with a high bonding precision in order to reach a high bonding accuracy, which is permitted by using the thermo-compression bonding method with flip-chip bonders.

Add-ons have first been assembled to the pillar-less side of the interposer, using adapted tooling in order to preserve the integrity of the pillars during the bonding process.

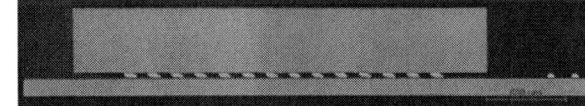

Figure 7.   Cross-section of a TIA bonded to a photonic interposer, using Au-Au thermocompression bonding

InFigure 7, a picture of a TIA assembled to the interposer using Au-Au thermo-compression bonding is shown. Since both the interposer and the TIA have Al pads, they both were stud bumped prior to thermo-compression bonding, in order to carry out Au-Au thermo-compression bonding and not Au-Al. For the optical components, i.e. VCSELs and PDs, it is only necessary to bump the interposer, as the components have Au pads.

Once the pillar-less side has been populated with components, components are assembled to the pillar side of the interposer. Since components are already present on one side of the interposer, specific tooling is used to shelter the components during the bonding process, while offering sufficient support to the areas where components are bonded.

For the assembly of the VCSELs to the interposer, a specific tilted bonding method has been developed and implemented (see [7]). Indeed, on the pillar-side of the interposer, grating couplers to enable coupling of the light emitted by the VCSELs into the Si photonics interposer are also integrated. The optimal coupling between the VCSEL beam and the grating coupler occurs at an angle of 10° [8]. The bonding of the VCSELs is still carried out using Au-Au thermocompression bonding, with Au stud bumps. The first step of this tilted bonding process is therefore to obtain a 10° angle between the stud bumps placed on the bond pads of the interposer.

Figure 8. Schematic of the tilted bonding of a VCSEL to a grating coupler

In Figure 8, a schematic of a VCSEL bonded with a given angle is shown. From this sketch, we see that for a 10° angle, we have

$$z22-z21=W_{VCSEL}*\tan(10°)$$

Where W_{VCSEL} is the distance between the pads of the VCSEL. The distance between the pads is here 110µm, therefore a height difference between the pads of 19µm is necessary to obtain a 10° angle. An appropriate set of stud bumping parameters was found and the height difference between pads could be controlled accurately. In Figure 9, SEM pictures of single and double stacked bumps after coining are shown.

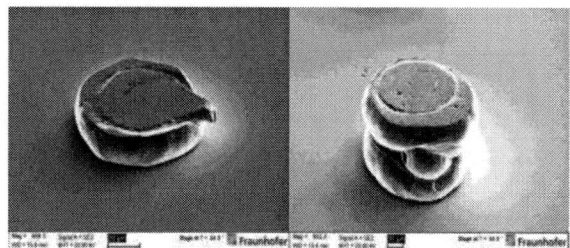

Figure 9. Au stud bumps after coining, single (left) and stacked (right)

After suitable stud bumping parameters have been found, VCSELs were bonded to such a set of stud bumps, using a tilted bonding tool in order to verify if the angle could be kept after bonding as well. In Figure 10, a cross-section of a tilted bonded VCSEL is shown.

Figure 10. Cross-section of a VCSEL bonded with a 10° tilt

As can be seen from the cross-section, the obatined bonding angle is very close to 10°, accounting for the inaccuracy of the measurement of the angle with the software of our optical microscope.

C. Assembly of the interposer to a substrate

The last step in assembly/packaging is the bonding of the interposer populated with components to a glass or Si substrate for testing purposes.

Figure 11 shows a picture of the interposer, with optical and electronic add-ons, assembled to a glass submount.

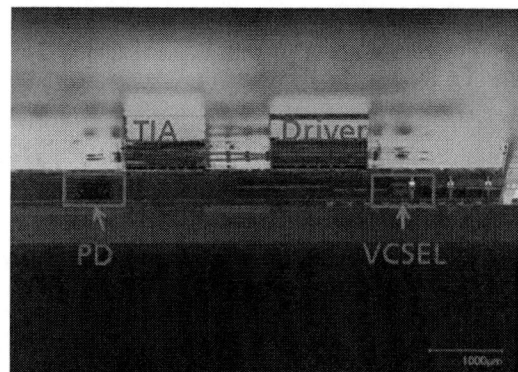

Figure 11. Picture of photonic interposer with electronic and optical add-ons, bonded to a glass substrate

In Figure 12, an overview cross-section of the interposer populated with components and assembled to a Si substrate is shown. Components on the bottom-side of the interposer cannot be shown in this cross-section, as they were located in a different plane. The various interconnect levels that are present are also recapitulated:

- The bumped Cu pillars realizing the interconnection between the interposer and the substrate
- The Au stud bumps interconnecting add-ons and interposer
- The optical and electrical TSVs interconnecting both sides of the interposer
- Gratings, connecting optical components to the interposer (not shown here)

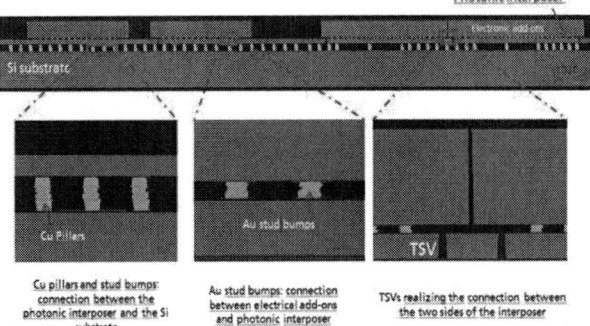

Figure 12. Cross-section of the interposer bonded to a Si substrate (top), together with a cross-section of Cu pillars realizing the connection from interposer to substrate (bottom left), a cross-section of Au stud bumps realizing the connection between an electronic add-on and the top side of the interposer (bottom middle) and a cross-section showing TSVs realizing the connection between the two sides of the interposer (bottom right).

V. DEMONSTRATOR EVALUATION

Once the assembly of the opto-electical components on the Si interposer was completed and the interposer itself was flip-chip assembled on a substrate we have performed some tests. In this section we will focus on two main tests: the characterization of one channel of the receiver part of the interposer (based on IMEC's TIA and Fraunhofer HHI's PD) and one channel of the transmitter part (IMEC's Driver and Vertilas's VCSEL). The entire chip has 12 channels for the transmitter and 12 separate channels for the receiving part, with a total throughput of 1.28Tbps. However, the entire capacity cannot be demonstrate at this time since the OPCB is not fabricated yet, thus we will present the obtained results for one TX and one RX channel. All opto-electrical devices have been developed within the PhoxTroT project.

Vertilas's high-speed VCSELs for data rates above 40 Gb/s are based on the short-cavity (SC) design. Two dielectric DBRs are used to shrink the cavity length and thus reduce photon lifetime resulting in a smaller damping of the electro-optic oscillator and it increases the resonance frequency. The intrinsic speed is impaired by parasitic elements which have been minimized in the final design. Most importantly the mesa diameter is shrunk and thus the junction capacitance around the optical cavity is reduced. Additionally, a smaller contact pad is implemented to push the parasitic bandwidth a little more, achieving optical output power up to 3.9 mW and the small signal modulation bandwidth reaches 18.5 GHz at room temperature and a side mode suppression ratio of 50dB. These components have been separately tested and successfully demonstrated an error-free BTB transmission at 56Gb/s.

IMEC's driver design is optimized for Vertilas's BJT VCSELs in a common-anode array, with a worst-case VCSEL operating current of 17mA. Every channel can be individually controlled via a serial peripheral interface (SPI). In this way, all channels can be optimized and it is e.g. possible to change the VCSEL wavelength across the array by controlling the bias current. For this 3D router optochip, it is required for the driver chip to operate without clock, so

that the FFE will be a fractional symbol spaced digital filter. The power consumption of 180mW per channel is obtained for an average load current of 13mA drawn from a 3V power supply. At 40Gb/s, this results in an energy efficiency of 4.4pJ/b. The input data signals are differential with 100Ω differential impedance. The minimum required input swing is 600mVpp differential while the lower speed digital inputs and outputs of the SPI interface are 2.5V CMOS compatible.

For the case of the 3D optoelectronic router, the VCSEL arrays had to be boned on top of the silicon grating couplers on the bottomside (waveguide side) of the optochip, while being driven by the electronics driver array on the top-side of the optochip, which requires electrical interconnection through electrical TSVs.

Figure 13 shows the voltage and optical output power versus the driving current at 20° C and 80° C as well as the optical spectrum and direct modulation frequency response. The device exhibits 1 mA threshold current with 50 Ω series resistance allowing very low electrical power consumption (<20 mW). The maximum optical power at the output exceeds 3.9 mW at 20° C and 1.5mW at 80° C. The VCSEL emits in a polarization stable TEM00 mode at all temperatures. The spectrum at 20°C shows a side mode suppression ratio greater than 50 dB (Figure 13 (b)). The device exhibits a modulation bandwidth as high as 18.5 GHz at 20° C, however, for the 40 Gb/s operation, an optical bandpass filter (OBPF) [11] is used to improve the signal-to-noise ratio (SNR) of the produced OOK intensity modulation before the detector. The filter exhibits a squared Lorentzian spectral shape with 3dB-bandwidth of 3.1 nm. Here we use the OBPF to enhance the limited bandwidth of the device, as described in detail in [12]. The OBPF is slightly red-shifted than the central wavelength of the laser (~0.75 nm-shift) so as to achieve optical equalization of the modulated spectrum and eventually enhance the eye-diagram performance in terms of signal-to-noise ratio (SNR). In particular, the direct modulation of the VCSEL causes the "1" intensity level to be blue-shifted than the "0" level because of the adiabatic frequency chirp induced during the cavity modulation process.

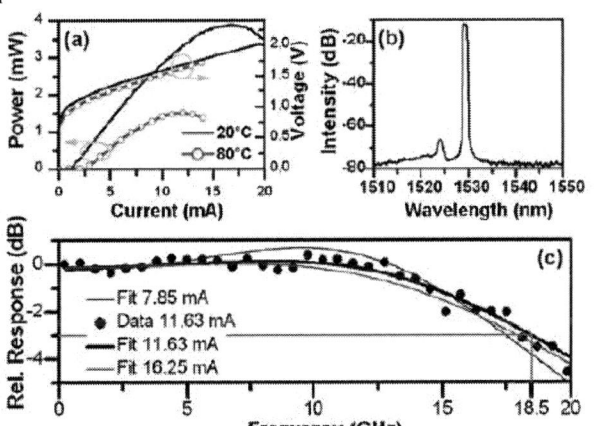

Figure 13. DC and RF VCSEL characterization. (a) L/I curve, (b) spectrum and (c) frequency response

The eye-diagrams and traces at 40 Gb/s are shown in Figure 14, as obtained before (a) and after (b) the OBPF. Optical temporal intensity before (c) and after (d) the filter, for the particular sequence "10100000100100001000 1000011011000 011 101110".

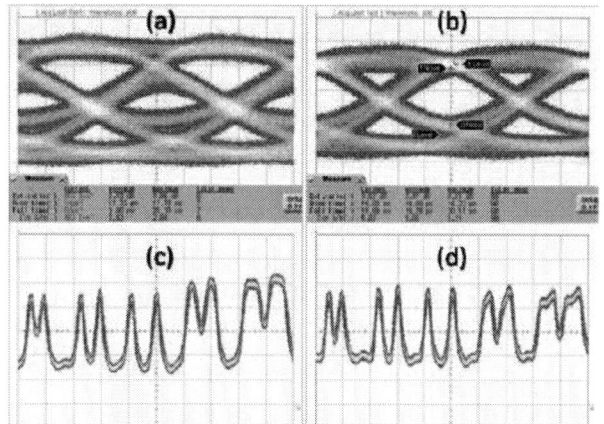

Figure 14. Eye-diagram and traces of one channel of the transmitter at 40 Gb/s before (a) and after (b) the OBPF

On the other hand, Fraunhofer HHI developed the PD arrays required for the receiver part our concept. The technology used for the fabrication is InP based and the final generation of PD shows the following characteristics: dark current below 10nA, crosstalk below -25dB, responsivity above 0.5A/W and a 3dB bandwidth of 33GHz.

IMEC's TIA array design targets the development of 40Gb/s TIA chips, optimized for Fraunhofer HHI's 1550nm PIN PDs. 40Gb/s TIAs are commercially available, featuring a very good sensitivity, but with a power consumption which is two times higher as IMEC's TIA. The TIA design is optimized for low power consumption and NRZ (non-linear) operation for Fraunhofer HHI's 40Gb/s PDs. Dedicated functionality for Received Signal Strength Indication, Gain Control and Offset Compensation improve the performance and/or simplify testing. Every channel can be individually controlled via a serial peripheral interface (SPI).

For this test, a continuous wave laser provided a 1550nm light source which was fed into a Mach-Zehnder modulator. The modulated light source passed through a variable attenuator and finally reached the photo diode on the test board using a manually aligned lensed fiber. An SHF 12100B pattern generator (clocked by an Anritsu MG3696B clock generator) directly drove the modulator. Eye diagrams were captured by connecting the 86118A remote sampling heads of a DCA 86100C oscilloscope to the differential outputs on the test board. For bit-error rate (BER) measurements, the TIA outputs were connected to an SHF 11100B error analyzer.

The BER was measured for various input powers. To eliminate the effect of extinction ratio (ER) on the performance, the input power was characterized by its optical modulation amplitude (OMA) rather than the average optical input power (P_{avg}): $OMA = 2\,P_{avg}\,(ER - 1)\,/\,(ER + 1)$. The BER and eye diagram plots for 40 Gb/s are shown in Figure 15 and Figure 16 respectively.

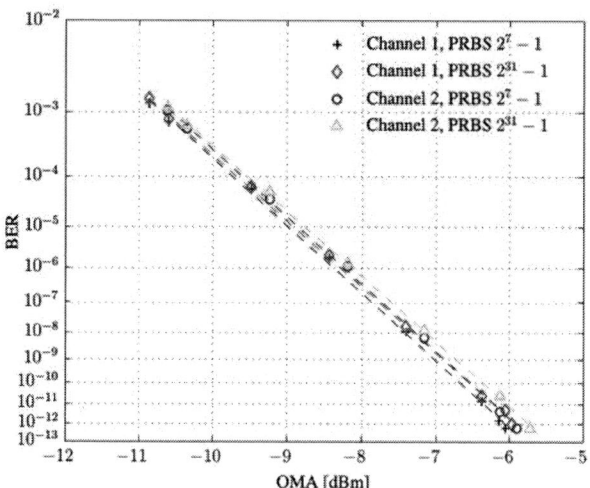

Figure 15. BER measurements of 2 adyacent channels of the the receiver at 40 Gb/s

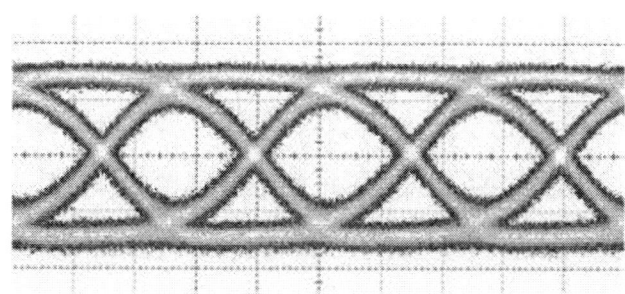

Figure 16. Eye-diagram of one channel of the receiver at 40 Gb/s at -3dBm OMA. Horizontal scale is 10 ps/div, vertical scale is 50 mV/div.

VI. CONCLUSION

Towards the development of the single mode 3D optoelectronic router a silicon photonic interposer was developed, which requires the development of Silicon Photonics layer, the VCSELs and Photodiodes, the electronic chips Drivers and TIAs, 3D integration techniques with thinned wafer and at least it requires the development of the assembly process and characterization of the technologies.

The concept, processing in the fabrication site of ams AG, assembly procedures and the successful characterization of the transceiver and receiver parts of the test vehicle are described. For the processing of the Si Photonics wafers with waveguides, ams AG had to develop a waveguide process module. Passive waveguides but also active waveguide switches have been successfully been processed and tested, including a 4x4 silicon switch following the Benes topology. The silicon photonic interposer on the other hand involves a series of complex fabrication procedures, towards a thinned optochip with optical and electrical TSVs, and dual metallization on both sides where electro-optical components have been assembled using thermos-comprehension bonding techniques.

ACKNOWLEDGMENT

The authors gratefully acknowledge the received funding for the PhoxTroT project from the European Union's Seventh Framework Programme for research, technological development and demonstration under grant agreement no 318240.

REFERENCES

[1] D. Thomson et al., "Roadmap on silicon phtonics", J. Opt., vol. 18, 2016, pp 1-20, doi: 10.1088/2040-8978/18/7/073003

[2] D. Liang et al., "Hybrid integrated platforms for silicon photonics", Materials, vol. 3, 2010, pp. 1782-1902, doi: 10.3390/ma3031782

[3] A. Y. Liu et al., "Quantum dot lasers for silicon photonics", Photon. Res., Vol. 3, No. 5, pp. B1-B9, doi: 10.134/PRJ.3.0000B1

[4] A. Moscoso-Mártiret al., "Hybrid Silicon Photonics Flip-Chip Laser Integration with Vertical Self-Alignment," in 2017 Conference on Lasers and Electro-Optics Pacific Rim, (Optical Society of America, 2017), paper s2069.

[5] https://phoxtrot.eu/

[6] J. Kraft, G. Meinhardt, K. Molnar, T. Bodner and F. Schrank, Electrical and optical Through Silicon Vias (TSVs) for high frequency photonic applications, ECTC 2016, pp. 2389-2394, doi: 10.1109/ECTC.2016.188

[7] H. Oppermann, "Arrangement of a substrate having at least one optical waveguide and an optical coupling point and an optoelectronic component and methods for making such an arrangement", German patent DE102013011581B4

[8] L. Brusberg et al., "Electro-optical circuit board with single-mode glass waveguide optical interconnects", Proceedings Vol. 9753, Optical Interconnects XVI, doi: 10.1117/12.2208103

[9] M. Moralis-Pegios et al, "A Programmable Si-Photonic Node for SDN-enabled Bloom Filter Forwarding in Disaggregated Data Centers", Proceedings Volume 10109, Optical Interconnects XVII; 101090X (2017) https://doi.org/10.1117/12.2252043

[10] D. Taillaert et al., "Grating couplers for coupling between optical fibers and nanophotonic waveguides", Jpn. J. Appl. Phys., Vol 45, No. 8A, 2006, pp. 6071-6077, doi: 10.1143/JJAP.45.6071

[11] Matsui, Y.; Mahgerefteh, D.; XueyanZheng; Liao, C.; Fan, Z.F.; McCallion, K.; Tayebati, P., "Chirp-managed directly modulated laser (CML)," IEEE Photonics Technology Letters, vol.18, no.2, pp.385,387, Jan. 15, 2006

[12] A. Malacarne, V. Sorianello, A. Daly, B.Kögel, M. Ortsiefer, C. Neumeyr, M. Romagnoli and A. Bogoni, Performance Analysis of 40-Gb/s Transmission Based on Directly Modulated High-Speed 1530-nm VCSEL, Photonics Technology Letters, vol 28, no. 16, 2016

Flip-Chip III-V-to-Silicon Photonics Interfaces for Optical Sensor

Yves Martin, Jason S. Orcutt, Chi Xiong, Laurent Schares, Tymon Barwicz, Martin Glodde, Swetha
Kamlapurkar, Eric J. Zhang, William M. J. Green
IBM T. J. Watson Research Center, Yorktown Heights, NY 10598
Victor Dolores-Calzadilla*, Ariane Sigmund, Martin Moehrle
Fraunhofer Heinrich-Hertz Institute, Germany
*Present: Institute for Photonic Integration, Eindhoven U. of Technology, Eindhoven 5600 MB, The Netherlands
ymartin@us.ibm.com

Abstract - **We demonstrate flip-chip solder assembly of InP chips on Silicon-Photonic (Si-Ph) substrates aimed at high volume manufacturing using typical microelectronic lead-free solders.**

In our show-case application, an InP die is both a light source and a detector in an integrated optical methane gas sensor that operates near 1.6mm. For high-resolution laser absorption spectroscopy sensing, a single-mode tunable laser is desired. We create an external cavity laser with InP as optical gain, butt-coupled to a Si-Ph external cavity, which incorporates the laser's frequency selective elements. For minimal reflection at the InP-Si interface, waveguides are angled to the facet, an index-matching medium is applied between the mating surfaces, and an anti-reflection coating designed for the index-matching medium is applied to the optical coupling facet of InP chip.

Sub-micron alignment accuracy is obtained without high-accuracy assembly tooling. Lithographically defined alignment features on both InP and Si components allow reproducible high-accuracy alignment. Interface throughput loss were measured to be as low as 1.4 dB, and interface reflections are more than 30dB smaller than main signal beams.

Keywords - Silicon & III-V photonics; heterogeneous assembly; flip-chip assembly; solder reflow;

I - INTRODUCTION

Heterogeneous photonics offers potential for high performance and low cost in integrated optics devices [1]. A cost-effective and high precision technology that combines III-V chips and Silicon Photonics (Si-Ph) substrates enables the fabrication of complex devices for optical communication and for integrated optical sensing. A range of nascent heterogeneous fabrication techniques [2,3,4] address the issue of coupling single-mode III-V light sources to silicon photonic components. Among them, butt-coupling can

provide high optical throughput and very wide spectral bandwidth suitable to the fabrication of high performance external cavity lasers [5-10].

In many butt-coupled demonstrations, the required high-alignment accuracy between single-mode waveguides from III-V chip to Si-Ph substrates was provided via high-accuracy alignment & bonding tools, often equipped with advanced image recognition systems. This low-throughput assembly technique brings high assembly cost and low scalability in volume production. In this paper, we demonstrate an assembly method where precision primarily resides in the components themselves, with the goal of making the chip placement a low precision operation. Highly optimized lithographic and etching steps, inherited from decades of innovation and development in micro-electronics, enable wafer-level fabrication of micro-alignment features on Si wafers that are key to the low-loss waveguide-to-waveguide interface.

Our bonding method includes the well-known lead-free solder joints that are prevalent in the microelectronic industry. Solder provides a four-fold bond & connectivity between InP and Silicon components: mechanical, electrical, thermal and optical. The thermal contact, in particular, is highly efficient through solder which is kept thin in our configuration. Heat dissipated in the InP lasing junction is effectively conducted through thin layers of InP and solder to the bulk silicon of the substrate where it is efficiently spread and removed. A thin solder layer also provides good mechanical stability even under variation of ambient temperature. Furthermore, we selected lead-free tin-based solder rather than gold-tin eutectic: tin-based solder is more developed and available in the microelectronic industry, the melting point is substantially lower, and the alloy is more malleable and less brittle than gold-tin thereby reducing fracture risks.

978-1-7281-1500-9/19 $31.00 © 2019 IEEE

II- SENSOR OVERVIEW

The motivation of this work was the fabrication of optical integrated spectrometers for the detection of traces of methane. They are to be deployed at natural gas production fields, as well as at gas storage and distribution sites, and serve as sensitive gas leak sensors. Details of the spectroscopy and applications are described elsewhere [11-13]. The heart of the sensor is a small (approx. 10 by 8 mm) Si-Ph component with a bonded InP chip, shown on Figure 1. The InP chip is butt coupled to waveguides on the Si-Ph substrate. One essential component is a wavelength tunable External Cavity Laser (ECL) having one reflector on one end of the InP chip (the high-reflectivity coated rear facet) and the

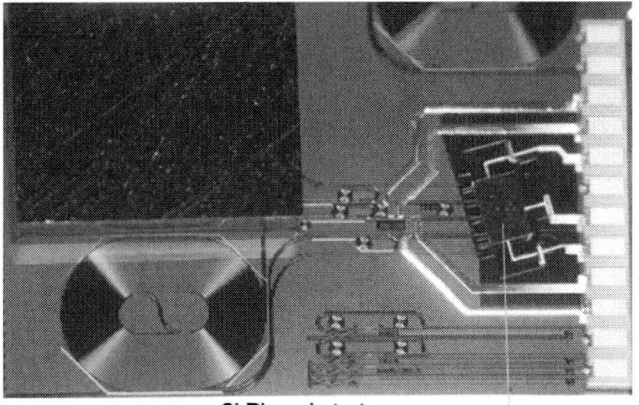

Si Ph. substrate

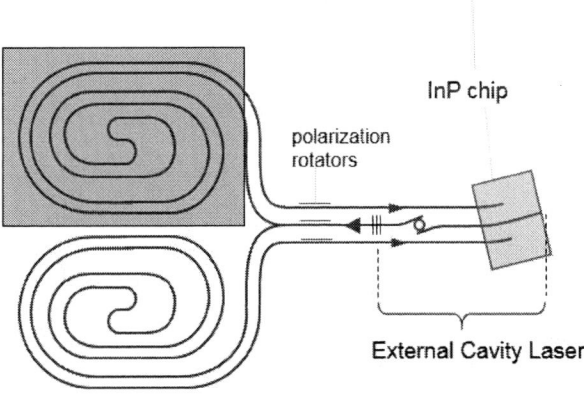

Figure 1. CSi-Photonic integrated spectroscopic sensor. Top is a view of the full sensor which measures approximately 10 x 8 mm. Bottom is a diagram depicting the essential Si-photonic waveguides and the InP chip

second one on the Si-Ph substrate (a Bragg-reflector). An intra-cavity tunable ring filter, depicted on Fig.1, is part of the ECL laser. The detailed operation of the ECL laser is the subject of a separate publication [12].

TE radiation emitted by the ECL is rotated [14] to TM, split, and directed into two long coiled waveguides at the surface of

the Si-Ph chip for the purpose of sensing ambient gas. Thereafter, the transmitted TM radiation is rotated back to TE and directed to two sensitive photodetectors built inside the InP chip. When the wavelength matches a resonance peak of ambient gas (methane gas in our application where λ_{abs} = 1.656 um), some of the evanescent radiation around the waveguides is absorbed by the gas and the resultant decreased amplitude is detected by the photodetectors. One of the coiled waveguides is enclosed in a sealed silicon cavity filled with methane, and serves as a reference cell to precisely tune the ECL wavelength at the peak gas absorption. Our InP chip was therefore built with two purposes. As active component of the ECL laser, it was designed to be a Reflective Semiconductor Optical Amplifier (RSOA), with one reflective facet coated with a high-reflective multilayer coating, and one facet coated with an anti-reflective coating. The active III-V stripe is forward biased and extends the full length of the chip. As a second purpose, two side-detectors are built in the same epitaxial layers as the main lasing strip, but extend only a small fraction of the full chip length. Without forward bias, radiation is absorbed over a few tens of microns and induces photo-currents.

Butt-coupling of radiation was selected for several reasons. Low coupling loss over broad wavelength range can be obtained when III-V and Si-Ph components are precisely aligned. Most importantly, our previous work had demonstrated the ability to precisely align and bond the two components in a way conducive to volume production [15,16]. For this purpose, lithographically defined mechanical stops are etched in both Si-Ph substrate and InP chip. External forces, either mechanically applied or via surface tension of solder when melted between substrate and chip, push the chip into precise X-Y-Z alignment against the substrate.

III- DESIGN OF InP AND SI-PH COMPONENTS

Figure 2 shows details of an InP chip, with several key elements indicated with arrows in Fig. 2.a. Four gold-coated electrical contact pads appear in yellow, as well as the contact strip which covers the emitter gain section of the InP chip, in the center and along the full length of the chip. Five brown recessed areas are built for mechanical positioning of the chip. The outer four recesses will land on top of standoffs (posts) formed on the Si-Ph substrate to provide alignment registration in the vertical (Z) direction. The central recess is deeper than the outer four recesses and is part of the lateral alignment feature that is described later. The three ridge waveguides correspond to the central IR emitter and to the two IR photodetectors on either side of it. To facilitate the separate biasing conditions needed for the RSOA and photodetector sections, electrical isolation trenches are etched to remove any p-doped or quantum well material below the depth of the standard ridge etch. Fig. 2.b is a cross section

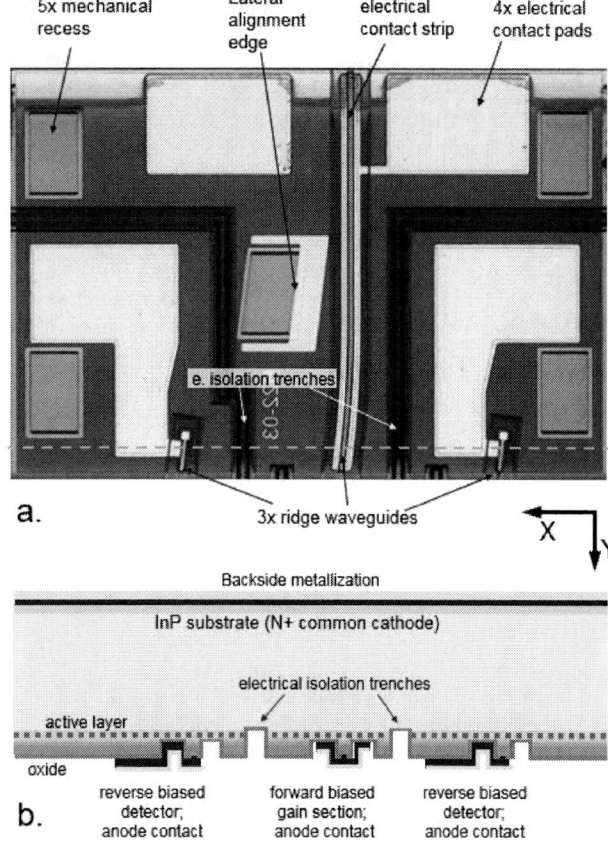

a.

5x mechanical recess | Lateral alignment edge | electrical contact strip | 4x electrical contact pads

e. isolation trenches

3x ridge waveguides

X
Y

Backside metallization

InP substrate (N+ common cathode)

electrical isolation trenches

active layer

oxide

b.

reverse biased detector; anode contact | forward biased gain section; anode contact | reverse biased detector; anode contact

Figure 2. a. Bottom view of the InP chip which is 750 x 500 um in size. Four electrical contact pads are visible, three of them are connected to the three P-doped anodes of the three ridge waveguides. Five brown areas correspond to recessed areas that will land on 5 standoffs (pedestals) of the Si-Ph substrate. The picture was inverted in the horizontal direction to match the views of the Si-Ph chip below.

b. cross-section diagram of the InP chip, at the dotted line location on a., that schematically depict several of the doped and electrical layers as well as ridges and isolation trenches at the bottom of the chip.

diagram (not entirely to scale) of the chip along the dotted line of Fig. 2.a. The 3 main components, forward biased gain region of the emitter, and reverse biased detectors are highlighted and built in close proximity to the active layer near the bottom surface of the chip.

Another essential feature visible on Fig. 2.a is the lateral alignment edge. The primary function of the trapezoidal-looking central recess is for lateral alignment (in X-direction) of the chip to the Si-Ph substrate. Precise X-alignment occurs when the chip's lateral alignment edge butts against the central standoff of the Si-Ph substrate. Spacing between this edge and the angled waveguides is highly accurate, since owing to lithography and process design: a single mask defines both the waveguides and the location of the lateral

alignment edge. The edge is angled to be parallel to the angled waveguide to allow for positional uncertainty of the front facet, defined by cleaving with a typical +/-10 µm variability.

Figure 3 shows the region of the Si-Ph substrate that receives the InP chip. Through semiconductor processing steps, three general types of features were created in the substrate for coupling with the InP chip and are highlighted on Fig. 3 & 4: mechanical structures for precise X-Y-Z alignment, solder pads for bonding and for surface-tension induced pulling forces during the melting phase, and optical waveguides with mode widening. The overall area that receives the InP chip has been recessed by nearly 20 µm below the top surface via

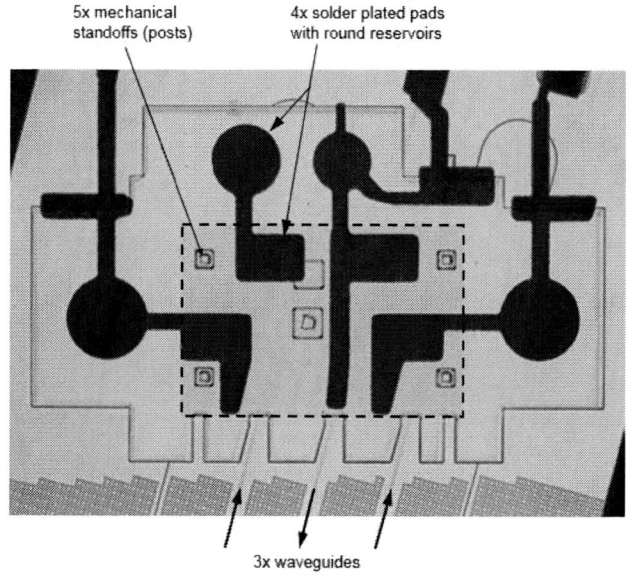

5x mechanical standoffs (posts) | 4x solder plated pads with round reservoirs

3x waveguides

Figure 3. View of the Si-Ph substrate. The final location of the InP chip has been delineated with a dotted rectangle.

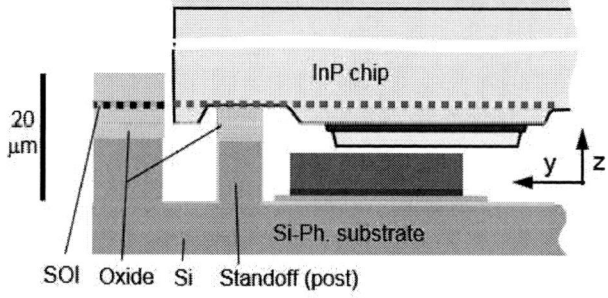

InP chip

20 µm

Si-Ph. substrate

y
z

SOI   Oxide   Si   Standoff (post)

Figure 4. Cross sectional diagram of substrate and chips, showing approximately scaled layers in the vertical direction and structures for mechanical alignment in the Y-Z directions. Dotted lines represent waveguides in both substrate (in the SOI layer) and in the InP chip.

978-1-7281-1500-9/19 $31.00 © 2019 IEEE   1062

etching, as shown on the cross sectional diagram of Fig. 4. Five standoffs (or posts) were left remaining at a height corresponding to the top Silicon layer in the original Silicon-On-Insulator (SOI) wafer. Four solder plated pads with round reservoirs were defined at the bottom of the recessed area and are electrically connected to lines at the top of the recess. Fig. 4 illustrates the vertical layers in both substrate and chip. Using the SOI layer in the substrate and the optical active layer in the InP chip as references and etch stops, the relative height of the waveguides in both components can be precisely set with an accuracy on the order of 0.1 μm or better. Fig. 4 illustrates Z-alignment of the two waveguides in dotted lines facing each other, after the InP chip is placed over the standoffs.

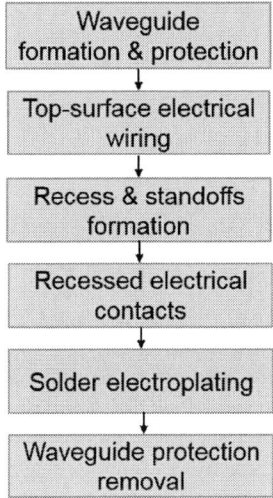

Figure 5. Overall process flow for the micro-fabrication of the Si-Ph substrate

Figure 5 gives the overall sequence for the fabrication of the Si-Ph substrate. Starting with an SOI wafer, the first steps involve 193nm UV lithography to define the sub-micron optical waveguides and associated components for ECL, beam splitting and polarization control. Low roughness and dimensional control of the waveguides is essential for low transmission loss and for minimizing etalon fringes [17]. Next steps in the process flow involved mid-UV lithography and thicker photoresist patterning, especially for deep RIE etching or for contact patterning at the bottom of the recess region. The recesses and standoffs are formed by first etching down to the buried oxide layer of the silicon-on-insulator substrate in the regions that standoffs are formed. The second, full depth, etch step is patterned to form the standoff and recess geometries. This process enables precise mechanical definition of the lateral alignment features to the limit of the mid-UV lithography alignment and CD control. The vertical alignment precision is limited by the timed etch required to offset the optimal alignment of the InP chip mode. Since the

vertical offset is only 100 nm, precise control of the height of the standoffs is possible. Electroplating involves deposition of an Under-Bump Metallization (UBM) layer of Nickel, and of lead-free solder. The final step removes a thin oxide protection layer from the coiled sensing waveguides.

Waveguide cross-sections have been optimized near the interface for both InP and Si-Ph chips to match the optical modes for minimum insertion loss. Calculated elliptical cross-sections of the modes are shown on Fig. 6. The resulting insertion loss as a function of mechanical alignment considerations is then shown in Fig. 7. The mode of the InP chip can be widened further to relax the alignment precision in the X direction by tapering the waveguides at the front facet, but this initial demonstration was designed to use the standard single-mode waveguide width of the InP chip.

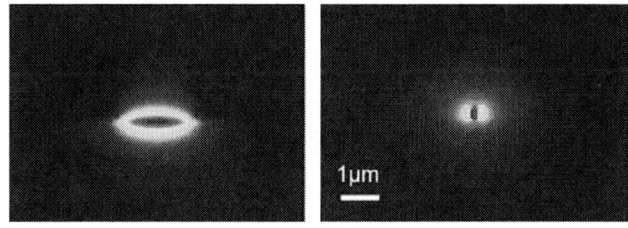

Figure 6. Left: Calculated optical mode in the InP chip, in the region near the coupling edge. Right: Calculated optical mode in the Si-Ph chip in the region near the coupling edge. The waveguide width and refractive index of deposited dielectrics have been optimized to minimize coupling loss to the InP chip.

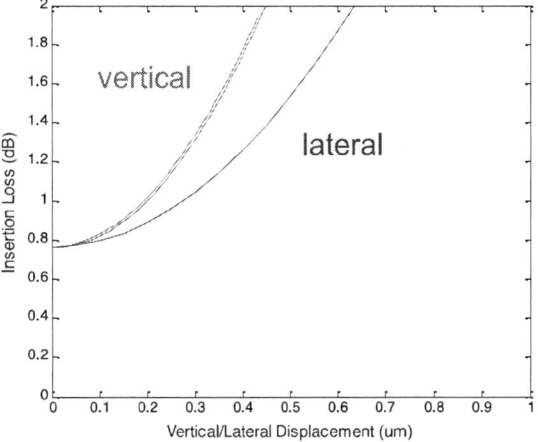

Figure 7. Calculated insertion loss between the InP and Si-Ph chips for the case of a fixed 0.5μm gap of refractive index 1.48 material in the direction of optical propagation (Y-direction) as a function of vertical (Z-direction) and lateral (X-direction) misalignment..

VI - ALIGNMENT & BONDING OF THE InP CHIP

Assembly of the InP on Si-Ph calls for picking and placing the flipped InP chip at a distance 10 to 20 µm from the desired butting position, in both X and Y. This placement position corresponds to having the five Si-Ph standoffs (visible on Fig. 3) being nearly centered in the InP landing regions (in brown on Fig. 2). In this position, the chip rests on the four outer standoffs and has free but limited motion range in X and Y, as illustrated on Fig. 4. During the solder melting phase, the plated solder on the Si-Ph substrate balls up and touches the metal pads of the InP chip, subsequently to wet the entire surface of the pad. Corresponding metal pads have been offset in both X and Y directions, as illustrated in Fig. 4 and Fig. 8. While in the melted phase, the solder surface tension pulls the chip against the alignment stops in both X and Y direction [18,19]. Fig. 8 shows the InP solder pads (in yellow) being offset in the negative X and negative Y directions. Hence, the overall pulling force and chip motion occur in the positive X and Y directions on the figure, until the chip comes into contact with X and Y alignment stops.

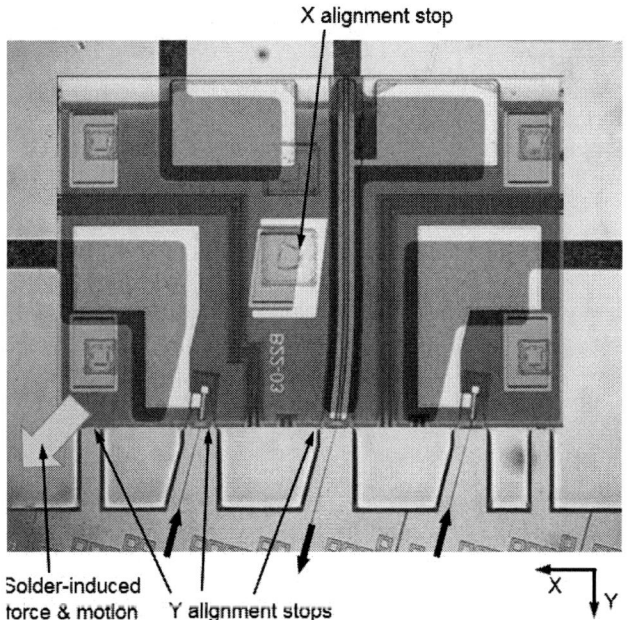

Figure 8. Composite picture of the 750x500 µm InP chip (rendered transparently) overlaid over the Si-Ph substrate with highlighted mechanical alignment stops.

The proper working of the solder surface tension forces hangs on an optimized balance between lateral and vertical forces. The vertical force is predominantly solder-induced, while weight accounts for only approximately 1% of the total force.

Two types of solder-induced forces account for the resulting vertical force:

a. the vertical component of the tension force, around the perimeter of solder pads, is a pulling downward oriented force.

b. the hydrostatic pressure within the liquid solder, can produce a relatively large vertical force. It can range from a large downward pulling force to a large upward pushing force, pending on the solder curvature according to the Young-Laplace equation.

Ultimately, the hydrostatic pressure is a strong function of the relative amount of plated solder. A shallow plated thickness leads to a large downward pulling force which creates a binding friction between the chip and standoffs and prevents chip motion. An excess plating induces a large pushing force that lifts the chip off the standoffs and fails the alignment in Z.

To broaden the solder plating window, we had introduced the concepts of solder reservoirs [20,21], seen as the round-shaped pads on Fig. 3. Hydrostatic pressure in the molten phase drives solder flow between reservoirs and chip-joining pads to equilibrate solder volumes and compensate for over or under plating. Meanwhile, the solder curvature sustained by the reservoirs stays relatively shallow and constant, and yields a desired small pushing force. Calculations were done according to the model shown in [20]. Fig. 9 illustrates the dependency of the vertical force on solder thickness, for different reservoir diameters. Selection of reservoir diameter and thickness were done for a small attractive vertical force on the order of 40 µN and a plating window around +/- 2 µm.

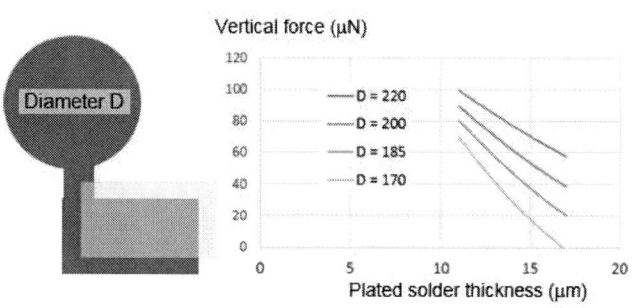

Figure 9. Calculated vertical force on the chip solder pad (in yellow) as a function of the solder thickness, for different solder reservoir diameters. The overall vertical force must be attractive (positive on the scale) and smaller than the lateral force, which is on the order of 60 µN

Our initial attempts to align chips via solder tension forces were hampered by unforeseen alloying effects between gold on the InP pads and solder on the Si-Ph substrate. A protective platinum barrier on the InP pads did not perform as expected, exposing thick Au to solder and resulting in solid Au-Sn intermetallic compounds formed before chip motion was induced. In our earlier work [15], a thin Au layer served as the wetting layer and we did not encounter this issue. As a temporary solution, we manually pushed the chips into X-Y alignment stops, and shortened the melting time of the solder to a few seconds to minimize the amount of intermetallic formation. The process was aided with the addition of a temporary adhesive gel between chip and substrate, and by loading the chip with an external weight. We are in the process of modifying the InP pad metallization to minimize the amount of gold and eliminate solid intermetallic formation, and thereby recover self-alignment via solder surface tension.

V - OPTICAL COUPLING AND RESULTS

Optical coupling was measured using modified Si-Ph substrates having additional waveguides and optical ports that were externally coupled to lensed fibers at the periphery of the Si-Ph substrates. One waveguide in particular allowed input of metered external light directly onto one of the photo-detector through one optical coupling between Si-Ph substrate and InP chip. A separate loop-back waveguide and two lensed

fiber couplings served to calibrate the lensed fiber input loss. Using an approximate photodetector efficiency of 0.9 A/W, estimated previously in actively-aligned component testing, we measured an optical transmission coefficient of about 0.73 which corresponds to a 1.4 dB loss through the InP to Si-Ph coupling, which is similar to earlier results [15].

Proper tuning of the ECL is also pending on minimal reflections at interfaces, going from the InP chip to the Si-Ph substrate. To this effect, angled waveguides were selected and designed for the interface to minimize specular reflections back into the InP waveguide. Furthermore, anti-reflection coating was deposited on the exiting surface of the InP chip and a 1.4 index adhesive was applied between the butting surfaces. Additional measurements indicated a minimized reflection down to nearly -40dB of the transmitted radiation. As a result, good single-mode ECL operation was obtained with side mode suppression ratio of 50dB. Wavelength tunability was obtained around a methane absorption line near 1650nm, and methane absorption peak was clearly resolved with the coiled sensing waveguide [12]. Figure 10 illustrates sub-milliwatt laser radiation going through a 30cm long exposed waveguide. Weak scattering at imperfect transitions between curved and straight portions of the waveguide, as well as a few scattering particles, serve as hallmark for the presence of the radiated light throughout the length of the waveguide.

1 mm

Figure 10. I-R microscopy of a 30cm coiled sensing waveguide of the gas sensor under operation.

VI - CONCLUSION

We demonstrate flip-chip assembly of an InP chip precisely aligned and soldered to a Si-Ph substrate for the fabrication of a spectroscopic sensor. Heterogeneous integrated components include a tunable External Cavity Laser, sensor waveguides, polarization rotators and splitters, and detectors, all operating at a wavelength around 1.6um. Measured optical throughput between butt-coupled InP chip and Si-Ph substrate indicated a modest loss of 1.4dB, and a small reflections (between -30 and -40dB) that enables ECL single mode operation with better than 50dB side-band suppression.

Our integrated optical gas sensor highlights a path to low cost mass production of spectroscopic sensors with small form factor and low power consumption. Butt coupling of waveguides between heterogeneous electro-optic components is demonstrated as a solution to an otherwise delicate and costly operation and can find application in a broad array of integrated optic products.

ACKNOWLEDGEMENT

The information, data, or work presented herein was funded in part by the Advanced Research Projects Agency-Energy (ARPA-E), U.S. Department of Energy, under Award Number DE-AR0000540. The views and opinions of authors expressed herein do not necessarily state or reflect those of the United States Government or any agency thereof. The authors would like to thank the staff at the IBM Microelectronics Research Laboratory and Central Scientific Services for their assistance in the fabrication of the Si-Ph components, as well as Theodore van Kessel, Levente Klein, Ramachandran Muralidhar, and Hendrik Hamann (all of IBM Research) for many discussions.

REFERENCES

1. T. Komljenovic et al., "Heterogeneous Silicon Photonic Integrated Circuits", J. LightwaveTechn. 2016
2. J.E. Bowers et al., "A Comparison of Four Approaches to Photonic Integration", OFC, 2017
3. Bowen Song et al., "3D integrated hybrid silicon laser", Opt. Express 2016
4. M.R. Billah et al., "Hybrid integration of silicon photonics circuits and InP lasers by photonic wire bonding", Optica 2018
5. A. J. Zilkie et al., "Power Efficient III-V/Silicon external cavity DBR lasers", Optics Express Vol.20, 2012
6. S. Tanaka et al., "High-output-power, single-wavelength silicon hybrid laser using precise flip-chip bonding technology", Optics Express Vol.20, 2012
7. S. Tanaka et al., "Four-Wavelength Silicon Hybrid Laser Array with Ring-Resonator Based Mirror for Efficient CWDM Transmitter", OFC/NFOEC Technical Digest 2013
8. J.H. Lee et al., "High power and widely tunable Si hybrid external-cavity laser for power efficient Si photonics WDM links", Optics Express Vol.22, 2014
9. J. Bovington et al., "III–V/Si Vernier-Ring Comb Lasers (VRCLs)", IEEE Quant.Elect.(23), 2017
10. B. Song et al., "High-Thermal Performance 3D Hybrid Silicon Lasers", IEEE Phot.Techn.Let. Vol.29, 2017
11. E.J. Zhang et al., "Methane absorption spectroscopy with a hybrid III-V silicon external cavity laser", CLEO: Science and Innovations (STh1B.2), San Jose CA, May 5-10, 2018.
12. Chi Xiong et al., "Silicon photonic integrated circuit for on-chip spectroscopic gas sensing", Proc. of Photonics West Conf., 2019
13. W.M. Green et al., "Silicon Photonic Gas Sensing", OFC 2019
14. W.D.Sacher et al., "Polarization rotator-splitters in standard active silicon photonics platforms", Optics Express Vol. 22, 2014
15. T. Barwicz et al., "Demonstration of Self-Aligned Flip-Chip Photonic Assembly with 1.1dB Loss and >120nm Bandwidth", 2016 Frontiers in Optics, OSA Technical Digest (online)), paper FF5F.3.
16. T. Barwicz et al., "Automated, high-throughput photonic packaging", Optical Fiber Technology 44, 24-35 (2018).
17. C.Xiong et al, "Correlation between optical return loss and transmission fringe noise in high-index contrast waveguides", CLEO (OSA) 2017, paper SW1N.5.
18. K.P. Jackson et al., "A High-Density, Four-Channel, OEIC Transceiver Module Utilizing Planar-Processed Optical Waveguides and Flip-Chip Solder-Bump Technology", J. Lightwave Techn. (12), 1994
19. J.-W. Nah, Y. Martin, S. Kamlapurkar, S.Engelmann, R.L. Bruce, and T.Barwicz, "Flip chip assembly with sub-micron 3D re-alignment via solder surface tension," in IEEE Proc. 2015 ECTC, pp.35-40.
20. Y. Martin et al., "Toward high-yield 3D self-alignment of flip-chip assemblies via solder surface tension", in IEEE Proc. of 2016 ECTC, pp.588-594.
21. Y. Martin et al, "Novel solder pads for self-aligned flip-chip assembly ", in IEEE Proc. of 2019 ECTC.

Extremely Low-profile Single Mode Fiber Array Coupler Suitable for Silicon Photonics

Mitsuharu Hirano, Akira Furuya, Hideki Machida, Koichi Koyama, Yasunori Murakami, and Kazunori Tanaka
Sumitomo Electric Industries, Ltd.
Yokohama, Japan
hirano-mitsuharu@sei.co.jp

Abstract—We report a novel, single mode optical fiber array coupler that couples the single mode fibers (SMFs) with grating couplers (GCs) on a silicon photonics chip and has D-shaped (DS) fibers. This DS fiber array coupler has obliquely polished tips of the SMFs that are fixed on a V-grooved glass, and a mirror plate is put on the polished tip surface of the fiber array coupler. Light that advances in the SMF is reflected by the mirror plate. The reflected light advances through the side face of the SMF, and couples with the GC on the silicon photonics chip. In the vicinity of the obliquely polished tips, the side face of the clad of the SMF is ground in such a way that the cross section of the SMF is formed into a D-shape to get good optical coupling with the GC. The DS fiber array coupler has extremely low-profile of 1.5 millimeters height above the silicon photonics chip surface. We fabricated the DS fiber array coupler, and measured coupling losses between this fiber array coupler and the GCs. The averaged additional optical coupling loss of 0.2dB due to the reflection structure is obtained for 93 coupling interfaces in the 11 fiber array couplers.

Keywords: silicon photonics, optical fiber couplers

I. INTRODUCTION

Silicon photonics technology has an advantage for fabrication of one-chip optical transceivers (TRXs) with parallel-multi-channelization such as parallel single mode (PSM) communication [1]. This is because silicon CMOS technology easily gives high-density integration of optical circuit for multi-channel transmitter and receiver in the one-chip optical TRX.

The parallel-multi-channelization requires multiple optical input and output interfaces on the TRX chip for multiple fibers [2, 3]. Grating couplers (GCs) fabricated on the TRX chip is one of promising optical interfaces for multiple single mode fibers (SMFs) [4, 5]. The light is input or output through the GCs on the surface of the TRX chip. The optical coupling to the SMF is simply carried out by butting the end face of the SMF to the GCs on the chip surface, so that wafer level optical testing of the TRX chip is easily carried out.

Optical coupling between the SMFs and the GCs in the parallel-multi-channelized TRX chip are usually achieved with arrayed fiber butting in a lump. Figure 1 shows a conventional coupling scheme between a fiber array coupler and the GCs. The polished end faces of the SMFs are directly connected with the coupling faces of the GCs as facing each other. The fibers stick up in the direction nearly vertical to the TRX chip surface, and then, the SMFs are bent toward the horizontal direction. In this coupling structure, the optical fiber array requires considerable space above the TRX chip. This space is determined by acceptable bending radiuses of the fibers. The SMFs usually require the bending radius more than 5 millimeters, and polarization maintaining fibers (PMFs) usually require the more. These bending radiuses often cause the difficulty in installing the fiber array coupler into small optical form factor TRXs such as OSFP or QSFP-DD. For example, QSFP-DD gives only the height around 3.8 millimeters above the TRX chip for installing the fiber array coupler.

In this paper, we report a novel fiber array coupler structure that has D-shaped fibers and a reflection mirror to give extremely low-profile. We fabricated this D-shaped (DS) fiber array couplers and evaluated their optical coupling losses with GCs.

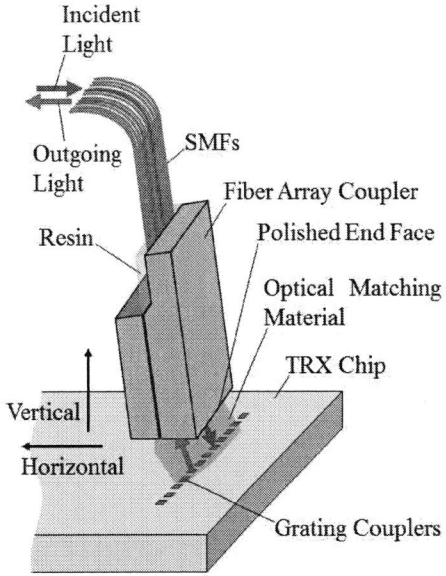

Figure 1. Conventional coupling scheme between a fiber array coupler and GCs.

II. STRUCTURE

A. Reflection Structure

Figure 2(a) shows a photo of the fabricated DS fiber array coupler. Figure 2(b) shows the schematic structure of the DS fiber array coupler. Twelve SMFs are arranged along V-grooved glass parallel to one another and fixed with resin. The tips of the SMFs are polished obliquely [6, 7]. An optical mirror plate is stuck onto the obliquely polished surface. The other end of the fibers are arranged into the MT connector to be suitable for PSM4 application.

Light entered from the MT connector propagates in the SMF, and is reflected by the mirror plate. The reflected light changes the traveling direction and exits from the side face of the SMF. On the contrary, light entered from the side face of the fiber with a suitable input angle is reflected by the mirror plate, and coupled to the SMF. The coupled light propagates in the SMF and exits from the MT connector end.

B. Optical Coupling

Figure 3 shows the coupling features between the DS fiber array coupler and the GCs on the TRX chip. Light entered from the end of the MT connector propagates in the SMF to the other end, and is reflected by the mirror plate and exits from the side face of the SMF. Then the light passes through optical matching material such as optical resin, which is provided between the surface of the TRX chip and the side face of the SMF. The optical matching material has a refractive index almost the same as that of the SMF. Then the light enters one GC on the TRX chip. On the contrary, light emitted from one GC enters the fiber array coupler from the side face. The light is reflected by the mirror plate and coupled with one fiber of the fiber array. The light propagates in the SMF and exits from the MT connector end.

GCs usually have the optimum distance between the SMF end and the GC for minimized the optical coupling loss. In this DS fiber array coupler structure, the minimum optical distance is determined by the distance between the core position and the side face position of the SMF due to the reflection structure. Generally, the extra optical distance from the optimum distance increases the optical coupling loss. As the side face of the SMFs of the DS fiber array coupler is ground in such a way that the cross section of the SMF is formed into a D-shape, the distance between the core position and the side face position of the SMF is enough short and the distance between the SMF end and the GC can be controlled and optimized.

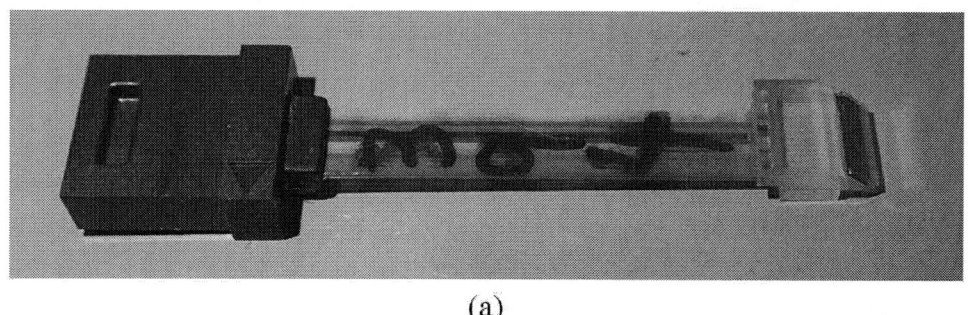

(a)

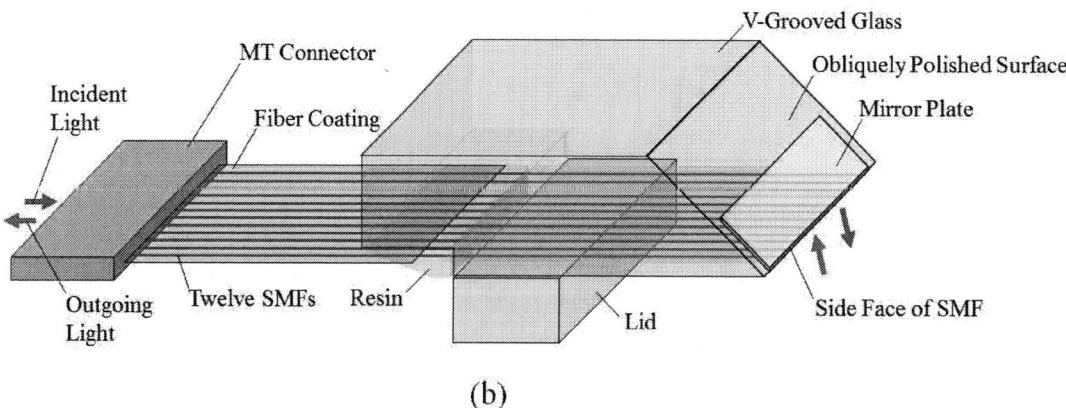

(b)

Figure 2. Structure of the DS fiber array coupler. (a) Photo of the DS fiber array coupler; (b) schematic structure of the DS fiber array coupler.

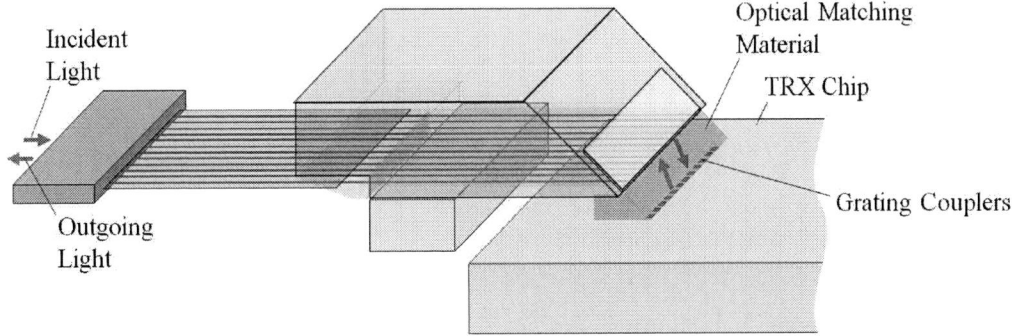

Figure 3. Coupling features between the DS fiber array coupler and the GCs on the TRX chip.

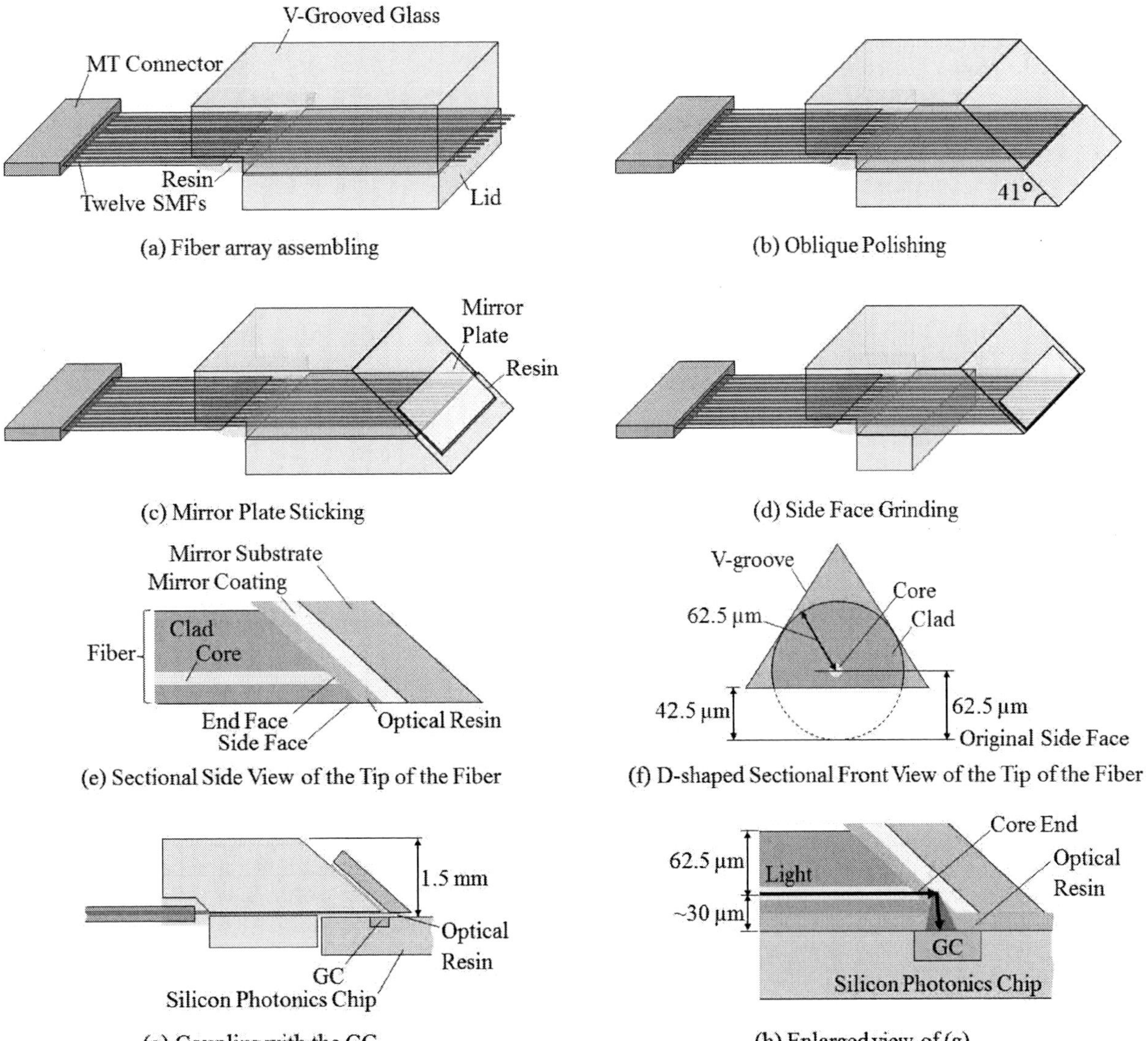

(a) Fiber array assembling

(b) Oblique Polishing

(c) Mirror Plate Sticking

(d) Side Face Grinding

(e) Sectional Side View of the Tip of the Fiber

(f) D-shaped Sectional Front View of the Tip of the Fiber

(g) Coupling with the GC

(h) Enlarged view of (g)

Figure 4. Manufacturing processes of the DS fiber array coupler. (a) Fiber array assembly; (b) oblique polishing: (c) mirror plate sticking; (d) side face grinding; (e) sectional side view of the tip of the fiber; (f) D-shaped sectional front view of the tip of the fiber: (g) coupling with the GC on the silicon photonics chip; (h) enlarged view of (g).

978-1-7281-1500-9/19 $31.00 © 2019 IEEE   1069

III. MANUFACTURING PROCESS AND USAGE

Figure 4 shows manufacturing process of the DS fiber array coupler. First, twelve SMFs with an MT connector are arranged along V-grooves in the glass. The V-grooves are prepared in 250-micrometer pitch. They are held by a glass lid and fixed with resin as shown Figure 4(a). Next, the tips of the SMFs, the V-grooved glass and the lid are simultaneously cut and polished obliquely as shown Figure 4(b). Oblique angle is selected to be 41 degrees, which fits to the optical input and output angle of the GC in this experiment. Then, as shown Figure 4(c), an optical mirror plate on which gold film was evaporated is stuck onto the obliquely polished surface using optical resin, whose refractive index is almost the same as that of the SMF. Generally, the edges of obliquely polished tips of the SMFs are fragile from external impact. The optical mirror plate also has a role to protect the edges from external impact. Finally, the side faces of the SMFs and a part of the lid are simultaneously ground off as shown Figure 4(d). Figure 4(e) and Figure 4(f) show the side sectional view and the front sectional view of the tip of the ground SMF, respectively. In the front sectional view, the cross section of the SMF is D-shaped to be ground off the side face of the SMF.

Figure 4(g) shows the example of the optical coupling between the DS fiber array coupler and a GC on the silicon photonics chip surface. The fiber array coupler is fixed above the GC on the silicon photonics chip with optical resin, whose refractive index is almost the same as that of the SMF. The height of the DS fiber array coupler is defined by the distance between the silicon photonics chip surface and the back side surface of the V-grooved glass. In our fabrication, extremely low-profile of 1.5 millimeters height is achieved.

GCs used in the experiment in this paper are designed to have the optimum coupling distance of 30 micrometers between the core end of the SMF and the GC, under the condition that optical matching material whose refractive index is almost the same as that of the SMF. The optical resin are filled between them as shown in Figure 4(g). At first, we measured the correlation between the extra distance from the optimum position and the excess coupling loss using conventional fiber array coupler. The result is shown in Figure 5. The graph shows that the extra distance increases the excess optical coupling loss.

Figure 4(h) shows the enlarged view of Figure 4(g). Since the radius of the SMF is generally 62.5 micrometers, the length between the core end of the DS fiber array coupler and the GC is, at least, about 30 micrometers longer than the optimum coupling distance of the GC. This extra distance would cause at least 0.7dB excess optical loss shown in Figure 5. Therefore, we grind off the clad of the SMF into 42.5 micrometers depth from the original side face as shown in Figure 4(f) in order to reduce the optical distance enough to reduce the expected excess optical loss.

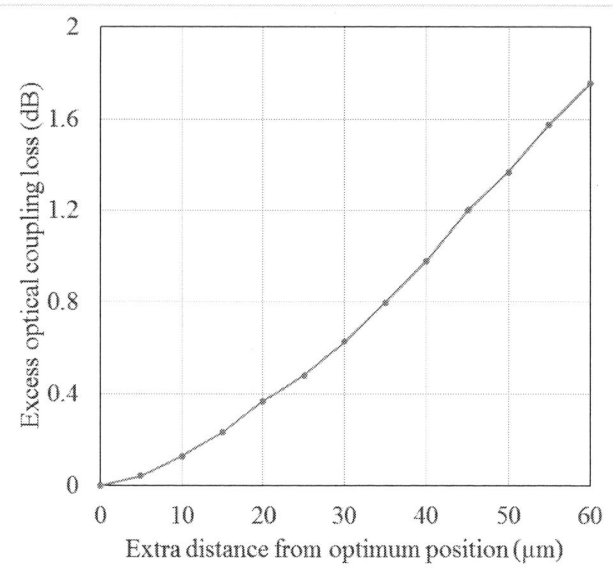

Figure 5. Correlation between the extra distance from the optimum position and the excess optical coupling loss using conventional fiber array coupler.

In the fabricated DS fiber array coupler, light passes through the ground surface of the SMF. However, since ground surface of the SMF is next to the optical matching resin, the low optical scattering loss would be expected.

IV. OPTICAL COUPLING LOSS

We made samples of the DS fiber array coupler, and measured optical coupling losses between these couplers and the GCs on the silicon photonics chip. Eleven samples were made. Each of them involved twelve SMF channels. A conventional fiber array coupler shown in Figure 1 was also prepared for comparison.

Test set-up for optical coupling loss evaluation is shown in Figure 6. Figure 6(a) and Figure 6(b) show the test set-up for the fabricated DS fiber array coupler and the conventional fiber array coupler, respectively.

In Figure 6(a), light from laser source having the wavelength of 1.3 micrometers is introduced into a SMF. The light passes a λ/4 plate, a λ/2 plate and an FC connector, and enters one channel of the fiber array coupler through the intermediary an MT connecter. The light is reflected by the mirror plate at the tip of the fiber and exits from the groundside face of the SMF. Then the light passes through matching oil having a refractive index almost the same as the SMF. The light enters one GC on the surface of the silicon photonics chip. The light entered in the GC propagates along the loop-shaped optical waveguide, reaches another GC at the end of the waveguide, and exits from this end GC. This end GC is prepared 750 micrometers apart from the entering GC. The light passes thorough the matching oil again and enters another channel of the fiber array coupler from the side face of the SMF. The light entering the fiber array coupler goes back through the MT connector and arrives at a photo detector. The intensity of the light is measured by the photo detector.

In Figure 6(b), light from laser source is introduced into a SMF. The light passes a λ/4 plate and a λ/2 plate, enters one channel of the conventional fiber array coupler through the intermediary an FC connecter, and exits from the end face of the SMF. Then the light passes through matching oil, and enters one GC on the surface of the silicon photonics chip. This silicon photonics chip is as same as one in the Figure 6(a). The light entered in the GC propagates along the loop-shaped optical waveguide, reaches another GC at the end of the waveguide, and exits from this end GC. The light passes thorough the matching oil again and enters another channel of the fiber array coupler from the end face of the SMF. The light entering the fiber array coupler goes back and arrives at a photo detector. The intensity of the light is measured by the photo detector.

In the evaluation test set-up in Figure 6(a) and Figure 6(b), it is able to move the fiber array couplers in parallel to the X axis, the Y axis and the Z axis and to rotate them around the X axis, the Y axis and Z axis. In addition, the λ/4 plate and the λ/2 plate is used controlling polarization of the light at the entering GCs. The position of the fiber array coupler and the GCs, and the polarization of the light at the entering the GCs are adjusted to get the best coupling in the each test set-up. By shifting the adjusting position in 250-micrometer pitch in the X direction, a plurality of the measured data were obtained for one sample.

In Figure 6(c), light from laser source is introduced into a SMF. The light passes a λ/4 plate, a λ/2 plate and an FC connector twice and arrives at a photo detector. The intensity of the light is measured by the photo detector for the reference.

We evaluated optical coupling loss of the each fiber array coupler as shown below.

First, the laser source output power is kept the same in the test set-ups of Figure 6(a), Figure 6(b) and Figure 6(c).

Then, in the test set-up of Figure 6(c), the light intensity at the photo detector was measured. This detected light include the optical loss effects of the optical components of the λ/4 plate, the λ/2 plate, and two FC connectors.

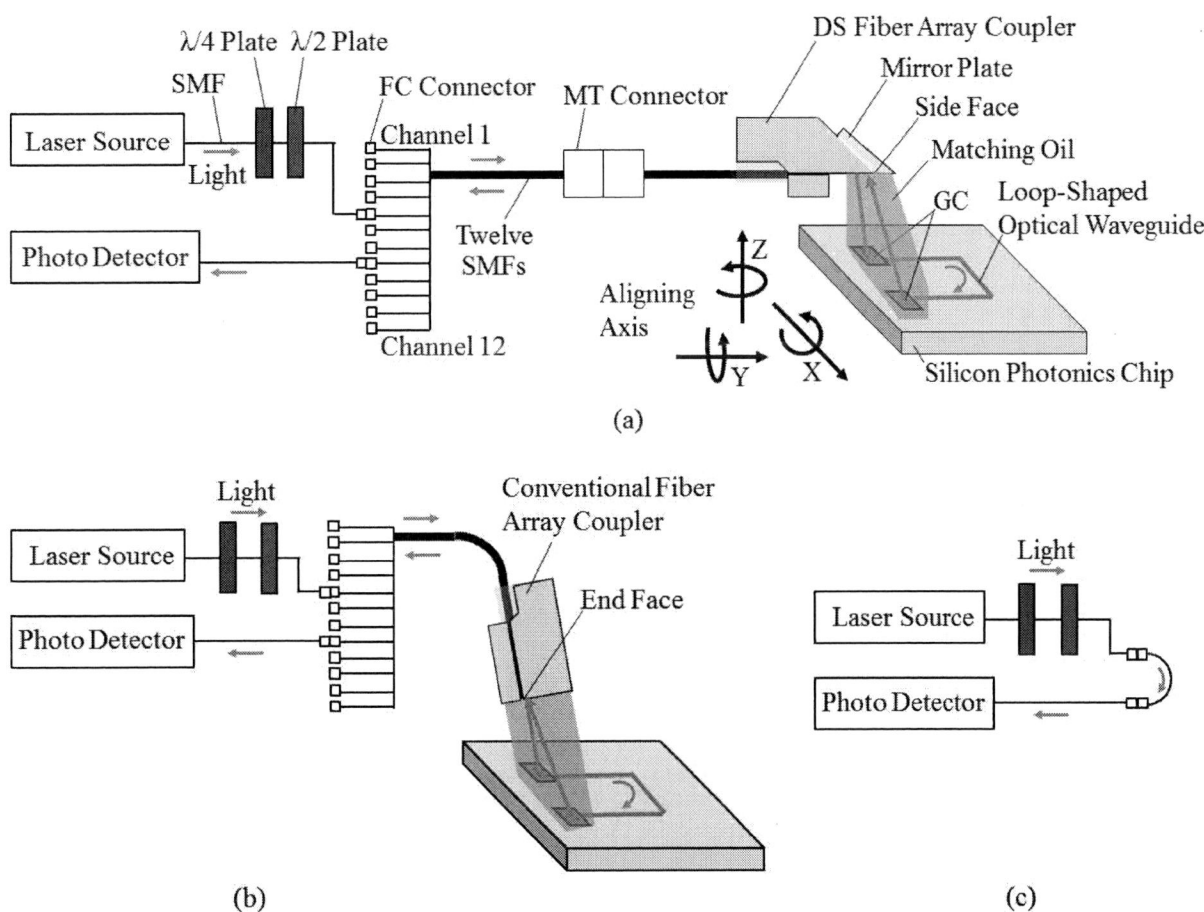

Figure 6. Optical coupling loss evaluation test set-up. (a) For DS fiber array coupler; (b) for conventional fiber array coupler; (c) for measurement of reference.

Then, in the test set-up of Figure 6(b), the light intensity at the photo detector was measured. This detected light include the effect of the optical components of the λ/4 plate, the λ/2 plate, two FC connectors, and two coupling losses between the GC and the conventional fiber array coupler. The light intensity ratio between the test set-up of Figure 6(b) and that of Figure 6(c) shows the twice of the coupling losses between the GC and the conventional fiber array coupler. Therefore, we can calculate the single optical coupling loss between the GC and the conventional fiber array coupler.

Next, in the test set-up of Figure 6(a), the light intensity at the photo detector was measured. This detected light include the optical loss effects of the optical components of the λ/4 plate, the λ/2 plate, two FC connectors, two MT connectors, and two coupling losses between the GC and the DS fiber array coupler. The light intensity ratio between the test set-up of Figure 6(a) and that of Figure 6(c) shows the twice of the summation of the MT connector coupling loss and the twice of the coupling losses between the GC and the DS fiber array coupler. By separately evaluating the MT connector coupling loss and deducting it from the light intensity ratio in the test set-up of Figure 6(a), we can calculate the single optical coupling loss between the GC and the DS fiber array coupler.

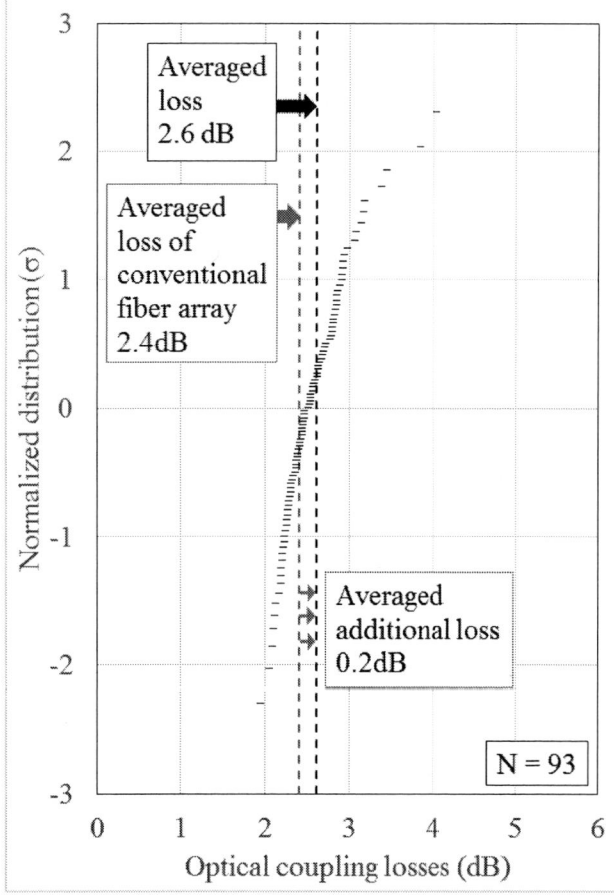

Figure 7. Normalized distribution of the single optical coupling losses between the SMFs of the DS fiber array couplers and the GCs on the silicon photonics chip.

The losses of SMFs in these evaluation test set-up were ignored because the sum of these losses is less than 0.01dB and is sufficiently small compared with other losses.

Figure 7 shows the single optical coupling losses between the SMFs of the DS fiber array couplers and the GCs on the silicon photonics chip for 93 measured points. The average of the optical coupling losses of 2.6dB is obtained. For the conventional fiber array coupler, the average of the optical coupling losses of 2.4dB is obtained. Consequently, the averaged additional optical coupling loss due to the novel reflection structure of 0.2dB is evaluated.

The origins of this additional optical coupling loss are considered to be composed of the transmission loss of the optical resin, the reflection loss by the mirror plate, the scattering loss at the end face and the side face of the fiber, and the beam shape deformation due to 41-degree-angled reflection. The separately evaluated optical loss of the reflection by the gold metal on the mirror plate is 0.12dB. Thus the losses by the other factors are 0.08dB and considered to be fairly small.

Though, some high coupling losses were observed in the tail of the distribution in Figure 7, we infer that these are due to some unintentionally formed structural imperfection such as included dust in the optical resin or small scratch on the reflection mirror.

V. MACHINING ACURRACY

We checked the machining accuracy in the side face grinding process. Figure 8(a) shows the perspective view of the fabricated DS fiber array coupler. Figure 8(b) shows the sectional front view of the red circled part of Figure 8(a). The difference between the actual ground depth and the target depth (42.5 micrometers) represents the machining error in Figure 8(b). The excess grinding error is shown as positive number. We checked the machining error of the eleven fiber array samples for twelve fibers each.

Figure 8(c) shows the machining errors. The errors were within plus or minus 5 micrometers. No outstanding machining error distribution was observed in the direction from the channel 1 to the channel 12.

By referring the data in Figure 4, the machining error within 5 micrometers gives additional optical loss of less than 0.04dB. This shows that the achieved machining accuracy is enough to give the low optical coupling loss.

VI. HEAT CYCLE TEST

We tested heat cycle endurance for two samples of the DS fiber array coupler. The heat cycle condition was from -40°C to 85°C and 60 cycles. After the heat cycle stress, we check the disconnection of the side-face ground SMFs by light passing test equipment.

No disconnection of the side-face ground SMFs was observed. The result of the heat cycle test shows the robustness of the ground part of the SMFs.

This result shows that the DS fiber array coupler structure with the side-face ground SMFs is applicable to the actual TRX usage.

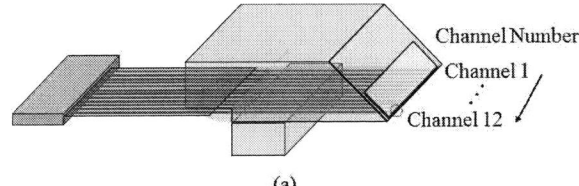

(a)

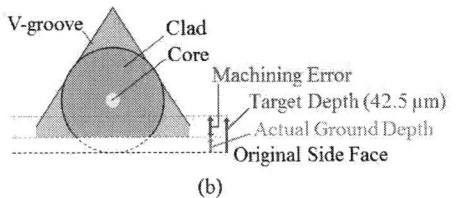

(b)

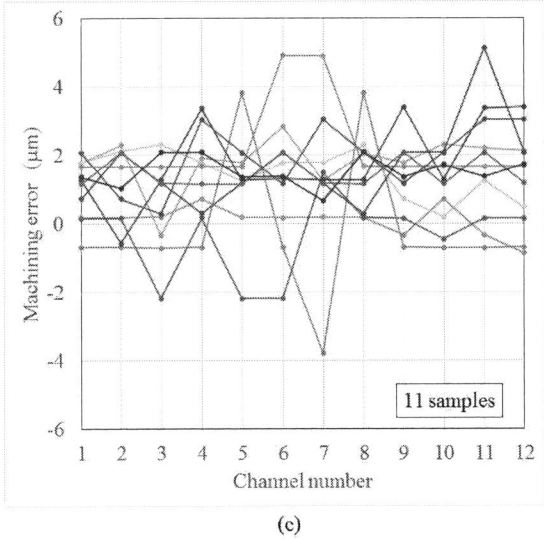

(c)

Figure 8. Machining errors. (a) Perspective view of the fabricated DS fiber array coupler; (b) sectional front view of the red circled part of Figure 8(a); (c) machining errors of the eleven samples for twelve fibers each.

VII. CONCLUSION

We developed a novel fiber array coupler and its fabrication process.

This fiber array coupler structure has the D-shaped (DS) SMFs, the tips of which are obliquely polished, and the mirror plate on the obliquely polished surface. The fiber array coupler has extremely low-profile of 1.5 millimeters. The grinding process of the side faces of the SMFs are applied to improve the optical coupling with the GCs.

We evaluated the characteristics for the fabricated DS fiber array couplers. The averaged additional optical coupling loss due to the novel reflection structure is 0.2dB, compared to that of the conventional fiber array coupler. The grinding fabrication process for the side faces of the SMFs has the machining errors within 5 micrometers and is precise enough to keep the low optical coupling loss. The ground part of the SMFs is robust to the heat cycle stress under the condition of 60 cycles from -40°C to 85°C.

These results show that the DS fiber array coupler is very promising for the application to the small optical form factor TRX with a silicon photonics chip.

REFERENCES

[1] Frédéric Boeuf, Sébastien Crémer, Enrico Temporiti, Massimo Ferè, Mark Shaw, Charles Baudot, Nathalie Vulliet, Thierry Pinguet, Attila Mekis, Gianlorenzo Masini, Herve Petiton, Patrick Le Maitre, Matteo Traldi, and Luca Maggi, "Silicon Photonics R&D and Manufacturing on 300-mm Wafer Platform," Journal of Lightwave Technology, vol. 34, no. 2, pp. 286-295, 2016.

[2] Lee Carroll, Jun-Su Lee, Carmelo Scarcella, Kamil Gradkowski, Matthieu Duperron, Huihui Lu, Yan Zhao, Cormac Eason, Padraic Morrissey, Marc Rensing, Sean Collins, How Yuan Hwang, and Peter O'Brien, "Photonic Packaging: Transforming Silicon Photonic Integrated Circuits into Photonic Devices," Applied Science, vol. 6, No. 12, pp. 426, 2016.

[3] Christophe Kopp, St´ephane Bernab´e, Badhise Ben Bakir, Jean-Marc Fedeli, Regis Orobtchouk, Franz Schrank, Henri Porte, Lars Zimmermann, and Tolga Tekin, "Silicon Photonic Circuits: On-CMOS Integration, Fiber Optical Coupling, and Packaging," IEEE Journal of Selected Topics in Quantum Electronics, vol. 17, no. 3, pp. 498-509, 2011.

[4] Siddharth Nambiar, Purnima Sethi, and Shankar Kumar Selvaraja, "Grating-Assisted Fiber to Chip Coupling for SOI Photonic Circuits," Applied Science, vol. 8, no. 7, pp. 1142, 2018.

[5] Dirk Taillaert, Frederik Van Laere, Melanie Ayre, Wim Bogaerts, Dries Van Thourhout, Peter Bienstman and Roel Baets, "Grating Couplers for Coupling between Optical Fibers and Nanophotonic Waveguides," Japanese Journal of Applied Physics, vol. 45, no. 8A, pp. 6071-6077, 2006.

[6] N. Hoppe, M. Haug, T. Polder, M. Félix Rosa, W. Vogel, P. Scheck, L. Rathgeber, D. Widmann, and M. Berroth, "Sealed and Compact Fiber Links to Integrated Photonics Using Grating Couplers," Proc. 67th Electron Components Technology Conference, pp.139, 2017

[7] Bradley Snyder, and Peter O'Brien, "Packaging Process for Grating-Coupled Silicon Photonic Waveguides Using Angle-Polished Fibers," IEEE Transactions on Components, Packaging, and Manufacturing Technology, vol. 3, no. 6, pp. 954-959, 2013.

Micro lens array assembly for optical organic substrate

Patrick Jacques, Richard Langlois, Élaine Cyr,
Alexander Janta-Polczynski, Paul Fortier
IBM Bromont
23 Boul. de l'Aéroport, Bromont, QC Canada
e-mail: pajacque@ca.ibm.com

Koji Masuda, Masao Tokunari, Hsiang-Han Hsu
Science & Technology,
IBM Research - Tokyo
Kawasaki, Kanagawa Japan
e-mail: masudak@jp.ibm.com

Abstract—A polymer lens attach process is developed on Optical Multi-Chip Modules (OMCM) with integrated optical waveguides and tested successfully in Deep Thermal Cycle (DTC) at -40°C/125°C for 1000 cycles. The lens attach is completed in a two step process: UV tack of the lens array using a high precision pick and place tool followed by the dispense of a viscous sidefill material along the lens array's periphery to mechanically secure it to the OMCM laminate. A screening protocol is developed based on manufacturing and reliability criteria for testing lens array UV tacking and sidefill adhesives. Manufacturing selection criteria include good adhesion and low infiltration in small gaps. Low infiltration under the lens array is necessary to prevent contamination of laminate optical features. For testing lens attach reliability, a list of critical process parameters are chosen including silicone and epoxy UV tacking and sidefill adhesives with a wide range of thermo-mechanical properties: Young's Modulus (E), Coefficient of Thermal expansion (CTE), glass transition temperature (Tg). Lens attach reliability performance is assessed by both visual inspection for mechanical integrity failures, and using lens to lens insertion loss optical test readouts.

Keywords-lens, optical, OMCM, waveguides, reliability

I. INTRODUCTION

The continued exponential growth in data communication as well as data processing is fueled by the increase in IoT devices, the popularity of social media, the deployment of 5G infrastructures and innovations in artificial intelligence technologies. We are only at the tip of the iceberg, and this growth is expected not only to continue, but accelerate as we move forward in time. This puts tremendous pressure on high performance computing, datacenters and interconnect networks to provide solutions for increased bandwidth. To meet these bandwidth requirements, signal speeds and density need to increase. Limitations in bandwidth density of electronics is driving initiatives towards bringing the conversion to optical signals closer and closer to the source. Co-packaging of optical engines within microelectronics modules is one level of integration that is gaining traction. The two primary optical technologies that are being considered for co-packaging are silicon photonics and VCSELs. Both technologies have their pros and cons, and in the end, it is most probable that both technologies will co-exist, finding the application spaces where they will shine. Silicon photonics provide benefits of higher levels of signal multiplexing, translating into higher bandwidths, as well as being able to efficiently send data over long distances. VCSELs, on the other hand, have been the workhorse of optical communication in the datacenters for a long time and offer a mature technology with unparalleled low cost and low power. The focus of the work presented here is to provide a packaging solution that allows the shortest possible distance between the electronic chips and multimode VCSEL based optical chips. Rather than attaching pre-packaged optical engines onto the microelectronic package, the individual components of the optical engines are directly mounted into the microelectronics package where the ASIC or CPU resides [1]. The optoelectronic chips (i.e. VCSELs and photodiodes) are flip chip mounted in close proximity to the ASIC chip, allowing for an extremely short electrical interconnection distance. The organic substrate to which the chips are attached contains embedded waveguides to route the optical signals from the optoelectronic chips to the edge of the substrate where the light is coupled to a lens/fiber system. The laser driver and transimpedance amplifier functionality is integrated into the ASIC chip. This packaging scheme allows us to achieve the highest possible bandwidth density for an ASIC chip using VCSEL based optics.

II. LENS ATTACH PROCESS DESCRIPTION

The OMCM reliability test vehicle laminate and lens array components are shown in Figure 1. The OMCM laminate is 50 mm X 50 mm in size and has 6 integrated waveguide sites, two of which are designed to receive/transmit lens to lens optical signals. The 4 other sites are designed for OE die to lens array optical transmission. The lens array is 5 mm x 3.5 mm x 1 mm thick and has 24 micro lenses at a pitch of 250 microns. It is made from polymer resin using an injection molding process. Two 1 mm x 1 mm x 0.5 mm thick wing-like structures are molded as part of the lens body and will be used for UV tacking of the lens array to the OMCM laminate. The wings are staggered to allow the lens arrays to be placed in a high density configuration on the OMCM laminate. The lens attach assembly process for qualification consists of attaching 8 lens arrays on the OMCM laminate using commercial high volume manufacturing tools. In order to achieve proper light coupling efficiency, the lens array needs to be attached within $\pm$ 5 μm of its targeted location [2]. This will ensure proper alignment of the lens array micro lenses with the laminate waveguide mirrors. The lens array attach process is completed in two steps: 1) Lens placement and UV tacking followed by 2) lens array sidefill. Figure 2 shows the lens UV tacking detailed process steps: Two dots of 1 mm in diameter and 100 microns high, of UV curable adhesive are dispensed on the laminate surface prior to lens array placement. The lens array is picked and placed using a high precision placement tool with UV light curing capability.

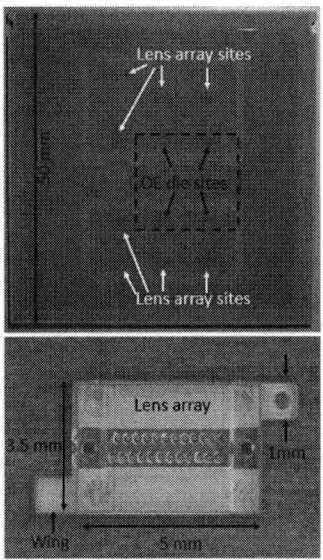

Figure 1. 50mmX50 mm OMCM laminate (top view). Polymer lens array (bottom view).

When the lens array is precisely aligned and placed on the laminate, the wings come into contact with the adhesive dots. To maintain the alignment, the placement tool's pick tip head maintains a light pressure on the lens array to prevent it from moving. The adhesive dots are then cured with UV light coming out of the placement tool pick tip head via an integrated light guide. The UV light goes through the 0.5 mm thick lens array wings before reaching the adhesive dots.

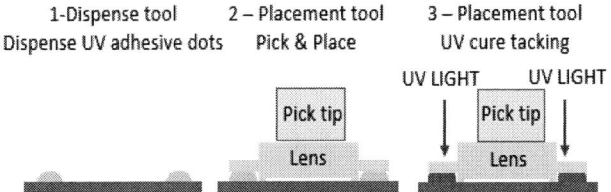

Figure 2. Lens UV tacking: Two adhesive dots are dispensed (left). Lens array is picked and placed (center) and held by pick tip while UV curing the adhesive dots through the lens array wings (right).

Lens array sidefill: This process is done using an automatic dispense tool with a high viscosity adhesive that is dispensed along the lens array's periphery. Figure 3 shows a lens array after UV tacking and after sidefill.

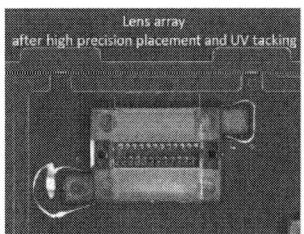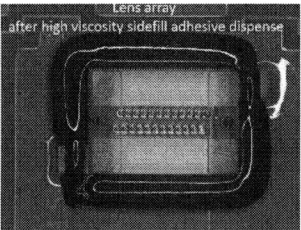

Figure 3. A lens array after UV tacking (left) and after sidefill (right).

After lens attach, placement accuracy can be measured using machine vision on the circular fiducials built-in the laminate cladding layer (Figure 4). The center position of the lens array alignment hole is measured relative to the center position of the laminate fiducial. This is done for both the left side and right side of the lens array. From this data a X, Y and offset angle theta can be calculated.

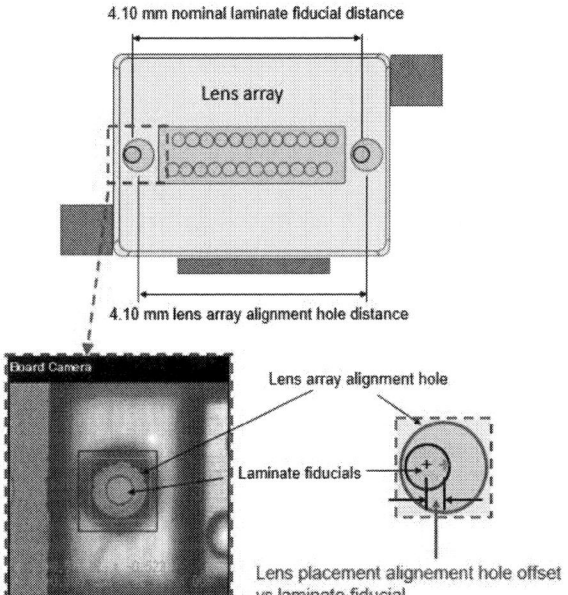

Figure 4. Lens array placement offset measured using laminates built-in fiducials and lens array alignment holes.

III. LENS ATTACH PROCESS PARAMETERS

A. Lens attach adhesive selection

Selection criteria for the UV tacking adhesive include that the material has a high viscosity so the adhesive will hold its round shape (non-collapsing) after dispense and will be able to contact the lens array wings which are 70 microns off the laminate surface once assembled. The adhesion to the laminate and lens array surfaces must be strong enough to survive handling until the sidefill process has been completed. The adhesive also needs to be UV curable to allow for short lens attach cycle time to enable high volume manufacturing. A number of adhesives were put through a series of shear, hardness, and dispense tests using OMCM hardware to determine the best UV tacking adhesives and lens attach assembly process parameters prior to sidefill and DTC reliability stresses. Both rigid epoxy and softer silicone adhesives were tested. For evaluation of viscosity, non-collapsing behaviour and wetting of the UV tacking adhesives, dots of different materials were dispensed on a laminate surface while diameter and height were measured over time. A shear test was developed using lens array material strips glued on a laminate surface for evaluation of adhesion and UV curing protocols for rigid epoxy adhesives. A hardness test was used on cured dots for evaluation of the softer silicone adhesives. Shear and hardness test results are shown in Table 1. The UV curing protocols used were based on the manufacturer's recommendations and process requirements.

Table 1. Shear and hardness results for testing adhesion and curing protocols.

Adhesive	UV Curing protocol	Shear strength (Lbf/in²)	Hardness (A) Cured adhesive dot
Epoxy A	1	175	NA
Epoxy B	2	178	NA
Epoxy C	2	153	NA
Epoxy D	2	490	NA
Epoxy E	2	155	NA
Silicone F	3	NA	11

Based on the results above, all adhesives shown were deemed viable for UV taking of lens arrays. The strongest adhesion at T0 (before reliability testing) being with adhesive D. Selection criteria for the sidefill adhesive are high viscosity to prevent infiltration and good adhesion. The adhesive must not flow under the lens array and reach the waveguide mirrors via the laser ablation holes [3] which are at 0.9 mm from the lens array's edge (Figure 5). The sidefill adhesive must also mechanically secure the lens array to the laminate and allow fiber connector plug in and removal during optical testing.

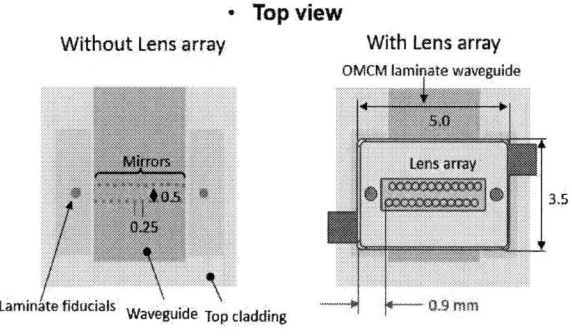

Figure 5. Shortest distance between lens array edge and laminate optical features (mirrors, laser ablation holes) is 0.9 mm.

Sidefill infiltration was evaluated using lens array material strips fixed on an OMCM laminate (Figure 6). One end of the strip is shimmed up to 15 microns while the other end rests directly on the laminate. This allows a variable infiltration gap between 0 to 15 microns. The actual lens arrays to laminate gap on functional samples is 10 microns. The sidefill material is dispensed, and infiltration is measured over time up to 30 minutes and again after

adhesive cure. Table 2 shows the smallest infiltration distance results for the 4 best sidefill materials tested. Only 2 candidates meet the 0.9 mm maximum infiltration requirement after dispense, 30 min wait time and adhesive cure. Figure 7 shows a top view of an actual dispensed test sample with silicone adhesive G. A sidefill infiltration distance of 0.25 mm is observed under the lens array material strip at a 10 microns gap. Infiltration distance depends not only on material viscosity but also the wetting affinity of the adhesive with the OMCM laminate and lens array material strip surfaces.

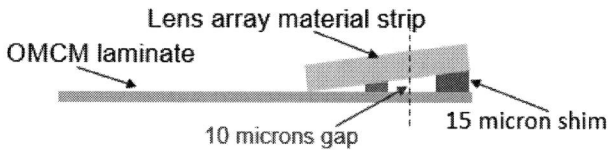

Figure 6. Sidefill material infiltration test set-up using 15 microns shim and lens array material strip glued on OMCM laminate.

Table 2. Sidefill infiltration measurements over time.

Delay dispense-measurement	Viscosity (Pa*s)	Infiltration distance (mm) over time (min)			
		5 min	10 min	30 min	After cure
Silicone G	304	0.05	0.07	0.12	0.25
Epoxy B	1200	0.25	0.28	0.4	0.4
Epoxy H	75	0.25	0.5	0.9	NA
Epoxy C	400	0.18	0.5	0.9	NA

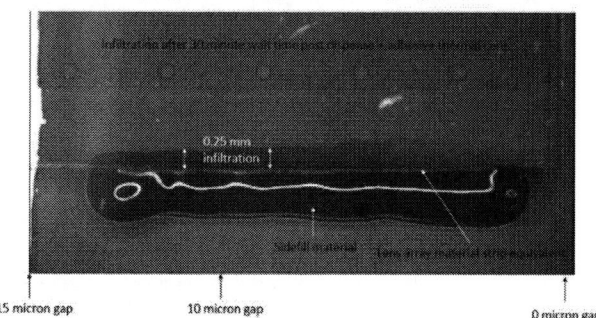

Figure 7. Actual sidefilled infiltration sample test for silicone G showing 0.25 mm of infiltration after dispense, 30 minutes wait time and adhesive thermal cure.

IV. DEEP THERMAL CYCLE RELIABILITY TEST CELLS

After the lens assembly and placement measurements, OMCM laminates were submitted to pre-conditioning (thermal shock of 5 cycles of -40°C to 60°C) followed by deep thermal cycle stresses of -40°C to 125°C for 1000 cycles. Microscope inspection for mechanical defects and/or lens to lens optical read-outs were taken after assembly (T0), after pre-conditioning and at 100, 250, 500 and 1000 DTC cycles. The goal is to maintain mechanical integrity of the lens arrays and OMCM laminate features including: laminate waveguides, mirrors, laser ablation cavities and lens array

978-1-7281-1500-9/19 $31.00 © 2019 IEEE

micro-lenses. These features will be under stress due to the CTE mismatch between laminate and lens array (Table 3)

Table 3. Thermo-mechanical properties of lens array and OMCM laminate

Component	Young's modulus (MPA)	CTE (PPM/K)
Lens array	1500-2000	80-100
OMCM laminate	900-2500	15-25

To evaluate the impact of these thermo-mechanical stresses, a list of process parameters, shown in Table 4 and 5 below, are chosen to be tested in a 22 test cell reliability DOE. Process parameters taken into account include : UV tacking adhesives, sidefill adhesives (including epoxy and silicone), sidefill dispense pattern, lens thermal pre-conditioning.

Table 4. List of UV tacking and sidefill adhesives and their distribution in the 22 reliability test cells.

UV tacking adhesive	No of cells	Sidefill Adhesive	No of cells
A (Epoxy)	5	A (Epoxy)	2
B (Epoxy)	6	B (Epoxy)	7
C (Epoxy)	2	C (Epoxy)	2
D (Epoxy)	2	D (Epoxy)	2
E (Epoxy)	4	E (Epoxy)	4
F (Silicone)	3	G (Silicone)	5
Total cells	22	Total cells	22

Table 5. Distribution of pre-conditioned lens arrays and sidefill dispense patterns in the 22 reliability test cells.

Lens thermal Pre-conditioning 180°C/30 min	No of cells	Dispense pattern	No of cells
Yes	12	Partial	5
No	10	Full	17
Total cells	22	Total cells	22

Lens thermal pre-conditioning is done prior to lens attach at a set profile temperature of 180°C for 30 minutes, as suggested by the lens array manufacturer. The objective is to relieve the internal stresses in the lens array which may have accumulated during the injection molding process. This could reduce the total stress seen by the lens array during reliability. To evaluate the effect of the lens array thermal pre-conditioning, lens array alignment hole pitch was measured by the manufacturer prior to thermal pre-conditioning, after thermal pre-conditioning and again after successive thermal excursions at 150°C (maximum sidefill cure temperature) and 250°C (BGA solder reflow). These lens array alignment hole pitch measurements (Table 6) show that after being heated during thermal pre-conditioning and cooled back down to room temperature, the lens array alignment hole pitch increases by 33 microns. No further

significant permanent changes in alignment hole pitch are measured at room temperature if the lens array is re-heated and cooled back down from peak temperatures of 150°C and 250°C.

Table 6. Lens array manufacturer measurements for alignment hole pitch following thermal pre-conditioning and successive thermal excursions.

Lens array alignment hole pitch measurement (mm)			
Initial	Pre-conditioning (180°C / 30 min)	150°C / 30 min	250°C (BGA reflow)
4.083	4.116	4.116	4.119

UV tacking and sidefill adhesive thermo-mechanical properties were chosen to cover a wide range of values as shown in Table 7. To maintain the precise alignment of the lens array during the manufacturing process of lens attach, more rigid (higher Young's modulus) and low shrinkage adhesives are favored. Requirements for passing DTC stresses are different, and could come in conflict with the manufacturing requirements. With a lower Young's modulus softer adhesives like E, F, G could absorb the reliability stresses through their elastic deformation. A lower Tg (Adhesive E) would spend more time above its Tg during thermal cycling (-40°C to 125°C) and be in a softer state (lower Young's modulus above Tg vs below Tg). A lower CTE (adhesives A, C) would better match the laminate's CTE and reduce stress at the laminate's surface while increasing the stress at the lens array's surface. A higher CTE (adhesives B, D) would better match the lens array's CTE and lower the stress at the lens array's surface while increasing the stress at the laminate's surface.

Table 7. Wide range of values for the thermo-mechanical properties of the lens attach adhesives.

Adhesives	CTE(ppm/°K)	Tg (°C)	E (MPA)
A (Epoxy)	10-20	100-160	3000-3500
B (Epoxy)	60-80	100-160	3000-3500
C (Epoxy)	10-20	100-160	3000-3500
D (Epoxy)	50-60	100-160	3000-3500
E (Epoxy)	50-60	20-50	2000-2500
F (Silicone)	200-300	NA	5-20
G (Silicone)	200-300	-50 to -25	3-10

Two different sidefill dispense patterns were developed for lens attach: full sidefill and partial sidefill (Figure 8). Full sidefill covers all the lens array's periphery. By covering more surface area this dispense pattern could lower the average stress per unit area on the lens array and OMCM laminate. Partial sidefill dispense pattern leaves the 4 lens array corners, as well as the short side, free of adhesive. Not attaching the lens array's corners to the OMCM could lower the peak stress at this vulnerable location caused by the different shrinkage and thermal expansion of the attached components during DTC.

Figure 8. Full sidefill (left) covers the entire periphery of the lens array. Partial sidefill (right) does not cover the lens array corners or short sides.

The lens attach process parameters listed in Tables 4 and 5 were mixed and matched in a 22 cell DOE and run through DTC reliability testing. One test cell consisting of 1 lens array attached on an OMCM laminate lens site. A total of 3 different OMCM laminates populated with these 22 test cells were assembled (Figure 9). After lens UV tacking, lens placement accuracy was measured, and the results showed an average offset of less than 3 microns, which is within expectation based on the placement tool's capability.

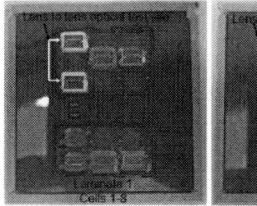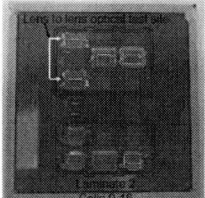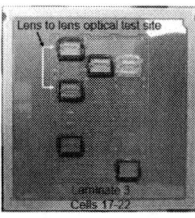

Figure 9. 3 different OMCM laminates populated with a total of 22 lens arrays. Each lens array site is a test cell.

V. DEEP THERMAL CYCLE RELIABILITY RESULTS

Following the 1000 DTC cycles only 3 cells out of 22 have successfully passed without showing any mechanical defects (microscope inspection at 40X). The other 19 cells have all shown early mechanical defects after only the first 100 DTC cycles read-out inspection (Table 8).

Table 8. DTC (-40°C/125°C) read-out microscope inspection at 40X for mechanical defects.

DTC reliability read-out	Total cells with mechanical defects
T0	0 / 22
After Thermal shock (-40°C/60°C)	5 cells / 22 Partial sidefill cells more affected
DTC 100 cycles	19 cells / 22 Light to severe mechanical defects
DTC 250 cycles	19 cells / 22 Severe mechanical defects
DTC 500 cycles	19 / 22 (3 good cells remain out of 22)
DTC 1000 cycles	19 / 22 (3 good cells remain out of 22)

A. DTC pre-condtioning (Thermal shock)

After DTC pre-conditioning (thermal shock) 5 out of 22 cells already start showing mechanical defects. Cells with partial sidefill are more affected and show cracks in the adhesive fillet as shown in Figure 10. Because of the smaller surface contact area , the partial sidefill fillet seems more vulnerable at this early stage of reliability stressing than the full sidefill fillet.

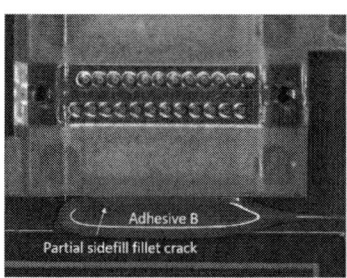

Figure 10. Inspection after thermal shock. Cracks found in partial sidefill fillets.

B. DTC 100 cycles and 250 cycles

After 100 DTC cycles, 19 out of the 22 cells are showing light to severe mechanical defects. Sidefill fillet cracks, and lens array corner cracks are shown in Figure 11 for a test cell using a lower CTE epoxy adhesive (adhesive A). This adhesive could be transferring too much stress to the lens body, which has a significantly higher CTE than the adhesive, resulting in lens array corner cracks.

Figure 11. Lens array corner cracks and sidefill fillet cracks found in test cell with low CTE adhesive A.

Other fail signatures, like laminate delamination, can be observed in Figure 12 for a test cell using a higher CTE epoxy adhesive (adhesive B). This adhesive could be transferring too much stress to the laminate, which has a significantly lower CTE than the adhesive, resulting in adhesive to laminate and laminate interlayer delamination. Figure 13 shows lens corner cracks specifically located at the wings and lens body connection. This signature is observed on a test cell using a combination of 2 adhesives: a higher Young's modulus Epoxy for UV tacking (adhesive B) and a thermally cured lower Young's modulus silicone adhesive as sidefill (adhesive G). In this case the more rigid UV tacking material located under the wing is transferring too much stress to the region where the wing connects to the lens body because the softer silicone sidefill adhesive is not able to offer enough mechanical support. Following the 250 DTC cycles, 19 out of 22 cells now only show severe mechanical defects. Review of the test cells data shows that any lens attach done using epoxy adhesives as UV tacking and/or sidefill material has resulted in early DTC fails. Also, lens arrays exposed to the 180°C/30 minutes thermal pre-conditioning prior to lens attach have the same failure rate as those that were not subjected to pre-conditioning.

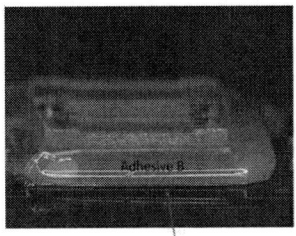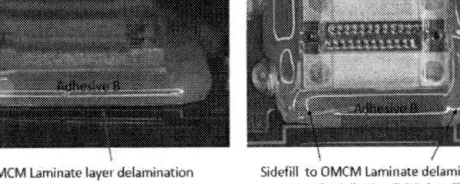

OMCM Laminate layer delamination
UV tack : Adhesive B (Higher CTE)
Sidefill : Adhesive B (Higher CTE)

Sidefill to OMCM Laminate delamination
UV tack : Adhesive B (Higher CTE)
Sidefill : Adhesive B (Higher CTE)

Figure 12. Laminate interlayer delamination (left) and adhesive to OMCM laminate delamination (right) observed in test cells using higher CTE adhesive B.

Lens array corner cracks
UV tack : Adhesive B (epoxy, Higher modulus)
Sidefill : Adhesive G (silicone, Lower modulus)

Lens array corner cracks
UV tack : Adhesive B (epoxy, Higher modulus)
Sidefill : Adhesive G (silicone, Lower modulus)

Figure 13. Lens array corner cracks located at the wing to lens array body connection. This signature is found in test cells using rigid epoxy adhesive B for UV tacking and softer silicone adhesive G for sidefill.

C. DTC 1000 cycles

The remaining 3 good test cells successfully completed 1000 DTC cycles with no mechanical defects seen after 40X microscope inspection as shown in Figure 14. These 3 test cells were assembled using lower Young's modulus silicone adhesive F and G for respectively UV tacking and sidefill. In this configuration the mechanical stresses induced by the DTC and the CTE mismatch of the lens to laminate assembly are relieved through the elastic deformation of both softer silicone lens attach adhesives.

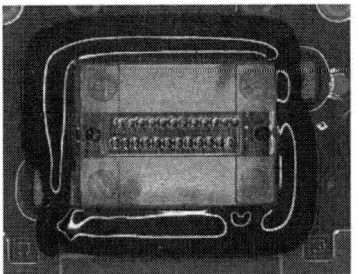

No mechanical defects
UV tack : Adhesive F (silicone, lower Young's modulus)
Sidefill : Adhesive G (silicone, lower Young's modulus)

Figure 14. One of 3 successful test cells showing no mechanical defects after 1000 DTC cycles. Softer silicone adhesive F is used for UV tacking, softer silicone adhesive G is used for lens array sidefill.

D. Optical lens to lens signal read-outs 1000 cycles

Optical lens to lens insertion loss read-outs were performed on specific laminate waveguide sites populated by two communicating lens arrays. These optical tests were done after DTC pre-conditioning (thermal shock), 100, 500 and 1000 DTC cycles. Lens to lens insertion loss optical measurements for 2 successful lens array test cells with no visual mechanical defects after 1000 cycles are shown in Table 9. Average insertion loss read-out for 24 channels are calculated relative to reference measurements taken after pre-conditioning. The 100 cycle optical test read-out data is disregarded because of suspected issues with the fiber to lens array connection. Taking this into account, relative measurements at 500 and 1000 cycles show no significant degradation during reliability stressing when compared to results after pre-conditioning.

Table 9. Lens to lens optical test read-out measurements during reliability testing.

Reliability Testing	Cycles	Lens to lens optical test results average insertion loss (dB) relative to pre-conditioning reference measurements
		Channel 1 to 24
Thermal shock (-40°C / 60°C)	5	Reference measurement
DTC - 40°C/125°C	100	+6.47 (disregarded)
	500	+1.72
	1000	-0.68

VI. CONCLUSION

A polymer lens array attach process on OMCM laminates using UV tacking and sidefill adhesive was developed to build a Deep Thermal Cycle reliability DOE. UV tacking consists of attaching the 5 mm X 3.5 mm X 1mm lens array to the OMCM laminate using two small 1 mm diameter dots of adhesives. The lens array is then sidefilled by dispensing a high viscosity adhesive along the lens array's periphery. The sidefill adhesive must not flow underneath the lens array 10 microns gap or it will contaminate the laminate optical features (mirrors). After being screened for good adhesion and low infiltration in small gaps with different test protocols, a list of 7 lens attach adhesives with both epoxies and silicones were chosen to build a 22 test cell DTC reliability DOE running at -40°C/125°C for 1000 cycles. Following the reliability stresses only the cells using a combination of soft silicones for lens UV tacking and sidefill adhesives successfully passed the microscope 40X inspection with no visual mechanical defects. These same cells using, only the soft silicone adhesive, also had no significant degradation in the optical signals after the reliability stresses. The softer silicone lens attach adhesives were able to absorb the reliability stresses through their elastic deformation. All test cells using epoxy adhesives for UV tacking and/or sidefill failed and showed severe mechanical defects at 250 DTC cycles. All epoxy lens attach adhesives proved to be too rigid. Lower CTE epoxy adhesives transferred too much stress to the higher CTE lens array which resulted in lens array corner cracks. Higher CTE epoxy adhesives transferred too much stress to the lower CTE laminate which resulted in different types of delaminations.

ACKNOWLEDGMENT

The authors would like to thank François Arguin, Evelyne Déragon, Lise Brault, Yan Thibodeau and Sébastien Gouin from the IBM opto-electronics team for their work and support in developing the lens attach process and optical test.

REFERENCES

[1] M. Tokunari et al., "Assembly and Demonstration of High Bandwidth-Density Optical MCM," in Proc. 65th Electronic Components and Technology Conf., San Diego, CA, May. 2015, pp. 799-803.

[2] K.Masuda et al., "High Density Micro-lens Array Connector for Optical Multi-chip Module," Proc. 66th Electronic Components and Technology Conf., Las Vegas, NV, June 2016, pp. 2317-2322.

[3] M. Tokunari et al., "Optoelectronic Chip Assembly Process of Optical MCM", in Proc. 67th Electronic Components and Technology Conf., Orlando, FL, June 2017, pp. 545-550.

Effects of In and Zn Double Addition on Eutectic Sn-58Bi Alloy

Shiqi Zhou, Yu-An Shen and Hiroshi Nishikawa

Joining and Welding Research Institute, Osaka
University
Osaka, Japan
charleszhou1992@gmail.com

Tiffani Uresti, Vasanth C. Shunmugasamy and
Bilal Mansoor
Mechanical Engineering Program, Texas A&M
University at Qatar
Doha, Qatar

Abstract— The effects of 0.5 wt. % In as well as 0.5 wt. % In and 1 wt. % Zn double (In & Zn) additions to eutectic Sn58Bi alloy on the microstructure and mechanical properties were studied before and after thermal aging. Newly designed In & Zn-added Sn58Bi alloy showed much finer microstructure than eutectic Sn58Bi and In-added Sn58Bi alloys. The elongation improvements of 36 % and 41 % before and after 1008 h aging were obtained in In & Zn-added Sn58Bi alloy compared to eutectic Sn58Bi alloy. In induced solid solution softening (SSS) effect on Sn phase was revealed by nanoindentation tests. A hardness decrease and a large creep displacement were obtained in both In- and In & Zn-added Sn58Bi. The effects of Zn and In were combined responsible for the elongation improvement of In & Zn-added Sn58Bi.

Keywords-Sn58Bi; Zinc; Indium; nanoindentation; hardnes; creep

I. INTRODUCTION

Recently, lead-free solder alloys have been widely used in the electronics industry. Solder alloys, such as Sn–Ag–Cu and Sn–Cu based alloys, have been firmly developed [1]. High reflow temperature can damage many temperature-sensitive components, such as LEDs. In addition, thermal warpage has become a serious issue for three-dimensional integrated circuit packaging and step soldering assembly. Therefore, low-temperature solder alloys are of great demand [2–4].

With a melting point of 139 °C, eutectic Sn58Bi (wt. %) alloy is considered potentially promising as a low-temperature lead-free solder alloy [3]. However, the brittleness of Sn58Bi alloy has to be solved. During solid-state thermal aging, a coarsened Bi grain can form owing to the Ostwald ripening of Bi particles [3]. In addition, because of the intrinsic brittleness of Bi, a coarsened microstructure deteriorates the mechanical properties [5]. It is reported that an approximately 3 wt. % Bi solid solution in the Sn phase can significantly increase the hardness of the Sn phase, thereby degrading the ductility of eutectic Sn58Bi alloy [6,7]. Refining the microstructure and reducing the hardness of Sn and Bi phases are of critical concerns in eutectic Sn58Bi alloy.

Some studies focusing on the two issues have been published. It is reported that the addition of Indium (In) to eutectic Sn58Bi alloy improves elongation effectively [3,6]. Since In solid solutes in Sn phase counteracts the strengthening effect of Bi solution, the softened Sn phase enhances the elongation of the solder alloy [6]. In addition, McCormack et al. found that a small amount of Ag addition improved the elongation of Sn55Bi alloy by three times, owing to the refined microstructure [8].

In our previous study, a refined microstructure in eutectic Sn58Bi with 0.5 and 1 wt. % Zn additions was achieved [9]. The diffusion of Sn and Bi atoms through the Sn–Bi phase boundaries was hindered by segregated Zn on Sn-Bi phase boundary during aging, thereby refining the microstructure. As a result, the elongation and ultimate tensile strength (UTS) showed higher values compared to eutectic Sn58Bi alloy before and after aging. This microstructure refinement method is considered very effective for eutectic Sn58Bi alloy. However, the elongation decreases during aging, because the Zn content decreases in the Bi phase because of Zn segregation, and, therefore, the effectiveness of SSS is predicted decreases.

In order to address this issue, in the present work, the effects of In & Zn additions on the microstructure and mechanical properties of eutectic Sn58Bi alloy were investigated. We designed Sn58Bi0.5In1Zn alloy. The authors believe further improved mechanical properties and thermal reliability will be achieved by combining the softening effect of In and microstructural refinement effct of Zn.

Additionally, based on the nanoindentation (NI) test result, the SSS phenomenon of In as well as In & Zn additions are discussed in detail to supplement our previous study [9]. Also, the creep behaviors of the Sn and Bi phases in each alloy were discussed based on the load–displacement curves (p–h curves) of the NI test.

II. EXPERIMENTAL

1 wt. % Zn (99.99 %) and 0.5 wt. % In (99.99 %) were alloyed with commercially available eutectic Sn58Bi alloy to produce new alloys, i.e., Sn58Bi0.5In and Sn58Bi0.5In1Zn. The alloying process included two steps. First, precise amounts of starting materials were melted in-house at 700°C for 5 h. Next, the newly synthesized alloys were subsequently remelted at 250 °C for 1 h before casting. A bar-shaped mold was used for casting. Some of these bar-shaped alloys were

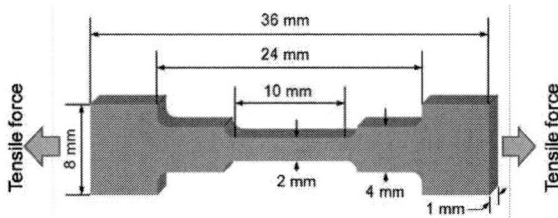

Figure 1 Schematic illustration of tensile test sample

then machined into tensile test specimens, as shown in Figure 1. The detailed alloy fabrication processes in this study were very similar to those described in our previous study [9].

The thermal reliability of the alloys was examined using a solid-state aging test for 504 h and 1008 h in an oil bath at 80 °C. The oil bath was used to prevent potential oxidation of the alloys during the high-temperature storage.

To quantify the magnitude of this microstructural change during thermal aging, the eutectic spacing and number of coarsened Bi grains were measured and calculated using an Image-pro Plus software. Three scanning electron microscope (SEM) images (dimensions: 250 μm × 200 μm) with 50 randomly selected eutectic spacings in each image were used to calculate the average eutectic spacing for each alloy. Ten SEM images (dimensions: 250 μm × 200 μm) were randomly chosen in each alloy for counting the coarsened Bi grains. A coarsened Bi grain normally has a rectangular shape [10]. In addition, we noticed that the size of a coarsened Bi grain in these 2D SEM images was usually larger than a 10 μm diameter circle. It was difficult to identify a coarsened Bi grain with a diameter smaller than 10 μm from a Bi phase in a eutectic structure; thus, we decided to base the counting on the two size categories: between 10 μm and 20 μm (small size) and larger than 20 μm (large size).

A field-emission SEM (FESEM, Hitachi SU-70) was used to observe the microstructure of the alloys and fracture surfaces after tensile tests. In addition, elemental composition and distribution were obtained using elemental-mapping, elemental-line and elemental-point analyses on a JEOL JXA-8530F field-emission electron probe microanalyzer (EPMA).

Tensile tests was performed using a Shimadzu Autograph AG-X machine at room temperature of 25 °C and under a strain rate of 0.0005/s.

NI test results were used to study the mechanical properties of individual Sn and Bi phases in different alloys. NI tests were conducted on a Hysitron Nano-Indenter TI Premier at room temperature on the as-cast alloys. A Berkovich indenter with a tip radius of 100 nm and a total included angle of 142.3° was used to perform the indents. The well-known Oliver–Pharr method was employed in the tests by applying a load function with a peak load of 300 μN and a loading time of 15 s. The indenter was then held at a peak load for 20 s before it was completely withdrawn from the specimen with 5 s unloading time. The peak load applied to the specimen was kept small to ensure that each indent was within a single phase. Data from the NI tests were analyzed

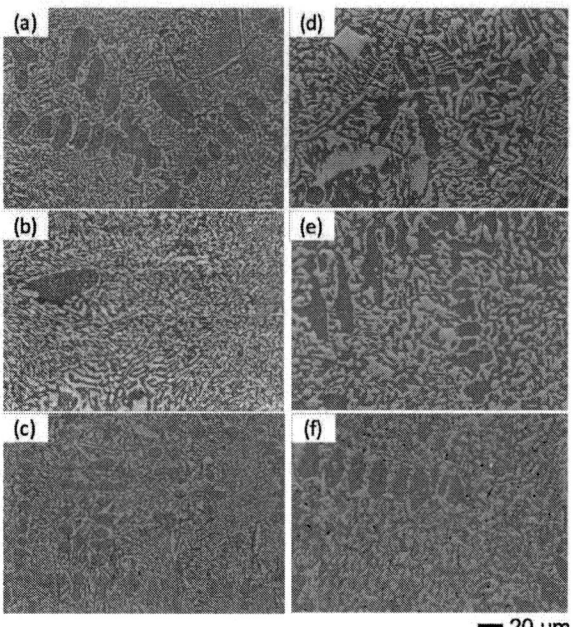

— 20 μm

Figure 2 Cross-sectional microstructure of as-cast (a) eutectic Sn58Bi, (b) Sn58Bi0.5In and (c) Sn58Bi0.5In1Zn. (d-f) are the corresponding microstructure after 1008 h aging.

and exported through the analysis tab, where a summary of information was obtained (Reduced Modulus, Hardness, etc.) for each indentation made on the sample.

III. RESULT AND DISCUSSION

A. Microstructure and Mechanical Properties Change Before and After Aging

Microstructures of as-cast eutectic Sn58Bi, Sn58Bi0.5In, and Sn58Bi0.5In1Zn alloys were shown in Figure 2a-c. All three alloys consisted of the bright-color Bi phase and dark-color Sn phase, which showed as interlocked lamellar structure. Additionally, some needle-shape Zn flakes were observed in Sn58Bi0.5In1Zn alloy. Sn58Bi0.5In1Zn contains an obviously finer microstructure than those of other two alloys. By quantify the fineness of microstructure, eutectic spacing was calculated, as shown in Table 1. By comparing with the eutectic spacing of Sn58Bi1Zn from our previous study [9], Sn58Bi0.5In1Zn had a similar value both before and after aging.

Figure 2d-f show the microstructure after thermal aging. The finer microstructure in the Sn58Bi0.5In1Zn alloy was

TABLE 1 EUTECTIC SPACING OF TESTED ALLOYS (μm)

Aging time (h)	Composition (wt.%)			
	Eutectic Sn58Bi	Sn58Bi1Zn	Sn58Bi0.5In	Sn58Bi0.5In1Zn
0	4.5	2.1	3.8	2.5
504	8.1	4.1	8.2	4.9
1008	8.2	4.2	9.1	4.0

TABLE 2 THE ELEMENT COMPOSITION IN SN AND BI
PHASES IN TESTED ALLOYS DURING AGING

| Phase | Element | Composition of each phase in Zn added Sn58Bi (wt. %) | | | |
| | | Sn58Bi0.5In | | Sn58Bi0.5In1Zn | |
		As-cast	1008h	As-cast	1008
Sn	Sn	94.00	93.52	93.46	92.75
	Bi	4.47	5.05	4.34	5.33
	Zn	-	-	0.66	0.53
	In	1.53	1.43	1.54	1.39
Bi	Sn	0.17	0.16	0.24	0.20
	Bi	99.83	99.84	99.37	99.79
	Zn	-	-	0.39	0.01
	In	0	0	0	0

Fracture mode was shown in Figure 4. Two distinguished phases with a cleavage fractured Bi phase and a bulge fractured Sn phase were observed in eutectic Sn58Bi. By contrast, in In- and In & Zn-added Sn58Bi, bulge fracture occupied almost entire fracture surfaces. After 1008 h aging, for eutectic Sn58Bi, the Sn and Bi phases in the fracture surfaces became much easier to distinguish, due to the coarsened microstructure. By contrast, the fractography of Sn58Bi0.5In1Zn remained unchanged after aging.

In order to reveal the mechanism of the increased elongation, we believe the study via a microcosmic demonstration is necessary. For eutectic Sn58Bi based alloy system, its microstructure was mainly composed of a binary eutectic interlocked Sn-Bi structure. Thus, the mechanical properties of individual Sn and Bi phases have great impacts on the mechanical performance of a whole solder bulk. However, the study in this aspect is rarely seen. Therefore, in this study, nanoindentation tests was implemented to exam the individual Sn and Bi phases of Sn58Bi0.5In and Sn58Bi0.5In1Zn alloy. The newly obtained hardness, creep displacement data sets were discussed and the possible reason for elongation improvement was proposed in the C section.

B. Zn Segragation on Sn-Bi Phase Boundary in Sn58Bi0.5In1Zn Alloy

The element distribution in Sn58Bi0.5In and Sn58Bio.5In1Zn after aging was shown in Figure 5. In remained as a solid solution in the Sn phase for both alloy. In contrast, Zn segregated on the Sn–Bi boundaries in Sn58Bi0.5In1Zn. The Zn concentration peaks on the Sn-Bi boundary as an evidence was detected by elemental-line analysis, which was shown in Figure 6, while In did not have this phenomenon.

To confirm this result, EPMA elemental-point result was conducted and shown in Table 2. In Sn58Bi0.5In, the compositions of Bi and In solid solute in Sn phase remained stable during thermal aging. In Sn58Bi0.5In1Zn, Bi, In, and Zn solid solutions in Sn phase remained constant during thermal aging as well. By contrast, the amount of Zn solid

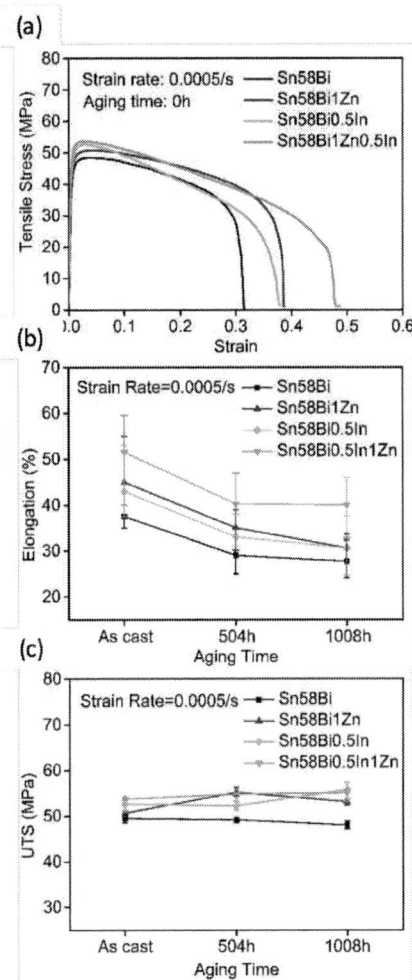

Figure 3 (a) SS curves of as-casted alloy. (b) elongation and (c) UTS before and after aging of tested alloys.

observed. A large volume fraction of coarsened lamellar structure was observed in both eutectic Sn58Bi and Sn58Bi0.5In alloy. Several coarsened Bi grains were found in eutectic Sn58Bi alloy after 1008 h aging, while no sign in Sn58Bi0.5In1Zn alloy. The eutectic spacing was measured and the results are shown in Table 1. During aging, the eutectic spacing grew continually for eutectic Sn58Bi and Sn58Bi0.5In alloy. In contrast, a slightly increased eutectic spacing was observed for Sn58Bi0.5In1Zn.

Because of the refined microstructure, Sn58Bi0.5In1Zn showed the highest elongation and UTS values among all the tested alloys throughout the aging period, as shown in Figure 3. Compared to eutectic Sn58Bi, Sn58Bi0.5In1Zn exhibited an elongation improvement of approximately 36 % before aging and 41 % after 1008 h aging. Sn58Bi0.5In and Sn58Bi0.5In1Zn had a similar UTS level at approximately 54 MPa. The UTS of eutectic Sn58Bi was the lowest. For reference, Sn58Bi1Zn and Sn58Bi0.5In had a similar mechanical performance.

978-1-7281-1500-9/19 $31.00 © 2019 IEEE

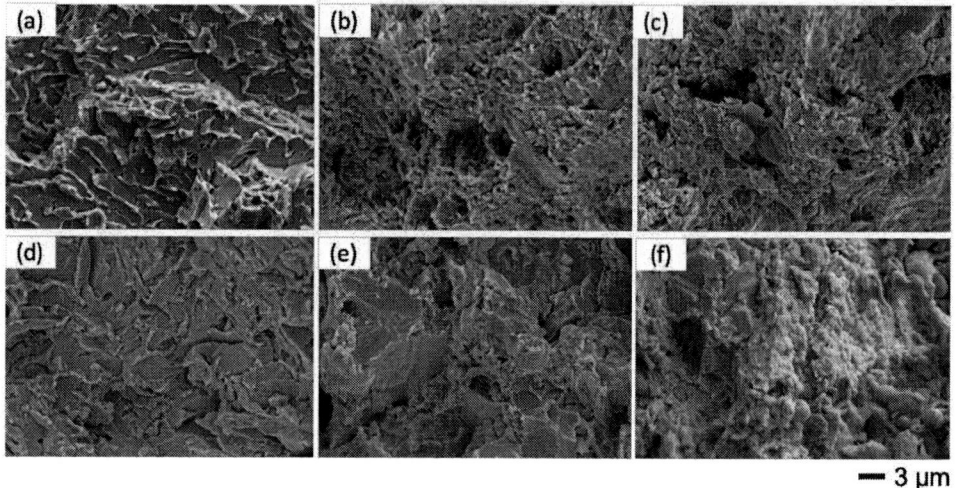

Figure 4 Top view of fracture surfaces of as-cast (a) eutectic Sn58Bi, (b) Sn58Bi0.5In and (c) Sn58Bi0.5In1Zn. (d-f) are the corresponding images of alloys thermal aged 1008 h.

solution in the Bi phase decreased. The Zn solution segregated to Sn–Bi phase boundary. This result was consistent with the outcome of our previous study.

C. SSS in Sn Phase Induced by In Solid Solution

Hardness result indicated that hardness of the Sn phase in Sn58Bi0.5In decreased dramatically to just below 0.5 GPa from 1.71 GPa in eutectic Sn58Bi, as shown in Table 3. This decreased Sn phase hardness was responsible for the elongation improvement of Sn58Bi0.5In bulk, which could be explained by the theory of SSS. The fundamental mechanism this phenomenon still remains unknown, but the study [6] assumed that the In solid solution in Sn phase

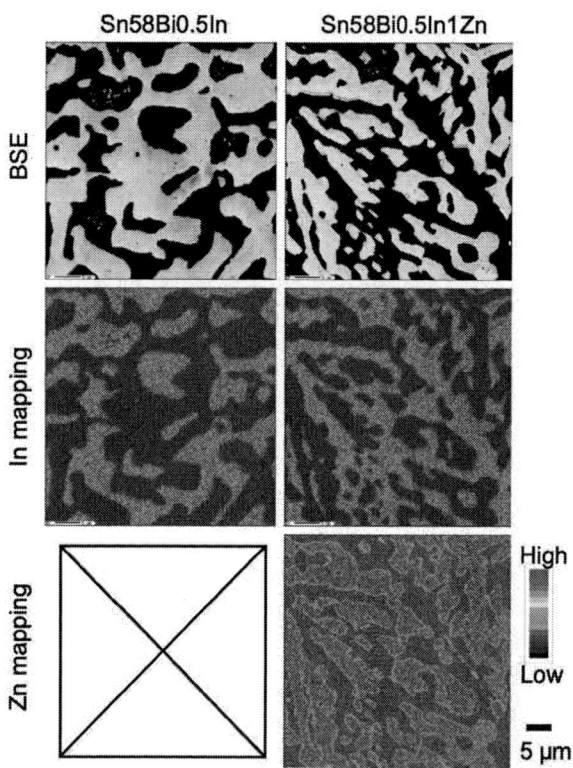

Figure 5 EPMA mapping of 1008 h thermal aged Sn58Bi0.5In and Sn58Bi0.5In1Zn.

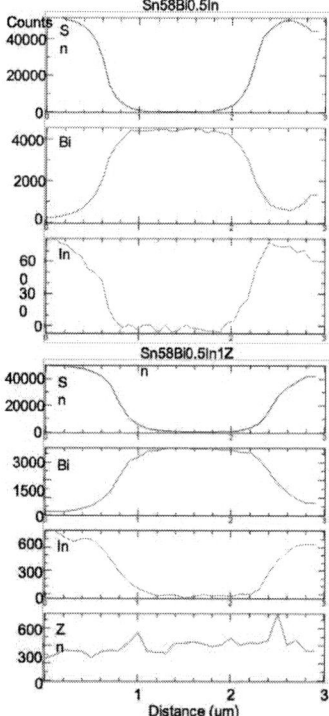

Figure 6 EPMA line analysis in a Sn58Bi0.5In and Sn58Bi0.5In1Zn by crossing two Sn-Bi phase boundaries.

TABLE 3 THE HARDNESS OF SN AND BI PHASES IN TESTED ALLOYS BEFORE AGING

Phase	Hardness (GPa)		
	Sn58Bi	*Sn58Bi0.5In*	*Sn58Bi0.5In1Zn*
Sn	1.71	0.45	0.61
Bi	0.95	0.41	0.75

TABLE 4 THE CREEP DISPLACEMENT IN SN AND BI PHASES IN TESTED ALLOYS BEFORE AGING

Phase	Creep displacement Δh (nm)		
	Sn58Bi	*Sn58Bi0.5In*	*Sn58Bi0.5In1Zn*
Sn	20.4	43.8	24.3
Bi	10.5	26.8	25.1

counteracts with the Bi phase and results in a reduced Sn phase hardness.

It has been reported that SSS can be manifested by yield strength decrease [11,12]. Thus, the relation between hardness and yield strength should be established. Yang at el. studied that for Sn−Bi-based alloy, the ratio of H_V and the yield strength δ_y is strongly follows a linear dependence [5], as shown in Equation (1).

$$H_V > 3 * \delta_y \qquad (1)$$

In Sn58Bi0.5In, because the hardness of the Sn phase decreased, the yield strengths based on Equation (1), therefore, decreased as well. As a result, the SSS induced by In solid solution on the Sn phases could be proved by the yield strength decrease. On the other hand, although in Sn58Bi0.5In1Zn, the hardness of both Sn and Bi phase was slightly higher than that in Sn58Bi0.5In, they were still lower than those in eutectic Sn58Bi. This means that the SSS effect by the In and Zn double addition was still effective in Sn58Bi0.5In1Zn.

In addition to the hardness discussion, it has been reported that the intrinsic mechanism of SSS is the interactions between substitutional solute atoms and dislocations, represented by the increased dislocation mobility [11]. Dislocation creep under a constant stress is a suitable mechanism to reveal dislocation mobility in a crystal [13,14]. Creep displacement Δh during a holding time of 20 s was tested, as shown in Table 4. A larger creep displacement means that the dislocation glides easier in the crystal. For Sn58Bi0.5In, the Δh of the Sn and Bi phases was the largest among the tested alloys, indicating that plastic deformation or dislocation movement became much easier in a In added alloy. On the other hand, the yield strengths of the Sn and Bi phases in Sn58Bi0.5In was the lowest among the four alloys. By comparing the hardness and the creep displacement of In- and In & Zn-added Sn58Bi, similar phenomena were obtained. Therefore, the creep displacement was directly related to the SSS, and the creep result served as another evidence of SSS in the studied alloy system.

It is considered that SSS was response for the elongation improvement of Sn58Bi0.5In and Sn58Bi0.5In1Zn. It is noted that, for Sn58Bi0.5In, the elongation decreased rapidly after 504 h aging, because the microstructure coarsened in the Sn58Bi0.5In after 504 h aging. The elongation was deteriorated even though the SSS provided by In solid solution was promising on the alloy. By contrast, owing to the fine microstructure controlled by Zn segregation on the Sn−Bi boundary, the SSS induced by In in Sn58Bi0.5In1Zn

can be maintained during aging. Because of the two different diffusion behaviors of In and Zn solutes during aging, the elongation and UTS performance of Sn58Bi0.5In and Sn58Bi0.5In1Zn were different. This result suggests that the effects of In and Zn were well combined in the In & Zn-added Sn58Bi alloy.

IV. CONCLUSION

In this study, Sn58Bi0.5In and Sn58Bi0.5In1Zn alloys were designed. The results show that the minor In and Zn addition has potential in solving the brittle issue of eutectic Sn58Bi alloy.

Compared to eutectic Sn58Bi, the Sn58Bi0.5In1Zn had finer microstructure, i.e., smaller eutectic spacing and fewer coarsened Bi grains, due to the Zn segregation phenomenon.

Elongation improvements of 36 % and 41 % were obtained in the Sn58Bi0.5In1Zn before and after 1008 h aging.

The softening of the Sn phase induced by the In solid solution was proposed in Sn58Bi0.5In and Sn58Bi0.5In1Zn. Quantitative hardness and creep data of individual Sn and Bi phases in each alloy were obtained.

The fine microstructure induced by the Zn addition as well as the In solid solution induced SSS were combined responsible for the elongation improvement and mechanical thermal reliability in Sn58Bi0.5In1Zn.

REFERENCES

[1] M. Abtew, G. Selvaduray, Lead-free Solders in Microelectronics, Mater. Sci. Eng. R Rep. 27 (2000) 95–141.

[2] K.M. Kumar, V. Kripesh, L. Shen, A.A.O. Tay, Study on the microstructure and mechanical properties of a novel SWCNT-reinforced solder alloy for ultra-fine pitch applications, Thin Solid Films. 504 (2006) 371–378.

[3] O. Mokhtari, H. Nishikawa, Correlation between microstructure and mechanical properties of Sn-Bi-X solders, Mater. Sci. Eng. A. 651 (2016) 831–839.

[4] E.E.M. Noor, N.M. Sharif, C.K. Yew, T. Ariga, A.B. Ismail, Z. Hussain, Wettability and strength of In-Bi-Sn lead-free solder alloy on copper substrate, J. Alloys Compd. 507 (2010).

[5] L. Yang, W. Zhou, Y. Ma, X. Li, Y. Liang, W. Cui, P. Wu, Effects of Ni addition on mechanical properties of Sn58Bi solder alloy during solid-state aging, Mater. Sci. Eng. A. 667 (2016) 368–375.

[6] X. Chen, F. Xue, J. Zhou, Y. Yao, Effect of in on microstructure, thermodynamic characteristic and mechanical properties of Sn-Bi based lead-free solder, J. Alloys Compd. 633 (2015) 377–383.

[7] L. Shen, P. Septiwerdani, Z. Chen, Elastic modulus, hardness and creep performance of SnBi alloys using nanoindentation, Mater. Sci. Eng. A. 558 (2012) 253–258.

[8] M. Mccormack, H.S. Chen, G.W. Kammlott, S. Jin, Significantly improved mechanical properties of Bi-Sn solder alloys by Ag-doping, J. Electron. Mater. 26 (1997) 954–958.

[9] S. Zhou, O. Mokhtari, M.G. Rafique, V.C. Shunmugasamy, B. Mansoor, H. Nishikawa, Improvement in the mechanical properties of eutectic Sn58Bi alloy by 0.5 and 1 wt.% Zn addition before and after thermal aging, J. Alloys Compd. (2018).

[10] O. Mokhtari, H. Nishikawa, Effects of in and Ni addition on microstructure of Sn-58Bi solder joint, J. Electron. Mater. 43 (2014) 4158–4170.

[11] Y. Huo, J. Wu, C.C. Lee, Solid solution softening and enhanced ductility in concentrated FCC silver solid solution alloys, Mater. Sci. Eng. A. 729 (2018) 208–218.

[12] V.P. Soldatov, V.D. Natsik, A.N. Diulin, G.I. Kirichenko, Low-temperature softening of β-tin single crystals on doping with substitutional impurities, Low Temp. Phys. 26 (2000) 160–168.

[13] X. Li, B. Bhushan, A review of nanoindentation continuous stiffness measurement technique and its applications, Mater. Charact. 48 (2002) 11–36.

[14] H. Ma, J.C. Suhling, A review of mechanical properties of lead-free solders for electronic packaging, J. Mater. Sci. 44 (2009) 1141–1158.

Microstructural Evolution in SAC+X Solders Subjected to Aging

Jing Wu, Jeffrey C. Suhling, Pradeep Lall
Center for Advanced Vehicle and Extreme Environment Electronics (CAVE[3]), and
Department of Mechanical Engineering
Auburn University
Auburn, AL 36849, USA
E-Mail: jsuhling@auburn.edu

Abstract—Aging effects are common in lead free solder joints within electronic assemblies that are exposed to isothermal environments for extended periods. Such exposures lead to evolution of the solder microstructure, which results in changes in the mechanical properties and creep behavior of the solder joints. These changes often lead to dramatic reductions in reliability of lead free electronic assemblies subjected to aging. In our recent investigations, we have been utilizing Scanning Electron Microscopy (SEM) to better understand aging induced degradations. In particular, our approach has been to monitor aging induced microstructural changes occurring within fixed regions in selected lead free solder joints and to create time-lapse imagery of the microstructure evolution. With such an approach, quantitative analysis of the microstructural changes can be performed, removing the limitations of many prior studies where aged and non-aged microstructures were taken from two different samples and could only be qualitatively compared.

In our recent paper presented at ECTC 2018, we illustrated the SEM approach for small fixed regions in SAC305 solder joint samples, and provided aging data for up to 1500 hours of aging. In the current study, we have extended this work to SAC_Q (SAC+Bi) alloys. In particular, microstructural evolution has been studied for several different regions from several different SAC_Q joints, and both short term (up to 100 hours) and long term (up to 2000 hours) aging of the joints have been performed. In all of the aging experiments, the microstructure evolutions were observed in solder joint samples exposed to isothermal conditions at T = 125 °C. The microstructures in several fixed regions of interest were recorded after predetermined time intervals of aging, which were 1 hour (up to 12 hours) and 10 hours (up to 100 hours) for the short term aging samples; and 250 hours (up to 2000 hours) for the long term aging samples. Using the recorded images and imaging processing software, the area and diameter of each IMC particle was tracked during the aging process. The measured data have been correlated with the results of our prior work on SAC305 microstructural evolution.

As expected, the quantitative analysis of the evolving SAC_Q microstructure showed that the particles coalesced during aging leading to a decrease in the number of particles. This caused an increase in the average diameter of the particles of slightly more than 100% for long term aging of 2000 hours. For SAC305, the average particle diameter was found to increase at three times the rate (increase of 300% after 2000 hours of aging). Thus, coarsening of IMC particles was greatly mitigated in the SAC_Q alloy relative to that observed in SAC305. Immediately after reflow solidification, Bismuth rich phases were present in the SAC_Q joints. During aging at T = 125 °C, the bismuth was observed to quickly go into solution both within the beta-Sn dendrites and in the intermetallic rich regions between dendrites. This resulted in solid solution strengthening of the lead free solder. It was also found that the aging-induced presence of bismuth in solution within the beta-Sn matrix provided an increased resistance to the Ostwald ripening diffusion process that coarsens the Ag_3Sn IMC particles. The combination of these two effects in the SAC+Bi alloy lead to greatly improved resistance to aging induced effects relative to the SAC305 solder alloy. Finally, we have compared the time dependent evolution of microstructure with the degradation in strength during aging for of the two solder alloys, and good correlations were observed.

Keywords: Solder, Aging, Microstructure, SEM, IMC, Intermetallic, Microstructural Evolution.

I. INTRODUCTION

The microstructures of popular SAC lead free solders are inherently unstable at relatively low temperatures, and continually evolve when utilized in electronic packaging assemblies. This evolution is primarily caused by diffusion phenomena resulting in coarsening of the intermetallic phases (Ostwald ripening). In addition, there can be coarsening of subgrains, breakdown of dendrite structures, as well as potential recrystallization at Sn grain boundaries. The microstructural changes lead to degradations of the constitutive and failure behaviors of lead free solder joints [1], and these combined effects are typically referred to as solder aging.

The recent literature contains a large selection of research on lead free solder aging. For example, aging leads to large reductions in bulk mechanical properties such as modulus and strength [2-5]. Other mechanical responses affected by aging include creep behavior [2-6], viscoplastic Anand model parameters [7-12], high strain rate mechanical behavior and drop reliability [13-14], ball shear strength [15], fracture behavior [16], uniaxial cyclic stress-strain curves and fatigue life [17-21], shear cyclic stress-strain curves and fatigue life [22-23], and thermal cycling reliability [6, 24-28]. In addition, tests on joints using nanoindentation have revealed aging induced degradations in modulus, hardness, and creep rate [29-33]. Several investigations [5, 12, 20-21, 31, 34] have also reported that the use of dopants (SAC+X), particularly bismuth (X = Bi), can help mitigate aging induced degradations in lead free SAC alloys.

978-1-7281-1500-9/19 $31.00 © 2019 IEEE

The effects of aging on the evolution of solder microstructure have been studied by several researchers, e.g. [35-54]. The primary changes are coarsening of the Ag_3Sn and Cu_6Sn_5 intermetallic compounds (IMCs) present in the eutectic regions between β-Sn dendrites. Several researchers [39, 43, 45-46, 50-54] have also proposed empirical models to describe the growth of IMC particles or layers as a function of aging temperature and aging time. In most studies on the effects of aging on solder microstructure, observations were made on two different solder joints (one non-aged and one aged). Comparisons made in this manner were qualitative in nature since the two microstructures were from different samples and could not be directly compared.

There are limited investigations that have involved the monitoring of evolving microstructure in a fixed region in a single solder sample [35, 40, 44, 51, 53]. Choi and Lee [35] made photographs of a growing intermetallic layer at the boundary between a Sn-Ag solder joint and a copper bonding pad. Ubachs, et al. [40] examined aging-induced solder microstructure evolution in a Sn-Pb sample, and compared their results to predictions made with two-dimensional diffusion simulations. Telang and coworkers [44] made before and after grain boundary maps in fixed regions in pure-Sn and Sn-Ag samples subjected to high temperature aging at 150 °C. Similarly, Kumar, et al. [51] and Shaym, et al. [53] observed changes to the IMC particles occurring within fixed regions in SAC105 and SAC305 solder joints exposed to aging at 150 °C for 100-300 hours. To obtain quantitative data and develop models, observations of a large set of IMC particles were made before and after aging in the latter study, and average particle size and subgrain size were measured and plotted [53].

Other changes in the SAC solder microstructure have also been observed to occur during aging, but they are not well characterized and their effects on mechanical behavior and reliability are not well understood. These include changes in size/shape of the β-Sn dendrites, growth of pure Sn subgrains within the β-Sn dendrites, migration of IMC particles to subgrain boundaries, changes in Sn grain orientation, etc.

In our recent research [55-57], we used Scanning Electron Microscopy (SEM) and Scanning Probe Microscopy (SPM) approaches to track microstructural evolution in small fixed regions in several different SAC305 joints. Results for both short term (50 hours) and long term (1500 hours) aging of the joints were presented, and quantitative data for the evolution of particle size and particle count were presented. In the current study, we have extended this work to SAC_Q (SAC+Bi) alloys. In particular, microstructural evolution has been studied for several different regions from several different SAC_Q joints, and both short term (up to 100 hours) and long term (up to 2000 hours) aging of the joints have been performed. In all of the aging experiments, the microstructure evolutions were observed in solder joint samples exposed to isothermal conditions at T = 125 °C. The microstructures in several fixed regions of interest were recorded after predetermined time intervals of aging, which were 1 hour (up to 12 hours)

and 10 hours (up to 100 hours) for the short term aging samples; and 250 hours (up to 2000 hours) for the long term aging samples. Using the recorded images and imaging processing software, the area and diameter of each IMC particle was tracked during the aging process. The measured data have been correlated with the results of our prior work on SAC305 microstructural evolution. In addition, we have compared the time dependent evolution of microstructure with the degradation in strength during aging for the two solder alloys.

II. EXPERIMENTAL PROCEDURE

The procedure reported in references [55-56] was used in this work. In brief, scanning electron microscopy was utilized to examine microstructural changes in lead free solder joints occurring due to aging. Both SAC305 and SAC_Q (SAC+Bi) solder joint samples were prepared using a nine-zone reflow oven and a typical lead free BGA temperature profile. Solder joint cross-sectional specimens were encapsulated in epoxy molds and polished using industry standard procedures. As shown in Figs. 1 and 2, small indentation marks were added to the cross-sections to facilitate subsequent locating of the same regions in the solder joints. Pyramidal idents were placed near the corners of the regions of interest using a nanoindentation system. As shown in Fig. 2, the indents were placed suitably away from the region of interest to minimize any influence on the microstructure evolution. In this case, the region of interest was approximately 60 x 40 µm, while the center-to-center spacing between the indentation marks was approximately 90 µm.

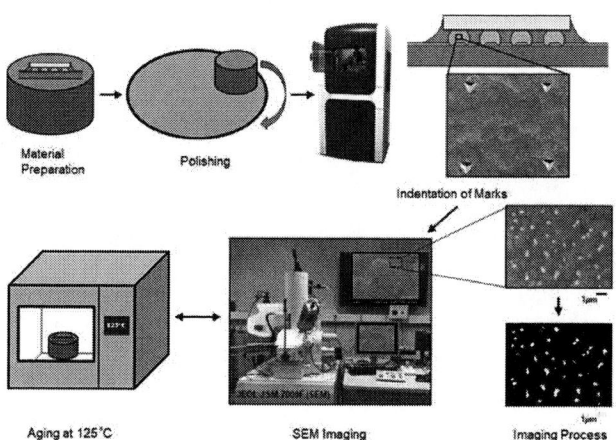

Figure 1. Flow Chart of Experimental Procedure
(Including JEOL JSM-7000F and Hysitron TI 950).

As discussed in refs. [55-56], the polished solder samples were aged at T = 125 °C for various durations. After each aging increment, the samples were placed in the SEM, and the regions of interest in the samples were located with the help of the indentation marks, and then digitally recorded. This process was repeated several times, yielding a photographic record of the microstructural evolution of each region. In the current work, several different regions from several different SAC_Q joints were examined, and both short term (up to 100 hours, with both 1 and 10 hour increments) and long term (up to 2000 hours with 250 hour increments) aging of the joints have been performed.

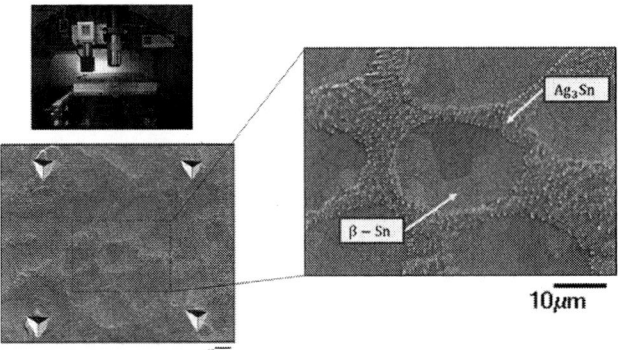

Figure 2. Example Region of Interest and Nanoindentation Markers.

III. RESULTS

A. Chemical Composition and Microstructure

The chemical composition of the SAC_Q alloy was explored with Energy Dispersive X-Ray Spectroscopy (EDX), and an average distribution of 3.3% bismuth (Bi), 3.41% silver (Ag), and 0.52% copper (Cu) was indicated. The chemical compositions of the two alloys are listed in 1 .

Table 1. Chemical Compositions of the Solder Alloys

Alloy	Sn	Ag	Cu	Bi
SAC305	96.50	3.00	0.50	0.00
SAC_Q	92.77	3.41	0.52	3.30

Typical microstructures for SAC305 and SAC_Q solder joints after reflow solidification are shown in Fig. 3. They typically consist of β-Sn dendrites surrounded by interdendritic eutectic regions incorporating a fine dispersion of Ag_3Sn and Cu_6Sn_5 intermetallic particles in β-Sn. The IMC particles provide pinning points that help mitigate and block dislocation movements. The inclusion of Bi in SAC_Q led to a dispersion of Bi rich phases, which appear a white regions in the SEM micrograph in Fig. 3(b) using back scattered electrons. As shown in the phase diagram in Fig. 4, bismuth has a high solid solubility in tin, which leads to an improved strength of the Sn matrix by the solid solution strengthening (SSS) mechanism. With an increase of temperature to approximately T = 50 °C, all 3.3% of the Bi will dissolve into solution with the Sn matrix. At T = 125 °C used for aging in this work, the solubility limit of Bi in Sn is approximately 17%.

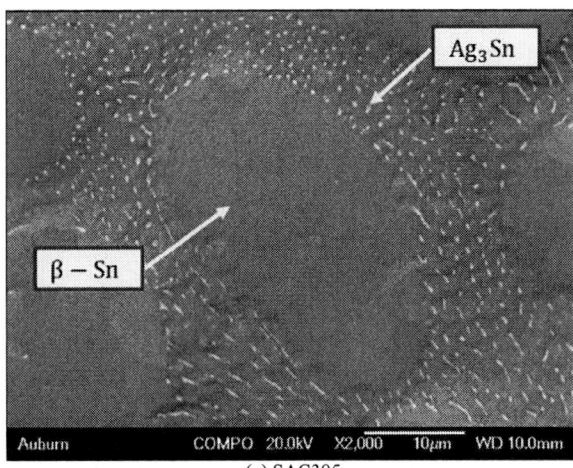

(a) SAC305

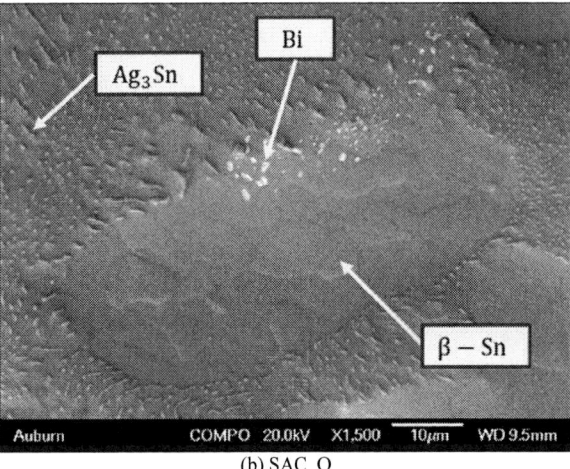

(b) SAC_Q

Figure 3. Microstructure of SAC305 and SAC_Q as Reflowed.

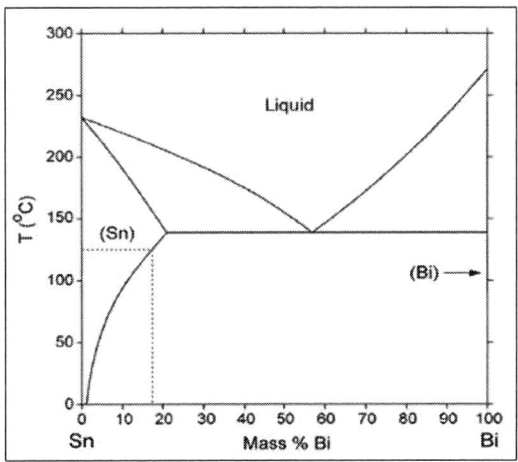

Figure 4. Sn-Bi Phase Diagram.
[http://www.metallurgy.nist.gov/]

B. Observations for Bismuth Rich Phases

Fig. 5 and Fig. 6 contain sets of images illustrating example microstructural evolutions of the bismuth rich phases in SAC_Q joints subjected to isothermal aging at T = 125 °C. As seen in Figure 4, the solid solution solubility of Bi increases from 1.8% at T = 25 °C, to about 17% at T = 125 °C. The images in Fig. 5(a) and Fig. 6(a) were taken immediately after reflow. They both exhibit a distribution of white Bi-rich phases throughout the lead free microstructure. EDS analysis of these white particles has confirmed that they are Bi rich phases with more than 80% Bi. During aging, the bismuth was observed to go into solution within the beta-Sn dendrites and also in the intermetallic rich regions between dendrites. The dissolving of the Bi-rich phases occurred in a short time. For example, the white phases in Fig. 5 disappeared in less than one hour of aging, while those in Fig. 6 were gone within 10 hours of aging. As discussed previously, this leads to strengthening of the solder during aging by the mechanism of solid solution strengthening.

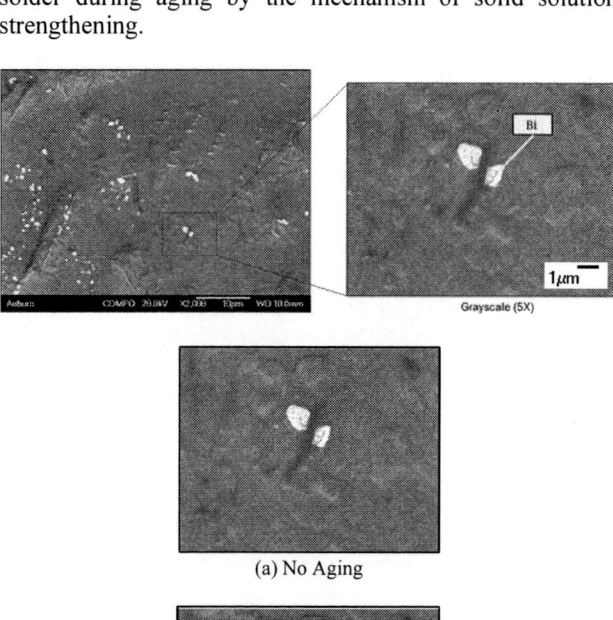

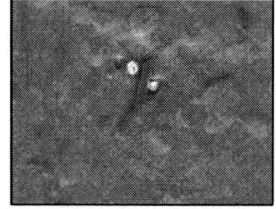

(a) No Aging

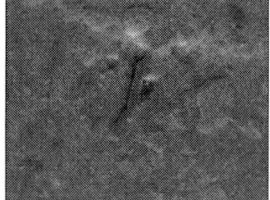

(c) 20 Minutes of Aging at 125 °C

(d) 40 Minutes of Aging at 125 °C

Figure 5. Dissolving of Bismuth Rich Phases During Aging for 40 Minutes.

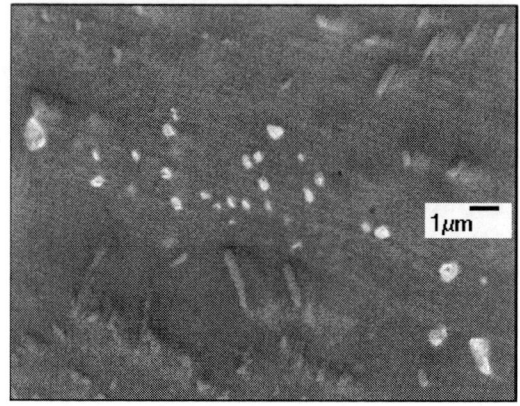

(a) No Aging

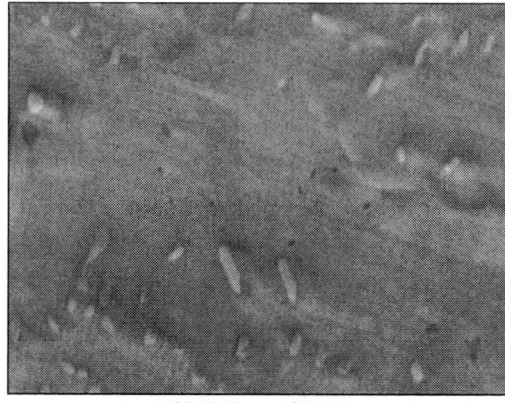

(b) 10 Hours of Aging

Figure 6. Dissolving of Bismuth Rich Phases During Aging for 10 Hours.

C. Observations for Short Term Aging

Figure 7 shows the observed microstructural evolution in one region of a SAC_Q sample for short term isothermal aging at 125°C for up to 12 hours. In this time lapse imagery, there is a one hour increment of aging between neighboring photos. As expected, some coarsening of the intermetallics occurred. This resulted in decreases in the number of IMC particles, and increases in the average particle size and average particle separation distance. In Fig. 7, blue circles show example positions where particles are growing, while red circles illustrate positions where the particles are decreasing in size and eventually disappear. Qualitatively, the changes observed in SAC_Q were fairly small relative to those found in SAC305 in our prior work [55-57]. Figure 8 contains zoomed-in high magnification views of the microstructure evolution seen in Fig. 7. In addition, Fig. 9 contains high magnification views of the evolution occurring in another region of a second joint. In this case, the total aging time was 100 hours, and photos were taken on 10 hour increments. In both Fig. 8 and Fig. 9, large aspect ratio particles tended to shift to more spherical shapes. There were also no significant changes observed in the size and shape of the dendrites.

978-1-7281-1500-9/19 $31.00 © 2019 IEEE

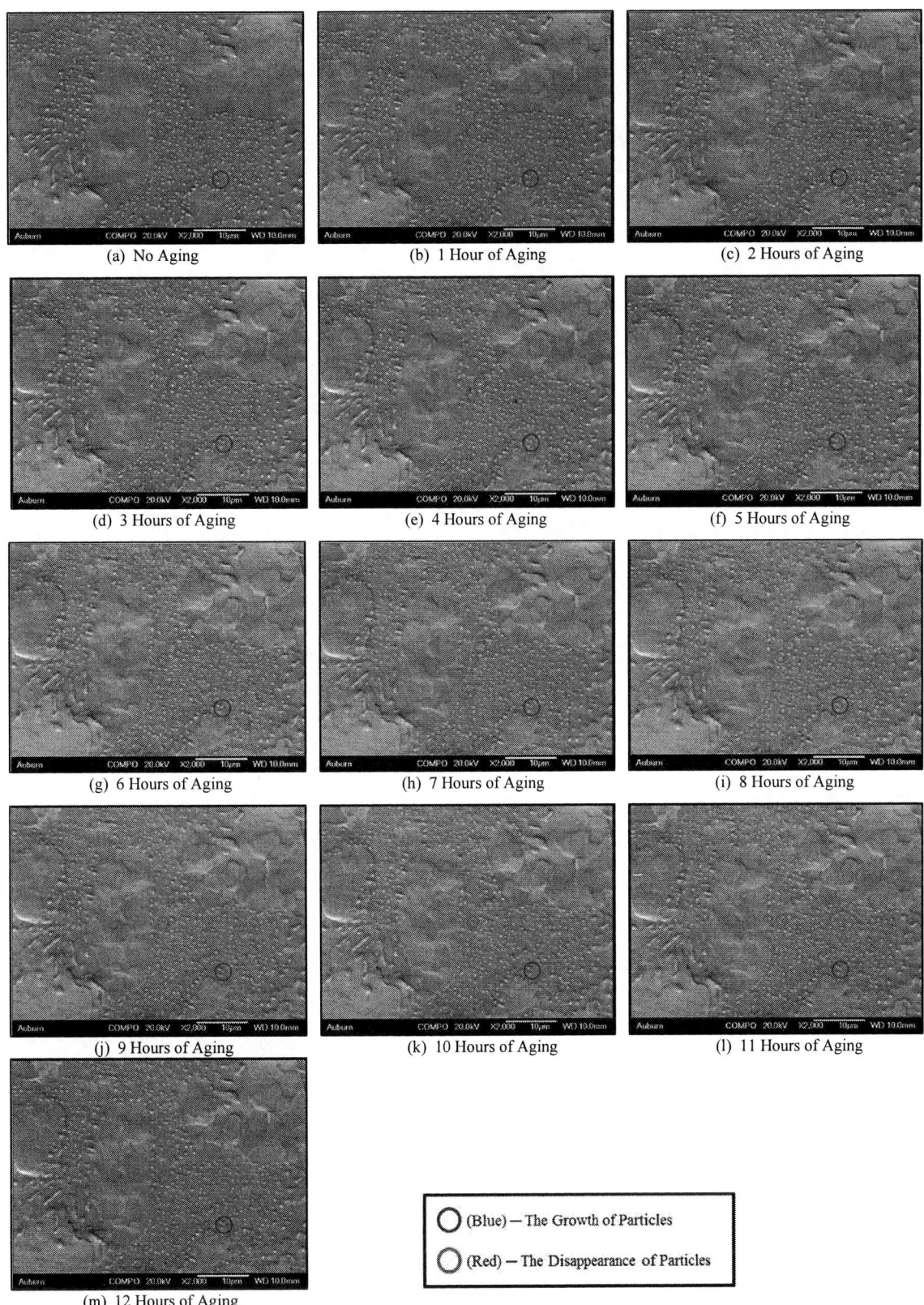

Figure 7. Example SAC_Q Short Term Microstructural Evolution (Aging at T = 125 °C).

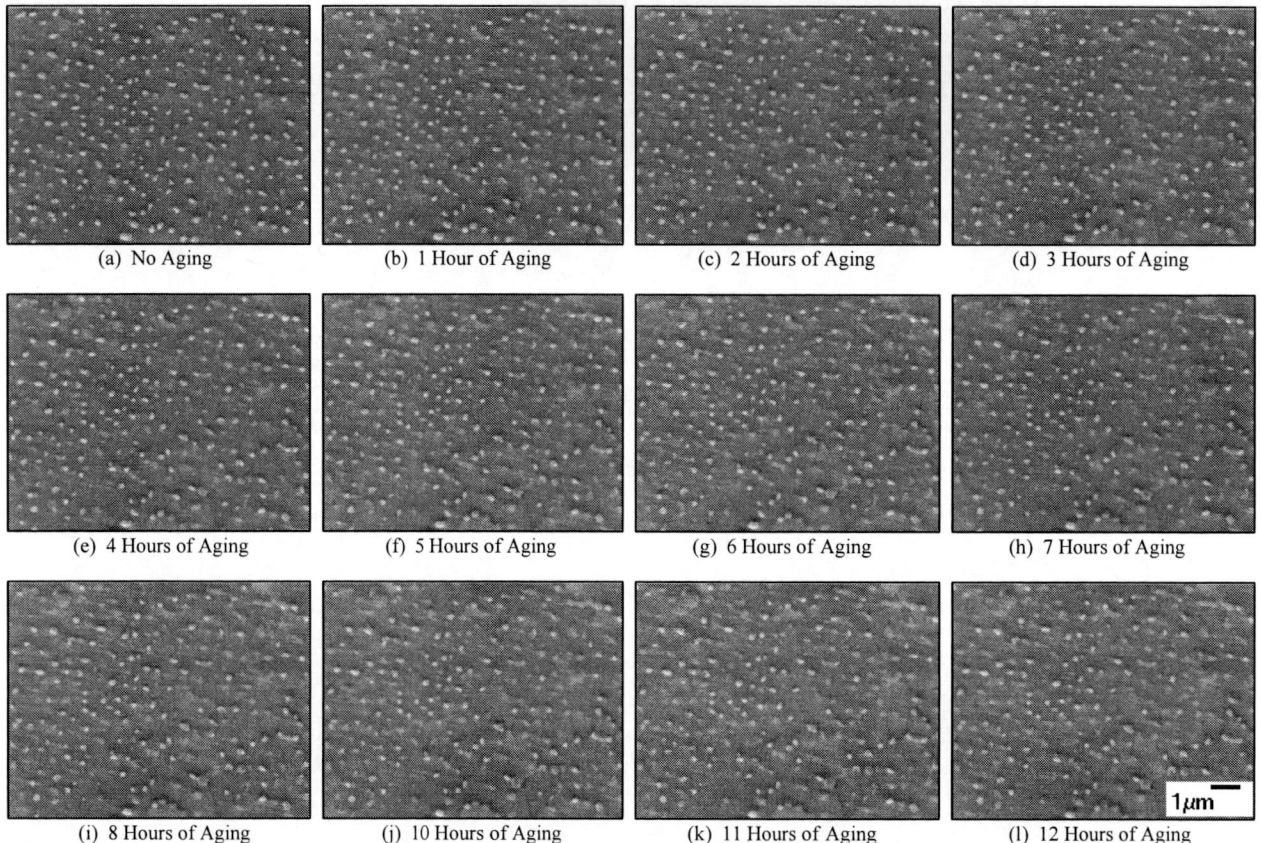

Figure 8. Example SAC_Q Short Term Microstructural Evolution over 12 Hours of Aging at T = 125 °C (Aging Increment: 1 Hour).

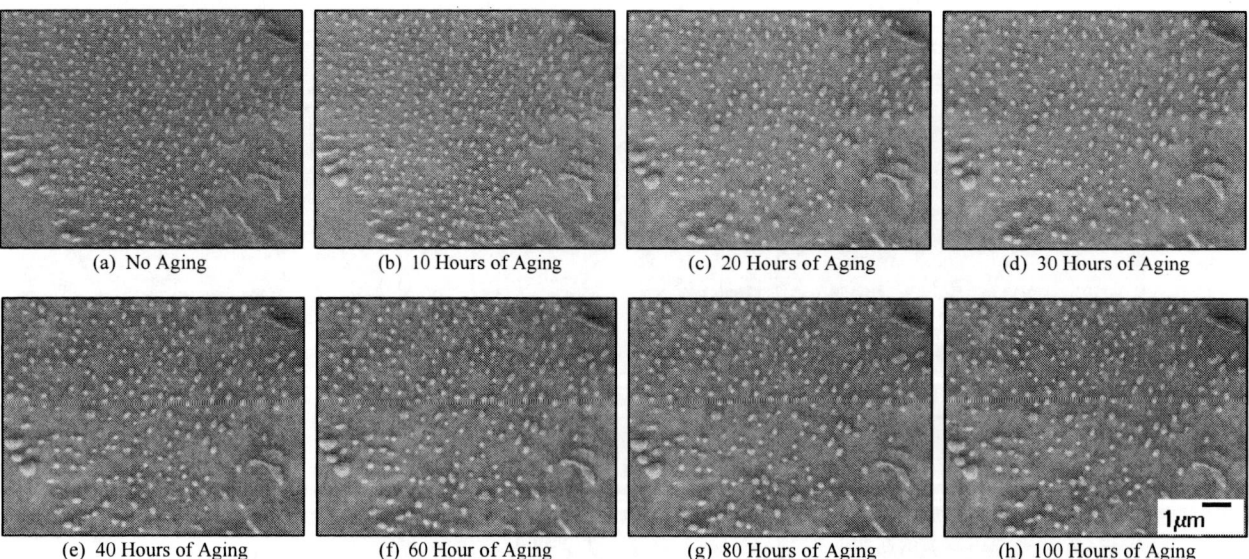

Figure 9. Example SAC_Q Short Term Microstructural Evolution over 100 Hours of Aging at T = 125 °C (Aging Increment: 10 Hours)

D. Observations for Long Term Aging

Figure 10 shows the observed microstructural evolution in one example interdendritic eutectic region from a SAC_Q sample subjected to long term isothermal aging at 125°C for up to 2000 hours. In this time lapse imagery, there is a 250 hour increment of aging between neighboring photos. Similar to our prior results for SAC305 in ref. [56], there is significant coarsening of the IMC particles during the first 250 hours of aging, while subsequent coarsening occurred at a very slow rate that is difficult to discern by casual observation.

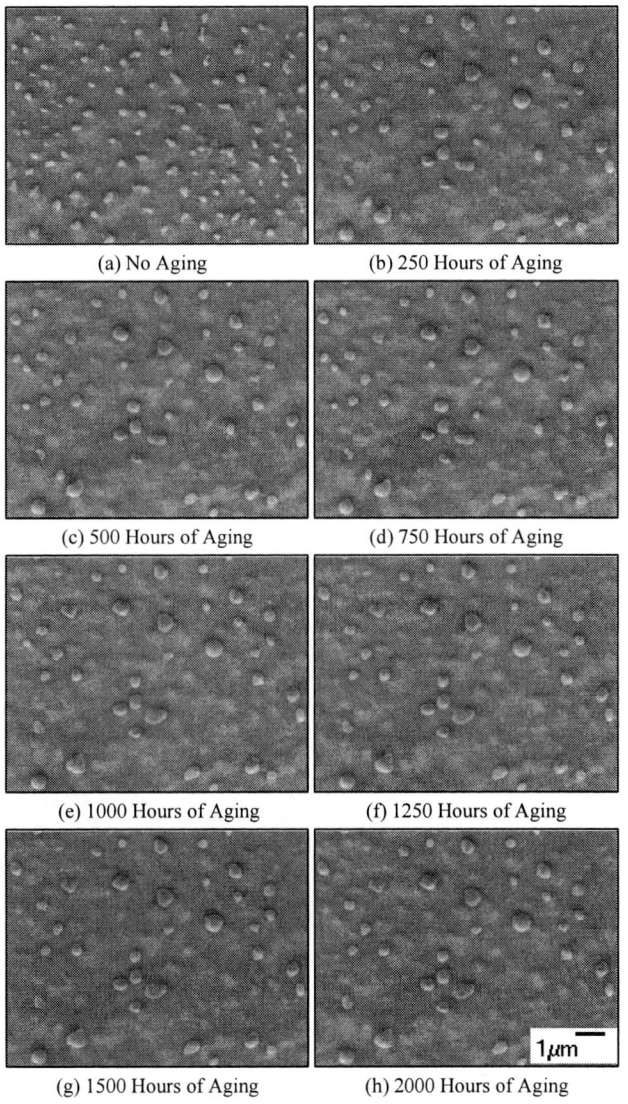

(a) No Aging   (b) 250 Hours of Aging

(c) 500 Hours of Aging   (d) 750 Hours of Aging

(e) 1000 Hours of Aging   (f) 1250 Hours of Aging

(g) 1500 Hours of Aging   (h) 2000 Hours of Aging

Figure 10. Example SAC_Q Long Term Microstructural Evolution over 2000 Hours of Aging at T = 125 °C (Aging Increment: 250 Hours)

E. Quantitative Analysis of Aging Induced Evolution

Quantitative analyses have been performed on the recorded SEM images to evaluate the size metrics for the IMC particle evolutions. The analyzed regions were typically chosen to be interdendritic regions with a heavy concentration of IMC particles. For example, the images shown in Fig. 8, were taken from a subregion of the images in Fig. 7, as shown below in Fig. 11. The area of each particle, the total area of all of the particles, and the total number of particles in each selected region and aging time were determined using image analysis software (ImageJ and Adobe Photoshop) and Matlab. These values were used to estimate the variation of the average particle diameter with aging time. For example, Fig. 12 shows the variation of the average particle diameter in SAC_Q for short term aging time up to 12 hours. In this plot, results for three different analyzed regions in the SAC_Q joint are shown, with one of them being the processed data from the images in Fig. 8. Similar results for three different SAC_Q regions with short term aging up to 100 hours and 10 hour measurement increments are shown in Fig. 13. In this case, one of the sets of data came from the images presented in Fig. 9. Finally, analogous results for three more SAC_Q regions in a joint subjected to long term aging up to 2000 hours and 250 hour measurement increments are shown in Fig. 14. In this case, one of the sets of data came from the images presented in Fig. 10. The red curves in Figs. 13, 14 and 15 indicate regression fits to the data presented in each plot using a two term exponential empirical model.

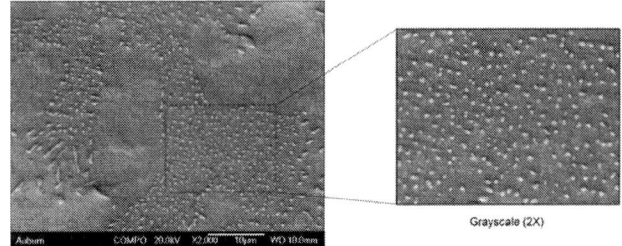

Figure 11. Sub-Region for Quantitative Analysis of Aging Effects on IMCs.

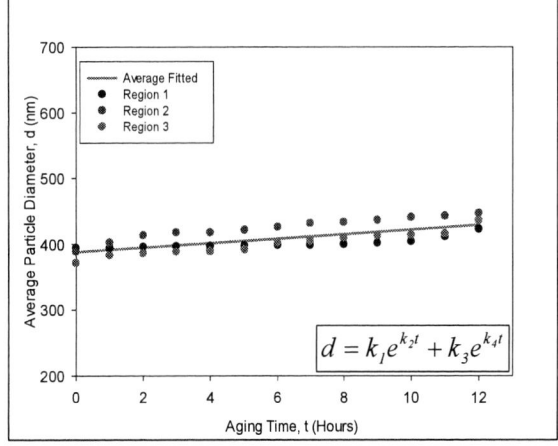

$$d = k_1 e^{k_2 t} + k_3 e^{k_4 t}$$

Figure 12. Variation of IMC Particle Diameter with Aging Time (Aging up to 12 Hours with 1 Hour Aging Increments).

The averaged data from Figs. 12-14 are plotted together in a single graph in Fig. 15 (includes both short term and long term aging data for SAC_Q). In this case, the particle diameter has been normalized using the initial value for each data set. Again, a two term exponential model has been used to fit the data. It is seen that the major changes in particle diameter occurred within the first 250 hours of aging (t < 250 hours). This matches our previous observations for the degradations in lead free solder mechanical behavior with aging [2-5, 29-33]. The IMC particle size continues to increase at a much slower but steady state rate for longer aging times (t > 250 hours). The results obtained from short and long term aging studies appear to be consistent with one another.

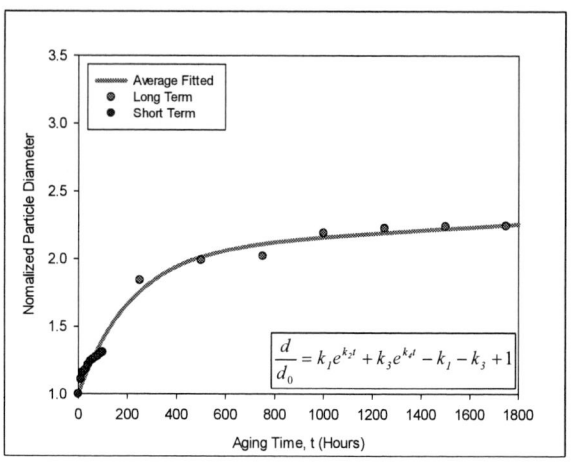

Figure 15. Variation of IMC Particle Diameter with Aging Time (Combined Short Term and Long Term Aging Data).

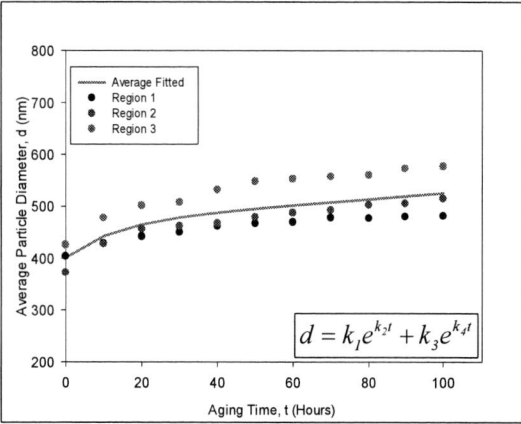

Figure 13. Variation of IMC Particle Diameter with Aging Time (Aging up to 100 Hours with 10 Hour Aging Increments).

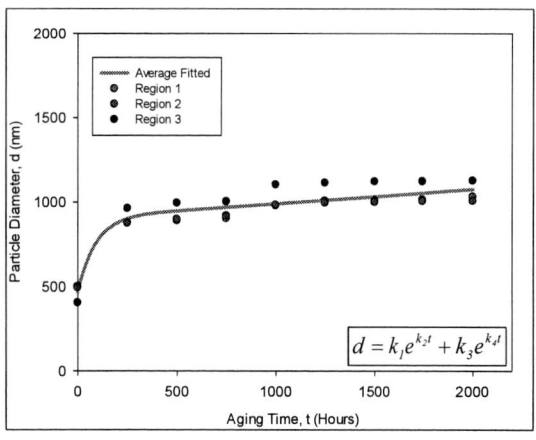

Figure 14. Variation of IMC Particle Diameter with Aging Time (Aging up to 2000 Hours with 250 Hour Aging Increments).

The quantitative analysis results reported in this work are based on coarsening phenomenon observed at the solder surface. The surface diffusion rates in the plane of the sample surface can be different than those in the sample bulk. The observations in question involve both bulk and surface diffusion phenomena. We are studying this further with additional experiments, and the results will be reported in a future publication.

F. Comparisons of IMC Evolution During Aging

As shown in Fig. 16, IMC particles tend to become rounder (spherical) with aging, and needle-shaped particles transition to more rounded shapes. During this diffusion based process, larger particles are produced at the expense of smaller particles through so-called Ostwald Ripening [58]. Such coarsening and coalescing of IMC particles during aging is known to play a critical role in the degradations of solder mechanical properties. IMC particles will pin and block the movement of dislocations. However, aging leads to both a smaller number of larger IMC particles, and increased spacing between the particles. This results in dislocations being able to pass more easily through the material, decreasing both the yield stress and strength. More detailed discussion can be found in references [55-58].

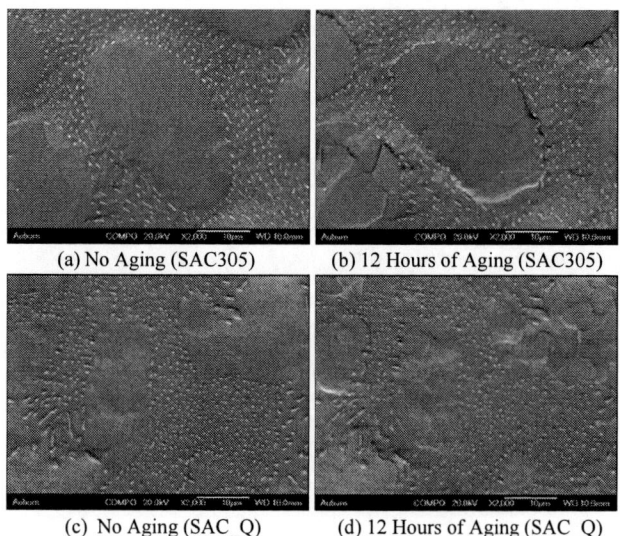

Figure 16. Aging Induced Coarsening of IMCs for SAC305 and SAC_Q (12 Hours of Aging at 125 °C).

During aging, intermetallic coarsening effects in SAC_Q are reduced relative to SAC305 because of adding bismuth. Figs. 17 and 18 show the coarsening of SAC305 and SACQ microstructures under similar aging conditions for aging at 125 °C for durations of 4 hours and 100 hours, respectively. Even though the samples were aged for only 4 hours, the microstructure of SAC305 in Fig. 17 coarsened significantly, while that for SAC_Q showed only slight changes. Similarly, the results in Fig. 18 visually demonstrate huge microstructural changes for SAC305, while only small changes in SAC_Q.

The observed variations of the average particle diameter for SAC305 and SAC_Q are compared in Fig. 19 for short term aging at 125 °C. In the first 12 hours of aging, the average IMC particle diameter increased by approximately 48.8% for SAC305 and only 9.6% for SAC_Q. A plot of the variations of the average normalized IMC particle diameter with aging time for both short term and long term aging of SAC305 and SAC_Q are shown in Fig. 20. This graph includes averaged data from this study for SAC_Q from Fig. 15, as well as all the data for SAC305 from reference [56]. For both alloys, the major changes in particle diameter occurred within first 250 hours of aging. After that point, the particle size was seen to increase at a slow but steady state rate. After 2000 hours of aging, the average IMC particle diameter increased by approximately 300% for SAC305 and approximately 120% for SAC_Q. Because they were formed using the same reflow profile, the initial IMC particles sizes were almost the same for the SAC305 and SAC_Q joints. After long term aging for 2000 hours, the average SAC305 particle diameter was almost twice that of SAC_Q.

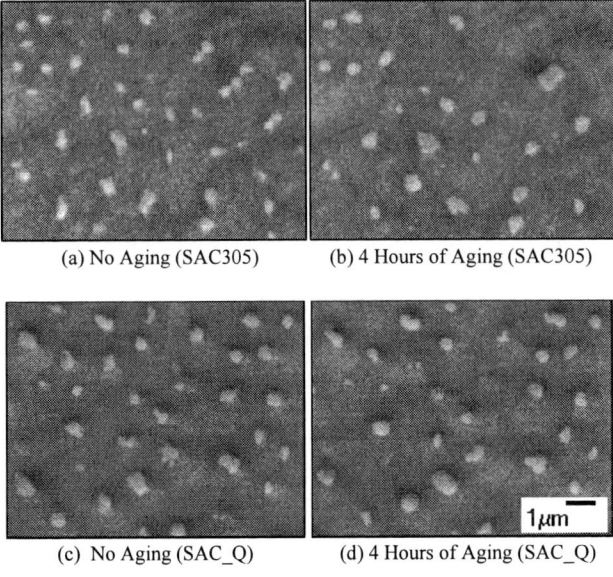

(a) No Aging (SAC305)   (b) 4 Hours of Aging (SAC305)

(c) No Aging (SAC_Q)   (d) 4 Hours of Aging (SAC_Q)

Figure 17. Aging Induced Coarsening of IMCs for SAC305 and SAC_Q (Aging at 125 °C for 4 Hours)

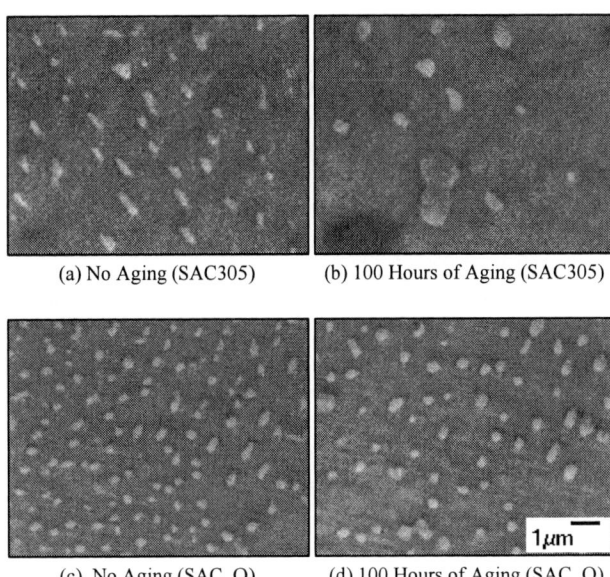

(a) No Aging (SAC305)   (b) 100 Hours of Aging (SAC305)

(c) No Aging (SAC_Q)   (d) 100 Hours of Aging (SAC_Q)

Figure 18. Aging Induced Coarsening of IMCs for SAC305 and SAC_Q (Aging at 125 °C for 100 Hours)

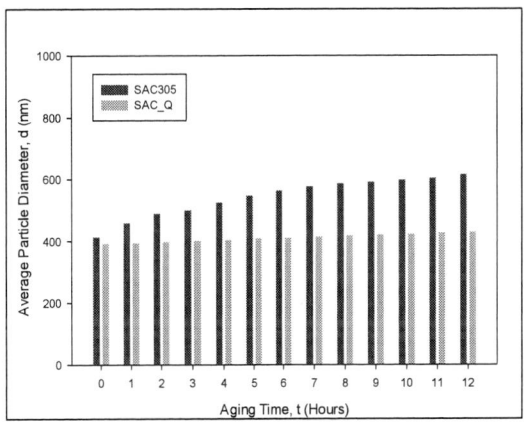

Figure 19. Variation of IMC Particle Diameter with Aging Time (SAC305 and SAC_Q; Short Term Aging up to 12 Hours).

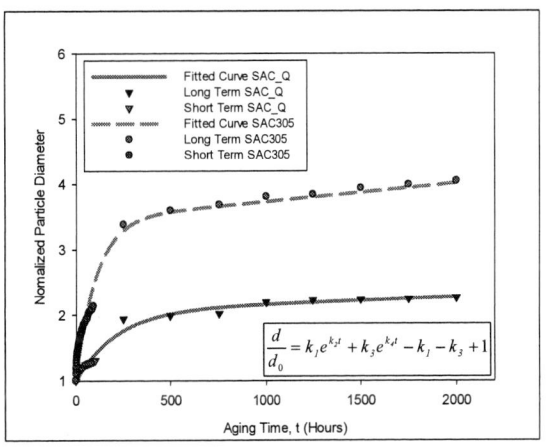

$$\frac{d}{d_0} = k_1 e^{k_2 t} + k_3 e^{k_4 t} - k_1 - k_3 + 1$$

Figure 20. Variation of IMC Particle Diameter with Aging Time (SAC305 and SAC_Q; Short Term and Long Term Aging).

Correlations between the reductions in ultimate tensile strength and the increase in IMC particle diameter are shown in Figs. 21 and 22 for SAC305 and SAC_Q, respectively. The mechanical property data for both alloys under 125 °C aging were taken from our prior work in reference [59]. It is clear that the evolution of the mechanical strength of the solder alloys is closely linked to variation in IMC particle diameter, and that the effects of aging are much greater for SAC305 than SAC_Q. The bismuth present as a dopant in SAC_Q made it relatively insensitive to aging induced degradations by a combination of the solid solution strengthening mechanism and a slower coarsening of the Ag_3Sn IMC particles. The aging induced presence of bismuth in solution within the beta-Sn matrix provided an increased resistance to the Ostwald ripening diffusion process that coarsens the IMC particles.

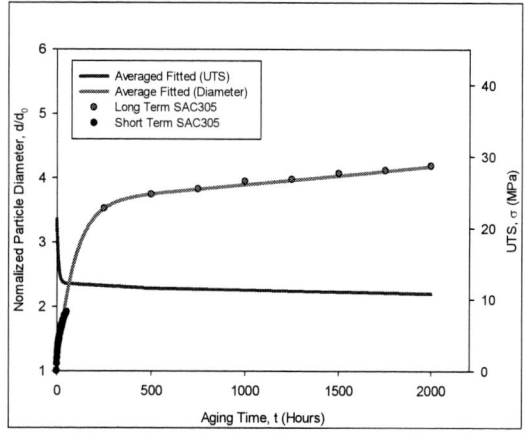

Figure 21. Variations of IMC Diameter and UTS with Aging (SAC 305).

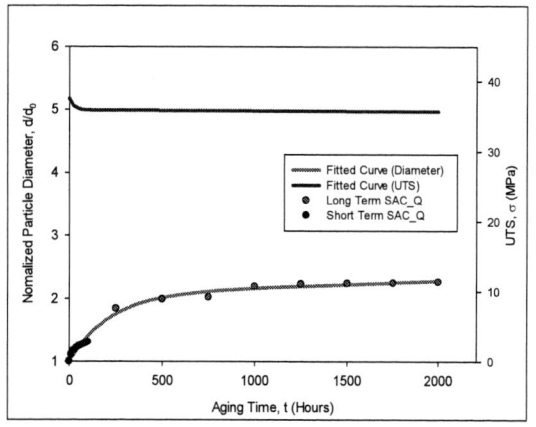

Figure 22. Variations of IMC Diameter and UTS with Aging (SAC_Q).

IV. SUMMARY AND CONCLUSIONS

In this investigation, Scanning Electron Microscopy has been utilized to examine aging induced microstructural changes occurring within SAC+Bi (SAC_Q) lead free solder. Microstructural evolution has been studied for several different regions from several different SAC_Q joints, and both short term (up to 100 hours) and long term (up to 2000 hours) aging of the joints have been performed. The size metrics for each IMC particle were tracked during the aging processes, and the results correlated to observed mechanical behavior changes.

As expected, bismuth rich phases were present in the SAC_Q joints after reflow solidification. During aging, the bismuth was observed to go into solution within the beta-Sn dendrites and in the intermetallic rich regions between dendrites. This leads to strengthening of the solder during aging by the mechanism of solid solution strengthening. In addition, it was observed that the coarsening of the Ag_3Sn intermetallic compounds was greatly mitigated in the SAC_Q alloy relative to that observed in SAC305. Quantitative analysis of the microstructures has shown that the average IMC particle diameter increased by approximately 9.6% for SAC_Q and 48.8% for SAC305 during the first 12 hours of aging. After 2000 hours of aging, the average IMC particle diameter increased by approximately 120% and 300% for the SAC_Q and SAC305 alloys, respectively. After long term aging for 2000 hours, the average SAC305 particle diameter was almost twice that of SAC_Q. Finally, we have compared the time dependent evolution of microstructure with the degradation in strength during aging for of the two solder alloys, and good correlation was observed.

ACKNOWLEDGMENTS

This work was supported by the US Navy, US Army, and the NSF Center for Advanced Vehicle and Extreme Environment Electronics (CAVE³).

REFERENCES

[1] Ma, H., and Suhling, J. C., "A Review of Mechanical Properties of Lead-Free Solders for Electronic Packaging," *Journal of Materials Science*, Vol. 44, pp. 1141-1158, 2009.

[2] Ma, H., Suhling, J. C., Lall P., Bozack, M. J., "Reliability of the Aging Lead-Free Solder Joint," *Proceeding of the 56th IEEE Electronic Components and Technology Conference*, pp. 849-864, San Diego, California, 2006

[3] Ma, H., Suhling, J. C., Zhang, Y., Lall, P., and Bozack, M. J., "The Influence of Elevated Temperature Aging on Reliability of Lead Free Solder Joints," *Proceedings of the 57th IEEE Electronic Components and Technology Conference*, pp. 653-668, Reno, NV, May 29-June 1, 2007.

[4] Zhang, Y., Kurumaddali, K., Suhling, J. C., Lall, P., and Bozack, M. J., "Analysis of the Mechanical Behavior, Microstructure, and Reliability of Mixed Formulation Solder Joints," *Proceedings of the 59th IEEE Electronic Components and Technology Conference*, pp. 759-770, San Diego, CA, May 27-29, 2009.

[5] Cai, Z., Zhang, Y., Suhling, J. C., Lall, P., Johnson, R. W., Bozack, M. J., "Reduction of Lead Free Solder Aging Effects Using Doped SAC Alloys," *Proceedings of the 60th IEEE Electronic Components and Technology Conference*, pp. 1493-1511, Las Vegas, NV, June 2-4, 2010.

[6] Zhang, J., Hai, Z., Thirugnanasambandam, S., Evans, J. L., Bozack, M. J., Sesek, R., Zhang, Y., Suhling, J. C., "Correlation of Aging Effects on Creep Rate and Reliability in Lead Free Solder Joints," *SMTA Journal*, Volume 25(3), pp. 19-28, 2012.

[7] Motalab, M., Cai, Z., Suhling, J. C., Zhang, J., Evans, J. L., Bozack, M. J., Lall, P., "Improved Predictions of Lead Free Solder Joint Reliability That Include Aging Effects," *Proceedings of the 62nd IEEE Electronic Components and Technology Conference*, pp. 513-531, San Diego, CA, May 30 - June 1, 2012.

[8] Motalab, M., Cai, Z., Suhling, J. C., Lall, P., "Determination of Anand constants for SAC Solders using Stress-Strain or Creep Data," *Proceedings of ITherm 2012*, pp. 910-922, San Diego, CA, May 30 - June 1, 2012.

[9] Motalab, M., Cai, Z., Suhling, J. C., Zhang, J., Evans, J. L., Bozack, M. J., Lall, P., "Correlation of Reliability Models Including Aging Effects with Thermal Cycling Reliability Data," *Proceedings of the 63rd IEEE Electronic Components and Technology Conference*, pp. 986-1004, Las Vegas, NV, May 28-31, 2013.

[10] Basit, M. M., Motalab, M., Suhling, J. C., and Lall, P., "The Effects of Aging on the Anand Viscoplastic Constitutive Model for SAC305 Solder," *Proceedings of ITherm 2014*, pp. 112-126, Orlando, FL, May 28-30, 2014.

[11] Basit, M. M., Ahmed, S., Motalab, M., Roberts, J. C., Suhling, J. C., and Lall, P., "The Anand Parameters for SAC Solders after Extreme Aging," *Proceedings of ITherm 2016*, pp. 440-447, Las Vegas, NV, June 1-3, 2016.

[12] Ahmed, S., Suhling, J. C., Lall, P., "The Anand Parameters of Aging Resistant Doped Solder Alloys" *Proceedings of ITherm 2017*, pp. 1416-1424, Orlando, FL, May 30 - June 2, 2017.

[13] Lall, P., Shantaram, S., Suhling, J., and Locker, D., "Effect of Aging on the High Strain Rate Mechanical Properties of SAC105 and SAC305 Leadfree Alloys," *Proceedings of the 63rd IEEE Electronic Components and Technology Conference*, pp. 1277-1293, Las Vegas, NV, May 28-31, 2013.

[14] Chiu, T. C., Zeng, K., Stierman, R., Edwards, D., and Ano, K., "Effect of Thermal Aging on Board Level Drop Reliability for Pb-Free BGA Packages," *Proceedings of the 54th IEEE Electronic Components and Technology Conference*, pp. 1256-1262, 2004.

[15] Lee, C. B., Jung, S. B., Shin, Y. E., and Chang, C. C., "Effect of Isothermal Aging on Ball Shear Strength in BGA Joints with Sn-3.5Ag-0.75Cu Solder," *Materials Transactions*, Vol 43(8), pp. 1858-1863, 2002.

[16] Deng, X., Sidhu, R. S., Johnson, P., and Chawla, N., "Influence of Reflow and Thermal Aging on the Shear Strength and Fracture Behavior of Sn-3.5Ag Solder/Cu Joints," *Metallurgical and Materials Transactions A*, Vol. 36A, pp. 55-64, 2005.

[17] Mustafa, M., Cai, Z., Suhling, J. C., and Lall, P., "The Effects of Aging on the Cyclic Stress-Strain Behavior and Hysteresis Loop Evolution of Lead Free Solders," *Proceedings of the 61st IEEE Electronic Components and Technology Conference*, pp. 927-939, Orlando, FL, June 1-3, 2011.

[18] Mustafa, M., Roberts, J. C, Suhling, J. C., and Lall, P., "The Effects of Aging on the Fatigue Life of Lead Free Solders," *Proceedings of the 64th IEEE Electronic Components and Technology Conference*, pp. 666-683, Orlando, FL, May 28-30, 2014.

[19] Fu, N., Suhling, J. C., Lall, P., "Cyclic Stress-Strain Behavior of SAC305 Lead Free Solder: Effects of Aging, Temperature, Strain Rate, and Plastic Strain Range," *Proceedings of the 66th IEEE Electronic Components and Technology Conference*, pp. 1119-1127, Las Vegas, NV, May 31 - June 3, 2016.

[20] Chowdhury, M. M. R., Fu, N.; Suhling, J. C., Lall, P., "Evolution of the Cyclic Stress-Strain Behavior of Doped SAC Solder Materials Subjected to Isothermal Aging" *Proceedings of ITherm 2017*, pp. 1369-1379, Orlando, FL, May 30 - June 2, 2017.

[21] Chowdhury, M. M. R., Hoque, M. A., Fu, N., Suhling, J. C., Hamasha, S., and Lall, P., "Characterization of Material Damage and Microstructural Evolution Occurring in Lead Free Solders Subjected to Cyclic Loading," *Proceedings of the 68th IEEE Electronic Components and Technology Conference*, pp. 865-874, San Diego, CA, May 29 - June 1, 2018.

[22] Mustafa, M., Cai, Z., Roberts, J. R., Suhling, J. C., Lall, P., "Evolution of the Tension/Compression and Shear Cyclic Stress-Strain Behavior of Lead-Free Solder Subjected to Isothermal Aging," *Proceedings of ITherm 2012*, pp. 765-780, San Diego, CA, May 30 - June 1, 2012.

[23] Mustafa, M., Suhling, J. C., Lall, P., "Experimental Determination of Fatigue Behavior of Lead Free Solder Joints in Microelectronic Packaging Subjected to Isothermal Aging," *Microelectronics Reliability*, Vol. 56, pp. 136-147, 2016.

[24] Zhang, J., Hai, Z., Thirugnanasambandam, S., Evans, J. L., Bozack, M. J., Zhang, Y., Suhling, J. C., "Thermal Aging Effects on Thermal Cycling Reliability of Lead-Free Fine Pitch Packages," *IEEE Transactions on Components, Packaging, and Manufacturing Technology*, Vol. 3(8), pp. 1348-1357, 2013.

[25] Hai, Z., Zhang, J., Shen, C., Snipes, E. K., Suhling, J. C., Bozack, M. J., and Evans, J. L., "Reliability Degradation of SAC105 and SAC305 BGA Packages Under Long-Term, High Temperature Aging," *SMTA Journal*, Vol. 27(2), pp. 11-18, 2014.

[26] Hai, Z., Zhang, J., Shen, C., Evans, J. L., Bozack, M. J., Basit, M. M., and Suhling, J. C., "Reliability Comparison of Aged SAC Fine-Pitch Ball Grid Array Packages Versus Surface Finishes," *IEEE Transactions on Components, Packaging, and Manufacturing Technology*, Vol. 5(6), pp. 828-837, 2015.

[27] Basit, M., Motalab, M., Suhling, J. C., Hai, Z., Evans, J. L., Bozack, M. J., and Lall, P., "Thermal Cycling Reliability of Aged PBGA Assemblies - Comparison of Weibull Failure Data and Finite Element Model Predictions," *Proceedings of the 65th IEEE Electronic Components and Technology Conference*, pp. 106-117, San Diego, CA, May 27-29, 2015.

[28] Zhao, C., Shen, C., Hai, Z., Basit, M. M., Zhang, J., Bozack, M. J., Evans, J. L., and Suhling, J. C., "Long Term Aging Effects on the Reliability of Lead Free Solder Joints in Ball Grid Array Packages with Various Pitch Sizes and Ball Arrangements," *SMTA Journal*, Vol. 29(2), pp. 37-46, 2016.

[29] Hasnine, M., Mustafa, M., Suhling, J. C., Prorok, B. C., Bozack, M. J., Lall, P., "Characterization of Aging Effects in Lead Free Solder Joints Using Nanoindentation," *Proceedings of the 63rd IEEE Electronic Components and Technology Conference*, pp. 166-178, Las Vegas, NV, May 28-31, 2013.

[30] Hasnine, M., Suhling, J. C., Prorok, B. C., Bozack, M. J., and Lall, P., "Exploration of Aging Induced Evolution of Solder Joints Using Nanoindentation and Microdiffraction," *Proceedings of the 64th IEEE Electronic Components and Technology Conference*, pp. 379-394, Orlando, FL, May 28-30, 2014.

[31] Hasnine, M., Suhling, J. C., Prorok, B. C., Bozack, M. J., and Lall, P., "Nanomechanical Characterization of SAC Solder Joints - Reduction of Aging Effects Using Microalloy Additions," *Proceedings of the 65th IEEE Electronic*

Components and Technology Conference, pp. 1574-1585, San Diego, CA, May 27-29, 2015.

[32] Hasnine, M., Suhling, J. C., Prorok, B. C., Bozack, M. J., and Lall, P., "Anisotropic Mechanical Properties of SAC Solder Joints in Microelectronic Packaging and Prediction of Uniaxial Creep Using Nanoindentation Creep," *Experimental Mechanics*, Vol 57(4), pp. 603-614, 2017.

[33] Ahmed, S., Hasnine, M., Suhling, J. C., and Lall, P., "Mechanical Characterization of SAC Solder Joints at High Temperature Using Nanoindentation," *Proceedings of the 67th IEEE Electronic Components and Technology Conference*, pp. 1128-1135, Orlando, FL, May 30 - June 2, 2017.

[34] Ahmed, S., Basit, M., Suhling, J. C., Lall, P., "Effects of Aging on SAC-Bi Solder Materials" *Proceedings of ITherm 2016*, pp. 746-754, Las Vegas, NV, May 30 - June 3, 2016.

[35] Choi, W. K., Lee, H. M., "Effect of Soldering and Aging Time on Interfacial Microstructure and Growth of Intermetallic Compounds between Sn-3.5Ag Solder Alloy and Cu Substrate," *Journal of Electronic Materials*, Vol. 29 (10), pp. 1207-1213, 2000.

[36] Ahat, S., Sheng, M., Luo, L., "Microstructure and Shear Strength Evolution of SnAg/Cu Surface Mount Solder Joint during Aging," *Journal of Electronic Materials*, Vol. 30 (10), pp. 1317-1322, 2001.

[37] Guo, F., Choi, S., Lucas, J. P., Subramanian, K. N., "Microstructural Characterization of Reflowed and Isothermally-Aged Cu and Ag Particulate Reinforced Sn-3.5Ag Composite Solders," *Soldering and Surface Mount Technology*, Vol. 13 (1), pp. 7-18, 2001.

[38] Miao, H., Duh, J., "Microstructure Evolution in Sn-Bi and Sn-Bi-Cu Solder Joints under Thermal Aging," *Materials Chemistry and Physics*, Vol. 71, pp. 255-271, 2001.

[39] Dutta, I., "A Constitutive Model for Creep of Lead-Free Solders Undergoing Strain-Enhanced Microstructural Coarsening: A First Report," *Journal of Electronic Materials*, Vol. 32, pp. 201-207, 2003.

[40] Ubachs, R. L. J. M., Schreurs, P. J. G., Geers, M. G. D., "Microstructure Evolution of Tin-Lead Solder," *IEEE Transactions on Components and Packaging Technologies*, Vol. 27 (4), pp. 635-642, 2004.

[41] Duan, L. L., Yu, D. Q., Han, S. Q., Ma, H. T., Wang, L., "Microstructural Evolution of Sn-9Zn-3Bi Solder/Cu Joint during Long-Term Aging at 170 °C," *Journal of Alloys and Compounds*, Vol. 381, pp. 202-207, 2004.

[42] Xiao, Q., Bailey, H. J., Armstrong, W. D., "Aging Effects of Microstructure and Tensile Property of Sn3.9Ag0.6Cu Solder Alloy," *Journal of Electronic Packaging*, Vol. 126(2), pp. 208-212, 2004.

[43] Allen, S. L., Notis, M. R., Chromik, R. R., Vinci, R. P., "Microstructural Evolution in Lead-Free Solder Alloys: Part I. Cast Sn-Ag-Cu Eutectic," *Journal of Materials Research*, Vol. 19 (5), pp. 1417-1424, 2004.

[44] Telang, A. U., Bieler, T. R., Lucas, J. P., Subramanian, K. N., Lehman, L. P., Xing, Y., Cotts, E. J., "Grain-Boundary Character and Grain Growth in Bulk Tin and Bulk Lead-Free Solder Alloys," *Journal of Electronic Materials*, Vol. 33(12), pp. 1412-1423, 2004.

[45] Dutta, I., Pan, D., Marks, R. A., Jadhav, S. G., "Effect of Thermo-Mechanically Induced Microstructural Coarsening on the Evolution of Creep Response of SnAg-Based Microelectronic Solders," *Materials Science and Engineering A*, Vol. 410-411, pp. 48-52, 2005.

[46] Xu, L., Pang, J. H. L., Prakash, K. H., Low, T. H., "Isothermal and Thermal Cycling Aging on IMC Growth Rate in Lead-Free and Lead-Based Solder Interface," *IEEE Transactions on Components and Packaging Technologies*, Vol. 28(3), pp. 408-414, 2005.

[47] Hao, H., Shi, Y., Xia, Z., Lei, Y., Guo, F., "Microstructure Evolution of SnAgCuEr Lead-Free Solders under High Temperature Aging," *Journal of Electronic Materials*, Vol. 37(1), pp. 2-8, 2008.

[48] Fix, A. R., Nuchter, W., Wilde, J., "Microstructural Changes of Lead-Free Solder Joints During Long-Term Ageing, Thermal Cycling and Vibration Fatigue," *Soldering and Surface Mount Technology*, Vol. 20(1), pp. 13-21, 2008.

[49] Seo, S., Kang, S. K., Shih, D., Lee, H. M., "The Evolution of Microstructure and Microhardness of Sn-Ag and Sn-Cu Solders During High Temperature Aging," *Microelectronics Reliability*, Vol. 49, pp. 288-295, 2009.

[50] Dutta, I., Kumar, P., Subbarayan, G., "Microstructural Coarsening in Sn-Ag-Based Solders and its Effects on Mechanical Properties," *Journal of Metals (JOM)*, Vol. 61, pp. 29-38, 2009.

[51] Kumar, P., Talenbanpour, B., Sahaym, U., Wen, C. H., Dutta, I., "Microstructural Evolution and Some Unusual Effects During Thermo-Mechanical Cycling of Sn-Ag-Cu Alloys," *Proceedings of the ITherm 2012*, pp. 880-887, San Diego, CA, May 30 - June 1, 2012.

[52] Kumar, P., Huang, Z., Chavali, S., Chan, D., Dutta, I., Subbarayan, G., Gupta, V., "A Microstructurally Adaptive Model for Primary and Secondary Creep of Sn-Ag-based Solders," *IEEE Transactions on Components, Packaging and Manufacturing Technology*, Vol. 2(2), pp. 256-265, 2012.

[53] Sahaym, U., Talebanpour, B., Seekins, S., Dutta, I., Kumar, P., Borgesen, P., "Recrystallization and Ag₃Sn Particle Redistribution during Thermomechanical Treatment of Bulk Sn-Ag-Cu Solder Alloys," *IEEE Transactions on Components, Packaging and Manufacturing Technology*, Vol. 3(11), pp. 1868-1875, 2013.

[54] Borgesen, P., "Microstructurally Adaptive Constitutive Relations and Reliability Assessment Protocols for Lead Free Solder," *SERDP Project WP-1752 Final Report*, 2015.

[55] Fu, N., Ahmed, S., Suhling, J. C., Lall, P., "Visualization of Microstructural Evolution in Lead Free Solders During Isothermal Aging Using Time-Lapse Imagery" *Proceedings of the 67th Electronic Components and Technology Conference*, pp. 429-440, Orlando, FL, May 30 - June 2, 2017.

[56] Ahmed, S., Wu, J., Fu, N., Suhling, J. C., Lall, P., "Quantification and Modeling of Microstructural Evolution in Lead Free Solders During Long Term Isothermal Aging" *Proceedings of the 68th Electronic Components and Technology Conference*, pp. 162-171, San Diego, CA, May 29 - June 1, 2018.

[57] Ahmed, S., Suhling, J. C., Lall, P., "Evaluation of Aging Induced Microstructural Evolution in Lead Free Solders Using Scanning Probe Microscopy," *Proceedings of ITherm 2018*, pp. 1062-1070, San Diego, CA, May 29 - June 1, 2018.

[58] Dieter, G. E., *Mechanical Metallurgy*, 3rd Edition, McGraw-Hill, 1986.

[59] Alam, M. S., Hassan, K. M. R., Suhling, J. C., Lall, P., "Investigation of the Effects of High Temperature Aging on the Mechanical Behavior of Lead Free Solders," *Proceedings of InterPACK 2018*, Paper InterPACK2018-8396, pp. 1-9, 2018.

2019 IEEE 69th Electronic Components and Technology Conference (ECTC)

Microstructure Signature Evolution in Solder Joints, Solder Bumps, and Micro-Bumps Interconnection in A Large 2.5D FCBGA Package during Thermo-Mechanical Cycling

[1]Arman Ahari, [1]Andy Hsiao, [1]Greg Baty, [2]Peng Su, and [1]Tae-Kyu Lee

[1]Portland State University, Portland, OR
[2]Juniper Networks, Sunnyvale, CA
taeklee@pdx.edu

Abstract— **Large body-size and heterogeneously integrated packages have become essential for high-performance computing applications. As an example, designs such as silicon interposer-based 2.5D packages have enabled the integration of high-performance silicon and memory in close proximity, greatly increasing the bandwidth and throughput of these devices. Within such a package, the interaction among the many sub-components and materials creates a complex thermo-mechanical response in the interconnections, which includes micro-bumps and C4 bumps. In addition, such components frequently require a high-layer count and high-thickness PCB, which creates a challenge for the reliability of the solder joints. As a result, the overall reliability of PCB assembly needs to be evaluated at every level of the interconnect. In this study, a large 2.5D flip chip package was subject to temperature cycling testing. This component was also attached to a PCB, and the entire assembly went through temperature cycling as well. Over the duration of testing, a series of microstructure evaluations were performed at the micro-bump, C4 bump, and solder joint level. Each analysis included polarized optical imaging, SEM (Scanning Electron Microscope), EBSD (Electron Backscatter Diffraction) and strain contour analysis. With these techniques, the methodology was able to not only observe the degradation and microstructure evolution, but was also able to reveal the damage by collecting high-resolution strain / stress distribution data at critical locations such as corner bumps and solder joints. These data provided insight into metallurgical processes that alter the grain structure of solder joints at different dimensions and locations, and ultimately the details of the failure mechanisms and processes.**

Keywords- 2.5D interposer, EBSD, solder bump, micro-bump, solder joint, heterogeneous integration.

I. INTRODUCTION

Given the recent evolution in advanced signal integration and demand for faster signal-speed, heterogeneous integration in electronic devices has reached the point of active commercialization. Along with new challenges for

multiple suppliers integrated into one device, the technological challenges have been largely addressed but still requires investigation, as the new deployment of these devices will generate new failure modes and new reliability challenges. Unlike the consumer market sector with handheld and portable electronic devices, the demands of high signal speed and huge capacity requires the components in telecommunication to move towards larger body sizes and more complex multi-die layered stacking structures with various types of interconnections. This also drives printed circuit board design to more complex configurations with denser Cu trace density and increase in board thickness.[1-3] As shown in Fig. 1, this multi-layered structure with different types and scales of interconnections, from solder joints to solder bumps and micro-bumps between the interposer and components, and the micro-bumps between the multi-layered high bandwidth memory component, needs a detailed assessment and monitoring to identify any instability or degradation which can potentially during field service. [4,5]

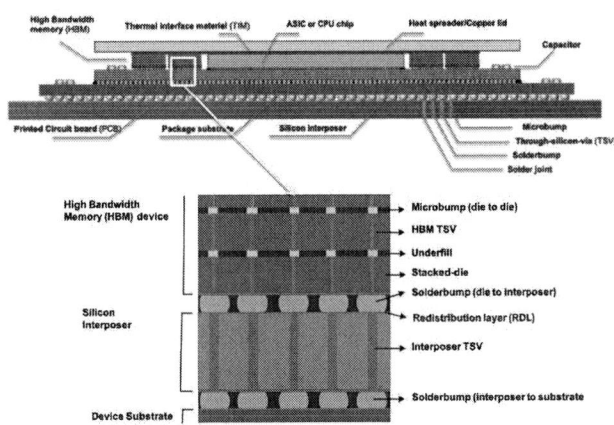

Figure 1. Schematic representation of (a) 3D package containing multiple devices and through silicon vias and (b) a schematically highlighted region at the interface between high band memory-interposer and substrate. [5]

978-1-7281-1500-9/19 $31.00 © 2019 IEEE   1099

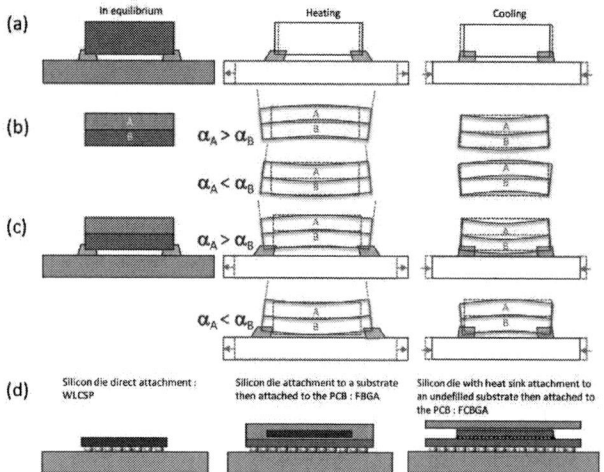

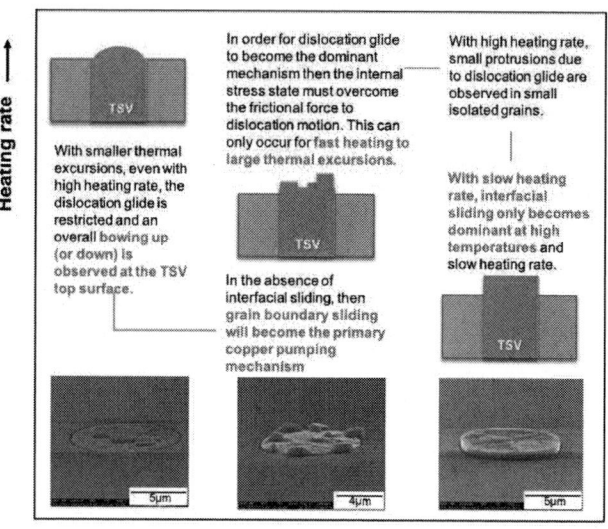

Figure 2. Effects of a temperature change on stress and strain in solder joints; (a) single component attached on ends, (b), warpage in a two layer assembly, (c) effect of bilayer warpage on stress in attachments on the end, and (d) practical geometries with these effects. [4]

Thermal cycling Temperature Range ⟶

Figure 4. Schematically shown TSV protrusion mechanisms correlated to heating rate and thermal cycling temperature range. [5]

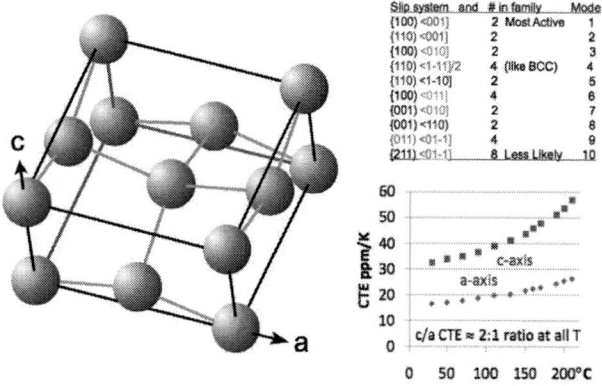

Figure 3. Sn lattice structure, main activated slip systems and thermal expansion ratio between c-axis and a-axis per temperature. [6]

The challenges in monitoring the stability of these interconnects is that the whole multi-layered structure is not homogeneous from the beginning, and is a continuously evolving structure, since the microstructure of the interconnects are evolving continuously, thus affecting the properties of the structure. Since the lead-free conversion, industry gained knowledge on solder and solder bump interconnection stability and shifting property changes with the support from academia. Experience-based and research-based studies on microstructure evolution at interconnection levels, combined with simulation studies, brought the whole industry to a comfort zone regarding interconnect reliability and performance. One of the identified main driving forces for instability and degradation mechanisms was the shear and

warpage-induced tension and compression force during thermal cycling due to thermal expansion mismatch as shown in Fig. 2. The accumulated thermal cycling test data gives confidence for certain package types and interconnect alloys which are critical for reliability assessment and risk mitigation. With the new multi-layered heterogeneous integration, the driving forces for stress and strain within these components have become more complex. Shown in Fig. 1, failure processes in a multi-layered structure are influenced not only by external driving forces, but also by internal factors. For example, the Sn-based interconnects contain the anisotropic expansion and slip systems, which are activated differently during thermal cycling. As shown in Fig. 3, the Sn lattice is a tetragonal lattice, which contains active slip systems based on the lattice orientation. In addition, the thermal expansion in the c-axis and a-axis direction is anisotropic, with a higher expansion in c-axis direction compared to the a-axis direction. With the miniaturized interconnections, the orientation of Sn-rich grains affects greatly the strain level of each solder bump or micro-bump, regardless of the external driving force. Another internal driving force is the through-silicon via (TSV) structure. The solder-bump or micro-bump diameter to the Cu-TSV diameter ratio is getting closer to one. This similarity in physical dimensions increases the impact of TSVs on the strain and stress behavior of solder bumps directly connected to them. Fig. 4 is a summary of Cu TSV's z-axis movement due to stress relaxation in TSV structures, as presented in a recent publication [5]. During thermal cycling, the thermal expansion mismatch between copper TSVs and the silicon dies creates large internal stresses in the heterogeneous structure, resulting in intrusion or protrusion of the TSV relative to the Si die. At low thermal excursions (i.e. small ΔT), no TSV protrusion will occur because insufficient stress has developed. At high temperatures and fast heating rates, dislocation glide will be the dominant mechanism. At moderate heating rates, moderate to high temperatures grain boundary sliding will be dominant. Finally, only at high temperatures and slow heating rates will interfacial sliding be dominant. Regardless of the amount the Cu TSV moves and regardless of the mechanism, the presence of TSV is a potential driving force which can affect the strain and stress state of the connected micro-bump. Given all mentioned internal driving force factors, the heterogeneously integrated multi-layered device will experience not only external force induced stress and strain, but also internally generated driving force-induced stress and strain. Thus, for effective reliability assessment all crucial factors need to be investigated. In this study, a large 2.5D flip chip package was selected for investigation. With a series of thermal cycling, microstructure evaluation was performed at the micro-bump, C4 bump, and solder joint level. Each analysis included polarized optical imaging, SEM (Scanning Electron Microscope), EBSD (Electron Backscatter Diffraction) and strain contour analysis.

II. EXPERIMENTAL PROCEDURE

Fig. 5 shows the selected component configuration. The 2.5D flip chip BGA (FCBGA) contains a large silicon die and four multi-layered components surrounding the large chip. Both the main chip and four multi-layered components are assembled on a large interposer. The FCBGA is assembled on a 140mil (3.6mm) thickness PCB. FCBGAs assembled on the PCB are subjected to thermal cycling then cross sectioned to reveal the solder joint, solder bump and micro-bump interconnect microstructure for optical, SEM and EBSD analysis. Two thermal cycling conditions were applied: 0°C to 100°C with a 10min ramp rate, 10min holding time and -40°C to 125°C with a 10min ramp rate, 10min holding time. Including the initial state where the FCBGA was not assembled to the PCB, three acceleration testing conditions were applied to the FCBGA samples, indicated as TC0, TC1 and TC2. TC0 is the initial state FCBGA which was not assembled to the PCB, TC1 is the board assembled FCBGA sample thermal cycled with 0°C to 100°C cycle to failure, TC2 is the sample with 0°C to 100°C cycle to failure then applied additional -40°C to 125°C thermal cycling for 600 cycles. Cross section samples were prepared to reveal multiple locations for solder joints, solder bump between the substrate, micro-bump between the interposer and main chip, and the micro bump between the interposer and the multi-

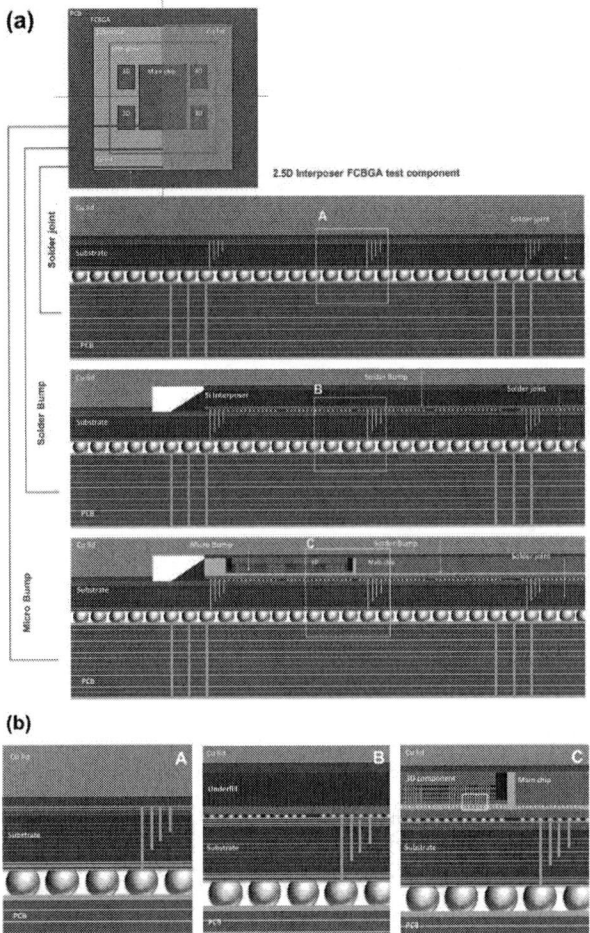

Figure 5. Test vehicle schematics. (a) cross section region at the corner of the FCBGA revealing the solder joints, solder bumps and micro-bumps. (b) Higher magnification schematics indicated in Fig.e 5(a) inlet boxes.

layered component. The solder joints at the FCBGA corner region and the solder bumps between the interposer and substrate are analyzed as indicated in Fig. 5. Cross sectioned samples are subjected to microstructure analysis using optical polarized images, EBSD for grain orientation scanning, and strain contour mapping.

III. RESULTS AND DISCUSSION

Fig. 6 shows polarized optical images of eight solder joints from the corner into the mid-section of the first row per TC0, TC1 and TC2 sample as indicated in Fig. 5(a). The red arrows mark the joints with partial and full crack propagation. As expected, the TC1 cross section joints have crack development in four solder joints at the package side interface. TC2 sample has additional -40°C to 125°C thermal

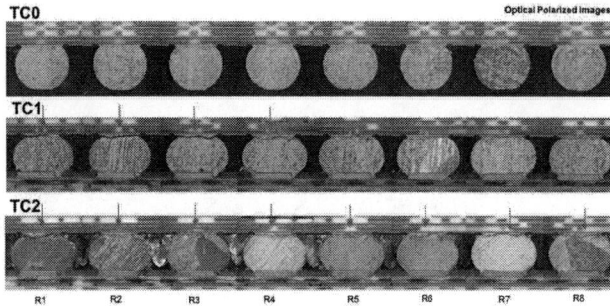

Figure 6. Eight solder joints optical polarized images from the corner into the mid-section of the first row per TC0, TC1 and TC2 sample.

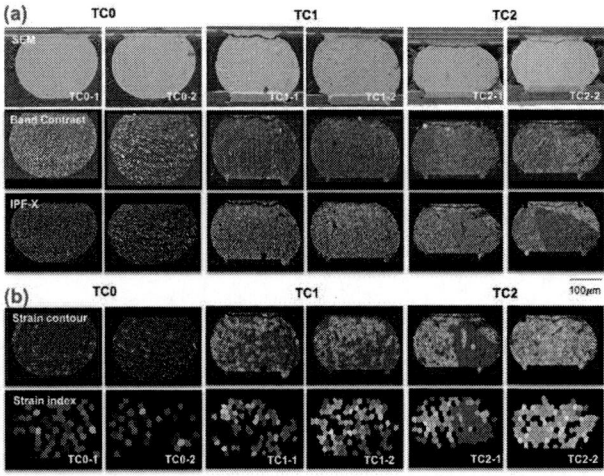

Figure 7. EBSD scanning per solder joint sample condition on selected joints. (a) SEM, band contrast, inverse pole figure (IPF) and (b) strain contour map and image processed strain index map.

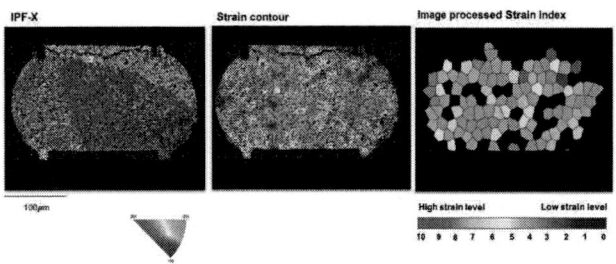

Figure 8. Example of EBSD IPF converting to strain contour map and strain index map.

cycling applied to the already 0°C to 100°C thermal cycled sample. More joints revealed the crack propagation with all joints containing a full crack propagation. The selected solder joints associated EBSD images are presented in Fig. 7. Two joints per condition were selected and presented. The SEM image, band contrast images, and inverse pole figures (IPF)

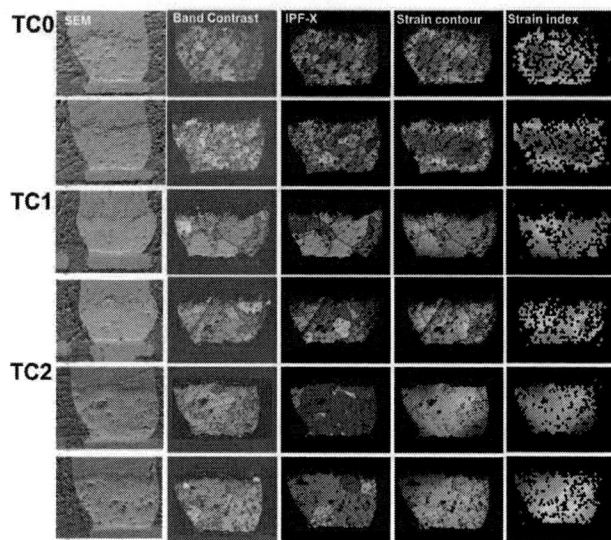

Figure 9. EBSD scans per solder bump sample condition on selected bumps. SEM, band contrast, inverse pole figure (IPF), strain contour map and image processed strain index map.

are shown. The band contrast image indicates a fine grain structure, but the IPF images show that all joints contain one to three oriented grain structures. The fine grain structure is due to the existence of eutectic regions, which surrounded the Sn grains. From EBSD images of each joint, a strain contour map can be developed. As shown in Fig. 8, each EBSD scan can construct an IPF image which indicates Sn grain orientations. Since each Sn grain orientation has a theoretically constructed Kikuchi pattern associated with the certain orientation, this can be compared to the measurement Kikuchi pattern per scanning location. The deviation can be back-calculated, which reveals the deteriorated angle per scanning location. The distortion is used to show the strain level at the localized scan region, which is the strain contour map. Although further calibration may be needed to improve accuracy of the absolute strain, the contour maps provide sufficient data on the highly strained regions within each joint. In addition, the strain contour map can be processed via image processing to an index map, which provides a quantitative and straightforward measurements of the overall strain levels within each joint. Fig. 7(b) shows the strain contour and index map of solder joints from each of the three conditions. Comparison of the strain levels from the three test conditions reveals that strain level increases with more thermal cycling. TC0 condition joints have mostly low intensity versus high intensity with TC2 joints. It is also shown that the intensity increase is detected at the package side interface region with thermal cycling and relative lower intensity once the crack is propagated, indicating a possible stress relaxation. The solder bumps, which are the interconnects between the substrate and the interposer, show

978-1-7281-1500-9/19 $31.00 © 2019 IEEE   1102

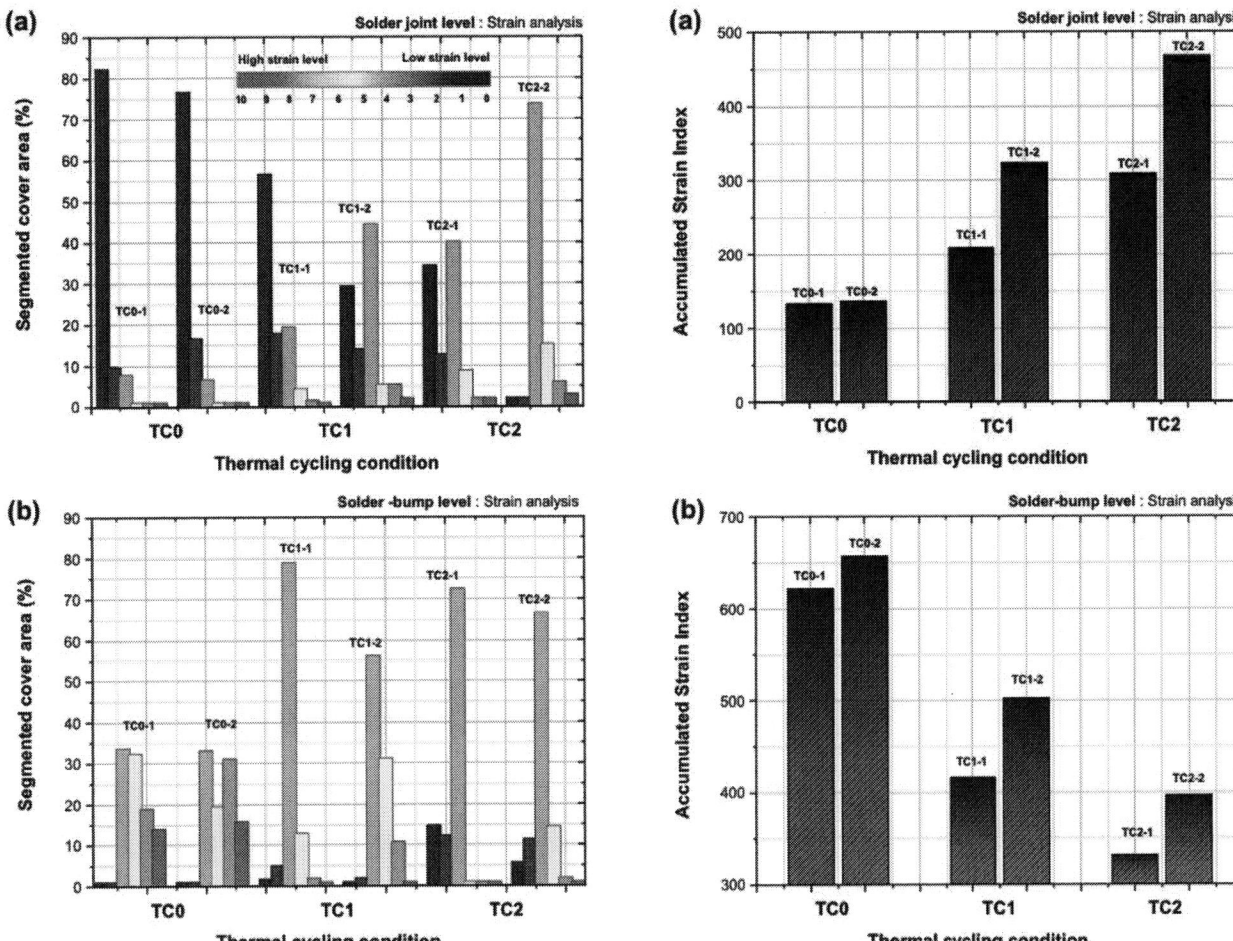

Figure 10. The intensity level segmented strain area plot. (a) segmented area per condition for solder joint sample. (b) for solder bump sample

Figure 11. The overall accumulated strain per joint can be plotted. (a) solder joint accumulate strain, and (b) solder bump accumulated strain amount.

different trends of strain evolution compared with the solder joints. As shown in Fig. 9, the band contrast and IPF images show continuous grain growth from thermal cycling condition T0 to TC1 and TC2. The associated strain contour map indicates a higher level of strain with TC0 sample solder-bumps, with a gradual decrease of strain intensity levels as more temperature cycles were applied. This trend is in the opposite direction of the solder joints. To convert the images to a potential quantitative measure, the accumulated strain index values were extracted from the image processed EBSD strain contour maps. Fig. 10 shows the segmented intensity plot. From the strain index images, the area percentage of specific strain levels were extracted and the area percentage were plotted per thermal cycling condition. Fig. 10(a) is the accumulated data for the selected solder joints in Fig. 7. The low intensity level indicated in blue were dominant at TC0 but reduced to a zero after TC2, with a constant increase of the mid-level intensity in green. The solder bumps on the other hand have a high intensity level at

TC0, and then the strain levels shifted to green and blue with thermal cycling. Since the color code indicates the strain level, the accumulated strain per joint can be also calculated by a simple scoring system. For examples, given the highest intensity ten and the low level of blue region an assigned score of one. The overall accumulated strain per joint can be plotted. Fig. 11(a) shows the accumulated strain within the solder joints, and Fig. 11(b) shows the accumulated strain in the solder bumps. As shown in Figure 11, the overall accumulated strain has a gradual increase with thermal cycling in solder joints but a gradual decrease in solder bumps. While the behavior of the solder joints appears to be logical, the data set from the solder bumps is somewhat curious. While this trend should not be interpreted as a reduction in risk of failure as more temperature cycles were added to the component, it does show the effects of grain size and crystallographic orientation on the strain level. The microstructure change could be moving these solder bumps to a lower energy state, before the eventual crack formation

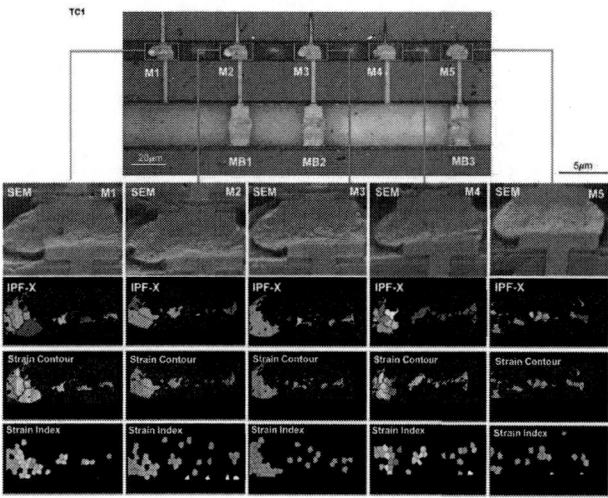

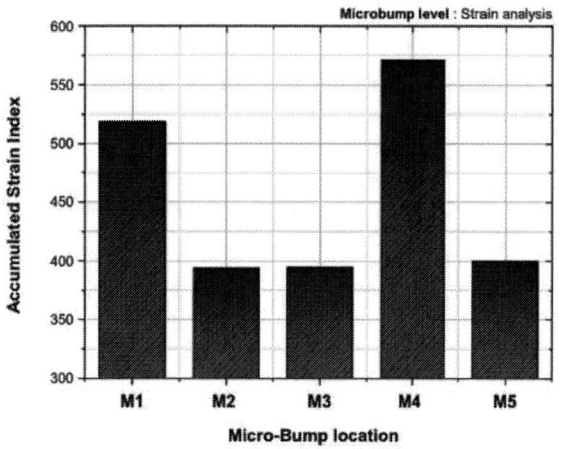

Figure 12. EBSD scanning per Micro-bump after TC on selected bumps. Location is indicated in Fig. 5(b). SEM, inverse pole figure (IPF), strain contour map and image processed strain index map.

Figure 13. The overall accumulated strain per micro-bump from M1 to M5.

at either the top or bottom interfaces. Further, strain levels alone do not indicate the risk of failure. Due to the highly anisotropic nature of the Sn grains, the yield and fracture strength is also anisotropic. For further investigation, the failure limit for each texture map should also be generated to provide an assessment of reliability risks. Another interesting observation is the TSV and micro-bump interaction associated with the strain development. As shown in Fig. 12, the micro-bumps at the multi-layered component are subjected for EBSD after TC1. The micro-bumps indicated also in Fig. 5(b) in the inlet box have different geometrical configurations. M1 and M4 do not have a micro-bump underneath the TSV, whereas M2, M3 and M5 have a directly connected micro-bump below. The strain contour map and strain index identify an interesting trend that where the M1

and M4, which has no micro-bump underneath have higher strain accumulation than the M2, M3 and M5 micro-bump. The accumulated strain index also reveals the same trend as shown in Fig. 13. This suggests that the residual strain primarily originates from the bottom side of the component, where the magnitude of thermal expansion is higher between features such as the silicon interposer and the organic substrate. The micro bumps between the first silicon and the interposer transfers this strain upwards and thus induces higher strains in the micro bumps above.

IV. SUMMARY AND CONCLUSION

Large body-size and heterogeneously integrated packages have become essential for high-performance applications. The overall reliability of PCB assembled large 2.5D FCBGA needs to be evaluated at every level of the interconnect. In this study, a large 2.5D flip chip package was subject to temperature cycling testing. Over the duration of testing, a series of microstructure evaluation was performed at the micro-bump, C4 bump, and solder joint level. Each analysis included polarized optical imaging, SEM (Scanning Electron Microscope), EBSD (Electron Backscatter Diffraction) and strain contour analysis. With these techniques, the methodology was able to not only reveal the damage and microstructure evolution, but also show the damage by collecting high-resolution strain / stress distribution data at critical locations such as corner bumps and solder joints. Using the EBSD pattern and profiles, strain contour maps can be extracted and used to identify the overall strained region and the accumulated strain per joint. A constant increase of accumulated strain was observed in solder joints and a stabilization of strain level was observed in solder-bump level. Further data collection and analysis is needed to help future understanding of the effects of grain microstructure change and strain distribution on the overall failure process of these solder joints.

REFERENCES

[1] P. Kumar, I. Dutta, Z. Huang and P. Conway, "Chapter 4: Microstructural and Reliability Issues of TSV", in *3D Microelectronic Packaging*, Y. Li and D. Goyal, eds., Springer Series in Advanced Microelectronics, Vol. 57, 2017, pp. 71-100 (ISBN: 978-3-319-44584-7).

[2] 5. I. De Wolf, K. Croes, O. Varela Pedreira, R. Labie, A. Redolfi, M. Van De Peer, K. Vanstreels, C. Okoro, B. Vandevelde, and E. Beyne, Microelectron. Reliab. **51**, 1856 (2011).

[3] H.K. Kim and K.N. Tu, Phys. Rev. B 53, 16027 (1996).

[4] Tae-Kyu Lee, Thomas Bieler, Choong-un Kim and Hongtao Ma, Fundamentals of Lead-Free Solder Interconnect Technology, Chapter 6, Springer, 169-210 (2015)

[5] H.Yang, Tae-Kyu Lee, L. Meinshausen and I.Dutta, Heating Rate Dependence of the Mechanisms of Copper Pumping in Through-Silicon Vias, J. Electron. Mater. 48 (1), 159-169 (2019).

[6] Bite Zhou, Quan Zhou, Thomas R. Bieler and Tae-kyu Lee, J. Electron. Mater. 44 (3), 895-908 (2015).

[7] T.-K. Lee, B. Zhou, T. Bieler, and K.-C. Liu, J. Electron. Mater. 41 (2), 273 (2012)

2019 IEEE 69th Electronic Components and Technology Conference (ECTC)

Long-Term Reliability of Solder Joints in 3D ICs under Near-Application Conditions

Omar Ahmed[1], Golareh Jalilvand[1], Hector Fernandez[1], Peng Su[2], Tae-Kyu Lee[3], and Tengfei Jiang[1]

[1] Department of Materials Science and Engineering & Advanced Materials Processing and Analysis Center, University of Central Florida, Orlando, FL 32816, USA
[2] Component Engineering, Juniper Networks, Sunnyvale, CA 94089, USA
[3] Department of Mechanical and Materials Engineering, Portland State University, Portland, OR 97201 USA

Abstract— **In this work, the long-term reliability of solder joints in 3D memory ICs (Integrated Circuits) was investigated. Three types of acceleration tests were carried out, where the temperature-time profiles were chosen to reflect near-operation conditions. For components after 500, 1000, and 1500 hours of testing, the formation and growth of intermetallic compounds (IMCs) in C4 bumps was inspected and analyzed using characterization techniques including scanning electron microscopy (SEM), energy dispersive x-ray spectroscopy (EDS), and Electron Backscatter Diffraction (EBSD). Different IMC growth behaviors were observed among the three different test conditions and discussed. The different behaviors between C4 bumps and μ-bumps were also evaluated, which can be traced to differences in the solder volume and the IMC/solder ratio between these two structures.**

Keywords-3D integration; Intermetallic compounds; Solder bumps; Micro-bumps; Reliability; Lead free solder.

I. INTRODUCTION

The stacking of multiple memory dies in three-dimensional memory ICs offers distinct advantages in integration density, speed, bandwidth, form factor, latency, and power savings [1-3]. While micro bumps (μ-bumps) are used to connect the multiple thin dies, Controlled Collapse Chip Connection (C4) is typically used to connect the die stack to a packaging substrate. Pb-free Sn based solder alloys are typically used for both the μ-bumps and the C4 bumps, due to environmental concerns and the favorable characteristics of Sn-based alloys, including enhanced electrical characteristics, excellent mechanical properties, and high thermal conductivity [4]. However, due to the mismatch of material properties, thermal stresses are inevitably generated in μ-bumps and C4 bumps, which raises thermo-mechanical reliability concerns during the fabrication and operation of the 3D ICs [5]. The reliability of the 3D ICs is further complicated by the formation and growth of intermetallic compounds (IMCs) due to the material reactions between solder and the contact pads. Kirkendall voiding and crack formation at critical interfaces has been observed, which can degrade device performance or even cause electrical failure [6-8].

In our previous work [9], we investigated the long-term reliability behaviors of μ-bumps in 3D memory ICs using non-standard reliability conditions to emulate real-life application conditions, where temperature profiles are often complex and irregular. Comparing to μ-bumps, which are about 30 μm in diameter, C4 bumps are much larger at about 100 μm in diameter and are expected to exhibit different reliability behaviors. Investigating the long-term reliability of C4 solder joints in a 3D memory IC, especially its difference from μ-bumps, is of both practical and scientific importance and is the focus of this work. The effect of material reactions and thermal stresses on the reliability of C4 bumps under near application conditions was studied and compared to that of μ-bumps. The results from this study will provide better insights regarding the long-term reliability of 3D memory IC devices under application conditions.

II. EXPERIMENTAL PROCEDURES

A. Description of the Test Vehicles

The 3D IC used in this study is illustrated in Fig. 1. The structure contains a stack of five thin dies connected by μ-bumps, and the substrate interfaces with the first level die by C4 solder bumps. The solder in these C4 bumps was made of Sn-Ag alloy. The under-bump metallization (UBM) on the substrate was ENIG (Electroless Nickel Immersion Gold), and a tri-layer ENIG/Cu/ENIG was used on the chip side. The average diameter and height of the C4 bump was 100 μm and 60 μm, respectively. The pitch among C4 solder bumps was 150 μm. The average diameter and height of a μ-bump was 30 μm and 32 μm, respectively, as described in our previous publication [9].

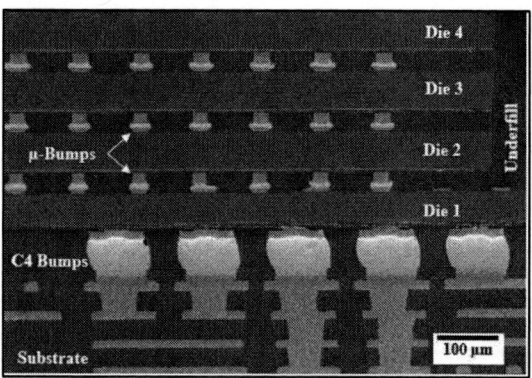

Figure 1. A typical 3D-ICs showing the structural components enabling 3D integration including C4 bumps and μ-bumps.

B. The Reliability Tests

Three types of long-term reliability tests were carried out for the test vehicles: High Temperature Storage (HTS), Sequential Test (SEQ), and Thermal Cycling with Long Dwell (TC-LD) test. Details of the experimental tests used in

978-1-7281-1500-9/19 $31.00 © 2019 IEEE

this study have been described previously [9] and only a short overview will be described here. The HTS test subjected the test vehicles continuously at 150°C (JEDEC standard JESD22-A103E Condition). The SEQ test subjected the test vehicles to 150°C for 95 hours, followed by 5 hours of thermal cycling (TC) between -55°C and 125°C. The TC portion had fixed heating and cooling rates of 15 °C/ minute (JEDEC JESD22-A104E Condition B) and a rate of 2 cycles per hour. Thus, for every 100 hours during the SEQ test, the test vehicles experienced 10 thermal cycles. The TC-LD test subjected the test vehicles to thermal cycling between -55°C and 125°C, with a long dwell time of 2.3 hours at each

temperature extremes. One cycle of the TC-LD test lasts 5 hours with heating and cooling rates fixed at 15 °C/ minute. Thus, for every 100 hours of TC test, the samples experienced 20 cycles between -55°C and 125°C. For each test, HTS, SEQ, and TC-LD, three test vehicles were removed from each test chamber at multiples of 500 hours for analysis. The temperature-time profiles of the three test conditions are shown in Fig. 2. The total number of hours that a test vehicle was held at 150°C and the total number of temperature cycles that a sample was subjected to is listed the Table 1.

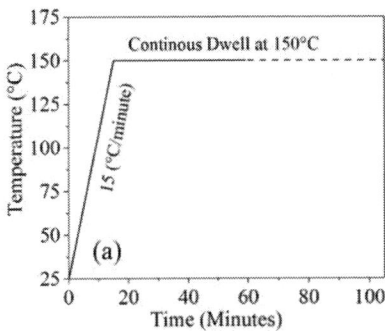 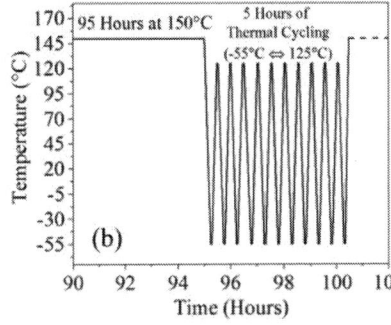 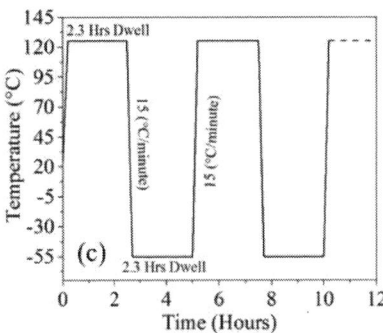

Figure 2.   Temperature-time profiles of reliability test conditions. (a) High temperature storage (HTS). (b) Sequential test (SEQ), where the components were subjected to an alternation of 95 hours of HTS at 150°C and 10 thermal cycles between 125°C and -55°C. Each thermal cycle lasted 30 minutes with a heating/ cooling rate of 15°C/ minute. (c) Temperature cycling with long dwell (TC-long dwell), where the components were thermally cycled between 125°C and -55°C with a 2.3 hours dwell at each temperature.

TABLE I.   TEST MATRIX, INCLUDING THE TOTAL NUMBER OF HOURS AND CYCLES FOR EACH TEST

Test Condition	Read Points	Total time at 150°C	Total No. of TC
HTS	500 hours	500 hours	0
	1000 hours	1000 hours	0
	1500 hours	1500 hours	0
SEQ	500 hours	475 hours	50
	1000 hours	950 hours	100
	1500 hours	1425 hours	150
TC-long dwell	500 hours	230 hour	100
	1000 hours	460 hour	200
	1500 hours	960 hour	300

C. Characterization

Samples removed at each read point were cross-sectioned and polished for further characterization. The morphology and distribution of the interfacial IMCs at the UBM-solder interface were examined by a field emission scanning electron microscope equipped with a secondary electron (SE) detector and a backscattering electron (BSE) detector. Elemental analysis was carried out using energy dispersive spectroscopy (EDS). Electron backscatter diffraction (EBSD) was also carried out to investigate the microstructure of the solder bumps.

III.   RESULTS

A. Zero Hour

Using EDS, the distribution of Cu, Ni, Sn, Si, and Ag elements in the C4 bumps was measured at time zero before the reliability tests. As shown in Fig. 3, a concentration of Cu can be seen at the location of the interfacial IMC, and there exists an overall distribution of Ag across the bulk of the solder. The identify and characteristics of the IMCs are discussed in the following sections.

B.   HTS

Representative micrographs from C4 bumps subjected to HTS test at 500, 1000, and 1500 hours are shown in Fig. 4. Based on the BSE image contrast, two IMCs can be distinguished in all of the images. IMC1, as denoted by the green arrows in Fig. 4b, was formed at the top and bottom interfaces between the solder and the UBM and has a scalloped shape. IMC2, denoted by the blue arrows in Fig. 4b, is seen as particles dispersed randomly throughout the solder.

For an HTS sample tested for 1500 hours, two EDS line-scans were made in the area labeled A in Fig. 4c. Fig. 5a is a magnified image of area A seen in Fig. 4c, where two scan paths, labeled *a* and *b,* are shown. For path *a*, the line scan started from the UBM, crossed the IMC1 layer, and stopped in the solder. Profiles for Ni, Cu, Ag, and Sn are plotted in Fig.5b. The Ni concentration started constant along the first 6.5 μm of scan and began to decrease after reaching the

UBM/IMC1 interface. Beyond the UBM/IMC1 interface, the concentration of Ni continued to decrease gradually, while the concentrations of Sn and Cu started to rise. At about 11.5 μm along the scan path, the Cu concentration started to decrease, which was accompanied by a rise in Sn concentration. At about 15 μm along the scan path, the scan entered the solder matrix and both the Ni and Cu concentrations had dropped to below 5 at.%. Between 6.5 μm and 11.5 μm along the scan path, the concentration of Sn was found to be 45.1 at.%, and the combined concentration of Ni and Cu was 54.9 at.%. Therefore, IMC1 is identified as the ternary intermetallic $(Cu_xNi_{1-x})_6Sn_5$.

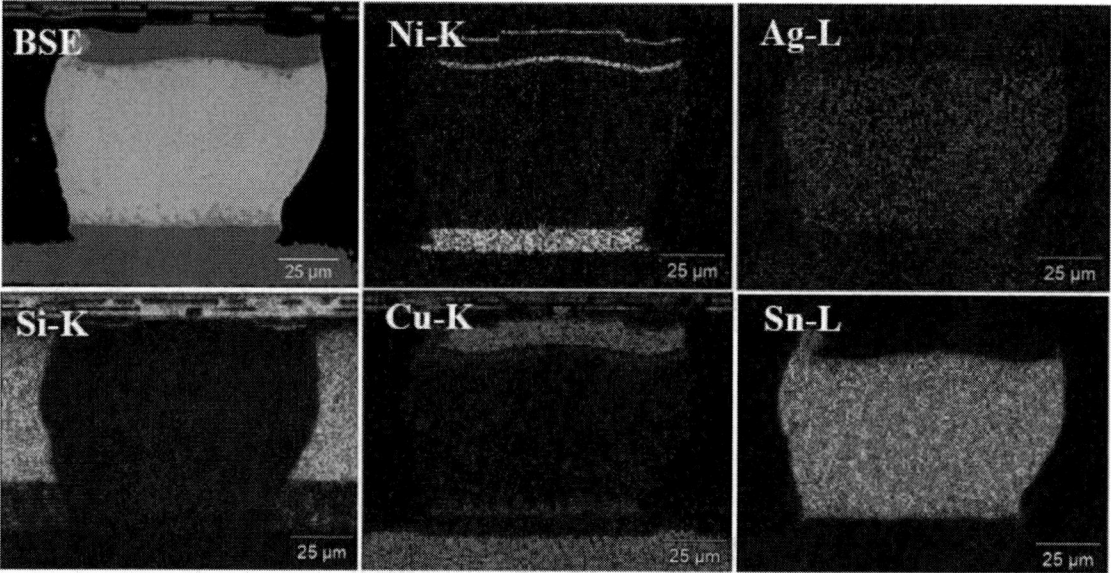

Figure 3. EDS map scan for an as recieved C4 bump showing the elemental distribution of all elements.

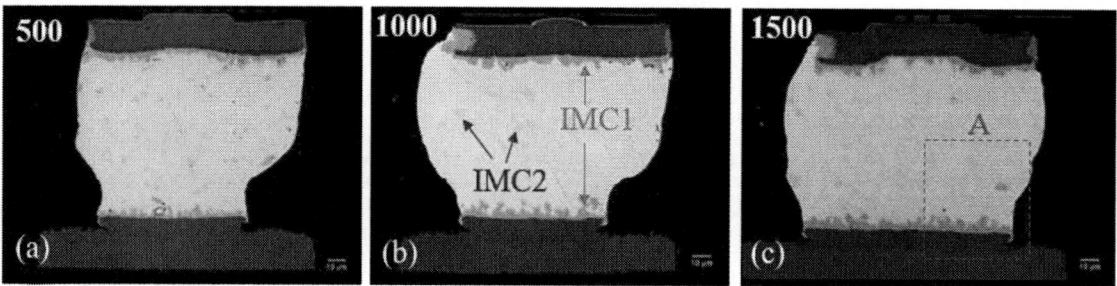

Figure 4. BSE images of C4 bumps subjected to HTS test for (a) 500 hours, (b) 1000 hours, and (c) 1500 hours.

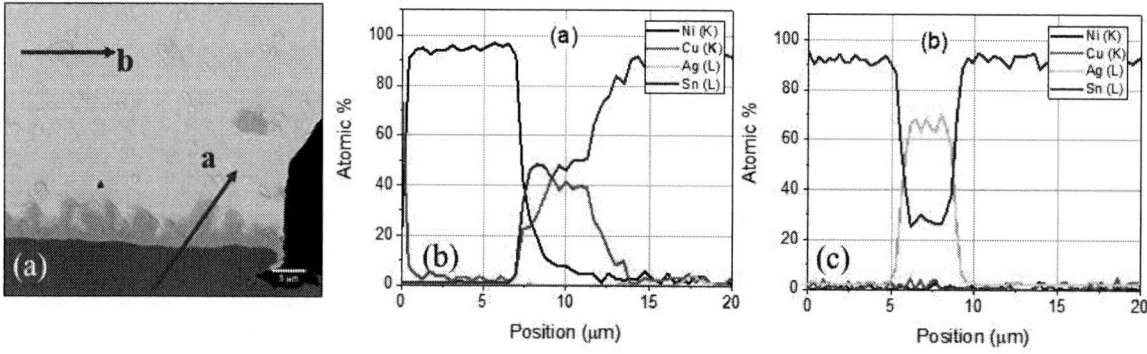

Figure 5. BSE image of a portion of a C4 bump after 1500 hours of HTS, showing the paths for EDS line scans and the corresponding line scan profiles for (b) Path a, and (c) Path b.

978-1-7281-1500-9/19 $31.00 © 2019 IEEE   1108

Scan path *b* crossed a particle of IMC2 and the concentration profiles for Ni, Cu, Ag, and Sn were obtained and plotted in Fig.5c. Within the Sn based solder matrix, which was the first 5μm along the scan path, the concentration of Sn remained constant. When the scan crossed the solder/IMC2 interface, Sn concentration dropped, accompanied by the rise of Ag concertation. Inside the IMC2 particle, the ratio of Sn/Ag remained mostly constant. At the plateau of the EDS profiles of Sn and Ag seen in Fig. 5c, Sn has an average concentration of 25.22 at.%, and Ag has a concentration 74.78 at.%. Therefore, IMC2 is identified as Ag_3Sn. This intermetallic is found to be randomly distributed across the solder matrix regardless of the test time. It is also observed from Fig. 4a-c that Ag_3Sn adopts a semi spherical morphology with varying particle sizes. This will be further discussed in the next sections.

C. SEQ

Cross-sectional BSE micrographs corresponding to samples subjected to SEQ test for 500, 1000, and 1500 hours are shown in Fig. 6. Similar to the case of HTS, two IMCs were formed and are identified as $(Cu_xNi_{1-x})_6Sn_5$ and Ag_3Sn. A scalloped layer of $(Cu_xNi_{1-x})_6Sn_5$ is seen at the UBM/solder interface. There also exists some spalling of $(Cu_xNi_{1-x})_6Sn_5$ into the solder, as indicated by the presence of $(Cu_xNi_{1-x})_6Sn_5$ particulates in the solder matrix away from the interface. The spherically shaped Ag_3Sn particles are also found to be dispersed randomly throughout the solder matrix. Particles of Ag_3Sn are also detected to have varying sizes. Although cracks were observed in the μ-bumps in the vicinity of IMCs as early as 1000 hours into the SEQ test [9], no cracks are seen at any location across the C4, including in the vicinity of interfacial IMCs even after 1500 hours of testing.

To further understand the influence of the SEQ test on the microstructure of C4 bumps, EBSD measurements were carried out. For a C4 bump subjected to SEQ test for 500 hours, as shown in Fig 7a, the band contrast revealed an average grain size of 3.2μm. The EDS mapping of Cu, Ag and Ni confirms the presence of $(Cu_xNi_{1-x})_6Sn_5$ at the interface and Ag_3Sn inside the solder. The associated EBSD inverse pole figures (IPF) and the strain contour map indicate that no specific localized strain was developed inside the solder bump. The C4 bumps subjected to 1500 hours of SEQ test show an evolved microstructure as seen in Fig. 7b. The average grain size increased to 10.3μm, indicating grain growth. The Cu and Ni EDS maps confirm the presence of $(Cu_xNi_{1-x})_6Sn_5$ and the EDS map of Ag indicates a reduced number of Ag_3Sn particles. The strain contour map also indicates a localized straining near the top interface, but no physical cracks were found.

D. TC-LD

For TC-LD test samples, representative BSE micrographs for the C4 bumps tested for 500, 1000, and 1500 hours are presented in Fig. 8. Similar to the observations for the HTS and SEQ samples, $(Cu_xNi_{1-x})_6Sn_5$ and Ag_3Sn IMCs are detected in the TC-LD samples. No cracks can be seen at any location in any of the C4 bumps, despite of the fact that the

TC-LD samples were subjected to the highest number of thermal cycles.

IV. DISCUSSION

A. Evolution of $(Cu_xNi_{1-x})_6Sn_5$

During reflow, solder reacts with the UBM and pad/trace that it comes into contact with, to form intermetallic compounds as reaction products. A thin layer of intermetallic is beneficial to the solder joint in providing a desirable metallurgical bond. Nevertheless, excessive intermetallic growth that occurs during service or thermal aging of the device often leads to detrimental mechanical consequences in the solder joints. The reliability of the solder joint, therefore, will be strongly affected by the properties of the intermetallics, including the composition, morphology, physical properties, and thickness. In this study, $(Cu_xNi_{1-x})_6Sn_5$ is the only intermetallic compound that was formed at the interface between the Sn solder and Ni-UBM, an observation that is in agreement with literature [10]. As can be seen from Fig. 5 (b), there exists a Ni and Cu concentration gradient across the IMCs layer (between 7.5 μm and 12.5 μm along the scan path). This suggests that IMC1 does not have a fixed composition. Instead, since Cu and Ni are soluble due to their similarities in crystal structure, atomic radius, and electronegativity, a ternary intermetallic compound of $Cu_xNi_{1-x})_6Sn_5$ was formed. The concentration gradient of Ni and Cu observed in Fig. 5b can be attributed to the diffusive process of IMC formation during testing [8, 11].

B. Evolution of Ag_3Sn Particles

The presence of Ag_3Sn IMCs has been observed in Sn-Ag based solders[8, 12]. The solubility limit of Ag in solid Sn is below 0.052wt.% at 150°C[12], therefore, excess Ag will precipitate as Ag_3Sn particles. The Ag_3Sn particles are found to be finely sized and randomly dispersed across the solder matrix after the reflow process [13]. During thermal aging and thermal cycling, the Ag_3Sn particles have been observed to coarsen, which leads to continued changes in creep, fracture, and thermomechanical fatigue characteristics of the solder. It is important to understand how the test conditions used in this study affect the morphological evolution of Ag_3Sn particles, because of the important role these particles play in the overall C4 bumps' strength and reliability.

In Fig. 9, the density of the Ag_3Sn particles for the three tests is plotted as a function of test duration. For each data point, the number of particles per area was obtained by counting and averaging over three SEM images. Fig. 9 clearly shows the reduction of density of Ag_3Sn particles as a function of increased test duration, indicating that the continuous reduction of the number of Ag_3Sn particles in the solder matrix is due to coarsening. This is also indicated by the reduced number of Ag_3Sn particles seen in Fig. 7. Therefore, all the test conditions used in this study has promoted the coarsening of Ag_3Sn. Nevertheless, the TC-LD test was found to be the most aggressive test in promoting coarsening of Ag_3Sn, due to the combined effect of thermal aging and thermal cycling. This can be seen in Fig. 9 by examining the slope of three curves for the three reliability tests. The slope was -3305 $(cm^{-2}.hr^{-0.5})$ for the TC-LD test, which was steeper than the slopes of -2470 $(cm^{-2}.hr^{-0.5})$ and -2389 $(cm^{-2}.hr^{-0.5})$ for the SEQ and HTS tests, respectively.

This finding indicates that thermal cycling induced stresses promote the coarsening of Ag_3Sn particles [14, 15].

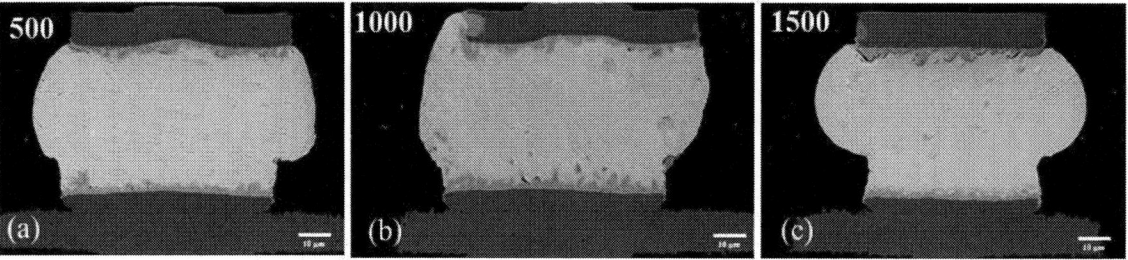

Figure 6. BSE images of C4 bumps subjected to SEQ test for (a) 500 hours, (b) 1000 hours, and (c) 1500 hours.

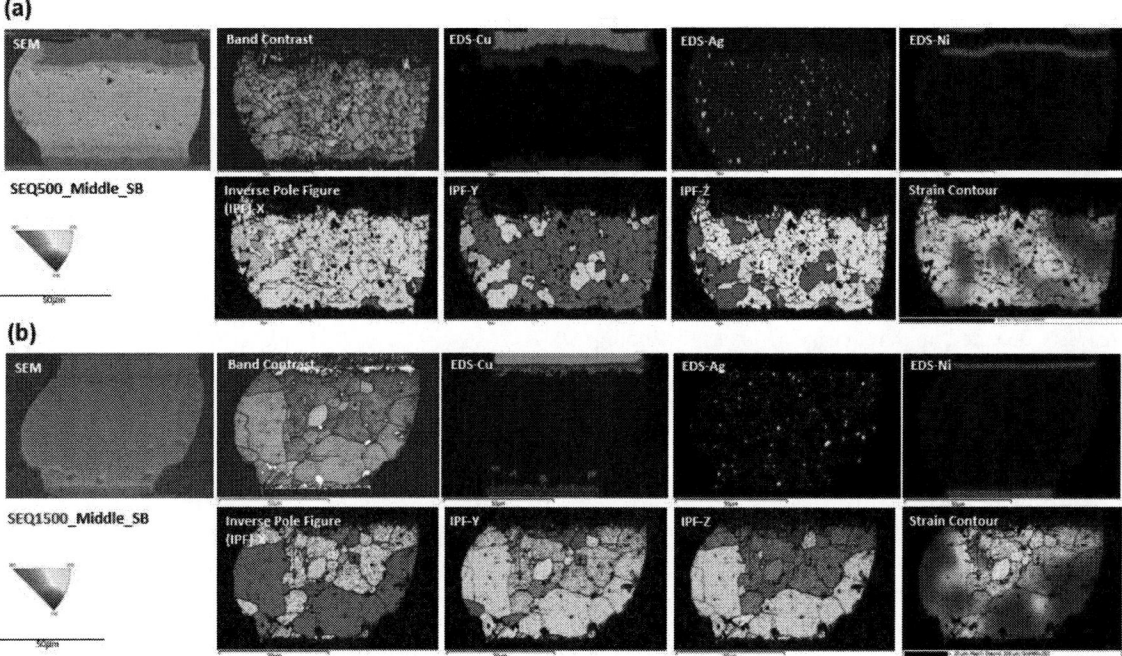

Figure 7. Microstructure analysis by EBSD for C4 bumps subjected to SEQ test for (a) 500 hours, and (b) 1500 hours.

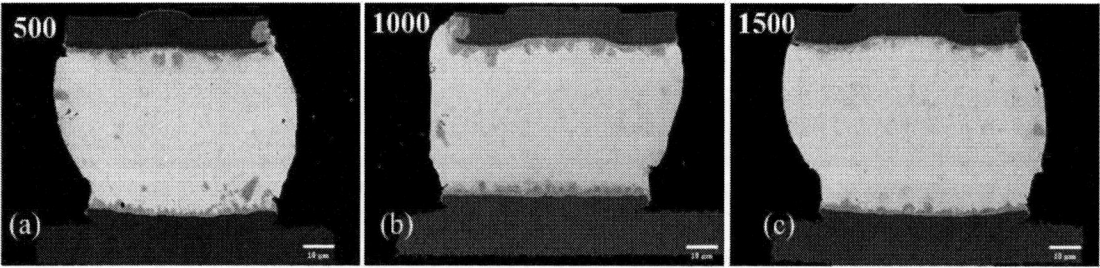

Figure 8. BSE images of C4 bumps subjected to TC-LD test for (a) 500 hours, (b) 1000 hours, and (c) 1500 hours.

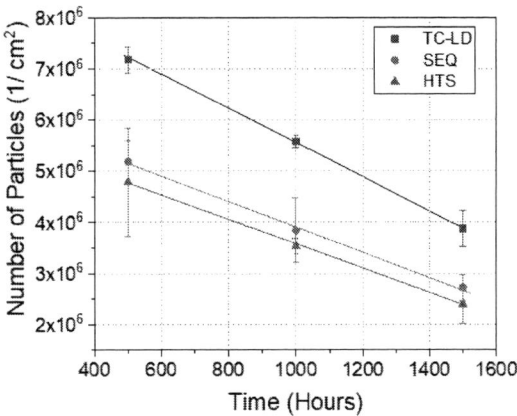

Figure 9. The evolution of Ag$_3$Sn particles in the three tests: HTS, SEQ, and TC-LD

C. Comparison of C4 Bump and μ-Bump Behavior under HTS, SEQ, and TC-LD Tests

In our previous study [9], we investigated the reliability aspects associated with the exposure of μ-bumps to the three test conditions: HTS, SEQ, and TC-LD. These test conditions were the same for the C4 bumps in this study. Therefore, a comparative analysis is presented in this section regarding the metallurgical, structural, and the reliability aspects of both μ-bumps and C4 bumps.

For the μ-bumps from all the test conditions and test periods, formation of two intermetallic compounds was observed. The first intermetallic was identified as Ni$_3$Sn$_4$ and was found adjacent to the Ni-UBM as a continuous layer across the width of the μ-bumps. The second intermetallic was identified as (Au,Ni)$_x$Sn$_y$. This latter intermetallic was found in the regions between Ni$_3$Sn$_4$ and the solder matrix, and adopted a scalloped morphology. Although some particles of Ag$_3$Sn were observed in the solder matrix, their presence was not significant. An interfacial intergranular crack was observed in a μ-bump subjected to SEQ testing for 1000 hours (100 cycles), located at the top interface of Ni$_3$Sn$_4$ and (Au,Ni)$_x$Sn$_y$. In μ-bumps after 1500 hours (150 cycles) of SEQ testing, cracks were observed at both the top and the bottom interfaces. Compared to cracks in the μ-bump tested for 1000 hours, the cracks had grown in the μ-bump after testing for 1500 hours. No crack was found in μ-bumps subjected to HTS and TC-LD for the same test duration. Therefore, the SEQ test was the most aggressive condition for μ-bumps to initiate the formation of cracks. Table 1 shows that the SEQ test subjected the test vehicles to more thermal cycles when compared to the HTS test and to longer aging at 150°C when compared to the TC-LD test. It was the combined effect of IMC growth during aging and thermal stress during thermal cycling that led to the observed cracking in the μ-bumps subjected to SEQ testing.

For the C4 bumps, as discussed earlier, the formation of (Cu$_x$Ni$_{1-x}$)$_6$Sn$_5$ at the solder/Ni-UBM interface and the formation of the finely dispersed Ag$_3$Sn particles across the solder matrix were observed. No crack was found in any C4

bump subjected to any of the three test conditions. Even the C4 bumps located at the corners of the package, a location known to be associated with large thermomechanical stresses, did not show any cracks anywhere in the vicinity of IMCs in any of the tested samples. The density of the Ag$_3$Sn particles was found to decrease in all test conditions, with a larger rate of reduction observed in C4 bumps subjected to TC-LD. No crack or void was observed in the solder. Therefore, the C4 bumps showed a more reliable behavior than the μ-bumps when being subjected to the same test conditions.

To understand the different metallurgical behaviors between C4 bumps and μ-bumps, a few aspects that distinguish these two components are discussed. The size of a typical μ-bump is much smaller than that of a C4 bump [16]. As a result, during thermal aging, the μ-bump will experience extensive IMC formation and may be entirely converted to IMCs, while a large volume ratio of solder will be retained in the C4 bumps. The resulting difference in the IMC to solder ratio was observed for the μ-bumps and the C4s subjected to different durations of SEQ tests, where nearly 60% of the μ-bumps were converted to IMCs, compared to less than 10% in the C4 bumps. As a direct consequence of this difference in IMC to solder ratio, the equivalent mechanical response of the μ-bumps will be dominated by that of the brittle IMCs, leading to their poor thermo-mechanical resistance and potentially shorter fatigue life, while the C4 bumps will largely retain their ductility despite of the formation of IMCs. Additionally, it is known that IMC formation is associated with a net volume shrinkage. For example, the formation of Ni$_3$Sn$_4$ between the Ni UBM and Sn in a μ-bump is associated with a large volume shrinkage of -11.4% [10, 16, 17]. The net volume shrinkage will impose considerable amount of stress on the μ-bumps and contribute to the formation and propagation of voids and cracks, which was observed in the μ-bumps subjected to SEQ tests. On the other hand, with ductility provided by the large solder volume in the C4 bumps, no cracks were observed in any of the test conditions.

V. SUMMARY AND CONCLUSION

In summary, we have conducted long-term reliability tests on 3D memory ICs subjected to three test conditions. In addition to the standard HTS test, TC-LD and SEQ test conditions were used to assess the reliability of the components under near application conditions. C4 bumps subjected to different test durations were characterized for all three tests by a variety of experimental methods. Formation of two IMCs was observed: the interfacial (Cu$_x$Ni$_{1-x}$)$_6$Sn$_5$, and the finely dispersed Ag$_3$Sn particles in the solder matrix. The concentration of (Cu$_x$Ni$_{1-x}$)$_6$Sn$_5$ was found to vary across the IMC layer thickness with the absence of any cracks in the vicinity of (Cu$_x$Ni$_{1-x}$)$_6$Sn$_5$. Coarsening of the Ag$_3$Sn particles was observed, which was found to be larger in TC-LD testing due to strain-induced effects. Grain growth was confirmed, and no cracks were observed in C4 bumps in any of the samples up to 1500 hours of testing.

Reliability responses of C4 bumps were compared to those of μ-bumps in the 3D memory IC of our previous study [9]. We showed that two intermetallic compounds, Ni_3Sn_4 and $(Au,Ni)_xSn_y$, were formed in the μ-bumps. Cracks along the interface between Ni_3Sn_4 and $(Au,Ni)_xSn_y$ were observed in some bumps after 1000 hours of SEQ testing. In the μ-bumps, the small solder volume and the high IMC/ solder ratio may induce sufficient thermomechanical stresses to generate microcracking during thermal cycling, as the effective mechanical properties of the μ-bumps could be dominated by those of the IMCs. In addition, and importantly, the volume shrinkage associated with IMC formation may have contributed greatly to the voiding and micro-cracking in the μ-bump. However, within the C4 bumps, the higher solder/IMC volume ratio may have mitigated these effects and thus less damage is observed.

Data from this study and our previous work have provided more insights on the long-term reliability of 3D IC devices. With proper mechanical design and material set selection, these components demonstrate robust reliability performance even during highly aggressive testing conditions.

ACKNOWLEDGMENT

The financial support by UCF Startup fund is gratefully acknowledged.

REFERENCES

[1] Chen, K.-N. and K.-N. Tu, Materials challenges in three-dimensional integrated circuits. MRS Bulletin, 2015. **40**(3): p. 219-222.

[2] Chen, Y., M.L. Huang, and N. Zhao. Microstructure and interfacial reactions of Sn-Au-Ag solder joints on Cu and Ni substrates. in 2016 17th International Conference on Electronic Packaging Technology (ICEPT). 2016.

[3] Pun, K.P.L., et al., Solid-state growth kinetics of intermetallic compounds in Cu pillar solder flip chip with ENEPIG surface finish under isothermal aging. Journal of Materials Science: Materials in Electronics, 2017. **28**(17): p. 12617-12629.

[4] Yoon, J.-W., et al., Mechanical reliability of Sn-rich Au–Sn/Ni flip chip solder joints fabricated by sequential electroplating method. Microelectronics Reliability, 2008. **48**(11): p. 1857-1863.

[5] Schubert, A., et al. Fatigue life models for SnAgCu and SnPb solder joints evaluated by experiments and simulation. in 53rd Electronic Components and Technology Conference, 2003. Proceedings. 2003.

[6] Lu, M., et al., Effect of Sn grain orientation on electromigration degradation mechanism in high Sn-based Pb-free solders. Applied Physics Letters, 2008. **92**(21): p. 211909.

[7] Zhang, H., et al. An effective method for full solder intermetallic compound formation and Kirkendall void control in Sn-base solder micro-joints. in 2015 IEEE 65th Electronic Components and Technology Conference (ECTC). 2015.

[8] Laurila, T., V. Vuorinen, and J.K. Kivilahti, Interfacial reactions between lead-free solders and common base materials. Materials Science and Engineering: R: Reports, 2005. **49**(1): p. 1-60.

[9] Ahmed, O., et al. Study of the Long Term Reliability of 3D IC under Near-Application Conditions. in 2018 IEEE 68th Electronic Components and Technology Conference (ECTC). 2018.

[10] Yang, T.H., et al., Effects of Aspect Ratio on Microstructural Evolution of Ni/Sn/Ni Microjoints. Journal of Electronic Materials, 2019. **48**(1): p. 9-16.

[11] Alam, M.O., et al., Effect of 0.5 wt % Cu in Sn−3.5%Ag Solder Balls on the Solid State Interfacial Reaction with Au/Ni/Cu Bond Pads for Ball Grid Array (BGA) Applications. Chemistry of Materials, 2005. **17**(9): p. 2223-2226.

[12] Shnawah, D.A., et al., Study on coarsening of Ag3Sn intermetallic compound in the Fe-modified Sn–1Ag–0.5Cu solder alloys. Journal of Alloys and Compounds, 2015. **622**: p. 184-188.

[13] Kanjilal, A., V. Jangid, and P. Kumar, Critical evaluation of creep behavior of Sn-Ag-Cu solder alloys over wide range of temperatures. Materials Science and Engineering: A, 2017. **703**: p. 144-153.

[14] Qi, L., et al., Effect of thermal-shearing cycling on Ag3Sn microstructural coarsening in SnAgCu solder. Journal of Alloys and Compounds, 2009. **469**(1): p. 102-107.

[15] Yin, L., et al., Recrystallization and Precipitate Coarsening in Pb-Free Solder Joints During Thermomechanical Fatigue. Journal of Electronic Materials, 2012. **41**(2): p. 241-252.

[16] Li, C.C., et al., Volume Shrinkage Induced by Interfacial Reaction in Micro-Ni/Sn/Ni Joints. Metallurgical and Materials Transactions A, 2014. **45**(5): p. 2343-2346.

[17] Li, C.C., et al. Volume shrinkage induced by interfacial reactions in micro joints. in 2013 IEEE International Symposium on Advanced Packaging Materials. 2013.

Experimental Investigation of the Correlation between a Load-Based Metric and Solder Joint Reliability of BGA Assemblies on System Level

Fabian Schempp[*‡], Marc Dressler[*], Daniel Kraetschmer[*], Friederike Loerke[*] and Juergen Wilde[‡]

[*]*Automotive Electronics*
Robert Bosch GmbH, 71701 Schwieberdingen, Germany
[‡]*Department of Microsystmes Engineering (IMTEK)*
University of Freiburg, Georges-Koehler-Allee 103, 79110 Freiburg, Germany
Email: Fabian.Schempp@de.bosch.com

Abstract—A fast and accurate method based on experimentally obtainable parameters for the solder joint reliability prediction under thermo-mechanical load can add significant value in the product development process. On system level, for products with over thousand components, the evaluation of every individual solder joint is not always feasible. This is first and foremost true for products with many large BGA components that have very high ball counts. In addition, especially for these components, board level lifetime test results cannot be easily transferred to the product due to system level effects like stress triaxiality. The CTE mismatch between PCB and housing leads to PCB warpage and complex solder joint loading conditions for bending sensitive components under temperature cycling. The work presented in this paper advances the state of the art in solder joint reliability prediction by using only experimentally obtainable parameters such as PCB strain, PCB curvature and component warpage as metric for the reliability prediction. With this metric, a fast pre-qualification for BGA assemblies on system level with easier accessible parameters is possible. In order to verify the method, two fine pitch BGA packages with SAC solder joints, identical footprint and lateral package dimensions but different warpage behavior over temperature are investigated. The number of cycles to failure of the solder joints are experimentally determined in passive temperature cycling test under varying PCB strains and curvatures by superimposing a 3P-bending load. In this bending test, well defined PCB strain and curvature can be applied on the components in order to investigate the influence of these loads on the solder joint rcliability. The correlation of component warpage, PCB strain and curvature with the solder joint reliability of these BGA packages is assessed.

Keywords-solder joint; BGA; thermomechanics; reliability;

I. INTRODUCTION

For the reliability assessment of solder joints of electronic components subjected to temperature cycling, usually board level tests are performed on freely expandable PCBs with accelerated temperature cycles or thermal shock cycles [1], [2], [3], [4]. Global CTE mismatches between components and PCB and local mismatches between solder and component as well as solder and PCB lead to thermomechanical stresses within the solder joint that cause cracking and thus failure of the joints.

To relate test results from accelerated tests to field con-

ditions Coffin and Manson proposed a model that correlates the strain range in the solder joint over a temperature cycle with the solder joint lifetime in the low cycle fatigue regime [5], [6]. With the assumption of a constant quotient of strain range and temperature delta, the change in lifetime between test temperature profiles and field profile can be correlated to the change in temperature delta. However, this assumptions does not hold true if board level results are transferred to system level where additional mechanical loads can be induced on the solder joints through PCB bending and stretching caused by a CTE mismatch between PCB and housing.

The finite element method is a widely accepted tool that is used to evaluate local damage related strain or energy based parameters for the reliability prediction [7], [8]. These local damage related parameters can be obtained in board and system level simulations. Together with experimental data, a correlation between damage related parameter and solder joint lifetime can be established. For complex systems with a great number of components with high I/O counts this procedure can be very tedious. A knowledge based pre-assessment of the components with easier accessible parameters would thus simplify the reliability engineering on system level in the product development process.

The aim of this work is to investigate and develop a method that allows the reliability predictions of Ball Grid Array (BGA) component solder joints on system level with parameters like strain and curvature on the PCB and warpage behavior of the component over temperature. This paper presents an experimental investigation of the influence of these parameters on the solder joint reliability of two BGA components. The components are tested under passive temperature cycles with superimposed mechanical load through 3P-bending. Numerous studies of board level bending tests exist for electronic components, however, with relatively high bending frequencies and targeted on the use of the components in portable electronic products [9]. A slow 4P-bending test as alternative for thermal cycling has been conducted by Vandevelde et al. for area array solder joints, however, under costant temperature [10], [11]. In the

978-1-7281-1500-9/19 $31.00 © 2019 IEEE

experiment presented in this paper, the mechanical load on system level, caused by the CTE mismatch between PCB and housing, is simulated by a bending load. The intensity of this bending load is relative to the temperature change in a slow temperature cycling test.

With this extensive experiment of a slow temperature cycle test with superimposed 3P bending the influence of parameters like PCB strain, PCB curvature and BGA warpage on solder joint reliability is investigated. 320 BGA components have been tested, 1920 daisy chains have been electronically monitored to generate a comprehensive database.

II. INVESTIGATED BGA COMPONENTS

Two BGA components that have the same lateral dimensions of 11 mm by 10.5 mm, the same SAC solder joint material and an identical footprint with 134 solder balls are investigated (cf. Fig. 5). The solder joints are solder mask defined on component side and non solder mask defined on PCB side. BGA1 has a 5 μm thinner solder mask than BGA2. BGA1 has one large silicon die whereas BGA2 has two small dies stacked on top of each other (cf. Fig. 8). The main difference between BGA1 and BGA2 is the respective warpage behavior. Component internal CTE mismatches between interposer, solder mask, die attach, silicon die and mold material cause the component to warp over temperature.

A. Component Warpage

White light interferometry measurements have been performed of the tested components both unsoldered and soldered on the PCB. White light interferometry is an optical and absolute measurement technique that allows the measurement of the out of plane deformation of the component over temperature. For this measurement both the unsoldered components and a PCB with soldered components are positioned on a plate that can be temperature regulated. Measurements have been conducted in a temperature range from −40°C to 125°C.

BGA1: 4 samples of BGA1 have been measured and the out of plane deformation was evaluated over the component diagonal on the mold compound. The raw measurement data has been filtered and a polynomial has been fitted to the data in a post processing step. In Fig. 1 the average of these 4 measurements is plotted at different temperatures. It can be seen that already at room temperature the component is not flat but has a height difference of approximately 50 μm. For the entire temperature range the component exhibits a cry like shape. In the soldered state the component is warped in the opposite direction as can be seen in Fig. 2. Also the warping is significantly reduced, which has been observed for other BGA components also by other researchers [12].

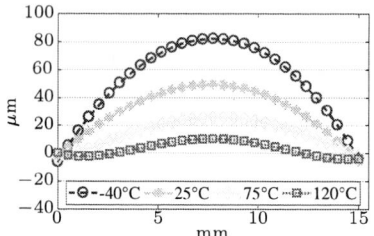

Figure 1. Out of plane deformation of BGA1 component over temperature.

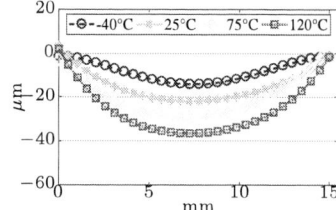

Figure 2. Soldered BGA 1 component and out of plane deformation at different temperatures measured over component diagonal.

The deformation displayed in the soldered state is the average of three measured components.

BGA2: The unsoldered BGA2 component changes at high temperatures from a cry to a smile like shape (Fig. 3). The warpage values at low temperatures are smaller compared to BGA1. BGA2 exhibits a smile like shape over the temperature range in the soldered state like BGA1 does (Fig. 4). The warpage values are, however, smaller than the ones measured for BGA1. The difference in warping of the components over temperature causes different loadings on the solder joints for the two components. Thus, different solder joint lifetime for the two components is expected under the same PCB load.

III. EXPERIMENTAL SETUP

For the temperature cycling test with superimposed bending the components are soldered on a 380 mm x 190mm, 8 layer PCB with a thickness of 1.6 mm in a two step reflow solder process. Components are assembled only on one side of the PCB. Fig 5 shows an x-ray image of the array after the solder process. The copper pads on the PCB cause the solder joints to have a slight oblong shape. Each PCB is equipped with in total 32 components. 4 components in 8 rows (cf. Fig. 6).

A. Three-Point Bending Test

In total 10 PCBs, five PCBs of each component, with 320 BGA components are tested under slow and passive temperature cycles and superimposed bending load in a temperature chamber. The PCBs are in a mechanical fixture without constraining the in plane movement. Thus, only the bending load and no stretching is applied. With the positioning of the components in 8 rows, the 4 components

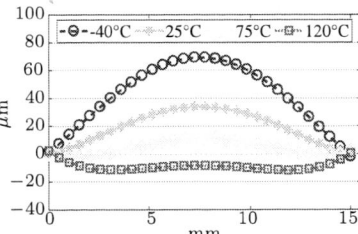

Figure 3. Out of plane deformation of BGA2 component over temperature.

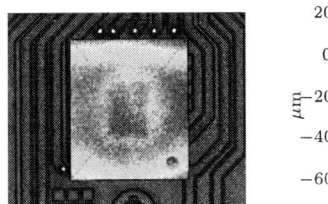 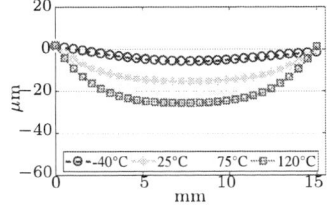

Figure 4. Soldered BGA2 component and out of plane deformation at different temperatures measured over component diagonal.

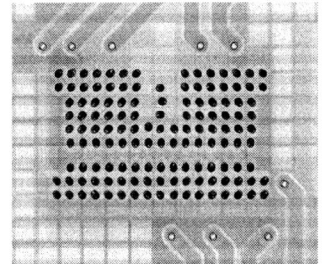

Figure 5. X-Ray image of BGA component.

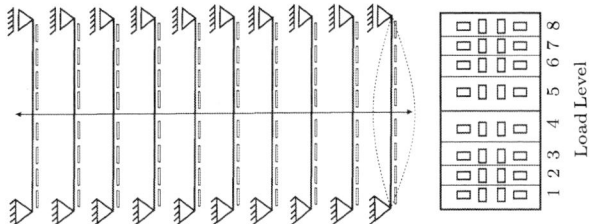

Figure 6. Left: 10 tested PCBs with mechanical boundary condition. Right: Schematic of test PCB with 32 components oriented parallel (middle columns) and orthogonal (left and right side of the PCB) to the bending direction.

in each row are subjected to a similar mechanical bending load. Under three point bending the strain and also curvature of the PCB increases towards the center. Thus, the influence of different strain and curvature on the solder joint reliability of the two components can be investigated. Two out of four components in each row are positioned in bending direction, whereas the other two components are orientated orthogonal to the bending direction with their long side (cf. Fig. 6).

Temperature Profile and Deflection: The temperature during the cycles is measured on the PCBs at nine different positions in the chamber. A $-40°C$ to $125°C$ test profile is run. The measured profiles of the sensors over one temperature cycle can be seen in Fig. 7. The chamber ventilation system, the mass of the mechanical construction for the fixture of the PCBs as well as the cables for the electrical online measurement of the solder joints cause a certain inhomogeneity of the temperature in the chamber. It is, however, made sure that at all positions in the chamber with a temperature sensor the corner temperatures are reached and a minimum of 20 min dwell time at high temperature ($125°C \pm 5K$) is guaranteed. Due to the inhomogeneity the maximum dwell time is 45 min. SAC relaxation rates are the highest at the beginning of the dwell period and decrease with time. Clech reported a 50 % stress decrease after 10 min and a 74 % stress decrease after 60 min for SAC PBGA assemblies [13]. It is thus assumed that the difference in creep between 20 min and 45 min is in a negligible magnitude.

The deflection of the PCBs is relative to the temperature change and thus imitates the mechanical load applied on the PCB on system level through a CTE mismatch between PCB and housing. The PCBs are deflected 1.1 mm/10 K such that the components are on the tension side at high temperature and on the pressure side at low temperature (cf. Fig. 6). At high temperature this results in to a maximum deflection of 11 mm and at low temperature of -7.2 mm.

B. Electrical Online Measurement

The tested components are special daisy chain test components. These components have an internal wiring in the copper layers of the interposer that directly connect adjacent solder joints. The internal connections can be seen as black lines between solder joints (circles) in Fig. 8. With the according wiring in the test PCB, the resistance of the 8 highlighted daisy chains in Fig. 8 can be electrically monitored during the temperature cycling test. As shown schematically in Fig. 9, a voltage is applied to the electric circuit containing the n solder joints of the daisy chain and the resistance value is measured. A fully cracked solder joint can be detected through an opening of the circuit. In this test a daisy chain consists of at least 2 solder joints. It is not possible to detect from the measurement which of the solder joints within the daisy chain actually had the 100 % crack. Thus, the daisy chains have been kept small, except for daisy chain 7, which consists exclusively of solder joints underneath the die shadow of BGA2, and daisy chain 8 with solder joints only in the center of the footprint.

Therefore, with the 8 daisy chains it can be investigated at which position within the footprint the solder joints fail first. Corner solder joints are monitored as well as solder joints both underneath the die shadow of BGA1 (daisy chain 5 and 6) and BGA2 (daisy chain 7). Daisy chain 8 in the center of the footprint is only monitored for components that

978-1-7281-1500-9/19 $31.00 © 2019 IEEE

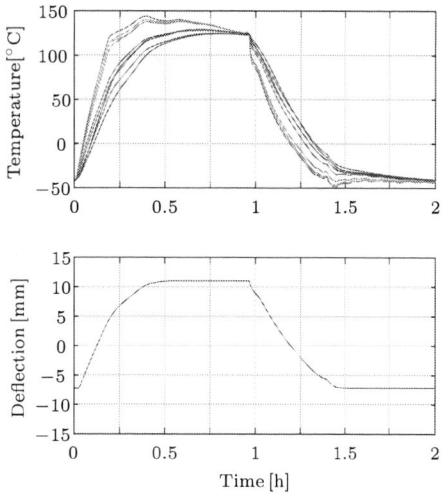

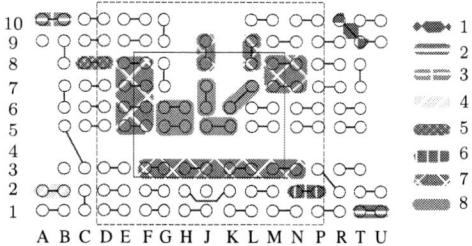

Figure 7. Top: Temperature profiles at different locations in the chamber. Bottom: Deflection distance in the center of the boards.

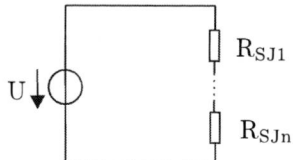

Figure 9. Schematic of electric circuit for electrical online measurements of daisy chain components.

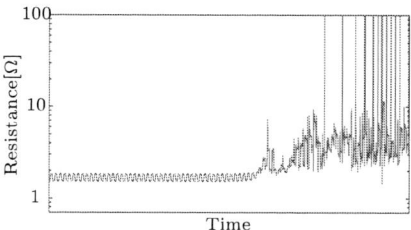

Figure 10. Resistance value over time for a daisy chain. Increase in resistance over $100\,\Omega$ indicates a cracked solder joint.

Figure 8. Internal daisy chain wiring of the BGA components. The highlighted solder joints show the 8 measurable daisy chains. The square with the dashed line shows the dimension of the silicon die of BGA1, the square with the solid line the die dimension of BGA2.

are orientated with their long side parallel to the bending direction due to limitations of the maximum number of measurement points. The same reason restricts the electrical online measurement to components on 8 out of the 10 test PCBs. A total of 1920 measurement points have been monitored over the 8 PCBs.

The resistance of a daisy chain is measured over the entire temperature cycle at a frequency of 0.1 Hz. An alternating resistance value between $1\,\Omega$ and $2\,\Omega$ due to temperature change can be seen for the electrically functioning joint in Fig. 10. The measurement at both high and low temperatures is necessary since a cracked solder joint might only cause an open circuit at a specific temperature. With the two point measurement performed in this test a significant change in the measured resistance can be seen only a few cycles before total failure, meaning $100\,\%$ crack of the solder joint. The failure criteria is defined to be met with the surpassing of a resistance value of $100\,\Omega$.

IV. BENDING LOAD

Depending on the position of the component on the test PCB, the solder joints will be subjected to different levels of strain and curvature. As described earlier, the deflection

of the PCBs is in relation to the temperature change in the chamber und thus resembles the conditions on system level on a housed PCB. The actual strain on the panels is measured at several positions with strain gauges in a quarter-bridge configuration. Over temperature, the total strain, consisting of thermal and mechanical strain, is measured. Strain gauges measurements on load level $1, 3, 4, 5, 6$ and 8 are available (cf. Fig 6). Strain values at the corner temperatures are listed in Table I only for load level $1 - 4$ for symmetry reasons. Also strain measurements are conducted at room-temperature deflecting the PCBs to the low temperature deflection (ltd) and high temperature deflection (htd) distance to measure the pure bending load without thermal strain. The curvature values are an approximation obtained from analytical calculation of a simply supported beam.

TABLE I. TOTAL STRAIN AND CURVATURE ON PANEL AT CORNER TEMPERATURES AT THE END OF DWELL TIMES.

Level	Strain $[\mu m/m]$ Measured $-40/125°C$	Measured RT : ltd/htd	Curvature $[1/m]$ Analytically RT : ltd/htd
1x = 8x 1y = 8y	$-1000/1600$ $-1100/1700$	$-60/70$ $15/-15$	$0.07/0.1$ $-$
2x = 7x 2y = 7y	$-$ $-$	$-$ $-$	$0.17/0.26$ $-$
3x = 6x 3y = 6y	$-1140/1750$ $-1100/1650$	$-200/240$ $30/-70$	$0.27/0.41$ $-$
4x = 5x 4y = 5y	$-1240/1900$ $-1100/1650$	$-320/385$ $40/-90$	$0.44/0.68$ $-$

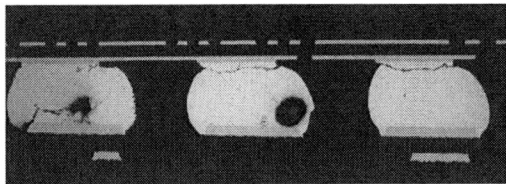

(a) U-R row 10.

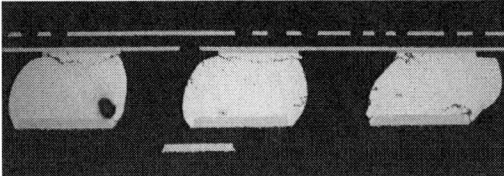

(b) C-A row 10.

Figure 11. Cross sections of BGA1 through center plane after 461 temperature cycles.

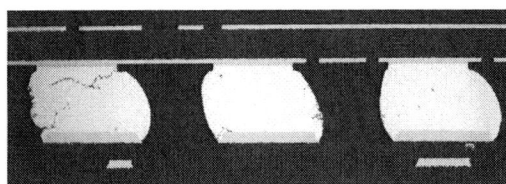

(a) U-R row 10.

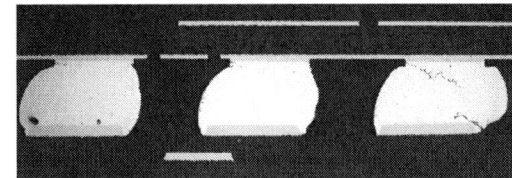

(b) C-A row 10.

Figure 12. Cross sections of BGA2 through center plane after 461 temperature cycles.

V. EXPERIMENTAL FINDINGS

A. Failure Mechanism

It is important to ensure for an accelerated temperature cycling test that the desired failure mechanism is triggered. Especially in a bending test different failure modes can be triggered through the additional bending load [14]. Besides solder joint cracks on component and PCB side, also pad cratering and delamination or cracks in the PCB copper traces could occur.

Cross sections of row 10 (compare Fig. 8) have been made for BGA1 and BGA2 of the components orientated parallel to the bending direction and positioned on the highest load levels 4 and 5. The two electrically not monitored test PCBs are removed after 461 temperature cycles and used for the cross sectioning. Displayed in Fig. 11 and 12 are the three outermost solder joints on both sides of row 10.

In all cross sections no other failure mode than solder joint cracks have been observed. Interestingly, the two adjacent solder joints to the corner joint of BGA1 show already 100 % cracks on component side in 3 cross section planes, whereas the corner solder joint has a significantly shorter crack that seems to extend over the diagonal of the solder joint. The cracks in the corner solder joints starts both form component and PCB side. The corner joint of BGA2 also shows the same crack growth behavior over the diagonal. The two adjacent solder joints, however, have only very little to no damage yet, making the corner solder joint the most likely, out of the solder joints displayed in Fig. 12, to fail first.

B. Solder Joint Lifetime Determination

For the evaluation of the detected electrical failures, components of BGA1 and BGA2 are grouped in catagories depending on the orientation of the component on the PCB, with the long side parallel to the bending direction or orthogonal (cf. Table II). With 4 PCBs of each BGA1 and BGA2 and 8 different load levels there are 64 components in each category. Only the respective daisy chains within the footprint of components on the same load level are finally grouped together for the determination of solder joint lifetime. Two components on one load level per category result over 4 PCBs in a maximum of 8 equally loaded daisy chains and thus 8 possible failures that can be used for statistical evaluation. However, during the test one electrically monitored PCB of BGA1 and BGA2 was removed for later cross sectioning. If daisy chains of the removed panels haven't failed the removed daisy chains are considered in the evaluation as right-censored data. This is also the case if daisy chains haven't failed after completion of the test. Only groups with at least 3 failures have been used for evaluation.

The daisy chain failures of each group have been used to fit a weibull distribution with the commercial software Weibull++ using a least squares method (cf. Fig. 13). With the weibull fit the characteristic life η, at which 63.2 % of the daisy chains of the specific group have failed, and the shape parameter β can be determined. Since at least 2 solder joints are in a daisy chain the number of cycles to failure cannot be assigned to one specific solder joint without the help of, for example, the finite element method to numerically determine the most loaded solder joint within a daisy chain.

TABLE II. CATEGORIES OF COMPONENTS FOR SOLDER JOINT EVALUATION.

	BGA1		BGA2	
Orientation	Parallel	Orthogonal	Parallel	Orthogonal
Category	1	2	3	4

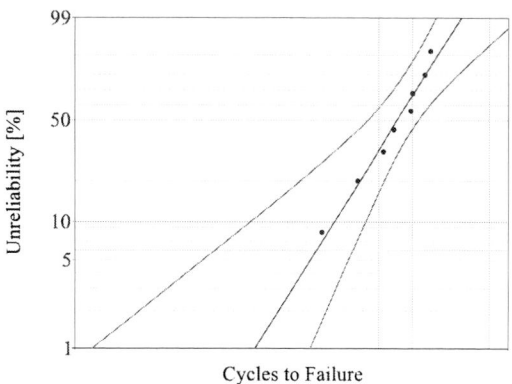

Figure 13. 2 parameter weibull fit [β=5.6, η] with 90 % confidence bounds of daisy chain 6 on load level 4 of category 1.

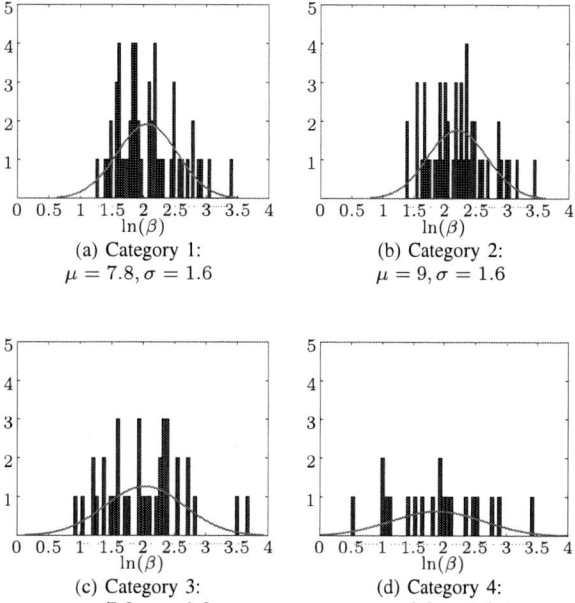

(a) Category 1: $\mu = 7.8, \sigma = 1.6$

(b) Category 2: $\mu = 9, \sigma = 1.6$

(c) Category 3: $\mu = 7.6, \sigma = 1.9$

(d) Category 4: $\mu = 6.6, \sigma = 2.1$

Figure 14. Lognormal distributions over the weibull formfactors.

The shape parameter β describes the failure rate of the daisy chains. Figure 14 shows histograms of the failure rates of the evaluated daisy chains in the 4 categories fitted with a lognormal distribution to determine a mean value μ. The values for of the mean β between 6.6 and 9 are in the region of common values for failure rates of BGA components on board level tests without bending [15]. Also the values of β obtained for the 4P-bending test of area array solder components by Vandevelde et al. at constant temperature are in a similar range [10]. Both the investigation of the failure mode and also the failure rate provide evidence that the test conditions with superimposed bending over passive temperature cycling did not trigger any unexpected effects.

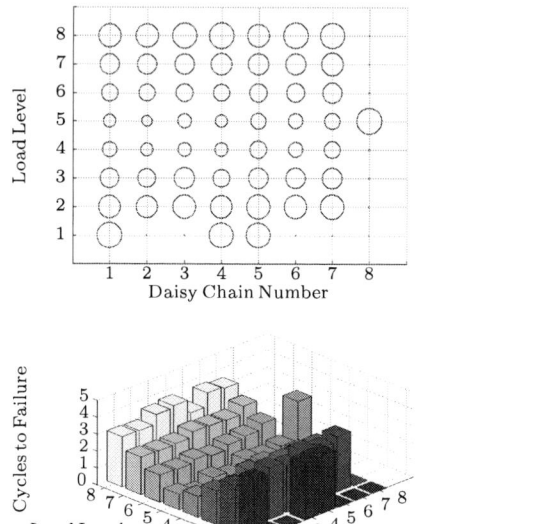

Figure 15. Category 1: Normalized cycles to failure of BGA1 components orientated parallel to bending direction. Circle size in top figure corresponds to cycles to failure. For exact daisy chain positions within the footprint of the component compare with Fig. 8 ($1 - 4$ corner area of footprint; $5, 6$ close to BGA1 die shadow; 7 BGA2 die shadow; 8 component center).

C. Impact of PCB strain, PCB curvature and Component Warpage on Solder Joint Lifetime

The normalized lifetime data for all categories in respect to load level and daisy chain numbers are plotted in Fig. 15-18. Two plots are used in these figures for the same data for better visualization. With these results the impact of PCB strain and curvature as well as component warpage on solder joint reliability is discussed.

Category 1: For the characteristic life of the daisy chains in category 1 a clear dependency on load level can be seen. Daisy chains of components on the highest loaded levels 4 and 5 fail first. For BGA1 the daisy chains $1 - 4$ right at or close to the corner solder joint fail before the daisy chains towards the center. A really significant jump in cycles to failure within the daisy chains on one load level can only be seen for daisy chain number 8 in the center of the component. With decreasing bending load towards the outside of the test PCB the lifetime increases continuously.

Category 2: BGA1 components orientated orthogonal to the bending direction still show a dependency to the load level, with daisy chains on high loaded levels in the center of the PCB failing first. However, this dependency is not as pronounced as for the components in category 1, orientated with the long side parallel to the bending direction.

Category 3: It can be immediately seen that significantly less daisy chains have failed for BGA2 at the end of the test. It can also be seen that daisy chains 2, 3 and

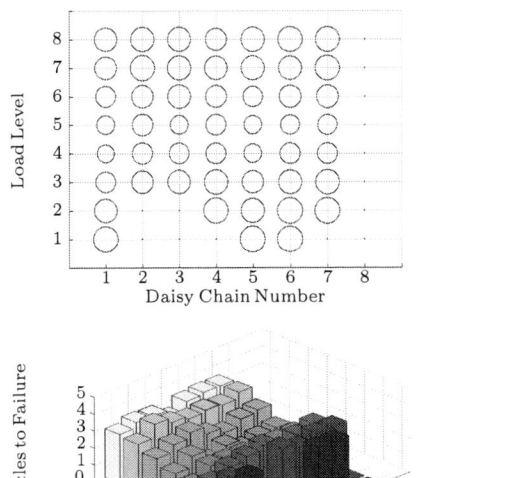

Figure 16. Category 2: Normalized cycles to failure of BGA1 components orientated perpendicular to bending direction.

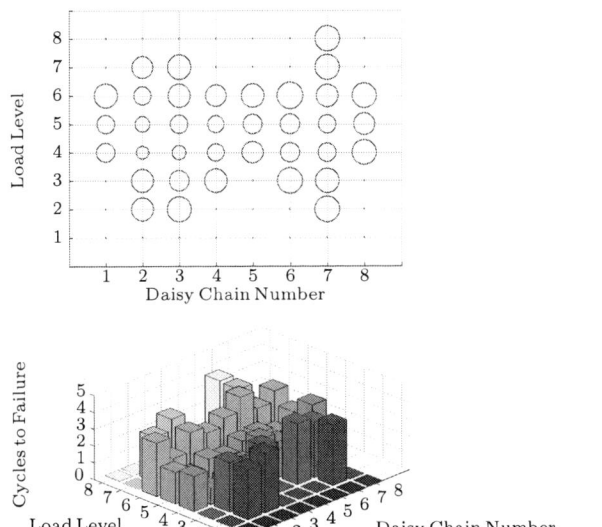

Figure 17. Category 3: Normalized cycles to failure of BGA2 components orientated parallel to bending direction.

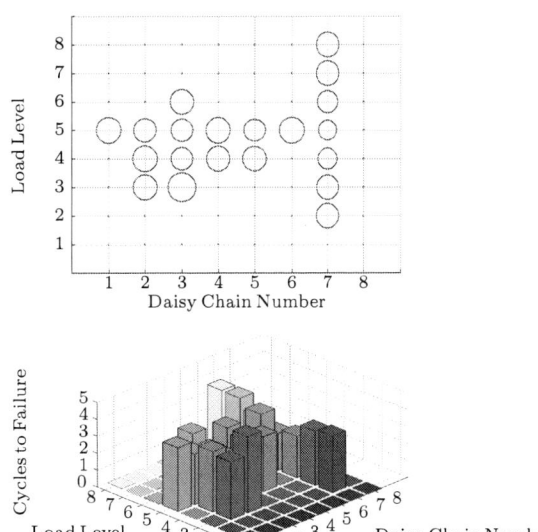

Figure 18. Category 4: Normalized cycles to failure of BGA2 components orientated perpendicular to bending direction.

Category 4: The fewest failures have been recorded for the daisy chains in category 4, BGA2 components orientated orthogonal to the bending direction. Similar trends can be seen as for the results in category 3. Also daisy chain 2 and 3 are higher loaded but the effect of the die shadow is clearly more pronounced as for the components in category 3 orientated parallel to the bending direction.

1) PCB Strain and Curvature: A dependency of PCB strain and curvature on solder joint lifetime can be seen for all categories. The most loaded solder joints are on load level 4 and 5 close to the center of the PCB, where the highest bending load occurs. As bending load decreases towards the outside of the PCB, lifetime increases. Under these test conditions, strain and curvature due to bending are coupled on the PCB and the impact on lifetime cannot be evaluated independently. By changing the orientation of the component on the PCB, both the effective strain and curvature applied on the component change. For both BGA1 and BGA2, components at high load levels fail earlier if oriented parallel to the bending direction. However, at low levels this clear difference cannot be observed anymore. The ratio of cycles to failure between category 1 and 2 is close to 1 for daisy chains on load level 1 and 8. This ratio increases with increasing anisotropy of the load towards the center of the PCB. The higher the PCB strain and the curvature, the higher the effect of component orientation.

2) Component Warpage: How component warpage influences solder joint reliability becomes apparent from comparing categories 1 and 2 with categories 3 and 4. Categories 1 and 2 with BGA1 have significantly more failures after

7 have more electrical failures than the other daisy chains in the footprint. Daisy chains 2 and 3 are the only daisy chains that include a corner solder joint. The fact that these fail earlier than daisy chain 1 or 4 does match the findings from the cross sections where the corner solder joint had a significantly longer crack than the adjacent solder joints (cf. Fig. 12). The failures of daisy chain 7 can be explained with the effect of the die shadow. The experimental results show a decrease of the load from the corner joints inwards, an increase underneath the die shadow and again a decrease in load towards the center of the component.

the end of the experiment than the categories of BGA2. The warpage measurements of the two components in Fig. 2 and Fig. 4 show that BGA1 exhibits both higher absolute warpage values and also a slightly greater delta warpage over temperature. Knowing that the same load is applied from the PCB, the difference in lifetime for BGA1 and BGA2 can be attributed to the different warpage behavior.

VI. CONCLUSION

The goal of the experiment was to investigate the influence of PCB strain, PCB curvature and component warpage behavior on the reliability of BGA solder joints. Within each of the 4 evaluated categories a clear dependency of solder joint lifetime on PCB strain and curvature can be seen. Comparing the results of the different categories, the impact of component warpage and also the impact of component orientation on the PCB is apparent. The results indicate that a metric based on PCB strain, PCB curvature and component warpage behavior could be developed for solder joint lifetime prediction of BGA components on system level.

With the significant amount of data gathered in this experiment the solder joint deformation based metric proposed by Schempp et al. will be further developed [16]. Also a correlation of local strain or energy based damage parameters with the experimental data will be established. This correlation will be used as a reference for the deformation based metric development.

Additionally, another passive temperature cycling test of the two BGA components will be conducted in housed PCBs to investigate the transferability of the board level bending test results to system level.

REFERENCES

[1] M. N. Tamin and N. M. Shaffiar, "Solder joint reliability assessment - finite element simulation methodology," *Springer International Publishing Switzerland*, vol. 37, 2014.

[2] J.-P. Clech, D. M. Noctor, J. C. Manock, G. W. Lynott, and F. E. Baders, "Surface mount assembly failure statistics and failure free time," *Proc. ECTC*, pp. 487–497, 1994.

[3] J.-P. Clech, "Acceleration factors and thermal cycling test efficiency for lead-free Sn-Ag-Cu assemblies," *Proc. SMTAI*, 2005.

[4] D. Pustan, "Belastungs- und Zuverlässigkeitsanalyse einer Ball-Grid-Array-Bauform: Von der Herstellung über den Einsatz in Kfz-Elektronik bis zum Ausfall im Test," *Dissertation, Albert-Ludwigs-Universität Freiburg im Breisgau*, 2011.

[5] S. S. Manson, "Thermal stress and low-cycle fatigue," *McGraw-Hill Book Company New York*, 1966.

[6] L. F. Coffin, "A study of the effects of cyclic thermal stresses on a ductile metal," *Trans. ASME*, vol. 76, pp. 931–950, 1954.

[7] A. Syed, "Accumulated creep strain and energy density based thermal fatigue life prediction models for snagcu solder joints," *Proc. ECTC*, no. 54, pp. 737–746, 2004.

[8] R. Darveaux, "Effect of simulation methodology on solder joint crack growth correlation," *Proc. ECTC*, 2000.

[9] R. Darveaux and A. Syed, "Reliability of area array solder joints in bending," *Proceedings of the SMTA*, pp. 313–324, 2000.

[10] B. Vandevelde, F. Vanhee, D. Pissoort, L. Degrendele, J. de Baets, B. Allaert, R. Lauwaert, R. Labie, and G. Willems, "Four-point bending cycling as alternative for thermal cycling solder fatigue testing," *Proc. EuroSimE*, pp. 1–5, 2016.

[11] B. Sabuncuoglu, F. Vanhee, G. Willems, B. Vandevelde, and D. Vandepitte, "Evaluation of fatigue behavior of lead-free solder joints in four-point bending test by finite-element modeling," *IEEE Trans. CPMT*, vol. 7, no. 12, pp. 1957–1964, 2017.

[12] D. Pustan, Z. Wenxin, and J. Wilde, "Experimental damage analysis and numerical reliability modeling of lead-free ball-grid-array second level interconnects," *Proc. ESTC*, pp. 1–6, 2010.

[13] J.-P. Clech, "Lead-free solder joint reliability," *Springer, Boston, MA*, 2007.

[14] R. Micheloni, G. Campardo, and P. Olivo, "Memories in wireless systems: Signals and communication technology," *Springer-Verlag Berlin Heidelberg*, 2008.

[15] E. Zukowski, "Probabilistische Lebensdauermodelle für thermomechanische Ermüdung von Lötverbindungen in CSP-Bauelementen," *Dissertation, Albert-Ludwigs-Universität Freiburg im Breisgau*, 2014.

[16] F. Schempp, M. Dressler, D. Kraetschmer, F. Loerke, and J. Wilde, "Introduction of a new metric for the solder joint reliability assessment of BGA packages on system level," *Proc. ECTC*, pp. 2192–2197, 2018.

Fatigue Life Prediction Model Development for Decoupling Capacitors

Krishna Tunga, Joseph Ross, Kamal Sikka and Bakul Parikh
IBM Corporation
Hopewell Junction, NY, USA
Email: ktunga@us.ibm.com

Abstract—There is an increasing need for a low cost solution to increase the speed and functionality of microprocessors. Decoupling capacitors mounted on the organic substrate close to the processor chip serve this purpose. Tin-Silver-Copper based solder materials are typically used to attach the capacitors to the substrate. Fatigue failure of the solder joints during field-use due to CTE mismatch between the capacitor and the substrate and due to proximity of the capacitor to the chip is a potential cause of concern. The solder joint materials exhibit highly rate dependent and microstructure dependent non-linear deformation behavior. The microstructure of the solder joints within a decoupling capacitor is very different from the microstructure of solder joints within a ball grid array package. Projecting fatigue life during field-use, where the temperature excursions experienced are generally less than 50 degrees centigrade, using a predictive model that was developed for a larger ball grid array solder joints or that was developed based on accelerated stress testing alone where the temperature excursions are much larger will lead to inaccurate life prediction. In this paper, we have developed a predictive model specifically for decoupling capacitors that can be used to determine the number of cycles for first fail and the mean number of cycles to failure during any thermal cycling conditions - ranging from harsh to benign - due to solder joint fatigue. Experimental data from test vehicles with 0306 and 0508 decoupling capacitors that were attached to the substrate using either SAC105 or SAC305 solder alloy material was used in conjunction with non-linear finite-element based mechanical modeling for fatigue life predictive model development.

Keywords: Decoupling capacitors, finite-element modeling, viscoplasticity, solder joint fatigue, MTLIC and acceleration factor

I. INTRODUCTION AND BACKGROUND

There is an increasing need for microprocessors that can function at higher speed and frequencies with lower noise and signal latencies without a significant addition to the cost. Decoupling capacitor has emerged as a low-cost solution that serves this purpose[1]. The effectiveness of a decoupling capacitor depends on its proximity to the chip. As shown in Fig. 1, decoupling capacitors are therefore typically placed on the substrate at various locations around the chip (option 1), below the substrate and chip shadow region (option 2), embedded within the substrate (option 3) [2], within the solder joint cage replacing the solder joints (option 4) [3] and within the chip back end of the line layers (option 5) [4]. The cost of implementing each of the options shown generally increases as the capacitors are placed closer to the chip. Option 1 is therefore the lowest cost alternative for increasing the microprocessor speed and functionality and is the focus of this paper.

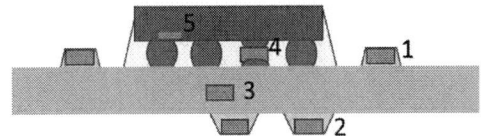

Figure 1. Location options for Decoupling Capacitor Placement

Solder alloys are used to attach the capacitors to the substrate (option 1 above). Environmental concerns have required the electronic industry to move away from using lead-based solder joints to lead-free solder joints with SnAgCu being the lead contender currently being used extensively in the industry. The electronics industry is also moving towards using organic substrates due to lower cost and to remain competitive. Organic substrates have a higher coefficient of thermal expansion (CTE), ranging from 12ppm/°C to 18ppm/°C, compared to the decoupling capacitors which are typically ceramic based and have a CTE ranging from 5ppm/°C to 12ppm/°C. The CTE mismatch between the capacitors and the organic substrate leads to increased stresses and strains within the solder joints connecting the capacitors to the substrate. When placed in close proximity to a large chip, the stresses and strains experienced by the solder interconnects within the capacitors increase considerably due to increased curvature of the substrate surface at regions closer to the chip. Failure of the capacitor due to fatigue of the solder interconnects during field-use due to thermal cycling is therefore a cause of concern. Field-use conditions are typically benign compared to the accelerated thermal cycling conditions used for qualifying an electronic package. Solder joint materials have a high homologous temperature and therefore exhibit significant time-dependent creep and a highly non-linear stress-strain deformation behavior [5][6] even at temperatures close to room temperature. Linearly extrapolating the failure rate observed during accelerated stress conditions to predict the life of the solder joints during more benign field-use conditions using temperature range alone could therefore lead to inaccurate prediction of solder joint failure risk. Norris-Landzberg [7], [8] type acceleration factor models are widely used in the industry to project field-use life using failure data from accelerated stress testing. Even though these models are non-linear, they are also limited in that they only take into account the temperature range and the thermal cycling frequency for life prediction. They don't consider additional factors [9] such as the complex non-linear behavior of solder materials, capacitor location, orientation, dimensions and high temperature dwell duration to provide a more accurate estimate of field-use life.

An extensive amount of work on predictive model development for SnAgCu based solder materials has been published in the literature [10]–[12]. However, most of the work is based on ball grid array (BGA) type solder joints that are used as interconnects between an active chip/package and

978-1-7281-1500-9/19 $31.00 © 2019 IEEE

a substrate/printed circuit board. The microstructure of the solder joints in a ball grid array package is generally very different from solder joints in a typical surface mounted capacitor. The models developed using BGA solder joint failure data therefore cannot always be used with confidence to predict the life of solder joints used in decoupling capacitors. There has been some work done in the recent past on predicting fatigue life of solder joints used in decoupling capacitors [13]–[15]. However, the study was limited to solder joints with higher silver content (SAC305/SAC405) or to solder joints that were exposed only to higher and more extreme temperature excursions seen during accelerated stress testing. Using such a model could lead to underprediction of field-use life thereby leading to a more conservative design than required.

In this work, a fatigue life predictive model is developed along with an acceleration factor model that can be used to predict the life of solder joints used in decoupling capacitors when subjected to any thermal cycling loading conditions – from harsh to benign. Application specific integrated circuit (ASIC) chips assembled on organic substrates were used to build test vehicles with decoupling capacitors. Two different types of decoupling capacitors – 0306 and 0508 – were assembled on an organic substrate using either SAC305 or SAC105 based solder alloy interconnects at various locations and orientations around the chip so as to introduce various levels of stresses and strains within the solder interconnects. These test vehicles were then subjected to three different temperature cycling conditions from harsh to benign: -55°C to -125°C (harsh), 0°C to 100°C (intermediate) and 25°C to 75°C (benign). Electrical integrity of the capacitors was monitored at regular intervals to determine the onset of failure. In parallel, the stresses, strains and plastic work experienced by the solder interconnects within the capacitors during the thermal cycling conditions were determined using non-linear Finite-element modeling. Experimental data in conjunction with simulation results was then used to develop a fatigue life prediction model that can be used to predict the life of decoupling capacitors when subjected to any given thermal cycling condition.

The experimental details including the test vehicle description, chip and capacitor assembly process and thermal stressing conditions will be outlined first. This will be followed by a presentation of the experimental results, modeling results and life prediction model development. The paper finally concludes with a summary of the key findings.

II. EXPERIMENTAL DETAILS

A. Test Vehicle Description

The description of the test vehicles with decoupling capacitors is shown in Table 1.

TABLE 1. TEST VEHICLE DESCRIPTION

Test Vehicle	Qty	Capacitor Type	Capacitor Quantity per Module	Solder Alloy
TV1	45	0306	8	SAC305
		0508	12	SAC305
TV2	45	0306	8	SAC105
		0508	12	SAC105

The test vehicle consisted of a 20mm x 20mm ASIC chip assembled on a 55mm x 55mm organic substrate. The substrates were baked, ashed and fluxed before chip and capacitor placement. Various 0306 and 0508 multi-terminal low inductance capacitors (MTLIC) with either SAC105 or SAC305 solder alloy were then placed at different locations and orientations around the chip on the substrate top surface as shown in Fig. 2. The 0306 capacitors spanned one of the two perpendicular orientations while the 0508 capacitors spanned one of the five different orientations as shown in the Fig. 2. The entire assembly was then reflowed to attach and bond the capacitors and the chip on the substrate surface. A 1 mm thick copper lid was adhered on the substrate surface along a 3 mm wide region around the periphery using silicone based seal as shown in Fig. 2. In addition, the center part of the lid was also adhered to the chip top surface using a thermal interface material.

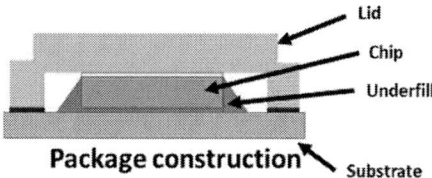

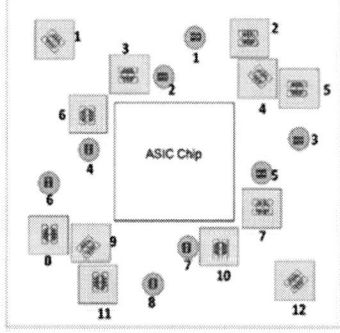

Figure 2. Test Vehicle Indicating the Placement of the 0306 (Blue Circles) and 0508 (Orange Squares) Capacitors

Both the 0306 and 0508 capacitor types have eight pads per capacitor. The 0508 capacitor was larger compared to the 0306 capacitor and had larger pads and larger post-reflow solder fillet size around the pads. The outer terminal pads of a capacitor were electrically connected though a daisy chain which allowed for monitoring the integrity and reliability of the capacitor during stress testing.

B. Accelerated Stress Testing

All the test vehicles were stressed by subjecting them to one of the three different thermal cycling conditions – from benign to harsh – as shown in Table 2.

TABLE 2. TEST VEHICLE STRESSING CONDITIONS

Test Vehicle	Qty	Thermal Cycling Condition
TV1	15	-55°C to 125°C (Harsh)
	15	0°C to 100°C (Intermediate)
	15	25°C to 75°C (Benign)
TV2	15	-55°C to 125°C (Harsh)
	15	0°C to 100°C (Intermediate)
	15	25°C to 75°C (Benign)

The actual thermal cycling profiles the test vehicles were subjected to in the thermal chambers were determined by measuring the temperatures using thermo-couples that were attached directly to the modules. The measured temperature profiles are shown in Fig. 3.

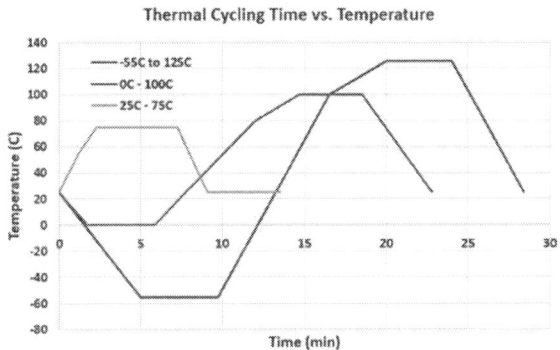

Figure 3. Thermal Chamber Temperature Profile

The test vehicles were taken out of the chamber at 50 or 100 cycle intervals initially, at 250 cycle intervals after 5000 cycles, at 500 cycle intervals after 15000 cycles and electrically tested, using a four point resistance measurement from the outer legs of the terminal through the body of the capacitor, to check for the integrity of the capacitors. A failure was deemed to have occurred when the daisy chain resistance increased by more than 20% of its initial value.

III. EXPERIMENTAL RESULTS

All test vehicles have completed over 40,000 cycles of accelerated stress testing. The number of cycles for first fail (FF) and the mean cycles to failure (MTTF) were determined assuming either a Weibull or a Lognormal distribution for the failure data. Plots showing a comparison of FF and MTTF for a given alloy and capacitor type along with 95% confidence intervals are shown in Fig. 4. Each data point in the plot represents a capacitor at a given location on the module. Both FF and MTTF are positively correlated for all four cases

shown. Capacitors that fail first can therefore be expected to also have a lower mean life to failure. Some of the outliers that were observed occurred for capacitors that were subjected to benign thermal cycling conditions where the variability in the FF and MTTF was higher compared to the other two conditions.

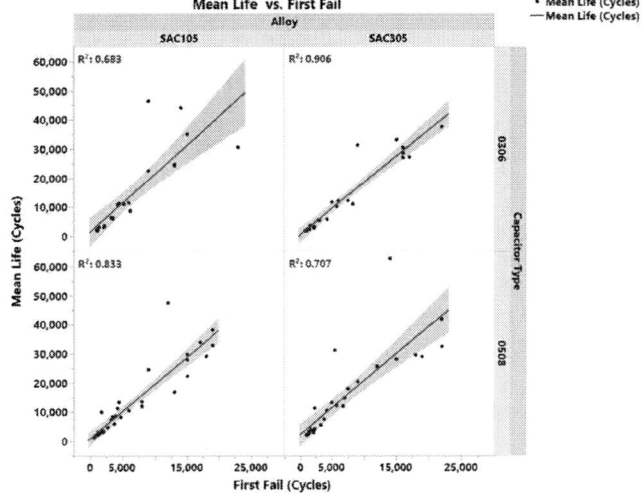

Figure 4. First Fail vs. Mean Life Correlation Plot

The variation of FF and MTTF as a function of thermal cycling temperature range are shown in Figs. 5 and 6 respectively. Each bar represents an average across all capacitor locations for a given capacitor type and solder alloy. The variability within each bar is due to the variability in life seen between capacitors placed at different locations on the module. As the temperature range of cycling decreases, the damage accumulated within the solder alloy decreases and this increases both FF and MTTF as seen in the Figures. Since, the damage accumulated within the solder is cumulative in nature, it keeps building up at different rates for capacitors placed at different locations and reflects as high variability in FF/MTTF as the number of cycles increases.

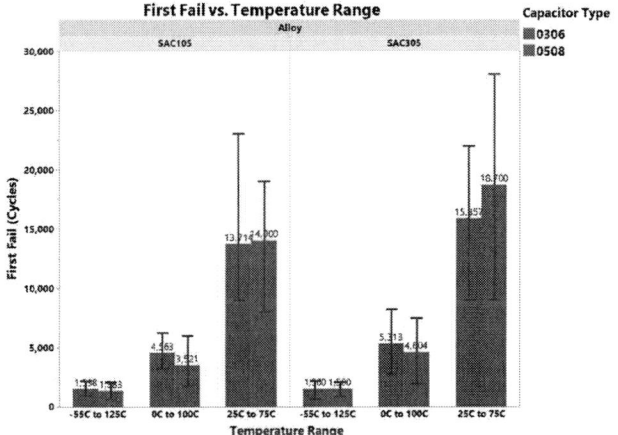

Figure 5. First Fail Data

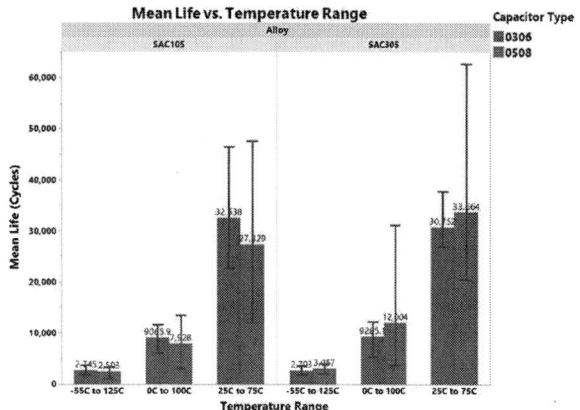

Figure 6. Mean Failure Life Data

To evaluate the effect of location on the capacitor life, both FF and MTTF were correlated with distance from the chip edge as shown in Fig. 7.

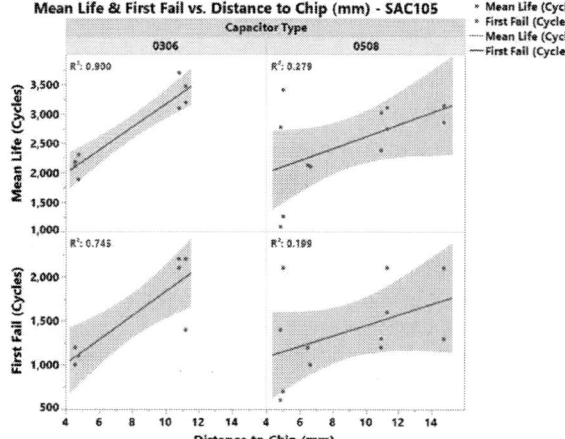

Fig. 7a)

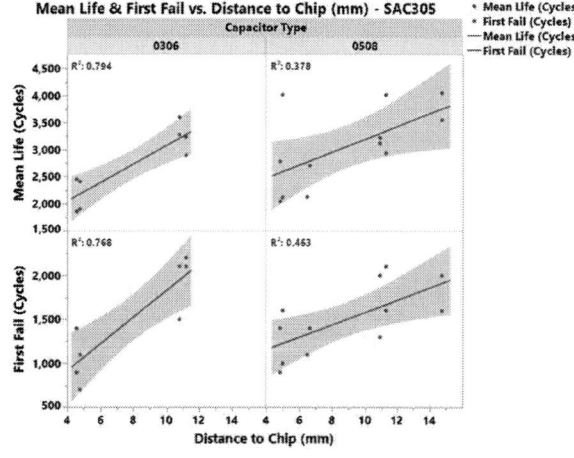

Fig. 7b)

Figure 7. Fatigue Life vs Distance from Chip with 95% Confidence Interval for a) SAC105 b) SAC305 Alloy

Experimental data for both solder alloys indicates a positive correlation between fatigue life (both FF and MTTF)

and distance from the chip edge. Due to high CTE mismatch between the chip and the substrate, the substrate surface close to the chip edge has a high degree of curvature. Capacitors placed close to the chip edge therefore experience higher strain levels and hence a lower fatigue life. For both solder alloys, the correlation between fatigue life and distance was higher for the 0306 capacitors compared to the 0508 capacitors. The twelve 0508 capacitors placed on the module spanned five different orientations. The fatigue life was therefore not just dependent on the distance from the chip edge but also on the orientation of the capacitor thereby leading to a lower correlation value with distance alone. This can also be seen from Figs. 5 and 6, where we see a higher variability in FF and MTTF for 0508 capacitor type compared to the 0306 capacitor type for both solder alloys and all three temperature excursions.

The failed samples were cross-sectioned to determine the exact failure location. All observed failures occurred within the solder material between the capacitor and the substrate. A cross-section of a typical failed location is shown Fig. 8.

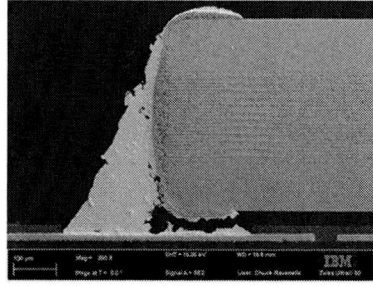

Figure 8. Typical Failure Site After Accelerated Thermal Cycling

IV. MECHANICAL MODELING

Solder joint failure during thermal cycling is known to be directly related to the stresses and strains the solder joint experiences during thermal cycling. The stresses and strains lead to damage within the solder that keeps on accumulating during thermal cycling. When the damage exceeds a threshold, which depends on the solder material, the solder joint fails leading to an electrical open. Finite element based mechanical modeling can be used to quantify the damage the solder joint experiences when exposed to repetitive thermal cycling conditions. Volume averaged plastic work per cycle [16], [17] and plastic strain range per cycle [18] are common metrics that have been used in the past to quantify the damage accumulated within the solder joint.

A finite element model of the ASIC module test vehicle with all the 0306 and 0508 capacitors was created using ANSYS v18. The nominal dimensions of the capacitors as provided by the vendor was used in the model. The dimensions of the solder joint fillet and thickness between the capacitor and the substrate were determined by cross-sectional analysis of an assembled module. The finite-element model and the meshing details for the 0306 and 0508 capacitors are shown in Fig. 9. The solder joints and the capacitor were modeled using VISCO107 linear elements and all the remaining materials within the module were modeled using SOLID186 quadratic elements.

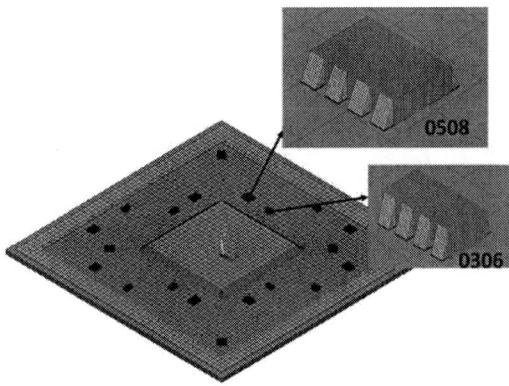

Figure 9. Finite-Element Modeling

The substrate, underfill and the seal materials were modeled as temperature dependent linear elastic materials. The material properties were determined using in-house testing tools. The silicon chip, copper lid and the ceramic capacitors were modeled as temperature independent linear elastic materials. The SAC305 and SAC105 solder alloys were modeled using Anand's viscoplastic model. Anand's viscoplastic constants for SAC305[19] and SAC105[20] solder already available in literature were used in the models. A stress-free temperature of 150°C was used for the entire module. The model was subjected to thermal cycling loading similar to what the test vehicles experienced in the thermal chamber as shown in Fig. 3. A typical plot of the plastic work accumulated per cycle within the solder joints located between the capacitor pads and the substrate is shown in Fig. 10. The maximum plastic work/cycle occurs within the outer solder pads. The maximum plastic work was also found to occur mostly at the solder/capacitor interface with the location changing to the solder/substrate interface under few cases. The plastic work accumulated per cycle was therefore volume averaged across these two interface regions as shown in Fig. 10 for all eight pads within a capacitor. The maximum value across both the regions and across all eight pads was chosen as being representative of the damage experienced by the capacitor.

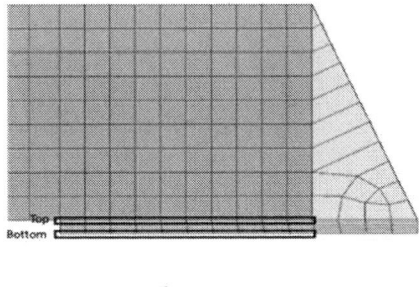

Figure 10. Top and Bottom Volume Averaged Regions with a Typical Contour Plot of Plastic Work/Cycle

The solder material being rate dependent, exhibits hysteresis during thermal cycling. It takes few thermal cycles for the plastic work accumulated per cycle to stabilize [21]. In our previous work [15], we have shown that it takes about three to five thermal cycles for the hysteresis behavior to stabilize. The volume averaged plastic work accumulated during the stabilized cycle was used for fatigue life predictive model development.

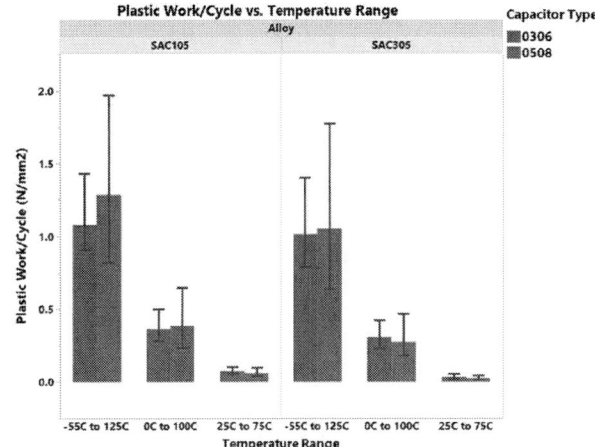

Figure 11. Plastic Work Dependency on Temperature Range and Solder Alloy

The dependency of plastic work/cycle for all capacitors of a given type as a function of solder alloy and thermal cycling temperature range is shown in Fig. 11. On an average, we can see that the plastic work accumulated during the -55°C to 125°C thermal cycling condition was found to be about 3X and 12X higher compared to the plastic work accumulated during the 0°C to 100°C and 25°C to 75°C thermal loading conditions respectively. We can also see that the location dependent variability in plastic work is higher for 0508 capacitors due to higher number of orientations spanned by this capacitor type. This trend confirms the location dependent variability seen in experimental data.

V. FATIGUE LIFE PREDICTION MODEL

Experimental data in conjunction with mechanical modeling results was used to develop a fatigue life predictive model for FF and MTTF for decoupling capacitors. Both FF and MTTF were plotted and non-linearly regressed against the plastic work accumulated per cycle for each decoupling capacitor. Separate plots were made for SAC105 and SAC305 based alloys and are shown in Figs. 12 and 13. The plots also show a 95% confidence interval around the fitted line.

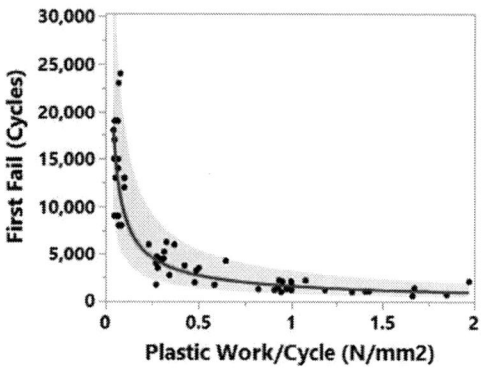

Fig. 12a) Initial Fail

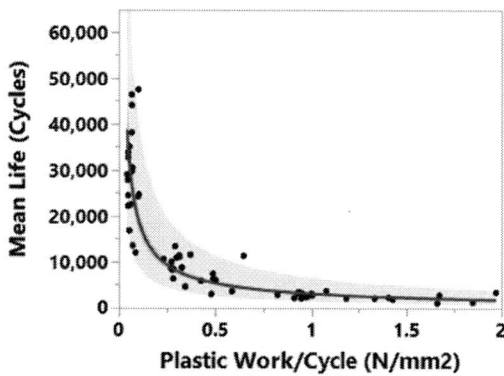

Fig. 12b) Mean Life to Failure

Figure 12. Predictive Model Fit for Capacitors with SAC105 Solder Alloy

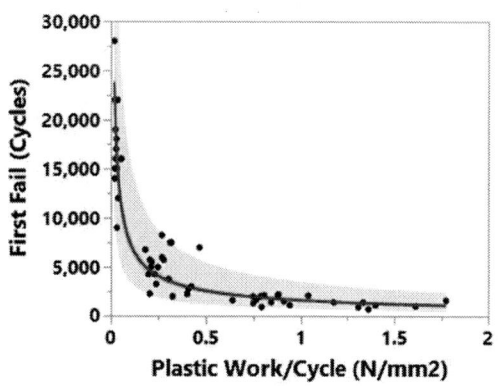

Fig. 13a). Initial Fail

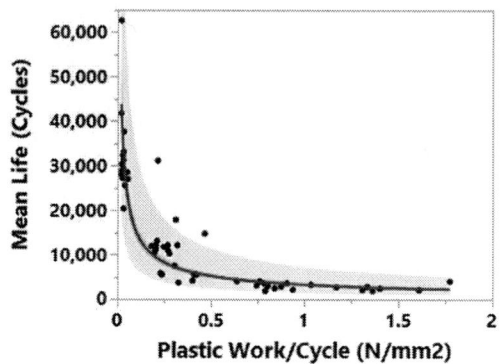

Fig. 13b) Mean Life

Figure 13. Predictive Model Fit for Capacitors with SAC305 Solder Alloy

Using the data from the fit, a fatigue predictive model for FF and MTTF can be obtained as shown in Equation 1.

$$\text{FF or MTTF} = A\Delta W^n \qquad (1)$$

Where, 'ΔW' represents the stabilized volume averaged plastic work accumulated per cycle during thermal cycling and 'A' and exponent 'n' are constants for the predictive model. The values for the two constants as a function of solder alloy composition, capacitor type and the failure level are listed in Table 3. Along with the mean values, the minimum and maximum values for the constants based on 95% confidence interval are also listed. The R^2 value for all the fitted equations are greater than 80% indicating a good fit between modeling results and experimental data.

TABLE 3. PREDICTIVE MODEL CONSTANTS

Alloy	Capacitor	Failure Level	R^2(%)	A	A_{min}	A_{max}	n	n_{min}	n_{max}
SAC105	306	Initial	88.6	1694	1380	2078	-0.798	-0.928	-0.668
SAC105	508	Initial	89.0	1538	1305	1810	-0.753	-0.846	-0.661
SAC105	0306 and 0508	Initial	88.3	1605	1415	1819	-0.765	-0.839	-0.691
SAC305	306	Initial	85.8	1656	1283	2137	-0.694	-0.822	-0.566
SAC305	508	Initial	89.0	1562	1302	1874	-0.661	-0.744	-0.577
SAC305	0306 and 0508	Initial	87.5	1605	1389	1856	-0.671	-0.739	-0.602
SAC105	306	Mean	94.1	3047	2589	3583	-0.908	-1.011	-0.805
SAC105	508	Mean	86.0	3032	2512	3659	-0.754	-0.860	-0.648
SAC105	0306 and 0508	Mean	87.7	3078	2689	3523	-0.800	-0.880	-0.720
SAC305	306	Mean	91.4	3039	2495	3696	-0.708	-0.807	-0.610
SAC305	508	Mean	82.4	3429	2744	4290	-0.643	-0.754	-0.532
SAC305	0306 and 0508	Mean	86.0	3268	2813	3793	-0.669	-0.743	-0.594

The constants along with the 95% confidence interval ranges are plotted in Figs. 14 and 15 for comparison. The constant A is indicative of the fatigue life in an average sense and the constant n is indicative of the sensitivity of fatigue life to plastic work/cycle. A higher value of magnitude n therefore signifies a higher rate of failure for a given plastic work/cycle.

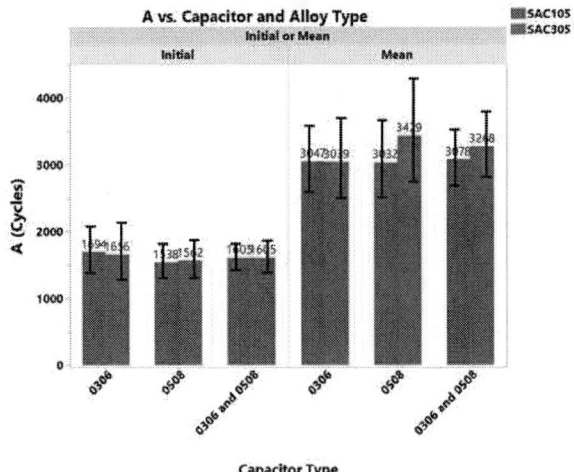

Figure 14. Variation of Predictive Model Constant 'A'

Since, MTTF is always higher than FF, the constant A used to determine MTTF is always higher as shown in Fig. 14. The constant A, however, was statistically similar and found to be not very sensitive to the capacitor type or solder alloy composition.

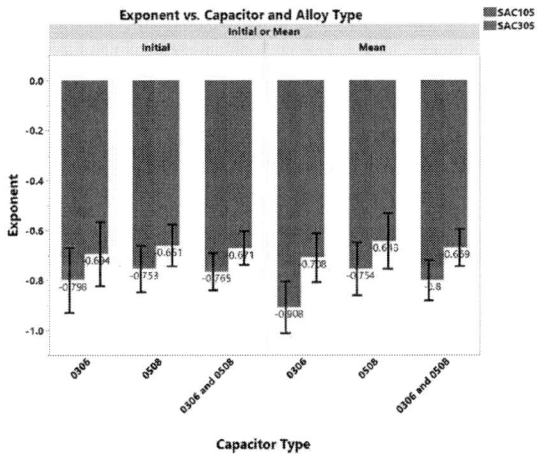

Figure 15. Variation of Predictive Model Exponent

As shown in Fig. 15, the exponential factor n shows a clear dependency on solder alloy. The constant n is higher for the SAC105 alloy indicating a higher rate of failure compared to the SAC305 alloy for a given plastic work accumulated within the solder. The constant n is also marginally higher for 0306 capacitor despite it being of a smaller size compared to 0508 capacitor. This could be due to the smaller pad size and fillet size for the 0306 capacitors which tends to increase the stresses experienced by the solder joints and more than compensates for the decrease in stresses due to smaller body size.

In our earlier work[15] we had determined predictive model constants based on data from -55°C /125°C and 0°C /100°C thermal cycling only. Fatigue data for thermal cycling between 25°C /75°C was not taken into account because it was not available. The values listed in Table 3 is therefore different from values we had published earlier. A comparison between our previously published work and current work is shown in

Table 4. For both solder alloys, the fitted constant A is comparable and within the 95% confidence limits listed in Table 3. However, the constant n is lower compared to our earlier published work suggesting that by not considering failure rates at lower temperature range excursions, a higher capacitor failure rate is predicted which might lead to an overly conservative design than warranted.

TABLE 4. COMPARISON TO EARLIER PUBLISHED WORK

Alloy	Constants	Earlier Published Work	This Work
SAC105	A	2860	3078
SAC105	n	-0.891	-0.800
SAC305	A	2875	3268
SAC305	n	-0.799	-0.669

VI. ACCELERATION FACTOR MODEL

Field-use conditions are characterized by thermal cycling conditions with temperature ranges that are typically under 50°C[22]. Determining the number of cycles to failure during field-use during the package qualification stage is therefore time consuming and not practical. An acceleration factor helps estimate the FF and MTTF during field-use conditions quickly using existing accelerated thermal cycling failure data. Norris-Landzberg[7] type acceleration factor equations are widely used in industry to estimate the number of cycles to failure during field-use. In addition to temperature cycling range, these equations also factor in the frequency of loading or switch on/off cycles during field-use. However, these equations cannot take into account the location, orientation and the dimensions of the capacitors for projecting failure. Finite-element based mechanical modeling offers an advantage in that it not only considers these additional parameters but can also simulate the exact temperature profile during accelerated testing and field-use and provide a more accurate estimate of failure rate during field-use.

Experimental data in conjunction with modeling results was used to determine the acceleration factor. In our earlier work [15] we had presented an Arrhenius type acceleration factor equation for MTTF. This was possible because the reciprocal of the plastic work/cycle had good linear correlation with the log of the MTTF. However, after including the failure data from thermal cycling between 25°C /75°C, we see that the linear correlation does not hold and Arrhenius type relationship breaks down as shown in Fig. 16.

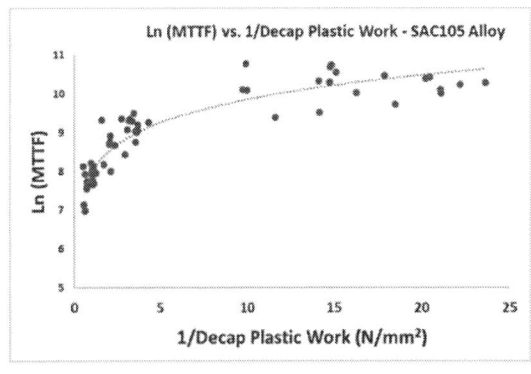

Fig. 16a) SAC105 alloy

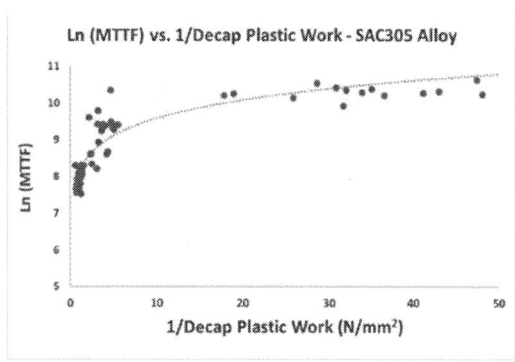

Fig. 16b) SAC305 alloy

Figure 16. Non-Linear Relationship between log(MTTF) and 1/Plastic Work/Cycle

Acceleration factor was therefore determined using a ratio of predicted life during accelerated thermal cycling and during field-use as shown below.

$$AF_{FF \text{ or } MTTF} = \frac{N_{ATC}}{N_{FU}} = \left[\frac{\Delta W_{ATC}}{\Delta W_{FU}}\right]^n \qquad (2)$$

Where, AF is the acceleration factor for FF or MTTF, N_{ATC} and N_{FU} represent the number cycles for FF or MTTF during accelerated thermal cycling and field-use conditions respectively. ΔW_{TC} and ΔW_{FU} represent the stabilized volume averaged plastic work accumulated per cycle during accelerated thermal cycling and field-use conditions respectively. The constant n shown in Equation 2 is the same constant listed in Table 3.

VII. SUMMARY AND CONCLUSIONS

Failure of the decoupling capacitors, mounted on an organic substrate, due to solder joint fatigue from thermal cycling is a cause of concern. Organic substrate based ASIC module test vehicles with decoupling capacitors were assembled. Two different types of capacitors – 0306 (smaller) and 0508 (larger) – were mounted on top of the substrate at different locations and orientations around the chip. The capacitors were attached using either SAC305 or SAC105 based solder alloy. The modules were then stress tested by subjecting them to one of the three different thermal cycling conditions for over 40,000 cycles or until the mean life to failure can be projected: -55°C to 125°C (harsh), 0°C to 100°C (intermediate) and 25°C to 75°C (benign).

Experimental data showed a positive correlation between FF and MTTF indicating that capacitors that fail first are also expected to have a lower mean life to failure. The fatigue life was found to be sensitive to location and orientation with the fatigue life being lower for capacitors that are located closer to chip edge.

Stabilized volume averaged accumulated plastic work per cycle as determined from Finite element modeling was used as a metric for solder joint damage. The plastic work accumulated in the solder during harsh thermal cycling condition was found to be about three times higher and twelve times higher compared to the plastic work accumulated during the intermediate and benign thermal cycling conditions respectively. Fatigue life prediction models for FF and MTTF were developed using modeling results in conjunction with experimental data. Separate prediction model constants for different capacitor types and solder alloys have been provided. Good correlation with an R-squared value greater than 80% was observed for all the fitted life predictive models. Models for SAC105 alloy had a higher exponent compared to SAC305 alloy indicating a higher failure rate for a given plastic work accumulated within the solder. It was also found that including failure data at lower temperature range excursions for the predictive model development is necessary to avoid over-predicting the failure rate and leading to an overly conservative design.

Acceleration factor model for projecting life during field-use has also been provided. The acceleration factor model is more comprehensive compared to the traditional Norris-Landzberg model in that it considers the capacitor location, orientation, dimensions and the exact temperature profile during accelerated testing and field-use to provide a more accurate estimate of failure rate during field-use.

ACKNOWLEDGMENTS

The authors would like to thank Eric Kastberg for scheduling and coordinating the thermal cycling activity for all the test vehicles.

REFERENCES

[1]   A. Sanna and G. Graziosi, "Optimization of on-package decoupling capacitors considering system variables," in *2018 IEEE 22nd Workshop on Signal and Power Integrity (SPI)*, 2018, pp. 1–4.

[2]   P. Muthana *et al.*, "I/O Decoupling in High Speed Packages Using Embedded Planar Capacitors," in *2007 Proceedings 57th Electronic Components and Technology Conference*, 2007, pp. 299–304.

[3]   C. Arvin, J. Audet, B. Quinlan, C. Reynolds, and B. Sundlof, "Multi terminal capacitor within input output path of semiconductor package interconnect," US20180012838A1, 2016.

[4]   T. Charania, A. Opal, and M. Sachdev, "Analysis and Design of On-Chip Decoupling Capacitors," *IEEE Trans. Very Large Scale Integr. Syst.*, vol. 21, no. 4, pp. 648–658, 2013.

[5]   R. S. Sidhu, X. Deng, and N. Chawla, "Microstructure characterization and creep behavior of Pb-free Sn-rich solder alloys: Part II. Creep behavior of bulk solder and solder/copper joints," *Metall. Mater. Trans. A Phys. Metall. Mater. Sci.*, vol. 39, no. 2, pp. 349–362, 2008.

[6]   S. T. Jenq, Y. S. Chiu, R. J. Lin, and Y. S. Lai, "The stress-strain relationship of Sn63Pb37 and SAC305 solder materials at elevated temperature condition," in *International Microsystems Packaging Assembly and Circuits Technology Conference, IMPACT 2010 and International 3D IC Conference, Proceedings*, 2010, pp. 1–4.

[7]   K. C. Norris and A. H. Landzberg, "Reliability of Controlled Collapse Interconnections," *IBM J. Res. Dev.*, 1969.

[8]   R. Dudek *et al.*, "Solder fatigue acceleration prediction and testing results for different thermal test- and field cycling

environments," in *Proceedings of the 5th Electronics System-Integration Technology Conference, ESTC 2014*, 2014, pp. 1–8.

[9] A. Syed, "Limitations of Norris-Landzberg equation and application of damage accumulation based methodology for estimating acceleration factors for Pb free solders," in *2010 11th International Thermal, Mechanical & Multi-Physics Simulation, and Experiments in Microelectronics and Microsystems (EuroSimE)*, 2010, pp. 1–11.

[10] J. Liu, W. Hu, H. Chen, and J. Wang, "A fatigue life model of BGA solder joints based on energy," in *2017 2nd International Conference on Reliability Systems Engineering, ICRSE 2017*, 2017, pp. 1–6.

[11] B. Metais *et al.*, "A viscoplastic-fatigue-creep damage model for tin-based solder alloy," in *2015 16th International Conference on Thermal, Mechanical and Multi-Physics Simulation and Experiments in Microelectronics and Microsystems, EuroSimE 2015*, 2015, pp. 1–5.

[12] K. Tunga and S. K. S. Sitaraman, "Predictive Model Development for Life Prediction of PBGA Packages with SnAgCu Solder Joints," *IEEE Trans. COMPONENTS Packag. Technol.*, vol. 33, no. 1, pp. 84–97, 2010.

[13] G. T. Ostrowicki, J. Williamson, V. Gupta, and S. P. Gurrum, "Thermal cycling reliability of lead free solder joints on multi-terminal passive components," in *Proceedings - Electronic Components and Technology Conference*, 2015, vol. 2015–July, pp. 127–134.

[14] J. M. Williamson, G. T. Ostrowicki, V. Gupta, S. P. Gurrum, and A. Zhang, "Impact of Lead Free Solder Joint Orientation on Multi-Terminal Passive Components during FCBGA Board Level Reliability," in *Proceedings - Electronic Components and Technology Conference*, 2016, vol. 2016–Augus, pp. 1112–1118.

[15] K. Tunga *et al.*, "Fatigue Life Predictive Model and Acceleration Factor Development for Decoupling Capacitors," in *Proceedings of the 17th InterSociety Conference on Thermal and Thermomechanical Phenomena in Electronic Systems, ITherm 2018*, 2018, pp. 1169–1176.

[16] R. Darveaux, "Effect of simulation methodology on solder joint crack growth correlation and fatigue life prediction," *J. Electron. Packag.*, 2002.

[17] S. Hamasha, A. Qasaimeh, Y. Jaradat, and P. Borgesen, "Correlation Between Solder Joint Fatigue Life and Accumulated Work in Isothermal Cycling," *IEEE Trans. Components, Packag. Manuf. Technol.*, vol. 5, no. 9, pp. 1292–1299, 2015.

[18] N. Fu, J. C. Suhling, and P. Lall, "Cyclic Stress-Strain Behavior of SAC305 Lead Free Solder: Effects of Aging, Temperature, Strain Rate, and Plastic Strain Range," in *2016 IEEE 66th Electronic Components and Technology Conference (ECTC)*, 2016, pp. 1119–1127.

[19] K. Mysore, G. Subbarayan, V. Gupta, and R. Zhang, "Constitutive and aging behavior of Sn3.0Ag0.5Cu solder alloy," *IEEE Trans. Electron. Packag. Manuf.*, vol. 32, no. 4, pp. 221–232, 2009.

[20] D. Bhate, D. Chan, G. Subbarayan, T. C. Chiu, V. Gupta, and D. R. Edwards, "Constitutive behavior of Sn3.8Ag0.7Cu and Sn1.0Ag0.5Cu alloys at creep and low strain rate regimes," *IEEE Trans. Components Packag. Technol.*, vol. 31, no. 3, pp. 622–633, 2008.

[21] S. K. Sitaraman, R. Raghunathan, and C. E. Hanna, "Development of virtual reliability methodology for area-array devices used in implantable and automotive applications," *IEEE Trans. Components Packag. Technol.*, vol. 23, no. 3, pp. 452–461, 2000.

[22] K. Tunga, "Experimental and theoretical assessment of PBGA reliability in conjunction with field-use conditions," 2004.

A Study Of Substrate Models And Its Effect On Package Warpage Prediction

Van-Lai Pham, Huayan Wang, Jiefeng Xu, Jing
Wang, and Seungbae Park
Department of Mechanical Engineering
State University of New York at Binghamton
NY, USA
e-mail: vpham3@binghamton.edu

Charandeep Singh
Corning Inc.
NY, USA

Abstract—In this work, a study of different substrate models on package warpage is performed. Three different substrate models are built: First, the package substrate is simplified and modeled as one effective layer. Second, a three-layer model is proposed with a top build-up layer, the middle that contains a low Coefficient of Thermal Expansion (CTE) core, and a bottom build-up layer. Third, a multi-layer laminate substrate that accounts for the complexity of copper trace, and the combination of polymer or non-metallic materials is considered. Package warpage among these models are compared and evaluated by both an analytical approach and Finite Element Analysis (FEA) in conjunction with the empirical data. The analytical and FEA results reveal that the three-layer model and multi-layer model could predict package warpage behavior in close approximation to the experimental results, whereas the one effective model provides an outlier quantity. A small amount of uncorrected warpage prediction may result in a large discrepancy of service life assessment of interconnected solder joints. The multi-layer model with detailed copper trace configuration is prohibitively expensive, while one effective layer could not represent correctly the major mechanical properties of the substrate; above all, the three-layer model is an optimal consideration and is recommended to have a proper FEA model for a more exact life prediction of solder interconnections.

Keywords- stack up substrate model; digital image correlation; finite element analysis; classical laminate theory, warpage prediction

I.   INTRODUCTION

Finite Element Analysis (FEA) has been utilized to help industry accelerate decision-making in the early product development phase to avoid production costs. An accurate FEA model helps in safe and reliable decision making. Warpage, which is mostly caused by CTE mismatch between the substrate and the package, is a critical issue for electronic packages. Therefore, warpage prediction has drawn industrial concerns. To obtain correct package warpage prediction, a FEA model should accurately depict the thermal mechanical properties of the substrate as well as package constituents. While the die package is solid, there are numerous effort that try to effectively model the substrate simply yet accurately [1]. A variety of simplified approaches has been discussed, however, the selection of a practical and reliable model to precisely predict the warpage is not straightforward. This is because each model has its own drawbacks [2, 3]. Apparently, warpage prediction that features the metal copper trace and non-metallic material

detail is superior; however it is expensive and unaffordable at most times. Considering the whole substrate as only one effective layer and that its properties are determined from measurement data is an oversimplified substrate model and that could provide imprecise warpage prediction due to deficient properties in the model. Reduction of the overall thickness of the microelectronics as well as the substrate is the ongoing trend. As the substrate layer thickness is reduced, the copper trace is prominent compared to the rest of the polymer or non-metallic material. Consequently, the overall mechanical properties of the substrate such as CTE and Young's modulus profoundly depend on the properties of copper. Modeling a substrate structure in the presence of complex pure copper trace and a mixture of other non-metallic composites at each laminate is excessively costly.

Package substrates are usually comprised of several laminate composites and a middle core layer. The substrate with low CTE and high Young's modulus of the core layer has been used widely in electronic packages because it is believed to reduce the overall package substrate CTE. Stiffness of core material usually is much higher than that of buildup laminate materials so that the core can assure the substrate rigidity mechanically, while stack up of buildup layers provides a conformity to the solder joint. Lower overall substrate CTE results in a lower level of stress imposed on the interconnections, and the subsequent enhancement of solder joint endurance.

One effective layer model has been commonly used [4-7]; however, there is an issue where the effective properties of the composite substrate are measured experimentally on the surface. With this assumption, although the structure inside the substrate has been changed, the influence on the effective Young's Modulus or CTE is minor because the result from experimental data was obtained from the top surface. Yuling et al. [8] has shown there is a noticeable strain discrepancy between the top surface and the midplane. Hence, the effective properties of the composite structure based on properties measured on the surface plane are irrelevant. Those may distort the material properties of the overall substrate package once the substrate is considered as a single layer.

Hence, it is not ideal to assume the whole package as only one effective model. On the other hand, it is more appropriate to consider it as separate three layers, which could preserve both the conformity of the buildup layers and the firmness of the core layer. In this study, various substrate model options are investigated with Finite Element Method for package warpage comparison. Each model has a different structure configuration. Evaluation of each model's accuracy to the warpage prediction is presented. First, the substrate is modeled as one effective layer. Second, the three-layer model is studied with the top build-up layer, the middle low CTE core, and the bottom build-up layer. Third, the multi-layer model includes detailed configuration of copper trace and mixture of other non-metallic composites at each laminate. The methodology to demonstrate the multi-layer substrate model is by employing copper randomization method with ANSYS[TM]. The proposed copper randomization method is to impose varied copper percentage at each laminate. Package warpage among these models is compared and evaluated along with the empirical data. Experiments for CTE of materials are accomplished using the digital image correlation (DIC) technique while those for the Elastic Modulus is achieved with Dynamics Mechanical Analyzer (DMA). The stereo-DIC system has been implemented to evaluate displacements and strains in a three-dimensional view.

DIC is a non-contact technique, capable of high-resolution and full field optical measurement, which can quantify both in-plane and out-of-plane displacement or strain by correlating the sequential images of the object at the deformed and undeformed stages. Applications of the DIC technique is used broadly in various engineering fields [9-15]. In the electronic packaging field, DIC method is desirable for package warpage characterization to study the reliability of package components or interconnected solder joints [16-21]. Available DIC data are used for initial correlation of the finite element model and analytical results. Based on analysis results, a precise substrate model that could provide more accurate warpage prediction is suggested.

II. SAMPLE PREPARATION/MATERIAL CHARACTERIZATION

The original structure of the bare die package is depicted as in Figure 1. This package consists of an organic substrate, underfill, controlled collapse chip connection (C4), and silicon die. This simple package structure is ideal for investigation of various substrate models to the overall package warpage.

Figure 1: The bare die package

Furthermore, the underfill and C4 bumps are treated as one effective layer, which is adequate in this study because the scope of this work focuses mainly on the substrate layer and its structure. The material properties of constituent parts within the package are characterized experimentally. The substrate structure is detailed as in Figure 2, which includes solder resist on the top and the bottom, numerous metal layers (M1-M10, and M11-M20) associated with Ajinomoto Build-up Film (ABF), and the middle core.

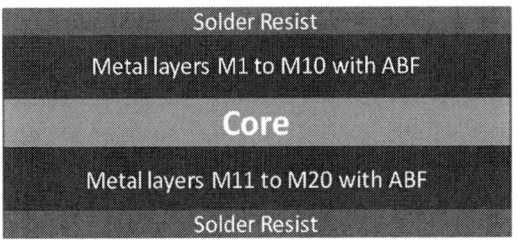

Figure 2: Schematic of the substrate

Each metal layer is a combination of a high percentage of copper and polymer material (ABF). In order to characterize the material properties of components for finite element analysis, the substrate was cut into strips with the size of 20 mm x 6 mm. To expose the core, the sample was ground off the build-up layer on both sides. To expose the build-up layer, the strips were ground off manually on one side of the build-up layer and the core. The strips after being ground off are shown in Figure 3.

The Figure 4 shows three different substrate models presented in this work: The one effective layer, the three major layers, and the multi-laminate layer.

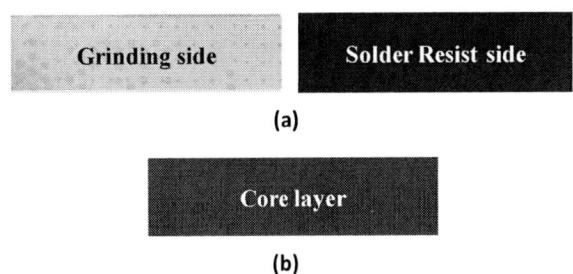

Figure 3: (a) Buidt-up layer after grinding off the core (Size: 20 mm x 6 mm x 0.4 mm) (b) Core layer after grinding off both sides (Size: 20mm x 6 mm x 0.82 mm).

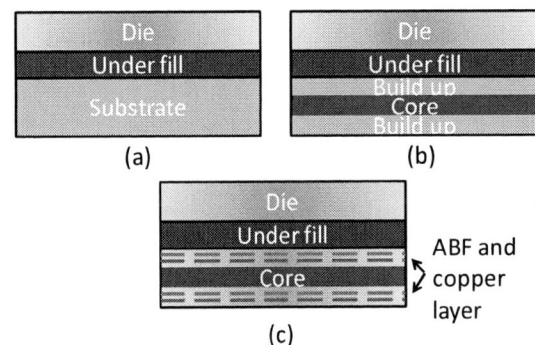

Figure 4: 3 Substrate package models investigated. (a) Package with one effective substrate. (b) Package model with three-layer substrate. (c) Package model with multi-layer substrate.

The DMA is utilized to conduct the experiments to characterize the Young's Modulus property of materials. The measurement results of the original package, the buildup and bottom layer, and the core layer are shown in Figure 5. The DMA test performed with 0.1% strain induced by 1 Hz in a three-point bending test. The temperature-dependent elastic modulus from the test was extracted and applied into FEA models.

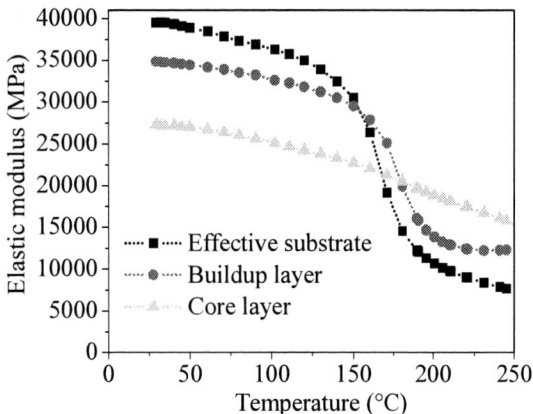

Figure 5: Young's Modulus for the substrate, buildup layer, and the Core

Along with the Elastic Modulus, CTEs of the original package, the buildup layer, and the core are concerned. A technique applied to characterize CTE properties of material in this work is the DIC system. A schematic setup that was used to measure deformation of the tested packages under thermal loading is illustrated as in Figure 6. The experimental setup consists of a correlation software (ARAMIS system), a heating chamber, and a pair of CCD cameras. Images at interested temperatures are captured via electrical triggered signals controled by the heating chamber. Through a clear glass window placed on the top chamber surface, the CCD cameras are able to observe the whole specimen clearly. It is recommended that the measurement field of view should be larger than the specimen size to avoid the effect of rigid body movement. The rigid body movement effect is usually caused by environmental factors such as the power chamber running or convection system blowing air inside the chamber.

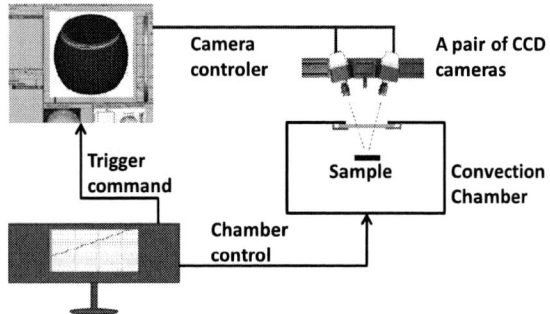

Figure 6: Schematic of the experimental setup for 3D DIC

Samples experienced surface treatment with random speckle patterns generated from high temperature resistant black and white paint before measurements. High contrast random patterns are desirable for the accurate correlation process. To measure the package deformation in excessive stretching stages or the resulting dome shape during thermal load, the stereovision system should be calibrated using a calibration panel with uniformly spaced markers to make sure the DIC can capture changes at extreme cases. Through the calibration process, intrinsic parameters of the stereoscopic system and the effect of optical imprecision are evaluated and compensated.

From the strain values measured by the DIC, CTEs of the buildup layer, the effective substrate, and the core layer are plotted as in Figure 7. To avoid the effect of uncorrected strain after the glass transition temperature, the thermal load profile for CTEs estimation is limited from room temperature to 150°C.

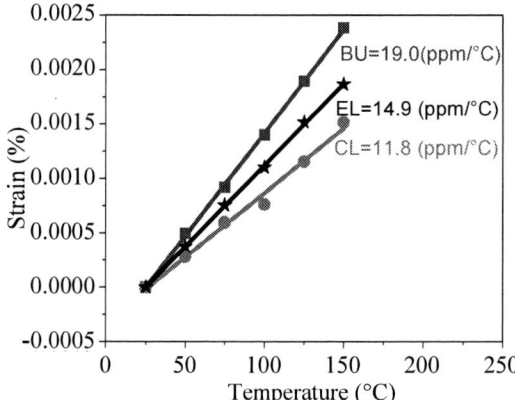

Figure 7: CTE characterization. (BU: Buildup Layer, EL: One Effective Layer, CL: Core Layer)

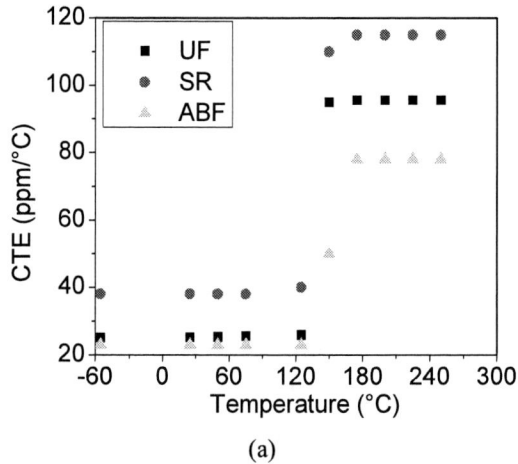

(a)

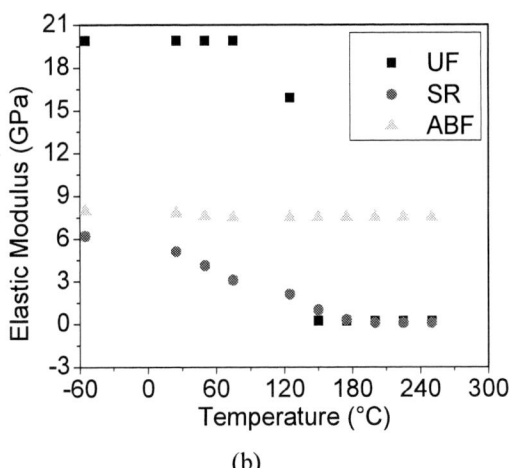

(b)

Figure 8: Properties of materials (UF: Underfill; SR: Solder Resist; ABF: Ajinomoto Buildup Film). (a) Coefficient of thermal expansion; (b) Elastic Modulus

The material properties of Underfill, Solder Resist, and Ajinomoto Buildup Film (ABF) is provided by the manufacturer and presented in Figure 8. The dimension of three models is summarized in the Table 1.

Due to the existence of the copper associated with ABF material at each laminate, the effective properties within each layer such as Young's modulus and CTE need to be estimated. All these are assumed to be in-plane properties. Figure 9 shows the copper percentage within each laminate. The comprised layer numbers of the substrate are named in the order from the die side downward to the bottom package.

TABLE 1: DIMENSION OF THREE PACKAGE MODELS

Model	Layers	Thickness (mm)	Area (mmxmm)
1 effective layer	Die	1.000	26x26
	Underfill	0.075	
	Substrate	1.652	75.6x70
3 major layer model	Die	1.000	26x26
	Underfill	0.075	
	Top build up	0.416	
	Bottom buildup	0.416	75.6x70
	Core	0.820	
Multi-layer model	Die	1.000	26x26
	Underfill	0.075	
	Solder Resist	0.021	
	9 x (Cu+ABF)	0.015	
	9 x ABF	0.025	
	Cu layer near	0.035	
	Core	0.820	75.6x70
	Cu layer near	0.035	
	9 x (Cu +ABF)	0.015	
	Solder Resist	0.021	

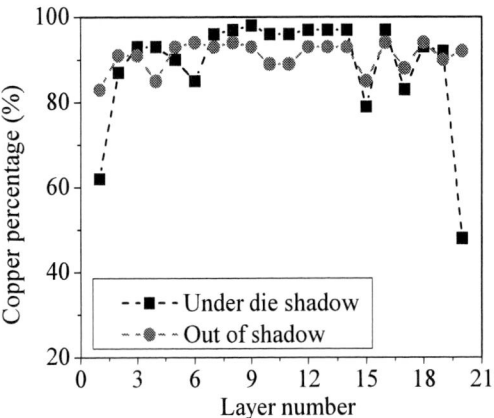

Figure 9: Copper percentage in each metal layer

III. FINITE ELEMENT ANALYSIS FOR VARIOUS SUBSTRATE MODELS

A. Substrate Modeled As One Effective Layer

The model with one effective substrate layer is popular due to its simplicity; the major material properties of the one effective layer, such as CTE and Elastic Modulus have been characterized by DIC and DMA.

B. Substrate Modeled As Three Major Layers: Top Buildup, the Core, and the Bottom Buildup Layer

The substrate with three major layers is a novel proposed model, which includes the soft buildup layers and firmer middle core. This model is believed to present precisely the mechanical properties of the whole substrate rather than the simple one effective substrate model while still preserving major properties of the complex substrate with copper trace and mixture of polymer material.

C. Substrate Modeled As Multi-layers

The ongoing trend of substrate structure contains a large number of copper traces. Copper with high modulus and high coefficient of thermal expansion becomes more prominent compared to the rest of the non-metallic material. Precisely modeling a substrate with detailed copper trace volume and geometry requires tremendous effort [3]; an alternate approach to account for the existence of copper is randomly generated copper location in the control of copper percentage. The percentage or the presence of copper rather than its location will determine the package warpage.

978-1-7281-1500-9/19 $31.00 © 2019 IEEE

Copper Randomization Method

This study proposes an approach that randomly generates the copper elements based on its fraction at each layer. The detail of the technique is described in Figure 10.

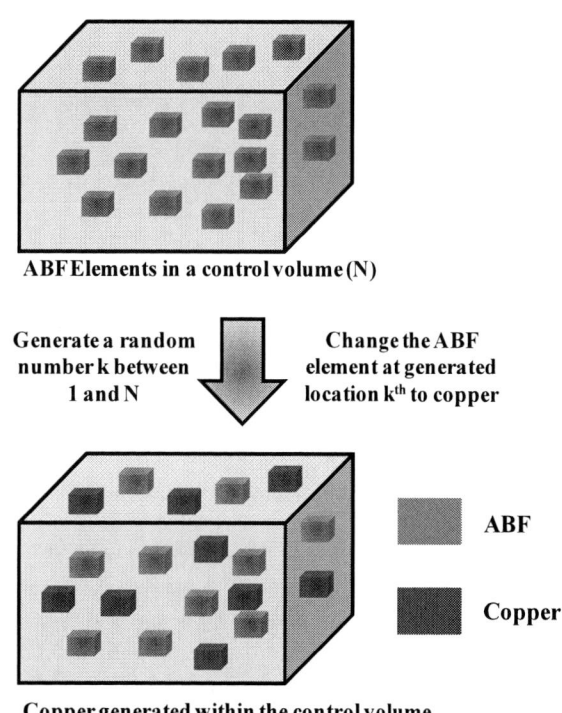

Figure 10: A control volume of a metal layer

Within a control volume, all elements are assigned to properties of ABF material, and the next step is to apply the random function provided by ANSYS to generate a number which indicates the position of that element in the control volume. Then switching that ABF element to copper is executed. Depending on the detail of the percentage of copper at each layer, the process is repeated until the copper reaches the requirement assumption of its presence within each layer.

Because the percentages of copper under the die shadow area and the amount of copper distributed outside the die shadow are different (as in Figure 9), the proposed randomization process has been done twice. One is for the control volume under the die shadow area and another is for the control volume out of the die shadow area. Figure 11 shows the copper randomization methodology implemented into the ANSYS at the first layer from the die. The laminate number is named in the order of the die downward to the bottom substrate. It can be seen that copper is obviously dominant to the ABF outside the die shadow area.

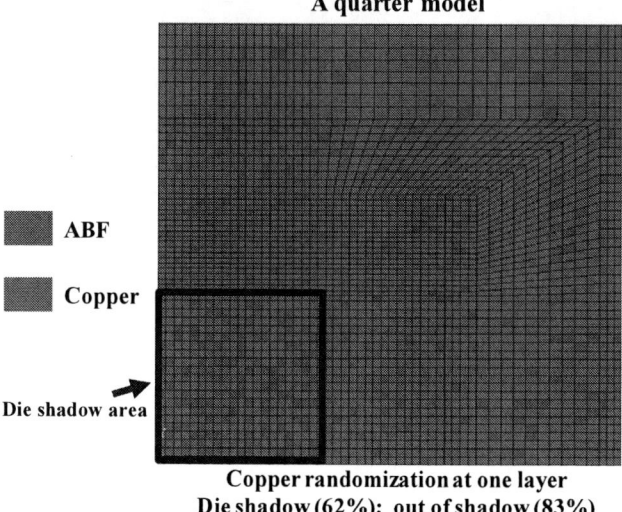

Figure 11: The copper randomization within a laminate

The mesh of three different substrate models is shown in Figure 12. The 3D quarter package was built for warpage analysis. The models include silicon die, effective underfill, and the substrate. The substrate is modeled in three different ways. The model was subjected to a thermal cycling load between 25°C to 250°C. Experimental results of material properties characterized by DIC and DMA were applied for three different models as discussed in Section II. For the three models, all constituents of the package have been considered as isotropic materials.

Accuracy Verification of the Proposed Copper Randomization Methodology

Copper has been randomly generated. Hence, the accuracy of the proposed copper randomization approach should be verified carefully so that it does not contaminate the predicted package warpage. Hence, the multi-layer model has experienced a consistent check through eight running cases with the thermal loading from 25°C to 250°C. Results at selected temperature stages for all cases are recorded as in Figure 13. It can be seen that the proposed randomization approach provides excellent consistency of package warpage estimation. In term of warpage concerns, it means that the package substrate modeled with the copper randomization method could entirely represent the complex, exact substrate structure.

978-1-7281-1500-9/19 $31.00 © 2019 IEEE

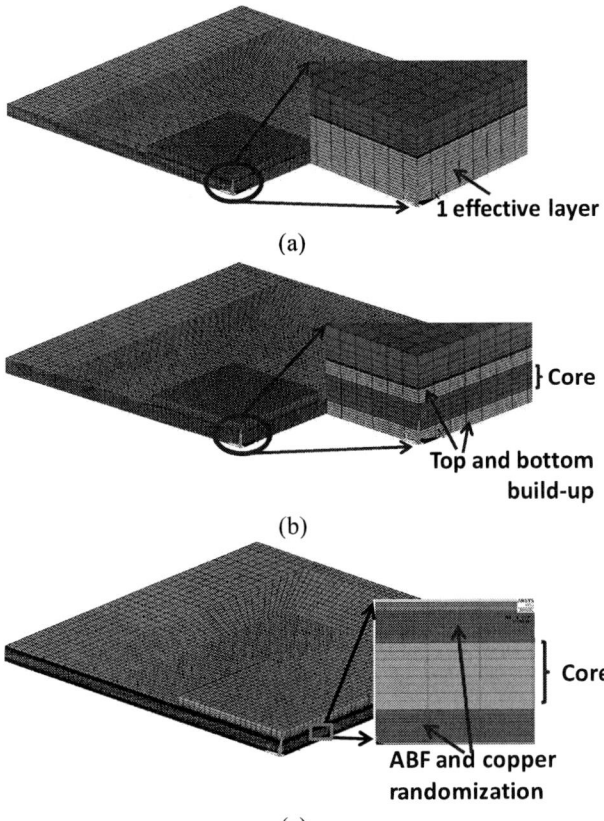

(a)

(b)

(c)

Figure 12: Package models for warpage investigation (a) One effective layer; (b) 3 major layers; (c) multi-layer model

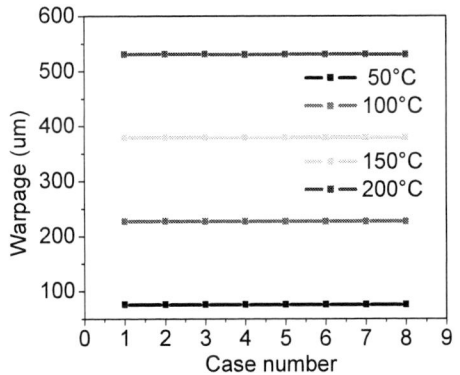

Figure 13: Warpage consistency of the proposed randomization method

IV. WARPAGE PREDICTION BY AN ANALYTICAL APPROACH

A. Effective Properties

For effective property of only one substrate layer, the Young's Modulus and CTE can refer from experimental result as in Figure 5 and Figure 7.

For three major layer substrate model, the properties of the buildup and the core layer are found in Figure 5 and Figure 7, similarly.

For the multi-layer model, it is required to derive the properties at each laminate due to the presence of copper and mixture of polymer or non-metallic material (Figure 9). The effective properties such as Young's Modulus and CTE are derived based on the rule of mixture. Park et al. confirmed that the in-plane properties are more appropriate to represent isotropic properties of a composite structure rather than the out-of-plane properties. [22, 23]. The rule of mixture (ROM) is constructed by assuming that the strain on the in-plane direction is uniform. The effective modulus, the coefficient of thermal expansion, and the Poisson's ratio can be given by derived equations. The schematic of the laminate structure is depicted as in Figure 14. Applying the ROM, the force that acts on the whole substrate $(\sigma_{x,y}A)$ is equal to the sum of individual force that acts on constituted layers ($\sigma_i A_i$). Based on this assumption, we have:

$$\sigma_{x,y}A = \sum \sigma_i A_i \qquad (1)$$

Where, A and A_i is the cross-section area of the whole structure and of each layer, respectively. From (1), the total stress $(\sigma_{x,y})$ can be represented by comprised parts (σ_i).

$$\sigma_{x,y} = \frac{\sum \sigma_i A_i}{A} = \sum \sigma_i c_i \qquad (2)$$

Applying Hook's Law:

$$\varepsilon_{x,y}E_{x,y}^{eff} = \sum E_i c_i \varepsilon_i \qquad (3)$$

In the condition that the strain is uniform, the strain over the whole laminate ($\varepsilon_{x,y}$) is the same as each constituent (ε_i), $\varepsilon_{x,y} = \varepsilon_i$, the equation (3) is simplified by:

$$E_{x,y}^{eff} = \sum E_i c_i \qquad (4)$$

Where, $E_{x,y}^{eff}$ is the effective Modulus of the overall structure, c_i and E_i is the volume fraction and the Young's modulus of the ith constituent, respectively.

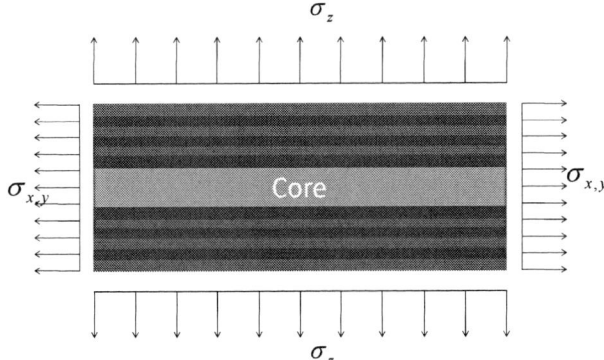

Figure 14: Schematic of a laminate structure

The equation (4) is well known as the rule of mixture for a composite structure. Applying the mixture rule to derive the effective Poisson's ratio ($\upsilon_{x,y}^{eff}$) and effective CTE ($\alpha_{x,y}^{eff}$).

$$\upsilon_{x,y}^{eff} = \sum \upsilon_i c_i \qquad (5)$$

$$\alpha_{x,y}^{eff} = \frac{\sum E_i \alpha_i c_i}{\sum E_i c_i} \qquad (6)$$

Where, υ_i and α_i are individual Poisson's ratio and CTE at the ith constitute layer, respectively. From the percentage of copper within each layer, the volume fraction of copper is known; therefore, the effective properties of the composite constituents are estimated.

B. Classical Laminate Theory

Classical laminate theory provides an efficient approach to understand the warpage behavior of a composite structure. Among the theories that can be applied to study the warpage behavior of laminate structure, classical laminate theory has been utilized and provided better data [22-25]. It is preferably applicable to this package. In order to apply the classical laminate theory, the constituents of the laminate are assumed as isotropic materials. At the mid-plane, due to extensional strains $\{\varepsilon^o\}$ and its curvatures $\{\kappa\}$, the laminate warpage is deduced. The resulting forces $\{N_\Lambda\}$ and moments $\{M_\Lambda\}$ are in the relation with midplane strain and curvatures as follows:

$$\begin{Bmatrix} \varepsilon^o \\ \kappa \end{Bmatrix} = \begin{bmatrix} A & B \\ B & D \end{bmatrix}^{-1} \begin{Bmatrix} N_\Lambda \\ M_\Lambda \end{Bmatrix} \qquad (7)$$

Where A, B and D are stiffness matrices, which can be calculated as follows:

$$[A] = \sum_{k=1}^{n} \left[Q_{ij}\right]_k (z_k - z_{k-1}) \qquad (8)$$

$$[B] = \frac{1}{2}\sum_{k=1}^{n} \left[Q_{ij}\right]_k (z_k^2 - z_{k-1}^2) \qquad (9)$$

$$[D] = \frac{1}{3}\sum_{k=1}^{n} \left[Q_{ij}\right]_k (z_k^3 - z_{k-1}^3) \qquad (10)$$

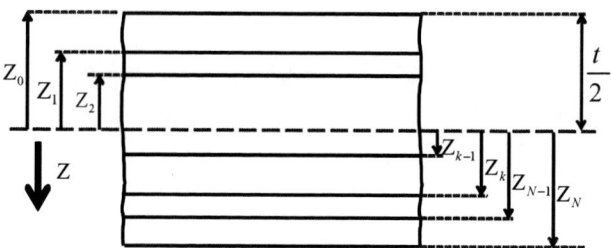

Figure 15: Geometry of a N-layered laminate

The classical laminate theory calculates the resultant forces and moments as follows:

$$\{N_\Lambda\} = \sum_{k=1}^{n} \left[Q_{ij}\right]_k \alpha_k \Delta T \begin{Bmatrix} 1 \\ 1 \\ 0 \end{Bmatrix} (z_k - z_{k-1}) \qquad (11)$$

$$\{M_\Lambda\} = \frac{1}{2}\sum_{k=1}^{n} \left[Q_{ij}\right]_k \alpha_k \Delta T \begin{Bmatrix} 1 \\ 1 \\ 0 \end{Bmatrix} (z_k^2 - z_{k-1}^2) \qquad (12)$$

For an isotropic material, $[Q]_k$ can be given as follows:

$$\left[Q_{ij}\right]_k = \frac{E_k}{1-\upsilon_k^2} \begin{bmatrix} 1 & \upsilon_k & 0 \\ \upsilon_k & 1 & 0 \\ 0 & 0 & \dfrac{1-\upsilon_k}{2} \end{bmatrix} \qquad (13)$$

At the midplane surface, the warpage or out of plane deformation can be obtained as follows [22]:

$$w = -\frac{1}{2}\left(\kappa_x x^2 + \kappa_y y^2 + 2\kappa_{xy} xy\right) \qquad (14)$$

Where, z_k is the distance from the middle plane to the laminate k position (Figure 15) and α_k is the CTE at that layer. For one effective substrate layer, the laminate theory can be applied into three layers: the die, the underfill, and the substrate. The three major layers of substrate model can be considered as five-layer laminate: the die, underfill, top buildup and bottom buildup layer, and the core. And the multi-layer model is a combination of 42 layers (as described in Table 1)

V. RESULTS AND DISCUSSION

A. Warpage Prediction by Analytical Solution

Based on effective properties generated from Section IV. A, the mechanical properties provided by manufacturers, and the application of the equation (15), the warpage for the laminate structure can be totally calculated. The warpage within the die shadow area is interested.

978-1-7281-1500-9/19 $31.00 © 2019 IEEE

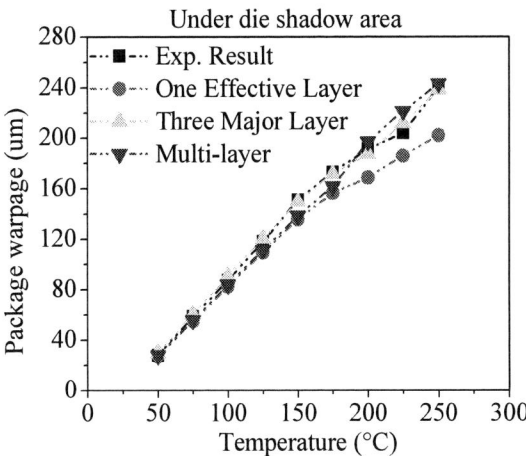

Figure 16: Analytical warpage result under die shadow area

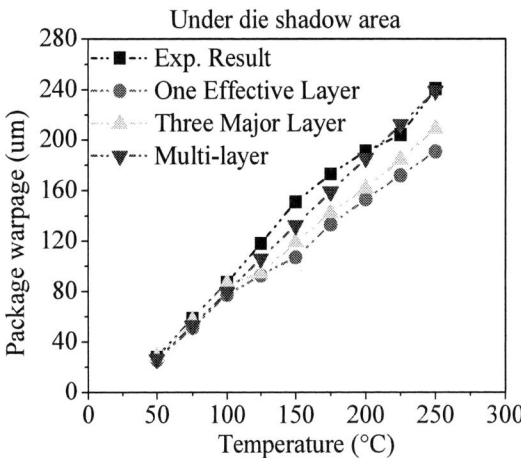

Figure 17: Package warpage behavior under die shadow area

The package warpage is presented as in Figure 16. It can be seen that, the package warpage level is proportional with the increasing of temperature. The warpage data from the three-layer model and multi-layer model provide good agreement with the experimental data, while the one effective layer gave a decent deviation from those data, especially at high temperature stages. Based on analytical results, the three-layer model can represent the exact, complex package structure. The following section will discuss about the results from finite element analysis.

B. Warpage Prediction by FEA Approach

The FEA results from three models are provided in Figure 17 and Figure 18 in a comparison with the experimental data. In order to correlate with analytical solution, warpage under the die shadow area is concerned. Warpage data were extracted along a diagonal of the package bottom within the die shadow area and along a diagonal of the whole package at different temperature stages. As can be seen in Figure 17, the warpage under the die shadow from the multi-layer model matched most excellently with the experimental result; next is the three-layer model, and the one effective model is the worst case, which gives a noticeable offset amount of warpage prediction. Among these models, the multi-layer substrate structure costs the most in terms of modeling effort, while modeling the other models is not a challenging task.

In both the analytical solution and FEA result, warpage prediction from the multi-layer model matches well with experimental results. Results from three-layer model are closer to the experimental data and the multi-layer model than the one effective layer. Moreover, in good agreement with the analytical solution, FEA results reveal that the three-layer model and multi-layer model have provided the package warpage prediction in close approximation to experimental results rather than only one effective model. The detail of relative warpage along the substrate diagonal among those models is presented in Figure 19. The reference stage is at room temperature.

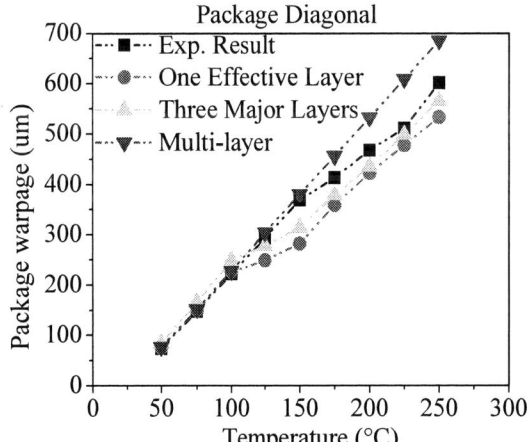

Figure 18: Warpage behavior along package diagonal

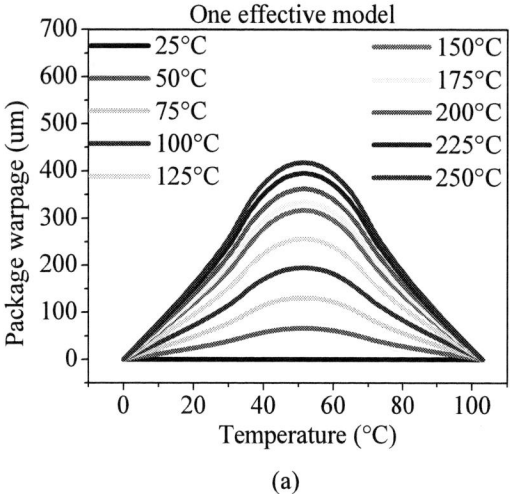

(a)

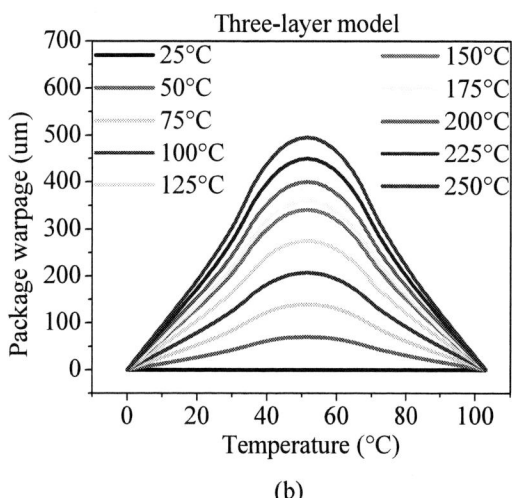

(b)

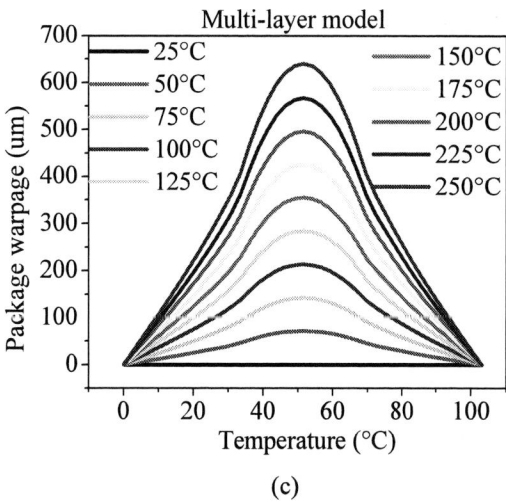

(c)

Figure 19: Warpage behavior at interested temperature stages. (a) Package with one effective substrate model. (b) Package with three-layer substrate model. (c) Package with multi-layer substrate model.

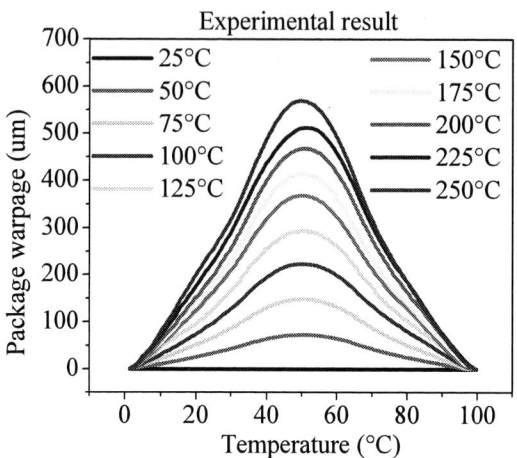

Figure 20: Warpage behavior at interested temperature stages from experimental results

The DIC measurements were conducted with the temperature profile that started from room temperature to 250°C. The heating ramp rate is set as low as 5°C; the warpage behavior of the package is observed at every 25°C. Image captured at 25°C was selected as the reference stage. Then, images were recorded consecutively in 25°C increments up to 250°C. The measurement results and the tendency of package warpage are depicted as in Figure 20. Clearly, it has same features as those exploited by FEA results. This has confirmed the data from the FEA model and proven the precision of the model. At all interested temperature stages, the FEA results from the three-layer model, and the multi-layer model always shows better warpage prediction than the one effective layer.

VI. CONCLUSION

This study proposes a comprehensive analysis of warpage prediction among package substrate models. There are three different substrate models under investigation: the simplified package substrate as one effective layer; the three-layer model with top build-up layer, the middle core with low CTE and high Young's Modulus, and the bottom build-up layer; and the multi-layer laminate substrate that considers the presence of copper traces. Both the analytical approach and Finite Element Analysis (FEA) have been utilized to examine package warpage among these models in a correlation with the empirical data. This research reveals that in terms of warpage, the substrate should be considered as three major layers because not only does it keep the FEA model simplified, but it maintains the foremost characterizations of the complex, full package substrate structure.

In conclusion, it is recommended that a FEA model of the package substrate should be considered as three layers for better warpage estimation or more exact board level solder joint prediction. The effect of various package substrate models on board level solder joint reliability is out of the scope of this work. However, it is an excellent future research topic as follow-up to this research.

978-1-7281-1500-9/19 $31.00 © 2019 IEEE

ACKNOWLEDGMENT

The authors would like to thank members from the Opto-Mechanics and Physical Reliability Lab at the Mechanical Engineering Department, State University of New York at Binghamton (SUNY Binghamton) for their valuable suggestions and supports.

REFERENCES

[1] M. Wang, and B. Wells, "Substrate Trace Modeling for Package Warpage Simulation," in 2016 IEEE 66th Electronic Components and Technology Conference (ECTC). 2016.

[2] N. Islam, A. Syed, T. Hwang, Y. Ka, W. Kang, "Issues in fatigue life prediction model for underfilled flip chip bump," in 2011 IEEE 61st Electronic Components and Technology Conference (ECTC). 2011.

[3] L.O. McCaslin, S. Yoon, H. Kim, S. K. Sitaraman, "Methodology for Modeling Substrate Warpage Using Copper Trace Pattern Implementation," IEEE Transactions on Advanced Packaging, 2009. 32(4): p. 740-745.

[4] S. Shao, Y. Niu, J. Wang, R. Liu, S. Park, H. Lee, G. Refai-Ahmed, L. Yip, "Comprehensive Study on 2.5D Package Design for Board-Level Reliability in Thermal Cycling and Power Cycling," in 2018 IEEE 68th Electronic Components and Technology Conference (ECTC). 2018.

[5] J. Hong, K. Choi, D. Oh, S. Park, S. Shao, H. Wang, Y. Niu, V.L. Pham, "Design Guideline of 2.5D Package with Emphasis on Warpage Control and Thermal Management," in 2018 IEEE 68th Electronic Components and Technology Conference (ECTC). 2018.

[6] J. Wang, Y. Niu, S. Park, "Modeling and design of 2.5 D package with mitigated warpage and enhanced thermo-mechanical reliability.," in 2018 IEEE 68th Electronic Components and Technology Conference (ECTC). 2018. IEEE.

[7] J. Xu, Y. Niu, S. R. Cain, S. McCann, H. H. Lee, G. Refai-Ahmed, and S. B. Park, 'The Expermental and Numerical Study of Electromigration in 2.5d Packaging', in 2018 IEEE 68th Electronic Components and Technology Conference (ECTC) (2018), pp. 483-89.

[8] Y. Niu, V. Pham, J. Wang, "An Accurate Experimental Determination of Effective Strain for Heterogeneous Electronic Packages With Digital Image Correlation Method.," IEEE Transactions on Components, Packaging and Manufacturing Technology, 2018. 8(4): p. 678-688.

[9] N. S. Ha, V.T. Le, N.S. Goo, J.Y. Kim, "Thermal strain measurement of austin stainless steel (ss304) during a heating-cooling process," International Journal of Aeronautical and Space Sciences, 2017. 18(2): p. 206-214.

[10] N. S. Ha, V.T. Le, and N.S. Goo, "Investigation of punch resistance of the Allomyrira dichtoloma beetle forewing," Journal of Bionic Engineering, 2018. 15(1): p. 57-68.

[11] V.T Le, N. S. Ha, T. Jin, N.S. Goo, J. Kim, "Thermal interaction of a circular plate-ring structure using digital image correlation technique and infrared heating system," Journal of Mechanical Science and Technology, 2016. 30(9): p. 4363-4372.

[12] T. Jin, N.S. Ha, V.T. Le, N.S. Goo, H. Jeon, "Thermal buckling measurement of a laminated composite plate under a uniform temperature distribution using the digital image correlation method," Composite Structures, 2015. 123: p. 420-429.

[13] N.S. Ha, V.T. Le, and N.S. Goo, "Investigation of fracture properties of a piezoelectric stack actuator using the digital image correlation technique," International Journal of Fatigue, 2017. 101: p. 106-111.

[14] C. Singh, S. Chaparala, C. Zhou, B. Zhang, S. Park, "Deformation of Display for Handheld Devices During Drop Impact," in Electronic Components and Technology Conference (ECTC), 2016 IEEE 66th. 2016. IEEE.

[15] C. Singh, S. Chaparala, and S.B. Park, "Measurement of Dynamic Response Parameters of an Underdamped System," in Dynamic Behavior of Materials, Volume 1. 2017.

[16] Y. Niu, J. Wang, and S. Park, "The complete packaging reliability studies through one digital image correlation system," in Electronic Components and Technology Conference (ECTC), 2017 IEEE 67th. 2017. IEEE.

[17] V.L. Pham, Y. Niu, J. Wang, C. Singh, S. Park, C. Zhong, S.W. Koh, J. Wang, S. Shao, "Experimentally Minimizing the Gap Distance Between Extra Tall Packages and PCB Using the Digital Image Correlation (DIC) Method," in 2018 IEEE 68th Electronic Components and Technology Conference (ECTC). 2018.

[18] Y. Niu, J. Wang, S. Shao, H. Wang, H. Lee, S. Park, "A comprehensive solution for electronic packages' reliability assessment with digital image correlation (DIC) method," Microelectronics Reliability, 2018. 87: p. 81-88.

[19] H. Wang, S. Shao, V.L. Pham, P. Shang, C. Zhong, S. Park, "Quantification of Underfill Influence to Chip Packaging Interactions of WLCSP," 2018(51920): p. V001T01A004.

[20] Y. Niu, H. Wang, and S. B. Park, 'A General Strategy of in-Situ Warpage Characterization for Solder Attached Packages with Digital Image Correlation Method', Optics and Lasers in Engineering, 93 (2017), 9-18

[21] Y. Niu, H. Wang, S. Shao, and S. B. Park, 'In-Situ Warpage Characterization of Bga Packages with Solder Balls Attached During Reflow with 3d Digital Image Correlation (Dic)', in 2016 IEEE 66th Electronic Components and Technology Conference (ECTC) (2016), pp. 782-88.

[22] S. Park, H.C. Lee, B. Sammakia, K. Raghunathan, "Predictive Model for Optimized Design Parameters in Flip-Chip Packages and Assemblies," IEEE Transactions on Components and Packaging Technologies, 2007. Vol. 30(2): p. 294-301.

[23] C. Singh, Y. Kim, and S. Park, "Comparative Study of Analytical Models to Predict Warpage in Microelectronics Packages," 2014(46590): p. V010T13A066.

[24] C. Singh, A Study of Warpage of Electronic Package and Effect of Reflow (State University of New York at Binghamton, 2014).

[25] R.M. Jones, "Mechanics of composite materials", Materials Science and Engineering Series, 2nd edition, CRC Press, 1998.

3D Fan-Out Package Technology with Photosensitive Through Mold Interconnects

Kentaro Mori, Soichi Yamashita, Takafumi Fukuda*, Masahiro Sekiguchi, Hirokazu Ezawa**, Shuzo Akejima

Toshiba Electronic Devices and Storage Corporation

Toshiba Development & Engineering Corporation*, Toshiba Memory Corporation**

33, Shinisogo-Cho, Isogo-ku, Yokohama 235-0017, Japan

kentaro3.mori@toshiba.co.jp

Abstract—Fan-out package technology for realizing high density and high integration is promising for 3D integration. This paper presents an innovative 3D fan-out packaging technology with photosensitive through mold interconnects. This is the first-time demonstration of 3D integration based on fan-out technology using a photosensitive mold compound material to provide high-aspect-ratio and fine-pitch through mold via openings, thereby advanced substrate-free packaging. 10×10 mm Si test dies with 100-µm thickness was prepared to confirm feasibility of the 3D integration packaging process. The chip design comprises over 2000 pads with a 200-µm pad pitch. After embedding the test die into a thick photosensitive molding compound, the lithography process conditions for shallow vias on the die and deep visa around the die were evaluated to realize a 3D integration packaging without conventional tall electroplated Cu pillars. The prototype package with photosensitive through mold interconnects shows a decent reliability over a 1000-cycle package-level thermal cycle test, a 1000-h high-temperature storage test, and a 96-h pressure cooker test. Thermal stress simulation has revealed that package warpage with the photosensitive material is small compared with that of a widely used epoxy molding compound for chip first fan-out wafer-level package. This study demonstrates future viability of the fan-out process using a photosensitive molding compound to enhance the versatility of 3D integration, thereby dramatically expanding the application field for fan-out packages.

Keywords- 3D fan-out wafer level packaging, Through mold interconnects, Photosensitive dielectric film

I. INTRODUCTION

Advanced packaging technology provides a powerful solution for system integrity with high performance, increased functionality, lower cost, and smaller footprint in a post Moore's era. Fan-out wafer-level packaging (FOWLP) is being a mainstream technology for 3D integration. Current 3D fan-out package technologies based on Package-on-Package (PoP) has already realized high-density interconnections for interfaces between logic and memory devices. Its PoP-based packages with epoxy molded substrate usually require tall electroplated Cu pillars, whose top surface has to be revealed by chemical mechanical polishing (CMP) of the mold surface to interconnect the two packages [1-3]. This process scheme has already been implemented in mass production for some mobile applications [1]. However, due to high process costs, many device package designers are reluctant to use this process as a 3D integration solution for many semiconductor device products and system modules. In particular, the thick

photoresist and CMP process required to form tall Cu pillars are high-cost processes.

Therefore, various methods for lowering costs and increasing density for through mold interconnects have been developed [4-7]. Table I shows various through mold interconnect technologies. In conventional PoP base package technologies, vias are drilled using laser ablation [5-6]. This is a cost-effective solution because there is no need for a thick photoresist and CMP process. However, epoxy molding compound (EMC) contains filler particles, so the size and spatial distribution of fillers dominate the sidewall quality in the drilled via, and random lack of filler on the sidewall makes it difficult to provide the via with a fine pitch. Vertical wire technologies utilizing wire bonding are expected to lower manufacturing costs and improve fine-pitch interconnections [7]. This technology does not require a thick photoresist process, but does require a CMP process because they must be formed before the epoxy molding process. Since the pad pitch is finer and the wire diameter is smaller, there may be one of the risk of wire flow issue during the epoxy molding process. If these conventional epoxy-based molding compounds are retained for the fan-out process, 3D integration packaging must rely on costly Cu pillars and wire bonding before molding, as well as laser drilling that cannot provide a fine pitch.

TABLE I THROUGH MOLD INTERCONNECT TECHNOLOGIES

TMI technology	Cu pillar	Laser via	Wire bonding	Photo via
Resin type	EMC	EMC	EMC	Photosensitivity
Thick photoresist	Need	No need	No need	No need
CMP process	Need	No need	Need	No need
Via pitch	Fine	Coarse	Fine	Fine
ref.	[4]	[4]	[4]	This work

This paper demonstrates for the first time 3D integration based on fan-out technology using a photosensitive mold compound material to provide high aspect ratio and fine-pitch through mold via openings, thereby improving on current substrate-free packaging and increasing flexibility as compared with conventional EMC. Unlike tall Cu pillar processes, the Cu electroplating in this process dispenses with the need for bottom-up via filling capabilities because Cu redistribution layers (RDLs) simply lie on the sidewalls of via openings, reducing costs and improving productivity.

Below, section II describes the value of the proposed 3D FOWLP technology using photosensitive mold material, and section III discusses the fabrication process, package reliability, and warpage management.

II. VALUE PROPOSITION

The proposed 3D FOWLP is characterized by using a photosensitive dielectric material instead of conventional non-photosensitive EMC to achieve 3D integration. The photosensitive dielectric is a hybrid organic material mainly composed of silicone. The following describes the merits of this material and the design flexibility and higher productivity realized by applying it to 3D FOWLP.

A. Material characteristics

- Photosensitive via opening: The photosensitive dielectric material enables fine-pitch vias and high flexibility for 3D hetero-integration technologies.
- Low viscosity: The material has very low viscosity, so it is possible to completely embed components with height and to flatten the surface.
- Fillerless: The sidewall quality on a photo via opening is smooth due to the fillerless nature of the material, allowing smaller via formation and finer pitch via.
- Thick material: The material can form a film with thickness exceeding 100 μm, and so can be used as a molding material.
- Film shape: The film material will be suitable for panel-level packaging processes.

B. Design flexibility and higher productivilty

When applying a material with the above characteristics to 3D integration, high design flexibility can be expected. Figure 1 shows an example of the design flexibility realized by using photosensitive dielectric material. In that figure, a thick photosensitive dielectric is laminated onto a substrate on which multi dies are mounted, allowing the surface to be completely flattened (Fig. 1(a)–(b)). The two lithography process options are available to form vias with different depths and sizes. One is a multiple exposure process for different depth vias using different designed photo-masks for each type of vias described in Fig. 1(c). In the other process, all via openings are formed by a single exposure using one mask with the exposure condition targeting the deepest via opening described in Fig. 1(d). In that case, the mask diameter of vias smaller than the deepest via must consider shrinkage due to excessive exposure doses. The one-time process is a cost-effective solution for higher productivity improvement.

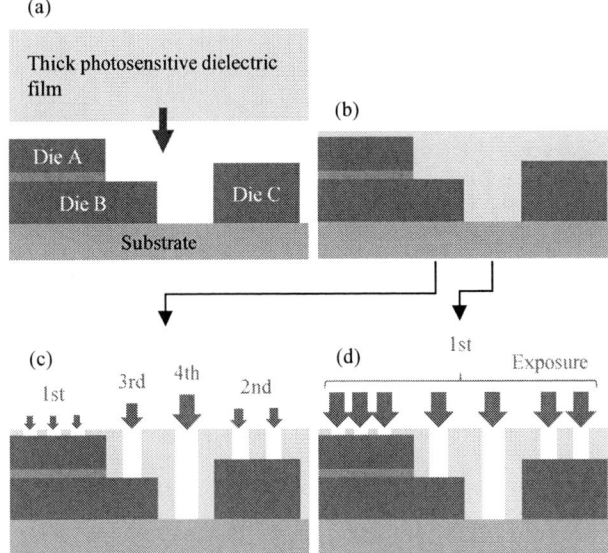

Figure 1 Exposure process using photosensitive dielectric material. (a) Multi dies mounting on base substrate, (b) Thick photosensitive dielectric film lamination and the surface flattening. (c) Multiple exposure process. (d) Single exposure process.

III. PROPOSED 3D FAN-OUT PACKAGE TECHNOLOGY

A. Fabrication process

Si test dies with (10 × 10 mm, 100 μm thickness) were prepared to confirm feasibility of the 3D integration packaging process. Figure 2 shows the fabrication process for the developed package. The die design comprises a 100 × 100 μm square Al bond pad with a 200-μm pad pitch and a 60 × 60 μm pad opening, and has over 2000 pads. The fan-out process was optimized for this demonstration of 3D integration. The die is mounted face up on the bottom substrate (Fig. 2(a)) and embedded in a more than 100 μm-thick low-viscosity photosensitive dielectric film by means of a vacuum lamination process (Fig. 2(b)). Considering viability of the packaging process in a large panel fan-out process, a dry film formed of the dielectric material was prepared. After curing the film to realize a thickness of 110 μm, the lithography process was executed to form openings in the film with 30-μm diameter and 10-μm depth on the Al pads of the embedded chip as shallow vias, and with 80-μm diameter and 110-μm depth on the Cu pads of the bottom substrate as deep vias. The 10-μm shallow vias were opened with an exposure dose of 400 mJ/cm^2, whereas the 110-μm deep vias were opened with an exposure dose of 1600 mJ/cm^2 (Fig. 2(c)). Cu RDLs were fabricated to connect Al bond pads on the test chip with Cu pads on the bottom substrate in the usual manner using Cu/Ti sputtering deposition, resist patterning, Cu electroplating, photoresist removal, and etching of sputter-deposited Cu and Ti (Fig. 2(d)). In the final step, a top solder resist was coated on the whole substrate. Cu plating of deep vias with aspect ratios less than 1.4 was confirmed. After singulation of the substrate into individual 15 × 15 mm packages, daisy chains

978-1-7281-1500-9/19 $31.00 © 2019 IEEE

connecting 2116 vias with 10-μm depth and 200-μm pitch on the chip and 768 vias with 110-μm depth and 300-μm pitch were checked to measure their electrical resistance.

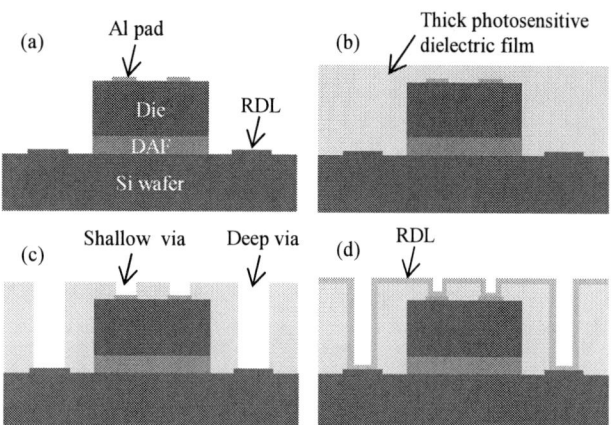

Figure 2 Fabrication process for developed package.
(a) Die mounting on Si wafer using die attach film (DAF). Thick photosensitive dielectric film lamination. (c) Via opening. (d) Metallization layer formation.

B. Photo via formation

Figure 3 shows the relation between exposure dose and top diameter of shallow and deep vias. In the small-dose region, there are delamination issues between the support substrate and the photosensitive dielectric material resulting from low adhesion due to insufficient exposure. The figure also shows that as the exposure amount increases, via diameters tend to be small. This is a reasonable result, because this material is a negative type in which light-irradiated regions do not dissolve in the development process. In this study, the mask diameter for shallow vias is 60 μm and that for deep vias is 100 μm. 400 mJ/cm^2 was adopted as the center exposure condition for forming shallow vias. It was confirmed that the actual via top diameter was 54 μm, meaning shrinkage from the mask diameter was 6 μm. Similarly, the center exposure condition for deep vias was 1600 mJ/cm^2, and the actual via top diameter and shrinkage were 80 μm and 20 μm, respectively. Figure 4 shows a photograph after via opening and confirms that both shallow and deep vias were via openings. In contrast, even if the exposure amount for shallow vias is the same as for deep vias (1600 mJ/cm^2), shallow vias can be formed with a larger shrinkage amount (27 μm). This shows that mask fabrication must consider the large shrinkage amount needed for via opening to establish a new process for simultaneously opening vias of different sizes. Figure 5 shows the relation between via diameter and aspect ratio (AP) at each resin thickness, demonstrating that vias with an AP ratio of 2.0 or more cannot be formed. Figure 6 shows a Cu wiring cross-section after plating, and confirms that deep vias of height 98.0 μm, top diameter of 80.0 μm, and AP ratio 1.2 connected to underlying pads, as did shallow vias with height 17.5 μm, top diameter 30.2 μm, and AP ratio 0.58. Figure 6 clearly shows that Cu metallization formed in both

shallow and deep vias on the Al pad on the embedded die and on the Cu pad on the substrate.

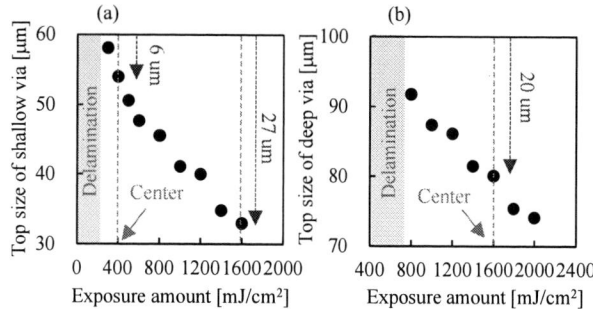

Figure 3 Relation between the exposure dose and via top diameter.
(a) Shallow vias. (b) Deep vias.

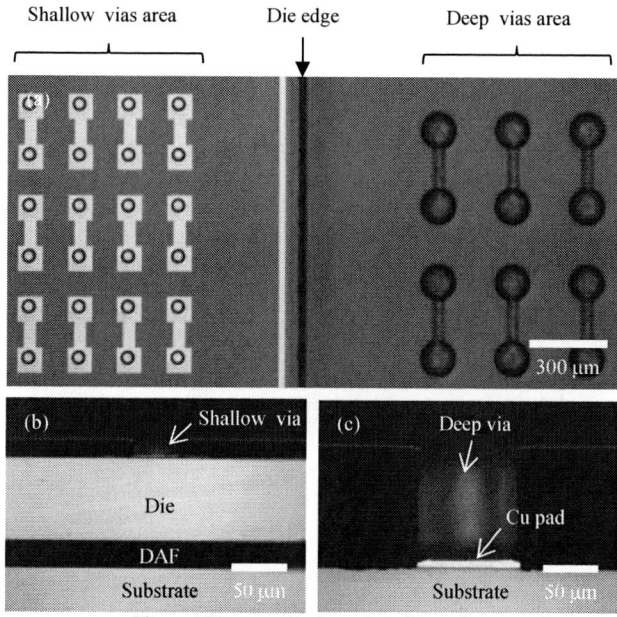

Figure 4 Photograph taken after via opening.
(a) Top view of the embedded die edge. (b) Cross-sections of shallow vias. (c) Cross-sections of deep vias.

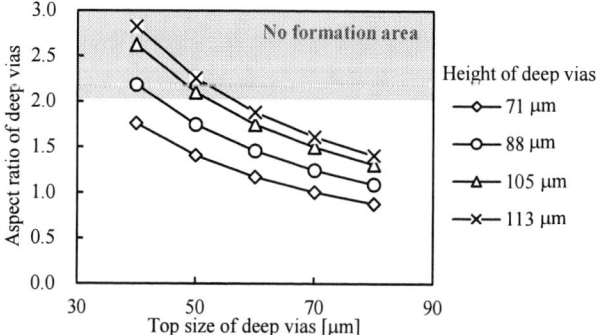

Figure 5 Relation between via diameter and aspect ratio at each resin thickness.

978-1-7281-1500-9/19 $31.00 © 2019 IEEE

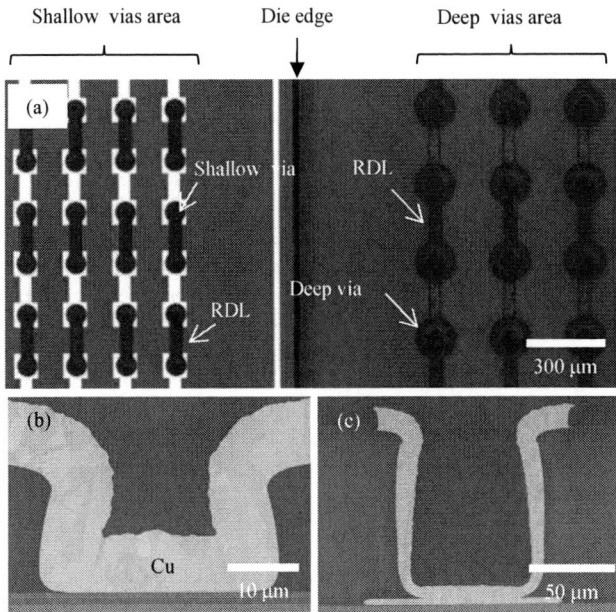

Figure 6 Photograph taken after Cu wiring formation.
(a) Top view of the embedded die edge. (b) Cross-sections of shallow vias.
(c) Cross-sections of deep vias.

C. Prototype package

Figure 7 shows a developed prototype package based on the fabrication process described in section A. In Fig. 7(a), a Si die is embedded face-up in the novel mold compound and Cu RDL interconnects the deep photo processed via with the bond pads on the Si die. The face up process will be afraid to remain a large step gap of the mold thickness around the Si die after curing, which could impact Cu RDL electrical. But, in Fig.7 (a), Δ h, indicated in the figure, is less than 3 µm and does not affect any electrical open/short yield. In Fig 7 (b), the 110 µm deep vias can be placed with a 150 µm pitch, which has confirmed that the present process has a great advantage for high-density 3D FOWLP compared with existing through mold interconnect technologies. In addition, In Fig 7 (c), deep vias can be placed at 100 µm from the die edge, greatly reducing the keep-out zone to place through mold vias for miniaturization of package size and higher pin counts of 3D stacked devices.

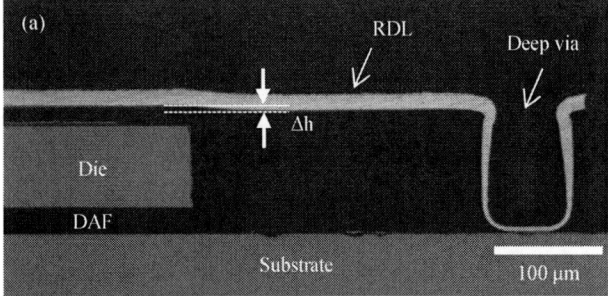

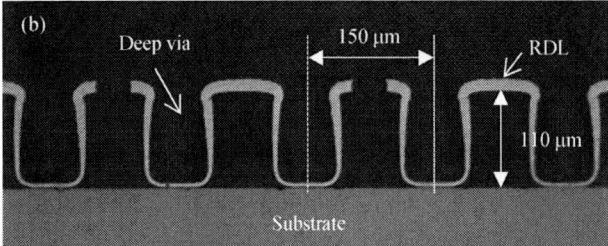

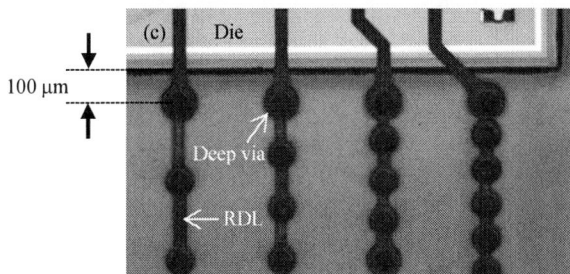

Figure 7 Photograph of a developed prototype package.
(a) Cross-sections of the embedded die edge. (b) Cross-sections of deep vias. (c) Top view of deep vias.

D. Package reliability

No electrical open and short failures were obtained for all measured daisy chains, confirming that the lithography process conditions for a photosensitive molding compound were well optimized for realizing a 3D integration package. Table II shows specifications of the prototype package, and Fig. 8 shows a top view of the package. The package is 15 mm x 15 mm in size, and the embedded test die is 10 mm. Reliability assessment was made on the packaged samples. Failure criteria was higher than 10% shift of the electrical resistance of the daisy chain. Table III shows the test conditions and results. Before the test, all package samples experienced the pretreatment under JEDEC level-3 conditions. The prototype package with photosensitive through mold interconnects have no failures after 1000-cycle package-level thermal cycle test, a 1000-h high-temperature storage test, and a 96-h pressure cooker test.

TABLE II SPECIFICATIONS OF THE PROTOTYPE

Die	Size (mm)		10 x 10
	Thickness (μm)		100 (including DAF)
Package	Size (mm)		15 x 15
	Thickness (μm)		835
Via	Shallow	Pad count	2,116
		Pad pitch (μm)	200
	Deep	Pad count	768
		Pad pitch (μm)	300

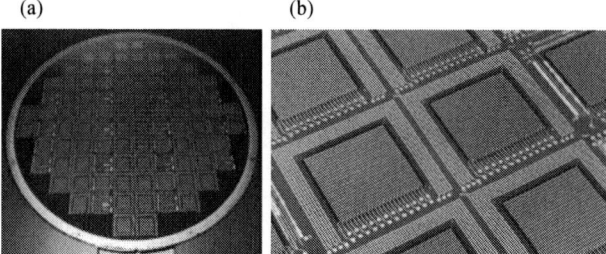

(a)   (b)

Figure 8 Photograph of the developed prototype package.
(a) 8-inch wafer. (b) Top view of the developed package.

TABLE III PACKAGE-LEVEL RELIABILITY TEST

Test Item (Test condition)	End point	Results
Temperature Cycling / TC (-55degC/125degC)	1000 cycles	pass
High Temperature Storage / HTS (150degC)	1000 h	pass
Pressure Cooker Test / PCT (121degC/100%RH)	96 h	pass

E. Warpage management

The novel developed material in the present study has characteristics of a higher CTE and a lower elastic modulus, compared to the conventional materials dedicated for a usual FOWLP. Thermal stress on Cu RDL due to its high CTE is usually regarded as a concern for its thermal cycling reliability. Therefore, thermal stress simulation was performed for the developed material and compared with a standard material for 3D FOWLP structures. A detailed 3D-solid package-level model including a silicon substrate, embedded die, shallow vias, deep vias, and RDL lines was employed to calculate warpage of the package and to know stress distribution around the vias due to a mismatch in CTE between the different type of materials to build the package model. Stress simulation was performed using ADVENTURECluster^{TM} finite element analysis software. In the simulation, all of the materials worked to be isotropic linear elastic. The effective stress-free temperature was defined as room temperature, due to the need for relatively comparing the stresses for the comparison of the stress induced by the two types of the molding materials with different curing temperatures. The stress simulation has disclosed how the novel developed material can be improved for more reliable 3D packaging.

The thermomechanical stress was calculated at the maximum and minimum temperatures in temperature-cycle testing. The package warpage was evaluated from the measurement along the package surface from its center to edge. Table IV and Fig. 9 show the 3D-solid package model and specific properties of the novel developed material and a standard material. This analysis addressed a 1/4 region with a package size of 15 mm. Figure 10 shows the measured warpage values for the two different molding materials at the developed and standard materials at −55 and 125 °C. It was found that warp displacement of the novel material (x) at −55 to 125 °C was about 47% less than that of the standard material (y).

TABLE IV GENERAL PROPERTIES OF THE DEVELOPED MATERIAL AND STANDARD MATERIAL

Item	Unit	Standard material	Developed material
Material type	-	Epoxy	Silicone
Young's modulus (25 degC)	GPa	22	0.095
Tg	degC	165	105
CTE (α1)	ppm / K	7	220

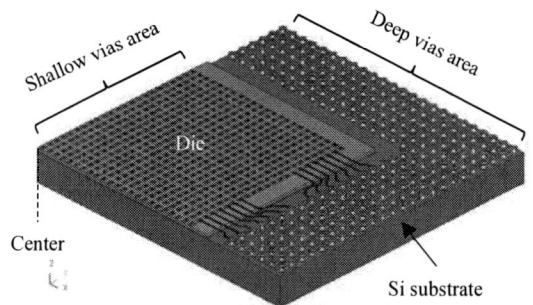

Figure 9 3D-solid package model.

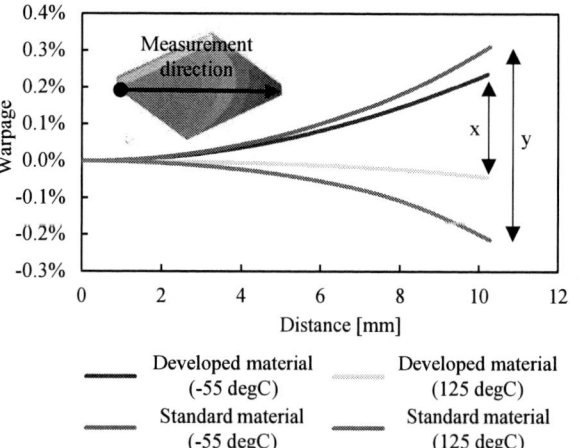

Figure 10 Warpage versus distance.

F. Material issues

Finally, material issues of the developed photosensitive material are described. Thermal stress analysis confirmed that the developed material with high CTE elongated in the vertical direction during thermal cycling. Tensile stress might be applied to Cu RDL lines along the sidewall of the vias and around the top corner of the vias opening, as shown in Fig. 11. Actual post-thermal cycle testing confirmed that tiny cracks were observed with their growth from the top edge of the vias opening in Fig.12, which can be predicted form the stress simulation results in Fig.11. Such Cu cracking may be problematic in products requiring high reliability or harsh testing environments. For more reliable packaging dedicated to automotive grade applications, the Cu cracks as observed in Fig.12 might be a reliability detractor. In the next stage of the development of the photosensitive molding materials, more CTE reduction will be a significant matter to alleviate the reliability concern while a lower modulus of the present material is kept as it is.

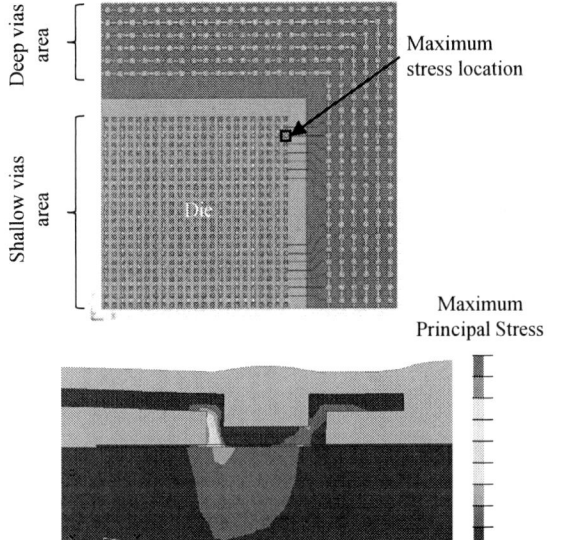

Figure 11 Distribution chart for principal stress.

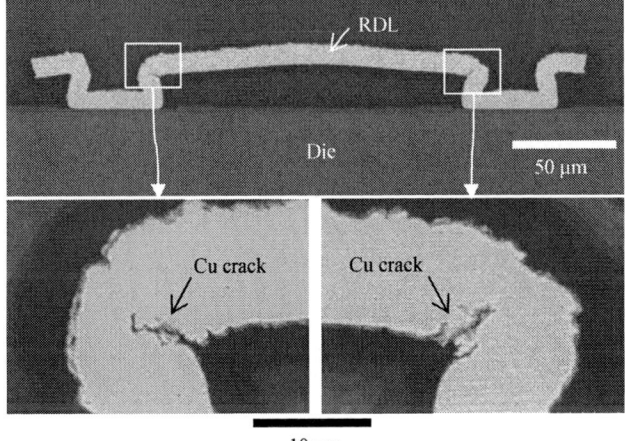

Figure 12 Test sample cross-section of after 1000 thermal cycles.

IV. CONCLUSION

An innovative 3D fan-out packaging technology with photosensitive through mold interconnects was presented. After a test die was embedded into a thick photosensitive molding compound, the lithography process conditions for shallow via on the die and deep via around the die were well optimized to realize 3D integration packaging without conventional tall electroplated Cu pillars. A prototype package with photosensitive through mold interconnects passed a 1000-cycle thermal cycle test, a 1000-h high-temperature storage test, and a 96-h pressure cooker test. In a thermal stress simulation, package warpage with the photosensitive material was less than that with a standard epoxy molding compound. This study revealed future viability of a fan-out process using a photosensitive molding compound to enhance 3D integration capability, thereby realizing a dramatic expansion of the application field for fan-out packages.

ACKNOWLEDGMENTS

The authors wish to acknowledge Shin-Etsu Chemical Co., Ltd. for material and process support. The authors would like to specially thank Yousuke Hisakuni and Daisuke Ando from Package Solution Technology Development for technical assistance. This study is based on results obtained from a project commissioned by the New Energy and Industrial Technology Development Organization (NEDO) to develop cross-sectoral technologies for IoT promotion.

REFERENCES

[1] Chien-Fu Tseng, et al., "InFO (Wafer Level Integrated Fan-Out) Technology," in Proc. IEEE 66th Electronic Components and Technology Conference (ECTC), Las Vegas, NV, USA, 2016, pp. 1-6.

[2] WonMyoung Ki, et al., "Chip Stackable, Ultra-thin, High-Flexibility 3D FOWLP (3D SWIFT® Technology) for Hetero-Integrated Advanced 3D WL-SiP," in Proc. IEEE 68th Electronic Components and Technology Conference (ECTC), San Diego, CA, USA, 2018, pp. 580-586.

[3] Feng-Cheng Hsu, et al., "3D Heterogeneous Integration with Multiple Stacking Fan-Out Package," in Proc. IEEE 68th Electronic Components and Technology Conference (ECTC), San Diego, CA, USA, 2018, pp. 337-342.

[4] Soon Wee Ho, et al., "Through mold interconnects for fan-out wafer level package," in Proc. IEEE 18th Electronics Packaging Technology Conference (EPTC), Singapore, 2016, pp. 51-56.

[5] Jinseong Kim, et al., "Application of through mold via (TMV) as PoP base package," in Proc. 58th Electronic Components and Technology Conference (ECTC), Lake Buena Vista, FL, USA, 2008, pp. 1089-1092.

[6] Hsu, Hsiangchen Chen, et al., "Reliability design and optimization process on through mold via using ultrafast laser," Polymers and Polymer Composites, Volume 26, Issue 1, 2018, Pages 1-8.

[7] Ivy Qin, et al., "Advances in Wire Bonding Technology for 3D Die Stacking and Fan Out Wafer Level Package," in Proc. IEEE 67th Electronic Components and Technology Conference (ECTC), Orlando, FL, USA, 2017, pp. 1309-1315.

Effects of the Materials Properties of Epoxy Molding Films (EMFs) on Fan-Out Packages (FOPs) Characteristics

Sangmyung Shin[1], Hanmin Lee[1], JunMo Kim[2], Tae-Ik Lee[2], Taek-Soo Kim[2], Youjin Kyung[3], Minsu Jeong[3], Kwangjoo Lee[3] and Kyung-Wook Paik[1]

1 Dept. of Materials Science and Engineering
Korea Advanced Institute of Science and Technology (KAIST)
Daejeon, Korea
2 Dept. of Mechanical Engineering
Korea Advanced Institute of Science and Technology (KAIST),
Daejeon, Korea
3 LG Chem R&D Center
Daejeon, Korea
e-mail: kwpaik@kaist.ac.kr

Abstract—Fan-Out Packages (FOPs) technology is actively researched because it enables high Input / Output (I/O) pads density, thin package, and cost reduction. Epoxy Molding Compound (EMC) materials play a major role in implementing FOPs. EMC acts as a substrate and also protects the chips from the outside. Recently Epoxy Molding Films (EMFs) have been introduced because EMFs are advantageous for large-area molding and can solve planarization problem. However, there are many problems before using at FOPs such as voids, delamination, warpage, and die shift. Voids and delamination occur when moldability is bad. Coefficient of Thermal Expansion (CTE) mismatch between the EMFs and the chips induces thermos-mechanical stress resulting in a warpage during the molding process. Furthermore, die shift from their original position can be also found by warpage, resin flow, and curing shrinkage. Therefore, EMFs materials properties and their effects on FOPs characteristics should be investigated. In this study, thermo-mechanical properties of EMFs were optimized by epoxy formulation to improve FOPs characteristics. Epoxy, silica contents, curing agent, and additives (coupling agent, carbon black) were investigated to produce low CTE and high Glass Transition Temperature (Tg) properties. Then, multiple 10 mm × 10 mm × 200 μm chips were molded with the size of 100 mm × 100 mm using 400 μm thick EMFs. The chip to chip distance could be reduced from 12 mm to 5.7 mm without voids and delamination. The warpage of EMFs were measured using 3 Dimensional - Digital Image Correlation (3D - DIC) method, and the effects of process conditions on the warpage of EMFs were also evaluated. And the die shift after molding was quantitatively measured assessed by optical microscopy. Finally, thermal cycle and 85 °C / 85 % RH tests were performed to evaluate the FOPs using optimized EMFs

Keywords-component; Fan-Out Panel-Level Packages, Epoxy Molding Films, Epoxy Formulation, Warpage, Die shift

I. INTRODUCTION

Fan-Out Packages (FOPs) technology is actively researched because it enables high Input / Output (I/O) pads density, thin package, and cost reduction. In accordance with these requirements, Chip Scale Package (CSP) was introduced. CSP has the advantage of having a small form factor and being able to be mass-produced in a wafer-level. However, as numbers of signal increase, Input / Output (I / O) pads density increases, resulting in fine pitch handling difficulties. In order to solve these problems, Fan-out types of packages are required. Fan-out packages (FOPs) can form I / O pads outside of chips through RDL (Re-Distribution Layer) process, which can increase I / O pad counts while preventing fine pitch issues. It also has a small form factor like CSP, and can reduce costs through a wafer-level process. Recently, research from wafer-level to panel-level is underway to reduce FOPs cost.

Fan-Out Panel-Level Packages (FOPLPs) use Epoxy Molding Compounds (EMCs) with compression molding instead of transfer molding. When FOPLPs are manufactured using conventional liquid and granule type EMCs, local non-uniform planarization problems occur. To solve this problem, Epoxy Molding Films (EMFs) have been newly introduced. [1] EMFs can not only improve the issue of non-uniform planarization, but also have the advantage of being suitable to be applied for large area molding. In this paper, material properties affected by the formulation of EMFs, warpage, die shift, and reliability of PLP (Panel Level Package) have been investigated.

II. EXPERIMENT

A. EMFs formulation

EMFs are composed of epoxy, curing agent, filler, and additives. Solid epoxy resin was added as a base resin of EMFs which enhanced Coefficient of Thermal Expansion (CTE), Glass Transition Temperature (Tg) of EMC. And to improve the thermo-mechanical properties of EMCs, an average of 5μm silica filler was also added. The curing agent was added to cure the epoxy. In addition, the coupling agent was also added as an additive to improve adhesion. Finally, carbon black was added for the laser marking.

CTEs were measured using a Thermo-Mechanical Analyzer (TMA), and the measurement was carried out at a rate of 5 °C per a minute with 100 mN load. The modulus and Tg were measured with a Dynamic-Mechanical Analyzer (DMA) of 100 mN load, amplitude 5 μm, 1 Hz frequency, 10 °C per a minute temperature elevation. And adhesion was measured with a die shear test. The viscosity was measured with a rheometer at 1000 Pa stress and 5 °C per a minute temperature elevation. The degree of cure was measured using the intensity of the epoxy peak using a Fourier-Transform Infrared Spectroscopy (FT-IR) equipment. The transmittance was measured in the visible region (300 nm to 800 nm) using an Ultraviolet / Visible (UV / VIS) spectrometer. The sheet resistances of the Panel Level packages (PLPs) was measured using a 4 point probe.

B. PLPs fabrication

The PLPs molding process using EMFs was shown in Figure 1. 40μm EMFs were fabricated by using a film coater and then laminated at 70 °C for 10 times to produce 400 μm EMFs sheet. Pre-lamination process was carried out at 110 °C to prevent the void capture. And then, using a hot press, the molding was performed at 110 °C with 3 MPa pressure for 15 minutes. Four metal spacers of 400 μm were used to control the thickness of the PLPs, and glass carrier was used as a molding carrier. Two-sided thermal release tape was used to attach the chip to the carrier. The Si bare chip was 10 mm × 10 mm × 200μm chip was used. The PLPs were made of 100 mm × 100 mm × 400 μm. The chip to chip distance was reduced from 12 mm to 5.7 mm. The SAT (Scanning Acoustic Tomography) measurement was performed in the C-scan mode to verify the void and the delamination of PLPs.

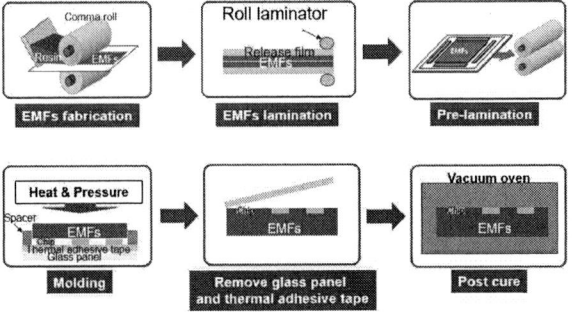

Figure 1. PLPs molding processes using EMFs

C. Warpage evaluation

The warpage of PLPs with 50 mm × 50 mm × 400 μm was measured using the 3 Dimensional -Digital Image Correlation (3D - DIC). In order to measure the warpage, sensing particles were sprayed on the PLPs. At first, warpage at room temperature was measured. And after PLPs were heated up to 250 °C with the rate of 22 °C per a minute by a heating plate, warpage at 250 °C was measured. Then, PLPs were cooled down to room temperature, and then warpage was measured again.

D. Die shift evaluation

For the die shift evaluation, PLPs were fabricated with 5 × 5 chips with the dimensions of 100 mm × 100 mm × 400 μm. The chip in the middle of the PLPs was designated as a reference assuming no die shift occurred. Die shift values were measured using the distance variation between the reference and other chips before and after the molding process. Distance between the chips was measured by using the panorama function of an optical microscopy.

E. Reliability test

100 mm × 100 mm × 400 μm PLPs were singulated into 15 mm × 15 mm × 400 μm PLP by a blade dicing. Blade dicing was performed at 20,000 rpm with 20 mm per a second velocity. In order to confirm the crack and delamination by CTE miss-match, the temperature cycle (T / C) test of 15 minutes holding at -55 °C and 15 minutes holding at 125 °C was performed. SAT measurement was performed every 250 cycles. To confirm the delamination due to moisture absorption, 85°C / 85% RH test was performed. Also, SAT measurement was performed every 250 hours.

III. RESULTS AND DISCUSSION

A. EMFs formulation

Thermo-mechanical properties were evaluated as the contents of thermoplastic resin decreased. As shown in Figure 2, when the contents of thermoplastic resin decreased, Tg, modulus increased and viscosity decreased. This is because the Epoxy Equivalent Weight (EEW) of the thermoplastic resin was so high that the Tg and modulus increased with decreased thermoplastic resin contents. In addition, as the thermoplastic resin had a high molecular weight, the minimum viscosity of the EMC decreased as the contents of the thermoplastic resin decreases. However, as shown in Figure 3(a), when the contents of thermoplastic resin was reduced to 2 g or less, film formability became problematic. This is due to the condensation of heat in the film coater process as the viscosity of the resin becomes lower.

Silica filler of 81 wt% was added to improve CTE, Tg properties. [2] And it was confirmed that there was no problem in the film formability, when the high amount of silica filler was added without adding thermoplastic resin as silica filler improved the viscosity of the resin as shown in Figure 3(b). The properties of CTE α1 14.8 ppm / °C, CTE α2 33.7 ppm / °C, Tg 221 °C and Modulus 14.4 GPa were obtained by adding 81 wt% silica filler to the epoxy resin without thermoplastic resin.

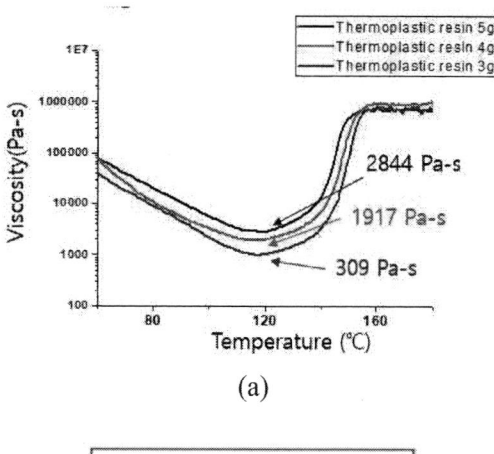

(a)

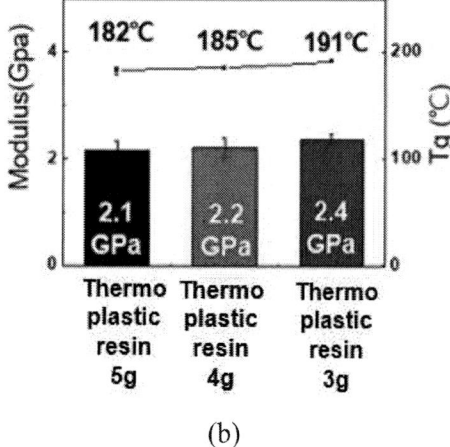

(b)

Figure 2. Comparison of the thermoplastic resin contents in terms of (a) viscosity and (b) modulus and Tg

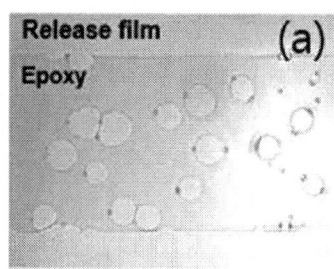

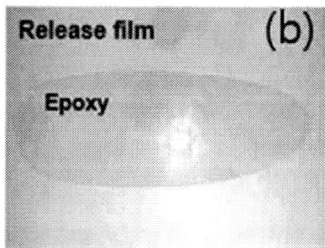

Figure 3. Film formability of (a) the resin without thermoplastic resin and (b) the resin with thermoplastic resin plus 81 wt% silica

The curing agent was added to cross-link the epoxy to form a network. The degree of cure after the film coater process according to the kind of curing agent was measured using a FT-IR. As a result, it was confirmed that curing agent B and C did not show pre-curing as shown in Figure 4(a) due to the low reactivity. As a result of the viscosity measurement, the C type curing agent showed the lowest minimum viscosity and broad U-curve in viscosity, which means better flowability as shown in Figure 4(b). Therefore, curing agent type C was the most suitable when reactivity and flowability were both considered.

The coupling agent was added to improve the adhesion between the chip and the EMFs. As a result, it was confirmed that adhesion increased as the coupling agent contents increased as seen in Figure 5(a). However, as the contents increased, it was confirmed that the thermo-mechanical properties were deteriorated due to the decrease of the Tg and modulus as shown in Figure 5(b). This was because the cross-linking density was reduced as the coupling agent had fewer epoxy groups per a monomer. The coupling agent was optimized as 1wt% considering the adhesion improvement and reduction of Tg.

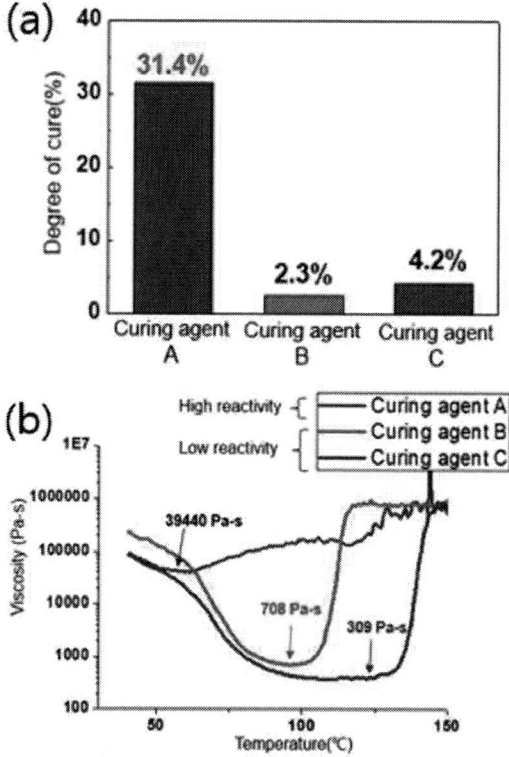

Figure 4. Comparison of the curing agent types in terms of (a) degree of cure and (b) viscosity

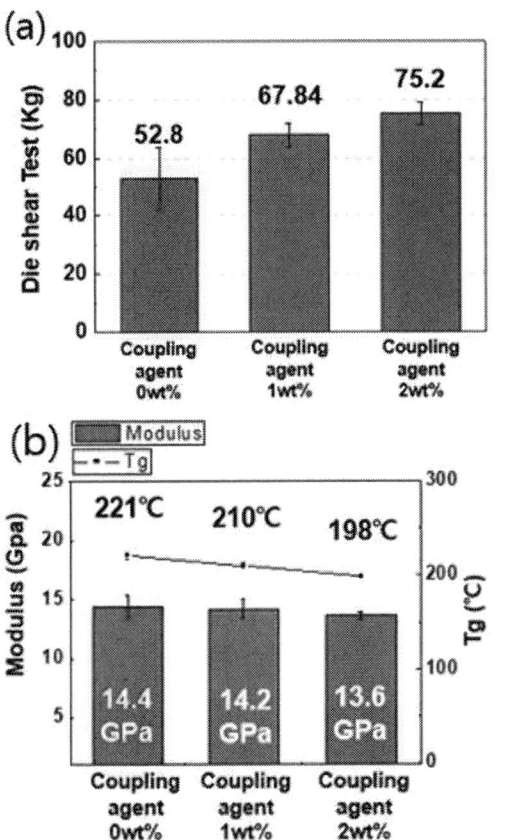

Figure 5. Comparison of coupling agent contents in terms of (a) adhesion and (b) modulus and Tg

Carbon black was added for the laser marking contrast. Visible light transmittance was measured by UV / VIS spectrometer. Figure 6 shows that transmittance was almost saturated with 0.25 wt% of carbon black. However, since the carbon black was conductive, the contents were needed to be minimized considering the insulating property. Therefore, the content of carbon clack was fixed with 0.25 wt%.

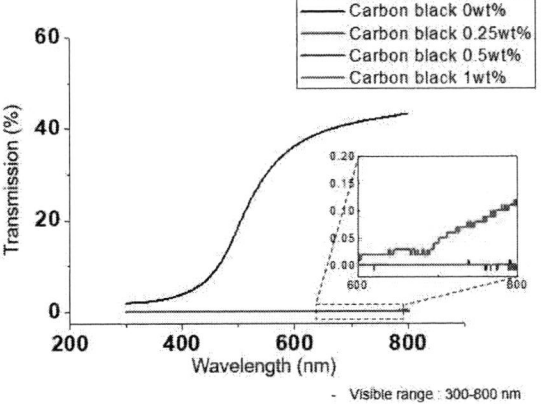

- Visible range : 300-800 nm

Figure 6. Comparison of carbon black contents in terms on the light transmission

B. PLPs fabrication

In order to fabricate 100 mm × 100 mm × 400 μm PLPs, EMFs were laminated to the chips to prevent void capture. Next, the molding was performed and the time was fixed at 15 min, the pressure was fixed at 3 MPa, and the molding was performed at 110 °C. The chip to chip pitch was reduced from 12 mm to 5.7 mm in Fig. 7. No voids or delamination were observed in PLPs as shown in Figure 7.

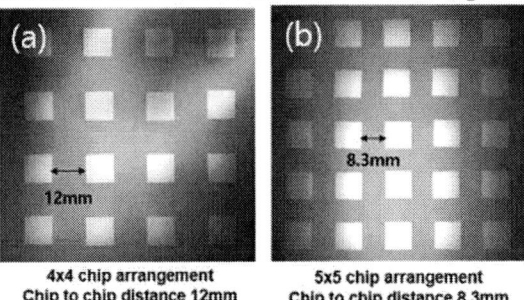

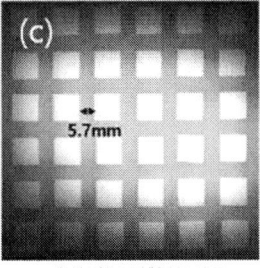

Figure 7. SAT images of PLPs with various chip arrangements of (a) 4x4, (b) 5x5, and (c) 6x6

C. Warpage evaluation

50 mm × 50 mm × 400 μm PLPs were post-cured for 2 hours at 170 °C, 200 °C, and 250 °C. As a result of the FT-IR measurement, it was confirmed that full-cure was observed from 200 °C. After post-cure at 200 °C and 250 °C, warpage measurement was performed using the 3D - DIC. Both PLPs decreased in warpage as temperature increased, which was due to a thermal relaxation. And the warpage increased when cooling to room temperature, because the thermal stress built up again. In Figure 8, the warpage at room temperature was lower at 200 °C, which was considered to be due to the small thermal stress build-up from process temperature reduction. [3]

978-1-7281-1500-9/19 $31.00 © 2019 IEEE

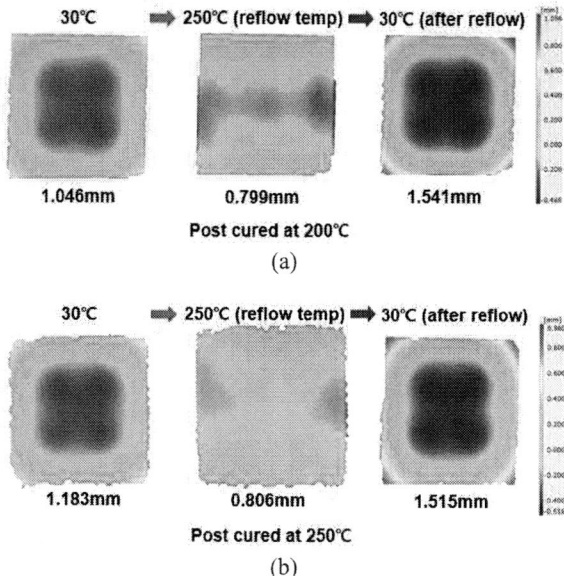

Post cured at 200°C

(a)

Post cured at 250°C

(b)

Figure 8. 3D - DIC images at 30°C and 250°C with post-cure temperature
of (a) 200°C and (b) 250°C

D. Die shift evaluation

Die shift was evaluated after molding process with a 5×5 chip structure. As shown in Figure 9, the shift of about -29.85 μm was observed in chips 18 mm away from the chip and the shift of about -48.71 μm was observed at a die 26 mm away from the die. Therefore, die shift occurred in the center direction of the chips. This is because the effect of warpage and curing shrinkage is greater than that of resin flow. [4]

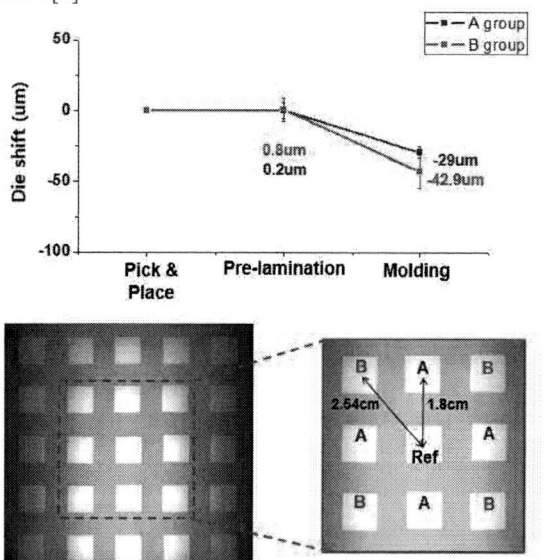

Figure 9. Die shift after molding measured with optical microscopy images

E. Reliability test

In the T / C test, SAT was measured to check crack and delamination after 0 cycle, 250 cycles and 500 cycles. Void and delamination did not occur as shown in Figure 10(a). After 85 °C / 85% RH for 0hr, 250 hrs, and 500 hrs, PLPs were measured with SAT to confirm delamination. As can be seen in Figure 10(b), no delamination occurred.

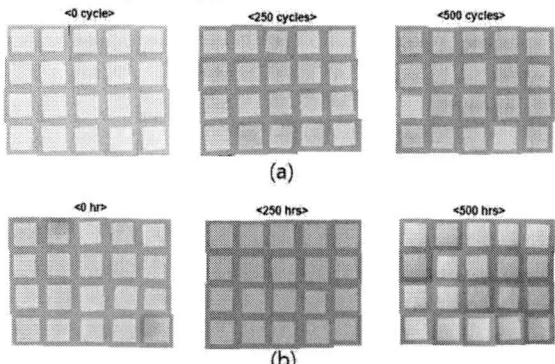

Figure 10. SAT images of singulated PLPs after (a) 500 cycles of T/C test
and (b) 500 hrs of 85°C / 85% RH condition

IV. CONCLUSION

In this study, epoxy, silica filler, curing agent, and additives were evaluated to make the optimized EMFs. Since thermoplastic resin had a high molecular weight and ded not contribute greatly to cross-linking, the content was minimized to increase the Tg, and decreased the viscosity. And 81 wt% of silica filler was added to improve CTE, Tg, and film formability. Then curing agent C type was added in terms of flowability and reactivity. The coupling agent contents were fixed as 1 wt% considering adhesion and Tg. From the viewpoint of transmittance, appropriate carbon black content was 0.25 wt%. Then 100 mm × 100 mm × 400 μm PLPs were fabricated with optimized EMFs and no voids or delaminations were observed in SAT measurement. Thereafter, the warpage was measured according to the post-cure temperature. The lower the post-cure temperature, the smaller the warpage was due to the less thermal-stress build-up. And die shift was also measured, and it was confirmed that warpage and curing shrinkage shift were more dominant than shift by a resin flow. Finally, T / C test 500 cycle and 85°C / 85% RH 500 hrs were performed. No chip crack and delamination were observed in all singulated PLPs in SAT measurement.

REFERENCES

[1] JongHo Park, HanMin Lee, SeYoung, Lee, Youjin Kyung, Jung Hak Kim, Kwangjoo Lee and Kyung-Wook Paik, "Fabrication and Characterization of Epoxy Molding Films (EMFs) for Wafer-Level and Panel-Level Fan Out Packages", Electronic Components and Technology Conference (ECTC), 2018 IEEE 68th, pp. 712 – 717

[2] HanMin Lee, SeYong Lee, SeMin Cho, YoungHyun Yu, JongHo Park and Kyung-Wook Paik, AStudy on Nano-sized Silica Contents and Size Effect in Non-Conductive Films (NCFs) for Ultra Fine-pitch Cu-pillar/Sn-Ag Micro-Bump Interconnection", Electronic Components and Technology Conference (ECTC), 2017 IEEE 67thm pp. 399 -404

[3] F.X. Che, David Ho, Mian Zhi Ding, Daniel Rhee MinWoo, "Study on Process Induced Wafer Level Warpage of Fan-Out Wafer Level Packaging", Electronic Components and Technology Conference (ECTC), 2016 IEEE 66th, pp. 1879 -1885

[4] Simo Yeon, Jeanho Park and Hye-Jin Lee, "Compensation Method for Die Shift Caused by Flow Drag Force in Wafer-Level Molding Process", Micromachines 2016, 7, 95 ; doi:10.3390/mi7060095

Mechanism of Moldable Underfill (MUF) Process for RDL-1ˢᵗ Fan-Out Panel Level Packaging (FOPLP)

Lin Bu, F. X. Che, Vempati Srinivasa Rao, Xiaowu Zhang

System in Package (SIP),
Institute of Microelectronics, A*STAR (Agency for Science, Technology and Research),
2 Fusionopolis Way, #08-02 Innovis Tower, Singapore 138634.
e-mail: bul@ime.a-star.edu.sg

Abstract— In order to achieve higher productivity and lower cost, Fan-out Panel Level Packaging (FOPLP) is being developed recently to increase throughput significantly compared to FOWLP. Gen III panel with a size of 650mm×550mm is investigated in this study. However, large panel size makes the issues (such as voids, and warpage) more severe than FOWLP. In the present paper, RDL-1ˢᵗ approach is used. One of the major advantages of RDL-1ˢᵗ process is to improve RDL process yield. This is due to chip-to-panel bonding is used before underfill/molding processes or modable underfill (MUF) process. MUF process has the advantage of simplified process steps. However, the challenge of moldability is increased with decreasing bump pitch and stand-off height. C-Scan or Thru-Scan test can obtain the void size, shape and location after molding. However, it is hard to analyze the mechanism of MUF process by only the testing results. Numerical simulation is a valuable way to track the transient information and make the whole process transparent. A large amount of fully populated chips with solder bumps are bonded onto the Gen-3 glass panel, which would demand high computational resources. More than 4000 bumps are located under each chip. Hence, it is difficult to use full model to construct the modeling. Instead, the simplified modeling is used to solve this issue. MUF process for liquid type EMC and sheet type EMC are simulated separately in the current study. The results show that packing pressure is one of the key factors for reducing voids for both types of EMC. Although lower molding speed is favorable in mold-1ˢᵗ process, it does not help in RDL-1ˢᵗ process. Staggered bump layout design and large pitch help to reduce voids for sheet type EMC.

Keywords-Fan-Out Panel Level Packaging (FOPLP), Modable Underfill (MUF), mold flow simulation, compression molding process.

I. INTRODUCTION

For Fan-out Wafer Level Packaging (FOWLP), wafer size limit was a big barrier for higher productivity and lower cost. In the recent years, fan-out panel-level packaging (FOPLP) was addressed to further reduce the cost. Fig.1 shows the cost reduction of die based on substrate size. Gen2 FOPLP can reduce cost by 2.5x, comparing to 200mm FOWLP.

For FOPLP, the redistribution layer (RDL) construction technologies can be divided into two categories. T. Braun et. al. [2-5] used PCB based technology with resin coated copper (RCC) materials. In our platform, thin film technology is used, which is a mature technology to achieve fine line/space requirement.

Gen III panel with a size of 650mm×550mm with using RDL-1ˢᵗ approach is investigated in this study. However, large panel size makes the issues more severe than FOWLP, i.e., voids & warpage. Hence, mold flow modeling on voids for FOPLP is investigated in the current study.

RDL-1ˢᵗ process starts with the chip-to-panel bonding process on the 2 layer RDL panel. Then panel level compression molding process for moldable underfill (MUF) is done subsequently. MUF process is one of the key processes in the packaging manufacturing. Good MUF process increases yield and benefits package reliability [6].

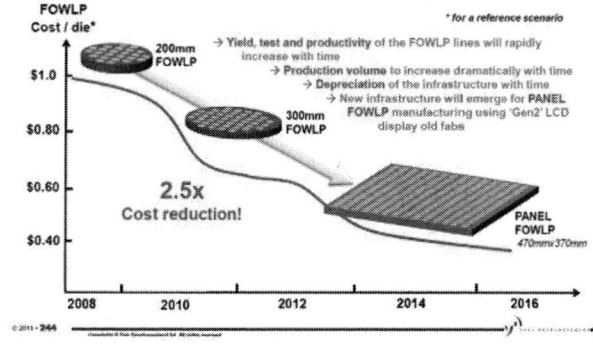

Figure 1.   Cost/die based on substrate size for Fan-Out packages [1]

II. TEST VEHICLE AND PACKAGE DESIGN

In the present study, a RDL-1ˢᵗ FOPLP approach is investigated. Package size is 15x15 mm², as shown in Fig.2. Chip with a size of 10x10 mm² locates at the center of the package. Signals from the die are routed out by 2 RDL layers with line width/line space 8um/8um. Total number of I/Os are more than 4000. The Gen3 panel size is 650x550 mm². Bottom cavity is used in molding process. Detailed package specifications are listed in Table I.

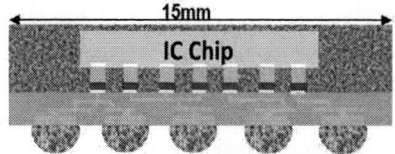

Figure 2.   RDL 1ˢᵗ package details

TABLE I. FOPLP PACKAGE SPECIFICATIONS

Test Vehicle	RDL-1st FOPLP
Target Applications	Mobile/Baseband/AP/2.5D
Package Size	15mm x 15mm
# of Chips/Pkg and Size	Single chip of 10mm x 10mm
RDL LW/LS	8um/8um (up to 2 layers)
# of I/Os	Up to 4000
Panel Size	650mm x 550mm (Gen 3)

In the special design, bumps are located in the central and peripheral of the chip. 4116 bumps are located under each chip, and structured mesh is used for each bump for the selected chips, as shown in Fig.3. Table II lists the bump details.

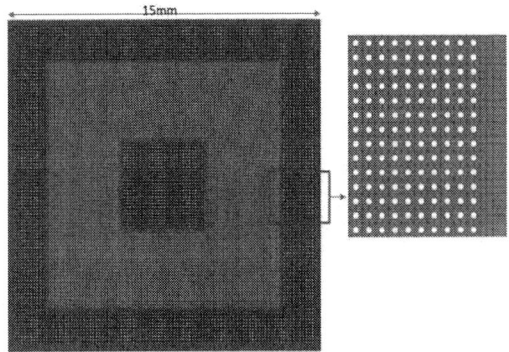

Figure 3. Mesh of the bumps

TABLE II. BUMP DETAILS

Micro-bump Pitch	100μm
Micro-bump stand-off height	50μm
Micro-bump Diameter	50μm
Micro-bump number	4116

III. MODELING DESCRIPTION

In this paper, computational fluid dynamics (CFD) simulation is used to understand the mechanism of voids issue. The boundary of epoxy molding compound (EMC) and air is tracked by volume of fluid method, which was proposed by Hirt and Nichols [7]. Volume of fluid (VOF) method is used to track the melting front of EMC material. Two different kinds of EMCs, liquid type and sheet type, are investigated. The liquid phase, EMC, is considered as incompressible material. The gas phase, air, is considered as compressible. Dynamic mesh technique is used to simulate the motion of bottom mold chase in the present study.

The modeling consists of the following assumptions.
1) 3D quarter panel model is used.

2) At the initial time, EMC is assumed to be fully melted.

For liquid type EMC, modeling conditions are: vent size is 3mm (width) x 15um (depth) and vent pitch is 8mm; packing pressure is 6MPa and vacuum pressure is 100Pa. A validated model is constructed, as shown in Fig.4. Table III lists the simulation DOE matrix for liquid type EMC. For the dispensing pattern, square dispensing pattern and strip pattern are modeled in simulation according to molding machine capability. Bottom cavity is used in the simulation.

Quarter model

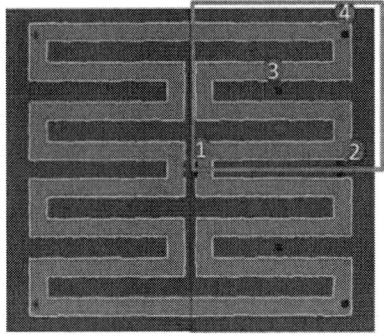

8 strips dispensing pattern

(a) top view

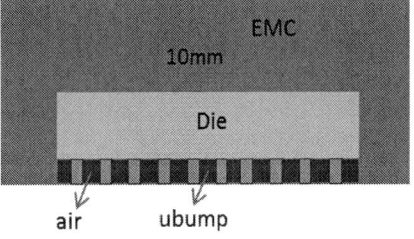

(b) cross section view

Figure 4. CFD model of panel with 8 strips dispensing pattern for liquid type EMC (a) top view (b) cross section view

TABLE III. DOE MATRIX FOR LIQUID TYPE EMC

Die thickness/mold thickness	0.25mm/0.4mm, 0.775mm/0.925mm
Dispensing pattern	4 strips, 8 strips, Rectangle
Molding speed	500um/s & 100um/s, 500um/s
EMC viscosity	EMC3 (7 Pa·s), EMC1 (94 Pa·s)

For sheet type EMC, modeling conditions are: vacuum pressure is 30Pa. A validated model is constructed, as shown in Fig.5. Table IV lists the simulation DOE matrix. Vent is not individually located at the panel edge as liquid type model. However, the outmost edges in the simulation are all opened as a big vent. The sheet edge to the utmost edge in

978-1-7281-1500-9/19 $31.00 © 2019 IEEE

the simulation is 40mm. For packing pressure, 1MPa, 5MPa and 10MPa are used in simulation.

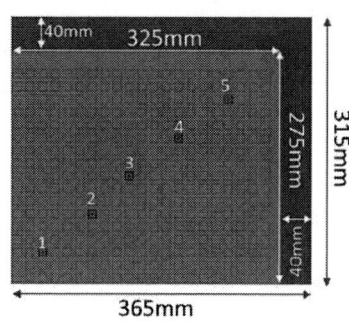

(a)   top view

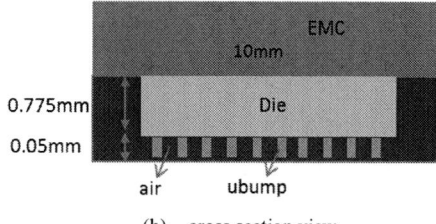

(b)   cross section view

Figure 5.   CFD model of panel with sheet type EMC (a) top view (b) cross section view

TABLE IV.   DOE MATRIX FOR SHEET TYPE EMC

Die thickness/mold thickness	0.775mm/0.925mm
Dispensing pattern	sheet type
EMC viscosity	264 Pa·s, 104 Pa·s
Packing pressure	1MPa, 5MPa, 10MPa

IV.   RESULTS AND DISCUSSION

A large amount of fully populated chips with solder bumps are bonded onto the Gen-3 glass panel, which would demand high computational resources. More than 4000 bumps are located under each chip. Hence, it is difficult to use fully populated model to construct the modeling for each bump. Instead, the simplified modeling is used to solve this issue. In the current study, only the investigated chips with labels use detailed mesh for each bump. For other chips, voids are fully occupied beneath the die. In the 4 packages model, 4 adjacent detailed packages are simulated, while in the 1 package model, only 1 detailed package is simulated. Location 1, 2D, 3D, 4D in 4 packages model are compared with the same location 1, 2, 3, 4 in the 1 package model. The condition of the two models are: 0.25mm die

thickness/0.4mm final mold thickness, 8 strips dispensing pattern, molding speed of 500um/s, EMC3. The results of the two models before packing are quite similar, as shown in Fig.6. Hence, the 1 package model is used in the subsequent studies.

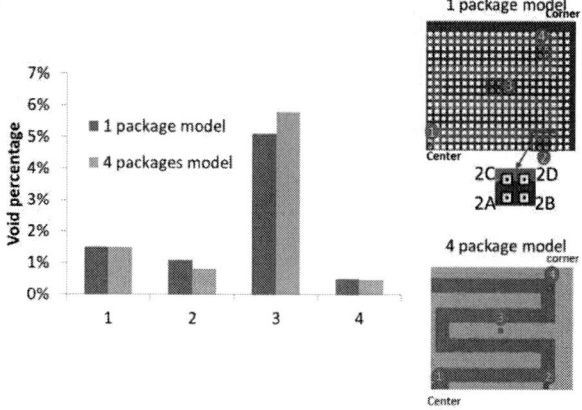

Figure 6.   Comparison of 1 package model and 4 package model

A.   Simulation results for liquid type EMC

The condition of the reference model for liquid type EMC is: 0.25mm die thickness/0.4mm final mold thickness, 8 strips dispensing pattern, molding speed of 500um/s, EMC3. Four typical die locations were selected: die1 at central and partially covered by EMC initially on the top, die2 and 3 uncovered, die4 fully covered by EMC on the top initially. Coordinate of each die central is listed in Table V.

TABLE V.   TYPICAL 4 DIE LOCATIONS

Die location number	Coordinate of die central /mm
1	(7.5,7.5)
2	(262.5,7.5)
3	(157.5,132.5)
4	(272.5, 227.5)

Fig.7 shows the void results of the reference model for liquid type EMC in the 4 locations. Usually the molding process can be separated into filling and packing two stages. In the filling stage, the process is controlled by molding speed; in the packing stage, the process is controlled by packing force. Void size is die location dependent. Die 3 shows large void before packing due to entrapped air during flow. Void size reduced significantly after packing.

978-1-7281-1500-9/19 $31.00 © 2019 IEEE   1154

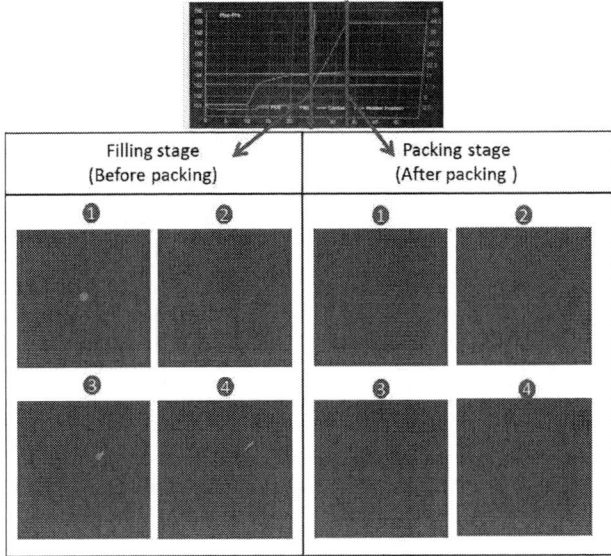

Figure 7.   Void results of reference model for liquid type EMC

Fig.8 shows the transient histories of EMC coverage beneath die1 to die4. EMC coverage beneath die at the initial state is 0. The sequence of EMC coverage beneath die is: die4, die1, die2, die3. Generally speaking, fully covered die at initial stage has a speedy coverage beneath die, and the uncovered die at the initial stage has a slow coverage beneath die.

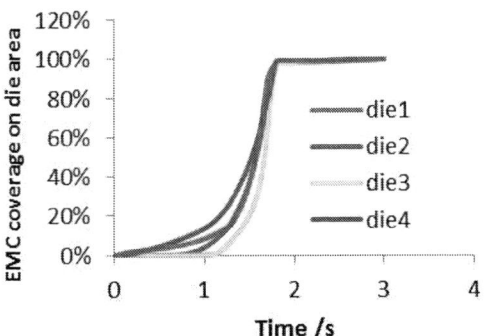

Figure 8.   EMC coverage beneath die1-die4

In this section, design parameters, i.e., the physical design parameters and process design parameters are optimized. The physical design parameters include die/mold thickness and copper pillar design etc. The process parameters, i.e., molding speed and dispensing pattern are also optimized.

Die/mold thickness is also a design parameter for warpage calculation. Combining the results of mold flow and warpage simulation, die/mold thickness will finally be optimized. Comparing to die and mold (250um/400um) in the reference model, thicker die and mold (775um/925um) is simulated, as shown in Fig.9. Thin die design leads to small void due to less entrapped air volume. By reducing die

thickness from 775um to 250um, the void size will decrease to 38% of the original one.

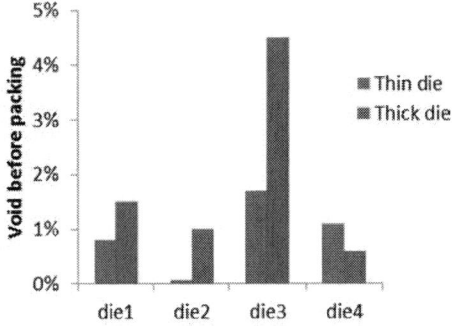

Figure 9.   Effect of die/mold thickness for liquid type EMC

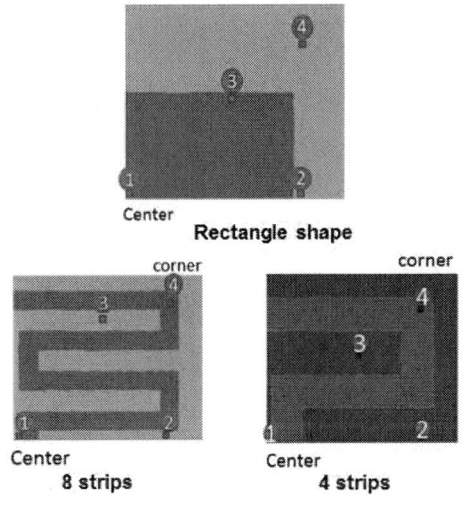

Figure 10. Three dispensing patterns for liquid type EMC

Fig.10 demonstrates three different dispensing patterns, i.e., rectangle shape, 8 strips and 4 strips. All the dispensing patterns are captured from real process. The volume of EMC material is the same and the initial molding height is adjustable in the simulation. Dispensing pattern has a significant effect on void size. Using optimized dispensing pattern helps to reduce entrapped air volume. 4 strips dispensing pattern leads to larger voids before packing, which is 3 times larger than 8 strips dispensing pattern and 9 times larger than rectangle dispensing pattern, as shown in Fig.11. Among three dispensing patterns, rectangle shape pattern is the best one. Rectangle dispensing pattern results in smaller void before packing and no void after packing. This is due to at the initial state, 4 strips and 8 strips dispensing pattern form the closed loop, a lot of air are entrapped in it, as shown in Fig.12.

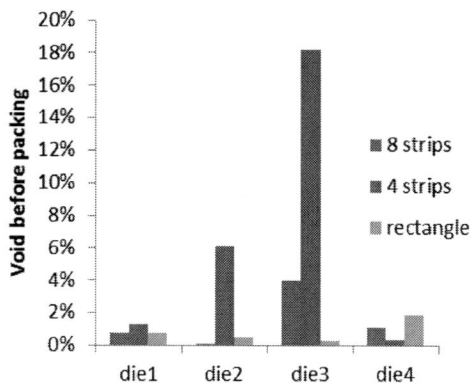

Figure 11. Effect of dispensing patterns for liquid type EMC

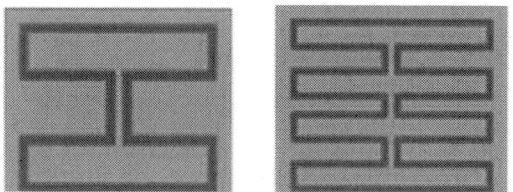

Figure 12. Closed loop for 4 strips and 8 strips dispensing pattern at initial state for liquid type EMC

High viscosity EMC leads to large void before packing. When the material viscosity changes from 94Pa·s to 7 Pa·s, void size reduces by 60%, as shown in Fig.13.

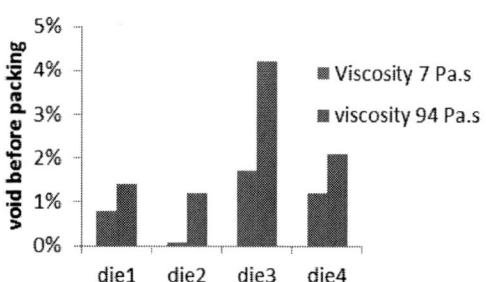

Figure 13. Effect of viscosity for liquid type EMC

High molding speed helps to reduce void size. By increasing the molding speed from 100um/s to 500um/s, the void size will decrease to less than 24% of the original one, as shown in Fig.14. For mold-1st process, usually lower speed at the final stage can help to reduce the drag force and flow induced die shift/sliding [8-10]. However, in the RDL-1st process, low molding speed does not help.

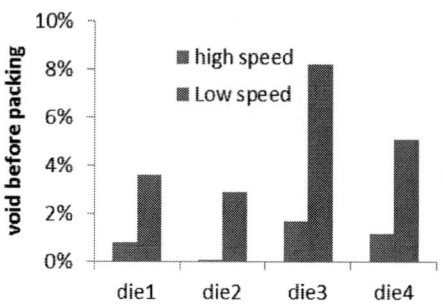

Figure 14. Effect of molding speed for liquid type EMC

Large packing force helps to reduce void size significantly. For 4 strips dispensing pattern, void size can be reduced by 86% with increasing the packing force from 3MPa to 6MPa, as shown in Fig.15. Large packing force is an effective way in reducing the voids but it is confined by machine limit.

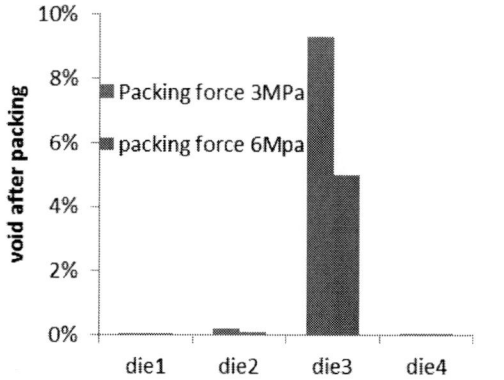

Figure 15. Effect of packing force for liquid type EMC

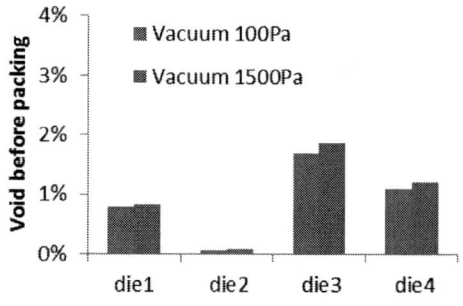

Figure 16. Effect of vacuum pressure for liquid type EMC

Vacuum level of 1500Pa results in slightly large void than vacuum level of 100Pa, as shown in Fig.16. Vacuum level can't reduce void size much due to flow resistance plays a much bigger role in the molding process. Fine pitch solder bumps induce a very high flow resistance, which makes the vacuum level insignificant.

978-1-7281-1500-9/19 $31.00 © 2019 IEEE

B. Simulation results for sheet type EMC

The viscosity of the sheet type EMC used in this study is temperature dependent, which is 264 Pa·s at 130°C and 104 Pa·s at 150°C. Molding temperature is usually fixed in the molding process. Hence, we use the viscosity of the material at different tempreature in the simulation so that appropriate temperature can be selected in the real process.

Larger viscosity results in larger void and void percentage in the packing pressure range from 1MPa to 10 MPa, as shown in Fig.17. Voids locate at the center of chip and at the edge of the panel. Void size is not sensitive to chip location for sheet type EMC molding process, as shown in Fig.18. This is due to sheet type EMC doesn't have the dispensing pattern effect. All the chips are fully covered by sheet type EMC on the top at the initial state.

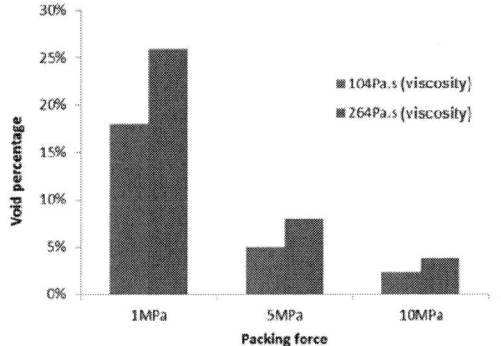

Figure 17. Effect of viscosity on void for sheet type EMC

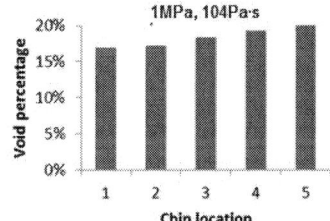

Figure 18. Effect of chip location on void for sheet type EMC

For sheet type EMC, packing pressure also plays a big role in reducing the voids. Increasing packing pressure helps to decrease void size and percentage significantly, as shown in Fig.19.

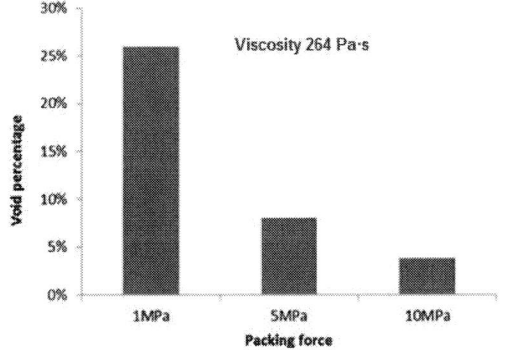

Figure 19. Effect of packing force on void for sheet type EMC

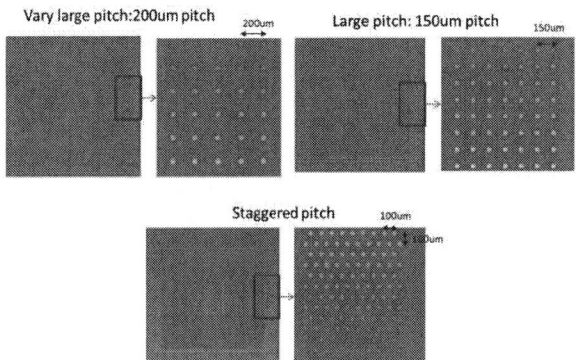

Figure 20. Bump design with different Cu pillar pitch

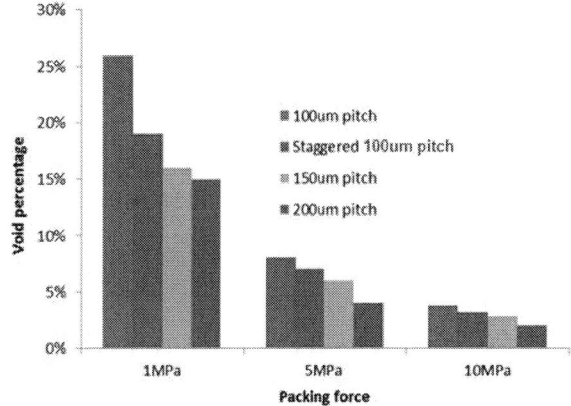

Figure 21. Effect of bump pitch on void

Fig.20 shows the three different designs such as 200μm pitch, 150um pitch and staggered 100um pitch. Staggered 100um pitch shows 30% smaller void size than 100μm pitch at 1MPa, as shown in Fig.21. With the same pitch, staggered design is more favorable in the molding process. Large pitch is also favorable to redcue void in the molding process.

V. CONCLUSION

1. From process point of view, dispensing pattern, packing force and molding speed play important roles in molding process.

   a) Dispensing pattern has significant effect on void size. Among three dispensing patterns, rectangle shape pattern is the best one. Rectangle dispensing pattern results in smaller void before packing and no void after packing. 4 strips dispensing pattern leads to larger voids before packing, which is 3 times larger than 8 strips dispensing pattern and 9 times larger than rectangle dispensing pattern.

   b) For both types of EMC, increasing packing force is an effective way for reducing voids. For 4 strips dispensing pattern, void size can be reduced by 86% with increasing the packing force from 3MPa to 6MPa for liquid type EMC.

   c) High molding speed help to reduce void size. By increasing the molding speed from 100um/s to

500um/s, the void size will decrease to less than 24% of the original one for liquid type EMC.

    d) Vacuum level of 1500Pa results in slightly large void than vacuum level of 100Pa for liquid type EMC.

2. From material selection point of view, low viscosity EMC helps to reduce void size. When the material viscosity changes from 94Pa·s to 7 Pa·s, void size reduces by 60% for liquid type EMC.

3. In design point of view, large Cu pillar pitch or staggered pillar layout design is recommended as it can reduce flow resistance. Thin die is recommended as it can entrap less air at the initial state. Thin die design leads to small void due to less entrapped air volume. By reducing die thickness from 775um to 250um, the void size will decrease to 38% of the original one for liquid type EMC.

ACKNOWLEDGEMENTS

This work is the result of a project initiated by Fan-Out Panel Level Packaging (FOPLP) Consortium. The authors greatly appreciate the members' participation in discussions and encouragement throughout the course of the project which makes this research possible.

REFERENCES

[1] Yole Development, "Convergence of FE, BE and FPD manufacturing technologies: a closer look," Jan 7th, 2012. http://www.i-micronews.com/news.

[2] T. Braun, K.-F. Becker, S. Voges, J. Bauer , "Challenges and Opportunities for Fan-out Panel Level Packing (FOPLP)," Proc.

International Microsystems, Packaging, Assembly and Circuits Technology Conference (IMPACT 9),Taibei, Taiwan, Oct 2014, pp. 154-157.

[3] T. Braun, K.-F. Becker, S. Voges, J. Bauer , "24"x18" Fan-out Panel Level Packing," Proc. Electronic Components and Technology Conference(ECTC 60), California, USA, May 2015, pp. 1077-1083.

[4] T. Braun, K.-F. Becker, S. Voges, J. Bauer , "Large Area Compression Molding for Fan-out Panel Level Packing," Proc. Electronic Components and Technology Conference(ECTC 59), Florida, USA, May 2014, pp. 940-946.

[5] T. Braun, K.-F. Becker, S. Voges, J. Bauer , "From Wafer Level to Panel Level Mold Embedding," Proc. Electronic Components and Technology Conference(ECTC 58), Nevada, USA, May 2013, pp. 1235-1242.

[6] Sheng Liu, Yong Liu, Modeling and simulation for microelectronic packaging assembly: manufacturing, reliability and testing, John Wiley & Sons, 2011.

[7] C. W. Hirt, B. D. Nichols, "Volume of Fluid (VOF) Method for the Dynamics of Free Boundaries," J. Comp. Phys., Vol.39, Jan, 1981, pp. 201–225.

[8] Lin Bu, Siowling Ho, Sorono Dexter Velez, Boyu Zheng, Ser Choong Chong, Booyang Jung, Taichong Chai, Xiaowu Zhang, "Investigation on decap shift and incomplete fill issues in the wafer level compression molding process," Proc. Electronics Packaging Technology Conference(EPTC 15), Singapore, 2013, pp.766-770.

[9] Lin Bu, Siowling Ho, Sorono Dexter Velez, Taichong Chai, and Xiaowu Zhang, "Investigation on Die Shift Issues in the 12-in Wafer-Level Compression Molding Process," IEEE Transactions on components, packaging and manufacturing technology, Vol.3, Oct 2013, pp.1647-1653.

[10] Lin Ji, Dexter Velez Sorono, Tai Chong Chai, and Xiaowu Zhang, "3-D Numerical and Experimental Investigations on Compression Molding in Multichip Embedded Wafer Level Packaging," IEEE Transactions on components, packaging and manufacturing technology, Vol.3, Apr 2013, pp.678-687.

Study of the Board Level Reliability Performance of a Large 0.3 mm Pitch Wafer Level Package

Bernd Waidhas, Jan Proschwitz, Christoph Pietryga, Thomas Wagner, Beth Keser
Intel Deutschland GmbH
Am Campeon 10-12, 85579 Neubiberg, Germany
bernd.waidhas@intel.com

Abstract—Board level reliability investigations have been performed on 36 mm² wafer level packages (WLP) with a 0.3 mm ball pitch. Three different solder ball alloys were included in the temperature cycling, thermal shock and mechanical shock test. In addition to daisy chain test vehicles to address the solder joint reliability, 28 nm die packaged with WLP were included in the assessment to check the stress impact on extreme low K dielectric (ELK) in the die back-end-of-line (BEOL). The tests have included selected studies on the influence of printed circuit board (PCB) thickness and usage of a board level underfill. WLP with SAC-Q solder ball show a significant improvement versus the industry standard SAC405 solder balls. In temperature cycling on board (TCoB), WLP's with SAC-Q (SAC405 with 3% Bi) achieve more than 2000 cycles without fail in temperature range from -40°C to 85°C on a 0.8 mm thick PCB, whereby first fails with SAC405 balls were observed above 800 cycles. All WLP's with LF35 (SAC125Ni) solder alloy fail before 1000 cycles. The failure mode in TCoB changes from solder joint fatigue for SAC405 and LF35 to Cu redistribution layer (RDL) cracks for SAC-Q. The 28 nm functional die packaged with WLP show no ELK crack or any other fail after 1000 cycles for SAC-Q alloy balls. The TCoB with SAC405 WLP pass 2000 cycles without fail when an underfill was applied. Non-underfilled WLP with SAC405 balls assembled on a thickness reduced 0.4 mm PCB show also a significant improvement with a first fail above 1900 cycles. WLP with SAC-Q and SAC405 passed mechanical shock (24 drops at 10k g).

Keywords-Packaging; Reliability; Wafer Level Package; Wafer Level Chip-Scale Package; 0.3mm pitch; solder ball alloy; temperature cycle on board; SAC405; SAC-Q

I. INTRODUCTION

Market driven performance increases in handheld applications drives feature density as well as system on chip integration in electronic systems. Advanced wafer technologies in combination with footprint optimized package solutions are used to fulfill the demand. Wafer level packages (WLP) have advantages in form factor, electrical and thermal performance. Therefore, they are widely used in mobile devices. However, the WLP does not include a substrate or leadframe, underfill or mold compound, so the mechanical protection of the die is reduced compared to other package types and there is a more direct interaction between die and the printed circuit board (PCB). Therefore

WLP board level reliability aspects like temperature cycling robustness and mechanical shock impact have to be carefully characterized. Optimization of design, material set and system construction can manage a suitable stress distribution for reliable WLP applications.

In our case, a 6.4 mm x 6.0 mm wafer level package with a 0.3 mm ball pitch was investigated for solder joint fatigue and ELK robustness. A two redistribution layer (2L-RDL) WLP with an under ball metallization (UBM) was chosen according to the targeted product requirements. A harder SAC-Q (SAC405 with 3% Bi) ball alloy for improved solder joint fatigue and a softer LF35 (SAC125Ni), the industry standard for drop test improved laminate packages, for reduced stress on package redistribution layers and the ELK were selected as well as the WLP industry standard SAC405 ball alloy. The parts were assembled on a 0.8 mm thick PCB. Temperature cycling on board -40°C to +85°C and mechanical shock 10k g were applied. Daisy chain die were used for in-situ resistance measurement to detect solder joint or redistribution layer cracks in the WLP. In addition, 28 nm functional die packaged with 0.3 mm pitch WLP are tested using typical board level reliability (BLR) stress tests to check for ELK cracks. For the functional die, a die metal stack failure analysis was done by a layer by layer top-down analysis (TDA), X-ray inspection and scanning acoustic microscopy (SAM). Further investigation of the SAC405 WLP were completed with reduced PCB thickness and underfill.

The paper will discuss the influence of the improvement measures on the board level reliability and the detected failure modes.

II. TEST VEHICLE DESCRIPTION

A wafer level package with two Cu redistribution layers (RDL) and an under-ball metallization (UBM) construction has been investigated. The package terminal ball grid array has a pitch of 0.3 mm. A 29 x 27 ball matrix in a 45° rotated arrangement with regards to the package edges is placed on a 6.4 mm x 6.0 mm die. There are some minor depopulations in the ball matrix reflecting real product constraints (Figure 1). The UBM and the solder balls are matched to 180 μm diameter. The package has a body thickness of 295 μm with a backside protection film.

There are two different silicon die used for the test vehicle package to address the different failure modes: one with a daisy chain (DC) connection for in-situ detection of interconnect fails during stress test by resistance monitoring and another has a full metal stack extreme low K (ELK) 28 nm technology die to investigate possible defects in die metal stack caused by the reliability test mechanical stress.

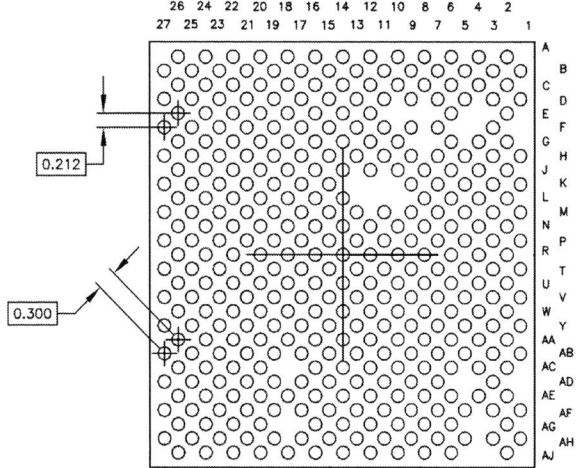

Figure 1. Top-view of test vehicle ball grid array used in this study.

Three solder ball alloys have been investigated. Besides the WLP industry standard SAC405, there was used a harder SAC-Q alloy (SAC405 with 3% Bi) for improved temperature cycling solder joint reliability and LF35 (SAC125Ni), a softer solder alloy, which is often used in FCCSP's for drop test robustness improvement. For WLP, it is used to reduce the stress on the fragile ELK die metal stack (Table 1).

TABLE I.   MECHANICAL PROPERTIES OF THE SOLDER BALL ALLOYS USED IN THIS STUDY.

Solder Ball Properties		Solder Ball Material		
		LF35	SAC405	SAC-Q
Young's Modulus	GPa	47.6	50.6	51.7
Tensile Strength	MPa	38	49	91
Elongation	%	48	61	37
CTE	ppm/°C	21.5	21	21

The positive influence of Bi content in SAC solders on TCoB reliability was reported in previous studies [1, 2, 3]

The WLP's are soldered on a PCB for the reliability tests. A 0.8 mm thick 10-layer FR4 board was used as the default test board stack-up. A thinner 0.4 mm thick 8-layer PCB, similar to smart watches or other small form factor device applications, was added to investigate the reliability improvement potential of a thinner, more flexible board. The ball pads on the PCB are non-solder mask defined with a copper pad diameter of 180 μm and an OSP surface protection.

A further variant of the investigation was using board level underfill (BLUF), a method widely used for handheld devices to improve the reliability against mechanical stress. In this investigation, the mechanical properties of the underfill were selected with focus on the 0.3 mm ball pitch WLP temperature cycling solder joint fatigue risk. Figure 2 shows the mechanical properties of the BLUF materials considered in simulation and normalized TCoB characteristic life prediction. Underfill material #3 with the CTE close to the solder ball material shows the highest improvement potential and was selected for the TCoB tests.

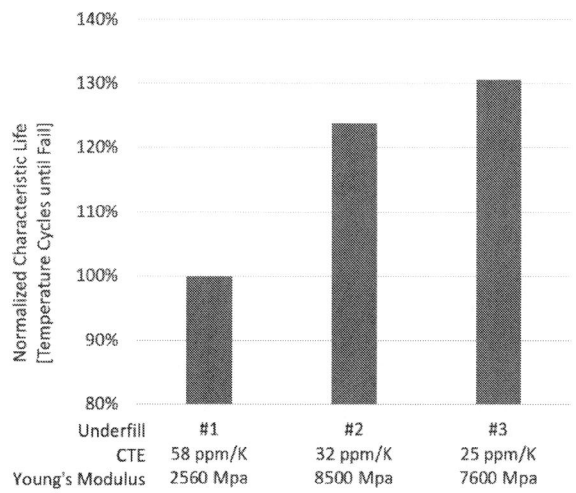

Figure 2. WLP test vehicle characteristic life at TCoB -40 / +125°C on 10-layer 0.8mm thick PCB with SAC 405 solder balls.

III.   BOARD LEVEL RELIABILITY STRESS TESTS AND FAILURE ANALYSIS METHODS

The reliability tests and analysis methods were determined to address and detect the potential failure modes generated by thermo-mechanical and mechanical stress on the WLP in the application environment: solder joint cracks of the WLP connections to the PCB and cracks or delamination in RDL stack or in the BEOL stack of the 28 nm ELK silicon die. Temperature cycling on board (TCoB), thermal shock and mechanical shock accelerated stress tests were applied to the test vehicles.

TCoB stress conditions were chosen according to IPC-9701 with -40°C to 85°C temperature range at 1 cycle per hour and a ramp of 15°C/min (Figure 3). The daisy chain WLP devices used for this test to investigate solder joint fatigue as well as cracks in the RDL or the connection of the RDL to the die pad by sensitive in-situ resistance measurements. The failure mode was confirmed by standard cross-section analysis or focused ion beam (FIB) cuts.

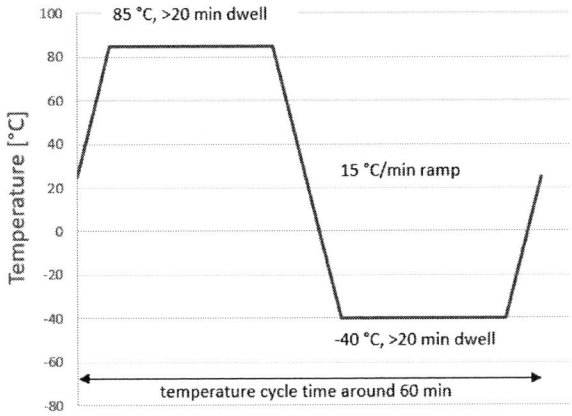

Figure 3.   TCoB cycle profile.

In addition to testing daisy chain test vehicles, temperature cycling of the 28 nm full metal stack devices soldered on the PCB were performed to identify any defects in the ELK BEOL stack. An electrical measurement analysis of these device was not possible. The devices are stressed until 1000 cycles followed by a SAM failure analysis then a top-down preparation with optical inspection. Figure 4 shows the principal flow. The WLP's are thinned to increase the resolution in SAM.

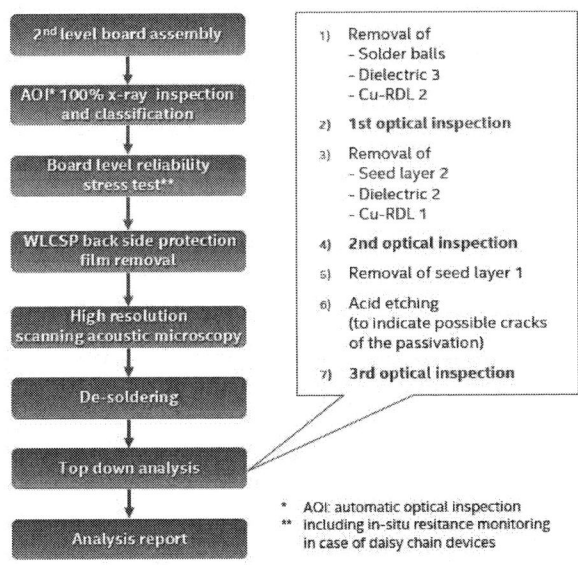

Figure 4.   Reliability stress test flow and failure analysis.

Furthermore, temperature shock test (liquid-to-liquid) is applied according to JESD22-A106 Condition D load with -65°C low temperature bath and +150°C high temperature bath (Figure 5) to check the robustness of the assemblies, especially the ELK layer stack of the 28 nm die under these harsh temperature change conditions. Analysis was done using SAM investigation and a top-down preparation

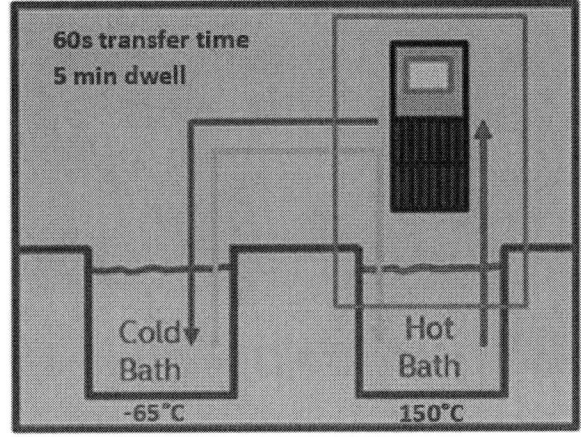

Figure 5.   Liquid-to-liquid thermal shhock test procedure

Mechanical shock tests (MS) have been applied on the daisy chain as well as on the 28 nm full metal stack die devices to simulate the mechanical stress impact, which occurs for example by an accidental drop of a hand-held device. This impact is supposed to generate brittle cracks in intermetallic phases of the solder at the UBM or PCB Cu pad, interconnection of RDL to die as well as to cracks or delamination in the RDL or die metal stack. The test was carried out according to JESD22-B104 with 10000 g acceleration at a half sine wave with a pulse width of 0.25 ms. The pulse is measured on top of the package. The test vehicles were stressed several times at the 6 orientations (+x, -x, +y, -y, +z, -z) each. The failure detection was done by resistance measurement before each direction change for the daisy chain devices in test and scanning acoustic microscopy and top-down preparation for the 28 nm full metal stack die test vehicles.

IV.   RESULTS

A.   Temperature Cycling on Board

LF35 solder ball devices fail first in the temperature cycling test as expected, due to the low tensile strength of this alloy. The first open in-situ resistance measurement was detected at 520 cycles and more than 95% of the 78 devices fail before 1000 cycles, and by 1111 cycles all devices have failed. The industry standard SAC405 solder ball test vehicles also show fails below 1000 cycles. The first failure was detected at 836 cycles followed by another 2 electrical opens before 1000 cycles. The devices with SAC-Q show a significant improvement in temperature cycling reliability. The first open was measured at 2612 cycles. The test was

978-1-7281-1500-9/19 $31.00 © 2019 IEEE

stopped after 3339 cycles with in total 5 fails. Figure 6 shows the TCoB reliability Weibull plots.

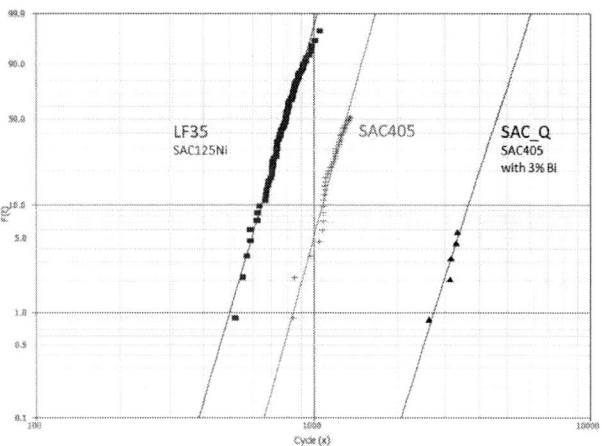

Figure 6. TCoB Weibull plots (-40°C / +85°C, 0.8mm thick 10L PCB), WLP with various solder ball alloys.

The cross section analysis of the defect parts confirms a typical solder joint fatigue crack failure mode for LF35 and SAC405 devices. The cracks are located in the solder joint bulk material. Conversely, the SAC-Q devices show a different failure mode. The weakest interface for these devices was observed at the interconnect between the RDL and the chip pad especially in regions near the device edge, where such interconnects are under or near to solder ball pads (Figure 8). This failure mode could be confirmed for all failed devices by C-SAM analysis and validated by FIB cut (Figure 9).

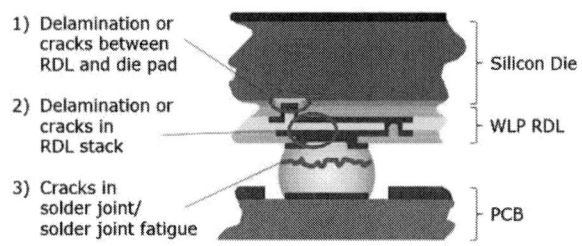

Figure 7. Failure mode classification.

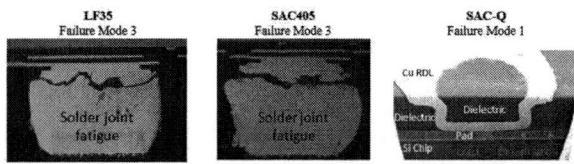

Figure 8. TCoB failure mode of WLP depending on solder ball alloy

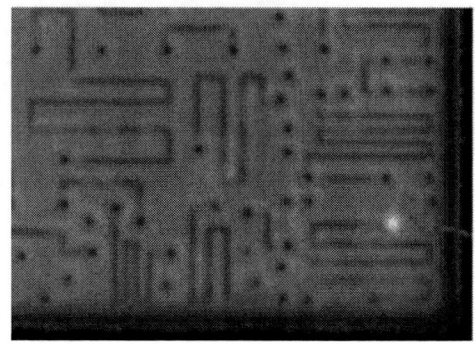

Figure 9. C-SAM analysis of WLP indicates a delamination/crack at a die pad.

The stress distribution as well as the location of weakest interface in the device is impacted by the die metal stack and design. Therefore, in addition to the investigation with simple metal stack daisy chain devices full metal stack 28 nm devices were stressed under the same condition to investigate the risk of damage in the RDL or BEOL stack for an active device as observed on the SAC-Q daisy chain devices. 72 devices each of SAC405 and SAC-Q are mounted on a 0.8 mm 10-layer PCB and exposed to 1000 temperature cycles and analyzed with top-down preparation and SAM inspection. For both solder balls, SAC405 and SAC-Q no delamination, dielectric or passivation crack was detected (Figures 10, 11 and 12).

Figure 10. SAM analysis of 28nm ELK WLP with SAC405 solder balls after 1000x TCoB -40°C/85°C.

 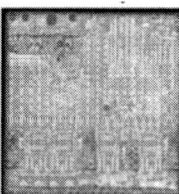 

Figure 11. SAM analysis of 28nm ELK WLP with SAC-Q solder balls after 1000x TCoB -40°C/85°C.

978-1-7281-1500-9/19 $31.00 © 2019 IEEE

After RDL2 removal	After RDL1 removal	After acid etch
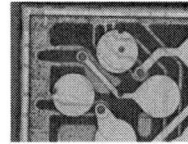	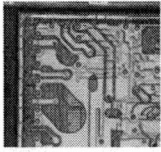	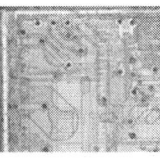

Figure 12. Top down analysis of 28nm ELK WLP with SAC-Q solder balls after 1000x TCoB -40°C/85°C.

B. Temperature Shock (Liquid-to-Liquid)

The harsh stress thermal shock (liquid-to-liquid) was used to reconfirm the device robustness of the full metal stack 28 nm device soldered on the 10-layer 0.8 mm thick PCB. 50 pcs each of the 3 solder ball alloy devices are exposed 75 cycles at -65°C and 150°C. The analysis was done with C-SAM inspection and top-down preparation. All 150 devices show no delamination or other damages (Figure 13).

LF35 solder balls	SAC405 solder balls	SAC-Q solder balls
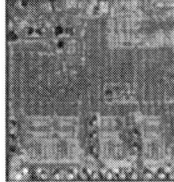	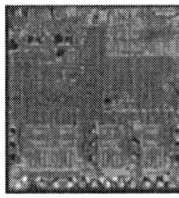	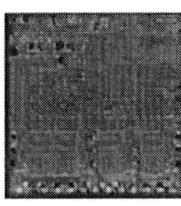

Figure 13. SAM analysis of 28nm ELK WLP after 75x LLTS -65°C/150°C.

C. Mechanical Shock

Daisy chain vehicles with the both higher risk to fail harder solder ball alloys SAC-Q and SAC405 were investigated with mechanical shock tests. The devices were stressed with shock impulses of 10000 g with a duration of 8 times in each of the 6 directions. There was no open detected during the stress test in-situ resistance measurement of the daisy chain connection.

28 nm full metal stack devices with SAC-Q solder balls were investigated with mechanical shock test to evaluate the functional active device reliability under these stress conditions. Analysis with C-SAM inspection and top-down preparation was done after 4 shock events in each of the 6 directions. No failure on the 78 devices in test was observed.

D. Influence of PCB Thickness on TCoB Performance

The board level reliability tests results discussed previously were generated with devices mounted on 0.8 mm thick 10-layer PCBs. The PCB construction and its stiffness has significant impact on the mechanical and thermo-mechanical stress induced into the solder joint connection, the RDL and the die. In some applications, thinner boards are used for small form factor devices like smart watches. In general, thinner board will have a positive effect on the device board level reliability. A 0.4 mm thick 6-layer PCB was used to investigate the improvement potential of a thinner board.

The WLP with the industry standard SAC405 solder balls have shown solder joint fatigue fails below 1000 cycles during TCoB stress on 0.8 mm thick PCB before. Daisy chain devices with this solder ball alloy were soldered onto the thinner PCB for TCoB. The electrical measurement results show a significant improvement with a first fail at 1927 cycles only. The Test was stopped after 2000 cycles with no additional fail.

E. Reliability Robustness Improvement by Usage of Board Level Underfill

The application of a board level underfill material to protect the solder joint connections by mechanical stress distribution and stress reduction in the critical interfaces is widely used in the industry. The underfill materials are often optimized to improve the drop impact reliability especially in handheld applications. Typical failure modes observed after mechanical drop or shock drop stress are cracks in brittle intermetallic phases of the interconnections or cracks in traces sometimes in combination with delaminations in the assembly layer stack [4]. In contradiction, solder joint fatigue crack is the typical failure mode for temperature cycling stress. Underfill with a coefficient of thermal expansion (CTE) near the solder material, a glass transition temperature (T_g) above the maximum temperature in the application and a high elastic-modulus (E) are recommended to improve the temperature cycling robustness. Underfill #3 properties (Figure 2) fit best. This material was used in the tests.

Devices with the both most promising solder ball materials from the previous tests, SAC405 and SAC-Q, were investigated in TCoB and LLTS test with board level underfill applied (Figure 14).

In TCoB test, the daisy chain device samples were stressed up to 2000 cycles. For each solder ball alloy, 78 devices were tested. There was no electrical fail detected in the daisy chain connection. Also no abnormalities were observed on the 46 devices for each of the both solder alloy 28 nm full metal stack devices after 1000 cycles.

978-1-7281-1500-9/19 $31.00 © 2019 IEEE

Figure 14. WLP with underfill.

50 pieces of SAC405 and 50 pieces of SAC-Q solder ball 28 nm devices were tested with LLTS. The analysis after 75 cycles show no delamination or other defects (Figure 15).

SAC405 solder ball   SAC-Q solder ball

Figure 15. SAM analysis of 28 nm ELK WLP with underfill #3 after 75x LLTS -65°C/150°C.

V. CONCLUSIONS

The board level reliability of a 36 mm² WLP with 0.3 mm ball pitch was tested with different solder ball alloys. The BLR stress impact on a 28 nm full metal stack ELK device was investigated. In addition, the influence of a reduced PCB thickness and a board level underfill was demonstrated.

The WLP test vehicle using the industry standard SAC405 solder ball alloy shows limitations in TCoB robustness. Solder joint fatigue fails are observed below 1000 cycles in temperature cycling test from -40°C to 85°C if the samples are soldered on a 10-layer 0.8 mm thick PCB. Defects in the 28 nm die metal stack are not detected in any of the performed stress tests using SAC405 solder balls in combination with the 2-layer RDL and UBM WLP architecture.

LF35 solder ball alloy has shown a significant reduction in the TCoB performance of the WLP used in the test condition of 0.8 mm thick 10-layer PCB and without a board level underfill. Such silver content reduced solder ball alloys, which are often used for flip-chip chip-scale packages (FCCSP) in drop sensitive applications, leads to early solder joint fatigue cracks when compared to harder solder ball alloys of WLP because of its small mechanical buffer zone for the CTE mismatch between the silicon die and the PCB.

The SAC-Q solder ball devices show positive results in TCoB, LLTS and mechanical shock test. The shift in TCoB failure mode from the desired solder joint fatigue to cracks in interface between Al pad of the die to RDL is in this case not critical because the first failure occurs more than factor 2 above the release criteria. The general stress impact from this harder solder ball material on the critical interfaces is manageable by suitable chip-package-board co-design solutions in WLP's similar to the test vehicle.

The positive influence of less stiff system constructions, like a reduced 0.4 mm PCB thickness, and the usage of an appropriate underfill on the board level reliability was demonstrated. In the tests, both measures show an improvement potential of factor 2 and more in TCoB reliability for the SAC405 WLP.

ACKNOWLEDGMENT

We would like to acknowledge Chooi Yan Tan and Yen Nee Lim from Intel Malaysia who helped assist in the failure analysis. Additionally, we would like to thank Stephan Stoeckl for the thermo-mechanical simulation studies that helped to guide these empirical studies. Furthermore, we would like to acknowledge Robert Capka and Adem Ali for execution of board level reliability tests as well as Lidia Parisoli for the test management.

REFERENCES

[1] Tak-Sang Yeung, Henry Sze, Keith Tan, Javed Sandhu, Chong-Wei Neo, Edward Law, "Material characterization of a novel lead-free solder material — SACQ", IEEE 64th Electronic Components and Technology Conference (ECTC), 2014, pp.518 - 522

[2] Wei Lin, Quan Pham, Bora Baloglu, Michael Johnson, „SACQ Solder Board Level Reliability Evaluation and Life Prediction Model for Wafer Level Packages", IEEE 67th Electronic Components and Technology Conference (ECTC), 2017, pp. 1058-1064

[3] Rey Alvarado, Beth Keser, Tong Cui, Ahmer Syed, Steven Xu, Brian Roggeman, "Multi DOE Study on 28nm (RF) WLP Package to Investigate BLR Performance of Large WLP Die with 0.35mm Ball Pitch Array", IEEE 67th Electronic Components and Technology Conference (ECTC), 2017, pp. 587-594

[4] Frank Kao, Zhi Hao Tseng, Chun Sheng Ho, Stan Chen, Chang-Yi Lan, Feng Lung Chien, "A study of board level reliability test with bump structure of WLCSP lead-free solder joints", 2007, International Microsystems, Packaging, Assembly and Circuits Technology, pp. 323-326

978-1-7281-1500-9/19 $31.00 © 2019 IEEE

Study of Board Level Reliability of eWLB (embedded wafer level BGA) for 0.35mm Ball Pitch

Kang Hai Lee, Yeow Kheng Lim, Seng Guan Chow, Kang Chen, Won Kyung Choi and Seung Wook Yoon
STATS ChipPAC Pte. Ltd., JCET Group
5 Yishun Street 23, Singapore 768442
seungwook.yoon@statschippac.com

NW Liu, Yenyao Chi and Benson Lin
Advanced Package Technology, Mediatek Inc.
Hsinchu City, Taiwan

Abstract

The number of WLP (Wafer Level Packages) used in semiconductor packaging has experienced significant growth since its introduction due to the small form factor and high performance requirements. With the advancement of fan-out wafer level packaging technology, it is more and more promising compared with fan-in WLP, because it can offer greater feasibility and flexibility for more I/Os, multi-chips, and system integration. But there are some restrictions in possible applications for Fan-In WLP or Fan-out WLP since global chip trends tend toward smaller chip areas with an increasing number of interconnects and better thermal performance. For wider applications of WLP additional development is needed to move past those restrictions.

In this study, board level reliability has been performed on eWLB (embedded wafer level BGA) FOWLP (Fanout Wafer Level Package) with 0.35mm ball pitch, with/without UBM, design factors (pad size, pad opening size etc.) with a comprehensive DOE study. The 0.35mm pitch eWLB test vehicles included a 7mm × 7mm package with over 280 balls. A Design of Experiment (DOE) study was reviewed which demonstrates improved Temperature Cycle on Board (TCoB) with thermo-mechanical simulation and experimental results of a daisy chain device. Several DOE test vehicles were prepared with multiple design variables. Daisy chain eWLB test vehicles were used for the TCoB (Temperature Cycle on Board) reliability study in JEDEC test conditions. Additionally, a JEDEC board level drop test was also carried out. With these parametric studies and reliability tests, the final test vehicle passed 1000 cycles TCoB in characteristic life time and also passed 200x drop test. Destructive analyses were performed to investigate potential structural defects and to conduct a failure mode study after reliability test.

Keywords-component; board level reliability, eWLB, FOWLP, thermo mechanical simulation

I. INTRODUCTION

In just one decade the hand phone has transformed from a simple communication device into more complex system integrating features that allow customers to use it as a multipurpose gadget. The carrier technology has jumped from 1G to 4G, LTE and 5G, changing at the rate of every two years and with room for potential growth for global adoption. Moving forward with this trend, packaging semiconductor devices for handheld electronics has become more challenging than ever before. The growing mismatch in interconnect gap, adding different functional chips for different features and application in similar system footprint, and package size reduction to increase battery size for extended usage has opened the window for innovative embedded package technology.

New and emerging applications in the consumer and mobile space, the growing impact of the Internet of Things (IoT), wearable electronics, automotive, 5G and mmWave applications and the complexities in sustaining Moore's Law have been driving many new trends and innovations in advanced packaging technology. The semiconductor industry now has to focus on density scaling and system level integration to meet the ever-increasing electronic system demands for performance and functionality as well as the reduction of form factor, power consumption and cost.

This paradigm shift from chip scaling to system-level scaling has been reinventing microelectronics packaging, driving increased system bandwidth and performance, and helping to sustain Moore's Law. Demand for maximum functional integration in the smallest and thinnest package will continue to growth with an order-of-magnitude requirement for lower cost and power consumption. The challenge for the semiconductor industry is to develop a disruptive packaging technology capable of achieving these goals.

To meet the above mentioned challenges, eWLB was developed to offer additional space for routing higher I/O chips on top of the Silicon (Si) chip area which was not possible in conventional WLP or WLB [1]. eWLB also offers comparatively better electrical, thermal and reliability performance at reduced cost with the possibility of addressing more Moore (decreasing technology nodes with low-k dielectrics in SoC) and more than Moore (heterogeneous

978-1-7281-1500-9/19 $31.00 © 2019 IEEE

integration of chips with different wafer technology as SiP solution in multi die or 3D eWLB approaches).

eWLB technology uses a combination of front- and back-end manufacturing techniques with parallel processing of all the chips on a wafer, which can greatly reduce manufacturing costs. Its benefits include a smaller package footprint compared to conventional leadframe or laminate packages, medium to high I/O count, maximum connection density, as well as desirable electrical and thermal performance. It also offers a high-performance and power-efficient solution for the wireless and mmWave markets [2].

eWLB (embedded Wafer Level BGA) Technology

eWLB technology addresses a wide range of factors. At one end of the spectrum are the packaging and testing costs. At the other end, physical constraints, such as its footprint and height, exist. Other parameters that were considered during the development phase included I/O density, which was a particular challenge for small chips with a high pin count; the need to accommodate SiP approaches, thermal issues related to power consumption and the device's electrical performance (including electrical parasitic and operating frequency) [3].

The obvious solution to the challenges was some form of WLP and there were two choices to choose from: Fan-out or Fan-in. FO-WLP is an interconnection system processed directly on the wafer and compatible with motherboard technology pitch requirements. It combines conventional front- and back-end manufacturing techniques, with parallel processing of all chips. There are three stages in the process. Additional fab steps create an interconnection system on each die, with a footprint smaller than the die. Solder balls are then applied and parallel testing is performed on the wafer. Finally, wafers are sawn into individual units, and are used directly on the motherboard without the need for interposers or underfill.

(a)

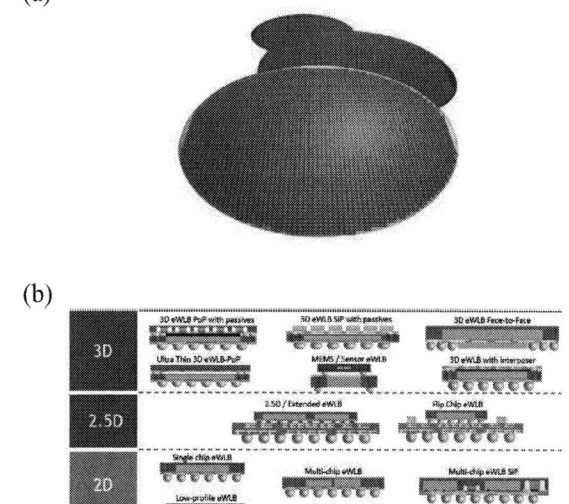

(b)

Figure 1. (a) Evolution of eWLB carrier size from 200 to 300 and 330mm (b) 2/2.5/3D eWLB packages.

The wafer level chip scale package (WLCSP) was introduced in the late 1990's as a semiconductor package wherein all manufacturing operations were done in wafer form with dielectrics, thin film metals and solder bumps directly on the surface of the die with no additional packaging [1]. The WLCSP's basic structure has an active surface with polymer coatings and bumps with bare silicon (Si) exposed on the remaining sides and back of the die. The WLCSP is the smallest possible package size since the final package is no larger than the required circuit area. Although WLCSP is now a widely accepted package option, initial acceptance was limited due to concerns with the Surface Mount Technology (SMT) assembly process and the fragile nature of the exposed silicon inherent in the package design. While assembly skills and methods have improved since the introduction of WLCSP, damage to the exposed silicon still remains a concern today. This is particularly true for advanced node products with fragile dielectric layers.

A very promising solution for mmWave packaging is the eWLB platform [1,4], based on an embedded device technology with fan-out redistribution. The thin-film redistribution layer (RDL) of the eWLB enables very flexible and highly customizable package designs. The length of the redistribution lines is within the range of the die size. eWLB has the ability to attain minimum interconnection length and excellent electrical performance. It is possible to achieve a 20-40% reduction in package size and over 50% volume reduction using eWLB as compared to other packaging solutions such as flip chip or wirebonding packages due to its slim and smaller form factor. For radio frequency (RF) and high-frequency microwave or mmWave devices, a significant improvement in overall device performance was illustrated by eWLB, showing less parasitic electrical interference.

II. EXPERIMENTAL RESULTS

A. Thermomechanical simulation DOE study

In this study, 3D full package finite element analysis (FEA) models were constructed due to the asymmetric nature in package structure, e.g. ball layout and die position. Table 1 shows the test vehicle package specification. Test vehicles were mounted on a 1.00mm thick printed circuit board (PCB) in accordance to JEDEC Specification JESD22-B111. For a light weight package, it was found that collapsed solder bump shape after reflow could be reasonably estimated using truncated-sphere theory as reported in [5]. All models were subject to the temperature cycling condition ranging from -40 to 125°C, with 15 minutes ramp and 10 minutes dwell times. Except for solder balls, all materials were assumed to be linear elastic as shown in Table 1. A constitutive model [6] was used to represent the in-elastic deformation behavior of the lead-free solders. Three nodes at the test board bottom surface were chosen to be fixed with Ux=Uz=0, Ux=Uy=Uz=0, and Uz=0 to prevent the rigid body movement. Solder fatigue life model can be generally expressed through a power law relationship as the following form:

$$N_f = \alpha(\Psi)^{-\beta}$$

N_f is the number of cycles to failure, Ψ is a damage parameter, α and β are empirical fitting parameters determined by least-squares regression analysis. In this study, the first failure lifetime is used as the solder fatigue life.

Equivalent creep strain or creep strain energy density accumulated per stable cycle can be chosen as the damage parameter for evaluating the solder joint fatigue life. Critical solder joint can be identified by the maximum value of the chosen damage parameter.

Table 1. Test Vehicle Package Specification

Item	Description
Package Type	eWLB
Package size	7 x 7 mm
Die Size	6.6 x 6.6 mm
Bump Count	150
Bump diameter/pitch	0.22/0.350 mm

Table 2. Material properties of eWLB package assembly

Constituent	Young's Modulus (MPa)	Poisson's Ratio	Tg (°C)	CTE1 / CTE2 (ppm/°C)
Die	131000	0.28	-	2.7
Molding Compound	23000	0.30	165	7.3 / 32
Dielectric Layer	3500	0.4	205	65
Copper	117000	0.35	-	17.3
PCB	22000	0.28	140	18
Solder	51700	0.35	-	20

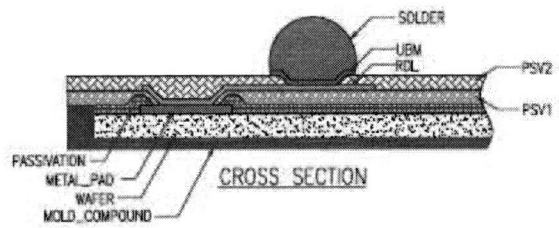

Figure 2. Cross-section Diagram of eWLB with 1L-RDL with UBM

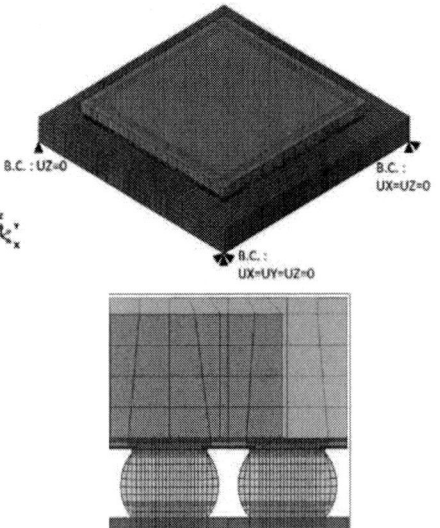

Figure 3. FEA Models of eWLB package and its close-up view of corner balls with refined meshes

A DOE study was carried out to investigate how to improve board level reliability with a parametric study. The temperature cycle profile was within the range of -40/+125°C, 15min ramp time, and 10 min dwell time. For studying of TCoB performance as a function of ball count, overlap between solder mask and ball pad, and RDL pad metal size. Additionally, a simulation study was carried out for 3 different package ball configurations, i.e.,283, 300 and 320 as shown in Table 3.

Table 3. DOE study details in design parameters

Feature	Leg 0	Leg 1	Leg 2	Leg 3	Leg 4	Leg 5	Leg 6	Leg 7	Leg 8
Ball Count	283	300	320	283	300	320	283	300	320
PSV2 Overlap size	25 um	25um	25 um	45 um	45 um	45 um	20 um	20 um	20 um
Pad metal size	240 um	240 um	240 um	240 um	240 um	240 um	190um	190um	190um

After simulation works, higher stresses were observed on the RDL pads' bottom sides whenever a die edge crossed over them as shown Fig. 4.

As shown in Table 3 below, TCoB performance improved with additional ball counts and higher PSV2 overlap. In addition, a smaller pad metal size showed minor improvement in TCoB performance too.

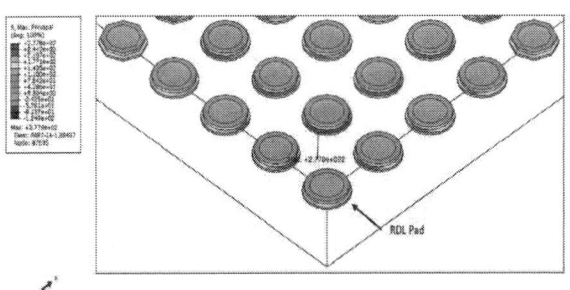

Figure 4. Max. principal stress plot of UBM and RDL Pad at -40°C @end of ramp down stage in 3rd Thermal Cycle

Table 4. Simulation results of TCoB life time of DOE study

	Feature	Leg 0	Leg 1	Leg 2	Leg 3	Leg 4	Leg 5	Leg 6	Leg 7	Leg 8
Critical Solder Joint	Predicted First Failure Life	449 Cycles	551 Cycles	581 Cycles	624 Cycles	816 Cycles	889 Cycles	652 Cycles	847 Cycles	927 Cycles
	Error Percentage	+7.2%	-	-	-	-	-			
	Relative Predicted First Failure Life	1.00	1.23	1.29	1.39	1.82	1.98	1.45	1.89	2.06

B. Daisychain ToCB and drop DOE study

After the simulation works, a final design was optimized with the addition of considerations regarding product design as well as the manufacturing process. A 2nd round DOE study was carried out with daisychain samples. In this DOE, 3 different test vehicles were fabricated as function of UBM and non-UBM, RDL pad size, different PSV opening size.

As shown in Fig 5 and 6, leg#1 with UBM showed the over 1000 cycle characteristic life time. And larger metal pad and PSV2 opening size also showed improvement in TCoB performance.

Table 5. DOE study of daisychain test vehicles

DOE	Leg#1	Leg#2	Leg#3
UBM/non-UBM	UBM	Non-UBM	Non-UBM
Pad metal size (um)	240	250	245
PSV2 opening size (um)	190	220	210

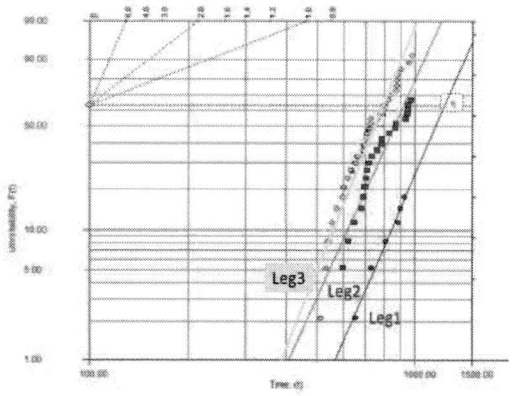

Figure 5. Weibull plot of TCoB reliability test results for Table 5.

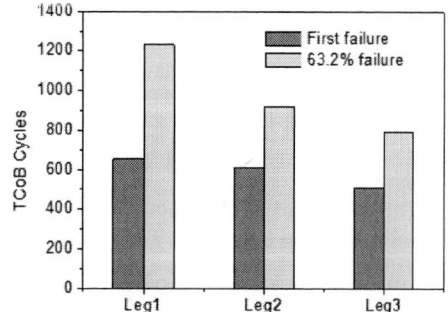

Figure 6. TCoB reliability test results for Table 5.

Fig.7 showed Leg#1 with UBM showing significant improvement of drop performance over non-UBM Leg#2,3. Showing over 400 drops characteristic life time. And larger metal pad and PSV2 opening size also showed the improved drop reliability performance. The trend of TCoB and Drop tests were similar with the DOE study.

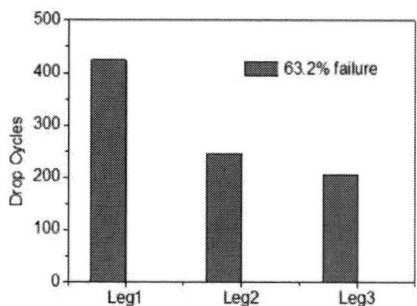

Figure 7. Drop reliability test results for Table 5.

III. CONCLUSION

Board level reliability studies have been performed on eWLB FOWLP with 0.35mm ball pitch, with/without UBM, design factors (pad size, pad opening size etc.) with comprehensive DOE study carried out in 3D finite element analysis and daisychain test vehicle for TCoB and drop reliability tests.

1. A three-dimensional finite element analysis was conducted to simulate the design parameter study in TCoB reliability. TCoB performance was improved with additional ball counts and higher solder mask overlap. Also smaller pad metal size showed improvement.
2. Daisychain test vehicles were fabricated and tested for TCoB and drop reliability as per DOE simulation work. The final test vehicle passed 1000 cycles TCoB in characteristic life time. It also passed the 200x drop test. Test vehicle with UBM showed significant improvement of characteristic life time.

Furthermore, factors such as superior high frequency electrical performance and the ability to enable heterogeneous integration such as the integration of passives like inductors/resistor/capacitor into the various thin-film layers, active/passive devices into the mold compound or encapsulation, and achievement of 3D vertical interconnections for new 3D SiP and 2.5D/3D packaging solutions, differentiate eWLB from other packaging technologies. eWLB technology provides a more holistic performance relevant to an increasingly broad range of emerging applications including automotive, 5G and mmWave applications.

978-1-7281-1500-9/19 $31.00 © 2019 IEEE

ACKNOWLEDGMENT

The authors would like to express their sincere appreciation to the eWLB NPI and R&D team of STATS ChipPAC Pte Ltd for their helpful advice and support of this work.

REFERENCES

[1] M. Brunnbauer, et al., "Embedded Wafer Level Ball Grid Array (eWLB)," Proceedings of 10th Electronic Packaging Technology Conference, 9-12 Dec 2008, Singapore (2008)

[2] Graham pitcher, "Good things in small packages," Newelectronics, 23 June 2009, p18-19 (2009)

[3] Seung Wook YOON, Meenakshi PADMANATHAN, Andreas BAHR, Xavier BARATON and Flynn CARSON, "3D eWLB (embedded wafer level BGA) Technology: Next Generation 3D Packaging solutions," San Francisco, IWLPC 2009 (2009)

[4] M. Brunnbauer, E. Fürgut, G. Beer, T. Meyer, H. Hedler, J. Belonio, E. Nomura, K. Kiuchi, K. Kobayashi, "An Embedded Device Technology Based on a Molded Reconfigured Wafer", 56th Electronic Components and Technology Conference (ECTC 2006), June 2006.

[5] K.N. Chiang and C.A. Yuan, "An Overview of Solder Bump Shape Prediction Algorithms with Validations", IEEE Trans on Adv Packag., Vol. 24, No. 2, pp. 158-162, May 2001 (2001)

[6] Chang, J., Wang, L., Dirk, J., and Xie, X., "Finite Element Modeling Predicts the Effects of Voids on Thermal Shock Reliability and Thermal Resistance of Power Device," Welding Journal, pp. 63-70, Mar. 2006 (2006)

2019 IEEE 69th Electronic Components and Technology Conference (ECTC)

Board Level Reliability Study of Fan-Out Single Die Package with 350um Bump Pitch

Chieh-Lung Lai, Gu-Yan Lin, Tz-Yuan Chao, Yih-Sin Chen, and Feng-Lung Chien
Siliconware Precision Industries Co., Ltd.
No. 19, Keya Rd., Daya, Taichung, Taiwan, R.O.C.
E-mail: chiehlunglai@spil.com.tw

Abstract—**350um bump pitch is trend for Fan-Out (FO) package and wafer-level-chip-scale package (WLCSP) to reduce package size. 220um or 230um ball grid array (BGA) ball is used for package with 350um bump pitch. Board level reliability (BLR) test is standard qualified method for new package design. BLR test items of FO package and WLCSP are comprised of temperature cycling test (TCT) and drop test. However, BGA ball size reduction will increase ball stress and worsen BLR test result, comparing to 250um BGA ball with 400um bump pitch. How to improve BLR performance of package with 350um bump pitch is key design character.**

In this study, Fan-out single die (FOSD) package is selected to do BLR study and is comprised of one redistribution layer (RDL), one under bump metal (UBM) layer, and two passivation layers. Fan-in die size is 6.6mm x 6.6mm embedded into 7mm x 7mm FOSD package. Polyimide (PI) is used for passivation layers. UBM pad size, BGA ball size, and 2nd passivation layer thickness (PI-2) are selected as key factors to improve BLR performance. Ball stress and UBM neck stress simulations are used to select FOSD package structure with best BLR performance. From stress simulation results, UBM size is the most important factor, and PI-2 thickness and BGA ball size are second and third ones. 5% UBM diameter increase can reduce 14% ball stress and 37% UBM neck stress. 30% PI-2 thickness increase can reduce 9% ball stress and 37% UBM neck stress. 5% BGA ball diameter increase can reduce 4% ball stress but increase 5% UBM neck stress. BLR tests are comprised of TCT condition G (-40~125℃) and drop test 1500G. From BLR test results, PI-2 thickness is the most important factor, and UBM size and BGA ball size are second and third ones. 30% PI-2 thickness increase can increase 28% Drop performance and 87% TCT performance. 5% UBM diameter increase can reduce 16% Drop performance but increase 19% TCT performance. 5% BGA ball diameter increase can reduce both 15% Drop performance and 7% TCT performance.

Keywords- 350um Bump Pitch; Fan-Out Package; Board Level Reliability; Temperature CyclingTest; Drop Test

I. INTRODUCTION

A FOSD package is comprised of one die that is embedded into molding compound, one or multiple PSV layers, one or multiple RDL layers which is fan-out metal lines from die area to compound area, UBM layer which is not necessary, and BGA balls, as shown in Figure 1. FOSD

is like 2.5D IC package, as in [1] and [2], has warpage issue during packaging process, but FOSD package does not use substrate during packaging process. FOSD package can directly mount on board by Surface-mount Technology (SMT). With the demand of more I/O and thinner package for mobile device, bump pitch and bump size are needed to be decreased.

Due to resource limitation of Outsourced Semiconductor Assembly and Test (OSAT) provider, PI or PBO PSV layer and Cu daisy chain (DC) are used to replace CVD PSV layer and Al pad, as shown in Figure 2. In this study, PI PSV is coated on bare Si wafer as insulation layer, and then Cu DC is plated on PI PSV layer.

In this study, 350um bump pitch design is to replace 400um bump pitch design and 220um diameter BGA ball is to replace 250um BGA ball, as shown in Figure 3. UBM diameter is changed from 250um to 220um, bump diameter is changed from 270um to 240um, and bump height is changed from 190um to 170um. Therefore, I/O density of 350um bump pitch package can increase to 1.306 times, and package thickness can decrease 20um, comparing to 400um bump pitch design. However, the stress on UBM, bump, and RDL will be increased with the decrease of UBM diameter and BGA ball diameter. How to keep the same stress for better BLR performance is a key for FOSD package design with 350um bump pitch. The FOSD package is comprised of one RDL layer, one UBM layer, and two PI layers. Fan-in die size is 6.6mm x 6.6mm and is embedded into 7mm x 7mm FOSD package.

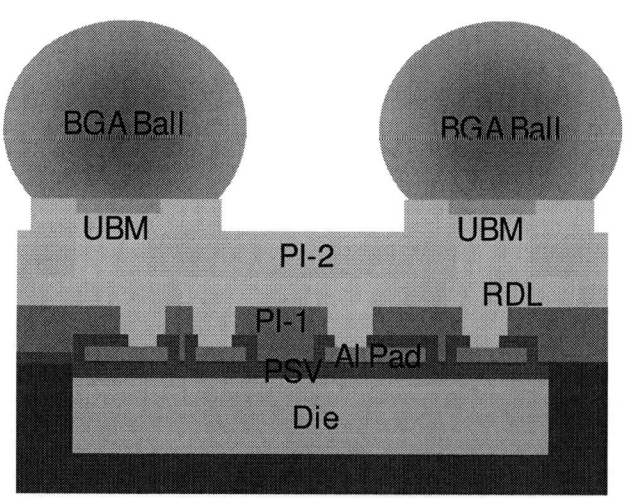

Figure 1. General FOSD Package

978-1-7281-1500-9/19 $31.00 © 2019 IEEE

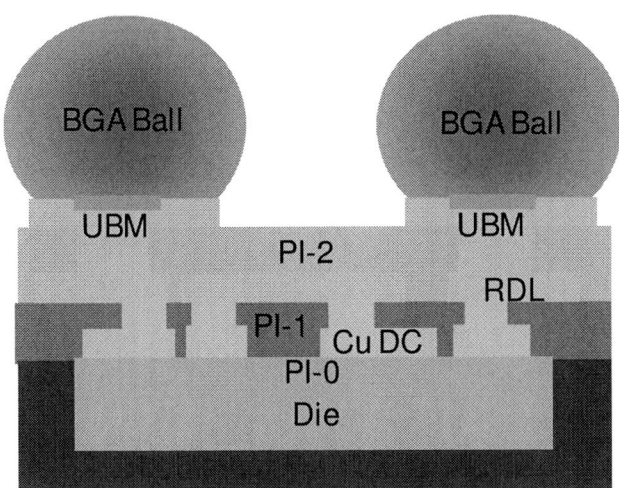

Figure 2. General FOSD DC Package

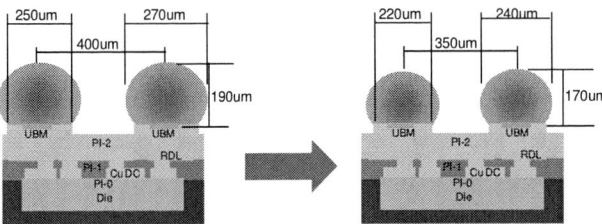

Figure 3. Bump Structure Comparison between Bump Pitch 400um and 350um

Table I Bump Dimension Comparison

	Original	New	Ratio (350/400um)
Bump Pitch (um)	400	350	0.88X
Bump Diameter (um)	270	240	0.89X
Bump Height (um)	190	170	0.89X
UBM Diameter (um)	250	220	0.88X
I/O Density	1 X	1.306 X	-
UBM Area	1 X	0.7744 X	-

II. PROCESS FLOW

FOSD is one of Fan-out Wafer Level Package (FOWLP) and is comprised of Reconstituted (Recon) process and WLCSP extending process which includes RDL process and Assembly-A (AA) process, as shown in Figure 4. Recon process, as shown in Figure 5, starts from 8 inch or 12 inch wafer mounting and wafer die sawing. Then 12 inch carrier is pasted with tape, and bond machine picks up and bond dies on tape. After die bonding, carrier with dies will mold with compound. After molding, carrier is removed and 12 inch compound wafer is formed. RDL process, as shown in Figure 6, starts from 1st PI layer, and following seed layer process, PR coating, PR exposure, PR development, copper plating, PR strip, and seed layer etching process are used to from RDL layer. Then 2nd PI layer is coated on RDL layer, and repeating RDL layer process is used to from UBM layer.

Finally, BGA balls are placed on UBM by flux and stencil, and reflow process is used to melt BGA balls onto UBM layer. There are two kinds of AA process, and they are AA process with Backside Lamination (BSL) and AA process without BSL. AA process with BSL and without BSL starts from lapping process or BSL process to make package thickness to meet design criteria, and then do laser marking to create package ID on compound or BSL surface. Final process is singulation to make compound wafer into single package form.

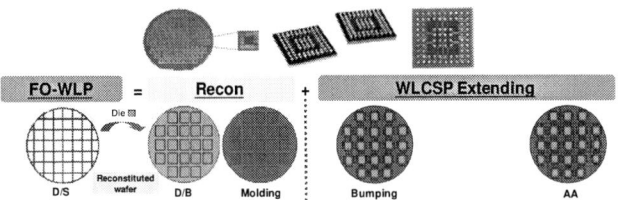

Figure 4. Process Flow of FOSD

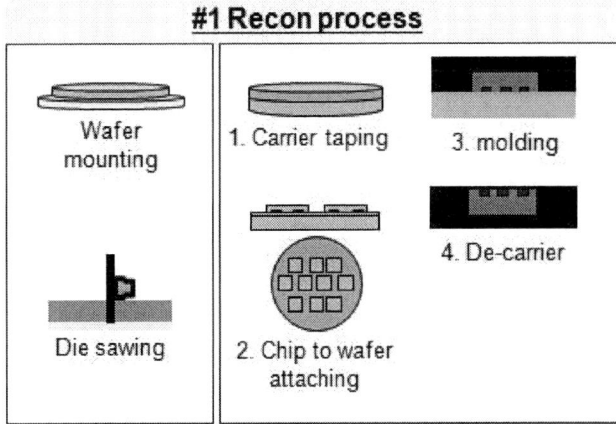

Figure 5. Process Flow of Recon Process

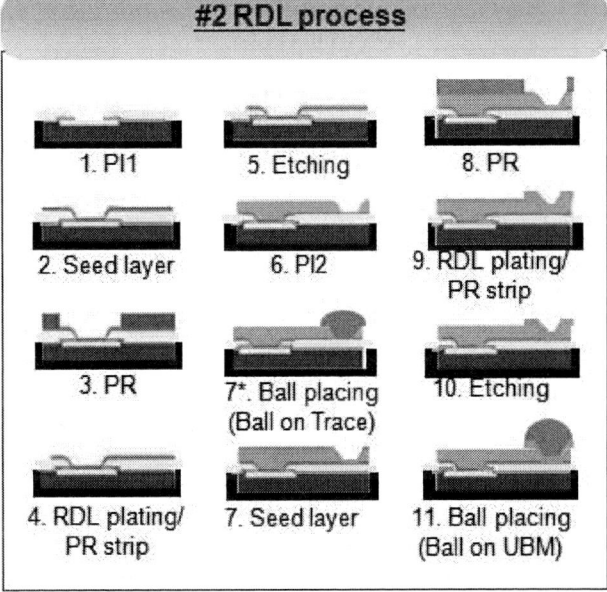

Figure 6. Process Flow of RDL Process

978-1-7281-1500-9/19 $31.00 © 2019 IEEE

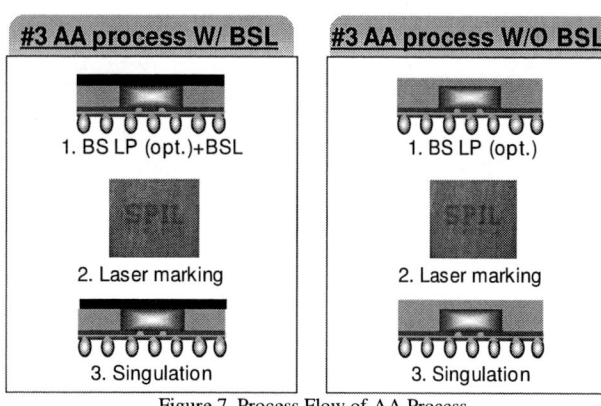

Figure 7. Process Flow of AA Process

III. STRESS SIMULATION

Generally, the BLR performance is highly related with BGA ball stress and UBM neck stress. With the demand of low bump pitch for thin package height and high I/Os, bump diameter decreasing will directly increase BGA ball stress and UBM neck stress at the same package design. High BGA ball stress will induce bump crack at package side or PCB side,, and high UBM neck stress will induce RDL crack during BLR, as shown in Figure 8 and Figure 9. Therefore, package design for lower BGA ball stress and UBM neck stress is important to keep or enhance BLR performance.

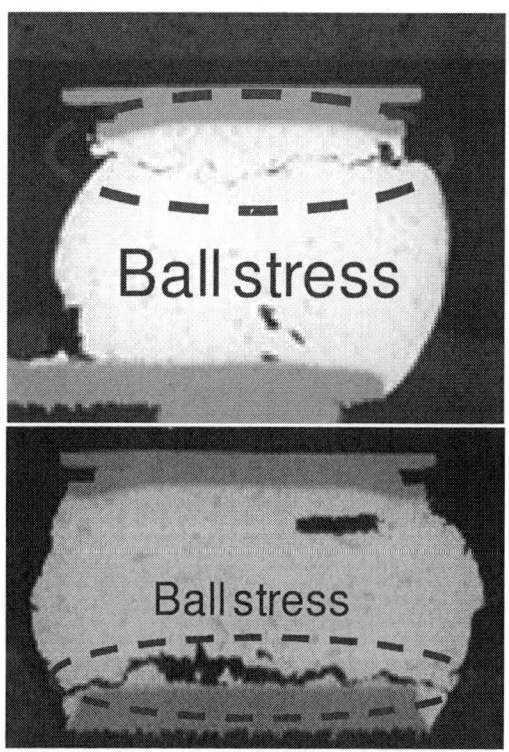

Figure 8. Failure Mode of High Ball Stress

Figure 9. Failure Mode of UBM Neck Stress

In order to optimize BLR performance of 350um bump pitch package, the models are built in ANSYS software by quarter setup for simulating its BGA ball stress and UBM neck stress at TCT tests. In this study, PI-2 thickness, UBM size, and BGA ball size are used to tune BGA ball stress and UBM neck stress. Passivation layer thickness is our first priority, because there is least change for package structure. Increase UBM size and BGA ball size will increase bump diameter and decrease space between bumps. Then bump bridge risk will increase at the same time. Therefore, thick passivation layer design is implemented at each simulation legs.

Stress simulation results are shown in Table II. The leg POR (Process of Record) is FOSD package with 400um bump pitch and 250um BGA ball, and is used as baseline which is passed BLR test. The leg 1~5 are FOSD packages with 350um bump pitch, and package size is changed from 6.8x6.8mm to 7x7mm comparing to the leg POR. The leg 1 is original package design with 350um bump pitch, and there is highest BGA ball stress and UBM neck stress. The leg 2 is to increase PI-2 thickness to 1.3X comparing to the leg 1. The leg 3 is to increase UBM pad size from 220um to 230um comparing to the leg 2. The leg 4 is to increase BGA ball size from 220um to 230um comparing to the leg 2. The leg 5 is to combine all change items of the leg 2~5, and is the most expensive leg. The bump diameter will be increased from 240um to 245um, but bump height will be decreased from 170um to 165um when UBM pad size increases from 220um to 230um. The bump diameter will be increased from 240um to 245um, and bump height will be increased from 170um to 180um when BGA ball size increases from 220um to 230um. The bump diameter will be increased from 240um to 250um, and bump height will be increased from 170um to 175um when both UBM pad size and BGA ball size increase. From simulation results, the lowest BGA ball stress results are at leg 5, the second lowest is at leg 3, the third lowest is at leg 4, and the fourth lowest is at leg 2. From BGA ball stress ratio results, UBM pad size increasing could get best stress reduction, PI-2 thickness increasing gets second one, and BGA ball size increasing gets third one. From UBM neck stress ratio results, PI-2 thickness increasing and UBM pad size increasing could both get best stress reduction, but BGA ball size increasing could worsen UBM neck stress.

Table II Stress Simulation Results

LEG	Bump Pitch (mm)	Package Size (um)	PI-2 Thickness (um)	UBM Size (mm)	Ball Size (mm)	Bump Height (um)	Bump Diameter (um)	PCB Pad Size (um)	Ball Count	Ball Stress Ratio	UBM Neck Stress Ratio
POR	0.4	6.8x8.8	1X	0.25	0.25	190	270	220	1X	1.00	1.00
1	0.35	7x7	1X	0.22	0.22	170	240	220	1.35X	1.15	2.12
2	0.35	7x7	1.3X	0.22	0.22	170	240	220	1.35X	1.06	1.75
3	0.35	7x7	1.3X	0.23	0.22	165	245	220	1.35X	0.92	1.38
4	0.35	7x7	1.3X	0.22	0.23	180	245	220	1.35X	1.02	1.8
5	0.35	7x7	1.3X	0.23	0.23	175	250	220	1.35X	0.90	1.42

IV. RESULT

Real BLR results are shown in Table III, and there are Drop test and Temperature cycle test in BLR tests. The Weibull plots of Drop test and TCT test are shown in Figure 10 and Figure 11. The failure mode of drop test of the leg 1~5 with 350um bump pitch is "solder crack at PCB site", and this is different from POR leg with 400um bump pitch which is "Die RDL crack". This difference shows 250um BGA ball strength is better than 220um and 230um BGA ball and RDL. The leg 2 of thick PI-2 shows best drop test performance. The leg 4 of thick PI-2 and big BGA ball and the leg 3 of thick PI-2 and large UBM pad show almost the same drop test performance and show second better performance between all legs with 350um bump pitch. The leg 5 of thick PI-2, large UBM pad, and big BGA ball shows third better drop test performance. The leg 1 of original design shows worse drop test performance. From above results, PI-2 layer plays the most important role at drop test and is like a shock absorber to protect RDL, UBM, and bump at package structure. On the contrary, large UBM pad and big BGA ball both worse drop test performance, although they can enhance package strength and reduce ball stress and UBM neck stress.

From Temperature cycle test results, the leg 3 of thick PI-2 and large UBM pad and the leg 5 of thick PI-2, large UBM pad, and big BGA ball show almost the same TCT performance and are best ones between legs with 350um bump pitch. The failure mode of the leg 3 and the leg 5 is solder crack at PCB site. The leg 2 of thick PI-2 shows second better TCT performance, and the leg 4 of thick PI-2 and big BGA ball shows third better one. The failure mode of the leg 2 and the leg 4 is solder crack at package site. The TCT performance trend of all legs is matched with UBM neck stress simulation results. Bigger BGA ball structure with original UBM pad worsens a little UBM neck stress and TCT performance between the leg 2~5. However, when large UBM pad design adds into package structure, BGA ball stress, UBM neck stress, and TCT performance are all improved, and the solder crack of TCT failure mode changes from package site to PCB site. The leg 1 shows worst TCT performance, and the TCT failure mode is Die RDL crack. This shows that PI-2 protection of original design is not enough, and there is highest UBM neck stress. From above BLR results, PI-2 layer is the most important factor at both Drop test and TCT test, UBM pad size is second one, and BGA ball size is third one.

Table III Real BLR Results

Leg	Purpose	Package Size (mm)	PI-2 Thickness	UBM Size (um)	Ball Size (um)	BLR Result								Simulation Result	
						Drop(1500G, 0.5 ms)				TCT (-40C ~125C)					
						1st fail	10% fail	63.2% fail	Failure Mode	1st fail	10% fail	63.2% fail	Failure Mode	Ball Stress Ratio	UBM Neck Stress Ratio
POR	Reference	6.8x6.8	1X	250	250	0.39X	0.71X	1.48X	Die RDL crack	>1.85Y	NA	NA	NA	1X	1X
1	Original Design	7x7	1X	220	220	0.39X	0.45X	1X	Solder crack at PCB side	0.42Y	0.53Y	1Y	Die RDL crack	1.15	2.12
2	PI-2↑	7x7	1.3X	220	220	0.36X	0.60X	1.28X	Solder crack at PCB side	1.10Y	1.29Y	1.87Y	Solder crack at Package side	1.06	1.75
3	PI-2↑ UBM ↑	7x7	1.3X	230	220	0.47X	0.54X	1.12X	Solder crack at PCB side	1.14Y	1.42Y	2.06Y	Solder crack at PCB side	0.92	1.38
4	PI-2↑ Ball size↑	7x7	1.3X	220	230	0.51X	0.54X	1.13X	Solder crack at PCB side	0.94Y	1.12Y	1.80Y	Solder crack at Package side	1.02	1.8
5	PI-2↑ UBM↑ Ball size↑	7x7	1.3X	230	230	0.29X	0.53X	1.11X	Solder crack at PCB side	1.40Y	1.56Y	1.90Y	Solder crack at PCB side	0.90	1.42

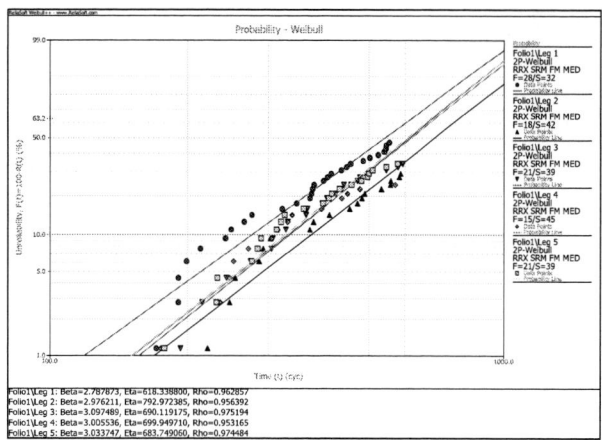

Figure 10. Drop Weibull Plot

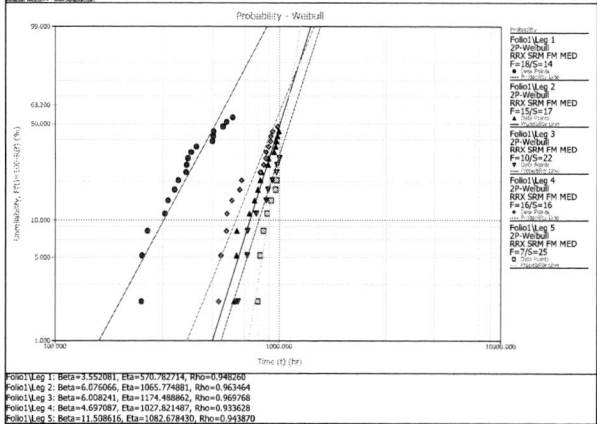

Figure 11. TCT Weibull Plot

V. CONCLUSION

In this study, FOSD package design factors from stress simulation results are successfully validated in BLR tests of FOSD package with 350um bump pitch, as shown in Table IV. From stress simulation results, UBM size is the most important factor, and PI-2 thickness and BGA ball size are second and third ones. 5% UBM diameter increase can reduce 14% ball stress and 37% UBM neck stress. 30% PI-2 thickness increase can reduce 9% ball stress and 37% UBM neck stress. 5% BGA ball diameter increase can reduce 4% ball stress but increase 5% UBM neck stress.

From BLR test results, PI-2 thickness is the most important factor, and UBM size and BGA ball size are second and third ones. 30% PI-2 thickness increase can increase 28% Drop performance and 87% TCT performance. 5% UBM diameter increase can reduce 16% Drop performance but increase 19% TCT performance. 5% BGA ball diameter increase can reduce both 15% Drop performance and 7% TCT performance. These BLR results are a little different from stress simulation results, but are matched with our preference. Because there is bump bridge

risk for both large UBM pad size and big BGA ball design, PI-2 thickness increase is the best design method for FOSD package with 350um bump pitch. For the gap between stress simulation and BLR tests, simulation model modification and design single factor experiment of larger UBM pad or Big BGA ball for FOSD package with 350um bump pitch to do BLR tests will be our future work to clarify this difference.

Table IV Package Design Factor Comparison

Factor	Drop 63.2% fail	TCT 63.2% fail	Stress Simulation	
			Ball Stress Ratio	UBM Neck Stress Ratio
Leg 1 Original Design	1X	1Y	1X	1X
PI-2 Thickness-1.3X	1.28X	1.87Y	0.91X	0.63X
UBM Pad Size-1.05X	0.84X	1.19Y	0.86X	0.63X
BGA Ball Size-1.05X	0.85X	0.93Y	0.96X	1.05X

REFERENCES

[1]. Chieh-Lung Lai, Hung-Yuan Li, Allen Chen, Terren Lu, "Silicon Interposer Warpage Study for 2.5D IC without TSV Utilizing Glass Carrier CTE and Passivation Thickness Tuning," 66th IEEE Electronic Components and Technol. Conf.(ECTC), May 31- June 3, 2016, pp.310-315, doi: 10.1109/ECTC.2016.164

[2]. Chieh-Lung Lai, Hung-Yuan Li, Sam Peng, Terren Lu, Stephen Chen "Warpage Study of Large 2.5D IC Chip Module", 67th IEEE Electronic Components and Technol. Conf.(ECTC), May 30-June 2, 2017, pp.1263-1268, doi: 10.1109/ECTC.2017.210.

The Analysis for Bump Resistance Improvement by Optimizing the Sputter Condition

M. S. Su*, C. N. Wang, Clair Tsai, T. L. Yang, Rolance Yang, W. C. Wu, C. S. Liu, J. M. Chiu, Y. F. Chen,
Ponder Pang, Harry Ku, Kirin Wang, C.H. Su, Steven Hsu, Calvin Lu, K. C. Liu, Marvin Liao.

Taiwan Semiconductor Manufacturing Company, Southern Taiwan Science Park, Tainan, Taiwan
Mail* : mssuq@tsmc.com

Abstract – **Along with the advanced technology development, the tinier polyimide opening (PIO) and bump size are required. However, it will make the bump resistance (Rc) become higher which is closely related to the chip probing (CP) performance. In addition, the additional pre-oven method may have limited effect to improve it especially for advanced tech node. Therefore, this paper proposes some alternative approaches for Rc improvement through optimizing pre-etching chamber condition during the sputter process. First of all, the backside cooling method to cool down the temperature of wafer to reduce the outgassing reaction on surface is introduced. Furthermore, a novel concept presented is to utilize the hybrid gas in pre-etching chamber for bump Rc improvement, whose purpose is to induce reduction-oxidation reaction (redox reaction) by other gas to remove the oxidation on wafer surface. The experiment is performed by a test vehicle with small PIO, and the Rc results show that presented methods can effectively improve bump resistance. In addition, two kinds of experimental polyimide material with different glass transition temperature coefficient (Tg) are used to verify the feasibility of proposed methods.**

Keywords – sputter, bump resistance, plasma, polyimide opening

NOMENCLATURE

K_e	Kinetic energy.
F	Electrical force.
q	Unit electric charge.
ε	Electrical field.
λ	Mean free path (MFP).
n	Particle density.
σ	Particle collision cross section area.

I. INTRODUCTION

Flip chip technique has become one of the most popular and widely used in IC assembling because of advantages such as smaller bump size and pitch. That is, a large number of tiny bumps used to be as I/O interconnections can be placed on a small chip [1]-[4]. However, as the CMOS technology goes into more advanced technology node in semiconductor industry, one of the concomitant challenges directly impacted to far-backend packaging engineering is the polyimide (PI) layer opening (PIO) size shrinkage, which affects the bump resistance (Rc) [5]-[8] during chip probing (CP) test. Besides PIO size, it is reported that PI material characteristics and the process condition during sputtering also play important roles on the bump electrical performance. In general, organic

compound redecoration on the PIO surface by polyimide outgassing during sputter pre-etching process [10] is a well-accepted suspected root cause for higher bump resistance [11]. Therefore, an extra oven baking process before sputter is adopted to reduce the outgassing of PI and further improve Rc. However, for advanced bumping technology, the ordinary oven baking process may have limited effect for Rc reduction.

To overcome the problem, the additional approaches to decrease Rc for advanced tech nodes, which are to directly optimize chamber condition in sputter pre-etching process are studied in this paper. The two techniques: backside cooling and hybrid gas plasma are explained how to attain better Rc performance. A flip-chip bump test vehicle with small PIO is adopted to evaluate the effeteness of these approaches. In addition, the sensitivity analysis between the Rc and hybrid gas concentration and the effect of proposed methods on different PI material are analyzed in this paper as well.

This reminder of this paper is organized as follows. In Section II, three proposed methods for Rc improvement are introduced step-by-step. Section III describes the experiment procedure, including the test vehicle properties, split condition, and the measurement technique. To verify the feasibility and effectiveness of the proposed approaches, the experimental results and discussion are analyzed in Section IV. Finally, the conclusion and future prospects are presented in Section V.

II. PROPOSED METHOD

In this section, the proposed approaches for Rc reduction by optimizing pre-etching chamber condition in sputter process are described step-by-step as following.

A. Wafer Backside Cooling

The plasma is generated by the collision. The electrons are accelerated in the electric filed which obtain extremely high kinetic energy, and the energy is transmitted when they collide with gas molecules or atoms. The reaction equation can be formulated as Eq. (1).

$$e^- + Ar \longrightarrow Ar^+ + 2e^- \tag{1}$$

The heat generated by plasma bombarded makes the wafer surface become more easily reacting with organic outgassing; thus, cooling down the temperature of wafer is a direct and helpful solution. Inserting cooling gas from wafer backside to take away the heat during the pre-etching process, which is shown as Fig. 1 [12].

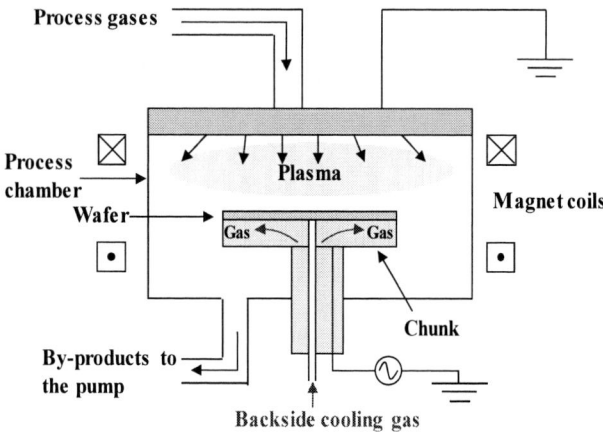

Fig. 1 Schematic diagram of a Pre-etching Chamber [12].

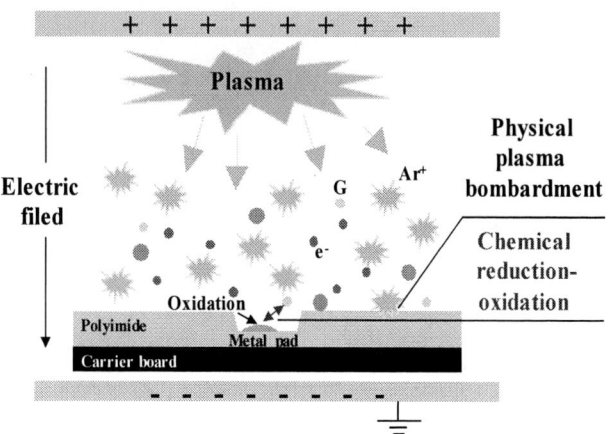

Fig. 2 Schematic diagram for Hybrid Gas Plasma.

B. Hybrid Gas Plasma

Inert gas, such as argon, is known to be a main source gas for plasma during pre-etching process. However, the effect of hybrid gas for bump resistance reduction is still unknown and lack of study. For example, it's discovered that the mixed gas can lead to other chemical reaction such as reduction-oxidation reaction (redox reaction) on the metal pad. Redox reaction is a chemical reaction in which the oxidation states of atoms are changed. It involves both a reduction process and a complementary oxidation process with electrons transferring [13]. In this way, the oxidation on the metal pad can be removed which can successfully and effectively improve the Rc. The schematic diagram for hybrid gas during plasma process is shown in Fig. 2, where the G stands for another gas studied in this paper. Furthermore, the different gas amount condition analysis for single and hybrid gas will be systematically investigated in this paper.

III. EXPERIMENTAL PROCEDURE

A. Test Experiment Condition

To match the requirement for advanced technology bump cell design, a flip-chip bump test vehicle with small PIO is adopted in this paper. Additionally, two kinds of experimental PI material defined as A-type and B-type with different glass transition temperature coefficient (Tg) are adopted. Tg stands for the temperature range where the polymer transfers from a hard and brittle "glassy" state into a soft state with increasing temperature [14]. In this way, the feasibility of proposed techniques on various PI type is verified to fulfill the request for developing advanced technology in the future. The experimental environment is summarized as in Table I. Furthermore, the combination with different split condition to verify and compare the effeteness for Rc improvement on each approach is listed in Table II, whose result is analyzed in the following section.

TABLE I
EXPERIMENTAL TEST VEHICLE CONDITION

Property	Value	Unit
Bump type	Flip chip	--
PIO	Below 30x30	um
PI layer material	A-type/B-type	--

TABLE II
EXPERIMENT SPLIT CONDITION

Condition	Extra oven	Backside cooling	Plasma gas type	PI type
Leg1	-	--	Ar only	A-type
Leg2	V			
Leg3	--	V		
Leg4			Hybrid	
Leg5		--	Ar only	B-type
Leg6		V	Hybrid	

B. Bump Rc Measurement by Kelvin Structure Bump

The bump resistance Rc is the key index to evaluate the effeteness for the improved methods, which is very relevant to the CP performance [6]. Generally speaking, Rc is estimated to be of the order of few mohm. The Kelvin structure is a well-known approach to monitor the resistance of single bump, which is drawn in Fig. 3 [11]. The three bumps are connected by the under metal layer, and a current I_{12} is drawn through from pins 1 to 2, and the voltage difference (V_{34}) between pins 3 and 4 is detected. Thus, the bump resistance can be determined by V_{34}/I_{12}.

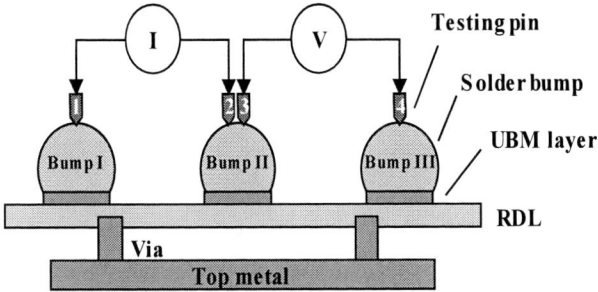

Fig. 3 Kelvin (4-wire) Resistance Measurement Structure [11].

C. Basic Formula of Plasma Energy

The Eq. (2) shows the kinetic energy formula for an electronic existing in a parallel plate electrical field, which is similar to the plasma environment in pre-etching chamber. It shows that the K_e is dependent on λ, which is called mean free path (MFP) meaning the average distance that a particle can move before colliding to another one.

$$K_e = F \cdot d = q \cdot \varepsilon \cdot \lambda \tag{2}$$

The MFP can be determined by (3), which is relevant to the particle density and collision cross section area. In this paper, the mixed gas is injected into the chamber so that the particle concentration will change and influence λ and K_e, which will be described in next section.

$$\lambda = \frac{1}{\sqrt{2}n\sigma} \tag{3}$$

IV. RESULTS AND DISCUSSION

In this section, Rc result is compared among different conditions, including the original one, improved by ordinary extra oven and proposed techniques. Moreover, the advanced sensitivity analysis about hybrid gas concentration is discussed as well.

A. Bump Resistance Rc Comparison

Firstly, Rc result of different methods is shown as Fig. 4. Leg1 means the original Rc performance without any improved approach, and thus it comes the worst. The traditional extra baking method can decrease the Rc effectively, because it can remove some outgassing during oven process. As for proposed techniques through optimizing the chamber condition, the backside cooling method denoted as Leg3 can dramatically reduce Rc, showing that that the wafer surface temperature plays an important role to better Rc due to less oxidation reaction. Lastly, for more aggressive improvement, the hybrid gas injected to the chamber approach is studied, marked as Leg4. Compared with Leg3, the mixed gas can make Rc reach better performance because of more oxidation is removed by the redox reaction induced by G gas. To sum up,

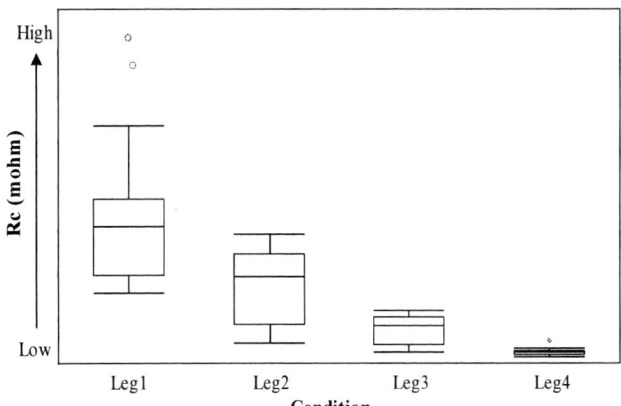

Fig. 4 Rc Results Comparison.

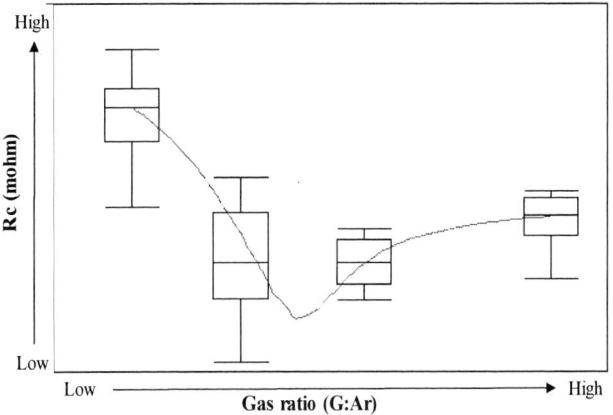

Fig. 5 Relationship between Rc and Gas ratio.

the experiment result shows that proposed methods provide alternative ways to improve bump resistance in advanced tech nodes if the ordinary method may have limited effects on it.

B. Hybrid Gas Concentration and Rc Analysis

Although the mixed gas can induce the redox reaction to remove the oxidation on the metal pad surface, there is still a limitation of its concentration in the chamber to guarantee a better Rc result. The sensitivity analysis between the hybrid gas concentration and Rc is drawn in Fig. 5, which performs a quadratic curve. In the beginning, the Rc decreases with gas ration (G:Ar) since more reduction-oxidation reaction happening during pre-etching process. Nevertheless, once the ratio gets too much, the effeteness becomes worse and Rc climbs up again. It's because that too much G gas occupying the chamber space which restricts the MFP resulting in that plasma cannot obtain enough kinetic energy to remove the outgassing on the wafer surface. As a consequence, an appropriate gas ratio of G:Ar is essential to achieve the best Rc performance.

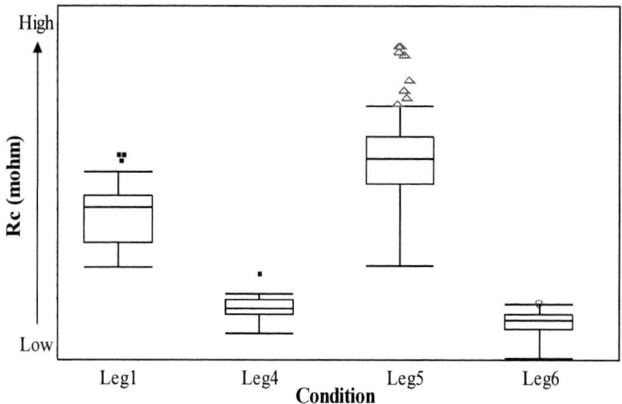

Fig. 6 Rc Result on Different PI Material.

C. Different Polyimide Material Effect Analysis

The experimental result discussed previously is tested on the A-type PI material, showing that proposed methods can effectively reduce Rc. In addition, to verify feasibility of them on different PI material, another B-type is executed and marked as Leg5 and Leg6 respectively, shown as in Fig. 6. Based on it, the original Rc without any improved method much higher on B-type PI material. However, through optimizing the pre-etching process condition, it can be significantly declined which ensures the CP performance. The experiment result shows no matter the PI type it is, proposed approaches can still achieve better Rc performance.

V. SUMMARY

To overcome the challenge of bump resistance rising because of the shrinking PIO in advanced technology, the Rc reduction approaches by optimizing the pre-etching chamber conditions have been propose in this paper. Starting from different aspects, the backside cooling and hybrid gas plasma techniques are introduced and verified the effect and feasibility. The experiments are executed on a test vehicle matching the bump cell design for the advanced tech node requirement. According to the experimental results, each approach can be helpful to decrease Rc and keep it below the specification to guarantee the follow CP test result. Moreover, the sensitivity analysis for mixed gas ration versus Rc is presented as well. Last of all, considering the potential request for different PI material, another PI material is utilized to verify the presented technique, showing that it can reach same effect on it. To sum, this paper presents some alternative approaches to improve Rc by directly adjusting chamber condition instead of ordinary extra oven process, which can meet the request for advanced technology development no matter the PIO or PI material properties in the future.

ACKNOWLEDGEMENT

This paper appreciates for TSMC APTS department's team cooperation and providing great support and research resource so as to achieve this paper presentation.

REFERENCES

[1] S.W. Liang, Y.W. Chang, C. Chen, Y.C. Liu, K.H. Chen, and S.H. Lin, "Geometrical Effect of Bump Resistance for Flip-Chip Solder Joints, Finite-Element Modeling and Experimental Results," *Journal of Electronic Materials*, vol. 35, no.8, pp. 1647-1654, Aug. 2006.

[2] M. Fendler, F. Marion, D. S. Patrice, V. Mandrillon, F. Berger, and H. Ribot, "Technological and electrical performances of ultrafine-pitch flip-chip assembly based on room-temperature vertical interconnection," *IEEE Trans. Compon. Packag. Manuf. Technol.*, vol. 1, no. 3, pp. 291-298, Mar. 2011.

[3] M. C. Hsieh, C. C. Lee, and L. C. Hung, "Comprehensive thermomechanical stress analyses and validation for various Cu column bumps in fcFBGA," *IEEE Trans. Compon., Packag. Manuf. Technol.*, vol. 3, no. 1, pp. 61-70, Jan. 2013.

[4] M.K. Md Arshad, U. Hashim, M. Isa, "Under Bump Metallurgy (UBM)-a Technology Review for Flip Chip Pachaging," *International Journal of Mechanical and Materials Engineering (IJMME)*, vol. 2, no.1, pp. 48-54, Jan. 2007.

[5] P. Lianto, K. J. Chui, B. Bhushan, H. M. C. Chua, L. Tang, B. S. S. C. Rao, X. Wang, A. L. Wu, Y. Gu, G. H. See, and A. Sundarrajan, "Under-Bump Metallization Contact Resistance (Rc) Characterization at 10-μm Polymer Passivation Opening," *IEEE Trans. on Compon., Packag., Manuf. Technol.*, vol. 7, no. 10, pp. 1592-1597, Aug. 2017.

[6] B. Tunaboylu, "Testing of Copper Pillar Bumps for Wafer Sort," *IEEE Trans. on Compon., Packag., Manuf. Technol.*, vol. 2, no. 6, pp. 985-993, Jun. 2012.

[7] S.W. Liang, T.L. Shao, C. Chen, C.C. Yeh, and K.N. Tu, "Relieving the current crowding effect in flip chip solder joints during current stressing," *J. Mater. Res.*, vol. 21, no. 1, Jan. 2006.

[8] B. J. Lwo, C.L. Teng, K.F. Tseng, T. Ni, S. Lu, "Contact Resistance of Microbumps in a Typical Through-Silicon-Via Structure," *Components IEEE Trans. Compon. Packag. Manuf. Technol.*, vol. 7, no. 1, pp. 27-32, Jan. 2017.

[9] J. Park et al., "Contact resistance of solder bump with low cost photosensitive polyimide for high performance SoC," in *Proc. IEEE Int. Rel. Phys. Symp. (IRPS)*, Apr. 2015, pp. CP.3.1–CP.3.2.

[10] J. Jun, I. Kim, M. Mayer, Y. N. Zhou, S. Jung and J. Jung, "A New Non-PRM Bumping Process by Electroplating on Si Die for Three Dimensional Packaging," *Materials Transactions*, vol. 51, no. 10, pp. 1887-1892, Aug. 2010.

[11] H.K. Cheng, S. P. Feng, Y. J. Lai, K. C. Liu, Y. L. Wang, T. F. Liu, and C. M. Chen, "Effect of polyimide baking on bump resistance inflip-chip solder joints," *Microelectron. Rel.*, vol. 54, no. 3, pp. 629-632, Mar. 2014.

[12] H. Xiao, Introduction to Semiconductor Manufacturing Technology, *US: Society of Photo Optical*, 2012.

[13] John Wiley & Sons Publishers, "Redox Reactions." Abstract retrieved Mar. 2019, from https://www.wiley.com/college/boyer/0470003790/reviews/redox/redox.htm.

[14] Polymer Science Learning Center, "The Glass Transition." Abstract retrieved Mar. 2019, from https://pslc.ws/macrog/tg.htm

[15] I. Mann, N. Meyer-Vernet, A. Czechowski, "Dust in the planetary system: Dust interactions in space plasmas of the solar system," *Physics Reports*, vol. 536, no. 1, pp. 1-40, Mar. 2014.

Hybrid prepreg conventional build up laminate for 112Gbit/s SerDes

Kwang Won Choi[a], Edmund Blackshear[a], Eric Tremble[a], David Stone[a], Jean Audet[b], Keiichi Hirabayashi[c]*

a. GLOBALFOUNDRIES US Inc., Hopewell Junction, NY 12533, USA

b. IBM Corporation, Bromont, Quebec, Canada

c. Shinko Electric Industries Co., LTD., Nagano-shi, Japan

* E-mail: kwangwon.choi@globalfoundries.com

Abstract—High speed SerDes standards demand higher speed signals for the next generation - 112Gbit/sec. The integrity of signals is affected strongly by the physical characteristics of the printed circuit substrate in a 112Gbit/s high speed device. The effect on circuit performance is primarily shown as insertion loss.

Insertion loss mainly consists of conductor loss and dielectric loss. In a build-up laminate substrate using stripline structures for high speed signals, parameters which can be varied to affect the insertion loss are dielectric properties, geometry, and Cu surface finish. Both conductor and dielectric loss benefit from coarse geometry. Coarse geometry for high speed conflicts with microelectronics trends for small size and high density. High speed laminate circuit design will require high wiring layer counts to enable necessary signals.

In this development, we designed, fabricated, and evaluated a flip chip laminate assembly using both low loss 33um particle filled and 65um low loss glass cloth reinforced dielectric. The latest generation low loss materials of both types were used. Cu surface roughness in the particle filled layers was 250nm Ra while the glass cloth reinforced (prepreg) layer Cu roughness was 200nm Ra. The configuration was a 65mm square PFCBGA(Plastic Flip Chip Ball Grid Array) laminate cross section consisting of four particle filled dielectric layers, four glass reinforced dielectric layers, two conventional core layers, four glass reinforces dielectric layers, four particle filled dielectric layers. The chip was approximately 25x25mm fully populated with flip chip bump at 150um pitch fabricated in 14 nm technology. Impedance and insertion loss characterization coupons were fabricated in all signal layers. Coupons were placed at controlled angle relative to glass cloth direction to enable evaluation of differential pair skew. In addition to high speed signal characterization, the reliability of high speed signal structures and mixed dielectric via stacks were evaluated.

Results indicate a feasible path to 112Gbit/sec HSS SerDes in a manufacturable highly complex (cost) configuration. Currently, the target speed is going to 116Gbit/sec.

Keywords- high performance; signal; insertion loss, substrate

I. INTRODUCTION

Serializer/Deserializer or SerDes is an industry standard interface protocol for sending and receiving signals commonly used to achieve high data-rates in data center communications. A single-channel in next generation SerDes can reach up to 112Gbit/s. The SerDes IO blocks can be integrated into the main logic chip or can be made as stand-alone chips connected to logic via a highly parallel interface. The main logic chip typically controls multiple SerDes communication channels simultaneously in high-speed data exchange. To achieve high-speed with required SerDes link counts, package substrate design requires a large form factor. There are many challenges in substrate materials and fabrication to fulfill the requirements for large form factor chips and laminate substrates[1]. Substrates are backbone parts that provide the microelectronics package with structural support and a form an electrical interface that allows signals to access the silicon device. We carefully considered both electrical performance and mechanical reliability to develop the package structure. The structure required high stiffness or high Young's Modulus for warpage control in the very large package, while maintaining an intermediate CTE to enable second (system board) level reliability[2]. The effect on circuit performance is primarily shown as insertion loss for high-speed communications and lower insertion loss is a major challenge[3]. As a consequence of the use of glass cloth reinforced dielectrics for high speed signals, differential pair skew became a significant issue which was managed by design.

Insertion loss, which consists of the sum of conductor loss, dielectric loss, radiation loss, and the leakage loss is the total loss of a high speed RF circuit. The conductor loss and dielectric loss are significant factors for the RF circuit. In general, thin (<167µm) circuit or package substrate's insertion loss are dominated more by the conductor loss while thick(>254µm) ones are dominated by dielectric loss[4]. Since the thickness of single layer in recent laminate technology is less than 167µm, we optimized the roughness of Cu surface in highest priority and set to 200~250nm range using proper Cu etchant for the prepreg layers. In general, foil roughness of Cu in prepreg is rougher than build up layer. After Cu surface process, the Cu surface roughness in the particle filled layers was 250nm Ra while the glass cloth reinforced (prepreg) layer Cu roughness was 200nm Ra. But, the Cu foil roughness at the bottom in the interface between Cu and dielectric still remains intact even after the Cu etching. So, foil roughness of Cu in prepreg is rougher than build up layer at the bottom interface where the Cu was initially plated. Smooth surface in copper conductor line is needed for faster electrical performances due to the skin effect, but too smooth surface could not give good adhesion between Cu and dielectric materials. So proper Cu roughness needs to be selected. Due to demand of larger package size, the thicker substrate is preferred to endure the reliability against larger thermal stress from body. Also, thicker substrate stripline layers show lower insertion loss.

The dielectric materials in substrate require lower Df and Dk for better electrical performances, lower values of these properties were prioritized for materials selection. While the selected materials satisfy these demands, good insulation reliability is also important for the thin layers of the substrate. In this paper we will mainly discuss about the challenges in substrate materials and challenges we encountered in developing a laminate structure fulfilling industry target requirements.

.

II. THE TECHNICAL CHALLENGES

The conventional sequential build up laminate substrate has been used for PFCBGA(Plastic Flip Chip Ball Grid Array) with a variety of layer stacks for many years due to its major benefit of high speed high performance[3][5]. In high speed signal devices, multi-story stacked vias are necessary to make the electron path shorter. The build up stacked via substrate could satisfy this technical need. Laser drilling of blind micro vias and subsequent copper filling has been used for the standardized manufacturing technique for these high density interconnects. Typically, 33um thickness of resin filled dielectric is popularly used to build the package substrate. However, the main bottle neck of this laminate is the insertion loss in case of high speed data transfer. To further reduce the insertion loss beyond the capability of a single layer dielectric, a skip layer design may be used. A skip layer means a layer of trace that has thicker dielectric thickness than other traces. Since thickness of dielectric layer in the same layer always same due to the inherent process of sequential film lamination process, it is hard to make a dielectric area of specific trace in the layer thicker than other area. To achieve thicker dielectric in a specific trace, two layers could be merged to enable. In a skip layer approach, 2 plies of dielectric are deposited, but plating is completely removed from one layer, resulting in effectively doubled dielectric thickness in the region immediately adjacent to the trace. To obtain impedance similar to a conventional wiring structure much coarser geometry is required. Significantly improved loss performance is obtainable using this approach. The coarser geometry and copper removed area create a zone which is copper poor compared to the surrounding substrate area, greatly increasing the risk of fatigue resin cracking near signals in a skip layer configuration.

To overcome this technical limitation, 65um of thick dielectric material based on glass cloth reinforced material is used in prepreg wiring layers. Since thicker conventional film has a technical limitation to form taller vias, 65um of thick prepreg is selected. The thick dielectric make the insertion loss smaller at 112Gbit/s speed of communication.

Since a large laminate substrate body induces higher mechanical stress at the laminate corners in next level assembly, and worse warpage at the module level, a thick laminate core with high young's modulus is frequently used[2]. The main role of the core is to use as back bone in the substrate during buildup lamination. The thick core may limit the electrical performance and utility of the package. Thin core may contribute to better electrical performance[5]. To satisfy both mechanical and electrical design targets, a single thick core structure may be replaced by a prepreg built up multilayer core. Wiring in thick prepreg buildup layers over a thin core base is an alternative approach to the single thick core solution.

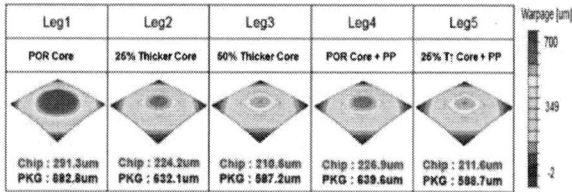

Fig. 1 Finite Element Analysis to estimate the core effect on warpage

To support the laminate core structure selection decision, we have preliminary simulation data to see the core effect on warpage. These three-dimensional computational models for PFCBGA type (a flip chip on top of a package substrate) packages are done using Finite Element Analysis. Fig. 1 shows the Finite Element Analysis result to estimate the core effect on warpage. Five simulation models from Leg1 to Leg5 have been built to simulate the core effect. The materials and form factors were taken from a successful predecessor package. Leg1 is a package type with a conventional sequential build up laminated package substrate. Leg2 and Leg3 were using 25% and 50% thicker laminate core material compares to the POR Leg1. Leg4 is a hybrid prepreg conventional structure mentioned in this study. The model is built using a thin core and prepreg layers are stacked after that, then the conventional build up layers are laminated. Compares to Leg4, Leg5 uses 25% thicker laminate core material. The magnitude of thermal warpage were ranked Leg3, Leg5, Leg2, Leg4, and Leg1 respectively. This simulation showed the thicker effective core of prepreg buildup reduces the warpage.

III. HYBRID PREPREG CONVENTIONAL STRUCTURE

A. Test Vehicle Design

The package substrate configuration was a 65mmx 65mm PFCBGA type design. The silicon chip on top was approximately 25x25mm fully populated with flip chip bump at 150um pitch fabricated in 14 nm technology. The TV was designed as an evaluation vehicle for 7nm ASIC packaging. We will use 14nm chip for the packaging test and move to a 7nm chip in the product or high volume manufacturing.

B. Laminate Structure

The package substrate or laminate's cross section was consisting of four particle filled dielectric layers, four glass reinforced dielectric layers, two conventional core layers (epoxy glass), four glass reinforces dielectric layers (Prepregs), four particle filled dielectric layers (Build up films). In describing the metal layers we call it 4-4-2-4-4 structure. (Table. 1, Fig. 2) (the central core dielectric layer has metal on both top and bottom)

Layer Name	Purpose	Material	Thickness(µm)
	Solder Mask	Epoxy Compound	18
FC9	Trace	Cu	15
	Dielectric	Particle Filled Build Up Film	33
FC8	Trace	Cu	15
	Dielectric	Particle Filled Build Up Film	33
FC7	Trace	Cu	15
	Dielectric	Particle Filled Build Up Film	33
FC6	Trace	Cu	15
	Dielectric	Particle Filled Build Up Film	33
FC5	Trace	Cu	20
	Dielectric	Glass Cloth Prepreg	60
FiC4	Trace	Cu	20
	Dielectric	Glass Cloth Prepreg	60
FiC3	Trace	Cu	20
	Dielectric	Glass Cloth Prepreg	60
FiC2	Trace	Cu	20
	Dielectric	Glass Cloth Prepreg	60
FiC1	Trace	Cu	25
	Core	Glass Cloth Prepreg	200
BiC1	Trace	Cu	25
	Dielectric	Glass Cloth Prepreg	60
BiC2	Trace	Cu	20
	Dielectric	Glass Cloth Prepreg	60
BiC3	Trace	Cu	20
	Dielectric	Glass Cloth Prepreg	60
BiC4	Trace	Cu	20
	Dielectric	Glass Cloth Prepreg	60
BC5	Trace	Cu	20
	Dielectric	Particle Filled Build Up Film	33
BC6	Trace	Cu	15
	Dielectric	Particle Filled Build Up Film	33
BC7	Trace	Cu	15
	Dielectric	Particle Filled Build Up Film	33
BC8	Trace	Cu	15
	Dielectric	Particle Filled Build Up Film	33
BC9	Trace	Cu	15
	Solder Mask	Epoxy Compound	18

Table. 1 Laminate Structure

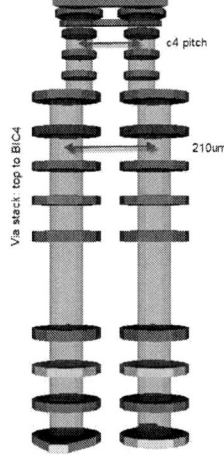

Fig. 2 Via Stacks in Laminate Structure

C. Test Coupon Design

C-1. Asymmetric Sripline Structure on Build Up Layer (FC8)

The main purpose of this test vehicle is to see signal perfomance of thick dielectric to minimize the insertion loss in the laminate. So aforementioned prepreg was mainly used in signal layers, but eventually finer wiring is required to enable fine pitch connection with C4 bump. So in the outer layers, conventional thin build up film was used to meet the finer pitch requirement. A skip layer approach was used to make effectively thick dielectric in the thin film area, the copper below signal traces were omitted. With this method, two dielectric area combined and it is possible to make asymmetrically thicker dielectric around signals. This assymetric stripline structure of trace enables thicker dielectric to reduce the insertion loss. So asymmetric stripline structure of test coupons were implemented in the North and South side of laminate area (Fig. 3). The idea of assymetric or low capacitance design has been visualized using pilot signal integrity simulation. Since the assymetric trace has lower capacity due to thicker dielectric area above the sinal trace, assymetric signal trace showed lower insertion loss (Fig. 4).

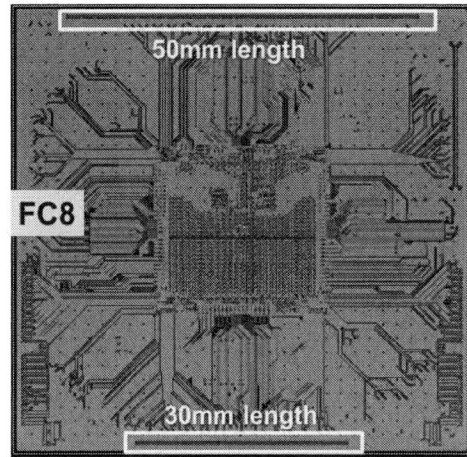

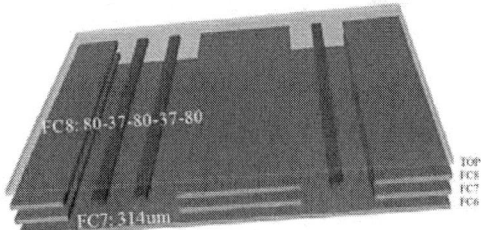

Fig. 3 Asymmetric Stripline Skip Layer Structures of test coupons in the build up in the FC8 layer

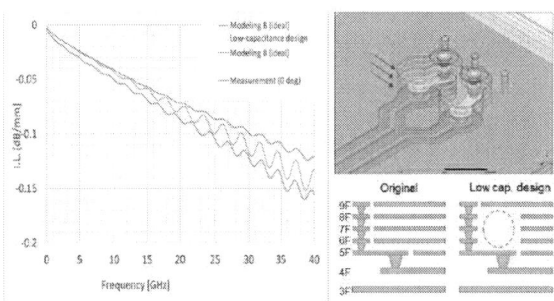

Fig. 4 Simulation on trace in 4F layer to compare the assymetric and the symetric signal trace

C-2. Symmetric Stripline Structures on Prepreg Layers (FiC2, FiC4, BiC2, BiC4)

One of the main purpose of this test vehicle is to see signal skew perfomance of build up layers in varioius condition including different angles and lengths in the prepreg wiring (Fig. 5, Fig. 6, and Fig. 7).

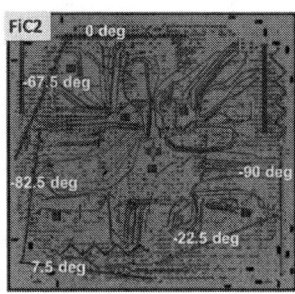 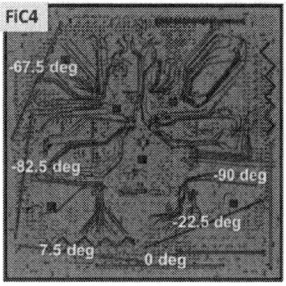

Fig. 5 Test Coupon to measure the insertion losses and impedances in various trace angles in the prepreg layers

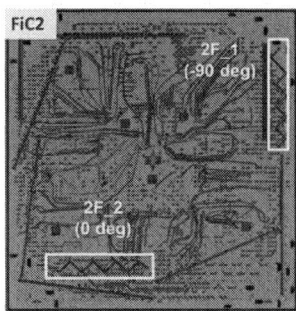 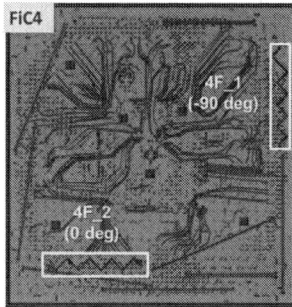

Fig. 6 Test Coupons to measure the insertion losses and impedances in zigzag shape of traces in the prepreg layers

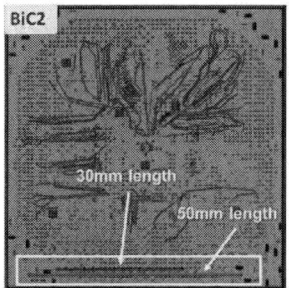 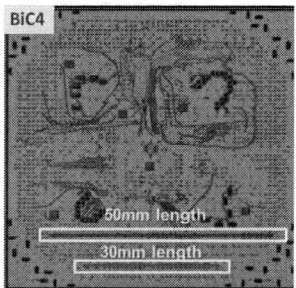

Fig. 7 Test Coupons to measure the insertion losses and impedances in the different trace lengths in the prepreg layers

C-3. Skew Test Coupons on Prepreg Layer (FiC4)

Prepreg dielectric was used in this test vehicle to characterize signal performance in thick dielectric. On the south side of the laminate, two types of test coupons were implemented. Straight line coupons designed orthogonal to glass cloth bundles are used to measure the insertion loss and worst case skew. Angled and zigzag line type test coupons measure the insertion loss and skew in trace at different angles. The zigzag type test coupon is especially to see the best case trace angle effect (45°) compares to orthogonal strain line (0°) in the prepreg layers. This horizontal strain line at 0° is aligned with one of the prepreg's glass cloth fiber orientatioins.

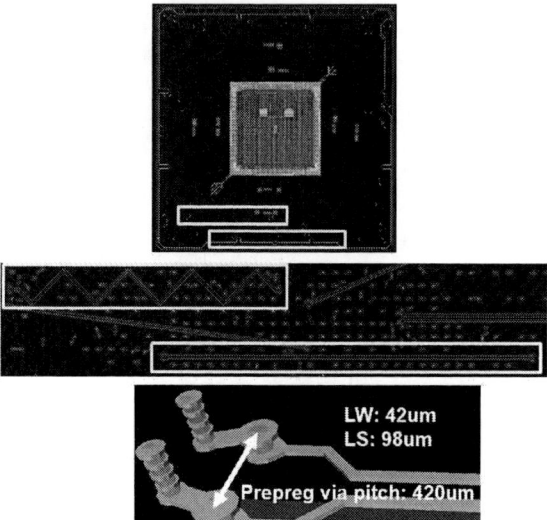

Fig. 8 Skew Test Coupons in the Prepreg layer

C-4. Insertion loss coupons with and without microvia stack on Prepreg Layer (FiC4)

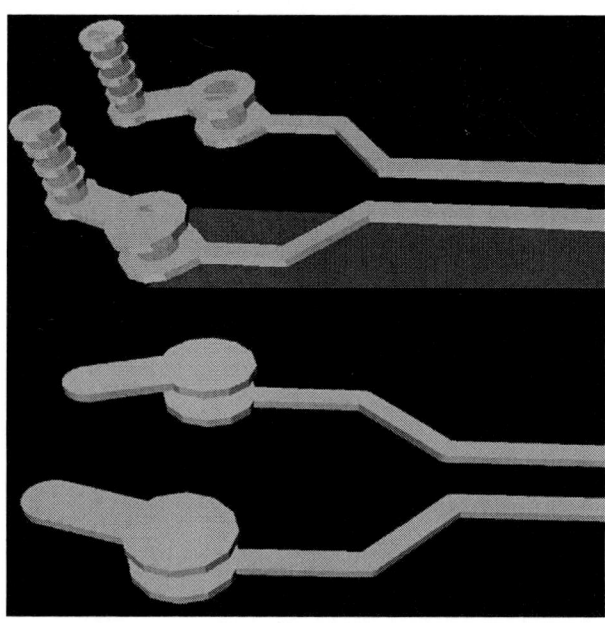

Fig. 9 Structure of Test Coupons with (Top)/without (Bottom) microvia stacks in the Prepreg layer

A laminate with removed microvia stacks was prepared to measure the insertion loss without microvia stacks in the Prepreg layer (Fig. 9 Bottom).

IV. CIRCUIT PERFOMANCE

Insertion loss and skew characterization coupons were fabricated in all signal layers. Coupons were placed at controlled angle relative to glass cloth direction to enable evaluation of differential pair skew. In addition to high speed signal characterization, the reliability and of high speed signal structures and mixed dielectric via stacks were evaluated.

A. Insertion Loss

In SerDes transmission, energy is lost from a transmitted signal as joule heating in the Cu trace and ground planes due to resistance and as heat in the surrounding dielectric caused by the material properties of dielectric. Resistive loss in Cu is caused by a combination of the geometry, plating characteristics, and surface roughness. Losses from the dielectric material are primarily caused by heating as a result of the movement of non-crosslinked polar groups on the polymer chains in the electric field[6][2]. Total loss is therefore mainly the sum of conductor loss and dielectric loss and usually expressed in decibels (dB). Since insertion loss is a key factor affecting high speed channel performance, it is an important parameter to measure when high speed device is developed.

A-1. Asymmetric Build Up vs Symmetric Prepreg

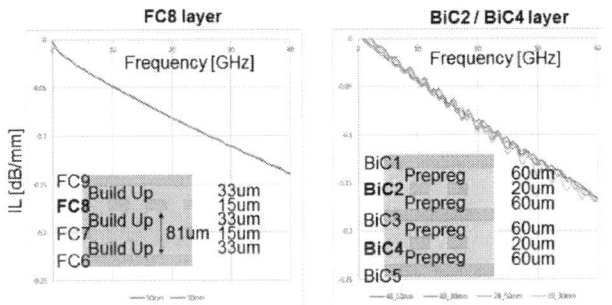

Fig. 10 Insertion loss at asymmetric stripline on FC8 in Fig. 3 vs symmetric strip line on BiC2 and BiC4 in Fig. 7

TABLE 2. INSERTION LOSS MEASUREMENTS OF TEST COUPONS IN FIG. 10

	FC8		BiC4		BiC2	
Length	50mm	30mm	50mm	30mm	50mm	30mm
10GHz	-0.0488	-0.0483	-0.0415	-0.0453	-0.0436	-0.0464
20GHz	-0.0816	-0.0483	-0.0857	-0.0937	-0.0900	-0.0967
28GHz	-0.1050	-0.1043	-0.1158	-0.1179	-0.1183	-0.1252
Imped ance Ω	95.1	92.5	94.3	93.5	99.5	98.4

Insertion loss at asymmetric stripline on FC8(Build Up Film), symmetric strip line on BiC2 and BiC4 (Prepreg) were measured and listed in the table 2 in the unit of dB/mm.

A-2. Insertion loss in traces in various angles

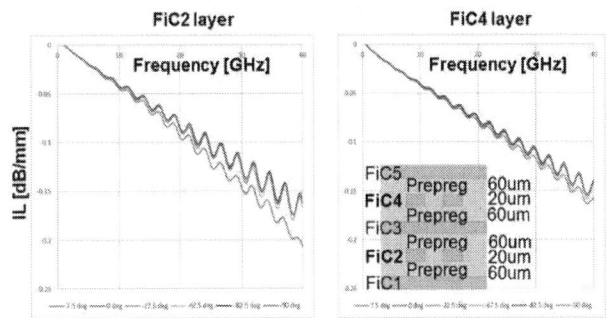

Fig. 11 Insertion loss in various angles of striplines on FiC2 and FiC4 in Fig. 5

TABLE 3. INSERTION LOSS MEASUREMENTS OF TEST COUPONS IN FIG. 11

Measurement in BiC2						
Degree	7.5°	0°	-22.5°	-67.5°	-82.5°	-90°
10GHz	-0.0427	-0.0449	-0.0419	-0.0429	-0.0421	-0.0432
20GHz	-0.0872	-0.0953	-0.0851	-0.0883	-0.0866	-0.0859
28GHz	-0.1258	-0.1392	-0.1231	-0.1278	-0.1254	-0.1274
Imped ance Ω	100.7	101.3	101.6	100.2	100.6	100.7
Measurement in BiC4						
Degree	7.5°	0°	-22.5°	-67.5°	-82.5°	-90°
10GHz	-0.0408	-0.0422	-0.0403	-0.0416	-0.0404	-0.0412
20GHz	-0.0773	-0.0809	-0.0757	-0.0776	-0.0760	-0.0778
28GHz	-0.1069	-0.1146	-0.1064	-0.1098	-0.1080	-0.1097
Imped ance Ω	96.4	96.3	97.3	96.3	96.2	96.6

Insertion loss at stripline in various angles on FiC2 and FiC4 were measured and listed in the Table 3. These insertion losses depend by the angles of differential lines relative to the glass cloth weave and they lead to skew effect.

A-2. Insertion loss in zigzag shaped traces

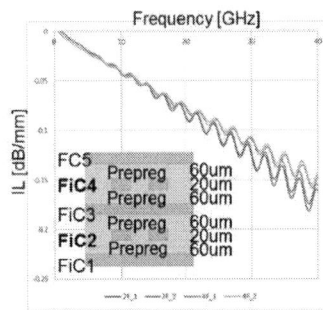

Fig. 12 Insertion loss in zigzag shaped traces in Fig. 6

TABLE 4. INSERTION LOSS MEASUREMENTS OF TEST COUPONS IN FIG. 12

Length	2F_1	2F_2	4F_1	4F_2
10GHz	-0.0428	-0.0422	-0.0418	-0.0408
20GHz	-0.0868	-0.0851	-0.0774	-0.0755
28GHz	-0.1261	-0.1237	-0.1077	-0.1065
Imped ance Ω	100.4	101.4	96.9	96.6

Insertion loss at stripline in various angles on FiC2 and FiC4 were measured and listed in the Table 4.

A-3. Microvia stacks effect on insertion loss in prepreg traces

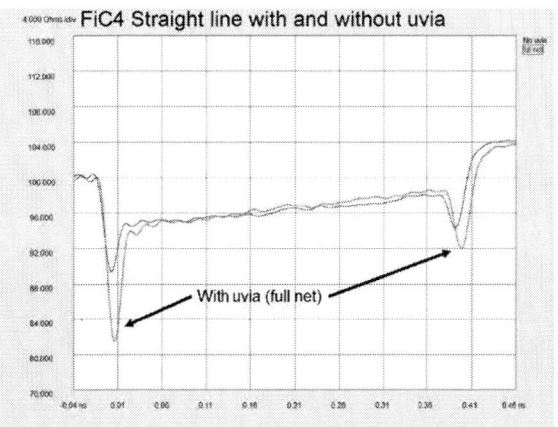

Fig. 13 Impedance comparison between with and without microvia stacks on coupons in Fig. 9 Top/Bottom

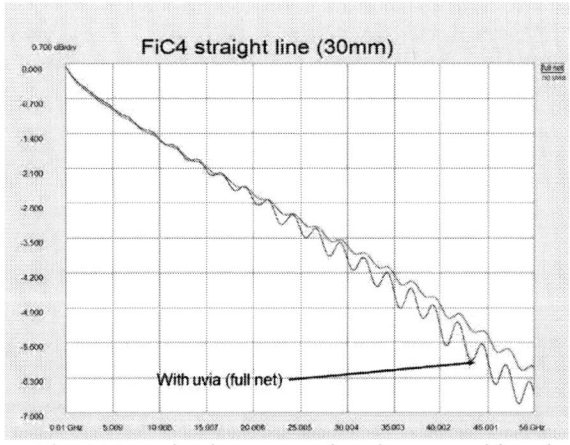

Fig. 14 Insertion loss comparison between with and without microvia stacks in coupons in Fig. 9

B. Skew

Differential pair signal skew is defined as the time of flight difference between a signal propagating down the 2 legs of the pair. Skew may be caused by trace length differences or differences in the impedance of the dielectric through which the signal propagates.

Due to usage of glass fiber and resin in the prepreg material, skew issues must be considered. Fig. 15 shows the distribution of glass fiber in resin. Due to the distribution of the glass fibers in prepreg, some areas glass fiber rich, while others are resin rich. As a result, there are impedance and insertion loss difference between Cu A trace and Cu B trace in Fig. 15. If these two trace form a differential pair, the signal performance varies by the composition of dielectric over which each signal travels. This skew depends by the angle of differential line relative to the glass cloth weave. For traces parallel or near parallel to a fiber bundle, it has a significant impact to the high speed signal such as 112Gbit/sec

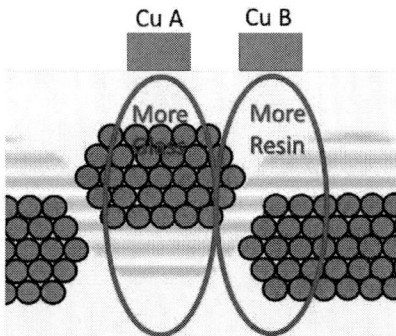

Fig. 15 Cross Section of prepreg shows distrubution of glass fiber in resin

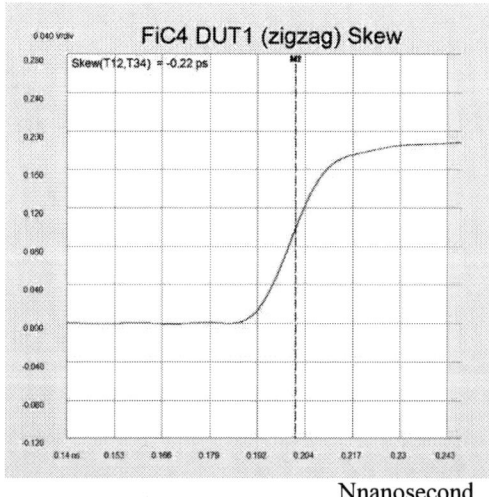

Nnanosecond

Fig. 16 Skew measurement from Test Coupon 4F_2 on FiC4 in Fig. 8

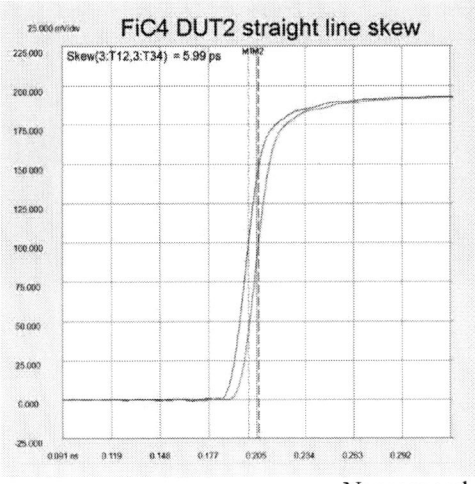

Nanosecond

Fig. 17 Skew measurement from Straight Line Test Coupon on FiC4 in Fig. 8

C. Quality and Reliability Result

Several Quality and Reliability tests have been done to access the quality and reliability of package substrate and the final package.

C-1. Coplanarity and Warpage Test

The land coplanarity at room temperature were measured using a non-contact optical inspector by the JEDEC plane of best fit (least squares method)[7]. The Fig. 18 shows that coplanarity values were relatively small compares to the large laminate size (65mmx65mm) and core thickness (200um). High stiffness due to glass fiber cloth in the prepreg plays major role to mitigate planar distortion during the hot temperature lamination and curing process. Warpage was measured at room temperature and 225°C using the Shadow Moiré measurement system. Fig. 19 and Fig. 20 show very flat warpage and thermal warpage. The thickness, material property of insulators above and below core were well balanced and Cu distribution in the overall insulator layers were well designed to mitigate substrate level warpage.

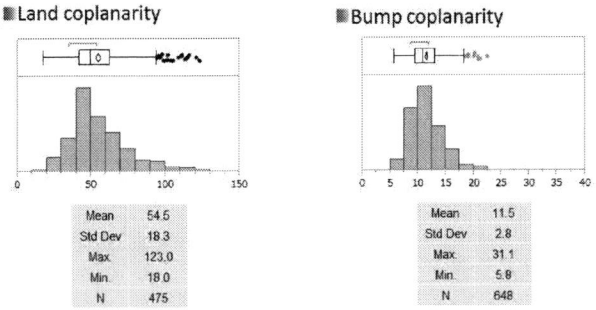

Fig. 18 Coplanarity in substrate level measured at room temperature

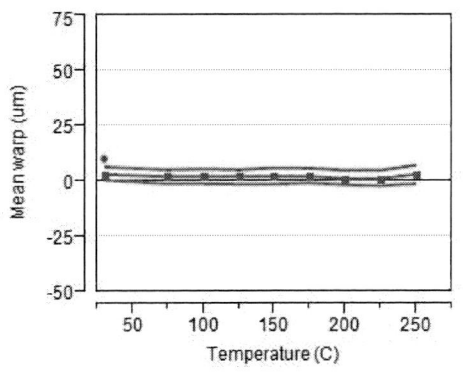

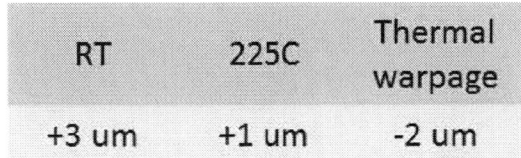

Fig. 19 Shape inversion plot

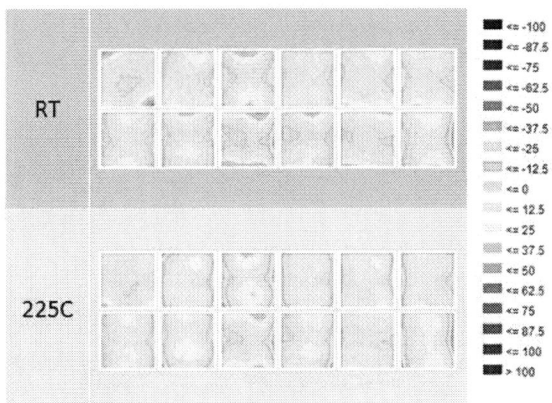

Fig. 20 Substrate level warpage shapes at room temperature and 225°C

C-2. Thermal Shock Test

The reliability of package substrate was tested at +150°C and -65°C condition. This test was done to see weakest area in package substrate. The temperature setting was similar to JEDEC Temperature Cycle C condition, but this test utilizes liquid medium instead of regular hot air chamber. Two containers with hot liquid (+150°C) and cold liquid (-65°C) were prepared. The samples were fully immersed into hot liquid, then they were fully immersed into cold liquid. This set of immersions is defined as 1 cycle. This extreme temperature cycling was done until 2000 cycles. With this liquid to liquid thermal shock test, the weakest area in the substrate could be found in an accelerated fashion. 10 substrate samples were used for this test.

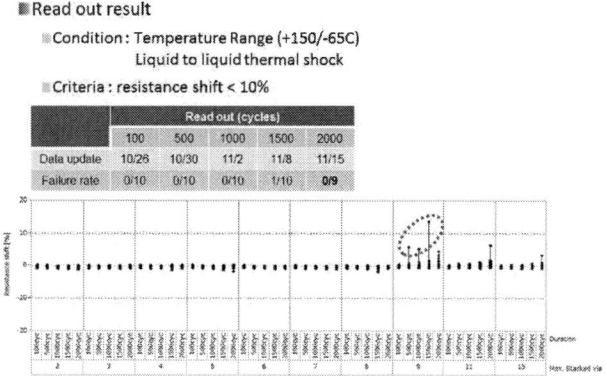

Fig. 21 Thermal Shock read out result

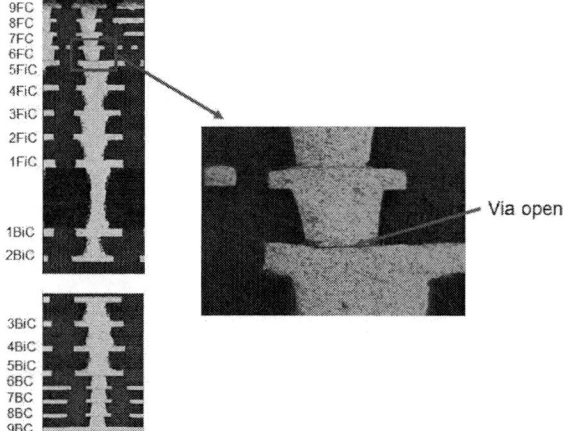

Fig. 22 Thermal Shock Failure Analysis Result

Fig. 21 shows the thermal shock result. The resistances in the test elements were not shifted until 1000cycles of thermal shock. But, after 1500cycles, the weakest area was found. Among all test elements in the substrate, the resistance in test element containing via stacks that connect all layers from top copper to core copper were shifted more than 10%. Fig. 22 shows a thin crack in the cross section. This is an expected result. The area that has a crack is an interface between build up film and prepreg dielectric. Usually, stacked via has the largest stress in the substrate and the interface between build up film and prepreg has more CTE mismatch than other interfaces. After the weakest region is detected using this thermal shock test, the drillability of the buildup film was modified to get better interface. 18 different DOE legs with different fine-tuned build up materials were prepared for good drill condition. Compares to the POR sample with the initial build up material, the leg#13 showed the best drill condition for the new build up material (Fig. 23).

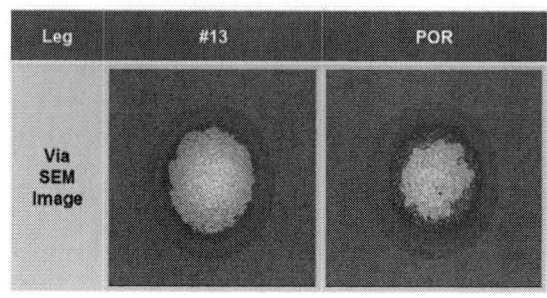

Fig. 23 CO_2 Laser drill evaluation results

Fig. 23 shows SEM images in top-bottom view after the via was formed using CO_2 laser drill. Leg 13 shows good results.

Mechanical tests were also conducted to ensure the adhesion strength.

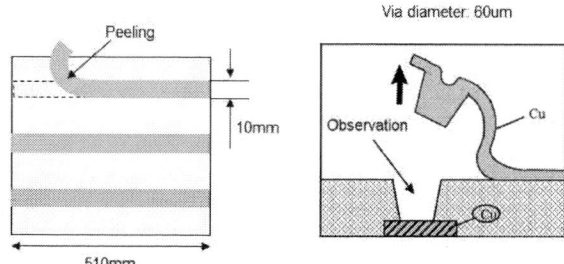

Fig. 24 mechanical peeling test to check adhesion strength

Fig. 24 shows the mechanical peeling test that conducted to ensure the adhesion strength of the via. The test was done on a package substrate sample with arbitrary pattern shown in Fig. 24. After the Cu was peeled off, the via bottoms were observed. The sample fails if any broken base plated Cu is visible and it passes if broken base plated Cu is not visible.

C-3. BHAST

The standard JEDEC BHAST(Biased Highly Accelerated Temperature and Humidity Stress) test was conducted[8]. The used condition was 130°C/85% Relative Humidity/ 3.7V Bias and 6 unit of substrate were used. For HAST, an open net is used as a test element. The initial resistance is very high as there is no connected conductor. If resistance starts to drop during the test, it means a leakage path is forming. Because Cu is not electrochemically stable, migration, clouding, dendrite and re-deposition are sometimes found in organic substrate stress. If the read out after BHAST reaches below the criteria at 1e+8Ω, it's considered a fail. All BHAST test elements in layer-layer, line-line in buildup film(BF), and line-line prepreg(PP) did not show any abnormality (Fig. 25).

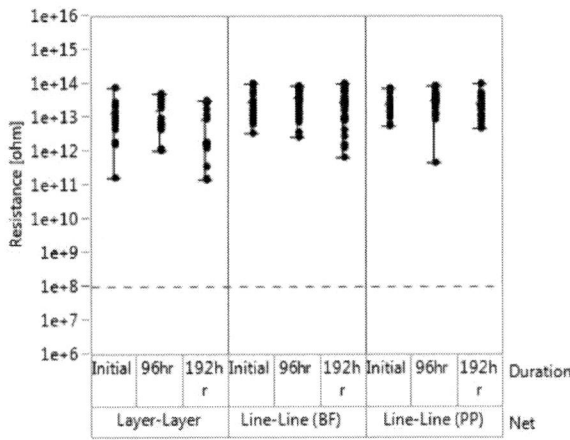

Fig. 25 BHAST result

C-4. T0 Assembly Test

The final 165 package modules were assembled as 2D PFCPBGA package type and read out were conducted after the final assembly. The 165 samples were fully passed the initial Assembly Test. The JEDEC assembly reliability tests are in progress.

V. CONCLUSION

In this study, a flip chip laminate assembly was designed and built using both 33um low loss particle filled and 65um low loss glass cloth reinforced dielectric. The integrity of signals is mainly affected by the insertion loss in the package substrate in a 112Gbit/s high speed application. Even though coarse geometry for high speed conflicts with microelectronics trends for small size and high density, coarse geometry was adopted because there is a benefit in insertion loss from this design. Coarse geometry was mainly implemented using thick prepreg dielectric, but layers in top and bottom need to have thin dielectric film to enable fine pitch interconnection. To implement coarse geometry in these thin film layers, skipped layers are used. Both coarse geometry from prepreg and skipped layer of thin film trace shows low insertion loss that enable high speed signal communication. However skew effect due to utilization of glass cloth should be evaluated for the trace in the prepreg layers. The skew measurements of lines at various angles show different results. In the case when the differential lines are parallel to the glass cloth weave orientation signal delay may vary between signals due to changes in dielectric constant as a result of glass fiber distribution in the prepreg. In cases where signals are at an angle to the glass cloth signal delays are similar. In contrast to non-fiber reinforced substrate, trace in glass cloth reinforced dielectric needs to be carefully analyzed and signal lines need to be oriented properly to mitigate the skew induced signal delay.

There were benefits from using glass fiber reinforced dielectric material. Even though the core was thin, the overall warpage and coplanatiry shows very good result. High young's modulus due to glass fiber cloth in the prepreg plays major role to mitigate planar distortion during the hot temperature lamination and curing process. Even in copper depleted areas dielectric resin cracking was eliminated. Insertion loss performance shows the expected benefit from the coarse geometry once via stacks effect are removed. Skew shown to be manageable by design.

Even though the transition interface between prepreg and build up layer consists of different dielectric materials, the result shows organic-organic adhesion was good, but attention is required to select matched CTE property against thermally induced mechanical stress.

Results indicate a feasible path toward meeting the 112Gbit/sec HSS SerDes insertion loss objective. Currently, the target speed is going to 116Gbit/sec. High cost of these approaches, may hinder their widespread adoption.

ACKNOWLEDGMENT

Authors would like to express very great appreciations to Gilles Poitras and Roxan Lemire at IBM Canada.

REFERENCES

[1] Z. Wu *et al.*, "Chip-Package Interaction Challenges for Large Die Applications," in *2018 IEEE 68th Electronic Components and Technology Conference (ECTC)*, 2018, pp. 656–662.

[2] E. Blackshear *et al.*, "Advanced laminate carrier module warpage considerations for 32nm pb-free, FC PBGA package design and assembly," in *2011 IEEE 61st Electronic Components and Technology Conference (ECTC)*, 2011, pp. 523–529.

[3] E. D. Blackshear *et al.*, "The evolution of build-up package technology and its design challenges," *IBM J. Res. Dev.*, vol. 49, no. 4.5, pp. 641–661, 2005.

[4] J. Coonrod, "The impact of final plated finishes on insertion loss for high-frequency PCBs," *Printed Circuit Design & Fab / Circuits Assembly, March*, p. 30~36, 2018.

[5] J. Wang, Y. Ding, L. Liao, P. Yang, Y. Lai, and A. Tseng, "Coreless substrate for high performance flip chip packaging," in *2010 11th International Conference on Electronic Packaging Technology High Density Packaging*, 2010, pp. 819–823.

[6] F. R. Eirich and H. F. Mark, "New Resins for Electrical Applications," *Trans. Am. Inst. Electr. Eng. Part III Power Appar. Syst.*, vol. 81, no. 3, pp. 183–185, Apr. 1962.

[7] JESD22-B108B, "Coplanarity Test for Surface Mount Semiconductor Devices," *Jedec Stand.*, 2010.

[8] JESD22-A110, "Highly-Accelerated Temperature and Humidity Stress Test (HAST)," *Jedec Stand.*, 2015.

PI/SI Analysis and Design Approach for HPC Platform Applications

Sungwook Moon, Chamin Jo, and Seungki Nam,
Foundry Business Division, Samsung Electronics Co. Ltd.
1-1 Samsungjeonja-ro, Hwaseong-si, Gyeonggi-do, South Korea
Email: {sw2013.moon, chanmin.jo, seungki.nam}@samsung.com

Abstract— **In this work, we introduce a comprehensive methodology for the power integrity/signal integrity analysis of SoC and silicon-based interposers in a HPC platform. By conducting a comprehensive investigation of the global chip-level PDN impact by electrical interactions between the chip, package, and board in the system, an efficient power delivery network was designed to meet HPC specifications. This was achieved by extracting long-term test scenarios (microsecond-order duration) to estimate voltage drop at the bumps and deployment of decoupling capacitors considering the optimal placement and numbers within an allowable cost budget. This analysis also yielded the benefit of reduction in power domains/ball count/package layers which further reduced costs. In addition to the Power Integrity, signal properties were analyzed under the allowed channel conditions (channel width and spacing, shielding lane placement). As a result, the channel design parameters are found to meet the criteria of eye opening including power noise and crosstalk effects. It was demonstrated that the electrical properties of the HBM2e IP are successfully operational up to 3.2Gbps.**

Index Terms—**High Performance Computing (HPC), Power Integrity, Signal Integrity, Simulation, Decoupling Capacitance Optimization, Power Delivery Network (PDN)**

I. Introduction

The recent explosive growth of advanced technology on artificial intelligence for processing big data is making fundamental changes in the information-centric society, causing an industry shift towards a big data-based working model [1]. To manage the complex problems of such mammoth data handling, a lot of data centers based on multilateral-interaction service are being established. These applications, called high-performance computing (HPC), require efficient computing algorithms for iterative operations and ultra-fast internal processing hardware with wide bandwidth [2], [3].

The HPC platform is configured with multi-chip configurations (main processing chip and external memory chips) on a silicon-interposer. In terms of design configuration, high-density thousand-order interconnects between a SoC and a High Bandwidth Memory (HBM) chip need to be manipulated for supporting wide bandwidth. The HBM chips need to be as close to the SoC as possible to minimize additional channel losses. In terms of design performance, a

HBM solution is suitable for making high-intensive training/inference operations. The power delivery network (PDN) of the system is of critical importance of such complex HBM2 operations and needs careful design and evaluation [4].

In order to implement high-power silicon chips for HPC applications, it is required to prepare design solutions for supporting high-power IPs with power consumption of a few hundred Watt and work properly for high-performance HBM with Tera B/s bandwidths. Thus, it is only possible to work properly, when optimally configured with combination of multiple decoupling capacitors covering a wide range of capacitance types. Also, power/signal design for enabling HBM interface becomes more challenging considering expected high power supply and marginal eye margins. Thus, it is required to suppress crosstalk between adjacent channels at high-power and high-speed operations in thousands of I/Os. Specifically, the significance of initial feasibility analysis for multichip placement becomes more important due to the large design complexity of interconnects on a Si-interposer. By early recognition of the physical limitations of design parameters, the solution exploration by sweeping a set of selected design parameters can allow to find a sweet spot in the huge possible solution space for PSI design optimization in HPC applications. The deficiency of initial design study at an early stage is possible to lead to significant cost increases when a silicon product is signed off and hence, it is of critical importance. Therefore, in this work we focus on design methodology on HBM2e IP enabling considering signal integrity and system-level power integrity considering the electrical properties of a Si-interposer.

II. HBM2e IP Enabling

A HBM2 approach with TB/s bandwidth is suitable for HPC applications with extremely fast internal processing speeds. Recently, HBM2e with 3.2Gbps data rate was introduced and it is almost 30% faster than existing HBM2 (2.4Gbps bandwidth). Thus, it is expected to have significantly more challenges in meeting signal and power quality specs compared to the HBM2 IP enabling.

A. Signal Integrity (SI) Enabling

As the shape of HBM2e is shorter and fatter than that of existing HBM2 and operating bandwidth is higher, the micro-bump distance between a logic chip and a HBM chip is nearly 1mm longer and the UI is shrunk by 30% compared to the

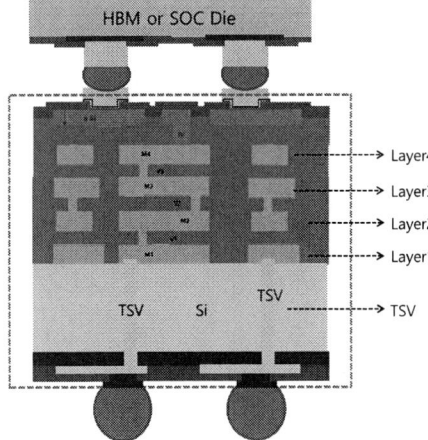

Figure 1. A cross-section image of Si-interposer for HBM2 I/F.

HBM2 specs. This makes the design of HBM2e more challenging due to the longer signal lengths involved.

Currently, the signal of HBM2 interface propagates on a Si-interposer. It consists of 4-metal layers, stacking vertically as shown in Fig. 1. Since Si-interposer uses mature silicon process, the geometry (width and height) of signal lines on a Si-interposer is relatively smaller than that of the conventional organic package substrate process. Thus, signal losses are higher when compared to the package process and this causes difficulty to maintain signal quality. In addition to the loss of the signal, the structural nature of parallel signaling in interposer causes higher crosstalk between adjacent lines. Therefore, it is of vital importance to design the stack-up structure of an interposer considering HBM2e SI characteristics.

1. Determination of Si-interposer stack-up (metal & dielectric height)

As described previously, the stack-up configuration of Si-interposer becomes an important factor to HBM2e signal quality. Since a stack-up is related to process cost, it is necessary to determine optimized stack-up through SI analysis among acceptable stack-up options. Currently, the stack-up configuration for HBM2 interface is option 0 in Table 1. Based on the following configuration, the best case was found to have the optimal signal quality by varying metal thickness, and height of dielectric layers through various simulation cases. The stack-up options used in simulation are shown in Table 1.

Table 1. Si-interposer Stack-up Options (Height)

Height [um]	M1	V1	M2	V2	M3	V3	M4
Option 0	A	B	A	B	A	B	A
Option 1	A x 1.25	B x 1.25	A x 1.25	B x 1.25	A x 1.25	B x 1.25	A x 1.25
Option 2	A x 1.5	B x 1.25	A x 1.5	B x 1.25	A x 1.5	B x 1.25	A x 1.5
Option 3	A x 1.75	B x 1.25	A x 1.75	B x 1.25	A x 1.75	B x 1.25	A x 1.75
Option 4	A x 2	B	A x 2	B	A x 2	B	A x 2
Option 5	A x 2	B x 1.67	A x 2	B x 1.67	A x 2	B x 1.67	A x 2
Option 6	A x 2.25	B x 1.67	A x 2.25	B x 1.67	A x 2.25	B x 1.67	A x 2.25
Option 7	A x 2.5	B x 1.67	A x 2.5	B x 1.67	A x 2.5	B x 1.67	A x 2.5
Option 8	A x 2.75	B x 1.67	A x 2.75	B x 1.67	A x 2.75	B x 1.67	A x 2.75

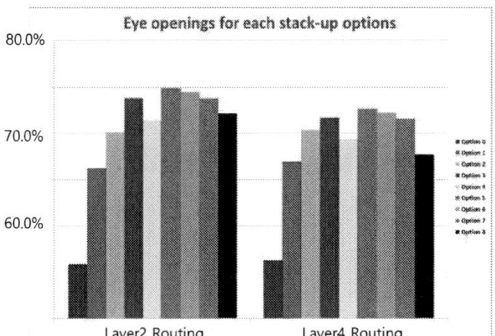

Figure 2. Simulation results

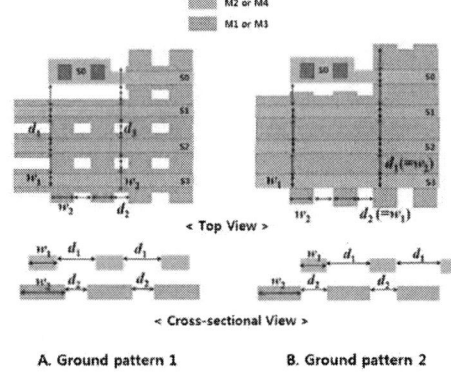

Figure 3. Ground patterns [5]

The SI analysis results corresponding to the stack-up options are shown in Fig. 2. In this work, option 5 was found to have the optimum characteristics and it was chosen as final configuration

2. Determination of Ground structure

The design of a Si-interposer has to satisfy the design rule of metal density based on the silicon fabrication process. The fabrication process does not allow for ground planes to be formed. This makes it necessary to ensure a high-quality ground for guaranteeing SI characteristics. Electrically, the resistance of the signal net is determined by the physical structure, but the capacitance is determined by the layout of the signal and ground. In order to reduce the signal loss, it is necessary to reduce the self-capacitance and decrease the coupling capacitance for suppressing crosstalk. Since the layout configuration that can satisfy both self and coupling capacitances is hard to achieve, the best configuration can be found through experimental analysis, considering transmission and crosstalk effects. In order to properly place the ground considering the density rule, two structures were analyzed as shown in Fig 3. Based on extensive analysis, it was found that the SI characteristics of placing the ground between the signals are better than the ground under the signal by comparing the trade-off between the signal loss and coupling effect as shown in Fig 4.

3. Signal routing guide for HBM2e Interface

978-1-7281-1500-9/19 $31.00 © 2019 IEEE

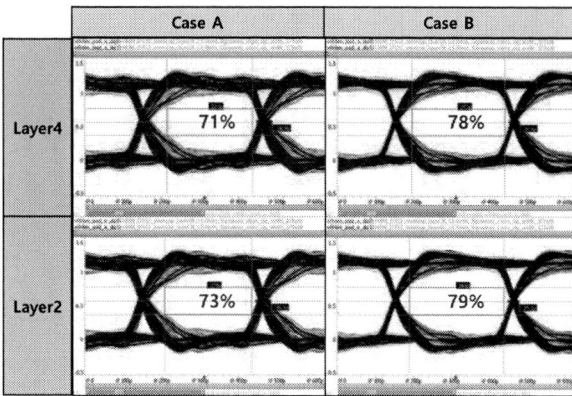

Figure 4. Comparison Eye openings for each Ground structure

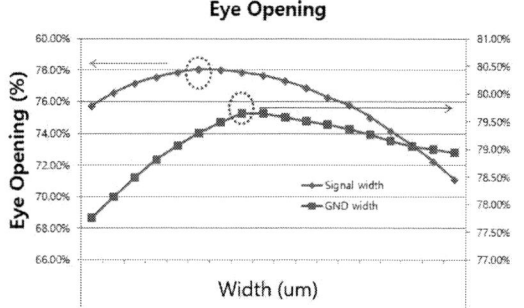

Figure 5. Simulation results for various signal and ground width

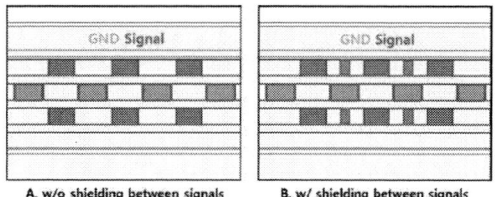

Figure 6. Ground shielding structure

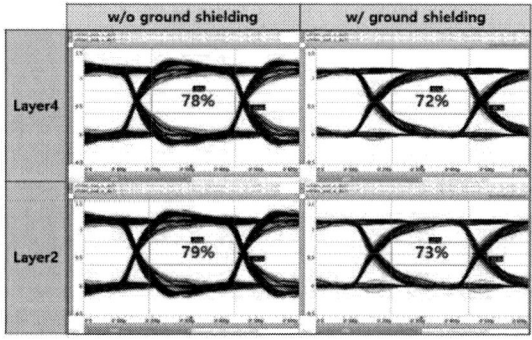

Figure 7. Ground shielding effect

A HBM signaling structure consists of total 8 channels, and each channel has about 200 signals, including 128 DQ signals and the others includes WDQS, RDQS, DM, DBI, ADD, CLK and so on. In addition, 128 DQ signals are divided into 32bit DQ blocks, and the remaining blocks and channel signals are configured in the same format as bump map. On the basis of the 32bit DQ block, all of the blocks are properly routed, and

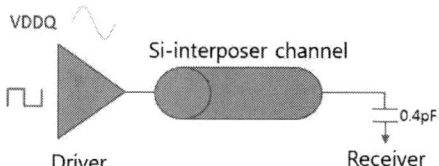

Figure 8. IO duty simulation

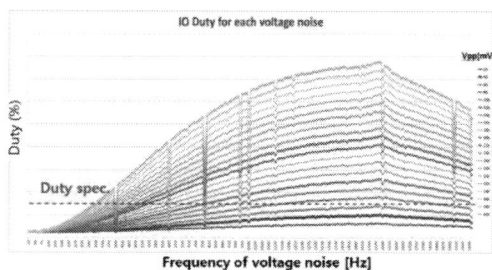

Figure 9. Voltage ripple spec. considering IO Duty

signal pitch is determined, considering the physical limitation of routing space and via insertion. The good candidates with parameter combination were found to ensure optimal SI characteristics through various analyses with determined signal width and space. Fig. 5-7 also helps in determining the requirement of ground shielding between signals.

B. Power Integrity Enabling

HBM2 in terms of power delivery has the same voltage level of 1.2V as SOC IO (VDDQ), HBM2 IO (VDDQ), and HBM2 Core (VDDC). HBM2e has also the same power domains as the HBM2. Thus, accurate analysis needs to be performed by considering the current and on-die capacitance of the VDDQ/VDDC, associated with the HBM2e. In both HBM2 and HBM2e, VDDQ and VDDC were designed to be separated. Especially, since the decoupling capacitance of the VDDC domain is larger than that of SOC IO VDDQ domain, it becomes more significant to analyze its PI characteristics considering the VDDC domain.

1. Determination of Voltage ripple spec. for VDDQ domain

In the process of IP development, it is necessary to determine on-die decoupling capacitance requirements in SOC IO (VDDQ) domain. It is possible to be determined by understanding the voltage noise requirements. Since the Vmin value of power (VDDQ/VDDC) related to HBM2e clearly is described in the JEDEC standard datasheet [6], there is no problem if the corresponding specifications are met to specified criteria, but the IP developer need to define the detailed voltage specifications for IO VDDQ domain. The specification in the VDDQ domain of HBM2e allows power noise for duty / jitter, conditions for tCH/tCL, and the decoupling capacitance value within the corresponding budget. Based on configuring IO scheme as shown in Fig. 8, simulation was performed where single-tone noise is induced to the IO by adjusting frequency and voltage ripple level. Finally, the allowable peak-to-peak voltage ripple (Vpp) was determined based on the analysis result as shown in Fig. 9.

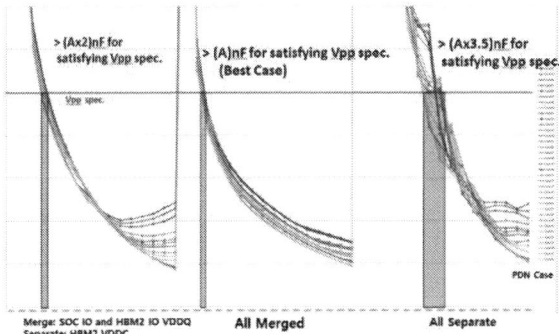

Figure 10. PDN structure for Vpp simulation

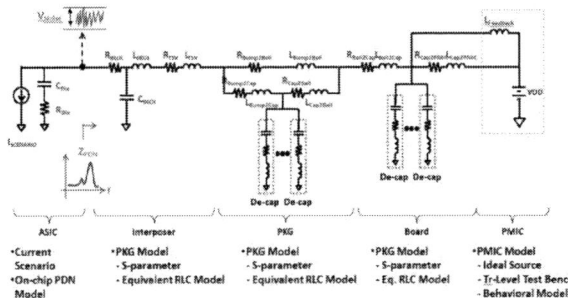

Figure 11. A simplified schematic for system-level PDN analysis at early design stage.

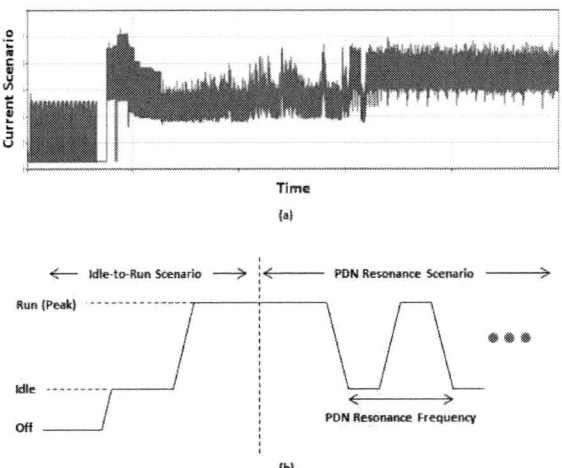

Figure 12. A current scenarios for system-level PDN analysis at early design stage: (a) vector-based scenario and (b) simplified worst case scenario

2. Determination of Decoupling capacitance for SOC IO VDDQ domain

Based on the previously determined Vpp specifications, voltage ripple is analyzed under the PDN cases as shown in Fig. 10. The number of maximum operations of IOs is assumed to operate simultaneously in 1-HBM DQ case when DBI option turns on. The analysis cases consider multiple scenarios like: (1) all power domains are merged (SOC IO VDDQ, HBM IO VDDQ and HBM core VDDC), (2) all power domains are individually separated, and (3) VDDQ

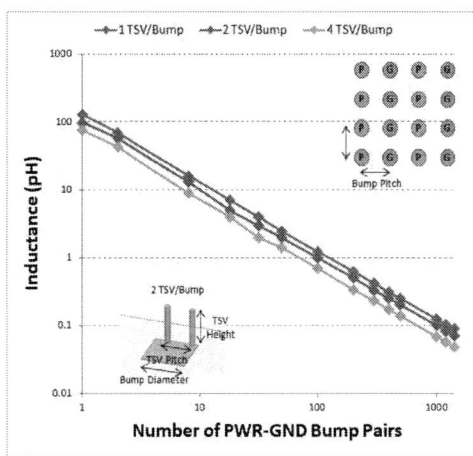

Figure 13. The effective inductance of TSVs for various number of C4 bumps.

domains are merged, but VDDC domain is separated. As a result, when all power domains are merged, voltage ripple was found to be the lowest at same PDN conditions. Although it depends on the characteristics of system-level PDN (Board/Package), it was confirmed that in most cases, the decoupling capacitance requirement for SOC IO (VDDQ) domain is larger than reference value (A) to satisfy Vpp specifications. This analysis was performed on 1-HBM, and when HBM2e connections are increased such as 2-HBM and 4-HBM, additional analysis is required to arrive at values for inserted decoupling capacitance.

III. HIGH POWER IP ENABLING

To design system-level PDN with qualified power integrity performance for high power systems, it is important to optimize system-level PDN and estimate target PDN parameters of each part, composing PDN system at early design stage. Fig. 11 shows a schematic for analysis and optimization of system-level PDN for core PDNs including CPU, NPU, and GPU at early design stage. To estimate initial parameters of the schematic, an initial model library was established based on model parameters extracted from the previous generation technologies. For on-chip capacitance estimation, the on-chip capacitance was extracted per unit area from the library and initial parameters of the new project were estimated using the target IP areas. For current scenario, one of the two different scenarios was applied as shown in Fig. 12. One is to use the extracted average power per cycle using on-chip tools such as PTPX, assuming that vector scenario in micro-second order including idle-to-run operations can be extracted from emulators at early design stage. The other is to use the simplified worst case scenario including idle-to-run operations or iterative operations at PDN resonance based on the initial estimated powers between idle mode and run mode. Using the simplified worst case scenarios, the effect of di/dt at idle-to-run transition and the effect of PDN resonance operation on voltage drop can be analysed, considering various system PDN conditions.

For an interposer, the main parasitic for core PDNs is the effective inductance of TSVs. The initial effective inductance

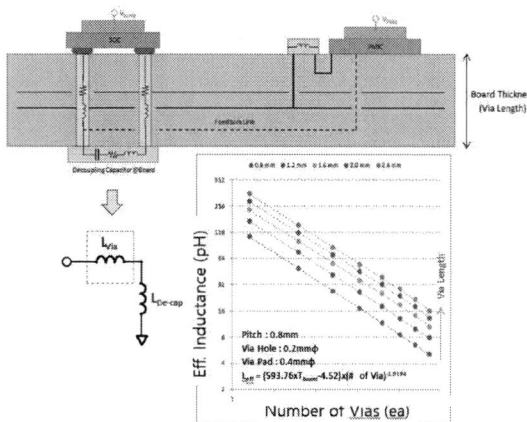

Figure 14. The simplified board model for the estimation of the needed number of balls and the effective inductance of boards for various via conditions.

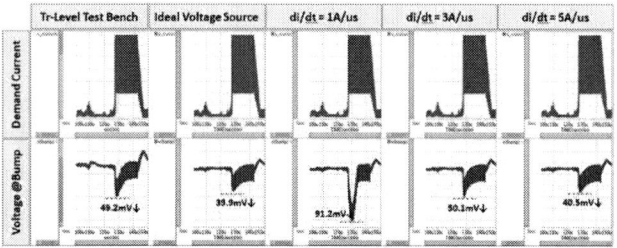

Figure 15. The effect of the current supply capability of PMIC on the voltage drop.

of TSVs can be estimated from the pre-extracted database for the effective inductance due to the number of TSVs as shown in Fig. 13. The required number of C4 bumps can be also estimated using the database from the extracted target inductance of an interposer by the optimization of system PDN at early design stage. For a package model, the effective inductance from C4 bumps to balls can be estimated based on the number of balls and thickness of packages. The required number of balls for packages can be estimated from the library of effective via inductance for various via conditions in boards and the target inductance of vias for connection between package balls and decoupling capacitors of boards. As shown in Fig. 14, the effective board inductance of boards consists of the inductance of vias to connect package balls to decoupling capacitors and effective inductance of decoupling capacitors when the number of decoupling capacitors is enough on boards. By extracting the target board inductance for the optimization of system PDNs at early design stage, the required number of balls can be estimated.

The other major factor to consider in PI analysis is PMIC. There are two main characteristics in PMICs. One is the feedback function to compensate the IR drop which can be modelled by feedback inductance and the other is the current supply capability. As shown in Fig. 15, the system-level voltage drop using real PMIC is larger than that using ideal voltage source due to lower current supply capability. The current supply capability of PMIC can be implemented using a Verilog-A model.

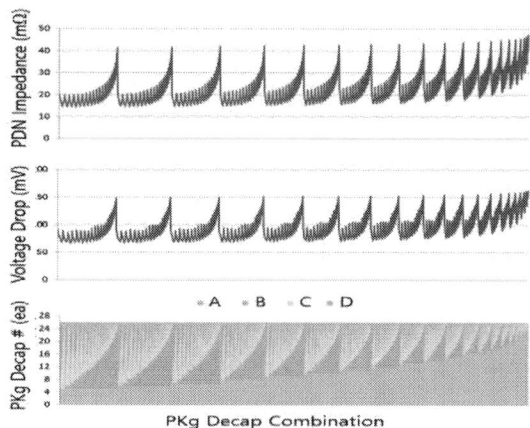

Figure 16. The PDN impedance and voltage drop for vario combinations of decoupling capacitors on package.

By using the simplified schematic of a system-level PDN and its estimated initial parameters, the optimum combination of decoupling capacitors on package can be extracted by the optimization of system-level PDN at early design stage. Fig. 16 shows the PDN impedance peak and voltage drops for various combinations of 26 package decoupling capacitors with 4 different values. As shown in Fig. 16, the PDN impedance peak and voltage drop have large variations due to the combinations of decoupling capacitors, so it is very important to perform the what-if analysis of system PDN at early design stage to estimate minimum system-level voltage drop performance when the optimum number and combination of decoupling capacitors are used.

IV. CONCLUSIONS

In summary, PI and SI design considerations on enabling HBM2e IP and Core IPs for HPC platform applications were suggested and based on extensive simulation analysis, HBM2e IP up to 3.2Gbps and core IPs with high power consumption were successfully operationalised. In conclusion, the PI/SI design for accommodating HPC platform requires a comprehensive strategy on working with composing HBM IP and core IPs on a compact package form and this work addressed all these challenges successfully by providing an efficient methodology to achieve the required goals.

ACKNOWLEDGEMENTS

The authors would like to thank Sumant Srikant and Jungil Son for their analysis support and technical discussion.

REFERENCES

[1] X.-W. Chen and X. Lin, "Big data learning: challenges and perspectives" *IEEE Access*, vol. 2, pp. 514-525, May 2014.
V. Kindratenko, P. Trancoso, "Trends in high-performance computing," IEEE Computing in Science & Engineering ,vol. 13, no. 3, pp. 92-95, Mar. 2011.

[2] A. Batra and A. Kumar, "High performance computing into cloud computing services," IEEE International Conference on Computing Sciences, pp. 172-175, 2012.

[3] D. Espino, G. Ares, M. Pedemonte, and P. Ezzatti, "Overview of HPC benchmarks in hybrid hardware platforms (CPUs+GPUs)," 2016 XLII Latin American Computing Conference, pp. 1-10, 2016.

978-1-7281-1500-9/19 $31.00 © 2019 IEEE

[4] M. Swaminathan, J. Kim, I. Novak, and J. P. Libous, "Power distribution networks for system-on-package: status and challenges," IEEE Transactions on Advanced Packaging, vol. 27, no. 2, pp. 286–300, 2004.

[5] Bo Pu, Jun So Pak, Chanmin Jo and Sungwook Moon, "Design of 2.5D Interposer in High Bandwidth Memory and Through Silicon Via for High Speed Signal", UBM DesignCon Conference, Santa Clara, CA, USA, Jan. 29-31, 2019.

[6] Standard, JEDEC (2013), High bandwidth memory (hbm) dram. JESD235.

2019 IEEE 69th Electronic Components and Technology Conference (ECTC)

PoP LPDDR5 (6.4 Gbps) NTODT and 1-tap DFE for Signal Integrity Enhancement

Sunil Gupta
Qualcomm Technologies, Inc.
5775 Morehouse Drive, San Diego, CA - 92121
sungupta@qti.qualcomm.com

Abstract— **LPDDR5 SI (Signal Integrity) enhancements are presented by using non-target DRAM termination and 1-tap DFE. LPDDR5 interface was running at 6.4 Gbps data rate, 0.5V VDDQ (TT) and VOH= ~300mV (WRITES). Mobile SoC-DRAM system in PoP (Package-on-Package) configuration was analyzed. Non-target DRAM termination in a dual-rank system mitigates the reflections coming to the target DRAM leading to improved SI, ~7% UI improvement was observed. 1-tap DFE (Decision Feedback Equalizer) is also employed to reduce the ISI (Inter-Symbol Interference) due to reflections, ~2ps UI improvement was observed. SI components (crosstalk and ISI) and VdIVW show that VdIVW having big impact on the system timing. Additionally, decreasing CIO and VdIVW show significant improvement in eye-apertures.**

Keywords—LPDDR5, Signal Integrity, crosstalk, ISI, Non-Target ODT, NTODT, DFE, PoP, SoC, Termination, Reflections, DRAM, VdIVW, TdIVW.

I. INTRODUCTION

This paper presents LPDDR5 (Low Power Double Data Rate 5) interface SI analysis. LPDDR5 is the latest standard in the evolution of the SoC-DRAM data transactions. It has maximum data transfer rate of 6.4 Gbps which is 1.5 times faster than the previous standard of LPDDR4X running at 4.266 Gbps. In order to enhance the SI of the LPDDR5 signals, two new features are employed, namely – 1) NTODT (Non-Target On-Die Termination) in a dual-rank system and 2) The use of 1-tap DFE. Both features help with mitigating the reflections. Detailed analysis of these two features is presented.

Background information on LPDDR5 basics is covered in section II. LPDDR is a parallel interface with single-ended data lanes and differential clock/strobe lanes. The primary SI degradation factors being reflections, ISI and single-ended data-lanes crosstalk. The two main SoC-DRAM configurations of PoP and External DRAM are described.

Section III covers the SI analysis framework, schematics of unterminated and terminated NTODT cases and the 1-tap DFE block diagram. It explains the LVSTL TX driver and the construction of 1-tap DFE. It also covers the SI/PI system timing budget components and the JEDEC standard parameters of the eye-aperture.

Section IV presents the results of the NTODT and 1-tap DFE impact on SI improvements. It also covers the impact of DRAM CIO (Input/Output Capacitance) and VdIVW eye-mask on eye-apertures. In-addition, SI components breakdown

into crosstalk, ISI and VdIVW_impact is presented. Finally, the key takeaways of LPDDR5 SI improvements are summarized in section V.

II. BACKGROUND

A. LPDDR5 PoP and External DRAM Configurations

Fig. 1a shows the PoP (Package-on-Package) block diagram where the SoC sits on the PCB and the DRAM sits on top of the SoC. PoP configuration is typically found in high-end smartphones. In this paper this configuration is analyzed.

Fig. 1b shows the external SoC-DRAM block diagram where the SoC and DRAM sit side-by-side on the PCB. This type of configuration is typically found in tablets and automotive systems.

Fig. 1a. SoC-DRAM PoP   Fig. 1b. SoC-DRAM External

B. LPDDR5 x16 and SoC with 4*x16 channels

Fig. 2 shows the schematic of x16 channel with two x8 data buses with CA (Command Address) bus in the middle.

Fig. 3 shows a SoC with 4*x16 channels on the periphery. Each x16 channel of the SoC is connected to a single/dual-rank DRAM. Generally, in a high-end system dual-rank DRAM is employed to get higher capacity.

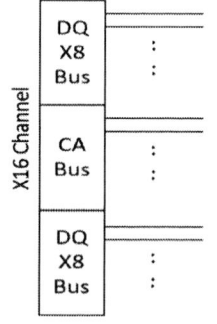
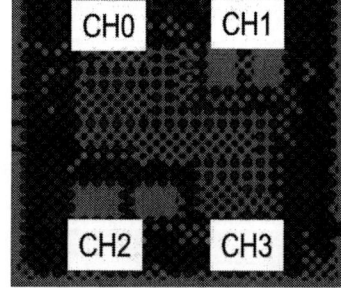

Fig. 2. x16 Channel   Fig. 3. SoC with 4*x16 channels

C. SoC-DRAM PoP System

Fig. 4 illustrates SoC-DRAM PoP system in detail. The package substrate with BGA attaches to PCB. The DRAM is

978-1-7281-1500-9/19 $31.00 © 2019 IEEE

attached to the SoC through the interposer. The SI (Signal Integrity) channel consists of SoC die and package, the interposer and the DRAM package and die. It does not go through the PCB whereas the PDN (Power Delivery Network) includes the PCB.

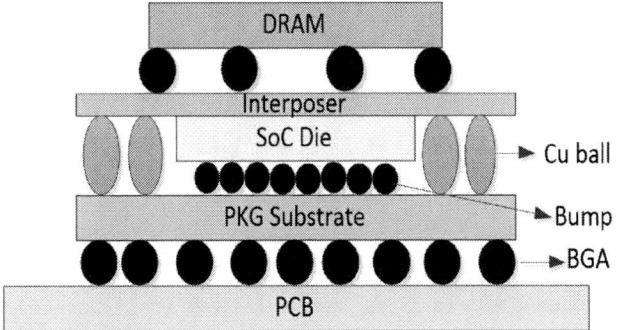

Fig. 4. Package-on-Package System.

III. SIGNAL INTEGRITY ANALYSIS FRAMEWORK

The signal integrity analysis was performed using the mobile SoC TX, SoC package, DRAM package and the DRAM RX. Writes from SoC to DRAM were analyzed to see the impact of NTODT and 1-tap DFE on SI eye-apertures.

A. SoC-DRAM Untermianted and Termianted NTODT DRAM

Two SoC-DRAM configurations in dual-rank system were analyzed. Fig. 5 illustrates the schematic of the target DRAM0 with unterminated NTODT DRAM1 as shown by "OFF" label. In this configuration, the reflections from the Non-Target DRAM1 degrade the signal and the eye-aperture at the target DRAM0.

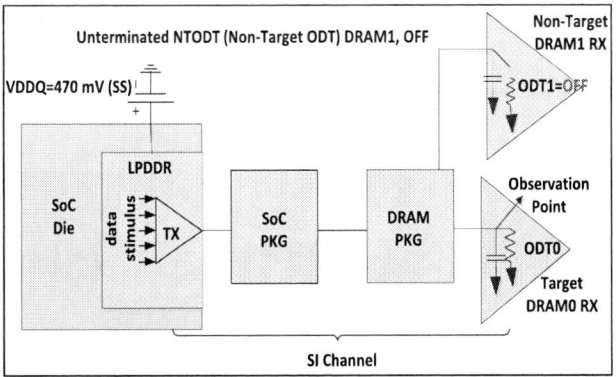

Fig. 5 Target DRAM0 with *Unterminated* NTODT DRAM1.

Fig. 6 illustrates the schematic of the target DRAM0 with terminated NTODT DRAM1 as shown by "ON" label. In this configuration, the reflections from the Non-target DRAM1 are mitigated by the presence of the on-die termination which leads to improved eye-aperture at the target DRAM0.

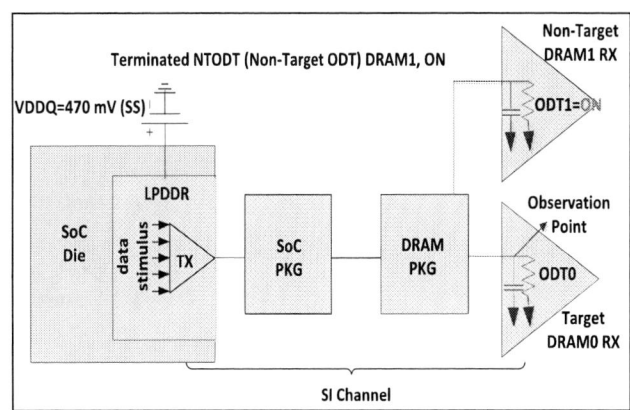

Fig. 6 Target DRAM0 with *Terminated* NTODT DRAM1.

B. SI Analysis Framework

The SI analysis framework consisted of the following.

1. Fig. 5 and Fig. 6 show the LPDDR5 SoC-DRAM systems analyzed.

2. VDDQ supply rail was at 470mV at SS (slow-slow) corner. VDDQ = 500mV at TT (Typ.-Typ.) corner.

3. VOH level (steady state value) of the received signal at the DRAM was ~280mV (0.6*VDDQ, SS).

4. WRITE traffic (SoC → DRAM) on 4*x16 channels was used with all channels being excited.

5. Dual-Rank DRAM in PoP configuration was used.

6. LPDDR5 interface was running at 6400 Mbps.

7. Data pattern for TX stimulus consists of PRBS (Pseudo Random Bit Sequence), SSO (Simultaneous Switching Output) and Victim/Aggressor sequences. This pattern represents a worst-case scenario.

8. IO IBIS models for SoC TX drivers were used.

9. HSPICE simulator was used for running the simulations.

10. DRAM RX load, during WRITES, consists of ODT in parallel with CIO.

11. SI effects of crosstalk, ISI and reflections, among others were included in the analysis.

12. PI (Power Integrity) effects were not considered since ideal supply was driving the IO circuits.

C. SoC LVSTL TX and DRAM RX Effective ODT

SoC employs LVSTL (Low Voltage Swing Terminated Logic) TX driver. The effective DRAM ODT is the ODT0 ∥ ODT1. Fig. 7 shows the SoC TX (PU) pull-up and (PD) pull-down segments and the DRAM loads. The channel consists of SoC and DRAM packages.

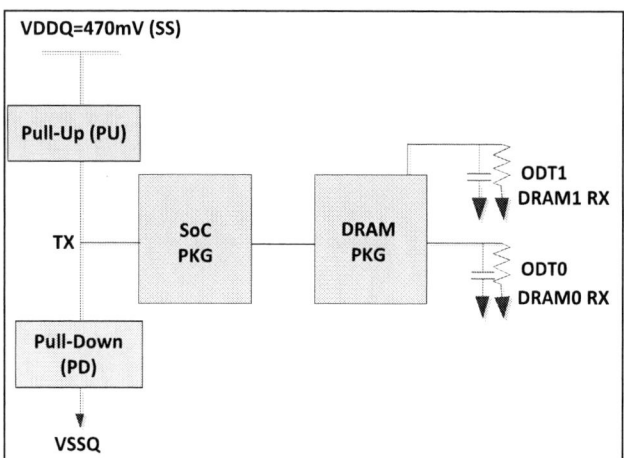

Fig. 7. LVSTL TX and DRAM ODT.

The TX PU code is matched to the RX ODT code in order to get JEDEC standard VOH levels. Table 1 shows the acceptable values of the ODT0 and ODT1 when the PU=60.

Table 1. ODT0 and ODT1, PU=60.

PU	ODT0, ohm (Target)	ODT1, ohm (Non-Target)
60	**60**	**Off (Open)**
	80	240
	120	**120**
	240	80
	Off (Open)	60

The symmetrical values of ODT0 = ODT1 are preferred as non-target DRAM can be in power-down mode for power savings.

If the PU=40, then the ODT0 and ODT1 values accordingly change to give effective ODT=40 ohm. The symmetric values in this scenario would be ODT0 = ODT1 = 80 ohms.

D. 1-tap DFE block diagram

LPDDR5 makes use of DFE to further improve the signal integrity by cancelling ISI due to reflections. Only 1-tap is employed to keep power dissipation in check due to stringent power requirements in mobile system.

CTLE (Continuous Time Linear Equalizer) is not employed since the PoP channel is short and as such the channel loss is minimal and would not justify the added cost of area and power dissipation.

Fig. 8 illustrates the schematic of the 1-tap DFE. X_i is the input to the DFE and Y_i is the output of DFE. The weighting factor 'W' is determined during the training process.

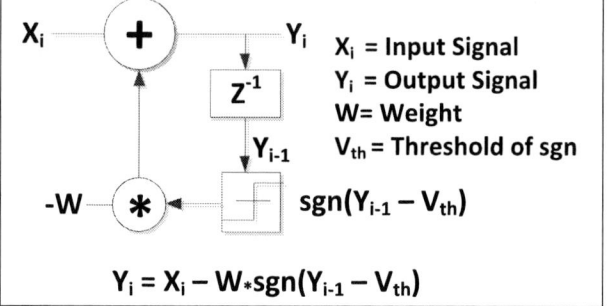

Fig. 8. 1-tap DFE.

E. SI/PI System Timing Budget and FOM (Figure-of-Merit)

The LPDDR SoC-DRAM link timing budget is divided into three main buckets, namely 1) SoC timing budget. 2) DRAM timing budget governed by JEDEC standard parameter TdIVW (Time of data Input Valid Window) and 3) SIPI timing budget. Fig. 9 illustrates the system timing budget in one UI (Unit Interval). Any time left over is the margin of the eye.

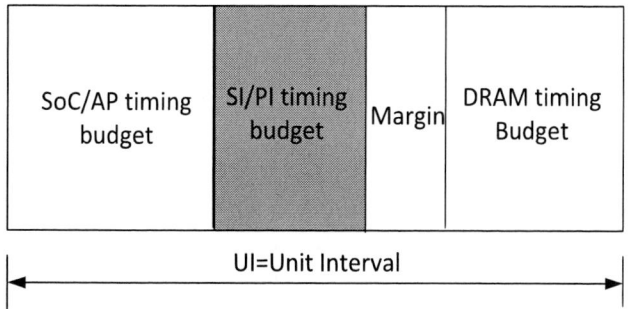

Fig. 9. System Timing Budget.

The SI/PI FOM is the eye-aperture with eye-mask window VdIVW = +/- 50mV from Vref/Vcent as defined by JEDEC standard. TdIVW2 is the DRAM RX timing window at VdIVW levels. TdIVW1 is the DRAM RX timing window at Vref. Fig. 10 shows the LPDDR5 one UI of data-eye with JEDEC standard TdIVW1, TdIVW2 and VdIVW parameters.

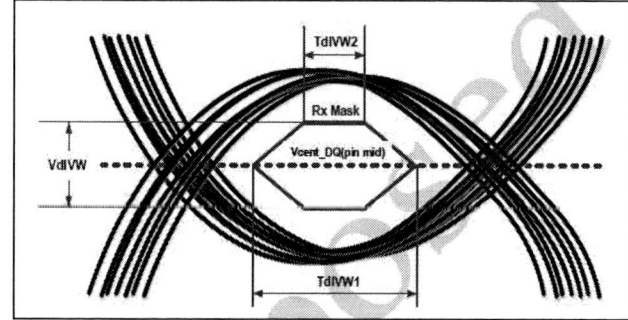

Fig. 10. JEDEC LPDDR5 RX eye-mask.

IV. RESULTS

This section presents the results of NTODT and 1-tap DFE impact on SI. These new capabilities in LPDDR5 greatly help with SI enhancements. In addition, effects of VdIVW and CIO were quantified on eye-apertures. Furthermore, SI components were isolated and quantified in terms of % of UI.

A. NTODT Impact on Signal Integrity

Results of NTODT impact are presented by comparing the pulse responses and eye-aperture comparisons. NTODT greatly helps when there is RDL (Re-Distribution Layer) trace in DRAM dies.

a. NTODT Unterminated vs Terminated Pulse Responses

Fig. 11 illustrates the pulse responses. The purple pulse is applied at the input of the SoC TX. Red pulse is the ideal RX response at DRAM0 when no reflections are coming from non-target DRAM.

Light blue pulse, case a, is the response at DRAM0 when non-target DRAM1 is unterminated and target DRAM0 is terminated by ODT.

Yellow pulse, case b, is the response at DRAM0 when both non-target and target DRAMs are terminated by 2*ODT. Yellow pulse shows lower reflections with smaller amplitudes in the tail compared to light blue. All waveforms are probed at the DRAM0 RX.

In fig. 11 *only*, effect of RDL was exaggerated to highlight the differences due to NTODT. In actual LPDDR5 system, the differences between case a and case b will be very subtle.

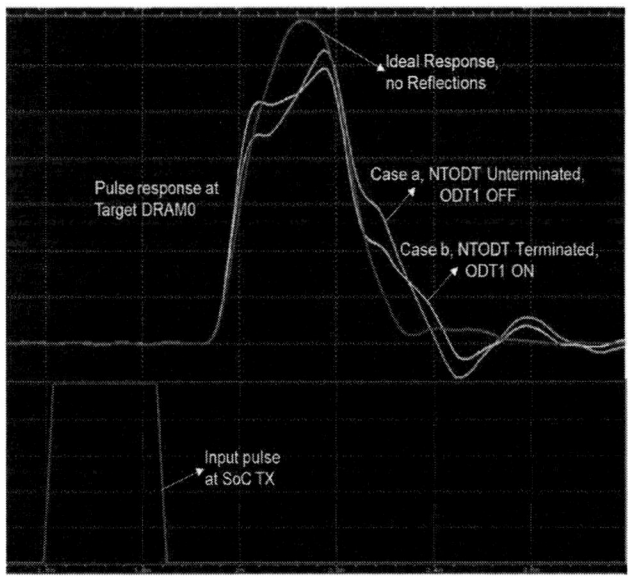

Fig. 11. Pulse response unterminated vs terminated NTODT.

b. Unterminated vs Terminated NTODT Eye Comparisons

Fig. 12, case a, illustrates the eye-diagram of the target DRAM0 bit (dq4) when the NTODT is unterminated (ODT0 = 60 ohm, ODT1 = open/OFF) in LPDDR5 system.

Fig. 12, case b, illustrates the eye-diagram of the same target DRAM0 bit (dq4) when the NTODT is terminated (ODT0 = ODT1 = 120 ohm, ODT1=ON) in LPDDR5 system.

X = eye-aperture when NTODT is unterminated.
Y = eye-aperture when NTODT is terminated.
Y > X.

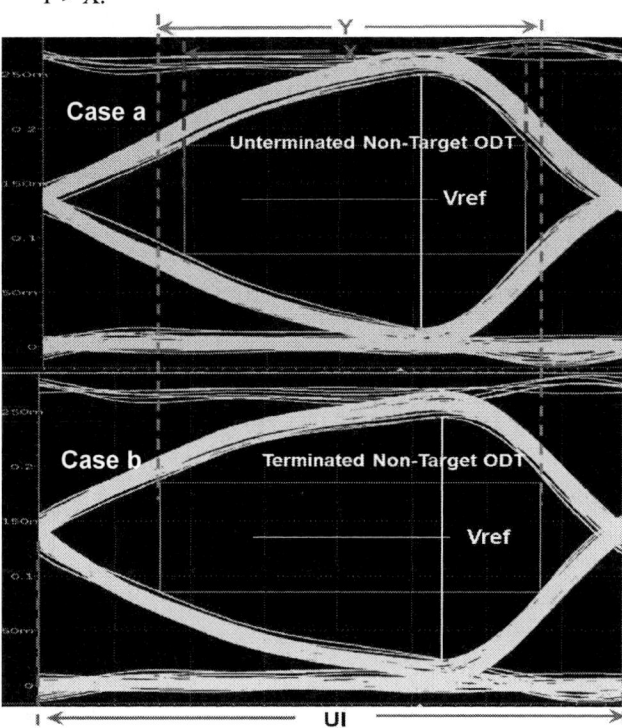

Fig. 12. Eyes of Unterminated vs Terminated NTODT.

c. Unterminated vs Terminated NTODT Eye-Apertures Comparisons of a Byte.

Fig. 13 shows the eye-apertures comparison for the target DRAM0 and the non-target DRAM1 of one byte of data in a dual-rank LPDDR5 system. Eye-apertures of target DRAM0 improve with termination of NTODT DRAM1.

On average, the eye-apertures improve ~7% UI going from ODT0=60 ohm and ODT1=open/OFF to ODT0 = ODT1 = 120 ohm. ODT0 = ODT1 is preferred as non-target DRAM can be in power-down mode.

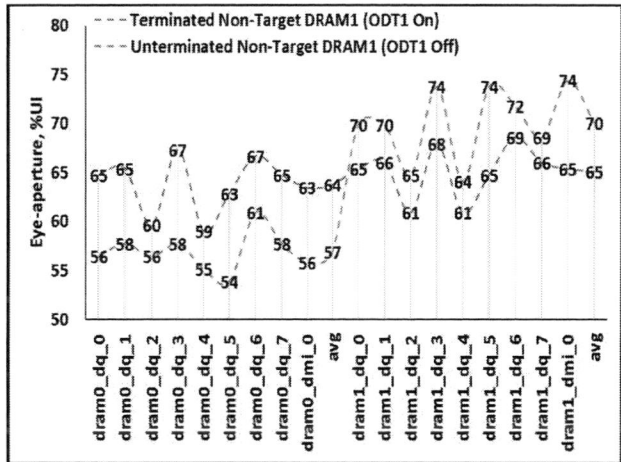

Fig. 13. Eye-apertures comparison of a Byte.

B. 1-tap DFE Impact on Signal Integrtiy

LPDDR5 makes the use of DFE to further improve the signal integrity. Only 1-tap is employed to keep power dissipation in check. Since the channel is short, it is sufficient to cancel the reflection based ISI. Dual-rank PoP SoC-DRAM system with NTODT enabled was investigated.

a. Eye Comparison with and without 1-tap DFE.

Fig. 14 A) shows the eye (green) without DFE with aperture = 108ps and B) shows the eye (yellow) with DFE with aperture = 110ps. Employing 1-tap DFE improves the aperture by about ~2ps (110ps – 108ps). The improvements vary depending upon the value of feedback weighting factor (W). In this case, "W" was adjusted to obtain the best improvement due to 1-tap DFE.

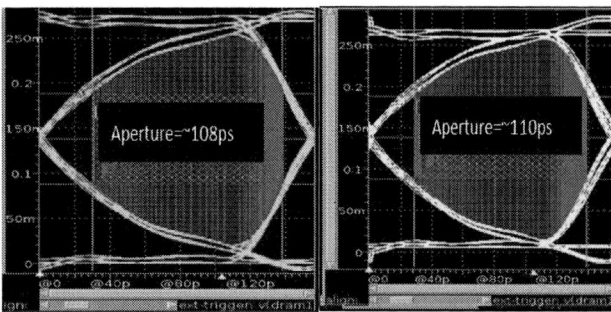

A) Eye without DFE   **B) Eye with DFE, width larger by ~2ps**

Fig. 14. Eye-apertures comparison of a Byte.

b. Time-domain Waveforms with and without 1-tap DFE

Fig. 15 shows the short duration time-domain waveform comparison of with and without 1-tap DFE. The Yellow waveform (with 1-tap DFE) shows sharper edges compared to Green waveform (without 1-tap DFE).

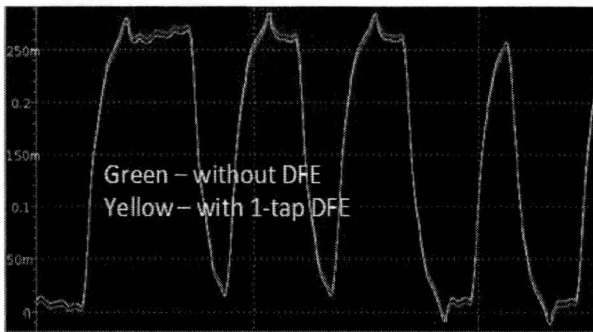

Fig. 15. Eye-apertures comparison of a Byte.

C. VdIVW and CIO Impact on Eye-Apertures

The impact of VdIVW and CIO on eye-aperture was examined in a dual-rank PoP SoC-DRAM system with NTODT enabled. VdIVW being the vertical height of the eye-mask window and CIO being the DRAM bit capacitive load. The values and ranges of VdIVW and CIO are specified by JEDEC standards committee and are lower in LPDDR5 compared to previous LPDDR4X standard.

a. Eye-apertures Increase with Decreasing VdIVW

Decreasing the VdIVW mask window improves the eye-aperture. This improvement is related to the sensitivity of the DRAM receiver and shows about ~5% UI improvement in eye-aperture for every 10 mV decrease in VdIVW mask.

Fig. 16 depicts this impact for VdIVW values varying from +/-50mV to +/-30mV. The graph shows the eye-apertures for one byte of data in dual-rank system.

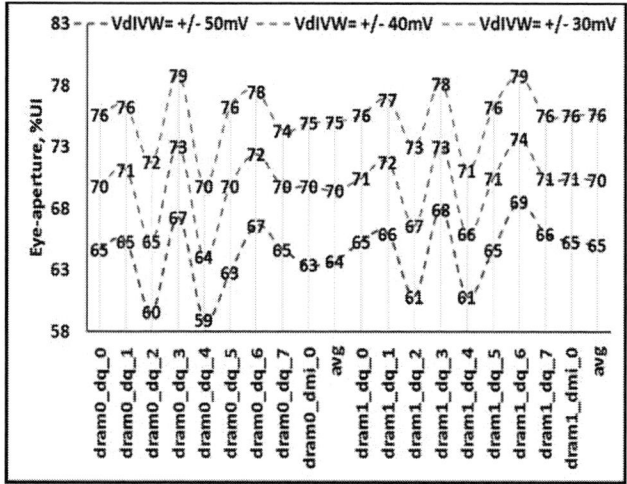

Fig. 16. Eye-apertures increase with decreasing VdIVW.

b. Eye-apertures Increase with Decreasing CIO

Decreasing CIO shows improvement in eye-apertures, since lowered CIO results in shorter charging/discharging time. Analysis results show improvements in eye-aperture of ~2% UI for every 0.1pf decrease in CIO.

Fig. 17 illustrates the impact of decreasing CIO from 0.9pf to 0.7pf on eye-apertures for one byte of data in a dual-rank system.

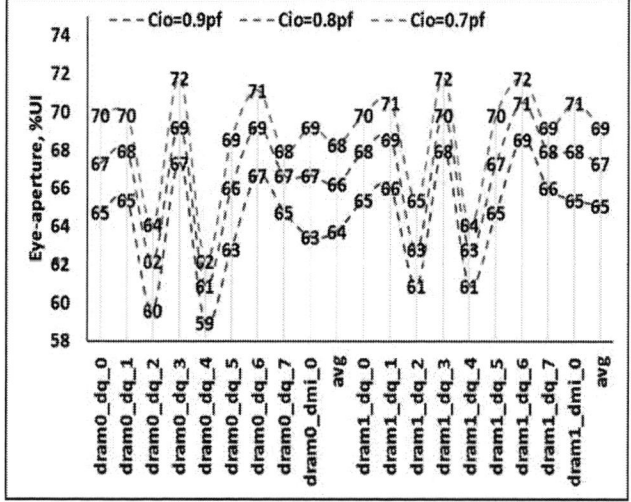

Fig. 17. Eye-apertures increase with decreasing CIO.

D. SI Components Breakdown and VdIVW Impact

The SI components and VdIVW impact of the dual-rank PoP SoC-DRAM were investigated with NTODT enabled. Fig. 18 shows the contribution of these elements in the eye-aperture, of one bit, as summarized below.

1. Crosstalk = ~3% (69.75% - 66.28%) of UI.

2. ISI = ~2% (72% - 69.75%) of UI.

3. VdIVW_impact = ~28% (99.6% - 72%) of UI.

VdIVW eye-mask encompasses DQ offset, Vref noise and signal noise. Thus, lowering VdIVW has a big impact on overall timing. In this analysis VdIVW=+/-50mV from Vref was used.

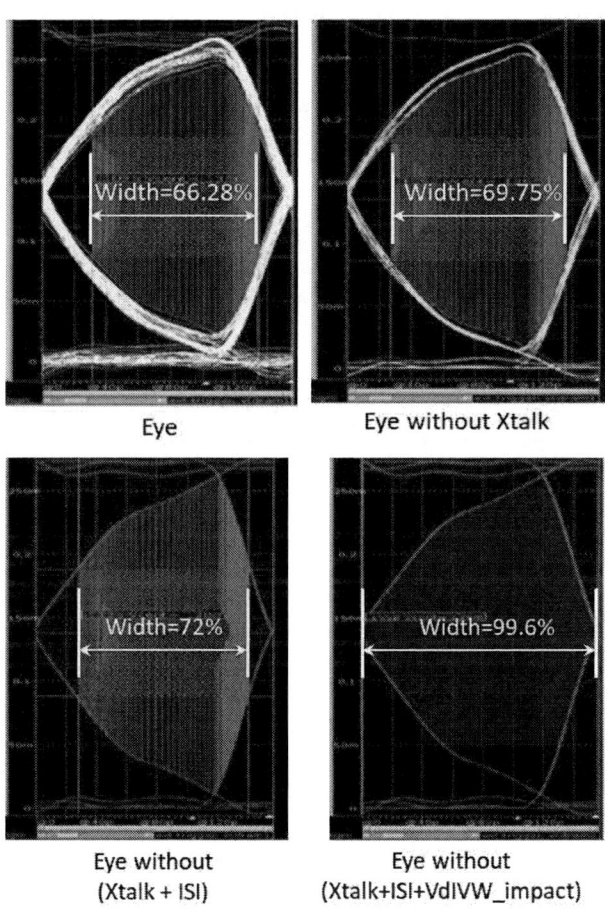

Fig. 18. SI components breakdown of eye-aperture.

These values correspond to the particular system analyzed and would be somewhat different for another system but at the same time it gives insights into the relative contributions of different components.

V. CONCLUSIONS

The key takeaways of this paper are as follows.

1. LPDDR5 running at 6.4 Gbps presents SI challenges due to decreasing unit interval and timing margins.

2. SI improvements were obtained on target/active DRAM by employing NTODT on passive DRAM to mitigate the reflections. ~7% UI improvement was observed in eye-apertures.

3. SI improvements were obtained by employing 1-tap DFE to further mitigate reflections. ~2ps improvement was observed.

4. Eye-aperture improvements were observed with decreasing values of VdIVW and CIO.

5. SI components (crosstalk and ISI) and VdIVW show that VdIVW having big impact on the system timing.

6. Only Writes (SoC → DRAM) data analysis was presented in this paper. Reads (DRAM → SoC) data analysis could also be performed like Writes and would benefit due to NTODT and DFE equalization features.

ACKNOWLEDGMENT

The author thanks LPDDR, IO, PCB and PKG groups for their assistance.

REFERENCES

[1] JEDEC Standard, LPDDR5, February 2019.

[2] E. Bogatin, Essentail principles of signal integrity, 2014.

[3] S. Gupta, SoC On-die Decap Optimization for PDN in LPDDR, EPEPS, San Jose, CA, October 2018.

[4] S. Gupta, LPDDR4X (3732 Mbps) DBI impact on SI/PI and Power, EPEPS, San Jose, CA, October 2017.

[5] TN_4003_DDR4_network_design_guide, Micron Technical Note, 2014.

[6] Understanding and characterizaing timing jitter, Tektronix, 2012

2019 IEEE 69th Electronic Components and Technology Conference (ECTC)

OpenCAPI Memory Interface signal integrity study for high-speed DDR5 Differential DIMM channel with standard loss FR-4 material and SNIA SFF-TA-1002 connector

Biao Cai[1], Jose Hejase[2], Kyle Giesen[2], Junyan Tang[2], Brian Connolly[2], KyuHyoun Kim[3], Daniel Dreps[2]
IBM Corporation
SCE[1], SG[2] and research[3]
Poughkeepsie United States
biaocai@us.ibm.com

Zhineng Fan, Rocky Huang, Luyun Yi, Qiaoli Chen, Yifan Huang, and Stephen Smith
Amphenol ICC
Research and development
Nashua, United States
zfan@amphenolacp.com

Abstract—**DDR5 Differential DIMM (DDIMM) is being defined in JEDEC and will be introduced to the market in 2020. DDR5 DDIMM uses OMI (OpenCAPI Memory Interface) as the host interface. On the DDIMM printed circuit board (PCB), the minimum data transfer rate per data differential pair over the OMI bus is 25.6Gbps. This is a significant data rate increase for DRAM modules over conventional single-ended data transferring DIMMs. For example, the DDR5 LRDIMM data transfer rate per pin is 3.2Gbps. Careful attention must be paid to the bill of materials of the DDIMM to control its cost towards general market acceptance of this new DIMM technology. As a result, it is desired to use standard loss FR-4 material to build the DDIMM PCB. Validating the DDIMM PCB wiring for the high-speed differential memory bus requires accurate high-speed link simulations. These simulations require accurate models representing differential wiring in the DDIMM PCB stack-up. The models must be built using not only representative physical dimensions but also accurate frequency dependent material properties obtained through PCB characterization. The Short Pulse Propagation (SPP) method will be used to extract PCB frequency dependent material properties. PCB suppliers usually have different Copper Clad Laminate (CCL) and prepreg material set selections largely due to the supplier relations. The test coupons of this study will be built using CCL and prepreg materials from two suppliers. System level differential memory bus simulation based on SPP characterization will be performed and the simulation results from different material/stack-up designs will be benchmarked.**

DDR5 DDIMM will be paired with the Storage Networking Industry Association (SNIA) SFF-TA-1002 high speed connector which differs significantly from the JEDEC RDIMM connector for improved electrical signaling characteristics. In this work, the SNIA high-speed connector to PCB interface design will be studied as it is of utmost importance for achieving good signal integrity. In addition to high speed signal integrity, the DDIMM PCB mechanical interaction with the SFF-TA-1002 connector is studied and it will be highlighted in this paper. The PCB mechanical outline design proposal is made to mitigate module/connector mechanical interference. To understand the PCB yield impact, DIMM PCB suppliers' process capabilities for critical feature/dimensions will be studied as well.

Keywords-CAPI (Coherent Accelerator Processor Interface), OMI (OpenCAPI Memory Interface); DDR5(Double Data Rate 5th-generation), DDIMM (Differential Dual In-line Memory Module), PCB (Printed Circuit Board), SNIA (Storage

Networking Industry Association), SNIA SFF-TA-1002 connector, signal integrity, mechanical interference

I. INTRODUCTION

IBM Power 8/9 memory buffer uses high speed proprietary differential memory interface (DMI) to communicate with processor. OpenCAPI memory interface (OMI) is an improved and open memory interface standard. JEDEC is defining the standard of DDR5 Differential DIMM (DDIMM) – an OpenCAPI interface buffered DIMM, where the memory buffer uses OMI, an open/agnostic protocol, to communicate with the processor. The Differential DIMM gets its name from the differential bus of the OMI channel. DDIMM brings benefits of lower latency, more bandwidth and fewer pins then DDR [1,2]. DDIMM will use high speed SNIA TA-1002 connector which is common to Gen-Z[1].

DDIMM is expected to be the low cost industry standard memory module[2]. The bill of material (BOM) cost reduction or avoidance opportunity, including Printed Circuit Board (PCB), should be explored during the early development stage. This paper studies various DDIMM PCB stack-up designs with the target to provide adequate OMI channel bandwidth to support 25.6Gbps data rate, which is expected to be challenging. To avoid the laminate material cost premium, standard FR-4 materials for JEDEC registered and load-reduced DIMM PCB are used as the baseline.

In addition to high speed signal integrity, this paper studies the DDIMM PCB mechanical interaction with the SNIA TA-1002 connector[1], which is significantly different than the conventional JEDEC DIMM connector.

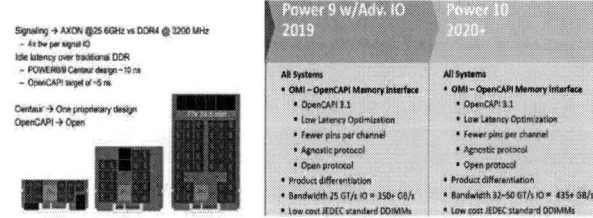

Figure 1. DDIMM and OMI features [1][2]

978-1-7281-1500-9/19 $31.00 © 2019 IEEE

II. DDIMM PCB MATERIAL AND STACK-UP

The conventional U/R/LR (Unbuffered/Registered/Load Reduced) DIMM PCB designs typically use tier 2 standard FR-4 material as shown in figure 2[3]. A low Dk/Df laminate is desired for high speed applications but the material cost is higher than that of the standard FR-4[4].

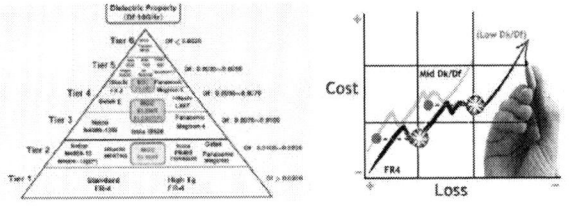

Figure 2. FR-4 laminate material cost increase with lowing Dk/Df [3][4]

A. Stack-ups for this study

To avoid a material cost premium, the laminate material selection for the DDIMM PCB starts from tier 2 of the pyramid shown in figure 2. The 1U DDIMM[5] PCB stack-up could be a 14 layer design and the differential OMI strip lines could be routed on layer 3 and 12 as shown in figure 3. To avoid a low volume custom manufacturing process cost premium, the physical design guidelines of the stack-up and OMI strip line are selected within the boundary of the High Volume Manufacturing (HVM) of the PCB suppliers.

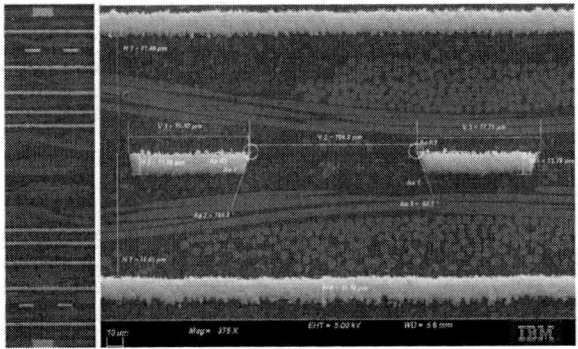

Figure 3. Typical 1U DDIMM PCB stack-up and OMI strip line of this study

Two primary PCB suppliers (A, B) for JEDEC DDR R/LR DIMM were selected for this study. Each PCB supplier has two stack-up designs under study. The laminate material of those stack-ups are from two major material suppliers (X,Y). Three of the four stack-ups are a standard tier 2 FR-4 material and one stack-up is hybrid using tier 3 materials for L3/12 OMI strip line dielectrics. Table 1 summarizes the stack-ups of this study, which all have satisfied the DDIMM module supplier's electrical requirement for clock and DDR nets. The hybrid stack-up has about a 10% price premium and is expected to have lower Dk/Df than the other stack-ups.

TABLE I. STACK-UPS FOR THIS STUDY

PCB mfger	material mfger	Laminate	Datasheet dielectric parameter @ 1GHz Resin content: 50%		STACKUP ID
			Dk	Df	
A	X	hybrid	3.4(4.3)	0.009(0.013)	Hybrid_A
	X	std FR4	4.3	0.013	std_AX
B	X	std FR4	4.3	0.011	std_BX
	Y	std FR4	4.4	0.014	std_BY

B. Reference stack-up

The reference for this study was selected from stack-ups of IBM production experience with tier 2 FR-4 material (table II). The simulation of the OMI channel topology (shown in figure 9) with the DDIMM PCB of this reference stack-up indicates adequate bandwidth to support a 25.6Gbps data rate (Detail will be covered in section IV of this paper).

III. SHORT PULSE PROPOGATION (SPP) TEST AND BROADBAND PERMITIVITY EXTRACTION

A. SPP test

This study uses the SPP test to verify the modeling parameters of PCB OMI wiring. The general procedure (figure 4a) of the SPP test can be found in IPC-TM-650 [6], with major steps including TDR, low frequency and TDT. The TDT step is where the test gets its name: short pulse is created by differentiating a step function and propagates through the transmission line pair under test. In this study, the setup shown in figure 5 launched a short pulse to the 3 cm and 10 cm Differential strip lines on the SPP coupon as shown in figure 4b. Please note IPC is in the process of revising the PCB signal loss test method and this study is following the current version.

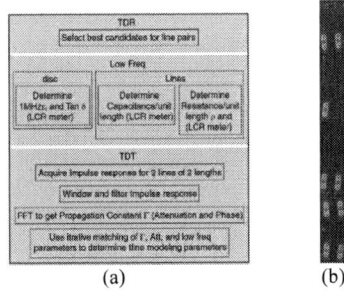

Figure 4. SPP test flow and coupon: (a) the general procedure of the SPP test (b) the SPP coupon for this study

After the signal processing of the captured short pulses to eliminate end effort, Fourier transform was performed to create a frequency domain response: for short strip line of length l_1 [$A_1(f)$, $\phi_1(f)$] and for long strip line of length l_2 [$A_2(f)$, $\phi_1(f)$]. The attenuation $Att(f)$ and phase constant $\beta(f)$ are computed, respectively, from a real part and an imaginary part of the propagation constant $\Gamma(f)$, as shown in formula [5-1] and [5-2].

$$\Gamma(f) = \alpha(f) + \beta(f) =$$
$$-\frac{1}{l1 - l2}\ln\left(\frac{A1(f)}{A2(f)}\right) + j\frac{\phi1(f) - \phi2(f)}{l1 - l2} \quad [5-1]$$

$$Att(f) = 20\log\left(e^{Re(\Gamma(f))}\right)$$
$$\beta(f) = Im(\Gamma(f)) \quad [5-2]$$

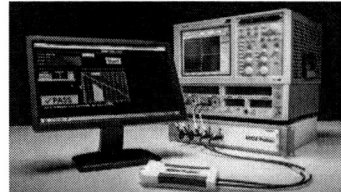

Figure 5.   Test setup to launch short pulse

The *Att(f)* measurements are plotted in figure 6. In the future, *Att(12GHz)* could be used to setup screen criteria for DDIMM PCB stack-up selection since it is close to the Nyquist frequency of the channel data rate.

The attenuation curve of the three standard FR-4 stack-ups are mostly overlapped and show more loss than the reference. By comparison, the hybrid stack-up has shown less loss then the reference. The measured effective Dk values at 20GHz are summarized in table II. The measured effective Dk value of the hybrid stack-up is very close to the reference while the three standard stack-ups have higher measured values than the reference.

Based on the SPP measurements, the hybrid stack-up has lower attenuation and similar effective Dk as compared to the reference. A DDIMM PCB with this stack-up is expected to have similar bandwidth as the reference and capable of supporting a 25.6Gbps data rate assuming everything else is equal. Since three standard FR-4 stack-ups have higher Dk and attenuation values than the reference, additional study is required to estimate whether the channel using DDIMM PCB of those stack-ups is capable to support a 25.6Gbps data rate.

TABLE II.   Dk EXTRACTION BASED ON SPP MEASUREMENT

STACKUP ID	PCB mfger	material mfger	Laminate	Measured Dk@20GHz
Hybrid_A	A	X	hybrid	4.0
std_AX	A	X	std FR4	4.3
std_BX	B	X	std FR4	4.4
std_BY	B	Y	std FR4	4.4
reference	C	Z	std FR4	3.9

B. Broadband extrapolation for complex permitivity

In this study, broadband extrapolation is performed by iterative modeling based on causally-enforced field solver (CZ2D)[7]. Multiple iterations are performed until the *att(f)* converged to measured values within specified tolerances for the entire frequency range. R(f), L(f), C(f), and G(f) per unit length values based on the actual cross sectional dimensions were calculated by CZ2D, as shown in formula [5-3]. Iteration enforces that the calculated $\Gamma(f)$ agrees with the measurement shown in figure 6. The complex permittivity was calculated by CZ2D as well, as shown in formula [5-4].

$$\Gamma(f) = \sqrt{(R + j\omega L)(G + j\omega C)} \quad [5-3]$$
$$\varepsilon_r(\omega) = \left(\frac{C(\omega)}{C1MHz}\right) \times \varepsilon_r1MHz$$
$$tan\delta(\omega) = \frac{\varepsilon_i(\omega)}{\varepsilon_r(\omega)} = \frac{G(\omega)}{\omega C(\omega)} \quad [5-4]$$

Due to the similarity of the measured Γ, broadband extrapolation was performed on one of the three standard FR-4 stack-ups – "*std_AX*" (figure 6). The calculated *Att(f)* values were compared with the measured values (figure 7). Broadband complex permittivity values are plotted in figure 8. The CZ2D calculated broadband permittivity matches with the SPP measurement within allowed tolerance. It is worthy to note the measured impedance value of the *std_AX* stack-up impedance coupons is average around 77 ohms as compared to 85 ohms target. Even though the coupons marginally comply to the manufacturing impedance control tolerance but limit this study to almost the low impedance corner. This limitation is planned to be addressed by follow up study.

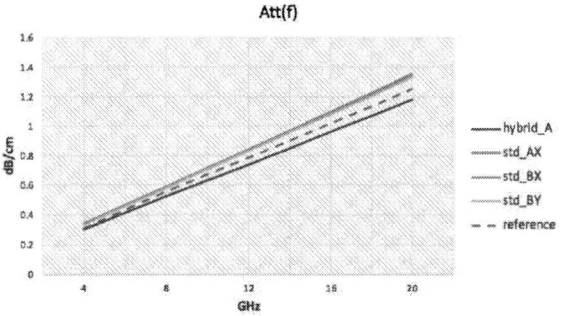

Figure 6.   Attenuation extraction based on SPP measurement

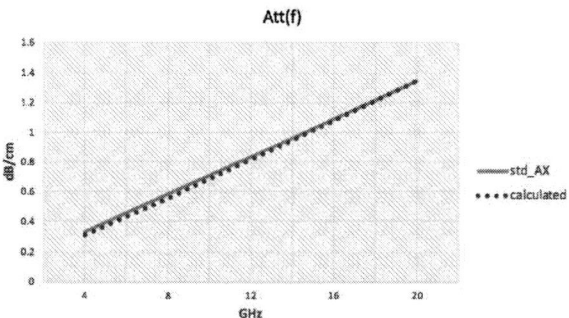

Figure 7.   Stack-up *std-AX* calculated *Att(f)* values comparing with measured values

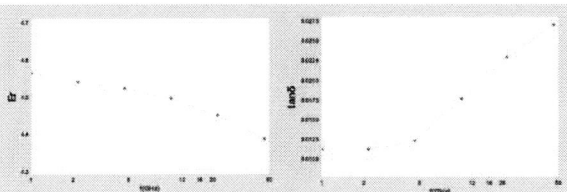

Figure 8. stack-up std-AX broadband complex permittivity

IV. OMI CHANNEL SIMULATION SETUP AND PROCEDURE

A. OMI channel topology and modeling

An OMI end-to-end channel planned to run at 25.6Gbps enabling data transmission between the CPU package and the memory buffer package was simulated in time domain (using IBM's HSSCDR high speed channel simulation tool) to show the channel performance under the different DDIMM PCB laminate material assumptions: SPP test material (table II *std_AX* stack-up) and reference material (table II reference stack-up). Prior to carrying out the simulations, channel s-parameters were generated by EM modelling of the different channel signaling sections.

In the communication channel, the signal travels between the chips on the CPU and memory buffer packages going through a combinations of vias, connectors, PCB differential trace wiring including the wiring in the pin area fields of connectors and package BGAs. Each component in the channel needs to be designed and optimized so that the signal integrity of the channel supports the bit error rate (BER) performance requirements of the communication bus. Figure 9 shows a sketch of the communication bus channel with its main components labelled.

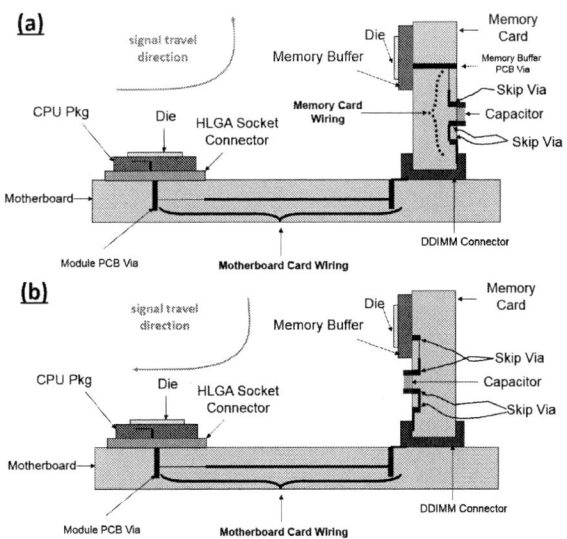

Figure 9. Sketch of the 25.6Gb/s OpenCAPI Memory Interface (OMI) bus channel: (a) CPU transmitting direction (b) CPU receiving direction.

The Hybrid Land Grid Array (HLGA) connector is used to link the CPU package with the motherboard. It maintains a two signal pin to one ground pin ratio for the OMI nets. A CPU package PCB via reflecting the same signal to ground ratio as the HLGA takes the signal from the surface wiring layer to an internal wiring layer. The active via length of the PCB via is 1.91mm. The HLGA and via signal/ground distribution was carefully planned to achieve good crosstalk isolation, ground referencing abundance and symmetry in order to maintain good signal integrity.

The dielectric material where the OMI bus resides in the motherboard is assumed to be an ultra-low loss material having a nominal loss of 0.754 dB/inch at 12.8GHz, Dk=3.26 and tanδ=0.003399 at 10GHz. The total length of the mother board wiring is 11.5 inches which includes 1.5 inches of wiring passing through the via field that is under the HLGA connector. This via field wiring is degrading to the signal integrity due to being in close proximity to anti-pad voids, other differential pairs, having to neck down the differential pair in tight wiring regions and being more sensitive to PCB layer mis-registration. This naturally results in impedance mismatch, additional crosstalk noise and worse differential to common mode conversion. The rest of the wiring on the mother board (10 inches) is in the open area wiring region which is designed to be 85 Ohms differential impedance while maintaining low cross-talks level due to good spacing properties.

A PCB via between the motherboard and the DIMM connector is designed to ensure that the differential impedance is close to 85 Ohms at the transition while the maintaining a low crosstalk level between the differential pairs. The active via length of this model also matches with the module PCB via's (1.91mm).

The DIMM connector along with its breakout traces on both the motherboard and the DIMM card are combined as a single model for optimization so that the best signal integrity properties can be achieved. The details and considerations of this modelling are presented and discussed in the following section.

A total of 1.5 inches of wiring using standard loss material properties is included in the DIMM card. Two different materials are considered for the DIMM card modelling: the reference material and the SPP test material (table I: std-AX). Note that both material considerations use slightly different cross-section assumptions. The material properties for the reference and the std-AX SPP test materials are presented in sections II and III. As is seen in the channel topology sketch, DC blocking capacitors are utilized on the surface of the DDIMM PCB. Their associated skip vias and capacitor-to-PCB interface are designed to meet the desired signal integrity properties. A representative model for memory buffer package is utilized at the non-CPU end of the channel. Most vias designed and assumed for the DIMM card are selected to be skip vias in order to capitalize on the fact that they are short and with no stubs thus improving signal integrity of the channel. The only signal plated through hole PCB via utilized is the memory buffer receiving direction via. In anyway, all vias are designed as much as possible to impedance match to 85Ohms while keeping within the limits of manufacturing and wiring constraints.

B. OMI channel DDIMM Connector modeling

This section focuses on the DDIMM connector design modelling as it is one of the most critical parts of the channel. The mother board PCB interface used for the DDIMM connector is shown in figure 10. Differential wiring having 85 Ohm impedance is implemented on the motherboard for the signal traces to connect to the connector signal pads. The direction which the signal trace breaks out from the signal pads is opposite for the two adjacent rows. This implementation not only reduces the row-to-row crosstalk but also creates more wiring space in the PCB layers below the connector. Ground vias and their pads are also tied to the connector ground pads between each differential signal pad group to increase the crosstalk isolation and create good return paths. A rectangular void is created on the first ground layer below the each connector differential signal pad pair to reduce capacitive impedance mismatches.

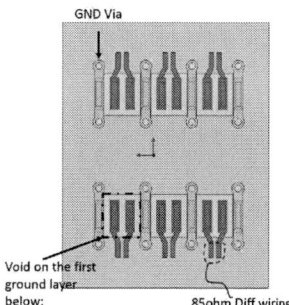

Figure 10.   Top view of the DIMM interface on mother board

The DDIMM PCB connector interface is shown in figure 11 where multiple characteristics have been implemented to improve the SI properties. As can be seen, window voiding is implemented on both sides of the PCB cross-section under all OMI bus differential signal pads except for the central two layers. This was done to get better impedance matching while transitioning from the DDIMM PCB to the connector. The two central layers must not be voided to avoid crosstalk between connector pads on the opposite sides of the DDIMM PCB. If the central two planes are not ground layers; ground islands must be created beneath the connector pads for isolation and return path purposes. Similarly to the DDIMM mother board interface, ground via and pad are used to isolate each differential signal group on the DDIMM PCB contact with the connector. Two ground vias are associated with each of the ground pads. The first ground via is placed at a short breakout distance from the ground connector contact pads to provide a short return path for the high speed signals. The second ground via is introduced into the ground connector contact pad for the purpose of pushing SI resonances away from frequency band of operation and is created due to redundant return paths.

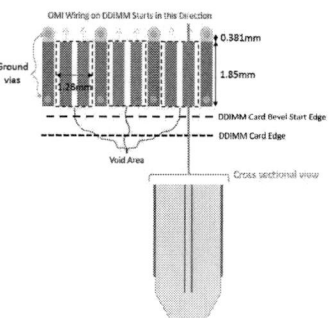

Figure 11.   Top view of the DIMM interface on mother board

V. OMI CHANNEL SIMULATION RESULT, ANALYSIS AND CONCLUSION

OMI channel simulations at 25.6Gbps were carried out in the frequency and time domains to show the channel performance when considering the different proposed material assumptions on the DDIMM PCB. The two sets of materials considered are the reference material (table II reference stack-up) and the SPP test material (table II *std-AX* stack-up).

Figures 12 and 13 show the frequency domain properties of the OMI channels for both signal directions (when the CPU transmits and receives) for the two PCB materials considered on the DDIMM PCB. The insertion loss and the cross talk levels at the operation frequency are also indicated on the figures. The insertion loss levels of both sets of channels are generally similar as are the crosstalk levels between DC and the Nyquist frequency of 12.8GHz. It is worthy to mention though that the reference channel insertion loss curves for CPU transmitting and receiving directions show more deviation than their SPP test material channel counterparts. This is opposite to what would have been expected if both DDIMM wiring traces were 85Ohms, it would have been expected to see more insertion loss deviation with the SPP test material scenario due to the vias having more capacitive impedance mismatch with the higher dielectric constant of the SPP test material. Having a lower impedance than planned with the SPP test material probably contributed to some impedance matching effect which decreased the insertion loss deviation. Whether this impedance matching effect is generalizable for different DDIMM wiring lengths or only specific to this example will be investigated in follow up studies.

Eye simulations were carried out at 25.6 Gbps and eye opening margins were collected at BER=10^{-15}. The eye simulation set up has built-in circuit jitter and noise in addition to having a variety of equalization settings to sweep through in search of the best results in terms of eye opening capabilities. Figures 14 and 15 show the eye diagrams of the channels that were evaluated. The reference channel when the CPU transmits has a horizontal eye opening of 29.1%UI and vertical eye opening of 58mVpk (zero to peak) while the test channel with DDIMM PCB using material properties obtained from the SPP measurements has a horizontal eye of 29.3%UI and vertical eye of 59mVpk. Additionally, the reference channel when the CPU receives has a horizontal eye opening of 23.4%UI and a vertical eye opening of 41mVpk (zero to peak) while the test channel with DDIMM PCB using material properties obtained from the SPP measurements shows a horizontal eye opening of 26.9%UI and vertical eye opening of 51mVpk. There is clearly an improvement seen with the SPP test material as opposed to the reference material. As a result, this confirms the previous observation made about how the SPP test material channels stand in a more favorable position from an insertion loss deviation perspective. The eye opening improvement is more pronounced when the signal level is higher at the memory buffer side of the channel –CPU receiving direction- (ie DDIMM side of the channel). This can be explained by the fact the rise time is still low at the buffer transmitting side, thus becoming more sensitive to the DDIMM wiring and the material assumptions there.

Based on the OMI bus channel eye simulation results, DDIMM PCB with stack-up *std_AX* is slightly better than the reference stack-up. This result supports the argument that tier 2 standard FR-4 material could be used to build DDIMM PCB, and, with careful design for each channel sections (section IV), should be capable to have adequate channel bandwidth to support a 25.6Gbps data rate. But before reaching the conclusion, we should note the limitation of this study that the test coupons are mostly at the low impedance corner (section III) and future study is planned to address this limitation (section VII).

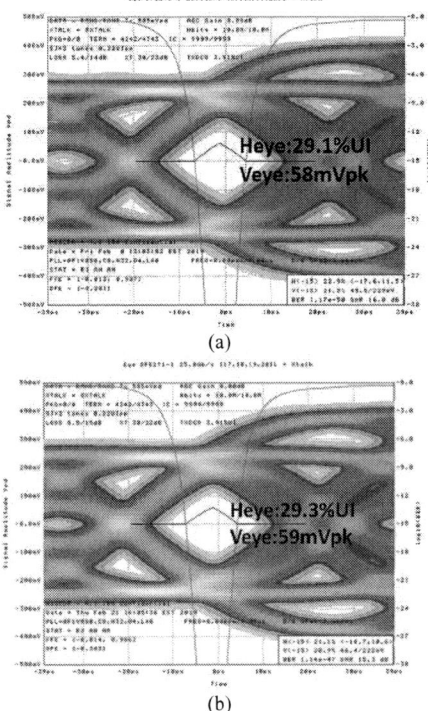

Figure 14. OMI bus channel eye simulation results when the CPU is transmitting: (a) reference case (b) test channel with material properties obtained from SPP measurements (std_AX).

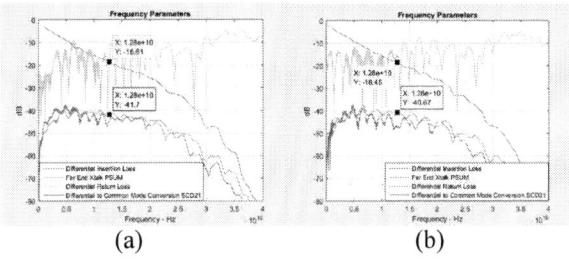

Figure 12. OMI bus channel frequency properties when the CPU is transmitting: (a) reference case (b) test channel with material properties obtained from SPP measurements (std_AX).

Figure 13. OMI bus channel frequency properties when the CPU is receiving: (a) reference case (b) test channel with material properties obtained from SPP measurements (std_AX).

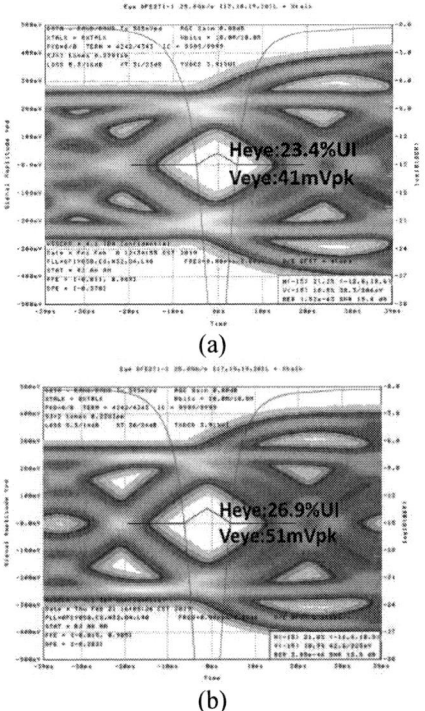

Figure 15. OMI bus channel eye simulation results when the CPU is transmitting: (a) reference case (b) test channel with material properties obtained from SPP measurements (std_AX).

VI. DDIMM PCB AND CONNECTOR MECHANICAL INTERACTION

The high insertion/extraction force of the conventional JEDEC U/R/LR DIMM presents challenge for server/storage system box-line manufacturing. DIMM insertion/extraction induced mechanical damage has been a known culprit of box-line DIMM fallout and even field failure. Starting from DDR4, the step and ramp PCB bottom edge has been specified in the JEDEC Module Outline (MO) [8] to reduce the insertion/extraction force and therefore mitigate the box-line DIMM insertion/extraction induced quality and reliability concerns. The mechanical interface of the DDIMM to the SNIA TA-1002 connector significantly differs from that of DDR U/R/LR DIMM (table III), the interaction study has been carried out to understand the insertion/extraction behavior and impact on contact reliability.

TABLE III. MODULE CONNECTOR MECHANICAL INTERFACE COMPARISION

	number of pin	pitch unit: mm	module bottom edge: step and ramp	module thickness unit: mm	height of alignment key unit: mm	latch: integrated with connector
DDR4 U/R/LR DIMM	288	0.85	Yes	1.4+/-0.1	3.85	Yes
DDR5 U/R/LR DIMM	288	0.85	Yes	1.27+/-0.1	3.85	Yes
DDR5 DDIMM	84	0.6	No	1.57+/-0.13	7	No

A. PCB component keepout and outline tolerance requirement

In general, DIMM has high component density. DDIMM is expected to have Power Management IC (PMIC) on the module [1]. PMICs and associated passives further increase the component density. The component placement challenge is most noticeable with the 1U DDIMM, the component area (gray area in figure 16) specified in the JEDEC DDIMM MO is expected to be completely filled with components.

During module insertion/extraction, the operator may not hold module evenly and tilting is expected. A common failure mode is the component (or latch) damage when the module bottom is caught by the latch, as shown in the upper right box of figure 16.

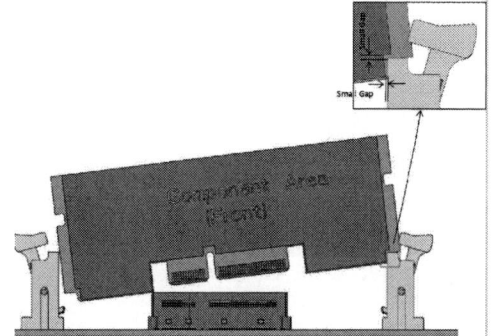

Figure 16. 1U DDIMM module titling during insertion or extraction

To avoid mechanical impact on components during DDIMM insertion/extraction, the JEDEC MO defines the component keepout (green area in figure 16). But due to the component placement challenges, the keepout area is minimized. The PCB edge +/-0.1mm tolerance [5] must be satisfied to mitigate the risk of component/latch mechanical interference. During PCB manufacturing, the edge outline is defined by router. Even computer numerical control (CNC) is common, but the typical tolerance is at +/-0.15mm. DDIMM manufacturers should drive the improvement of PCB suppliers' card edge routing capability.

B. Latch ejection study

A latch ejection study has been performed based on the JEDEC DDR5 DDIMM MO and Socket Outline (SO) proposal. Based on the current MO/SO, latch will not completely eject the DDIMM module from the connector; manually pulling DDIMM out of the connector is necessary. The extraction force of the SNIA TA-1002 connector is expected to be low, which was confirmed by a preliminary study that estimated the extraction force average around 13.4N (figure 17), only a third of the measured extraction force for DDR5 288 pin U/R/LR DIMM which is average around 38N.

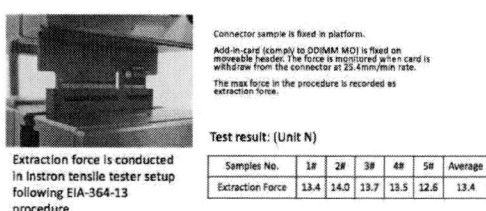

Figure 17. Preliminary DDIMM PCB extraction experiment

Ergonomic features on the latch keepout is an option on the current MO as shown in figure 18. In the future MO revision, this ergonomic feature might become mandatory to encourage the operator to hold the module on the edge to mitigate the "fat finger" failure mode – solder joint or component damage caused by an operator holding the DDIMM on the components during insertion/extraction.

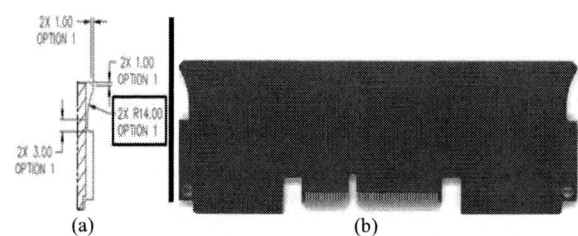

Figure 18. Ergonomic features to encourage operator to hold module on the edge during insertion/extraction [5]: (a) JEDEC MO latch keepout option (b) 1U DDIMM dummy PCB for dimension tolerance study

The latch ejection study also covers the rotation of the "one hand process". Even though the common operation is a "two hand process" but it is hard to perfectly synchronize the motion of two hands. In reality, the "two hand process" is composed of two "one hand processes". DDIMM module will have certain degree of rotation around the connector key as illustrated in figure 19. The TA-1002 connector housing is deep which limits the rotation angle. When the connector key resists the rotation, it could be damaged. The design optimization of the connector key is necessary to mitigate the risk of DDIMM module rotation induced mechanical damage.

Figure 19.   1U DDIMM module titling during insertion or extraction

C. Contact reliability with plated thru hole (PTH) under DDIMM PCB contact pad (gold finger)

A PTH under the DDIMM PCB contact pad (figure 11) is necessary for ground pins that separate adjacent OMI channels in order to control the cross talk (detail in section IV). A PCB with a pad on PTH has been in mass production, but has not been applied to connector contact pad.

The JEDEC proposal requires 200 um drill diameter PTH on the 380um contact pad. The survey responses from PCB suppliers indicate adequate drill alignment capability to mitigate contact pad breakout. The drilled hole will be copper plated with 18um minimum copper wall thickness. The plated hole will be filled with epoxy. After epoxy cure/grinding, the filled hole will be plated over with copper that at a subsequent process will be plated with Ni/Au. At the end of the process, on the 380 um contact pad, a dimple or bump with height around 1mil (or 25.4um) is expected.

The contact reliability will be impacted by the surface properties of the contact pad on the PTH. Critical surface properties could be impacted by the filler epoxy material/plated copper cap material properties, contact pad surface condition, dimple/bump size and location. Future study is planned to understand the contact reliability between the gold finger on the PTH and the TA-1002 connector contact beam.

D. PCB tie bar requriement

Removing the contact pad tie bar will reduced the contact parasitic capacitance which will improve the signal integrity. The PCB tie bar is necessary during the manufacturing for electrical Ni/Au plating. DDIMM 0.6mm contact pad pitch is smaller than that of conventional R/LR DIMM (table III). Smaller pitch leads to a thinner tie bar which leads to a smaller adhesion area to the laminate. When the contact beam wipes through tie bar, the tie bar could break and become the culprit

for electrical short. Mechanical bevel is the common process to remove the tie bar but is prone to creating a burr as shown in figure 20. In the future, an alternative process such as double etch back will be studied to mitigate the shorting risk of tie bar burr and peel off.

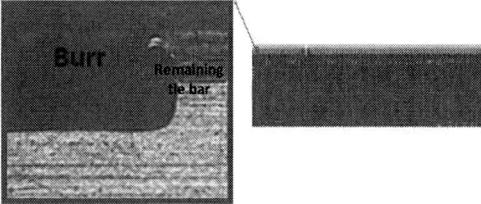

Figure 20.   Bevel process induce tie bar burr

VII.   FUTURE WORK

The SPP coupons for this study have measured impedance values average around 77 ohms while the target impedance is 85 ohms (section III). Even though those coupon samples are within the manufacturing tolerance for impedance but the study was limited to low impedance corner. The follow up study with improved impedance control is planned to address this limitation. In addition, the PCB supplier's process capability to control the strip line geometrical parameters will be studied to build impedance/degradation corner models.

The future study of DDIMM PCB and connector mechanical interaction will prioritize the contact reliability study of the contact interface between the DDIMM PCB contact pad on PTH and the SNIA TA-1002 connector contact beam.

ACKNOWLEGMENT

The SPP coupon construction analysis of this study was performed by Hongqing Zhang from IBM East Fishkill material characterization. The technical review was performed by Marie Cole and Zhaoqing Chen.

REFERENCES

[1]   Jeff Steuchli, "OpenCAPI a data-centric approach to server design", OpenPower Summit Europe, October 3-4, 2018

[2]   Kevin McIlvain, "Facilitating IP development for the OpenCAPI memory interface", OpenPower Summit Europe, October 3-4, 2018

[3]   George Dudnikvo, "PCB Materials and Technology for RF Microwave Application", IPC Printed Circuit Board Defense Roadmap, 2009

[4]   Steve Iketani and Brian Nelson, "Key Factors Influenciong Laminate Material Selection for Today's PCBs", The PCB design Magazine, September 2013

[5]   JEDEC DDR5 DDIMM Module Outline

[6]   IPC Test methods manual, "Test Methods to Determine the Amount of Signal Loss on Printed Boards"

[7]   Alina Deutsch and Roger Krabbenhoft, "Use of the SPP Technique to Account for Inhomogeneities in Differential Printed-Circuit-Board Wiring", SPI 2008

[8]   JEDEC 288 PIN DDR4 DIMM 0.85 MM Pitch Module Outline

Effectiveness of equalization and performance potential in DDR5 channels with RDIMM(s)

Nanju Na
Xilinx Inc.
San Jose, California, USA
e-mail: nanjun@xilinx.com

Hing "Thomas" To
Xilinx Inc.
San Jose, California, USA
e-mail: tto@xilinx.com

Abstract—**As data transport speed DDR5 supports jumps up to 6.4Gbps, JEDEC requires DFE at DRAM receive as a new equalization feature in evolving specifications. This paper investigates equalization behaviors of DFE and CTLE and their effectiveness at DDR5 operation speeds in varying channel configurations with RDIMM. Relative effects of DFE taps and potential of combined usage with CTLE are studied with eye opening simulations. While margin for operability is subject to device jitter budgets in TX and RX which still are evolving, eye opening margin at 1e-16BER with dual rank RDIMM appears enough for barely over 4Gbps in typical DDR4-like board routing setup and presumptive DDR5 device setup with full crosstalk.**

Keywords-component; memory interface; DDR5 equalization; RDIMM channel simulation; signal integrity

I. INTRODUCTION

The speed bins expected in DDR5 memory interface move to much higher transmit rates than previous DDR generations to meet demand for ever-increasing data bandwidth of memory interface while single-ended signaling scheme of data transfer is still intact. Its maximum data rate is set as high as 6.4GTS and there is a hint that even higher data rates is considered in future JEDEC specifications. To support signal integrity of higher speed signals, DFE (Decision Feedback Equalizer) as a new equalization feature is now required in DRAMs by JEDEC specification. In this paper equalization effectiveness and performance potential of DDR5 channels with RDIMMs are explored among wide range of channel configurations as an early application target of DDR5 for data center applications. Whereas 4-taps DFE is a requirement in JEDEC specification for DRAMs, the number of taps can vary over speed ranges of operations to be effective and DFE behavior with tap coefficients tuning differs in different channel configurations, say, with the number of ranks, routing designs in both main board and DIMM PCB, equalization features in host controller chips, whether 'READ' or 'WRITE' operation, etc. Host controller chip may also be designed with different equalization features, for example, CTLE (Continuous Time Linear Equalization) combined with varying number of DFE taps.

DDR5 interface can, like earlier DDR generations, operate in different configurations with DRAMs on DIMM(s) or on board in varying number of ranks. The paper investigates different equalization behaviors and effects on performance in various combinations of equalization setting over speeds with early focus of applications using RDIMM (Registered DIMM) channels as illustrated in Figure 1. In DDR5 RDIMM interface, command-address (CA) signals for host controller to RCD on RDIMM are operated in DDR speed on point-to-point interface requiring 4-tap DFE at RCD input similarly to DQ data interface and DIMM pins are defined with 1:1 signal to ground. Therefore, performance on host to RDIMM interface is rather limited by DQ signals with longer routing and more crosstalk than CA signals to RCD on RDIMM assuming similar routing in DQ and CA signals on main board. An early version of JEDEC reference raw card design was extracted to model signals in RDIMM for simulations in the following sections.

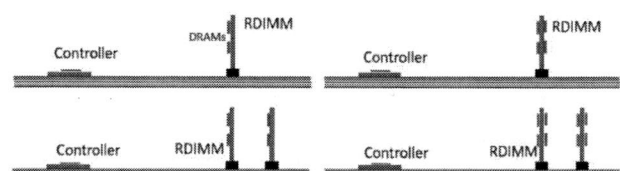

Figure 1. Channel topologies with RDIMMs - single rank per DIMM on left (1D1R, 2D1R) and dual rank per DIMM on right (1D2R, 2D2R)

II. CHARACTERIZATION OF EQUALIZATION METHODS

A. Pulse Responses on Channels and Effectiveness of DFE taps

JEDEC specifications now requires 4-tap DFE features in DDR5 DRAMs to help with degrading signal quality at higher data rates. Functional diagrams in Figure 2 illustrate working of DFE on data bits (DQ's) at WRITE operation in DRAM receivers. Figure 3 illustrates the channel simulation models which consist of main board routing and DIMM card routing. Board via models were generated using 3D full-wave tools for device pin-out patterns used in DDR4 channels and RDIMM routing was modeled from reference design extraction of early JEDEC raw card version. For higher crosstalk effect in typical board routing, vias were modeled for routing escape in lower PCB layer with device pins on top layer.

Electric field formed by charging and discharging of the channel lingers over longer time interval of a data bit and interferes with subsequently incoming signal bits.

978-1-7281-1500-9/19 $31.00 © 2019 IEEE

Efficiency of DFE is determined by how effectively the lingering field of prior signal bits can be canceled with taps setting for feedback.

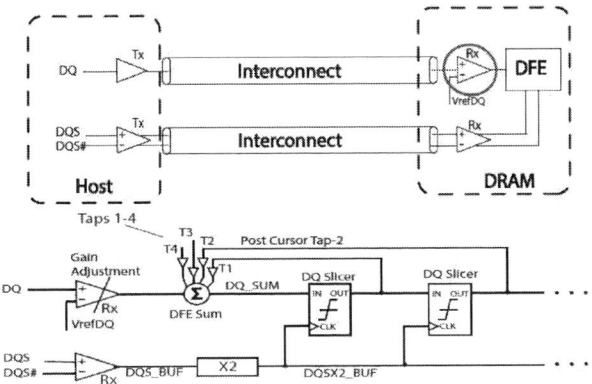

Figure 2. 4-tap DFE requirements on DQ signals at DRAM receive by JEDEC [1]

Single rank and dual rank channels illustrated in Figure 3 are examined for pulse responses varying in READ/WRITE direction. Figure 4 shows pulse responses on single rank channel for 4Gbps, 5.2Gbps and 6.4Gbps indicating lingering voltage trails at subsequent bit intervals, which are measures of DFE tap weights for feedback. More significant tap weights are expected with higher speeds as measures of voltage trails seen in Figure 4.

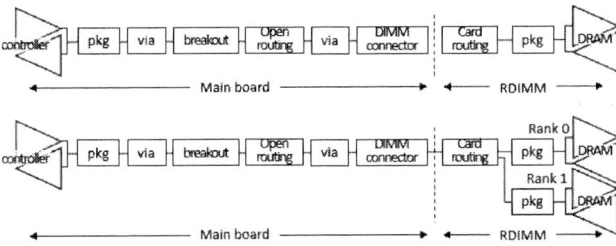

Figure 3. Channel configuration with single & dual-rank RDIMM systems

Note that voltage trails in the pulse waveforms are non-monotonic due to discontinuities in the channels. Pulse waveforms for dual rank interface in Figure 5 exhibit reflection behavior prominently differing between WRITE and READ directions with capacitance loading effect varying by signaling direction.

Single-rank and dual-rank interfaces were simulated with 6.4Gbps WRITE setup in which no crosstalk was applied from surrounding pins. Drivers and receivers were set identical in capacitance load of 1pF and impedance 40Ω throughout the tests in this study. The resulting eye diagrams are shown in Figure 6 to compare effect of 4-tap DFE equalization with eye opening with no equalization. While the effects of DFE equalization is clear on both channels, more prominent equalization effect is observed in dual-rank interface.

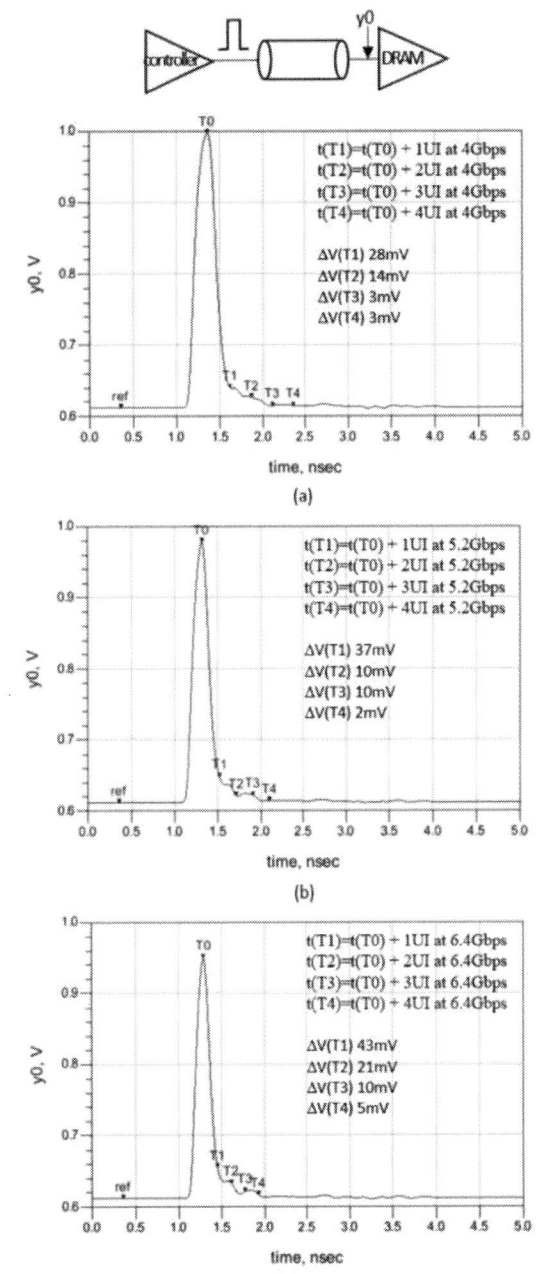

Figure 4. Pulse responses on single-rank system channel with pulse width for (a) 4Gbps, (b) 5.2Gbps and (c) 6.4Gbps; voltage trails $\Delta V(T)=V(T)-Vref$ as lingering field effect at post bit sampling times T1-T4 to offset with negative weights $-\Delta V(T)$ for DFE taps.

Note that the crossing levels are different before and after equalization, 835mV versus 795mV for single-rank interface and 1014mV versus 955mV for dual-rank interface. Vref will need to be tuned accordingly to expected new levels after equalization for DFE to be effective. Eye opening performance was tested on the interfaces sweeping in various equalization settings. Eye widths in the following tests were captured at crossing levels and at voltage offset levels in consideration of device RX sensitivity as illustrated in Figure 7.

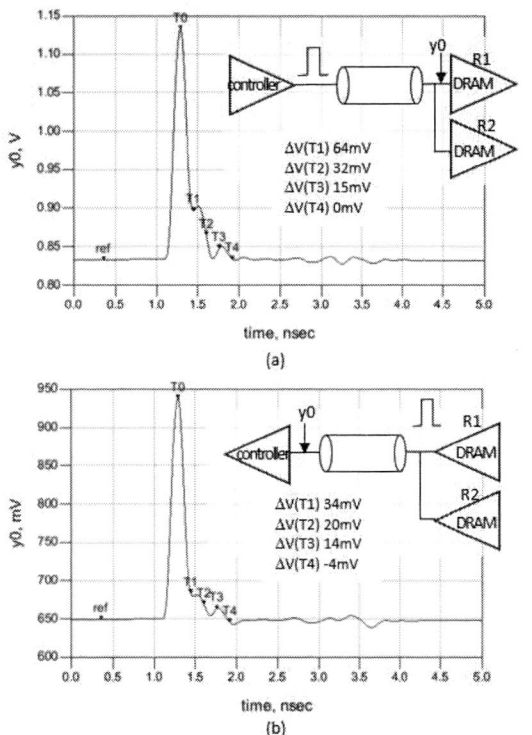

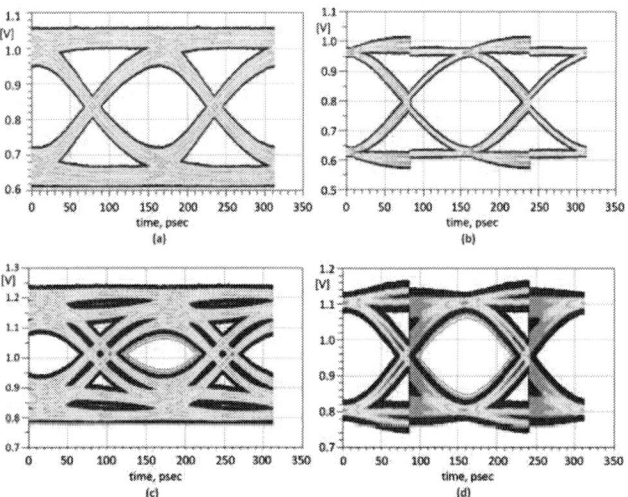

Figure 5. Pulse responses on dual-rank system channel with pulse width for 6.4Gbps in (a) WRITE data path and (b) READ data path; voltage trails at post bit unit intervals T1-T4

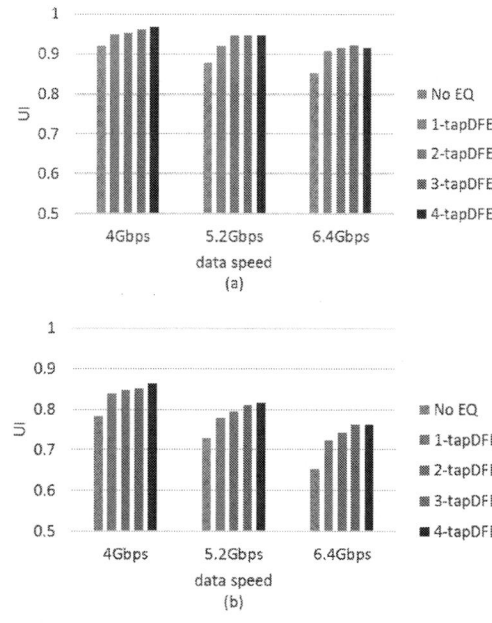

Figure 6. Eye opening at 6.4Gbps (no crosstalk from surrounding pins applied): on single-rank interface with (a) no equalization and (b) 4-tap DFE; on dual-rank interface with (c) no equalization and (d) 4-tap DFE

Figure 8 shows eye widths of single-rank channel with no crosstalk captured at crossing level as Vref and at 80mVpp over 4Gbps to 6.4Gbps with sweeping number of DFE taps. Note that all the eye data presented here were collected at 1e-16BER (bit error rate) for design requirement by JEDEC specifications.

Eye width measured at crossing level improved with the first tap but nearly no improvement is observed with the second and higher-order taps whereas those high-order taps appear to be relatively more effective when reading the eye width at 80mVpp though incremental. While 80mVpp was used for consistency in comparison among different speeds, smaller voltage mask levels are also considered for 5.2Gbps and 6.4Gbps in later tests assuming devices of higher speed grades would be capable of lower RX sensitivity.

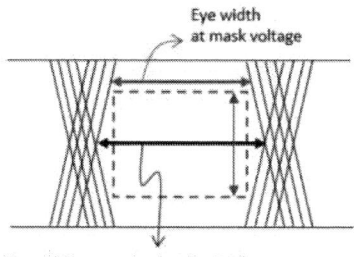

Figure 7. Eye opening as performance measure

Figure 8. Eye width of single-rank channel (no crosstalk) with varying number of taps in DFE (a) at crossing level and (b) at 80mVpp, 10^{-16} BER

In Figure 9, eye widths of 6.4Gbps on dual-rank interface for WRITE and READ are shown with sweeping number of DFE taps. The eye opening is notably reduced by more than 0.2UI at 6.4Gbps with dual-rank interface compared with single-rank interface. Without equalization, the channel performance in WRITE direction is significantly worse than in READ direction. The channel in WRITE, however, is shown to outperform with greater DFE effectiveness for which tap weights setting was predictable from pulse responses in Figure 5.

978-1-7281-1500-9/19 $31.00 © 2019 IEEE

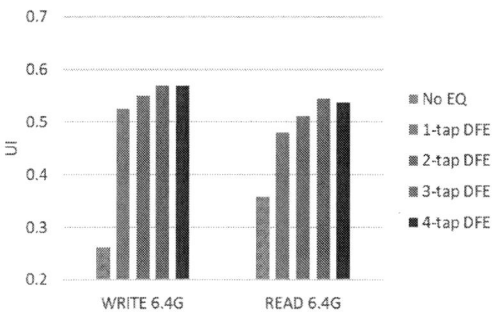

Figure 9. Eye width of dual-rank channel (no crosstalk) for 6.4Gbps at 80mVpp mask voltage in WRITE and READ directions, 10⁻¹⁶ BER

B. CTLE and effect of combined equalization

CTLE (continuous time linear equalization) feature has been employed in some DDR4 devices. Though CTLE is not part of requirements for DDR5 DRAMs in JEDEC specifications, it can be still utilized in existing circuits for DDR5. This section investigates equalization effect in combination of DFE and CTLE for DDR speeds. A CTLE for the test was chosen for the frequency response in Figure 10 with the peak gain 5.8dB at 2.5GHz which is best-tuned for 5.2Gbps among the test speeds. Single-rank interface is tested for equalization configurations in Figure 11 to investigate behaviors of CTLE and DFE at the speed of 5.2Gbps.

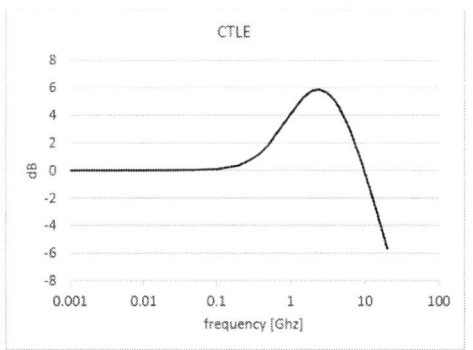

Figure 10. Frequency response of CTLE with peak gain 5.8dB at 2.5GHz chosen for test

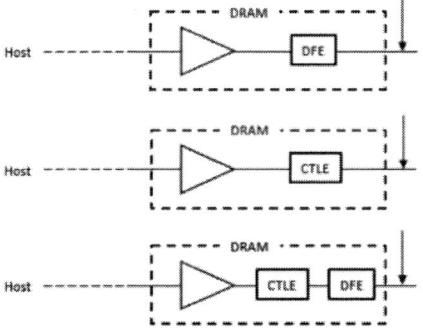

Figure 11. Cases of equalization setup for test

Single-rank interface was simulated for 5.2Gbps with 4-tap DFE alone and with CTLE alone. Eye diagrams in Figure 12 exhibit different equalization effects of two methods. DFE works to cancel field effect of previous bits with tap weights to correct DC level on incoming bit whereas CTLE works to improve rising and falling edges of the waveforms by boosting the high frequency components. By differing mechanisms, DFE tends to be more efficient with opening eye width and CTLE increasing eye height as exhibited in better-clustered eye traces on left and sharpened transition slopes on right of Figure 12. The effects of two methods on eye widths and heights are clear in sweeping test results in Figure 13 where eye width is greater with 4-tap DFE and eye height with CTLE. DFE could be more effective for devices with lower RX sensitivity and CTLE may work better for devices requiring higher RX sensitivity though it can vary from channel to channel as well as with device capability and loading in driver and receiver.

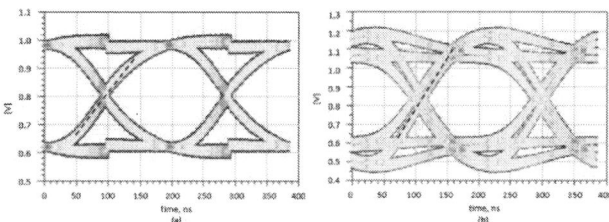

Figure 12. Eye opening behaviors of single-rank interface (no crosstalk) for 5.2Gbps with (a) 4-tap DFE alone and (b) CTLE alone.

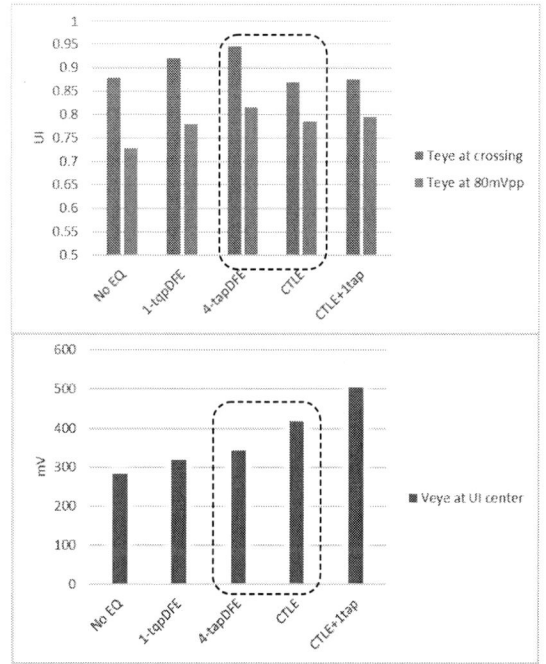

Figure 13. Eye opening with DFE vs CTLE in width (up) and height (down) for 5.2Gbps on single-rank interface (no crosstalk), 10⁻¹⁶ BER

978-1-7281-1500-9/19 $31.00 © 2019 IEEE   1211

III. EQUALIZATION PERFORMANCE WITH CROSSTALK

Crosstalk is the most significant contributor to jitter in memory interfaces. Crosstalk occurs mostly through vias on main board and DIMM card varying with via placements subject to device pin-out patterns and subject to escape routing layer(s) on board. Crosstalk in DIMM cards is relatively well-defined as the pin-out of DRAM(s) and reference designs for raw cards are specified by JEDEC. Crosstalk on the main board can vary with via placements for DIMM connector and via patterns from controller device pin-out. The channels of single-rank and dual-rank configurations were modeled with vias for controller pin-out of signal to ground 3:1 as illustrated in Figure 14 for crosstalk effect, and with RDIMM design extraction of early version of JEDEC raw card reference design. Single-rank and dual-rank interfaces were simulated for 6.4Gbps with full crosstalk from neighboring pins. Eye performance results with 4-tap DFE are shown in Figure 15. Comparing those with no crosstalk in Figure 6 using the same setup, it is observed that crosstalk cost eye widths by 45% for 6.4Gbps with 4-tap DFE for example. It demonstrates that crosstalk dictates the performance and limits prospect of speed potential.

Figure 16 shows effect of crosstalk on eye width at 80mVpp sweeping over speeds and equalization setup.

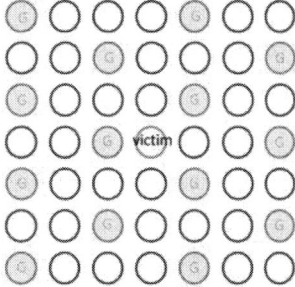

Figure 14. Controller pin-out with 3:1 signal to ground ratio for via crosstalk consideration

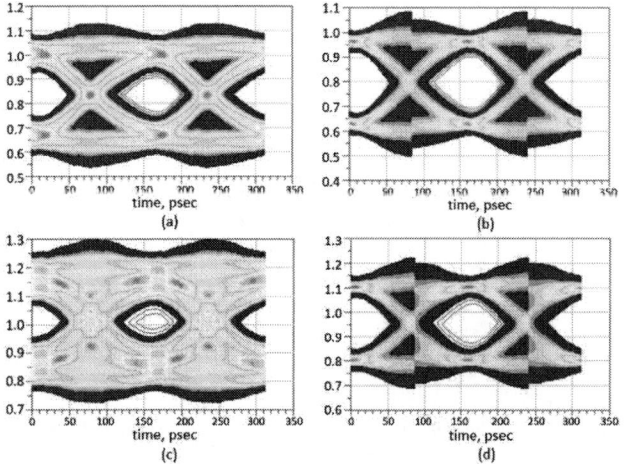

Figure 15. Eye opening at 6.4Gbps with crosstalk: on single-rank channel with (a) no equalization and (b) 4-tap DFE; on dual-rank channel with (c) no equalization and (d) 4-tap DFE

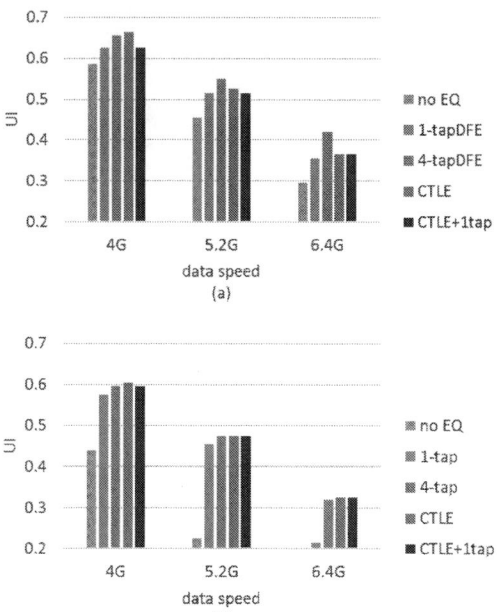

Figure 16. Eye width at 80mVpp with crosstalk (a) on single-rank channel and (b) on dual-rank channel WRITE, 10^{-16} BER

Performance degradation with crosstalk in Figure 16 (a) is rather abrupt with speed increase compared to those with no crosstalk in Figure 6 (b) for single rank interface. Degradation with crosstalk over speeds looks to be even more abrupt for dual rank interface in Figure 16 (b). Equalization effect of CTLE is comparable to 4-tap DFE on dual rank interface in this test. However, it may need broader range of channel analysis to draw a general conclusion.

Though improvement on DIMM card is limited with DRAM pin-out specified by JEDEC standards, further development of sophisticated via routing can still help reduce crosstalk on RDIMM card to some extent. Driver signaling control may help improve as well. As for main board factors, controller pin-out and board routing layer decisions with signaling density can affect interface performance. Note that the channels in the study were modeled for lower layer routing of 100-mil PCB to account for worse crosstalk from long via path with routing density assumption as typical in DDR4 interfaces.

A channel model was formed with a via model for 2:1 signal to ground controller pin-out as illustrated in Figure 17 for reduced crosstalk. The eye widths for 6.4Gbps captured at 80mV are compared for S/G ratios of 3:1 and 2:1 in Figure 18 where 10% to 12% eye width improvement is observed with 2:1 S/G over 3:1 S/G. 4-tap DFE appears more effective on single-rank channel whereas CTLE looks to be more effective on dual-rank channel in this setup. Even incremental improvements can be substantial where margin for driver and receiver device jitter is not enough. However, device pin-out choice of reduced S/G ratio is likely a cost increase, which may limit the option.

Figure 17. Via pattern with signal to ground ratio 2:1 in controller pin-out for reduced crosstalk test

Next, the eye performances with 4-tap DFE equalization on three DIMM configurations are sorted in Figure 19 as reference to JEDEC specification of 4-tap DFE requirement for DDR5 DRAMs.

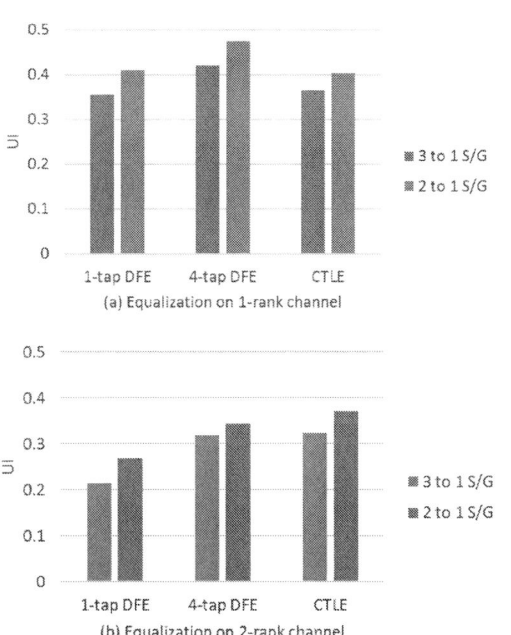

Figure 18. Eye opening at 80mVpp for 6.4Gbps with crosstalk on single-rank channel with controller pinout of (a) S/G ratio 3:1 and (b) S/G ratio 2:1

In Figure 19 and Figure 20, "1D1R" refers to a channel with controller to 1 DIMM on board configured in single rank, "1D2R" with controller to 1 DIMM configured in dual ranks and "2D1R" with controller to 2 DIMMs on board configured in single rank each with 1″ routing between the first DIMM (D0) and the second DIMM (D1). 1D2R and 2D1R configurations form dual rank operation interfaces. Eye widths for 5.2Gbps and 6.4Gbps in Figure 19 are shown with two different voltage mask levels in overlap. 80mVpp was used for

common reference of performance for three speeds at base.

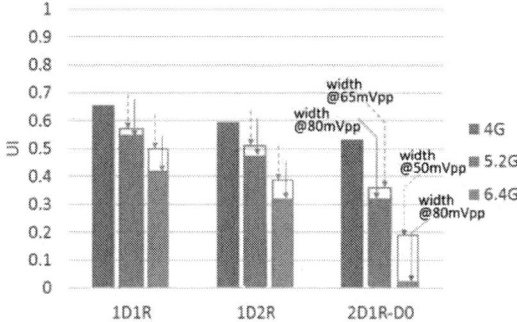

Figure 19. Eye opening with 4-tap DFE on 1 DIMM (1D1R & 1D2R) and 2 DIMM (2D1R) channels with 3:1 S/G controller pin-out; eye widths captured at 80mVpp (solid fill) for all speeds, at 65mVpp for 5.2Gbps (orange, no-fill) and at 50mVpp for 6.4Gbps (grey, no-fill).

Additional voltage mask levels 65mVpp for 5.2Gbps and 50mVpp for 6.4Gbps were used to explore an idea that devices of higher speed grades are assumed to be capable of lower RX sensitivity as in discussions in JEDEC specification ballots. As expected, eye opening degrades in dual rank systems with added loading compared to single rank (1D1R) and their degradation is much steeper over speeds. Also, it is noteworthy that eye width with 2 DIMMs of 1 rank at each DIMM is drastically closed toward 6.4Gbps.

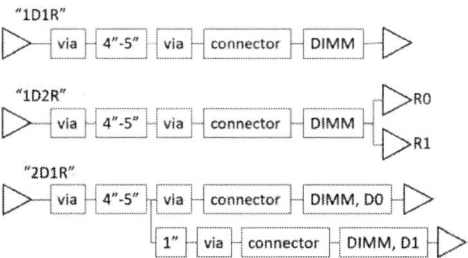

Figure 20. Channel configurations for tests in Figure 19.

IV. DISCUSSION AND CONCLUSION

As the analysis was conducted on a specific setup in device setting and DDR4 board routing, the results presented in this study may vary in some extent with device characteristics and setup variations along with board designs in real systems for DDR5. While 4-tap DFE is settled as a new equalization feature for DDR5 DRAM required by JEDEC specifications, CTLE utilized in DDR4 devices may continue to be implemented in combination with DFE in varying reasons. When considering DFE alone, 1-tap DFE looks likely enough for 4Gbps while incremental improvements from subsequent 3 taps in 4-tap DFE can still be substantial for 6.4Gbps. CTLE and DFE work differently on incoming waveforms, they exhibit different eye-opening behaviors. DFE tends to yield wider opening at crossing level or Vref assuming Vref training is accurate whereas CTLE tends to yield taller opening with sharpening transition edges. If receive circuits can afford lower RX sensitivity, DFE may work better whereas CTLE may be relatively beneficial for devices with higher RX sensitivity. Operable speeds estimated by simulation test results depend on driver and receiver budget of devices which are still in the working of JEDEC specifications. Operability in speeds can also differ with interface types with RDIMMs ranging in RDIMM ranks and the number of RDIMMs participating in configuration for which degradation is steep in higher speeds and multiple ranks/DIMMs. For example, if considering 0.25UI jitter budget for each of driver and receiver in the interface and lower RX sensitivity assumed for 5.2Gbps and 6.4Gbps, 1 DIMM channel may be viable up to 6.4Gbps if single rank interface and 5.2Gbps if dual ranks under those specific assumptions. If requiring eye opening at 80mVpp at all speeds, the interfaces appear to barely make 4Gbps according to the test data. Performance on a channel with 2 DIMMs of dual ranks each, not presented here, resulted to be detrimental even at 4Gbps with full crosstalk for the same loading and driving setup.

Since crosstalk is the most critical factor to eye closure, controlling crosstalk is a key design factor to improving speeds. As channels constructed in the study were chosen to represent a worse routing corner of typical DDR4 design, there are ways to improve for reduced crosstalk with PCB layer and routing choices. One option is to lower S/G ratio in pin-out definition at the cost of package increase and/or careful arrangement of signaling kinds to alleviate crosstalk effect. Other options may include tighter length control, routing layer for shorter via crosstalk with or without via back-drilling, etc. which require further analysis.

REFERENCES

[1] JEDEC DDR5 Full Spec Draft Rev0.5f
[2] JEDEC Standard JESD79-4B: DDR4 SDRAM
[3] JEDEC Standard No. 21C: DDR4 SDRAM Registered DIMM Design Specification

2019 IEEE 69th Electronic Components and Technology Conference (ECTC)

Inductive links for 3D stacked chip-to-chip communication

X. Sun, N. Pantano, S.W. Kim, G. Van der Plas and E. Beyne

Imec

Kapeldreef 75, Leuven, Belgium

sunx@imec.be

Abstract—**This paper focusses on an experimental inductive link study based on *S*-parameter measurements up to 67 GHz for 3D stacked chip to chip communication. This is complemented by a calibrated HFSS model for coupling between two chips that is then extended to stacks with up to eight chips. This approach allows for a complete analysis of the inductive link mechanisms between stacked chips and the impact of different parameters. Moreover, a circuit level assessment of the inductive link for a transmitter circuit has been performed using FinFET technology.**

Keywords-inductive link; chip-to-chip; stacked chip; N=2 chip staking; N=8 chip stacking; linear regime; non-linear regime; operating frequency; Self-resonance frequency;

I. Introduction

3D stacked chips are amongst the most attractive candidates for next generation high-performance large-system integration. Inductive links have been proposed as one of the interconnect solutions between 3D stacked chips because of their low fabrication cost, long communication distances, and the possibility to stack more than 3 chips. However, the electromagnetic coupling between the inductive link [1, 2, 3, 4] and nearby interconnects can cause voltage fluctuations that affect the interconnect performance. This can cause circuit malfunctioning and signal integrity problems. Therefore, it is essential to analyze and understand the coupling mechanisms of inductive links.

This paper focusses on the inductive link study for 3D stacked chip to chip communication and it is divided into 7 sections. First, a brief introduction of imec's hybrid bonding technology is given in section II.

In section III, chip to chip communication using inductive links for two stacked chips are discussed. Inductive link test structures were first processed respectively on two Si chips, subsequently imec's hybrid bonding technology was used to stack these two chips. The *S*-parameters of the test structures were measured and analyzed. Both linear and nonlinear inductive coupling has been observed. Meantime, HFSS models for two stacked chips were developed and compared with the measurements, the excellent agreement with measured data validates these models. It is demonstrated that the complex RF field distribution between the chips can be accurately represented by the HFSS model presented in this paper.

In section IV, the impact of different factors on the inductive link performance is investigated based on the validated HFSS model. Studied factors are inductor loop

size, Si chip thickness, Si substrate resistivity, as well as the impact of misalignment between the coupled inductors.

In a next step, the communication between eight stacked chips is investigated in section V. As expected, the amplitude of the signal decreased as the distance between two chips increased.

Finally, in section VI, a circuit level assessment of the inductive coupling for a transmitter circuit is described that is based on FinFET technology.

II. Hybrid Bonding Technology description

In 3D stacking integration, various technologies and techniques have been proposed and investigated for different system partitioning schemes [5]. Amongst the processing options for 3D SoC applications, hybrid bonding technology by directly joining BEOL layers from two substrates has emerged as a favorable option for high density interconnect applications. As shown in Figure 1, imec's hybrid bonding integration scheme uses scalable low temperature SiCN–SiCN dielectric bonding in combination with direct Cu–Cu bonding, resulting in excellent interconnect performance [6].

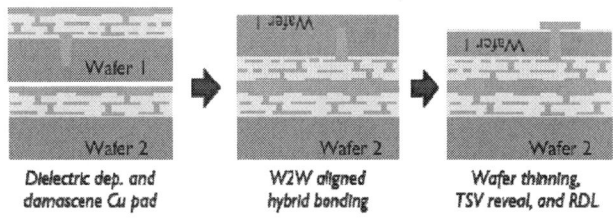

Figure 1. Schematic of the face-to-face hybrid wafer bonding process.

All chips use SiCN as the dielectric bonding medium and a unique Cu chemical mechanical polishing concept, which is developed to cope with scaled Cu pad dimension and thermal budget limitations for the targeted 3D applications. After wafer alignment and bonding at room temperature, the 3D stacked wafer was annealed first at 250°C to ensure the SiCN–SiCN bonding strength, followed by an annealing step at 350°C form stable Cu–Cu connections.

III. Inductive links for vertical communication between two chips

In this section, we focus on the inductive link study for vertical communication between two stacked chips using face-to-face hybrid bonding technology. As indicated in Figure 1, after the hybrid bonding, the top Si wafer was thinned down to a nominal thickness of 5 µm. Next, an

978-1-7281-1500-9/19 $31.00 © 2019 IEEE   1215

metal layer (RDLT) was processed on the backside of the thinned top chip.

The inductive link between the top chip and bottom chip was investigated through two inductors realized respectively on the metal layer RDLT (top chip) and layer M1B (bottom chip) without any direct electrical connection as shown in Figure 2. All communication between the two chips thus occurred through inductive coupling.

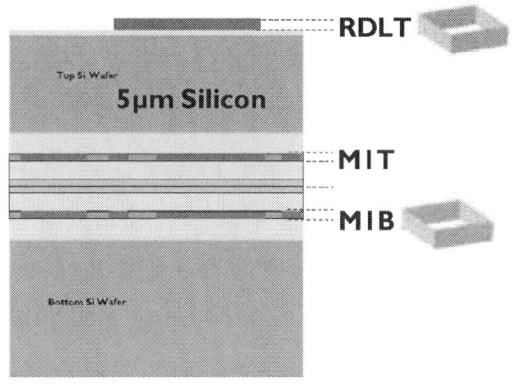

Figure 2. Inductive link test structures cross section for N = 2 chip stacks.

A. Inductive link test structures for N = 2 chip stacks

A top view of the inductive link test structure is shown in Figure 3. The image shows the inductor realized on the RDLT metal layer on the top chip, whereas the bottom inductor realized on the M1B layer of the bottom chip is less visible. The RDLT and M1B coil widths were 5 μm and 2 μm, respectively. The structure used for the HFSS modeling of the test structures is also illustrated in Figure 3.

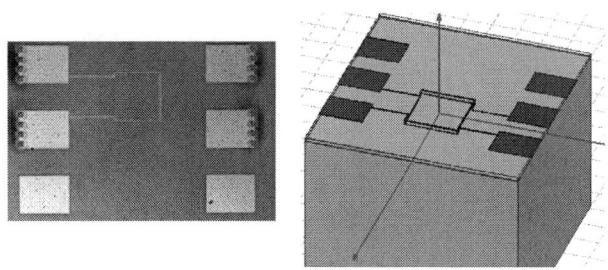

Figure 3. Inductive coupling test structures top view (N = 2 chip stacks). Left: Test structure image; right: setup of the HFSS model.

The RF measurements have been performed on two pairs of coupled inductors with different inductor inner dimensions, W, of 75 μm and 100 μm, respectively. The results are shown in Figure 4. Both linear and nonlinear inductive coupling regimes were observed. The inductive coupling increased from -60 dB to -15 dB in the linear region between 300 MHz and 30 GHz. Above 30GHz, it then starts to decrease in the nonlinear region.

However, below -50 dB, the inductive coupling is too low and can be neglected. This indicates that signals will need high frequency components to be transmitted by coupled inductors. The excellent agreement between the simulated and measured S_{12}-parameters validates the HFSS model for inductive link chip-to-chip communication.

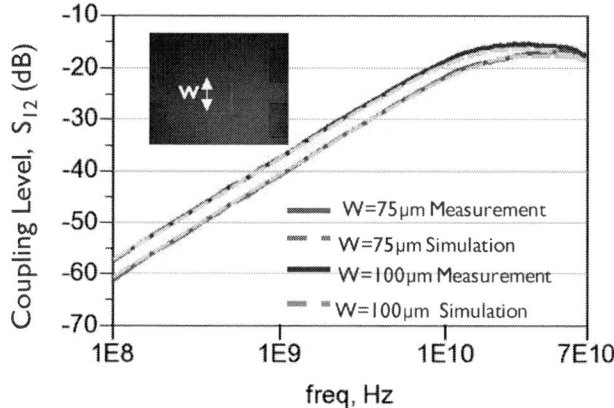

Figure 4. Inductive link RF measurements *vs.* HFSS simulation.

B. Inductive link coupling regimes for vertical chip-to-chip communication

To further understand the inductive coupling mechanisms, six pairs of coupled inductors with different inner dimensions from 75 μm to 200 μm have been simulated using the validated HFSS model. Simulated S_{12}-parameters for the inductive link between these coupled inductors are shown in Figure 5.

1) Inductive link in the linear regime

Below the self-resonance frequency, f_{SR}, inductive links stay in the linear regime and the self-inductance determines the inductive coupling level. As increasing the inductor inner dimension increases the self-inductance, it thus also increases the inductive coupling, however it also simultaneously reduces the f_{SR} as in Figure 5.

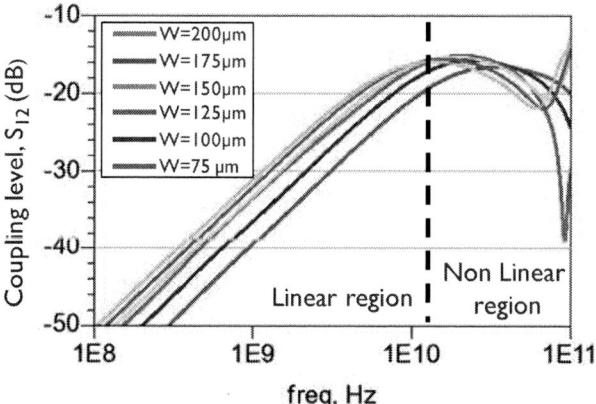

Figure 5. Impact of the width of the square inductor on the inductive link.

Figure 6 shows in the linear region below f_{SR}, increasing the inner dimensions of the inductor from 75 μm to 200 μm led to an increase of the coupling level from -40 dB to -31 dB at 1 GHz.

978-1-7281-1500-9/19 $31.00 © 2019 IEEE

For coupled inductors of the same size, the inductive coupling increased on average by about 18 dB from 1 GHz to 10 GHz.

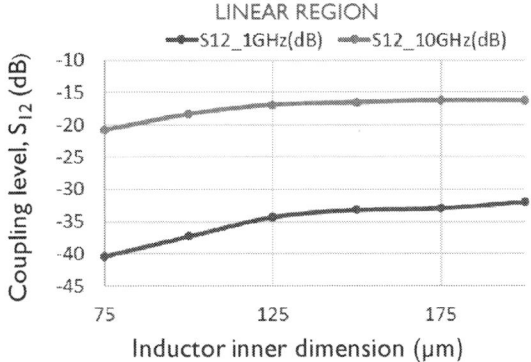

Figure 6. Inductive coupling level *vs.* inductor inner dimension and operating frequency in the linear region.

2) Inductive link in the nonlinear regime

In the nonlinear regime above the self-resonance frequency and beyond the operating bandwidth, the coupling of the inductive link was not proportional to the inductance, as shown in Figure 7. Hence, increasing the inductance was ineffective in improving the inductive coupling level. According to equation (1), increasing the operating frequency and the channel bandwidth f_{SR} required thus a reduction of the self-inductance L. However, reducing the self-inductance also decreased the inductive coupling in the linear region, thus a trade-off between the inductance and the operating bandwidth is needed.

$$f_{SR} = \frac{1}{2\pi\sqrt{LC}} \qquad (1)$$

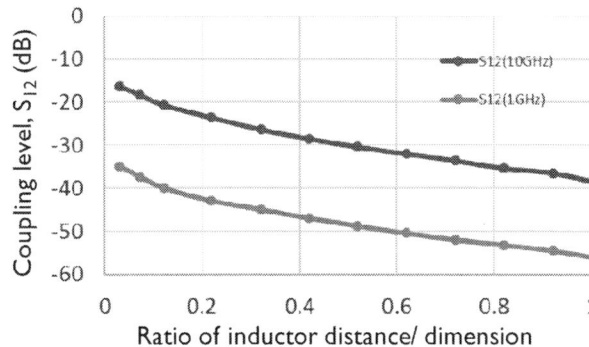

Figure 7. Inductive coupling level vs. inductor inner dimension in the nonlinear region.

IV. IMPACT OF DIFFERENT PHYSICAL PARAMETERS ON CHIP-TO-CHIP COMMUNICATION

In the previous section, it was shown that larger dimensions of the coupled inductors lead to higher inductive coupling between the chips within the channel bandwidth f_{SR}. Moreover, there are also other parameters that impact the behavior of inductive links, such as the thickness of the

Si chip, the misalignment between the chips, as well as the resistivity of the Si substrate of the chip.

A. Impact of chip-to-chip communication distance

Thinner the Si chip, shorter the communication distance. Thus thinner Si chips were found to lead to a better inductive link quality. Concretely, for an inductor inner dimension of 100 μm, the inductive coupling increased by 22 dB upon decreasing the Si chip thickness from 100 μm to 1 μm in the linear region as in Figure 8.

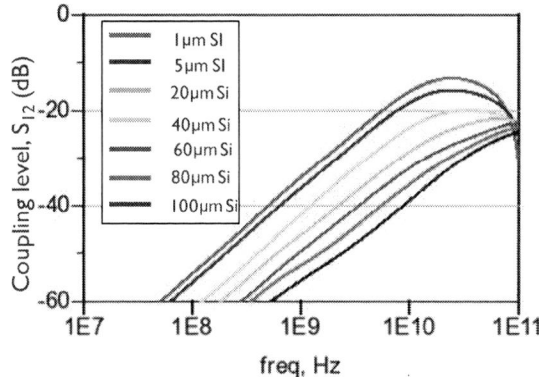

Figure 8. Coupling level of an inductive link as a function of the communication distance.

In a next step, the ratio between the distance of the coupled inductors and the inductor inner dimension was investigated. Figure 9 shows that the larger the ratio of the communication distance over the inductor inner dimension, the weaker was the inductive link. For a fixed distance, the dimension of the coupled inductors can be increased to lower the distance-dimension ratio, resulting in a larger coupling level in the linear region.

Figure 9. Coupling level of inductive link vs. the ratio between the communication distance and the inductor inner dimension.

B. Impact of the channel misalignment

The misalignment between the communication channels plays also a role for chip-to-chip communication. By misaligning the two channels equally in the x and y directions as in Figure 10, the effect of the misalignment was investigated for different Si chip thicknesses. With

978-1-7281-1500-9/19 $31.00 © 2019 IEEE

respect to thin Si chips, thicker Si chips were less influenced by the misalignment of the inductor loop between the top and bottom chips. As shown in Figure 11, for the 1-μm-thick Si chip, a misalignment of 10 μm led to a decrease of the inductive coupling by 4dB. However for the 50-μm-thick Si chip, a misalignment of 10 μm only led a decrease of the coupling level by 1dB. Hence, the thicker the Si chip, the smaller is the influence of the misalignment.

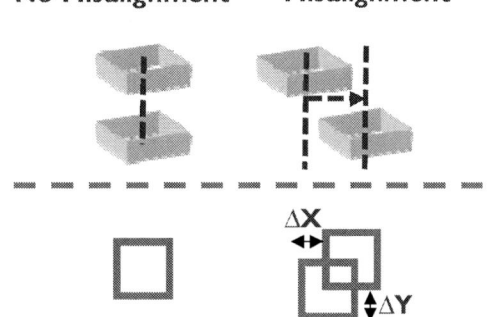

Figure 10. Misalignment for a single channel.

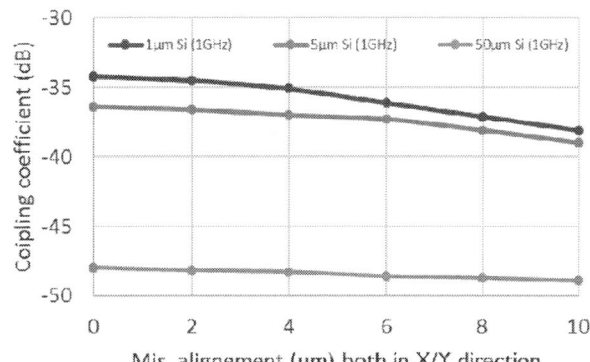

Figure 11. Inductive link vs. misalignment of the communication channel.

C. Impact of Si chip resistivity

The impact of the Si resistivity was investigated for values between in 0.1 Ωcm and 20 Ωcm. The simulation results in Figure 12 indicate that the Si chip resistivity did not affect the inductive link up to cut-off frequency that was dependent on the thickness of the Si chip. However, above the cut-off frequency, the coupling degraded much faster for the low-resistivity Si chip with a resistivity of 0.1 Ωcm with respect to the 20 Ωcm Si chip.

In an inductive coupling channel, eddy currents can be induced in the Si substrate by the generated magnetic field by coupled inductor, and this can partially counteract the magnetic field of the inductors and thus reduce the coupling strength between them. Above the cut off frequency, the eddy current becomes more dominate and thus lea to lower inductive coupling level.

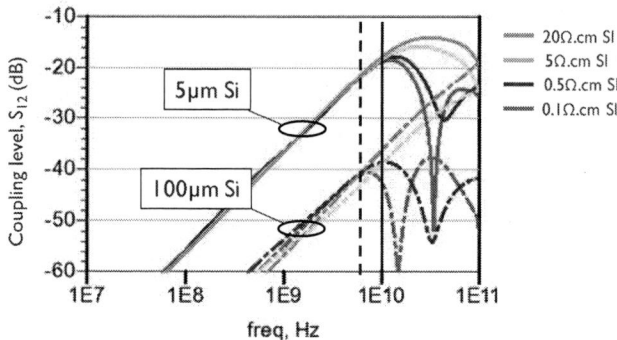

Figure 12. Inductive link coupling as a function of Si resistivity for an inductor with an inner dimension of 100 μm.

As shown in Figure 12, The cut-off frequency was 10 GHz for a 5-μm-thick Si chip; and 7 GHz for a 100-μm-thick Si chip.

V. INDUCTIVE COUPLING FOR VERTICAL COMMUNICATION AMONG 8 STACKED CHIPS

Furthermore, the communication between eight stacked chips was investigated. The binary-tree schema for vertical chip stacking is shown in Figure 13. Face-to-face bonding was used to stack two individual chips. Then, after thinning, backside oxide deposition, and chemical-mechanical polishing, back-to-back bonding was used to stack four chips. Finally, the same process sequence was repeated to stack all eight chips.

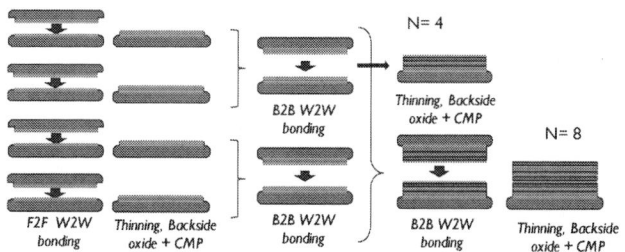

Figure 13. N=8 vertical stacked chips.

As the stacking of two chips is the basic step for the stacking of all eight chips, two coupled inductor coils were first processed on M1T (top chip) and M1B (bottom chip). As shown in Figure 14, good agreement was obtained between RF measurements and HFSS simulation, which validated the simulation model.

The model was then extended further to eight vertically stacked chips. At 10 GHz and for 5-μm-thick Si chips, a coupling of -18 dB was found between two adjacent bottom chips, whereas the coupling between the bottom and the top chips was -32dB. Figure 15 shows that the communication strength decreased as the distance between two chips increased, as expected.

978-1-7281-1500-9/19 $31.00 © 2019 IEEE   1218

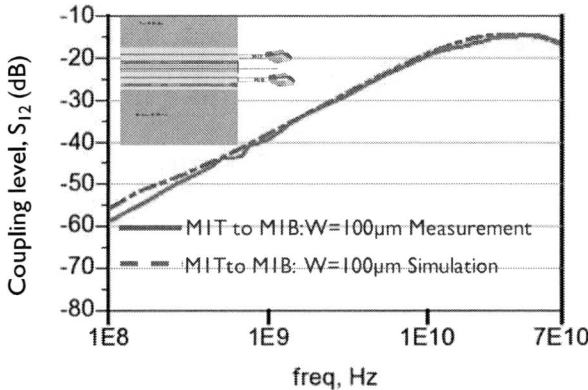

Figure 14. Inductive link for coupled inductors realized on M1T and M1B: measurement *vs.* HFSS simulation

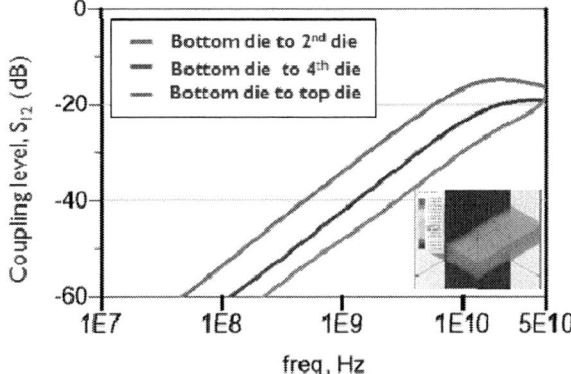

Figure 15. Inductive link for eight stacked chips.

VI. INDUCTIVE LINK CIRCUIT LEVEL ASSESSMENT

To evaluate the performance of chip-to-chip links by inductive coupling, a simulation-based circuit level assessment of a transmitter was performed. The schematic of the studied system is shown in Figure 16. The transmitter topology was similar to the one presented in [7] and was designed using FinFET technology. It consisted of a tri-state inverter connected to one branch of an inductor with the other branch connected to a capacitor. To transmit data, the capacitor is charged or discharged through the inductor. The current variations induce, by magnetic coupling, a voltage variation at the receiver side. This voltage variation needs then to be sampled and converted into a digital signal. In this paper, the receiver circuit was not included as it is beyond the scope of the work.

To avoid voltage ringing at the receiver side, the slew rate of the transmitter needs to be controlled. It can be tuned by varying the drive strength of the inverter or the size of the capacitor C_S. Decreasing the drive strength will reduce the ringing, but the peak current will also be affected which in turn affects the peak voltage at the receiver side.

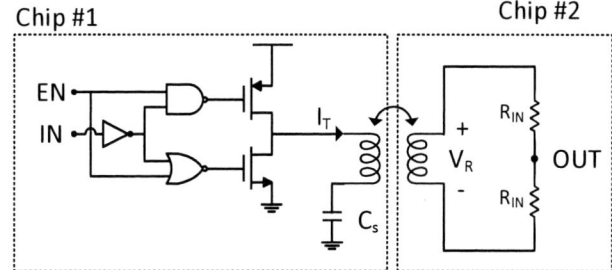

Figure 16. Circuit diagram representing the inductive link

For this reason, this is not the preferred option. On the other hand, adjusting the capacitor size has the advantage that the peak current remains almost constant and only the slew rate varies. To assess the impact of the capacitor, a stack of two chips was considered with a Si thickness of 5 µm. Figure 17 and Figure 18 show the simulation results of the impact of the capacitor size on the current I_T and the voltage V_R. As the size of the capacitor increased, the voltage ringing at the receiver was reduced. A capacitor of 500 fF was chosen as no ringing was then observed.

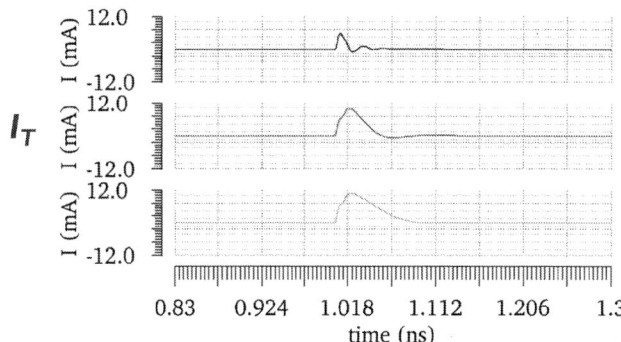

Figure 17. Variation of the current I_T as a function of the capacitor size.

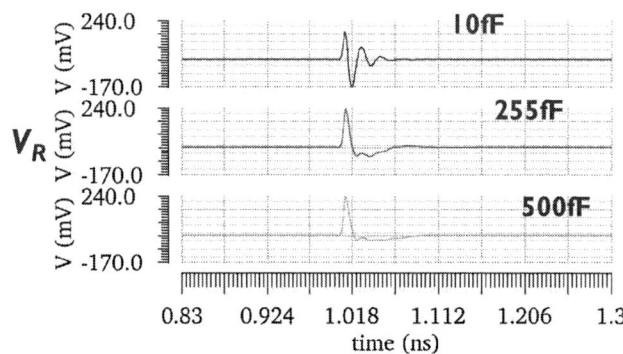

Figure 18. Variation of the voltage V_R as a function of the capacitor size.

Finally, the link was simulated for eight stacked Si chips. Each die had a Si thickness of 5 µm. A signal was transmitted from the bottom chip to the 2nd, 4th, and 8th chip, respectively. The simulation results are shown in Figure 19. As expected, the amplitude of the received signal decreased as the separation between the transmitting and the receiving chip increased. Moreover, oscillations could be observed at the intermediate (4th) and top chips (8th) due to intermediate inductors between transmitter and receiver. Nevertheless, a data rate of 1 Gbps could easily be achieved when taking into account the time required by the receiver to sense V_R. However, the technique requires large currents and a steep rise time of the signal, leading to large power usage.

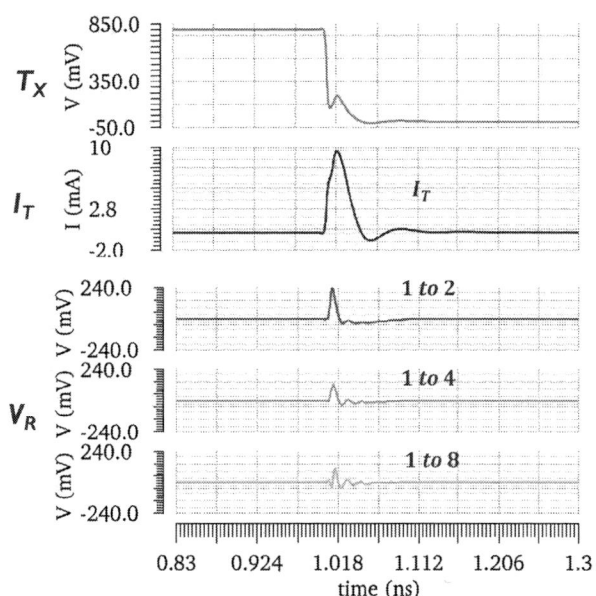

Figure 19. Simulation results for an inductive link between eight stacked Si chips.

VII. CONCLUSIONS

This paper focusses on an experimental inductive link study based on *S*-parameter measurements up to 67 GHz for 3D chip stacks. The experiments were complemented by a calibrated HFSS model for two stacked chips that was then extended to stacks of eight chips.

Frequency regions with both linear as well as nonlinear inductive coupling have been observed. The bandwidth of the inductive channel is defined by the cut-off frequency between the linear and non-linear regime, which is determined by the self-resonance f_{SR}. In the linear region, larger inductance values led to higher inductive coupling. Above the linear region, increasing the inductor dimension or its inductance value became ineffective. To further increase the operation frequency, the bandwidth of the inductive channel needed to be increased by decreasing the self-inductance value. However decreasing the self-inductance also decreased the inductive coupling in the linear region. Thus trade-offs need to made depending on the targeted application frequency. The impact of resistivity of the Si wafer (between 0.1 Ωcm and 20 Ωcm) on the inductive coupling was found to be negligible up to certain frequency, related to the Si wafer thickness.

It was further demonstrated that the inductive coupling technique can be used to transmit data at 1 Gbps through eight stacked Si chips.

REFERENCES

[1] I.A. Papistas et al., "Corsstalk noise effect of on-chip inductor links on power delivery networks," Proceedings of the IEEE International Symposium on curcuits and Systems, pp. 1938-1941, May 2016

[2] S. Han et al., "Performance improvement of resonant inductive coupling for wireless 3D IC interconnect," Proceedings of the IEEE International Symposium on Antennas and propagation, pp.1-4, july 2010

[3] K. Niitsu et al., " Interference from power/signal lines and to SRAM circuits in 65nm CMOS inductive-coupling link," Processdings of the IEEE Asian Solid-State Circuits Conference. Pp. 131-134, November 2007

[4] H. Ali et al., " Mathematical modeling of an inductive link for optimizing efficiency," IEEE symposium on Indutrial Electronics & Application. Vol.2, pp.831-855, 2009

[5] E. Beyne, "The 3-D Interconnect Technology Landscape," in IEEE Design & Test, vol. 33, no. 3, pp. 8-20, June 2016.

[6] E. Beyne et al., "Scalable, sub 2μm pitch, Cu/SiCN to Cu/SiCN Hybrid Wafer-to-wafer Bonding Technology" IEDM, 2017, pp. 729-732.

[7] N. Miura et al., "A 195-gb/s 1.2-W inductive inter-chip wireless superconnect with transmit power control scheme for 3-D-stacked system in a package," IEEE Journal of Solid-State Circuits, Vol: 4, Jan. 2006

System Co-Design of a 600V GaN FET Power Stage with Integrated Driver in a QFN System-in-Package (QFN-SiP)

Jie Chen, Xie Yong, Django Trombley, and Rajen Murugan

Texas Instruments Incorporated
Dallas, TX, USA
r-murugan@ti.com

Abstract — The adoption of GaN technology, in power electronic applications, is greatly facilitated by its inherent performance benefits as compared to its MOSFET counterparts. The ability to co-integrate controller and multiple passives, along with the GaN FETs on standard, cost-effectiveness, QFN-Module/SiP packaging technology, helps to further support GaN-based system proliferation. However, these benefits can easily be outweighed if system co-design practices, to minimize components interaction, are not adopted early in the design process. In this paper we present the electrical system co-design and measurement validation results of a high-performance 600V 70mΩ GaN transistor with integrated gate driver packaged in a 32-pin 8.00mm × 8.00mm QFN-SiP package. Due to the complex system-in-package (SiP) integration electromagnetic interactions between driver/FET, package, and PCB are exacerbated, compromising benefits of integration and system performance. We detail here how optimization of the system, was achieved through a coupled circuit-to-electromagnetic co-design modeling and simulation methodology. Laboratory measurements on a 600V driver-integrated GaN 70mΩ FET power stage for power conversion are presented that validate the integrity of the co-design modeling and simulation methodology.

I.   INTRODUCTION

The high breakdown voltage, high saturation velocity and superior thermal characteristic have made GaN device an ideal choice for high voltage, high power applications over its silicon counterpart [1-4]. Traditionally, GaN devices are packaged as discrete device and driven with separate driver, due to process technologies incompatibility. However, the discrete approach exacerbates electrical performance due to significant package and PCB interconnects parasitics power loops as have been demonstrated by multiple authors [5-8]. To overcome these challenges, an integrated architecture with driver and FET on same package is preferred. The integrated solution provides for minimization of electrical parasitics (viz. resistance and inductance) of the system (i.e. power stage + package + PCB) [9-10]. While beneficial, an integrated solution, if not design with appropriate considerations can lead to multiple challenges. Integrating the GaN FET with a driver in a single, cost-effective, QFN-SiP (system in package) packaging

technology, as demonstrated here, require multi-disciplinary design awareness and optimization. Design reliability, assembly, manufacturing, thermal, mechanical, and electrical challenges are some of the considerations to achieve desired performance. A detail discussion of the packaging reliability, manufacturing, design guidelines can be found in [7]. The work discussed here focuses on the system-level electrical co-design that were undertaken to achieving a high performance integrated solution.

As shown in Figure 1, in the integrated solution, the driver die and GaN FET together form a control loop with lower inductance, which is critical for high di/dt application. Also shown is the power loop which consists of GaN FET and interconnects on both the package and PCB. Co-design and co-optimization of these parasitics loops is critical to the design performance.

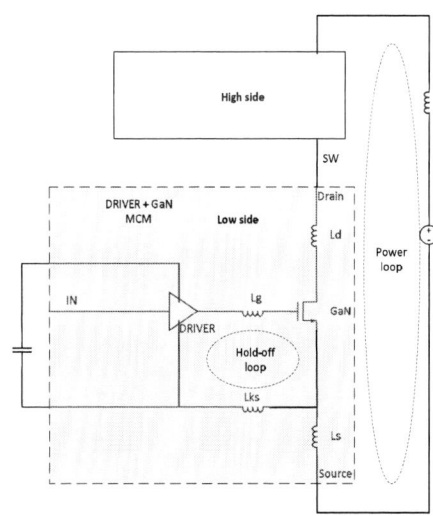

Figure 1.   Critical power and signal loops in the integrated solution.

After a brief overview of the integrated device in section II, the electrical parasitics co-extraction methodology developed for the package and PCB performance assessment is presented in section III. The system level transient analysis and measurement details of the power loop parasitics and system are presented in sections IV and V respectively.   System silicon characterization to modeling is discussed in Section VI.

978-1-7281-1500-9/19 $31.00 © 2019 IEEE

II. DEVICE DESCRIPTION

The 600V GaN power stage with integrated driver and protection device enables designers to achieve new levels of power density and efficiency in power electronics systems. The inherent advantages over silicon MOSFETs include ultra-low input and output capacitance, zero reverse recovery to reduce switching losses by as much as 80%, and low switch node ringing to reduce EMI [4]. These advantages enable dense and efficient topologies like the totem-pole PFC (power factor correction).

The device provides a smart alternative to traditional cascode GaN and standalone GaN FETs by integrating a unique set of features to simplify design, maximize reliability and optimize the performance of any power supply. Integrated gate drive enables 100V/ns switching with near zero V_{DS} ringing, <100 ns current limiting self-protects against unintended shoot-through events, Over-temperature shutdown prevents thermal runaway, and system interface signals provide self-monitoring capability.

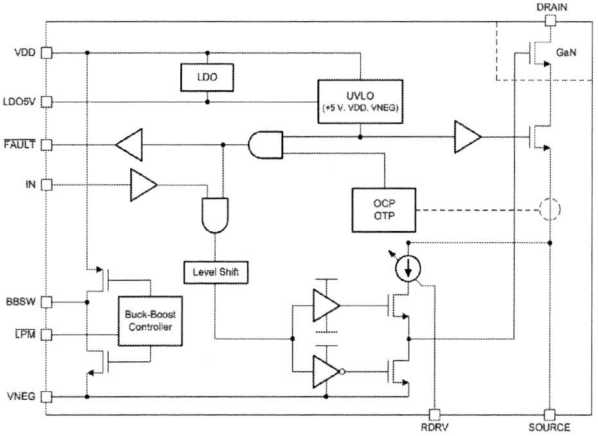

Figure 2. Functional block diagram of the integrated device.

Figure 2 above shows the functional diagram of the GaN power stage. The device utilizes a Direct Drive architecture to control the GaN FET. When the driver is powered up, the GaN FET is controlled directly with the integrated gate driver. Since it is a depletion mode device, the off-state gate voltage is -14V and the on-state is 0V. This architecture provides superior switching performance compared with the traditional cascode approach [6].

A comprehensive detail of the device can be found in [12]. The device includes numerous features to provide increased switching performance and efficiency in customers' applications while providing an easy-to-use solution. These include the Direct Drive architercture as described above, an internal inverting buck-boost

converter that generates a regulated negative voltage (V_{NEG}) for the turn-off supply of the GaN device, an internal auxiliary low-dropout (LDO) regulator to supply external loads (e.g. such as digital isolators for the high side drive signal, and fault detection schemes – namely over-current protection (OCP), over-temperature protection (OTP), and under voltage lockout (UVLO). It delivers 50% lower power losses in a totem-pole PFC compared with state-of-the-art silicon, FET-based boost power-factor converters. Lower power loss leads to high efficiency, which will be valued the most in high density industrial and consumer Power Supplies, multi-level converters, corporate computing and renewable energy applications.

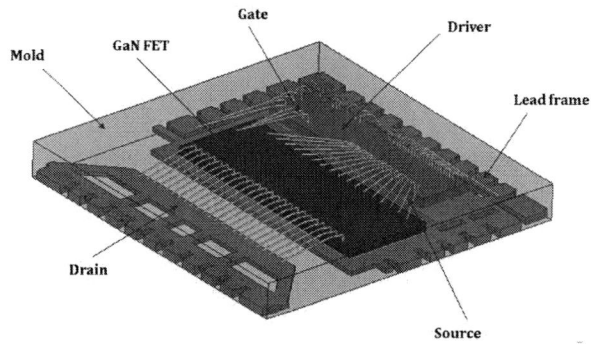

Figure 3. Device in an 8x8mm QFN-SiP package.

The device uses an 8x8mm QFN-SiP innovative package as shown in Figure 3 bove. For a comprehensive detail of the uniqueness of the QFN in system-in-package see reference [7]. As mentioned earlier, GaN devices are packaged as a discrete device and driven with a separate driver, because they are based upon different process technologies and may come from different manufacturers. Each package will have package bond wires and/or leads that introduce parasitic inductance. When being switched at high slew rates of 10s to 100s of volts per nanosecond, these parasitic inductances can cause switching loss, ringing, and reliability issues. To that end, the integration of the gate driver and GaN FET, in the same package, is expected to eliminate the common source inductance (CSI) and the gate loop inductance. Low gate loop inductance mitigates gate stress during turn-off and improves hold-off capability for GaN [11].

To evaluate the effect of parasitics on device performance and improve the designs of the package as well as the PCB, parasitic extraction is needed and a half bridge switching circuit needs to be simulated with the extracted models. The accuracy of parasitic inductance is critical and will be discussed in the next sections.

978-1-7281-1500-9/19 $31.00 © 2019 IEEE

III. EXTRACTION AND MODELING METHODOLOGY

To assess the electrical performance of the different package/PCB layout design iterations and their impact to the system, a coupled circuit-to-electromagnetic methodology was developed. Figure 4 below shows the high-level modeling flow for the assessment of the package and PCB performance. As per the flow, the package and PCB are designed with inputs from manufacturing/assembly rules and engineering/customer specs. RLGC (resistance, inductance, conductance, and capacitance) parasitics of the interconnect structures (bond wire, Leadframe and PCB routing) that connect the input circuit, driver circuit and the GaN FET were performed using 3D quasi-static electromagnetic solver. The solver combines MoM (method of moments) and FEM (finite element method) numerical extraction schemes to compute RLGC and corresponding LC mutual matrices. The RLGC matrices are then combined into a circuit model to represent the overall behavior of the whole interconnect structure.

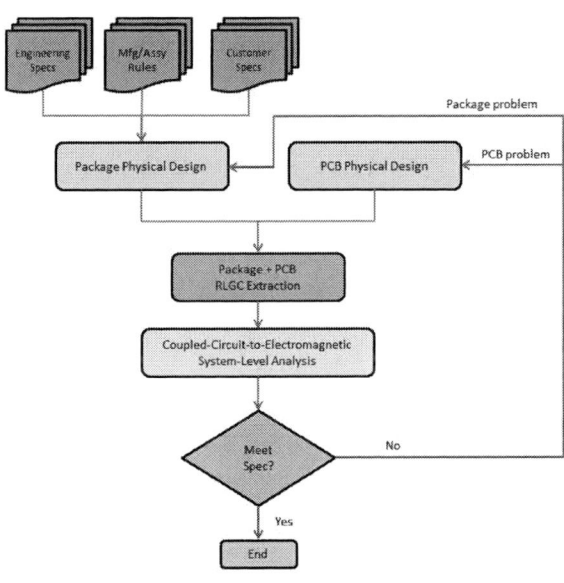

Figure 4.   Package/PCB Co-design and Co-analysis methodology.

The extracted models are then pushed through to the system-level circuit analysis step. In the system-level circuit analysis step, the time domain models of the package and PCB are coupled to the silicon/GaN FET transistor spice netlist and transient analysis was performed. If the requirements are met for the investigated figure of merits (ringing, slew rate, overshoot) the flow is exited. If the requirements are not met, then it iteratively looped back to the package and PCB and chip floor planning step and re-designs until requirements are met or exceeded. Implementation and accuracy of the core methodology has been demonstrated in other power electronics applications [13-15].

IV. SYSTEM LEVEL EXTRACTION & ANALYSIS

A. Package and PCB Extraction

For the transient system-level circuit analysis, the package and PCB time-domain Spice models are cascaded to the GaN FET Spice models. For accurate RLGC extraction, the package physical design was physically merged to the PCB physical design, as shown in Figure 5 below, before extraction was performed. As an example of the merged extraction process is shown for the LS FET (Figure 5). The parasitic power loop inductance from the PCB design and the inductance inside the low side device package are shown in Figure 1.

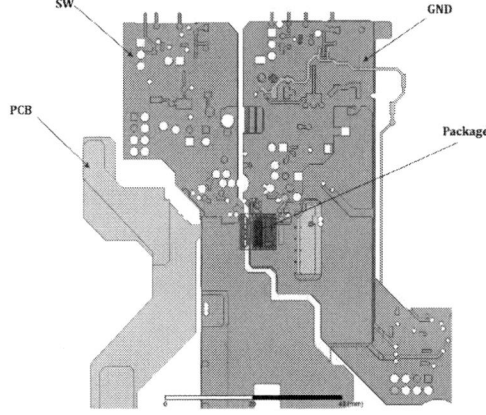

Figure 5.   Merging of package and PCB designs for extraction.

To extract the combined RLGC, ports are setup on the merged package and PCB in the 3D quasi-static extractor environment. Ports (source or sink) are located on the PCB and the wire tips of the package bond wires. The parasitics solver assumes equal potential within a defined port area. An understanding of the actual current path, through the system, is critical for realistic source/sink ports assignment if accurate parasitics extraction is required. Figure 6 below the ports set-up on the package and PCB for the combined RLGC model extraction.

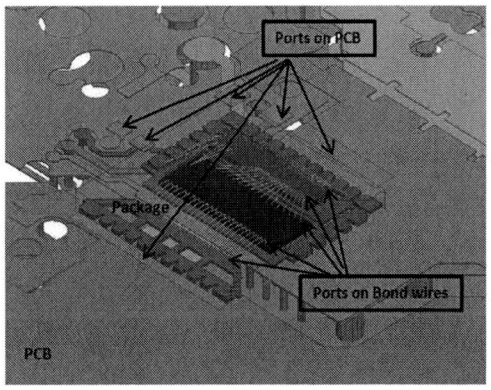

Figure 6.   Sink/source ports set-up on merged design.

Once the ports are set-up as per Figure 6, the parasitics extraction is performed. Figure 7 below shows the SPICE netlist representation of the merged package and PCB RLGC model. The model includes parasitics from the Drain lead frame and bond-wires, Gate bond-wires, source wires, together with the parasitics of the biasing/supply/control paths on the PCB. This Spice netlist representation is then cascaded to the GaN power stage circuit model, decaps model and biasing/supply/control circuit for the coupled circuit-to-electromagnetic system simulations. By probing the voltage ringing at the critical nodes on the circuit, contributions of each segment can be characterized. The package or PCB layout optimization decision can be made in multiple iterations when necessary in the system co-design and co-analysis process described in section III.

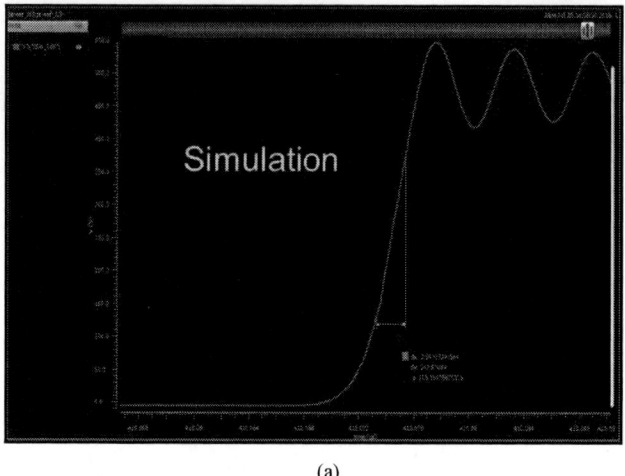

(a)

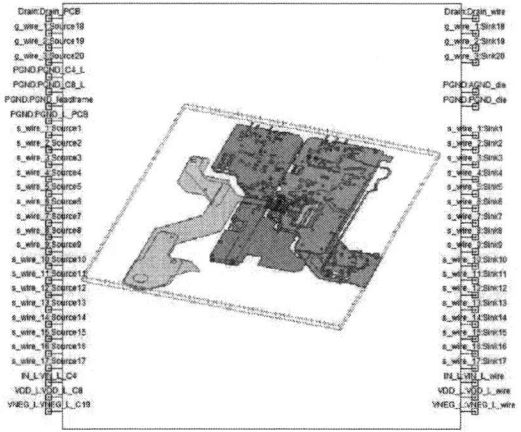

Figure 7. SPICE netlist represtation of the merged design with ports.

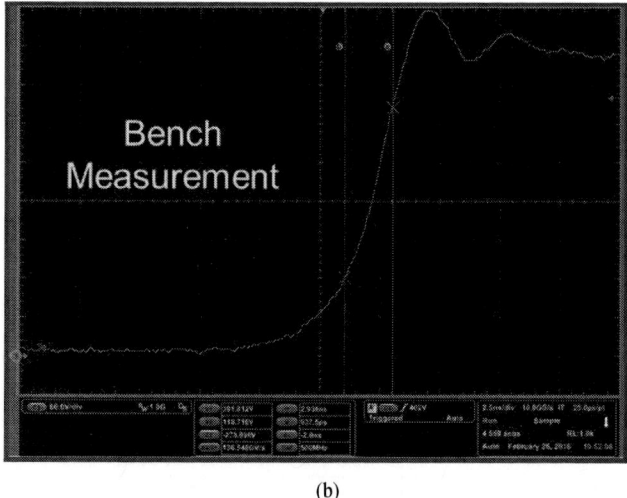

(b)

Figure 8. (a) Simulated and (b) Measured hard swtiching of SW node.

B. System-Level Analysis

When the GaN FETs are switching, the power loop current flows through the power loop inductance and causes ringing across the FETs. The ringing needs to be minimized to reduce voltage stress on the FETs. Figure 8(a) shows the simulated hard-switching transition of the switch (SW) node.

The parasitic inductances around the gate pull-down loop have important effects in holding the low-side device off while the high-side device is turning on and pulling the low-side drain towards the bus voltage. The current caused by the miller capacitance of the low-side flows through the gate loop. This may cause the low side device to turn on while it is expected in the off state and introduce a shoot through current. The shoot through current would cause additional loss in the system and may even damage the device if it is too high.

To find the effect of gate loop parasitics on the shoot through performance, the gate voltage holding off the low-side GaN FET is swept while the high-side is turning on. The gate voltage and drain current of the low side are monitored in the simulation as shown in Figure 9 below. A higher than usual peak current would indicate shoot-through. The result shows the shoot-through begins to appear when the hold-off voltage is around -11.5V. This result would help the designers to determine the hold-off voltage in early design stage. The hold-off voltage was designed to -14V to get a margin of 2.5V. The ability to perform predictive analyses with the package and PCB parasitics models is critical to assessing design robustness and performance early in the design process.

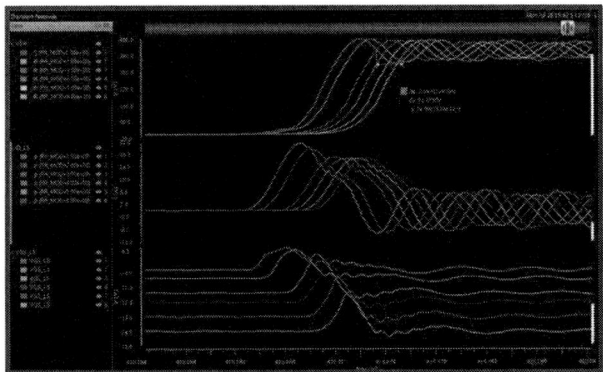

Figure 9.   SW node, low-side I_{ds} and V_{gs} while high-side turns on.

V.   MEASUREMENT DETAILS

To assess the accuracy of the 3D parasitics extractor, parasitics measurements were performed. Performance of the full wire-bond transition model (from lead-frame to die) can be tested and verified by measuring the RF parameters/figure of merits. Packaged parts were de-processed and de-capped to expose the lead-frame, wire-bond and silicon die. The measurement system was comprised of a HP4294A, DC manipulators and a calibration substrate to allow for corrections to the measurement. The calibration removes the stray cable/probe capacitance and inductance from the system. DC co-axial pins were used and care was taken to minimize the ground return path between the two RF pins otherwise the LCR meter will have issues balancing the bridge and virtual ground. The parameters were extracted using the series Ls model provided by the instrument. Figure 10 below shows the minimum set-up for the electrical characterization of the parasitics of the package.

Additionally to assess the accuracy of the system-level transient analysis, measurements were performed on the packaged device on an evaluation module (EVM). The packaged devices were tested in a half bridge configuration to correlate with the system level simulation in order to validate the extracted system-level models. Figure 11 below shows the silicon laboratory test bench and set-up.

Figure 11. Half bridge test setup to correlate the power loop and gate loop inductance.

VI.   DISCUSSIONS

Figure 12 shows the measured versus simulated inductance from 30MHz to 50MHz. As can be seen good correlation is achieved considering the difficulty and challenges with making sub nano-Henry (nH) measurements. Multiple measurement sets were performed and averaged.

Figure 10. Laboratory set-up to measure RLC parasitics of the package.

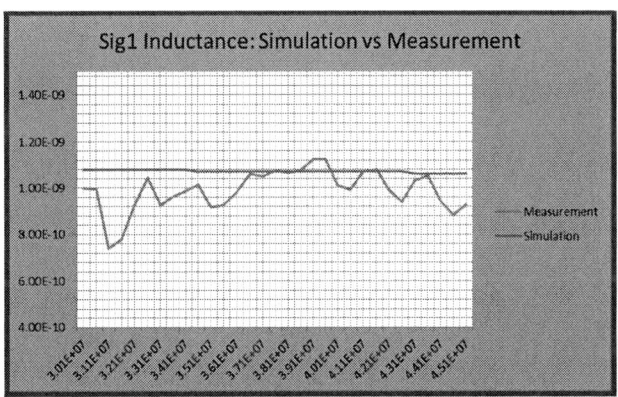

Figure 12. Measured vs Simulated Inductance for SIG1 signal.

As for SW waveform correlation, as shown in Figure 8 (a-b), good correlation was observed for the transient analysis. The SW node waveform showed good match in estimating the power loop inductance. The simulated ringing lasts longer showing lack of damping. This is due to the fact that ideal inductors were used in the extracted simulation. In reality the parasitic inductances are not ideal so the ringing is damped quickly.

It is difficult to correlate the gate-loop simulation result in Figure 9 since it is challenging to measure the short current pulse that only lasts for several nano seconds. However we can switch an unloaded half bridge in order to measure shoot-through. The current measured on the bus voltage supply can be expressed as:

$$I_{cc} = 2 \bullet f_{SW} \bullet Q_{OSS} \qquad (1)$$

where Q_{oss} is the output charge of each GaN FET, f_{sw} is the switching frequency. This current should be independent of the hold-off gate voltage. However as the gate voltage is moved up causing a shoot-through, I_{cc} increases due to the shoot-through current. Therefore a shoot-through can be detected by measuring the supply current of an unloaded half-bridge.

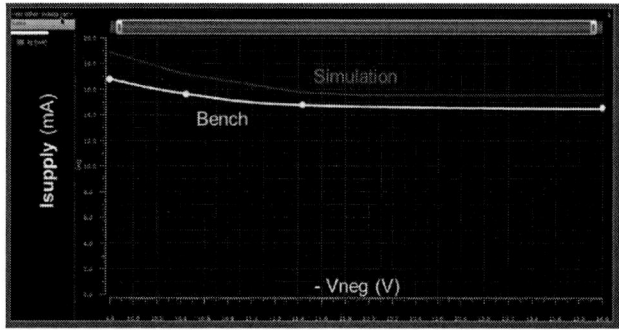

Figure 13. Supply current vs hold-off voltage while switching an unloaded half-bridge.

Figure 13 above shows the bus supply current when the hold-off gate voltage is swept. The simulation indicates shoot-through begins to happen at around -11.5V, which correlates well with the bench data thus shows the extracted gate loop matches the actual package parasitics. Note there is an offset between bench and simulation data since the output charge of the GaN model is slightly off due to GaN process variation.

VII. CONCLUSION

Power GaN FETs, due to their extremely low gate charge and output capacitance, can be switched at extremely high speeds with significantly reduced switching losses and improved efficiency compared to silicon FETs. The fast switching transients magnify the impact of parasitic inductance in device packaging and PCB due to Faraday's Law. As demonstrated here, by integrating the driver and GaN FETs in the same package, impact of common-source inductance and gate-loop inductance are greatly reduced. With proper PCB co-design additional parasitic inductance reduction can be achieved. Predictive co-modeling and co-analysis, at system-level, as shown here is critical for assessing design performance early in the design process for first time success.

ACKNOWLEDGMENT

The authors would like to thank Cetin Kaya for his ideas and valuable contributions in the system-level measurement of the shoot-through current.

REFERENCES

[1] Kikkawa, T. et al., "Current Status and Future Prospects of GaN HEMTs for High Power and High Frequency Applications", ECS Transactions. 50. 323-332., 2012.

[2] Travis J. Andersona et al., "GaN Power Devices – Current Status and Future Directions", Electrochem. Soc. Interface Winter 2018 volume 27, issue 4, 43-47.

[3] Amano, H et al., "The 2018 GaN Power Electronics Roadmap", Journal of Physics D: Applied Physics. 51. 10.1088/1361-6463/aaaf9d, 2018.

[4] Narendra M., "GaN FET module performance advantage over silicon", Texas Instruments, Inc. Application Note, SLYY071, 2015.

[5] Xie Y. and Brohlin P., "Optimizing GaN performance with an integrated driver", Texas Instruments White Paper, SLYY085, Feb 2016.

[6] Brohlin P. et al., " Direct-drive configuration for GaN devices", Texas Instruments White Paper, SLPY008A, 2018.

[7] Mishra D. et al., "Packaging Innovations for High Voltage (HV) GaN Technology", IEEE 67th Electronic Components and Technology Conference (ECTC), 2017.

[8] Lidow A. and Strydom J., "eGaN FET Drivers and Layout Considerations", "EPC White Paper, WP008, 2012.

[9] Reusch D., "Optimizing PCB Layout"., EPC White Paper, WP010, 2014.

[10] Letellier, Adrien et al., "Calculation of PCB Power Loop Stray Inductance in GaN or High di/dt Applications", IEEE Transactions on Power Electronics. 34. 612-623, 2018.

[11] Y. Xie, C. Kaya, P. Brohlin, "GaN HEMT Driver with Adjustable Slew Rate Control", Texas Intruments, Inc. TI Technical Leadership Conference, (TCL) 2016.

[12] LMG341xR070 600-V 70-mΩ GaN with Integrated Driver and Protection datasheet, Texas Instruments, Inc. http://www.ti.com/product/LMG3410R070.

[13] Goller J., Chen J. and Murugan R., "System-Level Crosstalk-Induced Efficiency Impact of DCDC Converter: Simulation to Measurement Correlation", APEC 2016, Long Beach California, March 21-24, 2016.

[14] Murugan R., Ai N. and C.T.Kao, "System-Level Electro-Thermal Analysis of RDS(ON) for Power MOSFET", IEEE SEMI-THERM33, March 13-17, 2017.

[15] Jie Chen et al., (2018), "System ElectroThermal Co-Design of a Zero-Drift Current-Shunt Monitor with Precision Integrated Shunt Resistor", International Symposium on Microelectronics: Fall 2018, Vol. 2018, No. 1, pp. 000193-00019.

Flexible probe for electrical neural signal recording

Sajay Bhuvanendran Nair Gourikutty and Lim Ruiqi

Institute of Microelectronics, A*STAR (Agency for Science, Technology and Research), 2 Fusionopolis Way, #08-02, Innovis, 117528, Singapore.
E-mail: nairs@ime.a-star.edu.sg

Abstract— **Implantable brain recording devices have provided a pivotal contribution in understanding the complex functioning of the human brain. However, these devices can be used to its fullest potential only if it can be implanted for a long period of time without compromising the functionality. In this paper, we report on an innovative idea of a flexible neural probe which can be implanted into brain tissue which causes a minimal immune reaction. The developed probe comprises of recording electrodes sandwiched between polyimide layers backed with bio dissolvable porosified silicon substrate. By employing different anodization recipes we are able to develop probes with porosities of 50%, 60% and 70%. It took 1-2 weeks for the porous silicon shank to fully dissolve in fluid Phosphate-buffered saline depending on the porosity and design used. We also report on recording electrodes made of platinum and platinum black which correspondingly exhibits an impedance of 700 kΩ and 6 kΩ at 1000 Hz. Lowering the impedance without changing the physical diameter of the electrodes will provide improved signal-to-noise ratio.**

Keywords-Neural probe, Porous silicon, MEMS, Electrical neural signal recording, immune reaction, Brain tissue response, Platinum black

I. INTRODUCTION

Neural recording from the human brain has demonstrated great potential in clinical studies and has vast possibilities in providing treatments for neurological conditions [1-3]. Neuro-technologies offer direct information on the neurological signal and neurochemical processes [4]. By developing a reliable system which can record and manipulate the activities inside the brain, it is possible to reestablish some of the functions for those patients who lost capabilities due to ailments or accidents. Recent developments show that by using high-density microelectrode arrays, the behavior of the body and its corresponding neural activity can be investigated.

For long term implants lifetime of the implanted probes which records quality neural signal is limited and therefore, it cannot meet the requirements of chronic recording medical applications. The main reason for the degraded neural signal recording quality is as a result of the immune reaction of the tissue. Irrespective of the method employed, the implantation of a device into the brain causes the blood-brain barrier (BBB) to be compromised and this will leads to several intricate cellular responses. Due to the implant, the tissue around the device may undergo unfavorable occurrences such as injury, strain due to mechanical impact, glial cell triggering and deterioration of neurons. Over time these deviations in the tissue behaviour around the implant adversely affect the sensitivity and lead to inaccurate recordings [5-9].

For a conventional rigid implantable probe, due to the mechanical property mismatch between the probe (~160 GPa for Si) and the cortical brain tissue (~100 KPa), damage of the tissue keeps happening as a result of relative micromotion between the tissue and rigid probe. This will ultimately cause the degradation of neural signal recording quality over time due to the formation of glial scar tissue and neuron degradation around the electrode. Even though the regeneration of healthy neurons is possible, the scar prevents the neuronal cells to occupy the regions proximal to the recording electrodes. By making the implanted probes flexible, the tissue immune response can be reduced and thereby the signal quality can be maintained. Figure 1 illustrates the tissue reaction in response to long term probe implantation.

Several attempts have been made to achieve a flexible probe by using materials such as parylene, PDMS, SU-8 [10-12]. However, making the probe flexible alone cannot achieve the objective, since the probe can hardly penetrate the brain tissue. So on top of flexibility, another layer which can assist the insertion of the probe is required. A number of materials have been investigated to employ as bio dissolvable coating to strengthen the recording probes such as silk and sugar, however, a thick coating is required to assist the implantation process [13]. Thick probes are not desirable since it displaces a big volume of tissue and also causes unnecessary implantation trauma. Porous silicon is an excellent substrate which can be employed as a candidate for strengthening the flexible implant. This material can provide enough strength during the implantation process and will degrade over time when it comes in contacts with bodily fluids. In addition to that, porous silicon does not cause any harm to the living tissue and exhibit a high degree of biocompatibility [14-15]. Another advantageous characteristic of this material is that it can be used to carry drugs since the substrate can act as a reservoir for drug due to the porous structure [16]. This property can be employed to rectify the scar caused due to the implantation process by gradually releasing the drug.

In this study, we describe the design, fabrication and testing of a flexible neural interface probe. In our approach, a flexible probe made of polyimide material strengthened by porous silicon shank is employed. The porous silicon shank functions as a mechanical stiffener which provides enough strength and facilitates smooth probe insertion process. Once implanted, the porous silicon shank will subsequently dissolve when it comes in contact with body fluids, leaving only the flexible polymer probe with the recording electrodes. This design will reduce the stiffness incompatibility between the implant and the brain tissue, thereby mitigating tissue response.

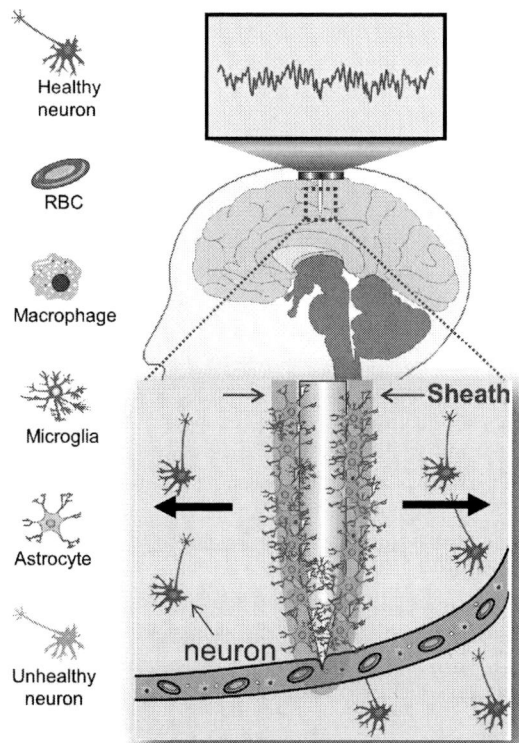

Figure 1. Demonstration of tissue reaction in response to long term probe implantation. Migration and degeneration of neuron followed by encapsulation of the implant with layers of astrocytes, macrophage and microglia causing the formation of a barrier.

II. MATERIALS AND METHODS

A. Design of the Probe

The designed probe consists of electrodes sandwiched between two polyimide layers (5 μm each) on top of 60 μm thick rigid porous silicon shanks. Although the probe is a single structure, it consists of two parts in terms of the materials used for the substrate. The vertical needle-like structure also called shank in which the electrodes are placed is made of porous silicon. This substrate is capable of degrading when it comes in contact with bodily fluids leaving only the metal sandwiched layers. The horizontal structure to which the shanks are connected namely the

backbone is made of normal silicon. Backbone is the structure used for die attach process on to a PCB or other 3D silicon platforms. It also has bond pads of the recording electrodes to which wire bonding is done.

To cater for diverse applications, the neural probes with a different number of shanks (1, 2, 4 and 18) were designed, shank lengths (2 mm, 3.5 mm and 5 mm) and different electrode placement pattern on the shank. The probes with shanks of 5 mm length are suitable for larger animals such as primates or to access the deep brain tissues of smaller animals. The probe backbone provides 20 bonding pads and the probe itself has a thickness of 70 μm consisting of 18 neural signal recording electrodes and 1-2 reference electrodes depending on the design. The electrodes are circular in shape and designed with a diameter of 20 μm to 30 μm. Figure 3d and 3e show the microscopic image of fabricated probes with 18 shanks having a shank length of 5 mm and 2 mm respectively.

B. Microfabrication

The neural probes were fabricated by employing silicon microfabrication methods. A highly doped p-type 8-inch silicon wafers were used for chip fabrication and the process starts with photolithography for defining the area for Deep reactive-ion etching (DRIE) to create probe outline trench. A highly anisotropic Bosch etching process is used to form the trench by etching the silicon with precise process control. A deep trench of 100 μm height is created to define the probe shape. Subsequently, the trench is filled with Si_3N_4 and undoped polysilicon deposition for protection during Si anodization process followed by Chemical-mechanical polishing (CMP) to flatten the uneven surface. Porosification of the silicon substrate is achieved by a process called Anodization. Here a single cell Anodization is done to introduce pores in silicon which consists of a Platinum cathode and Silicon Anode (Wafer) with HF solution as a medium. The mechanism of forming pores can be explained like this: A depletion region forms near the interface of HF-Silicon and the width of this depletion region depends on doping concentration in the wafer. When depletion region of two different pores come closer to each other there will be pinch off and Si etching is blocked in that area. This is the reason pores form during Anodization rather than complete Silicon etching. The depth of the anodization from the surface is controlled by varying the etching time. A 5 μm thick polyimide is coated on top of the Porosified substrate, patterned and cured. This polyimide layer acts as the bottom sandwiching layer for the electrodes once the substrate dissolves inside the tissue. Subsequently, the metal layer which forms the electrode, trace and bond pad is deposited on top of the polyimide layer by means of the evaporation process. A second 5 μm thick polyimide is coated on top of the metal layer and undergoes photolithography to define the opening for the recording electrodes. Lastly, the trenches are reopened using the DRIE process and the individual probes are released after wafer back grinding process. The fabrication process flow can be seen from Fig. 2.

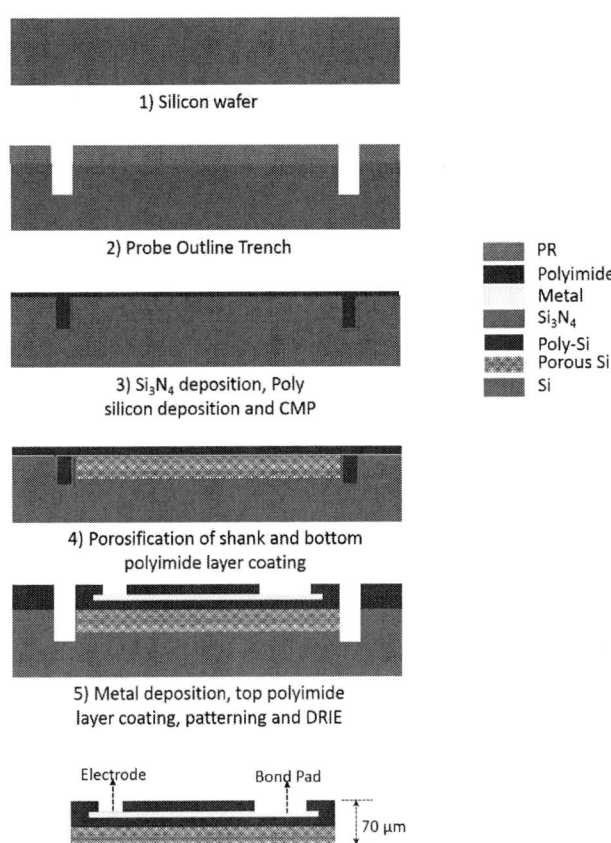

1) Silicon wafer

2) Probe Outline Trench

3) Si₃N₄ deposition, Poly silicon deposition and CMP

4) Porosification of shank and bottom polyimide layer coating

5) Metal deposition, top polyimide layer coating, patterning and DRIE

Electrode   Bond Pad
                          70 μm
6) Back grind and probe release

PR
Polyimide
Metal
Si₃N₄
Poly-Si
Porous Si
Si

Figure 2. The microfabrication process flow of Porous silicon probe.

C. Cytotoxicity test

Cell culture was used to evaluate the cytotoxic effect of the microfabricated neural probes as per ISO 10993 recommendation. NCTC Clone 929 Mouse Connective tissue cell line is used in this study because it is one of the recommended cell cultures that can be used and it is the one widely used for assessing medical devices. A cell suspension of approximately 10^5 cells per ml prepared in culture medium was transferred to each well of 6-well plates. The setup was then incubated at 37 degree Celsius until subconfluent monolayer was formed. The confluency and morphology of the cultures were verified by microscopic observation prior to the start of the test. All test were performed in triplicates. The sample is created by using 4g of probes to be tested in 20 ml of complete MEM10 cell culture medium. The extract from the sample was filtered by passing through a 0.2 μm membrane filter. The negative control sample is prepared by using one piece of 60 cm² HDPE material in 20 ml of extraction medium. And the positive control sample is made by adding 0.016 gram of Zinc sulphate in 20 ml of extraction medium. After incubation, the culture medium from the confluent cell layer was removed and replaced by the extract of the testing sample, positive and negative controls in each well respectively. The samples were then incubated for 48 hours at 37 degree Celsius and 5% CO_2 in a humidified incubator. After incubation, the cell culture plates were examined under the microscope.

D. Bio-dissolvability test

The bio-dissolvability of Porous silicon probes were evaluated by monitoring the degradation of the substrate over time in the fluid. To mimic the bodily environment Phosphate-buffered saline (1× PBS) was used. The probe to be tested were added to 60 mm petri dish containing 10 ml of PBS solution and placed inside the humidified incubator for periodic monitoring. The samples were inspected every two days by mounting it on to an upright motorized microscope (BX63, Olympus). Image capturing was performed with a 17.28 megapixel 14-bit CCD camera (Catalog# DP73, Olympus). The image acquiring process was performed with a 10× objective lens. For further analysis of sample, a 20× objective lens was used when required.

III. RESULTS AND DISCUSSION

A. Selective porosification

The designed neural probe becomes functional only if it can be assembled onto a PCB to access the recording electrodes. Therefore, the backbone of the probe comprising bond pads should withstand the die attach process and wire bonding to complete the assembly on to the PCB. To facilitate this, the probes are designed such way that the backbone is made of a rigid substrate, that is silicon and at the same time, the needle-like shanks are made of bio dissolvable porous silicon substrate which can dissolve gradually once it comes in contact with tissue. Therefore, a selective porosification is needed to ensure a rigid backbone and bio dissolvable shanks. This is made possible by a sequence of special microfabrication processes. Since the porosification is required across the whole thickness (60 μm) of the probe a longer anodization process is needed. Therefore, conventional masking layers such as silicon oxide are not sufficient since the oxide layer gets etched in HF. Here, we used a combination of silicon nitride and polysilicon to protect the backbone area from getting porosified.

To enable the release of the fragile probe at the end of the microfabrication, a deep trench is created to outline the probe from the beginning as shown in Fig. 3a. Subsequently, to avoid defects and to facilitate smooth downstream processes such as spin coating and deposition processes the trench is buried with silicon nitride and polysilicon (Fig. 3b). Lastly, right before the back grinding process to release the probes the trenches are reopened by means of DRIE. Figure 3c clearly shows the cross-sectional

SEM image of Porosified silicon in contrast to the normal silicon.

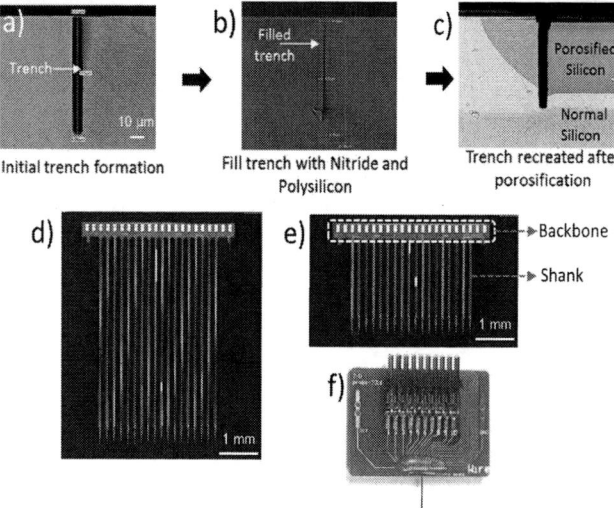

Figure 3. Probe fabrication and assembly a-c) Steps showing the probe outline trench formation, filling with silicon nitride and polysilicon and the formation of porous silicon after the anodization process. It can be clearly seen that the porousified silicon (darker region) in comparison to normal silicon. A fabricated 18 shank probe with a shank length of d) 5 mm e) 2 mm. f) A 5 mm single shank probe assembled on to a PCB for testing.

B. Characterization of electrodes

Due to microscopic dimensions of neural signal recording and stimulation electrodes, usually available probes have very high impedances. For recording single unit neuronal activities, it is essential to design the electrodes small and comparable to the size of the cell body. Electrodes with reduced impedance are developed here by modifying the surface by means of electroplating and creating a layer of platinum black. Platinum black provides a very large surface area due to uneven microscopic structures on its surface and therefore low impedance. The benefit of using platinum black is that it can reduce the impedance without impacting the physical size of the electrodes. Maintaining the small diameter of the electrode but lowering the impedance can assist stimulation studies as well by providing reduced stimulation voltages and reduced stimulation artefacts.

In this study, electrodes with three different electrode materials have been developed namely Gold (Au), Platinum (Pt) and Platinum Black (Pt Black). Gold and Platinum electrodes are made by the microfabrication process called metal evaporation followed by photolithography to pattern the metal layer. To create the neural probe with the platinum black electrode, an electroplating process is employed to add a layer of platinum black on to the preexisting platinum electrode. The platinum electrode to be blackened is made the cathode with respect to a large-area anode in the solution. A platinum mesh structure will act as anode and an aqueous solution of hexachloroplatinum is used as the electrolyte at room temperature. Platinum black was deposited in the presence of sonic agitation at ultrasonic frequency range to improve the stability of platinum black on to underlying layer. Figure 4a shows the microscopic image of a probe with platinum electrodes and Fig. 4b is made with platinum black electrodes.

Electrochemical Impedance Spectroscopy (EIS) is done for gold, platinum and platinum black electrodes whereby the frequency is scanned ranging from 0 to 10^5 Hz and impedance was measured. The test shows Pt Black electrodes exhibit low impedance (6 kΩ at 1 kHz) in comparison to gold and platinum electrodes (~700 kΩ at 1 kHz). These electrodes with platinum black can help to provide a better signal-to-noise ratio. Figure 4 c and 4d show the impedance spectroscopy of platinum and platinum black respectively. Even though the probes are developed for neural recording applications, electrodes modified to platinum black are also suitable for brain stimulation application since the electrodes larger surface area means lower impedance and higher charge injection ability [18-19].

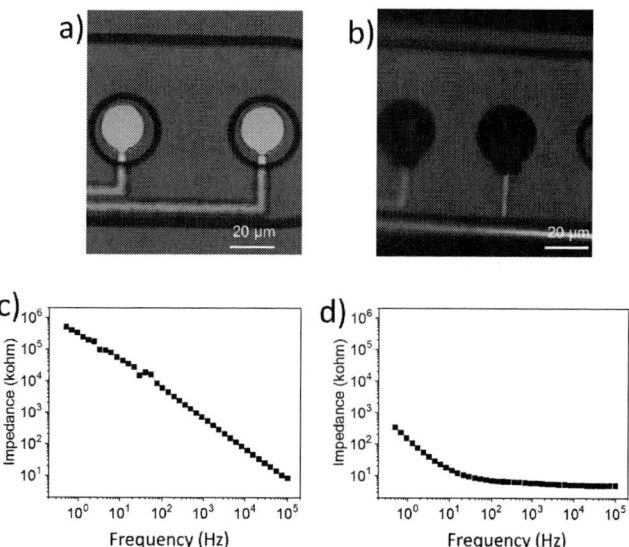

Figure 4. Microscopic image of neural probe electrode a) Platinum electrode b) Platinum black electrode. Electrochemical Impedance Spectroscopy (EIS) of the neural probe with c) Platinum electrode d) Platinum black electrode.

C. Cytotoxicity and cell viability

Since the device developed is an implantable device the effect of the material on to the living cells are very critical. To determine if the materials used for neural probe have an adverse effect on cell growth, a test called cytotoxicity is performed. The samples to be tested are too small in dimensions and therefore, a direct contact method is not practical. Here, an elution method is employed which is an assay designed to show the presence of toxic material eluted from the probe sample. This test is a biological evaluation

978-1-7281-1500-9/19 $31.00 © 2019 IEEE

that use tissue cells in vitro to observe the cell growth, reproduction and morphological effects by the designed probe. It was performed using NCTC Clone 929 Mouse connective tissue cell line. Figure 5a clearly shows the microscopic image of cells treated with the probe samples without affecting its morphology or growth. The cell culture was well maintained and proliferated in contact with probe samples indicating the probe materials have not toxic effect on the cells. Whereas, the positive control (Fig. 5b) shows the growth of the cell culture is substantially affected by the Zinc sulphate solution.

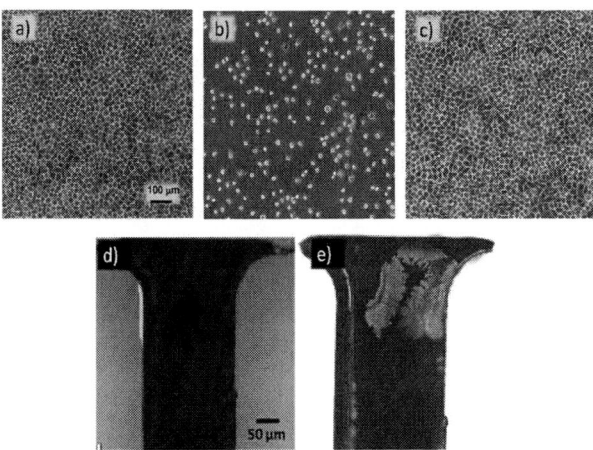

Figure 5. Cytotoxicity test conducted using NCTC Clone 929 Mouse connective tissue cell line. Microscopic image of cells treated with a) sample extracted from neural probe b) Positive control using Zinc sulphate solution c) Negative control using HDPE material. Microscopic images showing the degrading of the probe with a porosity of 50 % in Phosphate-buffered saline d) day 1 e) day 4.

D. Porosity and Bio-dissolvability

To achieve different levels of strength and dissolvability, porous silicon substrates with three different porosity were developed such as 50%, 60% and 70%. This is realized by developing special anodization recipes by varying current density ($20 \ J/cm^2$ to $64 \ J/cm^2$) and etching time (30 min–120 min). The degradation of the porous silicon can be explained like this - in an aqueous solution, the functional group SiH_x on the surface of the porous silicon substrate will get oxidized to soluble oxides. Once the substrate is degraded the byproduct mainly consists of various silicic acid [17]. The silicic acid is naturally occurring compound and it is not harmful to the body. A highly porous silicon substrate will degrade faster in comparison to the silicon with a low percentage porosity. Although high porosity is desirable for faster degrading and thereby reduced immune reaction to the brain tissue, it will increase the fragility of the substrate. It was observed that the substrate of the shanks was fully degraded in saline solution within 12, 8 and 6 days with the porosity of 50%, 60% and 70% respectively. Figure 5e shows the microscopic image of a

degrading porous silicon probe with a porosity of 50%. This test confirms the ability of the probe to maintain rigidity during the time of implantation and degrade over time evolving into a thin flexible recording device. Hence, mitigating the immune reaction of host tissue facilitating long term recording.

IV. CONCLUSIONS

We have presented a development of bio dissolvable neural probe which will provide minimal damage to brain tissue during long term implantation. The designed probe is made of metal electrodes sandwiched between polyimide layers and backed with a rigid porous silicon substrate. The probe is realized by CMOS compatible microfabrication process in which only the shank of the probe is Porosified while the backbone of the probe is silicon to facilitate the assembly process. Depending on the porosity of the probe, after tissue insertion, the shank of the probe will degrade within days leaving thin and flexible recording electrodes. As a result by minimizing the mechanical mismatch between the probe and brain tissue, immune responses of the host tissue are greatly reduced. We also investigated different types of electrode materials which can provide improved signal to noise ratio during neuron signal recording. Therefore, this novel approach is expected to substantially improve the long term recording of neural probes.

ACKNOWLEDGMENT

This work was supported by the Science and Engineering Research Council of A*STAR (Agency for Science, Technology and Research), Singapore under the grant number IAF311022.

REFERENCES

[1] Roitman, M. F., G. D. Stuber, P. E. Phillips, R. M. Wightman and R. M. Carelli. "Dopamine Operates as a Subsecond Modulator of Food Seeking." J Neurosci 24, no. 6 (2004): 1265-71.

[2] Cirrito, J. R., K. A. Yamada, M. B. Finn, R. S. Sloviter, K. R. Bales, P. C. May, D. D. Schoepp, S. M. Paul, S. Mennerick and D. M. Holtzman. "Synaptic Activity Regulates Interstitial Fluid Amyloid-Beta Levels in Vivo." Neuron 48, no. 6 (2005): 913-22.

[3] Collinger, J. L., B. Wodlinger, J. E. Downey, W. Wang, E. C. Tyler-Kabara, D. J. Weber, A. J. McMorland, M. Velliste, M. L. Boninger and A. B. Schwartz. "High-Performance Neuroprosthetic Control by an Individual with Tetraplegia." Lancet 381, no. 9866 (2013): 557-64.

[4] Kozai, T. D., A. S. Jaquins-Gerstl, A. L. Vazquez, A. C. Michael and X. T. Cui. "Brain Tissue Responses to Neural Implants Impact Signal Sensitivity and Intervention Strategies." ACS Chem Neurosci 6, no. 1 (2015): 48-67.

[5] Kozai, T. D., A. L. Vazquez, C. L. Weaver, S. G. Kim and X. T. Cui. "In Vivo Two-Photon Microscopy Reveals Immediate Microglial Reaction to Implantation of Microelectrode through Extension of Processes." J Neural Eng 9, no. 6 (2012): 066001.

[6] Holson, R. R., J. F. Bowyer, P. Clausing and B. Gough. "Methamphetamine-Stimulated Striatal Dopamine Release Declines

Rapidly over Time Following Microdialysis Probe Insertion." Brain Res 739, no. 1-2 (1996): 301-7.

[7] Holson, R. Robert, Russell A. Gazzara and Bobby Gough. "Declines in Stimulated Striatal Dopamine Release over the First 32 H Following Microdialysis Probe Insertion: Generalization across Releasing Mechanisms." Brain Research 808, no. 2 (1998): 182-189.

[8] Robinson, T. E. and D. M. Camp. "The Effects of Four Days of Continuous Striatal Microdialysis on Indices of Dopamine and Serotonin Neurotransmission in Rats." J Neurosci Methods 40, no. 2-3 (1991): 211-22.

[9] Kozai, Takashi D. Y., Zhanhong Du, Zhannetta V. Gugel, Matthew A. Smith, Steven M. Chase, Lance M. Bodily, Ellen M. Caparosa, Robert M. Friedlander and X. Tracy Cui. "Comprehensive Chronic Laminar Single-Unit, Multi-Unit, and Local Field Potential Recording Performance with Planar Single Shank Electrode Arrays." Journal of neuroscience methods 242, (2015): 15-40.

[10] Fernández, Luis J., Ane Altuna, Maria Tijero, Gemma Gabriel, Rosa Villa, Manuel J. Rodríguez, Montse Batlle, Roman Vilares, Javier Berganzo and F. J. Blanco. "Study of Functional Viability of Su-8-Based Microneedles for Neural Applications." Journal of Micromechanics and Microengineering 19, no. 2 (2009): 025007.

[11] Brien, D. P. O', T. R. Nichols and M. G. Allen. "Flexible Microelectrode Arrays with Integrated Insertion Devices." In Technical Digest. MEMS 2001. 14th IEEE International Conference on Micro Electro Mechanical Systems (Cat. No.01CH37090), 216-219, 2001.

[12] Chen, Y. Y., H. Y. Lai, S. H. Lin, C. W. Cho, W. H. Chao, C. H. Liao, S. Tsang, Y. F. Chen and S. Y. Lin. "Design and Fabrication of a Polyimide-Based Microelectrode Array: Application in Neural Recording and Repeatable Electrolytic Lesion in Rat Brain." J Neurosci Methods 182, no. 1 (2009): 6-16.

[13] Wu, F., L. W. Tien, F. Chen, J. D. Berke, D. L. Kaplan and E. Yoon. "Silk-Backed Structural Optimization of High-Density Flexible Intracortical Neural Probes." Journal of Microelectromechanical Systems 24, no. 1 (2015): 62-69.

[14] Herranz-Blanco, Bárbara, Laura R. Arriaga, Ermei Mäkilä, Alexandra Correia, Neha Shrestha, Sabiruddin Mirza, David A. Weitz, Jarno Salonen, Jouni Hirvonen and Hélder A. Santos. "Microfluidic Assembly of Multistage Porous Silicon–Lipid Vesicles for Controlled Drug Release." Lab on a Chip 14, no. 6 (2014): 1083-1086.

[15] Shahbazi, M. A., M. Hamidi, E. M. Makila, H. Zhang, P. V. Almeida, M. Kaasalainen, J. J. Salonen, J. T. Hirvonen and H. A. Santos. "The Mechanisms of Surface Chemistry Effects of Mesoporous Silicon Nanoparticles on Immunotoxicity and Biocompatibility." Biomaterials 34, no. 31 (2013): 7776-89.

[16] Sun, T., S. Merugu, W. M. Tsang, W. Park, N. Xue, Y. Liu, B. Han, G. Dawe and A. Y. Gu. "A Microfabricated Neural Probe with Porous Si-Parylene Hybrid Structure to Enable a Reliable Brain-Machine Interface." In 2016 IEEE 29th International Conference on Micro Electro Mechanical Systems (MEMS), 153-156, 2016.

[17] Anglin, Emily J., Lingyun Cheng, William R. Freeman and Michael J. Sailor. "Porous Silicon in Drug Delivery Devices and Materials." Advanced drug delivery reviews 60, no. 11 (2008): 1266-1277.

[18] Cogan, S. F. "Neural Stimulation and Recording Electrodes." Annu Rev Biomed Eng 10, (2008): 275-309.

[19] Long, Mingce, Jingjing Jiang, Yan Li, Ruqiong Cao, Liying Zhang and Weimin Cai. "Effect of Gold Nanoparticles on the Photocatalytic and Photoelectrochemical Performance of Au Modified Bivo4." Nano-Micro Letters 3, no. 3 (2011): 171-177.

Stretchable, Implantable Nanomembrane Biosensor for Wireless, Real-Time Monitoring of Hemodynamics

Robert Herbert and Woon-Hong Yeo*

George W. Woodruff School of Mechanical Engineering, Wallace H. Coulter Department of Biomedical Engineering,
Institute for Electronics and Nanotechnology,
Georgia Institute of Technology
Atlanta, GA, USA
*e-mail: whyeo@gatech.edu

Abstract— Unruptured cerebral aneurysms exist in as many as 6% of the population in the world, introducing risk of serious damage if not properly treated. Current gold standard of care involves invasive treatment via clipping or embolization in the targeted aneurysm sacs. Although post-treatment monitoring of blood flow is recommended, the existing practice using angiography is costly and invasive. Recently developed flow diverters, inserted in the region of a blood vessel and neck of a sac, offers lower procedural risks, while improving effectiveness. Even though some devices show a possible method of sensor integration for active monitoring of hemodynamics, they still require cumbersome tethering with an external data acquisition system. Here, we introduce a wireless, implantable, stretchable biosensor system, comprised of a miniaturized capacitive sensor and inductive coil, for integration into a batteryless functional device. Quantitative analysis of transient sensor signals allows accurate identification of a resonant frequency. The capacitive sensor is fabricated via aerosol jet printing of silver nanoparticles and polyimide. Analytical and experimental study optimizes the inductive coupling mechanism to achieve the maximum readout distance over 5.5 cm in air with a fabricated biosensor. The optimized wireless system and low-profile form demonstrates a 2.4 ratio of readout distance to cross-sectional area and possesses the most applicable performance among reported implantable, inductive coupling sensors. *In vitro* experiments include real-time, wireless monitoring of flow rates in a cerebral aneurysm model. Overall, the wireless system and stretchable biosensor provide a method to offer real-time monitoring of hemodynamics of blood flow in cerebral aneurysms with an abnormal focal dilation.

Keywords—Stretchable hybrid electronics, wireless monitoring, hemodynamics, aerosol jet 3D printing

I. INTRODUCTION

Weakened portions of blood vessel walls lead to an enlargement of the vessel wall and formation of an aneurysmal sac. Unruptured cerebral aneurysms occur in about 6% of the population [1]. Treatments often aim to minimize blood flow into the aneurysmal sac as untreated aneurysms include a risk of rupture and resulting death [2, 3]. Cerebral aneurysms occur in highly contoured and narrow blood vessels, and thus stretchable and flexible devices are required. However, standard treatments include the invasive procedure of surgical clipping or coiling, which is less widely applicable [4]. Recent progress in cerebral aneurysm treatments include flow diverters based on thin film nitinol offers a more promising approach [5, 6].

Both short-term and long-term follow-up care is still necessary to ensure treatment progress and identify potential risk of rupture [4]. Current follow-up monitoring techniques include angiography, an invasive and costly procedure. To improve the post-monitoring capability, implantable sensors have been investigated in our prior works by using a low-profile capacitive flow sensor [4, 6, 7]. Even though this study shows a feasibility of hemodynamics quantification, this method still needs wireless powering and telemetry for a fully implantable system. Widely used wireless interrogation for implantation primarily include radiofrequency (RF) communication and inductive coupling [8, 9]. Considering a narrow, highly contoured cerebral blood vessel, inductive coupling offers the ideal option since it does not require an electronic circuit. Some recent works have attempted the integration of rigid sensors and the inductive coupling method with medical stents [10, 11]. However, these devices show readout distances less than 3 cm in air, which cannot address cerebral aneurysm applications as cerebral aneurysms often occur 5 cm from the skin [12].

Here, we introduce an inductive coupling method that offers enhanced wireless telemetry for a flow-diverter and associated sensor system. Fig. 1 captures the overview of a newly proposed monitoring system for quantification of hemodynamics in cerebral aneurysms. For the restricted cerebral vessel, the miniaturized sensor coil has a specific dimension between 3.5 and 7.5 mm in diameter and up to 30 mm in length. A low-profile, capacitive flow sensor monitors blood flow into an aneurysmal sac without disrupting normal hemodynamics (Fig. 1A).

Aerosol jet printing (AJP) is investigated and applied for the multilayer, highly stretchable sensor. The wireless telemetry method employs two external coil antennas to record transient sensor signals from the implanted device (Fig. 1B and 1C). Quantitative analysis of the transient signals enables detection of sensor resonant frequency and flow changes (Fig. 1D). Optimization of relevant parameters in the inductive coupling between coils is performed by analytical and computational calculations. In vitro experimental study demonstrates the performance of the optimal sensor coils in wireless detection of resonant frequencies through air, saline, and meat. *In vitro* demonstrations display wireless monitoring of cerebral aneurysms and indicate potential for *in vivo* applications.

978-1-7281-1500-9/19 $31.00 © 2019 IEEE

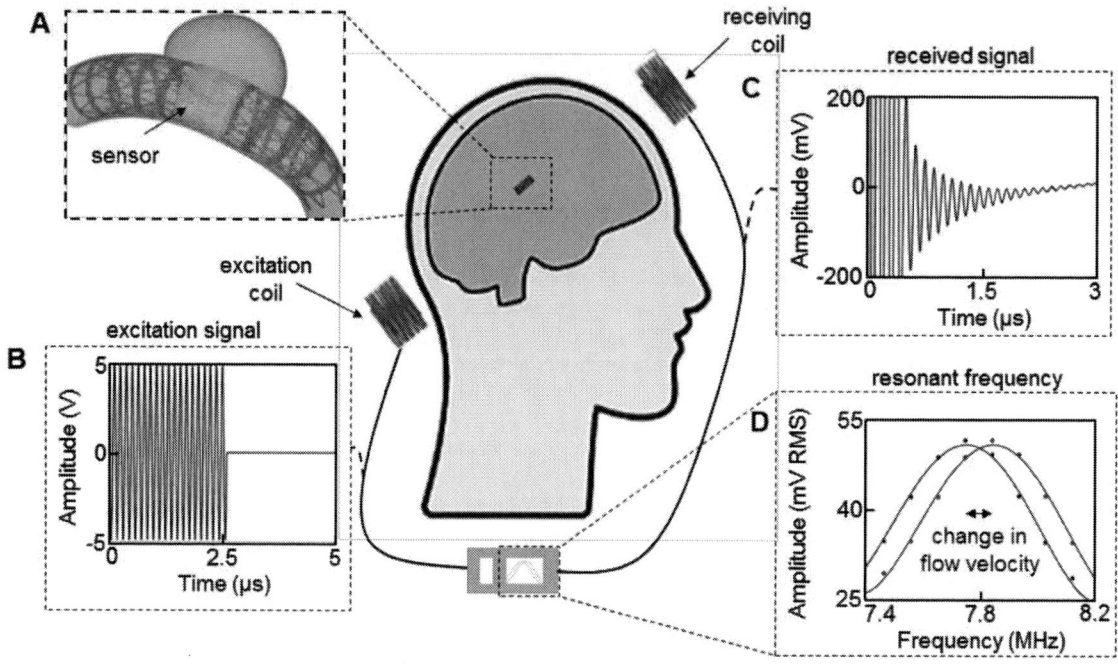

Figure 1. (A) Illustration of flow sensor implanted at cerebral aneurysm. (B) Excitation signal burst of specified frequency. (C) Signal at receiving coil showing transient response of sensor. (D) Frequency sweep showing shift in resonant frequency corresponding to a change in flow velocity.

II. AJP-ENABLED FABRICATION OF BIOSENSOR

The capacitive biosensor is fully fabricated via AJP and then transferred to silicone. Fig. 2A demonstrates the patterned, material deposition by AJP onto a sacrificial polymethyl methacrylate (PMMA) layer. The flow sensor consists of polyimide as the supporting layer for transfer and the dielectric layer. Silver nanoparticle (AgNP) layers form the capacitor's electrodes. Sensor capacitances vary between 60 and 80 pF and may be tuned via electrode area and dielectric thickness. In the present study, copper wire connects to the sensor electrodes via silver paste to enable attachment to an inductive coil.

Silicone elastomer encapsulates the sensor prior to conformal integration onto a medical stent (Fig. 2B). The resulting stent and sensor system maintain high stretchability and flexibility as confirmed by finite element analysis. The compact sensor pattern (Fig. 2C) enables stretchability up to 250% before plastic deformation. The sensor also achieves a 180° bending radius of 0.5 mm, indicating high flexibility. These results are validated via experimental stretching and bending while monitoring electrode resistance changes. The ultrathin form factor of the capacitive sensor prevents disruption to normal hemodynamics and enables a highly compliant device (Fig. 2D).

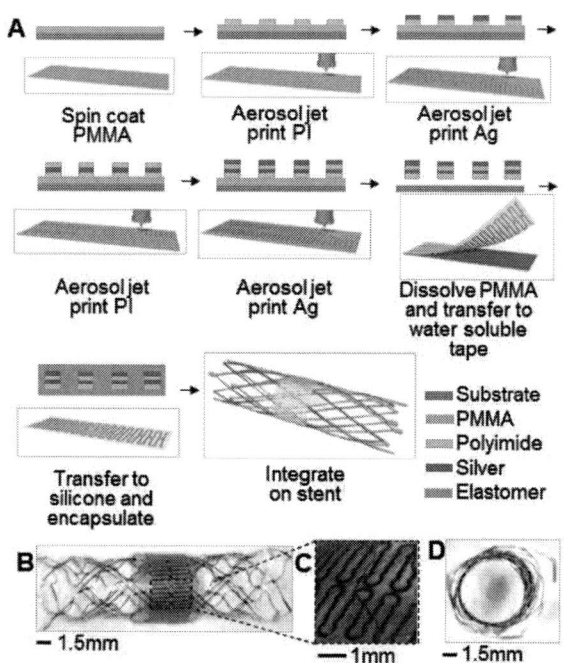

Figure 2. (A) Illustration of the AJP-enabled fabrication of capacitive flow sensor. (B) Nanomembrane sensor integrated with a medical stent. (C) Zoom-in view of the sensor pattern. (D) Cross-sectional view of the stent-embedded low-profile sensor.

978-1-7281-1500-9/19 $31.00 © 2019 IEEE

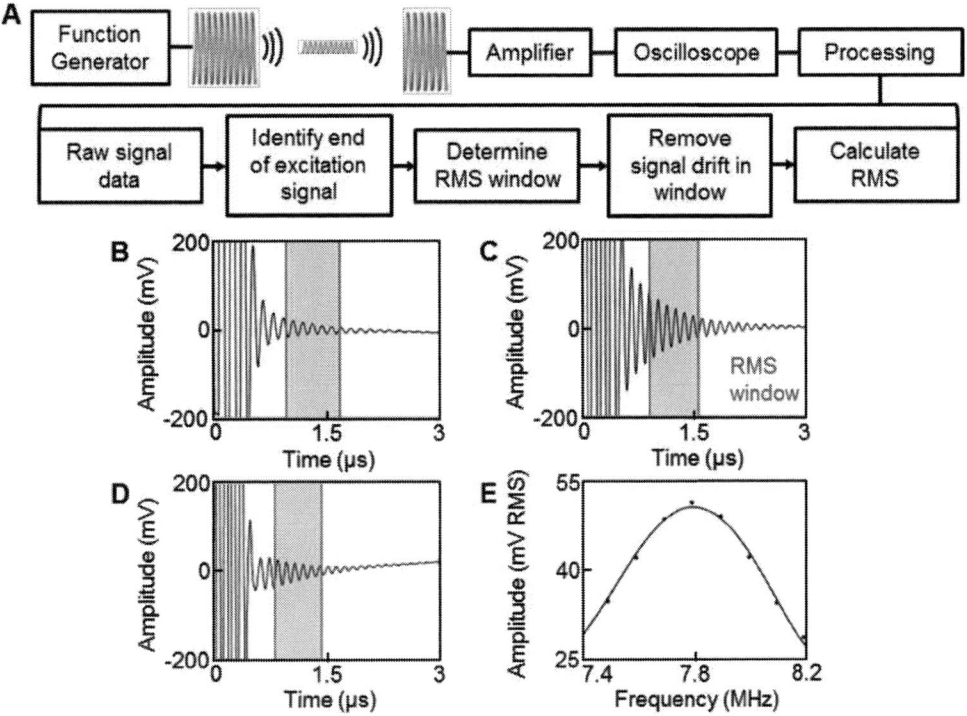

Figure 3. (A) Diagram of wireless system and resonant frequency monitoring. (B-D) Sensor transient signals and RMS windows for excitation frequencies (B) below resonance, (C) at resonance, and (D) above resonance. (E) Frequency sweep generated by analyzing transient signals.

III. WIRELESS TELEMETRY

A. System Design and Parameters

The wireless readout method, illustrated in Fig. 3A, applies inductive coupling principles between a sensor coil in a flow-diverter system and two external coils to record transient signals and measure sensor resonance. Sensor resonance is defined as:

$$f = 1/(2\pi\sqrt{LC}) \qquad (1)$$

This time-domain method, compared to the frequency domain method of observing impedance changes, has been proven to achieve longer readout distances [10]. During the measurement, three coils are axially aligned and the overall readout range is defined as the distance between the end of the sensor coil and the nearest external coil. An excitation external coil (8 AWG copper wire), connected to a function generator (33500B, Keysight), transmits a pulse of a given number of sine cycles at a specified frequency and amplitude. The readout coil, attached to a low-noise amplifier (Model ZFL-1000LN+, Mini-Circuits) and oscilloscope (Model TBS1052B, Tektronix), records the signals from the excitation coil and sensor circuit. The oscilloscope is triggered to record the transient signal by syncing it with the function generator.

A custom Matlab program processes transient sensor signals. First, the excitation pulse is identified to define an appropriate sampling window. Within this window, signal drift is removed and then the RMS of the signal is calculated. This process is repeated for each excitation frequency, generating a frequency sweep. Fig. 3B-3D displays sensor transient signals and RMS windows during a frequency sweep. Fig. 3B and 3D indicate the excitation signal is off resonance. The larger amplitude signal in Fig. 3C indicates the excitation frequency is near the sensor's resonant frequency. A curve is fitted to the RMS data and the maximum RMS value on the curve is taken as the sensor's resonant frequency (Fig. 3E).

In order to achieve functional interrogation distances, an analytical study is applied to optimize parameters, including coil length (l), coil diameter (d), and number of turns (N), as labeled in Fig. 4A, which was used to improve the wireless readout distance via inductive coupling. In this study, low resonant frequency is preferable to allow a higher magnetic field, while reducing tissue absorption from a sensor during wireless telecommunication, which enables a long readout distance [13]. Specific absorption rate (SAR) is a measure of safety when radio frequency is applied and is known to scale with frequency by $f^{2.15}$ [14]. Sensor inductance (L) is calculated in Equation (2) and includes correction coefficients as these offer more accurate estimates for the present coil dimensions of 3.5 mm diameter and 30 mm maximum length [15].

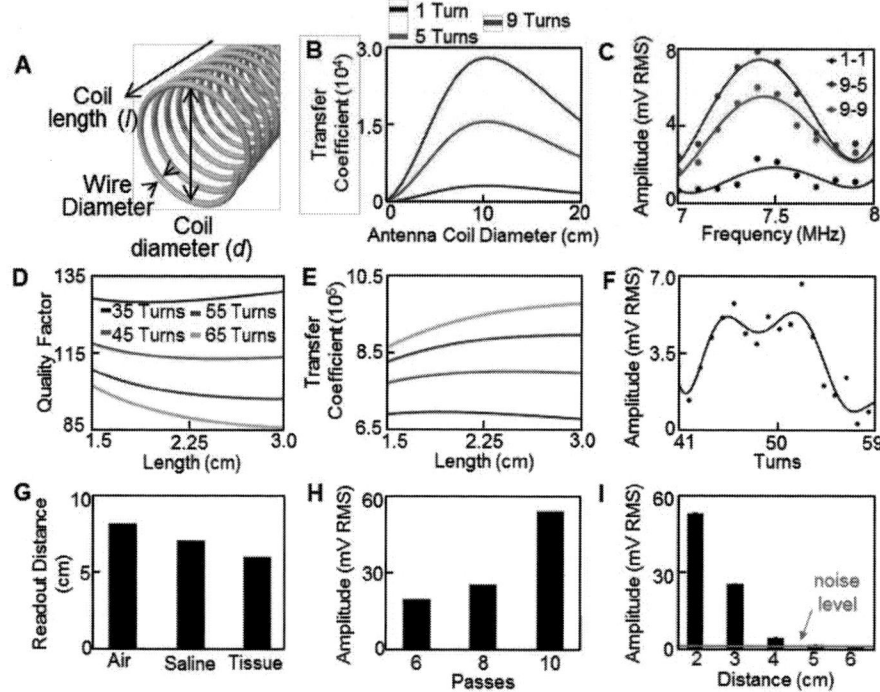

Figure 4. (A) Coil parameters. (B) Analytical transfer coefficient of antenna coil while varying number of turns and diameter. (C) Frequency sweep using a 9 turn excitation coil and 5 turn receiving coil provided the most distinct frequency sweep. (D) Analytical quality factor for sensor coil while varying number of turns and length. (E) Analytical transfer coefficient for sensor coil increases with more turns. (F) Experimental transient signal RMS values across a range of sensor coil number of turns. (G) Maximum readout distance of model sensor circuit in air, saline, and tissue. (H) Increasing the number of AJP passes for the silver nanoparticle electrodes increases signal amplitude at resonance. (I) Fabricated capacitive sensor achieves a redout distance of 5.5 cm in air.

The coefficient k_m is a mutual inductance correction for a round wire and k_s is a self-inductance correction [16, 17]. Inductance along with sensor capacitance (C) and resistance (R) define the sensor's quality factor (Q) according to Equation (3). Q plays a key role in determining readout distance as it is a measure of speed of response decay, power transfer efficiency at resonance, and bandwidth. A larger quality factor is desired to achieve a more detectable transient response. Mutual inductance (M) in Equation 4 and coupling coefficient (k) in Equation (5) also play key roles in power transfer and they are dependent on the separation distance (z) [18]. More efficient power transfer between the sensor and receiver coils improves the readout distance. Lastly, to quantify the impacts of quality and coupling on power transfer, overall transfer coefficient (Π), defined in Equation (6), and transfer efficiency (η) calculated with Equation (7) are defined [19].

$$L = \frac{\mu N^2 \pi d^2}{4l} - \mu \frac{d}{2} N(k_s + k_m) \tag{2}$$

$$Q = \frac{1}{R}\sqrt{\frac{L}{C}} \tag{3}$$

$$M = \frac{\mu \pi N_i N_e d_i^2 d_e^2}{16\sqrt{d_i^2 + z^2}} \tag{4}$$

$$k = \frac{M}{\sqrt{L_i L_e}} \tag{5}$$

$$\Pi = \frac{k\sqrt{Q_i Q_e}}{1 + k^2 Q_i Q_e} \tag{6}$$

$$\eta = \frac{k^2 Q_i Q_e}{1 + k^2 Q_i Q_e} \tag{7}$$

B. Antenna Coil Optimization

These analytical equations are applied to comprehensively design the wireless system. Antenna coil diameter and number of turns are studied according to coupling and transfer coefficients. Results indicate that the transfer coefficient between a sensor coil and antenna coil has a maximum value at a coil diameter based on separation distance (Fig. 4B). A coil diameter of 20 mm is used in the present study. Increasing turns enhances the transfer efficiency and is expected to improve readout range. However, a previous study employing a time-domain method between two coils stated that increasing antenna coil turns only increased received noise. To test the effects of improved transfer efficiency and increased noise, frequency sweeps are recorded while varying the number of turns for the excitation and receiving coil between 1 and 20. Results indicate that balancing the number of turns offers an improvement in performance (Fig. 4C). The optimal number of turns, determined by the maximum increase of signal RMS at resonance, for the present system is 9 for the excitation coil and 5 for the receiving coil. Increasing the coil diameter improved the readout range for single turn antenna coils, but 20 mm diameter, multi-turn coils more consistently achieved longer readout distances. Additionally, tuning the antenna coils increased sensor transient signals near resonance and provided a more consistent noise level across tested frequency ranges.

C. Sensor Coil Optimization

A similar procedure is performed to optimize the sensor coil parameters. Resonant frequency, quality factor, coupling, and power transfer efficiency based on sensor coils were investigated as they dictate system performance. Sensor coil parameters include the number of turns, length, and wire diameter, while the coil diameter is fixed to 3.5 mm for the cerebral aneurysm application; the coil pitch is determined by turns and length. Wire diameter, which determines the conductive cross-sectional area, indicates that larger was better, up to the skin depth phenomena, in inductive coupling, as a larger area reduces resistance and improves quality factor and transfer efficiency. In the analytical and experimental studies, a 100 μm-diameter wire is used. Fig. 4D shows sensor coil quality factor while varying length and number of turns.

Quality factor for a given sensor capacitance improves for fewer turns, while the variation due to length is less prominent. A lower sensor capacitance enhances quality at the cost of increasing resonant frequency. However, fewer turns negatively affects coupling and transfer efficiency (Fig. 4E), indicating that the number of turns may be balanced to achieve optimal performance. Moreover, maintaining a low resonant frequency adds a restriction on possible values of sensor inductance and capacitance. Experimental investigations of sensor parameters followed to test and validate the analytical calculation. Impacts of sensor characteristics are investigated through comparison of RMS responses at resonance while varying number of turns. An optimal number of turns is determined to be between 44 and 53 turns at 30 mm length (Fig. 4F). The values are collected at 2 cm distance with single-turn antenna coils (20 mm in diameter). Coils within this range shows higher RMS values at resonance and allows a longer readout distance. This finding can be explained analytically since more turns increases coupling and transfer coefficients, thus the induced power, but decreases quality. Resulting coils have analytical inductances between 1.5 and 1.9 μH.

D. Maximum Readout Distance

With these parameters, experimental studies identify the maximum wireless readout distance in three different conditions; sensor coils placed in air, saline or meat (beef). Fig. 4G shows the maximum readout distances with a ceramic capacitor. In air, the system achieves a working distance of 8 cm. Simulating implantation in a blood vessel by placing the sensor in saline solution reduces the distance to 7 cm. Lastly, encapsulating the coil in meat to simulate human tissue reduces the range to 6 cm. These distances indicate feasibility for cerebral aneurysm applications that require a readout distance of approximately 5 cm.

Using the AJP-fabricated flow sensors, the readout range is limited by the resistance of the individual AgNP electrodes. Increasing the number of print passes for the electrodes enables thicker silver traces and improves power transfer. Fig. 4H shows a significant increase in signal amplitude at resonance for increasing the passes from 6 to 10, correlating to a thickness increase of approximately 4 μm. With the 10 pass electrodes, a readout distance over 5 cm is achieved in air (Fig. 4I). With this distance, the ratio of readout distance to cross-sectional area, a measure of overall electrical and mechanical performance, is 2.4.

To ensure compatibility with safety limits, a computational model estimates the RMS magnetic field strength 1 cm away from the excitation coil. The proposed system shows an average strength of 0.78 A/m, well below the 1.6 A/m limit [13]. Moreover, the readout system uses bursts of excitation signals for less than a minute, whereas the safety limit is averaged over a 30-minute period. This result suggests that the magnetic field of our wireless system can be increased further to maximize the interrogation range, though closer examination is required.

IV. CEREBRAL ANEURYSM MONITORING

By using the fabricated biosensors and wireless system, in vitro experiments demonstrate the sensor's capability to monitor flow change in cerebral aneurysms. The low-profile sensor and stent are implanted in a PDMS mold of a cerebral aneurysm (Fig. 5A). The sensor is aligned at the aneurysm sac opening to detect flow changes into the sac. The low-profile form of the sensor (Fig. 5B) enables detection of flow into the aneurysm without disrupting normal hemodynamics. Deflection of the sensor is the primary mechanism of capacitance change. To wirelessly monitor flow changes, a 7.5 mm copper coil is externally connected to the capacitive sensor (Fig. 5C). Blood flow through the model is simulated by a pulsatile blood pump (Harvard Apparatus) with mean flow velocities varied from 0 to 0.35 m/s at 0.05 m/s intervals. In vivo intracranial blood flow velocities often exceed 0.5 m/s [20].

First, an LCR meter (B&K Precision) directly measures capacitance change of the flow sensor. The flow sensor captures the pulsatile form of the blood flow, as shown in Fig. 5D for a flow velocity of 0.35 m/s. Initial sensor capacitance is 61.53 pF and increases to 61.63 pF at the maximum flow velocity. The average change is demonstrated in Fig. 5E from 0 to 0.35 m/s. This increase is captured in Fig. 5F and yields a sensitivity of 0.29 pF/(m/s). Wireless monitoring is then performed to monitor average sensor capacitance by recording resonant frequency. Experimental tests indicate an initial resonant frequency of approximately 6.2 MHz. Continuously performing frequency sweeps in a defined range allows for continuous monitoring of flow velocity changes.

978-1-7281-1500-9/19 $31.00 © 2019 IEEE

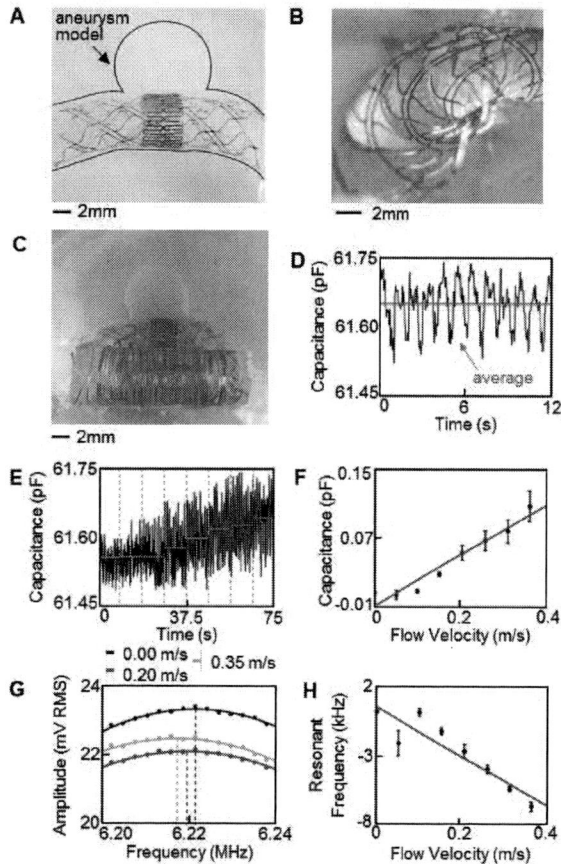

Figure 5. (A-B) Sensor and stent deployed in cerebral anuerysm mold. (C) Sensor connected to copper coil for wireless monitoring. (D) Sensor capacitance during pulsatile flow. (E) Average sensor capacitance increases with increasing mean flow velocity. (F) Sensor capacitance increases linearly with flow velocity. (G) Frequency sweeps for three flow velocities. (H) Decreased sensor resonant frequency as flow velocity increases.

Fig. 5G displays a set of frequency sweeps corresponding to three different flow velocities. A decrease in resonant frequency is observed, corresponding to the increase in capacitance, as flow is increased. Fig. 5H summarizes the recorded resonant frequency change across the tested flow velocities. A wireless sensitivity of -18.9 kHz/(m/s) is observed. This indicates a 6.5 kHz shift for a 0.1pF change, which correlates well with the analytical shift of 5.9 kHz.

V. CONCLUSIONS

This paper introduces a stretchable, implantable flow sensor, fabricated via AJP for monitoring cerebral aneurysm flow changes. The highly compliant, low-profile sensor is capable of detecting flow velocity changes into an aneurysm sac. A comprehensive study of wireless powering and communication system via analytical and experimental methods enables a functional readout distance of the sensor. Detection of resonant

frequency in the range of multi-centimeters with a sensor coil in air exhibits a markedly enhanced inductive coupling system that is now capable of addressing a variety of needs of wireless implantable sensors. Experimental studies of wireless flow monitoring demonstrate the capability of the system. A low-profile, implantable, wireless system in a single package will find various types of medical applications that require real-time, post-treatment monitoring of pressure or flow variation.

ACKNOWLEDGMENT

We acknowledge a grant from the Korea Institute of Industrial Technology, seed grant from the Georgia Tech Institute for Electronics and Nanotechnology and this work was performed in part at the Institute for Electronics and Nanotechnology, a member of the National Nanotechnology Coordinated Infrastructure, which is supported by the National Science Foundation (Grant ECCS-1542174).

REFERENCES

[1] J. Wardlaw and P. White, "The detection and management of unruptured intracranial aneurysms," Brain, vol. 123, pp. 205-221, 2000.

[2] J. R. Cebral, F. Mut, J. Weir, and C. M. Putman, "Association of hemodynamic characteristics and cerebral aneurysm rupture," American Journal of Neuroradiology, vol. 32, pp. 264-270, 2011.

[3] R. Deruty, I. Pelissou-Guyotat, C. Mottolese, and D. Amat, "Management of unruptured cerebral aneurysms," Neurological research, vol. 18, pp. 39-44, 1996.

[4] C. Howe, S. Mishra, Y.-S. Kim, Y. Chen, S.-H. Ye, W. R. Wagner, J.-W. Jeong, H.-S. Byun, J.-H. Kim, Y. Chun, ACS nano 2018, 12, 8706.

[5] M. Babiker, Y. Chun, B. Roszelle, W. Hafner, H. Farsani, L. Gonzalez, et al., "In vitro investigation of a new thin film nitinol-based neurovascular flow diverter," Journal of Medical Devices, vol. 10, p. 044506, 2016.

[6] Y. Chen, C. Howe, Y. Lee, S. Cheon, W.-H. Yeo, and Y. Chun, "Microstructured thin film nitinol for a neurovascular Flow-Diverter," Scientific reports, vol. 6, p. 23698, 2016.

[7] C. Howe, Y. Lee, Y. Chen, Y. Chun, and W.-H. Yeo, "An Implantable, Stretchable Microflow Sensor Integrated with a Thin-Film Nitinol Stent," in Electronic Components and Technology Conference (ECTC), IEEE 66th, 2016, pp. 1638-1643.

[8] R. Herbert, J.-H. Kim, Y. S. Kim, H. M. Lee, and W.-H. Yeo, "Soft material-enabled, flexible hybrid electronics for medicine, healthcare, and human-machine interfaces," Materials, vol. 11, p. 187, 2018.

[9] Y. Chen, Y.-S. Kim, B. W. Tillman, W.-H. Yeo, and Y. Chun, "Advances in Materials for Recent Low-Profile Implantable Bioelectronics," Materials, vol. 11, p. 522, 2018.

[10] D. S. Brox, X. Chen, S. Mirabbasi, and K. Takahata, "Wireless telemetry of stainless-steel-based smart antenna stent using a transient resonance method,"

IEEE Antennas and Wireless Propagation Letters, vol. 15, pp. 754-757, 2016.

[11] J. Park, J.-K. Kim, S. J. Patil, J.-K. Park, S. Park, and D.-W. Lee, "A wireless pressure sensor integrated with a biodegradable polymer stent for biomedical applications," Sensors, vol. 16, p. 809, 2016.

[12] H.-J. Kang, Y.-S. Lee, S.-J. Suh, J.-H. Lee, K.-Y. Ryu, and D.-G. Kang, "Comparative analysis of the mini-pterional and supraorbital keyhole craniotomies for unruptured aneurysms with numeric measurements of their geometric configurations," Journal of cerebrovascular and endovascular neurosurgery, vol. 15, pp. 5-12, 2013.

[13] J. C. Lin, "A new IEEE standard for safety levels with respect to human exposure to radio-frequency radiation," IEEE Antennas and Propagation Magazine, vol. 48, pp. 157-159, 2006.

[14] P. Röschmann, "Radiofrequency penetration and absorption in the human body: limitations to high-field whole-body nuclear magnetic resonance imaging," Medical physics, vol. 14, pp. 922-931, 1987.

[15] E. B. Rosa and F. W. Grover, "Formulas and tables for the calculation of mutual and self inductance.(Revised.)," Journal of the Washington Academy of Sciences, vol. 1, pp. 14-16, 1911

[16] F. W. Grover, "A Comparison of the Formulas for the Calculation of the Inductance of Coils and Spirals Wound with Wire of Large Cross-Section," Bureau of Standards Journal of Research, vol. 3, pp. 163-190, 1929.

[17] D. W. Knight, "Solenoid Inductance Calculation," http://www.g3ynh.info/zdocs/magnetics/Solenoids.pdf, 2013.

[18] P. T. Theilmann and P. M. Asbeck, "An analytical model for inductively coupled implantable biomedical devices with ferrite rods," IEEE transactions on biomedical circuits and systems, vol. 3, pp. 43-52, 2009.

[19] L. Rindorf, L. Lading, and O. Breinbjerg, "Resonantly coupled antennas for passive sensors," in Sensors, 2008 IEEE, 2008, pp. 1611-1614.

[20] S. Ferns, J. Schneiders, M. Siebes, R. van Den Berg, E. van Bavel, and C. Majoie, "Intracranial blood-flow velocity and pressure measurements using an intra-arterial dual-sensor guidewire," American journal of neuroradiology, vol. 31, pp. 324-326, 2010.

A Wearable Passive pH Sensor for Health Monitoring

Saikat Mondal, Saranraj Karuppuswami, Rachel Steinhorst, and Premjeet Chahal

Department of Electrical and Computer Engineering
Michigan Sate University
East Lansing, USA
chahal@egr.msu.edu

Abstract—In this paper, a miniaturized battery less passive transponder is designed and demonstrated for wireless pH monitoring. The proposed transponder is integrated into a wearable device for wireless real-time monitoring of salivary pH in patients. The transponder consists of two parts, the RF front end with a digital modulation circuitry and sensing electrodes for electrochemical detection. The potential difference between two custom fabricated electrodes is monitored for change in a range of pH from 4 to 9. The digital circuitry converts the sensor data into a bit sequence and provides the digital sensing data over the reflected backscattered signal. The fabricated pH electrode pair demonstrated a sensitivity of 49.5 mV/pH over the range from 4 to 9. The demonstrated sensor is not just limited to health monitoring but also can be used in monitoring pH in soil, food, chemicals, etc.

Index Terms—pH sensor, energy harvesting, wireless, passive.

I. INTRODUCTION

Wireless health monitoring applications has significantly altered traditional clinical diagnosis into a personalized health care management. The growth in population and increase in the number of diseases have prompted development of wearable technologies that perform real-time patient monitoring allowing timely medical attention [1]. Among the different monitored vital parameters, salivary pH has the potential to diagnose and monitor a number of common medical conditions such as dental caries, gingivitis, periodontitis, and gastroesophageal reflux disease (GERD) [2], [3]. Moreover, saliva is a convenient diagnostic medium, as the measurement is relatively less invasive than other bodily fluids.

The common pH monitoring technique is the electrochemical method in which a sensing electrode and a reference electrode creates a potential difference when dipped inside the liquid under test based on the pH level [4]. In literature, a number of active wireless techniques have been reported for monitoring pH which require a battery for operation [5], [6]. The active sensors are limited by battery life, bulky, and can turn into a serious health concern if leakage occurs. There are many recorded cases of battery ingestion which becomes a serious health concern most often when the battery becomes lodged in the esophagus, resulting in esophageal burns due to local generation of hydroxide [7].

Passive sensing technology is attractive due to the infinite life time, battery-free, and less power consumption require-

ments. In literature, batteryless passive wireless approaches using short range Inductor Capacitor (LC) resoanant tanks coupled with pH sensitive materials has been proposed in [8]–[10]. A wireless magnetoelastic sensor coated with pH sensitive polymers is reported in [11] for detecting pH using mass loading. Although these techniques are battery-less, they are limited by the read range and are bulky in size. For a robust real-time monitoring, a long range passive pH sensor is necessary. Furthermore, the conversion of the analog pH sensor data into digital domain allows the technqiue to be compatible with Radio Freqeuncy Identification (RFID) based sensing technology. In this format, each sensor is represented by an unique digital signature (ID) and transmits both the ID and the sensor information on demand. In our previous work [12], a harmonic approach is used to realize a long range passive transponder system for detecting pH using commercial electrodes. This work is analogous to our previous work and uses a custom designed electrode pair and a transponder for detecting pH in saliva.

In this paper, a custom designed Silver/Silver Chloride (Ag/AgCl) and Antimony/Antimony oxide (Sb/Sb_2O_3) based pH electrodes are evaluated for detecting pH. The change in pH changes the potential across the electrode pair and is read out using a transponder tag. The transponder consists of an RF front end with an antenna for long range transmission as well as energy harvesting and a digital circuitry for converting the analog pH sensing information to digital bit sequence. The antenna impedance is modulated according to the digital sensor information, which is read in the interrogator from the tag backscattered signal. The proposed wireless sensing technique can be easily adapted beyond wearable health care applications in multiple domains such as food monitoring, soil monitoring, chemical quality monitoring, etc.

II. DESIGN AND METHODS

The pH sensor tag consists of two parts: 1) pH sensor electrode pair, and 2) digital and RF circuit to convert the analog voltage signal into digital sequence and transmit it wirelessly. First, the design and fabrication process of the pH electrodes are described.

978-1-7281-1500-9/19 $31.00 © 2019 IEEE

A. pH sensor fabrication

The pH electrode pair consists of one Ag/AgCl electrode and another Sb/Sb_2O_3 electrode. Both the electrodes are fabricated on top of a 380 μm thick flexible Rogers PTFE (RT Duroid 5880) board. Intially, the copper metal on top of the substrate is patterned using standard photolithography and Cu etching process (1:4 by volume Sodium per sulphate: Distilled water) in order to create a bridge like interconnect for the electrodes. An inverse lift-off mask is patterned using Shipley S1813 negative photoresist on the substrate. A 60 nm thick intermediate Titanium (Ti) layer is sputter deposited first for better adhesion of the Ag with the copper followed by a 1 μm of Ag using a Ag sputtering target of 99.9% purity. After deposition, the redundant Ag layer is removed using a lift-off process by ultrasonication of the substrate in an Acetone bath for 5 minutes. Contamination of Ag layer is possible due to interdiffusion of Cu atoms from the underlying layer [13]. Additionally, as the Ag layer is very thin, irregularity in the deposition creates small pockets, through which the electrolyte interacts directly with the Cu layer, and thus can change the desired potential. Hence, a continuous layer of Cu backbone layer is avoided beneath the Ag electrode. Next, a layer of AgCl is formed on the exposed Ag electrode by immersing the substrate in a 0.1 M potassium chloride (KCl) solution with a 0.5 mm diameter Platinum (Pt) cathode dipped in it. 2 V is applied on the electrode for 50 s for chlorination of Ag. The substrate that contains the freshly prepared Ag/AgCl electrode is heated at 85^0 C for 15 minutes for uniform diffusion of Chloride ions into Ag. Following the AgCl deposition, the electrode is dip-coated with a protective Nafion layer by dipping in a 5% Nafion solution (Liquion-1105-MeOH) from Ion Power. After dip coating, the Ag/AgCl electrode is cured at 120^0 C for 1 hour to evaporate the methanol [14]. This dip coating is performed thrice to obtain a thick Nafion layer coating. Nafion is chosen because it is a selective ion exchange polymer allowing the H+ cation to pass through but prevents Cl- anion. The complete fabrication process of the electrodes is described in Fig. 1.

The copper backbone layer for Sb/Sb2O3 electrode is fabricated in a similar fashion. A 1 um thick Sb layer is deposited on top of a 60 nm Ti layer by sputtering. Unwanted Sb layer is removed from the substrate using lift-off process. The Sb metal is oxidized by heating at 480^0 C for 20 minutes. Once the Sb/Sb_2O_3 electrode is fabricated, it is used along with Ag/AgCl electrode as a combination electrode for pH measurement. The fabricated pH electrode pair is shown in Fig. 2. The electrodes are connected to the transponder through the copper pads.

B. Transponder system

The electronic circuit of the sensor transponder is simple and consists of two major components 1) RF front end with antenna for energy harvesting and 2) a digital circuit to convert the analog sensor signal into a digital bit sequence. A single antenna is used for wireless energy harvesting and digital modulation. The complete circuit can be impedance

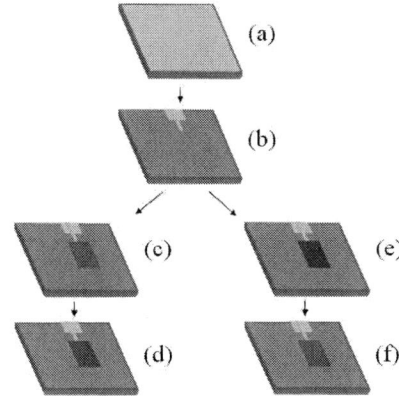

Fig. 1: Fabrication procedure of pH combination electrodes a) Cu plated duroid substrate; b) patterning of Cu layer to create Cu backbone; c) sputtering followed by lift-off of Ag metal; d) Chlorination of Ag layer; e) sputtering followed by lift-off of Sb metal; f) oxidation of Sb layer.

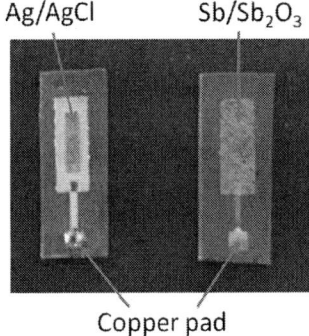

Fig. 2: Fabricated separate pH electrode pairs.

matched to any desired frequency through a resonating far field or near field antenna depending on the applications. The complete transponder circuit schematic is shown in Fig. 3. The transponder system consists of an RF front end with an energy harvesting unit. The energy harvesting unit harvests dc power from the incoming RF signal. A dc voltage regulator is used to regulate the voltage at a fixed 3 V required for powering the digital circuit of the transponder. A high frequency switch is connected across the antenna to modulate the antenna impedance according to the generated digital data sequence. The digital circuit of the transponder consists of a timing unit and an ADC. The timing unit generates the necessary clocks for the ADC and the ADC converts the analog sensor signal into digital data sequence. The RF system is intended to work at 900 MHz RFID frequency. Hence, a dipole antenna is designed at 900 MHz for receiving the RF energy at the tag.

III. RESULTS

Once the sensors are fabricated, they are electrically characterized to evaluate their performance in terms of electrode

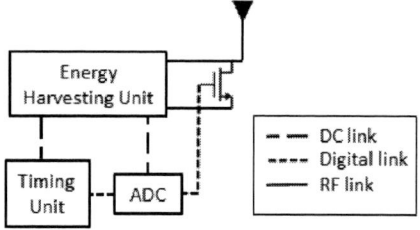

Fig. 3: The transponder system with RF front end.

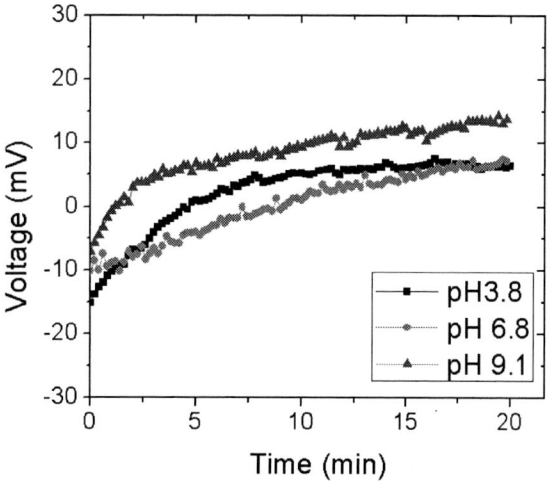

Fig. 4: Voltage stability of the fabricated Ag/AgCl electrode relative to a commercial Ag/AgCl electrode.

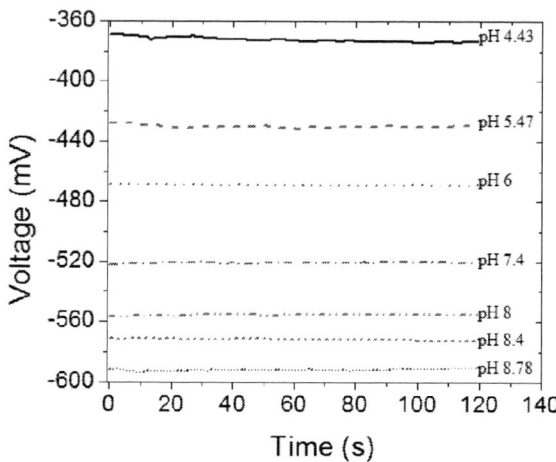

Fig. 5: Time response of fabricated electrodes at different pH levels.

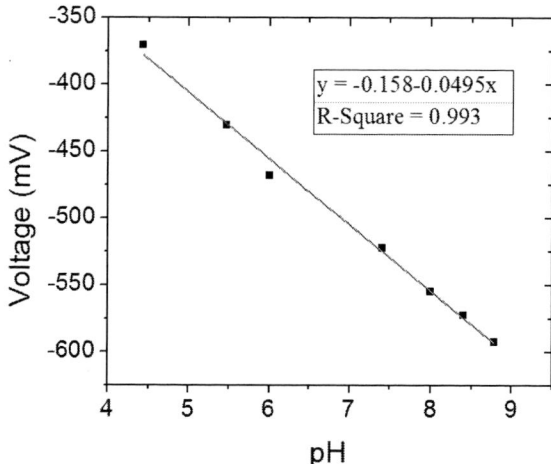

Fig. 6: Voltage response of the electrodes at different pH levels.

stability, repeatability, sensitivity, and response time. All the voltage measurements are performed at room temperature using a digital multimeter with an output impedance of 2 MΩ.

First the stability of fabricated Ag/AgCl electrode was measured relative to a commercial Ag/AgCl electrode. The measurement was performed for 20 minutes in three different pH buffer solutions starting from 3.8 to 9.1 as shown in Fig. 4. The maximum voltage variation was 25 mV for the entire 20 minutes period.

Once the stability of the fabricated Ag/AgCl electrode is verified, the voltage response of the Sb/Sb_2O_3 electrode is measured in reference to the fabricated Ag/AgCl electrode at different pH levels starting from 4.43 to 8.78. A stable voltage response is obtained with less than 10 mV of variation at maximum as shown in Fig. 5 within a timeframe of 120 seconds. The settling time for the electrodes in pH buffer solution is not shown in Fig. 5.

The voltage variation at different pH solution is shown in Fig. 6. Voltage variation is shown with linear approximation of 49.5 mV/pH at different pH levels ranging from 4.4 to 8.78. The R squared value of 0.99 indicates that the pH response is quite linear to the corresponding linear fit. Different pH buffer solutions are prepared by mixing potassium di-hydrogen phosphate salt with 0.1 M sodium hydroxide at different quantity.

Finally the repeatability and response time of the sensor electrodes are measured in different pH solutions with time as shown in Fig. 7. First the sensor electrodes are dipped into a pH solution of 4.6 and then it was transferred into a pH solution of 6.4. Finally the electrodes are transferred into a pH solution of 7.3. This process is repeated in a reverse order to ensure the repeatability of measurements from the electrodes. The electrodes are immersed in a single pH solution for 60 seconds before transferring to a different ph solution. Between the steps of pH solution transfer, the electrodes are washed in DI water for 6 seconds. As the voltage response is erratic during the washing period, the readings are ignored for that time.

First the digital section of the transponder is verified to be working under different conditions. The ADC converts the

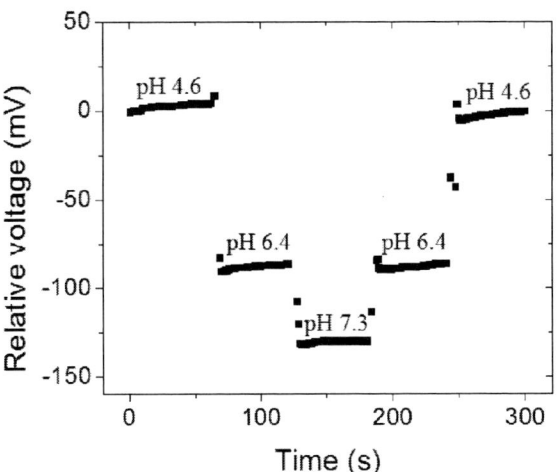

Fig. 7: Repeatability of the electrodes at different pH.

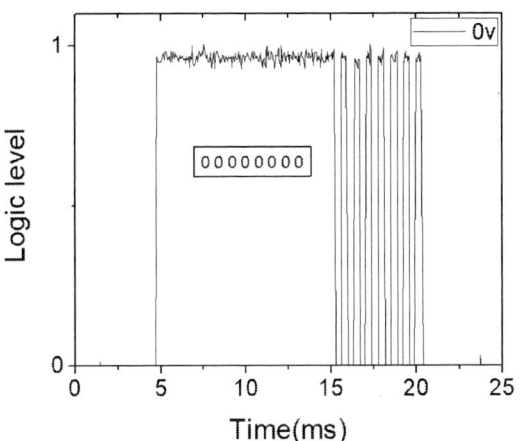

Fig. 8: ADC output bit sequence of 00000000 for 0 V input.

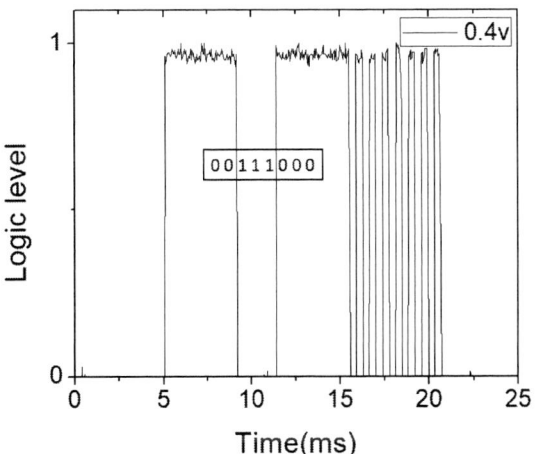

Fig. 9: ADC output bit sequence of 00111000 for 0.4 V input.

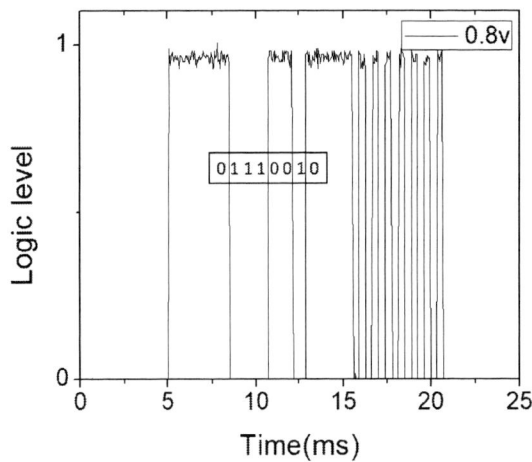

Fig. 10: ADC output bit sequence of 01110010 for 0.8 V input.

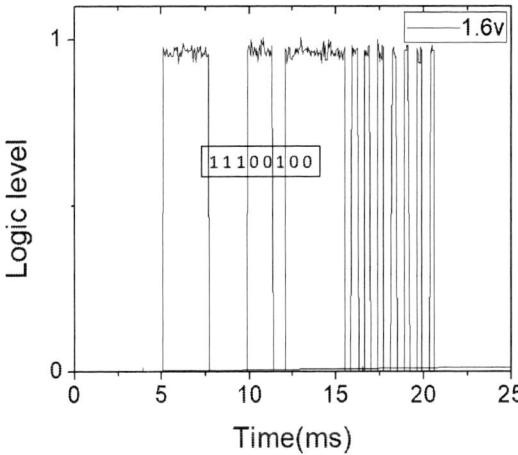

Fig. 11: ADC output bit sequence of 11100100 for 1.6 V input.

analog voltage signal into digital bit sequence. The analog reference high and low voltages are set as 1.8 V and 0 V, respectively, which means when the analog input voltage is 0 V, the 8 bit ADC output would be 00000000 and when the analog input voltage is 1.8 V, the ADC output would be 11111111. The output bit sequence is shown in Fig. 8, Fig. 9, 10, and 11 for different input voltage at 0 V, 0.4 V, 0.8 V, and 1.6 V respectively. However, the digital bits are flipped version of the actual bits due to inverse operation before transmission. The starting and ending extra bits before and after the sensor bits are included to track the sensor bits in a long digital data stream. The 8 bit ADC can provide 256 different digital sequence for 1.8 V analog input swing with a resolution of around 7 mV.

IV. WEARABLE PH SENSOR DESIGN

Apart from the RF front end, digital circuits and pH sensors, the transponder includes antenna to wirelessly receive RF

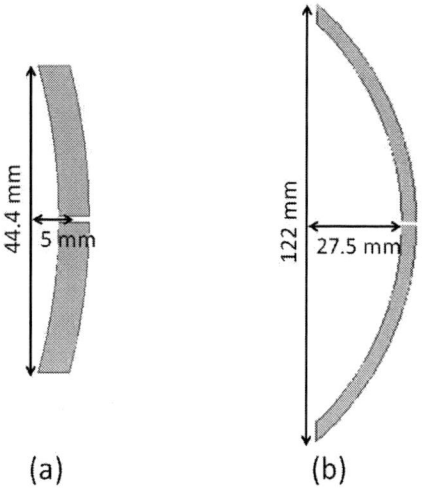

(a)   (b)

Fig. 12: Antenna dimensions for (a) inside human mouth and (b) in air medium.

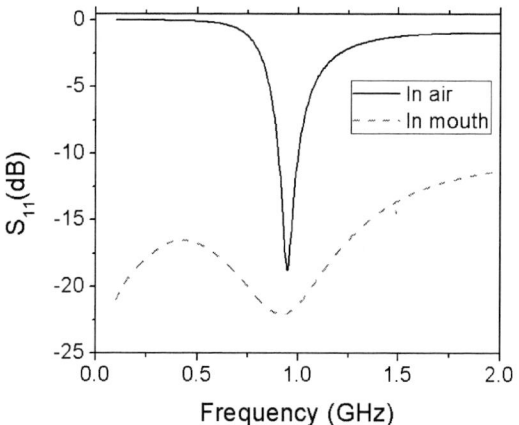

Fig. 13: Simulated reflection co-efficients for in air and in mouth antenna designs.

power and transmit back the sensor signals.

A. Antenna Design

When the mouthguard is to be placed within human mouth, the antenna needs to be changed due to loading of the antenna by the human body. A human head model was considered as described in [15] to design the antenna. Due to dielectric loading, the effective length of the antenna reduces resulting in a compact and small size antenna. The antenna dimensions for inside human mouth and in air medium are given as in Fig. 12. The simulation result for those antennas is shown in Fig. 13. From the reflection co-efficients it can be observed that the in air antenna is resonating at a fixed frequency. However, the simulation result of the in-mouth antenna shows apart from the resonance frequency, the reflection co-efficient is little at other frequencies due to loading effect by the human body.

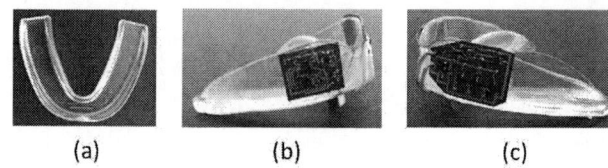

(a)   (b)   (c)

Fig. 14: The sensor tag integrated on mouthguard (a) bare mouthguard, (b) digital part of the transponder at the right side and (c) RF circuit of the transponder at the left side of the mouthguard.

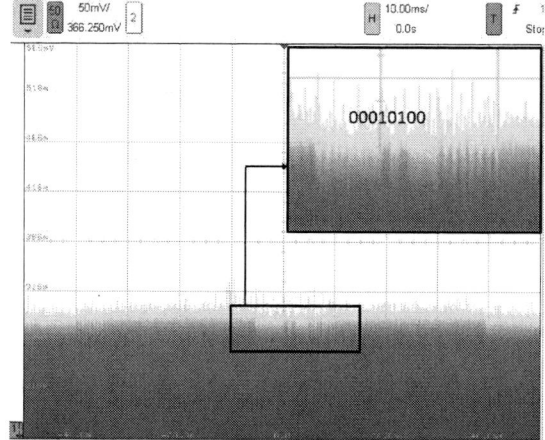

Fig. 15: Detected signal in Oscilloscope in buffer pH solution of 5.3.

B. Transponder Design

The complete transponder is shown in Fig. 14 from different side views in Fig. 14(b) and Fig. 14(c) and the top view of the bare commercial mouthguard in Fig. 14(a). The transponder consists of two circuit boards a) the digital part is visible at the right side and b) the RF part is visible at the left side of mouthguard in Fig. 14. The dipole antenna was placed at the bottom of the mouthguard along the two arms and the electrodes are mounted at the center. HSMS 2852 Schottky diode was used to make a six stage RF to DC converter. BF1105R MOSFET was used as RF switch as shown in Fig. 3.

The complete sensor was integrated with pH sensor electrodes and transponder with connected antenna resonating at 930 MHz. Each part of the whole sensor was integrated on a mouthguard and the back-scattered response from the tag was measured using an Oscilloscope at different pH solutions. When buffer pH solution of 5.3 was used, a digital sensor sequence of 00010100 was detected. Digital sensor sequence of 00111000 was received in buffer pH solution of 9.5. The antenna was designed to use in air medium. The antenna design can be further optimized for use within human mouth as proposed earlier. During measurement, the tag was kept at 8 cm from the reader, which was transmittiong 13 dBm of power at 930 MHz.

978-1-7281-1500-9/19 $31.00 © 2019 IEEE   1244

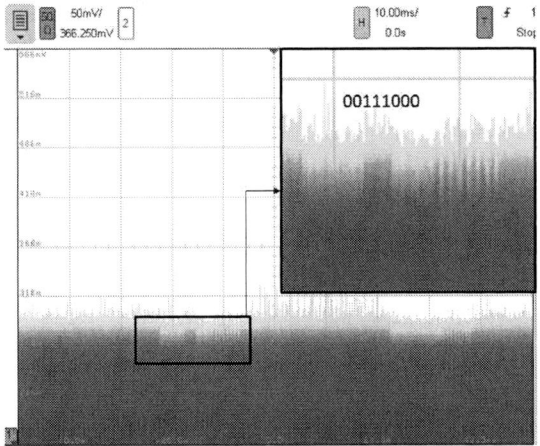

Fig. 16: Detected signal in Oscilloscope in buffer pH solution of 9.5.

V. CONCLUSION

The design and implementation of a complete ultra-low power wearable pH sensor is presented. The pH sensor harvests energy from the external RF power eliminating the use of battery within the package. The proposed pH electrodes exhibited linear response at different buffer solutions ranging pH from 4 to 9 with a sensitivity of 49.5 mV/pH. The wearable transponder system provided a digital response according to the pH. The RF switch modulated the antenna and thus perturbing the RF field according to the digital bit sequence. The complete digital circuit works at very low power of 70 μW at 3V, which is harvested using the RF energy harvester. The preliminary test results showed the pH transponder can work at different pH levels.

ACKNOWLEDGMENT

The authors would like to thank Mr. Brian Wright from MSU ECEshop for his help with silver and antimony sputtering. This work is supported by The Axia Institute.

REFERENCES

[1] S. Coyle, K.-T. Lau, N. Moyna, D. O'Gorman, D. Diamond, F. Di Francesco, D. Costanzo, P. Salvo, M. G. Trivella, D. E. De Rossi *et al.*, "BIOTEXBiosensing textiles for personalised healthcare management," *IEEE Transactions on Information Technology in Biomedicine*, vol. 14, no. 2, pp. 364–370, 2010.

[2] H. Cao, V. Landge, U. Tata, Y.-S. Seo, S. Rao, S.-J. Tang, H. F. Tibbals, S. Spechler, and J.-C. Chiao, "An implantable, batteryless, and wireless capsule with integrated impedance and pH sensors for gastroesophageal reflux monitoring," *IEEE Transactions on Biomedical Engineering*, vol. 59, no. 11, pp. 3131–3139, 2012.

[3] V. Gopinath and A. Arzreanne, "Saliva as a diagnostic tool for assessment of dental caries," *Archives of orofacial sciences*, vol. 1, pp. 57–59, 2006.

[4] B. E. Horton, S. Schweitzer, A. J. DeRouin, and K. G. Ong, "A varactor-based, inductively coupled wireless pH sensor," *IEEE Sensors Journal*, vol. 11, no. 4, pp. 1061–1066, 2011.

[5] J. Kim, S. Imani, W. R. de Araujo, J. Warchall, G. Valdés-Ramírez, T. R. Paixão, P. P. Mercier, and J. Wang, "Wearable salivary uric acid mouthguard biosensor with integrated wireless electronics," *Biosensors and Bioelectronics*, vol. 74, pp. 1061–1068, 2015.

[6] T. Arakawa, Y. Kuroki, H. Nitta, P. Chouhan, K. Toma, S.-i. Sawada, S. Takeuchi, T. Sekita, K. Akiyoshi, S. Minakuchi *et al.*, "Mouthguard biosensor with telemetry system for monitoring of saliva glucose: A novel cavitas sensor," *Biosensors and Bioelectronics*, vol. 84, pp. 106–111, 2016.

[7] T. Litovitz, N. Whitaker, and L. Clark, "Preventing battery ingestions: an analysis of 8648 cases," *Pediatrics*, vol. 125, no. 6, pp. 1178–1183, 2010.

[8] S. Bhadra, D. S. Tan, D. J. Thomson, M. S. Freund, and G. E. Bridges, "A wireless passive sensor for temperature compensated remote pH monitoring," *IEEE Sensors Journal*, vol. 13, no. 6, pp. 2428–2436, 2013.

[9] V. Sridhar and K. Takahata, "A hydrogel-based passive wireless sensor using a flex-circuit inductive transducer," *Sensors and Actuators A: Physical*, vol. 155, no. 1, pp. 58–65, 2009.

[10] W.-D. Huang, S. Deb, Y.-S. Seo, S. Rao, M. Chiao, and J. Chiao, "A passive radio-frequency pH-sensing tag for wireless food-quality monitoring," *IEEE Sensors Journal*, vol. 12, no. 3, pp. 487–495, 2012.

[11] Q. Y. Cai and C. A. Grimes, "A remote query magnetoelastic pH sensor," *Sensors and Actuators B: Chemical*, vol. 71, no. 1-2, pp. 112–117, 2000.

[12] S. Mondal, D. Kumar, and P. Chahal, "A Wireless Passive pH Sensor with Clutter Rejection Scheme," *IEEE Sensors Journal*, 2019.

[13] H. Suzuki, A. Hiratsuka, S. Sasaki, and I. Karube, "Problems associated with the thin-film Ag/AgCl reference electrode and a novel structure with improved durability," *Sensors and Actuators B: Chemical*, vol. 46, no. 2, pp. 104–113, 1998.

[14] P. Hashemi, P. L. Walsh, T. S. Guillot, J. Gras-Najjar, P. Takmakov, F. T. Crews, and R. M. Wightman, "Chronically implanted, nafion-coated Ag/AgCl reference electrodes for neurochemical applications," *ACS chemical neuroscience*, vol. 2, no. 11, pp. 658–666, 2011.

[15] F. Kong, C. Qi, H. Lee, G. D. Durgin, and M. Ghovanloo, "Antennas for intraoral Tongue Drive System at 2.4 GHz: design, characterization, and comparison," *IEEE Transactions on Microwave Theory and Techniques*, vol. 66, no. 5, pp. 2546–2555, 2018.

Novel Packaging Structure and Processes for Micro-TFB (Thin Film Battery) to Enable Miniaturized Healthcare Internet-of-Things (IoT) Devices

Bing Dang[1], Qianwen Chen[1], Leanna Pancoast[1], Yu Luo[1], Hongqing Zhang[2], Jae-woong Nah[1], John Knickerbocker[1], Andy Shih[3], Po Wen Cheng[3], Kai Liu[3], Mengnian Niu[3], and Simon Nieh[3]

[1]IBM T. J. Watson Research Center, Yorktown Heights, NY, 10514, USA
[2] IBM Systems, Hopewell Junction, NY, 12533, USA
e-mail: dangbing@us.ibm.com
[3] Front Edge Technologies Inc., 13455 Brooks Drive, Suite A, Baldwin Park, CA 91706
e-mail: simonnieh@frontedgetechnology.com

Abstract—**In this work, a novel packaging structure has been demonstrated for micro-TFB cells. Various semiconductor fabrication and assembly processes have been applied to thin substrate via formation and sealing, micro-TFB singulation and handling, as well as metallic sealing for hermetic battery packaging. Micro-TFB cells in the dimensions of 2.5mm x 2.5mm x 0.1mm have been fabricated and tested. A normalized capacity of ~ 190 µAh/cm² has been demonstrated.**

Keywords-Microbattery; Healthcare; IoT; Solid State Thin Film Battery; Medical Device; Implantable

I. INTRODUCTION

In the past five years, the application of IoT has extended to many fields with the number of IoT devices increasing exponentially. One example is self-powered wireless sensor networks, which drives a massive number of miniaturized devices deployed everywhere [1,2]. Another rapidly growing application for IoT micro-systems is the use of healthcare sensors and medical devices used for healthcare monitoring, providing new healthcare data for insights using analytics, diagnosis and tracking medication efficacy in everyday life [3,4]. The form factor of these devices varies significantly and a very small size in X-Y-Z dimensions is often required to power up miniaturized microsystems in wearable, embedded or implanted solutions. Commercially available batteries often do not meet the needs in these applications due to size or cost. As a result, the demand for new low-cost, ultra-thin, or ultra-small size rechargeable batteries is increasing [5,6].

Solid-State Thin Film Battery (TFB) technology and processes were first developed by Oak Ridge National Laboratory (ORNL) in 1996 [7-9]. It has become an attractive choice of power solution for healthcare and medical devices because of its high energy density, low self-discharge rate, low-profile, non-flammability and long cycle life. However, due to the sensitivity of Lithium to reaction in air and humidity, the packaging of TFB cells must be hermetic or near-hermetic. These packaging requirements are critical to enable shelf life of months to years for various healthcare products and their usage conditions. Meanwhile, the ultra-small form factor means that the seal width and encapsulation requirements are measured in microns to achieve highest energy density for these ultra-small batteries. Therefore, multiple research challenges need to be addressed to achieve the highest energy densities with 10's of microns level sealing solutions to support volume manufacturing of ultra-small size thin film batteries (TFB).

In this work, a novel battery sealing solution has been developed for TFB cells as small as 2.5mm x 2.5mm in X and Y size with thickness of <100um. First, various deposition approaches including PVD, plating, and injection soldering were studied for the sealing of vias as small as 50um through thin substrates. SEM cross-section analysis shows good via fill and coverage, which can enable hermetic electrical interconnections for cathode and anode terminals of the micro-TFB cells. After fabrication of the micro-TFB cells, programmable precision laser milling and cutting processes were developed to singulate large quantities of these cells. Finally, various polymer coatings and metal coatings were evaluated for the sealing and packaging of these ultra-small micro-TFB cells. The finished cells consist of sealed electrical terminals and are encapsulated with a hermetic or near-hermetic metal case. These demonstrated prototype solid-state micro-TFB cells have been tested and results showed a battery capacity of >12 µAh with more than 50 charging-discharging cycles.

II. CONVENTIONAL THIN FILM BATTERY DESIGN

Conventional TFB structure can be illustrated as shown in Figure 1. The overall TFB structure consists of a stack of thin films on a substrate with package on the top. Compared with conventional Li-Ion-Battery, the main advantage of TFB is its ultra-thin thickness and long cycle life. However, since the thicknesses of the cathode, anode as well as the electrolyte are limited, the capacity of a TFB cell is primarily determined by the X-Y dimensions. In 2001, W.C. West et al., reported the use of TFB for micro-size batteries and demonstrated a cell in the size of 50 µm and a normalized capacity of 65 µAh/cm² [7]. Since then, more research on the solid-state micro-size battery has been reported recently. In 2016, A. Kutbee et al., reported an enhanced normalized capacity of 147 µAh/cm² with a free-form flexible TFB using etch-back method [8]. These miniaturized power source solutions may be applied to

978-1-7281-1500-9/19 $31.00 © 2019 IEEE

various microsystems such as integrated MEMS sensors, implantable medical devices, and IoT devices.

Nevertheless, most of these studies were based on conventional TFB design using lateral thin film metal as terminal connections, which consumes a significant portion of the lateral area. In addition, the edge sealing area may account for a significant portion of battery area, which may largely reduce the available active battery area. Figure 2 shows that the percentage of sealing area for a 2.5mm x 2.5mm square TFB cell is plotted as a function of its edge sealing width. If the sealing width is 0.3mm, the percentage of sealing area relative to battery area exceeds 40% of the total battery area. Therefore, we propose the use of vertical vias as terminal connections and thin metallic sealing to reduce sealing width. These changes can significantly improve the lateral area efficiency in micro-TFB cells.

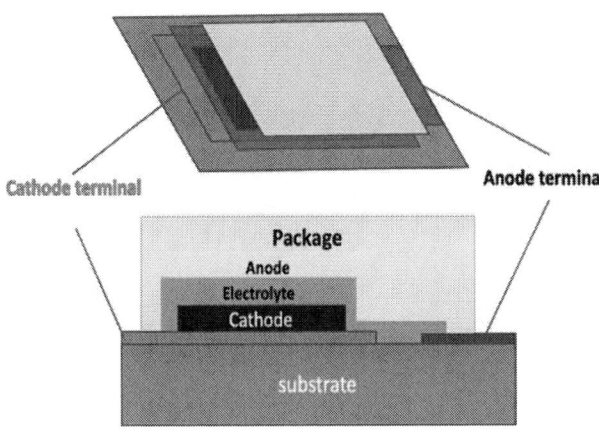

Figure 1. Schematic of a Conventional Solid State Thin Film Battery with lateral terminal connections [7-9]

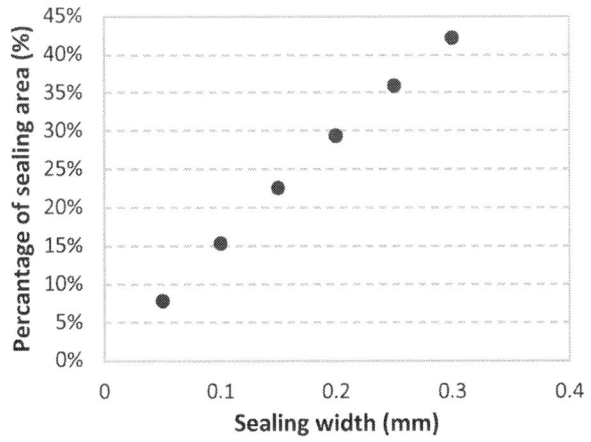

Figure 2. Percentage of sealing area over a TFB cell ~ 2.5mm x 2.5mm as a function of sealing width

III. DEVELOPMENT OF HERMETIC VIAS FOR MICROBATTERY TERMINAL CONNECTIONS

A. Via filling with Ag-paste

For micro-battery applications, the available active area is very limited and vertical interconnection is desirable for size reduction. Therefore, we first proposed the use of Ag-filled via through the thin-film substrate, as shown in Figure 3 since Ag paste is commonly used for via-filling in ceramic packaging substrates. Following stencil printing, the Ag-paste filled via is usually sintered at elevated temperature to remove organic paste materials and to permit silver particles to coalesce. Figure 4 shows an example of an as-sintered Ag-filled via. Characterization of the Ag paste filled vias by scanning electron microscopy (SEM) and lithium anode degradation indicated that such a via was not hermetic. Air and moisture could pass through the structure and degrade the battery structure. Unfortunately, it is well-known that lithium anode is sensitive to O_2, N_2 and H_2O and a via sealing solution is needed.

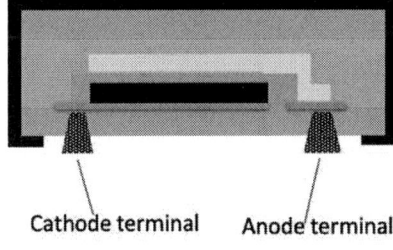

Figure 3. Schematic of a sealed solid state TFB with vertical via connections and metal packaging

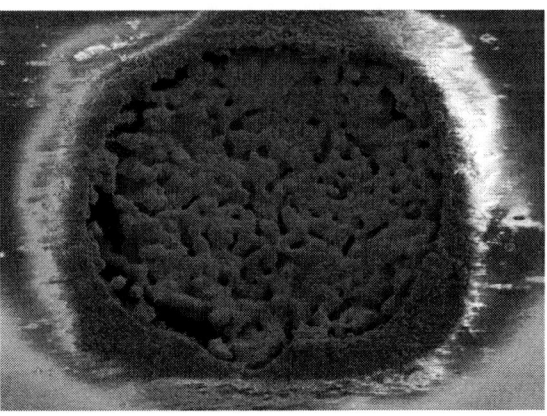

Figure 4. SEM image of an Ag-filled via shown high porosity

B. Via plug and capping with metal

We explored electroless plating, electroplating, as well as injection soldering as options to create a hermetic cap over the porous Ag-filled via (50 μm in diameter). Figure 5a shows a via after electroless plating. The advantage of electroless plating was that it does not require a seed layer. In this approach, the porous Ag via had been further filled and its surface covered by a densely packed metal cap. However, the coverage of the deposition was still limited because the electrochemical deposition only nucleates and grows on a properly activated metal surface. In addition, the sealing at the edge of a via was not consistent to ensure a hermetical via seal.

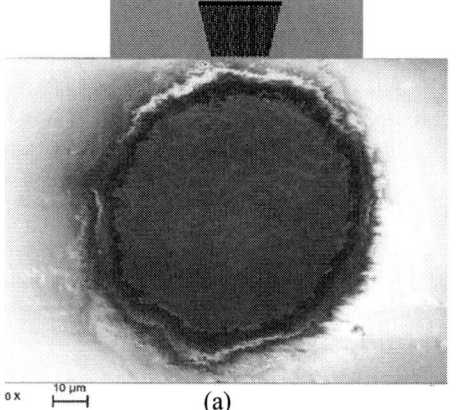

(a)

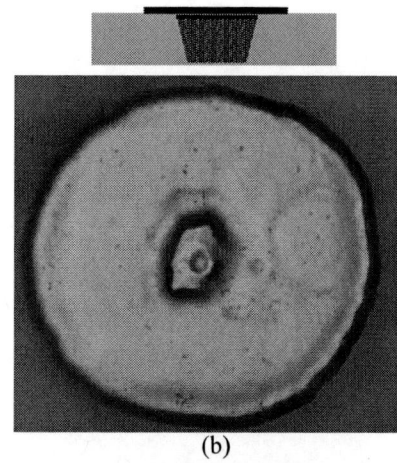

(b)

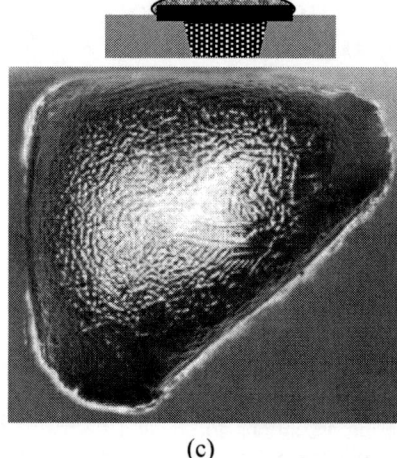

(c)

Figure 5. via with caps using various deposition approaches. (a) electroless plating (b) electroplating with an extended capturing pad (c) solder over an extended capturing pad in triangle shape

Figure 5b. shows an electroplated via with a larger coverage area. In this case, a thin layer of metal pad was first deposited as a seed layer using PVD to cover the via and define the plating area and thicker metal was then electroplated. Since the completed cap is larger than the via, much better sealing coverage was achieved. Figure 5c shows a soldered cap on top of the via. In this case, a capping pad was also first deposited and a controlled volume of solder was injected and reflowed to form a hermetic sealing cap on top of the via and sealing structure (Triangular in this demonstration).

Figure 6. shows the cross-section view of the sealed vias by electroplating and injected solder respectively. It can be seen that defect-free sealing caps had been formed with both methods although the Ag-filled via was porous. The plated sealing cap was smooth and flat, while the soldering approach tended to deposit a "dome shape" cap due to solder reflow. The molten solder has reacted and formed an intermetallic interface on top of the Ag filler in the via. Process enhancements have demonstrated a planar surface for either electroplating or injected solder structures.

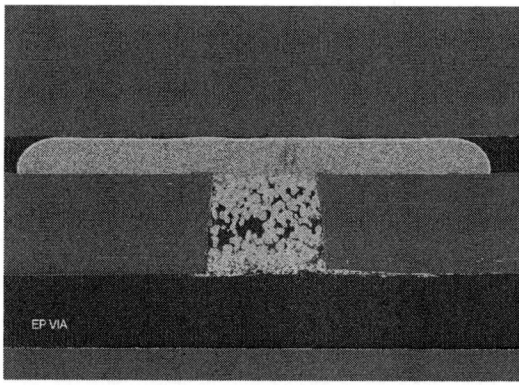

(a) Electroplated cap over via

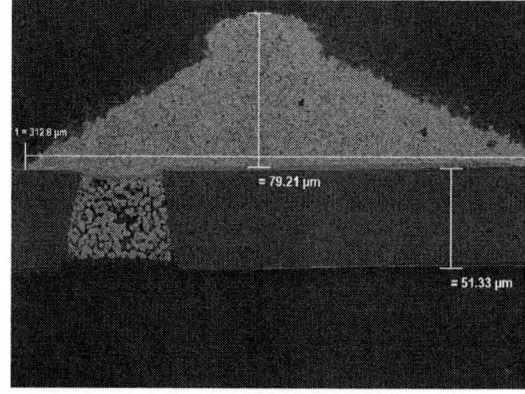

(b) Injected solder over via

Figure 6. Cross-section view of a sealed via with (a) electroplated pad (b) solder over via

Figure 6 also shows that the soldered cap tends to be much thicker because of the large solder volume. In order to control the total thickness of the micro-TFb cells, we studied various approaches and successfully demonstrated flattened solder cap over vias. Figure 7a shows an image of a TFB cell with flattened solder caps over via termicals. The 3D profile is illustrated in Figure 7b. and indicates that the total solder cap thickness is below 10 μm.

(a) a TFB cell with flattened solder cap over vias

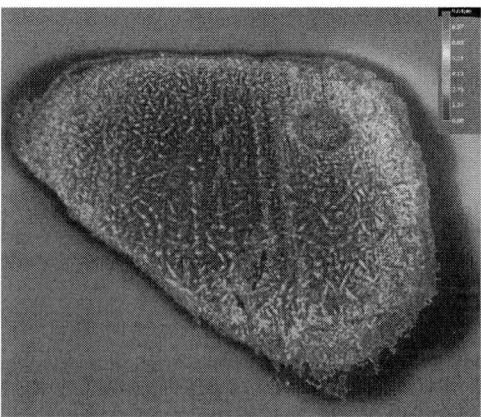

(b) 3D profile of a flattened solder cap over a via

Figure 7. Micrography of a flattened soldering cap in triangular ship over a via (thickness < 10 μm)

In order to test the effectiveness of the solder capped vias for hermetic sealing, dummy TFB cells have been fabricated and subjected to accelerated storage test. During the test, samples were stored in enclosed chambers with relative humidty of ~ 85% at temperature of 65°C. As mentioned previously, lithium anode is very sensitive to air and moisture. When an air or moisture leak occurs, the lithium metal near the vias can react with oxygen or H_2O and turn into dark color. Micrographs have been taken before and after the accelerated storage test to compare the lithium anode morphology. As shown in Figure 8, after 187 hours, only a small reduction of lithium (~ 2 μm) was observed near the sealed vias, which indicated that an effective sealing for the vias was demonstrated. Exposure in such an accelerated

test for 1 week is equivalent to a storage life of about 1 year under normal ambient storage condition.

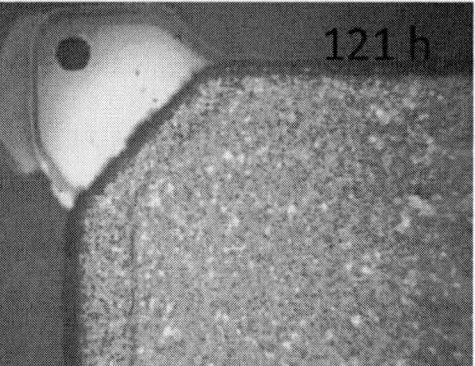

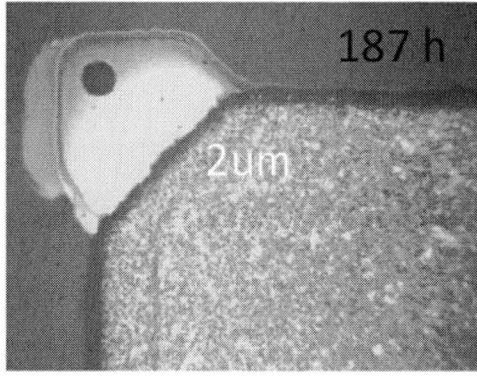

Figure 8. Micrograph of TFB cells with sealed vias connections after accelerated storage test 65°C/RH 85% for various periods of time.

IV. SINGULATION & PACKAGING OF MICRO-TFB CELLS

A. Singluation of TFB cells

In addition to vias, the edges of the micro-TFB cells need to be sealed. An array of solid-state TFB cells are usually fabricated on a thin film substrate panel. After the deposition of thin film battery layers and encapsulation, the micro-TFB cells must be singulated and carefully sealed and packaged since the edges of these cells may be prone to defects if laser parameters are not optimized. An example of defects created with laser milling or cutting is thermally induced defects,

978-1-7281-1500-9/19 $31.00 © 2019 IEEE

which may create a leakage path. Another challenge is the efficient handling of multiple small size TFB cells as they are only 2.5mm in x-y dimension and 100 μm in thickness. Therefore, we developed a special laser milling and cutting processes, which leaves tabs on each TFB cell to support handling before the final singulation. As shown in Figure 8, a precision-controlled laser milling and cutting program was used to first create a trench around a TFB cell without cutting the two small tabs. Various types of tabs or structures may be used depending on the cell design, as shown in Figure 9a and Figure 9b. Once a trench is formed, the edges of a TFB cell are exposed, either a controlled ambient or combined step for edge sealing can provide adequate time prior to subsequent process steps.

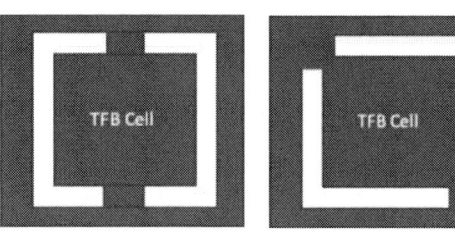

(a) center tab design   (b) corner tab design

Figure 9. Two examples of tab designs for handling during laser cutting

B. TFB cell edge sealing with metal coating

Figure 10 shows the photograph of a processed TFB panel with deposited metal coating. The high-magnification micrograph shows that the metal coating covers not only the top of a cell but also its sidewall.

Figure 10. TFB panels after metal deposition. (both top and sidewalls have been coated)

Figure 11 further shows detailed view of a singulated micro-TFB cell, which shows defect-free metal coating on the top and the sidewall. Electrochemical deposition was effective in the metal coverage in all directions.

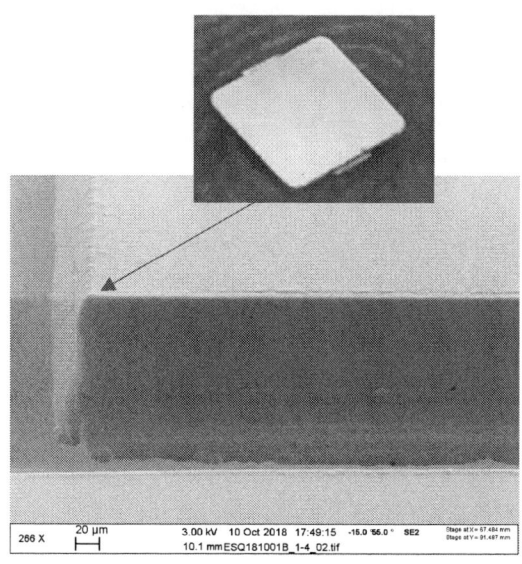

Figure 11. Micrograph and SEM image of singulated TFB cell (good sidewall coverage was demonstrated)

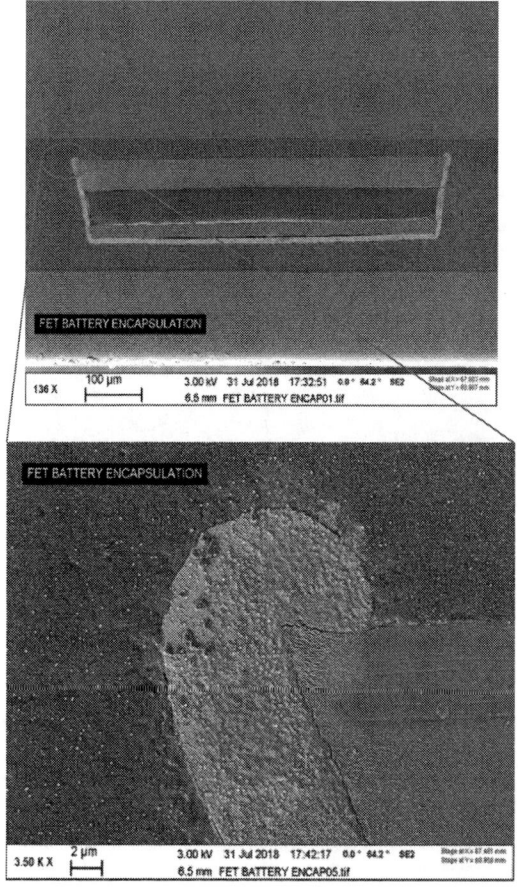

Figure 12. Cross-section image of metal sealed micro-TFB cell.

The Scanning Electron Micrograph (SEM) of cross-section of a sealed TFB cell is shown in Figure 12, which clearly demonstrates the completed coverage around a TFB cell. The metal coating thickness is 10um, which provides mechanical support and sealing of a TFB cell. Furthermore, because of electrochemical deposition, an extended metal coating is visible at the bottom of the cell.

V. MICROBATTERY TESTING

Upon the completion of micro-TFB fabrication, the cells have been tested with charging and discharging. Figure 13a shows the charging capacity as a function of time. A 2.5mm x 2.5mm micro-TFB cell was charged within 4 minutes. Figure 13b shows the discharging curve. It can be seen that the working voltage of such a micro-TFB ranges from 4.05V to 3.6V.

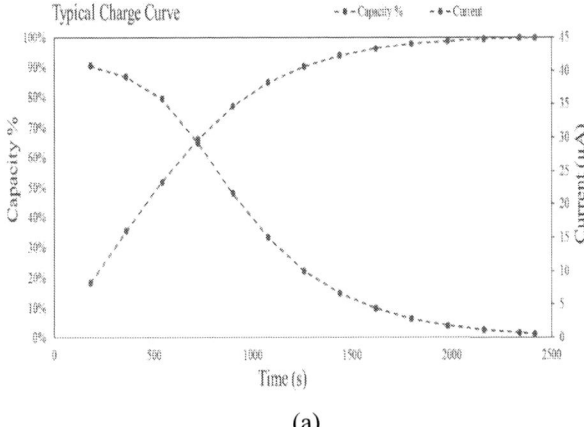

(a)

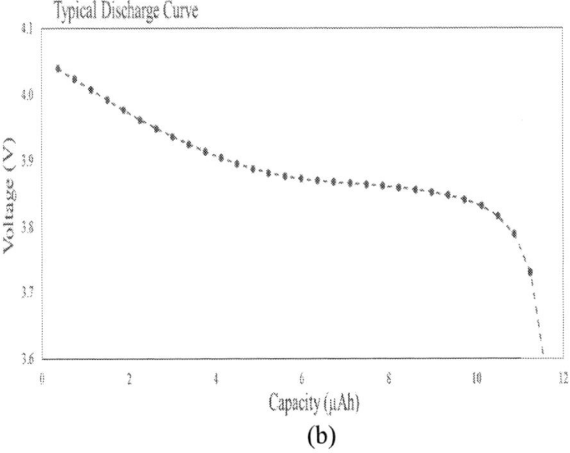

(b)

Figure 13. Example of (a) charging curve (b) discharging cure for a micro-TFB cell

VI. CONCLUSION

A novel packaging concept and structure have been proposed and demonstrated for solid-state micro-TFB (Thin Film Battery) cells. Vertical interconnection structure has been fabricated based on semiconductor metallization and bumping processes technologies. Metal coatings have been deposited for the sealing of all edges of TFB cells with batch plating processing. Finally. a normalized capacity of ~ 190 µAh/cm² has been successfully demonstrated for micro-TFB cells (2.5mm x 2.5mm x 0.1mm), which is higher than those reported in literature to the best of our knowledge. The resulted prototype micro-TFB cells may be used in miniaturized microsystems including wearable or implantable devices for various healthcare and medical applications.

ACKNOWLEDGMENT

This work has been conducted under a Joint Research Agreement between FET Inc. and IBM. The authors are grateful to Dr. Ajay Royyuru (VP. of Healthcare and Life Sciences) for the management support. Acknowledgements are also extended to technical staff of Centeral Scientific Service at IBM T. J. Watson Research Center.

REFERENCES

[1] A. Hande, T. Polk, W. Walker and D. Bhatia, "Self-Powered Wireless Sensor Networks for Remote Patient Monitoring in Hospitals,"*Sensors* 2006, *6*(9), 1102-1117

[2] R Torah1, P Glynne-Jones, M Tudor, T O'Donnell, S Roy and S Beeby, "Self-powered autonomous wireless sensor node using vibration energy harvesting," Measurement Science and Technology, Vol. 19, No. 12, 2008

[3] E. Sardini and M. Serpelloni, "Self-Powered Wireless Sensor for Air Temperature and Velocity Measurements With Energy Harvesting Capability," in *IEEE Transactions on Instrumentation and Measurement*, vol. 60, no. 5, pp. 1838-1844, May 2011.

[4] G. Yang *et al.*, "A Health-IoT Platform Based on the Integration of Intelligent Packaging, Unobtrusive Bio-Sensor, and Intelligent Medicine Box," in *IEEE Transactions on Industrial Informatics*, vol. 10, no. 4, pp. 2180-2191, Nov. 2014.

[5] P. Singh, Xiquan Wang, R. Lafollette and D. Reisner, "RF-recharged microbattery for powering miniature sensors," IEEE SENSORS, vol.1, pp. 349-352,Vienna, 2004.

[6] M. Nathan, "Microbattery Technologies for Miniaturized Implantable Medical Devices," Current Pharmaceutical Biotechnology, Vol. 11, No. 4, pp. 404-410(7), June 2010.

[7] B. Wang. J.B. Bates, F.X. Hat, B.C. Sales, R.A. Zuhr, and J.D. Robertson, *J. Electrochem. Soc.*. 143, 3203 (1996).

[8] B.J. Neudecker, N.J. Dudney and J.B. Bates, J. Electrochem. Soc., 147, 517 (2000).

[9] W. C. West, J. F. Whitacre, E. J. Brandon and B. V. Ratnakumar, "Lithium micro-battery development at the Jet Propulsion Laboratory," in *IEEE Aerospace and Electronic Systems Magazine*, vol. 16, no. 8, pp. 31-33, Aug

Screen Printed Temporary Tattoos for Skin-Mounted Electronics

Samuli Tuominen and Matti Mäntysalo
Faculty of Information Technology and Communication Sciences
Tampere University
Korkeakoulunkatu 3, P.O Box 692, FI33104
Tampere, Finland
matti.mantysalo@tuni.fi

Abstract—This paper focus on fabrication and analyzation of screen-printed temporary transfer electrical tattoos. Stretchable conductive traces and insulator layers are printed on a transfer tattoo paper. The screen-printed structures are electromechanically analyzed. Four different test structures were used to evaluate the sheet resistance and the quality of the printing process. The sheet resistance of the samples are 41 ± 6 mΩ/$\square$. The breaking point for the tensile test was 19 %. Finally, on-skin performance was evaluated by attaching a printed tattoo on a knee and backhand of a test person. The skin-mounted tattoo showed good electrical performance and conformability.

Keywords-stretchable electronics, temporary electrical tattoo, printed electronics

I. INTRODUCTION

Current wearable electronics is mainly based on rigid circuit boards. The circuit board defines the size, shape, and form factor of the unit, leading typically to bulky and clumsy wearable systems. Miniaturization of electronics has improved the user comfort, however, to reach the full potential of wearables, electronics hardware must become soft, light-weight, thin, conformable to the body and, especially, inexpensive to manufacture. Therefore, the electronics must be stretchable and mass-producible.

The development of functional inks has enabled the use of printing methods to manufacture electronics. So far electronics is mainly fabricated using substractive methods such as photolithography. Printed electronics, however, is an additive manufacturing method. Additive means that a small amount of functional ink is applied directly and accurately on a substrate, for example, by inkjet, screen, offset, or gravure printing technologies. The solution can contain functional materials such as metal flakes, oxides, polymer conductors, or semiconductive materials. Furthermore, recent development with silver nanowire [1], carbon nanotube (CNT) [2], and graphene inks [3], have enabled stretchable interconnections [4], sensor elements [5, 6], and circuits [7, 8].

Skin-mounted e-patch is an ideal form factor for remote monitoring of health metrics, and therefore, academics and industry have shown growing interest towards these technologies in recent years. Technology solutions have evolved from thick bandage-like structures currently on market to novel ultra-thin, soft, and transparent tattoo-like films. When mounting electronics on top of the skin, things like stretchability and conformability must be considered. In some places, human skin can stretch even more than 30 % [9]. However, in many wearable and skin-mounted applications it can be less [10]. This sets the strain target for skin mounted electronics. High conformability can be achieved with ultra-thin layers [11]. L. Wang et al. investigates the relationship between a membrane thickness and conformability [12]. Ameri et al. demonstrated that 510 nm and thinner films of PMMA conforms on a human skin [13]. This thickness is an extreme requirement for complete stack, which is hard to achieve together with a low resistance conductive elements requirement in wireless communication system. However, the situation is not as straight forward, and several parameters are determining the conformability of the system as presented in [12]. As an example, the maximum film thickness for fully conformable films depends from the film and membrane material.

The earliest electrical tattoo-like work was reported by Dae-Hyeong Kim et al. [14]. They also introduced a new term "epidermal electronics" to refer electronics that is like an epidermal layer of a human skin. This means that electronics should have skin like properties in terms of thickness, mechanical behavior (i.e. elastic modulus, bending), and density. Later Woon-Hong Yeo et al. [11] demonstrated a fabrication method and an electrical device based on poly(vinyl alcohol)(PVA) template that was used to transfer an electrical circuit on a skin.

Casson et al. [15] proposed inkjet printing of electrodes on to a transfer tattoo paper. This enabled transferring the electrodes on the skin by wetting the paper. However, they needed an adhesive sheet with a thickness of 10 µm to improve the adhesion between the skin and the electrodes. Ameri et al. [13] reported ultra-thin and transparent electrical tattoos based on the graphene. They proposed so thin structure that it conformed on a skin, and therefore, they did not need an additional adhesive. However, the sheet resistance of the device was relatively high, which might be suitable for some applications like electrodes and sensors. However, much lower sheet resistance values are needed for high-frequency components like antennas.

978-1-7281-1500-9/19 $31.00 © 2019 IEEE

This contribution presents a screen printed conductive and dielectric films on a temporary transfer tattoo template. The fabricated films are electromechanically characterized.

II. Materials and Methods

A. Materials

Substrate material in this research was temporary transfer tattoo paper from Silhouette [16]. Thermal properties of this particular paper was not available. Sanchez-Romaguera et al. investigated the heat stability of another tattoo paper using thermogravimetric analysis (TGA) [17]. According to their findings, the processing temperature should not exceed 135 °C. A baking test was done in order to estimate maximum annealing temperature of the paper. The pieces of the transfer tattoo paper were baked in oven at 100 °C, 120 °C, 140 °C, and 160 °C for 20 min. The paper did not show any visible color change with 100 °C and 120 °C. Minor and remarkable color changes were identified in samples annealed at 140 °C and 160 °C, respectively. Based on the work of others and these results, it was concluded that maximum annealing temperature for this tattoo paper is 20 min at 120 °C.

For conductive traces, a commercially available conductive silver-flake ink was used (CI-1036 by EMS) [18]. This ink is screen printable and stretchable. It has high conductivity, but it is not transparent. Furthermore, the estimated film thickness is higher than 10 µm, which may restrict the conformability with the skin. Despite of these limitations, this ink was selected due to the low processing temperature (120 °C) suitable for the selected substrate. Furthermore, the ink has high stertchability and conductivity suitable for RF antennas.

For cover layers, a UV-curable screen printable DI-7540 (from EMS) was selected [19]. This ink is stretchable and designed to function with the CI-1036 ink. The recommended film thickness when printed on top of the silver conductor is 36 µm in order to avoid pinholes [19]. The cover layer will increase the total thickness of the stack (temporary tattoo, conductive film, and dielectric film) and might decreases the conformability of the electrical tattoos.

B. Fabrication

Printing was done by using a TIC SCF-300 semi-automatic screen-printing machine. A polyester mesh screen was used in the printer. The mesh was attached to a 500×3000 mm^2 aluminum frame with a profile of 30×30 mm^2. The mesh count was 79 threads/cm, the thread diameter was 55 µm, and the mesh opening was 69 µm. The stretching angle of the mesh was 22.5°.

The screen printer was prepared according to instructions, and the conductive layers were printed first. After setting pressure values, snap-off, and alignment of squeegees, the conductive patterns were printed, dried, and annealed according to instructions (II. A.). Cover layer was printed on top of conductors after sheet resistance measurement. The cover layer was annealed according to manufacturer instructions.

C. Characterization

Four different geometries were designed and fabricated for electrical and electromechanical characterization as shown in Fig 1. These structures are named: i) Greek cross (GC), ii) bridge resistor (π), iii) big squares (BS), and iv) U-shape. The width of the lines is 1.0 mm and the size of the pads are approx. 9.0 mm^2. The length of the bridge resistor (π) under the test is 10 mm. The size of BS pattern is 900 mm^2. The U-shape pattern is previously used on strethcable electronics characterization. The total length of the U-shape pattern is 188.4 mm [10].

The sheet resistance of GC, π, and BS were determined using four point method that consists of a probe station and Keithley 2425 Sourcemeter. Sheet resistance of U-shape is determined using 2-point measurement setup. Each test card contained 6 GC, 6 π, 4 BS, and 5 U-shape patterns. Four test card were fabricated. The total number of test samples are listed in Table I.

TABLE I.   Number of Samples

	GC	π	BS	U	
Set 1	6	6	4	5	21
Set 2	6	6	4	5	21
Set 3	6	6	4	5	21
Set 4	6	6	4	5	21
	24	24	16	20	

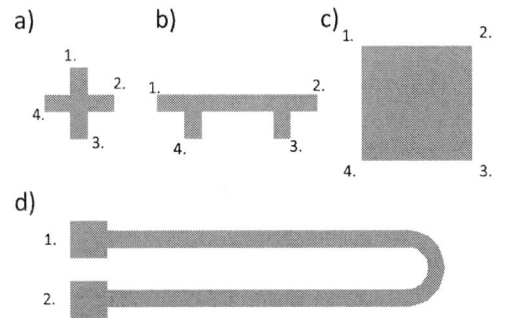

Figure 1. Sheet resistance test structures: a) Greek cross (GC), b) bridge resistor (π), c) big squares (BS), and d) U-shape.

Electromechanical characterization were performed by integrating a tensile test and a sourcemeter. Tensile tests were done with U-shape pattern and 2-probe method. Previously used Keithley 2425 Sourcemeter were combined with a LabView program via GPIB interface to store the data. Tensile tests were done with both an Instron 4411 and on-skin measurement setups.

The challenge with electromechanical characterization is that the printed temporary tattoo structure cannot be electromechanically characterized if not transferred onto a template or carrier. Thermoplastic polyurethane (TPU) is commonly used substrate material for printed stretchable electronics. We selected Epurex Platilon U4201 to act as a carrier material during the tests. The data logging frequency was set to 5 Hz.

In on-skin resistance measurement, the U-shape pattern was attached on a knee of a test person.

III. RESULTS AND DISSCUSSION

A. Sheet resistance measurement

Sheet resistance of thin film can be determined with different structures. In this work, different test methods and their suitability for computing sheet resistance are analyzed. At the beginning, the sheet resistance of GC structures were determined by measuring the sheet resistance from two positions. The results are shown in Fig. 2. As seen from the Fig., the average sheet resistance values between the Set 1-3 are almost identical, but the Set 4 showed clearly a higher sheet resistance value as well as a wider confidence interval (95 % CI) indicating also a larger variance. This means that Set 1-3 have quite uniform printing quality in both direction in x-y plane and between each other.

After this, the sheet resistance of different geometries were computed in order to determine the quality between different resistance patterns. Fig. 3. shows the sheet resistances of the GS, BS, π, and U-pattern in each set. The average between the samples as well as the variance are quite similar between the samples. However, the sheet resistance of Set 4 is clearly higher compared to Set 1-3 as

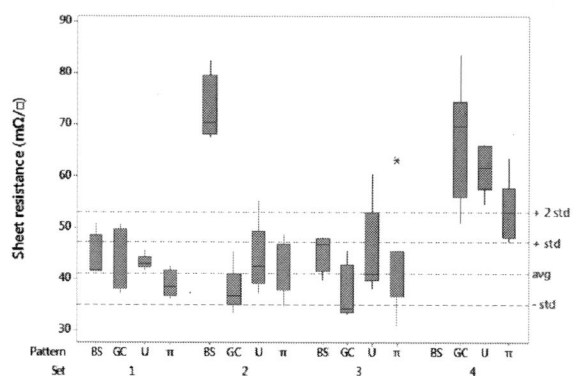

Figure 3. Sheet resistances of the BS, GC, U, and π patterns including average and standard deviations.

already noted. In addition, the Set 2 BS pattern seemed to have higher resistance compared to other patterns inside the Set 2. The higher resistance of Set 2 BS is most likely due to fabrication deviation inside the test card while the higher resistance in Set 4 is most likely due to process variance between the test cards. Potential explanation is the drying of the ink on the screen, which will partially block the holes of the screen, and therefore less material is transferred on the substrate. This emphasizes the need of screen maintenance during the fabrication process. The average sheet resistance of samples, excluding the resistances of Set 2 BS and Set 4, was 41 ± 6 m$\Omega/\square$. The sheet resistance values of Set 2 BS and Set 4 are more than two standard deviation higher than the average of other samples as seen from Fig 2. These numbers verifies the above statement about problems in ink transfer. Furthermore, the obtained sheet resistance values are comparable with the results achieved with same ink printed on a TPU (36 ± 5 m$\Omega/\square$) substrate [10].

B. Influence of the cover layer

In order to analyze the impact of the insulator layer, a dielectric cover layer was printed on top the Set 1 and Set 2 conductors, and cured with UV. Fig. 4. shows the histogram of the sheet resistance of non-covered and covered samples. It can be seen that printing and annealing of the cover layer

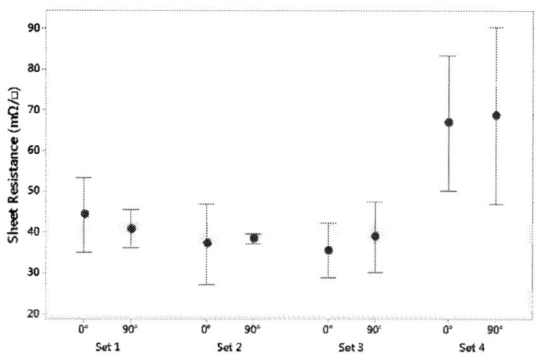

Figure 2. GC sheet resistance values of Set 1-4 in measured in 0° and 90°.

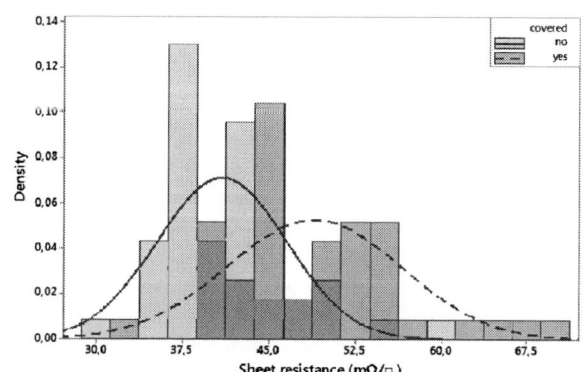

Figure 4. Sheet resistances of non-covered samples as a function of covered samples from Set 1 and Set 2.

increases the resistance and standard deviation of the sheet resistance. The average sheet resistance and standard deviation of non-covered and covered samples are 41 ± 6 m$\Omega/\square$ and 49 ± 8 m$\Omega/\square$, respectively. This equals an increase of 17 %, which is relatively large. Despite this increase, the sheet resistance is still on a relative low level compared to the results reported for graphene [3], CNT [2] and Ag-nanowires [1] as listed in Table II. This means that printed transfer tattoo could be used in wireless applications like antennas for RFID, WLAN or Bluetooth.

TABLE II. SHEET RESISTANCE OF DIFFERENT MATERIALS IN TEMPORARY TATTOO APPLICATIONS

Ink	Sheet resistance ($\Omega/\square$)	Ref
AgNW	50-100	[1]
CNT	20-700	[2]
Graphene	30 -100 k	[3]
Ag-flake	0.05	[this work]

C. Electromechanical performance

The electromechanical characterization was performed using Instron 4411 tensile strength tester as described earlier using U-shape pattern that were covered with insulator ink. Fig. 5. presents a resistance as a function of strain. The y-axis is normalized to initial resistance (R/R_0).

From Fig. 5. it can be seen that there is large fluctuation and sudden spikes in the resistance value at the beginning and at the end of the test. An exponential increase of the resistance and the fluctuation of the resistance at the end is expected due to the disconnection and fracture in a conductive path. Similar results are reported earlier in [10]. Furthermore, Fig. 5. shows some unexpected and undesirable fluctuation of the resistance at the beginning of the test. The resistance rapidly increased 50-100 times on these peaks. These spikes can be caused by the measurement setup itself, which is not ideal for transfer tattoo characterization. The general problem is that electrical tattoos are so thin that they need to mounted on a separate carrier / template in order to perform electromechanical tests. This will cause variation in

strain uniformity of the device under test. In ideal case, the mechanical properties of the carrier and the adhesion between the electrical tattoo and the carrier should be comparable with a human skin. If the mechanics and adhesion differs a lot, it causes uncertainly to the measurement setup since the stretching is actually happening though the carrier and not directly by moving clamps. In our tests, we used TPU carrier, which is not an ideal in order to mimic the human skin. This carrier was selected since it was previously used to analyze the performance of the stretchable ink itself. By selecting the same substrate, we can compare the results of electrical tattoo to directly printed results. This, however, will restricts the analysis only for the breaking point of the conductor.

As seen from the Fig. 5, the resistance starts to increase suddenly around the strain level of 16 %, and the complete failure occurs at the strain of 19 %. This is relatively small value compared to the results achieved by printing the same ink directly on TPU. J. Suikkola et al. received 74% elongation with the same ink printed directly on the TPU [10]. These results demonstrates that substrate material has remarkable impact on electromechanical behavior of the system. Furthermore, the results indicates to limit the use of presented electrical tattoo on body locations were the stretchability is limited to less than 15 %.

D. On-skin resistance measurements

Since the strain test with tensile test machine was limited

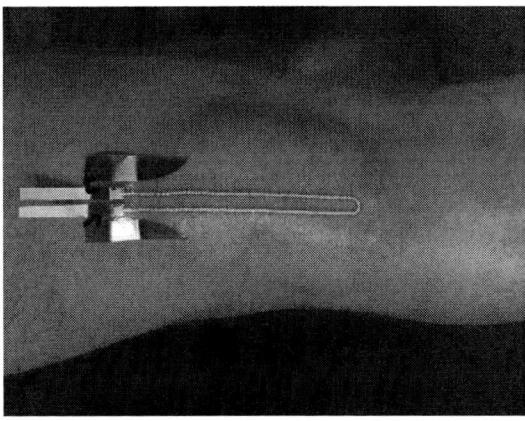

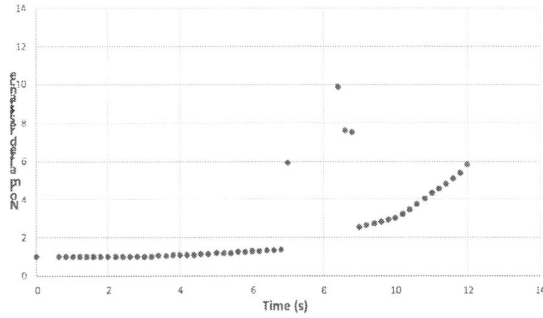

Figure 6. (top) image of an U-shape tattoo mounted on a skin and (bottom) a normalized resistance the test sample as a function of time when knee is bended.

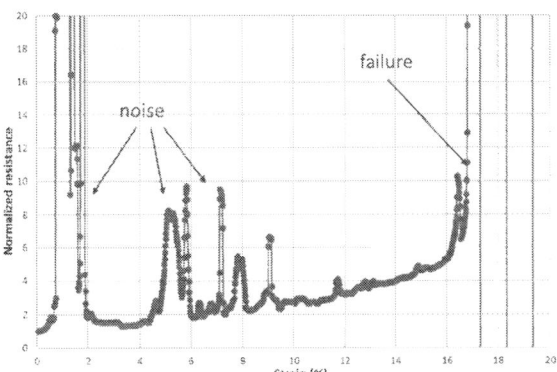

Figure 5. Normalized resistance of U-shape pattern from strain tests.

due the fact that electrical tattoo was mounted on TPU carrier, we performed on-skin resistance measurement. On-skin resistance measurement was done by attaching a U-pattern sample on the knee of a test person. This test method has other limitations and difficulties, which limited the number of test samples, and therefore, only couple of samples were successfully measured. Fig. 6 shows the normalized resistance as a function of time when the knee of the test person was bended. Fig. 6. shows much smoother resistance increase with less noise compared to electromechanical tests shown in Fig. 5, especially at the beginning of the test. This supports the previous hypothesis related to the challenges of the test methods and substrate interaction.

In Fig. 6., the normalized resistance increases up to 25 at the eight second mark. However, the conductor cannot be considered broken, since the resistance returns back to the original steady increase until it breaks after 12.4 seconds.

E. Conformability of the screen printed electrical tattoos

Finally, the conformability of the proposed electrical tattoos was observed by transferring a π-shape stack on the back of the hand of a test person. The conformability was visually observed. Fig. 7 shows an example of printed π-shape transfer tattoo on the backhand of a test person. In order to analyze the results in detail, we need to remind that the tattoo has a stack of materials. The silver conductors are printed on a top of a tattoo paper and blue color shows the cover layer insulating the conductor from the skin. The cover layer is not printed on top of the measurement pads. The full stack of the conductor and cover layer seem to conform on the skin at least partially. This can be seen from the bridge resistor location. The thickness of this stack is approx. 40 µm, which is relatively large, and needs additional adhesive film to be mounted on the skin. The stack on top of measurement pads is thinner, approx. 15µm, and therefore, the features of the skin are much clearer on the areas where only the silver is applied.

IV. CONCLUSIONS

In this work, we presented a fabrication of electrical temporary transfer tattoo by screen printing conductive and

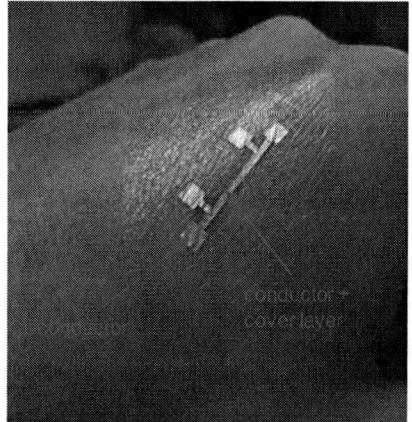

Figure 7. An example of printed tattoo stack mounted on hand.

insulating layers. A sheet resistance of 41 mΩ/□ was achieved. This is relative low value compared to reported values with CNT, AgNW, and graphene. Low sheet resistance is needed for high-frequency applications such as remote healthcare applications.

The elongation at break achieved during the electromechanical tests was 16 %, which is much smaller value compared to values reported by direct printing on 50 µm thick TPU film (74 %). Thus the proposed screen-printed temporary tattoo stack is recommend to use in low strain (10-15%) level applications.

ACKNOWLEDGMENT

This work is financially supported by Academy of Finland (Grants #288945 and #319408).

REFERENCES

[1] Silver Nanowire Coating Ink A30, Novarials Corporation, website. Available (accessed on 3.5.2018): http://www.novarials.com/ProductsAgNWCIA30.html.

[2] Technical Data Sheet Nink-1000, NanoLab Incorporation. Available (accessed on 8.5.2018): https://sep.yimg.com/ty/cdn/nanolab2000/Nink1000-TDS-2017.pdf?t=1511629666&.

[3] J. Li, et al, Efficient Inkjet Printing of Graphene, Adv. Mater. 2013, 25, 3985–3992

[4] Mahmoud Mosallaei, et al, Geometry Analysis in Screen-Printed Stretchable Interconnects, in IEEE Trans. on Comp., Pack. and Manuf. Technology, vol. 8, no. 8, pp. 1344-1352, Aug. 2018.

[5] T. Vuorinen, et al, Inkjet-Printed Graphene/PEDOT:PSS Temperature Sensors on a Skin-Conformable Polyurethane Substrate, Sci Rep. 2016 Oct 18;6:35289. doi: 10.1038/srep35289.

[6] A. Koivikko, et al, Screen-Printed Curvature Sensors for Soft Robots,in IEEE Sensors Journal, vol. 18, no. 1, pp. 223-230, Jan.1, 1 2018.

[7] H. Sirringhaus, et al "High-resolution inkjet printing of all-polymer transistor circuits", Science, vol. 290, pp. 2123-2126, Dec. 2000

[8] R. Shiwaku, et al, Printed organic inverter circuits with ultralow operating voltages, Adv. Electr. Mater., vol. 3, no. 5, p. 1600557, Mar. 2017

[9] A.J. Gallagher, et al, Dynamic tensile properties of human skin, 2012 IRCOBI Conference Proceedings - International Research Council on the Biomechanics of Injury, 12–14 September, 2012, Dublin, Ireland, pp. 494–502.

[10] J. Suikkola, et al, "Screen-printing fabrication and characterization of stretchable electronics, 2016, Scientific reports 6, 25784

[11] W. Yeo, et al, Multifunctional epidermal electronics printed directly onto the skin, Advanced Materials, Vol. 25, Iss. 20, 2013, pp. 2773–2778.

[12] L. Wang, N. Lu, Conformability of a thin elastic membrane laminated on a soft substrate with slightly wavy surface, Journal of Applied Mechanics, Transactions ASME, Vol. 83, Iss. 4, 2016.

[13] S. K. Ameri, et al, Graphene Electronic Tattoo Sensors, ACS Nano, Vol. 11, Iss. 8, 2017, pp. 7634–7641.

[14] D.-. Kim, et al, Epidermal electronics, Science, Vol. 333, Iss. 6044, 2011, pp. 838–843.

[15] A.J. Casson, R. Saunders, J.C. Batchelor, Five Day Attachment ECG Electrodes for Longitudinal Bio-Sensing Using Conformal Tattoo Substrates, IEEE Sensors Journal, Vol. 17, Iss. 7, 2017, pp. 2205–2214.

[16] Printable Tattoo Paper, Silhouette Europe, website. Available (accessed on 14.5.2018): http://www.silhouettecameoeurope.com/printable-tattoo-paper.html.

[17] V. Sanchez-Romaguera, et al., Towards inkjet-printed low cost passive UHF RFID skin mounted tattoo paper tags based on silver nanoparticle inks, Journal of Materials Chemistry C, Vol. 1, Iss. 39, 2013, pp. 6395–6402.

[18] Technical Data Sheet CI-1036, Engineered Materials Systems Incorporation, 2017.

[19] Technical Data Sheet DI-7540, Engineered Materials Systems Incorporation, 2014.

Thermoset polymers for bioelectronic interfaces - engineering of thermomechanical properties

A. Carolina Duran-Martinez
Bioengineering
The University of Texas at Dallas
Richardson, United States
Adriana.DuranMartinez@utdallas
.edu

Seyedmahmoud Hosseini
Chemistry
The University of Texas at Dallas
Richardson, United States
sxh144830@utdallas.edu

Daniel Del Nero
Electrical Engineering
The University of Texas at Dallas
Richardson, United States
Daniel.DelNero@utdallas.edu

Alexandra Joshi-Imre
Engineering Innovation
The University of Texas at Dallas
Richardson, United States
Alexandra.Joshi-Imre@utdallas.edu

Walter E. Voit
Materials Science and
Engineering
The University of Texas at Dallas
Richardson, United States
walter.voit@utdallas.edu

Melanie Ecker
Engineering Innovation
The University of Texas at Dallas
Richardson, United States
Melanie.Ecker@utdallas.edu

Abstract— **In this work, we present three compositions of thiol-click polymers engineered to soften at various degrees upon implantation into the living body. Dynamic mechanical analysis has been used to assess the softening process and the storage modulus under simulated physiological conditions. In detail, we present polymers that are engineered to soften under physiological conditions within 10 to 20 minutes from 1.7-2.5 GPa to 1.7 GPa, 800 MPa, or 20 MPa, dependent on their composition. Critical material parameters such as the glass transition temperature and the cross-link density can be tuned by the choice of monomers and their relative proportions within the pre-polymer solution, as well as by the specific curing conditions used in the polymerization process. Because these polymers are used as substrate material for bioelectronic interfaces, we also present their electrical properties under simulated physiological conditions. For this, electrochemical impedance spectroscopy was used to assess dielectric constant and electrical resistivity. Typical relative permittivity of our compositions ranges from 4 to 6, while volume resistivity ranges from 10E10 to 10E13 Ohm*m, both measured at 37 ° C temperature in phosphate buffered saline solution. We discuss the results in the perspective of building functional electrical circuits on the presented polymer substrates.**

Keywords— *thermoset polymers, tunable thermomechanical properties, bioelectronic interfaces, low relative permittivity*
Introduction

Bioelectronic interfaces are used to monitor and control bioelectronic activity of a living muscle or nervous tissue by connecting it to electronic circuitry. These interfaces need to be biologically, electrically and mechanically compliant to the living tissue. Biological compliance may be reached by using inert and non-toxic materials as components. The most studied materials are silicon, tungsten, polyimide, and parylene-C [1-6]. Mechanical compliance is achieved if no

harmful mechanical stress is exerted on the tissue, which is an aim particularly difficult for practical realization. One of the causes of early failure of devices has been attributed to the mechanical mismatch between the biomedical interface and the target tissue. This mechanical mismatch promotes a higher rejection of the implant by the body, which causes a series of acute and chronic foreign body response against the implanted device [7-11]. Designing the bioelectronic interface to be soft is an attempt towards improving the mechanical compliance. Silicone substrates have been used to prevent tissue damage by the mentioned mechanical mismatch [12, 13]; however, significant swelling of silicones like polydimethylsiloxane (PDMS) limits photolithography processes for device microfabrication [14] and lead to delamination of electrical functionalities [15, 16]. Our group at The University of Texas at Dallas has pioneered thermoset thiol-click polymers such as thiol-epoxies, thiol-enes and thiol-ene acrylates to be used as substrate material for soft bioelectronic interfaces[17-21]. The key benefit of using these polymers is the capability to tune thermomechanical properties, specifically the glass transition temperature. This allows for a controlled degree of softening upon implantation into the body. Softening polymer based bioelectronic interfaces maintain their rigidity during handling and implantation, and soften post implantation. The phenomenon of softening is the result of a small increase in temperature and plasticization caused by fluid uptake upon exposure to physiological conditions [22]. The polymer initial stiffness provides the benefit of facilitation of fabrication, handling and positioning of the device, and the capability of penetrate nervous tissue for penetrating probes. Once implanted, the polymer substrate will soften, becoming a more compliant material. Other groups are currently working in tuning the thermomechanical properties of softening polymer

978-1-7281-1500-9/19 $31.00 © 2019 IEEE

compositions due to their promising benefits in biomedical applications [8, 23-31]

In most applications, the bioelectronic interface operates by means of electrically driven microelectrodes, that are implanted to the immediate vicinity or even inside the target tissue, in order to provide the desired selective function. For example, microelectrodes typically need to be spaced closer than 100 micrometer to neuronal axons in order to record individual action potentials with appropriate signal to noise ratio[32], for microwire arrays, and an intersite of less than 50 μm for high density probes, such as tetrodes or micromachined silicon probes [33]. The softening degree of the polymers upon implantation can be tuned by the thiol – click polymer composition. Dynamic mechanical analysis (DMA) has been widely used to characterize polymer properties, such as storage modulus and glass transition temperature. Using this technique, we can understand and relate how the composition and concentration of monomers affect the thermomechanical properties including the modulus and glass transition temperature T_g of a softening polymer in dry and soaked conditions [34, 35]. Additionally, we can test on a same composition but different processing of the polymer to evaluate the mechanical properties of the product [36]. Finally, DMA can be used to quantify the softening process of the polymer while it is immersed in solution at isothermal conditions. Electrical properties are also studied in this paper as these polymers are intended to be used as substrate material for bioelectronic interfaces. As the thiol-click polymers soften due to water uptake and plasticization, this may lead to the formation of undesirable leakage current pathways. Electrochemical impedance spectroscopy (EIS) was used to obtain dielectric constant and leakage current test was conducted to measure electrical resistivity. Both studies were performed at simulated physiological conditions for the three studied compositions.

I. MATERIALS AND METHODS

A. Polymer synthesis

The monomers used in this study, as shown in Figure 1, are an alkene, 1,3,5-Triallyl- 1,3,5-triazine-2,4,6(1H,3H,5H)-trione (TATATO); an acrylate, tricyclo [5.2.1.02,6] decanedimethanol diacrylate (TCMDA); two thiols, trimethylolpropane tris(3-mercaptopropionate) (TMTMP), and Tris[2-(3-mercaptopropionyloxy)ethyl] isocyanurate (TMICN); and a photo initiator, 2,2- dimethoxy-2-phenylacetophenone (DMPA)(Fig. 1).

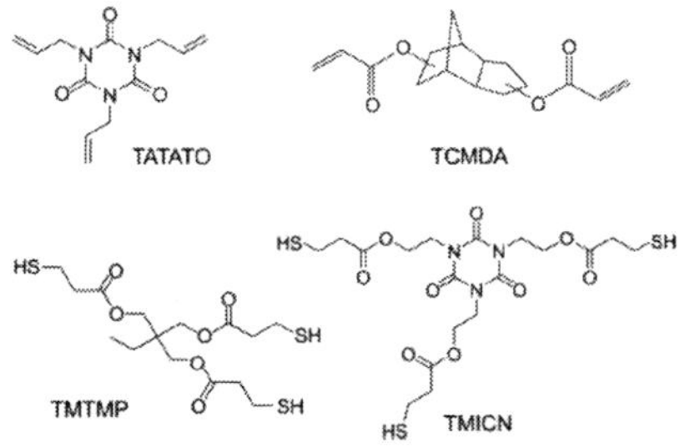

Figure 1. Chemical structures of representative monomers used for thiol-click polymers

TABLE I. MOLE FRACTIONS OF MONOMERS USED IN POLYMER FORMULATIONS

Composition	TATATO (alkene) [c]	TMICN (thiol) [c]	TMTMP (thiol) [c]	TCMDA (acrylate) [c]
TE	0.5	0.05	0.45	0
TEA1	0.345	0.345	0	0.31
TEA2	0.275	0.275	0	0.45

Only TMICN was purchased from Evans Chemetics LP (Teaneck, NJ), all other chemicals were purchased from Sigma Aldrich (St. Louis, MI). All the chemicals were used as received without further purification. Note that TATATO and TMICN have an Isocyanurate as a core, and that TMTMP has the same structure than TMICN, but it has carbon-carbon bonds with Sp3 hybridization as a core.

B. Sample preparatn for mechanical characterization

Three polymer compositions were studied based on their degree of softening at 37 °C, one thiol-ene (TE-Fully-Softening) and two thiol–ene/acrylates (TEA1-Semi-Softening and TEA2-Non-Softening), having a total of 0.1 wt% DMPA of total monomer concentration dissolved into the solution. Exact mole fractions are given in the table in Table I, which show a 1:1 ratio between the alkene and the thiol composition and a variable ratio of the acrylate, having no acrylate in one composition. Exposure to light was prevented by covering each polymer solution vial with aluminum foil. Monomers were mixed thoroughly by planetary speed mixing at 3000 rpm for 15 minutes. The polymer solutions were spin cast on pre-cleaned 75 mm × 50 mm type II soda lime glass 0215 Corning slides using a Laurell WS-650-8B spin coater. Spin parameters were chosen to achieve thicknesses of about 30 μm, resulting in an acceleration of 2000 rpm s⁻¹ and 45 seconds at 500 rpm (TE), 120 seconds at 800 rpm (TEA1), and 120 seconds at 700 rpm (TEA2). Afterwards, polymerization was accomplished at ambient temperature using an UVP CL-1000 crosslinking chamber with five overhead 254 nm UV bulbs for 30 seconds followed by curing under 365 nm light for 1 hour, for the composition effect study. Cured polymer samples were post-cured in a vacuum oven at for 12 hours to complete network conversion.

C. Sample preparation for dynamic mechanical analysis measurements

DMA samples were photolithographically defined in the UT Dallas Class 10,000 cleanroom facility as previously described [35, 37]. Spin coated SMP-on glass substrates were used as the starting material in the cleanroom. Low temperature silicon nitride (using PlasmaTherm-790 PECVD) was deposited to act as a hard mask for the following plasma etching processes in which the device outline/shape was patterned. Adjacently, the nitride hard mask was etched away in the 1:10 hydrofluoric acid dip. The delamination of the test devices from the glass slide was achieved by soaking in deionized water for about 20 minutes.

D. Mechanical characterization

Dynamic mechanical analysis was used as previously described [34, 35] to assess the thermomechanical properties of the three polymer formulations in the dry and in the wet state, and their softening behavior under different curing parameters.

In detail, RSA-G2 solids analyzer (TA Instruments) equipped with immersion fixture was used to measure the storage modulus (E') and the tan δ of SMP samples in tension mode. For all measurements, the frequency was set to 1 Hz, the oscillation amplitude was 0.275%, and temperature ramp was 2 K min^{-1}. For the dry-sample-test, the samples were measured in the range of 10 to 140 C, ensuring all glass transition temperatures are captured.

For the wet measurement, samples were positioned in the system, then soaked in phosphate buffered saline (PBS) at room temperature (RT) for 30 minutes, cooled down to 10 °C before DMA measurements were done between 10-80 °C. For the softening profile test, the samples were positioned in the immersion system. Measurements were performed immediately after adding PBS at room temperature in the immersion cell. The PBS was heated to 37 °C, followed by isothermal soaking of the polymer substrates for 60 minutes. This method is described in [34], and has been used previously to study the mechanical properties of softening polymers [35, 36].

E. Sample preparation for electrical characterization

Samples for electrical characterization were fabricated on clean glass slides as carrier substrate. The cleaning procedure of the glass slides were performed in a 10000-class clean room. A 1g/50mL solution of Alconox soap and deionized water (DIW) was prepared. Glass slides were rubbed with a wet soft cloth with the solution, rinsed with DIW and placed in 20-slide holder inside 1000 mL plastic beaker. The rinsed glass slides were sonicated between 10-15 minutes in Alconox solution, rinsed with water in ultrasound bath for 2 minutes, followed by an acetone rinse for 1-minute. A last rinse with DIW was followed by drying the glass slides with a Nitrogen gun. Cleaned samples were baked in an oven at 120°C for 5 min for dehydration.

A 30 nm titanium (Ti) layer, used as adhesion layer, followed by a 200 nm gold (Au) layer, used as electrode plate, were deposited by a CHA electron-beam evaporator. Adhesive tape was used to cover an approximately 1 cm wide strip at the edge of the gold-coated slide in order to mask for electrode contact area. Approximately 30-μm layer of monomer solution was spin-coated and polymerized using 30 s exposure to 254 nm UV light. The tape was removed, and polymerization was completed using 60 minutes exposure in 365 nm UV cross-linker oven followed by 18 hours vacuum oven treatment at 120 °C. An 8 cm long solid copper wire (26AWG PVC-insulated) was attached to the exposed Au layer, using conductive silver epoxy. The silver epoxy was cured at 120 °C for 10 minutes. The wire connection was insulated using Dow Corning 3140 curable coating, allowing it to cure at room temperature for at least 2 days. Sample thickness was measured with a DekTak XT Profilometer, using a range of 524 μm, a stylus force of 12.5 mg, a length of 20 mm during 90 s, having a resolution of 1.11 μm/pt and a speed of 333.33 μm/s.

F. Electrical characterization

Electrical properties under simulated physiological conditions were studied using I-V measurements and electrochemical impedance spectroscopy (EIS). The polymer samples were prepared for measurement by clamping a glass adapter onto the glass slides in a manner described in [38] [39], and filling it with warm (37 °C) PBS. The room temperature conductivity of PBS was 24.95 mS/cm and the average pH was 7.2. All measurements were performed at 37 °C. Relative permittivity was extracted from EIS data after 2 hours of soaking in PBS. Resistivity was measured in a separate I-V setup afterwards.

- **Electrochemical impedance spectroscopy**

A Gamry Reference 600+ potentiostat/galvanostat/ZRA instrument was used for EIS measurements. The samples were set up in a two-electrode configuration, where the "working electrode" was connected to the gold layer underneath the polymer, and the "counter electrode" was connected to a platinum wire placed inside the PBS. The "reference electrode" of the potentiostat was connected to the counter electrode. EIS was performed from 100 kHz to 10 mHz by applying a 10 mV rms AC voltage signal between working and counter electrodes. The samples were measured inside Faraday cage, placed inside a Fisher Scientific Isotemp oven at 37 °C.

Coating capacitance was obtained from impedance values in the 10-100 kHz frequency range, by fitting a leaky capacitor model (so-called paint model[40][39, 41, 42]). Relative permittivity was then calculated from the coating capacitance, using the measured thickness of each sample and the adapter's O-ring area of 3.46 cm^2.

- **Current-voltage measurements**

Current-Voltage (IV) measurements were performed using a Keithley 6482 dual channel pico-ammeter/voltage source instrument. The samples were set up in the same two-electrode arrangement described earlier for the EIS measurements (with the exception of the counter electrodes being stainless steel and that there is no reference electrode attached to the counter). A DC voltage staircase of 50 mV steps was measured between + 250 mV limits, cycling through three times. The resulting DC current at every voltage step was measured and recorded. The current versus voltage curves obtained this way, were fitted with a linear in order to determine leakage resistance of the polymer coatings.

All measurements were performed at 37°C, on samples soaked for approximately 1 day in 37°C PBS. The three cycles of IV measurements took approximately 18 hours to complete on each sample, and they resulted yielded very consistent results.

II. RESULTS AND DISCUSSION

Three compositions of thiol-ene and thiol-ene/acrylate polymers were used to demonstrate the effects of monomer composition and polymerization process parameters on the thermomechanical properties of the resulting polymer. Our thiol-click polymers were engineered to present various degrees of softening upon implantation into the living body. We have selected three compositions to represent various degrees of softening in simulated physiological conditions, respectively, in PBS at 37°C.

A. Thermomechanical properties

In order to tune thermomechanical properties we selected tri-functional thiol and alkene molecules and a bi-functional acrylate molecule to serve as monomers. Between two of them (TEA1 and TEA2) we varied the concentration of the selected acrylate (TCMDA), which affects its glass transition temperature. The next composition variation is the interchange of another thiol (TMTMP) instead of the acrylate (TCMDA), keeping a 1:1 stoichiometrically ratio of the thiols and the alkene to maximize the cross-linking chain reactions. We used the same curing parameters for all three-polymer compositions. TEA1 and TEA2 are composed of TATATO, the thiol TMICN, and the acrylate TCMDA. The Thiol-ene (TE) monomer possesses a combination of two thiols, TMICN and TMTMP , along with TATATO (Table 1). Figure 2 shows the storage modulus E' and tan delta for the different compositions in the dry (top) as well as in the wet (center) states. In addition, Figure 2 (bottom) displays the time it takes for each composition to reach their respective softening state. The polymers were designed to be in the glassy state at room temperature in dry conditions (Figure 2 top), but to soften to different degrees under simulated physiological conditions. Figure 2 (center) shows the DMA measurements of the plasticized polymers. The polymer composition TEA2 is close to its glassy state, TEA1 within the glass transition region, and TE almost reaches its rubbery state in PBS at 37°C, respectively.

The cross-link densities are slightly affected by the addition of a bi-functional acrylate in the polymer composition, which is reflected in the storage modulus in the glassy and rubbery regions, as seen in Figure 2. The bi-functional acrylate acts as a rigid chain extender, due to its geometric constrain and steric hindrance. Thus, the higher the acrylate content in the polymer composition, the higher the polymer constrain chain segment mobility, which subsequently increases the T_g when comparing TEA1 to TEA2. However, increasing the acrylate content creates a less homogeneous polymer matrix as acrylate can react with itself, acting as a chain extender. For this reason, the regions with more acrylate will be stiffer than those with less acrylate. This can be noticed by the flattering effect on the storage modulus curve, and a broadening in the tan delta curve with increasing acrylate content (also observed in [22, 36]).On the contrary, when comparing the geometry of the thiols monomers TMTMP and TMICN, one can see that TMTMP is less constrained since it has higher rotational freedom. Therefore, as more TMTMP replaces TMICN, the higher the polymer chain mobility and the lower the T_g.

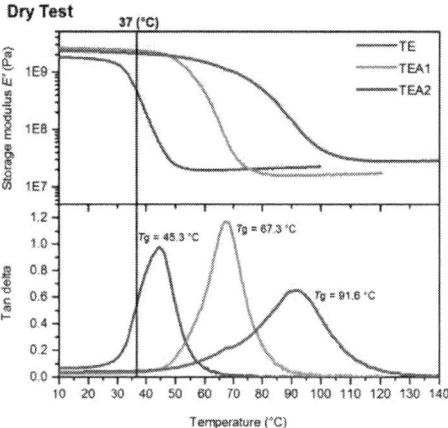

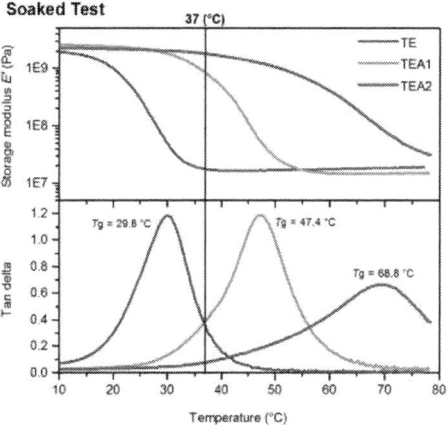

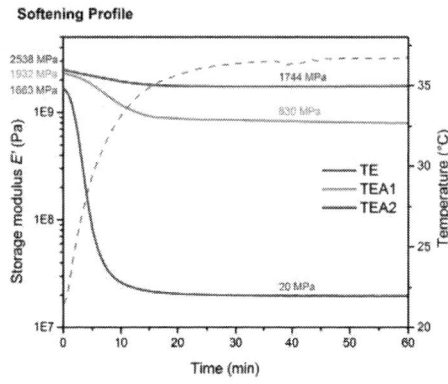

Figure 2. The effect of monomer composition on the thermomechanical properties. DMA dry (top) and wet (center) tests showing storage modulus and tan delta vs temperature of the three compositions TE, TEA1, and TEA2 and highlight 37 °C. Bottom graphic displays the. DMA softening profile.

When looking at the soaked measurements (Figure 2 center), it can be seen that the thermomechanical properties of the thiol-click polymers are affected by aqueous environments. This is due to the plasticization effect of water molecules in the polymer network. Water uptake increase free volume between the polymer chains, which also increase chain segment mobility and depresses the T_g as a consequence [43]. However, the water uptake should be controlled as swelling of the samples may lead to film delamination. The water uptake of these polymer compositions is less than 1.1% for TEA1 and TEA2 [36] and less than 3% for the TE polymer [22]. We observe a downshift in the T_g, of about 15.5-23°C after the samples were soaked in PBS, which is in line with samples that were tested previously [22, 35]. As mentioned earlier, we could observe that each polymer composition resulted in a different softening state at the target condition (PBS at 37 °C). TE has achieved its rubbery state, TEA1 is closer to its T_g and TEA2 just started to soften very gradually and is still in its glassy state. In Figure 2 (bottom), we show the softening dynamics of the polymer samples due to plasticization in PBS accompanied by temperature rise from room temperature to body temperature, as in [22, 35, 44]. We followed the DMA method described by Hosseini *et al.* [34] to measure the softening effect of polymers due to plasticization in simulated implantation to the body procedure conditions. It can be noticed that the polymer compositions achieve their equilibrium states after 10 to 20 minutes transition time, and that the softening is also dependent on the corresponding bath temperature and not only dependent on plasticization effects.

In Figure 3, we show the effect of polymerizing the same composition of a thiol-click polymer using various exposure times at high curing energy (254 nm), followed by an hour at 365 nm wavelength. Previous studies have shown that the curing time and wavelength affect the glass transition temperature, and can be used to determine the best polymerizing conditions [36]. Here, we have demonstrated that T_g fine-tuning can be achieved by controlling the exposure time of the polymer at high energy. Figure 3 top displays DMA curves of TE polymer after various exposure times at 254 nm followed by 1 h at 365 nm, respectively. It can be seen, that high energy curing between 30 s to 3 min does not affect the T_g noticeably, while increasing or decreasing the exposure time results in bigger T_g shifts. This is related to the efficiency of creating radicals in the material to activate the polymerization process. The speed of reaction is proportional to the created radicals. A lower wavelength corresponds to higher energy and thus, a higher impact will be caused to the exposed sample. For short exposure times, the higher energy increases the activation of radicals that initiate the polymerization process. However, for long exposure times, the sample surface will be damaged by the intense photon collisions, altering the composition of the polymer at the surface [35]. The higher the exposure time, the more damaged is the surface and the deeper the sample is affected. For this reason, exposure time at higher energy must be minimized. Thus, it is important to find a sweet spot

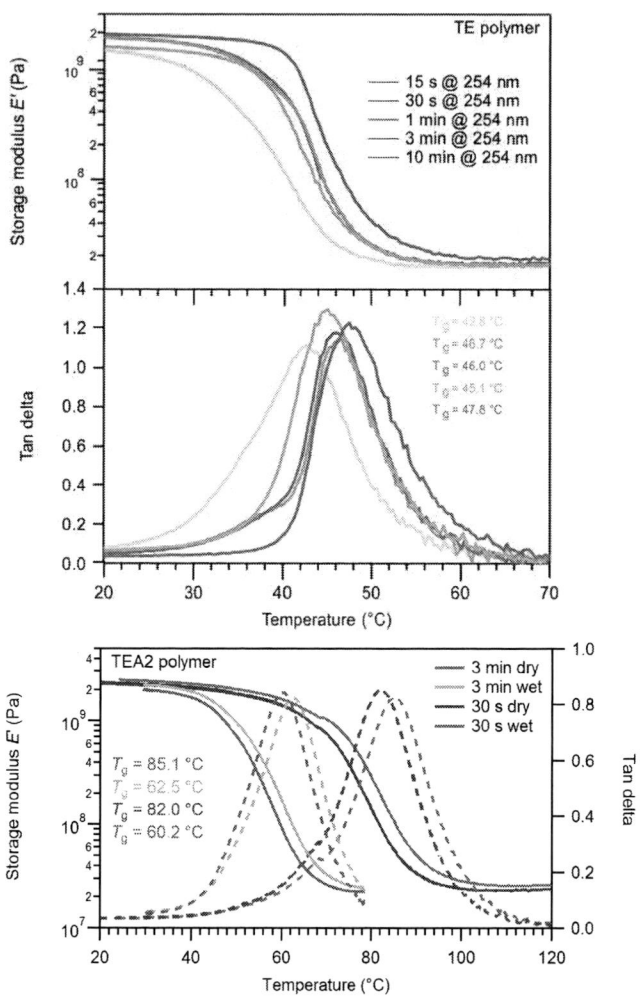

Figure 3. The effect of polymerization conditions. Storage modulus and glass transition temperature of dry TE after various times of exposure (top). TEA2 polymer storage modulus and glass transition temperature in dry and wet conditions at two UV-exposure times (bottom).

where the T_g can be shifted without damaging the polymer. Here, we have applied a limited high-energy exposure to lock the chain movement, followed by 1-hour exposure to a lower intensity to fully convert the network and allow for chain rearrangements, which reduce cure stresses. This will be the best approach to obtain uniform, low-stress thiol-click substrates with high conversion. Figure 3 bottom shows how a slight variation of 254 nm exposure from 30 seconds to 3 minutes does not only affect the dry T_g of the TEA2 polymer, but also the wet state by a shift of 3 and 2.5 °C, respectively.

B. Electrical properties

For implantable biomedical applications, it is critical to understand the electrical properties of thiol-click polymers in physiological conditions. Here, we have obtained relative permittivity and electrical resistivity in a simulated physiological environments at 37°C in PBS, and these are summarized in Table II. In general, we obtained very repeatable results for the resistivity of the soaked polymer compositions, and found narrow distributions with small standard deviations. As for the relative permittivity values, the TE polymer exhibited the highest relative permittivity at 5.8, while the TEA1 and TEA2 polymers were not significantly different from each other (4.8 and 4.7 respectively). The comparison between the three polymer compositions is visualized in Figure 4. The difference in relative permittivity values may be explained by the degree of softening (i.e. the storage modulus). As seen in Figure 2, the TE polymer is close to its fully-softened rubbery state, while the TEA1 and TEA2 compositions are closer to their glassy state at 37°C. In the viscoelastic region between glassy and rubbery states, polymers gradually loose secondary bonding between chain segments as the temperature increases. The softer the state of the material, the more freedom of movement the chain segments will have, and free volume increases due to material expansion. This, and the presence of water molecules are expected to be mainly responsible for the amount of polarization evoked by the applied alternating electric fields during EIS.

Electrical resistivity values show correlation with the storage modulus as well. The softer the state of the material, the lower its resistivity. We measured 44.7 GΩm, 852 GΩm and 3020 GΩm for the TE, TEA1 and TEA2 polymers respectively. We speculate that the polymers absorb some ions from the PBS, which contributes to DC conductivity, and that an increased degree of softening results in increased ion mobility.

Note, that resistivity values for the TE polymer were measured with the highest confidence, and the I-V curves were fitted for resistance with a coefficient of determination at least $R^2=0.997$. For the TEA1 polymer, we obtained fittings with $R^2=0.653-0.991$. In case of the TEA2 polymer, the measured leakage currents were so low that they were indistinguishable from alternative leakage pathways in our measurement setup, and for this reason, the resistivity of the TEA2 polymer is considered to be at least 3 TΩm, with a possibility of being higher.

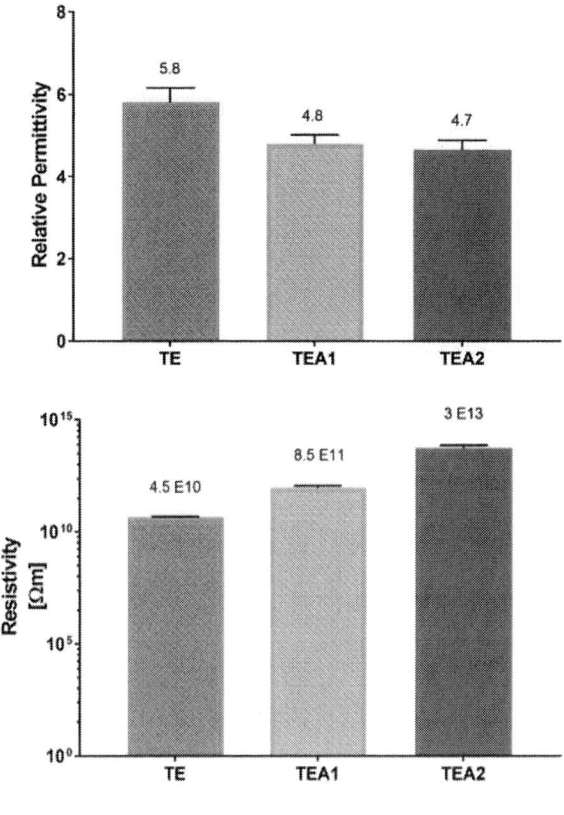

Figure 4. Electrical properties of PBS-soaked thiol-ene/acrylate polymers at 37°C. Average relative permittivity (top) and average resistivity (bottom).

III. CONCLUSIONS

Thiol-click polymers are a promising substrate for biomedical applications. Measurements in simulated biological environment show that a wide range of polymer moduli are achievable. Engineering the thermomechanical properties such as the modulus relies on the choice of polymer composition and the polymerization procedure. We show that in the case of thiol-click polymers, crosslink density and glass transition temperature (T_g) can be effectively tuned by selecting appropriate monomer compositions. Fine-tuning of the T_g for the same composition can be achieved by varying UV-curing exposure time and wavelength.

In addition, we have demonstrated that there is correlation between the dielectric properties and the softening state of the polymer. In order to fully understand this behavior, further investigation is needed.

TABLE II. ELECTRICAL PROPERTIES OF PBS-SOAKED THIOL-ENE/ACRYLATE POLYMERS AT 37°C

Composition	Relative permittivity			Resistivity (Ωm)		
	Mean	Stand. Dev.	n	Mean	Stand. Dev.	n
TE	5.801	0.348	10	4.47E10	5.45E9	10
TEA1	4.793	0.220	6	8.52E11	2.87E11	7
TEA2	4.65	0.231	7	3.02E13*	2.91E13	8

* Resistivity considered at least 3 TΩm due to equipment limitations

ACKNOWLEDGMENT

We express gratitude the University of Texas at Dallas for funding the Center for Engineering Innovation (CEI), which supported this research project. The work was partly done at the UT Dallas Cleanroom Research Laboratory. We thank our colleagues Vindhya Reddy Danda and Lisa Spurgin for their help with sample preparation.

REFERENCES

[1] Hassler, C., T. Boretius, and T. Stieglitz, *Polymers for Neural Implants (vol 49, pg 18, 2011)*. Journal of Polymer Science Part B-Polymer Physics, 2011. **49**(3): p. 255-255.

[2] Barrese, J.C., J. Aceros, and J.P. Donoghue, *Scanning electron microscopy of chronically implanted intracortical microelectrode arrays in non-human primates*. J Neural Eng, 2016. **13**(2): p. 026003.

[3] Wurth, S., et al., *Long-term usability and bio-integration of polyimide-based intraneural stimulating electrodes*. Biomaterials, 2017. **122**: p. 114-129.

[4] Wise, K.D., et al., *Microelectrodes, microelectronics, and implantable neural microsystems*. Proceedings of the Ieee, 2008. **96**(7): p. 1184-1202.

[5] Rousche, P.J., et al., *Flexible polyimide-based intracortical electrode arrays with bioactive capability*. Ieee Transactions on Biomedical Engineering, 2001. **48**(3): p. 361-371.

[6] Mercanzini, A., et al., *Demonstration of cortical recording using novel flexible polymer neural probes*. Sensors and Actuators a-Physical, 2008. **143**(1): p. 90-96.

[7] Barrese, J.C., et al., *Failure mode analysis of silicon-based intracortical microelectrode arrays in non-human primates*. J Neural Eng, 2013. **10**(6): p. 066014.

[8] Harris, J.P., et al., *Mechanically adaptive intracortical implants improve the proximity of neuronal cell bodies*. J Neural Eng, 2011. **8**(6): p. 066011.

[9] Polikov, V.S., P.A. Tresco, and W.M. Reichert, *Response of brain tissue to chronically implanted neural electrodes*. J Neurosci Methods, 2005. **148**(1): p. 1-18.

[10] Polikov, V.S., et al., *In vitro model of glial scarring around neuroelectrodes chronically implanted in the CNS*. Biomaterials, 2006. **27**(31): p. 5368-5376.

[11] Wyser, Y., C. Pelletier, and J. Lange, *Predicting and determining the bending stiffness of thin films and laminates*. Packaging Technology and Science: An International Journal, 2001. **14**(3): p. 97-108.

[12] Lacour, S.P., et al., *Flexible and stretchable micro-electrodes for in vitro and in vivo neural interfaces*. Medical & Biological Engineering & Computing, 2010. **48**(10): p. 945-954.

[13] Minev, I.R., et al., *Electronic dura mater for long-term multimodal neural interfaces*. Science, 2015. **347**(6218): p. 159-163.

[14] Koh, K.-S., et al., *Quantitative Studies on PDMS-PDMS Interface Bonding with Piranha Solution and its Swelling Effect*. Micromachines, 2012. **3**(2): p. 427.

[15] Chah, S., J. Noolandi, and R.N. Zare, *Undulatory delamination of thin polymer films on gold surfaces*. The Journal of Physical Chemistry B, 2005. **109**(41): p. 19416-19421.

[16] Guo, L., et al., *A PDMS-based integrated stretchable microelectrode array (isMEA) for neural and muscular surface interfacing*. IEEE transactions on biomedical circuits and systems, 2013. **7**(1): p. 1-10.

[17] Gutierrez-Heredia, G., et al., *Highly Stable Indium-Gallium-Zinc-Oxide Thin-Film Transistors on Deformable Softening Polymer Substrates*. Advanced Electronic Materials, 2017. **3**(10): p. 1700221.

[18] Ware, T., et al., *Smart Polymers for Neural Interfaces*. Polymer Reviews, 2013. **53**(1): p. 108-129.

[19] Garcia-Sandoval, A., et al., *Chronic softening spinal cord stimulation arrays*. Journal of neural engineering, 2018. **15**(4): p. 045002.

[20] Reeder, J.T., et al., *3D, Reconfigurable, Multimodal Electronic Whiskers via Directed Air Assembly*. Advanced Materials, 2018. **30**(11): p. 1706733.

[21] Ellson, G., et al., *Tunable thiol–epoxy shape memory polymer foams*. Smart Materials and Structures, 2015. **24**(5): p. 055001.

[22] Ecker, M., et al., *From softening polymers to multimaterial based bioelectronic devices*. Multifunctional Materials, 2018. **2**(1): p. 012001.

[23] Lendlein, A., et al., *Light-induced shape-memory polymers*. Nature, 2005. **434**(7035): p. 879-882.

[24] Hearon, K., et al., *Electron Beam Crosslinked Polyurethane Shape Memory Polymers with Tunable Mechanical Properties*. Macromolecular Chemistry and Physics, 2013. **214**(11): p. 1258-1272.

[25] Wojtecki, R.J. and A. Nelson, *Small changes with big effects: Tuning polymer properties with supramolecular interactions*. Journal of Polymer Science Part A: Polymer Chemistry, 2016. **54**(4): p. 457-472.

[26] Xie, T. and I.A. Rousseau, *Facile tailoring of thermal transition temperatures of epoxy shape memory polymers*. Polymer, 2009. **50**(8): p. 1852-1856.

[27] Xie, T., *Tunable polymer multi-shape memory effect*. Nature, 2010. **464**(7286): p. 267.

[28] Yakacki, C.M., et al., *Strong, tailored, biocompatible shape‐memory polymer networks*. Advanced functional materials, 2008. **18**(16): p. 2428-2435.

[29] Ge, Q., et al., *Multimaterial 4D printing with tailorable shape memory polymers*. Scientific reports, 2016. **6**: p. 31110.

[30] Ghosal, K., et al., *Electrospinning over solvent casting: tuning of mechanical properties of membranes*. Scientific reports, 2018. **8**(1): p. 5058.

[31] Qin, X., et al., *Tuning glass transition in polymer nanocomposites with functionalized cellulose nanocrystals through nanoconfinement*. Nano letters, 2015. **15**(10): p. 6738-6744.

[32] Frega, M., *Dissociated Neuronal Networks Coupled to Micro-Electrode Arrays Devices*, in *Neuronal Network Dynamics in 2D and 3D in vitro Neuroengineered Systems*. 2016, Springer. p. 9-29.

[33] Harris, K.D., et al., *Improving data quality in neuronal population recordings*. Nature neuroscience, 2016. **19**(9): p. 1165.

[34] Au - Hosseini, S.M., W.E. Au - Voit, and M. Au - Ecker, *Environmental Dynamic Mechanical Analysis to Predict the Softening Behavior of Neural Implants*. JoVE: p. e59209.

[35] Ecker, M., et al., *Sterilization of Thiol-ene/Acrylate Based Shape Memory Polymers for Biomedical Applications*. Macromolecular Materials and Engineering, 2017. **302**(2): p. 1600331.

[36] Do, D.-H., M. Ecker, and W.E. Voit, *Characterization of a Thiol-Ene/Acrylate-Based Polymer for Neuroprosthetic Implants*. ACS Omega, 2017. **2**(8): p. 4604-4611.

[37] Hosseini, S.M., et al., *Softening Shape Memory Polymer Substrates for Bioelectronic Devices With Improved Hydrolytic Stability*. Frontiers in Materials, 2018. **5**: p. 66.

[38] Murray, J.N., *Electrochemical test methods for evaluating organic coatings on metals: an update. Part I. Introduction and generalities regarding electrochemical testing of organic coatings*. Progress in Organic Coatings, 1997. **30**(4): p. 225-233.

[39] Loveday, D., P. Peterson, and B. Rodgers, *Evaluation of organic coatings with electrochemical impedance spectroscopy. Part 3: Protocols for testing coatings with EIS*. JCT coatingstech, 2005. **2**(13): p. 22-27.

[40] Murray, J.N., *Electrochemical test methods for evaluating organic coatings on metals: an update. Part III: Multiple test parameter measurements*. Progress in Organic Coatings, 1997. **31**(4): p. 375-391.

[41] Loveday, D., P. Peterson, and B. Rodgers, *Evaluation of organic coatings with electrochemical impedance spectroscopy*. JCT coatings tech, 2004. **8**: p. 46-52.

[42] Loveday, D., P. Peterson, and B. Rodgers, *Evaluation of Organic Coatings with Electrochemical Impedance*

Spectroscopy Part 2: Application of EIS to Coatings—Gamry Instruments. JCT CoatingsTech, 2004: p. 88-93.

[43] Ecker, M., et al., *From Softening Polymers to Multi-Material Based Bioelectronic Devices.* Multifunctional Materials, 2019. **2**(1): p. 012001.

[44] González-González, M.A., et al., *Thin Film Multi-Electrode Softening Cuffs for Selective Neuromodulation.* Scientific reports, 2018. **8**(1): p. 16390.

Direct Heterogeneous Bonding of SiC to Si, SiO₂, and Glass for High-performance Power Electronics and Bio-MEMS

Jikai Xu, Chenxi Wang*, Qiushi Kang, Shicheng Zhou, and Yanhong Tian

State Key Laboratory of Advanced Welding and Joining
Harbin Institute of Technology
Harbin 150001, China
E-mail: wangchenxi@hit.edu.cn

Abstract—Low-temperature direct bonding is an attractive technique for joining materials with high bonding strength. Silicon carbide (SiC) is a promising materials for the next generation power electronics and optoelectronic devices. However, it is difficult to realize heterostructures of SiC due to the excellent chemical inertness, ultrahigh hardness, and large mismatch in the lattice constant and coefficient of thermal expansion. To solve this technical difficulty, we developed a VUV/O₃ activated direct bonding method to achieve the integration of SiC on Si-based materials (i.e., Si, SiO₂, and glass). Hydrophilic surfaces have been obtained after the VUV/O₃ activation based on the water contact angle test. This change was mainly attributed to the increase of hydroxyl groups according to the Raman spectra. The direct bonding interfaces were observed by scanning electron microscopy. The nanoscale transition layers have also been confirmed by transmission electron microscopy. The UV-Vis transmittance spectra for SiC/glass pairs exhibited excellent optical visibility which demonstrated that the VUV/O₃ activated direct bonding method has a beautiful prospects for the application of high-performance bio-MEMS devices.

Keywords-silicon carbide; vacuum ultraviolet; direct bonding; heterostructure; interface

I. INTRODUCTION

Silicon carbide (SiC) has been recognized as one of the most promising candidates for the fabrication of the next generation power electronics and optoelectronic devices due to its excellent properties [1-3]. The wide band gap (2.2-3.3 eV) can make devices work at high temperature as high as 600 °C for a long time [4]. The high breakdown electric field ($0.8\text{-}3.0 \times 10^6$ V·cm⁻¹) which is 10 times of silicon can greatly reduce conduction loss, making SiC suitable for the fabrication of high-voltage and high-power devices [5]. The high thermal conductivity (4.9 W·cm⁻¹·K⁻¹) enables the heat generated by the devices to be quickly emitted from the substrates compared with traditional devices [6]. Meanwhile, it can miniaturize devices because no extra radiator is needed. Additionally, the high electron saturation velocity (2×10^7 cm·s⁻¹) also determines that SiC-based devices can achieve higher operating frequency and power density [7]. However, processing of SiC is very difficult due to the excellent inertness and ultrahigh hardness. Silicon-based materials (i.e., Si, SiO₂, and glass) is the most commonly used substrates in the electronic integrated circuits. It has the most mature production and processing technology. Therefore, the

combination of SiC and Si-based materials can exhibit all their respective advantages at the same time. Unfortunately, the integration of SiC on Si-based materials is an enormous challenge because of the large mismatch in the lattice constant and coefficient of thermal expansion (CTE). The CET of SiC is 4.5×10^{-6}·K⁻¹, while that of Si and SiO₂ are 2.6×10^{-6}·K⁻¹ and 0.56×10^{-6}·K⁻¹, respectively, at 25 °C [7-9].

Low-temperature direct bonding is an efficient technique for joining materials with high bonding strength [10-12]. After the surface activation, atoms on the surfaces are in the active states. Thus, atoms at the matting surfaces can diffuse to each other even at low annealing temperatures. Vacuum ultraviolet (VUV) is a short wavelength UV light [13]. It can clean surfaces more effectively owing to the high energy. In addition to getting atomic clean and smooth surfaces, covalent bonds on them can also be broke to form super-active surfaces. For these reasons, VUV/O₃ activated direct bonding method has been demonstrated feasibly to realize reliable Si/Si, glass/glass, and Si/glass direct bonding at 150 °C [14-16]. Nevertheless, the applicability of this bonding method still needs to be further investigated.

In this paper, we developed a VUV/O₃ activated method for the direct bonding of SiC to Si, SiO₂, and glass. This method is facile, universal, and cost-effective. Based on the water contact angle, Raman spectra, scanning electron microscopy (SEM) and transmission electron microscopy (TEM) analyses, hydrophilic surfaces and void-free interfaces have been confirmed. The excellent optical transmittance of the SiC/glass bonded pair has verified the possibility of the application in high-performance SiC-on-glass bio-MEMS devices.

II. EXPERIMENTAL

Double-side polished, n-type, and 2-in. (0001)-oriented 6H-SiC wafers were used in the experiments. The (100)-oriented Si and glass wafers were 3-in. and double-side polished. The SiO₂ layers with a thickness of 300 nm were grown on the 3-in. Si wafers via the thermal oxide. The thickness of SiC was 300 μm, while that of Si and glass were 500 μm. Then all the wafers were cut into 10 mm × 10 mm to be the bonded samples.

Due to the high energy of the VUV, it can react with oxygen (O₂) to produce ozone (O₃). Thus, we designed the VUV/O₃ activated bonding equipment to product the human health and to meet the requirements of the direct bonding process, as presented in our previous published papers. The whole VUV/O₃ activated direct bonding process includes four steps: firstly, put the Si and glass directly under the VUV. The distance between the samples and light source is about 2 mm. The VUV irradiation time is varied from 5 min to 25 min. Secondly, make the activated surfaces contact to form the pre-bonded pairs. Thirdly, drive the bubbles out with the manual force to form the perfect bonding. And then the pre-bonded samples are stored at atmospheric environment to obtain the maximum pre-bonding strength. Finally, anneal the pre-bonded pairs at 150-200 °C to further increase the bonding strength.

RESULTS AND DISCUSSION

A. Surface wettability

Wettability is an important parameter for the hydrophilic direct bonding. The more hydrophilic the surfaces are, the more favorable the formation of the direct bonding. Therefore, the wettability is performed by the water contact angle tester with a single side camera. Fig 1 showed the water contact angle images of SiC, Si, SiO₂, and glass substrates before and after the VUV irradiation.

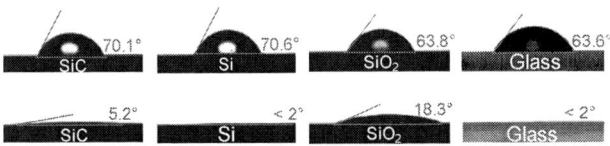

Fig 1. Water contact angle images of SiC, Si, SiO₂, and glass before and after the VUV irradiation

It was clear that all the samples to be bonded became very hydrophilic after the VUV/O₃ activation. The water contact angle of the bare SiC, Si, SiO₂, and glass samples were 70.1°, 70.6°, 63.8°, and 63.6°, respectively. And the water contact angle of VUV/O₃ activated SiC surface decreased to 5.2°, while that of Si, SiO₂, and glass substrates reduced to < 2°, 18.3°, and < 2°, respectively. These hydrophilic surfaces are very beneficial for the direct bonding. The reason for the surfaces became hydrophilic was attributed to the changes of surface chemical states. To further investigate this phenomena, we adapted the Raman spectra to study the surface chemical states before and after the VUV irradiation.

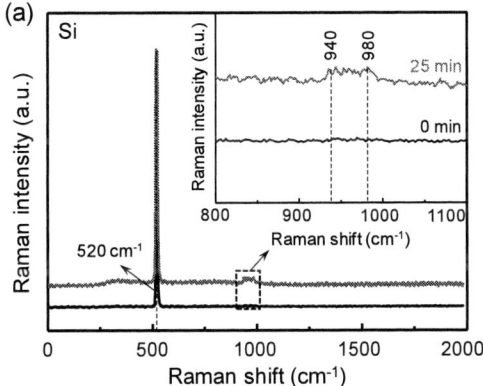

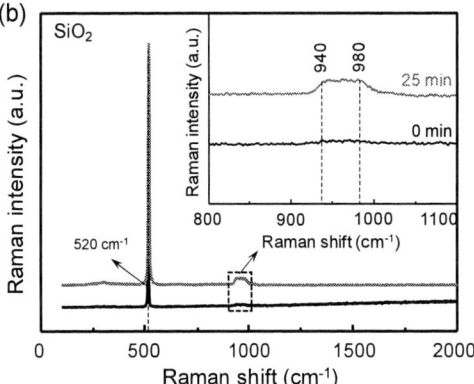

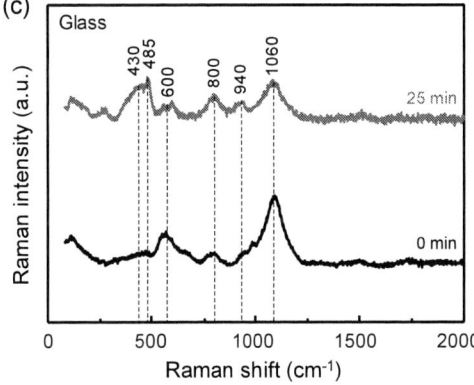

Fig 2. Raman spectra of (a) Si, (b) SiO₂, and (c) glass samples before and after the VUV irradiation.

Fig 2 showed the Raman spectra of Si, SiO₂, and glass before and after the VUV irradiation. Peaks at 940 cm⁻¹ and 980 cm⁻¹ are assigned to be hydroxyl (-OH) groups [9]. For the Si and SiO₂ substrates, the intensity of -OH groups has been obviously increased, as shown in Fig 2(a) and (b). Additionally, we can also find that the oxidation of Si substrate has been greatly improved. It was mainly due to the VUV can break the covalent bonds of O₂ to produce O₃ and oxygen radicals. Meanwhile, the covalent bonds on sample surfaces can also be destroyed and combined with the oxygen radicals to generate oxygen-enriched layers on the Si subsurface. For the glass substrates, the intensity of -OH groups has also been obviously increased. This indicated that hydrophilic glass surfaces have been obtained. Moreover, peaks at 430 cm⁻¹, 435 cm⁻¹, and 800 cm⁻¹ are reported to be large silica rings. Peaks at 600 cm⁻¹ and 1060 cm⁻¹ are attributed to be small silica rings [8-10]. After the VUV/O₃ activation, the intensity of large silica rings has improved, while that of small silica rings has greatly decreased, as shown in Fig 2(c). It was owing to the effects of VUV, which broke the covalent bonds on the glass surfaces and makes the small silica rings change to the large silica rings. And this change increased the total energy of the whole system, making the unstable surfaces generate. Therefore, atoms are easy to diffuse across the bonding interfaces even at low annealing temperatures.

B. Optimization of pre-bonding paremeters

Subsequently, we bonded the SiC to the Si-based materials at different VUV irradiation time to explore the optimum bonding parameters. We calculated the SiC/Si, SiC/SiO$_2$, and SiC/glass pre-bonding area at different VUV/O$_3$ activation times. And then choosing the parameters corresponding to largest bonding area as optimum activation time.

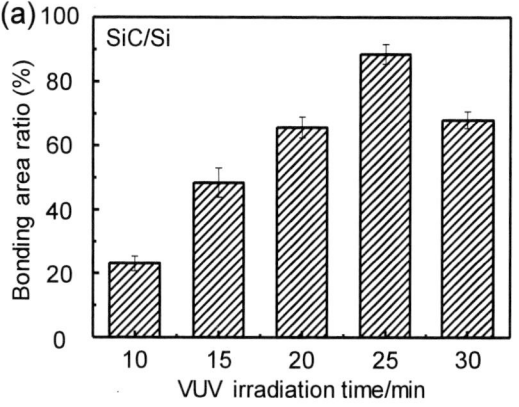

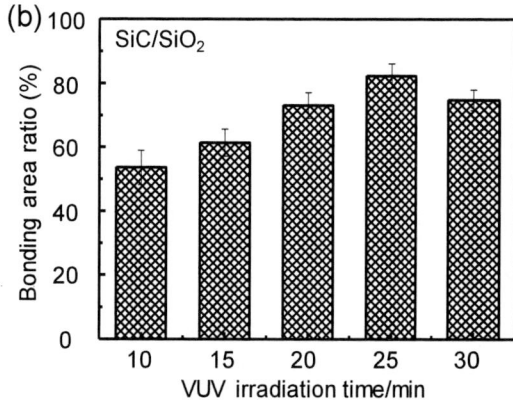

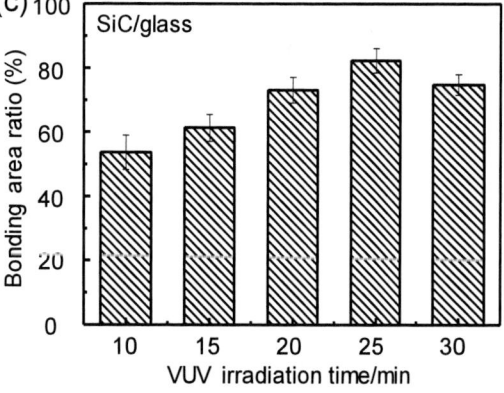

Fig 3. Bonding area ratio of VUV/O$_3$ activated (a) SiC/Si, (b) SiC/SiO$_2$, and (c) SiC/glass bonded pairs at different VUV irradiation time.

Fig 3 showed the bonding area ratio of the VUV/O$_3$ activated SiC/Si, SiC/SiO$_2$, and SiC/glass bonded pairs at different VUV irradiation time. With the increase of the VUV/O$_3$ activation time, the bonding area greatly increased.

As our previous studies, 10 min VUV irradiation can slightly reduce the surface roughness of SiC substrates. However, the wettability is not ideal for the direct bonding. The intensity of –OH groups on the sample surfaces still need to further improve to get the perfect bonding. Thus, the increased VUV irradiation time can lead to more bonding area ratio when the irradiation time varied from 10 min to 20 min. There are two main reasons for this change: one is the surface wettability increased after prolonging the activation time. On the other hand, the surface roughness of the samples to be bonded gradually reduced due to the influence of VUV. The maximum was realized when the VUV irradiation time reached to 25 min. All the bonding area ratio of SiC/Si, SiC/SiO$_2$, and SiC/glass pairs were exceeded 80%. However, when the VUV irradiation time ran up to 30 min, the bonding efficiency became to decrease. This was mainly due to the surface roughness of SiC, Si, SiO$_2$, and glass has increased under the effects of the VUV/O$_3$ activation. Therefore, we chose 25 min VUV irradiation as the optimum parameter for the subsequent bonding strength and bonding interface analyses.

C. Bonding strength and bonding interfaces

After the above exploration of the optimum pre-bonding parameters, 25 min VUV/O$_3$ activated SiC/Si, Si/SiO$_2$, and SiC/glass pre-bonded pairs were used during the annealing process. The pre-bonded samples were annealed at different temperatures (i.e., 100 °C, 150 °C, 200 °C, and 250 °C) for 24 h. The heating rate was 1 °C·min^{-1}. The tensile test results showed that the bonding strength gradually increased before 150 °C for the SiC/Si and SiC/glass bonded pairs. That meant the maximum bonding strengths were obtained when the annealing temperature was 150 °C. The corresponding bonding strengths for SiC/Si and SiC/glass were ~5.0 MPa and ~5.2 MPa, respectively. For the SiC/SiO$_2$ annealed pairs, the maximum bonding strength was got when the annealing temperature reached to 200 °C. And the corresponding bonding strength was ~9.7 MPa. The main reason for the degeneration of the SiC/Si bonding strength at 200 °C was mainly due to the interface water reacted with the Si substrate and produced hydrogen. And the hydrogen gas gathered at the bonding interface to produce bubbles, which reduced the bonding strength. Therefore, the ideal annealing temperatures for SiC/Si, SiC/SiO$_2$, and SiC/glass were 150 °C, 200 °C, and 150 °C, respectively.

To further inspect the bonding reliability, we also used the SEM and TEM to observe the bonding interfaces. The TEM samples were fabricated by the focused ion beam (FIB) technique. After the rigorous grinding and polishing, SEM images of the bonding interfaces were displayed in Fig 4(a), (e), and (i). We could find that all the bonding interfaces were tight and void-free. And Fig. 4(b), (f), and (j) showed the TEM samples for SiC/Si, SiC/SiO$_2$, and SiC/glass pairs fabricated by FIB. The interfaces still kept intact even after a long time bombardment of gallium ion beam. This also demonstrated that the VUV/O$_3$ activated direct bonded pairs

978-1-7281-1500-9/19 $31.00 © 2019 IEEE

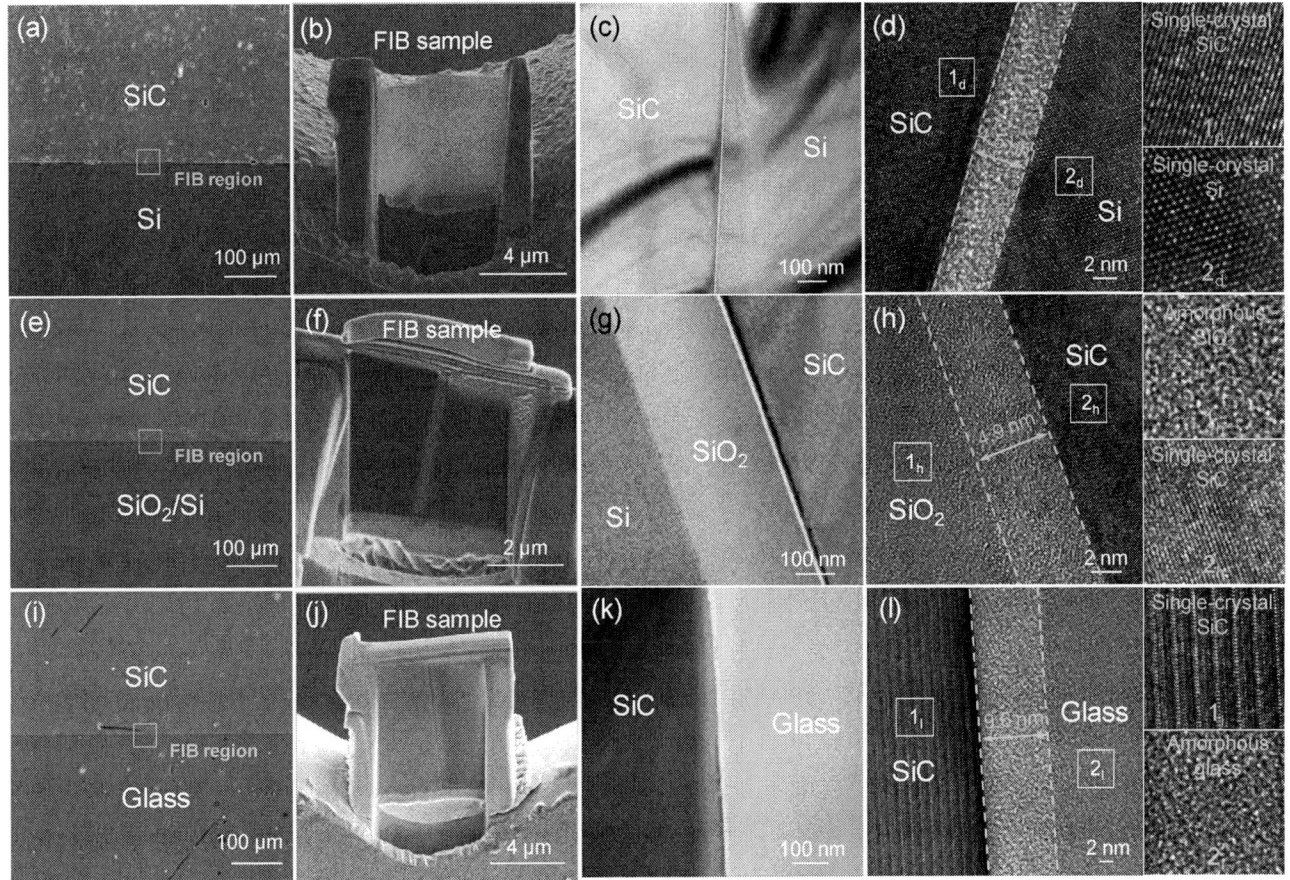

Fig 4. SEM and TEM images of (a)-(d) SiC/Si, (e)-(h) Si/SiO₂, and (i)-(l) Si/glass sample after annealing.

can withstand the thermal and mechanical stress during the microfabrication process. Fig 4(c), (g), and (k) exhibited the TEM images of the bonding interfaces under the low magnification. All of them were neat without cracking. It also could been seen that the color of the SiC substrates closed to the interfaces were slightly dark. That was caused by the thermal stress because of the large mismatch in the CTE. The high resolution transmission electron microscopy (HRTEM) images were also shown in the Fig 4(d), (h), and (l). And atomic boning has been realized confirmed by the TEM observation. The transition layers of SiC/Si, SiC/SiO₂, and SiC/glass bonded pairs were 4.5 nm, 4.9 nm, and 9.6 nm, respectively. The nanoscale layers will also promote the development of device miniaturization. Additionally, the power density will further improve because the excellent thermal conductivity of the SiC substrates can achieve the rapid heat dissipation via the tiny interfaces. Moreover, the SiC was defect-free and well oriented. This will lay a good foundation for the top-level SiC-based power electronic devices.

To detect the elemental composition across the SiC/Si, SiC/SiO₂, and SiC/glass bonding interfaces, the elemental line scanning analyses were used to study the interfaces, as shown in Fig 5. For the SiC/Si direct bonded pair, the

concentration of the C and O elements increased first from the SiC side while that of Si element decreased. Then, the concentration of each element has changed in the opposite trend, as presented in Fig 5(a). For the SiC/SiO₂ bonded pair, the C and Si elements have the same variation trend as the SiC/Si bonding interface. The O element has undergone a series of changes in peaks and valleys. It was mainly due to the SiO₂ substrates possessed a high O concentration. The O element will diffuse along the concentration gradient during the annealing process. Thus, the concentration of O element at the bonding interface closed to the SiO₂ decreased first while that near the SiC increased, as shown in Fig 5(b). For the SiC/glass bonding interface, it has the same changes as the SiC/SiO₂ pair, as displayed in Fig 5(c).

We could also find that carbon-enriched layers were formed at each bonded pairs. Based on the XPS depth analysis we have done in our previous publication, the carbon-enriched layer was induced by the influences of VUV/O₃ activation. This layer will also be beneficial for the heat dissipation. Another interesting phenomenon is that the concentration of C element on the Si substrate maintained at a high level (~80%). This was caused by the FIB samples have explored in the air before the TEM observation. Therefore, the organic contaminant were adsorbed on the

978-1-7281-1500-9/19 $31.00 © 2019 IEEE

FIB samples, causing a high concentration of C element on the Si substrate.

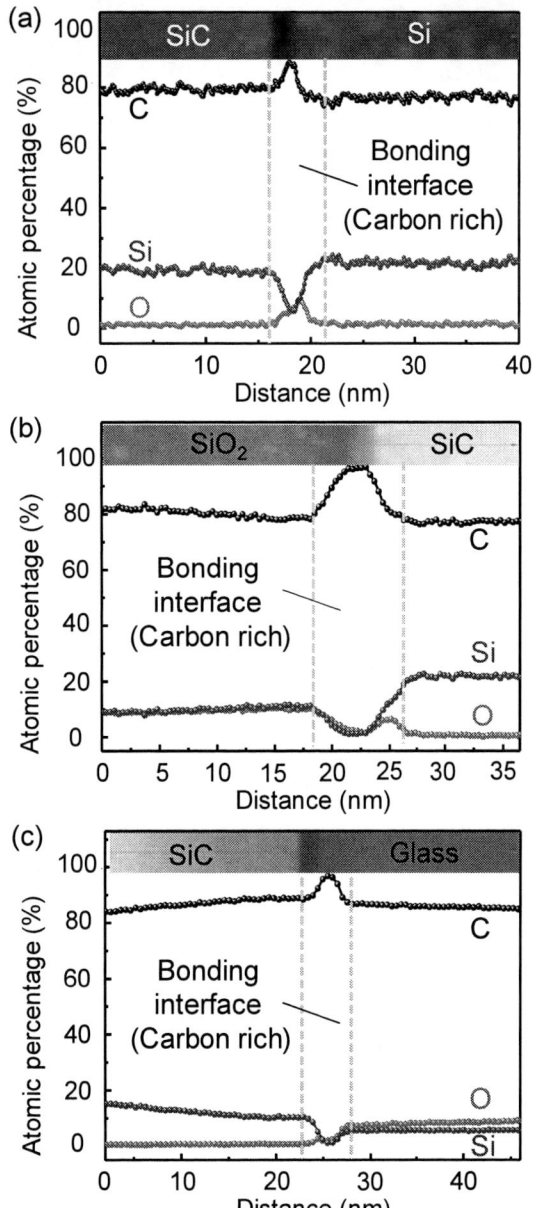

Fig 5. Line scanning of (a) SiC/Si, (b) SiC/SiO₂, and SiC/glass bonded samples across the bonding interfaces.

D. Transmittance test

The glass-on-SiC heterostructure was usually used for the application of bio-MEMS devices. The visibility is a very important parameter for biological devices. Therefore, the UV-Vis spectra of SiC/glass bonded pairs were performed, as shown in Fig. 3. Compared with the bare SiC substrates, the SiC/glass samples only degraded ~2.5% and ~0.7% at 200-400 nm and 600-1000 nm, respectively, even after annealing. The tiny bonding interface and excellent

transmittance has demonstrated that the VUV/O₃ activated direct bonding method has a beautiful prospects for the application of high-performance bio-MEMS devices.

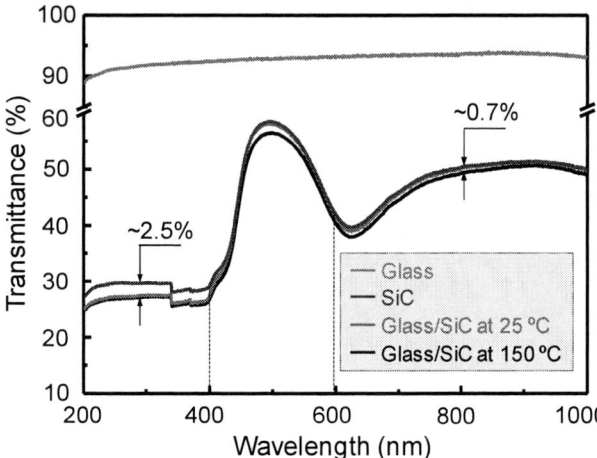

Fig 6. UV-Vis spectra of SiC and glass substrates at room temperature. The SiC/glass bonded pairs were also tested before and after annealing.

III. CONCLUSIONS

In summary, we developed a VUV/O₃ activated direct bonding method for SiC/Si, SiC/SiO₂, and SiC/glass. Based on the surface characterizations, VUV can lead to hydrophilic and smooth surfaces which are beneficial for the direct bonding at the suitable VUV irradiation time. Tight bonding interfaces with high bonding strength have been obtained according to our experimental results. Additionally, atomic bonding has also been realized according to the TEM observation. The nanoscale transition layer (< 10 nm) across the bonding interfaces can greatly promote the development of device miniaturization. Moreover, the excellent optical transmittance of the SiC/glass bonded pair also indicated that this bonding method is suitable for high-performance SiC-on-glass bio-MEMS devices.

ACKNOWLEDGMENT

This work was supported by the National Natural Science Foundation of China (Grant No. 51505106), the China Postdoctoral Science Foundation (Grant No. 2017M610207), and the Heilongjiang Provincial Postdoctoral Science Foundation (No. LBH-Z16074).

REFERENCES

[1] F. Mu, M. Uomoto, T. Shimatsu, Y. Wang, K. Iguchi, H. Nakazawa, Y. Takahashi, E. Higurashi, and T. Suga, "De-bondable SiC-SiC wafer bonding via an intermediate Ni nano-film", Appl. Surf. Sci., vol 465, Sep. 2018, pp. 591-595, doi:10.1016/j.apsusc.2018.09.050.

[2] F. Mu, R. He, and T. Suga, "Room temperature GaN-diamond bonding for high-power GaN-on-diamond devices", Scrip. Mater., vol 150, Mar. 2018, pp. 148-151, doi:10.1016/j.scriptamat.2018.03.016.

[3] F. Mu, Y. Morino, K. Jerchel, M. Fujino, and T. Suga, "GaN-Si direct wafer bonding at room temperature for thin GaN device transfer after epitxial lift off", Appl. Surf. Sci., vol 416, Apr. 2017, pp. 1007-1012, doi:10.1016/j.apsusc.2017.04.247.

[4] C. Wang, J. Xu, S. Guo, Q. Kang, Y. Wang, Y. Wang, and Y. Tian, "A facile method for direct bonding of single-crystalline SiC to Si, SiO_2, and glass using VUV irradiation", Appl. Surf. Sci., vol 471, Nov. 2018, pp. 196-204, doi:10.1016/j.apsusc.2018.11.239.

[5] J. Xu, C. Wang, D. Li, J. Cheng, Y. Wang, C. Hang, and Y. Tian, "Fabrication of SiC/Si, SiC/SiO_2, and SiC/glass heterostructures via VUV/O_3 activated direct bonding at low temperature", Ceram. Inter., vol 45, Oct. 2018, pp. 4094-4098, doi:10.1016/j.ceramint.2018.10.231.

[6] F. Yu, C. Hang, M. Zhao, H. Chen, "An interconnection method based on Sn-coated Ni core-shell power preforms for high-temperature applications", J. Alloys Compd., vol 776, Oct. 2018, pp. 791-797, doi:10.1016/j.jallcom.2018.10.267.

[7] Y.-M. Lin, C. Dimitrakopoulos, K. A. Jenkins, D. B. Farmer, H.-Y. Chiu, A. Grill, Ph. Avouirs, "100-GHz transistors from wafer-scale epitaxual graphene", Science, vol 05, Feb. 2010, pp. 662-663, doi:10.1126/science.1184289.

[8] J. Xu, C. Wang, T. Wang, Y. Wang, Q. Kang, Y. Liu, and Y. Tian, "Mechanisms for low-temperature direct bonding of Si/Si and quartz/quartz via VUV/O_3 activation", RSC Adv., vol 8, Mar. 2018, pp. 11528-11535, doi:10.1039/C7RA13095C.

[9] J. Xu, C. Wang, T. Wang, Y. Liu, and Y. Tian, "Direct bonding of silicon and quartz glass using VUV/O_3 activation and a multistep low-temperature annealing process", Appl. Surf. Sci., vol 453, May. 2018, pp. 416-422, doi:10.1016/j.apsusc.2018.05.109.

[10] C. Wang, J. Xu, X. Zeng, Y. Tian, C. Wang, T. Suga, "Low-temperature wafer direct bonding of silicon and quartz glass by a two-step wet chemical surface cleaning", Jpn. J. Appl. Phys., vol 57, Feb. 2018, pp. 02BD02, doi:10.7567/JJAP.57.02BD02.

[11] C. Wang, J. Xu, X. Qi, Y. Liu, Y. Tian, C. Wang, and T. Suga, "Direct homo/hetrogeneous bonding of silicon and glass using vacuum ultraviolet irradiation in air", J. Electrochem. Soc., vol 165, Jan. 2018, pp. H3093-H3098, doi:10.1149/2.0161804jes.

[12] R. He, M. Fujino, A. Yamauchi, Y. Wang, and T. Suga, "Combined surface acivated bonding technique for low-temperature Cu/dielectric hybrid bonding", ECS J. Solid State Sci. Technol., vol 5, Jan. 2016, pp. P419-P424, doi:10.1149/2.0201607jss.

[13] J. Xu, C. Wang, B. Wu, Y. Tian, and X. Qi, "Communication-Defect-free direct bonding for high-performance glass-on-$LiNbO_3$ devices", J. Electrochem. Soc., vol 165, Jan. 2018, pp. B727-B729, doi:10.1149/2.0871814jes.

[14] R. He, A. Yamauchi, and T. Suga, "Sequential plasma activation methods for hydrophilic direct bonding at sub-200 °C", Jpn. J. Appl. Phys., vol 57, Feb. 2018, pp. 02BD03, doi:10.7567/JJAP.57.02BD03..

[15] R. He, M. Fujino, A. Yamauchi, and T. Suga, "Novel hydrophilic SiO2 wafer bonding using combined surface-activated bonding technique", Jpn. J. Appl. Phys., vol 54, Mar. 2015, 030218, doi:10.7576/JJAP.54.030218.

[16] J. Xu, C. Wang, Y. Tian, B. Wu, S. Wang, and H. Zhang, "Glass-on-$LiNbO_3$ heterostructure formed via a two-step plasma activated low-temperature direct bonding method", Appl. Surf. Sci., vol 459, Aug. 2018, pp. 621-629, doi:10.1016/j.apsusc.2018.08.031.

Development of Flexible Hybrid Electronics
Using Reflow Assembly with Stretchable Film

Weifeng Liu, Ph. D., William Uy,
Alex Chan, Dongkai Shangguan, Ph.D.,
Advanced Manufacturing Engineering, FLEX Intl.
Milpitas, CA 95035
Email: weifeng.liu@flex.com

Andy Behr, Takatoshi Abe, Tomohiro Fukao
Panasonic Corporation,
Kadoma City
Osaka, Japan
Email: andy.behr@us.panasonic.com

Abstract—Flexible hybrid electronics (FHE) are manufactured by combining traditional circuit board fabrication and assembly processes with emerging printed electronics technology. By integrating surface mounted electronic components with printed stretchable conductive circuits and compliant/stretchable substrates these hybrid constructions have potential to revolutionize electronic assemblies used for Internet of Things (IOT), wearable, medical, wellness, automotive and aerospace markets. By employing FHE principles, designers can create heterogeneous electronic systems with unique form factors and functionality. These devices can conform to the curves of a human body or even be applied to the surface of or molded within an irregularly shaped mechanical structure. FHEs also offer the promise of light-weight and cost-effectiveness, scalable manufacturing.

The FHE industry remains in the early stages of development. A variety of design, material, assembly and reliability issues remain to be addressed. For example, the typical polymer based conductive pastes used for forming FHE circuit structures are not as conductive as the etched copper on traditional printed circuit boards (PCBs.) Additionally, most of these polymer-based conductive pastes are not readily solderable and the electrical interconnections formed with conductive adhesives in current FHE designs may not be as conductive or reliable as those formed with solder. Additionally, commercially available stretchable thermoplastic film substrates have relatively low thermal resistance and cannot withstand the current lead-free surface mount technology (SMT) reflow temperatures.

This paper discusses these challenges and presents an FHE manufacturing process utilizing a stretchable thermosetting polymer substrate, a combination of both screen printed stretchable conductive paste and etched copper structure, and the conventional SMT processes to create a functional proof of concept double-sided device integrating both active and passive components.

Key Words: Flexible Hybrid Electronics, FHE, Flexibility, Stretchability, Stretchable Hybrid Electronics, SHE, Wearable, IOT, Printed Electronics, Conductive Paste, Copper Clad Stretch, CCS.

INTRODUCTION

Flexible hybrid electronics (FHE) refers to a special category of electronic assemblies made by mounting components onto circuits formed using high-precision paste printing technologies [1]. By integrating semiconductor and passive components with printed pastes and flexible substrates, FHE has the potential to revolutionize the IoT, wearables and other industries by providing unique form-factor, lighter assemblies and lowering manufacturing costs by leveraging high-volume printing technology. With FHE, heterogeneous system integration can be achieved by forming input devices like printed sensors onto the same substrate as semiconductor components, passives and even output devices like haptics. These constructions can be pliable and may even conform to the curves of a human body or stretched into irregular shapes - all while preserving the full functionality of traditional electronic systems.

Electronic assembly manufacturing technology has continued to evolve for more than 60 years. The early generations of electronic assemblies were rigid, bulky constructions with components mounted on stiff, multi-layer printed circuit boards using tin-lead based solder. Later, flexible printed circuit (FPC) boards were developed for applications requiring static or dynamic bending. Future generations of electronic assemblies may be made entirely with printing or other additive manufacturing technologies. Not only may the circuit traces be formed by printing conductive paste, but the passives and active semiconductor devices may be printed using functional pastes. However, compared to conventional surface mount semiconductor components, the performance and functionality of the printed semiconductors are still far from meeting most application performance requirements. Conventional semiconductor components will continue to dominate the market in the foreseeable future.

Flexible hybrid electronics (FHE) technology is the "next generation" solution for making electronic assemblies that provide the benefits of using conventional semiconductor and passive component form-factors, while leveraging the capabilities of printed electronics to provide scalability, lower manufacturing costs and system flexibility. Table 1 provides a relative comparison between rigid PCBA, FPC and FHE.

Table 1. Qualitative Comparison of Typical Rigid PCBA, FPC and FHE.

	Rigid PCBA	FPC	FHE
Circuit Formation	Subtractive (Etching)	Subtractive (Etching)	Additive (Printing)
Conductor Material	Copper	Copper	Silver Paste
Conductivity	High	High	Low
Solderability	Yes	Yes	No
Substrate	Epoxy-Glass	Polyimide	PET or TPU
Resin Class	Thermoset	Thermoset	Thermoplastic
Substrate Pliability	Rigid	Bendable	Bendable (PET) Stretchable (TPU)
Temp. Tolerance	High	High	Low
Assembly Process	SMT	SMT	Conductive Adhesive (CA)

The FHE industry is still in the early stages of development and a variety of critical design, material, assembly and reliability issues remain to be addressed. For example, typical silver filled polymer based conductive pastes are not as conductive as the etched copper used to make PCBs. Additionally, these thermoplastic pastes are generally not solderable, so components must be mounted with electrically conductive adhesive (ECA). The electrical and mechanical interconnections formed with ECA are not usually as conductive or reliable as conventional SMT solder joints. Additionally, commercially available thermoplastic based films like PET and TPU have relatively low temperature resistance and cannot withstand the current solder reflow temperature regimens. As an alternative to conductive pastes, some researchers have investigated the use of serpentine or zig-zag trace designs with copper on stretchable substrates to create a "spring" effect to compensate for inherent inelasticity of copper [2].

To address these challenges, the FHE industry requires new substrate and circuit materials that are stretchable/conformable and can withstand conventional SMT assembly temperatures. As a "sketch to scale" provider of electronic manufacturing services, Flex considers it an attractive proposition to utilize existing high volume SMT infrastructure and equipment for FHE manufacturing. To accomplish this goal, the component mounting pads need to be solderable and the substrate needs to be compatible with a conventional reflow process. The formation of solderable copper pads on a high temperature stretchable substrate provides a potential solution. Therefore, our research approach was to make a flexible hybrid electronic device utilizing a combination of:

- A conformable and stretchable substrate capable of withstanding high processing temperature, especially at lead free reflow temperature (260°C maximum).
- A stretchable conductive high temperature resistant paste that can be printed on the substrate noted above to form flexible and stretchable conductive traces.
- Solderable bond pads for component assembly using standard solder pastes and reflow profiles.

This paper reviews the development of a process used to construct a proof of concept FHE device, utilizing a soft stretchable thermosetting polymer substrate capable of withstanding high reflow temperatures, thermosetting polymer based conductive paste to make stretchable traces for connecting surface mount components, and etched copper foil to form solderable pads. At the end of the process, functional FHE samples with integrated passive and active devices were fabricated and demonstrated.

MATERIALS

Stretchable polymeric materials used in FHE constructions should not only exhibit a low elastic modulus and high elongation, but also the ability to recover following bending, stretching, twisting and other deformation stresses that may be encountered during assembly or field use. Ideally these materials need to be durable, tolerate elevated temperatures and exhibit chemical resistance. Preferably, they will also be compatible with the other standard electronic materials and processes typically used in high volume circuit fabrication and SMT assembly processes. A variety of thermoplastic materials, such as thermoplastic polyurethane (TPU) films and thermoplastic polyester-based silver pastes have been used in stretchable hybrid electronics (SHE) constructions. However, these materials have some performance drawbacks, particularly in terms of permanent deformation (hysteresis), mechanical durability and temperature/humidity resistance [3].

Thermosetting resins predominate electronics manufacturing due to their broad range of desirable properties like heat tolerance, chemical resistance and electrical insulation. The substrate film and conductive paste used for this evaluation were made with a novel, non-silicone, stretchable, thermosetting resin system designed specifically for SHE applications that exhibits good stretchability, low hysteresis and high temperature resistance.

Thermosetting Stretchable Film

A liquid form of the resin noted above (the A-stage) is formed into a stretchable film using laboratory scale roll to roll coating, drying and curing equipment. This coating line can make films between 25 and 150 microns thick and approximately 300 mm in width. The film used for this study were 100 µm thick and cast on a 75 µm high-temperature polyester carrier. Because of the handling properties imparted from this film-on-carrier construction, the supported film is compatible with many "down-stream" assembly processes such as lamination, slitting and die cutting. By modifying the base resin formulation researchers can control mechanical properties such as modulus, stretchability, elongation recovery rate (stress relaxation) and hysteresis. A degree of stress relaxation property may be attractive for epidermal devices where skin comfort is a design priority. Because polymer decomposition occurs around 310°C, this film exhibits high temperature resistance. For example, the film does not degrade after floating in liquidous solder at 260°C for 60 seconds. The film properties are presented and compared to a representative thermoplastic polyurethane (TPU) film in Table 2.

Table 2. Stretchable Film Comparison.

Film Type	Developmental Film	Commercial TPU*
Resin Class	Thermosetting	Thermoplastic
Elongation Ratio at Break (%)	211.7	<1,000
Elastic Modulus (Mpa)	3.4	7.2
Stress Relaxation Value (%)	25.4	25.3
Hysteresis	0.3	5.9
Temperature Tolerance	>260° C	<150° C

*Representative thermoplastic polyurethane (TPU) material provided for comparison only.

Copper Clad Stretchable Film

Both PCB and FPC substrates are typically formed by laminating copper foil to one side, or more often, both sides of thermosetting polymer-based substrates. For rigid PCBs an epoxy/glass composition is commonly used. For flexible printed circuit (FPC), polyimide film is common. For this project a double-sided copper clad stretch (CCS) film construction was formed by laminating 12.5 μm (~1/3 oz.) electro-deposition (ED) copper foil to both sides of the 100 μm stretchable thermosetting film described above. The total thickness of the construction is 125 μm, with the substrate film at 100μm thickness and 12.5μm of copper foil on both sides.

Stretchable Conductive Paste

Interfacial adhesion between conductive elements and dielectric substrates is important for virtually all types of electronic circuitry, but perhaps nowhere quite as much as in the creation of stretchable circuitry. The use of compatible resin systems is a well-known approach to increase circuitry and substrate compatibility. An experimental conductive paste based on the stretchable thermoset resin chemistry noted previously was used in this evaluation. The electrical conductance of the cured conductive paste is

a result of percolation, by which the silver particles are into contact to generate electrical pathways at loadings above the percolation threshold [4]. The formulation consisted of silver particles, resin, solvent and additives blended to form a thixotropic paste designed for deposition via screen printing. When printing onto the thermosetting film substrate based on the same resin system, no surface preparation like abrasion or plasma cleaning was required. After printing, the paste was dried at 100°C for 10 minutes then cured at 170°C for 60 minutes. The properties of the conductive paste are presented in Table 3.

Table 3. Properties of Stretchable Conductive Paste.

Conductive Paste	Properties	Condition
Resin Class	Thermosetting Polymer	
Viscosity (Pa·s)	10	Paste
Elongation Ratio at Break (%)	380	Cured
Elastic Modulus (MPa)	7.0	Cured
Volume resistivity ($\Omega \cdot cm$)	1.0×10^{-3}	Cured

Module Design and Components

This proof of concept study employed a low power, Long Term Evolution (LTE) Bluetooth module design. This device was constructed using the following components:

- One LGA package (0.4mm pitch, 64 I/Os).
- One CSP package (0.35-0.45 mm pitch, 8 I/Os).
- Passive Devices (mostly 0603, 0402 and 0201).
- One SM connector.

MODULE MANUFACTURING PROCESS

This device was originally designed to be assembled using SMT soldered components on a rigid double-sided PCB. The experimental plan was to demonstrate an FHE design concept by converting the rigid PCB to a pliable construction using the stretchable substrate film, copper clad technology in combination with printed conductive paste, and reflow process to attach components.

Copper Circuit Formation

Copper structures including via holes, circuit traces, bond pads and land areas were formed using standard PCB manufacturing (patterning and etching) processes. To electrically connect both sides of the CCS, 100 μm via holes were drilled and plated using standard mechanical drilling equipment and copper plating processes. To form the circuit structures on both sides of the CCS, a dry-film photoresist was applied each side of the CCS. After applying a photo-tool, the film was imaged using high-intensity ultraviolet (UV) light and the exposed regions photoresist were cross-linked. The uncured photo resist was subsequently developed off. The exposed areas of copper were etched, the remaining photoresist stripped off exposing the resulting copper structures on both sides of the stretchable film. A simplified, single-sided copper circuit formation process is shown in Figure 1.

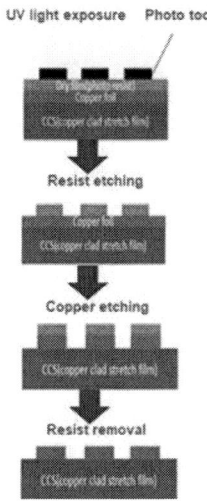

Figure 1: Circuit Formation with Copper Clad Stretch (CCS).

Device Construction

As this was a wholly new approach with many unknown factors, it was decided to move forward in two phases. The first leg was to make the circuitry using full copper clad technology; the second leg was to combine copper clad technology with thermoset stretchable conductive paste printing. Figure 2 shows the flow chart for both leg 1 and leg 2. This process started with manufacturing the copper clad stretchable film (CCS) as noted in the "Materials" section above. This was followed by drilling, plating, photolithographic patterning and etching the copper to form a double-sided interconnect circuit structure. In Leg 2, a hybrid approach was pursued by replacing some of the copper traces with screen printed stretchable conductive paste. The screen-printing process was added after the fabrication of double-sided copper interconnect. The addition of the printed conductive paste in leg 2 is highlighted with a red outline in Figure 2.

For the screen printing of the conductive paste, a screen made of stainless-steel wire with mesh count 250 was used. The wires were woven into special weaving structure, yielding a mesh 3 times thicker

than the wire diameter. The CCS parts were carefully placed top side up onto a tacky carrier film to avoid entrapping air bubbles between the circuit and the carrier. After printing the conductive paste on the top side, the parts went through a drying and curing process with the carrier sheet (drying for 10 minutes at 100℃, followed by curing for 60 minutes at 170℃.) Then, the parts were peeled from the carrier sheet. After this, paste printing was done on bottom side followed by the drying and curing process noted above. because conductive paste on the top side was fully cured, the printed patterns were not distorted by the printing and curing processes on the second side.

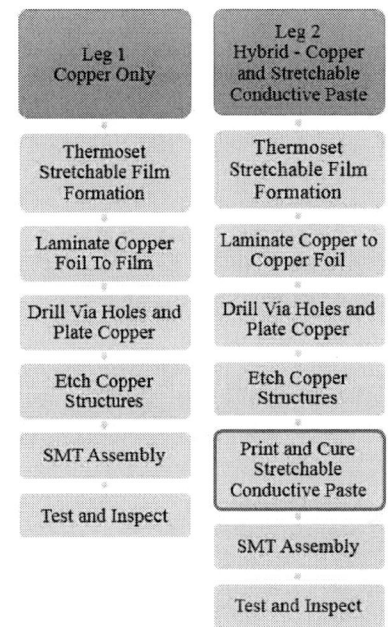

Figure 2: Manufacturing Process Flow for Leg 1 and Leg 2.

Module Assembly

The circuits were assembled using two classes of solder pastes. SnBiAg (melting temperature 138°C) and SAC305 (melting temperature 217°C) pastes were stencil printed on the pads, followed by component placement and reflow processes. The assembled modules were inspected under optical microscope and x-ray. The functioning of the module was demonstrated by powering-up the module with a coin cell battery and communicating with the device via in-house software installed in tablet computer. Shear testing was performed on the 0402 passive chip components to understand the solder joint strength with stroke speed 0.5mm/sec and shear height 30μm. Cross section was also performed to inspect the integrity of the solder joints. Figure 3 shows the solder reflow profiles for SnBiAg and SAC305 respectively.

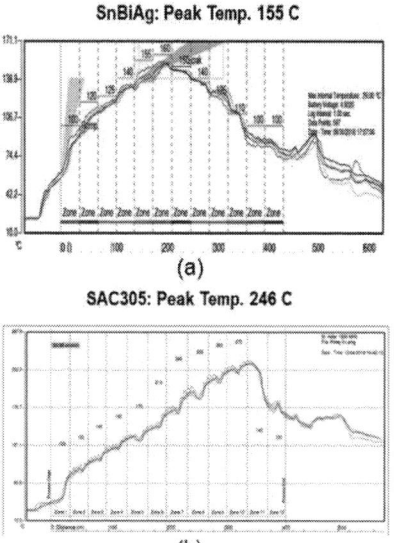

(a)

(b)

Figure 3: Reflow Profiles: (a)SnBiAg, (b) SAC305).

RESULTS AND DISCUSSIONS

Figure 4 shows the Leg 1 circuit panel made using CCS film and typical PCB fabrication process (drilling, plating, patterning, etching). 100μm diameter via holes were drilled and plated to connect the top and bottom circuitry.

Figure 4: Circuitry Formed on Copper Clad Stretch (CCS) Thermoset Film.

The components were assembled using both SnBiAg and SAC305 solder pastes. Figure 5 shows the assembled Leg 1 module using low temperature solder (SnBiAg). The solder joints for the passive devices were visually and optically inspected. The solder joints showed good wetting to the component terminals with acceptable contact angles and shapes. Figure 6 shows examples of passives mounted on the substrate.

Figure 5: Assembled Flexible Bluetooth Module.

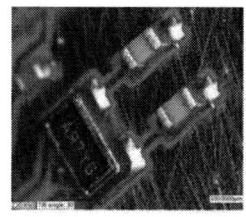 

Figure 6: Passive Components Mounted on the Substrate with Acceptable Solder Joints.

Figure 7 shows the cross sectioning on the assembled passives on a Leg 1 circuit for both SnBiAg and SAC305 assemblies. Both show acceptable fillet formation and good bonding with the copper pads.

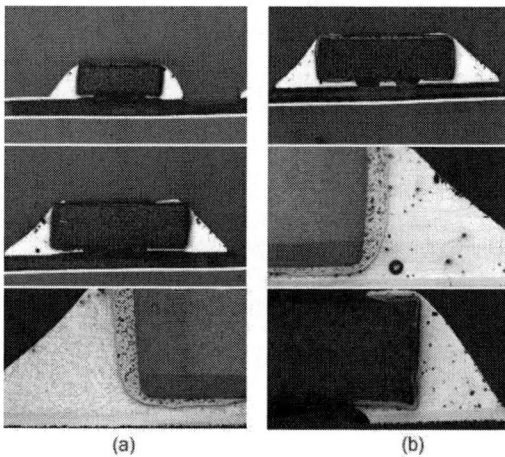

(a)   (b)

Figure 7: Cross Sections of Assembled Components: (a) SnBiAg, (b) SAC305.

Shear strength testing was performed on selected 0402 passive chip components. Figure 8 shows the shear strength data for SnBiAg and SAC305 respectively. Solder joints formed with both solder pastes show acceptable shear strength with the low-end strength values (95 percentiles) well above 1.2Kg, which are equal or better than the same passives assembled on the rigid boards [5]. Interestingly, the passives assembled with SnBiAg solder showed higher values than the SAC305 assemblies. One reason for this result might be the higher concentration of voids in the

978-1-7281-1500-9/19 $31.00 © 2019 IEEE   1276

SAC305 solder, which is known to decrease the solder joint strength. Alternatively, degradation of bonding strength between copper and substrate at the elevated SAC305 reflow temperatures might be another potential cause.

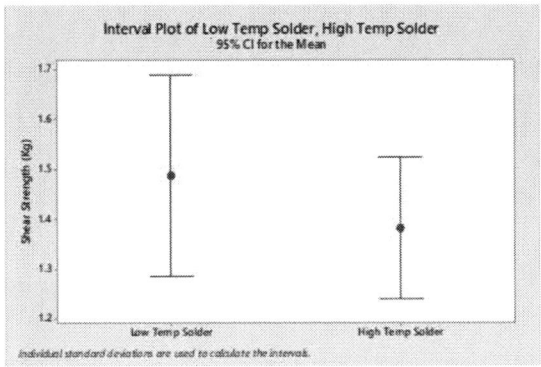

Figure 8: Shear Test of Solder Joints (SnBiAg and SAC305).

Figure 9 shows the optical images of the Leg 1 passives displaced after shear testing. For both SnBiAg or SAC305 assemblies, both copper pad delamination and solder joint crack were observed, with the former outnumbering the latter.

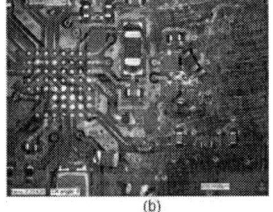

Figure 9: Post Shear Testing Images: (a) SnBiAg assembly, (b) SAC305 assembly.

To check the functionality of the Leg 1 devices, the fabricated modules were powered-up by a coin battery. An in-house software installed on a tablet computer was used to check the wireless communication with the module. A strong signal was received from the module at a decent distance. Not all the assembled modules were functioning. Further inspection and testing were done on the failed modules to understand the solder joint integrity and identify potential root causes for the failure.

One issue with assembling the Leg 1 circuits was the titling of the LGA components. This phenomenon occurred for both SnBiAg solder and SAC305 solder. It was suspected that the tilting may have caused the some LGA pads to be unsoldered, leading to the observed system failures. To confirm this, x-ray inspection was performed and one LGA component was manually separated from the substrate. Figure 10

(a) shows the x-ray image of the solder joint array under LGA package. Figure 10 (b) and (c) show the optical images of the Leg 1 substrate and LGA package respectively after prying. It was observed that approximately half of the LGA pads had not formed joints with the substrate pads.

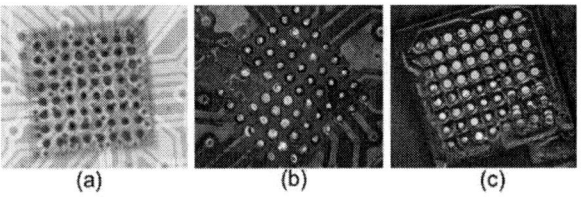

Figure 10: Partially Soldered Component: (a) X-ray, (b) Substrate, (b) Component.

To determine the cause of the tilting LGA components during the reflow process, the assembled Leg 1 modules were inspected closely. A bulge on the bottom side of the substrates, particularly underneath the LGA components was observed. Figure 11 (b) and (c) show the optical images of the substrate warpage underneath the LGA component. It was determined that this substrate warpage caused the tilting of the LGA components, resulting in the unsoldered LGA pads.

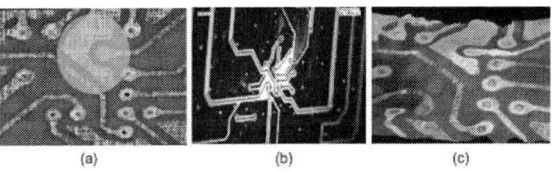

Figure 11: Substrate Bulge during Reflow: (a) Substrate before Assembly, (b) Substrate after Assembly, (c) Substrate after Assembly: 3D Mapping.

Substrate warpage is a common issue for electronic assembly and microelectronics packaging, caused by the CTE mismatch between building elements within the component or substrate [6, 7]. The CTE mismatch between the unfilled polymeric substrate film and the copper foil (Unfilled substrate film CTE 241 ppm/°C and copper CTE 17 ppm/°C), in combination with unbalanced copper on top and bottom sides of the substrate, caused the substrate bulge, as can be seen in Figure 12 (b) and (c).

To address the substrate bulge issue, several options were considered. A new circuitry could be designed with more balanced copper traces on both sides of the substrate. A fixture could be used during the reflow process to attach to the substrate and restrict bulging of the substrate. A third option was to use conductive paste instead of copper. The stretchable conductive paste is softer and more flexible than copper foil and

978-1-7281-1500-9/19 $31.00 © 2019 IEEE   1277

therefore, thermomechanical stresses due to CTE mismatch should be minimized. For this project, we attempted the latter approach by developing the hybrid manufacturing process (Leg 2) by combining conductive paste printing with copper clad technology.

Figure 12 shows the completed Leg 2 circuit with both copper and printed stretchable paste structures. The printed paste is mainly used as ground plane, and the copper traces are for power and signal.

Figure 12: Leg 2 Circuit with Both Copper and Thermosetting Stretchable Paste Traces.

A discoloration on the copper due to oxidation was observed. In a manufacturing environment, this phenomenon could be addressed with anti-oxidation coating on the copper pads and traces, which we plan to evaluate in later phase of the project. A quick test was performed to check the solderability of the oxidized copper pads. SnBiAg solder paste was printed and processed through the reflow process without mounting components. It appeared the SnBiAg solder paste fully wetted the copper pads as Figure 13 shows.

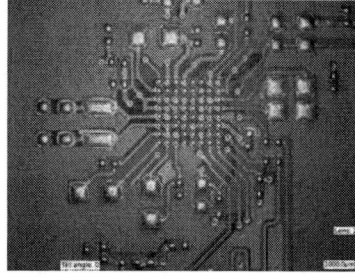

Figure 13: Solderability of Copper Pads after Conductive Paste Curing (170°C 60 minutes).

Figure 14 (a) shows the assembled Leg 2 modules using SnBiAg reflow process with a peak temperature of 155°C. Further inspection of the assembled modules revealed that the LGA components were soldered to the substrate with improved component topology and solder joint quality, as shown in Figure 14 (b) and (c). This confirms the initial conclusion and validates the process to use conductive paste to reduce substrate bulging and component tilting. Further analysis and optimization are being conducted.

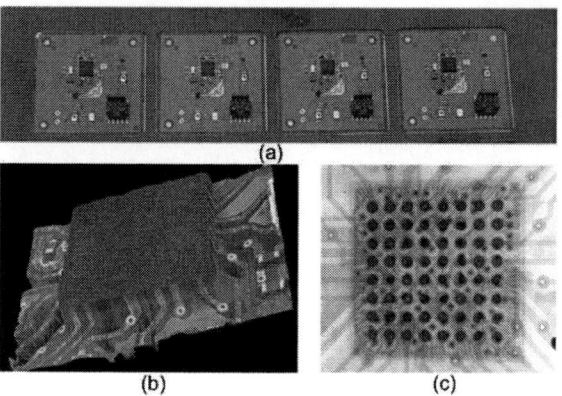

(a)

(b)   (c)

Figure 14: Assembled Module using SnBiAg: (a) Optical Image, (b) 3D Mapping, (c) X-ray.

SUMMARY AND CONCLUSIONS

Development of a proof of concept flexible hybrid electronics (FHE) device was presented in this paper. A hybrid manufacturing process utilizing a stretchable thermosetting polymer substrate which can withstand high reflow temperatures, printed thermosetting conductive paste to make flexible traces in combination with copper pads to enable soldering process using both SnBiAg and SAC305 solders. Functional devices with surface mounted passive and active devices are fabricated and demonstrated. Further development and evaluations are on-going

ACKNOWLEDGMENTS

The authors would like to acknowledge the contributions from Tuyen Nguyen, Tu Tran, Robert Penning, Dr. Jie Lian, Dr. Francoise Sarrazin, Dennis Willie of the Advanced Manufacturing Engineering (AME) Lab of Flex, Lipika Bhattacharya, Glenn Henriquez, Hien Tran, Arnold Andreas, Bruce Thompson, Raj Dhesikan of the company Innovation Lab.

REFERENCES

[1] E.Forsythe, B.Leever, "Flexible Hybrid Electronics Manufacturing 2016" (http://www.dtic.mil/dtic/tr/fulltext/u2/1029424.pdf).

[2] University of California – San Diego. "Building up stretchable electronics to be as multipurpose as your smartphone." Science Daily August 13, 2018.

[3] B. Song, F.Wu, K.Moon, CP Wong, R.Bahr, and M. Tentzeris, "Stretchable, Printable and Eclectically Conductive Composites for Wearable RF Antennas" IEEE ECTC 2018.

[4] J. Pan, M.Keif, J.Ledgerwood, X.Rong, and X.Wang, "Screen printing Fine Pitch Stretchable Silver Inks onto a Flexible Substrate for Wearable Applications," IMAPS Pasadena California 2018.

[5] T. Lentz, J. Bath, N. Pavithiran, "How Does Printed Solder Paste Volume Affect Solder Joint Reliability," SMTA International, Rosemont, Illinois, 2018.

[6] W.Liu, "Challenges of Organic Substrates from EMS Perspective," INEMI Workshop, Toyoma, Japan, 2013.

[7] Zh.Feng, J. Basani, D.Bernard, and E.Krastev, "Modern 2D X-ray Tackles BGA Defects," SMT, 2008.

978-1-7281-1500-9/19 $31.00 © 2019 IEEE   1278

Highly Compact RF Transceiver Module using High Resistive Silicon Interposer with Embedded Inductors and Heterogeneous Dies Integration

G. Pares, J.P. Michel, E. Deschaseaux, P. Ferris, A. Serhan, A. Giry

Univ. Grenoble Alpes, CEA, LETI, 38000 Grenoble, France

Gabriel.pares@cea.fr

Abstract

Silicon interposers are providing interesting alternatives to organic packages for the fabrication of complex system in package (SIP) modules in particular for RF application. Among the advantages of this technology are the capability to fabricate fine pitch redistribution layers and to embed inside the interposer some high quality passive elements very close to the active chips resulting in a highly integrated solution. To keep the technology at a reasonable cost these last add-on features need to be fabricated with no or minor additional process steps that the ones needed for the fabrication of the interposer itself.

In this work we present the design and the fabrication of a high resistive silicon interposer conceived for hosting a functional RF SIP transceiver including a Front End Module (FEM), a Dual Band Dual Mode Power Amplifier working at 400 and 900 MHz and some embedded planar and 3D inductors. The interposer consists in symmetrical stack with one thick level of copper RDL on each side of the 200 µm thin HR silicon substrate and connected with through silicon via-last (TSV-last). The inductors are built with no additional metallization level.

For the inductors development, a dedicated test vehicle was first designed to study different designs including planar spirals and 3D solenoids and torus. Simulation work was first carried out for dimensioning the structures to target inductance values in the range of 0.5 to 10 nH and high quality factor greater than 20. We present the process used for the fabrication, in particular the realization of the thick copper RDL layers on both sides of the interposer and the of the TSV-last module. Physical characterization are presented showing the integrity and the good control of the technology. DC and RF electrical measurements assess the performances achieved for the different inductor variants exhibiting low resistance, constant inductance up to 5 GHz and factor of quality higher than 30.

The functional RF transceiver SIP module will be then presented. The architecture is an extremely compact multi chip module composed of four actives CMOS chips (VGA, VCO, modulator and the PA) stacked by flip chip with Cu/SnAg µbumps on the HR silicon interposer and surrounded by a hundred of passive SMD components. With this technology the footprint of the module is reduced by a factor of 2 in comparison with the same module made on a µPCB substrate.

The signals and ground are fed through the thick copper RDL lines and the TSVs from LGA pads present on the backside of

the interposer for assembly on a test board. Different design configurations are made on the same wafer that include some version using only SMD inductors and some using 2.5 and 3D inductors in order to compare the corresponding performances. Other test structures are also present like filters and individual inductors for DC and RF characterization. We will finally present the resulting performances obtained from this SIP integration.

Introduction

With the increase of electronic systems complexity, particularly the I/O counts and the pitch reduction of the multiple ICs chips being integrated in a single module, silicon interposers are becoming an interesting option for complex IC System In Packages (SIP) [1, 2]. Such silicon interposer has a main function of interconnecting and redistributing the signal between the different chip's I/Os and to the external world. The use of existing CMOS fabrication lines to fabricate these interposers with BEOL interconnect levels allows to realize extremely dense routing levels that are not achievable today with other solutions like organic or ceramic substrates. Moreover it allows to embed thin-film devices such as MIM capacitors or spiral inductors which improves further the package density as well as the overall performances of the system thanks to closer proximity between the active chips and the passive elements. Today such silicon interposer technology have been adopted for large digital components like FPGA and GPU [3]. However, for RF applications silicon interposers are not commonly adopted for the moment due to several reasons. Firstly, some technological limitations exist that may induce high penalty on signal transmission such as the difficulty of fabricating thick metallization levels and the use of lossy thin dielectric materials for the insulation layers between metal levels or in the TSV [4]. Secondly, standard doped silicon substrate is a quite lossy material so high resistive silicon must be used instead. Thirdly, warpage issues caused by asymmetric structure of the interposer tend to limit the number of signal layers when decreasing the thickness of the interposer. Finally, RF components are relatively cost sensitive compared to higher end products so silicon interposer has to be economically competitive with other solutions like organic substrates.

In this paper our approach is to develop a silicon interposer platform to achieve a compact system in package for Professional Mobile Radio (PMR) applications, and taking advantage of this technology to embed high quality passive devices such as inductors while trying to overcome the above mentioned limitations.

3D inductors fabricated in HR silicon or glass substrates have already been published by several teams [5-8] in the past and recently at CEA-leti [9]. However, there are very few realization of a full functional SIP transceiver that embed these features along with flip chipped active ICs on to a highly compact interposer.

Our goals will be first to obtain an unprecedented integration level of the full RF system using tight design rules achievable simply with silicon interposer technology and targeting the lowest level of technological complexity. This implies the limitation of the number of metallization levels and the use of the same levels to fabricate the high-Q inductors.

The second objective is to adapt the technology to obtain the best performances in term of RF losses and inductor performances. For that, a thick copper RDL is developed for the metallization levels and the TSVs filling and to limit the signal losses, thick organic layers are employed for dielectrics and High Resistivity (HR) silicon is also used for interposer substrate.

System in package for RF transceiver

The objective is to achieve a RF SIP module (Figure 1) based on the proposed silicon interposer platform. The RF SIP includes a RF transmitter using three CMOS chips (Modulator, PLL, and VGA), a dual band reconfigurable power amplifier (PA) using a single SOI chip and the corresponding matching and decoupling networks.

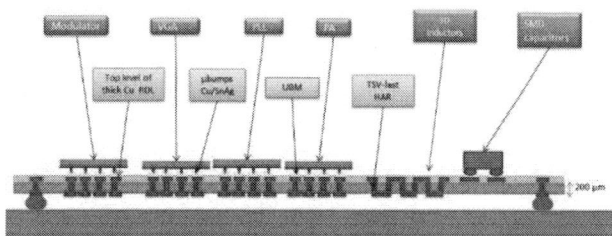

Figure 1: Cross section view of the RF SIP concept based on silicon interposer technology.

Metal routing and passive elements implementation need to be compatible with the same interposer technology while providing equivalent or better performances and reduced footprint compared to the classical organic-based MMIC module design approach.

Regarding the passive elements around 50 SMDs need to be assembled on top of the interposer consisting essentially in decoupling capacitors with high capacitance values.

In order to mitigate the complexity of the interposer fabrication it is important to limit the number of metal levels for routing. The goal is to have only one level on the top and one level on the bottom with TSVs for connection in between.

The interposer is then composed of the 3 main features (Figure 2).

-    One level of thick (10 μm) copper for routing realized with a semi-additive process and protected by a thick polymer passivation with opened contact pads covered with an Under Bump Metallization (UBM) for flip chip hybridization of the top chips and electrical tests.

-    Through silicon via (TSV) inside the High Resistivity Silicon substrate done from the backside after thinning (TSV-last approach)

-    One level of backside metallization (RDL) connecting the TSVs to the fan-out solder pads which are also equipped with a UBM interface.

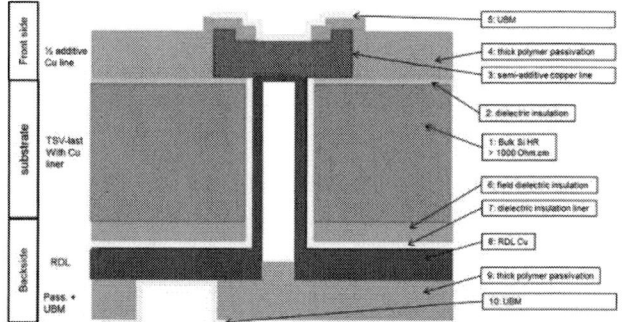

Figure 2: Cross section of the interposer technology with the different layers.

Fabrication of the Si-HR interposer

The technology used to fabricate the silicon interposer is described below with the process flow of Figure 3.

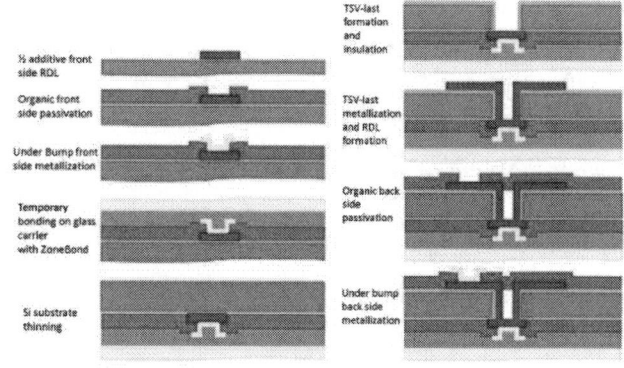

Figure 3: Process flow for interposer fabrication.

It consists first in the realization of a thick copper metallization level on top of a high resistivity 200 mm silicon substrate insulated by thermally grown field silicon oxide layer. This metallization is formed using a semi-additive electrochemical copper deposition of 10 microns of thickness. A passivation stack is then applied on top consisting on a thin SiN PECVD layer directly on the copper followed by a thick (11 μm) organic passivation using AL-X polymer from Asahi Glass. This passivation stack is opened by lithography to form the front side connecting pads which are subsequently metallized with an Under Bump Metallization (UBM) level of Ti/Ni/Au metal stack.

Secondly, the substrate is temporarily bonded on a glass carrier using BSI glue from Brewer Science and thinned by back grinding down to the final interposer thickness of 180 μm.

Third, the backside process is carried out with the etching of the TSVs by deep RIE Bosch process, the insulation from the substrate with the deposition of a SiON PECVD liner, the opening on the landing pads of the front side metal by RIE etch back and the metallization by semi-additive electrochemical copper deposition of 7 μm. Finally, the same passivation stack and UBM pads than for the front side are formed. The front side and backside stack are then perfectly symmetrical resulting to equilibrate the thermo-mechanical constraints of the two sides and to ensure that the final interposer has a low warpage.

This flow is fully compatible with the fabrication of the 3D inductors so they can be fabricated at the same time and with no additional steps. Figure 4 illustrates the realization of some 3D inductors with the front side and backside metallization connected with TSV-last inside the HR silicon substrate.

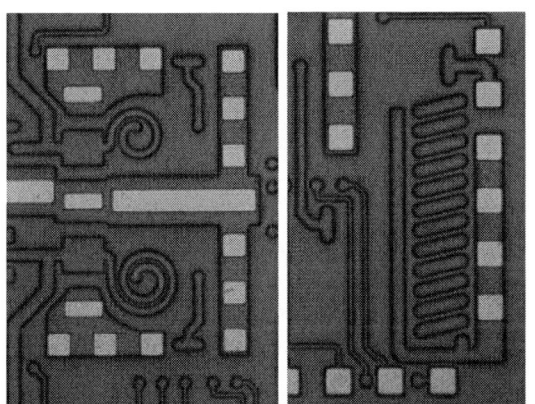

Figure 4: optical microscope views of embedded inductors in Si interposer; left: backside 2.5D spiral inductors: right: backside 3D inductor.

RDL module

In order to achieve the best performances for inductor elements copper metallization layers are required to be as thick as possible including inside the TSV structures to reduce the resistance. However there are some technological limitations to grow thick semi-additive ECD copper lines essentially linked to stress issues that can lead to delamination of the copper RDL on top of the underlying dielectric layers. This is particularly observed with the backside RDL formed on low temperature PECVD dielectrics. It has been observed that delamination issues

is likely to occur preferentially on large copper features and when the copper thickness is greater than 5 μm.

Also with thick copper lines, the passivation layer has to be thick enough and with a sufficient step coverage to cover the lines. An illustration is shown in Figure 6 where the thickness of the dielectric is less than 1 μm on the top corner of the line. In order to improve this situation a thicker and better planar coating layer of polymer has been set up to obtain a sufficient thickness over the copper lines. Another improvement on the process has been the addition of a thin SiN layer in between the copper and the Al-X to avoid the formation of an interfacial oxidized copper layer than can be observed in the Figure 5.

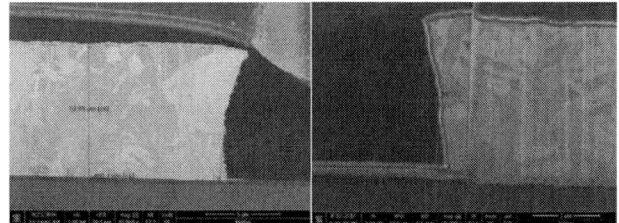

Figure 5: SEM cross section of a front side RDL copper line and AL-X passivation. (Left): insufficient AL-X coverage and presence of oxidized copper interface; (right): optimized passivation with better AL-X coverage and SiN underlayer.

TSV module

The insulation of the TSV inside the HR silicon substrate is achieved with the formation of a layer of SiOxNy deposited at low temperature (200°C) by plasma enhanced chemical vapor deposition (PECVD). This layer is etched back to open the bottom of the via and the resulting thickness profile from the top corner to the bottom corner varies respectively from 2 to 0.4 μm. This minimum thickness is sufficient to guaranty of good insulation of the TSV in the current range of our application.

Figure 6: SEM cross section of a TSV-last showing the copper metallization and the contact on the front side RDL.

The metallization is obtained by copper ECD growth on a dual sputtering and MOCVD Ti/Cu seed layer for better step coverage. The profile of the copper ECD is shown on the Figure 6 having the desired 10 μm thickness in the bottom corner region and for the backside RDL and with some little overgrowth of copper in the mouth of the TSV structure. Albeit this thick copper layer in the TSV and the RDL geometries, no delamination is observed on the interposer wafers, being effectively favorable for a very low resistive inductor element.

DC electrical characterization

The DC electrical characterization is performed on dedicated test structures including single line and comb/serpentine for in plane continuity and isolation, daisy chains and pseudo kelvin for vertical interconnections. Tests are performed after each level of metallization during the fabrication of the test vehicle wafers to ensure the quality and the reproducibility of the process.

For RDL copper line a typical value of 65 mOhm/mm is obtained for a 30 μm linewidth and a thickness of 7 μm.

For the TSV the kelvin resistance is typically in the range of 3-4 mOhm for one via and the yield of up to 160 TSVs daisy chain structures varies from wafer to wafer between 65 to 95%.

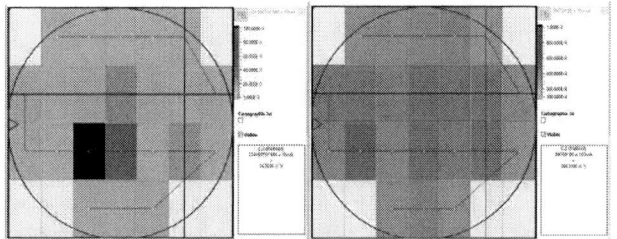

Figure 7: Wafer map on DC structures, left: TSV daisy chain of 160 vias with mean value of 1.8 Ohm; right: kelvin resistance of one TSV with mean value of 3.7 mOhm..

Power Amplifier design

Current Multimode Multiband Power Amplifier (MMPA) modules are mostly based on multi-chip module approach that uses GaAs technology for the PA core. Many research activities are currently engaged to push the cost and size reduction further by improving PA integration level and introducing new technologies and design techniques. The concept of reconfigurable PA appears as a promising solution to reduce the size and cost of the PA module. In addition, SOI CMOS and 3D package technologies offer new opportunities for advanced PA integration at reduced cost [11, 12].

Figure 8 shows the simplified block diagram of the SOI MMPA which was implemented in a 130nm RF SOI technology.

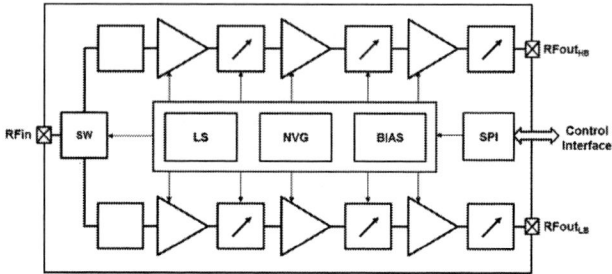

Figure 8: Block diagram of the PA.

The proposed MMPA address the broadband PMR frequency bands and output power specifications with optimized performances thanks to high efficiency SOI LDMOS devices and low loss tunable matching networks. As shown in Figure 1, the SOI PA is composed of several building blocks:

- The RF part which includes the PA line-up with an input switch (SW) that allows the selection of low-band (400MHz bands) or high-band paths (700-800MHz bands).
- A SPI interface which allows controlling the tunable components of the RF part.
- Bias circuits (BIAS) for the PA stages.
- Negative voltage generator (NVG) used for high power RF switch.
- Level shifter (LS) used to interface the SPI with the tunable RF components.

As shown in Figure 9, the SOI PA die includes the different amplification stages with several tunable capacitors, along with bias and control circuitry. The SOI die is flipped-chip on a dedicated SiHR interposer that includes 2D (L2D) and 3D (L3D) inductor designs and SMD capacitors for matching networks and decoupling networks respectively (Figure 10 and Figure 11). 2D and 3D inductors were co-designed with the SOI circuit in order to achieve a high performance MMPA module with reduced footprint.

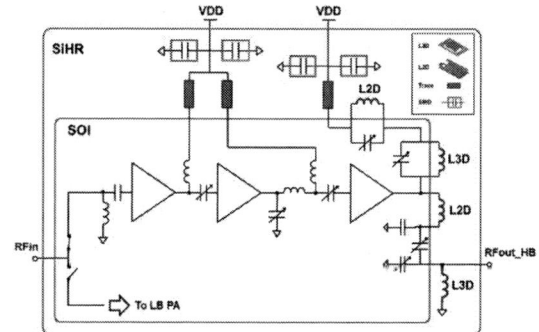

Figure 9: Schematic of the MMPA (HB part only).

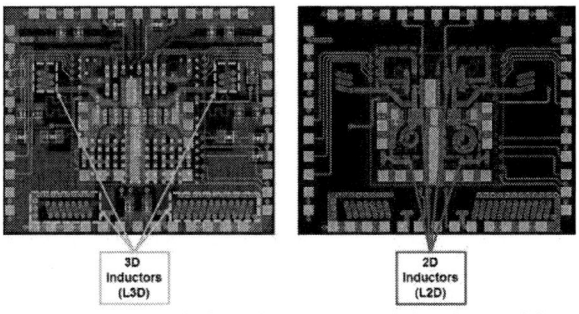

Figure 10: TOP (left) and BOTTOM (right) views of the PA module.

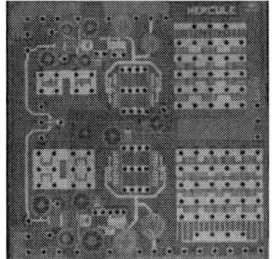

Figure 11: Picture of the SOI PA chip.

Functional interposer designs

The design of the full functional silicon HR SIP interposer was achieved and compared to an equivalent module realized on an advanced organic substrate as represented in the Figure 12. In this last case only the PA chip is flip-chipped, the FE module chips being die-attached and wire bonded on the laminate. Near one hundred of passive elements are placed close to the active chips (PA and FE module) with a minimum space à 200 µm. Depending on the electrical characteristics, the size varies from 0302 down to 01005.

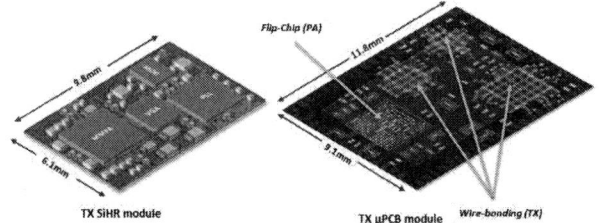

Figure 12: view of the comparative footprint between Si HR interposer and µPCB SIP.

Thanks to the tighter design rules of the silicon interposer technology, in particular for the vertical interco, the overall footprint is significantly reduced with a factor of 0.6 reduction compared to the organic solution.

Based on the development realized on the test vehicle and the RF performances measured, 2.5D and 3D inductors were designed in the Si HR functional interposer for matching network

elements of the PA. Another design with equivalent SMD inductors is also realized in order to have a direct comparison between the two configurations on the overall performance of the module. In addition, stand-alone inductors and LC structures are also designed a part in order to be tested individually and that precise characteristics can be extracted.

Functional interposer realization

The pictures of Figure 13 and Figure 14 illustrate the front side and the backside of the silicon interposer with the copper metallization in dark red and the UBM pads in yellow. For the front side the pads are both for the SMDs and the active chips stacking whereas for the backside they are for the assembly to the board.

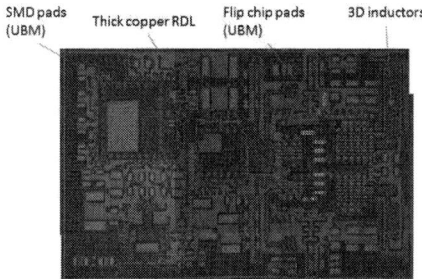

Figure 13: Picture of the front side of the functional Si-HR interposer.

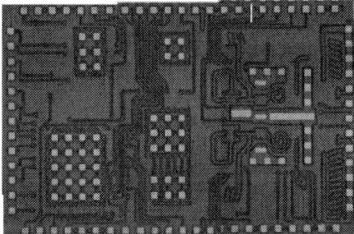

Figure 14: Picture of the Backside of the Si-HR interposer with copper RDL and UBM pads.

Passive elements integration

Depending on the specifications, some inductors are embedded inside the interposer itself and the others are SMD chips stacked on the front side whenever it provides performance benefits. Practically the inductors will be advantageously embedded in the interposer for inductance values lower or equal to 2 nH. For higher inductor values the size the 3D structure become larger than SMD ones. The main reason is that SMDs use magnetic cores that improve significantly the inductance density. Meanwhile some space can be further gained on the interposer by stacking for example SMD capacitor over 3D embedded inductors.

The inductors embedded inside the interposer are realized in 3D with a coil formed by the TSVs, the front side and backside metallization as shown in the RX tomography cross section of Figure 15 and exhibiting uniform thick copper liner inside the TSVs.

978-1-7281-1500-9/19 $31.00 © 2019 IEEE   1283

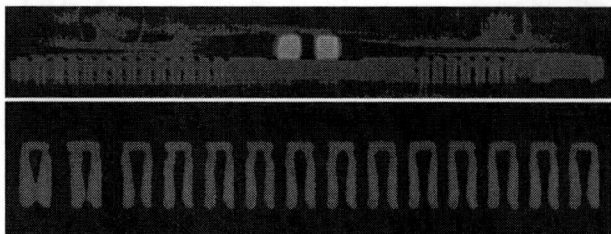

Figure 15: Picture of the RX cross-section of the interposer in the 3D inductors plan; top picture: view of the 3D inductors (L1, L2); bottom picture: detail of the TSVs.

3D inductor's RF characterization

The inductors designed on the interposer have been implemented on dedicated dies to verify their electrical behavior. S-parameters of the two types of inductors (2D spiral and 3D solenoid, see figure 4) have been performed up to 20 GHz with a VNA using GSG probes. De-embedding procedure have been used to remove the effects of contact pads. Inductance and quality factor values are extracted from the input admittance Y11 using the following equations:

$$L = \frac{Im\left(\frac{1}{Y_{11}}\right)}{2\pi f} \qquad (1)$$

$$Q = -\frac{Im(Y_{11})}{Re(Y_{11})} \qquad (2)$$

Figure 16 shows the inductance values and quality factors of the 3D embedded solenoids designed for the decoupling networks. 3D inductors formed between front side RDL, TSV and backside RDL with 3, 7 and 12 turns have been measured. As expected, the inductances values are 2 nH, 3.5 nH and 6 nH at 10 MHz. The quality factors of the smallest inductor reach 36 at 1.6 GHz.

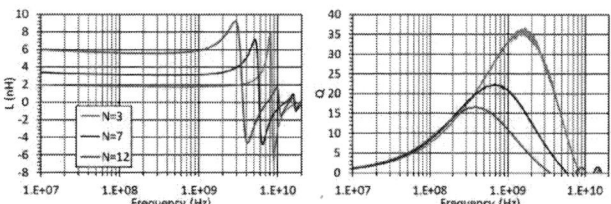

Figure 16: Inductance and quality factors of 3D solenoid inductors.

Figure 17 shows the inductance values and quality factors of the 2D spiral inductors designed for matching networks. This inductors with 1 and 2 turns have lowest inductance values of 1.15 nH and 2 nH at 10 MHz. The maximum quality factor for the 1 turn inductor is around 38.

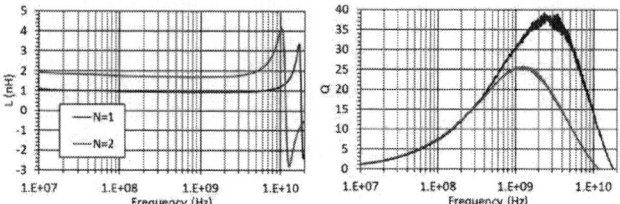

Figure 17: Inductance and quality factors of 2D spiral inductors.

The comparison between 2D and 3D inductors cannot be strictly performed because of the effects of the surrounding structures as ground planes and crossing vias. Nevertheless, trends can be observed for the 2 nH inductors (see Figure 18).

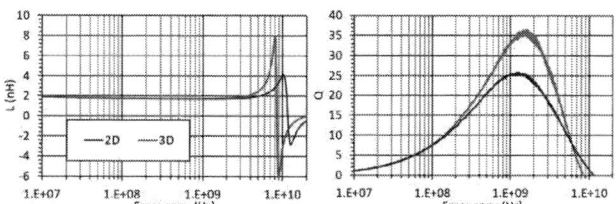

Figure 18: Inductance and quality factors of 2D spiral inductors.

For inductors with the same inductance value and almost the same size, the low frequency region of the quality factor indicates a similar resistance between these two inductors. But the quality factor reach highest values for the 3D inductor. Lowest proximity effects between turns for the 3D solenoid inductor can explain this. However, the resonance frequency is higher for the 2D spiral inductor, indicating lowest parasitic capacitive effects between turns and lowest losses from the substrate.

The 2D and 3D inductors realized on the SiHR interposer with high thickness metallization exhibits good quality factors. They are suitable to achieve a high performance MMPA module with reduced footprint and without assembly additional cost.

Chip stacking technology

For SIP integration, each of the four IC silicon chips needs to be flipped and stacked on the silicon interposer. To do so, a front end post-process µbumping module is implemented at wafer level for each die in CEA-Leti's 200 mm fabrication line. The µbump interco added on top of the aluminum I/O pads are composed of a copper pillar capped with SnAg solder bump. For the metallurgy of the bump, a nickel interlayer of 2 µm is added between the copper pedestal and the tin-silver solder to promote the formation of Sn-Ni intermetallic compounds (IMC) rather than Sn-Cu ones [9].

Functional interposer assembly

The Si interposer assembly process flow begins with printing some solder paste on the SMD pads through a screen mesh, then comes the placement and the soldering of the SMDs, then the

the ICs are flip-chiped and the submitted to reflow, finally the underfilling of the chips is achieved. Using this sequence the underfill dispense is then not critical with respect to the presence of the SMDs.

The final topology of the module is shown in Figure 19 where the chips that have been kept at the full silicon thickness of 730 μm are thicker than the SMD components. RX tomography analysis confirm the good connections between the chips and the interposer by mean of the μbumps interconnects.

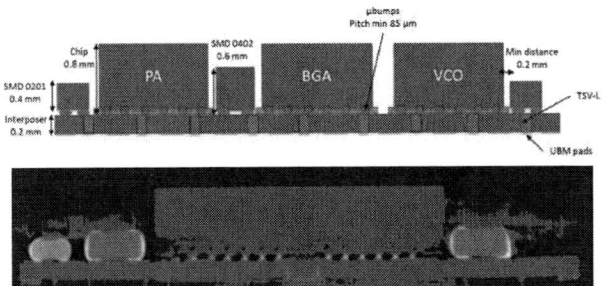

Figure 19: Picture of the topology of the SIP module; top picture: schematic cross section; bottom picture: RX cross-section showing the μbumps interconnections.

The top view of the module is shown in Figure 20 with the active chips presented before flip chip with their copper bumps interconnections. The very dense foot print yet obtained appears clearly particularly since the connections to the board are on the opposite side thanks to the presence of the TSVs avoiding to have connection pads on the front side.

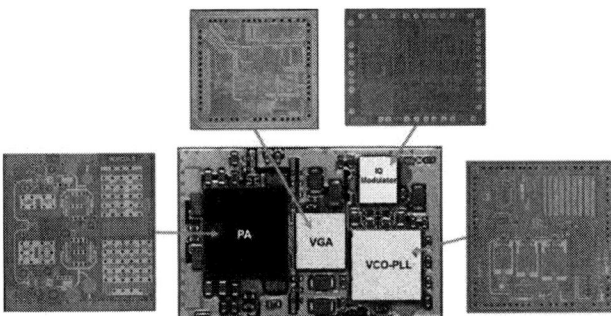

Figure 20: Picture of the SIP module after ICs and SMDs stacking.

Functional interposer testing

Figure 21 shows the silicon interposer with only the PA part and the associated SMDs for testing of the PA performances alone.

Figure 21: Pictures of the SOI PA die (left) and the assembly on the partial interposer mounted on the test board.

Output power performances of the MMPA with a 10MHz LTE test signal are summarized below for the different modes (3GPP and PMR) and bands (LB and HB) and an $ACLR_{EUTRA}$ of -37dBc.

Bands	Mode	P_{LIN} [dBm]
LB	3GPP	28
	PMR	29.5
HB	3GPP	28.5
	PMR	30

Testing of the functional interposer has not been possible at the time of the paper writing. Some results will be presented later.

Conclusions

A HR silicon interposer has been developed for professional mobile radio RF TX-RX application featured one thick RDL metallization on each side connected by TSV-last. With no additional process step 3D inductors are embedded in this interposer to be closely coupled to the active chips.

The resulting integration exhibit a significant form factor reduction (60%) compared to the equivalent module made on an organic substrate.

The 3D solenoids are characterized in RF up to 10 GHz for different variations of design. The obtained results exhibit quite good performances with a peak quality factor of 35 in the medium frequency range RF SIP and for inductance values in the range of 1-2 nH.

The footprint of the 3D inductors is equivalent to that of a SMD and with no additional cost with regard to the interposer fabrication. Moreover, we demonstrate that the tighter design rules achievable with such silicon interposer technology compared to organic solution allows a significant gain in the integration of a multichip SIP module.

Acknowledgments

This work was supported by the European commission in the frame of the CATRENE-CA118 project "Fully Integrated Transceiver for Next Generation Emergency Services" (FITNESS).

References

1. Parès, G., et al., *Full Integration of a 3D Demonstrator with TSV-First Interposer, Ultra-Thin Die Stacking and Wafer Level Packaging*, in *2013 63th Electronic Components and Technology Conference, ECTC*. 2013: Las Vegas, NV.

2. Franzon, P.D., W.R. Davis, and T. Thorolffson. *Creating 3D specific systems: Architecture, design and CAD*. in *Proceedings -Design, Automation and Test in Europe, DATE*. 2010.

3. Kim, N., et al. *Interposer design optimization for high frequency signal transmission in passive and active interposer using through silicon via (TSV)*. 2011. Lake Buena Vista, FL.

4. Lamy, Y.P.R., et al., *RF characterization and analytical modelling of through silicon vias and coplanar waveguides for 3D integration*. IEEE Transactions on Advanced Packaging, 2010. **33**(4): p. 1072-1079.

5. Tida, U.R., et al., *On the Efficacy of Through-Silicon-Via Inductors*. IEEE Transactions on Very Large Scale Integration (VLSI) Systems, 2015. **23**(7): p. 1322-1334.

6. Kim, J., et al. *High-Q 3D RF solenoid inductors in glass*. in *2014 IEEE Radio Frequency Integrated Circuits Symposium*. 2014.

7. Bontzios, Y.I., M.G. Dimopoulos, and A.A. Hatzopoulos. *Prospects of 3D inductors on through silicon vias processes for 3D ICs*. in *2011 IEEE/IFIP 19th International Conference on VLSI and System-on-Chip, VLSI-SoC 2011*. 2011.

8. Zhang, J., Indcutor with Patterned Ground Plane, 2008.

9. Pares, G., et al., *INDUCTORS USING 2.5D SILICON INTERPOSER WITH THICK RDL AND TSV-LAST TECHNOLOGIES*. International Symposium on Microelectronics, 2017. **2017**(1): p. 000072-000077.

10. Mondal, S., S.-B. Cho, and B.C. Kim, *Modeling and Crosstalk Evaluation of 3-D TSV-Based Inductor With Ground TSV Shielding Development of an ultra thin die-to-wafer flip chip stacking process for 2.5D integration*. IEEE Transactions on VLSI Systems. **vol. 25**: p. pp. 308-318.

11. Serhan, A. and e. al., *A Broadband High-Efficiency SOI-CMOS PA Module for LTE/LTE-A Handset Applications*. to be published in IEEE Radio Frequency Integrated Circuits Symposium (RFIC), 2019.

12. Giry, A., A. Serhan, and D. Parat, *Power Amplifier and Front-End Module Design in RF SOI Technology*. European Microwave Conference, Sept 2018. **Workshop WS-04, Madrid**.

Process Induced Wafer Warpage Optimization for Multi-chip Integration on Wafer Level Molded Wafer

Chen-Yu Huang, Daniel Ng, Hung-Ho Lee, Vito Lin, Chang-Fu Lin, C. Key Chung
Corporate R & D
Siliconware Precision Industries Co., Ltd
Taichung, 42881, Taiwan
chenyuhuang@spil.com.tw; keychung@spil.com.tw

Abstract—In this paper, the demonstration of test vehicle by two kinds of process flows noted as "C4 first" and "C4 last", which integrate chips on mold-based, Cu via wafer with glass carriers, are presented. Their warpage behavior during wafer-form integration will be experimentally and numerically evaluated, and also compared with wafer warpages of 2.5D assembly which applied Si interposer with TSV (through Si via). The C4-first flow is to attaching chips on wafers where the C4 bumps have been completed between mold layer and glass carrier. This flow is similar to 2.5D manufacturing process that the Si interposer was temporarily bonded on a carrier which can suppress the interposer warpage variation during reflow process. The temporary glue is required to protect the C4 bumps during chip on wafer procedures. In regarding to the cycle time and cost, the flow to complete C4 after chips attaching will be further studied. The processed induced u-bump (micro-bump) and u-pad (micro-pad) shift post jointing are observed to be larger than that in 2.5D flow. In the manufacturing process of molded wafer, the high CTE glass carriers are used to reduce CTE mismatch between molding compound and carrier. The significant wafer warpage changing from concave to convex shape was observed after chips attaching on the wafer. And the warpage will be further increased after underfilling and molding processes. To well predict and address the warpage trend, the finite element analyses are carried out to understand the process-induced warpage behaviors, and thus to select better material and process parameter. The key parameters affecting wafer warpages like material properties of molding compound, glass carrier and top-mold thickness are determined by finite element simulation. In light of the handing procedure of wafer form assembling equipment, the preferred wafer warpage shape is determined by experimental results. Finally, the test vehicle have been assembled with substrate as well. The chip module warpage in 2.5D structure is about double than that with molding interposer at 230°C and reversed direction in a convex shape, which is different from molding chip module.

Keywords- C4 first, C4 last, chip module, interposer

I. INTRODUCTION

In recent years, the explosive growth and demand for mobile, computing and networking devices are driving the development of heterogeneous integration technology, such as 2.5D (multi chips on Si interposer with TSV), FO-MCM (Fan-Out Multi Chip Module) and EMIB (Embedded Multi-die Interconnect Bridge), which can make more functionality integrated, provide high bandwidth data transportation and better power performance. The 2.5D packaging technology enables heterogeneous assembling of ASIC and memory dies, and currently becomes more mature and popularly applied in many products. However, there exists the process cost concern for TSV and Cu damascene layer fabrication at interposer front side and backside via revealing, especially for large-sized Si interposer. For the purpose to have cost reduction, lots of research works have proposed some integration technologies to be the alternative solutions for Si interposer with TSV. One structure similar to fan-out MCM and named RDL interposer package can enhance electrical performance by substituting Si interposer for polymer based, multilayered RDL with fine L/S to 1/1 um, thereby reducing signal loss and structural stress due to low modulus and high CTE of RDL [1]. Another type of electrical connection in Si-less interposer is performed by applying dual damascene layer and Si etching process, which has been used for both heterogeneous and homogeneous chip integration [2][3]. Further, the Si with fine pitch interconnections to be embedded in substrate and the 2.1D system in package that uses an organic interposer on substrate are getting notable attention because they can provide lower cost than Si interposer and a flexible way for heterogeneous integration [4][5]. Additionally, a novel package structure which integrates wide I/O DRAM, flash memory, logic die, two high density TPVs (Through Package Vias), and the two Si bridging dies by using wafer level molding and fan-out processes has been presented as well [6].

Apparently, most of these packages utilize fan-out plus thin film process, and materials like polyimide and molding compound to create an interposer structure. So these new proposed packing scenarios all possibly face the challenges of process-induced wafer warpages and handling problems [7-9]. Unlike conventional FCBGA (Flip Chip Ball Grid Array) using single Si chip to be attached on substrate, and 2.5D chip module with Si-based interposer, the side-by-side,

978-1-7281-1500-9/19 $31.00 © 2019 IEEE

multi-die chip module comprising high CTE materials such as underfill, molding compound, Cu and polyimide has a more complicated warpage behavior under thermal variation [1][2]. Besides, the risk of die shift and misalignment during micro bumps jointing is potentially higher than that in 2.5D flow, especially for the large sized chip module. There have been some studies focusing on the offset during die attaching on wafer [6][10].

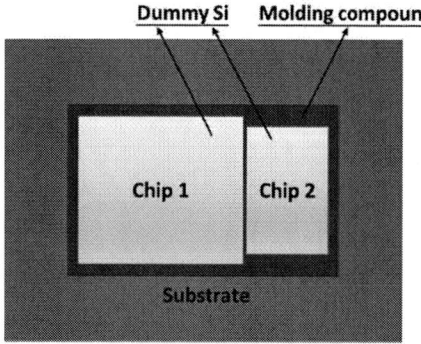

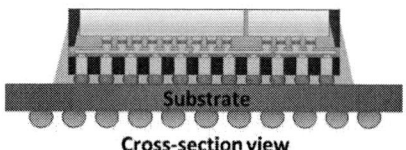

Figure 1. Top view and cross-section view of test vehicle

II. TEST VEHICLE INFORMATION

A. Package Structure

In this paper, a preliminary study using the molding compound-based interposer with single-layered RDL is assembled with two dummy Si chips bumped using Cu pillars and SnAg solder. From the package top view in Fig. 1, the dimension of chip 1 and 2 are 16 mm × 14 mm and 8 mm × 12 mm respectively, and both are grinded to 0.5 mm thick. The minimum u-bump pitch of 45 um and Cu pillar diameter of 25 um are designed in both dummy chips, which appear to be encapsulated by underfill and then molding compound around die edges, as drawn in cross-section view. The "chip module", defined as a structure covering the top dies to bottom interposer and C4 bumps, is 28 mm × 15 mm × 0.63 mm with a 150 um bump pitch and I/O count of 13,000 SnAg solder bumps. The die to die gap is around 100 um which is fully filled by underfill during u-bumps encapsulation. Similar to current 2.5D production flow, the multi-chip module was attached on a 3/2/3 substrate with 0.5 mm core thickness, then followed by underfill encapsulation, metal ring attaching and ball placement. The substrate used in this test vehicle has a size of about 1200 mm in area and thickness of 0.8 mm. The 0.6 mm thick metal stiffener ring with a 3 mm footprint width is attached on the substrate front side to constrain package's coplanarity. The effect of material selection in mold compound based chip module on the warpage behavior and assembly coplanarity is

investigated and compared with 2.5D product using Si interposer. In Fig 2(a), the SEM (Scanning Electron Microscope) image shows inside of the molding interposer with Cu post connected to C4 bumps, then to the substrate pre-solder after mass reflow. Fig. 2(b) presents the image of u-bumps after jointing with u-pads on a polyimide/RDL layer at front side of molding interposer. The u-bump structure is a Cu pillar with solder tip in a total height of 35 um and a diameter of 25 um. The measured stand-off height of micro bump after jointing is about 38 um.

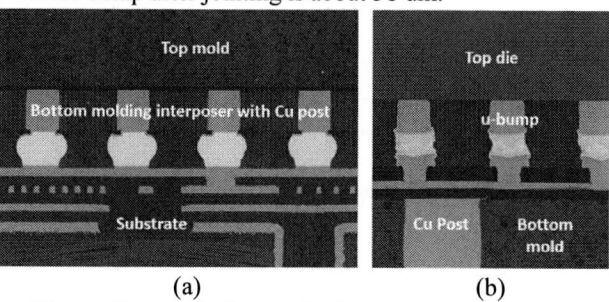

Figure 2. Cross-section images of package structure at (a) molding interposer with Cu post connected to substrate, (b) micro-bumps connected to molding interposer through RDL and micro-pads

B. Process flow

In this paper, the wafer form heterogeneous integration on molding interposer with through-mold Cu posts is performed. Two process flows are implemented in the early development stage. The naming of these two flows: "C4-first" and "C4-last" processes are defined by the process step of back-side C4 bumping completed before and after the top dies attaching on the molded carrier wafer. In the first step shown in Fig. 3(a), thin release layer needs to be coated on the glass carrier for de-bonding process requirement. Then the Cu posts with about 100 um height and 75 um diameter are fabricated on the carrier by applying conventional dry film lithographic patterning process. Next, the wafer-form liquid molding process is performed and sequentially followed by mold grinding process to planarize the molding compound surface, and to exposure Cu posts for electrical connection. The automatic optical inspection and grinded surface treatment are required before the first polyimide layer coating and via opening process. The Cu RDL is electroplated to about 5 um thick, and then coated with the second polyimide layer. For the purpose to joint with u-bumps at top dies, the micro pads were formed on the top of via openings which connect to Cu RDL.

In C4 last flow, the dummy dies are attached and reflowed on the mold-based wafer on glass carrier, which is followed by underfill encapsulation and wafer-form molding steps. Before backside C4 bumping process, the bottom glass carrier need to be removed from the wafer by laser de-bonding and the residual release layer on the wafer backside need to be cleaned as well. After that, the polyimide layer is coated on mold surface and then to be followed by via opening, UBM (Under Bump Metallization) patterning, SnAg solder electroplating and reflow to complete whole backside C4 bumping processes. When C4 last flow is conducted with assembling using real top dies, the C4

978-1-7281-1500-9/19 $31.00 © 2019 IEEE   1288

process yield become critical due to real die scrapping resulted from C4 bumping yield loss. In C4 first flow, the 2nd glass carrier is bonded on wafer front side using temporary glue after u-pad fabrication. The same C4 bumping procedures performed in C4 first flow is carried out on the wafer without top die attaching. Next, the 3rd glass carrier is bonded using thicker temporary glue to protect backside C4 bumps and followed by the 2nd carrier de-bonding to exposure the u-pads in previous steps. After completion of bottom wafer bumping at both sides, the top dies are jointed with u-pads, encapsulated by underfill and molding compound sequentially, and the 3rd glass carrier is de-bonded before singulation process. Different from TSMC's CoWoS (Chip on Wafer on substrate) process, die-last 2.5D wafer integration [11] is featured by attaching top dies after the bottom wafer process, as shown in Fig. 3(b). The thick Si interposer with TSVs is processed from front side u-pad bumping, 1st carrier bonding, backside via revealing process, backside C4 bumping process and then 2nd carrier bonding. Before the chip on wafer flow, the 1st carrier need to be de-bonded to appear front side u-pads. Further, the chip on wafer procedures are performed as mentioned before.

After thinning down and wafer sawing process, the molding chip module is mounted onto substrate and conducted with conventional FCBGA assembling processes. In above three process flows, only one carrier is required in C4 last flow. Two and three glass carriers are conducted in die-last 2.5D flow and C4 first flow respectively.

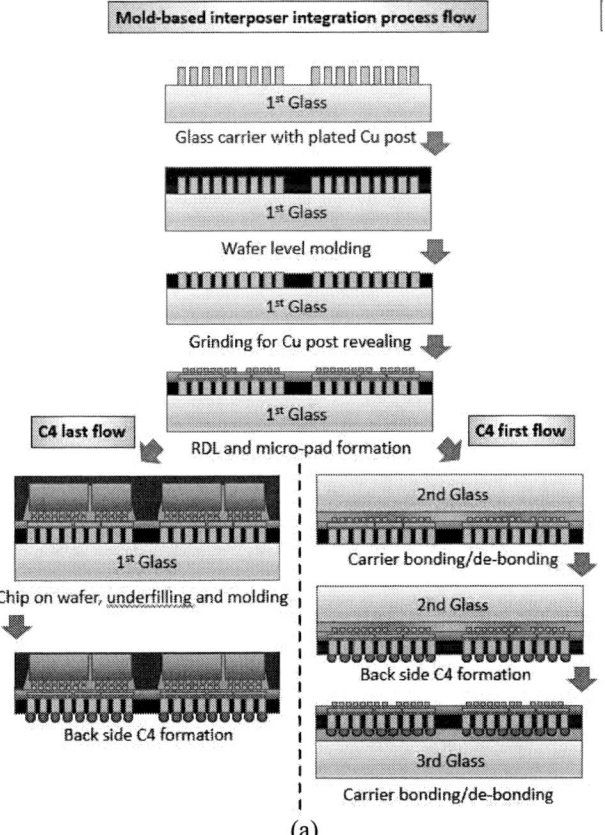

(a)

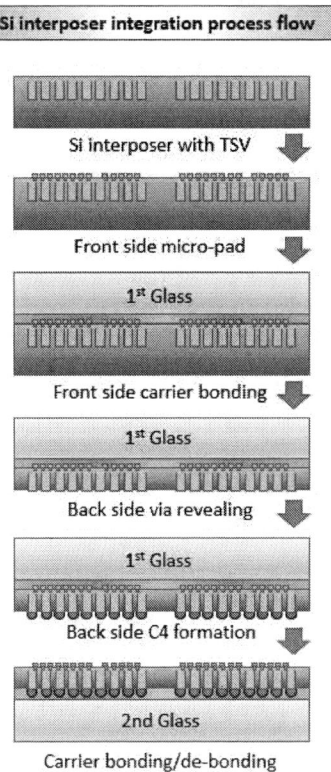

(b)

Figure 3. (a) C4 last and C4 first flows for chip integration on molding interposer; (b) die-last 2.5D flow for chip integration on Si interposer

III. METHODOLOGIES

A. Finite Element Analysis

In order to shrink development cycle time and the cost of materials used in wafer samples, FEM (Finite Element Method) using ANSIS code is implemented to simulate the warpage behavior of wafer form, chip module and package structures of multi-chip integration using molding interposer under thermal loadings. To avoid complexity and difficulty of creating mesh models with multi-scaling level, the equivalent material properties, which are calculated by considering the material's volume ratio, are employed in the simplified models of interposer with Cu via, bump/underfill layer and multi-layered substrate. In this paper, the effect of different molding compound, glass carrier with different CTE and modulus, mold thickness above top die on the process and thermal induced deformation are determined by FEM models. Two-time molding process is required from the wafer flow depicted in Fig. 3(a). The bottom molding compound covers the Cu posts and the top mold compound surrounds the top dies and underfill fillets. The material properties of mold compounds and glasses used in the wafer and package modeling are shown in table I. Four kinds of molding compounds and four kinds of glass carrier are investigated in the FEM model. A half-symmetric model is performed in wafer model to evaluate warpage behavior at different integration process steps which including u-pad wafer before and after top die attaching, wafer post underfill and molding compound curing.

TABLE I.   MATERIAL PROPERTIES USED IN FEM MODEL OF WAFER AND PACKAGE.

Material	Young's modulus (GPa)		CTE (ppm/K)		Tg (°C)
	under Tg	over Tg	under Tg	over Tg	
Mold A	19	1.6	12.4	38	150
Mold B	12	0.7	11	30	170
Mold C	17	1.8	8.1	25	160
Mold D	9	0.6	8.5	20	165
Glass A	70 ~ 80	na	3.8 ~ 4.8	na	na
Glass B	60 ~ 70	na	4.8 ~ 5.8	na	na
Glass C	60 ~ 70	na	5.8 ~ 6.8	na	na
Glass D	70 ~ 80	na	6.8 ~ 7.8	na	na

B. Warpage Measurement System

In the wafer fabrication processes, their warpages are measured by laser-optical system which can provide the 3D contours and values for further analysis. For monitoring chip module and substrate's deformation behavior versus temperature, the commercial shadow moiré system is applied to measure the out-of-plan deformation of specimens under thermal loading. The grating glass with 100 lines per inch and the phase-shifting function are used in the moiré system. Additionally, the pre-baking of the specimen is also required to eliminate the moisture absorbed in the sample to make sure the warpage results are not affected by moisture absorption. A temperature range from 25°C to 260°C is set up in one thermal cycling. The varied fringe patterns are recorded by camera and to be converted to warpage distributions and 3D contours.

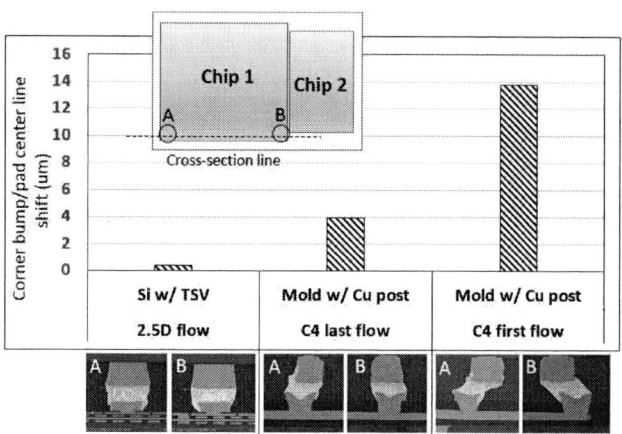

Figure 4. chip 1 corner u-bump to u-pad shift post chip-on-wafer process

IV.   RESULTS AND DISCUSSION

A. Micro-bump/pad shifting measurement

In this paper, the shift of u-bumps jointed to u-pads are measured on the wafers from different process flows shown in Fig. 3(a) and 3(b). From the top-view of chip module in Fig. 4, the u-joints at location A and B are picked to be observed along chip 1 longer edge. The u-bump pattern on the top die has lower expansion due to low CTE of Si. However, the u-pad pattern has higher expand rate under high temperature. Apparently, the SEM photos show the corner micro pads on molding interposer move outward after solder jointing. The shift is about 14um in the wafer of C4 first flow, but only minor shift observed in 2.5D flow that using Si top dies and interposer. This shift could be resolved by process parameters tuning or to re-design the relative position of micro bump and micro pad.

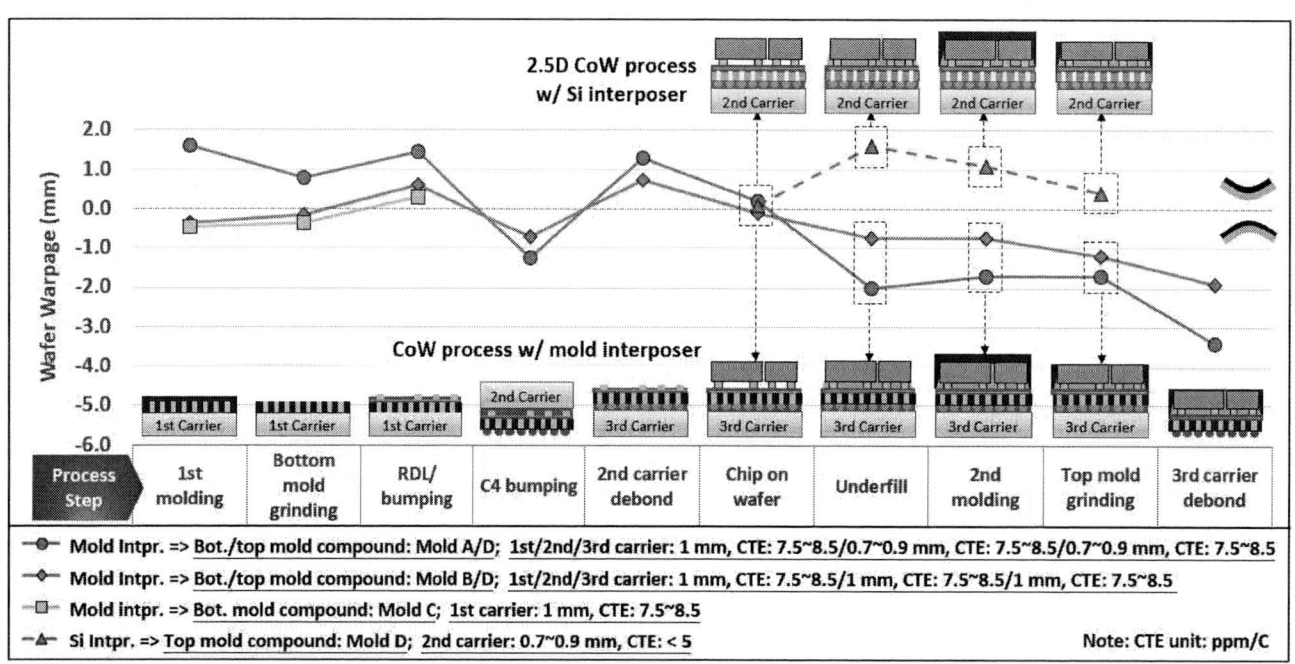

Figure 5. Wafer warpage trend in the process of using Si interposer (2.5D) and molding interposer (C4-first process)

B. Wafer Warpage Measurement Results

In this study, the wafer warpages in most of the process steps are measured by the tool using laser AF (Auto Focus) system for the purpose to figure out the process induced wafer deformation. As shown in Fig 5, the wafer warpage trend of chip on wafer process using silicon and molding compound based interposer are presented. The glass carrier with high CTE ranging from 7.5 ~ 8.5 ppm/C is applied for mold interposer handling to minimize the thermal mismatch between molding compound and glass. In the first molding (bottom mold) process, the wafer with mold A, which has higher modulus and CTE than mold B and C, has concave and concentric warpage of about 1.5 mm. After silicon attaching process, the wafer warpage change to convex shape because high CTE materials are all under center line of in-plane view. The wafer form molding process could slightly reduce convex warpage value, which is contributed from shrinkage of molding compound above top dies. In the final step of carrier de-bonding, the warpage increased to about 3.5 mm in a convex shape (top die upward), which is resulted from the removal of constrain from glass carrier.

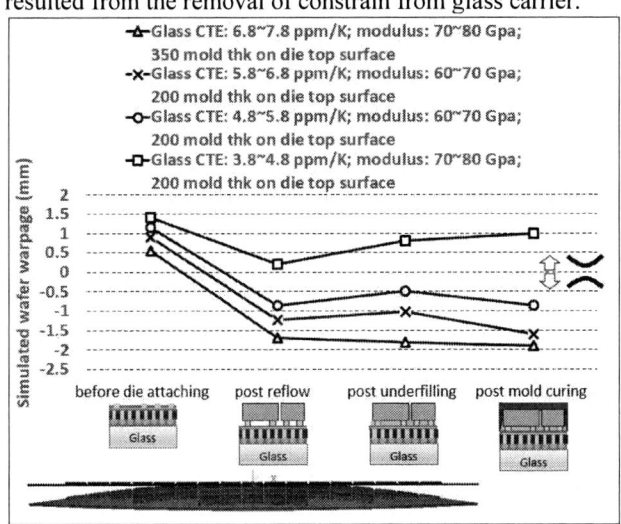

Figure. 6. Simulated wafer warpage with different glass CTE, modulus and mold thickness on top die in chip-on-wafer process

Material properties of glass carrier such as CTE, modulus, and mold thickness on top die post wafer form are studied by finite element simulation. Key process steps like wafer with completed u-pads, Si chip attaching on wafer, underfill and mold encapsulations are included in the simulated results as shown in Fig. 6. It reveals that the glass with the CTE range of 3.8 ~ 4.8 ppm/°C has higher wafer warpage right after u-pad bumping completion due to high CTE mismatch between glass and molding compound. Then the warpages are increased with the glass CTE increasing to 6.8 ~ 7.8 ppm/°C before die attaching. For these four glasses with different CTE ranges, the wafer warpage performs to be decreased and even to concave shape when the Si top dies are assembled with bottom wafer. In further process steps, the handing risk is getting higher for the wafer using high CTE glass which may suffer glass handing or crack risk induced by handing applied bending forces.

C. Chip module and package warpage

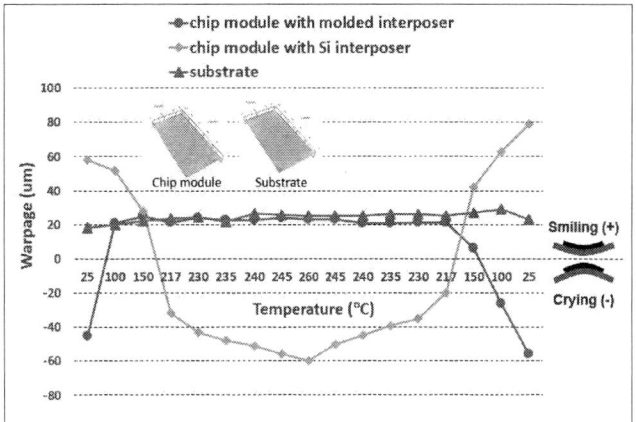

Figure 7. Chip module and bare substrate warpage with varied temperature by shadow moiré measurement.

As shown in Fig. 7, the chip module warpages with respect to temperature are measured by shadow moiré system. The 3D contours show the warpage shapes of the chip module using molding interposer and its substrate at 235°C are more consistent. For the chip module using Si interposer, its warpage performs from concave at room temperature to convex at high temperature as the top die upward. The chip module warpage in 2.5D structure is about double than that with molding interposer at 230°C and reversed direction in a convex shape.

	Condition	Molded interposer assembled on substrate	Si interposer assembled on substrate
Chip module warpage	at reflow temp.		
	at room temp.		
Substrate warpage	at reflow temp.		
	at room temp.		
Package warpage	at room temp.		

Figure 8. Chip module/ substrate/ package warpage configurations from reflow to room temperature

In order to further understand the thermal deformation behavior of chip module using Si and molding interposers assembled on substrates and their packages warpage, the decomposed warpage variation configurations under thermal loading are depicted in Fig. 8. Totally different warpage trends are found on the chip modules with Si and molding compound, but both have similar convex package warpage. Further, the package warpage are measured by ICOS tool and compared with simulation results as follow.

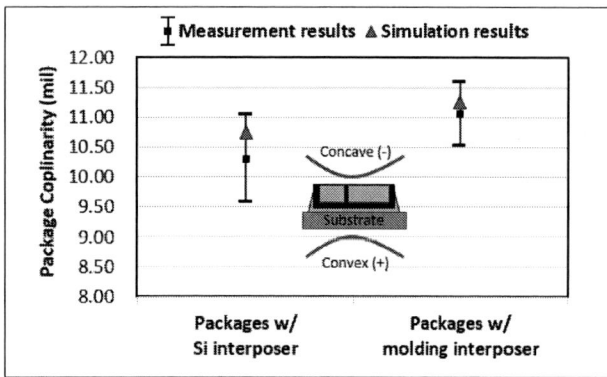

Figure 9. Experimental and simulated coplanarity of packages using Si interposer and molding interposer

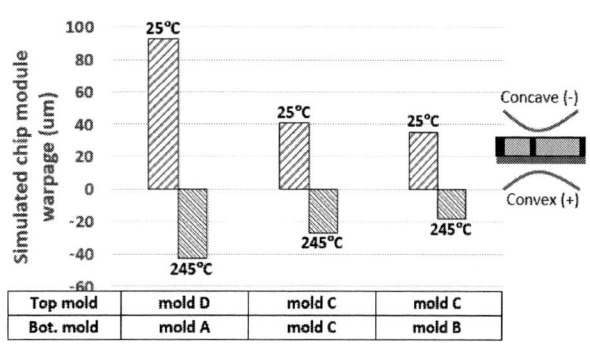

(a)

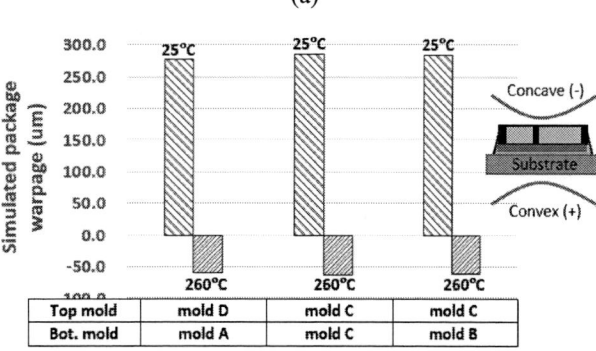

(b)

Figure 10. FEM simulation results of (a) chip modules using molding interposer; (b) packages using molding interposer

Fig. 9 shows the package coplanarity from experimental measurement and simulation results, which compares the effect from chip modules using Si interposer and molding interposer. From the results, the finite element models are shown to be able to correctly predict the package coplanarity post assembling, which appears the packages with molding interposer has higher convex deformation than that using Si interposer. The higher coplanarity is possibly resulting from the chip module warpage deformation from concave to convex shape during cooling process. The high stiffness from Si interposer and its warped trend after reflow makes the package convex value smaller. The FEM simulated chip module and package warpages at room and high temperature, which apply molding interposer, are presented in Fig. 10(a).

The mold A with high CTE and high young's modulus used for bottom molding compound has larger chip module warpage variation, especially at room temperature. However, the package warpages are not apparently affected by applying different molding compound combinations.

V. CONCLUSION

For multi-chip integration, the warpage behavior of the wafers, chip modules and packages applying molding and Si interposers have been studied with a combination of experimental and numerical approaches. The die corner u-bump to u-pad shift is found to be about 14um in the wafer from C4 first flow, 4um in the wafer from C4 last flow, but only minor shift observed in 2.5D flow that using Si top dies and interposer. For molding interposer as bottom wafer, the lower CTE mismatch between molding compound and glass carrier could present lower warpage after u-pad fabrication. After top dies attaching, the wafer warpage change to convex shape because high CTE materials are under Si chips. The chip module warpage variation using molding interposer is opposite of Si interposer. High CTE and high modulus molding compound in the bottom mold could induce higher warpage variation under thermal loading. The package coplanarity post assembling with molding interposer appears higher convex deformation. Conclusively, the molding interposer, chip module and its package has different warpage behavior compared with that using Si interposer, and material selection like molding compound and glass is critical for reducing wafer process risk.

REFERENCES

[1] K. L. Suk, S. H. Lee, J. Y. Kim, S. W. Lee, H, J, Kim, S. C. Lee, P. W. Kim, D. W. Kim and D. K. S. Oh, "Low cost si-less RDL interposer package for high performance computing applications," Proc. IEEE 68th Electronic Components and Technology Conf. (ECTC), 2018, pp. 64–69.

[2] F. X. Che, M. Kawano, M. Z. Ding, Y. Han, and S. Bhattacharya, "Study on low warpage and high reliability for large package using TSV-free interposer technology through SMART codesign modeling," IEEE Transactions on Components, Packaging and Manufacturing Technology, Vol. 7, No. 11, Nov. 2017.

[3] W. S. Kwon, S. Ramalingam, X. Wu, L. Madden, C. Y. Huang, H. H. Chang, C. H. Chiu, S. Chiu, and S. Chen, "Cost effective and high performance 28nm FPGA with new disruptive silicon-less interconnect technology (SLIT)," International Symposium on Microelectronics Fall 2014, vol. 2014, no. 1, pp. 000599-000605.

[4] R. Mahajan, R. Sankman, N. Patel, D. W. Kim, K. Aygun, Z. Qian, Y. Mekonnen, I. Salama, S. Sharan, D. Iyengar, and D. Mallik, "Embedded multi-die interconnect bridge (EMIB) a high density, high bandwidth packaging interconnect," Proc. IEEE 66th Electronic Components and Technology Conf. (ECTC), 2016, pp. 557–565, DOI: 10.1109/ECTC.2016.201.

[5] Y. Uematsu, N. Ushifusa, and H. Onozeki, "Electrical transmission properties of HBM interface on 2.1-D system in package using organic interposer," Proc. IEEE 67th Electronic Components and Technology Conf. (ECTC), 2017, pp.1944–1949, DOI: 10.1109/ECTC.2017.34.

[6] A. Podpod, J. Slabbekoorn, A. Phommahaxay, F. Duval, A. Salahouedlhadj, M. Gonzalez, K. Rebibis, R.A. Miller, G. Beyer and E. Beyne, "A novel fan-out concept for ultra-high chip-to-chip interconnect density with 20-um pitch," Proc. IEEE 68th Electronic Components and Technology Conf. (ECTC), 2018, pp. 370–377.

[7] S. Joblot, A. Farcy, N. Hotellier, A. Jouve, F. de Crécy, A. Garnier, M. Argoud, C. Ferrandon, J.P. Colonna, R. Franiatte, C. Laviron and S. Cheramy, "Wafer level encapsulated materials evaluation for chip on wafer (CoW) approach in 2.5D Si interposer integration," 2013 IEEE International 3D Systems Integration Conference (3DIC)

[8] H. Y. Li, A. Chen, S. Peng, G. Pan, and S. Chen, "Warpage tuning study for multi-chip last fan out wafer level package," Proc. IEEE 67th Electronic Components and Technology Conf. (ECTC), 2017, pp. 1384–1391.

[9] M. K. Shih, C. Hsu, Y. S. Chang, K. Y. U. Chen, I. Hu, T. Lee, D. Tarng, and C. P. Hung, "Warpage characterization of glass interposer package development," Proc. IEEE 67th Electronic Components and Technology Conf. (ECTC), 2017, pp. 1392–1397.

[10] V. S. Rao*, C. T. Chong, D. Ho, D. M. Zhi, C. S. Choong, S. L. PS, D. Ismael and Y. Y. Liang "Process and reliability of large fan-out wafer level package based package-on-package," Proc. IEEE 67th Electronic Components and Technology Conf. (ECTC), 2017, pp. 616–622.

[11] M. Ma, S. Chen, J. Y. Lai, T. Lu, A. Chen, G. T. Lin, C. H. Lu, C. H. Liu, S. L. Peng, "The development and technological comparison of various die stacking and integration options with TSV Si interposer," Proc. IEEE 66th Electronic Components and Technology Conf. (ECTC), 2016, pp. 336–342.

978-1-7281-1500-9/19 $31.00 © 2019 IEEE

Improved Structure for Package Substrates with Embedded Thin-Film Capacitor

Tomoyuki Akahoshi, Daisuke Mizutani
Fujitsu Laboratories Ltd.
10-1 Morinosato-Wakamiya
Atsugi-shi, Kanagawa, Japan
t. akahoshi@fujitsu.com

Kei Fukui, Seigo Yamawaki, Hidehiko Fujisaki
Fujitsu Interconnect Technologies Ltd.
36 Oaza Kitaowaribe
Nagano-shi, Nagano Japan

Manabu Watanabe, Masateru Koide
Fujitsu Advanced Technologies Ltd.
4-1-1 Kamikodanaka, Nakahara-ku
Kawasaki-shi, Kanagawa Japan

Abstract—**Embedding thin-film capacitors (TFC) in a package substrate is a technology aimed at improving the performance of power supply. The package substrate that we have developed and produced embeds TFCs into the core layer of a built-up substrate. Using this substrate promises to dramatically reduce the impedance on power supply lines and provide a stable power supply to the integrated circuit (IC) chip, resulting in improved performance of electronic devices. In this paper, we propose three improved structures for package substrates with embedded TFCs: the first is an asymmetric single-sided embedded TFC structure that uses the warpage-control layer; the second is a standalone embedded TFC structure that places solitary TFCs directly under the IC chip only; the third is a coreless embedded TFC structure that embeds TFCs in a coreless substrate. These technologies promise to help lower the cost of package substrates, increase their efficiency, and reduce their thickness.**

Keywords- thin film capacitor; package substrate; power integrity; impedance

I. INTRODUCTION

As the voltages of IC's have been reduced in recent years due to advances in CMOS technology, restricting the ripple voltage during IC operations has been an issue in order to improve the performance of electronic devices. Placing a decoupling capacitor near the IC chip is an effective way of restricting the voltage fluctuation, and this had been previously achieved by placing chip-type capacitors on the surface and bottom side of the package substrate. However, this approach has a drawback in that the power supply impedance is not reduced at high frequencies due to the effects of power line resistance and inductance, thereby causing an interference with the increase in IC performance. To counter this effect, we developed a package substrate with an embedded thin-film capacitor. After evaluating its electrical characteristics, mechanical characteristics, and reliability of this package substrate, we successfully created a product based on it [1]-[3].

Fig. 1 shows an image of the TFC material, Figs. 2–4 present an overview and cross-sectional photograph of the package substrate of the processor used in Fujitsu's SPARC M12 Unix server. Thin-film capacitors are thin films consisting of a high dielectric layer made of barium titanium oxide (BTO) less than 1 µm thick. Both sides of the high dielectric layer are coated with a nickel and copper layer. The capacitance per unit area is approximately 1 µF/cm^2 (supplier: TDK Corporation) [4].

This substrate with embedded TFCs is a built-up substrate of 26 layers in a 6-14-6 arrangement, with two TFC layers embedded in the core layer. As the TFCs are built into the core, multiple TFCs can be easily connected with plated through holes (PTHs) by mechanical drilling.

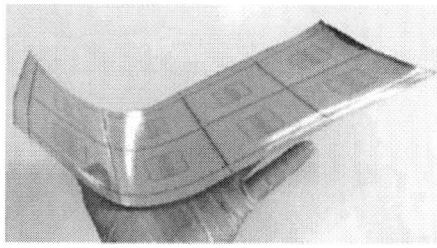

Figure 1. Photo of TFC material

Figure 2. Appearance of package substrate with embedded TFC

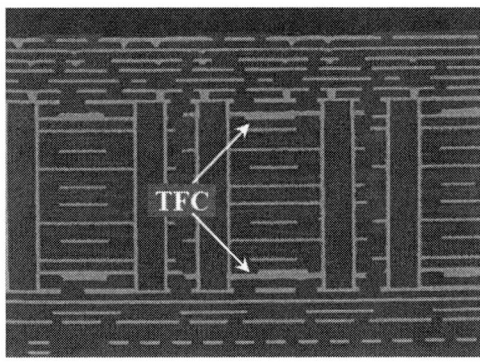

Figure 3. Cross section image of package substrate with embedded TFC

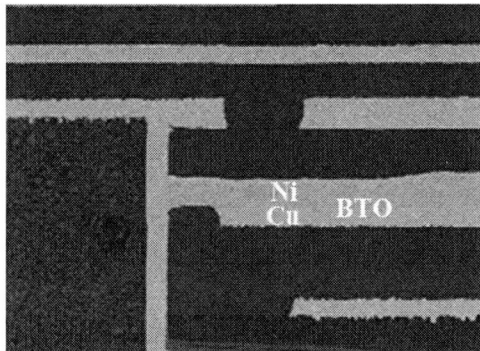

Figure 4. Enlarged view of connection area between PTH and TFC

The capacitance is determined by the areas facing the power planes and ground planes formed on both sides of the BTO layer. The TFCs' effects can be maximized by controlling the design factors, such as number of PTHs connecting between layers, their pitch, and their clearance. By controlling these factors, we were able to reduce the impedance in the range of 10–100 MHz, which has been difficult with conventional on-chip or on-package capacitors, and significantly reduced the power supply noise [1].

Power integrity technologies for the stable supply of power have been developed for high-end packages, such as those used in enterprise server CPUs and deep-learning GPUs not only to handle devices with low voltage, but also to accommodate traces for multiple power supplies that feed multi-core chips and multi-chip packages, on-package voltage regulators, etc. [5]-[7]. Furthermore, for consumer packages such as mobile phones and wearable devices, reducing the size and thickness at the same time can improve power supply characteristics.

In order to contribute to the further development of power integrity, we improved the substrates with embedded TFCs that can be a part of a power delivery network (PDN). The specific improvements are as follows:

(1) an asymmetric single-sided embedded TFC structure that uses the warpage-control layer,
(2) a standalone embedded TFC structure in which an individual TFC is placed directly under the IC chip, and
(3) a coreless embedded TFC structure in which the TFC is built into a coreless substrate.

These improvements are expected to ensure a stable supply of power and greatly contribute to cost reduction and creation of thinner package-substrates.

II. SINGLE-SIDED EMBEDDED TFC STRUCTURE

Currently, in addition to the conventional 1 $\mu F/cm^2$ capacitance per unit area, TFC film characterized by high volume type of 2 $\mu F/cm^2$ has become available. Thus, we have obtained two new options in addition to the conventional structure of $1\mu F / cm^2$ x 2 layers for the substrate structure. First is a capacitance doubling structure of $2\mu F/cm^2$ x 2 layers, and second is a $2\mu F/cm^2$ x 1 layer single-sided embedded structure. As the single-sided embedded structure can secure the same capacitance as the conventional type with one TFC, it is expected to reduce the manufacturing cost of the substrate. However, substrate warpage is a concern with this vertically asymmetric layer structure. To address this issue, we have developed a novel embedded TFC substrate manufacturing process with a warpage-control layer. The warpage-control layer was formed by plating copper on nickel foil. The advantage of the plating process is that its control of thickness is more precise than that of foil thickness. The thickness of the nickel foil is 13 µm, which is the same as the thickness of the TFC layer. The thickness of the copper plating is designed so that the thermal stress warpage is minimized using finite element analysis software Abaqus (Dassault Systems). The analysis model on 50 mm x 50 mm package substrate is shown in Fig. 5, and the analysis result is shown in Fig. 6. Forming a warpage-control layer with plating copper thickness can help optimize the substrate warpage. In this case, the copper thickness of the warpage-control layer is 17 µm.

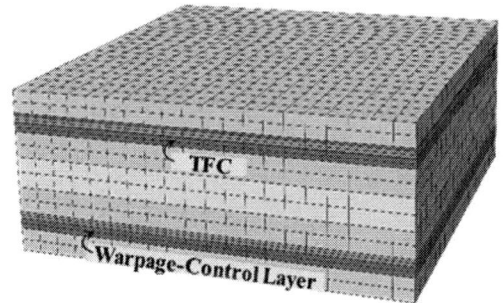

Figure 5. Overview of simulation model on finite element analysis

978-1-7281-1500-9/19 $31.00 © 2019 IEEE

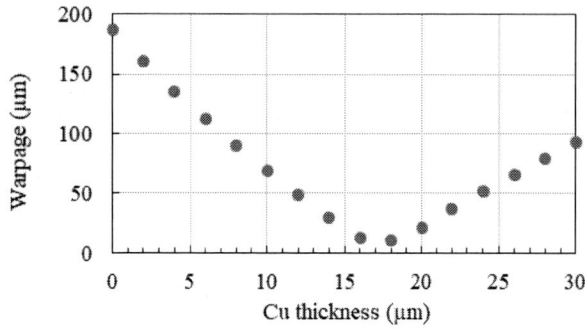

Figure 6. Simulation result with change of warpage control Cu thickness

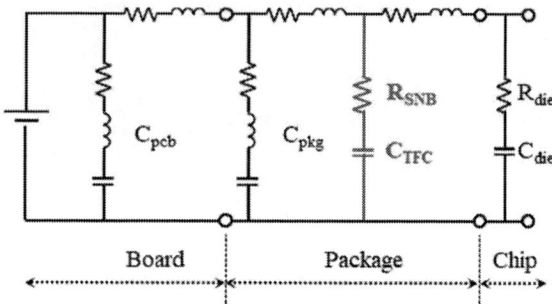

Figure 8. PDN model with RC snubber circuit using TFC

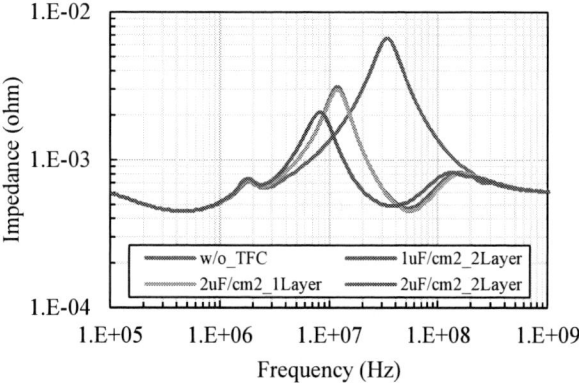

Figure 7. Simulation results of PDN model

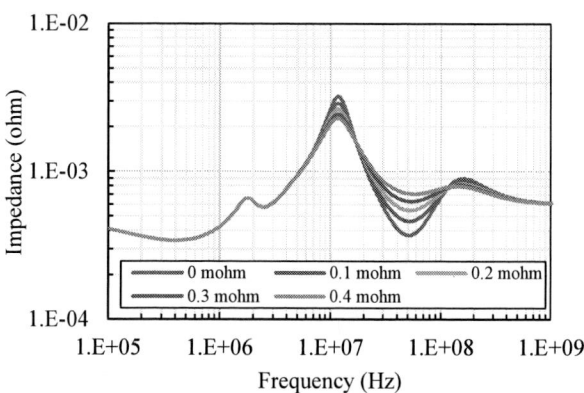

Figure 9. Simulation results with change of snubber resistance (R_{SNB})

A power delivery network (PDN) model was created to verify the electrical characteristics of the single-sided embedded TFC structure and the impedance curve was analyzed. For the analysis tool, Advanced Design System (Keysight Technologies Inc.) was used. The verification result is shown in Fig. 7. Compared with the case without TFC, the anti-resonance peak of the impedance of the embedded TFC substrate is less than half, shifting to the low frequency side. Furthermore, although the $1\mu F/cm^2$ x 2 layer and the $2\mu F/cm^2$ x 1 layer show almost the same behavior, the $2\mu F/cm^2$ x 1 layer has a slightly lower impedance. This difference is presumed to be the influence of resistance and inductance of PTH. As an application example using this embedded TFC structure, an in-package RC snubber circuit is proposed. The RC snubber circuit is a circuit technology that reduces resonance peak by applying resistance component in series to capacitance [8]. By embedding the TFC with only one layer, we can generate an in-package RC snubber circuit, as shown in Fig. 8, by forming a resistance in a vacant region by power/GND patterns and vias. As shown in Fig. 9, impedance reduction can be achieved by applying a slight resistance. In this way, with the development of the high-volume TFC and single-sided embedded TFC structure, doubling the capacitance is possible, along with other functions such as the RC snubber circuit in the package substrate.

III. STANDALONE EMBEDDED TFC STRUCTURE

Instead of embedding TFC in the entire substrate, we have newly developed a structure that embeds individual TFCs only in the required area under the IC chip. Fig. 10 shows the manufacturing process flow. A piece type TFC was placed in the uncured resin film laminated on the core layers which were formed by conventional processes. After the resin was cured, the surface conductor of TFC was patterned by etching process. Then, laminated sheets were covered both side of the substrate, and via-holes were formed by laser and mechanical drilling and plating processes. Finally, the build–up layers were formed by conventional semi-additive processes (laminating, laser drilling, and patterning).

Furthermore, not only capacitance increase but also relaxation of the alignment accuracy were realized by limiting the surface to be patterned to only the upper side. A comparison of the conventional double-sided patterning structure and the improved single-sided patterning structure is shown in Fig. 11. The region where the power plane and the GND plane face each other functions as a capacitance; thus, making the pattern only on one side leads to an increase in the facing area and capacitance. On the other hand, the PTHs of the power or the GND are inevitably arranged around the individual TFC, instead of the conventional alternate arrangement. Therefore, the influence of parasitic inductance should be considered. Tables I and II show the analysis results of verifying the electrical characteristics in these PTH arrangements. In the simulation model, PTH with a diameter of 0.15 mm was arranged at a 0.50 mm pitch in various TFC sizes. The capacitance and inductance were obtained from the impedance curve. As a result, the small TFC size showed better characteristics than the large one. Through the improved structure, large capacitance was obtained while the increase in inductance was suppressed. In the large TFC size, although the ratio seemed worse, the absolute value remained low. Therefore, the influence can be observed in the entire PDN model.

(a) Placing a TFC in the uncured resin

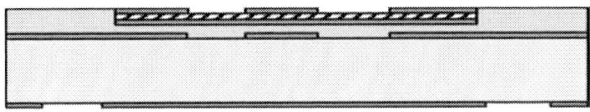

(b) Curing resin, patterning the surface of TFC

(c) Laminating films both side of substrate

(d) Fabricating vias by laser and mechanical drilling

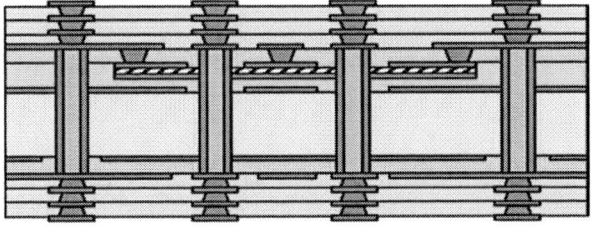

(e) Stacking up build-up layers

Figure 10. Manufacturing process flow of build-up substrate with standalone embedded TFC

(a) Conventional PTH arrangement in entire embedded TFC

(b) Improved PTH arrangement in standalone embedded TFC

Figure 11. Comparison of cross section structures
(red: power line, blue: GND line, green: BTO layer)

Table I. Analysis results of capacitance (nF)

TFC size	Conventional	Improved	Ratio
5 x 5 mm	282	521	185 %
10 x 10 mm	1,087	1,463	135 %
20 x 20 mm	4,222	4,808	114 %

Table II. Analysis results of inductance (pH)

TFC size	Conventional	Improved	Ratio
5 x 5 mm	1.81	3.94	218 %
10 x 10 mm	0.37	1.15	311 %
20 x 20 mm	0.08	0.45	563 %

As an application example using this standalone embedded TFC structure, an on-package integrated voltage regulator circuit is proposed. Normally, the voltage regulator circuits are composed of discrete component capacitors, inductors and driver ICs. Since the high performance IC chip with various driving voltage requires subdivision of the power supply line in the substrate, the higher impedance from the voltage regulator unit to the IC chip is a concern.

With the standalone embedded TFC structure, an inductor can be made using wiring layout in the TFC disappearance region such as the outer circumferential or bottom build-up regions. Through the combination of capacitance formed by TFC and inductance formed by wiring, LC circuits can be easily formed in the package substrate. Constructing the voltage regulators in the vicinity of the IC chip in the package substrate is possible, and this structure can contribute to the power integrity improvement.

IV. CORELESS EMBEDDED TFC STRUCTURE

In order to broaden the applications of embedded TFC substrates, we developed a coreless embedded TFC structure that achieves a thinner substrate. The coreless substrate is manufactured by a usual build-up process in which the insulator lamination and conductor pattern formation are repeated on the supporting panel. By laminating TFC instead of the insulator, the coreless embedded TFC structure is easily obtained.

The external and cross-sectional photographs of the manufactured test vehicle substrate are shown in Figs. 12-14. The substrate size is 29 mm x 29 mm, 0.6 μm thick BTO film is inserted between L2 and L3, and the power and GND lines are connected by a laser via with 60 μm diameter. Through a 30 μm thick insulation layer, the substrate thickness is suppressed to approximately 400 um even though 8 conductor layers are used.

Figure 12. Appearance of coreless substrate with embedded TFC

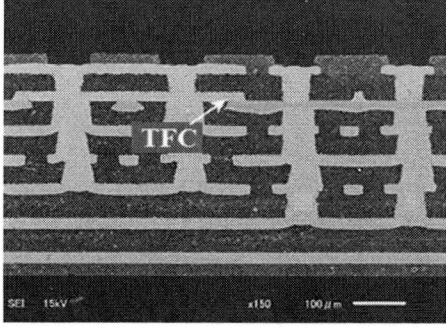

Figure 13. Cross section image of coreless substrate with embedded TFC

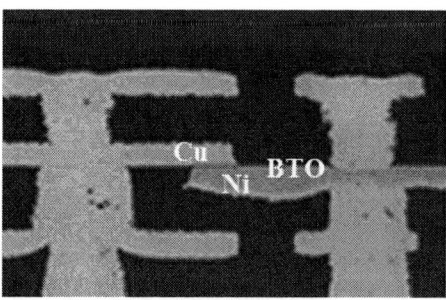

Figure 14. Enlarged view of TFC layer

Since the manufacturing process of the coreless substrate repeats the lamination and peeling steps from the support panel, the design of the TFC layer is easy to control and the design flexibility is high. In other words, when the TFC layer is embedded in the IC chip side close to the high density C4 bump, the inductance to the IC chip becomes low, but the capacitance also decreases. On the other hand, when it is embedded in the motherboard side with a low via density, the inductance up to the IC chip is high, but a large capacitance can be ensured. The conductor pattern can be designed according to the power supply characteristics.

Fig. 15 shows the measurement result of the impedance. Using Vector Network Analyzer E5061B (Keysight Technologies. Inc.), GS and SG probes were brought into contact between the power and the GND pad on the package surface, and the impedance was measured by a shunt-thru method. A photograph of the measurement system is shown in Fig. 16. The measurement frequency was 1 KHz to 3 GHz. Substrates with embedded TFC of 2 $\mu F/cm^2$ have lower impedance than those with embedded TFC of 1 $\mu F/cm^2$, and the measurement results of the capacitance in the 29 mm SQ substrate are 5.7 μF and 10.6μF, respectively. In addition, at frequencies higher than the resonance peak around 10 MHz, the substrates have a similar behavior, indicating that inductance is equal.

Fig. 17 shows AC voltage specification results measured by Impedance Analyzer 4294A (Keysight Technologies Inc.). The capacitance of TFC increased as the oscillation voltage level increased. This result is peculiar to Ferroelectric capacitors such as BTO, and designing a power supply line based on this result improves the accuracy of power integrity

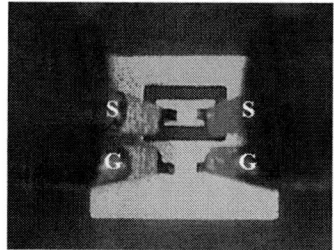

Figure 15. Photo of RF probes on surface layer

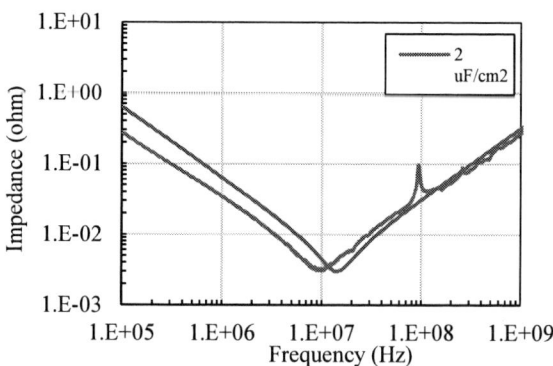

Figure 16. Impedance measurement results in frequency domain

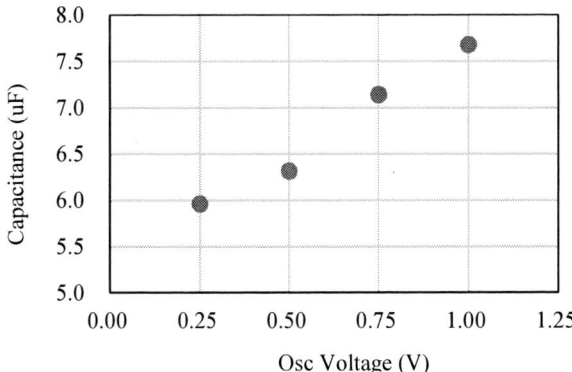

Figure 17. AC specification in coreless embedded TFC substrate

V. CONCLUSION

In this paper, we proposed three improved structures for package substrates with embedded TFCs, namely, single-sided embedded TFC structures, standalone embedded TFC structures, and coreless embedded TFC structures. Then we discussed the mechanical and electrical characteristics of each structure. By combining the proposed structures, placing the required capacitance in the relevant locations is possible, and placing the inductors in the empty locations facilitates the manufacture of RC or LC circuits. This technology can be applied not only to FCBGA package substrates for high-end CPUs, but also to optical-module substrates that require high-density components and power-supply design constraints, as well as DC-DC converters that control the voltages of nearby devices.

ACKNOWLEDGMENT

The authors would like to thank Hideaki Nagaoka at Fujitsu Laboratories Ltd. for supporting the mechanical stress analysis, and to Akiko Tsukada and Masaki Kirinaka at Fujitsu Interconnect Technologies Ltd. for assisting in the SI/PI design and simulations.

SREFERENCES

[1] T. Akahoshi, et al. "Development of CPU Package Embedded with Multilayer Thin Film Capacitor for Stabilization of Power Supply" IEEE 67th Electronic Components and Technology Conference (ECTC), pp. 179-184, 2017

[2] M. Koide, et al. "Development of Large-size CPU Package Structure Using Embedded Thin Film Capacitor Package Substrate" IEEE 67th Electronic Components and Technology Conference (ECTC), pp. 704-710, 2017

[3] M. Koide, et al. "Development of Large-size CPU Package Structure Using Embedded Thin Film Capacitor Substrate" IEEE CPMT Symposium Japan (ICSJ), pp. 159-162, 2017

[4] H. Tanaka, et al. "Embedded High-k Thin Film Capacitor in Organic Package" 10th IEEE Electronics Packaging Technology Conference(EPTC), pp.988-993, 2008

[5] R. Bertran, et al. "Voltage Noise in Multi-Core Processors: Empirical Characterization and Optimization Opportunities" 47th Annual IEEE/ACM International Symposium on Microarchitecture (MICRO), pp. 368-380, 2014

[6] K. Oi, et al. "Development of new 2.5 D package with novel integrated organic interposer substrate with ultra-fine wiring and high density bumps" IEEE 64th Electronic Components and Technology Conference (ECTC), pp. 348-353, 2014

[7] William J. Lambert, Michel J. Hill, K. Radhakrishnan, L. Wojewoda, and Anne E. Augustine "Package Inductors for Intel Fully Integrated Voltage Regulators" IEEE Transactions on Components, Packaging and Manufacturing Technology, vol. 6, No. 1, pp. 3-11, 2016

[8] Y. Iijima, M. Matsumura, and T. Sudo "Anti-Resonance Peak Damping of PDN Impedance by On-board Snubber Circuits" IEEE Electrical Design of Advanced Packaging and Systems Symposium (EDAPS), pp. 127-130, 2012

3D Packaging with Embedded High-Power-Density Passives for Integrated Voltage Regulators

Teng Sun, Robert G. Spurney, Atom Watanabe, P. Markondeya Raj, Himani Sharma, Rao Tummala
3D Systems Packaging Research Center
Georgia Institute of Technology
Atlanta, USA
tsun34@gatech.edu

Furukawa Yoshihiro
Nitto Denko Corporation
Osaka, Japan
yoshihiro.furukawa@nitto.com

Abstract— **Highly-integrated 3D voltage regulators (IVRs) for high-power applications are developed for emerging applications such as AI computing and server. With this 3D process integration, passive components such as inductors and capacitors are embedded into substrates and placed close to the chips, resulting in short power delivery networks (PNDs) and high power efficiency. High-density tantalum capacitors are integrated with high-density magnetic-core inductors to realize IVRs with module thickness around 0.7 mm. By incorporating high-permeability magnetic materials as the cores, the inductors achieved 20X improvement in inductance as compared to air-core inductors. The high inductance allows inductors to be designed with less number of windings, resulting in low component resistance of 5 mΩ. The integrated components have package-compatible terminals that are compatible with electrolytic plating process. The terminals allow them to be connected with low-resistance vias to further reduce parasitic losses and improve the power efficiency. Short PDNs and low-resistance interconnections and low-resistance components make the demonstrated IVRs ideal for high-power density computing applications with high efficiency low-impedance power delivery networks.**

Keywords-Magnetic composites; integrated buck converter inductors;

I. INTRODUCTION

Artificial Intelligence (AI) such as machine learning and deep learning requires the capability to process vast amounts of data in a short time. Tera Operations per Second (TOPs) is a common metric to evaluate the performance of AI computing. For servers with many AI chips, higher power densities is critical to supporting the ultra-high bandwidth. TOPs per watt (TOPs/W) extends the evaluation to capture both performance and efficiency. Today's voltage regulator architectures based on lateral or 2D power distribution will create major bottlenecks to power efficiency due to long power delivery networks (PDNs) [1]. As shown in Figure 1, current needs to travel a long distance from voltage regulator (VR) to power the IC chips. For high-power applications such as AI where tens or hundreds of current is necessary to power the AL chips, high DC resistance caused by the long PDNs inevitably lead to high DC loss and low power efficiency due to joule heating [2]. Additionally, the dissipated heat can also lead thermal challenges such as thermal warpage to the packages. PDNs with short interconnection length and low DC resistance are highly desired for voltage regulators.

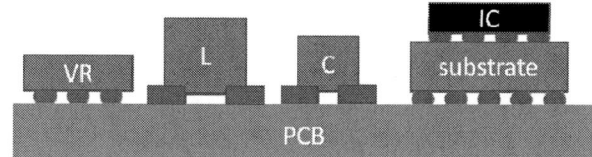

Figure 1: Cross section of a voltage regulator with 2D strucutre

In order to reduce the interconnection length, the passive components such as inductors (L) and capacitors (C) need to be stacked vertically and placed under the IC chips. Such structures have been demonstrated by TI and Murata. However, in the demonstrations, the components are still connected with solder which can lead to high resistance although the interconnection length is reduced [3, 4].

Voltage regulators with 3D integrated inductors (L) and capacitors (C) were demonstrated to reduce the interconnection length and resistance. Both the capacitors and inductors in this work have terminals that allow them to be electrically connected with electroplated low-resistance Cu vias instead of solder to further reduce the interconnection resistance. Figure 2 shows the cross section of the integrated voltage regulators (IVRs). The inductors are fabricated with panel-level process where the inductor layers can be used as the substrates to enable the integration of capacitors above the layers. The total thickness of the 3D LC network is ~0.7 mm which allows the integration of AI chips or other active components directly above them. With this 3D integration approach, the interconnection length is significantly reduced to provide lower parasitic losses and better power efficiency. Also, the ability to integrate a greater number of passive components due to savings in lateral area can be used to enhance the total power handling capability of the system.

978-1-7281-1500-9/19 $31.00 © 2019 IEEE

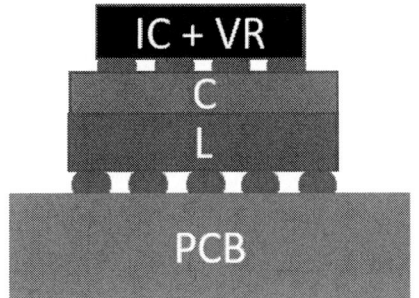

Figure 2: Cross section of a voltage regulator wit 3D structure

II. HIGH-DENSITY INDUCTOR

Inductors with high inductance density can be easily achieved with large number of windings. However, this inevitably leads large form factor and high DC resistance. A low resistance is critical for LC networks to mitigate joule heating and thermal stress problems. By incorporating advanced high-permeability magnetic composites as the cores, the inductors can achieve high-inductance density without need of large number of windings. Magnetic-core inductors with single winding turn were designed in this to achieve a low DC resistance of 5 mΩ. A panel-level process was developed to fabricate the designed inductors. The process flow is shown in Figure 3. A 5 μm ABF™ film was first laminated onto magnetic sheet. The film was used to promote adhesion of copper on the magnetic substrates. Copper windings were then formed onto the dielectric film. After copper formation, laser-drilled vias were formed in another magnetic sheet. A 5 μm dielectric film was then laminated onto the magnetic sheet with vias. The magnetic sheet was then laminated onto the copper windings to form magnetic-core inductor. The last step was to remove excess dielectric film above the copper windings.

Figure 4 (a) shows top view of the fabricated magnetic-core and air-core inductor arrays by using the panel-level process. Each inductor has a size of 10 mm^2 with a winding thickness of 100 μm. For the magnetic-core inductors, the copper windings are sandwiched between two 200 μm thick magnetic composites to improve the inductance. The total component thickness of magnetic-core inductors is around 460 μm. Figure 4(b) shows the measured inductance of both inductors by using impedance analyzer. As compared to air-core inductors with the same structure, the magnetic-core inductors show a 20x improvement in inductance.

This significant improvement is caused by the high-permeability magnetic composites. By engineering particles into flake morphology, the magnetic composites show a high in-plane permeability. To keep the high permeability upto 10 MHz, the flakes were fabricated with a thickness less than the skin depth to suppress eddy current losses. At 10 MHz, the skin depth of the magnetic flake is calculated as 8.2 μm by using (1).

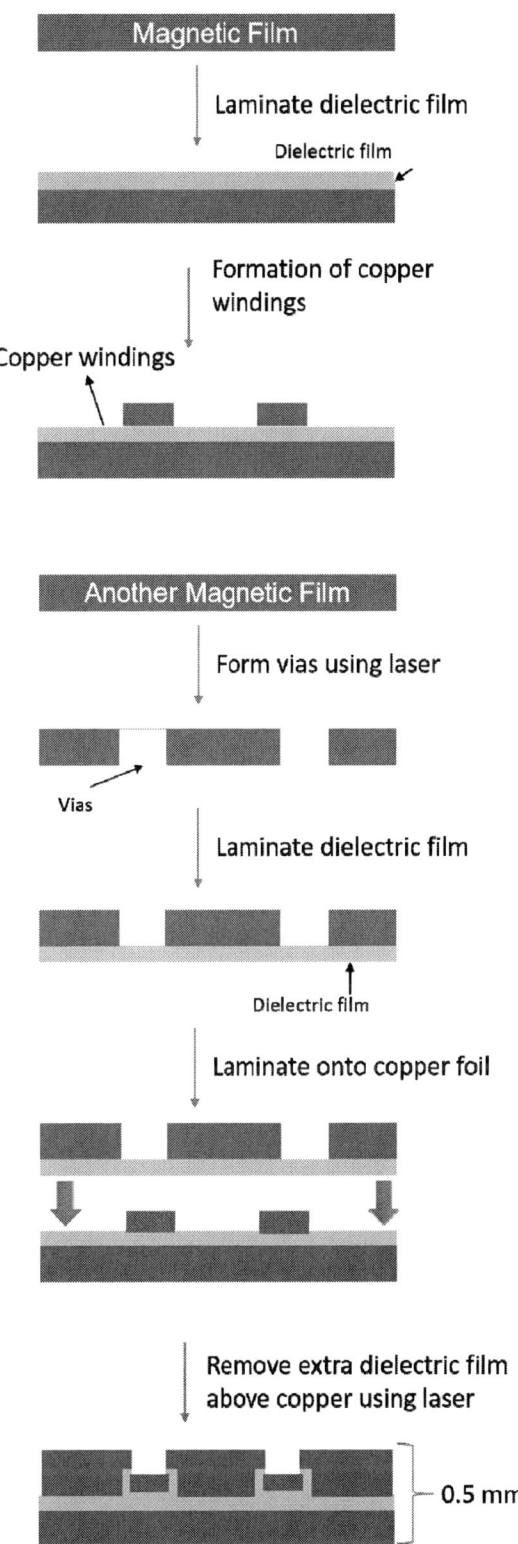

Figure 3: Fabrication process of magnetic-core inductors

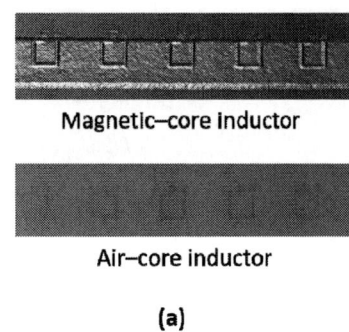

Magnetic–core inductor

Air–core inductor

(a)

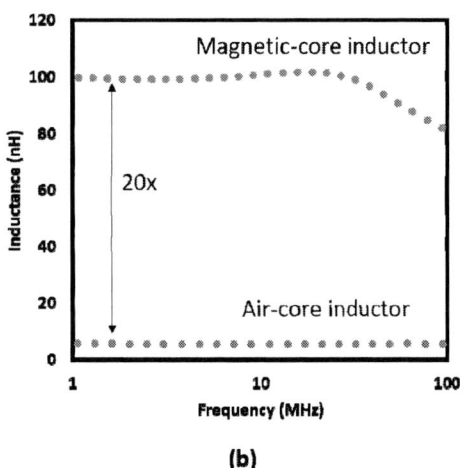

(b)

Figure 4: (a) Top views of magnetic-core and air-core inductors, (b) Measure inductance at different frequency, (c) Measured complex permeability of the magnetic composites

$$\delta = sqrt(\rho/(\pi*f*\mu_o*\mu_r)) \qquad (1)$$

where ρ is the resistivity of magnetic flake, f is the frequency, μ_o is the vacuum permeability and μ_r is the relative permeability of magnetic flake. Frequency-dependent permeability at different flake thickness is then calculated by using (2).

$$\mu = \mu_r*(2*\delta)/((1+i)*d)*tanh((1+i)*d)/(2*\delta)) \qquad (2)$$

where μ_r, δ and d are intrinsic permeability, skin depth and thickness of magnetic flake. Figure 5 shows the permeability spectrum of single magnetic flake at different thickness. For flake with 8.2 μm thickness, real permeability and imaginary permeability starts to drop beyond 10 MHz due to eddy current losses. By thinning the flake beyond the skin depth value, the flakes show improved real permeability and reduced imaginary permeability. This makes the flakes more suitable for 10 MHz applications.

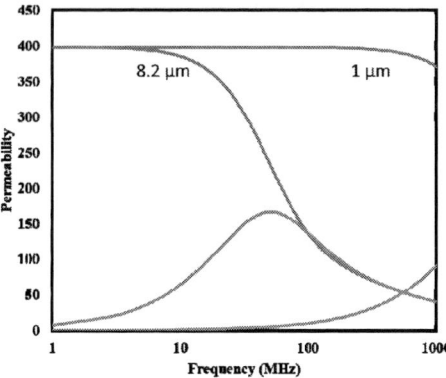

Figure 5: Calculated complex permeability of magnetic flakes with 8.2 μm and 1 μm thickness

III. LC INTEGRATION

With the magnetic-core inductors as substrates, high-density capacitors with small thickness profiles were then integrated onto the inductor substrates. In-house fabricated thin-film tantalum capacitors were used in the 3D LC network. The tantalum capacitors came with Au terminals which allow them to be electrically connected with inductors by using electroplated low-resistance Cu vias. The tantalum capacitors have a density of ~1 μF/mm^2 at a thickness of only 100 μm. Figure 6 shows the top view of fabricated Ta capacitors and measured capacitance at different frequencies.

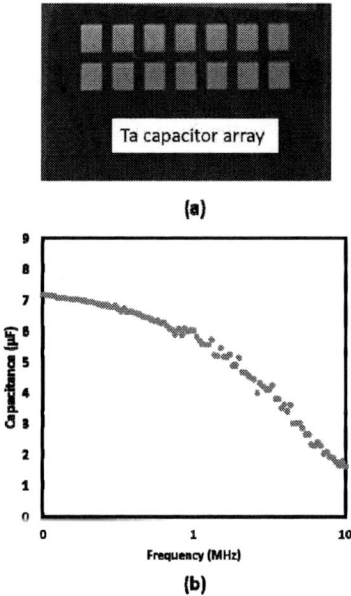

Figure 6: a) Top view of capacitor array consisting of 5 mm2 fabricated Ta capacitors and (b) Measured capacitance at different frequencies

The tantalum capacitors with Au terminal exhibit ~50 mΩ equivalent series resistance (ESR), thus helping to reduce power losses through the network and improving the AC filtering capability. As compared with MLCCs, the tantalum capacitors provide the benefit of showing highly

978-1-7281-1500-9/19 $31.00 © 2019 IEEE

stable capacitance over a wide range of temperatures and DC bias, especially at higher frequencies [5, 6]. This is due to the fact that the MLCCs rely on their ferroelectricity to obtain high-permittivity dielectrics, and thus high capacitance density, while tantalum capacitors rely on their high surface area. The ferroelectric dielectric material loses polarizability at increased electric fields or changing temperatures, resulting in unstable capacitance values [7]. The expected volumetric capacitance density of the ultra-thin tantalum capacitor technology is included in Figure 7. The tantalum capacitors show stable capacitance at various temperature and voltages.

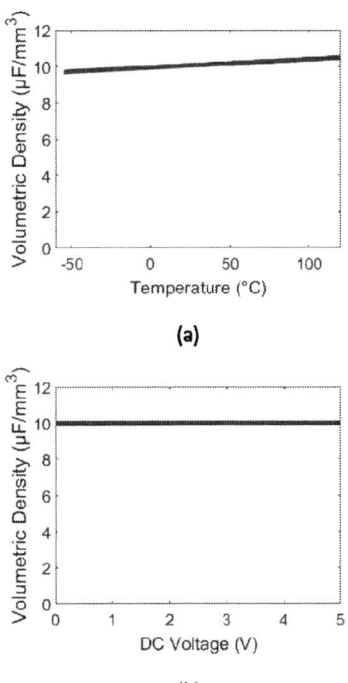

(a)

(b)

Figure 7: Effect of (a) temperature and (b) DC voltage on volumetric density of thin-film tantalum capacitors.

Thus, the type of capacitor technology used should be carefully selected based on system design needs. If the LC substrate is going to be used for a high-bandwidth CPU, then it makes sense to use the temperature-stable tantalum capacitors to maintain high capacitance despite localized heating caused by the silicon chip. On the other hand, if system efficiency is most critical then it may make more sense to use the low-impedance MLCCs.

An innovative process was developed to integrate the inductors and capacitors to form a 3D LC network. The process flow is shown in Figure 8. A die attach film was first laminated onto the inductor substrates. Capacitors were then picked and placed onto inductor substrates. By curing the die attach film, the position of capacitors are fixed. The capacitors were then embedded by molding compound to achieve a planar surface where active components can be assembled onto the surface. Laser-drilled vias are then

formed onto inductor and capacitor terminals by using UV laser. When fully filled with electroplated Cu, the vias provide low DC resistance and parasitic losses. Figure 9 and Figure 10 show the cross sections of integrated tantalum capacitors with air-core inductors and magnetic-core inductors. The two components are connected with Cu vias and re-distribution layer (RDL) to form LC networks. The total module thickness is 330 μm and 680 μm for LC network with air-core inductors and magnetic-core inductors respectively. For the LC network with magnetic-core inductors, the via depth is 330 μm which is 2X deeper than the LC networks with air-core inductors. The deep vias pose electro-plating challenges. As shown in Figure 10, the vias are not conformal vias. The electroplating process needs to be modified to address this challenge.

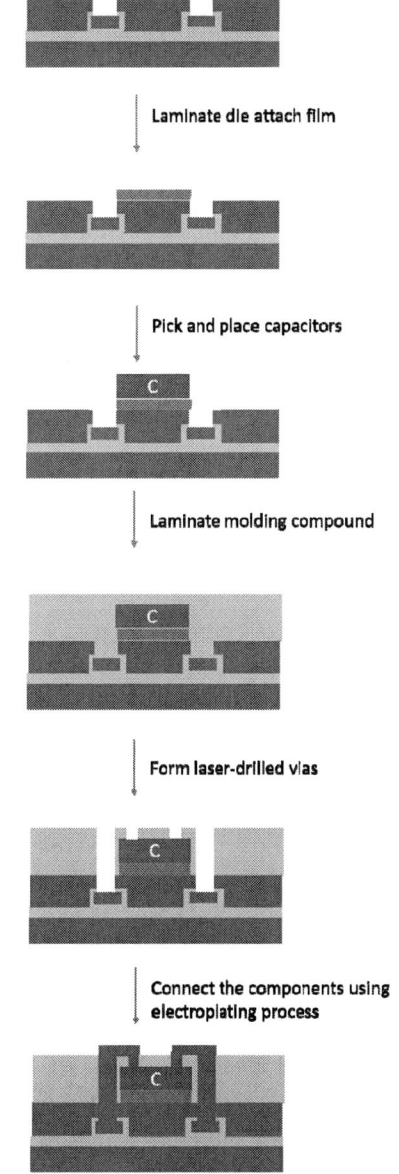

Figure 8: Integration process of 3D LC network

978-1-7281-1500-9/19 $31.00 © 2019 IEEE   1303

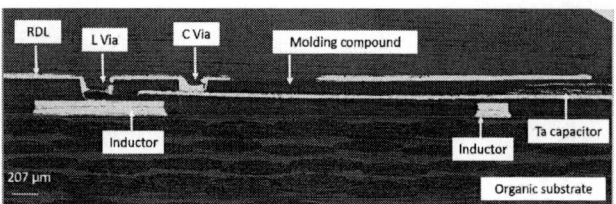

Figure 9: Cross section of substrate-integrated air-core inductors and Ta capacitors.

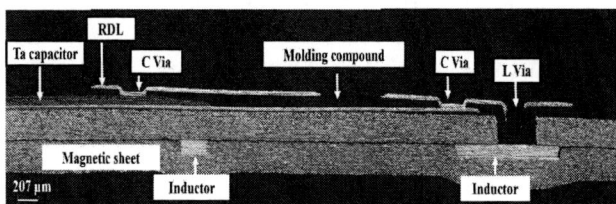

Figure 10: Cross section of substrate-integrated magnetic-core inductors and Ta capacitors

As illustrated in (3), hydrogen gas is generated during eletroless process. The hydrogen gas can be trapped inside the deep vias due surface tension [8]. Thus, copper is only deposited on top of vias and no copper is deposited on bottom.

$$Cu^{2+} + 2HCHO + 4OH^- = Cu + 2HCOO^- + H_2 + H_2O \quad (3)$$

By adding vibration during electroplating process, the trapped hydrogen gas can be eliminated resulting in copper inside the vias as shown in Figure 11.

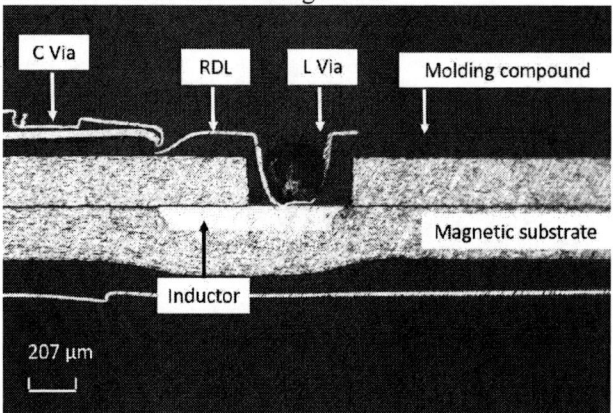

Figure 11: Cross section of magnetic-core inductor with copper inside the via

IV. LC CHARACTERIZATION

Electrical measurements were performed to the 3D LC networks with tantalum capacitors. Inductance and capacitance at 1 MHz were measured before and after the integration process by using LCR meter. The results are tabulated in Table 1. The performance variation for inductors and capacitors is around 10%. The small variation indicates that both components were not damaged during integration process. The increase in inductance after

integration resulted from the Cu vias and RDLs which act as part of copper windings to rise inductance.

Table 1: Measured inductance and capacitance at 1 MHz before and after integration.

	Before	after	Variation
Magnetic-core L	100 nH	109 nH	+11%
Ta capacitor	6 μF	5.5 μF	-8.3%

V. SUMMARY

Substrate-integrated 3D LC networks with short interconnection length and low interconnection resistance are developed and demonstrated improve the power efficiency of AI computing. Key enablers for this 3D LC networks are miniaturized embedded thin-film high-density inductors and capacitors.

Thin-film tantalum capacitors with high capacitance were used. The tantalum capacitors feature a capacitance density of ~1 μF/mm² at a thickness of only 100 μm. By incorporating high-permeability magnetic materials as the cores, the high-density inductors achieved 20x improvement in inductance as compared to air-core inductors. This improvement in inductance allows the inductors to be designed with less number of windings, resulting in low DC resistance of 5 mΩ. When the 3D LC networks are used in high-power applications, the low DC resistance mitigates joule heating and thermal stress problems in the LC networks.

A panel-level substrate-compatible process was developed to integrate the high-density inductors with high-density thin-film tantalum capacitors to form a 3D LC network. With the 3D integration approach, the LC networks are placed close to the AI chips to reduce interconnection length and improve the power efficiency. The fabricated inductors and capacitors are also compatible with electrolytic plating process, which allows them to be connected with low-resistance Cu vias to further improve the power efficiency.

Electrical measurements were performed with the fabricated 3D LC networks. Small variation in inductance and capacitance before and after integration indicates that the developed integration process does not degrade the component performance.

ACKNOWLEDGMENT

The research is funded by industry consortium program at Georgia Tech Packaging Research Center. The authors would like to give their sincere thanks to all the consortium member companies in supporting this research

REFERENCES

[1] C.-T. Wang, C.-L. Chen, J.-S. Hsieh, C.-H. Tsai, V. Chang, and D. Yu, Foundry WLSI Technology for Power Management System Integration, 2016.

[2] S. Naffziger, "Integrated Power Conversion Strategies across Laptop, Server and Graphics Products," in Power Supply On Chip, Madrid, 2016.

[3] "MicroSiP Power Modules," ed.

[4] Murata LXFC series. Available: http://www.murata.com/en-us/about/newsroom/news/product/capacitor/2009/1005b

[5] M. M. Strömme, G. A. Niklasson, K. Forsgren, and A. Hårsta, "A frequency response and transient current study of beta-Ta2O5: Methods of estimating the dielectric constant, direct current conductivity, and ion mobility," Journal of Applied Physics, vol. 85, 1999.

[6] S. Zedníček, J. Petržílek, P. Vanšura, M. Weaver, and C. Reynolds. A Basic Guide to AVX Conductive Polymer Capacitors. v8.5.

[7] Murata, "Chip Monolithic Ceramic Capacitor Electrical Characteristics Data - Part No. GRU152D80G105ME02," 2015.

[8] T. Bernhard, F. Brüning, R. Brüning, T. Sharma, D. Brown, S. Zarwell, et al., "The Impact of Hydrogen Gas Evolution on Blister Formation in Electroless Cu Films," Journal of Microelectronics and Electronic Packaging, vol. 12, pp. 86-91, 2015.

A Novel Panel Level Double Side Embedded Package for Small Size Power Devices

Kunpeng Ding[1], Zhichao Wu[2], Mian Huang[1], Bowei Zhang[3], Jian Cai[2*]

[1]Shenzhen Siptory Technologies Co., Ltd, Shenzhen 518057, China
[2]Department of Microelectronics and Nanoelectronics, Tsinghua University,
Beijing 100084, China
[3]Wuxi Sky Chip Interconnection Technology Co., Ltd, Wuxi 214000, China
*Email: jamescai@tsinghua.edu.cn

Abstract—For small size devices, fan-out chip-embedded technology has become an attractive and holistic packaging solutions with various advantages in comparison to conventional and mature technology. And high efficient and high reliability of the embedded package for small-size power devices has become a hot topic in industry. In this paper, a novel panel level double side-embedded packaged process (DS-Embedded) for small size power devices has been presented. The entire process can be performed on both sides of the carrier. In order to reduce cost and time-to-market, 2D and 3D finite element model were used to get a reliable package structure. The result shows that the thermomechanical fatigue life of the model after parameter optimization is much larger than the actual use requirements which was in good agreement with the experimental result. The actual reliability tests such as vibration, mechanical shock, shear test also prove the feasibility of the product. However, the failure problem caused by Ag corrosion and migration in the H3TRB experiment still needs further improvement.

Keywords-power device; DS-Embedded (Double Side-Embedded); thermal shock; reliability; Fatigue life

I. INTRODUCTION

The trend of electronic products is high-density integration and miniaturization. As Moore's Law becomes closer to the limit, electronic package solutions have increasingly become a key technology to implement these goals. This makes advanced package such as fan out technology play more and more important role in More than Moore era. Up to now, different fan-out technologies, which are favored by semiconductor fabs and OSATs, have been developed for different applications in the industry, institutes, and universities [1-2].

For small size power devices, wire bonding or flip chip assembly would be typical technologies for conventional and mature package solutions. Since last few years, embedding of semiconductor chips which is one of the fan-out technology has become an attractive and holistic packaging solutions for the power device [3-4]. This technology can be applied for most of the small-size power devices, including Schottky diode, FET, ESD chip, MOSFET, etc. Compared to wire-bonding or flip chip package solutions, embedded die technology has the advantages of faster switching speeds, lower cost, better heat dissipation, smaller form factor, lighter weight and simple process. And high efficient and high process yield of the embedded package process for small-size power devices has become a hot topic in industry.

In this paper, a novel panel level double side-embedded packaged (DS-Embedded) process for small size power devices has been presented and the packages were built. This panel level embedded technology combines the individual process of PCB and package. The carrier is a double-sided copper clad laminate which means that package can be done in the both sides of carrier. This novel process can not only reduce the residual stress during the process, but also greatly increase the production efficiency. At the same time, the packaging process can be adjusted according to specific interconnection requirements, such as the number of pins or the number of dies, and more package solutions suitable for the various small size power devices can be designed. This package technology has already been applied to the commercial production of small-sized devices.

However, for silicon-based power chip packages, there are many thermal mechanical reliability issues in embedded structure. The power chip packages long-term exposure to high ambient temperatures and operating temperatures may result in failures such as chip cracking, separation of dielectric and copper layers, cracking of metal traces due to the mismatch of coefficient of thermal expansion (CTE) among the constituents. These potential defects can lead to serious packaging quality problems and product failure. Therefore, it is necessary to study the thermomechanical reliability of the embedded package in the early stages of the product. In recent years, many research institutions around the world have used finite element methods to study the reliability of the embedded chip technology [5-7].

In order to reduce cost and time-to-market, in this paper, finite element simulation combined with actual thermal shock reliability experiments was used to identify the critical location of failure, optimize parameters of this product. And the effects of parameters such as thickness of the die, thickness of the Cu layer, thickness of Ag metallization layer and thickness of solder bonding layer on the thermo-mechanical reliability of the product were studied in the course of the technology development. The paper is organized as follows, firstly, the process flow of the embedded die technology is briefly descripted. Next, a 2D model was used to optimize the parameters of the package structure and the fatigue life was calculated with the 3D model. The predicted results of the simulation were in good

978-1-7281-1500-9/19 $31.00 © 2019 IEEE

agreement with the experimental results. Then, the optimized package of small-sized power device were performed for various reliability test verifications, such as vibration, mechanical shock, shear test and the 1000-hour reliability verification of H3TRB. The test results showed that the package has good reliability. However, several failures still occurred when the H3TRB test time exceeds 1000h. And it needs further improvement in subsequent research.

II. PACKAGING AND ASSEMBLY PROCESSES

The small size power device used in this paper was a bidirectional common capacitor electrostatic discharge protection device(ESD) produced by advanced silicon circuit technology with 3μm Ag pad on the top side and （0.5-2）μm Ag metal layer on the back side of the die. Figure 1 shows the top view of the chip. The chip's low clamping voltage and fast response speed make it ideal for embedded packaging technology.

Figure 1.   Top view of the chip with Ag pad

Figure 2 shows the whole process of panel level double side-embedded package. It starts with a carrier which has a panel size of 400mmx500mm. The carrier is a double-sided copper clad laminate, creatively using ultra-thin copper foil to bond with conventional PCB materials to develop a new package carrier. This package carrier has a double-sided copper foil and the copper foil can be separated after the packaging and assembly process so that the carrier can be reused in manufacture.

And conventional PCB processes have been used to realize the pattern as a base for chip placement. Then a layer of solder with a thickness of about （5-10）μm is plated on the copper base as a bonding layer between the silicon chip and the copper foil.

And then, the packaging processes are list as following: a 200μm×200μm silicon-based power chip is picked and placed on the carrier with an accuracy of ±25μm by a dedicated die bonder, after that the sheet molding material is processed through a large-sized plastic sealing device to complete the plastic molding of the chip.

The electrical interconnection in the package is achieved by PCB processes. Laser drilling, copper plating, line patterning and etching are used to complete the fabrication of the blind vias and traces. The finished vias on the top of the chip and the copper foil are 75μm and 200 μm, respectively. And the thickness of electroplated copper wall is about

10μm.The entire process can be performed on both sides of the carrier.

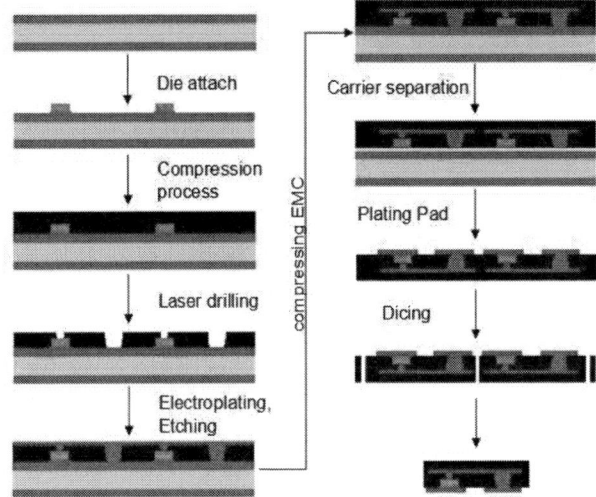

Figure 2.   Process flow of panel level double side embedded package

After the electrical interconnection process has been completed, PP resin and selected EMC material are laminated to complete the final via filling and molding. Finally, after completing the carrier plate separation, the final product processing and production is realized by the traditional back end of packaging processes such as singulation, testing, marking and packing. Figure 3 shows the completed package samples and a picture of the X-section which showing the electrical interconnection structure inside the package.

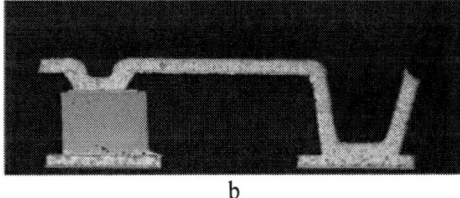

Figure 3.   The view of (a) packaged product and (b) cross section

978-1-7281-1500-9/19 $31.00 © 2019 IEEE

Finite element simulation was used to assist the design of the package and carry out the reliability theory analysis. And finished embedded packages require reliability tests to verify the rationality of the simulation by comparing the simulation results with the test results. Among them, thermomechanical reliability analysis which is based on the thermal cycle or thermal shock test is the most common in extensive studies.

III. SIMULATION MODELS

A. Model Structure

Finite element modeling was employed to simulate the stresses and strains of the package during thermal loading. This simulation model was based on the actual embedded package structure in the thermal shock test experiment. The models were simplified to improve simulation efficiency, including copper foil, copper pad, chip, plated solder for die bonding, metallized Cu vias, PP resin, and around these structures was epoxy molding compound (EMC). In order to get a more reliable product, the effects of the package structure parameters such as thickness of the die, the Cu layer, Ag metallization layer on the back of the chip and the solder bonding layer on the thermo-mechanical reliability of the product were studied in the course of the technology development. Figure 4 and 5 show the finite element model used for simulation. In order to obtain more accurate simulation results and reduce calculation time, the 2D model shown in Figure 4 is used for the optimization of the above parameters, and the 3D model shown in Figure 5 is used for the analysis of the failure hotspot and the calculation of the failure life. In order to display the internal structure of the package more intuitively, the EMC was hidden in Figure 5.

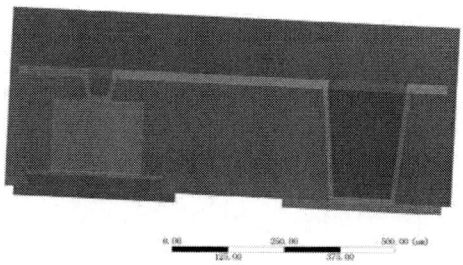

Figure 4. A view of 2D model

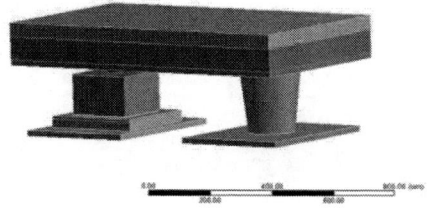

Figure 5. A view of 3D model

B. Loading Conditions

The thermal mechanical reliability test procedure and conditions were based on the JEDEC standard. According to thermal shock standard JESD22-A106B-01[8], test condition C was chosen as a reference for thermal mechanical reliability testing and simulation models, and the final test profile was a temperature range from -55℃ to +125 ℃ with a dwell time of 15 minutes and a conversion time of less than 10 seconds. Only one cycle begin from room temperature 25℃ was selected in the simulated temperature load. The temperature curve is shown in Figure 6 below. At the same time, in order to match the thermo-mechanical reliability test of the component-level package structure well, the boundary conditions of the model used in the simulation are as following. For the nodes in the bottom surface of the pad under the copper foils, the degree of freedom in the vertical direction was set to 0. It was considered that the pad is always in close contact with the test plane in the experiment.

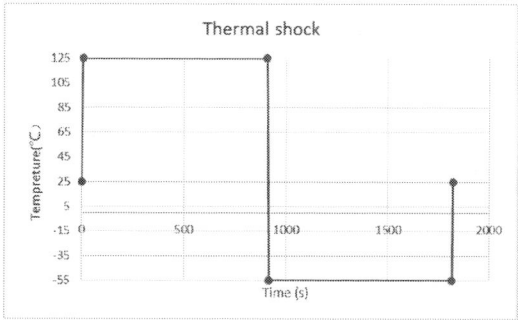

Figure 6. The view of temperature load curve

C. Material Characterization

The materials used in the model include Si, solder, Ag, electroplated copper, PP resin and EMC. In reliability test, the failure of embedded die package can be judged by means of electrical measurement, and this depends on whether there is local damage on the electrical interconnection structure. Therefore, in the processing of the simulation model, the calculation model of materials were simplified. Si power die, the molding material PP resin and EMC were treated as a linear elastic model, and the elastic modulus of EMC was temperature dependent. For the metal interconnect material solder, Ag, Cu were treated as an elastoplastic model. The specific material parameters are shown in Table 1. The materials' properties were derived from material suppliers and references in related research fields [9-10].

TABLE I. MATERIAL PROPERTIES

Material	Elastic Modulus (GPa)	CTE (ppm/℃)	Poisson's ratio	Yield (MPa)	Tangent modulus (MPa)
Cu	119	16.5	0.326	175	4500

978-1-7281-1500-9/19 $31.00 © 2019 IEEE

Material	Elastic Modulus (GPa)	CTE (ppm/℃)	Poisson's ratio	Yield (MPa)	Tangent modulus (MPa)
Ag	73.2	19.5	0.32	287	412.7
Solder	54.4	23.5	0.33	160	600
Die	190	4.1	0.3	-	-
PP	0.44	40 (T<170℃) 60 (T>170℃)	0.3	-	-
EMC	0.95 (0℃) 0.9 (50℃) 0.75 (100℃) 0.55 (150℃)	40 (T<140℃) 65 (T<140℃)	0.3	-	-

IV. RESULTS AND DISCUSSION

A. Failure Criteria

Figure 7 shows the von mise stress contour of the whole structure when cooling down from the maximum temperature 125℃ to the minimum temperature -55℃. From the results of the 3D model, it can be concluded that the thermal stress at the corner of Si power chip is the largest, which may cause cracking of the die. The risk of this failure may also exist in the actual package assembly process. And the von mise stress of the metal interconnect structure next to the chip such as the Ag pad on the chip and the solder plating layer is also very large, which may cause plastic deformation of the electrical interconnect layer. Figure 8 shows the plastic strain distribution of this package, it can be found that the plastic strain value at solder plating layer is the largest.

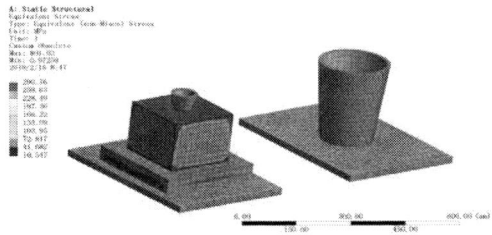

Figure 7. The von mise stress contour of 3D model

Figure 8. The contour of plastic strain distribution

B. Structral Parameter Optimization

Although the simulation results of the 3D model are more accurate, the process of the simulation also takes too

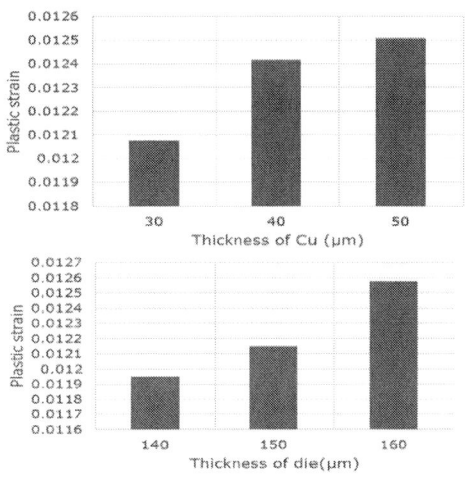

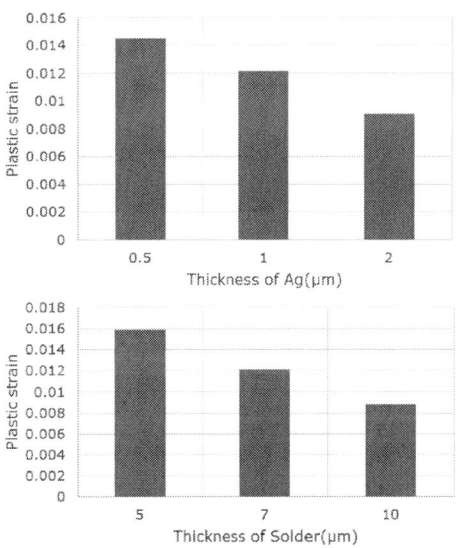

Figure 9. Influence of different structural parameters on plastic strain

much time. In order to improve the simulation efficiency, in this paper, simplified 2D model was used to study the influence of structural parameters on stress-strain behavior. The four parameters of the thickness of the metallized Ag layer on the back side of the chip, the thickness of the copper foil, the thickness of the chip, and the thickness of the solder layer were optimized. Figure 9 shows the simulation results of four factors, respectively. It is obviously that these four geometric factors have a significant influence on the thermomechanical plastic strain of the package. And with the decrease of Cu, die thickness and the increase of Ag and solder thickness, the plastic strain of the package undergoes a thermal shock was

978-1-7281-1500-9/19 $31.00 © 2019 IEEE

reduced. This means that the thermomechanical reliability of the packaged product can be well improved by adjusting the structural parameters.

However, excessively thick Ag layer may be easier to migrate and lead to failure, this will be mentioned in the fifth part. So in this paper, the Cu thickness of 30μm, Ag thickness of 2μm, silicon-based power chip thickness of 140μm, and the solder bonding layer thickness of 10μm were used to simulation the stress and strain of this package after thermal shock as a parameter-optimized structure. And 3D simulation model was used to get more accurate calculation results. The plastic strain of this optimized structure is 0.0117 as shown in Figure 10.

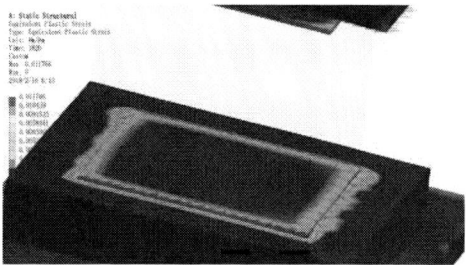

Figure 10. The plastic strain contour of 3D model

C. Fatigue Life

Finite element simulation can effectively predict the fatigue life of the package. And in the numerical analysis, the calculation of fatigue life is usually performed with the coffin-Mason relationship [11], as shown in (1).

$$N=(1/2)(\Delta\gamma / 2\varepsilon)^{1/\theta} \qquad (1)$$

ε represent the fatigue ductility coefficient, and for tough materials, $2\varepsilon= 0.65$, N represent the mean cycles to failure, $\Delta\gamma$ represent the equivalent inelastic shear strain range in each cycle, $\Delta\gamma =\sqrt{3}\Delta\alpha$, $\Delta\alpha$ represent the plastic strain range in each cycle, θ represent fatigue ductility exponent which is related to the temperature . The specific value of θ can be calculated by (2)

$$\theta= -0.442-6\times10^{-4}T_m+ 1.74\times10^{-2}\ln(1+f) \qquad (2)$$

Where:

T_m represent the average temperature of thermal cycle;
f represent cycle frequency, 1<f< 1000 cycles/day

For the experimental conditions in this paper, the T_{max} is 125℃ and the T_{min} is -55℃, T_m=35℃ can be deduced. And f=48, take them into (2) and calculate：θ = -0.395

The plastic strain of the solder-plated layer after one cycle of thermal shock is about 0.0117, substituted into (1), it can be inferred that the thermal shock fatigue life of the panel level double side-embedded package is 3251. The result is much greater than 1000 cycles, the fatigue life can meet the product's demand in the real market which means that such a structure has good thermo-mechanical reliability. And the reliability of this product in other aspects still needs to be verified by experiments.

V. RELIABILITY TEST

In order to obtain products that are more suitable for practical application reliability. It is necessary to perform more reliability tests on the parameter-optimized package. During the development of this product, the reliability under thermal mechanical test, mechanical shock and High Humidity High Temperature Reverse Bias(H3TRB) were tested. The thermal mechanical test was carried out with a thermal shock test chamber. Test conditions have been introduced in the previous section, while the H3TRB test was carried out in a Programmable temperature & humidity chamber. The experimental conditions were 5V bias and 85 ° C 85% RH. And mechanical reliability test was based on the drop test, the cycle shall be performed 3 times in each of orientation(X,Y,Z). The peak acceleration of the test is 1500g, the duration of pulse is 0.5ms.

The results after the test showed that there was no any failure after 1000 times thermal shocks. This result was consistent with the results obtained in the simulation. As for mechanical test, after the embedded package drop 3 times in each of orientation(X,Y,Z), there was also no failure of any product. However, after the H3TRB test, it was found through electrical measurement that there would be a certain proportion of package failure.

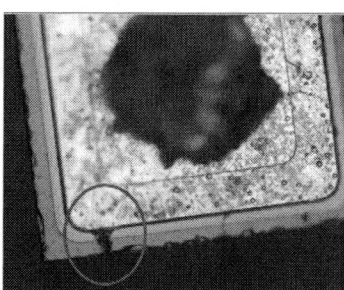

Figure 11. Optical microscope photo of the failure chip

Figure 11 shows an optical microscope photo of the failure chip. It can be found that there is a clear agglomerate at the edge of the metal pad.

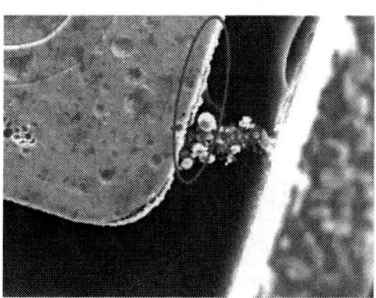

Figure 12. SEM view of the failure chip

The SEM picture of the failure package is shown in Figure 12. Through the result of energy spectrum, it is found that the area outside the pad as shown has Ag content and therefore the failure was probably due to Ag corrosion and migration in the pad. Also the obvious cracks can be observed at the edge of the Ag pad.

Metal corrosion and migration of chip pads in the high humidity high temperature reverse bias test has been mentioned in some device reliability studies [12-13], which becomes a reliability problem that must be paid attention to. The product in this paper used a chip with an Ag pad, when exposed to a bias of 5v, an electric field is generated in the electrical interconnect structure to power the migration of Ag ions. If there is delamination between the molding material and the chip, water vapor in an environment of 85% humidity will enter into the package through the gap which may corrode the silver pad to produce more ions. At the same time, the existence of the temperature field will accelerate the migration process and eventually lead to product failure.

In order to prevent this kind of phenomenon, one or more of the three factors of power, package hermeticity and environment should be changed. However, the power chip must have an electric field during normal operation and long-term exposure to high ambient temperatures and operating temperatures. Therefore, in this paper, the selected EMC material can be closely attached to internal interconnect structure which provides good hermeticity.

In addition, the risk of silver migration can also be reduced by reducing the thickness of the Ag layer on the top surface of the chip and reducing the total amount of silver metal. The fatigue life of this product has been improved by the new molding material A and reducing the thickness of the silver layer on the top side of the chip. It is increased from 500 hours to 1000 hours without failure at H3TRB test. However, 1000 hours later, there will still be sporadic failures in the package.

At the same time, due to the complex multi-physics effects of H3TRB test, simulation theory analysis and corresponding quantitative analysis have certain difficulties, which were not discussed in this paper .Subsequent research will focus on trying to study the package performance of this test through simulation models and further improve reliability of this product.

VI.   CONCLUSION

In this paper, a novel panel level double side-embedded packaged process (DS-Embedded) for small size power devices has been presented and the demonstrators were built. The entire process of this product which can be performed on both sides of the carrier can not only reduce the residual stress during the process, but also greatly increase the production efficiency

2D and 3D Finite element model combined with actual thermal mechanical reliability experiments was used to identify the critical location of failure, optimize parameters

of this product. And the four parameters of the thickness of the metallized Ag layer on the back side of the chip, the thickness of the copper foil, the thickness of the chip, and the thickness of the solder layer are optimized. The result showed that with the decrease of Cu, die thickness and the increase of Ag and Sn thickness, the plastic strain of the package undergoes a thermal shock reduction.

The parameter-optimized package structure successfully passed the various reliability tests, such as thermal shock test, drop test, 1000h H3TRB test. However, there will still be sporadic failures in the package after 1000h test due to the corrosion and migration of Ag. Subsequent research is needed to further improve the reliability of the test.

Acknowledgment

The authors would like to thank Shennan Circuits Company Limited, and the project team for technical support in technology development.

References

[1] Mori, Kentaro, et al. "Reliability of thin seamless package with embedded high-pin-count LSI chip." IEEE 60th Electronic Components and Technology Conference (ECTC), 2010, pp. 36-39.

[2] Yap, Daniel, et al. "Reliability of eWLB (embedded wafer level BGA) for Automotive Radar Applications." IEEE 67th Electronic Components and Technology Conference (ECTC), 2017, pp. 1473-1479.

[3] Han, Younggun, et al. "Process Feasibility and Reliability Performance of Fine Pitch Si Bare Chip Embedded in Through Cavity of Substrate Core." IEEE Transactions on Components, Packaging and Manufacturing Technology, vol. 5, (4), pp. 551-561, 2015.

[4] Lee, Doohwan, et al. "Fabrication of die embedded substrate and mechanical stress evaluation at active area of the embedded die." IEEE 10th Electronics Packaging Technology Conference(EPTC), 2008, pp. 224-229.

[5] Lu, Xiuzhen, et al. "Reliability analysis of embedded chip technique with design of experiment methods." 2005 International Symposium on Electronics Materials and Packaging. IEEE, 2005, pp. 43-49.

[6] Yang, Daoguo, et al. "Reliability modeling on a MOSFET power package based on embedded die technology." Microelectronics Reliability, vol. 50, (7), pp. 923-927, 2010.

[7] Liu, Yumin, et al. "Modeling characterization and reliability analysis of a power system in package." IEEE 61st Electronic Components and Technology Conference (ECTC), 2011, pp.731-739.

[8] JEDEC Standard JESD22-A106B-01, Temperature shock (Nov 2016).

[9] Wang, Fuliang, Yun Chen, and Lei Han. "Modeling and experimental study of the kink formation process in wire bonding." IEEE Transactions on Semiconductor Manufacturing, vol. 27, (1), pp. 51-59, 2014.

[10] Gao, Xiang, et al. "Thermo-mechanical reliability of copper-filled and polymer-filled through silicon vias in 3D interconnects." IEEE 63rd Electronic Components and Technology Conference(ECTC), 2013, pp. 2132-2137.

[11] Engelmaier, Werner. "Fatigue life of leadless chip carrier solder joints during power cycling." IEEE transactions on components, hybrids, and manufacturing technology, vol. 6, (3), pp. 232-237, 1983.

[12] Papadopoulos, Charalampos, et al. "The influence of humidity on the high voltage blocking reliability of power IGBT modules and means of protection."Microelectronics Reliability, vol. 88, pp.470-475, 2018.

[13] Barbieri, Thomas, et al. "Reliability Testing of SiC JBS Diodes for Harsh Environment Operation." PCIM Europe 2018; International Exhibition and Conference for Power Electronics, Intelligent Motion, Renewable Energy and Energy Management. VDE, 2018, pp. 1-5.

Chiplet Micro-Assembly Printer

Bradley B. Rupp, Anne Plochowietz, Lara S. Crawford, Matthew Shreve, Sourobh Raychaudhuri, Sergey Butylkov, Yunda Wang, Ping Mei, Qian Wang, Jamie Kalb, Yu Wang, Eugene M. Chow and Jeng Ping Lu
PARC, 3333 Coyote Hill Road, Palo Alto, CA 94304 USA
email: echow@parc.com

Abstract—A deterministic, directed, parallel electrostatic assembly and transfer process is being developed to arrange chips for electronics applications. Singulated chips 10 um – 200 um in size and initially in solution, are sorted, transported and oriented to form programmed patterns. The arrangements are then transferred to a final substrate with contact stamping or an electrostatic roller belt. Demonstrations achieved include automated parallel assembly, micrometer scale registration, heterogeneous integration, inch scale outputs, and basic functional circuits. The eventual goal is the ability to integrate millions of chiplets into systems with fine control over large areas to enable next generation electronic systems.

Keywords - assembly; printed electronics; chiplets; transfer; heterogeneous integration, flexible electronics, microLED

I. INTRODUCTION

Microelectronic devices are generally fabricated through intimate integration of thin films from highly specialized materials and processes in large facilities that are extremely expensive to develop and manufacture. This approach does not scale to different materials, processes or form factors, so assembly of singulated integrated circuit chips is the general approach used for building systems. Advanced assembly methods are needed to enable future electronic systems which will require increasingly more heterogeneous, fine-scale integration, more parts and complexity, more customization, and more large area capability. Example applications include flexible LED displays, large area sensor systems and custom, heterogeneous electronic and optical systems.

Current chip assembly approaches have different limitations [1-4]. Robotic pick and place tools have low throughput and cannot handle small chiplets. Parallel pick and place transfer processes such as with contact rubber stamps have fine registration, but need specialized chip fabrication with anchors, is not programmable, cannot perform heterogeneous integration in one step, and does not scale to continuous processes. Laser release processes have higher throughput potential, but are still ultimately limited by slow mechanical scanning, do not handle very small parts, and cannot vary orientation. Fluidic assembly approaches have the highest throughput potential but are generally not deterministic or controllable. We are using programmable directed electrostatic assembly of chips to attempt to address these limitations.

We aim to develop the capability to rapidly sort, place, and orient micro chiplets into custom positions and transfer them to a final substrate [5-7]. Laser printers use static charge templates to assemble billions of toner particles into custom patterns over meter scales, showing that electrostatics is a powerful method for manipulating small objects. To achieve deterministic electrostatic control, we use an active-matrix addressable two-dimensional array of electrodes as actuators to manipulate the chips. We previously showed open loop control methods for transporting and orienting chips, which could be used to perform the majority of the coarse assembly for large ensembles of chips. Results using closed loop feedback is the focus here, where chip positions and orientations are tracked and then the field pattern is altered to direct the chips to the target location. A human can be in the loop by controlling the chip with a computer mouse, or fully computer automated parallel assembly can be used.

In this paper we first describe in Section II the chiplet printer experimental setup for performing the assembly and transferring the arranged chips to a final substrate. In Section III we demonstrate a continuous feed chip assembly process with a 50 um pitch actuator array, a key step towards future large area applications. In Section IV results for small chips (<50 um) are presented, including heterogeneous and micrometer scale registration, using a 10 um pitch actuator array. Finally, functional electronic circuits fabricated with the microchip printer are summarized in Section V, showing basic LED circuits and diode characterization, suggesting the process does not alter the electrical properties of the chips.

II. EXPERIMENTAL SETUP

We have built protype systems to develop the basic processes for a microchip assembly printer. Singulated semiconductor chips in a dielectric fluid are used as "ink" in the printer. The chips can be pre-fabricated integrated circuit chips from standard foundries, as no special films or features are required on the chips for future assembly, though the ability to see the chip orientation is generally needed. Lithographically patterned charge patterns on the chips can be used to identify and guide chips, but is not needed or used in this work. Any compatible singulation process can be used. Silicon, glass, polymer, gallium arsenide, gallium nitride, and pure metal micro-objects have all been used in the system, as the dielectrophoretic and electrophoretic forces being used are very general.

A key component of the system (Figure 1) is the assembly region, where an electrode array generates dynamic electric field patterns. An active matrix addressable array, similar to the backplane of a display, was built by fabricating a thin film photo-switch array, and then addressing the array with a commercial video projector (Figure 2). A fully electronic active matrix could also be used for future systems, but the optically addressed versions are expedient for development as the video projector electronics enable addressing for different array pitches simply by changing the magnification.

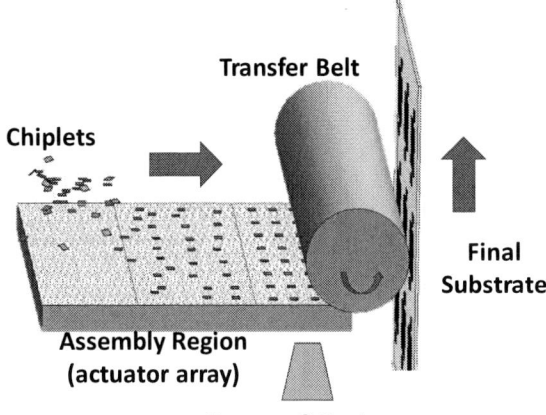

Figure 1. Microassembly system schematic. Chiplets dispersed in solution are directed by dynamic electric field patterns into custom arrangements, and then transferred to a final substrate.

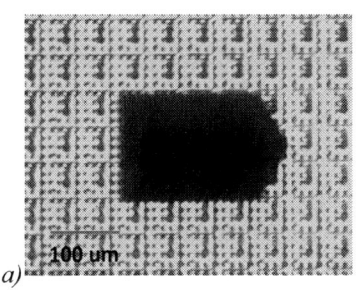

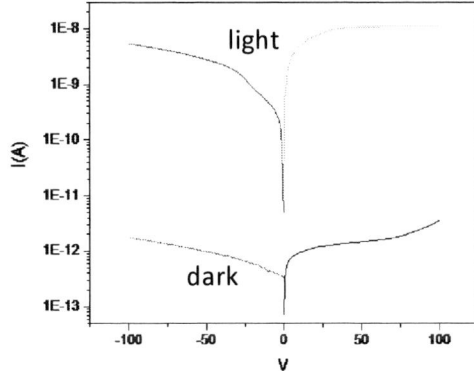

b)

Figure 2. Electrode array used to actuate the chips. a) Photo of a dummy silicon die (200 um x 150 um) on a 50 um pitch photo-switch electrode array. The chip has a pointed end to simplify orientation identification for the dummy chip. b) Measured current-voltage properties of an actuator pixel in light and dark conditions.

There are two system configurations. A 50 um pitch electrode array over a 4 cm^2 area, which uses up to 200 V actuation patterns, and 10 um pitch electrode array over a 1 mm^2 area which uses up to 30 V actuation patterns. The 50 um pitch array is transparent, enabling addressing and imaging from the bottom of the array. This leaves room on the top of the array for a roller based transfer system, consisting of an intermediate conductive belt that electrostatically picks up assembled chips row by row under the transfer nip using a non-contact process (Figure 3). The assembled chips can then be transferred again to a final substrate for subsequent electrical interconnect processing. For the 10 um pitch we are using a rubber stamp transfer process after drying the chips.

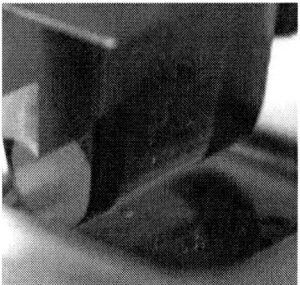

Figure 3. Microchip assembly printer setup with the transfer belt above the assembly region, with the cross process width of 2.5 cm.

III. CONTINUOUS ASSEMBLY AND TRANSFER

A continuous feed process is important to demonstrate because it shows the potential for generating large systems, where the final substrate is from a continuous roll and much larger than the assembly region. LED displays, sensor arrays, micro solar arrays, heterogeneous antennas or metamaterials, are all examples that could use such a capability.

We demonstrated a continuously printed array of 70 chips of size 50 um x 200 um (Figure 4) with single pixel registration (less than the actuator pitch). Chips in a reservoir area are automatically selected, transported and assembled into rows by the control system. All the chips are moving and being controlled in parallel. Before the transfer belt electrostatically picks up a row, the control system orients the chip. The final registration for the 70 chips after transfer was less than +/- 50 um and +/- 12°. The centroid registration was less than +/- 25 um before transfer (Figure 5, Figure 6), and the time for each row to assemble was ~60 seconds.

These results were expected based on the assembly control system parameters and transfer setup used. For applications which need finer orientation control, the camera imaging resolution and actuation algorithms could be improved. Centroid registration could be improved with finer actuator pitches and transfer gap control. Assembly time is currently limited by orientation time which can be improved with more advanced control algorithms and sensing. Basic translation speeds average ~2 mm/sec, but centimeter per second has been observed. Smaller chips are observed to travel faster than larger chips.

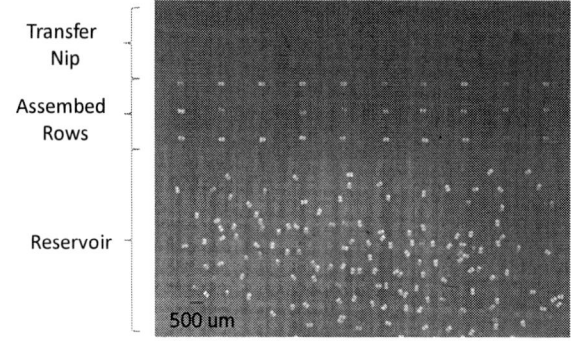

Transfer
Nip

Assembled
Rows

Reservoir

500 um

a)

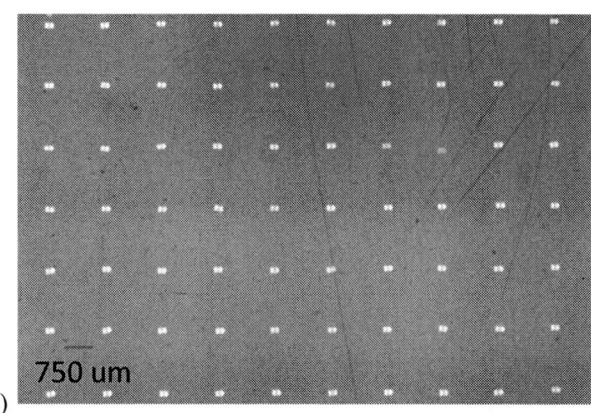

750 um

b)

Figure 4. Automated assembly and orientation control of 150 um Si chiplets (made by Sandia National Labs). a) View from under the array (see Fig 1 for viewing angle), showing the reservoir of chip as they are directed along the actuator array (up in the photo) to form rows near the transfer nip. b) A 7x10 array at 1.5 mm pitch on a final rubber substrate, after transfer from the intermediate electrostatic transfer belt.

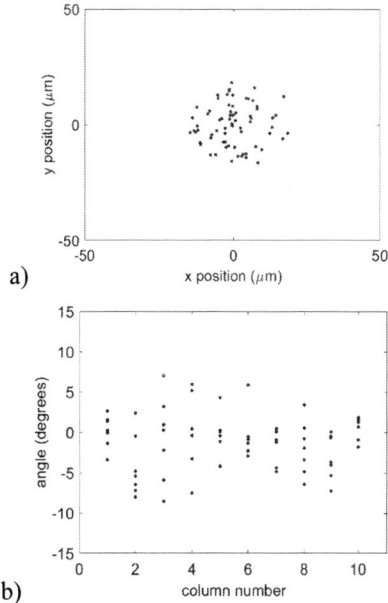

a)

b)

Figure 5. Registration of chips after assembly for a) xy centroid location and b) angle with respect to the intended placement.

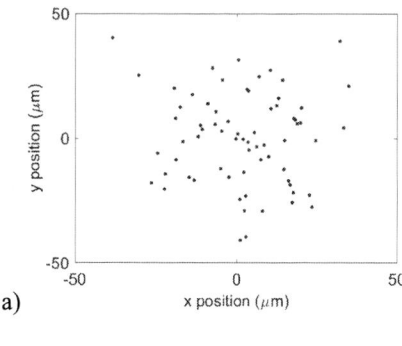

a)

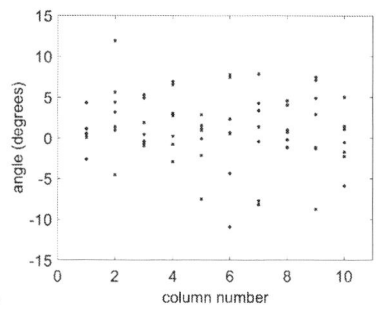

b)

Figure 6. Registration of chips after transfer to the final substrate a) xy centroid location and b) angle with respect to the intended placement.

IV. SMALL CHIPLETS

While small chiplets are generally cheaper to make than larger chiplets, they are particularly difficult for standard assembly methods. MicroLEDs can be 1-50 um and millions of transistors can fit on micrometer scale chips.

We used a 10 um pitch electrode array to manipulate small chiplets 10 um to 50 um in size, into a uniform array and heterogeneous configurations (Figure 7). These assemblies used human guided feedback to perform the assembly by moving the actuation pattern which moves the chips to the desired location. For transfer the fluid was removed and a planar rubber stamp was used to transfer the image, with a final registration error of < 1/5 of an actuator pixel (< 2 um for a 10 um pitch electrode).

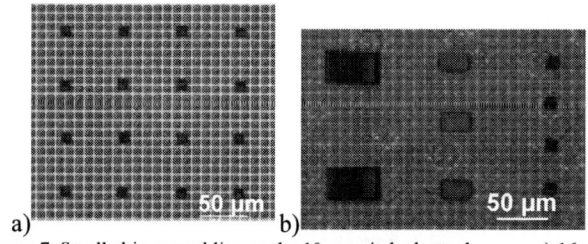

a)   b)

Figure 7. Small chip assemblies on the 10 um pitch electrode array. a) 16 chiplets of 15 um size arranged in an array, and b) heterogeneous assembly of chips of width 15 um, 30um and 50um.

978-1-7281-1500-9/19 $31.00 © 2019 IEEE   1314

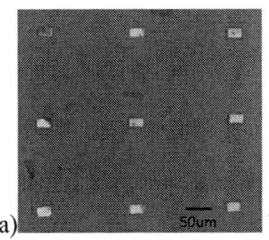 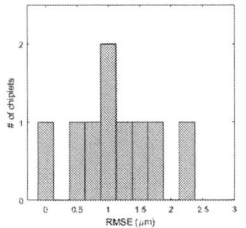

Figure 8. A 3x3 array of silicon chiplets 30x20x5 um a) after transfer to a final rubber substrate and b) measured xy centroid displacement, showing an RMS error of 1.1 +/- 0.7um. Rotation error was <1°.

V. ACTIVE CHIPS & INTERCONNECTS

Pre-fabricated electrical chiplets were successfully microassembled and transferred with the system. Ink jet printed silver lines were then used to electrically interconnect the chips and enable probing of the circuit. GaAs diodes were shown to have no change in IV measurements before and after the process (Figure 9) and nine GaN LEDs were successfully integrated into a "X" pattern (Figure 10, Figure 11) which emits light and can be power modulated to blink.

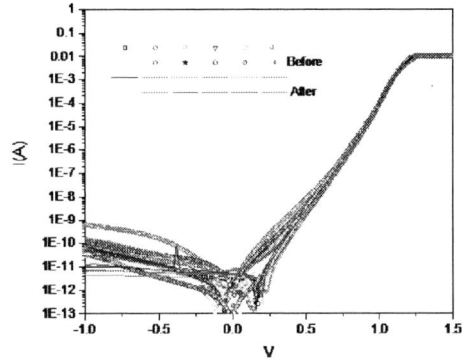

Figure 9. a) Measured IV curves for 12 GaAs photovoltaic diodes before and after the PARC electrostatic microassembly and transfer process showing coincident before and after measurements. Slight variations observed below the 1 nA reverse bias region are typical for such devices.

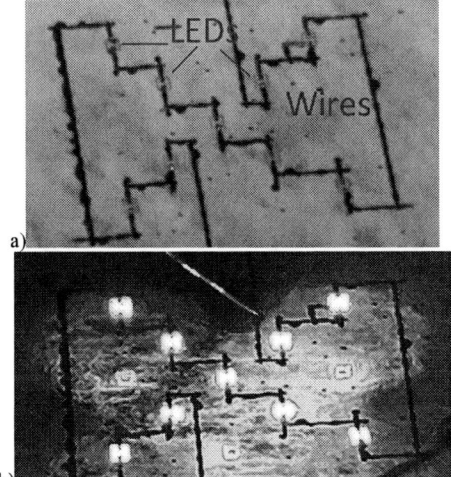

Figure 10. LED circuit demonstration using electrostatic microassembly and transfer of 9 commercial LEDs (~200um) followed by silver ink-jet printed interconnects. The LEDs a) off and b) on, emitting and blinking.

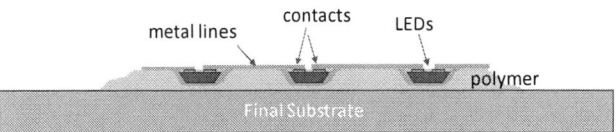

Figure 11. Cross section schematic for electrically interconnecting the printed LEDs.

VI. CONCLUSION

A microchip directed assembly and transfer process demonstrated parallel automated assembly with continuous feed transfer capability, small chip assemblies, and active circuits. Registration is less than the actuator pitch. The process could potentially serve as a general tool for integrating chiplets into heterogeneous, large area, custom systems.

ACKNOWLEDGMENT

We thank Julie Bert, Pat Maeda, Dave Biegelsen, Rene Lujan, Gregory Burton, Ion Matei, Dan Davies, Sai Nelaturi, Prakhar Jaiswal and Armin Volkel for early contributions. We thank Boston University for fabricating the small chips (<50 um) and Sandia National Labs for fabricating the GaAs diodes and performing the initial IV curves. The information, data, or work presented herein was funded in part by PARC, Xerox, the Defense Advanced Research Projects Agency (DARPA), and the Advanced Research Projects Agency-Energy (ARPA-E), U.S. Department of Energy. The views and opinions of authors expressed herein do not necessarily state or reflect those of the United States Government or any agency thereof.

REFERENCES

[1] S. Khan, L. Lorenzelli, and R. S. Dahiya, "Technologies for Printing Sensors and Electronics Over Large Flexible Substrates: A Review," IEEE Sensors Journal, vol. 15, no. 6, pp. 3164–3185, Jun. 2015.

[2] S. Biswas, M. Mozafari, T. Stauden, and H. O. Jacobs, "Surface Tension Directed Fluidic Self-Assembly of Semiconductor Chips across Length Scales and Material Boundaries," Micromachines, vol. 7, no. 4, p. 54, Apr. 2016.

[3] J. Yoon, S.-M. Lee, D. Kang, M. A. Meitl, C. A. Bower, and J. A. Rogers, "Heterogeneously Integrated Optoelectronic Devices Enabled by Micro - Transfer Printing," Advanced Optical Materials, vol. 3, no. 10, pp. 1313-1335, Oct. 2015.

[4] R. C. Y. Auyeung, H. Kim, S. Mathews, N. Charipar, and A. Piqué, "Laser additive manufacturing of embedded electronics," in Laser Additive Manufacturing, Elsevier, 2017, pp. 319–350.

[5] J. P. Lu, J. D. Thompson, G. L. Whiting, D. K. Biegelsen, S. Raychaudhuri, R. Lujan, J. Veres, L. L. Lavery, A. R.Völkel, and E. M. Chow, "Open and closed loop manipulation of charged microchiplets in an electric field", App. Phys. L., 105, 2014.

[6] E. M. Chow et al., "Micro-object assembly with an optically addressed array," in 2017 19th International Conference on Solid-State Sensors, Actuators and Microsystems (TRANSDUCERS), 2017, pp. 682–685.

[7] I. Matei, S. Nelaturi, J. P. Lu, J. A. Bert, L. S. Crawford, and E. Chow, "Towards printing as an electronics manufacturing method: Micro-scale chiplet position control," in 2017 American Control Conference (ACC), 2017, pp. 1549–155.

Effect of Intermetallic Compound Growth on Electromigration Failure Mechanism in Low-profile Solder Joints

H. Madanipour[1], Y.R. Kim[1], Choong-Un Kim[1,4]

N. Shahane[2], D. Mishra[2], T. Noguchi[3], M. Yoshino[3], and L. Nguyen[2]

[1]Department of Materials Science and Engineering, The University of Texas at Arlington
501 West First Street, Box 19031, Arlington, Texas 76019
[2]Texas Instruments, Inc., Santa Clara, CA 95052
[3]Texas Instruments, Inc., Hiji, Japan
[4]contact e-mail: choongun@uta.edu

Abstract

This paper describes the kinetic and microstructural mechanism of electromigration (EM) failure found in low-profile solder joints where EM and intermetallic phase formation compete for the same volume of Sn. The low-profile solder joint used in our study was made of 20-25um thick solder situated in between a Cu pillar and a Ni coated Cu lead frame (LF). The samples were EM tested in a temperature range of 140-170°C with the current densities varying between 35-45 KA/cm^2 in an oil bath to induce failure without Joule Heat induced artifacts. Our studies on EM failure kinetics and microstructural mechanism have produced two key findings. The first finding suggests that the EM diffusivity (Z*D) of diffusing species (Sn, Ni, Cu) in the solder matrix can be uniquely ranked from microstructural analysis, and it is estimated to be $(Z*D)_{Cu} > (Z*D)_{Sn} > (Z*D)_{Ni}$. This difference in EM diffusivity causes Cu-Sn and Ni-Sn intermetallic compounds (IMC) to develop in distinctively different manners under EM, leading to different EM failure mechanisms. The second finding is that EM in low-profile solder joints consists of multiple failure stages: a) with EM-related voiding in Sn dominating at lower temperatures; while b) thermally-induced IMC growth and invasion competes with EM-induced Sn voiding at high temperatures leading to the complete failure of each joint.

Keywords: low-profile solder joint; Cu-post, electromigration, Cu-Sn intermetallic compounds, Ni-Sn intermetallic compounds.

Introduction

Electromigration (EM) failure has long been the subject of extensive studies with its role in limiting long-term reliability of electronic devices. With increasing demands for devices operating at higher temperatures under higher current, the threat is especially significant for solder interconnects and becomes a major technical challenge in advanced electronic packaging. Previous studies [1]–[9] on EM failure in solder joints have provided many valuable findings in resolving such challenges, but there still exist a few unanswered questions of critical importance, especially for the next generation of packaging technology where the use of low-profile solder joints is essential. Unlike in conventional solder joints, such as in Ball Grid Array (BGA), low-profile joints use a limited amount of solder to a degree to convert the joint to a sizable fraction of

IMC during the reflow process. Since EM and IMC growth compete with each other for the same volume of solder and EM rate in IMC is considered to be negligibly small, EM failure in such joints is expected to show a unique kinetic mechanism fundamentally different from what is known. It is therefore essential to understand the kinetic interplay between the IMC growth and EM failure rate in such joints and, ideally, find material/structural parameters suppressing EM failure.

There are a few previous attempts of experimental assessment on the failure mechanism in low-profile solder [10]–[20]. A study conducted by Hsiao et al [18] is one of few limited studies on the subjects with results suggesting that EM failure in microjoints can be indeed complex due to IMC taking a significant fraction of the joint and its growth removing EM-prone Sn matrix. A similar conclusion was made in our own previous study in which immortality of the joint was seen in the low-profile joint when tested under moderate EM conditions due to a full conversion of Sn into Cu-Sn IMC before voiding by EM reached the critical point sufficient enough to result in failure[21]. The findings made in these studies present a strong implication that EM failure may not follow a simple activation energy behavior, where failure rate increases exponentially with temperature, because temperature affects both the EM and IMC growth rate in the same direction. They also suggest that a dramatic change in the failure rate and failure mechanism is possible even with a small change in interfacial microstructure (and thus IMC type and/or IMC growth rate) between the solder and Cu metallization. These postulations are plausible, but they are not easy to substantiate with difficulties in conducting EM testing with sufficient acceleration to induce failures without noises stemming from Joule Heat (JH) induced testing artifacts. The low-profile solder joints need to be EM tested at exceptionally high current densities in order to induce failure without relying on temperature in accelerating failure. In such a case, however, JH is difficult to manage without the use of special arrangement for heat dissipation, which probably was a cause of limited prior studies on the concerning subject.

Motivated by the need for properly addressing the complexity of EM failure in low-profile solder joint, we first developed a testing system where silicone based liquid coolant was circulated to manage JH and then conducted EM testing of samples with low-profile solder joints with a variation in dimension and interfacial structures. Our testing as well as microstructural analysis has yielded a number of anticipated results, which are indicative of a unique EM mechanism in the low-profile solder joint due to keen impact IMC growth on the failure process, confirming our hypotheses. Among various conclusions, this paper presents evidence leading to two key conclusions. The first conclusion is the confirmation of the anticipated mechanism that EM failure rate of the low-profile solder joint is determined by two competing kinetic factors, namely the growth rate of IMC (or conversion rate of IMC) and the EM voiding rate. One of the most striking evidences is found from the temperature dependence of EM failure rate because data suggests that samples would survive better at higher temperatures once temperature exceeds the critical point, above which IMC growth rate becomes dominant over the EM voiding rate. The microstructure of failed joints appears to reflect such a mechanism. In the case of joints without a Ni barrier, EM is found to add a significant assistance to the growth of Cu-Sn IMC more so at higher temperature and causes the growth to continue beyond the Sn solder, extending deep into the Cu post. As a result, samples failed at higher temperatures tend to show voiding activity more in the Cu post than the original Sn/Cu interface. The presence of a Ni diffusion barrier, on the other hand, is found to force the failure to develop within the solder layer with Ni preventing reaction between the Cu post and the Sn solder. Nevertheless, the similar kinetic interplay between the rate of IMC growth (Ni bearing IMC at this time) and the voiding affecting EM failure rate of the joint is found to exist. Interestingly, EM is found to have a negligible influence on the growth of Ni-Sn IMC; rather

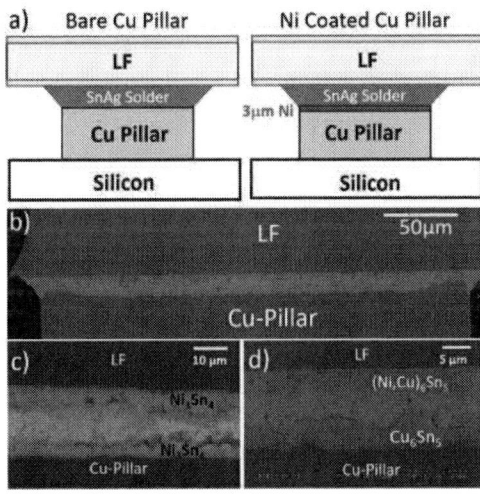

Figure 1. Schematic representation (a) and cross-sectional SEM micrographs showing the overall view of the assembly (b), the joint microstructure of N-post (Ni-coated) (c), and B-post (bare Cu) (d).

evidences suggested that Ni-Sn IMC growth occurred primarily by chemical diffusion. This result, which is our second key conclusion, physically implies that the EM diffusivity (a product of effective valence Z* and diffusivity D) of Ni in Sn matrix is negligibly small to a degree where the Ni-Sn IMC growth is predominantly governed by diffusion. This is unexpected behavior as Ni is known to exhibit high EM diffusivity[22]; however, consistency in our observations led us to speculate that low EM diffusivity of Ni in Sn may be a result of the stress-driven backflow that is likely to active in the low-profile solder joint in reducing EM. This paper presents

some of highlights evidencing these findings along with a discussion of the probable mechanisms.

Experimental Method and Sample Configuration

The samples used in our investigation are structurally similar to what was used in our previous study. In essence, they were based on a flip chip on LF structure with Pb-free SnAg solder alloys. As presented in Figure 1(a), where the schematic representations and cross-sectional SEM view of the joint are shown, the Cu-pillar is patterned on top of Si wafer and mated to the LF using electroplated lead-free SnAg alloys. The dimension of the Cu pillar is approximately 300µm in length, 110µm in width and 65µm in height. The mating electrode of the bump is the Cu-LF coated with Ni and Pd as the pre plated layer (Pd pre-plated LF). The solder joint is formed by by subjecting the assembly to a standard reflow condition. While the joint made of the bare Cu pillar was the focus of our study, we also prepared the samples with Cu pillar having ~3µm Ni finish to study the impact of the diffusion barrier on the EM mechanism. The distance between the top of the Cu-post and bottom of the LF before reflow is approximately 20-25µm, but due to formation of IMC phases both at the post and the LF interface, the actual solder layer thickness after the reflow is reduced to ~15-20 µm for both the assembly with the Ni coated post (N-post) and the bare post (B-post). As shown in Figures 1(c) and (d), where a cross-sectional SEM view of the N-post and B-post is compared, the growth of IMC phase on both interfaces, post/solder and solder/LF, causes the solder layer to be substantially reduced. The IMC phase found in the N-post is determined to be Ni_3Sn_4 both at the post and at the LF interface. In the case of the B-post, on the other hand, Cu_6Sn_5 phase is found to form at the post interface while it is the $(Ni,Cu)_6Sn_5$ phase found at the LF interface. The $(Ni,Cu)_6Sn_5$ phase formed at the LF interface is different from the same LF interface in N-post even if the same LF is used for both assemblies. The difference can be attributed to the availability of Cu atoms in the solder matrix originated from the Cu-pillar in the case of B-post with an absence of a Ni barrier in the pillar.

Table 1 A list showing the bath temperature used to conduct EM testing of samples at target temperatures under j=35KA/cm^2.

Sample	EM T, °C	Bath T, °C	δT, °C
Bare Cu	160	152	8
	152	144	8
	144	136.5	7.5
Ni coated Cu	160	151	9

In order to confine the EM damage to the target joint, we placed a number of supporting joints near the EM pillar and used them to provide distributed current paths to the joint under EM testing. This arrangement was used to make the test current flow from the LF to Cu-pillar, meaning that EM is directed from the pillar to the LF direction. Also developed for this study was a customized EM test system with oil bath to enable effective JH dissipation and maintenance of temperature uniformity and stability for the duration of EM testing. The temperature of the post during testing was monitored using the

978-1-7281-1500-9/19 $31.00 © 2019 IEEE

on-chip temperature sensor, and the temperature of ambient was adjusted to bring the post temperature to the target temperature. With the help of circulating oil, we were able to limit the temperature difference between the sample and oil to under 10°C even with a current density exceeding 35 KA/cm². Table 1 shows the bath temperature used to conduct EM testing of given samples at the target temperature.

Results

Characteristics of EM Failure Kinetics

As noted in our previous study [13], EM failure in low-profile solder joints would require extremely harsh EM loads in order to induce a sufficient level of EM damage before the full conversion of Sn into IMC. Samples were subjected to EM test at current densities ranging from 25 – 40 kA/cm² and temperatures ranging from 140-160°C. Figure 2, where the resistance change is plotted as a function of testing time, compares the representative failure characteristics of higher current density (35kA/cm², 144°C) versus lower current density at similar temperatures (27kA/cm², 140°C). As shown in Table I, JH was well controlled with help of custom-made setup at higher current densities. Therefore, any difference in the failure behaviors is related to the difference in EM mechanism not to JH level. Note the saturation of resistance in the case of the sample tested under lower current density. In the case of higher current density, on the other hand, the resistance does not show saturating behavior but continues to rise until failure. It seems that the failure develops in two-stage processes. In the first stage, the resistance increases slowly in a similar manner as in the case of lower current density but without saturation. After the first stage, it was followed by the second stage where the rate of resistance increase is much more rapid. All failed samples, irrespective of testing temperature and current densities, showed the same failure behaviors, indicating that the failure involved at least two stages of change in resistance shown in Fig.2.

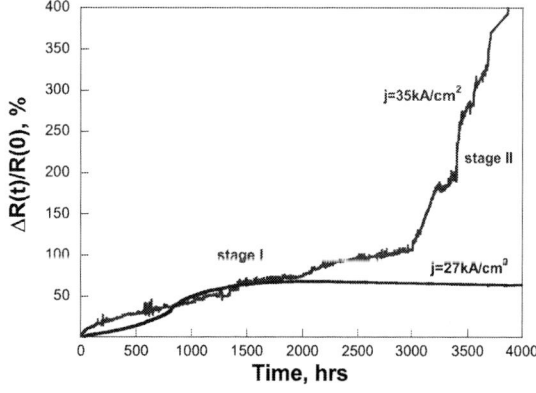

Figure 2. A plot showing the representative change in the sample resistance during EM testing found in present study in comparison with the previous one.

The first stage of failure shown in Fig. 2 can be related to the voiding in Sn phase prior to its conversion to IMC. The rate of resistance rise decreases with time because the growth of IMC results in the reduction of EM prone Sn phase in the joint.

With a sufficient EM load, the joint would trap a sizable amount of damage before the full conversion of Sn to Cu-Sn IMC takes place. Local JH in the voided area as well as availability of high concentration of EM induced vacancies may force IMC growth to continue towards the direction of the Cu pillar. This may lead to the ideal condition for extensive formation of a Kirkendall void inside of the IMC phase, that in turn induces more JH and vacancy flux[18]. Runaway failure can occur by this cascading propagation of EM damage, which may be responsible for the second stage of failure shown in Fig. 2. This mechanism seems reasonable and explains the kinetic behaviors of EM failure presented in Fig. 2. One of the most striking and surprising evidence for EM failure mechanisms involving multiple processes is shown in Fig. 3, where abnormal temperature dependence of failure kinetics seen in our study is presented.

Figure 3 shows the normalized mean time to failure (MTTF) of B-post samples under 35KA/cm² as a function of testing temperature. The MTTF shown in this plot is collected by fitting the time to failure of 17~20 samples to a log-normal

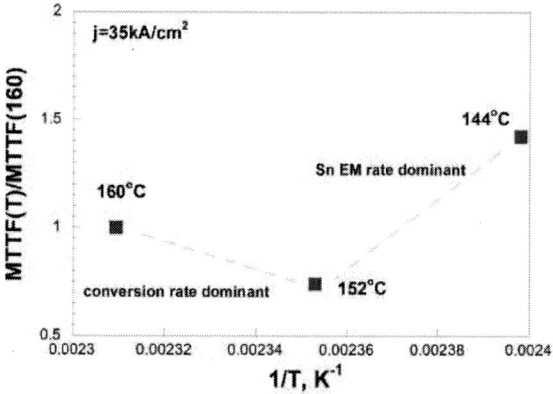

Figure 3. A plot showing the normalized MTTF of EM failure as a function of test temperature. More than 17 samples per each temperature was tested and MTTF was determined from log-normal statistics.

distribution, as shown in Fig. 4, where a normalized time to failure of each sample is plotted as a function of cumulative probability. The normalized time to failure is defined to be the time to reach 500% increase in resistance in reference to MTTF of samples tested at 160°C. The normalized MTTF used in Fig. 4 is in reference to the same MTTF, that is MTTF of samples tested at 160°C. Notice that the time to failure statistics fits very well to a log-normal distribution with the shape factor reasonably uniform across the samples failed at different temperatures. Note the fact that the MTTF in Fig. 3 does not follow a single activation energy relationship with temperature; rather, the data seems to suggest that some sort of abnormal process is involved in the failure and its role or influence changes at temperature near 152°C. It is almost tempting to attribute this behavior to the error in setting the test temperature because JH induced temperature change is often difficult to accurately compensate. However, as indicated in Table 1, temperature was well regulated and cannot be a source of the abnormality. We believe that it is rooted more to the physical mechanism, that is the EM failure kinetics being governed by two competing processes. The first is the kinetics of voiding by EM in Sn responsible for inducing critical damage, while the second is the rate of IMC growth in the joint. The factors affecting the kinetics of these two processes are not necessarily the same and scale differently with temperature. Specifically, the voiding kinetics is governed by the EM diffusivity of Sn atoms in Sn matrix $(Z^*D)_{Sn}$, while the IMC growth rate is governed more by either chemical diffusivity or EM diffusivity of Cu in Sn $(Z^*D)_{Cu}$. It is our belief that activation energy for voiding, $(Z^*D)_{Sn}$, is lower than the one for Sn conversion, $(Z^*D)_{Cu}$, and that the conversion rate is dominant at higher temperatures. The EM failure then occurs less readily than what is predicted from the failure rate at low temperatures (where voiding rate is more dominant). This seems to explain the data in Fig. 3 and also the microstructural failure mechanism detailed in the following section. It is important to note that the effect of failure governed by two competing processes would not be noticeable in the case of the conventional solder joint because excessively large dimensions of solder causes the effect of IMC formation to have negligible influences and failure to proceed without the competition.

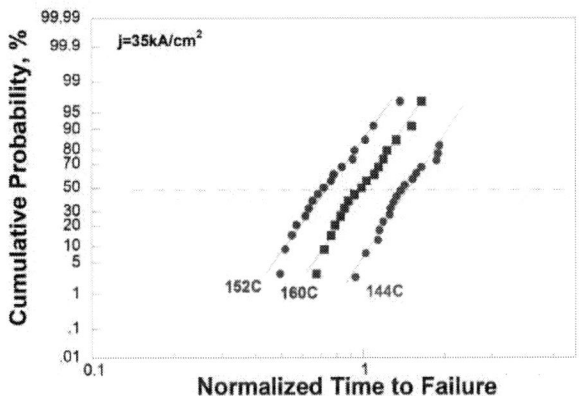

Figure 4. A plot showing the normalized time to failure of samples tested under 35kA/cm^2 at three different temperatures as a function of normal cumulative probability.

Microstructural Failure Mechanism

In order to better understand the failure mechanism active in the low-profile solder joints, especially the ones responsible for unique and unusual failure behaviors shown in Figs.2-4, a large number of failed samples tested at various conditions were subjected to microscopic characterization. This effort enabled us to find consistent features which are in support of the failure mechanisms proposed in our study.

Extensive characterization of the failed joint led to the conclusion that the two-stage failure kinetics in Fig. 2 illustrates the formation of initial EM damage at Sn phase in the joint and subsequent consumption of the Cu pillar by the growth of IMC. Samples with significant voiding both in the joint and Cu-Sn IMC after extended growth into Cu post pillar were found. Figure 5, where the cross-sectional SEM micrographs of three B-post samples after EM testing under 35KA/cm^2 at 144°C, 152°C, and 160°C are shown, present a representative example of the voiding in the joint (a) and in the Cu$_3$Sn IMC grown deep into Cu post (b, c). It can be seen that

the majority of voids trapped inside of the joint are formed at the original interface between Cu pillar and Sn solder.

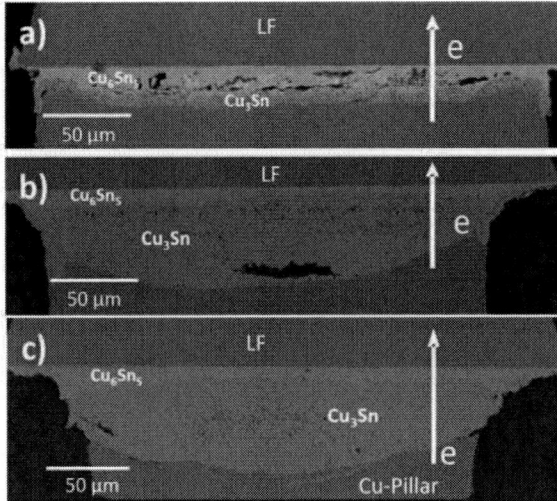

Figure 5. Cross-sectional SEM micrographs of three B-post samples after EM testing under 35kA/cm^2 at 144°C, 152°C, and 160°C. (a) voiding in the joint and (b,c) Cu$_3$Sn IMC grown deep into the Cu post. sample after EM testing under 35kA/cm^2 at 160°C.

This occurs because EM force directed from Cu pillar (cathode) to LF (anode) direction causes Sn to migrate toward LF side producing counter-flux of vacancy to accumulate at Sn/Cu post interface. Note also the exaggerated growth of Cu$_3$Sn IMC in Cu-pillar in Fig. 5-b, c. Regular interdiffusion cannot produce IMC growth to such an extent. It can happen only with assistance from EM. Some of the vacancies directed towards the Sn/Cu pillar interface by EM may leak into the Cu pillar and promote the rapid growth of Cu$_3$Sn phase with enhanced diffusivity of interdiffusing atoms. Although it is not to the same extent, enhanced growth of IMC phases in the direction of the Cu pillar is also evident in Fig. 5-a. It is difficult to generalize, but this extended growth of Cu$_3$Sn phase seems to appear more along the periphery of the Cu post, probably a result of surface diffusion being involved in its growth. It is also noted that the failure type in Fig.5-a, that is failure with more trapped voids in the joint, is more prevalent at lower temperature, which is in consistent with the failure mechanism we propose for unusual temperature dependence of EM failure kinetic in Fig.3.

The failure facilitated by Cu$_3$Sn phase growth into the Cu pillar becomes absent when the surface of Cu-pillar is coated with a Ni barrier (N-post). Figure 6 shows the representative microstructure of a failure site found in N-post samples tested under 35KA/cm^2 at 160°C. With a Ni diffusion barrier, the Cu pillar is well protected from the growth of IMC phase and the voiding area is confined only within the joint. Note that the voiding occurred along the original interface between Sn and Ni/Cu-pillar, which is caused by the collection of vacancies produced by Sn-EM during the early stage of testing. With a limited extent of EM testing conducted on N-pillar sample, it is somewhat difficult to determine if the Ni barrier provides better

978-1-7281-1500-9/19 $31.00 © 2019 IEEE   1319

protection against EM or not. The available test results along with microstructural analysis on the failure seem to suggest that the N-post would exhibit somewhat better EM resistance than B-post likely would, probably because the loss of Sn from the joint by the process of interdiffusion is prevented unlike in the case of B-post. However, we also note the possibility that the expected benefit may not exist due to a slow conversion of the joint into Ni_3Sn_4 IMC. In fact, a majority of the EM data we have collected so far supports the latter.

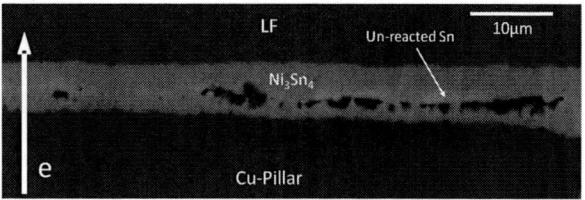

Figure 6. Cross-sectional SEM micrograph showing the voids trapped inside of the joint found in N-post sample after EM testing under 35kA/cm² at 160°C.

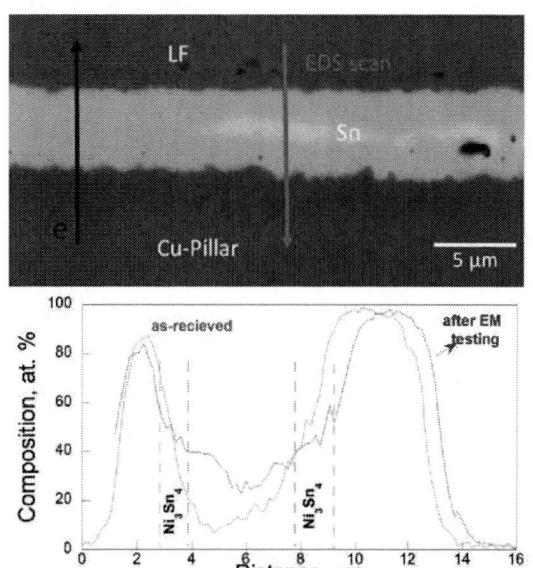

Figure 7. EDS scan profile of Ni across the N-post joint after 348 hrs of EM under 35KA/cm² at 160°C in comparison to the as-received.

As an attempt to characterize the kinetics of the joint conversion into Ni_3Sn_4 under an EM load, a number of samples were extracted after the elapse of prescribed testing time and were subjected to microstructural analysis. Our main objective was to determine the conversion rate, but it was discovered that the conversion (growth of Ni_3Sn_4) is not much assisted by EM as it is in the case of the B-post. We found evidence suggesting that Ni_3Sn_4 growth is directed more by chemical diffusion of Ni into Sn even with the presence of strong EM force. It is our conclusion that at least the relation of $(Z^*D)_{Sn}>(Z^*D)_{Ni}$ is true. Supporting evidence behind our conclusion is presented in Fig. 7, where the compositional profile scanned by EDS (electron energy dispersive spectroscopy) is compared for the case of as-

prepared and testing after 348 hrs under 35KA/cm² at 160°C. Notice that the growth of Ni_3Sn_4 phase proceeds without any dependence on the direction of EM, evidenced by the similar amount of its thickness both at anode (Ni coated LF) and cathode (Ni/Cu post). In the case when EM of Ni is active, Ni at cathode side should dissolve and migrate toward the LF to be deposited as Ni_3Sn_4 phase. This process results in thinning of Ni or Ni_3Sn_4 phase at the cathode and thickening of Ni_3Sn_4 at the anode as is schematically illustrated in Fig. 8-a[20], [24]–[26]. The Cu-Sn IMC growth in B-post was found to proceed in this manner[27]–[33]. On the other hand, when EM diffusivity of Ni in Sn is negligible or smaller than that of Sn, that is $(Z^*D)_{Sn}>(Z^*D)_{Ni}$ or $(Z^*D)_{Ni}\sim0$, then the Ni_3Sn_4 IMC phase would grow at an equal rate at both interfaces; this is the case schematically shown in Fig. 8-b. Experimental evidence clearly supports the small diffusivity of Ni in Sn, but it is not consistent with previous studies on Ni EM in Sn matrix. In fact, the EM diffusivity of Ni is reported to be considerably higher than that of Sn. While this discrepancy calls for further study, the results we gained so far suggest that the benefit of Ni used in N-post would be limited with a slower conversion rate of the joint than anticipated.

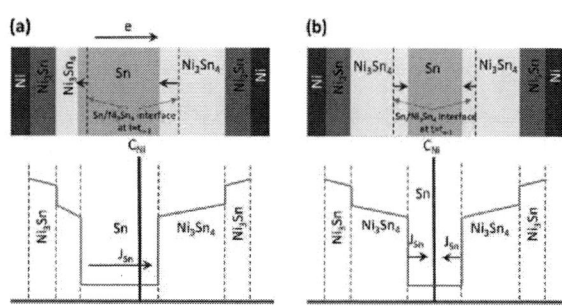

Figure 8. Schematic illustration of the Ni_3Sn_4 phase growth in case when a) $(Z^*D)_{Ni}>(Z^*D)_{Sn}$ and b) $(Z^*D)_{Sn}>(Z^*D)_{Ni}$ or $(Z^*D)_{Ni}\sim0$.

Discussion

Our study testifies the fact that the mechanism of EM failure in solder joints with a limited amount of Sn is indeed complicated and poses considerable theoretical and experimental challenges. Although we successfully manage to find the failure mechanism that seems to be consistent with and provides a reasonable explanation for the observations, we feel that there still exist a few observations that cannot be fully explained within the developed frame of understanding, which become a focus of ongoing investigation.

One of the persistently nagging questions has been the mechanism behind the unusually high resistance of low-profile solder joints in general against EM failure. An extensive literature survey on EM failure in solder joints led us to find a general trend that EM failure is developing far slower in low-profile solder joints than it is in joints with a larger solder volume. The mechanism developed in this study attributes the reason for a removal of EM-prone Sn phase and its eventual conversion to EM-immune IMC phase in the joint, causing the damage rate to slow down with the progression of EM. There

978-1-7281-1500-9/19 $31.00 © 2019 IEEE

is no doubt that such a mechanism is involved; however, it seems that there exists a secondary mechanism. In this respect, the observation made by Ouyang et al [16] is noteworthy because it reported a complete EM immunity of microjoint for a 3D package consisting of Ni coated 12 µm diameter Cu UBM with 6µm thick Sn rich solder alloys when tested under j~20KA/cm^2 at T=110°C. The immortality reported in this report occurs without full IMC conversion of Sn and authors relate it to the "Blech length" effect where EM force cease to exist due to a counteracting force from a stress gradient[34]–[37]. The stress gradient arises due to EM induced back-stress in the joint (compression at anode and tension at cathode), resulting in reduction in atomic flux J, namely

$$J = \frac{CD}{kT}\left(Z^*e\rho j - \frac{\Omega d\sigma}{dx}\right), \tag{1}$$

where ρ, Ω, and σ denote the resistivity of Sn, atomic volume, and hydrostatic pressure, respectively, while others have their usual meanings. The flux vanishes when EM force is completely counterbalanced by the back-stress gradient, leading to the well-known critical "Blech length" condition of

$$(jL)_c = \frac{\Omega\Delta\sigma}{Z^*e\rho} \tag{2}$$

where L represents the thickness of the solder joint. What eqs. (1)-(2) suggest is that the EM rate and thus EM failure rate would scale inversely with the solder thickness at a given current density. It is therefore conceivable that the bulk solder is inherently more prone to EM failure because the solder thickness is too large for IMC and its growth to add any beneficial influence but also for a stress gradient to develop in a sizable degree. On the other hand, a low-profile solder joint would benefit greatly from the back-stress effect for two reasons. The thinness of Sn is one reason but also, more importantly, restriction in stress relaxation is another factor that causes the back-stress effect to play a significant role in low-profile solder joint. In a bulk solder, the stress is limited at the yield stress beyond which stress is relaxed through plastic deformation carried out by dislocation, glide, or climb. Such stress relaxation demands the availability of the slip system toward the free surface. The relaxation should become increasingly difficult with decrease in solder thickness and aspect ratio (thickness/width). Therefore, with a lack of relaxing stress, the low-profile solder naturally produces higher back-stress and thus less EM flux at a given current density. In this respect, the solder joint used in our study is particularly unique because it has an extremely low aspect ratio, ~0.2, along with an extremely thin original Sn thickness, ~15µm. It is therefore reasonable to conclude that the EM failure kinetics measured in our study includes the EM suppression effect originated from the stress gradient. .

The back-stress effect may also explain our observation of small EM diffusivity of Ni in Sn. This result directly contradicts several measurements made in previous studies because those studies consistently report the EM diffusivity of Ni in Sn to be more than an order of magnitude higher than the EM diffusivity of Sn. In addition, dissolution and migration of the Ni barrier by EM is a common observation in bulk solder joints. The high back-stress active in our sample may provide answers to the question for why the reduction in Ni EM diffusivity is seen only in our study. EM of Sn is the very source

of stress generation, and Ni has to migrate under the stress environment created by Sn EM. In the case of when the stress gradient produces a far greater impact on the EM force acting on Ni than on Sn, such as by having a larger Ω of Ni, the EM force acting on Ni may become reduced or even ceases to exist. This causes the Ni to migrate only by the diffusion potential. This seems to be a reasonable mechanism that can explain the results we obtained but will require substantiation through an in-depth investigation. Some of the needed investigation is currently ongoing in our research team and our initial results, which favor the proposed theory, will be published in the near future.

Finally, it should be noted that the behavior shown in Fig. 4, non-classical dependence of EM failure on temperature, will vary sensitively not only with specifics of solder geometry (thickness, initial volume of Sn and IMC thickness, etc.) but also with testing current. It is our prediction that the temperature dependence would appear more normal when a thicker solder joint is used and/or when EM testing is conducted under higher current conditions than we used. In such cases, kinetics of Sn EM is fast enough to outcompete the IMC growth rate at all temperatures, causing the EM failure kinetics to follow what appears to be a single activation process. Ongoing studies may provide supporting evidence, yet the understandings made in this study may be sufficient to suggest the need for being careful in assessing EM reliability of low-profile solder joints.

Conclusions

The kinetic and microstructural mechanism of EM failure active in low-profile solder joint were investigated in our study by conducting EM testing of the flip-chip on the LF solder joint under highly accelerating current and temperature conditions. The resulting failure kinetics along with microstructural characteristics of the failure site reveals that the EM failure in low-profile solder joints proceeds by two competing kinetic processes: EM induced voiding in Sn and IMC growth. The first consequence of this mechanism is that the failure develops in a two-step sequence that is reflected in the resistance change, namely slow increase in resistance by initial voiding at Sn followed by rapid rise in resistance due to runaway failure by JH assisted EM and IMC growth. In addition, the two competing processes caused the EM failure to follow a highly unusual temperature dependence, that is that the fastest EM failure occurs at an intermediate temperature instead of at the highest temperature. EM-related voiding in Sn dominates at lower temperatures while thermally-induced IMC growth dominates at higher temperatures. These discoveries are found to be the cause of a unique mechanism, which is also supported by the microstructural mechanism. Finally, although the mechanism is presently unknown, our evidence suggests that the growth of Ni-Sn IMC (Ni$_3$Sn$_4$) is found to be unaffected by the presence of EM force, which is in contradiction with findings made in various previous studies.

Acknowledgments

This research is partially supported by Semiconductor Research Corporation (SRC 2089.001).

References

[1] K. N. Tu, H. Y. Hsiao, and C. Chen, "Transition from flip chip solder joint to 3D IC microbump: Its effect on microstructure anisotropy," *Microelectron. Reliab.*, vol. 53, no. 1, pp. 2–6, 2013.

[2] H. Huebner *et al.*, "Microcontacts with sub-30 μm pitch for 3D chip-on-chip integration," *Microelectron. Eng.*, vol. 83, no. 11–12, pp. 2155–2162, 2006.

[3] C. C. Lee, K. S. Kao, R. S. Cheng, C. J. Zhan, and T. C. Chang, "Reliability enhancements of chip-on-chip package with layout designs of microbumps," *Microelectron. Eng.*, vol. 120, pp. 138–145, 2014.

[4] O. M. Abdelhadi and L. Ladani, "Effect of Joint Size on Microstructure and Growth Kinetics of Intermetallic Compounds in Solid-Liquid Interdiffusion Sn3.5Ag/Cu-Substrate Solder Joints," *J. Electron. Packag.*, vol. 135, no. 2, p. 021004, 2013.

[5] M. Lee *et al.*, "Study of interconnection process for fine pitch flip cship," *Proc. - Electron. Components Technol. Conf.*, pp. 720–723, 2009.

[6] H. Y. Son, S. K. Noh, H. H. Jung, W. S. Lee, J. S. Oh, and N. S. Kim, "Reliability studies on micro-bumps for 3-D TSV integration," *Proc. - Electron. Components Technol. Conf.*, pp. 29–34, 2013.

[7] Y. M. Lin *et al.*, "Electromigration in Ni/Sn intermetallic micro bump joint for 3D IC chip stacking," *Proc. - Electron. Components Technol. Conf.*, pp. 351–357, 2011.

[8] S. Annuar, R. Mahmoodian, M. Hamdi, and K. N. Tu, "Intermetallic compounds in 3D integrated circuits technology: a brief review," *Sci. Technol. Adv. Mater.*, vol. 18, no. 1, pp. 693–703, 2017.

[9] K. N. Tu, "Recent advances on electromigration in very-large-scale-integration of interconnects," *J. Appl. Phys.*, vol. 94, no. 9, pp. 5451–5473, 2003.

[10] D. T. Chu *et al.*, "Growth competition between layer-type and porous-type Cu3Sn in microbumps," *Microelectron. Reliab.*, vol. 79, pp. 32–37, 2017.

[11] Y. A. Shen and C. Chen, "Effect of Sn grain orientation on formation of Cu6Sn5intermetallic compounds during electromigration," *Scr. Mater.*, vol. 128, pp. 6–9, 2017.

[12] C. Chen, T. Chiu, Y. Chiu, C. Lee, and K. Lin, "Intermetallics Current induced segregation of intermetallic compounds in three- dimensional integrated circuit microbumps," *Intermetallics*, vol. 85, pp. 117–124, 2017.

[13] C. E. Ho, P. T. Lee, C. N. Chen, and C. H. Yang, "Electromigration in 3D-IC scale Cu/Sn/Cu solder joints," *J. Alloys Compd.*, vol. 676, pp. 361–368, 2016.

[14] Y. C. Chan and D. Yang, "Failure mechanisms of solder interconnects under current stressing in advanced electronic packages," *Prog. Mater. Sci.*, vol. 55, no. 5, pp. 428–475, 2010.

[15] Y. Wang *et al.*, "Effect of intermetallic formation on electromigration reliability of TSV-microbump joints in 3D interconnect," *Proc. - Electron. Components Technol. Conf.*, no. May, pp. 319–325, 2012.

[16] F. Y. Ouyang, H. Hsu, Y. P. Su, and T. C. Chang, "Electromigration induced failure on lead-free micro bumps in three-dimensional integrated circuits packaging," *J. Appl. Phys.*, vol. 112, no. 2, 2012.

[17] C. Y. Liu, Y. C. Hsu, Y. K. Tang, E. J. Lin, and H. W. Tseng, "Effect of Cu solubility on electromigration in Sn(Cu) micro joint," *J. Appl. Phys.*, vol. 122, no. 9, p. 095702, 2017.

[18] Y. H. Hsiao, K. L. Lin, C. W. Lee, Y. H. Shao, and Y. S. Lai, "Study of electromigration-induced failures on Cu pillar bumps joined to OSP and ENEPIG substrates," *J. Electron. Mater.*, vol. 41, no. 12, pp. 3368–3374, 2012.

[19] H. Zhang, J. Li, and W. Zhu, "Electromigration in flip chip with Cu pillar having a shallow Sn-3.5Ag solder interconnect," *Proc. - 2018 19th Int. Conf. Electron. Packag. Technol. ICEPT 2018*, pp. 1653–1656, 2018.

[20] D. Goyal, P. Liu, and A. Overson, "A Comparison Study of Cu Dissolution Mechanism and Kinetics in Solder Joints under Electromigration and Extended Reflow," *Proc. - Electron. Components Technol. Conf.*, vol. 2018–May, pp. 440–447, 2018.

[21] M. Y. Kim, L. S. Chen, S. H. Chae, and C. U. Kim, "Mechanism of void formation in Cu post solder joint under electromigration," *Proc. - Electron. Components Technol. Conf.*, vol. 2015–July, pp. 135–141, 2015.

[22] S. N. Mei, J. Shi, and H. B. Huntington, "Diffusion and electromigration in lead alloys. I. Nickel as a mobile element," *J. Appl. Phys.*, vol. 62, no. 2, pp. 444–450, 1987.

[23] M. N. Bashir and A. S. M. A. Haseeb, "Improving mechanical and electrical properties of Cu/SAC305/Cu solder joints under electromigration by using Ni nanoparticles doped flux," *J. Mater. Sci. Mater. Electron.*, vol. 29, no. 4, pp. 3182–3188, 2018.

[24] C.-B. Lee, I.-Y. Lee, S.-B. Jung, and C.-C. Shur, "Effect of Surface Finishes on Ball Shear Strength in BGA Joints with Sn-3.5 mass%Ag Solder," *Mater. Trans.*, vol. 43, no. 4, pp. 751–756, 2005.

[25] P. Liu, C. Chavali, A. Overson, and D. Goyal, "Failure Mechanism and Kinetics Studies of Electroless Ni-P Dissolution in Pb-Free Solder Joints under Electromigration," *2017 IEEE 67th Electron. Components Technol. Conf.*, pp. 441–447, 2017.

[26] Y. W. Chang *et al.*, "A new failure mechanism of electromigration by surface diffusion of Sn on Ni and Cu metallization in microbumps," *Sci. Rep.*, vol. 8, no. 1, pp. 1–10, 2018.

[27] N. Mookam and K. Kanlayasiri, "Evolution of Intermetallic Compounds between Sn-0.3Ag-0.7Cu Low-silver Lead-free Solder and Cu Substrate during Thermal Aging," *J. Mater. Sci. Technol.*, vol. 28, no. 1, pp. 53–59, 2012.

[28] and C. A. H. K.-W. Moon, W.J. Boettinger, U.R. Kattiner, F.S. Biancaniello, "Experimental and Thermodynamic Assessment of Sn-Ag-Cu Solder Alloys," *J. Electron. Mater.*, vol. 29, no. 10, pp. 1122–1136, 2000.

[29] B. Chao, S. H. Chae, X. Zhang, K. H. Lu, J. Im, and P. S. Ho, "Investigation of diffusion and electromigration

parameters for Cu-Sn intermetallic compounds in Pb-free solders using simulated annealing," *Acta Mater.*, vol. 55, no. 8, pp. 2805–2814, 2007.

[30] Y. C. Chan, A. C. K. So, and J. K. L. Lai, "Growth-kinetic-studies-of-Cu–Sn-intermetallic-compound-and-its-effect-on-shear-strength-of-LCCC-SMT-solder-joints_1998_Materials-Science-and-Engineering-B.pdf," vol. 55, pp. 5–13, 1998.

[31] N. Mookam *et al.*, "Analysis of Low-Temperatur Intermetallic in Copper-Tin Couples," *Metall. Trans. A*, vol. 28, no. 1, pp. 857–864, 2012.

[32] J. Z. Mei, A.J. Sunwoo and J.W. Morris, "Analysis of Low-Temperatur Intermetallic in Copper-Tin Couples," *Metall. Trans. A*, vol. 23, no. 3, pp. 857–864, 1992.

[33] R. Labie, W. Ruythooren, K. Baert, E. Beyne, and B. Swinnen, "Resistance to electromigration of purely intermetallic micro-bump interconnections for 3D-device stacking," *2008 IEEE Int. Interconnect Technol. Conf. IITC*, pp. 19–21, 2008.

[34] I. A. Blech, "Electromigration in thin aluminum films on titanium nitride," *J. Appl. Phys.*, vol. 47, no. 4, pp. 1203–1208, 1976.

[35] M. Lu, D. Y. Shih, P. Lauro, and C. Goldsmith, "Blech effect in Pb-free flip chip solder joint," *Appl. Phys. Lett.*, vol. 94, no. 1, pp. 0–3, 2009.

[36] J. Lienig and M. Thiele, *Fundamentals of Electromigration- Aware Integrated Circuit Design.* 2018.

[37] R. G. Filippi *et al.*, "The effect of current density, stripe length, stripe width, and temperature on resistance saturation during electromigration testing," *J. Appl. Phys.*, vol. 91, no. 9, pp. 5787–5795, 2002.

Effect of Grain Orientation and Microstructure Evolution on Electromigration in Flip-Chip Solder Joint

Xing Fu
Science and technology on reliability physics and application of electronic component laboratory
Guangzhou, China
albertfuxing@sina.com

Bin Zhou
Science and technology on reliability physics and application of electronic component laboratory
Guangzhou, China

Yunfei En
Science and technology on reliability physics and application of electronic component laboratory
Guangzhou, China

Si Chen
Science and technology on reliability physics and application of electronic component laboratory
Guangzhou, China

Ruohe Yao
School of Electronics and Information South China University of Technology
Guangzhou, China

Abstract—Only bulk diffusion occurs in the solder joint of the monocrystalline structure without grain boundary diffusion. The migration direction of intermetallic compounds (IMCs) is mainly determined by grain orientation. Correspondingly, both bulk diffusion and grain boundary diffusion exist in the polycrystalline structure solder joints. The grain boundaries with random orientation and different lengths are distributed in the bumps, which can effectively inhibit the IMCs migration under the stressing of electron flow. The structure of lead-free solders mainly composed of Sn-based solder tends to nucleate cyclic twin structure with only two grains. The misorientation angle between the two grains is calculated to be a constant value of 57.2 degrees in this study. It is confirmed that the solder joints of the cyclic twin structure have a excellent inhibitory effect on electromigration failure.

Keywords-electromigration; cyclic twin; IMCs; orientation; EBSD

I. Introduction

With the development trend of electronic packaging to the direction of three-dimensional system in package, the bump size and the pitch in SiP and TSV have been continuously to 1 μm[1,2]. As a result, the current density in miniaturized interconnects increases dramatically. When the current density exceeds $10^4A/cm^2$, electromigration (EM) will take place to accelerate the interconnection degradation dominantly[3]. It becomes a vital issue for interconnect reliability in highly integrated SiP and TSV structures[4]. SnAgCu solder has basically replaced traditional lead-containing solders in the field of electronic products due to its excellent thermal properties. Many scholars have found that the micro-solders composed of tin-based solders have a crystallographic structure inside the body-centered tetragonal lattice structure, which makes the solder joints have obvious anisotropy in mechanical, electrical and diffusion properties[5-7]. The trajectory and migration mechanism of IMC migration under the influence of electronic wind is still unclear. In this study, the migration of IMC was characterized by backscattered electron image combined with backscatter diffraction pattern. The diffusion mechanism of IMCs in the bump under current was analyzed theoretically.

II. Experiment

The solder of the Sn-3.0-Ag-0.5-Cu was fabricated into bumps in the form of flip-chip bonding by reflow at a temperature of 240 °C. Figure 1 was the schematic map of a flip-chip samples. The upper and lower surfaces of the bumps were soldered to the copper pads for electrical and mechanical contact. The pads and solder balls have diameters of 420μm and 500μm, respectively. The thickness of the copper plating on the pad surface was designed to be 20 μm. A precision DC regulated power supply was used to keep the current value constant at 2.5A. The ion milling method was adopted by applying argon ion source bombardment to remove stress layers with the thickness of approximately 10μm. The IMCs migration and crystallographic orientation were characterized utilizing SEM and EBSD analysis.

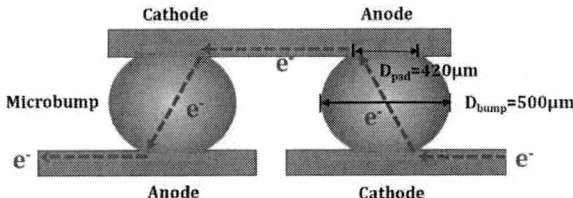

Figure 1 cross-sectional schematic map of a flip-chip samples

III. RESULTS AND DISSCUTION

A. Effect of Monocrytalline and Polycrystalline Structures on Electromigration

Figure 2 is the inverse pole figure and SEM topography of the Sn3.0Ag0.5Cu bumps loaded at a current of 2.5 A for 163.5 h. The current direction are shown by the arrows. The EBSD image in the left column is in one-to-one correspondence with the bumps represented by the SEM photographs in the right column. As can be seen from the inverse pole figures of 2a, 2c, 2e and 2g, the four solder bumps all exhibit a monocrystalline structure. The difference in color of the inverse pole figures indicate that the four single crystal structures have different grain orientations, respectively. In figure 2b, 2d, 2f and 2h, it can be found that the intermetallic compound layers at the upper cathode interfaces degradated from the original smooth state to uneven due to the continuous impact force during 163.5h as shown by the red dotted box in the figure2. The driving force of the rapidly IMCs migration at cathode regions were the rapid diffusion of Cu atoms forced by the electron wind. The electrical force acting on the Cu atoms was taken to be main driving force as illustrated in the equation below proposed by Huntington and Grone[8].

$$F_{em} = Z^* eE = (Z_{el}^* + Z_{wd}^*)eE$$

Here, e is the electron charge of the, E is the electric field ($E = \rho j$, where ρ is the resistivity, j is the current density), and Z^* is the effective charge number of EM. Z_{el}^* can be interpreted as the nominal valence of the diffusing ions ignoring the shielding effect. $Z_{el}^* eE$ is regarded as the direct force. Z_{wd}^* is a number of charges representing the momentum exchange. $Z_{wd}^* eE$ is called electron wind force, and it is commonly found to be much bigger than the direct force in electromigration.

In addition, an interesting phenomenon was clearly characterized in the last monocrystallin strcture bump. The above four bumps characterized in figure 2 were stressed at the same current conditions. It is interesting to note that fewer IMCs migrated in long trajectory inside the first to third bumps, while a striking amount of intermetallic compounds migration occured in the last bump. This indicated that the specific crystal orientation of the internal structure in the fourth bump caused a rapid migration of the intermetallic compound. In the monocrystally structure bumps, since only the bulk diffusion mechanism exists, the experiment showed that when a crystal orientation was just

favorable for the direction of electron flow, the migration of intermetallic compounds can be greatly accelerated.

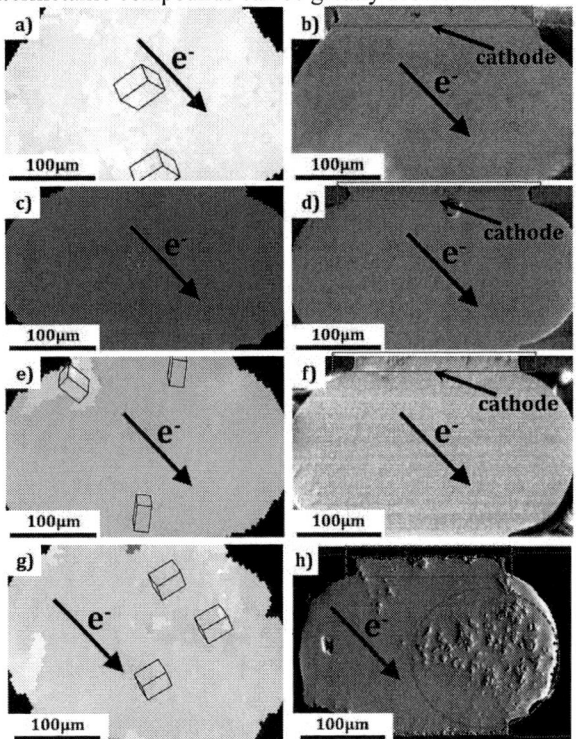

Figure 2 EBSD images and the corresponding SEM images of the microbumps with monocrystalline structure stressed at 2.5A for 163.5h at room temperature
a) c) e) g) inverse pole figures corresponding to b) d) f) h) SEM images

In the previous study, we concluded that the crystal structure inside the bump of the tin-based structure was a body-centered tetragonal structure, and had superior diffusivity along the c-axis compared to other axis [9]. The migration of the IMCs can be greatly accelerated when the electron flow is nearly parallel to the c-axis. The solute atoms in Sn-based micro-interconnects migrated rapidly when the c-axis direction is parallel to the electron flow trajectory.[10] Consequently, only bulk diffusion occurs in the solder joint of the monocrystalline structure without grain boundary diffusion. The migration direction of intermetallic compounds (IMCs) is mainly determined by grain orientation.

Figure 3 showed the EBSD images and the corresponding SEM images of the microbumps with polycrystalline structure stressed at 2.5A for 163.5h at room temperature. Analysis from the perspective of basic theory, bulk diffusion and grain boundary diffusion exist in the polycrystalline structure solder bumps simultaneously. It indicated that the orientation of each grain was random, and grain boundaries with different lengths were zigzag distributed in the bump. From the microstructure of SEM photographs, almost none significant intermetallic compounds migration occured inside the bump. Due to the impact of the electron impact, the interface of the cathode on the upper side of the bump is degraded to some extent. Comparing the analysis of the diffusion mechanism in the monocrystalline structure solder joint, we can draw conclusions that the polycrystalline structure can effectively inhibit the migration of intermetallic compounds due to its tortuous grain boundaries and uncertain crystal orientation.

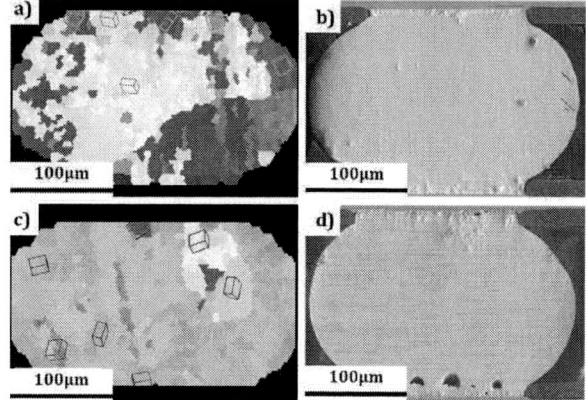

Figure 3 EBSD images and the corresponding SEM images of the microbumps with polycrystalline structure stressed at 2.5A for 163.5h at room temperature
a) c) inverse pole figures corresponding to b) d) SEM images

B. Effect of Cyclic Twin Structure Solder Joints on Electromigration

Figure 4 showed the EBSD images and the corresponding SEM images of the microbumps with cyclic twin structure stressed at 2.5A for 163.5h at room temperature. Figure 4a and 4d were the inverse pole figures corresponding to figure 4b and 4e SEM images. Figure 4c and 4f were the misorientation distribution figures corresponding to 4a and 4d EBSD images. Striking cyclic twin structure was characterized in the two solder joints, and the striking twin boundarys were formed across the bumps as shown by the red dotted line in the figure 4a and 4d. The misorientation angles of the two main grains were calculated to be 57.2 degrees, which has been also confirmed by L.P. Lehman [11].

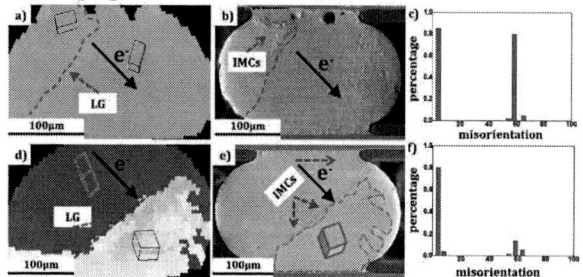

Figure 4 EBSD images and the corresponding SEM images of the microbumps with cyclic twin structure stressed at 2.5A for 163.5h at room temperature
a) d) inverse pole figures corresponding to b) e) SEM images
c) f) misorientation distribution figures corresponding to a) d) EBSD images

The results in figure 4 indicated that the cyclic twin grain boundaries tended to form long grain boundaries throughout the bump. It was also found that some intermetallic compounds migrated to grain boundaries and accumulated after 163.5 hours of current stressing as shown in figure 4b

and 4e, while no accumulated compounds were found inside the grains on both sides of the grain boundary. This is different from the case of intermetallic compounds migration in monocrystalline structure solder joints. The grain boundary diffusion played a major role in bumps with twin structure. We can concluded that the cyclic twin structure can inhibit the long-range migration of IMCs to a certain extent. It can do favor to inhibit the electromigration degradation in the Sn-based solder bumps. The angles of the two main grains' misorientation have been calculated and characterized to about 57.2 degrees as shown in the figure 4c or 4f. Consequently, in the bumps of the cyclic twin structure, when the extending direction of the long boundary is nearly perpendicular or the angle of the electrical current flow direction is larger, the diffusion direction of the atoms depends on the lattice orientation of β-Sn on both sides of the grain boundary as shown in figure 4. This is consistent with our previous research results. In another case, when the extension direction of the long crystal boundary is nearly parallel or the angle of the electron flow is smaller, the intermetallic compounds will migrate along the grain boundary. At this time, the lattice orientation on both sides of the grain boundaries is not conducive to the diffusion of each atom[9]. This is a very valuable research conclusion.

IV. CONCLUSION

(1) Only bulk diffusion occurs in the solder joint of the monocrystalline structure without grain boundary diffusion. The migration direction of intermetallic compounds (IMCs) is mainly determined by grain orientation.

(2) Both bulk diffusion and grain boundary diffusion exist in the polycrystalline structure solder joints. The grain boundaries with random orientation and different lengths are distributed in the bumps, which can effectively inhibit the IMCs migration under the stressing of electron flow.

(3) The structure of lead-free solder joints mainly composed of Sn-based solder tends to nucleate cyclic twin structure with only two grains. The misorientation angle between the two grains is calculated to be a constant value of 57.2 degrees in this study. It is confirmed that the solder joints of the cyclic twin structure have a excellent inhibitory effect on electromigration failure.

ACKNOWLEDGMENT

This work was supported by the "Thirteen Five" pre-research project and Key Laboratory Fund (Grant No.JAB1728200; No.JAB1728240), and

REFERENCES

[1] C. C. Lee, T. F. Yang, C. S. Wu, K. S. Kao, C. W. Fang, and C. J. Zhan, "Impact of high density TSVs on the assembly of 3D-ICs packaging", Microelectronic Engineering. J. 2013, 107(1):101-106.

[2] F. Y. Ouyang, W. C. Jhu, "Comparison of thermomigration behaviors between Pb-free flip chip solder joints and microbumps in three dimensional integrated circuits: Bump height effect", Journal of Applied Physics. J. 2013, 113(4):972.

[3] H. T. Chen, C. J. Hang, X. Fu and M. Y. Li, "Microstructure and Grain Orientation Evolution in Sn-3.0Ag-0.5Cu Solder Interconnects Under Electrical Current Stressing", Journal of Electronic Materials. J. 2015, 44(10):3880-3887.

[4] Cheng, H. En and R. S. Chen, "Interval optimal design of 3-D TSV stacked chips package reliability by using the genetic algorithm method", Microelectronics Reliability. J. 2014, 54(12):2881-2897.

[5] J. Q. Chen, K. L. Liu, J. D. Guo, H. C. Ma, S. Wei, and J. K. Shang, "Electromigration anisotropy introduced by tin orientation in solder joints", Journal of Alloys & Compounds. J. 2017, 703:264-271.

[6] S. Wei, H. Ma, J. Chen, and J. Guo, "Extreme anisotropy of electromigration: Nickel in single-crystal tin", Journal of Alloys & Compounds. J. 2016, 687:999-1003.

[7] J. Q. Chen, J. D. Guo, H. C. Ma, S. Wei, and J. K. Shang, "Cu6Sn5 intermetallic compound anisotropy introduced by single crystal Sn under current stress", Journal of Alloys & Compounds. J. 2017, 695:3290-3298.

[8] H. B. Huntington, A. R. Grone, "Current-induced marker motion in gold wires", Journal of Physics & Chemistry of Solids. J. 1961, 20(1):76-87.

[9] X. Fu, H. Chen, B. Zhou, S. Chen, R. Yao and X. He, "Effect of Crystal Boundary Character and Crystal Orientation on Electromigration in Lead-free Solder Interconnects with Cyclic Twin Structure," 2018 19th International Conference on Electronic Packaging Technology (ICEPT), Shanghai, 2018, pp. 701-703.

[10] L. P. Lehman, Y. Xing, T. R. Bieler, "Cyclic twin nucleation in tin-based solder alloys", 2010, pp.3546-3556.

[11] T. C. Huang, T. L. Yang, J. H. Ke, C. H. Hsueh, and C. R. Kao, "Effects of Sn grain orientation on substrate dissolution and intermetallic precipitation in solder joints under electron current stressing", Scripta Materialia. J. 2014, 80(2):37-40.

High Electromigration Lifetimes of Nanotwinned Cu Redistribution Lines

I-Hsin Tseng[1], Yu-Jin Li[1], Benson Lin[2], Chia-Cheng Chang[2] and Chih Chen[1]

Department of Materials Science and Engineering, National Chiao Tung University[1]
MediaTek Inc[2]

1001 Ta Hsueh Road, Hsin-Chu 30010, Taiwan
PT, MediaTek Inc., Hsinchu 300, Taiwan

chih@mail.nctu.edu.tw

Abstract— In this paper, we measured the electromigration lifetimes for nanotwinned copper (nt-Cu) and regular copper redistribution lines (RDLs) for 3D-ICs packaging. The width of the RDLs is 10μm in width and 5μm in thickness, and the lines were coated by polyimide. The average lifetime of the nt-Cu RDL copper line is 445 hours to reach 1.2 times of its initial resistance. However, it takes 160 hours to reach the same resistance increase for regular copper RDL line. The nt-Cu has better EM lifetime than regular copper.

INTRODUCTION

The requirements of commercial electronic products continue to increase and the package sizes continue to decrease. The redistribution lines (RDLs) in fan-out packages and three dimensional integrated circuit (3D IC) become narrower, resulting in the high current density in the RDLs. The three dimensional integrated circuits which can stack chips vertically and decrease RC-delay. As the scaling of large-scale ICs faced the physical limit, the IC industry is trying to follow the Moore's law to fulfill the needs of customers. By using the 3D IC, which not only reduce the form factor, but also enhance the power efficiency. The input/output pin numbers increasing while the size of the solder joints are shrinking. It is inevitably leads to higher current densities and operating temperatures in the joints. It raises severe reliability issues such as electromigration and thermomigration. Furthermore, the RDLs need carry higher current density than before. The redistribution layers [1]-[4] have become one of the priority researches. Lui, et al. [5] found that during the electromigration process, the damage occurred between the C4 bumps and the microbumps has a positive correlation between the electromigration and the joule heating effect, so that the wires locally reach the melting point of the copper, and the RDLs copper lines are melted and broken. Therefore, it is necessary to find a way to strengthen the line [6], such as covering the RDLs copper line with metal [7], improving the contact surface with Low K [8], or changing the microstructure in the copper line to enhance the properties.

As the volume of 3D ICs getting smaller, copper is the critical material in interconnects. Highly preferred oriented nanotwinned copper has been extensively studied. Many researched used sputtering to grow nanocrystalline copper (nt-Cu) [9] [10] and found it to have excellent mechanical strength and high ductility. Moreover, the electrical conductivity of nt-Cu is almost the same as that of bulk copper. Copper has low resistivity of 1.7 μΩ, high conductivity 401 W m^{-1} K^{-1}. Recently, In 2012, Hsiao et al reported electrodeposition of highly (111) preferred orientation of nanotwinned copper (nt-Cu) by direct current[11]. This new material may have better electric and mechanical properties and may meet the needs of the packaging industry.

It is reported that the nano-scale twins in the copper contribute to high resistance of the electromigration. Chen et al[12], using in-situ TEM to observed electromigration in nt-Cu. They found that the microstructure of copper has nano-scale twins help to improve resistance to electromigration, when copper atoms flow through the intersection of twin boundaries and grain boundaries (triple junction), the diffusion rate of copper atoms could be reduced by 10 times, and nanotwinned copper can effectively reduce the rate of electromigration.

Therefore, we use electroplated nt-Cu as the material of RDLs and observe destruction mechanism of electromigration tests. We also compare the electromigration lifetime of nt-Cu RDL lines to regular Cu.

EXPERIMENTAL

We produced samples with lithography and electrodeposition processes, as illustrated in Fig.1, After the sputtering of Ti/Cu seed layer on Si wafers with oxides, we dice the wafer into pieces in order to electroplate copper in our lab. we electroplated (111) prefer oriented nt-Cu by using periodic reverse electroplating on silicon substrates. The electroplating bath was high-purity $CuSO_4$ solution inclusive of 0.8M copper cations, 40 ppm hydrochloric acid, 100 g/ L sulfuric acid and the additive for nanotwin growth provided by Chemleaders, Inc. The condition of the electroplating is on time 4 A/dm^2 and reverse time -1 A/dm^2. We also make a sample without nano-scale twin structure as the controlled samples. The condition of the electroplating is 4 A/dm^2 by using direct current electroplating in the above bath with a commercial additive.

978-1-7281-1500-9/19 $31.00 © 2019 IEEE

After electroplating, we rinsed samples in acetone for 5 minutes to remove the photoresist, then rinse them in DI water for 30 seconds and etch copper seed layer by copper etching. Finally, A Cu line with 800 μm long, 5 μm thick and 10 μm wide was fabricated for electromigration tests. In order to avoid oxidation, in Fig. 2, low temperature polyimide (PI) was adopted to passivate the copper line and the curing temperature was at 230 °C for 1 hour.

Four-point probes are employed to monitor the resistance change during the electromigration tests of the RDL copper lines. The RDL copper line is stressed with a direct current on a 200 °C hot plate and the applied current density is 1×10^6 A/cm^2.

We examine orientation maps of copper lines to confirm the quality of nt-Cu by an EDAX electron back-scattered diffraction (EBSD) system, within the scanning electron microscope (JEOL 7800 FESEM). We observe the microstructure of RDL copper lines to figure out the void distribution by focused ion beam (FIB).

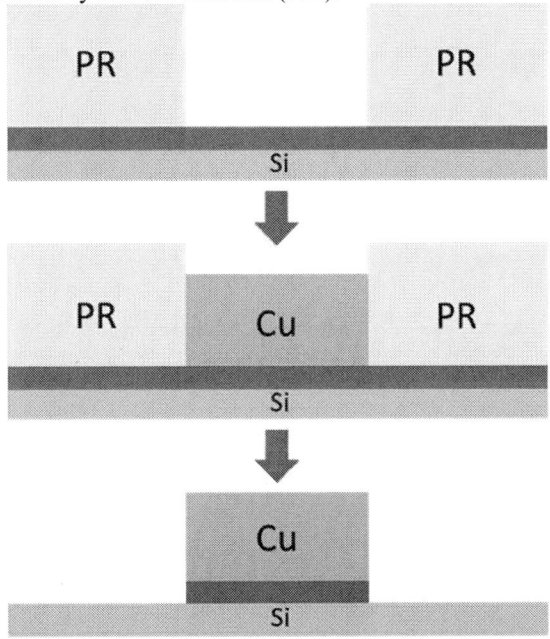

Figure. 1 Schematic process showing the fabrication procedure for the test structure of Cu lines.

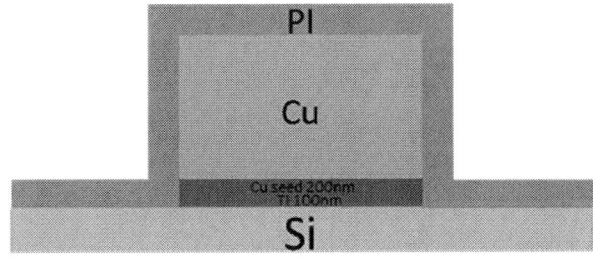

Figure 2 The cross-sectional schematic diagram of a RDL copper line for electromigration tests.

RESULTS AND DISCUSSION

In order to measure the resistance change in copper line, we designed a pattern with 800 μm copper line and seven probing pads. Figure 3 shows the layout for the EM test pattern. The top two pads are for applying current. The others pads can measure the line resistance in different line lengths.

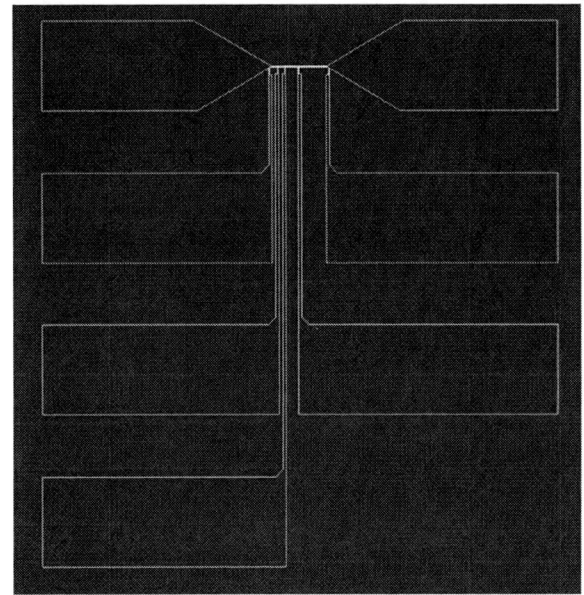

Figure. 3 The layout pattern for electromigration tests. Current was applied through the top two pads, and the other probing pads can be used to mesure the resistance change of various segment of the Cu line.

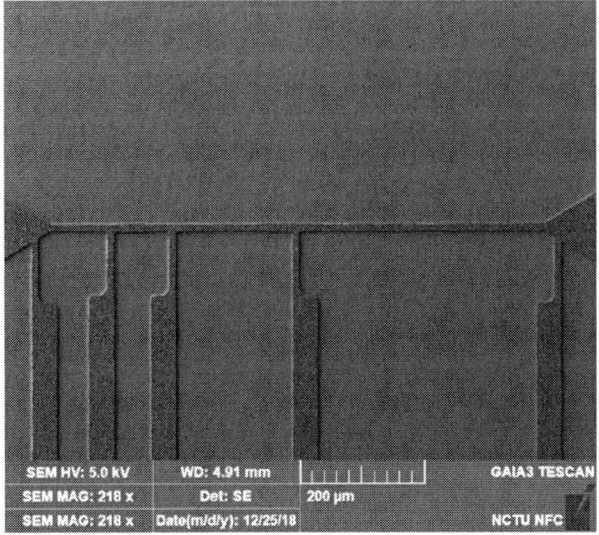

Figure.4 Plan-view SEM image for the fabricated EM layout.

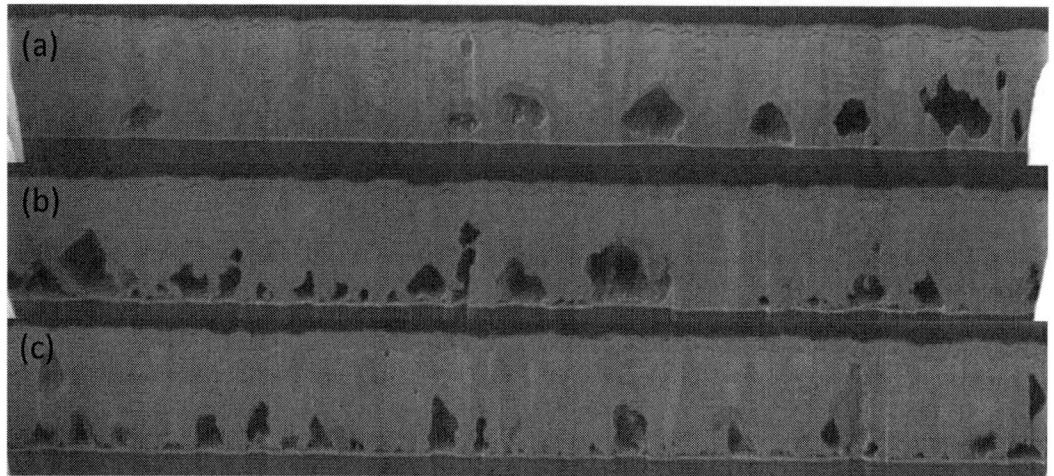

Figure 8 (a) The cross-section of cathode end of nt-Cu line (b) the cross-section of middle of nt-Cu line. (c) the cross-section of anode end of nt-Cu line

After electrodeposition of the Cu lines, we use focused ion beam (FIB) to observe the microstructure of RDLs copper line. Fig. 4, SEM image shows the top view of 800 μm length copper line. Fig. 5, shows the ion beam cross-section of nt-Cu. There are columnar grains with nano-scale twins in copper line. From Fig. 6, the XRD results shows that nt-Cu has highly (111) preferred orientation. Fig. 7, we use EBSD to obtain the orientation map of nt-Cu. We can see most of the grains are (111) orientation which is shown in blue color and the average grain size is about 1μm.

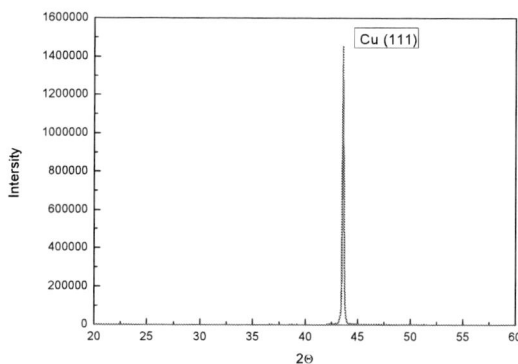

Figure. 6 XRD of the nt-Cu lines.

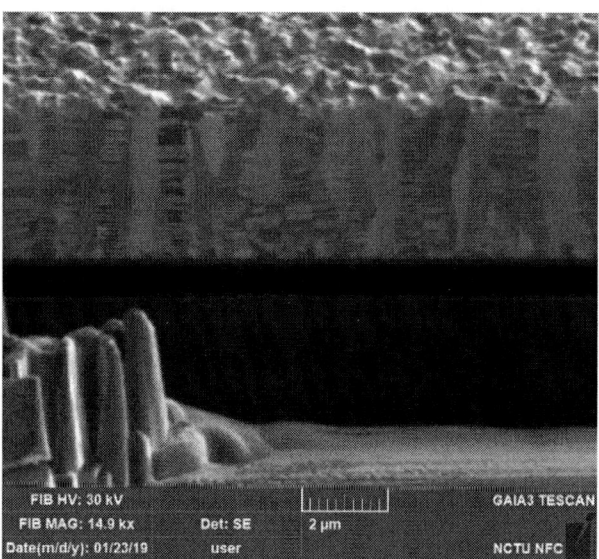

Figure. 5 Cross-sectional FIB image showing the columnar grains with nanotwins.

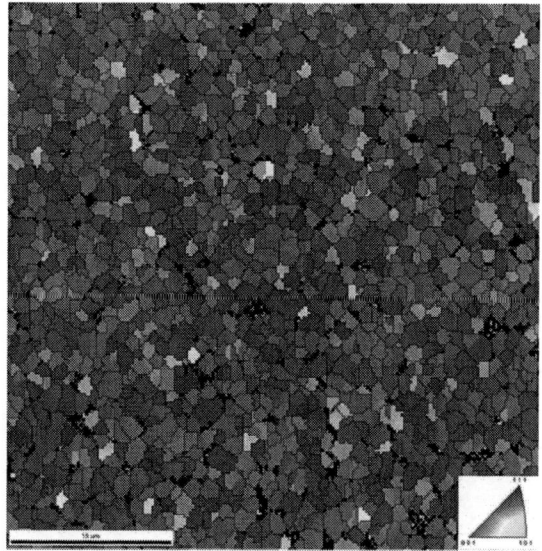

Figure.7 Plan-view EBSD image of nt-Cu lines.

After 20% resistance increase of its initial value of the copper line, we stop the electromigration test. We observe the microstructure change by FIB and try to observe failure mechanism. Fig. 8, shows the FIB cross-section of cathode end, middle and anode end of nt-Cu line for the nt-Cu RDL. Each cross-section is 50μm long. We can see that there are a lot of voids in the copper line. We figured out that there are some voids with columnar shape. We suggest that the reason is the columnar grain of nt-Cu. When copper atoms are left to be taken away by electromigration, forming a small void. There are many voids existing in the interface of the Cu seed and the electroplated Cu films. Moreover, the voids are scattering in the nt-Cu line.

We speculate that the voids are the cause of the increase in resistance. From the cross section, we can still observe the presence of nanotwinned copper. The temperature of the copper line is close to 230 °C because of joule heating during EM test for 445 hours, showing that nanotwinned copper has excellent thermal stability. Finally, we collect the data of resistance change from four point probes and we obtain the figure of resistance versus time. From Fig. 9, the resistance of nt-Cu line takes 445 h to increase 20% of the initial resistance.

We also fabricated the regular copper line for comparing with nt-Cu line. The samples were electroplated at commercial copper electroplating solution. Before the EM test, we examined the microstructure of regular copper line, as shown in Fig.10. There are no nanotwins in the Cu lines. On the contrary, there are small grains in it.

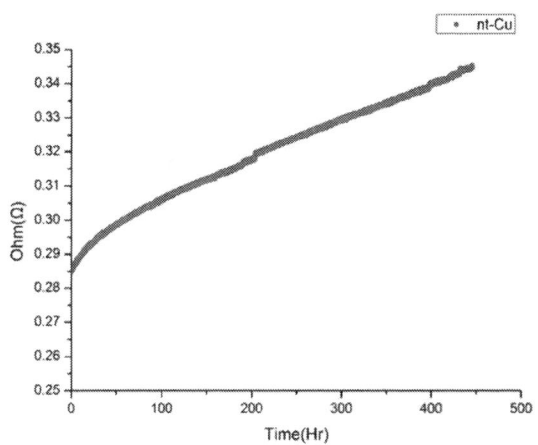

Figure. 9 The plot of resistance versus stressing time of nt-Cu

Figure. 10 FIB cross-section of regular copper line

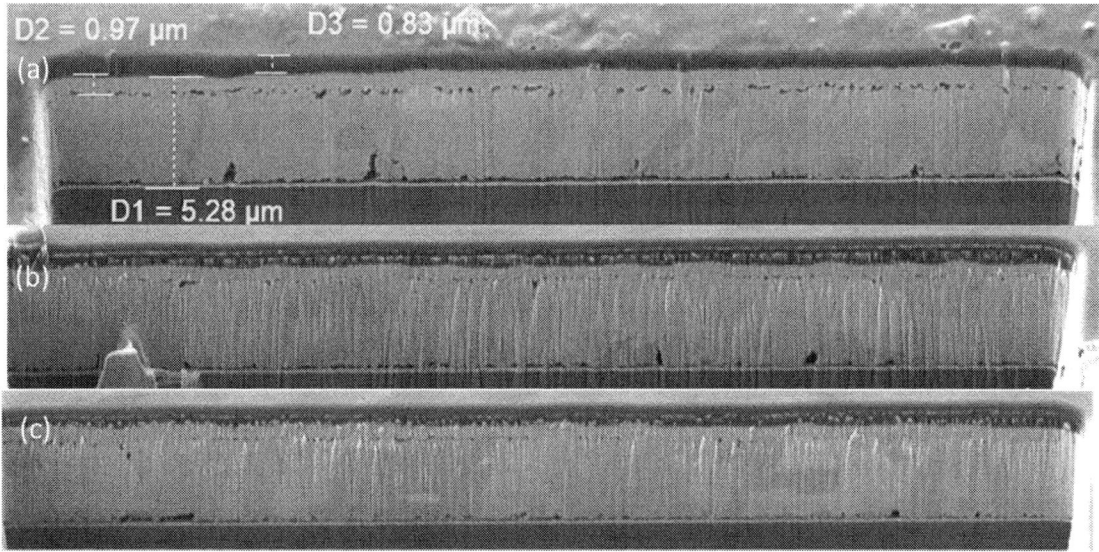

Figure 11, (a) The cross-section of cathode end of nt-Cu line (b) the cross-section of middle of nt-Cu line. (c) the cross-section of anode end of nt-Cu line

978-1-7281-1500-9/19 $31.00 © 2019 IEEE

According to the four-point measurement, we ended the EM test when the resistance increase exceed 20% of its initial resistance value. From Fig. 11, there are voids at the bottom of the regular copper line and on the interface of copper oxide. But the voids we observed were not large enough to increase 20 percent of its initial resistance. Because of the machine issue, it's hard to observe everywhere of 800 µm of copper line. We suggest that there are some large voids in the copper line and we are unfortunately to find them.

Polyimides (PI) are widely use in packaging industrial. PI are known for thermal stabiliy, good chemical resistance, excellent mechanical properties but PI cannot isolate the oxygen and moisture completely. So there were thick oxides after the EM tests.

From Fig.11, (a) the copper oxide is approximately 1µm thick after 160 hours. However, from Fig.8,(a) after 445 hours EM test, the copper oxide is approximately 0.6 µm thick. Tseng. Et al[13] reported that the oxidation rate of the highly (111)-oriented nt-Cu film was lower compared to the randomly-oriented Cu film. Therefore, the copper oxide in regular copper is thicker than that on the nt-Cu lines.

We collect the data of resistance changes from four point probes and we obtain the figure of resistance versus stressing time. From Fig.12, it takes 160 hrs to reach the resistance increase of 20% of the initial resistance. In the initial stage, the resistance increase slowly. However, the resistance increase rapidly around 150 h.

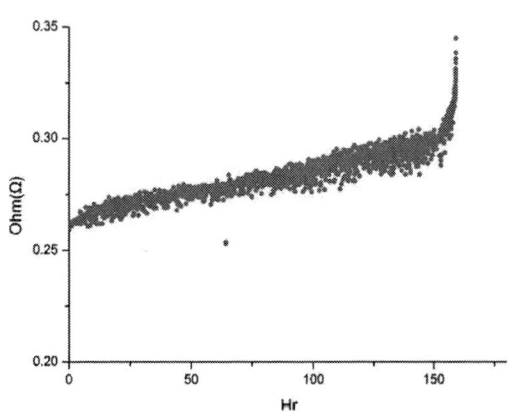

Fig.12 The plot of resistance versus stressing time of regular copper line

CONCLUSIONS

We have designed a test vehicle for measure the EM lifetime of RDLs copper line. We found that nanotwinned copper is highly resistant to electromigration. It takes 445 h to reach 20% increasing of its initial resistance for nt-Cu RDLs lines. However, the lifetime of regular RDLs copper line is only 160 h.

We found different ways of EM failures between nt-Cu and regular copper. There are lots of columnar-like voids scattered among the nt-Cu RDLs lines. The columnar voids grow slowly under high current stressing at high temperature. However, there are some large voids in the regular copper lines which leads to the rapid resistance increase.

Nt-Cu has many advantages such as high EM resistance, good thermal stability, low oxidation rate and excellent mechanical properties. With these excellent properties, nt-Cu RDLs line is a promising material to enhance the electromigration performance of high power chip in high-end electronic products.

ACKNOWLEDGMENT:

This work was financially supported by the "Center for the Semiconductor Technology Research" from The Featured Areas Research Center Program within the framework of the Higher Education Sprout Project by the Ministry of Education (MOE) in Taiwan. Also supported in part by the Ministry of Science and Technology, Taiwan, under Grant MOST-108-3017-F-009-003.

REFERENCES

[1] Ogawa, Ennis T., et al. "Electromigration reliability issues in dual-damascene Cu interconnections." IEEE Transactions on reliability 51.4 (2002): 403-419.

[2] Vairagar, A. V., et al. "In situ observation of electromigration-induced void migration in dual-damascene Cu interconnect structures." Applied physics letters 85.13 (2004): 2502-2504.

[3] Hu, C. K., et al. "Mechanisms for very long electromigration lifetime in dual-damascene Cu interconnections." Applied Physics Letters 78.7 (2001): 904-906.

[4] Vairagar, A. V., S. G. Mhaisalkar, and Ahila Krishnamoorthy. "Electromigration behavior of dual-damascene Cu interconnects—Structure,

width, and length dependences." Microelectronics Reliability 44.5 (2004): 747-754.

[5] Liu, Yingxia, et al. "Synergistic effect of electromigration and Joule heating on system level weak-link failure in 2.5 D integrated circuits." Journal of Applied Physics 118.13 (2015): 135304.

[6] Xu, Di, et al. "Nanotwin formation and its physical properties and effect on reliability of copper interconnects." Microelectronic Engineering 85.10 (2008): 2155-2158.

[7] Yang, C-C., et al. "Dependence of Cu electromigration resistance on selectively deposited CVD Co cap thickness." Microelectronic Engineering 106 (2013): 214-218.

[8] Lin, M. H., et al. "Electromigration lifetime improvement of copper interconnect by cap/dielectric interface treatment and geometrical design." IEEE Transactions on Electron Devices52.12 (2005): 2602-2608.

[9] Lu, Lei, et al. "Ultrahigh strength and high electrical conductivity in copper." Science 304.5669 (2004): 422-426.

[10] Li, Xiaoyan, et al. "Dislocation nucleation governed softening and maximum strength in nano-twinned metals." Nature 464.7290 (2010): 877.

[11] Hsiao, Hsiang-Yao, et al. "Unidirectional growth of microbumps on (111)-oriented and nanotwinned copper." Science 336.6084 (2012): 1007-1010.

[12] Chen, Kuan-Chia, et al. "Observation of atomic diffusion at twin-modified grain boundaries in copper." Science 321.5892 (2008): 1066-1069.

[13] Tseng, Chih-Han, K. N. Tu, and Chih Chen. "Comparison of oxidation in uni-directionally and randomly oriented Cu films for low temperature Cu-to-Cu direct bonding." Scientific reports 8 (2018).

This page intentionally left blank.

Non-destructive failure analysis of various chip to package interaction anomalies in FCBGA packages subjected to temperature cycle reliability testing

Vishnu V. B. Reddy, I. Charles Ume
Woodruff School of Mechanical Engineering
Georgia Institute of Technology
Atlanta, GA, US
e-mail: vishnuvardhan@gatech.edu

Jaimal Williamson, Luu Nguyen
WW Semiconductor (SC) Packaging
Texas Instruments Inc.
Dallas, TX, US
e-mail: jaimal@ti.com

Abstract— Failure analysis and root cause investigation of various package anomalies are key criteria to predict the failures, and thus the life of a device after reliability testing. Currently, quality assurance teams depend on less reliable electrical resistance probe testing and destructive methods such as cross-sectioning and microscopy, for the failure analysis. However, a reliable non-destructive test is preferred for failure analysis to provide more accurate and efficient data. Laser Ultrasonic and Laser Interferometer (LULI) Inspection System is a non-contact and non-destructive system that can be used for failure analysis of package anomalies. This system uses pulsed laser to generate ultrasound within the sample and a laser interferometer to sense the surface displacement response. To predict an anomaly at a location on the test sample, the surface displacement response at that location on the test sample will be compared against the response at the same location on the reference sample. For this research, original Flip-Chip Ball Grid Array (FCBGA) samples were used to evaluate various chip-to-package anomalies including inner and outer layer substrate discontinuities, second level joint imperfections (like voiding), and pad cratering in the printed circuit board. This paper presents promising results that can circumvent time consuming cross sectioning of substrate layers and laborious sample preparation consistent with general failure analysis tactics to facilitate root cause investigation of various package anomalies using the nondestructive LULI inspection technique.

Keywords-FCBGA; Laser Ultrasonic and Laser Interferometer inspection; failure analysis; chip package anomalies

I. INTRODUCTION

Cyclic thermal loads during packaging assembly process and service life are the major cause of failures of electronic packaging. Thermal expansion mismatch between various materials in a package lead to these failures. Delamination, cracking, which can manifest via electrical opens are the most common defects that need to be detected early to ensure a quality product. Thermal reliability test is a well-adopted qualification test to determine failure types and predict lifetime of component under cyclic thermal loads [1].

Non-destructive methods are preferred for failure analysis and root cause investigation of various package anomalies to preserve the test samples and to reduce the overall effort from destructive methods. Traditional automatic optical inspection systems cannot inspect the

solder connections after reflow, because the connections are hidden beneath the package and cameras cannot detect them through the body of the package. Most commonly used non-destructive techniques for identification of board level defects are X-rays, electrical curve tracing, and scanning acoustic microscopy (SAM). However, X-ray and acoustic microscopy can only provide non-optimal solutions for addressing inspection reliability and efficiencies. Although X-ray gives visual images to identify defects, there are significant limitations in both 2D and 3D X-ray. 2D X-ray is limited to top-down view which has poor accuracy of defects as the connections are underneath the components. While 3D X-ray overcomes this limitation, it takes significant processing time, up to 10 minutes per part, and prohibitive up-front investment cost [2]. Acoustic microscopy uses acoustic waves to measure and create visual images of variations in samples, allowing detection of delamination, voids and cracks in the connections between components and board. However, acoustic microscopy which has limited resolution requires coupling medium and cannot detect along the electronic package edge [2]. Electrical curve tracing is a simple and cheap method for preliminary analysis to predict the failure net based resistance measurement. However, this method is not accurate and cannot give exact location of failure site.

This paper presents a novel LULI failure analysis method for FCBGA test samples subjected to thermal reliability testing. LULI is a noncontact, nondestructive, fast, and high-sensitivity system. This automatic inspection system identifies defects at various levels of solder interconnections by analyzing electronic package vibration generated by the laser system [3-6]. LULI identifies precise defect locations and severity of defects [3][4]. This system is capable of high measurement throughput, with good accuracy at a relatively low cost [7].

FCBGA test samples were highly stressed after end-of-life conditions and qualification release point as subjected to board level reliability testing based on IPC 9701 test conditions. Thereafter, the samples were inspected using the LULI inspection system, where results correlated with electrical resistance test results. It was found based on a systematic process of mapping failing nets within the FCBGA package to electrical resistance results, that outputs of the LULI non-destructive technique demonstrated very good correlation between electrical resistance results and the failure region.

II. SYSTEM DESCRIPTION

The original image of the LULI system is shown in Figure 1. A Polaris II Nd:YAG laser source generates laser pulses with a repetition rate of 20Hz. Single pulsed beam from the source is split into two beams in multiplexer and then fed into two optical fibers. Focusing objectives at the end of fibers focus the laser on to the surface of the package to be tested. Laser pulses serve to excite the sample and generate ultrasound in the sample. The generated ultrasound from the excitation point propagate deep within the sample and create nano-amplitude vibrations in the sample [8]. Any anomaly at a location will alter amplitude and frequency of the ultrasonic waves (or vibrations) at that location. A fiber-coupled heterodyne interferometer is used to measure the transient out-of-plane surface displacement on surface of the sample. The interferometer is enhanced with an autofocus system to ensure optimal signal strength [9]. The vibration signal acquired by the interferometer is digitized and then averaged to improve signal-to-noise ratio (SNR). Two independent positioning stages are used to position the laser excitation point and interferometer detection point at required locations. A vision camera is used to locate the fiducials and determine the position and orientation of the sample on the fixture with respect to global coordinates. Major system parameters used in this research are listed in Table I.

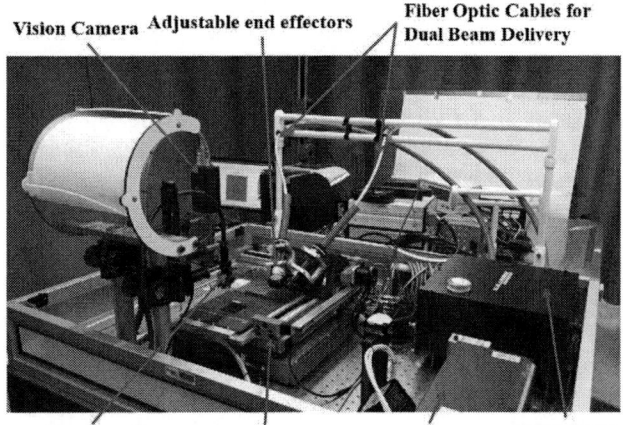

Figure 1. Original image of dual fiber array Laser Ultrasonic Inspection System.

TABLE I. LULI SYSTEM PARAMETERS

Laser wavelength	1064 nm
Pulse length	5ns
Pulse frequency (repetition rate)	20 Hz
Peak average pulsed laser power in each fiber	200 mW
Excitation laser spot size	6.14 mm^2
Interferometer sampling rate	50 MS/Sec
# of sample points considered for MCC	3000
# of signals averaged per inspection point	128

III. TEST SAMPLES DESCRIPTION

The test samples were supplied by Texas Instruments to test the LULI system's ability to identify various chip to package interaction anomalies. The physical test sample is shown in the Figure 2. These samples are Flip Chip Ball Grid Array (FCBGA) packages. The schematic of BGA solder ball locations is shown in the Figure 3. Results from two failure samples, designated as 5U1 and 6U2, which were subjected to board level thermal reliability testing are presented in this paper. One time zero sample, which was not subjected any reliability test, was considered as a reference sample for comparison with failed samples to determine potential failure sites.

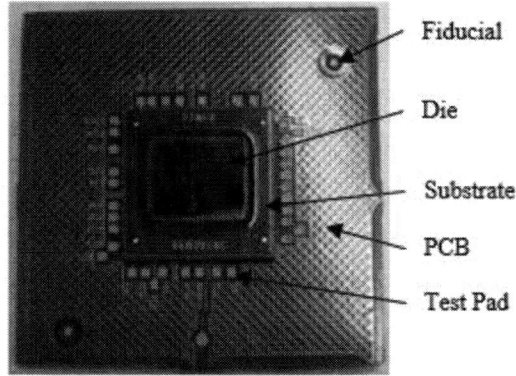

Figure 2. Original Image of the TI FCBGA test sample.

A. Reliability Test Condition

The two failed samples under consideration were subjected to thermal cycling testing for qualification requirements based on IPC 9701 test conditions. The samples were highly stressed after end-of-life conditions and qualification release point. In accordance with IPC 9701, a temperature cycle range of -40°C to +125°C was applied. The occurrence of the first event followed by nine additional events within 10% of the test time upon the first event recorded is defined as a failed sample. Test sample 5U1 failed at 2,011 thermal cycles and test sample 6U2 failed at 2,291 thermal cycles.

B. LULI Inspection Pattern

Interferometer displacement data should be collected at various inspection points for signal analysis and predicting the anomalies. The BGA solder ball layout of the test samples, as shown in Figure 3, is not a full area array. So, it is decided to have one interferometer inspection point at every alternative BGA solder ball location. Red line in Figure 3 indicates the die shadow. High noise in interferometer data collected from the top of underfill spread around the die was observed because of uneven/rough surface. Thus, the underfill spread area is omitted to have any inspection points. Inspection points on top of the die to collect the information about anomalies under the die shadow are represented with blue circles in Figure 3. Similarly, red circles are the inspection points to collect information under BGA substrate outside the die shadow. Interferometer data from the top of die and from top of BGA substrate other than die, collected and presented separately because the reflectivity of the laser light is different for different surfaces.

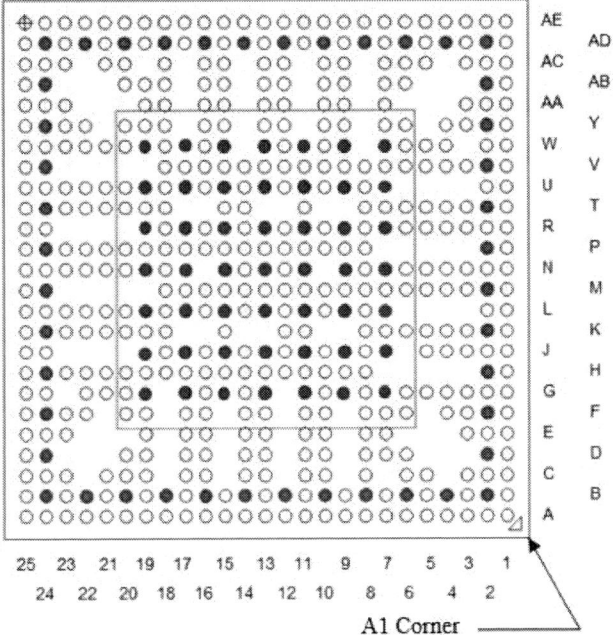

Figure 3. BGA solder ball layout in TI FCBGA package; Red line: Die shadow; Red circles: Inspection points to collect info under substrate - outside the die shadow; Blue circles: Inspection points on die top to collect info under the die.

IV. MODIFIED CORRELATION COEFFICIENT

Modified Correlation Coefficient (MCC) is a signal interpretation method based on the signal similarity, in which the solder bump quality is assessed by comparing the signal with a reference signal [10]. Figure 4 shows the basic idea of how this method works. MCC, as given in equation (1), is the correlation between the signal obtained from a detection point on the surface of a test sample and the reference signal obtained from the surface of the reference sample at the corresponding detection point. MCC calculation returns a value that lies between 0 and 1.

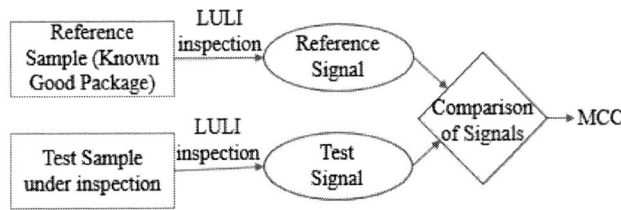

Figure 4. MCC method interpretation in LULI inspection.

$$MCC = 1 - \left(\frac{\sum_n (R_n - \bar{R})(A_n - \bar{A})}{\sqrt{(\sum_n (R_n - \bar{R})^2)(\sum (A_n - \bar{A})^2)}} \right)^2 \qquad (1)$$

Where
R_n: Reference Signal
$\bar{R}$: Average value of R_n
A_n: Test sample signal
$\bar{A}$: Average value of A_n
n: # of sampling points of the signal

The closer the MCC value is to 0, the more correlation there is between the test signal and reference signal. An MCC value of 0 means there is complete correlation that is the test package is exactly the same as the reference sample. A reasonably low MCC value close to 0 indicates that the test sample is good at the inspection point and does not contain any significant anomalies. A MCC value equal to 1 indicates that the signals are completely dissimilar. This means that the test sample is grossly defective and contains very high amount of anomalies at the location of inspection point. The major cause of the difference in the signals is the presence of anomalies in the die, die solder bumps, BGA substrate, vias in BGA substrate and BGA solder balls.

V. LULI RESULTS ON FAILURE TEST SAMPLES

LULI analysis was carried out on failed samples which are highly stressed after end-of-life conditions and results from two failed samples designated as 5U1 and 6U2 were presented in this paper. Experiments were carried out with the laser parameters listed in table I and on inspection points shown in figure 3. The total average laser power from the two fibers without damaging the die surface used for this study is 400mW. Separate sets of experiments were conducted to collect interferometer data on top of die and on peripheral inspection points (i.e. on BGA substrate outside the die). Correlation analysis is carried out using equation (1) at all inspection points and MCC values at each detection point were calculated and plotted in the form of 3D histograms as shown in figures 5 to 8. XY plane give the location of inspection points and z-axis gives the MCC value.

Figure 5 shows the MCC values on the top of die for failed sample 5U1. It is observed that the corner values are relatively more compared with the values in the middle. A thermally cycled sample typically fails at the corners due to the difference in coefficient of thermal expansion of various materials (especially between die and substrate and substrate and printed circuit board,). Based on the symmetry of MCC values, it is predicted that the corner solder balls would have failed, and the expected anomaly is a crack in BGA solder joints. Figure 6 shows the MCC values on the peripheral locations for the failed sample 5U1. All the MCC values are very low. BGA solder balls outside the die shadow are predicted not to show any significant anomalies or failures.

978-1-7281-1500-9/19 $31.00 © 2019 IEEE

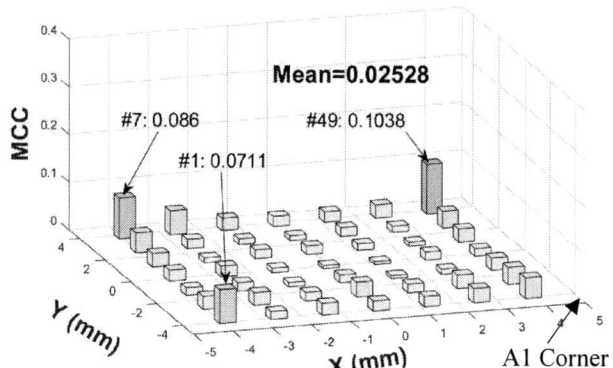

Figure 5. LULI results (MCC values) from the top of die for test sample 5U1 in 3D histogram format.

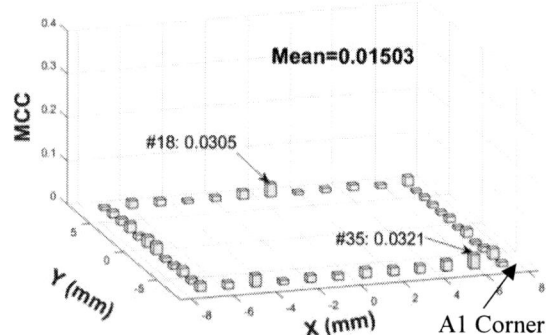

Figure 6. LULI results (MCC values) from the peripheral locations for test sample 5U1 in 3D histogram format.

Figure 7 shows the MCC values on the top of die for failed sample 6U2. It is observed that MCC values are high in almost all inspection points on top of die. Also, MCC values at the corner inspection points on top of the die are relatively higher than other points. From this it can be predicted that most of the BGA solder joints would have failed under die shadow and corner BGA solder joints under the die shadow would have the worst effects. Figure 8 shows the MCC values on the peripheral locations for the failed sample 6U2. All the MCC values are not as high as the values from the top of die. A25 corner has high MCC values. Thus, anomalies are expected to be present around A25 corner.

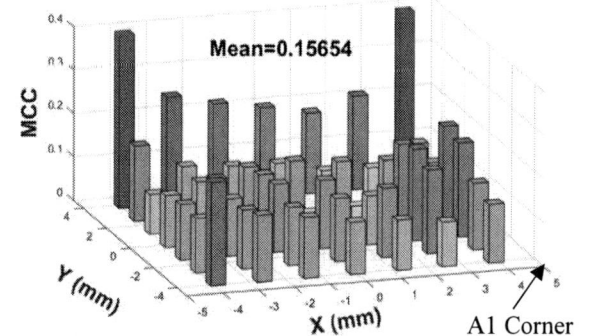

Figure 7. LULI results (MCC values) from the top of die for test sample 6U2 in 3D histogram format.

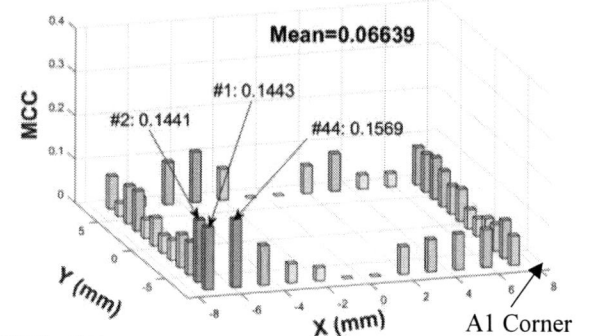

Figure 8. LULI results (MCC values) from the peripheral locations for test sample 6U2 in 3D histogram format.

VI. ELECTRICAL TEST RESULTS

During the board level reliability testing of FCBGA samples under discussion, the end of life was monitored by electrical testing. After every ~150 thermal cycles, the samples were pulled out of thermal cycling test chamber and electrical resistance was measured on test pads with appropriate net. Any abrupt increase in resistance between test pads based on the aforementioned criteria indicates the failure in the net connecting two test pads. However, it is not possible to know exact anomaly location or type in the failing net from electrical test.

From the electrical tests, sample 5U1 failed after 2011 thermal cycles and the failure net is between M11 − Y7 as shown in the Figure 9. Sample 6U2 was failed after 2291 thermal cycles and two nets have resulted in high electrical resistance. The failure nets are M11 − Y7 and Y20 − F20 as shown in Figure 10.

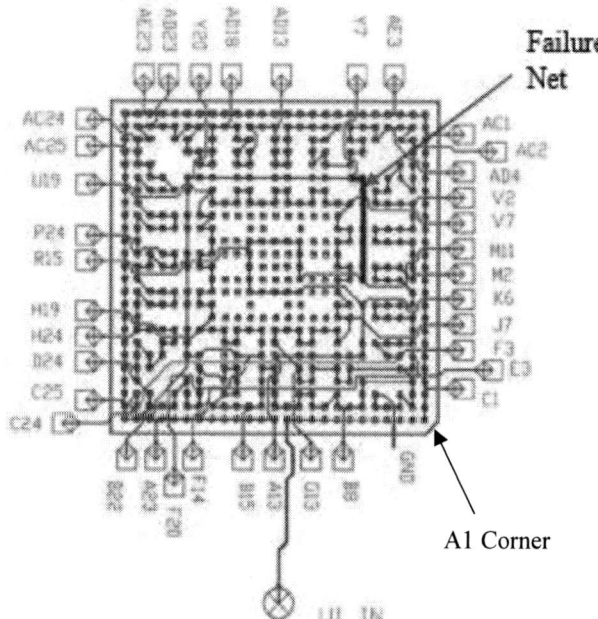

Figure 9. Failure net in sample 5U1 from electrical resistance test.

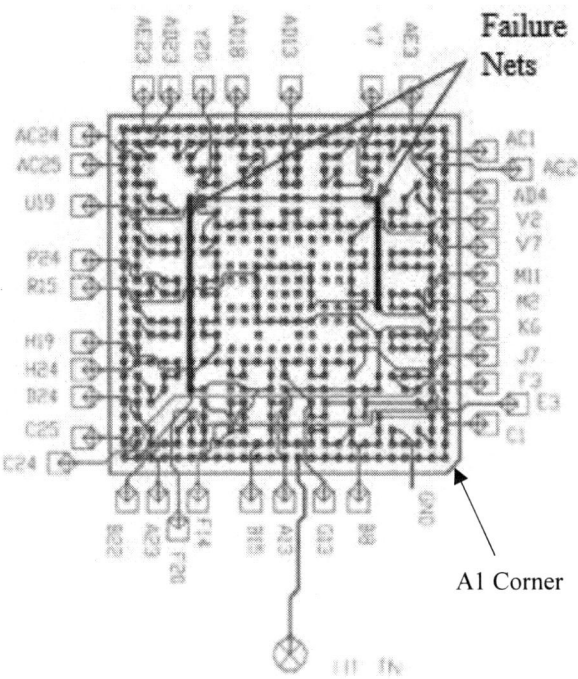

Figure 10. Failure net in sample 6U2 from electrical resistance test.

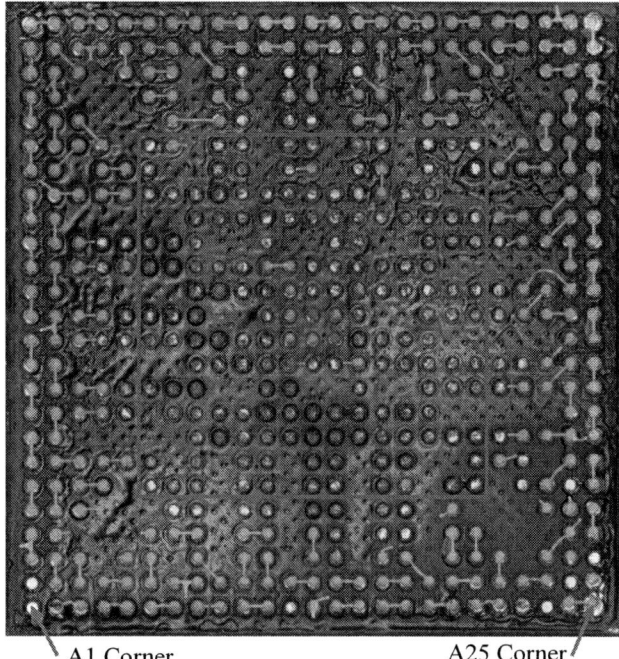

Figure 11. Dye and Pry test result on failure sample 6U2 (red border represents die shadow).

VII. DISCUSSION

LULI results from the failed sample 5U1 predict the anomalies on the corner BGA solder balls under the die shadow. Top right corner surrounding BGA solder ball Y6 has the highest MCC value of 0.1038 as given in the Figure 5. Electrical test on 5U1 also shows that the failed net is M11 – Y7 which is at the top right corner under die shadow as shown in the Figure 9. This validates LULI result that the top corner BGA solder balls under die shadow must have failed. LULI also predicts minor failures at top left and bottom left corners under the die shadow which would not have opened completely to result in high resistance in electrical test.

LULI results from the failed sample 6U2 predict anomalies almost everywhere under the die shadow in the sample as shown in Figure 7. However, electrical results show that only two nets M11 – Y7 and Y20 – F20 have failed. In order to further validate LULI results, dye and pry test was carried out on 6U2 failure sample. The dye and pry test result is shown in Figure 11. Dye was present on almost on all the solder joints under the die shadow, which means that BGA solder joints under the die shadow have partial to complete cracks. Thus, the LULI results are validated. It is also observed that most of the solder joint connections outside the die shadow are intact and no dye was penetrated. This confirms that the corner solder joints under the die shadow are the first connections to fail in thermal cycling testing.

To validate the high MCC values at the lower left corner (A25) from peripheral inspection points, Scanning Electron Microscopy (SEM) was carried out for image analysis, as shown in the Figure 12. It is observed from SEM image that the A25 corner solder balls have lower stand-off height and have been compressed completely. Therefore, it is concluded that lower stand-off height at the A25 corner produced high MCC values in that corner.

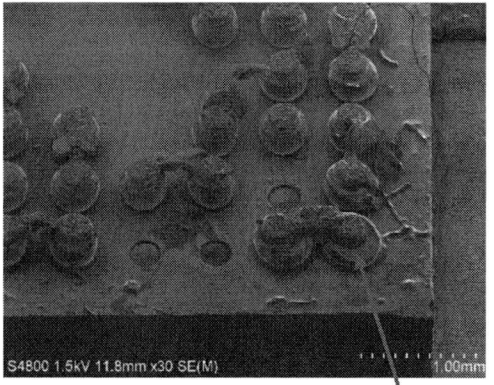

Figure 12. SEM image of A25 corner after Dye and Pry test.

VIII. CONCLUSION

This paper demonstrated LULI method to determine chip-to-package interaction anomalies. FCBGA samples which were subjected to temp cycle reliability testing were used for this research. The LULI results on two failed samples were presented and validated using electrical test results. LULI results on the failed sample 6U2 were further validated using dye and pry test and SEM analysis. Overall, the LULI method has proven to be a promising technique to determine chip-to-package anomalies. The major limitation of the LULI method is that it could not explicitly determine the type of the anomaly present inside the sample. Other non-destructive techniques such as SAM, X-ray have serious limitations in determining defects especially at BGA solder joints. Future plans include Gage Repeatability and Reproducibility (GRR) studies to confirm the consistency of waveform patterns generated from the LULI technique as well as optimization of inspection patterns for improved signal-to-noise ratio to allow better definition and accuracy of package anomaly. It is also envisaged to perform finite element simulation for thermal reliability modelling of FCBGA package to study insights of failure locations and severity of the failures.

ACKNOWLEDGMENT

The authors would like to acknowledge the financial support from Semiconductor Research Corporation (SRC Task ID: 2790.001). The authors would like to thank Texas Instruments Inc, for providing samples and conducting required failure analysis.

REFERENCES

[1] Flip Chip Ball Grid Array Package Reference Guide, Texas Instruments, Literature Number: SPRU811A, May 2005.

[2] Aryan, P.; Sampath, S.; Sohn, H., "An Overview of Non-Destructive Testing Methods for Integrated Circuit Packaging Inspection. Sensors," 2018, 18, 7.

[3] V. V. B. Reddy et al., "Non-destructive Inspection of Flip-Chip BGA Solder Ball Defects Using Two Laser Beam Probe Ultrasonic Inspection Technique," 2018 IEEE 68th Electronic Components and Technology Conference (ECTC), San Diego, CA, 2018, pp. 764-770.

[4] Vishnu V. B. Reddy et al., "Assessment of 2nd level interconnect quality in Flip Chip Ball Grid Array (FCBGA) package using laser ultrasonic inspection technique," SMTA International Conference Proceedings, Rosemont, IL, October 2018.

[5] L. Zhang, I. C. Ume, J. Gamalski, and K. P. Galuschki, "Detection of flip chip solder joint cracks using correlation coefficient, and auto-comparison analyses of laser ultrasound signals," IEEE Trans. Compon. Packag. Technol., vol. 29, no. 1, pp. 13-19. March 2006.

[6] Gong, et al., "Non-destructive Evaluation of Solder Ball Quality under Mechanical Bending Using Laser Ultrasonic Technique," SMTA Journal, Volume 28 Issue 3, 2015.

[7] A.M. Mebane et al., "Feasibility Studies and Advantages of Using Dual Fiber Array in Laser Ultrasonic Inspection of Electronic Chip Packages," IEEE Trans. Compon. Packag. Technol., DOi: 10.1109/TCPMT.2019.2896603.

[8] Scruby, C. B., and Drain, L. E., Laser ultrasonics: techniques, and applications. Bristol: Adam Hilger, 1990.

[9] T. W. Randolph, "Development of automated method of optimizing strength of signal received by laser interferometer," Ph.D. dissertation, Dept. Mech. Eng., Georgia Inst. Technol., Atlanta, GA, USA, 2009.

[10] Zhang, L., et al., "Detection of flip chip solder joint cracks using correlation coefficient and auto-comparison analyses of laser ultrasound signals," IEEE Trans. Compon. Packag. Technol. 29(1), 13–19 (2006).

Assessment of Accelerometer versus LASER for Board Level Vibration Measurements

V. Thukral, M. Cahu, J.J.M. Zaal, J. Jalink, R. Roucou, R.T.H. Rongen

NXP Semiconductors

Gerstweg 2, 6534AE, Nijmegen, The Netherlands

varun.thukral@nxp.com

Abstract—The ongoing trend to deploy ICs in more complex and harsher applications, entails precise evaluation of solder joint reliability of components subjected to vibration loads. For this, a good understanding of the PCB vibrational motion during a board level vibration test is essential. This can only be achieved by a well characterized vibration test setup. The vibration motion can be recorded by using a contact-based measurement approach, i.e. using an accelerometer, or a contactless measurement configuration, i.e. using a Laser Doppler Vibrometer (LDV).

This paper evaluates both measurement techniques by recording the PCB dynamic response, i.e. the resonance frequency and peak-to-peak displacement, in a board level vibration test set up. Bare and assembled printed circuit boards (PCBs) are investigated using different PCB form factors and package outlines (Wafer Level Chip Scale Package (WLCSP) and Ball Grid Array (BGA)), showing that LDV enables better lateral resolution and a more accurate measurement solution. Especially when the weight of the accelerometer cannot be neglected compared to the weight of the component on the PCB. An accelerometer is shown to perturb the PCB vibration motion.

It is found that depending upon the test objectives and PCB electronic system involved, both techniques can be used as complementary to one another. The accelerometer weight may give rise to substantial modification of vibration response which can be used to simulate the presence of a component on a bare PCB. In addition, both methods are expected to recognize the same trends when e.g. studying the environmental impact during vibration tests. Finally, the experimental observations are also confirmed using a Finite Element Model (FEM).

Keywords- board level reliability; vibration test; LASER Doppler Vibrometer, PCB dynamic response; modal analysis; Wafer Level Chip Scale Package

I. INTRODUCTION

Vibration phenomenon permeates through this entire universe. Its presence is ranging from celestial level to the subatomic level. In many applications, vibration is used deliberately to help the functionality. The vast range of such examples include MEMS based IC components, ultrasonic devices, aerospace equipment, musical instruments like guitar, ukulele, etc. However, vibration can also be devastating and hence, undesirable in some applications. It is said that all electronic assemblies are usually subjected to some type of vibration at some point in time [1]. It can occur either in its service life or during transportation, manufacturing and installation process itself.

The motion generated during a mechanical vibration loading event leads to oscillations of a PCB. Hence, the solder attachments of the component are agitated under some differential PCB strain load. As a result, integrity of IC components and its interconnection to the Printed Circuit Board (PCB) can be affected and in the worst case, functional failure in the product can occur [2]. Assessment of the solder joint lifetime under such vibrational loading is typically investigated in an accelerated life test [3-8]. However, it is shown that the experimental setup and the environmental conditions are important to define a correct vibration test setup [9,10]. In this paper, additional data is gathered to further collect the best-known techniques associated with the vibration test set up. Emphasis of this work will be primarily on understanding the impact of vibration characterization technique on the dynamic response of test boards containing distinct and small packaging constructions.

In general, accelerometers are employed to characterize the dynamic response of printed circuit boards during board level vibration tests. Measurement with accelerometers allow cheap and easy measurement test set up justification. However, accelerometers might influence the vibration motion of some electronic systems. Therefore, contact-less measurement methods are gaining renewed interest as alternate vibration measuring units. With its contactless measurement ability offering no additional mass in the vibration system, it is expected to measure more realistic vibration values when compared to that of an accelerometer. On the other hand, contactless techniques usually involve complex installation set ups.

978-1-7281-1500-9/19 $31.00 © 2019 IEEE

The objective of this paper is to investigate the potential influence of accelerometers and compare it to an alternative contact-less method like LASER Doppler Vibrometer (LDV). Initially, both accelerometers and LDV are mounted and focused respectively, at the shaker base plate to verify that the same results are obtained as the mass of the accelerometer can be neglected in comparison to the mass of the shaker. With both methods, the acceleration of the PCB is measured independently and finally combined. Then, different PCB form factors are evaluated, and vibration characteristics are gathered for board assemblies with components ($3x3mm^2$ to $12x12mm^2$) mounted on it. Finally, modal and harmonic analysis is performed using a finite element model (FEM) developed to mimic both measurement set up situations.

II. EXPERIMENTAL AND COMPUTATIONAL DETAILS

A. Experimental Test Set up

Test board layouts for different component types are designed according to the test board prescribed by the JEDEC Board Level Drop Test Method for handheld components, JESD22-B111A (image shown in Figure 1) [2]. The circuit board type will be referred to as Test board B111A (as mentioned in [11]).

The dynamic response of a PCB with and without components is recorded to assess the impact of the package to the resonance frequency and peak-to-peak displacement. Two package technologies (described in Table 1) are used as test vehicles.

TABLE 1: PACKAGE DETAILS

Package type	WLCSP	BGA
Package size [mm x mm]	3.4 x 3.0	12.0 x 12.0
I/O Pins	49	180
Pitch [mm]	0.40	0.50
Weigth [g]	0.02	0.50

The vibration modes of test boards are excited using a controlled shaker system environment supplying a swept sine vibration signal with constant acceleration of 5g. PCBs are mounted on to the shaker base plate using four custom made stand offs, see Figure 1. Finally, the PCB dynamic response is recorded and analyzed using a typical miniature IEPE-type piezoelectric accelerometer (0.5 gram) and is compared with the response measured from LDV.

Images and schematics for both test set ups are shown in Figure 1 measurements) and Figure 2.

The LDV uses a He-Ne LASER beam (633 nm) and measures the velocity based on the principle of the Michelson interferometer [13]. LDV enables direct access to the PCB dynamic behavior. The LDV type used in this study is capable of measuring velocity, displacement and acceleration. Here, it is used in acceleration measurement mode to allow for comparison with the accelerometer measurements. The LDV offers a velocity measurement range of 0 to 10m/s.

B. Numerical Equations

The PCB resonance frequency (f_0) is linked to the material properties, which can be described by the following equation from Steinberg [1]:

$$f_0 = \lambda \left(\frac{1}{a^2} + \frac{1}{b^2} \right) \sqrt{\frac{Eh^3}{12\rho(1-\upsilon^3)}} \quad (1)$$

where, λ is a constant depending upon clamping (among others), a and b are the length and width of the board respectively, E is the Young's modulus, h is the board thickness, ρ and υ refer to the density and Poisson's ratio respectively.

Figure 1: Vibration test set up at PCB level using LDV (a) and Accelerometer (b)

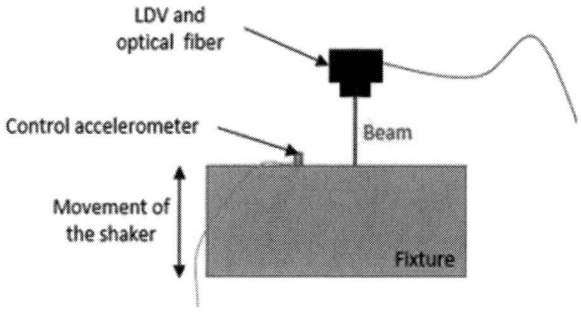

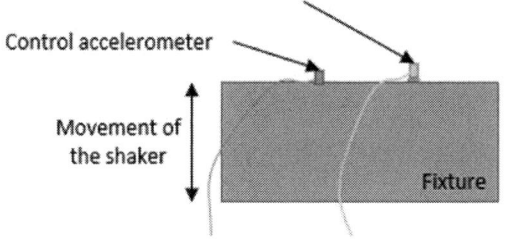

Figure 2: Vibration test set up schematic at fixture using LDV (a) and Accelerometer (b)

Then, assuming that the PCB acts like a single-degree-of-freedom system at its fundamental resonance mode, the peak-to-peak displacement at the first resonance frequency (f_0) can be determined using the following relation [11]:

$$d = \frac{a}{2\pi^2 f_0{}^2} \quad (2)$$

It shall be noted that this equation is subjected to the assumption that PCB motion can be approximated as sinusoidal in nature when it is excited at the first resonance mode. It can only be applied near the center of the PCB.

Relative motion between a vibrating board and its component produces strains in the solder joints. Therefore, another vibration characteristic, called relative displacement (δ) (see Figure 3) between the component and PCB (As defined in [1]) is used to better describe the stress experienced at solder joint levels. It is used as a measure to compare the lateral resolution of tested measurement methodologies.

Assuming a circuit board to be a flat square plate, with four edges of the PCB are simply supported and a uniform load is distributed, is being vibrated in a direction perpendicular to the plane of the plate (see Figure 4) [1].

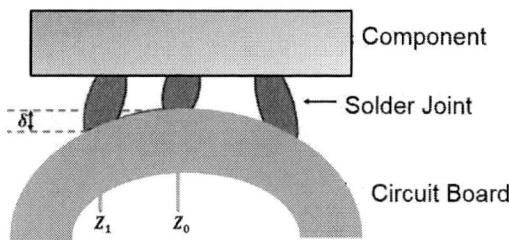

Figure 3: Relative displacement between a component and circuit board during vibration

Then, displacement at any PCB location ($Z(X,Y)$) excited at its fundamental resonance mode, can be simplified to the following expression (Equation 3) [1]:

$$Z(X,Y) = Z_0 sin\frac{\pi X}{a} sin\frac{\pi Y}{a} \quad (3)$$

Where Z_0 is displacement at the center of the PCB sizing a.

At PCB location, Y = a/2 (cross-section shown in Figure 3)

The above equation, the displacement Z1 at any point on the curve can be shown as below [1]:

$$Z_1 = Z_0 sin\frac{\pi X}{a} \quad (4)$$

$$\delta = Z_0 - Z_1 = Z_0\left(1 - sin\frac{\pi X}{a}\right) \quad (5)$$

The above equation (Equation 5) represents the out of plane stretch in the solder joints. It also represents deflection or deformation between a PCB and the edge of a component. It shall be noted that the Equation 5 conservatively assumes that there is no bending in the component body and the solder joint is loaded with tensile stress only. It is only an approximation whilst the bending moment induced in the solder joint is neglected [1]. Also, it assumes that the PCB is simply supported on all four sides. However, in actual boards are screwed at the four edges only.

C. *Simulation model set up*

Marc/Mentat software is used to numerically calculate the expected PCB resonance modes. The PCB and accelerometer are meshed into brick- and penta-shape elements. PCB material properties (described in Table 2) are measured using Dynamic Mechanical Analysis and Thermo-Mechanical Analysis. It is used as input for setting up the eigen modal analysis in Finite Element Modeling (FEM). These properties of the PCB are homogenized. Figure 4 shows the first PCB resonance mode calculated using FEM.

TABLE 2: TEST BOARDS DESCRIPTION

PCB details	Test Board B111A
PCB form factor [mm]	77 x 77
PCB thickness [mm]	1
PCB Material	FR4
Storage Modulus [GPa]	25
Number of Copper layers	10
PCB Weight [in grams]	16

Figure 4: Resonance mode of a Square PCB type with Accelerometer

III. MEASUREMENT COMPARISON ON VIBRATION SHAKER PLATE

Circuit boards are mounted on to a shaker base plate. For verification purposes, an experiment is performed by mounting and focusing both accelerometer and LASER respectively at the shaker base plate. This is done to confirm that both measurement methods generate the same value for the acceleration.

Figure 5 shows normalized vibration acceleration measurements of the shaker base plate. Such analysis allows easier dissection of deviations recorded by the monitoring or measurement sensor. Both measurements are within 1%, demonstrating that similar results can be expected when the weight of the accelerometer can be neglected.

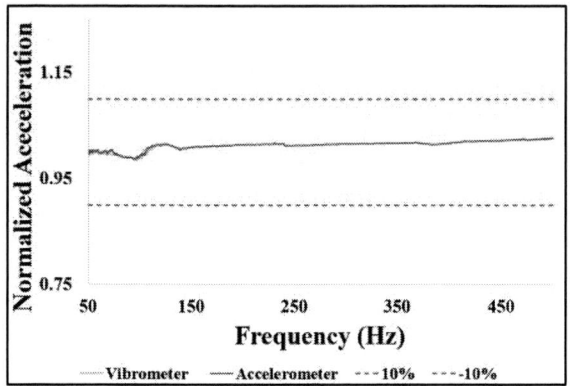

Figure 5: Normalized data at fixture: LDV vs Accelerometer

IV. PCB VIBRATION MEASUREMENT RESULTS & COMPARISON

A. Impact of measurement variations at bare PCB levels

Stress at solder joint level is depending upon the following PCB vibration parameters: relative displacement between a component and PCB, bending moment (function of component size, PCB peak-peak displacement, resonance frequency, form factor) and the amount of bending cycles that a board receives whilst vibrating at its fundamental resonance. Initially, bare PCB vibration responses are measured separately and or simultaneously with both accelerometer and LDV.

Figure 6 depicts the first resonance mode of the square PCB type. First, it is measured using an accelerometer placed at the center of the PCB. Results show similar values as presented in [11] and [14]. Then, LDV is pointed on top of the measuring accelerometer in order to assess the impact of mounting accelerometer. Finally, accelerometer is removed from the vibration system and measurements are performed using LDV only. Figure 6 depicts the measured vibration response from LDV.

The recorded resonance frequency and peak-peak displacement are significantly different for the two measurement methods, see Table 3. The resonance frequency and peak-peak displacement for B111A PCB layout, when measured using LDV is 415Hz ± 0.1 and 1.07mm ± 0.02. The resonance frequency when measured using LDV is higher than that of the resonance frequency measured by the accelerometer mounted on this circuit board. Also, peak-peak displacement of the PCB with LDV is larger when compared to 0.88mm ± 0.03 found using accelerometer.

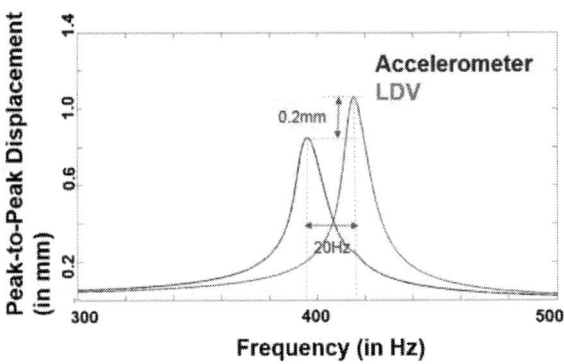

Figure 6: Resonance mode of Square PCB type: LDV vs Accelerometer

TABLE 3: VIBRATION CHARACTERISTICS OF BARE PCB

Measurement Methodology	Measured		FEM
	Peak-Peak Displacement [mm]	Resonance Frequency [Hz]	*Resonance Frequency* [Hz]
Accelerometer	0.88 ± 0.03	397 ± 1	435
LDV (with Accelerometer on PCB)	0.90 ± 0.01	395 ± 2	NA
LDV	1.07 ± 0.02	415 ± 0.1	448

Furthermore, FEM results (see Table 3) show a measurement difference of 18Hz, when comparing LDV to the accelerometer measurements. It exhibits good agreements with the experimental findings, showing a measurement difference of about 18Hz. It shall be noted that the resonance frequency prediction from FEM is slightly different when compared to the measured resonance frequency. It can be attributed to the assumption that the PCB material properties are homogenized during FEM analysis.

Both FEM and experimental results highlight the impact of using an accelerometer in the vibration measurement test set up. Deviations observed in peak-peak displacement can be explained using Equation 2. The mass of an accelerometer can lead to reduction of PCB resonance frequency. Contact less measurement ability of the LDV unit allows lower vibrating mass (see Equation 1). Thereby, achieving higher resonance frequency in the absence of accelerometer. System damping is also significantly affected by an accelerometer. Combination of larger peak-peak

displacement and smaller damping leads to a smaller bending cycle count. This is evident in measurements recorded using LDV only measurement configuration. Ultimately, these results underline the fact that depending upon the weight of an accelerometer and its weight relative to the weight of the PCB, accelerometer perturbs the PCB vibration motion.

A.2. Impact of measurement differences at bare PCB levels

In accordance with [2] and [11], a square shaped PCB construction is manufactured using a stiff PCB material. It amounts to a relatively smaller peak-to--peak displacement and a higher resonance frequency when compared to some other board geometries involving higher distance between the mounting screws and lower resonance frequency. This can cause higher measurement differences between LDV readings and accelerometer signal. Furthermore, the amount of error induced by the mass of an accelerometer can be larger for a light weight asymmetric (for instance, rectangular) board type when compared to the B111A PCB layout.

A customized rectangular board type is used to show the impact of PCB types on the measurement errors originating from an accelerometer weight. Rectangular PCB comprises of ten Copper layers sandwiched in between a more damped and less stiff PCB material when compared to the material used in a B111A PCB type. This reduces the PCB weight down to 8 grams only. It is half of the PCB weight measured for B111A type board. Then, the vibration response of this customized rectangular PCB is measured using both LDV and accelerometer.

LDV signal records higher peak-to-peak displacement (1.25mm ± 0.01) and resonance frequency (393Hz ± 0.6) when compared to the accelerometer measurements, showing 0.92mm ± 0.02 as peak-peak displacement and a resonance frequency of 360Hz ± 1. Expectedly, a larger alteration of vibration characteristics is observed for the rectangular PCB type when compared to the amount of deviation shown by a B111A PCB type. It is represented by deviation (in percentage), which is described as the difference in the response with LDV to the response with Accelerometer, normalized to the response with Accelerometer (shown in Figure 7). It highlights the bigger impact of accelerometer on the rectangular PCB type when compared to that of the B111A circuit board type.

978-1-7281-1500-9/19 $31.00 © 2019 IEEE

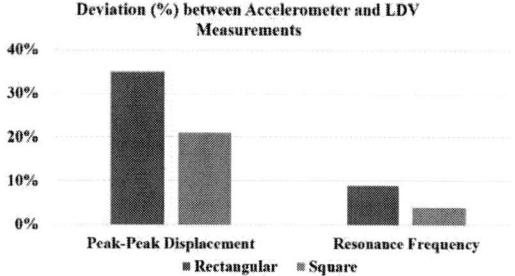

Figure 7: Influence of PCB type: Accelerometer vs. LDV

To summarize, depending upon the type of circuit board type that is required to be characterized, the measured vibration behavior by an accelerometer can be rather subordinate when compared to the LDV measurements. To reduce such large measurement errors allured by an accelerometer, the weight of an accelerometer shall be kept to bare minimum (e.g. in previous studies [7] and [9]), and preferably there shall not be any additional weight in the measurement system (e.g. by using contact less measurement techniques). Another way to reduce such large measurement errors is by designing circuit boards with large resonance frequency and weight. For example, in this study, B111A type board is preferred over a rectangular board type from precise measurement strategy standpoint. It also asserts the advantage that LDV measurement methodology enables more freedom in terms of board design. While, an accelerometer methodology is preferred for bulky and stiff PCB constructions.

A.2. Impact of measurement differences on the stress at Solder joint levels

Finally, the influence of measurement techniques on the stress at solder joint level is required to be understood in detail. The PCB flexing curvature at a component location can be better described using relative displacement between the component and PCB (described using Equation 5). B111A PCB layout for big EMC type package (see details in Table 1) is used for this study. Measurements using both LDV and accelerometer are performed at the footprint of solder joint locations.

When analyzing the flexure of PCB across the EMC component location, a relative displacement of about 0.026 mm is calculated for LDV signals. In contrast, accelerometer signal shows a relative displacement of about 0.016 mm. So, when examining the impact of the two-contested vibration measurement approaches, the accelerometer weight causes it to conservatively underestimate the relative axial stretch experienced by the solder joints.

These results assert the lateral resolution advantage that can be achieved using LASER. Contact less way of measurement enables better characterization of the PCB deformation when compared to an accelerometer. Accelerometer measurements might be limiting when measuring such small PCB deformations. It is due to the weight and occupied surface area by an accelerometer. This downside of accelerometers is well exposed when PCB deformation for small WLCSP packaging construction (see details in Table 1) is measured. The accelerometer is not capable of sensing differences between the center and corner of such a WLCSP component footprint.

B. Impact of measurement differences at Test Board Assembly levels

To understand the measurement variation at the PCB assembly level, the vibration motion of a B111A board carrying components (referred as assembled PCB) is measured. Figure 8, describes a vibration response difference obtained between a bare PCB and a WLCSP (see Table 1) component-based PCB assembly. It is measured using an accelerometer. Then, measurements are performed using LDV, focused on top of the component. The modification (see Figure 9) is captured by measuring a downward shift in resonance frequency and peak-peak displacement. Results are summarized in Table 4.

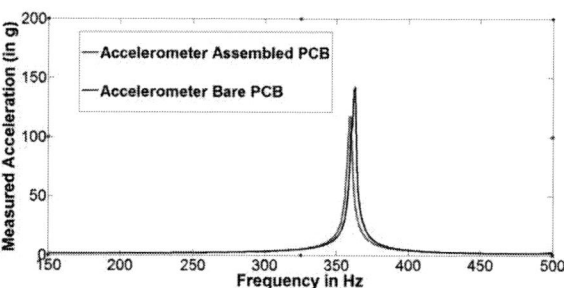

Figure 8: Bare PCB vs WLCSP Assembled PCB vibration response measured using Accelerometer

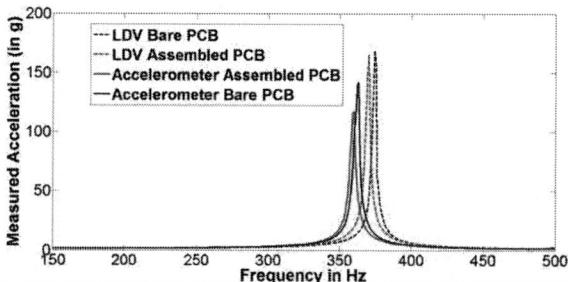

Figure 9: Resonance trends WLCSP-based PCB assemblies: LDV vs Accelerometer

978-1-7281-1500-9/19 $31.00 © 2019 IEEE

When comparing LDV measurements to the accelerometer readings on a WLCSP package-based PCB assembly, a difference of about two times is observed in resonance frequency. The observations are in line with the assumption that an accelerometer weight increases the overall mass of the vibrating body, causing lower resonance values when compared to that of the measured LDV signal (see Equation 1). Depending upon the package size, package style and package weight, the correction in the vibration signal may be significant. This has been demonstrated by comparing the response of two different assembled circuit boards in the subsequent section.

TABLE 4: VIBRATION CHARACTERISTICS OF ASSEMBLED BOARDS

Measurement Methodology	Resonance Frequency [Hz]	
	WLCSP based PCB Assembly	BGA based PCB Assembly
Accelerometer (Acc.)	344 ± 1	349 ± 1
LDV	369 ± 0.1	356 ± 0.1
Measurement Difference (LDV - Acc.)	25	7

B.1. Impact of measurements differences at different packaging constructions

Two distinct packaging constructions presented in Table 1 are tested to understand the significance of measurement modification induced by accelerometer weight. Figure 10 shows the impact of big BGA package and small WLCSP component on signal deviation observed between LDV and accelerometer measurements. Clearly, the package type is key in determining the overall deviation in PCB vibrational response.

Before comparing and comprehending the amount of impact that a measurement methodology can have on different PCB assemblies, it is required to understand the difference between vibration response determined for the WLCSP and BGA package-based PCB assemblies. The upward trend in resonance confirms that for larger packaging constructions, stiffness of the package is dominant over the mass of the package, causing an overall increase in resonance frequency. Whereas, for a smaller WLCSP component, increased stiffness due to silicon body is compensated by small weight of a WLCSP component (see Equation 1). Thereby, showing small impact on the resonant

frequency of a WLCSP based PCB system. Comparable results were observed in [11] for similar packages. FEM predictions (in [11]) showing turnover point for large BGA packages, matches well with the experimental observations presented in this paper.

When comparing LDV response to the Accelerometer response for a large BGA package PCB assembly (see Table 4), the mass of the component is comparable to the mass of the used accelerometer. Therefore, less deviation in accelerometer measurement is observed when contested with LDV measurements. It testifies the fact that when the weight or size of a component is not of the same order of magnitude as the mass of an accelerometer, the PCB vibration may not be measured correctly by the accelerometer.

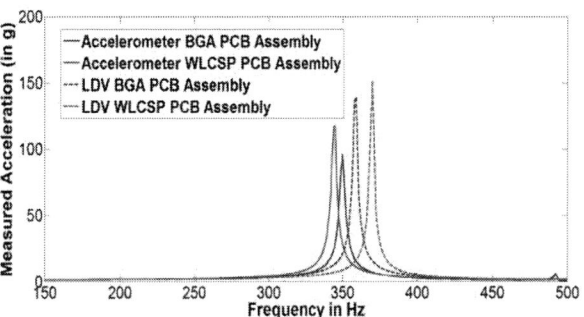

Figure 10: Influence of package type: Accelerometer vs LDV

V. CONCLUSIONS

The influence of two vibration measurement methods, LDV & Accelerometer, are compared on different PCB types and package constructions. Experimental test results exhibit the tendency of an accelerometer influencing both bare and assembled PCB vibration motion. Depending upon the PCB weight and stiffness, accelerometer weight can induce a significant deviation in the PCB vibration response. Several factors affecting this modification have been identified. Amongst others, a major factor includes the mass of an accelerometer used with respect to the weight of a moving PCB target. However, both measurement systems yield the same vibration characteristic trends with respect to the PCB vibration characteristics.

Next to the bare PCB vibration response analysis, the influence of package styles mounted on a B111A PCB type is also assessed. The deviation between the contact less LDV and accelerometer signal is essentially larger for a smaller WLCSP package PCB construction than that of a bigger BGA package PCB assembly. It is linked to the weight of an accelerometer which is greater than the weight of a WLCSP package, thereby inducing large measurement differences in WLCSP based PCB assembly.

To summarize, it is shown that the measurement method influences the PCB vibration response that is required to be measured. And it shall be avoided or minimized by keeping the accelerometer weight as low as possible. In order to measure precise vibration response for PCB assemblies carrying light weight WLCSP components, it is recommended to use accelerometers lighter than the weight of the accelerometer used in this study (0.5g). In such cases, it is advocated to either use very light weight accelerometers (<0.2g), weighing comparable to that of a small WLCSP component or use contact less measurement solutions. On the other hand, accelerometers can be used for measuring vibration response of PCB assemblies containing heavier components weighing more than the weight of the accelerometer itself.

ACKNOWLEDGEMENT

I would like to express heaps of gratitude to Peter Vullings and Kevin P. Hussey for providing the management support during this work.

REFERENCES

1. Steinberg, D.S., "Vibration Analysis of Electronic Equipment," 2nd ed, John Wiley, 1988

2. JEDEC Standard JESD22-B111a, Board Level Drop Test Method of Components for Handheld Electronic Products; Nov 2016

3. H.W. Zhang, Y. Liu, J. Wang and F.L. Sun "Effect of elevated temperature on PCB responses and solder interconnect reliability under vibration loading", Microelectronics Reliability 55 (2015), pp. 2391-2395

4. Y. Liu, F. Sun, H. Wang, Z. Zhou, Y. Qin "A Comparison of Two Board Level Mechanical Tests- Drop Impact and Vibration Shock", International Symposium on Advanced Packaging Materials (APM) 2011, pp. 199-203

5. H. Qi, S. Ganesan, J. Wu, M. Pecht, P. Matkowski and J. Felba, "Effects of Printed Circuit Board Materials on Lead-free Interconnect Durability", IEEE Polytronic 2005 – 5th International Conference on Polymers and Adhesives in Microelectronics and Photonics, pp. 140-144

6. P. Lall, G. Limaye, J.Suhling, M. Murtuza, b. Palmer, W. Cooper, "Reliability of Lead-free SAC Electronics Under Simultaneous Exposure of High Temperature and Vibration", 13th IEEE ITHERM Conference, 2012

7. C. Choi and A. Dasgupta, "Effect of Temperature on Vibration Durability of SAC305 Printed Wiring Assemblies", 13th IEEE ITHERM Conference, 2012

8. F.X. Che and J.H.L. Pang, "Study on Board Level Drop Impact Reliability of SnAgCu Solder Joint by Considering Strain Rate Dependent Properties of Solder", in IEEE Transactions on Device and Material Reliability, vol. 15, no.2, pp. 181-190, June 2015.

9. D. Iyengar, S.Shantaram, P. Gupta, D. Panchagade, J. Suhling, "Feature extraction and health monitoring using image correlation for survivability of lead-free packaging under shock and vibration," EurosimE 2008.

10. R. Roucou, J.J.M. Zaal, J. Jalink, R. de Heus, R. Rongen, "Effect of Environmental and Testing Conditions on Board Level Vibration" 2016 IEEE 66th Electronic Components and Technology Conference (ECTC), Las Vegas, NV, 2016, pp. 1105-1111

11. J. Jalink, R. Roucou, J.J.M. Zaal, J. Lesventes, R.T.H. Rongen, "Effect of PCB and Package Type on Board Level Vibration using Vibration Spectrum Analysis", 2017 IEEE 67th Electronic Components and Technology Conference (ECTC), Orlando, FL, 2017, pp. 470-475

12. X. Xing, G. Chen, M. Cheng, "Study on the failure behavior of BGA solder interconnections under fatigue loading", 2015 IEEE International Conference on Electronic Packaging Technology (ICEPT), pp. 729-733

13. M. Martarelli. Exploiting the Laser Scanning Facility for Vibration Measurements, Imperial College of Science, Technology & Medicine, University of London (2001).

14. V. Thukral, J.J.M. Zaal, R. Roucou, J. Jalink, R.T.H. Rongen, "Understanding the Impact of PCB Changes in the Latest Published JEDEC Board Level Drop Test Method", 2018 IEEE 68th Electronic Components and Technology Conference (ECTC), San Diego, CA, 2018, pp. 756-763

Effect of Process Parameters on the Long-Run Print Consistency and Material Properties of Additively Printed Electronics

Pradeep Lall[1], Nakul Kothari[1], Amrit Abrol[1], Jeff Suhling[1], Sudan Ahmed[1],
Ben Leever[2], Scott Miller[3]

[1]Auburn University, NSF-CAVE3 Electronics Research Center
Department of Mechanical Engineering, Auburn, AL 36849
[2]US Air Force Research Labs, Wright-Patterson AFB, OH
[3]NextFlex Manufacturing Institute, San Jose, CA
Tele: +1(334)844-3424; E-mail: lall@auburn.edu

Abstract— **Traditionally, the printed circuit assemblies have been fabricated through a combination of imaging and plating based subtractive processes involving use of photo-exposure followed by baths for plating and etching to form the needed circuitry on rigid and flexible laminates. Additive electronics is finding applications for fabrication of IoT sensors. The emergence of a number of additive technologies poses an opportunity for the development of processes for manufacture of flexible substrates using mainstream additive processes, which are now commercially available. Aerosol-Jet printing has shown the capability for printing lines and spaces below 10 μm in width. The Aerosol-Jet system supports a wide variety of materials, including nanoparticle inks and screen-printing pastes, conductive polymers, insulators, adhesives, and even biological matter. The adoption of additive manufacturing for high-volume commercial fabrication requires an understanding of the print consistency, electrical and mechanical properties. Little literature exists that addresses the effect of varying sintering time and temperature on the shear strength and resistivity of the printed lines. In this study, the effect of process parameters on the resultant line-consistency, mechanical and electrical properties has been studied. Print process parameters studied include the sheath rate, mass flow rate, nozzle size, substrate temperature and chiller temperature. Properties include resistance and shear load to failure of the printed electrical line as a function of varying sintering time and varying sintering temperature. Aerosol-Jet machine has been used to print interconnects. Printed samples have been exposed to different sintering times and temperatures. The resistance and shear load to failure of the printed lines has been measured. The underlying physics of the resultant trend was then investigated using elemental analysis and SEM. The effect of line-consistency drift over prolonged runtimes has been measured for up to 10-hours of runtime. Printing process efficiency has been gauged a function of process capability index (Cpk) and process capability ratio (Cp). Printed samples were studied offline using optical Profilometry to analyze the consistency within the line width, line height, line resistance and shear load to study the variance in the electrical and mechanical properties over time.**

Keywords-additive, printed-electronics, aerosol-jet, resistance, shear strength, line-consistency, process capability.

I. INTRODUCTION

Technologies for producing printed electronics have gained much popularity because of their low ramp-up time, simplified fabrication procedures, ability to fabricate varied complex functional devices and relatively high print resolutions in comparison to incumbent trace metallization processes. Advantages of printed electronics have driven grown of application area in a number of market segments. A recent reports point to the growth of printed electronics with applications, which span, displays, conductive inks, printed flexible sensors, lighting, and batteries [IDTechEx 2018]. The growth rate of flexible printed electronics has been high with reported growth in the neighborhood of 29-percent as recently as 2012 [BCC 2012]. The strong growth rate of flexible printed electronics is expected to continue with predicted growth rate of 14.9-percent between 2018-2023 [Markets 2018]. In recent times, the main market for printed electronics is found to be in the low-end devices with low fabrication costs in addition to customized electronics [Lupo 2013]. Manufacturing of application-specific electronics using existing printing infrastructure like flexography, gravure printing, offset lithography printing or screen-printing is prohibitively expensive because of the wastage of raw material and large-area print format.

The use of Aerosol Jet Printing allows for the use of multiple print heads, each capable of carrying different ink with different viscosities. The print tool-paths are driven by computer-aided design models. The process uses pressurized inert gas to focus a beam of aerosol (nano-particles) onto a substrate. The elimination of the need for masks and tooling for printing allows quick ramp-up without extensive setup costs associated with traditional existing techniques. The process can handle inks with varying viscosities ranging from 1 to 1000 cP, using either Ultrasonic atomization or Pneumatic atomization of the ink. In addition, the use of different printing standoff distance and sample orientation allows for printing on non-planar or curved surfaces. Previous researchers have published upon use of AJP for printing 3D conformal electronics with micron-scale features, micro-vias [Paulsen 2012, Krzeminski 2017]. The feasibility of using aerosol-jet printing has been demonstrated for a number of applications including passive microwave circuitry [Cai 2014] and quasi-optical terahertz filters [Oakley 2017]. A number of prior studies focus on the feasibility of the use of aerosol-jet printing on a number of substrates. Aerosol-jet printing has been studied for compatibility with first-level interconnects including wire bonding [Stoukatch 2012], for printing sensors for Internet of Things applications [Navratil 2018], printing on silicon solar cells [King 2009], making electrical contacts in SOI MEMS [Khorramdel 2018]. A number of material and process parameters may impact print process output including parameters of the atomization process, ink, substrate, and print-environment (Figure 1).

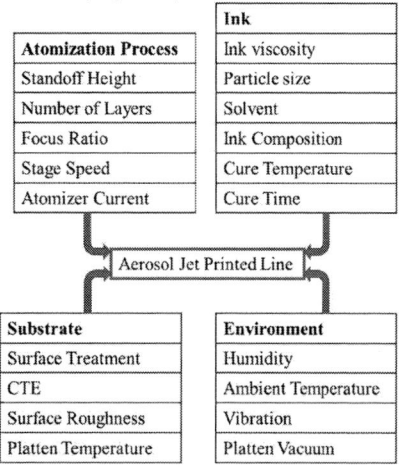

Figure 1: Parameters affecting AJP line properties

In this paper, the correlation of the mechanical properties of aerosol jet printed lines with process parameters has been studied. The reliability of the aerosol-jet printed traces has been correlated to the mechanical properties and electrical properties of the printed lines. The elastic modulus, the shear modulus or shear load to failure, resistivity of the printed line have been used for selection of suitable ink type, process parameters and sintering times and temperatures. Process parameters studied include sheath flow rate, ink flow rate, stage speed, standoff height, chiller temperature, platen temperature (Figure 1). Better understanding of the effect of each process parameter in the printed lines will allow users to select appropriate process parameters for design functional performance. In addition, the development of process-property relationships will provide visibility into the critical parameters, which require added level of attention for process-control. In addition, the process parameter drift over a long production run may be important for a high-volume production environment. The effect on the print process may be in the form of line consistency, and resistance – both of which have been quantified in this study. A procedure for finding process parameters that minimize the resistivity of printed lines has been developed. Effect of sintering time,

and sintering temperature on the mechanical and electrical properties of the printed lines has been quantified. Scanning electron micrographs have been used to understand the electrical and mechanical behavior of printed traces.

II. AEROSOL JET PRINTING

The Aerosol Jet Printer uses a combination of a pneumatic atomizer, an ultrasonic atomizer, a deposition head, a process control module (PCM) and a 3-axis linear and rotational motion driver for printing lines. The ultrasonic atomizer is used for inks with low viscosity and small solid particle sizes (< 100 nm). The pneumatic atomizer is required for inks with high viscosity (particle size up to 500 nm). Once the ink has been aerosolized in the atomizer, it is carried by the carrier gas, and aerodynamically focused in the print head into a concentrated beam and then projected onto the substrate. The deposited inks are sintered or cured based on properties of the material type and substrate. Sintering profiles are controlled though specification of the sintering temperature and sintering time.

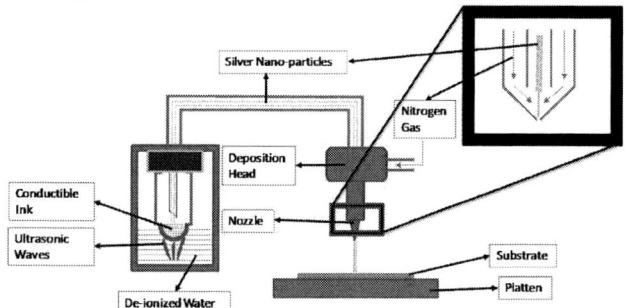

Figure 2: Principle of Ultrasonic Atomization

The width of the aerosol beam and thus the line width of the applied layers may be influenced by a number of parameters including among them the ratio of sheath gas to aerosol stream in the deposition head. This ratio is called the Focus Ratio. Increasing amounts of sheath gas results in a compression of the aerosol stream. Furthermore, line width increases with greater nozzle distance from the substrate. Gas flows and aerosol formation can be controlled via the process control module of the Aerosol Jet system. The print process can be separated into the following 4 stages [Secor 2018] including (1) Aerosol generation (2) Aerosol transport (3) Aerosol jet collimation (4) Impaction on substrate.

Optomec AJ-300 employs two-types of aerosol generators, an ultrasonic atomizer and a pneumatic atomizer. Depending on the viscosity of the ink, a suitable atomizer may be selected. Generally, an ink with higher viscosity (>30cP) is used with pneumatic atomizer and an ink with viscosity lower than 30cP may be used with ultrasonic atomizer. Ultrasonic atomizer has an ultrasonic bath underneath the vial containing ink. A piezoelectric actuator outputs ultrasonic waves, which travel upwards through deionized water as the medium, and sets up a capillary wave on the ink surface leading to the formation of ink droplets with a defined size distribution [Lang 1962; Peskin 1963]. The size of the droplet thus formed is dependent on the capillary wavelength, frequency, ink density and the ink surface tension [Rodes 1990]. The ink droplets are suspended in a solvent and cosolvent, which forms the ink. In order to prevent drying of the aerosol droplets, inks commonly have approximately 10-percent of low volatility cosolvent [Ha 2013]. Upon atomization, when the ink droplets meet a dry gas, the solvent evaporates [Ravindran 1982; Widmann 1997]. The closed enclosure that carries the atomized becomes saturated with solvent vapor in matter of seconds once the process has started.

Once the aerosol droplets are formed, they are carried to the deposition head in a plastic tube. It has been reported that, on an average it takes 10s for droplets generated to travel through the tube and reach the deposition head. During the process of droplet transport, some droplets collide with the inner walls of the tube and never reach the deposition head. Collison of the droplets with the inner walls of the tube is a source of losses attributed to the two mechanisms of gravitational settling and diffusion. Among the droplets being generated, larger sized droplets are susceptible to gravitational settling and smaller droplets are susceptible to diffusion. These mechanisms largely govern the output of the atomized ink on the substrate. When the droplets come out of the deposition nozzle, the aerosol droplets travel towards the substrate carried by the carrier gas. Smaller diameter droplets in the flow stream owing to their smaller mass may be carried away from the deposition axis by the sheath gas coming out of the nozzles. Increase in the sheath-gas flow rate increases the jet velocity minimizing spreading of the droplet stream. The distribution of the droplet size in the aerosol spray results in a line, which has a dense core region surrounded by a diffuse overspray. The flow rate, sheath gas flow rate, focus ratio, nozzle diameter, and ink droplet size determine the profile of the printed lines. In general, overspray is associated with larger diameter nozzles, low total flow rate, and a low sheath-to-flow rate ratio or low focus ratio. Further, a more stable size distribution of the droplet would result in printed lines with relatively lesser overspray and well-defined line edges. Well defines edges also result in more uniform cross section, resulting in lesser resistivity.

III. AEROSOL JET PRINTING INK

The ink used in this study is a silver nano-particle based ink. A detailed formulation of the ink is given in the Table 1.

TABLE 1: INK FORMULATION

Properties	Ag nano-particles
Solvent	Ethylene gycol
Sintering profile	200°C; 1hr
Form	Liquid
Particle size	70nm
Viscosity	15cP
Storage Temperature	20-25°C

Given the low viscosity of the ink, the ultrasonic atomizer was used for aerosol jet process. The ink has a nominal particle size of 70nm. Recommended sintering temperature of the ink is 200°C for 60 minutes in a conventional thermal chamber or oven. The ink has a density of 1.8 g/cm^3 and a shelf life of 6 months.

IV. IDENTIFICATION OF PROCESS PARAMETERS

Aerosol-Jet printing process parameters have been selected through the design of an experimental matrix and measurement of performance of the printed lines. Parameters controlled include, sheath gas flow rate, ink flow rate, focus ratio, stage speed, standoff height, nozzle diameter, chiller temperature, platen temperature are just the most important process parameters. In addition to the print parameters, a number of post-print parameters affect the performance of the printed line including the sintering profile quantified by the sintering time and sintering temperature.

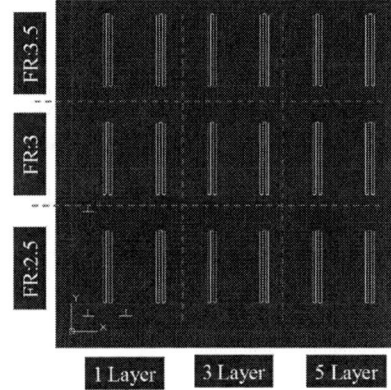

Figure 3: Print pattern and matrix for finding process parameters that minimize resistivity

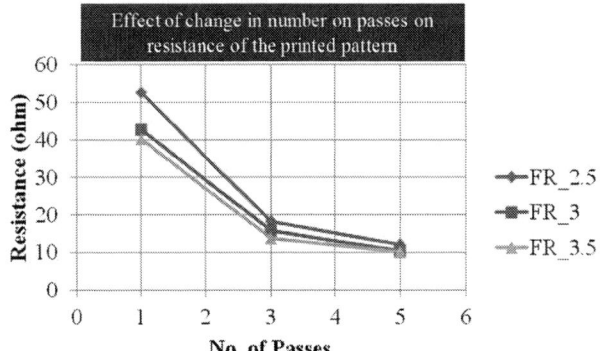

Figure 4: Effect of change in number on passes on resistance of the printed pattern

Once the lines are printed, and before the process of sintering, the printed lines are just an agglomeration of nanoparticles or nanoflakes. Exposure to heat through thermal sintering, photonic sintering or laser sintering causes the silver nano-particles to melt and form more coherent traces through surface melting and compaction of the particles. In addition to resistivity, sintering affects the mechanical properties like shear strength and elastic modulus of the line. In this study, the search for desirable process parameters has been guided by the configuration that provided the lowest resistivity. The effect of the sintering profiles has been quantified on the electrical properties including resistivity and mechanical properties, including the shear strength and the elastic modulus. Three different FR and three different number of passes or layers were printed and sintered at the stipulated print profile of 200°C for 60 minutes. First, lines with different ink flow rate and sheath gas flow rates were printed at a standoff height of 3mm from the substrate, chiller temperature of 20°C and stage speed ranging from 0.3 to 1mm per sec. The lines obtained were examined under an optical microscope to check for continuity. It was found that, an ink flow rate of 20sccm and sheath-gas flow rate ranging from 50sccm and above at a speed of 0.5mm per sec lead to continuous lines.

Schematic of the pattern to be printed was designed in a CAD modeling software and imported as a program file (g codes and m codes). The print pattern with varying FR and number of layers is shown in Figure 3. This test matrix of pattern was printed on to the large 2 inches by 3-inch substrate. The cleaning process of the substrate was set to be constant and not varied in the experimental matrix. Each substrate was cleaned with Acetone and Iso-propyl alcohol and blow dried with compressed air prior to printing. The AJ300 machine doors were kept closed at all times during the printing process to prevent any disruption in the aerosol jet during the printing process from airflow in the room.

TABLE 2:FINALIZED PROCESS PARAMETERS

Process Parameter	Value
Ink mass flow rate	20 sccm
Sheath gas flow rate	75 sccm
Stage Speed	0.5mm/s
Stand-off height	3mm
Nozzle diameter	150microns
Number of passes	5

The printed lines were probed on a resistance spectroscopy equipment to measure the resistance. The focus ratio was adjusted by only changing the sheath-gas flow rate, without changing the ink flow rate. The figure below is a graph of resistance vs number of layers. The resistance of the printed lines decreased with the increase in the number of print-passes. It was observed that improvement in resistance from 1-pass to 3-passes, for a FR of 3.5 was ~65% and the improvement from 3 passes and 5 passes was ~28%. The average resistance value found at 5-passes for all the FRs was found to be in the range of 10Ω to 14Ω, indicating that the gains in resistance value diminish beyond 5-passes. Thus, 5-passes were decided upon as the final number of layers. In addition, focus ratio of 3.5 was also observed to give the least resistance value in the graph as expected. The process parameters finalized that lead to least resistance are stated in Table 2. These parameters were used throughout the remainder of the study.

V. TEST MATRIX

Once the process parameters were finalized, a detailed study was conducted to understand the correlation of mechanical and electrical properties of the printed lines with variation in sintering times and temperatures. The process parameters of the additively printed traces have been correlated to the mechanical and electrical properties of these micro-scale printed lines. The measured data fills a void in literature for additively printed interconnects. Properties including elastic modulus, shear load to failure, resistivity are available for conventional electrical traces that are manufactured through subtractive printing process. The conventional traces are typically treated as homogenous linear isotropic materials – an assumption that breaks down for additively printed lines. Additively printed lines are instead a collection of spheres, or flakes depending on the ink formulation.

TABLE 3: VARYING SINTERING TEMPERATURE

Sintering Temperature	Sintering Time	Resistance	Shear Load to failure	Elastic modulus
200°C	60 min	X	X	X
240°C	60 min	X	X	X
280°C	60 min	X	X	X
300°C	60 min	X	X	X

TABLE 4: VARYING SINTERING TIME

Sintering Temperature	Sintering Time	Resistance	Shear Load to failure
200°C	60 min	X	X
200°C	105 min	X	X
20°C	150 min	X	X
200°C	180 min	X	X

Furthermore, when nano-sized particles are additively printed to form electrical traces that are much longer in length with respect to its width, with multiple print passes made, the resultant structure is anisotropic in nature. The material properties are dependent on the geometry of the resultant structure and the geometry is dependent on process variations. A test matrix has been developed to study the effect of varying sintering time and temperature on resistivity and f mechanical properties of the printed line. Table 3 and Table 4 show the test matrices for sintering temperature and sintering time respectively

VI. RESISTANCE MEASUREMENT

Patterns to be printed were designed in a CAD modeling software. Figure 5 shows the design of the resistive pattern used for quantification of resistance as a function of process parameters. The pattern shown in Figure 5 with 4 units was printed on the substrates, for the previously selected process parameters of ink mass flow rate, sheath gas flow rate, stage speed, standoff height, nozzle diameter, and number of passes.

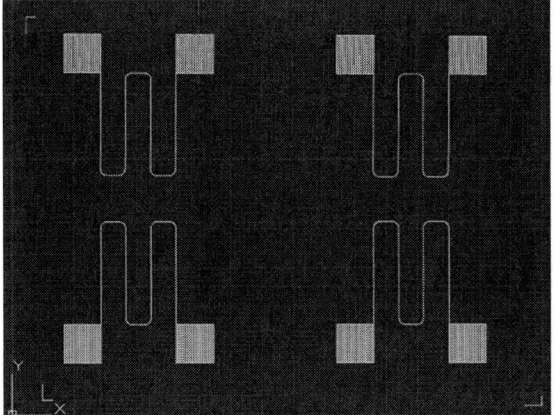

Figure 5: CAD of the pattern to be printed

In order to measure the resistance, aluminum based conductive epoxy mixed as per the stipulated ratio was deposited on the pads and a wire was connected using a conductivity epoxy mixed in the ratio of 1:2. The resistance was measured using a digital multimeter by probing the connected wires. An average of four resistance values obtained per print per condition has been reported for each reported resistance measurement in the paper.

VII. SHEAR LOAD TO FAILURE

Additive or subtractive 3D printed electronics may be subjected to shear and torsion loads during manufacturing, assembly or service. The print pattern shown in Figure 5 was printed and cured at the different sintering temperatures and times shown in Table 3 and Table 4.

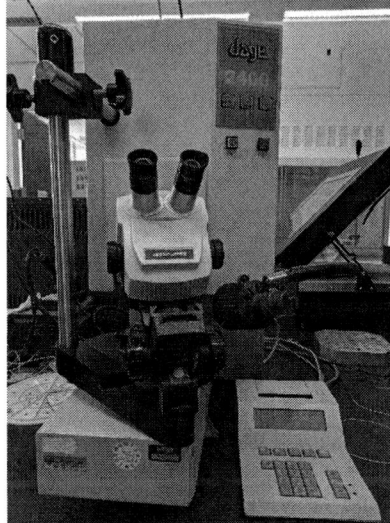

Figure 6: Dage Ball Shear Tester

A Dage shear tester shown in Figure 6 was used to apply a shear load using a knife edge. The failure load reported, once the transverse cross-section failed was reported. An average of 6 values was reported for each condition measured in the study.

VIII. MEASUREMENT OF ELASTIC MODULUS

Elastic modulus of the silver lines printed lines was experimentally measured using a nano-indentation technique. This technique involves using a pointed indenter to indent the surface of sample. The displacement of the indenter as a function of applied force is recorded as a function of time through the loading and the unloading cycle of the test. During the loading cycle, the indenter undergoes both elastic and plastic deformation. Several methods for computing the elastic modulus and hardness have been published. In this work, the Oliver-Pharr method has been used to compute the mechanical properties from the nano-indenter load displacement curves. The hardness, H is defined as the mean pressure under the indenter [Oliver 1992],

$$H = \frac{P_{max}}{A_c} \tag{1}$$

Where, P_{max} is the maximum applied force and is obtained from the load vs displacement graph, A_c is the projected area of the indenter tip. For a perfect Berkovitch indenter the contact area can also be represented as:

$$A_c = 24.56 h_c^2 \tag{2}$$

Where, h_c is the contact depth of the indenter tip. The slope of the unloading curve can be used to compute the elastic modulus using E_r, the reduced modulus of the specimen. For a Berkovitch indenter, the contact stiffness is given as follows:

$$S = \frac{2}{\sqrt{\pi}} E_r \sqrt{A_c} = \frac{dP}{dh} \tag{3}$$

Where, S is the contact stiffness. P and h are the indenter force and indenter height. Equation (3) can be rearranged to obtain,

$$\frac{1}{E_r} = \left(\frac{\pi}{4}\right)^{1/2} \left(\frac{1}{A_c}\right)^{1/2} \tag{4}$$

The measured elastic modulus has contributions from both the specimen and the indenter which are related as follows,

$$\frac{1}{E_r} = \frac{1 - v_s^2}{E_s} + \frac{1 - v_i^2}{E_i} \tag{5}$$

Where, v_s and v_i are the Poisson's ratio of the indenter and specimen. From equation above, the elastic modulus for thin films samples can be given as,

$$E_s = \left(\frac{1}{E_r} - \frac{1 - v_i^2}{E_i}\right)^{-1} \tag{6}$$

IX. EFFECT OF SINTERING TEMPERATURE ON RESISTIVITY

Figure 7 shows the graph of resistivity with respect to change in sintering temperature. The sintering time has been kept constant for all the measurements at 60-min of thermal sintering at specified temperature. Experimental measurements indicate that the resistivity decreases with the increase in sintering temperature.

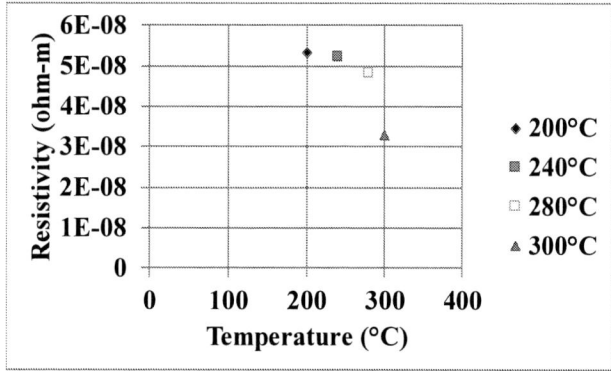

Figure 7: Resistivity Vs Sintering Temperature

The resistivity decreases or improves with increasing sintering temperature. In order to understand the physics of changes in resistivity, scanning electron microscopy images were taken to investigate the resistivity measurements (Figure 8). A comparison of the images from 200°C @1-hr with the images for 300°C @1-hr reveals that the nanoparticle in the printed ink are more clearly visible at lower sintering times. Nanoparticles are not discernable in sintering conditions of 300°C @1-hr. Increase in sintering temperature allows the nanoparticles to coalesce more readily resulting in a reduction in the resistivity of the printed traces. The trend is consistent for sintering temperature of 240°C, and 280°C sintering for 1-hour.

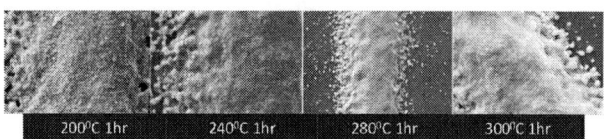

Figure 8: SEM images for varying sintering temperature and constant sintering time

X. EFFECT ON SINTERING TIME ON RESISTIVITY

The effect of sintering time has been measured by decoupling the effect of the sintering temperature in this part of the measurements. The sintering temperature has been kept constant at 200°C for the specified period of time. Figure 9 shows the graph of resistivity with respect to changes in sintering time. Experimental measurements indicate that the resistivity increases with increase in sintering time.

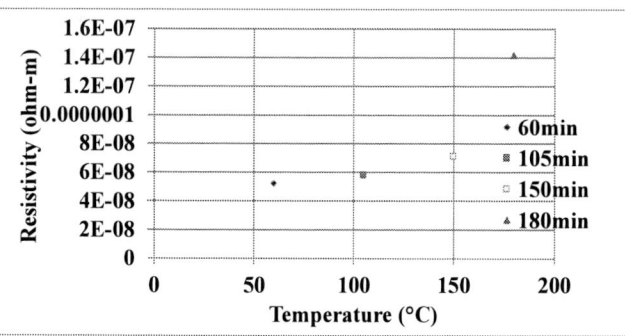

Figure 9: Resistivity Vs Sintering Time

Resistivity increases or degrades with the increase in sintering time for a specified sintering temperature. In order to understand the physics of changes in resistivity, scanning electron microscopy images were taken to investigate the resistivity measurements (Figure 10). A comparison of the images from 200°C @1-hr with the images for 200°C @3-hr reveals that the nanoparticle in the printed ink are more clearly visible at lower sintering times. Nanoparticles are not discernable in sintering conditions of 200°C @3-hr. Increase in sintering temperature allows the nanoparticles to coalesce more readily resulting in a reduction in the resistivity of the printed traces.

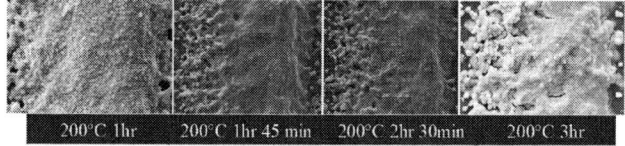

Figure 10: SEM imagery for varying sintering time

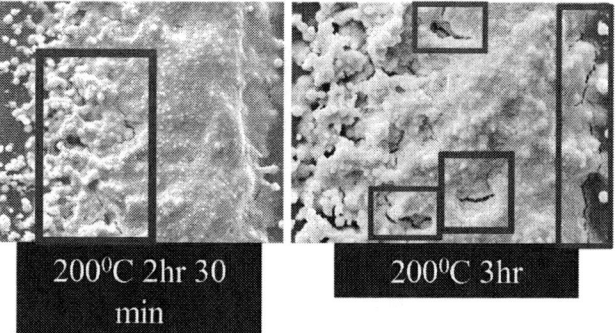

Figure 11: Appearance of micro-cracks

Higher sintering conditions also show appearance of micro cracks in the printed lines after sintering. The locations of microcracks are highlighted with blue-rectangles in Figure 11. The elimination of point-contact between particles results in a lower resistivity with the increase in sintering time. The trend is consistent for sintering temperature of 240°C, and 280°C sintering for 1-hour. The opposite trend of the increase in resistivity with the increase in sintering time is due to the appearance of micro cracks, which increase the current density in the printed-trace due to reduction in effective cross-sectional area. The occurrence of micro-cracks could be attributed to the evaporation of solvent and co-solvent that are deposited on the substrate prior to sintering. Ethylene glycol is the solvent in this ink, mixed with water and other co-solvents. The boiling point of the ethylene glycol as published in data sheets is 197.1°C. This residual chemical may undergo boiling when sintered for longer duration and escape out of the lines leaving cracks. A higher mean free path of electrons corresponds to higher resistivity because more scattering is observed as higher resistance. The presence of cracks would increase the mean free path of the electrons, increasing the resistance and in turn deteriorate the resistivity. Figure 8 show the scanning electron microscopy images for lines sintered for 1 hour and for increase in the sintering temperature.

XI. EFFECT OF SINTERING TEMPERATURE ON THE SHEAR LOAD-TO-FAILURE

A Dage shear tester with a knife-edge was used to find the shear load to failure for lines printed at varying sintering temperatures at a sintering time of 1-hr. Figure 12 shows the graph of shear load to failure with increase in sintering temperature.

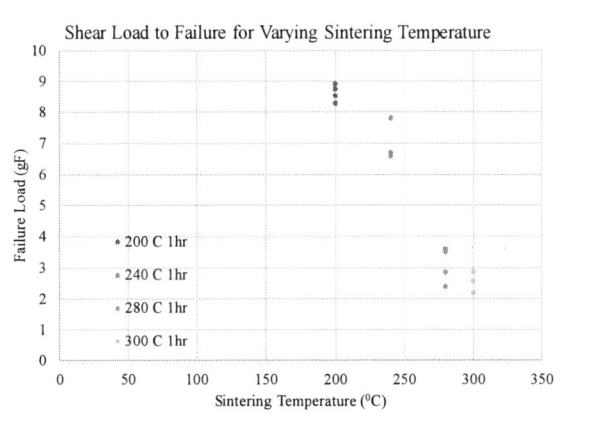

Figure 12: Shear Load to Failure Vs Sintering Temperature

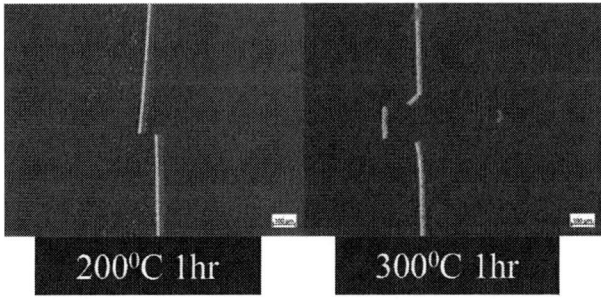

Figure 13: Sheared line for increasing sintering temperature

Experimental data indicates that the shear load-at-failure reduces with the increase in sintering temperature. The optical images of the sheared lines for increasing sintering temperature are shown in Figure 13. Sintering process is known to result in increase in nano-particle volume fraction, caused due to solvent evaporation. The sheared line for 200°C, 1-hour sintering condition shows a relatively ductile failure in comparison to the sheared line for 300°C, 1-hour sintering condition. When the printed line may be sintered for a prolonged period, the silver nano-particles coalesce resulting in making the structure more brittle, and a reduction in the ultimate tensile strength. This explains the transition from a ductile to brittle behavior of the printed lines shown in Figure 13.

XII. EFFECT OF SINTERING TIME ON THE SHEAR LOAD-TO-FAILURE

A Dage shear tester with a knife-edge was used to find the shear load to failure for lines printed at varying sintering time at a sintering temperature of 200°C. Figure 14 shows the graph of shear load to failure with increase in sintering time. Experimental data indicates that the shear load-at-failure reduces with the increase in sintering time. The optical images of the sheared lines for increasing sintering time are shown in Figure 15.

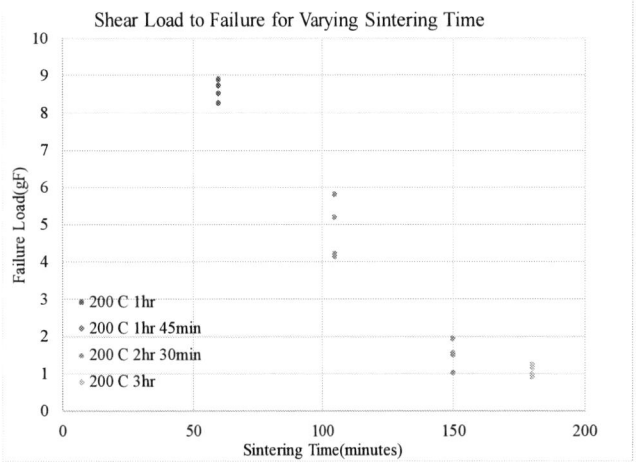

Figure 14: Shear Load to Failure Vs Sintering Time

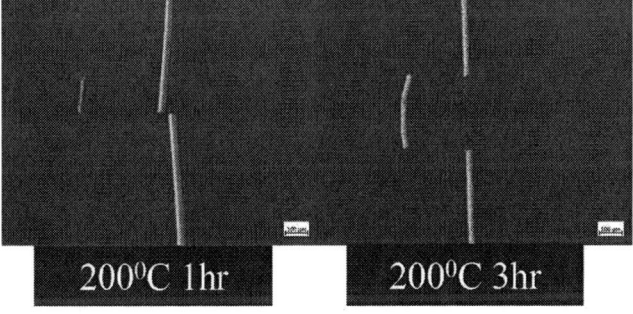

Figure 15: Sheared line for increasing sintering time

Sintering process is known to result in increase in nano-particle volume fraction, caused due to solvent evaporation. The sheared line for 200°C, 1-hour sintering condition shows a relatively ductile failure in comparison to the sheared line for 200°C, 3-hour sintering condition. When the printed line may be sintered for a prolonged period, the silver nano-particles coalesce resulting in making the structure more brittle, and a reduction in the ultimate tensile strength. This explains the transition from a ductile to brittle behavior of the printed lines shown in Figure 15.

XIII. EFFECT OF SINTERING TEMPERATURE ON ELASTIC MODULUS

Aerosol jet printed samples were printed, potted in resin and polished for the nano-indentation tests. A load of 20,000 micro-newtons was applied to the samples that resulted in an indentation depth of 920nm. A relatively high load was used to measure elastic modulus in the bulk region and avoid the influence of surface roughness.

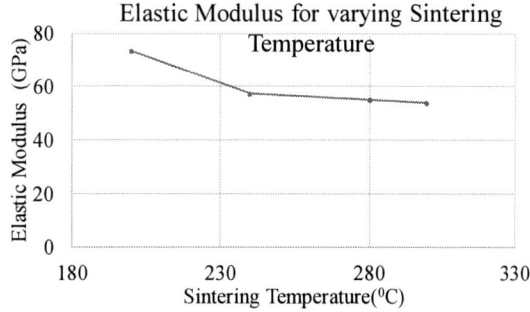

Figure 16: Elastic modulus for varying sintering temperature

Figure 16 shows the effect of varying sintering temperature on elastic modulus. Elastic modulus was found to decrease monotonically with the increase in sintering temperature from 200°C to 300°C for a sintering time of 1-hour. A 26-percent reduction in elastic modulus was found between the printed lines sintered at 300°C for 1-hr with respect to the printed lines sintered at 200°C for 1-hr. This reduction may be attributed to the degradation of grain boundaries at a higher temperatures.

XIV. LONG TERM PRINT CONSISTENCY

In order to investigate the long-term print consistency of the AJP process, a similar silver -nano particle-based ink was used with same nano-particle size and viscosity compatible with the an ultrasonic atomizer. For longer print durations, a bubbler for solvent replenishing was also used to maintain the viscosity of the ink. The results in this section are presented with and without the presence of bubbler. The drift in the process parameters was quantified and their effect on the dimensions of the printed lines studied. For the purpose of this study, the Aerosol jet printer was allowed to print continuously for a duration of 10 hours using the ultrasonic atomizer. Printing process consistency was gauged through the quantification of process capability index (C_{pk}) and

process capability ratio (C_p). Four main parameters, nitrogen or sheath gas flow rate (ShFR), carrier gas flow rate (CFR), ultrasonic atomizer current (UA_MAX) and platen heater, which influence the print quality, were recorded in-situ for the entirety of the tests conducted. Figure 17 shows the four process parameters recorded for a period of 10 hours. The control values were pre-defined before start of the 10-hour cycle count; these values are listed in Table 5. It can be seen from Table 5 how the response values correspond with the control values.

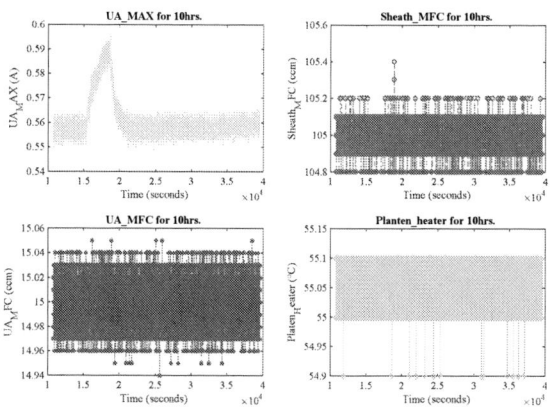

Figure 17: Variation in process parameters with bubbler

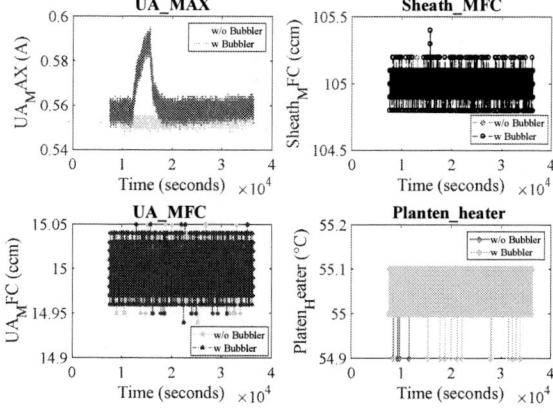

Figure 18: Comparison of the Process Parameters with and without Bubbler

TABLE 5: RESPONSE VALUES OF THE PROCESS PARAMETERS

Parameter	Control Values	Mean	STD Dev	Variance
ShFR	105	104.99	0.0564	0.0032
CFR	15	14.99	0.0127	1.62e-4
UA_MAX (mAmps)	0.56	0.561	0.0081	6.52e-5
Platen Heater(□)	55	55.003	0.0178	3.18e-4

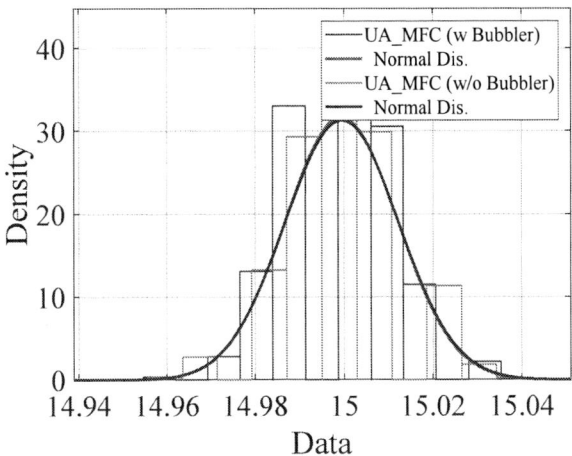

Figure 19: Probability Density Function of CFR fitted with Normal Distribution

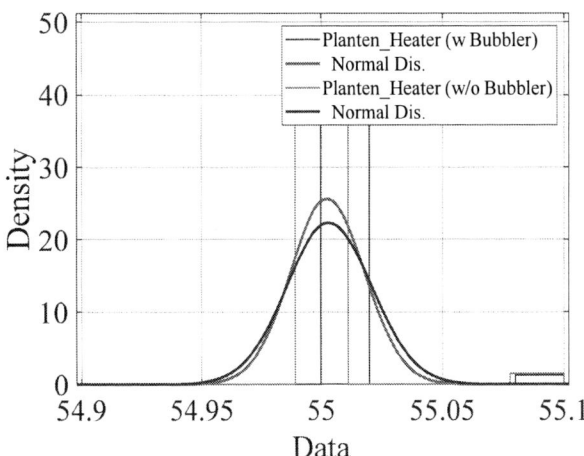

Figure 22: Probability Density Function of Platen Heater fitted with Normal Distribution

Probability density functions (PDFs) of the process parameters fitted with normal distributions are shown in Figure 19 through Figure 22 to assess the likelihood that the random values of the process parameters will be close to the actual set-value of the parameter. The data is shown with and without the use of bubbler. The variance in the CFR, ShFR, UA_MAX, Platen Heater Temperature, is lower with the use of a bubbler in all cases. The PDFs were computed for the full 10 hour using in-situ data. Figure 23 thru Figure 26 show the probability plots for the four process parameters where the distribution of the data in x is compared to the normal distribution. The data is plot against a theoretical normal distribution. For a normal distribution, the points should form an approximate straight line and any deviation from the straight line indicates deviations from normality. An example can be seen in Figure 25, which shows the probability plot for UA_MAX. The curvature can be attributed to the spike observed in the atomizer current data shown in top left of Figure 17.

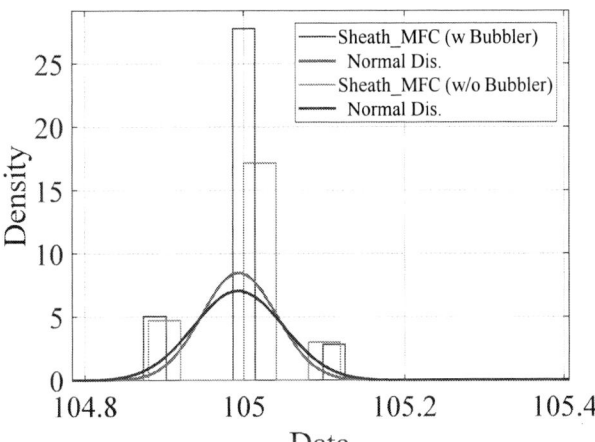

Figure 20: Probability Density Function of ShFR fitted with Normal Distribution

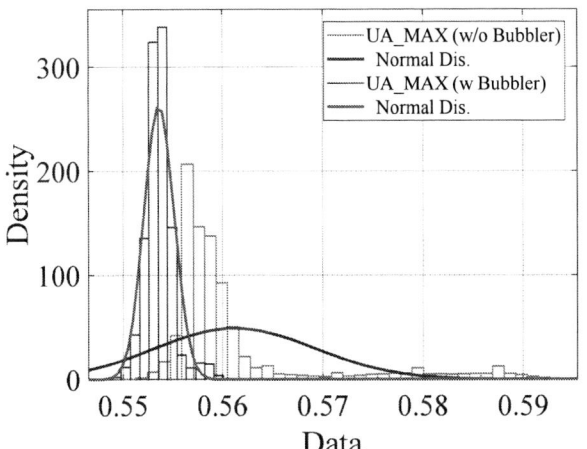

Figure 21: Probability Density Function of UA_MAX fitted with Normal Distribution

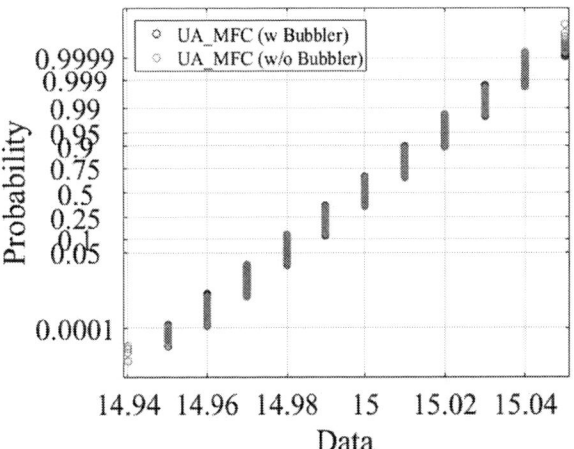

Figure 23: Accumulated Probability plot for CFR for a 10 hour run

978-1-7281-1500-9/19 $31.00 © 2019 IEEE

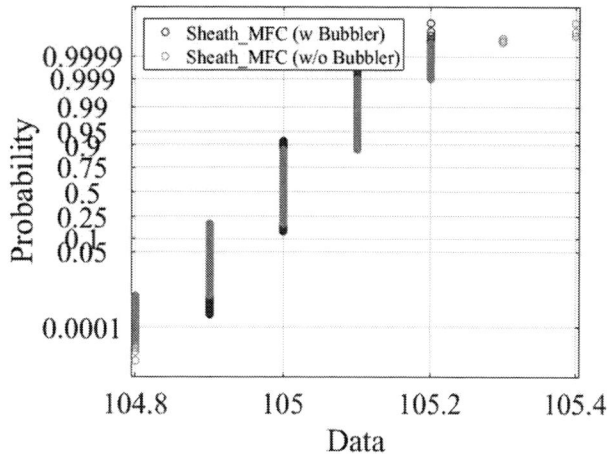

Figure 24: Accumulated Probability plot for ShFR for a 10 hour run

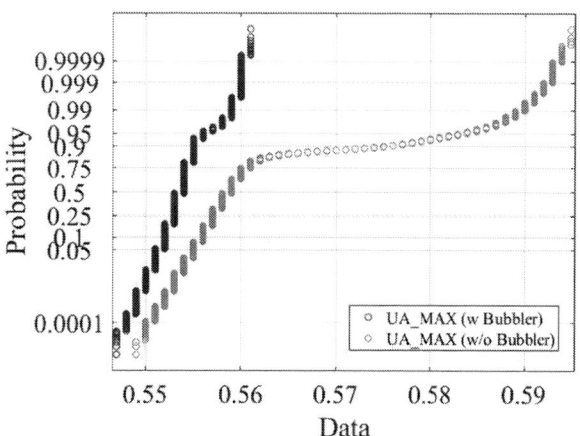

Figure 25: Accumulated Probability plot for UA_MAX for a 10 hour run

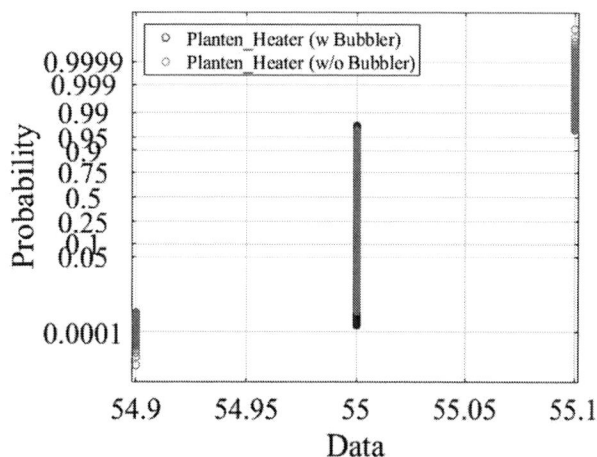

Figure 26: Accumulated Probability plot for Platen Heater for a 10 hour run

XV. DIMENSIONAL PARAMETERS

The printed dimension of the lines was used to compute the effect of print run on the dimensional consistency of the printed traces. Print line width and line height were computed by conducting surface-profilometry experiments whereas the pattern resistance was computed using a 4-point resistance method using a resistance probe.

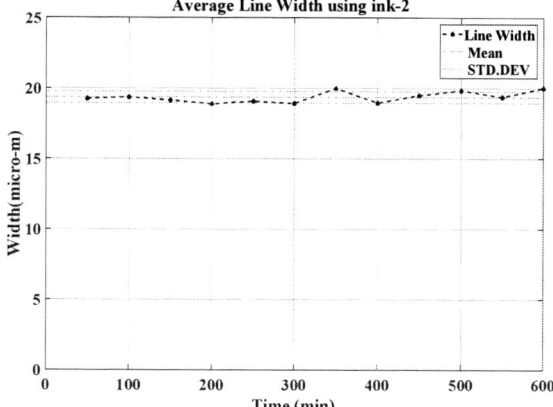

Figure 27: Average Line Width of the printed samples with Bubbler

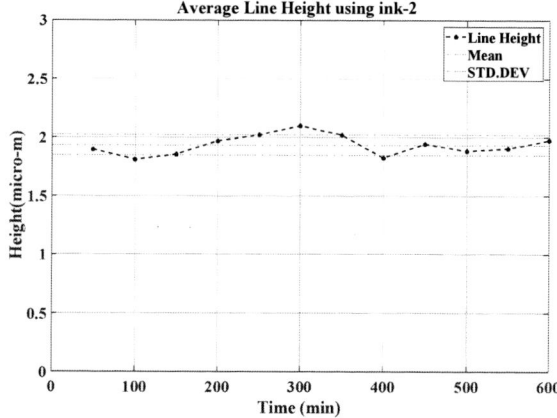

Figure 28: Average Line Width of the printed samples with Bubbler

Figure 27 and Figure 28 show that the average line width was between 19-20μm whereas the average line height was between 1.8-2μm. Over the course of the 10-hr print-period, the line dimensions show variations mainly along the height of the printed line. These variations do increase and decrease with time and the line height does not stay constant whereas the variations along the line width seem to be constant during first 300 minutes of printing but then drift positively or negatively as the print time increases.

XVI. PRINT RESISTANCE AS A FUNCTION OF RUN TIME

Several substrates were printed over the 10-hour period, where each print-run represents 50-min of print time. The printed pattern has been described about in the above sections. Each sample (glass slide) has 4-patterns, below shown is the print resistance as a function of operational time in Figure 29. Each data point on the resistance plots is an average of the 4-patterns (for each slide representing their respective time instances).

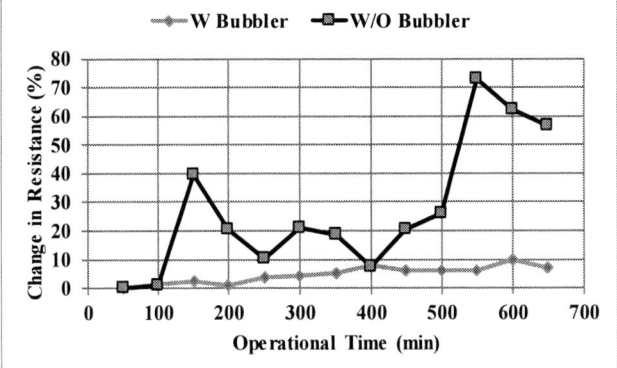

Figure 29: Print Resistance with and without bubbler

XVII. PRINT QUALITY AS A FUNCTION OF PRINT-TIME

Optical images were used to visualize any evidence of variation in the print quality as a function of print-time. It was also observed from the dimensional plots (Figure 27 and Figure 28) that line width tends to remain within +/-σ (one standard deviation mark) whereas the line height shows variance, which violates the one standard deviation limit.

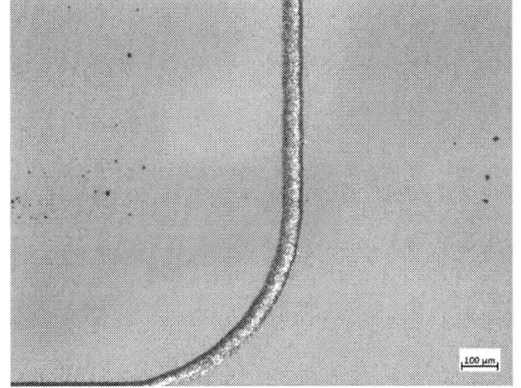

Figure 30: Print Quality of the loop feature from the pattern at 50th min mark

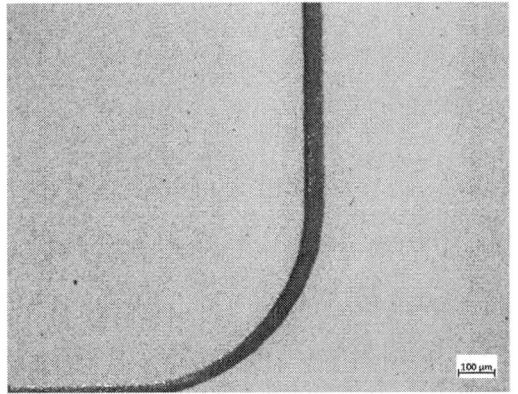

Figure 31: Print Quality of the loop feature from the pattern after 10th hour mark

Figure 30 and Figure 31 show a comparison of the print quality at the 50-min mark with that of the 10-hour print run mark. The prints appear comparable with the use of a bubbler and no significant change in the quality of the printed line is observed.

XVIII. STATISTICAL APPROACH

The 10-hour long print consistency was studied using a statistical approach to quantify process robustness. In this work, we showcase the potential of the four control variables including. ShFR, CFR, UA_MAX and Platen Heater and three response variables line-width, line-thickness, and line resistivity as a function of process capability index (C_{pk}) and process capability ratio (C_p). Two questions, which are answered using this approach, are - (a) How does one statistically characterize the printing process? (b) Is this process under control for reliable use?

$$Cp = \frac{(UpperSpecLimit) - (LowerSpecLimit)}{6\sigma} \quad (7)$$

$$Cpk_1 = \frac{(UpperSpecLimit) - (X_{50})}{3\sigma} \quad (8)$$

$$Cpk_2 = \frac{(X_{50}) - (LowerSpecLimit)}{3\sigma} \quad (9)$$

Where X_{50} is the mean value of the variable, and σ is the standard deviation. The equations listed above represent the process capability indexes (C_{pk}) and process capability ratio (C_p). The tolerances for the four process parameters were established using measurement of variance from the respective mean values.

TABLE 6: PROCESS CAPABILITY OF CONTROL AND PRINT PARAMETERS WITHOUT BUBBLER

	Platen	Sheath	UA	UA_Max	Resistance
Cp	1.0727	1.0656	1.0583	1.0963	0.767
Cpk1	1	1.01	1.011	1.096	0.744
Cpk2	1.01	1	1.01	1.1	0.732

TABLE 7: PROCESS CAPABILITY OF CONTROL AND PRINT PARAMETERS WITH BUBBLER

	Platen	Sheath	UA	UA_Max	Resistance	Height	Width
Cp	1.067	1.043	1.051	1.089	1.224	1.09	1.06
Cpk1	1.001	1	1.031	1.002	1.226	1.02	1.03
Cpk2	1.01	1.022	1	1	1.213	1	1.01

The main task of this approach is to find whether or not we can fit six standard deviations within this range to form a basis of a 6σ capable process. Three-sigma processes have a C_{pk} of 1 and six-sigma processes have a C_{pk} of 2. Process capability ratio and indexes have been established for CFR (UA_MFC). Table 6 and Table 7 show the computation for the process capability index (C_{pk}) and process capability ratio (C_p) for configuration with the bubbler and without the bubbler respectively. The values of the process capability index (C_{pk}) and process capability ratio (C_p) have been computed for both the control parameters and the print-performance parameters. A process capability of 1 indicates a 3-sigma process and a process capability of 2 indicates a 6-sigma process. Computations in Table 6 indicate that without bubbler, the resistance consistency is approximately a 2.125-sigma process. Computations in Table 7 indicate that with bubbler, the resistance consistency is approximately a 3.6-sigma process, height consistency is approximately 3-to-3.25-sigma process, and width consistency is a 3-to-3.25-sigma process.

XIX. SUMMARY AND CONCLUSIONS

In this paper, a detailed study of the effect of aerosol-jet process parameters on the resistance, shear strength, and elastic modulus has been presented. Experimental measurements indicate that the resistivity improved with the increase in the number of passes. Increase in sintering time at a constant sintering temperature was found to increase resistivity of the printed line resulting from the increase in number of cracks in the printed lines. Increase in the sintering temperature at a constant sintering time was found to decrease the resistivity of the printed line – which based on SEM analysis can be attributed to the decrease porosity of the printed line. Mechanical properties including the elastic modulus and shear strength decreased with the increase in sintering temperature. Line-consistency measurements of line-width, line height, and resistance over a continuous 10-hour print run are reported. Process measurements indicate that viscosity of the ink has a pivotal role in maintaining print consistency. Line width, line height, and resistance measurements exhibit much lower variance with the use of bubbler, which enables control over ink-viscosity during the 10-hour print run. In addition, computation of the process capability index (C_{pk}) and process capability ratio (C_p) indicate that without bubbler, the resistance consistency is a 2.125-sigma process. Further, the use of bubbler allows for a wider process window where, resistance consistency is a 3.6-sigma process, height consistency is 3-to-3.25-sigma process, and width consistency is a 3-to-3.25-sigma process.

ACKNOWLEDGMENTS

The project was sponsored by the NextFlex Manufacturing Institute under PC 3.3 Open Call Project titled – Additive and Semi-Additive Methods for High-Density Interconnects. This material is based, in part, on research sponsored by Air Force Research Laboratory under agreement number FA8650-15-2-5401, as conducted through the flexible hybrid electronics manufacturing innovation institute, NextFlex.

The U.S. Government is authorized to reproduce and distribute reprints for Governmental purposes notwithstanding any copyright notation thereon. The views and conclusions contained herein are those of the authors and should not be interpreted as necessarily representing the official policies or endorsements, either expressed or implied, of Air Force Research Laboratory or the U.S. Government.

REFERENCES

[1] BCC Research, Printed Electronics: The Global Market, https://www.bccresearch.com/market-research/information-technology/printed-electronics-market-ift066b.html, 2012

[2] Binder S, GlatthaarMand Rädlein E 2014 Aerosol Sci. Technol. 48 924

[3] Cai, F., Aerosol jet printing for 3-D multilayer passive microwave circuitry, 44th European Microwave Conference, pp. 512-515, 2014.

[4] Ha M, Zhang W, Braga D, Renn, MJ, Kim, CH and Frisbie CD 2013 ACS Appl. Mater. Interfaces 5 13198

[5] IDTECHX 2018: https://www.idtechex.com/research/articles/printed-electronics-key-trends-00015712.asp, Raghu Das, October 2018

[6] Khorramdel, B., A. Torkkeli and M. Mäntysalo, Electrical Contacts in SOI MEMS Using Aerosol Jet Printing, in IEEE Journal of the Electron Devices Society, vol. 6, pp. 34-40, 2018.

[7] King, B. H., M. J. O'Reilly and S. M. Barnes, "Characterizing aerosol Jet®multi-nozzle process parameters for non-contact front side metallization of silicon solar cells," 34th IEEE Photovoltaic Specialists Conference (PVSC), Philadelphia, PA, pp. 001107-001111, 2009.

[8] Krzeminski, J., Pads and microscale vias with aerosol jet printing technique, 2017 21st European Microelectronics and Packaging Conference (EMPC) & Exhibition, Warsaw, pp. 1-4, 2017

[9] Lang R J 1962 J. Acoust. Soc. Am. 34 6

[10] Lupo, D., W. Clemens, S. Breitung, K. Hecker, E. Cantatore, "Oe-a roadmap for organic and printed electronics" in Applications of Organic and Printed Electronics, US:Springer, pp. 1-26, 2013.

[11] Markets and Markets, Printed Electronics: https://www.marketsandmarkets.com/PressReleases/printed-electronics-market.asp, accessed, Dec 2018.

[12] Navratil, J., T. Rericha, R. Soukup and A. Hamacek, "Aerosol Jet Printed Sensor on Fibre for Smart and IoT Applications," 2018 41st International Spring Seminar on Electronics Technology (ISSE), Zlatibor, pp. 1-4, 2018.

[13] Oakley, C., A. Kaur, J. A. Byford and P. Chahal, Aerosol-Jet Printed Quasi-Optical Terahertz Filters, 67th ECTC, pp. 248-253, 2017.

[14] Oliver, W. C., G. M. Pharr, An improved technique for determining hardness and elastic modulus using load and displacement sensing indentation experiments, J. Mater. Res., vol. 7, no. 6, pp. 1564-1583, June 1992.

[15] Paulsen, J. A., M. Renn, K. Christenson and R. Plourde, "Printing conformal electronics on 3D structures with Aerosol Jet technology," Future of Instrumentation International Workshop (FIIW) Proceedings, Gatlinburg, TN, pp. 1-4, 2012.

[16] Peskin R L and Raco R J, J. Acoust. Soc. Am. 35 1378, 1963

[17] Ravindran P and Davis E J, J. Colloid Interface Sci. 85 278, 1982

[18] Rodes C, Smith T, Crouse R and Ramachandran G, Aerosol Sci. Technol. 13 220, 1990

[19] Secor, E., Principles of Aerosol Jet Printing, Flexible and Printed Electronics, 035002, 2018.

[20] Stoukatch et al., Evaluation of Aerosol Jet Printing (AJP) technology for electronic packaging and interconnect technique, 4th Electronic System-Integration Technology Conference, Amsterdam, Netherlands, pp. 1-5, 2012.

[21] Widmann J F and Davis E J, Aerosol Sci. Technol. 27 243, 1997

A Viscoplastic-Based Fatigue Reliability Model for the Polyimide Dielectric Thin Film

Yu-Chen Chang, Tz-Cheng Chiu
Department of Mechanical Engineering
National Cheng Kung University
Tainan, Taiwan
e-mail: tcchiu@mail.ncku.edu.tw

Yu-Ting Yang, Yi-Hsiu Tseng, Xi-Hong Chen
Central Development Engineering
ASE Group Chung-Li
Taoyuan, Taiwan

Abstract— In this study the fatigue characteristics of polyimide thin film used in redistribution interconnects were considered. Cyclic tension tests under various stress- and strain-ranges were performed to investigate the damage accumulation in the thin film. It was observed that, under cyclic strain loading, the deformation of the thin film is mainly governed by viscoelastic stress relaxation behavior, and damage of the thin film is insignificant. On the other hand, the thin film under cyclic stress loading exhibits a viscoplastic ratcheting response. Furthermore, the envelope of the strain response under stress-controlled cycling increases as the fatigue cycle increases. A phenomenological model was developed for characterizing the damage accumulation under cyclic stress condition. The model considers the viscoplastic response by superpositioning a power-law plastic model and a linear viscoelastic model for predicting the growth of the strain envelope. The fatigue prediction model can be applied for predicting the fatigue failure of the polyimide dielectric in the redistribution interconnects.

Keywords-viscoelastic; plastic; cycling; redistribution; fracture

I. INTRODUCTION

One of the current focuses of technology development in electronic packaging is the heterogeneous integration of microelectronic components through fan-out wafer-level package (FO-WLP). Various processing approaches including fabricating interconnects on reconstituted wafer (typically referred to as "die-first") and assembling die on processed interconnect structures (typically referred to as "die-last") have been implemented for the FO-WLP. Alternative integration approaches such as using panel as opposed to wafer-shaped carrier, stacking multiple FO-WLPs to form a single module were also developed. While there are many integration schemes implemented for the fan-out packages, a common feature of the fan-out design is that the traditional copper-clad laminate interconnect is replaced by the redistribution layer (RDL) interconnect consisted of Cu conductors embedded in polymeric dielectric thin films.

The common polymer resins used for RDL in fan-out packages include benzocyclobutene (BCB), polybenzoxazole (PBO) and polyimide (PI). These materials have also been used extensively in flip-chips and fan-in wafer-level packages as the passivation thin film for Si die surface

redistribution and bump structures. Among these polymer resins, PI has advantages of high mechanical compliance as well as good high-temperature stability, and is the most popular choice. While PI is clearly a favorable solution for die surface interconnect passivation, extending the technology into multilayered interconnect with submicron-pitch is complicated due to the increased risk of fatigue failure under board-level temperature cycling conditions. The typical failure modes of RDL under temperature cycling include Cu trace cracking, dielectric cracking [1, 2], and debonding of the polymer-Cu or polymer-die interface. These failures are driven mainly by thermal stresses related to the high coefficient of thermal expansion (CTE) of the thin film resin. With the expected introduction of chemical-mechanical polishing (CMP) processes for multilayered RDL planarization in the near future, these cracking and debonding failure modes would not only occur during the reliability phase, but will also likely to appear in the processing stage [3, 4]. Consequently, an important task in the continuous scaling of the interconnect technology is to characterize the fatigue response of the dielectric thin film such that the thermomechanical reliability of the redistribution interconnect can be accurately assessed and optimized.

The typical approach used for characterizing and predicting reliability related to the cracking and debonding failures is via a semi-empirical approach, e.g., [3, 5, 6]. In this approach, numerical thermoelastic simulations are conducted to extract stresses or strains at critical locations in the structure of interest, and the results are compared to experimental findings from design-of-experiments reliability evaluations to establish the empirical correlation between the stress or strain indices and the failure. Disadvantage of the semi-empirical prediction scheme is that the effects of process history and nonlinear material responses are confounded, and the underlying physical mechanisms causing the failure cannot be well defined. In order to capture the physics of failure accurately and to quantitatively predicting the fatigue reliability, a phenomenological model that describes the damage accumulation in the PI thin film under fatigue loading are necessary.

An important factor to be included in modeling the cracking involving the compliant PI thin film is the time-dependent inelastic deformation of the thin film and the related energy dissipation. Under fatigue cycling conditions, the energy dissipation from each cyclic loading can be attributed to the defect formation and growth in the polymeric thin film. In this study, the prediction model for the fatigue cracking of the polymeric thin film was established by experimentally characterizing the inelastic constitutive behavior of the thin film, applying the constitutive model to consider the cyclic loading response, and then validating with the fatigue experiments. The model can be combined with numerical models of the real RDL structures to estimate damage accumulation and fatigue reliability of the dielectric material and enable quick material selection for process development.

II. THE CONSTITUTIVE BEHAVIOR OF PI

Traditional photo-imageable dielectric polymers such as epoxy acrylate based solder mask deform in a linear-elastic fashion under typical use condition, and the corresponding failure is through brittle fracture. Contrast to these materials, PI exhibits ductile behavior and has high elongation at film fracture (at strain value around 20% or higher). Therefore, an important first step for characterizing the fatigue response of the PI thin film is to understand the mechanical behavior under monotonic load and establish the corresponding constitutive model.

For the evaluation of the mechanical response under monotonic load, uniaxial tensile test was first conducted to investigate the PI thin film of thickness of 5 μm. The specimen was prepared by coating PI on Si wafer, curing and then debonding by using HF. The thin film specimen was then mounted on a 100 μm-thick PI window frame and then mounted on an Instron 5565 tester with 500 N loadcell as shown in Fig. 1. The gauge length of the specimen is 20 mm, and the width is either 4 or 5 mm. Shown in Fig. 2 is the stress-strain curve of the PI thin film under a strain rate of 4.2×10^{-3} 1/s. It can be seen from Fig. 2 that the thin film exhibits significant nonlinear deformation before fracture. Additional experiments were also conducted to further examine the nonlinear behavior of the thin film at room and elevated temperatures. It was observed from the experimental results that the behavior of the PI thin film under the strain-rate range of 4.2×10^{-5} 1/s and 4.2×10^{-3} 1/s is relatively unchanged. Furthermore, when a tensile stress of 90 MPa was applied to the PI thin film and then immediately released, the strain recovery was significantly delayed, but still returned to zero eventually after one day. On the other hand, when a tensile load above 110 MPa was applied and then unloaded, the strain recovery was not complete and a permanent set was observed. Based on these observations, it is assumed that the strain of the PI thin film under loading can be superpositioned by the viscoelastic and the plastic responses, i.e.,

$$\varepsilon = \varepsilon^{\text{ve}} + \varepsilon^{\text{p}} \qquad (1)$$

where ε^{ve} and ε^{p} denote the viscoelastic and plastic strains, respectively.

For the viscoelastic behavior of the PI thin film, it is assumed that linear viscoelasticity applies, and the one-

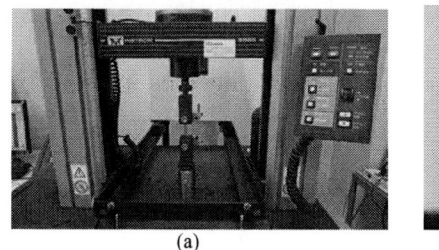

Figure 1.   Setup for PI thin film tensile experiment, (a) the tester, (b) zoom-in around the thin film specimen.

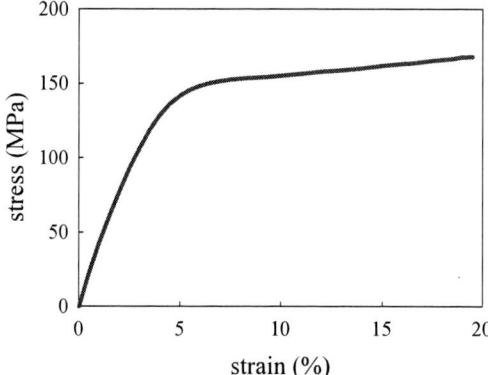

Figure 2.   The stress-strain relationship of the PI thin film.

dimensional stress-strain relationship can be written as a convolution integral given by

$$\sigma(t) = \int_{0^-}^{t} E(t-\zeta,T)\frac{d\varepsilon^{\text{ve}}(\zeta)}{d\zeta}d\zeta \qquad (2)$$

where σ denotes the one-dimensional stress, t denotes time, T denotes temperature, and E is the viscoelastic Young's relaxation modulus. By further assuming the PI being thermorheologically simple, the time-temperature correspondence principle is applicable for modeling the relaxation modulus. A master function for the cure-dependent viscoelastic relaxation modulus may be described by using the generalized Maxwell model (also referred to as the Prony series), given by

$$E(t,T) = E(\xi,T_{\text{T}}) = E_0\left[w_\infty + \sum_{i=1}^{N} w_i \exp\left(-\frac{\xi}{\tau_i}\right)\right] \qquad (3)$$

where ξ is the reduced or pseudo time, T_{T} is the reference temperature for the time-temperature correspondence, E_0 is the glassy modulus, w_i and τ_i are the weighting factor and relaxation time, respectively, for the i-th Maxwell element, and $w_\infty + \sum w_i = 1$. The relationship between the pseudo time and the real time is given by

TABLE I.   STRAIN-CONTROLLED FATIGUE CYCLING PROFILES

Case	Strain Range		Frequency
	ε_{min} (%)	ε_{max} (%)	f (Hz)
1	0	1	0.1
2	0	2	0.1
3	0	4	0.1
4	2	4	0.5
5	2	5	0.5
6	2	6	0.5
7	4	6	0.5
8	6	8	0.5
9	8	10	0.5
10	10	12	0.5

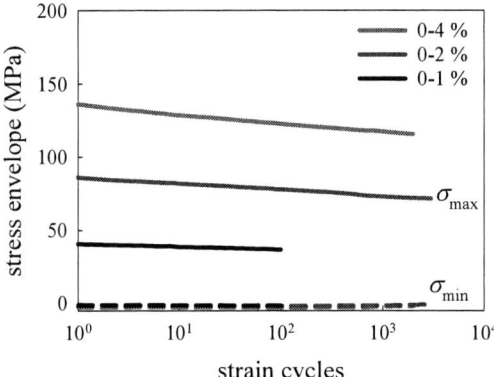

Figure 3.   Evolution of the stress envelop over cycles in the 0.1 Hz cyclic strain test.

$$\xi = \int_0^t a_{\mathrm{T}}(T(\zeta))d\zeta \qquad (4)$$

where a_{T} is the temperature shift function, and can be modeled by using the Williams, Landel, Ferry (WLF) equation, given by

$$\log_{10} a_{\mathrm{T}} = \frac{C_1(T - T_{\mathrm{T}})}{C_2 + (T - T_{\mathrm{T}})} \qquad (5)$$

In Eq. (5), C_1 and C_2 are the model constants. For determining the material constants of the generalized Maxwell model and the WLF shift function of the PI thin film, a frequency-sweep-and-temperature-step dynamic mechanical test under uniaxial oscillating condition was conducted by using a TA Instruments Q800 tester. A master curve was joined by shifting the frequency axis of the storage-modulus curve at each test temperatures. The generalized Maxwell model parameters including E_0, w_i and τ_i for the Prony series were obtained by fitting the master curve with the time-harmonic response obtained from Eq. (3) [7]. The model constants of the WLF equation (5) were

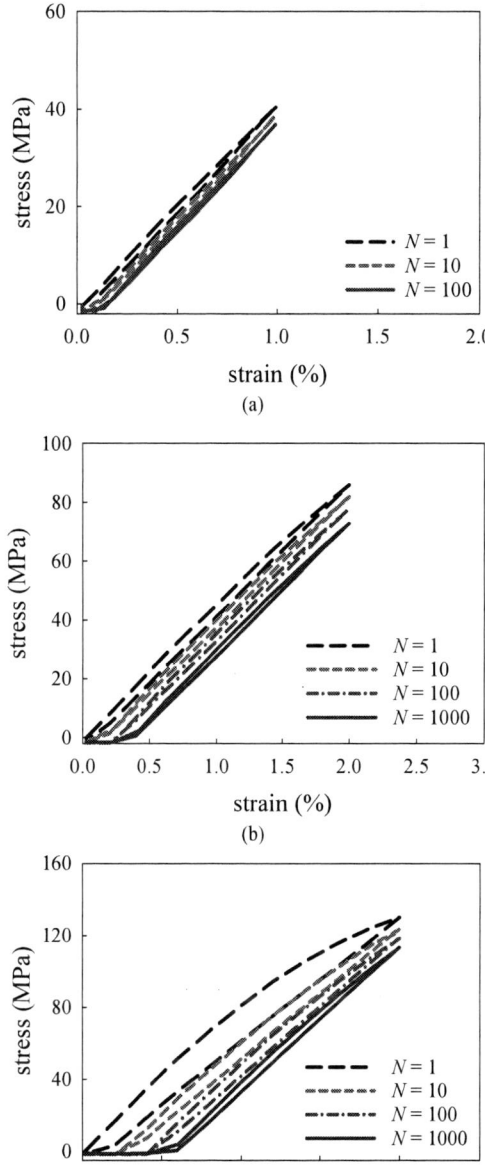

Figure 4.   Hysteresis loops at various cycles (N) in the 0.1 Hz strain-controlled cyclic tests, cyclic strain range: (a) 0-1, (b) 0-2%, (c) 0-4%.

determined by fitting to the shift factors used for constructing the master curve [7].

The plastic behavior of the PI thin film was modeled by using a simple power-law plastic model. For obtaining the related model constants, monotonic tensile, Viscoelastic contribution of the strain response under uniaxial tensile load was calculated numerically by using the generalized Maxwell model and then deducted from the experimentally measured responses of the PI thin film. The remaining part of the strain response was then fitted for the plastic model constants.

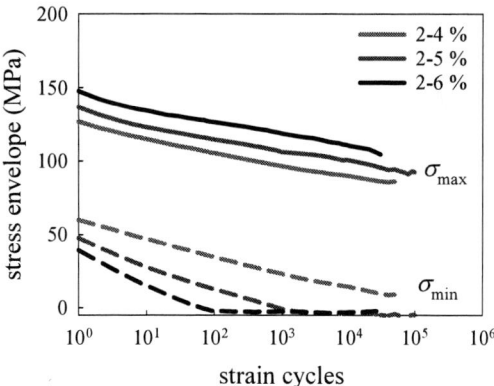

Figure 5. Evolution of the stress envelop over cycles in the 0.5 Hz cyclic strain test.

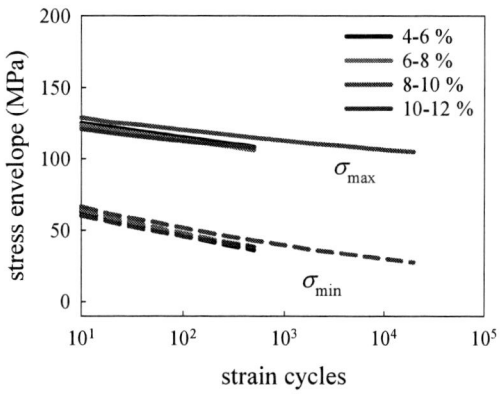

Figure 6. Effect of applied mean strain on the stress enevelopes over cycles in the 0.5 Hz cyclic strain test.

III. FATIGUE RESPONSES OF POLYIMIDE

The fatigue behaviors of thin films under cyclic loading conditions had been investigated extensively for metallic materials (e.g., for Cu [8-10]). The responses of metallic thin films under cyclic loading conditions and the related damage accumulations are mainly associated to the evolutions of the microstructural features. Fatigue models for these thin films are typically based either on the stress-cycle (S-N) curve (for high-cycle fatigue) or the Coffin-Manson equation (for low-cycle fatigue). These models may not be applicable to the polymeric thin films due to the distinct difference in their structures. In this study, both strain- and stress-controlled cycling tensile tests were conducted for characterizing the fatigue response of PI thin film. The magnitudes of the cyclic strain and stress were selected based on the monotonic test results as shown in Fig. 2 such that the strain- and stress-ranges are either within or beyond the linear limit of the monotonic response.

TABLE II. STRESS-CONTROLLED FATIGUE CYCLING PROFILES

Case	Stress Range		Frequency
	σ_{min} (MPa)	σ_{max} (MPa)	f (Hz)
1	5	60	0.0455
2	5	90	0.0294
3	5	110	0.0238
4	5	130	0.02

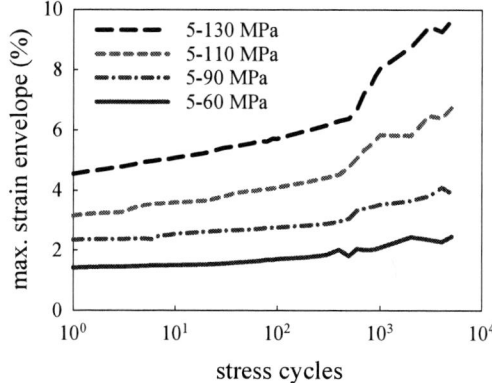

Figure 7. Maximum strain envelopes of cyclic stress experiments.

A. Response under Cyclic Strain Loading

For the strain-controlled fatigue experiments, the cyclic strain was applied in triangular waveform with strain range and frequency values given in Table I. In the first set of test conditions (Cases 1-3), the strain ranges applied (0 to up-to-4%) are within the linear regime of the monotonic stress-strain curve. Shown in Figs. 3 and 4 are the evolutions of the stress envelopes and the stress-strain curves at various fatigue cycles, respectively. It can be seen from Fig. 3 that the cyclic peak stress decreases as the number of cycle increases. Furthermore, the stress-strain responses of the 0-2% and 0-3% cases (Fig. 4a and 4b) exhibit minimal hysteresis; the area of the hysteresis loop is more obvious in the case of the 4% peak strain (Fig. 4c), but it still reduces rapidly as the number of fatigue cycle increases. From these results, it can be concluded that the irreversible inelastic response of the PI thin film under 4% peak strain is insignificant and the behavior of PI thin film is governed by viscoelastic stress relaxation.

For the second set of strain-controlled cyclic experiments (Cases 4-6 in Table I), the applied cyclic strain amplitudes are similar to the values in the first set, but the mean strain values are increased. The evolutions of the stress envelopes are shown in Fig. 5. It can be seen from Fig. 5 that the lower-bound stress relaxes at a faster rate when the applied strain range is higher. By further extending the applied cyclic strain into the nonlinear regime of the monotonic stress-strain response (Fig. 2), the stress envelopes for experiments with a 2%-strain amplitude (Cases 7-10 in Table I) are shown in Fig. 6. It can be seen from Fig. 6 that the stress relaxation responses are relatively independent of the applied cyclic mean strain when the mean strain is 5% or higher. Based on the results shown in Figs. 3-6, it may be concluded that, other than the initial plastic strain step, the general stress response of the PI thin film under cyclic strain condition is mainly governed by the viscoelastic relaxation. Consequently, the reliability failure is likely in the regime of high-cycle fatigue.

B. Response under Cyclic Stress Loading

The cyclic stress experiments were applied under a constant loading-and-unloading rate of 5 MPa/s. The corresponding stress cycle profiles are given in Table II. Shown in Fig. 7 are the envelopes of the maximum strains measured from every fatigue cycles. It can be seen from Fig. 7 that the maximum strain envelope increases monotonically as the fatigue cycle increases. For the cases of 5-60 MPa and 5-90 MPa cyclic stresses, the growth rate of the maximum strain envelope remains somewhat constant, which implies that the deformation is mostly viscoelastic creep deformation. On the other hand, for the 5-110 MPa and 5-130 MPa cases, the growth rate of envelope strain accelerates as the cycle number reach beyond 700. The accelerated strain increase may be attributed to the ratcheting deformation at higher stress levels, in addition to the viscoelastic creep.

IV. THE FATIGUE RELIABILITY MODEL

Based on the experimental observations described in Section III, the dominant deformation modes of PI thin film under cyclic loading include the viscoelastic stress relaxation and the non-recoverable plastic deformation. The superposition approach described in Eq. (1) is therefore applicable for considering the response of the PI thin film under cyclic loading. Again, by assuming the viscoelastic response is linear, the corresponding strain contribution can be expressed as

$$\varepsilon^{\text{ve}}(t) = \int_{0^-}^{t} J(t-\zeta)\frac{\mathrm{d}\sigma(\zeta)}{\mathrm{d}\zeta}\mathrm{d}\zeta \tag{6}$$

where $J(t)$ is the creep compliance. For simplicity, the viscoelastic behavior under cyclic loading condition is considered by using a two-element generalized Maxwell model ($i = 1$ in Eq. (3)), which is also known as the Zener model. The creep compliance of the Zener model can be written as

$$J(t) = \frac{1}{w_\infty E_0}\left[1 - w_1 \exp\left(-w_\infty t/\tau_1\right)\right] \tag{7}$$

where the material parameters E_0, w_∞, w_1 and τ_1 are given in Table III. By approximating the cyclic fatigue stress loading as a sinusoidal function, i.e.,

$$\sigma(t) = \sigma_m - \sigma_a \cos(\omega t) \tag{8}$$

TABLE III.   MATERIAL PARAMETERS OF THE VISCOELASTIC MODEL

Parameter	Value
E_0	4.19 GPa
w_∞	0.425
w_1	0.575
τ_1	5469 s

TABLE IV.   MATERIAL PARAMETERS OF THE PLASTIC MODELS

Parameter	Value
σ_0	90 MPa
C	8.56×10^{-4}
n	7.9
D	4.15×10^{-4}
m	0.34

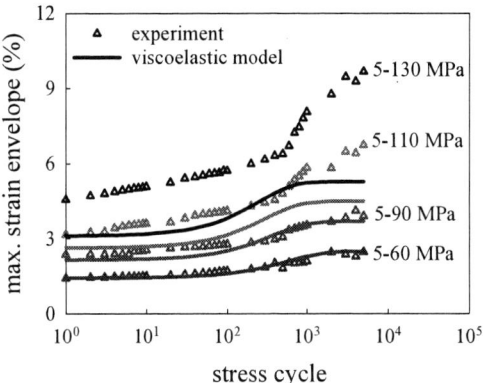

Figure 8. Viscoelastic model predictions compared to cyclic stress experimental data.

where σ_m and σ_a are the mean and amplitude stresses respectively, and ω is the angular frequency. By substituting Eqs. (7) and (8) into Eq. (6), the viscoelastic strain under cyclic loading can be expressed as

$$\varepsilon^{\text{ve}} = J(t)\left(\sigma_m - \sigma_0\right) + \frac{\sigma_a}{E_0 w_\infty}$$
$$-\frac{\sigma_a}{E_0\left(w_\infty^2 + \tau_1^2\omega^2\right)}\left[w_1\tau_1\omega\sin(\omega t)\right. \tag{9}$$
$$\left.+\left(w_\infty + \tau_1^2\omega^2\right)\cos(\omega t) + \left(\frac{w_1}{w_\infty}\right)\tau_1^2\omega^2\exp\left(-w_\infty t/\tau_1\right)\right]$$

The comparisons of the maximum strain envelopes obtained from Eq. (9) and the experimental results are shown in Fig. 8. It can be seen from Fig. 8 that the predictions obtained from Eq. (9) agree well to the experimental results for the cases of 5-60 MPa and 5-90 MPa stress cycling, but

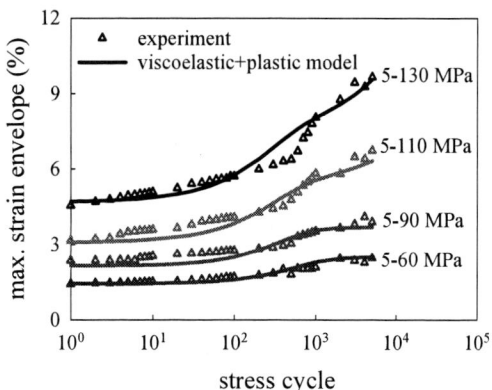

Figure 9. Combined viscoelastic-plastic model predictions compared to cyclic stress experimental data.

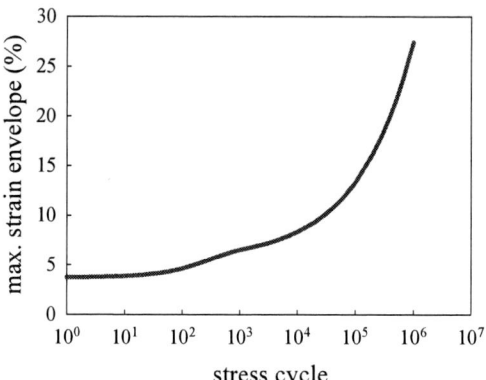

Figure 10. Evolution of maximum strain envelope for PI thin film under 0.021 Hz cyclic 0-120 MPa stress condition.

under-estimate the responses of the 5-110 MPa and 5-130 MPa stress cycling by as much as 50%. The agreement of the model prediction and the low-stress experimental data confirms that the viscoelastic creep deformation is the dominant mode of deformation under cycling stress in the linear regime of the constitutive behavior. For the fatigue loading with peak stress in the nonlinear regime, on the other hand, the plastic deformation should also be considered.

The plastic deformation of the PI thin film can be considered by adding the initial plastic step and the ratcheting plastic strain growth occurring in the subsequent stress cycles, i.e.,

$$\varepsilon^{p} = \varepsilon_0^{p} + \Delta\varepsilon^{p} \qquad (10)$$

where ε_0^{p} is the initial plastic strain and $\Delta\varepsilon^{p}$ is the subsequent ratcheting strain increment. The initial plastic strain can be modeled by a power-law relationship given by

$$\varepsilon_0^{p} = C\left(\sigma/\sigma_0\right)^{n} \qquad (11)$$

where C, n and σ_0 are the material parameters obtained from the monotonic stress-strain curve (Fig. 2), and are given in Table IV. The envelope of the subsequent plastic strain ratcheting growth is also considered by a simple power-law model, given by

$$\Delta\varepsilon^{p} = DN^{m(\sigma_{max}/\sigma_0)}, \quad \sigma_{max} > \sigma_0 \qquad (12)$$

where D and m are material parameters, N is the cycle number, σ_{max} is the cyclic peak stress, and $\sigma_{max} = \sigma_m + \sigma_a$. By combining Eqs. (9)-(12) and best-fitting to the experiment results, the values of D and m were obtained and are given in Table IV. The comparison of the combined viscoelastic-plastic model and the experimental data is shown in Fig. 9, from which it can be seen that the model prediction agrees well to the experimental data.

For predicting the fatigue reliability of the PI thin film, the strain envelope evolution model can be combined with the critical fracture strain to estimate the cycles to PI cracking failure. As an application example of the model, the evolution of envelope strain under a cyclic 0-120 MPa stress loading can be estimated by adding the viscoelastic contribution given in Eq. (9) and the plastic contribution given in Eq. (10), and is shown in Fig. 10. Assuming that the PI thin film ruptures at the strain value of 19%, fatigue failure would occur around 3.4×10^5 fatigue cycles.

V. CONCLUSIONS

The fatigue responses of PI thin film under both strain- and stress-controlled cyclic tensile experiments were investigated. Under strain-controlled fatigue condition, the thin film exhibited significant viscoelasticity relaxation. Furthermore, when a cyclic strain with peak value of 10% was applied, fatigue fracture did not occur at 1×10^5 cycles. Under stress-controlled cyclic loading with peak loading above 90 MPa, both viscoelastic creep and plastic ratcheting deformations were observed. A combined viscoelastic-plastic strain evolution model was developed for predicting the fatigue reliability of PI thin film. An obvious benefit of this strain evolution based phenomenological model is that this approach can readily be adopted to evaluate the fatigue reliability of other types of compliant polymeric thin films.

ACKNOWLEDGMENT

This study was partially supported by the Ministry of Science and Technology, ROC, under the grants MOST 104-2221-E-006-056-MY2, MOST 107-2221-E-006-138 and by ASE Group.

REFERENCES

[1] X. Fan, "Wafer level packaging (WLP): fan-in, fan-out and three-dimensional integration," Proceedings of the 11th international conference on thermal, mechanical and multiphysics simulation and experiments in micro-electronics and micro-systems, EuroSimE 2010, pp. 1-7.

[2] C. K. Yu, W. S. Chiang, N. W. Liu, M. Z. Lin, Y. H. Fang, M. J. Lin, B. Lin, and M. Huang, "A unique failure mechanism induced by chip

to board interaction on fan-out wafer level package," Proceedings of the 2017 IEEE international reliability physics symposium, IRPS 2017, pp. 4A-6.1-4A-6.4

[3] C. H. Yu, L. J. Yen, C. Y. Hsieh, J. S. Hsieh, Victor C. Y. Chang, C. H. Hsieh, C. S. Liu, C. T. Wang, K. C. Yee, and D. C. H. Yu, "High performance, high density RDL for advanced packaging," Proceedings of 68th electronic components and technology conference, ECTC 2018, pp. 587-593.

[4] C. Nair, H. Lu, K. Panayappan, F. Liu, V. Sundaram, and R. Tummala, "Effect of ultra-fine pitch RDL process variations on the electrical performance of 2.5D glass interposers up to 110 GHz," Proceedings of 66th electronic components and technology conference, ECTC 2016, pp. 2408-2413.

[5] M.-K. Shih, R. Chen, P. B. S. Chen, Y.-C. Lee, K. Y. U. Chen, I. Hu, T.-Y. Chen, L. T., E. Chen, E. Tsai, D. Tarng, and C. P. Hung, "Comparative study on mechanical and thermal performance of eWLB, M-series and fan-out chip last packages," Proceedings of 68th electronic components and technology conference, ECTC 2018, pp. 1670-1676.

[6] W. Wang, D. Zhang, Y. Sun, D. Rae, L. Zhao, J. Zheng, M. Schwarz, M. Shah, and A. Syed, "Study of polyimide in chip package interaction for flip-chip Cu-pillar packages," Proceedings of 68th electronic components and technology conference, ECTC 2018, pp. 1039-1043.

[7] T.-C. Chiu, B.-S. Lee, D.-Y. Huang, Y.-T. Yang, and Y.-H. Tseng, "Time- domain viscoelastic constitutive model based on concurrent fitting of frequency-domain characteristics," Microelectronics Reliability, vol. 55, 2015, pp. 2336-2344.

[8] X. J. Sun, C. C. Wang, J. Zhang, G. Liu, G. J. Zhang, and X. D. Ding, "Thickness dependent fatigue life at microcrack nucleation for metal thin films on flexible substrates," Journal of Physics D: Applied Physics, vol. 41, 2008, p. 195404.

[9] D. Wang, C. A. Volkert and O. Kraft, "Effect of length scale on fatigue life and damage formation in thin Cu films," Materials Science and Engineering: A, vol. 493, 2008 , pp. 267-273.

[10] Y. Hwangbo and J.-H. Song, "Fatigue life and plastic deformation behavior of electrodeposited copper thin films," Materials Science and Engineering: A, vol. 527, 2010, pp. 2222-2232.

Explicit FE Failure Prediction of Interfaces and Interconnect in Potted Electronics Assemblies Subject to High-g Acceleration Loads

Pradeep Lall[1], Kalyan Dornala[1], Ryan Lowe[2], John Deep[3]

[1] Auburn University
NSF-CAVE3 Electronics Research Center
Auburn, AL 36849
[2] ARA Associates, Littleton, CO 80127
[3] Air Force Research Laboratories, Eglin, FL 32542
Tele: 334-844-3424
E-mail: lall@auburn.edu

Abstract—Survivability of fine-pitch electronics at high-g loads requires the use of additional structural support and shock damping through the use of potting-materials. The potting compounds may serve additional functions including the ability to sustain thermo-mechanical loads and the high humidity in transport, storage and use environments. Failure of potted assemblies has been shown to be at the interface between the potting material and the printed circuit board. There is a dearth of computational tools to allow for the prediction of the initiation of damage and the progression of damage under high-g shock loads. Defense electronics and military systems have longer lifetimes in the neighborhood of 20-40 years and higher reliability requirements. New packaging architectures, which often push the edge of the envelope in terms of miniaturization, cannot be compared with the state-of-art systems and lack decades of historical data to provide robust proof of their survivability. Tools and techniques are needed to determine the failure envelopes for new component technologies for operation under high acceleration loads in current and next generation military systems. Component's survivability in a product is influenced by many factors including board construction, board size, board thickness, and component design rules. The same component may have large variance in shock survivability depending on the product implementation. The fracture properties and interfacial crack delamination of the PCB/epoxy interface was determined using three-point bend loading with a pre-crack in the epoxy near the interface. The fracture toughness and crack initiation of the three-epoxy systems was compared with the cure schedule and temperature. A finite element model framework was developed for a circular PCB with fine pitch BGA packages, which are encapsulated with potting material. The interface between the PCB and the potting compound was modeled using cohesive zone elements. The test assembly model was subjected to high-g mechanical shock loads up to 25,000g.

Keywords-High-g, mechanical shock, reliability, potting compounds, ball-grid arrays, explicit finite elements, cohesive zone models, fracture toughness, interface fracture.

I. INTRODUCTION

Electronics in military applications increasingly rely on the use of commercial off-the-shelf components for enabling critical electrical functions. Defense electronics may often be subjected to high-g acceleration loads in addition to extremes of temperature, humidity and prolonged storage. Consumer applications often require the components to survive the shock conditions specified in JESD22-B111A test standard, which specifies a g-level of 1500g. The JESD22-B111A test standard is often used to evaluate the drop performance of surface mount components for handheld electronics product applications in accelerated test conditions. Defense electronics, however, may experience g-levels significantly higher than consumer applications tested for handheld electronics. In addition, the electronic systems may be subjected to prolonged storage prior to deployment in field environments. Potting compounds and underfills are a supplemental restraint mechanism for augmenting the reliability in high-g environments and meeting the survivability and storage requirements [Lall 2016]. The final mechanical properties and interface properties of the underfills and the mold compounds may depend on the cure conditions and the materials of the encapsulated electronics. However, the diversity of component configurations, material sets, environmental and use conditions require the availability of computational tools and interface properties to allow for meaningful prediction of time-to-initiation of damage and time-to-failure once cracks have formed at the interfaces.

The effect of temperature, moisture on the toughness of for EMC and copper interfaces in microelectronics were presented by many researchers [Tran 2014; Xiao 2008]. However, there are no studies in open-literature on the interface properties of molding compounds as a function of cure conditions and their use for prediction of high strain rate reliability of electronics. In this study, the interface properties have been measured as a function of the cure condition. A notched bi-material specimen was prepared suitable for three-point bend test with the potting compound mated to PCB. A pre-crack was introduced right at the interface of PCB and epoxy to initiate the crack at the interface. Load-displacement curves and the critical load to initiate the crack at the interface were recorded from the three-point bend tests. Interface fracture properties such as the critical strain energy release rate and stress intensity factor were obtained. The measured properties have been ported to a computational cohesive-zone framework. The cohesive-zone model traction parameters based on the bi-linear traction separation law have been computed from the experiment measurements of the PCB-Epoxy interface. The cohesive-zone model has been integrated into an explicit finite-element model used to predict the occurrence of delamination at the potting compound-PCB interface in a circular assembly with fine-pitch ball-grid array components subjected to 25,000g. in addition, the damage progression under multiple high-g shock events at 25,000g shock was studied.

II. INTERFACE TOUGHNESS AND CZM IMPLEMENTATION

Previously, many researchers have computed the interfacial fracture parameters of bi-material interfaces under four-point bend loading for structural materials outside of electronic potting-compound applications [Charalambides 1989; Hutchinson 1987; Zhang 1997; Pozuelo 2009]. Damage accrual initiates from a primary crack that propagates to the interface, followed by immediate delamination. It is often argued that the distance of the initial crack to the interface may not be influential since the delamination occurs at the interface after the crack reaches the interface. Once the primary crack reaches the interface, and starts to deflect along the interface, the crack geometry would result in a notched bi-material three-point or four-point bending configuration during the crack initiation and crack growth phases of the specimen deformation. Prior researchers have shown that the strain energy release rate exhibits steady-state characteristics when the interfacial crack exceeds the thickness of the stiffer layer of the bi-material beam. A schematic representation of the bi-material interface crack in three-point bend loading is shown in Figure 1. Further, it is common to assume both linear-elastic and plane-strain conditions due to the very small-scale yielding ahead of the crack tip in the three-point bend test. For a very small h/L ratio of interface crack under steady state loading assuming plane strain conditions the fracture toughness G_{ss} is given by

$$G_{ss} = \frac{M^2(1-v_2^2)}{2E_2}\left(\frac{1}{I_2} - \frac{\lambda}{I_c}\right) \qquad (1)$$

Where G_{ss} is the energy release rate of the bend sample, $\lambda = E_2(1-v_1^2)/E_1(1-v_2^2)$; E and v are the Young's modulus and Poisson's ratio of the materials. M and I are the applied moment and moment of inertia per unit width. In addition, P is the total applied load, b the sample width and l the space between the two supports or gage length, we have, $M = Pl/4b$ and $I_2=h_2^3/12$ where h_2 is the thickness of the PCB. The stress intensity factor characterizing the interface favorable to delaminate is obtained using the following:

$$K_c = \left(\frac{G_c E_2}{1-v_2^2}\right)^{1/2} \qquad (2)$$

Where, K_c is the critical stress intensity factor, G_c is the critical energy release rate for delamination, E_2 is the elastic modulus of the second-layer, v_2 is the Poisson's ratio of the second layer. The epoxy potting compound used in this study

were two-part resins cured at 65°C for 1hr. The properties of the cured potting compound are shown in TABLE I.

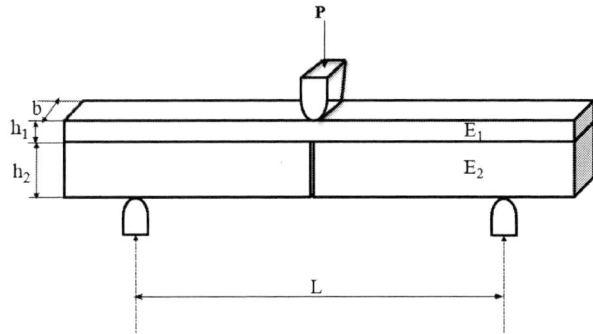

Figure 1. Three-point bend test schematic representation

TABLE I. Technical Data for the Cured Potting Material

T_g (°C)	Density (Kg/m³)	Elastic Modulus (GPa)	Max Elongation %	Hardness
60	1320	1.42	12	Shore D, 72

A pre-crack was defined at the interface of PCB and the epoxy by using an *Exacto* knife before curing. The three-point specimen has dimensions of 85mm (L) x 10mm (W) x 9mm (H) with gage length being 60mm. The specimen was loaded on the three-point bend fixture and load was applied at a rate of 2mm/min while acquiring the load-displacement data. A micro imaging setup was employed to capture the slow-motion video of the whole loading event. Figure 2, shows the images from micro video at various downward displacement of the loading fixture. The primary crack first extends towards the interface, i.e. perpendicular to the specimen surface, and then delaminates parallel to the interface creating a sub-interface crack between the solder mask and substrate of the PCB.

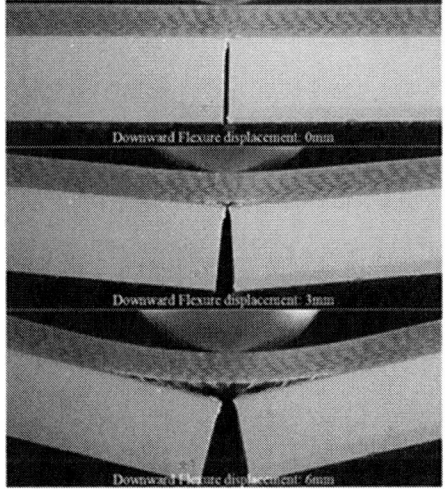

Figure 2: Delamination of Epoxy/PCB bi-material specimen under three-point bend loading

978-1-7281-1500-9/19 $31.00 © 2019 IEEE

Figure 3 shows the fracture tougness as a function of cure temperature for epoxy-A, epoxy-B, and epoxy-C potted PCB specimens. The fracture toughness has been computed for each of the measured samples for various cure conditions shown in Table 2, Table 3, and Table 4.

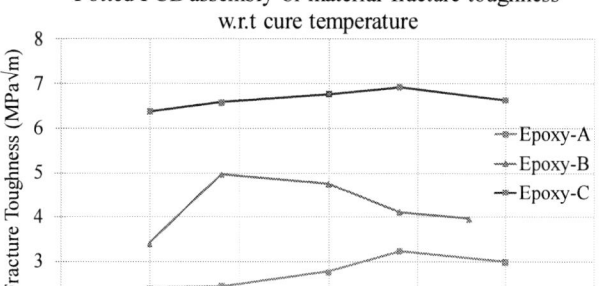

Figure 3: Fracture toughness vs Cure temperature for Epoxy-A/PCB, Epoxy-B/PCB, Epoxy-C/PCB bi-material interface

TABLE 2: EPOXY-A PCB SAMPLES FRACTURE TOUGHNESS CALCULATIONS

EpoxyA/PCB Samples	2Hr @ 45C		2Hr @ 75C		2Hr @ 95C	
	A1	A2	A1	A2	A1	A2
E_1 (Gpa)	2.81	2.81	2.81	2.81	2.81	2.81
E_2 (Gpa)	20.7	20.7	20.7	20.7	20.7	20.7
v_1	0.28	0.28	0.28	0.28	0.28	0.28
v_2	0.2	0.2	0.2	0.2	0.2	0.2
Span l(mm)	60	60	60	60	60	60
h_2 (mm)	1.71	1.71	1.71	1.71	1.71	1.71
h_1 (mm)	4	4.2	4.21	4.3	4.7	4.4
Width b(mm)	9.84	9.16	9.4	9.55	9.62	9.56
Critical Load (N)	41.6	51.4	47.4	57.5	61.2	62.6
Moment (N/mm)	63.415	84.170	75.638	90.314	95.426	98.222
G_c (KJ/m²)	0.204	0.362	0.293	0.419	0.475	0.497

Measurements indicate that the fracture toughness shows variation with respect to cure temperature for all three potting compounds. For epoxy-A, the peak fracture toughness value of 2.5 MPa.m$^{1/2}$ was observed at 25□, and a value of 3.2 MPa.m$^{1/2}$ was observed at 95□. In the case of epoxy-B, the fracture toughness value first increases then drops. The initial fracture toughness value was measured to be in the neighborhood of 3.4 MPa.m$^{1/2}$ at 25˚C cure, and 5 MPa.m$^{1/2}$ at 45˚C cure. Epoxy-B has higher fracture toughness in comparison with epoxy-A. No interfacial delamination was observed in any of the samples of Epoxy-C for all cure temperatures. The fracture toughness for epoxy-C/PCB specimens at different cure temperatures is shown in Table 4.

TABLE 3: EPOXY-B PCB SAMPLES FRACTURE TOUGHNESS CALCULATIONS

EpoxyB/PCB Samples	1Hr @ 45C		1Hr @ 75C		1Hr @ 115C	
	B1	B2	B1	B2	B1	B2
E_1 (Gpa)	1.42	1.42	1.42	1.42	1.42	1.42
E_2 (Gpa)	20.7	20.7	20.7	20.7	20.7	20.7
v_1	0.38	0.38	0.38	0.38	0.38	0.38
v_2	0.2	0.2	0.2	0.2	0.2	0.2
Span l(mm)	60	60	60	60	60	60
h_2 (mm)	1.71	1.71	1.71	1.71	1.71	1.71
h_1 (mm)	4.8	4.36	4.2	4.36	4.67	4.33
Width b(mm)	8.86	9.07	9.87	9.91	9.68	8.46
Critical Load (N)	95.4	85.6	98.7	94.5	78.4	68.4
Moment (N/mm)	161.51	141.57	150.00	143.04	121.49	121.28
G_c (KJ/m²)	1.312	0.984	1.092	1.004	0.738	0.721
K_c (MPa.m$^{1/2}$)	5.320	4.606	4.853	4.653	3.988	3.942

TABLE 4: EPOXY-C PCB SAMPLES FRACTURE TOUGHNESS CALCULATIONS

EpoxyC/PCB Samples	2Hr @ 45C		2Hr @ 75C		2Hr @ 95C	
	B1	B2	B1	B2	B1	B2
E_1 (Gpa)	0.315	0.315	0.315	0.315	0.315	0.315
E_2 (Gpa)	20.7	20.7	20.7	20.7	20.7	20.7
v_1	0.42	0.42	0.42	0.42	0.42	0.42
v_2	0.2	0.2	0.2	0.2	0.2	0.2
Span l(mm)	60	60	60	60	60	60
h_2 (mm)	1.71	1.71	1.71	1.71	1.71	1.71
h_1 (mm)	4.33	4.07	4.24	4.1	4.46	4.05
Width b(mm)	9.18	9.3	9.2	9.33	10.1	9.58
Critical Load (N)	143.8	149.7	157.7	145.8	168.2	158.9
Moment (N/mm)	234.97	241.45	257.12	234.41	249.80	248.80
G_c (KJ/m²)	2.012	2.010	2.365	1.907	2.331	2.124
K_c (MPa.m$^{1/2}$)	6.586	6.583	7.142	6.413	7.089	6.767

To compute the nominal normal strain at the interface of PCB and epoxy, using composite-beam theory the width of the epoxy is reduced to an equivalent width of the PCB. Figure 4 shows the schematic of the bi-material specimen under applied bending moment, M. Here E_1 and E_2 represents the elastic modulus of the PCB and epoxy materials respectively. For transforming the section into one made entirely of PCB, the transformation factor is given by n, where,

$$n = \frac{E_1}{E_2} \qquad (3)$$

The equivalent width of epoxy is equal to $b_2 = nb_1$. The loaction of the neutral axis from the bottom of the bi-material cross-section is computed by,

$$\bar{y} = \frac{\Sigma \bar{y} A}{\Sigma A} = \frac{(h_2/2)(h_2 b_2) + (h_2 + h_1/2)(h_1 b_1)}{h_1 b_1 + h_2 b_2} \qquad (4)$$

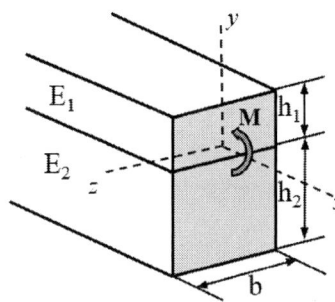

Figure 4: Sectional view of bi-material specimen under bending moment, M

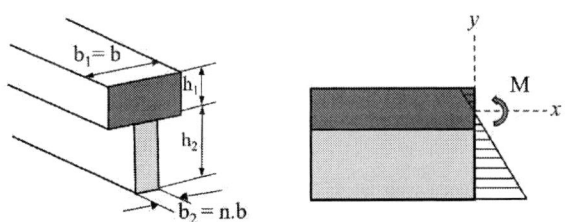

Figure 5: Equivalent width of epoxy (b2) reduced in terms of the PCB (b1) and the nominal normal state of stress

The equivalent width of the epoxy reduced in terms of PCB using the transformation factor and the nominal normal state of stress is shown in Figure 5. The moment of inertia about the neutral axis is given by

$$I_{NA} = \Sigma(I_A + Ad^2) \qquad (5)$$

$$I_{NA} = \left\{ \frac{b_1 h_1^3}{12} + b_1 h_1 [(\frac{h_1}{2}) + h_2 - \bar{y}]^2 \right\} + \qquad (6)$$

$$\left\{ \frac{b_2 h_2^3}{12} + b_2 h_2 [\frac{h_2}{2} - \bar{y}]^2 \right\}$$

Applying the flexure formula, the normal stress in the PCB at the epoxy-PCB interface of the bi-material three-point bend specimen is by,

$$\sigma_{PCB} = \frac{M(\bar{y} - h_2)}{I} \qquad (7)$$

The nominal strain at the contact point is given by,

$$\varepsilon = \frac{\sigma_{PCB}}{E_{PCB}} = \frac{M(\bar{y} - h_2)}{E_1 I} \qquad (8)$$

The constitutive response of the cohesive elements used in this paper is defined by bi-linear traction-separation law as shown in Figure 6, where T represents the interfacial strength, Δ_c is the critical separation, Δ_{fail} is the separation at failure, and the area under the curve, G_c, is the critical strain energy release rate. Under mixed-mode fracture, the properties required to define the bilinear traction-separation law are the three critical fracture energies G_{IC}, G_{IIC}, G_{IIIC}, the penalty stiffnesses K_1, K_2, K_3, and the interfacial strengths T, S_1, and S_2. In the present work, the shear strengths in the two orthogonal directions, S_1 and S_2 are assumed to be the same and are referred to as S. The initial response of the cohesive element is assumed to be linear until a damage initiation criterion is met. The penalty stiffness, K_i, of the bi-linear traction-separation law is defined as [Song 2008; Turon 2006, 2007]

$$K_i = \frac{T_i}{\Delta_i^c} \qquad (9)$$

where T is the interfacial strength, and Δ_i^c is the critical separation for damage initiation. Generally, the value of penalty stiffness is chosen to prevent interpenetration of the crack faces when closed and to prevent artificial compliance from being introduced into the model by the cohesive elements.

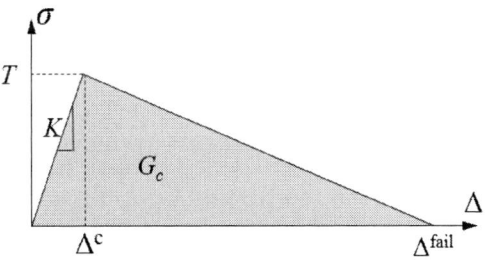

Figure 6: Bi-linear traction separation law

High values of penalty stiffness often may lead to convergence difficulties and unreasonably low values may result in interpenetration of the crack faces. Several guidelines have been proposed for obtaining the penalty stiffness of a cohesive element. The initial compliance of the bulk material is selected to be larger than the initial compliance of the cohesive element [Turon 2007]:

$$K_3 \geq \alpha. \frac{E_3}{t} \qquad (10)$$

Where α is a parameter much larger than 1, E_3 is the transverse elastic modulus of the material, t is the thickness of an adjacent laminate or the printed circuit board. The value of $\alpha = 50$ is used for the initial compliance equation, based on prior work by Turon [2006 and 2007]. In the present paper, the analysis assumes the same penalty stiffnesses for all modes, i.e., $K_1=K_2=K_3=K$. The length of the cohesive zone is computed from the following equation,

$$l_{cz} = E_2 \frac{G_{IC}}{T^2} \qquad (11)$$

Where, l_{cz}, is the distance from the crack tip at the interface to the point when the maximum cohesive traction is attained, G_{IC} is the critical energy release rate. Prior work by other researchers has demonstrated that the length of the cohesive zone is a material property for infinite-size geometries. In addition, E_2 is the transverse elastic modulus of the stiffer material. The length of the cohesive element where N_e is the number of elements is given by,

$$l_e = \frac{l_{cz}}{N_e} \qquad (12)$$

Where, N_e is the number of elements along the cohesive interface. More than three elements in the cohesive zone were used to accurately represent the fracture energy in the cohesive zone. Once the damage initiation criterion of the cohesive element was satisfied, the stiffness of the element was reduced. The following equation represents the softening response of the cohesive element.

$$\sigma_i = (1-d)K_i\Delta_i, \ i = 1,2,3 \qquad (13)$$

where d is the damage variable, which has the value d = 0 when the interface is undamaged, and the value d = 1 when the interface is fully fractured. Intermediate values are computed from the following equation,

$$d = \frac{\Delta_{fail}\left(\Delta_{max} - \Delta_c\right)}{\Delta_{max}\left(\Delta_f - \Delta_c\right)} \qquad (14)$$

Where the subscript "max" fail indicate the maximum separation during loading, "fail" indicates the separation at failure, and "c" indicates the separation at peak load. The cohesive zone finite element model approach was used to simulate the initiation and propagation of the crack at the interface of PCB and epoxy during bending loading. The analysis was performed in ABAQUS standard with definition of cohesive elements of COH3D8 type and 0.01mm layer thickness. The model geometry, boundary conditions and loading were made similar to that of experiment. Damage parameters such as fracture energy from the experiment was used to accurately represent the cohesive interface for crack evolution. An elastic traction separation type behavior was implemented in the model. The damage evolution is based on the steady state fracture energy acquired from the experiment which is 0.54 kJ/m². Other model input parameters for the cohesive elements include the definition of maximum stress damage tolerance which is 0.05. A 3D model of the epoxy/PCB three-point bend specimen is shown in Figure 7. Linear elastic FE material parameters were given to the bulk of PCB and epoxy potting material. The definition of cohesive elements at the PCB and epoxy interface is highlighted in the Figure 7.

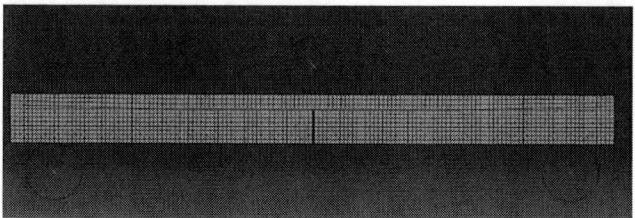

Figure 7: Highlighted cohesive elements and definition of pre-crack in epoxy elements at the interface

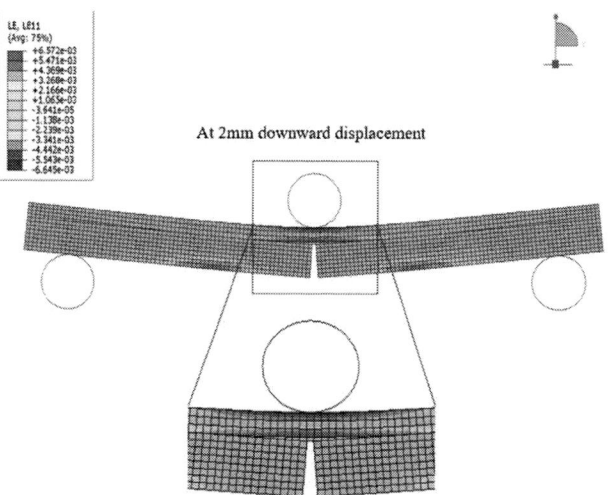

Figure 8: Stress contours at 2mm downward displacement

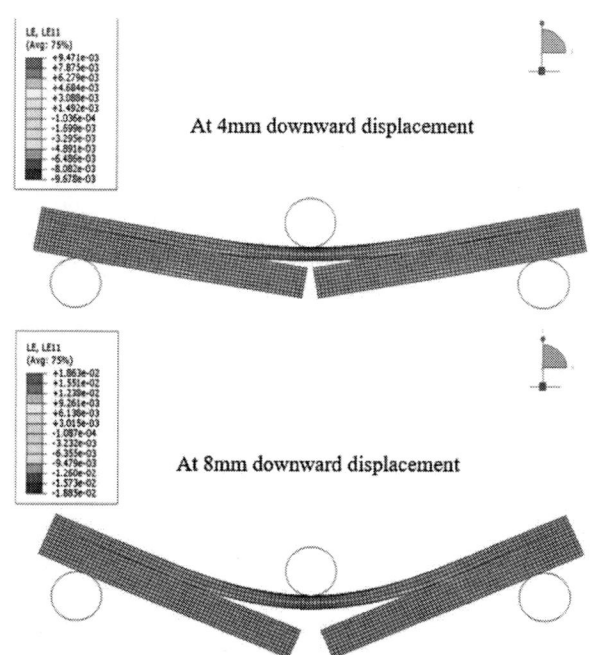

Figure 9: Stress contours at 4mm and 8mm downward displacement. Cohesive elements automatically deleted as they reach the damage limit

Stress contours and the flexure load versus the flexure extension parameters were extracted from the model output. Figure 8 shows the stress contours and the crack opening at 2mm downward displacement. Similarly, Figure 9 show stress contours at 4mm and 8mm downward displacement respectively. Once the cohesive elements reach the damage limit extension, they were deleted to represent the crack opening at the interface. A comparison between the FE model and experiment result for load versus displacement plot is shown in Figure 10. Once the model was calibrated against the experimental load-displacement curve and the strain energy release rate, the cohesive zone parameters can be applied in the actual shock model between the PCB and epoxy interface.

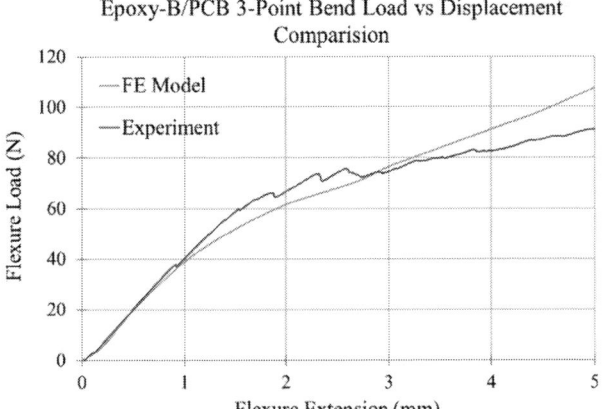

Figure 10: Load vs Displacement comparison between FE model prediction and experiment results

III. HIGH-G SHOCK DAMAGE MODELING

Explicit Finite element modeling was performed in ABAQUS explicit using cohesive zone elements, COH3D8 of 0.01mm thick layer between the PCB and epoxy bulk materials. The PCB was modeled with S4R elements, packages, underfill material and epoxy compound with C3D8R, and the solder interconnects were modeled as Timoshenko beam elements of type B31. The material property definitions required for model setup is shown in TABLE V.

TABLE V. MATERIAL PROPERTY DEFINITIONS FOR FEA

	Elastic Modulus (GPa)	Possion's Ratio	Density (Kg/m³)
Solder Ball (SAC305)	51	0.37	7500
Mold Compound	23.5	0.25	1650
Die	162	0.28	2329
Die Attach	2.76	0.35	7800
Epoxy Film	0.649	0.35	2100
BT Substrate	17.4	0.28	1800
Bare FR406	21.2	0.35	1850

To input boundary conditions in the finite element model, position tracking was employed by capturing the shock event at high frame rate using high-speed cameras. A snapshot of the drop and shock event at 25,000g shock is shown Figure 11. Velocity measurements of DMSA during the high-g shock event at 25,000g is derived from tracking the position of targets on the DMSA of drop base. The images generated from the high-speed camera were processed in a software called MotionPlus shown in Figure 12. The targets on the DMSA and on the drop table were tracked with respect to the reference co-ordinates and motion parameters like position, velocity and acceleration were generated. The velocity rate vs time plot is shown in Figure 13.

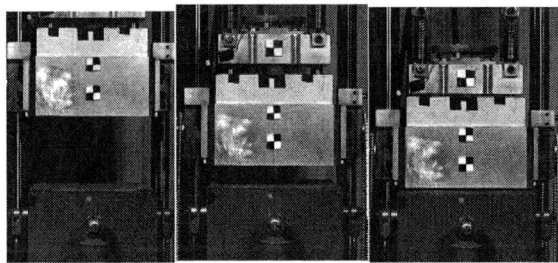

Figure 11: High-speed camera picture snaps of the drop table along with DMSA during 25,000g shock event.

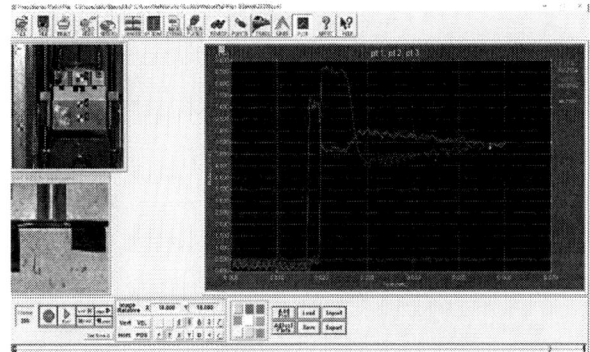

Figure 12: Motion plus software interface

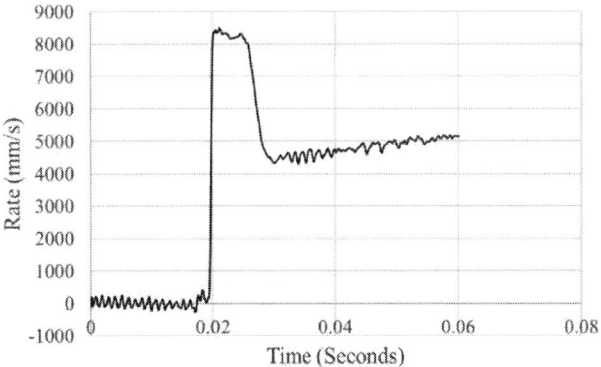

Figure 13: DMSA target velocity

The PCB along with the BGA components is shown in Figure 14. An epoxy potted test vehicle meshed model with the front side and the backside is shown in Figure 15. The epoxy model

978-1-7281-1500-9/19 $31.00 © 2019 IEEE   1371

showing the backside of the PCB with an annular masking similar to the actual test vehicle has been modeled. The constrained model along with the drop base and rigid floor is shown in Figure 16. A reference node has been defined beneath the rigid floor for application of constraints. Node-to-surface contact has been used for impact between the shock table and the floor. An event length of 5ms after impact has been simulated. Time history was monitored at a time-period of 0.1ms at the corner solder joints of all BGAs in the test assembly. A section-view of the potting encapsulated BGA package is shown in Figure 17. Cohesive elements layers have been defined between the PCB and epoxy interfaces on both the backside and front-side of the assembly.

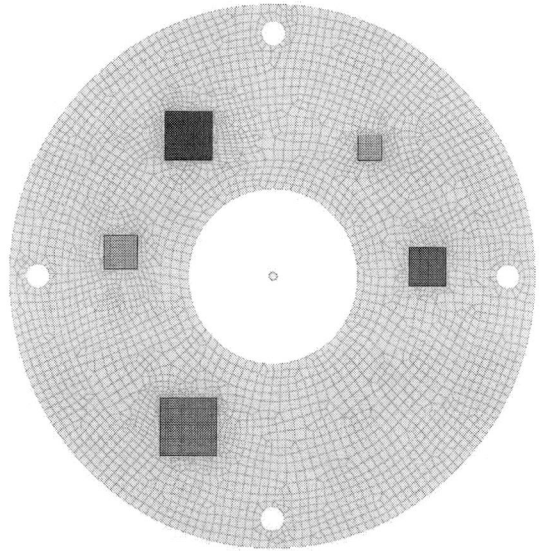

Figure 14: Circular PCB Model

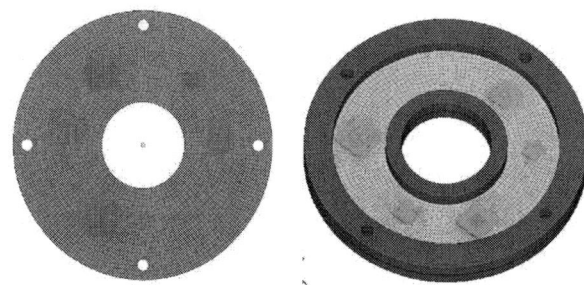

Figure 15: Potting encapsulated test vehicle

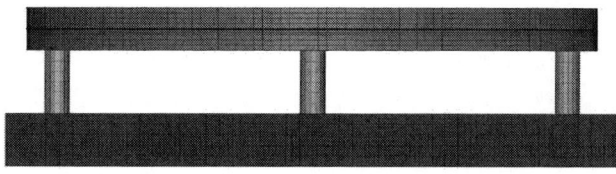

Figure 16: Transient dynamic model for potted assembly

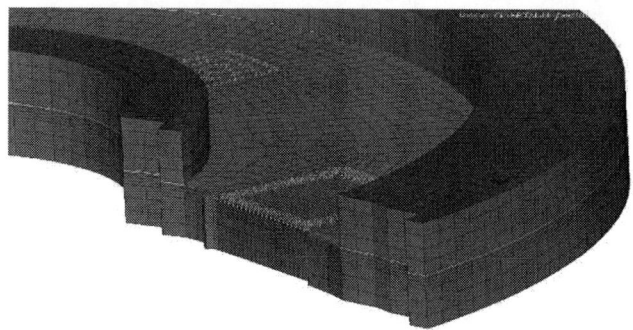

Figure 17: Section view of potted assembly along with cohesive elements definition

A. First Drop

With the interface properties and the cohesive zone defined between the PCB and epoxy layers the circular PCB model with the BGA components was simulated under the boundary conditions of 25,000g shock. Peak out-of-plane displacements and the status of the cohesive element deletion as the parameters reach the damage tolerance is reported. The out of plane displacement at 0ms just before the impact is shown in Figure 18. Figure 19 shows the peak displacement at 0.3ms time-step, which is 4.8mm. The delamination of the inner lip and the severity of the board deflection is evident through the progression of time. Figure 20 shows the model deformation at time-step of 0.5ms after impact – at which the inner lip of the assembly starts to delaminate completely and gets separated. The status of the cohesive element deletion when the elements reach damage criterion at initial timestep is shown in Figure 21.

Figure 18: Out-of-plane displacement of potted test assembly at 25,000g high-g shock at 0ms time step

Figure 22 shows the element deletion at the potting compound-PCB interface at 0.3ms after impact when a greater number of cohesive elements reach the damage state and the outer/inner lip delamination at the PCB and epoxy interface can be seen clearly. For the time step 0.5ms shown in Figure 23, the inner lip starts to separate from the assembly. High-speed cameras recording at 10,000fps were used to capture the transient dynamic event of potted test assemblies under 25,000g shock.

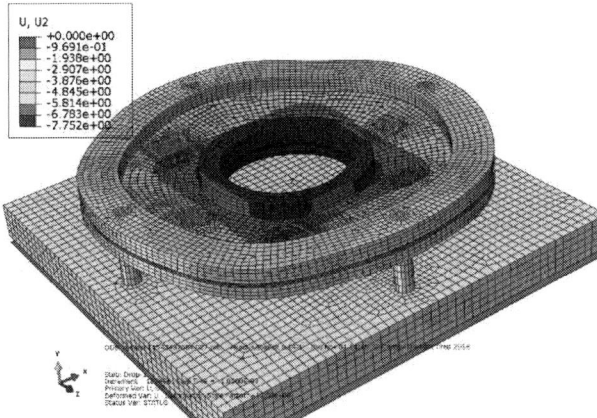

Figure 19: Out-of-plane displacement of potted test assembly at 25,000g high-g shock at 0.3ms time step

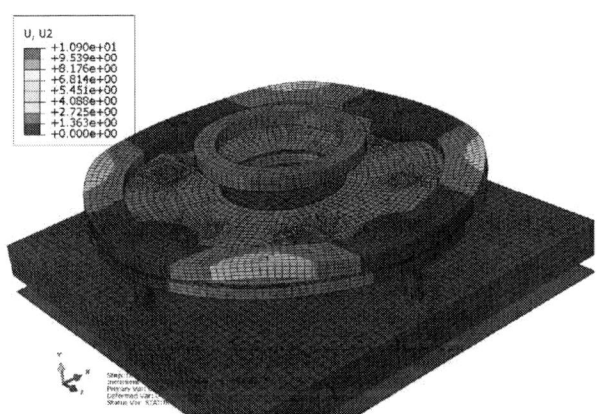

Figure 20: Out-of-plane displacement of potted test assembly at 25,000g high-g shock at 0.5ms time step

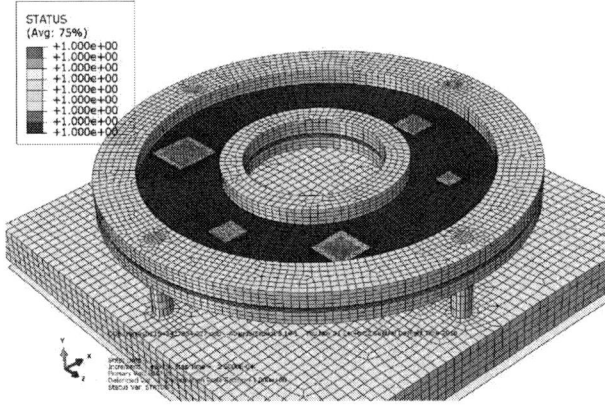

Figure 21: Status of element deletion under 25,000g high-g shock at 0ms time step

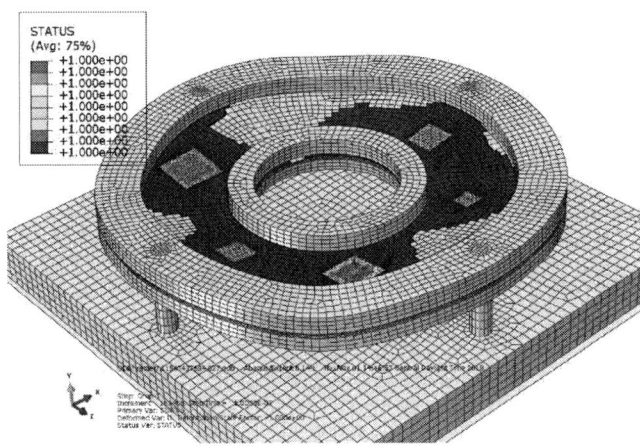

Figure 22: Status of element deletion under 25,000g high-g shock at 0.3ms time step

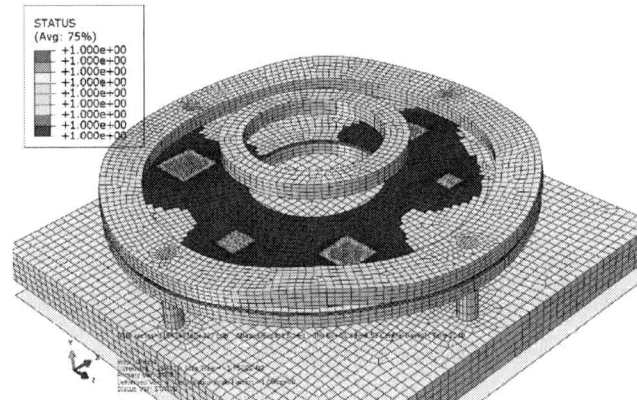

Figure 23: Status of element deletion under 25,000g high-g shock 0.5ms time step

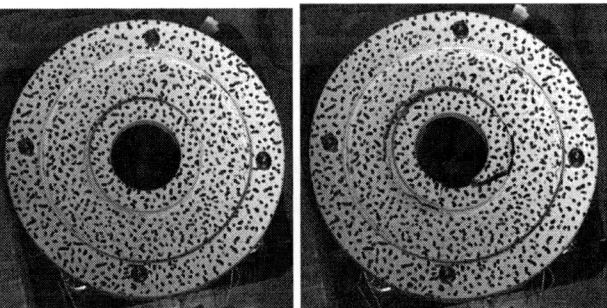

Figure 24: Epoxy potted test assembly under 25,000g shock. Just before impact snapshot on left and the after-impact snapshot on right

Figure 24 shows the snapshots from the high-speed imaging just before the impact and after the impact. The speckle patterns on the back-side of the PCB are for the digital image correlation analyses purposes. Figure 25 shows the delamination of the potting compound at the PCB and epoxy interface near the inner lip during the high-g shock event. The finite element model with the cohesive zone damage modeling has correctly predicted the delamination of the potting compound during shock when compared to the failure modes from the experimentation.

Figure 25: Delamination of potting compound at the PCB and epoxy interface near the inner lip under 25,000g shock

B. Damage Progression through Second Drop

The epoxy potting material failure through delamination was assessed under multiple drops using the state of damage from the first drop at 25,000g shock. The final damage state in the first drop with the deleted elements was the input for second drop simulation and this was implemented through Abaqus explicit using *import* scripting. The out-of-plane displacement at 0ms just before the impact for drop-2 is shown in Figure 26. We can identify the inner lip which was separated in the first drop was not imported in the second drop as it is not part of material response required for drop-2. Figure 27 shows the peak displacement at 0.3ms timestep which is 5.8mm. At time step 0.5ms shown in Figure 28 the separation between the PCB and epoxy is more severe for the second drop.

Figure 26: Drop-2, Out-of-plane displacement of potted test assembly at 25,000g high-g shock at 0ms time step

The final damaged state from the first drop simulation was imported as the first step in the second drop and Figure 29 shows the status of the cohesive elements in the initial timestep. The status of the cohesive element deletion at 0.3ms timestep is shown in Figure 30. At time step 0.5ms shown in Figure 31, over 70-percent of the cohesive elements have exceeded the damage criterion and delamination between the

PCB and epoxy potting is more pronounced. Figure 32 shows the image of the actual potted test assembly under 25,000g after second-drop. The potting material on the component side had shown failure through delamination at the solder mask interface.

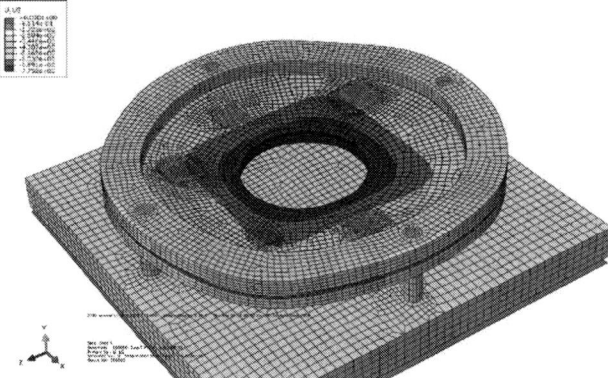

Figure 27: Drop-2, Out-of-plane displacement of potted test assembly at 25,000g high-g shock at 0.3ms time step

Figure 28: Drop-2, Out-of-plane displacement of potted test assembly at 25,000g high-g shock at 0.5ms time step

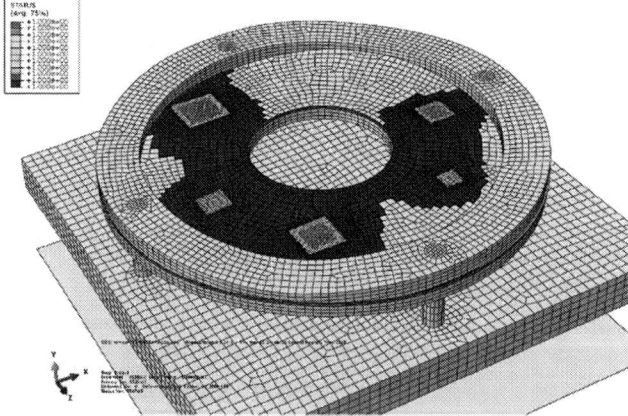

Figure 29: Drop-2, Status of element deletion under 25,000g high-g shock at 0ms time step

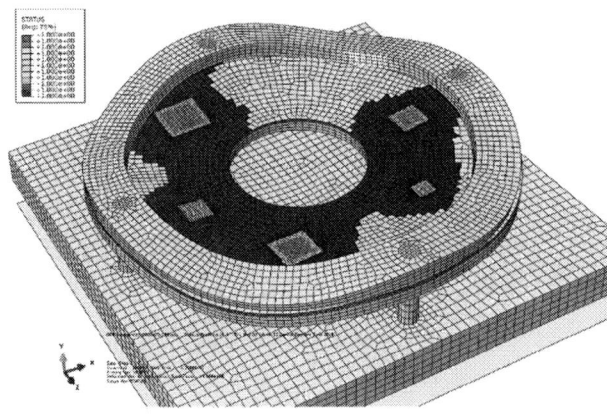

Figure 30: Drop-2, Status of element deletion under 25,000g high-g shock at 0.3ms time step

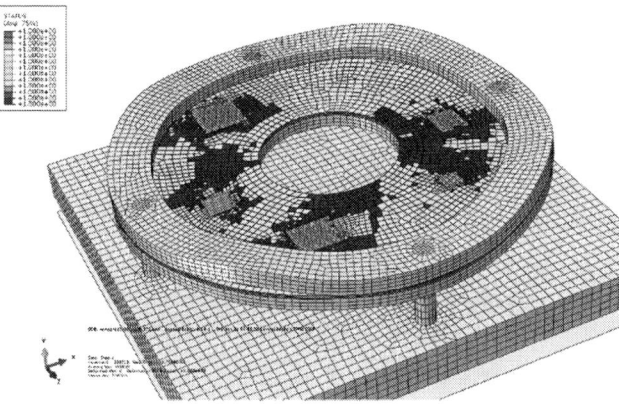

Figure 31: Drop-2, Status of element deletion under 25,000g high-g shock at 0.5ms time step

Figure 32: Delamination of epoxy potting after 2nd drop under 25,000g shock

Although the damaged potting material that had fractured and fallen off during drop-1 was only about 20-percent of the entire structure, the potting material that was still intact might show delamination, which is evident through the failure of the components underneath the epoxy in Figure 29.

IV. SUMMARY AND CONCLUSIONS

Interface properties characterization was done for the potting compound-PCB interface for three different epoxies using bi-material PCB/epoxy three-point bending specimens. A variety of cure conditions have been studied. The fracture toughness has been measured for various cure conditions of all three epoxy-systems studied in this paper. The measured values of fracture toughness have been used to compute the cohesive zone model parameters for implementation in an explicit finite element framework. The explicit finite element model framework has been used to predict the interface damage initiation and progression under high-g shock impact loading for repetitive shocks. Accrued damage in prior drops have been used a baseline for successive drops simulation in order to allow for damage progression after initiation of damage in earlier drops. The predictive power of the model has been quantified by modeling the three-point bend test of the notched bi-material specimen and the comparison of the load-displacement curves and the strain energy release rates with the experimental findings. Model predictions on circular daisy-chained printed circuit assemblies have been correlated with experimental data for successive drop-events. Model predictions indicate that the damage model for circular PCB with finepitch BGAs under high-G shock is able to predict the failure of the epoxy near the inner lip validating the observation found in high-G shock tests. The location of initiation of the delamination between the epoxy and PCB was found to be random but concentrated towards the inner lip of the circular PCB assembly.

ACKNOWLEDGMENT

The research presented in this paper has been supported by NSF-CAVE3 Electronics Center consortium-members.

REFERENCES

[1] Berman, M. S., "Electronic components for high-g hardened packaging," DTIC Document, 2006.

[2] Chao, N.H., J. A. Cordes, D. Carlucci, M. E. DeAngelis, and J. Lee, "The use of potting materials for electronic-packaging survivability in smart munitions," Journal of Electronic Packaging, vol. 133, no. 4, p. 41003, 2011.

[3] Charalambides, P. G., Lund, J., Evans, A. G., & McMeeking, R. M. (1989). A test specimen for determining the fracture resistance of bimaterial interfaces. Journal of applied mechanics, 56(1), 77-82.

[4] Haynes, A.S., Cordes, J.A., Krug, J., Thermomechanical Impact of Polyurethane Potting on Gun Launched Electronics," Journal of Engineering, vol. 2013, p. e148362, Nov. 2012.

[5] Hutchinson, J. W., Mear, M. E., & Rice, J. R. (1987). Crack paralleling an interface between dissimilar materials.

[6] Keith, R. E., "Potting Electronic Modules: A Report. NASA SP-5077," NASA Special Publication, vol. 5077, 1969.

[7] Kerlee, C., "Selecting encapsulants and potting compounds for electronic modules," in 1971 EIC 10th Electrical Insulation Conference, 1971, pp. 166–170.

[8] Lall P, Dornala K, Suhling J, Deep J. "Interfacial Delamination and Fracture Properties of Potting Compounds and PCB/Epoxy Interfaces Under Flexure Loading After Exposure to Multiple Cure Temperatures". ASME 2017 International Technical Conference and Exhibition on Packaging and Integration of Electronic and Photonic Microsystems.

[9] Lall P, Dornala K, Suhling J, Deep J. "Interfacial Delamination and Fracture Properties of Potting Compounds and PCB/Epoxy Interfaces Under Flexure Loading After Exposure to Multiple Cure Temperatures". ASME 2017 International Technical Conference and Exhibition on Packaging and Integration of Electronic and Photonic Microsystems.

[10] Lall P., Dornala K., Lowe R., and Foley J., "Survivability assessment of electronics subjected to mechanical shocks up to 25,000g," in 2016 15th IEEE Intersociety Conference on Thermal and Thermomechanical Phenomena in Electronic Systems (ITherm), 2016, pp. 507–518.

[11] Peng, H., et al., "Underfilling fine pitch BGAs," IEEE Transactions on Electronics Packaging Manufacturing, vol. 24, no. 4, pp. 293–299, Oct. 2001.

[12] Pozuelo, M., Cepeda-Jiménez, C. M., Chao, J., Carreño, F., & Ruano, O. A. (2009). Fracture toughness for interfacial delamination of Cr–Mo steel multilayer laminate. Materials Science and Technology, 25(5), 632-635.

[13] Tran, H.T., Shirangi, M.H., Pang, X. et al. "Temperature, moisture and mode-mixity effects on copper leadframe/EMC interfacial fracture toughness" Int J Fract (2014) 185: 115.

[14] Turon, A. Camanho, P. P., and Dávila, C. G., "A Damage Model for the Simulation of Delamination in Advanced Composites under Variable-Mode Loading," Mechanics of Materials, Vol. 38, No. 11, pp.1072-1089, 2006.

[15] Turon, A., Dávila, C. G., Camanho, P. P., and Costa, J., "An Engineering Solution for Mesh Size Effects in the Simulation of Delamination Using Cohesive Zone Models," Engineering Fracture Mechanics, Vol. 74, No. 10, pp. 1665-1682, 2007.

[16] Xiao, A., H. Pape, B. Wunderle, K. M. B. Jansen, J. de Vreugd and L. J. Ernst, "Interfacial fracture properties and failure modeling for microelectronics," 2008 58th Electronic Components and Technology Conference, Lake Buena Vista, FL, 2008, pp. 1724-1730. doi: 10.1109/ECTC.2008.4550213

[17] Zhang, J., & Lewandowski, J. J. (1997). Delamination study using four-point bending of bilayers. Journal of materials science, 32(14), 3851-3856.

978-1-7281-1500-9/19 $31.00 © 2019 IEEE

Numerical simulation on the formation process of metal droplets by pneumatic diaphragm drop-on demand technology

Kun Ma[a], Sheng Liu[a*], Zhiwen Chen[a,*], Li Liu[b], Hao Zheng[c], Yao Zhang[c]

[a]The Institute of Technological Sciences, Wuhan University, Wuhan, China
[b]School of Materials Science and Engineering, Wuhan University of Technology, Wuhan, People's Republic of China
[c]China Ship Development and Design Center, Wuhan, China
victor_liu63@126.com ,zwchen_lu@163.com

Abstract—In this paper, a theoretical analysis is carried out on the pneumatic diaphragm drop-on demand technology. The typical forming process of the droplet is analyzed, and the working principle of the pneumatic droplet injection is studied. The fluid calculation software FLUENT is used as the calculation platform to simulate the formation process of metal micro-droplets under gas pulse. Experiments demonstrate that this micro-jetting technology has a good performance on micron droplets generation and dispensing. And the experimental results agree well with the simulation results, which proves this simulation model and method is correct.

Keywords-micro-droplet jetting; pneumatic diaphragm drop-on-demand technology; numerical simulation; jetting characteristics

I. INTRODUCTION

Micro-droplet jetting technology is an attractive method for the production of solder particles , electronic packaging of MEMS and the rapid forming of metal parts owing to low-cost, low-waste, and simple process. Metal droplet injection technology has many advantages over traditional lithography and electroplating methods in terms of fast manufacturing time, low cost and no need for pickling[1]. Moreover, metal droplet injection technology is a non-contact direct deposition method, which deposited solder directly on the substrate to form bumps without masks and expensive devices[2].

Metal droplet injection technology mainly includes two modes: drop-on-demand (DOD) and continuous injection (CIJ)[3]. Continuous injection method is difficult to control the deposition position of the metal droplets, and the precision is not high, drop-on-demand ejection method can accurately control the size and deposition position of the droplets with high precision[4]. Drop-on-demand injection technology is usually divided into piezoelectric and pneumatic. The principle of piezoelectric type is that piezoelectric crystals vibrates under the effect of voltage pulses to extrude liquids to form droplets[5].However, piezoelectric materials typically operate at temperatures below 300 ° C. making them unsuitable

for use with high melting point materials. Therefore, in a high temperature environment, it is necessary to design complicated cooling devices, and the cost is relatively high[6]. The principle of the pneumatic metal droplet ejection technique is that actuator generate a gas pressure. The pressure in the injection chamber is changed to force droplets out of a nozzle[7]. As for the pneumatic type, it has no complex parts so that the equipment is easy to build and operate. It is important that the pneumatic type can endure high temperature [8]. Compared with pneumatic drop-on-demand technology, pneumatic diaphragm DOD technology has strong resistance to pressure disturbance due to its mental diaphragm actuator[9, 10]. Since the air pressure is driven directly by gas, the pressure change in the cavity has a great influence on the stable injecting of a single droplet. Therefore, the formation process of metal droplets should be emphasized in further study[11].

What's important of using DOD technology to weld small and complex interconnections are the generation, diffusion and wetting of solder droplets on substrates[12].In this paper, firstly, a pneumatic diaphragm drop-on demand experimental system is developed. Uniform droplets with good sphericity were obtained through the experimental. Experiments demonstrate that this micro-jetting technology has a good performance on micron droplets generation and dispensing. Secondly, the typical forming process of the droplet is analyzed, and the working principle of the pneumatic droplet injection is studied. The fluid calculation software FLUENT is used as the calculation platform to simulate the formation process of metal micro-droplets under gas pulse. Finally, the comparison between the test results and the simulation results reveals characteristics of the metal micro-droplet jetting process. And the experimental results agree well with the simulation results, which proves this simulation model and method is correct.

II. MODELING

A. Description of the Simulation Model

The morphology control of the formed droplets, is the key to determining whether a drop-on-demand device can be applied. If the design experiments are carried out directly on the injection devices of different structures, experimental repeatability and cost are relatively large. Therefore, simulation method can obtain the influence law of different parameters on the injection progress at low cost.

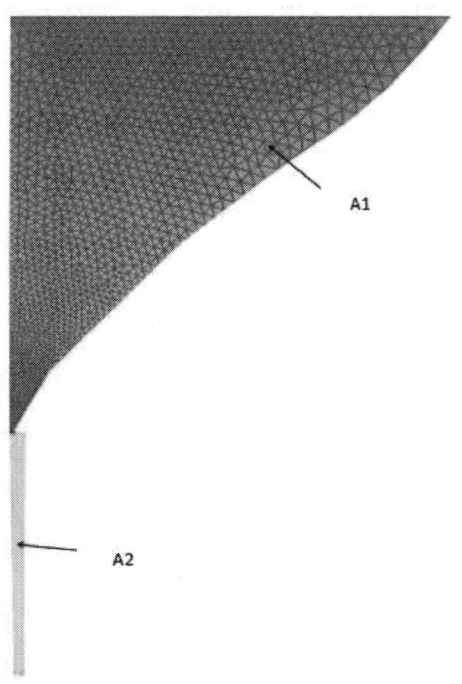

Figure1. Schematic diagram of 2D axisymmetric model

A two-dimensional axisymmetric model is established to simulate the injection process of metal droplets. Numerical analysis uses a simplified model to simulate the entire process. As shown in Figure1, the entire calculation area of the simulation model is divided into an internal flow field A1 and an external flow field A2. The internal flow field A1 consists of the air chamber, the helium and the nozzle of the injection device. The gas and liquid phases coexist in the internal flow field. Under the effect of the gas pulse, the gas and liquid phases flow at helium, nozzle, and the external flow field A2. External flow field A2 is the area where the liquid flow and the droplets fly. An unstructured grid was applied for the internal flow field A1, and a structured grid was used for the external flow field A2 to obtain accurate simulation results.

According to the principle of pneumatic metal droplet injection. Gas-liquid coupled two-phase flow model was applied to conduct the numerical simulation. This paper uses the VOF model (Volume of Fluid Model). Boundary condition was set as follows.The inlet condition as the pressure inlet (Pressure). Inlet). Boundary of the nozzle and the nozzle as the wall, and the boundary of the external flow field as the pressure outlet (Pressure Outlet). According to the principle of pneumatic injection, the inlet condition of the solution model is the pressure pulse function changing with time. The PISO algorithm for transient problem was applied to solve the problem. The second-order upwind style with higher computational precision was used to discretize the control equation.

III. EXPERIMRNTAL METHOD

A. Working Principle of the Equipment

The overall structure of the pneumatic diaphragm type metal droplet ejection printing system constructed in this paper is shown in Figure2.The core of the system is a pneumatic diaphragm type droplet ejection device, which completes storage, heating and melting metal forming materials. Driving energy of the spraying device is high-pressure gas. The environmental atmosphere protection is needed during the forming process, so that the pneumatic control subsystem is needed. From a single droplet to a three-dimensional entity, the injection device and the substrate are moved according to the planned path. A three-dimensional motion platform is needed, and the metal forming material needs to be heated and melted before spraying. Moreover, the temperature needs to be maintained near the working temperature, so that temperature control subsystem is needed. The progress of forming is completed on the substrate, and the temperature of substrate was kept constant during the forming process. The above components need to be organically combined under the control of the control system, so the driver board is required to directly control the various components of the system, such as the motion control of the three-dimensional motion platform, the on-off control of the pulse-driven air pressure, the temperature control, etc. Through the connection between the upper computer and the driver board, the monitoring of the working state of the system and the command control of the system are realized.

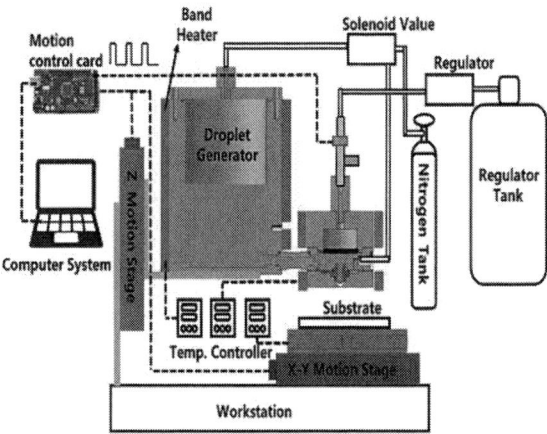

Figure 2. Schematic diagram of experimental system

Figure 3 shows the physical map of the equipment we designed. The spraying process of the metal droplet ejection device is mainly as follows. Firstly, the back pressure above the liquid storage tank is added to the stage where the liquid is about to fall. Secondly, a pulse signal is issued to open the electromagnetic valve, and the compressed gas enters the air chamber through the electromagnetic valve. At the same time, due to the presence of the exhaust pipe, the pressure in the air chamber rises and falls in a short time. The change of pressure causes the diaphragm to deform and recover in a short time, so that the droplets are squeezed out. The process monitor system which consists of a CCD

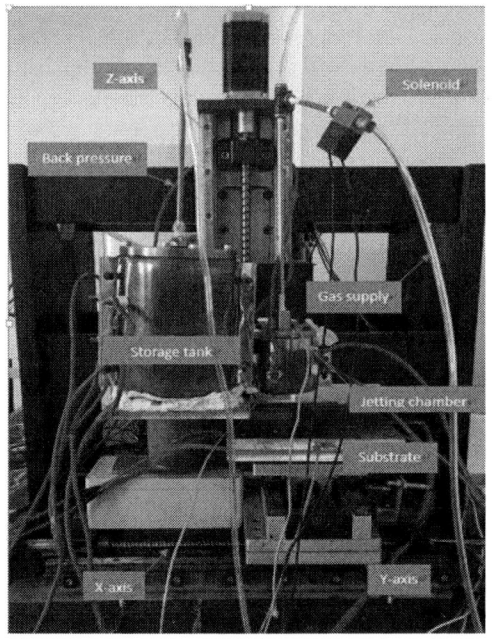

Figure3. Pneumatic diaphragm DOD micro-jet system

camera and an image acquisition card was used to observe the spraying and deposition process of droplets.

B. Experiment Procedure

The material, in this research, is Sn99.3CuO.7. Its properties are shown in table 1. When the solder melts into liquid, it has low viscosity to be able to spread out easily form the nozzle tip. Moreover, it has wide applications in electronic packages and electron industry.

TABLE1. PROPERTIES OF SPRAY MATERIAL

Spray material	Density $(kg \cdot m^{-3})$	Melting point /°C	Tensile strength /MPa	Kinematic viscosity /Pas
Sn$_{99.3}$Cu0.7	7400.5[a]	227	32	0.0013

In order to study the effect of different driving pressures on the droplet formation of a pneumatic metal droplet on-demand injection process, we obtained a sample by depositing metal droplets in glycerol, which ensures that the sample is spherical. The working parameters of injection process shown in table 2.

TABLE 2. WORKING PARAMETERS OF THE EXPERIMENT

Back pressure(MPa)	0
Nozzle diameter(mm)	0.15
Pulse width(ms)	35
Nozzle temperature(℃)	240
Driving pressure(MPa)	0.06,0.08,0.10,0.12,0.16,0.18

In order to analyze the formation mechanism of metal droplets during metal droplet spraying, driving gas pressure of 0.05 MPa was set to monitor the gas-liquid distribution and pressure field changes in the vicinity of the spraying process.

IV. RESULTS AND DISCUSSION

A. Effect of Drive Air Pressure

To verify the feasibility of the model we built, we compared the average diameter of the metal droplets simulated at different driving pressures with the average diameter obtained from the experiment. Figure4 indicated that the micro-droplets produced by our equipment had good sphericity. The uniform micro-droplets indicated the possibility of our equipment being used in electronic packaging. The comparison between simulation and experiment mean diameter is shown in Figure5 which shows the simulation results of the average diameter of the metal droplets at different air pressure are agreed well with the measured results. From Figure5, we can intuitively see that as the driving air pressure increases, the average diameter of the droplets increases.

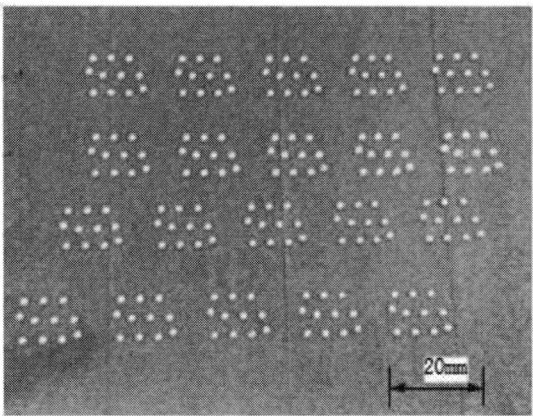

Figure4. Droplets deposited on the substrate

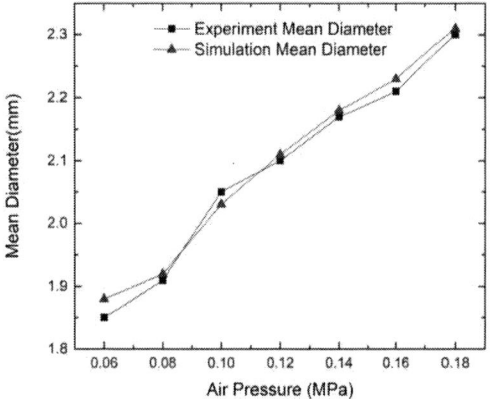

Figure5. The comparison between simulation and experiment result

B. Simulation Results of Pressure Field Variation

Figure6 and Figure7 show the formation of metal droplets during metal droplet ejection and the pressure field as a function of time. According to this, the whole process can be divided into four stages of jet-neck-fracture-flight.

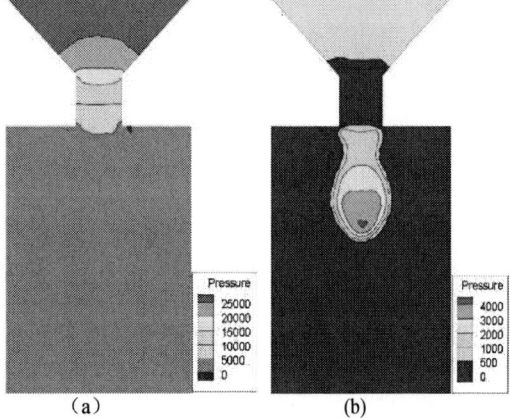

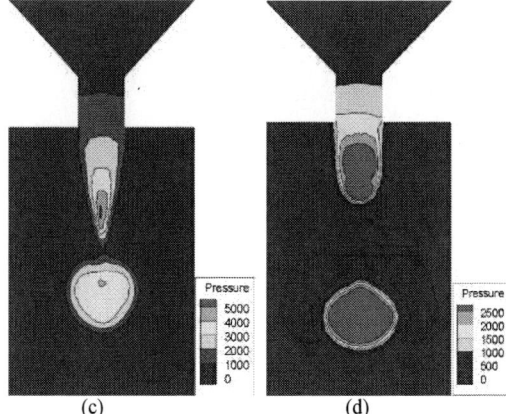

Figure6. Variation of pressure field during metal droplet injection, (a) pressure field of the jet phase, (b) pressure field of the necking stage, (c) pressure field of the fracture stage,(d) pressure field of the flight phase

The first stage of the metal droplet spraying process is the jet phase. During this stage, the pneumatic pulse is introduced into the discharge chamber, and the gas pressure in the chamber rises rapidly and acts on the liquid surface. Through the pressure contour (a) of the nozzle at this stage, it can be seen that the pressure decreases in turn from the storage chamber to the nozzle and then to the external flow field. The second stage of the injection process is necking stage, in which the droplets ejected from the nozzle produce a radial depression in the neck. The comparison between the pressure contour (a) and the pressure contour (b), we can find that the pressure in the storage chamber decreases. The reason for the pressure drop in the reservoir is because the gas pulse has been completed and the pressure in the reservoir is discharged through the vent. Fracture stage is the third stage in the micro-droplet injection process. The simulation results of pressure contour (c) show that the pressure in the storage chamber and nozzle is lower than that of the ejected liquid flow. The tail of the liquid flow has a tendency to recirculate, while the continuous necking causes the diameter of the liquid flow neck to decrease continuously, and the liquid droplet moves downward under the action of gravity. The final stream breaks in the neck, forming separate droplets. The fourth stage of the spraying process is the flight phase of the droplet, in which the droplets move downward under the action of inertia and gravity, and the droplets gradually condense into a spherical shape under the action of surface tension. The pressure in the reservoir is lower than the pressure at the nozzle.

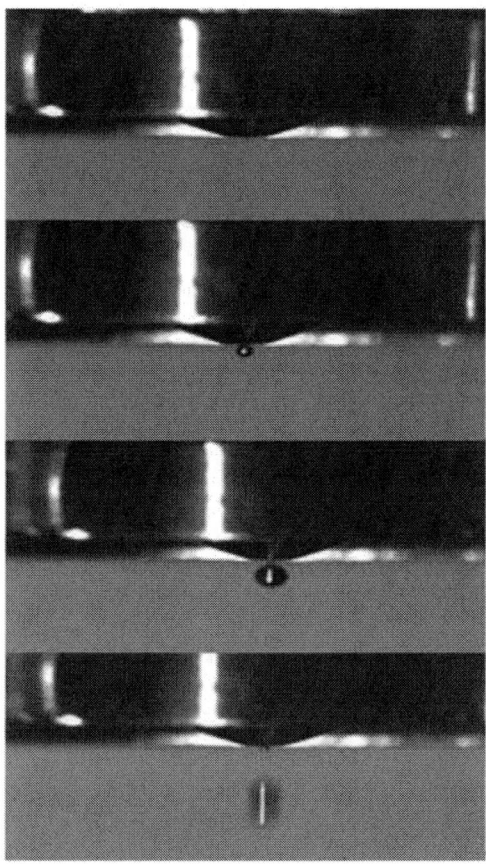

Figure6. The sequential images of droplets formation

V. CONCLUSIONS

In this paper, a two-dimensional axisymmetric model is proposed to simulate the spraying process of pneumatic metal droplets. Through the pneumatic diaphragm type metal droplet ejection printing system designed by our group, we studied the effect of different air pressure on the average diameter of the metal droplets. The formation process of metal droplets by pneumatic diaphragm drop-on demand technology was discussed as well. The main conclusions are as follows:

1) The simulation results are consistent with the experimental results of droplet pattern and droplet diameter in the process of single droplet formation. The accuracy of the model is verified by the experimental and simulation results. The size of the metal droplets increases with the increase of driving pressure.

2) The simulation results agreed well with the experimental observations. The whole process of droplets formation can be divided into four stages of jet-neck-fracture-flight.

3) The sequential images of droplets formation indicated that the generation of metal droplets by

our device was repeatable and controllable, can be applied to electronic packaging.

ACKNOWLEDGMENT

This work was supported by The Startup Foundation for Introducing Talent of Wuhan University (NO. 413100011).The experimental work was mainly carried out at the Research Center of Electronic Manufacturing and Packaging Integration in School of Power and Mechanical Engineering, Wuhan University.

REFERENCES

[1] J. Luo, L. H. Qi, J. M. Zhou, X. H. Hou, and H. J. Li, "Modeling and characterization of metal droplets generation by using a pneumatic drop-on-demand generator," *Journal of Materials Processing Tech*, vol. 212, no. 3, pp. 718-726, 2012.

[2] J. Luo, L. H. Qi, S. Y. Zhong, J. M. Zhou, and H. J. Li, "Printing solder droplets for micro devices packages using pneumatic drop-on-demand (DOD) technique," *Journal of Materials Processing Technology*, vol. 212, no. 10, pp. 2066-2073, 2012.

[3] M. Vaezi, H. Seitz, and S. Yang, "A review on 3D micro-additive manufacturing technologies," *The International Journal of Advanced Manufacturing Technology*, journal article vol. 67, no. 5, pp. 1721-1754, July 01 2013.

[4] M. Orme and R. F. Smith, "Enhanced Aluminum Properties by Means of Precise Droplet Deposition," (in eng), *Journal of manufacturing science and engineering: Transactions of the ASME*, vol. 122, no. 3, pp. 484-493, 2000.

[5] Q. Liu and M. Orme, "High precision solder droplet printing technology and the state-of-the-art," *Journal of Materials Processing Technology*, vol. 115, no. 3, pp. 271-283, 2001/09/24/ 2001.

[6] A. Amirzadeh, M. Raessi, and S. Chandra, "Producing molten metal droplets smaller than the nozzle diameter using a pneumatic drop-on-demand generator," (in English), *Experimental Thermal and Fluid Science*, vol. 47, pp. 26-33, May 2013.

[7] S. X. Cheng, T. Li, and S. Chandra, "Producing molten metal droplets with a pneumatic droplet-on-demand generator," *Journal of Materials Processing Tech*, vol. 159, no. 3, pp. 295-302, 2005.

[8] D. Liao, D. Su, S. Liu, X. Li, and H. Zhang, "Effects of drive air pressure, exhaust pipe length and deposition height on the morphology of Sn99.3Cu0.7 solder ball," in *2016 17th International Conference on Electronic Packaging Technology (ICEPT)*, 2016, pp. 1508-1511.

[9] C. Wang, D. Liao, S. Liu, X. Li, and H. Zhang, "Effect of droplet spacing & deposition height on Sn conductive lines by pneumatic diaphragm drop-on-demand technology," in *2016 17th International Conference on Electronic Packaging Technology (ICEPT)*, 2016, pp. 1512-1516.

[10] D. Xie, H. H. Zhang, X. Y. Shu, J. F. Xiao, and S. Cao, "Multi-materials drop-on-demand inkjet technology based on pneumatic diaphragm actuator," *Science China-Technological Sciences*, vol. 53, no. 6, pp. 1605-1611, Jun 2010.

[11] L. Suli, W. Zhengying, D. Jun, Z. Guangxi, W. Xin, and L. Bingheng, "Experimental Investigation of the Overlap Process in Metal Droplet Successive Deposition and Solidification," *Rare Metal Materials and Engineering*, vol. 46, no. 12, pp. 3645-3650, 2017/12/01/ 2017.

[12] L.-h. Qi, J. Luo, J.-m. Zhou, X.-h. Hou, and H.-j. Li, "Predication and measurement of deflected trajectory and temperature history of uniform metal droplets in microstructures fabrication," *The International Journal of Advanced Manufacturing Technology*, vol. 55, no. 9, pp. 997-1006, 2011/08/01 2011.

On Curing-induced Residual Stresses after Molding Processes: Mold Shrinkage, Chemical Shrinkage or Both?

Changsu Kim, Sukrut Phansalkar, Hyun-Seop Lee, and Bongtae Han*
Center for Life Cycle Engineering (CALCE)
Mechanical Engineering Department, University of Maryland
College Park, MD 20742, USA
bthan@umd.edu

Abstract— **Various definitions of shrinkage are reviewed: mold shrinkage, total chemical shrinkage and effective chemical shrinkage. Their relationship with the curing-induced residual stresses is discussed. Then, an advanced technique, called "dual configuration fiber Bragg grating (DC-FBG) technique", is employed to measure the effective chemical shrinkage of EMCs for two conditions: after removing the mold pressure before curing and while maintaining the pressure during curing. The results are discussed with the transfer molding and compression molding processes. A procedure to estimate the total chemical shrinkage from the mold shrinkage is also presented.**

Keywords- mold shrinkage, total chemical shrinkage and effective chemical shrinkage

I. INTRODUCTION

The volumetric shrinkage (or chemical shrinkage) of epoxy molding compounds (EMCs) occurs during curing. As the modulus evolves, the shrinkage produces stresses inside molded parts while curing, which results in "curing-induced" residual stresses. They can affect the reliability of components molded by EMCs.

The issue of "chemical shrinkage" is not new. Numerous testing methods have been developed for many decades, and some of them are practiced routinely to measure the intrinsic (or total) chemical shrinkage [1]. It is important to note, however, that not all of the chemical shrinkage contributes to the residual stresses simply because the chemical shrinkage occurs even before the gel point where polymers start to have mechanical strength [1, 2].

EMC manufactures often provide a shrinkage quantity called "mold shrinkage". Mold shrinkage is measured at room temperature, and it should be distinguished from the chemical shrinkage measured at the curing temperature.

In this paper, various definitions of shrinkage are reviewed first, and their relation to curing induced residual stress is discussed. Then, an advanced technique, called Dual Configuration Fiber Bragg Grating technique (DC-FBG), is described and implemented to measure the "effective chemical shrinkage" of EMCs with and without the applied pressure. A procedure to estimate the total chemical shrinkage from the mold shrinkage is also discussed.

II. MOLD SHRINKAGE AND CHEMICAL SHRINKAGE

A. Mold Shrinkage

The mold shrinkage is defined to be the difference between the dimensions of the mold and the molded part at room temperature [3]. It is intended to provide a dimensional change at room temperature caused by the mold process. This value should not be used to predict the residual stresses at the mold temperature simply because it contains the effect of the coefficient of thermal expansion (CTE) of the mold as well as the cured thermoset polymers.

In transfer molding, as in compression or injection molding, the difference between the dimensions of the mold and of the molded article produced therein from a given material vary according to the design and operation of the mold. It is affected by the size and temperature of the pot or cylinder and the pressure on it, as well as on mold temperature and molding cycle [4-6].

Typically, the mold shrinkage is measured using a standard cavity. As illustrated in Fig. 1, the mold shrinkage (%) is defined as

$$MS\ (\%) = \frac{D - d}{D} \times 100 \qquad (1)$$

where MS (%) is mold shrinkage in %; D is the dimeter of a mold at room temperature; and d is the diameter of a cured EMC at room temperature.

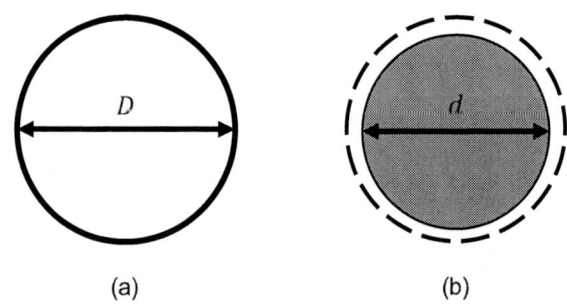

(a)   (b)

Fig. 1. Illustration of the dimensions of (a) cavity at room temperature and (b) molded EMC at room temperature

B. Chemical Shrinkage

Chemical shrinkage occurs as the distance between the molecules changes during crosslinking. The curing properties (chemical shrinkage and modulus) evolves non-linearly with time.

The behavior of a typical polymer system is illustrated in Fig. 2, where the x-axis shows an arbitrary time scale, and the y-axis represents the evolution properties ($\varepsilon_{ch}(t), E(t)$), normalized by the properties after curing is completed, ($\varepsilon_\infty, E_\infty$) [1]. The figure illustrates the gel point, t_{gel}, where the polymer starts to build mechanical strength [7-9]. As mentioned earlier, the chemical shrinkage does occur before the gel point. However, the modulus associated with the chemical shrinkage developed before t_{gel} is virtually "zero", and it does not contribute to the residual stresses. The chemical shrinkage after the gel point, is called the "*effective chemical shrinkage*" (denoted as ε_{ch}^{eff}) [1, 2].

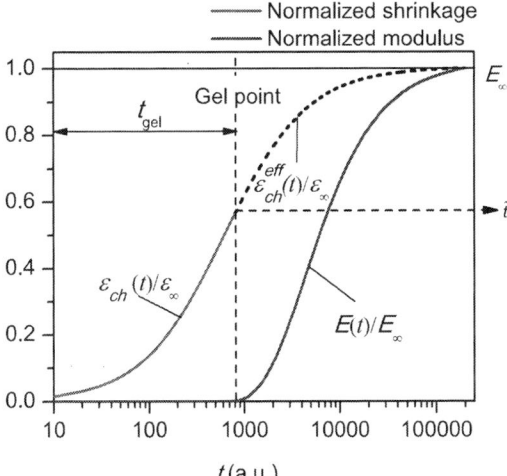

Fig. 2. Evolution of chemical shrinkage and modulus during the curing

The effective chemical shrinkage should be used for calculation of curing-induced residual deformations or stresses; otherwise, the magnitude of residual stresses can be significantly overestimated. If a polymer is elastic (e.g. silicone rubber), the residual stress can be calculated by using only the effective chemical shrinkage and the equilibrium modulus. If a polymer has a viscoelastic behavior after the gel point, however, the residual stress can be determined only when the evolution history of the effective chemical shrinkage, and the curing-dependent viscoelastic properties are known.

III. EFFECTIVE CHEMICAL SHRINKAGE MEASUREMENT WITHOUT AND WITH MOLD PRESSURE

A. Dual Configuration Fiber Bragg Grating (DC-FBG) technique [10]

As illustrated in Fig. 3, a fiber Bragg grating (FBG) sensor was embedded in a cylindrically-shaped EMC specimen. After the gel point, the effective chemical shrinkage of EMC compresses the FBG in the axial direction, and a Bragg Wavelength (BW) shift occurs. The BW shift, $\Delta\lambda_B$, after the gel point is illustrated in Fig. 4. The BW is continuously measured while the specimen cures. The effective chemical shrinkage is then determined from the governing equation that defines the relationship between the BW shifts and the effective chemical shrinkage. More details about the method can be found in [1, 2].

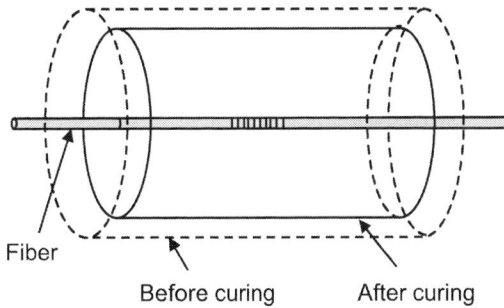

Fig. 3. Illustration of the volumetric shrinkage of an FBG sensor embedded polymer

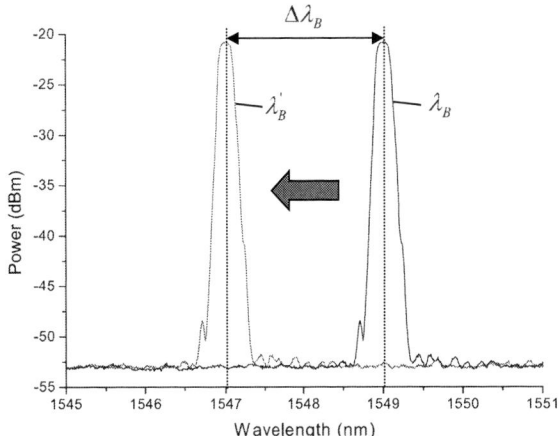

Fig. 4. Bragg Wavelength change due to the effective chemical shrinkage

B. Experimental Setup

The specimen is fabricated using a custom designed steel mold assembly, which is shown schematically in Fig. 5 [11, 12]. The inset shows a plunger that applies the required mold pressure to the specimen. The optical fiber (diameter of 125 μm) is inserted through a small through-hole drilled in the center of an EMC pellet. The vertical position of the Bragg grating is adjusted until it is placed in the middle of the pellet. The internal surfaces of the mold are treated with a release agent and thermal grease.

A large mechanical pressure has to be applied during curing. The required pressure (7 MPa) is achieved by the mechanical plunger. As illustrated in the inset of Fig. 5, the plunger was connected to the piston of a pneumatic cylinder. The diameter of the cylinder (31.75 mm) was much larger than the diameter of the plunger (8.9 mm). In this way, the curing pressure was achieved only by an air pressure of 0.54 MPa (78 psi), which was readily available in laboratories.

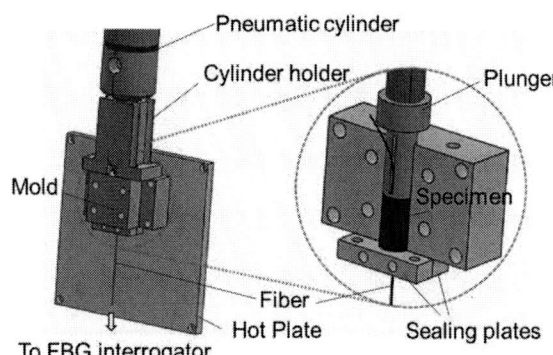

Fig. 5. Curing mold attached to the hot plate and the pneumatic cylinder.

C. Test Results

Two tests were conducted: (1) after removing the mold pressure before the gel point and (2) while maintaining the mold pressure until curing was completed. For the first experiment, the pressure was applied immediately after the specimen was heated to the pre-heating temperature where the EMC became viscous fluid. The pressure was removed as soon as the specimen reached the curing temperature, i.e., before the gel point.

The results from the first experiment are shown in Fig. 6, where the BW shift is shown as a function of time. It is important to note that the BW shift is zero for a short initial period. This is a clear indication that the FBG responded only after the gel point. The total BW shift was 507 pm, and the corresponding effective chemical shrinkage was 0.032%.

In the second test, the pressure was maintained until curing was completed. The second test results are shown in Fig. 7 (a). The total BW shift was 2.596 nm and the corresponding shrinkage was 0.167%. The BW change is nearly five times as high as the result of the first experiment. It was attributed to the hydrostatic creep deformation caused by the pressure after the gel point. The pressure was removed after curing was completed. The behavior is shown in Fig. 7 (b). A sudden increase of the BW shift is apparent, which represents the elastic recovery of EMC deformed by the pressure.

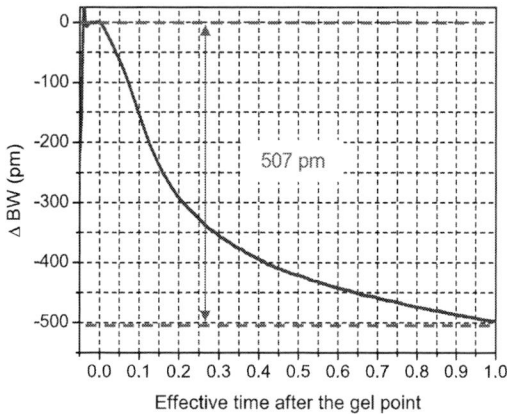

Fig. 6. Results from the test where the mold pressure was removed at the curing temperature

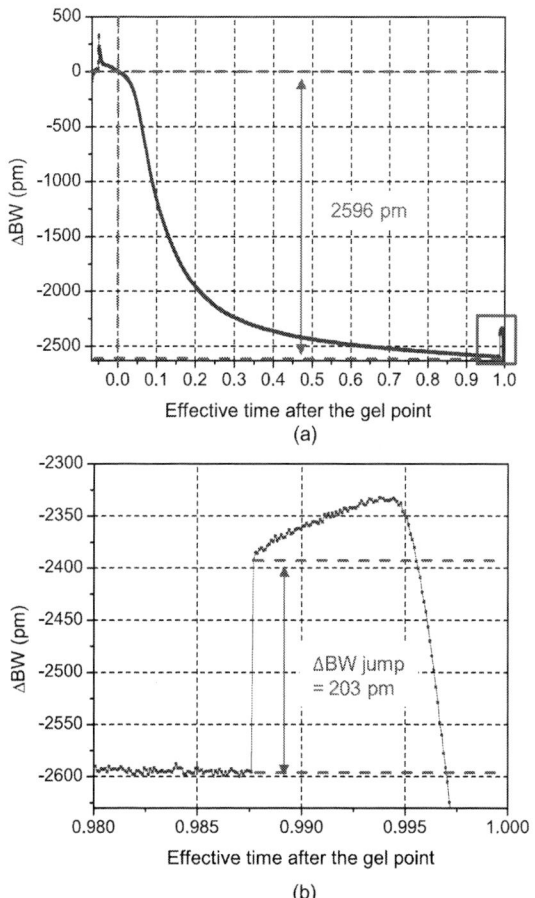

Fig. 7. Results from the test where the pressure was maintained during curing: (a) BW shift history and (b) zoom-in view marked by a box in (a).

IV. DISCUSSION

A. Transfer Molding vs Compression Molding

An important issue to consider for accurate prediction of the residual stresses is the pressure applied during molding processes. In the transfer molding process (Fig. 8), the charge (thermoset polymers) is transferred from the transfer pot to the cavities through channels called "runners", which connect the transfer pot to the gates of cavities. As a result, the pressure to fill the cavities is not transferred to the polymers after the gel point even when the pressure is maintained throughout the molding process. In this case, only the "effective chemical shrinkage" should be used for the residual stress prediction.

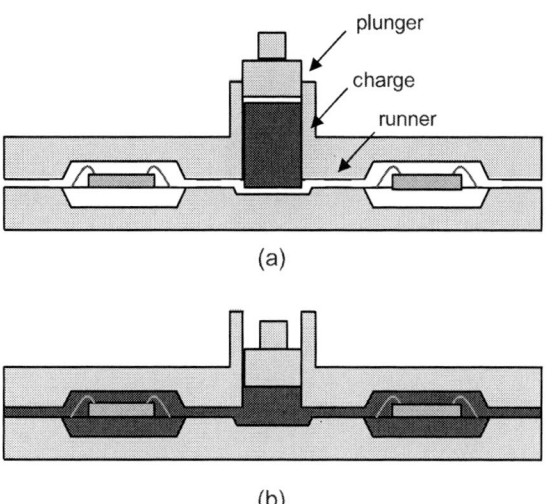

Fig. 8. Schematic diagram of the transfer molding (a) when charge is loaded, (b) after charge is transferred.

In the compression molding process (Fig. 9), however, the pressure to compress the charge is continuously applied to the polymers until the mold is opened. Unless the charge amount is precisely calculated and dispensed, the pressure causes the hydrostatic creep deformation of polymers after the gel point, and thus, the shrinkage caused by hydrostatic pressure should be considered together with the "effective chemical shrinkage" to determine the state of residual stresses correctly.

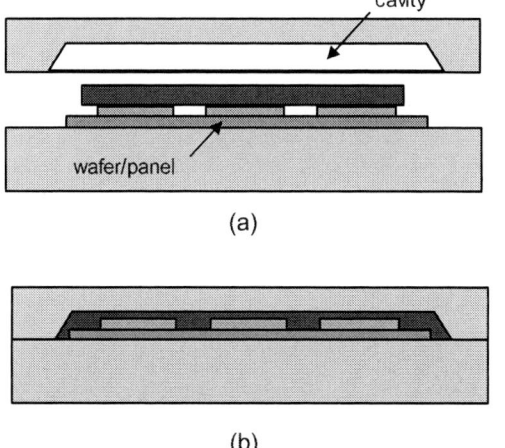

Fig. 9. Schematic diagram of the compression molding (a) before and (b) after the process.

B. Estimation of Total Chemical Shrinkage from Mold Shrinkage

The total chemical shrinkage can be estimated from the mold shrinkage. The mold cavity and the EMC during the mold process are shown schematically in Fig. 10. The mold cavity diameter at room temperature, D, is shown in (a). The mold cavity at curing temperature before and after charge is shown in (b). The mold cavity diameter at curing temperature, D', can be expressed as:

$$D' = D(1 + \alpha_m(T_c - T_r)) \qquad (2)$$

where α_m is the CTE of mold, and T_r and T_c are the room and curing temperatures, respectively.

The EMC diameter after curing at the curing temperature, D'', is shown in (c), and it can be expressed as:

$$D'' = D'(1 - \varepsilon_{ch}^{tot}) \qquad (3)$$

where ε_{ch}^{tot} is the total chemical shrinkage of EMC.

The EMC diameter at room temperature after cooling, d, is shown in (d), and it can be expressed as:

$$d = D''(1 - \alpha_1(T_g - T_r) - \alpha_2(T_c - T_g)) \qquad (4)$$

where T_g is the glass transition temperature of the cured EMC; α_1 and α_2 are the CTEs of cured EMC below and above T_g, respectively.

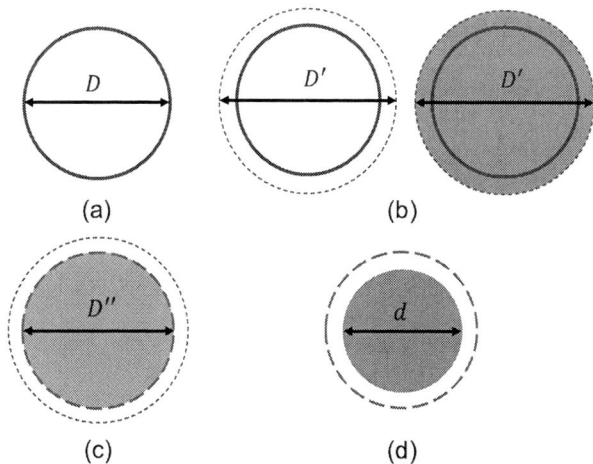

Fig. 10. Schematic illustration of mold cavity and EMC: (a) mold cavity at room, (b) mold cavity at curing temperature – before and after charge, (c) EMC diameter after curing at the curing temperature, and (d) EMC diameter at room temperature after cooling

From the above relationship, the total chemical shrinkage can be estimated from the mold shrinkage, MS, as:

$$\varepsilon_{ch}^{tot} = MS - \left\{ \alpha_1(T_g - T_r) + \alpha_2(T_c - T_g) - \alpha_m(T_c - T_r) \right\} \quad (5)$$

V. SUMMARY

Definitions of mold shrinkage, total chemical shrinkage and effective chemical shrinkage were reviewed. The dual configuration fiber Bragg grating (DC-FBG) technique was employed to measure the effective chemical shrinkage of EMCs for two conditions: after removing the mold pressure before curing and while maintaining the pressure during curing. As expected, the latter produced much higher shrinkage (nearly five times), which was attributed to the hydrostatic creep deformation after the gel point. The results were discussed with the transfer molding and compression molding processes. It was also shown that the total chemical shrinkage can be estimated from the mold shrinkage if the CTE of cured EMC is known.

REFERENCES

[1] Y. Wang, L. Woodworth, and B. Han, "Simultaneous measurement of effective chemical shrinkage and modulus evolutions during polymerization," *Experimental mechanics,* vol. 51, pp. 1155-1169, 2011.

[2] Y. Sun, B. Han, E. Parsa, and A. Dasgupta, "Measurement of effective chemical shrinkage and equilibrium modulus of silicone elastomer used in potted electronic system," *Journal of materials science,* vol. 49, pp. 8301-8310, 2014.

[3] ASTM, "Standard Test Method for Measuring Shrinkage from Mold Dimensions of Molded Thermosetting Plastics," vol. D6289-13, ed, 2013.

[4] F. Goffreda, A. Griff, L. Livinghouse, T. Walsh, and J. Scuralli, "Tool and Manufacturing Engineers Handbook Knowledge Base," *Society of Manufacturing Engineers,* 1998.

[5] L. P. Rector, S. Gong, T. R. Miles, and K. Gaffney, "Transfer molding encapsulation of flip chip array packages," in *Proceedings-SPIE the International Society for Optical Engineering,* 2000, pp. 760-767.

[6] S. Liu, G. Chen, and M. Yong, "EMC characterization and process study for electronics packaging," *Thin Solid Films* vol. 462, pp. 454-458, 2004.

[7] A. Spoelstra, G. Peters, and H. Meijer, "Chemorheology of a highly filled epoxy compound," *Polymer Engineering Science* vol. 36, pp. 2153-2162, 1996.

[8] Z. Zhang, E. Beatty, and C. P. Wong, "Study on the curing process and the gelation of epoxy/anhydride system for no‐flow underfill for flip‐chip applications," *Macromolecular Materials Engineering,* vol. 288, pp. 365-371, 2003.

[9] Z. Zhang, T. Yamashita, and C. P. Wong, "Study on the Gelation of a No‐Flow Underfill Through Monte Carlo Simulation," *Macromolecular Chemistry Physics,* vol. 206, pp. 869-877, 2005.

[10] Y. Sun, Y. Wang, Y. Kim, and B. Han, "Dual-configuration fiber Bragg grating sensor technique to measure coefficients of thermal expansion and hygroscopic swelling," *Experimental mechanics,* vol. 54, pp. 593-603, 2014.

[11] Y. Sun, H.-S. Lee, and B. Han, "Measurement of elastic properties of epoxy molding compound by single cylindrical configuration with embedded fiber Bragg grating sensor," *Experimental mechanics,* vol. 57, pp. 313-324, 2017.

[12] H. S. Lee, Y. Sun, C. Kim, and B. Han, "Characterization of Linear Viscoelastic Behavior of Epoxy Molding Compound Subjected to Uniaxial Compression and Hydrostatic Pressure," *IEEE Transactions on Components, Packaging Manufacturing Technology,* vol. 8, pp. 1363-1372, 2018.

Realistic Solder Joint Geometry Integration with Finite Element Analysis for Reliability Evaluation of Printed Circuit Board Assembly

Chun-Sean Lau
Western Digital Corporation
Plot 301A Persiaran Cassia Selatan 1
Taman Perindustrian Batu Kawan,
MK 13 Batu Kawan, Seberang Perai Selatan
14100, Penang, Malaysia
sean.lau@wdc.com

Ning Ye, Hem Takiar
Western Digital Corporation
951 SanDisk Dr, Milpitas, CA 95035

Abstract— Solder joint failure is a serious reliability concern in area array technologies, such as flip chip (FC), Plastic Ball Grid Array (PBGA), Fan-In and Fan-Out Wafer Level Packages (WLP) of advanced IC package. The selection of different substrate materials, solder material, molding compound, stacked dies structure, and laminate material could affect the solder joint stress-strain condition. It is therefore important to know the solder joint shape and standoff height accurately after the reflow process to estimate the reliability of solder joint assembly in three aspects: temperature cycling, mechanical shock, and vibration. A strategy for importing three-dimensional computed tomography (CT) data into a Finite Element based reliability evaluation is outlined. Three dimensional CT is a very fast, non-destructive automatic inspection machine. Moreover, with new version of CT scanning in high resolution, full solder geometry is reconstructed throughout the entire area array on printed circuit board assembly (PCBA). Finite Element Analysis (FEA) is used to calculate the accumulated plastic work per cycle for BGA packages on PCBA. The accumulated plastic work is then used to calculate the number of cycles to failure based on thermal fatigue life model of solder joints. FEA is also used to predict the damage index during shock and vibration event, and used to study mounting configurations and structural integrity of solder joints. The reliability results showed a good agreement with the experimental results based on two designs on new solid state drive (SSD) form factor. It was found that the cycles to failure and critical location among four corner joints match well with experimental results. From simulation results, it was also found that new design was much improved over old design. The methodology was extended to reliability evaluation for BGA packages such as FC controller, DDR SDRAM, and NAND packages on PCBA. Results demonstrate the excellent capability of the proposed integration tools for predicting the robustness of PCBA. The proposed approach greatly reduces reliability evaluation time, shortens the product life cycle development, and is more cost effective to address the reliability issues.

Keywords— Solder Joint Failure; Computed Tomography; Printed Circuit Board Assembly; Reliability of Solder Joint Assembly; Finite Element Anaysis; Solid State Drive

I. INTRODUCTION

As electronic packaging technologies continue to advance toward higher density, the use of area array type packages has gradually increased. Embedded and fan-out system in package (SIP) was the solution for higher density and miniaturization package. 2.5D ASIC and 3D IC was the solution for high performance package. Both future advanced IC packages utilized the area array platform to be mounted on printed circuit board (PCB). Solder joint failure is a common failure mechanism in area array technologies. The thermal expansion mismatch between the package and the board causes cyclic loading on the solder joints during temperature cycling and causes reliability issues. Also, rigidity plays a key role in PCBA in multiple aspects, especially in shock and vibration specification. With larger body size and weight per component, the solder joints have a higher chance of failure due to mechanical shock or vibration. Thus, knowing the solder joint shape and standoff height after the reflow process is important to determine the reliability of a solder joint assembly.

Automatic Optical Inspection (AOI) or Automated Visual Inspection (AVI) is an inspection process, widely used in surface mount technology (SMT). It evaluates the quality of finished goods with the help of visual information after their assembling sequences, i.e., paste printing, component placement, and soldering. Modern machines used in SMT lines such as paste printers, component placement machines, and etc. are capable of producing significantly better results than those required in normal standard specifications. However, to reach 6σ quality control, three-dimensional computed tomography (CT) systems are used as inline quality inspection appliances. The main advantage of these systems is their ability to detect failures under area array when the product has been assembled. Three dimensional CT is a very fast, non-destructive automatic inspection machine, which is currently used to check void percentage of area array, Head in Pillow (HIP), Non-Wet and other BGA related defects.

The motivation for investigating the effect that solder joint shape and standoff height have on the fatigue life is directly related to the quality of the soldering process. Since with high resolution of CT scanning, full solder geometry is reconstructed throughout the entire area array on printed circuit board assembly (PCBA). The data of solder joint shape and standoff height is transferred to the Finite Element based reliability evaluation, where it is subjected to a thermal cycle. With the aid of post-processing methodology which uses a solder fatigue method, the fatigue life of the solder joint can be estimated. This procedure can then be applied on PCB by systematically inputting the standoff height, solder shape and, other parameters to determine all BGA components influence on the predicted fatigue life and damage index.

In this paper, a strategy for importing CT data into a Finite Element based reliability evaluation is outlined. Then, Finite Element Analysis is used to calculate the accumulated plastic work per cycle for every BGA packages on PCBA. The accumulated plastic work is then used to calculate the number of cycles to failure based on thermal fatigue life model of solder joints and compared with experimental data with acceptable error. FEA is also used to predict the damage index during shock and vibration event and to study mounting configurations and structural integrity of solder joints. The methodology was extended to reliability evaluation for all BGA packages such as FC controller, DDR SDRAM, and NAND packages on PCBA.

II. COMPUTED TOMOGRAPHY INSPECTION

A. Description of the physical system

The three-dimensional computed tomography system used a gantry robot to move the assembled printed circuit board underneath an X-ray source to image the component joints that require inspection. The positioning of board was guided with the use of Computer Aided Design (CAD) data, which represented the outer layers of a printed circuit board's electrical design. The three-dimensional computed tomography system used classical tomosynthesis or laminography to create an image "slice," or image plane that will be distinct from other image planes on the object to be imaged. A slice will remove obstructions above or below the plane of focus so that only the regions of interest remain. Fig. 1 shows a classical laminography, a computed laminography (CL), and a CT image of solder contacts of a printed circuit board [1].

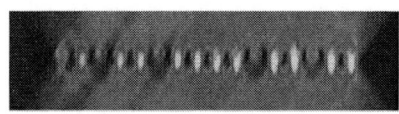

Classical laminography

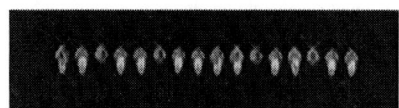

Computed laminography (CL)

Computed tomography (CT)

Figure 1: Solder joints of a printed circuit board [1]

The cross section obtained by classical laminography is blurred and only few details are visible. All solder joints are imaged clearly with a uniform quality throughout the slice by using CL. CT image the solder joints are clearly visible and free of artifacts in the center [1]. Our in-house system uses CT scanning in high speed and high resolution, as shown in Fig. 2. Moreover, full 3D image is able reconstructed for more accurate BGA inspection. Fig. 3 shows that example of 5 slices post reflow solder diameter for different NAND location.

Figure 2: Computed Tomography System

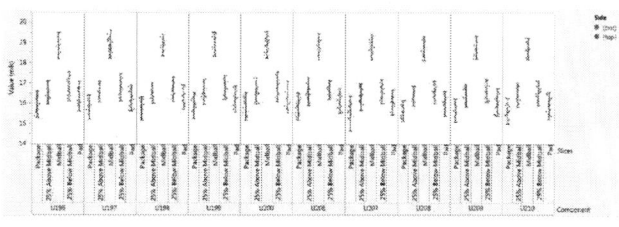

Remark:
Package= a; Pad= b ; Midball= c

Figure 3: 5 slices post reflow solder diameter for different NAND location

B. Solder joint shape calculation

Previously, many authors [2-3] determined solder ball geometry exclusively using surface evolver software tool based on energy method. In this study, post reflow solder diameter was measured and used to calculate the standoff height of the balls, with combination of truncated sphere theory. The function describing the shape of the solder joint was derived using the structure in Fig. 4. The standoff height of barrel shaped solder joint [2] could be calculated from (1) and (2) by knowing a, b, and c, where a is radius of solder resist opening; b is radius of PCB pad; c is radius of the middle of solder joint; t is thickness of PCB pad. Then, the final standoff height is defined as $(d + e + t)$. The volume of the truncated sphere by geometry considerations is given by (3).

$$d = \sqrt{c^2 - a^2} \tag{1}$$

$$e = \sqrt{c^2 - b^2} \tag{2}$$

$$V = \frac{\pi}{3}\left[d(3c^2 - d^2) + e(3c^2 - e^2)\right] \tag{3}$$

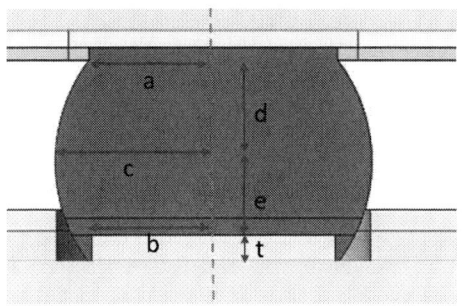

Figure 4: Truncated sphere model

C. Methodology

This section explains about the add-on inspection flow used for importing CT data into a Finite Element based reliability evaluation. The methodology flow is shown in Fig. 5. Finite Element based reliability evaluation is described in the order shown in next section.

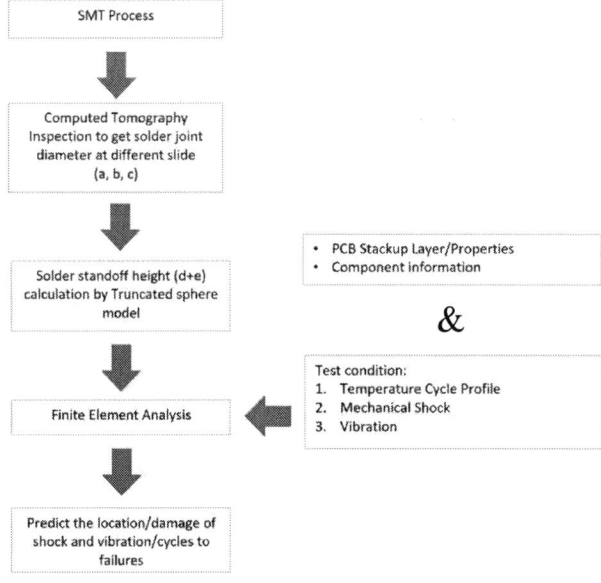

Figure 5: Methodology

III. FINITE ELEMENT ANALYSIS

A. Printed circuit board assembly

Fig. 6 shows two designs (Design A and B) on new solid state drive (SSD) form factor. Due to different copper trace underneath PCB, component structure and reflow process, the thermal response for every component are singular, causing the dissimilar solder joint formation. The cycles to failure, worst location among components, critical location among four corner joints and damage index for these two designs are compared with new proposed methodology. The model was built and export in ODB++.

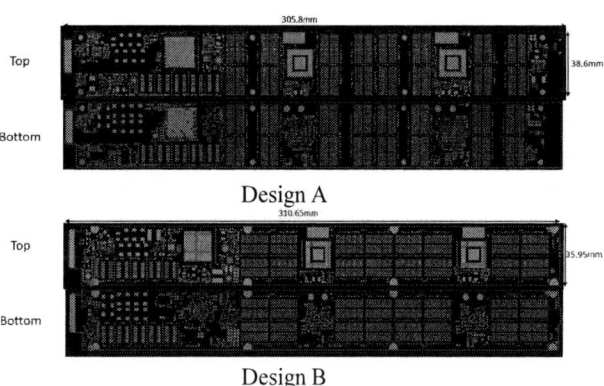

Figure 6: Solid State Drive (SSD) ruler form factor

B. Component and PCB information

There are 16 NAND packages, 3 FC controller, and 10 DDR SDRAM mounted on top and bottom of SSD ruler form factor. Other types of BGA package such as temperature sensor and microcontroller are also considered. Each design completed two cycles, top and bottom, of surface mount technology (SMT) process. The detail BGA component information are listed in Table 1. Solder joint diameters and standoff height are determined and as described in previous section.

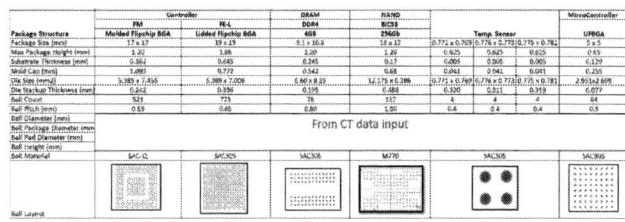

Table 1: Component Information

The equivalent material property based on 16-layer PCB was imported from ODB++ into Finite Element based reliability evaluation software, as shown in Table 2. The thickness, material type, and copper percentage of each layer used to generate the coefficients of thermal expansion (CTE) and Young's modulus for overall board behavior are shown in Table 2.

	Board Dimension: 306 x 39 mm [12.0 x 1.5 in]	CTExy: 16.786 ppm/C	Board Weight: 46.44 grams
	Board Thickness: 1.512 mm [59.5 mil]	CTEz: 38.625 ppm/C	Total Part Weight: 36.75 grams
	Density: 2.6471 g/cc	Exy: 32,114 MPa	Mount Point Weight: 0 grams
	Conductor Layers: 16	Ez: 4,264 MPa	Fixture Weight: 0 grams

Layer	Type	Material	Thickness	Density	CTExy	CTEz	Exy	Ez
1	SIGNAL	COPPER (24.9%) / COPPER-RESIN	36.8 micr..	3.5679	41.932	41.932	30,766	30,766
2	Laminate	Generic FR-4	57.7 micr..	1.9000	15.000	40.000	24,804	3,450
3	POWER	COPPER (87.3%) / COPPER-RESIN	19.5 micr..	7.9983	21.715	21.715	99,094	99,094
4	Laminate	Generic FR-4	73.2 micr..	1.9000	15.000	40.000	24,804	3,450
5	SIGNAL	COPPER (7.7%) / COPPER-RESIN	16.5 micr..	2.3467	47.505	47.505	11,932	11,932
6	Laminate	Generic FR-4	99.8 micr..	1.9000	15.000	40.000	24,804	3,450
7	SIGNAL	COPPER (10.7%) / COPPER-RESIN	15.2 micr..	2.5597	46.533	46.533	15,216	15,216
8	Laminate	Generic FR-4	76.2 micr..	1.9000	15.000	40.000	24,804	3,450
9	POWER	COPPER (87.5%) / COPPER-RESIN	15.2 micr..	8.0125	21.650	21.650	99,312	99,312
10	Laminate	Generic FR-4	75.7 micr..	1.9000	15.000	40.000	24,804	3,450
11	SIGNAL	COPPER (24.7%) / COPPER-RESIN	15.2 micr..	3.5537	41.997	41.997	30,546	30,546
12	Laminate	Generic FR-4	101.6 mic..	1.9000	15.000	40.000	24,804	3,450
13	POWER	COPPER (87.5%) / COPPER-RESIN	15.2 micr..	8.0125	21.650	21.650	99,312	99,312
14	Laminate	Generic FR-4	72.6 micr..	1.9000	15.000	40.000	24,804	3,450
15	POWER	COPPER (80.7%) / COPPER-RESIN	30.5 micr..	7.5297	23.853	23.853	91,866	91,866
16	Laminate	Generic FR-4	76.2 micr..	1.9000	15.000	40.000	24,804	3,450
17	POWER	COPPER (83.1%) / COPPER-RESIN	30.5 micr..	7.7091	23.076	23.076	94,494	94,494
18	Laminate	Generic FR-4	72.8 micr..	1.9000	15.000	40.000	24,804	3,450
19	POWER	COPPER (87.5%) / COPPER-RESIN	15.2 micr..	8.0125	21.650	21.650	99,312	99,312
20	Laminate	Generic FR-4	101.6 mic..	1.9000	15.000	40.000	24,804	3,450
21	SIGNAL	COPPER (24.7%) / COPPER-RESIN	15.2 micr..	3.5537	41.997	41.997	30,546	30,546
22	Laminate	Generic FR-4	75.7 micr..	1.9000	15.000	40.000	24,804	3,450
23	POWER	COPPER (87.5%) / COPPER-RESIN	15.2 micr..	8.0125	21.650	21.650	99,312	99,312
24	Laminate	Generic FR-4	76.2 micr..	1.9000	15.000	40.000	24,804	3,450
25	SIGNAL	COPPER (10.7%) / COPPER-RESIN	15.2 micr..	2.5597	46.533	46.533	15,216	15,216
26	Laminate	Generic FR-4	99.8 micr..	1.9000	15.000	40.000	24,804	3,450
27	SIGNAL	COPPER (7.7%) / COPPER-RESIN	16.5 micr..	2.3467	47.505	47.505	11,932	11,932
28	Laminate	Generic FR-4	73.2 micr..	1.9000	15.000	40.000	24,804	3,450
29	POWER	COPPER (87.3%) / COPPER-RESIN	19.5 micr..	7.9983	21.715	21.715	99,094	99,094
30	Laminate	Generic FR-4	57.7 micr..	1.9000	15.000	40.000	24,804	3,450
31	SIGNAL	COPPER (24.9%) / COPPER-RESIN	36.8 micr..	3.5679	41.932	41.932	30,766	30,766

Table 2: PCB Information

C. Test condition

The thermal cycling test (TCT) conditions applied in the present study are the same as JESD22-A104 Standard. The temperature profile in Fig. 7 shows a range from 0 to 100 °C at 40 min per cycle, with 10 min dwell time at the extreme temperatures and 10 °C/min ramp rate.

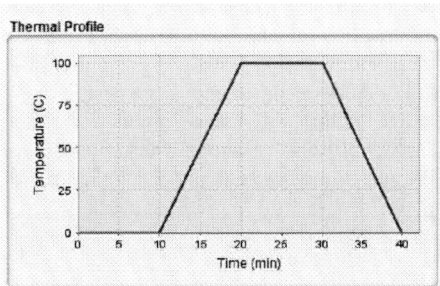

Figure 7: Temperature Cycle Profile

D. Importing solder joint geometry into a Finite Element based reliability evaluation

Normally, the first failure of solder fatigue is known to have occurred at the peripheral ring in the corner solder joints with the maximum distance to neutral point (DNP). Critical location among four corner joints was used to generate creep strain energy. Then, the creep strain energy used to predict cycles to failure. Many fatigue laws [4] have been developed to estimate the low-cycle fatigue life of a solder joint and classified as shown in Fig. 8.

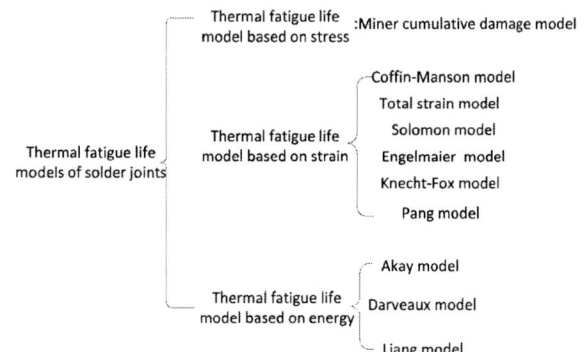

Figure 8: Thermal fatigue law of solder joints [4]

Fatigue laws were required to predict the number of cycles to fail when provided with either stress, strain or energy. Among these models, Darveaux's or Liang's model was widely accepted. Liang et al. [5] have developed a fatigue life prediction methodology that accounts for the geometry of the solder joint based on elastic and creep analyses. The fatigue life is calculated on an energy-based fatigue failure criterion and is shown in (4).

$$N_f = C(W_{ss})^{-m} \tag{4}$$

Where W_{ss} is the Energy density under the hysteresis loop of stress and strain, determined based on hysteresis loop; C and m are material constants.

Typically, failure mode related to mechanical shock can cause pad cratering issue and intermetallic fracture on solder joint. In this paper, simple spring mass approximation was used to predict the board deflection during a shock event. Finite Element based reliability evaluation utilizes an implicit transient dynamic simulation. A shock event with half-sine pulse of 100G/6ms is transmitted through the mounting points into the PCBA, as shown in Fig. 9. Two different mount points was applied to Design A and Design B, respectively, as shown in Fig. 10. The resulting PCBA strains are extracted from the simulation results and used to predict robustness under shock conditions.

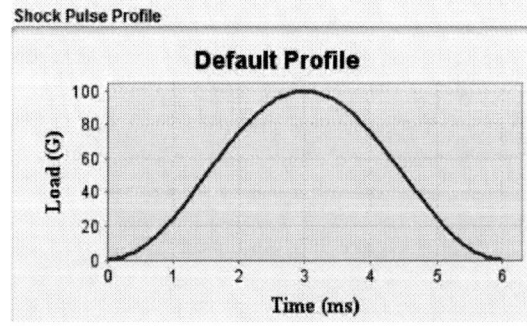

Figure 9: Mechanical Shock Profile

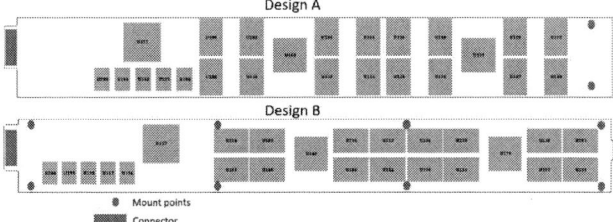

Figure 10: Mechanical Shock Profile

Vibration Fatigue is typically considered to be high cycle fatigue. The PCB displacement during vibration is modeled as a single degree of freedom system (spring mass approximation) using the natural frequency. Random vibration (0.012 G^2/Hz from 5 Hz to 1000 Hz) can be applied to Design A and Design B through the same mount points as shown in Fig. 10. Random vibration is a continuous spectrum of frequencies, and the profile is shown in Fig. 11. During vibration the board strain is proportional to the solder joint strains and therefore can be used to calculate the damage to failure. Fig. 12 shows that the summary of importing solder joint geometry into a Finite Element based reliability evaluation.

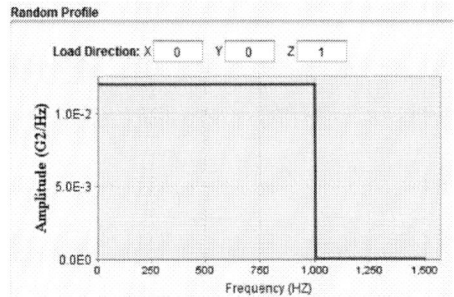

Figure 11: Vibration Profile

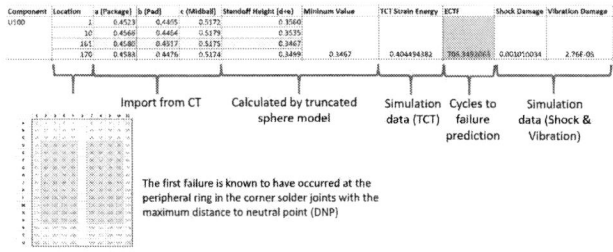

Figure 12: Importing solder joint geometry into a Finite Element based reliability evaluation

IV. RESULTS AND DISCUSSION

A. Validation

A cross section of a solder joint after SMT process was conducted to measure the standoff height. The definition of standoff height and location of NAND package are shown in Fig. 13. The measured values, plotted in Fig. 14, show inconsistence of standoff height. Then, the values are compared by solder joint shape calculation from CT data, as shown in Fig. 15. Both show similar trend and close correlation.

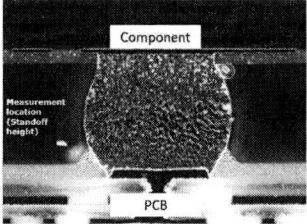

(a) Standoff height definition

(b) Location of NAND packages

Figure 13: Standoff height and location of NAND packages

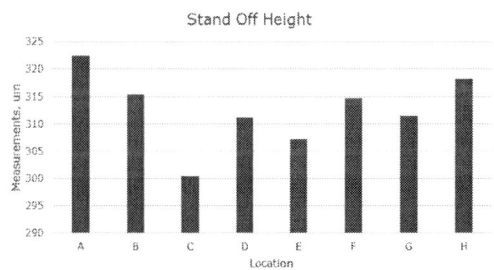

Figure 14: Measured standoff height values

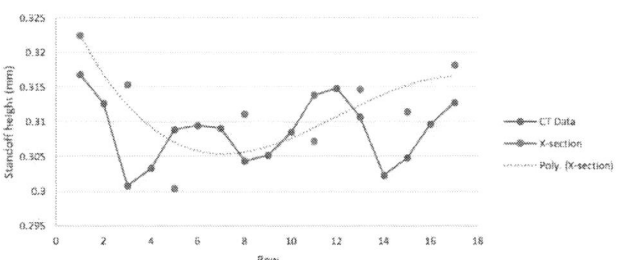

Figure 15: Comparison of CT and X-section data

Cross section analysis are one of many tools for approaching failures on a PCBA. Other techniques, like printed circuit board delayering or "dye-and-pry" (the use of a brightly colored dye to reveal any cracking or delamination on solder interfaces) might be more appropriate for a given failure mode, such as pad catering, delamination, and ball cracking. It is important, therefore, to validate the critical location among four corner joints. Fig. 16 shows an example of dye-and-pry for one NAND location. Good correlation between the critical locations with CT data showed that initial crack start from corner. Having developed a model that correlates well with the experimental data, the modeling methodology was used to analyze cycles to failure, worst location among components, critical location among four corner joints, and damage index of components.

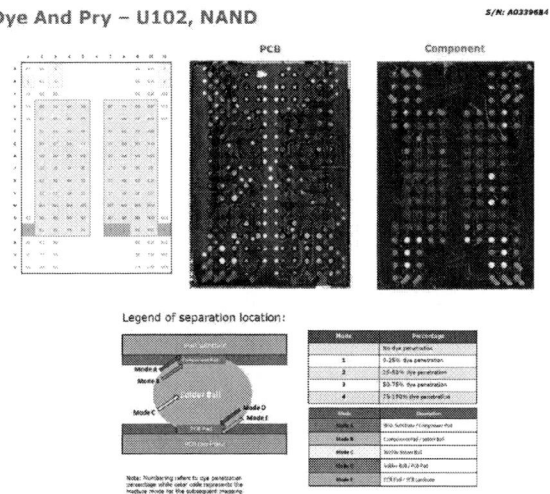

Figure 16: Dye-and-pry of NAND location

B. Analysis

Fig. 17 shows the top 10 worst locations for NAND packages and highlights the direction of initial crack start from corner location for Design A. The analysis shows that NAND packages on bottom side have low fatigue life compare to top side. Components on bottom side experiencing 2 times of SMT reflow compare to 1-time SMT reflow on top side, which accumulate more internal stress causing overall low fatigue life. Fig. 18 shows that cycles to failure for all NAND packages. The overall fatigue life of Design A was assumed as 1.0X cycles.

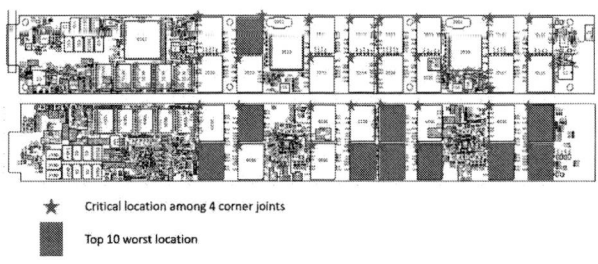

Figure 17: Top 10 worst locations for NAND packages (Design A)

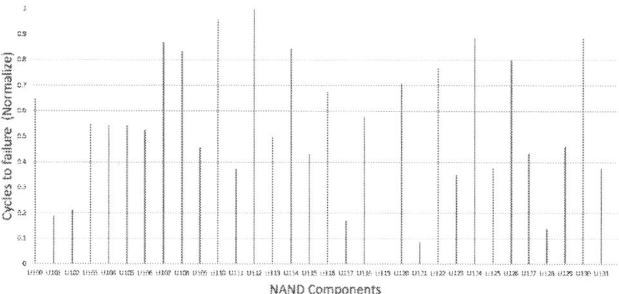

Figure 18: Fatigue life for NAND packages (Design A)

Some improvement was made to place the NAND packages horizontally, PCB routing design, and top and bottom reflow profile (double soak time and increase peak temperature by 5 °C). Besides, the die size of NAND package in Design B is smaller than Design A. Top 10 worst locations for NAND packages and the direction of initial crack start from corner location for Design B were highlighted, as shown in Fig. 19. 6 out of 10 worst locations are located on bottom side. Fig. 20 shows that cycles to failure for all NAND packages. The overall fatigue life is 1.6X cycles, which is was improved compared to Design A.

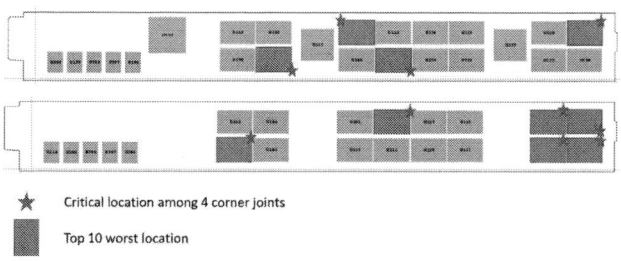

Figure 19: Top 10 worst location for NAND packages (Design B)

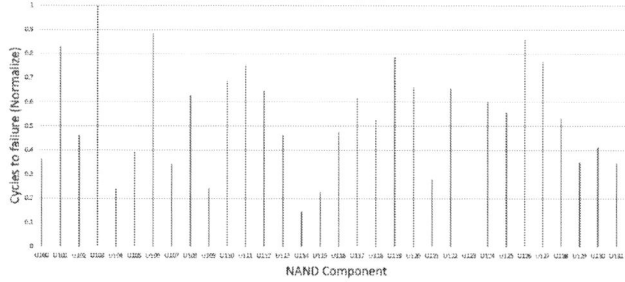

Figure 20: Fatigue life for NAND packages (Design B)

The methodology was extended to reliability evaluation for all BGA packages such as FC controller and DDR SDRAM. Top 5 worst locations for DDR packages and the direction of initial crack start from corner location for Design B are highlighted, as shown in Fig. 21. 3 out of 5 worst locations are located on bottom side. Fig. 22 shows cycles to failure for all DDR packages. The overall fatigue life is 2.1X cycles, which is longer compared to NAND packages.

DDR

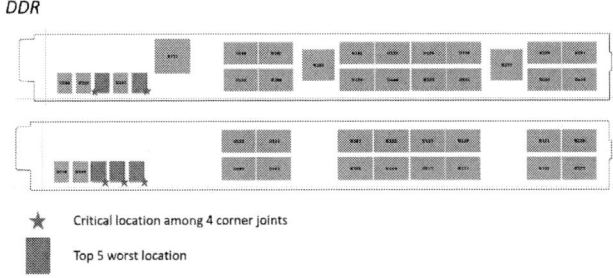

★ Critical location among 4 corner joints

▮ Top 5 worst location

Figure 21: Top 5 worst location for DDR packages (Design B)

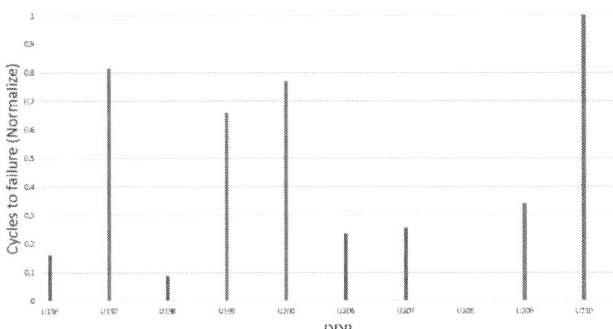

Figure 22: Fatigue life for DDR packages (Design B)

Besides, top worst location for FM Controller and the direction of initial crack start from corner location for Design B were highlighted, as shown in Fig. 23. Fig. 24 shows cycles to failure for all FM Controller. The fatigue life is 3.6X cycles, which is longer compared to DDR and NAND packages.

FM Controller

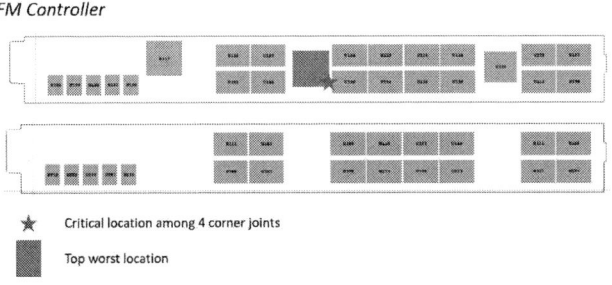

★ Critical location among 4 corner joints

▮ Top worst location

Figure 23: Top worst location for FM Controller (Design B)

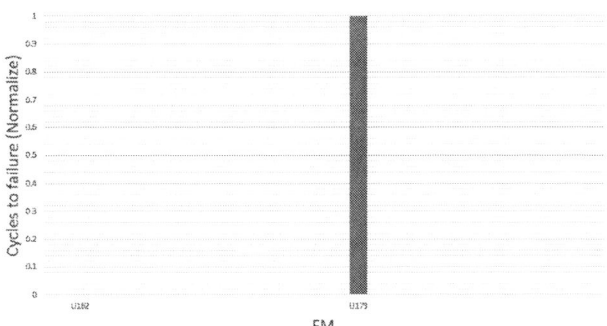

Figure 24: Fatigue life for FM Controller (Design B)

Table 3 shows overall cycles to failure of NAND packages for both designs. Design B showed the demonstration of smaller die size of NAND package and alternate reflow profile for top and bottom SMT. The fatigue life was improved around 60%. Moreover, Table 4 shows overall cycles to failure based on different type of BGA on Design B. The results clearly illustrates that different type of BGA component has different number of cycles to failure. Compared with experimental data, the predicted fatigue life of solder joints was within acceptable error. FM controller shows the best fatigue life on SSD ruler form factor. Further improvement is needed for NAND packages to achieve better reliability.

Type	Design A	Design B
Cycles to failure	1.0X	1.6X

Table 3: Overall NAND TCT

Type	NAND	DDR	FM	FE-L
Cycles to failure	1.6X	2.1X	3.6X	3.0X

Table 4: Overall TCT based on different type of BGA on Design B

Finite Element based reliability evaluation of shock and vibration issues is necessary to adequately capture the complex solder joint shape, mounting configurations, and response of PCBA. Shock damage index for Design A is much higher throughout the whole component on PCB by two order of magnitude compared to Design B, as shown in Fig. 25. With additional mount points in Design B, the shock damage index is significantly reduced. Besides, only components near the mount points in Design B indicated certain amount of damage index. In addition, vibration damage index for Design A is higher throughout the whole component on PCB by one order of magnitude compare to Design B, as shown in Fig. 26. The same observation has been made with vibration conditions that the additional mount points help reduce the damage index of components.

978-1-7281-1500-9/19 $31.00 © 2019 IEEE

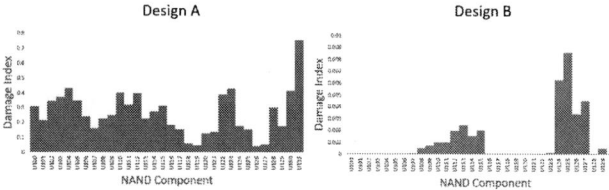

Figure 25: Shock damage index for Design A and Design B

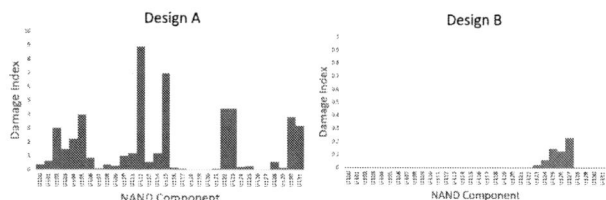

Figure 26: Vibration damage index for Design A and Design B

Displacement plot demonstrates critical board bending condition associated with shock and vibration. Fig. 27 shows the shock displacement for Design A and Design B. Fig. 28 shows the vibration displacement for Design A and Design B. Both plots show the mounting configurations played important role in bending displacement associated with shock and vibration. This approach greatly reduces product development cycle, and gives guideline to address the subject of Design for Reliability (DfR).

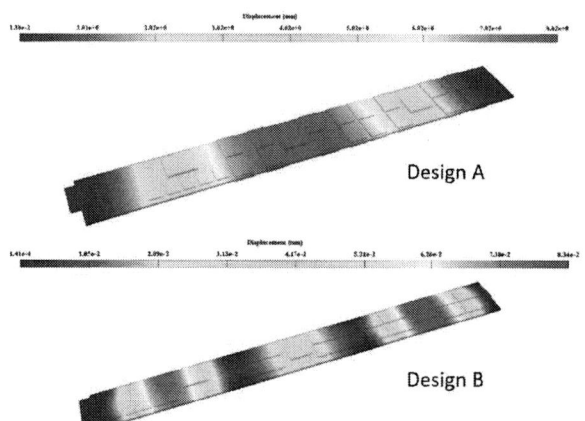

Figure 27: Shock displacement plot

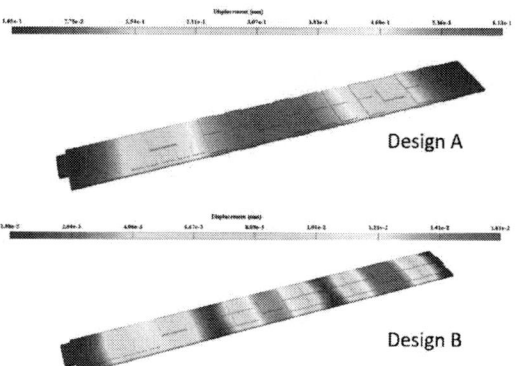

Figure 28: Vibration displacement plot

V. CONCLUSION

A strategy for importing CT data into a Finite Element based reliability evaluation is outlined in this study. Then, Finite Element Analysis is used to calculate the accumulated energy density for every BGA packages on PCBA. The accumulated energy density is then used to calculate the number of cycles to failure based on thermal fatigue life model of solder joints and compared with experimental data with acceptable error. The fatigue life was improved by 60% on Design B with a combination of smaller die size of NAND package and alternate reflow profile for top and bottom SMT. The methodology was extended to reliability evaluation for all BGA packages such as FC controller, DDR SDRAM, and NAND packages on PCBA. Further improvement is needed for NAND packages to achieve better reliability. Finite Element based reliability evaluation of shock and vibration condition is necessary to capture the complex solder joint shape, mounting configurations, and damage index of PCBA. Results demonstrate the excellent capability of the proposed integrated approach for predicting the robustness of PCBA in three aspects: temperature cycling, mechanical shock, and vibration. The proposed approach greatly reduces reliability evaluation time, shortens product development cycle, and is more cost effective to address reliability issues.

ACKNOWLEDGMENT

The authors would like to thank the Package Engineering Team and Reliability Team at Western Digital Corporation for their valuable technical support.

REFERENCES

[1] S. Gondrom, J. Zhou, M. Maisl, H. Reiter, M. Kröning, and W. Arnold, "X-ray computed laminography: an approach of computed tomography for applications with limited access," Nuclear Engineering and Design, vol. 190, pp. 141–147, 1999.

[2] S., B. Richard and N. Devendra, "Solder Joint Shape and Standoff Height Prediction and Integration with FEA-Based Methodology for Reliability Evaluation", Electronic Components and Technology Conference, February 2002, pp. 1739-1744, doi: 10.1109/ECTC.2002.1008345.

[3] H. A. Emeka and N. E. Ndy, "High temperature reliability of lead-free solder joints in a flip chip assembly" Journal of Materials Processing Technology, vol. 212, pp. 471–483, October 2012.

[4] X. Li, R. Sun and Y. Wang, "A review of typical thermal fatigue failure models for solder joints of electronic components" Materials Science and Engineering, vol. 242, 2017.

[5] W. W. Lee, L. T. Nguyen and G. S. Selvaduray, "Solder joint fatigue models: review and applicability to chip scale packages" Microelectronics Reliability, vol. 40, pp. 231-244, 2000.

Multi-physics modelling and Experimental Investigation – An original approach for Laser-dicing/grooving process optimization

Jeff. MOUSSODJI, Dominique DROUIN
3IT, Université de Sherbrooke,
Sherbrooke QC, Canada
jeff.moussodji@usherbrooke.ca,
dominique.drouin@usherbrooke.ca

Oswaldo CHACON, Francis SANTERRE
IBM Canada Ltd
Bromont QC, Canada
ochacon@ca.ibm.com, fsanterr@ca.ibm.com

Abstract—**The highly complex technology requirements of today's integrated circuits (ICs), lead to the increasingly use of several materials types such as metal structures, brittle dielectrics, porous low-k and ultra-low-k materials which are used in both front-end-of-line (FEOL) and back-end-of-line (BEOL) process for wafer manufacturing. In order to singulate chips from wafers, a critical laser-grooving process, prior to blade dicing, is used to remove these layers of materials out of the dicing street. The combination of laser-grooving and blade dicing allows to reduce the potential risk of induced mechanical defects such micro-cracks, chipping, on the wafer top surface where circuitry is located. Nevertheless, challenges related to unexpected drawbacks on process such as efficiency, quality and reliability still remain. To maximize control of this critical process and reduce its undesirable effects, numerical models of nanosecond laser pulsed and multi-stack material interaction have been developed. The modeling strategy using finite elements formalism is based on the convergence of two approaches, numerical and experimental Validation. To evaluate this interaction, several laser grooved samples were performed using IBM 14 nm technology node wafer and were correlated with finite elements modeling. Three different aspects were studied; phase change, thermo-mechanical and optical sensitive parameters. The numerical model makes it possible to simulate groove profile (depth, width, etc.) of a single pulse or multi-pulses on BEOL wafer material. Moreover, the heat-affected zone (HAZ) has been estimated as a function of laser operating parameters (power, frequency, spot size, defocus, speed, etc.). After modeling validation and calibration, a reasonable agreement between experiment and modeling results has been observed in terms of groove depth, width and HAZ.**

Keywords-Laser-dicing; nano-second pulsed laser; wafer back-end-of-line; multi-layer; multi-stacks, multiphysics modeling

I. INTRODUCTION

The wafer back-end-of-line (BEOL) materials currently face significant advanced changes as semiconductor technology nodes steps down to 45 nm, 22 nm, 14 nm, and more advanced nodes. This scale down trend in technology nodes (increased transistors and interconnect density), coupled with the rapid shrinking of integrated circuit critical dimensions have created great challenges on BEOL material such as thermo-mechanical properties and associated reliability [1]. Moreover, the wafer BEOL region mainly consists on brittle and porous materials such as low-k, ultra-low-k and interlayer dielectrics (ILDs), which can easily fail when the silicon wafer is exposed to high thermo-mechanical

stresses. These fragile materials are susceptible to induced damages from manufacturing processes, which could become a reliability concern [2, 3]. Thus, conventional techniques such as single diamond blade dicing process is not an adequate solution for high quality process and yield requirements due to the potential risk of mechanical defects on the BEOL layers. Therefore, alternative technique such as laser-grooving, prior to blade dicing is currently used to remove these layers of fragile materials out of the dicing street in the semiconductor industry [4]. Indeed, laser grooving is a powerful tool as it allows more benefits over traditional diamond blade process such as high flexibility and excellent groove profile quality on wafer BEOL materials. The combination of laser-grooving and blade dicing will reduce the potential risk of induced mechanical defects (micro-cracks, chipping etc.) on the wafer top surface where circuitry is located [4]. However, previous experimental observations have shown some drawbacks related to undesirable effects such as layer and sidewall delamination, nucleation of micro cracks, ILD damages, chipping induced by debris and recast which occur during the nanosecond laser grooving process [5, 6]. It seems therefore essential to have a fundamental understanding to address the different mechanisms involving in the nanosecond laser-grooving step. This study will allow for an optimum processing of fragile materials to reduce the thermo-mechanical damages and improve the reliability. From the production point of view, optimizing the laser process requires repetitive and time-consuming steps for designing trail runs, executing (hardware intensive) and analyzing results to find a process window that will eliminate or at least minimize the damages on BEOL fragile materials. While, numerical modeling could allow to investigate laser parameters impact on kerf without tedious experiments. However, previous numerical works were mostly focused on different mechanism surrounding laser-mater interaction such as ablation, welding, cutting or drilling a single layer sheet [7, 8]. In this context, the purpose of this paper is to investigate the potential implementation of a numerical tool, able to accurately predict physical phenomena involved in the laser-matter ablation in a production environment. This application could highlight the main physical mechanisms such as heat distribution in multilayered structure, the prediction of the vapor/liquid front, the HAZ area, shape and profile of the resulting grooves for a single or multiple pulses and for any custom laser pattering. To reach this goal, a numerical model describing the various processes related to the laser-matter interaction has been developed using a multilayered structure with different

978-1-7281-1500-9/19 $31.00 © 2019 IEEE

materials and associated physical properties. This original approach consisting of numerical and experimental approaches, allowed to investigate the interaction of the nanosecond pulsed laser and BEOL wafer materials. To evaluate this interaction, several laser-grooved samples were compared with finite element modeling, in which three different aspects, phase change, thermo-mechanical and optical sensitive parameters were considered. This modeling makes it possible to highlight a groove profile (depth, width, etc.) of a single pulse or multi-pulses on BEOL wafer materials. Moreover, the heat affected zone, and thermo-mechanical stress can be predicted as a function of laser operating parameters (power, frequency, spot size, defocus, speed, etc.).

This study will focus on short-pulsed regime in which a nanosecond laser is used to process kerf on wafer BEOL materials. In such process, a large number of mechanisms coexist; the thermal, the fluid and gas flow, the vaporization, the recoil pressure generated by the melting of the metal, the surface tension that leads to the Marangoni convection. Indeed, in the molted part of material, strong convection movements appear due to the Marangoni effect and could have a significant influence on the heat transfer. Its origin is related to the surface tension gradients induced by the gradients of temperature along the free surface [9, 10]. Moreover, this type of problem is strongly nonlinear since most thermo-physical parameters, such as thermal conductivity, fluid density, specific heat, latent heat are temperature dependent and present a more or less abrupt change in their values during the solid-liquid phase change of materials. For all these reasons, modeling nanosecond laser-matter interaction may be challenging.

As a first step, we present a design of experiment that was performed to assess the impact of critical parameters influencing the geometry of the groove crater. We will show how this method has been optimized to obtain best conditions for removing material of all structure while minimizing the HAZ and preserving the integrity of the structure, which was not exposed to laser. In the second step, a description of the modeling strategy and the most important parameters to improve the laser grooving, will be highlighted. The discussion around the correlation between experiment and modeling results in terms of groove depth, width and HAZ will be presented in the last section.

II. MOTIVATIONS

A typical nanosecond laser-grooving process consists of several nanosecond pulses at high frequency (f ~ 200 kHz) and displaced in space with a constant velocity V_{tr}. To maximize the spatial overlapping among pulses, the laser radiation is focused by a set of lenses, conferring to the beam the shape of an elongated ellipse (Fig. 1), along the grooving direction. This results in the formation of a deep crater, which can directly remove the BEOL materials (thicker than ~10 to15 um, depending on the technology) (Fig. 2).
The motivation to develop a numerical model for the laser beam in a kerf grooving application arises when one must go through a repetitive process of creating customized recipes for new die technology nodes. This is a time consuming and expensive process to meet adequate parameters. The goal of

this proposed modeling is to accelerate the convergence to set of parameters that can further be validated with experimental measurements.

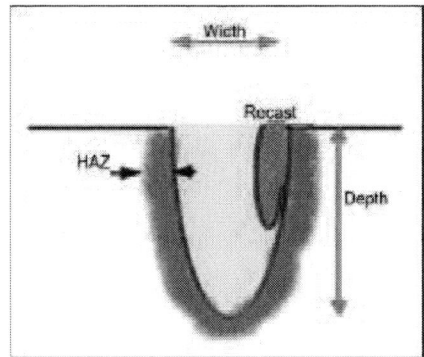

Figure 1 Laser grooving profile and HAZ characteristics

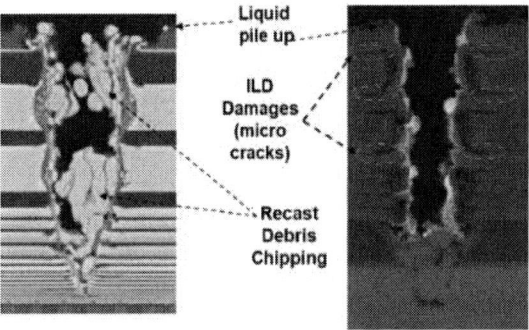

Figure 2 ILD Damages (post grooving)

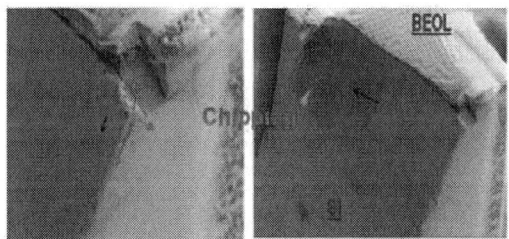

Figure 3 IC corner damage (post dicing)

For a new technology introduction, it is essential to find the best grooving conditions to:

- Reduce Heat Affected Zone (HAZ), which may generate inter layer damages (ILD) that may later expose to reliability problems (Fig. 2)

- Reduce recast material that is more fragile than BEOL materials. These recasts may generate damage on the die sidewall (cracks or corner chip out) during the blade dicing step that follows laser grooving. This can results on yield or reliability issues (Fig. 3)

III. EXPERIMENTAL APPOACH

To facilitate the development of numerical model, experimental data is a benefit not always available. However, the availability of experimental data speeds up the model creation process. The challenge then becomes, to mathematically represent laser beam characteristics close enough for a good correlation study and allow the team to build trust on this new development tool.

In addition, with the experimental data collected in the development phase, the team should be able to confirm and identify the main critical parameter set for the laser grooving process that have the most impact. This, combined with the measures from cross-section analysis, yields the correlation dataset needed.

A. Experimental data (DOE)

A design of experiment (DOE) investigating three critical parameters: Power, Feed Speed and Defocus are required to study the laser grooving process, as these parameters are inter dependent and affect other associated parameters such as laser frequency, step, overlap, FWHM, Spot size etc. A full set of experiments is the safest way to gain a good insight in our process. It was decided to perform the full 3^3 factorial matrix (27 different cells and cross-section images).

TABLE 1   CRITICAL PARAMETERS FOR LASER GROOVING APPLIED IN EACH CELL

Cell ID	Power	Feed Speed	Defocus
Cell 1	Low	High	A
Cell 2	Low	Med	B
Cell 3	Med	Med	B
Cell 4	High	Med	B
Cell 5	High	Low	C

Table 1 presents a summary of the most interesting cell results as well as the critical parameters for laser that were used during the grooving step. Cells 1 & 2 are the lowest power combinations while cell 3 represents the middle range power setting case. Cells 4 & 5 are the highest power setting case. In this experimental study, the groove profile (Depth and Width) represents the response variables as well as the HAZ and recast.

B. Major challenges for a laser groove process

The stack material thickness of dicing street and layer composition play a major role on the performance response in the laser grooving process. Uniform dicing street contents are not always possible due to design and test requirements for BEOL. For instance, laser grooves will need to penetrate different types of material (metal only, oxide only, or mixed metal oxide region). Indeed, laser processing different type of metals or oxides leads to different associated thermo-mechanical and optical responses due to the difference in

thermo-mechanical and optical properties of each material. Therefore, the laser groove recipes to apply will have to process effectively: the worst-case kerf region contents (the ones that need the most aggressive energy parameters), but also, minimize the impact on more fragile regions, which are close to these hard to groove regions. The choice of laser grooving parameters is then specific to die technology and will need to be customized (or at least confirmed) per wafer product.

For the IBM 14 nm technology node, the DOE experimental results yielded different but interesting results in the investigation for the best set of grooving parameters. On a given wafer, there are different material with different thermos-mechanical and optical properties on the dicing street to be grooved. For our study, we focus on regions with diametrically opposite behavior (with respect to the laser parameters) to enhance the difference in material properties as shown in Figure 4 with aluminum pads and copper lines (left) and oxide only region (right).

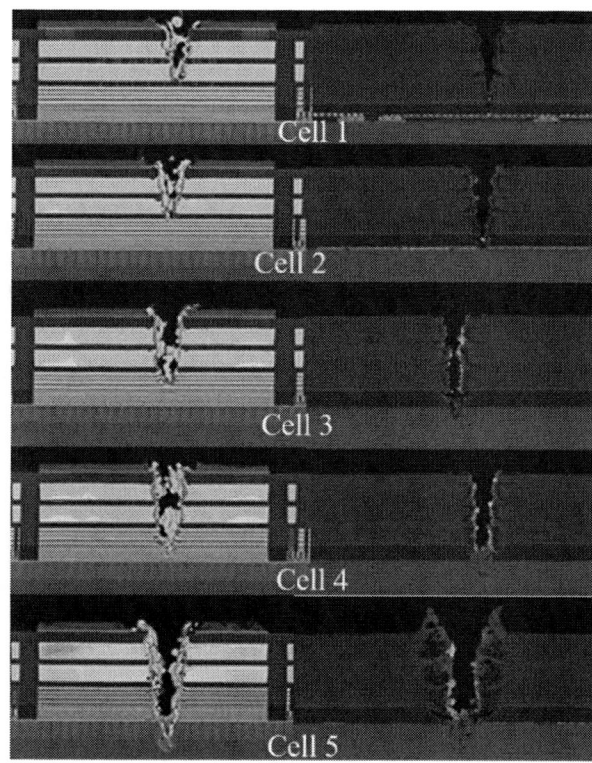

Figure 4. Experimental cross-section images of grooves in oxide only region (right) Vs Al/Cu region (left)

Qualitative analysis of the resulting groove profile in terms of depth, width , HAZ, and recasts after performing nanosecond laser grooving with different set of laser operating parameters in both regions have been illustrated in the cross-section of the Figure 4. These results allowed to monitor the evolution of each critical output function of the laser power, the feed speed and the defocus. This study will allow to identify the best set of laser operating parameters in this experimental investigation. The cross section of oxide only Figure 2 (right) and aluminum pads Figure 2 (left) illustrate the difference on the heat propagation and the groove shape between both regions.

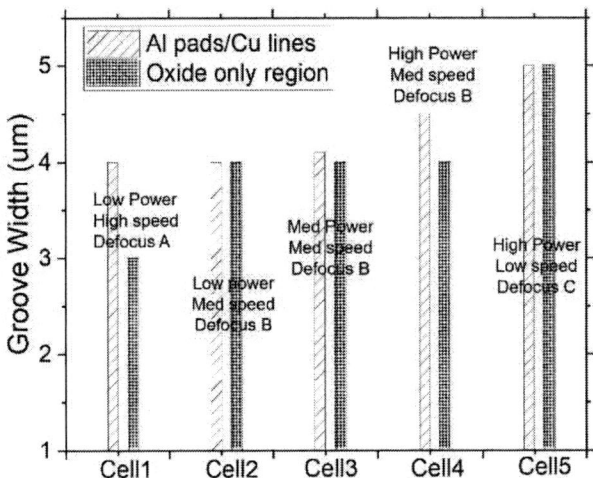

Figure 5 Measurement result of groove penetration depth in each cell in oxide only region Vs aluminum pads/Cu lines

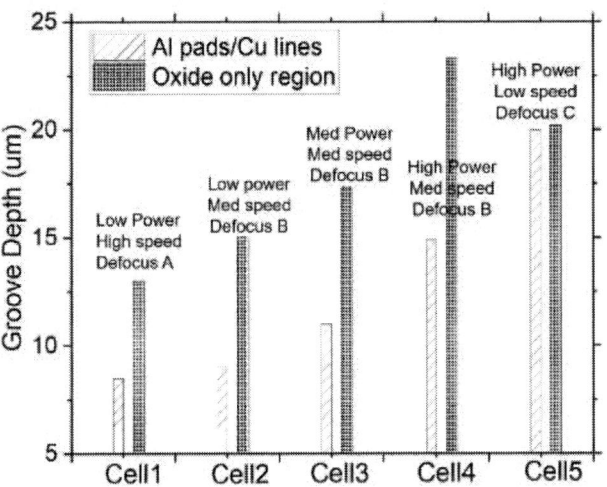

Figure 6 Measurement result of groove width in each cell in each cell in oxide only region Vs aluminum pads/Cu lines

Other points of interest are the extreme conditions for the parameter set range for cells 4 & 5 in Figure 4. We can see that there is a major difference in material behavior and some conclusions can be drawn on their performance for each region of interest. We can see that minimum parameter set values (Figure 4, cell 1) will not allow for a full material penetration of the laser beam on any of the regions of interest. At the other end, using extreme parameters set (Figure 4 cell 5), we can see that, this grooving parameter set allowed us to see a better material penetration in all regions but at the price of increasing the HAZ, especially in the oxide region that increase the risk of micro-crack /damage nucleation. In the experimental matrix, we have achieved the main goal of fundamental understanding the choice of grooving critical parameters and their effects on the thermo-mechanical behavior of BEOL materials. Experiential results show that the choice of laser grooving parameters is highly dependent on the material regions to process.

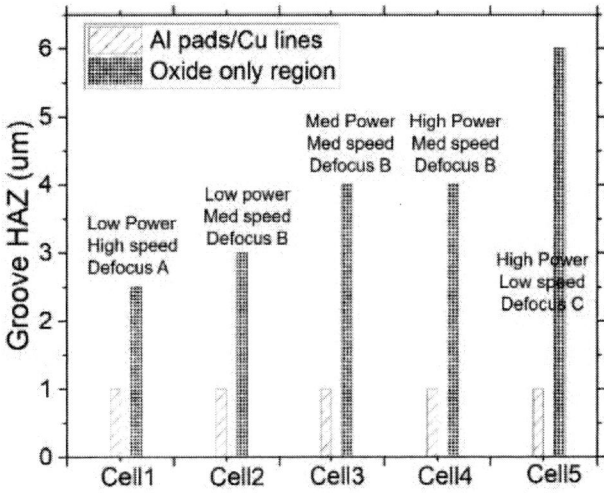

Figure 7 Measurement result of groove HAZ in each cell in oxide only region Vs aluminum pads/Cu lines

Measurements for a given set of parameters were taken on each region to compare laser grooving behavior as shown in the Figures 5, 6, 7 and 8. The Figures 5 & 6 illustrate the groove width and depth respectively of both aluminum pads and oxide only region in each cell, while the Figures 7 & 8 show the HAZ characteristic and the recast surface. These experimental data show the strong dependence of groove depth on the laser power especially in the case of cells 2, 3 and 4 (Fig. 6), where speed and defocus are set constant. However, the experimental results show that, the effect of speed and defocus on the resulting kerf characteristics should be taken into account. These experimental data allowed to select the best set of parameters that allows to process both aluminum pads and oxide only region. As we can see on the cross-section images of cells 1 & 2 in Figure 4, there is no penetration of the beam in the large metal region yielding incomplete grooving. As per the oxide region, we can see a marginally good groove as it penetrates the full layer stack. We can see here from the evolution of the parameter set in the DOE that we are approaching a solution that will be able to meet all BEOL conditions.

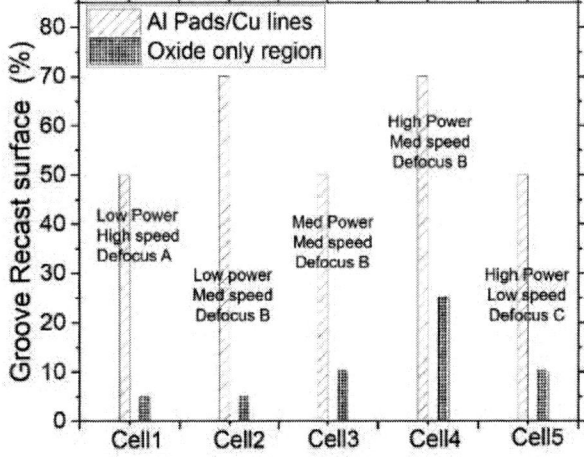

Figure 8 Measurement result of groove Recast surface in each cell for

Ideally, a uniform kerf BEOL material layout would benefit the laser grooving parameter optimization exercise but, in real life applications, it is not practical. A compromise parameter is required in order to successfully process the kerf regions. This search for the optimum parameter set may be long and tedious and therefore, justifying the development of numerical modeling.

IV. NUMERICA APPROACH

The following model has been implemented using finite element formalism with COMSOL multiphysics® Software. As mentioned above, the problem is multi-physics and requires the coupling of heat transfer and fluid flow modeling. The modeling strategy lies on a two-phase effort. In a first step, a 2D Cartesian approach allows a good approximation of groove profile and HAZ characterization. As a first approximation, to prevent mesh distortions responsible for poor accuracy, the free surface evolution is described by a moving mesh method, an Arbitrary Lagrangian Eulerian (ALE) method [11], by imposing the recoil pressure and the energy deposition as boundary conditions [12]. At each time step, a steady hyper-elastic problem is solved by propagating the moving boundary displacement throughout the domain to obtain a controlled mesh deformation. Specific boundary conditions are used to control these mesh displacements. As from the Lagrangian point of view, the normal component of the mesh velocity (u_{mesh}) at the interface is equal to the normal fluid velocity. For the lateral conditions, the mesh is free to move vertically. Concerning the lowest boundary, the mesh nodes are stationary. The validation and calibration of this light-matter interaction 2D modeling can be found in J. Moussodji previous work [13].

The more complex 3D approach, which includes more complex geometry, fluid displacement, recrystallization, recast and mechanical behavior is currently under development, but will drastically increase the computation time. In the following, a brief mathematical model development will be described, pointing out which particularities have been added regarding the standard modeling. Aspects concerning thermal, laser-grooving and micro fluidics will be briefly introduced.

A. Thermal problem

The properties of all the materials used in the model are temperature dependent. Materials with different thermal conductivity (thermal diffusivity) were modeled. The heat conductivity, the specific heat and mass density for metal and oxide as a function of temperature are given in [14, 15] and the third polynomial fitting of their thermal conductivity is given by:

$$k = B_0 + B_1T\ B_2T^2 + B_3T^3 \tag{1}$$

Where B_0, B_1, B_2, and B_3 are the polynomial fitting parameter and T the temperature. The theoretical equation of the specific heat of silicon, aluminum and copper is given as

$$C_p = A_0 + A_1T\ A_2T^2 + A_3T^3 \tag{2}$$

Where A_0, A_1, A_2, and A_3 are the polynomial fitting parameter and T the temperature. In this context, the distribution of heat within the material is governed by the heat equation.

$$\rho C_p \frac{\partial T}{\partial t} + \rho C_p u.\nabla T = \nabla.(k\ \nabla T) + Q_i \tag{3}$$

Where ρ is the mass density, C_p is the specific heat capacity, k the heat conductivity, T the temperature, u the fluid velocity vector and Q_i the volumetric heat source. The specific heat source taking into account the latent heat of fusion for the solid/liquid phase transition is given by

$$C_p = C_p^*(T) + \delta C_p * L_f \tag{4}$$

Where C_p is the temperature heat capacity L_f the latent heat of fusion. With ΔT the temperature range and T_{fusion} the melting point of material.

$$\delta C_p = \frac{1}{\Delta T\sqrt{\pi}}e^{-(\frac{T-T_{fusion}}{\Delta T})^2} \tag{5}$$

B. Laser interaction problem

In this work, two laser sources parameters are considered, consisting on two fluxes: the first flux is due to the absorption of lasers beam in semi-transparent materials and can be expressed according to a periodic Gaussian distribution as follows:

$$Flux_{laser} = \frac{Abs}{\pi(\frac{w(z)}{2})^2} * e^{-\frac{(x^2+((y-source\ (t))^2}{(\frac{w(z)}{2})^2}} \tag{6}$$

With P_{Laser} the laser power, Abs the equivalent absorptivity depending on the irradiated material, w (z) is the radius of the expanding laser beam given by the following expression

$$w(z) = w_0 * \sqrt{(1 + (\frac{z-zfoc}{w_0})^2)} \tag{7}$$

Where w_0 is the radius of focused laser beam and zfoc is the distance between working piece surface and the plan of concentrating lens. The Source (t), a function of time related to the laser step between two successive laser beams and defined as follows:

$$source(t) = overlap * \frac{T}{period} \tag{8}$$

The second flux term is related to the evaporation and is defined by:

$$Flux_{evap} = h * (T_{vap} - T) \tag{9}$$

Where h is a sufficient constant which allows assuming that, the interface temperature remains close to the vaporization temperature. The net mass flux leaving the interface is expressed as follows:

$$\dot{m} = Flux_{evap} * L_v \qquad (10)$$

The grooving speed is therefore evaluated by

$$v_{vap} = \frac{q_{vap}}{\rho L_v} \qquad (11)$$

Where ρ the ablated material mass density.

C. Micro-fluidic problem

To evaluate the temperature value in this process and to characterize the melt pool, the conservation of energy, and momentum, Navier-Stockes equation are solved

$$\frac{\partial \rho}{\partial t} + \nabla . \rho \boldsymbol{u} = 0 \qquad (12)$$

$$\begin{aligned}
\rho \left(\frac{\partial \boldsymbol{u}}{\partial t} + (\boldsymbol{u}.\nabla)\boldsymbol{u} \right) \\
= \nabla P + \nabla . [\boldsymbol{\eta}(\nabla \boldsymbol{u} + \nabla \boldsymbol{u}^T)] \\
+ \rho g \beta (T - T_m)
\end{aligned} \qquad (13)$$

Where $\boldsymbol{\rho}$ (kg/m3) is the material density, P (N/m2) is the pressure, $\boldsymbol{u}$ (m/s) is the molten metal velocity, β (1/K) is the thermal expansion coefficient, η is the viscosity, g is the gravity acceleration and T_m is the melting temperature. At the liquid gas interface, recoil pressure is mainly responsible for the grooving. Indeed, when the laser beam with w (z) as a radius illuminates the working piece surface, the fusion, then the vaporization of the localized zone quickly occur due the very high-localized energy density. The vaporization is often accompanied by a recoil pressure, which acts on the liquid, inducing the deformation and the appearance of the groove. Normal surface tension forces (Laplace forces) counterbalance the effect of this pressure. Near the irradiated surface, Marangoni effects influence the enlargement of the groove size. Leaning on [16], the recoil pressure is modeled by adjusting the Clausius-Clapeyron law with the re-condensation rate.

$$p_{recoil} = \frac{1 + \beta_R}{2} P_0 e^{\frac{L_v M}{R T_{vap}}(1 - \frac{T_{vap}}{T})} \qquad (14)$$

In this work, a simplified expression will be used as a first approach with N, a numerical parameter

$$p_{recoil} = N e^{\frac{r^2}{r_0^2}} \qquad (14)$$

As we can see in this approach, three phases are present:

- The vapor, which mainly directed upwards, following the equation 10. So, the interaction between vapor and liquid is assumed limited during the ablation process

- The recoil pressure in the equation 14, governs the liquid, which flows around the groove in complex movements, during the process. Only the gravity and the surface tension are the force acting on the liquid flow when the laser beam pulse switches off. Note, in this first modeling approach, the liquid flow is neglected during the cooling.

- The solid, which the initial state of the working piece assembly is at room temperature. Note that, each layer in the working piece assembly is supposed to be homogeneous and isotropic regarding thermal analysis. The mechanical stress related to the difference of mechanical properties like the coefficient of thermal expansion (CTE) of each layer involved in the assembly is ignored and other mechanical properties are subjected to have zero effect in this first modeling approach.

V. CALCULATION VS EXPERIMENT: RESULTS AND DISCUSSION

We explored the resulting grooving profile with respect to width and depth, HAZ characteristics, liquid-vapor front, recast and debris inside bulk for different materials. Using this method, simulations have been performed using critical laser operating parameters of cell 2, cell 3 and cell 4 in which feed speed and defocus have been considered non-variable (Table 1), in order to reach two goals. The first goal was to input experimental operating laser conditions such as (power, frequency, overlap, spot size, FWHM), which could determine acceptable groove profile and HAZ characteristics. The second goal was to monitor the transient evolution of the grooved region geometry, to be able to validate the accuracy of the numerical model with this type of condition.

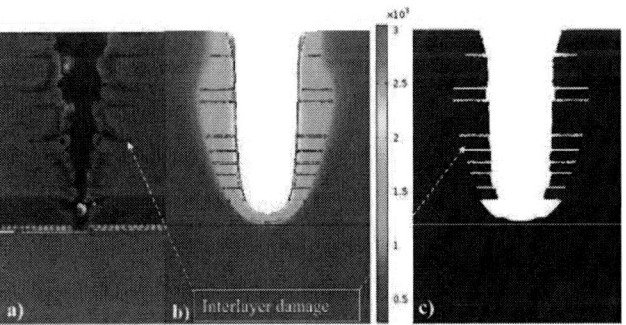

Figure 9 Cross section of cell 2 oxides only after nanosecond laser grooving; Comparison between a) the cross section image of the experimental groove, b) isthermal simulated groove taking into account the melt pool front without flowing, c)simulated groove assuming the melt pool is negligible. Fluence 5.3 J/cm2

To illustrate the laser power impact on groove geometry, HAZ, recast and the vapor liquid front, simulations using same laser operating parameters and same BEOL stacking have been performed. The model was designed to be able to input critical parameters of the laser grooving process and BEOL material multi-stack characteristics. In these simulations, both oxide only pads and aluminum pads/copper lines were used as samples to assess the modeling prediction versus experimental results. In the oxide only region, a relatively weak interlayer (ILD) material surrounds each dielectric layer. The aluminum The aluminum pads consist of a multilayered assembly of aluminum, copper, and oxide layers. Each layer is surrounded by the interlayer (ILD) material.

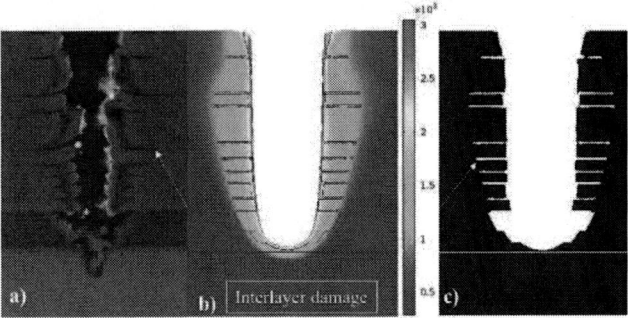

Figure 10 Cross section of cell 3 oxides only after nanosecond laser grooving; Comparison between a) the cross section image of the experimental groove, b) isthermal simulated groove taking into account the melt pool front without flowing, c)simulated groove assuming the melt pool is negligible. Fluence 7.03 J/cm2

Correlating experimental and numerical models was performed to calibrate and validate the precision of numerical application. For our application, a 10% deviation between experimental and modeling is acceptable for critical outputs such as depth, width, HAZ, lateral delamination appearance and recast surface approximation.

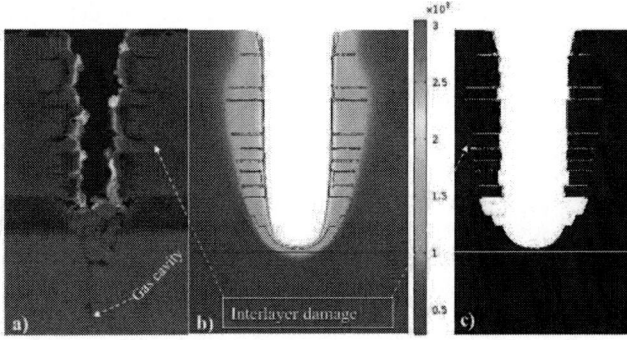

Figure 11 Cross section of cell 4 oxides only after nanosecond laser grooving; Comparison between a) the cross section image of the experimental groove, b) isthermal simulated groove taking into account the melt pool front without flowing, c)simulated groove assuming the melt pool is negligible. Fluence 8.4 J/cm2

It should be noted that, the calculation's grooving size is minimum when the melt pool zone is considered without flowing (b) and maximum when the melt pool part is neglected (c). The simulation results have been compared with experimental grooves of cell 2, cell 3 and cell 4 of Table 1. A good correlation between experiment and simulated grooves geometry has been found. Indeed, in the case of oxide only region, Figures 9 a, 10 a, and 11 a, present the experimental grooves shape for varying fluence of 5.3 J/cm2, 7.03 J/cm2 and 8.84 J/cm2 respectively, which have been correlated with the isotherm simulated groove shown in the Figures 9 b, 10 b and 11 b. In the same way, Figures 12 a, 13 a and 14 a, present experimental grooves in the case of aluminum pads, which

have been also compared with the simulated grooves Figures 12 b, 13 b, and 14 b.

The calculation results show that penetration depth and groove width response variables in both oxide only region and aluminum pads, are significantly impacted by the laser power (fluence), confirming the experimental observation (Fig. 5 & 6). However, the penetration depth response in the Figure 15, seems more important in the case of oxide only region than in the case of aluminum pads, whereas, in the Figure 16, the opposite is observed in the groove width

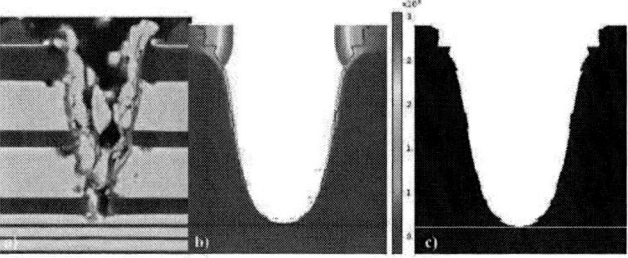

Figure 12 Cross section of cell 2 aluminum pads/Cu lines after nanosecond laser grooving; Comparison between a) the cross section image of the experimental groove, b) isothermal simulated groove taking into account the melt pool front without flowing, c) simulated groove assuming the melt pool is negligible. Fluence , 5.3 J/cm2

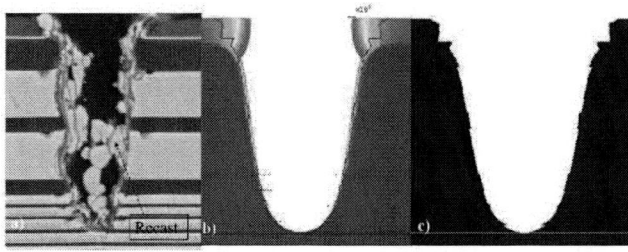

Figure13 Cross section of cell 3 aluminum pads/Cu lines after nanosecond laser grooving; Comparison between a) the cross section image of the experimental groove, b) isthermal simulated groove taking into account the melt pool front without flowing, c)simulated groove assuming the melt pool is negligible. Fluence 7.03 J/cm2

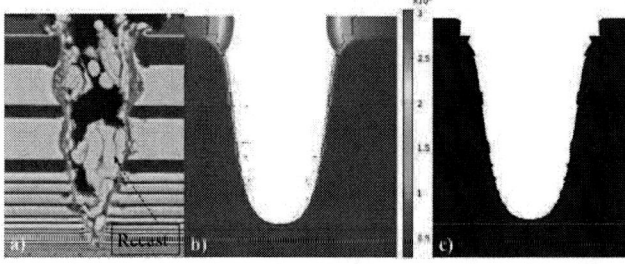

Figure14 Cross section of cell 4 aluminum pads/Cu lines after nanosecond laser grooving; Comparison between a) the section of real crater and a calculated one: b) taking into account the melt pool front without flowing, c) assuming the melt pool part is negligible. Same BEOL stack. Fluence, 8.4 J/cm2

These observations are verified in both experimental results and numerical results (Figures 14 &15). These results illustrate the impact of the BEOL material structure and material properties effect on the resulting groove profile. As demonstrated in previous works by Fabbro et al [17], the isotherms vertical inclination of the liquid surface in

978-1-7281-1500-9/19 $31.00 © 2019 IEEE   1402

simulation results increases with the power (fluence) leading to potential instabilities.

As also expected, in both structures, the temperature level increases with laser power, leading to a more intense vaporization phenomenon. It appears in Figures 15 & 16 that the penetration depth increases quicker than the width, which tends to stabilize at around 6.6 µm in the case of oxide only region, and at about 11.1 um for the aluminum pads. Indeed, the focal spot diameter is limited and energy will concentrate at the bottom of the groove increasing the depth rather than the lateral diffusion.

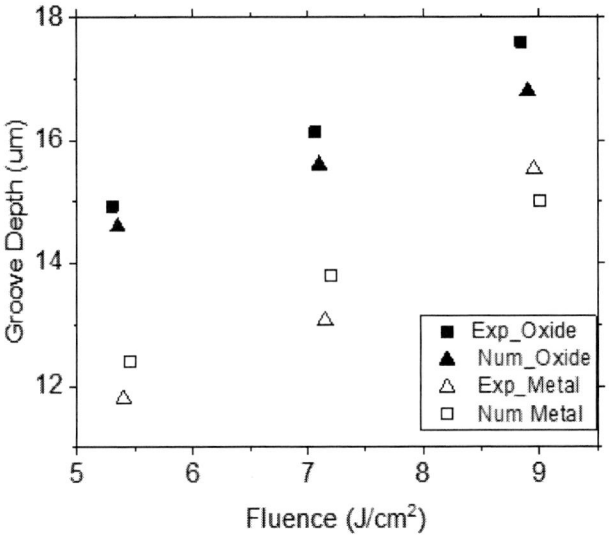

Figure 15 Comparison of simulated and experimental grooves depth for cells 2, 3 & 4 on oxides only regions and aluminum pads as a function of the laser power (fluence). a) 5.3 J/cm2, b) 7.03 J/cm2 and c) 8.84 J/cm2

On the other hand, simulation allows to qualitatively determine the extension of the HAZ around the crater as shown in figures 9, 10 and 11 in the case of oxide only regions. Indeed, experimental results, Figures 9 a, 10 a and 11 a, show a strong preponderance of ILD damages due to the interaction between nanosecond-laser beam and BEOL stack structures. These damages could be a concern as far as they could be the starting point of the delamination propagation observed during reliability testing phases. Several factors could explain these ILD damages; the formation of hot spots at oxide-ILD interfaces, the accumulation of thermo-mechanical stress and the thermo-mechanical sensitive properties of these ILD materials [3, 4, 5, 6]. The recoil pressure effect could also have a significant part in the appearance of these damages [16].

The concern on ILD damages has not clearly been observed in the case of aluminum pads in both experiments and simulations confirming the good agreement between experimental and modeling approaches. This result could be explained by the presence of aluminum and several copper layers and lines throughout the BEOL structure in the aluminum pads/copper lines. Indeed, aluminum and copper metals are excellent thermal conductors and behave like a heat pump. The consequence is the important lateral diffusion of heat leading to the decrease of the local energy density. The decrease of the local energy density could prevent the formation of hot spots and the accumulation of thermo-mechanical stresses leading to the reduction groove shape and HAZ characteristics as it is shown in Figures 11 c, 12 c, 13 c.

However further and finer studies are in progress to understand if micro cracks can originate from metal regions affected by laser grooving.

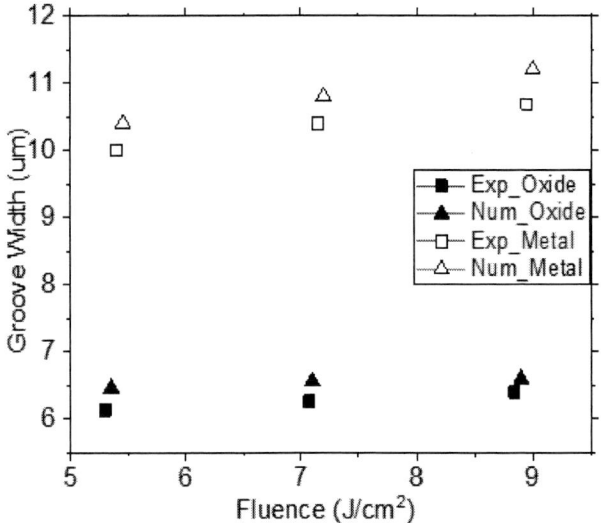

Figure 16 Comparison of simulated and experimental grooves width for cells 2, 3 & 4 on oxides only regions and aluminum pads as a function of the laser power (fluence). a) 5.3 J/cm2, b) 7.03 J/cm2 and c) 8.84 J/cm2

The reader should keep in mind that for this proposed modeling approach; the liquid flow and the phenomenon of recrystallization are neglected. That could explain why the gas cavity formed during groove crater collapse as observed in Figure 10 a in the experimental cross-section images of cell 4 (oxide only region), is not shown in simulated results. Similarly, the melt pool piling up on the groove top surface observed experimentally is not considered in this simulation.

A constant discrepancy between experimental results and numerical predictions exists due to the assumptions used in this modeling approach as well as the use of simplified multilayer structures in order to meet reasonable computation time and avoid solver convergence issues

VI. CONCLUSION

In this paper, two original approaches have been performed to investigate wafer BEOL laser grooving process. We show how thanks to an experimental approach, it is possible to optimize laser grooving process on the wafer BEOL materials, despite of time consuming and DOE processing. Moreover, a numerical model has been implemented to simulate a grooving process by laser beams in order to better understand the fundamental mechanisms at play and the material behavior when processing a BEOL device stack. This alternative way is a multiphysics model and consists of a full 2D laser beam simulation which ablates a multilayered BEOL structure. Both approaches have been applied on IBM 14 nm technology node and the conclusions show that simulation seems in agreement with experimental results. Although these preliminary results are very positive, work continues with other simulation structure, which will help to confirm the 2D model simulation accuracy. This proposed study is a first step toward implementing a quick and accurate assessment tool for design and debug of multiple laser grooving conditions with limited experiments and hardware in industrial application.

ACKNOWLEDGEMENT

This work has been performed within the framework and supported by the NSERC/IBM Industrial Research Chair in Smarter Microelectronics Packaging for Performance Scaling. The authors are grateful to Vincent Bruyère and his team from SIMTEC (Grenoble France) to their support for our modeling investigation as well as Richard Indyk (IBM East Fishkill, New York) for his collaboration in the experimental data phase

REFERENCES

[1] J.D. Meindl, J.A. Davis, P. Zarkesh-Ha et al. "interconnect opportunities for gigascale integration", IBM Journal of R&D, p 245-263 (2002).

[2] W. K. Chien, Y. A. Zhao, L. Zhang, et al., " Investigation and detection on a new BEOL dielectric failure mechanism at advanced technologies" Microelectronics reliability 81, p 368-372 (2018)

[3] W. ZhiJie, S. Wang, J. H. Wang et al, "300 mm Low K Wafer Dicing Saw Study" IEEE 6th International Conference on Electronic Packaging Technology, (2005)

[4] M. Fuegl, G. Mackh, E. Meissner et al. "Assessment of Dicing Induced Damage and residual Stress on the Mechanical end Electrical Behavior of Chips" IEEE Electronic Component & Technology Conference (2005)

[5] M. Cioban, T. Sinha, T. M. Shaw, "Back-end-of-line, (BEOL) Integrity Evaluation: A mixed-Mode Double cantilever Beam Test for Crackstop Strength Assessment" IEEE 68th Electronic Components and technology Conference (ECTC) p 467 - 475, (2018)

[6] M. rabie, N. A. Polomoff, M. K. Hassan, et al., "Innovative Design of crackstop Wall for 14 nm technology Node and Beyond" IEEE 68th Electronic Components and technology Conference (ECTC) p 460 - 466, (2018) ,

[7] P., Parandoush, A. Hossain, "A review of modeling and simulation of laser beam machining,"International Journal of Machine Tools & Manufacture 85, p 135-145 (2014)

[8] G. Galasso, M. Kaltendbacher, B. Karnamurthy et al., " Multiphysical modeling of nanosecond laser dicing on ultra-thin silicon wafers" IEEE, 15th International Conference on Thermal, mechanical and multi-physics Simulation and Experiments in Microelectronics and Microsystems, p 1-6, EuroSim (2014)

[9] K.-H. Leitz, B. Redlingshöfer., Y., Reg et al., " Metal Ablation with Short and UltraShort laser pulses" Physics Procedia 12, p 230-238 (2011)

[10] A. Otto, H. Koch, K.-H. Leitz et al., "Numerical Simulation –A Versatile Approach for better Undertanding Dynamics in Laser material processing" Physics Procedia vol 12, p 11-20 (2011).

[11] J. Donea, A. huerta, J.-Ph. Ponthot et al. " Arbitrary langrangian-Eulerian methods" Encyclopedia of computational Mechanics John Wiley & Sons, Ltd (2004).

[12] . Clerk Maxwell, A Treatise on Electricity and Magnetism, 3rd ed., vol. 2. Oxford: Clarendon, 1892, pp.68–73.

[13] J. Moussodji, O. Chacon, F. Santerre, D. Drouin "Laser Grooving of multi stack material modeling : Implementation of a high accuracy tool for laser-grooving and dicing application", Spie Photonics West Conference, Proceedings Vol: 10905, (2019)

[14] S.I. Abu-Eishah, "Correlation for the thermal conductivity of metals as a function of temperature," International Journal of Thermophysics, vol 22, N0 6 (2001)

[15] S.I. Abu-Eishah et al, "A new correlation for the specific heat of metals, metal oxides and metal fluorides as a function of temperature," lat. Am. Appl., vol 34, N0 4 p 257-265 (2004)

[16] K. Kheloufi, E. H. Amara, "Numerical investigation of the effect of some parameters on temperature field and kerf width in laser cutting process", Physics Procedia 39 p 872–880 (2012)

[17] R. Fabbro, S. Slimani, F. Coste et al., "Analysis of the various melt pool hydrodynamic regimes observed during CW Nd-YAG penetration laser welding", ICALEO Conference 2007, Orlando, USA

978-1-7281-1500-9/19 $31.00 © 2019 IEEE

Thermal Characteristics of Vertically-Integrated GaN/SiC-on-Si Assemblies: A Comparative Study

Kimmo Rasilainen*, Per Ingelhag[†], Peter Melin[†], Torbjörn M. J. Nilsson[‡],
Mattias Thorsell[*‡], and Christian Fager*

* Department of Microtechnology and Nanoscience, Chalmers University of Technology, SE-412 96 Gothenburg, Sweden
[†] Ericsson AB, SE-417 56 Gothenburg, Sweden
[‡] Saab AB, SE-412 89 Gothenburg, Sweden
e-mail: kimmor@chalmers.se

Abstract—This work investigates the thermal characteristics of a vertically-stacked, heterogeneously-integrated assembly intended for millimetre-wave communications systems. The assembly combines materials that enable the generation of high output power as well as high degree of integration for improved performance, and vertical integration of the different materials enables a compact footprint. Suitability of thermal solutions based on metal pillars, solder balls, and ball grid arrays (BGA) is investigated. Both ideal, fully-populated arrays of interconnects and partially-filled ones more suitable for practical implementations are considered using theoretical calculations and numerical thermal simulations. With the assumptions used, simulation results show that arrays of Cu pillars and large solder bumps with a pitch of $150\,\mu m$ provide good thermal performance also with a simplified grid and reduced number of interconnects. In the current geometry, the most important locations for the pillars and bumps are near the heat sources, and the use of a rim of interconnects around the assembly perimeter can reduce the temperature by several degrees — even when the majority of the other interconnects is focused beneath the heat sources.

Index Terms—Ball grid array (BGA), bumps, heterogeneous integration, microwave, pillars, silicon (Si), silicon carbide (SiC), solder, thermal analysis.

I. INTRODUCTION

Future-generation communications systems (5G and beyond) are expected to provide improved performance and functionality on both device and system levels, as well as to support a larger number of simultaneous users [1]. Technological solutions to achieve this include using millimetre-wave frequencies for broader bandwidths, and multi-antenna communications. Similar technologies can also be used in military applications.

Efficient implementation of these systems with high power levels and small size calls for a substantial degree of integration, which increases the significance of the assembly and packaging techniques used [2]. Additionally, the cost of the integrated systems should remain at an affordable level in both commercial and military solutions [3], [4]. One of the major design challenges in developing highly-integrated systems is to find materials that can handle the required power levels, and which enable an efficient heat transfer. An overview of various thermal challenges and some of their possible solutions is presented in, e.g., [5].

One concept that has the potential to solve these issues is heterogeneous integration of materials such as Si and III–V

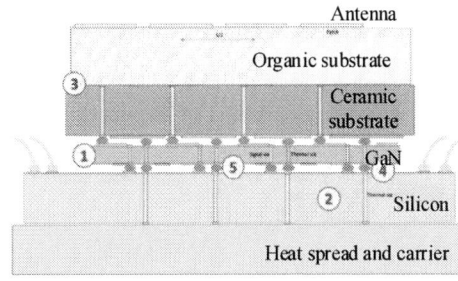

Fig. 1. Conceptual illustration of the envisioned integration approach. The present study does not consider the layers above the GaN/SiC chip.

compound semiconductors (see, e.g., [6]–[8]). As an example, combining the high power-handling capability of gallium nitride (GaN) and the high integration density and technological maturity enabled by silicon complementary metal-oxide semiconductor (Si CMOS) processes can be a suitable way of providing the necessary performance. In the Diverse Accessible Heterogeneous Integration (DAHI) project by the Defence Advanced Research Projects Agency (DARPA), the integration of, e.g., GaN and Si has been studied, but the focus has mainly been on placing or growing GaN transistors on Si substrates [8], [9]. An example of heterogeneous integration with vertically-stacked assemblies using Si and gallium arsenide (GaAs) can be found in [10].

From a technological point of view, application areas of heterogeneous integration include stacked integrated circuits (ICs), processors and other components, as well as some recent photonics applications [11]–[14]. As a generalisation, the end goal is to integrate, e.g., the ICs, RF components, and thermal assemblies on a system level [15]. Even though GaN-based technologies can withstand high power levels, there are nevertheless some fundamental limits for the suitability of particular cooling strategies [16].

Because of increasing integration and power level, it becomes more and more important to understand the effects caused by the resulting structural heating. On one hand, the heterogeneous nature of the materials involved in the structure means that they have different thermal parameters. This causes, e.g., differences in thermal expansion during the heating/cooling cycles under typical operating conditions,

978-1-7281-1500-9/19 $31.00 © 2019 IEEE

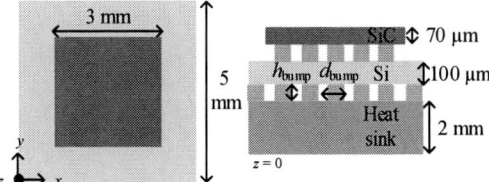

(a) Top and side view of the assembly and heat sources

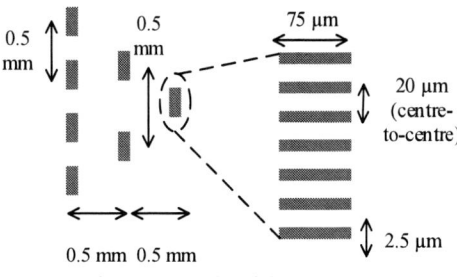

(b) Geometric details of the heat sources

Fig. 2. Illustration of the (a) studied assembly geometry and (b) the heat sources used. Dimensions of the thermal interconnects (d_{bump}, h_{bump}) are only indicative, and the drawing is not entirely to scale.

which can potentially reduce the lifetime and durability of the structure. On the other hand, structural heating also warms up the different electronic components, such as amplifiers, which can degrade their performance from nominal operating conditions. Proper characterisation of the combined electrothermal effects is therefore very significant (see, e.g., [17], [18]), but these aspects are not considered in this work.

In this study, we investigate the thermal properties of a heterogeneous stacked assembly. Fig. 1 presents a schematic illustration of the studied concept. The performance of different types of thermal interconnects are considered by varying their dimensions (height, diameter, pitch) and type (metal pillars, solder bumps, ball grid array [BGA]). In the simulations, effects of various "non-idealities" such as partially-filled arrays of interconnects and including a multi-layer PCB in the vertical thermal path are considered.

II. INVESTIGATED ASSEMBLY GEOMETRY

This work investigates the thermal characteristics of a stacked, heterogeneous GaN/SiC-on-Si assembly for a multi-antenna communications system at mm-wave (20–50 GHz) frequencies. In the study, numerical thermal simulations are carried out in the ANSYS Icepak software (version 19.1).

A. Chip Dimensions and Unit Cell

The assembly is visualised in Fig. 2(a), and it combines a 70-µm thick GaN/SiC chip and a 100-µm thick Si chip with vertical interconnects. These carry signals to the antennas and transfer heat downwards in the structure. Initially, the area of the SiC and Si chips is $3 \times 3\,\mathrm{mm}^2$ and $5 \times 5\,\mathrm{mm}^2$, respectively. In the present study, only the thermal effects in the interconnects and bulk substrates are considered, and the effects of the thin GaN layer are not taken into account.

TABLE I
THERMAL PROPERTIES OF THE MATERIALS USED IN THE STUDY.

Material	Density $\rho\,(\mathrm{kg/m}^3)$	Thermal conductivity $k\,(\mathrm{W/m \cdot K})$	Specific heat $C_p\,(\mathrm{J/kg \cdot K})$
Cu	8933	387.6	397[a]
Si	2330	180	770[b]
SiC	3210	490	677.8
Solder	8690	56.9	239
Underfill	1210	0.8	1172
FR4	1250	0.35	1300

[a] $C_{p,\mathrm{Cu}}(0, 77, 313, 1000)\,\mathrm{K} = (195, 195, 397, 397)\,(\mathrm{J/kg \cdot K})$
[b] $C_{p,\mathrm{Si}}(0, 77, 313, 1000)\,\mathrm{K} = (180, 180, 770, 770)\,(\mathrm{J/kg \cdot K})$

Table I summarises the material parameters assumed in the study (default values available in the Icepak software). As indicated in the footnote of Table I, the specific heat of Cu and Si is defined using a piecewise-linear temperature dependance. All other parameters are assumed to be constant.

A compact footprint is achieved through vertical stacking and using a small unit cell size of $5 \times 5\,\mathrm{mm}^2$. The idea is to have at least 16 antennas in the array (placed on a separate substrate above the GaN layer), with each unit cell driving four elements. This concept also scales up to larger antenna arrays. A fairly small unit cell size is needed as the inter-element spacing in the antenna array is proportional to the wavelength, which is between 15 and 6 mm from 20 to 50 GHz.

The present thermal study does not, however, consider the antenna properties nor the implementation of radio-frequency (RF) interconnects between the different chips. For this reason, components located above the GaN layer in the schematic of Fig. 1 are beyond the scope of the current work.

B. Heat Source Modelling

In the assembly model, a three-stage configuration of heat sources is used to model the intended high-power GaN electronics. The heat sources are modelled down to transistor finger level, and the resulting dimensions are given in Fig. 2(b). These can be considered feasible in terms of currently available GaN processes. Each transistor is assumed to dissipate 0.4 W of power (about 3 W per "unit heat source"), and the four chains of heat sources in Fig. 2(a) generate approximately 12 W of power. The heat source dimensions and power levels involved result in a localised heat flux of $36.4\,\mathrm{kW/cm}^2$. Such levels require careful consideration about suitable packaging options for proper thermal management to prevent very high localised hotspots, and to keep the GaN transistor channel temperature within its tolerance limits (150–170 °C) [19].

To study how to transfer the heat generated in the assembly downwards as effectively as possible, different configurations of thermal interconnects (metal pillars, solder bumps, ball grid arrays) with varying dimensions and distribution (pitch and patterning) are investigated in the following sections. Parametric studies help to identify possible thermal bottlenecks. The assembly is placed on a heat sink to further distribute the heat, and a comparison between ideal, solid-metal heat sinks

978-1-7281-1500-9/19 $31.00 © 2019 IEEE

TABLE II
DEFAULT DIMENSIONS OF THE INVESTIGATED INTERCONNECTS.

Case	Diameter (μm)	Height (μm)	Pitch (μm)	Notes
Cu pillars	75	45+25	150	$h_{Cu} + h_{solder}$
Solder bumps #1	60	30	100	
Solder bumps #2	100	75	150	
BGA CSP	300	(ca.) 200	500	

and multi-layer PCBs with thermal vias connected to a metal block is made.

III. THERMAL CHARACTERISTICS OF THE ASSEMBLY WITH DIFFERENT THERMAL INTERCONNECTS

Table II shows the default dimensions of the different interconnect types considered. These consist of Cu pillars (with a connecting solder layer), small and large solder bumps (subsequently called Bumps #1 and Bumps #2, respectively), and a BGA. The properties of the interconnects are initially investigated using parametrised calculations and then with thermal simulations.

A. Calculated Thermal Conductivity of the Interconnects

As a starting point for the analysis of different interconnect types, calculations are performed on the thermal conductivity of an individual bump/pillar/ball as a function of its diameter (d) and height (h), as well as of the pitch (p) between adjacent interconnects. Calculations use the material parameters of Table I, and of the three quantities (d, h, and p) two always have the default values of Table II when the third one is altered.

The thermal resistance of individual interconnects can be calculated as [20]:

$$R_{th} = \frac{h_{int}}{k \cdot A_{int}} = \frac{h_{int}}{k \cdot \pi \cdot (0.5 d_{int})^2}, \quad (1)$$

where A_{int}, h_{int} and d_{int} are the area, height and diameter of the interconnect, respectively, and k is the thermal conductivity. In the Cu pillar case, the thermal path consists of the metal pillar itself and of a thin solder layer, and the thermal resistance becomes $R_{th,tot} = R_{th,Cu} + R_{th,solder}$, with dimensions according to Table II.

Effective thermal conductivity (k_{eff}) over a given rectangular pitch area is

$$k_{eff} = \frac{h_{tot}}{R_{th} \cdot (0.5 p)^2}, \quad (2)$$

and the side of the rectangle equals half of the pitch between two adjacent interconnects. In (2), h_{tot} accounts for the total height of the interconnect (solder and possible Cu pillar).

Fig. 3 depicts the calculated thermal conductivities as a function of the different parameters. In terms of pitch and height (Fig. 3(a) and (b), respectively), the Cu pillars and Bumps #2 have better k_{eff}, whereas diameterwise the Cu pillars and Bumps #1 performs best (Fig. 3(c)). In Fig. 3(b), the thermal conductivity of the solder bumps becomes height-independent as h cancels out in these cases when applying

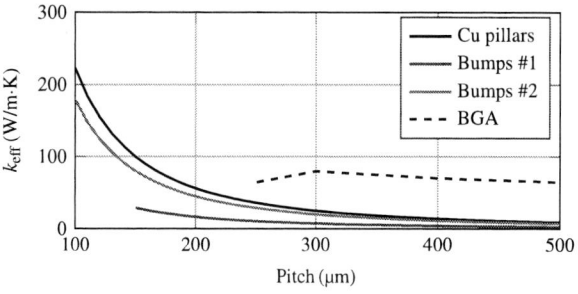

(a) Effective thermal conductivity as a function of pitch

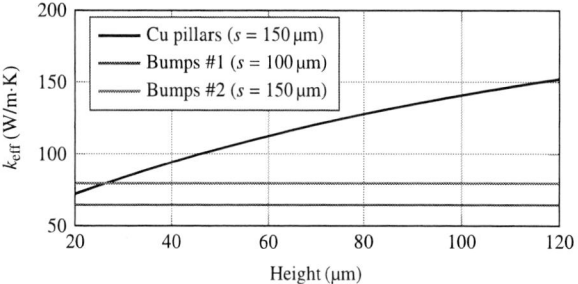

(b) Effective thermal conductivity as a function of height

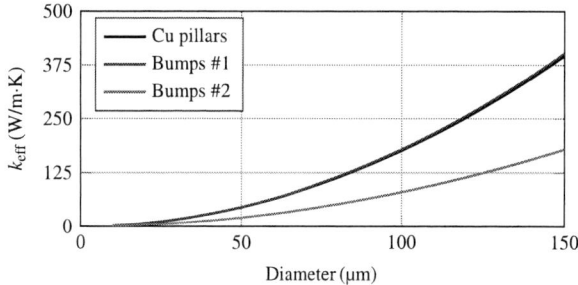

(c) Effective thermal conductivity as a function of diameter

Fig. 3. Calculated effective thermal conductivity of the pillars, bumps and BGA of Table II as a function of their (a) pitch, (b) height, and (c) diameter.

(1) into (2). Especially with larger diameters, Bumps #1 have better thermal conductivity than Bumps #2 and similar performance as the Cu pillars, but a key drawback is their small height (30 μm); higher bumps are better from a stress point of view [21].

The BGA shows a slightly different themal conductivity profile than the other three interconnects, but it should be noted that the calculations only have four data points (250, 300, 400, and 500 μm). Properties and dimensions of the BGA balls have been adapted from a commercially available packaging solution from Amkor Technology [22], and the ball size is pitch-dependent. As only discrete BGA diameter and height values are available, this interconnect type was not included in the analysis of the height and diameter cases in Figs. 3(b)–(c).

B. Simulated Thermal Properties of Ideal Interconnect Arrays

Thermal simulations in this study are done in two stages: 1) by considering ideal, fully-populated (regular) arrays of

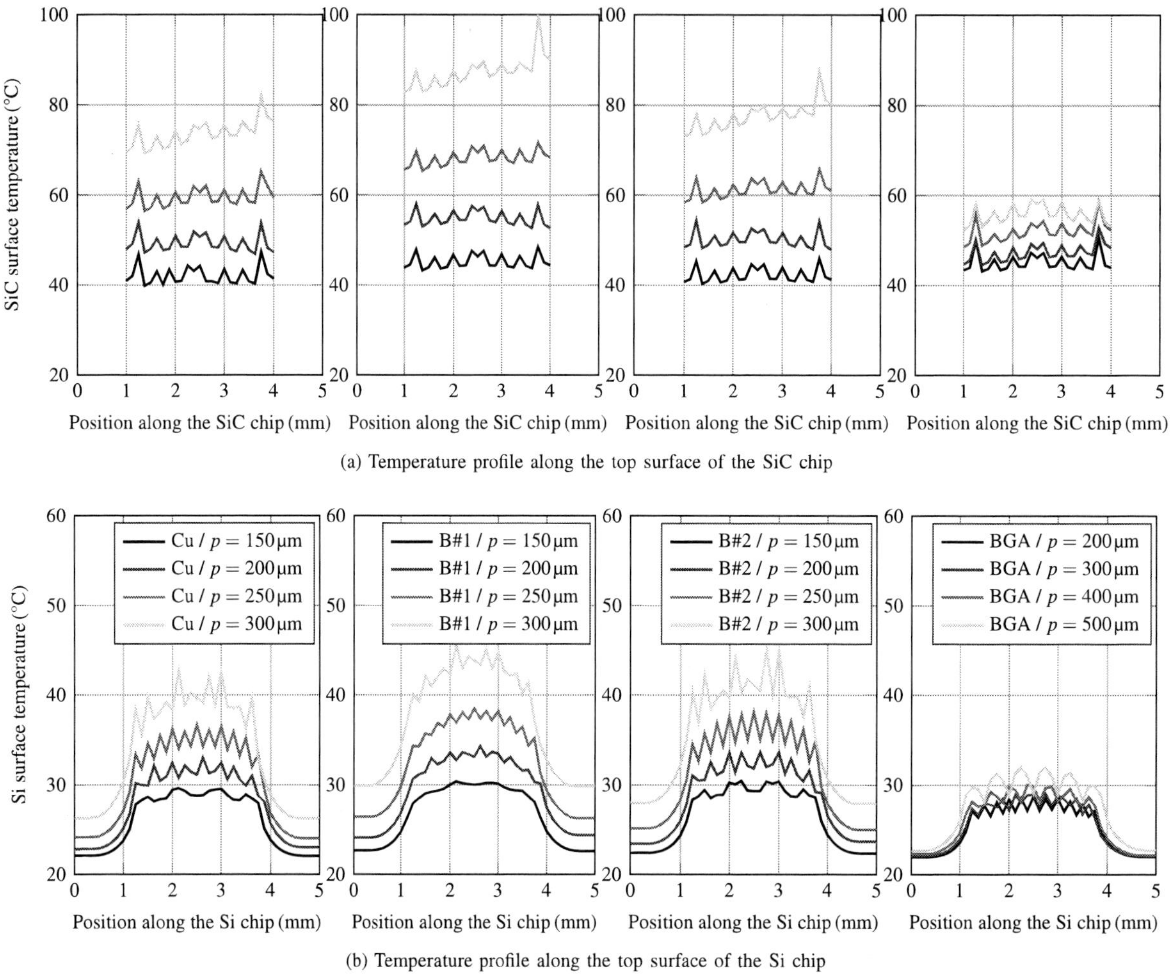

(a) Temperature profile along the top surface of the SiC chip

(b) Temperature profile along the top surface of the Si chip

Fig. 4. Temperature profiles along the top surfaces of the (a) SiC and (b) Si chips with different thermal interconnects assuming regular arrays and different values of pitch (p). The colouring scheme is the same in the upper and lower figures. Cu = Cu pillars, B#1 = Bumps #1, and B#2 = Bumps #2.

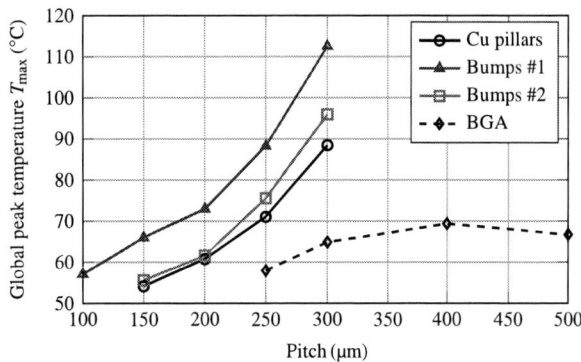

Fig. 5. Simulated global peak temperature ($T_{\max}$) in the assembly with ideal arrays of interconnects. The curve markers show the investigated pitch values.

interconnects, and 2) by performing a series of successive modifications (simplifications) on the arrays. Simplifications made in the second stage investigate how significant perfor-

mance degradation can be seen using a non-ideal (partially populated) interconnect array, with some unoccupied area reserved for placing various electronic components and chips.

Simulated pitch values for the different interconnects are 150 μm, 200 μm, 250 μm, and 300 μm (Cu pillars, Bumps #1, and Bumps #2), and 250 μm, 300 μm, 400 μm, and 500 μm for the BGA. An additional pitch of 100 μm was studied only with Bumps #1, as this value is not feasible for the diameter of the Cu pillars and Bumps #2. In all cases except for the BGA, the diameter and height remained constant with different pitches.

With the different interconnect cases, Fig. 4 shows the simulated temperature profiles along a linear trajectory at the top surface of the SiC and Si substrates with different pitch values. The resulting global peak temperatures ($T_{\max}$) are displayed in Fig. 5. In all cases, the y-directional cut intersects eight (4+4) heat sources at the outermost stage, and this shows a worst-case scenario in terms of local heat generation.

The temperature in the different surfaces increases with the pitch of the bumps/pillars/balls, as can also be expected. The effect is most pronounced in the SiC chip, where the

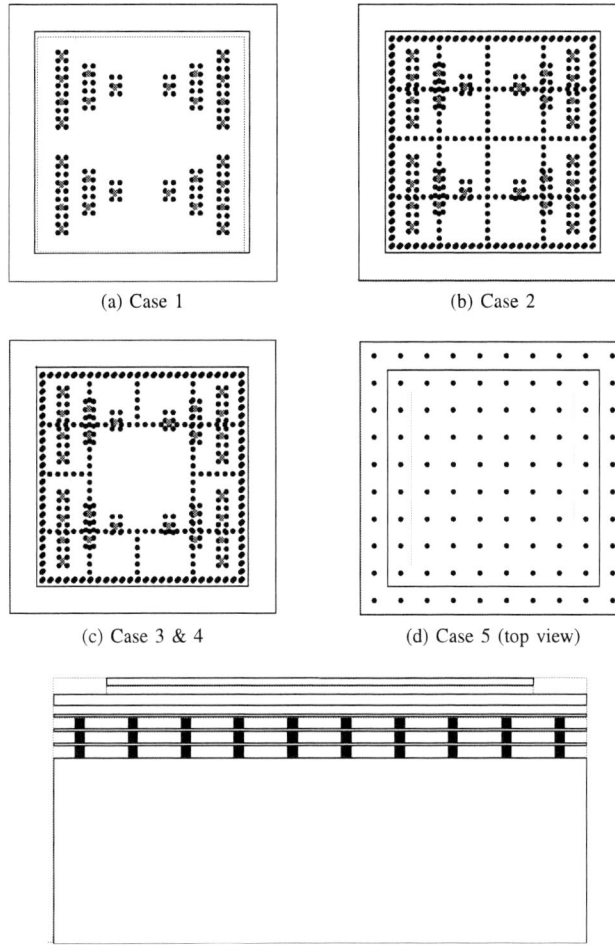

(a) Case 1   (b) Case 2

(c) Case 3 & 4   (d) Case 5 (top view)

(e) Case 5 (side view)

Fig. 6. Illustration of the principal modifications made to the interconnect arrays in Cases 1–5. The locations of the interconnects are marked in black, and the heat sources are shown in red. These drawings show the dimensions and pitch of Bumps #2, but the same concept is also used for the Cu pillars.

temperature difference between the densest and sparsest grid can be 30–40 °C or more. At the plane of the Si chip, heat distribution along the surface of the chip is more even, which reduces the temperature difference caused by varying the pitch.

In the case of Cu pillars and both bump variants, the 300-µm pitch shows a temperature increase towards the edge of the SiC chip (at around 4 mm). This effect, which is most evident with Bumps #1, is caused by a missing row of interconnects at that edge of the SiC chip; the pillar/bump diameter and the particular pitch value would have placed this row impractically close to the chip edge.

The obtained T_{max} results agree well with the behaviour of the thermal trajectories with different pitch values, and Cu pillars and Bumps #2 have similar T_{max} characteristics (especially with smaller pitch values). Being smaller in size than the other three interconnects, Bumps #1 enables the largest number of interconnects in all cases. However, the

small size also limits the vertical heat transfer capability, which results in peak temperatures consistently over 10 °C higher than with other interconnects. Even when the pitch is 100 µm, the peak temperature is higher than for Cu pillars and Bumps #2 with a 150-µm pitch.

The temperature rise of the BGA has a different trend than the pillars and bumps, which relates to changes in the ball dimensions (and hence thermal resistance) depending on the pitch. This can also be seen in the temperature profiles of Fig. 4, where the variation with different BGA pitch values is considerably smaller than with the other three interconnects. Even though the BGA does not have all too bad thermal performance, the large pitch values mean that the grid is inevitably sparse, and stress-related problems may occur.

C. Modifications Performed on the Interconnect Arrays

To transform the previously analysed fully-populated interconnect arrays into something a bit more feasible from a practical implementation point of view, the number and distribution of pillars are gradually modified. The main idea is to simplify the arrays by concentrating the pillars in locations where they are assumed to be most needed. These modifications are performed for Cu pillars and Bumps #2, assuming the pitch is 150 µm, and this value is also used for the added/removed interconnects. Fig. 6 illustrates the main details of the modifications. The array connecting the Si chip to the heat sink is also simplified by converting it to a chessboard-like pattern instead of a fully-populated one (the error of doing this can be assumed to be rather small).

The first modification (Case 1) is to increase size of the 3×3-mm^2 SiC chip to 4.5×4.5 mm^2, and to form the array of interconnects into rows located directly beneath the heat source. At the same time, the interconnect array at the Si/heat sink interface is converted into a chessboard-like pattern. Using a larger chip with a smaller number of interconnects is necessary to reduce the risk of a local hotspot in the proximity of the two centermost transistors (this effect already begins to manifest itself in the curves of Fig. 4).

In the second step (Case 2), a uniform rim of interconnects is placed along the perimeter of the SiC/Si interface. This enables an additional downward thermal path to compensate for possible problems with increased thermal resistance between the layers as the number of interconnects reduces. The third step (Case 3) removes the centremost part of the interconnects in the lower array to allow for more space for potential placement of electronics. Case 4 is a straightforward modification of Case 3, in which an underfill is infiltrated between the interconnects in the SiC/Si and Si/heat sink interfaces. The underfill is characterised using the default flip-chip underfill material in Icepak (parameters given in Table I).

The last modification (Case 5) is to replace part of the solid-metal heat sink with a stack of metal and dielectric layers, which are used to represent a PCB onto which the different chips can be assembled. In this assembly, the thickness of the Cu and dielectric layers is assumed to be 25 and 100 µm, respectively, and they are connected with an array of Cu vias

978-1-7281-1500-9/19 $31.00 © 2019 IEEE

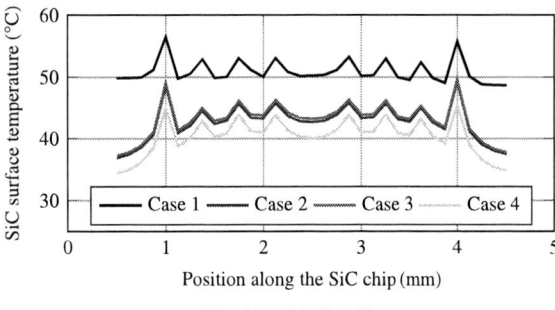

(a) SiC chip with Cu pillars

(b) Si chip with Cu pillars

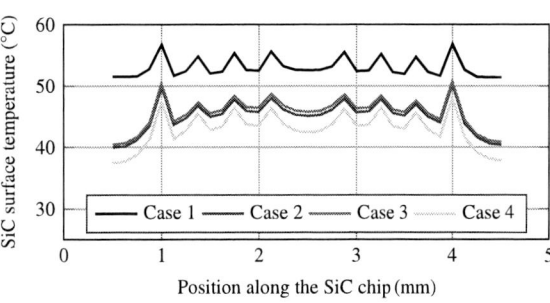

(c) SiC chip with Bumps #2

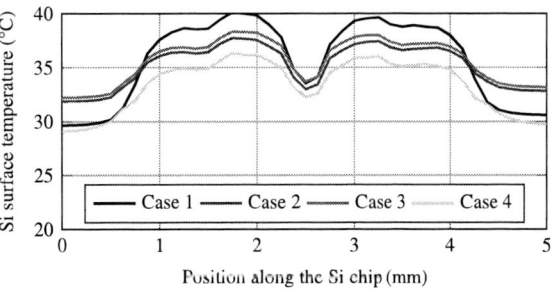

(d) Si chip with Bumps #2

Fig. 7. Temperature profiles on the SiC and Si chips obtained with Cu pillars and Bumps #2 using modified thermal interconnect distributions (Cases 1–4).

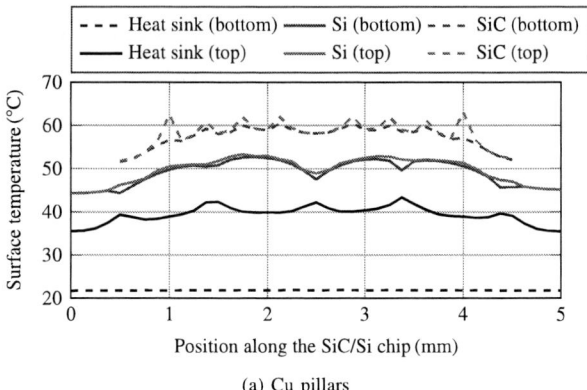

(a) Cu pillars

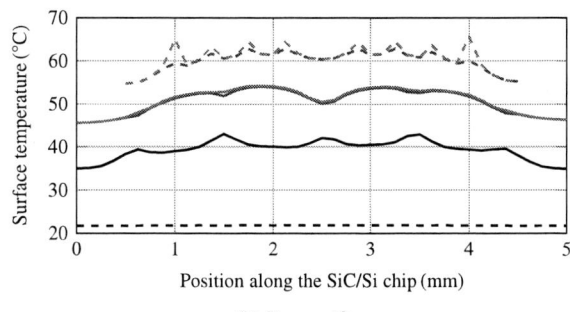

(b) Bumps #2

Fig. 8. Temperature profiles on the SiC and Si chips obtained with Cu pillars and Bumps #2 using a multi-layer PCB with thermal vias (Case 5).

TABLE III
TEMPERATURE EXTREMES IN DEFAULT AND MODIFIED ASSEMBLIES USING CU PILLARS AND BUMPS #2.

Case	T_{max} (°C) (global)	T_{max}/T_{min} (°C) (upper interconnect layer)
Cu pillars / Default	54.1	41.8 / 26.4
Bumps #2 / Default	55.5	43.6 / 26.5
Cu pillars / Mod 1	63.0	52.1 / 34.8
Cu pillars / Mod 2	59.1	44.3 / 30.0
Cu pillars / Mod 3	59.6	44.7 / 30.4
Cu pillars / Mod 4	52.8	41.8 / 28.5
Cu pillars / Mod 5	70.7	59.6 / 45.3
Bumps #2 / Mod 1	61.6	53.4 / 34.7
Bumps #2 / Mod 2	60.3	46.9 / 30.6
Bumps #2 / Mod 3	61.1	47.6 / 31.1
Bumps #2 / Mod 4	55.1	44.6 / 29.1
Bumps #2 / Mod 5	73.1	62.5 / 46.0

($d_{via} = h_{via} = 100\,\mu m$, pitch = $500\,\mu m$). The dielectric layers are assumed to be made of FR-4 material. The choice of via dimensions and dielectric material is made mainly to simplify the analysis rather than to describe any particular process or technology. Overall thickness of the combined heat sink-PCB structure (2 mm) is the same as in the the previous structure with a solid heat sink.

D. Simulated Thermal Properties of Modified Cases

Temperature profiles visualising the effects of the structural modifications are shown in Figs. 7 and 8. In the modified cases, the pitch of both the pillars and bumps is $150\,\mu m$, and the location of the thermal trajectory relative to the heat sources is the same as in Fig. 4. Three-dimensional temperature distributions in the pillars and bumps are presented in Fig. 9 for Cases 4 and 5, although in the former the centremost part of

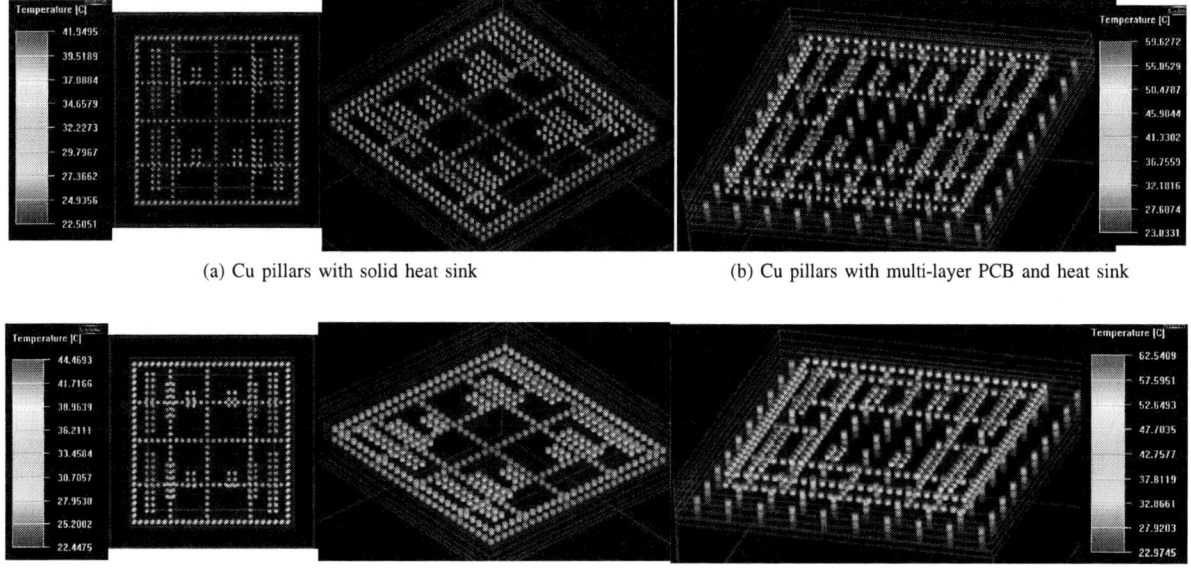

(a) Cu pillars with solid heat sink   (b) Cu pillars with multi-layer PCB and heat sink

(c) Bumps #2 with solid heat sink   (d) Bumps #2 with multi-layer PCB and heat sink

Fig. 9.   Temperature distributions in Cu pillars and Bumps #2 with solid heat sink (Case 4), and with a multi-layer PCB and heat sink (Case 5).

the lower interconnect array has not been removed (the effect of this difference is small). Table III shows the global peak temperatures and the temperature extremes in the interconnects (between SiC and Si chips) resulting from the modifications.

Comparison of the temperature profiles of Figs. 4 and 7 shows that with the larger SiC chip and more concentrated arrays of interconnects (Case 1), the temperature rises both in the SiC and Si substrates. In this case, a more significant increase occurs in the Si substrate. Increased distance between the two centremost heat sources can be seen as a local minimum in the Si thermal profile. Including a rim of pillars or bumps along the edge of the SiC substrate decreases its temperature by approximately $10\,°C$. The overall effect of the additional interconnects on the vertical heat transfer heavily depends on the thickness of the SiC chip, and this aspect can be a subject of future studies.

Partial removal of the interconnects in the interface between the Si chip and the heat sink does not affect the temperature profile at the chosen trajectory, as Cases 2 and 3 in the curves of Fig. 7 are more or less identical. At the plane of the lower interconnects, the heat is generally quite uniformly distributed and only a low number of heat sources is above the area of removed interconnects (when going upwards in the assembly). As a result, the global effects of introducing Case 3 are small. Including the underfill material in Case 4 slightly decreases the temperature along the trajectory compared to the results of Case 3, mainly caused by additional heat transfer through the underfill itself (although it is small, $k_{\text{underfill}} > k_{\text{air}}$).

Fig. 8 shows the simulated temperature profiles for the Cu pillars and Bumps #2 when part of the solid Cu heat sink is replaced with a multi-layer PCB structure. This modification

reduces the vertical heat transfer capability (thermal resistance increases), and as a result, the temperature increases in both the SiC and Si chips by up to $10\,°C$. For the heat sink, the 'top' curve is taken at the top surface of the uppermost metal layer, and the 'bottom' curve at the plane of the solid metal block. Across the stacked layers of metal and dielectric material, the vertical temperature drop can be as high as $20\,°C$.

IV. Observations and Discussion

Based on the studies performed in this work, the most important locations of thermal paths are directly below and near the heat sources. Instead of using rows of interconnects across the entire chip width, concentrating them below the sources reduces the consumed space, but as a result of this the peak interconnect temperature increases by several degrees. On the other hand, including a frame of pillars around the perimeter of the SiC/Si interface can be used to compensate for this increase. Thus, the frame clearly contributes to thermal management by introducing additional thermal paths for vertical heat transfer, but it is most likely also useful for mechanical robustness and structural integrity during operation considering the thermomechanical stress caused by structural heating and cooling.

With the most optimal thermal interconnect dimensions and spacing so far, the global maximum temperature of the assembly is approximately $60\,°C$ when using a solid metal heat sink, as seen in Table III. This applies to both Cu pillars and Bumps #2, and to the different structural modifications performed in Cases 1–4. However, when the heat sink is replaced by a more realistic multi-layer PCB approach in Case 5, the global maximum temperature becomes about $70\,°C$.

978-1-7281-1500-9/19 $31.00 © 2019 IEEE

Both of these temperatures are within practical limits for the materials and interconnects considered in the present study.

Throughout the study, the feasibility of manufacturing has been considered, and the modifications performed on the interconnect arrays have aimed at making the design more practical to implement. It is important to maintain a balance between having sufficiently many thermal interconnects for efficient heat transfer while still leaving a sufficient amount of unoccupied area for placement of various electronic components and systems in the assembly.

From the point of view of practical assembly and mounting, one topic for future research is the possible trade-off in the interconnect dimensions in terms of thermal performance and ease of fabrication. This is an important aspect to consider because, depending on the mounting process, particular "extreme" interconnect dimensions (e.g., very thin or very tall ones) can be challenging to produce in an accurate and repeatable way.

V. CONCLUSION

This study has investigated the suitability and thermal properties of metal pillar, solder bump, and ball grid array interconnects for high-power integrated GaN/SiC-on-Si assemblies. Parametric studies, calculations, and thermal simulations have revealed that Cu pillar and solder bump designs, both with a 150-μm pitch, have good and comparable thermal performance with the investigated structures and power levels, also with non-ideal configurations of interconnects and more realistic PCB designs. The present study provides improved understanding of the major thermal challenges in this type of design, and it can enable us to find an assembly whose operation is not thermally restricted. This will be a key aspect in the search for compact, integrated, and reliable systems for the high-performance communications needs of tomorrow. Future and ongoing work includes development of thermal test structures for experimental verifications, and studies on co-integration of thermal and RF/signal interconnects.

ACKNOWLEDGEMENT

This work was supported by the strategic innovation program "Smarter Electronic Systems", a joint effort by the Swedish Government Agency of Innovation Systems (VINNOVA), Formas, and the Swedish Energy Agency, through decision number 2017-01898.

REFERENCES

[1] J. G. Andrews *et al.*, "What will 5G be?" *IEEE J. Sel. Areas Commun.*, vol. 32, no. 6, pp. 1065–1082, Jun. 2014.

[2] I. Ndip and K.-D. Lang, "Roles and requirements of electronic packaging in 5G," in *Proc. 7th Electronic System-Integration Technology Conf. (ESTC)*, Dresden, Germany, Sep. 2018, pp. 1–5.

[3] K. K. Samanta and I. D. Robertson, "Surfing the millimeter-wave: Multilayer photoimageable technology for high performance SoP components in systems at millimeter-wave and beyond," *IEEE Microw. Mag.*, vol. 17, no. 1, pp. 22–39, Jan. 2016.

[4] H. H. Meinel, "Commercial applications of millimeterwaves: History, present status, and future trends," *IEEE Trans. Microw. Theory Techn.*, vol. 43, no. 7, pp. 1639–1653, Jul. 1995.

[5] S. V. Garimella *et al.*, "Thermal challenges in next-generation electronic systems," *IEEE Trans. Compon. Packag. Technol.*, vol. 31, no. 4, pp. 801–815, Dec. 2008.

[6] K. K. Samanta, "Designing and packaging wide-band PAs: Wideband PA and packaging, history, and recent advances: Part 1," *IEEE Microw. Mag.*, vol. 17, no. 10, pp. 35–45, Oct. 2016.

[7] ——, "Pushing the envelope for heterogeneity: Multilayer and 3-D heterogeneous integrations for next generation millimeter- and submillimeter-wave circuits and systems," *IEEE Microw. Mag.*, vol. 18, no. 2, pp. 28–43, Mar. 2017.

[8] D. S. Green, C. L. Dohrman, J. Demmin, Y. Zheng, and T. H. Chang, "A revolution on the horizon from DARPA: Heterogeneous integration for revolutionary microwave/millimeter-wave circuits at DARPA: Progress and future directions," *IEEE Microw. Mag.*, vol. 18, no. 2, pp. 44–59, Mar. 2017.

[9] Diverse Accessible Heterogeneous Integration (DAHI). Defense Advanced Research Projects Agency (DARPA). [Online]. Available: https://www.darpa.mil/program/diverse-accessible-heterogeneous-integration

[10] J. Zhou, J. Yang, and Y. Shen, "3D heterogeneous integration technology using hot via MMIC and silicon interposer with millimeter wave application," in *Proc. IEEE MTT-S Int. Microwave Symp. (IMS)*, Honolulu, HW, USA, Jun. 2017, pp. 499–502.

[11] Y. Zhang, Y. Zhang, and M. S. Bakir, "Thermal design and constraints for heterogeneous integrated chip stacks and isolation technology using air gap and thermal bridge," *IEEE Trans. Compon. Packag. Manuf. Technol.*, vol. 4, no. 12, pp. 1914–1924, Dec. 2014.

[12] Y. Zhang, T. E. Sarvey, and M. S. Bakir, "Thermal evaluation of 2.5-D integration using bridge-chip technology: Challenges and opportunities," *IEEE Trans. Compon. Packag. Manuf. Technol.*, vol. 7, no. 7, pp. 1101–1110, Jul. 2017.

[13] T. Komljenovic, D. Huang, P. Pintus, M. A. Tran, M. L. Davenport, and J. E. Bowers, "Photonic integrated circuits using heterogeneous integration on silicon," *Proc. IEEE*, vol. 106, no. 12, pp. 2246–2257, Dec. 2018.

[14] O. Marshall, M. Hsu, Z. Wang, B. Kunert, C. Koos, and D. V. Thourhout, "Heterogeneous integration on silicon photonics," *Proc. IEEE*, vol. 106, no. 12, pp. 2258–2269, Dec. 2018.

[15] R. Tummala, N. Nedumthakady, S. Ravichandran, B. DeProspo, and V. Sundaram, "Heterogeneous and homogeneous package integration technologies at device and system levels," in *Proc. Pan Pacific Microelectronics Symp. (Pan Pacific)*, Big Island, HI, USA, Feb. 2018, pp. 1–5.

[16] Y. Won, J. Cho, D. Agonafer, M. Asheghi, and K. E. Goodson, "Fundamental cooling limits for high power density gallium nitride electronics," *IEEE Trans. Compon. Packag. Manuf. Technol.*, vol. 5, no. 6, pp. 737–744, Jun. 2015.

[17] E. Baptista, K. Buisman, J. C. Vaz, and C. Fager, "Analysis of thermal coupling effects in integrated MIMO transmitters," in *Proc. IEEE MTT-S Int. Microwave Symp. (IMS)*, Honolulu, HW, USA, Jun. 2017, pp. 75–78.

[18] D. Gómez, C. Dufis, J. Altet, D. Mateo, and J. L. González, "Electrothermal coupling analysis methodology for RF circuits," *Microelectron. J.*, vol. 43, no. 9, pp. 633–641, Sep. 2012.

[19] M. Garven and J. P. Calame, "Simulation and optimization of gate temperatures in GaN-on-SiC monolithic microwave integrated circuits," *IEEE Trans. Compon. Packag. Technol*, vol. 32, no. 1, pp. 63–72, Mar. 2009.

[20] G. N. Ellison, *Thermal Computations for Electronics: Conductive, Radiative, and Convective Air Cooling.* CRC Press, 2011.

[21] C. Huang, Y. Liang, T. Li, G. Xiong, and S. Wu, "The FEM analysis of stress and strain in stacked solder bump under power load," in *Proc. 14th Int. Conf. Electronic Packaging Technology*, Dalian, China, Aug. 2013, pp. 587–590.

[22] "Wafer Level Chip Scale Packaging (WLCSP)," Amkor Technology, Tempe, AZ, USA, Data Sheet DS720I, 2017. [Online]. Available: www.amkor.com

Comprehensive Investigation on Warpage Management of FOPLP with Multi Embedded Ring Designs

Chang-Chun Lee[1], Yan-Yu Liou[1], Pei-Chen Huang[1], Fussen Hsu[2], Puru Bruce Lin[2], Cheng-Ta Ko[2], Yu-Hua Chen[2]

[1] Microsystems Mechanical Design & Reliability Analysis Laboratory,
Dept. of Power Mechanical Engineering, National Tsing Hua University, Hsinchu 300 , Taiwan, R.O.C.
[2] Dept. of New Business Development, Unimicron, Taiwan, R.O.C.
Phone: 886-3-5162410. Fax: 886-3-5722840, and E-mal: cclee@pme.nthu.edu.tw

Abstract—On the urgent multi-function requirements of mobile products and internet of things, a more than Moore's rule becomes a major concern in accordance with the scaling limitation of advanced devices. Accordingly, the architecture of fan out panel-level packaging (FOPLP) grows into the mainstream to meet the essentials of three-dimensional chip stacking and heterogeneous integration. Nevertheless, unfavorable warpage and residual stress issues resulted from the reconstruction and assembly processes of FOPLP are introduced, decreasing the assembly yield and subsequently long-term packaging reliability. To resolve the above-mentioned problem, this research proposes a 370 × 470 mm² carrier size FOPLP vehicle combined the novel designs of double embedded metal rings. The major geometrical characteristics of the foregoing metal rings are separately embedded in the multi-redistribution layers (RDLs) and epoxy molding compound (EMC) regions to control the process-induced warpage. On the basis of simulation-based factorial design, the key material selections, such as RDLs composed of photoimagable dielectric (PID) and EMC within the embedded metal rings, is investigated to significant restrain the induced warpage. Notably, a stiffer and a lower coefficient of thermal expansion (CTE) for PID are indicated to be an effective solution of miniaturizing the induced warp during the corresponding curing step. Consequently, a superior capability in warpage reduction after debonding process of panel level vehicles is acquired by means of the present simulation methodology and experimental validation.

Keywords-FOPLP packaging; chip-last process; finite element simulation; warpage bahavior; multi-embedded rings

I. INTRODUCTION

On the basis of development on smart devices with the concept of internet of things, the low consumption/high performance devices are continuously evolved to meet the requirement of Moore's law. However, the physical limitation of nano-scaled device is restricted the further expansion of More Moore. For this reason, more than Moore technologies has becomes the promising solutions to enhance the performance of heterogeneous device. The panel-based advanced fan-out packaging, Fan-Out Panel Level Packaging (FOPLP) have the superior area usage rate as compared with the wafer-based technologies [1-3]. Thus, the FOPLP-based techniques are regarded as the major solution for next-generation electronic packaging [4]. The fan-out packaging can be assorted into two main techniques, chip-first and chip-

last processes [5]. In view of the foregoing process, many researches is focused on the manufacture process and layout design of different fan-out packaging [6-15]. However, the heterogeneous integration on planar design is restricted by the scaling limitation of advanced packaging [16-18]. Therefore, the concept of three-dimensional integration is proposed to accomplish the high performance and multi-functional operation with high I/O density, such like the InFO (Integration Fan-Out, InFO) architecture evolved by TSMC [19]. In addition, the thermo-mechanical responses induced by the fabrication process and subsequent reliability test also play an important role on the yield of FOPLP vehicle. The reliability issue of FOPLP included die warp [20-22], die shift [22-23], and failure mechanism in multi-redistribution layers (RDLs) [24]. Among them, the warpage profile is the critical issue for the large size FOPLP and significantly influenced the assembly and solder reliability. In addition, novel material system, such like epoxy molding compound (EMC), is also widely adopted in the fan-out based packaging [25-35]. The foregoing material introduced the new resource of thermal deformation and the corresponding material selection and design rule is needed to achieve the better warpage control ability of FOPLP architecture. For those reasons, this study is mainly focused on the warpage estimation of FOPLP vehicle with narrow line width/spaces of 2μm/2μm.

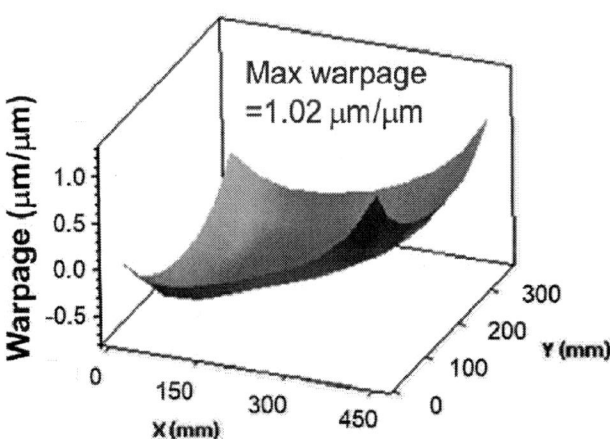

Figure 1. Experimental warpage measurement of FOPLP vehicle.

The feasibility of chip-last FOPLP vehicle is demonstrated and an actual warpage measurement is performed to investigate the warpage behavior of presented packaging (Referred to Fig. 1). However, the warpage profile of single packaging contained in entire panel is not fully examined. On the other hand, the material combination design effect on the FOPLP vehicle is also not estimated to further control the process-induced warpage issue. Furthermore, the multi-rings designs are embedded in the RDL and EMC regions to restrain the process-induced warpage in chip-last FOPLP process. The layout design of single packaging in FOPLP vehicle is shown as Fig. 2, the inner and outer pin area are separately designed with different pin density. In accordance with the cross-sectional view in Fig. 2, three dielectric layers is deposited on the temporary glass substrate. In this study, four kinds of photoimagable dielectric (PID) and EMC are separately adopted to restrain the process-induced warpage. Moreover, multi embedded rings design are also employed in PID and EMC regions to further inhibit the warpage issue of presented FOPLP vehicle.

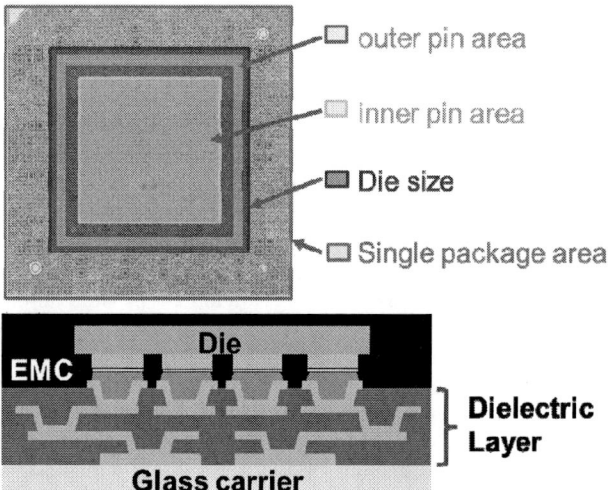

Figure 2. Layout design of FOPLP vehicle utilized in this study.

II. FINITE ELEMENT MODELING OF FOPLP PACKAGING VEHICLE

A. Equivalent FEA model of single FOPLP vehicle

The structure components and dimensions of presented FOPLP vehicle is shown as Fig. 3, the detailed RDL structure is modeled as equivalent multi-layer architecture using volume percentage of material composition. The inner and outer pin are also modeled by the same method. Under the assumption of plane strain condition, a half symmetric 2D FEA model of presented vehicle with multi rings embedded in PID and EMC regions is constructed as illustrated in Fig. 4, the constructed FEA model are contained 8100 four-node solid element, the x-directional displacements of symmetric axis of packaging is fixed, and

the y-directional displacement of the bottom of symmetric axis is also fixed to prevent the rigid body motion.

TABLE I. MATERIAL COMPARISON OF SEVERAL PID AND EMC UTILIZED IN THIS STUDY

Properties	PID	EMC
Young's modulus (GPa)	B > A > D > C	1 > 3 > 2 > 4
CTE(ppm/℃)	C > D > A > B	3 < 1 < 4 < 2
Curing Temperature (℃)	A > C > B > D	4 < 1 = 2 = 3

B. Fabrication process-oriented simulation flow of presented packaging vehicle

The process-oriented simulation flow is shown as Fig. 5, the RDL layer is electroplating on the glass carrier, then deposited the subsequent PID layer with embedded metal ring, the foregoing processes are repeated three times, and finished the lead layer to accomplish the RDL structures.

Accordingly, the silicon chip is bonded on the RDL structure enabled by the reflow process. The second metal ring is located on the outside of lead region. After molding process of EMC, the packaging vehicle is debonding from the glass substrate to finish the fabrication of single packaging vehicle.

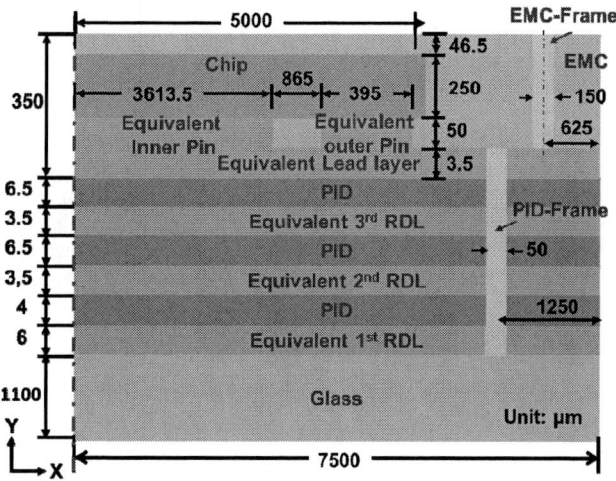

Figure 3. Structural dimensions of FOPLP vehicle integrated with multi-embedded ring designs in RDL and EMC regions.

III. RESULTS AND DISCUSSION

To achieve an estimation of warpage behavior during each process step, the normalized warpage profile on top surface of FOPLP vehicle are extracted and the corresponding physical mechanism are described in following sections. Therefore, the simulation-based parametric study is performed to examine the warpage control effects of different concerned material combinations, such as PID and EMC. The comparison of material characteristics of foregoing materials is listed as table. 1.

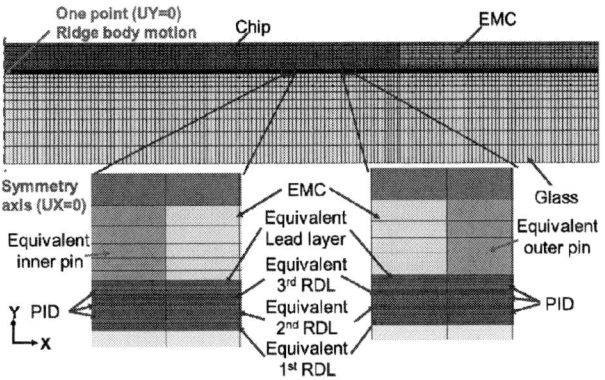

Figure 4. FEA model of FOPLP vehicle with corresponding boundary conditions adopted in warpage simulation.

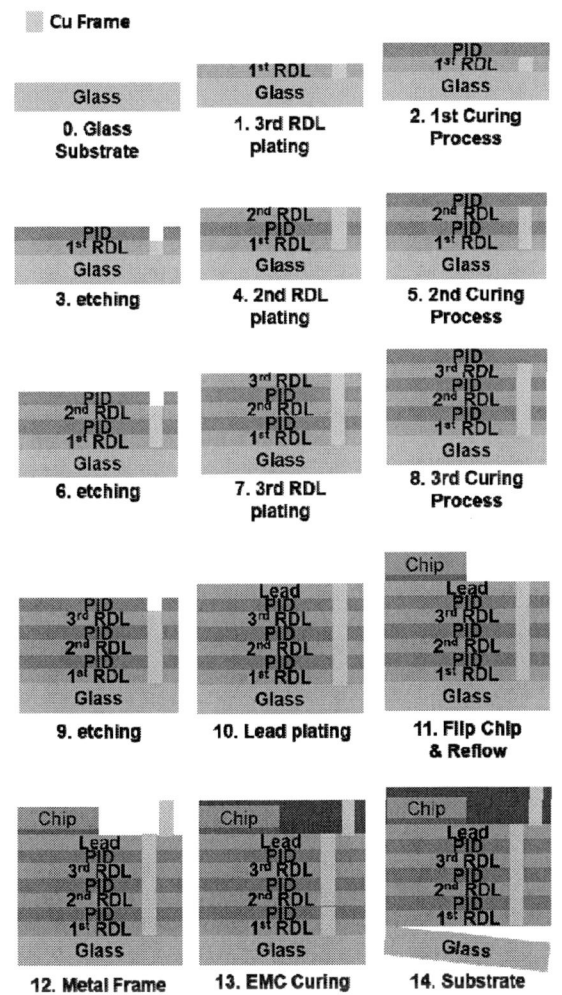

Figure 5. Fabrication process-oriented simulation flow of presented FOPLP vehicle.

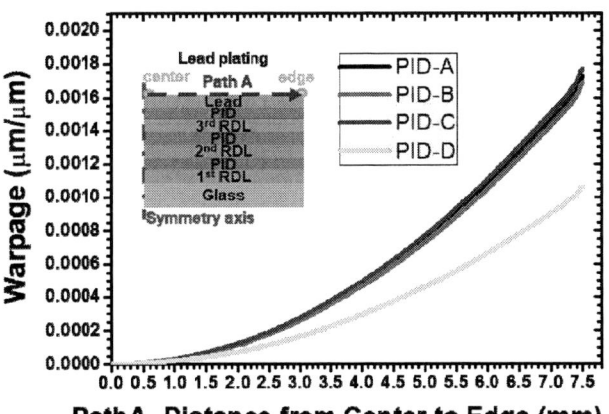

Figure 6. Warpage profile along top surface of single packaging after lead plating.

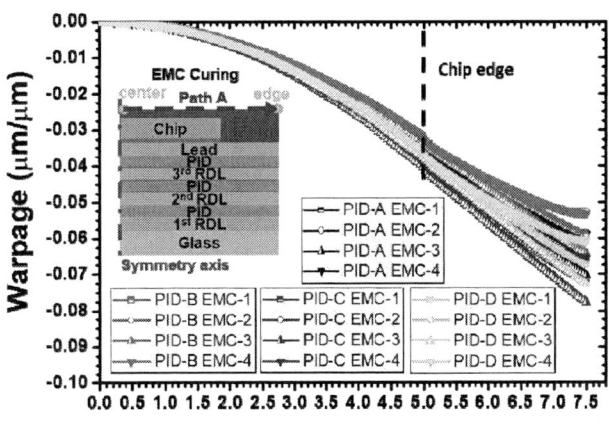

Figure 7. Warpage profile along top surface of single packaging after EMC curing.

A. Warpage profile after RDL process

The warpage profile of vehicle top surface after RDL process is shown as Fig. 6, four different types of PID material is adopted to observe the process induced warpage magnitude. The warpage profile of FOPLP packaging with type A, B, C PID shows significant difference as compared with the type D PID. This behavior could be attributed to the lowest curing temperature of type D PID among all PID materials. Thus, the coefficient of thermal expansion (CTE) mismatch between glass carrier and deposited RDL layer could be subsided when the type D PID is accomplished, inducing the low warpage after RDL process. In addition, the curing temperature difference between type A, B, C PID are not obvious, the warpage behavior is mainly determined by the Young's modulus and CTE of PID material itself at

978-1-7281-1500-9/19 $31.00 © 2019 IEEE

this time. For this reason, type B PID revealed the better warpage behavior among type A, B, C PID material.

B. Warpage profile after EMC curing process

In this session, four types of EMC material are further adopted in the parametric study to explore its influence on the warpage profile after EMC curing process. Thus, totally sixteen curve of warpage profile is shown in Fig. 7 when considering several types of PID and EMC materials. After EMC curing process, the warpage profile transferred from smile face type to cry face type, this phenomenon could be ascribed to the EMC thickness is almost 12 times as thick as RDL structure. Thus, the harsher CTE mismatch mechanism is induced after EMC process as compared with RDL process step. Accordingly, the change of warpage profile is dominated by the EMC process. The simulated results revealed that the type B PID integrated with EMC 2 shows the smoothest warpage profile after EMC curing process. This result showed that the stiffer PID material is conduced to restrain the thermal deformation induced by the CTE mismatch between EMC and glass carrier.

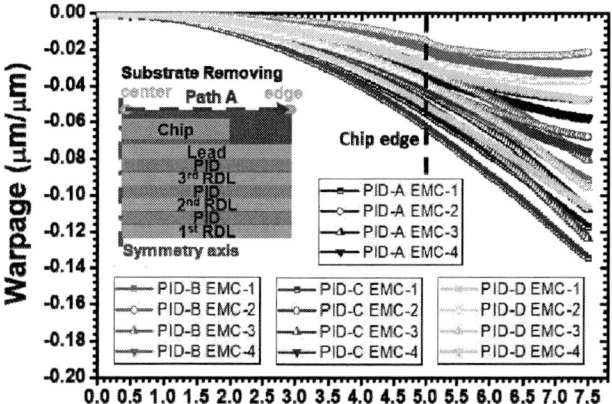

PathA- Distance from Center to Edge (mm)

Figure 8. Warpage profile along top surface of single packaging after debonding from temporary glass substrate, comparing totally sixteen combination with various PID and EMC materials.

C. . Warpage profile after FOPLP vehicle removed from glass carrier

Similar to the warpage curves represented in Fig. 8, the warpage profile after packaging removed from glass carrier shows the cry face type behavior. An almost flat warpage profile is observed on the fan-out region of EMC top surface and the corresponding normalized warpage value is being converged to -0.02 when the material selections of type B PID and EMC-2 are taken into account.

D. Warpage profile of FOPLP vehicle with multi-embedded rings in PID and EMC regions

After ensuring the best material combinations based on the simulated results presented in previous session, the multi-

rings design is further utilized to estimate the warpage restraint ability of copper ring. Comparison of warpage profile between presented FOPLP vehicle with and without PID & EMC embedded rings are shown in Fig. 9. It is found that the warpage restraint mechanism is observed on the local region of ring embedded in EMC region. This mechanism might be attributed to the micro level tiny dimensions of ring adopted in this study, indicting different warpage control behavior as compared with the meso-level ring design revealed in previous literatures [36-38]. Thus, the embedded ring design with present dimensions only expressed the local warpage control ability, not comprehensive warpage restraint capability among whole FOPLP vehicle.

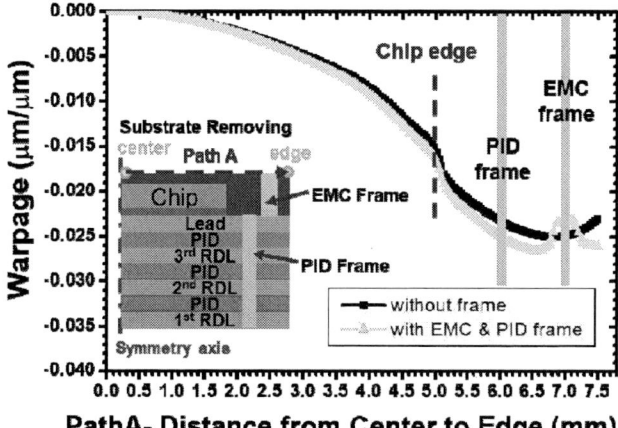

PathA- Distance from Center to Edge (mm)

Figure 9. Warpage profile of packaging vehicle without metal rings as compared with the vehicle with PID and EMC multi embedded rings.

IV. CONCLUSIONS

FOPLP technology has a high potential to increase the yield as compared with the wafer-based fabrication process. However, the process-induced warpage becomes a major reliability issue when the large panel size, complex material selection, and different CTE mismatch mechanism occurred in each process are taken into account. For those reasons, this study performed the process-oriented simulation to explore the warpage behavior of each critical process step in FOPLP architecture. Through the parametric study presented in this research, the appropriate material selection of PID and EMC for warpage restraint could be revealed. In addition, the multi-embedded copper rings are also adopted in the presented FOPLP vehicle to investigate its effect on warpage behavior. The presented results revealed that the material selections composed of the type B PID and EMC-2 shows the warpage control ability after EMC process and removed from glass. In addition, the multi-embedded rings also expressed the warpage restriction potential, but this behavior only represented in the local region of copper ring located at the EMC region. This phenomenon might be attributed the narrow structure dimensions of ring itself. The results presented in this study are beneficial in settling process-induced warpage issue during the development of FOPLP packaging in the future.

ACKNOWLEDGMENT

The authors would like to thank the Ministry of Science and Technology (MOST), Taiwan, R.O.C., for providing financial support under contract numbers MOST 105-2628-E-007-015-MY3 and MOST 107-2622-E-007-010-CC3.

REFERENCES

[1] F. X. Che , X. Zhang, and J. K. Lin, "Reliability study of 3D IC packaging based on through-silicon interposer (TSI) and silicon-less interconnection technology (SLIT) using finite element analysis," Microelectron. Reliab., vol. 61, pp. 64–70, 2016.

[2] R. McCleary, P. Cochet, T. Swarbrick, C. Sim, G. Singh, C. B. Yong, and A. K. Aung, "Panel Level Advanced Packaging," in Proc. 66th Electron. Compon. Technol. Conf. (ECTC), Las Vegas, NV, USA, pp. 25-30, 2016.

[3] H. D. Chang, D. Chang, K. Liu, H.S Hsu, R. F. Tai, H.C. Huang, Y. C. Lai, C. L. Lu, C. T. Lin, and S. Chiu, "Development and Characterization of New Generation Panel Fan-out (P-FO) Packaging Technology," in Proc. 64th Electron. Compon. Technol. Conf. (ECTC), Orlando, FL, USA, pp. 947–951, 2014.

[4] T. Braun, M. Töpper, S. Raatz, S. Voges, R. Kahle, V. Bader, J. Bauer, K.-F. Becker, T. Thomas, R. Aschenbrenner, and K.-D. Lang, "Opportunities and Challenges for Fan-out Panel Level Packaging (FOPLP)," SiP Global Summit, 2015.

[5] S. Chen, S. Wang, J. Hunt, W. Chen, L. Liang, G. Kao and A . Peng, "A Comparative study of a Fan Out Packaged Product : Chip First and Chip Last," in Proc. 66th Electron. Compon. Technol. Conf. (ECTC), Las Vegas, NV, USA, pp.1483-1488, 2016.

[6] T. Braun, K.-F. Becker, S. Voges, J. Bauer, R. Kahle, V. Bader, T. Thomas, R. Aschenbrenner, and K.-D. Lang, "24"x18" Fan-out Panel Level Packing," in Proc. 64th Electron. Compon. Technol. Conf. (ECTC), San Diego, CA, USA, pp. 940-946, 2014.

[7] K. Kikuchi, Y. Karasawa, Y. Nedzu, K. Takano, T. Yoshinobu, and T. Sugino, "Research of Fan-Out Panel Level Package (FOPLP) manufacturing process using single-sided adhesive tape," in Proc. CPMT Symposium Japan (ICSJ), Kyoto, Japan, pp. 83-86, 2017.

[8] M. Woehrmann, T. Braun, M. Toepper, and K.-D. Land, "Ultra-thin 50 um Fan-Out Wafer Level Package: Development of an innovative assembly and de-bonding concept," in Proc. 68th Electron. Compon. Technol. Conf. (ECTC), San Diego, CA, USA, pp. 675-681, 2018.

[9] K. Kikuchi, Y. Nedzu, and T. Sugino, "Warpage Analysis with Newly Molding Material of Fan-Out Panel Level Packaging and the Board Level Reliability Test Results," in Proc. 68th Electron. Compon. Technol. Conf. (ECTC), San Diego, CA, USA, pp. 973-978, 2018.

[10] T. Braun, S. Raatz, S. Voges, R. Kahle, V. Bader, J. Bauer, K.-F. Becker, T. Thomas, R. Aschenbrenner, and K.-D. Lang, "Large Area Compression Molding for Fan-out Panel Level Packing," 65th Electron. Compon. Technol. Conf. (ECTC), San Diego, CA, USA, pp. 1077-1083, 2015.

[11] T. Braun, K.-F. Becker, S. Voges, T. Thomas, R. Kahle, V. Bader, J. Bauer, R. Aschenbrenner, and K.-D. Lang, "Challenges and opportunities for Fan-out Panel Level Packing (FOPLP)," 9th Microsystems, Packaging, Assembly and Circuits Technology Conference (IMPACT), Taipei, Taiwan, pp. 154 – 157, 2014.

[12] K. Best, G. Singh, and R. McCleary, "Advanced packaging lithography and inspection solutions for next generation FOWLP-FOPLP processing," 17th International Conference on Electronic Packaging Technology (ICEPT), Shanghai, China, pp. 1090-1094, 2016.

[13] E. D. Silveira, S. Gurvinder, and M. Roger, "Advanced Packaging Lithography and Inspection Solutions for Next Generation FOWLP-FOPLP Processing," Device Packaging HiTEC, HiTEN, & CICMT, pp. 1-36, 2017.

[14] C.T Ko, H. Yang, J. H. Lau, M. Li, M. Li, C. Lin, J. W. Lin, T. Chen, L. Xu, C.-L. Chamg, J.-Y. Pan, H.-H. Wu, Q. X. Yong, N. Fan, E. Kuah, Z. Li, K. H. Tan, Y.-M. Cheung, E. Kai, J. Hao, R. Beica, M. Lin, Y.-H. Chen, Z. Cheng, K. S. Wee, J. Ran, C. Xi, S. P. Lim, and N.C. Lee, "Chip-First Fan-Out Panel-Level Packaging for Heterogeneous Integration," IEEE T Com. Pack. Man., vol. 99 pp. 1561-1572, 2018.

[15] M. C. Lu, "Warpage Management for Fan-Out Packaging Moving from Wafer Level to Panel Level," in Proc. 68th Electron. Compon. Technol. Conf. (ECTC), San Diego, CA, USA, pp. 360-367, 2018.

[16] G. Pares, C. Bouvier, M. Saadaoui, J. Mazuir, J. Noiray, K. Martinschitz, A. Planchais, and G. Simon, "3D embedded wafer-level packaging technology development for smart card SIP application," in Proc. 14th Electronics Packaging and Technology Conference (EPTC), Singapore, pp. 304-310, 2012.

[17] S. W. Yoon, P. Tang, R. Emigh, Y. Lin, P. C. Marimuthu, and R. Pendse, "Fanout Flipchip eWLB (embedded Wafer Level Ball Grid Array) Technology as 2.5D Packaging Solutions," in Proc. 63rd Electron. Compon. Technol. Conf. (ECTC), Las Vegas, NV, USA, pp. 1855-1860, 2013.

[18] K. Chen, J. A. Caparas, L. Chua, Y. Lin, and S. W. Yoon, "Advanced 3D eWLB-PoP (embedded Wafer Level Ball Grid Array - Package on package) technology," in Proc. Microsystems, Packaging, Assembly and Circuits Technology Conference (IMPACT), Taipei, Taiwan, pp. 96-100, 2015.

[19] S. C. Chong, D. H. S. Wee, V. S. Rao, and N. S. Vasarla, "Development of package-on-package using embedded wafer-level package approach," IEEE Trans. Compon. Pack. Manuf. Technol., vol. 3, no. 10, pp. 1654-1662,Oct. 2013.

[20] H. W. Liu, Y.-W. Liu, J. Ji, J. Liao, A. Chen, Y.-H. Chen, N. Kao, and Y.-C. Lai, "Warpage Characterization of Panel Fan-out (P-FO) Package," in Proc. 64th Electron. Compon. Technol. Conf. (ECTC), Orlando, FL, pp. 1750–1754, 2014.

[21] S. S. Deng, S.-J. Hwang, and H.-H. Lee, "Warpage prediction and experiments of fan-out wafer level package during encapsulation process," IEEE Trans. Compon. Packag. Manuf. Technol. vol. 3, no. 3, pp.452–458, 2013.

[22] T.H. Kuah, J. Y. Hao, J. P. Ding, Q. F. Li, W. L. Chan, S. C. Ho, H. M. Huang, and Y. J. Jiang, "Encapsulation challenges for wafer level packaging," in Proc. 11th Electron. Packag. Technol. Conf. (EPTC), pp. 903–908, 2009.

[23] G. Sharma, A. Kumar, V. S. Rao, S. W. Ho, and V. Kripesh, "Solutions strategies for die shift problem in wafer level compression molding," IEEE Trans. Compon. Packag. Manuf. Technol. vol. 1, no. 4, pp. 502–509, 2011.

[24] C.Y. Chou, T.-Y. Hung, S.-Y. Yang, M.-C. Yew, W.-K. Yang, and K.-N. Chiang, "Solder joint and trace line failure simulation and experimental validation of fan-out type wafer level packaging subjected to drop impact," Microelectron. Reliab., vol. 48, no. 8-9, pp. 1149–1154, 2008.

[25] L. Laplatine, E. Luan, K. Cheung, D. M. Ratner, Y. Dattner, and L. Chrostowski, "System-level integration of active silicon photonic biosensors using Fan-Out Wafer-Level-Packaging for low cost and multiplexed point-of-care

diagnostic testing" Sensors and Actuators B: Chemical vol. 273: pp. 1610-1617, 2018.

[26] C. Song, S. Kim, and S. Kim, "Fabrication and Characteristics of Spin-on Dielectric for Multi-level Interconnect in WLP," in Proc. 68th Electron. Compon. Technol. Conf. (ECTC), San Diego, CA, USA, pp. 1879-1884, 2018.

[27] M. Woehrmann et al., in Proc. 68th Electron. Compon. Technol. Conf. (ECTC), San Diego, CA, USA, pp. 675-681, 2018.

[28] C. Jones, S. Burgess, T. Wilby, and P. Densley, "UBM/RDL Deposition by PVD for FOWLP in High Volume Production," in Proc. 68th Electron. Compon. Technol. Conf. (ECTC), San Diego, CA, USA, pp. 1226-1232,2018.

[29] C. H. Yu, L. J. Yen, C. Y. Hsieh, J. S. Hsieh, Victor C. Y. Chang, C. H. Hsieh, C. S. Liu, C. T. Wang, KC Yee, and D. C. H. Yu, "High Performance, High Density RDL for Advanced Packaging," in Proc. 68th Electron. Compon. Technol. Conf. (ECTC), San Diego, CA, USA, pp. 587-593,2018.

[30] C. H. Lee, J. Su, X. Liu, Q. Wu, J.-W. Lin, P. Lin, C.-T. Ko, Y.-H. Chen, W.-W. Shen, T.-Y. Kou, S.-Y. Huang, Y.-M. Lin, K.-N. Chen, and A.Y. Lin, "Optimization of Laser Release Process for Throughput Enhancement of Fan-Out Wafer-Level Packaging," in Proc. 68th Electron. Compon. Technol. Conf. (ECTC), San Diego, CA, USA, pp. 1824-1829,2018.

[31] A. Lujan, "Comparison of Package-on-Package Technologies Utilizing Flip Chip and Fan-Out Wafer Level Packaging," in Proc. 68th Electron. Compon. Technol. Conf. (ECTC), San Diego, CA, USA, pp. 2089-2094,2018.

[32] S. H. You, S. Jeon, D. Oh, K. Kim, J. Kim, S.-Y. Cha, and G. Kim, "Advanced Fan-Out Package SI/PI/Thermal Performance Analysis of Novel RDL Packages," in Proc. 68th Electron. Compon. Technol. Conf. (ECTC), San Diego, CA, USA, pp. 1295-1301,2018.

[33] T. G. Lim, and D. H. S. Wee, "Electrical Design for the Development of FOWLP for HBM Integration," in Proc. 68th Electron. Compon. Technol. Conf. (ECTC), San Diego, CA, USA, pp. 2142-2148,2018.

[34] A. Martins, , M. Pinheiro, A. F. Ferreira, R. Almeida, F. Matos, J. Oliveira, H. M. Santos, M. C. Monteiro, H. Gamboa, and R. P. Silva, "Heterogeneous Integration challenges within Wafer Level Fan-Out SiP for Wearables and IoT," in Proc. 68th Electron. Compon. Technol. Conf. (ECTC), San Diego, CA, USA, pp. 1485-1492,2018.

[35] F. X. Che, D. Ho, M. Z. Ding, and D. R. MinWoo, "Study on Process Induced Wafer Level Warpage of Fan-Out Wafer Level Packaging," in Proc. 66th Electron. Compon. Technol. Conf. (ECTC), Las Vegas, NV, pp. 1879-1885, 2016

[36] C. K. Yu, W. S. Chiang, P. S. Huang, M. Z. Lin, Y. H. Fang, M. J. Lin, C. Peng, B. Lin, and M. Huang, "Reliability Study of Large Fan-Out BGA Solution on FinFET Process," in Proc. 68th Electron. Compon. Technol. Conf. (ECTC), San Diego, CA, USA, pp. 1623-1627, 2018.

[37] F. X. Che, D. Ho, and T. C. Chai, "Study on Warpage and Reliability of Fan-Out Interposer (FOI) Technology," in Proc. 68th Electron. Compon. Technol. Conf. (ECTC), San Diego, CA, USA, pp. 1-11, 2018.

[38] S. McCann1, H. H. Lee, R.-A. Gamal, T. Lee, and S. Ramalingam, "Warpage and Reliability Challenges for Stacked Silicon Interconnect Technology in Large Packages," in Proc. 68th Electron. Compon. Technol. Conf. (ECTC), San Diego, CA, USA, pp. 2345-2350, 2018.

Development of High Power and High Junction Temperature SiC Based Power Packages

Gongyue Tang*, Leong Ching Wai, Teck Guan Lim, Yong Liang Ye, Ravinder Pal Singh, Lin Bu, Boon Long Lau, Tai Chong Chai, Kazunori Yamamoto and Xiaowu Zhang
Institute of Microelectronics, A*STAR (Agency for Science, Technology and Research),
2 Fusionopolis Way, #08-02 Innovis Tower, Singapore 138634.
*Email: tangg@ime.a-star.edu.sg

Abstract— In this paper, a half-bridge leg single phase power package with SiC based MOSFETs (metal-oxide semiconductor field-effect transistors) on a AMB (active metal brazing) substrate was developed. The developed power package consists of a AMB substrate with specially designed cavities, four high power rated SiC chips, three types of customized copper clips forming the source and gate interconnects. High temperature die attach/solder material for the chip, substrate and copper clip bonding, and high temperature endurable encapsulation mould compound (EMC) for cavity and gap filling have been evaluated. The fabricated power packages were undergone the specified reliability assessments, i.e. unbiased highly accelerated stress test (HAST) test, thermal cycling (TC) test (-40~200°C), High temperature storage (HTS) test at 250°C and power cycling test (ΔT=150°C). 5 out of 5 samples passed the standard unbiased HAST test and HTS test for 500 hours. 4 out of 5 samples passed the TC test for 1000 cycles and 3 out of 5 samples passed the power cycling test for 10000 cycles. Electrical open failures were detected between clip 1, traces on the substrate and gate pads of the SiC chips. Interconnects between clip 2/3 and source pads of the SiC chips show good connections. Delamination between the clip 1 and sintered Ag were observed on the failed samples by the cross section failure analysis which is the potential root causes of electrical failure.

Keywords- power package; power inverter module; SiC chip; dual-side-cooling, wide band gap semiconductor.

I. INTRODUCTION

The needs of high temperature endurable and high performance power packages are increased dramatically with the evolution of the high power electronics devices for aerospace, automotive, solar panel, wind generator, and power grid applications, in which high power management, switching and conversion are critical [1, 2]. In general, power package, which consists of several power semiconductor devices (e.g., field-effect transistor (FET) dies and diodes) and assembled with different packaging technologies, provides mechanical support, electrical interconnection, protection and thermal management to the power semi-conductor devices. Correspondingly, the power package performance relies on the characteristics of the power semiconductor devices, the packaging technology and the package configuration. As one of the wide band gap (WBG) devices, SiC chips have excellent mechanical and thermal properties, as well higher critical breakdown field in comparison with silicon chips. Owning to its superior

properties over the conventional silicon based power package, the SiC chip based power package has attracted extended interests [3].

However, one of the key challenges of these wide band gap devices is the heat dissipation at the device junction area during operation when they deal with much higher power than that for the conventional silicon based power package. Conventional Si (silicon) based power devices have the limitation on the continuous operation temperature limit and it should be operated under 150°C normally and maximum 175°C at the junction, however, the wide band gap devices such as SiC still can be functional even above 250~300°C. In this point of view, the packaging technologies of the wide band gap devices with higher junction temperature than that of the Si power device is considered as the key limitation because the packaging process and materials of Si based power device is defined to meet the temperature up to 150~175°C only. For the case of SiC based high power devices with much higher junction temperature over 200~250°C, it is crucial to develop the new packaging technology [4, 5].

In this paper, a SiC device based power package capable for double side cooling scheme is designed and developed for high power, high performance application. The high power application is attributed to the customized interconnects and the use of paralleled high-power SiC devices in the developed package. High temperature endurable materials for the interconnections and encapsulation enables high temperature sustainability of the package. In order to develop this high power, high performance, with double side cooling capability, SiC chip based power package, the following works have been carried out in this study. Firstly, design analysis has been conducted to optimize the design of the proposed power package. Then investigation has been conducted to evaluate the high temperature endurable interconnection and encapsulation materials, as well the substrate surface finishing. After that, the optimized package has been fabricated and assembled. Lastly, the reliability assessment has been conducted for the developed power package and failure analysis has been conducted for the failed samples.

II. SiC BASED POWER PACKAGE DESIGN

The schematic structure of the proposed SiC power package is shown in Fig. 1, which presents the conceptual design of the proposed SiC power package with (a) single side cooling solution and (b) advanced dual side cooling solution. The proposed package mainly includes SiC power devices which are embedded in the ceramic based substrate, copper clips which are attached to SiC power devices and substrate to form the interconnects. A liquid cooled heat sink is attached to the bottom side of the power package with one layer of electrically isolated thermal interface material (TIM), forming a single side cooled power package (refer to Fig. 1(a)). Another liquid cooled heat sink can be added to the top side of the package with another layer of electrically isolated TIM to form the dual side cooled power package (refer to Fig. 1(b)).

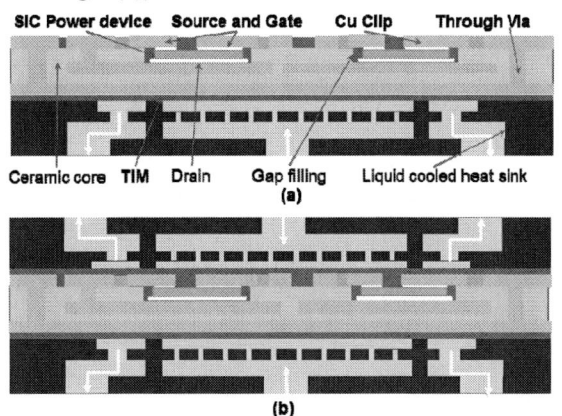

Figure 1. Conceptual design of the proposed SiC power package with (a) single side cooling solution and (b) advanced dual side cooling solution

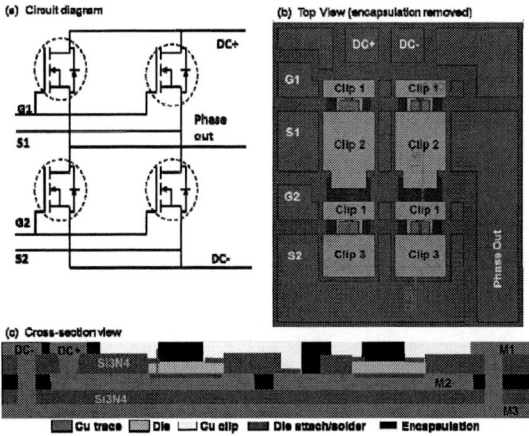

Figure 2. Design of the proposed SiC power package for (a) Circuit diagram of the single phase high power package with 4 SiC DMOSFETs (b) Top view of the proposed SiC power package with encapsulation removed (c) Cross section view of the proposed SiC power package

Fig. 2 shows detailed design of the proposed SiC power package for (a) Circuit diagram of the single phase high power package with 4 SiC MOSFETs (b) Top view of the proposed SiC power package with encapsulation remove (c) Cross section view of the proposed SiC power package. Fig. 3(a) presents the design of AMB substrate and Fig. 3(b) presents the design of the copper clips. As shown in Fig. 2, four discrete SiC power devices are embedded into AMB substrate (refer to Fig. 2 (b) and Fig. 2(c)) to form the power

package with a single-phase circuit (refer to Fig 2(a)). The designed power package is with small factor of 34.5mm x 29.5mm x 1.74mm and capable for double side cooling solution. In detail, four cavities are firstly designed in the AMB substrates of the power package (refer to Fig. 3(a)). The upper two cavities are with size of 5.1mm x 7.9mm and the lower two cavities are with size of 5.1mmx10.8mm. Then four commercial SiC power devices [7], each device is with size of 3.1mm x 5.9mm x 0.18mm, are embedded into the AMB substrate. Moreover, three types of customized metal clips with size of 2.0mm x 6.1mm x 0.25mm, 8.8mm x 6.1mm x 0.25mm and 4.92mm x 6.1mm x 0.25mm (refer to Fig. 3(b)) are bonded with the gate and source pads of the SiC chips and AMB substrate at the same level of the top metal layer of the substrate, and a flat surface on the top side of the power package is achieved, as such the advanced dual side cooling solution can be enabled.

The applied SiC power device is MOSFET (CPMF-1200-S040B) from CREE which is shown in Fig. 4. The implemented SiC power device is with drain source breakdown voltage up to 1200V and the typical rating of RDS(on) is 40mΩ. The device size is 3.1mm x 5.9mm x 0.18mm with backside metallization of Ni/Ag (0.8/0.6μm) and front side metallization of Al (4 μm).

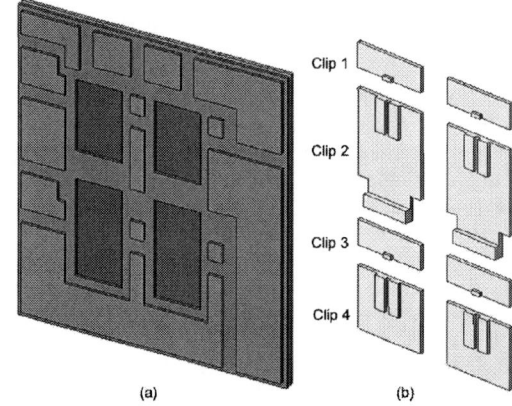

Figure 3. (a) Design of the AMB substrate and (b) design of the copper clips

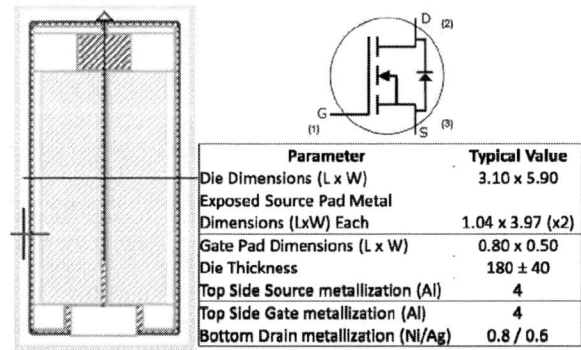

Parameter	Typical Value
Die Dimensions (L x W)	3.10 x 5.90
Exposed Source Pad Metal Dimensions (LxW) Each	1.04 x 3.97 (x2)
Gate Pad Dimensions (L x W)	0.80 x 0.50
Die Thickness	180 ± 40
Top Side Source metallization (Al)	4
Top Side Gate metallization (Al)	4
Bottom Drain metallization (Ni/Ag)	0.8 / 0.6

Figure 4. SiC MOSFET device (CPMF-1200-S040B) feature and its specifications

978-1-7281-1500-9/19 $31.00 © 2019 IEEE

III. MATERIAL EVALUATION AND SELECTION

Two critical materials for the proposed SiC power package are evaluated, one critical material is the interconnect bonding material and the other critical material is the encapsulation material.

Considering the high temperature application, silver (Ag) sintering material is selected as the bonding material, and shear test has been conducted to evaluate the bonding strength of the Ag sintering layer [8,9]. Two types of Ag sintering materials are introduced for evaluation and the shear test samples based on these two materials are built as shown in Fig. 5, which shows the images of the shear test samples for Ag sintering material evaluation.

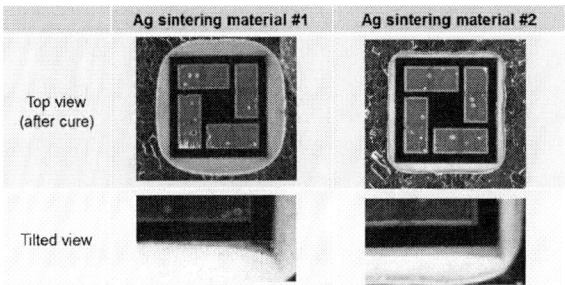

Figure 5. Image of the shear test samples for Ag sintering material evaluation

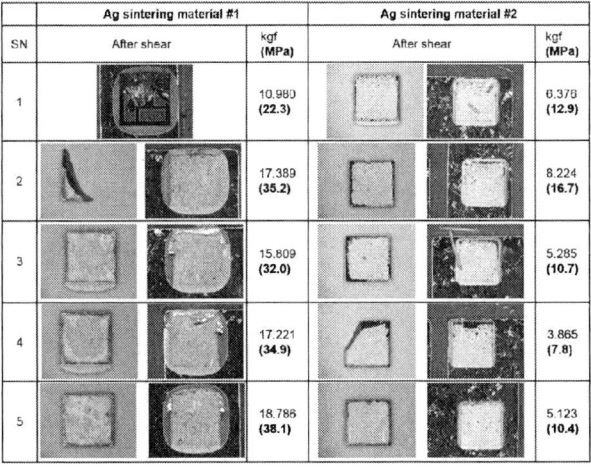

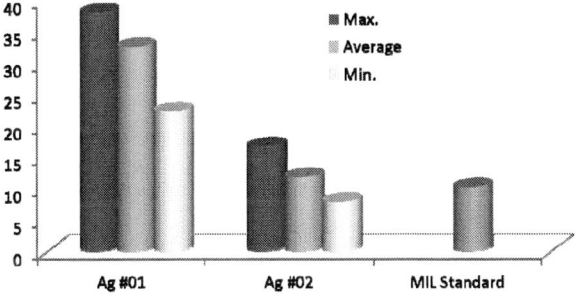

Figure 6. Shear test results of the Ag sintering material

The shear test results for Ag sintering material is shown in Fig. 6. It can be seen from Fig. 6 that the shear stress for Ag sintering material #01 is much higher than that for the Ag sintering material #02. While the average shear stress for Ag sintering#02 is marginally higher than the average shear stress requirements for the military standard. Therefore, the Ag sintering material #1 is selected as the bonding material to build the samples for final reliability assessment.

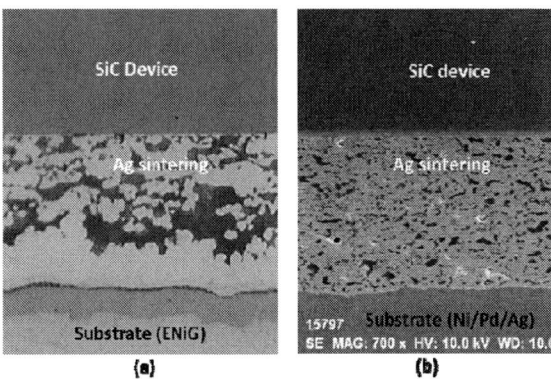

Figure 7. SEM images of the Ag sintering layer between SiC device and substrate with difference surface finishing after HTS for 100 hours for (a) substrate with ENIG surface finishing and (b) substrate with Ni/Pd/Ag surface finishing

In order to evaluate the bonding quality of the Ag sintering paste to the substrate with different surface finishes, two types of test vehicles have been built with two different substrates. One substrate is with electroless nickel immersion gold (ENIG) surface finish and the other substrate is with Ni/Pd/Ag multi-layer-metallization surface finish. These two types of test vehicles were subjected to high temperature storage (HTS) test at 250°C for 100hours. After 100hours, the samples were cross-sectioned and the quality of the Ag sintering layer is evaluated through the scanning electron microscopy (SEM). The evaluation results are shown in Fig.7, which presents the SEM images of the Ag sintering layer between SiC device and substrate with difference surface finishes after HTS for 100 hours for (a) substrate with ENIG surface finish and (b) substrate with Ni/Pd/Ag surface finish. It can be seen from the figure that delamination at bonding interface was observed for the test sample with ENIG surface finish due to the Ag diffusion into the Ag and formation of dense Au-Ag alloy layer. While there are no defects observed at the Ag sintering layer for sample with Ni/Pd/Ag surface finish, suggesting good bonding quality between the Ag sintering paste and the substrate with Ni/Pd/Ag surface finish can be achieved. Hence, the substrates with Ni/Pd/Ag surface finish is used to build the samples for final reliability assessment.

Besides the Ag sintering materials and substrate surface finish, different encapsulation materials have been evaluated as per their cavity filling capability and high temperature sustainability. The results are shown in Fig. 8, which presents the encapsulation material evaluation and selection through (a) cavity filling process and (b) high temperature storage (HTS) test (250°C, 100 hours). Form Fig. 8(a), it can be found that big voids underneath the copper clips are observed for the encapsulation material named CF#4, while on the other hand, the gaps can be fully filled with the encapsulation materials named CF#5, CF#6 and CF#8, indicating that encapsulation materials of CF#5, CF#6 and CF#8 are with good cavity filling capability, while CF#4 is not a suitable encapsulation material for this application. Furthermore, the samples with CF#5, CF#6 and CF#8 were subjected to high temperature storage (HTS) test at 250°C for 100hours, and the results are shown in Fig. 8(b). It can be seen from Fig. 8(b) that encapsulation materials CF#5 and CF# 6 are not suitable for such high temperature application (i.e. 250°C) while encapsulation material CF#8 is a good candidate for the proposed SiC power package

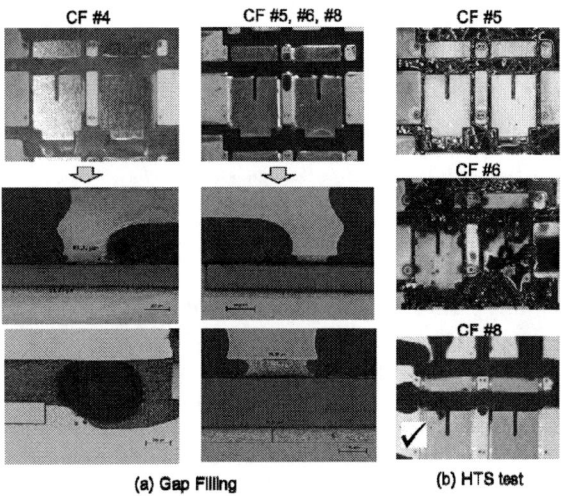

Figure 8. Encapsulation material evaluation and selection through (a) gap filling and (b) high temperature storage (HTS) test (250°C, 100 hours)

IV. PROCESS OPTIMIZATION AND ASSEMBLY FLOW OF THE SiC POWER PACKAGE

It is well known that the dispensing pattern of die attach is important to produce the complete filling under the chip, as such to form good quality interconnections between the chip and substrate. In this study, the dispensing pattern of the Ag sintering paste based die attach has been evaluated and the results are presented in Fig. 9, which shows the SiC chip to substrate attachment process for (a) dispensing pattern, (b) Image of die attach between chip and substrate and (c) X-ray image of die attach between the chip substrate. It can be seen from Fig. 9 that a dispensing pattern with a cross design (pattern #1) is not able to produce complete Ag sintering filling under the chip, while the dispensing pattern with an Asterisk design can produce the complete Ag sintering paste filling. Furthermore, the asterisk design based dispensing pattern with shortest horizontal line (i.e. pattern #4) produces the best fillet

coverage and complete fill of Ag sintering paste between the chip and substrate.

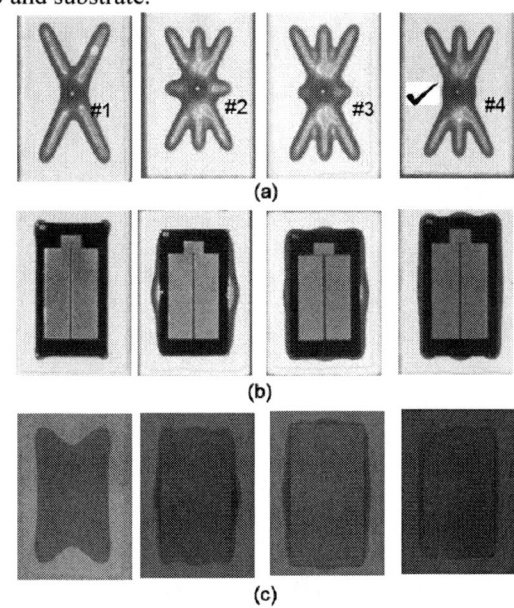

Figure 9. SiC chip to substrate attachment: (a) despensing pattern, (b) Image of die attach between chip and substrate and (c) X-ray image of die attach between the chip substrate

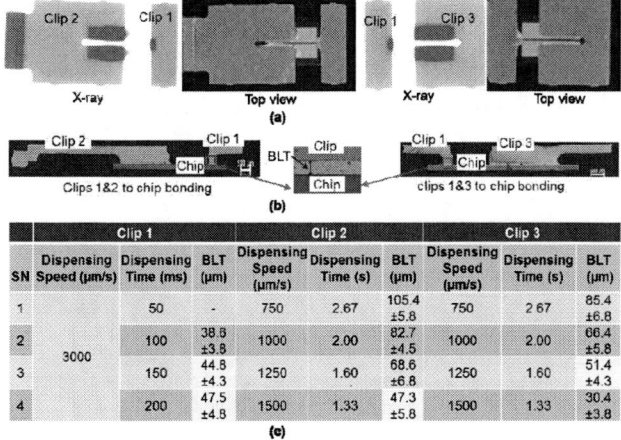

SN	Clip 1			Clip 2			Clip 3		
	Dispensing Speed (μm/s)	Dispensing Time (ms)	BLT (μm)	Dispensing Speed (μm/s)	Dispensing Time (s)	BLT (μm)	Dispensing Speed (μm/s)	Dispensing Time (s)	BLT (μm)
1	3000	50	-	750	2.67	105.4 ±5.8	750	2.67	85.4 ±6.8
2		100	38.8 ±3.8	1000	2.00	82.7 ±4.5	1000	2.00	66.4 ±5.8
3		150	44.8 ±4.3	1250	1.60	68.6 ±6.8	1250	1.60	51.4 ±4.3
4		200	47.5 ±4.8	1500	1.33	47.3 ±5.8	1500	1.33	30.4 ±3.8

(c)

Figure 10. Clip to SiC chip bonding process evaluation: (a) Image of the samples for clip to chip bonding process, (b) Cross-section image of the samples for clip to chip bonding process and (c) clip to chip bonding evaluation parameters and BLT results

Besides chip to substrate attachment process, clip to chip bonding process and the control of bond line thickness (BLT) are very challenging and it is also evaluated in this study. The evaluation and optimization results are presented in Fig. 10, which summarizes clip to SiC chip bonding process evaluation: (a) image of the samples for clip to chip bonding process, (b) cross-section image of the samples for clip to chip bonding process and (c) clip to chip bonding evaluation parameters and BLT results. In short, BLT meets the target value (i.e. 40μm – 60μm) for clip 1 when the

978-1-7281-1500-9/19 $31.00 © 2019 IEEE

dispensing speed is 3000 μm/s and dispensing time is 150ms or 200ms. BLT meets the target value (i.e. 40μm – 60μm) for clip 2 when the dispensing speed is 1500 μm/s and dispensing time is 1.33s. BLT meets the target value (i.e. 40μm – 60μm) for clip 3 when the dispensing speed is 1250 μm/s and dispensing time is 1.6s.

The assembly process flow for proposed package are shown in Fig. 11. As seen from Fig. 11, the assembly process starts from the AMB substrate with specially designed cavities. The major steps for the proposed power package assemble include [10]:

1. Silver sintering material dispensing on the bottom layer of the AMB substrate
2. SiC chip attachment and curing
3. Silver sintering material dispensing on SiC chip and top layer
4. Copper clip attachment and curing
5. Cavity filling and curing

As mentioned in the previous sections, Ag sintering material #01, substrate with Ni/Pd/Ag surface finish and encapsulation material #8 are identified for final samples. the picture of test vehicles for the developed SiC power packages is shown in Fig. 12.

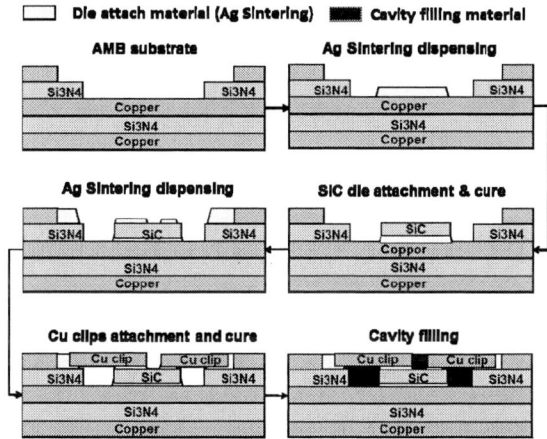

Figure 11. Optimized assembly process flow for the developed SiC power package

Figure 12. Picture of the assembled final SiC power package with Ni/Pd/Ag surface finishing and cavity fill material #08

V. RELIABILITY ASSESSMENT FOR THE DEVELOPED SIC POWER PACKAGE

The test vehicles of the developed SiC power packages are undergone the reliability assessment with specified testing conditions. The test items include the HAST, TC test, HTS test and PC test. The test conditions, methods and process for each test item are listed in Table I, including on Resistance measurement and C-SAM scan. The evaluation results show that the test vehicles are still functional after each step of reliability assessments as summarized in Table I. 5 test vehicles were tested for each reliability assessment item. It can be seen from Table I that all of the allocated test vehicles passed unbiased HAST test and HTS test at 250°C for 500hours. 4/5 test vehicles passed the TC test for 1000 cycles and 3/5 test vehicles passed power cycling test (ΔT=150°C) up to 10,000 cycles.

Table I: Summary of reliability assessment for developed SiC power package

Item	Samples	Conditions	Test Method	Process	Remark
HAST	5	Unbiased HAST JESD22-A110B 130°C / 85% 96 hours	CSAM inspection and DC Electrical test before & after	Rds measurement	5/5 passed
				C-SAM Before	
				Reliability Test	
				Rds measurement	
				C-SAM after	
TC	5	-40°C to +200°C, 15 min. dwell, 1000 cycles	Measure resistances before, at 100, 250, 500, 750, 1000 cycles	Rds measurement 100 cycle	4/5 passed
				Rds measurement 250 cycle	
				Rds measurement 500 cycle	
				Rds measurement 750 cycle	
				Rds measurement 1000 cycle	
				Rds measurement	
HTS	5	250 °C, 500 hours	Measure resistances before, at 100, 250, 500 hours	Rds measurement 100 hours	5/5 passed
				Rds measurement 250 hours	
				Rds measurement 500 hours	
				Rds measurement	
Power Cycling	5	ΔT = 150°C 10,000 cycles	CSAM inspection and DC Electrical test before & after	Rds measurement	3/5 passed
				C-SAM Before	
				Power Cycling test	
				Rds measurement	
				C-SAM after	

VI. FAILURE ANALYSIS

Failure analysis was conducted to the samples failed during the TC and power cycling test. At first, non-destructive electrical probing method was used to check the electrical connections between the clips, SiC chips and AMB substrate. Electrical open was detected between clip 1 and gate pad of SiC chip for the failed sample of the TC test. Electrical opens were detected between clip 1 and copper trace on the AMB substrate for the failed samples of the power cycles test. The Interconnections between clip 2, clip 3, top Cu pad and source/drain of SiC chip show good connections. Thus the clip 1 connection failures for all the failed samples are identified. Then, cross-section was conducted for all the failure samples and Fig. 13 shows cross section location along X-X and Y-Y directions on the failed power package. X-X direction is along the top clip 1. where the electrical opens were observed. Y-Y direction is along the right side of clip 1, 2 and 3. The failure mechanism can be identified by check the interconnection between clip 1, top Cu pad and gate of SiC chip.

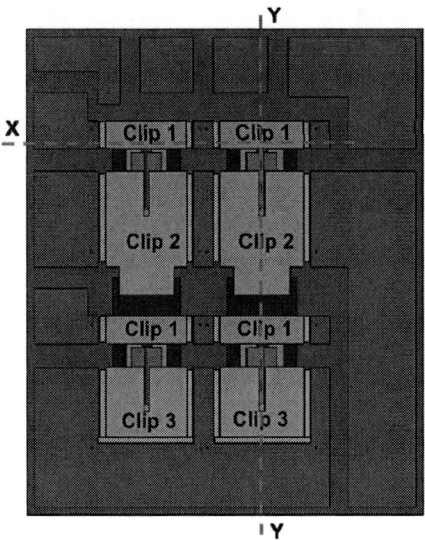

Figure 13. Cross section location along X-X and Y-Y directions.

Figure 14 shows the cross section of interconnection bottom right clip 1 to gate of SiC chip. It can be found that the Cu clip 1 is not connected with the gate of SiC chip at this location. The dispensed sintered Ag is not enough to form perfect interconnection which is the potential root causes of open failure under the tests. Figure 15 shows the cross section of SiC chip bonding on the middle Cu pad of the substrate with sintered Ag. It can be seen that good bonding sintered Ag was formed between the SiC chip and middle Cu pad. The sintered Ag is with porous structure. Figure 16 shows the cross section of interconnection of right clip 2 to middle Cu pad on substrate. Good sintered Ag bonding was formed between right clip 2 to middle Cu pad of substrate observed from the SEM picture.

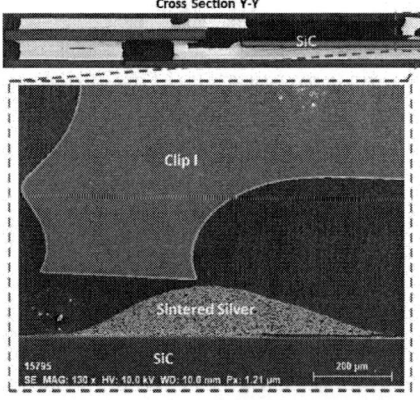

Figure 14. Cross section of interconnection bottom right clip 1 to gate of SiC chip along Y-Y direction.

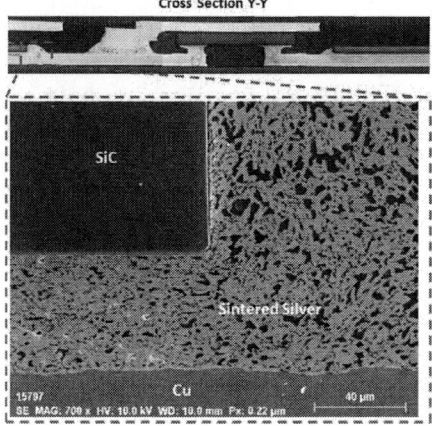

Figure 15. Cross section of SiC chip bonding on the middle Cu pad with sintered Ag along Y-Y direction.

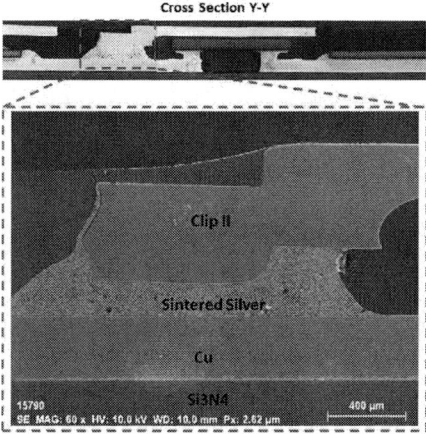

Figure 16. Cross section of interconnection of right clip 2 to middle Cu pad of substrate along Y-Y direction.

Figure 17 and 18 show cross section of interconnection of top left and right clip 1 to top Cu pad of substrate along X-X direction. It can be seen that delaminations were observed at the interface between the clip 1 and sintered Ag which is the potential root causes of electrical open failure under thermal cycling test. The sintered Ag was peeled off under the thermal cycling loading, suggesting the bonding of the Ag sintering to the Cu clips 1 need to be improved.

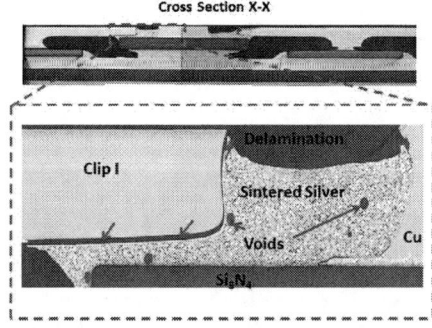

Figure 17. Cross section of interconnection of top left clip 1 to top Cu pad of substrate along X-X direction.

978-1-7281-1500-9/19 $31.00 © 2019 IEEE   1424

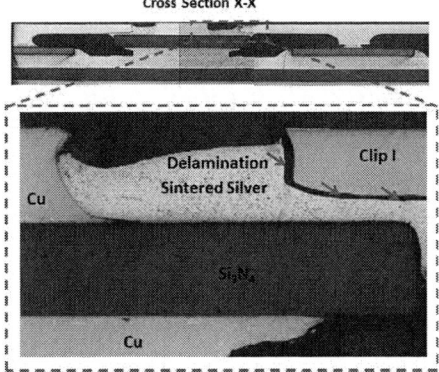

Figure 18. Cross section of interconnection of top right clip 1 to top Cu pad of substrate along X-X direction.

VII. SUMMARY

In this paper, a SiC device based power package with double side cooling capability is designed and developed for high power, high performance application. The developed power package mainly consists of high power rated SiC chips, customized copper clips and AMB substrates with specially designed cavities. The customized interconnects and the use of paralleled high-power SiC devices in the developed package enable its usage for high power applications. High temperature sustainability is obtained by utilizing the high temperature endurable materials for the interconnections and encapsulation. High performance is realized by applying the special substrate design and implementing the dual side liquid cooling solution. In addition, by embedding the chip inside the AMB substrate and replacing the wire-bond interconnections with the flatted copper clip interconnections, the developed power package is with low profile. High temperature endurable package materials (e.g. die attach and encapsulation material) have been evaluated. The developed power package has been fabricated and passed the specified reliability assessments, i.e. unbiased HAST, TC test (-40~200°C), HTS test at 250°C and power PC test (ΔT=150°C). 5 out of 5 samples passed the standard unbiased HAST test and HTS test for 500 hours. 4 out of 5 samples passed the TC test for 1000 cycles and 3 out of 5 samples passed the power cycling test for 10000 cycles. Electrical opens were detected between clip 1 and, copper trace of the substrate and gate pads of the SiC chips of the failed samples. Interconnects between clip 2/3 and source pads of the SiC chips show good connections. Delamination between the clip 1 and sintered Ag were observed on the failed samples by the cross section failure analysis which is the potential root causes of electrical failure.

ACKNOWLEDGMENTS

This study is supported by the power electronics packaging consortium project, and the members of the consortium include Fairchild, GLOBALFOUNDRIES, Kulicke & Soffa and Shin-Etsu. The author would sincerely thank the consortium members and the technical staffs in IME for their supports and contributions to this project.

REFERENCES

1. R. Khazaka, L. Mendizabal, D. Henry, and R. Hanna, "Survey of high-temperature reliability of power electronics packaging components", IEEE Trans. on Power Electronics, Vol. 30, No. 5 (2015), pp. 2456-2464.

2. Horio, M. et al. "New Power Module Structure with Low Thermal Resistance and High Reliability for SiC Devices." PCIM Europe. 2011, Proceeding CD, p.229-234.

3. Satoshi Tanimoto and Kohei Matsui, "High Junction Temperature and Low Parasitic Inductance Power Module Technology for Compact Power Conversion Systems," IEEE Trans. on Electron Devices, Vol. 62, No. 2 (2015), pp. 258-269.

4. Daniel Rhee Min Woo; Hwang How Yuan; Jerry Aw Jie Li; Lee Jong Bum; Zhang Hengyun, "Miniaturized Double Side Cooling Packaging for High Power 3 Phase SiC Inverter Module with Junction Temperature over 220°C," Proc IEEE 66th Electronic Components and Technology Conference (ECTC), Las Vegas, Nevada, May 2016, pp. 1190 – 1196.

5. Daniel Rhee M. W.; Hwang H. Y.; Jerry Aw J. L.; Ho S. L.; Lee J. B.; Zhang S. B.; Zhang H. Y.; Susai L. Selvaraj; Sorono D. V. and Ravinder P. S., "High power SiC inverter module packaging solutions for junction temperature over 220°C," Proc 16th Electronics Packaging Technology Conference (EPTC), Singapore, Dec. 2014, pp. 31-355.

6. G. Y. Tang, T. C. Chai, and X. W. Zhang, "Thermal Optimization and Characterization of SiC Based High Power Electronics Packages with Advanced Thermal Design", IEEE Trans-CPMT, 2018, DOI 10.1109/TCPMT.2018.2860998 (online)

7. CREE_ CPM2_1200_0040B datasheet, www.cree.com

8. Lee J. B.; Hwang H. Y.; Pan W. C.; Rhee M. W. Daniel, "Interfacial reaction and reliability of high temperature die attach solders for power electronics", Proc 17th Electronics Packaging Technology Conference (EPTC), Singapore, Dec. 2015, pp. 1-5.

9. Lee Jong Bum, Aw Jie Li, Hwang How Yuan, Pan Wei Chih, Yong Ling Xin and Rhee Min Woo Daniel, "Characterization of pressure-less Ag sintering on Ni/Au surface", Proc 17th Electronics Packaging Technology Conference (EPTC), Singapore, Dec. 2015, pp. 1-4.

10. Leong Ching Wai; Mian Zhi Ding; Gong Yue Tang, "High Temperature Endurable Die Attach Material for Power Electronics Package – Process Challenges", Proc 19th Electronics Packaging Technology Conference (EPTC), Singapore, Dec. 2017, pp. 1-5.

New Developments of Copper Plating Technology for Embedded Power Chip Packages Challenges

Yung-Da Chiu, Shiu-Chih Wang, David Tarng, An-Tai Wu, Allenyl Chen, Louis Chen and Chi-Tsung Chiu
Advanced Semiconductor Engineering (ASE) Inc., Kaohsiung, Taiwan
E-mail: StevenYD_Chiu@aseglobal.com

Abstract–Copper plating has been extensively employed in the fabrication of embedded packaging to reach high-density, high-speed, high performance electronic products. With through holes (TH) as well as blind via aspect ratios increase, development of a reliable plating technology is very important. When the depth of through hole was over 200um, it is difficult to fill without void by using direct current (DC) electroplating. In order to overcome this problem, organic additives were applied to cause faster copper deposition at the TH center rather than at the opening. Besides of that, the x-shape through hole was also developed due to its particular geometry form, which was beneficial for the copper bridge at the TH center. In this paper, the x-shape through hole with depth of 350 um was fully filled by DC electroplating within 2hrs. This filling ability development enable the embedded chip package design more flexibility.

I. INTRODUCTION

The content of advanced Embedded Active System Integration (a-EASI) was lead frame based single or multi-dice embedded in organic laminate material. The process of the embedding of thin chips into build-up layers was by the use of well-known substrate technology. Electrical contacts to the chips were achieved by laser-drilled and metallized microvias as schematically shown in fig. 1. Figure 2 was a-EASI cross section example of embedded power module which contained dice attached on lead frame.

The a-EASI lead-frame based die embedding application was mainly focused on the power packaging such as multichip for DC/DC converter modules, power device, metal-oxide-semiconductor field-effect transistor (MOSFET) and fast switching power, insulated gate bipolar transistor (IGBT) etc. Figure 3 showed the representative product of a-EASI.

The technology is to embed passive or active devices into substrate. The laser drilling via metallization was subsequently applied to form electric path from chip I/O pad to outside. The structure provides shorter interconnection for the fast electrical response application.

The embedded die package had several advantages such as miniaturization (at least 20~25% area reduction) and excellent electrical performance (>80 % reduction in resistance). The applied area includes multi chip power module, power discrete device, voltage regulator, analog IC or high thermal condition package etc.

As electronic devices become smaller and include more functions, the complexity of PCB design continues to expand.

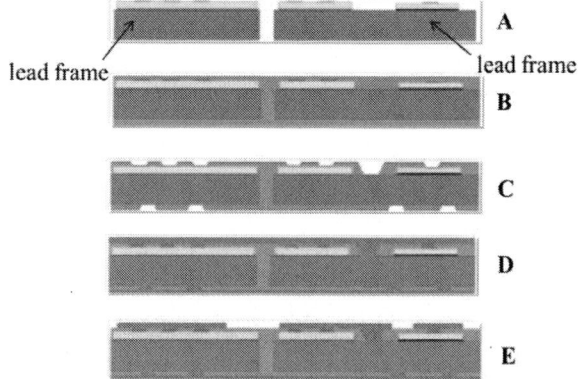

Figure 1: Process steps of chip embedding: (a) die attach, (b) polymer layer lamination, (c) laser drilling of vias to chip and substrate, (D) metallization of , (E) patterning

Figure 2: The cross-section of embedded module

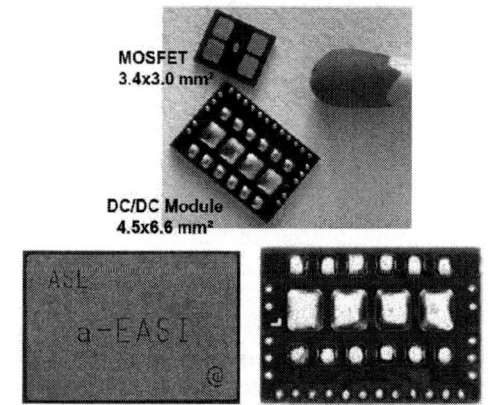

Figure 3: The examples of a-EASI product: Power device, MOSFET and DC/DC converter module.

The embedded packaging had several benefits as blow:
· Miniaturization & Design Flexibility
- Embedded chip enables more space for other components or shrinks overall solution
- Design flexibility shifts from 2D to 3D

• Improved Thermal & Electrical Thermal Performance
- Lower electrical & thermal resistivity in package improves power performance
• Improved Reliability and Mechanical Stability
- High mechanical system stability due to stable Cu interconnections

Through-holes (THs) and microvias played the electrical interconnection role in the embedded packaging to reach high-density, high-speed, high performance electronic products. With through holes as well as blind via aspect ratios increasing demands, development of a reliable plating technology was very important to meet the subsequent reliability test.
The filling of blind micro via has been widely studied; the so-called bottom-up filling was achieved attributable to the interaction between the accelerator and the suppressor [1-3]. However, there have been fewer studies of TH filling by copper plating. The through hole is a bit different from blind micro as there was no bottom area during copper plating. In order to overcome the lack of bottom, a strategy was utilized to cause the fastest copper deposition occurring at the TH center rather than at the opening, which means the current density distribution can be changed by organic additives along the TH. Otherwise, the result will tend to create a seam or void left in the TH as well as blind micro via seen in fig. 4 [9].

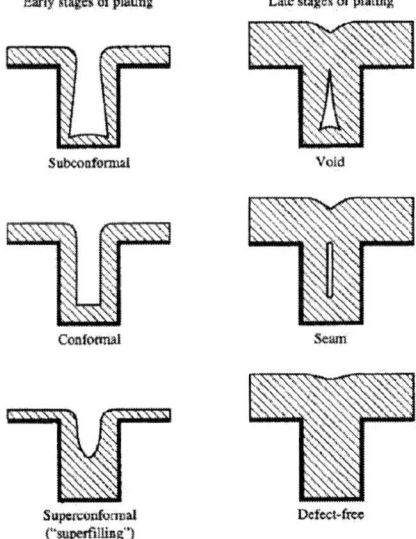

Figure 4: The three different filling model of micro via. [9]

During the plating process, the crosses sectional shape of the filled copper looked like a butterfly. Therefore, it was also referred to the butterfly technology (BFT) proposed by Wei-Ping Dow et al. [8]. Figure 5 showed the he butterfly plating steps.
With increasing the core thickness for IC substrates, it is extremely difficult for TH to fill without void formation and require a long plating time. For up to 200 μm substrate thickness, the drilled method was changed from mechanical drilling to laser drilling due to advantages such as faster drilling rates and higher quality [4, 5].

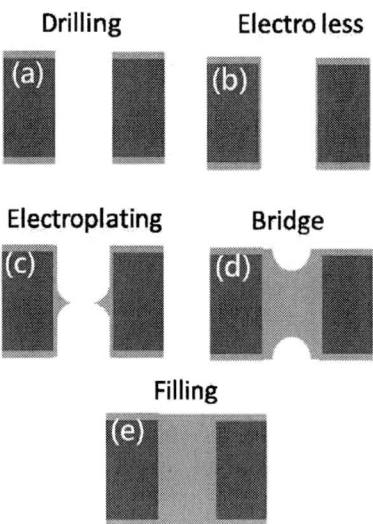

Figure 5: Illustration of the butterfly plating steps.

The drilled outcome of laser drilling for TH displayed an x-shape formation which was different from column-shape of mechanical drilling as seen in fig. 6.
Fortunately, this geometry characteristic was beneficial for the copper bridge at the TH center, which avoids the void left on the TH and reduces the plating time. Besides of that, the organic additives also play a significant influence on the plating results. The ratio between brightener (B), leveler (L) and carrier (C) must have an optimum value to reach synergistic effect for better filling performance during copper plating.

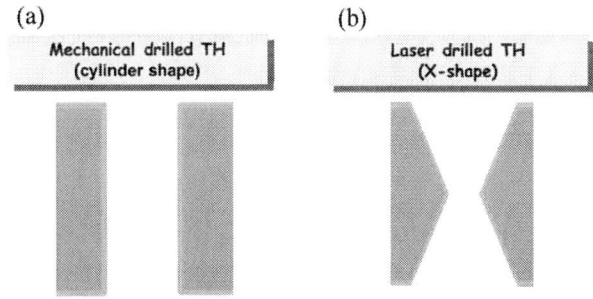

Figure 6: Diagram of (a) mechanical drilling and (b) drilled through holes.

II. RESULTS AND DISCUSSION

A. Plating formula testing

Figure 7 showed the galvanostatic measurements for different ratio additive addition in the plating bath. Here, the leveler and carrier were controlled at 4ml/L and 10ml/L respectively. Different amounts of brightener were mixed into the plating solution, which were 4 ml/L, 5 ml/L, and 6ml/L. From the potential monitor curves, the overpotential decreased with increasing the amount of brightener addition (shift to more positive potential) as seen in fig. 7(a~c). The depolarization results were caused from the synergetic interaction between the brightener and leveler.

The result may be caused by the unbalance between leveler and brightener during plating. The filling performance of these three conditions were also investigated and compared. Figure 9(a) showed poor filling results at lower brightener concentration which was due to weaker reduction rate in the center of through hole.

To overcome the issue of weak acceleration, the concentration of brightener was adjusted to 5ml/L and 6 ml/L respectively as shown in fig. 9(b,c). It was obviously found the filling performance improved dramatically with increasing the concentration of brightener.

To verify the entire filling results, an x-ray inspection had been employed for analysis these array. X-ray can penetrate the samples to explore the void or seam inside the x-shape column. The filling performance of plating condition with B/L=6/4 ml/L was better than the other two through the inspection of x-ray in fig. 9(d~f).

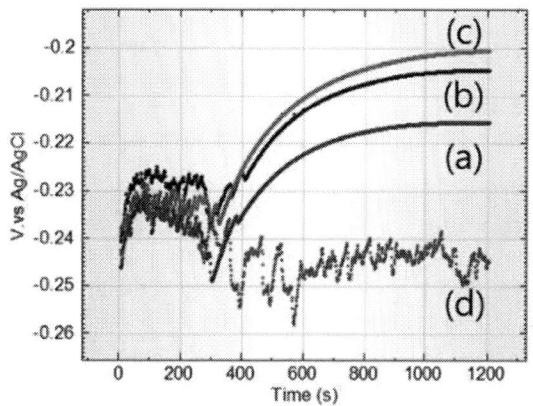

Figure 7: Polarization curves for Cu plating bath with 4ml/L leveler, 10 ml/L carrier and (a) 4 ml/L brightener (b) 5 ml/L brightener, (c) 6ml/L (d) 0ml/L brightener respectively.

From the fig. 8, we found the ratio between B/L played an important role during the plating. If the ratio of B/L was lower than 1.5, the filling ability will become poor. Besides of that, a lot of copper nodules also will be formed on the surface to reduce the reliability. According to previous works, Cl⁻ is a key additive in a copper plating solution. It will interact with brightener and leveler to exert their functions during the copper electrodeposition [6, 7].

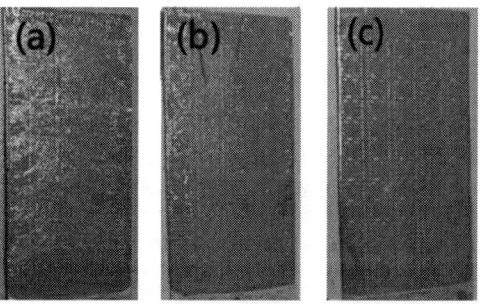

Figure 8: Top view image of test vehicle after electroplating. The ratios of B/L were (a) 4/4 ml/L, (b) 5/4 ml/L, and (c) 6/4 ml/L.

Figure 10 showed the influence of chloride ion concentration on filling performance. Increasing the chloride ion could markedly improve the void size to become small at fixed plating formula. Figure 10(c) even found filling results without voids left on the center of x-shape TH when the chloride ion increased to 90ppm.

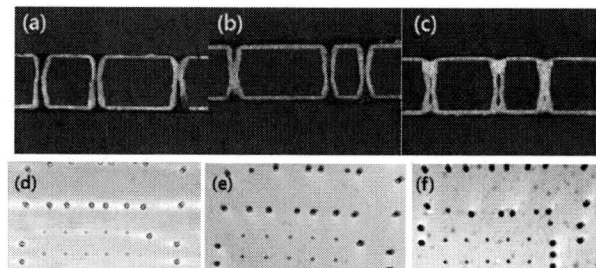

Figure 9: Cross-sectional images of x-shaped through hole after copper electroplating. The additive ratio of B/L was (a) 4/4 ml/L, (b) 5/4 ml/L, (c) 6/4 ml/L, (d~e) was the X-ray inspection corresponding to (a~c).

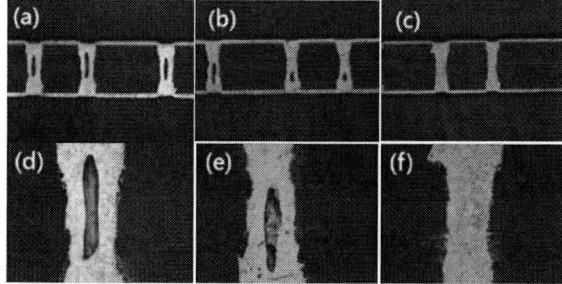

Figure 10: Cross-sectional images of x-shaped through hole after copper electroplating. The plating solution were composed of 3ml/L L, 10 ml/L C, 5ml/L B and (a)30ppm, (b)60ppm, (c,)90ppm, Cl⁻ respectively. (d~f) was the magnification corresponding to (a~c).

Thus, appropriate addition of chloride ion concentration and ratio between additives achieved a good filling performance for high depth x-shape TH.

B. Test vehicle for x-via reliability test

The prepared x-via test vehicle information was as below :
Top diameter: 120 +/-15 um
Via depth: 350 +/-20 um
Neck <= 50 % via diameter
Via space design from 75 to 300 um

The plating conditions were chosen from the previous optimum conditions in the follow-up experiments. The temperature cycle test (TCT), highly accelerated temperature humidity stress test (HAST), and high temperature storage test (HTST), were subsequently used to conduct reliability tests. Detail test conditions were presented in Table 1 followed the JEDEC standard and the scanning acoustic tomography (SAT) monitoring was preliminarily used to determine reliability pass or fail. Table 2 checked the total TCT results from SAT analysis. From the results of monitoring, showing that there was no abnormal appearance observed after TCT 500, 1000, 2000 cycles in fig. 11(a~c).

Table 1. Reliability test conditions

Reliability item	condition
TCT	-55°C/125°C, 500X , 1000X and 2000X SAT and Cross section
HAST	130°C/85%, 96h and 192h SAT and Cross section
HTST	150°C, 500h and 1000h SAT and Cross section

Table 2. Reliability test results for TCT.

Strip No	Via Space	TCT-500	TCT-1000	TCT-2000
		Result	Result	Result
6	75µm	30/30	30/30	30/30
5	300 µm	30/30	30/30	30/30
4	225 µm	30/30	30/30	30/30
3	120µm	29/29	29/29	29/29

* Failure quantity/Total sample quantity

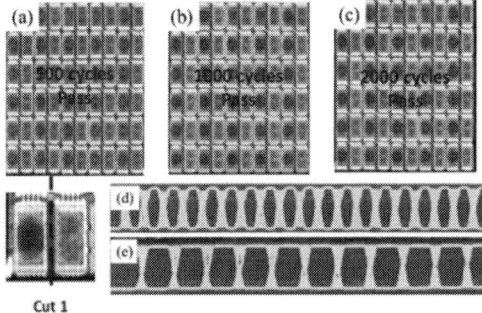

Cut 1

Figure 11: SAT image of TCT result of different cycle (a)500, (b)1000, (c) 2000 cycles and 2000 cycles SEM x-section image of x-via space of (d) 75 um, (e)300um

The cross section images of TCT 2000 cycles also did not displayed any delamination or crack for x-via space of 75um and 300 um respectively in fig. 11(d, e).The TCT 1000 cycles was also taken into inspect. In cut1 position, it was found slight delamination between polypropylene (PP) and lead frame around the corner. But the x-via still maintained its original geometric structure in cut 2 position as seen in fig. 12. The other two reliability tests were completed through HAST and HTST. From the fig. 13 and 14, SAT monitoring showed the samples were all passed. The results of test vehicle were summarized in Table 3 and 4.

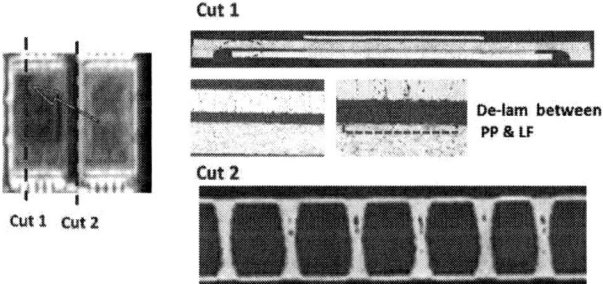

Figure 12: SAT image of TCT 1000 cycles and OM x-section images of position in cut1and cut2

Table 3. Reliability test results for HAST.

Strip No	Via Space	HAST 96	HAST 192
		Result	Result
6	75µm	30/30	30/30
5	300 µm	30/30	30/30
4	225 µm	30/30	30/30
3	120µm	29/29	29/29

* Failure quantity/Total sample quantity

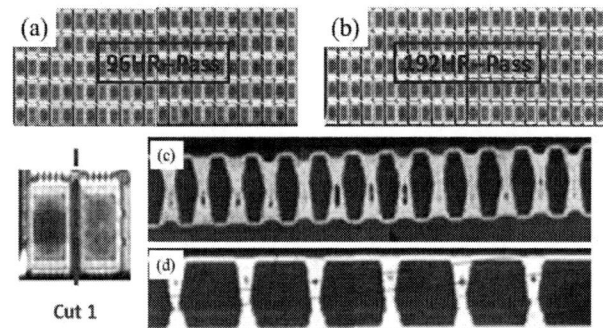

Figure 13: SAT image of HAST results of different time (a) 96 hr, (b)192 hr, and OM x-section image of x-via space of (a) 75 um, (b)300um.

Table 4. Reliability test results for HTST.

Strip No	Via Space	HTST 500	HTST 1000
		Result	Result
6	75µm	30/30	30/30
5	300 µm	30/30	30/30
4	225 µm	30/30	30/30
3	120µm	29/29	29/29

* Failure quantity/Total sample quantity

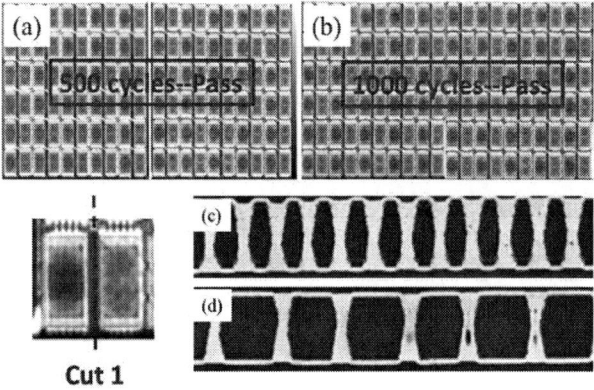

Figure 14: SAT image of HTST result of different cycle (a)500, (b)1000 cycles, and OM x-section image of x-via space of (a) 75 um, (b)300um.

From the cross section inspections, there were still small voids found in the x-via center in above reliability test conditions. But we did not find any delamination or crack among the x-vias. Through the calculation of residual void volume left in the whole x-via column. The residual void rate did not exceed 30 %.

978-1-7281-1500-9/19 $31.00 © 2019 IEEE   1429

C. Real sample for x-via reliability test

The real sample x-via conditions were as below
Top diameter: 150 +/-15 um
Via depth: 350 +/-20 um
Neck <= 50 % via diameter

The process of the embedding die into build-up layers was carried out by the use of PCB technology and the subsequent pre-condition of MSL was taken to confirm the x-via reliability test. The table 5 displayed the detail test condition. From the cut1 SEM cross section results, there was no delamination or crack observed between the connective interface of die and passive component as seen in fig. 15. The fully filling x-via was even observed in it. But the delamination phenomenon appeared in the cut2 position which may be caused by incomplete Ag sintering.

Figure 16 showed the other position for the x-via pattern. It was obviously found voids left in the hole. By the calculation, the residual volume rate of these observed voids were around 30%. Beside of that, there was no crack or deformation which was observed. The residual void size could be fine tuned by chemical or equipment parameters such as formula ratio and flow rate to improve its filling performance in the future work. Above experiments indicated the plating technology for deep through hole filling applied in the embedded package was feasible. The results showed that the deep through hole plating development will enable more space for more dice and components embedding, which facilitates the design more flexible to reach shorter interconnections, lower electrical and thermal resistivity for improving power performance. More detailed reliability tests such as TCT, HAST, and HTST were also arranged in the following work.

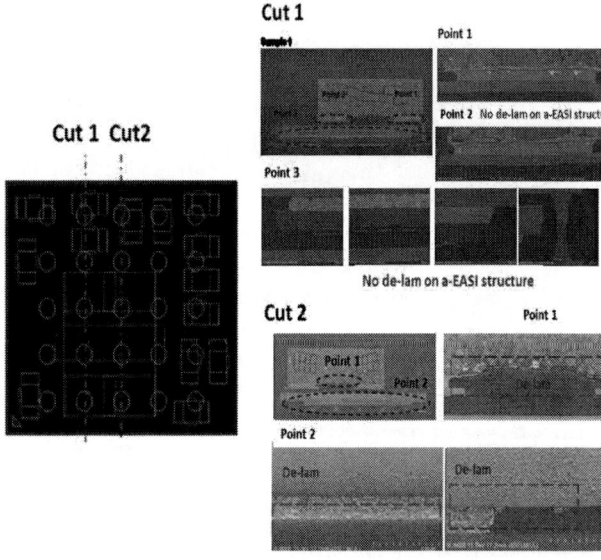

Figure 15: SEM x-section image of real sample after MSL precondition

Table 5. Reliability test conditions

Reliability item	condition
MSL	TCT X 5 Bake: 125°C / 24hr, Moisture soak: 168hr Reflow:265 +0/-5°C X 3 Cycle

Figure 16: OM x-section image of position in cut1 after MSL pre-condition

III. CONCLUSIONS

In this work, it was possible to fill laser drilled through hole up to a substrate thickness of 350 um with a diameter of 120/150 um by DC electroplating within 2hrs. This work demonstrated that the ratio between B/L must exceed 1.5; otherwise there will be a lot of Cu nodules spreading around the surface to cause the reliability issue.

The chloride ion concentration was also a critical factor for the filling performance. When the chloride ion concentration was higher than a specific value, normally between 60~90 ppm, the void size will become small, even appeared void free results.

Transferring the optimum parameters into test vehicle plating, the results showed that the filling ability was feasible to fulfill 350um x-via with the residual void rate within 30%. Reliability tests of the TCT, HAST, and HTST were also to ensure the integrity of the vehicle under specific conditions for a certain period of time. There was no delamination or crack observed among the x-via parts, which indicated the deep through hole plating development had the possibility applied in embedded package for more flexible design in the future.

ACKNOWLEDGEMENTS

The authors would like to thank the a-EASI Chi-Tsung Chiu team members for their support in making this work possible.

REFERENCES

[1] W. P. Dow, C. C. Li, Y. C. Su, S. P. Shen, C. C. Huang, C. Lee, B. Hsu, S. Hsu, "Microvia Filling by Copper Electroplating Using Diazine Black as a Leveler," Electrochim. Acta. 2009, 54: 5894-5901.

[2] R.Tenno, A. Pohjorantaz, "An ALE Model for Prediction and Control of the Microvia Fill Process with Two Additives," J. Electrochem. Soc. 2008, 155: D383- D388.

[3] M. Sugimoto, K. Yamaguchi, H. Kouzai, H. Honma, J. Appl. Polym. Sci. "Synthesis and practicability of novel additives for copper electroplating with semiconductor packaging." 2005, 96: 837-840.

[4] C. J. Moorhouse, F. J. Villarreal, H. J. Baker, and D. R. Hall, "Laser drilling of copper foils for electronics applications." IEEE Transactions on Components and Packaging Technologies. 2007, 30: 254-263.

[5] K. W. Eric, Gan, H.Y. Zhe1ng, and G.C. Lim, "Laser Drilling of Micro-Vias in PCB Substrates." IEEE Electronics Packaging Technology Conference. 2000, 321-326.

[6] J. G. Long, P. C. Searson, P. M. Vereecken, "Electrochemical Characterization of Adsorption-Desorption of the Cuprous-Suppressor-Chloride Complex during Electrodeposition of Copper." J. Electrochem. Soc. 2006, 153: C258-C264.

[7] V.D. Jovi´c, B.M. Jovi´c, "Copper electrodeposition from a copper acid baths in the presence of PEG and NaCl." J. Serb. Chem. Soc. 2001, 66: 935-952.

[8] W. P. Dow, H. H. Chen, M, Y, Yen, W. H. Chen, and K. H. Hsu, "Through-Hole Filling by Copper Electroplating," J. Electrochem. Soc., 2008, 155: D750-D757.

[9] P.C. Andricacos et al., "Damascene copper electroplating for chip interconnections," IBM J. Res. Devel, 1998, 42: 567-574.

Innovative Flip Chip Package Solutions for Automotive Applications

Tom Tang, Bo-Siang Fang, David Ho, B.H. Ma, Jensen Tsai, Yu-Po Wang

Siliconware Precision Industries Co., Ltd. No. 153, Sec. 3, Chung-Shan Rd. Tantzu
Taichung 427, Taiwan, R.O.C.
E-mail: tomtang@SPIL.com.tw

Abstract—Recently, the electronics industry is moving maturely on the mobile/tablet market. With the aggressive demand on self-driving car, the next fast growing market will be Automotive in the near future. Advance technology/packages are needed to provide ideal solutions for reliablitly and high electrical performance. Multi-function integration is also one of the critical requirements. To approach these requirements, innovative packages including Flip Chip Chip Scale Pakcage (FCCSP) and Flip Chip Ball Grid Array (FCBGA) are the potential solutions.

A key components of self-driving car are radar systems that basically detects the distance and the relative speed of the vehicles. In this paper, a FCCSP device was demostrated for the 77-GHz automotive radar application. To achieve low signal loss on the high frequency (77-GHz), coreless substrate and low dielectric tangent material were implemented, the impact of substrate trace routing and laser via design were also simulated and discussed. FCCSP could provide a cost-effective, reliable package solution for 77-GHz automotive radar systems.

Advance FCCSP and FCBGA also are the solutions to meet the increasing preformance requirements for the infotainment and Advanced Driver Assistance Systems (ADAS) applications in car. 2.5D IC is a platform for mono- or multi-functional integration. CPU, GPU, DRAM and main-broad might be all shrunken into one chip package. 2.5D IC is also a solution of high bandwidth, small form factor and multi-function integration. In this paper, another FCCSP device was demostrated for the power managerment system in car. Another FCBGA device with copper pillar bump also was developed, it not only provides the high computing performance but also passed the Automotive related qualification and is under production. Finally, this paper will sum up the progress of advanced Flip Chip packages-FCCSP and FCBGA for the Automotive application.

Keywords-automotive radar; flip chip chip scale (FCCSP) package; flip-chip ball grid array (FCBGA) package; Coreless substrates; Antenna in package (AiP).

I. INTRODUCTION

Electronic is the major driving force behind the evolution of automotive technology. Around eighty percent of automotive-related innovations are electronics, including automotive engines, safety and entertainment systems. Recently, ADAS and AI are new areas and driving forces for the growth of automotive electronics market. The IC in the car will continue to maintain two digital growth rates in coming years [1].

Figure 1 shows general data processing flow of the car in the near future, the sensors are for the data collection, and the micro control units control the sensors or other mechanical components and are for the initial data processing. Wire bond packages are the major packaging solutions for general sensors, such as image sensors and so on. Fan-out wafer level package (FO-WLP) and FCCSP are for car radar. The data collected by the sensors is transmitted to in-car computer for data analysis. Sometimes, computers in cars also needs to response very quickly for some emergencies in ADAS applications. Performance is a key requirement for in-car computing. FCBGA is the main package type. The 2.5D package is also the potential solution in AI applications. System in package (SiP) modules are the primary package type for connectivity, or communication.

Recently, lots of consumer electronics products have been designed and manufactured, especially for smartphones and wearable devices. Billions of devices have been built. Under intensive cost pressure and high performance requirements, innovative packaging technologies, including copper pillar bump, advanced substrate and materials, were developed and implemented. Since the production of consumer electronics is very large, there are huge production databases and knowledge bases. The automotive industry and some consumer electronics companies use this knowledge in the automotive IC manufacturing. Due to the huge production database, the results of adopting these new technologies are also positive. It also helps to save a lot of time and money in the development of automotive test vehicle.

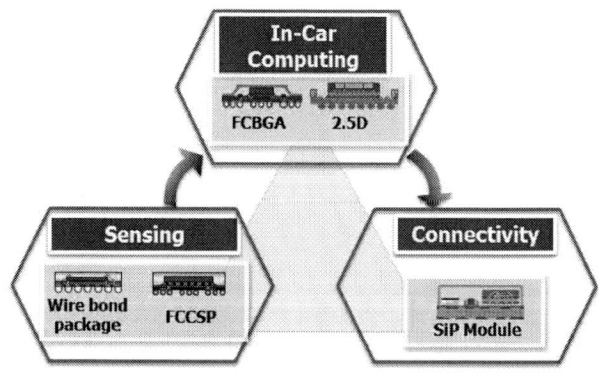

Figure 1. Packaging Solutions for Automotive Applications

II. CAR RADAR

Recently, one of the key innovations in automotive electronics are packaging solutions for ADAS, ADAS alerts drivers to potential hazards and problems to prevent collisions and improve vehicle safety. A key component of ADAS is automotive radar system, which basically detects the distance and relative speed of the vehicles [2, 3]. 24 GHz have been used in legacy automotive sensors for short-range radar, but according to spectrum specifications and standards from the European Telecommunications Standards Institute (ETSI) and the Federal Communications Commission (FCC), 24 GHz for automotive radar will be phased out and will not be available on January 1, 2022 both in Europe and the U.S. [4, 5]. Moving forward, new automotive radar implementations in the industry will likely shift to 77 GHz high frequency radar systems. The high-frequency signal has a shorter wavelength, which is more prone to signal mismatch and further loss of transmission signal. Therefore, the package structure, design and transmission trace quality requirements are high. Further study on the packaging solution for 77-GHz automotive radar systems is required.

A. FC-ETS for 77 GHz Radar

Coreless substrates with embedded trace technology definitely meet the high functionality requirements of automotive radar packages. Embedded trace technology provides superior fine-line capability and minimize transmission trace width and space variation [6]. Figure 2 shows the available packaging solutions for automotive radar applications, including flip chip embedded trace substrate (FC-ETS), conventional FCCSP, and Fan-out wafer level package (FO-WLP). Compare to conventional FCCSP substrate, coreless substrate technology eliminates the substrate cores and utilize build-up layers to interconnect chips and PCB boards. Not only does it bring a low z-height, but it also has good power integrity, short interconnects and less interconnection parasitic, which is very critical for high frequency applications.

The FO-WLP is another solution for 77-GHz automotive radar system. The FO-WLP has the shortest interconnection and is proposed to minimize transmission losses from the chip to PCB, particularly for 77-GHz radar. But the shortest interconnect also means the thinnest buffer layer, the coefficient of thermal expansion (CTE) mismatch between the chip and the PCB would lead to reliability challenges. Reliability enhancement and specific design are required for FO-WLP in automotive radar applications. FO-WLP technology is also expensive and only applied to luxury cars. The FC-ETS package utilizes the advantages of coreless substrate technology with excellent electrical and thermal performance, as well as ease of fabrication and reduced assembly costs. Table 1 shows a comparison between conventional FCCSP, FC-ETS and FO-WLP. The proposed low-cost solution of the FC-ETS package minimize losses of transition from chip to antenna, which also meets the requirements of 77-GHz automotive radar systems.

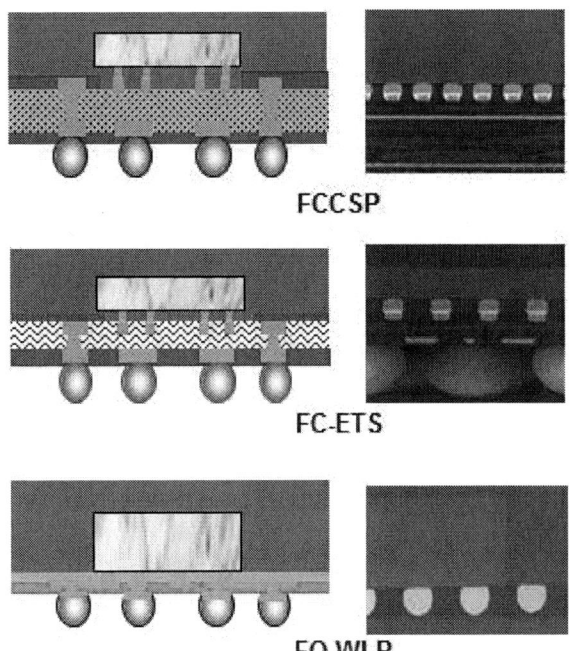

Figure 2. Packaging Solutions for Automotive Radar Systems

TABLE I. THE COMPARISON OF PACKAGING SOLUTIONS FOR AUTOMOTIVE RADAR SYSTEMS

	FC-ETS	FCCSP	FO-WLP
Body Size (mm²)	7*7	7*7	7*7
Die Size (mm²)	3.5*3.5	3.5*3.5	3.5*3.5
Thickness (mm²)	< 0.8	< 0.8	< 0.8
Line / Space (um)	10/10	25/25	10/10
Trace Routing Layer Count	2 Layers	2 Layers	1 Layer
Thermal	Better	Normal	Better
Electrical (dB) (insertion loss- S21 @ 77GHz)	-1.0~2.0	-2.0~3.0	-1.0~2.0

B. Losses of Signal Transition

Figure 3 shows a simulation model of return loss of differential pair transmission lines. It is performed by electromagnetic simulation software (ANSYS HFSS). Figure 4 shows the simulation results of insertion loss of FCCSP, FC-ETS and FO-WLP. The conventional FCCSP has a substrate core and has the largest insertion loss, but its loss is still less than 5dB. After integration with the antenna, its actual device marginally passed System level electrical test. The coreless FC-ETS has less insertion loss than conventional FCCSP and meets the requirement. The FO-WLP has the lowest insertion loss in the initial stage.

978-1-7281-1500-9/19 $31.00 © 2019 IEEE

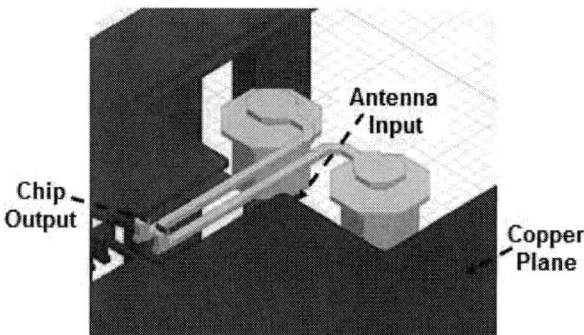

Figure 3. Simulation model

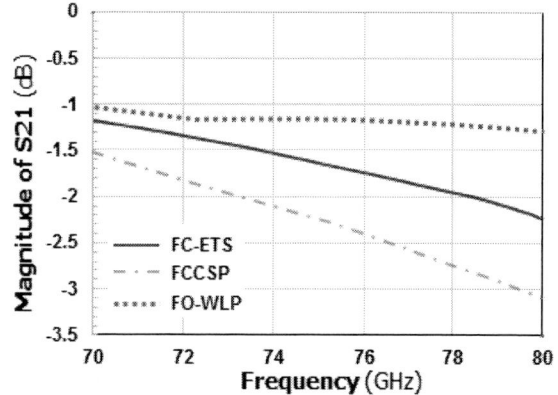

Figure 4. Simulation results of insertion loss of FCCSP, FC-ETS and Fan-out WLP

Design optimization and material selection are critical for the electrical performance of the FC-ETS, as shown in Figure 5. For the first stage, we only used normal prepreg (PP) and followed the general substrate deign rule, the performance of FC-ETS is slightly lower than the FO-WLP. However, after we changed the PP material from a normal one to a specific low loss material, its performance was improved. As shown in Figure 5, the FC-ETS (optimized) could achieved slightly better performance than FO-WLP.

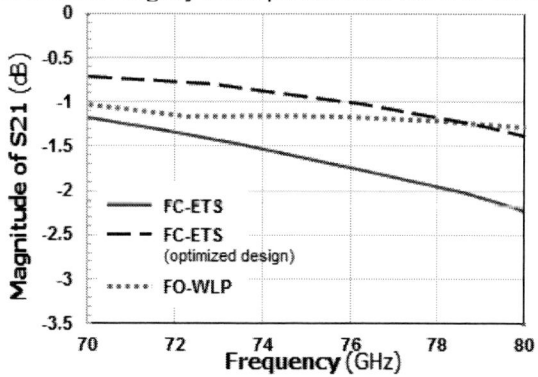

Figure 5. Simulation results of insertion loss among FCCSP, FC-ETS and Fan-out WLP

C. Impact of FC-ETS Design

The effects of FC-ETS design parameters, such as ball pad size, prepreg (PP) thickness, trace length, and the space between the trace and ground plane, are clearly demonstrated in this work. Figure 6 (a) ~ (d) show the effect of solder ball pad size, trace length and trace space on the frequency response.

The solder ball pad size has a large effect on the insertion loss, as shown on the Figure 6 (a). To achieve a short path length, solder ball pad on the laser via is used for this design. There are two parallel landing pads on the top of the laser via top and at the bottom of the laser via. These two parallel Cu plates have parasitical capacitance during the signal transmission. In addition, solder balls and laser via are typically inductive. Their parasitic capacitance could be adjusted by fine-tuning the Cu plate size to meet the requirement of inductance matching, which reduces return loss and improves insertion loss. In this case, when we shrink the size of the two Cu plates, it helps to reduce its parasitic capacitance, and then the insertion loss is reduced.

Figure 6 (b) shows the effect of PP thickness. The thin PP helps to slightly shorten the distance from the chip output to the antenna input. When we reduce the PP thickness of 20um. The insertion loss is reduced by around 0.1dB. But when we reduce the PP thickness of 40um, the effect of enhanced insertion loss is saturated.

Reducing the trace length could help to shorten the signal transmission distance from the chip output to the antenna input, normally, a shorter distance means lower insertion loss. However, impedance matching is also important for high frequency signal transmission, and impedance conversion is significant when the length of the transmission line exceeds 1/10 of the wavelength, especially when the frequency is up to 77 GHz, a 500um change in the length of transmission line induces an impedance shift, which causes signal reflection and increases insertion loss due to the energy conservation. This wavelength-dependent signal reflection is frequency dependent, so it causes the insertion loss to change sharply with frequency. As shown on the Figure 6 (c), when we shrink the trace length by 500um, it induces the mismatch and leads to larger and unstable insertion loss.

The characteristic impedance of the transmission line is needed to be carefully designed to avoid impedance mismatch of the circuit. It is usually related to the line width of the transmission line and its distance to the ground plane. The design and process control of these two dimensions should be optimized. As shown on the Figure 6 (d), we increased the space between the trace and ground plane by 20um, and the insertion loss associated to this impedance mismatch is close to 0.5dB, so precise control of the line pitch is required for 77 GHz signal transmission.

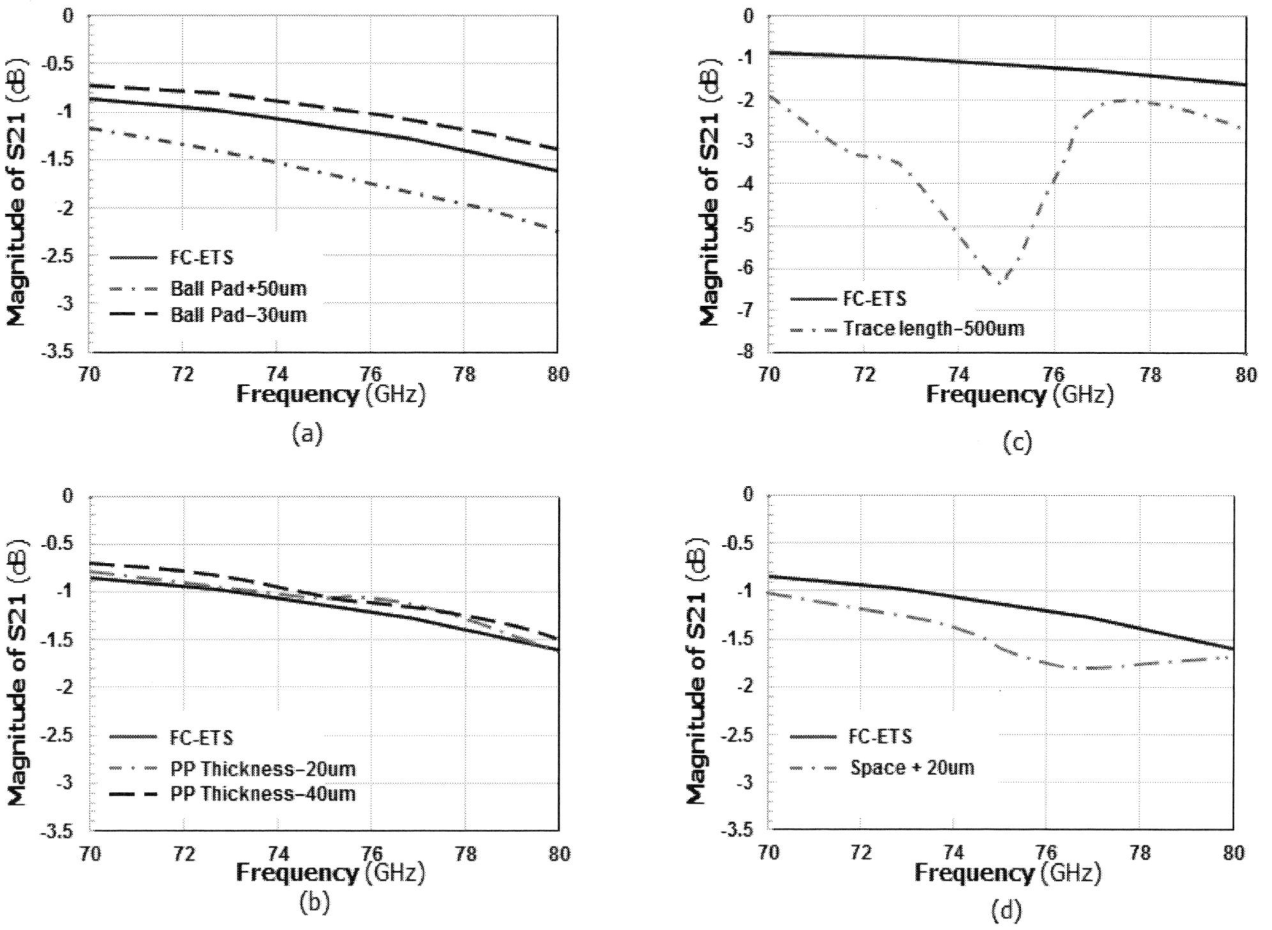

Figure 6. Impact of the FC-ETSSubstrate Design on the frequency response. (a) Solder ball pad diameter. (b)Prepreg thickness. (c) Trace length. (d) Space between trace and ground plate.

D. Antenna-in-Package

Antenna in Package (AiP) is being developed for the next generation of 77GHz automotive radar systems. As shown in Figure 7, FCCSP with landside die is used for AiP applications. Compared to conventional antenna-on-PCB board design, land side die structures can achieve shorter path from chip output to antenna input and reduce the transmission loss of high frequency signal. Compared to antenna-on-chip, AiP shows a good compromise between performance, integration and cost. The patch array on the top layer of substrate is one of the potential solutions for the antenna design of the AiP package because of its good planarity, high gain and orientation characteristics. In previous literature, many antenna arrays for AiP or millimeter waves were fabricated for high gain characteristics [7]. In general, hick dielectric layers in AiP are required to ensure antenna performance, such as radiation efficiency bandwidth and antenna gain, which are different from the requirements of conventional substrate designs of interconnects in packages. A typical AiP design uses a balanced substrates design that requires multiple substrate layers to pursue good antenna characteristics. As shown in Figure 7, two t unbalanced 7-layers metal substrates with thick cores are proposed and constructed, which have fewer substrate layers and lower than conventional balance designs.

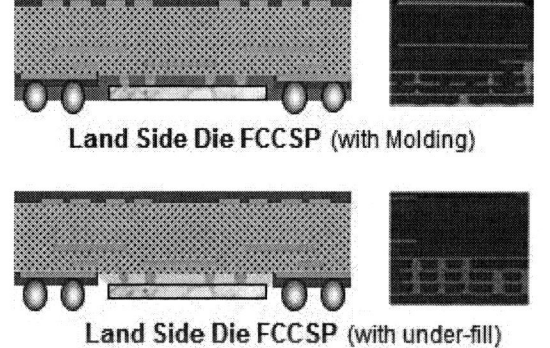

Figure 7. Two Types of Antenna-in-package

978-1-7281-1500-9/19 $31.00 © 2019 IEEE

III. IN-CAR COMPUTING

For entertainment systems in the car cabin, wire bond BGAs are typically used for audio and radio, or media centers within cars. Flip chip BGA is used for ADAS and infotainment, and flip chip BGA is the traditional one and is in production several years ago. The 2.5D package is the latest solution and is one of potential solutions to realize autonomous driving and maintaining general infotainment applications. But like previous technology transition, we can expect it to move to production in the next two or three years.

Figure 8 shows an example of the FCBGA. There is heat sink attached to help for the heat dissipation, and its size is about twenty-five to twenty-five millimeters. On the right side is a cross section of copper pillar bump with solder on board. Typically, in an actual ADAS application, there are two FCBGA packages on a single PCB board. It passed AEC Q100 grade two. Figure 8 is an example of a 2.5D package with a processor and four stacked memory ICs, normally, we attached these dies to a silicon interposer in wafer form, and then we sawed it and attached it to the substrate, the major challenge is alignment control. The package already passed AEC Q100 grade three.

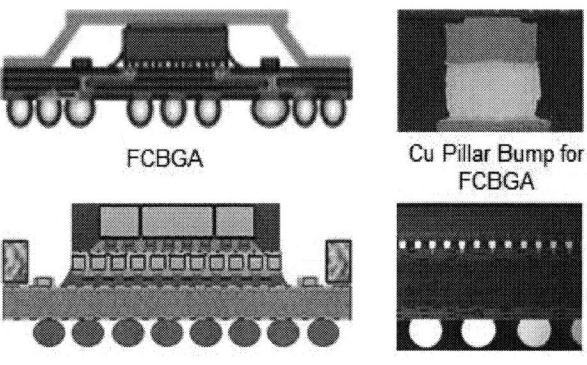

FCBGA

Cu Pillar Bump for FCBGA

2.5D Package

Figure 8. Packaging Solutions for In-car Computing

IV. CONNECTIVITY

The SiP module is used for the integration and integrates baseband, WIFI, Bluetooth RFIC and power management ICs. In general, EMI shielding is required to separate different function ICs. By using substrate and assembly technology, we can reduce the area by about 50% area by using SiP modules compared to normal PCB and side by side design. As shown in Figure 9, the WLCSP package is used for Power management applications, and the right side is a hybrid package for baseband application. We mounted all components on the substrate, then the metal lid was attached for t EMI shielding. This package already passed AEC Q100 grade three and is in production.

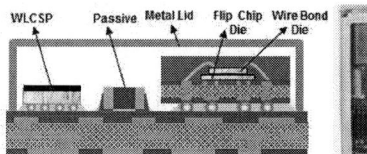

Figure 9. SiP Module for the Connnectivity

V. CONCLUSION

Currently, there are three major flip chip package trends in automotive applications, one is copper pillar flip chip packaging, which is especially suitable for automotive radar applications. The second trend is the high performance package, including flip chip BGA and 2.5D packages. The third is the integration, system in package module.

Depending on the application, there are three groups, the first group is traditional audio, radio and sensor applications. Wire bond packages are still dominant. It includes the lead-frame base and substrate base wire bond packages. The main trend is to use copper pillar flip chip packages, especially in automotive radar and audio applications. For the communication applications, SiP module is one of major package types. In order too separate the different function ICs, EMI shielding is also necessary. For in-car computing, performance is a major requirement. Currently, the Flip chip BGA is in production, and 2.5D package is under development. In the near future, these technologies will help fulfill the automotive IC package requirements.

REFERENCES

[1] ISLAM, Nokibul; HSIEH, Ming-Che; KEONTAEK, Kang. Advanced Packaging Need for Automotive In-cabin Application. In: 2017 IEEE 67th Electronic Components and Technology Conference (ECTC). IEEE, 2017. p. 1468-1472.

[2] HO, Cheng-Yu, et al. A 77GHz Antenna-in-Package with Low-Cost Solution for Automotive Radar Applications. In: 2018 IEEE 68th Electronic Components and Technology Conference (ECTC). IEEE, 2018. pp. 191-196.

[3] GOPPELT, Markus; BLÖCHER, H.-L.; MENZEL, Wolfgang. Automotive radar–investigation of mutual interference mechanisms. Advances in Radio Science, 2010, 8.B. 3: 55-60.

[4] FCC Report and Order – Radar services in the 76-81 GHz band, ET docket No. 15-26.

[5] ETSI EN 301 091 – Radar equipment operating inthe 76 GHz to 77 GHz range.

[6] TANG, Tom, et al. Challenges of flip chip packaging with embedded fine line and multi-layer coreless substrate. In: Electronic Components and Technology Conference (ECTC), 2015 IEEE 65th. IEEE, 2015. pp. 1219-1222.

[7] LU, Ying-Wei, et al. Mm-Wave Antenna in Package (AiP) Design Applied to 5th Generation (5G) Cellular User Equipment Using Unbalanced Substrate. In: 2018 IEEE 68th Electronic Components and Technology Conference (ECTC). IEEE, 2018. pp. 208-213.

Reliability of Laminated Bond Structure Using (Cu,Ni)/Sn TLP Bonding with Al Interlayer for High Temperature Power Electronics Packaging

Yanghe Liu, Shailesh N. Joshi, Ercan M. Dede
Electronics Research Department
Toyota Research Institute of North America
Ann Arbor, USA
Email: yanghe.liu@toyota.com

Abstract— **Future automotive power electronics will be using wide band gap (WBG) devices that requires high temperature operation of over 200°C and current solder material cannot be operated over 150°C. New bonding technology such as transient liquid phase(TLP) bonding whose melting point is over 400°C has shown as one of promising high temperature bonding. In this study, reliability of a new laminated bond structure using Al interlayer is investigated by subjecting the structure to thermal cycling from -40 to 200°C for 500 cycles. The new structure mitigates thermal stress due to CTE mismatch between the copper substrate and the power device. The results show that after 200°C of thermal storage for 100 h, complete consumption of tin was observed along with formation of (Cu,Ni)/Sn intermetallic compound(IMC). The same samples were thermal cycled from -40°C to 200°C and the confocal scanning acoustic microscopy(CSAM) results display consistent bonding quality. Then, micro-level crack was observed through a further Scanning Electron Microscopy(SEM) check on cross-section of sample. The root cause of the crack was identified as excessive thermal stress, especially normal stress, through numerical analysis. The study demonstrates the potential and challenges of the new laminated structure for low cost and reliable high temperature bonding material for power devices.**

Keywords- Power electronics, TLP bonding, thermal cycling, reliability

I. INTRODUCTION

Development of high power density compact power modules is essential to future advanced power electronics in electrified vehicles. Fig. 1 illustrates a schematic of cross-section of typical single side power module. The silicon(Si) power chip is bonded to substrate using die-attach material, typically tin (Sn) based solder. The common substrate materials are direct bonded copper (DBC) or direct bonded aluminum (DBA)[1, 2]. Pure copper (Cu) substrate is also a good candidate due to its excellent thermal and electrical conductivity which is utilized in this study. Heat was generated from chip and dissipated through die-attach layer to cooling structure below substrate. The bottlenecks of conventional silicon(Si) based power semiconductors include: high switching loss, junction temperature limitation, power density limitation etc. Wide bandgap (WBG) semiconductors, typically silicon carbide (SiC), are deemed as promising candidate of future power device in vehicle to achieve higher power densities and lower switching loss[3-5]. However, the preferable functional temperature of WBG

devices are over 200°C which is above operation temperature limit of current solder system. Thus, it's necessary to explore alternative high temperature die-attach technologies.

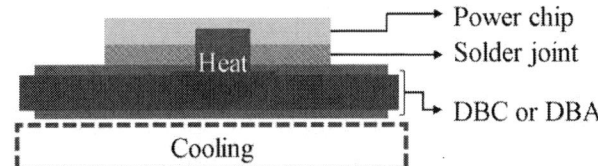

Figure 1: Typical single side power module die-attach

As one of the promising high temperature bonding technologies compatible with current industrial soldering process, transient liquid phase (TLP) sintering enables low fabrication temperature and high remelting temperature[6-9]. In a binary alloy TLP process, two metals of different melting point form their intermetallic compounds (IMC). The accomplished material has higher melting point than process temperature. E.g. Cu-Sn alloy melting point reaches over 400°C while the processing temperature is below 300°C. Other potential high temperature bonding options such as, Au-Sn bonding, Ag sintering, nano Cu sintering, etc. is not selected due to either high material cost or fabrication complexity. In this paper, the formed IMC composition is mainly (Cu, Ni)-Sn. Ni source, as a common surface plating material, is from chip and substrate metallization. The advantage of using complex form of (Cu,Ni)-Sn IMC is faster fabrication owing to rapid diffusion rate[10-12].

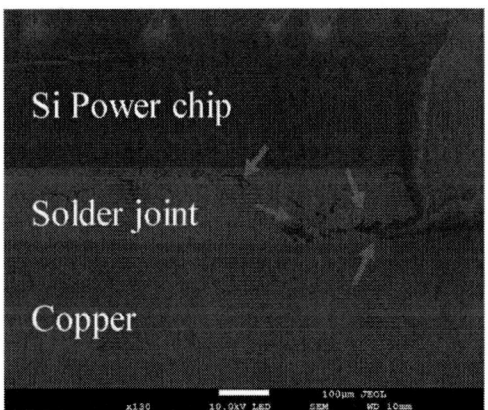

Figure 2: Typical die-attach solder joint failure

Table 1: Summary of high temperature bonding thermal cycling results [11-19]

Bonding material	Die	Substrate	Min. Temp.(°C)	Max. Temp. (°C)	No. of cycles
TLP Cu-Sn	Si	Cu	-55	200	20
TLP Ni-Sn	Si	DBC	-40	200	100
TLP Ag-Bi-Sn	SiC	DBC	-55	195	1000
Ag	SiC	DBC	-40	250	1000
Ag	Si	Si_3N_4	-55	165	1000
Ag	Si	Cu	-40	200	1000
Ag	Si, SiC	Alumina, DBC	-55	300	1000
Nano Cu	Al_2O_3	Cu-65wt% Mo	-40	200	1000
Core shell Cu-Sn TLP	DBC	Cu	-55	200	1000

Besides function of dissipating heat, die-attach material must exhibit high reliability under continuously changing environment. As temperature varies, excessive stress could occur at interface between layers of different materials owing to coefficient of thermal expansion (CTE) mismatch. Fig. (2) illustrates a solder failure caused from excessive thermal stress accumulated in thermal cycling test[13]. In this paper, chip material and substrate material are Si (α, 3 ppm/K) and nickel-plated Cu (α, 17 ppm /K), respectively, which has much higher CTE mismatch than die-attach of Si chip and DBC (α, 7ppm/K). Numerous investigations have been conducted on thermal cycling testing of high temperature bonding materials. Table 1 [14-22] illustrates a summary of thermal cycling results published by other research groups. Most of the researchers utilize DBC substrate which has less CTE mismatch concern. Those tests with Si chip and Cu substrate either seems to run only small number of cycles or use higher material cost Ag sintering.

This paper presents reliability testing results of a novel laminated TLP bonding followed by a stress analysis via finite element analysis. In order to handle large CTE mismatch between Cu substrate and Si chip, a thick aluminum(Al) core layer is sandwiched in bond line as a stress absorber[23]. This article is organized as follows. In section II, the laminated TLP bond structure using Al core layer and sample fabrication method are reviewed. Section III shows the thermal storage test results and analysis. Thermal cycling (-40°C/200°C) are demonstrated in Section IV. According to observations from thermal cycling tests, a thermal stress numerical analysis is conducted and detailed in Section V. Summary and our future work based on current observations are provided in Section VI.

II. LAMINATED TLP BONDING USING ALUMINUM INTERLAYER

A. Concept

In our previous work[23], we described a symmetrical laminate design of a TLP bond structure. Compared to conventional solder or other high temperature bonding, the proposed bond structure has at least three advantages including a high re-melting point > 400 °C, fast diffusion time to process completion, and thermal stress reduction capability from Al core layer. Considering brittle nature of material property of TLP IMC, addition of ductile Al layer as

buffer could mitigate thermal induced stress under changing temperature. Figure 3 illustrates the bonding structure.

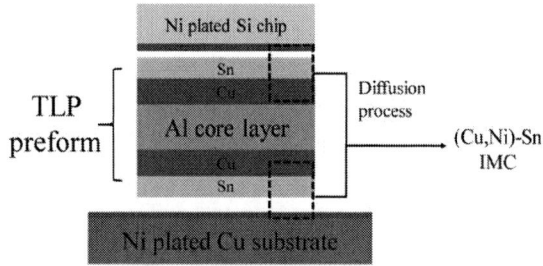

Figure 3: Laminated (Cu,Ni)-Sn TLP structure using Al interlayer

The commercially available Si IGBT chip as well as the TLP preform has ~12 mm × ~13 mm overall footprint area. The thickness of IGBT is ~200 μm with 1 μm thick Ni metallization. The total thickness of laminated bond structure is ~140μm with 100 μm thickness of Al core layer and 10 μm thickness of each Sn and Cu layer. The overall footprint area of Cu substrate is 18 mm × 35 mm with 3 mm total thickness. And note that there is a 10 μm Ni plating on the substrate. The prepared die-attach stack is positioned using a graphite fixture. Details of fabrication are illustrated in next subsection.

B. Fabrication

For preparation of TLP preform, Cu and Sn layers are uniformly electroplated on a 99.5% purity Al foil. Plasma cleaning are utilized for sample surface oxidation elimination to achieve low voiding bonding quality. Sample stack is aligned and clamped by our custom graphite fixture. To be closely compatible with conventional pressureless process, the compression force from fixture is less than 0.1 MPa. Although the low force may lead to some residual Sn in bond layer, the consumption of Sn can be fully completed with either longer fabrication time or post thermal storage as described in Section III. The assembled sample is mounted in SST1200 soldering station by Palomar Technologies Inc. After precleaning process by flushing the chamber with dry nitrogen, the sample is under vacuumed and then ramped up to 300 °C. At the peak temperature, 5%/95% hydrogen/nitrogen forming gas is injected and raise in-chamber pressure to 30 psi. After soaking for 1 min, sample is cooled down to room temperature while keeping the chamber oxygen free.

C. Quality check

Post fabrication of bonding, quality of sample is checked through CSAM a Sonix Echo VS Acoustic Microscope with a 110 MHz transducer. And a cross-sectional SEM and EDX analysis for microstructure confirmation of a duplicated sample is followed by. Fig. 4 illustrates the results of initial bond quality.

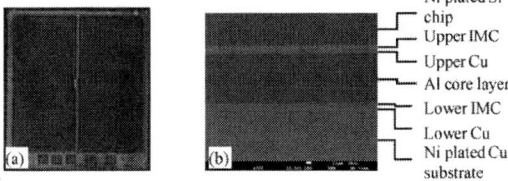

Figure 4: (a)C-SAM and (b)cross-sectional SEM image of laminated structure TLP bonding

Only minor void from CSAM was observed at upper right corner. A uniform bond compromising the expected layer structure is clearly displayed. No delamination was found along the bond line. In some duplicated samples, some residual Sn was found in bond layer, especially at near edge area. As explained above, this is mainly because of nearly pressureless fabrication and insufficient processing time for simulating current industrial solder reflow. The extra Sn will be further consumed in the following thermal storage. Details of the results will be introduced in Section III.

III. THERMAL STORAGE

Samples were subjected to thermal storage at 200°C for 100 h. During the process, thermocouple is directly attached to substrate surface to ensure stability of the condition. During the process, no chip cracking or obvious delamination was observed. Through C-SAM comparison between before and after thermal storage, only some negligible voids were identified at bottom side near edge area, shown in Fig. 5.

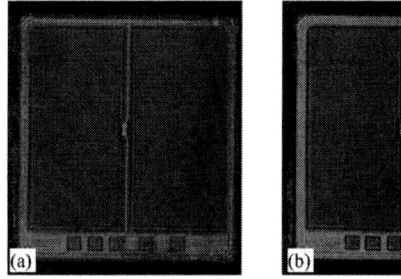

Figure 5: CSAM of (a) before and (b) after thermal storage test

The SEM and EDX results in Figure (6) indicate that TLP process is further completed after thermal storage. Before the process, existence of residual Sn was found at both upper and lower IMC layers. After the process, both Ni and Cu diffused forming IMC. In actual power card fabrication, a post heating for curing epoxy material is similar as the described process. If TLP bonding is adopted to mass production, this step could further guarantee the consumption of residue Sn.

IV. THERMAL CYCLING

After thermal storage test, samples were directly thermal cycled from -40 to 200°C for 500 cycles with dwell time of 30 min and ramping time of 15 min. Similarly, as thermal storage, directly attached thermocouple is utilized to ensure sample surface temperature. At every 100 cycles step, samples were taken out from chamber for CSAM quality recording. In this process, a die-attach solder sample using Cu substrate is added as a reference. The peak temperature of thermal cycling for solder sample is adjusted to 150°C which is commonly used in solder reliability test. The purpose is to compare difference of bond layer degradation between solder and our laminated TLP structure.

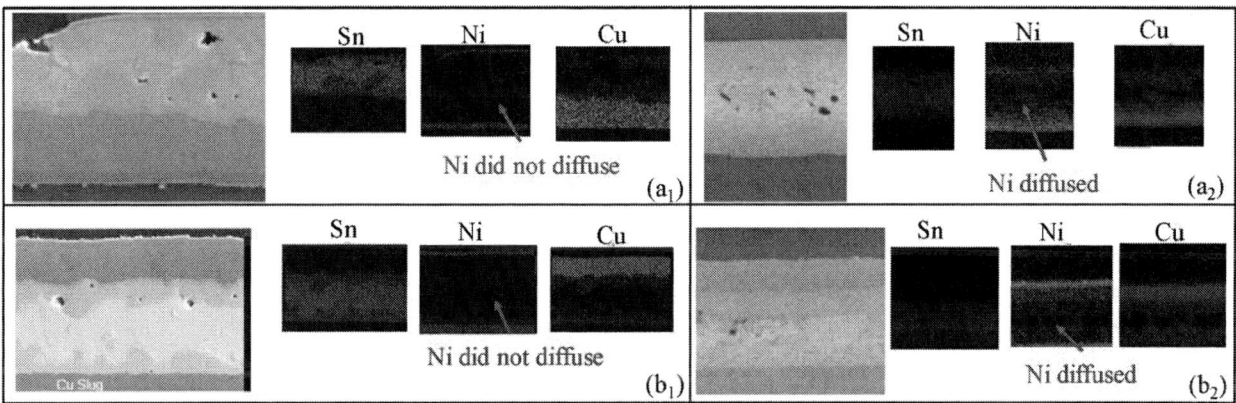

Figure 6: Cross-sectional SEM of upper IMC (a₁) before and (a₂) after thermal storage, lower IMC (b₁) before and (b₂) after thermal storage

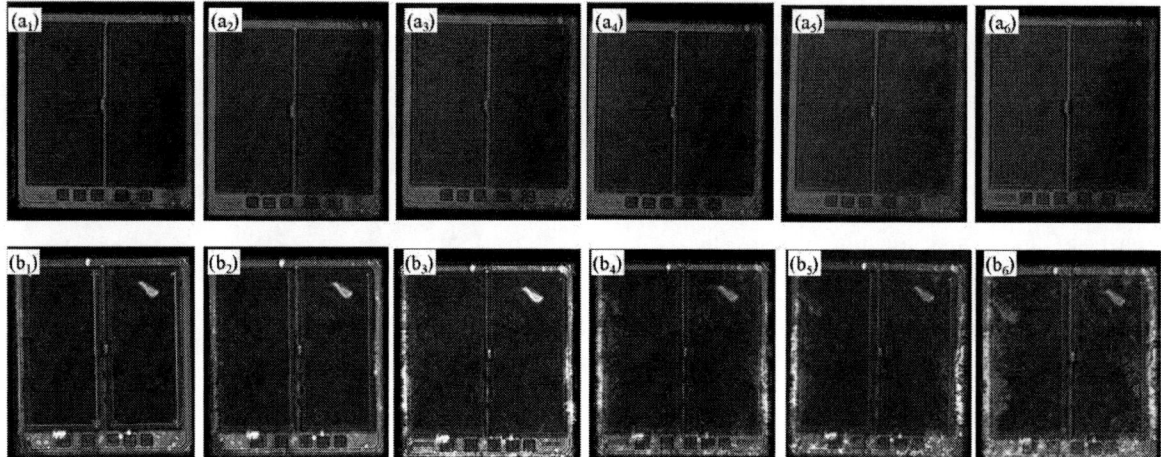

Figure 7: Thermal cycling CSAM of TLP and solder: (a₁) and (b₁) 0 cycles; (a₂) and (b₂) 100 cycles; (a₃) and (b₃) 200 cycles; (a₄) and (b₄) 300 cycles; (a₅) and (b₅) 400 cycles; (a₆) and (b₆) 500 cycles.

Fig. 7 depicts the CSAM results of TLP sample and solder sample of thermal cycling tests. TLP samples display a consistent bonding quality. No major delamination or voiding issue was found. However, slight difference of contrast and brightness needs to be noticed after 300 cycles. Right side area seems to become brighter which may indicate bond density change. For solder samples, the propagation of defects is obvious. Starting from 100 cycles, voiding area grows from edge side to inner side. The observations indicate that the laminated TLP structure with interlayer seems to be more capable of handling large thermal stress.

The post thermal cycling TLP sample was cross-sectioned followed by SEM analysis. In Figure (8), while lower IMC layer displays smooth bond with clear IMC layer, some cracks were found at upper IMC layer. It seems that these cracks initiated from edge and form a severe delamination which reflects the brittleness of IMC material. CSAM doesn't reflect the failure precisely which may be caused from the penetration of water into the failed area during CSAM. Note that this CSAM issue can be easily solved by sealing bond edge using epoxy. A root cause analysis of upper IMC failure via finite element analysis is described in section V.

v.19.2. A one-quarter symmetry model including the detailed bond composition for each scenario was created with a refined structured mesh. Mesh sensitivity tests were performed to ensure the accuracy of the results. For computational simplicity, all material properties are assumed to be temperature independent and isotropic. The plasticity of the IMC material is further neglected due to its known brittle nature. However, for ductile Al core layer, a non-linear plastic region was assigned to reflect enhanced stress absorption by plastic deformation of the Al layer. Two thermal cycles were simulated with creep behavior of materials ignored. The initial stress-free state is assumed to be at 22°C. Fig. 9 is the equivalent von-mises stress contour of upper IMC layer under thermal stress. Larger stress distributed along the 1mm edge. To avoid effect of stress singularity in simulation using maximum thermal stress for evaluation, the average stress of 1mm edge at both upper and lower IMC layers is replotted Fig. 10. It's obvious that the magnitude of maximum thermal stress at peak temperature of upper IMC is almost ~3X larger than lower IMC, which explains the experimental results above.

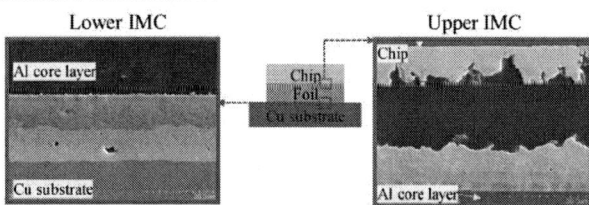

Figure 8: Cross-sectional view of TLP samples after 500 thermal cycles

V. NUMERICAL ANALYSIS

Thermal stress induced in the bond layer is evaluated through FEA analysis using commercial software ANSYS

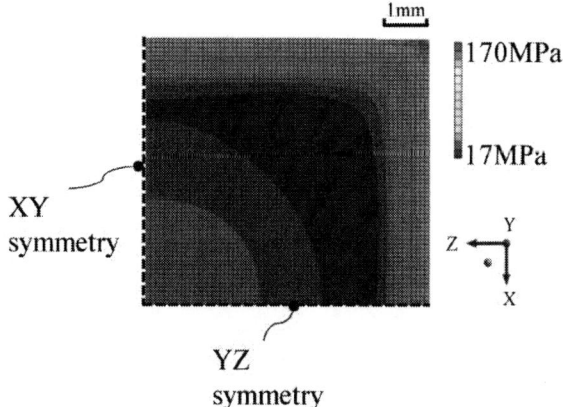

Figure 9: Upper IMC layer von-mises thermal stress

978-1-7281-1500-9/19 $31.00 © 2019 IEEE

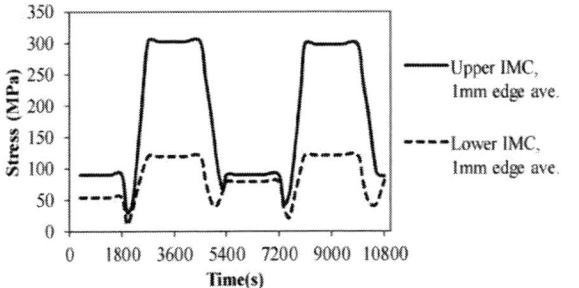

Figure 10: Average equivalent thermal stress of upper IMC layer and lower IMC layer

The upper IMC stress is further analyzed by replotting diagonal in-plane shear stress parallel to sample surface plane and normal stress perpendicular to surface plane, shown in Fig. 11. At near edge area, the behavior of the stress is mainly normal stress. While the shear stress is maintained at low value. This difference could lead to upper IMC bond layer splitting. And considering the brittleness of IMC layer, the crack is possibly caused by the excessive splitting thermal stress.

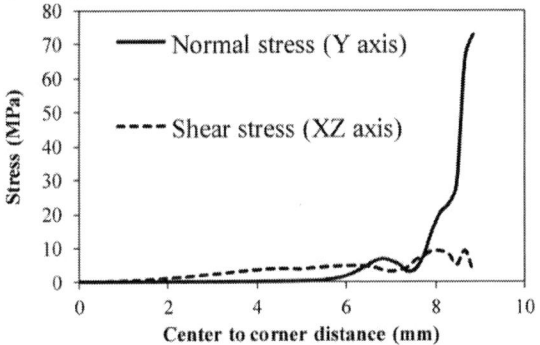

Figure 11: Upper IMC diagonal shear and normal thermal stress

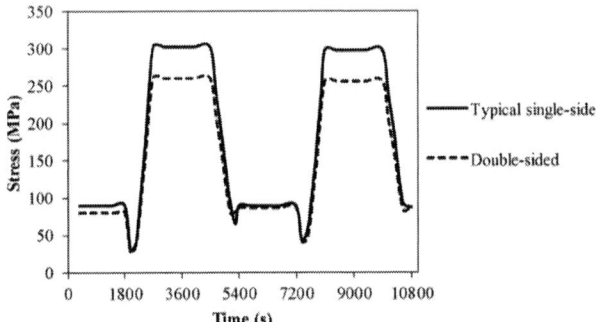

Figure 12: Average equivalent thermal stress of upper IMC layer of single sided packaging and double-sided packaging

The normal stress is mainly due to buckling of single sided Cu substrate. More symmetrical packaging method, e.g. double side packaging[24], could potentially mitigate normal stress. The main purpose of using double sided bonding was for cooling enhancement. However, in a symmetrical structure, normal stress from both sides can be mutually canceled out. Thus, chip deformation can be eased predictably. Based on the simulation results, Fig. 12 compares the upper IMC thermal stress of single-sided and double-sided packaging. More than 10% stress reduction are realized in this result. In the future, if large CTE mismatch substrate and chip are employed, we also need to take effect from proper packaging on thermal stress of bond layer into account.

VI. SUMMARY AND FUTURE WORK

This paper presents reliability test results of a novel laminated TLP structure using Al interlayer on large CTE mismatch Cu substrate and Si chip high temperature die-attach. The results show the potential of adopting this bonding technology to future high operation temperature power electronics. Meanwhile, challenges of mitigating thermal stress is also addressed by analyzing test results through numerical analysis. Effect on bond thermal stress from various packaging method will be further studied in our future work.

ACKNOWLEDGMENT

The authors would like to thank members at Toyota Motor Corporation (TMC) Japan for technical discussion and support. Also, the authors thank Innovative Circuits Engineering, Inc. for assistance in sample analysis.

REFERENCES

[1] Kanata, T., Nishiwaki, K., and Hamada, K., 2010, "Development trends of power semiconductors for hybrid vehicles," The 2010 International Power Electronics Conference-ECCE ASIA-, IEEE, pp. 778-782.

[2] Nozawa, S., Maekawa, T., Yagi, E., Terao, Y., and Kohno, H., 2010, "Development of new power control unit for compact-class vehicle," 2010 22nd International Symposium on Power Semiconductor Devices & IC's (ISPSD), IEEE, pp. 43-45.

[3] Bajwa, A. A., Qin, Y., Reiner, R., Quay, R., and Wilde, J., 2015, "Assembly and packaging technologies for high-temperature and high-power GaN devices," IEEE Transactions on Components, Packaging and Manufacturing Technology, 5(10), pp. 1402-1416.

[4] Baliga, B. J., 1989, "Power semiconductor device figure of merit for high-frequency applications," IEEE Electron Device Letters, 10(10), pp. 455-457.

[5] Hornberger, J., Lostetter, A., Olejniczak, K., McNutt, T., Lal, S. M., and Mantooth, A., 2004, "Silicon-carbide (SiC) semiconductor power electronics for extreme high-temperature environments," Aerospace Conference, 2004. Proceedings. 2004 IEEE, IEEE, pp. 2538-2555.

[6] Welch, W. C., Chae, J., and Najafi, K., 2005, "Transfer of metal MEMS packages using a wafer-level solder transfer technique," IEEE transactions on advanced packaging, 28(4), pp. 643-649.

[7] Greve, H., and McCluskey, F. P., 2013, "Transient Liquid Phase Sintered Joints for Power Electronic Modules," ASME 2013 International Technical Conference and Exhibition on Packaging and Integration of Electronic and Photonic Microsystems, American Society of Mechanical Engineers, pp. V001T001A004-V001T001A004.

[8] Greve, H., Moeini, S. A., McCluskey, P., and Joshi, S., 2015, "High Temperature Shear Strength of Cu-Sn Transient Liquid Phase Sintered Interconnects," ASME 2015 International Technical Conference and Exhibition on Packaging and Integration of Electronic and Photonic Microsystems collocated with the ASME 2015 13th International Conference on Nanochannels, Microchannels, and Minichannels, American Society of Mechanical Engineers, pp. V002T002A037-V002T002A037.

978-1-7281-1500-9/19 $31.00 © 2019 IEEE

[9] Greve, H., Moeini, S. A., McCluskey, F. P., and Joshi, S., 2016, "Microstructural Evolution of Transient Liquid Phase Sinter Joints in High Temperature Environmental Conditions," Electronic Components and Technology Conference (ECTC), 2016 IEEE 66th, IEEE, pp. 2561-2568.

[10] Yang, P.-F., Lai, Y.-S., Jian, S.-R., Chen, J., and Chen, R.-S., 2008, "Nanoindentation identifications of mechanical properties of Cu6Sn5, Cu3Sn, and Ni3Sn4 intermetallic compounds derived by diffusion couples," Materials Science and Engineering: A, 485(1-2), pp. 305-310.

[11] Labie, R., Ruythooren, W., and Van Humbeeck, J., 2007, "Solid state diffusion in Cu–Sn and Ni–Sn diffusion couples with flip-chip scale dimensions," Intermetallics, 15(3), pp. 396-403.

[12] Yoon, J.-W., Noh, B.-I., Kim, B.-K., Shur, C.-C., and Jung, S.-B., 2009, "Wettability and interfacial reactions of Sn–Ag–Cu/Cu and Sn–Ag–Ni/Cu solder joints," Journal of Alloys and Compounds, 486(1-2), pp. 142-147.

[13] Lu, H., Bailey, C., and Yin, C., 2009, "Design for reliability of power electronics modules," Microelectronics reliability, 49(9-11), pp. 1250-1255.

[14] Ikeda, H., Sekine, S., Kimura, R., Shimokawa, K., Okada, K., Shindo, H., Ooi, T., Tamaki, R., and Nagata, M., 2017, "Cu-Sn based joint material having IMC forming control capabilities," 2017 International Conference on Electronics Packaging (ICEP), IEEE, pp. 171-176.

[15] Yoon, S. W., Shiozaki, K., and Kato, T., 2014, "Double-sided nickel-tin transient liquid phase bonding for double-sided cooling," 2014 IEEE Applied Power Electronics Conference and Exposition-APEC 2014, IEEE, pp. 527-530.

[16] Shen, Z., Johnson, R. W., and Hamilton, M. C., 2015, "SiC power device die attach for extreme environments," IEEE Transactions on Electron Devices, 62(2), pp. 346-353.

[17] Suganuma, K., Asatani, N., Kimoto, K., Suetake, A., Zhang, H., Nagao, S., and Sugahara, T., 2017, "Ag sinter joining and wiring for high power electronics," 2017 IMAPS Nordic Conference on Microelectronics Packaging (NordPac), IEEE, pp. 147-150.

[18] Le Henaff, F., Greca, G., Salerno, P., Mathieu, O., Reger, M., Khaselev, O., Boureghda, M., Durham, J., Lifton, A., and Harel, J. C., 2016, "Reliability of Double Side Silver Sintered Devices with various Substrate Metallization," PCIM Europe 2016; International Exhibition and Conference for Power Electronics, Intelligent Motion, Renewable Energy and Energy Management, VDE, pp. 1-8.

[19] Watanabe, T., Nakajima, N., and Takesue, M., 2017, "Material design and process conditions of pressureless sintered silver for 200/-40 C thermal cycling reliability," PCIM Europe 2017; International Exhibition and Conference for Power Electronics, Intelligent Motion, Renewable Energy and Energy Management, VDE, pp. 1-4.

[20] Yu, F., Cui, J., Zhou, Z., Fang, K., Johnson, R. W., and Hamilton, M. C., 2017, "Reliability of ag sintering for power semiconductor die attach in high-temperature applications," IEEE Transactions on Power Electronics, 32(9), pp. 7083-7095.

[21] Ishizaki, T., Usui, M., and Yamada, Y., 2015, "Thermal cycle reliability of Cu-nanoparticle joint," Microelectronics Reliability, 55(9-10), pp. 1861-1866.

[22] Chen, H., Hu, T., Li, M., and Zhao, Z., 2017, "Cu@ Sn core–shell structure powder preform for high-temperature applications based on transient liquid phase bonding," IEEE Transactions on Power Electronics, 32(1), pp. 441-451.

[23] Liu, Y., Joshi, S. N., and Dede, E. M., 2018, "Novel Transient Liquid Phase Bonding for High-Temperature Automotive Power Electronics Systems," Submitted for Journal of Electronic Packaging, unpublished.

[24] Yoon, S. W., Glover, M. D., Mantooth, H. A., and Shiozaki, K., 2012, "Reliable and repeatable bonding technology for high temperature automotive power modules for electrified vehicles," Journal of Micromechanics and Microengineering, 23(1), p. 015017.

Silver Sintering on Organic Substrates for the Embedding of Power Semiconductor Devices

Alexander Schiffmacher, Lorenz Litzenberger, Juergen Wilde
Laboratory for Assembly and Packaging Technology
Department of Microsystems Engineering - IMTEK
University of Freiburg, Germany
alexander.schiffmacher@imtek.uni-freiburg.de

Till Huesgen, Vladimir Polezhaev
Electronics Integration Lab
University of Applied Science
Hoschschule Kempten, Germany

Abstract — **The requirements for power electronic assemblies are continuously increasing and are mainly driven by costs, functionality, and reliability. A novel and promising approach is the embedding of power semiconductor devices into PCB-materials. Benefits are the reduction in size and volume of the system. The embedding of semiconductor devices provides a high degree of miniaturization. Also printed circuit board technology in combination with the use of established processes apparently has the potential for low-cost manufacturing. Further functional advantages are the possibility to place passive components and peripheral circuits close to the switching devices, enabling shorter commutating paths. In consequence, they are expected to produce smaller parasitic effects caused by the package, which results in higher possible frequencies and reduced conduction and switching losses. However, there is a significant challenge regarding package design, processing, and materials selection to make use of this potential even at high operating temperatures. To address only one aspect, generally used materials, like epoxy-glass-substrates (FR4) and solder alloys like PbSnAg or SAC are not suitable for temperatures above 150 °C. This work will introduce and evaluate a concept for double-side Ag-sintered semiconductor chips, which are embedded between two organic high-temperature PCBs. A proof-of-concept will be presented by setting up a 30 kW (600 V, 50 A) power package as a demonstrator.**

Keywords—Power electronics; die attachment; silver sintering; electronic packaging; embedding technology; embedded power module; high performance substrates; system integration; PCB embedding; HELP-H

I. INTRODUCTION

The establishment of ceramic substrates in power electronics industry was driven by increased thermal durability, as well as, a more efficient heat distribution compared to conventional epoxy-glass-substrates. While the operating temperatures of power semiconductors can reach 175°C or more, the maximum operating temperature of common epoxy resin based printed circuits boards (PCBs) is limited to 150°C due to limited long term temperature stability [1, 2]. Apart from this, these materials are not compatible with meanwhile well-established processes like high pressure-assisted silver sintering. Previous investigations showed damage of epoxy-glass-substrates during die bonding due to too high pressures and too high process temperatures [3].

Despite the technical risks, there is a great interest in the development of new assembly concepts based on PCBs to provide highly integrated and cost-effective power electronic devices for electromobility. Embedding approaches are already in use for signal electronics and low to medium power applications but it is still challenging to apply them for higher power classes in many aspects. This work pursues the approach of using high-temperature stable benzoxazine-based PCBs, which were developed in a German-funded project "HELP" [2]. The suitability of these organic substrates, up to a temperature of 200 °C for 150 h, has already been presented in a previous work [3]. Moreover, the usually used solder on PCBs is replaced by silver sintering to improve the maximum operating temperature. Due to its high reliability, outstanding thermal performance (180…360 W/m*K) and high electrical conductivity (40…60 m/Ωmm²) silver sintering has great potential for bonding power electronic components. Furthermore, the elimination of the wire bonds by double-sided silver sintered semiconductor chips is of high attractiveness. The parasitic inductances are reduced by the compact design, which enables higher switching frequencies [2].

In order to accomplish sinter-embedded assemblies, the quality and reliability of sintered joints on these novel PCB-substrates were first systematically investigated. For this purpose, the principal process variables, pressure (P) and temperature (T) were systematically changed and the shear strength was evaluated. The achieved shear strengths were compared with values on standard substrates, e.g. direct copper bonded (DCB) ones. In addition, metallographic cross-sections and investigations using X-ray photoelectron spectroscopy (XPS) were performed. Based on these studies, the power semiconductor devices were embedded into the PCB material. For this purpose, a special sinter-lamination process was developed in order to form connections to the topside pads of gate and emitter on an IGBT electrically insulated from each other by silver sintering.

This project has been supported in the frame of the ECPE Joint Research Programme

978-1-7281-1500-9/19 $31.00 © 2019 IEEE

II. EXPERIMENTAL

A. Dummy chips and semiconductors

Silicon dummy chips with a sputtered backside metallization consisting of 50 nm titanium (Ti), 100 nm nickel (Ni) and 200 nm silver (Ag) were used for the preliminary examinations to obtain suitable conditions for silver sintering on PCBs. While Ti acts as an adhesion promoter and Ni is used as a diffusion barrier, the surface consists of Ag to form a bondable surface. Devices (IGBT IGC28T65T8M & diode SIDC14D60C8) from Infineon were used embedding. Electroless nickel immersion gold (ENIG) has been plated on the semiconductor top side pads to provide bondable surfaces. The die sizes and additional information are listed in Table I.

TABLE I.   PROPERTIES OF THE INVESTIGATED SILICON DEVICES

Device properties	Dummy chips	IGBT [a]	Diode [a]
Blocking voltage	n. a.	650V/ 50A	600V/ 50A
Metal Front-/Backside	none/TiNiAg	AlSiCu/NiAg	AlSiCu/NiAg
Die Size	2 x 2 mm²	6.57 x 4.2 mm²	4.6 x 3.05 mm²
Thickness	525 µm	80 µm	70 µm

[a.] As specified by the manufacturer.

B. Silver sintering pastes

A number of three silver sintering pastes were examined for the preliminary investigations in order to test various process parameters and their influence on the organic substrate. Two of these pastes consisting of silver microparticles were provided by a German manufacturer. They are further referred to low pressure (S-LP) and high pressure (S-HP) paste in reference to the applied process conditions. The second paste is from a Japanese manufacturer, which uses silver nanoparticles instead. It was developed to be sintered without pressure (S-NP).

C. Substrates

PCBs based on a benzoxazine resin system, which shows a thermal cycling capability of more than 1500 passive thermal cycles (-40/175 °C) were used for the investigations [2]. The thermal conductivity (λ) is improved by the addition of ceramic fillers, which also decrease the coefficient of thermal expansion (CTE) in z-direction to 38 ppm/K (Table II). The material promises significantly higher reliability compared to common epoxy-based materials and is further referred to HT-PCB.

TABLE II.   MATERIAL PROPERTIES OF THE USED PCB MATERIAL

Material properties [b]		Benzoxazin-based PCB
Transition temperature T_g (DSC)	°C	200
Coefficient of thermal expansion CTE (z) ($T < T_g$) ($T > T_g$)	ppm/K	38 210
CTE (x/y) ($T < T_g$) ($T > T_g$)		17/13 14/5
Thermal conductivity λ	W/mK	0.7

[b.] As specified by the manufacturer.

Shear tests results on benzoxazine-based PCBs were compared with standard substrates for power electronics. Table III shows the individual layers of the examined substrates and their thicknesses. The double line separates the substrates of the preliminary examinations from the substrates for the embedding of the 600V/50A IGBT half bridge module.

TABLE III.   MATERIAL PROPERTIES OF THE INVESTIGATED SUBSTRATES

Substrate	Description
	HT-PCB: Au ≈ 100 nm Ni ≈ 5 µm Cu ≈ 35 µm PCB ≈ 1.2 mm ...
	DCB (Al₂O₃): Au ≈ 50 nm Ni ≈ 5 µm Cu ≈ 300 µm Al₂O₃ ≈ 380 µm ...
	DCB (AlN): Ag ≈ 1 µm Cu ≈ 300 µm AlN ≈ 630 µm ...
	Copper: Ag ≈ 200 nm Cu ≈ 1 mm
	HT-PCB: (Top Board) Size: 50 x 40 mm² -------------------------------- 1450 µm thick 4-layer boards with 70 µm Cu thickness and thermal vias -------------------------------- **HT-PCB: (Bottom Board)** Size: 70 x 60 mm²

As shown in Table III, the assembly of the embedded module consists of two 4-layer boards where the semiconductors will be placed in between. A circuit layout for the top and bottom board was designed in order to build up half bridge circuits of IGBTs and diodes:

Top board:

- Contact pads for gate and emitter signals; placement of an external gate SMD resistor

- Inner shielding layer

- Connections by power traces for high-side and low-side of the IGBTs and diodes to the output and DC-

- Landing pads for flip-chip bonding and Cu strips

- Through-vias connecting the lands and the contact pads

Bottom board:

- Externally accessible contacts (DC+, DC-, out)

- Placement of Rogowski-coils for current measurement and decoupling capacitors

- Thermal vias for enhanced heat transfer

- Thermal contact pad for the heat sink

Additionally, holes for reference and alignment are located in the corners of both boards.

D. Silver sintering process

Initially, the paste was printed onto the substrates using a 100 µm thick stencil. The silicon chips were then placed in the sintering paste by a pick and placer. The micro-scaled silver sintering pastes were dried for 30 min at 130 °C respectively 160 °C before being sintering at elevated temperature (220-260 °C) and pressure (6-30 MPa). The nano-scaled silver sintering paste was sintered pressure-less at 175-220 °C for 60 min without pre-drying.

E. Embedding process sequence

The embedding process is split up into two individual assembly steps comprising the flip-chip die attachment by silver sintering and the silver sinter-lamination for embedding. Fig. 1 shows the sequence of fabrication of the demonstrator: First the flip-chip die attachment is done with the process described in the section before. For the chip placement, a flip-chip machine (Fine Placer) was used. This process results in a layer stack with approximately 3 µm Al, 1 µm Ni, and 0.1 µm Au. The top board is assembled with two IGBTs, two diodes, and two metallized copper strips, as shown in Fig. 1 (b).

Fig. 1 (c) shows the applied underfill, which supports the edges and the unfilled cavities of the chip. This mechanical support is necessary to avoid fracture of the chips in the subsequent sinter-lamination process that is shown in Fig. 1 (d). The stacking of the two printed circuit boards is done in a fixture by using pins to align the copper strips and the silicon chips of the top board to the applied sintering paste on the bottom board. A corresponding structured pre-preg sheet is placed between the boards to fill the cavities. Both boards are bonded together by simultaneous sintering of the Ag paste and laminating of the prepreg sheet while applying heat and pressure. This combined process results in the final stack displayed in Fig. 1 (f).

III. RESULTS

A. Shear strengths in comparison

To check for a suitable process window of silver sintered die attachments on the investigated high-temperature PCB (HT-PCB) materials, shear strengths on different substrate types were compared. Same process conditions were applied for the die attachment process to the substrates introduced in Table III. A comparison of the achieved shear strengths for a sintering time of 5 min at a temperature of 220 °C and a pressure of 6 MPa is shown in Fig. 2.

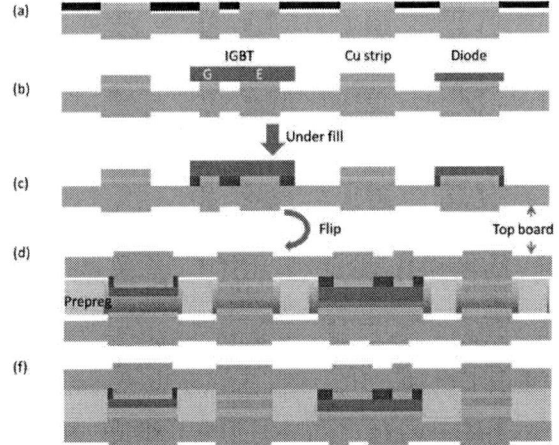

Fig. 1. Proposed embedding process for the fabrication of the power package. Fabrication sequence from top to bottom.

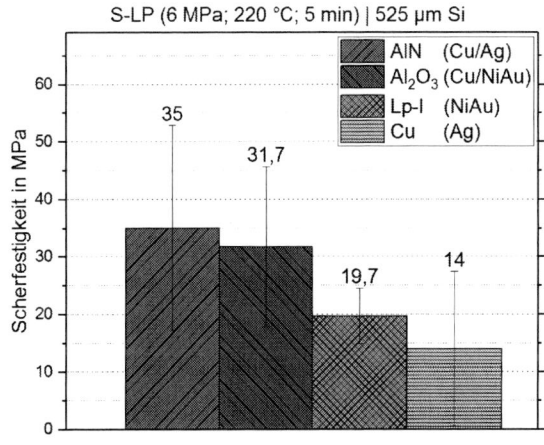

Fig. 2. Shear strength of silver sintered die attachments. Comparison of printed circuit boardswith standard power electronics substrates, e.g.DCB (15 measurements per value).

The findings can be summarized as follows:

- DCB substrates (Al_2O_3/AlN) achieve on average 69 % higher shear strengths compared to PCB substrates for equivalent process conditions.

- The lowest shear strength was found for pure copper substrates with Ag-metallization and amounts 71 % of the shear strength of PCBs.

- Dummy chips mounted on PCBs achieve intermediate shear strengths of approximately 20 MPa.

- The dominant failure mechanism of PCB mounted chips occurs on the interface between substrate metallization and the silver-sintered layer, while DCB-mounted chips also fail by fracture and at the interface between the silver sintered layer and the chip metallization.

These differences in shear strengths can be attributed partly to residual stresses which are formed after the layered composite material has cooled from the stress-free temperature (joining temperature) to room temperature during the manufacturing process. The influence of the adhesion will be discussed later. In order show a comparison here, the residual stresses in the die attachment were calculated from the center to the edge according to Suhir [4] (formulas 1 - 5) with the substrate thicknesses listed in Table III. The subscripts in the formulas represent the materials of the assembly according to Table IV. Furthermore k depicts the longitudinal compliance, α the coefficient of thermal expansion, λ the axial compliance, l half of the chip length, κ the interfacial compliance, υ the Poisson's ratio, E the Young's modulus, G the Shear modulus, D the flexural rigidity and t the thickness of the materials used. The effective material properties of the substrates were taken into account with the dimensions of the copper layers and neglecting the surface finish. Material properties from Table IV were applied for the calculation. Based on the previous experiments a chip size of $2 \times 2\ mm^2$, a chip thickness of 525 µm and a temperature difference (ΔT) of 200 K was used.

$$\tau_0(x) = k \cdot \frac{\Delta\alpha \cdot \Delta T}{\lambda \cosh kl} \cdot \sinh kx \qquad (1)$$

$$k = \sqrt{\frac{\lambda}{\kappa}} \quad ; \quad \Delta\alpha = \alpha_3 - \alpha_1 \qquad (2)$$

$$\lambda = \frac{1-\upsilon_1}{E_1 t_1} + \frac{1-\upsilon_3}{E_3 t_3} + \frac{(t_1+t_2+t_3)^2}{4(D_1+D_2+D_3)} \qquad (3)$$

$$\kappa = \frac{t_1}{3G_1} + \frac{2t_2}{3G_2} + \frac{t_3}{3G_3} \qquad (4)$$

$$G_i = \frac{E_i}{2(1+\upsilon_i)} \quad ; \quad D_i = \frac{E_i t_i^3}{12(1-\upsilon_i)} \qquad (5)$$

TABLE IV.   USED MATERIAL PROPERTIES FOR THE CALCULATION OF THE RESIDUAL STRESS

	Material	α in ppm/K	E in GPa
(1) Chip	Si	2.8	170
(2) Die-attachment	Ag	-	36
(3) Substrate	Al_2O_3	6.6	406
	HT-PCB	14.0	25
	Cu	16.5	115

The highest shear stress results for the copper substrate. This correlates with the lowest tested shear strength of the investigated substrate materials. For the DCB (Al_2O_3) substrate a comparable sum (Σ) of stress and strength can be observed compared to copper (see Table V). The lowest residual stress is calculated for the PCB substrate. According to the simplified hypothesis that the residual stress reduces the shear strength, the PCB substrate should achieve the highest shear strength. However, the measurement of the shear strength is contradictory to that. The most likely reason for this is the low adhesion of the sintered layer to the PCB substrate.

TABLE V.   MEASURED SHEAR STRENGTH AND CALCULATED RESIDUAL STRESS FOR A TEMPERATURE DIFFERENCE (ΔT) OF 200 K

	PCB	DCB (Al_2O_3)	Cu
$\tau_{\text{residual stress}}$	50 MPa	70 MPa	100 MPa
$\tau_{\text{shear strength}}$	20 MPa	32 MPa	14 MPa
Σ	70 MPa	102 MPa	114 MPa

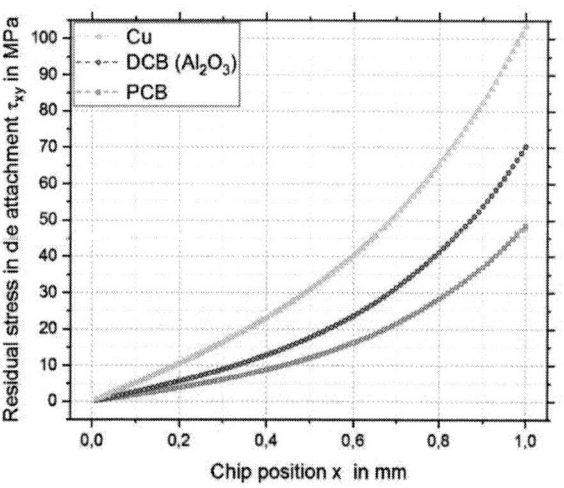

Fig. 3. Calculated residual shear stresses in the die attachment from chip center to edge of silver sintered silicon chips on different substrates.

All specimens tested for shear strength on the HT-PCB show insufficient adhesion to the substrate metallization (NiAu). This depends on the chemical condition of the surface, adsorbed layers and the microstructure [5]. Impurities, e.g. of organic origin or oxidation of the surface, can lead to adhesion problems. Therefore, the surface of the PCB substrate was examined with X-ray photoelectron spectroscopy (XPS). In Fig. 4 the atomic concentration of the investigated elements as a function of sputtering time is showed for the investigated sample.

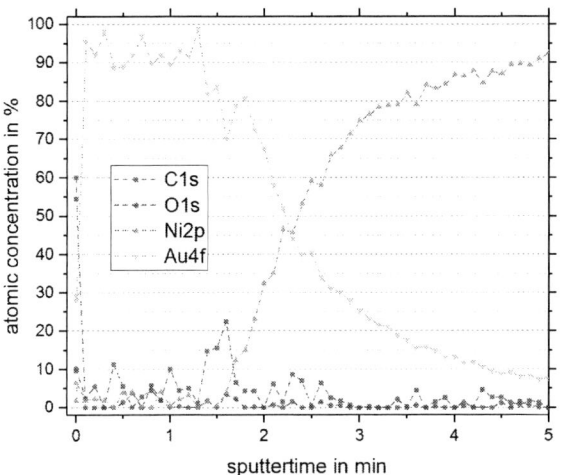

Fig. 4. Atomic concentration of the investigated elements as a function of sputtering time (X-ray source Al-Kα).

The atomic concentration of nickel on the surface amounts to 6.4 at.%. In a similar investigation, Blank et al. analyzed the shear strength of silver sintered chips on ENIG surfaces. They concluded that residual values of nickel should not exceed two at.% in order to achieve shear values higher than 20 MPa [6]. It should also be mentioned that the concentrations given should not be seen as absolute values, but rather as ratios compared to the other examined elements. Nevertheless, this could explain the comparatively low shear strengths achieved on the first batch of HT-PCBs. Also, these PCBs had a longer storage time. For this reason, freshly produced PCBs were used for further investigations. Also, we purposely selected an increased Au-finish of 100 nm.

B. Evaluation for higher process conditions

These freshly metalized PCBs with an increased Au-finish were used for investigations with process conditions of higher temperatures and higher pressures. While changing the temperature respectively the pressure, the other process conditions were kept constant. For investigating temperature influences a pressure of 20 MPa and a time of 5 min were selected. For evaluating sintering pressures a temperature of 230 °C and a time of 5 min were used. The results are shown in Fig. 5. No significant change in shear strength could be observed over the considered temperature range. This is consistent with the observation made in previous investigations [3].

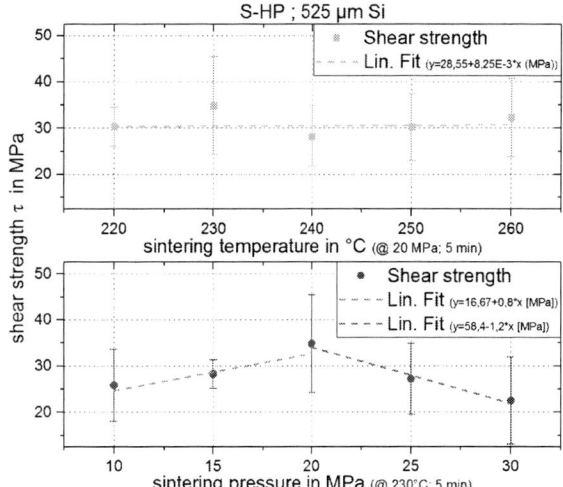

Fig. 5. Influence of the process temperature (220 to 260 °C) and the pressure (10 to 30 MPa) on the shear strength for a sintering time of 5 min (15 measurements per value).

However, this does not exclude the possibility of other temperature-induced changes to the sintered layer. The optimum process pressure for a 2-layer board was found to be in the region of 20 MPa. Below a sintering pressure of 20 MPa, the shear strength decreases by approximately 0.8 MPa per MPa of sintering pressure.

Above 20 MPa we found a decrease in shear strength by 1.2 MPa/MPa, while other examinations report an increase up to a process pressure of 50 MPa [7]. Examinations with Materialography were performed at this point. In the cross sections, damages of the dummy chips and the PCB substrates were detected. No deformation of the PCB was observed at pressures up to 15 MPa, whereas a slight imprint of about 2 μm was found for 20 MPa. At pressures of 30 MPa, the chip is deeply pressed into the PCB. In comparison to the original height of the metal pad (red line), a sinking in of 14 μm can be observed (Fig. 6). The process temperature of 230 °C is considerably higher than the glass transition temperature of HT-PCB material of 200 °C. This results in a significant reduction in Young's modulus and irreversible deformation. In combination with excessive pressures, serious damage to the PCB material can occur. Based on this observation, pressure levels above 20 MPa are not recommended for 2-layer boards without vias.

An influence of the sintering pressure can also be seen on the failure mechanisms (FM) of the sintered connections as listed below:

- FM (1): Fracture of the chip

- FM (2): Interface fracture on the chip side ↔ sintered layer

- FM (3): Interface fracture at the substrate ↔ sintered layer

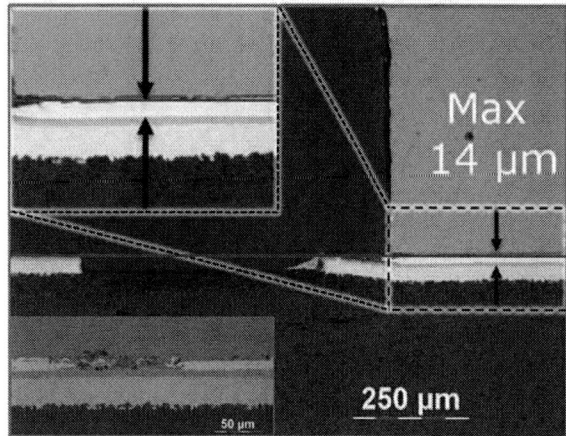

Fig. 6. Cross-sections trough sintered dummy chips sintered with 30 MPa.

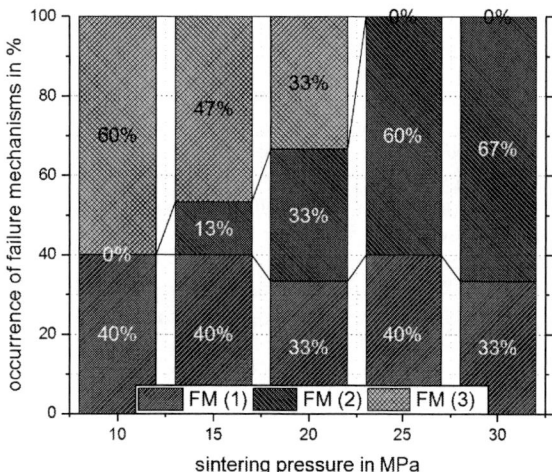

Fig. 7. Distribution of the failure mechanisms (FM) as a function of the sintering pressure (15 samples correspond to 100%).

Fig. 7 illustrates that initial adhesion weakness to the substrate metallization (60% at 10 MPa) switches to adhesion weaknesses for backside metallization of the chip (60% at 25 MPa) with increasing sintering pressure. At a pressure of 20 MPa, which achieved the highest shear strength, the occurrence of the failure mechanisms is of equal size when compared to each other. This is interpreted as an indication that a balance is reached, where the adhesion is approximate equally strong on both sides of the chip. The occurrence of chip fracture remains almost constant at 33-40 %.

One reason for the significant increase in adhesion weakness to the chip metallization above 20 MPa sintering pressure is the possible damage to chip and substrate. In a next step, additional specimens were sintered on PCBs with thermal vias under the chips. Cross-sections were made on diodes sintered with pressures of 10 MPa and 30 MPa. No damages to PCBs with vias (Fig. 8) could be observed. The results vindicate that vias under the chips improve the mechanical stability of printed circuit boards. In this case, higher sintering pressures can be used without damaging the board material.

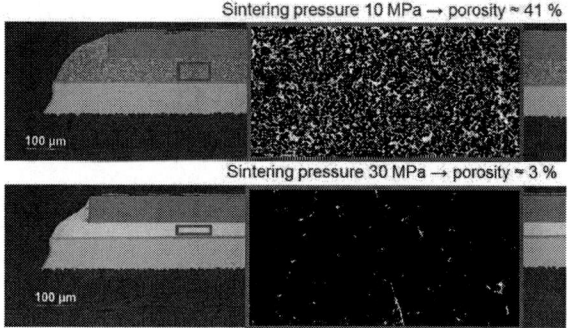

Fig. 8. Cross-sections through sintered diodes on PCBs with thermal vias sintered wit pressure of 10 and 30 MPa (Vias not visible).

The porosity was determined by taking microscope images of the sintered layers and converting them into binary images (Fig. 8). Subsequently, the area proportions of the black and white areas were determined and placed in relation to each other. For 10 MPa a porosity of 41.1% and for 30 MPa a porosity of 2.8% could be determined.

C. Pressureless silver sintering

The sintering paste containing nano-scaled silver particles was compared to the high-pressure paste of the previous investigations. Each paste was processed under the respective optimum conditions. For pressure-less applications long process times of one hour are necessary. The "nano paste" achieves shear strengths of 38 MPa on PCB before temperature cycling (Fig. 9). This value is almost maintained even after 250 passive temperature shock cycles (TS). Manufactured under low pressure and under significantly shorter time the shear strength is slightly lower before and after cycling. Summarizing, equivalent strengths can be achieved when the low-pressure paste and the high-pressure paste are processed under their respective optimum conditions.

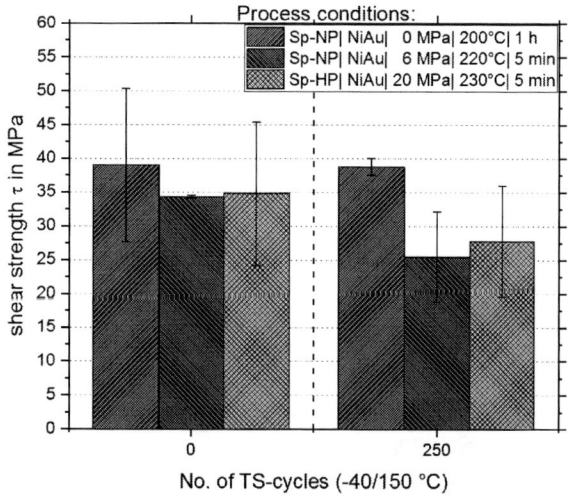

Fig. 9. Comparison of non/low-pressure silver sintering of nano scaled particles with optimum high-pressure sintering of micro particles. Shear strength of dummy chips on PCB after fabrication and after 250 passive temperature cycles.

978-1-7281-1500-9/19 $31.00 © 2019 IEEE

D. Evaluation of the embedding process

The embedding process was performed according to the concept outlined in Fig. 1. In the first step, the copper strips and the semiconductor chips were flip-chip sintered onto the top board with the parameters developed in the preliminary investigations (III B). The opening of the stencil printing mask for the paste application was reduced by 100 µm on each side to avoid electrical short circuits by smearing of the paste. Additionally, the resulting gap was under filled. A first functional test was applied at this stage to validate the blocking capabilities by setting a reverse bias of 200 V. Devices with a leakage current larger than 1 mA were considered as defective. Moreover, a current of 0.5 A was switched with a gate voltage of 15 V. A process yield of 100 % could be achieved in this first process step.

For the subsequent sinter-lamination process it was found that the thickness of the assembled components is essential for a successful process. For 10 samples, the average thickness of the Cu-pad, the sintered layer, and the IGBT amounted to 190 µm. Thicknesses of 175 µm for the diode and of 159 µm for the Cu-strip were determined. The quite clearly differing layer thicknesses result of the fact that the actual sintering was performed individually for each component. The sinter-lamination process had to compensate this difference in thickness by an increased layer thickness. As the prepreg requires a pressure of 3 MPa at 210 °C for 70 min, these parameters were used for the sinter-lamination process. Process conditions of 60 min at 200 °C without pressure showed already sufficient shear strengths for the nanoscaled sintering paste. Therefore this paste was selected for the sinter-lamination. Another big challenge was the selection of the thicknesses of the prepreg and the sintering paste on the bottom board. Starting with a layer thickness of 150 µm for the paste and 260 µm for the prepreg, no functional modules could be produced. Cross sections revealed that all chips were crushed to pieces by excessive pressure. As the pre-preg thickness was significantly lower than the chip stack, the bonding force first compressed the sintering interface at the chip before the pressure was shared between all interfaces. Consequently, the local and temporary overpressure led to a fracture of the devices. For a second attempt, a paste thickness of 100 µm and a thickness of 350 µm for the prepreg were chosen. With this design, a process yield of approximately 50 % functional switches could be achieved.

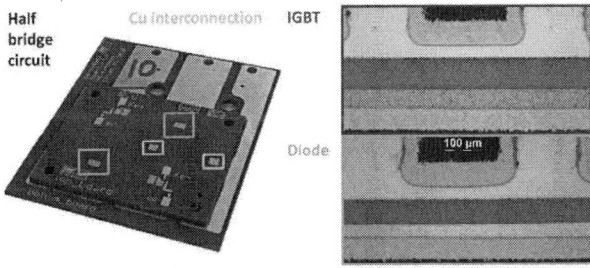

Fig. 10. Picture of a half-bridge demonstrator module (left) and cross sectional view of an IGBT and a diode in the module (right).

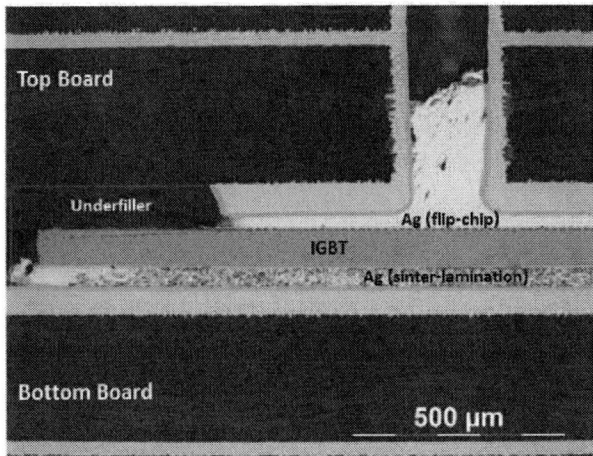

Fig. 11. Cross-sections through a embedded on IGBT with thermal vias with the introduced sinter-lamination process.

Cross sections of the sintered device are shown in Fig. 10 and in detail in Fig. 11. The functionality of the power package was proven by electrical characterization. The static forward characteristics for the diode and the IGBT are shown in Figures 12 and 13. The blocking was proven up to 600 V. Also double-pulse tests with a collector-emitter-voltage of 280 V and a current of 90 A were successfully performed.

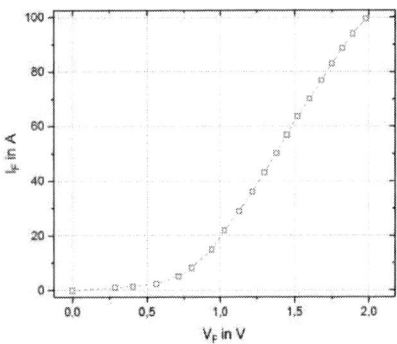

Fig. 12. Static forward characteristics of an embedded diode measured at 25 °C.

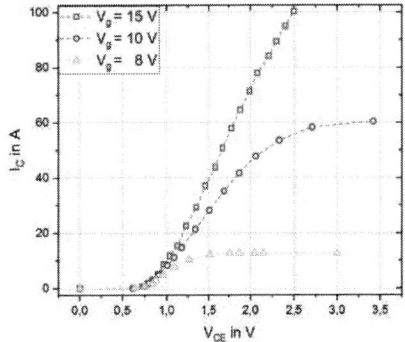

Fig. 13. Static forward characteristics of an embedded IGBT for different gate voltages measured at 25 °C.

IV. CONCLUSION

In this work, the silver sintering process was successfully transferred to high temperature printed circuit boards. Furthermore, the execution and potential of a sinter-lamination concept were demonstrated. The following points summarize the main outcome:

- Although the lowest residual shear stresses were calculated for die attachments of PCBs, the shear strength was lower than on DCB-substrates. This finding could be attributed to the diffusion of nickel and contamination of the gold surface using XPS. In consequence, the Au thickness was increased and higher process pressures were used to achieve higher shear strengths (III A).

- In the context of this work stable process conditions for silver sintering on printed circuit board materials could be proven. In addition, irreversible damage above 20 MPa was seen for 2-layer PCBs without vias. Significantly higher sinter pressures can be used when vias are present under the chips. Then a damage of the PCBs could not be determined in the examined range up to 40 MPa. As a result, layers with low porosity ($\approx 3\%$) can be produced (III B).

- For a pressure-less process with nano-scaled silver particles high shear strength was achieved, which lasted even after 250 passive temperature shock cycles. This silver nano-paste is particularly suitable for a sinter-lamination process (III C).

- The presented embedding process with double-sided silver sintered semiconductor chips was evaluated and the process yield could be increased significantly. The new process presented for embedding is based on already established technology and machines Therefore it can be practicable in existing production lines. Functional demonstrators were successfully tested. In the future, the process has to be improved with regard to uniform layer thicknesses. (III D).

ACKNOWLEDGMENT

The authors gratefully thank the ECPE Joint Research Programme for financing und supporting this research project. Further thanks go to the industrial partners Robert Bosch GmbH, Infineon Technologies AG, Heraeus Deutschland GmbH & Co. KG and Schweizer Electronic AG who all together provided support.

REFERENCES

[1] R. Ratchev, C. Mager, M. Guyenot, A. Khoshamouz, T. Gottwald, and S. Kreuer, "Hochtemperatur-Leiterplattenmaterialien für die Leistungs-elektronik unter erhöhter Belastung," Proc. EBL 2018 – 9. DVS/GMM-Conf., Fellbach, Germany, 2018, pp.59-65.

[2] M. Guyenot, D. Maas, R. Ratchev, A. Khoshamouz, T. Gottwald and S. Kreuer, "New failure Mechanism in High Temperature Resin Materials," *2018 IEEE 68th Electronic Components and Technology Conference (ECTC)*, San Diego, CA, 2018, pp. 1238-1244.

[3] A. Schiffmacher, L. Litzenberger, J. Wilde, V. Polezhaev and T. Huesgen, "Power electronic assemblies on Printed Wiring Boards Mounted by Silver Sintering," 2018 7th Electronic System-Integration Technology Conference (ESTC), Dresden, 2018, pp. 1-6.

[4] Tsai, M., Hsu, C. and Han, C. (2004). A note on Suhir's solution of Thermal Stresses for a Die-Substrate Assembly. Journal of Electronic Packaging, 126(1), p.115.

[5] M. Y. Wang, Y. H. Mei, R. Burgos, D. Boroyevich and G. Q. Lu, "Effect of substrate surface finish on bonding strength of pressure-less sintered silver die-attach," 2018 International Conference on Electronics Packaging and iMAPS All Asia Conference (ICEP-IAAC), Mie, 2018, pp. 50-54.

[6] T. Blank *et al.*, "Low temperature silver Sinter Processes on ENIG Surfaces," *CIPS 2016; 9th International Conference on Integrated Power Electronics Systems*, Nuremberg, Germany, 2016, pp. 1-6.

[7] Y. Zhao, P. Mumby-Croft, S. Jones, A. Dai, Z. Dou, Y. Wang, F. Qin, "Silver sintering die attach process for IGBT power module production," 2017 IEEE Applied Power Electronics Conference and Exposition (APEC), Tampa, FL, 2017, pp. 3091-3094.

[8] T. Herboth, M. Guenther, A. Fix and J. Wilde, "Failure mechanisms of sintered silver interconnections for power electronic applications," in Proc. IEEE Electronic Components and Technol. Conf. (ECTC), Las Vegas, May 28-31, 2013, pp. 1621-1627.

[9] E. Möller, A. Bajwa, E. Rastjagaev, and J. Wilde, "Comparison of new die-attachment technologies for power electronic assemblies," in Proc. IEEE Electronic Components and Technol. Conf. (ECTC), Lake Buena Vista, May 27-30, 2014, pp. 1707-1713.

High temperature resistant packaging technology for SiC power module by using Ni micro-plating bonding

Kohei Tatsumi[1], Isamu Morisako[1], Keiko Wada[1], Minoru Fukuomori[1], Tomonori Iizuka[1], Nobuaki Sato[2], Koji Shimizu[2], Kazutoshi Ueda[2], Masayuki Hikita[3], Rikiya Kamimura[4], Naoki Kawanabe[5], Kazuhiko Sugiura[6], Kazuhiro Tsuruta[6], Keiji Toda[7]

[1]IPS Research Center/Gradulate School,Waseda University,2-7 Hibikino, Wakamatsu-ku,Kitakyushu-shi,Fukuoka,808-0135 JAPAN   E-mail: tatsumi.kohei@waseda.jp

[2]Mitsui High-tec Inc., 2-10-1 Komine, Yahatanishi-ku, Kitakyushu-shi, Fukuoka, 807-8588 JAPAN

[3]Department of Electrical and Electronic Engineering, Kyushu Institute of Technology
1-1 Sensui-cho, Tobata-ku, Kitakyushu-shi, Fukuoka, 804-8550 JAPAN

[4]Kitakyushu Foundation for the Advancement of Industry, Science and Technology,
1-103 Hibikino-kita, Wakamatsu-ku, Kitakyushu-shi, Fukuoka, 808-0138, JAPAN

[5] R&D Center WALTS Co.,LTD. Research Center for 3D Semiconductors #310 1963-4Higashi, Itoshima-shi, Fukuoka, 819-1122JAPAN

[6]DENSO Corporation, 500-1Minamiyama, Komenoki, Nisshin-shi, Aichi, 470-0111 JAPAN

[7]TOYOTA Motor Corporation, 543, Kirigahora, Nishihirose-cho, Toyota, Aichi, 470-0309 JAPAN

Abstract—**There is an increasing expectation to incorporate silicon carbide (SiC) as inverter power modules for hybrid electric vehicles (HEVs) and electric vehicles (EVs). In order to maximize the performance of SiC devices, new packaging technologies that can realize high-temperature heat resistance by replacing solder joint or Al wire bonding, have been strongly demanded. In order to meet these demands, we have developed a new interconnection technology named Nickel micro-plating bonding (NMPB), that enables the interconnection in a narrow gap between electrodes of SiC devices and substrates via our newly designed lead frame, whose lead surface is formed into chevron shape. The plating bath for NMPB is a sulfamic acid bath consisting of nickel sulfamate and several additives, and was specifically prepared to plate narrow areas at the bonding interface without plating defects. It was found that when columnar crystal grains grow from both facing electrode surfaces, a strong bond without defects such as micro-voids at the interface is obtained. The bonding strength of NMPB was confirmed by shear test to be higher than that of Pb free solder joints and not to deteriorate even after high temperature storage (HTS) tests at 250°C for 1000hrs and after 1000cycle thermal cycle tests (TCTs, 250°C/-45°C). The NMPB was applied to the manufacture of one leg SiC inverter power module using two pairs of SiC MOS-FETs and SBDs, which were interconnected with a newly designed lead frame for double sided cooling structure. After molding resin copper heat spreaders were formed on the outer surfaces of both sides of the NMPB leads by additive method. Our newly developed SiC power module showed stable I-V characteristics over 250°C and lower switching loss. The reliability of the modules was confirmed by TCTs and power cycle tests.**

Keywords- Silicon carbide; Ni; Micro-plating; NMPB; SBD; MOS-FET; Inverter; Cu lead; EV; Power module;

I. INTRODUCTION

To reduce the size and increase the efficiency of inverter mounted in hybrid vehicles or electric vehicles, research on increasing the output power density using technology such as silicon carbide (SiC) devices is in progress, resulting in an increase in the demand for highly heat-resistant packaging technologies [1-4]. While studies have been conducted on heat-resistant interconnecting materials that can replace aluminum bonding wire or solder materials [5-7], various problems have been encountered.

To realize the high temperature reliable interconnection, the high temperature resistant materials need to be selected and connected at relatively lower temperature. Fig. 1 summarizes the relationship between the bonding temperature and the melting point of the materials for the interconnection being used or studied. Ultrasonic bonding can be applied to join high melting point materials at relatively low temperature, but its application is limited to the materials of wire or ribbon shape.

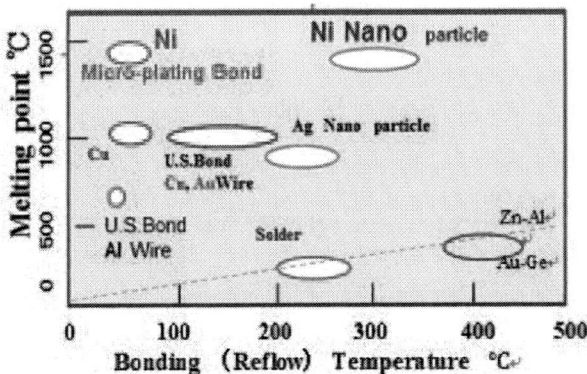

Fig. 1 Relationship between the bonding temperature of interconnection materials and the respective melting point of the material

We have been developing Ni micro-plating bonding (NMPB), which is processed around 55°C for the materials of melting points over 1400°C [8-13]. Since Ni exhibits excellent corrosion resistance, thermal stability and relatively low electrical resistivity, Ni was selected as a plating connection metal.

Fig.2 shows a sectional view of a double-sided cooling type Si-IGBT power module for inverters currently used in the latest HEV [14]. Based on this double-sided cooling type structure, we fabricated a high temperature heat resistant power module by replacing solder or Al wire interconnection with NMPB. The SiC chips are sandwiched between two lead frames and directly bonded to the chevron shaped portion of the lead by NMPB, which enables the interconnection in a narrow gap by Ni electro-plating as illustrated in Fig.3.

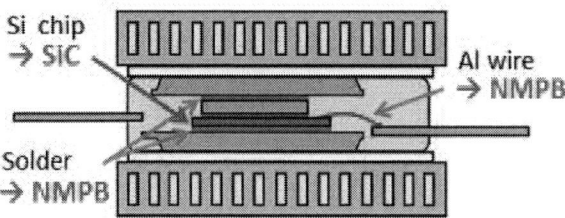

Fig,2 Double-sided cooling Si IGBT power module

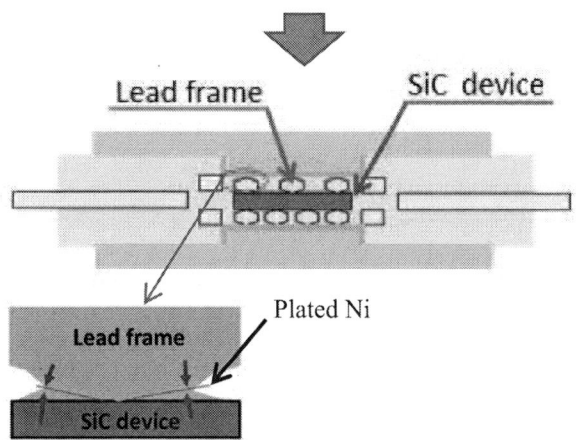

Fig.3 Double-sided cooling SiC power module interconnected by using NMPB

In our previous studies [8-11], we have identified the optimum angles of the chevron shape of lead frames and the NMPB conditions for the joining. Fig.4 (a) shows the cross section of the joining of Cu lead and SiC electrode observed by an optical microscope. Fig. 4(b) and (c) show an example of the distribution of the crystallographic orientation of Ni plating formed in the gap between Cu lead and SiC electrode obtained by SEM-EBSP mapping [11]. These figures indicate the <110> preferred columnar grain growth. As a result of forming texture of these columns, the bond interface consists of inherently so-called twist boundary, having low boundary energy and being stable, compared with tilt boundary or general boundary. We can apply this NMPB method for lead-free, high-temperature-resistant, corrosion-free power device interconnection, which is processed at a lower temperature around 55°C.

In this study, first, SiC schottky barrier diodes (SBDs) were mounted on a TO type package by NMPB and their reliability was evaluated by HTS and TCTs. Then, using our newly designed Cu lead frames for NMPB, one-leg power modules

mounted with two pairs of MOSFET and SBD were fabricated, and their I-V and switching characteristics and high temperature reliability were investigated.

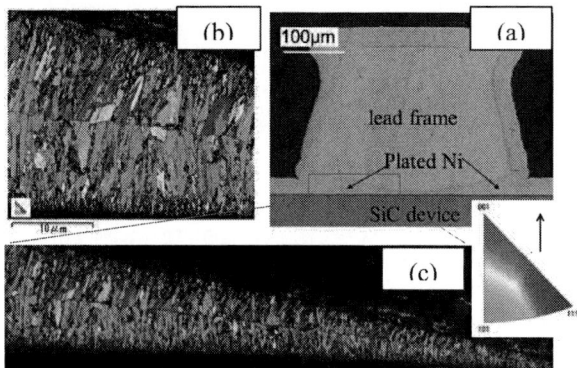

Fig.4 Optical microscope image of a cross section of bond formed by NMPB (a), SEM-EBSD mapping of plated Ni around the interface (b,c) [Ref. 11].

II. EXPERIMENTAL

(1) NMPB process

The plating for the NMPB process used a sulfamic acid bath with some additive agents. The plating solution was specially prepared by Japan Pure Chemical Co.,Ltd. for high speed plating in the narrow gap between Cu leads and chip electrodes. The bath temperature was controlled at 55°C and the typical plating current density was $1.5 A/dm^2$. Fig. 5 shows a schematic view of the plating bath, where a stacked set of SiC chips sandwiched with Cu lead frames was connected by electrolytic plating. The electrodes to be connected are all simultaneously bonded in the plating bath.

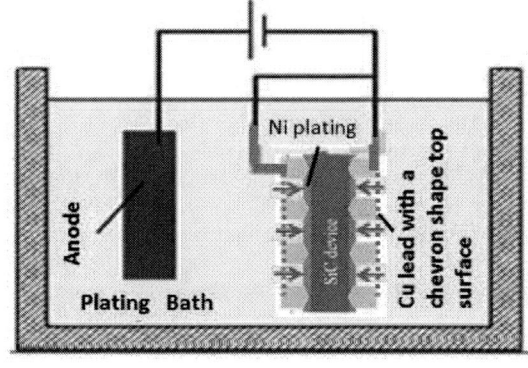

Fig.5 Schematic representation of NMPB process

The microstructure of the Ni plating was observed by optical microscope and SEM (Hitachi-SU5000). The electron backscattering diffraction (EBSD/ Oxford NordlysMax2) analysis was performed to examine the crystallographic orientation of the growth of Ni plating.

(2) Lead frame and SiC chips

As for bond reliability evaluation, commercially available TO264 lead frames and Cu plates (8mmx8mm, t=1mm) were used as plating substrates. Metalized SiC dummy chips (5mmx5mm, t=035mm) and SiC-SBDs (6mmx6mm, t=0.35mm) were used for various evaluation. For prototyping of 1-leg power module, a set of lead frames were designed in which two pairs of SiC-MOSFET(4.08mmx7.35mm) and SiC-SBD(6mmx6mm) were sandwiched from both sides of the chips. In order to form a circuit of the module, additional lead frames are prepared for electrically connecting the lead frame terminals.

(3) Shear strength measurement, TCT and power cycle tests

To evaluate the shear strength of the chip bonded to the substrate or lead frame, a bond tester (Nordson DAGE 4000plus) was used. TCT evaluation was performed between -45 °C to 250°C using fully automated TCT equipment (Espec, TSA-73ES). Fig.6 shows a typical TCT temperature cycle pattern used. For comparison, chips bonded by using commercially available Pb-free solder were tested in the same manner as NMPB samples, where the temperature cycle was changed from -45/250°C to -45/150°C. Power cycle test was performed by using Power Tester 1500A, Mentor Graphics with cold plate equipped. The test cycle pattern is shown in Fig. 15.

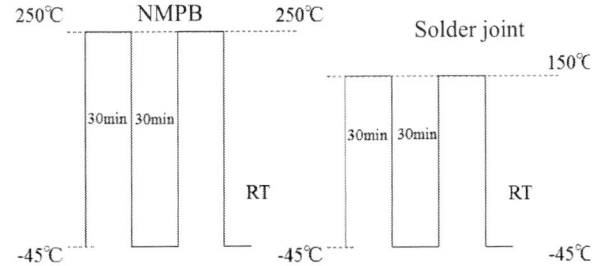

Fig.6 Thermal cycle in TCT for NMPB (left) and solder joint (right)

(4) I-V and switching characteristic evaluation

In order to evaluate the I-V characteristic, a conventional curve tracer (Iwatsu, CS-3300) was used. For high temperature operation tests, the power module was heated in an oven at temperature up to 250°C and the I-V characteristics were evaluated in site In oder to confirm the high-temperature operation due to internal heating of the MOS-FET, the surface of the Cu lead was exposed by removing the mold resin on one side of the module, and the surface temperature was measured with an infrared thermo viewer(R550-Pro, Nippon Avionics Co., Ltd.) .

III. RESULTS AND DISCUSSIONS

(1) Cu lead connection by NMPB and bond reliability

Fig. 7 shows a schematic view of cross section of a bond set of SiC chip and Cu plate formed via chevron shaped Cu leads by using NMPB for shear strength evaluation.
In order to evaluate the reliability of the bonds formed by NMPB, the shear tests were performed after each certain number of TCT cycles. The measured values were compared with those obtained for Pb-free solder joint.

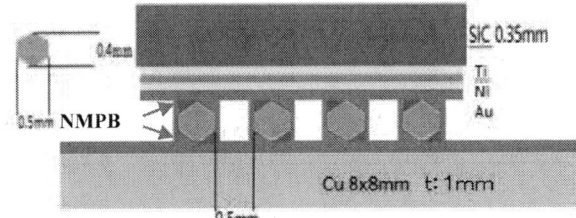

Fig.7 Cross sectional image of a bond set formed by NMPB using Cu leads for shear strength evaluation

Fig.8 shows the results of shear strength variation as a function of TCT cycle number, where the advantage of NMPB connection is confirmed clearly, showing the higher shear strength value than that of Pb-free solder joint for any TCT cycles. The shear fracture mode was inside of the chip in all cases for NMPB, whereas in case of the solder joint the fracture occurred inside of the solder or at the interface.

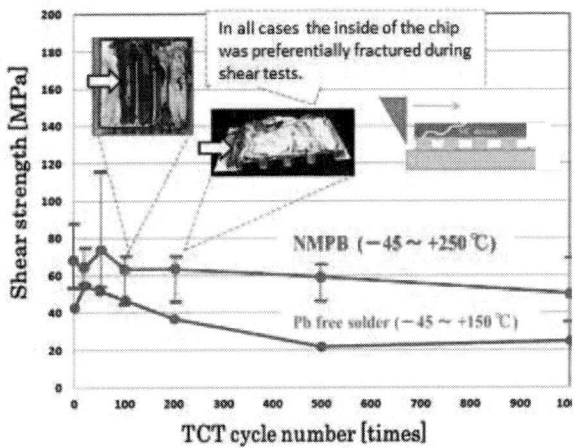

Fig.8 Shear strength as a function of TCT cycle number

(2) Inverter module

Fig.9 shows the schematic outline and cross-section of our NMPB module for inverter one-leg system. The circuit diagram is shown in Fig.10. The lead frames were designed for NMPB process and manufactured by etching and press forming. The entire bonding process for one leg module assembling is performed basically in one process simultaneously. A bond set of lead frames and chips are molded with heat-resistant molding resin by compression molding equipment. Fig.11(a) and 11(b) show a schematic image of lead frame interconnection by NMPB and an example of a cross section of optical microscope image of one-leg NMPB module with molding resin respectively. Heat spreaders can be formed to this module by high speed copper electroplating or other additive methods like Cu cold spray as needed. Fig.12 shows each process of prototyping of one leg inverter module using NMPB lead frame and an example of an appearance of the module with heat spreaders formed by copper electroplating.

(3) I-V characteristics and high temperature operation of power modules

First, the I-V characteristics of the power module were evaluated at room temperature, 150 °C and 250°C, using curve tracer, where the modules were heated in an electric oven.

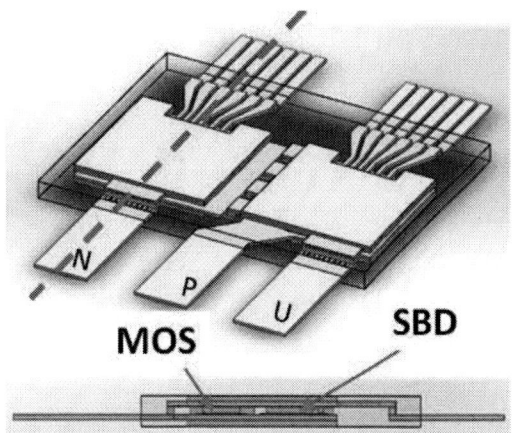

Fig.9 Schematic outline and cross section of NMPB module

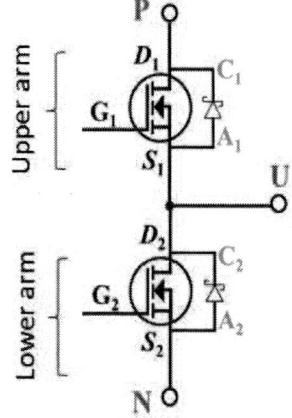

P, U, N: output leads, G, D, S: gate, drain, source for MOSFET and SBD

Fig.10 Circuit diagram for one-leg inverter module

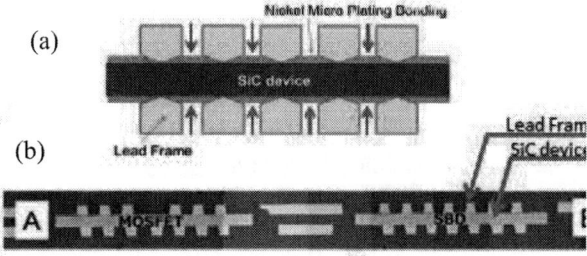

Fig.11 (a) Cross sectional schematic image of lead frame interconnection by NMPB (b) Optical micrograph of cross section of one-leg NMPB module with molding resin

Fig.12 Appearance of each process of prototyping of one leg inverter module using NMPB lead frame

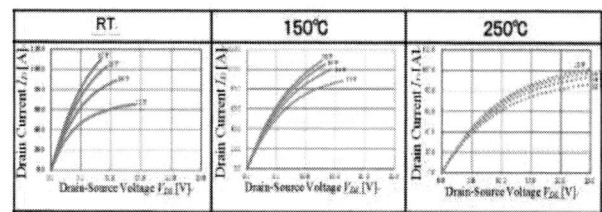

Fig. 13 I-V Characteristics of NMPB power module at at RT, 150°C, 250°C.

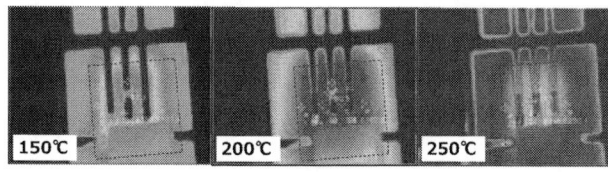

Fig.14 Thermographic images of NMPB power module at RT, 150°C, 250 during the operation of MOSFET

As shown in Fig.13, we can see the normal behavior in I-V characteristic even at 250°C, which indicates that the NMPB bonding and lead frame system is effective for high temperature operation of power modules.

In oder to confirm the high-temperature operation of the device during self-heating by the MOS-FET driving, the surface of the Cu lead was exposed by removing the mold resin on one side of the module, and the surface temperature of the chip portion was measured by an infrared thermo viewer. Fig. 14 shows thermographic images of NMPB power module at RT, 150°C, 250°C. The module with NMPB interconnections was found to exhibit normal behavior even under high temperature switching operation.

(4) Power cycle tests for various packaging types of power modules

Power cycling tests focus on the degradation of our developed module using NMPB (a), compared with that of conventional module packages (b) (c). The ΔT is generated through the self-heating effects from a Drain-Source current flowing. In the conventional module package (b) the same SiC devices as those mounted on our developed module are interconnected with high-temperature solder and thick Al wires. The commercial SiC module (c) is packaged with Pb free solder and thick Al wires. The power cycle test results for the modules (a), (b) and (c) are shown in Fig.15. The power cycle test condition is shown in the figure. The commercial SiC module (c) was tested under the condition of $T_j$max =150°C, ΔT=100°C, as the low melting temperature solder is used for the interconnection. Our developed module shows a power cycle life of 120,000 cycles or more even under the condition of $T_j$max = 250 °C and ΔT=150 °C, indicating extremely excellent high temperature heat resistance.

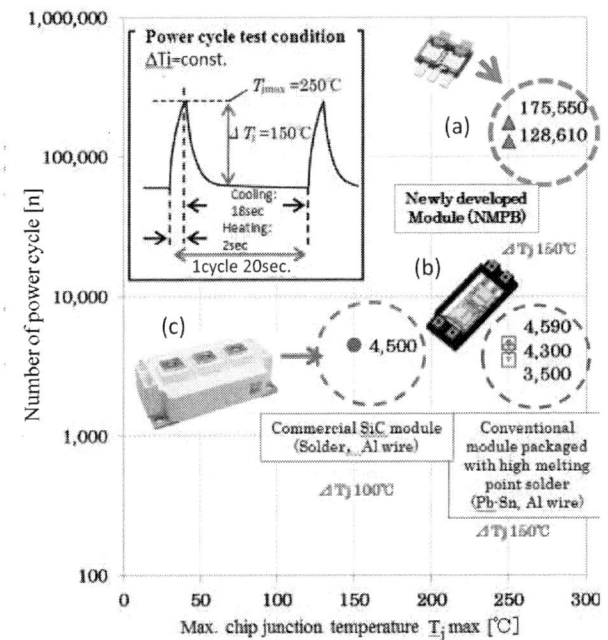

Fig. 15   Number of power cycles to failure for newly developed module (a),   conventional module package (b), and commercial SiC module package (c)

(5) Switching characteristics of the module and inverter

Switching waveforms were measured with an input voltage of V_{DC}=650V and a current of I_D=100A for our newly developed   module by using NMPB (a) and conventional module package (b). The configuration of the modules is illustrated in Fig. 16. Fig.17 shows an example of turn-off waveforms for NMPB module (a) and conventional module package (b).The surge voltage in NMPB module was found to be reduced by 40%. Furthermore, switching loss was calculated from the waveforms of turn on and turn off, and as a result, the loss of our developed NMPB module was estimated to be 57% of the conventional package. This can be related to the fact that the parasitic inductances of the

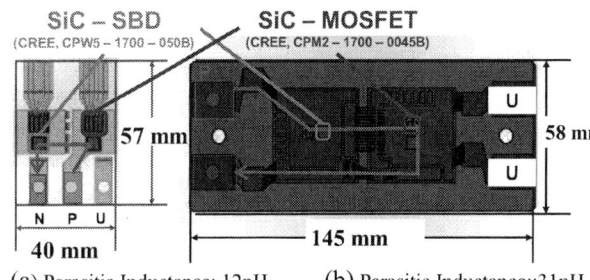

(a) Parasitic Inductance: 12nH   (b) Parasitic Inductance::31nH

Fig.16 Configuration of NMPB module (a) and conventional module package (b). [Ref. 15].

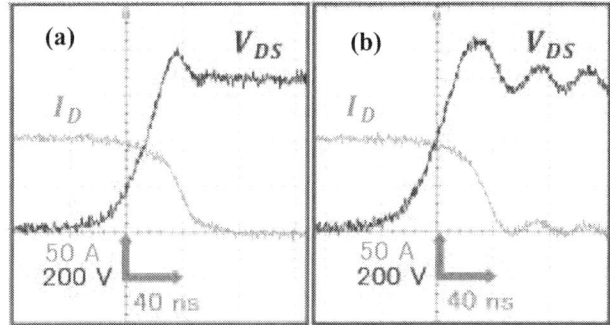

Fig.17 Turn-off waveforms for NMPB module (a) and conventional module package(b) [Ref. 16].

developed module and the conventional module are measured as 12nH and 31nH, respectively[16]..

Furthermore, we fabricated prototype inverter for automobile using developed module technologies[17]. Fig. 18 shows the appearance and size of the inverter. The modules of double-sided cooling type shown in Fig.12 are sandwiched between two cold plates. Currently we are further evaluating inverter characteristics. By applying these development technologies, it is considered that the Si-IGBT inverter used in existing EV or HEV can be greatly miniaturized.

Fig. 18 Prototype inverter unit using NMPB SiC power modules [Ref. 17]

IV. CONCLUSIONS

The SiC high temperature resistant module was fabricated as a prototype by using NMPB technology that can securely bond the narrow gap between the SiC device and the specially designed lead frame. Through the evaluation of the reliability and characteristics of the module fabricated by using NMPB, the followings were clarified.

1.The bond reliability revealed that there was no deterioration in the TCT (250 ° C./-45° C.) for 1000 cycles or more and in the power cycle test (Tjmax 250 ° C., $\Delta T = 150$ ° C.) up to more than 100,000 cycles.

2. The normal operation of the module at high temperature due to the environmental temperature of 250 ° C. and the chip s elf-heating (250 ° C. heating) in the module was confirmed.

3. The switching loss of our developed module was calculated from the waveforms of turn on and turn off, and as a result, the loss of the module was estimated to be 57% of the conventional SiC module package.

ACKNOWLEDGMENT

This work was supported by the Council for Science, Technology and Innovation (CSTI), Cross-ministerial Strategic Innovation Promotion Program (SIP), and "Next-generation power electronics/ Research and development of packaging technology for high temperature resistant SiC module of automobile of next-generation SiC power electronics" (funding agency: NEDO)

REFERENCES

[1] F. Xu, T. J. Han, D. Jiang, L. M. Tolbert, F. Wang, J. Nagashima, S. J. Kim, S. Kulkarni, and F. Barlow, "Development of a SiC JFET-Based Six-Pack Power Module for a Fully Integrated Inverter", *TRANSACTIONS ON POWER ELECTRONICS*, pp. 1464-1477, 2013.

[2] T. Ishikawa, Y. Tanaka and T. Yatsuo, "SiC power devices for HEV/EV and a novel SiC vertical JFET", *IEDM 2014*, pp. 222-224, 2014.

[3] H.A. Mantooth, M. M. Mojarradi, and R. W. Johnson, "Emerging capabilities in electronics technologies for extreme environments. Part I high temperature electronics," *IEEE power Electronics Society Newsletter*, pp. 9-14, 2006.

[4] L. Coppola, D. Huff, F. Wang, R. Burgos, and D. Boroyevich, "Survey on high temperature packaging materials for SiC-based power electronics", *Proc. PESC, Orlando, FL*, pp. 2234-2240, 2007.

[5] M. Knoer and A. Schletz, "Power Semiconductor Joining through Sintering of Silver Nanoparticles, Evaluation of Influence of Parameters Time, Temperature and Pressure on Density, Strength and Reliability", *International Conference on Integrated Power Electronics CIPS2010*, pp. 16-18, 2010.

[6] S. Sakamoto,S. Nagao and K. Suganuma, "Thermal fatigue of Ag flake sintering die-attachment for Si/SiC power devices", *J. Mater. Sci: Mater Elctron*, pp. 593-2601,2013.

[7] V. R. Manikam, K. A. Razak, K. Y. Cheong, "Reliability of sintered Ag80–Al20 die attach nanopaste for high temperature applications on SiC power devices", *Microelectronics Reliabitiy*, pp.473-480, 2013.

[8] K, Tatsumi ELECTRODE CONNECTION STRUCTURE AND ELECTRODE CONNECTION METHOD, US.Patent 9601448B2

[9] N. Kato, A. Shigenaga, K. Tatsumi, " High temperature resistant packaging for SiC power devices using interconnections formed by Ni micro-electroplating", *Material Science Forum*, Vols 778-780,pp.1110-1113, 2014.

[10] N. Kato, S. Hashimoto, T. Iizuka, and K. Tatsumi, "High-Temperature-Resistant Interconnections Formed by Using Nickel Micro-plating and Ni Nano-particles for Power Devices" *Trans, JIEP* Vol.6, No.1, pp.87-92, 2013.

[11] Kohei Tatsumi, et.al., "Development of Packaging Technology for High Temperature Resistant SiC Module of Automobile Application, "*Proc. 67th Electronic Components and Technology Conference (ECTC 2017)*p1316-1321

[12] S. Terashima, Y. Yamamoto, T.Uno and K. Tatsumi, "Significantreduction of wire sweep using Ni plating to realize ultra fine pitch wire bonding", *Proceedings of the 52nd Electronic Components and Technology Conference*, pp.891– 896,2002

[13] T.Ando, K.Tatsumi, Y.Ohno, S.Shimizu, Y.Kudo, and T.Fujitsu, "Plating micro bonding used for Tape Carrier Package:",*Proc. NIST/IEEE VLSI PACKAGING WORKSHOP*, 10.12,1993.

[14] Y. Sakamoto, "Assembly technologyies of double-sided cooling power modules", *Denso Technical Review*, Vol.16, pp46-56,2011

[15] Akihiro Imakiire, Masahiro Kozako, Masayuki Hikita, Kohei Tatsumi,Masakazu Inagaki, Tomonori Iizuka, Hiroaki Narimatsu Nobuaki Sato, Koji Shimizu, Kazutoshi Ueda, Kazuhiko Sugiura, Kazuhiro Tsuruta, Makio Iida, Keiji Toda, "Thermal Characteristic Evaluation and Transient Thermal Analysis on Next-generation SiC Power Module at 250 ºC *Proc." 10th International Conference on Integrated Power Electronics Systems(CIPS)*2018,pp.174-179

[16] Tomoya Itose, Akihiro Kawagoe,Akihiro Imakiire, Masahiro Kozako, Masayuki Hikita, Kohei Tatsumi, Tomonori Iizuka, Kazuhiko Sugiura, Kazuhiro Tsuruta, Nobuaki Sato, Koji Shimizu, Kazutoshi Ueda, Keiji Toda, "Evaluation of Switching Loss of SiC Power Module using Ni Micro Plating Bonding" *Proc. Denki Gakkai Sangyo Oyo Bumontaikai 2018 IEEJ* 1-9, I,37-38 (in Japanese)

[17] Akihiro Kawagoe ,Tomoya Itose, Akihiro Imakiire, Masahiro Kozako, Masayuki Hikita, Kohei Tatsumi, Tomonori Iizuka,,Isamu Morisako, Nobuaki Sato, Koji Shimizu, Kazutoshi Ueda, Kazuhiko Sugiura, Kazuhiro Tsuruta, Keiji Toda, "Development and evaluation of SiC inverter using Ni micro plating bonding power module" *Proc. International Workshop on Integrated Power Packaging （IWIPP）2019*, to be published

Pb-free, High Thermal and Electrical Performance Driven Die Attach Material Development for Power Packages

Byong Jin Kim, Dong Su Ryu, HyeongIl Jeon, Muhammad Hadhari Hazellah,
Weng Tuck Chim, JinYoung Khim,
Amkor Technology Inc.,
Amkor Technology Korea, Inc., 150, Songdomirae-ro, Yeonsu-gu, Incheon 21991, Korea
byongjin.kim@amkor.co.kr

Abstract— Most power packages still use a high-lead (Pb) solder paste because the high-Pb solder paste has the advantage of no re-melt during the printed circuit board (PCB) reflow process as well as good wettability and high reliability performance. There are a few material candidates that could meet the reliability compliance to AEC-Q100/-Q101 Grade 0 and provide thermal and electrical performance comparable to lead solder. Silver (Ag)-sintered material is one of candidates. However, the different types of Ag-sintered materials available in the market have not been proven to meet the required quality level and have not been validated for workability.

This paper investigated Ag-based materials to replace a high-lead (Pb) solder paste and checked associated processing and reliability testing for power discrete packages. Evaluation was conducted from material screen-out by workability check, epoxy coverage and void-free performance. In addition, automotive reliability had to be confirmed with the process workability. Electrical/thermal performance were simulated with material and localized void ratio and confirmed for performance comparable or better than the current high-Pb material. The new material is expected to achieve void-free and high thermal/electrical performance and address the environmental issue with Pb.

Keywords— *Power discrete package, High Pb solder paste, Ag sintered material, Pb-free solution, Void-free solution, package reliability*

I. INTRODUCTION

Lead (Pb)-free and environmentally friendly materials have been used for die attach and clip attach in integrated circuit (IC) packages for many years, but most power packages still use a high-Pb solder paste as an interconnect material for high power and high reliability application. The major reason has been the lack of alternatives to replace a high-Pb solder paste that has advantage of no re-melt during the printed circuit board (PCB) reflow process as well as good wettability and high reliability performance[1]. Even though there are many good advantages with High Pb material, there are increasing demands to exit Pb material use and find alternative materials in the near term to address environment issues[2,3]. Figure1 shows the representative power discrete package with a copper (Cu) clip interconnection. There are three contact areas as interconnections between die bottom and leadframe, clip and die top and clip and leadframe (LF) which are mostly using high Pb material. They are the contact areas to be studied with silver (Ag) sintered material.

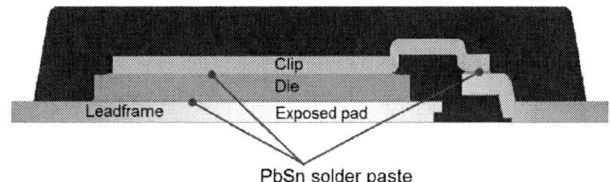

Figure 1. Schematic power package (Cross section view).

Along with environmental issues, higher reliability requirement and higher performance are being driven by new applications like battery-operated systems, renewable energy applications and high-power consuming cloud/network services[4-6]. These new drivers need to resolve the issues of solderball and voids which inherently exist with high Pb paste. Figure2 shows the issues of solder voids and solderballs occurring when high-Pb material is applied. A solder void is created during the reflow process when flux squeezes out. It can be minimized by reflow profile adjustment or vacuum reflow but it is impossible to achieve void-free attachment. These voids are known to degrade thermal and electrical performance. Solderballs occur randomly during reflow process. The problem results from not being able to estimate where it sits down and how it is critical enough to cause electrical migration.

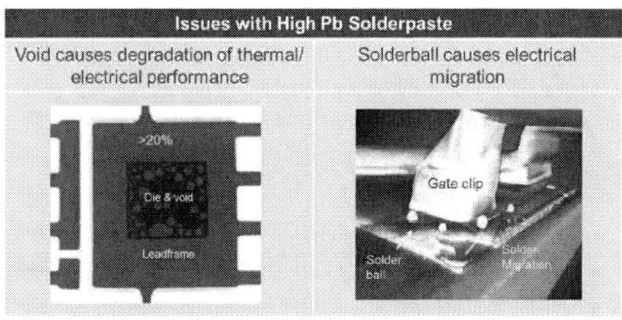

Figure 2. Voids and solderball issue with high-Pb material.

In this paper, Ag based materials have been studied to check an alternative material comparable with high Pb paste in terms of process workability, thermal/electrical performance and confirmed reliability. Several types of Ag sintered materials were screened out by epoxy coverage, die shear strength and void under clip or die and finally confirmed by automotive reliability testing. Also, thermal and electrical performance were simulated and checked to find out dominant factors to determine appropriate materials. Micro structure analysis was executed to understand the material bond between Ag material to leadframe and Ag material to die metallization.

II. EXPERIMENT AND RESULTS

A. Experiment to screen out Ag-sintered material candidates

The test vehicle for the evaluation consists of die size a 11.5mm² with 9-mm² bond pad opening area, A clip interconnection like Figure1. Leadframe thickness of 0.20mm (C19210 type) and clip thickness of 0.15mm (C19210 type) were applied. The assembly process is shown in Figure 3 where the major differences between high Pb and Ag-sintered material are die attach, clip attach and reflow processes. After the Ag-sintered material was dispensed on the leadframe for die attach, die bonding area for clip attach and leadframe post area, die attach and clip attach was performed in sequence and oven-cured together.

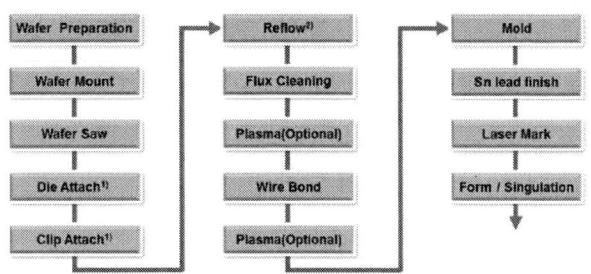

1) PbSn or Ag sintered material dispensing.
2) Oven cure (30min/150°C & 2hr 220°C) instead of reflow for Ag sintered material

Figure 3. Assembly process.

Material candidates were selected by different viscosity, modulus of elasticity and different thermal/electrical properties which could impact workability and reliability. Table 1 shows a comparison of the material candidates. All sintered material candidates have higher thermal conductivity than existing epoxy and high Pb solder paste. Thermal conductivity is different based on Ag content ratio and generally accounts for around 85~90% while volume resistance is almost the same or comparable with high-Pb material. The viscosity of Ag-sintered materials is significantly lower than high-Pb materials and could impact the workability. So, process verification will be confirmed through epoxy coverage for die and clip and die shear tests.

Table 1. Ag sintered material candidates.

Adhesive type	Epoxy	Solder Paste	Ag Sintered Paste			
	Standard	95Pb5Sn	A	B	C	D
Viscosity at 25°C at 5.0 RPM (kcP)	9.5	130~230	17	18	34.7	8
Modulus at 25°C (Gpa)	2.9	-	9.1	5.5	19	11.8
Modulus at 260°C (Gpa)	0.09	-	1.9	0.43	13	1.5
Thermal Conductivity (W/mK)	2~10	23	170	75	110	110
Volume resistance uOhm.m	-	0.2	0.2	0.3	0.2	0.2

Evaluation check items to screen out material candidates and verify process workability and reliability are:

1. Visual inspection to check paste coverage and bleed-out

2. Die shear test to check workability under wet condition

3. Bond line thickness (BLT) to meet consistent & required thickness

4. Void under clip and die by X-ray

5. Reliability: MSL1, TCx1000, HTS 1000h and uHAST 500h

The sample build performed using the process flow described in Figure 3 with four material candidates.

Figure 4 shows die shear test result. This was done under wet condition. Criteria is recommended have more than 2.5kgf(min) but material D shows less than that. The other three materials all are above 2.5kgf(min).

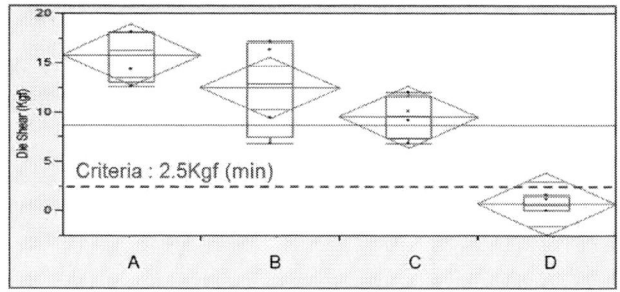

Figure 4. Die shear test.

Figure 5 shows bond line thickness (BLT) under the die before and after curing. It should be above 15µm, but material C is less and out of spec. However, BLT is considered quite stable and consistent because there is no significant difference before and after curing.

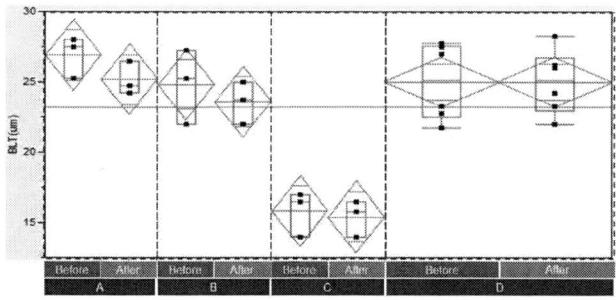

Figure 5. Bond line thickness before and after curing.

Figure 6 shows epoxy coverage and void level under the clip and die after curing. All material candidates were optimized for dispensing process first and checked for attach condition by X-ray. Epoxy coverage was manageable for both die and clip attach as well as lead post area. But material A shows a snake-like void under the clip different from the other candidates. Figure 7 shows a cross-section view after the molding process to confirm BLT condition and any abnormality. As shown, material C shows delamination between die top and clip which is suspicious of low adhesion under molding temperature

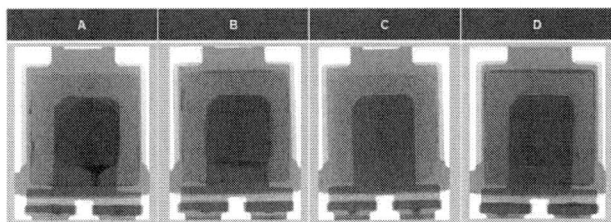

Figure 6. X-ray top view image after die/clip attach and curing.

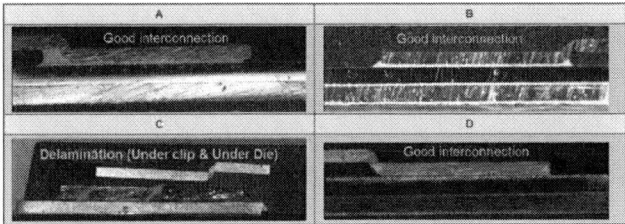

Figure 7. Cross section view.

B. Validation build and reliability test

Based on 1st sample build result, material C was excluded from further evaluation because it failed in BLT control and happened delamination during assembly process. So, 3 material candidates moved to 2nd sample build for reliability test. Reliability test was arranged under automotive conditions like Moisture Sensitivity Level (MSL) 1, unbiased Highly Accelerated Stress Test (HAST) 500 hours, High Temperature Storage (HTS) 1000 hour and Thermal Cycle (TC) 1000 cycle. Scanning Acoustic Tomography (SAT) analysis was performed after precondition MSL1 to check delamination. Electrical test was done to confirm any Ag-sintered material crack under die, clip and post area after reliability test. Fig. 8 shows SAT image where material D has a lot of delamination on clip top and leadframe area. Additional electrical testing confirmed this was an electrical failure.

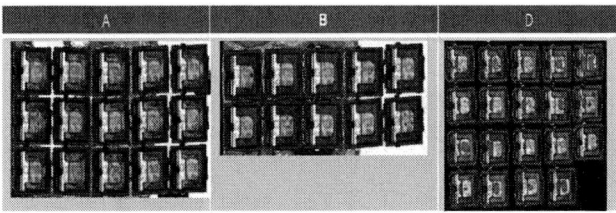

Figure 8. SAT images.

Table 2 is a summary of evaluation result after long term reliability test. Material A was failed due to electrical failure after TCx1000 and snake void not acceptable. Material C was failed due to BLT control and delamination. Material D was failed at MSL1 due to delamination and electrical failure. Material B only has no issue with workability and attach condition and passed all reliability test including precondition MSL1.

Table 2. Summary of evaluation result.

Sintering paste	Q'ty	Void (X-ray)	Die Shear	X-Section	MSL1	TCx1000
A	22	Snake Void	Excellent	Good	Good	Fail
B	22	Good	Excellent	Good	Good	Good
C	22	Good	Good	Delamination	N/A	N/A
D	22	Good	Not Good	Good	Fail	N/A

Figure 9 shows a X-ray view of material B by each process step and showing no abnormality in epoxy coverage for die and clip attach. There is no void under clip and die which is the major achievement compared with high-Pb material. Figure 10 shows a cross-sectional view after 1000-hr HTS to check any degradation of Ag material and BLT uniformity. BLT under clip is around 20 μ m and under clip is around 30 μ m and both are consistent. No degradation after long temperature storage is shown.

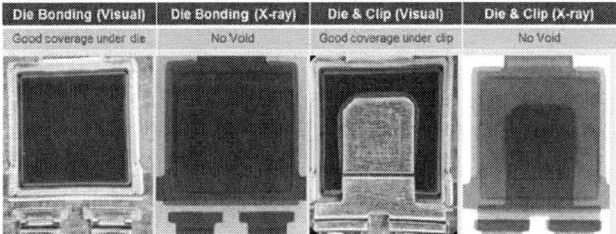

Figure 9. Sample view of each process.

Figure 10. Cross-sectional view after 1000 hours HTS.

C. Material Characterization

High-Pb materials are known to provid high thermal and high electrical performance compared with conventional epoxy die/ clip attach material as well as high reliability capabilities. Ag-sintered vs. high-Pb material should show no differences in electrical performance because their volume resistance is almost same as shown in Table 1. The Ag-sintered material has no detectable voids but voids are very common in high-Pb materials. To confirm this material difference, simulation was studied by intentional material volume change from 0% to 20% less. Figure 11 shows the electrical path on package structure and resistance result by material volume change. In this simulation, material was set up 0%, 5% and 20% less and checked. Naturally, less volume causes higher resistance by its volume gap. That is, the Ag-sintered material could have better consistent in electrical performance considering the Pb material has inherent and inconsistent voids. Actual sample tests were checked to compare material B and high Pb. It was measured from clip top to LF pad by forward voltage (mV) which was expected to check material resistance between LF to die bottom and die top to clip top at the same time. The other path to measure focused on the material on the post between the LF and clip.

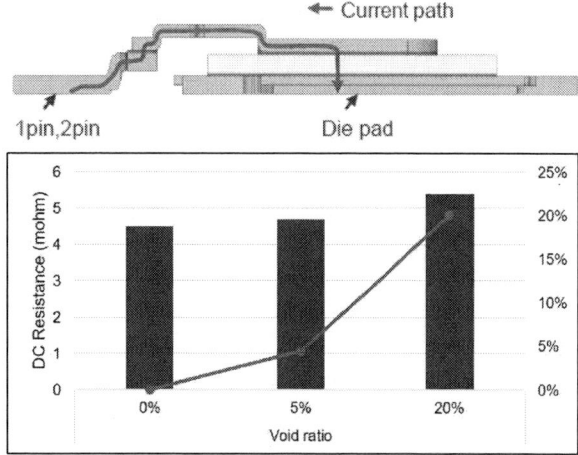

Figure 11. DC Resistance simulation by material volume change.

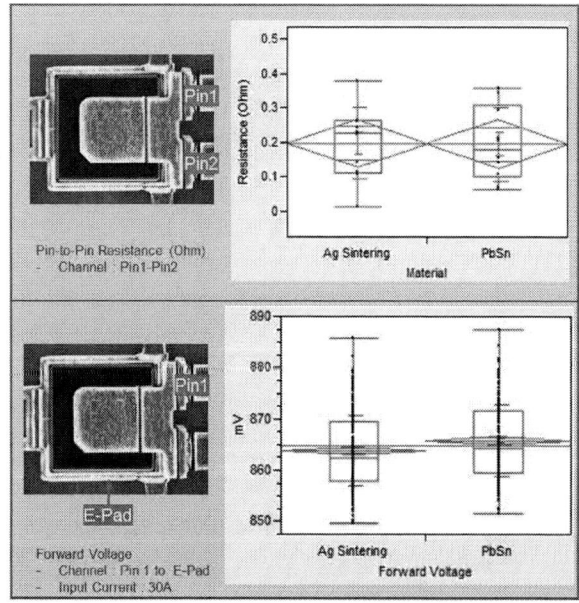

Figure 12. Sample test and measurement.

Figure 12 shows the actual sample test and its measurement. The Ag-sintered material showed comparable performance with high-Pb material. Unfortunately, the void effect was not detected because it is difficult to make the sample with intentional voids as well as measure variance by high resistance.

Thermal performance between high-Pb and Ag-sintered materials was also confirmed. Quite a difference in terms of package level was expected because material thermal characterization is significantly different based on datasheet values in Table 1. However, die and clip attach materials are not dominant factors and turned out to be less than a 1% improvement with the Ag-sintered material. Similar to the electrical analysis, thermal performance was further studied with the assumption of high-Pb void from 0% to 20% less. Depending on void level, thermal performance degraded up to 5.5% as shown in Figure 13.

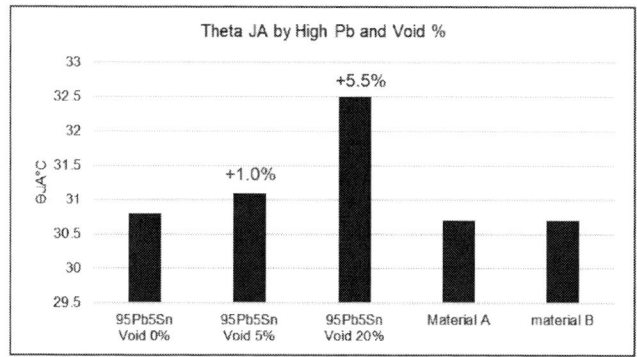

Figure 13. Thermal simulation by material and void.

D. Microstructure Analysis

The evaluation silicon (Si) wafer has surface metallization for solder wettability like die top with aluminum/nickel/gold (Al/Ni/Au) and die bottom with titanium/nickel/gold (Ti/Ni/Au). This was applied to the Ag-sintered material evaluation using the same terms for comparison and checked for intermetallic compound (IMC) and its microstructure after TCx1000.

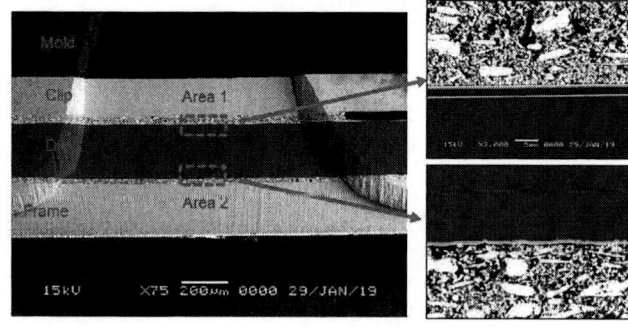

Figure 14. Cross-sectional view to check IMC layer

There could be two different IMC layers between die top and die bottom where new Ag sintered material was placed in this package structure. The IMC layer was confirmed by cross section view shown at Figure 14. Bond line thickness of die top and die bottom area were almost same and no clear IMC layer shows. The detailed analysis was carried out with SEM/EDX to check the IMC layer. Figure 15 is high resolution SEM image and EDX analysis at three points for die top area. Die top metallization is Al/Ni/Au and those elements would be diffused into Ag layer and supposedly make a specific compound during the curing process. But no IMC layer was detected.

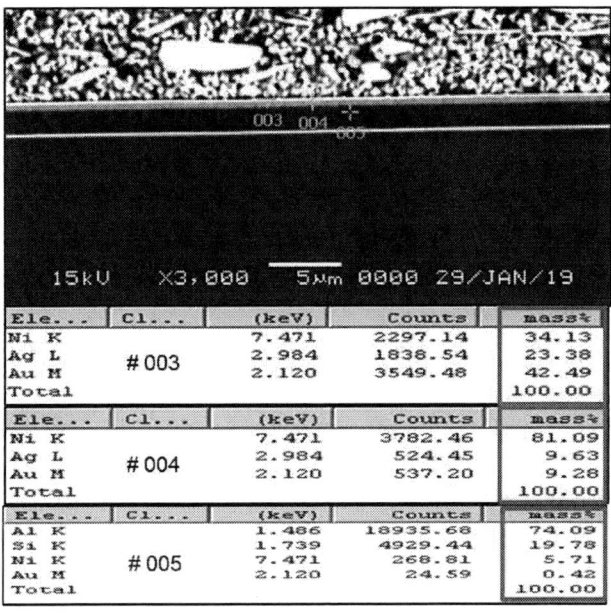

Figure 15. EDX at die top IMC layer.

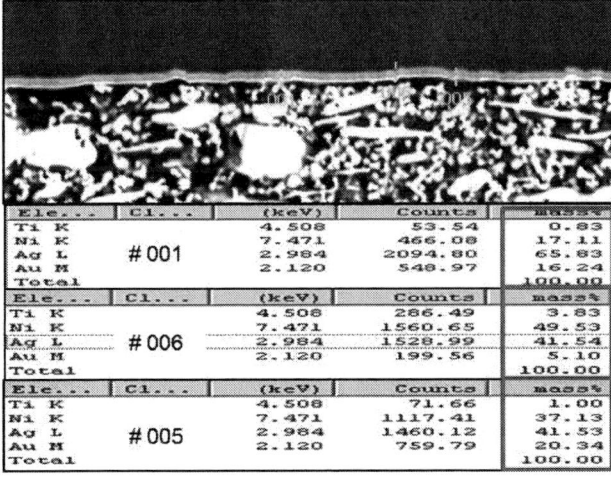

Figure 16. EDX at die bottom IMC layer.

The elements and their relative amount were identified by EDX. Point 003 which are far from die top metallization showed even distribution of Ag, Ni and Au. But as it goes down to die top layer, Point 004 contained Ni rich area with small amount of Ag and Au. Point 005 near Al passivation layer had no trace of Ag material. Figure 16 shows the interface of die bottom area and Ag-sintered material. Die bottom metallization is Ti/Ni/Au and no IMC layer was

detected, the same as die top area. This result is reasonable considering the first two metal element, Ni/Au, are the major contributors of interaction with Ag material. Point 1 is the layer near to Ag paste surface and 65.8% Ag detected. Point 6/ Point 5 are lower than Point 1 and show less Ag and more Ni than Point 1.

III. Conclusion

Most power packages still use a high-Pb solder paste as an interconnect material for high power and high reliability application. This is allowable under exemption conditions, but it is not intended for continuous use. Therefore, an alternative material escape the exemption time and criteria. Alternative materials should be compatible with current Pb material in terms of reliability and high thermal/ electrical performance. This paper has studied Ag sintered material to check any possibility of Pb material replacement from process workability to reliability and check thermal/ electrical performance by simulation and actual measurement. Finally, microstructures were checked to see inter-metallization between Cu to Si wafer surface metallization of Ni/Au. The findings from this study are:

1. Ag-sintered materials under development were tested in terms of process workability and reliability compatibility to high Pb material.

2. One of Ag-sintered materials passed process workability and automotive reliability requirement (AEC100) and confirmed high possibility to replace current Pb material.

3. Electrical and thermal characterization of the chosen Ag material was studied by simulation and actual measurement. Its characterization was confirmed compatible with high-Pb material. Considering that the Pb material has inherent solder void and solderball issues, the Ag material could be consistent and better in electrical and thermal performance but was not proven here.

4. Intermetallic compound and its layer were analyzed by SEM/EDX. Regardless of die top and bottom metallization, an IMC layer was not detected with Cu clip as well as Cu leadframe.

© 2019, Amkor Technology, Inc. All rights reserved.

Reference

[1] Sandeep Menon; Elviz George; Michael Osterman; Michael Pecht, "High lead solder (over 85 %) solder in the electronics industry: RoHS exemptions and alternatives", Journal of Materials Science: Materials in Electronics, 2015 Vol 26, pp 4021–4030.

[2] Kenny Chiong; HongWen Zhang; Sze Pei Lim, "High lead solder failure and microstructure analysis in die attach power discrete packages", 2016 IEEE 18th Electronics Packaging Technology Conference (EPTC), 2016

[3] Iver E. Anderson; Stephanie Choquette; Kathlene Thomas Reeve; Carol Handwerker, "Pb-free solders and other joining materials for potential

replacement of high-Pb hierarchical solders", 2018 Pan Pacific Microelectronics Symposium (Pan Pacific), 2018, pp.1-11.

[4] Byong Jin Kim; Dong Su Ryu; Jia Yunn Ting; HyeongIl Jeon; Weng Tuck Chim; Nathan Whitchurch, "High Reliability Power Package Design for Automotive Requirements", 2018 19th International Conference on Electronic Packaging Technology (ICEPT), 2018, pp. 1209-1212.

[5] Kay Hofmann; Christian Herold; Menia Beier; Josef Lutz; Jens Friebe, "Reliability of discrete power semiconductor packages and systems —

D2Pak and CanPAK in comparison," 2013 15th European Conference on Power Electronics and Applications (EPE), 2013, pp. 1-10.

[6] Xin Li; Xu Chen; Guo-Quan Lu, "Effect of Die Attach Material on Performance and Reliability of High Power Light-Emitting Diode Modules", Proceedings of Electronic Components and Technology Conference, 2010.

An RDL-First Fan-Out Panel-Level Package
for Heterogeneous Integration Applications

Yu-Min Lin[1,2], Sheng-Tsai Wu[1], Chun-Min Wang[3], Chia-Hsin Lee[2,4], Shin-Yi Huang[1], Ang-Ying Lin[1], Tao-Chih Chang[1], Puru Bruce Lin[3], Cheng-Ta Ko[3], Yu-Hua Chen[3], Jay Su[4], Xiao Liu[4], Luke Prenger[4] and Kuan-Neng Chen[1,2*]

[1]Electronic and Optoelectronic System Research Laboratories, Industrial Technology Research Institute (ITRI), Hsinchu, Taiwan
[2]National Chiao Tung University, Hsinchu, Taiwan
[3]Unimicron Technology Corporation
[4]Brewer Science Taiwan/Brewer Science, Inc.
*Tel: 886-3-5712121 ext. 31558, Fax: 886-3-5724361, Email: knchen@mail.nctu.edu.tw

Abstract—

Technologies of Fan-out panel-level packaging (FOPLP) are studies in this paper. First, the warpage control of a molded panel is a crucial problem for FOWLP technology development. In this paper, finite element analysis (FEA) is applied to study the influence of back end of the line (BEOL) process-induced warpage, as well as characterization for simulation, and investigation of each single process. In our process, a liquid release material is coated onto a 370 mm x 470 mm glass carrier. After baking, three layers of redistribution layer (RDL), passivation, and Cu leads are fabricated on the panel with coating, exposing, developing, lithography, and electroplating processes. Silicon test chips with a size of 10 mm x 10 mm and micro solder bumps with a pitch of 90μm are thinned down to 150 μm. Test chips are then flip-chip bonded onto glass carrier with pre-bond and reflow proces. After panel molding, a laser debonding method, another key technology advancement, is utilized for panel debond. Debond performance, which is directly related to laser parameter and panel-level package (PLP) structure, is critical. After debonding, the molded panel is cleaned, followed by dicing and OSP coating processes, and then the electrical performance of the interconnection is evaluated. Reliability tests at the component level, such as pre-condition, thermal cycling test (TCT), and unbiased HAST (uHAST), are performed. The demonstration of RDL-first PLP technology without interposers proves its great potential in heterogeneous integration applications.

Keywords- Fan-out panel-level packaging; FO-WLP; Process development; finite element analysis (FEA); warpage

I. INTRODUCTION

Goals of packaging development include low cost, high packaging density, and performance improvement, while maintaining the reliability of circuits. New assembly technologies and structure solutions for high density integrated circuit (IC) packaging have been developed, such as flip-chip bonding and 2.5D/3D IC packaging. A typical 2.5D/3D IC structure includes a top chip, silicon interposer, organic substrate, and printed circuit board. Silicon interposers play a key role in 2.5D/3D IC packaging and can allow a high-density fine line pattern [1-3]. The concept of flip-chip bonding technology is that a Si chip is aligned in a face-down configuration and then flip-chip bonded onto a Si substrate or organic circuit board. The high frequency performance, high pin count, and thermo-mechanical reliability using this method is better than other interconnection methods because the length of the connection path is minimized. Performance, packaging cost, and throughput drive the preference of using flip-chip bonding over the conventional wire bonding. On the other hand, to reduce total system and device cost, the concept of heterogeneous integration is proposed and developed. Fan-out wafer-level packaging, FOWLP, one of the new heterogeneous integration technologies, is based on chips that are embedded-inside epoxy molding compound (EMC) and then high-density RDL and solder balls are fabricated on the wafer surface to produce a reconstructed wafer [4-8]. These modules produced by FOWLP could be bonded onto an organic bismaleimide-triazine (BT) substrate or a circuit board directly without interposer and through silicon via (TSV) fabrication. Therefore, it has the advantages of lower-cost processing, thinner packaging, shorter signal transmission path, higher I/O channel, and flexible assembly collocation. Moreover, fan-out panel-level packaging (FOPLP) has been further developed in order to increase the throughput and potentially lower the cost compared with FOWLP [9-16]. However, substrate handling is the most complex process in FOPLP technology. There are so many issues that need to be solved. In this study, we present a redistribution layer (RDL)-first FOPLP technology and the fabrication of a three-layer-dielectric 370 mm x 470 mm-size panel-level fan-out structure. Control of the warpage of a 370 mm x 470 mm molded panel, is a key issue for FOPLP technology development.

In this paper, finite element analysis (FEA) of FOWLP is applied to study the influence of back end of the line (BEOL) process-induced warpage, as well as characterization for simulation, and investigation of each single process. Parameters such as chip thickness, molding thickness, chip number, and coefficient of thermal expansion (CTE) of molding compound are considered. Finally, a warpage mechanism with prediction for optimizing the reduction of panel warpage is proposed. With an optimized design, corresponding parameters can be applied to process flow of FOPLP.

In our process, a liquid release material is coated onto a 370 mm x 470 mm glass carrier. After baking, three layers of RDL, passivation, and under bump metallization (UBM) are fabricated on the panel with coating, exposing, developing, lithography, and electroplating processes. Silicon test chips are flip-chip bonded onto glass carrier with reflow process. After wafer molding and laser debonding, the molded panel is cleaned, followed by dicing and organic solderability preservative (OSP) coating processes, and then separated chips are carried out to evaluate the electrical performance of interconnection. Reliability test of component level, such as pre-condition, thermal cycling test (TCT), and unbiased HAST (uHAST), are performed. The demonstration of RDL-first PLP technology without interposers proves its great potential in heterogeneous integration applications. With conventional flip-chip assembly flow, it is also suitable for OSAT industry.

In this paper, we present the heterogeneous integration technology and fabrication of RDL-first FOPLP with a face-down method. The process flow details, finite element analysis, evaluation of release layer, and reliability test are presented in the following sections.

II. PROCESS FLOW INTRODUCTION

Figure 1 presents an RDL-first FOPLP process flow. Firstly, a 370 mm x 470 mm glass panel with 1.2 mm thickness was prepared. A release layer was coated on the panel as shown in Figure 1(a). Liquid laser release paste could be slide-coated as a thin film with a 150 nm thickness and then baked to form a thin solid film. Related research integrated with wafer form have been announced in previous researches [17-18]. Diffirent laser release materials were used for comparison in the study. In order to fabricated Cu RDL, a Ti/Cu seed layer was deposited on the release layer by a conventional sputtering method. Then, a polymer material with 8 μm thickness was used as the 1st passivation layer to protect and isolate the metal layer. After the process was repeated 4 times, 3 layer RDL/ passivation layers and Cu lead were fabricated on glass carrier as shown in Fig. 1(b)-Fig. 1(d). An OSP was formed on the surface of the Cu lead to protect the copper until soldering. After OSP coating, the panel was diced into strips. Top chips were bonded on to strip glass carrier then a plasma treatment system was used to clean the surface of Cu lead and test chips. Micro solder humps could melt and eutectic soldering with Cu lead formed a electric interconnection after reflow process.

Subsequent processes, shown in Fig. 1(i), are strip panel molding process and post-curing. Molding compound could cover the chip surface and glass panel and also seal the chip gap and bump space to protect the micro solder bump through the molded underfill (MUF) technology. Next, a laser debond process was preformed to decompose the release layer. After laser debond, RDL3 was exposed and protected with an additional OSP coating. Finally, test chips (15 x 15 mm^2) were strip diced from the strip .

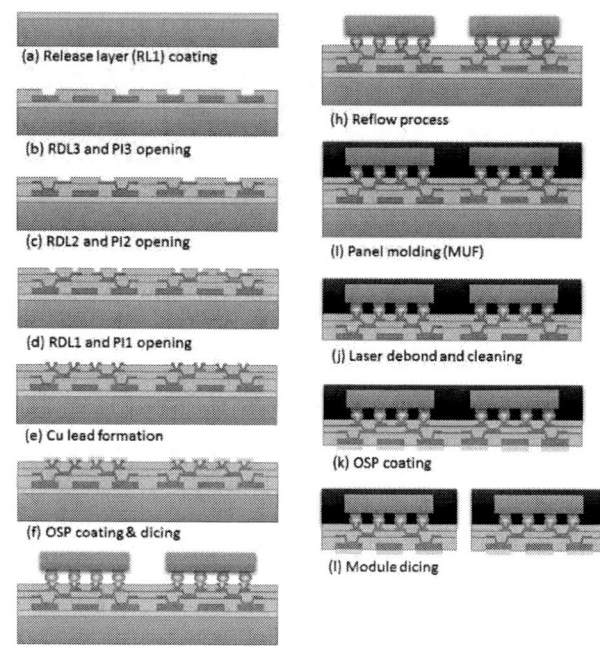

(a) Release layer (RL1) coating

(b) RDL3 and PI3 opening

(c) RDL2 and PI2 opening

(d) RDL1 and PI1 opening

(e) Cu lead formation

(f) OSP coating & dicing

(g) Prebonding

(h) Reflow process

(I) Panel molding (MUF)

(j) Laser debond and cleaning

(k) OSP coating

(l) Module dicing

Figure 1. Process flow of the RDL-first fan-out process structure

III. MATERIAL SELECTION FOR LASER RELEASE PROCESS

A laser release process is considered as the favorite mechanism for temporary bonding and debonding among thin wafer handling technologies, which can be attributed to advantages of high throughput and low stress generated on the device wafer during separation. Since the laser release process is associated with laser ablation of a release layer material, selection of the release layer becomes crucial for safety of separation of the device wafer from the carrier.

For RDL curing, reflow process, as well as chemical resistance to harsh semiconductor chemicals, thermal stability up to 250°C for hours is a common requirement of laser release layer. In this study, a laser release layer was synthesized with a thermal stability up to 350°C and also had desired absorption of laser energy from the targeted wavelengths in order to handle panel level RDL first process. Figure 2 shows the thermal decomposition (T_d) under N_2 and the thermal stability under N_2 for 1 hour. Table 1 shows the k (absorbance) of release layer. Moreover, Figure 3 shows coating thickness of laser release layer A exceeds 600 nm, which is thick enough to block most energy of laser beam as too much laser energy traveling through the laser release layer brings uncertainties to device wafer.

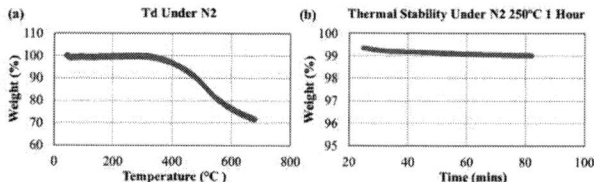

Figure 2. Thermogravimetric analysis (TGA) indicated Td of laser release layer A is up to 350°C with weigh loss only 0.27% at 250°C for 1 hour.

TABLE I. ABSORPTION OF LASER RELEASE LAYER AT VARIOUS WAVELENGTH

Wavelength	308 nm	343 nm	355 nm
k	0.36	0.26	0.24

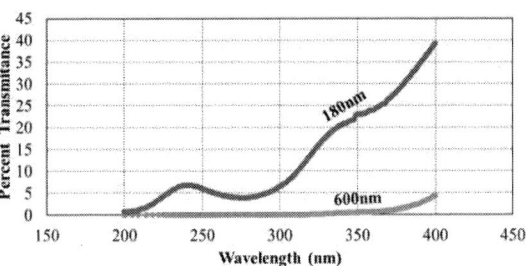

Figure 3. Percent Transmittance Data of laser release layer A

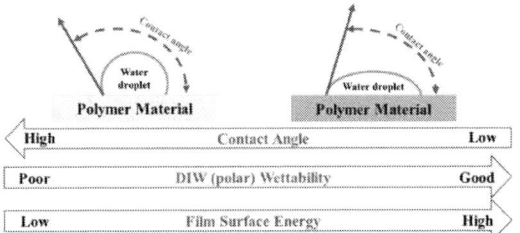

Figure 4. water contact angle and surface energy relationship

In addition to thermal stability, adhesion of laser release layer to the deposited metal film like titanium is crucial as warpage of bonding pair can cause delamination at interface [19-20]. As plasma treatment is commonly seen to control adhesion of two adjoined surfaces, by varying plasma treatment conditions it is possible to adjust wettability of given surface. The most important characteristic factor that affects interfacial interactions is surface energy. The surface energy of plasma-treated polymers can easily be examined by measuring the contact angles. Fig. 4 illustrates relationship of contact angle with surface energy.

To understand the changes made to the surfaces during plasma treatment, a laser release layer around 600 nm in thickness was coated on a 4-inch wafer prior to oxygen plasma with power ranging from 50 to 150 watts with a gas flow rate of 50 sccm and was analyzed. Table II summarizes contact of each wafer in 0, 3 and 7 days after plasma treatment, which indicates oxygen plasma is effectively to increase surface energy of release layer A.

Subsequently, a 100 nm titanium and 200 nm copper layer were deposited on the coated wafer before adhesion test method ASTM 3359 was utilized to determine adhesion of titanium and copper on laser release layer A. With 12 cuts in hash mark pattern made through the film to the substrate, pressure-sensitive tape was applied over the cut and then removed to assess adhesion qualitatively on a 0 to 5 scale.

The adhesion test result on table III indicated adhesion of release layer A with titanium increased significantly after oxygen plasma treatment. However, treatment for longer than 20 seconds with power of 150 watts seems to suggest at higher power there is a modification of the surface chemistry. Figure 5 shows roughness of wafer 1 at 12.2 nm, wafer 3 to

reach 20.1 nm and wafer 4 down to 8.9 nm due to removal of polymer surface. With treatment showing improvement of adhesion, wafer 3 condition was selected to carry on integration processes.

TABLE II. CONTACT ANGLES OF SELECTED LASER RELEASE LAYER A

Wafer	Gas Type	Power (watts)	Time (s)	Pressure (mtorr)	Gas Flow Rate (sccm)	Contact angle		
						0 day	3 days	7 days
1			Untreated			52.6	NA	NA
2	O₂	150	5	100	50	22.2	38.3	41.4
3	O₂	150	10	100	50	18.5	35.9	38.9
4	O₂	150	20	100	50	18.3	36.8	39.9
5	O₂	50	5	100	50	24.5	36.9	42.4
6	O₂	50	10	100	50	21.7	33.3	45.2
7	O₂	50	20	100	50	20.4	38.6	47.1

TABLE III. RESULT OF TAPE-PEEL INTERPRETATION (ASTM 3359) OF LASER RELEASE LAYER A

Wafer	Gas Type	Power (watts)	Time (s)	Pressure (mtorr)	Gas Flow Rate (sccm)	Cross-cut result
1			Untreated			0B
2	O₂	150	5	100	50	2B
3	O₂	150	10	100	50	5B
4	O₂	150	20	100	50	4B
5	O₂	50	5	100	50	3B
6	O₂	50	10	100	50	5B
7	O₂	50	20	100	50	5B

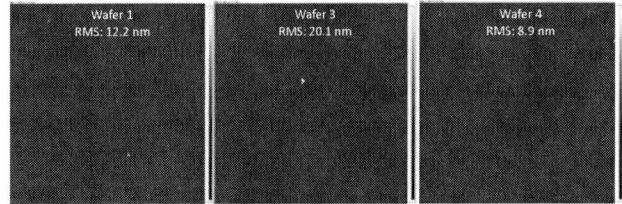

Figure 5. Roughness of release layer A after oxygen plasma treatment

IV. TEST VEHICLE DESIGN AND RDL PROCESS FLOW

To study the RDL-first fan-out panel level package, a panel size of 370 mm x 470 mm is applied to develop for increasing productivity. Figure 6(a) shows the panel and strip design. There are 5 strips in total inside the panel. The strip size is standard CA74 (240 mm x 74 mm), including 60 package units, as depicted in Figure 6(b). Figure 6(c) shows the body size of package unit is 15 mm x 15 mm, and the corresponding chip size iss 10 mm x 10 mm, which resulted in a chip-to-package ratio of 44.4%.

The initial fabrication was performed with a bare glass panel utilized as a temporary carrier. In order to minimize the residual stress and warpage control of panel, a glass panel with a CTE of ~8-8.5 ppm/°C with a thickness of 1.1 mm was chosen. At first, the glass carrier was coated with a releasing layer for debonding in future. After coating the release layer, a seed of Ti/Cu bi-layer was sputtered. The liquid photoresist (LPR) was slot coated onto the seed layer and then exposed by a proximity aligner which the minimum resolution was 8 μm. An aqueous developer of 2.38 wt% TMAH was adopted to develop the exposed LPR.

978-1-7281-1500-9/19 $31.00 © 2019 IEEE

The electrolytic copper deposition was applied later on and the approximate trace thickness of copper was 6 μm by utilizing Enthone's chemical solutions. The metal layer of RDL3 was completely formed by the following processes of LPR stripping and Ti/Cu etching. The copper thickness was measured by 3D laser scanning microscope, Keyence, and the sampling size was 35 points.

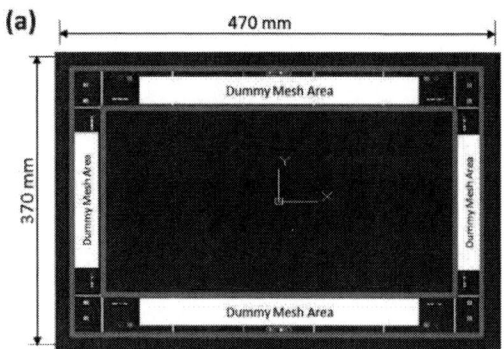

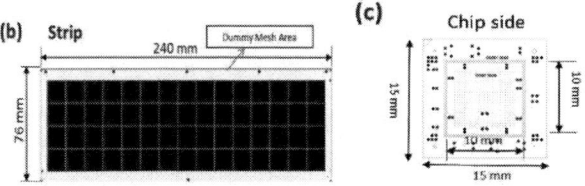

Fig.6 (a) Panel design for panel size of 370 mm x 470 mm, (b) Strip design for strip size of 240 mm x 76 mm and (c) Package design for chip size of 10 mm x 10 mm and package size of 15 mm x 15 mm.

The inter-metal dielectric was adopted with a photo-imageable-dielectric (PID) material and coated onto metal layer of RDL3. The photo-via was formed by exposure and development processes which were same with LPR lithography. For better interconnection quality, the plasma was utilized to remove the scum at the bottom of photo-via (Via23). Each RDL layer was formed by repeated the process of Ti/Cu seed layer to Ti/Cu etching, and PID material was also coated onto the RDL layer to finish each via layer. The designed redistribution layers consists of four metal layers and three PID layers are depicted in Figure 7.

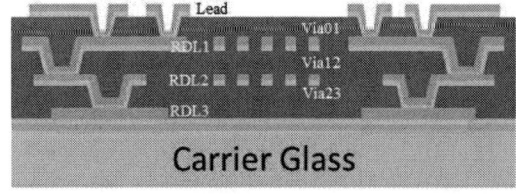

Fig. 7 Cross-section of the designed structure.

The specific layout of 8 μm/8 μm line width/space was designed on two-middle layer of RDL2 and RDL1. Figure 8(a) demonstrates the optical microscope picture of 8 μm

line width/space after LPR stripping and Ti/Cu etching in Figure 8(b) of RDL1. The trace width before etching was around 9.2 μm, and became 8.3 after Ti/Cu etching process. The etching rate of downward is about 0.9, which is good for line-width well control. Table IV shows the different RDL layer of Cu ratio of occupied panel area, indicating the warpage contribution is corresponding with the ratio.

The following PID was covered prior to the RDL layer and processed by a "hard bake" curing step at 250°C in 3 hr by N₂ oven. Figure 9 schematically showed the optical microscope picture of 17 μm via-opening in the layer of Via01. N₂ atmosphere is required while post-curing since O_2 may react with PID during the process of polymerization and degrade the mechanical properties, such as CTE and resistance of water absorption. Consequently, Sn-Ag solder bumping is made on the top lead layer with under bump metallurgy (UBM) stack-up and the minimum bump pitch was 90 μm with chip thickness of 150 μm.

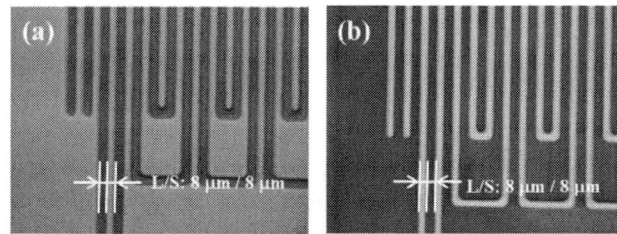

Fig 8. The optical microscopy pictures of 8 μm line width/8 μm line space (a) after LPR stripping and (b) after Ti/Cu etching of RDL1 layer.

Table IV. Each RDL layer of Cu ratio of occupied panel area

Layer	Cu ratio of panel
RDL3	29%
RDL2	52%
RDL1	51%
Lead	26%

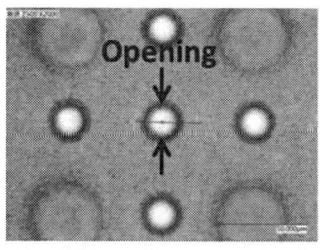

Fig.9 The optical microscopy picture of 17 μm via-opening in the layer of Via01.

978-1-7281-1500-9/19 $31.00 © 2019 IEEE

V. TEST CHIP BONDING PROCESS

The thin Si chip has an array of microsolder bumps with 90 μm pitch, which were fabricated with a bump structure of Cu/Ni/SnAg. The surface of test chip and strip panel are cleaned by plasma treatment. After that, test chips are bonded on strip glass panel in a subsequent die bonding process. The die bonding process used a different method. The new bonding method and traditional flip-chip bonding are described as shown in Figure 10 and 11, respectively.

There are some issues in traditional flip-chip bonding assembly process. Figure 10 shows that a process flow and eutectic soldering concept using a traditional assembly process. Paste type flux is coated on the substrate and then die attached on the substrate with adherence of flux. As the strip panel goes through the reflow oven, full melting occurrs between micro solder bump and UBM, causing self-alignment to occur. Flux could prevent oxidation ocurred during reflow process, but high density interconnection and bump structure in a bigger chip size make it difficult to remove flux residue. Figure 11 explains the new assembly process, where there are streamlined steps: pre-bond and formic acid reflow. At first, to remove organic residue and contamination, a plasma treatment is applied. It also could improve surface energy and soldering wettability. Partial melting occurs between micro solder bump and UBM in the pre-bond process. After reflow, oxidation is also removed as formic acid reflow is applied. Full eutectic melting occurs between micro solder bump and UBM. Eutectic soldring leads to self-alignment mechanism and excellent alignment accuracy $\leq \pm 1 \mu m$. The new assembly process does not use flux and could skip the flux cleaning process. It is cleaner on bonding interface and substrate surface.

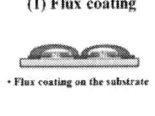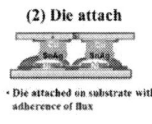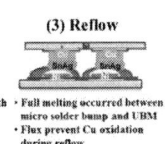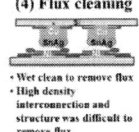

Figure 10. Traditional assembly process

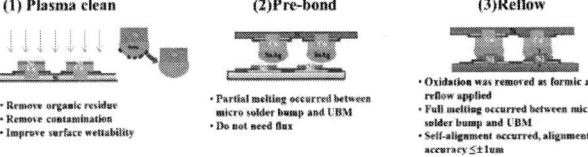

Figure 11. New assembly process: Pre-bond + formic acid reflow

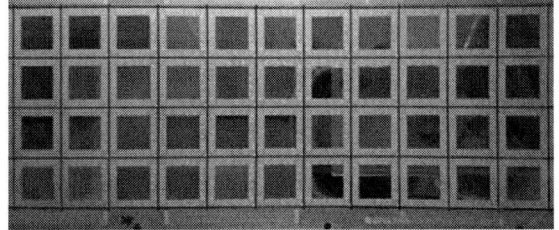

Figure 12. Assembly images of bonded strip panel

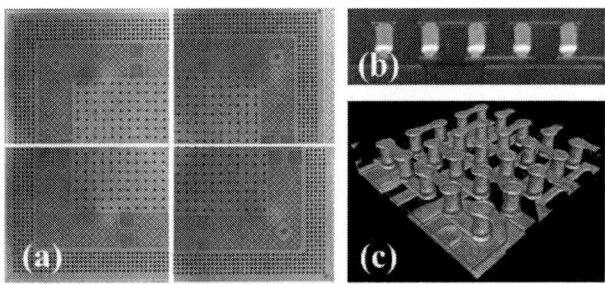

Figure 13. X-ray inspection images after reflow process (a) 2D X-ray; (b)2D x-ray; (c) 3D x-ray

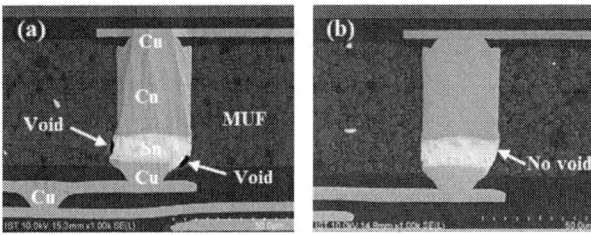

Figure 14. Inspection images after bonding process: (a) void appeared as no plasma treatment; (b) No void issue after plasma treatment

Figure 12 shows a strip panel that has finished the die bonding process. A nondestructive 2D x-ray and 3D x-ray inspection is used to check the bonding alignment and solder shape as shown in Figure 13. These x-ray inspection results reveal good bonding alignment accuracy and no appearance of solder shorting. In order to check the bonding interface and bonding quality after the molding process, cross-sectional SEM photos are analyzed for micro solder bumps and interconnection with destructive crosssection inspection. Figure 14(a) shows a void appeared between the MUF and solder. If Cu lead is not treated by plasma system, there will be some organic residues on Cu lead, which can lead to void or delamination during the panel molding process. No void was found as the suface had sufficient plasma treatment, as shown in Figure 14(b).

VI. FINITE ELEMENT ANALYSIS FOR WARPAGE

In general, the warpage control of a molded wafer/panel is a crucial problem for fan-out package technology development. In our previous investigation, we have used finite element analysis (FEA) and three dimensional solid model to study the influence of BEOL processes and their effect on induced warpage, and the simulation agrees with experimental results [20]. Figure 15 illustrates an opposite warpage situation as the chip number embedded in the molded wafer has a great difference. If a big chip pitch is applied, there are fewer chips that can be embedded in the wafer. This molded wafer will exhibit a concave shape. In contrast, with fine chip pitch, a large number of chips could be embedded in the molded wafer and then the wafer warp will reveal a convex shape. A comparison of various chip thicknesses, molding thicknesses, and chip numbers were collected for their impact on wafer warpage, as shown in Figure 16. A volume ratio of silicon/EMC is used to replace the diffcrence of chip thickness, molding thickness, and chip number. This relationship of volume ratio of silicon/EMC versus wafer warpage reveal a exponential curve. As the volume ratio decrease and becomes close to 0.6, the convex wafer warpage will transition to concave wafer warpage. This is an interesting discovery. Measurement warpage of our molded wafer also agrees and fits the trend. If the relationship of fan-out package can be defined, it would be advantageous to help control the warpage during the process.

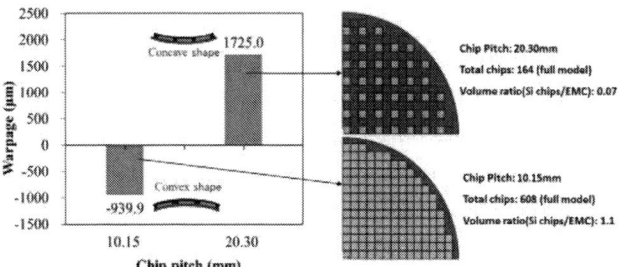

Figure 15. Chip number effect for wafer molding

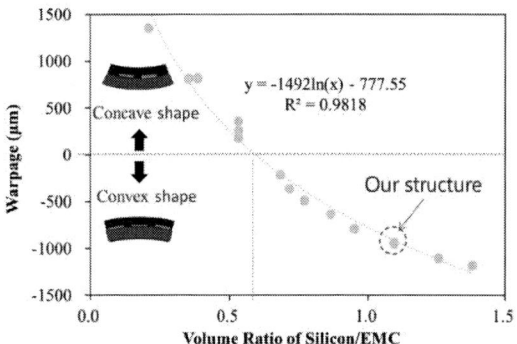

Figure 16. Volume ratio of silicon/EMC versus wafer warpage

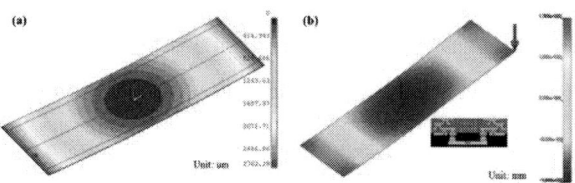

Figure 17. warpage comparison of molded strip panel: (a)warpage of simulation; (b) Measurement results

Based on the simualation results of wafer-level package, the same concept is also applied to a strip panel to monitor the warpage situation. Figure 17(a) is simulation results showing a warp of 2.74 mm. Measurement results of a real molded strip panel is 2.84 mm and explained in Figure 17(b). The FEA results are very close to the measurement value and reveal a predictable effect.

VII. RELIABILITY TEST

Test modules were conducted with pre-condition test (MSL3), temperature cycling test (TCT), and unbiased HAST (uHAST) as reliability tests. Related test items and detailed test conditions are listed in Table V.

Test modules were evaluated and the electrical performance of initial contact resistance of interconnection using with a probe measuring equipment. After pre-condition and TCT/ uHAST, test modules were evaluated again.

All these test modules passed each reliability test. Their testing results with daisy chain interconnection after using pre-condition test, TCT, and uHAST are illustrated in Figure 17(a) and (b), respectively. There is almost no difference before and after the reliability test. However, some TCT samples are found to have some voids around the solder surface, just like as Figure 14(a) shown in cross-sectioncal inspection, probably due to unsufficient plasma treatment on surface. It might slightly increase the contact resistance but does not lead to failure

TABLE V. RELIABILITY TEST ITEM AND CONDITION

Test Item	Test condition
Pre-condition for Reliability Test (MSL3)	Baking (125˚C, 24 hrs)→ Soaking (30˚C/60%RH, 192 hrs)→ Reflow * 3
Temperature Cycle test (TCT)	-55˚C - 125˚C, 1000 cycles, Dwell Time=5 min, Ramp rate =15˚C/min
Unbiased HAST (uHAST)	Temperature: 130°C, Humidity: 85%R.H Pressure: 33.3 Pa, Duration: 96hours

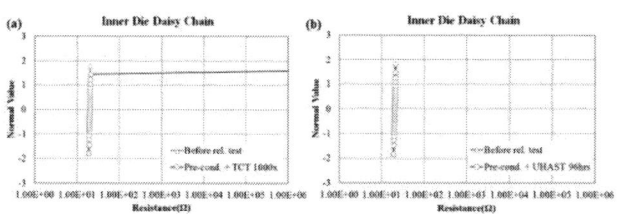

Figure 17. Reliability test results: (a) Pre-cond + TCT; (b) Pre-cond + uHAST

VIII. CONCLUSION

In this study, the demonstration of RDL-first PLP technology without interposers proves its great potential in heterogeneous integration applications. Finite element analysis with a three dimensional solid model has been established, as well as the evaluation of the influence of volume ratio and their effect on induced warpage was characterized, where the simulation agrees with experimental results. In the future, this finite element method could be further used to identify more material properties and define the dimensions of a robust package.

ACKNOWLEDGMENTS

Authors greatly appreciate panel molding supported from Towa Corp., Ltd.

REFERENCES

[1] H. Jihwan et al, "Fine Pitch Chip Interconnection Technology for 3D Integration," *Electronic Components and Technology Conference*, 2010, pp. 1399-1403.

[2] J. U. Knickerbocker et al, "Three-Dimensional Silicon Integration," *IBM Journal of Research and Development*, 2008, Vol. 52, pp. 553-569.

[3] A. L. Manna et al, "Challenges and improvements for 3D-IC intergation using ultra thin (25μm) devices," *Electronic Components and Technology Conference*, 2012 , pp. 532-536.

[4] S. W. Yoon, P. Tang, R. Emigh, Y. Lin, P. C. Marimuthu and R. Pendse, "Fanout flipchip eWLB (embedded Wafer Level Ball Grid Array) technology as 2.5D packaging solutions," *2013 IEEE 63rd Electronic Components and Technology Conference*, Las Vegas, NV, 2013, pp. 1855-1860.

[5] J. Osenbach, S. Emerich, L. Golick, S. Cate, M. Chan, S.W. Yoon, Y.J. Lin& K. Wong, "Development of exposed die large body to die size ratio wafer level package technology," *2014 IEEE 64th Electronic Components and Technology Conference (ECTC)*, Orlando, FL, 2014, pp. 952-955.

[6] Hong-Da Chang, David Chang, Kenny Liu, H.S Hsu, Rui-Feng Tai, Hsiao-Chun Huang,Yi-Che Lai, Chang-Lun Lu, Chun-Tang Lin, Steve Chiu, "Development and characterization of new generation panel fan-out (P-FO) packaging technology," *2014 IEEE 64th Electronic Components and Technology Conference (ECTC)*, Orlando, FL, 2014, pp. 947-951.

[7] F. X. Che, D. Ho, M. Z. Ding and X. Zhang, "Modeling and design solutions to overcome warpage challenge for fan-out wafer level packaging (FO-WLP) technology," *2015 IEEE 17th Electronics Packaging and Technology Conference (EPTC)*, Singapore, 2015, pp. 1-8.

[8] C. F. Tseng, C. S. Liu, C. H. Wu and D. Yu, "InFO (Wafer Level Integrated Fan-Out) Technology," *2016 IEEE 66th Electronic Components and Technology Conference (ECTC)*, Las Vegas, NV, 2016, pp. 1-6.

[9] Cheng-Ta Ko et al., "Chip-First Fan-Out Panel-Level Packaging for Heterogeneous Integration," *2018 IEEE 68th Electronic Components and Technology Conference (ECTC)*, San Diego, CA, 2018, pp. 355-363.

[10] Jinyoung Kim et al., "Fan-Out Panel Level Package with Fine Pitch Pattern," *2018 IEEE 68th Electronic Components and Technology Conference (ECTC)*, San Diego, CA, 2018, pp. 52-57.

[11] Siddharth Ravichandran et al., "2.5D Glass Panel Embedded (GPE) Packages with Better I/O Density, Performance, Cost and Reliability than Current Silicon Interposers and High-Density Fan-Out Packages," *2018 IEEE 68th Electronic Components and Technology Conference (ECTC)*, San Diego, CA, 2018, pp. 625-630.

[12] H. Liu, T. Lin, F. Hou, Fengchen, H. Zhang and L. Cao, "Process Development and Failure Analysis for Chip-Last Panel Level Fan-out Packaging (PLP)," *2018 19th International Conference on Electronic Packaging Technology (ICEPT)*, Shanghai, 2018, pp. 129-134.

[13] A. Ostmann, F. Schein, M. Dietterle, M. Kunz and K. Lang, "High Density Interconnect Processes for Panel Level Packaging," *2018 7th Electronic System-Integration Technology Conference (ESTC)*, Dresden, 2018, pp. 1-5.

[14] Tanja Brau et al., "Recent Developments in Panel Level Packaging," *2018 International Wafer Level Packaging Conference (IWLPC)*, San Jose, CA, 2018, pp. 1-7.

[15] J. H. Lau, "Redistribution-Layers (RDLS) for Fan-Out Panel-Level Packaging," *2018 International Wafer Level Packaging Conference (IWLPC)*, San Jose, CA, 2018, pp. 1-7.

[16] Keita Gunji and Toshihisa Hibarino, "Cost-effective Capacitive Testing for Fine Pitch RDL First," *2018 IEEE 20th Electronics Packaging Technology Conference (EPTC)*, Singapore, Singapore, 2018, pp. 561-564.

[17] Wen-Wei Shen, Yu-Min Lin, Sheng-Tsai Wu, Chia-Hsin Lee, Shin-Yi Huang, Hsiang-Hung Chang, Tao-Chih Chang, and Kuan-Neng Chen, " Warpage Characteristics and Process Development of Through Silicon Via-Less Interconnection Technology," *Journal of Nanoscience and Nanotechnology*, Vol. 18, 2018, pp.1-8.

[18] Wen-Wei Shen, Yu-Min Lin, Hsiang-Hung Chang, Tzu-Ying Kuo, Huan-Chun Fu, Yuan-Chang Lee,Shu-Man Lee, Ang-Ying Lin, Shin-Yi Huang, Tao-Chih Chang, Alvin Lee, Jay Su, Baron Huang, Dongshun Bai, Xiao Liu, and Kuan-Neng Chen, "Process Development and Material Characteristics of TSV-Less Interconnection Technology for FOWLP," *2017 IEEE 67th Electronic Components and Technology Conference (ECTC)*, Orlando, FL, 2017, pp. 35-40.

[19] Chia-Hsin Lee et al., "Optimization of Laser Release Process for Throughput Enhancement of Fan-Out Wafer-Level Packaging," *2018 IEEE 68th Electronic Components and Technology Conference (ECTC)*, San Diego, CA, 2018, pp. 1824-1829.

[20] Yu-Min Lin et al., "An RDL-First Fan-out Wafer Level Package for Heterogeneous Integration Applications," *2018 IEEE 68th Electronic Components and Technology Conference (ECTC)*, San Diego, CA, 2018, pp. 349-354.

978-1-7281-1500-9/19 $31.00 © 2019 IEEE

High Yield Precision Transfer and Assembly of GaN μLEDs using Laser Assisted Micro Transfer Printing

G. Ezhilarasu, A. Hanna, S. S. Iyer
Center for Heterogeneous Integration and Performance
Scaling, Department of Electrical and Computer
Engineering, University of California Los Angeles
Los Angeles, CA90095
goutham93@g.ucla.edu

Ajit Paranjpe
Veeco Instruments Inc.,
145 Belmont Dr. Somerset, NJ 08873
AParanjpe@veeco.com

Abstract—**Rapid developments in GaN based μLED mass transfer & assembly have been driven by the demand for high resolution, bright and efficient displays for various solid-state lighting applications. There has however been a roadblock for the commercialization of this technology due to the poor transfer yields attained and high processing costs. The Laser Lift-Off (LLO) process used to release the μLEDs from their native substrate (sapphire) is non-trivial as it can easily crack the chips. In this work, we propose a new μLED transfer and assembly process based on adhesive bonding using a laser debondable thermoplastic polyimide (HD3007) that can potentially achieve transfer yields >99%. The LLO process is also done more reliably by using mechanically supported μLEDs which helps to attain nearly 100% LLO yield.**

Keywords-GaN μLEDs; mass transfer; transfer yield; Laser-Liftoff; display assembly

I. INTRODUCTION

In recent years there has been a growing interest in building high resolution displays using micro Inorganic Light Emitting Diodes (μLEDs) as an alternative to conventional OLED or LCD displays [1]. The reasons for this are threefold: (1) μLEDs have narrow emission spectrum and can support high operating current densities with a large internal quantum efficiency thus attaining much higher illumination brightness and quality, (2) they are more resistant to environmental conditions such as humidity and temperature & (3) have longer operational lifetimes, typically > 100,000 hours [1], [2] & [3].

Inorganic μLED stacks are based on a GaN or InP material system and are typically grown on lattice matched substrates at very high temperatures (>600°C) [4]. The growth substrates such as c-plane sapphire in the case of GaN have very poor thermal conduction ($k_{sapphire}$~23 W/m°C at R.T.) and hence high-power operation of the devices on them is limited by thermal roll-over [5]. High resolution, fast refresh displays use an active matrix of pixels which requires the integration of individual μLEDs with compact high transconductance transistors and low leakage capacitors which are most readily fabricated on Silicon [6].

To overcome these issues, as fabricated & singulated μLEDs on growth substrates are to be released and then assembled onto the final packaging substrate, in most cases a Si backplane [7]. Due to the small size and thickness of the individual device (<100μm side, <7μm thickness), sequential pick and place for assembly is not practical. Hence a mass transfer technique is used where in all the devices from the growth wafer are released and transferred to the target wafer using a high yield (typically >99%) wafer-level process [8]. Before releasing from the growth substrate, the fabricated μLEDs are attached to a temporary carrier by bonding (metallic or adhesive) or stiction (Van der Waals forces as in elastomers) for accurate registration [9]. If the devices are chemically released, they could also be left anchored to the growth substrate by using polymers like photoresist so that a stamp could be used to pick up selected arrays of the released μLEDs for final assembly [10].

The release process of the μLEDs from the growth substrate is done by either chemical or optical means. In the case of InGaP (red or yellow) μLEDs, the release is typically done by chemically wet etching an underlying sacrificial layer such as $Al_{0.97}Ga_{0.03}As:Si$ in hydrofluoric acid (HF) without affecting the overlying device layers [10]. This release process has shown to be high yield and can be done with relative ease. For high quality GaN μLEDs grown on c-plane sapphire, simple wet release techniques are not possible due to the chemical stability of GaN and sapphire [11]. Such devices are instead released by an optical process called Laser Lift-Off (LLO) [12]. In LLO of GaN, a high-power UV or Q-switched solid-state IR pulsed laser (pulse width ~ 10ns) is shone from the backside of the polished sapphire substrate (see Figure 1).

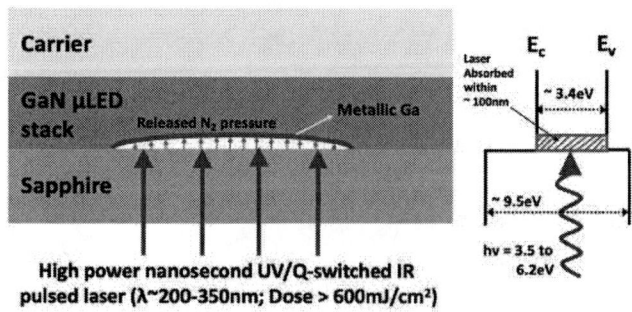

Figure 1. Schematic of the LLO process along with the band diagram of the GaN/sapphire interface

The laser is not absorbed by sapphire due to the high bandgap and hence passes through with little attenuation until it reaches the interface where it is absorbed by the GaN layer within a few hundreds of nm, refer [12]. The high energy (dose >600mJ/cm^2) of the absorbed light induces a local hotspot with a temperature >800°C with causes spontaneous decomposition of GaN into liquid Ga and N$_2$. The vapor pressure of released N$_2$ induces enormous stresses on the overlying GaN device and can cause pitting or crack formation on the surface which can fully damage the device causing yield reduction or degrade its performance by increasing leakage currents (see Figure 1) [13]. After the laser exposure, the sample can be heated to the melting point of Ga (~30°C) causing the release of the devices to the temporary carrier. Any metallic residue can then be cleaned using dilute acid treatment [14] and surface dry-etched to remove surface damage and incipient cracks.

The released µLEDs, held on the temporary carrier, are now selectively picked up using a stamp and then assembled onto the final target wafer using a process known as transfer printing [15]. Several approaches for the stamp pick-up of µLEDs have been explored in literature and are summarized in Table 1.

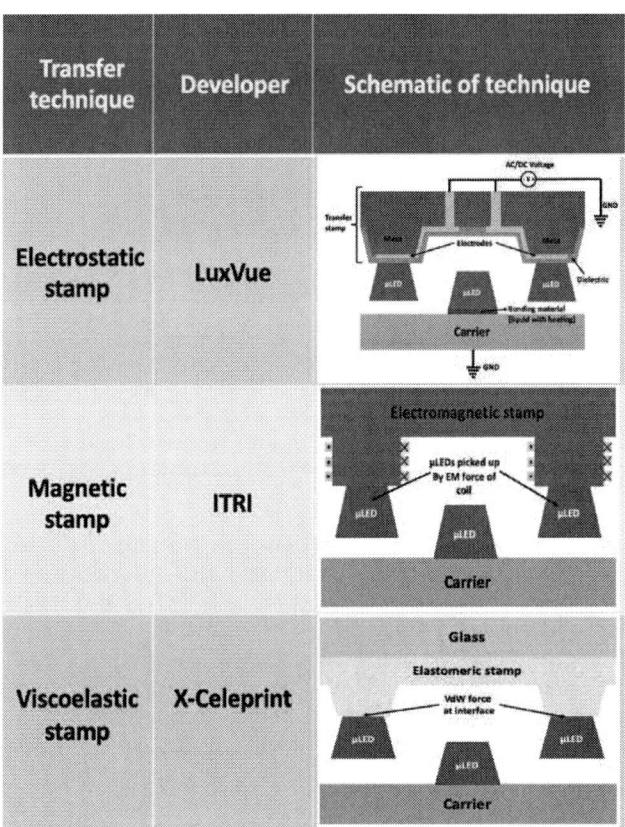

Table 1. Summary of various stamp-based transfer-printing techniques in literature

Although all the transfer approaches mentioned do work in principle, their commercial viability for high yield assembly onto a Silicon backplane is doubtful. Transfer approaches based on electrostatic forces, developed by LuxVue [16], uses a transfer head with an applied high voltage AC or DC to electrostatically pick up µLEDs from a carrier for transfer. The technique is based on the electrostatic gripper action [17] wherein an applied voltage to an electrode on the transfer head capacitively induces charges on the conductive surface of a µLED, separated from the electrode by a thin dielectric (see table 1), to attract it. The electrostatic force attracting the µLED is sensitive to the applied voltage, thickness & dielectric constant of the separating dielectric and most importantly, the presence of any air gaps or particles between the head and the µLED [16]. The complexity of operation and yield issues especially for warped wafers make this approach less attractive. Transfer printing using electromagnetic stamp techniques can only handle relatively large µLED pitches and require modification of the devices to make them magnetically attractive in addition to similar problems mentioned for the electrostatic case. Mass transfer using viscoelastic stamps (PDMS) rely only on Van der Waals (VdW) forces between the stamp and µLED [18]. They are hence not very reliable in holding the devices in place during the mechanical transfer process from the temporary carrier to the target wafer. Some µLEDs could shift significantly or even be completely lost from their original positions, in random locations of the stamp, and this in turn could reduce transfer yields to <99% across the wafer unless tedious fine tuning of process conditions is done. Even if double printing of µLEDs [18] can be used to allow circuit redundancy to account for the yield loss, the high cost per µLED would significantly push up the total cost of the manufactured display. Also, the viscoelastic property of the elastomeric stamp, that causes it to exhibit peel-rate dependent interfacial adhesion, is often used for the transfer printing [19]. When the pulling speed of the stamp is increased, the adhesion of the stamp to the µLED increases and vice-versa. So, a higher pull rate is used during the device pick-up process from the temporary carrier and a lower pull rate is used for the final printing. As the dependence of the pull-off force (for delamination of stamp from a surface) versus the pulling speed, as discussed in [19], is highly non-linear, using such a property for the transfer printing process requires significant tuning of process parameters. This could be difficult in a process setting where components of different sizes and materials are to be transfer printed at high yield.

In this work, we propose a new wafer-level µLED transfer and assembly process using adhesive bonding that can potentially attain >99% transfer yields. A thermoplastic, laser-debondable Polyimide based adhesive (HD3007) is used to attach the µLEDs to a temporary glass carrier before the LLO process. The same adhesive, deposited on a lithographically patterned glass wafer (see section II), is used to transfer print selected arrays of programmably debonded µLEDs from the temporary carrier to a pre-fabricated Silicon backplane for testing and evaluation. As strong adhesive bonding [20], instead of weaker electrostatic/electromagnetic or VdW forces, is used for the transfer printing, registration errors for the assembled devices can be kept to a minimum. Before the release process, the µLEDs are also supported using a 5-10µm thick electroplated Ni layer. This electroplated support layer serves to protect the device from

978-1-7281-1500-9/19 $31.00 © 2019 IEEE

the high stresses involved during the LLO process thus preventing device cracking or surface pitting. We have thus demonstrated nearly a 100% LLO yield (see section III).

II. PROPOSED PROCESS FLOW

The process flow for transfer printing pre-fabricated GaN μLEDs (4" device wafer from manufacturer) from c-plane sapphire wafer to the final target Si backplane is given in Figure 2. Liftoff is used to terminate the p and n contacts of the μLED with gold which will be needed for the final SLID bonding to the indium bumped Si backplane. A seed layer of Ti(20nm)/Cu(300nm) is then sputtered on the wafer followed by patterning of a thick layer (~10μm) of AZ4620 resist. Semi-additive plating is then used to form the Ni support layer which is needed for the LLO process as described in Section I. The same Ni layer is used as a hard mask to dry etch the exposed GaN in a $BCl_3/Cl_2/Ar$ ICP to singulate the GaN devices. The wafer is then bonded to a 4" temporary glass carrier coated with a thin layer (~2-5μm) of the laser-debondable Polyimide adhesive HD3007 using a SUSS Bonder SB6. The samples can then be sent for LLO. The temporary glass carrier with the released devices can then be cleaned with a dilute acid to remove any residual Ga, followed by dry etching of the adhesive between the μLEDs using an O_2 plasma. A lithographically patterned glass carrier with a layer of adhesive (HD3007) on it is now used to selectively pick up and assemble the devices on the final Si backplane/test wafer.

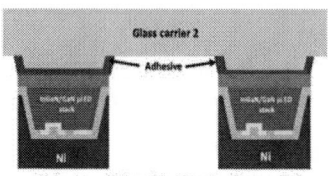
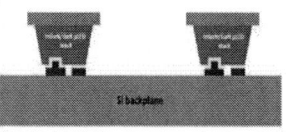

Figure 2. Proposed process flow for μLED transfer assembly using adhesive bonding

III. EXPERIMENTALS

A. Fabrication of samples for LLO yield testing

For the initial testing of LLO yield and quality, blanket 4" GaN wafers (no functional devices) were used. A layer of Ti(20nm)/Au(200nm) was used as seed layer. Patterned AZ4620 (~10μm) was then used to perform semi-additive plating to form the Ni support layer. Two different thicknesses of this layer were formed: 5μm and 10μm. The Ni layer is then used to dry etch about 7μm of the exposed GaN using a $BCl_3/Cl_2/Ar$ ICP to singulate GaN islands, see Figure 3. The Ni support layer on some samples is removed to perform a comparison. The samples are then bonded to a 4" glass wafer using the adhesive HD3007 and sent for LLO, see Figure 3.

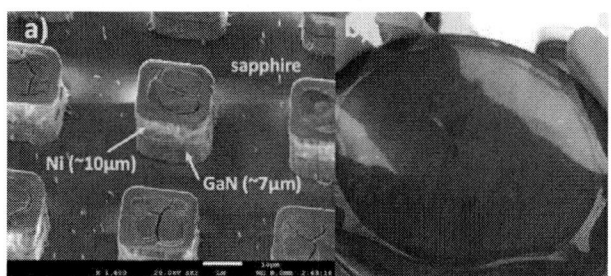

Figure 3. (a) GaN islands on sapphire after singulation. (b) Bonded sample for LLO

B. Evaluation of LLO quality and yield

After performing LLO, the glass carrier with the released GaN islands are cleaned with dilute HCl to remove any residual Ga. Any adhesive between the islands is also etched using O_2 plasma. The samples are then optically mapped (imaged) throughout the 4" area using a Nanotronics nSpec® PS microscope to detect for the presence of any visual cracks on the GaN islands, see Figure 4.

For further evaluation of the LLO quality, an AFM map of the GaN surface is also taken for the samples with and without the support layer.

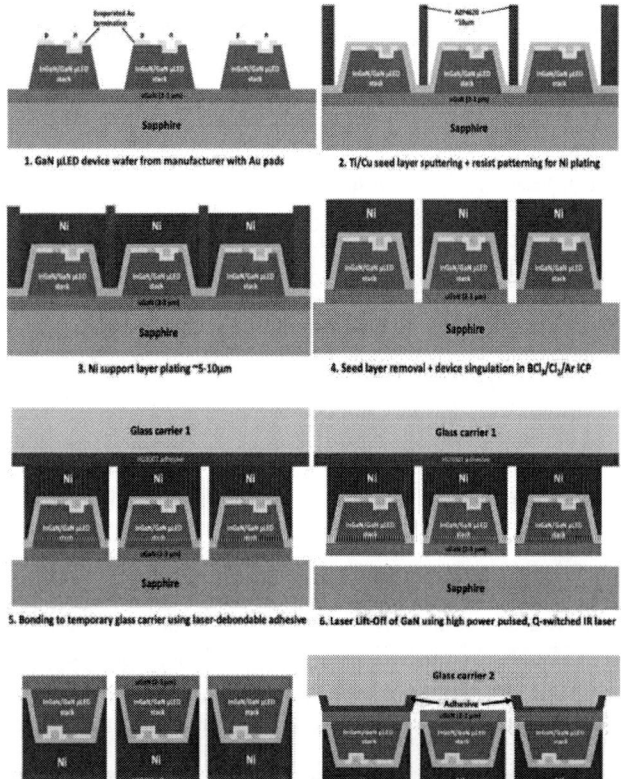

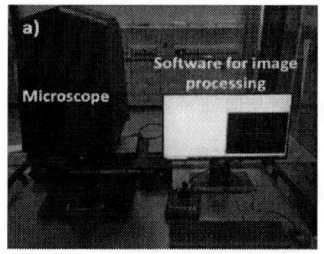

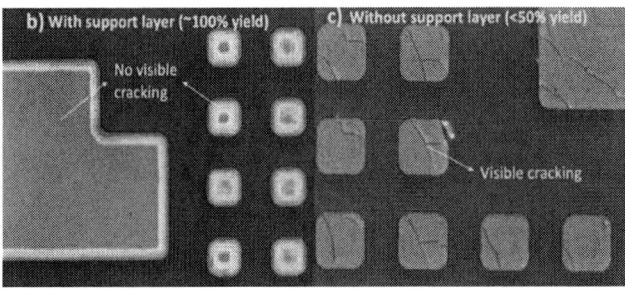

Figure 4. (a) nanotronics nSpec® PS microscope setup. Optical image of GaN islands, on the 4" temporary carrier, (b) with ~5μm support layer & (c) without support layer

IV. RESULTS AND DISCUSSIONS

A. Optical mapping of GaN islands on temporary glass carrier after LLO

The results of the optical mapping of the 4" temporary glass carrier with GaN islands is given in Figure 5 for 3 cases: Samples with no support layer, 5μm and 10μm support layers. We find that for samples with the 5μm and 10μm support layers, we get nearly 100% yield of crack free GaN islands throughout the area of the 4" wafer. For the unsupported samples however, > 50% of the GaN islands have developed surface cracks. These results thus demonstrate the effectiveness of the Ni support layer in preventing the stress induced GaN cracking due to LLO process.

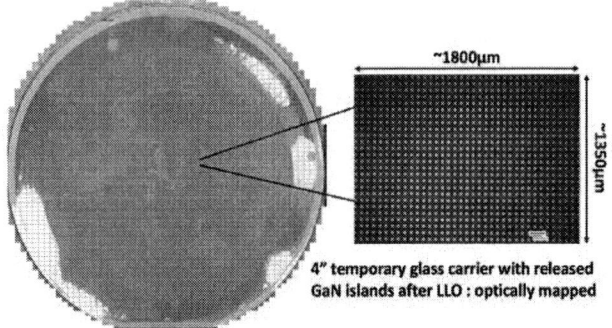

Figure 5. Results of optical mapping of 4" temporary glass carrier with GaN islands after LLO

B. AFM profile of GaN island surface after LLO

The results of the AFM profile of the GaN island surface after LLO with a 5μm support layer and without one is given in Figure 6. For the sample with the 5μm Ni support layer, we find that the RMS roughness of the surface is ~ 6.5nm. The surface is thus relatively smooth with the presence of some shallow (<3nm) surface pits and no surface cracks. The sample with no support layer however shows considerable surface cracking as can clearly be seen from the AFM profile.

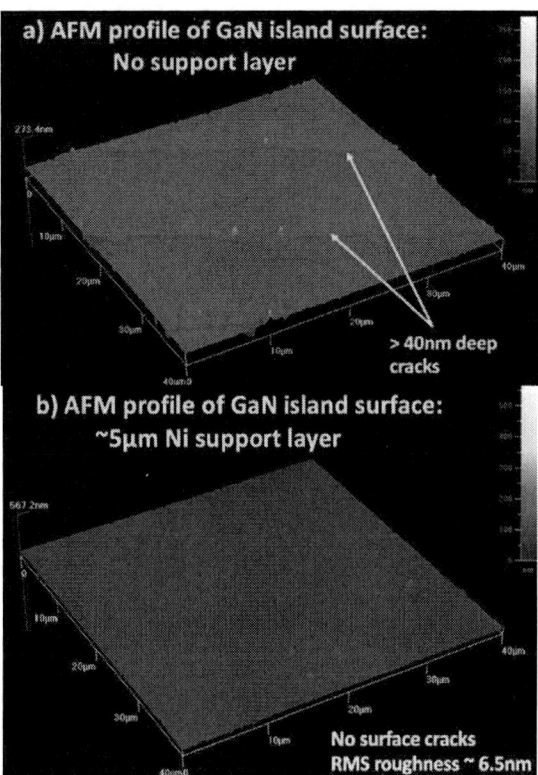

Figure 6. AFM surface profile of released and singulated GaN island (a) without a support layer and (b) with a 5μm Ni support layer

V. CONCLUSION AND FUTURE WORK

In summary, we have proposed a new process for high yield transfer printing of GaN based μLEDs from sapphire to a Si backplane that uses adhesive bonding. The release process of the GaN μLEDs from sapphire involves an optical process called LLO which can quite often damage the devices by inducing surface pits and cracks. To prevent this problem, we have successfully demonstrated the use of an electroplated Ni support layer on top of the GaN device (island) to relieve some of the stress during LLO, thus preventing the formation of surface cracks and significant pitting. This in turn allows us to achieve nearly 100% LLO yield, in terms of undamaged devices, throughout a 4" wafer area. Further analysis of the device damage needs to be done by taking a Photoluminescence map of the wafer. Currently, the rest of the process, including the selective transfer of the released μLEDs and final assembly of the devices on the Si backplane is under development.

ACKNOWLEDGMENTS

The authors gratefully acknowledge the contributions of UCLA CNSI and Nanolab cleanroom staff. We would like to thank Jay Lee and Aris Bernales from DISCO for their help with sapphire polishing and LLO. We would also like to thank Ron Legario of Hitachi DuPont MicroSystems™ for providing the adhesives used in this work. We are also very grateful to Habib Hichri from SUSS MicroTec for helping us with the laser debonding process. This work was supported by the UCLA CHIPS Consortium and the UC system.

REFERENCES

[1] T. Wu et al., "Mini-LED and Micro-LED: Promising Candidates for the Next Generation Display Technology," Appl. Sci. 2018, 8, 1557; doi:10.3390/app8091557.

[2] H. X. Jiang and J. Y. Lin, "Nitride micro-LEDs and beyond – a decade progress review," Opt. Express 21, A475-A484 (2013).

[3] V. W. Lee, N. Twu, I. Kymissis, "Micro-LED Technologies and Applications," Information Display 32(6), 16-23 (2016).

[4] R. F. C. Farrow et al. "Thin Film Growth Techniques for Low-Dimensional Structures," Springer-Verlag US 1987, ISBN: 978-1-4684-9147-0.

[5] J. Sun et al., "Thermal Management of AlGaN-GaN HFETs on Sapphire Using Flip-Chip Bonding With Epoxy Underfill," IEEE Electron Device Letters, Vol. 24, No. 6, June 2003.

[6] D. Armitage, I. Underwood, S. T. Wu, "Introduction to Microdisplays," John Wiley & Sons Ltd 2006, ISBN: 978-0-4708-5281-1.

[7] F. Templier, "GaN-based emissive microdisplay: A very promising technology for compact, ultra-high brightness display systems," Journal of the Society for Information Display, Vol. 24, Issue 11, Nov 2016, Pages 669-675.

[8] J. Yoon et al., "Heterogeneously Integrated Optoelectronic Devices Enabled by Micro-Transfer Printing," Adv. Optical Mater. 2015, 3, 1313-1335.

[9] S. H. Lee, S. Y. Park and K. J. Lee, "Laser Lift-Off of GaN Thin Film and its Application to the Flexible Light Emitting Diodes," Proc. of SPIE, Vol. 8460, 846011-1.

[10] Sang-il Park et al., "Printed Assemblies of Inorganic Light-Emitting Diodes for Deformable and Semitransparent Displays," Science 325, 977 (2009).

[11] C. H. Ko at al., "Photo-enchanced chemical wet etching of GaN," Material Science and Engineering B96 (2002), 43-47.

[12] T. Ueda, M. Ishida and M. Yuri, "Separation of Thin GaN from Sapphire by Laser Lift-Off Technique," 2011 Jpn. J. Appl. Phys. 50 041001.

[13] Y. Sun et al., "Properties of GaN-based light-emitting diode thin film chips fabricated by laser lift-off and transferred to Cu," Semicond. Sci. Technol. 23 (2008) 125022 (4pp).

[14] L. Li et al., "Heterogeneous Integration of Microscale GaN Light-Emitting Diodes and Their Electrical, Optical, and Thermal Characteristics on Flexible Substrates," Adv. Mater, Technol. 2018, 3, 1700239.

[15] A. Carlson et al., "Transfer Printing Techniques for Materials Assembly and Micro/Nanodevice Fabrication," Adv. Mater. 2012, 24, 5284-5318.

[16] A. Bibl et al., "Method of Transferring a Micro Device," LuxVue Technology Corporation, US Patent No. 8333860 B1, Dec. 18 2012.

[17] K. Asano, F. Hatakeyama and K. Yatsuzuka, "Fundamental Study of an Electrostatic Chuck for Silicon Wafer Handling," IEEE Transactions On Industry Applications, Vol. 38, No. 3, May/June 2002.

[18] C. A. Bower et al., "Emissive displays with transfer-printed assemblies of 8μm x 15μm inorganic light-emitting diodes," Photonics Research, Vol. 5, No. 2, April 2017.

[19] H. Cheng et al., "A Viscoelastic Model for the Rate Effect in Transfer Printing," Journal of Applied Mechanics, Vol. 80 / 041019-1, July 2013.

[20] P. Ramm, J. Lu, M. Taklo, "Hanbook of Wafer Bonding," 2012 Wiley-VCH Verlag GmbH & Co. KGaA, ISBN: 978-3-5273-2646-4.

High-Density Flexible Substrate Technology with Thin Chip Embedding and Partial Carrier Release Option for IoT and Sensor Applications

Kai Zoschke, Piotr Mackowiak, Ha-Duong Ngo,
Christian Tschoban, Carola Fritsche, Kevin
Kröhnert, Thorsten Fischer, Ivan Ndip

Fraunhofer Institute for Reliability and
Microintegration
Gustav-Meyer-Allee 25, 13355 Berlin
e-mail: kai.zoschke@izm.fraunhofer.de

Klaus-Dieter Lang

Technical University of Berlin
Gustav-Meyer-Allee 25, 13355 Berlin
e-mail: klaus-dieter.lang@izm.fraunhofer.de

Abstract— **The paper describes the fabrication of high-density flex circuits based on wafer level redistribution technology. The systems are build up by sequential processing of polyimide layers and semi-additive structured metal layers on a temporary carrier wafer. With the final removal of the carrier only 20-50 µm thick flexible circuit layers are generated. The technology allows multi-layer routing with up to three levels having line pitches of 30 µm and 35 µm vias in staggered configuration using the standard process by photo structuring of the polymer. With an advanced process, line pitches of 14 µm with vias of 10 µm diameter in stacked configuration are possible using laser structuring of the polymer. As additional features, thin active ICs can be embedded into the flex layers and with a special release sequence a partial rigidness of the flex can be obtained. A 24 GHz radar module was fabricated as first demonstrator of the described flex technology featuring two embedded only 20 µm thin transceiver ICs per module. Due to a partial carrier release a rigid-flex configuration was obtained, which allows the antenna arrays of the module to be tilted relative to the center part of the module where the transceiver ICs and IOs are located. A vibration and stress sensor module was fabricated as second demonstrator with two only 20 µm thin stress and acceleration sensor chips integrated into a flex circuit. The flex with the acceleration sensor overhangs the rigid part of the module, which allows the detection of oscillations with high sensitivity.**

Keywords: Flexible Substrates, Thin Chip Embedding, Chip in Flex, High-Density Wiring, Muil-Layer RDL, Laser Structuring, Laser Release

I. Introduction

The trend for devices being interconnected with each other to efficiently collect, analyze and process data is increasing steadily and known as the internet of things (IoT). The idea of IoT exists for all important application domains like consumer, commercial, industrial and infrastructure and relies on powerful sensing/acting functionalities combined with both, data processing and transmission. Scenarios like smart home, smart health care, smart transportation, smart manufacturing, etc. offer fantastic opportunities to improve the quality, sustainability and efficiency in wide areas of the daily life. [1-3]

In order to enable such IoT scenarios powerful technologies in all areas such as identification, communication, networking, software, data processing, energy storage as well as hardware implementation are required. Especially packaging technologies are requested to provide cost effective solutions for heterogeneous integration of systems with high performance and high degree of miniaturization. [4-7]

In this manuscript, we present the high-density flexible substrate technology from Fraunhofer IZM as one possible technology platform for integration of sensor systems or sub-systems, which are related to IoT applications. The technology enables the integration of high-density routing features as well as thin active IC components into a polymer foil. In contrast to other already published approaches [8-12] the flex technology allows combination of various features such as fabrication of vertical through connections, high density multi-layer wiring, ultra-flat profiles with total flex thickness of 50-60 µm and embedded ultra-thin ICs with 20 µm thickness as well as partial rigidness. The manufacturing processes are streamlined to require no special preparation processes for the thin chips, just grinding and polishing, and to use established technologies from the wafer level packaging portfolio, such as semi-additive metal deposition and polymer structuring by direct lithography or laser ablation.

II. Multi Layer Flex Technology based on Wafer Level Redistribution and Laser Assisted Debonding

This chapter explains the basic high-density flex technology, which is available since several years at Fraunhofer IZM. This basic technology relies on the well-established wafer level redistribution processes used for fabrication of multi-layer routing structures on up to 300 mm active or passive silicon wafers. The isolation layers of the multi-layer wiring are created by lithographic structuring of photosensitive polymer precursors and their subsequent thermal cure. The wiring layers are realized by semi-additive structuring of metal, which is preferably Cu or Au. As pad metallizations additional Ni+Au, Ni+Cu and solder finishes can be used.

978-1-7281-1500-9/19 $31.00 © 2019 IEEE

Mandatory for the creation of reliable flex circuits is the proper material selection for the polymer as well as a capable release technology of the polymer/metal multi-layer stack from the surface of the process wafer. As dielectric material, the polymer LTC9300 from Fujifilm was chosen. After the evaluation of different release mechanisms from silicon like mechanical substrate removal or usage of anti-sticking layers a laser assisted release of the polymer/metal multi-layer stack through a transparent carrier wafer was depicted to be the most promising approach. Figure 1 shows the schematic flow for the standard high-density flex fabrication process based on processing wafers of glass with a thickness below 1 mm.

In a first step, a release layer of polymer is deposited on the glass surface. The release layer is only temporary present and required for the final detach of the multi-layer stack from the glass carrier. It will not be part of the final flex circuit. In a second step, the backside pads of the later flex circuits are generated by semi-additive metal structuring. The sequence includes the sputtering of an adhesion and seed layer followed by photo resist lithography and electroplating to the desired metal. Finally, the resist is removed and the seed layer and adhesion layer are etched. As an abnormality, it has to be pointed out, that the electroplating steps for pad metallizations involving two metals like Ni+Au or Ni+Cu have to be performed in reverse order and that no solder finish can be used for the backside pads. A further important fact is that the sputtered adhesion and seed layer remain present below the backside pads. They have to be removed later after the flex is detached from the carrier wafer.

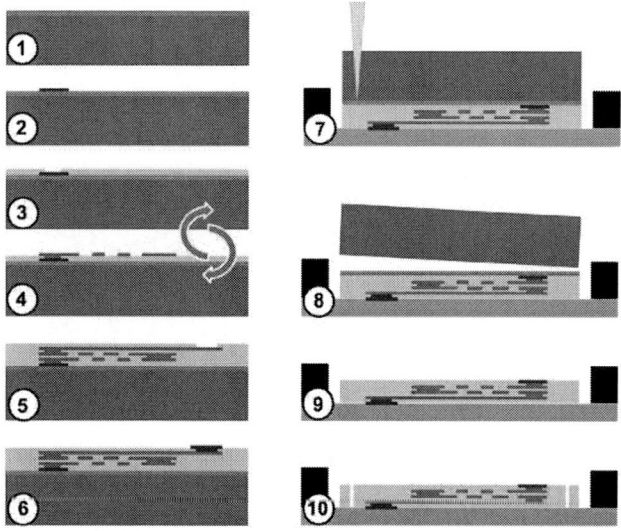

Figure 1: high-density flex technology based on standard wafer level multi-layer redistribution and laser assisted release

After the formation of the backside pads, the first polymer dielectric layer is deposited by spin coating, lithography and cure as shown in step 3 of Figure 1. The lithography step defines the vias to the backside pads. In the following step 4, the first wiring layer is deposited by semi-additive metal structuring. In most cases copper is used as wiring material, gold can be used as alternative.

The multi-layer wiring stack is now created by the sequential repetition of the steps 3 and 4. The maximum number of routing layers, which can be processed is limited by the warpage of the carrier wafers, which is increasing with each additional layer. Root cause for the warpage is the CTE mismatch between the carrier wafer and the deposited materials. Reduction of metal load and thicknesses of both, metal and polymer layers as well as introduction of interruptions in the polymer layers (like scribe lines) can reduce the warpage issue. Up to now, a multi-layer stack with four routing layers and a total thickness of 60 μm could be processed with the given technology.

After finishing the multi-layer stack with deposition and structuring of the final polymer layer in step 5, the topside IO pads are created by using again a semi-additive metal structuring technology. Different to the backside pads also solder finishes or solder bump like interconnects can be realized here.

The process wafer is now flipped onto a dicing tape on film frame carrier in step 7 to enable the detach of the multi-layer flex. With the backside of the glass wafer facing up a laser can be focused on the release layer and a full area laser exposure of the release layer is executed. The exposure is done using a 248 nm excimer laser implemented into a Süss ELP300 laser stepper system. The system features a high-speed x/y stage with guides the wafer under the exposure spot of 6.5×6.5 mm^2 size so that the entire wafer can be exposed. Due to the shoot repetition rate of 50 Hz a whole 300 mm wafer can be exposed within 60 s. [13]

During the laser exposure, the release layer absorbs the laser radiation completely within a layer thickness of several 100 nm behind the glass surface. Above a certain threshold the energy intake into the polymer caused by the laser exposure leads to a physical damage of the material at molecular level meaning the destruction of the materials molecular bonds. As a result, the remaining adhesion of the material to the glass wafer is close to zero, so that the glass carrier can be detached easily as shown in step 8 of Figure 1. [13] As a helpful item, the tensile stress in the multi-layer flex supports a smooth release of the glass wafer. As discussed before, almost the full thickness of the release layer remains on the backside of the multi-layer stack whereas the glass wafer is nearly free of material. A proper cleaning procedure from carbonized polymer residues allows the re-use of the carrier wafers.

As indicated in step 9 of Figure 1 the release layer and the sputtered adhesion layer, which is still present at the back side of the pads, have to be removed finally. Both cleaning tasks are performed in a RIE plasma etcher using a two-step plasma treatment. In the first stage the release layer is removed using a standard recipe for polymer etching until the pads are revealed and the sputtered adhesion layer is accessible. At this point the second stage is executed using a plasma recipe for metal etching to remove the Ti containing adhesion layer. The two plasma steps are run in one sequence in the same chamber.

The singulation of the multi-layer stack in step 10 finalizes the fabrication process. The samples can now be picked and sorted or shipped on the dicing tape.

For the standard process as described in Figure 1 the design rules as shown in Figure 2 are defined. In case that maximum three internal routing layers have to be realized, the minimum line width and spacing is set to 12 µm. Due to the via fabrication by direct lithography of the negative photosensitive acting polymer precursors the via diameter mustn't be smaller than 35 µm. The relatively large via diameter is required to ensure a safe and repeatable via formation at every location on the wafer even where unfavorable constellations of increased topography cause thickness variations or were reflections from neighboring features or rough metal surroundings cause undesired stray light, which hinders a proper via definition.

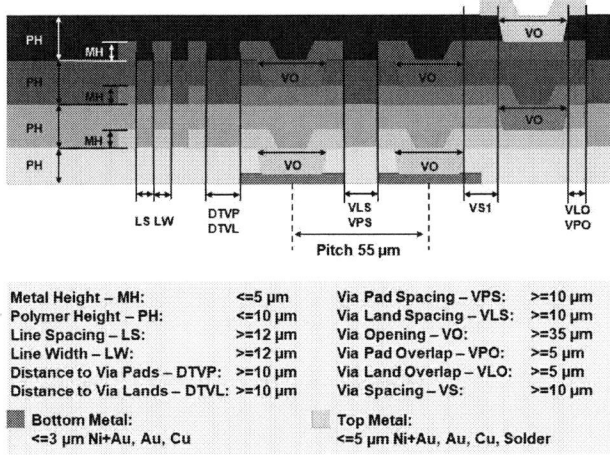

Metal Height – MH:	<=5 µm	Via Pad Spacing – VPS:	>=10 µm
Polymer Height – PH:	<=10 µm	Via Land Spacing – VLS:	>=10 µm
Line Spacing – LS:	>=12 µm	Via Opening – VO:	>=35 µm
Line Width – LW:	>=12 µm	Via Pad Overlap – VPO:	>=5 µm
Distance to Via Pads – DTVP:	>=10 µm	Via Land Overlap – VLO:	>=5 µm
Distance to Via Lands – DTVL:	>=10 µm	Via Spacing – VS:	>=10 µm

Bottom Metal:
<=3 µm Ni+Au, Au, Cu

Top Metal:
<=5 µm Ni+Au, Au, Cu, Solder

Figure 2: design rules for standard high density flex technology with three routing layers and staggered vias

As shown in Figure 2, the vias in successive layers mustn't be at the same location, but need to be realized with a lateral offset. This staggered via configuration is required to avoid the formation of larger topographies. Due to the via size, the following metal layer does not fully fill the via holes. The metal with a maximum thickness of 5 µm covers the via holes more or less conformal meaning that significant valleys remain at the locations of the via. If the next via would be located at the same position, a deeper opening would be the result. These deeper openings are no problem in general, but require an adjusted lithography process for the exposure and development of the photo resist for the feature definition of the next metal layer. However, the fine pitch wiring is defined with the same mask and the common processing of small features in thinner resist and large features in thicker resist is limited. Based on that, the via-in-via (or stacked) constellation is not allowed in following layers. Instead of this, the parameter "via spacing (VS)" defines a minimum distance of 10 µm to be kept for vias with common signal. The required distance between vias of different signals is defined by the parameters "via land spacing (VLS)" and "via pad spacing (VPS)", which are also both 10 µm. Together with the other parameters for metal overlaps a minimum IO pad pitch of 55 µm can be enabled.

Figure 3 shows an image of a flex with three internal routing layers. The upper cross sectional view at the right

side shows the typical staggered via configuration as discussed before. The vias here have a opening size of 50 µm and a larger distance compared to the given minimum design rules in Figure 2. The lower cross cut on the right side in Figure 3 shows the metallization lines. In this example the line width is 15 µm and the pitch is 30 µm.

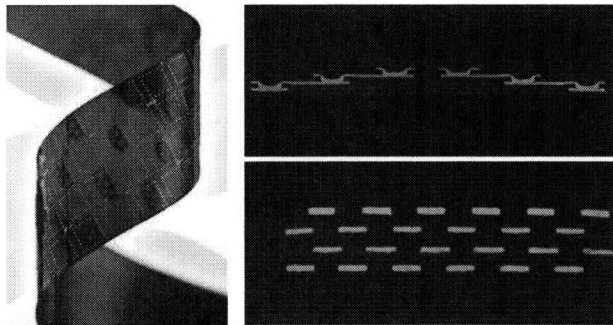

Figure 3: flex substrate after release from carrier (left), cross sectional view of staggered vias (right/top) and multi-layer wiring (right/bottom)

In order to show the robustness of the process and also the capability of the laser assisted detach approach the resistance of via chains was exanimated before and after detach from the glass process wafer at five different locations on the wafers. For this evaluation a test design with three internal routing layers and top metal pads was processed. Certain wafers were taken out of the process already after the second and also after the third routing layer were processed. Other wafers finished the complete process.

Based on that, it was possible to check the integrity and stability of the via connections stepwise. Wafers taken out after finalization of the second metal layer were used the measure the chains with 64 via 1◇2 connections before and also after release from the glass wafer. Wafers taken out after the third metal layer was completed were used to measure chains with 64 via 1◇2 connections and 64 via 2◇3 connections before and after release of the glass wafer. At the fully processed wafers the longest chains with 64 via 1◇2 connections and 64 via 2◇3 connections and 64 via 3◇4 connections could be measured. The diagram in Figure 4 shows the result of these examinations. Measurements taken before release of the glass carrier wafer are indicated with "bcr". Those measurements taken at the released flex circuits are indicated with "acr". As can be seen, no significant change in resistance is introduced by the laser release operation. The resistance of the single via 1◇2 chain approximately doubles and triples with the addition of each of the corresponding 64 via 2◇3 and via 3◇4 connections, which was the expectation. The slightly different absolute values occurring with same chain configurations at the different wafer positions are caused by slight metal height deviations over the wafer, which were not optimized during the fabrication of these test wafers. The image in Figure 4 shows an example of a via chain test structure with all three via test fields present. The photograph was taken from a wafer, which was fully processed including the final pad metallization.

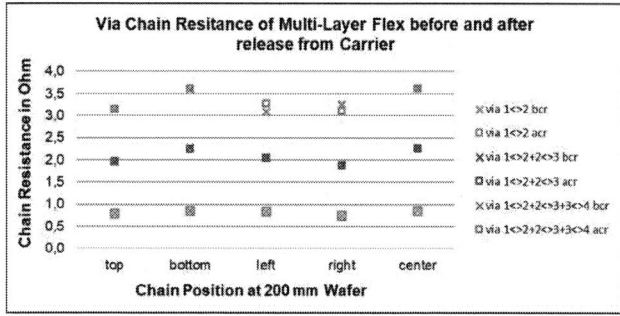

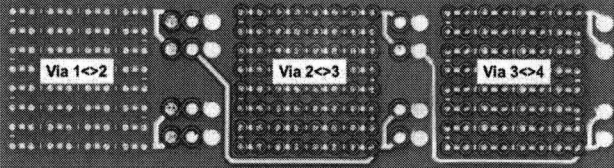

Figure 4: measurements of via chain resistance at different positions on a 200 mm wafer before and after release from carrier (top), via chain test structure (bottom)

III. MULTI LAYER FLEX TECHNOLOGY - OPTION FOR INCREASED ROUTING DENSITY

As discussed in the previous chapter, the photo structuring of the polymer vias is the limiting factor for the further shrink of the feature sizes in the standard high-density flex process. A reduction of the via size towards 10 μm diameter would allow also to decrease the minimum feature sizes for the wiring itself since the lateral dimension lay within a closer range and are more easy to handle for the photo resist lithography. Furthermore, with decreased via diameter an almost complete filling of the vias during the electroplating of the next metal layer could be possible, which could enable stacked via configurations.

Based on the given advantages of smaller vias, a new structuring method was evaluated and developed, which is schematically shown on the right side of Figure 5. In a first step the polymer layer is spin coated and cured to give a final thickness of 10 μm. In the second step, a 1 μm thick layer of aluminum is sputtered and a photo resist lithography is done to define openings in the aluminum layer. Wet etching against the rest mask creates the openings in the aluminum layer. After removal of the resist the aluminum layer represents a hard mask with defined openings were the vias have to be created (step 3). Following, an excimer laser is used to ablate the cured polymer out of the openings in the hard mask. (step 4) For this process the same laser as for the flex release operation is used. The wafer is stepped under the exposure spot so that the entire wafer area is exposed. Only the openings in the hard mask define were the polymer is abated down into Z direction until the landing pad in the previous metal layer is reached. Depending on the exposure dose, several shoots are required to fully ablate the hole into the 10 μm thick polymer. [14, 15] In the regions were the hard mask is closed the laser is reflected so that the underlying polymer is not damaged.

After the laser ablation is done, the hard mask is removed wet chemically revealing the polymer layer with the structured via holes. (step 5) As reference, a schematic flow of the original via structuring process using direct lithographic patterning of the photosensitive polymer is shown on the left side of Figure 5.

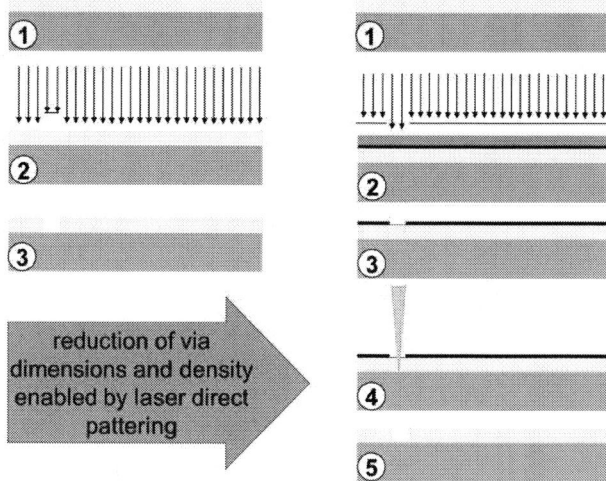

Figure 5: schematic process flow of via structuring using photosensitive polymers (left) and by laser ablation (right)

Based on processing of several test layouts using the new via structuring technology, the design rules as shown in Figure 6 were extracted. As expected, the 10 μm vias can be almost completely filled with copper during the realization of the following metal layer. Due to that, no severe topographies are created by the vias and the "via spacing" (VS) parameter could be set to zero so that true via-in-via (stacked via) configurations can be allowed over the full multi-layer stack. The minimum metal line width and spacing was set to 7 μm with the first release of this technology. A further reduction is under evaluation. Together with reduced values for overlaps and via land / pad distances a minimum IO pad pitch of 27 μm can be addressed with the new via technology.

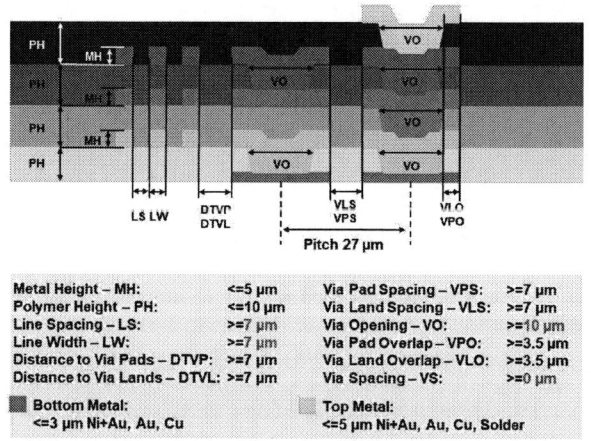

Figure 6: design rules for advanced high density flex technology with three routing layers and stacked vias

The images in Figure 7 show results from test layouts, which were processed with the new technology. The left image shows a FIB cut through a triple layer via-in-via structure. The maximum diameter of the vias is 10 µm. The right image shows a fine pitch Cu wiring structure with a thickness of 5 µm. The finest line pitch is 14 µm with 7 µm line width and 7 µm line spacing accordingly.

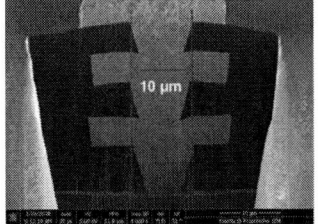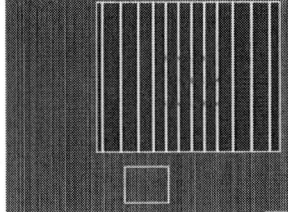

Figure 7: example of via stack through three polymer layers created by laser structuring and copper plating (left), high density routing with 7 µm line width and 7 µm line spacing (right)

IV. MULTI LAYER FLEX TECHNOLOGY - OPTION FOR EMBEDDING OF THIN ICS

As a further extension of the standard technology it was evaluated how thin ICs can be integrated inside the flex multi-layers. This allows the packaging of active components with small form factor and high-density connections and routing features for the fabrication of sub-systems with high complexity, which can be mounted to low cost substrates or carriers. The IC integration flow for thin ICs is designed as an option to the existing standard flex process and is schematically shown in Figure 8. The process sequence can be insert into the standard flow after the first polymer layer or after the first metal layer is fabricated, which is indicated in step 1 of Figure 8. Now, as shown in step 2, alignment marks and depots of die bond adhesive are structured onto the surface the polymer layer. The alignment marks are created from a thin sputtered metal layer, which is structured by lithography and wet etching. The die bond adhesive is deposited by spin coating followed by photo resist lithography and dry etching against the resist mask. Ideally, the die bond adhesive depots are laterally slightly larger than the dices to be bonded. In our case, the adhesive layer is a thermoplastic polymer with a thickness in the range of 3-5 µm.

Following, in step 3 ultra-thin ICs are aligned and fixed onto the adhesive depots using a thermo-compression type die bonding process utilizing a flip chip bonder. The desired total thickness of the ICs is <=20 µm. The thin ICs are prepared by wafer back grinding + dry polishing and subsequent mechanical dicing. The process temperature for the placing is approx. 250 °C, one cycle for the placing requires a time of approximately 20 s.

As shown in step 4, after the ICs are bonded, a thick polymer layer with target thickness of approx. 25 µm is coated over the wafers with the placed ICs. The polymer is direct structured by photo lithography to open vias on top of each chip IO pad. Those vias above the chip IOs can have a diameter down to 40 µm. If a bottom metal layer is present

and has to be connected, vias can also be opened next to the chips above prepared landing pads in the bottom RDL. However, due to the thicker material next to the chips larger via diameters in the range of >60 µm have to be considered here.

In step 5 a semi-additive metal structuring process is run to form the wiring layer which connects the chip IOs as well as (if present) the bottom routing layer. It has to be taken into account, that the large via opening for the connection of the bottom routing layer will not be fully filled with metal. Instead of this, larger dimples remain, which need to be planarized further with processing of following polymer layers.

In the example given in Figure 8, only one following polymer layer is processed, which is again directly structured by photo lithography. (step 6) A pad metallization process by semi-additive structuring finishes the simplified process flow. (step 7)

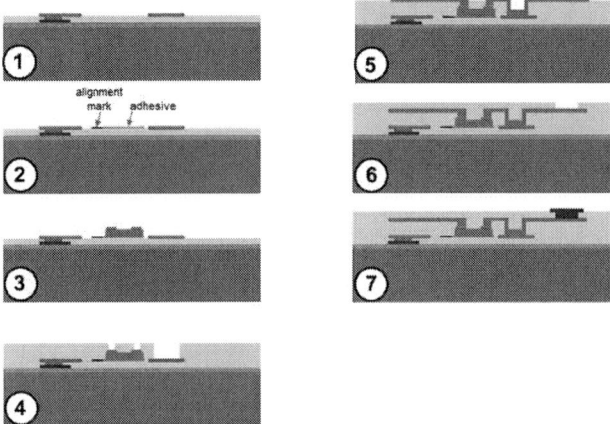

Figure 8: schematic process flow for the embedding of thin ICs into the flex build-up layers

V. MULTI LAYER FLEX TECHNOLOGY - OPTION FOR PARTIAL CARRIER RELEASE

The partial carrier release option allows to keep the carrier locally as permanent rigid substrate of the final circuit. To enable that option the carrier wafer has to be covered with a light blocking layer at the backside in those areas where the carrier should remain attached to the flex multi-layer, as schematically sown in step 1 of Figure 9. The light blocking layer can be deposited by sputtering of a thin metal such as Ti, Cr or Al and structured by wet etching. The pattern can be defined by lithography with referenced alignment to the flex multi-layer at the front side of the glass wafer.

Following, the areas with the light blocking layers of the glass carrier wafer have to be separated from areas without light blocking layer by mechanical dicing. It has to be ensured, that the flex multi-layer is not diced through here. The cut ends in the release layer as indicated in step 2.

Now, in step 3 the standard full area laser exposure can be run. The partial presence of the light blocking features prevents the flex multi-layer to be released from the glass

978-1-7281-1500-9/19 $31.00 © 2019 IEEE

here. On the contrary, in the regions without the blocking features the glass can be detached as shown in step 4. The already described cleaning process in the regions with the removed glass finishes the process finally (step5).

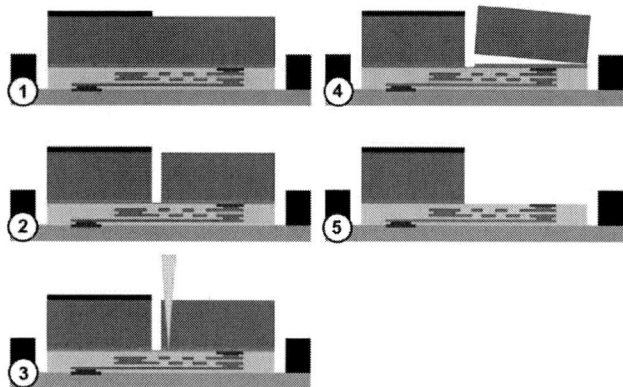

Figure 9: schematic process flow for the partial release of the rigid carrier

VI. USE CASE 1: 24 GHz RADAR SENSOR IN RIGID-FLEX TECHNOLOGY WITH EMBEDDED TRANSCEIVER ICS

A. Sensor concept and design

The various demands of different applications on frequency modulated continuous wave (FMCW) radar sensor systems in automotive and industrial environment are becoming more and more challenging. For the needed modularity, a highly integrated sensor system is inevitable. Additionally, in order to achieve a high spacial resolution of less than 10 cm, a high bandwidth is needed.

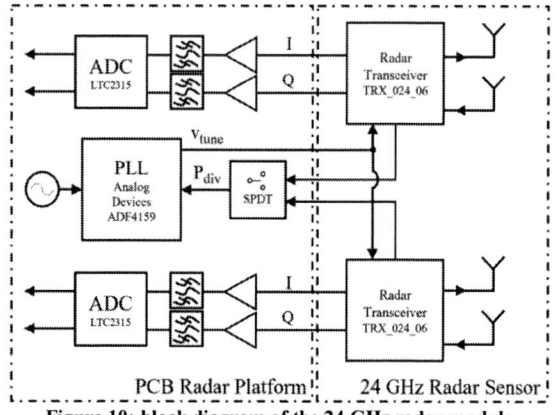

Figure 10: block diagram of the 24 GHz radar module

Figure 10 shows a block diagram of the developed radar module, which meets these criteria. The radar sensor consists of two ultra-thin 24 GHz radar transceiver monolithic microwave integrated circuits (MMICs) TRX_024_06 from Silicon Radar embedded in the thin film RDL stack on a glass carrier. The MMICs can be controlled and operated separately. The input and output signals (RF) of each MMIC are connected to 4x4 slot antenna arrays. Each array consists

of a single transmitting antenna element and 15 receiving elements resulting in a total peak gain of 11 dBi. The antennas arrays are also implemented in the same multi-layer stack-up with rigid glass carrier underneath and are located next to the positions were the MMICs are embedded. Between the antenna regions and the MMIC build-ups the carrier will be partially removed in order to obtain a bendable region and the possibility of a flexible alignment of the transmission plane of the antennas and thus of the detection area of the radar system. This provides an universally applicable radar sensor that can be assembled on different shapes for many applications. The transceiver MMICs provide complex in-phase and quadrature (IQ) intermediate frequency (IF) signals for signal processing. The radar sensor is mounted on a printed circuit board (PCB) radar platform. Together with the phase-locked-loop (PLL) ADF4159 from Analog Devices the FMCW radar signal is generated by the transceiver MMIC's voltage controlled oscillator (VCO). The output frequency of this VCO is specified from 23.2 GHz to 26.3 GHz, which covers the required wide bandwidth. On the radar platform the IQ signals are amplified, bandpass filtered and finally sampled by Analog Devices' 12 bit ADC LTC2315.

B. Sensor manufacturing

The sensor modules were fabricated according to the process flows shown in Figure 1 and Figure 8. Since the sensor modules need no backside contacts, only a non-structured polyimide layer was created on top of the release layer. The formation of backside pads as shown in step 2 of Figure 1 was left out.

In the next steps, according to step 2 in Figure 8, alignment marks and adhesive depots were deposited onto the non-structured polyimide base layer. A microscopic image of a die bonding site with alignment marks and adhesive depot can be seen in the upper left image of Figure 11. According to steps 3 and 4 in Figure 8 the thin ICs were now placed and subsequently over coated with polyimide which was directly structured by lithography to open the IO pads. The upper right image in Figure 11 shows an example of an embedded transceiver IC with the structured polymer layer after cure.

In the following processes a copper wiring layer was fabricated by semi-additive structuring. With that, the IOs of the ICs were connected and linked to the antenna structures, which were realized with the same metal layer. The bottom left image in Figure 11 shows an embedded transceiver chip with the copper wiring connecting its peripheral IOs. The image in the bottom right corner of Figure 11 shows a lower magnification of the chip position after processing of the passivation polymer layer and deposition of NiAu pads for the later connection of the radar modules to PCBs by soldering.

Figure 12 shows a cross sectional view of one embedded thin transceiver die at the location of the IO pads. As can be seen, the die including BEOL has only a thickness of approx. 20 µm. The total thickness of the build-up layers including the IC is 50 µm.

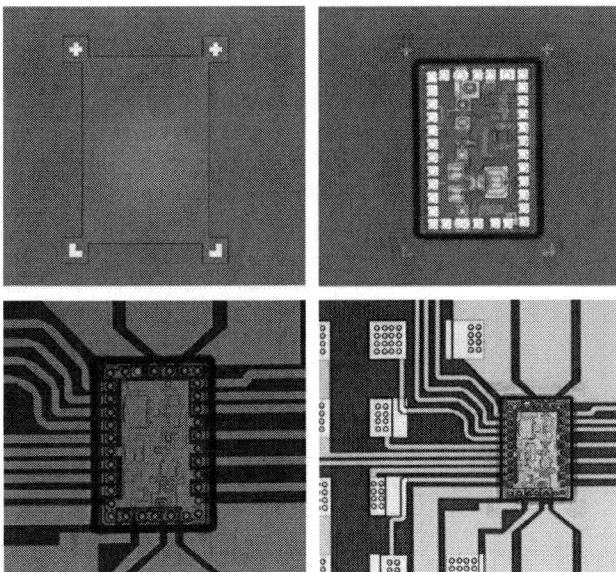

Figure 11: embedding and connection of thin transceiver ICs in flexible multi-layer substrate, die bonding adhesive (top left), placed IC with opened vias in covering polymer layer (top right), routing connected to IC pads (bottom left), smaller view with final NiAu pads of module (bottom right)

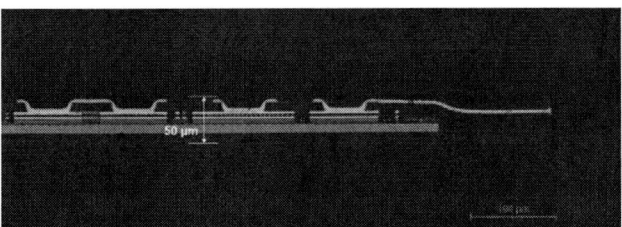

Figure 12: cross sectional view of IO pad connections to 20 μm thin transceiver IC embedded in polymer layers

In the final step, processing wafers were singuated by mechanical dicing. In order to create the required bendable zones for tilting of the antenna arrays, the carrier was removed partially at two areas along the full Y extension of the modules. According to Figure 9, the partial carrier removal was enabled by the local deposition of a light blocking layer followed by a soft dicing process for cutting only the carrier glass, but not the flex build-up layers along the zones were the carrier should be released. Now, after the laser exposure, thin stripes of the carrier could be removed to create the bending zones.

Figure 13 shows one 24 GHz radar module after full processing. The transceiver ICs are embedded in the center part of the module. That area also hosts the IOs of the module for its soldering to the PCB radar platform. The two antenna arrays are located to the left and right side. Due to the local removal of the rigid carrier, the antenna arrays can be deflected in a wide range around the horizontal (normal) position. The glass carriers in the rigid zones have a thickness of 700 μm. The multi-layer build up with the embedded ICs and also the bendable zones have a thickness of 30-50 μm

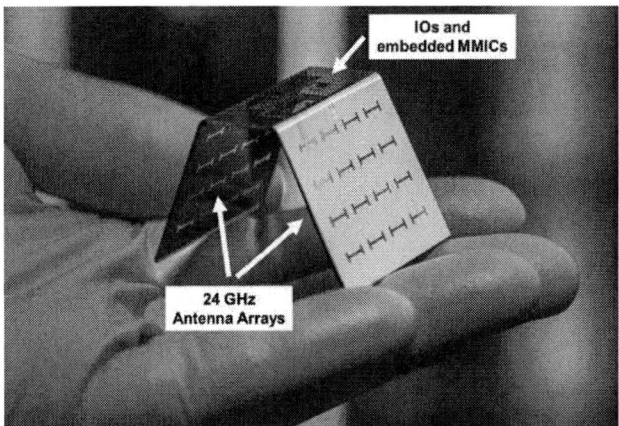

Figure 13: 24 GHz radar sensor with embedded transceiver ICs after singulation from wafer and partial substrate release

C. Functional test and characterization

For the measurement setup, the radar sensor module was connected to XYLINX's Basys3 FPGA board, which is using the FPGA Artix 7. This board sets the control voltages for the ICs and performs the radar signal processing of the IF signals. In order to obtain a triangular modulated output signal in a frequency range from 23.3 GHz to 23.5 GHz, the tuning voltage vtune of the PLL, shown in Figure 14, was a triangular signal from 1.3 V to 1.9 V. The transceiver MMIC provides the RF output signal divided by 32 for the PLL. For functional testing this RF frequency was measured with a N9030A PXA signal analyzer from Keysight (formerly Agilent). The measured spectrum is shown in Figure 15. Using this data the RF output frequency is in a range from 26.85 GHz to 27.1 GHz. This means the output frequency of the radar MMIC's VCO is shifted by approximately 3.6 GHz. That is because the transceiver MMIC was designed for bond wire assembly so the parasitic inductances were taken into account, which are missing in this embedded setup with connection by thin film metal lines. Consequently, the antennas do not transmit or receive any signal because they were designed for the original frequency range. Although no operating module could be shown, the functionality of the embedded MMICs could be proven. The to low inductance of the connection can be compensated by layout change of the wiring metallization and is under preparation.

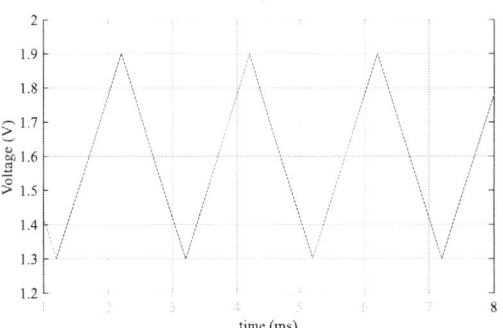

Figure 14: control voltage vtune provided by the PLL

978-1-7281-1500-9/19 $31.00 © 2019 IEEE

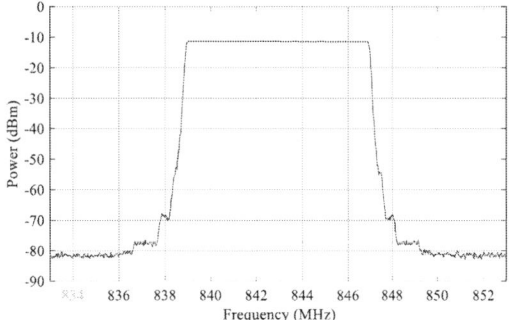

Figure 15: divided output spectrum of the radar sensor

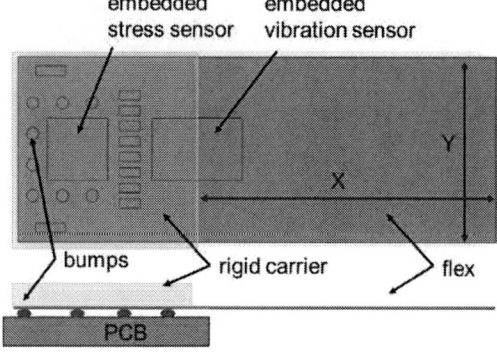

Figure 16: schematically drawn sensor concept with top view and cross section

VII. USE CASE 2: ACCELERATION AND VIBRATION SENSOR MODULE FOR CONDITION MONITORING IN RIGID-FLEX TECHNOLOGY WITH EMBEDDED ICs

A. Sensor concept and design

For condition monitoring of mechanical tools are typically two parameters used – vibration (acceleration) and mechanical stress. Knowing these parameters helps to optimize process performance and to increase the lifetime or identify preventive maintenance actions of equipment. Cracks, mechanical defects might have direct influence on the vibration frequency of moving components. Stresses could lead to many issues, like delamination, adhesion and offset issues in electronic devices. It is great demand of control and measure them.

The sensor system has been designed for the detection of both, lateral stress and acceleration (Z-direction). It features two sensor chips, which were designed and manufactured at Fraunhofer IZM, Berlin. The system concept is shown in Figure 16. Both sensor chips are extremely thinned and embedded into the polymer layers of a multi-layer rigid-flex circuit. The stress sensing device is a 2.37x2.37 mm2 silicon die embedded on a location where the rigid carrier remains. The only 20 µm thin silicon device hosts 4 implanted piezo-resistors. Stress in a tool where the sensor module is mounted on is transferred onto the rigid carrier and causes a lateral stress field in the silicon die, which can be sensed by the piezo resistors. As the carrier glass has limited stiffness and the silicon plate is very thin the concept is very sensitive to any change of the residual mechanical stress of the linked equipment.

The acceleration sensor has a size of 3.04x2.04 mm2 and consists of a micro beam with 4 implanted piezo resistors at both ends. As indicated in Figure 16, half of the die is located above the rigid carrier and the other half is only surrounded by the embedding polymer. Thus, half of the sensing beam on the chip can bend and can be deflected free around edge of the carrier. Any vibration of the tool leads to a bending of the silicon beam and a related value change of the piezo resistors. Due to a Wheatstone bridge arrangement of the resistors, the deflections can be detected very precisely and up to high frequencies.

The resonance behavior of the system could be adjusted by the size of the overhanging flex foil. The table below in Figure 17 shows the simulated dependency of the resonance frequency of the flex length X and its width Y with half of the silicon chip overhanging the rigid carrier as sown in Figure 16. The simulations were done with a 3D model using ANSYS.

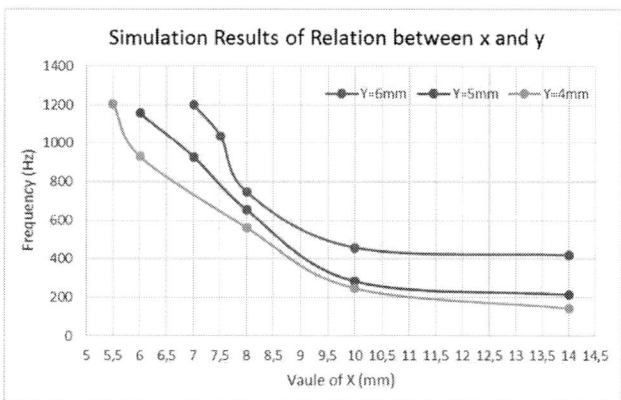

Figure 17: resonance frequency as function of the flex length X and width Y

B. Sensor manufacturing

The sensor modules were manufactured using a similar process sequence as for the previously described radar sensor. First, a polymer base layer was deposited onto the glass carriers followed by deposition and structuring of alignment marks and adhesive depots. Now, the 20 µm thin sensor dices were placed and over coated with an approx. 25 µm thick polymer layer. The polymer was structured directly by lithography and was cured subsequently. The top left and top right images in Figure 19 show the stress sensor chip and the vibration sensor chip after opening of the vias in the embedding polymer layer.

For the connection of the devices and re-routing of their IOs one semi-additive structured wiring layer of copper was used. The bottom images in Figure 19 show the two sensor chips after the fabrication of the wiring layer.

For passivation of the routing layer, a third polymer layer was deposited, structured and cured. As final IOs, pads were deposited by semi-additive structuring of NiAu. Finally, the devices were singulated and the carriers were partially removed as explained in chapter 5.

Figure 19 shows the final sensor modules with half of the vibration sensor inside a thin flex foil overhanging the rigid carrier. The rigid part of the module has a size of 6x7 mm^2 and hosts the embedded stress sensor as well as 10 solder balls for second level interconnection of the modules to PCBs.

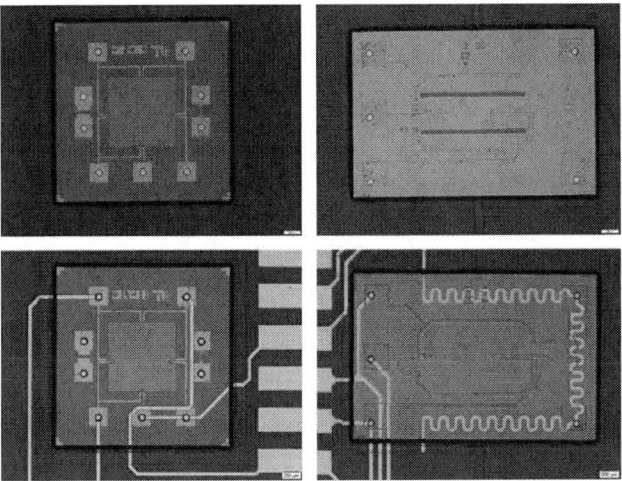

Figure 18: stress and vibration sensor after via formation in embedding polymer layer (top) and after wiring formation (bottom)

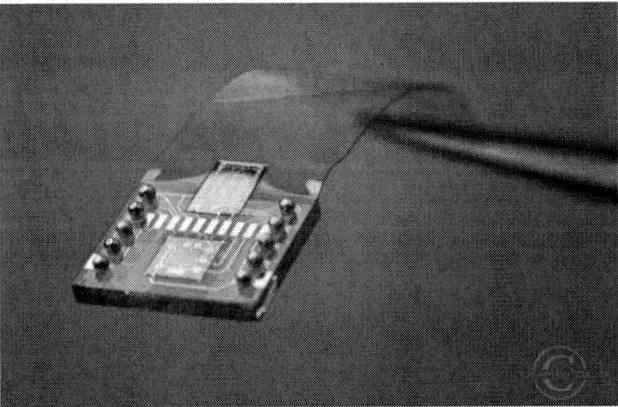

Figure 19: stress / vibration sensor module with thin embedded ICs after singulation from wafer and partial substrate release

C. Functional test and characterization

For evaluation the modules have been clamped carefully and mounted on a shaker, which can be adjusted in frequency and amplitude of the vibration. With the moment of inertia of the foil beam it starts to move as the shaker is excited. With the deflection of the beam the resistivity of the implanted piezo resistors change and this voltage drop can be detected by the attached microcontroller. The setup was intended to simulate a use of the sensor in a rotary pump, which runs with 50 Hz. The sampling rate of the microcontroller was set to 500 Hz in order to satisfy the Nyquist criteria. Figure 20 shows the ADC-signal recorded

by the microcontroller. One can clearly see the sinus excitation of the shaker with its 50 Hz frequency.

The system shows a good response to the excitation of the shaker, which proves the sensor concept and enables to retrofit condition monitoring of mechanical tools.

During the qualification test, also the functionality of the stress sensor could be approved by stress intake using a manual setup. However, an adequate test environment including defined device fixation and reproducible stress application to the device could not be established until the finalization of this manuscript.

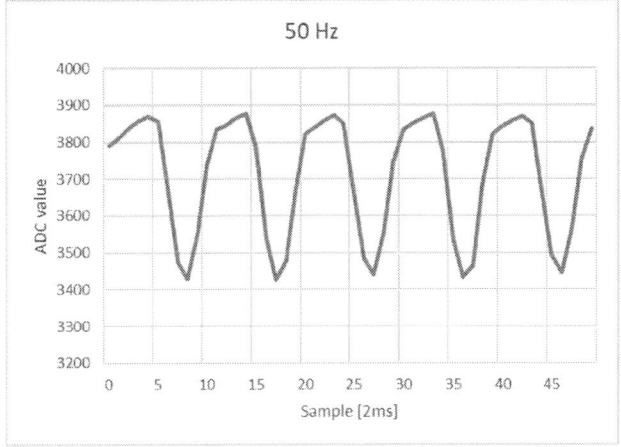

Figure 20: ADC signal of the microcontroller with 2 ms sampling rate

VIII. CONCLUSION

The presented high-density flex technology with thin chip embedding and partial carrier release option is a versatile base for system integration and packaging scenarios in the IoT domain. The technology offers a standard and an advanced process scheme with high and very high routing densities over up to three levels as well as front and backside contacts with a total flex thickness of up to 50-60 μm. The standard process scheme allows 30 μm line pitch with staggered vias of 35 μm diameter. The advanced process scheme allows reduced via diameters of 10 μm in stacked configuration and a minimum pitch of 14 μm with 7 μm line width and spacing. Further reduction of these values are in preparation.

The thin chip embedding option relies on back ground and polished ICs, which are die bonded in face-up configuration onto the first polymer layer of the flex build-up and which are subsequently over coated and thus embedded with the next polymer. The IOs of the ICs are connected by vias through the embedding polymer. The flex processing wafer can be removed completely or only partial, which gives the option to create fully flexible or rigid-flexible circuits.

The technology was approved by two use case demonstrators. The first demonstrator is a 24 GHz radar sensor module with two only 20 μm thick transceiver ICs embedded into the multi-layer routing. The module also includes two 24 GHz antenna arrays, which are realized in the same flex stack-up and are located next to each transceiver chip. The ICs have shown full functionality, but their working frequency was shifted 3,6 GHz up due to reduced inductance of the routing layers in the flex compared to connection by wire bond loops for which the ICs were designed. A redesign with adapted routing providing higher inductance can easily fix this problem and is in preparation.

The second demonstrator is a module for condition monitoring and hosts two only 20 µm thin silicon based stress and vibration sensors. The module has a rigid part were the IOs are located and were the stress sensor is embedded. In the flex part the vibration sensor is located so that the deflection of the flex causes an optimal agitation of the sensor. A 50 Hz vibration could be well detected when the module was mounted to a shaker, which was oscillating with that frequency.

The principal functionality of both demonstrators has shown the capability of the presented flex technology and its high versatility for the implementation of different application scenarios.

ACKNOWLEDGMENT

The authors thank all employees from Fraunhofer IZM, Berlin and Technical University of Berlin who were involved in that project.

REFERENCES

[1] P.V. Dudhe ; N.V. Kadam ; R. M. Hushangabade ; M. S. Deshmukh, *"Internet of Things (IOT): An Overview and its Application"*, 2017 International Conference on Energy, Communication, Data Analytics and Soft Computing (ICECDS), 1-2 Aug. 2017, Chennai, India, pp. 2650-2653

[2] X. Jia, J. Wang, Q. He, *"IOT Business Models and Extended Technical Requirements"*, IET International Conference on Communication Technology and Application (ICCTA 2011), 14-16 Oct. 2011, ISBN 978-1-84919-470-9

[3] https://en.wikipedia.org/wiki/Internet_of_things

[4] D. Bandyopadhyay, J. Sen, *"Internet of Things: Applications and Challenges in Technology and Standardization"*, Wireless Personal Communications 58(1), pp. 49 – 69

[5] W. Chen, *"Heterogeneous integration for IoT Cloud and Smart Things a Roadmap for the future"*, 2016 International Symposium on

3D Power Electronics Integration and Manufacturing (3D-PEIM), 13-15 June 2016, Raleigh, NC, USA

[6] A. Tai-Ying Lin, J. Lee, D. Lee, C.-C. Chen, *"The Development of IC Packaging under the Internet of Things Standards"*, 2016 11th International Microsystems, Packaging, Assembly and Circuits Technology Conference (IMPACT), 26-28 Oct. 2016, Taipei, Taiwan, pp. 209-211

[7] R. Beica, *"Enabling information age through advanced packaging technologies and electronic materials"*, 2018 Pan Pacific Microelectronics Symposium (Pan Pacific), 5-8 Feb. 2018, Waimea, HI, USA

[8] C. Landesberger, N. Palavesam, W. Hell, A. Drost, R. Faul, H. Gieser, D. Bonfert, Bock, C. Kutter, *"Novel processing scheme for embedding and interconnection of ultra-thin IC devices in flexible chip foil packages and recurrent bending reliability analysis"*, 2016 International Conference on Electronics Packaging (ICEP), 20-22 April 2016, Sapporo, Japan, pp. 473-478

[9] W. Christiaens, E. Bosman, J. Vanfleteren, UTCP: *"A Novel Polyimide-Based Ultra-Thin Chip Packaging Technology"*, IEEE Transactions on Components and Packaging Technologies (Volume: 33 , Issue: 4 , Dec. 2010), pp. 754-760

[10] M.-U. Hassan, C. Schomburg, C. Harendt, E. Penteker, J. N. Burghartz, *"Assembly and Embedding of Ultra-Thin Chips in Polymers"*, 9-12 Sept. 2013, Grenoble, France

[11] T.-Y. Kuo, Z.-C. Hsiao, Y.-P. Hung, W. Li, K.-C. Chen, C.-K. Hsu, C.-T. Ko, Y.-H. Chen, *"Process and characterization of ultra-thin film packages"*, 2010 5th International Microsystems Packaging Assembly and Circuits Technology Conference, 20-22 Oct. 2010, Taipei, Taiwan

[12] J. Wolf, J. Kostelnik, K. Berschauer, A. Kugler, E. Lorenz, T. Gneiting, C. Harendt, Z. Yu, *"Ultra-thin Silicon Chips in Flexible Microsystems"*, ECWC 13, 13th Electronic Circuits World Convention, Nürnberg, DE, May 7-9, 2014, pp.1-5

[13] K. Zoschke, M. Wegner, T. Fischer, K.-D. Lang, „*Temporary Handling Technology by Polyimide based Adhesive bonding and Laser assisted de-bonding"*, 6th Electronics System-Integration Technology Conference, September 13 – 16, 2016, Grenoble, France

[14] M. Woehrmann, O. Wuensch, K.-D. Lang, R. Gernhard, K. Hauck, K. Kroehnert, K. Zoschke, N. Juergensen, M. Toepper, T. Braun, H. Hichri, M. Arendt, *"New Excimer Laser-Based Dual Damacene Process For High I/O Applications with Ultra-Fine Line Routing"*, Süss Report 2016, pp 4-10

[15] K. Zoschke, K.-D. Lang, *"Evolution of Structured Adhesive Wafer to Wafer Bonding enabled by Laser Direct Patterning of Polymer Resins"*, 18th Electronics Packaging Technology Conference, 30th Nov – 3rd Dec 2016, Singapore, pp. 223-228

Advance Embedded Packaging For Power Discrete device

Jia ren Huo,Guan qiang Song,Jing Jiang
Product Research and Development
Wuxi Sky Chip Interconnection Technology co.,LTD
shenzhen, China
huojiar@scc.com.cn,
songgq@scc.com.cn,king@scc.com.cn

Tao Jun Wang,Ling wen Kong
R & D Department
Wuxi Sky Chip Interconnection Technology co.,LTD
Shenzhen,China
wangjt@scc.com.cn,konglw@scc.com.cn

Abstract—Power devices are developing towards high performance, small size, modularization and integration. These devices need not only high thermal performance, but also complex circuit layout. However, power devices of traditional wiring technology are unable to meet the requirements. In order to solve this problem, a new kind of advanced and extensible power package device is developed by the embedded packaging technology, which utilizes porous heat dissipation and conduction, with low resistance and good thermal performance.

The reliability of power devices after moisture adsorption and reflow soldering is a key problem to be overcome in project development. In order to verify the problem, DFN5×6 embedded package is used to replace the traditional lead package test, which uses finite element software to build the three-dimensional model as to analyze the distribution of relative moisture and thermal stress in embedded package during re-welding. According to the size of the simulation model to manufacture the experimental products, which proves that there's no product failed after verifying MSL3 through the electrical property test. This packaging method has excellent low resistance property and will be more and more used in power devices.

Keywords- power separated device, embedded package, internal resistance, thermal stress

I. INTRODUCTION

On the one hand, with the continuous upgrading of technology and the increasing demand for high voltage and high power, modern power semiconductor devices demand higher performance, faster speed and smaller size. With the development of multi-chip connected package, modularization is an inevitable trend. On the other hand, with the continuous expansion and deepening of the application field, the IC driven power can achieve higher efficiency, better control functions, simpler peripheral layout design, so high integration has become a very important direction of development. Power discrete devices (Figure 1) are developing towards miniaturization, high power and high efficiency.

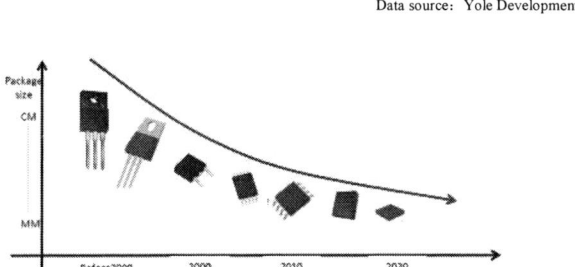

Data source: Yole Development

Figure 1. The packaging size of power discrete devices decreases year by year

Power modules are used in almost all power industrial products. Different requirements in various application fields require each part of power modules to be improved and innovated. Traditional wire bonding and copper Clip processes are difficult to achieve higher performance, faster speed, smaller size, multi-chip packed connection and modularizations. In the future, the packaging technology of power semi-conductor will develop towards more excellent embedded package.

II. PROCESS INTRODUCTION

A new advanced expandable power packaged device is developed by using embedded packaging technology. This technology uses copper plated interconnection to replace traditional wire-making and copper Clip welding process, which uses porous heat dissipation and conduction to realize chip Top layer heat dissipation. It has low resistance, low parasitic capacitance and inductance, with good thermal performance. Figure 2 below is the product structure diagram of embedded packaging, and Figure 3 is the product appearance diagram of embedded packaging.

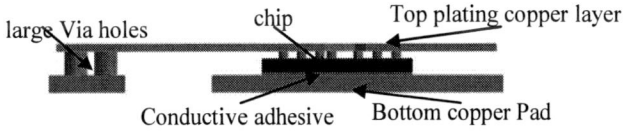

Figure 2. Structural chart of embedded packaging products

Figure 3. The external view of embedded packaging products

Figure 2 shows the product structure after hiding the plastic sealing materials, and the chip Face Up is placed. The chip is fixed on the copper pad at the bottom and connected with large and small Via on the top of the chip. The large Via holes are interconnected without filling the holes due to the limitation of process capability. The small Via holes are interconnected by filling holes, and small Via holes can improve the heat dissipation of the chip and reduce the internal resistance of the package.

Table I and Figure 4 is the comparison between single-side Via-hole plan and double-side Via-hole schemes on the internal resistance of conventional wire bonding and Clip welding package plan. Compared with the conventional TO-220 package plan, the maximum internal resistance of the single-side Via hole plan reduces 1.28MΩ of the space, with 85.3% decrease, as to meet the demands of high performance, small size, modularization and integration of power devices. Figure 5 shows the embedded package technology of Roadmap.

TABLE I.   THE RESULTS OF INTERNAL PACKAGE RESISTANCE COMPARISON

/	Single Via PAK	Double Via PAK	Typical DPAK
Model			
Resistance(mΩ)	0.23	0.22	1.5
/	Typical D2PAK	Typical To220	LFPAK
Model			
Resistance(mΩ)	0.8	1.0	0.2

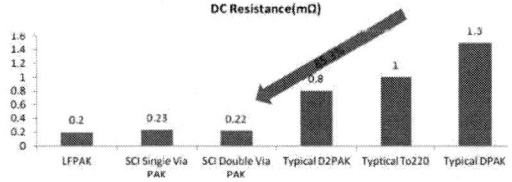

Figure 4. The comparison diagram of different internal package resistances

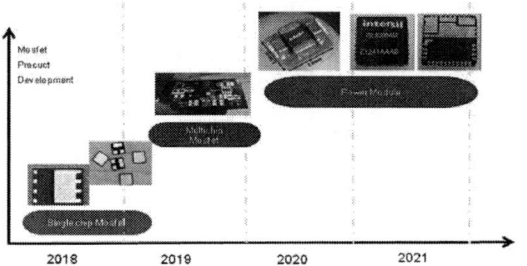

Figure 5. The embedded packaging technology of roadmap

III. MODELING AND SIMULATION

A. Heat Conduction and Moisture Diffusion

For most physical and chemical diffusion behaviors, the Fick's second law is followed. The heat conduction equation is [1]:

$$\frac{\partial y}{\partial t} = k/\rho C \times \text{div}（T）$$

In above formula, T is temperature, k is thermal conductivity, ρ is density and C is specific heat. Both moisture diffusion and heat conduction follow the same governing equation in mathematical form. It only needs to replace the temperature and thermal diffusion coefficient in the heat conduction equation to the moisture concentration and moisture spread coefficient. Temperature T in heat conduction and moisture concentration C in moisture diffusion are their driving forces respectively, so the equation of moisture conduction [2] is:

$$\frac{\partial C}{\partial t} = \nabla \cdot D \times \nabla C$$

C is the moisture, D is the moisture diffusion coefficient, and the corresponding characteristic parameters of moisture diffusion heat conduction are shown in Table II below.

TABLE II.   THERMAL-MOISTURE ANALOGY FOR MOISTURE DIFFUSION ANALYSIS IN FINITE ELEMENT ANALYSIS [3-4]

Features	Heat Conduction	Moisture Diffusion
Field variable	Temp. T/℃	Relative humidity
Density	ρ/（Kg/m³）	1
Thermal conductivity	k[W/(m·K)]	D·C_{sat}/[kg/(s·m)]
Specific heat capacity	C[(J/(kg·K)]	C_{sat}/(kg/m³)

According to JESD51-3 standard, DFN5X6 packaged MOS-8Pin devices are selected as analysis subject to construct PCB, and the dimension is 76.2×114.3mm×1.6mm. The circuit layer is a thermal equivalent model based on JEDEC standard, and its size is 50mm *50mm *0.071mm. Because the volume of Top copper plating layer on chip accounts for a large proportion, it has a great influence on the heat, and the model is not symmetrical, so the 1:1 three-dimensional model is used for the analysis. Figure 6 is the overall structure of DFN5X6 packaging model, and Figure 7 is the local structure of DFN5X6 packaging model.

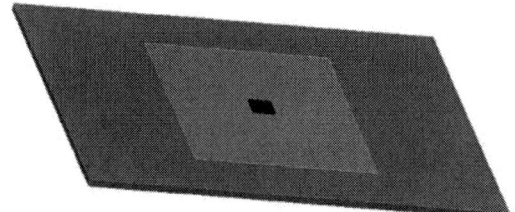

Figure 6. Integral structure of DFN5X6 package model

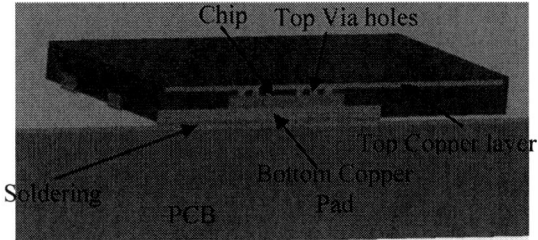

Figure 7. Local structure of DFN5X6 package model

B. Boundary Condition

The transient thermal analysis and steady-state structure analysis module in the finite element analysis software is used to simulate the moisture diffusion and stress in the re-welding process of plastic package devices. According to the moisture sensitivity rating Table [5], after selecting 30℃/60%RH/168H moisture condition, stimulate the moisture diffusion before re-welding. Using the transient thermal simulation model, the material properties of moisture diffusion are shown in Table III [6-7], and those of thermal diffusion and thermal stress are shown in Table IV.

TABLE III. PROPERTIES OF MOISTURE DIFFUSION MATERIALS

Material	30℃/60%RH		
	C_{sat}/ ($Kg\cdot m^{-3}$)	D/ ($mm^2\cdot S^{-1}$)	CME/(mm^3/mg)
Silicon	1×10^{-6}	1×10^{-6}	/
Solder paste	1×10^{-6}	1×10^{-6}	/
Copper	1×10^{-6}	1×10^{-6}	/
Plastic package materials	7.81	3.13×10^{-7}	0.222

TABLE IV. THERMAL DIFFUSION AND THERMAL STRESS MATERIAL PROPERTIES

Material category	Density(Kg/M^3)	Isotropic Thermal Conductivity ($W/m.℃$)	Specific Heat ($J/Kg.℃$)
Conductive adhesive	3.7	2.5	-
Copper	8300	401	385
FR-4	1840	0.38@X,0.38@Y, 0.30@Z	-
Plastic package materials	1550	0.65	-
Silicon	2330	124	702
Tin	10900	35	139

Material category	Coefficient of thermal expansion (/℃)	Young's Modulus (Pa)	Poisson's Ratio
Conductive adhesive	4.0×10^{-5}@0℃ 1.5×10^{-4}@120℃	4.41×10^9@-65℃ 3.93×10^9@25℃ 2.0×10^9@150℃ 3.03×10^8@250℃	0.3
Copper	1.80×10^{-5}	1.10×10^{11}	0.34
FR-4	1.25×10^{-5}@X 1.14×10^{-5}@Y 8.2×10^{-5}@Z	2.04×10^{10}@X 1.84×10^{10}@Y 1.5×10^{10}@Z	0.11@XY 0.09@YZ 0.14@XZ
Plastic package materials	1.80×10^{-5}@0℃ 4.80×10^{-5}@160℃	9.0×10^9@25℃ 1.20×10^7@250℃	0.35
Silicon	2.46×10^{-6}@20℃ 3.61×10^{-6}@250℃	/	/
Tin	37	1.20×10^9	0.3

Material attributes are derived from supplier's datasheet.

Temperature loading in re-welding process is different from the actual plate temperature. The simulation is based on measuring the actual PCB surface temperature. The set temperature and the actual measured temperature are measured in 13 load steps as shown in Table V.

TABLE V. THE LOADING STEPS OF TEMPERATURE LOADING

Step	Time (S)	Setting Temp. (℃)	Actual plate temp. (℃)
1	30	120	50
2	60	140	75
3	90	160	105
4	120	180	135
5	150	190	150
6	180	200	170
7	210	220	185
8	240	250	200
9	270	260	220
10	300	258	230
11	330	cooling	220
12	360	cooling	170
13	390	cooling	125

IV. SIMULATION RESULTS

A. Moisture simulation results

Figure 8, and Figure 9 are the moisture distribution of the plastic package material and the device cross-section moisture distribution after the device passes through the humid environment of 30 C/60% RH/168H. The results show that the plastic sealing materials on the displayer edge is basically saturated, while the middle of the device is still in a dry state. Because the device has a large copper layer and moisture can't diffuse through copper.

Chips and coppers are basically in a dry state, because the moisture diffusivity D. C_{sat} of plastic sealing material is much larger than that of chips and copper layers. Figure 10 shows the trend of moisture diffusion in the dry region of the device. The curve shows that the moisture diffusion becomes very slow in the later stage because of the copper layer protection in the region. It takes a long time for the device to reach moisture saturation.

978-1-7281-1500-9/19 $31.00 © 2019 IEEE   1487

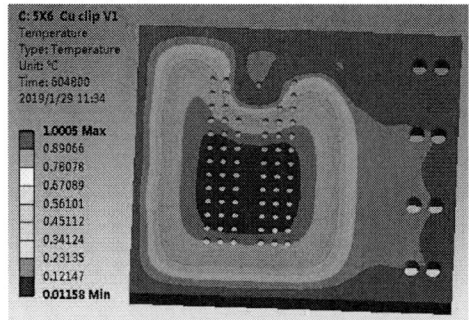

Figure 8. Moisture distribution of device's plastic package materials

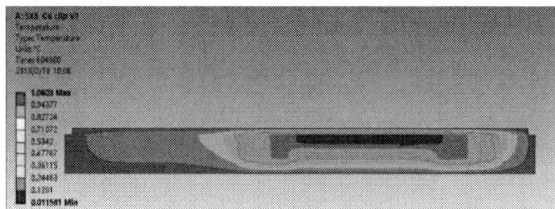

Figure 9. Moisture distribution of device's section

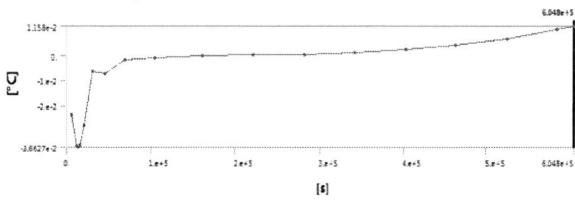

Figure 10. Moisture distribution trend in the drying area of plastic package materials

Figure 11 shows the simulation results of the moisture stress after absorption at 30 C/60% RH/168H. The maximum stress occurs on the Via porous copper column with a value of 142 Mpa. Figure 12 shows the moisture stress distribution at the interface of the chip after moisture diffusion. The maximum stress of the chip appears at the contact point between the chip and the Via hole, and the maximum stress is 107.5 Mpa. The interface between the chip and the plastic sealing material is the weak area of the device. If the bonding force between the two interfaces is not enough, the interface extends from the position of Via hole and begins to have de-lamination and cracks. With the increase of reflow temperature, the interfacial vapor pressure of de-lamination and fracture has a significant effect, even the phenomenon of "popcorn".

Figure 11. Simulation results of moisture stress

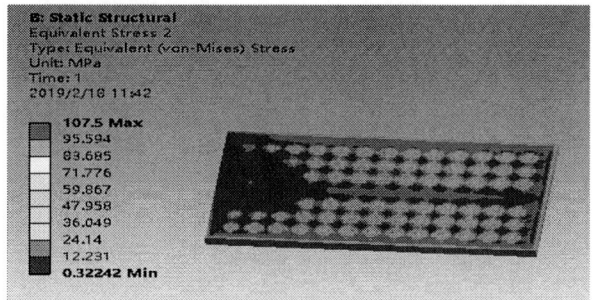

Figure 12. Moisture Stress Result on Chip Surface

The results of moisture simulation show that the interface area between chip and plastic packaging material is still in a relatively dry state after the device passes through 30 ℃ /60%RH/168H. The vapor density at the interface is relatively low, while the vapor pressure generated during over-current welding is relatively low. Therefore, the vapor pressure at the chip interface is not considered at the moment.

B. Thermal Stress Simulation

Figure 13, Figure 14, Figure 15 and Figure 16 show the distribution of equivalent stress on the chip surface at 50℃, 150℃, 230℃, and 125℃ respectively. Equivalent thermal stress is positively correlated with temperature, and the maximum stress occurs at 230℃ during reflow process. The maximum equivalent thermal stresses on the chip surface at 50℃, 150℃, 230℃, and 125℃ are 0.252 Mpa, 3.76 Mpa, 10.85 Mpa and 0.028 Mpa respectively. At different temperatures, the maximum equivalent stress occurs at the interface between the chip and the Via hole. The main reason is that the material expansion coefficient at the two ends of the interface is quite different. The thermal conductivity of the chip, copper Pad, electroplated copper layer and welding layer is higher than that of the plastic sealing material, and the corresponding material has faster thermal conductivity. There is a large thermal gradient between these materials and the plastic sealing material.

By comparing the thermal and moisture stress distributions in Figure.13 to 16 with those in Figure 12, the distributions are very similar, both of which occur at the interface between the chip and the Via hole. These stresses are too concentrated on the interface between the chip and the Via hole, which leads to the easy stratification of the interface.

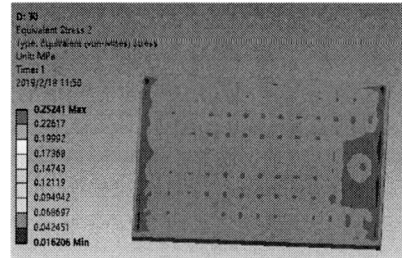

Figure 13. Equivalent Stress of PCB Chips at 50℃

978-1-7281-1500-9/19 $31.00 © 2019 IEEE   1488

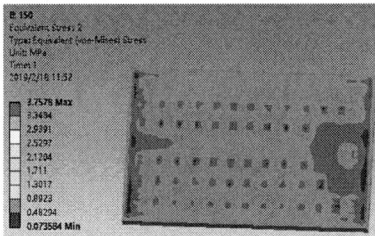

Figure 14. Equivalent Stress of PCB Chips at 150℃

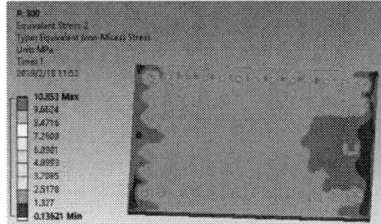

Figure 15. Equivalent Stress of PCB Chips at 230℃

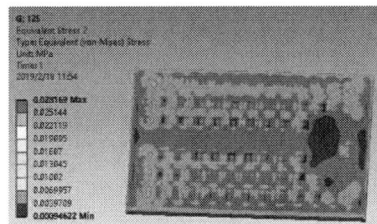

Figure 16. Equivalent Stress of PCB Chips at 125℃

V. TEST MANUFACTURE

A. Process flow

Table Ⅵ shows the process flow of embedded package. The first step is to make Patten L1 circuit layer of bottom Pad on the Carrier. The second step is to fix the chip on the copper Pad with wire glue on Pattern L1.The third step is to press the plastic sealing material and copper foil on the Carrier which has been crystallized. The fourth step is to process small Via holes and large Via holes on the chip by laser drilling process; the fifth step is to deposit copper and electroplating on the laser Via holes, fill the small Via holes, and make the pattern L2 layer. The sixth step is to press the insulation material in Pattern L2 to insulate the pattern L2 layer from the environment and fill in the large Via hole. In the seventh step, remove the carrier layer, and the device Pad is exposed and tin plated on the Pad. Finally, the whole product board will be marked and cut into DFN5*6 size.

TABLE VI.   PROCESS FLOW OF EMBEDDED PACKAGE

Process	Cross section
Pattern L1	
Die attach	
Insulator resin Lamination	
Pattern L2	
Removed Carrier	

B. Finished products

Figure 17 is the front view of the product, Figure 18 is the back view of the product, Figure 19 is the X-Ray drawing of the product, Figure 20 is the 45 degree X-Ray drawing of the product, Figure 21 is the X-Ray drawing of the side of the product, and Figure 22 is the slice figure of the product, and the overall thickness of the product is 0.7mm.

Figure 17. Product front view

Figure 18. Product back view

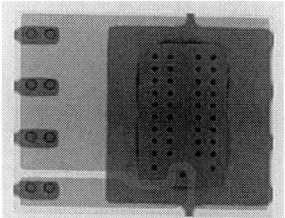

Figure 19. X-Ray drawing of product overview

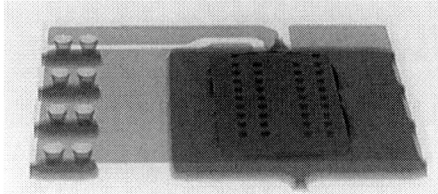

Figure 20. X-Ray drawing of product 45°

Figure 21. X-Ray drawing of product profile

Figure 22. Product slice figure

C. Verification result of product reliability

Table Ⅶ shows that according to the JESD22-A113E MSL3 test plan, 44pcs were tested. The samples are baked at 125℃ for 24H to remove moisture in the device, and are placed at 30℃ under 60%RH condition after 168H, and the plastic sealing materials of the device absorb moisture according to the standard requirements. The product is subjected to lead-free reflow three times after moisture absorption, and the sample is subjected to thermal shock. After the completion, compare the test to check if the product electrical performance is invalid and changed, and confirm whether the product has been invalid through slice analysis and SEM analysis.

TABLE VII. MSL3 TEST PLAN

Test items	The number of tests	Basic standard	Test means	Test methods
MSL 3	44	JESD22 -A113E	Electrical properties test, EM, slice	1.First bake the plate at 125℃ for 24h; 2.MSL3 level，30℃，60%RH，168H; 3.Lead-free reflux, 260℃, three times

Table Ⅷ compares the overall test data of 44pcs products before and after passing MSL3. Figure 23 shows the Rdson distribution before and after MSL3. All products have not failed, which can be judged that the electrical properties of

samples after MSL3 are OK. Figure 24-25 is the slice condition after MSL3. Figure 26, 27 and 28 are the SEM analysis diagrams of Via holes, on-chip and conductive adhesives after MSL3. From the slice and SEM analysis, it is confirmed that there are no gaps and stratification problems, so it can be judged that the sample meets the requirements of MSL3.

TABLE VIII. TEST DATA RESULTS AFTER MSL3

Table	Test Items	GSTH	BDSS	IGSSF	IGSSR	RDSON
Before MSL3	Average	1.720	34.13	44.72	6.74	1.248
	Minimum	1.702	33.95	33.27	4.15	1.137
	Maximum	1.779	34.31	66.93	9.20	1.387
After MSL3	Average	1.706	34.24	41.80	6.60	1.258
	Minimum	1.679	34.12	28.95	3.73	1.113
	Maximum	1.747	34.42	64.95	12.71	1.376

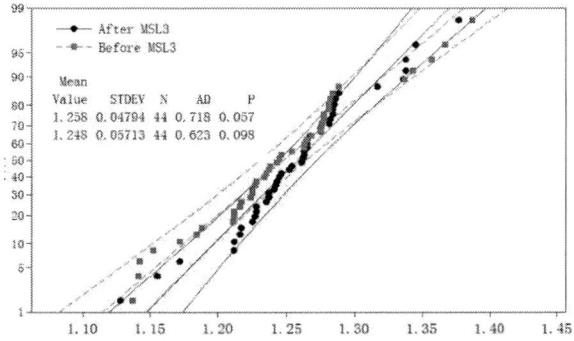

Figure 23. Rdson Distribution before and after MSL3

Figure 24. Small Via hole slice figure after sample MSL3

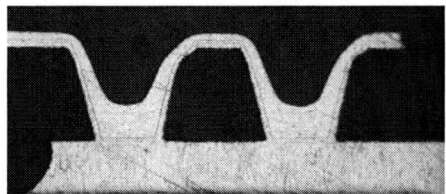

Figure 25. Large Via hole slice figure after sample MSL3

Figure 26. Via hole SEM

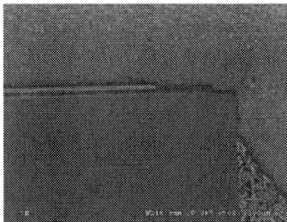

Figure 27. Chip corner SEM

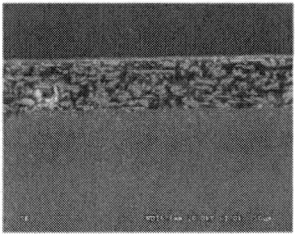

Figure 28. SEM Diagram under Conductive Adhesive

VI. SUMMARY AND CONCLUSIONS

A three-dimensional finite element model of DFN5X6-8Pin package device is established. The thermal and stress distributions of the plastic package device are obtained by simulating the process of moisture diffusion and reflow. It was confirmed that the moisture and thermal stresses were concentrated on the interface between the chip and the Via hole.

In the process of reflow, the stress decreases first and then increases. When the moisture stress and thermal stress

in the reflow exceeds the yield strength or fracture strength of plastic sealant, cracks and cracks will appear on the interface. When the cracks extend to a certain extent, the device will fail [8]. From the simulation results, the moisture equivalent stress and thermal equivalent stress don't reach the fracture strength of the material. Samples are fabricated according to the simulation plan. Samples were hygroscopically welded at 30℃, 60% RH and 168H. The electrical test performance of the sample hasn't been failure, and there's no crack and de-lamination through sample slice analysis and SEM analysis, so it is confirmed that the sample meets the requirements of MSL3 grade.

[1] WONG E H,KOH S W,LEE K H,et al.Advanced moisture diffusion modeling &characterization for electronic packaging[C]//Proe of Electronic Components and Technology conf,Difornia,2002;1297-1303.

[2] Chen X,Zhao S F,Zhai L.Moisture absorption and diffusion characterization of molding compound.ASME Journal of Electronic Packaging,2005,127:460-465.

[3] I Wong E H,Rajoo R, Koh S W,et al.The mechanics and impact of hygroscopic swelling of polymeric meterials in electronics packaging.ASME Journal of Electronic Packaging,2002,124:122-126.

[4] Wong E H,Rajoo R, Koh S W,et al.The mechanics and impact of hygroscopic swelling of polymeric meterials in electronics packaging.ASME Journal of Electronic Packaging,2002,124:122-126.

[5] Zhang Linchun, Grade Evaluation of Hygroscopic Sensitivity of Green Plastic-encapsulated IC [J]. Electronics Process Technology, 2005（6）: 29-32.

[6] LAM T F.FEA simulation on moisture absorption in PBGA packages under various moisture pre-conditioning [C]//2000 50nd Electronic Components and Technology Conference.Las Vagas,USA:IEE,2000.

[7] Ye Anlin, Qin Liancheng, Kang Xuejing, the Effect of Moisture Diffusion and Moisture Stress on the Reliability of Laminated Packages[J]. Electronic Components and Materials, 2008,27（1）: 69-73.

[8] VAN DRIEL W D,VAN GILS M A J,VAN SILFHOUT R B R,et al.Prediction of delamination related IC & Packaging reliability problems[J].Microelectron Reliab,2005,45(9/10/11):1633-1638.

2019 IEEE 69th Electronic Components and Technology Conference (ECTC)

Large Panel Size Bonder with High Performance and High Accuracy

Hubert Selhofer, Andreas Mayr, Hugo Pristauz
Besi Austria GmbH
Radfeld, Austria
hubert.selhofer@besi.com

Abstract—**Driven by massive cost down demands fan-out packaging is facing now a transition from wafer to panel level. To address the market needs a panel level fan-out bonder has to provide high productivity exceeding 7000 UPH as a key requirement. On the other hand embedded multi-die interconnect bridge packaging (EMIB) requires very high placement accuracy with a roadmap pointing to sub-micron level.**

This paper presents a panel level die-attach machine suitable for substrates up to 730 mm x 920 mm (GEN 4.5). The bonder provides global placement accuracy of 2 μm @ 3σ and local placement accuracy of 1 μm @ 3σ at 6000 UPH, the bonder's productivity can be scaled up to 7000 UPH. The machine concept is based on a building block approach and therefore can be configured for various fan-out pick and place processes, for silicon bridge embedding, thermal compression, and other processes.

Keywords-die attach; advanced packaging; fan-out packaging; panel level packaging; EMIB; SiP

I. INTRODUCTION

One of the major drivers for the industry to convert to Panel Level Packaging (PLP) is the ever present high pressure to lower the overall package costs. This goes hand in hand with the demand for thinner packages with higher performance at the same time.

Fan Out Wafer level packaging (FO-WLP) is an established technology fulfilling the latter requirements. Persistent pressure to reduce package cost forces the industry to go to large panel sizes. The players in PLP production are predicting a cost down by a factor of 2 [1].

This is the reason why we see PLP on the roadmap of nearly every outsourced semiconductor assembly and test (OSAT) company. As an example, Powertech Technology (PTI), which started Fan Out Panel Level Packaging (FO-PLP) production in small volumes already in 2017, recently announced the construction of a new plant for high volume production in 2020 [3].

As FO-PLP manufacturing is also an enabler for high performance packages at a lower cost compared to other technologies like 3D integration, we see IDMs like Samsung and Intel develop packages supporting their front-end chip business. In 2018 Samsung Electro Mechanics (SEMCO) announced that they switched production of the application processor unit (APU) for its Galaxy Watch to FO-PLP using their SiP-ePoP technology [2].

Intel´s heterogeneous integration roadmap, based on their Embedded Multi-Die Interconnect Bridge process (EMIB), will lead to die sizes much larger than 16x16mm. Running the EMIB process on panel level will lower the packaging costs but will require a large area die bonder with highest accuracy.

Currently FO-PLP is mainly used for mobile applications. But the market expects a dramatic increase of FO-PLP by 2020 or 2021 [2], due to 5G, AI, autonomous driving and server requirements as well as demand for modularity and high-speed data processing.

Yole Développement reported in its "Status of Panel Level Packaging 2018" report a growth rate of 79% CAGR and a market volume of almost US$279 million until 2023 [1].

FOPLP MARKET FORECAST (2018-2023)

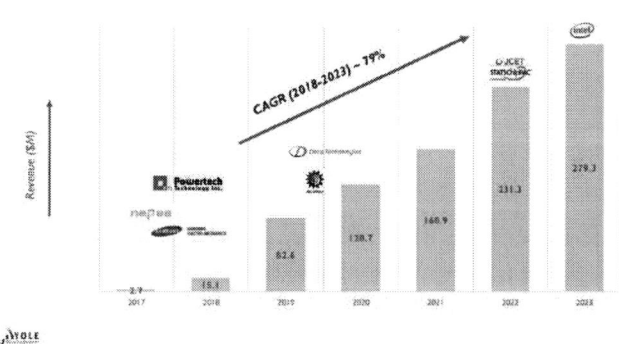

Figure 2. PLP market forecast [1].

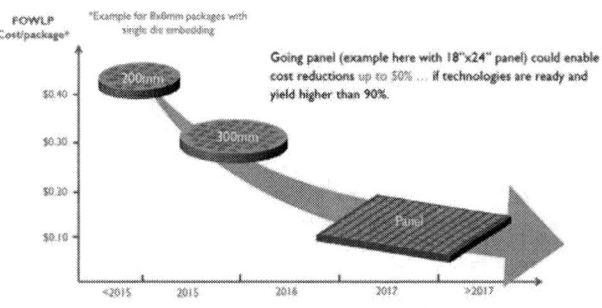

Figure 1. Fan-Out cost evolution analysis [1].

978-1-7281-1500-9/19 $31.00 © 2019 IEEE

As yet there is no standardization in panel size, and thus there exists on the market a wide variety of formats and dimensions. The sizes vary between 510 x 415 mm² (SEMCO), 510 x 515 mm² (Intel), 600 x 600 mm² (ASE/ DECA) and in some cases even up to 730 x 920 mm² (GEN 4.5).

II. THE NEW PANEL SIZE PLATFORM

Besi's PLP Die Bonder is ready to support panels up to 730 x 920 mm² and is based on a modular machine concept. A die attach cell with a dual gantry pick & place system can be linked with a broad variety of process modules to fulfill different customer requirements:
- flip-/non-flip module
- heated bond head and heated panel chuck
- thermo compression bond head
- multi die handling with auto tool exchange
- ISO 3 clean room kit
- slide fluxer
- tape & reel input
- large die kit (up to 75x75 die size)
- thin die ejector (> 50μm die thickness)
- multi nozzle bond head and die provider

To support enhanced accuracy at sub-micron level on panel area a disruptive approach for the metrology of an advanced gantry system has been chosen.

III. ADDITIONAL METROLOGY SYSTEM

During production a machine will be in different temperature states. Usually the machine is started in cold state, and then the induced heat from the motors will heat up the gantry system until it theoretically reaches a thermal steady state. However, the thermal steady state will see phases of cooling down (operator intervention, material loading) and heating up again. Thermal changes will cause mechanical deformations in the gantry system, especially in a large gantry system for panel size applications.

Conventional gantries use encoders and scales that are placed on the gantry beams and therefore suffer from the thermal impact of the motors. Additionally those encoders cannot detect any of those thermally induced mechanical deformations and therefore the accuracy performance is vastly degraded.

The UPH requirements for die attach machines necessitate the use of high acceleration and speed settings. The dynamical effects on the gantry beams (deformations like bending and torsion, cross influence between left and right gantry) are also mostly invisible to a conventional encoder system and that makes dynamics the second major cause for inaccuracies.

To overcome those effects we use a commercially available additional metrology and advanced motion control system which adds a second high precision encoder system having its reference point coupled to the substrate area. The design gives an additional position input mechanically and thermally decoupled from the gantry beam. Fig. 3 shows the basic principle behind this method.

In addition the gantry system is equipped with water cooling in order to minimize heat flow from the motor drives into the gantry structures, and subsequently by convection into the metrology beams [4]. Such systems have not yet been used for die bonding applications and specifically not for dual head gantry systems.

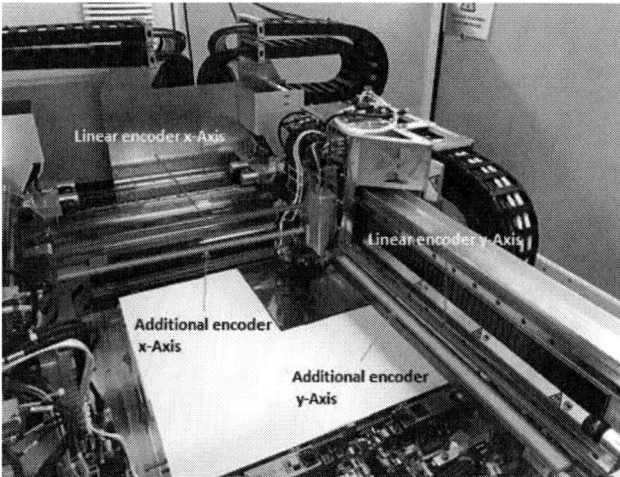

Figure 3. Additional metrology system.

IV. PRODUCTIVITY

Units per hour (UPH) are a key metric for machine productivity. Unfortunately there is no universal definition for machine UPH available since this number strongly depends on process parameters as well as the geometry of substrate and component.

Our UPH calculation is based on a real world reference product as benchmark. It does not include substrate and component wafer change time, but realistic process requirements (100 ms bonding delay, 50 ms transfer delay, and 0.3 mm over-travel distance) are taken into account.

This reference product consists of a 650 mm x 650 mm substrate with 2500 bonding positions on a mesh grid populated over the whole substrate. A 4-point substrate alignment is done once after substrate loading and clamping. Every component is ejected, flipped, transferred to the bond head, measured at an upward looking camera system and bonded.

The machine currently achieves a sustainable maximum movement speed of 2 m/s with an acceleration of 50 m/s². To use such motion parameters water cooling is required to prevent the powerful iron core linear motors from overheating.

In this benchmark the machine achieves 6400 UPH according to the aforementioned UPH calculation method. The performance is not hindered by any throttling due to overheating.

V. LOCAL ACCURACY

To assess the machine capability for different accuracy requirements the terms "local" and "global" accuracy are used. Within this section we will define local accuracy and the test procedure we use to measure it. Global accuracy will be covered in the subsequent section.

Local accuracy is a useful concept for the case, where an alignment mark (fiducial) is visible on the substrate for every bonding position and where the component has to be placed with defined offset relative to those fiducials.

Our test procedure uses a single glass chip which is placed on a precisely marked glass plate at multiple positions. The procedure further includes all steps which are usually required to bond a real component, i.e. picking a component, moving the component on the bond head to the upward looking camera, placing the substrate camera over the fiducial on the substrate, moving the die on the bond head to the substrate target position, and moving down in z, all this including the actual bonding process with slow travels and delays. After placing the component on the glass plate (see Fig. 4) the offset of the glass chip to the ideal position is measured with a relative post bond inspection (PBI), i.e. comparing the position of the glass die fiducials with those of the glass plate. The measurement error of the PBI has been estimated to be normally distributed with zero mean and a standard deviation of circa 45 nm.

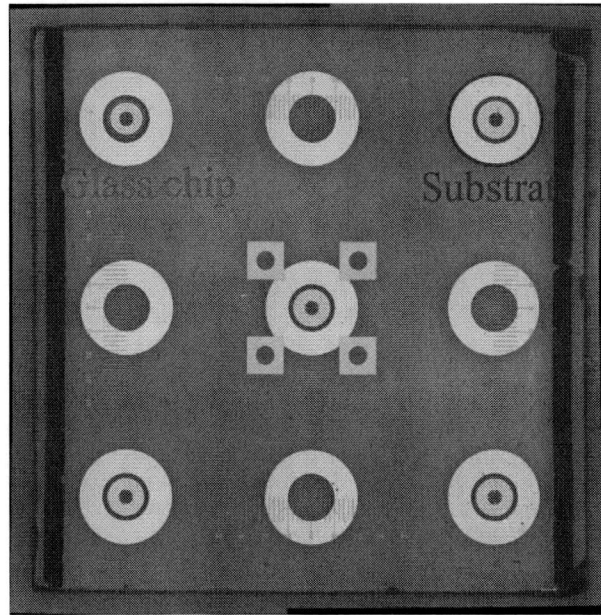

Figure 4. Glass chip "bonded" on substrate ready for the PBI.

Fig. 5 shows the result of a test run, where a 10 by 10 matrix with a pitch of 32 mm x 36 mm has been consecutively bonded for 150 times. The test was done using the left gantry only with motor parameters matching a 6k UPH (i.e. 3k UPH in single gantry mode).

The measured standard deviation in single gantry mode for x placement was 173 nm (cmk 1.85 at ±1 μm tolerance) and for y placement 147 nm (cmk 2.25 at ±1 μm tolerance).

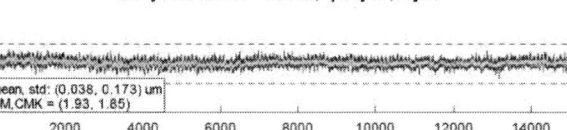

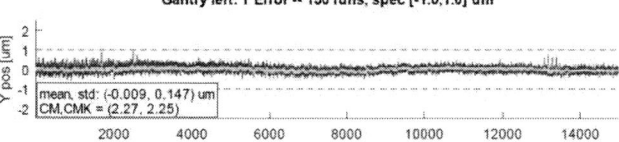

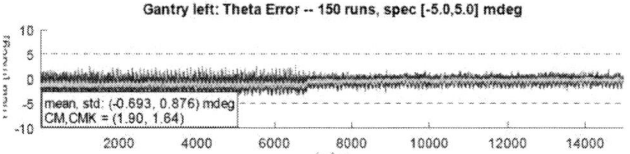

Figure 5. Local accuracy over 15000 bonds in single gantry mode.

As expected the same test in dual gantry mode shows some cross influence between left and right gantry, mainly in x placement. In Fig. 6 the first 50 measurements belong to the left gantry, the next 50 measurements belong to the right gantry, and this setup is repeated for the entire graph. The placements were done concurrently.

The observed cross influence depends on the exact timing of the gantry movements and a proper synchronization of both gantries can improve the result in x placement.

The measured standard deviation in dual gantry mode for x placement was 254 nm (cmk 1.25 at ±1 μm tolerance) and for y placement 144 nm (cmk 2.29 at ±1 μm tolerance).

The latter results prove that the standard deviation of the local accuracy is less than

$$\sigma_L = 260 \text{ nm.} \tag{1}$$

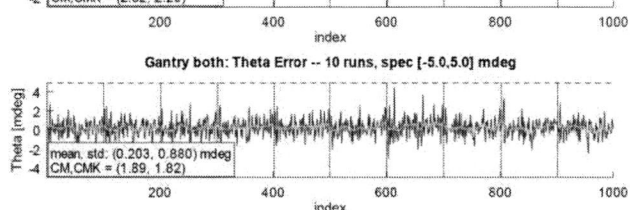

Figure 6. Local accuracy over 1000 bonds in dual gantry mode.

VI. REPEATABILITY

One important criterion for a stable production machine is repeatability meaning that the machine is capable to achieve the same accuracy over many hours operating in a reasonably stable environment (including controlled temperature, floor vibrations, and humidity) that is comparable to typical clean room conditions.

Fig. 7 shows the repeatability of 250 runs (i.e. 25000 bonds) in single gantry mode. The x-axis represents a bond location and the y-axis refers to the placement error on this bond location as a boxplot.

The results support the data presented in Fig. 5.

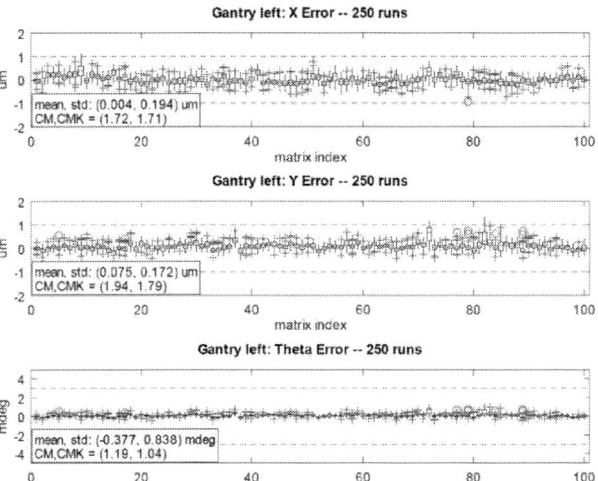

Figure 7. Repeatability and error distribution per bonding position.

VII. REPRODUCIBILITY AND THERMAL DRIFT

Once local accuracy and repeatability are accounted for and compliant with target specifications, the final critical influence on machine performance and subsequently on production yield is the reproducibility (stability) of the machine when challenged with typical workflow interruptions that occur on production equipment. To assess the impact of operator interventions, material change times, or production stops the (thermal) drift between the machine's cold state and the thermal steady state was analyzed.

As a first test the machine was allowed to cool down for 30 minutes after a thermally stable production run and then the dual gantry test run was restarted. Fig. 8 shows the result of this test. As in Fig. 6 the first 50 measurements belong to the left and the consecutive 50 measurements belong to the right gantry. Both gantries show a drift of about 4 µm in x direction during the first run. The second run (index 101 to 200) starts with a substrate alignment and an alignment of the upward looking camera with the substrate camera. That clearly compensates the drift that occurred during run 1. It is noteworthy to highlight that this procedure can amend the drift. During the second run there is still some transient thermal drift of approximately 1 µm present. Run 3 shows about 0.5 µm drift, and run 4 is relatively thermally stable.

On the Y axis much less drift is observed with the first run showing about 1 µm and the second run is relatively steady.

The overall drift observed in this experiment was about ±5.5 µm. This drift can either be compensated by a cyclical substrate adjust and camera realignment or a short warm-up cycle prior to production commencement.

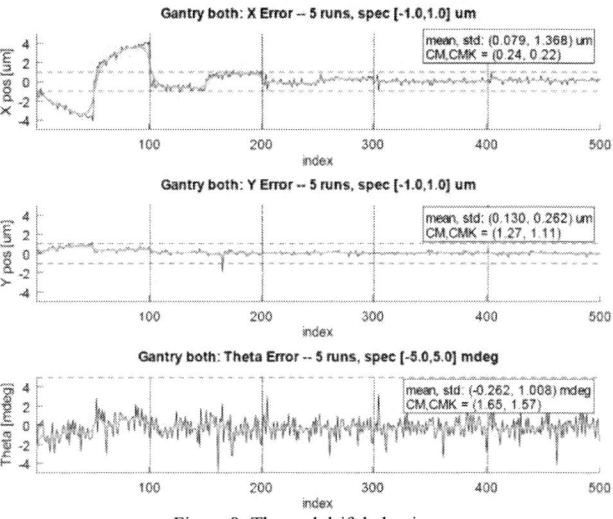

Figure 8. Thermal drift behavior.

As a second test for the thermal settling behavior a machine in cold state was used and performed 1000 consecutive 3-point adjusts on the substrate spanning 320 mm in x and 288 mm in y. Fig. 9 shows the drift of the measured substrate center according to the adjust. Thermal duty is quite different to the first test but still the range of the drift matches Fig. 8 and the thermal equilibrium is reached quickly.

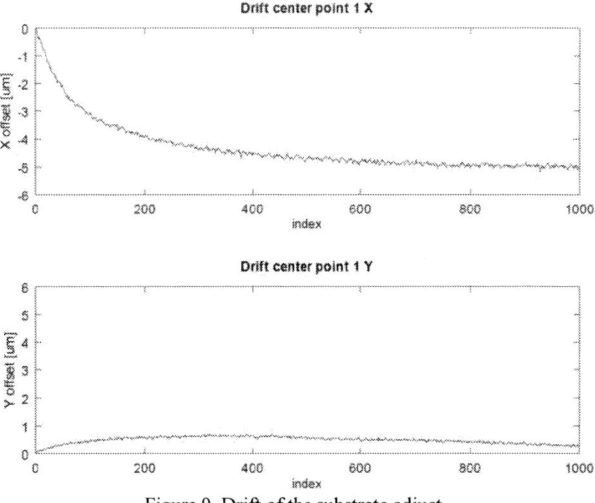

Figure 9. Drift of the substrate adjust.

VIII. GLOBAL ACCURACY

The second accuracy metric for which we present empirical data is "global" accuracy. It is tailored to assess machine performance for processes where only global fiducials on the substrate are available but no local fiducials at the bonding positions. This is for example the case for processes where the machine will build an artificial "wafer" by placing the components on a regular grid (matrix). The global alignment marks define the position where to populate the matrix, but the accuracy within the bonded matrix is completely determined by the machine's capability to find the bonding positions on the substrate in a reproducible way without the positional information of local fiducials.

Our test procedure uses a glass plate with fiducials on a regularly spaced grid, though this is not a hard requirement. To allow for inaccuracies on this test plate or the way the plate is handled and placed in the machine the following procedure is used:

1. Insert glass plate at roughly 0 degree
2. Measure all fiducials $m_0(i,j)$
3. Insert glass plate at roughly 180 degree
4. Measure all fiducials $m_{180}(i,j)$

It is strongly recommended to use a glass plate that was not used as reference for the calibration of the machine. Otherwise the 0 degree measurement is heavily biased and gives the result of a pure repeatability test. In other words the machine will find all fiducials on the exact same position where the calibration procedure measured them and the result will be overly optimistic very close to zero since the test run will confirm the "ground truth" of the calibration. On the other hand during the 180 degree measurement run any error on the glass plate would be registered roughly twice. This can be explained by the following example: assume there is just one gross outlier on an otherwise perfect glass plate that is used for the calibration and the test run. This outlier would clearly cause a bad spot in the calibration data. When doing a measurement in this "bad spot" of the calibration with the rotated glass plate the test run will register the deviation of the calibration from the now correct position of the fiducial. Additionally the bad fiducial in the rotated position will now show up as an error since the calibration data on this spot is correct. Therefore this bias will influence the proposed metric and the result will be worse than the true performance and thereby giving an upper bound.

Let $g(i,j)$ be the position of the fiducial with row and column index (i, j) on the glass plate with respect to a Cartesian coordinate system on the glass plate. The position of the fiducial (i, j) inside the machine is then given by

$$p_0(i,j) = R(\alpha) \cdot g(i,j) + t_0, \qquad (2)$$

where $R(\alpha)$ denotes a rotation matrix with angle α and t_0 is a translation vector. Similarly for the rotated glass plate with angle $(180 + \beta)$ degree we can write

$$p_{180}(i,j) = R(180+\beta) \cdot g(i,j) + t_{180}. \qquad (3)$$

We model the overall error of the machine's measurement by two additional terms: let $e_{MC}(x, y)$ be the systematic error of the machine at the gantry position (x, y)

including the error introduced by an imperfect calibration (mapping) at the position (x, y). The second error term is denoted by e_S that models the remaining stochastic error and includes at least the measurement error of the vision system and the placement error of the placement process. In this model e_S is by virtue of the central limit theorem a normally distributed random variable. Using this notation we can define the measurement results at both target angles by

$$m_0(i,j) = p_0(i,j) + e_{MC}(p_0(i,j)) + e_S(i,j) \qquad (4)$$

and once again similarly

$$m_{180}(i,j) = p_{180}(i,j) + e_{MC}(p_{180}(i,j)) + e_S(i,j) \qquad (5)$$

The measurement inaccuracy of the vision system is known to have zero mean and a standard deviation of about 30 nm which is therefore a lower bound for the standard deviation of the stochastic error [4].

Both formulas can now be used to derive an estimator for the machine error. For the moment let us assume an ideal machine, i.e. $e_{MC}(x, y) = 0$. Then the measurements m_0 at 0 degree and m_{180} at 180 degree can be mathematically fitted using for example the Levenberg-Marquardt algorithm (LMA) to solve a least squares problem with three degrees of freedom (rotation and translation in x and y) [5]. In other words we mathematically rotate the glass plate back to the angle and position of the 0 degree measurement and compare results.

Let this isometric transformation computed by the LMA be denoted by $T(\gamma, tx, ty)$, the least squares result is a maximum likelihood estimator for γ, tx, ty in this model. In our assumption the residuals defined by

$$r = m_0 - T(\gamma, tx, ty) \cdot m_{180} \qquad (6)$$

will have a distribution with zero mean and standard deviation of about $\sqrt{2}$ times the standard deviation of e_S. The key observation is that the exact positions of the fiducials on the glass plate $g(\cdot, \cdot)$ included in the terms p_0 and p_{180} will cancel out (in a least squares sense). This makes the method independent of the accuracy of the test glass plate. Fig. 10 shows a quiver plot of the residuals in (6).

This chain of thought is the motivation to use the residuals as estimator for the machine error. Any deviation from the expected stochastic error is necessarily an artifact of the machine and its calibration.

Using this procedure we observed a standard deviation of 395 nm in x direction and 309 nm in y direction.

The test glass plate covered about half of the working area in x and y, but the test gave similar results in all four quadrants of the working area. Therefore we can derive an estimate of the standard deviation of the global measurement error

$$\sigma_{GM} = 400 \text{ nm.} \qquad (7)$$

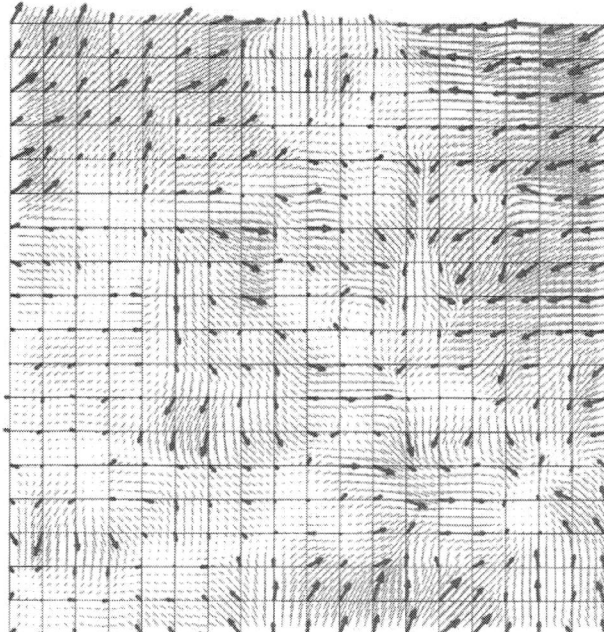

Figure 10. Quiver plot of the residuals.

It remains to link this estimate of the measurement error to the achievable placement precision. Since we have established the capability of the machine to move with the substrate camera to a perfect grid with respect to the global alignment marks, we need to consider the additional error introduced by the bonding process. The previous section on local accuracy gives an upper error bound that includes the measurement error of the local fiducial with the substrate camera and the complete placement process.

For global accuracy we could "virtually" move with the substrate camera to the expected local fiducial position w.r.t. the global fiducials, do a "virtual" measurement and end up with a local accuracy process. We consider these two steps to be independent and therefore we can add the variation of the total global error by

$$\sigma_L{}^2 + \sigma_{GM}{}^2 = \sigma_T{}^2. \tag{8}$$

So the standard deviation of the total global placement error in dual gantry mode can be computed using (1), (7), and (8)

$$260^2 + 400^2 = \sigma_T{}^2, \tag{9}$$

which gives σ_T = 477 nm. This conveniently fulfills a ±2 µm @ 3σ requirement.

IX. ROADMAP

Though the current prototype already offers high UPH and excellent accuracy, the design of the machine needs the flexibility to improve in both areas. To ensure the fitness for future challenges we present a short roadmap outlook.

One approach to achieve higher productivity is to increase the maximum speed of the y axis to 3 m/s or more

and to ramp the acceleration to 80 m/s². Using the reference product mentioned before the time for the long y stroke from the picking position to the bonding position is reflected one-to-one in the cycle time. Making this movement faster will enable 7000 UPH and more for the reference product.

But the largest UPH improvement clearly is moving to parallel die transfer [4]. Replacing the bond head, the flipper, and the eject system with a multi-nozzle system with $N > 1$ nozzles scales nicely since it reduces the number of necessary long y movements, that are in the critical path of productivity. Instead of N long movements to place N dies only one long movement and $N-1$ short movements between neighboring bond positions are required. After placing the die on the first nozzle only $N-1$ small movements from bonding position to bonding position are required. This paves the path to a throughput of 20k and more.

On the accuracy side theta is gaining more and more interest for Heterogeneous Integration (HI). Improving theta is necessary to achieve high x/y accuracy for large dies even in the corners of the die. This goal can be tackled by improving the rotational drive train with higher resolution encoders and an optimized mechanical design and extending the calibration procedures to take theta into account.

For applications where local accuracy in the nanometer range is paramount the machine concept can be extended to a direct metrology approach described in [4, 6]. This system is in prototype stage and targeting ±200 nm @ 3σ.

X. CONCLUSION

The new platform was developed with flexibility and scalability in terms of productivity and accuracy as primary targets. UPH can already reach up to 6400 with a clear roadmap to go even further.

Local accuracy was demonstrated to be better than ±1 µm @ 3σ, and global accuracy is at ±2 µm @ 3σ even at high productivity settings.

ACKNOWLEDGMENTS

The authors sincerely thank Wolfgang Vögele and Volker Groll for their hard work on the machine, numerous discussions, creativity, data analysis and support.

REFERENCES

[1] Santosh Kumar: "Status of Panel Level Packaging 2018", Yole Développement, April 25, 2018

[2] Richard (KwangWook) Bae: Samsung Electromechanics (SEMCO), Interviewed by Santosh Kumar (Yole Korea), 2018

[3] Julian Ho, Jessie Shen, DIGITIMES, 26 September 2018

[4] H. Pristauz, A. Mayr, S. Behler, "Disruptive Developments for Advanced Die Attach to Tackle the Challenges of Heterogeneous Integration", at IEEE 68th Electronic Components and Technology Conference 2018, San Diego

[5] R. Fletcher, "Practical methods of optimization: Unconstrained optimization", vol. 1, Wiley, Chichester, 1980.

[6] "Vorrichtung und Verfahren zum Montieren von Bauelementen auf einem Substrat" ("Device and method for mounting components on a substrate"), patent CH 713 732 A2, 2018.

Advances in high speed plating for vertical glass panel fine-line plating

Christian Dunkel, Herbert Ötzlinger, Tetsuya Onishi, and Raoul Schroeder
Semsysco GmbH
Karolingerstr. 7c
Salzburg, Austria
raoul.schroeder@semsysco.com

Abstract— The future of computing is well known to involve large amounts of data being collected, computed, and transferred wirelessly. Under the hood, this requires the CPU, memory, sensors, and transfer units to be packaged very closely to each other. While FOWLP (fan-out wafer level packaging) and other WLPs are the standard for the mobile, server, and autonomous automotive applications, significant cost reduction can be achieved when performing PLP (panel level packaging) due to the economy of scales.

The main hurdle has been to provide a path from typical line/space dimensions of 20/20µm first to 10/10µm, then 5/5µm and finally to 2/2µm in size. Especially for wet processing, there are many hurdles to be faced, especially when it comes to plating speed and uniformity. One of the obstacles is that the panel typically has to be handled vertically, and cannot be rotated to achieve optimal uniformity, which can drive the costs up, as more time needs to be spent to put down overburdened films that have to be polished and etched.

To overcome this issue for electrochemical deposition of metals, we report the successful scaling of high speed, extremely uniform plating technology from horizontal wafer plating to vertical panel plating. The patented high speed technology plating setup consists of a plate that normalizes the electric field very efficiently over the entirety of the plate to within 5%, whereas the electrolyte injection system within the same plate allows for a rapid exchange of depleted and fresh plating electrolyte even at very high plating speeds.

In fact, the replenishment is so close to the substrate surface, that we can effectively minimize the surface boundary layer to below the thickness where the exchange of the electrolyte occurs effortlessly. With this setup, there is an additional advantage that is often overlooked for high speed plating solutions: the role of the organics and chloride ions. In traditional plating systems, the accelerator especially requires the presence of chloride ions to function properly. At high plating speeds, however, we also require large electrical fields to drive the currents, and this electric field impedes the chloride more and more from reaching the equally negatively charge cathode. Our system overcomes this by direct injection close to the surface and therefore improves the plating speed of all electrolytes.

As the system scales very easily, we can manufacture systems with substrate sizes of 510x515mm², 600x610mm², 620x750mm² (GEN 3.5) with relative ease of the underlying engineering work, but even systems with sizes of over 1m², e.g. GEN 5 or GEN 6 should be achievable. With blanket uniformities around two percent, and patterned substrate uniformities at five to seven percent in range at up to 20 ASD, the question of how to economically metallize panel substrates for next generation PLP has been solved, finally allowing wide-spread adoption of this technology.

Panel-level packaging, fine line plating, electroplating, high density interconnects, glass panels

I.

II.   Introduction

During the rise of the CMOS semiconducting industry, aluminum (Al) was the metal of choice for contacting transistors and logic elements. This metal was always deposited as a physical vapor deposition (PVD), either by evaporation or sputtering. PVD is an easily controllable process but it has the disadvantage of needing a vacuum, showing relatively slow deposition speeds. The biggest hindrance in the late 90s was that aluminum had a relatively low conductivity and therefore the high currents induced by higher and higher clock speeds became a heating issue.

In 1997, IBM published results for the electrochemical formation of copper on thin liner materials [1], namely a barrier

layer (Ti) and a thin copper (Cu) seed layer. The barrier layer was the same that was already used before depositing Al, but the newly added thin Cu seed layer was necessary for electroplating, since any uneven addition of Cu would dramatically change the current density locally, giving uneven results. The barrier layer was even more necessary than before because Cu easily dissolves into the silicon (Si). [2] Even with the added steps, the benefits of this new process was so massive due to the superior conductivity and possibility to quickly deposit thick layers without material going to waste on the shadow masks or photoresist. The physical improvement was one of the key factors in holding up Moore's law in the 2000s, both for CPUs, GPUs, and memory.

Beyond the world of CMOS manufacturing, Cu electroplating was introduced in general into the field of electronics manufacturing. It was back in the early nineties that IBM also tried to use Cu as a contact metal in the early nineties, but there it faced some very hard manufacturing difficulties that proved difficult to overcome. [3] The PCB industry, however, was quite successful in adopting this technology, as they were driven to a level of miniaturization and higher integration new to that field back then – the push today for integration on panels is the next logical culmination of that. The Cu electrolyte baths and their additives for uniform plating were quite similar back then and still are today. [4] Even better, through a study of the morphology of differently deposited metallic layers, it showed the possibility of developing short loop tests as a stand-in replacement for much more complicated full device electrical tests. [5] Finally, the unprecedented boost of processing power, enabled in part by the Cu electroplating revolution in CPU and memory, especially the possibility to reduce dies in size while increasing their frequencies, meant that the packaging technology thus far was starting to fall behind and no longer met all the necessities for the new generations of CMOS silicon:

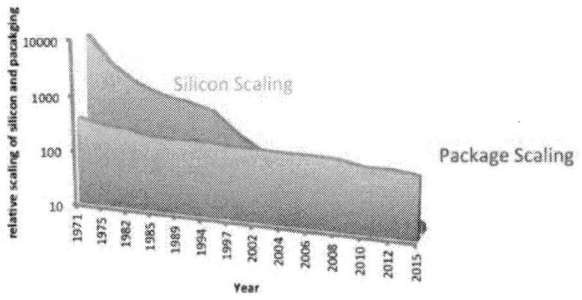

Figure 1. The dramatic difference of critical dimensions for silicon technologies compared to packaging scaling. [6]

This led to the development of the so-called wafer level packaging (WLP), which allowed the redistribution of contact pins from the high density layout of a small current generation die to the much less dense PCB packages of the time. This allowed very large I/O counts and communication speeds, while allowing the die shrinkage that enables yield improvements and cost cutting at the same time. [7-8] – Fan-Out Wafer Level Packaging was born (FOWLP). Thinking further, 2.5D and 3D integration, using strategies similar to PCBs with embedding, albeit at a smaller technology node, could be realized. To achieve thick layers, full filling of holes and equal thickness of lines and pads at very different sizes, new plating technologies with advanced prerequisites for both the hardware and the electrolytes as well as additives were developed. This culminated for instance in a CMOS compatible formation of through silicon VIAs. [9]

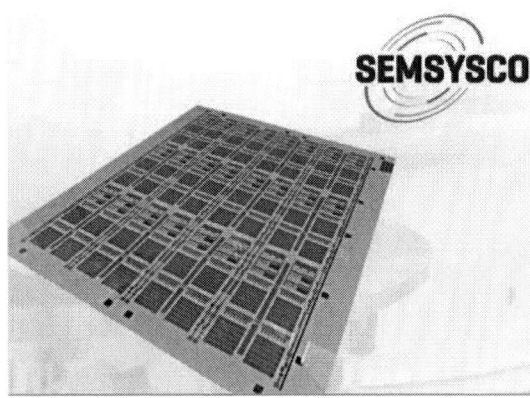

FIGURE 2. LARGE TEST VEHICLE OF RDL PACKAGES AVAILABLE IN PANEL LEVEL PACKAGING

As we see the market penetration of more and more WLPs used in computing, especially mobile computing, the search to reduce cost in packaging begins. Since WLP only consists of metal lines on silicon that move the interconnects to new, further out areas, using silicon, especially the large areas necessary for typical packages, can become a major cost factor. As several players, namely OSATs, old LCD factories and PCB manufacturers, are aiming to enter the panel level packaging market, electroplating with similar uniformities and speeds as on wafers become a crucial part, along with the uniform exposure and development of the photoresist that guides the RDL structures. Mastering this to get down to 5/5µm and 2/2µm line/spaces for glass or mould compound panels will enable a large part of WLP to transition to PLP at a lower cost. Transitioning to 450mm wafers would also reduce cost, but the 450mm tool install base in the world is very small. [10] This will narrow the gap to 0.8/0.8µm lines/spaces commercially used for wafer level packaging. Applying this kind of technology node to glass or compound panels is typically called "Fine Line Plating" as the key technology driving lateral integration to lower cost points and smaller scales, and keeping Moore's law alive at least for system components. By keeping plating speeds higher than 5 ASD, going up to 20 ASD for smaller open areas, while keeping the uniformity range generally below 10%, preferably going down to 5% by using clever design rules, enables the transition to PLP with the best possible costing economy, avoiding, e.g., planarization afterwards as much as possible.

III. EXPERIMENTAL

The plating equipment used for these experiments are part of the standard Semsysco plating platforms, Triton-type tools for single wafer and VHS-P –type tools for single or dual panel plating. Both wafers and panels, including photoresist patterning as well as PHV deposition of the barrier and seed layer, were provided by the Fraunhofer Institute IZM in Berlin and Dresden. Plating baths were off the shelf ready to use mix (VMS) or in-house mixed CuSO4 solution and H2SO4 available from the major vendors, who provided us directly with the organics. Plating baths have been kept within tight spec while plating using CVS measurements, the temperature typically at +/- 1°C, bath agitation at flow rates +/- 2% and organics within the supplier recommended values. Analysis of the panels was performed with stereoscopic microscopes for plating thickness and roughness

978-1-7281-1500-9/19 $31.00 © 2019 IEEE

analysis, as well as FIB cuts to analyze for voids and crystalline structures.

For the experiment, both the horizontal (wafer) as well as the vertical chambers (panel) were equipped with the so-called high speed plate. This plate serves to overcome many of the problems with electrolytic metal deposition at high currents and is the center piece of hardware allowing the deposition speeds necessary for cost-efficient packaging plating at the uniformities required to avoid costly follow-up processing downstream.

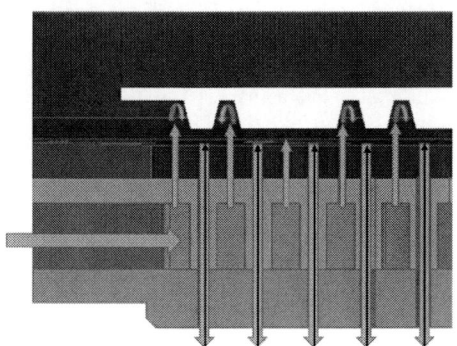

FIGURE 4. THE FUNCTIONALITY OF THE HIGH PRESSURE PLATE. UNIFORM LIQUID FLOW (LIGHT GRAY ARROWS UP), UNIFORM ELECTRIC FIELD (THIN BLACK ARROWS UP) AND BREAKING OF THE BOUNDARY LAYER (180° CURVED ARROWS)

Our panel level plating setup is fundamentally the same as the wafer type setup, which is important for the ease of transition of the users of these chambers. Essentially, the majority of process parameters gathered on wafers can be transferred to panels. This means that the technology can be demoed on wafers when customers do not have the whole panel line set up in its entirety yet. There are only two differences, that do not affect the overall plating process:

1. Large panels are held vertically in a chamber with two anodes, allowing for double sided plating or dual plating of single side substrates
2. Due to the impossibility of rotating a panel like a wafer, the panel runs through a 2D agitation pattern instead to achieve uniform plating uniformities.

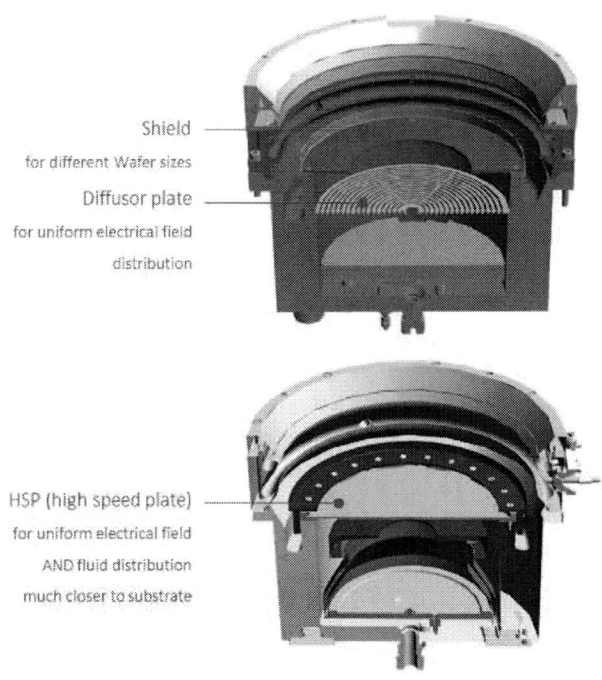

FIGURE 3. SINGLE WAFER FOUNTAIN PLATING CHAMBER COMPARED TO A HIGH-SPEED PLATING CHAMBER BELOW.

As visible in Fig. 3 the main difference between a normal fountain plating chamber is the presence of the high speed plate close to the wafer surface. It serves three purposes:

1. Injection of fresh chemical not at the anode, but close to the wafer surface

2. Breaking the electrolytic boundary layer.

3. Uniform liquid flow to the surface

4. Uniform electrical field close to the surface

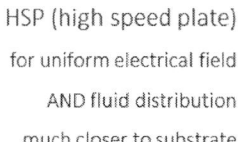

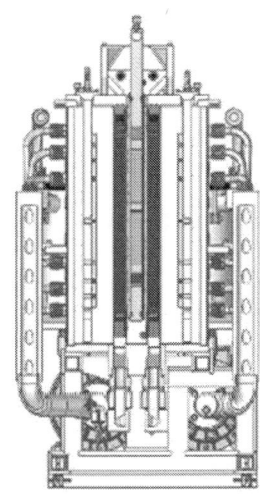

FIGURE 5. SINGLE OR DUAL SUBSTRATE VERTICAL HIGH-SPEED PLATING CHAMBER FOR RECTANGULAR OR SQUARE PANELS.

IV. RESULTS AND DISCUSSION

One of the most crucial measurements when moving towards high speed plating is to ensure that it does not affect the crystalline structure and size of the polycrystalline Cu. Electroplating is well known for its advantage in achieving large crystals upon deposition, which yields improved mechanical and electrical properties. High deposition speeds often lead to smaller crystals, so an important first step was to verify that the injection of fresh electrolyte at the surface allows the continuous growth of the crystals because deposition is not starved of Cu in the boundary layer.

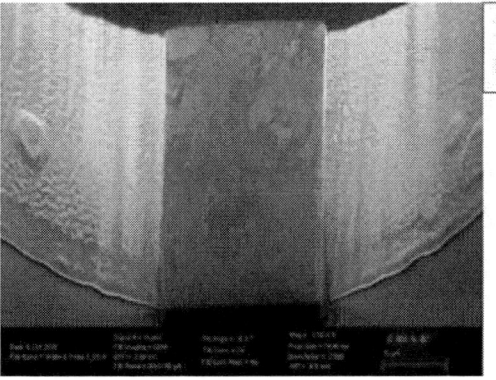

FIGURE 6A. FIB CUT OF 2MICRONS/MIN CU PLATED BUMP

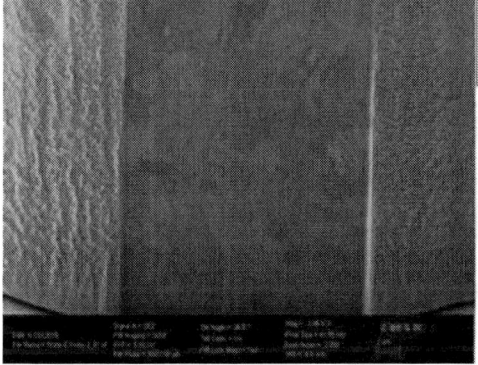

FIGURE 6B. FIB CUT OF 6MICRONS/MIN CU PLATED BUMP

In Fig. 6 a-b, we see that increasing the plating speeds does not affect crystalline structure as the deposition never reaches the Cu starved regime.

One of the critical parameters in increasing plating speeds, apart from having to control the uniformity better, is the performance of the electrolyte bath itself, specifically the organics to fill all features properly, and other components that they need to function, especially chloride Cl⁻. Normally, Cl⁻ is refreshed through the normal recirculation of the bath. [11] At higher currents, the voltage between the anode and the cathode, however, also becomes higher and this potential inhibits the free moving of chloride to the cathode, as it has the same charge. Especially with the flow in a fountain plater originating at the anode, this can become a severe problem. In our layout (cf. Fig 3 and 5), however, Cl⁻ is directly injected to the substrate surface at the cathode, which ensures that the accelerator works flawlessly even at elevated currents up to, but not limited to 20 ASD

(4.4μm/min). No new chemistry is required, at most, the organics have to be fine-tuned to the process. This allows the extension of current process results with known chemistries to higher and higher plating speed just by the introduction of a different hardware (Fig 7).

FIGURE 7. FLAT PROFILE OF A BUMP PLATED AT 20 ASD (4.4 MICRON/MIN) WITH REGULAR CHEMISTRY)

The final critical step to achieve all the process needs to the fine line packaging plating processes is to allow pads and lines with very different dimensions to be co-deposited with same or very similar heights. The plating electrolyte manufacturers have spent many years in tailoring three component system that allow the deposition of such features next to each other with limited impacts to the uniformity; they achieve this by improving accelerator, suppressor and leveler as well as how they interact with each other and the substrate.

These systems are often fine-tuned for lower deposition speeds, due to issues like the aforementioned Cl- depletion at the cathode for high plating currents. In our final experiment, we therefore looked into how far we could push this system in our plating setup. We indeed found that through direct injection of the plating electrolyte near the substrate surface, high speed plating was possible without compromising co-planar, but differently sized structures (Fig 7.), all the while achieving excellent uniformities of less than 4% in range on plated features. (Fig. 7b)

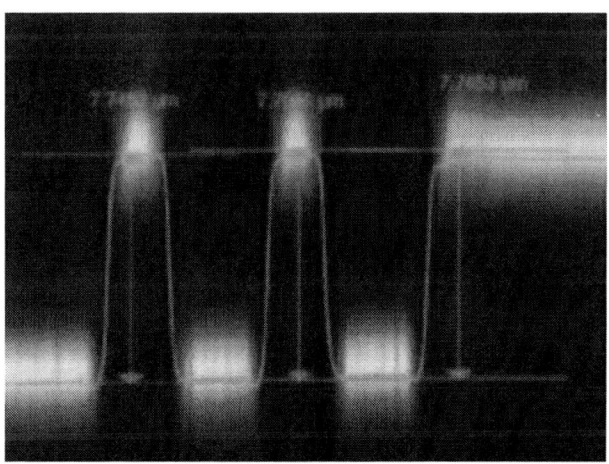

7,80	7,00	8,00	7,90	8,00	7,60	7,80	7,80	7,80	7,60
7,90	7,60	7,60	7,80	7,70	7,70	7,60	7,90	7,50	7,60
7,80	7,80	7,80	7,60	7,70	7,60	7,70	7,70	7,50	7,60
7,90	7,80	7,60	7,70	7,60	7,70	7,60	7,70	7,70	7,50
8,00	7,90	7,60	7,70	7,60	7,60	7,50	7,60	7,50	7,60
8,00	7,80	7,70	7,60	7,70	7,60	7,80	7,60	7,70	7,50
8,00	8,00	7,70	7,80	7,70	7,60	7,50	7,70	7,50	7,60
8,00	8,00	7,60	7,70	7,80	7,70	7,60	7,60	7,60	7,80
7,70	7,80	7,80	7,70	7,60	7,70	7,80	7,60	7,80	7,90
7,90	7,50	7,90	7,90	7,70	7,80	7,60	7,90	7,70	7,60

Uniformity of 3.2 % range/2*average

FIGURE 7. (A) FINE LINES (8 MICRONS) PLATED AT HIGH SPEED NEXT TO 120MICRONS PAD. (B) WITH A UNIFORMITY IN RANGE OF LESS THAN 6.8%.

V. CONCLUSIONS

With the need for cost optimization, we see a push for fine line panel level packaging throughout the industry. As part of providing the solution for the need to have a low-cost, high speed, and high uniformity plating process, we adapted our patented high speed plating system from our single-wafer to a vertical panel system. The unique hardware setup of this system allows a largely identical plating setup, but vertically handled with a 2d agitation system instead of substrate rotation. These two differences are masked by the overall process, allowing to move directly from wafer to panel plating without major adjustments of the chemistry or the process parameters. Due to the direct electrolyte injection at the substrate surface, high deposition speeds can be achieved without alteration to the electrolyte itself; at most, fine-tuning of the organic system is needed. With plating speeds of more than 5 ASD for large open area RDL structures and up to 20 and more ASD for bumps, while keeping uniformities below 10% and less than 5% while following some design rules, panel level plating is now a key enabler to driving advanced packages onto larger substrates, enabling further cost reductions.

REFERENCES

[1] D. Edelstein, J. Heidenreich, R. Goldblatt, W. Cote, C. Uzoh, N. Lustig, P. Roper, T. McDevitt, W. Motsiff, A. Simon, J. Dukovic, R. Wachnik, H. Rathore, R. Schulz, L. Su, S. Luce, and J. Slattery, *"Full Copper Wiring in a Sub-0.25mm CMOS ULSI Technology,"* Technical Digest, IEEE International Electron Devices Meeting, 1997, p. 773.

[2] P. C. Andricacos, C. Uzoh, J. O. Dukovic, J. Horkans, and H. Deligianni, *"Damascene Copper Electroplating for Chip Interconnections,"* IBM J. Res. & Dev. 42, 1998, 567–574.

[3] B. Luther, J. F. White, C. Uzoh, T. Cacouris, J. Hummel, W. Guthrie, N. Lustig, S. Greco, N. Greco, S. Zuhoski, P. Agnello, E. Colgan, S. Mathad, L. Saraf, E. J. Weitzman, C. K. Hu, F. Kaufman, M. Jaso, L. P. Buchwalter, S. Reynolds, C. Smart, D. Edelstein, E. Baran, S. Cohen, C. M. Knoedler, J. Malinowski, J. Horkans, H. Deligianni, J. Harper, P. C. Andricacos, J. Paraszczak, D. J. Pearson, and M. Small, Proceedings of the 10th International IEEE VLSI Multilevel Interconnection Conference, 1993, p. 15.

[4] Nichols RJ, Bach CE, Meyer H: *The effect of three organic additives on the structure and growth of electrodeposited copper: An in-situ scanning probe microscopy study",.* Berichte der Bunsen-Gesellschaft, 1993

[5] Nichols RJ, Beckmann W, Meyer H, Batina N, Kolb DM: *An in situ scanning tunneling microscopy study of bulk copper deposition and the influence of an organic additive."* J Electroanut Chem 330, 1992, pp. 381-394.

[6] S. S. Iyer, *"Heterogeneous Integration for Performance and Scaling",* IEEE Transactions on CPMT, 2006, pp. 975-984

[7] H. Hedler, T. Meyer, B. Vasquez *"Transfer Wafer Level Packaging",* US Patent US6727576B2, 2001

[8] M. Brunnbauer, T. Meyer, G. Ofner, K. Mueller and R. Hagen, *"Embedded Wafer Level Ball Grid Array (eWLB),"* 2008 33rd IEEE/CPMT International Electronics Manufacturing Technology Conference (IEMT), Penang, 2008, pp. 1-6.

[9] P. S. Andry, C. Tsang, E. Sprogis, C. Patel, S.L. Wright, B.C. Webb, L.P. Buchwalter, D. Manzer, R. Horton, R. Polastre, J. Knickerbocker, *"A CMOS-compatible process for fabricating electrical through-vias in silicon,"* 56th Electronic Components and Technology Conference 2006, San Diego, CA, 2006, pp. 7

[10] T. Braun, K.-F. Becker, S. Raatz, V. Bader, J. Bauer, R. Aschenbrenner, S. Voges, T. Thomas, R. Kahle, K.-D. Lang, *"From fan-out wafer to fan-out panel level packaging,"* 2015 European Conference on Circuit Theory and Design (ECCTD), Trondheim, 2015, pp. 1-4.

[11] M. Palazzola, N. Dambrowsky and S. Kenny, *"Via Filling: Challenges for the Chemistry in the Plating Process",* IPC APEX EXPO 2013, pp 32

Study of the properties of AlN PMUT used as a wireless power receiver

Dan Gong[1], Shenglin Ma[1*]
[1]Department of Mechanical & Electrical Engineering,
Xiamen University
Xiamen, Fujian, China
E-mail: mashenglin@xmu.edu.cn

Yihsiang Chiu[2], Hungping Lee[3], Yufeng Jin[2]
[2]Peking University Shenzhen Graduate School,
Shenzhen, China
[3]J-Metrics Technology, Shenzhen
E-mail: yfjin@pku.edu.cn

Abstract—This paper mainly designed an aluminum nitride based piezoelectric micro-machined ultrasonic transducer (AlN PMUT) as a wireless power receiver. AlN PMUT is a stacked layers structure of Si/SiO₂/Mo/AlN/Al about 4µm/1µm/0.1µm/2µm/1µm in thickness from bottom to up suspended by a cavity measuring in 0.496mm×0.506mm. Fabrication and characterization are conducted as a wireless power receiver. According to the wireless power transmission experiment, AlN PMUT can generate a peak voltage about 38.6mV when it receives an input sonic wave with a sound pressure of 2×10^4Pa.

Keywords-AlN PMUT; wireless power receiver; characterization

I. INTRODUCTION

AlN PMUT can be used as wireless power receiver and many applications will benefit from it, such as counterfeit banknote detection, micro-robots [1][2], IoT device communication [3], wearable devices [4], and activation of biomedical implants [5]. In 2017, Emad Mehdizadeh and Gianluca Piazza proposed a chip-level wireless power transmission with AlN PMUT and proved that thicker aluminum nitride was helpful for improving wireless power transmission efficiency [6]. In 2018, Xinxin Liu et al contrasted properties of PMUTs with square structure and circular shape and found used as a receiver [7]. Ibrahim, Ahmed demonstrated a power transmission efficiency of 0.65% is achieved with PMUT with 20mm³ in the size as a receiver, which is distanced about 3mm from the transmitter (TX) in the oil environment [8]. In this paper, a square AlN PMUT with 400µm×400µm in size is proposed, and its properties as a wireless power receiver is tested.

II. STRUCTURAL DESIGN

Fig.1 is the schematic diagram of the designed AlN PMUT, which is a stacked layers structure of Si/SiO₂/Mo/AlN/Al from bottom to up suspended by a cavity. When excited by incident acoustic waves, it vibrates, leading to stress in AlN film and further converted electrical charges due to the piezoelectric effect. It can be used to store electrical energy, which is significantly enhanced if it stimulated at its resonance frequency. Specifications of designed PMUT are summarized in Tab.1 and based on this device, a 3×3 PMUTs array is constructed with a spacing of 220µm .

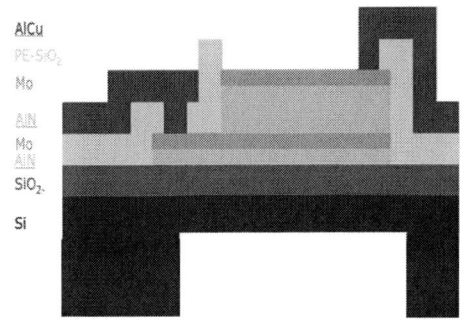

Figure 1. The schematic diagram of AlN PMUT.

TABLE I. PARAMETERS OF ALN PMUT

Parameters	Value
PMUT size	400µm×400µm
Suspended Si film thickness	2-5µm
Backside cavity size	0.496mm×0.506mm
AlN layer thickness	2µm
SiO₂ layer thickness	1µm
Mo layer thickness	100nm
Al/Cu layer thickness	1µm

In order to obtain the resonant frequency and impedance of PMUTs, simulation was performed using COMSOL simulation software. Fig. 2 shows the first order vibrating mode and the relative frequencies of PMUT at different locations in the 3×3 PMUTs array. Fig. 3 shows the simulated impedance curve of PMUTs at different locations in the 3×3 PMUTs array. A1, A2 and A3 in Fig. 2 and Fig. 3 are the three positions of the edge of the PMUTs array, and A5 is the middle position of the PMUTs array. The simulation results are summarized in Tab.2 and it has a resonant frequency around 500kHz. In order to better evaluate the property as a wireless power receiver, the power transmission link is simulated with COMSOL software. A sinusoidal sonic wave is input, which the resonant frequency is based on each PMUT and the sound pressure amplitude is variable. Fig. 4 shows the excited voltage wave when the sonic wave reach at PMUT is about 2×10^4Pa, it can be found a peak-to-peak value of 37.5mV.

978-1-7281-1500-9/19 $31.00 © 2019 IEEE

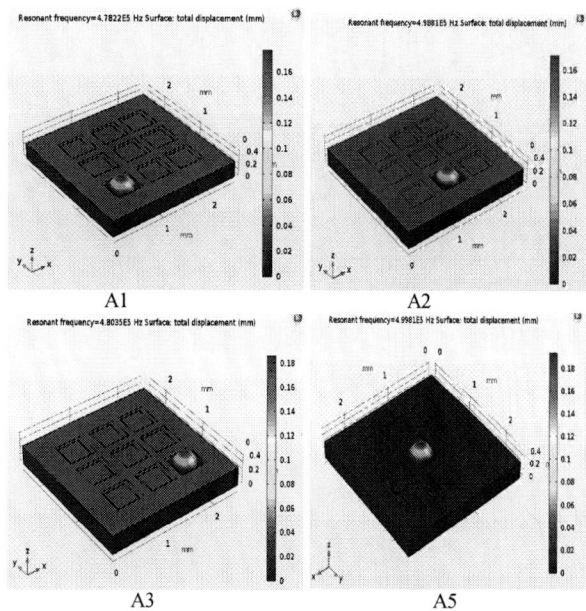

Figure 2. The first order vibrating mode and the relative frequencies of PMUT at different locations in the 3×3 array.

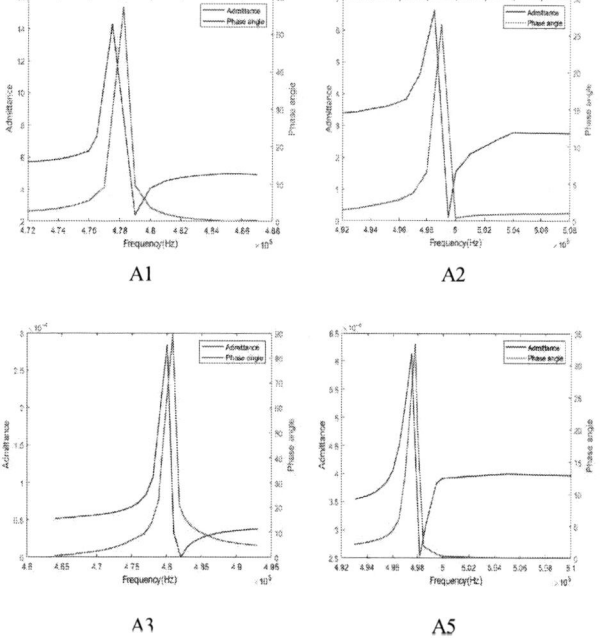

Figure 3. Simulated impedance curve of PMUT at different locations in the 3×3 array.

TABLE II. PMUT RESONANCE FREQUENCY

Parameters	Value			
	A1	*A2*	*A3*	*A5*
Resonant frequency (kHz)	478.22	498.80	480.35	498.81
Electromechanical coupling coefficient (%)	0.833	0.878	1.32	0.760

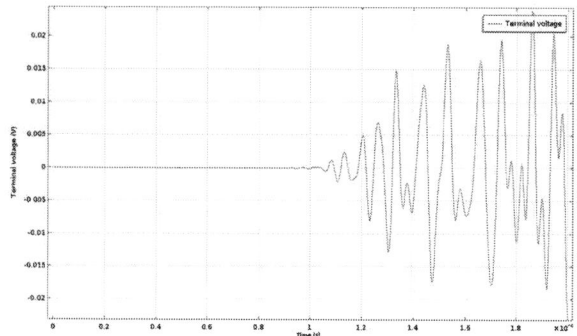

Figure 4. The excited voltage wave when it receives sonic wave input.

III. FABRICATION

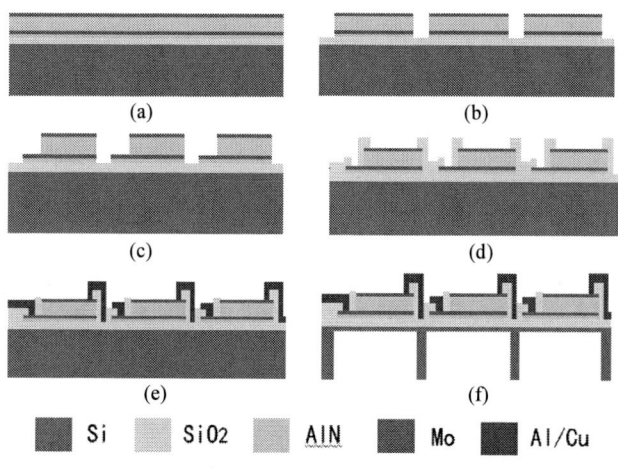

Figure 5. Process for AlN PMUT.

The AlN PMUT is fabricated on a Si substrate. Firstly, 1μm of silicon dioxide (SiO₂) is deposited on the front side, which is used as an insulating layer and also as a stress buffering layer due to CTE mismatch in the stacked layers. Secondly, in the same vacuum environment,100nm AlN, 100nm Mo, 2μm AlN and 100nm Mo are deposited on silicon dioxide in sequence as showed in Fig. 5(a). Thirdly, ICP process is used to etch AlN/Mo/AlN/Mo to define the square AlN PMUT as showed in Fig. 5(b). And then 100nm Mo and 2μm AlN are etched to expose the bottom Mo layer as showed in Fig. 5(c). Fourthly, 1μm SiO₂ is deposited on to form an insulating layer and contact vias are opened for the upper and lower electrodes as showed in Fig.5(d). A layer of Al/Cu is deposited and patterned by wet etching to form electrode as showed in Fig. 5(e). Finally, deep silicon ions etching (DRIE) is used to form cavities behind every PMUT device as showed in Fig. 5(f).

Based on the process, AlN PMUT is fabricated. Fig. 6 shows an optical microscope image of a single PMUT. Fig. 7 shows SEM photo of cross-section of a 3×3 PMUTs array and the remaining Si film is about 2μm in thickness. Fig.8(a) shows the optical photo of 3×3 PMUTs array captured on the backside, and Fig.8(b) shows the optical photo of packaged 3×3 PMUTs array.

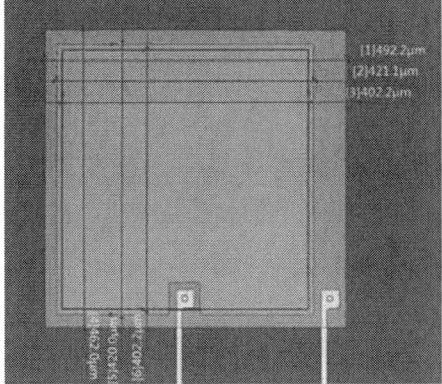

Figure 6.   Optical photo of a single PMUT.

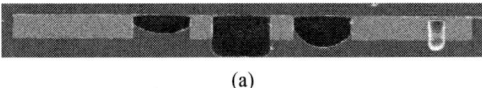

(a)

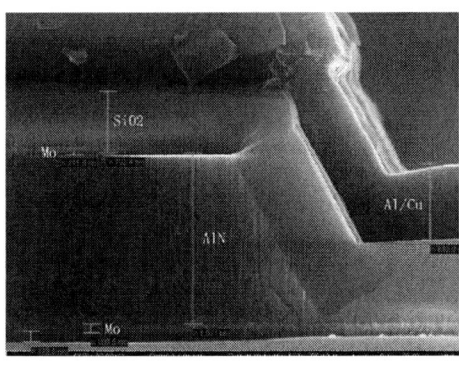

(b)

Figure 7.   SEM photo of cross-section of a 3×3 PMUTs array:(a) global view; (b) close-up of stacked layers of Si/SiO₂/Mo/AlN/Al.

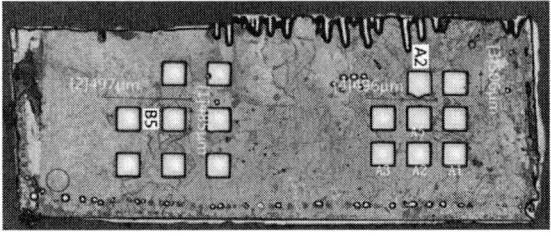

(a)

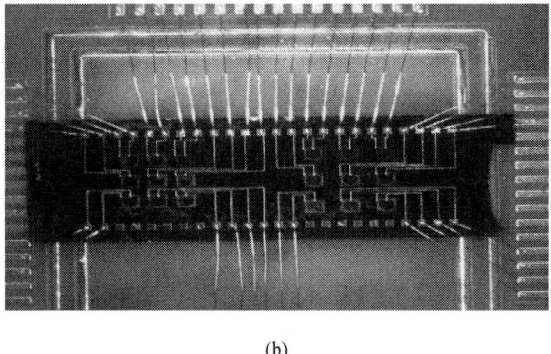

(b)

Figure 8.   Optical photo of a PMUT 3×3 PMUTs array: (a) optical photo of 3×3 PMUTs array captured on the backside, (b) photo of packaged 3×3 PMUTs array.

IV.   CHARACTERIZATION

A. AlN characteristics

X-ray diffraction (XRD) is used to analyze the grain orientation of AlN and the test result is shown in Fig. 9. It can be found that AlN has a strong peak at 36° and the intensity is above 10,0000, and that means it's the preferred c-axis (002) orientation.

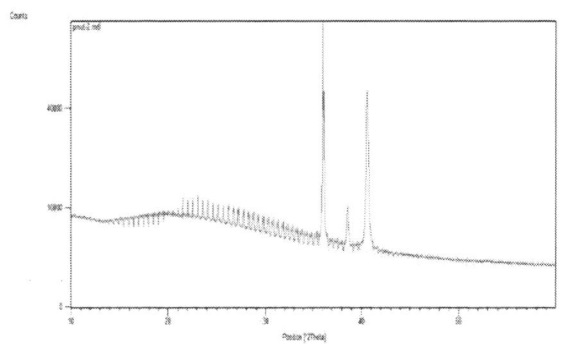

Figure 9.   Grain orientation of AlN.

Impedance analyzer (4294A) is utilized to obtain the impedance curve of the fabricated AlN PMUT and the test result is shown in Fig.10. The resonance frequency and electromechanical coupling coefficient are summarized in Tab.3. Compared with the simulated results, a consistent between them can be found.

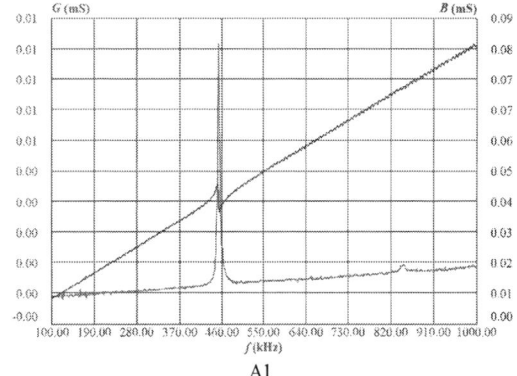

A1

978-1-7281-1500-9/19 $31.00 © 2019 IEEE

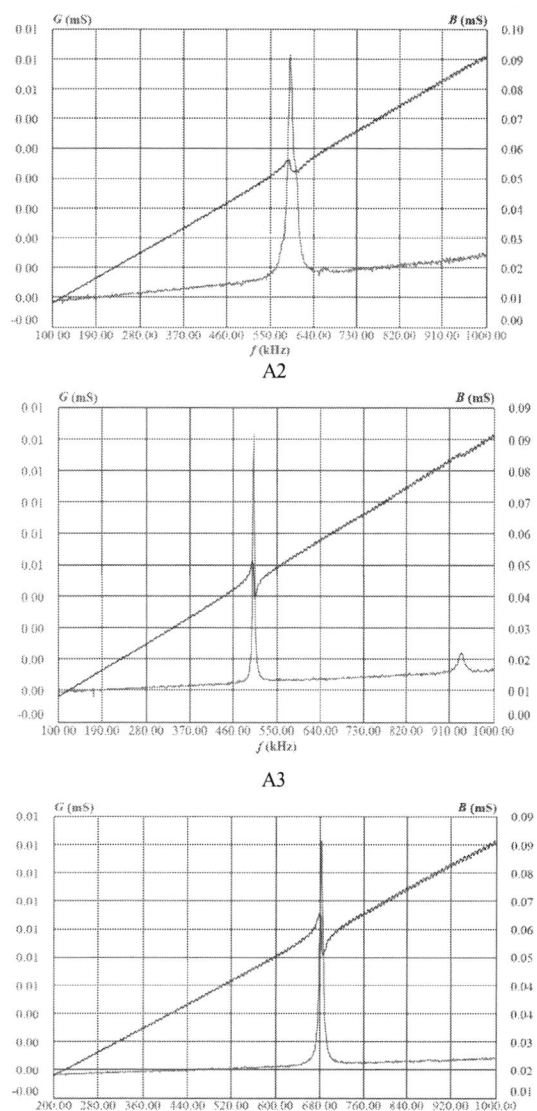

Figure 10. Resonance frequency test results for different positions of the PMUTs array with nine small cavities.

TABLE III. TEST RESULTS

Parameter	Value			
	A1	*A2*	*A3*	*A5*
Measured result (kHz)	451.44	591.11	502.13	672.59
Tested Electromechanical coupling coefficient (%)	4.699	5.973	4.523	3.025

B. Wireless power transmission experiment

In order to verify its property as a wireless power receiver, a power transmission link is constructed as showed in Fig.11. It includes a signal generator (33600A Series), RF power amplifier, preamplifier (MODEL SR560), standard hydrophone, power supply, oscilloscope (DSOX3054A), mechanical fixtures. In the experiment, the PMUT device is fixed with mechanical fixture facing to the emitter distanced about 12.7cm, immersed in oil tank as showed in Fig.12. The

signal generator emits a sinusoidal AC voltage with an amplitude of 500mV and 7 cycles (the frequency range is adjusted according to the resonant frequency of the PMUT under test). The power amplifier amplifies the input signal and excites the standard hydrophone, emits an ultrasonic wave transmitting to the PMUT, the electrical signal excited by AlN PMUT is obtained and displayed by the oscilloscope. Fig.13 show the tested output voltage of PMUTs in the 3×3 array. Excluding the effect of 5 times magnifications by preamplifier, the peak value of excited voltage is summarized in Tab.4.

Compared with simulation results, an agreement can be found. The test result is smaller than the simulation result. The reason may be that the boundary condition which is completely consistent with the test condition cannot be set in the simulation process, and there is a certain error. It is also possible that the purity of the transmission medium during the test and the transmission line has a certain interference with the sound pressure of the transmitter output. Tab.5 summarized the performance of reported PMUT as a wireless power receiver in recent years. It can be found that the PMUT studied in this paper has a higher transmission efficiency.

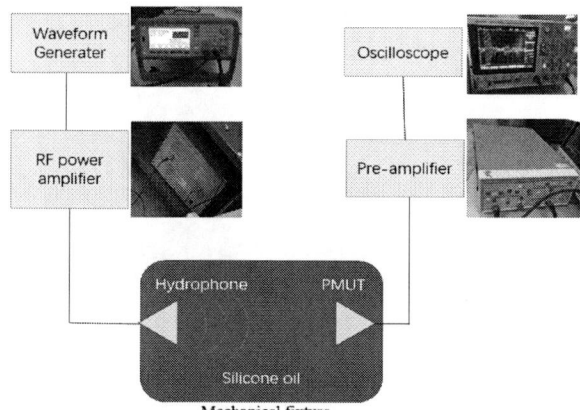

Figure 11. A power transmission link for evaluating PMUT property as wireless power receiver.

Figure 12. Photo during the experiment for evaluating PMUT property as wireless power receiver.

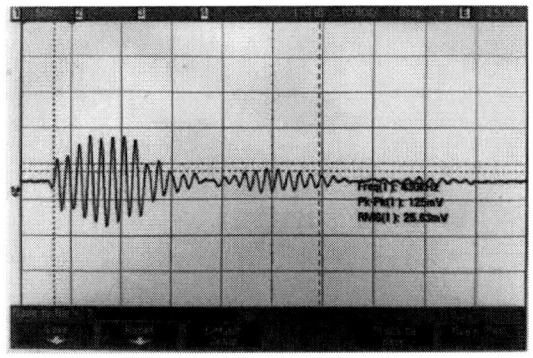

（a）

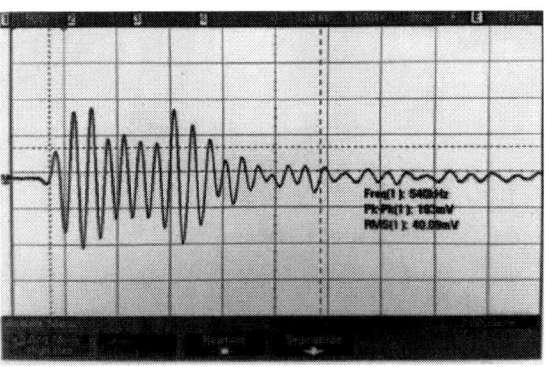

（b）

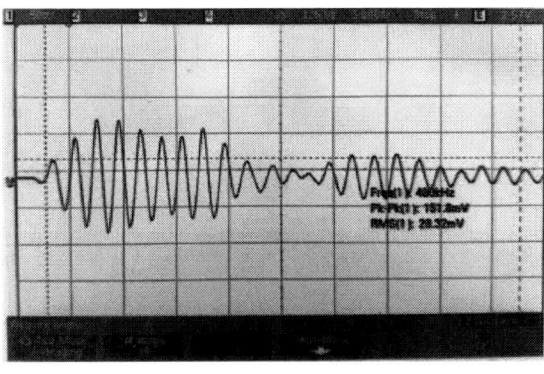

（c）

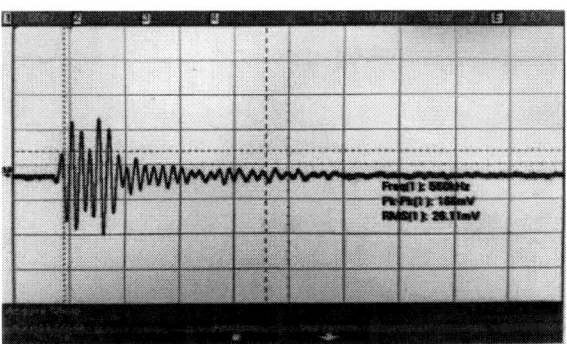

（d）

Figure 13. Tested output voltage of PMUT in the 3×3 array: (a), (b), (c) test results of PMUT output voltage for edge A1, A2, A3, respectively; (d) test results corresponding to intermediate PMUT A5.

TABLE IV.   THE PEAK VALUE OF EXCITED VOLTAGE AT PMUT

Parameter	Value			
	A1	*A2*	*A3*	*A5*
Peak-to-peak value (mV)	25	38.6	30.36	33.2

As showed in Fig.13, the output voltage signal is fluctuations with 7 cycles of sinusoidal AC input signal. It is very likely that the sound pressure reflected by the sidewalls in the device causes the signal enhancement of the PMUT.

TABLE V.   REPORTED PERFORMANCE OF ALN PMUT AS A RECEIVER

Parameter	Value			
	This work	*[9]*	*[7]*	*[8]*
Years	2019	2017	2018	2018
Affiliation	Xiamen University, Xiamen, China	Carnegie Mellon University, Pittsburgh, USA	Zhejiang University, Zhejiang, China	Pennsylvania State University, University Park, PA, USA
Size	400μm×400μm	200um×200μm	550μm (Square)	1.2mm×1.2 mm
Piezoelectric material	AlN	AlN	AlN	PZT
Resonant frequency	500kHz	3.2MHz	400 kHz	1.6MHz
Transmitter to receiver distance	12.7cm	5mm	28cm	13.3mm
Transmission medium	Silicone oil	Plastic material	Air	Castor oil
Input voltage	500mV	350mV	3.3V	500mV
Acoustic pressure of Input sonic wave reaching at PMUT(esti mated value)	$2×10^4$ Pa	$1.692×10^4$ Pa	$1.685×10^4$ Pa	$2.3×10^3$ Pa
Output voltage	38.6mV	29dB(49.8 mV)	7.5mV	3.8mV

V. CONCLUSION

In this paper, a 3×3 AlN PMUTs array is designed, fabricated, and characterized as a wireless power receiver. AlN PMUT is a stacked layers structure of $Si/SiO_2/Mo/AlN/Al$ about 4μm/1μm/0.1μm/2μm/1μm in thickness from bottom to up suspended by a cavity measuring in 0.496mm×0.506mm. According to X-ray diffraction (XRD) analysis, the fabricated AlN film has the preferred C-axis (002) orientation. Through the wireless power transmission experiment, it can generate a peak voltage about 38.6mV when it receives an input sonic wave with a sound pressure of $2×10^4$ Pa.

ACKNOWLEDGMENT

This work is supported by National Natural Science Foundation of China under Grant No.61404112, Collaborative Innovation Center of High-End Equipment Manufacturing in FuJian Province, is jointly conducted with the School of Microelectronics of Peking University and Shenzhen Maochen Limited Company.

REFERENCES

[1] T. Okada, S. Guo, F. Qiang, and Y. Yamauchi, "wireless microrobot with two motions for medical applications," in 2012 ICME International Conference on Complex Medical Engineering (CME), pp. 306-311, 2012.

[2] M. Vahabi, E. Mehdizadeh, M. Kabganian, and F. Barazandeh, "Modelling of a novel in-pipe microrobot design with IPMC legs," Proc. Inst. Mech. Eng. Part J. Syst. Control Eng., vol. 225, no. 1, pp. 63-73, Feb. 2011.

[3] A. S. Rekhi, B. T. Khuri-Yakub, and A. Arbabian, "Wireless Power Transfer to Millimeter-Sized Nodes Using Airborne Ultrasound," IEEE Trans Ultrason Ferroelectr Freq Control, vol. 64, no. 10, pp. 1526-1541, Oct. 2017.

[4] C. M. Nguyen et al. "Wireless Power Transfer for Autonomous Wearable Neurotransmitter Sensors," Sensors, vol. 15, no. 9, pp. 24553-24572, Sep. 2015.

[5] D. Ahn and M. Ghovanloo, "Optimal Design of Wireless Power Transmission Links for Millimeter-Sized Biomedical Implants," IEEE Trans. Biomed. Circuits Syst., vol. 10, no. 1, pp. 125-137, Feb. 2016.

[6] E. Mehdizadeh and G. Piazza, "Chip-Scale Near-Field Resonant Power Transfer via Elastic Waves," J. Microelectromechanical Syst., vol. PP, no. 99, pp. 1-10, 2017.

[7] X. Liu et al., "A High-Performance Square pMUT for Range-finder," IEEE 13th Annual International Conference on Nano/Micro Engineered and Molecular Systems (NEMS), IEEE, pp.106-109, 2018.

[8] A. Ibrahim, M. Meng, and M. Kiani, "A Comprehensive Comparative Study on Inductive and Ultrasonic Wireless Power Transmission to Biomedical Implants," IEEE sensors journal, 18(9): 3813-3826, 2018.

[9] E. Mehdizadeh and G. Piazza, "AlN on SOI pMUTs for ultrasonic power transfer," IEEE International Ultrasonics Symposium (IUS), IEEE, pp.1-4, 2017.

A Sequential Finite Volume Method / Finite Element Analysis of a Power Electronic Semiconductor Chip

Mario Gschwandl,[1] Peter Filipp Fuchs,[1] Thomas Antretter,[2] Martin Pfost, [3] Ivaylo Mitev,[1] Tao Qi,[4]
Thomas Krivec,[4] Angelika Schingale[5] and Michael Decker[5]

[1] *Polymer Competence Center Leoben GmbH, Roseggerstrasse 12, 8700 Leoben, Austria*
mario.gschwandl@pccl.at*

[2] *Institute of Mechanics, Montanuniversitaet Leoben, Franz-Josef-Strasse 18, 8700 Leoben, Austria*

[3] *Chair of Energy Conversion, TU Dortmund University, Emil-Figge-Strasse 68, 44227 Dortmund, Germany*

[4] *Austria Technologie & Systemtechnik Aktiengesellschaft, Fabriksgasse 13, 8700 Leoben, Austria*

[5] *CPT Group GmbH, Siemensstraße 12, 93055 Regensburg, Germany*

Abstract—The shift of the automotive industry towards e-mobility results in a strong demand for highly reliable power electronics. A major goal in their design is to improve the thermal management of all components. Most commonly power electronics are subject to high temperature loads, either internally generated by an active part (semiconductor) or externally applied. Depending on the materials used, such as metals, polymers, etc., thermo-mechanical stresses will arise and promote different failure mechanisms. The complexity of the loading situation, especially in the case of internally generated loads, calls for a sequential approach, consisting of a Finite Volume Method (FVM) and a Finite Element Analysis (FEA) for the lifetime assessment of these components. Using this methodology, the highly complex temperature distribution of any power package can be determined. Consequently, accurate results for the thermo-mechanical stress situation from chip to power packages are deduced and critical spots are identified. Based on the obtained stress fields, an enhanced lifetime assessment of power packages can be performed. The proposed methodology is validated on a standard TO-263 package for a short circuit loading scenario.

Keywords-power electronics; electro-thermo-mechanical simulation; finite element analysis; finite volume method;

I. Introduction

The shift of the automotive industry towards e-mobility results in a strong demand for highly reliable power electronics. Most commonly power electronic packages are subject to high temperature loads, either internally generated by an active part - e.g. a semiconductor [1] - or externally applied - e.g. during manufacturing or in service [2]. Depending on the materials used, such as metals, polymers, etc., thermo-mechanical stresses will arise and promote different failure mechanisms. Hence, a major goal is the improvement of the thermal management of all components by a proper material selection, material development, design, etc. ([3], [4], [5], [6]). Thus, knowledge of the applied materials as well as understanding of the designs of these systems is absolutely necessary. This can be facilitated by virtual prototyping methods and virtual testing methods; taking advantage of

a variety of different numerical approaches. However, each approach is offering specific benefits and restrictions. Due to the complexity of the loading situation of power packages, especially in the case of internally generated loads [7], a single numerical approach might not deliver adequate results. Hence, a sequential approach is needed to obtain precise results.

This research work is considering surface-mounted power packages, with a semiconductor chip as an active heat source encapsulated by a polymeric material. The proposed approach employs a combination of Finite Volume Method (FVM) analysis and a Finite Element Analysis (FEA). The FVM analysis is used to calculate the locally varying electrical material data of the active semiconductor device and the FEA is hereafter used for an electro-thermal (ET) and a thermo-mechanical (TM) analysis.

Using this methodology, the highly complex temperature distribution of power packages can be determined. Consequently, accurate results for the TM stress situation from chip to power packages are deduced and critical spots can be identified. Based on the obtained stress fields, a lifetime optimization of power packages can be performed.

II. Sequential Simulation Strategy

The research work is mainly focusing on surface-mounted power packages with a silicon semiconductor chip, nonetheless it is transferable and scalable to any device employing a silicon die.

The main advantage of the introduced methodology is the simple link of FVM and FEA. An existing tool based on FVM is used to calculate the temperature field generated by the active component for the entire power package. Generally, power packages exhibit a complex geometry, thus the transfer of the entire results, the temperature field, to a FEA domain (for a precise thermo-mechanical assessment [8]) is difficult and time consuming. In this work however, the FVM simulation is used to calculate the spatially resolved electrical material properties, such as the specific electrical resistivity of the chip's depletion zone for a defined loading scenario. As a consequence, for a defined loading case just the calculated electrical properties of the silicon semiconductor need to be transferred. Moreover, due to the simple geometry of a silicon chip - most commonly a rectangle - the data transfer is less complicated and fast.

Finally, using the FVM input an ET-FEA simulation is conducted and the resulting temperature distribution is then used in a consecutive TM-FEA simulation to compute the precise stress situation due to the internally generated loading of the package.

A. Electro-Thermal Finite Volume Simulation

The FVM offers an efficient computation with short processing times. The main advantage of the FVM is that an unstructured mesh can be used [9]. Thus, FVM is commonly used to solve electro-thermal problems.

For this approach an already existing 3-D numerical simulator based on the FVM is used [7]. This tool is used for two specific purposes: First, to calculate the material properties of the chip's depletion zone; delivering the spatially resolved specific electrical resistivity of the depletion zone for a given load case. Second, the FVM tool is already verified experimentally, so the calculated temperature distribution of the power packages is used for validating the ET-FEA results.

The electrical and thermal material data are taken from former measurements [2] and literature [7]. For the verification of the proposed framework a D2PAK (TO-263) is used. The FVM model considers the entire chips as well as the bondwires and the encapsulation material. For the demonstration of the proposed methodology a short-circuit loading case is considered, where thermal loads are induced into the package by active heating of the semiconductor die. To represent this loading scenario a highly resistive depletion zone is delivered by the FVM tool. The chip is modelled with a top layer metallization out of aluminum, with a thickness of 5 μm. Below the metallization the active layer, where the depletion zone is located, is found. This layer is 40 μm thick and all the heat will be generated in this area. Below, a 235 μm thick bulk silicon layer is found.

B. Transfer Finite Volume to Finite Element

As mentioned above the FVM and FEA require different discretization strategies. While FVM allows for an unstructured mesh, such as Manhattan-like geometries [7], FEA requires a structured mesh - see Figure 1a and 1b. The shared nodes depicted therein are commonly used for the FVM, as they allow for an easy discretization. However, they are not permitted in the FEA.

An exemplary FVM mesh of the active layer within the FVM tool is depicted in Figure 2. The outer boundary of the chip is displayed as well as the mesh within the depletion

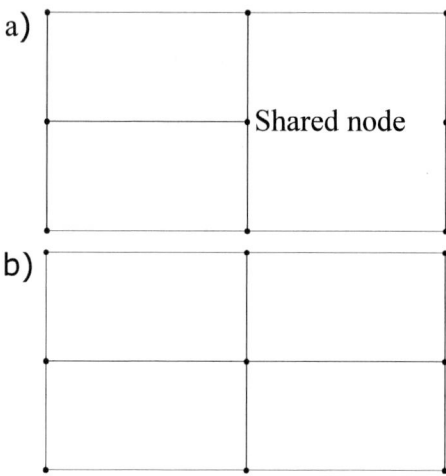

Figure 1. a) Unstructured mesh of FVM b) Structured mesh of FEA

zone (red area). The remaining area between the border of the chip and the depletion zone is defined as an inactive zone which will be considered as bulk silicon with a low electrical conductivity so no current will flow through this area. As depicted, the mesh is highly unstructured, as the discretization level is finer at the outer boundary of the depletion zone. Furthermore, numerous shared nodes are present. They are invalid in a FEA.

To this end, a fast and effective mapping tool is created using Matlab (R2018a,The MathWorks Inc., Natick, USA), which transfers the electrical material parameters of the active layer of the chip from the FVM to the FEA domain. This tool is clustering the FVM-elements of the active layer using equally spaced rectangular elements - see Figure 3. The overlaid pattern (blue lines) represents the clusters for the mapping procedure. Each cluster represents a hexahedral FEA-element in the following ET simulation. The level of discretization for the mapping procedure can be adjusted to the needs of the simulations and the complexity of the circuit inside the active layer of the chip. At this stage the

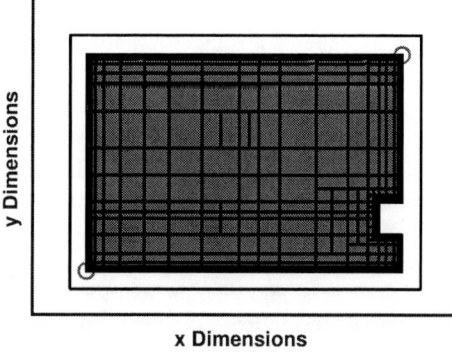

Figure 2. Exemplary FVM Mesh of used for an electro-thermal simulation in the FVM tool

978-1-7281-1500-9/19 $31.00 © 2019 IEEE

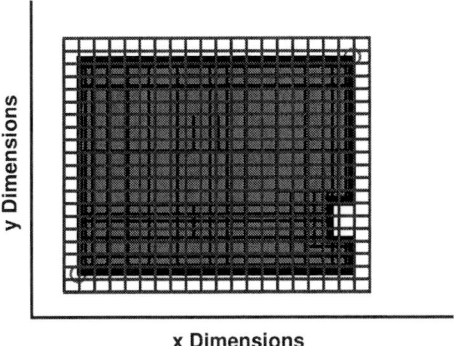

Figure 3. Mapping Clusters for an exemplary FVM mesh

discretization is defined equally along the x and y direction.

Each FVM-element is assigned a specific electrical resistivity ρ - see Figure 4. This spatial distribution of the specific electrical resistivity is an output generated from the FVM tool. For the exemplary case - Figure 4 - ρ is constant in the center of the chip and drops to a lower level at the boundary of the chip exhibiting a totally inactive area just before the borders of the chip.

For the mapping procedure each cluster is processed one after another. For the calculation of the specific electrical resistivity of a cluster all FVM-elements which intersect the area or lie totally within the processed cluster are considered - see Figure 5. Using this approach it is guaranteed that, the area of all considered FVM-Elements equals always the area of the processed cluster. For intersecting elements the intersecting area is calculated, whereas for FVM-elements which lie entirely inside the cluster the entire area of the FVM element is considered. Finally, the resistivity of each cluster is calculated as follows:

$$\rho_{\text{FEA}} = \frac{1}{A_{\text{FEA}}} \sum_{i=1}^{n} \rho_{\text{FVM}} A_{\text{FVM}}, \quad (1)$$

where ρ_{FEA} and A_{FEA} are the specific electrical resistivity and the area of the newly generated Finite Element, ρ_{FVM}

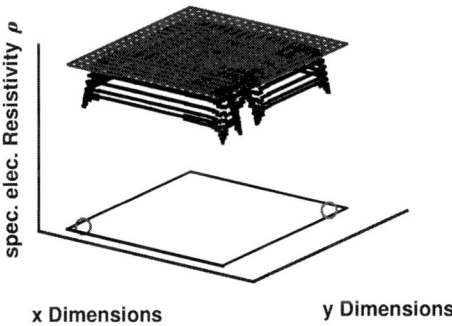

Figure 4. Specific electrical resistance of the exemplary FVM-mesh, indicating a high local resistivity at the boarders of the depletion zone.

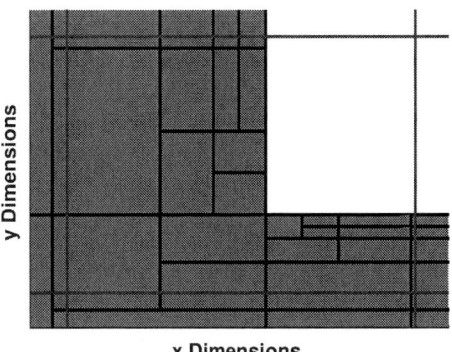

Figure 5. Zoom-in for one FEA-cluster (blue rectangle) of the mapping process. The red areas are FVM-Elements which either lie totally within (middle) the cluster or intersect the cluster (left side). The white area is considered as inactive part of the depletion zone.

and A_{FVM} are the specific electrical resistivity and the intersecting area of a FVM-element within the calculated cluster and n is the total number of FVM-elements intersecting this cluster.

The calculated data is then transferred to the input format of the FEA software - Abaqus (2017, Dassault Systemes Simulia Corp., Providence, RI, USA) - using an already developed parsing tool [10]. As a result, a ready-to-use FEA-model of the chip is provided; containing the top metallization layer as well as the mapped active layer and the non-active bulk silicon layer below.

C. Electro-Thermo-Mechanical Finite Element Simulation

In the finite element regime an electro-thermo-mechanical simulation is carried out, which is split into two parts: (1) a coupled thermal-electrical analysis to compute the temperature distribution induced via self-heating and (2) a thermo-mechanical simulation using the a-priori computed temperature field as an input parameter. Finally, the arising stress can be evaluated out of the thermo-mechanical simulation results.

The mapping tool provides a regular mesh using hexahedral elements. However, the thermo-mechanical simulation requires a different mesh. Due to the complex geometry of the other parts of the power package (encapsulation, bondwires, etc.) a tetrahedral mesh is applied for the electro-thermal simulation. Afterwards the parts are combined using a node merging strategy. The chip with the hex-mesh is then added to the complete package using tie-constraints [11]. This meshing strategy is valid for an electro-thermal analysis, but will not be precise for a thermo-mechanical assessment where a full tetrahedral mesh enabling a fine resolution in all directions is more appropriate. Since both FEA-simulations are carried out in Abaqus, a seamless interpolation between the two meshes inside the same software is guaranteed.

Coupled Electro-Thermal Analysis: By using a coupled electro-thermal simulation approach the temperature and electrical potential is calculated simultaneously at each node. For the calculation of the heat generated by the active part of the semiconductor chip the Joule-heating effect is used. The generated power P can be calculated according to Joule's first law, see Equation (2).

$$P = I^2 R, \qquad (2)$$

with I being the current and R the resistivity of the conductor. Hence, Joule heating occurs when electrical current is converted to dissipated energy - heat - whilst flowing through a conductor [11]. The resistance R is defined as:

$$R = \rho \frac{l}{A}, \qquad (3)$$

where ρ is the specific electrical resistivity, l the length of the conductor and A the cross-sectional area of the conductor. Thus, using the locally resolved specific electrical resistivity gained from the FVM-simulation, the heat generation can be described in detail and spatially resolved. The FEA software however, requires the electrical conductivity κ as input parameter which is defined as the inverse of the specific electrical resistivity:

$$\kappa = \frac{1}{\rho}. \qquad (4)$$

For the analysis κ is assumed as isotropic and independent of the electrical field. Moreover, the heat conduction is considered as transient, whereas electrical transient effects are neglected. Furthermore, it is assumed that all dissipated energy is converted into heat, indicating a Joule heating fraction of 1.0 [11].

The design of the simulation model for the (ET) FEA and a zoom of the chip area for ET-FEA simulation are depicted in Figure 6 & 7. As mentioned above, the different mesh strategies are clearly visible. The load (current) is applied via the right terminal and the attached bond ribbon. At the underside of the package the electrical potential is set to zero, representing the drain of the package. For the demonstration load case a current of 120 A is applied for a time span of 10 μs at an ambient temperature of 150°C - the parameters are taken in accordance to the data sheet. The locally resolved specific electric resistivity is depicted

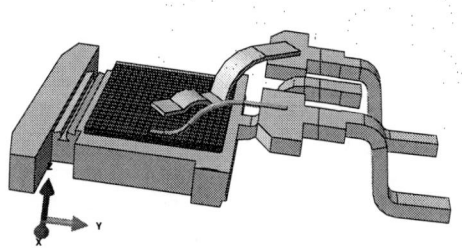

Figure 6. FEA-model design (ET-simulation) of D2PAK without the polymeric encapsulation

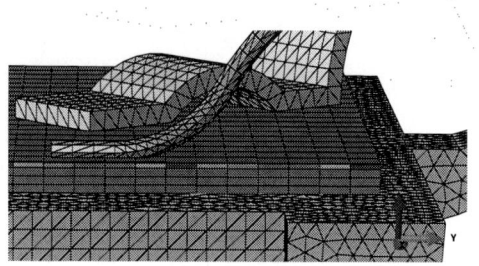

Figure 7. FEA-model zoomed in at the chip area for the electro-thermal simulation without the polymeric encapsulation

in Figure 8. All values are normalized to the maximum value of the specific electrical resistivity. As seen, a very homogeneous distribution of the resistivity is yielded. This means, that throughout the entire active area of the chip all power is dissipated, which simulates the heat generation during a short circuit behavior.

The bondwire gate (left terminal) is not attached in an electrically conductive manner to the chip's depletion zone. Furthermore, the ambient temperature is set to 150°C. At this stage, the systems are considered as adiabatic and no heat convection to the surrounding air is considered. Due to the very short time span convection will not influence the results.

Thermo-Mechanical Simulation: For the mechanical analysis of the power package a mesh with tetrahedral elements with quadratic shape functions is used. This allows a fine discretization of the highly complex geometries within the power packages, such as bondwires. A zoom of the chip area of model for the TM-FEA simulation is depicted in Figure 9. The material parameters for the mechanical properties, such as Young's Modulus, Coefficient of Thermal Expansion (CTE), etc., are taken from literature [7] and for the polymeric parts from previous measurements [2]. For the stress evaluation during the internal loading via the semiconductor, the temperature distribution resulting from the ET-FEA simulation is used as a boundary condition. As in the ET-

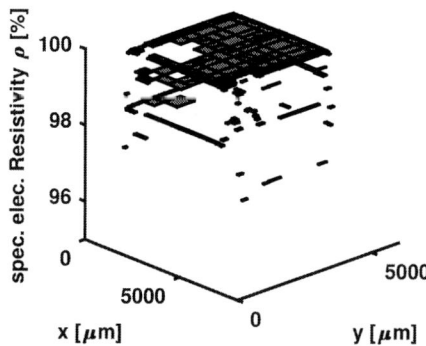

Figure 8. Locally resolved specific electrical resistivity for the applied short circuit load case. The values are normalized to the maximum specific resistivity.

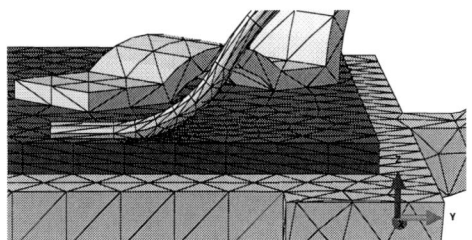

Figure 9. FEA-model zoomed in at the chip area for the thermo-mechanical simulation without the polymeric encapsulation

Figure 11. Zoom of the temperature distribution computed by electro-thermal FEA simulation at the bond ribbon area. As the heat is only present in the active part of the depletion zone and the metallization on top.

FEA simulation, the starting ambient temperature is defined as 150°C. Additionally, it is assumed that the power package is attached via solder bumps to a printed circuit board (PCB). For this stage the PCB and the solder bumps are not considered, nonetheless the bottom of the package is restrained according to this set-up. Moreover, convection phenomena are neglected, which should not interfere with the results given the short times under investigation.

III. RESULTS & DISCUSSION

With the calculated locally resolved electrical conductivity the electro-thermal FEA yields a homogeneous temperature distribution at the active area of the chip - see Figure 10. Due to the instantaneous electrical loading all heat is dissipated within the chip's active layer and the metallization layer on top. As depicted in Figure 11, the heat is generated solely in the active layer and spreads minimally. This setup is generating a high temperature gradient from the die area to the encapsulating polymer.

Comparing the results of the electro-thermal analysis of FEA and FVM a minor difference in the temperature development is visible - see Figure 12. However, compared to an electro-thermal simulation using literature values [12] for the electrical conductivity and a homogeneous active layer, a clear benefit of the mapped approach is visible. Moreover, the definition of the electrical properties of silicon is challenging, due to its high dependency on the structure and the doping concentration [12], [13]. Using the FVM tool upfront enables the clear calculation of the local resistivity

for a specific load case thus improving the quality of the electro-thermal FEA results. The difference between the heat development predicted by the FVM-tool and the FEA analysis is resulting from the current level of discretization. For the methodology validation the chip was partitioned into 400 elements, averaging the locally distributed specific electric resistivity to a certain extent. Increasing the level of detail for the mapping procedure should reduce the difference of the results of the two methods, however significantly increasing the computation time.

Using these results the thermo-mechanical stresses can be computed in the subsequent thermo-mechanical FEA simulation. As indicated by the temperature field the stresses are concentrated in the area of the active layer and metallization of the chip. At the interface of the metallization layer to the active area of the chip stresses (Mises) of 85 MPa appear. Considering the small thickness of the metallization layer (5 μm) this stress is, according to literature, already close to the yield stress of thin aluminum metallization layers. As indicated by Alpern et al. [14] a 3.5 µm thick metallization layer has a yield stress of 100 MPa. Considering the thickness of the present metallization of 5 µm the yield stress will be below 100 MPa and thus even closer to the stress magnitudes obtained from the thermo-mechanical simulation. According to the data-sheet of the package, this loading scenario of the chip shall be endured by the package

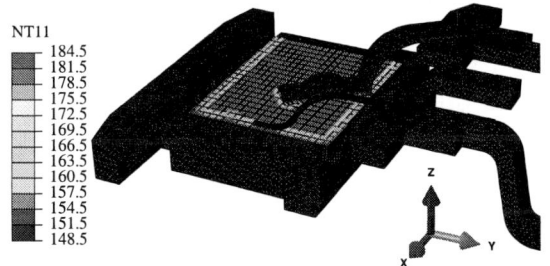

Figure 10. Results of the temperature distribution computed by electro-thermal FEA simulation. As depicted a homogeneous heat distribution through out the active area of the depletion zone is yielded.

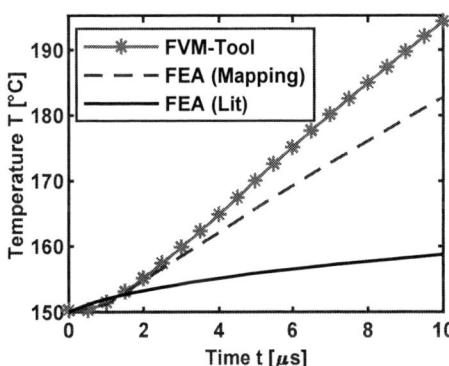

Figure 12. Comparison of the heat development using the FVM tool, ET-FEA simulation with mapping and literature values.

978-1-7281-1500-9/19 $31.00 © 2019 IEEE   1513

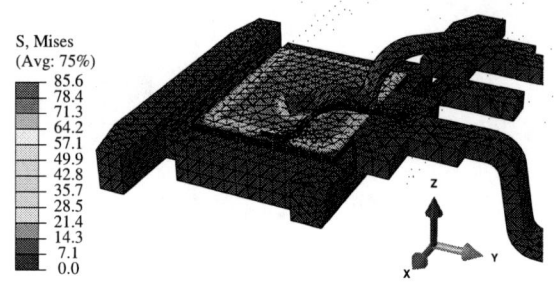

S, Mises
(Avg: 75%)
- 85.6
- 78.4
- 71.3
- 64.2
- 57.1
- 49.9
- 42.8
- 35.7
- 28.5
- 21.4
- 14.3
- 7.1
- 0.0

Figure 13. Results of the thermo-mechanical FEA simulation. As depicted the main stress is arising in the metallization layer.

once. According to the simulation results this requirement is satisfied.

IV. CONCLUSION

Using this methodology, the highly complex temperature distribution of a power package can be determined in a cost-effective manner. The transfer of the locally resolved electrical parameters from the FVM domain to the FEA domain is fully functional and the computed results of the electro-thermal FEA including the mapped data align well with the FVM tool's results. Consequently, accurate results for the TM stress situation from chip to power packages are deduced and critical spots are identified. Based on the obtained stress fields, an enhanced lifetime assessment of power packages can be performed.

ACKNOWLEDGMENT

The research work was performed within the K-Project "PolyTherm" at the Polymer Competence Center Leoben GmbH (PCCL, Austria) within the framework of the COMET-program of the Federal Ministry for Transport, Innovation and Technology and the Federal Ministry for Digital and Economic Affairs with contributions by by the Montanuniversitaet of Leoben, TU Dortmund University, AT&S Austria Technologie & Systemtechnik Aktiengesellschaft and by Continental Automotive GmbH. Funding is provided by the Austrian Government and the State Government of Styria.

REFERENCES

[1] O. Muscato and V. Di Stefano, "An energy transport model describing heat generation and conduction in silicon semiconductors," *Journal of Statistical Physics*, vol. 144, no. 1, pp. 171–197, 2011.

[2] M. Gschwandl, P. F. Fuchs, I. Mitev, M. Yalagach, T. Antretter, T. Qi, and A. Schingale, "Modeling of manufacturing induced residual stresses of viscoelastic epoxy mold compound encapsulations," in *Electronics Packaging Technology Conference (EPTC), 2017 IEEE 19th*. IEEE, 2017, pp. 1–8.

[3] R. W. Johnson, J. L. Evans, P. Jacobsen, J. R. Thompson, and M. Christopher, "The changing automotive environment: high-temperature electronics," *IEEE Transactions on Electronics Packaging Manufacturing*, vol. 27, no. 3, pp. 164–176, 2004.

[4] A. L. Moore and L. Shi, "Emerging challenges and materials for thermal management of electronics," *Materials Today*, vol. 17, no. 4, pp. 163–174, 2014.

[5] M. Morak, P. Marx, M. Gschwandl, P. Fuchs, M. Pfost, and F. Wiesbrock, "Heat dissipation in epoxy/amine-based gradient composites with alumina particles: A critical evaluation of thermal conductivity measurements," *Polymers*, vol. 10, no. 10, p. 1131, 2018.

[6] R. J. McGlen, R. Jachuck, and S. Lin, "Integrated thermal management techniques for high power electronic devices," *Applied thermal engineering*, vol. 24, no. 8-9, pp. 1143–1156, 2004.

[7] M. Pfost, C. Boianceanu, H. Lohmeyer, and M. Stecher, "Electrothermal simulation of self-heating in dmos transistors up to thermal runaway," *IEEE Transactions on Electron Devices*, vol. 60, no. 2, pp. 699–707, 2013.

[8] P. Cova and N. Delmonte, "Thermal modeling and design of power converters with tight thermal constraints," *Microelectronics Reliability*, vol. 52, no. 9-10, pp. 2391–2396, 2012.

[9] O. Kolditz, *Computational Methods in Environmental Fluid Mechanics*. Springer, Berlin, Heidelberg, 2002, ch. Finite Volume Method, pp. 173–190.

[10] M. Gschwandl, P. Fuchs, K. Fellner, T. Antretter, T. Krivec, and T. Qi, "Finite element analysis of arbitrarily complex electronic devices," in *Electronics Packaging Technology Conference (EPTC), 2016 IEEE 18th*. IEEE, 2016, pp. 497–500.

[11] *Abaqus 2017 Documentation*, Dassault Systmes Simulia Corp., Providence, RI, USA, 2017.

[12] J. Y. Seto, "The electrical properties of polycrystalline silicon films," *Journal of Applied Physics*, vol. 46, no. 12, pp. 5247–5254, 1975.

[13] L. Weber and E. Gmelin, "Transport properties of silicon," *Applied Physics A*, vol. 53, no. 2, pp. 136–140, 1991.

[14] P. Alpern, P. Nelle, E. Barti, H. Gunther, A. Kessler, R. Tilgner, and M. Stecher, "On the way to zero defect of plastic-encapsulated electronic power devices—part i: Metallization," *IEEE Transactions on Device and Materials Reliability*, vol. 9, no. 2, pp. 269–278, 2009.

Failure Life Prediction of Wafer Level Packaging using DoS with AI Technology

P. H. Chou[1], H.Y Hsiao[1] and K.N. Chiang[1,*]

[1,*] Advanced Microsystem Packaging and Nano-Mechanics Research Lab, Taiwan
*Dept. of Power Mechanical Engineering, National Tsing Hua University, Hsinchu 300, Taiwan, R.O.C.
Phone: 886-3-5742925. Fax: 886-3-5745377, and E-mail: knchiang@pme.nthu.edu.

Abstract— In regard to the design of new electronic packaging structure, e.g., wafer level packaging (WLP), one needs to consider many design factors that could affect the reliability characteristics of electronic package. Before mass production, the electrical packaging has to pass the reliability test. Thermal cycling test (TCT) is one of the standard reliability tests and has been commonly used in electrical packaging industry. To ensure the new products pass the thermal cyclic test is the critical issue in the electronic packaging industry. To react to the rapid growth of electronic components, it is imperative to shorten the development time and cycles of electronic packaging, design-on-experiment (DoE) method is quite time consuming and very costly, finite element method (FEM) based design-on-simulation technology could be used as a feasible development methodology for the reliability assessment and reliability prediction of electronic package. After the FEM model, mechanics theory, simulation procedures and reliability behavior, etc. being verified by experiments, simulation can be treated the same as the experiment if simulation can consistently get the results that similar to experiment. Simulation can build a set of reliability results database of different geometric structures of the WLP and provided this database for machine learning. Machine learning is an automatic analysis method which could learn a regression model from the database for predicting reliability cycles of package instantly. In this paper, the thermal cycling test (TCT) reliability of different geometry structures of Wafer Level Packaging is discussed through verified simulation technology. The database generated by FEM simulation is analyzed by artificial neural network (ANN), the training result shown ANN can predict the reliability of unknown structure of WLCSP in an accurate range.

Keywords—FEM simulation, WLP, Machine learning, Artificial neural network, Regression model

I. INTRODUCTION

Recently, artificial intelligence is applied extensively in various applications. Machine learning [1] automatically learns accurate trends from complex data through logistic algorithms to assist user making a decision. Machine learning is mainly used in prediction and classification. Whether the input data corresponding to output variables had been known will determine whether the machine learning is classified as supervised learning or unsupervised learning.

In 1943, McCulloch and Pitts [2] proposed an artificial neural network logic calculation method which is a multi-layer neural networks. The method is a concept of brain self-thinking learning imitating human nervous system processing information. As shown in Fig. 1.

This study focused on supervised learning, using FEM, empirical reliability prediction equation conjunction with artificial neural network for predicting reliability life of WLCSP, artificial neural network is designed to learn and establish a regression model between input design variable and output reliability life.

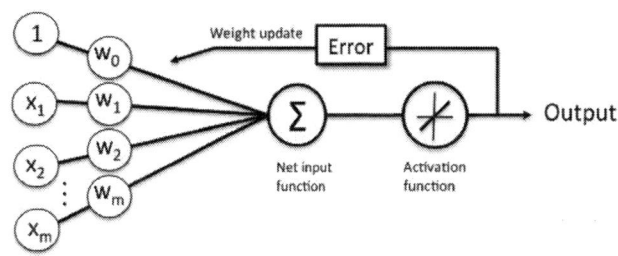

Fig. 1 Schematic of artificial neural network.

II. BOARD LEVEL RELIABILITY TEST

For solder ball type of WLCSP package, the typical failure of solder joint (Fig. 2) under thermal cycling loading (Fig. 3) may generate main crack in the corner of solder ball. The accumulated strain of solder joint eventually causes fatigue failure. In this section, dimensions of all the four test vehicles are first introduced, including the solder ball size and the dimension of the structure. The JEDEC standard for thermal cycling test condition G is selected as the testing condition.

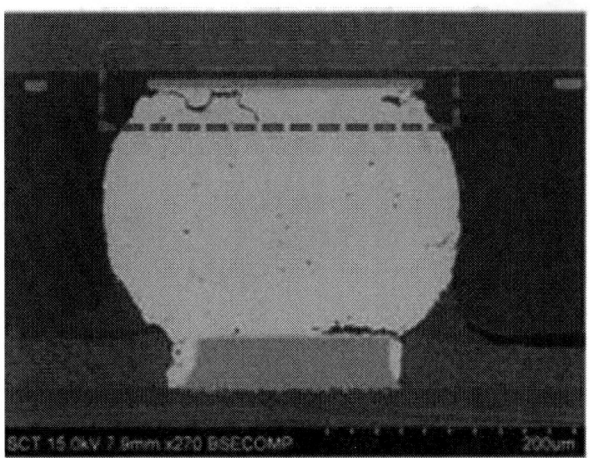

Fig. 2 WLCSP upper pad solder joint crack failure

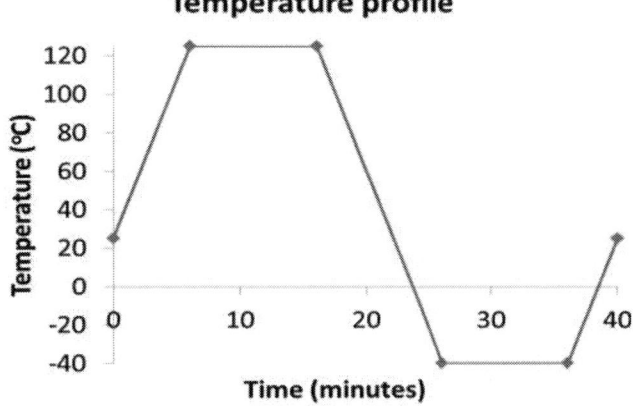

Fig. 3 Thermal cycling test

TABLE I. DIMENSIONS OF TEST VEHICLE 1 & 2

Material	TV1(mm)	TV2(mm)
Si Chip	5.3x5.3x0.33	4x4x0.33
Cu RDL	0.26x0.008	0.26x0.008
UBM	--	0.24x0.0086
Cu Pad	0.22x0.025	0.22x0.025
SBL1	5.3x5.3x0.0075	4x4x0.0075
SBL2	0.01	0.004
PCB	10.6x10.6x1	8x8x1
Low-k	5.3x0.005	4x0.005

TABLE II. DIMENSIONS OF TEST VEHICLE 3 & 4

Material	TV3(mm)	TV4 (mm)
Si Chip	4x4x0.33	6x6x0.33
Cu RDL	0.2x0.008	0.25x0.0065
UBM	0.19x0.0086	--
Cu Pad	0.22x0.025	0.24x0.04
SBL1	4x4x0.008	6x6x0.008
SBL2	0.0075	0.0065
PCB	8x8x0.0086	12x12x1
Low-k	4x0.005	--

TABLE III. DIAMETER, PITCH & COUNTS OF SOLDER BALL

Unit: mm	Ball diameter	Ball pitch	Ball counts
TV1	0.25	0.4	121
TV2	0.25	0.4	100
TV3	0.20	0.3	144
TV4	0.25	0.4	196

A. Dimension of Test Vehicles

All of the four selected test vehicles are WLCSP. The first selected test vehicle [3], TV1 (Test Vehicle 1), is with a packaging dimension of 5.3x5.3x0.33mm, Test vehicle II (TV II) and test vehicle III (TV III) both with a packaging size of 4x4x0.33mm [4]. TV4, with larger chip size 6x6x0.33mm [5]. Other dimensions and solder ball information of these WLCSP are shown in Table I, II and III.

B. Thermal Cycling Test

The thermal cycling test condition used in this research follows JEDEC standard JESD22-A104D condition G [6] (temperature ranges from -40°C to 125°C). The ramp rate is 16.5°C / min. with 10 min dwell time on low/high temperature stages.

C. Finite Element Models

Two-dimensional FEM models are established using ANSYS® to simulate four WLCSP packages with SAC305

978-1-7281-1500-9/19 $31.00 © 2019 IEEE

solder joints. To ensure the simulation quality and accuracy of reliability life prediction, this study applied a fixed mesh size at the key part of the solder joint and this methodology has been addressed in Tsou's [7] paper. Only mesh size of solder joints near the maximum distance to neutral point (DNP) is fixed with a critical mesh size, others are coarse mesh. As shown in Fig.4. The height and width of the fixed mesh size are 12.5μm and 7.5μm respectively. The solder joint geometries are generated by the Surface Evolver program [8], which is based on an energy-based algorithm. In this program, the reflow profile of solder joints can be accurately predicted based on its surface tension, internal force, pad sizes and volume.

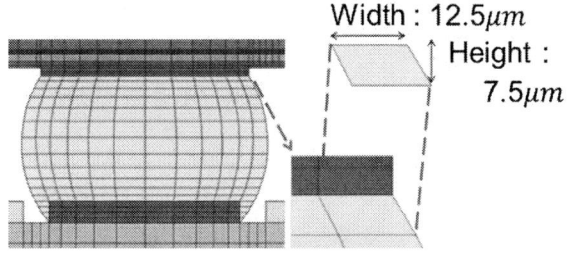

Fig. 4 Controlled mesh size

Fig.5 is the schematic of FEM model of WLCSP structure. Material and geometry of test vehicle 1 to 4 (TV1 - TV4) are shown in Table I, II and III. Since the structures are symmetric, half of the structure along diagonal is modeled.

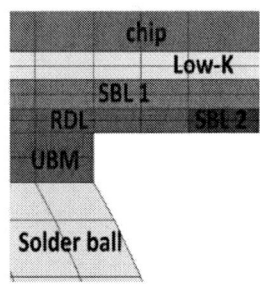

Fig.5 FE Mesh at the Solder Corner of WLCSP

D. Material Properties

The FEM model is built by the chip, two layers of stress buffer layer (SBL), UBM pads, low-k, copper RDL, solder ball, solder mask, and the printed circuit board. The material properties are listed in Table IV. The Young's modulus of the solder joint is temperature dependent and listed in Table V and the SAC solder joint's stress-strain curve [9] is shown in Fig. 6.

The element types selected in this research are PLANE182 and PLANE42. Both element types are four nodes element.

SAC solder joint using PLANE182 which has a better convergence performance for large deformation. Other materials are set PLANE42.

TABLE IV. MATERIAL PROPERTIES

Material	Young's modulus (GPa)	Poisson's Ratio	CTE (ppm/oC)
Solder joint	Temperature dependent	0.35	25
Silicon chip	150	0.28	2.62
Copper	68.9	0.34	16.7
Low-k	10	0.16	5
Solder mask	6.87	0.35	19
PCB	18.2	0.19	16

TABLE V. YOUNG'S MODULUS OF SAC SOLDER AT DIFFERENT TEMPERATURE

Temperature	Young's modulus (GPa)
233K	45.74
253K	42.22
313K	31.66
353K	24.62
398K	16.70

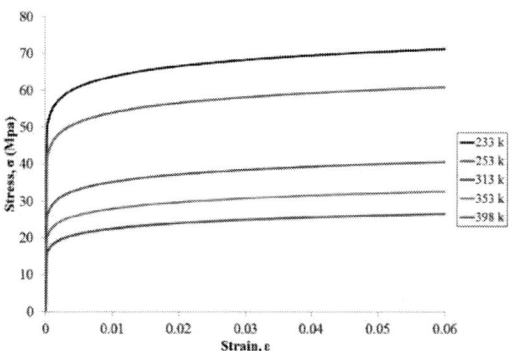

Fig. 6 Stress-strain curve for SAC solder joint [9]

E. Boundary Conditions

Since the electronic packages are symmetric, only half of the structure along diagonal are modeled. Therefore, x-directional displacement is fixed on every node which is located on the y symmetric line. Moreover, all degree of freedom of node at the neutral point of PCB is fixed for avoiding rigid body motion.

The FEM model boundary constrain is shown in Fig.7.

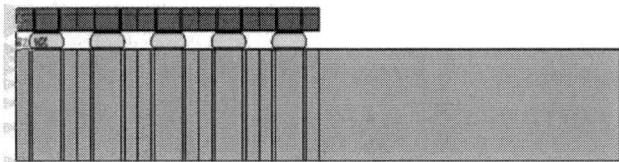

Fig.7 FE Model Boundary Constrains

F. Life Prediction Model

The life prediction models used in this study is the Coffin-Manson strain based model [10]. The Coffin-Manson empirical equation is widely used by many researchers, the equation is as below:

$$N_f = C(\Delta\varepsilon_{eq}^{pl})^{-\eta} \tag{1}$$

Where is mean cycle to failure, C and η are empirical constants. For SAC solder joint C and η are 0.235 and 1.75 [11].

In this research, the stabilized maximum incremental equivalent plastic strain after eight cycles is obtained through FEM analysis by ANSYS® and substituted into Coffin-Manson equation to predict the fatigue life of the solder joint.

G. Life Prediction Results

After finishing finite element analysis, the stable equivalent plastic strain is substituted into the Coffin-Manson strain based model (1). Fatigue prediction life is shown in Table VI, results shown that the simulation results are quite matched with experiment data. After validation, FEM simulation can replace experiment for generating a large number of reliability data for different WLCSP structures.

TABLE VI. RELIABILITY LIFE OF WLCSP

Test vehicle	Experiment data (life cycle)	FEM Coffin-Manson strain based model (life cycle)	Difference
TV1	318	295	7.23%
TV2	1013	970	4.24%
TV3	876	723	5.71%
TV4	904	876	3.10%

III. MACHINE LEARNING

Machine learning algorithm provides self-learning ability, an automatic analysis capability using the big data to learn and establish the regression model. This study applies an artificial neural network algorithm executed in Python. With the help of the regression model, we could assess the reliability of the package instantly.

A. Establishment of Dataset

Once the thermal cycling test reliability for WLCSP is simulated through FEM and validated by experiment. Different geometry structure of reliability prediction dataset could be created by FEM and trained by artificial neural network algorism. As shown in Table VII, FEM life prediction of three different design parameters (upper/lower pad sizes and die thickness) had been established.

Table VII FEM LIFE PREDICTION RESULTS

Parameter				Parameter			
Upper Pad Size	Lower Pad Size	Chip Thickness	FEM Life Prediction	Upper Pad Size	Lower Pad Size	Chip Thickness	FEM Life Prediction
0.22	0.20	0.33	956	0.24	0.20	0.40	903
0.22	0.22	0.33	870	0.24	0.22	0.40	824
0.22	0.24	0.33	737	0.24	0.24	0.40	749
0.24	0.20	0.33	1006	0.26	0.20	0.40	912
0.24	0.22	0.33	914	0.26	0.22	0.40	839
0.24	0.24	0.33	825	0.26	0.24	0.40	772
0.26	0.20	0.33	1,020	0.22	0.20	0.26	995
0.26	0.22	0.33	943	0.22	0.22	0.26	941
0.26	0.24	0.33	860	0.22	0.24	0.26	789
0.22	0.20	0.40	888	0.24	0.20	0.26	1,065
0.22	0.22	0.40	784	0.24	0.22	0.26	1,006
0.22	0.24	0.40	674	0.24	0.24	0.26	908
				0.26	0.20	0.26	1,132
				0.26	0.22	0.26	1,036
				0.26	0.24	0.26	944

B. Artificial Neural Network

Artificial neural network logic is shown in Fig. 8, one could enter the data to the input layer then the data will be calculated in the hidden layer. If there are more neurons in the hidden layer or more hidden layers, nonlinearity would be more significant. However, these adjustments would cause much difficulty of model learning. After integrated calculating, results would be displayed in the output layer.

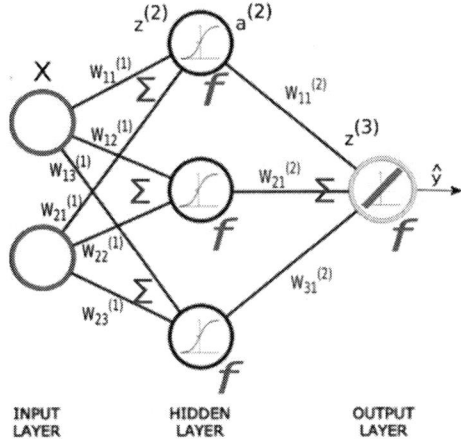

Fig. 8 Schematic of artificial neural network.

In the hidden layer, a_i^ℓ is the i_{th} activation element of the l_{th} layer. Except of a_0 is bias, every a_i is equal to input value times weight "w" and plus bias "b" as shown in (2):

$$a_i^l = \sum_{i=1}^{n} w_{ji}^l x_i + b_i^l \qquad (2)$$

Before entering the next layer, the original value of a_i is substituted into sigmoid function (activation function) for transforming to nonlinear behavior. As the result, we could get $a_i^{(\ell+1)}$ and enter the next layer.

$$a_i^{l+1} = \phi(a_i^l) \qquad (3)$$

Where the activation function as shown in (4).

$$\phi(Z) = \frac{1}{1+e^{-z}} \qquad (4)$$

IV. RESULT AND DISCUSSION

In this section, the structure of artificial neural network trained in Python and comparison of results between FEM life prediction and regression model are discussed. To verify the effectiveness of the regression model, extra testing data will be given and compared with FEM results.

A. Artificial Neural Network Trained in Python

In order to find the best hidden layer structure, we let the input data trained in Python. After being trained, we chose one layer of the hidden layer including 3 neurons as shown. Besides,

the function matrix of whole neural network is shown in (5) and (6).

$$\begin{bmatrix} a_4 \\ a_5 \\ a_6 \end{bmatrix} = \phi \left\{ \begin{bmatrix} -7.752 & -7.995 & -12.524 \\ 1.814 & -28.457 & -11.286 \\ 73.115 & -35.382 & 8.091 \end{bmatrix} \times \begin{bmatrix} a_1 \\ a_2 \\ a_3 \end{bmatrix} + \begin{bmatrix} -3.543 \\ 9.695 \\ -9.733 \end{bmatrix} \right\} \qquad (5)$$

$$[a_7] = [567.772 \quad 548.316 \quad 350.039] \times \begin{bmatrix} a_4 \\ a_5 \\ a_6 \end{bmatrix} + [298.458] \qquad (6)$$

Where a_1, a_2, a_3 are input parameter, a_4, a_5, a_6 are parameters of neuron in hidden layer and a_7 is output prediction result.

The regression model can fit the dataset well. Besides, global error is 0.72% and maximum local error does not exceed 3.26%.

For verifying the regression model trained by artificial neural network that can predict the reliability life of any design data within the design range of these three design parameters. Given a design parameter data in the range of three parameters but not in the training data. In Table VII, it is clear to see that the prediction result between FEM and regression model trained by artificial neural network are very close.

Table VII. COMPARISON OF FEM AND ANN PREDICTIONS

Parameter				Machine learning	
Upper pad size	Lower pad size	Die thickness	Finite Element Method life prediction	ANN	Difference
0.25	0.21	0.3	1015	1015.43	0.04%

V. CONCLUSION

For predicting the reliability of WLCSP, the FEM simulation results with fixed methodology and critical mesh size of the model and using Coffin-Manson strain based model had been verified with experimental results. FE simulation methodology proposed in this research can substitute experiment and use for generating a massive dataset for machine learning.

With an appropriate set of hidden layers, the artificial neural network regression model can have a good result even for small databases, 27 sets of data. Moreover, the regression model had been verified by extra testing data which shown ANN regression model has a good agreement with FE simulation result. In other words, the regression model trained by artificial neural network can predict the reliability of package in different geometry structure of WLCSP very well.

ACKNOWLEDGMENT

The authors would like to thank the National Tsing Hua University for supporting this research.

REFERENCES

[1] E. Alpaydim, "Introduction to Machine Learning", MIT Press, 2014.

[2] W. S. McCulloch, W. Pitts, "A logical calculus of the ideas immanent in nervous activity", The bulletin of mathematical biophysics, Vol.5, Issue 4, pp 115–133, 1943.

[3] M. C. Hsieh and S. L. Tzeng, "Solder joint fatigue life prediction in large size and low cost wafer-level chip scale packages," IEEE Electronic Packaging Technology (ICEPT), pp. 496–501, 2015.

[4] M. C. Hsieh, "Modeling correlation for solder joint fatigue life estimation in wafer-level chip scale packages", International Microsystems, Packagings, Assembly and Circuits Technology Conference (IMPACT), pp. 65–68, Oct. 2015.

[5] B. Rogers and C. Scanlan, "Improving WLCSP Reliability Through Solder Joint Geometry Optimization"z International Symposium on Microelectronics, vol. 2013, no. 1, pp. 546-550, 2013.

[6] JEDEC Standard, JESD22-A104D, Temperature cycling.

[7] C. Y. Tsou, T. N Chang, K. C. Wu, P. L. Wu and K.N. Chiang, "Reliability Assessment using Modified energy based model for WLCSP Solder Joints", ICEP2017, Yamagata, Japan, April 19-22, 2017.

[8] Chiang K. N. and Yuan C. A., "An Overview of Solder Bump Shape Prediction Algorithms with Validations," Advanced Packaging, vol.24, no.2, pp.158-162, May, 2001.

[9] J. Chang, L. Wang, J. Dirk, and X. Xie, "Finite Element Modeling Predicts the Effects of Voids on Thermal Shock Reliability and Thermal Resistance of Power Device", Welding Journal , Vol. 85, pp. 63-70, 2006.

[10] L. F. Coffin, "A study of the effects of cyclic thermal stress on a ductile metal," Transactions ASME, Vol. 76, pp. 931-950, 1954.

[11] Y. J. Xu, L. Q. Wang, F. S. Wu, W. S. Xia, H. Liu, "Effect of Interface Structure on Fatigue Life under Thermal Cycle with SAC305 Solder Joints," IEEE Electronic Packaging

Technology International Conference, pp, 959-964, Aug 11-14, 2013.

Thermal Cycling Simulation and Sensitivity Analysis of Wafer Level Chip Scale Package with Integration of Metal-Insulator-Metal Capacitors

Yi Zhou[1], Liangbiao Chen[2], Yong Liu[2]*, Suresh Sitaraman[1]

[1]Department of Mechanical Engineering, Georgia Institute of Technology
Atlanta, GA, USA

[2]ON Semiconductor
South Portland, ME, USA
*corresponding author. E-mail: Yong.Liu@onsemi.com

Abstract— Wafer-level chip scale package (WLCSP) under temperature cycles could be subjected to reliability issues due to thermal mismatch of different materials. Various factors could contribute to WLCSP thermo-mechanical performance, such as bond pad and solder material, die dimensions, as well as the locations of packaging components. In this paper, a novel sensitivity analysis technique is integrated with ANSYS® APDL to improve the thermal cycling performance of a WLCSP package with metal-insulator-metal (MIM) capacitors. In the first example, a finite-element model with details of passivation and metal layers were constructed in ANSYS® APDL. In the model, a total of 30 different variables were defined. Then the model was imported into an optimization software optiSLang to perform sensitivity analysis under thermal cycling loadings. Two runs of sensitivity analyses were carried out, each with 100 designs of experiments generated by space-filling Latin hypercube sampling technique. During sensitivity analysis, optiSLang was able to run 4~8 designs in parallel for faster results. In the first run, the individual contribution of all the parameters to the passivation stresses was examined through correlation coefficients and coefficients of prognosis. The results of the first 100 designs revealed that the Young's modulus of isolation material in MIM structure is a dominating parameter for the high thermal stress. Specifically, dielectric material Si_3N_4 had a much higher modulus than SiO_2 and was adopted as the isolation material in the MIM capacitor, causing higher stress. Due to the dominance of isolation modulus, the contribution of the other parameters was not visible. Therefore, the isolation modulus was fixed in the second run of sensitivity analysis. Then the three most important parameters were identified for both low and high isolation modulus. In the second example, a sensitivity analysis was performed by considering only the three critical parameters using a 3D parametric model to capture the stress patterns and magnitudes better. Through sensitivity analysis using the ANSYS® parametric model in optiSLang, we concluded that the methodology could quickly identify the key parameters for the high thermal stress in the WLCSP and could also provide potential solutions to mitigate the cracking issues.

Keywords- sensitivty analysis; stochastic sampling; WLCSP; MIM capacitor; thermal cycling; isolation cracking

I. INTRODUCTION

Fan-in wafer level chip scale package (WLCSP) is a mature packaging technology widely used in semiconductor products, especially in mobile applications [1]. The integrated passive components such as capacitors are more and more required in WLCSP due to increasing electronic system's functionality [2-3]. The capacitors in ICs typically use a metal-insulator-metal (MIM) structure. Such structure may add uncertainty to the reliability of the WLCSP due to the thermal mismatch of different materials under thermal cycling conditions [4]. Evaluation of MIM parameters, such as stacking geometries and choice of insulation materials, may be needed in addition to other factors that are contributing to the thermo-mechanical performance of WLCSP (i.e., bond pad and solder material).

Finite-element analysis (FEA) can be a useful tool to study the thermos-mechanical performance of WLCSP with MIM capacitors. Although it is ideal to examine all the potential parameters using FEA, it is also much challenging to construct the Designs of Experiments (DoE) by full factorial sampling approach, as the number of simulation legs grows exponentially with increasing design parameters. Therefore, one critical practice is to select only a few key parameters. The key parameters, however, are unknown for a new product or a new failure mechanism. To identify the key parameters a large parameter set, sensitivity analysis may be used with stochastic sampling techniques such as space-filling Latin hypercube sampling [5]. The stochastic-based sensitivity analysis has been applied to handle complex problems in microelectronics packaging with a large number of parameters [6-8]. However, its effectiveness on thermal-mechanical modeling of WLCSP under thermal cycling has not yet been sufficiently examined.

In this paper, we demonstrate the applications of sensitivity analysis on a 7×7, 0.4-mm-pitch WLCSP with the integration of MIM capacitors, as shown in Fig. 1. After the package was mounted on PCB and subjected to thermal cycling tests, the abnormal failure mode was observed. Usually the first failure should appear at the corner solder joints due to longer distance from neutral point [1,9]. However, for the package tested, the first failure always took place at the A4 solder joint, a central one in row A, as shown in Fig. 1. One possible reason may be that MIM capacitors are evenly distributed in the WLCSP at A4 location. Previous studies also indicated that MIM structure could have high thermal stress due to CTE mismatch of different layers [10]. Therefore, it is necessary to perform a sensitivity analysis of varying design parameters, including those related to MIM capacitors, such as the thickness and types of insulation materials.

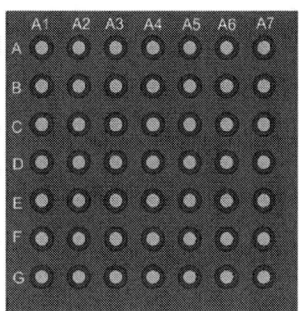

(a) Bottom view of 7x7 WLCSP

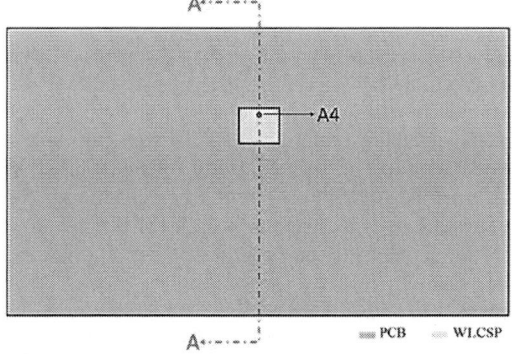

(b) 7x7 WLCSP mounted on PCB

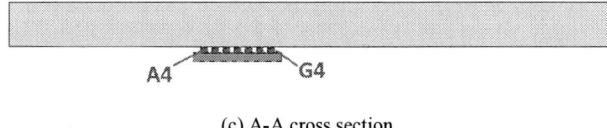

(c) A-A cross section

Figure 1. (a) Bottom view of 7x7 WLCSP; (b) WLCSP mounted on a 39×216.6 ×1.575 mm PCB; (c) A-A cross section of the package. Thermal cycling tests were performed after PCB mounting. The failure occurred at A4 location.

This paper is organized as followings: Section I gives an introduction of current work in the industry. Section II gives an overview of stochastic-based sensitivity analysis and the basic concepts of prognosis and correlation coefficients. Section III provides an example of sensitivity analysis with 30 parameters based on 2D finite-element models. Typical simulation results are shown, followed by two rounds of sensitivity analysis in order to identify the key parameters for different scenarios. As the second example, sensitivity analysis of a full 3D model with a reduced number of parameters is discussed in Section IV. The discussion and conclusions are given in Section V.

II. OVERVIEW OF STOCHASTIC SENSITIVITY ANALYSIS

A. Stochastic Sampling Methods

To perform a global sensitivity analysis, the Designs of Experiments (DoE) must be generated through either deterministic sampling schemes or stochastic schemes. The often-used deterministic sampling scheme is the full or fractional factorial design mainly based on a regular arrangement of the samples [11]. For full factorial design, the number of samples increases exponentially with increasing parameter dimension. For fractional factorial design, the number of the samples also depends on the investigating level: a high investigating level leads to a large number of design points. With a low investigating level, the design points can be reduced, but the model prediction quality will be lower.

Alternatively, stochastic or random sampling schemes can be applied, such as Monte Carlo Simulation (MCS). In MCS, random samples are generated independently in the given design space, assuming that the design variables follow a uniform distribution. With random sampling, the DoE size can be potentially reduced for a large parameter set. However, MCS sampling does not cover the design space adequately if the number of total design points is too small, which is often referred as to design holes or clusters [12]. The problem can be overcome by Latin Hypercube Sampling (LHS) [5]. A comparison between MCS and LHS is illustrated in Fig. 2. Based on LHS scheme, a standard Latin Hypercube approach has been developed to minimize of the undesired correlation. An advanced Latin Hypercube Sampling (ALHS) is also available to reduce the correlation errors by stochastic evolution strategies [13]. The approach of ALHS is recommended if the number of input variables is less than 50.

In the software package optiSLang used in this paper for sensitivity analysis, both deterministic and stochastic schemes are available.

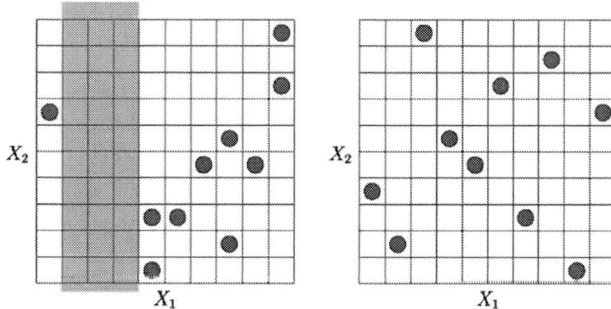

Figure 2. Comparison of two stochastic sampling schemes for a small number of design points. Left: Monte Carlo Simulation (MCS) with holes. Right: Latin Hypercube Sampling (LHS) without holes.[12]

B. Coefficients of Correlation and Prognosis

After a sampling technique is chosen and DoEs are generated, sensitivity analysis can be performed. Several statistical indices can be used to identify the key parameters, such as the correlation coefficient of and the coefficient of

978-1-7281-1500-9/19 $31.00 © 2019 IEEE

Prognosis (CoP). The correlation coefficient can be estimated from a given sampling set as follows [12]:

$$\rho(X,Y) \approx \frac{1}{N-1} \frac{\sum_{i=1}^{N}(x_i - \hat{\mu}_X)(y_i - \hat{\mu}_Y)}{\hat{\sigma}_X \hat{\sigma}_Y} \quad (1)$$

where ρ is the linear correlation coefficient, N is the number of samples; x_i and y_i are the ith sample value of variable X and Y, respectively; $\hat{\mu}$ and $\hat{\sigma}$ are the mean value and standard deviation of the sample value. This quantity ρ ranges from -1.0 (perfect negative correlation) and 1.0 (perfect positive correlation). One drawback of using correlation coefficientt is that it only describes the linear relationship between two variables. To better under the model quality and parameter importance, the coefficient of prognosis (CoP) can be used, as [12]:

$$CoP = 1 - \frac{SS_E^{Prediction}}{SS_T} \quad (2)$$

where $SS_E^{Prediction}$ the sum of squared prediction errors and $SS_T = \sum_{i=1}^{N}(y_i - \mu_Y)^2$ is the total variation. The prediction errors are estimated based on cross validation of sample points and are not dependent on regression and interpolation models [13]. The CoP value for each parameter can range from 0 to 1.0. The closer the CoP to 1.0, the more important of the parameter. The overall CoP value for the model can be used to indicate the quality of the prediction. Higher CoP value means higher model prediction quality.

In general, to perform a sensitivity analysis, one could follow the following steps: 1). build the parametric model; 2) define inputs and responses; 3) choose a DoE generation scheme; 4) solve all the DoEs; 5) evaluate the correlation and CoP values for variable sensitivities. In this study, the parametric model is built using ANSYS® APDL, and the sensitivity analysis is performed by the optimization program optiSLang which is built in ANSYS®.

III. EXAMPLE 1: SENSITIVITY ANALYSIS OF 2D STRESS SIMULATION

A. Model Description

In this example, the failure mode of 7x7 WLCSP shown in Fig. 1 is studied by finite-element analysis and then sensitivity analysis. To locate the failure spot on the package, a cross section through A4 and G4 was simulated (see Fig. 1). Figure 3 shows a typical mesh of the cross section with PCB. The stacking details of WLCSP at A4 location are illustrated. It can be seen that there are MIM capacitor structures between the solder pad and silicon die at A4 location, which is the major difference between A4 and the other balls. The MIM capacitor structure is composed of three layers of metals and four layers of isolation materials. The metal layers that are closer to the solder pad are labeled as Metal3 and Metal2; they were modeled as a series of pads. IMD4 is the isolation layer between solder pad and Metal3. In this analysis, all vias between layers were not modeled.

The size of a typical PCB is 39×216.6 ×1.575 mm. The thermal cycling condition is -40 ~125°C, 2 cycles per hour. A total of four cycles were run.

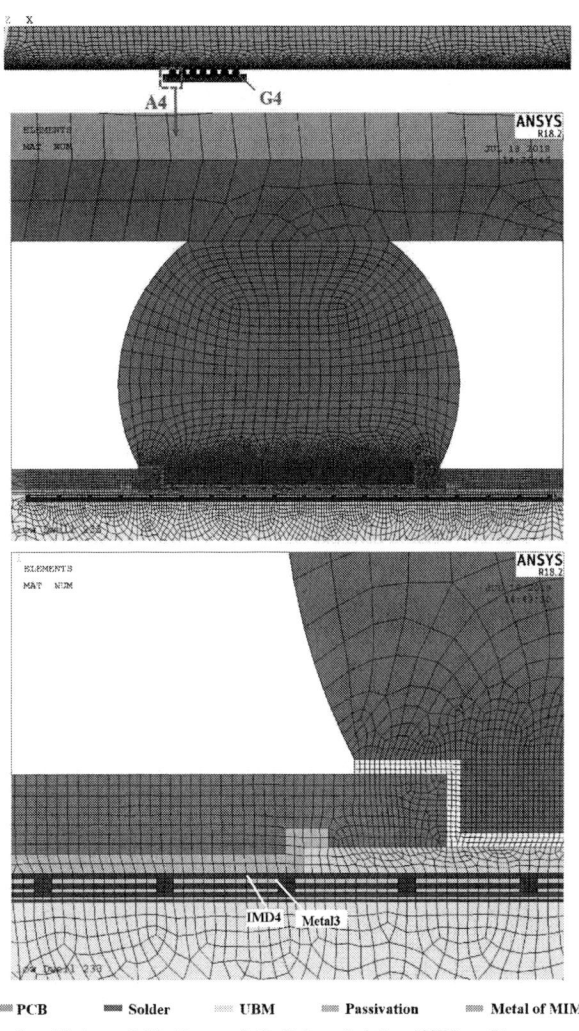

Figure 3. 2D Model of the WLCSP at A4 location with meshing details. There are no metals of MIM underneath other balls

Table I-III provides the material properties that were used in the FEA simulation. The FR4 material, silicon, silicon nitride, oxide passivation, polyimide and UBM are all considered as linear elastic materials. The UBM layer is nickel-based metal with flash gold on the top and a plating seed layer and adhesion metal. In the FEA model, the UBM is simplified as a single layer with its material properties made up of a combination of nickel, gold, plating seed metal and adhesion metal. For the solder type of SAC, Anand's model is use, and the parameters are shown in Table II. The copper board pad on PCB was considered as bilinear material, and the bilinear material properties are presented in Table III. The solder pad and the metal in MIM capacitors structure is Al-0.5Cu, and its bilinear material properties are also presented in Table III.

TABLE I. YOUNG'S MODULUS AND POISSON'S RATIO OF THE SIMULATION MATERIALS

Component	Young's modulus (MPa)	Poisson's ratio	CTE ppm/c
Die	161000	0.26	2.6
Board pad	117000	0.33	16.12
Solder ball	See Table II	0.37	21.9
FR4 Board	E_x=25420 G_{xz}=4971 E_y=25420 G_{yz}=4971 E_z=11000 G_{xy}=11450	v_{xy} = 0.11 v_{xz} = 0.39 v_{yz} = 0.39	$Alp_{x,y}$=14 Alp_z=45 (≤180°C) $Alp_{x,y}$=16 Alp_z=220 (>180°C)
Polyimide	3500	0.35	35
Pad	68900	0.33	20
UBM	124500	0.299	15
Passivation	155000	0.205	1.7
MIM Metals	68900	0.33	20
Isolation	314000	0.33	4

TABLE II. EXPERIMENTALLY DETERMINED AND FITTED ANAND'S MODEL CONSTANTS FOR SOLDER ALLOY [14]

Description	Symbol	Value
Initial value of s	s_0	1.3 MPa
Activation energy	Q/R	9000 K
Pre-exponential factor	A	500 /s
Stress multiplier	ξ	7.1
Strain rate sensitivity of stress	m	0.3
Hardening coefficient	h_0	5900 MPa
Coefficient for deformation resistance saturation value S	$\hat{S}$	39.4 MPa
Strain rate sensitivity of saturation value	n	0.03
Strain rate sensitivity of hardening coefficient	a	1.4

TABLE III. YIELD STRESS AND TANGENT MODULUS OF COPPER AND ALUMINUM PAD

Material	Yield Stress (MPa)	Tangent Modulus (MPa)
Copper	70	700
Al0.5Cu	200 @ 25 C 164.7 @ 125C	300@ 25 C 150@ 125 C

B. Simulation Results

Prior to sensitivity analysis, it is essential to examine the simulation results of the parametric model based on typical model parameters. Figure 4 compares the experimental failure location and the stress contours of isolation layers (assumed as silicon nitride material) under A4. It can be seen that higher stress appears in the isolation layers. The maximum isolation stress of 852.05 MPa is higher than the silicon nitride flexural strength of 689MPa. Therefore, the thermal cycling could result in cracking at the isolation and damage the MIM capacitors, which is consistent with the physical test data.

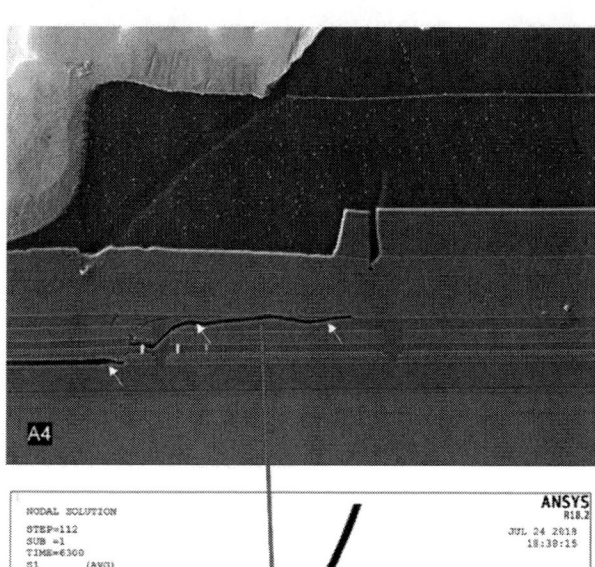

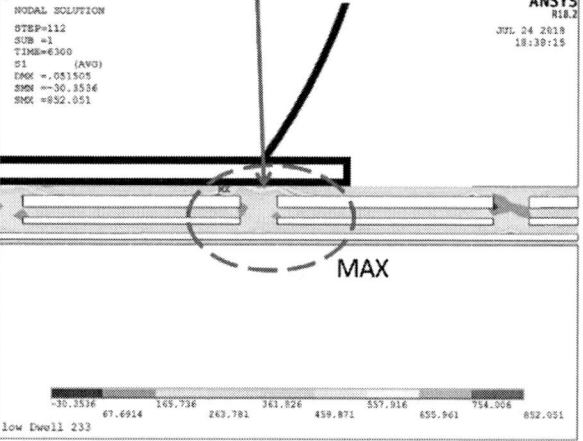

Figure 4 A comparison of experimental results with 2D simulation results under A4 solder joint. Maximum tensile stress is 852 MPa.

C. Sensitiivty Analysis Setup

After the 2D model was validated, sensitivity test was carried out to understand how the package geometry and material parameters impact the isolation stress in WLCSP. There are two ways to perform sensitivity analysis with optiSLang. One is to use its plugin in ANSYS® Workbench where the workflow has been built and no additional coding is needed. However, the simulation may be slow due to frequent communication between optiSLang, Workbench, and ANSYS® APDL. The other way is to use the stand-alone optiSLang to integrate ANSYS APDL directly, which requiring a script. The latter method is used in this paper.

In the sensitivity analysis, we considered 30 different parameters. The parameters include the geometry details of the WLCSP design such as each layer thickness, MIM capacitor size, and different solder final collapse heights. In general, the parameters related to the MIM capacitors structure were given a range of ±25%, and others were given ±10%. In optiSLang, exceptions can be applied to parameters that were constrained by the geometry. The material property variation was also taken as input parameters. In this study, three different solder materials were tested, including two SACs, and one high-Pb. Two solder pad materials were used in the simulation, Al-0.5Cu and pure copper. Furthermore, the isolation material's Young's modulus was another critical parameter. Depending on the material used (i.e., SiO2 or Si3N4), Young's modulus was set to be 69~340GPa. To study the isolation cracking, the output of the sensitivity test was set to be the first principal stress of the isolation layers. To speed up the sensitivity analysis, 4~8 designs were run in parallel.

D. Results of Sensitivityy Analysis

We first run 100 DoE designs with different input parameter values using the Latin Hypercube Sampling method.89 designs succeeded (out of 100) in the sensitivity run, while the other 11 designs failed because the parameters did not meet the geometry constraints. Figure 5 presents the correlation coefficient plot of the input parameters versus the insulationn first principle stress based on the successfull designs. The plot shows the correlation coefficients between important input parameters and the isolation tensile stress (which is called the model response). As seen from the Fig. 5, Young's modulus of the isolation material is the dominating factor and hasa positive linear relationship, with acorrelationn coefficient of 0.927. There is very little correlation between the isolation stress and the other parameters.

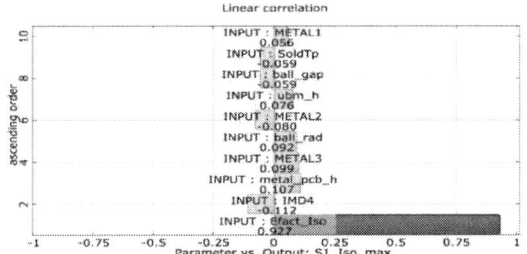

Figure 5. Correlation coefficient plot of first-run sensitivity analysis. The dominating factor is found to be Young's modulus of isolation material (labeled by Efact_Iso).

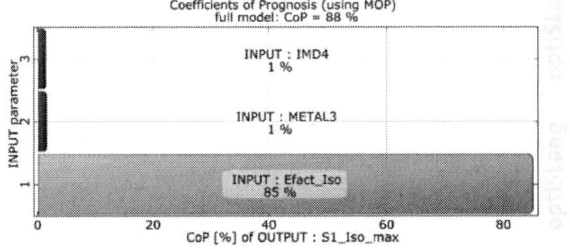

Figure 6. Coefficients of Prognosis (CoP) of first-run sensitivity analysis. The most important parameter is identified as Young's modulus of isolation material (labeled by Efact_Iso).

To examine the model quality based on the 89 design points, the values of CoP were reported in Fig. 6. The overalll CoP for the full model is 88%, which indicates a good model prediction. The three most important parameters identified are Young's modulus of isolation material, the thickness of metal layer Metal3 (the first metal layer under

solder pad), and the thickness of the isolation layer IMD4 (between solder pad and metal 3).Thee Young's modulus of isolation makes an 85% contribution to the maximum isolation tensile stress. The contribution of the other two parameters is less than 2%. Both the correlation plot and CoP plot indicate that among all input parameters, Young's modulus dominates the contribution to the maximum tensile stress of isolation materials.

To understand the effects of other parameters, two typical isolation materials were chosen, which are silicon nitride and silicon dioxide. Another round of sensitivity analysis was then carried out for each isolation material. Young's Modulus is chosen as 314 GPa for silicon nitride and 69 GPa for silicon dioxide. The other input parameters were kept the same as the first run. Figure 7 shows the correlation plot for the case of silicon nitride (high-modulus isolation). The results indicate that the solder pad type is an important parameter, with a negative correlation coefficient of -0.525. This means that Cu pad (type 2) yields less stress than Al-0.5Cu (type 1). Other important parameters for high isolation stress are large Metal3 thickness (correlation coefficient=0.447) and small IMD4 thickness (correlation coefficient=-0.314). The CoP plot in Fig. 8 is consistent with the correlation coefficient plot regarding the ranking of important parameters. Among the 50 designs, the isolation tensile stress difference could be up to 51%, as shown in Table IV.

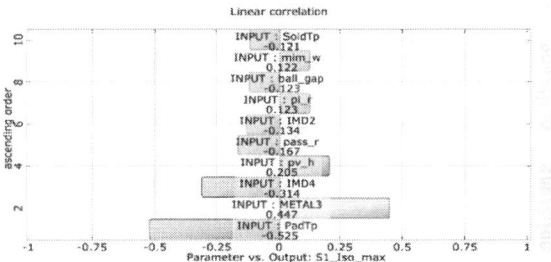

Figure 7. Correlation coefficient plot of 2nd run sensitivity analysis, with isolation Young's modulus = 314 GPa.

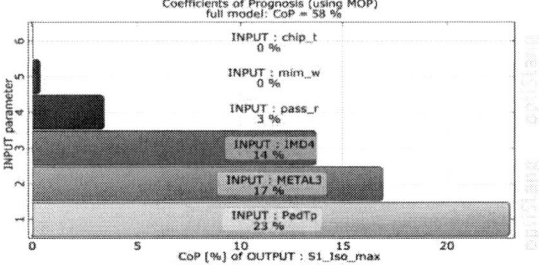

Figure 8. Coefficients of Prognosis (CoP) plot of 2nd run sensitivity analysis, with isolation Young's modulus = 314 GPa

TABLE IV. COMPARISON OF BEST CASE AND WORST CASE IN SENSITIVITY ANALYSIS WITH ISOLATION YOUNG'S MODULUS = 314 GPA

Parameter/output	Worse Case	Best Case	Difference
Pad Type	Al-0.5 Cu	Cu	/
Metal3 thickness	1.1 um	0.84 um	-23.6%
IMD4 thickness	0.81 um	1.11 um	+37%
Isolation S1, MPa	1396.0	683.1	-51.0%

Similarly, Figure 9 and Figure 10 present the correlation and CoP plots for the case of silicon dioxide (low-modulus isolation material). It can be seen that the ranking of the correlation coefficient plot is slightly different from the CoP plot. High correlation coefficients were found for solder pad type, PCB thickness (PCB_t), IMD1/2 thickness and IMD4 and metal-3 thickness. However, the CoP plot gives the same ranking of important parameters as the case of Si_3N_4, with solder pad type, metal3 thickness, and IMD4 thickness as the three most important parameters. The CoP value is 75% for the whole model, which is better than the low modulus case. Depending on the values of key parameters, the isolation tensile stress difference could be up to 52.8%, as shown in Table V.

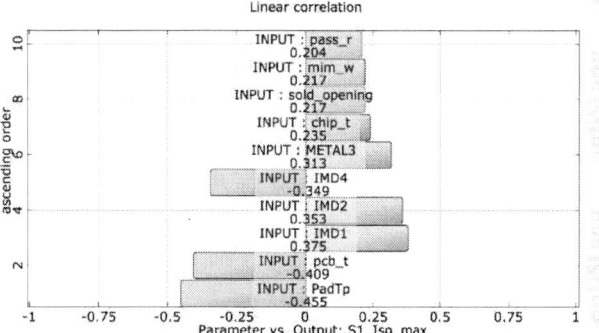

Figure 9. Correlation coefficient plot of 2nd run of sensitivity analysis, with isolation Young's modulus = 69 GPa.

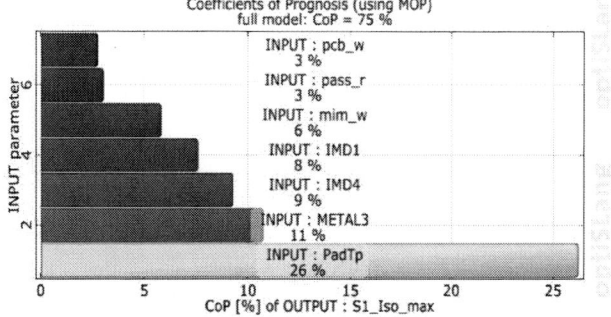

Figure 10. Coefficients of Prognosis (CoP) plot of run 2 sensitivity analysis, with isolation Young's modulus = 69 GPa

TABLE V. COMPARISON OF BEST CASE AND WORST CASE IN SENSITIVITY ANALYSIS WITH ISOLATION YOUNG'S MODULUS = 69 GPA

Parameter/output	Worse Case	Best Case	Difference
Pad Type	Al-0.5 Cu	Cu	/
Metal3	1.2 um	0.82 um	-32.0%
IMD4	0.37 um	0.47 um	+27.0%
Isolation S1	409.8	193.4	-52.8%

IV. EXAMPLE 2: SENSITIVITY ANALYSIS OF 3D STRESS SIMULATION

A. 3D Model Based on Reduced Parameter Set

The 2D model could help identify the key parameters based on the sensitivity analysis of a large parameter set. In the second example, we built a 3D model that represents the package better. The model is shown in Fig. 11, where the quarter model is used to simplify the simulation. Based on the first example, we already learned that there are three most parameters for isolation stress: solder pad type, metal-3 thickness, and IMD4 thickness from Fig. 8. The importance of their effects will be verified by using the 3D model. In this example, the high-modulus isolation material (Si_3N_4) is chosen for the sensitivity analysis. Due to the small number of parameters, we used level-3 full factorial sampling method, which results in 18 designs (the value of pad type is either 1 or 2, representing Al-0.5Cu pad or Cu pad, respectively). Both the IMD4 and Metal3 thickness have a range of 0.5 um to 2.0 um. The other material properties and geometries are the same as the 2D finite element model. To speed up the sensitivity analysis, 4~8 designs were run in parallel.

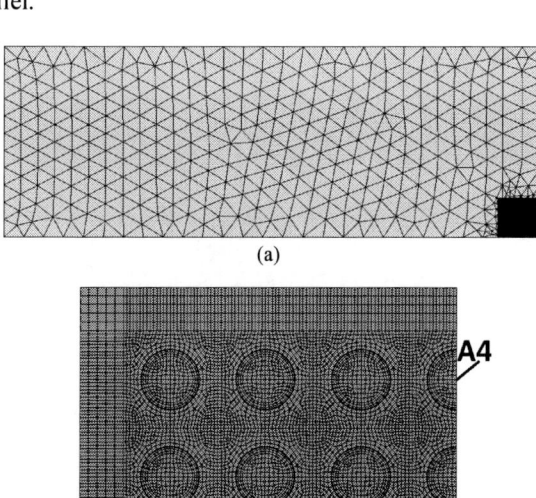

(a)

(b)

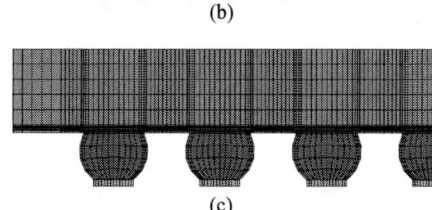

(c)

Figure 11. (a) Top view of 3D quarter model with PCB; (b) 3D WLCSP model, top view; (c) 3D WLCSP model, side view

978-1-7281-1500-9/19 $31.00 © 2019 IEEE

B. Results of Sensitivity Analysis Based on the 3D Model

Figure 12 shows a typical isolation tensile stress distribution under A4 according to the 3D stress simulation model. The maximum stress location matches well with the experimental tests. The cross-sectional stress distribution in Fig. 12b also agrees with the 2D stress simulation.

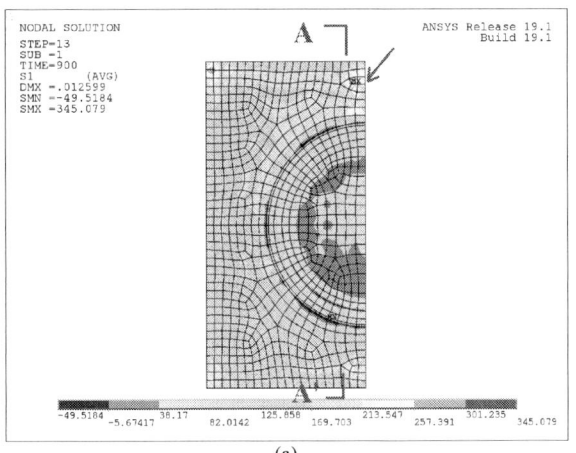

(a)

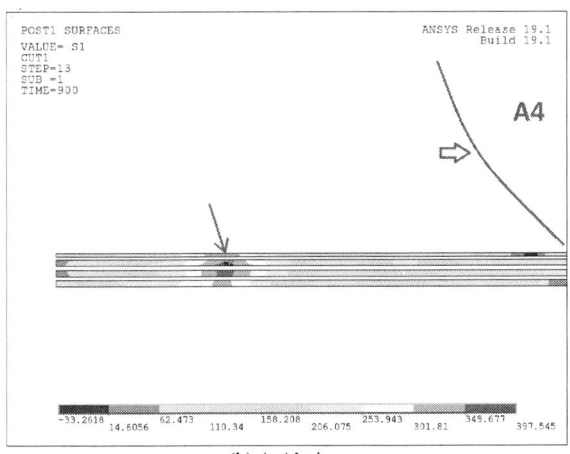

(b) A-A' view

Figure 12. Typical isolation stress distribution based on 3D stress simulation (a) Top view; (b) cross-sectional view through A-A' plane

Figures 13-14 provides the correlation plot and CoP plot for design points of the 3D model, respectively. It can be seen that 3D isolation stress almost purely depends on the metal3 thickness. Recall that the 2D isolation stress is also dependent on solder pad type and IMD4 thickness (see Fig. 7 and Fig. 8). In the 3D example, a perfect CoP value of 100% is obtained.

The response surface plot for isolation stress is given in Fig. 15, showing a nonlinear relationship between metal3 thickness and isolation stress. Unlike 2D stress simulation, changing IMD4 thickness and switching solder pad type does not change the isolation stress much in the 3D cases considering. The highest isolation stress obtained is 614 MPa when metal3 thickness is 2 um, which is closed to Si3N4 tensile strength. Using 0.5-um metal-3 thickness yields stress of 400 MPa, about 35% stress reduction. Therefore, the 2D

sensitivity analysis may be valuable for an initial estimation, and a 3D sensitivity analysis may be required to verify the 2D findings.

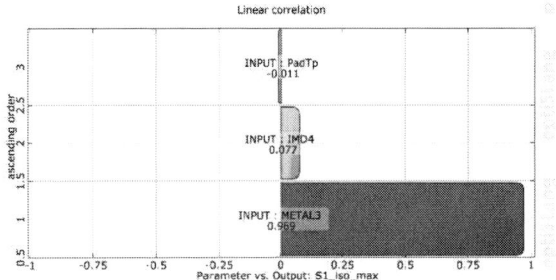

Figure 13. Correlation coefficient plot for 3D isolation tensile stress versus the input design parameters

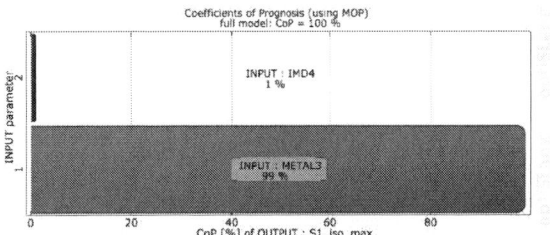

Figure 14. Coefficient of prognosis (CoP) for 3D isolation stress versus the most important design parameters

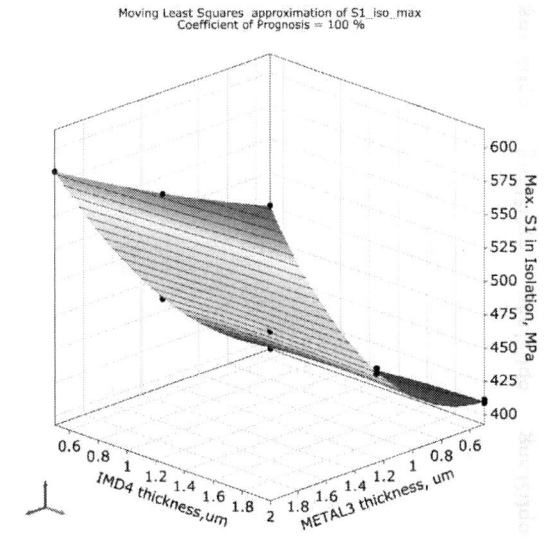

Figure 15. 3D response surface of isolation stress versus the two most important parameters

V. DISCUSSION AND CONCLUSION

This paper implemented both stochastic and deterministic sensitivity analysis to understand the failure modes of WLCSP with MIM capacitors. The FEA simulation results indicate that the failure of the 7×7, 0.4 mm pitch WLCSP in the thermal cycle test was due to the presence of the MIM

978-1-7281-1500-9/19 $31.00 © 2019 IEEE

capacitors structure. The silicon nitride layer for MIM structure has much higher stress than other components. The maximum stress of isolation appears in the layer closest to the solder pad, where the cracking could initiate. The simulation results are consistent with the experimental findings. To understand the contribution of material properties and geometry parameters to the failure mode, stochastic sensitivity tests with a large number of parameters were carried out using the advanced LHS built in optiSLang. The sensitivity tests clearly showed that higher isolation Young's modulus leads to higher stress in isolation, but did not provide sufficient information on other parameters. To better reveal the contribution of the other parameters, an additional sensitivity analysis was performed for two selected isolation materials, from which the three most important parameters were identified. Based on the first example, we built a 3D model to represent the package stress betters but used only three parameters in the sensitivity analysis. The 3D results provided a better estimation of isolation stress, and the sensitivity analysis showed that reducing MIM metal thickness could greatly reduce the isolation stress.

Based on the two examples, we learned that the primary challenge of sensitivity analysis for studying packaging thermal-mechanical performance lies in the building of a proper parametric model with different parameters. For 3D FEA analysis, the computational time is more expensive than 2D. To make sensitivity analysis more effective, stochastic-based sampling method and running multiple designs in parallel are recommended.

ACKNOWLEDGMENT

The authors would like to thank the mobile team of ON Semiconductor for their great support on this project.

REFERENCES

[1] S. Qu and Y. Liu. Wafer-Level Chip-Scale Packaging: Analog and Power Semiconductor Applications. Springer, 2016.

[2] P.J. van der Wel, J. F. Dijkhuis, C. Chanlo, H.K.J. ten Dolle, L.C.M. van den Oever and R.J. Havens, "RF Characterisation and Process Control for Passive Integration Components", Proceedings of 57th Electronic Components and Technology Conference,2007, pp. 1855-1860.

[3] A. den Dekker, A. van Geelen, P. van der Wel, R. Koster and E. C. Rodenburg, "Passi4: The next Technology for Passive Integration on Silicon," 2007 Proceedings 57th Electronic Components and Technology Conference, Reno, NV, 2007, pp. 968-973. doi: 10.1109/ECTC.2007.373914

[4] M. Hsieh, "Finite Element Analyses for Critical Designs of Low-Cost Wafer-Level Chip Scale Packages," IEEE Transactions on Components, Packaging and Manufacturing Technology, vol. 4, no. 3, pp. 451-458, March 2014.

[5] McKay, M., R. Beckman, and W. Conover, "A comparison of three methods for selecting values of input variables in the analysis of output from a computer code," Technometrics, vol. 21, pp. 239-245, 1979.

[6] S. Tandon, E. Liu, T. Zahner, S. Besold, W. Kalb and G. Elger, "Transient thermal simulation of high power LED and its challenges," 2017 18th International Conference on Thermal, Mechanical and Multi-Physics Simulation and Experiments in Microelectronics and Microsystems (EuroSimE), Dresden, 2017, pp. 1-8.

[7] R. Schlegel, A. Müller, R. Niemeier and P. J. Gromala, "Parameter identification for interface delamination processes in molded electronic packages," 2016 17th International Conference on Thermal, Mechanical and Multi-Physics Simulation and Experiments in Microelectronics and Microsystems (EuroSimE), Montpellier, 2016, pp. 1-4.

[8] P. Gromala, B. Muthuraman, B. Öztürk, K. M. B. Jansen and L. Ernst, "Material characterization and nonlinear viscoelastic modelling of epoxy based thermosets for automotive application," 2015 16th International Conference on Thermal, Mechanical and Multi-Physics Simulation and Experiments in Microelectronics and Microsystems, Budapest, 2015, pp. 1-7

[9] T.-K. Lee, T. Bieler, T., C. Kim, H. Ma. Fundamentals of Lead-Free Solder Interconnect Technology: From Microstructures to Reliability. Springer, 2018

[10] Y. Liu, Y. Liu, S. Irving and T. Luk, "Thermal stress simulation in the metal-insulator-metal (MIM) wafer fabrication process," 2008 58th Electronic Components and Technology Conference, Lake Buena Vista, FL, 2008, pp. 1067-1072

[11] R. Myers, and D. C. Montgomery. Response Surface Methodology (2 ed.). John Wiley &Sons, Inc, 2002

[12] Most, T. and J. Will, "Metamodel of Optimal Prognosis - an automatic approach for variable reduction and optimal metamodel selection," In Proc. Weimarer Optimierungs- und Stochastiktage 5.0, Weimar, Germany, November 20-21, 2008

[13] D.E. Hungtington and C. S. Lyrintzis, "Improvements to and limitations of Latin hypercubesampling," Probabilistic Enginerring Mechanics, vol. 13, pp. 245-253, 1998

[14] T.O. Reinikainen, P. Marjamäk, and J.L. Kivilahti, "Deformation Characteristics and Microstructural Evolution of SnAgCu Solder Joints," EuroSime Conference Proc., Germany, Apr. 2005

Effect of Time-dependent Bulk Modulus on Reliability Assessment of Automotive Electronic Control Unit

Hyun Seop Lee and Bongtae Han*
Mechanical Engineering Department
University of Maryland, College Park, MD 20742, USA
bthan@umd.edu

Przemyslaw Gromala
Automotive Electronics
Robert Bosch GmbH, 72703 Reutlingen, Germany
przemyslawjakub.gromala@de.bosch.com

Abstract — Time-dependent bulk modulus of EMCs is measured using an embedded fiber Bragg grating (FBG) sensor while subjecting EMCs to hydrostatic pressure. The time-dependent behavior of the bulk modulus, especially around the glass transition region, is discussed. Then, an LVE based numerical modeling is employed to quantify the effect of time-dependent bulk modulus on the internal stresses of encapsulated automotive ECU units subjected to the solder reflow profile. The results are compared with the case of the time-independent bulk modulus, and the effect on the reliability assessment is discussed.

Keywords- Epoxy molding compound, hydrostatic pressure, bulk modulus.

I. INTRODUCTION

Epoxy molding compounds (EMCs) are thermosetting polymers filled with a large quantity of silica particles. They have been widely used to protect active devices in semiconductor packaging.

It has been well known that EMCs show viscoelastic behaviors. The stress inside the packages is directly related to the viscoelastic properties of EMC. The EMC behavior has been described typically by the linear viscoelastic (LVE) model.

The constitutive law of the linear viscoelasticity is [1]:

$$s_{ij}\left(t\right) = 2\int G\left(t-\tau\right)\frac{\partial e_{ij}\left(\tau\right)}{\partial \tau}d\tau$$
$$\sigma_{kk}\left(t\right) = 3\int K\left(t-\tau\right)\frac{\partial \varepsilon_{kk}\left(\tau\right)}{\partial \tau}d\tau \tag{1}$$

where s_{ij} and e_{ij} are the deviatoric stress and strain tensors; $G(t)$ and $K(t)$ are the time-dependent shear modulus and bulk modulus; and, σ_{kk} and ε_{kk} are the dilatational stress and strain tensors.

In the linear viscoelasticity regime, two elastic constants in Eq. (1) have the following relationships with two other elastic constants as:

$$K(s) = \frac{E(s)G(s)}{9G(s)-3E(s)}, \quad G(s) = \frac{E(s)}{2\left(1+v(s)\right)} \tag{2}$$

where $E(s)$ and $v(s)$ are the transformed time-dependent Young's modulus and Poisson's ratio to the Laplace plane. Therefore, two of the four constants in Eq. (2) have to be measured as a function of time experimentally for predictive modeling.

The Young's modulus is the easiest to measure among four constants with standard testing apparatuses. However, due to the complexity involved in measuring elastic constants other than the Young's modulus, the bulk modulus is often assumed to be "time-independent" [2-4]. This assumption is based on the fact that the traditional viscosity theories based on the shear motion [3]. With this assumption, only Young's modulus and the Poisson's ratio at room temperature has to be measured for viscoelasticity model. With the measured Poisson's ratio and Young's modulus, the bulk modulus can be approximated as:

$$K(t) \approx K = \frac{E(0)\big|_{T=20°C}}{3\left(1-2v(0)\big|_{T=20°C}\right)} \tag{3}$$

It is important to note that the assumption of "time-independent" bulk modulus implies that the bulk modulus becomes temperature-independent if the EMC behavior follows the thermo-rheological simplicity (TRS) [5, 6]. Using the constant K value, the time-dependent shear modulus can be calculated using:

$$G(s) = \frac{3K \cdot E(s)}{9K - E(s)} \tag{4}$$

Recently, the Fiber Bragg Grating (FBG) sensor has been used to characterize the bulk modulus of EMC [7-12]. The results clearly indicated that the bulk modulus has a strong time-dependent behavior, especially around the glass transient region: i.e., the assumption of time-independent bulk modulus is not valid over a large temperature range.

One of the advanced electronics that have been adopted in automotive technologies to enhance user interfaces is an automotive electronic control unit (ECU), which is an embedded system that controls an electrical system or subsystems in a transport vehicle. In a conventional ECU, a protective metal case has been used to ensure reliability under harsh environmental conditions. Recently, the EMC has been adopted to replace the metal case. The EMC technology reduced a manufacturing cost significantly, yet the presence of a large amount of outer EMC can increase the stresses of ECUs during the transfer molding process and operations. The temperatures of the manufacturing process and the operating conditions of ECUs are higher than the glass transition temperature of the EMC, and thus, the time-dependent bulk modulus should be considered in stress prediction unless its effect is negligible.

In this paper, the time-dependent bulk modulus of EMCs is measured using an embedded fiber Bragg grating (FBG) sensor by subjecting EMCs to hydrostatic pressure. After discussing the bulk modulus behavior at various temperatures, an LVE based numerical modeling is employed to quantify the effect of time-dependent bulk modulus on the internal stresses of encapsulated automotive ECU units subjected to the solder reflow profile as well as the automotive operating conditions. The results are compared with the case of the time-independent bulk modulus, and the effect on the reliability assessment is discussed.

II. BACKGROUND: BULK MODULUS MEASUREMENT USING FBG SENSOR METHOD

The FBG sensor method was originally developed to characterize the curing behavior of advanced semiconductor packaging materials.[7-9] More recently, the method was extended to measure the temperature-dependent elastic properties of EMC. [12, 13] This section reviews the basic concept of the method, the procedure to prepare the EMC specimen, and the test setup for hydrostatic pressure.

A. Governing Equation

Fig. 1 shows the specimen configuration with FBG embedded in the center and general loading condition of axial pressure, P_1 ., and a radial pressure, P_2 . An analytical solution for the stress distribution in the fiber and polymer under the generalized plane strain condition is available in the literature, and the detailed description of the solution can be found in [12].

The BW shift can take the following form [12]:

$$\Delta\lambda = \Pi\left(E_p, \nu_p, \beta\right) \qquad (5)$$

where β is the ratio between the radius of the polymer (specimen) and the radius of the fiber, E_p and ν_p are the Young's modulus and Poisson's ratio of the polymer; Π is a nonlinear function that can be expressed explicitly as a function of the configuration (β) and the properties of polymer and the optical fiber [12].

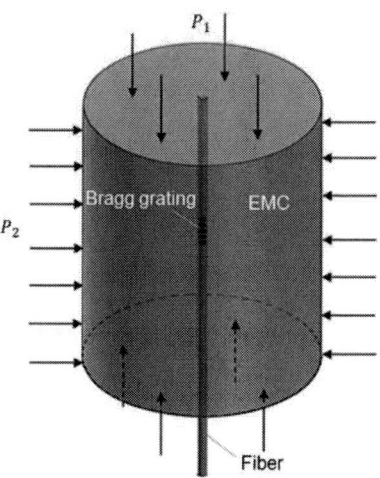

Fig. 1 Specimen configuration and general loading condition [13]

The BW shift is a function of any two of four elastic constants of the polymer with given loading and configuration. Two constants can be determined from the BW shifts measured from two independent experiments with different loading condition. In order to avoid the complex non-linear regression with two unknown parameters, a sequential procedure was proposed in Ref. [12], where Young's modulus was determined first from the results of uniaxial testing, and the bulk modulus was subsequently determined from the results of hydrostatic pressure testing.

The time-dependent properties can be determined from the BW shift measured as a function of time from the following inverse functions:

$$E_p(t) = \Pi^{-1}\left(\Delta\lambda(t), \beta\right);$$
$$K_p(t) = \Pi^{-1}\left(\Delta\lambda(t), \beta\right) \qquad (6)$$

where E_p and K_p are Young's modulus and bulk modulus of the polymer, respectively. The descriptions of the inverse functions can be found in Ref. [12].

B. Measurement of Time-dependent Bulk Modulus

The specimen for the measurement is fabricated using a custom designed stainless-steel mold assembly, which is shown schematically in Fig. 2. The inset shows a plunger with a slit for the optical fiber that applies the required pressure to the specimen while holding the optical fiber at the center. The optical fiber with diameter of 125 μm and 5 mm of FBG is paced through the curing mold and a small hole in the center of an uncured EMC pellet.

The internal surfaces of the mold are treated with a release agent and the uncured pellet is wrapped with a thin Teflon tape layer before the curing process starts. The purpose of these pre-processing is to ensure easy separation of the specimen from the mold after curing as well as no constraint from the mold walls during curing which will affect the curing condition of the specimen related to the signal quality.

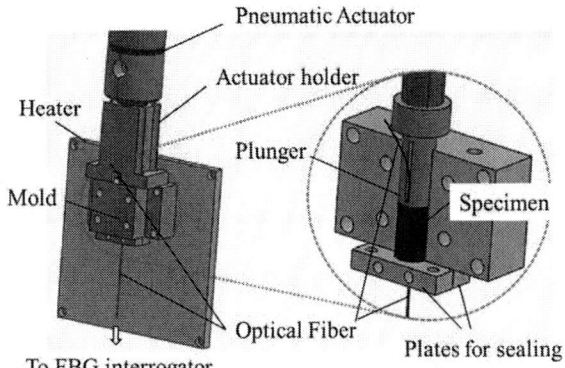

Fig. 2 Mold assembly to fabricate the EMC specimen [13]

A large mechanical pressure has to be applied during curing. The required pressure (7 MPa) is achieved by the mechanical plunger. More detailed procedure for EMC specimen preparation including the curing pressure application can be found in Ref. [12].

A small test chamber is designed to accommodate the required pressure of the gas to provide hydrostatic loading to the fabricated specimen as shown in Fig. 3. The specimen is placed inside the chamber which has a slightly larger internal diameter than the specimen so the gas can fill the gap and apply the hydrostatic pressure to the specimen.

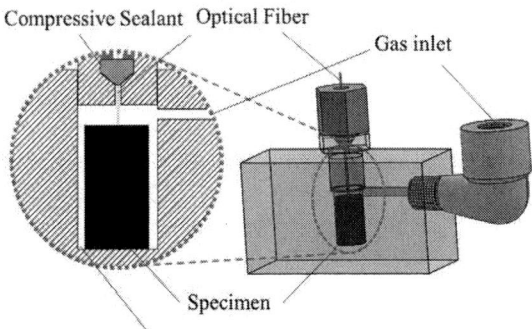

Fig. 3 Setup for the hydrostatic test [13]

The chamber has to be sealed completely so a special sealing system is used as schematically shown in Fig. 3. More detailed information of the setup and the procedure to measure time dependent Young's and bulk modulus can be found in [13].

The specimen was tested from 25 °C to 235 °C. The temperature interval between measurements was 20 °C below and above the glass transition temperature, and 5 °C over the glass transition range. Two different pressures of 6.9 MPa and 1.4 MPa were applied for the temperatures below and above the glass transition temperature, respectively.

The representative data obtained from hydrostatic creep testing at two different temperatures at which large time-dependent behavior under hydrostatic loading was observed are shown in Fig. 4. The time-dependent bulk moduli at various temperatures are shown in Fig. 5.

III. EFFECT OF BULK MODULUS ON ECU RELIABILITY

A. ECU Model

An overmolded ECU package is used to investigate the effect of the time dependent bulk modulus. The stress model for the outer mold and also DPAK package inside the ECU is investigated. Several failure modes are known in the DPAK molded inside an ECU. They include delamination between molding compound and copper lead frame, wire bond failure, and solder joint failure [16]. One of the most common failure modes is wire lift-off in the power electronic device [17-19]. Fig. 6 (a) illustrates the interfacial delamination between the gate Al wire and Cu lead inside a molded ECU after a series of thermal cycling test. The interface between an aluminum wire bond and metal pad shown in Fig. 6 (b) is selected to investigate the effect of the bulk modulus.

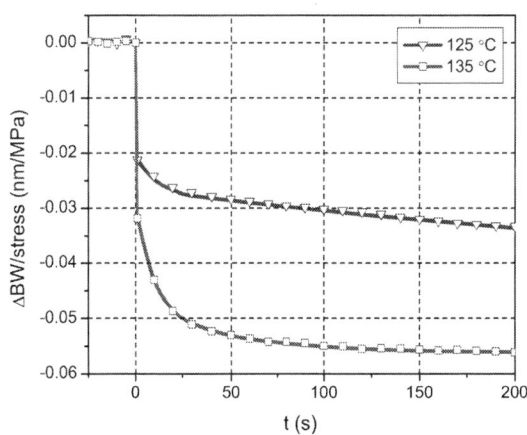

Fig. 4 BW change normalized by applied hydrostatic stress at representative temperatures with large time dependent behavior

The schematic of the cross-section of the molded ECU package used in the analysis is shown in Fig. 7.[20] The validity of the model was verified by Moiré interferometry in Ref. [20].

In the FEM modelling, the shear modulus and bulk modulus are required as an input data. The master curve of the bulk modulus was obtained from the data in Fig. 7. The shear modulus obtained from the previous data [13] was utilized in this study. The bulk modulus and shear modulus are fitted by the Prony series in (Eq.(7)). The final master

978-1-7281-1500-9/19 $31.00 © 2019 IEEE

curves for time-dependent shear and bulk modulus are shown in Fig. 8.

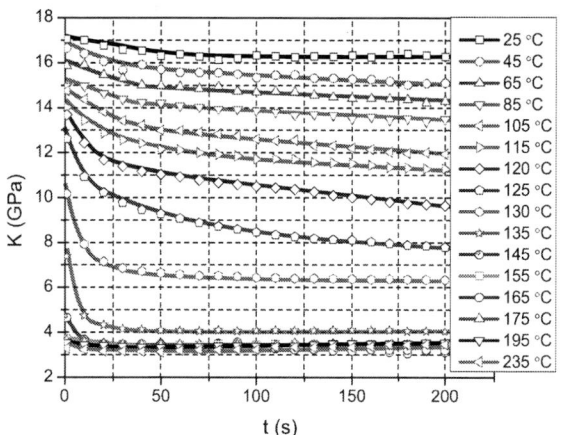

Fig. 5 Time and temperature dependent bulk modulus

The modelling result using these master curves will be compared to the master curves with time-independent bulk modulus assumption, which are shown in Fig. 9. As mentioned in the previous section, time-independent bulk modulus also implies temperature-independent bulk modulus as the master curves for linear-viscoelastic model will be used with shift factors assuming thermo-rheological simplicity (TRS) of the material. Isochronous bulk modulus difference with and without constant bulk modulus assumption is shown in

$$K(t) = K_\infty + \sum_{i=1}^{20} K_i \exp\left(-\frac{t}{\tau_i}\right)$$

$$G(t) = G_\infty + \sum_{i=1}^{20} G_i \exp\left(-\frac{t}{\tau_i}\right)$$

(7)

B. Result and Analysis

The DPAK package model shown in Fig. 11 was subjected to a thermal loading (heating from 25 to 240 °C) with two different bulk moduli: (1) temperature-dependent bulk modulus and (2) constant bulk modulus.

Normal stresses at the wirebond interface were analyzed to investigate the effect of the lift-off failure. The maximum normal stress was determined first with the temperature-dependent bulk modulus, and then it was compared with the case of the constant bulk modulus. When the constant K was used, the maximum normal stress at the interface increased by 9.2 %.

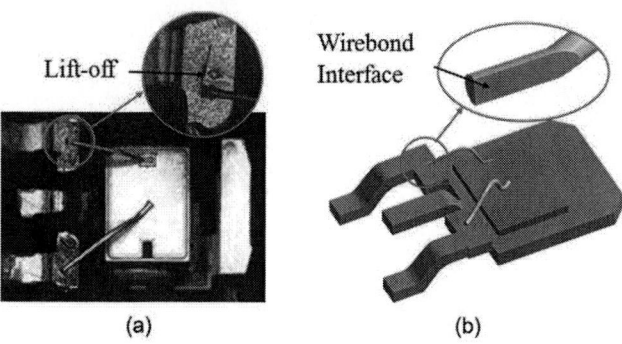

Fig. 6 (a) Wirebond "lift-off" failure in a DPAK, and (b) Al wire bond interface of interest [20]

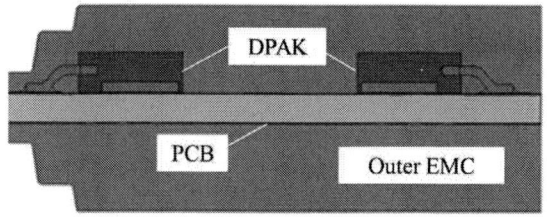

Fig. 7 Cross-sectional view of the FEA model to simulate the presence of the outer mold [20]

The same thermal loading condition was applied to the DPAK molded by an outer EMC as shown in Fig. 9. For the comparison purpose, different properties with and without the constant bulk modulus assumption was applied to the molding compound of DPAK.

The stresses of the bonding wire for two cases are shown in Fig. 12. The maximum normal stress at the wirebond interface increased by 41.2% when the constant bulk modulus was used. It is attributed to the fact that DPAK EMC was constrained excessively with the outer mold.

The maximum principal stresses of the DPAK EMC for both cases are shown in Fig. 13. They also increased approximately 3 times in both tension and compression when the constant bulk modulus was assumed. It is worth noting that the difference of the maximum principal stress of the DPAK EMC is comparable with the amount of relaxation of bulk modulus.

The above results clearly show that the DPAK EMC with a constant bulk modulus can constraint the package excessively, which results in significantly overestimated stresses in a package.

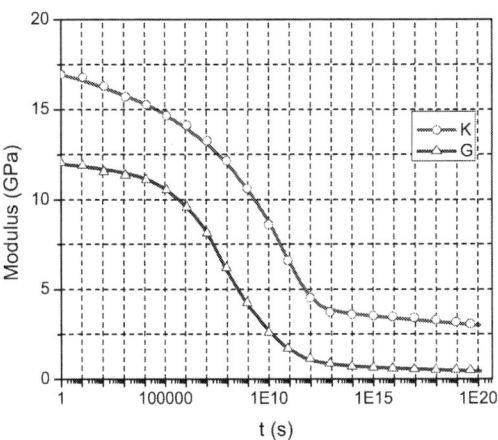

Fig. 8 Master curves of time-dependent bulk and shear modulus

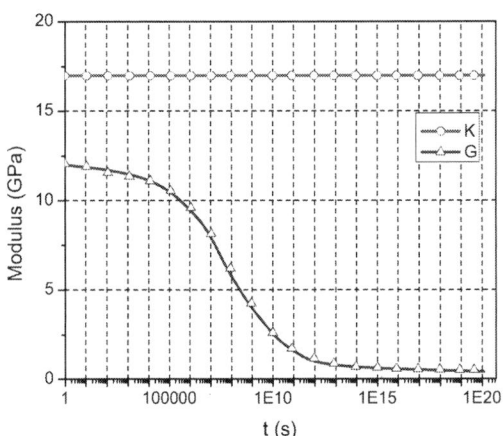

Fig. 9 Master curves with time-independent bulk modulus

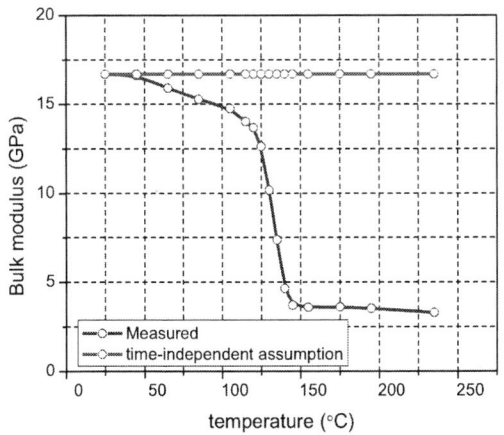

Fig. 10 Isochronous bulk modulus as a function of temperature

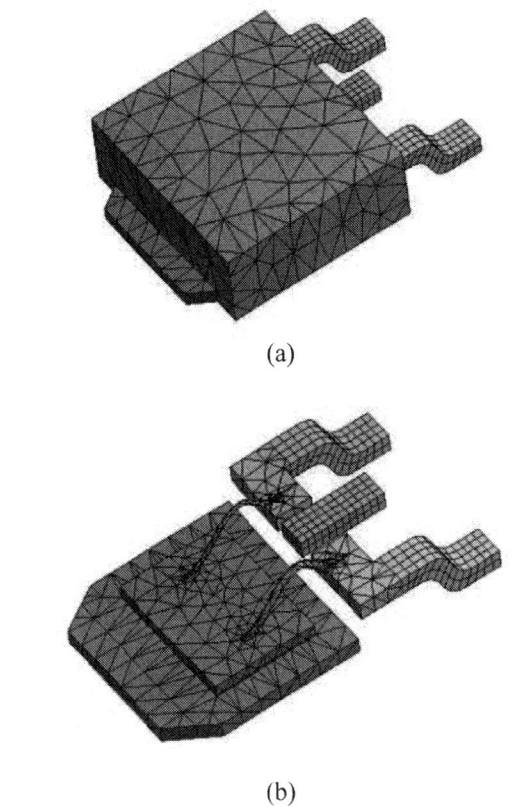

Fig. 11 FEA models of DPAK in ECU unit shown (a) with EMC (b) without EMC

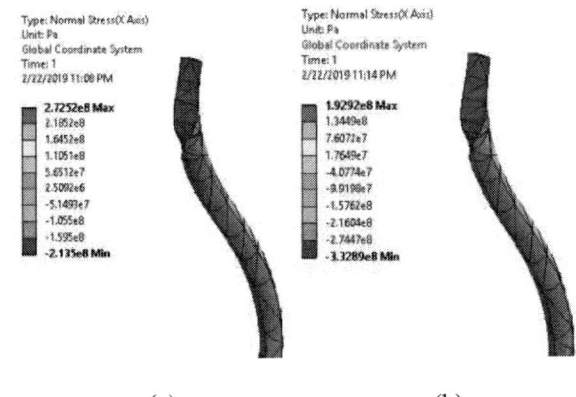

Fig. 12 Stresses of Al wire of overmolded DPAK with (a) constant bulk modulus (b) temperature-dependent bulk modulus

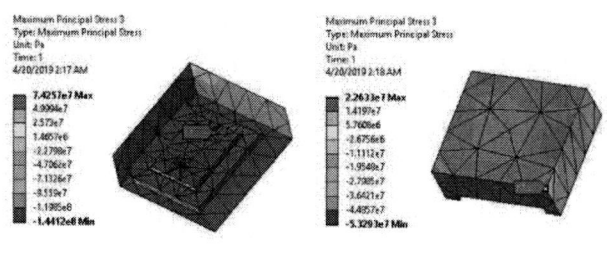

(a)   (b)

Fig. 13 Maximum principal stress on DPAK EMC with (a) constant bulk modulus and (b) temperature dependent bulk modulus

IV. CONCLUSION

The temperature and time-dependent bulk modulus of EMCs was measured by subjecting an EMC specimen to hydrostatic pressure. A fiber Bragg grating (FBG) sensor was embedded in the specimen, and it measured the deformations caused by hydrostatic pressure as a function of time. The results showed the significant time-dependent behavior of bulk modulus, especially around the glass transition temperature. The experimental data were utilized in LVE based numerical modeling to quantify the effect of time-dependent bulk modulus on the internal stresses of DPAK package in automotive ECU units. The results showed that the constant bulk modulus overestimated the wire-lift force significantly due to the extra-constraints.

Reference

[1] A. Amirkhizi, J. Isaacs, J. McGee, and S. Nemat-Nasser, "An experimentally-based viscoelastic constitutive model for polyurea, including pressure and temperature effects," *Philosophical magazine,* vol. 86, pp. 5847-5866, 2006.

[2] D.-L. Chen, P.-F. Yang, and Y.-S. Lai, "A review of three-dimensional viscoelastic models with an application to viscoelasticity characterization using nanoindentation," *Microelectronics Reliability,* vol. 52, pp. 541-558, 2012.

[3] H. F. Brinson and L. C. Brinson, *Polymer engineering science and viscoelasticity*: Springer, 2008.

[4] J. de Vreugd, K. Jansen, A. Xiao, L. Ernst, C. Bohm, A. Kessler, H. Preu, and M. Stecher, "Advanced viscoelastic material model for predicting warpage of a QFN panel," in *Electronic Components and Technology Conference, 2008. ECTC 2008. 58th,* 2008, pp. 1635-1640.

[5] R. Maksimov and É. Plume, "Predicting the creep of unidirectional reinforced plastic with thermorheologically simple structural components," *Mechanics of Composite Materials,* vol. 18, pp. 737-744, 1983.

[6] A. Sarhadi, J. H. Hattel, and H. N. Hansen, "Precision Glass Molding: Validation of an FE Model for Thermo☐Mechanical Simulation," *International Journal of Applied Glass Science,* vol. 5, pp. 297-312, 2014.

[7] Y. Wang, L. Woodworth, and B. Han, "Simultaneous measurement of effective chemical shrinkage and modulus evolutions during polymerization," *Experimental mechanics,* vol. 51, pp. 1155-1169, 2011.

[8] Y. Sun, Y. Wang, C. Jang, B. Han, and K. Choi, "Generalized Hybrid Modeling to Determine Chemical Shrinkage and Modulus Evolutions at Arbitrary Temperatures," *Experimental mechanics,* vol. 53, pp. 1783-1790, 2013.

[9] Y. Sun, Y. Wang, Y. Kim, and B. Han, "Dual-Configuration Fiber Bragg Grating Sensor Technique to Measure Coefficients of Thermal Expansion and Hygroscopic Swelling," *Experimental mechanics,* vol. 54, pp. 593-603, 2014.

[10] Y. Sun, B. Han, E. Parsa, and A. Dasgupta, "Measurement of effective chemical shrinkage and equilibrium modulus of silicone elastomer used in potted electronic system," *Journal of Materials Science,* pp. 8301-8310, 2014.

[11] D. Karalekas, J. Cugnoni, and J. Botsis, "Monitoring of process induced strains in a single fibre composite using FBG sensor: A methodological study," *Composites Part A: Applied Science and Manufacturing,* vol. 39, pp. 1118-1127, 2008.

[12] Y. Sun, H.-S. Lee, and B. Han, "Measurement of elastic properties of epoxy molding compound by single cylindrical configuration with embedded fiber Bragg grating sensor," *Experimental Mechanics,* vol. 57, pp. 313-324, 2017.

[13] H. S. Lee, Y. Sun, C. Kim, and B. Han, "Characterization of Linear Viscoelastic Behavior of Epoxy Molding Compound Subjected to Uniaxial Compression and Hydrostatic Pressure," *IEEE Transactions on Components, Packaging and Manufacturing Technology,* vol. 8, pp. 1363-1372, 2018.

[14] N. Tanaka, Y. Okabe, and N. Takeda, "Temperature-compensated strain measurement using fiber Bragg grating sensors embedded in composite laminates," *Smart materials and structures,* vol. 12, p. 940, 2003.

[15] X. Tao, L. Tang, W.-c. Du, and C.-l. Choy, "Internal strain measurement by fiber Bragg grating sensors in textile composites," *Composites Science and Technology,* vol. 60, pp. 657-669, 2000.

[16] P. Gromala, A. Palczynska, and B. Han, "Prognostic approaches for the wirebond failure prediction in power semiconductors: A case study using DPAK package," in *2015 16th International Conference on Electronic Packaging Technology (ICEPT),* 2015, pp. 413-418.

[17] P. McCluskey, "Reliability of power electronics under thermal loading," in *Integrated Power Electronics Systems (CIPS), 2012 7th International Conference on,* 2012, pp. 1-8.

[18] L. Yang, P. A. Agyakwa, and C. M. Johnson, "Physics-of-failure lifetime prediction models for wire bond interconnects in power electronic modules," *IEEE Transactions on Device and Materials Reliability,* vol. 13, pp. 9-17, 2013.

[19] J. Onuki, M. Koizumi, and M. Suwa, "Reliability of thick Al wire bonds in IGBT modules for traction motor drives," *IEEE Transactions on Advanced Packaging,* vol. 23, pp. 108-112, 2000.

[20] B. Wu, D.-S. Kim, B. Han, A. Palczynska, A. Prisacaru, and P. J. Gromala, "Hybrid Approach to Conduct Failure Prognostics of Automotive Electronic Control Unit Using Stress Sensor as In Situ Load Counter," *IEEE Transactions on Components, Packaging and Manufacturing Technology,* pp. 1-11, 2018.

2019 IEEE 69th Electronic Components and Technology Conference (ECTC)

Thermal and mechanical simulations for Fan-Out Wafer-Level Packaging technology: introduction of a "solder heatsink"

Jean-Philippe Colonna	Loic Marnat	Mathilde Cartier	Gabriel Pares	Dominique Noguet
CEA-Leti	CEA-Leti	CEA-Leti	CEA-Leti	CEA-Leti
Université Grenoble Alpes	Université Grenoble Alpes	Université Grenoble Alpes	Université Grenoble Alpes	Université Grenoble Alpes
Grenoble, France	Grenoble, France	Grenoble, France	Grenoble, France	Grenoble, France
Jean-philippe.colonna@cea.fr	Loic.marnat@cea.fr	Mathilde.cartier@cea.fr	Gabriel.pares@cea.fr	Dominique.noguet@cea.fr

Abstract—**The present study focuses on a 3D module composed of two stacks of 8 silicon chiplets interconnected with copper Through Molding Vias and ReDistribution Layer realized in Fan-Out Wafer-Level Packaging technology. The thermal management of such 3D module has been identified as a critical issue. The proposed multi-strata and multi-chiplet package is studied by means of numerical simulations with a focus on its thermal and thermo-mechanical behavior. After a description of the FO-WLP technology, a first part is dedicated to thermal simulations. After identifying the thermal impact of several parameters, we propose two thermal improvements: 1/ the addition of a thin graphite heatspreader on top of the 3D module. 2/ Solder a large area of the 3D module on a highly thermally conductive substrate in order to improve the bottom heat removal, as it is done in power module packaging. The second part is dedicated to the thermo-mechanical study of the 3D module and more precisely to the comparison between a ball grid array and the proposed thermal solution with the "central solder heatsink".**

Keywords – Fan-Out Wafer-Level-Packaging (FO-WLP), 3D stacking, thermal simulations, thermo-mechanical simulations, thermal management, co-design, co-integration.

I. INTRODUCTION

Fan-Out Wafer-Level-Package (FO-WLP) has received growing attention recently due to its high 3D heterogeneous integration capability, small footprint, thin modules and high number of interconnects (I/Os). It also covers an increasing range of applications, from logic (application processor) to RF (radar modules) [1, 2]. However, these applications may require dense stacking of thin silicon layers that can lead to thermal management issues. Indeed, thin silicon dies will exacerbate the hot spot effect due to limited spreading capabilities and the molding compound will contribute to thermally insulate the stacked dies.

The objective of this work is to study the proposed multi-strata and multi-chip compact package by means of numerical simulations with a focus on thermal management. The first part of this paper will describe the used FO-WLP technology with a focus on the key process parameters. The next part will be dedicated to thermal simulations with a multi-scale model. Critical parameters will be analyzed and

identified. Then, thermal solutions will be proposed. The last part of the paper will be dedicated to the thermo-mechanical study of the 3D module and the impact of the proposed thermal solutions.

II. FO-WLP TECHNOLOGY

The considered multi-strata and multi-chip FO-WLP process flow is described in figure 1.

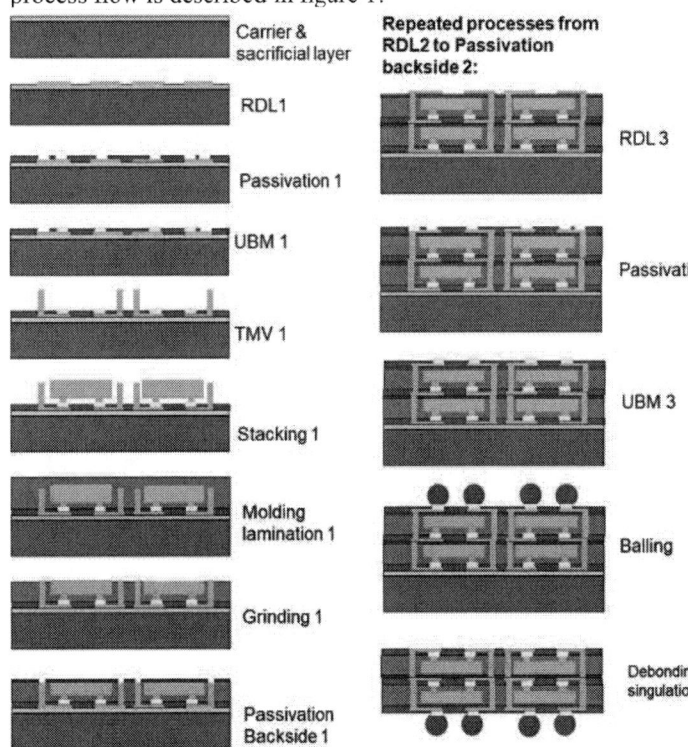

Fig. 1. Multi-strata and multi-chip FO-WLP process flow.

The process flow from fig. 1 can be described by the following steps:
- RDL: copper ReDistribution Layer from 2 to 10 μm in thickness.
- Passivation: patternable dielectric polymer – thickness of 10 μm maximum.
- UBM: Under Bump Metallization Ti/Ni/Au.

978-1-7281-1500-9/19 $31.00 © 2019 IEEE

- TMV : Through Molding via of 50-60 µm in diameter, up to 120 µm.
- Stacking of thinned dies with µ-bumps – die thickness from 40 to 120 µm, µ-bumps thickness of 30 µm minimum in order to allow molding-underfilling in the following lamination step.
- Molding lamination of a 200 µm thick film – also acts as underfill.
- Grinding of the molding layer in order to resume electrical contacts on TMV – grinding of silicon dies depending on the desired die thickness.
- Passivation backside: optional, depending on die thickness - same process as previous passivation.

These process steps are then repeated for stack 2.

RDL 3, passivation 3 and UBM 3 are used for balling. The stack is finally debonded and singulated.

The key process steps of this flow are the realization of the TMV and the lamination of the molding layer.

The TMV fabrication is based on semi-additive processes, which include seed layer deposition, lithography of a thick dry film (with exposure and development), Copper Electro-Chemical Deposition (ECD), resist stripping and seed layer etching. The aspect ratio for TMVs of 50-60 µm in diameter is about 2:1, that means a height of about 100-120 µm.

The molding lamination material has low stress specifications; it also acts as underfill for the µ-bumps of the Si dies and allows a Si coverage above 90%.

The resulting 3D module composed of two layers of eight silicon dies stacked on top of each other is presented in figure 2.

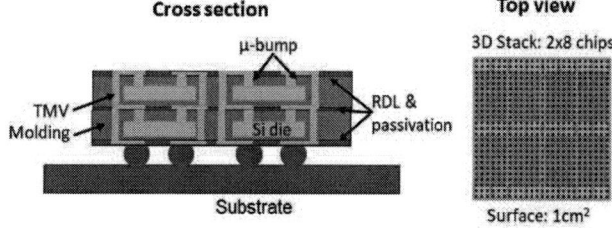

Fig. 2: Cross-section and top views of a 3D stacked multi-die module.

III. THERMAL SIMULATIONS

A. Model description

In this work, the commercial multi-physics finite element software, COMSOL v5.3, has been used. The module is complex and multiple elements such as RDL, integrated circuits dimensions, molding material and dimensions or the external assembly board may affect its thermal stability. To assess the impact of all these elements, two simplified models are defined considering two scales: firstly, the complete module with external board and secondly, a refined model focusing on the behavior of two stacked ICs among the molded module (1/8 of the complete module).

The first simplified model is assessing the module at system level. It is used to determine equivalent boundary conditions at the top and bottom interfaces (called H_{top} & H_{bot}) of the 3D chip, as shown in figure 3, to be used in the next more refined model.

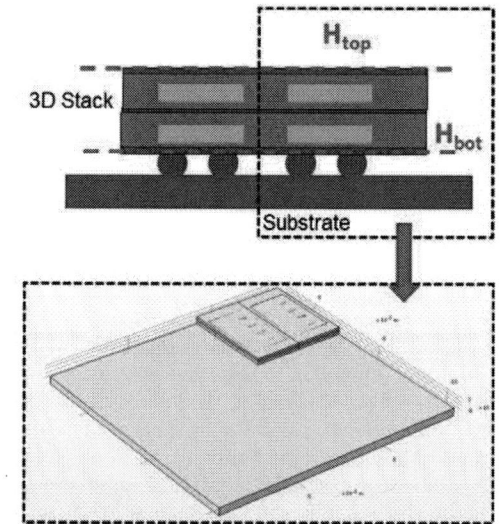

Fig. 3: Simplified model 1 with homogenized substrate and interconnects for ¼ of structure (4 Si dies).

The considered characteristics of the layers from top to bottom are:
- Simplified 3D Stack with Si dies and molding only with k_{Si}=130 W/m.K & $k_{molding}$=1 W/m.K (no RDL, vias & copper pillars considered);
- Homogenized solder balls: 150 µm thick, diameter 200 µm, pitch 300 µm k_Z=20, k_{XY}=0.05;
- Bottom substrates investigated:
- Small organic substrate (27x27 mm², thickness 400 µm) with k_Z=0.6, k_{XY}=40.
- Large organic substrate (45x45 mm², thickness 800 µm) with k_Z=0.6, k_{XY}=40.
- Large alumina substrate (45x45 mm², thickness 1 mm) with k_{Al2O3}=30.

The homogenization has been done according to the procedure proposed in [3].

Boundary conditions around and above this structure are considered with a convection of 15 W/m².K. A standard default tetrahedral mesh is used.

According to this model, bottom boundary condition can be extracted with heat transfer coefficients (HTC) for the bottom of the module corresponding to the three considered bottom substrate. The extracted coefficients are:
- H_{bot} = 180 W/m².K for a small organic substrate.
- H_{bot} = 400 W/m².K for a large organic substrate.
- H_{bot} = 460 W/m².K for a large alumina substrate.

These HTC are then used in a second more refined model. The considered 1/8 of the 3D module including two silicon dies with four heat sources each, molding compound, RDL, TMV and µ-bumps is presented in figure 4.

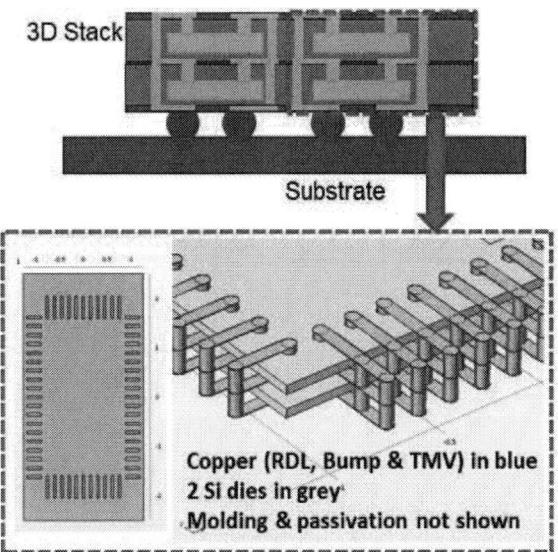

Fig. 4: Second refined model of 1/8 structure of the 3D module.

The heat sources or hot spots are shown in figure 5. In order to model more precisely these hotspots, the default fixed thermal conductivity of silicon has been replaced by a temperature dependent thermal conductivity $k_{Si}(T)$ taken from [4].

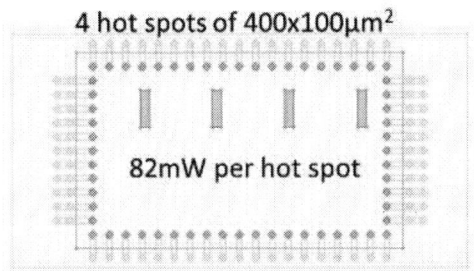

Fig. 5: Top view of one die with close RDL and via geometry and the location of the four 400x100 µm² heat sources areas.

B. First thermal results

In this second model, the implemented heat sources correspond to 5.25 Watt for the sixteen dies in whole 3D chip. The first results with this power input in steady state and the different bottom HTC found with the previous model are given in figure 6.

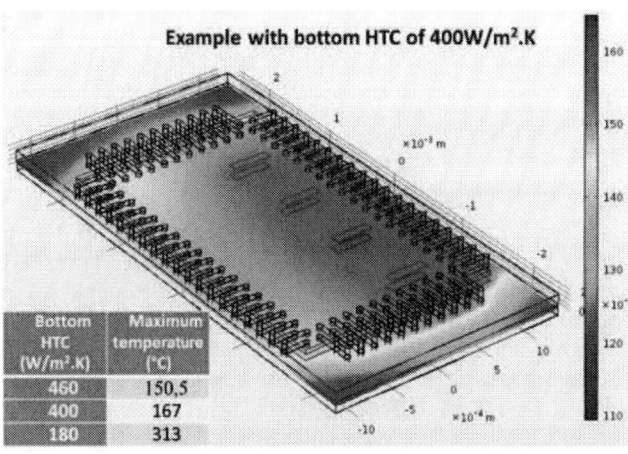

Bottom HTC (W/m².K)	Maximum temperature (°C)
460	150,5
400	167
180	313

Fig. 6: Simulated thermal map of the 1/8 structure with the maximum temperature related to the three considered bottom HTC.

Although the model is not calibrated by experimental measurements and cannot be considered as predictive, high maximum temperatures are observed meaning that a power of 5.25 W seems difficult to handle in a standard package with an organic substrate.

Four technological parameters have been varied within the range allowed by our FO-WLP technology to assess the variation on the maximum temperature within the 3D chip (see results in fig. 7):

- RDL thickness (t_{RDL}) ranging from 2 to 10 µm.
- Si die thickness (t_{si}) from 40 to 90 µm.
- Molding thickness under Si die (t_{mold}): 2 to 60 µm.
- µ-bump thickness (t_{bump}) from 2 to 60 µm.

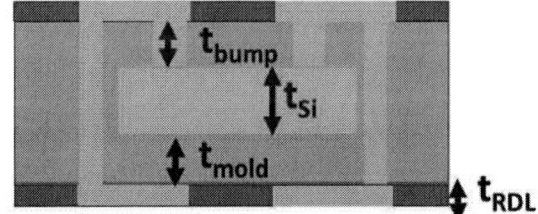

Silicon thickness (µm)	Maximum temperature (°C)	Molding thickness (µm)	Maximum temperature (°C)
40	167	2	166,2
70	161	30	166,8
90	159	60	167,4

Bump thickness (µm)	Maximum temperature (°C)	RDL thickness (µm)	Maximum temperature (°C)
2	167,6	2	166,8
30	166,9	5	166,9
60	166,2	10	167,0

Fig. 7: Influence of the FO-WLP technological parameters on maximum 3D chip temperature.

978-1-7281-1500-9/19 $31.00 © 2019 IEEE

One can see that the only parameter having a visible influence on the maximum temperature is the die thickness. Nevertheless, it has a significantly lesser influence than the bottom HTC.

C. Thermal improvements overview

Clearly, thermal improvements have to focus on the top and bottom HTC. For the considered device, the space constraint forbid the addition of a heatsink or a copper lid on top of the 3D module. So, concerning the top HTC, it is proposed to add a 70 µm thick graphite heatspreader on top of the 3D module [5]. This graphite heatspreader is provided with a Thermal Interface Material (TIM) of 25 µm of soft silicone, which has the advantage of being completely neutral in terms of mechanical stress [6]. Furthermore, these graphite sheets can be patterned and co-integrated with interconnects in an inter-tier position in a 3D package [7, 8, 9]. In order to improve the bottom HTC, it is proposed to solder a large area of the 3D module on a highly conductive substrate or a highly conductive part of the substrate, as it is done for power module packaging [10]. This solution, called "solder heatsink" has the advantage to maintain the stack thickness. Its mechanical impact will be studied in the last part of this paper.

D. Graphite Heatspreader performance

The thermal efficiency of the top graphite heatspreader is studied using the first model of ¼ of the structure. Two additional layers were added on top of the 3D stack and the substrate:

- A silicone layer of 25 µm with $k_{Silicone}$=0.8 W/m.K
- A graphite layer of 70 µm with k_Z=15 and k_{XY}=800 W/m.K.

Fig. 8a shows the geometry of the additional silicone and graphite layers in the first model (cross-section view) compared to a Scanning Electron Microscopy (SEM) picture of a previous experimental integration of graphite with a silicone TIM [5, 6, 9]. Fig. 8b shows the 3D view of the updated first model with the additional silicone and graphite layers spread all over the bottom substrate area.

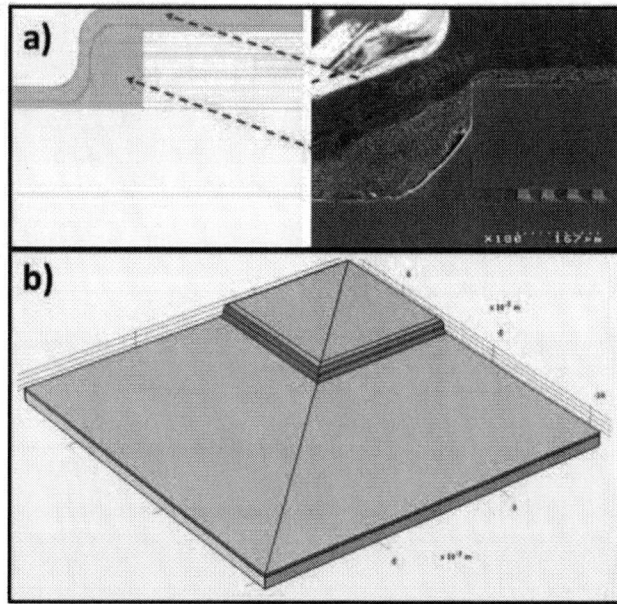

Fig. 8: (a) cross-section and (b) 3D views of the additional silicone and graphite layers on the simplified model 1 (¼ of structure).

The simulations were run with two substrates: the small and the large organic substrate. The new extracted HTC with graphite heatspreader are for the small organic substrate:

- H_{top} = 95 W/m².K
- H_{bot} = 115 W/m².K.

And for the large organic substrate:

- H_{top} = 170 W/m².K
- H_{bot} = 360 W/m².K.

The second refined model run with these new HTC gives a temperature reduction of 20°C for the small organic substrate and 29°C for the large organic substrate thanks to the addition of the graphite heatspreader. Indeed, the heat is still extracted through the substrate backside, but with the graphite sheet, part of it (up to 40%) goes up in the graphite, spreads over its entire surface before going back down into the substrate. This explains why the temperature reduction is more important with the larger substrate.

E. Solder heatsink performance

To evaluate the thermal impact of the solder heatsink, the first model (¼ of the structure without graphite layer) is used with two modifications (see fig.9):

- A large organic substrate with an additional 100 µm thick copper foil (k_{Cu}=400 W/m.K) on top of it.
- Two studied detailed interconnects between 3D module and substrate:
  - Solder balls only (considered as a reference) with k_{Solder}=50 W/m.K.
  - Smaller peripheral solder balls associated to a large central heat sink.

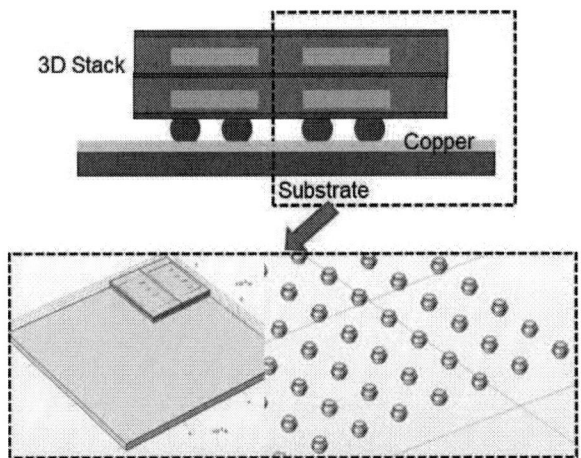

Fig. 9: Cross-section and 3D views of model 1 (¼ of structure) with detailed solder balls & additional copper foil on the substrate.

The reference structure is composed of an array of 576 solders balls (diameter 200 μm / pitch 400 μm) versus the structure with a continuous central solder heatsink of 7.4x8.2 mm² surrounded by an array of 400 smaller solder balls of diameter 130 μm / pitch 300 μm. The thermal simulation results are given in fig. 10 for the reference case and in fig. 11 for the central solder heatsink (the same temperature scale is used). They show temperature maps at the top silicon die surface (top picture) and at the interface between the module and the solder balls (bottom picture).

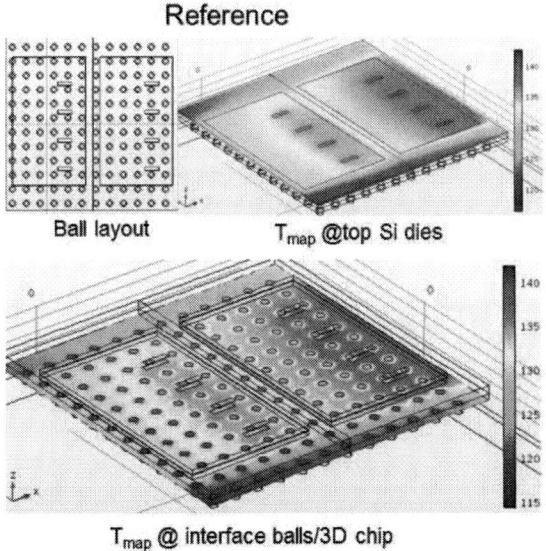

Fig.10: 2D temperature distribution for the reference case (only solder balls).

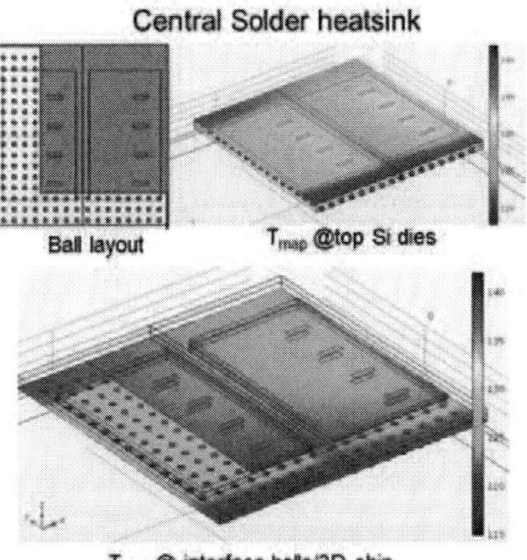

Fig.11: 2D temperature distribution using the solder heatsink.

A temperature reduction of 15°C is achieved thanks to the "central solder heatsink".

Again, these simulations are non-calibrated and the results cannot be considered as predictive. Furthermore, a direct comparison between the temperature reduction from the graphite heatspreader and the solder heatsink should not be done as those results were not extracted from the same models. Nevertheless, the graphite solution seems more efficient especially when associated with large substrates and does not affect the potential number I/Os of the module and relaxe routing constraints on the bottom substrate.

IV. THERMO-MECHANICAL SIMULATIONS

In this part, simulations aim at comparing several structures from a mechanical point of view but only on a qualitative way. Indeed, simplified structures and non-calibrated material properties are used with simplified physics.

A. Model description

The simplified structure is composed of three layers as shown in figure 12:
- Simplified 3D Stack with Si dies and molding only (no RDL, vias and μ-bumps),
- Detailed solder balls,
- FR4 substrate.

978-1-7281-1500-9/19 $31.00 © 2019 IEEE

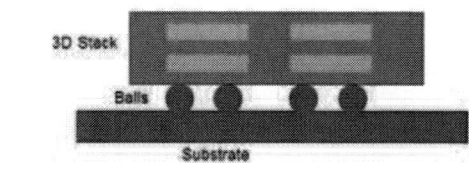

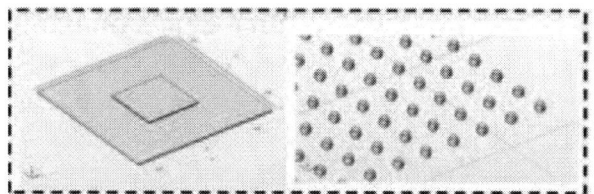

Fig. 12: Cross-section and 3D views of the simplified thermo-mechanical model.

The material properties used in the simulations are given in table I. The mechanical properties considered are the Young's Modulus E (in GPa), the Poisson ratio ν (no unit) and the Coefficient of Thermal Expansion, CTE (in ppm/K). The glass transition temperature Tg (in °C) is also considered for the molding polymer. The polymer properties come from vendor's datasheet and the other materials from COMSOL library.

	E (GPa)	ν	CTE (ppm/K)	Tg (°C)
Si	170	0.28	2.6	-
Molding	3.3	0.3	20	180
FR4	22	0.15	18	-
Cu	120	0.34	16.5	-
Solder	10	0.4	21	-

Table I: Material properties used in the thermos-mechanical simulations.

In the following, we consider the stress coming from CTE mismatch during assembly of the 3D module on the substrate (reflow @ 200-250°C). Theoretically, the structure is considered stress free at reflow temperature and cooled down to 20°C. However, considering molding polymer, the structure will be considered stress free above the Tg temperature. This assumption means that above Tg, polymers are supposed to be soft and able to yield. All materials are assumed to have linear and elastic behavior. To verify this hypothesis, the Von Mises stress criterion is calculated in the solder balls (assumed to be the weakest point of the structure) and compared with the typical yield stress for solder, which is around 220 MPa (from COMSOL library).

B. Reference case

In this section, the silicon dies are 70 μm thick and the molding thickness under the silicon dies is 30 μm. Solder balls of 130 μm in diameter and 400 μm pitch are used. The geometry of one ball is detailed in figure 13. The FR4 substrate is 400 μm thick for an area of 2.7x2.7 cm².

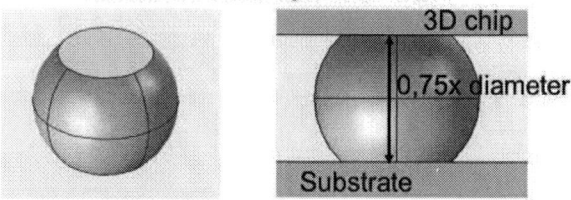

Fig. 13: Solder ball morphology and location between 3D chip and substrate.

Fig. 14 and 15 present the Von Mises Stress in MPa of ¼ of the 3D module, at the interface between bottom dies and molding for fig. 14 and in the solder balls for fig. 15.

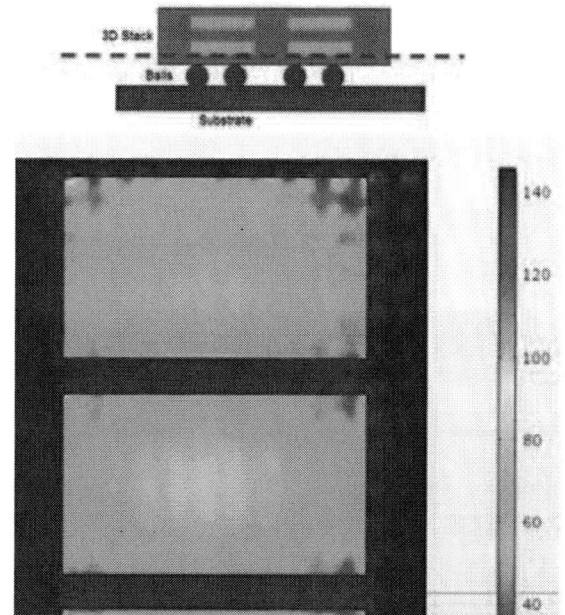

Fig. 14: Von Mises stress at Si/molding interface for the reference case.

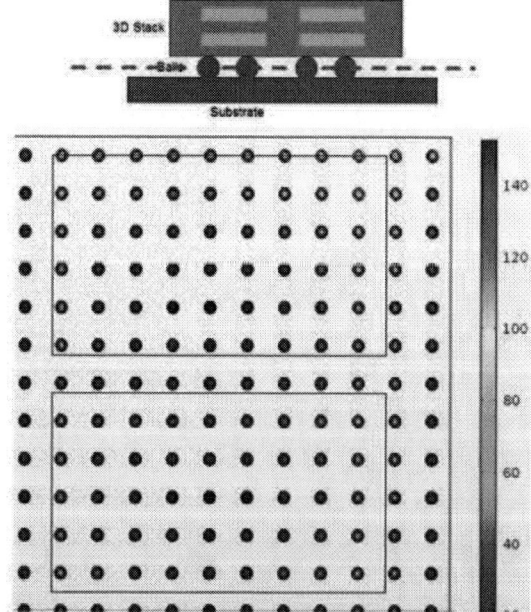

Fig. 15: Von Mises stress in solder balls for the reference case.

The maximum stress values for the solder balls are lower than the 220 MPa threshold, which validates our hypothesis. One can also notice that contrarily to bulk silicon, the stress distribution is different with maxima localized at the silicon/molding edges and not at the corners of the whole structure. Furthermore, the mechanical stress in a 3D module with silicon dies and molding is significantly better than a bulk silicon chip (simulation not shown here). This is expectable as the CTE of the molding is far closer to the CTE of the substrate as compared to the CTE of silicon. This may help to get more robust module regardless to their size contrarily to large bulk Si chip by adjusting the ball grid map at the module design level. In addition, it allows the use of smaller solder balls, which means more addressable I/Os with such FO-WLP based module than with a bulk silicon chip.

C. Solder heatsink

In this last section, the mechanical impact of the "central solder heatsink" is investigated and compared to the previous reference case using only larger ball grid array. The central heatsink is surrounded with smaller solder balls of 130 µm in diameter and a pitch of 300 µm (same as the thermal simulation). The Von Mises stress criterion is presented in fig. 16 with and without heatsink. The stress maps in the solder volume show no major difference, with similar stress values. On the other hand, the stress maps in the silicon plane clearly show the impact of the central heatsink with its footprint showing slightly higher stress values in the 3D module. It can be noted that stresses are remaining in an acceptable range and thermally much more efficient. Therefore, it can be concluded that the solder heatsink can be realized without major changes mechanically speaking.

V. CONCLUSION

The aim of such module is to drastically enhance the number of functions realized on a given area thanks to the 3D packaging. It has been shown that dense stacking of thin silicon may lead to high thermal hot spots due to limited spreading. Four small hot spots in each die have been considered to emulate active regions in a module composed of sixteen chips for a total 5.25 W power consumption. This can be representative of radar or 5G applications. Two seamless thermal solutions namely graphite heatspreader and solder heatsink have been presented and assessed in the frame of the complex multi-chip and multi-strata 3D module.

These solutions have shown efficient results with about 20°C reduction in the maximum temperature each.

The mechanical stability of these thermal solutions applied to such heterogeneous 3D module have been assessed. It reveals it can provide degrees of freedom for the thermal management. Thus, the position and size of the balls/heat sink; the Si dies size and position considering hot spots can and must be taken into account at the module design level. A co-design and co-integration process may lead to an optimized thermal distribution limiting the maximal temperatures while maintaining the mechanical stability of 3D module.

REFERENCES

[1] J. Bock and R. Lachner, "SiGe BiCMOS and eWLB packaging technologies for automotive radar solutions," in 2015 IEEE MTT-S International Conference on Microwaves for Intelligent Mobility (ICMIM), 2015, pp. 1–4.

[2] Hagelauer, M. Wojnowski, K. Pressel, R. Weigel, and D. Kissinger, "Integrated Systems-in-Package: Heterogeneous Integration of Millimeter-Wave Active Circuits and Passives in Fan-Out Wafer-Level Packaging Technologies," IEEE Microwave Magazine, vol. 19, no. 1, pp. 48–56, Jan. 2018.

[3] F. de Crécy, "A simple and approximate analytical model for the estimation of the thermal resistances in 3D stacks of integrated circuits," in 18th International Workshop on THERMal INvestigation of ICs and Systems, pp. 1–6, 2012.

[4] C. J. Glassbrenner and G. A. Slack, "Thermal Conductivity of Silicon and Germanium from 3°K to the Melting Point," Phys. Rev., vol. 134, no. 4A, pp. A1058–A1069, May 1964.

[5] R. Prieto et al. "Thermal measurements on flip-chipped System-on-Chip packages with heat spreader integration" in proceedings of 31th SEMITHERM Symposium, San Jose CA, march 2015.

[6] R. Prieto et al. "Thermo-mechanical assessment of copper and graphite heat spreaders for compact packages" in 22nd International Workshop on THERMal INvestigation of ICs and Systems, pp. 19-22, 2016.

[7] S. Snyder, et al., "Thermally enhanced 3 dimensional integrated circuit (TE3DIC) packaging" in Electronic Components and Technology Conference (ECTC), Orlando, USA, pp. 601–608, 2014.

[8] K. Matsumoto, H. Mori, and Y. Orii, "Cooling from the bottom side (laminate (substrate) side) of a threedimensional (3D) chip stack," in IEEE International 3D Systems Integration Conference (3DIC), 2014.

[9] J-P. Colonna et al. "Carbon-based patterned heat spreaders for thermal mitigation of wire bonded packages" in 23rd International Workshop on THERMal INvestigation of ICs and Systems, pp. 1-6, 2017.

[10] J. Zhang et al. "The improvement of soldering process of new power module packaging material with large soldering area", Electronic Packaging Technology (ICEPT), 2014.

Solder layout

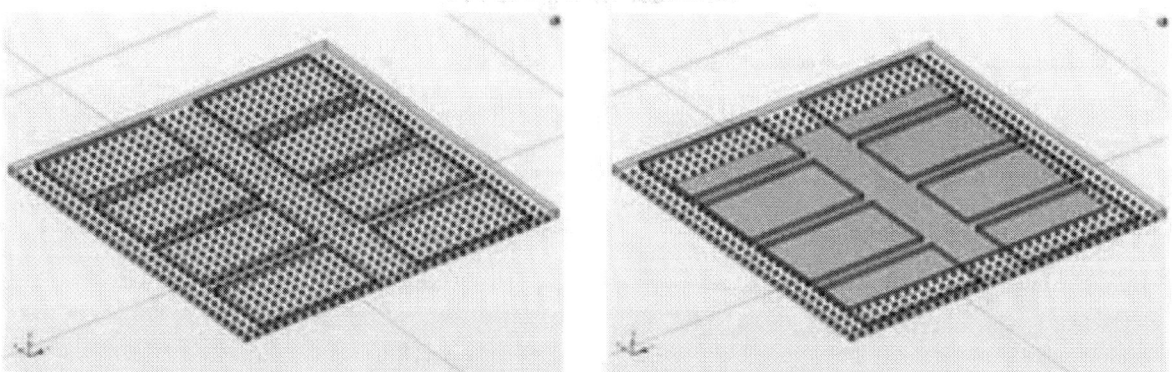

Solder stress map 1/4

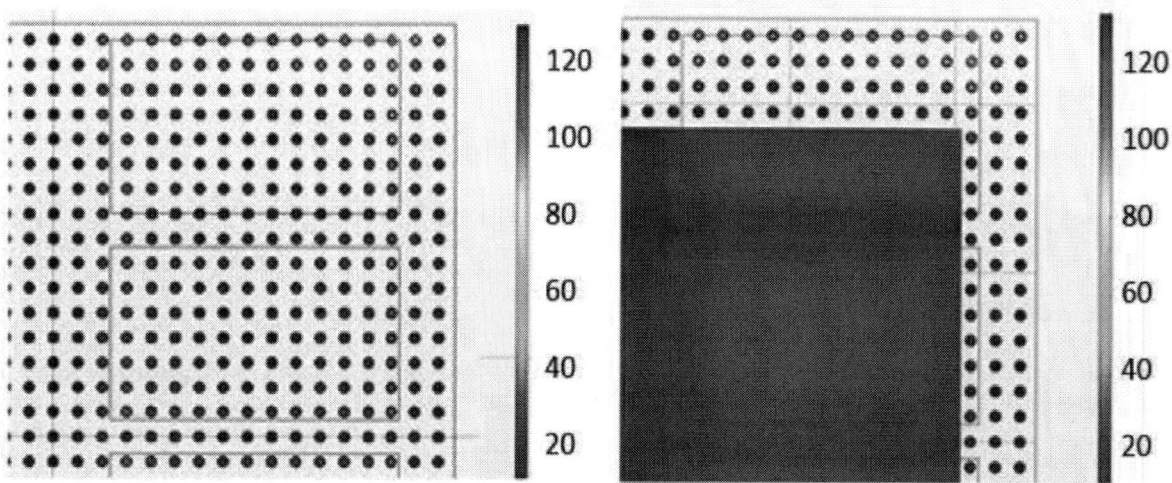

Silicon stress map

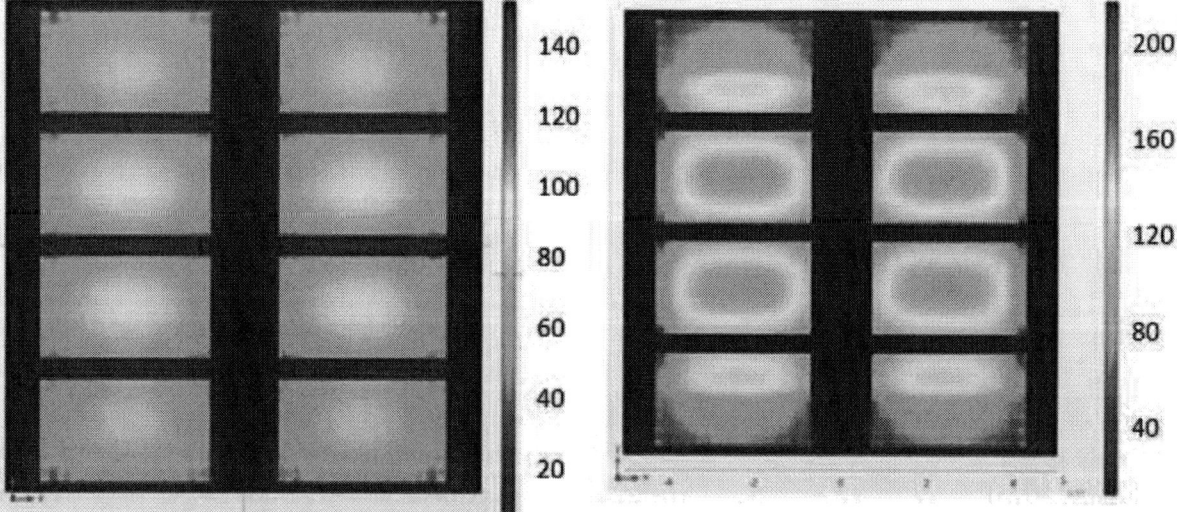

Fig. 16: Von Mises stress maps in solder balls and in silicon/molding interface with (right) or without (left) central solder heatsink.

Wafer Level Warpage Modelling and Validation for FOWLP Considering Effects of Viscoelastic Material Properties under Process Loadings

Zhaohui Chen [1*], Xiaowu Zhang [1*], Sharon Pei Siang Lim [1], Simon Siak Boon Lim [1], Boon Long Lau [1], Yong Han [1], Ming Chinq Jong [1], Songlin Liu [2], Xiaobai Wang [2], Yosephine Andriani[2]

1 Institute of Microelectronics, A*STAR (Agency for Science, Technology and Research), 2 Fusionopolis Way, #08-02 Innovis Tower, Singapore 138634

2 Institute of Materials Research and Engineering, A*STAR (Agency for Science, Technology and Research), 2 Fusionopolis Way, #08-03 Innovis Tower, Singapore 138634

* Email: chenz@ime.a-star.edu.sg, xiaowu@ime.a-star.edu.sg

Abstract— Wafer level warpage model was developed for the 12 inch mold-first FOWLP wafer with model change technique. The sequential process loadings of post-mold curing (PMC) (150 °C), warpage correction (150 °C and pressure), dielectric curing (190 °C), seed layer PVD (160 °C) and blank Cu electroplating were applied. The time and temperature dependent material properties of EMC and dielectric were considered in the model. The warpages of the 12 inch mold-first FOWLP wafer after the sequential processes were measured by the Nikon surface profiler and correlated with the predicted wafer warpage results. The predicted warpage results have good correlation with the experimental results. The warpage of FOWLP wafer was reduced after warpage correction. However, the warpage will increase after dielectric coating and curing. After PVD process, warpage will increase further. The viscoelastic material properties of EMC and dielectrics are critical for the warpage prediction under the process loadings. Only considering elastic material properties, the warpage behavior cannot be predicted. The developed wafer level warpage model considering time and temperature dependent material properties can be used for the warpage optimization and control for FOWLP development.

Keywords-Warpage correction, Viscoelastic, 12 inch mold-first FOWLP, Wafer level warpage

I. INTRODUCTION

Fan-out wafer level packaging (FOWLP) is becoming as a promising technology for the consumer electronic products with demands of heterogeneous integration. However, there are several technical problems, issues and challenges for the FOWLP, such as severe processing warpage, redistribution line (RDL) distortion and cracking due to the die protrusion, multi-layer RDL delamination and cracking, and solder joint reliability for large package size [1-2]. The wafer level warpage control is critical for the 12 inch FOWLP processing. Large wafer level warpage will hinder the processes moving on.

In this paper, wafer level warpage model considering time and temperature dependent material properties of epoxy molding compound (EMC) and dielectrics with model change technique was developed for the 12 inch mold-first FOWLP wafer under the typical FOWLP sequential fabrication processes like the post-mold curing (150 °C), warpage correction (150 °C and pressure), dielectric curing (190 °C), seed layer PVD (160 °C) and blank Cu electroplating. The time and temperature dependent material properties of EMC and dielectric were characterized and considered in the model. The warpage of fabricated 12 inch mold-first FOWLP wafer after post mold curing, during warpage correction, after warpage correction, after dielectric curing; after seed layer deposition, and after Cu plating was measured and correlated with the predicted results from the developed warpage simulation model.

II. TEST VEHICLE

The mold-first FOWLP was designed with 15 mm ×15mm package size and 8 mm ×8 mm chip size. The 12 inch FOWLP wafer is with thicknesses of 0.8 mm for 0.2 mm chip and 0.9 mm for 0.775 mm chip, respectively. Figure 1 shows the 12 inch mold-first FOWLP wafer of TV2 after pick and place of chips and after compression molding. The warpages of the 12 inch mold-first FOWLP wafer after the sequential processes of post mold curing, warpage correction, dielectric curing, seed layer PVD deposition and blank Cu electroplating were measured by the Nikon surface profiler.

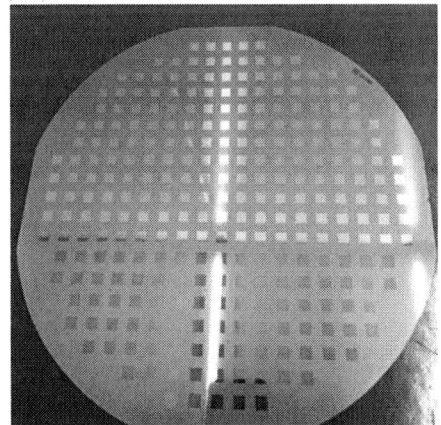

(a) After pick and place

978-1-7281-1500-9/19 $31.00 © 2019 IEEE

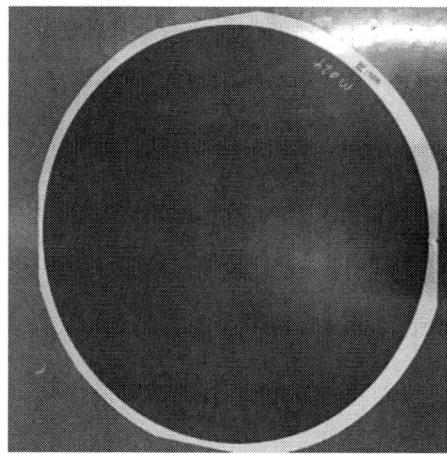

(b) After molding

Figure 1. 12 inch mold-first FOWLP wafer of TV2: (a) after pick and place of chip; (b) after compression molding.

TABLE I. DIMENSIONS OF TEST VEHICLE

	TV1	TV2
Chip size	8 mm×8 mm×0.775 mm	8 mm×8 mm×0.2 mm
Package size	15 mm×15 mm	15 mm×15 mm
Wafer size	12 inch	12 inch
Wafer thickness	0.8 mm	0.9 mm

III. WAFER LEVEL WARPAGE SIMULATION MODEL

Figure 2 shows the one quarter 12 inch mold-first FOWLP wafer warpage simulation model for TV1 under process loadings. The silicon side is facing to the rigid plate. The sequential process loadings of post mold curing (150 °C), warpage correction (150 °C and pressure), dielectric curing (190 °C), seed layer PVD (160 °C) and blank Cu electroplating were applied. Contact relation is built up between wafer and rigid plate, the wafer is contacted with rigid plate and pressure was applied on the top surface of silicon and EMC when the wafer is going through wafer correction process. After warpage correction the rigid plat was moved away and pressure was released. After that, the wafer went through the other process. The model change technique was used for the sequential processes simulation. The element of dielectric, seed layer, and blank Cu layer was deactivated when doing the warpage correction simulation and activated when doing the dielectric curing, seed layer deposition and Cu plating process simulation.

Figure 2 shows the one quarter 12 inch mold-first FOWLP wafer warpage simulation model for TV2 under process loadings. The EMC side is facing to the rigid plate. The sequential process loadings of post mold curing (150 °C), warpage correction (150 °C and pressure) were applied. Contact relation is also built up between wafer and rigid plate, the wafer is contacted with rigid plate and pressure

was applied on the top surface of EMC when the wafer is going through the wafer correction process. After warpage correction the rigid plat was moved away and pressure was also released. The curing shrinkage of EMC was also considered for both model which has significant contribution to the FOWLP warpage after post mold curing of EMC.

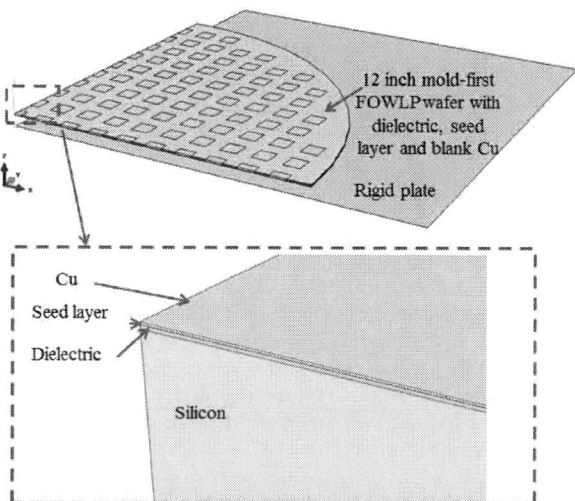

Figure 2. One quarter 12 inch FOWLP wafer warpage model of TV1 for post mold cure, warpage correction, dielectric curing, seed layer deposition, and Cu plating.

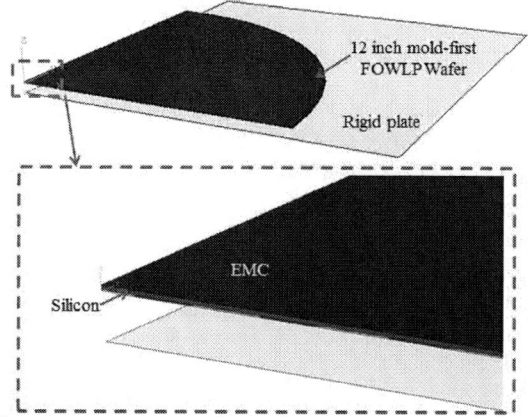

Figure 3. One quarter 12 inch FOWLP wafer warpage model for TV2 for post mold cure, warpage correction.

IV. MATERIAL PROPERTIES CHARACTERIZATION

Time and temperature dependent viscoelastic modelling provide the process time effects and stress relaxation history. A sinusoidal stress is applied to determine the complex modulus by dynamic mechanical analysis (DMA). The temperature or the frequency of the loading are often varied, which can lead to variations in the complex modulus. The constitutive equation for an isotropic viscoelastic polymer material can be expressed as:

$$\sigma_{ij} = \int_0^t 2G(t-\tau)\frac{de_{ij}}{d\tau}d\tau + \delta_{ij}\int_0^t K(t-\tau)\frac{d\Delta}{d\tau}d\tau \quad (1)$$

where σ_{ij} are the components of the Cauchy stress, δ_{ij} are Deviatoric strain components, Δ is the volumetric part of the strain, $G(t)$ and $K(t)$ are the shear and bulk modulus functions, t and τ are current and past time, and δ_{ij} are the components of the unit tensor. The Maxwell model, which consists of a linear elastic Hookean spring in series and a linear viscous Newtonian damper can be used in modelling of complicated linear viscoelastic behaviour of the polymer materials. For the generalized Maxwell model, the shear and bulk modulus functions can be expressed as a Prony series:

$$G(t) = G_\infty + \sum_{i=1}^n G_i e^{(-t/\tau_i)} \quad (2)$$

$$K(t) = K_\infty + \sum_{i=1}^n K_i e^{(-t/\tau_i)} \quad (3)$$

where G_∞ and G_i are the shear elastic modulus, K_∞ and K_i are the bulk elastic modulus, τ_i are the relaxation times for the various Prony series components. The time dependent stress-strain relations are typically extended to include temperature dependence by using the principle of time-temperature superposition and the empirical William-Landel-Ferry (WLF) shift function. The WLF formulation has the form:

$$Log_{10}(\alpha_T) = Log_{10}\left(\frac{t}{t_r}\right) = \frac{C_1(T-T_r)}{C_2+(T-T_r)} \quad (4)$$

where α_T is the shift factor, C_1 and C_2 are material constants, T is the temperature, T_r is the reference temperature [3-6].

The strip sample was prepared for the both EMC and dielectric. The EMC was molded by compression molding to achieve 1 mm thick 12 inch wafer. After post-mold curing at 150 °C, the 12 inch EMC wafer was diced to obtain 40 mm×10 mm×1 mm strip samples. The dielectric was coated on the 12 inch silicon wafer. After bake and curing at 190 °C, the 12 inch silicon wafer with dielectric was diced to obtain 40mm×10mm×1mm strip samples. The thin film dielectric was released from the silicon substrate. The viscoelastic materials properties of EMC, dielectric were characterized by the dynamic mechanical analysis (DMA) testing. Prony series was obtained by curve fitting of the mater curve. The DMA testing can also identify the temperature of glass transition (T_g) of the polymer materials. The coefficient of thermal expansion (CTE) of EMC was characterized by the thermo-mechanical analyzer (TMA) testing.

Figure 4 shows the storage modulus and loss modulus of EMC tested by DMA. The elastic modulus at 30 °C is 18.5 GPa and the glass transition temperature is 163 °C. Figure 5 shows the storage modulus and loss modulus of dielectric tested by DMA. The elastic modulus at 30 °C is 0.92 GPa and glass transition temperature is 240 °C. The details of the viscoelastic characterization and Prony series by curve fitting can be found in reference [7-8].

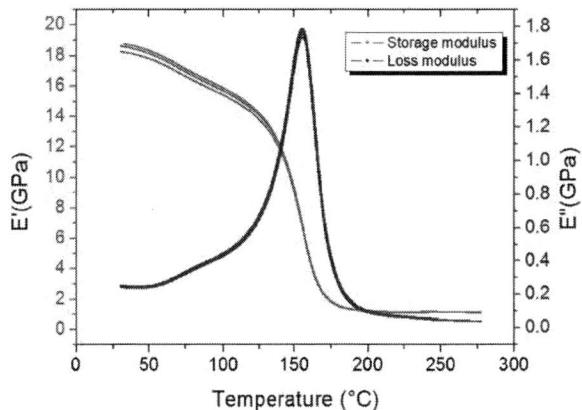

Figure 4. Storage modulus and loss modulus of EMC tested by DMA.

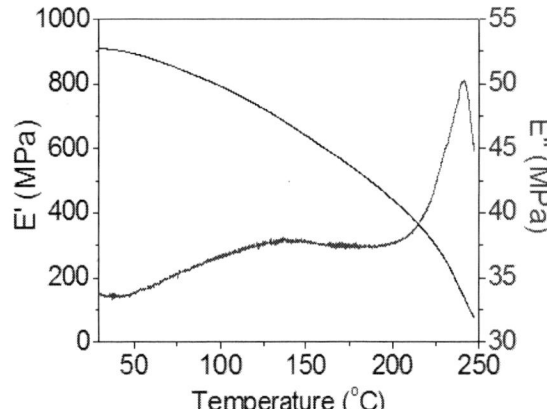

Figure 5. Storage modulus and loss modulus of dielectric tested by DMA.

Material properties used for the 12 inch FOWLP wafer level warpage simulation are listed in Table II. The viscoelastic material properties of EMC and dielectric were considered in the model.

TABLE II. MATERIAL PROPERTIES USED FOR WAFER LEVEL WARPAGE SIMULATION

Material	Young's Modulus (GPa)	Poisson Ratio	CTE (ppm/°C)	Tg (°C)
Si	131	0.3	2.8	-
EMC	18.5/1.2	0.3	9/18	163
Dielectric	0.92/0.1	0.3	80/227	240
Cu	117	0.34	17	-
Ti	116	0.34	8.9	-

978-1-7281-1500-9/19 $31.00 © 2019 IEEE

V. RESULTS AND DISCUSSION

Figure 6 shows the simulated warpage contour after different process steps of 12 inch mold-first FOWLP for TV1. The warpage of mold-first FOWLP wafer of TV1 after post mold curing is 0.971mm. The wafer becomes flatten under 150 °C temperature and pressure loadings of the warpage correction. The warpage is 0.107 mm during the warpage correction process. The warpage reduces to 0.332 mm after the warpage correction. The warpage becomes 0.732 mm after coating 6 μm dielectric and curing. The warpage of the wafer becomes 0.805 mm and 0.72 mm after seed layer deposition and 0.3 μm blank Cu plating.

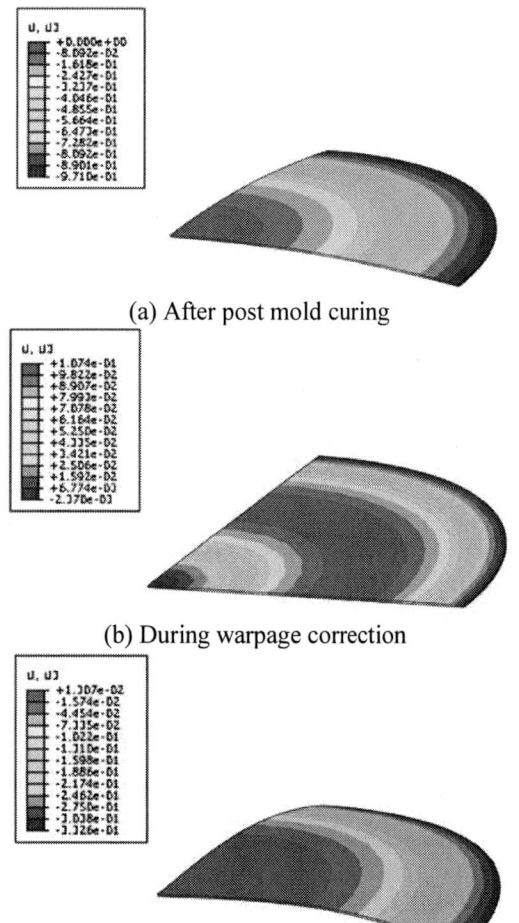

(a) After post mold curing

(b) During warpage correction

(c) After warpage correction

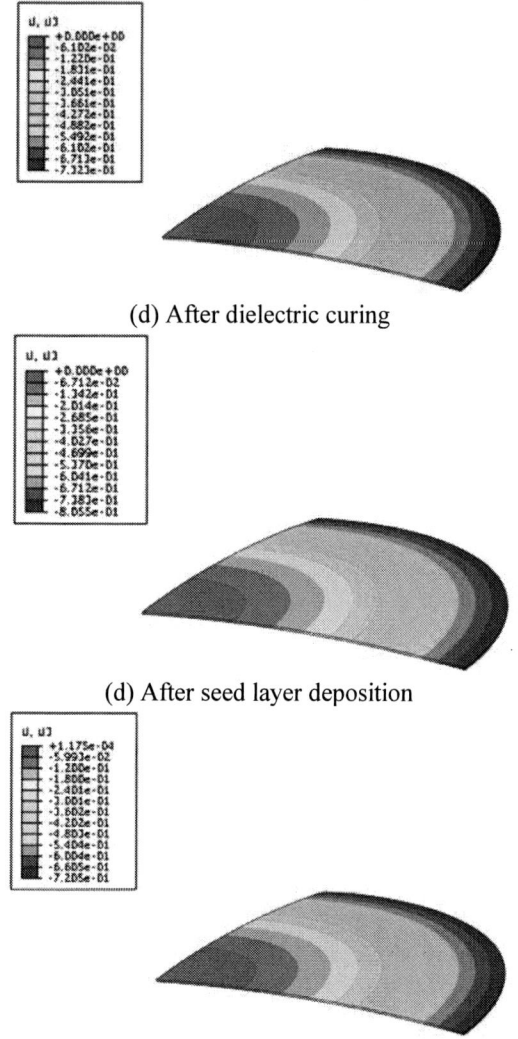

(d) After dielectric curing

(d) After seed layer deposition

(d) After blank Cu plating

Figure 6. TV1 Simulated warpage contour after different process steps: (a) After post mold curing; (b) During warpage correction; (c) After warpage correction; (d) After dielectric curing; (d) After seed layer deposition; (d) After blank Cu plating.

The warpage of fabricated TV1 FOLWP wafer after post mold curing, during warpage correction, after warpage correction, after dielectric curing; after seed layer deposition, and after blank Cu plating was measured. The experimental warpage data is shown as Figure 7. The warpage after post mold curing is 0.77mm. The warpage reduce to 0.43 mm after the warpage correction. The warpage becomes 1.07 mm after coating 6 μm dielectric and curing. The warpage of the wafer becomes 1.06 mm and 0.78 mm after seed layer deposition and blank Cu plating. The simulated warpage is consistent with the measured warpage data. With the viscoelastic material properties of EMC and dielectric, the warpage behaviour of 12 inch FOWLP wafer for TV1 under process loadings can be captured. Warpage simulation model under process loadings considering viscoelastic material properties was validated by the experimental data.

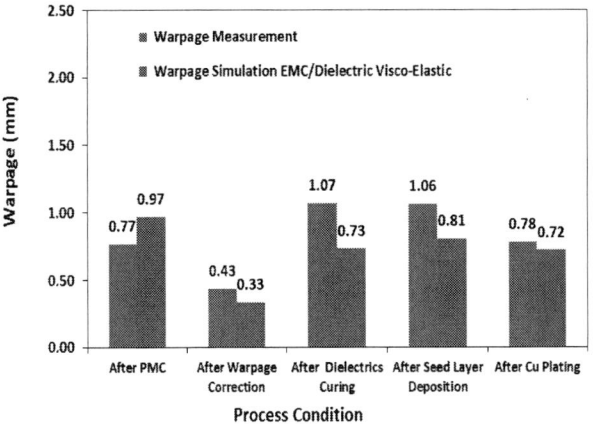

Figure 7. Measured and simulated wafer level warpage after different process steps of TV1.

The simulated warpage of TV1 FOWLP wafer after post mold curing, during warpage correction, after warpage correction, after dielectric curing, after seed layer deposition, and after blank Cu plating with elastic and viscoelastic material of EMC and dielectric was compared in Figure 8. It can be found that the wafer level warpage is overestimate with elastic material properties and only using the elastic material properties, the warpage behaviour cannot be simulated under the process loadings especially the warpage correction process.

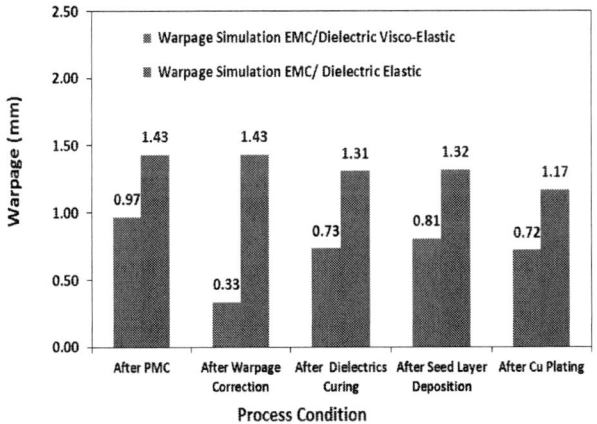

Figure 8. Simulated wafer level warpage of TV1 after different process steps with elastic and viscoelastic material properties of EMC and dielectric.

Figure 9 shows simulated warpage contour of TV2 wafer after different process steps of after post mold curing, during warpage correction, and after warpage correction. The warpage of 12 inch mold-first FOWLP wafer of TV2 after post mold curing is 2.64 mm which is higher than TV1. The reason is that the stiffness was reduced and the ratio of EMC to silicon was increased by changing the chip thickness from 0.775 mm to 0.2 mm. The warpage also becomes flatten under 150 °C temperature and pressure loadings of the warpage correction. The warpage is 0.026 mm during the warpage correction process. The warpage reduce to 1.48 mm after the warpage correction.

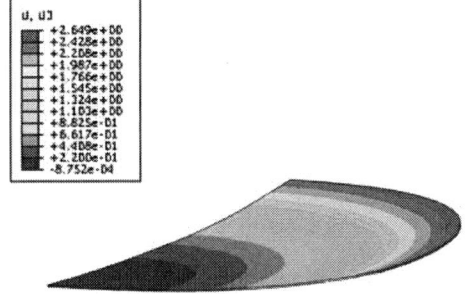

(a) After post mold curing

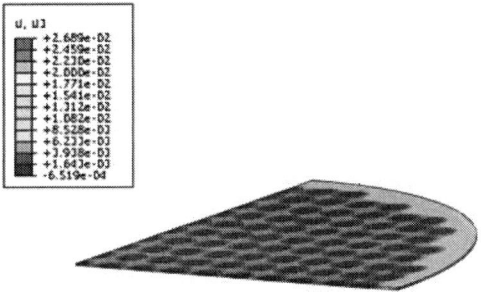

(b) During warpage correction

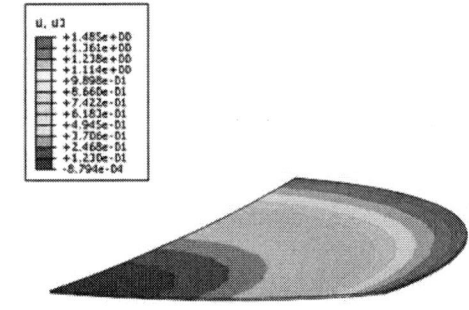

(c) After warpage correction

Figure 9. TV2 simulated warpage contour after different process steps: (a) After post mold curing; (b) During warpage correction; (c) After warpage correction.

The warpage of fabricated TV2 wafer after post mold curing and after warpage correction was measured. The experimental warpage data is showing as Figure 10. The warpage after post mold curing is 2.4 mm. The warpage reduces to 1.6 mm after the warpage correction. The simulated warpage is consistent with the measured warpage data. It is confirmed that with viscoelastic material properties of EMC the warpage behaviour of 12 inch FOWLP wafer under warpage correction loading can be captured. Warpage simulation model under warpage correction loadings considering viscoelastic material properties was validated by the experimental data.

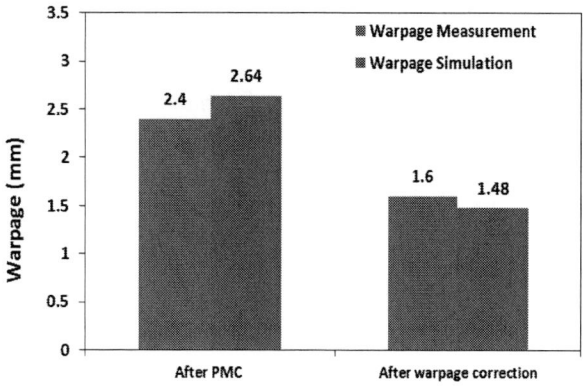

Figure 10. Measured and simulated wafer level warpage after PMC and after warpage correction of TV2 FOWLP wafer.

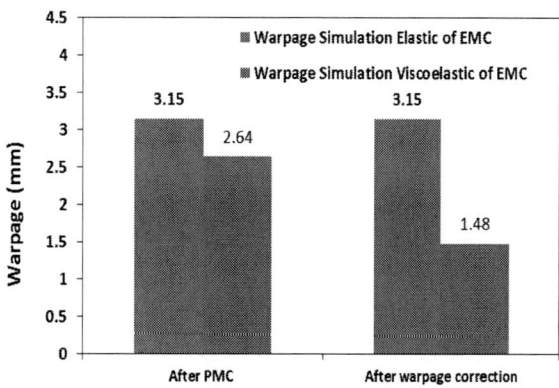

Figure 11 Simulated wafer level warpage of TV2 after different process steps with elastic and viscoelastic material properties of EMC.

The simulated warpage of TV2 wafer after post mold curing and after warpage correction with elastic and viscoelastic material of EMC was compared in Figure 11. It can be found that the wafer level warpage is overestimate with elastic material properties. Only use the elastic material properties, the warpage behaviour cannot be simulated under warpage correction process.

VI. CONCLUSIONS

The development of validated wafer level warpage modelling for 12 inch mold-first FOWLP under different processes loadings considering effects of viscoelastic material properties of EMC and dielectric has made a few significant achievements. Some of the important results of the paper are summarized below:

(1) Warpage of 12 inch mold-first FOWLP wafer predicted by the developed model considering viscoelastic material properties of EMC and dielectric have good correlation with the experimental results;

(2) The warpage of 12 inch mold-first FOWLP wafer can be reduced by after warpage correction. However, the warpage will increase after dielectric coating and curing. After PVD process, warpage of wafer will increase further.

(3) The viscoelastic material properties of EMC and dielectrics are critical for the warpage prediction under the FOWLP process loadings, especially the warpage correction. Only considering elastic material properties, the warpage behaviour cannot be predicted under the process loadings and the warpage is overestimated.

ACKNOWLEDGMENT

This work is funded by Science and Engineering Research Council (SERC), A*STAR through the Industry Alignment Fund: Pre-Positioning (IAF-PP; SERC Grant no. 16224 00012). The authors are grateful to IME staffs as well as IMRE staffs who had contributed and made this work possible.

REFERENCES

[1] Zhaohui Chen, Boon Long Lau, Zhipeng Ding, Eva Leong Ching Wai, Beibei Han, Lin Bu, Hyun-Kee Chang, Tai Chong Chai, "Development of WLCSP for Accelerometer Packaging with Vertical CuPd Wire as Through Mold Interconnection (TMI)," Proc. 68th Electronic Components and Technology Conference (ECTC), San Diego, USA, June 2018.

[2] Vempati Srinivasa Rao, Chai Tai Chong, David Ho, Ding Mian Zhi, Chong Ser Choong, Sharon Lim PS, Daniel Ismael and Ye Yong Liang, "Process and Reliability of Large Fan-Out Wafer Level Package Based Package-on-Package," Proc. 67th ECTC, Orlando, FL, USA, May 2017, pp. 615-622.

[3] Nusrat J. Chhanda, Jeffrey C. Suhling, Pradeep Lall, "Implementation of a viscoelastic model for the temperature dependent material behavior of underfill encapsulants," 13th IEEE ITHERM Conference, 2012, pp. 269-272.

[4] Huiqiang Shen, F. Qin, W. Wu, G. Xia, "Effect of viscoelastic behavior of EMC on predicting QFN fatigue life," 16th International Conference on Electronic Packaging Technology, 2015, pp.1242-1246.

[5] Nusrat J. Chhanda, Jeffrey C. Suhling, and Pradeep Lall, "Experimental Characterization and Viscoplastic Modeling of the Temperature Dependent Material Behavior of Underfill Encapsulants," ASME 2011 Pacific Rim Technical Conference and Exhibition on Packaging and Integration of Electronic and Photonic Systems, MEMS and NEMS: Volume 2, 2011, Paper No. IPACK2011-52209, pp. 749-761.

[6] Zhaohui Chen, Qin Zhang, Feng Jiao, Run Chen, Kai Wang, Mingxiang Chen, and Sheng Liu, "Study on reliability of application specific LED package by thermal shock testing, failure analysis and fluid-solid coupling thermo-mechanical simulation," IEEE

Transactions on Components, Packaging and Manufacturing Technology, 2012 2(7) pp.1135-1142.

[7] Yosephine Andriani, Xiaobai Wang, Songlin Liu, Zhaohui Chen, Xiaowu Zhang, "Thermomechanical and Viscoelastic Properties of Dielectric Materials Used in Fan-Out Wafer-Level Packaging," 2018 20th Electronics Packaging Technology Conference.

[8] Xiaobai Wang, Yosephine Andriani1, Songlin Liu, Zhaohui Chen, Xiaowu Zhang, "Dynamic Mechanical Analysis and Viscoelastic Behavior of Epoxy Molding Compounds Used in Fan-Out Wafer-Level Packaging," 2018 20th Electronics Packaging Technology Conference.

Ultra-thin Package Board Level Drop Impact Modeling and Validation

Shu-Shen Yeh*, P. Y. Lin, M. C. Yew, W. Y. Lin, K. C. Lee, C. C. Yang, J. H. Wang, P. C. Lai, C. K. Hsu, Shin-Puu Jeng

Taiwan Semiconductor Manufacturing Company,
No.6, Creation Rd. II, Hsinchu Science Park, Hsinchu, Taiwan (R.O.C.) 30077
*ssyehf@tsmc.com

Abstract—**Board level reliability during drop impact is a major concern for electronic packages. The impact force generated as the casing strikes the ground can cause electronic device failures in handheld products. The full drop testing procedure is costly and time-consuming due to complex sample preparation and test set-up procedures. Failure analysis also requires significant manpower to conduct. Therefore, an impact modeling method to predict the results of board level drops is highly desirable.**

We propose a dynamic modeling approach to describe the transient response of the package during impact based on an Input-G loading method with an implicit solver algorithm. The dynamic response of an ultra-thin package is obtained experimentally using a drop tester with accelerometer, strain gauge, and resistance monitor. For the Input-G method, the acceleration response of the impact pulse is then converted into a velocity form, and is taken as the loading input to a finite element (FE) model in this paper. The time dependent PCB strain, spectrum analysis, and modal analyses are used to correlate with the FE model and used to understand drop impact behaviors. Spectra of impact pulse and PCB dynamic strains are obtained by using the Fast Fourier Transform (FFT) technique. The extracted bending shape and frequency are consistent with the modal analysis results. The bending shape is mainly determined by the first mode. Knowledge about the spectra of PCB dynamic responses are required to understand the bending characteristics, which affect the package drop reliability. This FE model provides an accurate and reliable way to understand failure physics, and to help to achieve service life improvements in early development stage.

Keywords- Board level drop reliability, Input-G, Implicit method, fcCSP & Fan-out Package

I. INTRODUCTION

Consumer preferences for electronic products with smaller sizes, lighter weight, and multiple functionalities have resulted in the development of many handheld and portable devices with enhanced capabilities. These products are generally realized using ultra-thin packages. However, the impact force generated between the casing and the ground in the event that the product is accidentally dropped is transmitted to the electronic package through the printed circuit board (PCB), and may result in failure of the solder joints. Accordingly, increasing the board level drop reliability is an essential concern in improving the survivability of portable devices during mishandling or accidental drops [1-2].

The board level reliability of electronic packages is generally evaluated using in situ drop tests since, compared to other fast mechanical loading test methods, drop tests generate an impact closer to the actual loading condition experienced by the board in real-world shock or drop events [3].

There are several studies about the board level drop reliability of electronic products [4-10]. These studies focus on the solder joint failure mechanism (e.g., the crack location and characteristics) under the JEDEC standard. However, detailed investigations into the wider effects of the impact force on the strain and frequency response of a package and its constituent components remain lacking.

The present study proposes a comprehensive drop impact finite element (FE) modeling approach based on the velocity profile obtained in practical impact tests. The validity of the modeling approach is demonstrated by comparing the simulation results for the strain and frequency response of the PCB under various drop conditions with those obtained experimentally via in-situ drop tests. The validated model is then used to investigate issues with practical concern, including the relationship between the thermal interface material (TIM) coverage area and the stress induced in molded underfill (MUF), the effect of the underfill location on the crack risk of the solder ball gate array (BGA), and the effect of the drop acceleration and duration on the strain induced in PCB and substrate.

II. EXPERIMENTAL SETUP AND PROCEDURES

The simulation results from various models require validation through experimental results. Fig. 1 shows the typical setup of a board level drop tester [11-13]. The PCB/package is mounted with the drop tester components facing downwards on a drop table with fixing bolts. The drop table falls down along the guiding rods and hits the striking surface. The accelerometer and strain gauge are individually mounted on the drop table and the center of the PCB. A multi-channel real-time (in-situ) electrical monitoring system is used to measure the input acceleration of the drop table and strain at center of PCB. The impact force generated from the strike is transmitted to the PCB test board via the fixing bolts. Different impact pulses are achieved by adjusting the combination of drop height, striking surface/material and drop table. To achieve impact specifications, a fabric is placed on the rigid base to absorb the residual impact wave.

JESD22-B110A drop test condition B is adopted in this study. The input pulse is described by a half-sine wave, and the pulse is calculated by Eq. (1).

$$G(t) = G_0 \sin \frac{\pi}{\tau} t \qquad (1)$$

Where G_0 is the magnitude of peak acceleration and τ is pulse duration. Specific G_0 level and τ pulse duration can be obtained by adjusting drop height and strike surface with the fabric on the rigid base.

The experimental and theoretical impact pulses with 2,900G with 0.3ms duration are compared in figure 2. The difference between them is due to the friction on guiding rods and other unknown tester parameters. We use the experimental impulse as a loading input for dynamic modeling approach to describe the transient response of the package during impact.

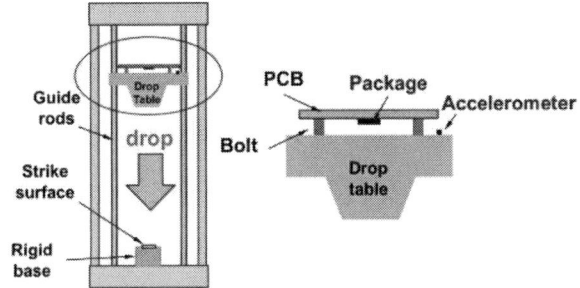

Figure 1. Typical experimental setup for board level drop test.

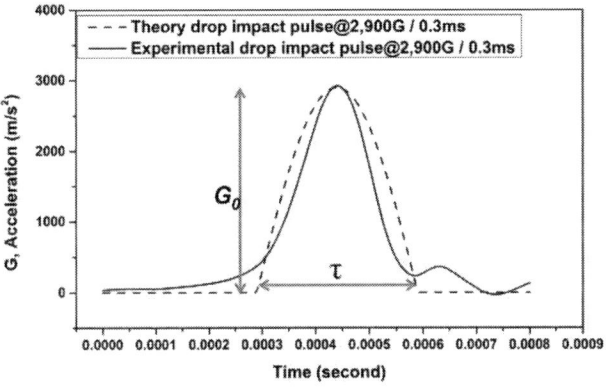

Figure 2. Comparison of experimental and theoretical impact pulse profile.

III. MODELING TECHNIQUES FOR BOARD LEVEL DROP TEST

There are three methods for modeling verification, i.e., PCB dynamics responses with strain, specturm analysis and modal analysis comparison. A three-dimensional FE model is developed for the following modeling study.

A. PCB Dynamic Responses

For drop modeling evaluation, a three-dimensional FE model is established using ANSYS software [14], which provides convenient tools for calculating the dynamic simulation. For a flip chip chip-scale package, the model is comprised of seven main components: the molded underfill, system-on-chip die, substrate/core, BGA, PCB, board level underfill (BL-UF), and bolts. When constructing the model, the geometry is meshed using 8-node solid elements. Because of the symmetry of the wafer form, only a quarter of the

wafer is used in the analysis with the load and boundary conditions.

As shown in Fig. 3, an Input-G (input acceleration) method is developed in this study. For this method, the impact pulse or input acceleration needs to be obtained beforehand, and will be fixed and input to the bolts of PCB subassembly directly as a boundary condition. In here, the input acceleration is obtained from the experimental data in Fig. 2. The data is applied as an input loading on the bottom surface of the bolts. The profile of time-dependent velocity (Fig. 4) can be integrated from the impact pulse (to velocity profile first), and then is directly applied to support bolts as boundary conditions of implicit Input-G model. This velocity form is integrated from Eq. (1), and it is described by Eq. (2) and Eq. (3).

$$V(t) = -\frac{G_0 \tau}{\pi} \cos \frac{\pi}{\tau} t - \frac{G_0 \tau}{\pi} \qquad (2)$$

$$V_{initial} = -2 \frac{G_0 \tau}{\pi} \qquad (3)$$

In order to reduce the huge computation amount of the drop simulation, the implicit Input-G method was proposed by Luan *et al.* [15]. This Input-G method in FE model ignores the drop table, screw and strike surface of rigid base, and the impact pulse is applied as an input acceleration loading at bolt fixed at the PCB hole. Because of the simplicity of the implicit Input-G method, the computing time is much reduced when compared with conventional drop analysis such as LS-Dyna. Therefore, the Input-G method is widely used in drop dynamic simulations.

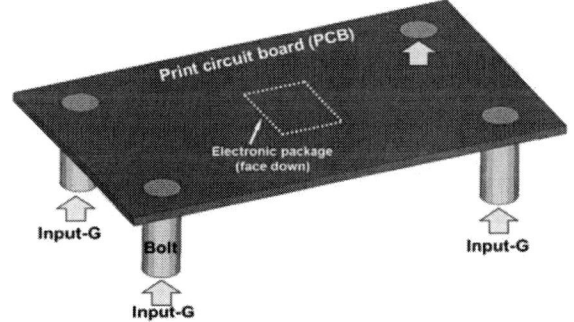

Figure 3. Schematic of Input-G method setup.

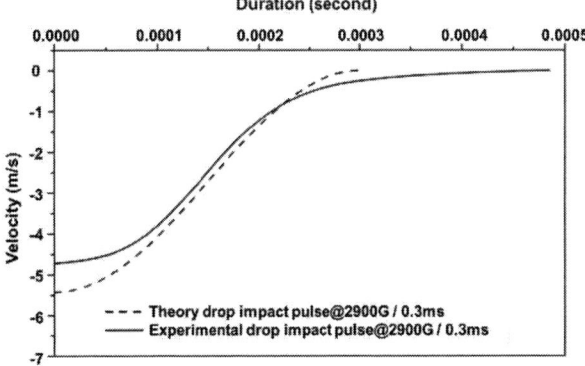

Figure 4. Comparison of theoretical and experimental drop velocity profiles.

978-1-7281-1500-9/19 $31.00 © 2019 IEEE

Before we compare the simulation and experiment, we must highlight that the strain gauge location is also the key to determining PCB strain. In general, there is a high degree of strain variation along the packaging edge. Then, along the PCB length direction, the PCB strain becomes more stable and changes the direction from the positive strain of tension to the negative strain of compression. This is due to the free vibration of the PCB after the initial impact force. As shown in Fig. 5, the results of the time-dependent dynamic strains in PCB length direction are predicted based on the Input-G method. It can be seen that the simulation results obtained from the model not only correlate well with the experimental measurement, but also shows that the amplitude and frequency of PCB vibration are almost identical between the model and the experiment.

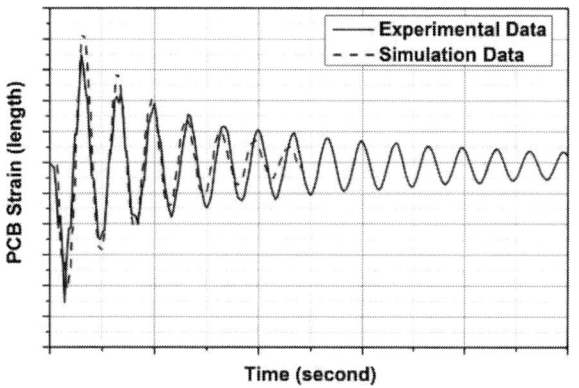

Figure 5. Comparison of experimental and simulated PCB strain in length direction.

B. Spectrum Analysis

Experimental spectrum analysis is performed using fast Fourier transform (FFT) technique from PCB dynamic strains. The bending modes and frequencies described in the spectrum of PCB dynamic strains are then compared with the values calculated by ANSYS modal analysis show in the next session. The time domain changes to frequency domain can be described as Eq. (4) and Eq. (5):

$$F(f) = \Im[f(t)] \tag{4}$$

$$F(f) = \int_{-\infty}^{\infty} f(t) e^{-j\omega t} \, dt \tag{5}$$

Where $F(f)$ is frequency domain. $f(t)$ is the time domain through the Fourier transform calculation, j is the complex number. ω is frequency, and t is time.

Theoretically, responses of a package to an impulse describe a system's characteristics, i.e. the natural frequencies. In this paper, the drop impact occurs during a very short duration (~0.3ms) which is close to an ideal impulse. Therefore, the PCB characteristics can be extracted from the dynamic responses of PCB strains through FFT.

Fig. 6 shows the spectrum of PCB dynamic strains. The peak in the spectrum graph is from the drop/bending mode of PCB. Obviously, the first mode is the most dominant with the frequency of 956Hz in the length direction. The first mode has the highest energy, and its frequency corresponds to the highest peak in the spectrum. Frequency domain analysis is the most comprehensive, quick method and much clearer for frequency study.

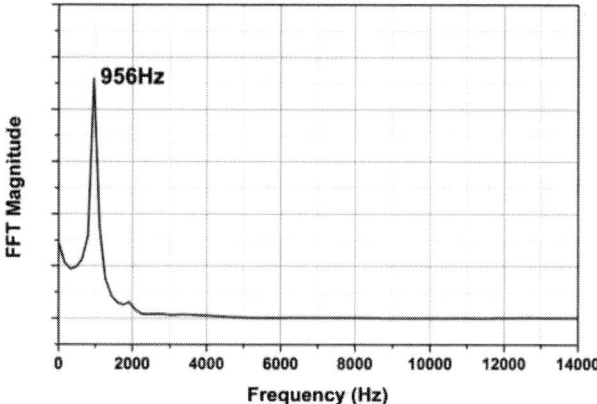

Figure 6. Experimental impact pulse in frequency domain under JEDEC standard of 2,900G/0.3ms.

C. Modal Analysis Comparison

Modal analysis is performed using finite element modeling. The dominant bending mode shapes of the first three modes are shown in Fig. 7. The calculated frequency values are 959Hz, 1026Hz and 1302Hz, respectively. It can be seen that the first mode is consistent with 1st peak frequency in the spectrum analysis. Therefore, from the 1st mode shape, the maximum deflection during drop impact is excited on the PCB center and leads to a large strain of in-length direction induced around the package.

The simulation work conducted by Luan et al. [10] show that the maximum t bending moment occurs at the region near the center of PCB. Thus, the maximum degree of peeling stress is expected when the component is situated there.

By using the Input-G method, the simulation results not only exhibit good agreement in time-domain analysis, but also show excellent correlation between spectrum analysis and modal analysis. The simulation facilitates a thorough understanding of PCB bending characteristics, and more importantly, the impact model we develop in this paper is successfully applied on package design optimization for to improve reliability.

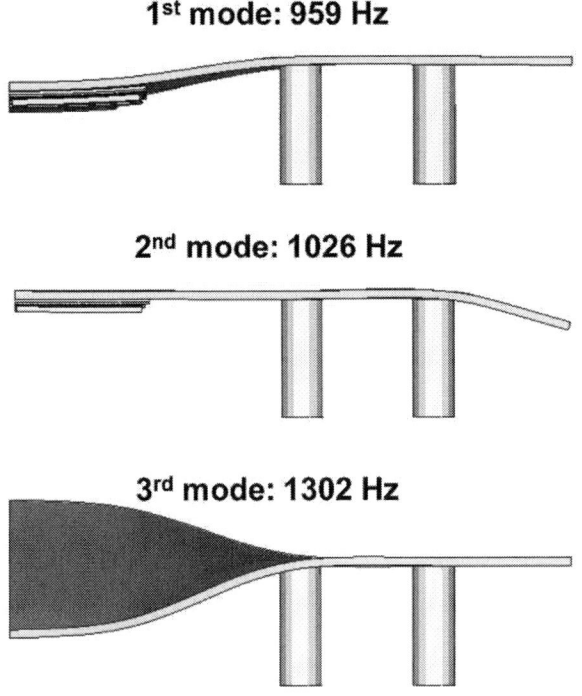

1st mode: 959 Hz

2nd mode: 1026 Hz

3rd mode: 1302 Hz

Figure 7. Analysis results of the first 3 modes.

IV. APPLICATION OF DROP IMPACT MODELS

A. Molding Compound Crack Risks Assessment on fcCSP

Fig. 8 presents a fcCSP package. This state-of-art packaging technology has many advantages, including a short interconnect length and a high degree of reliability for mobile applications. These advantages are achieved by using thermosetting polymers such as a molded underfill (MUF) or an epoxy molding compound (EMC). These materials have advantages for the packaging of electronic products, including low cost, high throughput, and excellent reliability. However, the high stress at the SoC die corner (see Fig. 9) induced during drop testing has a critical effect on the reliability of the package. A detailed understanding of the MUF high stress location can be derived from the aforementioned drop modeling approach. Furthermore, a different coverage of thermal interface material (TIM) is evaluated as shown in Fig. 10. A higher TIM coverage reaches a lower MUF P1 stress due to the ability of TIM to absorb energy, and therefore reduces the MUF crack risk.

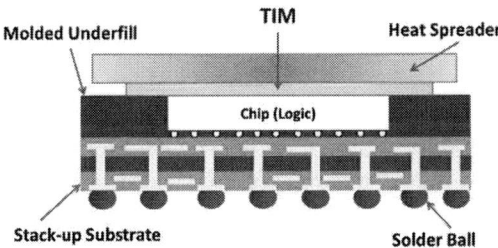

Figure 8. Schematic of a typical fcCSP.

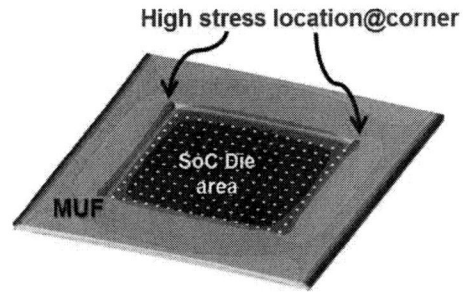

Figure 9. Contour of MUF P1 stress. High stress concentrates at corners.

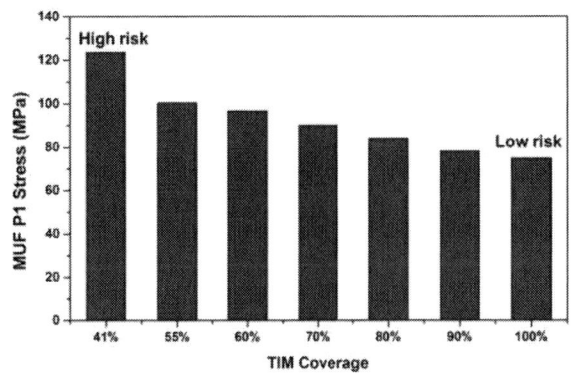

Figure 10. Relationship between TIM coverage and MUF P1 stress under the drop condition of 5,000G/0.2ms.

B. Solder Joint Reliability and Failure Mode

A second further series of simulations is also applied to multi-chip module (MCM) package. As shown in Fig. 11, the combination can be the flip chip chip scale package (fcCSP) MCM or fan-out MCM [16]. Both of these configurations are currently intensively discussed in industry for their use in high performance computing (HPC) products. In general, these configurations can be 1-SoC+4-DRAMs or 1-SoC+8-DRAMs. In this paper, we consider 1-SoC+4-DRAMs for evaluation.

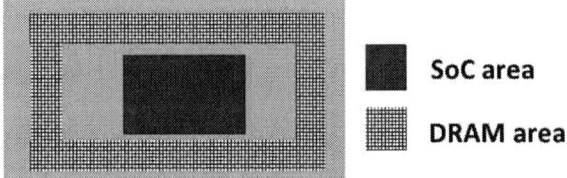

Figure 11. A typical multi-chip module.

As shown in Table 1, the DRAM BGA and package BGA joint risk on fcCSP MCM are evaluated under JEDEC standard drop condition of 2,900G/0.3ms. In the BGA crack risk assessment, the fine FE model has two-layer elements across the 25 µm interface thickness. The Anand constitutive model is used for the solder in this study. The accumulated equivalent plastic strain (EPS) in the top and bottom interface layer elements are compared as in Table 1. Besides, the

978-1-7281-1500-9/19 $31.00 © 2019 IEEE   1553

element size effect on the simulation result is not significant due to the use of the volume-averaging method.

TABLE I. SOLDER JOINT RELIABILITY AND FAILURE MODES WITH AND WITHOUT UNDERFILL

Drop Condition		2,900G / 0.3ms		
DoE Leg		Leg-1	Leg-2	Leg-3
DRAM underfill		x	x	v
Board level underfill		x	v	v
DRAM BGA	Max P1 Stress, MPa	210	230	50
	EPS (Top)	11%	12%	1.2%
	EPS (Bottom)	13%	15%	0.8%
	EPS Contours	See Fig. 12	See Fig. 12	See Fig. 12
Package BGA	Max P1 Stress, MPa	215	62	60
	EPS (Top)	52%	0.6%	0.6%
	EPS (Bottom)	55%	0.9%	0.7%
	EPS Contours	See Fig. 13	See Fig. 13	See Fig. 13
Drop Testing Status		DRAM corner BGA Package corner BGA Failed	DRAM corner BGA Failed	Passed

Three different DoE legs are considered in Table 1, including (Leg-1) no DRAM underfill (UF) + no board level UF; (Leg-2) no DRAM UF + with board level UF; (Leg-3) with DRAM UF + with board level UF. The results can be summarized as follows:

Leg-1: the DRAM BGA and package BGA both have a high P1 stress during the initial impact, and exhibit a high equivalent plastic strain after drop vibration (see Figs. 12 and 13). For both BGA layers, the solder joint fails along a 45-degree plane as a result of the fatigue induced by the free vibration of the PCB following the initial impact. Moreover, the BGA solder layers fail at the DRAM corner and package corner, respectively, due to the lack of underfill protection in both cases.

Leg-2: the DRAM BGA has a higher crack risk than the package BGA due to the absence of DRAM UF. Consequently, as shown in Table 1, the BGA solder layer fails along a 45-degree plane at the DRAM corner.

Leg-3: underfill is applied at both the DRAM BGA and the package BGA. Consequently, the risk of solder joint failure is reduced. The results presented in Table 1 confirm that the underfill prevents fatigue-induced failure in both BGA layers.

The results presented in Table 1 confirm that the underfill should be applied at both the DRAM level and the board level in order to improve the drop reliability of fcCSP and fan-out MCM packages. This dynamic response of PCB has great influence on solder joints due to higher order vibrating modes. From modeling evaluation, solder joint crack risk from fcCSP MCM to fan-out MCM can be

optimized to meet the requirement level to ensure 100% yield on board level drop reliability.

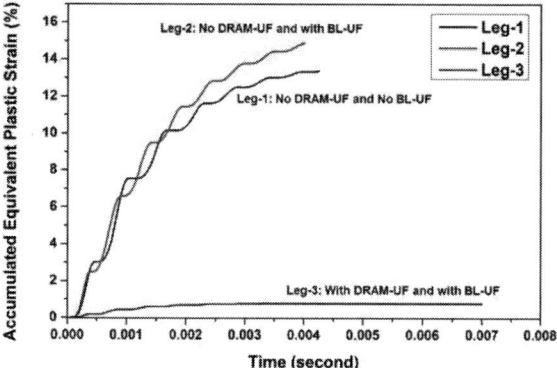

Figure 12. Accumulated equivalent plastic strain of DRAM BGA from different packages.

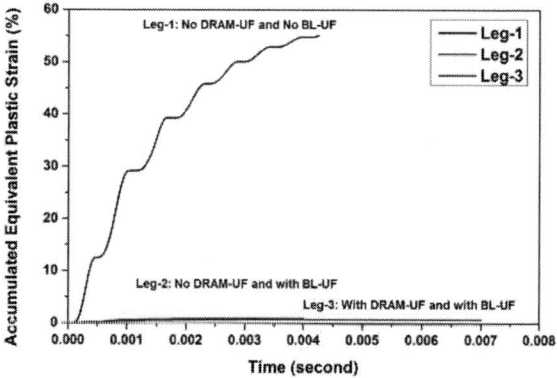

Figure 13. Accumulated equivalent plastic strain of BGA from different packages.

C. *Drop Acceelration versus the Drop Duration Effect*

A third series of simulations is also conducted to examine the relationship between drop acceleration and drop duration. A MCM package is used for evaluation.

Table 2 shows the maximum strain on substrate and PCB under various board level drop conditions. It can be seen that 5,000G (0.2ms) reaches a very high risk on substrate strain in the length direction with the level of around 0.7% during the drop, resulting in actual substrate cracking happening after real drop testing. However, when the drop condition is 2,900G (0.3ms), it is observed that the maximum substrate strain is around 0.4% and all samples pass the real drop testing, indicating low risk. Therefore, the strain of substrate failure criterion is defined on 0.7% in this case study.

The lower JEDEC drop condition of 1,500G (0.5ms) is also evaluated, and this study finds that both substrate and PCB strain are comparable to the higher condition of 2,900G (0.3ms) due to similar areas under drop profile (Fig. 14). These results are very useful for engineers in conducting testing, as the actual drop duration is hard to control for very short drop times.

It is challenging to precisely control short duration with haevy load during drop test. As shown in Table 2, the maximum strains of substrate and PCB increase significantly, if the drop duration deviates slightly from 0.3ms to 0.35ms with a heavy load of 2900G. Therefore, the conditon of 0.5ms duration with smaller 1,500G is recommended, as such a test condition is easier to control.

TABLE 2. MAXIMUM STRAIN MONITORED ON SUBSTRATE AND PCB UNDER DIFFERENT DROP CONDITION

BLR Drop Input Condition	1,500G (0.5ms)	2,900G (0.30ms)	2,900G (0.35ms)	5,000G (0.20ms)
Max Substrate strain (Length)	0.4%	0.4%	0.5%	0.7%
Max PCB strain (Length)	0.25%	0.25%	0.3%	0.45%
Drop testing status	Low Risk	Low Risk	Moderate Risk	High Risk

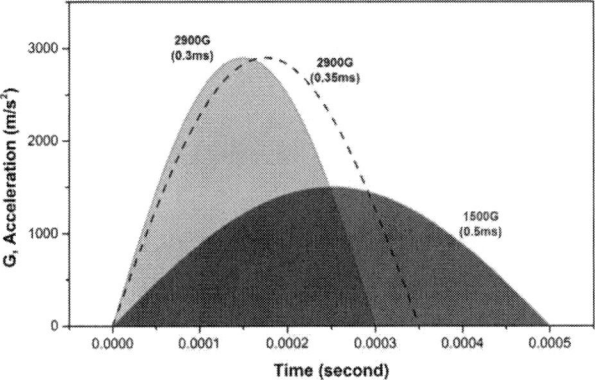

Figure 14. Acceleration versus drop duration under different drop conditions.

V. CONCLUSION

We demonstrate a simple and efficient approach for simulating the strain and frequency response of ultra-thin packages under the realistic drop conditions. The experimental acceleration values are inputs for the boundary condition in finite element (FE) Input-G model. The validity of this modeling approach is demonstrated by comparing the time-dependent strain and frequency responses. The model has been successfully applied to various practical issues in board level impact tests, such as (1) the relationship between the TIM coverage area and the MUF strain, (2) the effect of the underfill arrangement on the BGA solder stress and failure mode, and (3) the dependence of the PCB and substrate strains on the acceleration and time duration properties of the impact event. This approach provides an accurate and reliable means to evaluate the response of ultra-thin electronic packages under typical drop conditions. It is a useful tool for identifying potential risks to improve service life in early development stage.

ACKNOWLEDGMENTS

The authors would like to acknowledge Dr. De-Dui Marvin Liao for their support as well as the TSMC APTS (Advanced Packaging Technology and Service) teams for their efforts and contributions to this paper.

REFERENCES

[1] T. T. Mattila, J. Simecek, and J. Kivilahti, "Failure Modes of Solder Interconnections under Mechanical Shock Loading at Elevated Temperatures," Electronics System Integration Technology Conference, Dresden, Germany, Sept. 5-7, 2006, pp. 195-202.

[2] J. J. M. Zaal, W. D. van Driel, F. J. H. G. Kessels, and G. Q. Zhang, "Verification of Drop Impact Simulations Using High-Speed Camera Measurements," in Proc. IEEE Electronic Compoments and Technology Conf. (ECTC), Lake Buena Vista, FL, USA, May 27-30, 2008, pp. 2149–2155.

[3] Y. Liu, F. L. Sun, F. J. H. D. Kessels, W. D. van Driel, and G. Q. Zhang, "The Effects of Response Features on Failure Modes of Board Level Drop Impact Test," 11th International Conference on Electronic Packaging Technology & High Density Packaging, Xi'an, China, Aug 16-19, 2010, pp. 984-988.

[4] John Lau et al., "Reliability of Fan-Out Wafer-Level Packaging with Large Chips and Multiple Re-Distributed Layers," in Proc. IEEE Electronic Compoments and Technology Conf. (ECTC), San Diego, CA, USA, May 29-June 30, 2018, pp. 1574–1582.

[5] C. Y. Chou, T. Y. Hung, S. Y. Yang, M. C. Yew, W. K. Yang, and K. N. Chiang, "Solder joint and trace line failure simulation and experimental validation of fan-out type wafer level packaging subjected to drop impact," Microelectronics Reliability, no. 48, 2008, pp. 1149-1154.

[6] C. Y. Chou, T. Y. Hung, C. J. Huang, and K. N. Chiang, "Development of Empirical Equations for Metal Trace Failure Prediction of Wafer Level Package Under Board Level Drop Test," IEEE Transactions on Advanced Packaging, vol. 33, no. 3, 2010, pp. 681-689.

[7] T. Y. Tee, J. Luan, and H. S. Ng, "Development and Application of Innovational Drop Impact Modeling Techniques," in Proc. IEEE Electronic Compoments and Technology Conf. (ECTC), Lake Buena Vista, FL, USA, May 31-June 3, 2005, pp. 504-512.

[8] E. S. Ibe, K. I. Loh, J. E. Luan, and T. Y. Tee, "Effect of Unfilled Underfills on Drop Impact Reliability Performance of Area Array Packages," in Proc. IEEE Electronic Compoments and Technology Conf. (ECTC), San Diego, CA, USA, May 30-June 2, 2006, pp. 462-466.

[9] F. X. Che, and John H. L. Pang, "Fatigue Reliability Analysis of Sn–Ag–Cu Solder Joints Subject to Thermal Cycling," IEEE Transactions on Device and Materials Reliability, vol. 13, no. 1, 2012, pp. 36-49.

[10] Luan, J.E., Tee, T.Y., Pek, E., Lim, C.T., Zhong, Z.W., "Modal Analysis and Dynamic Responses of Board Level Drop Test," 5th EPTC Conference Proc., Singapore, 2003, pp. 233-243.

[11] JEDEC Standard JESD22-B111, Board Level Drop Test Method of Components for Handheld Electronic Products, 2003.

[12] JEDEC Standard JESD22-B104-B, Mechanical Shock, 2001.

[13] JEDEC Standard JESD22-B110, Subassembly Mechanical Shock, 2001.

[14] ANSYS Version 19.0, Ansys, Inc., Canonsburg, PA, 2018.

[15] J. E. Luan, T. Y. Tee, E. Pek, C. T. Lim, Z. Zhong, and J. Zhou, "Advanced numerical and experimental techniques for analysis of dynamic responses and solder joint reliability during drop impact," IEEE Transaction on Components and Packaging Technologies, Vol. 29, pp,449-456, 2006.

[16] S. S. Yeh, P. Y. Lin, K. C. Lee, J. H. Wang, W. Y. Lin, M. C. Yew, P. C. Lai, C. K. Hsu, C. C. Yang, and S. P. Jeng, "An Integrated Warpage Prediction Model Based on Chemical Shrinkage and Viscoelasticity for Molded Underfill," in Proc. IEEE Electronic Compoments and Technology Conf. (ECTC), San Diego, CA, USA, May 29–Jun 01, 2018, pp. 249–254.

978-1-7281-1500-9/19 $31.00 © 2019 IEEE

Flexible Graphene-Glass Fiber Composite Film with Ultrahigh Thermal Conductivity and Mechanical Strength as Highly Efficient Thermal Spreader Materials

Xiaoliang Zeng, Linlin Ren, Rong Sun, Jianbin Xu
Center for Advanced Material Research
Shenzhen Institutes of Advanced Technology,
Chinese Academy of Sciences
Shenzhen, China
E-mail: rong.sun@siat.ac.cn

Ching-Ping Wong
School of Materials Science and Engineering
Georgia Institute of Technology
Atlanta, United States
E-mail: cp.wong@mse.gatech.edu

Abstract—Most commercial thermal spreader materials have satisfactory thermal conductivity but inferior mechanical strength and flexibility. Therefore, how to simultaneously improve ultrahigh thermal conductivity and excellent mechanical properties for heat spreaders is a still under challenge. Here, we demonstrated that the use of graphene oxide and glass fiber can fabricate graphene-glass fiber composite film with ultrahigh thermal conductivity, high mechanical strength, and flexibility, by bar-coating method. The one-dimensional glass fibers were chose as the scaffolds, which reinforce the mechanical strength. The in-plane thermal conductivity of the graphene-glass fiber composite film reaches 952±104 W/mK, while the mechanical strength is up to 106.9 MPa. In addition, the composite film shows excellent flexibility. The bar-coating method is easily scaled up for wide applications. It is believed that the graphene-glass fiber composite film can be used as high-performance heat spreaders in high-power devices for highly efficient thermal management.

Keywords- Thermal spreader materials, graphene, glass fiber, thermal conductivity, mechancial strength

I. INTRODUCTION

High-degree integration, ligh-weight, high-frequency and high performance is the future development trend for modern electronic devices. However, these trend will results in serious heat accumulation in electronic devices during their working process. How to effectively remove redundant heat is crucial to ensure reliability and even lifetime for electronic devices [1]. Thermal spreader materials are essential ingredients of thermal management by quick removing local hot spot of electronic devices in the x-y direction [2]. Thermal spreader materials are typically categorized into metallic, ceramic, carbon-based, and their composite heat spreaders materials. The metallic based heat spreaders contain as copper [3], aluminium [4], copper–tungsten or copper–moly alloy. However, they have the thermal conductivity below 400 W/m K. On the other hand, composite heat spreader composite heat spreaders including ceramic [5, 6], metallic, and carbon composites have thermal conductivity below 1,200 W/m K. Interestingly, plates have the thermal conductivity in the range from 500 to

5,000 W/m K,and heat pipes have the thermal conductivity above 5,000 W/m K [7].Surprising, the oscillating flow heat spreaders possess thermal conductivity up to 120,000 W/m K [8]. Up to now, thermal spreader materials have been used widely in various commercial electronic devices, such as touch panels, smart phones, and light emitting diode lamps. Among all thermal spreader materials, carbon-based thermal spreader materials have important roles due to their excellent ability of thermal [9-11]. For example, the thermal conductivity of carbon nanotubes (CNTs) is in the range of of ~3000– 3500 W/m K [12, 13]. The room temperature thermal conductivity of graphene is calculated to be 5300 W/m K [14], which much higher than umper limit of bulk graphite (2000 W/m K) [15]. Therefore, the carbon based thermal spreader materials, especially graphene based thermal spreader materials[16], have attracted significant attention in recent years [9, 17]. Modern electronics that greatly develop day by day require high-performance heat spreader materials in order to meet the designed demand.

Although single CNT nanotube shows great potential application as heat spread for high-power electronic devices, the strong phonon scattering leads to low thermal conductivity of CNT film, so that CNTs are not considered to be heat spreader materials [18]. In market, the most used carbon heat spreader materials are graphitized polymide and graphite papers. Their thermal conductivity ranges from 100 to 1000 W/mK, which cannot satisfy demands of the future eletonic devices. Although the graphitized polyimide papers possess the thermal conductivity as high as 1950 W/mK, the graphitization at 3000 °C leads to high cost of the graphitized polyimide paper. On the other hand, most work in the area of thermal interface materials has focused on only thermal conductivity. Actually, besides high thermal conductivity, excellent mechanical properties are important for the thermal spreader materials, especially good flexibility for the next generation flexible electronics. Unfortunately, most commercial thermal spreader materials have satisfactory thermal conductivity but inferior mechanical strength and flexibility. Some researchers have devoted to improving both the thermal conductivity and mechanical strength [19]. For example, Q.Q. Kong et al fabricated a novel graphene-carbon fiber composite films that have the potentioal application in heat spreaders [20]. These composite films have ultra-high in-plane thermal conductivity of 977 W/mK. The function of carbon fiber is not only reinforcing the mechanical strength of the paper, but also avoiding the re-stacking of the single-layer graphene. However, the mechanical tensile strength is still low (15.3 MPa) and cannot meet the requirement. In our previous repot, we reported on a kind of boron nitride based heat spreaders with excellent mechanical strength (125.2 MPa). However, the in-plane thermal conductivity of the boron nitride films is only 6.9 W/mK [21]. G. Chao et al fabricated graphene films with high thermal conductivity of 1940 W/mK and superflexible [22]. However, its mechanical strength is only 60 MPa, and the preparation processing is too complex. G. Xin et al reported a kind of graphene fibers with superhigh thermal conductivity (1290 W/mK) and excellent mechanical strength (1080 MPa) by tunning the size of the graphene oxide and forming an interacted compact microstrure [23]. The thermal conductivity is up to, and the tensile strength reaches, which indicates that realizing simultaneously high mechanical and superior thermal conductivity is feasible. However, this approach can only prepare graphene fiber, how to simultaneously improve ultrahigh thermal conductivity and excellent mechanical properties for heat spreaders is a still under challenge.

Here, we demonstrated that the use of graphene oxide and glass fiber can fabricate graphene-glass fiber composite film with ultrahigh thermal conductivity, high mechanical strength, and flexibility, by bar-coating method. The one-dimensional glass fibers were chose as the scaffolds, which reinforce the mechanical stiffness and flexibility. The graphene-glass fiber composite film possesses in-plane thermal conductivity of 952±104 W/mK, while the mechanical strength is up to 106.9 MPa. In addition, the composite film shows excellent flexibility. The bar-coating method is easily scaled up for wide applications. It is believed that the graphene-glass fiber composite film can be used as high-performance heat spreaders in high-power devices for highly efficient thermal management.

II. EXPERIMENTAL SECTION

A. Materials

Graphene oxide nanosheets (GO) dispersion was provided by Shanxi Institute of Coal Chemistry (Shanxi Province, China). Borosilicate electrical grade glass fiber commonly known as E-glass fiber (glass style: E-106, plain weave, count: 56×56 ends/in, thickness: 0.0015 in) was purchased from Shengzhen Sailong Fiberglass Company Ltd, China. 3-aminopropyltriethoxysilane (98%) was obtained from J&K Scientific (GmbH, Germany). Ultrapure water (18.25 MΩ) was produced in the laboratory using Ulupure (UPHeI-20T). All other chemicals were used as received.

B. Fabrication of graphene-glass fiber composite film

The graphene-glass fiber composites films were prepared by the combination of bar-coating and thermal treatment. The glass fiber weave was firstly put on polyethylene terephthalate (PET) release film, so that the fabricated film is

flast can easily peel off the release film after dyring. The GO dispersion (2.0 mg/mL) was coated on the glass fiber weave by bar-coating method. The obtained GO-glass fiber composite film was dried at 80 °C for 24 h, followed by peeling off from the release film to obtain free-standing GO-glass fiber composite film. The composite film was then thermally treated at 500 °C, 750 °C, 1000 °C, and 1200 °C, respectively, in Ar atmosphere for 2 h to obtain the graphene-glass fiber composite film.

C. Characterization

The morphology and the thickness of GO was characterized with an atomic force microscope (Dimension ICON, Bruker Corporation, USA) in ScanAsyst mode. The morphology of the composite film was examined by a microscope (DM2700 H, Leica, Germany) and field emission scanning electron microscopy (FE-SEM, Nova NanoSEM 450, FEI, USA), working in the secondary electrons mode at 5 kV. The in-plane thermal diffusivity of the composites was measured by the laser flash method using LFA 467 (Nano-flash, NETZSCH, Germany) at 30 °C. The in-plane thermal conductivity was calculated by

$$K = \alpha \cdot C_p \cdot \rho \qquad (1)$$

where α is the thermal diffusivity, ρ is the density of composites which was calculated by weighing method. C_p is the specific heat capacity obtained by differential scanning calorimetry (DSC, TA Q2000, USA) with the sapphire method. Tensile tests were carried out on a universal texting machine (RGM-4000, REGER, China) at room temperature according to American Society for Testing and Materials (ASTMs) standard D822-09. The crosshead speed was set at 12.5 mm/min. The test samples were cut into the shape of $120 \times 10 \times 0.007$ mm^3. At least 5 samples were tested ensure the accuracy of the data. The temperature distribution image of the composite film was recorded by an infrared thermograph (FLIR E30, FLIR Systems, Inc., USA). The ambient temperature was approximately 20 °C.

III. RESULTS AND DISCUSSION

Figure 1 shows the representative AFM image of GO by drop-coating a drop of dilute solution onto a new cleaved silicon sheet. The individual GO sheets have an average lateral size of ~ 5.0 μm and a thickness of ~1.0 nm. The thickness of single-layer graphene is 0.34 nm. However, there are many functional groups on its surface, which results in a enhanced measured hight, as agreed well with the previous report [24].

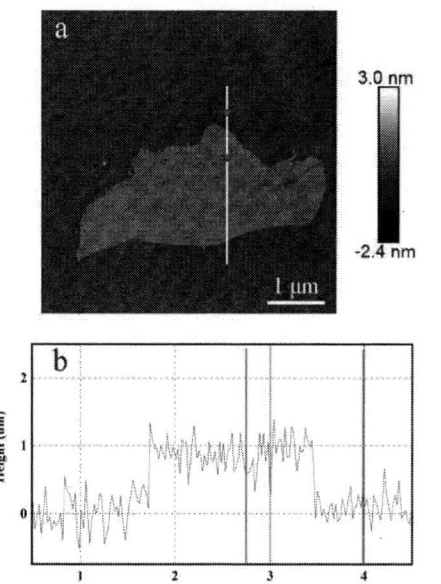

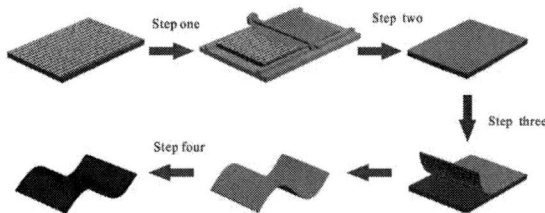

Figure 1. Morphology of GO nanosheets. (a) AFM image of GO nanosheets. (b) Corresponding AFM cross section.

Figure 2. Preparation process of the GF-RGO composite films by bar-coating. Step 1: Spreading the GO gels out on PET release film. Step 2: Drying the wet paper at room temperature. Step 3: Peeling off the GF-GO film from PET release film. Step 4: Thermal annealing at high temperature.

Figure 2 schematically illustrates the fabrication process of GF-RGO composite films. To achieve freestanding GF-RGO composite film, the graphene oxide (GO) gels were spread out on glass fiber mats which were put on PET substrate by an automatic bar-coater (step 1). This preparation process can be prepared GF-RGO composite films with large area and freestanding. The wet GF-GO films were dried overnight at 80 °C (step 2). The GF-GO films were then peeled off from the commercially available PET release substrate when dried (step 3). Finally, thermal annealing at high temperature leads to the formation of GF-RGO composite films with high thermal conductivity and excellent mechanical properties.

To improve the interaction between glass fibers and GO, surface modification of glass fibers was carried out by using functional silanes. The silanes were chosen since they enable a covalent bonding of the modified glass fiber surface to the GO. To confirm the surface modification of glass fiber, Fourier-transformed infrared (FTIR) spectroscopy was used to compare the original glass fibers and modified glass fibers. Figure 3 shows the FTIR spectra of the original glass fibers, and modified glass fiber. The unmodified glass fibers show a broad characteristic bands at the range of 900–1200 cm^{-1},

corresponding to Si–O–Si vibrations. For functionalized glass fibers, the peaks related to the $-CH_2$ groups appear at 2951 and 2861 cm^{-1}. The peaks at 3357 and 1567 cm^{-1} correspond to the $-NH_2$ group. In addition, C–N stretching vibration peak at 1388 cm^{-1} are also observed. On the basis of these observations, we confirm that 3-aminopropyltriethoxysilane was covalently bonded on the glass fiber surface.

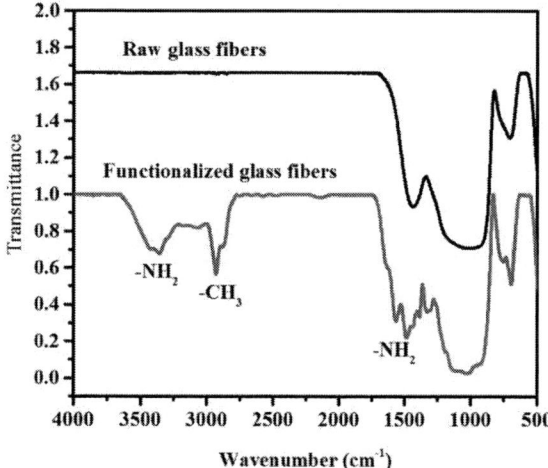

Figure 3. FTIR spectra of the raw glass fibers and functionalized glass fibers.

Figure 4a shows the optical image of the freestanding GF-RGO composite films fabricated by the facile scalable bar-coating method. The size of the obtained film is 20 cm × 20 cm with the thickness of 40 μm. We can also fabricated GF-RGO composite films with various thicknesses by bar-coating and drying repeat several times. The fabricated GF-RGO composite film shows excellent flexibility, because it can be easily folded to be a craft (Figure 4b). In addtion, the GF-RGO composite film exhibits good mechanical strength that it can hold a weight of 500 g steadily (Figure 4c). Figure 4d shows the glass fiber mats which consists of many glass fibers. After bar-coating graphene, the space between glass fiber bundles was filled with grapheen, as shown in Figure 4e. In addition, the high-magnification SEM image indicates good compatibility between graphene and glass fibers (Figure 4f).

978-1-7281-1500-9/19 $31.00 © 2019 IEEE   1558

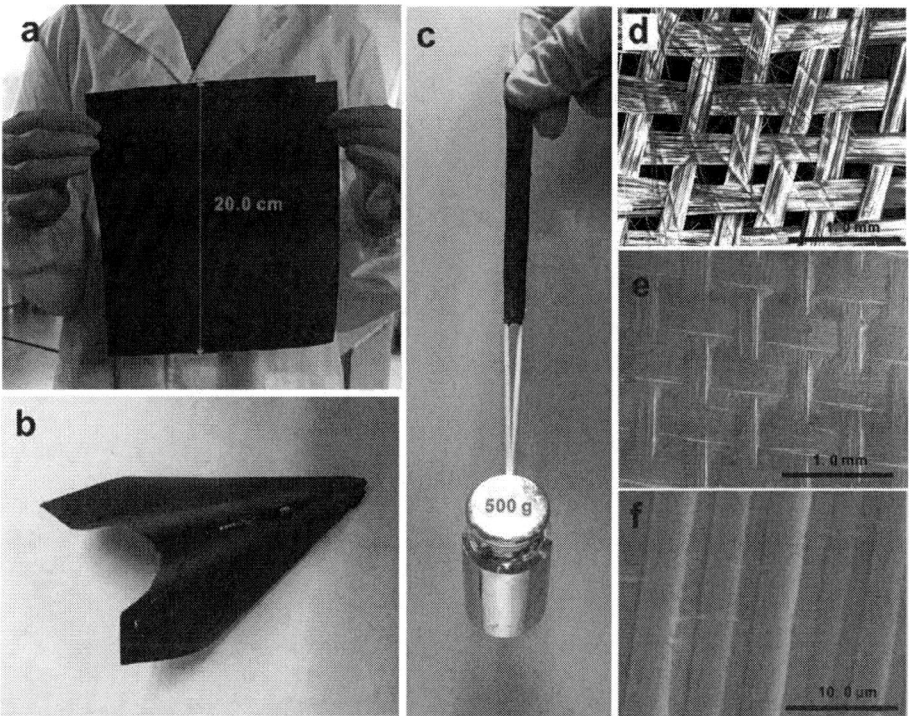

Figure 4. Morphology of the GF-RGO composite films. (a) Optical image of large-are and freestanding GF-RGO composite film with the size of 20 cm ×20 cm and thickness of 40 μm. (b) Optical image of a handmade craft using GF-RGO composite film. (c) Optical image of GF-RGO composite stretching weight of 500 g. (d) SEM image of glass fiber mats. (e) SEM image of GF-RGO composite film. (f) High magnification SEM image of GF-RGO composite film.

Figure 5 shows the in-plane and out-of-plane thermal conductivity of the GF-RGO composite films annealed at different temperatures. The in-plane thermal conductivity of the GF-GO composite film is only 45±4.9 W/mK. However, after thermal annealing, the thermal conductivity increases with the increase in the annealed temperature. When the annealed temperature reaches 1200 °C, the in-plane thermal conductivity is as high as 976±107.7 W/mK, as shown in Figure 5a. The high in-plane thermal conductivity of the GF-RGO composite film originates from fewer defects of RGO sheets. As demonstrated previously [25], thermal conduction of graphene is dominated by phonon diffusion from lattice vibrations of its covalent sp^2 bonding network. However, the covalent sp^2 bonding network was broken in GO structure, leading to substantial phonon scattering. The high temperature annealing process can heal the defects of GO, as demonstrated previously [22]. The out-of-plane thermal conductivity of the GF-RGO composite films was also tested as shown in Figure 5b. The out-of-plane thermal conductivity is two-magnitude lower than in-plane thermal conductivity. For example, the out-of-plane thermal conductivity of the GF-GO composite film is only 0.56 ±0.08 W/mK, because of the anisotropic structure of graphene. The carbon atoms in graphene are bonded covalently in a hexagonal lattice at in-plane direction, but bonded non-covalently at out-of-plane direction. However, the thermal annealing also leads to increase of the out-of-plane thermal conductivity. When the annealed temperature reaches 1200 °C, the out-of-plane thermal conductivity is up to 1.31±0.2 W/mK.

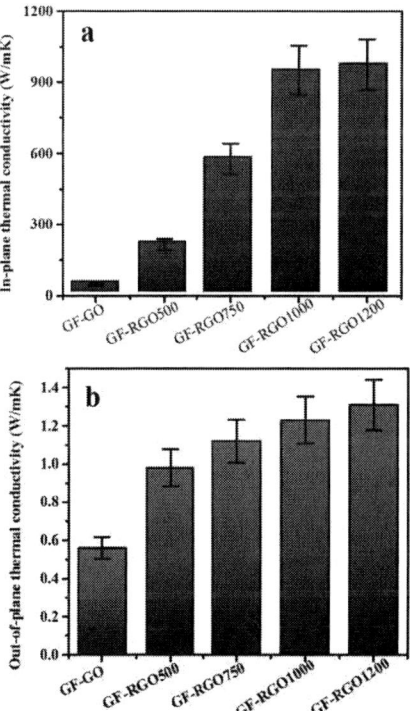

Figure 5. Thermal conductivity of the GF-RGO composite films annealed at different temperatures. (a) In-plane and (b) out-of-plane thermal conductivity.

In order to investigate the mechanical properties of the GF-RGO composite films, we performed tensile tests, determined tensile strengths, strain and Young's modulus. Figure 6 shows the mechanical properties of the GF-RGO composite films annealed at different temperatures. The typical stress-strain curves for films are exhibited in Figure 6a. The films show almost a linear behaviour in the elastic regions. Compared with GF-GO composite films, the GF-RGO composite films exhibit enhanced mechanical behaviors. After being annealed at 500 °C, the tensile strength of the GF-RGO500 increased to 75.5±6.9 MPa, while the tensile strength of the GF-GO composite film is 62.1±6.2 MPa, as exhibited in Figure 6b. Further increasing the annealed temperatures leads to the increase in the tensile strength of the GF-RGO composite film. The tensile strength reaches the maximum value of 106.9±9.8 MPa at the annealed temperature of 1000 °C. However, the tensile strength decreases to 82.9±8.1 MPa, at the annealed temperature of 1200 °C. This is possible due to that the annealed temperature of 1200 °C leads to the part thermal decomposition of glass fibers. Moreover, the enhancements in tensile strength take no negative effect on the ultimate elongation, as shown in Figure 6c. The ultimate elongation for the GF-GO composite film is 0.68%, but it increases to 0.89 % (increased by 30.9%), indicating a significantly enhanced toughness of the GF-RGO composite film. As expected, the Young's modulus increases with the annealed temperatures. The Young's modulus for the GF-GO composite film is 6.5±0.58 GPa, and increase to 12.8±1.3 GPa for the GF-RGO 750 film. The Young's modulus has slight increase to 11.1±1.2 GPa, when the annealed temperature is up to 1200 °C. We suggest that these enhancements in mechanical properties originate from the two factors: reinforcement of glass fiber and healing the defect of GO by annealing. The glass fibers are widely used as reinforced agents for polymers [26-29], due to their excellent mechanical properties, low cost, and good thermal resistance. In addition, thermal annealing is a widely used approach to reducing the defects of GO and improving the crystalline of the r-GO, resulting in the formation of r-GO with enhanced mechanical, thermal, and electrical properties [23, 30].

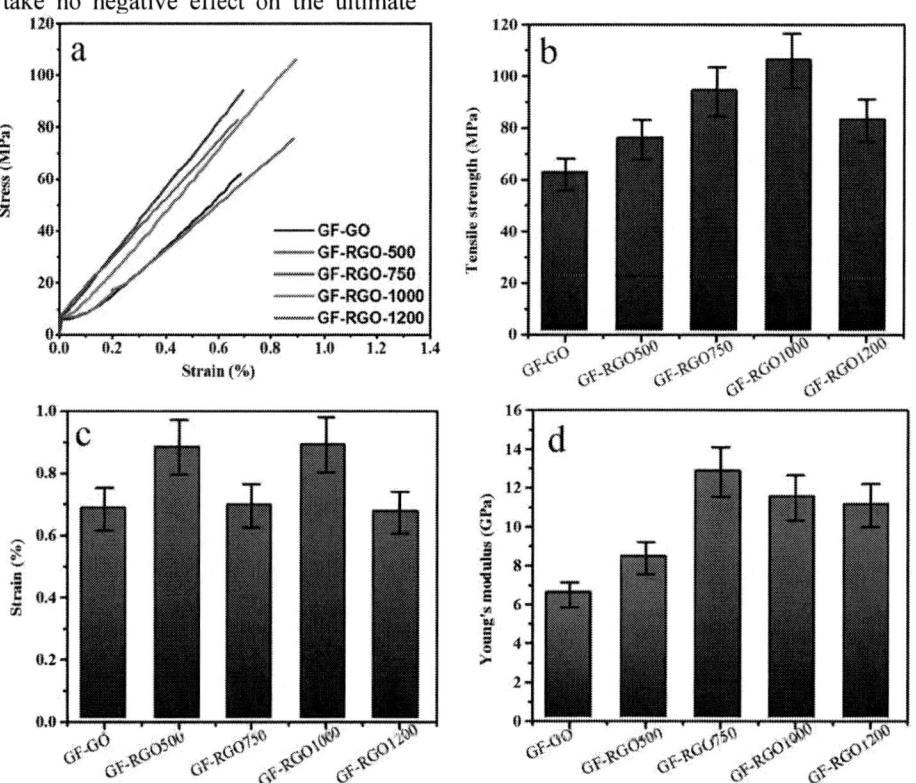

Figure 6. Mechanical properties of the GF-RGO composite films annealed at different temperatures. (a) Stress-strain curves, (b) tensile strength, (c) strain, and (d) Young's modulus of the GF-RGO composite films at the tensile mode.

Table 1. Comparison of in-plane thermal conductivity in GF-RGO composite films and previously reported graphene films.

No.	Samples	Thermal Conductivity W/mK	Tensile strength MPa	References and year
1	Pure graphene film	1940±113	60.0	2017[22]
2	Graphene/nanofibrillated cellulose	5.73	116.8	2017[31]
3	Pure graphene film	1390 ± 65	77.7	2015[32]
4	Pure graphene film	1219	10.2	2015[33]
5	Pure graphene film (Thickness: 1.0 μm)	2100±300	112.0	2017[34]
6	Pure graphene film (Thickness: 8.0 μm)	490±50	101.0	2017[34]
7	Graphene/carbon fiber	997	15.3	2014[20]
8	Pure graphene film	1434	58.20	2014[35]
9	Pure graphene film	524 ± 36	614.0	2015[36]
10	Pure graphene film	3200	78 .0	2018[37]
11	GF-RGO composites film	952±104	106.9	This work

In general, the quality of individual graphene and interfaces between graphenes play crucial role in the thermal conductivity and mechanical strength. High-quality graphene sheet and tight interfacial interaction between the sheets will be beneficial to high thermal conductivity and mechanical strength. Several authors have been reported the fabrication of graphene based films through chemical oxidation, thermal reduction and following graphitization at elevated temperatures, resulting in products with enhanced thermal conductivity and mechanical strength [31-37]. Unfortunately, simultaneously achieving high thermal conductivity and excellent mechanical strength using the facile method is still a key challenge. Graphene film prepared by assembling graphene crystals with large size into fibers possesses the thermal conductivity as high as 1940 ± 113 W/ mK [22]. However, the relatively poor interface interaction lowers its mechanical strength. In addition, this high-temperature thermal annealing (3000 °C) makes the grapheen films expansive and limits their large-scale applications. On the other hand, the strong interfacial interaction between graphene and nanofibrillated cellulose endows the composite films with high mechanical strength (116.8 MPa) [31].

IV. CONCLUSION

In summary, we demonstrated that the use of graphene oxide and glass fiber can fabricate graphene-glass fiber composite film with ultrahigh thermal conductivity, high mechanical strength, and flexibility, by bar-coating method. The one-dimensional glass fibers were chose as the scaffolds, which reinforce the mechanical stiffness and flexibility. The in-plane thermal conductivity of the graphene-glass fiber composite film reaches as high as 952±104 W/mK, while the mechanical strength is up to 106.9 MPa. In addition, the composite film shows excellent flexibility. The bar-coating method is easily scaled up for wide applications. It is believed that the graphene-glass fiber composite film can be used as high-performance heat spreaders in high-power devices for highly efficient thermal management.

ACKNOWLEDGMENT

However, a number of graphene defects and low graphene loadings lead to extremely low thermal conductivity (5.73 W/mK). Y. Guo et al. reported a blade-coating approach to prepare freestanding graphene with both high thermal conductivity (2100±300) and mechanical strength (112.0 MPa) for the thickness of 1.0 μm. However, the mechanical strength and thermal conductivity decrease with increased thickness, because of the introduced defects. M. Zhang et al. fabricated a kind of graphene with the tensile strength of 614 ± 12 MPa, by conttrolling the GO chemical groups and the microstructures of the films, but the fabricated thermal conductivity is only 524 ± 36 W/mK. For our GF-RGO composite film paper, glass fibers ensure the composite film high mechanical strength. Furthermore, the enhanced interfacial interaction between GO and glass fibers boost the mechanical strength of the GF-RGO composite films. Annealing at high temperature greatly reduces the amount defects, which leads to the formation of high-quality graphene sheets. The three factors lead to its superior thermal conductivity and mechanical strength to previously reported graphene films.

The authors would like to acknowledge the financial support from National Key R&D Program of China (No.2017YFB0406000), Frontier Sciences Key Research Program of the Chinese Academy of Sciences (No.QYZDY-SSWJSC010), National Natural Science Foundation of China (No. 51603226), National and Local Joint Engineering Laboratory of Advanced Electronic Packaging Materials (Shenzhen Development and Reform Committee 2017-934), R&D Funds for basic Research Program of Shenzhen (Grant No. JCYJ20150831154213681), and Science and Technology Planning Project of Guangdong Province, China (No. 2017A010106005).

REFERENCES

[1] A. L. Moore and L. Shi, "Emerging challenges and materials for thermal management of electronics," *Materials Today,* vol. 17, pp. 163-174, May 2014.

[2] X. C. Tong, "Materials and Design for Advanced Heat Spreader and Air Cooling Heat Sinks," in *Advanced Materials for Thermal Management of Electronic Packaging*, ed New York, NY: Springer New York, 2011, pp. 373-420.

[3] R. H. Horng, H. L. Hu, R. C. Lin, L. S. Tang, C. P. Hsu, *et al.*, "Cup-shaped copper heat spreader in multi-chip high-power LEDs application," *Optics Express*, vol. 20, pp. A597-A605, September 2012.

[4] K. Loutfy, "Aluminum-Diamond Metal-Matrix Heat Spreaders for GaN Devices," *Microwave Journal*, vol. 61, pp. 44-46, Junly 2018.

[5] S. Mateti, K. Yang, X. Liu, S. Huang, J. Wang, *et al.*, "Bulk Hexagonal Boron Nitride with a Quasi-Isotropic Thermal Conductivity," *Advanced Functional Materials*, vol. 28, p. 1707556, May 2018.

[6] L. Fu, T. Wang, J. Yu, W. Dai, H. Sun, *et al.*, "An ultrathin high-performance heat spreader fabricated with hydroxylated boron nitride nanosheets," *2d Materials*, vol. 4, Junly 2017.

[7] J. Li and L. Lv, "Experimental studies on a novel thin flat heat pipe heat spreader," *Applied Thermal Engineering*, vol. 93, pp. 139-146, January 2016.

[8] A. J. Robinson, J. Colenbrander, R. Kempers and R. Chen, "Solid and Vapor Chamber Integrated Heat Spreaders: Which to Choose and Why," *Ieee Transactions on Components Packaging and Manufacturing Technology*, vol. 8, pp. 1581-1592, September 2018.

[9] W. Feng, M. Qin and Y. Feng, "Toward highly thermally conductive all-carbon composites: Structure control," *Carbon*, vol. 109, pp. 575-597, November 2016.

[10] Y. Zhang, M. Edwards, M. K. Samani, N. Logothetis, L. Ye, *et al.*, "Characterization and simulation of liquid phase exfoliated graphene-based films for heat spreading applications," *Carbon*, vol. 106, pp. 195-201, September 2016.

[11] A. A. Balandin, "Thermal properties of graphene and nanostructured carbon materials," *Nature Materials*, vol. 10, pp. 569-581, August 2011.

[12] E. Pop, D. Mann, Q. Wang, K. E. Goodson and H. J. Dai, "Thermal conductance of an individual single-wall carbon nanotube above room temperature," *Nano Letters*, vol. 6, pp. 96-100, January 2006.

[13] P. Kim, L. Shi, A. Majumdar and P. L. McEuen, "Thermal transport measurements of individual multiwalled nanotubes," *Physical Review Letters*, vol. 87, November 2001.

[14] A. A. Balandin, S. Ghosh, W. Bao, I. Calizo, D. Teweldebrhan, *et al.*, "Superior Thermal Conductivity of Single-Layer Graphene," *Nano Letters*, vol. 8, pp. 902-907, March 2008.

[15] J. H. Seol, I. Jo, A. L. Moore, L. Lindsay, Z. H. Aitken, *et al.*, "Two-Dimensional Phonon Transport in Supported Graphene," *Science*, vol. 328, pp. 213-216, April 2010.

[16] S. H. Bae, R. Shabani, J. B. Lee, S. J. Baeck, H. J. Cho, *et al.*, "Graphene-Based Heat Spreader for Flexible Electronic Devices," *Ieee Transactions on Electron Devices*, vol. 61, pp. 4171-4175, December 2014.

[17] P. H. Lee, W. M. Tu and H. C. Tseng, "Graphene Heat Spreaders for Thermal Management of InGaP/GaAs Collector-Up HBTs," *Ieee Transactions on Electron Devices*, vol. 65, pp. 352-355, January 2018.

[18] K. H. Baloch, N. Voskanian and J. Cumings, "Controlling the thermal contact resistance of a carbon nanotube heat spreader," *Applied Physics Letters*, vol. 97, August 2010.

[19] Y. Chen, X. Hou, R. Kang, Y. Liang, L. Guo, *et al.*, "Highly flexible biodegradable cellulose nanofiber/graphene heat-spreader films with improved mechanical properties and enhanced thermal conductivity," *Journal of Materials Chemistry C*, vol. 6, pp. 12739-12745, December 2018.

[20] Q.-Q. Kong, Z. Liu, J.-G. Gao, C.-M. Chen, Q. Zhang, *et al.*, "Hierarchical Graphene–Carbon Fiber Composite Paper as a Flexible Lateral Heat Spreader," *Advanced Functional Materials*, vol. 24, pp. 4222-4228, March 2014.

[21] X. Zeng, L. Ye, S. Yu, H. Li, R. Sun, *et al.*, "Artificial nacre-like papers based on noncovalent functionalized boron nitride nanosheets with excellent mechanical and thermally conductive properties," *Nanoscale*, vol. 7, pp. 6774-6781, February 2015.

[22] L. Peng, Z. Xu, Z. Liu, Y. Guo, P. Li, *et al.*, "Ultrahigh Thermal Conductive yet Superflexible Graphene Films," *Advanced Materials*, vol. 29, p. 1700589, May 2017.

[23] G. Xin, T. Yao, H. Sun, S. M. Scott, D. Shao, *et al.*, "Highly thermally conductive and mechanically strong graphene fibers," *Science*, vol. 349, pp. 1083-1087, September 2015.

[24] G. Lian, C.-C. Tuan, L. Li, S. Jiao, Q. Wang, *et al.*, "Vertically Aligned and Interconnected Graphene Networks for High Thermal Conductivity of Epoxy Composites with Ultralow Loading," *Chemistry of Materials*, vol. 28, pp. 6096-6104, 2016/09/13 2016.

[25] A. A. Balandin, "Thermal properties of graphene and nanostructured carbon materials," *Nature Materials*, vol. 10, p. 569, 2011.

[26] L. Riaño, L. Belec, J.-F. Chailan and Y. Joliff, "Effect of interphase region on the elastic behavior of unidirectional glass-fiber/epoxy composites,"

Composite Structures, vol. 198, pp. 109-116, 2018/08/15/ 2018.

[27] M. Sahin, S. Schlögl, G. Kalinka, J. Wang, B. Kaynak*, et al.*, "Tailoring the interfaces in glass fiber-reinforced photopolymer composites," *Polymer,* vol. 141, pp. 221-231, 2018/04/11/ 2018.

[28] J.-H. Kim, D.-J. Kwon, P.-S. Shin, Y.-M. Beak, H.-S. Park*, et al.*, "Interfacial properties and permeability of three patterned glass fiber/epoxy composites by VARTM," *Composites Part B: Engineering,* vol. 148, pp. 61-67, 2018/09/01/ 2018.

[29] H. Mahmood, L. Vanzetti, M. Bersani and A. Pegoretti, "Mechanical properties and strain monitoring of glass-epoxy composites with graphene-coated fibers," *Composites Part A: Applied Science and Manufacturing,* vol. 107, pp. 112-123, 2018/04/01/ 2018.

[30] Z. Xu, H. Sun, X. Zhao and C. Gao, "Ultrastrong Fibers Assembled from Giant Graphene Oxide Sheets," *Advanced Materials,* vol. 25, pp. 188-193, 2013.

[31] N. Song, S. Cui, D. Jiao, X. Hou, P. Ding*, et al.*, "Layered nanofibrillated cellulose hybrid films as flexible lateral heat spreaders: The effect of graphene defect," *Carbon,* vol. 115, pp. 338-346, 2017/05/01/ 2017.

[32] P. Kumar, F. Shahzad, S. Yu, S. M. Hong, Y.-H. Kim*, et al.*, "Large-area reduced graphene oxide thin film with excellent thermal conductivity and electromagnetic interference shielding effectiveness," *Carbon,* vol. 94, pp. 494-500, 2015/11/01/ 2015.

[33] S.-Y. Huang, B. Zhao, K. Zhang, M. M. F. Yuen, J.-B. Xu*, et al.*, "Enhanced Reduction of Graphene Oxide on Recyclable Cu Foils to Fabricate Graphene Films with Superior Thermal Conductivity," *Scientific Reports,* vol. 5, p. 14260, 2015.

[34] Y. Guo, C. Dun, J. Xu, J. Mu, P. Li*, et al.*, "Ultrathin, Washable, and Large-Area Graphene Papers for Personal Thermal Management," *Small,* vol. 13, p. 1702645, 2017.

[35] G. Xin, H. Sun, T. Hu, H. R. Fard, X. Sun*, et al.*, "Large-Area Freestanding Graphene Paper for Superior Thermal Management," *Advanced Materials,* vol. 26, pp. 4521-4526, 2014.

[36] M. Zhang, Y. Wang, L. Huang, Z. Xu, C. Li*, et al.*, "Multifunctional Pristine Chemically Modified Graphene Films as Strong as Stainless Steel," *Advanced Materials,* vol. 27, pp. 6708-6713, 2015.

[37] N. Wang, M. K. Samani, H. Li, L. Dong, Z. Zhang*, et al.*, "Tailoring the Thermal and Mechanical Properties of Graphene Film by Structural Engineering," *Small,* vol. 14, p. 1801346, 2018.

Highly Thermal Conductive and Electrically Insulated Graphene Based Thermal Interface Material with Long-term Reliability

Nan Wang
SHT Smart High Tech AB
Gothenburg, Sweden
Nan.wang@sht-tek.com

Lilei Ye
SHT Smart High Tech AB
Gothenburg, Sweden
lilei.ye@sht-tek.com

Ya Liu
Department of Microtechnology and Nanoscience
Chalmers University of Technology
Gothenburg, Sweden
yaliu@chalmers.se

Johan Liu
Department of Microtechnology and Nanoscience
Chalmers University of Technology
Gothenburg, Sweden
johan.liu@chalmers.se

Shujing Chen
School of Automation and Mechanical Engineering
Shanghai University
Shanghai, China
csj427@i.shu.edu.cn

Abstract—Thermal management in high power devices are becoming more and more challenging due to the high density packaging as well as dramatically increased transistor integration. The conventional TIMs that are widely used in the microelectronic industry today are experiencing more and more stress due to their limited thermal performance and poor reliability. Composed by particle laden polymer matrix, conventional TIMs have thermal conductivity (K) values in the range of 1-5 W/mK, and such values can be even lower for electrically insulated TIMs. Conventional TIMs also suffer from severe pump-out and dry-out failures, which brought great threat to the performance and lifetime of the electronic devices. Here, we address these problems by utilizing a highly thermally conductive, electrically insulated and reliable graphene enhanced TIMs (I-GTs). Composed by vertical-aligned graphene layers, I-GTs provide a direct heat pathway from top to bottom, which enables superfast heat dissipation at through-plane direction. The highest bulk thermal conductivity of the conductive body at the through-plane direction is over 1000 W/mK, which is 100 times higher than conventional TIMs. I-GT also possesses good flexibility and can be easily compressed over 100% at Z direction upon small applied pressures. Therefore, fully contact between two surfaces can be achieved by using I-GT as gap fillers. The minimum thermal resistance measured for I-GTs reaches about 30 Kmm²/W, which is much lower than most of the conventional TIMs. To ensure fully electrical insulation, a smooth and soft adhesive layer with a thickness of few microns was coated on the surface of I-GT. The breakdown voltage of I-GT reaches up to 950 V. Thermal cycling test shows the highly stable nature of I-GT. The good compressibility and elasticity of I-GT ensures continued proper TIM contact with substrates, which counteracts the effect of internal stress induced by the mismatch of coefficient of thermal expansion (CTE) during temperature cycling. In addition, the I-GT has the advantages

of low density and good maintainability. The resulting I-GTs thus offers new solutions for addressing thermal management issues in high power electronics and other systems with small form-factor.

Keywords- graphene, reliability, insulation, compressibility, bonding strength

I. INTRODUCTION

The development of electronic industry towards miniaturization and higher performance leads to enormous thermal management challenges. To minimize adverse effects of such challenges both on the user's health and device's reliability and performance, efficient thermal management in high power density electronic products becomes highly essential. As a key element in thermal management, thermal interface materials (TIMs) act as a bridge to link the hotspot with cooling system to bring down the junction temperature of power modules in electronic devices [1]. Today, the widely used TIMs in the electronics industry are composed of high thermal conductive filler enhanced polymer matrix, Such formulas have the great advantages of high reliability, cost efficient and ease to use [2], [3]. However, their thermal conductivity (K) at through-plane direction is usually limited within 10 W/mK, and in most cases, is about 4 or 5 W/mK. Such low thermal conductivity of the conventional TIMs is strongly related to the large thermal contact resistance formed between individual particle fillers of TIMs [4]. Therefore, it is highly essential to develop new solutions to address the fast

978-1-7281-1500-9/19 $31.00 © 2019 IEEE

growing thermal management issues in modern power devices.

Many studies have been carried out to solve above problems. For example, increasing the amount of fillers in polymer matrix is the most common strategy to improve thermal performance of TIMs [5]. However, high filler contents can inevitably lead to the increase of material's hardness and make the composites difficult to conform to surface roughness, which results in relatively high contact thermal resistance. Moreover, the substantial filler particles can sacrifice the mechanical strength of polymer resin and increase the density of composites [6], [7]. Therefore, it becomes highly essential to investigate new approaches for improving thermal conductivity of TIM whilst maintaining low contact thermal resistance.

Here, a highly thermally conductive, electrically insulated, compressible and light-weighted graphene enhanced TIMs (I-GTs) were developed to achieve above goals. The conductive body of I-GT is composed by vertical-aligned graphene layers, which can provide a continuous highway for fast heat dissipation at the vertical direction. Therefore, excellent thermal properties can be achieved along the heat flow direction in I-GT. The highest thermal conductivity of the conductive body was found to be up to 1000 W/mK at the vertical direction. Such a value is over 100 times higher than conventional TIMs. I-GT also possesses good flexibility and can be easily compressed over 100% at Z direction upon small applied pressures ($\leqslant$ 400 KPa). The minimum thermal resistance measured for I-GTs is about 30 Kmm^2/W, which is much lower than most of the conventional TIMs. To ensure fully electrical insulation, a smooth and soft adhesive layer with a thickness of 4 μm was coated on the surface of I-GT. The breakdown voltage of I-GT reaches up to 950 V. Thermal cycling test shows the highly stable nature of I-GT. The good compressibility and elasticity of I-GT ensures continued proper TIM contact with substrates, which counteracts the effect of internal stress induced by the mismatch of coefficient of thermal expansion (CTE) during temperature cycling. In addition, I-GT has the advantages of low density and good maintainability. The resulting I-GTs thus offers new solutions for addressing thermal management issues in high power electronics and other systems with small form-factor.

II. Experiments

I-GTs were fabricated and provided by SHT Smart High Tech AB. Thermal conductivity of I-GTs at through-plane direction and thermal resistance of I-GT between two copper substrates were measured by LFA 447 instrument. Compression ratios of I-GT at different pressures and thermal resistance of I-GTs were tested by INSTRON ASTM. Bonding strength of I-GT was measured by a shear tester (Dage 4000 Bondtester, Nordson). Thermal reliability of I-GT was evaluated at the temperature range of -20°C to

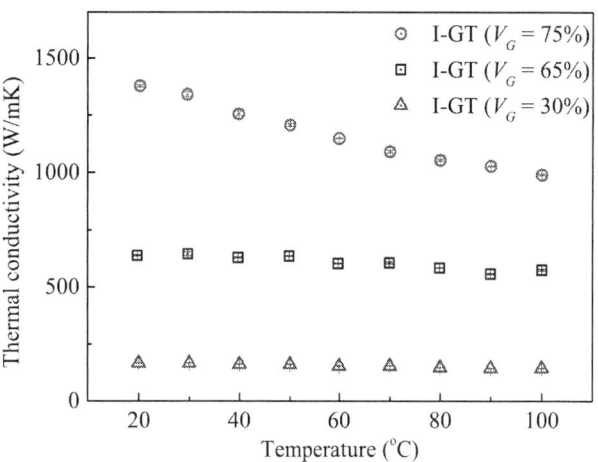

Figure 1. thermal conductivity changes of I-GT with different V_G proportions at the through-plane direction upon different temperatures.

125°C in a thermal cycling oven upon nitrogen atmosphere. Heat dissipation performance of I-GTs at the through-plane direction was shown by using a home-made testing platform. A heater and a heat sink equipped with a cooling fan were used to create temperature gradient. I-GTs was mounted between the heater and heatsink. The heater's working voltage is 23 V (power 3.57 W). Temperature change on the surface of heater was recorded by using a thermal imaging camera (Therm CAM PM595 NTSC) after the system reaches to the equilibrium state. A thin graphite coating layer was applied on the surface of heater to improve the accuracy of temperature recorded by IR camera.

III. Results and Discussion

The graphene volumatic ratios (V_G) play a key role on determine the through-plane thermal conductivity and contact thermal resistance of I-GTs. Here, I-GT samples with varied V_G ratios (30%, 65%, and 75%) were fabricated and tested. As illustrated in Figure 1, the raising of V_G ratios from 30% to 65% led to significant increase of through-plane thermal conductivity of I-GT samples from 167 ± 9 W/mK to 637 ± 31 W/mK. Further increase of V_G to 75% showed the highest thermal conductivity (1379 ± 138 W/mK). Such value is over 100 times and over 3 times higher than conventional TIMs and copper, respectively. The superior thermal properties of I-GTs are related to its unique structure composed by vertically aligned graphene layers.

Contact thermal resistance induced by surface roughness plays a key role in determining the final heat dissipation performance of TIMs. Therefore, good compressibility is highly essential for TIMs to decrease the contact thermal resistance. Figure 2a shows the change of deformation behavior of I-GTs at different V_G proportions. For I-GT with a large V_G proportion above than 65%, it showed a linearly increase of the compression ratio when the applied pressure increased, and the material recovered to its original thickness completely after the pressure releasing. The

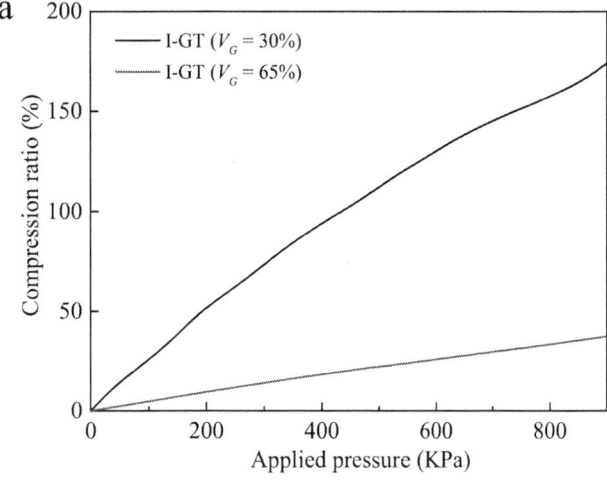

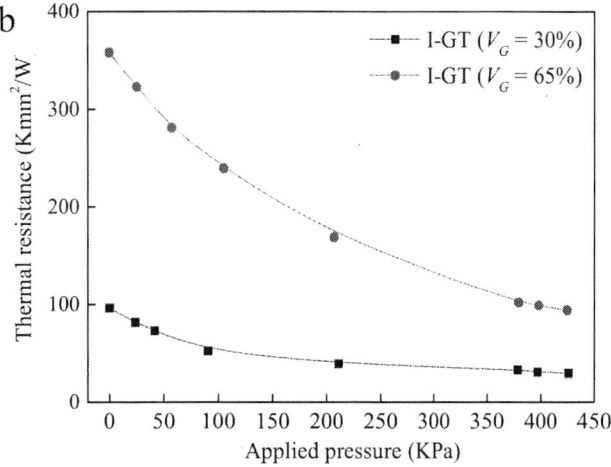

Figure 2. (a) Compression ratio changes of I-GT with different V_G proportions as a funtion of pressures. (b) Thermal resistance change of I-GT with different V_G propotions at different pressures

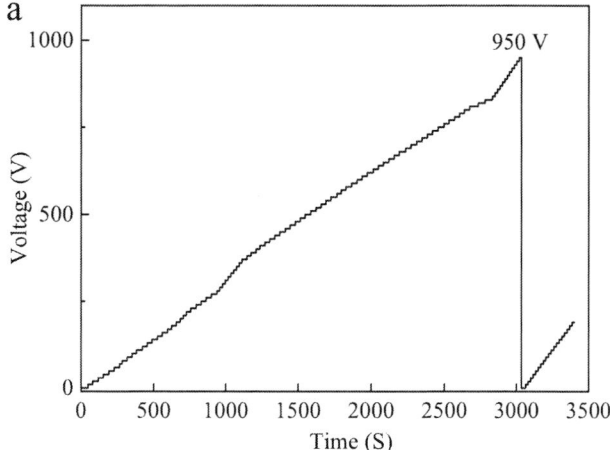

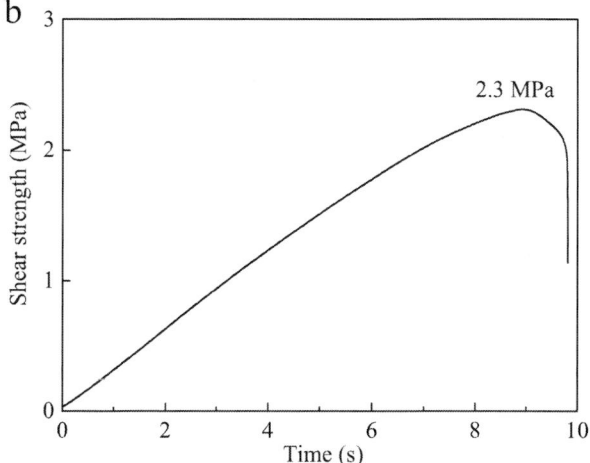

Figure 3. (a) Breakdown voltage of the I-GT film. (b) Shear strength curve of the I-GT film.

results showed that the composite is highly elastic at the through-plane direction. However, electronic devices have limitations on the applied pressures for TIM bonding (400 KPa) to prevent the functional components from damages. The compression ratio of I-GT samples with a large V_G proportion above than 65% is less than 20% within 400 KPa, which is difficult to reduce thermal contact resistance at interfaces. As a comparison, I-GT ($V_G = 30\%$) exhibits high softness and shows a large compression ratio around 100% in the pressure range of 400KPa. After releasing the pressure, the sample thickness can be recovered partially till to 80% of original thickness. We believe that the 20% of permanent structural deformation of I-GT ($V_G = 30\%$) was caused by the spontaneously bending and folding processes of the vertical graphene structure in the first pressing.

Thermal resistance of I-GT samples with different V_G proportions were measured by using ASTM D5470 equipment. The thickness of all measured samples is 2 mm. The applied pressure was varied in the range of 450 KPa. As shown in Figure 2b, the I-GT sample with a large V_G

proportion of 65% exhibited a large thermal resistance value of 358 Kmm²/W before pressing, which was about 4 times higher than that of I-GT ($V_G = 30\%$) (96 Kmm²/W). With the applied pressure increased, both samples showed a significant decrease of thermal resistance. When pressure reached to 425 KPa, the resistance values of I-GT ($V_G = 65\%$) and I-GT ($V_G = 30\%$) are 94 Kmm²/W and 30 Kmm²/W, respectively. The small thermal resistance value of I-GT ($V_G = 30\%$) in the measured pressure range was attributed to the high softness of I-GT structure which enabled the composite to decrease the contact thermal resistance significantly at interfacial areas.

To avoid potential risks of short circuit and contaminations, the surface of I-GT film needs to be protected by insulation layers. To meet this goal, a thermal plastic polymer was coated on the surface of I-GT. It has the advantages of highly electrical insulation, strong bonding and good elasticity. The thickness of the protective layer is 4μm. Figure 3a shows the breakdown voltage of the I-GT film. The maximum breakdown voltage that the I-GT film

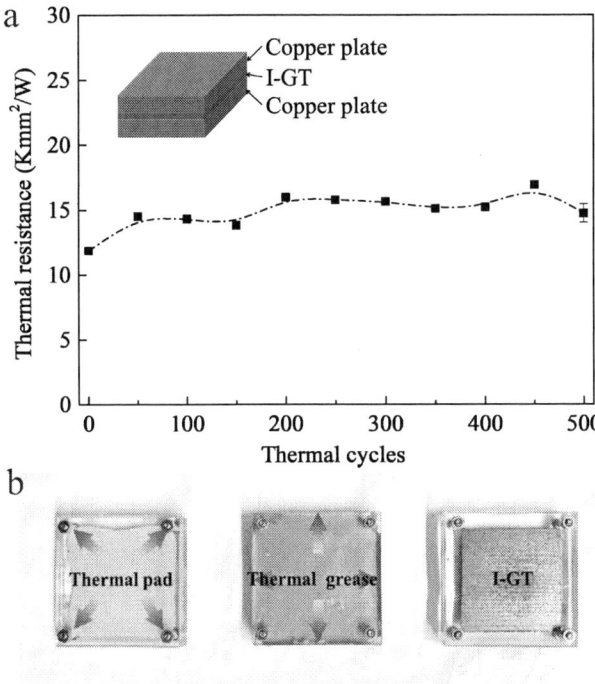

Figure 4. (a) Thermal reliability curve of I-GT in 500 cycles of thermal cycling tests. (b) Pump-out demonstrators of different thermal interface materials, including thermal pad, thermal grease and I-GT.

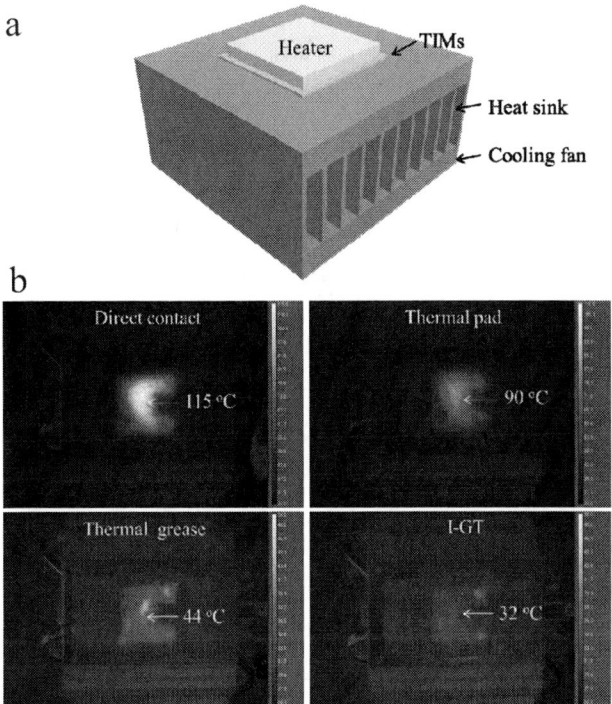

Figure 5. (a) The sketch of the home-made thermal testing platform (b) IR images of the working temperature of the heater by using different TIMs, including direct contact, thermal pad, thermal grease and I-GT.

can stand reaches to 950V. By increasing the temperature above than 100°C, the protective layer can form bonding with variety of substrates, including ceramic, metal, glass, and polymer. Figure 3b shows the bonding strength of the I-GT between two pieces of copper substrates. Shear strength was measured by using a shear tester. The maximum shear strength that the protective layer can provide is about2.3 MPa, which reaches the same level of mechanical performance as the commonly used acrylic double-side tapes.

Thermal reliability of the I-GT was investigated further in a thermal cycling oven at the temperature range of -20°C to 125°C. To simulate the actual application environment, the I-GT sample with a thickness of 200 μm was used as a TIM layer between two pieces of copper plates (shown in the insert image of Fig. 5a). Thermal resistance was measured by using laser-flash equipment. Figure 4a shows the thermal resistance variation of FSCs in total 500 thermal cycles. Notably, the thermal resistance remained constant around 15 Kmm²/W. It shows the excellent thermal-mechanical stability of I-GT during the temperature cycling test. The good stability of I-GT is mainly attributed to the flexible nature of the material which benefits on absorbing the inner stress caused by the mismatch of coefficient of thermal expansion (CTE). In addition, the unique vertical graphene structure of I-GT benefits on solving the pump-out and dry-out failures existing in most of common TIMs. As shown in Figure 4b, thermal pad and thermal paste exhibited obvious extrusion phenomenons upon pressing. In electronic devices,

repeated powering up and down processes during the usage can also generate a similar effect as the above motions which lead to the increase of thermal resistance and reduce the lifetime of the power modules. However, with the same pressure, I-GT did not show any structural change at the in-plane directions. The main structural deformation of I-GT was only taken place at the vertical directions in the forms of bending and folding of the vertical graphene structures. This advantage of I-GT can benefit improving the long-term thermal reliability of power devices and thereby improving their lifetime.

Light weight is another important advantage of I-GT material compared to most of commercial TIMs, The density of I-GT (VG = 30%) was measured to be 0.6 g/cm³, It is only half of the common paste TIMs and also one-fifteenth of copper.

A thermal testing platform was built to demonstrate the heat dissipation performance of I-GTs. Two types of commercial TIMs, including thermal pad with a thermal conductivity of 3 W/mK and thermal grease with a thermal conductivity of 2 W/mK, were also used as comparisons. A control group shows the heat dissipation situation of the thermal testing platform without using any TIMs. As shown in Fig. 5a. All tested TIMs were added between a heater and a heatsink equipped with a cooling fan. The operating power of the heater and cooling fan were 3.57 W and 0.57 W respectively. The working temperatures of the heater were recorded by using a thermal imaging camera when the testing platform reached to the equilibrium state. As shown in Fig. 5b, the control group that heater had a direct contact with the heatsink presented the highest surface temperature of 115°C. Obvious decreases on the working temperature were detected in the rest three samples. In cases of thermal pad and thermal grease, the working temperatures of heater were stabilized at 90°C and 44°C respectively. The much decreased working temperature in the thermal grease sample was attributed to its much lower contact resistance than the thermal pad sample. For the I-GT sample, it showed the lowest working temperature compared to the rest three samples, which was only 32°C. It is 83°C less than that of the control group. The temperature difference between the heater and ambient was only about 7° due to the fast heat dissipation property of I-GTs.

IV. CONCLUSIONS

In summary, I-GT was developed as a novel TIM for addressing thermal management issues in many high power density devices. I-GT has the advantages of light-weight, electrically insulation, good compressibility and high thermal conductivity. The developed I-GT demonstrated superior thermal-mechanical stability in long-term reliability study which is highly essential for prolonging the lifetime of electronic devices. The resulting I-GTs thus offers new solutions for addressing thermal management issues in high power electronics and other systems with small form-factor.

ACKNOWLEDGMENT

We thank for the financial support from the Swedish Foundation for Strategic Research (SSF) (No SE13-0061), Swedish National Board for Innovation (Vinnova) Graphene SIO-Agenda Program, Formas program on graphene enhanced composite as well as from the Production Area of Advance at Chalmers University of Technology, Sweden. Thanks for the financial support from the Ministry of Science and Technology of China (No: 2017YFB040600) and Chinese national Science foundation project (No: 51872182)

REFERENCES

[1] R. Prasher, "Thermal Interface Materials: Historical Perspective, Status, and Future Directions," *Proc. IEEE*, vol. 94, no. 8, pp. 1571–1586, Aug. 2006.

[2] R. S. Prasher, J. Shipley, S. Pistic, P. Koning, and J. Wang, "Thermal Resistance of Particle Laden Polymeric Thermal Interface Materials," *J. Heat Transf.*, vol. 125, no. 6, pp. 1170–1177, Nov. 2003.

[3] R. S. Prasher, P. Koning, J. Shipley, and A. Devpura, "Dependence of Thermal Conductivity and Mechanical Rigidity of Particle-Laden Polymeric Thermal Interface Material on Particle Volume Fraction," *J. Electron. Packag.*, vol. 125, no. 3, pp. 386–391, Sep. 2003.

[4] R. S. P. Amit Devpura Patrick E.Phelan, "Size Effects on the Thermal Conductivity of Polymers Laden with Highly Conductive Filler Particles," *Microscale Thermophys. Eng.*, vol. 5, no. 3, pp. 177–189, Jul. 2001.

[5] S. Kemaloglu, G. Ozkoc, and A. Aytac, "Properties of thermally conductive micro and nano size boron nitride reinforced silicon rubber composites," *Thermochim. Acta*, vol. 499, no. 1, pp. 40–47, Feb. 2010.

[6] Y. Xu, D. D. L. Chung, and C. Mroz, "Thermally conducting aluminum nitride polymer-matrix composites," *Compos. Part Appl. Sci. Manuf.*, vol. 32, no. 12, pp. 1749–1757, Dec. 2001.

[7] S.-Y. Fu, X.-Q. Feng, B. Lauke, and Y.-W. Mai, "Effects of particle size, particle/matrix interface adhesion and particle loading on mechanical properties of particulate–polymer composites," *Compos. Part B Eng.*, vol. 39, no. 6, pp. 933–961, Sep. 2008.

2019 IEEE 69th Electronic Components and Technology Conference (ECTC)

Further enhancement of thermal conductivity through optimal uses of h-BN fillers in polymer-based thermal interface material for power electronics

Han Jiang[1], Han Zhou[1], Stuart Robertson[1,2], Zhaoxia Zhou[2], Liguo Zhao[1], Changqing Liu[1]

[1] Wolfson School of Mechanical, Electrical and Manufacturing Engineering,
Loughborough University, Loughborough LE11 3TU, UK
[2] Loughborough Materials Characterisation Centre, Department of Materials,
Loughborough University, Loughborough LE11 3TU, UK
Email: C.Liu@lboro.ac.uk

Abstract—Due to the demand of miniaturization and increasing functionality in power electronics, thermal dissipation becomes a challenging problem for thermal management and reliability. To enable effective heat transfer across the interconnect interfaces, thermal interface materials (TIMs) are required. Electrically insulating TIMs are primarily polymer-based composites which use conductive fillers to enhance thermal conductivity (TC). In this study, the optimal hybrid filler constituents, achieved through mixing spherical and platelet h-BN particles with different ratios, in polymer-based TIM was predicted using finite element (FE) simulations. The underpinning mechanisms of the variation in TC of the TIMs were analyzed from the temperature distribution patterns and micro heat flux paths. Results showed that with the same total volume fraction of h-BN, mixed spherical and platelet h-BN fillers of a certain ratio can further improve the thermal properties of the TIMs compared with those with spherical or platelet h-BN particles alone.

Keywords - Power electronics; Thermal interface materials; h-BN; Thermal conductivity; Finite element simulation

I. INTRODUCTION

In the field of power electronics, most of the challenges arise from the thermal management, which is critical to the performance, lifetime and reliability of electronic devices [1]. Due to the demand of miniaturization and increased functionality, modern power electronics require higher integration of components and higher switching frequencies. The increase in switching frequency leads to increased power losses and dissipation of the generated loss heat becomes a challenging problem due to the physical integration of the system [1-3]. Although the components are usually made of solid materials with high TC, the contact surfaces of heat resources and heat sinks are rough, and the thermal resistance is huge due to the low effective contacting area. TIMs are thus introduced to minimize the thermal resistance between the two surfaces [4, 5].

For some applications, TIMs need to be highly thermal conductive but electrically insulating, polymer-based composites with improved TC are required. In general, the addition of thermally conductive inorganic fillers is indispensable for achieving high TC, and the filler type, size, shape, loading level and spatial arrangement have strong influences on the TC of polymer composites. Typically, heat conductive fillers with both high TC and electrically insulating property, such as alumina, boron nitride and aluminum nitride can be used [3, 6]. Notably, TC is not the only property that should be considered. It is widely recognized that TC relies greatly on the formation of continuous conductive network. Network formation, however, usually takes place at high filler volume levels. This can lead to poor processability, poor mechanical properties and high cost. Research effort is still needed to obtain better performance at lower filler volume. For different shapes of fillers, composites filled with high-aspect-ratio fillers usually exhibit higher TC even at a relatively lower filler content [7]. However, there are still obvious gaps between fillers of a single type, leading to deterioration in heat transfer of the adhesive. Meanwhile, the use of more or larger size particles can significantly decrease the processability and mechanical properties of the adhesive, especially for the high-aspect-ratio fillers, because of the increased friction between filler and matrix and the cut-off effect of the particles on the polymer matrix. Fillers with spherical morphology are favorable for lowering down the friction, but could also be less efficient in increasing TC. Hybrid filler system offers a better choice to combine the advantages of different fillers, as this not only can form better continuous network at lower filler content, but also improve vertical through-plane TC for some platelet filler composite systems, e.g., adding spherical fillers to BN/polymer composites to disrupt the alignment of platelet fillers.

Since the ways of material compositions are almost infinite, it is imperative to find approaches to limit the number of trial experiments by predicting the properties of the candidate materials. Theoretical analysis and numerical modelling play important roles in finding the optimal combination of material compositions [3]. Among the several methods, finite element (FE) modelling offers an opportunity to calculate composite TC accurately by taking realistic composite morphology into consideration, which is barely possible for analytical micromechanical models. In addition, FE simulation can predict the detailed distribution of temperature and heat flux, which is also difficult to achieve through experimental methods. Therefore, FE method can help to elaborate the influences of different filler constituents on the TC of TIMs, by simulating the conduction process quantitatively and enables the design of the filler geometries, constituents and distributions for a desired performance. There has been some research focusing on the composite TC calculation by FE method [8, 9], but most of them studied

978-1-7281-1500-9/19 $31.00 © 2019 IEEE 1569

composites filled with single type of filler, few models considered hybrid filler systems (fillers with different shapes).

In this study, the optimal design of polymer-based TIM, filled with mixed spherical and platelet h-BN particles, for power electronics with enhanced TC was conducted through three-dimensional FE modeling and experimental characterization. Based on the realistic composite morphology observed by SEM, the effect of hybrid filler constituents (mixed spherical and platelet h-BN particles with different ratios) on the TC of polymer-based TIM was studied. Carefully designed algorithms were proposed to construct the geometric models. Further, the underpinning mechanisms of the variation in TC of the TIMs were analyzed through the temperature distribution patterns and micro heat flux paths. Results showed that with the same total volume fraction of h-BN, mixed spherical and platelet h-BN of a certain ratio can improve the thermal properties of the TIM compared with those filled with spherical or platelet h-BN particles alone.

II. METHODS

A. Preparation of the TIM Samples

In this work, a two-component epoxy adhesive (EPO-TEK® 353ND, Epoxy Technology Inc.) was used as the polymer matrix, and the mix ratio of the two components is 10:1 by weight. Hexagonal boron nitride (h-BN) powders were kindly supplied by Momentive Performance Materials. Two types of the h-BN powders, PT110 and PTX25, were used as the fillers. PT110 are single-crystal platelets with a mean particle size of 45 μm, while PTX25 are spherical agglomerates with a mean diameter of 25 μm. The morphologies of the h-BN particles observed by SEM (JSM-7100F, JEOL) are shown in Fig. 1. All the materials mentioned above were used directly without further treatment.

Fig. 1. Morphologies of the h-BN particles: (a) PT110; (b) PTX25.

The TIM samples were prepared by mixing the fillers with the polymer matrix. The mixtures were blended uniformly using an automatic planetary gravity mixer (SpeedMixer™ DAC 150 FVZ-K, FlackTek Inc.). All the prepared TIM samples contained 75 wt.% epoxy matrix of the same composition and 25 wt.% fillers which were mixtures of PT110 and PTX25 with different ratios. The prepared TIM samples were designated as A1, A2, A3 and A4, A5 and A6, which have a PT110- PTX25 weight ratio of

5:0, 4:1, 3:2, 2:3, 1:4 and 0:5, respectively, as shown in Table I.

For the microstructure observation of the cured TIM samples under service, 3M adhesive tape was used to form a depressed area (10 mm × 10 mm) on the standard microscope slides and a small amount of the ICA was scraped over the area by a stainless-steel squeegee at an angle of 45°. Then the tape was peeled off, and the slides with printed TIM samples were kept at 150 °C for 1 hour in an electrically heated drying oven (MINO, Genlab Ltd.). The cross-sectional fracture morphologies of the as-obtained TIM samples imaged by SEM are presented in Fig. 2, from which it can be seen that h-BN platelets are evenly distributed in the resin matrix and their planes tend to align along the direction of the substrate plane due to the external force applied when scraping the samples on the glass substrate.

TABLE I. COMPOSITION OF TIM SAMPLES

Samples	Filler (25 wt.%)		Epoxy matrix
A1	PT110	25 wt.%	
	PTX25	0 wt.%	
A2	PT110	20 wt.%	
	PTX25	5 wt.%	
A3	PT110	15 wt.%	
	PTX25	10 wt.%	
A4	PT110	10 wt.%	75 wt.%
	PTX25	15 wt.%	
A5	PT110	5 wt.%	
	PTX25	20 wt.%	
A6	PT110	0 wt.%	
	PTX25	25 wt.%	

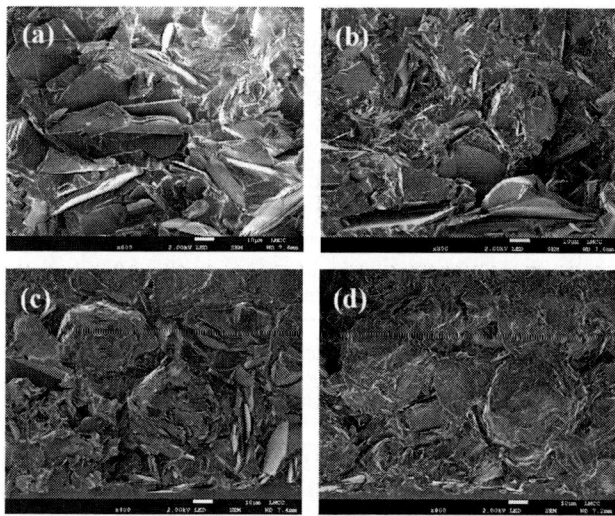

Fig. 2. Cross-sectional fracture morphologies of the TIM samples: (a) A1 (5:0); (b) A2 (4:1); (c) A5 (1:4); (d) A6 (0:5).

B. Finite Element Model

The simulations were performed with MATLAB and COMSOL Multiphysics software. The simulation process can be seen from Fig. 3. Based on the realistic morphologies observed by SEM, the geometric models of TIM samples with different filler constituents were constructed in MATLAB. As shown in Fig. 4, the micro-sized h-BN platelets were simplified as wafers with a radius of 22.5 μm and a thickness of 2 μm. The spatial orientation was controlled by the central coordinates (x, y, z) of the platelets. Due to the actual alinement of the h-BN platelets in the polymer matrix, the angle between the slice and the XOY plane, θ, was set as a random number between -10° and 10°. In order to reduce the amount of calculation, the rotation angle φ was set as 0° in this study. The location and size of the spherical fillers were controlled by their central coordinates (x', y', z') and the radius $(r' = 12.5$ μm). The length of the studied TIM cubes equals to 200 μm, and all the values of the central coordinates are randomly produced within the range of the cubes.

The random distribution of the particles in the resin matrix was realized by coding in MATLAB. The conductive particles were created one by one, without overlapping with each other and intersecting with the boundary of the TIM cube. Particles that did not meet the requirements were deleted and regenerated. Based on the experimental observations, the geometric models of the six TIM samples, designated as B1, B2, B3, B4, B5 and B6, were constructed, with the same volume of constituents as sample A1 to A6, correspondingly (see Table I). TIM cubes were attached with copper pads (with a thickness of 0.25 μm) on their upper and lower surfaces. The created geometric models and their grid partition details (polymer matrix excluded) can be seen in Figs. 5 and 6.

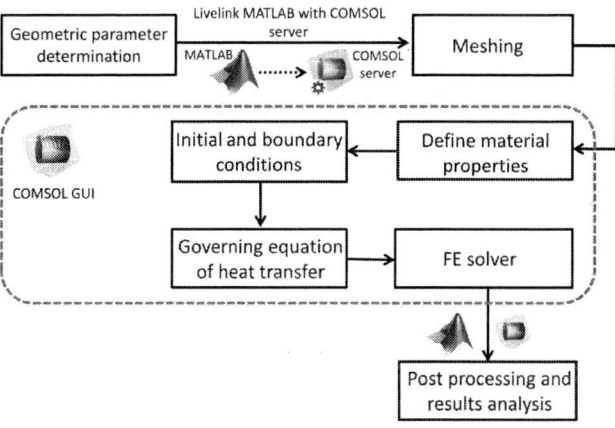

Fig. 3. Modeling process of this work.

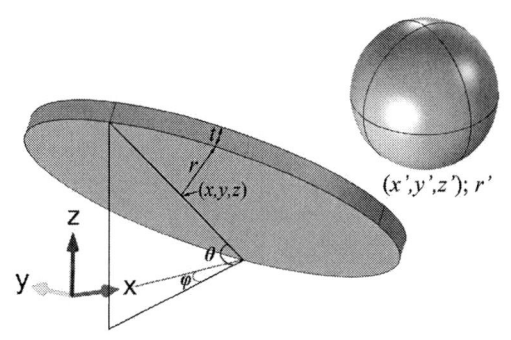

Fig. 4. Schematic diagram of the geometric models of the platelet and spherical conductive particles.

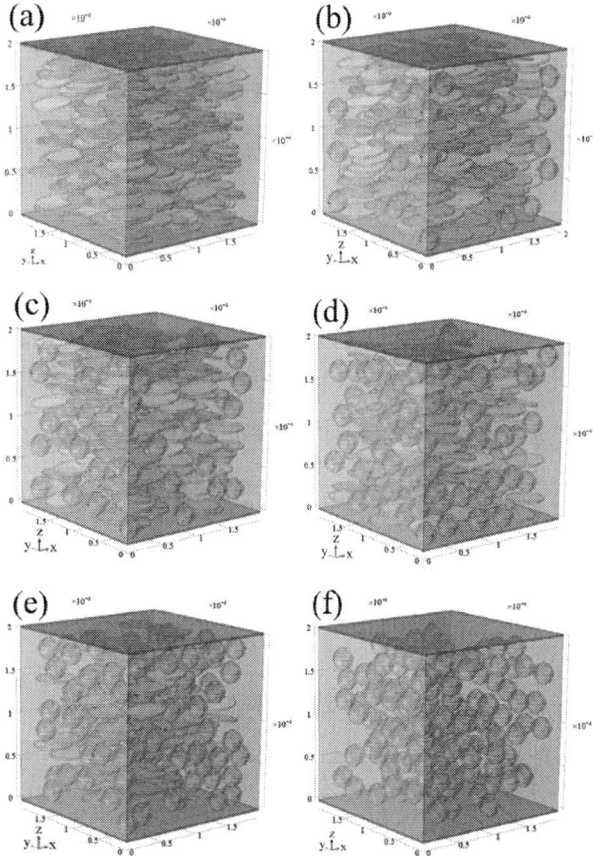

Fig. 5. 3D geometric models of B1 ~ B6: (a) B1 (5:0); (b) B2 (4:1); (c) B3 (3:2); (d) B4 (2:3); (e) B5 (1:4); (f) B6 (0:5).

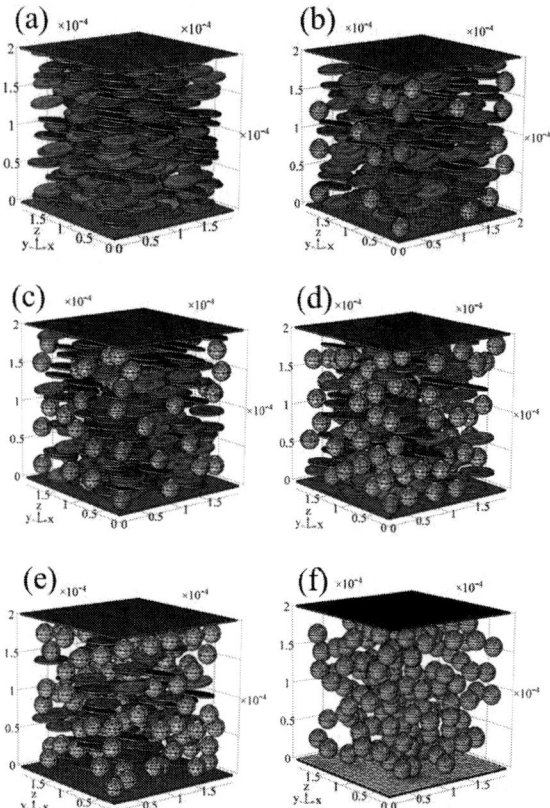

Fig. 6. Element grind partitions (polymer matrix excluded) of B1 ~ B6:
(a) B1 (5:0); (b) B2 (4:1); (c) B3 (3:2); (d) B4 (2:3); (e) B5 (1:4); (f)
B6 (0:5).

The geometric models after meshing were then imported
into COMSOL Multiphysics. Material parameters used in
this work are listed in Table II. The setting of the physical
field was depicted in Fig. 7. An initial heat flux (10000 W/m²)
was applied to the upper substrate and flowed through to the
bottom substrate. The temperature of the lower surface, T_1,
was set as 50 °C. The governing equation for heat transfer in
the present study can be presented as:

$$\rho C_p \frac{\partial T}{\partial t} + \nabla \cdot (-k \nabla T) = q, \qquad (1)$$

where q is the heat flux, ρ is the density, C_p is the specific
heat capacity and k is the thermal conductivity.

TABLE II.   MATERIAL PARAMETERS USED IN FE MODELS [10]

Material Properties	h-BN	Cu	Epoxy Matrix
Density (g/cm³)	2.15	8.93	1.18
Thermal Conductivity (W·m⁻¹·K⁻¹)	300	386	0.5
Specific Heat Capacity (J·kg⁻¹·K⁻¹)	793.71	385	1400

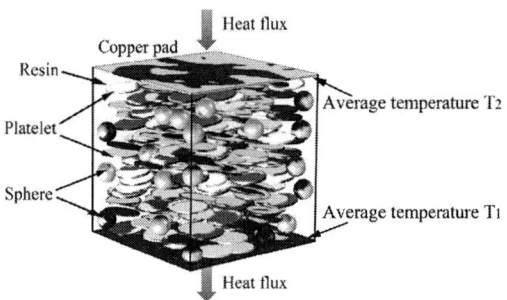

Fig. 7. Schematic diagram of heat transfer in the TIM cube.

III. RESULTS AND DISCUSSION

A. Inhomogeneous Temperature and Heat Flux Distributions

The detailed patterns of temperature and heat distribution
can be calculated through modelling. Fig. 8 shows the
morphology and distribution of the isothermal surfaces in the
TIM cubes. In general, the temperature differences between
the top and bottom surfaces are small due to the selected
micron size of the cubes. There is also a trend that the
distance between isothermal surfaces becomes larger
gradually with the increase of the ratio of spherical fillers. In
addition, the internal temperature distribution is
inhomogeneous and varies with the different filler
constituents. The unevenness depicts the shape of the
particles, and the temperature at the edge of the filler tends to
be higher.

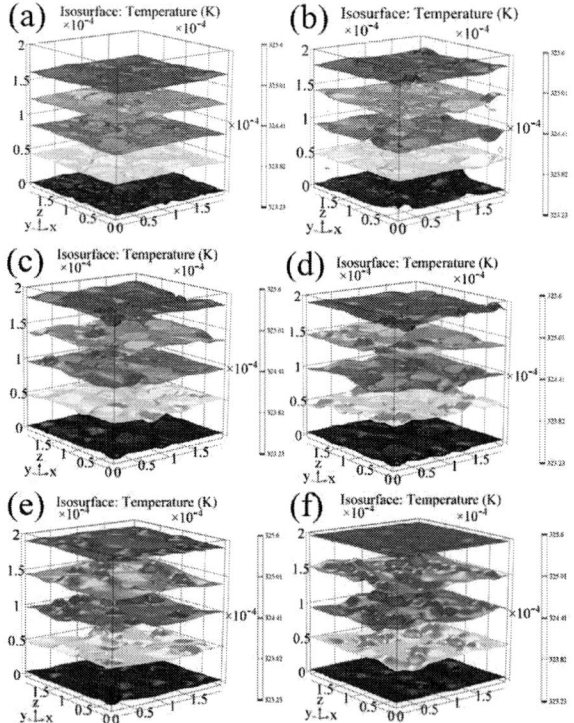

Fig. 8. Morphology and distribution of isothermal surfaces in B1 ~ B6: (a)
B1 (5:0); (b) B2 (4:1); (c) B3 (3:2); (d) B4 (2:3); (e) B5 (1:4); (f) B6 (0:5).

978-1-7281-1500-9/19 $31.00 © 2019 IEEE

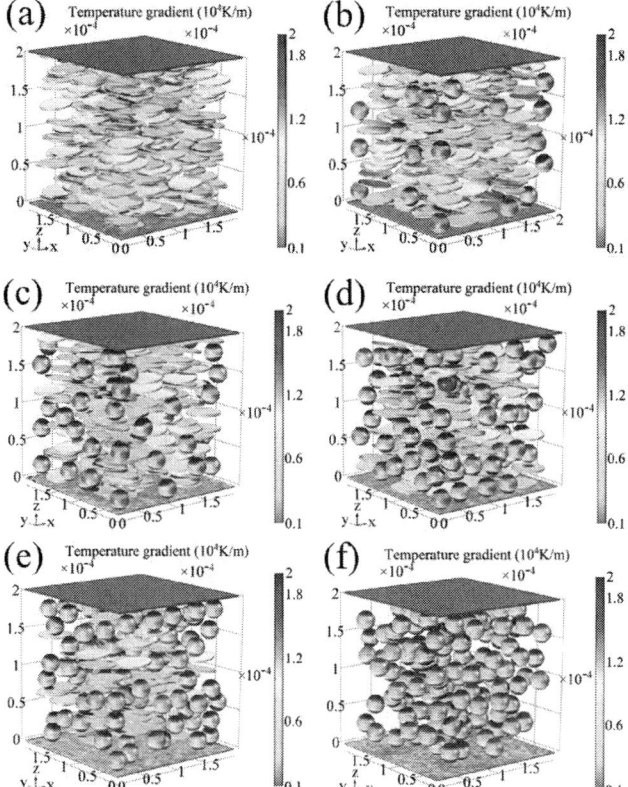

Fig. 9. Distribution of temperature gradient in B1 ~ B6: (a) B1 (5:0); (b) B2 (4:1); (c) B3 (3:2); (d) B4 (2:3); (e) B5 (1:4); (f) B6 (0:5).

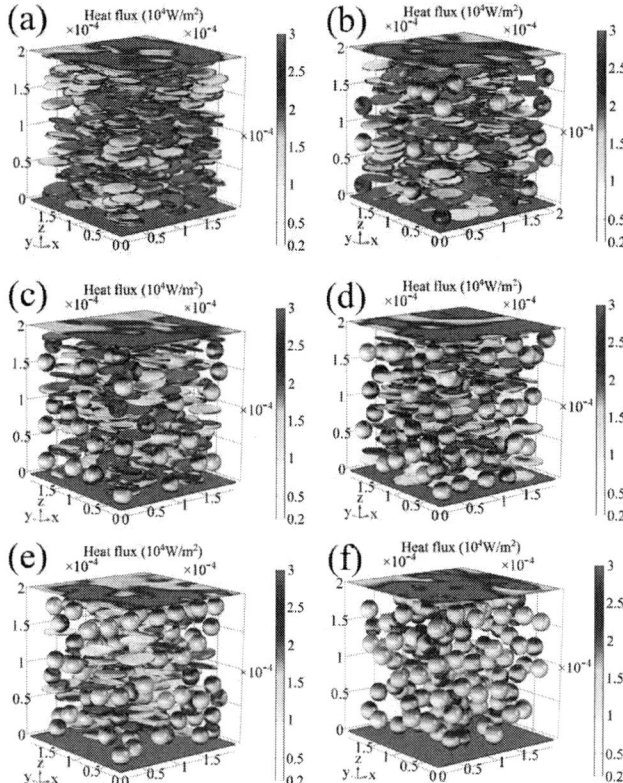

Fig. 10. Distribution of heat flux in B1 ~ B6: (a) B1 (5:0); (b) B2 (4:1); (c) B3 (3:2); (d) B4 (2:3); (e) B5 (1:4); (f) B6 (0:5).

The distribution of temperature gradient is shown in Fig. 9. Similar to Fig. 8, the internal temperature gradient is also inhomogeneous and varies with the different filler constituents of the TIM cubes. However, compared with the relatively small overall temperature differences of the top and bottom surfaces, the localized temperature change is quite conspicuous and the value can even reach 2×10^4 K/m, which also provides evidence for the possible performance impairment of TIM. Additionally, the temperature change at the edge of the particles tends to be more significant, especially for the contact surface between filler and epoxy matrix where the heat flows through. Furthermore, temperature gradient in the spherical particles tends to be more apparent than platelets because of their larger contacting area.

Fig. 10 presents the actual heat flux distribution inside the TIM cubes, and the uneven color patches indicate that the conduction of heat is highly preferable for a continuous thermal network. The distribution of the heat flux on the top surface of the cubes is uneven and changes when the filler constituent changes; while the temperature gradient distribution of them is fairly even and remains unchanged despite of the change of fillers as shown in Fig. 9. The homogenous distribution of temperature gradient corresponds to the initial even heat flux applied on the top surface, while the pratical heat flux distribution changes immediately due to its preferable conduction.

B. Relationship between Thermal Conductivity and Hybrid Filler Constituents

From the effective heat flux and temperature obtained from simulation, the heat conductivity k_e can thus be calculated by the following formula based on Fourier's law:

$$k_e = \frac{q'_z h}{T_2 - T_1},\qquad (2)$$

where q_z' is the average heat flux along z axis, h is the length of the TIM cubes along z axis, and T_2 and T_1 are the average temperature of the top surface and bottom surface of the sample, respectively.

As h and T_1 were set as constants, q_z' and T_2 are the variables that determine the thermal conductivity. The values of q_z' and T_2 of different TIM cubes are listed in Table III.

TABLE III.   VALUES OF THE VARIABLES

Variables	B1	B2	B3	B4	B5	B6
q_z' (W·m^{-2})	34805	96710	72266	77046	30286	10766
T_2 (K)	326.12	325.89	325.80	325.75	325.71	325.64

978-1-7281-1500-9/19 $31.00 © 2019 IEEE

Consequently, the thermal conductivity can be calculated, as shown in the histogram Fig. 11. Compared with those cubes filled wth fillers of single shape, mixed spherical and platelet h-BN of a certain ratio can obviously improve the thermal conductivity, although the total volume fraction of h-BN is the same. In this study, the thermal conductivity reaches 7.1 $W{\cdot}m^{-1}{\cdot}K^{-1}$ when the ratio of spherical particles to platelets is 4:1, more than three times that of pure platelets. The thermal conductivities of TIM cubes filled with pure platelets and spheres are 2.3 $W{\cdot}m^{-1}{\cdot}K^{-1}$ and 0.9 $W{\cdot}m^{-1}{\cdot}K^{-1}$, respectively, and both are lower than those for cubes filled with hybrid fillers.

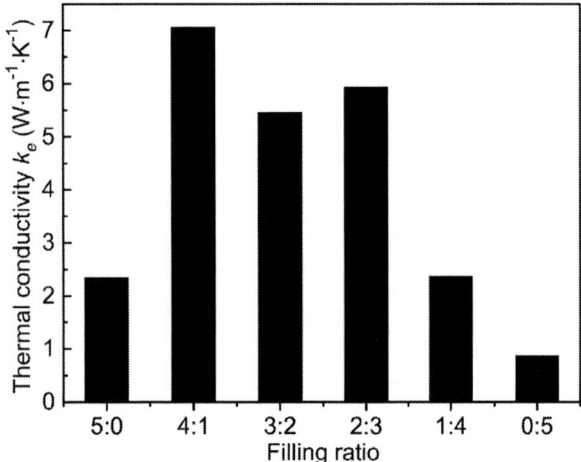

Fig. 11. Thermal conductivity of TIM cubes B1~B6.

IV. CONCLUSION

In this work, the optimal design of polymer-based TIMs filled with mixed spherical and platelet h-BN particles for power electronics with enhanced TC was conducted through three-dimensional FE modeling and experimental characterization. Based on the morphologies observed by SEM, the effect of hybrid filler constituents (mixed spherical and platelet h-BN particles with different ratios) on the TC of polymer-based TIM was observed. Exquisitely designed algorithms were proposed to construct the geometric models. Furthermore, the underpinning mechanisms of the variation in TC of the TIMs were analyzed through the temperature distribution patterns and micro heat flux paths. Results showed that with the same total volume fraction of h-BN, mixed spherical and platelet h-BN of a certain ratio can

further improve the thermal properties of the TIM compared with those filled with spherical or platelet h-BN particles alone, which provides a reference for the selection of the hybrid combination of multiple types of filler constituents.

ACKNOWLEDGEMENTS

This research is supported by EPSRC Underpinning Power Electronics 2017 - heterogeneous integration (HI) project (EP/R004501/1), and China Scholarship Council (201806150013). The authors are grateful to Momentive Performance Materials for supplying h-BN powder samples, and Huntsman Corporation for chemical samples. Dr. David Hutt, Dr. David Whalley, and Shanda Wang in Loughborough University are also acknowledged for their help with the experiments.

REFERENCES

[1] M. Rashid, Power Electronics Handbook, 4th ed. Butterworth-Heinemann, 2018.

[2] J. Hansson, C. Zandén, L. Ye and J. Liu, "Review of Current Progress of Thermal Interface Materials for Electronics Thermal Management Applications," Proc. 16th International Conference on Nanotechnology, IEEE Press, Aug. 2016, Japan, pp. 371-374.

[3] H. Chen, V. V. Ginzburg, J. Yang, Y. Yang, W. Liu, et al. "Thermal Conductivity of Polymer-Based Composites: Fundamentals and Applications," Progress in Polymer Science, vol. 59, Mar. 2016, pp. 41-85.

[4] K. M. F. Shahil and A. A. Balandin, "Graphene-Multilayer Graphene Nanocomposites as Highly Efficient Thermal Interface Materials," Nano Letters, vol. 12, Jan. 2012, pp. 861-867.

[5] C. I. Idumah and A. Hassan, "Recently emerging trends in thermal conductivity of polymer nanocomposites," Rev Chem Eng, vol. 32, Mar. 2016, pp.413-457.

[6] S. N. Leung, "Thermally conductive polymer composites and nanocomposites: Processing-structure-property relationships," Composites Part B, vol. 150, May 2018, pp. 78-92.

[7] Y. Tao, Z.Yang, X. Lu, G. Tao, Y. Xia, et al. "Influence of filler morphology on percolation threshold of isotropical conductive adhesives (ICA)," Sci China Technol Sci, vol. 55, 2012, pp. 28–33.

[8] K. Ramani and A. Vaidyanathan, "Finite element analysis of effective thermal conductivity of filled polymeric composites," J Compos Mater, vol. 29, 1995, pp. 1725–1740.

[9] X. Li, X. Fan, Y. Zhu, J. Li, J. M. Adams, et al. "Computational modeling and evaluation of the thermal behavior of randomly distributed single-walled carbon nanotube/polymer composites," Computational Materials Science, vol. 63, 2012, pp. 207-213.

[10] S. N. Leung, M. O. Khan, E. Chan, H. E. Naguib, F. Dawson, et al. "Synergistic effects of hybrid fillers on the development of thermally conductive polyphenylene sulfide composites," Journal of Applied Polymer Science, 2012, pp. 3293-3301.

Author Index

Aasmundtveit, Knut E 141
Abe, Takatoshi 1272
Abhijit, Dasgupta 498
Abrami, Avner 270
Abrol, Amrit 370, 1347
Agarwal, Amit 718
Ahari, Arman 1099
Ahasan, Kawkab 1860
Ahmed, Omar 1106
Ahmed, Sudan 1347
Ahn, Geun-Sik 197
Akahoshi, Tomoyuki 1294
Akashi, Takahiro 1022
Akazawa, Miyuki 94
Akejima, Shuzo 1140
Akiyama, Kentaro 1641
Alam, Arsalan 277
Alam, Mohammad S. 1815, 1958
Albertinetti, Andrea 2091
Albrecht, John 948
Alexandre, Giry 1279
Alhendi, M. 1946
Alhendi, Mohammed 768, 1581
Ali, Muhammad 960
Alizadeh, Azar 1581
Allain, Fabienne 1622
Allouti, Nacima 569
Alvarez, Claudio 1672
Amandine, Jouve 225
Amano, Takeru 2042
Ambat, Rajan 515
Ambhore, Pranav 620, 800, 1605
Amiran, Johnny 995
An, Sang-Ho 204
Anai, Kei 76
Andersson, Rickard 1870
Andreini, Antonio 2194
Andriani, Yosephine 1543

Antoniou, Antonia 655
Antretter, Thomas 1509, 2029
Apriyana, Anak Agung Alit 1735
Araki, Hitoshi 346
Araki, Naoko 1002
Araki, Noritoshi 175
Arayama, Chika 1022
Argoud, Maxime 569
Arnaud, Lucile 569, 1926
Arnold, Kim 340
Aschenbrenner, R. 861
Aslani Amoli, Nahid 249, 1939
Assous, Myriam 1622
Audet, Jean 1179
Aue, Maximilian 1883
Auer, Benedikt 1789
Aumont, Christophe 569
Aygun, Kemal 667
Ayssar, Serhan 1279
Ayukawa, Michael 2117
Azizi, Arad 1970
Azzopardi, Stephane 2173
Baelmans, Martine 126
Bajwa, Adeel Ahmed 620
Bakir, Muhannad S. 1803
Balaraman, Devarajan 163
Banerjee, Deepayan 1770
Barth, Maximilian 868
Bartl, Ulf 855
Barut, Atila 825
Barwicz, Tymon 528, 1060
Baty, Greg 1099
Bauwelinck, Johan 1052, 1757
Bea, J.C 1047
Becker, Karl-Friedrich 861
Bécu, Stéphane 168
Bedjaoui, Messaoud 995
Bedoin, Alexis 535

Behr, Andy	1272
Beica, Rozalia	14, 903
Beigné, Edith	1926
Bejugam, Vinith	47
Bellaredj, Mohamed	1672, 1939
Bellaredj, Mohamed L. F.	249
Beltritti, Jérôme	569
Bengsch, Sebastian	1883
Berger, Frédéric	569
Bernstein, Gary H.	2072
Bertheau, Julien	340
Bex, Pieter	340, 607, 674
Beyer, Gerald	340, 607, 1035, 2206
Beyne, Eric	126, 340, 437, 607, 674, 1035, 1215, 2206
Bharath, Krishna	2180
Bhattacharya, Surya	587, 917
Bhuvanendran Nair Gourikutty, Sajay	1227
Bicer, Mehmet	1622
Bilgen, Halim	1926
Billard, Christophe	1680
Biscarrat, Jérôme	168
Bito, Jo	896
Blachier, Denis	168
Blackshear, Edmund	1179
Blattau, Nathan	2103
Blecker, Ken	505
Boddu, Vijaya	2180
Bolkhovsky, Vladimir	1611
Bonam, Satish	2156
Boo, Hyunpil	277
Borgesen, P.	1946
Borgesen, Peter	768, 1581
Bouillard, Boris	428
Bowrothu, Renuka	695, 983, 2085, 2337
Bozano, Luisa	417
Brandl, Elisabeth	1789
Brandstätter, Birgit	1789
Braun, Tanja	363, 861, 861
Bravin, Julian	1789
B. Reddy, Vishnu V.	1333
Bruckner, Gudrun	878
Bruderer, Alex	726
Brun, Jean	995
Bu, Lin	1152, 1419, 1735
Butylkov, Sergey	1312
Bylund, Maria	1870
Byong Jin, Kim	1457
Cahu, M.	1339
Cai, Biao	1200
Cai, Jian	69, 1306, 2234
Cain, Stephen R.	2150, 2331
Campos, Didier	569
Cao, Liqiang	2318
Cao, Xinpei	753
Capecchi, Simone	910
Caplet, Stéphane	535
Carlsson, Mats	968
Cartier, Mathilde	1535
Castagné, Laetitia	1575
Castan, Clément	225
Cecchetto, Luca	2194
Chacon, Oswaldo	1396
Chahal, Premjeet	113, 948, 1240, 1687
Chai, Tai Chong	21, 1419
Chambion, Bertrand	535
Chan, Alex	1272, 1902
Chan, Chin-Wei	1751
Chandrasekhar, Arun	2180
Chang, Chia-Cheng	758, 1328
Chang, Chieh-Lin	14
Chang, Grace	1595
Chang, K.C.	763
Chang, Keng Tuan	41
Chang, Kuo-Chin	397
Chang, Megan	2079
Chang, Tao-Chih	1463

Chang, Victor C. Y. 688
Chang, Yu-Chen 1359
Chang Chien, Chien-Lin 461
Chao, Tz-Yuan 1170
Charbonnier, Jean 569, 1622
Charles, Matthew 168
Chausse, Pascal 569
Che, Fa Xing 842
Che, F. X. 1152, 2126
Chen, Allenyl 1426
Chen, Chen 661
Chen, Cheng-Chih 1826
Chen, Chih 642, 758, 1328
Chen, Chi-Jen 7
Chen, Chi-Yuan 289
Chen, Chuantong 474
Chen, C.S. 931, 1595
Chen, Dao-Long 1710, 1902
Chen, Fang-Cheng 594
Chen, J.H. 1693
Chen, Jie 1221, 1653
Chen, Jing 2061
Chen, J. Y. 700
Chen, Kang 1165
Chen, KarenYU 1710
Chen, Kuan-Neng 1463
Chen, Kuan-Ta 1704
Chen, Liangbiao 1521
Chen, Louis 1426
Chen, Ming-Fa 594
Chen, Qianwen 1246
Chen, Qiaoli 1200
Chen, Rui 249, 1939
Chen, S. 1629
Chen, Shujing 1564
Chen, Si 1324
Chen, Sihai 2186
Chen, Simon 700

Chen, S.M. 931
Chen, Tangsheng 1842
Chen, Tianfang 1983
Chen, Tony 14, 325, 903
Chen, Wei-Han 1877
Chen, Xi-Hong 1359
Chen, Xu 1889
Chen, Yan-Hao 1729
Chen, Y. F. 1175
Chen, YH 903
Chen, Y. H. 14
Chen, Yih-Sin 1170
Chen, Y.N. 1595
Chen, Yu-Hua 1413, 1463
chen, zhaohui 1543
Chen, Zhaoqing 1989
Chen, Zhiwen 63, 850, 1377
Chen, Zihao 917
Cheng, Po Wen 1246
Cheng, Shau-Fei 1877, 1877
Cheng, Ta-Chien 910
Cheng, Wei-Yuan 1877, 2009
Chenhsiu, Sung 1933
Chéramy, Séverine 569
Cheramy, Severine 225, 1926
Cherman, Vladimir 126
Cheung, Y. M. 14, 903
Chew, Ly May 87
Chew, Nam Piau 1735
Chi, Yenyao 1165
Chiang, Jack 2079
Chiang, K.N. 1515
Chiang, W.S. 493
Chien, Feng-Lung 1170
Chien, Feng Lung 1635
Chien, Han 800
Chien, Jason 1902, 2079
Chim, Weng Tuck 1457

Ching, Eva Wai Leong 917
Chiou, Wen-Chih 594
Chiu, Chi-Tsung 1426
Chiu, Jason 763
Chiu, J. M. 1175
Chiu, Julia 453
Chiu, Ryan 700
Chiu, Steve 2009
Chiu, Tz-Cheng 1359
Chiu, Yihsiang 1503
Chiu, Yung-Da 1426
Chiu, Yu-Shan 235
Cho, Cheng-Lin 258
Cho, Jae Kyu 910
Cho, NamJu 289
Cho, Sung-Il 204
Cho, Youngsang 300
Choi, Heejung 300
Choi, Hyun-Seok 204
Choi, Jinu 753
Choi, Kwang-Seong 197
Choi, Kwang Won 1179
Choi, TaeJin 2278
Choi, Won Kyung 968, 1165
Chong, Ser Choong 21, 191
Choong, Chong Ser 2126
Chou, P. H. 1515
Chow, Eugene M. 1312
Chow, Justin H. 785
Chow, Seng Guan 1165
Chowdhury, Md Mahmudur 792
Christie, Leroy 2349
Chu, Fu-Cheng 977
Chuang, Chun-Hsiang 258
Chuang, Oscar 2009
Chuang, Wallace 2067
Chuh, Erich 2180
Chung, C. Key 7, 1287

Chylak, Bob 620
Cibié, Anthony 168
Colonna, Jean-Philippe 168, 1535
Colosimo, Tom 620
Con, Celal 2219
Connolly, Brian 1200
Coquand, Rémi 1622
Coudrain, Perceval 168, 569
Crawford, Lara S. 1312
Cromwell, Kevin 1883
Crump, Cameron 948
Cui, Xiaole 1983
Cyr, Élaine 1074
Daeumer, Matthias A. 1970
Dahl, David 2240
Dahlbäck, Robin 968
Dalmia, Sidharth 294, 954, 1666
Dang, Bing 1246
Danovitch, David 467, 2117, 2252
Das, Rabindra 1611
David, Leslie 498
Day, Doug 1933
Decker, Michael 1509
Dede, Ercan M. 1437
Deep, John 1366
Dehe, Alfons 855
De Heyn, Peter 1757
Del Nero, Daniel 1258, 1848
Deloffre, Emilie 1926
DeProspo, Bartlet 334, 1588
DeProspo, Bartlet H. 924
Deschamps, Jerôme 535
Desmaris, Vincent 1870
Desmet, Andres 1757
De Vos, Joeri 1035
De Wolf, Ingrid 126
Dhandapani, Karthik 819
Dias, Rajen 163

Di Cioccio, Léa	168
Dincau, Brian	1860
Ding, Guifu	707
Ding, Kunpeng	1306
Ding, Qian	1897
Ding, Xuanyi	2186
Docanto, Manuel	1611
Dohi, Kazuhiro	2162
Dolores-Calzadilla, Victor	1060
Dorduncu, Mehmet	825
Dornala, Kalyan	1366
Dreissigacker, Marc	861
Dreps, Daniel	1200
Dressler, Marc	1113
Drouin, Dominique	1396, 2252
Du, Ke	69
Dubis, Monique	87
Dunkel, Christian	1498
Duran-Martinez, Adriana Carolina	1258
Durgun, Ahmet C.	667
Dutoit, Denis	569
Dwarakanath, Shreya	718
Ecker, Melanie	1258
Economou, Manthos	486
Edouard, Deschaseaux	1279
Eichhammer, Yann	1052
Eid, Aline	896
Ekstrom, Noah	753
EL Amrani, Abderrahim	2252
Eleouet, Raphaël	569
Elger, Gordon	2324
El-Mekki, Zaid	1035
Elmogi, Ahmed	1757
En, Yunfei	1324
England, Luke	600
Enomoto, Tetsuya	352
Eom, Yong-Sung	197
Erhart, Andreas	1833

Escoffier, René	168
Eto, Motoki	175
Evans, John	2309
Evans, John L.	792
Evertsen, Rogier	423
Exbrayat, Yorrick	569
Ezawa, Hirokazu	1140
Ezhilarasu, Goutham	277, 1470
Fager, Christian	1405
Fan, Nelson	14, 903
Fan, Xuejun	806
Fan, Zhineng	1200
Fana, Jilei	81
Fang, Bo-Siang	1432, 1704
Fang, Runiu	2168
Fang, Sheng-Po	1647, 1809
Fang, T.J.	931
Fang, Y.H.	493
Farcy, Alexis	569, 1926
Farrugia, M-L	479, 777
Fasoli, Andrea	417
Feng, Qingming	878
Fernandez, Hector	1106
Fernandez, Maïlys	535
Fernandez-Zelaia, Patxi	2349
Fettke, Matthias	47, 210
Feuchter, Michael	2029
Filipp Fuchs, Peter	2029
Finn, Daragh	453
Fischer, Thomas	855
Fischer, Thorsten	1475
Fisher, Daniel	600
Fisher, Timothy	1605
Fisher, Timothy S	277
Fitzgerald, Padraig	1660
Flaim, Tony	1722
Fleischman, Martin	811
Fortier, Paul	1074

Fortin, Clément 306
Fountain, Gill 628, 1041
Fournel, Frank 225
Fowler, Michelle 363
Franiatte, Rémi 225, 1622
Franieck, Erick 811
Fraschke, Mirko 218
Friedmann, T. A. 648
Friedrich, Georg 47, 210
Fritsche, Carola 1475
Fu, Haley 318
Fu, Xing 1324
Fuchs, Peter Filipp 1509
Fuguet Tortolero, César 569
Fujimagari, Junichiro 1641
Fujinaga, Tetsushi 358, 1865
Fujisaki, Hidehiko 1294, 1952
Fujiwara, Atsushi 1641
Fukuda, Takafumi 1140
Fukui, Kei 1294
Fukuomori, Minoru 1451
Fukushima, Takafumi 264, 1047
Furuya, Akira 1067
Gagnon, Pascale 306, 1744
Galbraith, Christopher 1611
Gao, Guilian 628, 1041
Garnier, Arnaud 569
Gaschet, Christophe 535
Geissler, Christian 855
George, Jinto 2117
Gerber, Mark 1902
Gernhardt, Robert 363
Ghannam, Ayad 1789
Ghosh, Tamal 2156
Giesen, Kyle 1200
Gillot, Charlotte 168
Gjokaj, Vincens 948
Glodde, Martin 1060

Goemare, Charlotte 1870
Goggin, Ray 1660
Goller, Bernd 855
Gong, Dan 1503
Goodelle, Jason 28
Goorsky, Mark 1605
Gordon, Seth 2309
Gore, Aaron 453
Gore, Brandon T. 726
Gorrell, Robin 330
Goto, Yoshio 101
Gottardi, Mathilde 569
Gottwald, Thomas 726
Goumans, L. 777
Gourvest, Emmanuel 428
Graap, Pascal 861
Graham, Samuel 1977
Green, Ryan B. 1782
Green, William M.J. 1060
Gromala, Przemyslaw 811, 1529
Gschwandl, Mario 1509, 2029
Gu, Han 243, 1916
Guarino, Lucrezia 2194
Guerrero, Alice 340
Gueugnot, Alain 569
Guevara, Gabe 628
Guidoni, Luca 1735
Guo, Huaixin 1842
Gupta, Sunil 1194, 2097
Gupte, Omkar 1028
Guthmuller, Eric 569
Guthrie, Bill 600
Hagn, Josef 954
Hah, Jinho 1977, 2349, 2359
Hai, Joe 486
Hajjar, Jean-Jacques 1660
Hama, Hiroki 550
Hamasha, Sa'd 792

Hamasha, Sa'd	2309
Han, Bongtae	811, 1382, 1529
Han, Jeong Sam	2246
Han, Kwangwoo	707
Han, Yong	21, 1543
Hanada, Tadahiko	2112
Hanisch, Anke	628
Hanna, Amir	277, 579, 800, 1470
Haque, Mohammad Aminul	2073
Harrison, Todd	1653
Hasegawa, Yasuo	101
Hashimoto, Keika	346
Hashmi, Mohammad	1770
Hassan, KM Rafidh	1815, 1958
Haumesser, Paul-Henri	168
Hayashi, Kazutaka	712
Hayashi, Toshihiko	1641
Hazellah, Muhammad Hadhari	1457
He, Eric	700
He, Jiangling	135
He, Peng	2022
He, Quanfeng	1716
He, Xuanke	119
Heinig, Andy	314
Heisig, Stephen J	270
Hejase, Jose	1200
Helbig, Stephan	855
Helou, Assaad	405
Henrion, Yann	1926
Henry, David	535
Henry, M. David	648
Hensley, Dale	2073
Herbert, Robert	1233
Hernandez, Natalie	2103
Hernandez, Selene	753
Herrmann, Matthias	855
Hester, Jimmy	896
Hikita, Masayuki	1451

Hillman, Craig	2103
Hirabayashi, Keiichi	1179
Hirano, Mitsuharu	1067
Hirose, Masakazu	2162
Hirt, Etienne	868
Ho, Bin-En	1693
Ho, Cheng-Yu	977
Ho, David	1432
Ho, David Soon Wee	917
Ho, Soon Wee	21
Hoang, Tim Tri	667
Hoelck, Ole	861
Hokari, Ryohei	1764
Honda, Kazutaka	446, 740
Hong, Xuan	753
Hook, Michael David	2219
Hooshmand, Nasrin	1588
Hopsch, Fabian	314
Hoque, Mohd Aminul	792
Horibe, Akihiro	1921
Hoshino, Hitoshi	437
Hosseini, Seyedmahmoud	1258
Hou, Xinnan	69
Hou, Zhuangzhuang	1716
Houston, Paul	2349
Hsiao, Andy	1099
Hsiao, Hsiang-Yao	21
Hsiao, H. Y.	1515
Hsiao, Yu-Hsiang	461
Hsieh, Chia-Ping	2009
Hsieh, Jeng-Shien	688
Hsieh, Ming-Che	289
Hsieh, Ricky	977
Hsieh, Tsun-Lung	41
Hsieh, Yi-Chen	1751
Hsu, C.C.	1595
Hsu, Che-Ming	461, 600
Hsu, Chieh-Hao	397

Hsu, Chih Chung 318
Hsu, Chih-Hsun 7
Hsu, Chung-Yi 2200
Hsu, C.K. 931
Hsu, C. K. 1550
Hsu, F.C. 931
Hsu, Fussen 1413
Hsu, Hsiang-Han 1074
Hsu, Ian 289, 493
Hsu, Po-Ning 642
Hsu, Steven 397, 763, 1175
Hsu, Y.N. 1693
Hu, Ian 1710
Hu, Yang 2234
Hu, Yougen 243, 1916
Hu, Yuan 277
Huang, Baron 1722
Huang, Chen-Yu 1287
Huang, Chih-Yi 41
Huang, Dick 700
Huang, Dinos 1710
Huang, G.C. 1595
Huang, H.L. 1595
Huang, Mian 1306
Huang, Mingliang 2022
Huang, Mingliang L. 1774, 2036
Huang, Pei-Chen 1413
Huang, P.S. 493
Huang, Rocky 1200
Huang, Shih-Ya 688
Huang, Shin-Yi 1463
Huang, Yifan 1200
Huesgen, Till 1443
Hung, Han-Tang 1729
Hung, Mi-Chun 41
Huo, Jia Ren 1485
Huo, Yongjun 150
Huynh, Michael 628

Hwang, Jisoo 300, 682
Hwang, Jung Woo 733
Hwang, Kihyun 636
Hwang, Kyo-sung 330
Hwang, Lih-Tyng 1751, 2200
Hwang, Taejoo 614
Hwangbo, Seahee 695, 983, 2085, 2337
Hyun, Sangjin 636
Iacovo, Serena 607, 2206
Ihori, Atsuhito 1865
Iida, Kenji 1952
Iizuka, Tomonori 1451
Im, Yunhyeok 300
Inaba, Takayuki 1952
Inamdar, Adwait 811
Ingelhag, Per 1405
Inoue, Fumihiro 437, 607, 2206
Inoue, Junishi 556
Irwin, Randall 277
Ishigure, Takaaki 550
Islam, Nokibul 325
Itawi, Ahmad 1575
Iwai, Toshiki 1952
Iwamoto, Hayato 1641
Iyer, S. S. 277
Iyer, Subramanian 620, 1470
Iyer, Subramanian S. 543, 579, 800, 1605, 2225
Jacquemond, Achille 264
Jacques, Patrick 1074
Jain, Ritesh 2180
Jalilvand, Golareh 1106, 1909
Jalink, J. 1339
Jamieson, Geraldine 1035
Janek, Florian 868
Jang, Joohee 636
Jangam, SivaChandra 543, 620
Jangam, Siva Chandra 800
Jani, Imed 1926

Janta-Polczynski, Alexander	1074
Jarecki, Robert	648
Jayabalan, Jayasanker	587
Jean-Philippe, Michel	1279
Jemaa, Salwa Ben	1744
Jeng, Shin-Puu	931, 1550
Jeon, HyeongIl	1457
Jeong, James	300
Jeong, leeseul	197
Jeong, Minsu	1146
Jeong, Se Young	733
Jhong, Ming-Fong	41
Ji, Hongjun	183
Jiang, Don-Son	1704
Jiang, Han	1569
Jiang, Jing	1485
Jiang, Tengfei	1106, 1909
Jiang, Yih-Jenn	7
Jin, Yufeng	1503, 1983, 2016, 2061, 2168
Jo, Chanmin	1188
Jo, Jung-Lae	76
Jo, Paul K.	1803
John Akkara, Francy	2309
Johnson, Leonard	1611
Joly, Pierre	535
Jong, Ming Chinq	1543
Joo, Jiho	197
Joo, Kisu	733
Jordan, Matthew B.	648
Joshi, Rahul	806
Joshi, Shailesh N.	1437
Joshi-Imre, Alexandra	1258, 1848
Jourdon, Joris	1926
Juang, Jing Ye	642
June Rebibis, Kenneth	437
Jung, Jin-San	204
Jung, Kwang-Ho	2290
Jung, Seung-Boo	2290

Jung, Seung-Yoon	283
Kabir, Mohammed	1870
Kahle, Ruben	861
Kalappurakal Thankappan, Kannan	2225
Kalb, Jamie	1312
Kalnitsky, Alex	1595
Kalyanam, Huthasana	2180
Kam, Nicholas	2219
Kamimura, Rikiya	1451
Kamlapurkar, Swetha	528, 1060
Kanagawa, Naoki	1022
Kandanur, Sashi	1588
Kaneko, Junichi	1933
Kang, Kuo-Chang	600
Kang, Minsoo	1977
Kang, Pilkyu	636
Kang, Qiushi	1266
Kannan, Jenefa	334
Kao, C. Robert	235, 1729, 2258
Kao, Feng	700
Kao, Hsuan-Ling	258
Karim, Karim S.	2219
Karlheinz, Bock	498
Karsten, Meier	498
Karuppuswami, Saranraj	113, 1240
Katagiri, Shunsuke	1009
Kathaperumal, Mohan	718
Kathaperumal, Mohanalingam	1796, 2112
Kathaperumal, Mohananlingam	334
Katkar, Rajesh	1041
Kavle, Pravin	931
Kawanabe, Naoki	1451
Kawano, Masaya	1996
Kaynak, Mehmet	218, 942
Ke, Chang-Bo	410
Ke, C.N.	1595
Keith Newman, Keith	806
Kelly, Mike	163

Kencana, Sagung Dewi 2067
Kennes, Koen 607
Kenney, Christopher 2072
Kerepesi, Peter 218, 942
Kersjes, Sebastiaan 1789
Keser, Beth 1159
Khazaka, Rabih 2173
Khim, Jin Young 1457
Khinda, G.S. 1946
Khurana, Gaurav 924
Kida, Tsuyoshi 1009
Kidera, Nobutaka 712
Kilger, Thomas 855
Kim, Changsu 1382
Kim, Choong-Un 1316
Kim, Dogeun 937
Kim, Dongsu 1647
Kim, Dong wook 204
Kim, Gahui 937, 2246
Kim, Hae-In 983, 2085
Kim, Haein 2337
Kim, Jaechoon 614
Kim, Ji-Hye 2272
Kim, Ji-Min 204
Kim, Jong Heon 35, 563
Kim, Jong-Hoon 1860
Kim, Ju hyeon 197
Kim, Jung Hak 197
Kim, Junghwa 300
Kim, JunMo 1146
Kim, Kilsoo 614
Kim, Kyoung-Tae 1809
Kim, KyuHyoun 1200
Kim, Nam Chul 35, 563
Kim, Nam-Seog 2246
Kim, Seokho 636
Kim, Soon-Wook 2206
Kim, Taehun 614

Kim, Taehwan 614
Kim, Taek-Soo 1146
Kim, Taeyeong 636
Kim, Yi-Ram 1316
Kim, Yoon-Hyun 733
Kim, Young-Cheon 2246
Kim, Young Ho 35
Kim, Youngja 2349
Kim, Young-Ja 2359
Kino, Hisashi 264
Kintaka, Kenji 556
Kirchner, Lisa 1722
Klengel, Robert 175
Klengel, Sandy 175
Klingler, Hannes 1789
Knickerbocker, John 270, 1246
Ko, Cheng-Ta 14, 903, 1413, 1463
Ko, T. 688
Kobayashi, Naoki 1599
Kodama, Shoichi 1002
Kohl, Paul A. 249
Koide, Masateru 1294
Kojima, Ryoji 1933
Kokash, M.Z. 1946
Kolbasow, Andrej 47, 210
Kong, Yuechan 1842
Kothari, Nakul 1347
Koyama, Koichi 1067
Koyama, Toshinori 1599
Koyama, Yutaro 346
Koyanagi, Mitsumasa 1047
Kozlovsky, William J. 726
Kraetschmer, Daniel 1113
Kraft, Jochen 1052
Krishna, Bhogaraju Sri 2324
Krivec, Thomas 1509
Kröhnert, Kevin 1475
Krumbein, Ulrich 855

Ku, Harry	1175, 1595, 1693
Ku, Terry	1
Kuah, Eric	14, 903
Kuang, Jiameng M.	2036
Kubo, Atsushi	334, 718, 924
Kubsch, Timo	210
Kudo, Hiroshi	94
Kudo, Tomoya	1015
Kuechenmeister, Frank	910
Kulick, Jason	2072
Kulterman, Ron W.	318
Kumar, Deepak	1687
Kumazawa, Yune	1009
Kuo, C.C.	1595
Kuo, C.H.	1595
Kuo, Hung-Chun	41
Kuo, Kuei Hsiao (Frank)	1635
Kuo, Ping-Jui	600
Kuo, Yu-Lin	2067
Kurihara, Kazuma	1764
Kurosaka, Seigo	474
Kurz, Helmut	218, 942
Kwon, Odal	55
Kwon, Yong Tae	35, 563
Kyung, Youjin	1146
Labarbera, Christine	2186
Lai, Chia-Chu	1704
Lai, Chieh-Lung	1170
Lai, Hsin-Cheng	1877
Lai, P. C.	1550
Lai, T.M.	931
Lai, Yen-Kun	397
Lall, Pradeep	370, 505, 792, 1087, 1347, 1366, 1815, 1958
Lambert, Renée	1611
Lambrecht, Joris	1757
Lan, Jia-Shen	2144
Lang, Klaus-Dieter	861, 1475, 1853
Langlois, Richard	1074

Larsson, Andreas	141
Lasfargues, Gilles	535
Lattard, Didier	569, 1926
Lau, Boon Long	1419, 1543
Lau, Chun Sean	1387
Lau, John	903
Lau, John H.	14
Laugier, Maxence	225
Lauser, Simone	515
Lavrik, Nickolay V.	2073
Le, Thanh Long	2173
Le, Wen-Kai	410
Leblanc, Alexandre	2117
Lee, Bob	2079
Lee, Bongsub	628, 1041
Lee, Chang-Chi	600
Lee, Chang-Chun	1413, 2009
Lee, Chang Woo	35, 563
Lee, Chia-Hsin	1463
Lee, Chin C.	150, 2302
Lee, Choong-Jae	2290
Lee, Chul-Hee	197
Lee, Chul Hyo	35
Lee, HanMin	1146, 2278
Lee, Heeseok	300
Lee, Heesok	682
Lee, Hohyung	2343
Lee, Hoi-jin	682
Lee, Hung-Ho	1287
Lee, Hungping	1503
Lee, Hyeong Gi	204
Lee, Hyun-Seop	1382
Lee, Hyun Seop	1529
Lee, Jae Cheon	35
Lee, Jeffrey ChangBing	1826
Lee, Joungphil	614
Lee, Jun Kyu	35, 563
Lee, Kang Hai	1165

Lee, K.C.	931
Lee, K. C.	1550
Lee, Kwang-Hee	197
Lee, Kwangjoo	197, 1146
Lee, Kyuha	636
Lee, Kyu Jae	733
Lee, NC	903
Lee, N. C.	14
Lee, Ning-Cheng	2186
Lee, Rick	1635
Lee, Ricky	14, 903
Lee, Sangil	2349
Lee, Seok-hyun	937
Lee, Seung Jae	733
Lee, SeYong	2278
Lee, Sung Hyuk	35
Lee, Tae-Ik	1146
Lee, Tae-Kyu	1099, 1106
Lee, Yisang	1047
Lee, Yuh-Zheng	1877
Leever, Ben	370, 1347
Legalland, Corinne	569
Li, Gang	81, 746
Li, Guanglin	243
Li, Hong Yu	1735
Li, Ji	135
Li, Jiahui	2003
Li, Jiaxiong	2296
Li, Junjie	661
Li, Kunkun	1983
Li, Ming	14, 903
Li, Mingyu	183
Li, Na	1983
Li, Ping	28
Li, Wen-Yang	7
Li, Yu-Jin	758, 1328
Li, Yu Jin	642
Li, Zhang	14, 903

Liang, Qi	661
Liang, Shui-Bao	410
Liang, Xianwen	746
Liao, Guanglan	661
Liao, Kuo-Hsien	1902
Liao, Marvin	1175, 1595, 1693
Liao, Siyuan	81
Lii, Mirng-Ji	397
Lii, M.J.	763
Lim, Francis Chee Peng	968
Lim, Jun Su	204
Lim, Ruiqi	1227
Lim, Sharon Pei Siang	1543
Lim, Simon Siak Boon	21, 1543
Lim, Sze Pei	14, 903
Lim, Teck Guan	917, 1419
Lim, Yeow Kheng	1165
Lim, Yew Kheng	968
Lim, Yu Dian	1735
Lim Sharon, Pei Siang	191
Lin, Ang-Ying	1463
Lin, Benson	493, 758, 1165, 1328
Lin, C.H.	931
Lin, Chang-Fu	7, 1287
Lin, Cheng Ping	924
Lin, Curry	14
Lin, Gu-Yan	1170
Lin, Marc	14
Lin, M.J.	493
Lin, M.Z.	493
Lin, Puru Bruce	1413, 1463
Lin, P. Y.	1550
Lin, P.Y.	931
Lin, Stanley	289
Lin, Tiesong	2022
Lin, Tong-Hong	896, 960
Lin, Vito	1287
Lin, W. Y.	1550

Lin, Yi-Hang	931
Lin, Yi-Sheng	461
Lin, Yu-Min	1463
Lin, Zhibin	453
Liou, Yan-Yu	1413
Litzenberger, Lorenz	1443
Liu, Canyu	63, 850
Liu, Changqing	63, 850, 1569
Liu, Chan-Yuan	1902
Liu, Chun-Chen	1693
Liu, C. S.	1175
Liu, Fuhan	334, 718, 924, 1796, 2112
Liu, Handa	135
Liu, Hao-Chun	397
Liu, Huan	1983, 2168
Liu, Hui	1897
Liu, Johan	1564
Liu, Kai	1246
Liu, K.C.	1595, 1693
Liu, K. C.	1175
Liu, Li	63, 850, 1377
Liu, Liyuan	2054
Liu, Meng-Hsiang	579
Liu, M.S.	931
Liu, N.W.	493
Liu, NW	1165
Liu, Penglin	1897
Liu, Ping	628
Liu, Sheng	850, 1377
Liu, Shengfa	63
Liu, Songlin	1543
Liu, Weidong	28
Liu, Weifeng	1272, 1826
Liu, Xiao	1463, 1722
Liu, Ya	1564
Liu, Yanghe	1437
Liu, Yingia	1716
Liu, Yong	1521

Lo, ChangHo	1826
Lo, I-Fang	1751
Lo, Jeffery	14, 903
Lo, Penny	14, 903
Lodermeyer, Johannes	855
Loerke, Friederike	1113
Loh, Wei Keat	318
Lombard, Marc	535
Lombardi, Jack	1581
Lombardi, J.P.	1946
Lord, David	453
Lowe, Ryan	1366
Lu, Calvin	1175, 1693
Lu, Chun-Lin	1697
Lu, JengPing	1312
Lu, Tan	1916
Lu, Tao	2054
Lu, Tian	2072
Lu, Ying-Wei	1704
Luo, Bin	890
Luo, Daojun	2054
Luo, Jiangbo	707
Luo, Yandong	579
Luo, Yu	1246
Luu, Thi-Thuy	141
Luu Trung Duong, Pham	834
Ma, B.H.	1432
Ma, H.T.	1629
Ma, Kun	1377
Ma, Li	28
Ma, Lulu	806
Ma, Shenglin	1503
Ma, Shuying	28
Ma, Xiao	410
Macaisa, Dexter	2117
Machida, Hideki	1067
Mackowiak, Piotr	1475
Madanipour, Hossein	1316

Madenci, Erdogan	825
Maeda, Toru	1933
Maehara, Masataka	1641
Maetani, Shinji	1002
Mahajan, Ravi	667
Maier, Dominic	855
Makita, Toshiyuki	924
Mandal, Rathin	834
Manepalli, Rahul	1588
Mansoor, Bilal	1081
Mantysalo, Matti	1252
Marchack, Nathan	528
Maria, Winkler	498
Marnat, Loic	1535
Martin, Letz	726
Martin, Yves	528, 1060
Martina, Manuel	726
Martineau, Donatien	2173
Maslyk, Dan	753
Massey, John P.	363
Masuda, Koji	1074
Masuda, Yuki	346
Matsukawa, Daisaku	352
Matsumoto, Keiji	417
Matthias, Jost	726
Maune, Holger	726
Mavinkurve, A.	479
Mavinkurve, Amar	777
Mayer, Michael	2219
Mayr, Andreas	1492
McCann, Scott	2150, 2331
McFarlane, Nicole	2073
Mei, Ping	1312
Meiler, Josef	942
Melin, Peter	1405
Melkote, Shreyes	2349
Mellen, Jon	453
Mercier, Denis	1680

Meth, Jeffrey	785
Meunier, Philippe	1789
Miao, Min	1983, 2168
Michailos, Jean	569
Michel, Jean-Philippe	1680
Michihiro, Toshiaki	2042
Miki, Shota	1599
Miller, Andy	340, 437, 1035, 2206
Miller, Scott	370, 1347
Milton, Basil	55
Min, Fan-Yu	600
Min, Kyung Deuk	2290
Ming Chinq, Jong	587
Minoret, Stéphane	569, 1622
Mirkarimi, Laura	628, 1041
Mishra, Dibyajat	1316
Missinne, Jeroen	1757
Mitchell, Nicholas C	2134
Mitev, Ivaylo	1509
Miura, Seiya	101
Miyazawa, Risa	1921
Miyazawa, Yoshinori	1952
Mizutani, Daisuke	1294, 1952
Moehrle, Martin	1060
Moeller, Berthold	437
Mogera, Umesh	620
Mogera, Umesha	1605
Mohan, Kashyap	655
Mohd. Ghazali, Mohd. Ifwat	1687
Mohd Ghazali, Mohd Ifwat	113
Momozawa, Aya	334
Mondal, Saikat	113, 1240, 1687
Montmayeul, Brigitte	225
Moon, Kwangjin	636
Moon, Kyoung-Sik	157, 1977, 2134, 2140, 2296, 2349
Moon, Kyoung-Sik (Jack)	2359
Moon, Seok Hwan	197
Moon, Sungwook	1188

Mori, Daichi	1933
Mori, Hiroyuki	417, 1921
Mori, Ken-Ichiro	101
Mori, Kentaro	1140
Mori, Kiyoharu	1047
Mori, Takahiro	1921
Morikawa, Yasuhiro	1865
Morisako, Isamu	1451
Motobe, Takeharu	352
Motoyoshi, Makoto	1047
Mourier, Therry	569
Mourier, Thierry	1622
Moussodji Moussodji, Jeff	1396
Mrozek, Pawel	628, 1041
Mu, Fengwen	989
Mudrick, John P.	648
Muehlbauer, Franz-Xaver	855
Muga, Karthik	1035
Müller, Ernst	868
Murakami, Yasunori	1067
Murayama, Takahide	1865
Murray, Bruce T.	1970
Murtagian, Gregorio	1028
Murugan, Rajen	1221, 1653
Murugesan, Murugesan	1047
Mydlak, Mathias	726
Na, Hoonjoo	636
Na, Nanju	1208
Nachiappan, Vivek Chidambaram	587
Nagai, Koji	1599
Nagamatsu, Tatsuo	1933
Nah, Jae-woong	528, 1246
Nahalingam, Kirthika	1666
Nair, Chandrasekharan	334, 924
Nakamura, Ai	1047
Nakamura, Eiji	417
Nakamura, Takuya	1641
Nakamura, Tomonori	1933

Nakayama, Tomoki	550
Nakazaki, Fukino	550
Nam, Ju Hyun	563
Nam, Seungki	1188
Narayanan, Rajeev	270
Naseem, Sadia	2079
Nauroze, Syed Abdullah	119
Ndip, Ivan	1475
Nedumthakady, Nithin	1588
Nemeth, Csaba	811
Neumeyr, Christian	1052
Ng, Daniel	1287
Ng, Eric	14
Ngo, Ha-Duong	1475
Nguyen, Hoang-Vu	141
Nguyen, Luu	655, 1316, 1333
Nguyen, Thong	1889
Nieh, Simon	1246
Nilsson, Torbjörn M. J.	1405
Nishikawa, Hiroshi	1081, 2003
Niu, Mengnian	1246
Niu, Yuling	819
Noguet, Dominique	1535
Nolmans, Philip	674
Nonaka, Toshihisa	446, 740
Noriki, Akihiro	2042
Oates, Daniel	1611
Oberndorff, P.	479
Ogawa, Tsuyoshi	446, 740
Ogura, Nobuo	972
Oh, Dan(Kyung Suk)	614
Oh, KwangSeok	163
O'Halloran, G.M.	479
O'Halloran, Orla	777
Ohba, Takayuki	1002
Ohde, Christian	106
Ohkubo, Tomohiro	1641
Oi, Kiyoshi	1599

Öjefors, Erik 968
Okada, Kazuya 1015
Okamoto, Daichi 718, 2112
On, JY 1710
Onishi, Tetsuya 726
Onitake, Shigeo 726
Oo, Aung Kyaw 968
Oppermann, Hermann 1052
Oprins, Herman 126
Orcutt, Jason S. 1060
Ortega, Carlos 2072
Osborn, Tyler 453
Ötzlinger, Herbert 1498
Owens, N. 479
Pacot, Guilhem 1870
Paeck, Marcus 1853
Paik, Kyung-Wook 283, 1146, 2022, 2213, 2266, 2272, 2278
Palmer, Jordan 2186
Palys, Anna 47
Pan, Jhih-Yuan 14
Pan, Ke 2343
Pan, Ponder 1693
Pancoast, Leanna 1246
Pang, Ponder 1175
Panigrahy, Asisa Kumar 2156
Panikkanvalappil, Sajanlal 1588
Pantano, Nicolas 674, 1215
Pantouvaki, Marianna 1757
Papapolymerou, John 948
Paradis, Etienne 2252
Paranjpe, Ajit 1470
Pares, G. 1279
Pares, Gabriel 1535, 1622
Parikh, Bakul 1121
Park, Gyu-Tae 2246
Park, Hyun Ho 733
Park, Jae-Hyeong 2213
Park, Jongcheol 2213

Park, JoonYoung 163
Park, Sang Yong 35, 563
Park, S.B. 2150, 2331
Park, Seungbae 1130, 2343
Park, SooIn 2278
Park, Yong-Jin 204
Park, Yong Sung 204
Park, Young-Bae 937, 2246
Parker, David 428
Parthasarathy, Srivatsan 1660
Paul, Jens 910
Pei, Yu 2234
Pei Siang, Sharon Lim 587
Peng, lan 607, 2206
Peray, Patrick 535
Petzold, Matthias 175
Pfost, Martin 1509
Pham, Van-Lai 1130
Pham, Vanlai 2150, 2331, 2343
Phansalkar, Sukrut 1382
Philip, Pierre-Emile 569
Phommahaxay, Alain 340, 437, 607, 2206
Pierre, Ferris 1279
Pietryga, Christoph 1159
Plant, Jason 1611
Plochowietz, Anne 1312
Podpod, Arnita 340, 437
Polezhaev, Vladimir 1443
Poliks, Mark D. 768, 1581, 1946
Ponthenier, Fabienne 569
Posthill, John 1041
Pozzobon, Fiorella 2194
Premerlani, Romeo 726
Prenger, Luke 1463
Prisacaru, Alexandru 811
Pristauz, Hugo 1492
Proschwitz, Jan 1159
Pu, Han-Ping 688

Pu, Li	1716
Puligadda, Rama	1722
Pulugurtha, Markondeya Raj	960
Pulugurtha, P. Raj	1300
Qi, Tao	1509
Qian, Zhiguo	667
Qiang, Song Guan	1485
Qiao, Y.Y.	1629
Qin, Ivy	55
Raad, Peter	405
Raghavan, Nagarajan	834
Rahim, Kaysar	800
Raj, Anto	2309
Raj, P. Markondeya	718, 972
Rajagoapal, Varun	334
Ramon, Hannes	1757
Rao, Vempati Srinivasa	842, 1152, 2126
Rasilainen, Kimmo	1405
Rastogi, Ravi	1611
Ravichandran, Siddharth	726, 1796
Ravinder, Pal Singh	1419
Raychaudhuri, Sourobh	1312
Raynaud, Christine	1680
Rebhan, Bernhard	218, 942
Refai-Ahmed, Gamal	2150, 2331, 2343
Ren, Chao	1977, 2296
Ren, Linlin	1556
Ren, Qin	1996
Ribière, Céline	569, 1622
Richter, Theresia	515
Rivera, Katie	600
Robertson, Stuart	63, 1569
Rogers, Jeff	270
Rolland, Emmanuel	225
Romano, Giovanni	569
Romero, Gilles	569
Rongen, Rene T.H.	479
Rongen, R.T.H.	777, 1339
Ross, Joseph	1121
Roucou, R.	1339
Rovitto, Marco	2091
R. Tummala, Rao	718
Rudolph, Catharina	628
Rupp, Bradley B.	1312
Ruzicka, Klaus	868
Ryu, Dong Su	1457
Ryu, Hyodong	2246
Sakai, Taiji	1952
Sakakibara, Shiori	352
Sakaue, Takahiko	76
Sakuma, Katsuyuki	270
Sakuyama, Seiki	1952
Salahoueldhadj, Abdellah	340
Salcedo, Javier	1660
Saleem, Amin	1870
Saleh, Rafat	868
Sammakia, Bahgat G.	1970
Sanchez, Juliet	753
Sanchez, Loïc	225
Santerre, Francis	1396
Sarangapani, Murali	2048
Sato, Muneyuki	1865
Sato, Nobuaki	1451
Sato, Yoichiro	712
Savage, Eric	2117
Saxena, Antra	1770
Scevola, Daniel	569
Schares, Laurent	1060
Scheller, Britta	106
Schellkes, Eckart	2067
Schempp, Fabian	1113
Schiffmacher, Alexander	1443
Schiffres, Scott N.	1970
Schingale, Angelika	1509
Schischka, Jan	175
Schmid, Maximilian	2324

Schmitt, Wolfgang 87
Schneider-Ramelow, Martin 861
Schroeder, Raoul 1498
Schulze, Gary 55
Schulze, Sebastian 218, 942
Schumann, Todd 695, 1647, 1809
Schuster, Christian 2240
Schutt-Aine, Jose 1889
Schwarz, Mark 392, 819
Schwenk, Erika 87
Seefisch, Henning 2284
Segal, Julie 2072
Segaud, Roselyne 569
Sekhar, Vasarla Nagendra 842
Sekiguchi, Masahiro 1140
Selhofer, Hubert 1492
Selvanayagam, Cheryl 834
Serebreni, Maxim 2103
Shaddock, David M. 768
Shah, Aashish 55
Shah, Ujash 1605
Shahane, Ninad 1316
Shakoorzadeh, Niloofar 800
Shambach, William 2186
Shang, Jintang 522, 890
Shangguan, Dongkai 1272
Shapiro, Dmitri 1611
Sharma, Himani 1300, 1588
Sharon, Gil 2103
Sharon Lim, Pei Siang 21
Sheikhi, Roozbeh 150
Sheikhnejad, Ommeaymen 243
Shelton, Douglas 101
Shen, Yu-An 1081, 2003
Shi, Aihua 884
Shi, Tielin 661
Shibahara, Hiromi 1933
Shibasaki, Yoko 2112

Shibata, Daisuke 2112
Shie, Kai Cheng 642
Shigetou, Akitsu 235
Shih, Andy 1246
Shih, Meng-Kai 1710, 1902
Shika, Seiji 1009
Shim, Ji Ni 563
Shim, Moo-Sup 197
Shimada, Sawako 1015
Shimatsu, Takehito 989
Shimazu, Takayuki 1764
Shimizu, Kan 1641
Shimizu, Koji 1451
Shin, SangMyung 1146, 2278
Shin, Youngmin 300, 682
Shiraiwa, Tomio 163
Shirley, Tim 543
Shoji, Hideaki 163
Shoji, Yu 346
Shreve, Matthew 1312
Shumarayev, Sergey Yuryevich 667
Shunmugasamy, Vasanth 1081
Sidorov, Victor 1052
Sierra-Suarez, Jonatan A. 648
Sigl, Alfred 855
Sigmund, Ariane 1060
Sikka, Kamal 1121
Silberer, Gerald 942
Simmons, Jacob C. 1970
Simon, Gilles 569
Singh, Chrandeep 1130
Singh, Shiv Govind 2156
Sirbu, Bogdan 1052
Sitaraman, Suresh 1521
Sitaraman, Suresh K. 249, 785, 1939
Sitaraman, Suresh K 382
Sivapurapu, Sridhar 249, 1939
Sivasubramony, Rajesh S. 1581

Sivasubramony, Rajesh Sharma 768
Slabbekoorn, John 340, 607
Sleeckx, Erik 340, 437, 607, 2206
Smet, Vanessa 655, 972, 1028, 1796
Smith, Stephen 1200
So, R. 14
So, Raymond 903
Soares, Francisco 1052
Son, Ho-Young 2246
Son, Kirak 937, 2246
Song, Bo 157, 2134, 2140
Song, Changming 2234
Song, Euseok 614
Soon-Wook, Kim 1215
Sorensen, Eric 543
Sosa, Ramón A. 655
Souriau, Jean-Charles 1575
Southard, Arthur 1722
Sover, Raanan 294
Spurney, Robert G. 1300
Sridhar, Sharath 2309
Srinivasan, Sriram 2180
Stegmann, Tamira 87
Steinhorst, Rachel 1240
Stephan, Tino 175
Stewart, Benjamin G 382
Stoffel, Nancy C. 768, 1946
Stone, Bill 392
Stone, David 1179
Stucchi, Michele 1035
Su, An-Jhih 1
Su, C.H. 1175
Su, Jay 1463
Su, Ming-Sin 1175
Su, Peng 1099, 1106
Su, Sinan 792, 2309
Su, Zhaoxi 890
Subbiah, Nilavazhagan 878

Sueoka, Kuniaki 1921
Suga, Tadatomo 989
Suganuma, Katsuaki 474
Sugiura, Kazuhiko 1451
Suhard, Samuel 437, 607
Suhling, Jeff 505, 1347
Suhling, Jeffery 2309
Suhling, Jeffrey C. 792, 1087, 1815, 1958
Sulkis, Michael 1977
Sun, Hongyu Y. 2036
Sun, Rong 81, 243, 746, 1556, 1916
Sun, Teng 1300
Sun, Xiao 1215
Sun, Yangyang 392
Sun, Yunna 707
Sun, Yunting 707
Sung, Yun Hyun 35
Surillo, Emanuel 334
Suryoatmojo, Heri 1751
Susumago, Yuki 264
Suzuki, Akiyoshi 1865
Suzuki, Takuya 1009
Suzuki, Yui 2162
Suzuki, Yuya 1015
Swaminathan, Madhavan 249, 1672, 1939
Syed, Ahmer 392, 819
Sylvestre, Julien 1744
Symonds, Ken 2103
Tai, Jui-Feng 7
Tak, Coen 2029
Takahashi, Noriyuki 264
Takano, Takamasa 94
Takiar, Hem 1387
Tal, Sharon 294
Talebbeydokhti, Pouya 294, 954, 1666
Tamura, Akira 1952
Tan, Chuan Seng 1735
Tan, KH 325

Tan, Kim Hwee	14, 903
Tanaka, Kazunori	1067
Tanaka, Masaya	94
Tanaka, Tetsu	264
Taneda, Hiroshi	1599
Tang, Gongyue	1419
Tang, Junyan	1200
Tang, Tom	1432
Tang, Zirong	661
Tani, Daisuke	1933
Tan Swee Seng, Eric	2048
Tao, Jing	1735
Tao, Mian	14, 903
Tao, Qi	2029
Tao, Wang Jun	1485
Tarng, David	1426, 1710
Tasi, Mike	1704
Tatsumi, Kohei	1451
Tehrani, Bijan	896
Tekin, Tolga	1052
Tentzeris, Manos M.	119, 896, 960, 972
Teoh, Kristie	1028
Tetsuya, Onishi	1498
Teutsch, Thorsten	47, 210
Thai, Trang	294, 954
Theil, Jeremy	628, 1041
Theng, Chih-Han	758
Theuss, Horst	855
Thirugnanasambandam, Sivasubramanian	2309
Thomas, Tony	505
Thorsell, Mattias	1405
Thukral, Varun	1339
Tian, Yanhong	1266
Tissier, Pierre	1622
To, Hing "Thomas"	1208
Toda, Keiji	1451
Toepper, Michael	1853

Tokunari, Masao	1074
Tollefsen, Torleif A	141
Tolunay Wipf, Selin	942
Tomikawa, Masao	346
Tomita, Yasunari	1022
Tomohiro, Fukao	1272
Topsakal, Erdem	1782
Tremble, Eric	1179
Trombley, Django	1221
Tsai, Chung-Hao	1
Tsai, Clair	1175
Tsai, Jensen	700, 1432
Tsai, Mike	700
Tsai, Sheng-Han	397
Tsao, Pei-Haw	763
Tschoban, Christian	1475
Tseng, I-Hsin	758, 1328
Tseng, Yi-Hsiu	1359
Tseng, Yu-Chou	1902
Tsfati, Yossi	954
Tsunoda, Masatoshi	2042
Tsuruta, Kazuhiro	1451
Tu, K N	642
Tu, K. N.	1716, 2003
Tummala, Rao	334, 655, 1028, 1300, 1588, 1796
Tummala, Rao R.	924, 960, 972, 2112
Tunga, Krishna	1121
Tuominen, Samuli	1252
Turcotte, Eric	467
Tutunjyan, Nina	1035
Twiefel, Jens	2284
Ueda, Kazutoshi	1451
Ueno, Keiko	446, 740
Ueno, Kenichi	2162
Ueta, Chiho	1015
Ume, I. Charles	1333
Uomoto, Miyuki	989
Ura, Shogo	556

Uresti, Tiffani	1081	von Waechter, Claus	855
Utano, Tetsuya	1933	Wada, Keiko	1451
Uy, William	1272	Wagner, Juergen	855
Vadimas, Verdingovas	515	Wagner, Thomas	1159
Vaisband, Boris	543, 579, 1605, 2225	Wai, Leong Ching	21, 1419
van Borkulo, Jeroen	423	Waidhas, Bernd	1159
Van Campenhout, Joris	1757	Wan, Weikang	2318
Vandendaele, William	168	Wan, Zhe	579
Vandeneynde, Aurélie	535	Wang, Chang-Ning	1175
Van der Plas, Geert	674, 1215	Wang, Chen-Chao	41, 977
van der Stam, Richard	423	Wang, Chengqian	28
van Haare, Niek	1789	Wang, Chenxi	1266
Van Huylenbroeck, Stefaan	1035	Wang, Chuei-Tang	688
Vanjari, Siva Rama Krishna	2156	Wang, Chun-Min	1463
van Olst, E.	479, 777	Wang, Daixing	2016
van Soestbergen, M.	479, 777	Wang, Huayan	1130, 2150, 2331, 2343
Van Steenberge, Geert	1757	Wang, Hui	243
Varga, Edit	2072	Wang, Huiying	707
Vasarla, Nagendra Sekhar	2126	Wang, J. H.	1550
Vélard, Rémi	569	Wang, Jin	69
Velenis, Dimitrios	674	Wang, Jing	1130, 2150, 2343
Venkataraman, Srikrishnan	2180	Wang, Jiunn Jie	1635
Venugopal, Archana	405	Wang, Kirin	1175, 1595, 1693
Verdonck, Patrick	2206	Wang, Lejun	392
Veres, Agnes	811	Wang, Liyuan	1983
Verhelst, Marian	674	Wang, Nan	1564
Verrun, Sophie	1622	Wang, Peiren	135
Viehweger, Kay	1833	Wang, Qian	69, 1312, 2234
Vijayakumar, Swathi	1666	Wang, Qidong	2318
Vinci, Andrea	569	Wang, Rung-De	1693
Viswanath, Ram	2180	Wang, Shiu-Chih	1426
Vitello, Dario	2091	Wang, Tai-Jui	1877
Vivet, Pascal	569, 1926	Wang, Wei	392, 819, 2016
Vladimirova, Kremena	168	Wang, Xiangy-Yu	1996
Vobl, Matthias	855	Wang, Xiaobai	1543
Voges, Steve	363, 861	Wang, Xin	2234
Voit, Walter	1848	Wang, Xinying	1889
Voit, Walter E.	1258	Wang, Xueqiao	2140

Wang, Yan 707
Wang, Yang 135
Wang, Yen Neng 1635
Wang, Yiteng 972
Wang, Y.P. 1629
Wang, Yu 1312
Wang, Yu-Cheng 1693
Wang, Yunda 1312
Wang, Yu-Po 700, 1432
Wang, Zhijie 819
Watanabe, Atom 924, 960, 1300
Watanabe, Atom O. 972
Watanabe, Manabu 1294
Watanabe, Naoki 924
Watanabe, Osamu 1933
Watanabe, Takuro 1764
Watariguchi, Shigeru 1047
Webb, Bucknell 270
Weber, Y. 479
Weerawarne, Darshana L. 1581
Weerawarne, D.L. 1946
Wei, Cheng 410
Wei, Tiwei 126
Weichart, Johannes 1833
Weichart, Jüergen 1833
Weir, Terence 1611
Weiss, Thomas 306
Weng, Chen-Yuan 600
Weng, I-An 1729
Wen Kong, Ling 1485
Werner, Thomas 628
Widiez, Julie 168
Wietstruck, Matthias 218, 942
Wigger, Benedikt 868
Wijewardena, Kanishka 1687
Wilde, Juergen 878, 1113, 1443
Williamson, Jaimal 1333
Wipf, Christian 942

Wohrmann, Markus 363
Wöhrmann, Markus 861, 1853
Wolfberger, Archim 2029
Wong, Chee Wei 277
Wong, Ching-Ping 81, 243, 746, 1556, 1916, 2296, 2349
Wong, CP 157, 2134
Wong, C. P. 1977, 2140, 2359
Wong, Nelson 55
Wong Chin Yeung, Jason 2048
Workman, Thomas 628
Wu, An-Tai 1426
Wu, C.M.L. 1629
Wu, Dapeng 968
Wu, Fan 157, 2134
Wu, Hsing-Hui 14
Wu, Jiaqi 2302
Wu, Jing 1087
Wu, Jui-Yang 2258
Wu, Mei-Ling 2144
Wu, Sheng-Tsai 1463
Wu, W. C. 1175
Wu, Yang 2022
Wu, Y.H. 1693
Wu, Zhichao 1306
Wu, Zhongming 486
Wurz, Marc Christopher 1883
Xi, Cao 14, 903
Xiang, Gengzhao 135
Xiang, Hui 63
Xiangyu, Wang 587
Xiao, Hui 2054
Xiao, Zhiyi 884
Xie, Dongji 486
Xie, Hong 28
Xie, Yong 1221
Xiong, Chi 1060
Xiong, Yaoxu 243
Xu, Hui 55

Xu, Iris	14, 903
Xu, Jianbin	1556
Xu, Jiefeng	1130, 2150, 2331, 2343
Xu, Jikai	1266
Xu, Xirui	884
Xue, Mei	2318
Yadav, M.	1946
Yadav, Manu	768
Yagi, Hidekazu	1933
Yahyaei-Moayyed, Farzaneh	2180
Yakabe, Sho	1764
Yalagach, Mahesh	2029
Yamada, Shuhei	1796
Yamada, Takashi	175
Yamamoto, Kazunori	842, 1419, 2126
Yamashita, Soichi	1140
Yamauchi, Sinichi	76
Yamawaki, Seigo	1294
Yamazaki, Noriyuki	352
Yang, C. C.	1550
Yang, Chen-Tsai	1877
Yang, Cheol-Woong	2246
Yang, Chi-Hau	2200
Yang, Fan	850
Yang, Henry	14, 903
Yang, Jenn-Ming	2258
Yang, Ming	2022
Yang, Rolance	1175
Yang, S.B.	1595
Yang, Sean	1729
Yang, Tilo H.	235
Yang, T. L.	1175
Yang, Xiaobing	28, 884
Yang, Yang	1983, 2168
Yang, Yong	1716
Yang, Yong-suk	330
Yang, Yu-Hsiang	811
Yang, Yu-Ting	1359

Yang, Zhuoqing	707
Yao, Ruohe	1324
Ye, Chen	522
Ye, Lilei	1564
Ye, Ning	1387
Ye, Yong Liang	1419
Yee, Kuo-Chung	1
Yeh, Meng-Kao	1697
Yeh, Shu-Shen	1550
Yen, Yee-Wen	2067
Yeo, Woon-Hong	1233
Yess, Kim	340
Yew, M.C.	931
Yew, M. C.	1550
Yi, Luyun	1200
Yildiz, Ömer Faruk	2240
Yin, Liang	768
Yin, Xin	1052
Yook, Jongmin	1647
Yoon, Dal-Jin	2266, 2272
Yoon, Gil-Sang	197
Yoon, Seung Wook	325, 968, 1165
Yoon, Yong-Kyu	695, 983, 1647, 1809, 2085, 2337
Yoshida, Shu	1009
Yoshihiro, Furukawa	1300
Yotsuyanagi, Hiroko	352
Young Suk, Kim	1002
Youssef, Toni	2173
Yu, C.K.	493
Yu, C.T.	931
Yu, Daquan	28, 884
Yu, Doug C.H.	594
Yu, Douglas	1, 688
Yu, Hai-Yang	235
Yu, Jambo	28
Yu, Ji-In	204
Yu, Shiang-Hwua	1751
Yu, Ta-Jen	289

Yuan, K.S.	1595
Yue, Xiang	522
Zaal, J.J.M.	1339
Zachariah, Ashwin Varkey	768
Zarr, Scott	1611
Zeb, Gul	467
Zeng, Qinghua	2061
Zeng, Xiaoliang	1556
Zhang, Baotan	81
Zhang, Bowei	1306
Zhang, Dongxiao	63
Zhang, Eric J.	1060
Zhang, Hong	1722
Zhang, Hongqing	1246
Zhang, Jianfeng	890
Zhang, Jincan	1983
Zhang, Shuye	2022
Zhang, Wenwu	183
Zhang, Xiaowu	1152, 1419, 1543
Zhang, Xin-Ping	410
Zhang, Xuefeng	392
Zhang, Yao	1377
Zhang, Zheng	474
Zhao, Cong	2309
Zhao, Liguo	1569
Zhao, Lily	392
Zhao, Lixin	69
Zhao, N.	1629
Zhao, Peng	1735
Zhao, Tao	81, 746, 1916
Zhao, Weiwei	183
Zhao, Xiuchen	1716
Zhao, Yang	2234
Zheng, Fengxia	28
Zheng, Hao	1377
Zheng, Ting	1803
Zhengyang, Qian	264
Zhou, Bin	1324
Zhou, Han	1569
Zhou, Jianwen	746
Zhou, Jie-Ying	410
Zhou, Min-Bo	410
Zhou, Shicheng	1266
Zhou, Shiqi	1081, 2003
Zhou, Yi	249, 1521, 1939
Zhou, Zhaoxia	63, 850, 1569
Zhu, Pengli	243, 746, 1916
Zhua, Pengli	81
Zhuo, Qizhuo Zhuo	753
Zoberbier, Ralph	106
Zoschke, Kai	1475
Zou, Lin	1774
Zou, Yichao	884
Zuber, Fabien	535
Zullino, Lucia	2194
Zussy, Marc	535

IEEE
445 Hoes Lane
Piscataway, NJ 08854-4141

ISBN 978-1-7281-1500-9